FIFTH EDITION

GAS PURIFICATION

Gulf Publishing Company
Houston, Texas

FIFTH EDITION

GAS PURIFICATION

Arthur L. Kohl
Richard B. Nielsen

FIFTH EDITION

GAS PURIFICATION

Copyright © 1960, 1974, 1979, 1985, 1997 by Gulf Publishing Company, Houston, Texas. All rights reserved. Printed in the United States of America. This book, or parts thereof, may not be reproduced in any form without permission of the publisher.

Gulf Publishing Company
Book Division
P.O. Box 2608 □ Houston, Texas 77252-2608

10 9 8 7 6 5 4 3

Library of Congress Cataloging-in-Publication Data

Kohl, Arthur L.
   Gas purification. — 5th ed. / Arthur Kohl and Richard Nielsen.
     p. cm.
   Includes bibliographical references and index.
   ISBN 0-88415-220-0
   1. Gases—Purification. I. Nielsen, Richard (Richard B.) II. Title.
TP754.K6 1997
665.7—dc21
                                                               96-52470
                                                                   CIP

Contents

Preface, vii

Chapter 1
Introduction, 1

Chapter 2
Alkanolamines for Hydrogen Sulfide and Carbon Dioxide Removal, 40

Chapter 3
Mechanical Design and Operation of Alkanolamine Plants, 187

Chapter 4
Removal and Use of Ammonia in Gas Purification, 278

Chapter 5
Alkaline Salt Solutions for Acid Gas Removal, 330

Chapter 6
Water as an Absorbent for Gas Impurities, 415

Chapter 7
Sulfur Dioxide Removal, 466

Chapter 8
Sulfur Recovery Processes, 670

Chapter 9
Liquid Phase Oxidation Processes for Hydrogen Sulfide Removal, 731

Chapter 10
Control of Nitrogen Oxides, 866

Chapter 11
Absorption of Water Vapor by Dehydrating Solutions, 946

Chapter 12
Gas Dehydration and Purification by Adsorption, 1022

Chapter 13
Thermal and Catalytic Conversion of Gas Impurities, 1136

Chapter 14
Physical Solvents for Acid Gas Removal, 1187

Chapter 15
Membrane Permeation Processes, 1238

Chapter 16
Miscellaneous Gas Purification Techniques, 1296

Appendix, 1374

Index, 1376

Preface

The first four editions of *Gas Purification* were authored by Arthur L. Kohl and Fred C. Riesenfeld. Mr. Riesenfeld died shortly after publication of the fourth edition in 1985. His considerable technical contributions and warm friendship will be sorely missed.

The present team of authors has endeavored to completely overhaul and update the text for publication as this fifth edition of *Gas Purification*. Three new chapters have been added to cover the rapidly expanding fields of NO_x control (Chapter 10), absorption in physical solvents (Chapter 14), and membrane permeation (Chapter 15). All other chapters have been expanded, revised, and rearranged to add new subject matter, delete obsolete material, and provide increased emphasis in areas of strong current interest. Examples of major additions to existing chapters are the inclusion of new sections on liquid hydrocarbon treating (Chapter 2), Claus plant tail gas treating (Chapter 8), biofilters for odor and volatile organic compound (VOC) control (Chapter 12), thermal oxidation of VOCs (Chapter 13), and sulfur scavenging processes (Chapter 16).

Because of the growing importance of air pollution control, the coverage of gas purification technologies that are applicable in this field, such as SO_2, NO_x, and VOC removal, has been expanded considerably. On the other hand, the use of ammonia for H_2S and CO_2 removal and the removal of ammonia from gas streams represent technologies of decreasing importance, primarily because of the declining use of coal as a source of fuel gas. Discussions of these two subjects have, therefore, been combined into a single chapter (Chapter 4).

Organization of the text represents a practical compromise between an arrangement based on unit operations or process similarities and one based on impurities being removed. Thus, Chapters 12 and 15 cover the operations of adsorption and membrane permeation, respectively, and the use of these technologies for the removal of a wide variety of impurities; while Chapters 7 and 10 cover single impurities (SO_2 and NO_x, respectively) and their removal by a number of different processes. Consideration is also given to the industrial importance of the technologies in the allocation of chapters; as a result, two chapters (Chapters 2 and 3) are devoted to the use of amines for the removal of H_2S and CO_2, while only one, rather short chapter (Chapter 6) covers the use of water for the absorption of gas impurities of any type.

The aim of this book is to provide a practical engineering description of techniques and processes in widespread use and, where feasible, provide sufficient design and operating data to permit evaluation of the processes for specific applications. Limited data on processes that were once, but are no longer commercially important, are also presented to provide an historical perspective. Subject matter is generally limited to the removal from gas streams of gas-phase impurities that are present in relatively minor proportions. The removal of discrete solid or liquid particles is not discussed, nor are processes that would more appropriately be classified as separation rather than purification.

A generalized discussion of absorption is provided in Chapter 1 because this unit operation is common to so many of the processes described in subsequent chapters. Discussions of other unit operations employed in gas purification processes, such as adsorption, catalytic conversion, thermal oxidation, permeation, and condensation are included in the chapters devoted to these general subjects.

No attempt has been made to define the ownership or patent status of the processes described. Many of the basic patents on well-known processes have expired; however, patented improvements may be critical to commercial application. In fact, a number of important proprietary systems are based primarily on the incorporation of special additives or flow system modifications into previously existing processes.

Technical data are normally presented in the units of the original publication. Practically, this means that most U.S. data on commercial operations are given in English units, while foreign and U.S. scientific data are presented in metric or, occasionally, SI units. To aid in the conversion between systems, a table of units and conversion factors is included as an appendix.

The assistance of many individuals who contributed material and suggested improvements is gratefully acknowledged. Thanks are also due to the companies and organizations who graciously provided data and gave permission for reproducing charts and figures. The number of such organizations is too large to permit individual recognition in this preface; however, they are generally identified in the text as the sources of specific data. We particularly acknowledge with appreciation the generous support of the Fluor Daniel Corporation in the preparation of this edition, and to the following Fluor Daniel personnel: Joseph Saliga, for providing much of the new data in Chapters 7 and 10 on SO_2 and NO_x removal processes; Michael Potter, for input on sulfur conversion processes; Paul Buckingham, for work on the physical solvent chapter; and David Weirenga, for assistance with the chapter on permeation processes. Other significant contributors to this fifth edition are Ronald Schendel, consultant, who provided data for Chapters 8 and 15 on sulfur conversion technology and permeation processes; John McCullough, Proton Technology, who supplied information for the discussion of amine plant corrosion in Chapter 3; Robert Bucklin, consultant, who provided detailed information on sulfur scavenging processes for Chapter 16; and Dr. Carl Vancini, who drafted the review of the Stretford process for Chapter 9. Finally, we wish to express gratitude to our families: Evelyn, Jeffrey, and Martin Kohl; and Theresa and Michael Nielsen for their support and patience during the preparation of this book.

Arthur L. Kohl
Richard B. Nielsen

FIFTH EDITION

GAS PURIFICATION

Chapter 1
Introduction

DEFINITIONS, 1

PROCESS SELECTION, 2

PRINCIPLES OF ABSORPTION, 6

    Introduction, 6
    Contactor Selection, 6
    Design Approach, 12
    Material and Energy Balance, 13
    Column Height, 15
    Column Diameter, 27
    General Design Considerations, 31

REFERENCES, 35

DEFINITIONS

Gas purification, as discussed in this text, involves the removal of vapor-phase impurities from gas streams. The processes which have been developed to accomplish gas purification vary from simple once-through wash operations to complex multiple-step recycle systems. In many cases, the process complexities arise from the need for recovery of the impurity or reuse of the material employed to remove it. The primary operation of gas purification processes generally falls into one of the following five categories:

1. Absorption into a liquid
2. Adsorption on a solid
3. Permeation through a membrane
4. Chemical conversion to another compound
5. Condensation

Absorption refers to the transfer of a component of a gas phase to a liquid phase in which it is soluble. Stripping is exactly the reverse—the transfer of a component from a liquid phase in which it is dissolved to a gas phase. Absorption is undoubtedly the single most important operation of gas purification processes and is used in a large fraction of the systems described in subsequent chapters. Because of its importance, a section on absorption and basic absorber design techniques is included in this introductory chapter.

Adsorption, as applied to gas purification, is the selective concentration of one or more components of a gas at the surface of a microporous solid. The mixture of adsorbed components is called the adsorbate, and the microporous solid is the adsorbent. The attractive forces holding the adsorbate on the adsorbent are weaker than those of chemical bonds, and the adsorbate can generally be released (desorbed) by raising the temperature or reducing the partial pressure of the component in the gas phase in a manner analogous to the stripping of an absorbed component from solution. When an adsorbed component reacts chemically with the solid, the operation is called chemisorption and desorption is generally not possible. Adsorption processes are described in detail in Chapter 12, which also includes brief discussions of design techniques and references to more comprehensive texts in the field.

Membrane permeation is a relatively new technology in the field of gas purification. In this process, polymeric membranes separate gases by selective permeation of one or more gaseous components from one side of a membrane barrier to the other side. The components dissolve in the polymer at one surface and are transported across the membrane as the result of a concentration gradient. The concentration gradient is maintained by a high partial pressure of the key components in the gas on one side of the membrane barrier and a low partial pressure on the other side. Although membrane permeation is still a minor factor in the field of gas purification, it is rapidly finding new applications. Chapter 15 is devoted entirely to membrane permeation processes and includes a brief discussion of design techniques.

Chemical conversion is the principal operation in a wide variety of processes, including catalytic and noncatalytic gas phase reactions and the reaction of gas phase components with solids. The reaction of gaseous species with liquids and with solid particles suspended in liquids is considered to be a special case of absorption and is discussed under that subject. A generalized treatment of chemical reactor design broad enough to cover all gas purification applications is beyond the scope of this book; however, specific design parameters, such as space velocity and required time at temperature, are given, when available, for chemical conversion processes described in subsequent chapters.

Condensation as a means of gas purification is of interest primarily for the removal of volatile organic compounds (VOCs) from exhaust gases. The process consists of simply cooling the gas stream to a temperature at which the organic compound has a suitably low vapor pressure and collecting the condensate. Details of the process are given in Chapter 16.

PROCESS SELECTION

The principal gas phase impurities that must be removed by gas purification processes are listed in **Table 1-1.**

Selecting the optimum process for removing any one or combination of the listed impurities is not easy. In many cases, the desired gas purification can be accomplished by several different processes. Determining which is best for a particular set of conditions ultimately requires a detailed cost and performance analysis. However, a preliminary screening can be

Table 1-1
Principal Gas Phase Impurities

1. Hydrogen sulfide
2. Carbon dioxide
3. Water vapor
4. Sulfur dioxide
5. Nitrogen oxides
6. Volatile organic compounds (VOCs)
7. Volatile chlorine compounds (e.g., HCl, Cl_2)
8. Volatile fluorine compounds (e.g., HF, SiF_4)
9. Basic nitrogen compounds
10. Carbon monoxide
11. Carbonyl sulfide
12. Carbon disulfide
13. Organic sulfur compounds
14. Hydrogen cyanide

made for the most commonly encountered impurities by using the following generalized guidelines.

Hydrogen sulfide and carbon dioxide removal processes can be grouped into the seven types indicated in **Table 1-2**, which also suggests the preferred areas of application for each process type. Both absorption in alkaline solution (e.g., aqueous diethanolamine) and absorption in a physical solvent (e.g., polyethylene glycol dimethyl ether) are suitable process techniques for treating high-volume gas streams containing hydrogen sulfide and/or carbon dioxide. However, physical absorption processes are not economically competitive when the acid gas partial pressure is low because the capacity of physical solvents is a strong function of

Table 1-2
Guidelines for Selection of H_2S and CO_2 Removal Processes

Type of Process	Acid Gas H_2S	CO_2	Plant Size	Partial Pressure	Sulfur Capacity
Absorption in Alkaline Solution	A	A	H	L	H
Physical Absorption	A	A	H	H	H
Absorption/Oxidation	A	—	H	L	L
Dry Sorption/Reaction	A	—	L	L	L
Membrane Permeation	A	A	L	H	L
Adsorption	A	A	L	L	L
Methanation	—	A	L	L	—

Notes: A = Applicable, H = High, L = Low; dividing line between high and low is roughly 20 MMscfd for plant size, 100 psia for partial pressure, and 10 tons/day for sulfur capacity.

partial pressure. According to Christensen and Stupin (1978), physical absorption is generally favored at acid gas partial pressures above 200 psia, while alkaline solution absorption is favored at lower partial pressures. Tennyson and Schaaf (1977) place the boundary line between physical and chemical solvents at a somewhat lower partial pressure (i.e., 60–100 psia) above which physical solvents are favored. They also provide more detailed guidelines with regard to the preferred type of alkaline solution and the effect of different acid gas removal requirements. The absorption of hydrogen sulfide and carbon dioxide in alkaline solutions is discussed in detail in Chapters 2, 3, 4, and 5. Chapter 14 covers the use of physical solvents.

Membrane permeation is particularly applicable to the removal of carbon dioxide from high-pressure gas. The process is based on the use of relatively small modules, and an increase in plant capacity is accomplished by simply using proportionately more modules. As a result, the process does not realize the economies of scale and becomes less competitive with absorption processes as the plant size is increased. McKee et al. (1991) compared diethanolamine (DEA) and membrane processes for a 1,000 psia gas-treating plant. For their base case, the amine plant was found to be generally more economical for plant sizes greater than about 20 MMscfd. However, at very high acid-gas concentrations (over about 15% carbon dioxide), a hybrid process proved to be more economical than either type alone. The hybrid process, which is not indicated in **Table 1-2**, uses the membrane process for bulk removal of carbon dioxide and the amine process for final cleanup. Membrane processes are described in Chapter 15.

When hydrogen sulfide and carbon dioxide are absorbed in alkaline solutions or physical solvents, they are normally evolved during regeneration without undergoing a chemical change. If the regenerator offgas contains more than about 10 tons per day of sulfur (as hydrogen sulfide), it is usually economical to convert the hydrogen sulfide to elemental sulfur in a conventional Claus-type sulfur plant. For cases that involve smaller quantities of sulfur, because of either a very low concentration in the feed gas or a small quantity of feed gas, direct oxidation may be the preferred route. Direct oxidation can be accomplished by absorption in a liquid with subsequent oxidation to form a slurry of solid sulfur particles (see Chapter 9) or sorption on a solid with or without oxidation (see Chapter 16). The solid sorption processes are particularly applicable to very small quantities of feed gas where operational simplicity is important, and to the removal of traces of sulfur compounds for final cleanup of synthesis gas streams. Solid sorption processes are also under development for treating high-temperature gas streams, which cannot be handled by conventional liquid absorption processes.

Adsorption is a viable option for hydrogen sulfide removal when the amount of sulfur is very small and the gas contains heavier sulfur compounds (such as mercaptans and carbon disulfide) that must also be removed. For adsorption to be the preferred process for carbon dioxide removal, there must be a high CO_2 partial pressure in the feed, the need for a very low concentration of carbon dioxide in the product, and the presence of other gaseous impurities that can also be removed by the adsorbent. Typical examples are the purification of hydrogen from steam-hydrocarbon reforming, the purification of land-fill effluent gas, and the purification of ammonia synthesis gas. Adsorption processes are described in detail in Chapter 12.

Two processes predominate for water vapor removal: absorption in glycol solution and adsorption on solid desiccants. These two processes are quite competitive and, in many cases, either will do an effective job. In general, a dry desiccant system will cost more, but will provide more complete water removal. For large-volume, high-pressure natural gas treating, glycol systems are generally more economical if dew-point depressions of 40° to 60°F are sufficient. For dew-point depressions between about 60° and 100°F, either type may prove more economical based on the specific design requirements and local operating costs. For dew-point depressions consistently above 100°F, solid desiccant processes are generally specified, although the use of very highly concentrated glycol solutions for attaining water dew points as low as −130°F is gaining favor. Solid desiccant processes are also preferred for very small installations where operating simplicity is a critical factor. Gas dehydration by absorption in liquids is covered in Chapter 11, and by adsorption using solid desiccants in Chapter 12.

Fewer options have been commercialized for the removal of sulfur dioxide and nitrogen oxides than for hydrogen sulfide and carbon dioxide. The predominant process for flue gas desulfurization (FGD) for large utility boiler applications is wet scrubbing with a lime or limestone slurry. This process can provide greater than 90% sulfur dioxide removal. Where lower removal efficiency can be tolerated, spray drying and dry injection processes have been found to be more economical. Commercially proven processes for the removal of nitrogen oxides from the flue gas of large boiler plants are currently limited to selective catalytic reduction (SCR) and thermal reduction processes. The SCR process is the only one capable of 90%-plus NO_x removal efficiency. Further details on the selection of FGD and nitrogen oxide removal processes are provided in Chapters 7 and 10.

Volatile organic compounds (VOCs) can be removed from gases by at least five types of processes: thermal incineration, catalytic incineration, carbon adsorption, absorption in a liquid, and condensation. Preliminary guidelines for selection of these are given in **Figure 1-1**, which is based on the data of McInnes et al. (1990). For typical VOC concentrations in the range of 100 to 1,000 ppmv, only thermal incineration can provide 99% removal efficiency. Of course, feed concentration and removal efficiency are not the only factors to be considered. For example, if energy consumption is a significant factor, catalytic incineration may be preferable to thermal incineration because it operates at a lower temperature and therefore requires less heat input. If chemical recovery as well as removal is required, a process other than incineration must be specified. A detailed discussion of factors to be considered in selecting the best VOC control strategy is given by Ruddy and Carrol (1993). Specific VOC removal processes are described in several chapters, Chapter 12—adsorption processes, Chapter 13—catalytic incineration and thermal incineration processes, Chapter 15—membrane permeation processes, and Chapter 16—condensation and absorption/oxidation processes.

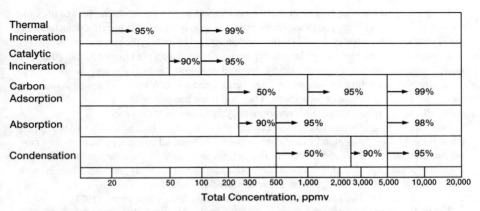

Figure 1-1. VOC removal efficiency of applicable processes vs. VOC concentration in feed gas. *Data of McInnes et al. (1990)*

PRINCIPLES OF ABSORPTION

Introduction

Absorption, as applied to gas purification processes, can be divided into the following general classifications based on the nature of the interaction between absorbate and absorbent:

1. *Physical Solution.* In this type of process the component being absorbed is more soluble in the liquid absorbent than other components of the gas stream, but does not react chemically with the absorbent. The equilibrium concentration of the absorbate in the liquid phase is strongly dependent on the partial pressure in the gas phase. An example is the absorption of hydrogen sulfide and carbon dioxide in the dimethyl ether of polyethylene glycol (Selexol Process). Relatively simple analytical techniques have been developed for designing systems of this type.
2. *Reversible Reaction.* This type of absorption involves a chemical reaction between the gaseous component being absorbed and a component of the liquid phase to form a loosely bonded reaction product. The product compound exhibits a finite vapor pressure of the absorbate which increases with temperature. An example is the absorption of carbon dioxide into monoethanolamine solution. Analysis of this type of system is complicated by the nonlinear shape of the equilibrium curve and the effect of reaction rate on the absorption coefficient.
3. *Irreversible Reaction.* In this type of absorption the component being absorbed reacts with a component of the liquid phase to form reaction products that can not readily be decomposed to release the absorbate. An example is the absorption of hydrogen sulfide in iron chelate solution to form a slurry of elemental sulfur particles. The analysis of systems involving irreversible reactions is simplified by the absence of an equilibrium vapor pressure of adsorbate over the solution, but becomes more complex if the irreversible reaction is not instantaneous or involves several steps.

Contactor Selection

The primary function of the gas absorption contactor is to provide an extensive area of liquid surface in contact with the gas phase under conditions favoring mass transfer. Contactors normally employ at least one of the following mechanisms: (1) dividing the gas into small bubbles in a continuous liquid phase (e.g., bubble cap trays), (2) spreading the liquid into thin films that flow through a continuous gas phase (e.g., packed columns), and (3) forming the liquid into small drops in a continuous gas phase (e.g., spray chambers). All three types of contact are employed in gas purification absorbers. They are interchangeable to a considerable extent, although specific requirements and conditions may favor one over the others.

Countercurrent contactors can also be categorized as staged columns, which utilize separate gas and liquid flow paths in individual contact stages; differential columns, which utilize a continuous contact zone with countercurrent flow of gas and liquid in the zone; and pseudo-equilibrium columns, which combine essentially countercurrent flow of gas and liquid streams with discrete stages. A simplified guide to the selection of gas-liquid contactors based on this categorization is presented in **Table 1-3** which is derived from the work of Frank (1977).

Table 1-3 is generally applicable for stripping columns as well as absorbers, although additional parameters may need to be considered. Bravo (1994) points out that biological or

Table 1-3
Selection Guide for Gas-Liquid Contactors

Conditions of Application	Staged Columns Perforated, or Valve Trays	Staged Columns Bubble Cap or Tunnel Trays	Differential Columns Randomly Packed	Differential Columns Systematically Packed	Pseudo-Equilibrium Downcomer-less	Pseudo-Equilibrium Disc and Donut
Low pressure (<100 mm Hg)	2	1	2	3	0	1
Moderate pressure	3	2	2	1	1	1
High pressure (>50% of critical)	3	2	2	0	2	0
High turndown ratio	2	3	1	2	0	1
Low liquid rates	1	3	1	2	0	0
Foaming systems	2	1	3	0	2	1
Internal tower cooling	2	3	1	0	1	0
Solids present	2	1	1	0	3	1
Dirty or polymerized solution	2	1	1	0	3	2
Multiple feeds and sidestreams	3	3	1	0	2	1
High liquid rates (scrubbing)	2	1	3	0	3	2
Small diameter columns	1	1	3	2	1	1
Columns with diameter 3–10 ft	3	2	2	2	2	1
Large diameter columns	3	1	2	1	2	1
Corrosive fluids	2	1	3	1	2	2
Viscous fluids	2	1	3	0	1	0
Low ΔP (efficiency no concern)	1	0	2	2	0	3
Expanded column capacity	2	0	2	3	2	0
Low cost (performance no concern)	2	1	2	1	3	3
Available design procedures	3	2	2	1	1	1

Notes:
Rating key: 0 - Do not use
    1 - Evaluate carefully
    2 - Usually applicable
    3 - Best selection
Staged columns: Tray columns with separate liquid and vapor flow paths.
  Common types: Bubble cap, sieve, valve.
  Proprietary types: Angle, Uniflux, Montz, Linde, Thorman, Jet.
Differential columns: True countercurrent flow of gas and liquid.
  Randomly packed: Raschig rings, saddles, slotted rings, Tellerettes, Maspac.
  Systematically packed: Flexipac, Goodloe, Hyperfil, Sulzer, Glitch Grid.
Pseudo-equilibrium stages: Countercurrent flow of gas and liquid with discrete trays.
  Downcomerless trays: Perforated, Turbogrid, Ripple.
  Low pressure drop trays: Disc and donut, shower deck.
Special devices (not rated in table):
  Venturi scrubber, turbulent contact absorber, marble bed absorber, horizontal spray chamber, cocurrent rotator.
Based on data of Frank, 1977

inorganic fouling of trays or packing may occur when volatile organic compounds (VOCs) are steam stripped from water; however, he concludes that neither type of column has an advantage when fouling occurs. In another paper, Bravo (1993) describes methods to avoid and contend with the fouling of packing.

Gas purification absorbers often operate with liquids that contain suspended solid particles. A detailed review of techniques and design issues involved in making a vapor/liquid mass transfer device operate with solid particles in the solvent is given by Stoley and Martin (1995). They rank mass transfer equipment for such service from most suitable to least suitable as

1. grid-structured packing
2. baffle trays (e.g., shed trays, disc-and-donut trays, side-to-side trays)
3. dual-flow trays (e.g., downcomerless perforated trays with large openings)
4. tab trays (e.g., fixed tabs, jet tabs)
5. sieve trays (with downcomers)
6. bubble-cap trays
7. third generation random packings (e.g., Glitsch CMR)
8. second generation random packings (e.g., Glitsch Ballast Plus Rings) or smooth-surface structured packings
9. aggressive-surface structured packings
10. first generation random packings (e.g., Raschig rings)
11. complex trays (e.g., film or valve trays)
12. mist eliminator pads or wire packings

As a further guide to the selection of absorbers, the relative costs of six types of tray columns and ten types of column packings are presented in **Table 1-4** (Blecker and Nichols, 1973). Generalized comments on the nature and fields of application for tray, packed, and spray contactors follow.

Tray Columns

Tray columns (also called plate columns) are particularly well suited for large installations; clean, noncorrosive, nonfoaming liquids; and low-to-medium liquid flow applications. Tray columns are also preferred when internal cooling is required in the column. Cooling coils may be installed directly on individual trays or liquid can readily be removed at one tray, cooled, and returned to another tray. Perforated trays (also called sieve trays) are widely used because of their simplicity and low cost.

The formerly popular bubble-cap design is now used primarily for columns requiring a very low liquid flow rate, although structured packing is being used as a replacement for bubble-cap trays in many such applications. A number of special tray designs have been developed, including valve, grid, and baffle types to overcome some of the limitations of simple perforated and bubble-cap trays. Valve trays have been particularly popular because they permit operation over a wider range of flow rates than simple perforated trays without the high liquid holdup of bubble-cap trays. Examples of proprietary designs are the Koch Flexitray, Glitsch Ballast Tray, and Nutter Float Valve Tray.

Conventional bubble-cap, perforated, and valve trays operate as crossflow contactors in which the liquid flows horizontally across the tray and contacts gas flowing vertically

Table 1-4
Relative Costs of Columns

Relative Costs of Tray Columns (for equal diameter and height)

Bubble Cap	1.00
Koch Kascade	1.243
Plate Tray	0.842
Sieve Tray	0.874
Turbogrid	0.855
Valve Tray	0.911

Relative Costs of Column Packings (installed cost for equal volumes of packing, $/cu. ft in 1973 dollars)

	1-in. Dia.	2-in. Dia.
Berl saddles, stoneware	13.10	—
Berl saddles, steel	26.30	—
Berl saddles, stainless steel	32.80	—
Intalox saddles, ceramic	13.30	10.40
Pall rings, polypropylene	36.90	26.30
Pall rings, stainless steel	13.60	9.80
Raschig rings, stoneware	6.30	4.38
Raschig rings, stainless steel	15.70	10.90
Raschig rings, steel	12.60	8.79
Tellerettes, HD polyethylene	26.30	—

Data of Blecker and Nichols, 1973

upward through openings in the tray. After traversing the tray, the liquid flows into a downcomer, which conveys it to the tray below. Downcomers typically occupy 5 to 20% of the column cross-sectional area.

In countercurrent trays, which are also available but less popular than crossflow types, the liquid flows from one tray to the next lower tray as free falling drops or streams. Examples of countercurrent trays include perforated (Dual Flow), slotted (Turbogrid), and perforated-corrugated (Ripple). The trays are reasonably efficient, but lack flexibility because tray holdup and operating characteristics are highly dependent on gas and liquid flow rates.

Baffle or shower deck tray columns also approximate countercurrent contactors. These trays are nonperforated horizontal or slightly sloped sheets, each of which typically occupies slightly more than half of the tower cross-sectional area. The liquid flows off the edge of one tray as a curtain of liquid or series of streams and falls through the gas stream to the tray below. Typically, the trays are half moon in shape on alternate sides of the column, or disc and donut designs with centrally located discs that are slightly larger than the openings in donut-shaped trays located above and below them. Baffle trays are used for extremely dirty liquids when highly efficient contact is not required and for heat exchanger duty—particularly the quenching of hot, particle-laden gas streams. Photographs of typical commercial trays are shown in **Figure 1-2.**

10 Gas Purification

Flexitray valve tray   Standard Flexitray valve

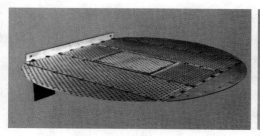

Sieve tray   Bubble-cap tray

Figure 1-2. Typical commercial trays. *Courtesy of: Koch Engineering Company, Inc.*

Packed Columns

Packed columns are gaining favor for a wide range of applications because of the development of packings that offer superior performance, as well as the emergence of more reliable design techniques. The most commonly used packing elements are packed randomly in the column. Non-random ordered (or structured) packings were originally developed for small scale distillation columns to handle difficult separations. Their use has recently expanded, however, and ordered packings are now offered by several companies for large scale commercial applications. The current availability of performance data and rational design procedures makes the use of ordered packing worth considering for cases requiring high mass transfer efficiency and low pressure drop.

Packed contactors are most frequently used with countercurrent flow of liquid and gas. However, in special cases they are used in a crossflow arrangement with the liquid flowing down through a bed of packing while the gas flows horizontally, or in cocurrent flow with liquid and gas flowing in the same direction. Cocurrent contactors using structured packing elements similar to in-line mixers are used for gas purification applications when a single contact stage is sufficient; for example, when an irreversible reaction occurs. They have the advantage of operating with much higher gas velocities than countercurrent designs without being subject to flooding problems.

Introduction 11

As compared to tray columns, packed columns are generally preferred for small installations, corrosive service, liquids with a tendency to foam, very high liquid/gas ratios and applications in which a low pressure drop is desired. Their use in larger sizes appears to be increasing, and there is also a growing use of packing to replace trays where an improvement in column performance is required.

Typical random packing elements are illustrated in **Figure 1-3** and one example of an ordered packing is shown in **Figure 1-4**.

Spray Contactors

Spray contactors are important primarily when pressure drop is a major consideration and when solid particles are present in the gas, for example, atmospheric pressure exhaust gas streams. They are not effective for operations requiring more than one theoretical contact stage or a close approach to equilibrium. Since the surface area for mass transfer in a spray chamber is directly proportional to the liquid flow rate, it is common practice to recycle the absorbent to increase absorption efficiency.

Types of equipment classified as spray contactors include countercurrent spray columns, venturi scrubbers, ejectors, cyclone scrubbers, and spray dryers. The use of spray dryers as absorbers is of particular interest in the removal of sulfur dioxide from hot flue gas (see Chapter 7).

Figure 1-3. Examples of random packing elements. *Courtesy of Koch Engineering Company, Inc.*

12 Gas Purification

Figure 1-4. Typical structured packing (Intalox). *Courtesy of Norton Chemical Process Products Company*

Design Approach

The design of countercurrent absorbers normally involves the following steps: (1) selection of contactor, including type of trays or packing, based on process requirements and expected service conditions; (2) calculation of heat and material balances; (3) estimation of required column height (number of trays or height of packing) based on mass transfer analysis; (4) calculation of required column diameter and tray or packing parameters based on gas and liquid flow rates and hydraulic considerations; and (5) mechanical design of the hardware. The steps are not necessarily performed in the above order and may be combined or reiterated in the design procedure. In the design of spray contactors, steps 3 and 4 are replaced by design calculations that define the configuration and operating parameters of the liquid breakup and separation equipment. For cocurrent contactors selecting and sizing the mixing elements are the principal design tasks.

The key data required for the design of absorbers are the physical, thermal, and transport properties of the gases and liquids involved; vapor/liquid equilibrium data; and, if chemical reactions are involved, reaction rate data. Configuration data on the trays or packing are, of course, also required. Appropriate data are included, when available, for processes described in subsequent chapters.

The design of absorbers (and strippers) typically involves a computer-assisted, tray-by-tray, heat- and material-balance calculation to determine the required number of equilibrium stages, which are then related to the required number of actual trays by an estimated tray efficiency. More recently, a non-equilibrium stage model has been developed for computer application which considers actual trays (or sections of packing) and performs a heat and material balance for each phase on each actual tray, based on mass and heat transfer rates on that tray.

To facilitate the use of computers in the design of absorbers, Kessler and Wankat (1988) have converted a number of commonly used correlations to equation form. These include

O'Connell's overall tray efficiency correlation (1946), Fair's flooding correlation for sieve tray columns (1961), Hughmark and O'Connell's correlation relating to pressure drop of gas through a dry tray (1957), Fair's correlation for tray weeping (l963), and Eckert's correlation for flooding in a packed tower (1970A).

A number of commercially available software programs that include absorber design routines are listed in the *CEP 1997 Software Directory* (*Chem. Eng. Prog.*, 1997). A packed tower design program for personal computers, which includes correlations for predicting the efficiency and capacity of high efficiency structured packings, is described by Hausch and Petschauer (1991). Detailed reviews of commonly used design procedures for absorption operations are presented in several texts and articles including those of Edwards (1984), Fair et al. (1984), Zenz (1979), Treybal (1980), Kohl (1987), and Diab and Maddox (1982). A brief summary of the principal design equations and correlations is presented in the following sections.

Material and Energy Balance

Figure 1-5 is a simplified diagram of a countercurrent absorption column containing either trays or packing. In order to work with constant gas and liquid flow rates over the length of the column, solute-free flow rates and mole ratios (rather than mole fractions) are used in material balance equation 1-1.

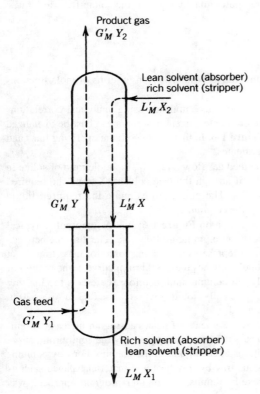

Figure 1-5. Material balance diagram for countercurrent contactor.

14 Gas Purification

$$G_M'(Y - Y_2) = L_M'(X - X_2) \tag{1-1}$$

Where: G_M' = solute-free gas flow rate, lb-mole/h ft
 L_M' = solute-free liquid flow rate, lb-mole/ h ft
 X = mole ratio solute in the liquid phase
 $= x/(1 - x)$, where x = mole fraction
 Y = mole ratio solute in the gas phase
 $= y/(1 - y)$, where y = mole fraction

Rearranging equation (1-1) gives

$$Y = L_M'/G_M'(X - X_2) + Y_2 \tag{1-2}$$

which is the equation of the operating line. The line is straight on rectangular coordinate paper and has a slope of L_M'/G_M'. The coordinates of the ends of the operating line represent conditions at the ends of the column, i.e., X_2, Y_2 (top) and X_1, Y_1 (bottom).

Absorber design correlations are not always based on solute-free flow rates and mole ratios. The original Kremser (1930) and Souders and Brown (1932) design equations, for example, are based on the lean solvent rate and the rich gas. When the solute concentrations in gas and liquid are low, x is approximately equal to X, and y is approximately equal to Y. In addition, the total molar flow rates are approximately equal to the solute-free flow rates. In these cases equation 1-2 simplifies to

$$y = L_M/G_M (x - x_2) + y_2 \tag{1-3}$$

which is easier to use because equilibrium data are usually given in terms of mole fractions rather than mole ratios.

Although absorber designs can be effectively accomplished using analytical correlations and computer programs, the performance of countercurrent absorbers can best be visualized by the use of a simple diagram such as **Figure 1-6.** In this figure both the operating lines and equilibrium curve are plotted on X,Y coordinates.

Typically the known parameters are the feed gas flow rate, G_M', the mole ratio of solute in the feed gas, Y_1, the mole ratio of solute (if any) in the lean solvent, X_2, and the required mole ratio of solute in the product gas, Y_2. The goal is to estimate the required liquid flowrate and, ultimately, the dimensions of the column.

Two possible operating lines have been drawn on **Figure 1-6;** line A represents a typical design, and line B represents the theoretical minimum liquid flow rate. The distance between the operating line and the equilibrium curve represents the driving force for mass transfer at any point in the column. Since line B actually touches the equilibrium curve at the bottom of the column, it would require an infinitely tall column, and therefore represents the limiting condition with regard to liquid flow rate. Typically liquid flow rates (or L/G ratios)—20 to 100% higher than the minimum—are specified.

Frequently absorption is accompanied by the release of heat, causing an increase in temperature within the column. When this occurs it is necessary to modify the equilibrium curve so that it corresponds to the actual conditions at each point in the column. For tray columns this can best be accomplished by a rigorous tray-by-tray heat and material balance such as one proposed by Sujata (1961). For packed columns, a computer program approach was

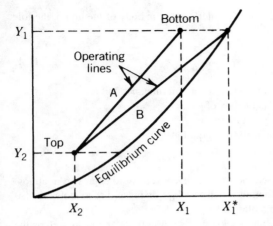

Figure 1-6. Operating line-equilibrium curve diagram for absorption column.

developed by von Stockar and Wilke (1977) which operates by dividing the column into an arbitrary number of segments. These authors also developed a shortcut method that does not require a computer (Wilke and von Stockar, 1978). More recently, computer programs have been developed that calculate heat and material balances around both the gas and liquid phases on each actual (not theoretical) tray or each selected slice in a packed column (Vickery et al., 1992; Seader, 1989; Krishnamurthy and Taylor, 1985A, B).

The overall distribution of heat release between the liquid and gas streams is determined primarily by the ratio of the total heat capacities of the two streams, $L_M C_q / G_M C_p$, where L_M is the flow rate of the liquid, G_M is the flow rate of the gas, C_q is the heat capacity of the liquid, and C_p is the heat capacity of the gas. When the ratio is high (over about 2), the liquid carries the heat of reaction down the column, the product gas leaves at approximately the temperature of the liquid feed, and the product liquid leaves at an elevated temperature determined by an overall heat balance. Typically the feed gas cools the outgoing liquid somewhat, resulting in a temperature bulge within the column. When the ratio is low (below about 0.5), the product gas carries essentially all of the heat of reaction out of the column. For ratios close to 1.0, the reaction heat is distributed between the liquid and gas products, both of which may leave at a temperature well above that of the incoming streams.

Column Height

Packed Columns

The concept of absorption coefficient, which is the most convenient approach for packed column design, is based upon a two-film theory originally proposed by Whitman (1923). It is assumed that the gas and liquid are in equilibrium at the interface and that thin films separate the interface from the main bodies of the two phases. Two absorption coefficients are then defined as k_L, the quantity of material transferred through the liquid film per unit time, per unit area, per unit of driving force in terms of liquid concentration; and k_G, the quantity transferred through the gas film per unit time, per unit area, per unit of driving force in terms of pressure. Since the quantity of material transferred from the body of the gas to the interface must equal the quantity transferred from the interface to the body of the liquid, the following relationship holds:

$$N_A = k_G(p - p_i) = k_L(c_i - c) \qquad (1\text{-}4)$$

Where: N_A = quantity of component A transferred per unit time, per unit area
p = partial pressure of A in main body of gas
p_i = partial pressure of A in gas at interface
c = concentration of A in main body of liquid
c_i = concentration of A in liquid at interface

Any consistent set of units may be used; however, it is convenient to express p in atmospheres and c in pound moles per cubic foot, in which case k_G is expressed as lb moles/(hr)(sq ft)(atm) and k_L as lb moles/(hr)(sq ft)(lb moles/cu ft).

The use of equation 1-4 for design requires a knowledge of both k_G and k_L as well as the equilibrium relationship and the interfacial area per unit volume of absorber. Although these factors can be estimated for special design cases, it is more practical to use overall coefficients which are based on the total driving force from the main body of the gas to the main body of the liquid and which relate directly to the contactor volume rather than to the interfacial area. These overall coefficients, K_Ga and K_La, are defined as follows:

$$N_A a\, dV = K_G a(p - p_e)\, dV = K_L a(c_e - c)\, dV \qquad (1\text{-}5)$$

Where: a = interfacial area per unit volume of absorber
p_e = partial pressure of A in equilibrium with a solution having the composition of main body of liquid
c_e = concentration of A in a solution in equilibrium with main body of gas
V = volume of packing

The overall coefficients are related to the individual film coefficients as follows:

$$\frac{1}{K_G a} = \frac{1}{k_G a} + \frac{H}{k_L a} \qquad (1\text{-}6)$$

$$\frac{1}{K_L a} = \frac{1}{k_L a} + \frac{1}{H k_G a} \qquad (1\text{-}7)$$

where H is Henry's law constant, p_i/c_i, or, in cases where Henry's law does not hold, $(p_i - p_e)/(c_i - c_e)$.

The use of overall coefficients is strictly valid only where the equilibrium line is straight over the operating region. However, because of their convenience, they are widely used for reporting test data, particularly on commercial equipment, and are therefore very useful for design.

In order to apply absorption coefficient data to the design of commercial columns, it is necessary to consider the changes in liquid and gas compositions that occur over the length of the column. This involves equating the quantity of material transferred (as indicated by gas- or liquid-composition change) to the quantity indicated to be transferred on the basis of the absorption coefficient and driving forces and then integrating this equation over the length of the column. For the individual film coefficients, this results in the following expression for column height (Sherwood and Pigford, 1952):

$$h = G'_M \int_{p_2}^{p_1} \frac{p_{BM} \, dp}{k'_G a (P-p)^2 (p-p_e)} = \frac{L}{\rho_L} \int_{c_2}^{c_1} \frac{dc}{k_L a (c_e - c)} \tag{1-8}$$

Where: h = height of packed zone, ft
G'_M = superficial molar mass velocity of inert gas, lb moles/(hr)(sq ft)
p_1 = partial pressure of solute in entering gas, atm
p_2 = partial pressure of solute in leaving gas, atm
P = total pressure of system, atm
p = partial pressure of solute in main gas stream, atm
p_e = partial pressure of solute in equilibrium with main body of solution, atm
p_{BM} = log mean of inert gas pressures
$k'_G = k_G (p_{BM}/p)$ = special mass-transfer coefficient which is independent of gas composition, lb moles/(hr)(sq ft)(atm)
L = liquid flow rate, lb/(hr)(sq ft)
ρ_L = liquid density, lb/cu ft (assumed constant)
c_1 = solute concentration in liquid leaving bottom of column, lb moles/cu ft
c_2 = solute concentration in liquid fed to top of column, lb moles/cu ft
c = solute concentration in main body of liquid, lb moles/cu ft
c_e = concentration of solute in liquid phase in equilibrium with main body of gas, lb moles/cu ft

The equations may be integrated graphically by a method developed by Walker et. al (1937) that is also described by Sherwood and Pigford (1952). Simplified forms of these equations have been developed which are much more readily used and which are sufficiently accurate for most engineering-design calculations. Two of these forms are particularly adapted to the low gas and liquid concentrations that are frequently encountered in gas purification. These equations assume that the following conditions hold:

1. The equilibrium curve is linear over the range of concentrations encountered (and therefore overall coefficients can be used).
2. The partial pressure of the inert gas is essentially constant over the length of the column.
3. The solute contents of gaseous and liquid phases are sufficiently low that the partial pressure and liquid concentration values may be assumed proportional to the corresponding values when expressed in terms of moles of solute per mole of inert gas (or of solvent).

In terms of the overall gas coefficient and gas-phase compositions, the tower height can be estimated by equation 1-9:

$$h = \frac{G_M}{K_G a P} \int_{y_2}^{y_1} \frac{dy}{y - y_e} \tag{1-9}$$

or, where the overall liquid absorption coefficient is available, the column height may be calculated in terms of liquid-phase compositions:

$$h = \frac{L_M}{\rho_M K_L a} \int_{x_2}^{x_1} \frac{dx}{x_e - x} \tag{1-10}$$

In equations 1-9 and 1-10, y and x refer to the mole fractions of solute in the gas and liquid streams, respectively, and L_M and ρ_M represent the molar values of liquid flow rate and density, i.e., lb moles/(hr)(sq ft) and lb moles/cu ft. The subscript 1 refers to the bottom of the column, subscript 2 to the top of the column, and subscript e to the equilibrium composition with respect to the main body of the other phase. The other symbols have the same significance as in the previous equations. In general, it is preferable to employ the overall gas-film coefficient when the gas-film resistance is predominant, and the overall liquid coefficient when the principal resistance to absorption is in the liquid phase.

Equations 1-9 and 1-10 may be solved by relatively simple graphical integration. However, a further simplification, which can frequently be employed, is the use of a logarithmic mean driving force in the rate equation rather than graphical integration. This can be shown to be theoretically correct where the equilibrium curve and operating line are linear over the composition range of the column. The equations then reduce to

$$h = \frac{G_M(y_1 - y_2)}{K_G a P(y - y_e)_{LM}} \tag{1-11}$$

or

$$h = \frac{L_M(x_1 - x_2)}{\rho_M K_L a(x_e - x)_{LM}} \tag{1-12}$$

where $(y - y_e)_{LM}$ and $(x_e - x)_{LM}$ are equal to the logarithmic mean of the driving forces at the top and bottom of the column. Although not theoretically correct, the logarithmic mean driving force is often used to correlate $K_G a$ values for systems where the equilibrium curve is not a straight line and even for cases of absorption with chemical reaction. This greatly simplifies data reduction but can lead to serious errors. In general, the procedure is useful for comparing similar systems within narrow ranges of liquid composition and gas partial pressure.

A considerable amount of data on absorption-column performance is presented in terms of the "height of the transfer unit" (HTU), and design procedures based on this concept are preferred by many because of their simplicity and similarity to plate-column calculation methods. The basic concept which was originally introduced by Chilton and Colburn (1935) is that the calculation of column height invariably requires the integration of a relationship such as (from equation 1-9)

$$\int_{y_2}^{y_1} \frac{dy}{y - y_e}$$

The dimensionless value obtained from the integration is a measure of the difficulty of the gas-absorption operation. In the above case, it is called the number of transfer units based on an overall gas driving force, N_{OG}, and equation 1-9 can be reduced to

$$h = \frac{G_M}{K_G a P} N_{OG} \tag{1-13}$$

The HTU for this case (based on an overall gas-phase driving force) is then defined as

$$H_{OG} = \frac{h}{N_{OG}} = \frac{G_M}{K_G a P} \qquad (1\text{-}14)$$

Since N_{OG} is dimensionless, H_{OG} will have the same units as h. Similarly, for the overall liquid case:

$$H_{OL} = \frac{h}{N_{OL}} = \frac{L_M}{\rho_M K_L a} \qquad (1\text{-}15)$$

As in the calculation of column height from $K_G a$ or $K_L a$ data, it is theoretically correct to use a logarithmic mean driving force when both the equilibrium and operating lines are straight. For this case, the number of transfer units (overall gas) may be calculated from the simple expression:

$$N_{OG} = \frac{y_1 - y_2}{(y - y_e)_{LM}} \qquad (1\text{-}16)$$

This equation may be combined with the equilibrium relation:

$$y_e = mx$$

and the material-balance expression:

$$L_M (x_1 - x) = G_M (y_1 - y)$$

to eliminate the need for values of y_e. The resulting equation which was proposed by Colburn (1939) is given below:

$$N_{OG} = \frac{\ln\left[\left(1 - \frac{mG_M}{L_M}\right)\left(\frac{y_1 - mx_2}{y_2 - mx_2}\right) + \frac{mG_M}{L_M}\right]}{1 - (mG_M / L_M)} \qquad (1\text{-}17)$$

Where: N_{OG} = number of overall transfer units
m = slope of equilibrium curve dy_e/dx
x_2 = mole fraction solute in liquid fed to top of column
y_1 = mole fraction solute in gas fed to bottom of column
y_2 = mole fraction in gas leaving top of column
G_M = superficial molar mass velocity of gas stream, lb moles/(hr)(sq ft)
L_M = superficial molar mass velocity of liquid stream, lb moles/(hr)(sq ft)

It will be noted that the parameter mG_M/L_M appears several times in equation 1-17.

This parameter is called the stripping factor, S, and its reciprocal, L_M/mG_M, is called the absorption factor, A. The absorption factor is used in a number of popular techniques for the design of both packed and tray absorbers. It can be considered to be the ratio of L_M/G_M, the slope of the operating line, to m, the slope of the equilibrium line. Plots of equation 1-17, which can be used for estimating the required number of transfer units, are given in several publications (e.g., Treybal, 1980; Perry and Green, 1984).

Alternative equations and graphical techniques have been developed to calculate N_{OG} for other design conditions (Colburn, 1941; White, 1940). A summary of useful design equations for transfer-unit calculations is presented by Sherwood et al. (1975).

The HTU concept can also be employed for analysis of the contributions of the individual film resistances although, in general, the individual absorption coefficients are preferred for basic studies. Values of N_{OG} are particularly useful for expressing the performance of equipment in which the volume is not of fundamental importance. In spray chambers, for example (see Chapter 6), the effectiveness of the equipment is more a function of liquid flowrate and spray nozzle pressure than of tower volume. The use of volume-based absorption coefficients for such units is quite meaningless.

An approach frequently used by vendors to describe the mass transfer efficiency of packing is the "height equivalent to a theoretical plate" (HETP) which is defined as follows:

HETP = Height of packed zone/Number of theoretical plates achieved in packed zone

In this approach the number of theoretical plates required is estimated as described in the next section for tray columns, and this number is simply multiplied by the HETP value given for the packing employed to obtain the required packing height. The HETP concept is not theoretically correct for packed columns, in which contact is accomplished by differential rather than stagewise action; however, it is very easy to use for column design. For the special case of parallel equilibrium and operating lines (i.e., $mG_M/L_M = 1$), HETP and HTU are equal.

The calculation of packed column height by these techniques requires a knowledge of the overall absorption coefficient (e.g., K_Ga), the height of a transfer unit (e.g., H_{OG}), or the height equivalent to a theoretical plate (HETP) and estimation of these values is usually the most difficult column design task. Although some success has been achieved in predicting packed-column, mass-transfer coefficients from a purely theoretical basis (e.g., Vivian and King, 1963), the use of empirical correlations and experimental data represents the usual design practice. Test or operating data relating to absorption coefficients are therefore given whenever feasible for processes described in subsequent chapters. Examples of K_Ga values for a number of gas absorption operations are presented in **Table 1-5**. Data for a variety of packings operating under similar conditions are given in **Table 1-6**. The values given in this table are calculated for the absorption of carbon dioxide in dilute sodium hydroxide solution by assuming zero equilibrium vapor pressure of carbon dioxide over the solution and using a log-mean partial pressure over the length of the column.

Generalized correlations for estimating the individual mass transfer coefficients have been proposed by Onda et al. (1968), Bolles and Fair (1982), and Bravo and Fair (1982). These correlations cover commonly used packings such as Raschig rings, Berl saddles, Pall rings, and related configurations. Correlations for structured packings have been developed by Bravo et al. (1985) for Sultzer BX (gauze) packing, and by Spiegel and Meier (1987) for Mellapak (sheet metal) packing. Fair and Bravo (1990) suggest that the Bravo et al. (1985) correlation can be used for sheet metal as well as gauze packing by using a ratio of interface area/packing area of less than 1.0, and they provide a simple method of estimating the ratio.

A computer model that makes use of correlations, such as those referred to above for the individual mass transfer coefficients, to predict the actual performance of small sections of

Table 1-5
Typical K$_G$a Values for Various Absorbate/Absorbent Systems

	KGa, lb moles/(hr) (ft3) (atm)					
	Absorbent					
	Water			Aqueous Solutions		
Absorbate	A	B	C	Solution	A	B
CO_2			0.07	4% NaOH	2.0	1.5
H_2S			0.4	4% NaOH	5.92	4.4
SO_2	2.96	2.2	0.32	11% Na_2CO_3	11.83	8.93
HCN	5.92	4.4				
HCHO	5.92	4.4				
Cl_2	4.55	3.4	0.14	8% NaOH	14.33	10.8
Br_2				5% NaOH	5.01	3.7
ClO_2	4.4	4.4				
HCl	18.66	14.0	16.0			
HBr	5.92					
HF	7.96	6.0				
NH_3	17.30		13.0	Dilute acid	13.0	13.0
O_2			0.007			

Notes:
A = data for #2 plastic Super Intalox packing, gas velocity 3.5 ft/s, liquid rate 10 gpm/sq ft. (ASHRAE Handbook, 1988)
B = data for #2 plastic Super Intalox packing, gas velocity 3.5 ft/s, liquid rate 4 gpm/sq ft. (Strigle, 1994)
C = data for 1.5 in Intalox Saddles, conditions not stated. (Eckert et al., 1967)

packing in a column has been proposed by Krishnamurthy and Taylor (1985B). The approach is based on one they originally proposed for tray columns (1985A) and does not involve the concepts of HTU or HETP; in fact, the attainment of equilibrium is assumed to occur only at the gas/liquid interface and not in the products of a theoretical stage. In this rate-based model, separate material balances are made for gas and liquid phases in each packing section; these are coupled by interface mass transfer rates which must be equal in each phase at the interface.

Tray Columns

A commonly used design concept for tray columns is the "theoretical tray." This concept is based on the assumption that, with a theoretically perfect contact tray, the gas and liquid leaving will be in equilibrium. Although this assumption does not exactly represent the operation of any actual tray (where much of the gas will not even come in contact with the leaving liquid), it greatly simplifies the design procedure, and the departure of actual trays from this ideal situation can be conveniently accounted for by an expression known as "tray efficiency."

Table 1-6
Typical K_Ga Values for Various Packings

	K_Ga, lb moles/(hr) (ft3) (atm)		
	Packing Material		
Packing	metal	plastic	ceramic
#25 IMTP	3.42		
#50 IMTP	2.44		
#70 IMTP	1.74		
1-in. Pall Rings	3.10	2.64	
2-in. Pall Rings	2.18	2.09	
3.5-in. Pall Rings	1.28	1.23	
#1 Hy-Pac Packing	2.89		
#2 Hy-Pac Packing	2.06		
#3 Hy-Pac Packing	1.45		
#1 Super Intalox Packing		2.80	
#2 Super Intalox Packing		1.92	
#3 Super Intalox Packing		1.23	
1-in. Intalox Saddles			2.82
2-in. Intalox Saddles			1.88
3-in. Intalox Saddles			1.11
1-in. Raschig Rings			2.31
2-in. Raschig Rings			1.63
3-in. Raschig Rings			1.02
Intalox Snowflake Packing		2.37	
Structured Packings			
Intalox 1T	4.52		
Intalox 2T	3.80		
Intalox 3T	2.76		

Notes: 1. Conditions: inlet gas—3.5 ft/s, 1 mol% CO2 in air; feed liquid—10 gpm/ft2, 1 N NaOH, 75°F; NaOH conversion less than 25%.
    2. IMTP, Hy-Pac, Super Intalox, Intalox, and Intalox Snowflake are trademarks of the Norton Company.
Source: Strigle (1987, 1994)

The number of theoretical trays required for absorption can be determined simply by stepping off trays on a diagram similar to **Figure 1-6**. An example of this procedure is shown in **Figure 11-32** for water absorption in triethylene glycol (TEG). In this case the coordinates are lb water/MMscf for the gas phase and lb water/lb TEG for the liquid phase. A modification of this technique has been proposed by Rousseau and Staton (1988) for strippers and absorbers employing chemical solvents. The key features are the use of $y_A/(1 - y_A)$ as the ordinate and f_A as the abscissa where

A = solute
y_A = mole fraction of A in the gas phase
f_A = fractional saturation of the reactive component of the solution with A

With this coordinate system the operating line is straight. The equilibrium curve may be based on actual data for the specific system at column operating conditions or may be approximated on the basis of related data. Rousseau and Staton outline steps for estimating equilibrium curves based on the Henry's law constant for unreacted component A in the liquid and the equilibrium constant for the chemical reaction of A with the reactive component.

The graphical procedure is very useful for preliminary studies to establish the minimum flow rates for absorption and stripping operations, and for estimating the number of ideal stages (theoretical trays) required once design flow rates have been set. The number of ideal stages can be converted to actual trays by applying an appropriate tray efficiency.

Analytical procedures that closely resemble those employed for calculating the number of transfer units have also been developed for tray columns. A particularly useful equation suggested by Colburn (1939) for the case of low solute concentration and a straight equilibrium line is

$$N_P = \frac{\ln\left[\left(1 - \frac{mG_M}{L_M}\right)\left(\frac{y_1 - mx_2}{y_2 - mx_2}\right) + \frac{mG_M}{L_M}\right]}{\ln(L_M/mG_M)} \qquad (1\text{-}18)$$

where N_P = number of theoretical plates and the other symbols have the same meaning as in equation 1-17.

As noted in the preceding section, the parameter mG_M/L_M represents the ratio of the slope of the equilibrium curve to the slope of the operating line and is called the stripping factor, S. This factor and its reciprocal, the absorption factor, A, normally vary somewhat over the length of the column due to changes in all three variables. Kremser (1930) proposed defining A in terms of the lean solution and feed gas flow rates as follows:

$$A = L_{M0}/mG_{M(N+1)} \qquad (1\text{-}19)$$

The fractional absorption of any component, C, by an absorber of N theoretical plates is given by the following equation, often referred to as the Kremser (or Kremser-Brown) equation:

$$(y_{N+1} - y_1)/(y_{N+1} - y_0) = (A^{N+1} - A)/(A^{N+1} - 1) \qquad (1\text{-}20)$$

Where: y_{N+1} = mole fraction C in the inlet gas
y_1 = mole fraction C in outlet gas
y_0 = mole fraction C in equilibrium with lean solution
L_{M0} = lean solution flow rate, moles/hr
$G_{M(N+1)}$ = feed gas rate, moles/hr
N = number of theoretical plates in the absorber
m = K = y/x at equilibrium (assumed constant over the length of the column)

The Kremser equation is useful in the preliminary design of plate columns for physical absorption processes, such as the dehydration of natural gas with glycol solutions (see Chapter 11) and the absorption of CO_2 and H_2S in nonreactive solvents (see Chapter 14).

As mentioned previously, the number of actual plates in an absorber is related to the number of theoretical plates by a factor known as the "plate efficiency." In its simplest definition, the "overall plate efficiency" is defined as "the ratio of theoretical to actual plates required for a given separation." For individual plates, the Murphree vapor efficiency (Murphree, 1935) more closely relates actual performance to the theoretical-plate standard. It is defined by the following equation:

$$E_{MV} = \frac{y_p - y_{p+1}}{y_{pe} - y_{p+1}} \qquad (1\text{-}21)$$

Where: y_p = average mole fraction of solute in gas leaving plate
y_{p+1} = average mole fraction of solute in gas entering plate (leaving plate below)
y_{pe} = mole fraction of solute in gas in equilibrium with liquid leaving plate

Murphree plate efficiency values can be used to correct the individual steps in graphical analyses of the number of plates required. The overall efficiency, on the other hand, can only be used after the total number of theoretical plates has been calculated by a graphical or analytical technique. When operating and equilibrium lines are nearly parallel, the two efficiencies can be considered to be equivalent. Under other conditions they may vary widely.

A third version of the plate efficiency concept is the Murphree Point Efficiency, which can be defined as the Murphree efficiency at a single point on a tray. The point efficiency is the most difficult to use but is the most useful in theoretical analysis of tray performance. The Murphree vapor tray efficiency and point efficiencies on the tray are related primarily by the degree of mixing that occurs on the tray. The two are equal if mixing is complete; while the tray efficiency can be appreciably higher than the point efficiency if no mixing occurs. Actual trays fall between the two extremes.

A computer model relating point and tray efficiencies is described by Biddulph (1977). In this model the calculations for a tray are started at the outlet weir, where the liquid composition is known, and move progressively through thin slices of the liquid against the liquid flow to the inlet weir. At each increment, the liquid composition and temperature, and the gas composition above the point are calculated, based on an assumed point efficiency for each component and the gas composition below the tray at that point. An eddy diffusion model is used to define mixing in a comparison of the computer simulation with actual commercial plant data from a distillation column.

For simple physical absorption, the principal factors affecting tray efficiencies are gas solubility and liquid viscosity, and a correlation based on these two variables has been developed by O'Connell (1946). His correlation for absorbers is reproduced in **Figure 1-7**. Unfortunately, other factors such as the absorption mechanism, liquid depth, gas velocity, tray design, and degree of liquid mixing also influence tray efficiency, so no simple correlation can adequately cover all cases. A more detailed study of bubble tray efficiency has been made by the Distillation Subcommittee of the American Institute of Chemical Engineers (1958). *The Bubble Tray Design Manual* resulting from this work provides a standardized procedure for estimating efficiency which takes the following into account:

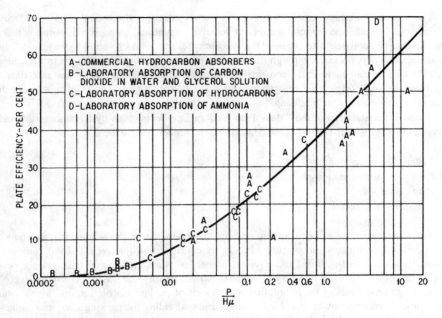

Figure 1-7. Correlation of overall plate efficiencies for commercial and laboratory absorbers; H = Henry's law constant in atm/(lb mole/cu ft), P = total pressure in atmospheres, and μ = liquid viscosity in centipoises. *From O'Connell (1946)*

1. The rate of mass transfer in the gas phase
2. The rate of mass transfer in the liquid phase
3. The degree of liquid mixing on the tray
4. The magnitude of liquid entrainment between trays

Properly designed sieve trays are generally somewhat more efficient than bubble-cap trays. A simplified approach for predicting the efficiency of sieve trays is given by Zuiderweg (1982) who presents a series of correlations defining their overall performance. The Zuiderweg study relies heavily on data released by Fractionation Research, Inc. (FRI) on the performance of two types of sieve trays (Yanagi and Sakata, 1981).

More recent studies aimed at developing models for predicting stage efficiencies include those of Chen and Fair (1984), Tomcej and Otto (1986), and Tomcej et al. (1987). The objective of the Tomcej et al. study is to provide a technique for designing trays for selective absorption. Specifically, the absorption of H_2S and CO_2 in amines is considered. The H_2S has a much higher tray efficiency because its rate of reaction in the liquid phase is faster. The approach makes use of a nonequilibrium stage model in which a parametric analysis is used to estimate tray efficiencies for the individual components. The calculated efficiencies are found to be strong functions of kinetic rate parameters and operating variables such as the gas velocity and the interfacial area and dispersion height generated on the tray. In one example, with 30% DEA solution operating at a pressure of 260 kPa, the CO_2 tray efficiency ranges from 7.5–9.1%, while the H_2S tray efficiency is relatively constant at about 42%.

The preceding discussion is based on using the concepts of "theoretical trays" and "tray efficiencies" to estimate the total number of actual trays required for a given absorption task. An alternative approach is to consider the mass transfer rate on each actual tray by modeling material and energy transfer through the interface between gas and liquid on the tray. Seader (1989) presents an historical perspective and generalized description of the rate-based approach for modeling staged separations and suggests that "the advantages of this approach can usher in a new era for modeling."

A detailed description of the "Mass Transfer Rate" model is given by Krishnamurthy and Taylor (1985A) who list the equations describing the model as follows:

1. Material balance equations
2. Energy balance equations
3. Rate equations
4. Equilibrium relations

Since the mass transfer occurring on an actual tray depends on the tray design, the model uses detailed information about column and tray configurations, as well as fluid compositions, flow rates, diffusivities, and physical properties. Mass and energy balances are performed around each phase on every actual tray. Krishnamurthy and Taylor (1985B) also propose a rate-based model for simulation and design of packed distillation and absorption columns. The packed tower model is based on simply dividing the packing zone into a number of sections (e.g., 10 for a typical absorber) around which the mass and energy balances are performed.

The rates of mass and energy transfer between phases are calculated based on gas and liquid film coefficients and concentration and temperature driving forces. Both thermal and chemical equilibria are assumed to exist only at the gas-liquid interface. The liquid film mass transfer coefficient is adjusted, if necessary, for chemical reactions occurring in the liquid phase by use of an enhancement factor (as defined in the next section). An absorption column simulator, which uses the rate-based approach, is described by Sardar et al. (1985). They demonstrate its predictive capabilities against operating data from a number of commercial plants employing various amines to remove H_2S and CO_2 in both tray and packed towers. The use of the rate-based design method to evaluate the performance of two amine plants is described by Vickery et al. (1992).

Effect of Chemical Reactions

A chemical reaction of the solute with a component in the liquid phase has the effect of increasing the liquid-film absorption coefficient over what would be observed with simple physical absorption. This results in an increase in the overall absorption coefficient in packed towers or an increase in tray efficiency in tray towers.

With very slow reactions (such as between carbon dioxide and water) the dissolved molecules migrate well into the body of the liquid before reaction occurs so that the overall absorption rate is not appreciably increased by the occurrence of the chemical reaction. In this case, the liquid film resistance is the controlling factor, the liquid at the interface can be assumed to be in equilibrium with the gas, and the rate of mass transfer is governed by the molecular CO_2 concentration-gradient between the interface and the body of the liquid. At the other extreme are very rapid reactions (such as those of ammonia with strong acids) where the dissolved molecules migrate only a very short distance before reaction occurs. The location of the reaction zone (and the value of the absorption coefficient) will depend primarily upon the diffusion rate of reactants and reaction products to and from the reaction zone, the concentration of solute at the interface, and the concentration of the reactant in the body of the liquid. However, since the distance that the solute must diffuse into the liquid is extremely small compared to the distance that it would have to travel for simple physical absorption, a high liquid-film coefficient is observed, and, in many cases, the gas-film resistance becomes the controlling factor.

Since the effect of chemical reaction is to increase the liquid film coefficient, k_L, over the value it would have in the absence of chemical reaction, k_L^o, a common approach is to utilize the ratio, k_L/k_L^o, in correlations. This ratio is called the enhancement factor. Both k_L and k_L^o are affected by the fluid mechanics, but fortunately their ratio, E, has been found to be relatively independent of these factors. It is primarily a function of concentrations, reaction rates, and diffusivities in the liquid phase.

The theoretical evaluation of absorption followed by liquid-phase chemical reaction has received a great deal of attention although the results are not yet routinely useful for design purposes. Early studies of serveral reaction types were made by Hatta (1929, 1932) and Van Krevelen and Hoftijzer (1948). This work has been expanded by more recent investigators to cover reversible and irreversible reactions, various reaction orders, and reaction rates from very slow to instantaneous. Important contributions have been made by Perry and Pigford (1953), Brian et al. (1961), Gilliland et al. (1958), Brian (1964), Danckwerts and Gillham (1966), Decoursey (1974), Matheron and Sandall (1978), and Olander (1960). The application of the theory to specific gas purification cases has been described by Joshi et al. (1981) (absorption of CO_2 in hot potassium carbonate solution), and by Ouwerkerk (1978) (selective absorption of H_2S in the presence of CO_2 into amine solutions).

Stripping in the presence of chemical reaction has been considered by Astarita and Savage (1980), Savage et al. (1980), and Weiland et al. (1982). In general, it is concluded that the same mathematical procedures may be used for stripping as for absorption; however, the results may be quite different because of the different ranges of parameters involved. It is always necessary to consider reaction reversibility in the calculation of stripping with chemical reaction.

It is beyond the scope of this introductory discussion to present even a listing of the numerous mathematical equations developed to correlate the effects of chemical reactions on mass transfer. Detailed equations and examples of their application are presented in comprehensive books on the subject by Astarita (1967), Danckwerts (1970), and Astarita et al. (1983).

Column Diameter

Packed Columns

The diameter of packed columns filled with randomly dumped packings is usually established on the basis of flooding correlations such as those developed by Sherwood et al. (1938), Elgin and Weiss (1939), Lobo et al. (1945), Eckert (1970A, 1975), Kister and Gill (1991), Robbins (1991), and Leva (1992). According to Fair (1990), the currently used correlation for packed tower pressure drop prediction—commonly called the Generalized Pressure Drop Correlation (GPDC)—should be attributed to Leva (1954). Other investigators have developed minor improvements. A generalized correlation for estimating pressure drop in structured packings is presented by Bravo et al. (1986). The Eckert (1975) version, which is the basis for the approach given by Strigle (1994), is widely used and is therefore included here.

The Eckert correlation is shown in **Figure 1-8.** The Y axis is called the Flow Capacity Factor and the X axis the Relative Flow Capacity. The flow capacity factor includes a packing factor, F, which is a characteristic of the packing configuration. Leva (1992) provides a simple procedure for calculating packing factor values for any non-irrigated, randomly dumped packing for which non-irrigated pressure drop data are available. However, for most packings acceptable packing factor values are available from the vendor or the open literature. Typical values are listed in **Table 1-7.**

It is normally considered good practice to design for a gas rate that gives a pressure drop of less than about 0.4 inches of water per foot of packing. At high L/G ratios (over about 20), which are encountered in many gas purification absorbers, the pressure drop may exceed the above value but the gas rate should not exceed 85% of the rate that results in a pressure drop of 1.5 inches of water per foot of packing as predicted from **Figure 1-8.** Systems that tend to foam should be operated to give a low pressure drop (e.g., 0.25 in./ft) and vacuum systems may require an even lower pressure drop to minimize overall column pressure drop.

Maximum liquid flow rates recommended by Strigle (1987, 1994) for typical packing sizes and low viscosity liquids are as follows:

Tray Columns

Most tray column design procedures are based on limiting the gas velocity through the available column cross section (A_n) to a value that will not cause flooding or excessive

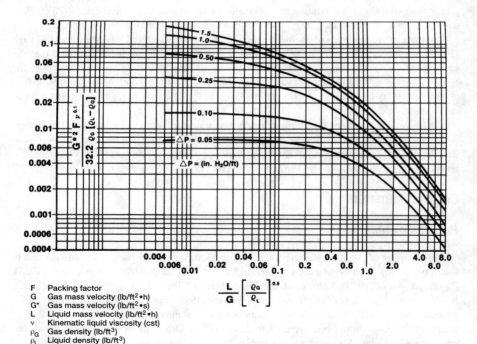

F   Packing factor
G   Gas mass velocity (lb/ft²•h)
G* Gas mass velocity (lb/ft²•s)
L   Liquid mass velocity (lb/ft²•h)
ν   Kinematic liquid viscosity (cst)
ρ_G Gas density (lb/ft³)
ρ_L Liquid density (lb/ft³)

Figure 1-8. Generalized pressure drop correlation for packed towers. *From Strigle (1994)*

Packing Size, in.	Maximum Liquid Loading, gpm/ft²
¾	25
1	40
1½	55
2	60
3½	125

Table 1-7
Value of the Packing Factor, F, for Packed Tower Pressure Drop Correlation

Packing	Material	¼	⅜	½	⅝	¾	1	1¼	1½	2	3	3½
Super Intalox Saddles	C						60		30			
Super Intalox Saddles	P						40		28	18		
Intalox Saddles	C	725	330	200		145	92		52	40	22	
Intalox Saddles	P						33		21	16		
Raschig Rings	C	1600	1000	580	380	255	179	125	93	65	37	
Raschig Rings, ¹⁄₃₂ in.	M	700	390	300	170	155	115		65	45		
Raschig Rings, ¹⁄₁₆ in.	M			410	300	220	144	110	83	57	32	
Berl Saddles	C	900		240		170	110		65	45		
Pall Rings	P				95		55		40	26		17
Pall Rings	M				81		56		40	27		18
Tellerettes	P						35			24		17
Maspac	P						32			24		21
IMTP Packing	M			51			41		24	18	12	12
Intalox Snowflake	P	Diameter 3.7 in., height 1.2 in., F = 13										
Hy-Pak Packing	M						43		26	18	15	
Jaeger Tri-Packs	P						28		16			
Jaeger VSP	M						32		21		12	

Notes: C = Ceramic, P = Plastic, M = Metal, Data compiled from Strigle (1994), Eckert (1970A), Norton Company (1990), Jaeger Products, Inc. (1990), and ASHRAE Handbook (1988)

entrainment. The exact column diameter, tray spacing, and design of column internals are then established on the basis of the liquid and gas properties, flow rates, and special system requirements. The vapor-velocity limitation is usually established from a correlation of the general form:

$$U = K_v \sqrt{\frac{\rho_L - \rho_G}{\rho_G}} \qquad (1\text{-}22)$$

Where: U = allowable superficial vapor velocity, ft/sec
ρ_L = liquid density, lb/cu ft

ρ_G = gas density, lb/cu ft
K_v = empirical constant

This equation was originally proposed by Souders and Brown (1934), on the basis of an analysis of the frictional upward drag of the moving gas stream upon suspended liquid droplets. A number of other factors such as plate design and spacing have been found to affect entrainment; however, equation 1-22 is still widely used as an empirical expression by adjusting K_v to the tray conditions. Typical values of K_v based on a correlation proposed by Fair (1963, 1987) are given in **Table 1-8.** The Flow Parameter, F_v, is defined by the following equation:

$$F_v = \frac{L}{G}\sqrt{\frac{\rho_G}{\rho_L}} \quad (1\text{-}23)$$

Where: L = liquid flow rate, lb/sec
G = gas flow rate, lb/sec

Table 1-8
Typical Design Values of K_v for Sieve, Bubble-Cap, and Valve Plates

Plate Spacing, in.	Kv – When Flow Parameter, Fv is:		
	0.01	0.1	1.0
6	0.15	0.14	0.065
9	0.18	0.17	0.070
12	0.22	0.20	0.079
18	0.30	0.25	0.095
24	0.39	0.33	0.13
36	0.50	0.42	0.15

Based on correlation of Fair (1963, 1987)

The values given are for a liquid with a surface tension, σ, of 20 dynes/cm. The calculated gas velocity, U, may be corrected for other surface tension values by multiplying by the correction factor $(\sigma/20)^{0.2}$. The correlation provides a means for estimating the maximum allowable gas velocity for all types of plate columns subject to the following restrictions:

1. The system is low or nonfoaming.
2. Weir height is less than 15% of the tray spacing.
3. Sieve plate perforations are ¼-in. or less in diameter.
4. The ratio of bubble-cap slot, sieve tray hole, or valve tray full opening area, A_h, to the active tray area, A_a, is 0.1 or greater.

The key column areas involved in the correlation are

A_a = Active area, the area on the tray actively involved in gas/liquid contact, typically the column cross section less two downcomers, sq. ft
A_h = Total slot, perforated, or open valve area on plate, sq. ft
A_n = Net area for vapor flow, typically the column cross section less one downcomer (used for calculating U), sq. ft

Fair (1987) points out that when the ratio of A_h/A_a is smaller than 0.1, jetting occurs because of the high velocity of gas through the tray openings. Fair suggests that the calculated allowable velocity be corrected as follows:

A_h/A_a	U_c/U
0.10	1.00
0.08	0.90
0.06	0.80

where U is the allowable velocity as calculated by the above equations, and U_c is the corrected allowable velocity.

A more accurate (and somewhat more complicated) correlation for predicting entrainment flooding on sieve and valve trays is proposed by Kister and Haas (1990). Their correlation is also derived from the original work of Souders and Brown, but provides a modified approach for determining K_v. They introduce, as an important parameter in the correlation, the clear liquid height at the froth-to-spray-regime transition, and suggest that it be calculated by an equation attributed to Jeronimo and Sawistowski (1973).

General Design Considerations

Packed Columns

Practical guidelines for random-packed tower design are given by Coker (1991) as follows:

1. The ratio of the diameter of the column to the packing should be at least 15:1.
2. Because of deformability, plastic packing is limited to an unsupported height of 10–15 feet, and metal to 20–25 feet.
3. Liquid redistributors are required every 5–10 tower diameters for rings, and at least every 20 feet for all types of dumped packing.
4. The number of liquid streams provided by the feed distributor should be 3–5 per square foot in towers larger than 3 feet in diameter.

Efficient liquid and gas distribution is a key requirement for high performance packed columns. The packed bed will normally cause the liquid flow to spread throughout the column as it flows downward; however, this results in a loss of effective column height.

A detailed study of liquid and gas distribution in commercial packed columns is described by Moore and Rukovena (1986). They conclude that the importance of high performance distributors increases as the stage count per bed increases. Liquid and gas flow rates as well as packing type or size have little or no effect. Practical guidelines for selecting, designing, and installing packed column distributors are given by Bonilla (1993).

Perry et al. (1990) define the following three basic measures of liquid distribution quality:

1. Distribution density (number of liquid streams)
2. Geometric uniformity of distribution points
3. Uniformity of liquid flow in the individual streams

On the basis of available test data they constructed a decision tree to aid in the selection of the distributor type for various column services and operating conditions. The resulting guide indicates that trough-type distributors are generally preferred (over orifice-pan, multipan 2-stage, or spray) for typical absorbers and strippers. The multi-pan 2-stage (MTS) distributor is a high-performance system designed for low liquid rate conditions (less than about 5 gpm per square foot). The development and application of the MTS distributor is described by Killat and Perry (1991).

For structured packing, Shah (1991) ranks liquid distribution at the top of the list of potential trouble spots. He points out that orifice distributors are more generally used for relatively small columns and are not recommended for fouling service or for liquid containing solids. Notched-trough distributors are particularly well suited for fouling service and are also used for large diameter columns, but are extremely sensitive to levelness. Generally, 4 to 7 distribution points are used per square foot (45–75/sq. m) of tower cross sectional area. Details of the design of packed tower internals are given by Chen (1984). Excellent standardized distributors for liquid feeding, and packing support plates that provide effective gas distribution are generally available from major packing suppliers.

Tray Columns

Items to be considered in the design of tray columns include

1. Type of tray (e.g., sieve, valve, or bubble-cap)
2. Tray spacing
3. Number and size of openings for gas flow
4. Dimensions of active area
5. Number of passes
6. Size and location of downcomers and weirs

These items are covered in detail in texts on the subject such as Treybal (1980), Van Winkle (1969), Bolles (1963), Fair (1963), and *Perry's Chemical Engineers' Handbook* (Perry and Green, 1984). Design data and procedures have also been published for specific tray column types. Sieve trays are reviewed by Chase (1967), Economopoulos (1978), and Barnicki and Davis (1989); valve trays by Bolles (1976); and slotted sieve trays by Smith and Delnicki (1975).

A comprehensive algorithm for designing sieve tray towers is presented by Economopoulos (1978). More recently, Barnicki and Davis (1989) authored a two-part article about sieve tray column design, including multipass trays, flow regime effects, and practical cost-effective standardizations. These authors divide column design into four tasks: (1) determining the approximate configuration of each tray, (2) selecting a common diameter for the column and dividing the column into zones of trays with the same number of passes and uniform active tray areas, (3) assigning hole areas for each tray based on pressure drop, structural, and flow regime limitations, and (4) checking each tray for excessive entrainment, entrainment flooding, downcomer backup, and weeping. Typical guidelines for the design of sieve plate columns as compiled by Barnicki and Davis (1989) are given in **Table 1-9**.

Table 1-9
Sieve Tray Design Guidelines

Tray Spacing:

Column Diameter, ft	<3	3–5	5–6	6–12	13–24
Tray Spacing, in.	6–12	18–24	24–30	30–36	35–48

Weir Height:

Should not exceed 15% of tray spacing
Froth regime: 1–4 in. (2 in. is normal)
Spray regime: >¼ in. (½ to ¾ in. is normal)

Downcomer Clearance:

Minimum: ½ in. less than weir height (¼–½ in. is normal)

Hole Diameter:

Typical: 3/16–½ in.

Plate Thickness:

Hole Diameter, in.	Plate Thickness/Hole Diameter	
	Stainless Steel	Carbon Steel
3/16	0.43	1.0
¼	0.32	0.75
⅜	0.22	0.50
½	0.16	0.38

Weir Loading:

Typical: less than 96 gpm/ft

Pressure Drop:

Maximum, 1.5–3.0 in. of liquid for vacuum
8.0–10.0 in. of liquid for one atm or higher

System Derating Factor:

	Factor
Nonfoaming	1.0
Moderate foaming (e.g., absorbers, amine and glycol regenerators)	0.85
Heavy foaming (e.g., amine and glycol absorbers)	0.75

Based on compilation of Barnicki and Davis (1989)

34 *Gas Purification*

The design of downcomers for sieve plate columns is reviewed by Biddulph et al. (1993). These authors present the following four rules of thumb for downcomer sizing based on years of experience:

1. Use a velocity of 1.6 ft/s (0.5 m/s) for liquid flow under the downcomer (based on unaerated liquid).
2. Use the same velocity for liquid flow under the downcomer and liquid flow on the tray to assure a smooth entry.
3. Keep the head loss due to the underflow clearance to no more than 1.0 to 1.5 inches of hot liquid.
4. Allow adequate residence time in the downcomer for the disengagement of vapor; 3 seconds for a nonfoaming system and 6 seconds for a foaming system.

Spray Contactors

Spray contactors can be categorized into two basic types: (1) preformed spray, which includes countercurrent, cocurrent, and crosscurrent spray chambers; spray dryers; cyclonic spray devices; and injector venturis, and (2) gas atomized spray, which consists primarily of venturi scrubbers. Many commercial spray systems use more than one type of spray contactor and often combine sprays with trays or packing.

The correlations developed for predicting the performance of tray and packed towers are not generally applicable to spray contactors because of fundamental differences in the contact mechanism, particularly with regard to "a," the effective area for mass transfer. In spray contactors, the contact area is related more to the number and size of droplets in contact with the gas stream at any time than to the configuration or volume of the contact chamber. Since, in most spray devices, these values are determined primarily by the liquid flowrate and the pressure drop across either the spray nozzles or the venturi throat, it is not surprising that attempts have been made to correlate spray system performance with power consumed in the operation. Such a correlation was originally proposed by Lunde (1958), and a plot that includes his data is reproduced in Chapter 6 (**Figure 6-18**). Roughly, the correlation indicates that to realize two transfer units (overall gas), for example, the total amount of power required is

Contactor	hp/1000 scfm
Venturi scrubbers	2.0
Crosscurrent sprays with mesh	1.0
Spray cyclones	0.5
Packed tower (3-in. Intalox Saddles)	0.2
Spray tower	0.1

Theoretical correlations have been developed for predicting mass transfer rates for both the gas and liquid phases with droplets of known size. Unfortunately these correlations are of little value for design because the droplet size is highly variable and uncertain with commonly used equipment. The problem of design is further complicated by backmixing in the gas phase, which is significant in most spray chambers.

Venturi scrubbers, ejectors, and most spray chambers are, at best, single stage contactors. Cocurrent contactors also fall into this category. The theoretically ideal performance of such units is to produce gas and liquid products that are in equilibrium; actual hardware can only approach this goal. Normally contactors of this type are used when the equilibrium vapor pressure of the absorbate over the product liquid is extremely low and can be neglected. Under these conditions, the equation for the number of transfer units reduces to

$$N_{OG} = \ln (1/(1 - E)) \tag{1-24}$$

where E = absorption efficiency, expressed as a fraction.

In accordance with equation 1-24, removal efficiency and N_{OG} are related as follows:

Removal Efficiency, %	N_{OG}
90	2.3
95	3.0
99	4.6

The N_{OG} values can be used to extrapolate test or operating data on a spray contactor to other systems or conditions. For example, if one spray unit provides 90% removal (2.3 transfer units), it can be expected that two identical units in series will provide about 99% removal (4.6 transfer units). Since commercial spray systems are widely variable, the development of a generalized design approach based on fundamentals is quite difficult. As a result, spray contactors are usually designed on the basis of previous experience with similar systems.

REFERENCES

American Institute of Chemical Engineers, 1958, *Bubble Tray Design Manual, Prediction of Fractionation Efficiency,* NY: AIChE.

American Society of Heating, Refrigerating, and Air-Conditioning Engineers, 1988, *ASHRAE Handbook Equipment Volume,* Chapt. 11, "Industrial Gas Cleaning and Air Pollution Control."

Astarita, G., and Savage D. W., 1980, *Chemical Engineering Science,* Vol. 35, p. 649.

Astarita, G., 1967, *Mass Transfer with Chemical Reactions,* Elsevier, Amsterdam.

Astarita, G., Savage, D. W., and Bisio, A., 1983, *Gas Treating with Chemical Solvents,* John Wiley & Sons, NY.

Barnicki, S. D., and Davis, J. F., 1989, *Chem. Eng.*, Vol. 96, No. 10, October, pp. 140 and 141, November, p. 202.

Biddulph, M. W., 1977, *Hydro. Process.*, Vol. 56, No. 10, October, p. 145.

Biddulph, M. W., Thomas, C. P., and Burton, A. C., 1993, *Chem. Eng. Progr.*, Vol. 89, No. 12, December, p. 56.

Billet, R., 1989, *Packed Tower Analysis and Design,* Ruhr University, Bochum, Germany.

Blecker, H. G., and Nichols, T. M., 1973, *Capital and Operating Costs of Pollution Control Equipment Modules, Data Manual,* Vol. 2., EPA-R5-73-023b, July, PB-224536.

Bolles, W. L., 1963, "Tray Hydraulics—Bubble Cap Trays," Chapt. 14 in *Design of Equilibrium Stage Processes,* B. D. Smith, Ed., McGraw-Hill, New York, NY.

Bolles, W. L., 1976, *Chem. Eng. Prog.*, Vol. 72, No. 9, September, pp. 43–49.

Bolles, W. L., and Fair, J. R., 1982, *Chem. Eng.*, Vol. 89, July 12, p. 109.

Bonilla, J. A., 1993, *Chem. Engr. Prog.*, Vol. 89, No. 3, March, p. 47.

Bravo, J. L., 1993, "Effectively Fight Fouling of Packing," *Chem. Eng. Prog.*, Vol. 89, No. 4, April, p. 72.

Bravo, J. L., 1994, "Design Steam Strippers for Water Treatment," *Chem. Eng. Prog.*, Vol. 90, No. 12, December, p. 56.

Bravo, J. L., and Fair, J. R., 1982, *Ind. Eng. Chem., Process Des. Dev.*, Vol. 21, No. 1, p. 162.

Bravo, J. L., Rocha, J. A., and Fair, J. R., 1986, *Hydro. Process.*, Vol. 65, No. 3, March, p. 45.

Bravo, J. L., Rocha, J. A., and Fair, J. R., 1985, *Hydro. Process.*, Vol. 64, No. 1, January, p. 91.

Brian, P. L. T., 1964, *AIChE J.*, 10:5.

Brian, P. L. T., Hurley, J. F., and Hasseltine, E. H., 1961, *AIChE J.*, Vol. 7, p. 226.

Chase, J. D., 1967, *Chem. Eng.*, Vol. 1, Part 1, July 31, pp. 105–106, Part 2, Aug. 28, pp. 139–146.

Chem. Eng. Prog., 1997, 1997 CEP Software Directory, Supplement to January 1997 issue, Vol. 93, No. 1.

Chen, G. K., 1984, *Chem. Eng.*, Vol. 91, No. 5, March 5, p. 40.

Chen, H., and Fair, J. R., 1984, *Ind. Eng. Chem. Process Des. Dev.*, Vol. 23, Part 1, p. 814, Part 2, p. 820.

Chilton, T. H., and Colburn, A. P., 1935, *Ind. Eng. Chem.*, Vol. 27, p. 255.

Christensen, K. G., and Stupin, W. J., 1978, *Hydro. Process.*, Vol. 57, No. 2, February, p. 125.

Coker, A. K., 1991, *Chem. Eng. Prog.*, Vol. 87, No. 11, November, p. 91.

Colburn, A. P., 1939, *Trans. Am. Inst. Chem. Engrs.*, Vol. 35, p. 211.

Colburn, A. P., 1941, *Ind. Eng. Chem.*, Vol. 33, p. 459.

Danckwerts, P. V., 1970, *Gas Liquid Reactions*, McGraw-Hill Book Company, NY.

Danckwerts, P. V., and Gillham, A. J., 1966, *Trans. Inst. Chem. Engrs.*, London, Vol. 44, p. T42.

Decoursey, W. J., 1974, *Chem. Eng. Sci.*, Vol. 29, p. 1867.

Diab, S. and Maddox, R. N., 1982, *Chem Eng.*, Vol. 89, Dec. 27, p. 38.

Eckert, J. S., Foote, E. H., Rollison, L. R., and Walter, L. F., 1967, *Ind. Eng. Chem.*, Vol. 59, p. 41.

Eckert, J. S., 1970A, *Chem. Eng. Prog.*, Vol. 66, No. 3, February, p. 39.

Eckert, J. S., 1970B, *Oil and Gas J.*, Vol. 60, Aug. 24, p. 39.

Eckert, J. S., 1975, *Chem. Eng.*, Vol. 82, April 14, p. 70.

Economopoulos, A. P., 1978, *Chem Eng.*, Vol. 85, Dec. 4, p. 109.

Edwards, W. M., 1984, "Mass Transfer and Gas Absorption," Sec. 14 in *Perry's Chemical Engineers' Handbook*, 6th ed., McGraw-Hill, NY.

Elgin, J. C., and Weiss, F. B., 1939, *Ind. Eng. Chem.*, Vol. 31, p. 435.

Fair, J. R., 1987, "Distillation," Chapt. 5 in *Handbook of Separation Technology*, R. W. Rousseau, Ed., John Wiley & Sons, NY.

Fair, J. R., 1990, *Chem. Eng. Prog.*, Vol. 86, No. 12, December, p. 102.

Fair, J. R., 1963, Chap. 15 in *Design of Equilibrium Stage Processes*, B. D. Smith, Ed., NY: McGraw-Hill Book Company, Inc.

Fair, J. R., Steinmeyer, D. E., Penney, W. R., and Brink, J. A., 1984, "Liquid-Gas Systems," Sec. 18 in *Perry's Chemical Engineers' Handbook,* 6th ed., McGraw-Hill, NY.

Fair, J. R., 1961, *Petro/Chem Eng.*, Vol. 33, No. 19, October, p. 45.

Fair, J. R., and Bravo, J. L., 1990, *Chem. Eng. Prog.*, Vol. 86, No. 1, January, p. 19.

Frank, O., 1977, *Chem. Eng.*, Vol. 84, No. 6, March 14, pp. 111–128.

Gilliland, E. R., Baddour, R. F., and Brian, P. L. T., 1958, *AIChE J.*, Vol. 4, p. 223.

Hatta, S., 1932, *Technol. Repts.*, Tohoku Univ., Vol. 10, p. 119.

Hatta, S., 1929, *Technol. Repts.*, Tohoku Univ., Vol. 8, p. 1.

Hausch, G. W., and Petschauer, F. J., 1991, "P.C. Based Packed Tower Design Program," presented at the AIChE Summer National Meeting, Pittsburgh, PA, Aug. 20.

Hughmark, G. A., and O'Connell, H. E., 1957, *Chem. Eng. Prog.*, Vol. 53, No. 3, March, p. 127.

Jaeger Products, Inc., 1990, *General Catalog 100,* JP1 2/90 5M.

Jeronimo, M. A. da S., and Sawistowski, H., 1973, *Trans. Inst. Chem. Engineers,* (London), Vol. 51, p. 265.

Joshi, S. V., Astarita, G., and Savage, D. W., 1981, *Transport with Chemical Reactions, AIChE Symposium Series No. 202,* Vol. 77, p. 63.

Kessler, D. P., and Wankat, P. C., 1988, *Chem. Eng.*, Vol. 95, No. 13, Sept. 26, p. 72.

Killat, G. R., and Perry, D., 1991, "A High Performance Distributor for Low Liquid Rates", presented at AIChE Annual Meeting, Nov. 17–22, Los Angeles, CA.

Kister, H. Z., and Gill, D. R., 1991, *Chem. Eng. Prog.*, Vol. 87, No. 2, February, p. 32.

Kister, H. Z., and Haas, J. R., 1990, *Chem. Eng. Prog.*, Vol. 86, No. 9, September, p. 63.

Kohl, A. L., 1987, "Absorption and Stripping," Chap. 6 in *Handbook of Separation Process Technology,* R.W. Rousseau, Ed., John Wiley & Sons, NY.

Kremser, A., 1930, *Natl. Petroleum News,* Vol. 22, May 21, p. 48.

Krishnamurthy, R., and Taylor, R., 1985A, *AIChE Journal,* Vol. 31, pp. 449–465.

Krishnamurthy, R., and Taylor, R., 1985B, *Ind. Eng. Chem. Process Des. Dev.*, Vol. 24, No. 3, p. 513.

Leva, M., 1954, *Chem. Eng. Prog. Symp. Series,* No. 10, Vol. 50, p. 51.

Leva, M., 1992, *Chem. Eng. Prog.*, Vol. 88, No. 1, January, p. 65.

Lobo, W. E., Friend, L., Hashmall, F., and Zenz, F., 1945, *Trans. Am. Inst. Chem. Engrs.*, Vol. 41, p. 693.

Lockett, M. J., 1986, *Distillation Tray Fundamentals,* Cambridge University Press, NY.

Lunde, K. E., 1958, *Ind. Eng. Chem.*, Vol. 50, No. 3, March, p. 293.

Matheron, E. R., and Sandall, O. C., 1978, *Am. Inst. Chem. Engrs. Journal,* Vol. 24, No. 3, p. 552.

McInnes, R., Jelinek, S., and Putsche, V., 1990, *Chem. Eng.* Vol. 97, No. 9, September, p. 108

McKee, R. L., Changela, M. K., and Reading, G. L., 1991, *Hydro. Process.*, Vol. 70, No. 4, April, p. 63.

Moore, F., and Rukovena, F., 1986, "Liquid and Gas Distribution in Commercial Packed Towers," presented at 36th Canadian Chemical Engineering Conference, Paper 23b, Oct. 5–8.

Murphree, E. V., 1935, *Ind. Eng. Chem.*, Vol. 17, p. 747.

Norton Co., 1990, Akron, OH, *Bulletin 1SPP-1R, 2M-150014302-5/90.*
O'Connell, H. E., 1946, *Trans. Am. Inst. Chem. Engrs.*, Vol. 42, p. 741.
Olander, D. R., 1960, *Am. Inst. Chem. Engrs. Journal,* Vol. 6, No. 2, p. 233.
Onda, K., Takeuchi, H., and Okumoto, Y., 1968, *Journal ChE,* Japan, Vol. 1, No. 1, p. 56.
Ouwerkerk, C., 1978, *Hydro. Process.*, Vol. 57, No. 4, p. 89.
Perry, R. H., and Pigford, R. L., 1953, *Ind. Eng. Chem.*, Vol. 45, p. 1247.
Perry, D., Nutter D. E., and Hale, A., 1990, *Chem. Eng. Prog.*, Vol. 86, No. 1, January, p. 30.
Perry, R. H., and Green, D. W., Eds., 1984, *Perry's Chemical Engineer's Handbook,* 6th ed., McGraw Hill, NY.
Robbins, L. A., 1991, *Chem. Eng. Prog.*, Vol. 87, No. 5, May, p. 87.
Rousseau, R. W., and Staton, J. S., 1988, *Chem. Eng.*, Vol. 95, No. 10, July 18, p. 91.
Ruddy, R. N., and Carroll, L. A., 1993, *Chem. Eng. Prog.*, Vol. 89, No. 7, July, p. 28.
Sardar, H., Sivasubramanian, M. S., and Weiland, R. H., 1985, "Simulation of Commercial Amine Treating Units," *Proc. Laurance Reid Gas Conditioning Conference,* University of Oklahoma, March 4–6, Norman, OK.
Savage, D. W., Astarita, G., and Joshi, S., 1980, *Chem. Eng. Sci.,* Vol. 35, p. 1513.
Seader, J. D., 1989, *Chem. Eng. Prog.*, Vol. 85, No. 10, October, p. 41.
Shah, G. C., 1991, *Chem. Eng. Prog.*, Vol. 87, No. 11, November, p. 49.
Sherwood, T. K., and Pigford, R. L., 1952, *Absorption and Extraction,* 2nd ed. NY, McGraw-Hill Book Company, Inc.
Sherwood, T. K., Pigford, R. L., and Wilke, L. G., 1975, *Mass Transfer,* NY, McGraw-Hill Book Company, Inc.
Sherwood, T. K., Shipley, G. H., and Holloway, F. A. L., 1938, *Ind. Eng. Chem.*, Vol. 30, p. 765.
Smith, V. C., and Delnicki, W. V., 1975, *Chem. Eng. Prog.*, Vol. 71, No. 8, August, pp. 68–73.
Souders, M., and Brown, G. G., 1934, *Ind. Eng. Chem.*, Vol. 26, p. 98.
Souders, M., and Brown, G. G., 1932, *Ind. Eng. Chem.*, Vol. 24, p. 19.
Spiegel, L., and Meier, W., 1987, *Int. Chem. Eng. Symposium Series,* Vol. 104, p. A203.
Stoley, A. W., and Martin, G. R., 1995, "Subdue Solids in Towers," *Chem. Eng. Prog.,* Vol. 91, No. 1, January, p. 64.
Strigle, R. F., Jr., 1994, *Packed Tower Design and Applications,* Gulf Publishing Co., Houston, TX.
Strigle, R. F., Jr., 1987, *Random Packings and Packed Towers,* Gulf Publishing Co., Houston, TX.
Sujata, A. D., 1961, *Hydro. Process.*, Vol. 40, No. 12, December, p. 137.
Tennyson, R. H., and Schaaf, R.P., 1977, *Oil & Gas J.,* Jan. 11, p. 78.
Tomcej, R. A., and Otto, F. D., 1986, "Improved Design of Amine Treating Units by Simulation using Personal Computers," presented at the World Congress III of Chemical Engineering, Sept. 21–25, Tokyo, Japan.
Tomcej, R. A., Otto, F. D., Rangwala, H. A., and Merrell, B. R., 1987, "Tray Design for Selective Absorption," *Proc. Laurence Reid Gas Conditioning Conf.,* University of Oklahoma, Norman, OK.

Treybal, R. E., 1980, *Mass Transfer Operations,* 3rd ed., McGraw-Hill, New York, NY.

Van Krevelen, D. W., and Hoftijzer, P. J., 1948, *Chem. Eng. Prog.*, Vol. 44, p. 529.

Van Winkle, M., 1969, *Distillation,* McGraw-Hill, New York, NY.

Vickery, D. J., Adams, J. T., and Wright, R. D., 1992, "The Effect of Tower Parameters on Amine Based Gas Sweetening Plants," *Proc. 42nd Annual Laurance Reid Gas Conditioning Conference,* March 2–4, University of Oklahoma, Norman, OK.

Vivian, J. E., and King, C. J., 1963, *Modern Chemical Engineering,* Vol. 1., Edited by A. Acrivo, NY: Reinhold Publishing Corp.

von Stockar, U., and Wilke, C. R., 1977, *Ind. Eng. Chem. Fundam.,* Vol. 16, No. 1, p. 44

Walker, W. H., Lewis, W. K., McAdams, W. H., and Gilliland, E. R., 1937, *Principles of Chemical Engineering,* 3rd ed., NY: McGraw-Hill Book Company, Inc.

Weiland, R. H., Rawal, M., and Rice, R. G., 1982, *Am. Inst. Chem. Engrs. J.,* Vol. 28, No. 6, p. 963.

White, G. E., 1940, *Trans. Inst. Chem. Engrs.*, Vol. 36, p. 359.

Whitman, W. G., 1923, *Chem. & Met. Eng.*, Vol. 29, p. 147.

Wilke, C. R., and von Stockar, U., 1978, "Absorption" in *Encyclopedia of Chemical Technology,* Vol. 1, 3rd edition, Kirk-Othmer, Eds., Wiley, NY.

Yanagi, T., and Sakata, M., 1981, 90th National AIChE Meeting, Symp. 44, Houston, TX, April.

Zenz, F. A., 1979, "Design of Gas Absorption Towers," Sect. 3.2 in *Handbook of Separation Techniques for Chemical Engineers,* P.A. Schweitzer, Ed., McGraw-Hill, NY.

Zuiderweg, F. J., 1982, *Chem. Eng. Sci.*, Review Article No. 9, Vol. 37, No. 10, p. 1441.

Chapter 2

Alkanolamines for Hydrogen Sulfide and Carbon Dioxide Removal

BACKGROUND, 41

BASIC CHEMISTRY, 42

SELECTION OF PROCESS SOLUTION, 48

    Monoethanolamine, 49
    Monoethanolamine-Glycol Mixtures, 50
    Diethanolamine, 50
    Diglycolamine, 51
    Diisopropanolamine, 53
    Methyldiethanolamine, 53
    Mixed Amines, 54
    Sterically Hindered Amines, 56
    Amine Concentration, 56

FLOW SYSTEMS, 57

    Basic Flow Scheme, 57
    Water Wash for Amine Recovery, 58
    Split-Stream Cycles, 59
    Cocurrent Absorption, 60

DESIGN DATA, 62

    Acid Gas-Amine Solution Equilibria, 62
    Amine Solution Vapor Pressures, 91
    Heats of Reaction, 91
    Physical Properties, 98

PROCESS DESIGN, 103

Design Approach, 103
Computer Programs, 110
Tray Versus Packed Columns, 111
Column Diameter, 112
Column Height, 113
Absorber Thermal Effects, 120
Stripping System Performance, 123
Simplified Design Procedure, 133
Commercial Plant Operating Data, 144
Organic Sulfur Removal by Amine Solutions, 151

AMINE TREATMENT OF LIQUID HYDROCARBONS, 156

Process Description, 156
Design Data, 157
LPG Treater Operating Conditions, 165
Amine Solution Flow Rates and Composition, 165
Absorber Designs, 166
Auxiliary Systems, 171
Removal of COS from LPG by Amines, 173

REFERENCES, 174

BACKGROUND

Credit for the development of alkanolamines as absorbents for acidic gases goes to R. R. Bottoms (1930), who was granted a patent covering this application in 1930. Triethanolamine (TEA), which was the first alkanolamine to become commercially available, was used in the early gas-treating plants. As other members of the alkanolamine family were introduced into the market, they were also evaluated as possible acid-gas absorbents. As a result, sufficient data are now available on several of the alkanolamines to enable design engineers to choose the most suitable compound for each particular requirement.

The amines that have proved to be of principal commercial interest for gas purification are monoethanolamine (MEA), diethanolamine (DEA), and methyldiethanolamine (MDEA). Triethanolamine has been displaced largely because of its low capacity (resulting from higher equivalent weight), its low reactivity (as a tertiary amine), and its relatively poor stability. Diisopropanolamine (DIPA) (Bally, 1961; Klein, 1970) is being used to some extent in the Adip process and in the Sulfinol process (see Chapter 14), as well as in the SCOT process for Claus plant tail gas purification (see Chapter 8). However, methyldiethanolamine (MDEA) is gradually displacing DIPA in these applications. Although MDEA was described by Kohl and coworkers at Fluor Daniel (Frazier and Kohl, 1950; Kohl, 1951; Miller and Kohl, 1953) as a selective absorbent for H_2S in the presence of CO_2 as early as 1950, its use in industrial processes has only become important in recent years. A somewhat different type of alkanolamine, 2-(2-aminoethoxy) ethanol, commercially known as Diglycolamine (DGA), was first proposed by Blohm and Riesenfeld (1955). This compound couples the stability and reactivity of monoethanolamine with the low vapor pressure and hygroscopicity of diethylene glycol and, therefore, can be used in more concentrated solutions than monoethanolamine.

In addition to simple aqueous solutions of the previously mentioned alkanolamines, proprietary formulations comprising mixtures of the amines with various additives are widely used. Formulated solvents are offered by: Dow Chemical Company (GAS/SPEC), UOP (and/or Union Carbide Corp.) (Amine Guard and UCARSOL), Huntsman Corporation (formerly Texaco Chemical Company) (TEXTREAT), and BASF Aktiengesellschaft (Activated MDEA). Some of Dow's GAS/SPEC and UOP's Amine Guard formulations are basically corrosion inhibited MEA and DEA solutions. However, the most significant development in formulated solvents is the advent of tailored amine mixtures. These are usually based on MDEA, but contain other amines as well as corrosion inhibitors, foam depressants, buffers, and promoters blended for specific applications. They can be designed to provide selective H_2S removal, partial or complete CO_2 removal, high acid gas loading, COS removal, and other special features (Manning and Thompson, 1991; Pearce and Wolcott, 1986; Thomas, 1988; Meissner and Wagner, 1983; Meissner, 1983; Niswander et al., 1992).

A different class of acid gas absorbents, the sterically hindered amines, has recently been disclosed by EXXON Research and Engineering Company (Anon., 1981; Goldstein, 1983; Sartori and Savage, 1983). These absorbents, some of which are not alkanolamines, use steric hinderance to control the CO_2/amine reaction. Several different solutions are offered under the general name of Flexsorb solvents.

Typical ethanolamine gas-treating plants are shown in **Figures 2-1, 2-2a, 2-2b,** and **2-3. Figure 2-1** is a photograph of a unit treating natural gas at high pressure to pipeline specifications using an aqueous diethanolamine solution (S.N.P.A.–DEA process). **Figures 2-2a** and **2-2b** depict a large gas treating complex (4 × 540 MMscfd Improved Econamine gas treating trains) located in Saudi Arabia which uses Diglycolamine as the solvent. **Figure 2-3** depicts another natural gas-treating plant using Diglycolamine.

BASIC CHEMISTRY

Structural formulas for the alkanolamines previously mentioned are presented in **Figure 2-4.** Each has at least one hydroxyl group and one amino group. In general, it can be considered that the hydroxyl group serves to reduce the vapor pressure and increase the water solubility, while the amino group provides the necessary alkalinity in water solutions to cause the absorption of acidic gases.

Amines which have two hydrogen atoms directly attached to a nitrogen atom, such as monoethanolamine (MEA) and 2-(2-aminoethoxy) ethanol (DGA), are called primary amines and are generally the most alkaline. Diethanolamine (DEA) and Diisopropanolamine (DIPA) have one hydrogen atom directly attached to the nitrogen atom and are called secondary amines. Triethanolamine (TEA) and Methyldiethanolamine (MDEA) represent completely substituted ammonia molecules with no hydrogen atoms attached to the nitrogen, and are called tertiary amines.

The principal reactions occurring when solutions of a primary amine, such as monoethanolamine, are used to absorb CO_2 and H_2S may be represented as

Figure 2-1. High-pressure natural gas-treating plant using diethanolamine solution (S.N.P.A.-DEA process). *Courtesy of The Parsons Corp.*

Ionization of water:

$$H_2O = H^+ + OH^- \tag{2-1}$$

Ionization of dissolved H_2S:

$$H_2S = H^+ + HS^- \tag{2-2}$$

Hydrolysis and ionization of dissolved CO_2:

$$CO_2 + H_2O = HCO_3^- + H^+ \tag{2-3}$$

Protonation of alkanolamine:

$$RNH_2 + H^+ = RNH_3^+ \tag{2-4}$$

Carbamate formation:

$$RNH_2 + CO_2 = RNHCOO^- + H^+ \tag{2-5}$$

(*text continued on page 46*)

Figure 2-2a. Shedgum, Saudi Arabia, gas treating complex. 4 × 540 MMscfd Improved Econamine trains using DGA on right, four Claus plants with incinerator stacks on left. For each Improved Econamine train, contactor is on right, regenerator on left. From left, air cooler sequence is regenerator condenser, lean DGA cooler No. 1, lean DGA cooler No. 2, and contactor side cooler. *Courtesy Fluor Daniel*

Figure 2-2b. Individual 540 MMscfd Improved Econamine gas treating train at Shedgum, Saudi Arabia. Contactor on right, regenerator with four vertical thermosyphon reboilers on left. *Courtesy Fluor Daniel*

Figure 2-3. High-pressure gas-treating plant using Diglycolamine solution (Fluor Econamine process). *Courtesy of Fluor Daniel*

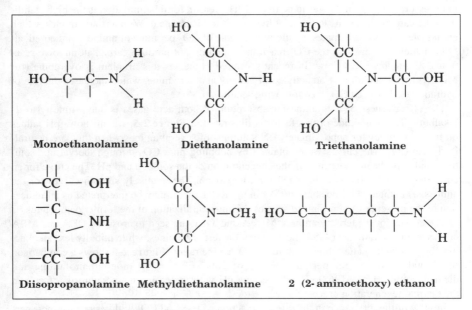

Figure 2-4. Structural formulas for alkanolamines used in gas treating.

(*text continued from page 43*)

Reactions 2-1, 2-3, 2-4, and 2-5 account for the principal species present in aqueous alkanolamine treating solutions. These species are the unionized molecules H_2O, H_2S, CO_2, and RNH_2 and the ions H^+, OH^-, HS^-, HCO_3^-, RNH_3^+, and $RNHCOO^-$. Alternative reaction paths can, of course, be postulated which relate to the same species.

Additional reactions may occur to produce species other than those listed, but these are not considered important in the basic absorption/desorption operation. Examples of such minor reactions are the dissociation of bisulfide to produce sulfide ions, the dissociation of bicarbonate to produce carbonate ions, and the reaction of carbon dioxide with some amines to produce nonregenerable compounds. Additional details with regard to chemical reactions involved in the absorption of H_2S and CO_2 are given in a subsequent section of this chapter entitled "Acid Gas—Amine Solution Equilibrium Correlations."

Although reactions 2-1 through 2-5 relate specifically to primary amines, such as MEA, they can also be applied to secondary amines, such as DEA, by suitably modifying the amine formula. Tertiary amine solutions undergo reactions 2-1 through 2-4, but cannot react directly with CO_2 to form carbamates by reaction 2-5.

The equilibrium concentrations of molecular H_2S and CO_2 in solution are proportional to their partial pressures in the gas phase (i.e., Henry's law applies) so reactions 2-2, 2-3, and 2-5 are driven to the right by increased acid gas partial pressure. The reaction equilibria are also sensitive to temperature, causing the vapor pressures of absorbed acid gases to increase rapidly as the temperature is increased. As a result it is possible to strip absorbed gases from amine solutions by the application of heat.

If the reaction of equation 2-5 is predominant, as it is with primary amines, the carbamate ion ties up an akanolammonium ion via equation 2-4 and the capacity of the solution for CO_2 is limited to approximately 0.5 mole of CO_2 per mole of amine, even at relatively high partial pressures of CO_2 in the gas to be treated. The reason for this limitation is the high stability of the carbamate and its low rate of hydrolysis to bicarbonate. With tertiary amines, which are unable to form carbamates, a ratio of one mole of CO_2 per mole of amine can theoretically be achieved. However, the CO_2 reactions which do not produce carbamate involve reaction 2-3, which is very slow. In recently offered processes this problem is overcome (for MDEA) by the addition of an activator, typically another amine, which increases the rate of hydration of dissolved CO_2 (see following section).

The effectiveness of any amine for absorption of both acid gases is due primarily to its alkalinity. The magnitude of this factor is illustrated in **Figure 2-5,** which shows pH values on titration curves for approximately 2N solutions of several amines when they are neutralized with CO_2. The curves were obtained by bubbling pure CO_2 through the various solutions and periodically determining the concentration of the solution and pH. The curve for an equivalent KOH solution is included for comparison. The relatively smooth curves for the amines, as compared to the sharp breaks in the KOH curve, may be interpreted as an indication of the presence of non-ionized species during neutralization of the former compounds.

The curves for the tertiary amines, MDEA and TEA, are seen to cross the DEA and MEA curves at a mole ratio near 0.5 indicating that the tertiary amines, while initially less alkaline, may be expected to attain higher ultimate CO_2/amine ratios. **Figure 2-6** shows a comparison of pH values versus temperature curves of 20% solutions of monoethanolamine and diethanolamine (Dow, 1962). The decreasing pH with increasing temperature is a factor in the thermal regeneration process.

In view of the difference in the rates of reaction of H_2S and CO_2 with tertiary amines, partially selective H_2S absorption would be expected with these compounds. The kinetics of

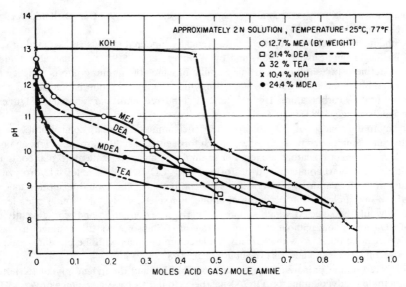

Figure 2-5. Titration curves showing pH during neutralization of ethanolamine and KOH solutions with CO_2.

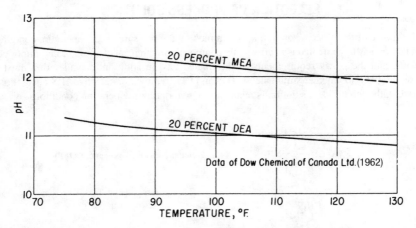

Figure 2-6. pH values of aqueous mono- and diethanolamine solutions (technical grade).

H_2S and CO_2 absorption into aqueous solutions of MDEA has been studied by a number of investigators (Savage et al., 1981; Ouwerkerk, 1978; Blanc and Elgue, 1981). Savage et al. (1981) found that although the rate of H_2S absorption could be thermodynamically predicted, the rate of CO_2 absorption, measured experimentally, appreciably exceeded that predicted on the basis of thermodynamic considerations, and they concluded that MDEA apparently acts as a base catalyst for hydration of CO_2. More recently, investigators have found that the rate of CO_2 absorption in MDEA can be significantly increased by the addition of relatively low

48 Gas Purification

concentrations of primary or secondary amines (Polasek et al., 1990; Campbell and Weiland, 1989; Katti and Wolcott, 1987; Vickery et al., 1988).

The chemistry of acid gas reactions with sterically hindered amines is discussed in some detail by Sartori and Savage (1983) and by Weinberg et al. (1983). A sterically hindered amine is defined structurally as a primary amine in which the amino group is attached to a tertiary carbon atom or a secondary amine in which the amino group is attached to a secondary or tertiary carbon atom. Typical sterically hindered amines are shown in **Figure 2-7** (Sartori and Savage, 1983).

The key to the concept of CO_2 absorption by such amines is that, by control of the molecular structure, amines can be synthesized which form either a stable carbamate ion, an unstable carbamate ion, or no carbamate ion. For example, by an appropriate molecular configuration an unstable carbamate would be formed with CO_2 which is readily hydrolyzable, resulting in the formation of bicarbonate as the end product. This would result in a theoretical ratio of one mole of CO_2 per mole of amine. For selective H_2S absorption, a molecular structure would be selected which suppresses carbamate formation and, consequently, the rate of CO_2 absorption, without affecting the rate of H_2S absorption. It is claimed that better selectivity can be obtained with sterically hindered amines than with the presently used tertiary or secondary alkanolamines (Weinburg et al., 1983).

A hindered amine prepared from tertiary butylamine and diethylene glycol (N-tertiary butyl diethylene glycolamine or TBGA) has been found to have advantages over MDEA with regard to selectivity, acid gas loading, and product gas purity. However, the material is still too expensive for general commercial use (Cai and Chang, 1992).

SELECTION OF PROCESS SOLUTION

The choice of the process solution is determined by the pressure and temperature conditions at which the gas to be treated is available, its composition with respect to major and minor constituents, and the purity requirements of the treated gas. In addition, consideration must, of course, be given to whether simultaneous H_2S and CO_2 removal or selective H_2S absorption is desired. Although no ideal solution is available to give optimum operating conditions for each

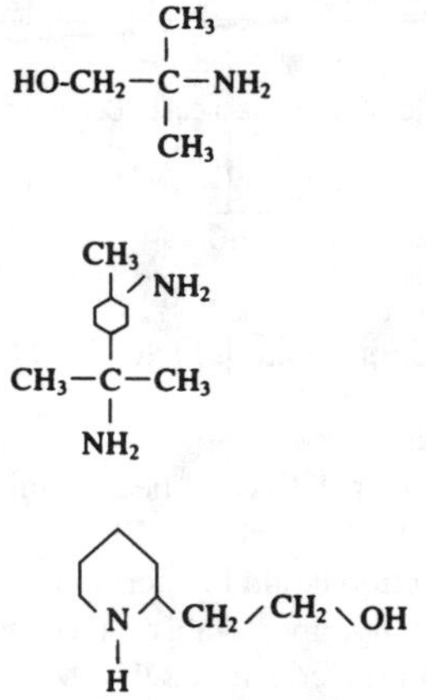

Figure 2-7. Examples of sterically hindered amines. (*Sartori and Savage, 1983*)

Monoethanolamine

Aqueous monoethanolamine solutions, which were used almost exclusively for many years for the removal of H_2S and CO_2 from natural and certain synthesis gases, are rapidly being replaced by other more efficient systems, particularly for the treatment of high-pressure natural gases. However, monoethanolamine is still the preferred solvent for gas streams containing relatively low concentrations of H_2S and CO_2 and essentially no minor contaminants such as COS and CS_2. This is especially true when the gas is to be treated at low pressures, and maximum removal of H_2S and CO_2 is required. The low molecular weight of monoethanolamine, resulting in high solution capacity at moderate concentrations (on a weight basis), its high alkalinity, and the relative ease with which it can be reclaimed from contaminated solutions (see Chapter 3) are advantages, which in many cases more than counterbalance inherent disadvantages. Among the latter, the most serious one is the forma-

case, sufficient data and operating experience with several alkanolamines are on hand to permit a judicious selection of the treating solution for a wide range of conditions. In many cases, process requirements can be met by a number of different amines (or other processes) and an economic analysis is required. A comparison of alkanolamines used for gas purification, based on selected physical properties and approximate cost, is shown in **Table 2-1**.

Table 2-1
Physical Properties of Alkanolamines

Property	MEA*	DEA*	TEA*	MDEA*	DIPA*	DGA**
Mol. weight	61.09	105.14	149.19	119.17	133.19	105.14
Specific gravity, 20/20°C	1.0179	1.0919 (30/20°C)	1.1258	1.0418	0.9890 (45/20°C)	1.0550
Boiling point, °C						
760 mmHg	171	decomp.	360	247.2	248.7	221
50 mmHg	100	187	244	164	167	—
10 mmHg	69	150	208	128	133	—
Vapor pressure, mmHg at 20°C	0.36	0.01	0.01	0.01	0.01	0.01
Freezing point, °C	10.5	28.0	21.2	−21.0	42	−9.5
Solubility in water, % by weight at 20°C	Complete	96.4	Complete	Complete	87	Complete
Absolute viscosity, cps at 20°C	24.1	380(30°C)	1,013	101	198(45°C)	26(24°C)
Heat of vaporization, Btu/lb at 1 atm	355	288(23 mm) (168.5°C)	230	223	184.5	219.1
Approximate cost, $/lb***	0.59	0.60	0.61	1.40	—	0.93

Notes:
 *Data of Union Carbide Chemicals Company (1957) except for pricing.
 **Data of Jefferson Chemical Company, Inc. (1969) except for pricing.
 ***Kenney (1995). Prices are for bulk sales. Add $0.10 per pound for drum sales.

tion of irreversible reaction products with COS and CS_2, resulting in excessive chemical losses if the gas contains significant amounts of these compounds. Furthermore, monoethanolamine solutions are appreciably more corrosive than solutions of most other amines, particularly if the amine concentrations exceed 20% and the solutions are highly loaded with acid gas. This feature limits the capacity of monoethanolamine solutions in cases where high partial pressures of the acid gases would permit substantially higher loadings. However, several systems, using effective corrosion inhibitors, reportedly overcome these limitations. Such systems include Dow Chemical Company's GAS/SPEC FT-1 technology, which is suitable for CO_2 removal in ammonia and hydrogen plants, as well as from sweet natural gas streams (Dow, 1983), and UOP's Amine Guard Systems (Butwell et al., 1973, 1979; Kubek and Butwell, 1979). In general, corrosion inhibitors are effective in CO_2 removal systems, permitting MEA concentrations as high as 30% to be used. However, they have not proven to be reliable in preventing corrosion with CO_2/H_2S mixtures.

Another disadvantage of MEA is its high heat of reaction with CO_2 and H_2S (about 30% higher than DEA for both acid gases). This leads to higher energy requirements for stripping in MEA systems. Finally, the relatively high vapor pressure of monoethanolamine causes significant vaporization losses, particularly in low-pressure operations. However, this difficulty can be overcome by a simple water wash treatment of the purified gas.

Monoethanolamine-Glycol Mixtures

Mixtures of monoethanolamine with di- or triethylene glycol, as first described by Hutchinson (1939), were once used extensively for simultaneous acid-gas removal and dehydration of natural gases. This process, commonly known as the glycol-amine process, has as its principal advantages the features of simultaneous purification and dehydration and somewhat lower steam consumption when compared to aqueous systems. Furthermore, glycol-amine solutions can be stripped almost completely of H_2S and CO_2, resulting in the capability of producing extremely high purity treated gas. However, the glycol-amine process has a number of drawbacks which have seriously limited its usefulness. Probably the most important of these is the fact that, in order to be effective as a dehydrating agent, the water content of the solution has to be kept at or below 5%, requiring relatively high reboiler temperatures. At these temperatures rather severe corrosion occurs in the amine to amine heat exchangers, the stripping column, and, under certain operating conditions, the reboiler. The only practical solution to the corrosion problem is the utilization of corrosion-resistant ferrous alloys or nonferrous metals. Another undesirable feature of the glycol-amine process is a high vaporization loss of the amine. Furthermore, because of the very low vapor pressure of the glycol, a contaminated glycol-amine solution cannot be reclaimed by simple distillation as is possible with the aqueous system. Finally, hydrocarbons, especially aromatics, are substantially more soluble in glycol-amine than in aqueous amine solutions. This feature is of major importance if the acid gas is to be further processed in a Claus type sulfur plant, as the presence of high molecular weight hydrocarbons usually leads to rapid catalyst deactivation and production of discolored sulfur. As a result of these limitations and problem areas, the glycol-amine process is no longer considered competitive. Details of this process are discussed in earlier editions of this text.

Diethanolamine

Aqueous solutions of diethanolamine (DEA) have been used for many years for the treatment of refinery gases which normally contain appreciable amounts of COS and CS_2, besides

H_2S and CO_2. As discussed in Chapter 3, secondary amines are much less reactive with COS and CS_2 than primary amines, and the reaction products are not particularly corrosive. Consequently, diethanolamine and other secondary amines are the better choice for treating gas streams containing COS and CS_2. The low vapor pressure of diethanolamine makes it suitable for low-pressure operations as vaporization losses are quite negligible. One disadvantage of diethanolamine solutions is that the reclaiming of contaminated solutions may require vacuum distillation. Another disadvantage of DEA is that DEA undergoes numerous irreversible reactions with CO_2, forming corrosive degradation products, and for that reason, DEA may not be the optimum choice for treating gases with a high CO_2 content (see Chapter 3).

Application of diethanolamine solutions to the treatment of natural gas was first disclosed by Bertheir (1959) and later described in more detail by Wendt and Dailey (1967), Bailleul (1969), and Daily (1970). This process, which is commonly known as the S.N.P.A.-DEA process, was developed by Societe Nationale des Petroles d'Aquitaine (S.N.P.A.)[1] of France in the gas field at Lacq in southern France. S.N.P.A. recognized that relatively concentrated aqueous diethanolamine solutions (25 to 30% by weight) can absorb acid gases up to stoichiometric molar ratios as high as 0.70 to 1.0 mole of acid gas per mole of DEA, provided that the partial pressure of the acid gases in the feed gas to the plant is sufficiently high. If the regenerated solution is well enough stripped when returned to the absorber and the operating pressure is high, purified gas satisfying pipeline specifications can be produced. The presence of impurities such as COS and CS_2 is not injurious to the solution. Under normal operating conditions, DEA decomposition products are removed quite easily by filtration through activated carbon. In general, diethanolamine solutions are less corrosive than monoethanolamine solutions unless corrosive decomposition products from side reactions build up in the solution (see Chapter 3).

As a result of S.N.P.A.'s experience in Lacq, the S.N.P.A.-DEA process has been widely used for the treatment of high-pressure natural gases with high concentrations of acidic components, especially if COS and CS_2 are also present in appreciable amounts. Beddome (1969) reported that in 1969 the S.N.P.A.-DEA process predominated for the recovery of sulfur from natural gas in Alberta, Canada. Comparative operating data for mono- and diethanolamine systems, as reported by Beddome (1969), for typical Canadian gas-treating plants are shown in **Table 2-2**. Although not stated in the article, it is assumed that all plants were operating at a pressure of about 1,000 psig, which is typical for Canadian operation.

Diglycolamine

The use of aqueous solution of Diglycolamine, 2-(2-aminoethoxy) ethanol, was commercialized jointly by the Fluor Corporation (now Fluor Daniel), the El Paso Natural Gas Company, and the Jefferson Chemical Company Inc. (now the Huntsman Corporation) (Holder, 1966; Dingman and Moore, 1968). The process employing this solvent has been named the Fluor Econamine process. The solvent is in many respects similar to monoethanolamine, except that its low vapor pressure permits its use in relatively high concentrations, typically 40 to 60%, resulting in appreciably lower circulation rates and steam consumption when compared to typical monoethanolamine solutions. A comparison of operating data for glycol-monoethanolamine and Diglycolamine solutions in a commercial installation, which treats natural gas containing 2 to 5% total acid gas at a pressure of 850 psig is shown in **Table 2-3,** Holder (1966).

---

[1]Now Societe Nationale Elf Aquitaine (Production) (SNEAP).

Table 2-2
Comparative Operating Data for MEA and DEA Systems

Gas Plant	A	B	B	C	D
Feed gas composition					
Mole % H_2S	2.1	7.1	7.1	2.4	16.5
Mole % CO_2	0.7	5.9	5.9	4.9	8.0
Solvent					
(% active reagent in water solution)	18% MEA	15% MEA	24% SNPA-DEA	22.5% DEA	27.5% SNPA-DEA
Solvent circulation					
Moles amine per mole acid gas	1.8	2.5	1.3	1.5	1.0
Gallons solvent per mole acid gas	74	123	68	84	44
Reboiler steam					
lb steam/gal solvent	1.0	1.2	1.5	1.2	1.0
lb steam/mole acid gas	74	148	72	101	44

Source: Beddome (1969)

Table 2-3
Comparison of Typical Operating Data of MEA-DEG and DGA Systems

	MEA-DEG	DGA
Gas volume, MMscfd	121.2	121.3
Solution rate, gpm	714	556
Reboiler steam, lb/hr	50,700	40,100
Solution loading, scf acid gas/gal	4.0	5.5
H_2S in treated gas, grain/100 scf	0.25	0.25
CO_2 in treated gas, Mol%	0.01	0.01

Source: Holder (1966)

DGA has proven to be very effective for purifying large volumes of low pressure (~100–200 psig) associated gas in Saudi Arabia. DGA is particularly useful for such applications because it can operate at high ambient temperatures and can produce sweet gas (<¼ grain H_2S/100 scf) at moderate pressures. Information on commercial applications of the Fluor Econamine process has been presented by Dingman (1977), Mason and Griffith (1969), Huval and van de Venne (1981), Bucklin (1982), and Weber and McClure (1981). Comparison of the process with systems using MEA solutions indicates some capital and operating cost savings, as well as improved operation at relatively low pressures (Huval and van de Venne, 1981). An additional advantage is partial removal of COS by the DGA solution. Furthermore, steam distillation can be used to recover a substantial portion of DGA from the degradation products resulting from reactions of DGA with CO_2 and COS (see Chapter 3). In 1996, Diglycolamine solutions were being used in more than 100 plants (Kenney, 1996).

Diisopropanolamine

Diisopropanolamine (DIPA) has been used in the ADIP and Sulfinol processes, both licensed by the Shell International Petroleum Company (SIPM). In the Sulfinol process, diisopropanolamine is used in conjunction with a physical organic solvent, and a more detailed discussion of this process is given in Chapter 14. The ADIP process, which employs relatively concentrated aqueous solutions of diisopropanolamine, has been described by Bally (1961) and by Klein (1970). It has been widely accepted, primarily in Europe, for the treatment of refinery gases and liquids which, besides H_2S and CO_2, also contain COS. It is claimed that substantial amounts of COS are removed without detrimental effects to the solution. Furthermore, diisopropanolamine solutions are reported to have low regeneration steam requirements and to be noncorrosive (Klein, 1970). SIPM has applied the ADIP process to the selective absorption of H_2S from refinery gas streams (Abe and Peterzan, 1980) and, as part of the SCOT process, to selective absorption of H_2S from Claus plant tail gas (see Chapter 8). However, SIPM is gradually replacing DIPA with MDEA in both of these applications. A theoretical study of the absorption kinetics involved in the selective absorption of H_2S in DIPA has been presented by Ouwerkerk (1978). Equations for mass transfer with chemical reaction are utilized in the study to develop a computer program which takes into account the competition between H_2S and CO_2 when absorbed simultaneously.

Methyldiethanolamine

Selective absorption of hydrogen sulfide in the presence of carbon dioxide, especially in cases where the ratio of carbon dioxide to hydrogen sulfide is very high, has recently become the subject of considerable interest, particularly in the purification of non-hydrocarbon gases such as the products from coal gasification processes and Claus plant tail gas. The early work at the Fluor Corp. (now Fluor Daniel) showed that tertiary amines, especially methyldiethanolamine, can absorb hydrogen sulfide reasonably selectively under proper operating conditions involving short contact times (Frazier and Kohl, 1950; Kohl, 1951; Miller and Kohl, 1953). A study by Vidaurri and Kahre (1977), in which selective absorption with several ethanolamines was investigated in a pilot and commercial plant, demonstrated that purified gas containing as little as 5 parts per million of hydrogen sulfide could be obtained with absorption of only about 30% of the carbon dioxide contained in the feed gas. The most selective solvent was methyldiethanolamine, although other amines also showed some selectivity.

Additional information on selective H_2S absorption with MDEA or MDEA-based solutions is presented by Pearce (1978), Crow and Baumann (1974), Goar (1980), Blanc and Elgue (1981), Sigmund et al. (1981), Dibble (1983), Robinson et al. (1988), and Katti and Langfitt (1986A, 1986B). The papers by Sigmund et al. and Dibble describe Union Carbide Corporation's proprietary process using MDEA-based solutions under the trade name of UCARSOL HS Solvents. These solvents are claimed to be more selective than conventional MDEA and DIPA solutions and, consequently, more economical with respect to energy consumption. A comparison of UCARSOL with DIPA for recovering H_2S from Claus plant tail gas (after hydrogenation) is shown in **Table 2-4**. The paper by Robinson et al. provides data on a gas treating plant that was converted from DEA to GAS/SPEC CS-3 Solvent—an

Table 2-4
UCARSOL HS 101 Solvent vs. DIPA in Claus Tail Gas Cleanup Unit

	27% DIPA	50% UCARSOL HS
Circulation Rate, gpm	76	28
CO_2 Slippage, %	84	95
H_2S Content in Recycle Stream to Claus Unit, %	35	66
Reboiler Steam Consumption, Mlbs/Hr	4,469	2,284
Steam Cost @ $5.50/Mlbs, $/Year	212,400	108,500
Savings with UCARSOL HS, $/Year	—	103,900

Source: Dibble (1983)

MDEA-based selective solvent formulation offered by Dow Chemical, USA. The papers by Katti and Langfitt also relate to the Dow GAS/SPEC selective solvents and report on the development and use of an absorber simulator to predict plant performance.

The data from many studies indicate that, with proper design, selective solvents can yield H_2S concentrations as low as 4 ppmv in the treated gas while permitting a major fraction of the CO_2 to pass through unabsorbed. Because of its low vapor pressure, MDEA can be used in concentrations up to 60 wt% in aqueous solutions without appreciable evaporation losses. Furthermore, MDEA is highly resistant to thermal and chemical degradation, is essentially noncorrosive (see Chapter 3), has low specific heat and heats of reaction with H_2S and CO_2 and, finally, is only sparingly miscible with hydrocarbons.

Mixed Amines

MDEA is also rapidly increasing in importance as a nonselective solvent for the removal of high concentrations of acid gas, particularly CO_2, because of its low energy requirements, high capacity, excellent stability, and other favorable attributes. Its principal disadvantage is a low rate of reaction with (and therefore absorption of) CO_2. The addition of primary or secondary amines, such as MEA and DEA, has been found to increase the rate of CO_2 absorption significantly without diminishing MDEA's many advantages (Polasek et al., 1990; Campbell and Weiland, 1989; Katti and Wolcott, 1987). The kinetics of CO_2 absorption into mixtures of MDEA and DEA has been studied by Mshewa and Rochelle (1994). They measured the rates of absorption and desorption of CO_2 in a 50 wt% solution of MDEA over a wide range of temperatures and partial pressures. The results were used with literature values of DEA reactions to develop a model for CO_2 absorption in DEA and mixtures of DEA and MDEA. The model predicts that the overall gas phase coefficient for CO_2 absorption in a solution containing 40% MDEA and 10% DEA is 1.7 to 3.4 times greater than that for CO_2 absorption in a 50% MDEA solution under typical absorption column conditions. A commercial process using this phenomenon was disclosed by BASF Aktiengesellschaft and described by Meissner (1983) and by Meissner and Wagner (1983). Mixed amine processes containing MDEA are now offered by several licensors.

The BASF Activated MDEA process employs a 2.5 to 4.5 M MDEA solution containing 0.1 to 0.4 M monomethylmonoethanolamine or up to 0.8 M piperazine as absorption activators (Bartholome et al., 1971; Appl et al., 1980). The activators apparently increase the rate of hydration of CO_2 in a manner analogous to the activators used in hot potassium carbonate solutions (see Chapter 5) and thus increase the rate of absorption. The process can be operated with one or two absorption stages, depending on the required gas purity. In one single-absorption stage version, which is suitable for bulk CO_2 removal from high pressure gases, the rich MDEA solution is regenerated by simple flashing at reduced pressure. In a two-stage version, when essentially complete CO_2 removal is required, a small stream of steam-stripped MDEA solution is used in the second stage.

The comparative capacities of MDEA and MEA for CO_2 recovery in an absorption/flash process are illustrated by **Figure 2-8** (Meissner and Wagner, 1983). If it is assumed that equilibrium is attained in both the absorption and stripping steps and that isothermal conditions are maintained, the maximum net capacity is simply the difference between equilibrium concentrations at the absorption and stripping partial pressures. A net CO_2 pickup of 30 vol/vol (0.297 mole/mole) is indicated for a 4.5 molar MDEA solution by flashing from a CO_2 partial pressure of 5 bar (72.5 psia) to one bar (14.5 psia) at 70°C (158°F). By comparison, a 4.1 molar MEA solution provides a net pickup of only 5 vol/vol for the same pressure change at a somewhat lower temperature of 60°C (140°F). This promoted MDEA process is particularly useful when CO_2 is present at high partial pressures, as either no steam or only a small amount of steam is required for regeneration.

Another useful feature of the MDEA-based, mixed amine systems is that the formulation can be varied to meet specific site requirements. Vickery et al. (1988) describe how the

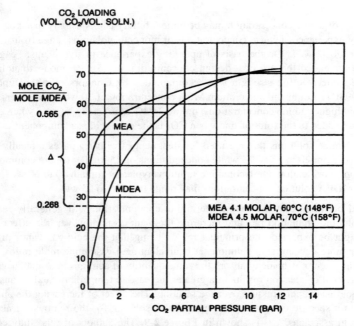

Figure 2-8. CO_2 solution isotherms in MEA and MDEA (*Meissner and Wagner, 1983*). *Reprinted with permission from* Oil & Gas Journal, Feb. 7, 1983. Copyright Pennwell Publishing Company.

selectivity for H_2S over CO_2 can be adjusted, as required, by blending a primary or secondary amine with a tertiary amine, such as MDEA, in just the right proportions.

Sterically Hindered Amines

Although sterically hindered amines are not necessarily alkanolamines, their characteristics as gas purification agents are sufficiently similar to those of the alkanolamines to be included in this chapter. A family of solvents based on hindered amines is licensed by Exxon Research and Engineering Co. under the broad designation of Flexsorb Solvents. The processes have been described in some detail by Goldstein (1983), Weinberg et al. (1983), and Chludzinski and Wiechart (1986). The hindered amines are used as promoters in hot potassium carbonate systems (Flexsorb HP); as components of organic solvent/amine systems with characteristics similar to Shell's Sulfinol process (Flexsorb PS); and as the principal agent in aqueous solutions for the selective absorption of H_2S in the presence of CO_2 (Flexsorb SE and SE+). Each system makes use of a different sterically hindered amine with a specifically designed molecular configuration. On the basis of pilot and commercial plant experience, substantial savings in capital and operating cost are claimed for this technology. As of 1994, it was reported that 32 Flexsorb plants were operating or in design (Exxon Research and Engineering Co., 1994).

Amine Concentration

The choice of amine concentration may be quite arbitrary and is usually made on the basis of operating experience. Typical concentrations of monoethanolamine range from 12 wt% to a maximum of 32 wt%. On the basis of operating experience in five plants, Feagan et al. (1954) recommended the use of a design concentration of 15 wt% monoethanolamine in water. The same solution strength was recommended by Connors (1958). Dupart et al. (1993A, 1993B) recommend a maximum MEA concentration of 20 wt%. However, it should be noted that higher amine concentrations, up to 32 wt% MEA, may be used when corrosion inhibitors are added to the solution and when CO_2 is the only acid gas component.

Diethanolamine solutions that are used for treatment of refinery gases typically range in concentration from 20 to 25 wt%, while concentrations of 25 to 30 wt% are commonly used for natural gas purification. Diglycolamine solutions typically contain 40 to 60 wt% amine in water, and MDEA solution concentrations may range from 35 to 55 wt%.

It should be noted that increasing the amine concentration will generally reduce the required solution circulation rate and therefore the plant cost. However, the effect is not as great as might be expected, the principal reason being that the acid-gas vapor pressure is higher over more concentrated solutions at equivalent acid-gas/amine mole ratios. In addition, when an attempt is made to absorb the same quantity of acid gas in a smaller volume of solution, the heat of reaction results in a greater increase in temperature and a consequently increased acid-gas vapor pressure over the solution. The effect of increasing the amine concentration in a specific operating plant using DGA solution for the removal of about 15% acid gas from associated gas is shown in **Figure 2-9**. The authors of this study concluded that the optimum DGA strength for this case is about 50 wt%. The effect of the increasing amount of DGA at higher concentrations is almost nullified by the decreasing net acid gas absorption per mole of DGA (Huval and van de Venne, 1981).

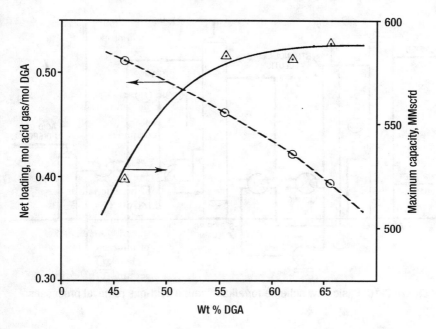

Figure 2-9. Effect of DGA concentration on maximum plant capacity and net solution loading in a large commercial plant (*Huval and van de Venne, 1981*)

FLOW SYSTEMS

Basic Flow Scheme

The basic flow arrangement for all alkanolamine acid-gas absorption-process systems is shown in **Figure 2-10**. Gas to be purified is passed upward through the absorber, countercurrent to a stream of the solution. The rich solution from the bottom of the absorber is heated by heat exchange with lean solution from the bottom of the stripping column and is then fed to the stripping column at some point near the top.

In units treating sour hydrocarbon gases at high pressure, it is customary to flash the rich solution in a flash drum maintained at an intermediate pressure to remove dissolved and entrained hydrocarbons before acid gas stripping (see Chapter 3). When heavy hydrocarbons condense from the gas stream in the absorber, the flash drum may be used to skim off liquid hydrocarbons as well as to remove dissolved gases. The flashed gas is often used locally as fuel. A small packed tower with a lean amine wash may be installed on top of the flash drum to remove H_2S from the flashed gas if sweet fuel gas is required (Manning and Thompson, 1991).

Lean solution from the stripper, after partial cooling in the lean-to-rich solution heat exchanger, is further cooled by heat exchange with water or air, and fed into the top of the absorber to complete the cycle. Acid gas that is removed from the solution in the stripping column is cooled to condense a major portion of the water vapor. This condensate is continually fed back to the system to prevent the amine solution from becoming progressively

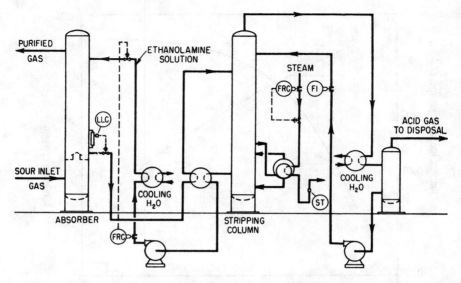

Figure 2-10. Basic flow scheme for alkanolamine acid-gas removal processes.

more concentrated. Generally, all of this water, or a major portion of it, is fed back to the top of the stripping column at a point above the rich-solution feed and serves to absorb and return amine vapors carried by the acid gas stream.

Many modifications to the basic flow scheme have been proposed to reduce energy consumption or equipment costs. For example, power recovery turbines are sometimes used on large, high-pressure plants to capture some of the energy available when the pressure is reduced on the rich solution. A minor modification aimed at reducing absorber column cost is the use of several lean amine feed points. In an arrangement described by Polasek et al. (1990), most of the lean solution is fed near the midpoint of the absorber to remove the bulk of the acid gas in the lower portion of the unit. Only a small stream of lean solution is needed for final clean-up of the gas in the top portion of the absorber, which can therefore be smaller in diameter.

A modification that has been used successfully to increase the acid gas loading of the rich amine (and thereby decrease the required solution flow rate) is the installation of a side cooler (or intercooler) to reduce the temperature inside the absorber. The concept has proved particularly useful for DGA plants operating in Saudi Arabia where air cooling is used (Huval and van de Venne, 1981). The optimum location for a side cooler is reported to be the point where half the absorption occurs above and half below the cooler, which results in a location near the bottom of the column (Thompson and King, 1987).

Water Wash for Amine Recovery

The simplest modification of the flow system in **Figure 2-10** is the inclusion of a water wash at the top of the absorber to reduce losses of amine with the purified gas. If acid gas condensate from the regenerator reflux drum is used for this purpose, no draw-off tray is required because it is necessary to readmit this water to the system at some point. It should be noted however, that this condensate is saturated with acid gas at regenerator condenser operating conditions and that this dissolved acid gas will be reintroduced into the gas stream if the water is used "as is" for washing. If the gas volume is very large, compared to the amount of wash water, this may be of no consequence. However, if calculations indicate that the quantity of acid gas so introduced is excessive, a water stripper can be included in the process. Alternatively, a recirculating water wash with a dedicated water wash pump can be utilized. This design uses a comparatively small wash water make-up and wash water purge.

A water wash is used primarily in monoethanolamine systems, especially at low absorber operating pressures, as the relatively high vapor pressure of monoethanolamine may cause appreciable vaporization losses. The other amines usually have sufficiently low vapor pressures to make water washing unnecessary, except in rare cases when the purified gas is used in a catalytic process and the catalyst is sensitive even to traces of amine vapors.

The number of trays used for water wash varies from two to five in commercial installations. Experience has indicated that an efficiency of 40 or 50% can be expected per tray under typical absorber operating conditions. From this, it would appear that four trays would be ample to remove over 80% of the vaporized amine from the purified gas and, incidentally, a major portion of the amine carried as entrained droplets in the gas stream.

It is probable that an even greater tray efficiency is obtained in the water wash section of the stripping column. However, because of the higher temperature involved, the amine content of the vapors entering this section may be quite high. Four to six trays are commonly used for this service.

Split-Stream Cycles

A flow modification which has been proposed for aqueous amine solutions to reduce the steam requirement is shown in **Figure 2-11.** The split-stream scheme, in which only a portion of the solution is stripped to a low acid-gas concentration, has been applied to several gas-purification processes, including the Shell tripotassium phosphate process and the hot

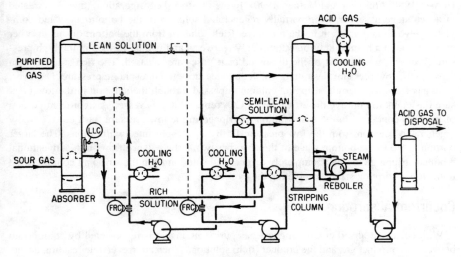

Figure 2-11. Split-stream flow modification for amine plants.

potassium carbonate process (see Chapter 5), and was first disclosed by Shoeld (1934). The rich solution from the bottom of the absorber is split into two streams, one being fed to the top of the stripping column and one to the midpoint. The top stream flows downward countercurrent to the stream of vapors rising from the reboiler and is withdrawn at a point which is above the inlet of the second portion of the rich solution. The liquid withdrawn from the upper portion of the stripping column is not completely stripped and is recycled back to the absorber to absorb the bulk of the acid gases in the lower portion of the absorber column. The portion of solution, which is introduced near the midpoint of the stripping column, flows through the reboiler and is very thoroughly stripped of absorbed acid gases. This solution is returned to the top of the absorber where it serves to reduce the acid gas content of the product gas to the desired low level. In this system, the quantity of vapors rising through the stripping column is somewhat less than that in a conventional plant. However, the ratio of liquid to vapor is lower in both sections because neither carries the total liquid stream.

The obvious drawback of this process modification is that it appreciably increases the initial cost of the treating plant. The stripping column is taller and somewhat more complex, and the two streams require separate piping systems with two sets of pumps, heat exchangers, and coolers. Commercial units utilizing a system of this type have been described by Bellah et al. (1949) and by Estep et al. (1962).

A simplified form of the split-stream cycle consists of dividing the lean solution before introduction into the absorber into two unequal streams. The larger stream is fed to the middle of the absorber, while the smaller stream is introduced at the top of the column. In cases where gases of high acid-gas concentration are treated, this scheme may be more economical than the basic flow scheme, as the diameter of the top section of the absorber may be appreciably smaller than that of the bottom section. Furthermore, the lean-solution stream fed to the middle of the absorber may not have to be cooled to as low a temperature as the stream flowing to the top of the column, resulting in reduction of heat exchange surface.

A split flow cycle specifically designed for the removal of CO_2 from high pressure gas streams with promoted MDEA solutions is shown in **Figure 2-12**. This and several other flow schemes are offered by BASF for use with their activated MDEA process (Gerhardt and Hefner, 1988; Meissner and Hefner, 1990). In the illustrated configuration, the gas is treated in a two stage absorber using partially regenerated solution in the bottom stage and completely regenerated solution in the top stage. Rich solution from the bottom of the absorber passes through a hydraulic turbine for energy recovery, and is then flashed in the high pressure flash unit where most of the dissolved inert gases are released. The rich solution then flows to the low pressure flash unit, which operates close to atmospheric pressure. Hot overhead gas from the thermal stripping column is passed through the solution in the low pressure flash tank to improve the efficiency of CO_2 removal in this vessel. A significant portion of the CO_2 contained in the rich solution is stripped in the low pressure flash unit. Partially regenerated solution from the low pressure flash step is split into two portions. The larger portion is fed to the bottom stage of the absorber; while the balance flows to a conventional reboiled stripping column. Completely regenerated solution from the stripping column is fed to the top absorption stage, completing the cycle.

Cocurrent Absorption

With cocurrent absorbers, the highest gas purity attainable is represented by equilibrium between the product gas and the product (rich) solution. When an irreversible reaction occurs in the liquid phase, the equilibrium vapor pressure of acid gas over the solution is negligible

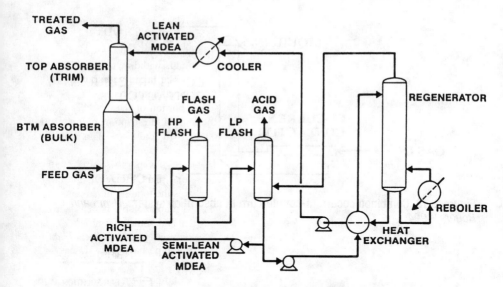

Figure 2-12. Activated MDEA split flow configuration.

and cocurrent contactors can yield high removal efficiencies. With amines, however, the rich solution usually has a significant acid gas vapor pressure, and a cocurrent contactor cannot normally be used as the sole absorption device. In some cases, however, a cocurrent contactor can be used in conjunction with a countercurrent unit to provide improved results.

Isom and Rogers (1994) evaluated several possible flow arrangements for incorporating an SMV high efficiency cocurrent contactor into an existing gas treating system to increase the H_2S removal efficiency. They developed a rate-based modeling method to predict the performance of combination systems that contain both a cocurrent contactor and a countercurrent unit. The existing countercurrent unit treated 10 MMscfd of gas containing 2.0% H_2S and 5.0% CO_2, and produced gas containing 1,000 ppm H_2S and 2.7% CO_2 using 66 gpm of 40% MDEA solution. The studies indicated that the optimum configuration would be that shown in **Figure 2-13** with the same 66 gpm of solution fed to the countercurrent absorber and 74 gpm recycled through the cocurrent unit. With this arrangement the outlet gas composition was determined to be 608 ppm H_2S and 2.5% CO_2.

A very similar configuration, in which the cocurrent contactor is in the form of a heat exchanger, was proposed by Kohl and Bechtold (1952). This concept has found only limited application to date. However with the trend toward more concentrated solutions, higher acid gas loadings, and closer approach to equilibrium with the sour gas, it may be worth reconsideration, particularly in areas such as the Middle East where cooling water is not available and lean amine temperatures obtained by air cooling are relatively high. In this circumstance, a substantial fraction of the heat of reaction can be removed by heat exchange upstream of the amine contactor. The concept is illustrated in **Figure 2-14.** A related concept is the use of a side cooler near the bottom of the absorber as previously described (see Basic Flow Scheme section). In both of these designs, cooling reduces the rich amine temperature, permitting higher rich solution loadings.

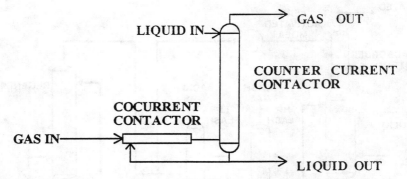

Figure 2-13. Combined cocurrent-countercurrent absorption system. (*Isom and Rogers, 1994*)

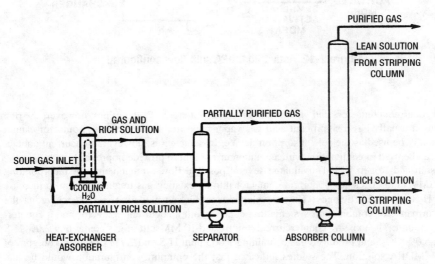

Figure 2-14. Flow system employing heat exchanger for initial gas contact.

DESIGN DATA

Acid Gas-Amine Solution Equilibria

The relationship between the concentration of acid gas in an amine solution and its partial pressure in the gas phase at equilibrium is probably the most important item of data required for the design of treating plants. The relationship may be referred to as gas solubility or vapor-liquid equilibrium (VLE). The concentration in the liquid phase is normally reported as moles acid gas per mole of amine (mole/mole or mol/mol). Since this value varies with the partial pressure (or more precisely with the fugacity) of the acid gas, temperature, type of amine, amine concentration in the solution, and nature and concentration of other components in the solution, the amount of data required to cover all possible conditions is enormous. Although significant data gaps still exist, numerous VLE studies have been conducted and reported in literature. Particularly noteworthy are the extensive publications of Dr. A. E. Mather and coworkers at the University of Alberta, Edmonton, Canada.

The acid gas solubility data presented in the following sections are generally limited to conditions near those most commonly encountered in commercial systems. References are provided to sources of additional data and to correlations that have been developed for predicting VLE relationships in the absence of specific data. Charts and tables of experimental data are useful for preliminary studies; however, correlations are needed for interpolating and extrapolating data to specific conditions, and are required for computer-based amine system design programs.

Monoethanolamine

Since MEA was one of the first ethanolamines used for gas treating and is still widely used, a large amount of VLE data has been published covering MEA solutions of CO_2 and H_2S. The magnitude of this effort can be appreciated by inspection of **Table 2-5,** which lists most of the papers presenting experimental data on the subject.

Much of the early work was conducted with dilute (~15 wt%) MEA solutions because such solutions were commonly used in commercial plants at the time since higher concentrations were considered too corrosive. With the advent of corrosion inhibited solutions and a better understanding of corrosion mechanisms, more concentrated solutions have become popular. This is reflected in the recent VLE data, which typically covers both 15 and 30 wt% solutions. More data are provided for MEA than for the other amines because of its widespread and long time commercial use. Also, many of the conclusions for MEA, such as the general effects of temperature, amine concentration, and the presence of other acid gases are also applicable to other amines.

Figures 2-15 through **2-28** and **Tables 2-6** and **2-7** present data on the solubility of CO_2, H_2S, and mixtures of the two acid gases in MEA solutions. Most of the data are for 2.5N (approximately 15 wt%) and 5.0N (approximately 30 wt%) solutions. **Figure 2-17,** which gives data for H_2S in 15.3 wt% MEA, includes curves calculated by the Kent-Eisenburg correlation, which is discussed later.

Figures 2-26 and **2-27** show the effect of temperature on the vapor pressures of CO_2 and H_2S respectively for various acid-gas/amine mole ratios. The curves are nearly straight lines on the log P versus 1/T coordinates, which aids in extrapolation to other temperatures. The plots also provide an indication of the heat of reaction, which, in accordance with the Clausius-Clapeyron equation, is proportional to the slope of the lines. The decreasing slope with increasing concentration of acid gas in the solution indicates that the heat of reaction decreases as more acid gas is absorbed.

Figure 2-28 shows the effect of increasing the concentration of MEA on the vapor pressure of CO_2 at various acid gas to amine mole ratios and at a temperature of 77°F. Increasing the MEA concentration increases the CO_2 vapor pressure at the same mole ratio. From a practical standpoint, this means that the quantity of solution required does not decrease in inverse proportion to the amine concentration. For example, with 10 psia CO_2 partial pressure in the feed gas, and equilibrium at 77°F, the maximum solution capacity is about 0.8 moles CO_2/mole amine in a 10 wt% MEA solution and 0.6 moles/mole in a 40 wt% solution.

Diethanolamine

Sources of data for the solubility of acid gases in DEA solutions are given in **Table 2-8.** Typical data are given in **Figures 2-29** through **2-35.** The principal charts covering the indi-

Table 2-5
Sources of Solubility Data for CO_2 and H_2S in Aqueous MEA Solutions

Reference	MEA Concentration kmol/m3	Acid Gas
Mason and Dodge, 1936	0.5, 3.0, 5.0, 9.5, 12.5	H_2S
Reed and Wood, 1941	2.5	CO_2
Riegger et al., 1944	0.57, 0.94, 1.34, 1.78, 2.53, 3.22, 3.82	H_2S
Lyudkovskaya and Leibush, 1949	0.5, 2.0, 5.0	CO_2
Leibush and Shneerson, 1950	0.93, 2.5	mixtures
Atadan, 1954	2.5, 5.0, 10.0	CO_2
Atwood et al., 1957	0.83, 2.5, 3.3, 5.0	H_2S
Muhlbauer and Monaghan, 1957	2.5	mixtures
Jones et al., 1959	2.5	mixtures
Goldman and Leibush, 1959	1.0, 2.0, 2.5, 5.0	CO_2
Murzin and Leites, 1971	0.5, 1.0, 2.0, 2.5, 3.4	CO_2
Lee et al., 1974B	2.5, 5.0	CO_2 and H_2S
Lee et al., 1975	5.0	mixtures
Lee et al., 1976A	2.5, 5.0	H_2S
Lee et al., 1976B	5.0	mixtures
Lee et al., 1976C	1.0, 2.5, 4.0, 5.0	CO_2
Lawson and Garst, 1976	2.5, 5.0	mixtures
Nasir and Mather, 1977	2.5, 5.0	CO_2 and H_2S
Isaacs et al., 1980	2.5	mixtures
Maddox et al., 1987	2.5	CO_2 and H_2S
Shen and Li, 1992	2.5, 5.0	CO_2
Murrieta-Guevara et al., 1993	2.5, 5.0	CO_2 and H_2S
Jou et al., 1995A	5.0	CO_2

Note: In some of the references, MEA concentrations are given as weight percent; these have been converted to approximate values of $kmol/m^3$; other references give the concentration as normality, N, which is essentially the same as $kmol/m^3$. For MEA (molecular weight 61.09) the conversions for 2.5 and 5.0 $kmol/m^3$ are approximately 15.2 and 30.2 wt%.

vidual acid gases, CO_2 and H_2S in DEA solutions (**Figures 2-29** and **2-31**), are smoothed curves based on the data of Maddox and Elizondo (1989); These curves appear to be in reasonable agreement with most previous data such as those of Lee et al. (1972, 1973A, 1973B) and Lal et al. (1980, 1985).

Diglycolamine

Data on the solubility of CO_2 and H_2S in DGA are somewhat limited, but are adequate for design, particularly when used in conjunction with recently developed VLE correlations. **Figures 2-36** through **2-39,** which are from Dingman et al. (1983), provide data for a 65 wt% DGA solution at 100°F and 180°F. The curves depict CO_2 partial pressures in the gas versus CO_2/DGA mole ratios in the liquid for several H_2S/DGA mole ratios, and H_2S partial

(*text continued on page 79*)

Alkanolamines for Hydrogen Sulfide and Carbon Dioxide Removal 65

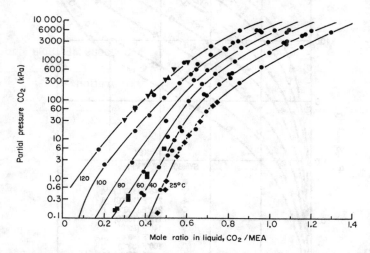

Figure 2-15. Solubility of CO_2 in 2.5 N MEA solution. ♦, *Muhlbauer and Monaghan (1957) (25°C);* ■, *Murzin and Leites (1971) (60°C);* ▼, *Jones et al. (1959) (120°C);* ▲, *Reed and Wood (1941) (120°C);* • *and smoothed curves, Lee et al. (1976C). Chart from Lee et al. (1976C)*

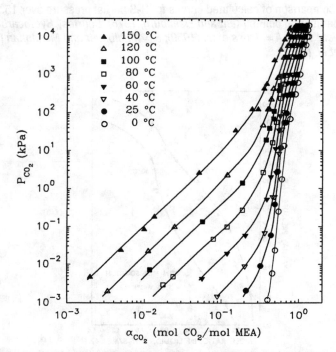

Figure 2-16. Solubility of CO_2 in 30 wt% MEA solution (*Jou et al., 1995A*). *Reprinted with permission from the Canadian Journal of Chemical Engineering, Copyright 1995, Canadian Society for Chemical Engineering*

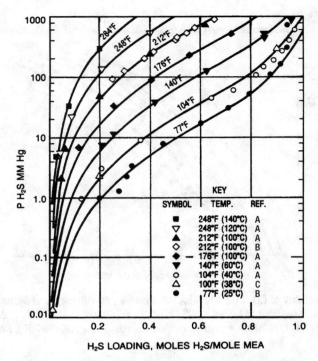

Figure 2-17. Comparison of calculated curves for H_2S partial pressure over 15.3 wt% MEA with experimental points *(Kent and Eisenburg, 1976). Courtesy Hydrocarbon Processing, Feb. 1979. A = Jones et al. (1959); B = Muhlbauer and Monaghan (1957); C = Atwood et al. (1957)*

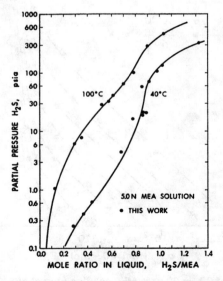

Figure 2-18. Solubility of H_2S in 5.0 N MEA solution *(Lee et al., 1974B). Reprinted with permission from the Canadian Journal of Chemical Engineering, Copyright 1974, Canadian Society for Chemical Engineering*

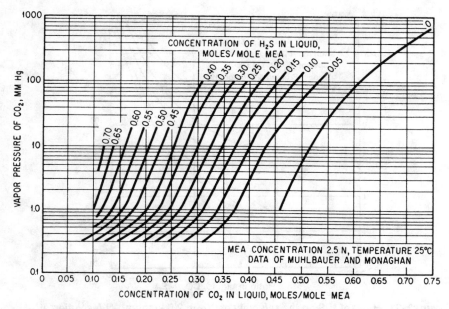

Figure 2-19. Effect of dissolved hydrogen sulfide on vapor pressure of CO_2 over 2.5 N monoethanolamine solution at 25°C. *Data of Muhlbauer and Monaghan (1957)*

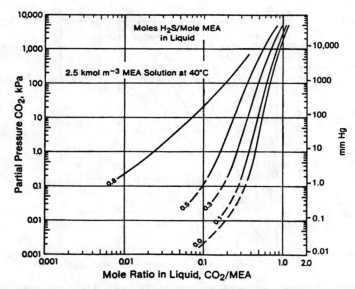

Figure 2-20. Effect of H_2S on the solubility of CO_2 in 2.5 kmol/m³ MEA solution at 40°C. (*Lal et al., 1980*)

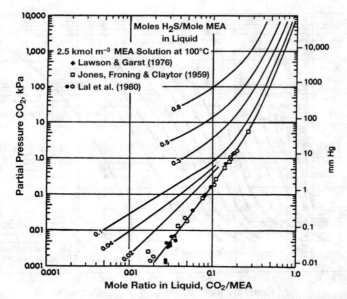

Figure 2-21. Effect of H_2S on the solubility of CO_2 in 2.5 kmol/m³ MEA solution at 100°C. (*Lal et al., 1980*)

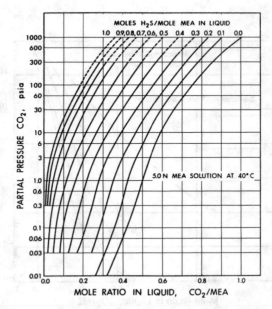

Figure 2-22. Effect of H_2S on partial pressure of CO_2 in 5.0 N MEA solution at 40°C (*Lee et al., 1975*). *Reproduced with permission from Journal of Chemical and Engineering Data, Copyright 1975, American Chemical Society*

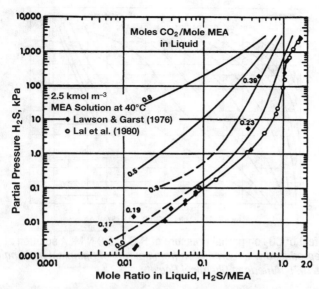

Figure 2-23. Effect of CO_2 on the solubility of H_2S in 2.5 kmol/m³ MEA solution at 40°C. (*Lal et al., 1980*)

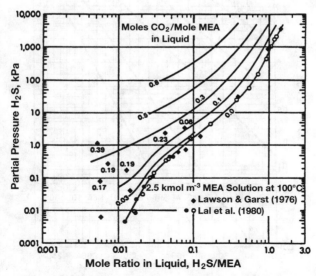

Figure 2-24. Effect of CO_2 on the solubility of H_2S in 2.5 kmol/m³ MEA solution at 100°C. (*Lal et al., 1980*)

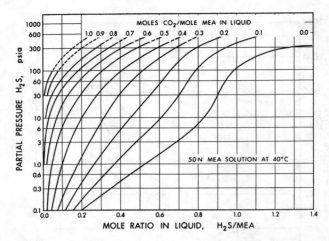

Figure 2-25. Effect of CO_2 on partial pressure of H_2S in 5.0 N MEA solution at 40°C. (*Lee et al., 1975*). *Reproduced with permission from Journal of Chemical and Engineering Data, Copyright 1975, American Chemical Society*

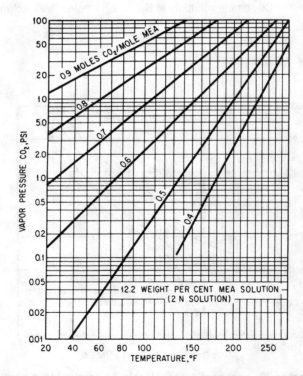

Figure 2-26. Effect of temperature on CO_2 vapor pressure for various CO_2 concentrations in 2 N monoethanolamine solution.

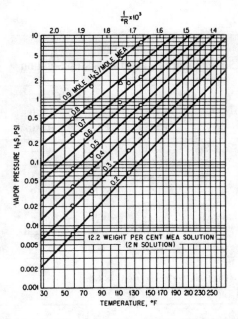

Figure 2-27. Effect of temperature on vapor pressure of H_2S for various H_2S concentrations in 2 N monoethanolamine solution.

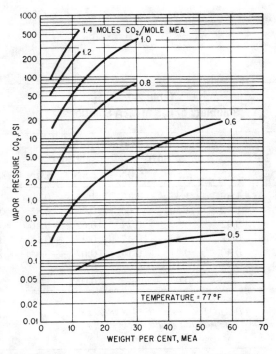

Figure 2-28. Effect of amine concentration on CO_2 vapor pressure.

Table 2-6
Solubility of CO_2 in 30 wt% MEA Solution

25°C		60°C		100°C		120°C	
P_{CO_2}	Mole Ratio	P_{CO_2}	Mole Ratio	P_{CO_2}	Mole Ratio	P_{CO_2}	Mole Ratio
19,936	1.25	19,893	1.11	19,812	0.94	17,723	0.86
9,973	1.17	9,959	1.03	9,871	0.86	9,770	0.78
2,996	1.04	2,977	0.88	2,899	0.71	2,804	0.64
297	0.81	282	0.64	376	0.59	422	0.47
55.1	0.65	34.1	0.57	39	0.42	47	0.35
2.8	0.54	2.01	0.44	1.43	0.17	2.3	0.12
0.06	0.44	0.06	0.20	0.14	0.057	0.098	0.025
0.0021	0.21	0.0043	0.056	0.0072	0.012	0.0020	0.0033

Note:
P_{CO_2} = partial pressure of CO2 in kPa (1 psia = 6.8948 kPa)
Source: Jou et al., 1995A

Table 2-7
Solubility of H_2S in MEA Solutions

	Mole Ratio H_2S/MEA							
	25°C		60°C		100°C		120°C	
P_{H_2S}	2.5N	5.0N	2.5N	5.0N	2.5N	5.0N	2.5N	5.0N
2,000	1.58	1.39	1.33	1.15	1.16	0.99	1.02	0.89
1,000	1.30	1.12	1.56	1.00	0.99	0.85	0.89	0.75
100	0.99	0.92	0.89	0.77	0.62	0.46	0.45	0.35
10	0.82	0.72	0.55	0.42	0.23	0.15	0.15	0.11
1.0	0.47	0.40	0.21	0.14	0.07	0.04	0.05	—
0.316	0.29	0.23	0.12	0.07	0.04	—	0.02	—
0.1	0.16	—	0.07	—	0.03	—	—	—

Note:
1. P_{H_2S} = partial pressure of H_2S in kPa (1 psia = 6.8948 kPa).
2. Values represent smoothed data from test results.
Source: Lee et al., 1976A

Table 2-8
Sources of Solubility Data for CO_2 and H_2S in Aqueous DEA Solutions

Reference	DEA Concentration kmol/m³ (wt%)	Acid Gas
Bottoms, 1931	(50%)	CO_2 and H_2S
Mason and Dodge, 1936	0.5, 2.0, 5.0, 8.0	CO_2
Reed and Wood, 1941	2.5	CO_2
Leibush and Shneerson, 1950	0.97, 2.0	H_2S and mixtures
Atwood et al., 1957	(10, 25, 50%)	H_2S
Murzin and Leites, 1971	0.5, 1.0, 2.0, 5.0, 8.0	CO_2
Lee et al., 1972	0.5, 2.0, 3.5, 5.9	CO_2
Lee et al., 1973A	0.5, 5.0	H_2S
Lee et al., 1973B	0.5, 2.0, 3.5, 5.0	H_2S
Lee et al., 1974A	2.0, 3.5	mixtures
Lal et al., 1980	2.0	H_2S, CO_2, and mixtures
Lal et al., 1985	2.0	H_2S, CO_2, and mixtures
Kennard and Meisen, 1984	1.0, 2.0, 3.0	CO_2
Maddox et al., 1987	0.5, 2.0	CO_2
Maddox and Elizondo, 1989	(20, 35, 50%)	H_2S and CO_2
Ho and Eguren, 1988	(5.3, 35, 50, 77.5%)	CO_2 and mixtures

Note: The DEA concentrations are generally stated in the same units as the reference, except that kmol/m3 is used instead of normality. For DEA (molecular weight 105.14) the conversions for 1, 2, and 5 kmol/m³ are approximately 10.4, 20.4, and 49.6 wt%.

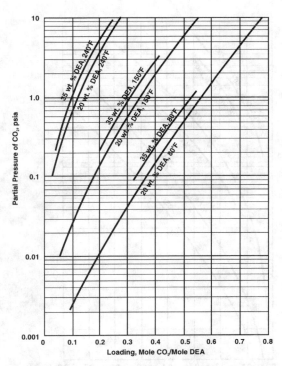

Figure 2-29. Solubility of CO_2 in aqueous DEA solutions. *Data of Maddox and Elizondo (1989)*

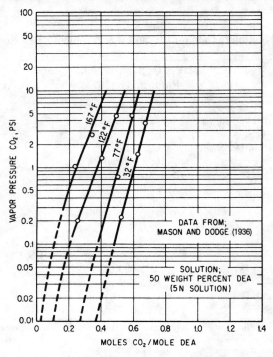

Figure 2-30. Vapor pressure of CO_2 vs. CO_2 concentration in 5 N diethanolamine solution.

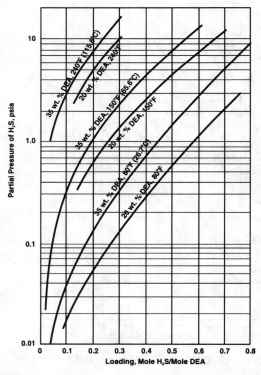

Figure 2-31. Solubility of H_2S in aqueous DEA solutions. *Data of Maddox and Elizondo (1989)*

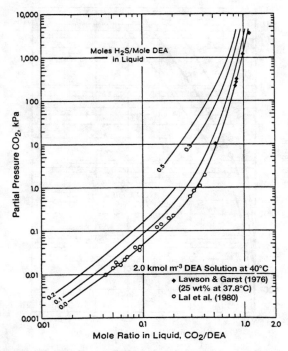

Figure 2-32. Effect of H$_2$S on the solubility of CO$_2$ in 2.0 kmol/m^3 DEA solution at 40°C. (*Lal et al., 1980*)

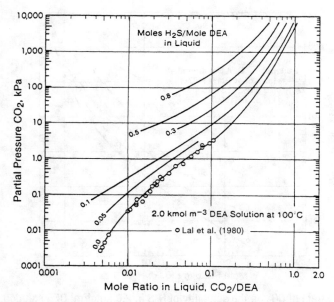

Figure 2-33. Effect of H$_2$S on the solubility of CO$_2$ in 2.0 kmol/m^3 DEA solution at 100°C. (*Lal et al., 1980*)

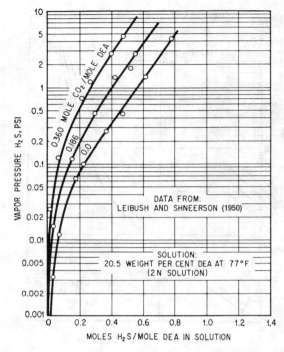

Figure 2-34. Effect of CO_2 on vapor pressure of H_2S over 2 N diethanolamine solution containing both CO_2 and H_2S.

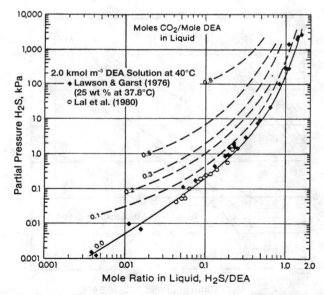

Figure 2-35. Effect of CO_2 on the solubility of H_2S in 2.0 kmol/m³ DEA solution at 40°C. (*Lal et al., 1980*)

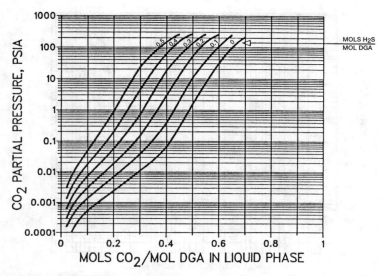

Figure 2-36. CO_2 partial pressure curves for the Diglycolamine agent - H_2S - CO_2 - H_2O system at 100°F, 65/35 amine-water weight ratio. (*Dingman et al., 1983*)

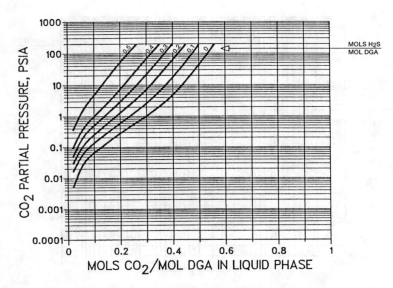

Figure 2-37. CO_2 partial pressure curves for the Diglycolamine agent - H_2S - CO_2 - H_2O system at 180°F, 65/35 amine-water weight ratio. (*Dingman et al., 1983*)

78 Gas Purification

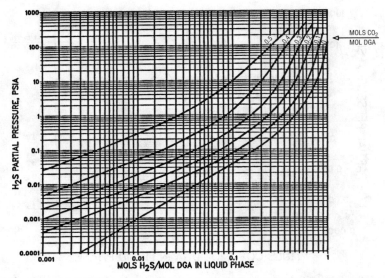

Figure 2-38. H_2S partial pressure curves for the Diglycolamine agent - H_2S - CO_2 - H_2O system at 100°F, 65/35 amine-water weight ratio. (*Dingman et al., 1983*)

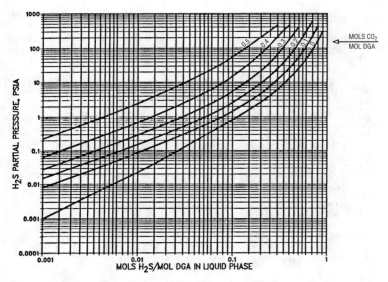

Figure 2-39. H_2S partial pressure curves for the Diglycolamine agent - H_2S - CO_2 - H_2O system at 180°F, 65/35 amine-water weight ratio. (*Dingman et al., 1983*)

(*text continued from page 64*)

pressures as a function of the H_2S/DGA mole ratio for several CO_2/H_2S mole ratios. **Figure 2-40** provides partial pressure versus mole ratio data for the two acid gases alone in a 60 wt% DGA solution at 50°C and 100°C. The curves in this figure are based on data published by Martin et al. (1978). Other sources of VLE data for DGA solutions are Christensen et al. (1985) and Maddox et al. (1987).

Diisopropanolamine

VLE data on DIPA solutions with CO_2 and H_2S are given by Isaacs et al. (1977A and B). The first of these references provides data on the acid gases independently and is the basis for **Figure 2-41,** which shows the partial pressure of the acid gases versus mole ratio in the liquid for 2.5 M diisopropanolamine at 40°C and 100°C. **Figures 2-42** and **2-43,** which are from Isaacs et al. (1977B), show the effects of one acid gas on the other in 2.5N DIPA at 40°C.

Methyldiethanolamine and Triethanolamine

MDEA and TEA are both tertiary amines and have somewhat similar properties and applications. Although TEA was the first to be used commercially, MDEA has become much

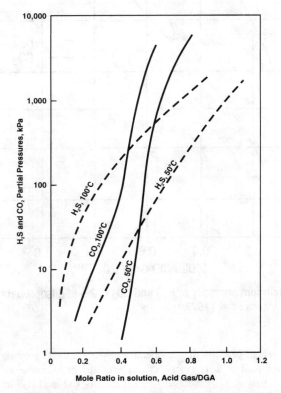

Figure 2-40. Solubility of CO_2 and H_2S in 60 wt% DGA. *Data of Martin et al. (1978)*

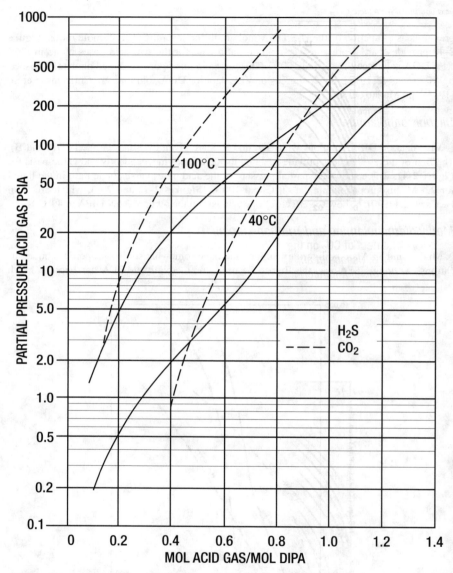

Figure 2-41. Equilibrium solubility of H_2S and CO_2 in 2.5 M diisopropanolamine solution. *Data of Isaacs et al. (1977A)*

more important in recent years, both as a selective absorbent for H_2S in the presence of CO_2 and for the bulk removal of acid gases. As a result, current literature on MDEA is much more extensive than that on TEA. Sources of VLE data for CO_2 and H_2S in MDEA solutions are listed in **Table 2-9**.

Alkanolamines for Hydrogen Sulfide and Carbon Dioxide Removal 81

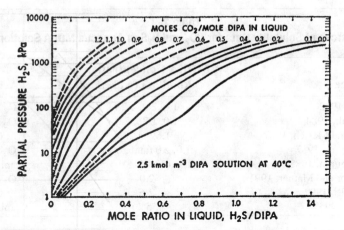

Figure 2-42. Effect of CO_2 on the partial pressure of H_2S over a 2.5 kmol m^{-3} DIPA solution at 40°C, (*Isaacs et al., 1977B*). *Reprinted with permission from the Canadian Journal of Chemical Engineering, Copyright 1977, Canadian Society for Chemical Engineering*

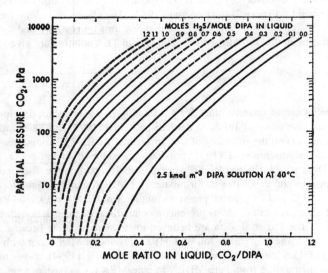

Figure 2-43. Effect of H_2S on the partial pressure of CO_2 over a 2.5 kmol m^{-3} DIPA solution at 40°C. *Isaacs et al. (1977B). Reprinted with permission from the Canadian Journal of Chemical Engineering, Copyright 1977, Canadian Society for Chemical Engineering*

Table 2-9
Sources of Solubility Data for CO_2 and H_2S in Aqueous MDEA Solutions

Reference	MDEA Concentration kmol/m3 (wt%)	Acid Gas
Jou et al., 1982	1.0, 2.0, 4.28	CO_2 and H_2S
Bhairi et al., 1984	1.0, 1.75, 2.0	CO_2 and H_2S
Chakma and Meisen, 1987	1.69, 4.28	CO_2
Maddox et al., 1987	1.0, 2.0 (and 20%)	CO_2 and H_2S
Ho and Eguren, 1988	(23, 49%)	mixtures
MacGregor and Mather, 1991	2.0	CO_2 and H_2S
Shen and Li, 1992	2.6	CO_2
Jou et al., 1993	(35%)	mixtures

Note:
The MDEA concentrations are generally stated in the same units as the reference, except that $kmol/m^3$ is used instead of normality. For MDEA (molecular weight 119.17) the conversions for 1, 2, and 4.28 $kmol/m^3$ are approximately 11.8, 23.5, and 50 wt%.

Data on the solubility of acid gases in 4.28N MDEA at various temperatures are given in **Figures 2-44** and **2-45** for CO_2 and H_2S respectively. The solubility of H_2S in 1.0N MDEA is described by **Figure 2-46** (Jou et al., 1982). The effects of H_2S on the solubility of CO_2 and of CO_2 on the solubility of H_2S are shown in **Figures 2-47** and **2-48,** respectively. Limited data on the solubility of CO_2 and H_2S in TEA solution are given in **Figures 2-49** and **2-50.**

Mixed Amines

The commercial use of mixed amines for gas treating is a very recent development, and only limited data have been published to date. Li and Shen (1992) and Shen and Li (1992) give experimental data on the solubility of CO_2 in several MDEA-MEA mixtures containing 30 wt% total amine. Austgen et al. (1991) provide some solubility data for CO_2 in aqueous mixtures of MDEA with MEA and DEA. Jou et al. (1994) report on the distribution of CO_2 between the aqueous and vapor phases for mixtures of MDEA and MEA at temperatures of 25, 40, 80, and 120°C, and CO_2 partial pressures ranging from 0.001 to 19,930 kPa.

Figure 2-51 (from Jou et al., 1994) presents a comparison of their data with those of Li and Shen (1992). The Jou et al. data are believed to be more accurate because the Li and Shen results do not agree with earlier Jou et al. (1982) results for MDEA nor with the data of Austgen et al. (1991). **Figure 2-52,** which is also from Jou et al. (1994), shows the effect of changing the amine mixture from pure MDEA to pure MEA (at a constant total amine concentration of 30 wt% and a constant CO_2/amine mole ratio of 0.1) on the partial pressure of CO_2 at several temperatures.

(text continued on page 87)

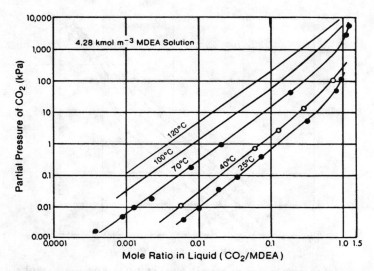

Figure 2-44. Effect of temperature on the solubility of CO_2 in 4.28 kmol/m³ MDEA solution (*Jou et al., 1982*). Reprinted with permission from Industrial and Engineering Chemistry, Process Design and Development, Vol. 21, No. 4. Copyright 1982, American Chemical Society

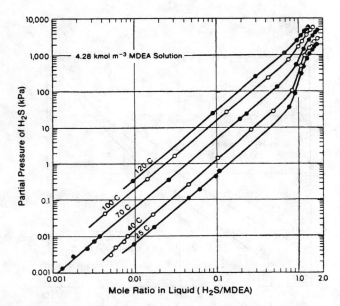

Figure 2-45. Effect of temperature on the solubility of H_2S in 4.28 kmol/m³ MDEA solution (*Jou et al., 1982*). Reprinted with permission from Industrial and Engineering Chemistry, Process Design and Development, Vol. 21, No. 4. Copyright 1982, American Chemical Society

84 Gas Purification

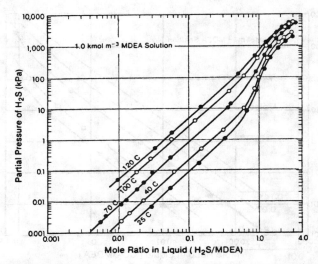

Figure 2-46. Effect of temperature on the solubility of H_2S in 1.0 kmol/m³ MDEA solution (*Jou et al., 1982*). *Reprinted with permission from Industrial and Engineering Chemistry, Process Design and Development, Vol. 21, No. 4. Copyright 1982, American Chemical Society*

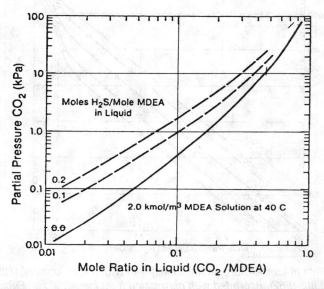

Figure 2-47. Effect of H_2S on the solubility of CO_2 in 2.0 kmol/m³ MDEA solution at 40°C. (*Jou et al., 1981*)

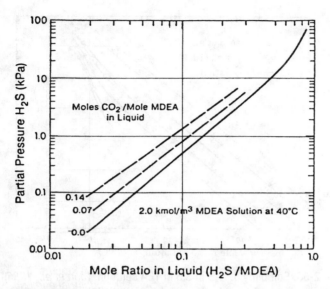

Figure 2-48. Effect of CO_2 on the solubility of H_2S in 2.0 kmol/m³ MDEA solution at 40°C. (*Jou et al., 1981*)

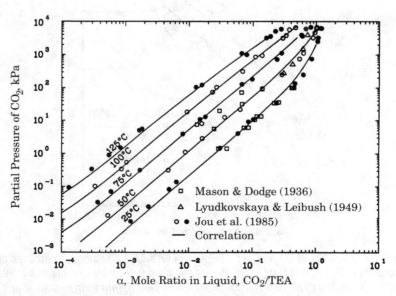

Figure 2-49. Solubility of CO_2 in 5.0 M TEA solution. (*Jou et al., 1985*)

86 *Gas Purification*

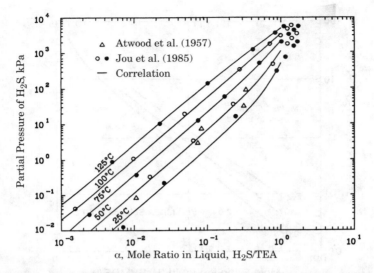

Figure 2-50. Solubility of H_2S in 3.5 M TEA solution. (*Jou et al., 1985*)

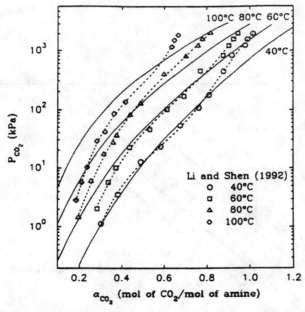

Figure 2-51. Solubility of CO_2 in 6 wt% MEA + 24 wt% MDEA solution. *Solid lines are interpolation of Jou et al. (1994) data. Points and dotted lines are Li and Shen (1992) data. Chart from Jou et al. (1994). Reproduced with permission from Industrial and Engineering Chemistry Research, Copyright 1994, American Chemical Society*

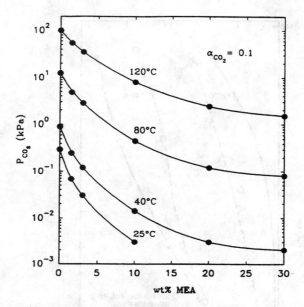

Figure 2-52. Partial pressure of CO_2 as a function of composition at a mole ratio (moles CO_2 per mole MEA) of 0.1 for mixtures of MEA and MDEA with a total amine concentration of 30% (*Jou et al., 1994*). *Reproduced with permission from Industrial and Engineering Chemistry Research, Copyright 1994, American Chemical Society*

(*text continued from page 82*)

Sterically Hindered Amines

Commercial application of hindered amines to gas treating was pioneered by Exxon Research and Engineering Company as their Flexsorb system. Much of the design data are considered proprietary; however, a limited amount of VLE data has been published. A comparison of the solubility of CO_2 in 3 M MEA and 3 M 2-amino-2 methyl-1-propanol (AMP), a hindered amine, at 40°C and 120°C is given in **Figure 2-53** (Sartori et al., 1994). More detailed information on the solubilities of CO_2 and H_2S in 2 and 3 M AMP are reported by Roberts and Mather (1988). Smoothed curves based on their data for CO_2 and H_2S in 2.0 M AMP are shown in **Figure 2-54**.

Acid Gas-Amine Solution Equilibrium Correlations

Considerable progress has been made in the development of generalized correlations for predicting phase equilibrium data for CO_2 and H_2S in aqueous amine solutions. The first practical and widely used model was proposed by Kent and Eisenberg (1976). Like all important subsequent models, the Kent Eisenberg correlation is based on defining the chemical reaction equilibria in the liquid phase. The key reactions identified are

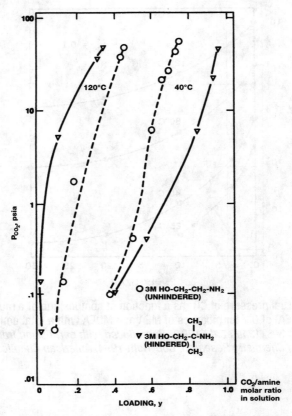

Figure 2-53. Comparison of VLE curves for CO_2 in MEA and AMP at 40°C and 120°C (*Sartori et al., 1994*). Reproduced with permission of the Royal Society of Chemistry, Cambridge, UK.

protonation of amine:

$$H^+ + RR'NH = RR'NH_2^+ \tag{2-6}$$

carbamate formation:

$$RR'NH + HCO_3^- = RR'NCOO^- + H_2O \tag{2-7}$$

hydrolysis of carbon dioxide:

$$H_2O + CO_2 = H^+ + HCO_3^- \tag{2-8}$$

dissociation of water:

$$H_2O = H^+ + OH^- \tag{2-9}$$

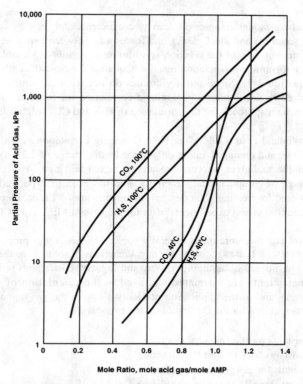

Figure 2-54. Solubility of CO_2 and H_2S in 2.0 M aqueous AMP. (*Roberts and Mather, 1988*)

dissociation of bicarbonate ion:

$$HCO_3^- = H^+ + CO_3^{2-} \tag{2-10}$$

dissociation of hydrogen sulfide:

$$H_2S = H^+ + HS^- \tag{2-11}$$

dissociation of bisulfide ion:

$$HS^- = H^+ + S^{2-} \tag{2-12}$$

solution of carbon dioxide:

$$P_{CO_2} = H_{CO_2}(CO_2) \tag{2-13}$$

solution of hydrogen sulfide:

$$P_{H_2S} = H_{H_2S}(H_2S) \tag{2-14}$$

The Kent-Eisenberg model assumes all activity coefficients and fugacity coefficients to be 1.0 (i.e., ideal solutions and ideal gases), and forces a fit between experimental and predicted values by treating two of the reaction equilibrium constants as variables. The reactions so treated are the amine dissociation reaction (equation 2-6) and the carbamate formation reaction (equation 2-7). Since tertiary amines do not form carbamates, a modified approach is required in developing a generalized correlation for these amines. Jou et al. (1982) describe such an approach for the correlation of H_2S and CO_2 solubilities in aqueous MDEA solutions.

According to Weiland et al. (1993), the Kent-Eisenberg correlation provides a good fit between experimental and predicted values only in the loading range of 0.2 to 0.7 moles acid gas per mole of amine, and gives inaccurate results for mixed acid gases. However, it has the important advantage of computational simplicity, and has been incorporated into several computer models used for treating plant design. A comparison of VLE data predicted by the Kent-Eisenberg correlation and experimental data for the system MEA-H_2O-CO_2 is given in **Figure 2-17.**

A more rigorous, and therefore more generally applicable model was proposed by Deshmukh and Mather (1981). It uses the same chemical reactions in solution as the Kent-Eisenberg correlation, but, instead of assuming activity and fugacity coefficients to be unity, values for these coefficients are estimated and used in the calculation of liquid phase equilibrium constants and in the application of Henry's law to the gas-liquid equilibrium. The basic elements required for the Deshmukh-Mather model are

1. equilibrium constants for the chemical reactions
2. Henry's law constants for CO_2 and H_2S in water
3. fugacity coefficients for gas phase components
4. activity coefficients for all species in the solution

Sufficient data and methods exist to permit reasonable estimates to be made for items 1, 2, and 3. The approach has, therefore, been taken to accept these estimates, and to adjust the interaction parameters used in estimating the activity coefficients so that the final calculated equilibrium values match experimental data. In Deshmukh and Mather's original publication, a rather cumbersome method was used to solve the system of equations. Chakravarty (1985) proposed a simpler technique, which greatly reduces computation times. The improved model has been used for estimating VLE data in an absorption system simulation model developed by Sardar and Weiland (1985) and by Weiland et al. (1993) for evaluating and condensing a large amount of published data for MEA, DEA, DGA, and MDEA. The interaction parameters developed by Weiland et al. provide a sound basis for estimating VLE data for the most important commercial amines over a wide range of conditions.

In a related study, Li and Mather (1994) used Pitzer's excess Gibbs energy equations (Pitzer, 1991) to predict VLE data for the MDEA-MEA-H_2O-CO_2 system using interaction parameters determined from experimental data for MDEA-H_2O-CO_2 and MEA-H_2O-CO_2 systems. Li and Mather's presentation provides insights into concentrations of various ionic and molecular species in the liquid phase when an acid gas is dissolved into a mixed amine solution. **Figure 2-55,** for example, shows how the concentrations of all key species vary with increasing CO_2/amine mole ratio in a solution containing 10 wt% MEA and 20 wt% MDEA at 40°C.

The most sophisticated, and probably the most accurate model available at this time was proposed by Austgen et al. (1991). This model is based on the electrolyte-NRTL model of

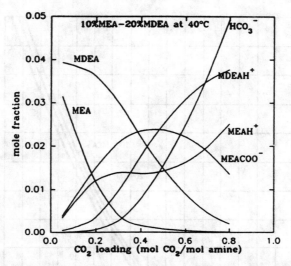

Figure 2-55. Liquid-phase concentration of a CO_2-loaded 10 wt% MEA + 20 wt% MDEA aqueous solution at 40°C (*Li and Mather, 1994*). *Reprinted with permission from Industrial and Engineering Chemistry Research, Copyright 1994, American Chemical Society*

Chen and Evans (1986). Austgen et al. used the model to successfully correlate experimental VLE data for the systems MEA-MDEA-H_2O-CO_2 and DEA-MDEA-H_2O-CO_2. The model is thermodynamically rigorous, but computationally complex, and has not been widely accepted for design applications.

Amine Solution Vapor Pressures

Data on the total vapor pressure of amine solutions as a function of temperature and amine concentration are necessary for the design of stripping columns and reboilers. Such data are presented in **Figure 2-56** for MEA solutions, **2-57** for DEA solutions, **2-58** for DGA solutions, **2-59** for TEA solutions, **2-60** for DIPA solutions, and **2-61** for MDEA solutions. These charts can be used to determine the boiling point of an amine solution as a function of concentration and pressure; however, they do not show the composition of the vapor phase. Vapor-liquid equilibrium composition charts for MEA and DGA solutions at selected pressures are given in Chapter 3. Additional data on amine solution vapor-liquid equilibrium can usually be obtained from the manufacturers.

Heats of Reaction

Data on the heats of reaction of amines with the acid gases are necessary for the generation of individual tray and overall vessel heat balances for the absorber and stripper, the calculation of the amount of steam needed in the reboiler, and the estimation of heat duties of heat exchange equipment in the plant. As noted previously, the heats of reaction are not constants for each amine and acid gas, but generally decrease as the acid gas concentration in

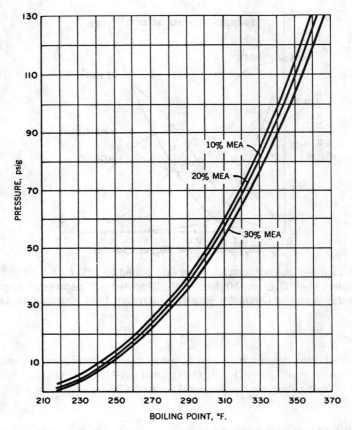

Figure 2-56. Vapor pressure of aqueous MEA solutions as a function of temperature and concentration. (*Dow, 1962*)

the solution increases. This effect is shown in **Figure 2-62** from Li and Mather (1994) for 30 wt% MEA, 30 wt% MDEA, and several mixtures of the two amines. The heat of reaction (or enthalpy of solution) for CO_2 in MEA is almost constant at 85 KJ/mol of CO_2 until a mole ratio of about 0.5 is reached (corresponding to formation of the carbamate) and then drops off rapidly. The heat of reaction of CO_2 and MDEA solution, on the other hand, starts at a lower value (about 62 KJ/mol of CO_2) and starts dropping immediately, but at a lower rate than MEA, so that both have an enthalpy of solution for CO_2 of about 30 KJ/mol at a mole ratio of 1.0 mole CO_2/mole amine.

The heat of reaction of H_2S is generally lower than that of CO_2 for the same amine and decreases with concentration in the solution, but does not show the sharp break typical of CO_2 in MEA. In the design of absorbers and strippers it is common practice to use an integrated heat of reaction representing the total heat released (per mole of each acid gas) over the solution composition change from lean solution to rich solution (or vice versa in the stripper).

A comprehensive study of the enthalpies of solution of CO_2 and H_2S in amines was conducted under the sponsorship of the Gas Processors Association (GPA) at Brigham Young

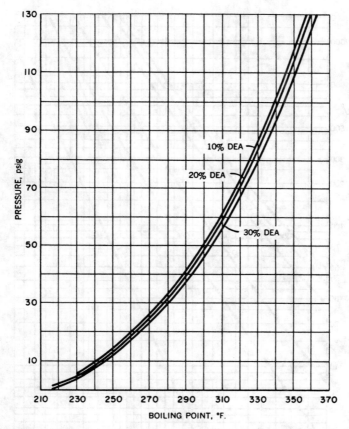

Figure 2-57. Vapor pressure of aqueous DEA solutions as a function of temperature and concentration (*Dow, 1962*)

University. The results of these studies are presented in a series of research reports. Research reports RR-85 (Christensen et al., 1985), RR-102 (Merkley et al., 1986), and RR-108 (Helton et al., 1987) provide data on CO_2 in aqueous diglycolamine, methyldiethanolamine, and diethanolamine solutions, respectively. Reports RR-114 (Van Dam et al., 1988) and RR-127 (Oscarson and Izatt, 1990) give similar data on H_2S in diethanolamine and methyldiethanolamine, respectively.

In general, the results of the GPA studies indicate that the enthalpy of solution for CO_2 or H_2S in any of the amines studied is essentially constant with solution loading up to the saturation point for the acid gas in the solution under the conditions of the tests. The enthalpy of solution was also found to be independent of acid gas partial pressure within the range studied. Temperature and amine concentration in the solution were found to have varying effects on the different systems. These effects are indicated by the following equations developed in the studies to represent the experimental data:

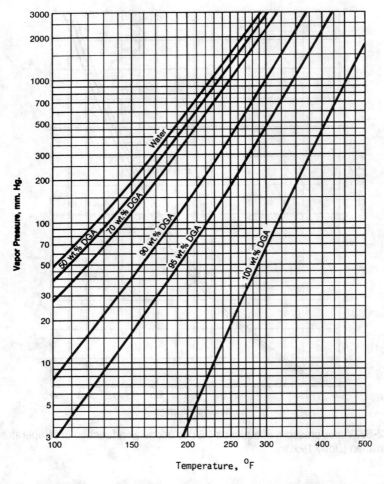

Figure 2-58. Vapor pressure of aqueous DGA solutions as a function of temperature and concentration. (*Jefferson Chemical Co., 1969*)

CO_2 in DGA solutions for T = 60 to 300°F, P_{CO_2} = 22.6 to 162.6 psia, and w = 10 to 60 wt% DGA (Christensen et al., 1985):

H (Btu/lb CO_2) = −1.8 w −790
(range of H: −810 to −830 Btu/lb CO_2) (2-15)

CO_2 in MDEA solutions for T = 60 to 300°F, P_{CO_2} = 22.6 to 212.6 psia, and w = 20 to 60 wt% MDEA (Merkley et al., 1986):

H (Btu/lb CO_2) = −0.9438 w − 0.6764 T − 400.76
(range of H: −460 to −660 Btu/lb CO_2) (2-16)

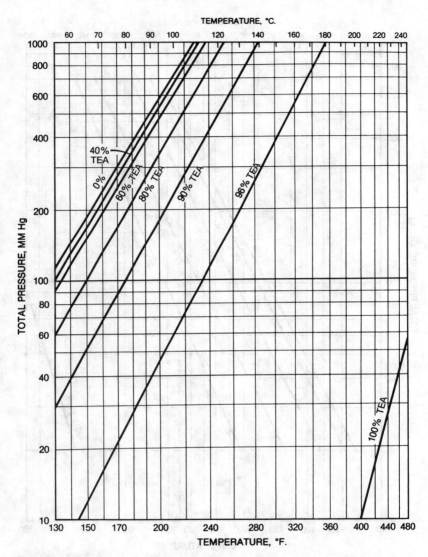

Figure 2-59. Vapor pressure of TEA solutions as a function of temperature and concentration. (*Dow, 1981*)

CO_2 in DEA solutions for T = 80 to 260, P_{CO_2} = 12.6 to 162.6 psia, and M = 2.0 to 5.0 molar DEA (Helton et al., 1987):

H (Btu/lb CO_2) = –0.6389 T –621.4
(range of H: –675 to –810 Btu/lb CO_2) (2-17)

96 Gas Purification

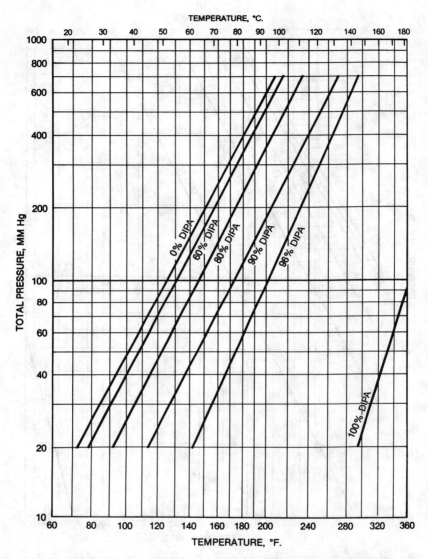

Figure 2-60. Vapor pressure of DIPA solutions as a function of temperature and concentration. (*Dow, 1981*)

H_2S in DEA solutions for T = 80 to 160°F, P_{H_2S} = 12.6 to 260 psia, and M = 2.0 to 5.0 molar DEA (Van Dam et al., 1988):

H (Btu/lb H_2S) = 0.1853 M − 0.6870 T −435.2
(range of H: −462 to −643 Btu/lb H_2S) (2-18)

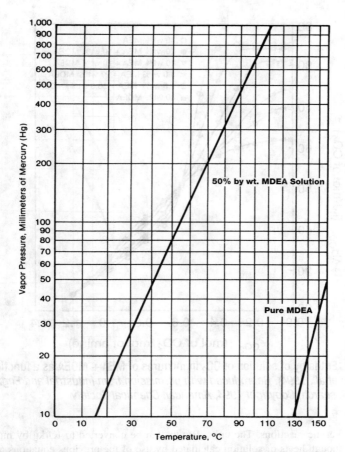

Figure 2-61. Vapor pressure of MDEA solutions as a function of temperature and concentration. (*Union Carbide, 1985*)

H_2S in MDEA solutions for T = 80 to 260°F, P_{H_2S} = 12.6 to 162.6 psia, and w = 20 and 50 wt% MDEA (Oscarson and Izatt, 1990):

$$H \text{ (Btu/lb } H_2S) = -1.24 \, w - 0.8311 \, T - 307$$
(range of H: −350 to −573 Btu/ lb H_2S) (2-19)

In equations 2-15 through 2-19, H is the enthalpy of solution (Btu/lb acid gas), w is the weight percent amine in the solution, T is the temperature (°F), M is the molarity of the amine solution, and P is the partial pressure of acid gas (psia). The enthalpies are actually integral values representing all heat released in bringing the solution acid gas content from zero to the final loading (moles acid gas per mole amine). However, since the enthalpies of solution were found to be constant with loading for conditions covered by the previous equations, differential and integral values are the same in this case. The negative enthalpy values

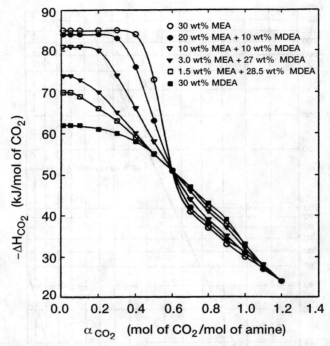

Figure 2-62. Enthalpy of solution of CO_2 in mixtures of MEA + MDEA as a function of loading (*Jou et al., 1994*). *Reproduced with permission from Industrial and Engineering Chemistry Research, Copyright 1994, American Chemical Society*

indicate exothermic reactions. The units, Btu/lb, can be converted to KJ/kg by multiplying by 2.325. Although heats of solution calculated by use of the previous equations are in fair agreement with other published values, other investigators have generally found heats of reaction to decrease with increasing acid gas loading. Examples of differential enthalpy of solution data on CO_2 and H_2S in MEA, DEA, and MDEA from several sources are presented in **Table 2-10**. Approximate average integrated enthalpy values for typical commercial plant absorber conditions are given in **Table 2-11**.

Physical Properties

Figure 2-63 gives the specific gravity of six alkanolamines (at 20°C relative to water at 20°C) as a function of the amine concentration in water. The effects of temperature on the specific gravity (or density) of MEA, DEA, DGA, and ADIP (DIPA) solutions are shown in **Figures 2-64** through **2-67**. For amines not included in **Figures 2-64** through **2-67**, the specific gravity at temperatures other than 20°C can be approximated from the Figure 2-63 value by assuming a specific gravity vs. temperature curve slope similar to those of the other amine solutions. Recent data on the densities of binary mixtures of water with MEA, DEA, and TEA over the full range of compositions and over the temperature range of 25°C to 80°C are given by Maham et al. (1994).

Table 2-10
Differential Enthalpy of Solution for H_2S and CO_2 in MEA, DEA, and MDEA Solutions

Mole Ratio Acid Gas/Amine	Differential Enthalpy of Solution, KJ/Mol Acid Gas				
	MEA (2.5N)		DEA 3.5 N	MDEA (1–4.28N)	
	H_2S	CO_2	CO_2	H_2S	CO_2
0.1	—	—	—	41.3	60.9
0.2	48.5	85.4	76.3	—	—
0.3	—	—	—	40.7	—
0.4	47.6	66.0	65.4	—	—
0.5	—	—	—	39.0	54.3
0.6	46.3	50.7	50.3	—	—
1.0	24.6	29.5	32.4	26.1	33.7
1.2	16.8	23.1	27.3	19.7	—

Sources: MEA data, Lee et al. (1974B); MDEA data, Jou et al. (1982); DEA data, Lee et al. (1972)

Table 2-11
Approximate Integral Heats of Solution for Absorption of H_2S and CO_2 in Alkanolamine Solutions

Amine	Integral Heat of Solution, Btu/lb Acid Gas	
	H_2S	CO_2
MEA	615	825
DEA	510	700
DGA	675	820
MDEA	520	575
TEA	430	465
DIPA	475	720

Notes: Values are approximate averages from several sources; based on total heat released when acid gas is absorbed from a mole ratio of 0 to about 0.4 moles acid gas per mole of amine at 100°F with typical commercial amine concentrations.

Figures 2-68 through **2-71** give the viscosities of MEA, DEA, DGA, MDEA, and DIPA (ADIP) as a function of temperature. Heat capacity data for the same amines are given in **Figures 2-72** through **2-77**. **Figure 2-77** shows the effect of acid gas loading on the specific heat of aqueous amine solutions. It is based on MEA data, but is believed to be approximately valid for other amines.

The freezing points of aqueous solutions of the six most commonly used alkanolamines are depicted in **Figure 2-78**. It is interesting to note that all have low freezing points in the concentration range of about 50 to 80 wt%. This favors the use of concentrated solutions in regions where low temperatures are encountered, which, from a practical standpoint, favors amines, such as DGA, which are commonly used in a concentrated form.

(text continued on page 103)

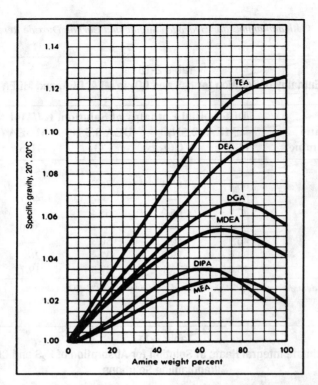

Figure 2-63. Specific gravity of aqueous alkanolamine solutions. (*GPSA, 1994*)

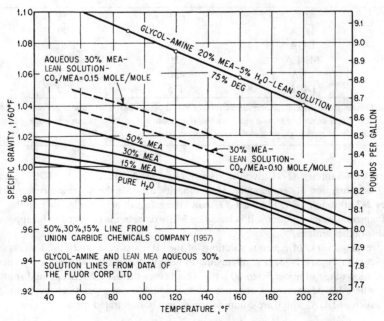

Figure 2-64. Specific gravity of monoethanolamine solutions.

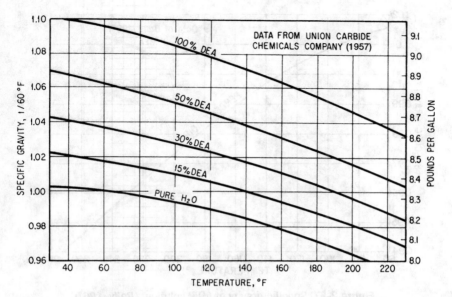

Figure 2-65. Specific gravity of diethanolamine solutions.

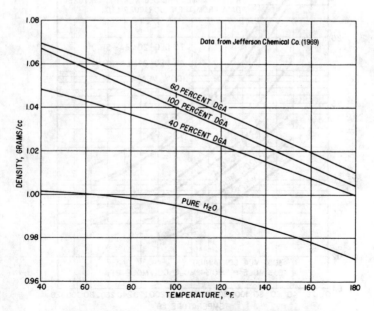

Figure 2-66. Density of Diglycolamine solutions.

102 Gas Purification

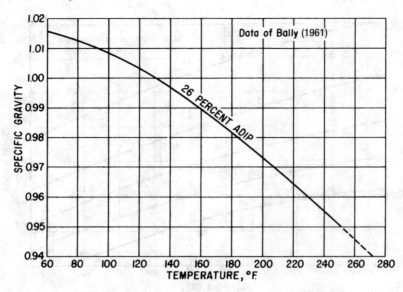

Figure 2-67. Specific gravity of ADIP solution. (*Bally, 1961*)

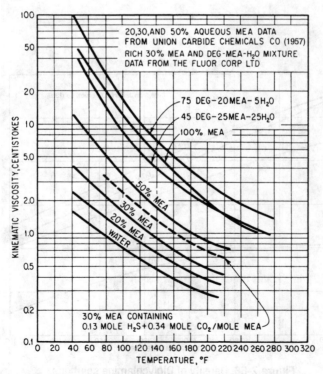

Figure 2-68. Viscosity of monoethanolamine solutions.

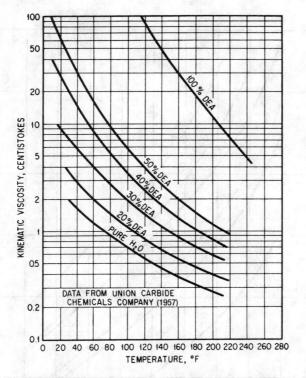

Figure 2-69. Viscosity of diethanolamine solutions.

(*text continued from page 99*)

The physical properties of aqueous solutions of the sterically hindered amine, 2-amino-2 methyl-1-propanol (AMP) are described by Xu et al. (1991). Additional amine physical property data are available from manufacturers and process licensors. The volume of such data is so large that it would be impractical to include it all in this text.

PROCESS DESIGN

Design Approach

The design of amine plants centers around the absorber, which performs the gas purification step, and the stripping system which must provide adequately regenerated solvent to the absorber. After selecting the amine type and concentration, as discussed in a previous section, key items which need to be determined by the designer are the required solution flow rate; the absorber and stripper types (tray or packed), absorber and stripper heights and diameters; and the thermal duties (heating and cooling) of all heat transfer equipment. Various

(*text continued on page 108*)

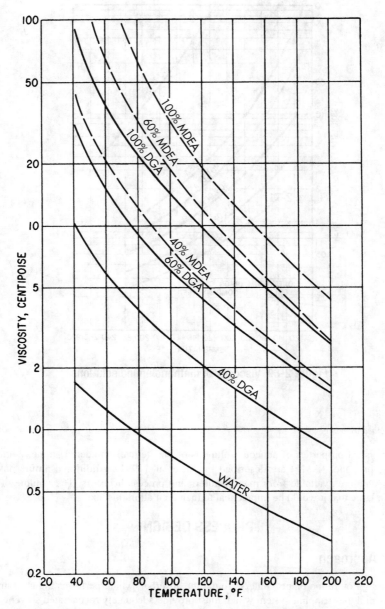

Figure 2-70. Viscosity of Diglycolamine and methyldiethanolamine solutions. *MDEA data from Teng et al. (1994); DGA data from Jefferson Chemical Co. (1969)*

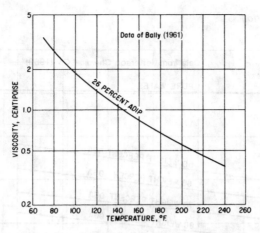

Figure 2-71. Viscosity of ADIP solution. (*Bally, 1961*)

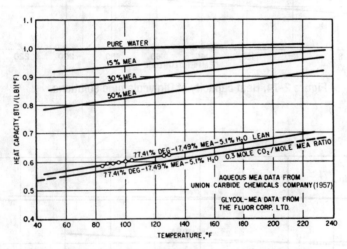

Figure 2-72. Heat capacity of monoethanolamine solutions.

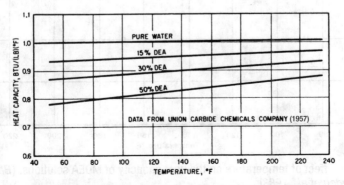

Figure 2-73. Heat capacity of diethanolamine solutions.

106 Gas Purification

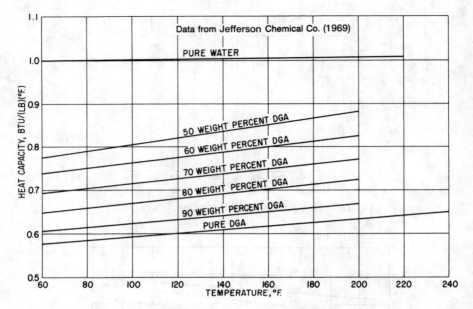

Figure 2-74. Heat capacity of Diglycolamine solutions.

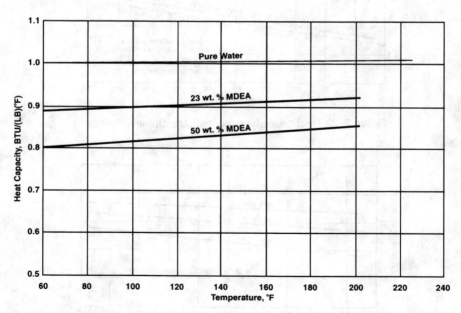

Figure 2-75. Effect of temperature on the heat capacity of MDEA solutions. (*Based on data of Hayden et al., 1983*)

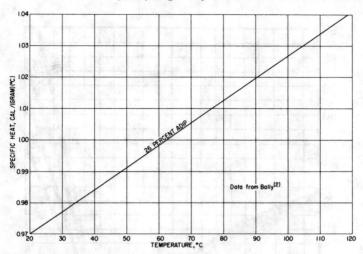

Figure 2-76. Specific heat of ADIP solution. (*Bally, 1961*)

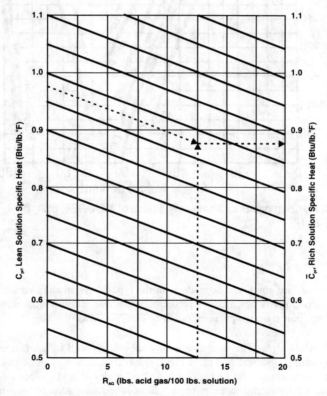

Figure 2-77. Effect of acid gas loading on the specific heat of aqueous amine solutions. *Data from Fluor Daniel (1995)*

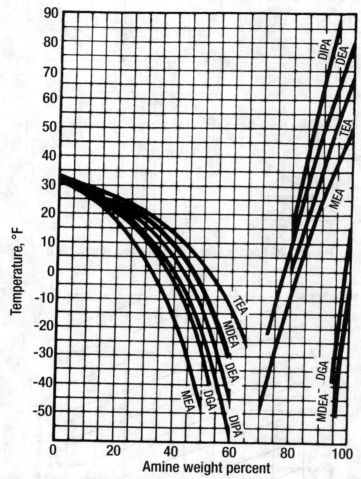

Figure 2-78. Freezing points of aqueous solutions of MEA, DEA, DGA, DIPA, MDEA, and TEA. (*GPSA, 1994*)

(*text continued from page 103*)

design approaches are employed depending on the type of gas treating application and the precision desired. Approaches used for predicting the required solution flow rate and establishing basic equipment design parameters can be categorized as follows:

1. *Rule of Thumb Designs:* This approach works surprisingly well for applications similar to existing operating systems. A typical MEA plant, for example, can be designed on the basis of an assumed net solution loading (e.g., 0.33 moles of acid gas per mole of amine), and an assumed steam consumption (e.g., 1 lb of steam per gallon of solution circulated) based entirely on experience correlations. Another rule of thumb is that about 20 sieve trays

will be required in both the absorber and the stripper. The above "rules" are applicable only if the unit is intended to provide essentially complete removal of acid gas, the amine solution concentration is relatively low (12–20 wt%), and the partial pressure of the acid gas relatively high (over about 1 psia). Pure rule of thumb designs are used only for "quickie" hand calculation estimates of conventional systems.

2. *Approach to Equilibrium Method:* This technique works only for nonselective absorption. It is based on the premise that the theoretical maximum rich solution loading is represented by equilibrium with the feed gas at absorber bottom conditions. Since this theoretical maximum cannot be attained in practical equipment, it is customary to assume that an approach of 75–80% will be attained. This approach can be considered in terms of either a rich solution with an acid gas vapor pressure 75–80% of the acid gas partial pressure in the feed gas, or a rich solution with an acid gas loading 75–80% of the loading in equilibrium with the feed gas. Both calculation methods are in use; however, the latter method tends to be more conservative (i.e., results in a higher design solution flow rate).

When both H_2S and CO_2 are present, they affect each other's vapor pressure and this must be taken into account. Normally, it is assumed that both acid gases are absorbed completely from the feed gas. Carbon dioxide is a stronger acid in solution than hydrogen sulfide and is therefore held more tightly by the amine. As a result, the H_2S/CO_2 ratio in the equilibrium vapor over the rich solution is usually higher than in the feed gas, and the H_2S vapor pressure is often the controlling factor in determining the 75–80% approach to equilibrium.

The approach to equilibrium method is normally used together with selected "rule of thumb" correlations and rigorous thermal calculations in the design of nonselective treating systems. Details of this technique are provided in subsequent sections of this chapter.

3. *Equilibrium-Based Tray Efficiency Techniques:* This is the standard technique for the design of absorbers and strippers for nonreactive systems. A theoretical stage is considered in which the liquid and gas phases attain equilibrium, then the performance of this theoretical stage is adjusted to represent a real tray by the use of a correction factor called the tray efficiency. With nonreactive systems, the tray efficiency can be correlated by consideration of gas and liquid properties, tray design, and flow dynamics. However, with amine plants, where chemical reactions occur in the liquid, the correlation of tray efficiency is much more complex because the reactions affect both the equilibrium relationships and the rate of absorption. Similar problems occur with the HETP concept for packed towers, which also uses an equilibrium-based theoretical plate. Nevertheless, equilibrium based tray efficiency methods are sometimes used because they are convenient, adaptable to graphical analysis, and amenable to either manual or computer calculations.

4. *Rate-Based Approach:* This approach is based on analyzing the mass and heat transfer phenomena occurring on an actual tray (or section of packing) rather than on a "theoretical" tray or in a packing height equivalent to a theoretical plate. The basic procedure was developed by Krishnamurthy and Taylor (1985A and B) for tray and packed towers, respectively, operating with nonreactive systems. The method considers each actual tray individually, and is based on separate mass and heat balances for each phase, which are solved simultaneously with mass and energy rate equations on the tray. Because of the large number of simultaneous equations to be solved, the rate-based approach is applicable only to computer calculation.

The problem becomes even more complex when chemical reactions occur in the liquid phase. The reactions affect both the vapor-liquid equilibrium and the rate of mass transfer. However, correlations have been developed to predict vapor-liquid equilibria for amine-acid gas systems (as discussed in a previous section), and to predict mass transfer rates with chemical reaction based on an enhancement factor to include the reaction effects (see Chapter 1). These have been used by several investigators to develop rate-based, nonequilibrium stage computer models for amine process absorbers and strippers. An example of such a model is described in detail by Sardar and Weiland (1985). An alternative approach, proposed by Tomcej et al. (1987), uses a sophisticated nonequilibrium stage model to determine tray efficiency values for CO_2 and H_2S absorption in alkanolamine solutions. Rate-based, nonequilibrium stage models have been used in commercial computer programs.

Computer Programs

Several computer programs have been developed for designing and simulating amine plants. Development of the Dow Chemical amine plant simulator is described by Katti and Langfitt (1986A and B). The absorber simulator includes models to predict vapor-liquid equilibria, reaction kinetics, hydrodynamics, and the effect of reactions on mass transfer. Thermal parameters included in the model are the heats of reactions, heat transfer between phases, and latent heats of evaporation and/or condensation. According to Katti and Langfitt (1986B), the Dow simulator on average solves a set of 200 coupled nonlinear equations per pass through the column.

Computer programs which have been reported to be commercially available for use in designing amine plants are

Program	Licensor	Reference
AMSIM	D. B. Robinson & Associates, Ltd. Alberta, Canada	Tomcej et al. (1987) Zhang et al. (1993)
APM	Oklahoma State University, Stillwater, OK	Vaz et al. (1981)
Gas Plant	Taylor, Weiland & Associates	Vickery et al. (1988)
TSWEET	Bryan Research & Engineering Bryan, TX	Bullin and Polasek (1982); King et al. (1985); Polasek et al. (1992)

These programs were included in a list published by Manning and Thompson (1991). Some may no longer be available. Also, most of the commercial process plant simulation programs (PRO/II, Hysim, Aspen Plus) have amine plant simulation capability. Of all the commercial amine plant simulators, TSWEET is probably the most popular.

A common problem with many of the commercially available programs is their inability to make an accurate prediction of the lean amine solution composition and therefore the product gas purity. As a result, it is good practice to check stripper performance predicted by a computer program with correlations derived from actual plant experience when available. For applications where little or no plant data are available, such as selective absorption systems, the computer models are the only practical design tools. Furthermore, the models are continually being upgraded, so deficiencies noted in previous versions may be resolved in more recent editions.

Tray Versus Packed Columns

Although bubble-cap trays and raschig ring packings were once commonly used in amine plant absorbers and strippers, modern plants are generally designed to use more effective trays (e.g., sieve or valve types) and improved packing shapes (e.g., Pall rings or high-performance proprietary designs). Very high-performance structured packing is seldom used for large commercial gas treating plants because of its high cost and sensitivity to plugging by small particles suspended in the solution.

The choice between trays and packing is somewhat arbitrary because either can usually be designed to do an adequate job, and the overall economics are seldom decisively in favor of one or the other. At this time, sieve tray columns are probably the most popular for both absorbers and strippers in conventional, large commercial amine plants; while packed columns are often used for revamps to increase capacity or efficiency and for special applications.

Tray columns are particularly applicable for high pressure columns, where pressure drop is not an important consideration and gas purity specifications can readily be attained with about 20 trays. Packing is often specified for CO_2 removal columns, where a high degree of CO_2 removal is desired and the low efficiency of trays may result in objectionably tall columns. Packing is also preferred for columns where pressure drop and possible foam formation are important considerations. Packing should not be used in absorbers treating unsaturated gases that can readily polymerize (propadiene, butadiene, butylene, etc.) as gum formation can lead to plugging of the packing. Also, packing should not be used in treating gases containing H_2S which are contaminated with oxygen because of the potential for plugging with elemental sulfur. General factors affecting the choice between tray and packed towers are discussed in Chapter 1.

Table 2-12 gives a comparison of trays and packing based on an analysis by Glitsch, Inc. and presented by Gangriwala (1987). In the table, valve trays on 24-in. spacing are assigned capacity and efficiency indexes of 100, and various other column internals are assigned comparative values. The data show, for example, that a column can be modified to give 132% of the original capacity at the same efficiency by converting from valve trays at 24-in. spacing to #3 Cascade Mini Rings. Alternatively, the efficiency of a trayed column can be increased at essentially the same (or slightly higher) capacity by converting to #2 or #2.5 Cascade Mini Rings. Conventional random packings (Ballast Rings) do not show any significant capacity/efficiency advantages over trays; however, they may offer other advantages such as a low pressure drop. Although the table covers only Glitsch products, it should be noted that other vendor products in the same categories show very similar performance characteristics.

The effect of contactor design on the selectivity of amine solutions for absorbing H_2S in the presence of CO_2 has been studied by Darton et al. (1987). They conclude that selectivity, as represented by the ratio of overall mass transfer coefficients for H_2S and CO_2, is about the same (100) for trays and 2-in. Pall rings in a MDEA contactor operating at atmospheric pressure. Their work indicates that even higher selectivities should be possible with alternative contactor designs, such as cyclones, centrifuges, and cocurrent gas/liquid flow tubes.

Vickery et al. (1988) compared valve trays with 2-in. steel Pall rings for treating high-pressure natural gas with 50% MDEA solution. The results of this analysis, based on GAS-PLANT software, are given in **Table 2-13**. These results, which indicate a much higher selectivity for Pall rings than for valve trays, do not appear to agree with those of Darton et al. (1987); however, the two studies were based on widely different operating pressures, and other parameters may also have differed substantially.

Table 2-12
Performance Comparison of Trays and Packing

	Capacity Index	Efficiency Index
Valve Trays		
24-in. spacing	100 (1)	100 (1)
18-in. spacing	83	133
30-in. spacing	114	80
Conventional Random Packing (Ballast Rings)		
1.5-in.	83	123
2-in.	91	100
3.5-in.	121	88
High-Performance Random Packing (Cascade Mini Rings)		
#2	98	143
#2.5	109	120
#3	132	100
Structured Packing (Gempak)		
4A	88	385
3A	109	268
2A	125	188
1A	167	109

Note:
1. Basis for comparison, valve trays at 24-in. spacing = 100.
Source: Gangriwala (1987)

Column Diameter

After establishing the liquid and gas flow rates, the column operating conditions, and the physical properties of the two streams, the required diameters of both the absorber and stripping column can be calculated by conventional techniques. For packed towers, correlations of the type proposed by Sherwood et al. (1938), and later modified and improved by Elgin and Weise (1939), Lobo et al. (1944), Zenz and Eckert (1961), Kister and Gill (1991), and others, have proven to be satisfactory for amine solutions. Pressure drop and flooding data for proprietary packing designs are available from the manufacturers. Additional information on the design of packed towers is given in Chapter 1, and the subject is covered in detail by Strigle (1994). It is usually necessary to use a conservative safety factor in conjunction with published packing correlations because of the possibility of foaming and solids deposition in gas treating applications.

The determination of tray column diameters is also discussed in Chapter 1 and covered in detail in standard chemical engineering texts such as *Perry's Chemical Engineer's Handbook* (1963) and the *Handbook of Separation Process Technology* (Fair, 1987). Data on proprietary tray designs are normally supplied by the manufacturers.

Table 2-13
Trays vs. Packing in Selective Treating with 50% MDEA Solution

	Field Data	Calculated Data	
	14 Valve Trays	14 Valve Trays	30 ft of Pall Rings (2″)
Feed Gas			
Pressure, psia	915	915	915
CO_2, vol%	4.0	4.0	4.0
H_2S, vol%	0.4	0.4	0.4
Product Gas			
H_2S, ppmv	8–12	11.2	2.57
CO_2, vol%	1.8–2.2	1.85	3.05
CO_2 slip, %	45–55	45.1	75.2
Temperature Bulge			
Location		mid-tower	mid-tower
Temperature,°F		171	146
Rich Solvent			
Moles CO_2/mole amine		0.597	0.273
Moles H_2S/mole amine		0.1094	0.1094
Moles acid gas/mole amine		0.706	0.382

Source: Vickery et al., 1988

Figure 2-79 is a highly simplified chart for estimating the required diameters for tray-type amine plant contactors. Similar charts are provided by Maddox (1985). Manning and Thompson (1991) suggest using the Souders-Brown equation (Chapter 1, equation 1-22) with an empirical constant of 0.25, reducing the gas velocity by 25 to 35% to avoid jet flooding and by 15% to allow for foaming. They also suggest limiting the liquid velocity in the downcomers to 0.25 ft/sec.

The stripping column diameter can be determined by the same procedures as used for the absorber. Maddox (1985) provides approximate diameter (and height or length) requirements for the stripping column and other regeneration system vessels as a function of the amine solution flow rate.

Column Height

Experience-Based Height Determination

Column heights for amine plant absorbers and strippers are usually established on the basis of experience with similar plants. Almost all installations that utilize primary or secondary amines for essentially complete acid gas removal are designed with about 20 trays (or a packed height equivalent to 20 trays) in the absorber. In bulk acid gas removal applications, experience has shown that if a 20-tray column is supplied with sufficient amine so that the rich solvent leaving the absorber has an acid gas loading that is 75 to 80% of the equilibrium value, then the amine on the upper 5 to 10 absorber trays is very close to equilibrium with the

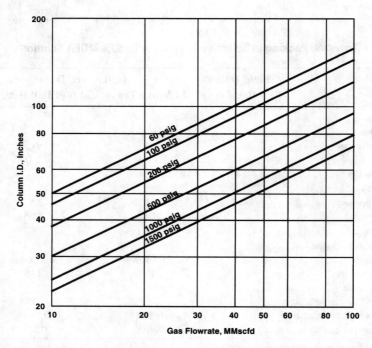

Figure 2-79. Approximate diameter required for tray-type amine plant absorbers. (*Khan and Manning, 1985*)

H_2S in the treated gas leaving these trays. Therefore, in these circumstances, the H_2S content of the treated gas is independent of the absorber design and depends only on the lean amine temperature and the amine regenerator performance. Absorbers with 20 trays can usually meet all common treated gas CO_2 specifications; however, more than 20 trays may be required if CO_2 in the treated gas is to be close to equilibrium with the lean amine. Therefore, in applications such as synthesis gas treating, where it is advantageous to reduce the CO_2 content of the treated gas to very low levels, absorbers containing more than 20 trays or the equivalent height of packing are often specified.

As previously noted, in typical 20-tray absorbers, the bulk of the acid gas is absorbed in the bottom half of the column, while the top portion serves to remove the last traces of acid gas and reduce its concentration to the required product gas specification. With sufficient trays and amine, the ultimate purity of the product gas is limited by equilibrium with the lean solution at the product gas temperature. When water washing is necessary to minimize amine loss (e.g., with low-pressure MEA absorbers), two to four additional trays are commonly installed above the acid gas absorption section. A high efficiency mist eliminator is recommended for the very top of the absorber to minimize carryover of amine solution or water.

Stripping columns commonly contain 12 to 20 trays below the feed point and two to six trays above the feed to capture vaporized amine. Split stream plants obviously require more trays. One unit, for example, employs a total of 33 trays in the stripping column with chimney trays at the base and at the seventeenth tray from the bottom (Bellah et al., 1949). The less volatile amines, such as DEA and MDEA, require fewer trays above the feed point to achieve adequate recovery of amine vapors. Typical DEA and MDEA stripping columns use two to four trays, while MEA systems use four to six trays above the feed point. Equilibrium conditions alone would indicate that the above numbers are overly conservative; however, the trays above the feed point serve to remove droplets of amine solution, which may be entrained by foaming or jetting action, as well as amine vapor.

Overall Gas Absorption Coefficients

Early studies on the absorption of CO_2 and H_2S were reported in terms of the overall gas absorption coefficient (K_Ga) and the Murphree vapor phase tray efficiency (E_{MV}) for packed and tray columns, respectively. Both approaches were based on the assumption that the absorption rate is controlled by the vapor/liquid equilibrium, which is an oversimplification when a chemical reaction occurs in the liquid. However, the approaches are very simple to use and can be helpful when comparing data for the same amine and acid gas at similar operating conditions.

Early experimental studies of the rate of absorption of CO_2 and H_2S in alkanolamines in packed towers were reported by Cryder and Maloney (1941), Gregory and Scharmann (1937), Wainwright et al. (1952), Benson et al. (1954, 1956), Teller and Ford (1958), Leibush and Shneerson (1950), Shneerson and Leibush (1946), and Eckart et al. (1967). Packed column performance data on the absorption of H_2S in MDEA solutions in the presence of CO_2 were presented by Frazier and Kohl (1950) and Kohl (1951). Much of the early work on acid gas absorption in alkalis and amines was reviewed by Danckwerts and Sharma (1966), who proposed design procedures based on fundamental concepts.

Most of the recent studies on the development of design techniques for packed towers have been based on the use of an enhancement factor to account for the effect of chemical reaction on the liquid phase mass transfer. The empirical approach using K_Ga based on overall gas/liquid equilibria is losing favor, but still occasionally used. Strigle (1994), for example, describes the use of K_Ga values for amine plant design. He provides K_Ga values for the absorption of CO_2 by NaOH solution and notes that a 3 N MEA solution will produce a K_Ga for CO_2 absorption about twice that of 1 N NaOH under the same conditions; the K_Ga for CO_2 absorption is about 40% of that for H_2S absorption under the same conditions; and the K_Ga for absorption into DEA solution is only about 50 to 60% of that for MEA solution of the same normality. The reported K_Ga values must be used with extreme caution because of the strong effects of variables such as acid gas partial pressure on the mass transfer coefficient.

Theoretical Stages and Stage Efficiencies

The theoretical stage approach is still employed in several design procedures. It is particularly useful for the design of plants for CO_2 removal where a high removal efficiency is desired, and for the design of MEA stripping columns, where high tray efficiencies are encountered. Gagliardi et al. (1989), for example, suggest that contactor and strippers be sized by graphically determining the required number of theoretical stages, and then applying appropriate tray efficiency or HETP values. They base their design procedure on Air Products & Chemicals, Inc.'s (APCI) extensive experience with MEA plants that remove CO_2 from high-pressure hydrogen and synthesis gas.

APCI plants operate with CO_2 partial pressures from 5 to 75 psia, CO_2 removals to less than 100 ppmv, and MEA concentrations up to 32%. The systems are designed with reflux ratios of 1.0 to 1.2 moles H_2O per mole of CO_2 in the stripper overhead; lean solution load-

116 Gas Purification

ings of 0.15 to 0.2 mole CO_2 per mole amine (0.16 to 0.18 preferred); and a final rich solution loading of 0.45 to 0.55 mole/mole. Under these conditions, the absorber requires 3 to 4 equilibrium stages, and the stripping column requires 6 to 18 theoretical stages. Typical tray efficiencies and HETP values suggested by Gagliardi et al. (1989) are given in **Table 2-14**. Detailed operating data from performance tests conducted on an ammonia plant MEA CO_2

Table 2-14
HETP and Tray Efficiency Ranges for CO_2 Removal Plants Operating on Hydrogen or Synthesis Gas

	Service	Type of Packing	Typical HETP Ranges, ft
HETPs			
	Absorber	1½-in. high eff. rings	10
	Absorber	2-in. high eff. rings	12
	Absorber	Saddle packing	10–16
	Stripper	1½-in. high eff. rings	2–3
	Stripper	2-in. high eff. rings	3–4
	Stripper	Saddle packing	3–6
Tray Efficiencies			
	Absorber		15–30%
	Stripper Column		<70%

Source: Gagliardi et al., 1989

removal unit are given in **Table 2-15**. The data were taken before and after the absorber and stripper columns were converted from trays to packing. The results indicate overall vapor tray efficiencies of 18% in the absorber and 70% in the stripper before the conversion, and HETP values of 8.1 ft in the absorber and 2.1 ft in the stripping column after the conversion.

Figure 2-80 shows a typical absorber tray diagram for the absorption of CO_2 in MEA solution. This figure is based on actual plant data from a 16 bubble-cap tray absorber treating atmospheric pressure flue gas for CO_2 recovery. Because of the low values of solution loading involved, the equilibrium line is almost coincident with the x-axis and is not shown. A pseudo-equilibrium line (dashed) has been drawn to represent actual gas and liquid compositions from each tray. The plate efficiencies in this column vary from about 14% in the bottom of the column to slightly over 16% at the top.

Figure 2-81 shows an approximate tray diagram for a column stripping CO_2 from 17 wt% MEA solution. The equilibrium curve is based on an extrapolation of available vapor pressure data. The concentration in the liquid is expressed as mole fraction CO_2 relative to both water and monoethanolamine because the water content of the solution varies between the feed point and the reboiler. The assumed conditions for the stripping operation are: (a) a pressure of 24 psia and a temperature of 240°F at the reboiler, and (b) 20 psia and 208°F at the top of the column. As can be seen, eight theoretical trays and sufficient steam to produce 2.1 moles H_2O per mole of CO_2 leaving the stripping section result in a lean solution containing 0.14 mole CO_2 per mole MEA from the reboiler. Because of the shape of the equilibrium curve, additional trays would be of little value in reducing the required reflux ratio. The

Table 2-15
Performance of Ammonia Plant MEA CO_2 Removal System with Tray and Packed Columns

	Tray Columns	Packed Columns (3)
MEA Solution Concentration, wt%	31.5	31.1
CO_2 Removal, tons/day	1,018	1,124
Lean Loading, moles/mole	0.18	0.18
Rich Loading, moles/mole	0.51	0.53
MEA Circulation Rate, gpm	2,450	2,350
Regenerator Heat Duty, MMBtu/hr	123.7	131.1
CO_2 Leakage, ppmv	502	27
Absorber		
Pressure drop, psi	9.3	5.4
Temp. rise, °F	53	50
Bottom temp., °F	204	166
Active height, number of trays or feet of packing	20 trays	31.5 ft
No. of theoretical stages (1)	3.60	3.88
Tray efficiency, % or HETP, ft	18%	8.1 ft
Stripper		
Pressure drop, psi (2)	4.5	2.5
Bottom pressure, psig	12.0	10.0
Bottom temp., °F	249	245
Reflux ratio, moles water vapor/mole CO_2	1.5	1.2
Active height, number of trays, or feet of packing	17 trays	31.5 ft
No. of theoretical stages	12.8	15.0
Tray efficiency, % or HETP, ft	70%	2.1 ft

Notes:
1. Includes inlet gas sparger in absorber bottoms (one theoretical stage).
2. Includes overhead piping and condenser.
3. Conversion from tray to packed columns included other minor system changes.
Source: Gagliardi et al. (1989)

performance indicated by the diagram of **Figure 2-81** is typical of stripping columns containing 12 to 16 trays below the solution feed point, indicating overall average tray efficiencies for stripping CO_2 from MEA in the 50–67% range.

Rigorous Column Design Approaches

The previous discussions cover column design procedures based primarily on empirical data. Such procedures have proven adequate for plants designed essentially for complete removal of acid gases from gas streams because an overly conservative design, with a few extra trays, can only improve performance. This is not true for selective absorption, however, because too many trays can destroy selectivity; while too few can cause the production of

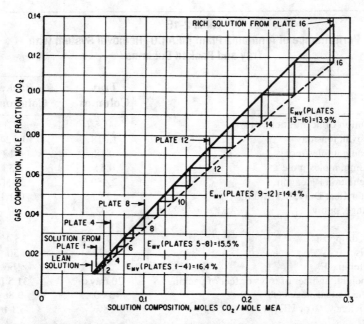

Figure 2-80. Graphical analysis of plate-efficiency data for CO_2 absorption with 14.5% aqueous monoethanolamine in an atmospheric pressure bubble-cap column. *Data of Kohl (1956)*

off-spec gas. As a result, the expanding usage of selective absorption systems has forced designers to develop more sophisticated column design techniques. This development work has generally started with attempts to accurately model the phenomena occurring at the gas-liquid interface. A key parameter in this modeling is the enhancement factor, E, which is defined as the ratio of the actual liquid phase mass transfer coefficient, k_L, to the mass transfer coefficient that would be experienced under the same conditions if no chemical reaction occurred in the liquid, k_{L_o}.

$$E = k_L/k_L^o \tag{2-20}$$

Equation 2-20 can be used to calculate k_L after k_L^o and E are determined. The value of k_L^o is readily estimated by the use of conventional correlations for physical absorption, which take into account the system hydraulics and physical conditions at the interface. The value of E is more difficult to determine because it requires information on both the rate and order of the reaction involved; however, a considerable amount of such data is becoming available. Both k_L and k_L^o are, of course, affected by fluid mechanics, but fortunately their ratio, E, is relatively unaffected. Comprehensive discussions of the theory of mass transfer with chemical reaction are given in texts by Astarita (1967), Danckwerts (1970), and Astarita et al. (1983) and will not be presented here.

Tomcej et al. (1987) proposed a nonequilibrium stage model to simulate the performance of real stages in amine contactors using mass transfer rates to calculate acid gas tray efficiencies. The model was the basis for the AMSIM simulation program in which individual com-

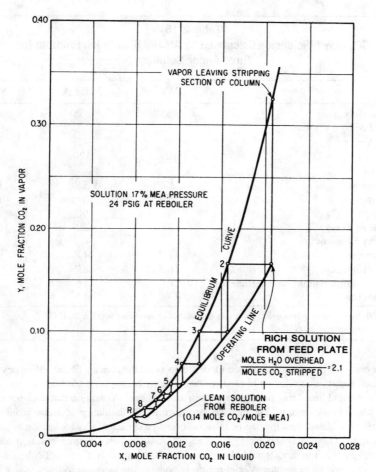

Figure 2-81. Calculated plate diagram for stripping of CO_2 from aqueous monoethanolamine solutions.

ponent stage efficiencies are calculated based on kinetic rate constants, interfacial area, tray hydraulics, operating conditions, and physico-chemical properties of the system. In the original model (Tomcej et al., 1987), the bulk amine concentration was used to calculate enhancement factors.

Rangwala et al. (1989) modified the AMSIM model to use interface amine concentrations rather than bulk concentrations to calculate enhancement factors for CO_2 absorption. They then used the modified model to calculate tray efficiencies for the absorption of CO_2 in solutions of DEA, MDEA, and MEA. Selected results of this study are given in **Table 2-16**. Since details of the tray design and operating conditions are not given, the efficiency values cannot be used for general design purposes; however, the results are of interest in showing the significant variations in efficiency over the length of a column and the wide differences between the three amines studied.

Table 2-16
CO₂ Tray Efficiencies Calculated by AMSIM Simulation Program for Three Amine Solutions

	DEA	MDEA	MEA
Inlet Pressure, psig	465	955	820
Amine Feed			
wt% Amine	35	45	18
Temp.,°F	100	120	110
Gas Feed			
CO_2, mole %	2.7	3.8	1.16
H_2S, mole %	15.3	0	1.43
Temp.,°F	90	90	100
Number of Trays	20	20	20
CO_2 Tray Efficiency			
Top tray	12.0	4.8	14
Max. efficiency	16.3	8.7	45
Tray No. from bottom, max. efficiency	19	16	15
Bottom tray	13.0	4.6	37

Note: Efficiencies are approximate values as calculated by modified version of AMSIM.
Source: Rangwala et al. (1989)

The recent trend in column simulation models is to avoid the concept of tray efficiency entirely. It is considered more relevant to predict what actually occurs on a tray than to predict how closely the gas and liquid tray products approach an equilibrium condition that does not actually exist anywhere in the column. The new rate approach to the simulation of amine plant contactors (or strippers) determines the degree of separation on each actual tray (or section of packing) by considering (1) material and energy balances, (2) mass and energy transfer rate models, (3) vapor-liquid equilibrium models, and (4) reaction rate effects models. Design of columns by this approach is entirely by computer. A brief review of some of the literature follows.

The basic rate approach for non-reactive systems has been described in detail by Krishnamurthy and Taylor (1985A, B). The approach was modified to cover CO_2 and H_2S absorption in alkanolamines by Cornelissen (1980), and alkanolamine regeneration by Weiland and Rawal (1980). Further improvements and comparisons with plant data were made by Sardar and Weiland (1984, 1985). Use of the rate-based model for amine blends and promoted amine solutions is described by Campbell and Weiland (1989) and Vickery et al. (1988). Tomcej (1991) proposed a model that extended the Krishnamurthy and Taylor (1985A, B) method by the addition of an unsteady-state, finite-difference mass transfer model to define the concentration profiles of absorbing and reacting species in the liquid. Tomcej (1992) proposed an improvement to his 1991 model which provides a more accurate representation of the concentration gradients in the liquid phase as it flows across a tray.

Absorber Thermal Effects

The absorber acts as both a reactor and a heat exchanger. Considerable heat is released by the absorption and subsequent reaction of the acid gases in the amine solution. A small amount of heat may also be released (or absorbed) by the condensation (or evaporation) of water vapor. To avoid hydrocarbon condensation the lean solution is usually fed into the top of the absorber at a slightly higher temperature than that of the sour gas, which is fed into the bottom. As a result, heat would be transferred from the liquid to the gas even in the absence of acid gas absorption. The heat of reaction is generated in the liquid phase, which raises the liquid temperature and causes further heat transfer to the gas. However, the bulk of the absorption (and therefore heat generation) normally occurs near the bottom of the column, so the gas is first heated by the liquid near the bottom of the column, then cooled by the incoming lean solution near the top of the column.

When gas streams containing relatively large proportions of acid gases (over about 5%) are purified, the quantity of solution required is normally so large that the purified gas at the top of the column is cooled to within a few degrees of the temperature of the lean solution. In such cases essentially all of the heat of reaction is taken up by the rich solution, which leaves the column at an elevated temperature. This temperature can be calculated by a simple heat balance around the absorber since the temperatures of the lean solution, feed gas, and product gas are known, and the amount of heat released can be estimated from available heat of solution data. A typical temperature profile for an absorber of this type is shown in **Figure 2-82.** This profile is for a glycol-amine system; however, very similar profiles have been observed for MEA and DGA plants.

The temperature "bulge" is a result of the cool inlet gas absorbing heat from the rich solution at the bottom of the column, then later losing this heat to the cooler solution near the upper part of the column. The effect is similar to that of preheating air and fuel to a burner with the combustion products to increase the temperature in the flame zone. The size, shape, and location of the temperature bulge depend upon where in the column the bulk of the acid gas is absorbed, the heat of reaction, and the relative amounts of liquid and gas flowing

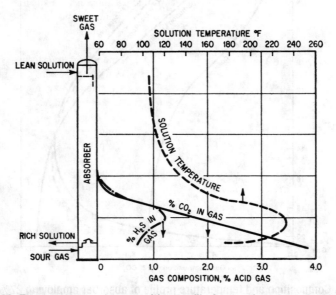

Figure 2-82. Temperature and composition profile for treating plant absorber handling a gas stream containing a high concentration of acid gas.

through the column. In general, for CO_2 absorption, the bulge is sharper and lower in the column for primary amines, broader for secondary amines, and very broad for tertiary amines, which absorb CO_2 quite slowly and also have a low heat of solution.

Since heat is transferred from the hot liquid to the cooler gas at the bottom of the column and in the opposite direction near the top, the temperature profiles for gas and liquid cross each other near the temperature bulge. This effect is shown in **Figure 2-83,** which gives computer generated curves reported by Sardar and Weiland (1985) for an absorber treating 840 psig natural gas containing 7.56% CO_2 and a trace of H_2S with a 27 wt% DEA solution. The temperature bulge is unusually broad because the solution is allowed to attain a very high loading and very little CO_2 absorption occurs in the bottom quarter of the column.

The effect of the liquid/gas ratio on the bulge is illustrated in **Figure 2-84.** The diagram depicts the absorption of CO_2 in a solution of MDEA. As the flow rate of amine is decreased, the temperature bulge increases in magnitude and moves up the column. The curves are based on the data of Daviet et al. (1984) for a small plant treating natural gas. Operating conditions for the three runs are summarized in **Table 2-17.**

When the feed gas contains very little acid gas, the quantity of solution required may be so small relative to the gas that the gas leaving the contact zone will carry more of the reaction heat than will the liquid. In the extreme case illustrated in **Figure 2-85,** the rich solution is

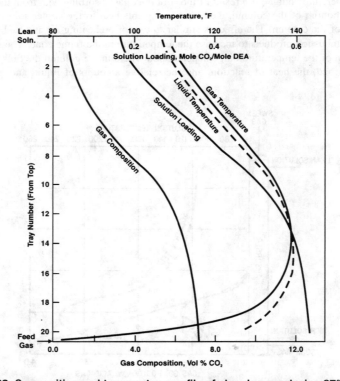

Figure 2-83. Composition and temperature profile of absorber employing 27% DEA solution to absorb CO_2 from high pressure natural gas based on computer simulation. *Data of Sardar and Weiland* (1985)

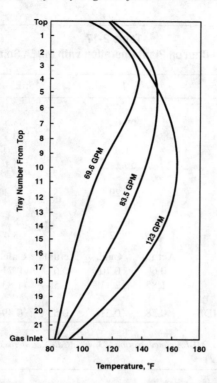

Figure 2-84. Temperature profiles for 21-tray absorber treating 31 MMscfd of natural gas with various flow rates of 33% MDEA. *Based on data of Daviet et al. (1984)*

cooled to approximately the temperature of the incoming gas before it leaves the column, and essentially all of the heat of reaction is taken out of the column by the product gas stream, which raises the temperature at the top of the column. This type of operation can cause problems with product gas purity because the high temperature at the top of the column adversely affects H_2S equilibrium. An empirical correlation used for estimating how much of the heat of reaction leaves with the product gas and how much with the rich solution is described later in this chapter in the section entitled "Simplified Design Procedure."

Stripping System Performance

In conventional stripping operations, heat is supplied to the column by steam or by a heat medium within tubes in the reboiler. Sufficient heat must be supplied to: (a) provide sensible heat to raise the temperature of the feed solution to the temperature of the lean solution leaving the reboiler, (b) provide sufficient energy to reverse the amine-acid gas reaction and dissociate the amine acid-gas compounds, and (c) provide the latent and sensible heat required to convert reflux water into steam which serves as the stripping vapor.

The quantity of stripping vapor required depends upon the solution purity needed to achieve the required product gas purity, the stripping column height, the nature of the solution, the ratio of CO_2/H_2S in the rich amine solution, and the regenerator operating pressure.

Table 2-17
Typical Data on Plant Operation with MDEA Solution

Run Number	1		2		3	
Lean Amine						
Flow rate, gpm	69.6		83.5		123	
Temperature, °F	97		100		120	
H_2S, gr/gal	1.0		1.0		0.5	
CO_2, gr/gal	50.1		54.1		57.8	
Inlet Gas						
Flow rate, MMscf/h	1.29		1.30		1.26	
Temperature, °F	84		85		92	
H_2S, ppmv	50		58		55	
CO_2, vol%	3.52		3.47		3.48	
Product Gas	Actual	Calc.	Actual	Calc.	Actual	Calc.
H_2S, ppm	~0.6	0.70	~0.6	0.71	<0.1	0.39
CO_2, %	1.85	2.11	1.58	1.83	1.13	1.18
Rich Solution						
mole acid gas/mole MDEA	0.58	0.51	0.55	0.49	0.45	0.49

Notes:
Absorber pressure: 800 psi; MDEA concentration: 33 wt%; Calculation by TSWEET program.
Source: Daviet et. al (1984)

The stripping vapor that passes out of the column with the acid gas is normally condensed and returned to the column as reflux. The ratio—moles water in the acid gas from the stripping column to moles acid gas stripped—is commonly referred to as the "reflux ratio" and is used in design as a convenient measure of the quantity of stripping vapor provided.

Typical reflux ratios in commercial columns range from 3:1 to less than 1:1. In general, aqueous monoethanolamine solutions require the highest reflux ratio (typically 2:1 to 3:1), DGA and DEA can be operated with appreciably lower ratios, and MDEA requires the least reflux. Experience with DGA solutions indicates that a reflux ratio of 1.5 mole water per mole acid gas is usually satisfactory for adequate stripping. For MDEA, reflux ratios ranging from 0.3 to 1.0 mole water per mole acid gas have been reported to be satisfactory (Dingman, 1977). The effect of reflux ratio on the H_2S content of purified gas from an MDEA plant is shown in **Figure 2-86** from Vidaurri and Ferguson (1977). For this particular plant, raising the lean solution temperature from 99°F to 129°F caused a major increase in the purified gas H_2S content, but increasing the reflux ratio in the stripping column above about 0.7 mole water vapor per mole of acid gas had little effect on performance at either lean solution temperature.

The operation of a typical stripping column can be visualized by inspection of **Figure 2-87**, which shows composition and temperature profiles for a packed column used for stripping CO_2 from MEA solution. The figure is based on data presented by Sardar and Weiland (1985) and represents a column packed with 37 feet of 2-inch Pall rings and operating at 12.8 psig. The curves were developed by a computer model, which was checked against actual plant data for a unit of this design. Feed to the column consists of 20 wt% MEA solution with a CO_2 loading of 0.5 mole/mole. The rich solution is fed into the column four feet below the top

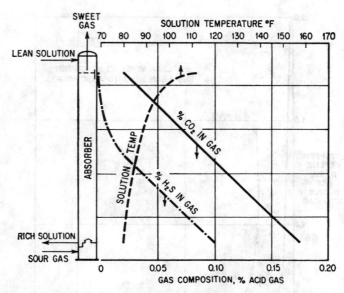

Figure 2-85. Temperature and composition profile of gas-treating plant absorber handling a gas stream containing a low concentration of acid gas.

of the packing at a temperature of 221°F. Reflux water is fed to the top of the packing at 130°F, and is somewhat hotter than the solution feed when the two are mixed. The resulting mixture heats up rapidly to more than 240°F in about 10 feet of packing, then heats up very slowly until it reaches the reboiler where it is finally heated to about 248°F.

The solution CO_2 loading drops rapidly in the column immediately below the feed point, then decreases more slowly until the bottom of the column is reached, and finally exits the reboiler with a CO_2 loading of 0.12 mole/mole. As would be expected, the liquid and vapor composition curves have the same shape. Above the feed point the vapor phase is about one-third CO_2 and two-thirds water vapor, representing a reflux ratio of 2:1 mole H_2O/mole CO_2. These results can be compared to **Figure 2-81,** which also shows vapor leaving the stripping section of an MEA CO_2 unit with a ratio of about 2:1 mole H_2O/mole CO_2.

The stripping of CO_2 from MEA solutions is aided by raising the temperature of the operation. This may be done by either increasing the pressure as described by Reed and Wood (1941) and Reed (1946), or by reducing the water content of the solution by adding a high boiling compound such as glycol. A comparison of the effect of raising the temperature by increasing the pressure with the effect of diethylene glycol addition is shown in **Figure 2-88.** As these curves were derived from plant data in which reflux ratio, number of stripping trays, and other variables were not necessarily constant, they can be taken only as an indication of expected performance.

The reason that CO_2 stripping from MEA solution increases with increased reboiler pressure can be explained on the basis of the effect of temperature on vapor pressures. The vapor pressure of CO_2 over an amine solution generally increases with temperature more rapidly than does the vapor pressure of the water/amine mixture over the same solution. As a result, stripping to the same mole fraction CO_2 in the vapor phase means stripping to a lower mole fraction in the liquid when the pressure (and thus the boiling point) are raised. It should be

126 Gas Purification

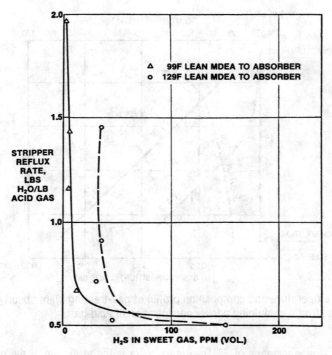

Figure 2-86. Effect of stripper reflux on H_2S in sweet gas. Data are from a refinery unit treating gas containing 8.5% H_2S and 1.4% CO_2 with 15% MDEA. The unit was operated to permit over half the CO_2 to pass through the contactor unabsorbed (*Vidaurri and Ferguson, 1977*)

noted that this phenomenon is peculiar to CO_2 in MEA and will not necessarily occur with other amines nor with H_2S in MEA.

In reports of operating data, it is common practice to express the heat requirements for solution stripping in terms of pounds of steam per gallon of circulated solution. This value is closely related to reflux ratio, but is also dependent on the temperatures of the rich solution and reflux entering the column and the temperature of the lean solution leaving the regenerator. Typical values for pounds of steam per gallon of solution range from less than 1 to 1.5 or even 2 lb/gal. In most cases a steam rate of about 1 lb/gal. is sufficient to obtain satisfactory treated gas purity when mono- or diethanolamine solutions are used.

The effect of steam flow rate on the stripping of H_2S from MEA solutions is indicated by the data of Buskel (1959) and Estep et al. (1962), which are plotted in **Figure 2-89**. The Buskel data were obtained from a plant treating 1,000 psig natural gas containing 11.0% H_2S and 6.0% CO_2 with a 20% MEA solution. The Estep et al. data are from a unit using 14.4% MEA solution to treat natural gas containing 19.9% H_2S and 1.84% CO_2.

The stripping of H_2S is aided by the presence of CO_2 in the amine solution. Fitzgerald and Richardson (1966A, B) provide a very valuable correlation of residual H_2S versus the H_2S/CO_2 ratio in the feed gas as a function of the steam rate for MEA plants treating high pressure natural gas. The results of this study are summarized in **Figures 2-90, 2-91,** and **2-92**.

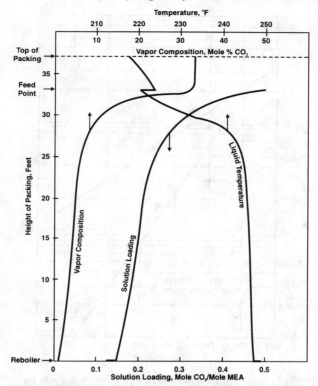

Figure 2-87. Temperature and composition profiles in a packed column used for stripping CO_2 from MEA solutions. *Based on data from Sardar and Weiland (1985)*

Figure 2-90 shows the plant data collected by Fitzgerald and Richardson. The curves on this figure clearly indicate that for each feed gas H_2S/CO_2 molar ratio, the lean solution H_2S content (grains/gallon) approaches an asymptotic lower limit, independent of the amount of stripping steam used (moles of reflux water/mole of acid gas). All amines exhibit this behavior. See, for example, the MDEA solution data presented in **Figure 2-86,** which shows that the H_2S content of the treated gas (an indication of the lean amine H_2S loading) reaches an asymptotic lower limit independent of the amount of stripping steam.

Figure 2-91 is a crossplot of **Figure 2-90.** As **Figure 2-91** indicates, the lean MEA solution loading (grains of H_2S/per gallon solution) at any given stripping steam ratio is a strong function of the feed gas H_2S/CO_2 ratio. These lean amine data are for an MEA solution containing 15.3 wt% MEA and for an amine regenerator with a bottoms temperature of 252°F (about 15 psig in the bottom of the regenerator). The data also assume that the lean/rich exchanger approach temperature (lean amine out minus rich amine in) is about 40°F. The dashed line in **Figure 2-91** indicates the stripper operating conditions required to produce ¼ grain treated gas (4 ppmv H_2S) at 900 psig with lean amine at 110°F. Less stripping steam is required for feed gases with a low H_2S/CO_2 molar ratio, and more stripping steam is required when the acid gas is predominantly H_2S. The dashed line can be used to estimate the treated gas H_2S content at other operating pressures. For example, at 450 psig the dashed line would denote an H_2S content of ½ grain H_2S or 8 ppmv.

128 Gas Purification

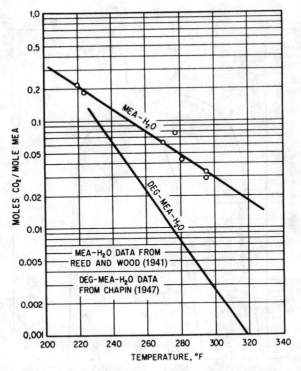

Figure 2-88. Effect of temperature on degree of stripping obtainable for monoethanolamine solutions. Temperature increased for aqueous MEA solution by increasing regenerator operating pressure. (*Reed and Wood, 1941; Chapin, 1947*)

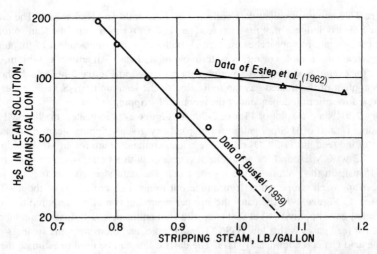

Figure 2-89. Effect of stripping steam rate on H_2S residual in lean monoethanolamine solutions.

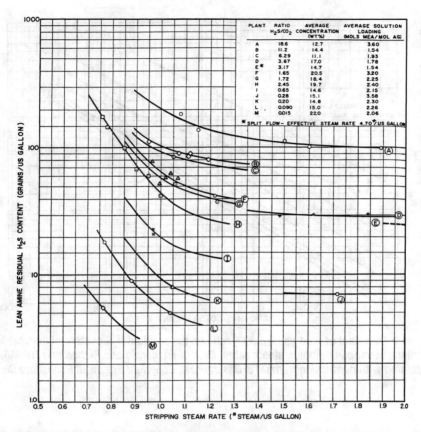

Figure 2-90. Plant data showing how stripping steam rate affects residual H_2S content in lean MEA solutions at various feed gas H_2S/CO_2 ratios. (*Fitzgerald and Richardson, 1966A, B*). Reprinted with permission from Oil & Gas Journal, Copyright Pennwell Publishing Co.

Figure 2-92 shows the lean amine CO_2 loading as a function of the feed gas H_2S/CO_2 ratio and the amount of stripping steam. At feed gas H_2S/CO_2 ratios above 1.0, the lean MEA solution CO_2 loading is a strong function of the feed gas H_2S/CO_2 ratio while, for a given stripping steam rate, the lean MEA solution CO_2 loading approaches an asymptotic limit for H_2S/CO_2 ratios less than 1. The residual CO_2 loadings are based on a fixed regenerator operating temperature of 252°F, which corresponds to a regenerator bottoms pressure of about 15 psig (Fitzgerald and Richardson, 1966B). These data can be approximately corrected for different regenerator operating pressures by use of the data of Reed and Wood (1941) as presented in **Figure 2-88.** The Reed and Wood chart shows stripping to about 0.1 mole CO_2/mole MEA (643 grains/gallon) at 252°F, which is consistent with **Figure 2-92** at a steam rate of slightly over 1.1 lb/gallon.

Figure 2-91 indicates that the MEA solution residual H_2S content decreases by a factor of ten or more as the feed gas H_2S/CO_2 ratio decreases from 10 to 0.1 at a constant stripping

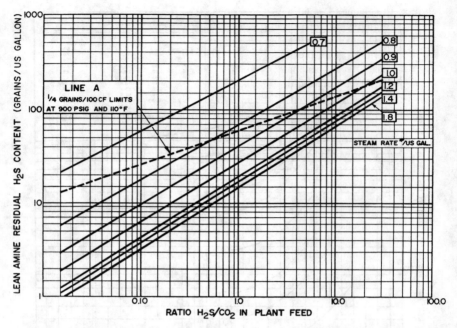

Figure 2-91. Effect of feed gas H_2S/CO_2 ratio on the residual H_2S content of lean MEA solutions at various stripping steam rates. *Data of Fitzgerald and Richardson (1966A). Reprinted with permission from Oil & Gas Journal, Copyright Pennwell Publishing Co.*

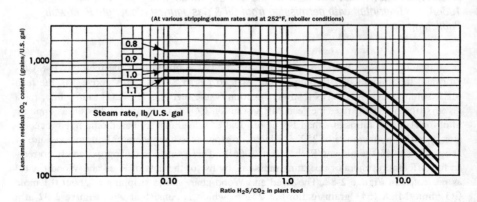

Figure 2-92. Effect of feed gas H_2S/CO_2 ratio on the CO_2 content of lean MEA solutions at various stripping steam rates. *Data of Fitzgerald and Richardson (1966B). Reprinted with permission from Oil & Gas Journal, Copyright Pennwell Publishing Co.*

steam ratio (lb steam/gal. solution). As noted by Fitzgerald and Richardson (1966A, B), unstripped CO_2 in the solution flowing through the stripping column and reboiler contributes to H_2S stripping by increasing the H_2S vapor pressure at a given H_2S loading. This effect of dissolved CO_2 on the vapor pressure of H_2S is clearly shown in equilibrium charts such as **Figure 2-23**. The effect is beneficial in that it reduces the lean amine solution H_2S content. However, the reverse effect can occur in the contactor, where the lean amine residual CO_2 content tends to raise the H_2S vapor pressure and increase the H_2S content of the treated gas. Fortunately, the net effect is normally beneficial because CO_2 stripping continues into the reboiler, and the vapor leaving the reboiler contains both CO_2 and H_2S in equilibrium with the boiling lean solution. Under the same stripping conditions, the partial pressure of either acid gas in the vapor would be less when both acid gases are present than it would be if only the one gas were present. Since the H_2S partial pressure in the reboiler vapor represents its equilibrium vapor pressure, it can be expected that the equilibrium vapor pressure of H_2S over the same solution at the contactor feed temperature will also be lower than it would be if only H_2S had been in the reboiler vapor.

The following discussion provides a chemical equilibrium explanation as to why the presence of another acidic component (such as CO_2) increases the vapor pressure of H_2S over an amine solution. Stripping of H_2S from a primary amine solution can be represented by the following equation:

$$RNH_3^+ + HS^- = RNH_2 + H_2S \qquad (2\text{-}21)$$

where RNH_2 and H_2S are the dissolved molecular forms of the two compounds. Molecular H_2S is then released into the gas phase. At equilibrium, the relationship between the concentrations of each reactant in solution is given by the following equation:

$$\frac{(H_2S)(RNH_2)}{(RNH_3^+)(HS^-)} = K_1 \qquad (2\text{-}22)$$

The concentration of RNH_3^+ ions must be equal to the sum of the negative ions to maintain electrical neutrality, and if both H_2S and CO_2 are present in a lean MEA solution, the principal negative ions are HS^- and carbamate ions, $RNHCO_2^-$. The amount of unreacted amine in solution is equal to the total amine concentration minus the sum of the concentrations of HS^- and $RNHCO_2^-$. Equation 2-22 can be rearranged to a more useful form by using these relationships and by defining L_{CO_2} and L_{H_2S} as the CO_2 and H_2S loadings in the solution in moles acid gas per mole amine and $[RNH_2]_o$ as the total moles of amine, ionized and un-ionized.

$$(H_2S) = \frac{L_{H2S}(L_{H2S} + L_{CO2})K_1}{(1 - L_{H2S} - L_{CO2})}[RNH_2]_o \qquad (2\text{-}23)$$

Equation 2-23 is approximate since it assumes that CO_2 and H_2S are present in solution only as HS^- and carbamate ions; however, it illustrates some valid general effects. Since the equilibrium partial pressure of H_2S in the gas is proportional to the concentration of molecular H_2S in the solution (Henry's Law), the equation indicates that the vapor pressure of H_2S increases with both H_2S and CO_2 loadings in the solution. The enhanced stripping of H_2S in the presence of dissolved CO_2 can, therefore, be attributed to the common ion, RNH_3^+, produced by both acid gases in reacting with an amine. Both primary and secondary amines react with CO_2 to form carbamates, which are difficult to decompose. Therefore, for these amines a significant amount of CO_2 may be present at the bottom of the regenerator, and enhanced stripping of H_2S may be realized. High regenerator pressures may also enhance H_2S stripping when dissolved CO_2 is present by forcing the release of additional CO_2 with its accompanying equilibrium concentration of H_2S.

The "common ion" effect has also been used to advantage in both the Sulften and Flexsorb SE+ Claus tail gas processes. In both of these processes, acids are added to enhance amine solution stripping (Dibble, 1985; Ho et al., 1990). Sulften is an MDEA-based process; while the Flexsorb SE+ process uses a hindered amine. With both processes, Claus plant tail gas H_2S emissions are reduced from 250 to less than 10 ppmv (Tragitt et al., 1986; Ho et al., 1990; Dibble, 1985). In the Sulften process, adding sufficient phosphoric acid to protonate, 2 to 13% of the amine is claimed to be the most beneficial form of the process (Dibble, 1985). Phosphoric acid addition also tends to increase the CO_2 slip (Dibble, 1985, Cordi and Bullin, 1992). As CO_2 absorption in amine solutions is thought to be base catalyzed, it is likely that the lowering of the solution pH when acid is added improves CO_2/H_2S selectivity, which further enhances the benefits of acid addition. Data showing the effect of phosphoric acid addition on the treated gas H_2S content are summarized in **Table 2-18.** The information in the table is for a pilot plant treating gas that contains 1.5% H_2S and 30% CO_2 with a solution containing 50 wt% MDEA (Dibble, 1985).

Acid addition is not beneficial for all amines. Dibble (1985) reports that acid addition actually increases the H_2S content of the treated gas for both TEA and DIPA, but is beneficial for MDEA, DEA/MDEA mixtures, and DEA. Many acids are evidently suitable for lowering the treated gas H_2S content. Dibble (1985) and Ho et al. (1990) state that phosphoric, hydrochloric, acetic, and formic acid, among others, are beneficial. This would imply that heat-stable salts inadvertently formed by the reaction of strong acids with amines could also enhance amine solution stripping and lower the treated gas H_2S content; however, no studies confirming this effect have been reported.

Less data are available on the performance of DEA strippers than MEA. Smith and Younger (1972) surveyed 24 DEA plants in Western Canada. Detailed operating data were not reported, but the following general conclusions were reached:

Table 2-18
Effect of Phosphoric Acid Addition on the H_2S Content of Treated Gas from MDEA Absorber

% H_3PO_4*	Mols H_3PO_4 Per Mol MDEA	% Protonation	ppmv H_2S In Offgas
0	0	0	52
0.1	0.0024	0.49	12.5
1	0.0245	4.9	1
2	0.04956	9.9	2
5	0.1278	25.6	100

*Based on the weight of the solution.
Source: Dibble (1985)

1. Specification gas (either <0.25 or <1.0 grain/100 scf) can be met as long as the H_2S content of the lean solution is less than 40 grains/gallon; however, the H_2S content of the gas does not appear to vary with solution H_2S content below 40 grains/gallon.
2. Specification gas can be made with solution flow rates equivalent to a range from 1 to 2 moles amine per mole of acid gas as long as the solution is adequately regenerated.
3. DEA strips more readily than MEA.
    a. At high $H_2S:CO_2$ ratios regenerated DEA solutions contain 0.1 to 0.3 wt% CO_2 compared to 1.1 to 1.3 wt% CO_2 in lean MEA solutions.
    b. At low $H_2S:CO_2$ ratios regenerated DEA solutions contain less than about 0.9 wt% CO_2 compared to 2.2 to 3.5 wt% CO_2 in lean MEA solutions.
4. Solution strengths from 20 to 30 wt% DEA can be used with no apparent change in treating ability.
5. Most of the plants use stripping steam rates of about 1.2 lb/gal.

Simplified Design Procedure

The simplified design procedure is based on the "approach to equilibrium" method (described earlier) to determine the amine circulation rate, plus a series of enthalpy balances to determine temperatures and heat duties. The calculation procedure involves four steps:

1. Estimating the lean amine loading. This is accomplished by the use of plant data and correlations from plants operating with the same amine and similar acid gas compositions.
2. Determining the heat and material balance around the contactor for the case of equilibrium at the contactor bottom. This is a trial and error calculation to establish the rich solution loading and rich solution temperature corresponding to equilibrium between either one of the acid gases (H_2S or CO_2) in solution and the feed gas.
3. Determining the heat and material balance around the contactor for the case of a 75% approach to equilibrium at the contactor bottom. This represents an adjustment of the step 2 results to provide a reasonable factor of safety and provides the basis for the actual plant design.
4. Designing the regenerator system. These calculations consist of enthalpy balances around regenerator system components based on flow rates and compositions developed in step 3 and empirical correlations of stripping column performance.

Estimating the Lean Amine Loading

The design procedure begins by establishing the lean amine temperature and the lean amine loading (moles acid gas/mole amine). The lean amine temperature is determined by the temperature of the available cooling medium and also by the requirement that the lean amine be at least 10°F hotter than the feed gas hydrocarbon dew point. The lean amine acid gas loading is more difficult to determine and is generally estimated on the basis of experience with similar systems.

The stripping of H_2S and CO_2 from amine solutions is discussed in the previous section entitled "Stripping System Performance." For the case of H_2S and CO_2 removal from natural gas with MEA solutions, the Fitzgerald and Richardson correlations (**Figures 2-90, 2-91, and 2-92**) can be used directly. **Figure 2-92** is based on a regenerator bottom temperature of 252°F, equivalent to a pressure of about 15 psig. Values of the lean solution CO_2 loading, estimated from **Figure 2-92**, can be extrapolated to other stripper pressures (and temperatures) by the use of the Reed and Wood correlation, **Figure 2-88.** Unfortunately, correlations similar to those generated for MEA by Fitzgerald and Richardson have not been developed for other amines, and lean solution loading estimates are usually based on the limited plant operating data that are available.

DEA is a weaker base than MEA. As a result, the same stripping conditions will produce a lower lean solution loading in DEA than in MEA solutions. Alternatively, a similar level of stripping can be accomplished with less steam consumption by the DEA system. In fact, improved steam economy is one reason many plants have been converted from MEA to DEA. A survey of 24 DEA plants, conducted by Smith and Younger (1972), suggests that 1.2 lb steam/gallon of solution is adequate to produce a lean solution containing less than 40 grains H_2S/gallon (about 0.007 mole H_2S/mole DEA) under a variety of DEA plant operating conditions.

As with MEA, carbon dioxide is the principal acid gas remaining in DEA solution. Smith and Younger found typical CO_2 loadings of lean DEA solution to be in the range of about 0.01 to 0.03 mole CO_2/mole DEA with high ratios of $H_2S:CO_2$ in the feed gas, and to be less than about 0.09 with low $H_2S:CO_2$ ratios. Other investigations report CO_2 loading much less than 0.09 mole CO_2/mole DEA, even at very low $H_2S:CO_2$ ratios. Butwell and Perry (1975), for example, describe two DEA plants treating natural gas with $H_2S:CO_2$ ratios of only 0.04 to 0.07 that produced lean DEA loadings in the range of about 0.01–0.02 mole CO_2/mole DEA.

DGA is a primary amine similar to MEA with regard to basicity. As a result, the correlations developed for MEA stripping can be used to provide a first approximation of lean solution loadings for DGA solutions. Additional approximate lean solution loading values for DEA and DGA are provided in **Table 2-19.** Little data have been published on the stripping of DIPA, and it is suggested that lean solution loadings estimated for DEA be used in the absence of more specific data. For material balance purposes, lean MDEA loadings can be assumed to be zero, as MDEA solutions are very readily stripped. For example, lean MDEA solution loadings less than 0.004 mole acid gas/mole amine have been reported by Dupart et al. (1993A, B).

Determining the Rich Solution Equilibrium and Design Loadings

After the lean amine temperature and lean solution loading are determined, the designer should use the following steps to determine the required circulation rate. Refer to **Figure 2-93** for information on the amine contactor calculation envelope and to **Table 2-20** for a definition of the variables used in these calculations.

1. Establish the lean amine concentration and loading, L_L (moles acid gas/mole amine), and break the loading down into L_{L,H_2S} (moles of H_2S/mole amine) and L_{L,CO_2} (moles of CO_2/mole amine).
2. Tabulate the feed gas conditions, including the following:
   a. Feed gas flow rate, M_F (moles/hr) and W_F (lb/hr).
   b. Feed gas inlet temperature, T_F (°F).
   c. Feed gas inlet pressure, P_F (psia).
   d. Feed gas composition (mol %).
   e. Water content, M_{F,H_2O} (moles/hr) and W_{F,H_2O} (lb/hr).
   f. Latent heat of water in the feed gas, h_F (Btu/lb).
3. Set the lean amine temperature, T_L (e.g., $T_F + 10$).

Table 2-19
Approximate DEA and DGA Lean Solution Loadings for Various Operating Conditions

		Stripping Steam		Regenerator Bottom Temp., °F	L/R Exchanger Approach, °F	Feed Gas CO_2/H_2S	Lean Solution Loading		
Amine	Wt% Amine	lb steam/ gal sol'n	mol H_2O/ mol AG				mol AG/ mol amine	mol CO_2/ mol amine	mol H_2S/ mole amine
DEA	~20	~1.2	—	—	—	>10	—	<0.07	<0.01
DEA	~20	~1.2	—	—	—	<10	—	<0.04	<0.01
DEA	20	1.0	1.7	240	—	0.5	—	<0.01	<0.01
DEA	20	1.2	3.2	230	30	0.4	—	<0.01	<0.01
DEA	25	0.9	0.4–0.6	—	—	0.3	—	—	0.004–0.011
DEA	20–22	1.1	—	230	56	1.9	—	—	0.005–0.006
DEA	20	1.0	1.7	240	—	0.5	—	0.01	0.005
DGA	60	1.5	1.5	250	30	1.0	0.10	—	—
DGA	54	—	3.7	250	67	1.1	—	0.04	0.01
DGA	67	1.6	—	271	42	2.9	—	0.01	Trace
DGA	55	1.1	1.4	246	54	34	—	0.09	Trace
DGA	62	1.1	1.1	247	20	4.0	—	0.06	0.01

136 *Gas Purification*

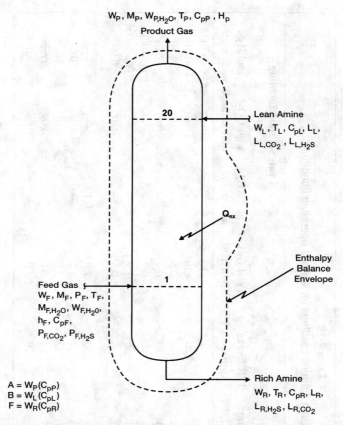

Figure 2-93. Heat and material balance schematic for amine contactor design.

4. Calculate the partial pressures of the CO_2 and H_2S in the feed gas (P_{F,CO_2} and P_{F,H_2S}) in psia.
5. Calculate the total acid gas components in the feed gas, $M_{F,AG}$ (moles/hr) and $W_{F,AG}$ (lb/hr); the ratio of CO_2 to H_2S, R; and the total heat of reaction, Q_{RX}. Q_{RX} is usually calculated on the basis that all of the acid gas is absorbed.
6. Estimate the rich amine solution temperature, T_R, at the contactor bottom (e.g., $T_L + 30$).
7. Next, use the VLE chart data in this chapter or in other references (GPSA, 1987; Dow, 1962; Maddox, 1985) in conjunction with the following steps to determine the vapor pressure of CO_2 and H_2S above the rich solution.
   a. Assume a rich solution acid gas loading, L_R (for example, $L_R = 0.6$ moles acid gas per mole amine is a reasonable starting point for most conventional MEA, DEA, or DGA plants).
   b. Calculate the rich solution H_2S and CO_2 loadings, L_{R,H_2S} and L_{R,CO_2}:

$$L_{R,H_2S} = \frac{R(L_{L,H_2S}) + L_R - L_{L,CO_2}}{R+1} \qquad (2\text{-}24)$$

$$L_{R,CO_2} = L_R - L_{R,H_2S} \qquad (2\text{-}25)$$

Table 2-20
Nomenclature for Simplified Design Procedure Equations

C_p = Heat capacity, Btu/(lb)(°F)
h = Latent heat of water vapor, Btu/lb
L = Solution loading, mole acid gas/mole amine
M = Gas flow rate, total, mole/hr
M' = Gas flow rate, dry basis, mole/hr
MW = Molecular weight, lb
P = Pressure or partial pressure, psia
P^o = Equilibrium vapor pressure over a solution, psia
Q_{RX} = Total heat of reaction, Btu/hr
Q_{RC} = Regenerator condenser duty, Btu/hr
Q_{LR} = Lean/rich heat exchanger duty, Btu/hr
Q_{RE} = Reboiler duty, Btu/hr
R = Acid gas ratio in feed gas, mole CO_2/mole H_2S
T = Temperature, °F
W = Mass flow, total, lb/hr
W' = Mass flow, dry basis, lb/hr

Subscripts:
A = Acid gas stream at reflux drum outlet
AG = Acid gas
B = Solution at regenerator bottom
C = Stripper overhead condensate
F = Feed gas entering absorber
L = Lean solution entering absorber
LX = Lean solution exiting lean/rich heat exchanger
P = Product gas leaving absorber
R = Rich solution leaving absorber
REF = Reference value
SO = Acid gas stream at stripper overhead

c. Use VLE chart data and the following equation (GPSA, 1987; Lee et al., 1973C) to determine the CO_2 and H_2S partial pressures at the estimated rich amine solution temperature, T_R, based on published data at T_1 and T_2:

$$\ln P_R^o = \ln P_1^o + \left[\frac{\frac{1}{T_1} - \frac{1}{T_R}}{\frac{1}{T_1} - \frac{1}{T_2}}\right] \ln\left(\frac{P_2^o}{P_1^o}\right) \qquad (2\text{-}26)$$

Equation 2-26 is based on the fact that the vapor pressure of acid gas above the rich amine solution is inversely proportional to the logarithm of the absolute temperature. Use of this equation is usually necessary as the VLE chart data cover a limited number of temperatures.

d. Adjust the estimated rich solution loadings appropriately until the calculated vapor pressure of one of the two acid gases is equal to its partial pressure in the feed gas. One of the two acid gases, usually H_2S, will be controlling. When the vapor pressure of one of the acid gases above the rich solution corresponds to the partial pressure of the same acid gas in the feed gas, the equilibrium solution loading at the estimated rich amine temperature has been determined.

8. Next, estimate the split of the heat of reaction, Q_{RX}, between the sweet gas product and the rich amine solution. The split of the heat of reaction is controlled by the ratio of A, the product of the mass flow and the specific heat of the product gas, and B, the product of the mass flow and the specific heat of the lean amine solution. If A/B is less than one, all the heat of reaction leaves the amine contactor with the rich amine solution. If the ratio is greater than 1, some of the heat of reaction leaves with the product gas:

138 *Gas Purification*

$$A/B = W_P(C_{pP})/W_L(C_{PL}) \tag{2-27}$$

Figure 2-94 summarizes a correlation based on plant data which can be used to estimate the split of the reaction heat between the treated gas and rich amine streams (Fluor Daniel, 1995).

Under most circumstances, all the heat of reaction is contained in the rich amine stream leaving the bottom of the contactor. However, when the acid gas content of the feed gas is low and/or the amine concentration of the lean amine is high, some of the heat of reaction can leave with the treated gas product stream. For example, Holder (1966) and Dingman and Moore (1968) report that temperature breakthrough occurred, i.e., the treated gas temperature was higher than the lean amine temperature, when a 60 wt% DGA solution was used to treat a feed gas containing less than 1 to 2 mol% acid gas.

a. The reaction heat split is determined by first calculating the acid gas pick-up, AGPU, in moles of acid gas removed from the feed gas per mole of amine:

$$AGPU = L_{R,H_2S} + L_{R,CO_2} - L_L = (L_{R,H_2S} - L_{L,H_2S})(1 + R) \tag{2-28}$$

b. Then calculate the lean amine circulation rate, W_L (lb/hr), using the following equation:

$$W_L = (M_{F,AG})(MW_{(AMINE)})/(AGPU)(100)(wt\% \text{ AMINE}) \tag{2-29}$$

c. Obtain the product gas rate, W_P (lb/hr), assuming that all the acid gas components are removed from the feed gas. Estimate the water content of the product gas, W_{P,H_2O} (lb/hr), using Raoult's law and basing the calculations on the water vapor pressure at the lean amine temperature and the mole fraction of water in the lean amine. Using steam tables, find the latent heat of water in the product gas, h_P (Btu/lb).

d. Obtain the heat capacity of the product gas, C_{pP} (Btu/lb°F), including the water. Heat capacities should be average values. In this case, C_{pP} is an average over the product gas temperature, T_P, and a selected reference temperature. It is usually advantageous to choose the feed gas temperature, T_F, as the reference temperature, as this simplifies the enthalpy balance around the amine contactor.

e. Multiply the mass flow rate of the product gas, W_P, by its average heat capacity, C_{pP}, to obtain A.

f. Determine the heat capacity, C_{pL}, of the lean amine solution at its operating temperature. Again, C_{pL} is an average over the lean amine temperature and the selected reference temperature.

g. Multiply the lean solution heat capacity, C_{pL}, by the lean amine circulation rate, W_L, to obtain B.

h. Calculate the ratio A/B.

i. If A/B is less than 1.0, set the product gas temperature, T_P, the same as the incoming lean amine solution temperature, T_L, and go to Step 9. If A/B is greater than 1, use the following procedure and **Figure 2-94** to determine the heat of reaction split between the treated gas and the rich amine.

  i. Set the rich amine solution temperature, T_R, equal to the feed gas temperature, T_F.

  ii. Assuming no heat of reaction, compute the product gas temperature as follows:

$$T'_P = (B/A)(T_L - T_F) + T_F \tag{2-30}$$

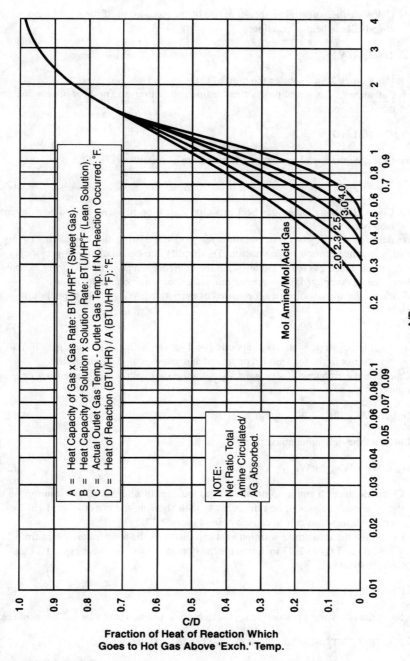

Figure 2-94. Correlation for determining the outlet gas temperature for amine contactors. *Data from Fluor Daniel (1995)*

140 Gas Purification

iii. Calculate the potential temperature rise of the gas, D (°F), if the product gas were to absorb all the heat of reaction:

$$D = Q_{RX}/A \qquad (2\text{-}31)$$

iv. Use **Figure 2-94** to obtain the ratio of C/D.

v. Calculate C, the actual product gas temperature minus the no-heat-of-reaction product gas temperature:

$$C = (C/D)(D) \qquad (2\text{-}32)$$

vi. Calculate the actual product gas temperature as follows:

$$T_P = T'_P + C \qquad (2\text{-}33)$$

vii. Check that the product gas heat capacity, C_{pP}, used in Steps 8d and 8e is approximately correct.

9. Next determine the rich amine solution temperature exiting the contactor, T_R, by an enthalpy balance around the contactor. To simplify the calculations, as previously discussed, use the feed gas temperature, T_F, as the reference temperature. However, before this balance can be completed, it is necessary to make some additional calculations:

a. Calculate the amount of water condensed or evaporated, W_{H_2O}, lb H_2O per hour:

$$W_{H_2O} = W_{F,H_2O} - W_{P,H_2O} \qquad (2\text{-}34)$$

If W_{H_2O} is negative, water is evaporated and leaves with the product gas. If W_{H2O} is positive, water in the feed gas is condensed and leaves with the rich amine.

b. Calculate the enthalpy loss or gain, Q_{H_2O}, associated with water condensing or evaporating:

$$Q_{H_2O} = W_{F,H_2O}(h_F) - W_{P,H_2O}(h_P) \qquad (2\text{-}35)$$

c. Calculate the rich amine mass flow, W_R, lb/hr:

$$W_R = W_L + W_{F,AG} + W_{H_2O} \qquad (2\text{-}36)$$

d. Calculate the rich amine solution acid gas loading, lb acid gas/100 lb solution, so that the solution loading can be referenced to data showing the effect of acid gas loading on rich amine solution heat capacity. See **Figure 2-77**.

e. Determine the lean amine solution heat capacity over the temperature range from T_R to T_F. Next use **Figure 2-77** to determine the rich amine solution heat capacity, C_{pR}.

f. Next, calculate F:

$$F = W_R(C_{pR}) \qquad (2\text{-}37)$$

g. Solve the following equation for T_R, the rich amine solution contactor exit temperature:

$$T_R = [B(T_L - T_F) - A(T_P - T_F) + Q_{RX} + Q_{H_2O}]/F + T_F \qquad (2\text{-}38)$$

Equation 2-38 is based on an enthalpy balance around the amine contactor as shown in **Figure 2-93**. In this balance, the reference temperature is selected as the feed gas temperature, T_F.

10. Check to see that T_R assumed in Step 6 matches T_R calculated by equation 2-38. If the temperatures match, the rich solution equilibrium composition and temperature have been correctly determined. Go to Step 13 to determine the actual design balance.
11. If T_R calculated by equation 2-38 is less than the T_R assumed in Step 6, the lean amine circulation rate can be reduced. Assume a new T_R between T_R calculated and T_R assumed, and repeat Steps 6 through 12 until a match is obtained.
12. If the calculated T_R is greater than the T_R assumed in Step 3, the circulation rate is too low and must be increased. Assume a new T_R between T_R calculated and T_R assumed, and repeat Steps 6 through 12 until a match is obtained.
13. The design heat and material balance is obtained after a converged rich amine solution loading and temperature have been determined. Set the design rich solution loading to 75% of the equilibrium loading and repeat Steps 8 and 9 to determine the design T_R and rich and lean amine flows:

$$L_{R(des.)} = 0.75 \, (L_{R,H_2S} + L_{R,CO_2}) \tag{2-39}$$

If the temperature of the product gas is high enough to affect its H_2S content (assuming equilibrium with the lean solution) consider reducing the amine solution concentration. Also, 75% of the equilibrium rich solution loading may be higher than desired as highly loaded rich amine solutions can be very corrosive. Refer to the maximum recommended rich amine loadings in Chapter 3 for guidance on maximum loadings.

Amine Regenerator Calculations

Amine regenerator design calculations that can be performed readily without a computer are based on a series of enthalpy balances. The basis for these balances is empirical data on the amount of steam required to strip the rich amine. These data are usually available in the form of: 1) the pounds of stripping steam required to strip one gallon of rich solution or; 2) the moles of water vapor per mole of acid gas in the regenerator overhead just upstream of the regenerator condenser. Since the water vapor is normally condensed and returned to the regenerator as reflux, the moles of water vapor per mole of acid gas leaving the regenerator is often called the "reflux ratio." There is a direct relationship between these two empirical measures of amine regenerator energy input.

Most rich amines can be adequately stripped with 0.9 to 1.2 pounds of steam per gallon of rich solution. See **Figure 2-91,** which is based on empirical MEA performance data collected by Fitzgerald and Richardson (1966A, B). Implicit in the data of Fitzgerald and Richardson is the assumption that the lean/rich exchanger approach (lean amine temperature out minus rich amine temperature in) is about 40°F. When the energy requirement is stated in terms of reflux ratio, the empirical data indicate that supplying enough reboiler energy to generate about 1 to 3 moles of reflux water per mole acid gas is usually sufficient to adequately strip most rich amine solutions. Use of either type of empirical stripping data is straightforward. The following calculation strategy is based on the availability of plant performance data based on pounds of stripping steam per gallon of rich solution, as there are more data of this type in the literature. Results can be qualitatively checked by calculating the moles of reflux water per mole acid gas.

The enthalpy balances required to size an amine regenerator are indicated in **Figure 2-95.** The balances begin with a direct calculation of the reboiler duty using a selected value for pounds of steam per gallon of rich solution based on **Figure 2-91.** This is followed by a calculation of the lean/rich exchanger duty. An enthalpy balance around the regenerator is then

142 Gas Purification

used to determine the condenser duty. The temperature and composition of the regenerator overhead stream just upstream of the condenser is then calculated by an enthalpy balance around the condenser.

Before starting these calculations, it is necessary to set the amine regenerator pressure profile. First the regenerator reflux drum pressure, P_A, is set. This pressure is usually determined by the pressure required to deliver the acid gas to a sulfur plant and/or the pressure required to provide the desired amine regenerator bottom temperature. The minimum regenerator reflux drum pressure is usually about 5 psig, with an upper pressure limit of about 15 or 20 psig depending on the amine and the maximum allowed operating temperature for that amine. See Chapter 3 for recommended maximum operating temperatures for various amines. For design purposes, a pressure allowance of 5 psi is reasonable for the pressure drop from the regenerator bottom to the regenerator reflux drum. In most circumstances, as indicated in **Figure 2-95,** the amine regenerator pressure is controlled by a pressure control valve located on the acid gas line leaving the regenerator reflux drum. A pressure allowance also has to be included for this valve.

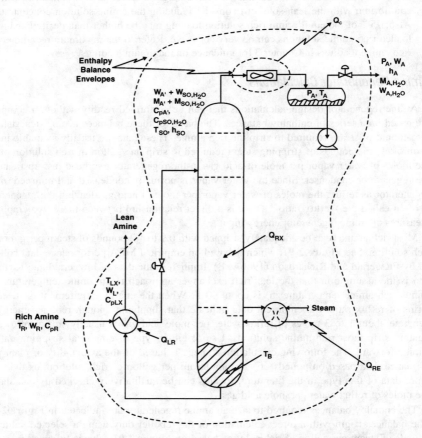

Figure 2-95. Enthalpy balances required to size an amine regenerator.

After the regenerator pressure profile and the reflux drum pressure are established, the reflux drum temperature, T_A, is fixed. If the acid gas is to be fed to a Claus sulfur plant, the reflux drum temperature should be minimized consistent with the available cooling medium temperature since water contained in the acid gas is detrimental to sulfur plant performance. For air cooling, an approach of 20°F is common, while a typical approach for cooling water would be in the range of 10°F.

After fixing the pressure profile and key temperatures, the required reboiler and lean/rich exchanger duties are calculated. The reboiler duty, Q_{RE}, is calculated directly from the stripping steam requirement obtained from **Figure 2-91** and the rich amine flow determined in the absorber design calculation outlined earlier. The steam latent heat is calculated based on the temperature/pressure conditions of condensing steam in the reboiler. This temperature is fixed by either the maximum allowable steam operating temperature or a minimum 20°F reboiler approach temperature (steam saturation temperature minus reboiler outlet temperature). A maximum steam operating temperature of 300°F has been recommended by several authors. See the discussion of recommended maximum steam and hot oil operating temperatures in Chapter 3.

In determining the lean/rich exchanger duty, Q_{LR}, it is first necessary to determine the regenerator bottoms temperature, T_B. This temperature is fixed by the lean amine composition, the lean amine type, and the regenerator pressure profile. As previously noted, a pressure drop of about 5 psi from the reflux drum to the stripper bottoms should be allowed. Available aqueous amine solution vapor pressure data are then used to determine the stripper bottoms temperature, T_B. See **Figures 2-56, 2-57, 2-58, 2-59, 2-60,** and **2-61** for vapor pressure/temperature data for MEA, DEA, DGA, TEA, DIPA, and MDEA, respectively.

After determining the regenerator bottoms temperature, the temperature of the lean amine leaving the lean/rich exchanger is estimated. A commonly used rule of thumb is that the difference between the lean amine outlet temperature, T_{LX}, and the rich amine inlet temperature, T_R, should be about 40°F. In addition, the rich amine outlet temperature should not be higher than about 210 to 220°F as excess heat recovery contributes little towards amine stripping (i.e., above a certain temperature, recovered heat generates steam at the feed tray, which is much less effective for amine stripping than steam generated by the reboiler). Having set the lean amine temperature, the lean/rich exchanger duty can be calculated with the following equation:

$$Q_{LR} = (W_L)C_{pL}(T_B - T_{LX}) \tag{2-40}$$

W_L is the lean amine circulation determined in the amine contactor calculation, and the lean amine specific heat is an average calculated over the T_B to T_{LX} temperature range.

Next the total acid gas flow (dry basis), M_A' (mole/hr) and W_A' (lb/hr), are calculated. Then steam tables are used to determine the latent heat of water vapor at the reflux drum outlet conditions, h_A, and to calculate the water content of the acid gas, M_{A,H_2O} (mole/hr) and W_{A,H_2O} (lb/hr). Finally, the total flow of acid gas plus water vapor leaving the reflux drum, M_A (mole/hr) and W_A (lb/hr), are determined.

Sufficient data are now available to perform an enthalpy balance around the regenerator to determine the regenerator condenser duty, Q_{RC}. The heat of reaction, Q_{RX}, is from the amine contactor design calculation:

$$\begin{aligned}Q_{RC} = &\ Q_{RE} - Q_{RX} + (W_R)C_{pR}(T_R - T_{REF}) - (W_A)C_{pA}(T_A - T_{REF}) \\ &+ W_{A,H_2O}(h_A) - (W_L)C_{pL}(T_{LX} - T_{REF})\end{aligned} \tag{2-41}$$

144 Gas Purification

In equation 2-41, T_{REF} is any suitably chosen reference temperature, and the specific heats are averages for the temperatures between T_{REF} and the relevant process temperature.

Next, an enthalpy balance around the condenser is used to determine the composition and temperature of the overhead stream just upstream of the condenser. The amount of water in this stream is given by the following equation:

$$W_{SO,H_2O} = \{Q_{RC} - W_A'C_{pA}'(T_{SO} - T_A) + (W_{A,H_2O})h_A\}/\{C_{pSO,H_2O}(T_{SO} - T_A) + h_{SO}\} \tag{2-42}$$

The subscript "SO" refers to the stripper overhead, i.e., the stream leaving the regenerator column. See **Figure 2-95**. T_{SO} and several terms in this equation, which are functions of T_{SO}, are not known. Neglecting relatively small terms, equation 2-42 can be simplified to

$$W_{SO,H_2O} = \{Q_{RC} + (W_{A,H_2O})h_A\}/h_{SO} \tag{2-43}$$

With an initial estimate of h_{SO}, a first approximation of W_{SO,H_2O} can be calculated by use of equation 2-43. Steam tables can then be used to determine T_{SO} and a more accurate value of W_{SO,H_2O} can be calculated by repeated direct substitution into equation 2-42. When determining T_{SO}, about 2 psi should be allowed for the pressure drop through the condenser and associated piping to the reflux drum.

When these calculations are completed, equipment sizing data can be generated from the results. For example, the regenerator diameter is determined by conventional column design correlations (see Chapter 1) based on the lean amine flowrate and the vapor flow generated by the reboiler duty.

Commercial Plant Operating Data

The availability of operating data on a full size plant, operating under similar conditions, is extremely helpful in the design of a new installation. Although many papers present information on commercial plants, most do not provide complete and consistent sets of operating data. This is particularly true for proprietary processes. However, an appreciable amount of useful plant data is available for the commonly used amines. Obviously, only a small portion of such material can be included in this text, and the reader is referred to the original publications for more complete information.

Monoethanolamine Plants

Design and operating data from two large MEA plants are summarized in **Table 2-21**. Case A is a plant designed primarily to remove H_2S from 200 psig natural gas and produce a sweet gas containing less than ¼ grain per 100 scf. Although no data are given on CO_2 removal, it is expected that essentially all of the CO_2 in the feed gas would also be absorbed in a plant of this type. Case B is a plant intended to recover CO_2 from flue gas at atmospheric pressure. Two 23-ft high-packed beds are used in a single tall absorber to provide efficient CO_2 removal with less pressure drop than an equivalent tray column. Additional operating data on an MEA plant using tray and packed towers are given in **Table 2-15** in the previous section of this chapter entitled "Tray Versus Packed Columns."

Operating data from a large test unit used to absorb CO_2 from ammonia synthesis gas are given in **Table 2-22** (Butwell et al., 1979). The test runs were made to obtain data for the

Table 2-21
Operating Data for Aqueous MEA Gas Treating Plants

Plant	A	B
Gas Feed Rate, MMscfd	50 (each absorber)	72
Solution Flow rate, gpm	~70–100 (each absorber)	~2,000
Solution Concentration, wt% MEA	17	10–18
Feed Gas Analysis		
H_2S, grains/100 scf	160–180	0
CO_2, vol%	0.3–0.4	10–15
SO_2, ppmv	0	10–100
Outlet Gas Analysis		
H_2S, grains/100 scf	0.02–0.3	0
CO_2, vol%	—	0.1–3.0
SO_2, ppmv	0	1–5
Lean Solution Loading		
H_2S, mole/mole	—	0
CO_2, mole/mole		0.062
Rich Solution Loading		
H_2S, mole/mole	—	0
CO_2, mole/mole		0.415
Absorber		
Number of Columns	5	1
Diameter, ID ft	7	14.5
Height, ft	68	133
Internals	23 trays (total)	two 23-ft beds of polypropylene saddles
Water Wash	3 trays	2 trays
Pressure, psig	200	0
Stripper		
Number of Columns	2	1
Diameter, ID, ft	7	12.5
Total trays	20	18
Wash trays	—	2
Pressure, psig	12	5–10
Bottom temp.,°F	250	245
Steam to reboiler, lb/gallon of solution	1.1–1.6 (max)	1.13

Notes:
Plant A, CO2 recovery from flue gas (Arnold et al. 1982).
Plant B, Natural gas treating plant (Carney 1947).

design of an MEA-based Amine Guard III system. The objective was to operate at the highest possible acid gas loading to minimize the reboiler heat requirement. By operating at a relatively high temperature (which increases the rate of CO_2 absorption) and using a large number of trays (30), the tests achieved an 87% approach to equilibrium at the bottom of the absorber (vs. 75–80% commonly used for the design of lower temperature, 20 tray columns).

Table 2-22
Operating Data from Amine Guard III MEA CO_2 Removal Test Unit in an Ammonia Plant

Constant conditions	
Absorber Pressure, psig	395
Number of Trays	30
Stripper Pressure, psig at bottom	12
Number of trays (solution feed to top tray)	17
Reboiler heat input, Btu/lb mole CO_2	~60,000
Reboiler pressure, psig	12
Stripper top pressure, psig	8
Feed Gas, % CO_2	17.8
Product Gas, ppmv CO_2	100

	Test Number		
Test Variables	1	2	3
Solution			
MEA concentration, %	43.6	35	42.4
Lean solution flow rate, gpm	2,750	2,750	2,550
Lean solution temp., °F	120	112	110
Lean solution loading, mole/mole	—	0.15	0.15
Rich solution temp., °F	215	206	222
Rich solution loading, mole/mole	—	0.47	0.44
Feed Gas Temp., °F	212	227	216
Stripper Conditions			
Inlet solution temp., °F	237	235	237
Reboiler temp., °F	247	244	244
Stripper overhead temp., °F	207	207	206
Ammonia Production			
Short tons/stream day	1,100	1,110	1,125

Source: Butwell et al. (1979)

Additional data on the operation of aqueous MEA plants, including one with a split flow cycle, are given by Estep et al. (1962). Operating data on Glycol-Amine plants, which employ a mixture containing MEA and DEG, are not included here because the process is considered obsolete; however, such data are included in previous editions of this book and are available in the literature (Kohl and Blohm, 1950).

Diethanolamine Plants

Diethanolamine is frequently used for purifying refinery gas streams because of its resistance to COS, which reacts nonregenerably with MEA. In refinery operations, the purified gas is often used for fuel within the refinery, so a high degree of purity is not required. Typical performance data for aqueous DEA plant absorbers are presented in **Table 2-23**. It will be noted that gas purities from 5 to 26 grains $H_2S/100$ scf are reported. Although such purities are often considered acceptable, the trend is to require greater sulfur removal efficiency in new refinery treating plants.

Operating data for an aqueous DEA plant in high pressure natural gas service have been presented by Berthier (1959), and are summarized in **Table 2-24** as plant A. The data were obtained in the early phases of development of the S.N.P.A.-DEA process at Lacq and are not quite representative of the current process which uses substantially lower solution circulation rates (see **Table 2-2**). Vaz et al. (1981) used the Berthier data as a check against a calculation algorithm called Amine Process Model (APM). Two of the calculated values are included in **Table 2-24** to provide a more complete data summary. Many other operating data values were also calculated by the model, and in general, the calculated results agreed very well with the actual data.

Table 2-24 also includes data on two DEA plants described by Butwell and Perry (1975). Plant B is designed to treat natural gas containing very little H_2S at a pressure of 900 psig. Plant C is a relatively low pressure unit (220 psig) treating gas containing a high percentage of acid gas (7% CO_2, 0.5% H_2S). All three DEA plants produced gas containing about ¼ grain H_2S per 100 scf or less.

Diglycolamine Plants

Operating data from two DGA plants are given in **Table 2-25**. Both plants treat natural gas containing a high percentage of acid gas and are able to produce gas containing 0.25 or less grains H_2S per 100 scf. Plant A is a small unit with an absorber which operates at a pressure of only 140 psig. It has a more difficult treating job than Plant B because of the low absorber pressure and higher $H_2S:CO_2$ ratio, but has more trays in both the absorber (25 vs. 20) and the stripper (21 vs. 18) and operates with a higher reboiler temperature (255 vs. 250°F). The two sets of data on Plant B were obtained several years apart.

Huval and van de Venne (1981) describe several DGA plants in Saudi Arabia treating gas streams containing 3–8 vol% H_2S and 8–14 vol% CO_2 at contactor pressures as low as 115 psig. The plants produced sweet gas containing 1–2 ppm H_2S and less than 100 ppm

Table 2-23
Performance Data for Typical Aqueous Diethanolamine Plant Absorbers Used to Remove H_2S from Refinery-Gas Streams

Absorber	Pressure, psig	Temperature, °F	H_2S Grains/100 scf In	Out
16 trays (1)	101	66	3,196	15
26 ft of 3-in. rings (1)	225	140	1,490	26
2-in. rings (1)	250	140–155	260	6
51-ft packing (2)	150	95–100	1,500	5
30 ft of ¾-in. rings (3)	175	125–130	2,500	15

Sources: 1. Reed and Wood (1941), 2. Love (1941), 3. Anon (1946)

Table 2-24
Operating Data for Aqueous Diethanolamine Plants Treating Natural Gas

Plant	A (1)	B (2)	C (2)
Gas Feed Rate, MMscfd	35.5	2.92	452
Solution Flow rate, gpm	1,540	188	106–111
Solution Composition, wt% DEA	20	28.7	28.9
Feed Gas Analysis			
H_2S, vol%	15.0	0.054	0.5
CO_2, vol%	10.0	1.3	7.0
COS, ppm	300	—	—
Outlet Gas Analysis			
H_2S, gr/100 scf	0.28	0.01	0.05
CO_2, ppmv	20	150	200
COS, ppmv	0	—	—
Lean Solution Loading, mole/mole			
H_2S	0.0219	—	—
CO_2	0.0063	—	—
Total acid gas	0.0282	0.01	0.02
Rich Solution Loading, mole/mole			
H_2S	0.423(^{3})	—	—
CO_2	0.274(^{3})	—	—
Total acid gas	0.697(^{3})	0.50	0.44
Absorber			
Number of Trays	30	20	20
Pressure, psig	1,000	900	220
Lean Solution Temperature,°F	—	102	95
Rich Solution Temperature,°F	124.5	109	133
Gas Inlet Temperature,°F	—	72	48
Gas Outlet Temperature,°F	—	108	81
Stripper			
Number of Trays	20	31	20
Stripper Top Tray Pressure, psig	25	9.8	19
Reboiler Temperature,°F	272	241	241
Steam to Reboiler, lb/gal. Solution	0.995	0.85	—

Sources: 1. Berthier (1959), 2. Butwell and Perry (1975), 3. Calculated values from Vaz et al. (1981)

CO_2. They also removed about 90% of the COS and a small fraction of the mercaptans present in the feed gas. Data on the organic sulfur removal in one of these plants are given in **Table 2-26.** The absorbers operate at unusually high temperatures (e.g., lean amine temperatures as high as 150°F). It is possible that this high temperature favors COS removal (by increasing the rate of the hydrolysis reaction) and inhibits mercaptan removal (by reducing equilibrium solubility).

Table 2-25
Operating Data for Aqueous DGA Plants Treating Natural Gas

Plant	A	B-1	B-2
Gas Feed Rate, MMscfd	6	121	83
Solution Flow rate, gpm	172	556	426
Solution Concentration, wt% DGA	50	60	56.5
Feed Gas Analysis			
H_2S vol%	5.48	0.5–1.25	0.74
CO_2 vol%	6.52	1.5–3.75	3.46
Outlet Gas Analysis			
H_2S gr/100 scf	0.205	<0.25	0.01–0.05
CO_2 vol%	—	<0.01	0.0093
Lean Solution Loading			
H_2S, gr/gallon	2		~21
CO_2 mole/mole amine	0.04		0.09
Rich Solution Loading			
H_2S mole/mole amine	0.09		0.06
CO_2 mole/mole amine	0.18		0.33
Absorber			
No. of trays	25		20
Pressure, psig	140		550
Lean Solution Temp.,°F	120		88
Rich Solution Temp.,°F	156		171
Stripper			
No. of trays, stripping	21		16
No. of rays, reflux	4		4
Pressure, psig	13		7.25
Reboiler temperature,°F	255		250

Sources: Plant A data from Harbison and Dingman (1972), Plant B-1 data from Holder (1966), Plant B-2 data courtesy Fluor Daniel (1995)

Methyldiethanolamine and Mixed Amine Plants

Ammons and Sitton (1981) describe an extensive series of tests conducted on a commercial natural gas treating plant using aqueous MDEA to selectively remove H_2S in the presence of a much larger concentration of CO_2. The approximate results of eight of the tests are presented in **Table 2-27**. Although complete operating data are not given, the results are of interest in showing the effects of MDEA concentration, solution flow rate, absorber temperature, and the number of active trays on the amount of CO_2 that passes through the absorber unabsorbed (slippage). The results can be summarized as follows:

CO_2 slippage increases with

1. Increased MDEA concentration (compare tests 1 and 2).
2. Decreased solution flow rate (compare tests 2 and 3, and also tests 6 and 8).

Table 2-26
Organic Sulfur Removal by DGA Solution

	COS	CH$_3$SH	C$_2$H$_5$SH
Sour Gas, ppmv			
Test A	44.1	85.2	28.4
Test B	54.2	78.1	33.5
Test C	54.4	85.0	29.9
Sweet Gas, ppmv			
Test A	4.4	63.3	27.5
Test B	4.2	74.5	29.4
Test C	7.4	75.6	23.6
Average Removal, %	90	14	12

Notes:
1. *Tests A, B, and C were conducted about 1 week apart on the Berri gas plant in Saudi Arabia.*
2. *Inlet gas averaged about 6.8% H$_2$S.*
3. *Solution contained 50–55% DGA.*

Source: Huval and van de Venne (1981)

Table 2-27
Operating Data from a Commercial MDEA Selective Natural Gas Treater

Gas Feed Rate, MMscfd				80				
Absorber Pressure, psig				940				
Feed Gas Temperatures, °F				90				
Feed Gas Analysis								
H$_2$S, ppm				40–60				
CO$_2$, vol%				3.83				
Test Number	1	2	3	4	5	6	7	8
Feed Tray	10	10	10	10	10	10	20	10
Amine Concentration, % MDEA	39	53	53	49	49	46	46	43
Amine Feed Temperature, °F	117	117	116	111	132	114	114	116
Amine Circulation Rate[1]	I	I	L	I	I	I	I	H
CO$_2$ Slippage %[2]	61	70	75	70	65	70	53	60

Notes:
1. *L = low amine circulation rate (1 pump), I = intermediate amine circulation rate (2 pumps), H = high amine circulation rate (3 pumps). Actual flow rates not reported.*
2. *H$_2$S in product gas less than 1 ppm unless stripping is inadequate. At minimum amine solution flow rate and minimum number of trays, a steam rate of 1.2 lb/gallon caused H$_2$S content of product gas to be 7 ppm, but increasing steam rate to 1.44–1.56 lb/gallon produced gas containing less than 1 ppm H$_2$S at all absorber conditions.*

Source: Table based on data reported by Ammons and Sitton (1981)

3. Decreased temperature (Compare tests 4 and 5).
4. Decreased number of trays (Compare tests 6 and 7).

Only 8 of 18 tests are included in the table. In all but one of the 18 tests (with minimum solution flow rate and 10 trays) the H_2S content of the product gas was less than 1 ppm. However, even in this one test, increasing the steam rate to the reboiler from 1.2 to 1.44–1.56 lb steam per gallon of solution at the same absorber conditions resulted in a return to less than 1 ppm H_2S product gas. Production of a product gas containing less than 1 ppm H_2S was maintained at solution feed temperatures as high as 145°F, although CO_2 slippage declined as the solution temperature increased (Ammons and Sitton, 1981).

Detailed operating data on a plant using an aqueous mixture of MDEA and DEA are given by Harbison and Handwerk (1987). The use of an amine mixture was not intentional in this case, but apparently resulted from carryover from a DEA plant or the inadvertent use of DEA as makeup to the MDEA unit. Operating data taken on the system during a 24 hour test run are summarized in **Table 2-28**. During the test period the solution contained 21.8% MDEA and 4.2% DEA. The results of the test showed the plant to be capable of producing gas containing only about 0.002 vol% H_2S (i.e., 99% removal) while removing only about 32% of the CO_2. Vickery et al. (1988) used the Harbison and Handwerk plant data to verify the performance of the GASPLANT-PLUS flow sheet simulator with the AMCOLR rate model for amine columns. The model duplicates actual plant performance reasonably well, and predicts that pure (25%) MDEA will provide greater selectivity (more CO_2 in the product gas) and purer product gas, with regard to H_2S, than a mixed amine of the type actually used in the plant.

Diisopropanolamine Plants

Operating data for three ADIP process plants, which employ aqueous DIPA solutions, are provided in **Table 2-29**. These plants operate at pressures from 59 to 360 psig on gases from high-temperature oil processing units which usually contain COS and CS_2. Data on COS removal are given for Plant 1, and indicate that the ADIP process removes 50% of the COS present in the feed gas. With regard to H_2S removal, the product gas purity varies from 2 ppm for Plant 1, which treats 350 psig gas with an $H_2S:CO_2$ ratio of 1:11, to 100 ppm for Plant 3, which treats a gas containing 15.6% H_2S and no CO_2 at a pressure of only 59 psig (Klein, 1970).

Organic Sulfur Removal by Amine Solutions

This section covers the removal of COS, CS_2, and light mercaptans from gas streams by amine solutions. These are the principal organic sulfur compounds normally encountered in fuel and synthesis gases. The removal of organic sulfur compounds from liquid hydrocarbons is discussed in the next section. The presence of the above components (and many other reactive species) in a gas to be treated raises two questions: (1) How much, if any, of the material will be removed during the treating operation? and (2) will the impurities cause deterioration of the amine solution? The question of solution deterioration by reaction with various gas impurities is discussed in detail in Chapter 3; this discussion is concerned primarily with removal of carbonyl sulfide, carbon disulfide, and mercaptans from the gas by amine solutions.

**Table 2-28
Operating Data for Plant Using a Mixed Aqueous Amine Solution Containing MDEA and DEA**

Gas Feed Rate, MMscfd	5.935
Solution Flow Rate, gpm	76.4
Solution Composition	
MDEA, wt%	21.8
DEA, wt%	4.2
Feed Gas Analysis	
CO_2, vol%	19.20
H_2S, vol%	2.078
Product Gas	
CO_2, vol%	14.38
H_2S, vol%	0.022
Lean Solution Loading	
CO_2, mole/mole amine	0.02
H_2S, mole/mole amine	0.004
Rich Solution Loading	
CO_2, mole/mole amine	0.545 (0.482[1])
H_2S, mole/mole amine	0.187 (0.157[1])
Absorber[2]	
Diameter, I.D., ft	3
Solution Feed Tray (from bottom)	10
Pressure, psia	134
Lean Amine Temp., °F	118
Rich Amine Temp., °F	139
Sour Gas Temp., °F	94
Bulge Temp., °F	145
Bulge Max. Temp. Tray	2
Effective Heat of Reaction	
Contactor, Btu/lb acid gas	514
Stripper, Btu/lb acid gas	512

Notes:
1. Values in parentheses are based on gas analysis.
2. Absorber contained single pass valve trays with 2-in. weir height and 24-in. spacing.
Source: Harbison and Handwerk (1987)

Carbonyl Sulfide (COS) and Carbon Disulfide (CS_2)

COS and CS_2 are discussed together because they usually occur together and have quite similar chemical characteristics (e.g., they both undergo hydrolysis with water to form H_2S and CO_2). However, COS is much more common and has been the most widely studied. It is therefore covered in greater depth.

COS and CS_2 removal from gases by amine solutions may involve one or more of the following mechanisms:

Table 2-29
Operating Data of ADIP Plants

Plant	1	2	3
Gas feed, cu ft/hr	700,000	85,000	1,200,000
H_2S, percent	0.5	10.4	15.6
CO_2, percent	5.5	2.5	—
COS, ppm	200	—	—
Absorber pressure, psig	350	280	59
Absorber temp., °F	104	95	104
No. of trays in absorber	25	20	15
Outlet gas:			
H_2S, ppm	2	10	100
COS, ppm	100	—	—
Steam, lb/lb acid gas removed	1.3	1.8	2.3
Power, KWH per ton of acid gas removed	14	15	6

Notes:
Plant 1: Synthesis gas from oil gasification unit.
Plant 2: Gases from catalytic cracking unit.
Plant 3: Offgases from gas oil hydrodesulfurizer.
Source: Klein (1970)

1. Hydrolysis (i.e., reaction with water) to form H_2S and CO_2 which may subsequently be absorbed.
2. Direct reaction with the amine to form a relatively stable compound (which may or may not be regenerable).
3. Physical solubility in the solution.

Hydrolysis is believed to be the primary mechanism when significant quantities of COS or CS_2 are removed from gas streams by amine solutions. Direct reaction can also be important, particularly when a nonregenerable reaction product is made; in fact, it is the primary reason MEA is considered unsuitable for treating gases containing COS and/or CS_2. Physical solubility is generally very low, but to the extent that it does occur, it aids removal efficiency, and acts as a first step in the other two mechanisms.

The hydrolysis reactions are

$$COS + H_2O = CO_2 + H_2S \tag{2-44}$$

$$CS_2 + 2H_2O = CO_2 + 2H_2S \tag{2-45}$$

The hydrolysis reactions occur to some extent in all aqueous amine solutions; however, the efficiency of removal by this means depends on the amine basicity, the temperature, and the time of contact, with hydrolysis increasing with increases in all three parameters.

With MEA solutions, COS reacts by both hydrolysis and direct reaction. Unfortunately, the direct reaction between MEA and COS results in the formation of a nonregenerable prod-

154 *Gas Purification*

uct, and the rate is high enough to make MEA unsuitable as an absorbent for gases with high concentrations of COS. DEA also undergoes both hydrolysis and direct reaction with COS. As with MEA, the direct reaction results in the formation of a nonregenerable compound; however, the rate of formation is extremely slow, so DEA can be used effectively to treat gas streams containing COS. According to Butwell et al. (1982), only about 2% of the COS in the feed gas stream reacts irreversibly with DEA; while overall COS removal efficiencies of 70–80% are attainable, primarily by hydrolysis. They claim that even higher levels of COS removal can be attained with DEA solutions by eliminating the competition of H_2S and CO_2 for reaction sites; i.e., contacting the gas with lean DEA solution in a second stage after H_2S and CO_2 have been removed in the first stage. Additional data on the reactions of COS with MEA and DEA are provided by Pearce et al. (1961).

DIPA is reported to react with COS even more slowly than DEA (Danckwerts and Sharma, 1966). Aqueous DIPA solutions can therefore be used for treating gases containing COS without serious deterioration. According to Klein (1970), DIPA, as used in the ADIP process, is suitable for partial removal of COS from gas streams. The Sulfinol process, which uses a mixture of DIPA and Sulfolane, reportedly provides very complete COS removal (see Chapter 14).

DGA is a primary amine which is chemically related to MEA and also reacts quite rapidly with COS. Fortunately, the main product of the reaction decomposes under conditions encountered in a DGA solution reclaimer allowing DGA solutions to be used for gases containing COS. Furthermore, the process can be operated to enhance COS hydrolysis and thereby provide effective COS removal. As previously mentioned, two key factors in COS hydrolysis are temperature and contact time. Two key features of the Fluor Daniel Improved Econamine Process, which is claimed to provide efficient COS removal, are operating at elevated temperatures and providing adequate contact time. Huval and van de Venne (1981) report on experience with 5 large DGA plants operating in Saudi Arabia. These plants operated with lean amine temperatures as high as 150°F. One typical unit averaged about 90% COS removal (see **Table 2-26**).

Singh and Bullin (1988) examined the kinetics of the reaction between COS and DGA over a temperature range of 307 to 322°K (93 to 120°F). They concluded that the reaction is controlled by hydrolysis and that the DGA has a catalytic effect on the hydrolysis reaction. The reaction was found to follow a second order rate equation; first order in COS and first order in DGA. A competing reaction occurs to form N,N'bis (hydroxyethoxyethyl) thiourea (BHEEU) which, fortunately, reacts with water at high temperatures to regenerate DGA.

The absorption of COS in aqueous solutions of MDEA was studied by Al-Ghawas et al. (1988). They concluded that the overall reaction between COS and MDEA is given by the following equation:

$$R_3N + H_2O + COS = R_3NH^+ + HCO_2S^- \qquad (2\text{-}46)$$

They performed laboratory tests to measure the rate of COS absorption in MDEA and used a model based on the penetration theory to calculate the kinetic rate constant for equation 2-46. The results should be of value in rigorous design calculations to predict the fraction of COS in a feed stream that will be absorbed in a commercial MDEA contactor.

Rahman et al. (1989) conducted an experimental study to determine the products of reaction between COS and several different amines. The protonated amine thiocarbamate salt was detected in the reaction products of COS with MEA, DEA, and DGA. The DIPA thiocarbamate salt could not be detected, possibly because the DIPA-COS reaction was too slow to provide enough reaction product for detection by the method employed. The tertiary amines, MDEA and dimethylethanolamene (DMEA), did not give any evidence of reaction with COS in these tests. The second order rate constants for reactions between COS and MEA, DEA, and DIPA have been reported to be 16, 11, and 6 $\text{Mol}^{-1}\ \text{s}^{-1}$, respectively at 25°C. (Danckwerts and Sharma, 1966; Sharma and Danckwerts, 1964; Sharma, 1965).

Mercaptans

Mercaptans are substituted forms of H_2S in which one of the hydrogen atoms is replaced by a hydrocarbon group. They have the general formula RSH, and their properties are governed to a large extent by the length of the hydrocarbon chain, R. Like H_2S they have acidic properties, but because of the hydrocarbon radical, they are much weaker acids than H_2S. Mercaptans behave less like acids and more like hydrocarbons as the hydrocarbon chain length increases.

Because of their acidic properties, mercaptans can react with alkalies to form mercaptides; however, the salts are weakly bonded and readily dissociated. As a result, the solubility of mercaptans in alkaline amine solutions tends to be higher than would be expected on the basis of simple physical solubility, and increases with increased alkalinity while decreasing with increased temperature. Since the partial pressure of mercaptans in most gases is very low and the quantity of solution flowing is based on the more reactive acid gases (H_2S and CO_2), the percent removal of mercaptans in most amine plants is small.

According to Butwell et al. (1982), the approximate removal efficiencies by MEA and DEA plant solutions are as follows:

Methy mercaptan   45–55%
Ethyl mercaptan   20–25%
Propyl mercaptan   0–10%

Huval and van de Venne (1981) describe several large DGA plants in Saudi Arabia treating gas streams containing 3–8% H_2S and 8–14% CO_2 at contact pressures as low as 115 psig. The plants produced sweet gas containing 1–2 ppmv H_2S and less than 100 ppmv CO_2. They also removed about 90% of the COS and a small fraction of the mercaptans present in the feed gas. Data on organic sulfur removal in one of these plants are given in **Table 2-26**. The absorbers operated at unusually high temperatures with lean amine temperatures as high as 150°F and much higher (but unreported) temperatures within the contactors. It is possible that these high temperatures favored COS removal (by increasing the rate of reaction), but inhibited mercaptan removal by reducing the equilibrium solubility.

Harbison and Dingman (1972) describe a small DGA plant that was reported to remove about 98% of the mercaptans from a natural gas stream. This unit operated with a very low rich solution acid gas loading (0.27 mole acid gas per mole amine), a moderate contactor bottom temperature (rich solution 156°F), a tall absorber (25 perforated trays), and a concentrated amine solution (50% DGA), all of which would be expected to enhance mercaptan removal efficiency. It should be noted, however, that the 98% efficiency value is based on the absorber inlet and outlet gas analyses. If the mercaptan removal efficiency is based on the amount of mercaptan in the feed gas to the absorber and the amount found in the acid gas stream, a removal efficiency of only about 45% is calculated. Kenney et al. (1994) claim that DGA solutions can achieve mercaptan removal levels as high as 90%.

Fortunately, mercaptans do not appear to react with any of the amines to form nonregenerable compounds. Rahman et al. (1989) looked for possible reactions between methyl mercaptan and MEA, DEA, DGA, DIPA, and MDEA and found none.

AMINE TREATMENT OF LIQUID HYDROCARBONS

Process Description

Amine-based liquefied petroleum gas (LPG) and gas treating processes are similar in that both involve contacting a low density hydrocarbon phase (liquid or gas) with a heavier, immiscible liquid phase (aqueous alkanolamine solution). Carbon dioxide, hydrogen sulfide, and carbonyl sulfide in the hydrocarbon phase are transferred to the aqueous phase where they react with the amine. Spent amine solution is removed from the contactor, regenerated, and recycled. In gas treating units, the gas is usually (but not always) the continuous phase because of the large volumes of gas involved and the limited gas flow capabilities of flooded columns. In LPG treaters, however, the volumetric flowrate of the hydrocarbon phase is relatively low, although usually larger than that of the amine, and designs with either phase continuous may be considered.

According to Tse and Santos (1993), the choice of amine as the continuous phase and LPG as the dispersed phase offers two advantages:

1. The total surface area for mass transfer is maximized when the dispersed phase has the larger flow rate.
2. Droplet residence times and mass transfer rates are increased when LPG is dispersed into the higher viscosity continuous amine phase.

Using amine as the continuous phase also increases the LPG treater amine residence time and provides more time for the operator to intervene if the interface level controller fails. This minimizes the chance of hydrocarbon breakthrough causing a major sulfur plant upset.

The advantage of using LPG as the continuous phase is that the LPG Amine Treater capacity (gpm of LPG/ft^2 of contactor area) is substantially higher when LPG is the continuous phase. Therefore, substantial capacity increases can often be achieved by a revamp which makes LPG the continuous phase. However, when LPG is the continuous phase, the height of packing equivalent to a theoretical stage is greater because the mass transfer area is less and the droplet residence time is decreased as previously noted. The vast majority of LPG treaters have amine as the continuous phase. And, unless otherwise noted, this is assumed in the following discussion.

LPG contactors are typically countercurrent columns containing 1.5 or 2.0-in. stainless steel or ceramic random packing. As the LPG flows upward through the bed, the droplets coalesce and mass transfer efficiency decreases. The maximum effective bed height is generally thought to be about 12 ft (Strigle, 1994), and bed heights in the range of 8 to 12 ft are typically used. When more than 12 ft of packing are required, it is necessary to use multiple beds and to collect and redisperse the LPG entering each bed. Specially designed disperser/support plates are used beneath each bed to collect and redisperse the LPG and provide support for the packing.

Most LPG treating applications require no more than three beds of packing. In some cases, e.g., removal of carbon dioxide alone, a single bed of packing is adequate. Single stage contact can be accomplished by the use of a single countercurrent packed bed or by the use of an efficient cocurrent contactor. Cocurrent contactors used for LPG treating usually consist of an eductor or in-line mixer followed by a phase separator.

Figure 2-96 depicts a typical LPG treating system. In this case the LPG feed is the condensed product of a refinery debutanizer column. The LPG is pumped on level control from the debutanizer column overhead accumulator and flows to a countercurrent liquid-liquid contactor which contains three beds of random packing. The LPG enters the bottom of the column, while filtered lean amine is fed to the top of the column at a controlled, constant flowrate.

The LPG feed is distributed evenly and formed into droplets by injection into the continuous amine phase at the bottom of the column. The lean amine is distributed across the top of the packing where it joins the continuous amine phase. The density difference between the two phases causes the dispersed LPG to flow upward through the continuous amine phase. The LPG/amine interface is maintained above the top bed and the amine distributor by an interface level controller which controls the rate of discharge of rich amine from the bottom of the contactor.

The treated LPG leaves the contactor and flows to a gravity settler or coalescer where entrained droplets of amine are removed from the LPG. **Figure 2-96** shows a gravity settler which also includes a LPG water wash system. Washing the treated LPG with water improves the recovery of entrained amine and also removes dissolved amine from the LPG. As shown in **Figure 2-96,** make-up wash water is added on flow control; while the water-amine purge stream is discharged on level control. A portion of the water-amine stream from the settler is recycled to the LPG stream entering the water wash mixer. Usually the wash water flow rate is about 25% of the LPG flow. The combined water-LPG stream flows through a static mixer or other mixing device and then to the gravity settler. If required, the LPG from the settler may be further processed in other treating units, such as an extractive Merox unit to remove mercaptans. The operating pressure of the entire LPG treating system is controlled by a pressure controller located downstream of the final LPG treating unit.

An alternate design, depicted in **Figure 2-97,** is often used when only one theoretical contact stage is required. In this application, LPG and amine are contacted in either an eductor or a static mixer and the intimate mixture of the two liquid phases is sent to a settling tank where the two phases are separated. The rich amine is discharged from the settler on level control and lean amine is added on flow control. A relatively large flow of rich amine is recycled through the eductor to assure an adequate surface area for transfer of CO_2 from the LPG to the amine. A single stage contactor system is commonly used for CO_2 removal from natural gas plant liquids (NGL) when hydrogen sulfide is not present and very high efficiency carbon dioxide removal is not required (Bacon, 1972; Perry, 1977, B; Honerkamp, 1975).

Design Data

Liquid-Liquid Equilibrium Data for H_2S and CO_2

The development of rigorous design methods for LPG treaters has been hampered by the lack of liquid-liquid equilibrium (LLE) data for amines and sour hydrocarbon liquids. Early designs were based on empirical rules of thumb and field experience with existing units operating under similar conditions.

(text continued on page 160)

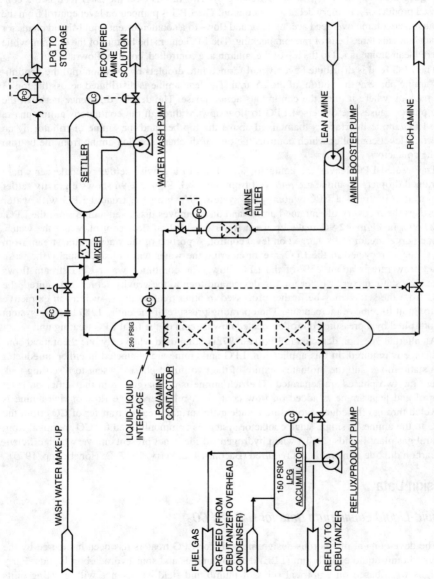

Figure 2-96. Typical LPG amine treating system with gravity settler and LPG water wash.

Alkanolamines for Hydrogen Sulfide and Carbon Dioxide Removal 159

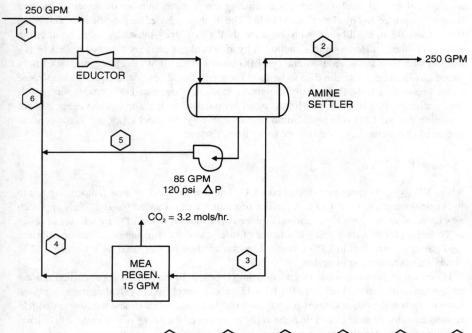

	1	2	3	4	5	6
	SOUR H.C. PRODUCT	SWEET H.C. PRODUCT	RICH MEA TO REGEN.	LEAN MEA FROM REGEN.	RICH MEA RECIRC.	SEMI LEAN MEA RECIRC
HYDROCARBON, Mol/hr.	1406.11	1406.11				
CO_2	3.10	NIL	4.026	0.926	22.814	23.740
H_2O			353.889	353.889	2005.371	2359.260
MEA			18.515	18.515	104.918	123.433
TOTAL, Mol/hr.	1409.21	1406.11	376.430	373.330	2133.103	2506.433
LBS. PER HR.	60472	60336	7678	7542	43511	51053
MW	42.91	42.91	20.40	20.20	20.40	20.37
SP. GR.	0.483	0.483	1.006	1.006	1.006	1.006
GPM	250	250	15	15	85	100
PRESSURE, PSIA	435	445	445	565	565	565
TEMP., °F	55	55	55	90	55	60

Figure 2-97. Single-stage liquid hydrocarbon CO_2 removal system with material balance. (*Bacon, 1972*)

(*text continued from page 157*)

An approach for generating approximate LLE data, which could be used with conventional graphical or analytical methods for estimating the required number of stages or packing height, was suggested by Honerkamp (1975). The method is based on the principle that when two phases are in equilibrium with the same third phase, they must also be in equilibrium with each other. In Honerkamp's method, a hypothetical gas phase is assumed to be in equilibrium with both the aqueous amine and LPG phases. This allows the large amount of published gas-liquid equilibrium data to be used for liquid-liquid applications. For example, the vapor pressure of CO_2, P_{CO_2}, above the amine solution is estimated by extrapolating available amine-carbon dioxide gas-liquid equilibrium data to the appropriate range. Then, assuming that the LPG is in equilibrium with the same hypothetical vapor phase, the concentration of CO_2 in the LPG is estimated using Raoult's law:

$$X_{CO_2} = P_{CO_2}/VP_{CO_2} \tag{2-47}$$

where VP_{CO_2} is the vapor pressure of pure CO_2 liquid at the LPG treating temperature (up to the critical temperature of CO_2). A similar procedure was proposed for H_2S. This approach was found to underestimate the carbon dioxide concentration in the LPG by a factor of about 6. To correct for this error and add a factor of safety, Honerkamp suggested that equilibrium acid gas concentrations in LPG estimated by this method be increased by a factor of 12 to obtain values to be used in design.

Holmes et al. (1984) modified the Honerkamp approach to improve its accuracy and adapt it to computer simulation. They used the Kent-Eisenberg correlation to calculate the vapor pressures of carbon dioxide and hydrogen sulfide over the amine solution, and the Soave-Redlich-Kwong equation of state to correlate the vapor pressures of the acid gases over the LPG phase. Equating the two correlations for each acid gas produces an expression that relates the concentration of the acid gas in the amine solution to its concentration in the LPG phase at equilibrium. This expression can then be used to generate liquid-liquid equilibrium curves and, with conventional column design techniques, to estimate the required number of theoretical stages.

If additional precision is desired, the basic approach of Honerkamp and Holmes could be further improved by using a more rigorous model for predicting the acid gas-amine solution equilibrium (e.g., the Deskmukh-Mather (1981) or Austgen et al. (1991) correlation coupled with the Peng-Robinson equation of state to define the acid gas-LPG equilibrium.

Holmes et al. (1984) describe the use of their LPG/amine equilibrium model in the TSWEET process simulation program to design LPG treaters. The TSWEET program, which was originally written for gas treatment plants, is based on rigorous tray-by-tray calculations. A comparison of the calculated results with actual plant operating data is given in **Table 2-30**. The operating data for plants 1 through 7 were originally reported by Honerkamp (1975). Although the measured product LPG acid gas concentrations are limited in number and precision, the results indicate that the calculation method provides a reasonable prediction of plant performance. Additional operating data for LPG treaters using random packing are reported by Tse and Santos (1993).

The use of the Holmes et al. calculation method to evaluate contactor design alternatives is described by Fleming et al. (1988). They considered MEA, DEA, and MDEA solutions for reducing the carbon dioxide concentration in 50 gpm of LPG from 7.7 mole % to 0.16 mole %. Both packed columns and static mixer/coalescer systems were evaluated. As a result of the study, a single-stage static mixer/coalescer was selected. Operating data showed the unit to be capable of reducing the CO_2 concentration to 0.10 mole % using 70 gpm of 25% DEA.

Table 2-30
Comparison of LPG Treater Operating Data with LLE Calculations

Case	1		2		3	
	Plant Data	Calculated	Plant Data	Calculated	Plant Data	Calculated
Hydrocarbon						
Rate, gpm	100	103	780	780	208	210
T,°F	50–60	55	90	90	77–89	83
P, psig	500	500	535	535	835	835
CO_2 in, ppm	10,000	10,000	2,030	2,023	33,000	33,000
H_2S in, ppm	—	—	Trace	2.7	—	—
CO_2 out, ppm	0	0.056	<100	0.42	Trace	0.4
H_2S out, ppm	—	—	—	0.01	—	—
Amine	MEA	MEA	MEA	MEA	MEA	MEA
wt%	20	20	15	15	20	20
Flow, gal/min	10.8	10	34	34	54	50
Ideal Stages**	—	3	—	2	—	2
Lean Loading*	0.10	0.108	0.10	0.13	0.10	0.11
Rich Loading*	0.44	0.44	0.28	0.32	0.53	0.53
Contactor	three 12-ft sections of packing		two 12-ft sections of packing		18 ft of packing	
Packing Type	1-in. Intalox		—		—	
Contactor, Diam., in.	30		—		48	

Case	4		5		6	
	Plant Data	Calculated	Plant Data	Calculated	Plant Data	Calculated
Hydrocarbon						
Rate, gpm	265	273	42	44	21	22
T,°F	—	80	80–90	85	—	85
P, psig	305	305	—	200	265	265
CO_2 in, ppm	2,460	2,450	5,000	5,020	nil	—
H_2S in, ppm	270	270	—	—	1,080	1,080
CO_2 out, ppm	Trace	1.2	100	47	—	—
H_2S out, ppm	nil	5.9	—	—	nil	9.2
Amine	DEA	DEA	MEA	MEA	MEA	MEA
wt%	20	20	8	8	5	5
Flow, gal/min.	20	20	5.3	5.3	5	5
Ideal Stages**	—	2	—	1	—	1
Lean Loading*	0.07	0.05	0.10	0.10	0.07	0.07
Rich Loading*	0.30	0.27	0.44	0.43	0.13	0.13
Contactor	two 8-ft sections of packing		10 perforated trays		jet eductor-mixer	
Packing Type	1½″ ceramic					
Contactor Diam., in	66					

(*table continued on next page*)

Table 2-30 (Continued)
Comparison of LPG Treater Operating Data with LLE Calculations

Case	7 Normal		7 Maximum		8	
	Plant Data	Calculated	Plant Data	Calculated	Plant Data	Calculated
Hydrocarbon						
Rate, gpm	250	250	200	205	105	105
T, °F	55	55	55	55	100	100
P, psig	435	435	435	435	385	385
CO_2 in, ppm	2,260	2,200	6,675	6,630	7,260	7,260
H_2S in, ppm	—	—	—	—	—	—
CO_2 out, ppm	nil	0.90	138	74	660	158
H_2S out, ppm	—	—	—	—	—	—
Amine	MEA	MEA	MEA	MEA	DEA	DEA
wt%	15	15	15	15	30	27
Flow, gal/min.	100	110	100	93	14	14
Ideal Stages**	—	1	—	1	—	1
Lean Loading*	0.20	0.20	0.38	0.38	0.03	0.03
Rich Loading*	0.22	0.22	0.46	0.45	0.25	0.32
Contactor	jet eductor-mixer		jet eductor-mixer		static mixer	
Packing Type	—		—		—	
Contactor Diam., in.	—		—		—	

Notes:
*Loadings are mole of acid gas per mole of amine. A lean solution loading of 0.1 mol/mol was assumed if no information was available.
**Equilibrium stages used in the absorber.
Source: Holmes et al. (1984)

Experimental data on the distribution of H_2S between an aqueous MDEA solution and liquid propane have recently been reported by Carroll et al. (1993). A small portion of this data is reproduced in **Table 2-31**. The results indicate that for solution loadings less than 1.0 mole H_2S/mole MDEA, there is a linear relationship between the logarithm of the amine loading and the logarithm of the mole fraction H_2S in the propane. The relationship is

$$\delta \log x / \delta \log R = 2.4 \tag{2-48}$$

Where: x = mole fraction H_2S in liquid propane.
R = moles H_2S/mole MDEA in the aqueous solution.

This relationship indicates that an order of magnitude change in the solution loading results in more than two orders of magnitude change in the concentration of H_2S in the propane.

Data on the distribution of CO_2 between phases in the system carbon dioxide-propane-3M MDEA are reported by Jou et al. (1995B). The vapor-amine solution equilibrium data were found to closely match predictions of the Deshmukh-Mather (1981) correlation. The vapor-liquid equilibrium for the system CO_2-propane appeared to fit the Peng-Robinson (PR) equation of state.

Table 2-31
Three Phase Equilibrium in the System H_2S—Propane—3M MDEA (Aqueous)

Temp °C	Gas Phase		Aqueous Amine Phase		Liquid Propane Phase
	Total Pressure, KPa	H_2S Partial Pressure, KPa	Mole H_2S/Mole Amine	Mole Propane/ Kg Solvent	Mole Fraction H_2S
25	960	0.154	0.065	0.019	0.00003
	966	1.03	0.166	0.017	0.00022
	966	2.91	0.286	0.014	0.00073
	995	18.8	0.625	0.012	0.00528
	1,050	89.9	0.909	0.010	0.0179
40	1,380	0.152	0.050	0.020	0.00003
	1,382	1.87	0.157	0.018	0.00031
	1,420	9.17	0.335	0.015	0.00178
	1,430	46.2	0.670	0.012	0.0111
	1,490	118.6	0.847	0.011	0.0278

Source: Carroll et al. (1993)

Mutual Solubility of Amines and LPG

Carroll et al. (1992) conducted a comprehensive study of phase equilibria in the system propane-3M MDEA solution including vapor-liquid, liquid-liquid, and vapor-liquid-liquid equilibria. Data on propane solubility in the amine solution are of particular interest because dissolved hydrocarbons represent a loss of valuable product, and, when stripped from solution with the acid gases, can affect performance of a downstream sulfur conversion unit. The study showed that the solubility of propane in 3M MDEA is low, but more than twice its solubility in pure water under the same conditions. For example, at 100°F (37.8°C) and a propane pressure of 10,000 kPa (1,450 psia), the mole fraction propane in the 3M MDEA solution was found to be about 0.0006. This compares to a mole fraction of propane in pure water of about 0.00025 under the same conditions.

Both Veldman (1989) and Stewart and Lanning (1991, 1994A, B) report solubility data for MEA, DEA, and MDEA in hydrocarbons. Both sets of data show that at a given wt% amine concentration, MEA is the most soluble, followed by MDEA, and then by DEA. Stewart and Lanning (1991, 1994A, B) report considerably higher amine solubilities in hydrocarbon than Veldman (1989). Since the data of Carroll et al. (1992) for the solubility of MDEA in propane closely match the Veldman data, the latter are reproduced in **Figure 2-98**. This figure can be used to estimate the residual solubility of amines in LPG after water washing and to estimate the required water wash make-up rate.

Distribution of Organic Sulfur Compounds

In addition to H_2S and CO_2, LPG and related light hydrocarbon streams may contain COS, CS_2, mercaptans, and other organic sulfur compounds. The amounts of each present depend on the concentration in the original plant feed material and the hydrocarbon processing steps. Unfortunately, the distribution of organic sulfur compounds cannot be predicted accurately by commonly used equations of state, such as that of Soave-Redlich-Kwong (SRK). The use

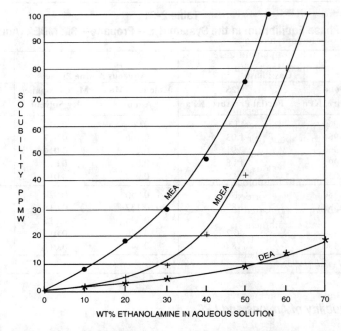

Figure 2-98. Solubility of ethanolamines in saturated hydrocarbons at 25°C. (*Veldman, 1989*)

of interaction parameters to correct SRK predictions for sulfur compounds in hydrocarbon systems is discussed by Goel et al. (1982).

Examples of sulfur compound distributions in natural gas liquids are given by Goel et al. (1982) and by Petty and Naeger (1987). The reported distributions are quite different, but generally show the following:

1. Almost all of the H_2S goes to the ethane product with a small amount appearing in the propane.
2. The COS also divides between the ethane and propane products; however, most is associated with the propane.
3. Methyl mercaptan goes primarily to the butane product, but significant amounts show up in the propane and even the ethane streams.
4. Ethyl mercaptan is split between the butane and gasoline fractions with most appearing in the butane stream.
5. Propyl and higher mercaptans appear primarily in the gasoline.

Amine-type liquid hydrocarbon treaters can be designed to remove both H_2S and COS efficiently, but mercaptans, which are much weaker acids than H_2S, do not react significantly with amines. However, they can be made to react with stronger alkalies, such as sodium hydroxide.

LPG Treater Operating Conditions

The operating temperatures and pressures of amine-type LPG treaters must be maintained within narrow limits to maintain the hydrocarbon in the liquid state, minimize hydrocarbon and amine entrainment, and optimize amine-hydrocarbon separation. Veldman (1989) recommends that the lean amine temperature be controlled so that the viscosity is around 2 centipoise at the amine/LPG interface to assure effective phase separation. The relationship between amine solution temperature and viscosity for several amines is illustrated in the previous section entitled "Physical Properties." As the data indicate, operating temperatures must be greater than about 100°F for typical MEA, DEA, and MDEA solutions to meet the 2 centipoise viscosity requirement. Changela and Root (1986) report that low amine operating temperatures (60–70°F versus a design of 110°F) increased the viscosity of the lean amine solution at one plant to the point that a significant amount of LPG was entrained in the rich amine. This problem was corrected by increasing the lean amine temperature above 110°F. A similar problem was reported by DuPart and Marchant (1989) when low temperatures led to excessive amine entrainment in the LPG product. In this latter case, improved control of both the amine and LPG inlet temperatures reduced amine losses.

As previously noted, the operating temperature must be below the LPG vaporization temperature throughout the treating system. The designer must carefully evaluate the pressure profile of the entire LPG treating system to ensure that an adequate safety margin is maintained. Maximum and minimum LPG and amine operating temperatures should be considered in making this evaluation and it may be desirable to provide some means of temperature control to ensure that neither the amine nor the LPG are too hot or too cold. The effect of the heat of reaction on the LPG bubblepoint should also be considered—particularly when treating a very sour LPG. Typical design margins between the LPG bubblepoint pressure and the LPG treater operating pressure are often 100 psi or greater, and the difference between the LPG bubblepoint pressure and the minimum operating pressure in the LPG treating system is set by most designers at 50 psi or more. For example, Changela and Root (1986) describe a gas plant liquids treater where bubblepoint hydrocarbon liquids from a surge drum with a design operating pressure of 330 psig were treated at 500 psig. In this case the margin between the operating and bubblepoint pressures in the LPG treater was about 170 psi.

Amine Solution Flow Rates and Composition

Rigorous calculation of the minimum required amine flow rate, the maximum product purity attainable, and the relationship between amine flow rate and number of theoretical trays requires accurate liquid-liquid equilibrium data. As previously discussed, these data can be estimated by the approximate method of Honerkamp (1975) or the more precise method of Holmes et al. (1984).

The maximum LPG purity possible represents equilibrium with the lean amine solution. Since equilibrium can only be approached in actual equipment, it is necessary to regenerate the amine to acid gas concentrations below the levels which would be in equilibrium with the desired LPG product. For the case of CO_2 removal from NGL with MEA, Honerkamp (1975) recommends that the lean amine be regenerated to less than 0.1 mole CO_2/mole amine.

The theoretical minimum amine flow rate occurs when the rich amine leaving the contactor is in equilibrium with the entering LPG. However, to provide an adequate driving force for mass transfer over the entire column, it is necessary to use an actual flow rate well above the minimum. As a result, the rich solution loading is always below the equilibrium value. Typical recommended rich solution loading values for MEA, DEA, and DGA are given in **Table 2-32**. These rich amine loading recommendations should be used with caution as the minimum amine circulation requirement is usually set by the approach to equilibrium at the bottom of the contactor. The calculation methods outlined by Honerkamp (1975) and Holmes et al. (1984) can be used to set the approach to equilibrium. Setting the approach to equilibrium at the bottom of the contactor at 60 to 70% usually provides a conservative design margin. **Table 2-32** also lists the amine concentrations recommended by several investigators.

Table 2-32
Recommended Solution Conditions for LPG Treating with Alkanolamines

	Reference	MEA	DEA	DGA
Amine Concentration wt%	A	5–20	25–35	50–70
	B	<18	—	—
	C	10–18	20–30	<40
Rich Solution, Moles Acid Gas/ Mole Amine	A	0.3–0.4	0.35–0.65	—
	B	<0.4		
	D	<0.5 (CO_2)		

References:
A. Holmes et al. (1984); B. Bacon (1972); C. Dupart and Marchant (1989); D. Honerkamp (1975)

Once the lean and rich amine solution loadings have been selected, the required solution flow rate can be calculated by a simple material balance around the contactor. The required number of theoretical stages can be estimated using graphical or analytical techniques described in many standard texts (e.g., Treybal, 1963; Perry, 1963).

If a graphical method is used, the material balance establishes the operating line, and the required number of theoretical stages can be stepped off between the operating line and the equilibrium curve. If too many theoretical stages are indicated (i.e., more than about 3), a lower acid gas loading in the rich solution should be assumed and the design procedure repeated.

In addition to the restrictions on amine concentration and rich solution loading, the volumetric flow rate ratio of the continuous phase (amine) to the dispersed phase (LPG) should be more than about 1:30 to ensure good mass transfer. Conservative design would call for a volumetric ratio of amine to LPG of 1:15 or greater.

Absorber Designs

Random Packed, Countercurrent Columns

Packing Selection and Bed Height. The use of packing increases mass transfer in the LPG contactor by increasing the residence time of the LPG droplets, reducing the possibility of back-mixing, and renewing the droplet surface film (Tse and Santos, 1993). Either metal or ceramic packings may be used. If amine is the continuous phase and if metal packing is used, the packing should be wetted by the amine phase first when the plant is initially commissioned or after a turnaround (Tse and Santos, 1993; DuPart and Marchant, 1989; Strigle, 1994).

According to Honerkamp (1975), typical random packing sizes for LPG treating are 1 to 1½ in. with 1 in. being the most common size. Tse and Santos (1993) and Perry (1977A) state that the most common packing sizes for LPG treating are 1½ in. to 2 in. Strigle (1994) states that for all liquid-liquid treating applications the maximum packing size should be 2 in. with 1½ in. the more common choice. Tse and Santos (1993) and DuPart and Marchant (1989) present an equation which indicates that random packing used for LPG treating should be ½ in. or greater. Also, according to Tse and Santos (1993) and Strigle (1994), the maximum packing size should be no greater than one-eighth the contactor diameter. All of these criteria are normally satisfied by the use of 1- to 2-in. packing.

As the LPG flows up through a bed of random packing, the LPG droplets tend to coalesce and mass transfer efficiency is lost. Accordingly, there is a maximum effective bed height. If the packing height requirement exceeds the maximum effective bed height, it is necessary to divide the bed into two or more separate packed sections. The LPG flowing into the upper bed is collected and redispersed to obtain good mass transfer. Honerkamp (1975) noted that LPG contactors with 16 ft of random packing performed as well as contactors with 36 ft of packing. While this suggests that individual bed heights should be limited to less than 16 ft, Honerkamp recommended that individual beds of random packing be restricted to a maximum height of 8 ft for LPG treating service. Strigle (1994) recommends a maximum random packing bed height of 12 ft for all liquid-liquid contacting applications. For LPG treaters, DuPart and Marchant (1989) also recommend that the maximum packing height per bed be limited to no more than 12 ft, since deeper beds result in a loss of efficiency. While the recommendations differ somewhat, 12 ft is probably a reasonable estimate of the maximum effective bed height for LPG treating with typical 1½ to 2-in. random packing.

Each 8 to 12 ft bed of packing is approximately equivalent to one theoretical stage. When more than one theoretical stage is required, it is necessary to use multiple beds with effective dispenser/support plates for each bed. Most liquid hydrocarbon treating requirements can be accomplished with three or less theoretical stages; therefore, one to three beds of packing are commonly used in commercial installations.

Contactor Diameter. It has been common practice in the industry to base the diameter of packed LPG/amine contactors on the superficial velocity of the two liquid phases combined together. There is considerable variation in design superficial velocity recommendations. Perry (1977A, B) recommends sizing LPG contactors based on 15 gpm/ft^2. Perry's recommendation, which is based on natural gas plant liquids, is supported by DuPart and Marchant (1989). Russell (1980) suggests a superficial velocity of 12 gpm/ft^2, while Changela and Root (1986) and Veldman (1989) recommend 10 gpm/ft^2. *The GPSA Engineering Data Book* (1987) recommends 20 gpm/ft^2. A design superficial velocity of 15 gpm/ft^2 is probably appropriate for natural gas plants because there is usually a large difference between the density of the LPG and amine phases, the amine and the LPG are usually clean, and hydrocarbon liquid production typically declines with time. In a refinery, 10 gpm/ft^2 would seem to be the proper choice because refinery capacity usually increases through a progressive series of revamps, the amine is often contaminated with particulates and surface active compounds, and the LPG usually has a higher gravity than gas plant liquids. In any case, design of a new LPG treater should be based on a design superficial velocity of 15 gpm/ft^2 or less.

The flooding correlation of Crawford and Wilke (1951) should be used with caution when evaluating LPG treaters. Strigle (1994) suggests that LPG treater design be based on 12% of the flooding velocity given by Crawford and Wilke, as flooding in LPG treaters can occur at 20% of the flooding velocity predicted by that correlation.

Distributors and Dispersers. If amine is the continuous phase, LPG must be evenly dispersed at the bottom of each additional bed of random packing. Usually this is accomplished with a ladder type LPG distributor beneath the bottom bed and a separate disperser/support plate beneath each additional bed of packing. The LPG ladder distributor holes should be pointed in the upward direction to minimize LPG entrainment in the rich amine (Tse and Santos, 1993). The disperser/support plate is designed so that the LPG pools below the plate and flows up into the random packing through orifices in the plate. Amine, which is the continuous phase, flows through downcomers in the disperser/support plate. The number of downcomers required is determined by the amine flowrate (DuPart and Marchant, 1989; Strigle, 1994). See **Figure 2-99** for details. Although separate devices can be used for packing support and LPG dispersion, Strigle (1994) and DuPart and Marchant (1989) recommend the use of a combined disperser/support plate, as the support plate commonly used for random packing can adversely affect the LPG dispersion. The merits of several disperser-packing support arrangements are reviewed by DuPart and Marchant.

According to Honerkamp (1975), the design of the hydrocarbon disperser/support plates between packed beds is critical. Excessive LPG velocity through the plate orifices can produce emulsions while too low a velocity causes poor distribution of the LPG. LPG and amine distributor and LPG redistributor design criteria are reviewed by DuPart and Marchant (1989), Tse and Santos (1993), and Strigle (1994). According to Tse and Santos (1993), LPG velocities in the ladder type distributor and in the disperser/support plate orifices should be less than 1.25 ft/sec. Russell (1980) recommends that the number and size of the orifices be based on a velocity of 1.0 ft/sec. Disperser/support plate LPG orifice design velocities should be in the range of 1.0–1.25 ft/sec (Russell, 1980; Strigle, 1994). Russell also states that operating at jet velocities below 0.5 ft/sec can lead to LPG entrainment in the rich amine. Orifice velocities above 1.25 ft/sec can lead to emulsion formation and amine carryover in the LPG (Perry, 1977A, B; Russell 1980). Orifice diameters should

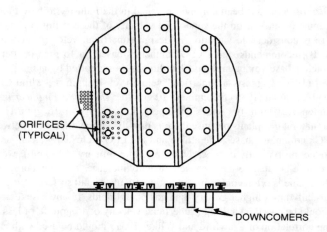

Figure 2-99. Disperser support plate. (*Strigle, 1994*)

be 0.19 to 0.25 in. (Strigle, 1994). Larger orifices produce irregular, non-uniform droplets that reduce mass transfer; while smaller orifices can lead to emulsion formation (DuPart and Marchant, 1989).

The amine distributor is usually a ladder type consisting of a series of parallel tubes fed by a central pipe. The orifices in the parallel tubes are pointed down into the top packed bed and are sized based on the amine velocity not exceeding 170 ft/min. (Strigle, 1994; DuPart and Marchant, 1989).

Although almost all LPG treaters have amine as the continuous phase with the interface controller above the top bed, some have LPG as the continuous phase with the interface below the bottom bed of packing. With the interface at the bottom, the amine is dispersed above each bed. Strigle (1994) describes LPG contactor internals for this interface control arrangement.

Minimizing Entrainment. Adequate space must be provided below the LPG distributor to prevent hydrocarbon entrainment in the rich amine leaving the LPG contactor. Honerkamp (1975) recommends 10 minutes amine residence time as measured from the LPG distributor to the bottom of the contactor. This residence time requirement should be coupled with a minimum distance requirement where the distance is measured from the bottom of the LPG distributor to the LPG contactor bottom tangent line. Tse and Santos (1993) report on an LPG contactor where this distance was 10.5 feet. This distance is somewhat greater than normal. A conservative criterion for rich amine residence time at the bottom of the LPG contactor is 10 minutes as measured from the bottom of the LPG distributor to the bottom tangent line or eight ft, whichever is greater.

Amine entrainment in the LPG is minimized by providing adequate LPG residence time above the normal LPG/amine interface level. Tse and Santos (1993) cite a case where the LPG residence time, as measured from the interface to the top of the contactor, was 8 to 10 minutes. This is quite conservative. A more reasonable criterion is believed to be the greater of 8 feet or 5 minutes LPG residence time as measured from the LPG/amine interface to the LPG contactor top tangent line. This assumes that an external gravity settler or coalescer is provided for the LPG product stream as shown in **Figure 2-96**.

The above criteria are based on maintaining the combined phase superficial velocity below 15 gpm/ft^2. If LPG entrainment in the rich amine is a problem, a shallow bed of small packing or wire mesh to coalesce the LPG can be added below the LPG distributor. Similarly, a shallow bed of packing or wire mesh can be added above the amine/LPG interface to reduce amine entrainment in the LPG.

LPG contactors using random packing have poor response to varying hydrocarbon flows (Honerkamp, 1975). Widely varying flow rates can also aggravate amine carryover. Changela and Root (1986) outline a technique where this problem was solved by recycling some of the treated LPG back to the LPG booster pump to maintain constant LPG flow through the treater. McClure and Morrow (1979B) also describe the use of recycle to maintain constant LPG flow.

Structured Packing

Sulzer and Koch have pioneered the use of structured packing for LPG treating with amines. As of 1994, Sulzer had 5 operating units in Europe, and Koch had 11 operating units in North America, 9 in gas plant service and 2 in refinery service (Rogers, 1994). For structured packing, the height of a theoretical stage is 6 to 8 ft, depending on the LPG specific gravity (Rogers, 1994). The structured packing is provided as a single bed; multiple beds are required for random packing. A single bed is possible because disperser plates are located between sections of structured packing in alternating stacks. By providing the packing as a single bed, the required bed height and contactor tangent to tangent distance can be reduced and more theoretical stages can be achieved for a given vessel height. Structured packing also has significantly higher capacity than random packing. Recent designs have achieved superficial velocities of 55–60 gpm/ft^2 based on the combined flow of both phases (Rogers, 1994). These designs had the interface controller located in the bottom of the treater and amine was the dispersed phase. At superficial velocities this high with LPG as the continuous phase, it may be necessary to install coalescing pads on both the rich amine and treated LPG streams to limit amine and LPG entrainment. Units revamped with structured packing are reported to have achieved highly efficient H_2S removal (Copper Strip test results: #1A) and to have met pipeline CO_2 specifications (Rogers, 1994). Although it is claimed that structured packing is more resistant to plugging than random packing, structured packing is more difficult to clean, and full flow filtration of the lean amine entering the LPG contactor is recommended.

Sieve Trays

While the vast majority of LPG treaters utilize random packing, sieve trays are occasionally used. However, Honerkamp (1975), Fleming et al. (1988), and DuPart and Marchant (1989) report that sieve trays are less efficient than random packing for LPG treating. Sieve tray operating data for LPG treating are given in **Tables 2-30** and **2-33**. **Table 2-33** provides extensive data for three different treating units.

Design criteria for sieve trays in LPG treating service are reviewed by Tse and Santos (1993). Average LPG velocities through the sieve tray holes should be from 0.5 to 1.0 ft/sec. The holes are set on either square or triangular pitch and are usually ⅛ to ¼ in. diameter. Tray spacing varies from 0.5 to 2 ft with 1.5 to 2 ft being more typical. Tse and Santos (1993) cite Laddha and Degaleesan (1978) for examples of detailed sieve tray rating calculations. Sieve trays have been reported to have limited turndown in LPG treating service (Honerkamp, 1975; Bacon, 1972). If LPG flows vary significantly, it may be desirable to provide for LPG recirculation as suggested by Changela and Root (1986).

Cocurrent Contactors

Honerkamp (1975) and Bacon (1972) reported on the design and operation of a single stage contactor using an eductor. **Figure 2-97** gives the flow diagram for this plant with a material balance and operating conditions. In this design, the amine was used as the driving fluid and experienced a 120-psi pressure drop across the eductor; the LPG experienced a slight pressure increase. Operating conditions were 445 psig at 55°F. The single-stage eductor-mixer treated LPG with 15 wt% MEA. The treated LPG attained a purity of 10 ppmw CO_2 when the rich MEA loading was 0.22 moles CO_2/mole MEA. The settler downstream of the eductor was designed for 30 minutes hydrocarbon liquid residence time, and the interfacial area provided for amine-hydrocarbon liquid settling was about 0.58 ft^2/gpm (equivalent to 2,500 gal hydrocarbon/day/ft^2). Honerkamp (1975) notes that the mass transfer efficiency of single stage contacting devices is relatively insensitive to varying LPG flows, while both sieve trays and random packing are affected by turndown.

Table 2-33
Operating and Design Conditions for Three Sieve Tray LPG Treaters

Plant Number	1	2	3
Inlet LPG			
Flow rate, gpm	445 (700)	102 (219)	671
Temperature, °F	60	90	110
Pressure, psia	510 (600)	225	165
Acid Gas In			
H_2S, mol%	trace	0.45	0.3
CO_2, mol%	1(0.66)	nil	nil
Acid Gas Out			
H_2S, ppm	nil	50	<4
CO_2, mol%	<0.075	nil	nil
Amine Solution			
Amine type	GAS/SPEC® CS-1M (MEA)	DEA	MDEA
Concentration, wt%	35 (15)	20	40
Circulation rate, gpm	80 (50)	47 (56)	60
Lean loading, mol/mol	0.001	0.01	0.002
Rich loading, mol/mol	0.22	0.21	0.1
Contactor Type	Sieve Tray	Sieve Tray	Sieve Tray
Column diameter, ft	6	4.5	9.5
No. of trays	20	15	20
Tray spacing, in.	18	18	30
Hole dia., in.			$3/16''$
Open hole space per tray, ft^2			2.123
Downspout area, ft^2			6.05 (8.54% of tower area)
Combined flow, gpm/ft^2	18.4 (26.5)	9.36 (17.2)	10.3

Note:
Parentheses ( ) indicate design conditions.
Source: Tse and Santos (1993)

Auxiliary Systems

Water Wash

LPG/amine treaters can have significant amine losses. Veldman (1989) states that amine losses for well designed and operated LPG treaters due to amine entrainment and solubility amount to about 0.02 lb of amine per barrel of LPG treated. Washing the LPG with a dilute amine-water solution removes entrained amine from the LPG and reduces the concentration of amine dissolved in the LPG because amine in the hydrocarbon phase establishes equilibrium with the dilute water wash phase. **Figure 2-96** includes a typical water wash system.

Design criteria for LPG water wash systems are reviewed by Stewart and Lanning (1991, 1994A, B). For a gravity settler as shown in **Figure 2-96,** they recommend LPG and wash water residence times of 15 and 20 minutes, respectively. They also recommend that the water make-up rate be set to produce a wash water purge to the amine system containing about 3 wt% amine. Although solubility data such as those of Veldman (1989) can be used to estimate amine solubility losses in the LPG, it is probably better to use an experience factor, such as on 0.02 lb of amine per barrel of LPG feed (about 100 ppmw), as this figure includes amine entrainment. Note that poorly designed systems with high superficial velocities or with inadequate LPG residence times in the LPG amine contactor can have amine losses as high as 500 ppmw (Veldman, 1989). While there is considerable variation, typical practice is to set the volumetric ratio of wash water to LPG at 1:4, although Stewart and Lanning (1991, 1994B) report on a liquid ethane treater where the wash water was only 6 to 7% of the LPG volumetric flow. See **Table 2-34** for details.

Table 2-34
Water Wash System Operating Conditions for Recovering
MDEA from Liquid Ethane

Ethane rate, BPSD	22,000	27,000
Water-wash rate, gpm	44	44
Water-wash amine content, wt%	0.3	0.4
Water-wash make-up, gpm	3.5	3.5
Water-wash rate, % of ethane flow	7	6

Source: Stewart and Lanning (1991, 1994B)

LPG Settlers and Coalescers

It is generally desirable to use an LPG coalescer or settler downstream of the LPG/amine contactor to remove entrained amine from the LPG. A settler is usually a horizontal vessel that uses gravity to separate two liquid phases. Baffles to prevent channeling are sometimes provided. A coalescer is a vessel that contains elements, such as wire mesh pads, which can agglomerate small liquid droplets into larger droplets which can then be easily separated by gravity (Brown and Wines, 1993). A coalescer is smaller than a settler and costs less; however, the elements can plug if the liquid is dirty. The performance and size of the coalescer depend on the design of the coalescing elements which are usually proprietary items. Although it is difficult to provide generalized guidelines for sizing coalescers, elements used for LPG/amine applications must be resistant to attack by the amine. They are usually made of stainless steel, polyester, or polypropylene. Fiberglass is not suitable, as amine degrades the fiberglass. Also, the performance of the coalescing element cannot depend on a hydrophilic coating because surface active amine contaminants can "disarm" the treated surface and reduce its coalescing efficiency (Brown and Wines, 1993).

Coalescers are lower in cost, but gravity settlers have the advantage of simplicity and reliability of operation. Settlers are usually designed on the basis of residence time. Honerkamp (1975) reports that a settler downstream of a single stage eductor was designed for a 30-minute hydrocarbon retention time with a hydrocarbon-amine interfacial area of 0.58 ft^2/gpm (equivalent to 2,500 gal/(day)(ft^2) used to set the settler dimensions. Separators downstream of contactors, using either random packing or sieve trays, can be smaller than those following mixers or eductors because the degree of agitation and, therefore, the formation of fine, entrainable droplets is less in the packed or tray columns. Stewart and Lanning (1991, 1994A, B) recommend LPG residence times of 15 minutes for a gravity settler downstream of a packed bed LPG treater.

Amine Filters

Figure 2-96 shows an optional amine filter on the lean amine solution flowing to the LPG/amine contactor. Although many LPG treaters do not have dedicated mechanical or activated carbon filters, their use should be considered. Bacon (1972) recommends full flow mechanical and activated carbon lean amine filters for LPG treaters. Russell (1980) reports particulate plugging of the random packing beds used in LPG treaters and recommends full flow amine filtration to minimize plugging. DuPart and Marchant (1989) also recommend both full flow mechanical and activated carbon filtration of the amine flowing to LPG treaters. As noted by both DuPart and Marchant (1989) and Strigle (1994), failure to remove impurities can reduce the capacity of the LPG treating system.

Removal of COS from LPG by Amines

A common specification for LPG is the Copper Strip corrosion test. This test indicates the level of elemental sulfur and H_2S by the degree of darkening of a strip of pure copper. A #1A Copper Strip test result corresponds to 1–2 ppmw or less H_2S in the LPG. COS in dry LPG is not detectable by the Copper Strip test (Bacon, 1972; Perry, 1977A, B). However, if the LPG comes in contact with water, e.g., during shipment in a tank car containing water, the COS hydrolyzes to H_2S and CO_2 and the LPG can fail the Copper Strip test upon retesting (Holmes et al., 1984; Perry, 1977A, B). The COS hydrolysis reaction is

$$COS + H_2O = H_2S + CO_2 \qquad (2\text{-}49)$$

Because of this possible reaction, COS removal during amine LPG treating is desired. DGA, MEA, and possibly DIPA can react with COS during the amine treating operation so that a #1A Copper Strip test result can be achieved even if the LPG is wet. DGA is reported to reduce COS levels from 4,500 ppmw to less than 2 ppmw (McClure and Morrow, 1979A, B, C). The DGA reacts with COS to form a degradation product that can be converted back to DGA by thermal reclaiming. MEA reacts irreversibly with COS and, when MEA is used for LPG treating, the treated LPG can pass the #1A Copper Strip test. However, MEA is not preferred for simultaneous CO_2, H_2S, and COS removal from LPG because of the amine loss caused by the irreversible reaction with COS (McClure and Morrow, 1979C). DIPA is also reported to be capable of removing COS to several ppmw (S.I.P.M., 1979). Mick (1976) reports that DEA can only remove COS to 90–135 ppmw. Bacon and Pearce (1985) report that MDEA can remove only 33% of the COS in LPG and cannot remove COS to levels low enough to pass the #1A Copper Strip test.

Although neither MDEA nor DEA can remove COS to levels low enough to pass the Copper Strip test, use of a formulated caustic wash downstream of the amine/LPG treater is reported to remove COS to levels significantly less than 1 ppmw (Bacon and Pearce, 1985). The composition of the formulated caustic may be based on technology outlined by Johnson and Condit (1952) in which MEA is added to the caustic prewash upstream of an extractive Merox unit to achieve COS removal. According to Johnson and Condit, the MEA concentration in the caustic must be maintained at about 1.5 wt% or higher to achieve nearly complete COS removal (<0.04 ppmw COS reported as S). The caustic concentration must be greater than 3.0 wt% for the reaction to occur, and the volumetric ratio of the formulated caustic to the LPG flow should be 10:1. When both MEA and caustic are present, the MEA is not consumed by reaction with the COS.

REFERENCES

Abe, T., and Peterzan, P., 1980, *Oil and Gas J.*, March 31, p. 139.

Al-Ghawas, H. A., Ruiz-Ibanez, G., and Sandall, O. C., 1988, "Absorption of Carbonyl Sulfide in Aqueous Methyldiethanolamine," presented at the Spring National Meeting of the AIChE, New Orleans, LA, March 1–10.

Ammons, H. L., and Sitton, D. M., 1981, "Operating Data from a Commercial MDEA Treater," *Proceedings of the 1981 Gas Conditioning Conference,* University of Oklahoma, OK.

Anon., 1946, *Petrol. Refiner,* Vol. 25, No. 10, p.121.

Anon., 1981, *Chem. & Eng. News,* Sept. 7, p. 58.

Appl, M., Wagner, U., Henrici, H. J., Kuessner, K., Volkamer, K., and Fuerst, E., 1980, "Removal of CO_2 and/or H_2S and/or COS from gases containing these constituents," Canadian Patent No. 1,090,098.

Arnold, D. S., Barrett, D. A., and Isom, E. H., 1982, "CO_2 Can Be Produced from Flue Gas," *Oil & Gas J.*, Nov. 22, p. 130.

Astarita, G., 1967, *Mass Transfer with Chemical Reaction,* Elsevier, Amsterdam.

Astarita, G., Savage, C. W., and Basio, A., 1983, *Gas Treating with Chemical Solvents,* Wiley, N.Y.

Atadan, E. M., 1954, "Absorption of Carbon Dioxide by Aqueous Monoethanolamine Solutions," Ph.D. thesis, University of Tennessee, Knoxville, TN.

Atwood, K., Arnold, M. R., and Kindrich, R. C., 1957, *Ind. Eng. Chem.*, Vol. 49, September, p. 1439.

Austgen, D. M., Rochelle, G. T., and Chen, C. -C., 1991, "Model of Vapor-Liquid Equilibria for Aqueous Acid Gas-Alkanolamine Systems," 2. "Representation of H_2S and CO_2 Solubility in Aqueous MDEA and CO_2 Solubility in Aqueous Mixtures of MDEA with MEA or DEA," *Ind. Eng. Chem. Res.*, Vol. 30, p. 543.

Bacon, K. H., 1972, "Liquid Treating," *Proceedings of the Gas Conditioning Conference,* University of Oklahoma, Norman, OK.

Bacon, T. R., and Pearce, R. L., 1985, "Energy Conservation and H_2S Selectivity With COS Removal," paper presented at Petro Energy '85, Sept. 16–20.

Bailleul, M., 1969, *Gas Processing Canada,* Vol. 61, No. 3, pp. 34–38.

Bally, A. P., 1961, *Erdöl und Kohle,* Vol. 14, pp. 921–923.

Bartholome, E., Schmidt, H. W., Friebe, J., 1971, "Process for Removing Carbon Dioxide from Gas Mixtures," U.S. Patent 3,622,267.

Beddome, J. M., 1969, "Current Natural Gas Sweetening Practice," paper presented at 19th Canadian Chemical Engineering Conference, October.

Bellah, J. S., Mertz, R. V., and Kilmer, J. W., 1949, *Petrol. Refiner,* Vol. 28 June, p. 154.

Benson, H. E., Field, J. H., and Jimeson, R. M., 1954, *Chem. Eng. Progr.*, Vol. 50, July, pp. 356–364.

Benson, H. E., Field, J. H., and Haynes, W. P., 1956, *Chem. Eng. Progr.*, Vol. 52, p. 33.

Berthier, P., 1959, *Science et Technique,* Vol. 81, January, pp. 49–55.

Bhairi, A., Mains, G. J., Maddox, R. N., 1984, "Experimental Measurements of Equilibrium Between CO_2 or H_2S and Ethanolamine Solutions," paper presented at AIChE National Meeting, Atlanta, GA, March 11–24.

Blanc, C., and Elgue, J., 1981, *Hydro. Process.*, Vol. 60, No. 8, August, p. 111.

Blohm, C. L., and Riesenfeld, F. C., 1955, U.S. Patent 2,712,978.

Bottoms, R. R., 1930, U.S. Patent 1,783,901; Re 1933, 18,958.

Bottoms, R. R., 1931, *Ind. Eng. Chem.*, Vol. 25, May, pp. 501–504.

Brown, R. L., Jr. and Wines, T. H., 1993, "Improve suspended water removal from fuels," *Hydro. Process.*, December, pp. 95–100.

Bucklin, R. W., 1982, *Oil and Gas J.*, Nov. 8, p. 204.

Bullin, J. A., and Polasek, J. C., 1982, "Selective Absorption with Amines," Proceedings of the 61st Annual GPA Convention, Dallas, TX, March 15–17, pp. 86–90.

Buskel, C., 1959, *Oil and Gas J.*, Vol. 57, Nov. 30, pp. 67–70.

Butwell, K. F., Hawkes, E. N., and Mago, B. F., 1973, *Chem. Eng. Prog.*, February, p. 57.

Butwell, K. F., Kubek, D. J., and Sigmund, P. W., 1979, "Amine Guard III," *Chem. Eng. Prog.*, February, p. 75.

Butwell, K. F., Kubek, D. J., and Sigmund, P. W., 1982, "Alkanolamine Treating," *Hydro. Proc.*, March, pp. 108–116.

Butwell, K. F., and Perry, C. R., 1975, "Performance of Gas Purification Systems Utilizing DEA Solutions," *Proceedings 1975 Gas Conditioning Conference,* University of Oklahoma, Norman, OK.

Cai, R.-X., and Chang, H.-G., 1992, "Selective Hydrogen Sulfide Absorption in Hindered Amine Aqueous Solution," *Journal of Natural Gas Chemistry,* No. 2, p. 175.

Campbell, S. W., and Weiland, R. H., 1989, "Modeling CO_2 Removal by Amine Blends," paper presented at the AIChE 1989 Spring National Meeting, Houston, TX, April 2–6.

Carney, B. R., 1947, *Oil and Gas J.*, Vol. 46, Aug. 30, pp. 56–63.

Carroll, J. J., Jou, F.-Y., Mather, A. E., and Otto, F. D., 1992, "Phase Equilibria in the System Water-Methyldiethanolamine-Propane," *AIChE J.*, Vol. 38, No. 4, pp. 511–520.

Carroll, J. J., Jou, F.-Y., Mather, A. E., and Otto, F. D., 1993, "The Distribution of Hydrogen Sulfide Between an Aqueous Amine Solution and Liquid Propane," *Fluid Phase Equilibria,* Vol. 82, pp. 183–190.

Chakma, A. and Meisen, A., 1987, "Solubility of CO_2 in Aqueous Methyldiethanolamine and N,N-Bis (hydroxyethyl) Piperazine Solutions," *Ind. Eng. Chem. Res.*, Vol. 26, No. 12, pp. 2461–2466.

Chakravarty, T., 1985, "Solubility Calculations for Acid Gases in Amine Blends," Ph.D. Thesis, Clarkson University, Potsdam, NY.

Changela, M. K., and Root, C. R., 1986, "NGL Product Treating for Acid Gas Removal," *Laurance Reid Gas Conditioning Conference Proceedings,* University of Oklahoma, Norman, OK.

Chapin, W. F., 1947, *Petrol. Refiner.*, Vol. 26, No. 6, pp. 109–112.

Chen, C. C., and Evans, L. B., 1986, "A Local Composition Model for the Excess Gibbs Energy of Aqueous Electrolyte Systems," *AIChE J.*, Vol. 32, pp. 444–454.

Chludzinski, G.R., Stogryn, E. L., and Weichert, S., 1986, "Commercial Experience with Flexsorb SE Absorbent," presented at the 1986 Spring AIChE National Meeting, New Orleans, LA.

Christensen, S. P., Christensen, J. J., and Izatt, R. M., 1985, "Enthalpies of Solution of CO_2 in Aqueous Diglycolamine Solutions," GPA Research Report 85, Project 821, Brigham Young University, Provo, UT, May.

Connors, J. S., 1958, *Oil and Gas J.*, Vol. 56, March 3, pp. 100–120.

Cordi, E. M. and Bullin, J. A., 1992, "Kinetics of Carbon Dioxide and Methyldiethanolamine with Phosphoric Acid," *AIChE J.*, Vol. 38, No. 3, March, p. 455–460.

Cornelissen, A. E., 1980, "Simulation of Absorption of H2S and CO2 with Aqueous Alkanolamines in Tray and Packed Columns," *Trans. Inst. Chem. Engrs.*, Vol. 58, p. 242.

Crawford, J. W., and Wilke, C. R., 1951, "Extraction Column—Correlation for Large Packings," *Chem. Eng. Prog.*, Vol. 47, No. 8, p. 423.

Crow, J. H. and Baumann, J. C., 1974, *Hydro. Process.*, Vol. 53, No. 10, October, p. 131.

Cryder, D. S., and Maloney, J. O., 1941, *Trans. Am. Inst. Chem. Engrs.*, Vol. 37, October, pp. 827–852.

Dailey, L. D., 1970, *Oil and Gas J.*, Vol. 68, May 4, pp. 136–141.

Danckwerts, P. V., 1970, *Gas-Liquid Reactions,* McGraw-Hill, NY.

Danckwerts, P. V., and Sharma, M. M., 1966, "The Absorption of Carbon Dioxide Into Solutions of Alkalies and Amines," *Chem. Eng.*, October, pp. 244-280.

Darton, R. C., Hoek, P. J., Spaniks, J. A. M., Suenson, M. M., and Wijn, E. F., 1987, "The Effect of Equipment on Selectivity in Amine Treating," *European Federation of Chemical Engineering (EFCE) Publication Series,* No. 62, pub. by EFCE, Amarousion, Pefki, Greece, pp. A323–A345.

Daviet, G. R., Sundermann, R., Donnelly, S. T., and Bullin, J. A., 1984, "Switch to MDEA raises capacity," *Hydro. Process.*, May, pp. 79–82.

Deshmukh, R. D., and Mather, A. E., 1981, "A Mathematical Model for Equilibrium Solubility of Hydrogen Sulfide and Carbon Dioxide in Aqueous Alkanolamine Solutions," *Chemical Engineering Science,* Vol. 36, pp. 355–362.

Dibble, J. H., 1983, "UCARSOL Solvents for Acid Gas Removal," paper presented at Petroenergy '83, Houston, TX, Sept. 14.

Dibble, J. H., 1985, "Absorbent formulation for enhanced removal of acid gases from gas mixtures and processes using same," European Patent Publication Number 0134948, Application No. 8417586.4, March 27.

Dingman, J. C., and Moore, T. F., 1968, "Gas Sweetening With Diglycolamine," *Proceedings of the 1968 Gas Conditioning Conference,* Univ. of Oklahoma, Norman, OK.

Dingman, J. C., 1977, "Gas Sweetening with Diglycolamine Agent," paper presented at Third Iranian Congress of Chemical Engineering, Shiraz, Iran, Nov. 6–11.

Dingman, J. C., and Moore, T. F., 1968, *Hydro. Process.*, Vol. 47, July, pp. 138–140.

Dingman, J. C., Jackson, J. L., Moore, T. F., and Branson, J. A., 1983, "Equilibrium Data for the H₂S-CO₂-Diglycolamine-Water System," paper presented at 62nd Annual Gas Processors Assoc. Convention, San Francisco, CA, March 14–16.

Dow, 1962, *Gas Conditioning Fact Book,* Dow Chemical Co.

Dow, 1981, *The Alkanolamines Handbook,* Dow Chemical Co.

Dow Chemical USA, 1983, "GAS/SPEC FT-1 Technology," Technical Bulletin.

DuPart, M. S., and Marchant, B. D., 1989, "Natural Gas Liquid Treating Options and Experiences," *Laurance Reid Gas Conditioning Conference Proceedings,* University of Oklahoma, Norman, OK.

Dupart, M. S., Bacon, T. R., and Edwards, D. J., 1993A, "Part 1—Understanding corrosion in alkanolamine plants," *Hydro. Process.*, April, pp. 75–80.

Dupart, M. S., Bacon, T. R., and Edwards, D. J., 1993B, "Part 2—Understanding corrosion in alkanolamine plants," *Hydro. Process.*, May, pp. 89–94.

Eckart, J. S., Foote, E. H., Rollison, L. R., and Walter, L. F., 1967, *Ind. Eng. Chem.*, Vol. 59, February, pp. 41–47.

Elgin, J. C., and Weiss, F. B., 1939, *Ind. Eng. Chem.*, Vol. 31, April, pp. 435–445.

Estep, J. W., McBride, J. T., and West, J. R., 1962, *Advances in Petroleum Chemistry and Refining,* Vol. 6, Interscience Publishers, NY, pp. 315–466.

Exxon Research and Engineering Co., 1994, "Flexsorb Solvents," in *Gas Processes '94, Hydro. Process.*, April, p. 80.

Fair, J. R., 1987, "Distillation," in *Handbook of Separation Process Technology,* R.W. Rousseau, editor, John Wiley and Sons, NY, NY.

Feagan, R. A., Lawler, H. L., and Rhames, M. H., 1954, *Petrol. Refiner,* Vol. 33, June, p. 167.

Fitzgerald, K. J., and Richardson, J. A., 1966A, "How Gas Composition Affects Treating Process Selection," *Hydro. Process.*, Vol. 45, No. 7, pp. 125–129.

Fitzgerald, K. J., and Richardson, J. A., 1966B, "New correlations enhance value of monoethanolamine process," *Oil and Gas J.*, Oct. 24, pp. 110–118.

Fleming, K. B., Spears, M. L., and Bullin, J. A., 1988, "Design Alternatives for Sweetening LPGs and Liquid Hydrocarbons With Amines," SPE paper 18231, presented at 63rd Annual Soc. Petrol. Eng. Conference, Houston, TX, Oct. 2–5, pp. 521–527.

Fluor Daniel, 1995, unpublished data.

Frazier, H. D. and Kohl, A. L., 1950, "Selective Absorption of H₂S from Gas Streams," *Ind. Eng. Chem.*, Vol. 42, November, pp. 2282–2292.

GPSA (Gas Processors Suppliers Assoc.), 1987, *Engineering Data Book, Tenth Edition,* Gas Processors Suppliers Association, 6526 East 60th St., Tulsa, OK.

GPSA (Gas Processors Suppliers Association) 1994, *Engineering Data Book, 10th Edition, Revised, Vol. II,* Tulsa, OK.

Gagliardi, C. R., Smith, D. D., and Wang, S. I., 1989, "Strategies to Improve MEA CO₂-Removal Detailed at Louisiana Ammonia Plant," *Oil and Gas J.*, March 6, pp. 44–49.

Gangriwala, H. A., 1987, "Packed Columns in Acid Gas Removal," *European Federation of Chemical Engineering (EFCE) Publication Series,* No. 62, pub. by EFCE, Amarousion, Perki, Greece, pp. B89–B99.

Gerhardt, W., and Hefner, W., 1988, "BASF's Activated MDEA: A Flexible Process to Meet Specific Design Conditions," presented at the 1988 AIChE Ammonia Safety Symposium, Denver, CO, Aug. 22–24.

Goar, B. G., 1980, "Selective Gas Treating Produces Better Claus Feed," *Proceedings of the 1980 Gas Conditioning Conference,* University of Oklahoma, Norman, OK.

Goel, R. K., Chen, R. J., and Elliot, D. G., 1982, "Data Requirements in Gas and Liquid Treating," *Gas Processors Assoc. Proceedings,* pp. 185–193.

Goldman, A. M., and Leibush, A. G., 1959, "Solubility of Carbon Dioxide in Aqueous Solutions of Monoethanolamine in the Temperature Range 75–140°C," Trudy Gasudarst, *Nauch.-Issledov Praekt. Inst.-Azot. Prom.*, Vol. 10, pp. 54–82.

Goldstein, A.M., 1983, "Commercialization of a New Gas Treating Agent," paper presented at Petroenergy '83 Conference, Houston, TX, Sept. 14.

Gregory, L. B., and Scharmann, W. G., 1937, *Ind. Eng. Chem.*, Vol. 29, May, pp. 514–519.

Harbison, J. L. and Dingman, J. C., 1972, "Mercaptan Removal Experiences in DGA Sweetening of Low Pressure Gas," *Proceedings of the 1972 Gas Conditioning Conference,* University of Oklahoma, Norman, OK.

Harbison, J. L., and Handwerk G. E., 1987, "Selective Removal of H_2S Utilizing Generic MDEA," *Proceedings of the 1987 Gas Conditioning Conference,* University of Oklahoma, Norman, OK.

Hayden, T. A., Smith, T. G. A., and Mather, A. E., 1983, *Journal of Chemical & Engineering Data,* Vol. 28, p. 196.

Helton, R., Christensen, J. J., and Izatt, R. M., 1987, "Enthalpies of Solution of CO_2 in Aqueous Diethanolamine Solutions," GPA Research Report 108, Project 821, Brigham Young University, Provo, UT, May.

Ho, B. S., and Eguren, R. R., 1988, "Solubility of Acidic Gases in Aqueous DEA and MDEA Solutions" presented at AIChE Spring National Meeting, New Orleans, LA, March 6–10.

Ho, W. S., Sartori, G., and Stogryn, E. L., 1990, "Absorbent Composition Containing a Severely Hindered Amine Mixture With Amine Salts and/or Aminoacid Additives for the Absorption of H_2S from Fluid Mixtures Containing CO_2," US Patent 4,961,873, Oct. 9.

Holder, H. L., 1966, "Diglycolamine—a promising new acid-gas remover," *Oil and Gas J.*, May 2, pp. 83–86.

Holmes, J. W., Spears, M. L., and Bullin, J. A., 1984, "Sweetening LPG's with Amines," *Chem. Eng. Prog.*, May, pp. 47–50.

Honerkamp, J. D., 1975, "Treating Hydrocarbon Liquids with Amine Solutions," *Gas Conditioning Conference Proceedings,* University of Oklahoma, Norman, OK.

Hutchinson, A. J. L., 1939, U.S. Patent 2,177,068.

Huval, M., and van de Venne, H., 1981, "DGA Proves out as a Low Pressure Gas Sweetener in Saudi Arabia," *Oil & Gas J.*, Aug. 17, p. 91.

Isaacs, E. E., Otto, F. D., and Mather, A. E., 1977A, "Solubility of Hydrogen Sulfide and Carbon Dioxide in an Aqueous Diisopropanolamine Solution," *Journal of Chemical and Engineering Data,* Vol. 22, pp. 71–73.

Isaacs, E. E., Otto, F. D., and Mather, A. E., 1977B, "The Solubility of Mixtures of Carbon Dioxide and Hydrogen Sulphide in an Aqueous DIPA Solution," *Canadian Journal of Chemical Engineering,* Vol. 55, pp. 210–212.

Isaacs, E. E., Otto, F. D., and Mather, A. E., 1980, "Solubility of Mixtures of H_2S and CO_2 in a Monoethanolamine Solution at Low Partial Pressures," *J. Chem. Eng. Data,* Vol. 25, pp. 118-120.

Isom, C. R., and Rogers, J. A., 1994, "Utilizing High Efficiency Co-Current Contactors to Optimize Existing Amine Treating Units," *Proceedings of the 1994 Laurance Reid Gas Conditioning Conference,* University of Oklahoma, Norman, OK.

Jefferson Chemical Co., 1969, *Gas Treating Data Book,* Jefferson Chemical Co. (now Huntsman Chemical Co.).

Johnson, A. B., and Condit, D. H., 1952, "Removal of Carbonyl Sulfide from Liquefied Petroleum Gas," US Patent No. 2,594,311, April 29.

Jones, J. H., Froning, H. R., and Claytor, E. E., Jr., 1959, *J. Chem. and Eng. Data,* Vol. 4, January, pp. 85–92.

Jou, F.-Y, Lal, D., Mather, A. E., and Otto, F. D., 1981, "Solubility of Acid Gases in MDEA Solutions," paper presented at CGPA Meeting, Nov. 12, Calgary, Alberta.

Jou, F.-Y., Mather, A. E., and Otto, F. D., 1982, "Solubility of H_2S and CO_2 in Aqueous Methyldiethanolamine Solutions," *Ind. Eng. Chem. Process Des. Dev.,* Vol. 21, pp. 539–544.

Jou, F.-Y., Otto, F. D., and Mather, A. E., 1985, "The Solubility of H_2S and CO_2 in Triethanolamine Solutions," in *Acid and Sour Gas Treating Processes,* (S.A. Newman, editor) p. 278–288, Gulf Publishing Co., Houston, TX.

Jou, F.-Y., Carroll, J. J., Mather, A. E., and Otto, F. D., 1993, "The Solubility of Carbon Dioxide and Hydrogen Sulfide in a 35 wt% Aqueous Solution of Methyldiethanolamine," *Can. J. Chem. Eng.,* Vol. 71, pp. 264–268.

Jou, F.-Y., Otto, F. D., and Mather, A. E., 1994, "Vapor-Liquid Equilibrium of Carbon Dioxide in Aqueous Mixtures of Monoethanolamine and Methyldiethanolamine," *Ind. & Eng. Chem. Res.,* Vol. 33, No. 8, pp. 2002–2005.

Jou, F.-Y., Mather, A. E., and Otto, F. D., 1995A, "The Solubility of CO_2 in a 30 Mass Percent Monoethanolamine Solution," *Can. J. of Chem. Eng.,* Vol. 73, February, pp. 140–147.

Jou, F.-Y., Mather, A. E., Otto, F. D., and Carroll, J. J., 1995B, "Experimental Investigation of the Phase Equilibrium in the Carbon Dioxide-Propane-3M MDEA System," *I&EC Research,* Vol. 34, p. 2526.

Katti, S. S., and Langfitt, B. D., 1986A, "Development of a Simulator for Commercial Absorbers Used for Selective Chemical Absorption Based on a Mass Transfer Rate Approach," presented at the 65th Annual Convention Gas Processors' Assoc., San Antonio, TX, March 10–12.

Katti, S. S. and Langfitt, B. D., 1986B, "Effect of Design and Operating Variables on the Performance of Commercial Absorbers Used for Selective Chemical Absorption," presented at the AIChE National Meeting, New Orleans, LA, April 6–10.

Katti, S. S., and Wolcott, R. A., 1987, "Fundamental Aspects of Gas Treating with Formulated Amine Mixtures," paper presented at the AIChE National Meeting, Minneapolis, MN.

Kennard, M. L., and Meisen, A., 1984, "Solubility of Carbon Dioxide in Aqueous Diethanolamine Solutions at Elevated Temperatures and Pressures," *J. Chem. Eng. Data,* Vol. 29, pp. 309–312.

Kenney, T. J., 1995, (Huntsman Chemical), personal communication, Oct. 18.

Kenney, T. J., 1996, (Huntsman Chemical), personal communication, March 14.

Kenney, T. J., Khan, A. R., Holub, P. E., and Street, D. E., 1994, "DGA Agent Shows Promise for Trace Sulfur Compound Removal from Hydrocarbon Streams," Presented at the 1994 GRI Sulfur Recovery Conference, May 15–17, Austin, TX.

Kent, R. L., and Eisenberg, B., 1976, "Better Data for Amine Treating," *Hydro. Process.*, Vol. 55, No. 2, February, pp. 87–90.

Khan, M., and Manning, W. P., 1985, "Practical Designs for Amine Plants," paper presented at Petro Energy Workshop, Houston, TX, September.

King, R. L., Spears, M. L., and Bullin, J. A., 1985, "Design Alternatives for Sweetening Liquids with Amines," Petroenergy Conference, Houston, TX, Sept. 17.

Kister, H. S., and Gill, D. R., 1991, "Predict Flood Point and Pressure Drop for Modern Random Packings," *Chem. Eng. Prog.*, Vol. 87, No. 2, February, p. 32.

Klein, J. P., 1970, *Oil and Gas Int.*, Vol. 10, September, pp. 109–112.

Kohl, A. L. and Blohm, C. L., 1950, "Technical Aspects of Glycol-Amine Gas Treating," *Petrol. Engr.*, Vol. 22, June, p. C-37.

Kohl, A. L., 1951, "Selective H_2S Absorption—A Review of Available Processes," *Petrol. Processing*, Vol. 6, January, pp. 26–31.

Kohl, A. L., and Bechtold, I. C., 1952, "Extraction of High Acidic Concentrations from Gases," U.S. Patent 2,607,657.

Kohl, A. L., 1956, "Plate Efficiency with Chemical Reaction—Absorption of Carbon Dioxide in Mononethanolamine Solutions," *AIChE J.*, Vol. 2, June, p. 264.

Krishnamurthy, R. and Taylor, R., 1985A, "A Nonequilibrium Stage Model of Multicomponent Separation Processes," *AIChE Journal*, Vol. 31, pp. 449–455 and 456–465.

Krishnamurthy, R., and Taylor, R., 1985B, "Simulation of Packed Distillation and Absorption Columns," *Ind. Eng. Chem. Process Des. Dev.*, Vol. 24, No. 3, p. 513.

Kubek, D. J., and Butwell, K. F., 1979, "Amine Guard Systems in Hydrogen Production," paper presented at AIChE National Meeting, April 1–5.

Laddha, G. S. and Degaleesan, T. E., 1978, *Transport Phenomena in Liquid Extraction*, McGraw-Hill, NY., NY.

Lal, D., Isaacs, E. E., Mather, A. E., and Otto, F. D., 1980, "Equilibrium Solubility of Acid Gases in Diethanolamine and Monoethanolamine Solutions at Low Partial Pressures," *Proceedings of the 1980 Gas Conditioning Conference*, University of Oklahoma, Norman, OK.

Lal, D., Otto, F. D., and Mather, A. E., 1985, "The Solubility of H_2S and CO_2 in a Diethanolamine Solution at Low Partial Pressures," *Can. J. Chem. Eng.*, Vol. 63, pp. 681–685.

Lawson, J. D., and Garst, A. W., 1976, "Gas Sweetening Data: Equilibrium Solubility of Hydrogen Sulfide and Carbon Dioxide in Aqueous Monoethanolamine and Aqueous Diethanolamine Solutions," *J. of Chem. and Eng. Data*, Vol. 21, No. 1, pp. 20–30.

Lee, J. I., Otto, F. D., and Mather, A. E., 1972, "Solubility of Carbon Dioxide in Aqueous Diethanolamine Solutions at High Pressures," *J. Chem. Eng. Data*, Vol. 17, pp. 465–468.

Lee, J. I., Otto, F. D., and Mather, A. E., 1973A, "Solubility of Hydrogen Sulfide in Aqueous Diethanolamine Solutions at High Pressures," *J. Chem. Eng. Data*, Vol. 18, pp. 71–73.

Lee, J. I., Otto, F. D., and Mather, A. E., 1973B, "Partial Pressures of Hydrogen Sulfide in Aqueous Diethanolamine Solutions," *J. Chem. Eng. Data*, Vol. 18, p. 420.

Lee, J. I., Otto, F. D., and Mather, A. E., 1973C, "Design Data for Diethanolamine Acid Gas Treating Systems," *Gas Processing Canada,* March–April, pp. 26–34.

Lee, J. I., Otto, F. D., and Mather, A. E., 1974A, "The Solubility of Mixtures of Carbon Dioxide and Hydrogen Sulfide in Aqueous Diethanolamine Solutions," *Can. J. Chem. Eng.*, Vol. 52, pp. 125–127.

Lee, J. I., Otto, F. D., and Mather, A. E., 1974B, "The Solubility of H_2S and CO_2 in Aqueous Monoethanolamine Solutions," *Can. J. Chem. Eng.*, Vol. 52, pp. 803–805.

Lee, J. I., Otto, F. D., and Mather, A. E., 1975, "Solubility of Mixtures of Carbon Dioxide and Hydrogen Sulfide in 5.0 N Monoethanolamine Solutions," *J. Chem. Eng. Data,* Vol. 20, pp. 161–163.

Lee, J. I., Otto, F. D., and Mather, A. E., 1976A, "Equilibrium in Hydrogen Sulfide-Monoethanolamine-Water System," *J. Chem. Eng. Data,* Vol. 21, pp. 207–208.

Lee, J. I., Otto, F. D., and Mather, A. E., 1976B, "The Measurement and Prediction of the Solubility of Mixtures of Carbon Dioxide and Hydrogen Sulfide in 2.5 N Monoethanolamine Solution," *Can. J. Chem. Eng.*, Vol. 54, pp. 214–219.

Lee, J. I., Otto, F. D., and Mather, A. E., 1976C, "Equilibrium Between Carbon Dioxide and Aqueous Monoethanolamine Solutions," *J. Appl. Chem. Biotechnol.*, Vol. 26, pp. 541–549.

Leibush, A. G. and Shneerson, A. L., 1950, *J. Appl. Chem.,* U.S.S.R., Vol. 23, No. 2, pp. 145–152.

Li, M.-H. and Shen, K.-P., 1992, "Densities and Solubilities of Solutions of Carbon Dioxide in Water + Monoethanolamine + N-Methyldiethanolamine," *J. Chem. Eng. Data,* Vol. 37, p. 288.

Li, Y.-G. and Mather, A. E., 1994, "Correlation and Prediction of the Solubility of Carbon Dioxide in a Mixed Ethanolamine Solution," *Ind. Eng. Chem. Res.*, Vol. 33, pp. 2006–2015.

Lobo, W. E., Friend, L., and Skaperdas, G. T., 1944, *Ind. Eng. Chem.*, Vol. 34, July, pp. 821–823.

Love, F. H., 1941, *Petrol. Engr.*, Vol. 13, November, pp. 31–32.

Lyudkovskaya, M. A., and Leibush, A. G., 1949, *J. Appl. Chem.*, U.S.S.R., Vol. 22, No. 6, pp. 558–567.

MacGregor, R. J., and Mather, A. E., 1991, "Equilibrium Solubility of H_2S and CO_2 and Their Mixtures in a Mixed Solvent," *Can. J. Chem. Eng.*, Vol. 69, pp. 1357–1366.

Maddox, R. N., 1985, *Gas Conditioning and Processing, Vol. 4, Gas and Liquid Sweetening,* Campbell Petroleum Series, 1215 Crossroads Blvd., Norman, OK.

Maddox, R. N., Bhairi, A. H., Diers, J. R., and Thomas, P. A., 1987, "Equilibrium Solubility of Carbon Dioxide or Hydrogen Sulfide in Aqueous Solutions of Monoethanolamine, Diglycolamine, Diethanolamine and Methyldiethanolamine," Research Report RR-104 to Gas Processors Association, Tulsa, OK, March.

Maddox, R. N., and Elizondo, E. M., 1989, "Equilibrium Solubility of Carbon Dioxide or Hydrogen Sulfide in Aqueous Solutions of Diethanolamines at Low Partial Pressures," GPA Research Report 124, Project 841, Oklahoma State University, Stillwater, OK, June.

Maham, Y., Teng, T. T., Hepler, L. G., and Mather, A. E., 1994, *Journal of Solution Chemistry,* Vol. 23, No. 2, p. 195.

Manning, F. S., and Thompson, R. E., 1991, *Oilfield Processing of Petroleum, Vol. One: Natural Gas,* Pennwell Publishing Co., Tulsa, OK.

Martin, J. L., Otto, F. D., and Mather, A. E., 1978, "Solubility of Hydrogen Sulfide and Carbon Dioxide in a Diglycolamine Solution," *Journal of Chemical and Engineering Data*, Vol. 23, pp. 163–164.

Mason, J. R., and Griffith T. E., 1969, *Oil and Gas J.*, June 5, p. 67.

Mason, J. W., and Dodge, D. F., 1936, *Trans. Am. Inst. Chemical Engrs.*, Vol. 32, No. 1, pp. 27–48.

McClure, G. P., and Morrow, D. C., 1979A, "MALAPROP™—An Amine Process for Removal of COS from Propane," *Gas Conditioning Conference Proceedings*, University of Oklahoma, Norman, OK.

McClure, G. P., and Morrow, D. C., 1979B, "MALAPROP process removes COS," *Hydro. Process.*, May, pp. 231–233.

McClure, G. P., and Morrow, D. C., 1979C, "Amine process removes COS from propane economically," *Oil and Gas J.*, July 2, pp. 106–108.

Meissner, R. E., 1983, "A Low Energy Process for Purifying Natural Gas," *Proceedings of the 1983 Gas Conditioning Conference*, University of Oklahoma, Norman, OK.

Meissner, H., and Hefner, W., 1990, "CO_2 Removal by BASF's Activated MDEA Process," presented at the 1990 European Conference on Energy Efficient Production of Fertilizers," Bristol, UK, Jan. 24–25.

Meissner, R. E., and Wagner, U., 1983, *Oil and Gas J.*, Feb. 7, pp. 55–58.

Merkley, K. E., Christensen, J. J., and Izatt, R. M., 1986, "Enthalpies of Solution of CO_2 in Aqueous Methyldiethanolamine Solutions," GPA Research Report 102, Project 821, Brigham Young University, Provo, UT, September.

Mick, M. B., 1976, "Treating Propane for Removal of Carbonyl Sulfide," paper presented at the 55th Annual GPA Meeting, March 24.

Miller, F. E., and Kohl, A. L., 1953, "Selective Absorption of Hydrogen Sulfide," *Oil and Gas J.*, Vol. 51, April 27, pp. 175–183.

Mshewa, M. M., and Rochelle, G. T., 1994, "Carbon Dioxide Absorption/Desorption Kinetics in Blended Amines," *Proceedings of the 44th Annual Laurance Reid Gas Conditioning Conference*, University of Oklahoma, Norman, OK, Feb. 27–March 2, p. 251.

Muhlbauer, H. G., and Monaghan, P. R., 1957, *Oil and Gas J.*, Vol. 55, April 29, pp. 139–145.

Murrieta-Guevara, F., Robolledo-Libreros, E., and Trejo, A., 1993, "Gas Solubility of Carbon Dioxide and Hydrogen Sulfide in Mixtures of Sulfolane and Monoethanolamine," *Fluid Phase Equilibrium*, Vol. 86, pp. 225–231.

Murzin, V. I., and Leites, I. L., 1971, "Partial Pressure of Carbon Dioxide Over Dilute Solutions of Aqueous Aminoethanol," *Zhur. Fiz. Khim.*, Vol. 45, pp. 417–420.

Nasir, P., and Mather, A. E., 1977, "The Measurement and Prediction of the Solubility of Acid Gases in Monoethanolamine Solutions at Low Partial Pressure," *Can. J. Chem. Eng.*, Vol. 55, pp. 715–717.

Niswander, R. H., Edwards, D. J., DuPart, M. S., and Tse, J. P., 1992, "A More Energy Efficient Product for Carbon Dioxide Separation," *Proceedings of the 42nd Annual Laurance Reid Gas Conditioning Conference*, University of Oklahoma, Norman, OK, March 2–4.

Oscarson, J. L., and Izatt, R. M., 1990, "Enthalpies of Solution of H_2S in Aqueous Methyldiethanolamine Solutions," GPA Research Report 127, Project 821, Brigham Young University, Provo, UT, August.

Ouwerkerk, C., 1978, *Hydro. Process.*, Vol. 57, No. 4, April, p. 89.

Pearce, R. L., 1978, "Hydrogen Sulfide Removal with Methyldiethanolamine," paper presented at 57th Annual GPA Convention, New Orleans, LA, March 20–22.

Pearce, R. L., Arnold, J. L., and Hall, C. K., 1961, *Hydro. Process.*, Vol. 40, August, pp. 121–129.

Pearce, R. L., and Wolcott, R. A., 1986, "Basic Considerations in Acid Gas Removal," presented at the AIChE Annual Meeting, New Orleans, LA, April 6–10.

Perry, C. R., 1977A, "Treating Systems for Ethane Recovery Plants," *Gas Conditioning Conference Proceedings,* University of Oklahoma, Norman, OK.

Perry, C. R., 1977B, "Several treating options for ethane recovery plants," *Oil and Gas J.*, May, pp. 76–79.

Perry, J. H., ed., 1963, *Chemical Engineers' Handbook,* 4th Edition, McGraw-Hill, NY, NY.

Petty, L. E., and Naeger, R. J., 1987, "NGL Treating—A Review of Today's Practices," *Gas Processors Assoc. Proceedings,* pp. 29–32.

Pitzer, K. S., 1991, "Ion Interaction Approach Theory and Data Correlation," in *Activity Coefficients in Electrolyte Solutions,* 2nd ed., Pitzer, K. S., Ed., CRC Press, Boca Raton, FL, Chapter 3, Appendix I.

Polasek, J. C., Donnelly, S. T., and Bullin, J. A., 1990, "The Use of MDEA and Mixtures of Amines for Bulk CO_2 Removal," presented at the AIChE National Meeting, Orlando, FL.

Polasek, J. C., Iglesias-Silva, G. A., and Bullin, J. A., 1992, "Using Mixed Amine Solutions for Gas Sweetening," *Proceedings of the 71st GPA Annual Convention,* pp. 58–63.

Rahman, M. A., Maddox, R. N., and Mains, G. J., 1989, "Reactions of Carbonyl Sulfide and Methyl Mercaptan with Ethanolamines," *Ind. Eng. Chem. Res.*, Vol. 28, No. 4, pp. 470–475.

Rangwala, H. A., Tomcej, R. A., Xu, S., Mather, A. E., and Otto, F. D., 1989, "Absorption of Carbon Dioxide in Amine Solutions," paper 56b presented at the 1989 Spring National AIChE Meeting, Houston, TX, April 2–6.

Reed, R. M., 1946, U.S. Patent 2,399,142.

Reed, R. M., and Wood, W. R., 1941, *Trans. Am. Inst. Chemical Engrs.*, Vol. 37, June 25, pp. 363–383.

Riegger, E., Tartar, H. V., and Lingafelter, E. C., 1944, *J. Am. Chem. Soc.,* Vol. 66, No. 12, pp. 2024–2027.

Roberts, B. E., and Mather, A. E., 1988, "Solubility of CO_2 and H_2S in a Hindered Amine Solution," *Chemical Engineering Communications,* Vol. 64, pp. 105–111.

Robinson, A., Bradley, V., and Daughtry, J., 1988, "Formulated Amine Solvent Improves Natural Gas Treating Process," *Proceedings of the 1988 Laurance Reid Gas Conditioning Conference,* University of Oklahoma, Norman, OK.

Rogers, J. A., 1994, Koch, Inc., Personal Communication, May 3.

Russell, R. M., 1980, "Liquid-liquid contactors need careful attention," *Oil and Gas J.*, Dec. 1, pp. 135–138.

S.I.P.M, 1979, "Shell treats LPG," *Hydrocarbon Processing,* October, p. 85.

Sardar, H., and Weiland, R. H., 1984, "A Non-equilibrium Stage Approach to the Design and Simulation of Gas Treating Units," paper presented at the AIChE National Meeting, San Francisco, CA, Nov. 25–30.

Sardar, H., and Weiland, R.H., 1985, "Simulation of Commercial Amine Treating Units," *Proceedings of the Laurance Reid Gas Conditioning Conference,* University of Oklahoma, Norman, OK.

Sartori, G., and Savage, D. W., 1983, *I.&E.C. Fundamentals,* Vol. 22, p. 239.

Sartori, G., Ho, W. S., Thaler, W. A., Chludzinski, G. R., and Wilbur, J. C., 1994, "Sterically Hindered Amines for Acid Gas Absorption," in *Carbon Dioxide Chemistry: Environmental Issues,* edited by J. P. and C.-M. Pradier, Pub. by the Royal Society of Chemistry, Cambridge, U.K.

Savage, D. W., Funk, E. W., and Astarita, G., 1981, "Selective Absorption of H_2S and CO_2 into Aqueous Solutions of Methyldiethanolamine," paper presented at AIChE Spring Meeting, Houston, TX, April 5–9.

Sharma, M. M., and Danckwerts, P.V., 1964, "Absorption of Carbonyl Sulfide in Amines and Alkalies," *Chem. Eng. Sci.,* Vol. 1, pp. 991–92.

Sharma, M. M., 1965, *Kinetics of Reaction of Carbonyl Sulfide and Carbon Dioxide with Amines and Catalysis by Brönsted Bases of the Hydrolysis of COS,* Trans. Faraday Soc., Vol. 61, p. 681.

Shen, K.-P., and Li, M.-H., 1992, "Solubility of Carbon Dioxide in Aqueous Mixtures of Monoethanolamine with Methyldiethanolamine," *J. Chem. Eng. Data,* Vol. 37, pp. 96–100.

Sherwood, T. K., and Holloway, F. A. L., 1940, *Trans. Am. Inst. Chem. Engrs.,* Vol. 36, Dec. 25, p. 39.

Sherwood, T. K., Shipley, G. H., and Holloway, F. A. L., 1938, *Ind. Eng. Chem.,* Vol. 30, July, pp. 765–769.

Shneerson, A. L. and Leibush, A. G., 1946, *J. Appl. Chem.* U.S.S.R., Vol. 19, No. 9, pp. 869–880.

Shoeld, M., 1934, U.S. Patent 1,971,798.

Sigmund, P. W., Butwell, K. F., and Wussler, A. J., 1981, *Hydro. Process.,* Vol. 60, No. 5, May, p. 118.

Singh, M., and Bullin, J. E., 1988, "Determination of Rate Constants for the Reaction Between Diglycolamine and Carbonyl Sulfide," presented at Spring National Meeting of AIChE, New Orleans, LA, March 1–10.

S.I.P.M., 1979, "Shell treats LPG," *Hydro. Process.,* October, p. 85.

Smith, R. F., and Younger, A. M., 1972, "Tips on DEA Treating," *Hydro. Process.,* July, pp. 98–100.

Stewart, E. J., and Lanning, R. A., 1991, "A Systematic Technical Approach To Reducing Amine Plant Solvent Losses," *Laurance Reid Gas Conditioning Conference Proceedings,* University of Oklahoma, Norman, OK.

Stewart, E. J., and Lanning, R. A., 1994A, "Part 1—Reduce amine plant solvent losses," *Hydro. Process.,* May, pp. 67–81.

Stewart, E. J., and Lanning, R. A., 1994B, "Part 2—Reduce amine plant solvent losses," *Hydro. Process.,* June, pp. 51–54.

Strigle, R. F., Jr., 1994, *Packed Tower Design and Applications,* Gulf Publishing, Houston, TX.

Teller, A. J., and Ford, H. E., 1958, *Ind. Eng. Chem.,* Vol. 50, August, p. 1201.

Teng, T. T., Maham, Y., Hepler, L. G., and Mather, A. E., 1994, *Journal of Chemical & Engineering Data,* Vol. 39, No. 2, p. 290.

Thomas, J. C., 1988, "Improved Selectivity Achieved with UCARSOL Solvent," *Proceedings of the Laurance Reid Gas Conditioning Conference,* University of Oklahoma, Norman, OK, March.

Thompson, R. E., and King, C. J., 1987, "Energy Conservation in Regenerated Chemical Absorption Processes," *Chem. Eng. Process.*, Vol. 21, p. 115–129.

Tomcej, R. A., 1991, "Rate-Based Model for Industrial Gas Sweetening Operations," paper presented at the 41st Canadian Chemical Engineering Conference, Vancouver, British Columbia, Oct. 6–9.

Tomcej, R. A., 1992, "Two Dimensional Model for Acid Gas Mass-Transfer in Commercial Amine Contactors," paper presented at the AIChE Spring National Meeting, New Orleans, LA, March 29–April 2.

Tomcej, R. A., Otto, F. D., Rangwala, H. A., and Morrell, B. R., 1987, "Tray Design for Selective Absorption," Proceedings of the Laurance Reid Gas Conditioning Conference, University of Oklahoma, Norman, OK, March 2.

Tragitt, G. N., Armstrong, T. R., Bourdon, J. C., and Sigmund, P. W., 1986, "SULFTEN System commercialized," *Hydro. Process.*, February, pp. 27–29.

Treybal, R. E., 1963, *Liquid Extraction,* 2nd ed., McGraw-Hill, NY.

Tse, J., and Santos, J., 1993, "The Evaluation and Design of LPG Treaters," *Laurance Reid Gas Conditioning Conference Proceedings,* University of Oklahoma, Norman, OK.

Union Carbide, 1985, Brochure F-47637B: "N-Methyl Diethanolamine for Gas Treating," 5/85/3M.

Union Carbide Chemicals Company, 1957, *Gas Treating Chemicals, Vol. 1.*

Van Dam, R., Christensen, J. J., Izatt, R. M., and Oscarson, J. L., 1988, "Enthalpies of H_2S in Aqueous Diethanolamine Solutions," GPA Research Report 114, Project 821, Brigham Young University, Provo, UT, May.

Vaz, R. N., Maines, G. J., and Maddox, R. N., 1981, "Ethanolamine Process Simulated by Rigorous Calculation," *Hydro. Process.*, Vol. 81, No. 4, April, p. 139.

Veldman, R., 1989, "How to Reduce Amine Losses," paper presented at Petroenergy '89, Oct. 23–27.

Vickery, D. J., Campbell, S. W., and Weiland, R. H., 1988, "Gas Treating with Promoted Amines," *Proceedings of the 38th Laurance Reid Gas Conditioning Conference,* University of Oklahoma, Norman, OK, March 7–9.

Vidaurri, F. C., and Kahre, L. C., 1977, *Hydro. Process.*, Vol. 56, November, pp. 333–337.

Vidaurri, F. C., and Ferguson, R. G., 1977, "MDEA Used in Ethane Purification," *Proceedings of the 1977 Gas Conditioning Conference,* University of Oklahoma, Norman, OK.

Wainwright, H. W., Egelson, G. C., Brock, C. M., Fisher, J., and Sands, A. E., 1952, U.S. Bur. Mines, Rept. Invest., No. 4891, October.

Weber, S. and McClure, G., 1981, *Oil and Gas J.*, Vol. 79, No. 23, pp. 160–163.

Weiland, R. H., and Rawal, M. Y., 1980, "Stripping of Carbon Dioxide from Rich Monoethanolamine Solutions," *Proceedings of Chemeca '80,* Melbourne, Australia.

Weiland, R. H., Chakravarty, T., and Mather, A. E., 1993, "Solubility of Carbon Dioxide and Hydrogen Sulfide in Aqueous Alkanolamines," *Ind. Eng. Chem. Res.*, Vol. 32, pp. 1419–1430.

Weinberg, H. N., Eisenberg, B., Heinzelmann, F. J., and Savage, D. W., 1983, "New Gas Treating Alternatives for Saving Energy in Refining and Natural Gas Processing," paper presented at 11th World Petroleum Congress, London, England, Aug. 31.

Wendt, C. J., and Dailey, L. W., 1967, *Hydro. Process.*, Vol. 46, October, pp. 155–157.

Xu, S., Otto, F. D., and Mather, A. E., 1991, "The Physical Properties of Aqueous AMP Solutions," *J. Chem. Eng. Data,* Vol. 36, pp. 71–75.

Zenz, F. A. and Eckert, R. A., 1961, *Petrol. Refiner.*, Vol. 40, February, pp. 130–132.

Zhang, D., Tazzaghi, M., and Ng, H.-J., 1993, "Modeling of Acid Gas Treating with Blended Amines," *Proceedings of the 72nd GPA Annual Convention,* pp. 76–82.

Chapter 3
Mechanical Design and Operation of Alkanolamine Plants

INTRODUCTION, 188

AMINE PLANT CORROSION, 188
   Background, 188
   Relationship of Process Conditions to Corrosion, 189
   Wet Acid Gas Corrosion, 191
   Amine Solution Carbon Steel Corrosion, 199
   Erosion-Corrosion, 213
   Cracking of Carbon Steel in Amine Service, 216
   Corrosion Inhibitors, 223
   Chloride Attack of Stainless Steel in Amine Service, 224

FOAMING, 224
   Causes of Foaming, 224
   Symptoms and Characteristics of Foaming Systems, 225
   Prevention of Foaming, 226

CHEMICAL LOSSES, 231
   Volatility Losses, 231
   Entrainment, 232
   Solution Degradation, 232

PURIFICATION OF DEGRADED SOLUTIONS, 242
   Amine Solution Mechanical Filtration, 242
   Activated Carbon Filtration, 250

Thermal Reclaiming, 255
Ion Exchange for Amine Solution Purification, 264
Electrodialysis for Amine Solution Purification, 264

NONACIDIC-GAS ENTRAINMENT IN SOLUTION, 265

REFERENCES, 269

INTRODUCTION

One of the reasons that alkanolamine processes have become the predominant choice for both refinery gas treating and natural gas purification is their comparative freedom from operating difficulties. Nevertheless, several factors can result in undue expense and cause difficulty in the operation of alkanolamine units. Chief among these, from an economic standpoint, are corrosion and amine loss. Other operating difficulties which occasionally limit the capacity of a plant for gas purification include foaming and plugging of equipment. In many cases, operation can be significantly improved by daily monitoring of key plant operating variables and by proper control and design of the treating plant.

AMINE PLANT CORROSION

Background

The most serious operating problem encountered with alkanolamine gas purification plants is corrosion, and, as would be expected, this problem has been given the widest attention. Several theories have been advanced to explain corrosion mechanisms, patents describing measures to eliminate or alleviate corrosion have been issued, and numerous papers have been published. Based on this appreciable amount of information and experience, the corrosion phenomena observed in a large number of plants operating under a wide variety of conditions can be reasonably explained, and certain guidelines can be established to minimize corrosion.

Amine systems are subject to corrosion by carbon dioxide and hydrogen sulfide in the vapor phase, in the amine solution, and in the regenerator reflux, and by amine degradation products in the amine solution. In refineries, amine systems often suffer from corrosion by several agents not generally found in natural and synthesis gases, namely ammonia, hydrogen cyanide, oxygen, and organic acids, some of which tend to accumulate in certain parts of the refinery amine system.

Amine units must be designed to overcome these special problems. This chapter reviews both the causes of corrosion problems and possible solutions. It describes the locations within amine units where the various agents cause corrosion, discusses the corrosion mechanisms in these places, and reviews the design practices and preventive measures required to mitigate corrosion. These measures include control of velocities and impingement, process configuration changes, process control strategies, selective use of corrosion resistant materials, sidestream amine solution purification, and the choice of amine.

Relationship of Process Conditions to Corrosion

Figure 3-1 depicts a typical alkanolamine treating unit. The feed gas, containing either CO_2 or H_2S or a mixture of both acid gases, flows into the bottom of a trayed or packed column where it contacts an amine solution. Acid gas components are removed from the gas by chemical reaction with the amine. The purified gas is the overhead product while the rich amine solution flows from the bottom of the contactor on level control to a rich amine flash drum. In the flash drum, the rich amine is flashed to a lower pressure to remove dissolved and entrained hydrocarbons. The rich amine then flows on level control from the flash drum through the lean/rich amine heat exchanger and on to the amine regenerator. In the regenerator, the acid gas components are stripped from the solution using heat supplied by the regenerator reboiler. Acid gas components are the amine regenerator overhead product, and lean amine solution is the bottom product. The hot lean amine from the regenerator is heat exchanged with the rich amine, pumped to the contactor operating pressure, and cooled before entering the contactor.

As **Figure 3-1** indicates, most of the equipment and piping in an alkanolamine plant is constructed of carbon steel. In fact, it is possible, in most cases, to build an alkanolamine plant entirely of carbon steel by keeping amine regenerator operating temperatures low and by minimizing the amine solution concentration and acid gas loading (moles acid gas/mole amine). However, as shown in **Figure 3-1**, it is common practice to construct certain sections of an amine plant with stainless steel or other alloy. Selective use of corrosion resistant alloys permits operation at higher amine concentrations and acid gas loadings, allows better stripping of the lean amine solution and improved treating, reduces corrosion in susceptible areas, and improves process economics. Locations marked in bold on **Figure 3-1** show where carbon steel is typically replaced by stainless steel. Areas marked with a dotted line show where either process or mechanical design modifications, or corrosion resistant materials may be required depending primarily upon the composition of the gas being treated. To comprehend these choices, it is necessary to understand why carbon steel corrodes in amine plants.

In the absence of inhibitors, carbon steel corrodes in aqueous solutions by an electrochemical mechanism. The anodic half reaction is the oxidation of iron to ferrous ion:

$$Fe = Fe^{2+} + 2e^- \tag{3-1}$$

The cathodic half reaction is the reduction of some form of hydrogen from the +1 oxidation state to the element:

$$H^+ + e^- = H° \tag{3-2}$$

Most of the hydrogen atoms, $H°$, combine to form molecular hydrogen, H_2. However, in some circumstances, atomic hydrogen migrates into the metal lattice.

The net chemical reaction is the sum of the two half reactions:

$$Fe + 2H^+ = 2H° + Fe^{2+} \tag{3-3}$$

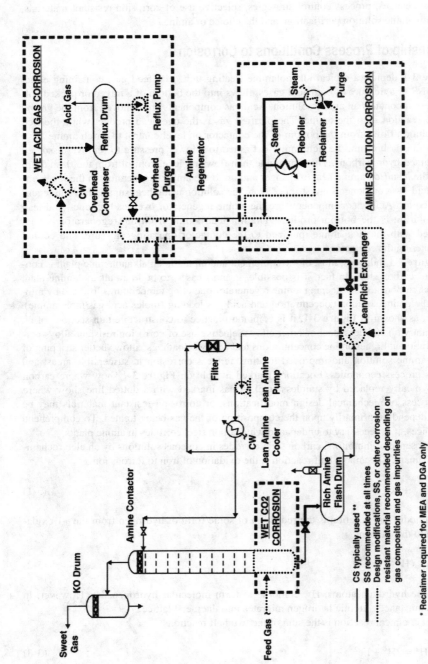

Figure 3-1. Flow diagram of a typical amine plant showing principal areas of corrosion and suggested materials of construction.

This is an irreversible reaction and, consequently, its rate is affected only by temperature and the concentrations of the species on the left side of the equation. Therefore, the rate of reaction 3-3 (and of corrosion) increases with increasing hydrogen ion concentration (i.e., decreasing pH), with increasing temperature, and, because the reaction is electrochemical, with increasing conductivity of the liquid medium.

Figure 3-1 indicates the principal areas where corrosion can occur in alkanolamine gas purification plants. As indicated in this figure, corrosion in amine plants can be divided into two broad categories:

1. *Wet Acid Gas Corrosion* of carbon steel is the reaction of CO_2 and H_2S with iron in an aqueous environment where little or no amine is present.
2. *Amine Solution Corrosion* is the corrosion of carbon steel in the presence of aqueous amine.

Wet Acid Gas Corrosion

Aqueous acid gas solutions occur in the overhead section of the amine regenerator and in the bottom of an amine contactor if the feed gas is water-saturated. Metal surfaces in these sections of the amine plant may, therefore, be contacted with aqueous acid gas solutions containing little or no amine. See **Figure 3-1**.

Mechanism of Wet CO_2 Corrosion

If there is a separate aqueous phase, and if the only acid gas present is carbon dioxide, the CO_2 will dissolve in the water and partially ionize to form a weak acid according to reaction 3-4:

$$CO_2 + H_2O = HCO_3^- + H^+ \tag{3-4}$$

The increase in hydrogen ion concentration from reaction 3-4 accelerates corrosion by reaction 3-3. As would be expected from reactions 3-3 and 3-4, the rate of corrosion increases with increased CO_2 concentration in the water (or increased CO_2 partial pressure in the gas phase).

As corrosion proceeds, ferrous ions produced at the corrosion site by reaction 3-3 react with bicarbonate ions in solution to form ferrous carbonate, which is essentially insoluble and precipitates as scale. Scale deposition reduces the rate of corrosion, but unfortunately the ferrous carbonate formed is quite porous and provides only limited corrosion protection.

Hydrogen ions are removed from solution at the iron surface by reaction 3-3. This causes the solution adjacent to the corroding surface to become less acid (more alkaline), thereby reducing the rate of corrosion. However, the evolution of gases or other actions that enhance liquid turbulence counter the build-up of alkalinity by bringing fresh acid to the metal surface.

With a weakly ionized acid, such as carbonic acid, the large concentration of dissolved, but un-ionized carbon dioxide in the water provides a reservoir of reactive molecules, which can produce additional hydrogen ions by reaction 3-4 to replace those used up by the corrosion reaction. With a strongly ionized acid, on the other hand, an extremely low acid concentration is needed to produce the same pH as the carbonic acid solution, and depletion of hydrogen ions at the corroding surface can reduce the corrosion rate. Such an effect may explain the higher rates of corrosion observed for CO_2 solutions than for strong acid solutions at the same bulk solution pH values (Berry, 1982).

The specific acid anion can also affect the corrosion rate. Some acids, such as H_2S, react with iron to form a relatively dense, non-porous scale that can significantly reduce continued corrosion. Other acids, such as oxalic, tend to increase solution corrosivity due to the formation of soluble iron chelates (Rooney et al., 1996).

An alternative explanation for the high rate of corrosion of iron in contact with CO_2-containing water is described by de Waard and Lotz (1993). In the proposed mechanism, carbonic acid, H_2CO_3, is formed by hydration of dissolved CO_2. It is postulated that the carbonic acid molecule participates directly in the corrosion reaction by accepting an electron from the corroding iron to form HCO_3^- and elemental hydrogen in accordance with the following equation:

$$H_2CO_3 + e^- = H° + HCO_3^- \tag{3-5}$$

The rate of corrosion by this mechanism would, of course, increase with increased carbon dioxide partial pressure and be affected by temperature.

Whether the primary mechanism for the corrosion of iron by carbon dioxide and water involves the acceptance of electrons from hydrogen ions or from molecular carbonic acid, the overall reaction is the same, i.e.:

$$Fe + 2CO_2 + 2H_2O = Fe^{2+} + 2HCO_3^- + 2H° \tag{3-6}$$

Figure 3-2 illustrates the effects of CO_2 partial pressure and temperature on the rate of corrosion of carbon steel by water saturated with CO_2. De Waard and Lotz (1993) have converted the data of **Figure 3-2** into the nomograph of **Figure 3-3**, which can be used to predict the corrosion of carbon steel by aqueous carbon dioxide solutions.

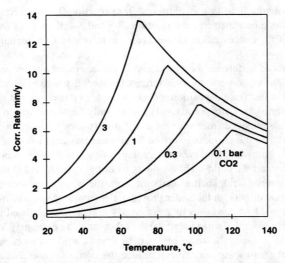

Figure 3-2. Effect of temperature on carbon steel corrosion at various CO_2 partial pressures. (*De Waard and Lotz, 1993*)

Figure 3-2 shows that the corrosion rate at any given partial pressure increases with temperature at moderate temperatures, but reaches a maximum as the temperature is further increased. De Waard and Lotz (1993) attribute this effect to the formation of a more protective film, either $FeCO_3$ or Fe_3O_4, at higher temperatures. The reduction of corrosion rate at elevated temperatures is accounted for in the nomograph by the inclusion of a scale factor from 0.1 to 1, which is used as a multiplier to reduce the corrosion rate read directly from the chart in the higher temperature region.

The nomograph of **Figure 3-3** can be used to provide an estimate of the corrosion rate in the regenerator overhead and the contactor bottoms of amine plants when CO_2 is the only acid gas present. De Waard and Lotz (1993) also discuss the use of correction factors to account for the effects of corrosion product films, pH, system pressure, system geometry, glycol or methanol, crude oil, inhibitors, and flow velocities on corrosion rates.

Mechanism of Wet H_2S Corrosion

If the acid gas includes hydrogen sulfide, the principal product of corrosion is ferrous sulfide, which is very insoluble, and forms a weakly adherent and somewhat protective film. The overall corrosion reaction is

$$Fe + H_2S = FeS + 2H° \tag{3-7}$$

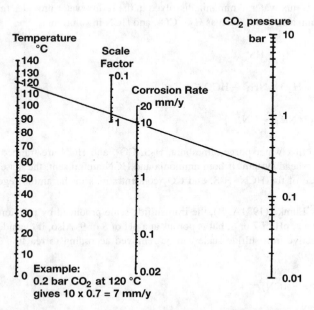

Figure 3-3. Corrosion nomograph for carbon steel in contact with wet CO_2. (*De Waard and Lotz, 1993*)

Iron sulfide is more protective than iron carbonate, and, if the acid gas contains sufficient H_2S, a protective sulfide film can be formed. If contaminants which disrupt this iron sulfide film are not present, and if the mechanical design minimizes erosion-corrosion, wet acid corrosion of exposed carbon steel surfaces will be limited. On the other hand, if the gas is predominantly CO_2, wet acid gas corrosion will occur and protective measures are required. As might be expected, there is disagreement on the exact CO_2/H_2S ratio above which protective measures are required. API Recommended Practice 945 suggests that wet CO_2 corrosion occurs when the acid gas contains 95% or more CO_2 (API, 1990).

Mechanism of Wet Acid Gas Corrosion Due to NH_3 and HCN

In refineries, the amine regenerator overhead system is often affected by wet acid gas corrosion due to the combined presence of ammonia, CO_2, and H_2S. Wet acid gas corrosion of the overhead system is accelerated if HCN is also present (Ehmke 1981A, B). In these circumstances, wet acid gas corrosion due to H_2S and CO_2 can occur when the CO_2 content of the acid gas is less than 90%. In fact, substantial corrosion can occur in the total absence of CO_2 if sufficient HCN and ammonia are present.

Ammonia and HCN are found in gas streams from delayed cokers, visbreakers, and fluid catalytic cracking units (FCCUs). Ammonia, HCN, and H_2S are also produced during hydrotreating and hydrocracking operations. When gases from these units are treated, the ammonia is readily absorbed by aqueous alkanolamine solutions. HCN, which is a weak acid and very water-soluble, is chemically absorbed by alkanolamine solutions and is released, along with the absorbed ammonia, in the amine regenerator.

In the regenerator overhead condenser, the gaseous ammonia and HCN are reabsorbed in the condensed reflux water. Ammonia dissolved in the reflux water provides the alkalinity to absorb and retain acid gases, such as H_2S, CO_2, and HCN in solution.

$$NH_3 + H_2S = NH_4^+ + HS^- \tag{3-8}$$

$$NH_3 + CO_2 + H_2O = NH_4^+ + HCO_3^- \tag{3-9}$$

$$NH_3 + HCN = NH_4^+ + CN^- \tag{3-10}$$

Without a reflux water purge, ammonia, H_2S, CO_2, and HCN are trapped in the amine regenerator overhead system. If both ammonia and HCN are present, the net effect is a substantial increase of the HCN, H_2S, and CO_2 concentrations in the amine regenerator overhead system.

As noted by Ehmke (1981A, B), the iron sulfide scale produced by reaction 3-7 provides some protection at pH of 7 or 8, but is porous at a pH of 8 or 9. Also, if cyanide ions are present, the protective iron sulfide scale can be removed according to reaction 3-11 (Ehmke, 1981A, B, 1960):

$$FeS + 6CN^- = Fe(CN)_6^{4-} + S^{2-} \tag{3-11}$$

Removal of the protective iron sulfide scale by reaction 3-11 is pH dependent, the rate increasing as the pH increases (Ehmke, 1981A, B). This is one of the few corrosion reactions where corrosion increases with increasing pH. As reactions 3-7 and 3-11 indicate, the higher the concentrations of H_2S and CN^-, the greater the rate of corrosion.

If CO_2 is trapped in the amine regenerator overhead system by ammonia, corrosion can result by reaction 3-6.

Reactions 3-6 and 3-7 produce atomic hydrogen, H^0. Under normal circumstances, the atomic hydrogen would combine at the metal surface to form molecular hydrogen. However, steel surface poisoning agents, such as sulfide and cyanide anions, prevent this recombination, and a significant fraction of the atomic hydrogen may migrate into the metal lattice (Fontana and Greene, 1967). Most of the atomic hydrogen passes completely through the steel and forms molecular hydrogen on the opposite surface. However, if the atomic hydrogen encounters an inclusion or a subsurface discontinuity in the base metal, it becomes trapped and recombines to form molecular hydrogen. As more and more molecular hydrogen is trapped at these locations, the pressure builds up past the yield strength of the steel. High molecular hydrogen concentrations at these locations can lead to hydrogen blistering/hydrogen-induced cracking (HIC). In areas of high stress, such as weld heat affected zones, HIC may propagate in a planar manner, through the metal wall thickness. This type of cracking is referred to as stress oriented hydrogen induced cracking (SOHIC). Also, atomic hydrogen dissolved in the steel lattice may embrittle areas of hard microstructures in carbon steel base metal or weldments. High atomic hydrogen concentrations in carbon steel can lead to sulfide stress cracking (SSC) of these hard areas. A detailed review of these cracking mechanisms is presented later in this chapter.

Amine units treating gas from FCCUs are particularly susceptible to low temperature hydrogen attack because the gas from these units can have a high cyanide content. According to Neumaier and Schillmoller (1955), hydrogen attack should be expected whenever the organic nitrogen compounds in the FCCU feed are greater than 0.05 wt%. Wash or reflux water with a blue color (Prussian blue) after oxidation by air indicates that cyanide-induced corrosion is taking place (Ehmke, 1981A, B). A simple plant test to detect cyanide using a dilute ferric chloride solution has been described by Neumaier and Schillmoller (1955) and by Ehmke (1981A, B).

Prevention of Wet CO_2 and H_2S Corrosion

As shown in **Figure 3-1,** wet CO_2 corrosion can occur in either the bottom of the amine absorber or the amine regenerator overhead system. Wet CO_2 corrosion can also occur in the upper section of the absorber above the feed tray when the absorber is designed for selective H_2S removal. However, if the amine solution completely wets all the exposed carbon steel surface, it greatly raises the pH of the condensate and reduces its corrosiveness as long as cyanide is not present. In cases where the acid gas is 95% CO_2 or greater, an amine spray in the regenerator overhead has been recommended to minimize carbon steel corrosion (API, 1990; Gutzeit, 1986; Ballard, 1966). Sufficient amine should be injected so that the reflux water contains 0.5 wt% amine (Gutzeit, 1986). In most refinery amine systems, wet CO_2 corrosion of the amine regenerator overhead system is not an issue because the acid gas is predominantly H_2S, and the carbon steel in the amine regenerator overhead system is protected by an iron sulfide film. However, an amine spray should be considered for amine regenerators in hydrogen plants where CO_2 is the only acid gas present—although an amine spray in the regenerator overhead may be counterproductive if HCN is trapped in the regenerator overhead system. See the prior discussion on the effect of HCN on wet acid gas corrosion for more information.

Wet CO_2 corrosion can also be a problem in hydrogen plant amine absorbers. In the bottom of the amine absorber, corrosion due to high CO_2 content of the acid gas can be prevented by introducing the gas through a distributor immersed in the amine absorber sump liquid (Dupart et al., 1993B). The distributor orifices should be immersed in the rich amine solution so that the emerging gas causes amine solution to splash and wet the carbon steel surface. If the distributor breaks or is installed above the solution level, wet CO_2 can rapidly attack the shell of the absorber and its internal elements. An alternative solution for protecting the carbon steel metallurgy in the bottom of the absorber from wet CO_2 corrosion is to drill a series of weep holes around the perimeter of the bottom tray support ring. Amine flowing down through the holes will wet the absorber walls and protect the carbon steel from attack by the CO_2-rich gas. An advantage of this technique in comparison to the use of a submerged sparger is that it minimizes gas entrainment in the rich amine leaving the absorber. If the absorber treating a CO_2-rich gas has carbon steel trays, the underside of the bottom tray may be attacked by acid gas until the tray collapses, whereupon the second tray may be attacked, and fail in turn. As shown in **Figure 3-1,** all the trays can be saved if the bottom tray is stainless steel. If the mechanical design of either the bottom of the absorber or the amine regenerator overhead system is such that amine does not wet all exposed carbon steel surfaces, piping should be stainless steel and all equipment should be either made of or lined with stainless steel.

Another factor that should be considered for high CO_2 content sour gases is minimizing high velocity acid gas vapor impingement on carbon steel surfaces. Dupart et al. (1993B) report one instance where impingement of a CO_2-rich gas led to corrosion in the bottom of an amine contactor. This problem was corrected by modifying the feed gas distributor to minimize impingement of the acid gas on the amine contactor wall.

Corrosion by wet H_2S occurs primarily by sulfide stress cracking (SSC), hydrogen-induced cracking (HIC), and stress oriented hydrogen-induced cracking (SOHIC). Control measures to mitigate these types of corrosion are discussed in a separate section of this chapter entitled "Cracking of Carbon Steel in Amine Service."

Preventing NH_3 and HCN-Based Corrosion

Several methods have been employed to reduce ammonia- and cyanide-induced corrosion in the amine regenerator overhead system. These methods include

- A reflux water purge to reduce the concentration of ammonium cyanide and bisulfide in the amine regenerator overhead. See **Figure 3-1**.
- A water wash upstream of the amine treating system to remove both ammonia and HCN from the feed gas.
- Injection of either ammonium or sodium polysulfide upstream of the amine regenerator or into the amine regenerator overhead system to convert the cyanide iron to thiocyanate.
- Stripping of corrosive components from the amine regenerator reflux.
- Use of corrosion resistant materials in the amine regenerator overhead system.
- Combinations of the above.

In general, the most economic method of reducing NH_3- and HCN-based corrosion is to modify the process conditions to permit the use of carbon steel. Elimination of HCN and ammonia upstream of the amine treating unit is usually the most effective solution. A brief review of each of these corrosion prevention methods follows:

Continuous Reflux Purge. Ammonia and cyanide build-up in the amine regenerator overhead system can be reduced and corrosion controlled with a continuous purge of amine regenerator reflux water. See **Figure 3-1**. A reflux purge is effective if the cyanide and ammonia concentrations in the sour gas are low (Neumaier and Schillmoller, 1955). If the cyanide content of the reflux water or upstream wash water is less than 100 ppmw, cyanide-induced blistering (HIC) of carbon steel will be minimal (Neumaier and Schillmoller, 1955). To minimize wet H_2S corrosion problems (SSC, HIC, SOHIC), the cyanide content of the wash water should be below 20 ppm (NACE, 1994B). On the basis of guidelines developed for hydroprocessing unit air coolers, carbon steel can be used in the regenerator overhead condenser and downstream piping if the ammonium bisulfide concentration in the reflux water is less than 2–3 wt% and if exchanger tube and piping velocities are kept below 20 feet per second (Piehl, 1975). Inhibitor injection (ammonium polysulfide) may be required for the amine regenerator overhead system if the cyanide concentration is too high and, in refineries, a standby inhibitor injection system is often provided (Neumaier and Schillmoller, 1955). If a reflux purge is used to control the NH_3 and cyanide content of the reflux water, it may be desirable to avoid the use of return bends in either air- or water-cooled overhead condensers (Piehl, 1975).

The disadvantage of an amine regenerator reflux purge is that there may be very costly amine losses because amine concentrations in the reflux water can be between 0.5 and 2.0 wt% depending on amine entrainment, the type of amine, and the number of water wash trays in the amine regenerator (Bucklin and Mackey, 1983). Also, the purge rate required to reduce cyanide and ammonia concentrations below corrosive levels is difficult to determine and control. Although amine losses can usually be reduced by installing a demister pad in the amine regenerator, some losses will occur, and corrosive conditions may exist even with a continuous reflux purge. For these reasons, when both ammonia and cyanide are present, amine units are typically designed with corrosion resistant alloys as shown in **Figure 3-1**. The amine regenerator is usually all 304L SS or is lined with type 304L integral cladding or weld overlay from two trays below the feed tray up to and including the upper head. The overhead piping is usually 304L SS, and titanium tubes are frequently used in the overhead condenser. Ehmke (1981A, B) notes that aluminum has been used successfully for the overhead condenser if pH, chloride ion content, impingement, and velocity are controlled. Carbon steel is frequently utilized for the reflux accumulator; however, a conservative corrosion allowance is necessary and HIC-resistant material is sometimes used. The reflux pump casing and impeller should be 316 SS. All of the reflux piping is typically 304L SS.

Upstream Water or Caustic Wash. Caustic washing was recommended by Polderman and Steele (1956) and used by Norris and Clegg (1947) for removal of formic and acetic acids from gas streams entering amine treating units. Since both formic and acetic acids are stronger acids than either H_2S or CO_2, they will replace the sulfide (or carbonate) salt. However, as noted by Polderman and Steele (1956) and Ehmke (1981A, B), caustic washing cannot remove HCN from gas streams because HCN is a weaker acid than H_2S or CO_2 and will be displaced from the caustic solution by either of these two acid gases.

Since HCN is more soluble in water than either H_2S or CO_2, water wash has been used to remove hydrogen cyanide from gas streams (Kelley and Poll, 1953; Neumaier and Schillmoller, 1955; Polderman and Steele, 1956). **Table 3-1** indicates that a single stage water wash alone is relatively ineffective in removing ammonia and cyanide (Neumaier and Schillmoller, 1955; Ehmke, 1981A, B). This means that additional protective measures are required, including a back-up inhibitor injection system (ammonium polysulfide) for the amine regenerator overhead system and an amine regenerator reflux purge (Neumaier and Schillmoller, 1955).

Table 3-1
Removal of Corrosive Components from an FCCU Gas Stream by a Single Stage Water Wash Followed by Amine Contacting

Constituents	Total Constituents (Pounds per Day)	Removed by Water Wash (Pounds per Day)	(Percent)	Absorbed by Diethanolamine Solution (Pounds per Day)	(Percent)	Balance to Gas Plant (Pounds per Day)	(Percent)
Ammonia	7,400	6,300	85	700	10	400	5
Sulfides	86,000	21,100	24	64,200	75	700	1
Carbonates	32,000	4,000	12	21,500	67	6,500	20
Cyanides and Thiocyanates	2,000	1,000	50	400	20	600	30
Phenols	7,100	1,750	25	100	1	5,250	74
Aliphatic Acids	6,500	1,850	28	1,600	25	3,050	47
TOTAL	141,000	36,000	25	88,500	63	16,500	12

Source: Neumaier and Schillmoller (1955)

Upstream Ammonium Polysulfide Wash. Corrosion due to cyanide and ammonium bisulfide can be greatly reduced by washing the gas upstream of the amine absorber with an ammonium polysulfide solution (Ehmke, 1981A, B). In the upstream wash, the ammonia is dissolved in the wash water while the ammonium polysulfide reacts with CN^- to form thiocyanate ion, SCN^-, by reaction 3-12. Thiocyanic acid, HSCN, is a strong acid and is more water-soluble than HCN. Ehmke (1981A, B) presents data that show that ammonium polysulfide in the upstream wash can reduce CN^- concentrations to less than 10 ppmw, which is low enough to prevent both HIC and SSC (Neumaier and Schillmoller, 1955; NACE, 1994B).

$$S_x^{-2} + CN^- = S_{x-1}^{-2} + SCN^- \tag{3-12}$$

Ammonium polysulfide is preferred over sodium polysulfide because it reacts faster with CN^- and does not raise the pH of the wash water (Ehmke, 1981A, B). Design and operating guidelines for ammonium polysulfide wash systems are provided by Ehmke (1981A, B). Design principles are reviewed by Bucklin and Mackey (1983) who recommend the use of a multistage counter-current contactor with 8 to 10 actual trays. The water wash system described in this reference uses stripped sour water containing 50 ppm NH_3 to reduce the ammonia in the washed gas to 0.1 ppm. Ammonium polysulfide is added to the wash water to convert the cyanide to thiocyanate. Additional polysulfide wash design and operating data are provided by Carter et al. (1977), Penderleith (1977), Lynch (1982), and Donovan and Laroche (1981). With an upstream ammonium polysulfide wash to remove both ammonia and HCN, the amine regenerator overhead system can be all carbon steel.

Regenerator Reflux Water Stripping. Neumaier and Schillmoller (1955) outline another method for eliminating corrosion by ammonia and cyanide in the amine regenerator overhead system. In this method the amine regenerator reflux stream is directed to a dedicated sour water stripper. The ammonia and HCN are then stripped from the reflux water in the sour water stripper. The stripped reflux water is then returned to the amine regenerator. Since cyanide and ammonia are not returned to the amine regenerator, there is no build-up of these components in the amine regenerator overhead and the entire amine regenerator overhead system can be made of carbon steel.

Amine Solution Carbon Steel Corrosion

Causes of Amine Solution Carbon Steel Corrosion

Amine solution corrosion of carbon steel can be influenced by a number of factors, including

- High operating temperatures
- High rich and lean amine loadings (moles acid gas/mole amine)
- The ratio of CO_2 to H_2S in the acid gas
- Amine solution contaminants including amine degradation products and heat-stable salts
- Amine solution concentration
- Amine type

Amine solution corrosion is most significant in the hot bottom section of the regenerator. It also occurs in the line connecting the contactor level control valve to the rich amine flash drum, in the lean/rich amine exchanger (rich solution side), and in the piping from the rich amine flash drum level control valve to the regenerator. See **Figure 3-1**. Amine solution corrosion in these areas can be minimized by selecting appropriate alloy materials of construction and by process modifications.

The general effects of process conditions and materials of construction on corrosion rates are indicated in **Table 3-2**. Cases A and B represent laboratory tests. Case A was conducted with pure MEA solution containing no acid gases, and all alloys performed well. For Case B, carbon dioxide was bubbled through the solution during the test, and severe corrosion of carbon steel occurred. Cases C through F are based on corrosion coupons installed in operating gas treating plants. In general, monel and the 300 series stainless steels performed well in all plants, although some pitting occurred in the 304 and 316 stainless steel samples that were exposed to DEA plant stripper conditions for 483 days. These conditions also proved too severe for the carbon steel coupons, which were completely destroyed. Aluminum did not prove suitable for any of the conditions tested.

Amine-Acid Gas Carbon Steel Corrosion Mechanisms

Pure amines and mixtures of only water and amines are not corrosive because they have low conductivity and high pH (Dupart et al., 1993A, B). However, rich amine solutions, which have high conductivity and a pH significantly lower than lean amine solutions, can be quite corrosive.

The effects of acid gas loading and temperature on solution pH are illustrated in **Figure 3-4**. The curves in this figure show that, for a 15% MEA solution, the pH decreases as either the acid gas loading or the temperature increases. They also show that, at equal loading, CO_2 has a greater ability than H_2S to reduce the pH; i.e., CO_2 is a stronger acid. It is interesting to

Table 3-2
Corrosion Rates of Metals in MEA and DEA Solutions

Case	A	B	C	D	E	F
Amine	MEA	MEA	MEA	MEA	MEA	DEA
Concentration, wt%	20	20	90–95	15–20	15	11–15
CO_2	0	Sat.	0	present	present	—
H_2S	0	0	0	0	present	10–50 gr/gal
Temp, °F	240	240	338–374	180–200	230	225–231
Test duration, days	21	21	36	100	270	483
Service	Lab. Test	Lab. Test	Reclaimer	Refinery Absorber	Nat. Gas Treater	Stripping Column
Indicated Corrosion Rate, mils per year						
Monel	1.0	3.0	0.3	1.6	1.3	2.1
302 and 304 SS	<1.0	nil	0.5	<0.1	<0.1	<0.1(2)
316 SS	1.0	<1	0.4	—	<0.1	—
410 SS	1.0	nil	8.0	—	—	0.1(3)
Aluminum 2S and 3S	—	—	(1)	—	—	(1)
Mild steel	1.0	103	3	1.4	5.4	(1)
Cast iron	—	—	—	8.2	2.1	17

Notes:
1. 0.031" thick specimens completely destroyed.
2. Pitted to a maximum depth of 2 mils.
3. Pitted to a maximum depth of 13 mils.
Source: Lang and Mason (1958)

note that extrapolation of the curves for CO_2-containing MEA solutions to typical reboiler temperatures indicates that near-neutral or even slightly acid conditions may occur.

Several mechanisms have been proposed for amine-acid gas corrosion. Riesenfeld and Blohm (1950, 1951A, B) were the first to note that significant amine corrosion was usually associated with the evolution of acid gases from the rich amine solution. Based on this observation, Riesenfeld and Blohm stated that amine solution carbon steel corrosion was due to the presence of the acid gases themselves. For example, carbon dioxide can be evolved from rich amine solutions according to reactions 3-13 and 3-14:

$$R_3NH^+ + HCO_3^- = R_3N + H_2O + CO_2 \tag{3-13}$$

$$R_2NH_2^+ + R_2NCO_2^- = CO_2 + 2R_2NH \tag{3-14}$$

The acid gases can then react directly with exposed carbon steel in the presence of water to form iron carbonate according to reaction 3-6 or 3-15:

$$Fe + H_2O + CO_2 = FeCO_3 + 2H° \tag{3-15}$$

This overall equation does not explain the detailed reactions at the point of metal corrosion where the released hydrogen may originate from H^+, H_2CO_3, or alkanolammonium ions.

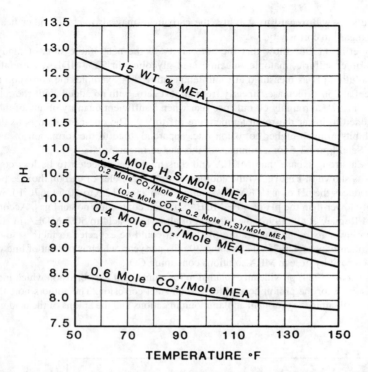

Figure 3-4. Effects of acid gas loading, acid gas composition, and temperature on the pH of 15 wt% MEA solution, Reproduced with permission from *Gas Conditioning and Processing, Vol. 4, Gas and Liquid Sweetening,* copyright 1985, Campbell Petroleum Series. (*Maddox, 1985*)

The iron carbonate produced by these reactions is only slightly soluble and forms a film over the active metal surface, which offers limited protection from further corrosion. Similar corrosion reactions occur with H_2S; however, the iron sulfide film covering the active metal surface is much more protective than iron carbonate, and the iron sulfide film resists further corrosion.

Attributing corrosion to simple acid gas attack explains several observed corrosion phenomena. For example, primary amines, such as monoethanolamine (MEA) and Diglycolamine (DGA), are more corrosive than secondary and tertiary amines because in amine systems employing primary amines, which are difficult to strip, high concentrations of amine-acid gas salts are present in the hottest areas of the process. Conversely, methyldiethanolamine (MDEA), a tertiary amine, is easily stripped of both CO_2 and H_2S. Therefore, it is less corrosive because the bulk of the acid gas is evolved from solution at a lower temperature.

Another possible mechanism for the high corrosivity of lean MEA solutions containing CO_2 involves the presence and behavior of carbamate. Austgen et al. (1991) conclude that carbon dioxide retained in partially stripped MEA solutions is almost entirely in the form of carbamate ions. An appreciable amount of monoethanolammonium carbamate does not decompose under stripper conditions and remains in solution. Since it resembles so-called heat-stable salts with regard to stability, the carbamate may also act like other heat-stable salts with regard to corrosivity.

Rooney et al. (1996) show that the corrosivity of the heat-stable salts of many acids (acetic, formic, sulfuric, malonic, succinic, and glycolic), in 50% MDEA, correlates well with the solution pH as measured at room temperature. For example, the corrosion rate for carbon steel at 250°F increases linearly from almost zero, with no added acid (room temperature pH = 11.57), to roughly 60 mils per year when a sufficient quantity of any of the above acids is added to lower the room temperature pH to 9.9. The one exception is oxalic acid, which exhibits a much higher corrosion rate, apparently due to the formation of a chelate with iron. See **Figure 3-5** for a summary of these experiments.

MEA is a stronger base than MDEA and thus might be expected to be less corrosive. However, as shown in **Figure 3-4** the addition of only 0.2 moles of carbon dioxide per mole of MEA reduces the pH of a 15% MEA solution from about 12.5 to 10.5 (at 70°F), which is appreciably lower than the pH of 50% MDEA solution containing no acid gas. According to the data of Rooney et al. (1996), the addition of almost any acid to 50% MDEA in sufficient quantity to lower its room temperature pH to 10.5 could be expected to increase its rate of corrosion of carbon steel at 250°F, to about 30 mils per year, which is not out of line with the observed corrosivity of lean MEA solutions containing CO_2.

Kosseim et al. (1984) provide an explanation for amine-acid gas corrosion which includes a plausible source for the proton needed for carbon steel corrosion. The authors note that acid gases react with amines to form alkanolammonium cations and the anions of the acid gases:

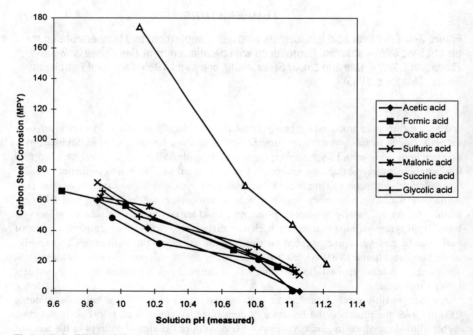

Figure 3-5. Effect of the pH of various heat-stable salt solutions on carbon steel corrosion at 250°F. (*Rooney et al., 1996*)

$$H_2S + R_3N = R_3NH^+ + HS^- \tag{3-16}$$

$$CO_2 + R_3N + H_2O = R_3NH^+ + HCO_3^- \tag{3-17}$$

$$CO_2 + 2R_2NH = R_2NH_2^+ + R_2NCO_2^- \tag{3-18}$$

Alkanolammonium ions ($R_2NH_2^+$ and R_3NH^+) are acids in that they can provide protons for the corrosion reaction. The corroding metal (iron in this case) will react with the most abundant acid in the solution. In amine solutions, the alkanolammonium ion is much more concentrated than the hydrogen ion. Thus, reaction 3-3 can be rewritten in the following form:

$$Fe + 2R_3NH^+ = Fe^{2+} + 2H° + 2R_3N \tag{3-19}$$

Reaction 3-19 implies that corrosion rates should increase in proportion to the concentration of alkanolammonium ions, and generally this is true. Richer solutions are more corrosive than leaner, other things being equal. Also, undegraded lean MEA solutions are more corrosive than undegraded lean diethanolamine (DEA) solutions, because MEA, the stronger base, retains a higher concentration of alkanolammonium cations and, therefore, forms a more corrosive solution.

Reaction 3-19, like reaction 3-3, is irreversible; therefore, its rate cannot in principle be affected by the concentrations of its products. In particular, heat stable anions such as the oxalate ion and amine degradation products that form complexes with ferrous ions should not affect the corrosion rate, yet experience shows that these contaminants often do aggravate corrosion (Rooney et al., 1996). The reason probably is that the deposition of a protective film of iron compounds is prevented by removal of iron atoms from the carbon steel surface by complexing agents. Furthermore, those complexing agents that associate with ferric as well as ferrous ions are very likely to inhibit the active-to-passive transition by preventing the formation of an insoluble ferric oxide passive film. For these reasons, certain heat stable salt anions, amine carbamate anions, and amine degradation products may strongly affect corrosion rates, although they may not appear in the overall corrosion reactions.

Uncontaminated solutions of tertiary amines such as MDEA are generally not corrosive whatever the acid gas. According to API 945, solutions of most amines are not corrosive if the ratio of hydrogen sulfide to carbon dioxide is above roughly 5/95, because the corrosion reaction leads to formation of a protective sulfide film (API, 1990). The most corrosive combinations appear to be those of primary or secondary amines with carbon dioxide.

Although it is widely believed that the acid gases are primarily responsible for the corrosion of carbon steel by amine solutions, the exact mechanisms have not been established unequivocally. As indicated by reactions 3-2, 3-5, and 3-19, there are several possible sources of protons that may accept electrons from elemental iron to cause corrosion and release elemental hydrogen. It is possible, of course, that more than one of the proposed species is active, depending on process conditions. Further fundamental work is needed to firmly establish the primary mechanisms and to clarify the effects of other factors such as solution contaminants, chelate formation, and scale deposition.

Effect of Amine Loadings and Temperature

Figure 3-6 shows the effect of CO_2 loading on carbon steel corrosion (Fochtman, 1963). At a given amine concentration, corrosion increases as the CO_2 loading increases. Through

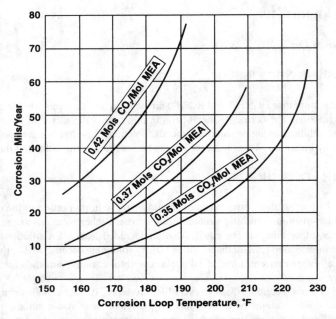

Figure 3-6. Results of pilot plant tests showing the effect of temperature and acid gas loading on carbon steel corrosion. *Reproduced with permission from Chemical Engineering, Vol. 70, No. 2, copyright McGraw Hill Companies, Inc.* (Fochtman et al., 1963)

experience, general guidelines have evolved on the maximum acceptable lean and rich amine loadings. **Table 3-3** summarizes recommended lean and rich solution loadings obtained from various sources. As might be expected, there are considerable variations in these recommendations. One reason for these differences is that the recommended maximum loadings ignore the inhibition of corrosion by H_2S. Therefore, these recommendations usually err on the conservative side.

Carbon steel corrosion due to high lean amine loadings can be limited by controlling the stripping operation in the amine regenerator. For a given stripping system (i.e., fixed number of trays, pressure, amine type, and concentration, etc.), the primary factor affecting the degree of stripping is the amount of stripping vapor (steam). The quantity of steam stripping vapor is usually expressed in terms of the reflux ratio (the mole ratio of water vapor to acid gas in the gas phase leaving the stripping column) or the weight of steam fed to the reboiler per unit volume of rich amine solution (lb/gallon or kg/m^3). Fitzgerald and Richardson (1966A, B) provide guidelines for estimating the amount of steam required (lb steam/gallon rich MEA solution) as a function of the H_2S/CO_2 ratio in the feed gas. See Chapter 2 for a more detailed discussion of amine stripping system design.

Two methods of controlling the degree of lean amine stripping are in common use. The first uses flow ratio control to set the reboiler heat medium mass flow at a fixed value in relation to the rich amine flow (kg of steam per m^3 of solution or lb of steam per gallon of rich amine solution). See **Figure 3-7**. Most rich amines can be adequately stripped using between 110 and 133 kg of steam per m^3 of rich amine solution (0.9 to 1.1 lb of steam per gallon)

Table 3-3
Recommended Maximum Amine Solution Concentrations and Maximum Rich and Lean Amine Solution Loadings

Amine	Maximum Wt% Concentration	Mol Acid Gas/Mol Amine Maximum Lean Loading	Maximum Rich Loading	Reference
MEA	15–20	0.10–0.15	0.30–0.35	Dupart et al., 1993B
MEA	15–20	0.08–0.12	0.35–0.40	Dingman et al., 1966
MEA*	18	0.13	—	Montrone and Long, 1971
MEA	18	—	0.3–0.4	Ballard, 1966
MEA	10–20	—	0.25–0.45	Butwell et al., 1982
MEA	<20	<0.10*	<0.4	Hall and Polderman, 1960
DEA	25–30	0.05–0.07	0.35–0.40	Dupart et al., 1993B
DEA	20–30	—	0.33–1.0	Smith and Younger, 1972
DEA	20–30	—	0.77–1.0	Wendt and Daily, 1967
DEA	20–40	—	0.50–0.85	Butwell et al., 1982
MDEA	50–55	0.004–0.010	0.45–0.50	Dupart et al., 1993B
DGA	50–65	—	0.25–0.45	Butwell et al., 1982
DIPA	20–40	—	0.50–0.85	Butwell et al., 1982

Notes:
*For CO2 only.

(Wendt and Daily, 1967). The second control method uses the temperature between the top regenerator tray and the regenerator overhead condenser to reset the reboiler heat medium flow. See **Figure 3-8**. This controls the moles of reflux water per mole of acid gas leaving the amine regenerator overhead because, at a fixed regenerator operating pressure, the temperature above the top tray is directly related to the mol% water in the acid gas leaving the top tray. Usually, a reflux ratio of 1.0–2.0 moles of water per mole acid gas is adequate to strip most amines (Dupart et al., 1993B).

As **Figure 3-6** illustrates, high rich amine loadings can also lead to excessive corrosion. **Figure 3-9** depicts a control strategy used to limit rich amine loadings and control corrosion (Dingman and Moore, 1968). As shown in **Figure 3-9,** in a typical amine absorber most of the heat of reaction is released in the bottom section of the tower. If there is too little amine in relation to the amount of acid gas, the temperature bulge moves up the column. If there is excessive amine, the temperature bulge moves to the bottom of the column and amine acid gas loadings are low. Using the temperature near the middle of the amine absorber to reset the amine flow, as depicted, for example, in **Figure 3-9,** maintains a relatively constant rich amine loading. This minimizes the chance of severe corrosion due to temporary overloading of the rich amine solution and also minimizes lean amine pumping and rich amine stripping costs. For this control scheme to work the temperature bulge must be at least 15 to 20°F, so it is not suitable for an absorber treating a gas stream with a low acid gas content.

There is general agreement that concentrated amine solutions with low acid gas loadings are less corrosive than less concentrated solutions with higher loadings when it is necessary to absorb more acid gas. Butwell (1968), for example, recommends that amine concentrations be

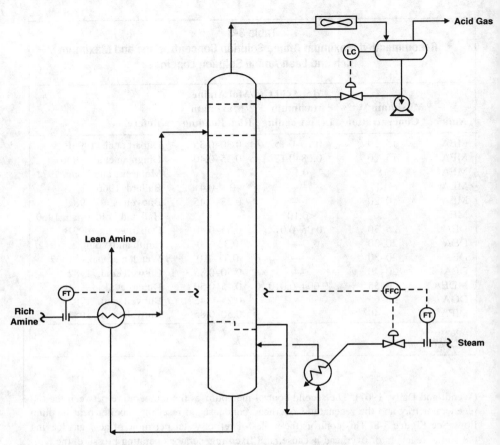

Figure 3-7. Use of flow ratio control to control amine stripping.

increased in preference to increasing amine loadings when it is necessary to absorb more acid gas. Although more concentrated amine solutions are more difficult to strip, they are preferred as long as the regenerator reboiler has enough capacity to regenerate the solution. **Figure 3-6** also shows that at constant amine acid gas loading, carbon steel corrosion increases as the temperature increases. To minimize corrosion, operating and reboiler heat medium temperatures are generally limited to maximum values. **Table 3-4** summarizes various recommendations. In most cases, it is desirable to limit the reboiler steam temperature to about 150°C (300°F).

Effect of the H_2S/CO_2 Ratio

The guidelines in API 945 (API, 1990) suggest that solutions of most amines are not corrosive if the ratio of hydrogen sulfide to carbon dioxide is above about 5/95, because when sufficient H_2S is present a protective iron sulfide film is formed according to reaction 3-20:

$$Fe + H_2S = 2H° + FeS \qquad (3\text{-}20)$$

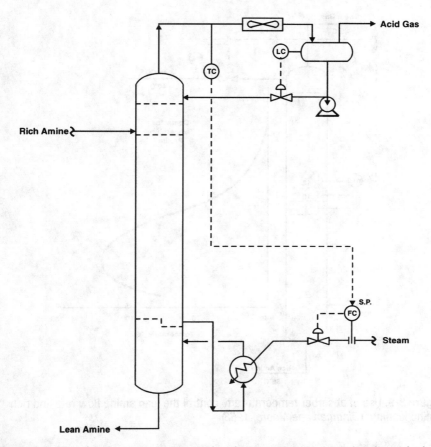

Figure 3-8. Use of amine regenerator overhead temperature to control amine stripping.

The ferrous sulfide film is sufficiently adherent and nonporous to prevent further corrosion as long as erosion-corrosion is prevented by good mechanical design, and impurities which remove the protective sulfide film, such as cyanide, are not present. The lack of corrosion at high H_2S/CO_2 ratios has been exploited in the SNPA-DEA process where acid gas loadings of 0.77 to 1.0 moles of acid gas per mole of amine have been achieved (Wendt and Dailey, 1967). High acid gas loadings might also be achieved with other amines at high H_2S/CO_2 ratios; however, no confirming data have been published. In the absence of better data, it is recommended that the guidelines of Dupart et al. (1993B) as summarized in **Table 3-3** for maximum rich solution loadings be followed even though they ignore the effect of the H_2S/CO_2 ratio.

Several studies have been published on the effect of the H_2S/CO_2 ratio on the corrosion of carbon steel by amine solutions (Froning and Jones, 1958; MacNab and Treseder, 1971; Lang and Mason, 1958; Dow, 1962). The results of corrosion tests reported by Froning and Jones (1958) are summarized in **Figures 3-10** and **3-11.** Both charts show the rate of corrosion of carbon steel in boiling MEA solutions with nitrogen gas containing CO_2 and/or H_2S

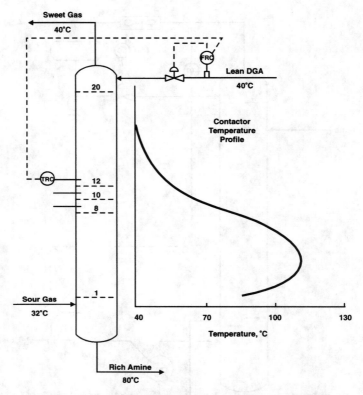

Figure 3-9. Use of absorber temperature to control the lean amine flow rate and rich amine loading. (*Dingman and Moore, 1968*)

bubbling through the liquid. The tests were conducted in stainless steel vessels, presumably at atmospheric pressure. **Figure 3-10** presents data for tests with a 10% MEA solution contacted by a nitrogen/H_2S mixture. The highest corrosion rate occurred with a gas containing 0.01 to 0.05% H_2S. The corrosion rate dropped rapidly as the H_2S content of the gas was increased, and appeared to level off at a low value (less than 1 mil/year) at H_2S concentrations above about 5% in the gas.

Figure 3-11 covers corrosion tests in boiling 10 and 20% MEA solutions contacted with gas containing both H_2S and CO_2. The results are quite similar to those of **Figure 3-10,** indicating that the corrosion rate is controlled primarily by the H_2S content of the gas and that 5% or more H_2S effectively inhibits corrosion regardless of the CO_2 content of the gas (up to at least 33% CO_2).

Since these tests were conducted with amine solutions at their boiling points, the vapor phase would be primarily steam, and neither the partial pressures of the acid gases nor the acid gas concentrations in the test solutions can be accurately estimated. The plant location most closely simulated by the test conditions is the stripping column, where the pressure is near atmospheric, the temperature is near the solution boiling point, and the vapor phase is primarily steam with some acid gas. However, the H_2S/CO_2 ratio in the acid gas is not con-

Table 3-4
Recommended Maximum Process and Heat Medium Temperatures

Amine	Maximum Process Temp, °C (°F)	Maximum Heat Medium Temperature(1), °C (°F)		Reference
		Hot Oil	Steam	
All	124 (255)	—	149 (300)	Dupart et al., 1993A
MEA	127 (260)	—	149 (300)	Dingman et al., 1966
MEA	121 (250)	—	149 (300)	Connors, 1958
MEA	127 (260)	—	141 (285)	Ballard, 1966
MEA/DEA	116–127 (240–260)	<230 (<450)	—	Ballard, 1986A, 1986B
MEA	—	188–227 (370–440)	—	Carlson et al., 1952
MEA	121 (250)	—	135 (275)	Hall and Polderman, 1960

Notes:

1. Fired heater reboilers should be designed to keep the tube wall temperature below 300°F which limits the heat flux to 6,500 to 8,500 Btu/(hr)(ft2) (Manning and Thompson, 1991; Ballard, 1966). Bacon (1987) recommends that the fired heater tube skin temperature be limited to a maximum of 350°F and heat flux should be limited to 6,000 to 8,000 Btu/(hr)(ft2). These constraints may require inlet ferrules to restrict heat flux at the fired inlet.

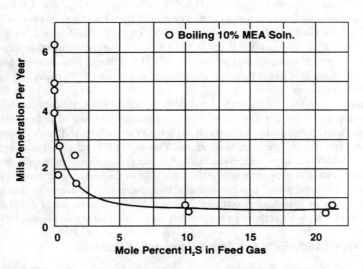

Figure 3-10. Corrosion of carbon steel in an H_2S-N_2-MEA-H_2O environment. (*Froning and Jones, 1958*)

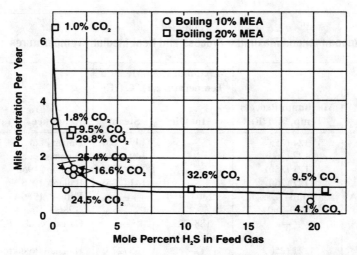

Figure 3-11. Corrosion of carbon steel in an H_2S-CO_2-N_2-MEA-H_2O environment. (*Froning and Jones, 1958*)

stant over the length of the stripping section and is equal to that in the feed gas only at the top. Nevertheless, the ratio of H_2S to CO_2 found to minimize corrosion in the Froning and Jones tests has been used for guidance in the design of MEA plants. Assuming that the curve of **Figure 3-11** is valid for all CO_2 concentrations tested (up to 32.6% CO_2) and that at least 5% H_2S is required to inhibit corrosion, an H_2S/CO_2 ratio of at least 5/32.6 or 1/6.5 would be needed. If it is assumed that the same curve would hold for an acid gas mixture of 5% H_2S and 95% CO_2, then one could conclude that the H_2S/CO_2 ratio should be at least 5/95. The observation in API Recommended Practice 945 (API, 1990) that corrosion has been least severe in low pressure amine plants that remove only H_2S or remove mixtures in which H_2S is at least 5% of the acid gas is based on the Froning and Jones (1958) paper and industrial plant experience (Gutzeit, 1994).

Corrosion tests reported by Riesenfeld and Blohm (1950, 1951A, B) confirm the conclusion of Froning and Jones that there is no correlation between the rate of carbon steel corrosion and the H_2S/CO_2 ratio when this ratio is higher than about 5/95 and that the presence of H_2S inhibits carbon steel corrosion. MacNab and Treseder (1971) have also investigated the effect of the H_2S/CO_2 ratio on amine solution carbon steel corrosion. Their results, which are based on tests with diisopropanolamine (DIPA or ADIP) solutions, do not support the conclusions of Froning and Jones. It is possible that this discrepancy is due to the use by MacNab and Treseder of sealed glass containers for the corrosion tests. Silica dissolved from glass containers is known to inhibit amine solution corrosion (Froning and Jones, 1958).

Effect of Heat-Stable Salts and Amine Degradation Products

Carbon dioxide and hydrogen sulfide are weak enough acids that their reactions with amines are thermally reversible. Acids that are sufficiently strong that their reactions with amines are not thermally reversible are called heat-stable acids, and products of their reactions with amines are heat-stable salts. If heat-stable acids enter an amine unit or are generated in the amine solution by reaction with trace amounts of oxygen or by thermal degradation of the amine, heat-stable salts can accumulate in the solution.

Heat-stable salts have several sources. In refineries, FCCU gases may contain traces of formic, oxalic, and acetic acids. Traces of oxygen in various refinery gas streams (e.g., FCCU, delayed coker, vacuum unit, vapor recovery system), air leaking into gas gathering systems which are operated at subatmospheric pressure, and oxygen in unblanketed amine storage tanks and sumps can react with the amine to form carboxylic acids and with H_2S to form elemental sulfur and thiosulfate. In refinery systems, elemental sulfur can then react with cyanide to form thiocyanate.

Heat-stable salts reduce the acid gas removal capacity of the amine solution by tying up a portion of the amine. The presence of heat-stable salts can also increase the corrosivity of amine solutions (Dupart et al., 1993B). Such salts are corrosive because they lower the amine solution pH, increase solution conductivity, and may also act as chelating agents, dissolving the protective film covering the base metal (Rooney et al., 1996). It is also possible that some of the weaker heat-stable acids, such as formic acid, vaporize in the amine regenerator to release the free acid, which could then react with exposed carbon steel (McCullough and Nielsen, 1996). Amine-CO_2 degradation products, some of which are strong chelating agents, may also contribute to amine solution corrosion by removing protective oxide or sulfide films (Polderman et al., 1955A, B; Chakma and Meisen, 1986). While it is generally agreed that heat-stable salts and amine degradation products contribute to amine solution corrosion, there is no definitive explanation of the corrosion mechanism. In fact, it is likely that several factors, including lowering of the amine solution pH and chelating effects, contribute to carbon steel corrosion by heat-stable salts and amine degradation products (Rooney et al., 1996). See **Figure 3-5** and the discussion on amine-acid gas carbon steel corrosion mechanisms for more information. Corrosion due to heat stable salts can be controlled by amine reclaiming and/or the addition of soda ash or caustic soda to neutralize the acids involved.

Amine Reclaiming. The operation of sidestream purification units (reclaimers) makes it possible to maintain a constant concentration of active amine in the treating solution and prevent the accumulation of corrosive heat-stable salts and amine degradation products. Commercial techniques used to reclaim amine solutions include distillation under vacuum; atmospheric or higher pressure distillation; ion exchange; and electrodialysis. Atmospheric or higher pressure distillation can only be used for MEA and DGA, which are primary amines. Secondary amines (DEA and DIPA) and MDEA, a tertiary amine, must be reclaimed by vacuum distillation, ion exchange, or electrodialysis because these amines decompose at atmospheric distillation temperatures. Design and operating guidelines for MEA thermal reclaimers are provided in several references: Hall and Polderman (1960), Blake and Rothert (1962), Blake (1963), Dow (1962), and Jefferson Chemicals (1963). DGA reclaiming is reviewed by Kenney et al. (1994), ion exchange by Keller et al. (1992), and electrodialysis by Union Carbide (1994) and Burns and Gregory (1995). Reclaiming of secondary and tertiary amines is usually on a contract basis, while primary amines are reclaimed as a part of normal operation. Amine reclaiming should be considered when the heat stable salt content is greater than 10% of the active amine concentration (Dupart et al., 1993B). A detailed review of amine reclaiming techniques is presented later in this chapter.

Heat Stable Salt Neutralization. Soda ash (or caustic soda) is often added to DEA and MDEA solutions to neutralize heat-stable salts, and there is considerable plant evidence that this is an effective means of reducing corrosion (Smith and Younger, 1972; Butwell et al., 1982; Liu and Gregory, 1994; Burns and Gregory, 1995; Liu et al., 1995; Rooney et al., 1996). Adding soda ash reduces amine solution corrosiveness by raising the solution pH. Soda ash addition may also reduce corrosion by preventing the release of weaker acids such as formic acid during amine regeneration (McCullough and Nielsen, 1996). Although soda ash addition can reduce corrosion, the amount that can be added is limited because solids will eventually be precipitated, plugging equipment and piping. However, solids precipitation and equipment plugging can be avoided if soda ash addition is combined with amine solution reclaiming using either batch distillation, ion exchange, or electrodialysis. Soda ash addition is particularly attractive for secondary and tertiary amines like DEA and MDEA since these amines cannot be reclaimed during normal operation. Therefore, for these amines, soda ash addition can be used to control corrosion until a contract reclaimer arrives at the plant site.

According to Scheirman (1973A, B), soda ash should first be added to DEA solutions when the heat stable salt concentration reaches 0.5 wt%. Nearly 20 wt% sodium salts can be tolerated before any solids precipitate. Potassium carbonate can also be used to neutralize heat stable salts and has the advantage of being about 25% more soluble by weight than sodium compounds (Scheirman, 1973A, B).

For MDEA solutions, Liu and Gregory (1994) recommend that soda ash should be added to keep the amine heat stable salts concentration below 2 wt%. The MDEA solution should be reclaimed when the total heat stable salt anion content reaches 4 wt%. Rooney et al. (1996) have also investigated caustic soda neutralization of MDEA solutions containing heat stable amine salts. They recommend that soda ash addition be used to keep the heat stable amine salts level below 0.5 wt%. In addition, they recommend that individual heat stable amine salt anions be kept below the following maximum levels:

Heat-Stable Amine Salt Anion	Maximum ppmw
oxalate	250
formate, glycolate, malonate, sulfite, or sulfate	500
acetate or succinate	1,000
thiosulfate	10,000

According to Rooney et al., the amount of soda ash that can be added is limited to a maximum of about 10% of the total MDEA concentration before solution viscosity and solids precipitation problems occur.

Effect of Amine Type

It is well known that the choice of amine affects corrosion (Dupart et al., 1993A, B). Primary amines like MEA and DGA are more corrosive than secondary amines like DEA and DIPA. In turn, DIPA and DEA are more corrosive than tertiary amines like MDEA. As noted by DuPart et al. (1993A, B) several investigators have shown that all amines are equally non-corrosive when no acid gas is present (MacNab and Treseder, 1971; Lang and Mason, 1958; Froning and Jones, 1958; Blanc et al., 1982A, B). Therefore, differences in corrosion cannot be due to the amine alone. Several explanations have been offered to account for the effect of amine type on corrosion. For example, MDEA differs from MEA, DEA, DIPA, and DGA in that it does not form amine-CO_2 degradation products. However, investigations by Polderman et al. (1955A, B) for MEA and by Chakma and Meisen (1986) for DEA suggest that although amine-CO_2 degradation products contribute to corrosion, they are not the primary cause. As noted by DuPart et al. (1993A, B), it is possible that the ability of primary and secondary amines to form carbamates according to reaction 3-18 may account for the differences in corrosion. Perhaps the more basic amines such as MEA are more corrosive due to the presence of undecomposed salts such as alkanolammonium carbamate and the resultant concentration of alkanolammonium ions in the hottest sections of the amine regenerator (see equation 3-19). Also, less basic amines such as DEA are easier to strip than MEA, while MDEA is very easily stripped. Consequently, DEA and MDEA alkanolammonium ion concentrations are low in the bottom section of the amine regenerator where the amine solution is the hottest. Therefore, these less basic amines are less corrosive than primary amines such as MEA and DGA.

Erosion-Corrosion

Erosion-corrosion is caused by high amine solution velocities, solution turbulence, and impingement of gas and amine on metal surfaces. Erosion-corrosion removes the protective iron sulfide, carbonate, or oxide film protecting the piping and equipment from corrosion. Areas that are subject to erosion-corrosion include the piping from the amine contactor pressure let-down valve to the rich amine flash drum, the piping from the rich amine flash drum level control valve to the amine regenerator, and the lean amine pump (see **Figure 3-1**). Other areas that are affected include heat exchanger tubes near the inlet nozzle, the amine contactor near the sour gas inlet, and the amine regenerator near the rich amine inlet. Erosion-corrosion is aggravated by dirty amine solutions containing suspended solids. Erosion-corrosion can be reduced by choosing the correct materials of construction and mechanical design details which minimize impingement, reduce turbulence, and lower amine solution velocities. Amine solution mechanical filtration to remove suspended solids also reduces erosion-corrosion. Several authors recommend that suspended solids levels be kept below 0.01 wt% (Lieberman, 1980; Hall and Polderman, 1960). Mechanical filtration of amine solutions is reviewed in more detail later in this chapter.

Erosion-Corrosion of Piping

API 945 recommends designing both lean and rich amine carbon steel piping for velocities less than 1.8 m/sec (6 ft/sec) (API, 1990). **Table 3-5** summarizes piping velocity recommendations from several sources. As noted, recommended velocities range from 0.9 to 1.8 m/sec (3 to 6 ft per sec). Although there is no published research to support these recommendations, it is thought that API 945 represents good practice because it is an industry consensus document. Sheilan and Smith (1984) and Dingman et al. (1966) recommend the use of seamless pipe and long-radius elbows to reduce amine piping erosion-corrosion. Sheilan and Smith also suggest that threaded connections or socket weld fittings be avoided.

Erosion-Corrosion of Heat Exchangers and Reboilers

Ballard (1966) and Dingman et al. (1966) recommend the use of multiple inlets and outlets to reduce corrosion of kettle and horizontal thermosyphon reboilers. See **Figure 3-12**.

Table 3-5
Recommended Maximum Amine Solution Piping Velocities for Rich and Lean Amine Solutions

Amine	Velocity, m/sec (ft/sec) (Note 1)				Reference
	Rich Solution		Lean Solution		
All	1.8	(6)	1.8	(6)	API, 1990
All	0.9	(3)	0.9	(3)	Campbell, 1981
MEA	0.9	(3)	0.9	(3)	Dingman et al., 1966
All	1.5	(5)	—		Dupart et al., 1993B
DEA (Note 2)	1.0	(3.3)	—		Dailey, 1970
DEA (Note 2)	0.9	(3.0)	—		Smith and Younger, 1972
DGA	1.5	(5)	1.5	(5)	Seubert and Wallace, 1985
All	0.6–1.5	(2–5)	0.6–1.5	(2–5)	Manning and Thompson, 1991

Notes:
1. *Figures also generally apply to heat exchanger tubing. However, Manning and Thompson (1991) recommend a maximum tubeside velocity of 2 to 4 ft/sec for lean/rich exchangers where rich amine is on the tubeside.*
2. *Recommended velocity applies only to rich DEA flowing through the tubeside of a lean/rich exchanger. Dailey (1970) and Smith and Younger (1972) report that severe corrosion occurred in the tubes of a lean/rich exchanger when the tubeside velocity exceeded 5.25 ft/sec.*

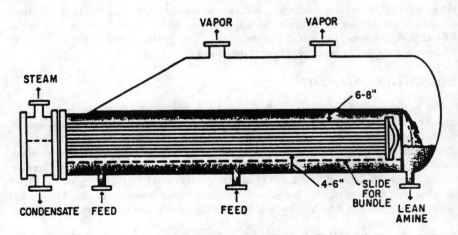

Figure 3-12. Recommended kettle reboiler dimensions and inlet and outlet piping configuration to minimize carbon steel corrosion. *Reproduced with permission from Hydrocarbon Processing, April 1966.* (*Ballard, 1966*)

Sheilan and Smith (1984) recommend perforated inlet baffles for heat exchangers to prevent corrosion due to impingement. Ballard (1966), Dingman et al. (1966), and Connors (1958) suggest designing reboilers with liberal disengaging space to minimize violent boiling and resulting erosion-corrosion. All recommend removal of tube rows to form a "V" or "X" in existing installations where violent boiling is a problem. See **Figure 3-13.** Ballard (1966) recommends designing exchangers and reboilers with square pitch to facilitate cleaning and reduce erosion-corrosion. Connors (1958), Dingman et al. (1966), and Smith and Younger (1972) advise keeping amine velocities in carbon steel heat exchangers below 0.9 m/sec (3 ft/sec). However, this guideline may be too conservative because the velocity recommendations for piping (1.8 m/sec or 6 ft/sec) summarized in **Table 3-5** should apply to heat exchangers as well. Ballard (1966) advises that the reboiler bundle should be covered with 15 to 20 cm (6 to 8 inches) of liquid to prevent localized drying of exposed tubes and overheating. He also recommends locating the reboiler tube bundle about 15 cm (6 inches) above the bottom of the reboiler shell to allow free circulation of the amine through the tube bundle. See **Figure 3-12** for details. The reboiler steam control valve should be located on the reboiler inlet, not the condensate outlet, to prevent high localized temperatures due to steam side reboiler flooding and resulting corrosion. As shown in **Figure 3-1,** the lean/rich

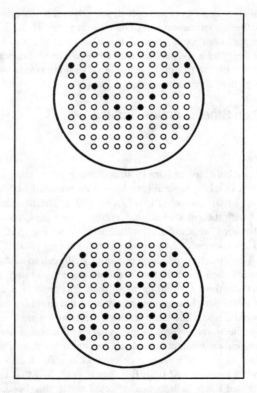

Figure 3-13. Recommended tube removal pattern to reduce vapor blinding in kettle or horizontal thermosyphon reboilers. (*Dingman et al., 1966*)

amine exchangers should be located upstream of the rich amine level control valve to minimize acid gas evolution inside the exchangers, and the rich amine should be on the tube side. If the lean/rich amine exchangers are stacked, rich amine should flow up through the tubeside of the bottom exchanger to the tubeside of the upper unit.

Erosion-Corrosion of Lean Amine Pumps

To minimize turbulence and erosion-corrosion of the lean amine pump impeller and casing, Sheilan and Smith (1984) recommend a minimum of 8 to 9 pipe diameters of straight pipe upstream of the pump suction. As shown in **Figure 3-1**, lean amine pumps should be located downstream of the lean/rich amine exchanger because the hot lean amine solution leaving the regenerator is often near its boiling point at the elevation corresponding to the lean amine pump suction. Placing the lean amine pump downstream of the lean/rich exchanger ensures that the lean amine is subcooled and, therefore, less subject to gas evolution when it enters the pump. Cooling the lean amine solution also raises the solution pH (see **Figure 3-4**) and makes the solution less corrosive (see **Figure 3-6**).

Erosion-Corrosion of Pressure Let-Down Valves

To reduce erosion-corrosion of pressure let-down valves downstream of absorbers, Graff (1959) recommends the use of carbon steel bodies with type 316 SS internals and stellited trim when the valve pressure drop is above 7 to 14 bar (100 to 200 psi). Scheirman (1976) recommends carbon steel globe body valves with stellited 316 SS internals, but also suggests that valves be selected with the maximum feasible valve body size in order to minimize the amine velocity through the valve body.

Cracking of Carbon Steel in Amine Service

Background

Four carbon steel cracking mechanisms in alkanolamine gas treating units have been identified. Reviews of these cracking mechanisms have been provided by Merrick (1989), Buchheim (1990), Gutzeit (1990), and in API 945 (API, 1990). The first three cracking mechanisms are associated with the entry of atomic hydrogen into the carbon steel lattice. These three cracking mechanisms are known as sulfide stress cracking (SSC), hydrogen-induced cracking (HIC), and stress-oriented hydrogen induced cracking (SOHIC). All three of these cracking mechanisms require the production of atomic hydrogen in an aqueous-H_2S solution. While there is no established lower H_2S concentration limit, industry practice has been to assume that aqueous solutions containing more than 50 ppmw H_2S can lead to cracking (NACE, 1994B). In the vapor phase, a commonly used threshold for SSC is an H_2S partial pressure of 0.34 kPa (0.05 psia). The three cracking mechanisms are distinguished from each other by what entraps the atomic hydrogen inside the metal lattice, whether it recombines to form molecular hydrogen, the orientation and features of the resulting cracks, and the corrective measures required to minimize each type of cracking. The fourth mode of cracking is alkaline stress corrosion cracking (ASCC). It is thought that ASCC is caused by a film rupture mechanism. Stressed areas such as heat-affected zones slip, breaking the passive film and exposing bare steel, which corrodes to form cracks. The passive film reforms, but residual stresses cause the film to rupture again, leading to more corrosion. Repetition of this process leads to cracking. A summary of each cracking mechanism follows. Examples of each cracking mechanism are provided in API 945, Appendix A (API, 1990).

SSC (Sulfide Stress Cracking)

Carbon steel is embrittled by atomic hydrogen dissolved in the metal lattice. In the heat-affected zones adjacent to welds there are often very narrow hard zones combined with regions of high residual tensile stress that may become embrittled to such an extent by dissolved atomic hydrogen that they crack. **Figure 3-14** shows sulfide stress cracking originating at a heat-affected zone of a weld (API, 1990). SSC is directly related to the amount of atomic hydrogen dissolved in the metal lattice and usually occurs at temperatures below 90°C (194°F) (Gutzeit, 1990). SSC is also dependent on the composition, microstructure, strength, and residual and applied stress levels of the steel (Buchheim, 1990). SSC has been found in attachment and seam welds in the amine regenerator overhead system, in the bottom of the amine absorber, in the top of the amine regenerator column, and on the rich side of the lean/rich amine exchanger (Gutzeit, 1990). These locations suggest that SSC is due mainly to wet acid gas corrosion. See **Figure 3-1**. This form of cracking can generally be prevented by limiting the carbon steel weld metal hardness to less than 200 Brinell (BHN) and by restricting the steel tensile strength to less than 621 MPa (90 ksi) (NACE, 1994B; 1987). Post weld heat treatment (PWHT) is beneficial in mitigating SSC because it reduces hardness and relieves stresses (Merrick, 1989; Buchheim, 1990).

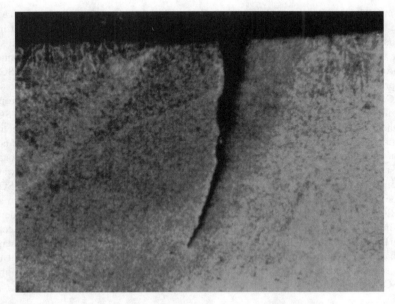

Figure 3-14. Sulfide stress cracking (SSC) in an existing hardened heat-affected zone on a weld. (*API, 1990*)

HIC (Hydrogen-Induced Cracking)

When hydrogen atoms dissolved in carbon steel meet a non-metallic inclusion, e.g., a sulfide or oxide particle, a slag inclusion, a lamination, or other discontinuity, they often combine irreversibly to form molecular hydrogen. The molecular hydrogen, unlike atomic hydrogen, cannot escape; therefore, it accumulates and builds up high pressure inside the metal. Eventually, the pressure causes the metal-inclusion interface to separate, resulting in cracking or blistering. The blisters are parallel to the steel surface because the carbon steel laminations or inclusions are typically elongated parallel to the carbon steel surface when the steel is rolled during manufacture. **Figure 3-15** shows HIC blistering in a steel pipe. HIC rarely occurs in product forms other than plate or plate products.

HIC is also called hydrogen blistering cracking or stepwise cracking. HIC depends on steel cleanliness and composition. It has been found primarily in the bottom of absorber towers, in the amine regenerator overhead system, and in the top section of the amine regenerator tower (Gutzeit, 1990). These locations suggest that the principal cause of HIC is wet acid gas corrosion. HIC may be avoided by using specially manufactured "clean steel" plates that are more HIC-resistant than conventional carbon steel. HIC-resistant carbon steel is made by ladle treating with either calcium or a rare earth metal for residual sulfide-inclusion shape control. A recently published NACE International Technical Committee Report reviews the manufacturing and test methods for HIC-resistant steel (NACE, 1994A). Since hydrogen induced cracking depends on the cleanliness of the carbon steel and its method of manufacture, HIC cannot be prevented by PWHT (Buchheim, 1990).

SOHIC (Stress-Oriented Hydrogen-Induced Cracking)

As in HIC, SOHIC is caused by atomic hydrogen dissolved in the carbon steel lattice combining irreversibly to form molecular hydrogen. The molecular hydrogen collects at imperfections in the metal lattice, just as in HIC. However, due to either applied or residual stresses, the trapped molecular hydrogen produces micro-fissures which align and interconnect in the through-wall direction. SOHIC can propagate from blisters caused by HIC, SSC, and from prior weld defects (Gutzeit, 1990; Buchheim, 1990). For example, **Figure 3-16** shows SOHIC propagating from a blister caused by HIC, and **Figure 3-17** shows SOHIC propagating from SSC (API, 1990). However, neither HIC nor SSC are preconditions for SOHIC (Buchheim, 1990). In amine systems, SOHIC has been found mostly in the upper section of the amine regenerator tower, in the amine regenerator overhead system, and in the bottom section of the absorber below the bottom tray (Gutzeit, 1990). As with HIC, these locations suggest that the primary cause of SOHIC is probably atomic hydrogen produced by wet acid gas corrosion (see **Figure 3-1**). PWHT improves the resistance of carbon steel to SOHIC, but does not totally eliminate it (Buchheim, 1990). In recent years, many users have specified HIC-resistant carbon steels, with PWHT, for SOHIC resistance. However, under very corrosive laboratory conditions even HIC-resistant steels have been shown to be susceptible to SOHIC (Cayard et al., 1994). Therefore, carbon steel plate clad with austenitic stainless steel has been used to eliminate the risk of SOHIC. Since SOHIC is most prevalent in the amine regenerator overhead system, cladding this area, as shown in **Figure 3-1,** can prevent both SOHIC and HIC.

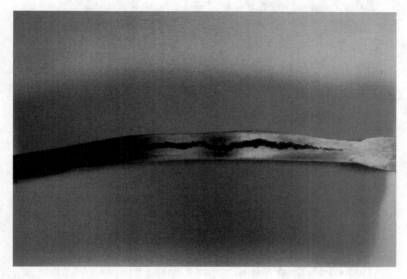

Figure 3-15a. Hydrogen surface blisters in steel pipe caused by H_2S service. (*API, 1990*)

Figure 3-15b. Splitting and bulging in carbon steel plate due to the growth of an embedded blister. (*API, 1990*)

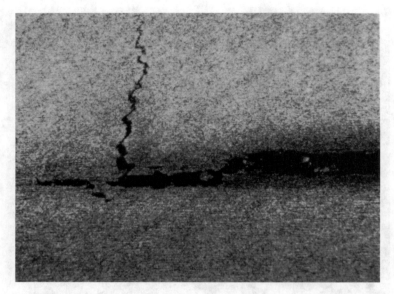

Figure 3-16. Crack extension from a hydrogen blister. (*API, 1990*)

ASCC (Alkaline Stress-Corrosion Cracking)

As noted earlier, it is thought that ASCC is caused by a film rupture mechanism. In areas of high residual stress, such as heat affected zones, slip causes the passive film to break, exposing bare steel, which corrodes to form cracks. If the corrosion rate is greater than the rate of passive film formation, the protective film will not reform, and pitting or some other form of localized corrosion will occur. However, if the passive film forms faster than the metal corrodes, the protective film will be restored. Repetition of this process results in alkaline stress corrosion crack growth. ASCC is the most common cracking mechanism in alkanolamine gas treating plants. It can occur in plants treating CO_2, H_2S, or mixtures of both acid gases (Richert et al., 1987; 1989). **Figure 3-18** shows ASCC in the vicinity of a weld.

ASCC in amine gas treating plants was first reported in 1951 by a NACE Committee (Schmidt et al., 1951). In 1953, Garwood reported ASCC in MEA plants treating natural gas. Cracks were found in heat exchanger heads, amine absorbers, piping, and amine regenerators. The cracks were intergranular and oxide-filled. Since the cracking occurred only in highly stressed, heat affected zones, PWHT was recommended to eliminate ASCC. As a result of this early work, the industry adopted a general policy of PWHT for all piping and equipment (except storage tanks) in contact with amine above a certain temperature. Depending upon the company and the amine, the selected temperature varied between 38 and 93°C (100 and 200°F) with 66°C (150°F) being a common choice (Richert et al., 1987; 1989).

In 1982, Hughes reported ASCC in non-PWHT carbon steel equipment in an MEA refinery unit. ASCC had occurred in welds in contact with amine at temperatures ranging from 53 to 93°C (127 to 200°F); although no cracking had occurred in PWHT welds operating at temperatures as high as 155°C (311°F). Hughes (1982) concluded that PWHT for all carbon steel piping and equipment in amine service would eliminate ASCC.

Mechanical Design and Operation of Alkanolamine Plants **221**

Figure 3-17a. Stress-oriented hydrogen-induced cracking (SOHIC) from pre-existing sulfide stress cracking (SSC). (*API, 1990*)

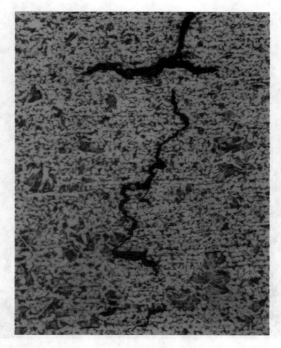

Figure 3-17b. Higher magnification view of the crack tip shown in Figure 3-17a. (*API, 1990*)

222 Gas Purification

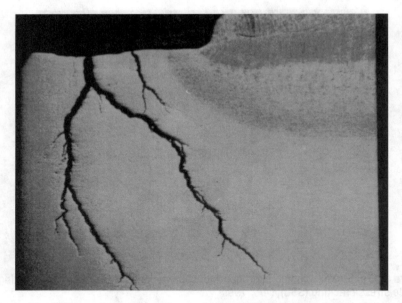

Figure 3-18a. Alkaline stress-corrosion cracking (ASCC) in the vicinity of a weld. (*API, 1990*)

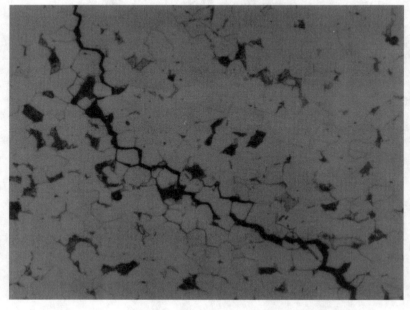

Figure 3-18b. Higher magnification view of the crack tip shown in Figure 3-18a. (*API, 1990*)

In 1984, an amine LPG treater ruptured, causing an explosion and fire that killed 17 people (McHenry et al., 1986; 1987). Although the disaster was apparently caused by a combination of HIC and SOHIC, this disaster and reports of extensive cracking in other amine systems (Gutzeit and Johnson, 1986; Anon, 1985) initiated investigations of carbon steel cracking in amine units. These investigations, which revealed many instances of ASCC, were undertaken by NACE Group Committee T-8 on Refining Corrosion in cooperation with the API.

Results of an industry survey by this group were reported in 1989 by Richert et al. This survey included 294 amine units, 272 of them in refineries, and most of them MEA and DEA units. This survey indicated that cracking occurred primarily in MEA service. Nearly every crack was associated with a weld that had not been PWHT. Cracks occurred in every type of equipment and at temperatures as low as ambient. However, no conclusive correlation was found between cracking and most process variables, including source of gas, amine concentration or acid gas loading, use of filters or reclaimers, use of corrosion inhibitors, type of carbon steel, or addition of caustic to neutralize heat-stable salts. ASCC was found to be generally independent of the H_2S/CO_2 ratio. However, all amine plants with less than 1 mol% H_2S in their feed experienced cracking. This suggests that at least 1 mol% H_2S in a predominantly CO_2 stream has some tendency to inhibit ASCC. This conclusion was confirmed by later laboratory investigations (Schutt, 1988; Parkins and Foroulis, 1988; Lyle, 1988). The survey data could not be used to demonstrate that ASCC was temperature dependent because most of the equipment operating at higher temperatures was PWHT. However, the experience documented by Garwood (1953) suggests that ASCC increases with increasing temperature when equipment and piping are not PWHT. Since 98% of the cracks reported in the survey occurred in carbon steel welds that had not been PWHT or at nozzles where PWHT is difficult, PWHT of all carbon steel piping and equipment in amine service was judged to be the single most effective measure to prevent ASCC. Later, additional data on DEA and DIPA gas treating units, which included data on ASCC below 66°C (150°F) for both DEA and DIPA, confirmed that PWHT is required to prevent ASCC in plants using these amines (Bagdasarian et al., 1991).

A report by the Southwest Institute (SwRI, 1989) and a related article by Lyle (1988) point out that the NACE survey includes data from only 16 natural gas treating units out of a total of 294 units of all kinds. These references state that a more comprehensive survey was made for natural gas treating units, but gave the results of the survey only in brief summary, and did not state how many units had been surveyed. Both reports did, however, give detailed results of a laboratory study, which concluded that ASCC of carbon steel was inhibited by the presence of H_2S. Also, according to both reports, ASCC in refineries occurs predominantly in lean amine solutions; whereas, ASCC cracking in natural gas plants occurs primarily in rich amine solutions.

Corrosion Inhibitors

Corrosion inhibitors are often classified as cathodic inhibitors, which inhibit reaction 3-2, anodic inhibitors, which inhibit reaction 3-1, and oxidizing passivators, which are discussed in the next paragraph. Cathodic and anodic inhibitors are often adsorbed on the corroding metal, like filming amines, or plated out on it, like arsenic and antimony. They have been recommended and patented at various times, but none has had much commercial success in amine units (API, 1990).

Oxidizing passivators are considerably stronger oxidizing agents than hydrogen ion, and they operate by shifting the potential of the steel to a more positive value where reactions 3-1 and 3-2 do not occur. Instead, an oxidation passivator converts the iron on the surface to the trivalent state, i.e., to a form of ferric oxide that is very adherent and protective. This oxide is called the passive film. Such inhibitors work well in units removing carbon dioxide only, reducing the corrosion rate practically to nil, and they are the only effective inhibitors for MEA in the absence of hydrogen sulfide (API, 1990). Conversely, they are destroyed by hydrogen sulfide.

While oxidizing passivators work extremely well when they are properly maintained, they have several drawbacks. First, regular solution analyses are required. Second, the passivator concentration must be maintained within specified limits. Third, they must be protected against impurities which destroy them, including hydrogen sulfide and large amounts of iron corrosion products, both soluble and insoluble; and fourth, when they fail, they often permit local attack, or even aggravate it. Finally, many of them contain toxic heavy metals, which makes disposal difficult and expensive.

Chloride Attack of Stainless Steel in Amine Service

Impurities such as chloride gradually build up in amine systems until a steady-state concentration is reached. Since most amine systems contain some stainless steel, it is of interest to know what chloride levels can cause pitting of stainless steels in an amine environment. Limited information is available in the literature. Experiments reported by Seubert and Wallace (1985) indicate little or no pitting tendencies with 304 SS exposed to DGA solutions containing up to 4,000 ppm chloride. Based on these experiments, the maximum acceptable chloride level for DGA plants containing type 304 SS was set at 1,000 ppm.

Halides have also been found to contribute to crevice corrosion and stress corrosion cracking in stainless steel heat exchanger plates in lean/rich amine plant plate exchangers. The corrosion occurred under ethylene-propylene-diene monomer (EPDM) rubber gaskets that contained significant concentrations of chlorine and bromine left over from the curing process. The problem was resolved by replacing the heat exchanger packs with 316 SS plates and peroxide-cured EPDM gaskets with a maximum total halogen concentration of 200 ppm (Hay et al., 1996).

FOAMING

Foaming of alkanolamine solutions is probably the most common operating problem in amine treating units. It is most frequently encountered in the contactor, but may also occur in the stripping column. Foaming may result in excessive amine losses, off-specification product gas, and reduced operating rates, and can also be responsible for the production of off-specification, dark sulfur if foam is carried over into the Claus sulfur plant. Lieberman (1980), Smith (1979A, B), Bacon (1987), Ballard and von Phul (1991), Manning and Thompson (1991), McCullough and Nielsen (1996), Thomason (1985), and Ballard (1966, 1986A, B) review contaminants that can cause amine solution foaming, and summarize plant operating practices and troubleshooting techniques that minimize amine plant foaming problems. A summary of amine plant foaming causes, symptoms, and remedies follows.

Causes of Foaming

Foaming in an amine unit is caused by solution contaminants since uncontaminated alkanolamine solutions will not form a stable foam. Common solution contaminants known to cause foaming are condensed hydrocarbons and acidic amine degradation products formed in the amine circuit. Feed gas sources that can contribute to or cause foaming include coalesced lubricating oil aerosols and surface active agents (e.g., water-soluble pipeline corrosion inhibitors and many well treatment compounds). Materials introduced into the amine circuit during operation and maintenance (such as excessive antifoam agents, phosphoric acid from acid-washed activated carbon, chemicals and contaminants in make-up water, and chemicals used in the manufacture of cotton filter elements) can also cause foaming (Pauley et al., 1988, 1989A, B; Pauley, 1991; Bacon, 1987; Ball and Veldman, 1991). Suspended finely-divided solids, e.g., iron sulfide or pipeline mill scale, do not cause foaming, but rather tend to stabilize foam and further aggravate any tendency to foam (Pauley, 1991). Pauley et al. (1988, 1989A, 1989B) present experimental data showing that liquid hydrocarbons, low and high molecular weight organic acids, and acidic amine degradation products increase both the amine solution foaming tendency and foam stability. Specific causes of foaming include the following:

1. Water-soluble surfactants in the feed gas (e.g., well treating compounds, pipeline corrosion inhibitors) which lower the amine solution surface tension. Excessive antifoam can also cause foaming.
2. Liquid hydrocarbons (e.g., entrained compressor lubricating oil in the feed gas or hydrocarbon condensation within the amine absorber).
3. Particulate contaminants (e.g., mill scale, FeS corrosion products, rust) contained in the feed gas or produced within the amine treating unit. Solids such as FeS do not cause foaming, but concentrate at the liquid/gas interface and stabilize the foam by increasing the surface viscosity and thereby retarding film drainage (Pauley, 1991).
4. Oxygen contamination of the feed gas or amine unit (usually at the amine sump or amine storage tank) and reaction of the amine with oxygen to form carboxylic acids and amine heat stable salts. Dissolved iron can catalyze the reaction of amine with oxygen to form carboxylic acids (Pauley, 1991).
5. Feed gas contaminants, such as carboxylic acids, which react with the amine to form heat stable salts.
6. Activated carbon that has been washed with phosphoric acid or that naturally contains phosphorus in the form of leachable phosphates (Pauley, 1991; Bourke and Mazzoni, 1989).
7. Contamination of the amine unit with greases and oils during a turnaround.
8. Amine filter elements that have been washed with surfactants or contaminated with oils during manufacture (e.g., string wound cartridge filters made with virgin cotton, but containing cottonseed fragments).
9. Contaminants in the amine plant make-up water such as boiler feed water treating chemicals and corrosion inhibitors (Smith, 1979A, B; Ballard, 1986A, B).

Symptoms and Characteristics of Foaming Systems

An amine plant subject to foaming will exhibit various characteristics including the following:

1. High amine losses and amine carry-over into downstream units
2. Reduced acid gas removal efficiency and failure to meet product gas specifications
3. High or erratic differential pressure measurements across the absorber and/or the stripper
4. An amine solution that is opaque and contaminated with suspended solids
5. More than 10% of the amine in the form of heat-stable salts

6. Failure of an amine solution foaming test. Bacon (1987) recommends that the foam height and break time as measured by the standard Dow foam test be less than 200 ml and 5 seconds, respectively. Smith (1979A, B) describes laboratory and field foaming tests
7. High operating costs (power, steam, and filtration costs)
8. Instrument taps plugged with particulates, solids accumulation in the regenerator overhead condenser, and reduction in amine filter cycle length (Lieberman, 1980)

Prevention of Foaming

Summary of Foam Prevention Techniques

Foaming can be reduced or controlled by proper care of the amine solution. The following techniques reduce amine solution contamination and minimize foaming:

1. A properly designed feed gas inlet separator and filter should be provided. A feed gas coalescer should be considered for feed gas streams contaminated with compressor lubricating oil and other finely dispersed aerosols (Ball and Veldman, 1991; Pauley, 1991). A properly sized slug catcher should be provided if slugs can accumulate in the feed gas line.
2. A feed gas water wash should be considered when the feed gas stream is severely contaminated with carboxylic acids or water-soluble, surface-active contaminants (Bacon, 1987). A feed gas water wash can also remove aerosols and ultra-fine particles (Ball and Veldman, 1991).
3. Mechanical and activated carbon filtration of the amine solution. A 10 micron mechanical filter on a 10 to 20% amine solution slipstream is usually sufficient; however, it is good practice to use the smallest micron rating that has an acceptable run time between filter element changes. Usually mechanical and activated carbon filtration of a 10 to 20% slipstream is sufficient. However, full stream filtration may be required for Claus plant tail gas units and for MDEA and DEA units treating feed gas streams with low H_2S/CO_2 ratios.
4. Onsite or offsite amine solution reclaiming to remove heat-stable salts and amine degradation products. No more than 10% of the amine should be tied up as heat-stable salts (Bacon, 1987).
5. Caustic addition to neutralize heat-stable salts to mitigate corrosion and thereby reduce iron sulfide formation (Bacon, 1987).
6. Temperature difference control of the lean amine solution feed to the absorber to ensure that the lean amine is 10 to 15°F warmer than the feed gas.
7. A properly sized rich amine flash drum to remove entrained and dissolved hydrocarbons.
8. Liquid hydrocarbon skimming facilities in the absorber sump, the rich amine flash drum, the regenerator sump, and the amine regenerator overhead accumulator (Bacon, 1987).
9. New plants and old plants that have undergone a major turnaround are often contaminated with oils, greases, welding fluxes, and corrosion inhibitors. A hot caustic wash (2 to 5 wt% caustic soda) followed by a hot condensate wash can remove these impurities and help prevent foaming.
10. The minimum contact temperature in the absorber should be greater than 50 to 60°F to ensure that high amine solution viscosity does not initiate foaming (Ballard, 1966).
11. Either batch or continuous antifoam addition (as a last resort).

A detailed review of foam prevention techniques follows.

Control of External Contamination

Foaming problems can usually be minimized by limiting the access of external contaminants such as compressor lubricating oil, well treating compounds, and pipeline corrosion inhibitors to the alkanolamine solution. A combination of a slug catcher, inlet filter/separator, and a gas-liquid coalescer installed in series in the feed gas line is the best way to prevent external contaminants from entering the system with the feed gas and can substantially reduce amine filtration costs and improve plant performance. According to Pauley and Perlmutter (1988), conventional inlet separators have difficulty in removing aerosols 3 microns or less in diameter, while specialized, high-efficiency gas/liquid coalescers can remove compressor lubricating oil droplets as small as 0.001 micron (Pauley, 1991). It is claimed that removal of these small aerosol particles can, in some cases, substantially reduce foaming problems in amine plants.

Prevention of Hydrocarbon Condensation and Liquid Hydrocarbon Accumulation

If the feed gas is near its hydrocarbon dewpoint, hydrocarbon condensation inside the contactor and related foaming can be avoided by keeping the temperature of the lean amine solution about 10 to 15°F above that of the gas in the amine contactor. This is usually accomplished by using a temperature difference controller to control a bypass around the lean amine cooler.

Amine contactors and regenerators should be supplied with differential pressure indication and alarm to provide an early warning of foaming problems. When the differential pressure (in feet of water) is 40 to 50% of the height between the instrument taps, the tower is usually foaming and amine is being carried over into downstream units (Lieberman, 1980).

Heavy hydrocarbons in the feed gas to the absorber can be dissolved or entrained in the rich amine. In systems with absorber pressures above about 100 psig, it is generally appropriate to use a rich amine flash drum to flash off hydrocarbon gases entrained or dissolved in the rich amine solution. **See Figure 3-1.** Rich amine flash drums are usually equipped with a skimming device to remove any liquid hydrocarbons that accumulate. Ball and Veldman (1991) recommend that the rich amine flash drum be designed for at least 20 to 30 minutes of liquid retention when half full. For refinery applications, where the risk of hydrocarbon contamination is high, they recommend that the flash drum operate at as low a pressure as possible, i.e., 5 to 10 psig, which would require pumps to transfer the rich amine to the regenerator. Manning and Thompson (1991) recommend 10- to 15-minute residence times for two-phase separators (gas/amine solution) and 20 to 30 minutes for three-phase flash tanks (gas/oil/amine solution). Bacon (1987) recommends liquid residence times ranging from 5 to 30 minutes depending on the type and degree of mixing between the hydrocarbon and amine. Long residence times and low pressure operation are probably appropriate when treating gas streams containing heavy hydrocarbons, i.e., many refinery applications. Shorter flash drum residence times and higher flash drum operating pressures can be considered when the gas being treated is relatively lean as in many plants treating natural gas.

Removal of Foam-Causing Impurities from Solution

Three techniques are commonly employed to remove impurities from gas treating plant amine solutions: Mechanical filtration, activated carbon adsorption, and thermal reclaiming (distillation). Additional, occasionally used solution purification techniques are settling, ion exchange, and electrodialysis.

Fine particles and sludge are removed from amine solutions by settling or filtration. Settling is quite effective; however, large vessels are required to provide sufficient time for the very small particles (usually iron sulfide) occurring in amine solution to separate by gravity. Filtration can be very difficult because the iron sulfide, which is formed as a corrosion product or enters the plant with the feed gas, is hard to remove. Designs based on etched disk or sintered metal fiber elements have proved effective for large amine flowrates, although they are expensive. Simpler, less expensive equipment such as bag and cartridge type filters are often used, particularly for the smaller amine flow rates.

High molecular weight organic molecules, such as some amine degradation products, can be removed from amine solutions by adsorption on activated carbon. A properly designed activated carbon system should treat a 10 to 20% slip stream of the circulating solution. Hot rich amine tends to liberate acid gas in response to the pressure drop across the mechanical and activated carbon filters. The liberated gas forms bubbles around the carbon granules and pockets of acid gas within the carbon filter that inhibit adsorption from the liquid (Leister, 1996). Therefore, the preferred location for carbon adsorbers is on the cooled lean amine. Activated carbon systems usually include mechanical filters both upstream and downstream of the carbon bed. The upstream filter minimizes plugging of the bed by fine particles, and the downstream unit catches particles of carbon that may be released from the bed.

Solution reclaiming by distillation (thermal reclaiming) effectively removes all non-volatile species including particulate matter, heat-stable salts, and high molecular weight organic compounds. Unfortunately, it is not equally effective for all amines, and is most commonly used for MEA and DGA solutions, which can be vaporized at pressures slightly above atmospheric without decomposing. Secondary and tertiary amines generally require vacuum distillation to avoid serious degradation. Since reclaimers designed to operate under vacuum are appreciably more complex, they are seldom incorporated into plant systems. When vacuum reclaiming is employed, it is usually performed by an outside contractor.

Ion exchange and electrodialysis processes have been developed to purify solutions that cannot readily be reclaimed by distillation. However, both processes are capable of removing only ionized species (such as heat-stable salts) and are ineffective against non-ionized organic compounds such as many amine degradation products. Since foaming problems are more apt to be caused by organic compounds than by heat-stable salts, ion exchange and electrodialysis are not common remedies for foaming.

Detailed descriptions of filtration, activated carbon adsorption, thermal reclaiming, ion exchange, and electrodialysis processes are given in a subsequent section of this chapter entitled "Purification of Degraded Solutions."

Use of Antifoam Agents

Foaming can in many cases be controlled by the addition of foam inhibitors (commonly called antifoams). The most widely used foam inhibitors are either silicone compounds or high-boiling alcohols such as oleyl alcohol or octylphenoxyethanol. The silicones are commercially available either as water emulsions or in their pure form. In amine systems, silicones are generally preferable to the high-boiling alcohols.

Usually antifoams are added batchwise when needed. Ballard (1986A, B) states that typical batch dosage levels are in the range of 5 to 20 ppm; whereas, Meusburger and Segebrecht

(1980) recommend antifoam batch addition concentrations of 10 to 300 ppm. Manning and Thompson (1991) recommend the following antifoams for amine systems:

- Dow Corning DB-100 Antifoam Compound
- Dow Corning DB-31 Antifoam Emulsion
- Tretolite VEZ D-83
- Tretolite D-95
- Natco DF-971
- Exxon Corrects-It for chlorides
- Union Carbide SAG Antifoams (GT-101, 102, 301, or 302)

Figure 3-19 shows typical equipment for batch addition of an antifoam agent. Separate antifoam injection points should be provided for the amine contactor and for the regenerator. The amine regenerator is usually provided with an antifoam pot on the suction of the regenerator reflux pump, while the amine contactor is often protected with an antifoam pot on the suction of the lean amine pumps. However, if the carbon filter is downstream of the lean amine pumps, it may be necessary to locate the antifoam pot downstream of the carbon filter and rely on the amine flow control valve to mix the antifoam with the amine solution. Alternatively, a chemical feed pump can be used to pump the antifoam directly to these locations. Although antifoam addition is usually on a batch basis, it may be necessary in some circumstances to add antifoam continuously.

An antifoam that works well in one plant may be ineffective or even promote foaming in another. Also, an antifoam that has worked satisfactorily in the past may gradually lose its effectiveness and have to be replaced with another product. When time is available, the suitability of an antifoam should be tested before it is used. Antifoam test procedures are described by Meusberger and Segebrecht (1980), Pauley et al. (1988, 1989A, B), and Smith

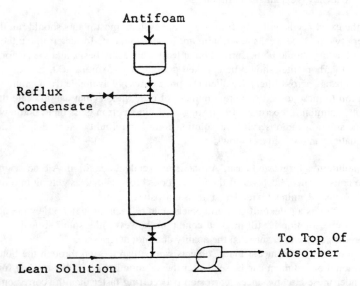

Figure 3-19. Pot for batch addition of antifoam agent. (*Meusberger and Segebrecht, 1980*)

(1979A, B), and are also available from amine vendors. Use of antifoam should not be regarded as a permanent solution to plant foaming problems. Once foaming is under control, the source of trouble should be investigated and the basic problem corrected.

Amine Solution Monitoring Program

Foaming of amine solutions and corrosion of amine units are best controlled by a comprehensive amine solution monitoring program. Pauley (1991) suggests that the following amine solution tests be performed by an outside laboratory:

- Organic acids (liquid chromatography)
- Heat-stable salts (ion exchange)
- Amine concentration (gas chromatography and alkalinity)
- Soluble metal content (atomic absorption)
- Iron sulfide content
- Liquid hydrocarbon content
- Water content

Bacon (1987) recommends the following tests:

- Lean and rich amine solution loadings
- Total heat-stable salts—free vs. total amine
- Anion assay—type and amount of anions
- Cation assay—type and amount of cations
- Soluble iron content of solution
- Gas chromatography—compare amine concentration against alkalinity. Identify and quantify degradation products
- Foaming tendency—extent and duration

While the exact needs will vary from system to system, most plants should run these tests about every two to three weeks at a minimum. In addition, tests for determining the lean and rich amine loading should be performed on at least a biweekly basis, and the performance of the activated carbon filter should be checked periodically (foaming test, turbidity, pressure drop, color change across filter). Detailed test procedures are available from amine vendors.

In addition to these tests, there are a number of observations that can indicate the quality of the amine solution. Lieberman (1980) suggests filling a clean glass quart bottle with unfiltered amine solution and observing the solution color and turbidity. According to Lieberman, solution quality can be rated as follows:

1. Amine solution is bright and clear. Amine is in excellent condition. A blue or green tinge is of no consequence; however, if the amine looks brownish, air is getting into the system and oxidizing the amine. Oxidized amine is corrosive.
2. Amine solution is a pale, dull gray and objects can be seen through bottle without difficulty. The amine solution is still in good condition; however, the solution quality should be monitored carefully to ensure that the quality does not deteriorate.
3. Amine solution is translucent black, objects can be barely seen through the bottle, and a small amount of sediment can be seen after the solution has been left standing for 10 minutes. Under these circumstances, corrosion is occurring faster than the corrosion products can be removed. The source of contamination (feed gas contaminants, high heat-stable

salts, inadequate amine solution regeneration, etc.) should be identified and corrective actions should be implemented.

4. Amine solution is opaque black, objects cannot be seen through the bottle, and substantial sediment has settled out of the solution after ten minutes. In this case, the amine system is probably becoming severely fouled with suspended solids and immediate corrective action is required.

CHEMICAL LOSSES

Volatility Losses

Amine volatility losses are usually not significant. DEA, DIPA, DGA, and MDEA all have very low vapor pressures at typical amine contactor and regenerator operating conditions. However, MEA has a substantially higher vapor pressure than other amines, and volatility losses in low pressure MEA contactors can be significant. In some cases, a water wash is provided for low pressure MEA contactors (See Chapter 2). **Figure 3-20** can be used to estimate amine volatility losses.

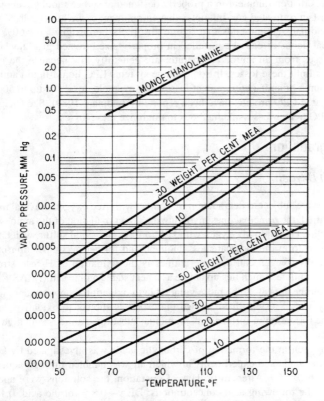

Figure 3-20. Vapor pressure of monoethanolamine and diethanolamine over aqueous solutions. *Chemetron Corporation and Texaco Chemical*

Entrainment

In many cases, the major cause of amine loss is entrainment (Veldman, 1989). Entrainment losses are caused either by inefficient mist extraction or by foaming and subsequent carry-over of solution. This problem can be minimized by the use of efficient mist elimination equipment and the application of foam inhibitors. Mist elimination equipment includes an efficient treated gas knockout drum downstream of the amine contactor as well as the installation of mist elimination equipment (chevron mist eliminators and wire and fiber mesh pads) inside the amine contactor or treated gas knockout drum. Pauley (1991) recommends fiber mesh pads rather than a wire mesh pad or chevron mist eliminator because fiber mesh pads can capture aerosol particles that are less than 3 microns in diameter. Wire and fiber mesh pads should, however, be used with caution in amine contactors treating gas streams that contain olefins or traces of oxygen because the mesh can gradually plug with elemental sulfur or polymers formed by the reaction of olefinic gas molecules with each other or with oxygen. In these circumstances, chevron mist eliminators should be considered because they are more resistant to plugging.

Entrainment losses from an amine absorber vary considerably depending on the mechanical design of both the upper section of the absorber and the mist elimination device. Veldman (1989) states that entrainment in a properly designed absorber should average less than 0.5 lb amine/MMscf of treated gas, but notes that entrainment of well over 3 lb/MMscf is not uncommon. Veldman recommends that amine absorbers be designed for 60% or less of flooding and that demister pads be installed in properly sized treated gas knockout drums. Entrainment losses from an amine regenerator are generally low because the regenerator water wash trays limit these losses. In some amine systems LPG liquid-liquid amine treaters are the major source of amine entrainment losses. For example, losses of up to 500 ppmw of amine in the treated LPG product have been reported (Veldman, 1989). See Chapter 2 for a discussion of LPG treater losses and methods to minimize these losses.

Solution Degradation

Reaction with Oxygen

Alkanolamines are subject to degradation by contact with free oxygen. Several mechanisms have been identified, the principal ones involving the direct oxidation of the amines to organic acids and the indirect reaction of oxygen with H_2S to form elemental sulfur, which then reacts with the amines to form dithiocarbamates, thiourea, and further decomposition products. A third route whereby oxygen can degrade amines is the oxidation of H_2S to stronger acid anions such as thiosulfate, which ties up amine as a heat stable amine salt (HSAS).

Monoethanolamine appears to be more vulnerable to oxidation than secondary and tertiary amines. Hofmeyer et al. (1956) have shown that MEA is subject to oxidative deamination that results in the formation of formic acid, ammonia, substituted amides, and high molecular weight polymers.

The role of oxygen in the formation of carboxylic acids was investigated by Blanc et al. (1982A, B). They conducted experiments in which air was bubbled through aqueous amine solutions at 194°F (90°C). After thirty days of operation, the solutions were analyzed and found to contain the following acid concentrations: DEA—0.8% formic acid, 0.15% oxalic acid, and 0.02% acetic acid; MDEA—0.3% formic acid; MEA—2.8% formic acid (the MDEA and MEA apparently were not analyzed for oxalic and acetic acids).

Blanc et al. (1982A, B) also determined that thiosulfate is formed by the reaction of H_2S and O_2 in the presence of amine. The experiment was performed with DEA solution in an autoclave at 194°F (90°C). The initial gas phase contained air at a partial pressure of 58 psia and H_2S at a partial pressure of 29 psia. After one hour the amine solution was found to contain 1.4 wt% thiosulfate ion. When the same experiment was performed with pure water instead of amine solution, only 200 ppm thiosulfate was formed. The same investigators analyzed solutions from operating DEA and MDEA plants for evidence of oxidation. They were able to identify acetic, propionic, formic, and oxalic acids as well as thiosulfuric acid. The concentrations of total amine salts of these acids in solution samples from nine plants ranged from 0.24 to 3%.

Oxygen incursion into the amine unit and degradation of the amine can be minimized by blanketing the amine sump and storage tank with nitrogen or fuel gas. It is also good operating practice to monitor the oxygen content of certain streams which can be contaminated with oxygen (e.g., FCC, Delayed Coker, and Vacuum Column fuel gas streams).

Irreversible Reaction with CO_2

Most of the commercial amines react in the presence of carbon dioxide to form degradation products. The degradation reactions of DEA involving CO_2 are catalyzed by (but do not consume) CO_2; while in other amine-CO_2 degradation reactions, CO_2 is consumed (Kim and Sartori, 1984; Kim, 1988). Degradation products can reduce amine solution absorption capacity, increase solution viscosity, increase solution foaming tendency, and in some cases contribute to amine plant corrosion.

MEA reacts with CO_2 to form a substituted imidazolidone that later hydrolyzes to produce a diamine and release the CO_2. See equations 3-21 through 3-24. Both the imidazolidone and the diamine degradation products can be removed from MEA solutions by thermal reclaiming (Polderman et al., 1955). CO_2 catalyzes a series of DEA degradation reactions that form a variety of degradation products, including high molecular weight polymers (Polderman and Steele, 1956; Kim and Sartori, 1984; Hsu and Kim, 1985). The degradation products of DEA and CO_2 are best removed by vacuum distillation.

Kim (1988) has shown that diisopropanolamine (DIPA) reacts with CO_2 to form an oxazolidone which, due to steric interference, is the sole degradation product of DIPA and CO_2. Diglycolamine (DGA) reacts with CO_2 to form a urea (Dingman, 1977). The reaction of DGA with CO_2 can be reversed by thermal reclaiming. Vacuum distillation is required to remove the degradation product of DIPA and CO_2. MDEA, which does not form a carbamate, is not degraded by carbon dioxide (Blanc et al., 1982A, B). Since a new carbon-nitrogen bond appears in all the amine-CO_2 degradation products, it is likely that carbamate formation is an essential step in these reactions. Those amines in which carbamate formation is inhibited by steric effects are probably resistant to CO_2 degradation.

The presence of CO_2-amine degradation products does not, in general, impair the absorption characteristics of the free amine contained in the solution as long as the concentration of free amine is kept at a constant value. However, as previously noted, accumulation of large amounts of amine degradation products results in increased treating solution viscosity and a consequent decrease of absorption efficiency. Also, several of the degradation products may be corrosive, particularly the polymeric degradation products of DEA, and could contribute to the general corrosivity of the solution.

The amine-CO_2 degradation reactions are relatively slow, but do occur at a significant rate under the conditions prevailing in the regeneration section of a purification plant. According to Meisen and Kennard (1982), the extent of these reactions can be limited by avoiding elevated temperatures. Reboiler heat flux should be limited, and amine circulation through the reboiler should be kept high. Amine regenerator operating temperatures may be limited by the need to minimize amine degradation (Polderman and Steele, 1956). Kim and Sartori (1984) have also shown that the degradation reactions of DEA depend on the CO_2 solution loading (i.e., the equilibrium CO_2 partial pressure above the rich solution, and the amine concentration), but are not affected by the presence of H_2S. This suggests that these reactions may impose additional limitations on the rich amine solution CO_2 loading and the amine solution concentration for MEA, DGA, DIPA, and DEA. MDEA is not affected by high CO_2 loadings because there are no CO_2-MDEA degradation reactions.

Irreversible Reaction of Monoethanolamine (MEA) with CO_2. Polderman et al. (1955A, B) first investigated the reaction of CO_2 with monoethanolamine to form amine degradation products. The reaction mechanism proposed by Polderman et al. (1955A, B) is essentially identical to that proposed later by Kim and Sartori (1984) for DEA and by Kim (1988) for DIPA. The reactions begin with the formation of the carbamate ion:

$$HOCH_2CH_2NH_2 + CO_2 = HOCH_2CH_2NHCO_2^- + H^+ \quad (3\text{-}21)$$

Monoethanolamine carbamate then condenses to form oxazolidone–2:

$$HOCH_2CH_2NHCO_2^- = \underset{\text{Oxazolidone-2}}{\underbrace{\begin{array}{c} CH_2 - CH_2 \\ | \quad\quad | \\ O \quad\quad NH \\ \diagdown \diagup \\ C \\ \| \\ O \end{array}}} + OH^- \quad (3\text{-}22)$$

Oxazolidone-2 then reacts with another molecule of monoethanolamine, yielding 1-(2-hydroxyethyl)-imidazolidone-2. Although this molecule contains organic nitrogen, it has no acid gas absorbing capacity and contributes no basicity to the solution (Polderman et al., 1955A, B).

$$\underset{\text{oxazolidone-2}}{\begin{array}{c} CH_2 - CH_2 \\ | \quad\quad | \\ O \quad\quad NH \\ \diagdown \diagup \\ C \\ \| \\ O \end{array}} + HOCH_2CH_2NH_2 = \underset{\text{1-(2-hydroxyethyl)-imidazolidone-2}}{\begin{array}{c} CH_2 - CH_2 \\ | \quad\quad | \\ HOCH_2CH_2N \quad NH \\ \diagdown \diagup \\ C \\ \| \\ O \end{array}} + H_2O \quad (3\text{-}23)$$

The substituted imidazolidone then hydrolyzes to CO_2 and N-(2-hydroxyethyl)-ethylenediamine, otherwise known as hydroxyethylenediamine or HEED.

$$\underset{\text{1-(2-hydroxyethyl)-imidazolidone-2}}{\text{HOCH}_2\text{CH}_2\text{N}\underset{\underset{\underset{\text{O}}{\|}}{\text{C}}}{\overset{\text{CH}_2\text{—CH}_2}{\diagup\hspace{-0.5em}\diagdown}}\text{NH}} + \text{H}_2\text{O} = \underset{\text{N-(2-hydroxyethyl)-ethylenediamine (HEED)}}{\text{HOCH}_2\text{CH}_2\text{NHCH}_2\text{CH}_2\text{NH}_2} + \text{CO}_2 \quad (3\text{-}24)$$

The hydrolysis of the substituted imidazolidone to the diamine releases the previously reacted CO_2 and restores part of the lost alkalinity and acid gas absorption capacity of the solution. However, because the ethylenediamine derivative is a stronger base than monoethanolamine, its sulfide and carbonate salts are more difficult to regenerate, and a significant portion of the diamine remains unregenerated. Although there are no published evaluations of the kinetics of these reactions, it appears that high regenerator operating temperatures, high equilibrium partial pressures of CO_2, and corresponding high solution loadings favor reactions 3-21 and 3-22, which in turn lead to increased HEED formation through reactions 3-23 and 3-24. Therefore, these contaminants do not significantly contribute to corrosion in properly maintained MEA solutions. In most circumstances, the formation of these MEA-CO_2 degradation products is limited and, if formed, they can be easily removed by side stream reclaiming (Polderman et al., 1955A, B).

CO_2 Catalyzed Degradation of DEA. Polderman and Steele (1956) first determined that irreversible reactions of DEA with CO_2 are responsible, in part, for DEA solution degradation. While Polderman and Steele identified some of the principal degradation products, Kim and Sartori (1984) first determined a reaction mechanism and derived a rate expression for the reaction kinetics that is consistent with experimental data. Their work demonstrates that CO_2 catalyzes a series of reactions that lead to the degradation of DEA. Kim and Sartori (1984) also proposed the existence of high molecular weight polymeric diethanolamine degradation products that were later characterized by Hsu and Kim (1985).

As shown by Kim and Sartori (1984), the irreversible reactions of diethanolamine with CO_2 are analogous to the reactions of monoethanolamine. The degradation reactions are probably initiated by the formation of the carbamate ion:

$$(\text{HOCH}_2\text{CH}_2)_2\text{NH} + \text{CO}_2 = (\text{HOCH}_2\text{CH}_2)_2\text{NCO}_2^- + \text{H}^+ \quad (3\text{-}25)$$

Diethanolamine carbamate then condenses to form 3-(2-hydroxyethyl)oxazolidone-2 (HEO):

$$(\text{HOCH}_2\text{CH}_2)_2\text{NCO}_2^- = \text{HOCH}_2\text{CH}_2\text{N}\underset{\underset{\underset{\text{O}}{\|}}{\text{C}}}{\overset{\text{CH}_2\text{—CH}_2}{\diagup\hspace{-0.5em}\diagdown}}\text{O} + \text{OH}^- \quad (3\text{-}26)$$

3-(2-hydroxyethyl) oxazolidone-2 (HEO)

HEO then reacts with another molecule of diethanolamine to release CO_2 and form N,N,N'-tris(2-hydroxyethyl)ethyldiamine (THEED):

$$HOCH_2CH_2N\underset{\underset{O}{\overset{\|}{C}}}{\overset{CH_2 - CH_2}{\diagdown \diagup}}O + (HOCH_2CH_2)_2 NH = (HOCH_2CH_2)_2 NCH_2CH_2NHCH_2CH_2OH + CO_2 \quad (3\text{-}27)$$

3-(2-hydroxyethyl) oxazolidone-2 (HEO)   N, N, N'-tris(2-hydroxyethyl) ethyldiamine (THEED)

Since the CO_2 that reacts in reaction 3-25 is released in reaction 3-27 by the formation of THEED, there is no net consumption of CO_2. Some of the THEED then slowly condenses with itself to form N,N'-bis(2-hydroxyethyl)piperazine (BHEP):

$$(HOCH_2CH_2)_2 NCH_2CH_2NHCH_2CH_2OH = HOCH_2CH_2N\overset{CH_2 - CH_2}{\underset{CH_2 - CH_2}{\diagup \diagdown}}N CH_2CH_2OH \quad (3\text{-}28)$$

N, N, N'-tris(2-hydroxyethyl) ethyldiamine (THEED)   N, N'-bis(2-hydroxyethyl)piperazine (BHEP)

N,N'-bis(2-hydroxyethyl)piperazine (BHEP), which has been identified in commercial diethanolamine solutions, is basic and capable of absorbing both H_2S and CO_2. Therefore, its formation results in only a partial loss of acid gas removing capacity.

Polderman and Steele (1956) were the first to identify 3-(2-hydroxyethyl)oxazolidone-2 (HEO) and N,N'-bis(2-hydroxyethyl)piperazine (BHEP). Hakka et al. (1968) first identified N,N,N'-tris(2-hydroxyethyl)ethyldiamine (THEED). These degradation products have also been identified by Kennard and Meisen (1980, 1983, 1985), Meisen and Kennard (1982), Kim and Sartori (1984), and by Blanc et al. (1982A, B).

Kim and Sartori (1984) note that HEO is the initial product of DEA degradation. THEED appears after an induction period and gradually becomes the major degradation product. The THEED concentration builds up to a maximum and then declines with time. BHEP forms after a long induction period. See **Figure 3-21** for details.

Material balances by Kim and Sartori (1984) indicated that a fourth degradation product, presumably a polymeric material, which was not detectable by gas chromatography, began to form about the same time as BHEP. These observations led Kim and Sartori to suggest that DEA degrades sequentially to HEO, THEED, BHEP, and finally to polymeric degradation products. These polymeric degradation products were identified in further studies by Hsu and Kim (1985).

Additional tests by Kim and Sartori (1984) demonstrated that these CO_2–DEA degradation reactions were independent of the H_2S concentration and were not initiated by thermal degradation of the amine at the temperatures of the experiment (120°C). They also established that the reaction kinetics were consistent with equations 3-25 through 3-28. Kim and Sartori also showed that the DEA degradation rate was roughly proportional to the CO_2 par-

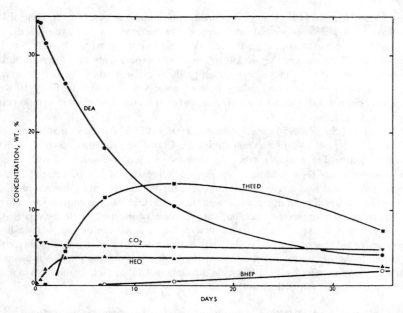

Figure 3-21. Rates and products of DEA degradation at 120°C. (*Kim and Sartori, 1984*)

tial pressure. Since there was no net consumption of CO_2, and since the presence of CO_2 was required for the reactions to take place at the conditions of the experiment (120°C), Kim and Sartori characterized CO_2 as a catalyst in DEA degradation reactions.

There are inconsistencies in the literature regarding the impact of DEA-CO_2 degradation products on amine plant corrosion. Blanc et al. (1982A, B) state that DEA-CO_2 degradation products do not contribute to amine plant corrosion. However, Chakma and Meisen (1986) demonstrate that DEA solutions contaminated with specific DEA-CO_2 degradation products are indeed corrosive, and cite the results of other investigators who confirm their results. What distinguishes these two investigations is that the corrosion experiments of Blanc et al. were performed in an H_2S atmosphere; whereas the experiments of Chakma and Meisen were performed in a CO_2 environment. Both investigators used synthetic solutions and, as noted by Chakma and Meisen (1986), the synthetic solutions of Blanc et al. (1982A, B) did not contain several of the major DEA degradation products. While the exclusion of major degradation products may have affected the results of Blanc et al., it is likely that the use of a CO_2 versus an H_2S environment also influenced the conclusions of Blanc et al. The tests of Chakma and Meisen may underestimate the corrosivity of degraded DEA solutions as the synthetic solutions used in their experiments did not contain the polymeric degradation products that have been identified by Hsu and Kim (1985). It is likely that these polymeric degradation products are excellent complexing agents for both ferrous ions, which are moderately insoluble in gas treating solutions, and for ferric ions, which are otherwise very insoluble in alkaline solutions. In fact, some degraded DEA solutions can dissolve up to 0.6 wt% ferric ion (McCullough, 1994). This suggests that degraded DEA solutions can dissolve passive ferric oxide films and, therefore, contribute to amine plant corrosion.

Meisen and Kennard (1982) suggest that the most important measures to control DEA degradation are to keep reboiler and heat exchanger temperatures at the lowest practical levels, to maximize circulation through the reboiler, and to operate with relatively low DEA concentrations. Furthermore, they found in laboratory experiments that filtration through activated charcoal did not remove degradation products of DEA and CO_2. This observation is contrary to the experience of other investigators, notably that reported by Perry. However, Perry (1974) did not distinguish CO_2-amine degradation products from other solution contaminants and degradation products.

Polderman and Steele (1956) and Polderman et al. (1955A, B) report that under certain conditions, the degradation of diethanolamine by CO_2 is significantly faster than that of monoethanolamine. For example, when an aqueous solution containing 20 wt% of either amine saturated with CO_2 was heated for 8 hours at a pressure of 250 psig and 259°F, 22% of the diethanolamine was converted, while practically no conversion of monoethanolamine was observed. In actual plant operation, lower rates of diethanolamine conversion have been observed compared to monoethanolamine. This seeming inconsistency with the reported laboratory results may be explained by the fact that the vapor-liquid equilibrium relationships of the systems, monoethanolamine-CO_2 and diethanolamine-CO_2, are such that in the reboiler (the point of highest temperature) of a plant using diethanolamine, the solution is essentially free of CO_2. On the other hand, appreciable amounts of CO_2 are left in monoethanolamine solutions at the same point of the plant.

Irreversible Reaction of Diisopropanolamine (DIPA) with CO_2. Kim (1988) demonstrated that the sole degradation product of DIPA and CO_2 is 3-(2-hydroxypropyl)-5-methyl-2-oxazolidone (HPMO) according to the following reaction, where diisopropanolamine carbamate condenses to form HPMO:

$$(CH_3CHCH_2)_2 \overset{OOH}{NCO_2^-} = CH_3\overset{OH}{CHCH_2}\underset{\underset{O}{\overset{\diagdown}{C}}\underset{\diagup}{\overset{\parallel}{}}O\underset{\diagup}{\overset{\diagdown}{}}\underset{CH_3}{\overset{\diagup}{CH}}}{-N-CH_2} + OH^- \qquad (3\text{-}29)$$

diisopropanolamine
carbamate (DIPA)

3-(2-hydroxypropyl)-5-methyl-2-oxazolidone (HPMO)

DIPA does not undergo further reaction to form polymeric degradation products as does DEA. Kim (1988) attributes the inhibition of these further reactions to the steric hindrance of the reaction of HPMO with DIPA. Although the degradation reactions of DIPA are limited to HPMO, DIPA degrades severely in the presence of CO_2. According to Butwell et al. (1982), operating plant DIPA solutions have HPMO concentrations up to 20%, and continuous reclaiming is required in CO_2 removal service. Since HPMO has no acid gas removal capacity, CO_2 degradation represents a serious loss of treating capacity.

Degradation of Methyldiethanolamine (MDEA) by CO_2. Blanc et al. (1982A, B) state that there are no MDEA-CO_2 degradation products. Since MDEA is a tertiary amine, it cannot form a carbamate ion and this may be the reason that it is not degraded by carbon dioxide.

Diglycolamine Reaction with CO_2. Diglycolamine (DGA) reacts with CO_2 to form a urea, N,N'-bis(hydroxyethoxyethyl) urea (BHEEU) (Kenney et al., 1994; Dingman, 1968;

1977). BHEEU has no acid gas removal properties; therefore, it reduces the solution treating capacity (Butwell et al., 1982). However, the presence of BHEEU does not, in general, impair the performance of DGA solutions as long as the BHEEU concentration is not allowed to increase to levels that affect the solution viscosity. Urea formation is probably due to the reaction of CO_2 with Diglycolamine to form a carbamate and the subsequent reaction of Diglycolamine carbamate with another Diglycolamine molecule to form BHEEU. See reactions 3-30 and 3-31. Note that in these reactions R denotes $HOCH_2CH_2OCH_2CH_2^-$:

$$RNH_2 + CO_2 = RNHCO_2^- + H^+ \tag{3-30}$$

$$RNHCO_2^- + RNH_2 = R-\overset{H}{\underset{|}{N}}-\overset{O}{\underset{\|}{C}}-\overset{H}{\underset{|}{N}}-R + OH \tag{3-31}$$

N,N'bis(hydroxyethoxyethyl) urea (BHEEU)

BHEEU formation is reversible at the temperatures used in a reclaimer, so that nearly all of the BHEEU is recoverable as DGA (Kenney et al., 1994; Dingman, 1977).

Irreversible Reactions with COS and CS_2

The reaction of carbonyl sulfide with alkanolamines, as applicable to gas-purification operations, has been studied by Pearce et al. (1961), Berlie et al. (1965), and Orbach and Selleck (1996). Pearce et al. found that the reactions between carbonyl sulfide and monoethanolamine are essentially analogous to those of CO_2 with monoethanolamine as shown in equations 3-21 through 3-23 except that they take place readily at ambient temperatures. In addition to oxazolidone and imidazolidone, the presence of N,N'-bis(hydroxyethyl) urea was reported. Orbach and Selleck conducted experiments in a continuous bench scale pilot plant simulating an absorption–regeneration cycle. When contacting essentially pure carbonyl sulfide with a 20 wt% monoethanolamine solution, they found rapid disappearance of alkalinity. A compound identified as 2–oxazolidone was isolated from the solution. The same experiment conducted with a 35 wt% diethanolamine solution indicated no loss of alkalinity with time, and no degradation product could be found in the solution. The results from this study are shown graphically in **Figure 3-22.** These data are in agreement with the finding of Pearce et al., who also report essentially no degradation of diethanolamine by carbonyl sulfide, both in laboratory experiments and in field tests.

In commercial plants all of the carbonyl sulfide present in the feed gas does not react with monoethanolamine. Pearce et al. (1961) and Berlie et al. (1965) report that the major portion of the COS undergoes hydrolysis, forming H_2S and CO_2, and only about 15 to 20% of the COS reacts irreversibly with the monethanolamine. The addition of strong alkalis, such as sodium carbonate or sodium hydroxide, to monoethanolamine solutions reduces losses due to reaction with COS substantially, probably by increasing the rate of COS hydrolysis (Pearce et al., 1961). Although monoethanolamine losses can be reduced by this method, use of diethanolamine is preferred if the gas to be treated contains appreciable amounts of COS.

According to Butwell et al. (1982), only about 2% of the COS in the feed gas to a typical DEA plant reacts irreversibly with the amine, while overall COS removal efficiencies of 70 to 80% are attainable, primarily by hydrolysis. DIPA is reported to react with COS even

240 Gas Purification

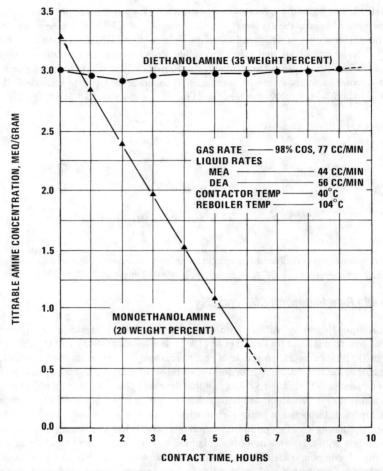

Figure 3-22. Effect of COS on mono- and diethanolamine. *Data of Orbach and Selleck (1996). Courtesy of Fluor Daniel*

more slowly than DEA (Danckwerts and Sharma, 1966). Aqueous DIPA plants are therefore considered suitable for treating gases containing COS.

DGA is a primary amine, and, like MEA, reacts rapidly with COS. However, the main reaction product can be regenerated under the conditions attained in a DGA solution reclaimer. As a result, DGA can be used to treat gases containing COS with only minor net amine loss. Since DGA plants can be operated with a high contactor temperature, which enhances COS hydrolysis, the overall COS removal can be quite high. Huval and van de Venne (1981) report COS removal efficiencies of 90% for DGA plants operating in Saudi Arabia. The kinetics of the COS-DGA reaction have been examined by Singh and Bullin (1988). The main product of the COS-DGA reaction is N,N'-bis(hydroxyethyl) urea (BHEEU), although some thiourea is also formed (Kenney et al., 1994).

Rahman et al. (1989) examined the reactions of COS with several amines. The protonated amine thiocarbamate was found in the reaction products of COS with MEA, DEA, and DGA. It was not found with DIPA, but this was attributed to a very slow rate for the COS-DIPA reaction. No evidence of reaction was found with the tertiary amines MDEA and dimethyethanolamine (DMEA), and it is generally assumed that tertiary amines do not react irreversibly with COS.

Carbon disulfide reacts with primary and secondary amines, first forming substituted dithiocarbamates and, subsequently, thiocarbamides (thioureas):

$$2(HOCH_2CH_2)NH + CS_2 = (HOCH_2CH_2)_2 \overset{\overset{S}{\|}}{N}CSH_2N(CH_2CH_2OH)_2 \quad (3\text{-}32)$$

diethanolamine   diethanolammonium N,N-bis-
                       (2-hydroxyethyl)dithiocarbamate

$$(HOCH_2CH_2)_2 \overset{\overset{S}{\|}}{N}CSH_2N(CH_2CH_2OH)_2 = (HOCH_2CH_2)_2 \overset{\overset{S}{\|}}{N}CN(CH_2CH_2OH)_2 + H_2S$$

diethanolammonium N,N-bis-   N,N,N',N'-tetrakis(2-hydroxyethyl)-   (3-33)
(2-hydroxyethyl)dithiocarbamate   thiocarbamide

A laboratory study of the reactions of carbon disulfide with monoethanolamine, diethanolamine, and diisopropanolamine has been reported by Osenton and Knight (1970). The results of this study indicate that all three amines form dithiocarbamic acid salts quite rapidly. While the salts formed from diethanolamine and diisopropanolamine appear to be thermally stable, the monoethanolamine salt is less stable and forms one mole of oxazolidone 2-thione for each mole of monoethanolamine. The end product of the CS_2-DGA reaction is reported to be bis(hydroxyethoxyethyl)thiourea (Kenney et al., 1994).

The MEA plant tests reported by Pearce et al. (1961) showed that, under the operating conditions investigated, more than half of the carbon disulfide was released as COS in the solution stripper, with the balance being hydrolyzed to H_2S and CO_2. This would indicate that the rate of reaction of monoethanolamine with carbon disulfide is slow and that amine losses resulting from this reaction are not serious.

Tertiary amines do not react with carbon disulfide.

Reactions with Other Gas Impurities

The irreversible reactions of amines with O_2, CO_2, COS, and CS_2 are discussed in the preceding sections. The principal other potentially reactive gas phase impurities which may enter the absorber are organic acids (such as formic or acetic acids), HCl, HCN, SO_2, NH_3, and mercaptans. All of the acidic compounds form salts with the alkaline amines. If the acids are stronger than CO_2 and H_2S, their amine salts are not efficiently decomposed under the stripping conditions (which are designed to decompose the amine-CO_2 and amine-H_2S salts), and they build up in the solution as heat-stable amine salts (HSAS). This, of course, lowers the capacity of the solution for CO_2 and H_2S, and can lead to other operational difficulties such as corrosion and foaming. Caustic soda or soda ash is often added to the solution to tie up the acids as sodium salts and release the amine; however, sodium salt build up is also undesirable, and an upper limit of 10 wt% is commonly set. Techniques for removing sodium salts from amine solutions are discussed in the next section.

Hydrogen cyanide, HCN, is also an acid, and as such, is absorbed in the alkaline amine solution. Unlike most other acids, however, it is weaker than CO_2, and is usually released from the amine solution in the stripper without degrading the amine. However, if both H_2S and oxygen are present, the HCN can react to form thiocyanate, which is the anion of a strong acid and forms a heat-stable salt.

Sulfur dioxide is a relatively strong acid and reacts rapidly with amines to form sulfite. However, the sulfite does not generally remain in solution as a heat-stable salt because of its high reactivity. For example, SO_2 in solution reacts rapidly with H_2S by the Claus reaction (which is catalyzed by liquid water) to form elemental sulfur and polysulfides. Zero valent sulfur can, in turn, disproportionate to form thiosulfate, which is a heat-stable anion. The sulfite may also be oxidized to sulfate which forms a heat-stable amine salt.

Ammonia is a common impurity in gases derived from coal. It is not an acid, but is readily absorbed by amine solutions because of its extremely high solubility in water. It does not react chemically with any of the amines, and is generally expelled in the stripper. However, its presence can cause operating problems such as corrosion in the stripper overhead system (see prior section on wet acid gas corrosion).

Mercaptans are related to H_2S. They are slightly acidic and are, therefore, absorbed to some extent by the alkaline amine solutions. Rahman et al. (1989) looked for possible irreversible reactions between mercaptans and MEA, DEA, DIPA, and MDEA and found none.

PURIFICATION OF DEGRADED SOLUTIONS

As previously discussed, amine solutions are degraded by reaction with CO_2, oxygen, organic sulfur compounds, and other gas impurities to form heat-stable salts and amine degradation products. These contaminants cause corrosion and lower the treating capacity of the amine solution. In commercial plant operations, the concentration of amine decomposition products should not be allowed to exceed about 10% of the active amine concentration. Once amine contamination has reached this level, contaminants should be removed by solution purging, solution change-out, or amine solution purification.

The operation of side-stream purification units makes it possible to maintain a constant concentration of active amine in the treating solution and prevent the accumulation of undesirable degradation products. Techniques used to purify amine solutions include distillation under vacuum or low pressure (thermal reclaiming), ion exchange, electrodialysis, and adsorption. Thermal reclaiming is the most commonly used process; however, both ion exchange and electrodialysis have become more common in recent years. Mechanical filtration and adsorption using activated carbon are very useful to control foaming; however, their application for solution purification is generally limited to the removal of solid particles (by filtration) and the removal of high-boiling or surface active organic compounds (by adsorption).

Table 3-6 provides a generalized comparison of mechanical filtration, activated carbon adsorption, thermal reclaiming, ion exchange, and electrodialysis for purifying amine solutions.

Amine Solution Mechanical Filtration

Suspended solids and surface-active contaminants can be removed from the treating solution by continuous mechanical and activated carbon filtration of a side stream. For large systems, Ball and Veldman (1991) recommend continuous filtration of a minimum of 10 to 15%

Table 3-6
Comparison of Amine Solution Purification Techniques

	Mechanical Filtration	Activated Carbon	Distillation (thermal reclaiming)	Ion Exchange	Electrodialysis
Application	Removal of particles and sludge	Removal of high mol. wt. and polar organics	Removal of all solids and nonvolatile species	Removal of ionized impurities	Removal of ionized impurities
Operating principle	Filtration	Adsorption (usually with filtration)	Vaporization of volatile species (amine, water, etc.)	Ions are captured by ion exchange resin	Ions are moved by electricity from amine to waste soln.
Limitations	Removes only particulate matter	Does not remove HSAS	Energy intensive. Some amines require vacuum	Does not remove nonionic species. Best for low salt conc.	Does not remove nonionic species. Best for moderate to high salt concs.
Waste products	Filter sludge, filter bags, and cartridges (hazardous)	Spent carbon and filter waste products (hazardous)	Reclaimer bottoms, containing salts, nonvolatile organics and some amine (hazardous)	Dilute aqueous stream containing removed ions plus excess regeneration and rinse water	Brine containing removed ions
Volume of wastes	Low	Low	Low	High	Moderate
Amine recovery	High	High	Moderate (about 85–95%)	High (about 99%)	High (about 98%)
Amine feed requirements	None (lean)	Prefiltered lean	HSAS neutralized	Cool, lean, hydrocarbon and particulate free	Cool, lean, hydrocarbon and particulate free; HSAS neutralized
Special requirements	None	None	Fuel gas or high temp. heat source	Regeneration chemicals	D.C. power supply
Operating mode	Semi-continuous; periodic clean-out required	Semi-continuous; periodic carbon replacement required	Batch or on-line; mobile units available	Semi-continuous, batch or on-line, mobile units available	Continuous on-line, or batch, mobile units available

Note: HSAS = Heat-stable amine salts.
Source: Data on Ion Exchange, Distillation, and Electrodialysis from Burns and Gregory (1995)

of the circulating solution. Most mechanical amine filters and the downstream activated carbon filter are located downstream of the lean amine pump and lean amine cooler. Usually a 10 to 20% slipstream of lean amine is filtered. See **Figure 3-1**. Filtering the cold, lean solution is preferred because, if the feed gas contains H_2S, filter maintenance is safer at this location because H_2S is not present in lethal concentrations. Also, the activated carbon filter is more effective on the lean side because contaminants are less soluble and more surface-active at lower temperatures (Pauley, 1991; Bacon, 1987).

However, in systems using DEA, MDEA, or MDEA mixtures to treat gases with low H_2S/CO_2 ratios, the most effective location for both the mechanical and activated carbon filter may be downstream of the rich amine flash tank and upstream of the lean/rich exchanger (Smith and Younger, 1972; Perry, 1974). With MDEA and DEA, amine degradation products cannot be removed by online slipstream reclaiming, and polymeric amine degradation products, which can chelate iron molecules, can build up in the amine solution. These amine degradation products can dissolve the protective iron sulfide film to form iron chelates which can, in turn, react with H_2S in the feed gas to form insoluble iron sulfide particles in the amine contactor. For MDEA and DEA solutions or mixed solutions containing MDEA and another amine, this can lead to plugging of the amine contactor and regenerator trays and other piping and equipment in contact with the rich amine solution (Smith and Younger, 1972; Perry, 1974). Placing the mechanical filter on the rich amine side minimizes plugging of the lean/rich exchanger and amine regenerator with iron sulfide particles.

Several types of mechanical filters have been used to remove solids and viscous liquids from alkanolamine solutions. These include string-wound cartridge, cotton sock, molded cellulose, pressed cloth, proprietary cartridge, bag, stacked paper, precoat, etched disk, and sintered metal fiber filters. Guidelines for selecting filters are provided by Ballard and von Phul (1991), Fabio (1992), and Pauley (1991). Mechanical filters are generally divided into two types: surface or depth type filters. Surface filters collect the particulate contaminants on the surface of the filter media. Depth filters trap particles inside their structure. Depth filters are usually used in relatively dirty services such as amine filtration as they have more dirt holding capacity than surface filters. Since depth filters trap suspended solids within the filter pure structure, they usually cannot be reused and must be disposed of when they are spent. Depth filters are either flexible (e.g., string-wound cartridges which deform as the differential pressure increases) or rigid (e.g., proprietary polypropylene cartridge filters). Rigid-depth filters have either a uniformly sized or a graded structure. The openings in graded structure depth filters become progressively smaller; whereas, depth filters with a uniformly sized structure have the same size openings throughout the filter media matrix (Ballard and von Phul, 1991). In amine service, depth filters with a graded structure usually have greater dirt holding capacity at a given micron rating than depth filters with a uniform structure.

Filter performance is usually quoted in terms of either a "nominal" or an "absolute" rating, e.g., 5 microns absolute or 5 microns nominal. Nominal ratings are assigned by the manufacturer and are not necessarily indicative of performance unless the manufacturer provides the removal efficiency in addition to the micron rating. For example, a filter with a nominal rating of 5 microns may capture only 30% of the particles that size, while another filter with a nominal 5 micron rating may capture 90%. Therefore, nominal ratings that do not include capture efficiency do not permit the comparison of one filter with another. Some filters have an absolute rating. An absolute rating is the diameter of the largest hard spherical particle that will pass through the filter. Most commercial amine filters do not have an absolute rating and even those filters that have an "absolute" or a "nominal" rating are not tested with an industry standard test. Given this lack of standardization and the variety of filter designs that are available, it is difficult to compare and select filters for amine service. However, most filters used in amine service have a nominal or absolute rating between 5 and 10 microns. Also, mechanical filtration of a 10 to 20% amine solution slipstream is usually provided both upstream and downstream of an activated carbon filter. The large mechanical filter upstream of the activated carbon filter is usually designed to remove smaller particles (e.g., 5 micron), and the smaller downstream filter is rated for larger particles, (e.g., 10 micron) (Calgon, 1986; Keaton and Bourke, 1983). Rating the downstream filter for a larger particle size ensures that particles bypassing the upstream filter will not plug the smaller downstream filter.

When the lack of an industry-wide filtration standard is coupled with a tendency to purchase filters on the basis of price, it is not surprising that there is a general tendency to undersize amine filters (Ballard and von Phul, 1991). A severely undersized filter that requires frequent filter element change out can seriously impact operator acceptance and lead to long-term operability and maintenance problems. Therefore, the designer is urged to size amine filters conservatively, as a filter that is too large is far superior to one that is too small. **Table 3-7** summarizes recommendations from various sources. The relative merits of various filter selections follow.

String-Wound Cotton Cartridge Filters

String-wound cotton cartridge filters are probably the most common filter used for amine solution filtration since they have lowest replacement cost. However, string-wound cotton cartridge filter changeouts may require considerable operating labor when these filters are

Table 3-7
Amine Mechanical Filter Sizing Criteria from Various Sources for Several Filter Types

Filter Type	Rating, micron(4)	Slipstream %	Max. ΔP, psi	Flux gpm/ft2(1)	Filter Life, months	Source
String-wound cotton	—	17%	—	0.65	0.25–2.0	Scheirman, 1973A
String-wound cotton	—	—	—	1–3	—	Fabio, 1992
String-wound polypropylene	5µN	5–100	30	1–3	—	Clark, 1996
Precoat, pressure leaf type	<1µN	12%	—	<1.0	0.25–0.50	Scheirman, 1973A
Precoat, vertical tubular	<1µN	—	—	<2.0	0.5–1.0	Scheirman, 1973A
Sintered metal fiber	10µA	10–15	30	4.2	2–8 hrs (3)	Clark, 1996
Etched disk	10µA	8–16	30	4.2 (2)	2–8 hrs (3)	Farrell, 1990; Clark, 1996

Notes:
1. Based on external surface.
2. Each element has 6 ft2 of external surface. There are a maximum of 7 elements (42 ft2) per housing. Each housing typically filters 175 gpm (4.2 × 42).
3. Typical time between automatic backwash cycles.
4. "A" denotes absolute rating. "N" denotes nominal rating.

used in amine systems with a circulation rate greater than 500 gpm. String-wound cotton cartridge filters also tend to deform when subjected to pressure and heat differentials and unload fines back into the amine solution. String-wound cotton cartridge filters made from recycled material should be avoided because they may contain oils and surfactants which can cause foaming (Pauley, 1991; Bacon, 1987). According to **Table 3-7,** conservative design would indicate that string-wound cotton cartridge filters with a nominal 5 to 10 micron rating be sized on the basis of 1 to 2 gpm/ft^2 of external surface. String-wound cotton cartridges used for amine service should be made from virgin cotton, be surfactant free, and contain no residual cottonseed particles.

Proprietary Cartridge Filters

Major filter vendors supply proprietary filter cartridges that are specifically designed for amine service. These filters are dimensionally stable under increasing differential pressure because they are made from extrudable polymers such as polypropylene. Proprietary cartridge filters often have an "absolute" rating because the thickness of the polymer can be carefully controlled, and they usually have higher capacities (in terms of gpm/ft^2 of external surface) than string-wound cartridge filters. Proprietary filter cartridges are usually made of specially engineered materials and often incorporate unique design features to increase dirt holding capacity and throughput. For example, the radial pleated cartridge filter depicted in **Figure 3-23** has a distinctive pleated design which provides 33 ft^2 of filter surface for every ft^2 of cartridge external surface. Due to these special design features, it is virtually impossible to compare one proprietary filter cartridge to another without plant test or operating data.

Pauley (1991) and Pauley et al. (1988, 1989A, B) review the merits of proprietary cartridge filters. Suggestions for specifying, selecting, operating, and maintaining cartridge filters intended for amine service are found in papers by Ballard and von Phul (1991) and Ballard (1986A, B). Factors that should be considered in selecting an amine filter include

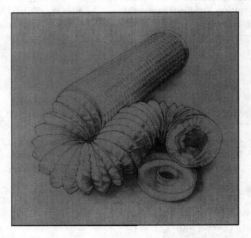

Figure 3-23. Proprietary meltblown polypropylene microfiber filter cartridge for amine service. Radial pleated design provides 33 ft^2 of filter surface per ft^2 of external cartridge surface. *Courtesy 3M Filter Products*

chemical resistance, dirt holding capacity, the optimum flow rate through the cartridge, the maximum allowable operating temperature, and the maximum allowable pressure differential before changeout is required.

Precoat Filters

Precoat filters in amine filtration service use a diatomaceous earth or perlite filteraid precoat (typically about 10 pounds of filteraid per 100 square feet of filter surface area) and a body feed of 1 pound of filteraid per pound of solids removed. They usually filter a 10% slipstream, and depending on the type of filter aid, can remove particles as small as 1 micron. Precoat filters were widely used in large amine systems (greater than 1000 gpm circulation) in the 1970s. However, precoat filters require significant operator attention, and amine losses can be substantial unless they are carefully operated. For these reasons, precoat filters are no longer widely used for amine filtration.

Two types of precoat filters (tubular and filter leaf) have been used for amine filtration. Typical sizing criteria and performance data are summarized in **Table 3-7**. A review of experience with several precoat filters in DEA plants located in Canada and the United States has been presented by Scheirman (1973A, B). Combined use of filter aid and activated carbon in a precoat filter of the pressure leaf type has been reported by Dailey (1970). Such a system removes suspended solids and undesirable dissolved compounds in one operation. However, Smith and Younger (1972) report that precoat filters using both activated carbon and filteraid are not as effective as a precoat filter combined with a dedicated activated carbon filter.

Bag Filters

Fabio (1992) discusses some of the advantages of bag filters for amine mechanical filtration. **Figure 3-24** depicts a proprietary bag filter. This filter has 35 separate filtration layers

Figure 3-24. Proprietary polypropylene bag filter for amine service. Each bag has multiple layers with internal bypasses which provide 32 ft^2 of filtration surface per bag. *Courtesy 3M Filter Products*

separated by internal bypasses which let unfiltered amine solution flow to the next layer. It is claimed that the multiple layers of filtration media provide 15 times more capacity than a string-wound cartridge and more than 7 times the capacity of a conventional bag filter (Anon., 1989). Since one bag filter is equivalent to 15 string-wound cartridge filters, considerable operating labor savings can be achieved during a filter element changeout. Ease of replacement can enhance operator acceptance, and disposal costs are also often less for bag filters due to their comparatively smaller volume.

Etched Disk and Sintered Metal Fiber Filters

Either cartridge or bag filters are usually acceptable for small to medium amine systems; however, larger systems (1,000 gpm or greater amine circulation or more than about 100 LT/day of sulfur production) may require considerable operating labor per filter element changeout. Operator acceptance of cartridges and bag filters in these larger systems is also often problematic. For these reasons, some operators of large systems have installed etched disk or sintered metal fiber filters. **Figure 3-25** depicts a fully automated etched disk filter in a large Canadian gas plant. This 350 gpm filter consists of two parallel filter housings each containing seven 25-gpm, 10-micron, etched disk elements. These totally automated systems require little or no operator attention.

Figure 3-25. 350-gpm automated Vacco Etched Disc amine filter in a Canadian gas processing plant. Two 175-gpm filter housings (7 elements per housing) are located on either side of the control panel. A solution decant tank is to concentrate filtered particulates located on the right. *Courtesy PTI Technologies, Newbury Park, CA*

Etched disk filter elements in amine service are usually rated for 10 microns absolute. Each 10 micron rated element has 6 ft² of external surface and, in amine service, fluxes range from 0.5 to 5 gpm/ft² depending on the amine solution particulate loading and size (Farrell, 1990; Clark, 1996). A typical flux is 4.2 gpm/ft² or 25 gpm of amine solution per element (Clark, 1996). An etched disk filter housing can contain as many as seven elements and can normally filter 175 gpm of amine solution. An etched disk filter element is depicted in **Figure 3-26a.** Amine solution flows through the elements, and particles collect on the ele-

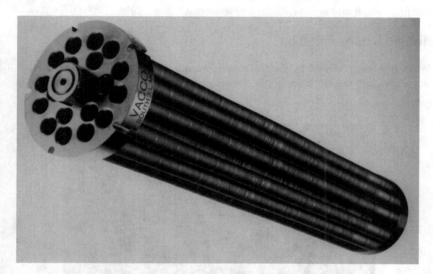

Figure 3-26a. Vacco Etched Disc element. *Courtesy PTI Technologies, Newbury Park, CA*

ment surface. The etched disk filter elements also trap a substantial fraction of smaller particles in the cake that forms on the element's outer surface. As the thickness of the particulate cake increases, the differential pressure across the filter rises. When the differential pressure reaches 30 psi, the filter is automatically backflushed. First, the amine is forced out of the filter housing by a water flush which reduces amine losses in the backflush cycle. After the water flush, the system is then automatically backflushed with high pressure (150 to 350 psig) nitrogen, steam, or preferably fuel gas. Sludge is collected in a decanter and removed periodically with a vacuum truck. The decanting drum minimizes amine losses by separating recoverable amine from the sludge particles. See **Figure 3-25.** Since particulates are collected on the surface of the element, the elements are relatively immune to plugging and have a 40-year design life. With element life this long, etched disk filters can be installed in rich amine service with minimal risk of operator exposure to lethal concentrations of H_2S. An etched disk filter for amine service (known as the Vacco Etched Disc filter) is manufactured by PTI Technologies, Inc. of Newbury Park, CA.

Sintered metal fiber filters are virtually identical in design and operation to etched disk filters. See **Figure 3-25** and the previous discussion. Sintered metal fiber filter elements consist of layers of metal fibers which have been mechanically compressed and then sintered together. Typical sintered metal fiber element life in amine service is 18 months (Clark,

250 Gas Purification

1996). The shorter life of sintered fiber elements in comparison to the etched disk element is compensated for by a significantly lower cost per element. Sintered metal fiber elements in amine service are typically rated at 10 microns absolute (Clark, 1996). Since these systems are completely automated, operator acceptance problems are avoided. A sintered metal fiber element housing for amine filtration service is depicted in **Figure 3-26b.** PTI Technologies of Newbury Park, CA, manufactures sintered metal fiber filters for amine service, while Pall Well Technologies, Inc. of East Hills, NY, makes similar automatic sintered metal powder and sintered metal fiber filters for amine service. Pall also markets an automatic backflush amine filter which uses non-metallic filter elements (Cathcart, 1996).

Activated Carbon Filtration

When foaming is caused by surface-active contaminants or dissolved or emulsified high molecular weight organic compounds, these agents may be eliminated by passage of the

Figure 3-26b. Sintered metal fiber elements for amine filtration. *Courtesy PTI Technologies, Newbury Park, CA*

solution through a bed of activated carbon as described by Fife (1932). Not all activated carbons are suitable for amine filtration. Bourke and Mazzoni (1989) recommend the use of a hard, fine mesh, steam-activated carbon having a broad range of pore sizes. They also recommend that the activated carbon have a low phosphorus content since leachable phosphates can cause foaming. According to Pauley (1991), activated carbon with an iodine number ranging from 900 to 1,100 is effective in removing amine degradation products, while carbons with a molasses number greater than 200 are more efficient in removing liquid hydrocarbons. (Both the iodine and molasses numbers are measures of the pore area available for adsorption.) Gustafson (1970) recommends a simple field test to judge the effectiveness of a given activated carbon in reducing foaming. Two quart jars are partially filled with amine solution, 1 wt% activated carbon is added to one jar, and both jars are then shaken vigorously. Comparison of the foam volume and duration of foaming indicates the degree of foam reduction that can be achieved with that activated carbon.

While it is generally agreed that activated carbon filtration should be considered for all amine units, it is highly recommended for secondary and tertiary amines (DEA, MDEA, and DIPA), which cannot be thermally reclaimed online like DGA and MEA. The preferred location for the activated carbon filter is on the lean side of the circulating system downstream of the lean amine cooler. The activated carbon is more effective on the lean side and, if H_2S is present, safety concerns are not an issue when the carbon is replaced. However, there are some circumstances when it may be advisable to locate the carbon filter on the rich side. See the previous discussion on mechanical amine filtration for details. In filtration systems where the carbon filter is located on the rich side there is a tendency for gas pockets to form inside the carbon filter (Leister, 1996). These gas pockets can reduce the effectiveness of the carbon filter and, in rich amine filtration systems where H_2S is present, care should be taken to continuously vent the carbon filter to a downstream location to prevent flash gases from being a safety hazard during filter changeouts as well as from accumulating and sealing off a portion of the filter (Bacon, 1987).

Table 3-8 summarizes activated carbon filter design data. As indicated in this table, a variety of activated carbons (8 × 30 mesh, 5 × 7 mesh, 4 × 10 mesh, etc.) derived from coal or wood have been used for amine filtration. Activated carbon filters using the finer carbon grades (8 × 30 mesh) are the most common type and are usually designed for a flux of 2 to 4 gpm/ft^2 with a 10 foot bed height (15 to 20 minutes contact time). Filters using the coarser carbon grades (5 × 7 mesh and 4 × 10 mesh) are designed for fluxes as high as 10 gpm/ft^2 and usually have a minimum bed depth of 5 feet. All of these carbon filters operate in a downflow mode and, in most cases, a 10 to 20% slipstream of lean amine is filtered. Small amine systems may use an activated carbon cartridge element.

Per **Table 3-8,** the recommended size of the slipstream filtered by activated carbon varies from 1% to 100% (full flow filtration). Early activated amine filters built in the 1970s were often designed for a 1 to 2% amine solution slipstream. Experience demonstrated that these filters were undersized, and most activated carbon filters are now designed for a 10 to 20% slipstream (Bourke and Mazzoni, 1989). However, in some applications, e.g., Claus plant tail gas treaters, activated carbon filters may be designed for full flow filtration. Typical activated carbon bed life is about 6 months to a year.

The activated carbon is replaced when the pressure drop across the bed exceeds design values, when foaming or turbidity tests indicate that the carbon is spent, or when there is no color change in the amine solution flowing through the filter. Recommended test procedures are available from the activated carbon vendors. Ballard (1986A, B) suggests that better activated carbon filtration is obtained when the amine temperature is in the range of 120 to

Table 3-8
Amine Activated Carbon Filter Sizing Criteria from Various Sources

Carbon Mesh Size	Amine % Slipstream	Flux, gpm/ft2 (1)	Carbon Bulk Density, lb/ft3	Carbon Type	Carbon Bed Contact Time min.	Carbon Bed Height, ft.	Source
8 × 30	>10	4	—	Calgon SGL	~20	10	Calgon, 1986
—	10–20	<4	—	—	>15	—	Bright and Leister, 1987
8 × 30	10	0.6(2)	30	Calgon SGL	120(2)	10	Keaton and Bourke, 1983
—	10–20	2–4	—	—	>15	—	Bourke and Mazzoni, 1989
8 × 30	10–20	<4	—	Coal Based	—	—	Manning and Thompson, 1991
—	>1	2–5	—	—	—	>10	Scheirman, 1973A
12 × 28	~2	0.5	—	—	—	>6	Gustafson, 1970
4 × 10	10–100	10	—	Lignite	—	5	Burns, 1996
4 × 10	10–100	<15	—	Wood	—	>5	Perry, 1974
5 × 7	2–100	2–10	16–18	—	20	—	Ballard, 1986A, B
5 × 7	10–25	—	30	Coal Based	—	—	Manning and Thompson, 1991

Notes:
1. Flux is the superficial velocity (gpm/ft2 of cross-sectional area) in the downflow direction.
2. Calgon recommended a design based on 4 gpm/ft2 flux and 20 minutes contact time, but client opted for more conservative design.

150°F. Below 120°F the amine solution viscosity may lead to a high pressure drop across the carbon bed; at temperatures above 150°F the adsorptive capacity of the carbon is reduced.

8 × 30 Mesh High-Density Activated Carbon Filter

Most activated carbon filters use a coal-based, 8 × 30 mesh activated carbon to treat a 10 to 20% amine solution slipstream. Bed heights are typically 10 feet with a flux of 2 to 4 gpm/ft^2 (contact time 15 to 20 minutes). Bright and Leister (1987) recommend that an 8 × 30 mesh activated carbon filter have a 5 micron mechanical prefilter to remove suspended solids from the amine solution and to prevent plugging of the activated carbon pores. The activated carbon filter should be followed by a 10 micron afterfilter to remove carbon fines which might exit the carbon filter. Many 8 × 30 mesh commercial activated carbon filters are designed more conservatively. For example, Keaton and Bourke (1983) review the design of an activated carbon system used to filter a 10% lean DEA solution slipstream. The 42 gpm slipstream of 25 wt% DEA solution is taken from the lean amine pump discharge. The activated carbon filter is a 10-ft diameter by 11-ft carbon steel vessel with a 316 stainless steel liquid collection system containing 20,000 pounds of Calgon SGL 8 × 30 mesh granular activated carbon. Surface loading is 0.6 gpm of amine per square foot and the carbon contact time is 120 minutes.

Figure 3-27a can be used to estimate the clean bed pressure drop for 8 × 30 mesh activated carbon. Since the 8 × 30 mesh activated carbon filter has an upstream mechanical filter,

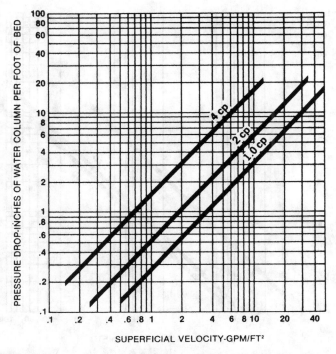

Figure 3-27a. Clean bed pressure drop of ethanolamine solution through Calgon SGL 8 × 30 mesh carbon. (*Calgon, 1986*)

254 Gas Purification

the carbon is usually spent before the maximum allowable pressure drop of 30 psi above the clean bed pressure drop is reached (Leister, 1996). Although steam regeneration has been used on occasion, in almost all cases the carbon beds are not regenerated since steam is not effective in removing high-boiling contaminants (Pauley, 1991). When steam regeneration is used, typical carbon bed life is 4 months after the first cycle, 7 weeks after the second cycle, and 3 to 4 weeks after the third cycle. Low pressure steam can be used for regeneration, but must be superheated to 500°F to remove high-boiling contaminants (Schierman, 1973A, B). Activated carbon filters are often supplied as skid mounted units by carbon vendors such as the Calgon Carbon Co. of Pittsburgh, PA, and, for larger systems, vendors will often assume responsibility for carbon replacement and disposal.

4 × 10 Mesh High-Density Activated Carbon Filter

The Perry Equipment Co. of Mineral Wells, TX, has marketed a 4 × 10 mesh activated carbon amine filter since the 1970s. Early versions of this filter utilized a wood-based, low-density activated carbon and were designed for a flux of 10 gpm/ft^2 with a minimum bed height of 5 feet (Perry, 1971; 1974; 1980). Recent designs utilize a 4 × 10 mesh, high-density, coal-based activated carbon (Burns, 1996). The filters are designed for a flux of 10 gpm/ft^2 and the bed depth is 5 feet. **Figure 3-27b** can be used to estimate the clean bed pressure drop for 4 × 10 mesh activated carbon. Usually the carbon is replaced when the differential pressure across the

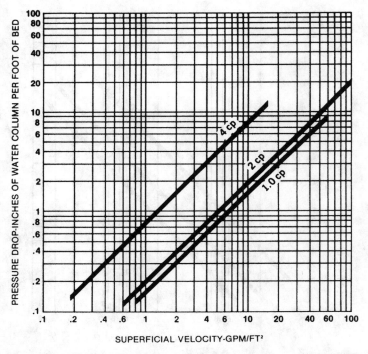

Figure 3-27b. Clean bed pressure drop of ethanolamine solution through Calgon LPA 4 x 10 mesh carbon. (*Calgon, 1986*)

bed exceeds 40 psi. Upstream and downstream mechanical filtration are recommended when the filter is used on a 10 to 20% slipstream. When the filter is used for full flow filtration, the carbon filter acts as a mechanical filter, and no mechanical prefilter is required.

Thermal Reclaiming

High-boiling degradation products of the amines, as well as suspended solids, can be removed from amine solutions by distillation of a small sidestream, usually 0.5 to 2% of the main flow. This stream is withdrawn from the solution leaving the regenerator reboiler and fed to a small steam-heated or direct fired kettle or reclaimer. For monoethanolamine and Diglycolamine, the reclaimer usually operates at the amine still pressure. **Figure 3-28** depicts a typical installation for reclaiming MEA. Reclaiming at the amine still-operating pressure allows the reclaimer vapor product to be used directly for reboiling the still and minimizes capital cost by eliminating the need for a separate condenser. Atmospheric distillation has been used for reclaiming MEA, but is less common than operation at the amine still pressure. Secondary and tertiary amines like DIPA, DEA, and MDEA are thermally reclaimed under vacuum because these amines degrade at the temperatures corresponding to atmospheric or higher pressure operation.

In the operation of thermal reclaimers, high-boiling organic compounds and salts concentrate in the reclaimer kettle. The boiling point of the mixture, and thus the kettle temperature, rises with increased concentration of these high boiling components. This temperature rise can cause an increased decomposition rate of both the amine and the high-boilers, resulting

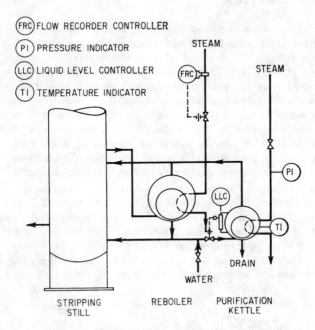

Figure 3-28. Purification system for aqueous monoethanolamine solutions operating at amine still pressure.

in additional amine loss, coking of heat transfer surfaces, and the formation of volatile decomposition products that can contaminate the amine vapor product. High salt concentration can also cause the precipitation of solid crystals that contribute to reclaimer fouling and plugging problems.

These considerations generally limit the degree of concentration attainable in the reclaimer bottoms product. Simmons (1991) recommends that the discharged residue contain sufficient amine to be a viscous liquid on cooling—about as viscous as crude oil. Amine remaining in the discharged material represents an unavoidable loss.

Techniques that have been developed to minimize problems associated with high residue concentration and high reboiler temperatures include 1) vacuum operation to reduce the amine boiling point, 2) use of stripping vapor (such as steam or dilution water) in the reclaimer kettle to reduce the amine partial pressure, and 3) addition of alkali to neutralize strong acids and liberate the bound amine and/or form inorganic salts that are molten at reclaimer temperatures.

MEA and DGA thermal reclaimers normally operate at atmospheric or at the treating plant still pressure and can easily be installed as permanent adjuncts to the treating unit. Thermal reclaimers for higher molecular weight amines normally require vacuum, resulting in a more complex and more costly installation. As a result, such units are frequently provided by a contract organization, either as a mobile unit brought in as required, or as a central processing plant. Contract plants, which process the amine from many treating units, can use a more sophisticated design, including the capability for continuous rather than batch operation.

Thermal Reclaiming of MEA

In plants using aqueous monoethanolamine solutions, purification is effected by semi-continuous distillation as shown in **Figure 3-28.** Blake and Rothert (1962), Blake (1963), Dow (1962), Hall and Polderman (1960), and Jefferson Chemicals (1963) provide guidelines for the operation and design of MEA reclaimers. In MEA reclaiming, sodium carbonate or hydroxide is added, if necessary, to liberate the amine from the heat-stable acid salts and to minimize corrosion. If carbon steel tubes are utilized in the reclaimer, Jefferson Chemicals (1963) recommends adding sodium carbonate to the kettle before commissioning the reclaimer to minimize corrosion. If the reclaimer has stainless tubes, low-chloride content sodium carbonate should be used to minimize the possibility of chloride stress corrosion cracking. A soda ash mix tank with a hard-piped connection to the reclaimer should be provided so that neutralizing solution can be added during the reclaiming cycle.

The amount of sodium carbonate or caustic required is based on the amine solution heat-stable salts concentration. One mole of amine tied up as a heat-stable salt (usually reported as the weight percent of bound amine based on the total solution) can be neutralized with one mole of caustic or one-half mole of sodium carbonate. Neutralizing the heat-stable salts with a strong base allows the bound amine to be liberated from the solution and raises the pH and reduces the corrosivity of the solution in the reclaimer. The reaction of sodium carbonate with heat-stable salts can be represented by equations 3-34 and 3-35 where R_3NH^+ represents the heat-stable salt cation:

$$CO_3^{2-} + R_3NH^+ = R_3N + HCO_3^- \qquad (3\text{-}34)$$

$$HCO_3^- + R_3NH^+ = R_3N + CO_2 + H_2O \qquad (3\text{-}35)$$

In the first reaction, the carbonate anion liberates one mole of amine per mole of carbonate anion. In the second, another mole of amine is liberated by the one mole of bicarbonate anion produced by the first reaction. One mole of sodium carbonate, therefore, neutralizes two moles of bound amine.

After the initial charging of the reclaimer kettle with amine solution, fresh solution diluted with water is fed at a slow rate on level control, and steam is added on flow control. At steady state, the amine solution is brought to a concentration where the equilibrium vapor leaving the reclaimer has the same amine content as the feed. As shown in **Figure 3-29,** if the solution fed to the purification kettle contains 20% (by weight) MEA in water, at atmospheric pressure the solution in the kettle has to be concentrated to an MEA content of 68% (by weight) in order to obtain an MEA concentration of 20% (by weight) in the vapor. At steady state, the reclaimer

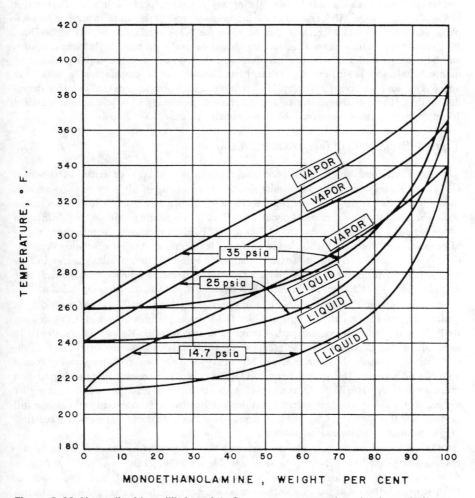

Figure 3-29. Vapor-liquid equilibrium data for aqueous monoethanolamine solution at various pressures. (*Dingman et al., 1966*)

operating temperature and amine concentration in the kettle can be varied by diluting the lean amine reclaimer feed with water. Water dilution is used to minimize MEA degradation and improve recoveries at high still operating pressures. Use of amine still reflux as dilution water has the advantage of maintaining the amine system water balance.

As the distillation continues, the temperature in the kettle increases because of the accumulation of solids and high-boiling constituents. As the temperature rises, the steam flow control valve opens, and the steam pressure in the kettle is gradually increased to a maximum of about 100 psig. The addition of lean amine feed is stopped when the temperature in the kettle reaches about 290°F. However, distillation is continued for a short time with the addition of water as liquid feed or as steam to remove as much of the residual MEA as economically possible. Additional caustic soda can be added to the reclaimer if tests indicate that more amine can be recovered from the still by neutralizing any remaining heat-stable salts. During this phase of operation, the MEA content of the kettle vapor product should be monitored by sampling, and steam or water injection stopped when the MEA content drops to 1 wt% (Dow, 1962). The kettle is then cleaned and recharged, and the cycle is repeated. This simple system can be used with aqueous monoethanolamine solutions because the vapor pressure of monoethanolamine is sufficiently high to permit distillation at temperatures at which the amine does not decompose thermally. High still operating pressures can lead to amine degradation. Dow (1962) recommends atmospheric pressure reclaiming when still pressures exceed 10 psig; however, most commercial MEA reclaimers operate at 15 to 20 psig.

Thermal Reclaiming of Diglycolamine (DGA)

Thermal reclaiming of Diglycolamine from degraded solutions, as described by Dingman (1977) and Kenney et al. (1994), is quite similar to reclaiming of MEA by semi-continuous steam distillation. A diagram of a DGA reclaimer is shown in **Figure 3-30**. The reclaimer should be sized for a sidestream of about 1 to 2% of the circulating solution, and distillation is conducted at a kettle temperature of 360 to 380°F. High reclaiming temperatures maximize reclaimer throughput, while lower operating temperatures minimize solution degradation. When steam is available, it is usually sparged in below the tube bundle. As seen in **Figure 3-31,** at a temperature of 360°F and at a pressure of 20 psia, the vapor from the reclaimer contains about 50 wt% Diglycolamine. This is returned directly to the stripping column.

DGA reacts with CO_2 to form bis(hydroxyethoxyethyl)urea (BHEEU) and with CS_2 to form bis(hydroxyethoxyethyl)thiourea. While DGA reacts preferentially with COS to form BHEEU, some bis-hydroxyethoxyethyl thiourea is also formed (Kenney et al., 1994). In most circumstances, BHEEU is the predominant Diglycolamine solution degradation product. Initially, as reported by Dingman and Moore (1968), DGA was reclaimed under vacuum with caustic addition. However, Mason and Griffith (1969) discovered that the degradation reactions to form BHEEU could be reversed without caustic addition by operation at higher temperature. Later it was demonstrated that the reclaimer could be operated at the amine still pressure and that the COS, CS_2, and CO_2 degradation reactions could be reversed at reclaiming temperatures without caustic addition.

DGA reclaimer control is based on reversing these reactions by maintaining reclaimer operation at a fixed temperature. As shown in **Figure 3-30,** steam, typically 300 psig, is added on flow control, lean DGA solution is added on level control, and condensate (or DGA stripper reflux) is added to maintain temperature control. Since the predominant degradation product is BHEEU, sodium carbonate or caustic soda is added to DGA reclaimers only when solution analysis indicates that heat-stable salts are present. Normally, caustic

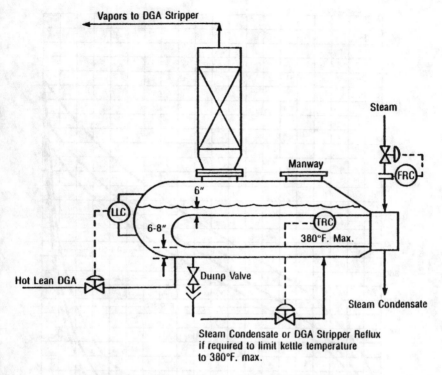

Figure 3-30. Diglycolamine reclaimer. (*Dingman, 1977*)

addition is not required for DGA reclaiming. The reclaimer should be run to keep the BHEEU concentration at a satisfactory level, typically less than 5 wt% of the total solution. Kenney et al. (1994) provide a detailed description of DGA reclaimer operation.

Mechanical Design of MEA and DGA Reclaimers

Design of DGA and MEA reclaimers is quite similar. Since there is little acid gas present in the solution, the kettle is usually constructed of carbon steel, although stainless steel tubes are often used. Square tube pitch with generous spacing is used for ease of cleaning and to keep vapor from sweeping liquid away from tube surfaces. Dow (1962) recommends that heat flux values for MEA reclaimers not exceed 5,000 Btu/ft^2/hr. DGA reclaimer design is based on a heat flux of 8,000 Btu/ft^2/hr for carbon steel tubes and 12,000 Btu/ft^2/hr for stainless steel. Kenney et al. (1994) recommend that the heat flux be less than 9,000 Btu/ft^2/hr for DGA reclaimers with carbon steel tubes.

According to **Figure 3-30,** the liquid level should be controlled so that the tubes are covered with at least 6 inches of liquid, and there should be a minimum of 6 to 8 inches of clearance between the reboiler bundle and the bottom of the kettle to provide settling space for sludge and to minimize corrosion of the bottom tubes. Others recommend greater clearances. Jefferson Chemicals (1963) recommends a bottoms clearance of 10 to 15 inches for MEA

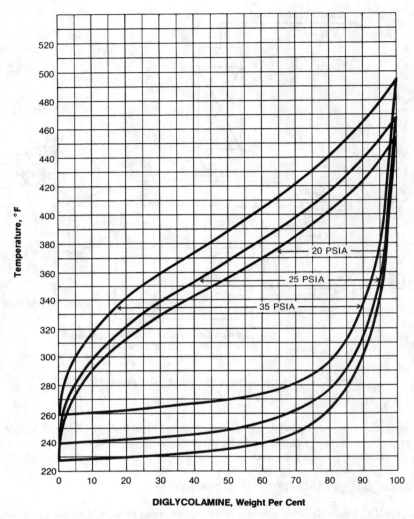

Figure 3-31. Vapor-liquid equilibrium data for aqueous Diglycolamine solutions at various pressures. (*Texaco Chemical, 1982*)

reclaimers, while Kenney et al. (1994) recommend 8 to 12 inches for DGA. Since the reclaimer sludge is quite viscous, the dump valve should be a minimum of 3 to 4 inches (pipe size) to prevent plugging of the drain line (Jefferson Chemicals, 1963; Kenney et al., 1994). In many jurisdictions, the reclaimer sludge is regarded as a hazardous waste. Providing a dedicated underground sump for reclaimer sludge storage can facilitate disposal and minimize operator exposure.

A manway should be provided at the top of the kettle to provide access so that the kettle can be hosed down at the end of the reclaiming cycle. To minimize corrosion, steam spargers should be designed with deflector plates so that the steam is directed away from the tube bundle and reclaimer shell (Jefferson Chemicals, 1963). Block valves should be provided on the vapor return line to the regenerator and on the reclaimer liquid feed lines so that the reclaimer can be isolated from the still for cleaning.

Entrainment must be minimized, or contaminated solution will return to the still. Disengagement space is provided inside the kettle to minimize entrainment. The rule of thumb is that the top of the reclaimer bundle should be at the kettle centerline. Another rule of thumb is that the reclaimer liquid residence time based on the flow of amine to the reclaimer should be 50 to 100 minutes (Blake and Rothert, 1962). Also, as shown in **Figure 3-30,** a short packed column on the reclaimer vapor outlet can be used to further minimize entrainment. Jefferson Chemicals (1963) recommends a 3- to 5-foot section of column filled with any common type of packing as shown in **Figure 3-30,** while Blake (1963) recommends a separate KO drum. A positive displacement type level transmitter should be used instead of a differential pressure level transmitter since changes in liquid gravity during reclaiming can affect the latter type and possibly expose reclaimer tubes and cause corrosion (Shaban et al., 1988).

Thermal Reclaiming of Secondary and Tertiary Amines

DEA, DIPA, and MDEA can be thermally reclaimed by distillation under reduced pressure. Operating conditions vary, but solution temperatures should be less than 400°F to prevent degradation of the amines and maximize recovery (Simmons, 1991). For example, typical operating conditions for thermal reclamation of DIPA are 50 to 100 mm Hg absolute at 350°F (Butwell et al., 1982).

When they are reclaimed, secondary and tertiary amine solutions are usually handled on a contract basis. Contract reclaiming can be either on-site, with a portable reclaimer, or off-site. The advantage of contract reclaiming is usually the use of a more sophisticated reclaiming unit operated by highly experienced operators. Contract reclaimers can be either batch or continuous. Continuous operation usually has the advantage of higher amine recovery since, with proper design, operating temperatures can be low. Millard and Beasley (1993) describe the design of a commercial amine reclaiming unit. Key design features are 1) continuous operation, 2) using waste liquid recycled through the tubes of a direct fired heater as the heat source for vaporizing the products (amine and water), 3) feeding fresh amine plant solution into the heated recycle liquid at a recycle-to-feed ratio of 40:1, 4) flashing amine and water vapor from the hot mixture, 5) discharging a portion of the remaining liquid as waste since required to maintain a constant inventory, 6) limiting the temperature rise of the liquid in the fired heater tubes to 25°F to minimize further decomposition, and 7) using a special fired heater design which limits heat transfer to the convective section.

A simple vacuum reclaimer used for reclaiming monoethanolamine-diethylene glycol solution is depicted in **Figure 3-32.** This installation consists of a small steam-heated or direct-fired kettle. Proper heat densities to avoid excessive skin temperatures have to be used in direct-fired kettles. The feed to the kettle is separated from the main solution stream at the reboiler outlet; overhead vapors are condensed, collected in an accumulator, and returned to the plant system. This vacuum reclaimer design, which was used in the now obsolete glycol-amine process, may not be suitable for those tertiary and secondary amines where the amine degradation products have nearly the same vapor pressure as the amine. In these instances, the reclaimer design is more complex. In the case of diisopropanolamine (DIPA), the principal degradation product, 3-(2-hydroxypropyl)-5-methyl-2-oxazolidone (HPMO), has nearly the same vapor pressure as DIPA. The reclaimer, which operates under vacuum (50 mm Hg,

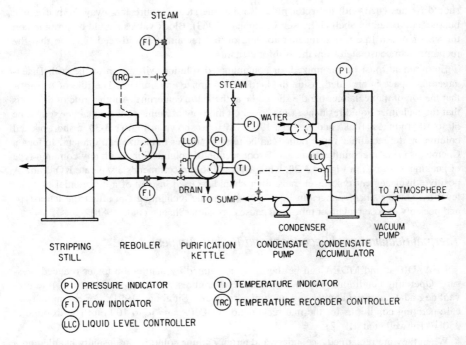

Figure 3-32. Monoethanolamine-diethylene glycol-water solution purification system at reduced pressure.

350°F), typically has 14 trays and utilizes direct steam sparging. There is no heat recovery, and the recovered DIPA is pumped back into the amine system (Butwell et al., 1982).

Electrodialysis and ion exchange compete with vacuum distillation when reclaiming MDEA, DEA, or DIPA solutions. However, vacuum reclaiming of these amines has the advantage of removing non-ionizable degradation products, such as those formed by the reaction of CO_2 with amines. These degradation products cannot be removed by either ion exchange or electrodialysis. **Table 3-6** summarizes the advantages and disadvantages of vacuum distillation in comparison to ion exchange and electrodialysis.

Miscellaneous Thermal Reclaiming Methods

A thermal reclaiming technique for purifying contaminated diethanolamine solutions that uses molten salt technology has been disclosed (Anon., 1963). This system has been developed for reclaiming solutions that treat refinery gases containing small amounts of acids. In order to maintain the activity of the diethanolamine solutions in such cases, continuous addition of sodium hydroxide is necessary to neutralize the amine-acid heat-stable salts. This results in a buildup of sodium salts, requiring periodic disposal of the solution. Adding potassium salts to the contaminated solution lowers the melting point of the salt mixture sufficiently to permit distillation of the diethanolamine at essentially atmospheric pressure without thermal decomposition or precipitation of salt crystals.

A schematic flow diagram of the unit is shown in **Figure 3-33**. A solution of 48 wt% potassium hydroxide is added to the contaminated diethanolamine before it enters the top of a packed tower where it is contacted with superheated steam rising from the reboiler. Essentially all of the amine and water is vaporized, and the product is withdrawn at a point below the packed section into a condenser, which is held at a slightly lower pressure than that prevailing in the reboiler. A steam jet is used to maintain the required pressure. The salts drop into the reboiler and melt at 420 to 440°F, and any residual diethanolamine is flashed off. The temperature of the diethanolamine-water vapor withdrawn is 300 to 375°F. The temperatures in the column and reboiler are sufficient to vaporize practically all the amine without decomposition. Heat is supplied by circulating oil at 600°F through the reboiler tubes. Recovery of 90% of reusable diethanolamine is claimed.

Another proposed thermal reclaiming method consists of withdrawing a small sidestream of the ethanolamine solution, adding sodium or potassium hydroxide, and then removing most of the water by distillation. After this, the solution, containing 5% to 10% of water, is mixed with an alcohol—preferably isopropyl alcohol—which causes precipitation of the sodium or potassium salts of volatile organic and inorganic acids. After filtration of these salts, the alcohol is removed from the mixture by distillation, and the amine is returned to the main solution stream (Paulsen et al., 1955).

Thomason (1985) described tests of a novel reclaiming technique in which DEA was reclaimed by sparging both CO_2 and low pressure steam into a reclaimer kettle that contains DEA. At 280°F and atmospheric pressure, reclaiming was successful, but part of the DEA was converted to TEA.

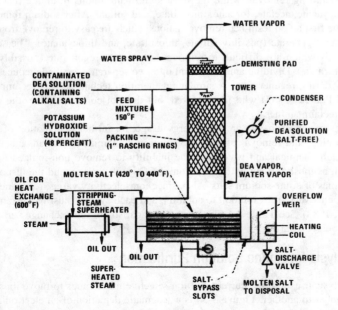

Figure 3-33. Diethanolamine reclaimer using molten salt bath.

Ion Exchange for Amine Solution Purification

Ion exchange is the reversible exchange of ions between a liquid and a solid in which there is no substantial change in the solid. Ion exchange removal of a typical heat-stable salt anion, Cl^-, from an amine by an anion exchange resin is illustrated in the following equation where parentheses denote ions bound to the ion exchange resin:

$$(OH^-) + R_3NH^+ + Cl^- = (Cl^-) + R_3NH + H_2O \tag{3-36}$$

The resin is then regenerated with caustic:

$$(Cl^-) + OH^- = (OH^-) + Cl^- \tag{3-37}$$

Similarly, undesirable cations (e.g., Na^+) can be removed by exchange with H^+ on a cation exchange resin. One of the first commercial applications of ion exchange for amine reclaiming was described by Morgan and Klare (1977). A side stream of the circulating amine solution, which was contaminated with sodium chloride, flowed downward through a bed of a strong base anion exchange resin and the chloride ion was replaced with hydroxide ions. The hydroxide ions reacted with CO_2 in the treating process, causing the precipitation of sodium bicarbonate crystals which were collected in the amine filter. A similar application is described by Bacon et al. (1986). In both of these applications, ion exchange was used as a temporary measure to clean up contaminated amine solutions.

Keller et al. (1992), Yan (1993), and Cummings et al. (1991) describe further developments in the use of ion exchange for amine reclaiming. This commercial technology, called the HSSX process, has a cation resin bed for removing sodium ions followed by two anion resin beds, each containing a different resin. According to Keller et al., sodium is usually the only cation requiring removal, while the most significant anions in amine reclaiming are chloride, thiocyanate, acetate, formate, thiosulfate, and sulfate. After sodium removal in the cation bed, the first anion resin bed removes anions which are easy to remove from solution, but difficult to regenerate (possibly sulfate, thiosulfate, and thiocyanate). The second bed removes anions which are difficult to remove, but easy to regenerate (possibly chloride, acetate, and formate). Dividing anion removal into two separate resin beds reduces chemical consumption during regeneration. A similar regeneration technique was recommended by Prielipp and Pearce (1957) who investigated, but did not commercialize, the use of ion exchange to reclaim amines.

Advantages of ion exchange are low energy usage, no further degradation of the amine during reclaiming, and removal of only the solution contaminants. Disadvantages include potentially high chemical and water use, the inability to remove non-ionized degradation products such as those formed by the reaction of CO_2 with amines, and the disposal of significant volumes of dilute sodium salts. Also, since ion selectivity varies, resin and regeneration requirements vary from solution to solution, and each system must be custom designed (Bacon et al., 1988). See **Table 3-6** for a further comparison of ion exchange versus vacuum distillation or electrodialysis.

Electrodialysis for Amine Solution Purification

Electrodialysis uses a direct current and ion-selective membranes to move ions from one solution chamber to another. **Figure 3-34** is a schematic depiction of an electrodialysis cell. The cell consists of a stack of alternating cation (C) and anion (A) selective membranes located between two electrodes. Commercial electrodialysis cells can contain 100 or more

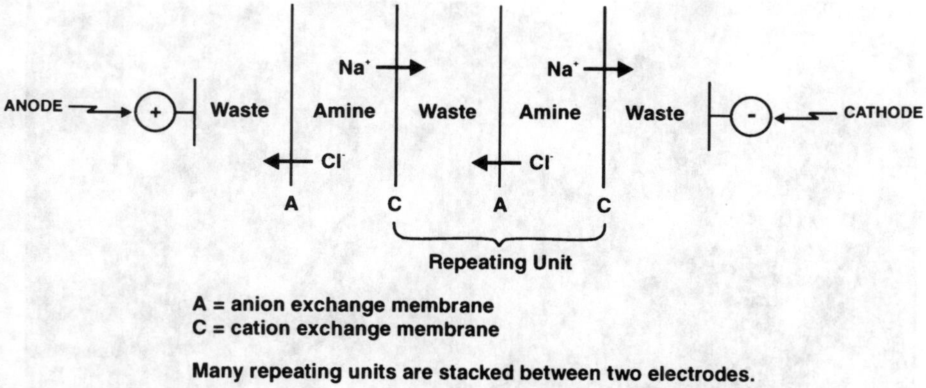

Figure 3-34. Schematic of electrodialysis cell for amine reclaiming.

A = anion exchange membrane
C = cation exchange membrane
Many repeating units are stacked between two electrodes.

stacked membranes. When a current is applied across the two electrodes, the cations move toward the negatively charged cathode, and the anions move toward the positively charged anode. Anions, such as Cl^-, can penetrate the anion exchange membrane, A, but not the cation exchange membrane, C. Cations, such as Na^+, can penetrate the cation exchange membrane, C, but not the anion exchange membrane. The net effect is to remove both cations and anions from the impure feed stream and collect them in a concentrated brine waste stream. Hydrogen and oxygen are produced at the anode and cathode, respectively. Stacking the membranes increases the ratio of the ions separated to the oxygen and hydrogen produced and lowers energy consumption.

Union Carbide has commercialized the use of electrodialysis to reclaim amines (Union Carbide, 1994; Gregory and Cohen, 1988; Burns and Gregory; 1995). A mobile, commercial unit is depicted in **Figures 3-35a** and **3-35b.** Commercial units reportedly process a 5- to 15-gpm slipstream of filtered (<10 micron), lean, cool amine which has been neutralized with caustic to liberate the amine tied up as heat-stable salts. It is claimed that a mobile unit can remove over 3000 lb/day of heat-stable salts.

Bacon et al. (1988) claim that electrodialysis requires less capital than ion exchange and that operating costs are competitive with ion exchange. Also, Bacon et al. state that electrodialysis produces a small volume of concentrated brine for disposal while ion exchange produces a high-volume, dilute salt solution as a purge stream. The disadvantage of electrodialysis (and ion exchange) in comparison to vacuum distillation is that non-ionic contaminants, such as amine-CO_2 degradation products, are not removed. In comparing ion exchange and electrodialysis, Keller et al. (1992) claim that electrodialysis requires higher electric power consumption to remove some anions, and that the salt purge stream may contain significant amounts of amine, which can adversely impact the waste water treatment system. They also claim that electrodialysis is most efficient when the heat-stable salt content of the amine is high and, therefore, when the amine solution is more corrosive. **Table 3-6** summarizes the advantages and disadvantages of electrodialysis in comparison to vacuum distillation and ion exchange.

NONACIDIC-GAS ENTRAINMENT IN SOLUTION

In certain operations, especially when acid gas removal is carried out at high pressure, appreciable amounts of nonacidic gases are carried by the solution from the contactor to the

266 Gas Purification

Figure 3-35a. Mobile electrodialysis unit for amine reclaiming. *Courtesy Union Carbide Corporation*

Figure 3-35b. Interior of mobile electrodialysis unit for amine reclaiming. *Courtesy Union Carbide Corporation*

regeneration section of the plant. This is particularly undesirable if the acid gases are intended to be used further, as for instance, for the production of dry ice or elemental sulfur. The nonacidic gas may be carried both in solution and as entrained bubbles (or as drops of liquid hydrocarbon). Mechanical carryover of nonacidic gases can be minimized by proper design of the bottom section of the contactor. Both splashing and the free fall of liquids through the vapor space (which result in froth in the bottom of the contactor) should be avoided by the installation of a properly designed downcomer, and the outlet line should include a vortex breaker. Even when the contactor bottom is properly designed, it is in most cases impossible to eliminate totally mechanical carryover. In addition, the solubility of most nonacidic gases in ethanolamine solutions becomes appreciable at high pressures. Provisions, therefore, must be made to separate these gases from the solution after it leaves the contactor and before it enters the regenerating section. Depending on the operating conditions of the plant and the purity requirements of the acid gas, a rich amine flash drum is often installed downstream of the rich solution level control valve and upstream of the lean-to-rich solution heat exchanger. To provide a maximum of vapor-disengaging area, horizontal vessels are frequently used.

The gases evolved in the rich amine flash drum contain acid gas in varying concentrations. This acid gas can be recovered by contacting the flashed gases with a small stream of lean amine solution in a small column installed at the top of the flash drum (see Chapter 2).

The solubility of methane in rich, aqueous MEA and glycol-monoethanolamine solutions is presented in **Figure 3-36** for two temperatures and varying pressures (Fluor Daniel, 1996).

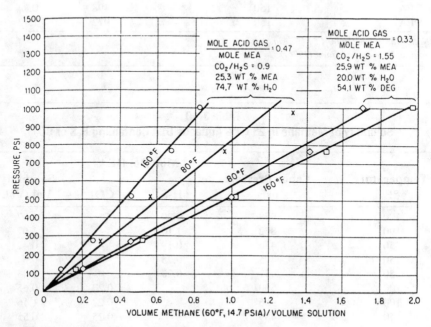

Figure 3-36. Solubility of methane in aqueous monoethanolamine and monoethanolamine-diethylene glycol-water solutions containing acid gas. *Courtesy Fluor Daniel*

Table 3-9
Solubility of Methane and Ethane in Aqueous MEA and DEA Solutions

Amine Type	wt% Amine	Temperature °F	Vapor Pressure-psia Methane	Vapor Pressure-psia Ethane	Solubility Vol./Vol.
MEA	15	100	954	—	1.55
MEA	15	150	498	—	0.73
MEA	15	150	993	—	1.37
MEA	15	100	—	491	0.92
MEA	15	100	—	868	1.19
MEA	15	150	—	501	0.74
MEA	15	150	—	955	1.18
DEA	25	100	510	—	0.76
DEA	25	100	968	—	1.42
DEA	25	150	511	—	0.68
DEA	25	150	982	—	1.28
DEA	25	200	516	—	0.61
DEA	25	200	920	—	1.19
DEA	25	100	—	480	0.91
DEA	25	100	—	868	1.19
DEA	25	150	—	498	0.74
DEA	25	150	—	972	1.13

Note: Gas volume at 60°F, 14.7 psia.
Source: Lawson and Garst (1976)

Table 3-10
Solubility of Methane in 25 wt% Aqueous DEA Containing H_2S or CO_2

Temperature °F	Vapor Pressure psia	Mol/Mol DEA H_2S	Mol/Mol DEA CO_2	Solubility Vol./Vol.
100	973	0.20	-	1.27
100	862	1.10	-	1.07
150	456	0.19	-	0.47
150	495	0.25	-	0.54
150	951	0.20	-	1.03
100	500	-	0.27	0.65
100	956	-	0.27	1.16
150	489	-	0.25	0.55
150	970	-	0.25	0.97

Note: Gas volume at 60°F, 14.7 psia.
Source: Fluor Daniel (1996)

This chart can be used in estimating the amount of methane (or dry natural gas) which will be dissolved in a treating solution and the quantity which will remain after an equilibrium flash separation. Data on the solubility of methane and ethane in aqueous monoethanolamine and diethanolamine solutions, either free of acid gas or partially saturated with H_2S or CO_2 were reported by Lawson and Garst (1976). Selected values from this study are shown in **Tables 3-9** and **3-10**. It should be noted that hydrocarbon gases are appreciably more soluble in amine solutions containing no dissolved acid gases than in solutions containing either one or both of the acid gases.

REFERENCES

Anon., 1963, *Chem. Eng.*, March 4, Vol. 70, pp. 40–41.

Anon., 1985, Minutes of NACE Group Committee Group T-8, March 27.

Anon., 1989, *What a Dirt Bag!* 3M Company Brochure 70-0702-2502-7.

API (American Petroleum Institute), 1990, "Avoiding Environmental Cracking in Amine Units," *API Recommended Practice 945, 1st Edition*, August, American Petroleum Institute, Washington, D.C.

Austgen, D. M., Rochelle, G. T., and Chen, C. C., 1991, "Model of Vapor-Liquid Equilibria for Aqueous Acid Gas-Alkanolamine Systems, 2. Representation of H_2S and CO_2 Solubility in Aqueous MDEA and CO_2 Solubility in Aqueous Mixtures of MDEA with MEA or DEA," *Ind. Eng. Chem. Res.*, Vol. 30, p. 543.

Bacon, T. R., 1987, "Amine Solution Quality Control Through Design, Operation, and Correction," *Proceedings 1987 Gas Conditioning Conference,* Univ. of Oklahoma, Norman, OK.

Bacon, T. R., Bedell, S. A., Niswander, R. H., Tsai, S. S., and Wolcott, R. A., 1988, "New Developments in Non-Thermal Reclaiming of Amines," *Laurance Reid Gas Conditioning Conference Proceedings,* Univ. of Oklahoma, Norman, OK.

Bacon, T. R., Krohn, J. V., Lewno, J. A., and Wolcott, R. A., 1986, "Alternative Economic Solutions for Amine Reclaiming," paper presented at the Sixty-Fifth Gas Processors Association, March 10–12, San Antonio, TX.

Bagdasarian, A. J., Shargay, C. A., and Coombs, J. W., 1991, "Stress Corrosion Cracking of Carbon Steel in DEA and 'ADIP' Solutions," *Materials Performance,* May, pp. 63–67.

Ball, T., and Veldman, R., 1991, "Improve Gas Treating," *Chem. Eng. Prog.*, January 4, pp. 67–72.

Ballard, D., 1966, "How to Operate an Amine Plant," *Hydro. Process.*, Vol. 45, No. 4 (April), pp. 137–144.

Ballard, D., 1986A, "Cut Energy/Chemical/Corrosion Costs in Amine Units," *Proc. of The Sixty-Fifth Annual Convention (Gas Processors Assoc.),* Tulsa, OK, pp. 223–236.

Ballard, D., 1986B, "Techniques to Cut Energy/Corrosion/Chemical Costs in Amine Units," *Proceedings 1986 Gas Conditioning Conference,* Univ. of Oklahoma, Norman, OK.

Ballard, D., and von Phul, S. A., 1991, "Cut Filtration Costs by 80%," *Chem. Eng. Prog.*, May, pp. 65–68.

Berlie, E. M., Estep, J. W., and Ronicker, F. J., 1965, *Chem. Eng. Prog.*, Vol. 61, April, pp. 82–85.

Berry, W. E., 1982, "Effect of CO_2 on Corrosion of Line Pipe," paper presented at the American Gas Assoc. Transmission Conference, Chicago, IL, May 17–19.

Blake, R. J., 1963, "Why Reclaim Monoethanolamine Solutions?" *Oil and Gas J.*, Sept. 9, pp. 130–134.

Blake, R. J. and Rothert, K. C., 1962, "Reclaiming Monoethanolamine Solutions," *Gas Conditioning Conference Proceedings,* Univ. of Oklahoma, Norman, OK.

Blanc, C., Grall, M., and Demarais, G., 1982A, "The Part Played by Degradation Products in the Corrosion of Gas Sweetening Plants Using DEA and MDEA," *Laurance Reid Gas Conditioning Conference Proceedings,* Univ. of Oklahoma, Norman, OK.

Blanc C., Grall, M., and Demarais, G., 1982B, "Amine-degradation products play no part in corrosion at gas sweetening plants," *Oil and Gas J.*, Nov. 15, pp. 128–129.

Bourke, M. J., and Mazzoni, A. F., 1989, "The Role of Activated Carbon in Gas Conditioning," *Laurance Reid Gas Conditioning Conference Proceedings,* Univ. of Oklahoma, Norman, OK, pp. 137–158.

Bright R. L., and Leister, D. A., 1987, "Gas treaters need clean amines," *Hydro. Process.*, December, pp. 157–158.

Buchheim, G. M., 1990, "Ways to deal with wet H_2S cracking revealed by study," *Oil and Gas J.*, July 9, pp. 92–96.

Bucklin, R. W., and Mackey, J. D., 1983, "Sulfur, Pollution and Corrosion Management in a Modern Refinery," paper presented at the 1983 AIChE Summer National Meeting in Denver, CO.

Burns, D., 1996, Perry Equipment Corporation, personal communication, May 16.

Burns, D., and Gregory, R. A., 1995, "The UCARSEPR® Process For Online Removal of Non-Regenerable Salts From Amine Units," *Laurance Reid Gas Conditioning Conference Proceedings, Univ. of Oklahoma,* Norman, OK.

Butwell, K. F., 1968, "How to Maintain Effective MEA Solutions," *Hydro. Process.*, Vol. 47, No. 4, pp. 111–113.

Butwell, K. F., Kubek, D. J., and Sigmund, P. W., 1982, "Alkanolamine Treating," *Hydro. Process.*, March, pp. 108–116.

Calgon Carbon Corporation, 1986, *Purification of Amines With Granular Activated Carbon,* Sales Brochure 23-61d.

Campbell, J. M., 1981, *Gas Conditioning and Processing,* 4th Edition, Campbell Petroleum Series, Inc., Norman, OK.

Carlson, E. C., Davis, G. R., and Huysak, K. L., 1952, "Corrosion in CO_2-H_2S—Amine System," *Chem. Eng. Prog.*, Vol. 48, No. 7, pp. 333–336.

Carter, W. A. P., Rodgers, P., and Morris, L., 1977, "Gas Desulphurization by the Stretford Process and the Development of a Process to Treat Stretford Effluent," *Proc. of the Symposium on Treatment of Coke-Oven Gas,* McMaster Univ., Hamilton, Ontario, Canada, May 26.

Cathcart, N., 1996, Pall Advanced Separations Systems, personal communication, May 17.

Cayard, M. S., Kane, R. D., Horvath, R. J., Prager, M., 1994, "Large-Scale Wet H_2S Performance: Evaluation of Mechanical, Metallurgical and Welding Variables," Second International Conference on Interaction of Steels with Hydrogen in Petroleum Industry Pressure Vessel and Pipeline Service, Vienna, Austria, October.

Chakma, A., and Meisen, A., 1986, "Corrosivity of Diethanolamine Solutions and Their Degradation Products," *Ind. Eng. Chem. Prod. Res. Dev.*, Vol. 25, No. 4, pp. 627–630.

Clark, R. C., 1996, PTI Technologies Inc., personal communication, April 25.

Connors, J. S., 1958, "Aqueous-amine acid-removal process needn't be corrosive," *Oil and Gas J.*, March 3, pp. 100–110.

Cummings, A. L., Veatch, F. C., Keller, A. E., Thompson, J. C., and Severson, R. A., 1991, "Process for Monitoring and Controlling an Alkanolamine Reaction Process," US Patent No. 5,162,084.

Dailey, L. W., 1970, "Status of SNPA-DEA," *Oil and Gas J.*, May 4, pp. 136–141.

Dancwerts, P. V., and Sharma, M. M., 1966, "The Absorption of Carbon Dioxide Into Solutions of Alkalies and Amines," *Chem. Eng.*, October, pp. 244–280.

de Waard, C., and Lotz, U., 1993, "Prediction of CO_2 Corrosion of Carbon Steel," Paper No. 69 presented at Corrosion/93, NACE.

Dingman, J. C., 1968, "Gas Sweetening with Diglycolamine Agent," *Laurance Reid Gas Conditioning Conference Proceedings,* Univ. of Oklahoma, Norman, OK.

Dingman, J. C., 1977, "Gas Sweetening with Diglycolamine Agent," paper presented at Third Iranian Congress of Chemical Engineering, Shiraz, Iran, Nov. 6–10.

Dingman, J. C., Allen, D. L., and Moore, T. F., 1966, "Minimize Corrosion in MEA Units," *Hydro. Process.*, Vol. 45, No. 9, pp. 285–290.

Dingman, J. C. and Moore, T. F., 1968, "Gas Sweetening With Diglycolamine," *Gas Conditioning Conference Proceedings,* Univ. of Oklahoma, Norman, OK.

Donovan, J. J., and Laroche, K. J., 1981, "Two Approaches to Effluent Treatment of a Coke-Oven Gas Desulphurization Process," paper presented at 64th Chemical Conference and Exhibition, Chemical Institute of Canada, May 31–June 3.

Dow, 1962, *Gas Conditioning Fact Book,* The Dow Chemical Co., Midland, MI.

Dupart, M. S., Bacon, T. R., and Edwards, D. J., 1993A, "Part 1—Understanding corrosion in alkanolamine gas treating plants," *Hydro. Process.*, April, pp. 75–80.

Dupart, M. S., Bacon, T. R., and Edwards, D. J., 1993B, "Part 2—Understanding corrosion in alkanolamine gas treating plants," *Hydro. Process.*, May, pp. 89–94.

Ehmke, E. F., 1960, "Hydrogen Diffusion Corrosion Problems In a Fluid Catalytic Cracker and Gas Plant," *Corrosion,* May, pp. 116–122.

Ehmke, E. F., 1981A, "Polysulfide stops FCCU corrosion," *Hydro. Proc.*, July, pp. 149–155.

Ehmke, E. F., 1981B, "Use Ammonium Polysulfide to Stop Corrosion and Hydrogen Blistering," paper #59, NACE, Corrosion/81, Ontario, Canada, April 6–10.

Fabio, D. G., 1992, "The 3M Bag Filter—The Cost Cutting Problem Solver," *Proceedings 1992 Laurance Reid Gas Conditioning Conference,* Univ. of Oklahoma, Norman, OK.

Farrell, L. T., 1990, "Summary Report on Amine Filtration," Vacco Industrial Filter Systems (now PTI Technologies, Newbury Park, CA) Report 97343-66, Jan. 25.

Fife, H. R., 1932, U.S. Patent 1,944,122.

Fitzgerald, K. J., and Richardson, J. A., 1966A, "How Gas Composition Affects Treating Process Selection," *Hydro. Process.*, Vol. 45, No. 7, pp. 125–129.

Fitzgerald, K. J., and Richardson, J. A., 1966B, "New correlations enhance value of monoethanolamine process," *Oil and Gas J.*, Oct. 24, pp. 110–118.

Fluor Daniel, 1996, unpublished data.

Fochtman, E. G., Langdon, W.M., and Howard, D. R., 1963, "Continuous Corrosion Measurements," *Chem. Eng.*, Vol. 70, No. 2, pp. 140–142.

Fontana, M. G., and Greene, N. D., 1967, *Corrosion Engineering,* McGraw-Hill, NY, NY.

Froning, H. R., and Jones, J. H., 1958, "Corrosion of Mild Steel in Aqueous Monoethanolamine," *Ind. & Eng. Chem.*, Vol. 50, No. 12, December, pp. 1737–1738.

Garwood, G. L., 1953, "What to do about Amine Stress Corrosion," *Oil and Gas J.*, July 27, 1953, pp. 334–340.

Graff, R. A., 1959, "Corrosion in Amine Type Gas Processing Units," *Ref. Eng.*, March, pp. C-12–C-14.

Gregory, R. A., and Cohen, M. F., 1988, "Removal Of Salts from Aqueous Alkanolamine Using An Electrodialysis Cell With Ion Exchange Membrane," European Patent No. 286,143.

Gustafson, K. J., 1970, "Conditioning Gas-Treating Liquids With Activated Carbon," *Proceedings 1970 Gas Conditioning Conference,* Univ. of Oklahoma, Norman, OK.

Gutzeit, J., 1986, "Refinery Corrosion Overview," in *Process Industries Corrosion—The Theory and Practice,* Moniz, B. J. and Pollock, W. I, editors, National Assoc. of Corrosion Engineers, Houston, TX, p. 184.

Gutzeit, J., 1994, consultant, personal communication.

Gutzeit, J., 1990, "Cracking of Carbon Steel Components in Amine Service," *Materials Performance,* September, pp. 54–57.

Gutzeit, J., and Johnson, J. M, 1986, "Stress corrosion cracking of carbon steel welds in amine service," *Materials Performance,* Vol. 25, No. 7, July, pp. 18–26.

Hakka, L. E., Singh, K. P., Bata, G. L., Testart, A. C., and Andrejchyshyn, W. M., 1968, "Some Aspects of Diethanolamine Degradation in Gas Sweetening," Paper presented at Canadian Natural Gas Processing Association meeting, May 9.

Hall, G. D., and Polderman, L. D., 1960, "Design and operating tips for ethanolamine gas scrubbing systems," *Chem. Eng. Prog.*, Vol. 56, No. 10, pp. 52–58.

Hay, M. G., Baron, J. J., and Moffat, T. A., 1996, "Elastomer Induced Crevice Corrosion and Stress Corrosion Cracking of Stainless Steel Heat Exchanger Plates in Sour Amine Service," paper presented at Corrosion 96, The NACE International Annual Conference and Exposition, Paper No. 389, Denver, CO, March 24–29.

Hofmeyer, B. G., Scholten, H. G., and Lloyd, W. G., 1956, "Contamination and Corrosion in Monoethanolamine Gas Treating Solutions," paper presented at National Meeting of American Chemical Society, Dallas, TX, April 8–13.

Hsu, C. S., and Kim, C. J., 1985, "Diethanolamine (DEA) Degradation under Gas-Treating Conditions," *Ind. Eng. Chem Prod. Res. Dev,* Vol. 24, pp. 630–635.

Hughes, P. G., 1982, "Stress Corrosion Cracking in an M.E.A. Unit," *Proceedings UK National Corrosion Conference,* November, pp. 87–91.

Huval, M., and van de Venne, H., 1981, "DGA Proves out as a Low Pressure Gas Sweetener in Saudi Arabia," Oil and Gas J., Aug. 17, p. 91.

Jefferson Chemicals, 1963, "Monoethanolamine Reclaiming Part II, Design Considerations," *Hydro. Process. and Refinery,* Vol. 42, No. 10, p. 225.

Jefferson Chemical Company, Inc., 1969, *Gas Treating Data Book.*

Keaton, M. M., and Bourke, M. J., 1983, "Activated carbon system cuts foaming and amine losses," *Hydro. Process,* Vol. 62, No. 8, August, p. 71.

Keller, A. E., Kammiller, R. M., Veatch, F. C., Cummings, A. L., and Thompsen, J. C., 1992, "Heat-Stable Salt Removal from Amines by the HSSX Process Using Ion Exchange," *Laurance Reid Gas Conditioning Conference Proceedings,* Univ. of Oklahoma, Norman, OK.

Kelley, A. E., and Poll, H. F., 1953, "Double-duty Gas Plant," *Petro. Proc.*, January, pp. 55–59.

Kennard, M. L., and Meisen, A., 1980, "Control DEA Degradation," *Hydro. Process.*, Vol. 59, No. 4, April, pp. 103–106.

Kennard, M. L., and Meisen, A., 1983, "Gas Chromatographic Technique for Analyzing Partially Degraded Diethanolamine Solutions," *J. of Chromatography,* Vol. 267, pp. 373–380.

Kennard, M. L., and Meisen, A., 1985, "Mechanisms and Kinetics of Diethanolamine Corrosion," *Ind. Eng. Chem Fundam.*, Vol. 24, pp. 129–140.

Kenney, T. J., Khan, A. R., Holub, P. E., and Street, D. E., 1994, "DGA Agent Shows Promise for Trace Sulfur Compound Removal from Hydrocarbon Streams," paper presented at the 1994 GRI Sulfur Recovery Conference, May 15–17, Austin, TX.

Kim, C. J., 1988, "Degradation of Alkanolamines in Gas-Treating Solutions: Kinetics of Di-2-propanolamine Degradation in Aqueous Solutions Containing Carbon Dioxide," *Ind. Eng. Chem. Res.*, Vol. 27, No. 1, pp. 1–3.

Kim, C. J., and Sartori, G., 1984, "Kinetics and Mechanism of Diethanolamine Degradation in Aqueous Solutions Containing Carbon Dioxide," *Int. J. of Chem. Kinetics,* Vol. 16, pp. 1257–1266.

Kosseim, A. J., McCullough, J. G., and Butwell, K. F., 1984, "Treating Acid & Sour Gas: Corrosion-Inhibited Amine Guard ST Process," *Chem. Eng. Prog.*, October, pp. 64–71.

Lang, F. S., and Mason, J. F., 1958, "Corrosion in Amine Gas Treating Solutions," *Corrosion*, Vol. 14, No. 2, pp. 105t–108t, February.

Lawson, J. D., and Garst, A. W., 1976, *J. of Chem. and Eng. Data,* Vol. 21, No. 1, pp. 30–32.

Leister, D. A., 1996, Calgon Carbon Corporation (personal communication), May 30.

Lieberman, N. P., 1980, "Amine appearance signals condition of system," *Oil and Gas J.*, May 12, pp. 115–120.

Liu, H. J., and Gregory, R. A., 1994, "Union Carbide Amine Management Program," paper presented at NACE Corrosion/94, March 3, Baltimore, MD.

Liu, H. J., Dean, J. W., and Bosen, S. F., 1995, "Neutralization Technology To Reduce Corrosion From heat-stable Amine Salts," paper presented at Corrosion in Gas Treating, Corrosion 95, NACE International Conference and Corrosion Show, Orlando, FL, March 29.

Lyle, F. F., Jr., 1988, "Stress Corrosion Cracking of Steels in Amine Solutions Used in Natural Gas Treatment Plants," Paper No. 158, Corrosion/88, NACE, St. Louis, March 21–25.

Lynch, P. A., 1982, *Iron and Steel Engineer,* December, p. 29.

MacNab, A. J., and Treseder, R. S., 1971, "Materials requirements for a gas treating process," *Materials Performance,* Vol. 10, No. 1, pp. 21–26.

Maddox, R. N., 1985, *Gas Conditioning and Processing, Vol. 4, Gas and Liquid Sweetening,* Campbell Petroleum Series, Norman, OK.

Manning, F. S., and Thompson, R. E., 1991, *Oilfield Processing of Petroleum, Vol. 1: Natural Gas,* Pennwell Publishing Co., Tulsa, OK.

Mason, J. R., and Griffith, T. E., 1969, "Conversion of a Sweetening Solution from MEA to DGA," *Gas Conditioning Conference Proceedings,* Univ. of Oklahoma, Norman, OK.

McCullough, J. G., 1994, consultant, personal communication.

McCullough, J. G., and Nielsen, R. B., 1996, "Contamination and Purification of Alkaline Gas Treating Solutions," paper No. 396, NACE, Corrosion 96, Denver, CO, March 24–29.

McHenry, H. I., Shives, T. R., Read, D. T., McColskey, J. D., Brady, C. H., and Purtscher, P. T., 1986, "Examination of a Pressure Vessel that Ruptured at the Chicago Refinery of the Union Oil Company on July 23, 1984," National Bureau of Standards Report NBSIR 86-3049, March.

McHenry, H. I., Read, D. T., and Shives, T. R., 1987, "Failure analysis of an amine-absorber pressure vessel," *Materials Performance,* August, pp. 18–24.

Meisen, A., and Kennard, M. L., 1982, "DEA degradation mechanism," *Hydro. Process.,* Vol. 61, No. 10, October, pp. 105–108.

Merrick, R. D., 1989, "An Overview of Hydrogen Damage to Steels at Low Temperatures," *Materials Performance,* February, pp. 53–55.

Merrick, R. D., and Bullen, M. L., 1989, "Prevention of Cracking in Wet H_2S Environments," paper presented at CORROSION/89, paper #269, NACE, Houston, TX.

Meusberger, K. E., and Segebrecht, E. W., 1980, "Foam Depressants for Gas Processing Systems," *1980 Gas Conditioning Conference Proceedings,* Univ. of Oklahoma, Norman, OK.

Millard, M. G., and Beasley, T., 1993, "Contamination Consequences and Purification of Gas Treating Chemicals Using Vacuum Distillation," *Laurance Reid Gas Conditioning Conference Proceedings,* Univ. of Oklahoma, Norman, OK.

Montrone, E. D., and Long, W. P., 1971, "Choosing Materials for CO_2 Absorption Systems," *Chem. Eng.,* Vol. 78, January, pp. 94–97.

Morgan, C., and Klare, T., 1977, "Chloride Removal from DEA by Ion Exchange," *Gas Conditioning Conference Proceedings,* Univ. of Oklahoma, Norman, OK.

NACE International Publication 8×194, 1994A, "Materials and Fabrication Practices for New Pressure Vessels Used in Wet H_2S Refinery Service," NACE, June, Houston TX.

NACE International, Material Requirement MR0175-94, 1994B, Standard Material Requirements, "Sulfide Stress Cracking Resistant Metallic Materials for Oilfield Equipment," NACE, Houston, TX.

NACE International, Refinery Practice RP0472-87, 1987, Standard Recommended Practice, "Methods and Controls to Prevent In-Service Cracking of Carbon Steel (P-1) Welds in Corrosive Petroleum Refining Environments," NACE, Houston, TX.

Neumaier, B. W., and Schillmoller, C. M., 1955, "Deterrence of Hydrogen Blistering at a Fluid Catalytic Cracking Unit," *Proceedings of the API Division of Refining,* Vol. 35, No. 3, pp. 92–109.

Norris, W. E., and Clegg, F. R., 1947, "Investigation of a Girbotol Unit Charging Cracked Refinery Gases Containing Organic Acids," *Petroleum Refiner,* November, pp. 107–109.

Orbach, H. K., and Selleck, F. T., 1996, "The Effect of Carbonyl Sulfide on Ethanolamine Solutions," unpublished paper, by permission of Fluor Daniel.

Osenton, J. B., and Knight, A. R., 1970, "Reaction of Carbon Disulphide with Alkanolamines used in the Sweetening of Natural Gas," paper presented at the Canadian Gas Processing Assoc., Fourth Quarterly Meeting, Calgary, Nov. 20.

Parkins, R. N., and Foroulis, Z. A., 1988, "Stress corrosion cracking of mild steel in monoethanolamine solutions," *Materials Performance,* Vol. 27, No. 1, January, pp. 19–29.

Pauley, C. R., 1991, "Face the Facts About Amine Foaming," *Chem. Eng. Prog.*, July, pp. 33–38.

Pauley, C. R., Hashemi, R., and Caothein, S., 1988, "Analysis of Foaming Mechanisms in Amine Plants," paper presented at AIChE Summer Meeting, Denver, CO, Aug. 22–24.

Pauley, C. R., and Perlmutter, B. A., 1988, "Texas plant solves foam problems with modified MEA system," *Oil and Gas J.*, Feb. 29, pp. 67–70.

Pauley, C. R., Hashemi, R., and Caothein, S., 1989A, "Analysis of Foaming Mechanisms in Amine Plants," *Proceedings 1989 Laurance Reid Gas Conditioning Conference,* Univ. of Oklahoma, Norman, OK.

Pauley, C. R., Hashemi, R., and Caothein, S., 1989B, "Ways to Control Amine Unit Foaming Offered," *Oil and Gas J.*, Dec. 11, pp. 67–75.

Paulsen, H. C., Holtzelaw, J. B., and McNamara, T. P., 1955, U.S. Patent 2,701,750.

Pearce, R. L., Arnold, J. L., and Hall, C. K., 1961, *Hydro. Process.*, Vol. 40, August, pp. 121–129.

Penderleith, Y., 1977, "Stretford Plant Operating Experience at DOFASCO," *Proc. of the Symposium on Treatment of Coke-Oven Gas,* McMaster University, Hamilton, Ontario, Canada, May 26.

Perry, C. R., 1971, "Filtration Method and Apparatus," U.S. Patent #3,568,405, March 9.

Perry, C. R., 1974, "Activated Carbon Filtration of Amine and Glycol Solutions," *Proceedings 1974 Gas Conditioning Conference,* Univ. of Oklahoma, Norman, OK.

Perry, C. R., 1980, "Activated Carbon Filtration of Amine and Glycol Solutions," *Laurance Reid Gas Conditioning Conference Proceedings,* Univ. of Oklahoma, Norman, OK.

Piehl, R. L., 1975, "Survey of Corrosion in Hydrocracker Effluent Air Coolers," Paper No. 5, presented at Corrosion/75, Ontario, Canada, April 14–18.

Polderman, L. D., and Steele, A. B., 1956, "Why diethanolamine breaks down in gas treating service," *Oil and Gas J.*, Vol. 54, July 30, pp. 206–214.

Polderman, L. D., Dillon, C. P., and Steele, A. B., 1955A, "Degradation of monoethanolamine in natural gas treating service," *Oil and Gas J.*, Vol. 53, May 16, pp. 180–183.

Polderman, L. D., Dillon, C. P., and Steele, A. B., 1955B, "Why MEA Solution Breaks Down in Gas Treating Service," *Oil and Gas J.*, Vol. 54, No. 2, pp. 180–183.

Prielipp, G. E., and Pearce, R. L., 1957, "Removal of Non-Volatile Acids from Diethanolamine Gas Scrubbing Solutions by Ion Exchange," *Gas Conditioning Conference Proceedings,* Univ. of Oklahoma, Norman, OK.

Rahman, M. A., Maddox, R. N., and Mains, G. J., 1989, "Reactions of Carbonyl Sulfide and Methyl Mercaptan with Ethanolamines," *Ind. Eng. Chem. Res.*, Vol. 28, No. 4, pp. 470–475.

Richert, J. P., Bagdasarian, A. J., and Shargay, C. A., 1987, "Stress corrosion cracking of carbon steel in amine systems," *Materials Performance,* January, 1988, pp. 9–18.

Richert, J. P., Bagdasarian, A. J., and Shargay, C. A., 1989, "Extent of stress corrosion cracking in amine plants revealed by survey," *Oil and Gas J.*, June 5, pp. 45–52.

Riesenfeld, F. C., and Blohm, C. L., 1950, "Corrosion Problems in Gas Purification Units Employing MEA Solutions," *Pet. Ref.*, Vol. 29, No. 4., pp. 141–150.

Riesenfeld, F. C., and Blohm, C. L., 1951A, "Corrosion in Amine Gas Treating Plants," *Pet. Ref.*, Vol. 30, No. 2., pp. 97–106.

Riesenfeld, F. C., and Blohm, C. L., 1951B, "Corrosion Resistance of Alloys in Amine Gas Treating Systems," *Pet. Ref.*, Vol. 30, October, pp. 107–115.

Rooney, P. C., Dupart, M. S., and Bacon, T. R., 1996, "Effect of heat-stable Salts on Solution Corrosivity of MDEA-Based Alkanolamine Plants," *Proceedings Laurance Reid Gas Conditioning Conference*, Univ. of Oklahoma, Norman, OK.

Scheirman, W. L., 1973A, "Diethanolamine Solution Filtering and Reclaiming in Gas Treating Plants," *1973 Gas Conditioning Conference Proceedings*, Univ. of Oklahoma, Norman, OK.

Scheirman, W. L., 1973B, "Filter DEA Treating Solution," *Hydro. Process.*, Vol. 52, August, pp. 95–96.

Scheirman, W. L., 1976, "Operating Experience With Amine Absorber Level Control Valves," *Gas Conditioning Conference Proceedings*, Univ. of Oklahoma, Norman, OK.

Schmidt, H. W., Gegner, P. J., Heinemann, G., Pogacar, C. F., and Wyche, E. H., 1951, *Corrosion*, Vol. 7, No. 9, p. 295.

Schutt, H. U., 1988, "New Aspects of Stress Corrosion Cracking in Monoethanolamine Solutions," *Materials Performance*, December, 1988, pp. 53–58.

Seubert, M. K., and Wallace, G. D., Jr., 1985, "Corrosion in DGA Treating Plants," paper #159, Corrosion/85, NACE, Boston, MA, March 25–29.

Shaban, H., Abdo, M. S. E., and Lal, D. P., 1988, "Failure of a MEA reclaimer tube bundle due to corrosion," *Corr. Prev. and Control.*, August, pp. 96–99.

Sheilan, M., and Smith, R. F., 1984, "Hydraulic-flow effect on amine plant corrosion," *Oil and Gas J.*, Nov. 19, pp. 138–140.

Simmons, C. V., 1991, "Reclaiming Used Amine and Glycol Solutions," *Laurance Reid Gas Conditioning Conference Proceedings*, Univ. of Oklahoma, Norman, OK.

Singh, M., and Bullin, J. E., 1988, "Determination of Rate Constants for the Reaction Between Diglycolamine and Carbonyl Sulfide," paper presented at the Spring National Meeting of AIChE, New Orleans, LA, March 1–10.

Smith, R. F., 1979A, "Curing foam problems in gas processing," *Oil and Gas J.*, July 30, pp. 186–192.

Smith, R. F., 1979B, "Foam problems and remedies for gas processing solutions," *Proceedings 1979 Gas Conditioning Conference*, Univ. of Oklahoma, Norman, OK.

Smith, R. F., and Younger, A. H., 1972, "Tips on DEA treating," *Hydro. Process.*, July, pp. 98–100.

SwRI, 1989, "An Investigation of Amine-Induced Stress Corrosion Cracking of Steels in Natural Gas Treatment Plants," Final Report SwRI Project No. 06-1202, prepared by the Southwest Research Institute, San Antonio, TX.

Texaco Chemical (now Huntsman Chemical), 1982, *Diglycolamine Agent,* Texaco Chemical Brochure TEX 4038-182.

Thomason, J., 1985, "Reclaim gas treating solvent," *Hydro. Process.*, April, pp. 75–78.

Union Carbide, 1994, Sales Brochure, *The Union Carbide Corporation UCARSEP Process.*

Veldman, R., 1989, "How to Reduce Amine Losses," paper presented at PETROENERGY 89, Oct. 23–27.

Wendt, C. J., and Dailey, L. W., 1967, "Gas Treating: The SNPA Process," *Hydro. Process.*, Vol. 46, No. 10, pp. 155–157.

Yan, T. Y., 1993, "Clean Up of Ethanolamine to Improve Performance and Control Corrosion of Ethanolamine Units," U.S. Patent No. 5,268,155.

Chapter 4
Removal and Use of Ammonia in Gas Purification

INTRODUCTION, 279

    Types of Coal Gas, 279
    Sources of Ammonia, 280
    Coal Gas Impurities, 281
    Sources of Sour Water, 282

BASIC DATA, 283

REMOVAL OF AMMONIA FROM GASES, 292

    Types of Processes, 296
    Water Scrubbing, 297
    Sour Water Stripping Process, 302
    Ammonium Salt Production, 308
    Phosam and Phosam-W Processes, 311
    Chevron WWT Process, 314

USE OF AMMONIA TO REMOVE ACID GASES, 318

    Basic Chemistry, 319
    Process Descriptions, 321
    Design and Operation, 326

REFERENCES, 326

INTRODUCTION

In previous editions of this book the removal of ammonia from gas streams and the use of ammonia to remove H_2S and CO_2 were covered in separate chapters. The two subjects have been combined into a single chapter in this edition because (1) the technologies are closely interrelated, (2) the use of aqueous ammonia to absorb H_2S and CO_2 is declining relative to other, more efficient processes, and (3) the subjects are of most interest with regard to coal gas purification, which is not currently an expanding field.

Types of Coal Gas

Table 4-1 lists the principal types of gases produced from coal. Producer gas, water gas, and carburetted water gas represent technology that was widely practiced during the first half of this century to provide fuel gas for residential and industrial use. The availability of abundant supplies of natural gas during the 1950s led to the abandonment of most of these types of coal gasification units in the United States, although their use has continued in other parts of the world. The production of coke-oven gas (COG) continued to grow in the United States during the 1950s and 1960s and is still a major operation worldwide.

The development of processes to produce low- and medium-Btu gas using large pressurized gasifiers received considerable attention during the 1970s and 1980s, but was de-emphasized in the 1990s when oil and natural gas supplies appeared to be ample. Limited development and commercial activities on large, pressurized, oxygen-blown gasifiers are continuing. This type of gasifier is considered to have high potential for future applications to provide fuel for combined cycle power plants and feed gas for the manufacture of ammonia, synthetic natural gas (SNG), and other products.

Table 4-1
Principal Types of Gas Derived from Coal

Gas Type	Process Conditions	Approx. HHV (Btu/scf)
Producer Gas	Air blown, moving bed, atmospheric pressure	150
Water Gas	Cyclic process—air to heat, steam to gasify, moving bed, atmospheric pressure	300
Carburetted Water Gas	Similar to water gas with oil added during steam blowing cycle	500
Coke-Oven Gas (COG)	By-product of coal coking process	300
Low-Btu Gas	Air blown; moving, entrained or fluidized bed; usually elevated pressure operation	150
Medium-Btu Gas	Oxygen blown; moving, entrained, or fluidized bed; elevated pressure operation	300

Sources of Ammonia

During the gasification, carbonization, and thermal treatment of coal, liquid petroleum, shale oil, and tar sands products, a portion of the nitrogenous material contained in the feed is converted to volatile compounds that appear in the gaseous products. The principal nitrogen compounds that have been identified in such gases are ammonia, cyanogen, hydrogen cyanide, pyridine and its homologues, nitric oxide, and free nitrogen. Although the nitrogen present in the fuel is the primary source of these compounds, small amounts of atmospheric nitrogen may also contribute to the presence of nitrogen compounds in the gas stream.

The distribution and concentration of nitrogen compounds in the gaseous products for a given fuel vary over a wide range depending on operating conditions of the gasifier, reactor, or coke oven. The principal operating variables influencing the distribution of nitrogen compounds are temperature, time at temperature, and quantity of steam or oxygen used. For coal carbonization, temperature has the most pronounced effect. For example, the fraction of the nitrogen contained in coal that is converted to ammonia varies from about 2% at a carbonization temperature of 400°C (750°F), which is typical for low-temperature carbonization processes, to 10–15% at 900°C (1,650°F) or higher, the range of typical high temperature carbonization installations.

The approximate distribution of nitrogen in the products under normal, high-temperature coking conditions is shown in **Table 4-2.** Typical concentrations of nitrogen compounds in gases from coal carbonization processes are given in **Table 4-3.** Detailed discussions of the effect of coal carbonization conditions on nitrogen distribution are presented by Kirner (1945) and Hill (1945).

Table 4-2
Distribution of Nitrogen in Carbonization Products

Products	Distribution
In the coke	30–50 percent
As ammonia in the gas	10–15 percent
As cyanide compounds in the gas	1–2 percent
In the tar	1–3 percent
As free nitrogen in the gas	Balance

Table 4-3
Typical Concentrations of Nitrogen Compounds in Coal Gases

Compound	Volume %
Free nitrogen	0.5–1.5
Ammonia	1.1
Hydrogen cyanide	0.10–0.25
Pyridine bases	0.004
Nitric oxide	0.0001

Coal Gas Impurities

Ammonia, hydrogen sulfide, and carbon dioxide are major impurities in coke oven gas (COG). In addition to these three components, COG often contains carbon disulfide, carbonyl sulfide, hydrogen cyanide, organic acids, pyridine, phenol, and other impurities, which can cause problems with conventional amine plants. The presence of large quantities of ammonia in these gases naturally led to consideration of its use for removal of H_2S and CO_2, and several ammonia-based coke oven gas purification processes have been developed and commercialized.

Processes that use aqueous ammonia to remove H_2S and CO_2 are still offered for coke-oven gas purification, and many such plants are in operation in the U.S. and Europe; however, it appears that the trend for new operations is toward the use of other absorbents. Other absorption processes that may be applicable for COG purification include the Takahax, Stretford, Vacuum Carbonate, Potassium Carbonate, and Sulfiban (MEA) processes. These processes can be designed to avoid serious adverse effects of trace impurities in the gas, and generally provide somewhat higher H_2S removal efficiency than ammonia scrubbing. The processes are described in detail in other chapters (see index).

Typical concentrations of major components and impurities in gases derived from coal are given in **Table 4-4**. The principal impurities removed by coal-gas purification processes are hydrogen sulfide and ammonia. These compounds are undesirable because they can cause corrosion, plugging, and ultimately, air pollution. In addition, both H_2S and NH_3 are relatively valuable chemicals, and their recovery and conversion to useful products, such as elemental sulfur and ammonia, can be of significant economic value. This was particularly true

Table 4-4
Typical Compositions of Raw Gas Streams from Coal Gasifiers and Coke Ovens

Component	Lurgi	Oxygen-Blown Gasifiers					COG
		Koppers Topsek	Texaco	BGC/ Lurgi	Shell		
C_1, C_2	3.6	—	0.3	4.2	—		28.6
H_2	16.1	18.7	29.8	26.4	30.0		59.0
CO	5.8	43.4	41.0	46.0	60.3		6.0
CO_2	11.8	6.1	10.2	2.9	1.6		1.3
H_2S	0.5	0.6	1.0	1.0	1.2		0.4
COS	trace	0.1	0.1	0.1	0.1		—
N_2	0.1	0.9	0.7	2.8	3.6		1.3
Ar	—	—	0.1	—	1.1		—
H_2O	61.8	30.2	17.1	16.3	2.0		25.2
NH_3	0.3	—	0.2	0.3	0.1		0.7
HCN	—	—	—	—	—		0.09

Oxygen-blown gasifier data from Simbeck et al. (1983) based on Illinois #6 coal containing 2.6–3.7% sulfur and 1.1–1.4% nitrogen.
Coke-oven gas (COG) data from Bloem et al. (1990); coal composition not given.

before the advent of synthetic ammonia, when coal gas constituted the largest source of fixed nitrogen.

Carbon dioxide is also a major impurity in coal gas. Although it is not generally necessary to remove CO_2 from gases used as fuels, partial removal is sometimes desirable to improve the heating value. Complete CO_2 elimination is required for gases undergoing processing at very low temperatures, for example, in coke-oven gas purification to provide hydrogen for ammonia synthesis. The removal of other impurities such as HCN, COS, organic acids, and pyridine may also be required when the concentration is high enough to cause pollution or operating problems in downstream systems.

As indicated by the data of **Table 4-4,** gases produced by oxygen-blown gasifiers generally contain appreciably less ammonia than coke oven gas. As a result, the use of aqueous ammonia has not been developed as a gas purification technique for such gases. Gases produced by air-blown gasifiers and by the thermal treatment of petroleum streams (including shale oil and tar sands liquids) are also relatively low in ammonia and are not considered to be appropriate candidates for ammonia-based scrubbing processes.

Sources of Sour Water

In refinery catalytic cracking operations, oxygen-blown coal gasifiers and similar high temperature processes that produce a gas stream containing both ammonia and hydrogen sulfide, the gas leaving the high temperature step is typically quenched with water or cooled by indirect heat exchange and then scrubbed with water. The resulting "sour water" contains essentially all of the ammonia, but normally contains only a small portion of the H_2S and CO_2 contained in the gas stream. It also contains water soluble impurities such as organic acids and phenols. Considerable development work has gone into the processing of sour water. Because its processing constitutes an essential step in the overall gas purification scheme, and ammonia is always one component, key processes for sour water treatment are described in the Ammonia Removal section of this chapter.

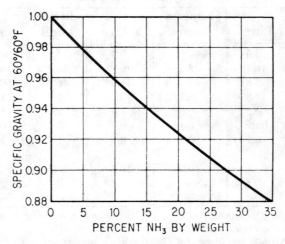

Figure 4-1. Specific gravity of aqueous ammonia solutions. *Data of Ferguson (1956)*

BASIC DATA

The specific gravity of aqueous ammonia solutions is given in **Figure 4-1.** Vapor pressures of ammonia over aqueous ammonia solutions and vapor-liquid equilibria for such solutions at the atmospheric pressure boiling points are presented in **Figures 4-2** and **4-3.** Vapor-liquid equilibria for the system ammonia-water above about 200°F up to the critical region are given by Sassen et al. (1990). At moderate temperatures, near ambient, dilute solutions of ammonia in water at pH 6 to 10 obey Henry's law, which may be expressed as $H = 4092/T - 9.70$ (Dasgupta and Dong, 1986). In this equation H is the Henry's law constant (M atm^{-1}) and T is the absolute temperature (°K).

The vapor pressures of H_2S, CO_2, and NH_3 over aqueous solutions containing these compounds have been extensively studied. Early investigators included Lohrmann and Stoller (1942), Pexton and Badger (1938), Badger and Silver (1938), Badger (1938), Badger and Wilson (1947), Dryden (1947), and van Krevelen et al. (1949).

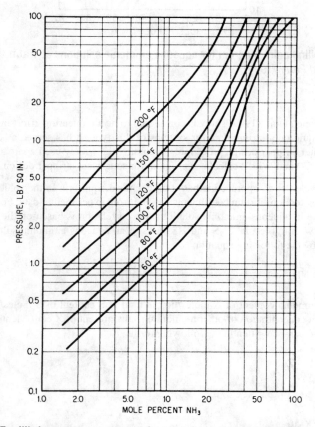

Figure 4-2. Equilibrium vapor pressure of ammonia over aqueous solutions. *From Perry (1941)*

284 *Gas Purification*

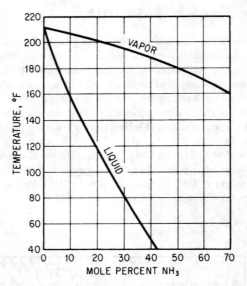

Figure 4-3. Boiling point diagram for aqueous ammonia solutions at 1 atm total pressure. *From Perry (1941)*

The van Krevelen et al. (1949) study is considered to be a pioneering effort in the correlation of vapor-liquid equilibrium data for systems of this type. It has served as the basis for several sour water stripper design procedures, including the widely used approach described by Beychok (1967) and Wild's calculator program for sour water stripper design (1979).

The van Krevelen et al. approach is based on chemical equilibria in the solution with the numerical values for equilibrium coefficients derived from experimental data. For the case of hydrogen sulfide in aqueous ammonia solutions, the correlation is based on the equilibrium relationship between molecular H_2S and HS^- ions in an aqueous ammonia solution, which can be defined by the following equation

$$NH_3 + H_2S \rightleftharpoons NH_4^+ + HS^- \tag{4-1}$$

For a total ammonia content of A moles/liter, a total H_2S content of S moles/liter, and an equivalent anion concentration of other ammonium salts present of Z equivalents/liter, equations 4-2 and 4-3 can be written in the following forms:

$$(NH_3) = A - S - Z \tag{4-2}$$

$$(HS^-) = S \tag{4-3}$$

$$(NH_4^+) = S + Z \tag{4-4}$$

An equilibrium coefficient K can be expressed as follows:

$$K = \frac{(NH_4^+)(HS^-)}{(NH_3)P_{H_2S}} = \frac{(S+Z)S}{(A-S-Z)P_{H_2S}} \tag{4-5}$$

from which the vapor pressure of H_2S is

$$P_{H_2S} = \frac{(S+Z)S}{(A-S-Z)K} \text{ mm Hg} \tag{4-6}$$

and from Henry's law,

$$P_{NH_3} = \frac{A-S-Z}{H_{NH_3}} \text{ mm Hg} \tag{4-7}$$

Van Krevelen et al. propose the following equation for estimating Henry coefficients for solutions of the type commonly encountered in H_2S absorption:

$$-\log\frac{H_{NH_3}}{H_0} - 0.025(NH_3) \tag{4-8}$$

Where: H_0 = Henry coefficient for ammonia in pure water
(NH_3) = concentration of free ammonia

The equilibrium coefficient K is not a true constant and varies with the concentration of dissolved salts in the solution. The authors found that this variation can be expressed as a function of the ionic strength (I):

$$I = \frac{1}{2}\Sigma C_i Z_i \tag{4-9}$$

Where: C_i = concentration of a given ion
Z_i = corresponding valency, by the equation:

$$\log K = a + 0.089I \tag{4-10}$$

Where a has the following values:

t, °C	a
20	−1.10
40	−1.70
60	−2.19

In aqueous solutions containing only ammonia and H_2S, I equals S, the total H_2S concentration, and equation 4-10 becomes

$$\log K = a + 0.089S \tag{4-11}$$

Van Krevelen et al. (1949) also proposed correlations to calculate the vapor pressures of components in carbon dioxide-ammonia-water and hydrogen sulfide-carbon dioxide-ammonia-water systems and, in general, results obtained by use of the correlations agree well with experimental data. However, it is important to note that the van Krevelen et al. correlations apply only to ammonia-rich solutions, and are further limited to restricted ranges of acid gas/ammonia ratios and to relatively low temperatures and pressures.

Edwards et al. (1975, 1978A, B) established a molecular-thermodynamic correlation for calculating vapor-liquid equilibria in aqueous solutions containing one or more volatile electrolytes, with special attention to the ternary systems, ammonia-carbon dioxide-water and ammonia-hydrogen sulfide-water. Their 1978 correlation was shown to give results in satisfactory agreement with the limited data then available for temperatures from 0° to 170°C (32° to 338° F) and ionic strengths of about 6 molal (equivalent to total concentrations of the electrolytes between 10 and 20 molal).

Maurer (1980) compared results calculated on the basis of the van Krevelen et al. (1949), Edwards et al. (1978A, B), and Beutier and Renon (1978) correlations with several sets of experimental data. The average deviation of the calculated partial pressures ranged from 10 to 50% depending upon which correlation and which set of experimental data were compared. Kawazuishi and Prausnitz (1987) updated values for some dissociation equilibrium constants and Henry's constants used in the Edwards et al. (1978A, B) correlation with the result that it predicted vapor pressure values in good agreement with the more recent experimental data of Muller (1983).

Because of the importance of sour water processes in the development of substitute natural gas supplies and in refinery operations, the Gas Processors Association (GPA) and American Petroleum Institute (API) supported an extensive program to obtain and correlate design data in this field. In a joint GPA and API-sponsored project, Wilson (1978) developed a program called SWEQ (Sour Water Equilibrium). According to Newman (1991), the SWEQ model is used in several process simulators (e.g. PROCESS, CHEMSHARE, HYSIM, and COADE) and is valid in the temperature range of 70° to 300°F. Newman published a series of charts based on the SWEQ program, which are convenient to use for the preliminary design of sour water systems. Four of his charts representing typical absorption and stripping temperatures are reproduced as **Figures 4-4, 4-5, 4-6** and **4-7.**

After development of the original SWEQ correlation, additional experimental data were obtained by Wilson et al. (1982) and Owens et al. (1983) under GPA sponsorship. The subject is reviewed by Wilson et al. (1985) who present new data on the partial pressure of sour water components at temperatures from 100° to 400°F; and pressures up to 1,000 psia. Sample data are given in **Table 4-5.** Wilson et al. (1985) point out that at temperatures above 300°F, the measured partial pressures of NH_3, H_2S, and CO_2 diverge significantly from values predicted by the earlier correlation (Wilson, 1978). They also found the Henry's law coefficients for the inert gases to be the same in sour water as in pure water.

The GPA program culminated with the development of a new model and computer program called GPSWAT (GPA Sour Water Equilibria), which extends the applicability of the precursor SWEQ program from 68°– 284°F to 68°– 600°F and extends the pressure range to 2,000 psia. This program may be purchased from the Gas Processors' Association.

(text continued on page 291)

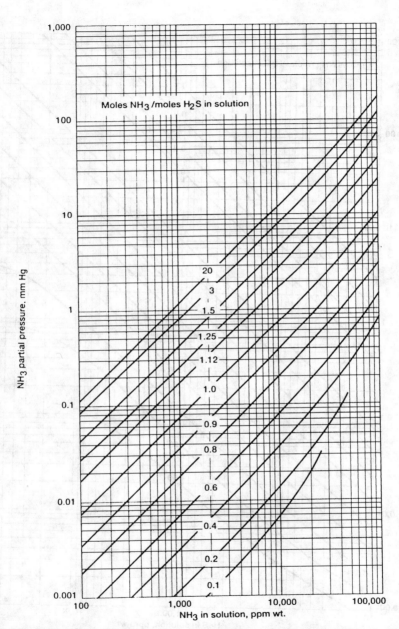

Figure 4-4. Ammonia equilibria at 100°F. *Data of Newman (1991). Reproduced with permission from Hydrocarbon Processing*

288 *Gas Purification*

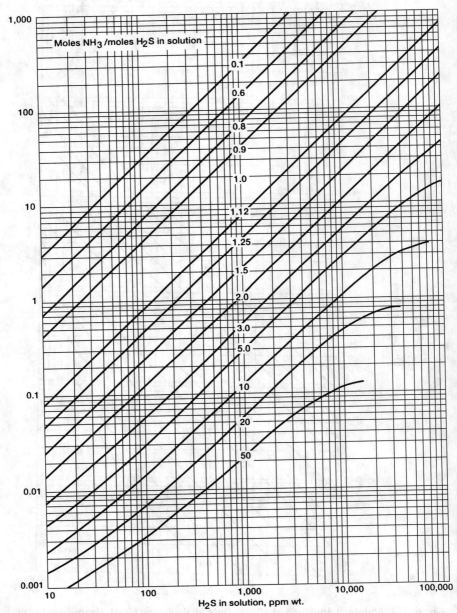

Figure 4-5. Hydrogen sulfide equilibria at 100°F. *Data of Newman (1991). Reproduced with permission from Hydrocarbon Processing*

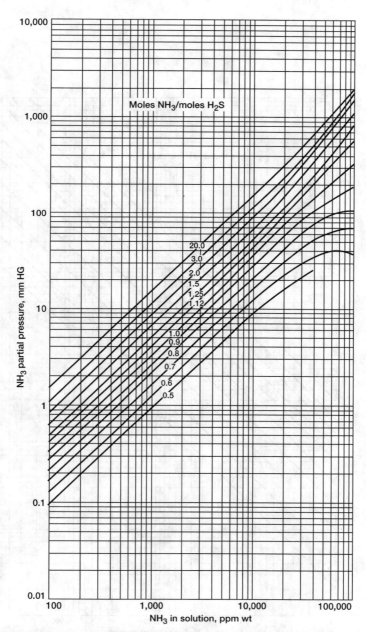

Figure 4-6. Ammonia equilibria at 240°F. *Data of Newman (1991). Reproduced with permission from Hydrocarbon Processing*

290 *Gas Purification*

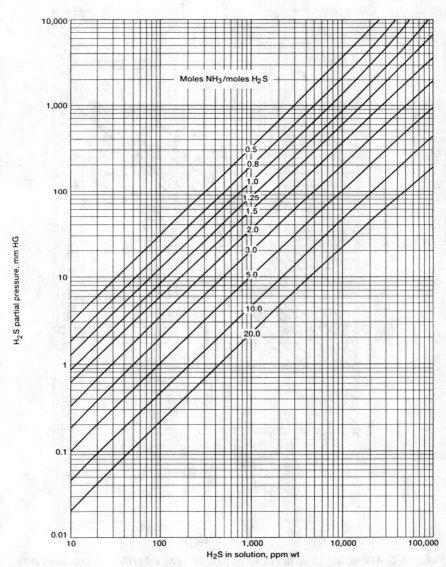

Figure 4-7. Hydrogen sulfide equilibria at 240°F. *Data of Newman (1991). Reproduced with permission from Hydrocarbon Processing*

Table 4-5
Measured Vapor-Liquid Equilibrium Data for Sour Water with Inerts at Elevated Temperatures and Pressures

Temperature and Component	Concentration in Liquid, Mole %	Measured Partial Pressure, psia at total pressure of:		
		33.4 psia	500 psia	1,000 psia
100°F				
NH_3	2.11	0.28	0.34	0.36
CO_2	0.27	0.022	0.015	0.012
H_2S	0.53	0.16	0.14	0.15
H_2O	97.01–97.09	0.78	0.93	0.93
inerts	0–0.08	32.16	498.61	998.6
200°F				
NH_3	2.05–2.02		2.40	2.75
CO_2	0.25–0.26		1.61	1.75
H_2S	0.51–0.52		2.37	2.34
H_2O	97.15–97.13		11.21	11.20
inerts	0.04–0.07		482.4	981.6
400°F				
NH_3	2.13			32.7
CO_2	0.45			222.8
H_2S	0.84			124.5
H_2O	96.51			239.8
inerts	0.07			380.2

Notes: Inerts are CO, N_2, CH_4, and H_2 for 100°F and 200°F runs; N_2, CH_4, and H_2 for 400°F run. Data from Wilson et al. (1985)

(text continued from page 286)

Vapor-liquid equilibria of typical coke-oven gases and liquids obtained in ammonia scrubbers (indirect process) are shown in **Figure 4-8**.

Differential and integral heats of solution of liquid ammonia in water are given in **Tables 4-6** and **4-7**.

The heat of reaction evolved in the saturator during absorption of gaseous ammonia in sulfuric acid, based on heat-of-formation data, is shown in equation 4-12:

$$2NH_3(g) + H_2SO_4(aq) \rightarrow (NH_4)_2SO_4(aq) - \Delta H = 86{,}000 \text{ Btu/lb mole} \tag{4-12}$$

In a typical process where ammonia is reacted with sulfuric acid, this heat of reaction is augmented by the heat of dilution of sulfuric acid from 60° Bé (77.6%) to 7% which amounts to 15,000 Btu/lb mole. The total heat effect is, therefore, 101,000 Btu/lb mole of ammonium sulfate.

292 Gas Purification

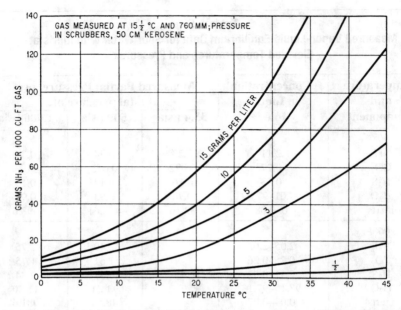

Figure 4-8. Vapor-liquid equilibria of typical coke-oven gases and liquids obtained in ammonia scrubbers (indirect process). *From Pacific Coast Gas Association, Gas Engineers' Handbook (1934)*

In processes where ammonia is reacted with phosphoric acid, the heat evolved in the formation of diammonium phosphate from ammonia and phosphoric acid may be estimated from equation (4-13):

$$2NH_3(g) + H_3PO_4(aq) \rightarrow (NH_4)_2HPO_4(aq) - \Delta H = 75,000 \text{ Btu/lb mole} \qquad (4\text{-}13)$$

The density, viscosity, and vapor pressure of aqueous ammonium sulfate solutions and the solubility of ammonium sulfate in water can be estimated from the nomogram presented in **Figure 4-9**. To obtain the density, viscosity, and vapor pressure, the temperature on the t scale is selected and aligned with the concentration on the C scale. The intersects of the line of the p, γ, μ, and d scales give the values for vapor pressure, viscosity, and density of the solution. The solubility of ammonium sulfate in water is obtained by aligning the temperature (on the t scale) with the saturation point S and reading the solubility on the C scale. The values obtained by the nomogram agree very closely with published data.

REMOVAL OF AMMONIA FROM GASES

It is generally necessary to remove ammonia, hydrogen cyanide, and pyridine bases from coal gases prior to industrial or domestic use to meet purity requirements of downstream systems. In addition, ammonia and the pyridine bases are relatively valuable chemicals and their recovery as by-products can be economically attractive when significant quantities are produced. Before the advent of synthetic ammonia processes, by-product ammonia from coal

Table 4-6
Differential Heats of Solution of Liquid Ammonia
(Btu per pound of ammonia dissolved)

Concentration, wt%	Heat of solution	Concentration, wt%	Heat of solution	Concentration, wt%	Heat of solution	Concentration, wt%	Heat of solution
0	347.4	11	302.8	21	253.8	31	197.6
1	343.8	12	298.2	22	248.4	32	191.9
2	340.2	13	293.6	23	243.0	33	186.1
3	336.6	14	289.0	24	237.6	34	180.4
4	333.0	15	284.4	25	232.2	35	174.6
5	329.4	16	279.4	26	226.4	36	168.1
6	325.0	17	274.3	27	220.7	37	161.6
7	320.6	18	269.2	28	214.9	38	155.2
8	316.2	19	264.2	29	209.2	39	148.7
9	311.8	20	259.2	30	203.4	40	142.2
10	307.4						

Concentration, wt%	Heat of solution
41	135.0
42	127.8
43	120.6
44	113.4
45	106.2
46	99.0
47	91.8
48	84.6
49	77.4
50	70.2

Source: Perry (1941)

**Table 4-7
Integral Heats of Solution of Liquid Ammonia**

Concentration, wt%	Btu/lb mixture	Btu/lb NH$_3$
0	0	358.0
10	34.4	343.8
20	65.7	328.5
30	92.5	308.2
40	108.2	270.0
50	109.4	218.8
60	101.9	169.7
70	84.8	121.1
80	60.7	75.8
90	31.9	35.5
100	0	0

Source: Perry (1941)

carbonization and gasification constituted the most important source of fixed nitrogen. At present, by-product ammonia accounts for only a small percentage of the ammonia produced, and there is a growing trend with small- and medium-sized installations to burn rather than recover ammonia removed from gas streams. Svoboda and Diemer (1990) suggest that ammonium sulfate production is particularly uneconomical in Europe because the acid soils in much of the region preclude the use of sulfate-containing fertilizer.

Since coal gasification has been practiced for many years, it is not surprising that the literature covering the removal of ammonia and other impurities is voluminous and repetitious. No attempt is made in this chapter to discuss all processes that have been developed; instead, detailed descriptions are presented for processes in current use, with emphasis on those that appear to be expanding in application. Comprehensive reviews of earlier technologies applied to coal gas purification in the 1940s and 1950s are given by Hill (1945), Wilson (1945), Wilson and Wells (1948), Bell (1950), and Key (1956). Updates covering coke oven gas and effluent treatment technology as of the mid to late 1970s are provided by Grosick and Kovacic (1981) and Massey and Dunlap (1975).

The proposed construction of very large oxygen- (or air-) blown coal gasifiers to provide feed for combined cycle power plants or for the synthesis of fuels and chemicals is a relatively recent development. Many more such plants have been designed and studied than actually built, and, with the exception of the large Lurgi gasifier installation in South Africa, plants built to date have been aimed primarily at process development and demonstration. The designs typically include a water wash (or quench) step immediately downstream of the gasifier to clean and cool the gas. The resulting sour water contains ammonia, hydrogen sulfide, and other soluble impurities. In most gasification plant designs, the sour water undergoes phenol extraction, steam stripping, and finally biological oxidation before disposal. Vapors from the sour water stripper are generally processed for ammonia recovery. Detailed flow diagrams of typical water processing systems for coal gasification plants are given in the United States DOE *Coal Conversion Systems Technical Data Book* (1982).

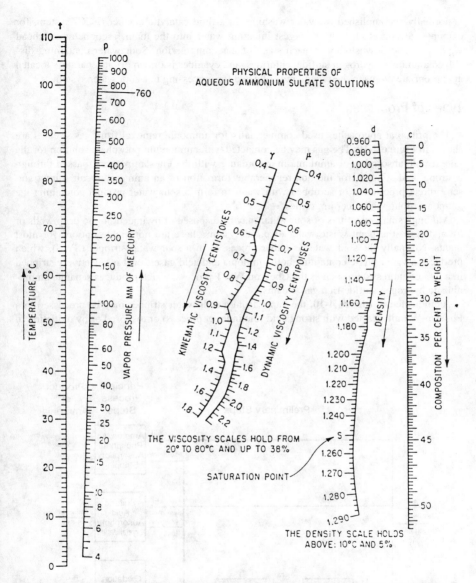

Figure 4-9. Nomogram for estimating density, viscosity, and vapor pressure of aqueous $(NH_4)SO_4$ solutions and the solubility of ammonium sulfate in water. *Data of Tans (1958)*

Petroleum refineries handling heavy crudes, tar sands, or shale oil typically encounter ammonia in the gas streams from catalytic crackers, cokers, and other high temperature operations. It is considered important to remove the ammonia and associated impurities as soon as possible after production to avoid corrosion and plugging of downstream equipment. This is normally accomplished by water washing. In a fluid catalytic cracker (FCC) system, for example, Strong et al. (1991) suggest injecting water into the main fractionator overhead vapor line and downstream of each stage of gas compression. Sour water containing dissolved ammonia, hydrogen sulfide, chlorides, and cyanides is drawn off at separators located at appropriate points in the circuit and collected for processing in a sour water system.

Types of Processes

The principal approaches used commercially for ammonia removal from gas streams are shown in **Figure 4-10.** The gas may be contacted with ammonium phosphate solution for the regenerable absorption of ammonia in accordance with the Phosam process, passed through a strong acid solution for the non-regenerable formation of an ammonium salt (direct and semi-direct processes), or scrubbed with water to form a sour water product containing the ammonia along with other impurities.

Although small quantities of sour water may be disposed of by injection into deep wells or by adding to the plant waste water disposal system, these are not viable options for most plants. Normally the sour water is either processed in a sour water stripper (SWS), which produces a vapor phase containing both ammonia and acid gases, or it is selectively stripped in a two-column system such as the Chevron WWT process, which produces separate ammonia and hydrogen sulfide-rich gas streams.

As indicated in **Figure 4-10,** the vapor from a sour water stripper may be processed in a Phosam-W unit, reacted with strong acid (the indirect process) or disposed of by oxidation in

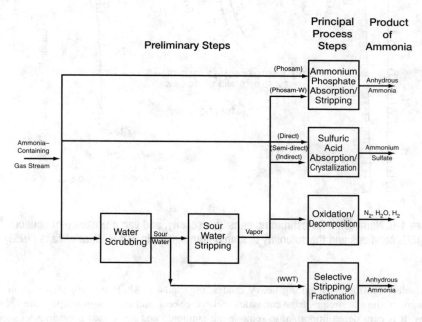

Figure 4-10. Basic steps in ammonia removal processes.

an incinerator or sulfur plant or by catalytic decomposition to nitrogen and hydrogen. The ammonia-rich vapor produced in the Chevron WWT process is typically purified to produce anhydrous ammonia as shown in **Figure 4-10;** however, it may also be combusted if ammonia recovery is not economical.

The terms "direct," "indirect," and "semi-direct" are used primarily in conjunction with coal-gas purification. In the direct process, the hot gas from the coke ovens is passed directly through concentrated sulfuric acid to produce ammonium sulfate. The process is fraught with operational difficulties and is no longer of commercial interest. In the indirect process, which is the oldest of the three, all of the ammonia is removed from the coke-oven gas by water washing; while in the semi-direct process, the gas is first cooled to condense out tar and a small volume of ammonia-containing water, then reheated and reacted with sulfuric acid as in the direct process. Both the indirect and semi-direct processes are currently used in coking plants. The processes identified in **Figure 4-10** are described in the following sections.

Water Scrubbing

Coke-Oven Gas Processing

The basic processes used in handling coke-oven gas have not changed greatly in many years. Typically, hot dirty gas from the coke ovens undergoes the following steps prior to ammonia and acid gas removal:

1. The gas is precooled by direct contact with a large volume of a mixture of coal tar and weakly ammoniacal aqueous solution which is sprayed directly into the collection main. During this operation the gases are cooled to about 75° to 100°C (167° to 212°F), and most of the fixed ammonium salts (typically about 30% of the ammonia originally present in the gas) as well as a major quantity of the tar are removed. The liquid, which is known as the "flushing liquor," is recycled to the collecting main after most of the tar has been decanted. A portion of the flushing liquor is continuously withdrawn from the cycle, combined with other liquid streams containing relatively low concentrations of ammonia (the so-called "weak ammoniacal liquor") and further processed for recovery of the ammonia. The liquid removed from the cycle is replaced by condensate from subsequent coolers.
2. The gas is further cooled to a temperature of 28°–30°C (82°–86°F) in primary coolers, which may be direct or indirect. Direct coolers are towers in which the gas is cooled by direct contact with a countercurrent stream of weakly ammoniacal solution. Packed towers were formerly used for this service, but naphthalene buildup on the packing caused excessive maintenance, and all recent installations have employed spray towers. The hot solution leaving the bottom of each tower is passed through water-cooled coils and recycled to the top of the tower. A portion of the circulating liquid is continuously withdrawn and added to the flushing liquor. Indirect coolers are shell-and-tube heat exchangers with cooling water flowing through the tubes. Condensate from indirect coolers, which contains some light tar and about 40% of the ammonia originally present in the coal gas, mostly as free ammonia, is collected in a decant tank. The tar is separated and the aqueous phase is added to the circulating flushing liquor.
3. The cooled coke-oven gas is normally passed through an electrostatic precipitator to remove fine droplets of tar, then compressed in an exhauster to a pressure of about 2 psig (Svoboda and Diemer, 1990).

The sequence of these steps may vary depending upon the overall process employed. When the ammonia is to be removed by water scrubbing prior to acid gas removal, the gas must be cooled again to remove the heat of compression. The required secondary cooler is typically installed in the bottom of the ammonia wash tower. A diagram of the water wash ammonia removal process, including the stripping section, is shown in **Figure 4-11;** typical compositions for key flow streams, as reported by Svoboda and Diemer (1990), are given in Table 4-8.

Coke-oven gas from the exhauster, at a temperature of 45°–55°C (113°–131°F), is cooled in the secondary cooler to about 28°C (82°F), then enters the absorption section of the vessel. A portion of the rich solution from the absorber is cooled and recycled over the bottom packed section to remove the heat of reaction and provide a high liquid flow rate in this zone. Cooled excess flushing liquor, which is a dilute solution of ammonia, is fed into the column in the lower third of the vessel. Water from the free ammonia stripper is fed into the top of the absorber. This water still contains fixed ammonia, but since this form of ammonia has a negligible ammonia vapor pressure, it does not significantly affect the gas purity attainable. According to Svoboda and Diemer (1990), this type of washer typically reduces the ammonia content of coke-oven gas from 200–500 g/100 scf to 2–7 g/100 scf.

Rich solution from the bottom of the wash tower is preheated by heat exchange with stripped water then fed to the top of the free ammonia stripper, which operates with direct steam injection at the bottom. Stripped water from this column is cooled and recycled to the top of the wash column. Excess stripped water, which still contains fixed ammonia, is processed in the fixed ammonia stripper before disposal. Caustic soda (or lime) is used to

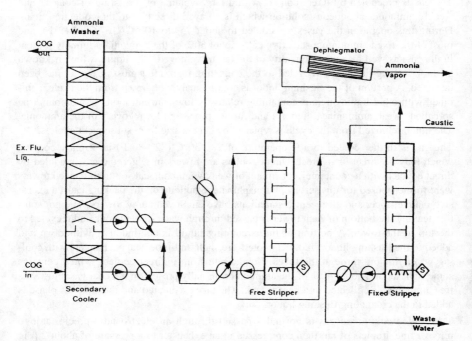

Figure 4-11. Coke-oven gas, water-wash flow diagram. *From Svoboda and Diemer (1990)*

Table 4-8
Typical Stream Compositions for Water-Wash Process for Ammonia Removal from Coke-Oven Gas

Component	Process Stream		
	Excess Flushing Liquor g/l	Rich Solution from NH_3 Scrubber g/l	Ammonia Still Vapor vol. %
NH_3, fixed	2.5–5.0	4.0	—
NH_3, free	1.0–2.5	0.9	38.7
H_2S	0.2–0.4	0.6	1.7
HCN	0.1–0.3	0.6	1.4
CO_2	0.2–0.5	1.6	2.5
H_2O	balance	balance	55.9

Data of Svoboda and Diemer (1990)

react with nonvolatile acids in the liquid feed to the fixed ammonia stripper to release the ammonia and allow it to be steam-stripped from the solution. The design of stripping systems is discussed in a later section entitled "Sour Water Strippers."

Overhead vapors from both the free and fixed ammonia strippers pass through a dephlegmator (partial condenser) and are then available for further processing to recover or dispose of the ammonia and acid gases. Processes for handling the overhead gas product include

1. Reaction with strong acid to produce ammonium sulfate (or phosphate) salts and a stream of acid gas (the indirect process)
2. Phosam process to produce separate streams of pure ammonia and acid gases
3. Catalytic destruction of ammonia in the presence of H_2S to produce a single gas stream containing N_2, H_2, H_2S, and other components, which can be recycled to the coke-oven gas stream ahead of the desulfurization unit (Svoboda and Diemer, 1990)
4. Oxidation of both the ammonia and H_2S in a Claus Plant or incinerator to produce a tail gas containing N_2, H_2O, and varying amounts of SO_2

Ammonia Absorption in Water

The rate of absorption of ammonia has been studied extensively. A list of key investigations is given in **Table 4-9**. There is general agreement that the operation is gas film-controlled due to the high rate of reaction of dissolved ammonia with water. A wide range of data on the rate of absorption of ammonia in water in packed towers has been presented by Fellinger (1941). Examples of his data, in terms of heights of overall transfer units based on the gas driving force (H_{OG}), are given in **Figures 4-12** and **4-13**. Although the charts show that the height of a transfer unit (H_{OG}) increases with gas rate (G), indicating poorer performance, the increase in H_{OG} is not proportional to the increase in G; therefore, the absorption coefficient also increases. (See Chapter 1 for equations defining H_{OG}.)

Table 4-9
Investigations on the Rate of Absorption of Ammonia in Towers

Authors	Column diameter in.	Packed height in.	Packing material	Gas	Solvent	Flow rate, lb/(hr) (sq ft) Liquid	Flow rate, lb/(hr) (sq ft) Gas	Temp., °F	Remarks
Kowalke, Hougen, and Watson (1925)	16	46	Spray tower	Air	Water	30–750	6–150	60	5 "Vermorel" nozzles
	16	41	Wood grids	Air	Water	21–670	19–240	68–110	
	16	41	No. 1 stoneware*	Air	Water	30–780	15–240	60–110	
	16	41	No. 2 stoneware†	Air	Water	30–780	15–240	60–110	
	10	19–31	Rings, 1-in.	Air	Water	570–830	55–530	77	Data of Borden and Squires
Sherwood and Holloway (1940)	10	19–31	Rings, 1-in.	Air	Water	660–710	65–700	54	Data of Doherty and Johnson
	10	19–31	Rings, 1-in.	Air	0.5–4.5 N H_2SO_4	1,520–1,850	210	77	
	10	19–31	Rings and berl saddles, ½-in., 1-in.	Air	3.5–4.5 N H_2SO_4	75–480	3,300		Data of Withers
Dwyer and Dodge (1941)	12	48	Rings, ½-in., 1-in., 1½-in.	Air	Water	100–1,000	100–1,000	74–88	
Fellinger (1941)	20	9, 17, 25¼, 25¼, 23¾	Rings, ⅜-in., ½-in., 1-in., 1½-in., 2-in.	Air	Water	500–4,500	200–1,000		
	20	20½, 22, 25	Berl saddles, ½-in., 1-in., 1½-in.	Air	Water	500–4,500	200–1,000		
	20	26	Triple spiral tile, 3-in.	Air	Water	500–4,500	200–1,000		
Parsly, Molstad, Cress, and Bauer (1950)			No. 6295 drip-point grid	Air	Water	1,900–15,000	100–1,000		
Pigford and Pyle (1951)	31.5	26, 52	Spray tower	Air	Water	285–900	230–800	6	Sprayco 5-B nozzles

300 *Gas Purification*

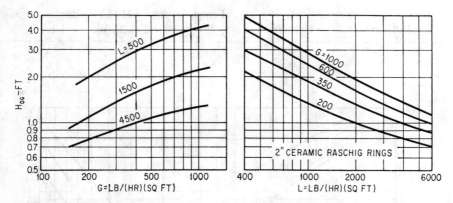

Figure 4-12. Values of H_{OG} for absorption of ammonia in water (packed tower, 2-in. ceramic raschig rings). *Data of Fellinger (1941)*

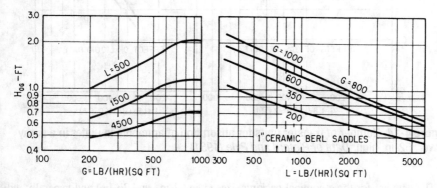

Figure 4-13. Values of H_{OG} for absorption of ammonia in water (packed tower, 1-in. ceramic berl saddles). *Data of Fellinger (1941)*

Rates of absorption of ammonia in water in countercurrent spray towers have been determined by Kowalke et al. (1925) and Pigford and Pyle (1951). Data obtained by Pigford and Pyle are shown in **Figure 4-14** for a 52-in. high spray tower. In this case the data are given in terms of the number of transfer units because spray column performance is not proportional to height. The indicated number of transfer units for the system is almost directly proportional to the liquid rate because the area for mass transfer increases almost linearly with flow rate in spray units.

Mass transfer data on specific packings and gas/solvent systems of interest can often be obtained from vendors. For example, Jaeger Products, Inc. (1990) gives the following data for the absorption of ammonia by water using Jaeger Tri-Packs, hollow spherical-shaped packings made of injection molded plastic:

G lb/hr ft²	L lb/hr ft²	Temp. °F	HTU, inches 2 & 3½-in.	1-in.
512	1,024	68	8.4	5.6
512	4,096	68	5.4	3.6

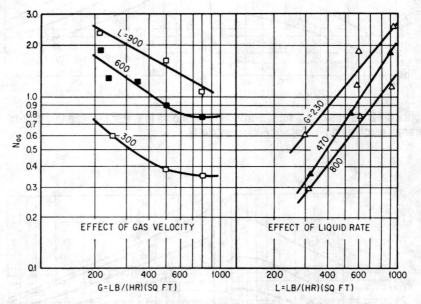

Figure 4-14. Number of transfer units for absorption of ammonia in water in a spray tower, 52-in. high. *Data of Pigford and Pyle (1951)*

The data are for three nominal packing sizes (1-in., 2-in., and 3½-in.) and generally indicate a performance considerably better than comparably sized ring or saddle-type packings. Packing (or tray) performance can also be predicted from correlations based on fundamental data. Available correlations are discussed in Chapter 1.

Sour Water Stripping Process

Sour water strippers are commonly used in refineries and other operations where a stream of water is produced containing ammonia and hydrogen sulfide. In most cases the sour water also contains carbon dioxide and a number of trace impurities. The principal object of the stripper is to remove the ammonia and hydrogen sulfide, typically to levels below 50 ppmw NH_3 and 10 ppmw H_2S, so that the water can be reused or disposed of as normal wastewater.

Carbon dioxide is not considered to be a noxious impurity in the water; nevertheless, it is completely removed with the ammonia and hydrogen sulfide during the stripping operation. The presence of phenol in the sour water may require the inclusion of a separate phenol extraction step in the water treatment process, while the presence of strong or nonvolatile acids can usually be handled in the sour water stripping system by the addition of caustic soda. The caustic soda displaces ammonia by forming stable salts with the acids. As a result the product water contains a small concentration of sodium salts, but no significant quantity of ammonia.

Process Description

A flow diagram of a typical sour water stripping system is shown in **Figure 4-15.** The sour water feed is passed through a separator where floatable light oil and heavy sludge are removed. It is next collected in a feed storage tank, which serves to smooth out flow rate and composition changes and, after preheating, is fed to the top of the stripping column. Sieve tray columns are probably the most popular, although other types of trays and packed columns are also used successfully. Both the water-cooled overhead condenser/reflux system and the steam-heated reboiler system shown in the flow diagram are optional items. The overhead vapor stream can sometimes be used without condensation as feed to a sulfur plant or incinerator, and stripping vapor can be provided by direct steam (or inert gas) injection; however, economics and practical considerations usually favor the arrangement shown for all but very small installations.

Trends in the design of sour water stripping systems as noted by Won (1982) include

1. Using less steam and more trays to accomplish a given stripping requirement to reduce operating costs
2. Using reboilers instead of direct steam injection to conserve clean condensate and reduce effluent volumes
3. Using an overhead condenser to provide reflux, particularly when the vapor stream is fed to a sulfur plant
4. Using controlled caustic injection to free fixed ammonia

An American Petroleum Institute (API) survey (Gantz, 1975) also noted a trend toward the use of tray columns instead of packed columns. About 75% of the plants surveyed used trays, and all units built after 1970 used trays.

Design Approach

A detailed procedure for designing sour water strippers is outlined by Beychok (1967). The Beychok approach is a simple tray-by-tray calculation that starts with a preliminary material balance around the tower and assumed values for the number of trays, the column top pressure, the individual tray pressure drop, and the stripping steam flow rate leaving the top tray. The tray-by-tray analysis begins at the top tray with an estimate of the temperature based on the partial pressure of water vapor in the gas phase. An initial value is assumed for the NH_3/H_2S mole ratio in the liquid on the top tray, then this ratio is adjusted as required until the composition of the gas and liquid phases leaving the tray are in equilibrium. A material balance around the top (first) tray gives the composition of vapor coming from the second tray.

The total pressure at the second tray is estimated by assuming a pressure drop of about 0.50 psi per theoretical tray (based on 0.25 psi per actual tray and 50% tray efficiency) and the above procedure repeated. The tray-by-tray analysis is continued down the column until the liquid leaving a tray meets the product water specification with regard to ammonia and hydrogen sulfide content. If necessary, the entire process is repeated with different initial

304 Gas Purification

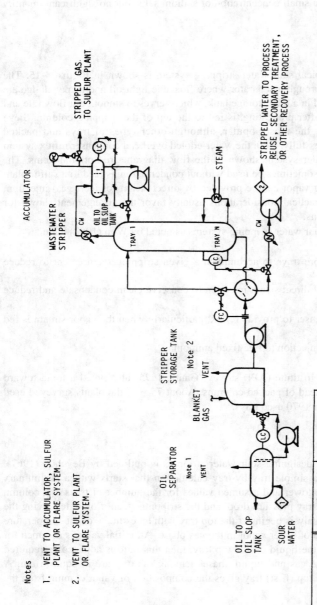

Figure 4-15. Flow diagram of sour water stripper system. *From U.S. DOE, Coal Conversion Systems Data Book (1982)*

assumptions until good agreement is obtained and optimum design conditions are established. Composition and temperature profiles calculated by Beychok for a sour water stripper operating with feed to the top tray and no reflux are shown in **Figure 4-16**.

Wild (1979) adapted the Beychok design procedure to a hand calculator. His sample calculation is of interest in defining typical design and performance parameters for a commercial sour water stripper. The design is for a stripper to process 100,000 lb/hr of sour water containing 0.73 wt% H_2S and 0.556 wt% NH_3 and produce water containing 5 ppm and 20 ppm of H_2S and NH_3, respectively. The initial design assumptions are a stripping steam rate at the top of the column of 10.2% of the feed (10,200 lb/hr), a column top pressure of 30 psia, and a pressure drop per theoretical tray of 0.45 psia. Results of the calculation, which are reported to compare favorably with the results from a commercial computer program, are summarized in **Table 4-10**.

The effect of the quantity of stripping steam and the number of theoretical trays on the amount of ammonia remaining in the bottoms liquid of a typical stripper is shown in **Figure 4-17** (Gantz, 1975). For comparison, actual data from a refinery installation (questionnaire No. 34 of the API survey) are included on the chart. The actual column, which contains 15 ft. of 3-in. raschig rings, appears to behave as predicted for a 4.7 theoretical tray column, indicating a height equivalent to a theoretical tray (HETP) of 3.2 ft. In most cases studied, HETPs for 3-in. raschig rings averaged 2.5 to 3.5 ft and overall tray efficiencies ranged between 30 and 50% (Gantz, 1975).

The API-sponsored survey of more than 80 sour water stripping operations found that the Beychok calculation model correlates satisfactorily with observed results down to an ammo-

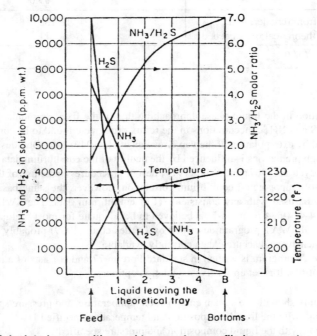

Figure 4-16. Calculated composition and temperature profile in sour water stripper. *From Beychok (1967). Used with permission from John Wiley & Sons, New York*

Table 4-10
Sour Water Stripper Design Calculations; Input Data and Results

Material Balance Stream	Pounds per hour		
	H_2O	NH_3	H_2S
Sour water feed	98,714.00	556.00	730.00
Tower bottoms	98,292.43	1.97	0.49
Overhead vapors	10,200.00	1,740.82	1,451.06
Reflux	9,778.43	1,186.79	721.55
Tail gas	421.57	554.03	729.51
Temperatures, °F			
Feed		199	
Reflux drum		184	
Bottoms		259	
Pressures, psia			
Tower top		30	
Reflux drum		27.2	
Pressure drop per theoretical tray		0.45	
Reboiler		35.00	
Steam rate from reboiler, lb/hr		18,158	
Number of theoretical trays needed		11	

Data of Wild (1979)

nia concentration in the liquid phase approaching that of the fixed ammonia (Gantz, 1975). Measured H_2S and NH_3 concentrations in the reflux drum were found to be considerably less than predicted by extrapolation of the van Krevelen et al. (1949) data. Obviously, the accuracy of the design predictions is influenced by the quality of the equilibrium data used.

Both the Beychok (1967) and Wild (1979) design procedures make use of the van Krevelen et al. (1949) vapor-liquid equilibrium correlation; however, the same basic tray-by-tray approach can be used with any consistent set of equilibrium data. The SWEQ correlation (see **Figures 4-4, 4-5, 4-6,** and **4-7**) is believed to be adequate for most preliminary designs; however, the GPSWAT program described by Gerdes et al. (1989) probably represents the most comprehensive and accurate data currently available.

The GPSWAT program is written in standard Fortran 77 and consists of a major program and 17 subroutines. It is set up for five calculation options:

1. Equilibrium flash, with the stream composition, temperature, and pressure specified
2. Bubble point, with the liquid composition and temperature specified
3. Bubble point, with the liquid composition and pressure specified
4. Dew point, with the gas composition and temperature specified
5. Dew point, with the gas composition and temperature specified

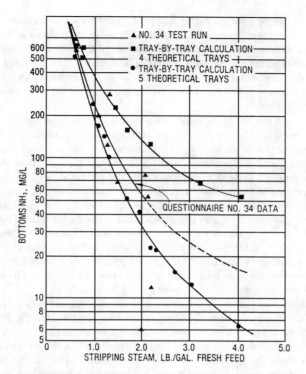

Figure 4-17. Effect of stripping steam and number of trays on sour water stripper performance. *From Gantz (1975). Reproduced with permission from Hydrocarbon Processing*

In each case, the unspecified temperature, pressure, and composition will be calculated. The pH can also be calculated, or a specified pH can be imposed on the calculation in which case the required amount of caustic will be calculated. The GPSWAT model is useful in predicting the composition of sour water that will result from a high pressure gas/water contact or in designing a sour water stripping system using either manual or computerized column design techniques.

Plant Operations

The sour water stripper system represents a very simple separation process, which generally operates in a predictable and straightforward manner. The three main problems encountered in many installations are: (1) the presence of fixed ammonia, (2) the presence of phenols, and (3) corrosion.

Ammonia is fixed, so it is not readily vaporized in the stripper, by the presence of nonvolatile or strong acids in the feed stream. The acids may be weak (e.g., benzoic acid) or strong (e.g., hydrochloric acid). One source of acid is the oxidation of sulfur compounds in the sour water to form, for example, thiosulfuric acid. This source can be minimized by preventing contact of the sour water with air at any point in its gathering and processing.

Fixed ammonia can be released during stripping by the addition of caustic soda. A study of caustic addition by Bomberger and Smith (1977) indicates that the addition of caustic does not interfere with hydrogen sulfide removal provided the caustic is added in an amount equivalent (on a molar basis) to the fixed ammonia in solution. They conclude that it is more efficient to add the caustic at the top of the column with the feed stream than to add it lower in the column in practice, an excess of caustic is added near the bottom to avoid interference with hydrogen sulfide stripping.

One problem with phenol in the sour water is that a column designed specifically to remove a high percentage of the ammonia and hydrogen sulfide is capable of removing only a relatively small fraction of the phenol. This is due to the low volatility of phenol, not to a low tray efficiency for phenol removal. Won (1982) studied the individual tray efficiencies for ammonia, hydrogen sulfide, hydrogen cyanide, and phenol in sour water strippers and found that ammonia and phenol have relatively high tray efficiencies (about 70%) while hydrogen sulfide, hydrogen cyanide, and a phenolic mixture have low efficiencies (about 20 to 40%). He used commercial column performance data published by the API which generally showed that 8- to 10-tray (actual) columns removed only about one-half the phenol when operated at steam rates of 1.0 to 2.7 lb/gal. If this level of phenol removal is not adequate, the column can be designed specifically for phenol removal, or a phenol extraction system can be installed to treat the sour water feed before it enters the stripper. The latter approach is generally preferred if significant amounts of phenol are present.

Sour water strippers are generally constructed of carbon steel and only minor corrosion is reported in the tower, trays, and feed-to-bottoms heat exchangers (Gantz, 1975). Materials experience and corrosion problems in sour water strippers are discussed in two API reports (1974, 1976). Significant corrosion has been encountered in the overhead systems of refluxed towers when the overhead systems were constructed of carbon steel. The API reports indicate that titanium is very resistant to corrosion in such situations.

Ammonium Salt Production

The Direct Process

This process, which was proposed by Brunck (1903), is designed to eliminate the necessity of recovering ammonia from aqueous solutions before its conversion to ammonium sulfate. The hot gas leaving the retorts or coke-ovens is kept at a temperature above its water dew point and passed directly through concentrated sulfuric acid. During absorption of the ammonia, an appreciable portion of the tar is also removed from the gas; this results not only in serious contamination of the ammonium sulfate, but also in acid degradation and contamination of tar. In addition, ammonium chloride contained in the gas is decomposed by the concentrated sulfuric acid, and the hydrogen chloride evolved causes extremely severe corrosion of the equipment. Some of the difficulties of the process can be alleviated by installation of very elaborate tar-separation systems; others, however, have proved to be almost insurmountable, and, as a result, the process has only been used in a few installations. The relative merits of the process as compared with the indirect and semi-direct methods are discussed in some detail by Ohnesorge (1910, 1923).

The Indirect and Semi-direct Processes

The indirect process requires that the ammonia in the gas first be removed by a water wash operation as described in the previous section entitled "Coke-Oven Gas Processing."

The stripper overhead vapor from this process is then passed through a "saturator" containing sulfuric acid to produce ammonium sulfate crystals.

The semi-direct process, which was developed by the Koppers Company, Inc. (1907) and is now licensed by ICF Kaiser Engineers, is a combination of the direct and indirect processes. A schematic flow diagram is shown in **Figure 4-18**. Gas evolved in the coke ovens is first treated in a manner similar to that described in the "Coke-Oven Gas Processing" section, steps 1 through 3.

After flowing through the electrostatic tar precipitator and exhauster, the gas flows through a reheater where its temperature is raised above its dew point, usually to about 60°C (180°F). The reheated gas (which may also contain the overhead vapor from an ammonia still) flows to the saturator where the ammonia is removed by reaction with a strong acid, in most cases sulfuric acid. The ammonia-free gas, which still contains most of the originally present weakly acidic gases, passes to the light-oil recovery and desulfurization operations.

Absorption of ammonia in acid (typically sulfuric acid) is carried out in a vessel called the "saturator." A diagrammatic view of a typical unit is shown in **Figure 4-19**. A discussion of different saturator designs is presented by Otto (1949). Saturators are usually lead- or stainless steel-lined, although other materials such as Monel or acid-resistant refractories may also be used. A concentration of approximately 5 to 7% free acid is maintained in the saturator liquid by continuous addition of 60° Bé (77.6%) sulfuric acid. Periodically, usually every 24 hr, the liquid level in the saturator is raised and the free-acid content is increased to 10 to 12%. This is done to dissolve crystals of ammonium sulfate which have accumulated on the vessel walls

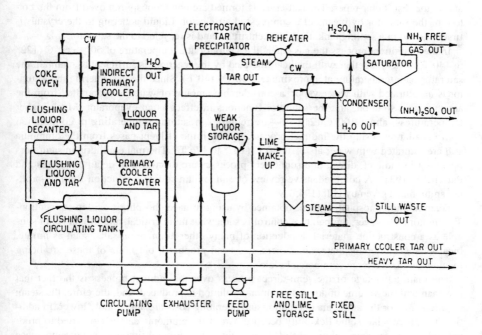

Figure 4-18. Diagrammatic flow sheet of semi-direct ammonia-removal process.

310 Gas Purification

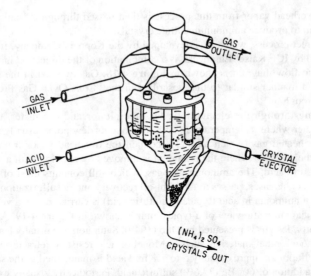

Figure 4-19. Saturator for production of ammonium sulfate.

and in the distributor pipes. The salt crystals formed are continuously removed from the bottom of the saturator by means of a compressed-air ejector. Liquid adhering to the crystals is first removed in a settling tank, then in a centrifuge, and is recycled to the saturator.

In the semidirect process the gas enters the saturator at a temperature of 50° to 60°C (122° to 140°F) which, together with the heat liberated by the reaction, is sufficient to maintain the saturator-bath temperature at approximately 60°C (140°F). Since the gas entering the saturator is unsaturated with water vapor, a considerable amount of water is evaporated from the aqueous acid solution, and the saturator functions in effect as an evaporator. Much higher temperatures—about 100°C (212°F)—are required in saturators operating in conjunction with installations using the indirect process, in which the effluent gases from the ammonia still are saturated with water at 75° to 80°C (167° to 176°F). Thermal equilibria in saturators operating in indirect, direct, and semidirect processes are discussed in detail by Terres and Patscheke (1931). A comprehensive review of ammonium sulfate production from by-product ammonia is given by Hill (1945).

Diammonium phosphate may be obtained by using furnace phosphoric acid in conventional saturators. However, temperature control is somewhat more critical, and Monel is not suitable as a construction material. Production of many other ammonium salts, such as nitrate, bicarbonate, monophosphate, and others, has been proposed, but none of these are being manufactured on a large commercial scale.

The main advantage of the semi-direct process over the indirect process is the fact that less than half the volume of aqueous ammonia solution is produced, and, therefore, the steam requirements for the ammonia distillation and operating costs are appreciably lower. In addition, the process has some flexibility because part of the ammonia can be converted to products other than the ammonium salts of strong acids. Furthermore, first cost and ground-space requirements are less than for installations using the indirect process, and, finally, smaller ammonia losses are incurred.

Phosam and Phosam-W Processes

The Phosam technology was initially developed by USX Corporation to remove ammonia from coke-oven gas and recover it as pure anhydrous ammonia. It is called the Phosam process when applied to coke plants (i.e., removal of ammonia from coke-oven gas, removal of ammonia from the overhead vapor stream of an ammonia liquor distillation column, or removal of ammonia from the overhead stream of a deacidifier column ahead of a Claus plant). It is designated as the Phosam-W process when used for any other application (i.e., chemical plants, coal gasification systems, shale oil processing plants, and petroleum refineries). The W in Phosam-W is related to its use for the treatment of sour water (Busa, 1992).

The technology is covered by U.S. Patents 3,024,090, 3,186,795, and 3,985,863 as well as patents in Canada, Germany and, Great Britain. Aristech Chemical Corporation is the proprietor of the Phosam and Phosam-W processes, and USX Engineers and Consultants, Inc. (UEC) has exclusive licensing rights worldwide. Detailed descriptions are given in UEC sales literature and in papers by Rice and Busa (1984, 1985) and Busa and Cole (1986).

Work on the Phosam Process began in the late 1950s and the first full-scale commercial plant was started up at USS Clairton Works in 1968. By 1994, twenty-four different companies worldwide had installed or licensed the Phosam process, while ten companies had installed or licensed the Phosam-W version.

Process Description

Typical flow diagrams for the Phosam and Phosam-W processes are given in **Figures 4-20** and **4-21.** It should be noted that the processes are identified by application, not flow arrangement, and the illustrated diagrams are not the only possible systems for either version. For example, when ammonia is removed from coke-oven gas by water washing and stripping,

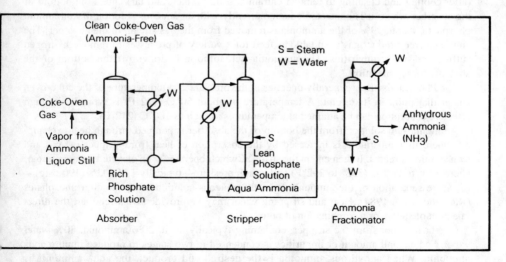

Figure 4-20. Flow diagram of typical Phosam process in a coke plant. *From USX Engineers and Consultants, Inc. (1991)*

312 Gas Purification

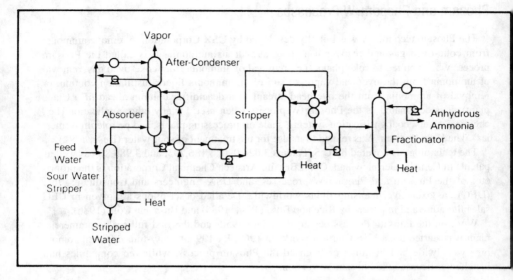

Figure 4-21. Flow diagram of Phosam-W process. *From USX Engineers and Consultants, Inc. (1991)*

the Phosam absorber would be installed to scrub ammonia from the stripper overhead vapor as illustrated for the Phosam-W case.

Referring to **Figure 4-20,** which represents a typical Phosam plant, the coke-oven gas (after cooling and cleaning to remove entrained solids, water, and tar) enters the bottom of the absorber where it is contacted countercurrently with an aqueous solution of ammonium phosphate. About 99% of the ammonia is removed from the gas stream in the absorber. The ammonia-free product gas is suitable as feed for a variety of processes to remove hydrogen sulfide and other impurities. The ammonia-rich solution is drawn off the bottom of the absorber for regeneration.

The Phosam absorber generally operates at the pressure and temperature of the coke-oven gas at that point in its circuit. A temperature of about 50°C (122°F) is typical, although effective absorption can be attained at temperatures as high as 100°C (212°F).

Ammonia-rich solution from the bottom of the absorber is pumped through heat exchangers where its temperature is increased by the absorption of heat from stripper off-gas and lean solution streams. It then enters the stripper, which operates at an elevated temperature of about 170° to 200°C (338° to 392°F) and a high pressure—typically 1,300 kPa (190 psig)—to assure separation of the ammonia from the aqueous solution of ammonium phosphates (Rice and Busa, 1984). Heat and stripping vapor may be provided by a reboiler, the direct injection of steam, or a combination of both.

Overhead vapor from the stripper, containing typically about 20% ammonia, 80% water vapor, and a small amount of impurities is condensed in two stages to produce impure aqua ammonia. When anhydrous ammonia is the desired end product, the aqua ammonia is pumped to a fractionator which separates ammonia and water. The fractionator is typically operated at a high enough pressure to permit the ammonia overhead vapor to be condensed at the temperature of the available cooling water. Part of the liquid ammonia is used as reflux and the balance is taken off as a salable product.

The Phosam-W process shown in **Figure 4-21** differs from the system of **Figure 4-20** primarily in the front-end configuration. Sour water from the refinery or other non-coke-oven operation is processed in a conventional sour water stripper. The stripped water can be reused in the plant or further processed for disposal. The vapor phase, containing, for example, 10% NH_3, 10% H_2S and CO_2, and the balance steam with traces of HCN and volatile organics, is fed into the Phosam absorber. Absorber conditions for this case are approximately 90°–100°C (194°–212°F) and 35–70 kPa (5–10 psig). About 95–99% of the ammonia is removed from the vapor phase passing through the absorber.

Vapor leaving the top of the absorber is next passed through an after-condenser where a small portion of the water vapor is condensed. The resulting small stream of sour water is recycled to the sour water stripper. Rich solution from the absorber is recycled around the bottom of the absorber to maximize its loading with ammonia. The net rich solution product is fed to the regeneration stage which is similar to that of the Phosam process.

Process Chemistry

The process uses a conventional absorption-regeneration cycle in which a thermally unstable compound is formed during absorption and decomposed by an increase in temperature and the presence of stripping vapor during regeneration. The chemistry is simple and can be represented approximately by the conversion of monoammonium phosphate to diammonium phosphate as follows:

$$NH_3 + (NH_4)H_2PO_4 = (NH_4)_2HPO_4 \tag{4-14}$$

In practice, the reactions are not carried to completion in either direction. The mole ratio NH_3/HPO_4 shifts from the range of 1.2–1.4 in the lean solution to 1.7–2.0 in the rich solution (Rice and Busa, 1984). In effect the dissolved salt has the formula, $(NH_4)_nH_{3-n}PO_4$, where n varies from less than 1.4 in the lean solution to over 1.7 in the rich solution. The concentration of salt in the circulating solution is maintained well below the limit of solubility of the existing salt form at all points in the circuit to avoid the formation of solid salt crystals.

Since phosphoric acid is essentially nonvolatile, thermal decomposition of diammonium phosphate in solution yields only ammonia gas in the vapor phase. The phosphate ion is extremely stable and is not altered in the process; therefore, no blowdown is required and only a small makeup of phosphoric acid is needed to compensate for unavoidable mechanical losses. The small amount of acid gases, acidic organics, and neutral compounds that are co-absorbed with the ammonia are either recycled to the absorber or reacted with caustic soda to prevent corrosion and ammonia product contamination in the fractionator.

Performance

Ammonia produced by the process will readily meet commercial or fertilizer grade specifications (99.5% minimum ammonia content, 0.2–0.5% wt. water, 5 ppm max. oil) for which there is a ready market. The fractionation system has the capability of producing an even purer ammonia when a market exists for such a product.

In plants where the hydrogen sulfide is ultimately converted to elemental sulfur, removal of ammonia from the acid gas stream greatly improves the operation and economics of the sulfur plant. This is because ammonia tends to form compounds that plug the catalyst beds and extra air is required to completely oxidize any ammonia in the feed gas. The presence of ammonia in the feed to a sulfur plant also requires the use of high efficiency burners to assure the complete destruction of ammonia before all oxygen is consumed by hydrogen sulfide.

Capital and Operating Costs

According to USX Engineers and Consultants, Inc. (1991), a Phosam Process plant is competitive with other available processes for removing ammonia from coke-oven gas, including ammonium sulfate production and ammonia destruction by incineration, and is a more compact system than either of these alternatives. They have provided the data given in **Table 4-11** as an aid to estimating projected operating costs. Data are not provided for Phosam-W installations because they are normally very site specific.

Chevron WWT Process

A process for treating ammonia-containing sour water generated during petroleum refining operations has been developed by the Chevron Research and Technology Company. It is anticipated that the process will also be applicable to the high-ammonia sour water produced in shale oil and tar sands processing plants. In this concept, which is called the Chevron WWT (waste water treatment) process, relatively pure streams of ammonia, acid gas, and stripped water are produced. According to the developer, approximately one new WWT plant per year is licensed; as of mid-1992 approximately 20 plants were operating worldwide. Descriptions of the process are given by Annessen and Gould (1971), Leonard et al. (1984, 1985), Busa and Cole (1986), and, more recently, by Bea and Wardell (1989).

Conventional sour water stripping systems use a single column and produce a vapor stream containing both ammonia and hydrogen sulfide. This mixed vapor stream may be fed to a Phosam unit, passed through a strong acid ammonia absorber, fed to a sulfur recovery unit, or incinerated. The WWT process also accomplishes sour water stripping, but uses a two-column stripping system that separates the ammonia and acid gases.

Table 4-11
Operating Requirements for Phosam Process Plants

Chemicals and Utilities	Per kg NH_3
H_3PO_4 Makeup (as 100%)	0.0075 kg
NaOH (as 100%)	0.01 kg
Steam (250 psig)	10–11 kg
Electricity	0.22 kW
Cooling Water (per °C temp rise)	3.5–5.0 m^3
Labor for Operation and Analytical Control	
Men per shift	less than 2

Data provided by USX Engineers and Consultants, Inc. (1991)

The production of separate ammonia-rich and acid gas-rich streams offers the following advantages:

1. Removal of ammonia from feed to a conventional sulfur recovery plant reduces sulfur plant operating problems and improves economics.
2. Ammonia-free hydrogen sulfide can be used as feed to a sulfuric acid plant.
3. The ammonia stream can be converted to a salable byproduct or burned without producing SO_2.

Process Description

As shown in **Figures 4-22** and **4-23,** the Chevron WWT process incorporates the following steps:

1. Degassing and feed storage
2. Acid gas stripping
3. Ammonia stripping
4. Ammonia purification and recovery

Plants have been designed to handle sour water feeds containing 0.3 to 6.0 wt% NH_3 and 0.3 to 10 wt% H_2S and can be designed to handle even higher concentrations of these two ingredients as well as high levels of CO_2 (Bea et al., 1989). When large volumes of dilute sour water are to be processed, it is often economical to use a preconcentrator ahead of the

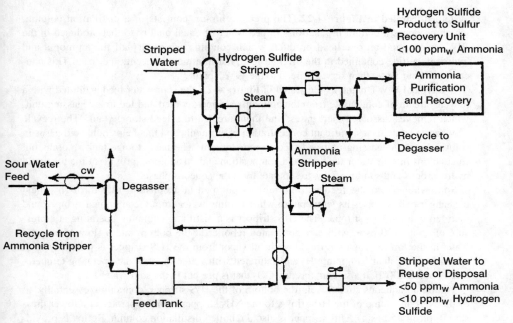

Figure 4-22. Flow diagram of Chevron WWT process. *From Bea et al. (1989)*

316 Gas Purification

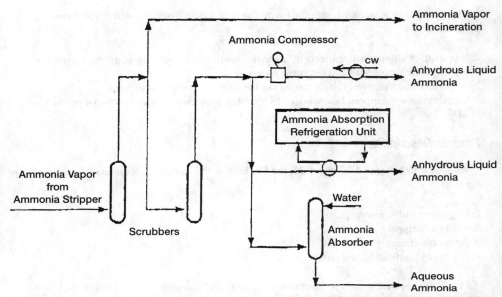

Figure 4-23. Chevron WWT process; ammonia recovery options. *From Bea et al. (1989)*

system illustrated in **Figure 4-22.** The preconcentrator consists of a distillation column, which produces a bottom product of stripped water of equal quality to that produced in the WWT plant, and an overhead product, which contains essentially all the ammonia and hydrogen sulfide contained in the original feed, but in a more concentrated form. This concentrated sour water then serves as feed to the WWT plant.

In a typical WWT plant, as illustrated in **Figure 4-22,** the sour water feed is diluted with a recycle stream of condensate from the ammonia stripper; cooled and fed to a degassing unit, which removes dissolved inert gases; and then directed to a feed storage tank. The recycle condensate contains a significant concentration of ammonia and this helps hold acid gases in solution in the degassing unit and feed storage system. The feed storage tank smooths out fluctuations in sour water feed rate and composition and, in addition, provides time for liquid hydrocarbons and solid particles to separate from the aqueous phase.

Sour water from the feed storage tank is pumped at a constant rate through a heat exchanger, which raises its temperature while cooling water from the ammonia stripper, and then fed into the H_2S stripper. The H_2S stripper is a distillation column, operating at a pressure of about 100 psig, with a steam-heated reboiler at the bottom and a flow of stripped water at the top to provide reflux. Overhead vapor from the H_2S stripper is primarily H_2S, containing less than 100 ppm NH_3 and saturated with water vapor at the operating temperature of about 100°F. It also contains any CO_2 that is present in the sour water feed.

The liquid aqueous stream from the bottom of the H_2S stripper, containing essentially all of the NH_3 and some of the H_2S, flows to the NH_3 stripper, which operates at a lower pressure (typically 50 psig). This stripper is also a reboiled distillation column. Reflux is provided by recycling water condensed from the overhead vapor stream. Part of this condensate is recycled to the sour water feed to the plant. The vapor stream, which is primarily ammonia saturated with water vapor at the condenser temperature, is sent to the ammonia disposal or recovery system.

The liquid product from the bottom of the ammonia stripper is clean water containing less than 50 ppm NH_3 and less than 10 ppm H_2S. After passing through a heat exchanger, where it is cooled while heating the sour water feed stream, and after further cooling, if necessary, the water is suitable for reuse in refinery processes or for disposal with other waste waters.

The vapor product from the ammonia stripper can be processed by one of the options shown in **Figure 4-23** for recovery or disposal of ammonia. When the quantity of ammonia produced or the local ammonia market make the production of salable ammonia uneconomical, the ammonia-rich vapor can be incinerated by using it as fuel in a process furnace or boiler. To prevent air pollution it may be necessary to scrub the ammonia-rich stream with water before burning it. This generates a small amount of sour water, which can be recycled to the plant feed stream.

The production of anhydrous or aqueous ammonia requires that the ammonia stripper overhead stream be further purified to remove H_2S. Two steps of water scrubbing are required for the production of anhydrous ammonia, while the production of aqueous ammonia may be accomplished with either one or two scrubbing steps depending on the desired product purity. Typical product specifications for the Chevron WWT process are given in **Table 4-12**.

Process Economics

Plant investment and utility requirements for a typical 200 gpm sour water WWT plant are given in **Table 4-13**. In many cases the higher investment cost and utility requirements of the ammonia recovery option are more than offset by the byproduct value of the ammonia product. The disposition of the ammonia product from 15 WWT plants was reported by Bea et al. (1989) to be as follows: seven utilized ammonia incineration; four produced ammonia for industry, and agriculture; and four used the ammonia to produce nitric acid.

**Table 4-12
Typical Product Specifications: Chevron WWT Process**

	Hydrogen Sulfide	Stripped Water	Anhydrous** Ammonia	Anhydrous Ammonia	Aqueous Ammonia
Form	Gas	Liquid	Gas	Liquid	Liquid
Ammonia, ppm	<100	<50	—	—	—
Hydrogen Sulfide, ppm	—	<10	<1500	<5	<2
Water, Wt %	0.1	—	1.5	0.4	75*
Temperature, °F	100	100–200*	100	100	100
Pressure, psig	25–100*	—	40	200	35

*As required
**For incineration
From Bea et al. (1989)

Table 4-13
Chevron WWT Process Economics; Ammonia Incineration vs. Ammonia Recovery

	NH$_3$ Incineration	NH$_3$ Recovery
Investment, $	5,100,000	7,100,000
Utilities		
150 psig Steam, lb/hr	45,700	46,100
Cooling Water, gpm	4,600	5,100
Electric Power, kW	74	278
Deaerated Condensate, gpm	20	20
	Design Basis	
Sour Water, Rate, gpm	200	
Ammonia		
Wt %	2.5	
Tons/day	30	
Hydrogen Sulfide		
Wt %	5.0	
Tons/day	60	

From Bea et al. (1989)

Leonard et al. (1984) present a comparison of the economics of using the WWT process to produce ammonia versus the use of a sour water stripper with the product gas going to a sulfur recovery plant. The study is based on plants processing 285 gpm of sour water containing 3.1 wt% ammonia (47 tons/day) and 6.2 wt% H_2S (94 tons/day). They conclude that the investment cost for a sour water stripper system would be about $1 million (12%) more than for a WWT system when the extra costs to enable the sulfur and tail gas plants to handle the ammonia are added to the base cost of the SWS system. Operating costs are about $120,000 per year higher for the WWT system assuming credits of $2.6 million per year for ammonia sold (at $165/ton), and $1.4 million per year for ammonia burned to produce steam in the sulfur plant. Clearly, generalizations cannot be made with regard to the most economical approach; each case requires a site-specific evaluation.

The recent development of Claus plant sulfur furnace burners that can handle acid gas streams containing significant percentages of ammonia can improve the economics of simple sour water stripper-sulfur recovery unit systems. Such burners are available from LD Duiker, B. V. of Holland (1990) and Lurgi Corporation (Fischer and Kriebel, 1988). See Chapter 8 for additional information on sulfur plant burners.

USE OF AMMONIA TO REMOVE ACID GASES

The use of ammonia to remove hydrogen sulfide and carbon dioxide from gas streams has declined in recent years; however, the process is still used to desulfurize coke-oven gas in a number of installations. Ammonia-based hydrogen sulfide removal processes are offered by the Krupp Wilputte Corporation (1988), Davy-Still Otto (1992), and Mitsubishi Kakoki Kaisha, Ltd. (Fumio, 1986).

Processes have also been developed which use aqueous ammonia solutions to remove carbon dioxide from synthesis gas. The use and relative economics of ammonia-based carbon dioxide processes as of the late 1950s have been described by Mullowney (1957). Ammonia has the advantage for such applications of being essentially unaffected by the presence of carbonyl sulfide, carbon disulfide, hydrogen cyanide, and hydrogen sulfide. However, the process is more complex than those based on nonvolatile alkaline absorbents and is not capable of providing very low levels of carbon dioxide in the product gas. As a result, the process is seldom considered for new commercial applications. A detailed discussion of the history of ammonia use for acid gas removal and descriptions of many processes that are no longer of industrial significance are given in previous editions of this book and in the *Chemistry of Coal Utilization* (Lowry, 1945), including the first and second supplements (Lowry, 1963; Elliott, 1981).

Basic Chemistry

The reactions occurring in the system comprised of ammonia, hydrogen sulfide, carbon dioxide, and water can be represented by the following equations:

$$NH_3 + H_2O = NH_4OH \tag{4-15}$$

$$NH_3 + H_2S = NH_4HS \tag{4-16}$$

$$2NH_3 + H_2S = (NH_4)_2S \tag{4-17}$$

$$2NH_3 + CO_2 = NH_2COONH_4 \tag{4-18}$$

$$NH_3 + CO_2 + H_2O = NH_4HCO_3 \tag{4-19}$$

$$2NH_3 + CO_2 + H_2O = (NH_4)_2CO_3 \tag{4-20}$$

$$NH_2COONH_4 + H_2O = (NH_4)_2CO_3 \tag{4-21}$$

$$(NH_4)_2CO_3 + H_2S = NH_4HCO_3 + NH_4HS \tag{4-22}$$

$$(NH_4)_2S + H_2CO_3 = NH_4HCO_3 + NH_4HS \tag{4-23}$$

$$NH_4HS + H_2CO_3 = NH_4HCO_3 + H_2S \tag{4-24}$$

Van Krevelen et al. (1949) have shown that under equilibrium conditions the ionic species NH_4^+, HCO_3^-, NH_2COO^-, and $CO_3^=$, as well as undissociated NH_3, are present in aqueous solution in measurable quantities. The same authors state that under most conditions H_2S is present in the form of HS^- ions. When the pH of the solution is below 12 the sulfide ion concentration is negligible, and even at pH 12 this ionic species amounts to only 0.1% of the fixed hydrogen sulfide.

The following equations present heats of reaction (based on published heats of formation) for the four reactions most predominant in processes using ammonia for the removal of hydrogen sulfide and carbon dioxide from gases:

$$2NH_3(g) + H_2S(g) \rightarrow (NH_4)_2S\ (aq) - \Delta H = 39{,}600\ Btu/lb\ mole\ (77°F) \quad (4\text{-}25)$$

$$NH_3(g) + H_2S(g) \rightarrow (NH_4)HS\ (aq) - \Delta H = 34{,}800\ Btu/lb\ mole\ (77°F) \quad (4\text{-}26)$$

$$2NH_3(g) + CO_2(g) + H_2O(l) \rightarrow (NH_4)_2CO_3(aq) - \Delta H = 73{,}200\ Btu/lb\ mole\ (77°F) \quad (4\text{-}27)$$

$$2NH_3(g) + CO_2(g) \rightarrow NH_4CO_2NH_2\ (aq) - \Delta H = 61{,}100\ Btu/lb\ mole\ (77°F) \quad (4\text{-}28)$$

The absorption of ammonia into water is quite rapid and governed almost entirely by the gas-film resistance; in fact, this is the classical system for chemical engineering studies of gas-film resistance. The rate of absorption of hydrogen sulfide into aqueous ammonia solutions is also rapid although it is dependent upon the ammonia concentration. In the presence of an adequate concentration of ammonia at the interface, it is probable that the rate of this absorption is also governed by the gas-film resistance. On the other hand, absorption of carbon dioxide into water or weak alkaline solutions is considered typical of liquid-film controlled systems, not because its gas-film resistance is any lower than that of H_2S and ammonia, but because its liquid-film resistance is very much greater. The net result is that when gases containing H_2S, ammonia, and CO_2 are contacted with water, the ammonia and H_2S are absorbed much more rapidly than the CO_2 and this difference can be accentuated by operating under conditions which reduce the gas-film resistance or increase the liquid-film resistance.

The absorption of CO_2 into water and dilute alkaline solutions is hindered by a slow chemical reaction which is required to convert the dissolved carbon dioxide molecules into the more reactive ionic species. In effect, the CO_2 molecules are absorbed until some molecules are removed by the hydration reaction. The efficiency of CO_2 absorption can, therefore, be improved by turbulence in the liquid film (this aids diffusion of unreacted molecules into the body of the liquid) and by an extended hold-time of liquid in the absorption zone (this provides for the continuous reaction of CO_2 molecules that do enter the liquid phase). These conditions can be met, for example, by a tall, packed column operating at a relatively high liquid-flow rate or by a liquid-filled column through which bubbles of gas are made to pass.

The hydrogen sulfide and ammonia-absorption rates can be increased by inducing turbulence in the gas phase at the interface, a condition that requires a high relative velocity between gas and liquid. This can be achieved by the use of high-pressure spray nozzles, which produce a much higher relative velocity than can be realized with gravity flow devices. If maximum selectivity for H_2S and ammonia is desired, the use of spray column in combination with relatively short contact-time is indicated.

An additional factor that bears on the selectivity of the process is the fact that, once in solution, CO_2 is an appreciably stronger acid than H_2S and, under equilibrium conditions, the process would actually be expected to be selective for CO_2. If the ammonia is added with the wash water, this is indeed possible providing that a sufficiently tall column is used. When the ammonia enters with the feed gas, however, it is absorbed near the bottom of the column and, because very little absorption of CO_2 can occur in pure water, additional column height will have little effect on the selectivity or on the total amount of acid gases absorbed.

With proper design and operation, most of the H$_2$S can be eliminated from the gas in a selective absorber. Removal of H$_2$S, CO$_2$, and HCN from the ammonia solution in a stripping column located ahead of the ammonia-distillation column permits completely separate processing of the ammonia and the acid gases. This serves to prevent some difficulties in the operation of the saturator or, if strong ammoniacal liquor is produced, results in much higher purity of the crude liquor. Finally, selective absorption of H$_2$S yields an acid-gas stream of high H$_2$S concentration, which is desirable in the further processing to sulfur or sulfuric acid.

Selective H$_2$S-removal processes do not result in complete elimination of H$_2$S and, if the treated gas is intended for use as a domestic fuel or in synthesis processes, a final purification step is required. The degree of H$_2$S removal depends on several operating variables, but it appears that elimination of about 90% of the H$_2$S is the maximum that can be attained economically. Substantial amounts of hydrogen cyanide are also removed in the selective absorber.

Process Descriptions

A basic flow diagram of the process for removing H$_2$S from coke-oven gas using ammonia solution is shown in **Figure 4-24.** In the process the gas passes through a hydrogen sulfide scrubber and an ammonia scrubber in series. Stripped water is fed to the top of the

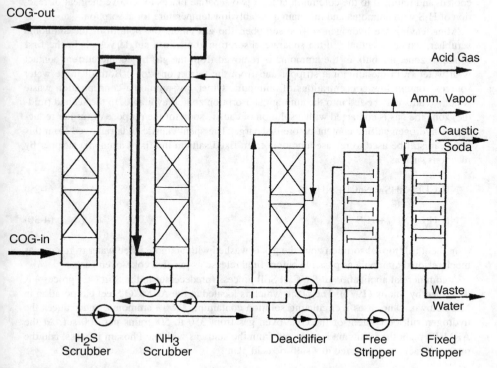

Figure 4-24. Basic flow diagram for coke-oven gas desulfurization with aqueous ammonia.

ammonia scrubber where it absorbs ammonia from the gas. The resulting ammonia solution is then used as the alkaline absorbent for H_2S in the hydrogen sulfide absorber. The rich solution from this unit, containing ammonium sulfide in solution, is fed to a deacidifier which decomposes the ammonium sulfide to produce a hydrogen sulfide-rich vapor and an ammonia-rich liquor. Part of the ammonia-rich liquor is fed to an intermediate point in the hydrogen sulfide absorber and the surplus is fed to an ammonia stripper where ammonia vapors are steam stripped from the solution. Caustic soda is normally added to the ammonia stripper or to a separate fixed-ammonia stripper to release ammonia that is present as the salt of a nonvolatile acid. The various processes differ in the details of heat exchange, recycle streams, wash steps, hardware designs, and process conditions; however, the basic operations and chemical reactions are essentially the same.

Still Otto Process

A more detailed diagram of the hydrogen sulfide and ammonia scrubber systems of the Still Otto process is shown in **Figure 4-25** (Davy/Still Otto, 1992). In this arrangement the ammonia-rich deacidified water is fed to the hydrogen sulfide scrubber near the bottom to minimize absorption of carbon dioxide; while the wash water product from the ammonia scrubber, which contains much less ammonia, is fed to the top of the scrubber to maximize the total amount of H_2S removed. The liquid flowing down through the H_2S scrubber is removed, cooled, and returned to the column in at least two separate places to remove the heat of reaction of H_2S with ammonia and maintain a suitably low temperature for absorption.

After leaving the hydrogen sulfide scrubber, the gas enters the bottom of the ammonia scrubber, which is actually three separate absorption zones in a single vessel. In the first (bottom) zone, the bulk of the ammonia is removed from the gas by countercurrent contact with water from the ammonia stripper and from the third (top) zone. Both of these water sources contain low concentrations of ammonia. Other low-ammonia-concentration waste water streams may be fed into the ammonia absorption zone if available. In the second (middle) zone the gas is contacted with a solution of caustic soda for the purpose of further reducing its hydrogen sulfide content before discharge. The spent caustic soda removed from this zone can still be used to release ammonia from fixed salts in the fixed ammonia stripper by reactions such as

$$NH_4Cl + NaHS = NaCl + NH_4HS \tag{4-29}$$

$$NH_4HS = NH_3 + H_2S \tag{4-30}$$

In the third (top) absorption zone the gas is washed with ammonia-free water to remove as much residual ammonia as possible before final release as purified coke-oven gas.

A commercial application of the Carl Still process (predecessor to the Still Otto process) is described by Snook (1977). The plant, which is located at the ARMCO Steel Corporation in Middletown, Ohio, uses a circulating solution containing 2 wt% ammonia, and reduces the hydrogen sulfide content of the coke-oven gas from 350 to 50 grains per 100 scf. At the ARMCO plant the ammonia is removed from the acid gas with the Phosam process, and the remaining acid gases are fed to a sulfuric acid plant.

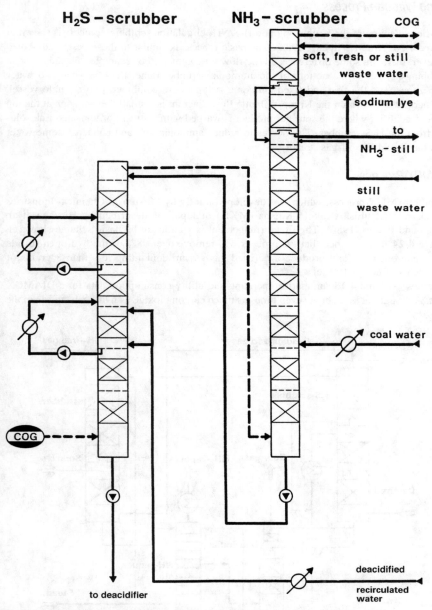

Figure 4-25. Diagram of Still Otto process hydrogen sulfide and ammonia scrubber system. *Davy/Still Otto (1992)*

Krupp Wilputte Process

A flow diagram of the Krupp Wilputte H_2S/NH_3 circulation scrubbing process is shown in **Figure 4-26** (Krupp Wilputte, 1988). The basic process is similar to those described above; however, there are some differences in the flow arrangement. For example, the caustic soda scrubbing zone is at the bottom of the ammonia scrubber rather than between two water wash zones, and the free-and fixed-ammonia strippers are combined into a single vessel. Another detail shown on the Krupp Wilputte flow diagram is a small contact zone at the top of the deacidifier where the sour gas product is washed with a stream of unheated rich solution from the H_2S scrubber. This serves to reduce ammonia loss and condense some water from the deacidifier off-gas stream.

DIAMOX Process

The DIAMOX process, which was developed jointly by Mitsubishi Chemical Industries (MCI) and Mitsubishi Kakoki Kaisha (MKK) in Japan, is described by Hiraoka et al. (1977) and Fumio (1986). The flow arrangement is similar to the basic scheme shown in **Figure 4-24.** It is claimed that the process will remove over 98% of the hydrogen sulfide from coke-oven gas and produce a purified gas stream containing less than 8 grains of hydrogen sulfide per 100 scf.

Tables 4-14 and **4-15** summarize operating and utility consumption data for a DIAMOX plant. Although there is about seven times as much carbon dioxide as hydrogen sulfide in the

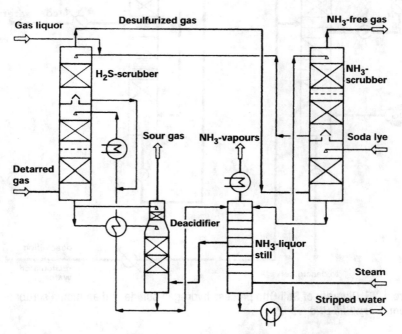

Figure 4-26. Krupp Wilputte H_2S/NH_3-circulation scrubbing process for desulfurization and ammonia removal. *Krupp Wilputte Corp. (1988)*

COG treated by this plant, the data show that the DIAMOX process selectively removes the hydrogen sulfide. Operating data from the MCI Kurosaki plant in **Table 4-14** show the amount of ammonia, hydrogen sulfide, and carbon dioxide present in the rich and lean liquors as a function of absorption liquor temperature and percent hydrogen sulfide removed from the COG. The data indicate that the lower the operating temperature, the greater the percent of desulfurization. Equal concentrations of hydrogen sulfide and carbon dioxide exist in the rich and lean solutions indicating that hydrogen sulfide is selectively removed over carbon dioxide. Only about 30% of the hydrogen cyanide present in the COG is absorbed during desulfurization and hydrolyzed into ammonia and other by-products.

The acid gas produced contains more than 45 vol% hydrogen sulfide with the balance being carbon dioxide and a small amount of other impurities such as 0.1 vol% hydrogen cyanide, .01 vol% ammonia, and 1.1 vol% aromatics. Consequently, the acid gas is a suitable feed for a typical Claus unit. Operating problems such as plugging and corrosion associated with larger concentrations of these impurities are avoided. As an alternative, the acid gas is a suitable feed for a sulfuric acid plant.

Table 4-15 shows the utility requirements for the DIAMOX process for two levels of COG desulfurization. The DIAMOX process does not require any chemicals since the absorption liquor is generated from the ammonia present in the COG.

Table 4-14
DIAMOX Operating Data from MCI Kurosaki Plant

Absorption Liquor Temp., °F	% Removal H_2S	Solution Analysis, Grains/U.S. Gal.					
		NH_3		H_2S		CO_2	
		Rich	Lean	Rich	Lean	Rich	Lean
90	96.2	473	473	118	12.3	117	2.3
98	91.1	436	417	108	14.6	123	2.3
104	86.0	418	386	102	16.4	106	5.2

Data from Hiraoka (1977)

Table 4-15
Utilities for DIAMOX Process for Two COG Desulfurization Levels
(Basis: 500 grains H_2S/100 SCF at inlet, 90 MMSCFD capacity)

Utility \ H_2S at Absorber Outlet	8 Grains/100 SCF	30 Grains/100 SCF
L.P. Steam, Lb/Hr	34,180	20,950
Cooling Water, GPM	2,910	1,590
Boiler Water, GPM	7	33
Electric Power, kW	1,000	570

Data from Hiraoka (1977)

Design and Operation

A simulation model for an ammonia-based coke-oven gas desulfurization plant is described by Bloem et al. (1990). The model was developed within the capabilities of Aspen-Plus, a widely used software package for chemical process flowsheet simulation. The two scrubbers were simulated on the basis of six theoretical stages for the hydrogen sulfide absorber and five stages for each of the two ammonia scrubbers. All important chemical reactions were included and equilibrium assumed. The difference in the absorption rates of carbon dioxide and hydrogen sulfide was corrected by adjusting the vaporization efficiency of carbon dioxide. The model was verified and adjusted on the basis of experimental data. It has been used to optimize the performance of a large coke-oven gas desulfurization plant and reduce sulfur emissions.

Corrosion problems in ammonia-based desulfurization plants are confined primarily to the elevated temperature zones, and are aggravated by the presence of HCN in solution and by the use of more concentrated ammonia (and acid gas) solutions. According to Massey and Dunlap (1975), the absorber and most heat exchangers are constructed of aluminum-killed carbon steel. Commercially pure aluminum is used for the deactifier vessel with the internals fabricated out of Haveg reinforced phenolic resin. Hastelloy is used for steam spargers, hot-solution circulation pumps, and high-temperature heat exchangers. Ammonia stripping columns are typically fabricated out of cast iron. Ceramics can be used for accessories with solutions containing high concentrations of ammonia and HCN.

REFERENCES

American Petroleum Institute, 1974, "1972 Survey of Materials Experience and Corrosion Problems in Sour Water Strippers," API 944, November.

American Petroleum Institute, 1976, "A Study of Variables That Effect the Corrosion of Sour Water Strippers," API 948, May.

Annessen, R. J., and Gould, G. D., 1971, *Chem. Eng.* Vol. 78, March 22, p. 67.

Badger, E. H. M., 1938, *J. Soc. Chem. Ind.* London, Vol. 57, No. 112.

Badger, E. H. M., and Silver, L., 1938, *J. Soc. Chem. Ind.* London, Vol. 57, No. 110.

Badger, E. H. M. and Wilson, D. S., 1947, *J. Soc. Chem. Ind.* London, Vol. 66, No. 84.

Bea, D. M., Wardell, J. R., and Law, C. V., 1989, "The Chevron WWT Process, A Proven Treatment Method for Sour Water," presented at the Petro-Safe '89 Conference, Houston, TX, Oct. 4.

Bell, J., 1950, *Coke and Gas,* Vol. 12, June, pp. 206–209 and 214.

Beutier, D., and Renon, H., 1978, *Ind. Eng. Chem. Process Design and Development,* Vol. 17, p. 220.

Beychok, M. R., 1967, *Aqueous Wastes from Petroleum and Petrochemical Plants,* John Wiley and Sons, NY.

Bloem, W. B., Boender, W., and Toebes, B. W., 1990, *La Revue de Metallurgie,* CIT, February, p. 129, (in English).

Bomberger, D. G., and Smith, J. H., 1977, *Hydrocarbon Processing,* July, p. 157.

Brunck, 1903, British Patent 8,287.

Busa, J. V., 1992, UEC, USX Engineers and Consultants, Inc., personal communication, Feb. 5.

Busa, J. V., and Cole, T. P. Jr., 1986, "Removing Ammonia Using the Phosam-W Process," reprint from the May–June issue of *Sulfur,* British Sulphur Corporation, Ltd, London, England.

Dasgupta, P. K., and Dong, S., 1986, *Atmospheric Environment,* Vol. 20, No. 3, p. 565.

Davy/Still Otto, Pittsburgh, PA, 1992, personal commumication "Still Otto Process," July 6.

Dryden, J. G. C., 1947, *J. Soc. Chem. Ind.*, London, Vol. 66, No. 59.

Dwyer, O. E., and Dodge, B. F., 1941, *Ind. Eng. Chem.*, Vol. 33, pp. 485–492.

Edwards, T. J., Maurer, G., Newman, J., and Prausnitz, J. M., 1978A, *AIChE Journal,* Vol. 24, No. 6, November, p. 966.

Edwards, T. J., Newman, J., and Prausnitz, J. M., 1978B, *Ind. Eng. Chem. Fundam.*, Vol. 17, No. 4, p. 264.

Edwards, T. J., Newman, J., and Prausnitz, J. M., 1975, *AIChE Journal,* Vol. 21, No. 2, March, p. 248.

Elliott, M. A., 1981, *Chemistry of Coal Utilization,* Second Supplementary Volume, John Wiley & Sons, NY.

Fellinger, L., 1941, Sc. D. Thesis in Chemical Engineering, Massachusetts Institute of Technology.

Ferguson, W. C., 1956, *Lange's Handbook of Chemistry,* 9th ed., McGraw-Hill Book Company, NY., p. 112.

Fischer, E., and Kriebel, M., 1988, "Boosting Throughput in Claus Plants," presented at the ACHEMA 1988, June 6, Frankfurt (Main)—FRG.

Fumio, S., 1986, "Diamox Process" in *Process Handbook 3,* Mitsubishi Kakoki Kaisha, Ltd. Kawasaki, Japan.

Gantz, R. G., 1975, *Hydrocarbon Processing,* May, p. 83.

Gerdes, K., Johnson, J. E., and Wilson, G. M., 1989, "Application of GPSWAT to Sour Water Systems, Design and Engineering Problems," presented at the 68th Annual GPA Convention, San Antonio, TX, March 13.

Grosick, H. A., and Kovacic, J. E., 1981, "Coke-Oven Gas and Effluent Treatment," Ch. 18 in *Chemistry of Coal Utilization,* Second Supplementary Volume, M. A. Elliott, Ed., John Wiley and Sons, NY.

Hill, W. H., 1945, *Chemistry of Coal Utilization,* Ed. H. H. Lowry, John Wiley & Sons, Inc. NY., Ch. 13.

Hiraoka, H., Tanaka, E., and Sudo, H., 1977, "DIAMOX Process for the Removal of H_2S in Coke Oven Gas," in *Symposium on Treatment of Coke Oven Gas,* McMaster University Press, McMaster University, Hamilton, Ontario, Canada.

Jaeger Products, Inc., 1990, "Plastic Jaeger Tri-Packs," Product Bulletin 600, JPI 3M/11-90.

Kawazuishi, K., and Prausnitz, J. M., 1987, *Ind. Eng. Chem. Res.*, Vol. 26, No. 7, p. 1482.

Key, A., 1956, *Gas Works Effluents and Ammonia,* 2nd ed., The Institution of Gas Engineers, London.

Kirner, W. R., 1945, *Chemistry of Coal Utilization,* H. H. Lowry Ed., John Wiley & Sons, Inc., NY., Ch. 13.

Koppers, H., 1907, U.S. Patents 846,035 and 862,976 (Re. 12,971 in 1909).

Kowalke, O. L., Hougen, O. A., and Watson, K. M., 1925, *Chem. & Met. Eng.* Vol. 32, No. 10, pp. 443–446 and No. 11, pp. 506–510.

Krupp Wilputte Corporation, 1988, brochure *Desulfurization of Coke Oven Gas, Recovery of Sulfur and Sulfuric Acid,* KWC 6.441 e.

LD Duiker b. v., 1990, brochure *Combustion Equipment for Sulfur Recovery Plants,* from LD Duiker b. v., Bovendijk 39, 2295 RV Kwintsheul—Holland.

Leonard, J. P., Haritatos, N. J., and Law, D. V., 1984, *Chem. Eng. Progress,* Vol. 10 Vol. October, p. 57.

Leonard, J. P., Haritatos, N. J., and Law, D. V., 1985, Ch. 26 in *Acid and Sour Gas Treating Processes,* S. A. Newman, Ed., Gulf Publishing Company, Houston, TX.

Lohrmann, H., and Stoller, P., 1942, *Arch. Bergbauliche Forsch.* Vol. 3, p. 43.

Lowry, H. H. (editor), 1945, *Chemistry of Coal Utilization,* John Wiley and Sons, Inc., NY.

Lowry, H. H. (editor) 1963, *Chemistry of Coal Utilization,* First Supplemental Volume", John Wiley and Sons, Inc., NY.

Massey, M. J., and Dunlap, R. W., 1975, "Assessment of Technologies for the Desulfurization of Coke Oven Gas," presented at the American Institute of Mechanical Engineers, Ironmaking Conference, Ontario, Canada, April 13.

Maurer, G., 1980, "On the Solubility of Volatile Weak Electrolytes in Aqueous Solutions," Ch. 7, *ACS Symposium Series 133,* Am. Chem. Soc., Washington, D.C.

Muller, G., 1983, Dissertation, University of Kaiserslautern, (cited in Kawazuishi and Prausnitz, 1987).

Mullowney, J. G., 1957, *Petrol. Refiner* Vol. 36, December, p. 149.

Newman, S. A., 1991, *Hydrocarbon Processing,* Part 1—September, p. 145, Part 2—October, p. 191.

Ohnesorge, O., 1910, *Stahl u. Eisen,* Vol. 30, pp. 113–116.

Ohnesorge, O., 1923, *Brennstoff-Chem.*, Vol. 4, pp. 118–122.

Otto, H., 1949, Am. Inst. Min. Met. Eng., Proc. Blast Furnace, Coke Oven and Raw Material Conference, Vol. 8, pp. 50–60.

Owens, J. L., Cunningham, J. R., and Wilson, G. M., 1983, "Vapor-Liquid Equilibria for Sour Water Systems at High Temperature," Research report RR-65, Gas Processors Association, Tulsa, OK, April.

Pacific Coast Gas Association, 1934, *Gas Engineers Handbook,* McGraw-Hill Book Company, NY., p. 443.

Parsly, L. F., Molstad, M. C., Cress, H., and Bauer, L. G., 1950, *Chem. Eng. Progr.*, Vol. 46, pp. 17–19.

Perry, J. H. (Ed.), 1941, *Chemical Engineer's Handbook,* 2nd ed., McGraw-Hill Book Company, NY, pp. 404, 2542, 2544.

Pexton, S., and Badger, E. H. M., 1938, *J. Soc. Chem. Ind.*, London, Vol. 57, p. 106.

Pigford, R. L., and Pyle, C., 1951, *Ind. Eng. Chem.*, Vol. 43, p. 17–19.

Rice, R. D., and Busa, J. V., 1984, *Chem. Eng. Progress,* Vol. 10, October, p. 61.

Rice, R. D., and Busa, J. V., 1985, Chapter 28 in *Acid and Sour Gas Treating Processes,* S. A. Newman, Ed., Gulf Publishing Company, Houston, TX.

Sassen, C. L., van Kwartel, R. A. C., van der Kooi, H. J., and de Swaan Arons, J., 1990, *J. Chem. Eng. Data,* Vol. 35, No. 2, p. 140.

Sherwood, T. K., and Holloway, F. A. L., 1940, *Trans. Am. Inst. Chem. Engrs.*, pp. 3621–36.

Simbeck, D. R., Dickenson, R. L., and Oliver, E. D., 1983, *Coal Gasification Systems; a Guide to Status, Applications and Economics,* EPRI AP-3109, Project 2207 final report, June.

Snook, R. D., 1977, "The Carl Still Coke Oven Gas Desulfurization Plant at the ARMCO Steel Corp. No. 3 Coke Plant" in *Symposium on Treatment of Coke Oven Gas,* McMaster University Press, McMaster University, Hamilton, Ontario, Canada.

Strong, R. C., Majestic, V. K., and Wilhelm, S. M., 1991, *Oil Gas J.,* Oct. 7, p. 86.

Svoboda, K. P., and Diemer, P. E., 1990, *Iron and Steel Engineer,* December, p. 42.

Tans, A. M. P., 1958, *Ind. Eng. Chem.*, Vol. 50, No. 6, pp. 971–972.

Terres, E., and Patscheke, G., 1931, *Gas-u Wasserfuch,* Vol. 74, pp. 761–764, 792–799, 810–814, 837–841.

U.S. Department of Energy, 1982, *Coal Conversion Systems Technical Data Book,* HCP/T2286-o1/5, Sections VF 50.1 and 51/2.

USX Engineers and Consultants, Inc. (UEC), 1991, Bulletins No. 2-01, "Aristech Phosam Process," and 2-12, "Aristech Phosam-W Process."

Van Krevelen, D. W., Hoftijzer, P. J., and Huntjens, F. Y., 1949, *Rec. Trav. Chim.*, Vol. 68, p. 191.

Wild, N. H., 1979, *Chem. Engineering,* Feb. 12, p. 103

Wilson, G. M., 1978, *A New Correlation of NH_3, CO_2, and H_2S Volatility Data from Aqueous Sour Water Systems,* American Petroleum Institute, Publication 555, Washington D.C.

Wilson, G. M., Gillespie, P. C., and Owens, J. L., 1985, Ch. 27 in *Acid and Sour Gas Treating Processes,* S. A. Newman, Ed., Gulf Publishing Company, Houston, TX.

Wilson, G. M., Owens, J. L., and Cunningham, J. R., 1982, "Vapor-Liquid Equilibria for Sour Water Systems with Inert Gases Present," Research report RR-52, Gas Processors Association, Tulsa, OK, April.

Wilson, P. J., 1945, *Chemistry of Coal Utilization,* Ch. 32, H. H. Lowry, Ed., John Wiley & Sons, NY.

Wilson, P. J., and Wells, J. H., 1948, *Blast Furnace Steel Plant,* Vol. 36, p. 806, 961.

Won, K. W., 1982, "Component-Dependent Stage Efficiencies for Sour Water Stripper Calculations," Presented at 1982 Spring AIChE National Meeting, Anaheim, CA, June 6–10.

Chapter 5

Alkaline Salt Solutions for Acid Gas Removal

INTRODUCTION, 330

ABSORPTION MECHANISMS, 331

ABSORPTION AT ELEVATED TEMPERATURE, 334

    Hot Potassium Carbonate (Benfield) Process, 334
    Catacarb Process, 363
    Flexsorb HP Process, 369
    Giammarco-Vetrocoke Process, 371

ABSORPTION AT AMBIENT TEMPERATURE, 378

    Carbon Dioxide Absorption in Alkali-Carbonate Solutions, 378
    Seaboard Process, 381
    Vacuum Carbonate Process, 383
    Vacasulf Process, 392
    Tripotassium Phosphate Process, 393
    Sodium Phenolate Process, 396
    Alkacid Process, 397
    Caustic Wash Processes, 401

REFERENCES, 410

INTRODUCTION

A prime requirement for absorptive solutions to be used in regenerative CO_2 and H_2S removal processes is that any compounds formed by reactions between the acid gas and the solution must be readily dissociated. This precludes the use of strong alkalies; however, the salts of these compounds with weak acids offer many possibilities, and a number of processes have been developed which are based on such salts. Typically the processes employ an aqueous solution of a salt containing sodium or potassium as the cation with an anion so selected that the resulting solution is buffered at a pH of about 9 to 11. Such a solution, being alkaline in nature, will absorb H_2S and CO_2 (and other acid gases), and, because of the buffering action of the weak acid present in the original solution, the pH will not change rapidly as the acid gases are absorbed. Salts that have been proposed for processes of this type include sodium and potassium carbonate, phosphate, borate, arsenite, and phenolate, as well as salts of weak organic acids.

The major commercial processes that have been developed for hydrogen sulfide, carbon dioxide, and mercaptan absorption using aqueous solutions of sodium or potassium compounds are discussed in this chapter. Several of the processes described are no longer commercial, but brief descriptions are included because of their historical significance. The principal technologies which are still employed are (1) processes based on hot potassium carbonate solutions, which are used for the removal of carbon dioxide (and sometimes hydrogen sulfide) from high pressure gas streams (the solutions normally contain a promoter to enhance the rate of carbon dioxide absorption); (2) processes based on absorption in ambient temperature sodium or potassium carbonate solutions with vacuum regeneration, which are used primarily for the removal of hydrogen sulfide from coke-oven gas; and (3) processes based on ambient temperature absorption into solutions containing free caustic, which are used to remove mercaptans and small traces of carbon dioxide or hydrogen sulfide from gases.

ABSORPTION MECHANISMS

As pointed out in Chapter 1, the occurrence of a chemical reaction in the solution has the effect of increasing the liquid phase absorption coefficient over that which would be observed with simple physical absorption. This increase can be quantified in terms of an enhancement factor: the ratio of the actual absorption coefficient with reaction to the absorption coefficient which would be expected under the identical conditions if no reaction occurred. The prediction of enhancement factors for various classes of reactions is quite complex (see Astarita et al., 1983) and requires a knowledge of the reaction path and rate as well as liquid phase physical properties. As the reaction rate increases, the enhancement factor, and thus the liquid phase absorption coefficient, also increases. When the reaction rate is extremely fast, the liquid phase absorption coefficient can be high enough to make the gas phase mass transfer resistance controlling.

When hydrogen sulfide is absorbed into an alkaline solution, it can react directly with hydroxyl ions by a proton transfer reaction:

$$H_2S + OH^- = HS^- + H_2O \tag{5-1}$$

This reaction is extremely rapid and can be considered instantaneous in comparison with diffusion phenomena (Savage et al., 1980).

Since hydrogen sulfide is absorbed more rapidly than carbon dioxide by aqueous alkaline solutions, partial selectivity can be attained when both gases are present. The data of Garner et al. (1958) indicate that selectivity is favored by short gas-liquid contact times and low temperatures. Commercial applications of selective absorption based on short residence time contact are described by Hohlfield (1979) and Kent and Abid (1985).

Carbon dioxide is a slightly stronger acid in solution than hydrogen sulfide. Its ionization constant for the first step ionization to H^+ and HCO_3^- is approximately 4×10^{-7} at 25°C compared to 1×10^{-7} for the corresponding hydrogen sulfide ionization. As a result, under conditions of extended gas-liquid contact where equilibrium is approached, carbon dioxide can displace previously absorbed hydrogen sulfide. This phenomenon is used commercially in the processing of Kraft paper mill liquors.

The chemical reaction of absorbed carbon dioxide with alkaline carbonate solutions takes place through two parallel mechanisms: (1) direct formation of HCO_3^- by reaction of CO_2 with the hydroxyl ion and (2) reaction of CO_2 with water followed by dissociation of carbonic acid. According to Astarita et al. (1981), the predominant mechanism at pH > 10 involves the direct reaction of dissolved CO_2 with OH^-:

$$CO_2 + OH^- = HCO_3^- \quad \text{(fast)} \tag{5-2}$$

$$HCO_3^- + OH^- = CO_3^= + H_2O \quad \text{(instantaneous)} \tag{5-3}$$

At pH < 8 the principal mechanism is based on the hydration of dissolved CO_2 to form carbonic acid followed by reaction of the carbonic acid with OH^-:

$$CO_2 + H_2O = H_2CO_3 \quad \text{(slow)} \tag{5-4}$$

$$H_2CO_3 + OH^- = HCO_3^- + H_2O \quad \text{(instantaneous)} \tag{5-5}$$

In the pH range of interest for commercial operations, pH > 8, the mechanism involving the direct reaction of carbon dioxide to form bicarbonate ions (reaction 5-2) predominates (Astarita et al., 1981).

Although the overall reaction of carbon dioxide with solution components results in the conversion of carbonate to bicarbonate, the local reaction rate is determined by the concentration of hydroxyl ions as indicated by reaction 5-2 (Tseng et al., 1988):

$$\text{reaction rate [g mol/(liter)(sec)]} = k_{OH}(CO_2)(OH^-) \tag{5-6}$$

The value of the second order rate constant, k_{OH}, can be estimated by the following equation suggested by Astarita et al. (1983):

$$\log_{10} k_{OH} = 13.635 - 2.895/T + 0.08\ I \tag{5-7}$$

Where: $T = °K$
I = Ionic strength of the solution

Equations 5-6 and 5-7 indicate that the rate of reaction of carbon dioxide can be increased by increasing the CO_2 concentration, the hydroxyl ion concentration, or the temperature.

According to Astarita (1967), the reaction rate of carbon dioxide in carbonate-bicarbonate solution is not fast enough at room temperature to enhance the absorption rate appreciably over that of physical mass transfer. At temperatures above about 318°K (113°F) the reaction rate is sufficiently high to enhance the mass transfer rate significantly, but even at temperatures as high as 378°K (221°F) the reaction rate is not high enough to be considered instantaneous (Savage et al., 1980).

The relatively low rate of absorption of carbon dioxide in carbonate-bicarbonate solutions has been an incentive for research on rate-increasing additives. Many such materials have been discovered and are usually referred to as promoters, activators, or catalysts. Ferrell et al. (1987) provide the following list of materials found to increase the rate of carbon dioxide absorption and the original references: formaldehyde, methanol, phenol, ethanolamines (Killeffer, 1937), arsenious acid (Roughton and Booth, 1938), glycine (Jeffreys and Bull, 1964), and the enzyme carbonic anhydrase (Alper et al., 1980). Promoters known to be used in commercial processes include diethanolamine, sterically hindered amines, glycine, and arsenious oxide.

In early theories to explain the rate enhancement effects of promoters it was assumed that arsenious acid and organic amines operated through different mechanisms. Roughton and Booth (1938) concluded that arsenious acid acts as a homogeneous catalyst that increases the rate of the key carbon dioxide reaction (reaction 5-2). A "shuttle" mechanism was proposed by Shrier and Danckwerts (1969) for amine type promoters at low temperatures. In this mechanism the carbon dioxide reacts rapidly with dissolved amine by a second order reaction of the following type:

$$CO_2 + RR'NH = RR'NCOO^- + H^+ \qquad (5\text{-}8)$$

Even with small amine concentrations in the solution, reaction 5-8 is significantly faster than reaction 5-2 causing reaction 5-8 to proceed near the gas-liquid interface. The carbamate ion then diffuses into the bulk of the liquid where equilibrium is re-established by reversal of reaction 5-8. Carbon dioxide released by the reverse reaction is consumed by reaction 5-2, and the resulting free amine can diffuse back to the interface to react with additional carbon dioxide.

Astarita et al. (1981) proposed a general mechanism to cover both arsenious acid and amine promoters involving the following two steps:

$$CO_2 + \text{promoter} = \text{intermediate} \qquad (5\text{-}9)$$

$$\text{intermediate} + OH^- = HCO_3^- + \text{promoter} \qquad (5\text{-}10)$$

The relative rate of equation 5-10 represents the key difference between the two types of promoters. With arsenious acid, the step indicated by equation 5-10 is very rapid and takes place immediately after equation 5-9 at the gas-liquid interface. With amines at moderate temperatures, reaction 5-10 is believed to take place primarily in the bulk of the liquid. At the higher temperatures of desorption, the reaction may be fast enough for amines to act as homogeneous catalysts (Astarita et al., 1981). Savage et al. (1984) indicate that the rate-promotion effect of amine in carbonate solutions can be described quite well in terms of homogeneous catalysis.

Sartori and Savage (1983) compare the rate promotion capabilities and the effects on vapor liquid equilibria of a conventional amine (DEA) and a sterically hindered amine. In the sterically hindered amine, bulky substituent groups are placed adjacent to the amino nitrogen site. This causes the hindered amines to form carbamates of intermediate to low stability. The low stability of the hindered amine carbamate is credited with the observed faster absorption rate and higher CO_2 loading at commercial operating conditions for carbonate solutions promoted with hindered amines compared to solutions promoted with DEA.

ABSORPTION AT ELEVATED TEMPERATURE

Hot Potassium Carbonate (Benfield) Process

This process was developed by the U.S. Bureau of Mines, at Bruceton, Pennsylvania, as part of a program on the synthesis of liquid fuel from coal. Research on CO_2 removal was conducted with the objective of reducing the cost of synthesis-gas purification by designing a process that would take maximum advantage of the synthesis-gas conditions; i.e., high CO_2 partial pressure and high temperature. A flow sheet of the basic process is shown in **Figure 5-1,** and a photograph of a large commercial plant is shown in **Figure 5-2.** The original process was described in considerable detail by publications of Benson and coworkers (Benson et al., 1954; Benson et al., 1956; The Benfield Corp., 1971).

The hot potassium carbonate process was developed further during the 1970s by Benson and Field, who conducted much of the original work at the U.S. Bureau of Mines, and many improvements were made. Among these, the development of an amine activator (DEA) for the potassium carbonate solution, resulting in substantial lowering of capital and operating costs and higher treated gas purity, is probably the most important (The Benfield Corporation, 1971). Major improvements were also made in energy economy through the recovery of internal heat (Clayman and Clark, 1980; Baker and McCrea, 1981; Grover and Holmes, 1987; Bartoo and Ruzicka, 1983), and the process has been demonstrated to be suitable for partially selective removal of hydrogen sulfide in the presence of carbon dioxide (Astarita, 1967).

Recent improvements to the process include the use of high efficiency packing in both the absorber and regenerator columns and the development of further improved activators thereby reducing capital and operating costs and resulting in higher treated gas purities when

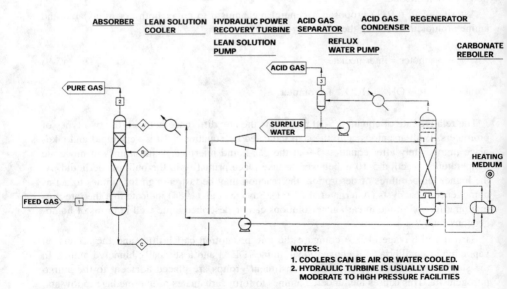

Figure 5-1. Flow diagram of hot potassium carbonate process for the absorption of CO_2, split-stream configuration. A = cooled lean solution; B = main lean solution stream; C = rich solution; 1 = feed gas; 2 = purified gas; 3 = acid gas. (*UOP, 1993*)

Figure 5-2. Two Benfield trains for natural gas purification. *Courtesy of UOP and Natural Gas Corporation of New Zealand*

compared with the original amine activator (Bartoo et al., 1991). To date, the Benfield process is practiced in more than 600 plants worldwide for the removal of carbon dioxide and hydrogen sulfide from ammonia synthesis gas, crude hydrogen, natural gas, town gas, and others (UOP, 1992B). The process is licensed under the name of "The Benfield Process" by UOP, Tarrytown, New York.

The applicability of potassium carbonate to CO_2 removal has been known for many years. A German patent, granted as early as 1904, described a process for absorbing CO_2 in a hot solution of potassium carbonate and then stripping the solution by pressure reduction without additional heating (Behrens and Behrens, 1904). Williamson and Mathews (1924) studied the rate of absorption of CO_2 in potassium carbonate solution and found that increasing the absorption temperature to 75°C (167°F) from 25°C (77°F) greatly increased the rate of absorption. The work of the U.S. Bureau of Mines, however, constitutes a major contribution, in that it resulted in the development of an economically commercial process. A patent covering one aspect of this work was issued to Benson and Field in Great Britain (1955). This patent describes the use of potassium carbonate solution as an absorbent at temperatures near its atmospheric boiling point and its regeneration by flashing and steam stripping.

As a result of the high-absorber temperature, the steam, which other regenerative processes require to heat the solution to stripping temperature, is not required in the hot potassium carbonate system. In addition, the need for heat-exchange equipment between the absorber and stripper is eliminated. The high temperature also increases the solubility of potassium bicarbonate, thus permitting operation with a highly concentrated solution.

Process Description

As can be seen from the flow diagram, **Figure 5-1,** the process is extremely simple. In the split-stream process shown, a portion of the lean solution from the regenerator is cooled and fed into the top of the absorber, while the major portion is added hot at a point some distance below the top. This simple modification improves the purity of the product gas by decreasing the equilibrium vapor pressure of CO_2 over the portion of solution last contacted by the gas. A somewhat more complex scheme termed "two-stage," has also been used for applications in which more complete CO_2 removal is required (see **Figure 5-3**). In this modification, the main solution-stream is withdrawn from the stripping column at a point above the reboiler so that only a portion of the solution passes down through the bottom of the stripping column to the reboiler. Since this portion of the solution is regenerated by the total steam supply to the stripping column, it is thoroughly regenerated and is capable of reducing the CO_2 content of the gas to a low value. The main solution-stream is fed into the midpoint of the absorber, while the more completely regenerated portion is fed at the top.

Packing is used almost exclusively today in Benfield units in preference to trays. Metal packing is the standard, although ceramic and plastic rings are found in some units. Neither ceramic nor plastic packings are recommended for hot potassium carbonate service because plastic rings have a tendency to melt or deform at the high operating temperatures, and the ceramic packing can become brittle and deteriorate with time causing small pieces to break off and damage other areas of the unit. Both carbon steel and stainless steel types are common depending on the particular processing conditions. Choice of an appropriate packing can lead to optimization in grassroots facilities and increased capacity or debottlenecking in revamped units. See Chapter 1 for details on packing types and sizes.

The basic Benfield flow scheme, without any heat conservation features, typically has a net heat duty in the range of 45,000 to 50,000 Btu/lb-mole CO_2. Several modifications of the basic flow patterns have been used, aimed primarily at greater heat economy and higher product gas purity. Examples of modified flow schemes are shown in **Figures 5-3, 5-4, 5-5,** and **5-6.** The Benfield LoHeat process, illustrated in **Figure 5-3,** uses low level heat, which would otherwise be lost in a solution cooler or in an overhead condenser, to satisfy part of the regeneration heat requirement. Hot lean solution exiting the regenerator is reduced in

Alkaline Salt Solutions for Acid Gas Removal 337

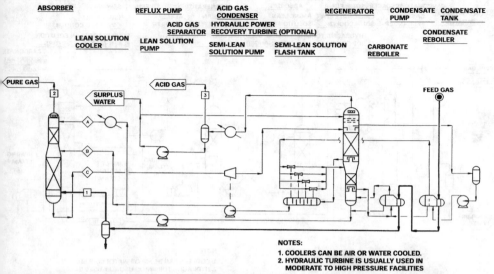

Figure 5-3. Flow diagram of two-stage Benfield LoHeat process with internal steam generation. A = lean solution; B = semi-lean solution; C = rich solution; 1 = feed gas; 2 = purified gas; 3 = acid gas. (*UOP, 1993*)

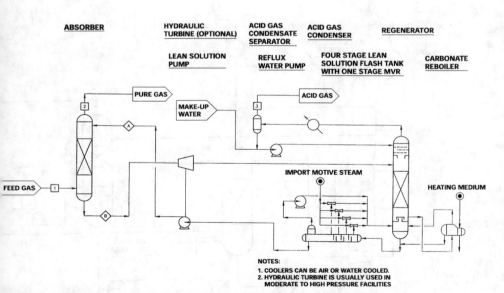

Figure 5-4. Flow diagram of Benfield LoHeat process with both steam ejectors and mechanical vapor recompression. A = lean solution; B = rich solution; 1 = feed gas; 2 = purified gas; 3 = acid gas. (*UOP, 1993*)

338 Gas Purification

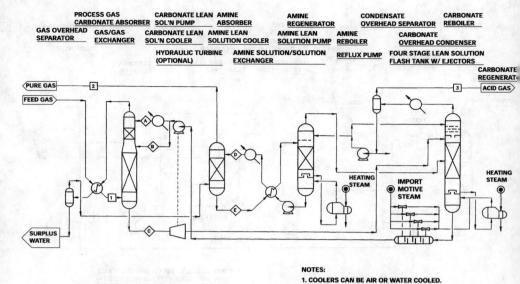

Figure 5-5. Flow diagram of Benfield Hi-Pure process with LoHeat system. A = cooled lean solution; B = main solution stream; C = rich solution; D = lean amine; E = rich amine; 1 = feed gas; 3 = acid gas. (*UOP, 1993*)

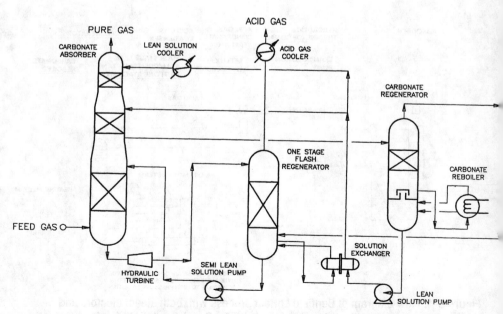

Figure 5-6. Flow diagram of Enhanced LoHeat Benfield process with one-stage rich solution flash. (*UOP, 1993*)

pressure to produce steam. Most LoHeat units use multiple stages to increase the overall energy efficiency by 10 to 15% (Bartoo et al., 1991). Each stage operates about 2 psi below the preceding stage. The flashed steam is then recompressed using a steam ejector or mechanical compressor and re-injected into the base of the regenerator (Grover and Holmes, 1987). As described by Baker and McCrea (1981), usable heat recovered from gas and liquid streams may reduce outside energy requirements by as much as 60% and, in some cases, result in the export of low pressure steam. It should be noted, as shown in **Figure 5-3,** that waste heat reboiling of internal reflux water to generate motive steam for the LoHeat ejector system via a condensate reboiler can further reduce the net energy requirements.

Whether the LoHeat process is used or not, the amount of stripping steam or gross regeneration energy required for solution regeneration is approximately the same. However, in a Benfield unit that utilizes the LoHeat process, part of the gross energy requirement is generated internally through the flash heat recovery. Hence, the total external energy requirement, or net regeneration energy, is reduced. Typical LoHeat energy requirements depend on the process configuration, but usually range from 30,000 to 35,000 Btu/lb-mole CO_2 when steam ejectors are used.

One important LoHeat option is the Benfield hybrid LoHeat scheme. As shown in **Figure 5-4,** this LoHeat system utilizes a combination of steam ejectors and a mechanical vapor recompressor (MVR) (Grover and Holmes, 1987). Typically, a multi-stage flash tank is employed with the first few stages operating on steam ejectors and a final stage using an MVR. The MVR allows for larger recompression ratios, thereby allowing a deeper flash and hence increased energy savings. Typical heating requirements for a hybrid LoHeat system will range from 25,000 to 28,000 Btu/lb-mole CO_2.

Another modification described by Benson and Parrish (1974) is the HiPure process, which is capable of producing a treated gas containing less than one part per million of H_2S and less than 50 parts per million of CO_2. The ability to remove these compounds to such low levels makes it an excellent choice for purification of natural gas to pipeline purity. The Benfield HiPure process has also been used in large liquefied natural gas (LNG) facilities where extremely low product specifications for CO_2, H_2S, COS, and mercaptans are required.

This process, as shown in **Figure 5-5,** uses two independent, but compatible circulating solutions in series to achieve high purity combined with high efficiency. The process gas is first contacted with normal hot potassium carbonate followed by contact with aqueous amine solution. The hot potassium carbonate serves to remove the bulk of the acid gases, while final purification is achieved with the second solution. The two solutions are regenerated separately in two sections of a regenerator with the stripping steam leaving the lower section of the regenerator being re-used in the upper section. The two systems are thermally integrated by using waste heat from the amine circuit to provide a portion of the regeneration heat in the carbonate circuit. The combined heat required for the two solutions is generally lower than that for a conventional hot carbonate system. Although the capital cost of the HiPure unit is somewhat higher than that of a normal Benfield unit due to the additional equipment required, the savings in heat energy and the ability to produce high purity product gas make this process quite attractive.

An extension of the conventional Benfield process, termed the "Enhanced LoHeat Benfield Process," has been developed (Grover, 1987). As shown in **Figure 5-6,** the thermal energy required for regeneration is substantially reduced as a result of a one or two-stage rich solution flash step, which serves as the driving force for a major portion of the regeneration. The balance of the regeneration energy is supplied via conventional thermal stripping. The bulk of the rich solution is regenerated in the flash tank and recycled back to the bottom

340 Gas Purification

section of the absorber where about two-thirds of the CO_2 contained in the feed gas is removed. The top section of the absorber receives thermally regenerated solution, which is used to remove the remaining CO_2 and produces a treated gas having the desired low level of impurities. Low level waste heat within the process is used to enhance the rich solution flash. Typical heating requirements depend on whether a one- or two-stage rich solution flash is chosen, but usually range from 18,000 to 25,000 Btu/lb-mole CO_2. The approximate heat requirement for several versions of the Benfield process is shown in **Figure 5-7** as a function of the acid gas partial pressure.

Basic Data

A large amount of comprehensive physical data on the potassium carbonate-potassium bicarbonate-carbon dioxide-water and potassium carbonate-potassium bicarbonate-potassium bisulfide-carbon dioxide-hydrogen sulfide-water systems is available in literature (Benson et al., 1954; Benson et al., 1956; Tosh et al., 1959; Tosh et al., 1960; Allied Chemical Corp., 1961; Bocard and Mayland, 1962). Some typical data are given below; however, for complete information, the reader is referred to the original sources.

The effect of temperature and percentage conversion to bicarbonate on the solubility of the salts in the system, potassium carbonate bicarbonate, has been determined by Benson et al. (1954), and their data are presented in **Figure 5-8,** together with data from other literature

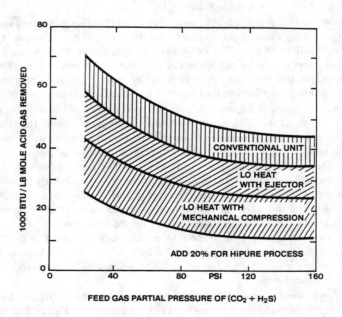

Figure 5-7. Approximate heat requirements of Benfield process systems as a function of acid gas partial pressure in the feed gas. (*Bartoo, 1984*)

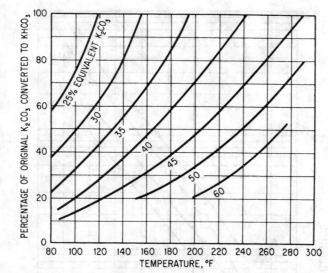

Figure 5-8. Effect of temperature and percentage conversion to bicarbonate on solubility of K_2CO_3 plus $KHCO_3$. Lines represent solubility limits for given salt concentrations (measured as equivalents of K_2CO_3). *Forty, 50, and 60% lines based on data of Benson et al. (1954); other lines based on data of Perry (1950), Seidell (1940), and Hill and Hill (1927)*

sources (Perry, 1950; Hill and Hill, 1927; Seidell, 1940). Lines on the chart represent the conditions under which crystals of potassium bicarbonate begin to precipitate for varying potassium carbonate concentrations. At 240°F, for example, the 60% solution can be converted to only about 30% bicarbonate without the formation of a precipitate. A 50% solution can be 50% converted; and a 40% solution can theoretically be converted 100%. On the basis of these data, it is concluded that a 40% equivalent concentration of potassium carbonate is about the maximum that can be used for the treating operation without precipitation occurring, and a 30% solution is considered to be a reasonable design value for most applications.

If cooling of the solution occurs at any point in the system, even 30% potassium carbonate may be too great a concentration. On the basis of commercial-plant experience with units treating natural gas, Buck and Leitch (1958) recommend 30% potassium carbonate equivalent as a maximum concentration. They found no appreciable effect on the absorptive capacity of the solution when its concentration was reduced to as low as 20%.

The equilibrium vapor pressure of CO_2 over a solution containing the equivalent of 20% potassium carbonate as a function of conversion to bicarbonate, based on the data of Tosh et al. (1959), is presented in **Figure 5-9**. These authors, who investigated vapor/liquid equilibria for 20, 30, and 40% equivalent potassium carbonate solutions, found that the CO_2 equilibrium vapor pressure remains practically the same for the range of 20 to 30%. This in effect confirms the observation of Buck and Leitch (1958) for commercial operations. The experimental CO_2 vapor pressure data were used by Tosh el al. (1959) as the basis for calcu-

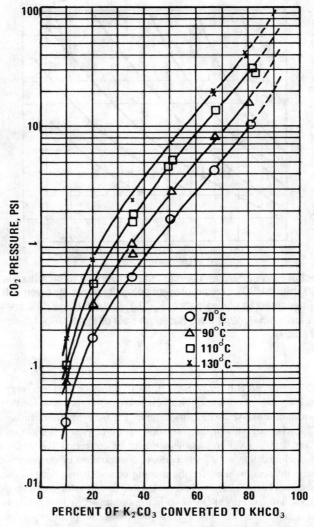

Figure 5-9. Equilibrium vapor pressure of CO_2 over 20 percent equivalent potassium carbonate solution. (*Tosh et al., 1959*)

lating the equilibrium constant, K, for the three solution concentrations according to the expression:

$$K = (KHCO_3)^2/(K_2CO_3)P_{CO_2} \tag{5-11}$$

with $KHCO_3$ and K_2CO_3 expressed in gram moles per liter and P_{CO_2} in mm Hg. K was found to be constant at a given temperature for each degree of conversion for the 20 and 30%

potassium carbonate solutions. From the values of K, the equilibrium vapor pressure can be calculated for any conversion within the given range of solution concentrations. **Table 5-1** shows average K values (arithmetic mean of all experimental data points) for 20 and 30% solutions (Tosh et al., 1959).

Table 5-1
Average Values of K for 20 and 30 percent K_2CO_3 Solutions

Temperature °C	K, 20% Solution	K, 30% Solution
70	0.042	0.058
90	0.022	0.030
110	0.013	0.017
130	0.0086	0.011

Source: Data of Tosh et al., 1959

The equilibrium vapor pressure of water over a solution containing the equivalent of 20% potassium carbonate as a function of conversion to bicarbonate is shown in **Figure 5-10**. This chart is also based on data obtained by Tosh et al. (1959). Here, again, there is not much difference in the vapor pressure of water for 20 and 30% equivalent potassium carbonate concentrations.

Additional vapor-liquid equilibrium data based on published information and experimental work have been reported by Bocard and Mayland (1962). Other physical data on the potassium carbonate-potassium bicarbonate-carbon dioxide system are shown in **Figures 5-11 to 5-14**. The data of Bocard and Mayland (1962) have been converted to a series of nomographs by Mapstone (1966).

The potassium carbonate-potassium bicarbonate-potassium bisulfide-carbon dioxide-hydrogen sulfide-water system has been studied extensively by Tosh et al. (1960) and Field et al. (1960) of the U.S. Bureau of Mines. It was found that it is necessary that the gas to be treated contain some carbon dioxide in addition to hydrogen sulfide for successful use of this system in a gas treating operation.

If hydrogen sulfide were the only acid gas to be absorbed, the following reactions would occur:

$$K_2CO_3 + H_2S = KHCO_3 + KHS \tag{5-12}$$

$$2KHCO_3 = CO_2 + H_2O + K_2CO_3 \tag{5-13}$$

Since with each absorption-regeneration cycle some carbon dioxide would be lost from the system, all of the potassium in the solution would eventually be converted to potassium bisulfide which was found to be essentially non-regenerable according to the following reaction (Tosh et al., 1960):

$$2KHS = K_2S + H_2S \tag{5-14}$$

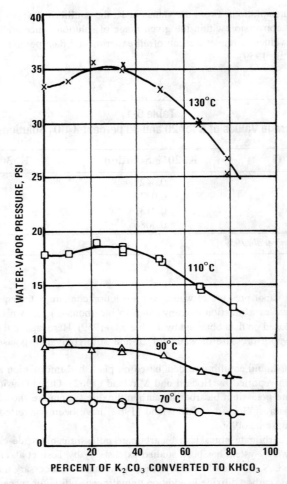

Figure 5-10. Equilibrium vapor pressure of water vapor over 20% equivalent potassium carbonate solution. (*Tosh et al., 1959*)

Equilibrium pressures of hydrogen sulfide, carbon dioxide, and water over solutions of 30% equivalent potassium carbonate are shown in **Figures 5-15** to **5-21**. The term "equivalent potassium carbonate" means that all potassium is assumed to be present as the carbonate. Values of the equilibrium constant:

$$K_1 = (P_{H_2S})(KHCO_3)/(P_{CO_2})(P_{H_2O})(KHS) \qquad (5\text{-}15)$$

are given by Tosh et al. (1960) for a 30% equivalent solution, two-thirds converted to potassium bicarbonate and potassium bisulfide and are shown in **Table 5-2**. Estimated heats of reaction of hydrogen sulfide with 30 or 40% potassium carbonate solutions of 11 and 22 Btu

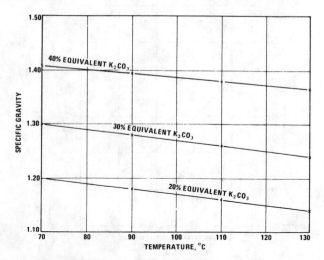

Figure 5-11. Specific gravity of potassium carbonate solutions. (*Allied Chemical Corporation, 1961*)

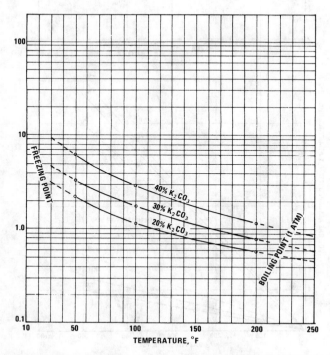

Figure 5-12. Viscosity of potassium carbonate solutions. (*Allied Chemical Corporation, 1961*)

346 *Gas Purification*

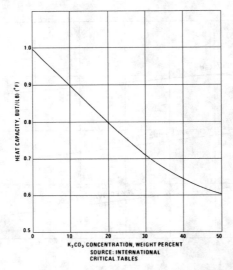

Figure 5-13. Heat capacity of potassium carbonate solutions.

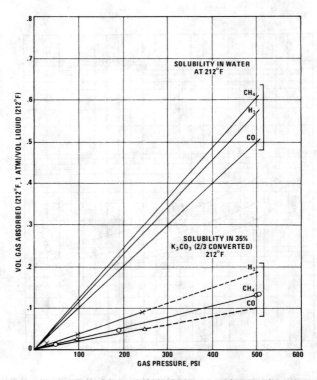

Figure 5-14. Solubility of CH_4, CO, and H_2 in 35% potassium carbonate solution and water. (*Field et al., 1960*)

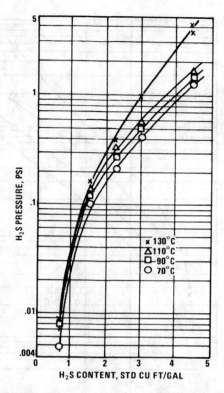

Figure 5-15. Equilibrium vapor pressure of H_2S over 30% potassium carbonate solution. (*Tosh et al., 1960*)

per cu ft of hydrogen sulfide (at 0°C and 760 mm of mercury) have been reported by Tosh et al. (1960).

Design and Operating Data

As a result of the comprehensive pilot-plant studies of the U.S. Bureau of Mines, data are available for the design of non-activated hot-potassium carbonate-process plants. Three series of pilot-plant tests on the hot potassium carbonate process have been reported. In the first (Benson et al., 1954), the absorber consisted of a 4-in. diameter, schedule 80 steel pipe, packed with 0.5-in. raschig rings to a depth of about 9 ft. In the second (Benson et al., 1956), the absorber consisted of a 6-in., schedule 80 pipe, packed to a height of 30 ft with 0.5-in. porcelain raschig rings. The regenerator for the first series consisted of a 6-in. diameter, schedule 40 steel pipe, packed to a depth of 4.75 ft with 0.5-in. raschig rings, and, for the second series, an 8-in. diameter, schedule 40 pipe was used, packed to a height of 25 ft with the 0.5-in. rings. Heat losses from the pilot plants were minimized by adequate insulation and use of steam tracing.

While the first two pilot-plant studies were concerned primarily with carbon dioxide removal, the third study was conducted to investigate the suitability of the hot potassium car-

348 Gas Purification

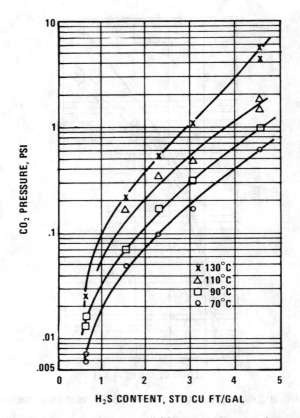

Figure 5-16. Equilibrium vapor pressure of CO_2 over 30% potassium carbonate solution containing H_2S. (*Tosh et al., 1960*)

bonate process for the removal of hydrogen sulfide and carbon dioxide (Field et al., 1960). The equipment used was arranged for single-stage split-stream, or two-stage split-stream operation. The absorber consisted of a 6-in. diameter, schedule 80 pipe and the regenerator of an 8-in. diameter, schedule 40 pipe. An additional regenerator consisting of a 10-ft packed section of 6-in. diameter pipe was used in the two-stage operation. The absorber contained two sections of packing: a lower section packed to a height of 24 ft, 8 in. with ½-in. porcelain raschig rings and an upper section packed to a height of 3 ft, 10 in. with ¼-in. porcelain raschig rings. This upper section was later increased in length to 10 ft. The split-stream was fed to the top of the upper section while the main stream entered the absorber above the lower packed section. The main regenerator was packed to a height of 25 ft with ½-in. raschig rings.

Commercial-plant operating data have been reported for a relatively small non-activated hot potassium carbonate plant treating natural gas containing about 7.5% CO_2. The plant was designed to remove 50% of the carbon dioxide from 8.5 MM scf/day of feed gas; it employed perforated trays in both absorber and stripping columns (Buck and Leitch, 1958).

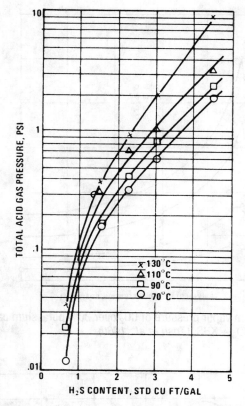

Figure 5-17. Equilibrium vapor pressure of H_2S and CO_2 over 30% potassium carbonate solutions. (*Tosh et al., 1960*)

Absorption. Absorption coefficients are presented for pilot-plant runs with the 4-in. absorber (Benson et al., 1954), and, although a complete variable study was not made, certain trends are apparent. Fortunately, solution and gas concentrations, temperatures, and liquid/gas ratios are in the range typically encountered in commercial-column design so that the data are quite useful.

All of the overall gas coefficients tabulated by Benson et al. (1954) for runs in the 4-in. column with hot 40% K_2CO_3, are presented in **Figure 5-22** as $K_Ga/L^{2/3}$ versus partial pressure of CO_2. The plot is intended to show the effect of CO_2 partial pressure on the absorption coefficient; $L^{2/3}$ is introduced with K_Ga to minimize the spread of points due to different liquid-flow rates. As can be seen, the points appear to fall on two straight lines on the semi-logarithmic coordinates. The first portion of the curve, which extends from 0.0 to 0.7 atm CO_2 partial pressure, has a slope which indicates $K_Ga/L^{2/3}$ to be proportional to $e^{-2.6p}$, where "p" is the partial pressure of CO_2 in atmospheres; while the second portion of the curve, which extends from a CO_2 partial pressure of 0.7 atm to the maximum value tested (5 atm), has a much smaller negative slope. For comparison purposes, a curve of data from Shneerson and Leibush (1946), on the absorption of CO_2 in 5 M monoethanolamine in a very small packed

350 Gas Purification

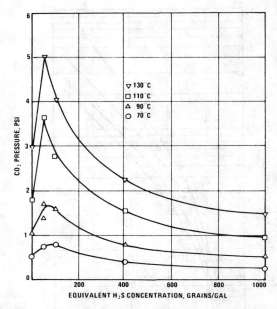

Figure 5-18. Equilibrium vapor pressure of CO_2 over 30% potassium carbonate, one-third converted to $KHCO_3$ + KHS. (*Tosh et al., 1960*)

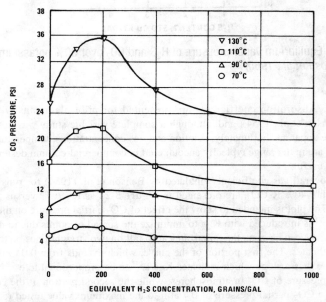

Figure 5-19. Equilibrium vapor pressure of CO_2 over 30% potassium carbonate solution, two-thirds converted to $KHCO_3$ + KHS. (*Tosh et al., 1960*)

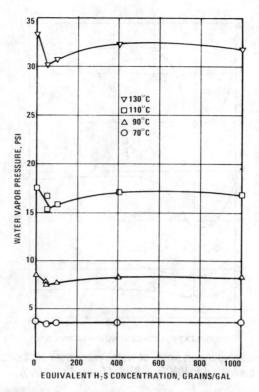

Figure 5-20. Equilibrium vapor pressure of water vapor over 30% potassium carbonate solution, one-third converted to KHCO$_3$ + KHS. (*Tosh et al., 1960*)

column, is also included (see Chapter 2). Over the partial-pressure range covered (0.0 to 0.5 atm), the points are very well correlated by a line plotting the following equation:

$$K_Ga \propto e^{-3.5p} \tag{5-16}$$

which is surprisingly close to the relationship for absorption in hot potassium carbonate solution over the same range.

The pronounced effect of the partial pressure of CO$_2$ on K_Ga is the result of the chemical reaction in the liquid phase and illustrates the difficulty of using K_Ga to correlate absorption when the primary resistance is in the liquid phase. The occurrence of a reaction causes an increase in the liquid film coefficient over that which would be observed with physical absorption alone. The degree of liquid film coefficient enhancement is a complex function of concentrations, reaction rates and diffusivities in the liquid phase. Discussions of enhancement factors for various generalized types of chemical reactions are given in several texts (Sherwood and Pigford, 1952B; Astarita, 1967; Danckwerts, 1970; Astarita et al., 1983).

The effects of DEA concentration, temperature, and percent conversion of potassium carbonate to bicarbonate on the overall liquid film mass transfer rate, K_L, are indicated by the

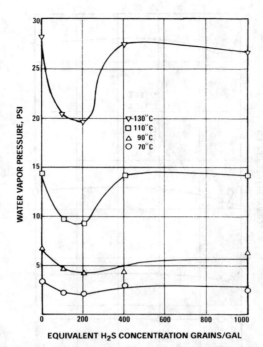

Figure 5-21. Equilibrium vapor pressure of water vapor over 30% potassium carbonate solution, two-thirds converted to $KHCO_3$ + KHS. (*Tosh et al., 1960*)

Table 5-2
Values of K_1 for 30% Equivalent K_2CO_3,
Two-Thirds Converted to $KHCO_3$ and KHS

Equivalent H_2S Concentration, Grains/gallon	K_1 110°C	K_1 130°C
100	0.12	0.39
200	0.12	0.41
400	0.09	0.39
1,000	0.09	0.37

data of **Table 5-3**. The table provides approximate values derived from charts presented by Tseng et al. (1988), which depict the results of experiments conducted with single-sphere absorbers. These investigators proposed a model for estimating the rate of absorption of carbon dioxide in potassium carbonate solutions containing DEA as an activator, based on the assumption that a zwitterion intermediate ($R_1R_2N^+HCOO^-$) is formed. Results of the modeling studies indicated that the CO_2-DEA reaction rate is controlled by the rate of formation of

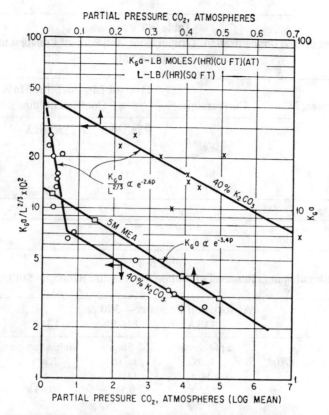

Figure 5-22. Effect of CO_2 partial pressure on absorption coefficients for hot potassium carbonate and monoethanolamine solutions. *Potassium carbonate data from Benson et al. (1954) for 4-in. diameter column packed with 9 ft of ½-in. ceramic rings; monoethanolamine data from Shneerson and Leibush (1946) for 1-in. column packed with 5- to 6-mm glass rings*

the zwitterion intermediate for carbonate conversions below 30% and by the rate of abstraction of a proton from the zwitterion for carbonate conversion greater than 30%.

A proposed methodology for the design of isothermal packed towers covering the specific case of CO_2 absorption in a hot aqueous solution of potassium carbonate is presented by Joshi et al. (1981). They compare the pilot plant data provided by Benson et al. (1954) with predicted values, and conclude that, in this case, equations based on a bimolecular irreversible reaction provide a satisfactory approximation of the rate enhancement due to chemical reaction.

Typical results from the pilot-plant test reported by Field et al. (1960) for simultaneous absorption of hydrogen sulfide and carbon dioxide are shown in **Table 5-4**. The absorption pressure was 300 psig and the regeneration pressure slightly above atmospheric. No attempt was made to remove hydrogen sulfide selectively.

Table 5-3
Effect of DEA Concentration, Temperature, and Percent Conversion on CO_2 Absorption Rate

Temperature, °C	Conversion, %	Overall Liquid-Film Mass Transfer Coefficient, cm/min.		
		0% DEA	2% DEA	5% DEA
50	20	4	12	25
50	50	2	8	12
90	20	35	90	180
90	50	20	40	90

Data of Tseng et al. (1988)

Table 5-4
Simultaneous Removal of Carbon Dioxide and Hydrogen Sulfide

	(Absorption Pressure—300 psig)			
Run:	115A	126A	131C	136A
Type of Flow:	Split Stream	2-Stage	Single Stream	2-Stage
Feed Gas: Rate, ft³/Hr*	1,262	1,214	1,185	583
CO_2, vol. %	11.5	11.35	11.0	11.9
H_2S, vol. %	2.65	0.51	0.88	10.0
Treated Gas:				
CO_2, vol. %	0.6	0.3	0.8	0.07
H_2S, vol. %	0.111	0.002	0.035	0.056
Solution:				
Rate, gal/hr.				
Main stream	34	33	42	40.5
Split stream	17	9	—	12
K_2CO_3, %	34	38	35	20
Temp., °C				
Main stream	114	109	109	102
Split stream	95	91	—	74
Capacity, ft³/gal				
CO_2	2.6	3.2	2.9	1.3
H_2S	0.565	0.152	0.24	1.15
Regeneration efficiency:				
ft³ acid gas/lb steam	6.8	5.7	4.7	5.0
Percent removal:				
CO_2	94.6	97.4	93.7	99.6
H_2S	95.5	99.7	96.4	99.6

*At 0°C and 760 mm Hg, dry
Source: Data of Field et al. (1960)

The relationship between the degree of hydrogen sulfide removal and that of carbon dioxide removal, based on pilot-plant test results with hydrogen sulfide concentrations in the feed gas ranging from 0.2 to 5%, is shown in **Figure 5-23**. Again, no attempt was made to remove hydrogen sulfide selectively (Field et al., 1960). The pilot-plant tests indicate that when properly designed, the hot potassium carbonate process should be suitable for the production of pipeline specification gas.

More recent pilot plant data are available from the Coal Gasification/Gas Cleaning Test Facility at North Carolina State University (Ferrell et al., 1987). The acid gas removal system includes a gas absorption column, one or more flash tanks for intermediate pressure reduction, and a packed stripping column operated with a reboiler. Both of the columns are packed with 0.25-in. ceramic Intalox saddles. The project included the development of a system model and simulation program covering absorption, flashing, and stripping.

Typical operating data from the University of North Carolina program are given in **Table 5-5**. The table shows the data for two runs operating at approximately the same conditions except for the presence of a DEA promoter in the solution of the second run. The results indicate that the addition of the promoter increased the efficiency of removal of all contaminants (CO_2, H_2S, and COS), but the effects on CO_2 and COS removal were most pronounced. Some general conclusions of the overall study were

1. The model predicted gas flow rates and compositions in close agreement with measured values.
2. Heat effects are small. Calculations based on isothermal operation in the absorber, flash tank, and stripper gave accurate results.

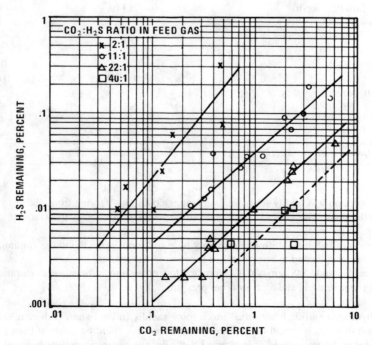

Figure 5-23. Relationship of H_2S and CO_2 in treated gas. (*Field et al., 1960*)

Table 5-5
Pilot Plant Data on Coal Gas Purification with Hot Potassium Carbonate

	Run A	Run B
Solvent Composition, %		
Water	80	75
Potassium carbonate	20	20
DEA	0	5
Solvent flow rate, L/min.	3.3	2.91
Sour gas flow rate, mol/h	798	674
Absorber		
Pressure, kPa	1,790	1,754
Packed height, m	2.16	2.18
Liquid inlet temp. °C	102	101
Gas inlet temp. °C	19	25
Flash tank		
Pressure	533	549
Liquid inlet temp.	94	95
Stripper		
Pressure, kPa	134	133
Packed height, m	4.33	4.33
Liquid inlet temp, °C	98	91
Steam flowrate, Kg/h	17.6	15.9
Steam temp. °C	172	173
Feed Gas Composition, mol %		
CO_2	23.4	23.8
H_2S	0.276	0.279
COS	0.0099	0.0132
Removal, percent		
CO_2	91.4	99.2
H_2S	98.7	99.7
COS	81.5	95.8

Source: Data of Ferrell et al. (1987)

3. CO_2 absorption was highly sensitive to operating conditions varying from 38.1% to 98.4% during the tests. The effect of H_2S and other species on the mass transfer rate of CO_2 appeared to be small.
4. Absorption of H_2S was controlled primarily by equilibrium at the top of the column.
5. COS removal appeared to be primarily a function of temperature.
6. Selective H_2S and COS removal is favored by low potassium carbonate concentration, low absorber temperature, and high absorber pressure.

Since hydrogen sulfide is absorbed much more rapidly in potassium carbonate solutions than carbon dioxide, it would be expected that the process could be made at least partially selective for hydrogen sulfide. It is claimed that selective hydrogen sulfide removal can be obtained by special design (The Benfield Corporation, 1971). Rib et al. (1982) report on the operation of a two-stage Benfield system used in a demonstration plant for selectively removing sulfur compounds from coal gas. The total acid-gas content of the gas varied from 6.4 to 17.1% CO_2, 2,600 to 3,200 ppmv H_2S, and 106 to 216 ppmv COS. In order to obtain high selectivity the unit was operated at 85°C (185°F), and the effects of lean and semi-lean solution flow and reboiler steam rates were studied for different feed gas compositions. The results of this work showed that, under the operating conditions investigated, over 90% of the H_2S and about 40% of the COS could be removed, with about 70% of the CO_2 retained in the product gas.

Contaminants present in gas streams in relatively small concentrations, such as carbonyl sulfide, carbon disulfide, mercaptans, thiophene, hydrogen cyanide, ammonia, and sulfur dioxide, are removed by hot potassium carbonate solution to various degrees. Parrish and Neilson (1974) report laboratory and commercial scale data obtained in purifying natural gases and gas from a pressurized Lurgi gasifier.

Carbonyl sulfide is hydrolyzed almost quantitatively to hydrogen sulfide and carbon dioxide and better than 99% removal can be achieved. Carbon disulfide is hydrolyzed in two steps—first to carbonyl sulfide and hydrogen sulfide, followed by hydrolysis of the carbonyl sulfide to hydrogen sulfide and carbon dioxide. The two-step reaction is slower than the simple hydrolysis of carbonyl sulfide with the result that only about 75 to 85% carbon disulfide removal is obtained under normal operating conditions.

Mercaptans, which are slightly acidic, react with the hot potassium carbonate solutions, forming mercaptides. Since the acidity of mercaptans decreases with their molecular weight, less complete removal of higher molecular weight mercaptans is obtained than of the lower mercaptans, i.e., methyl and ethyl mercaptan. However, the lower molecular weight mercaptans normally constitute the bulk of those present in gas streams. Removal efficiencies of up to 92% have been reported with a HiPure flow pattern. Thiophene does not react chemically with potassium carbonate, and any removal of this compound could only be attributed to physical solubility in the carbonate solution.

Ammonia is readily absorbed by potassium carbonate solutions. Sulfur dioxide and hydrogen cyanide, both acidic compounds, are also readily absorbed by hot carbonate solutions. After absorption, these compounds react further forming a variety of compounds, such as sulfates, thiosulfates, thiocyanates, polysulfides, and elemental sulfur, which accumulate in the solution.

Typical operating data from a commercial absorber are presented in **Table 5-6.** Contact in this absorber is by means of perforated trays. Unfortunately, the number of trays is not specified so that tray efficiencies cannot be calculated from the data. However, since this has been shown to be a liquid-film controlled system, tray efficiencies would be expected to be quite low. Data reported from other installations have indicated tray efficiencies (Murphree vapor) to be on the order of 5% for absorption of CO_2 in hot potassium carbonate solutions.

Regeneration. As shown in **Figure 5-9,** the CO_2 concentration in the solution decreases rapidly with decreased partial pressure of CO_2 over it. Because of this, solution regeneration can be carried out most effectively at a very low pressure. Benson et al. (1956) employed 2 to 10 psig and found the 2 psig regeneration pressure more favorable. Because of the strong dependence of CO_2 concentration in the solution on partial pressure, an appreciable portion of the regeneration is effected when the pressure is reduced over the rich solution leaving the absorber. From ⅓ to ⅔ of the absorbed carbon dioxide is released during this flashing operation.

Table 5-6
Operating Data for Commercial Perforated-Tray Absorber

Absorber Data	1	2	3	4	5
Feed-gas rate, MMSCF/day	1.13	2.60	4.32	5.80	8.56
Lean-solution rate, gpm	95.0	95.9	95.9	84.2	117.5
Gas composition, percent CO_2:					
Inlet	7.2	7.8	7.7	7.6	7.7
Outlet	0.4	2.0	3.0	2.7	4.0
Lean-solution characteristics:					
Specific gravity	1.23	1.27	1.26	1.205	1.275
Equivalent K_2CO_3 concentration, wt%		28.9	28.0	30.0	27.7
Free K_2CO_3 concentration, wt%		18.5	16.8		16.7
$KHCO_3$ concentration					
(as equivalents of K_2CO_3), wt%		10.4	11.2		11.0
CO_2 concentration in solution, cu ft/gal		3.0	3.3		3.2
Rich-solution characteristics:					
$KHCO_3$ concentration					
(as equivalent K_2CO_3), wt%		13.8	16.7		17.5
CO_2 concentration in solution, cu ft/gal		4.0	4.8		5.1
Percent of K_2CO_3 reacted		47.6	59.5		63.1
Percent of inlet CO_2 absorbed	95.0	68.3	62.7	68.6	48.5

Source: Buck and Leitch, 1958

The heat of reaction of carbon dioxide with the hot potassium carbonate solution is relatively small (about 32 Btu/cu ft of CO_2) (Benson et al., 1954), and theoretically, this heat would not need to be supplied in the regeneration step if only a simple flash were used and cooling of the solution were allowed to occur. Since a simple flash does not give adequate regeneration, it is necessary to provide stripping vapor in the regeneration column, and this is done by evaporating water in the reboiler. Essentially all of the heat added goes to the vaporization of water, and this heat must be removed from the system in the overhead condenser of the regenerator. A small amount of additional heat must, of course, be added to make up for heat losses from vessels and lines and to provide latent heat of vaporization for any water evaporated into the gas stream in the absorber.

The effect on regeneration efficiency of altering the operation to give increased solution-carrying capacity is illustrated (**Figure 5-24**) for two product-gas purity conditions. This figure is based on operations of the U.S. Bureau of Mines pilot plant, in which the solution-capacity changes were effected by varying the degree of stripping of the lean solution. As can be seen, attempting to increase the solution-carrying capacity to more than about 4 ft^3 CO_2/gal resulted in a decreased efficiency for both product purity conditions. Comparison of the two curves shows that operating to produce a 0.6% CO_2 product gas instead of a 2% CO_2 product results in a steam-consumption increase of about 30%. The tests illustrated were made with split-stream flow at an absorption pressure of 300 psig and with a feed gas containing 20% CO_2 and 80% nitrogen.

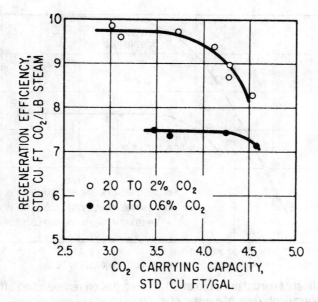

Figure 5-24. Effect of product gas purity and CO_2-carrying capacity of solution on regeneration efficiency (split stream at 300 psig and 12,000 SCF/hr). *Data of Benson et al. (1956)*

Since operation of the hot potassium carbonate process is based primarily on the difference in the solubility of CO_2 at high partial pressure (absorber conditions) and that at low partial pressure (stripper conditions), it would be expected to be more efficient with increased CO_2 partial pressure in the feed gas. The magnitude of this effect for the pilot-plant operations is shown in **Figure 5-25,** which includes curves for two outlet-gas compositions.

Guidelines for the selection of the most advantageous flow system and equipment, with respect to feed-gas composition, treated gas purity required, and heat consumption, have been presented by Benson and Field (1960). A convenient chart, developed by Bartoo (1984), is shown in **Figure 5-26.** An example design calculation for a hot-potassium-carbonate plant has been given by Maddox and Burns (1967).

Operating Problems. In the initial tests made by the U.S. Bureau of Mines, severe corrosion of carbon steel was encountered, especially where the conversion to bicarbonate was high or where carbon dioxide and steam were released by pressure reduction (Bienstock and Field, 1961). Potassium dichromate was found to be an effective corrosion inhibitor, and 0.2% was used in the solution for subsequent CO_2-absorption tests. This material is recommended as a corrosion inhibitor only for plants handling sulfur-free gas. Concentrations of 1,000 to 3,000 ppm have been found satisfactory for commercial installations.

Since H_2S and possibly other sulfur compounds rapidly reduce the chromate ion, the use of this inhibitor is uneconomical in the presence of appreciable quantities of such impurities. In addition to increasing operating costs by destroying the inhibitor, the reduction reaction results in the formation of insoluble precipitates, which cause erosion of equipment, fouling of heat exchanger surfaces, and other operating difficulties. Fortunately, H_2S itself appears to

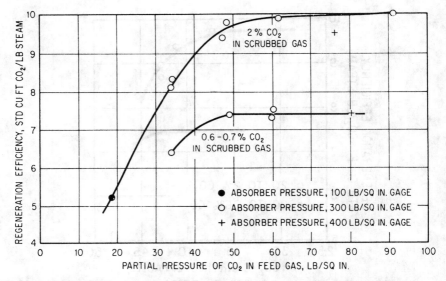

Figure 5-25. Effect of partial pressure of CO_2 in feed gas on regeneration efficiency (split-stream operation). *Data of Benson et al. (1956)*

inhibit CO_2 corrosion somewhat so that carbon steel can still be used for most of the equipment in plants handling gas containing both CO_2 and H_2S.

It has been shown in the laboratory, pilot plants, and in commercial plants, that small amounts (0.01–2.0 wt.%) of metavanadate salts are not only highly effective in inhibiting corrosion on steel surfaces in potassium carbonate systems, but also retain their solubility and anti-corrosion effectiveness in the presence of CO_2 and H_2S (Bienstock and Field, 1965). The metavanadate salts also have the additional advantage that they tend to increase the rate of CO_2 absorption in the absorber column.

In commercial units today, it is preferred to use the alkali metal metavanadates, particularly potassium metavanadate (KVO_3). If desired, vanadium pentoxide (V_2O_5) can be added directly to the potassium carbonate solution, forming potassium metavanadate by reaction with potassium carbonate. For the corrosion inhibitor to be effective, the plant must be designed so that exposed carbon steel surfaces are wetted by the inhibited solution. In order to maintain the metavanadate in its fully oxidized (pentavalent) state, the licensors of the Benfield process recommend that potassium nitrite be injected into the solution on a carefully controlled basis (Sorell, 1990).

The vanadate inhibits corrosion of carbon steel by forming a tightly bonded protective iron-oxide scale on the surface. This is effective at very low concentrations of H_2S; however, at moderate concentrations (about 100–1000 ppm) a much less protective scale containing iron sulfide is formed. This can result in pitting corrosion of carbon steels. The occurrence of this problem in a large gas purification plant and its solution by the use of stainless steel weld overlay has been described by Ferguson and Stutheit (1991). At very high H_2S concentrations a thick iron sulfide scale is formed, which can provide some protection against further corrosion without the use of a vanadate inhibitor; however, the H_2S concentration and other operating requirements to minimize corrosion in such plants are not well characterized.

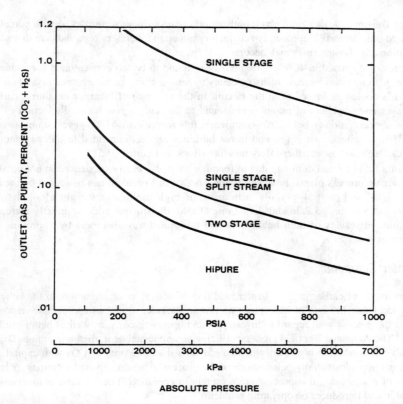

Figure 5-26. Guide for selecting process scheme for hot potassium carbonate plants. (*Bartoo, 1984*)

Stainless steels (Types 304 and 316) are recommended for the portions of the plant subject to particularly corrosive conditions. According to Sorell (1990), the licensors of the Benfield process specifically recommend that the following items be made of 300 series stainless steel:

- Solution circulating pumps
- Letdown hydraulic turbine
- Cladding and internals of regenerator shell above top bed (including top head)
- Top two feet of packing in each bed
- All solution check valves, throttling valves, and control valves
- Piping from rich solution letdown valve to regenerator
- Reboiler tubes, tube sheets, baffles, and tie rods
- Acid gas separators, coolers, and piping
- Overhead condensers
- Reflux pump and piping
- All demisters
- Cladding of feed gas separator

In the Bureau of Mines pilot plant, both absorber and stripper were packed with porcelain rings, and this material is apparently satisfactory although laboratory tests indicate that some dissolution of porcelain rings may occur.

As previously stated, a 40% solution has been found to be too concentrated for commercial use in some cases because of its tendency to form a slurry of bicarbonate crystals if the solution is cooled at any point in the circuit. In the presence of bicarbonate slurry, carbon steel cases and impellers of pumps were found to last only a few hours. Reduction of the solution concentration to below 30% eliminates this severe erosion; however, stainless steel (type 316) impellers, case rings, and throat bushings are recommended in pumps handling the K_2CO_3 solution as a further safety measure (Buck and Leitch, 1958).

Column flooding and dumping due to foaming have also been experienced in hot potassium carbonate process plants; however, it appears that this problem can be solved by the use of additives such as polyglycols, silicones, or high molecular weight alcohol foam inhibitors. According to Palo and Armstrong (1958), the silicone additive greatly increases the column efficiency of both bubble-cap and perforated-tray absorbers by improving the bubbling action.

Activated Solutions

As pointed out earlier in this chapter, addition of activators or promoters to the solution which accelerate the rate of absorption of carbon dioxide, results in appreciable improvement of the process with regard to investment and operating costs, as well as better product quality. The Activated Benfield process, utilizing an amine activator, diethanolamine (DEA), is capable of producing high-purity treated gas (less than 500 ppm of CO_2 with capital and operating costs substantially below those of the process using unactivated solutions. A large number of plants using this process are presently in operation. The activator is inexpensive and stable and introduces no operating problems.

Recently, a new organic activator has been found to be very effective for use in the Benfield process. This activator has demonstrated significantly higher absorption rates than DEA, particularly at low partial pressures of CO_2. The activator also exhibits excellent chemical stability with respect to degradation by CO_2, heat, and oxygen (Bartoo et al., 1991). Improved vapor-liquid equilibria have led to reduced circulating inventories and/or energy reduction compared to DEA alone. Use of the activator can result in lower capital costs and reduce energy consumption in grassroots plants, lower CO_2 product levels, increase capacity, or lower regeneration energy in existing units.

Economics

Cost data on the hot potassium carbonate process published since its first disclosure (Mullowney, 1957; Eickmeyer, 1958; Katell and Faber, 1960; Benson, 1961) are of limited value because of substantial changes in engineering and construction costs during the past years and, more importantly, because of the rapid development of technology. The process was originally thought to be most economical for bulk removal of carbon dioxide and, in some cases, hydrogen sulfide, with a final purification step, using, for example, an ethanolamine solution. However, recent developments have shown that product gas of high purity can be obtained economically with hot carbonate solutions alone. The economics of the process are primarily determined by the overall heat utilization in each specific case, and generalizations are not particularly useful.

Catacarb Process

The Catacarb process, which was disclosed by Eickmeyer (1962), is licensed by Eickmeyer and Associates of Prairie Village, Kansas. For most applications the Catacarb process utilizes a catalyzed hot potassium carbonate solution; however, potassium borate solutions are used for the removal of hydrogen sulfide in the absence of carbon dioxide (Gangriwala and Chao, 1985). The solutions contain undisclosed additives that catalyze absorption and desorption of acid gases, particularly carbon dioxide. The additives, which include a corrosion inhibitor, are claimed to have no effect on reformer or methanation catalysts that the purified gas may pass through downstream of the Catacarb absorber (Morse, 1968).

Process Description

The basic flow scheme of the Catacarb process is similar to that of the Benfield and other hot potassium carbonate based processes. Several versions of the process are available, aimed at providing improved product gas purity and/or greater heat economy. Gangriwala and Chao (1985) describe three primary flow arrangements: (1) single-stage design, in which the gas is scrubbed countercurrently by the lean solution in an absorber and the rich solution is regenerated in a simple reboiled stripping column; (2) split-cooled design, in which a portion of the lean solution is cooled and sent to the top of the absorber to provide a higher purity treated gas; while the bulk of the lean solution is sent hot to the middle of the absorber; and (3) two-stage design, in which two distinct stages of both absorption and stripping are used.

A flow diagram of the two-stage design is shown in **Figure 5-27**. In this arrangement, a small portion of the solution is thoroughly stripped in the bottom section of the regenerator and used in the top section of the absorber to assure production of a high purity treated gas. A larger portion of the circulating solution is less completely regenerated by contact with stripping vapors in the top section of the stripper, and this semi-lean solution is used for bulk removal of acid gas in the bottom portion of the absorber. This arrangement provides both high purity product gas and good heat economy, but results in a somewhat higher plant investment.

In addition to the three basic flow arrangements and minor variations required for special situations, the Catacarb process is available with "low-heat" technology. In the most frequently used low-heat concept, the hot solution from the regenerator is flashed to a lower pressure; the flashed vapors (primarily steam with some acid gas) are compressed by use of a steam jet or mechanical compressor; and the hot compressed vapors are fed into the regenerator to augment stripping vapor from the reboiler. A flow diagram of a single-stage design incorporating this type of low-heat technology is shown in **Figure 5-28**.

Although its principal application is the removal of relatively high concentrations of carbon dioxide from gas streams, the Catacarb process is also suitable for hydrogen sulfide removal from natural gas, yielding a product gas containing less than ¼ grain of H_2S per 100 cu ft (Eickmeyer, 1971). If required, the carbon dioxide content of hydrogen or ammonia synthesis gas can be lowered to 300 ppmv (Morse, 1968). The process is also capable of removing trace amounts of other sulfur-containing gases such as COS, CS_2, and RSH.

Additives

Promoters. The process utilizes two types of additive formulations—organic and inorganic—to meet specific treating needs. The proprietary organic formulations, which consist of

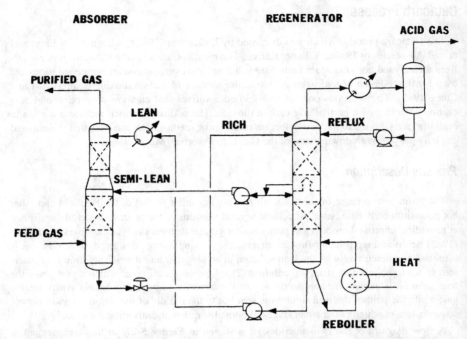

Figure 5-27. Catacarb process two-stage design. (*Gangriwala and Chao, 1985*)

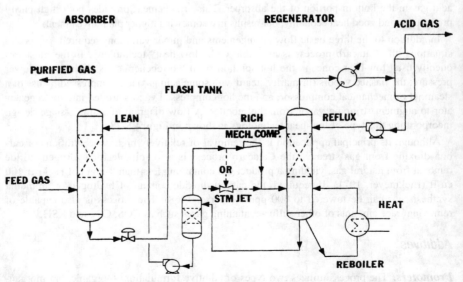

Figure 5-28. Catacarb process single-stage, low heat design. (*Gangriwala and Chao, 1985*)

amine promoters to improve mass transfer rates and corrosion inhibiting components, are typically used in the following applications (Eickmeyer & Associates, 1992):

- Ammonia synthesis gas
- Natural gas purification
- Hydrogen production
- Mole sieve regeneration gas
- Liquefied natural gas

Figure 5-29 indicates the effectiveness of the catalyst content on solution activity relative to a pure potassium carbonate solution. Gangriwala and Chao (1985) report that the proportion of the catalyst components may be altered to increase H_2S removal capacity or remove H_2S in the absence of CO_2.

Inorganic formulated catalysts are used in oxidizing environments where amines and other organic components may be degraded by oxygen present in the feed gas. Commercial applications for inorganic catalysts include (Eickmeyer and Associates, 1992)

- Ethylene oxide recycle gas
- N_2O production
- Vinyl acetate monomer recycle gas
- Autoclave recycle O_2
- CO_2 from flue gas

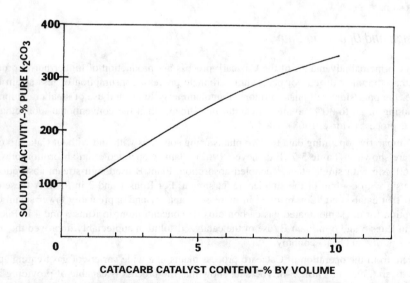

Figure 5-29. Effect of Catacarb catalyst concentration on relative solution activity for CO_2 absorption. Activities are based on observed overall liquid film absorption coefficient (*Eickmeyer, 1962*). Reprinted with permission from *Chemical Engineering Progress*, copyright 1962, American Institute of Chemical Engineering

366 Gas Purification

The inorganic catalyst formulations have very low vapor pressure, which is of particular importance in ethylene oxide plants where downstream ethylene oxide catalysts could be poisoned by contamination with trace amounts of organics. Inorganic catalyst formulations have been successfully commercially applied since the early 1980s. As of July 1992, 54 Catacarb plants were in operation using the organic catalyst and 10 plants using inorganic catalysts. More than 100 Catacarb plants purifying 10 billion scfd of feed gas have been built in 27 countries (Eickmeyer & Associates, 1992).

Corrosion Inhibitors. CO_2 removal applications use a standard vanadium-based Catacarb corrosion inhibitor. However, effective use of the inhibitor requires oxidation by continuous injection of a small amount of air (Scott et al., 1987). The inhibitor forms a protective film on metal surfaces that prevents corrosion due to CO_2 attack, thereby permitting carbon steel to be used in most of the plant, except in areas of high erosion. The vanadium itself does not combine with the iron to form the protective layer. Instead, its presence in the pentavalent state provides the oxidation potential necessary to form a passive, tightly adhering magnetite layer on the iron surface. The magnetite layer provides the corrosion protection and maintenance of the desired pentavalent vanadium concentration keeps the magnetite layer intact. Isolated corrosion problems associated with existing plants are believed to have occurred due to improperly prepassivated process equipment exposed to wet CO_2 and erosion resulting from high solution velocities. Mechanical filtration and activated carbon treating are reported to aid in maintaining a clean solution (Scott et al., 1987).

If H_2S is present in small amounts, the process requires an undisclosed inhibitor to prevent corrosion. No corrosion inhibitor is required when H_2S is present in large quantities since H_2S forms an iron sulfide film over the carbon steel surfaces, which in turn prevents further corrosion.

Design and Operating Data

The principal advantages of the Catacarb process are production of higher purity product gas than the uncatalyzed hot potassium carbonate process, consumption of less steam than either the potassium carbonate or monoethanolamine systems, and use of smaller equipment (resulting in 20 to 30% savings in capital investment) than the conventional monoethanolamine process (Morse, 1968).

Comparative operating data for two plants using solutions with and without Catacarb catalyst are shown in **Table 5-7** (Eickmeyer, 1962). Plant A operated at an absorption pressure of 360 psig with single stage, uncooled absorption. Plant B used split-stream absorption at 300 psig with cooling of the smaller liquid stream. For Runs 1 and 2 in Plant A these data show that catalyst addition resulted in increased capacity and appreciably lower steam consumption for the same treated gas carbon dioxide concentration. In Runs 3 and 4 in Plant A and in Runs 5 and 6 in Plant B, use of the catalyzed solution appreciably improved the treated gas purity and heat economy.

Data from the operation of Catacarb process plants in a wide variety of gas treating applications are given in **Table 5-8**. The data show that the process is capable of providing high efficiency removal of CO_2, H_2S, COS, and RSH from gas streams.

Table 5-9 shows the result of a design study comparing the energy requirements of three possible low-heat designs with a conventional Catacarb system. All of the low-heat concepts required less reboiler steam and correspondingly less cooling water than the conventional

Table 5-7
Comparison of Hot Potassium Carbonate and Catacarb Process Performance

Run no.	1	2	3	4	5	6
Plant	A	A	A	A	B	B
Catalyst content, %	0.0	6.8	0.0	7.0	0.0	5.0
Gas rate, MSCF/Hr	405.0	509.0	464.0	472.0	1,193.0	1,193.0
Inlet gas CO_2, %	23.4	22.8	22.9	23.0	15.2	14.8
Outlet gas CO_2, %	1.0	1.1	2.3	0.6	1.5	0.6
Liquid circulation, gpm	860.0	982.0	810.0	815.0	1,360.0	1,220.0
Steam, lb/MCF CO_2	169.0	121.0	169.0	149.0	218.0	167.0

Source: Eickmeyer, 1962

Table 5-8
Catacarb Process Operating Experience in a Variety of Acid Gas Removal Applications

Plant Product	Ammonia	Ethylene Oxide	Natural Gas	Natural Gas	Liquefied Natural Gas
Year of Construction:	1982	1989	1964	1980	1968
Removal Capacity:					
$CO_2 + H_2S$:					
(ST/DAY)	2,157	824	20	609	1,263
Process Gas Conditions:					
Pressure, psig	445	270	820	630	645
Feed Gas:					
Component, % Vol:					
CO_2	17.7	3.5	6.7	8.3	1.4
H_2S	—	—	2.1	1.03	0.79
COS	—	—	—	0.16	—
RSH	—	—	125 ppm	—	—
Treated Gas:					
Component, ppmv:					
CO_2	1,500	2,000	760	500	20
H_2S	—	—	1.5	20	2.3
COS	—	—	—	20	—
RSH	—	—	1.5	—	—

Source: Eickmeyer & Associates, 1992

Table 5-9
Catacarb Process Design Summary for CO_2 Removal in a 1000 Std.T/Day NH_3 Plant

CASE:	Conventional Design	Low-Heat Design (See Note)		
		Case A	Case B	Case C
Feed Gas:				
Pres., psig:	396	396	396	396
Temp., °F:	399	310	338	307
Flow, lb-mol/hr: (Dry)	12,400	12,400	12,400	12,400
CO_2, % Dry:	17.8	17.8	17.8	17.8
H_2O, lb-mol/hr:	5,068	3,007	4,963	2,901
Treated Gas:				
Pres., psig:	386	386	386	386
Temp., °F	180	180	180	180
CO_2, ppmv:	500	500	500	500
CO_2 Removed:				
Pres., psia:	20	20	20	20
Dry Rate, Std. T/day:	1,160	1,160	1,160	1,160
Purity, % Dry:	98.5+	98.5+	98.5+	98.5–
Utilities:				
Reboiler Duty, MMBtu/hr:	84.9	35.5	72.6	33.3
Steam, lb/hr:	—	32,400	—	—
Power, kW:	800	850	800	2,020
Cooling Duty, MMBtu/hr:	93.3	44.0	81.0	42.1
(To Water Balance)				
Efficiency, 1,000 Btu/lb-mole CO_2	38.7	30.5	33.1	15.6

Note: Low-heat design incorporates steam recompression and heat exchange techniques to increase energy efficiency.
Source: Eickmeyer & Associates, 1992

arrangement. However, Case A required 32,000 lb/hr of other steam and Case C required an additional 1,220 kw of power (presumably for vapor compression).

Wiberg et al. (1982) evaluated Catacarb process options in demonstrating techniques for attaining an optimum economic design. They considered a two-stage Catacarb unit for a 1,000 ton/day ammonia plant without a low-heat system and with a system employing a single flash and vapor compression. For the case where the steam cost is $4/1,000 lb and power is $.03/kwh, the use of an optimum flash pressure (12.5 psia) and vapor compression resulted in a small (2.2%) increase in plant cost, but a significant (10.6%) reduction in annual utilities' cost. For this case the payout period for the increased investment was only about 3 months. Obviously the economic benefit of this low-heat scheme would decrease as the cost of power increases or the cost of reboiler steam decreases.

Flexsorb HP Process

Process Description

The Flexsorb HP Process is a promoted hot potassium carbonate process developed and licensed by Exxon Research and Engineering Company. Four commercial plants using the process were reported to be in operation in 1992 (Exxon, 1992). The process is similar, with regard to flow schemes and general applications, to the activated Benfield and Catacarb processes. It is characterized by the use of a sterically hindered amine as the activator.

A sterically hindered amine has a bulky alkyl group attached to the amino group. Structurally, it is defined as either a primary amine, in which the amino group is attached to a tertiary carbon atom, or a secondary amine, in which the amino group is attached to a secondary or a tertiary carbon atom. In general, only aliphatic and cycloaliphatic amines are useful in gas treating applications. Aromatic amines are not sufficiently basic. In addition to the amino and alkyl groups, the compound must contain a hydrophilic group, such as hydroxy or carboxylic, to assure solubility in water and low volatility. Examples of typical sterically hindered amines are 2-amino-2-methyl-1-propanol (AMP) and 1,8-p-menthanediamine (MDA) (Sartori et al., 1987).

According to Exxon (1991), the use of hindered amines in hot potassium carbonate solution gives both higher working capacity and higher mass transfer rates than conventional amine promoters. The capacity effect is shown in **Figure 5-30** (Sartori and Savage, 1983). The curves in this figure represent vapor-liquid equilibria for carbon dioxide over unpromoted, DEA-promoted, and hindered amine-promoted solutions of potassium carbonate at 120°C. The maximum usable (cyclic) capacity of a solution is the difference between the equilibrium loading at the partial pressure of CO_2 in the feed gas and the equilibrium loading at CO_2 partial pressure conditions existing at the bottom of the stripper. For a partial pressure cycle of 0.01 psia to 200 psia, the data show the hindered amine promoted solution to have a cyclic capacity about 20–30% higher than that of the unpromoted or DEA promoted solutions.

The effect of hindered amine promoters on the rate of CO_2 absorption into a potassium carbonate solution has been evaluated using a single sphere absorber (Sartori and Savage, 1983) and a pilot scale packed column (Weinberg et al. 1983). The single sphere absorber tests showed that unpromoted potassium carbonate provides a mass transfer enhancement in the range of 10–30 times that of simple physical absorption. The addition of a conventional amine activator (DEA) further increases the mass transfer rate by a factor of about 3, and the substitution of a hindered amine for the DEA gives an additional increase by a factor of about 2.

The pilot scale absorption tests generally corroborated the single sphere tests. At an average solution loading of 0.3 mole CO_2/mole initial K_2CO_3, a temperature of about 200°C, and a pressure of 200 psia, the hindered amine-promoted solution provided an overall mass transfer coefficient for CO_2 slightly over twice that of the ethanolamine-promoted solution. The advantage was even higher at solution loadings of about 0.5 mole CO_2/mole initial K_2CO_3 and above.

Both the high mass transfer rate and the high equilibrium capacity are attributed to the low stability of carbamates formed by the reaction of carbon dioxide and hindered amines. Carbamates are formed at the gas-liquid interface with both conventional and hindered amine promoters. However, the hindered amine carbamates decompose rapidly, releasing fresh amine and thereby maintaining a high amine concentration at the interface, which results in a high rate of CO_2 absorption.

The rapid decomposition of carbamate and the high alkalinity of the hindered amines are believed to cause the observed capacity increase because the free amine acts as additional

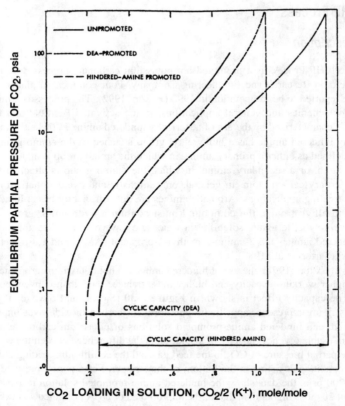

Figure 5-30. Comparison of vapor-liquid equilibrium curves for CO_2 over unpromoted, DEA-promoted, and hindered amine-promoted solutions of hot potassium carbonate (*Sartori and Savage, 1983*). Reprinted with permission from Industrial Engineering Chemical Fundamentals, Vol. 22, copyright 1983, American Chemical Society

base in the solution. With conventional amines, such as DEA, the stable carbamate ties up available amine in an essentially nonreactive form, minimizing the capacity enhancement relative to unpromoted potassium carbonate solution (Say et al. 1984).

Operating Problems

Corrosion problems and required control methods appear to be quite similar in the Flexsorb HP and other amine activated hot potassium carbonate processes. During pilot plant tests the corrosion rate was monitored by the use of corrosion coupons and Corrosometer probes. In many of the tests a small amount of H_2S, ranging from 0.1 to 0.5 mole %, was present in the feed gas. This amount proved adequate to form a protective sulfide layer that inhibited corrosion of carbon steel equipment. In tests without H_2S or inhibitor, severe corrosion was observed (Say et al. 1984).

Alkaline Salt Solutions for Acid Gas Removal **371**

For commercial plants removing only carbon dioxide or carbon dioxide plus low concentrations of hydrogen sulfide, the use of a vanadium base corrosion inhibitor is recommended. For high H_2S concentrations, the vanadium inhibitor is unsuitable; however, the formation of hydrogen sulfide scale permits operation without an inhibitor. Carbon steel construction is applicable for most equipment. The hindered amines used as promoters have low volatility, good stability, and low foaming tendency at plant operating conditions (Exxon, 1991).

Process Economics

An economic comparison of the Flexsorb HP Process with a conventional amine-promoted hot potassium carbonate process has been made by Goldstein et al. (1984). The basis of comparison was a 20 MM scfd hydrogen plant with a 250 psia absorber. Assumed gas compositions were 19.9 mole % CO_2 for the feed and 0.2 mole % CO_2 for the product. The calculated results showed a 19% higher solution circulation rate and 11% higher reboiler duty for the conventional amine-promoted hot potassium carbonate design.

Giammarco-Vetrocoke Process

The use of sodium and potassium arsenite solutions for the absorption of carbon dioxide at elevated temperatures and pressures was first licensed by Giammarco-Vetrocoke of Venice, Italy, in the 1950s. The introduction of improved, nontoxic organic-promoter-based processes has led to the replacement of arsenite-based solutions in many existing plants without any loss of capacity or performance. Giammarco-Vetrocoke currently proposes use of the nontoxic organic-promoted hot potassium carbonate solutions for new plants. However, most of the nearly 300 Giammarco-Vetrocoke plants operating worldwide are based on the arsenite solution. Giammarco reports approximately 190 plants operating in Europe, 80 in Asia and the balance in Africa, America, and Australia (Giammarco-Vetrocoke, 1992B). The arsenite-based process has found practically no acceptance in the United States, probably because of the toxicity of the arsenite solution, which requires special precautions in handling. However, one large installation treating high pressure natural gas has operated in the United States (Anon., 1960; Jenett, 1962). The Giammarco-Vetrocoke process for hydrogen sulfide removal, which is based on the use of aqueous solutions of alkali arsenite and arsenate, is described in Chapter 9.

Basic Chemistry

The general mechanisms for the absorption of CO_2 and H_2S in alkaline solutions and the effects of activators on CO_2 absorption are discussed in an earlier section of this chapter entitled "Absorption Mechanisms." The following discussions are specific to the Giammarco-Vetrocoke processes.

Organic Activators. CO_2 absorption into unpromoted aqueous alkaline solutions is believed to occur via the following reactions:

$$CO_2 + H_2O = HCO_3^- + H^+ \qquad (5\text{-}17)$$

$$CO_3^{2-} + H_2O = HCO_3^- + OH^- \qquad (5\text{-}18)$$

$$CO_3^{2-} + CO_2 + H_2O = 2HCO_3^- \qquad (5\text{-}19)$$

The reaction rate for the overall reaction 5-19 is determined by the sum of the rates of 5-17 and 5-18, where 5-18 represents the rate-controlling step. The Giammarco-Vetrocoke process uses a glycine activator, which increases the rate of absorption by providing an alternative reaction path for CO_2, forming glycine carbamate as shown by reaction 5-20:

$$H_2N\ CH_2\ COO^- + CO_2 = {}^-OOC\ NH\ CH_2\ COO^- + H^+ \tag{5-20}$$

At high temperatures and in the presence of OH^-, the carbamate is hydrolyzed and the activator is restored by the following reaction:

$$^-OOC\ NH\ CH_2\ COO^- + H_2O = H_2N\ CH_2\ COO^- + HCO_3^- \tag{5-21}$$

Reactions 5-20 and 5-21 take place continuously so that glycine acts as a CO_2 carrier. The CO_2 introduced into the liquid phase reacts quantitatively with potassium carbonate according to reaction 5-19.

Giammarco-Vetrocoke has also developed a dual activated solution. By adding an amine, glycine's capacity to act as a CO_2 carrier is greatly improved. The amine's activation mechanism is similar to that of glycine in forming a carbamate:

$$R_2NH + CO_2 = R_2NCOO^- + H^+ \tag{5-22}$$

$$R_2NCOO^- + H_2O = R_2NH + HCO_3^- \tag{5-23}$$

Since glycine's reaction rate is much higher than that of the amine, glycine carbamate formation prevails. Giammarco-Vetrocoke has confirmed the higher efficiency obtained using a glycine-activated solution compared to a DEA-activated solution. In a comparison of two plants the glycine-activated solution achieved an 18% increase in plant load using less steam (Giammarco-Vetrocoke, 1992B).

In the dual-activated system it is believed that only part of the amine reacts with CO_2, forming amine carbamate, leaving most of the amine available to catalyze the hydrolysis of glycine carbamate. The use of an amine also allows the glycine content to be significantly reduced since the formation and hydrolysis of glycine carbamate occurs rapidly. This results in an important reduction in the CO_2 vapor pressure in the rich solution due to the presence of the additional base. It is claimed that the dual-activated solution has a lower CO_2 vapor pressure, a higher regenerator efficiency, and a higher CO_2 absorption rate than the mono-activated solution. It is also claimed that the new activated solutions are clean, stable, and nontoxic (Giammarco-Vetrocoke, 1992B).

Arsenite Solutions. Addition of essentially stoichiometric proportions of arsenic trioxide to aqueous sodium or potassium carbonate solutions results in a marked increase in the rate of absorption and desorption of carbon dioxide, as compared with conventional carbonate solutions. **Figure 5-31** illustrates this phenomenon by comparing, qualitatively, the rate of absorption of carbon dioxide at 1 atm partial pressure and room temperature in 40% potassium carbonate and in a typical solution used in the Giammarco-Vetrocoke process (Riesenfeld and Mullowney, 1959). The effects of the more rapid absorption and desorption are appreciable savings in regeneration heat, reduction in equipment size, and production of treated gas of higher purity than is possible with ordinary hot carbonate solutions.

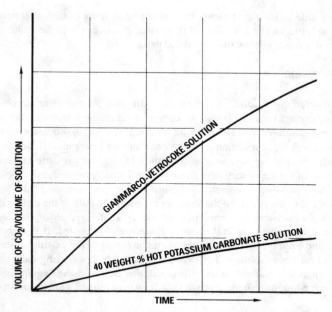

Figure 5-31. Comparative rate of absorption of CO_2 in 40% potassium carbonate and typical Giammarco-Vetrocoke arsenical solution. (*Riesenfeld and Mullowney, 1959*)

The chemical reactions occurring during absorption and desorption of carbon dioxide can be symbolized by the following equations:

Absorption:

$$6CO_2 + 2K_3AsO_3 + 3H_2O = 6KHCO_3 + As_2O_3 \tag{5-24}$$

$$CO_2 + K_2CO_3 + H_2O = 2KHCO_3 \tag{5-25}$$

Desorption:

$$6KHCO_3 + As_2O_3 = 2K_3AsO_3 + 6CO_2 + H_2O \tag{5-26}$$

$$2KHCO_3 = K_2CO_3 + H_2O + CO_2 \tag{5-27}$$

It is claimed that the addition of arsenic trioxide not only increases the rate of carbon dioxide absorption, but also the carrying capacity of the solution. Jenett (1962) reports that an arsenite solution of 30% equivalent potassium carbonate content, regenerated with 0.45 lb of steam per gallon, has about 25% more carrying capacity for carbon dioxide than a potassium carbonate solution of the same equivalent concentration. The carbon dioxide content of the gas treated with the arsenite solution is substantially lower than that of the gas treated with the carbonate solution. These observations indicate that arsenic trioxide increases the rate of hydration of carbon dioxide to carbonic acid in the absorption step and also the shift of pH

374 Gas Purification

toward the acid side in the regeneration step, resulting in more complete expulsion of the absorbed carbon dioxide. The net effect of these two phenomena is higher solution capacity and lower carbon dioxide content of the purified gas.

Process Description

Organic Activated Solutions. A typical flow diagram for the Giammarco-Vetrocoke process with organic activated hot potassium carbonate scrubbing solution is shown in **Figure 5-32**. The process illustrated uses two-pressure-level regeneration to provide good heat economy and split flow solution circulation to assure complete acid gas removal. Solution from the bottom of the high pressure regenerator is flashed in the low pressure regenerator then pumped to the top of the upper section of the absorber. Partially regenerated solution from the middle of the high pressure regenerator is flashed in the upper section of the low pressure regenerator then fed to the middle of the absorber to absorb the bulk of the carbon dioxide.

Organic activators achieve their maximum efficiency at elevated absorption temperatures since, under such conditions, the carbamate dissociation is promoted. It is therefore necessary to modify the equipment configuration when converting some existing systems to activated potassium carbonate solution. For example, plants which were originally designed for MEA require removal of the rich solution/lean solution heat exchange equipment, and may also require conversion from single-stage to split-flow operation to assure product gas purity.

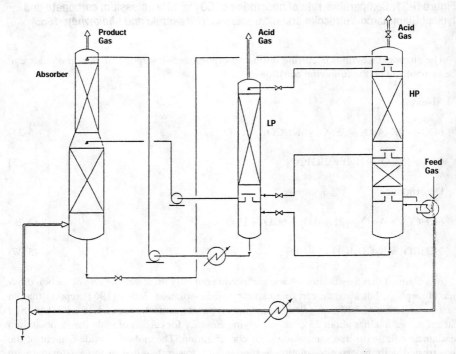

Figure 5-32. Giammarco-Vetrocoke process flow scheme with split-stream absorption and two-pressure-level regeneration. (*Giammarco-Vetrocoke, 1992B*)

Solution changeover may also provide the opportunity for a regenerator revamp to the two-pressure-level configuration shown in **Figure 5-32**. Giammarco-Vetrocoke reports a reduction in energy consumption as high as 40%, a capacity increase of as high as 25%, and a reduction in CO_2 slippage when the two-pressure-level regeneration system is used (Giammarco-Vetrocoke, 1992B).

Arsenite Activated Solutions. Typical flow diagrams for the Giammarco-Vetrocoke arsenite-based process are shown in **Figures 5-33** and **5-34**. The flow scheme shown in **Figure 5-33** illustrates the version of the process in which steam is used for solution regeneration. This arrangement is used primarily for carbon dioxide removal from synthesis gas or crude hydrogen when the gas to be treated is hot and inexpensive steam is available. The flow is quite conventional; the hot gas is washed countercurrently with the solution in the absorber which usually is a packed column, although trays can also be used. The rich solution flows first to a flash drum where a portion of the carbon dioxide is removed by pressure reduction. The partially stripped solution is then heated by exchange with the lean solution before entering the regenerator where it is stripped of its remaining carbon dioxide by steam rising in the column. The regenerator is also a packed column provided with a reboiler for supply of heat. The carbon dioxide leaving the top of the regenerator is cooled; the condensed water is collected in a reflux drum and recycled to the regenerator, while the cold carbon dioxide is vented to the atmosphere.

The lean solution leaving the bottom of the regenerator is first cooled by heat exchange against the partially stripped solution and then further cooled before being returned to the absorber. The second cooling step is optional, depending on the gas purity required.

Air regeneration of the arsenite solution is illustrated in **Figure 5-34**. The rich solution leaving the absorber is heated by steam before entering the flash drum, where part of the carbon dioxide is flashed off. The partially stripped solution then flows to the top of the regenerator where final stripping is obtained by countercurrent contact with a stream of preheated and water saturated air. The mixture of carbon dioxide and air leaving the top of the regenerator is washed with cold water in the dehumidifier in order to recover heat and condense water contained in the gas stream. The cooled mixture of air and carbon dioxide is vented to the atmosphere, while the water flows from the dehumidifier to the gas presaturator. In this manner, maximum heat economy is achieved.

Process Operation

Operating data for a large arsenite-based plant treating natural gas at high pressure have been reported by Jenett (1962). The feed gas to the unit contained a small amount of hydrogen sulfide, which was removed by an arsenate-arsenite solution (see Chapter 9) prior to carbon dioxide removal. The potassium arsenite solution was regenerated by air stripping. Operating data for this plant are shown in **Table 5-10**.

The entire plant is constructed from carbon steel, and no evidence of corrosion has been reported. Also, no discharge of toxic material to the atmosphere with the regeneration air has been observed. Economic comparisons of the Giammarco-Vetrocoke arsenite-based process with the hot potassium carbonate and ethanolamine processes have been presented by Riesenfeld and Mullowney (1959) and by Jenett (1962). These comparisons show a definite economic advantage of the Giammarco-Vetrocoke process over the other two processes with respect to both capital investment and operating costs. However, the comparisons do not take into account toxicity concerns, the more recently developed activated potassium carbonate solutions, and design improvements in the ethanolamine process and, therefore, are of limited value.

376 *Gas Purification*

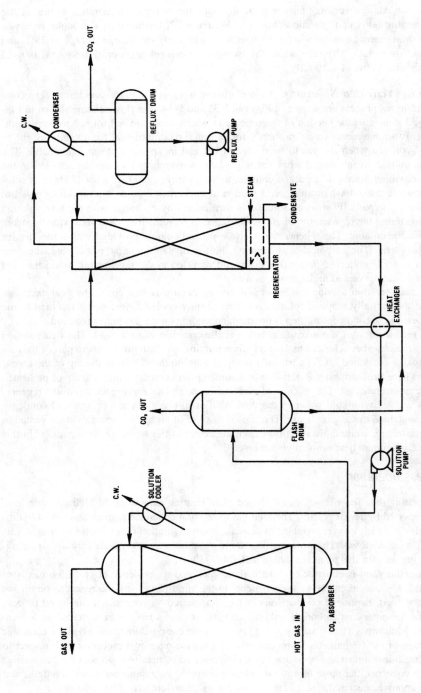

Figure 5-33. Typical flow diagram of Giammarco-Vetrocoke CO_2 removal process with steam regeneration.

Alkaline Salt Solutions for Acid Gas Removal 377

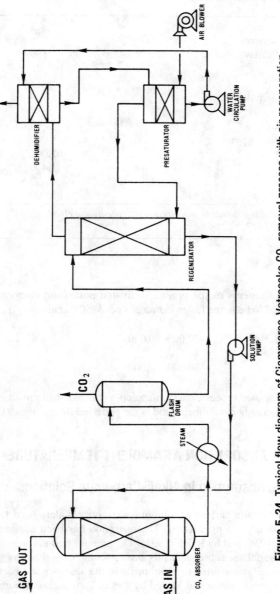

Figure 5-34. Typical flow diagram of Giammarco-Vetrocoke CO_2 removal process with air regeneration.

Table 5-10
Operating Data for Giammarco-Vetrocoke Plant for Natural-Gas Treating

Number of absorbers	3
Number of regenerators	6
Inlet gas, MMSCF/day	180
CO_2, %	28
H_2S, gr/100 SCF	2.5
Outlet gas, MMSCF/day	130
CO_2, %	2.0
H_2S, gr/100 SCF	<0.25
Solution circulation, gpm	8,000
Steam, lb/hr	65,000
Power, kW	1,400
Fuel, MMSCF/day	3.8
Absorber Pressure, psig	1,000

Note: CO_2 removal was followed by dehydration with triethylene glycol.
Source: Jenett, 1962

Operating requirements of the organic-activated potassium carbonate version of the process are reported to be in the following ranges per MSCF acid gas:

Regeneration heat, Btu	30,000–80,000
Power, kWh	1–2
Cooling, Btu	30,000–50,000

For the organic-activated process, the Giammarco-Vetrocoke dual pressure regeneration system results in low energy requirements and is the preferred flow scheme (Giammarco-Vetrocoke, 1992A).

ABSORPTION AT AMBIENT TEMPERATURE

Carbon Dioxide Absorption in Alkali-Carbonate Solutions

Sodium and potassium carbonate solutions were once widely used for CO_2 absorption from flue gases in dry ice production. The operation has been adequately described by Quinn and Jones (1936) and by Reich (1931). The primary reason for removing CO_2 was to provide a raw material for subsequent processing. In the CO_2-recovery process, the alkali carbonate was partially converted to bicarbonate in the absorber and back to the carbonate again in the regenerator, which was heated by steam. Two packed absorbers in series were often used because of the low rate of CO_2 absorption. Major drawbacks of the process were low CO_2 recovery efficiency and the high regeneration steam requirement. The majority of modern CO_2 plants employ aqueous monoethanolamine to remove CO_2 from flue gas (see Chapters 2 and 3).

Sodium carbonate (with free hydroxyl due to excess caustic) has also been used to remove the last traces of carbon dioxide from hydrogen (or other gases) from which carbon dioxide has been removed by the use of a relatively inefficient process or to provide a high purity product from a gas stream containing trace quantities of CO_2 (or other acid gases). In this operation, the sodium carbonate in the solution is formed by the reaction of CO_2 with free hydroxyl, and the alkalinity is maintained at a high level by the periodic addition of fresh caustic. No attempt is made to regenerate the solution, which is discarded or used elsewhere, when its alkalinity is reduced to the point where it is no longer effective for CO_2 removal. Processes of this type are described in a later section entitled "Caustic Wash Processes."

Design and Operating Data

The vapor pressure of carbon dioxide over sodium carbonate bicarbonate solutions can be related by the following equation presented by Harte et al. (1933):

$$P_{CO_2} = \frac{137 f^2 N^{1.29}}{S(1-f)(365-t)} \tag{5-28}$$

Where: N = sodium normality
f = fraction of total base present as bicarbonate
t = temperature, °F
P_{CO_2} = equilibrium partial pressure of CO_2, mm Hg
S = solubility of CO_2 in water under a pressure of 1 atm, moles CO_2/liter

Values for S are available in standard texts. Seidell (1940) gives the values listed in **Table 5-11**. When free hydroxide is present, the vapor pressure of CO_2 over the solution is so low that it can usually be neglected.

Numerous studies have been made of CO_2 absorption in sodium carbonate, bicarbonate, and hydroxide. An excellent review of the early work is presented by Sherwood and Pigford (1952A). For the carbonate bicarbonate region, the data of Furnas and Bellinger (1938) provide an equation relating the overall gas-film coefficient with packing characteristics and liquid rate, as follows:

$$K_G a = C'L^{1-n} a_d n \tag{5-29}$$

Table 5-11
Solubility of CO_2 in Water with Changing Temperature

Temperature	S, moles/liter
15°C (59°F)	0.0455
25°C (77°F)	0.0336
45°C (113°F)	0.0215
75°C (167°F)	0.0120

Source: Seidell (1940)

Where: $K_G a$ = overall gas-film coefficient, lb moles/(hr)(cu ft)(atm)
L = liquid rate; lb/(hr)(sq ft)
a_d = surface area of packing (dry), sq ft/cu ft
C′ and n = constants

At 77°F, a sodium carbonate molality of 0.5, and with 20% of the sodium in the form of the bicarbonate, the constants for equation 5-28 are as in **Table 5-12**. At temperatures above 77°F, the absorption coefficient increases rapidly. An increase to 130°F, for example, almost doubles the coefficient.

Table 5-12
Constants for Equation 5-29

Packing	C′	n	a_d(sq ft/cu ft)
Raschig rings, ⅜-in.	1.53×10^{-4}	0.56	148
Raschig rings, 1-in.	0.81×10^{-4}	0.36	58
Berl saddles, 1-in.	1.00×10^{-4}	0.42	79

In the carbonate-hydroxide region, the rate of absorption of CO_2 is found to be much higher than in the carbonate-bicarbonate region. It is theorized that this is because the molecular carbon dioxide can react directly with hydroxyl ions to form the carbonate ion and that this reaction proceeds rapidly in strongly basic solutions. Useful design data on the absorption of carbon dioxide by sodium hydroxide solutions have been presented by Tepe and Dodge (1943). The data were obtained in an absorber of 6-in. diameter packed with 36 in. of 0.5-in. carbon raschig rings. These observers found that $K_G a$ varies with sodium-ion normality and percentage conversion to carbonate as shown in **Figure 5-35**. They propose that $K_G a$ values taken from the chart be adjusted to other liquid rates and temperatures by the use of the following relation:

$$K_G a \, \alpha \, L^{0.28} T^6 \qquad (5\text{-}30)$$

where T is the absolute temperature and L is the liquid rate.

Tepe and Dodge found $K_G a$ to be essentially independent of gas rate, a condition which would normally indicate that the liquid film was controlling absorption. However, the exponent relating the effect of liquid rate is not as high as would be expected in a simple liquid-film-controlled absorption. As shown in the figure, the absorption coefficient increases with increased sodium hydroxide concentrations up to about 2 N and then decreases. The decrease is presumably due to the higher viscosity of more concentrated solutions—a phenomenon also observed for alkanolamine solutions.

The absorption of carbon dioxide in potassium hydroxide and carbonate solutions is generally similar to that of the corresponding sodium compounds. However, Spector and Dodge (1946) found that slightly larger coefficients are obtained for KOH than for NaOH. The absorption of carbon dioxide in hot potassium carbonate solutions is discussed in an earlier section of this chapter.

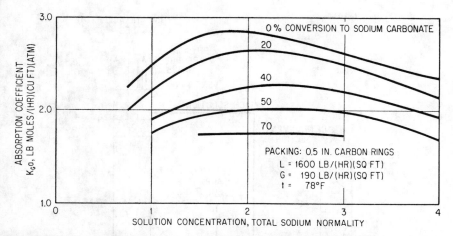

Figure 5-35. Absorption of CO_2 in sodium hydroxide-sodium carbonate solutions. *Data of Tepe and Dodge (1943)*

Seaboard Process

The Seaboard process, which was introduced in 1920 by the Koppers Company, Inc. (now part of ICF Kaiser Engineers of Pittsburgh, PA), was the first commercially applied regenerative liquid process for H_2S removal. The process is based on a dilute sodium carbonate solution absorbent with air stripping, and is thoroughly described by Sperr (1921). The Seaboard process has been largely displaced by the Vacuum Carbonate process, also developed by Koppers and now offered by ICF Kaiser. The principal advantage of the Seaboard Process is its simplicity. Its chief drawbacks are (a) the occurrence of side reactions caused by the introduction of oxygen in the air-regeneration step and (b) the disposal of the foul air containing H_2S. The process is capable of H_2S-removal efficiencies of 85 to 95% in single-stage plants.

As of 1992, ICF Kaiser Engineers reports two operating Seaboard plants in the United States. Environmental problems associated with the air regeneration step have restricted its use (ICF Kaiser Engineers, 1992).

Process Description

A flow sheet of a typical Seaboard-process installation is presented in **Figure 5-36**. The circulating solution normally contains 3.0 to 3.5 wt % sodium carbonate. This is used to wash the gas in a countercurrent absorber column and is regenerated in a separate actifier column by a countercurrent flow of low-pressure air. The principal chemical reaction in the process is as follows:

$$Na_2CO_3 + H_2S = NaHCO_3 + NaHS \tag{5-31}$$

According to Gollmar (1945A), the solution flow rate is between 60 and 150 gal/1,000 scf of gas, depending upon the H_2S and CO_2 concentrations in the gas. For typical coke-oven

382 Gas Purification

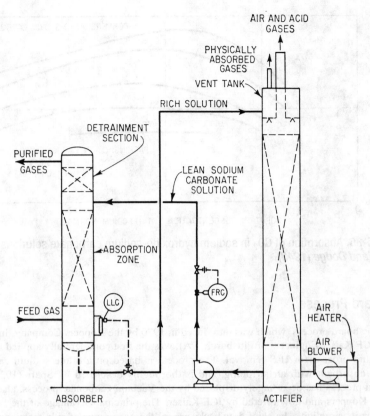

Figure 5-36. Flow diagram of Seaboard gas-purification process.

gas containing 300 to 500 grains of H_2S/100 scf and 1.5 to 2.0% CO_2, the solution flow rate will be in the range of 60 to 80 gal/1,000 scf, representing a carrying capacity of around 50 grains/gal. For higher H_2S concentrations and higher than normal CO_2, particularly the latter, solution flow rates up to 150 gal/1,000 scf may be required.

The air rate required for solution regeneration is usually from 1.5 to 3.0 times that of the gas, depending upon the H_2S removal efficiency desired. The actifier exhaust gas, therefore, contains the H_2S removed at a concentration of about one-half that in the feed-gas stream. Since such concentrations of H_2S are extremely objectionable, it is necessary to either vent the air through a tall stack or, preferably, to use it as combustion air for some other plant operation. Disposal by combustion offers advantages, including conversion of H_2S to the less objectionable SO_2, and better dispersal because of the high temperature and velocity of the combustion-stack gases.

Plant Operation

As shown by the flow diagram, the process is quite simple. In some cases, plants have been built with a single tall tower, half of this being used for absorption and the other half for stripping. Principal auxiliary equipment items are the air fan, solution pump, and solution and air heaters. Since air is required in large quantities for regeneration, it is important that the actifier column be designed for very low pressure drop to minimize horsepower requirements.

The use of air as stripping vapor has the disadvantage of promoting oxidation in the system when air is used; approximately 5% of the hydrogen sulfide that is absorbed is oxidized to thiosulfate. As sodium thiosulfate is not regenerable by simple stripping, buildup of this salt results in a loss of solution activity, and it becomes necessary to replace a portion of the solution periodically.

Hydrogen cyanide is also absorbed by the sodium carbonate solution and, to a considerable extent, is oxidized by the oxygen in the air used for stripping according to the following reaction:

$$2NaHS + 2HCN + O_2 = 2NaSCN + 2H_2O \tag{5-32}$$

Although it has been reported that the spent Seaboard process solutions have some value because of their NaSCN content, in most cases their disposal is a serious problem from the water pollution standpoint.

Vacuum Carbonate Process

The use of vacuum distillation for regeneration of the alkali-carbonate solutions used in H_2S absorption was a development of the Koppers Company, Inc., now part of the ICF Kaiser Engineers of Pittsburgh, PA (Sperr and Hall, 1925; Powell, 1941). The process was an outgrowth of the early Seaboard process (which used air reactivation) and offered the advantage over this process of recovering the H_2S in a concentrated, usable form. The use of vacuum reduced the steam requirement to about one-sixth that required for steam-stripping at atmospheric pressure (Gollmar, 1945B). The first installation of the Vacuum Carbonate process was in Germany in 1938 and used a potassium carbonate solution. Plants constructed in the United States have in general utilized sodium carbonate solutions. The process has been primarily applied to coke-oven gas streams, which generally contain 300 to 500 grains H_2S/100 scf. Coke-oven gas also contains HCN and other impurities which may cause difficulties with other H_2S-absorption systems. **Figure 5-37** depicts a Vacuum Carbonate-process installation in a large coke-oven plant.

Although the Vacuum Carbonate process is still commercially available, in recent years it has been replaced in many coke-oven gas treating installations by processes based on the oxidation of hydrogen sulfide to elemental sulfur in the liquid phase (see Chapter 9).

Process Description

A simplified process flow diagram is shown in **Figure 5-38.** The impure gas is contacted with dilute sodium carbonate in a packed countercurrent absorber. Rich solution is passed to the top of an actifier column, where it is regenerated by vacuum distillation. The regenerated solution is then pumped from the bottom of the actifier through a solution cooler and back to the absorber. Actifier overhead gases, which consist of H_2S, HCN, CO_2, and water vapor, pass through condensers and a vacuum pump system. In the process as illustrated in **Figure 5-38,** the heat required for activation is provided by low-pressure steam in a reboiler at the base of the actifier.

Figure 5-37. Vacuum Carbonate (hot-activation) gas-purification plant. *Koppers Company, Inc.*

In a modification of the process aimed at reducing steam consumption (Gollmar, 1949), a major portion of the heat is supplied to the solution in the actifier by heat exchange with a source of low level waste heat. Such heat is available in most coke-oven plants in the form of "flushing liquor," which is circulated through the hot gas-collecting mains and reaches a temperature of 75° to 80°C (167° to 176°F) (Gollmar and Hodge, 1956). Solution from the bottom of the actifier is pumped through heat exchangers, where it is heated by the flushing liquor. It then flows back to the lower portion of the actifier where stripping vapors are flashed off as a result of the relatively low boiling temperature in the vacuum actifier.

The original form of Vacuum Carbonate process (shown in **Figure 5-38**) was designed to remove approximately 90% of the H_2S. In order to comply with increasingly stringent environmental regulations, the Koppers Company developed the two-stage process to attain about 98% H_2S removal without excessive steam consumption (Grosick and Kovacic, 1981).

A flow diagram of Koppers' two-stage process is shown in **Figure 5-39.** Coke-oven gas enters the bottom of the absorber and is contacted countercurrently with semi-lean solution in the bottom (primary) stage then with lean (activated) solution in the top (secondary) stage. This arrangement produces high purity product gas without the requirement of completely stripping the entire circulating solution stream.

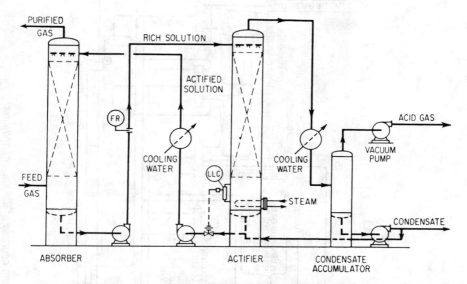

Figure 5-38. Simplified flow diagram of Vacuum Carbonate process.

The activator is also divided into two stages. The top (primary) stage provides partial stripping of the highly loaded solution from the primary absorber stage, and the bottom (secondary) stage provides very complete stripping of the smaller solution stream from the secondary absorber stage. Partially stripped solution from the primary activator stage is mixed with lightly loaded solution from the secondary absorber stage to form the semi-lean solution fed to the primary absorber stage.

The flow diagram of **Figure 5-39** includes a Koppers heat recovery system to minimize steam consumption. The high vacuum in the activator reduces the boiling point of the solution so that low temperature heat sources can be used to generate stripping steam. In the system shown, lean solution from the bottom of the activator is used as the cooling medium in the inter- and after-condensers and is also heated by heat exchange with flushing liquor before it is flashed back into the activator. Additional heat can be added by the use of live steam in a solution heater as shown in the diagram and by heat exchange with hot wash oil (not shown) (Koppers, 1978).

Another concept for reducing steam consumption in the Vacuum Carbonate process has been used successfully in a Ukraine coking plant (Redin et al., 1986). The plant originally used heat exchange between a recirculated stream of lean solution from the activator bottom and hot coke-oven gas as the primary source of heat to the activator. The modification consisted of substituting essentially pure water condensate for salt solution in the heat transfer loop. In the revised flow system, the condensate is heated by heat exchange with coke-oven gas then flashed to activator pressure. A portion of the water is vaporized and used as stripping steam; the remainder, which is cooled by the flashing step, is recirculated through the heat exchanger. Water condensed from the acid gas stream is used as makeup to replace the amount flashed off.

Since pure water has a lower boiling point and higher heat capacity than the salt solution used for absorption, it is capable of extracting significantly more heat from the hot coke-

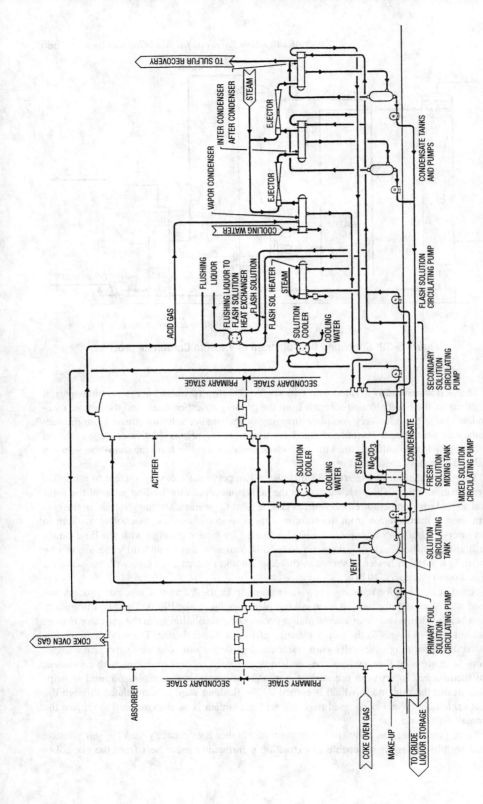

Figure 5-39. Koppers' two-stage Vacuum Carbonate process. *Diagram courtesy of ICF Kaiser Engineers*

oven gas. In the example given, the temperature of the liquid entering the coke-oven gas heat exchanger was reduced about 5°C (9°F), and steam consumption of the unit was reduced from about 15–20 LT/h to 0–5 LT/h. An additional benefit claimed for the process modification is a reduction of corrosion in the heat transfer loop (Redin et al., 1986).

Basic Data

Principal reactions occurring in the Vacuum Carbonate process are

$$Na_2CO_3 + CO_2 + H_2O = 2NaHCO_3 \tag{5-33}$$

$$Na_2CO_3 + H_2S = NaHS + NaHCO_3 \tag{5-34}$$

$$Na_2CO_3 + HCN = NaCN + NaHCO_3 \tag{5-35}$$

Side reactions, which may occur if air contacts the solution, include equations 5-36 and 5-37:

$$2NaHS + HCN + O_2 = 2NaSCN + 2H_2O \tag{5-36}$$

$$2NaHS + 2O_2 = Na_2S_2O_3 + H_2O \tag{5-37}$$

The CO_2 absorption-desorption reaction (equation 5-33) is not completely reversed in the stripping operation; thus, a considerable proportion of the soda ash remains as sodium bicarbonate, and very little net CO_2 removal is realized. An analysis of the actified solution from an operating plant, **Table 5-13,** has been given by Kurtz (1955).

The high values for NaSCN and $Na_2S_2O_3$ given in **Table 5-13** are not necessarily typical, as these will depend upon the extent to which oxygen is admitted to the system as well as on the age of the solution. In the plant from which the solution shown in **Table 5-13** was taken, a specific gravity of 1.125 at 60°F was adopted as the maximum allowable in the actified solution. The solution was discarded when dissolved solids increase to the point where this specific gravity was attained—usually after 6 to 8 months operation.

The active ingredient of the solution can be considered to be the Na_2CO_3, which provides alkalinity for the absorption of acid gases. The reactions between acid gas and Na_2CO_3 are not complete, however, and can only approach the equilibrium condition corresponding to the gas and liquid composition. When H_2S is absorbed by sodium carbonate solution, NaHS and $NaHCO_3$ are formed by the reaction given in equation 5-34. The vapor in equilibrium with such a solution thus contains both H_2S and CO_2 in accordance with the decomposition pressure of these two compounds. Carbon dioxide is the stronger acid and is held more firmly; however, its rate of absorption is much lower. As previously stated, the low rate of absorption of carbon dioxide is believed to be due to a slow chemical reaction that must occur before combination of CO_2 and Na_2CO_3 can take place. The rate-controlling reaction is most probably that of the dissolved molecular CO_2 with OH^- ions to form HCO_3^- ions. Hydrogen sulfide does not require hydration to form an acid and can, therefore, react rapidly with alkaline solutions.

Litvinenko (1952) studied the absorption of H_2S in solutions of sodium and potassium carbonate on a laboratory scale and found that at low H_2S partial pressures (1% H_2S in nitrogen at 1 atm total pressure) the gas film presented the major resistance. Solutions containing 5%

Table 5-13
Vacuum Carbonate Process Solution Analysis

Constituent	Concentration, g/liter
Na_2CO_3	20.25
$NaHCO_3$	25.38
NaHS	0.67
$Na_2S_2O_3$	9.48
NaSCN	66.62
Total solids	163.86 g/liter
Specific gravity (60°F)	1.103
Boiling point (30.38 in. Hg)	103.0°C (217.4°F)

Source: Kurtz (1955)

Na_2CO_3 and 15% K_2CO_3 were studied. Although the potassium carbonate produced slightly higher absorption coefficients, results with both solutions were similar, K_Ga was found to vary as $G^{0.8}$ (in a 13-mm ID column packed with 5-mm glass rings) and as $L^{0.34}$ (in a 9.5-mm ID wetted-wall column). These results indicate that the H_2S reaction rate is fast enough to make the liquid film resistance small relative to the overall resistance. Similar conclusions were reached by Garner et al. (1958).

When both H_2S and CO_2 are present in the gas stream contacting a sodium carbonate solution, a higher percentage of the H_2S will be absorbed (in a tower of reasonable height) because of the difference in rates of reaction. Unfortunately, however, the CO_2 that is absorbed will not be completely stripped and the bicarbonate concentration of the solution will gradually build up until a steady-state condition is attained. At this point, the quantity of CO_2 absorbed (which decreases with increased $NaHCO_3$ concentration) will equal the quantity stripped in the desorption step. The net effect of increased CO_2 concentration in the gas (with the same H_2S concentration) is thus to increase the $NaHCO_3$ content of the circulating solution, thereby decreasing its absorptive capacity and resulting in an increased solution circulation requirement.

Plant Design and Operation

Absorption. Packed absorbers are generally employed in the Vacuum Carbonate process. In one commercial unit (plant A of **Table 5-14**), the packing is in the form of hurdles made up of numerous small wood slats (West Virginia spruce) (Kastens and Barraclough, 1951). Ceramic saddles are generally used in Koppers' two-stage process absorbers (Koppers, 1978). Plants that have been described are used for treating coke-oven gas, which contains 250 to 500 grains $H_2S/100$ cu ft and 1½ to 3% CO_2. This type of gas is usually treated at a relatively low pressure (below 25 psig).

According to Smith (1953), up to 93% of the HCN and 5 to 7% of the CO_2 are also removed with the H_2S. When the process is operated to attain very high H_2S-removal efficiency, more CO_2 is also removed. Since the absolute quantity of H_2S absorbed does not increase materially, the resulting acid-gas stream then contains a higher percentage of CO_2. Operating data on the absorption step in two commercial installations are presented in **Table 5-14**.

Table 5-14
Vacuum Carbonate Process Absorber Data

Absorber	Plant A (Kastens and Barraclough, 1951)	Plant B (Kurtz, 1955)
Gas Rate, SCF/hr	2,300,000	1,000,000, avg. / 1,300,000, max.
Gas composition:		
H_2S, grains/100 SCF	300	406 avg.
H_2S, percent	0.48	0.64
HCN, percent	0.13	
Solution rate, gpm	533	375–400
Absorber pressure, psig	2	15–20
Purification efficiency, % removed:		
H_2S	93	85 avg.
HCN	85–90	

Desorption. The solution is stripped in a column called the "actifier," which is operated at a pressure of about 2.0 to 2.5 psia. Steam is used as stripping vapor and is generated by boiling the solution in the reboiler at the base of the actifier. Because of the reduced pressure, boiling occurs at a temperature of about 140°F (60°C). Both wood hurdle-packed and bubble-plate columns have been employed for actifying. In plant B of **Table 5-14,** the actifier is equipped with 15 stainless-steel bubblecap trays.

Since the reboiler heat can be supplied at a very low level, it is frequently possible to utilize waste heat from other plant streams. As previously mentioned, one such source of heat is the flushing liquor which is circulated in the hot-gas mains of coke-oven plants.

The quantity of heat required for regeneration depends to a considerable extent upon the degree of H_2S removal desired. In plant A of **Table 5-14,** flushing liquor is used as the source of heat. In plant B, it is reported (Kurtz, 1955) that 13,000 to 15,000 lb/hr of steam are used for reboiling at the base of each actifier. Since 1 to 1.3 MMscf/hr of gas are treated per unit, the steam requirement is about 12.0 lb/Mscf of gas purified.

Acid gas leaving the actifier of plant B contains 5 to 9% HCN. This is removed by a water wash prior to combustion of the H_2S for use in sulfuric acid manufacture. Acid-gas analyses for plants A and C are presented in **Table 5-15.**

At plant A, the HCN that is removed in the Vacuum Carbonate process is recovered as a by-product. In most other plants, this substance is either removed and burned (to permit the H_2S to be utilized in conventional acid plants), or it is destroyed in a specially designed sulfuric acid plant, which utilizes the H_2S.

Operating Problems. Minor constituents in coke-oven gas, particularly naphthalene, can cause difficulty in the operation of Vacuum Carbonate process plants. Smith (1953) recommends that the process be limited in application to gases containing not more than 2 grains naphthalene/100 scf; 1.5 grains ammonia (including pyridine)/100 scf; and 1 grain tar fog/100 scf.

Coke-oven gas normally meets these specifications after it has passed through benzene washers. If the Vacuum Carbonate process is located ahead of the benzene washers, which is

Table 5-15
Analysis of Acid Gas From Plants Using Vacuum Carbonate Process

Gas	Gas content, % Plant A (*Kastens and Barraclough, 1951*)	Plant C (*Smith, 1953*)
H_2S	55	70
HCN	15	10
CO_2	25	15
N_2, etc.	5	5

sometimes done to improve HCN yield, special precautions are necessary to minimize difficulties caused by naphthalene.

Naphthalene is only very slightly soluble in the sodium carbonate solution; however, a small amount is absorbed by the solution when it contacts the coke-oven gas. Absorbed naphthalene is stripped from the solution in the actifier and carried by the acid-gas stream to the condensers. The naphthalene concentration in the acid-gas stream will thus be a function of the initial concentration in the coke-oven gas. If this concentration is too high, condensation will occur when the acid gas is cooled. Since naphthalene condenses as a solid, this can cause plugging of condensers and other equipment.

Ammonia dissolves readily in the Na_2CO_3 solution and is also stripped in the actifier. Cooling the actifier effluent gas causes the condensation of water, which readily absorbs ammonia from the acid-gas stream. The ammonia dissolved in the condensate increases the solubility of H_2S and, as this is returned to the system, the overall H_2S-removal efficiency of the process is reduced. Pyridine behaves in a similar manner; however, the quantity of this compound is usually too small to have an appreciable effect on process efficiency.

Corrosion is generally not a major difficulty with Vacuum Carbonate plants because of the relatively low temperatures employed. Mild steel is the primary material of construction, although, as previously mentioned, wooden and ceramic packings and stainless-steel actifier trays are also used. Stainless steel is also used for minor components of the vacuum and other pumps, thermometer wells, and instrument diaphragms (Smith, 1953).

One problem common to all processes using liquids for the purification of coke oven gases or other coal derived gases is the handling of contaminated waste streams within the framework of existing environmental pollution control regulations. The principal offender in coke-oven gas is hydrogen cyanide, which is readily soluble in most solvents and is quite reactive with the chemicals used as active treating agents. The reaction products such as cyanides, thiocyanates, and iron-sulfur-cyanide compounds have to be eliminated before liquid waste streams can be disposed of in sewage systems. A considerable effort has been made to develop economical methods for acceptable waste disposal and some are in industrial use. A good review of such methods, including those used in vacuum carbonate units, has been presented by Massey and Dunlap (1975).

Operating Requirements. Comparative operating data for a Koppers two-stage Vacuum Carbonate plant and a single-stage Vacuum Carbonate plant, both with Koppers heat exchange systems to conserve steam, are presented in **Table 5-16** (Koppers, 1978). The

Table 5-16
Comparative Operating Data for a Koppers Two-Stage Vacuum Carbonate Plant and a Single-Stage Vacuum Carbonate Plant with Koppers Heat Exchange Systems

General Design Data	Two-Stage	Single-Stage
Coke-Oven Gas Processed, scfd	60,000,000	60,000,000
Average H_2S in Inlet Gas, grain/100 scf	500	500
Average H_2S in Outlet Gas, grain/100 scf	10	50
Average H_2S Removal, %	98	90
Cooling Water Temperature, °F	80	80
Vapor Temperature Leaving Vapor Condensers, °F	100	100
Coke-Oven Gas Temperature to Vacuum Carbonate Plant, °F	90	90
Coke-Oven Gas Pressure to Vacuum Carbonate Plant, Inches W.C.	23	20
Utility and Chemical Requirements		
150 psig Steam Required, lb/day		
Actifier Reboiler	—	—
Jet Ejectors	514,000	389,000
Fresh Solution Tank Agitator and Misc.	3,000	3,000
Total	517,000	392,000
City Water, gpd		
Cooling Tower Make-Up	360,000	312,000
Electric Power, kWH/day		
Solution Circulating Pumps	2,400	1,700
Flash Solution Pump	3,600	3,150
Condensate Pumps	30	30
Flushing Liquor Pump-Additional Power	580	580
Cooling Tower Fans	3,840	3,500
Cooling Tower Pump	3,600	3,360
Chemical Treatment Pumps	10	10
Lighting	30	25
Total	14,090	12,355
Chemicals, lb/day		
Sodium Carbonate	530	530
Cooling Tower Chemicals	335	290
Operating Labor		
Operator—Man Hours/Day	6	6
Laboratory Technician—Man Hours/Day	2	2

Source: Koppers (1978)

comparison is based on the treatment of 60,000,000 scfd of 90°F coke-oven gas containing 500 grains/100 scf H_2S. The single-stage plant is designed for 90% H_2S removal and the two-stage plant for 98% removal. The principal differences in operating requirements are about 32% higher steam and 14% higher electric power for the two-stage process.

Vacasulf Process

The Vacasulf process, offered by the Krupp Wilputte Corporation in the United States and GKT in Germany, is closely related to the Vacuum Carbonate process. It uses a potassium carbonate solution for H_2S absorption and vacuum regeneration. A simplified flow diagram of one application of the Vacasulf process, in which a portion of the coke-oven gas is treated at elevated pressure, is shown in **Figure 5-40.** A plant of this design was started up in Germany in March, 1991. Design and operating data for this plant are given in **Table 5-17.**

The low-pressure scrubber uses polyethylene packing and operates at essentially atmospheric pressure. It reduces the H_2S content of the gas to less than 10 grain/100 scf. The high-pressure scrubber operates at a pressure of 12 bar and is followed by a sodium hydroxide scrubber to reduce the H_2S content of the final product to 0.08 grain/scf.

Rich solution from the low-pressure scrubber is fed to the midpoint of the high-pressure scrubber because it is capable of absorbing additional H_2S at the H_2S partial pressure existing in the bottom section of the high-pressure scrubber. Cooled lean solution from the regenerator is split into two streams—one is fed to the top of the low-pressure scrubber, and the other to the top of the high-pressure unit. The total rich solution stream from the high-pressure scrubber is preheated and fed into the stripper. Heat for solution regeneration is provided by hot water, which enters the reboiler at 90°C and is returned to the water heater at 60°C. Alternatively, a portion of the heat may be provided by heat exchange with coke plant flushing liquor.

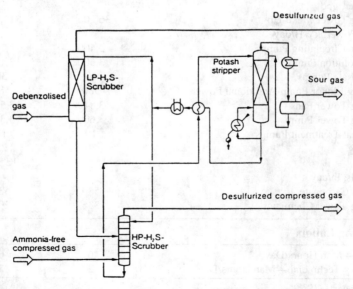

Figure 5-40. Simplified flow diagram of Vacasulf process with high- and low-pressure scrubbers and potassium carbonate solution. (*Krupp Wilputte, 1992*)

Table 5-17
Vacasulf Process Design and Operating Data

	Units	Actual Average Values	Contract Values
Coke-oven gas capacity			
-low pressure	MMSCFD	10–25	26.8
-high pressure	MMSCFD	27–50	53.6
Pressure loss			
LP-H_2S-scrubber	mbar	10–12	<15
H_2S-content inlet scrubber	gr/100 SCF	165	280
H_2S-content outlet			
-LP scrubber	gr/100 SCF	<10	21
-HP scrubber	gr/100 SCF	<0.08	0.08
HCN-removal	%	90	
CO_2-removal	%	10	
Consumption Figures			
Heat duty for desorption	MMBTU/h	16.0	22.0
KOH (50% by wt.)	lb/h	45	82.7
Electric power	kW	515	515
Cooling water	gpm	1,761	2,364

Source: Krupp Wilputte (1992)

The presence of HCN and O_2 in the coke oven gas results in the formation of nonregenerable potassium compounds in the circulating absorbent. Consequently, a portion of the scrubbing solution must be periodically purged and replaced (GKT, 1992; Krupp Wilputte, 1992).

Tripotassium Phosphate Process

The use of a tripotassium phosphate solution for H_2S removal was introduced by the Shell Development Company (Rosenstein and Kramer, 1934). The process has been displaced by alkanolamine and other processes for refinery and natural-gas purification and is no longer offered or supported by Shell (Shell Development Co., 1992).

From a historical viewpoint, the principal advantages of the process were nonvolatility of the active component in the solution, insolubility in hydrocarbon liquids, and nonreactivity with COS and other trace impurities. These advantages made the tripotassium phosphate process suitable for high-temperature applications, the treatment of liquid hydrocarbons, and the purification of refinery or synthesis gases. In common with the other alkali-salt processes, the tripotassium phosphate process offered some selectivity for H_2S in the presence of CO_2. This selectivity gave the phosphate process an economic advantage over monoethanolamine or diethanolamine systems when it was desired to remove H_2S with a minimum extraction of CO_2 from gas streams containing both at CO_2/H_2S ratios exceeding about 4:1.

Process Description

The absorption of hydrogen sulfide by tripotassium phosphate may be represented by equation 5-38:

$$K_3PO_4 + H_2S = K_2HPO_4 + KHS \tag{5-38}$$

The basic flow-cycle used is similar to that of alkanolamine processes and other heat-regenerative systems. Where a high degree of gas purification is required, a double-stream system such as that shown in **Figure 5-41** is suggested.

If no CO_2 is present, a 40 to 50% (by weight) tripotassium phosphate solution is used; however, if CO_2 is present, a 35% solution is generally used to avoid precipitation of bicarbonate. In the split-stream cycle, a portion of the solution is passed through a second stripping zone, and aqueous condensate is added to this portion after final stripping. This more completely regenerated (and more dilute) stream is fed into the top of the absorber to provide final cleanup of the gas stream. Higher purity is attainable with the dilute solution because, for a given H_2S/K_3PO_4 ratio, the vapor pressure of H_2S is lower over solutions containing less tripotassium phosphate.

Design and Operating Data

The equilibrium vapor pressure of hydrogen sulfide over a 50% tripotassium phosphate solution at several temperatures is shown in **Figure 5-42** which is based on the data of Rose-

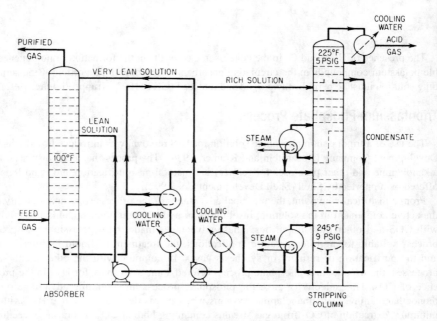

Figure 5-41. Flow diagram of tripotassium phosphate process, split-flow cycle.

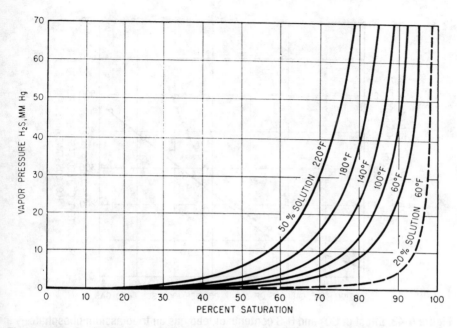

Figure 5-42. Equilibrium vapor pressure of hydrogen sulfide over 50 and 20% tripotassium phosphate solutions. *Data of Rosebaugh (1938)*

baugh (1938). One isotherm for a 20% solution is also included to point out the effect of dilution on the H_2S vapor pressure.

Detailed data on a pilot-plant study of the applicability of tripotassium phosphate solutions to the selective absorption of H_2S in the presence of CO_2 have been presented by Wainwright et al. (1953). Results of this study with regard to the effect of CO_2 and H_2S on solution circulation rate requirements are shown in **Figure 5-43.** As would be expected, an increase in either the CO_2 or H_2S content of the feed gas increases the quantity of 35% potassium phosphate solution required to make a product gas containing 25 grains H_2S/100 scf. The data were obtained with a column made of 10-in. standard iron pipe packed with 15 ft of 1-in. porcelain raschig rings and operated at 300 psig pressure. All runs were made with relatively high CO_2/H_2S ratios in the feed gas. The effect of varying this ratio on the quantity of H_2S absorbed per gallon of solution is shown in **Figure 5-44,** which also includes data on an Alkacid "dik" solution for comparison.

Wainwright et al. (1953) found that their initial solution, which contained 43.7% tripotassium phosphate, could not be used on gases of high CO_2 content because of the formation of potassium bicarbonate, which precipitated and plugged the absorber packing. The solution was therefore diluted to 32 to 35%, and no further difficulty was encountered as a result of precipitation.

Reactivation was accomplished in a column also made from 10-in. steel pipe and packed with 15 ft of 1-in. porcelain raschig rings. The reboiler pressure was generally maintained at about 3 psig. Steam consumption averaged about 1 lb/gal of circulating solution. Because of relatively high heat losses from the pilot-scale equipment, it was estimated that the quantity

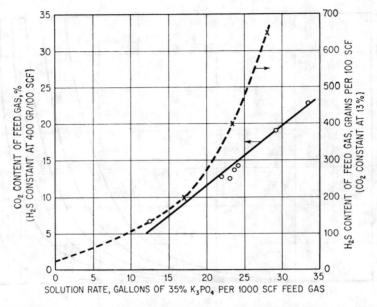

Figure 5-43. Effect of CO_2 and H_2S contents of feed gas on tripotassium phosphate—solution rate required to produce exit gas containing 25 grains H_2S/100 SCF. Absorber of 10-in. ID packed with 15 ft of 1-in. porcelain raschig rings; solution regenerated with steam, approximately 1 lb/gal. *Data of Wainwright et al. (1953)*

of steam actually used for reactivation was somewhat lower, about 0.9 lb/gal of solution. Increasing the steam flow rate by 20% was found to have only a slight effect on the degree of stripping. It has been reported (Liedholm, 1958) that commercial plants operating at about the same conditions have substantially lower steam consumption than that observed by Wainwright et al. (1953). This higher stripping efficiency is believed to be due to (a) more efficient stripping columns, (b) more nearly optimum solution circulation rates, and (c) less heat loss from large-scale equipment.

A commercial plant which removes H_2S and CO_2 from a stream of light hydrocarbons in the liquid state has been described by Zahnstecher (1950). Although this is not gas purification, the regeneration step is identical whether the absorber treats liquid or gaseous hydrocarbons. The stripping column consists of a 2 ft, 6-in.-diameter unit with 10 bubble cap trays installed directly on top of a steam-heated reboiler. With a solution-circulation rate of 25 gpm, approximately 1.5 lb of steam are required per gallon of solution circulated. The regenerator operates at a pressure of 2.5 psig and a temperature of about 226°F. During regeneration, the solution concentration is reduced from 0.187 mole H_2S and 0.640 mole CO_2 to 0.090 moles H_2S and 0.493 mole CO_2 per mole of K_3PO_4. In this plant, the feed to the absorption unit (liquid phase) contains 0.50 mole percent CO_2 and about 0.4 mole percent H_2S.

Sodium Phenolate Process

The sodium phenolate process is also of more interest for historical reasons than for practical applications as it has been largely supplanted by other processes. The process, which

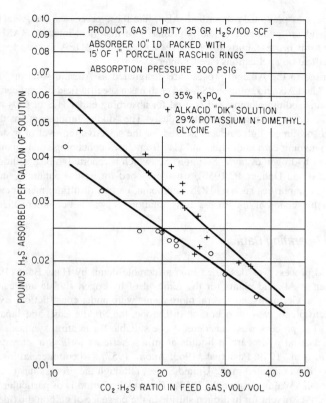

Figure 5-44. Effect of CO_2/H_2S ratio on capacity of K_3PO_4 and Alkacid "dik" solutions. Data of Wainwright et al. (1953)

was developed by the Koppers Company, Inc. (now part of ICF Kaiser Engineers), employs a fairly concentrated solution of sodium phenolate in a conventional heat-regenerative flow cycle. The solution contains approximately three moles of NaOH (120 g) and two moles phenol (188 g) per liter (Shreve, 1945). Its capacity for H_2S is quite high, up to about 35 scf/cu ft for an extremely sour gas (Powell, 1938). The basic disadvantages of the process are the low efficiency of H_2S removal (on the order of 90%) and the high steam consumption. The steam consumption can be cut almost in half by the use of a two-stage stripping system; however, this complicates the plant design considerably. Phenolate plants have also reportedly encountered corrosion difficulties.

Alkacid Process

The Alkacid (Alkazid) process, which was developed in Germany by I.G. Farbenindustrie (Bähr and Mengdehl, 1935), could be classified as three separate processes in that three different absorption solutions are used. However, all of the process variations use a solution of the salt of a strong inorganic base and weak organic nonvolatile acid and all use a conventional heat-regenerative cycle, such as employed by the tripotassium phosphate and sodium phenolate processes previously described. Also, like these processes, the Alkacid process is no longer considered competitive, and is primarily of historical interest. BASF, which formerly licensed this process, currently proposes an activated MDEA process for a wide variety of gas purification applications (see Chapter 2).

Solutions used in the Alkacid process are designated as Alkacid solution "M," Alkacid solution "dik," and Alkacid solution "S"; and each has a specific field of application. The "M" solution contains sodium alanine and is used for absorbing either H_2S or CO_2 when present alone, or for absorbing both gases simultaneously. The "dik" solution contains the potassium salt of diethyl- or dimethylglycine and is used for the selective removal of hydrogen sulfide from gases containing carbon dioxide and also from gases containing small quantities of carbon disulfide or hydrogen cyanide. Solution "S," which is reported to be a solution of sodium phenolate (Reed and Updegraff, 1950), was developed for gases containing an appreciable amount of other impurities such as HCN, ammonia, carbon disulfide, mercaptans, dust, and tar. However, the version of the process using solution "S" was not commercialized.

Design and Operating Data

The Alkacid process was described in considerable detail by Hans Bähr (1938). At that time a number of Alkacid plants for the removal of hydrogen sulfide and carbon dioxide were operating in Germany and several more plants were under construction. By 1959, more than 50 Alkacid plants were in operation in Europe, the Middle East, and Japan (Leuhddemann, 1959). The process was considered to be suitable for treating synthesis gases, water gas, natural gas, and hydrocarbon liquids at atmospheric as well as at elevated pressures (Leuhddemann et al., 1959; Pasternak, 1963; Anon., 1957). No commercial installations are known to have been operated in the United States, although the process has been studied on a pilot-plant scale (Wainwright et al., 1953). The "dik" solution is of particular interest as it is a more selective solvent for hydrogen sulfide in the presence of carbon dioxide. A comparison of the selectivity of the two solvents, based on the data of Leuhddemann et al. (1959), is shown in **Figure 5-45**. These data were obtained by shaking the "dik" and "M" solutions at ambient temperatures with pure hydrogen sulfide and carbon dioxide and measuring the volume of gas dissolved per volume of solution every minute, until saturation was reached. Vapor pressures of hydrogen sulfide and carbon dioxide over Alkacid "dik" and "M" solutions were also reported by Leuhddemann et al. (1959).

The U.S. Bureau of Mines studied the process to evaluate the Alkacid "dik" solution as a selective absorbent for hydrogen sulfide in the presence of carbon dioxide, and the results were compared with those of other absorbents tested in the same apparatus. The solution capacity of the Alkacid solution tested has been compared with that of the 35% tripotassium phosphate solution in **Figure 5-44**, which demonstrates the effect of the CO_2/H_2S ratio on the absorptive capacity of both solutions. It will be noted that the Alkacid solution takes up appreciably more H_2S particularly at low CO_2/H_2S ratios. Although its capacity is higher, the Alkacid solution was found to be somewhat less selective than tripotassium phosphate.

An indication of the capacity of the Alkacid solution for absorbing H_2S is shown in **Figure 5-46**, which presents the vapor pressure of H_2S over Alkacid solutions containing two concentrations of CO_2. This chart is based on U.S. Bureau of Mines data (Wainwright et al., 1953) and represents results obtained with a solution of the following composition and characteristics:

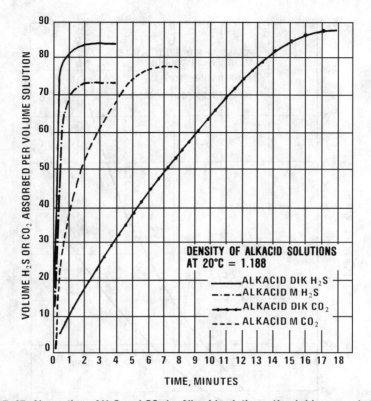

Figure 5-45. Absorption of H_2S and CO_2 by Alkacid solutions. (*Leuhddemann et al., 1959*)

Total nitrogen, %	2.60
Sulfated ash, %	19.26
NH_3, nitrogen %	0.01
Solids, %	29.2
Specific gravity	1.138

Another pilot-plant investigation comparing the selectivity of Alkacid "dik" with that of diethanolamine (DEA) and methyldiethanolamine (MDEA) was presented by Pasternak (1963). The results showed that, under the conditions of the test, the "dik" solution was most selective, followed by MDEA and DEA. Also, the "dik" solution showed the highest capacity for hydrogen sulfide for comparable partial pressures of hydrogen sulfide and carbon dioxide in the feed gas. It should be noted that no attempt was made in this study to attain complete hydrogen sulfide removal, the maximum being on the order of 70%. Another interesting result was that the steam requirement for solution regeneration was appreciably lower for the "dik" solution than for the two ethanolamines.

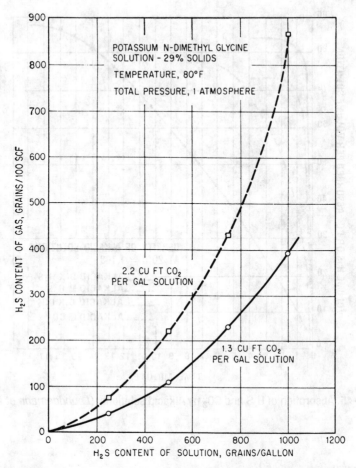

Figure 5-46. Vapor pressure of H_2S over potassium N-dimethylglycine solutions containing CO_2. *Data of Wainwright et al. (1953)*

The "dik" solution was reported to be quite stable although some aging and loss of selectivity was observed, especially when the gas to be treated contained HCN and oxygen (Pasternak, 1963).

Operating results from an Alkacid plant using "M" solution for the removal of hydrogen sulfide and carbon dioxide from natural gas are shown in **Table 5-18** (Leuhddemann et al., 1959). It has been claimed that the Alkacid solutions are relatively noncorrosive, and the equipment requires no special materials of construction except in a few locations which are particularly subject to wear (Bähr, 1938). Patents have been obtained on methods of avoiding corrosion (Farbenindustrie, 1935). Reportedly, it was German practice to use aluminum and special alloys for the hot-solution pumps and lines, the reactivator, and the reboiler (Peck, 1938).

Table 5-18
Removal of H_2S and CO_2 with Alkacid "M"

Feed gas volume, SCF per hour	650,000
Pressure, psia	85
Temperature, °F	78
H_2S, vol. percent	0.5
CO_2, vol. percent	3.0
Treated Gas:	
H2S, grains/100 SCF	4.4
CO_2, vol. percent	0.04
Steam, lb/lb CO_2	4

Source: Leuhddemann et al. (1959)

Caustic Wash Processes

Sodium (or potassium) hydroxide solution is a very effective, but nonregenerable absorbent for CO_2 and H_2S, forming stable salts such as sodium carbonate and sodium sulfide. Processes using caustic solutions are useful for removing trace amounts of these impurities from gas streams. In such systems it is common practice to recycle the absorbent solution to make maximum use of the caustic, and add fresh sodium hydroxide as required. As a result, the absorption actually takes place in a salt solution containing a small amount of free hydroxide.

Caustic solutions are also capable of absorbing mercaptans. These are monosubstitution products of hydrogen sulfide with the generic formula RSH. R is any hydrocarbon group, including aliphatic, aromatic or cyclic, and saturated or unsaturated. In most gas purification applications, R is a light aliphatic hydrocarbon radical, typically methyl or ethyl. Like their parent compound, H_2S, mercaptans are weak acids and form salts with strong alkalies. These salts, called mercaptides, are decomposed by heat, releasing the mercaptan and regenerating the hydroxide. When contacted with air, the dissolved mercaptans are subject to oxidation, forming disulfides that can be separated from the caustic. Both the thermal and oxidative regeneration techniques have been used with caustic scrubbing in mercaptan removal processes.

Carbon Dioxide Removal

Sodium hydroxide, in aqueous solution with sodium carbonate, is occasionally used to remove the last traces of carbon dioxide from hydrogen, or other gases, where the bulk of the carbon dioxide has been removed by a more economical, but less efficient regenerative process. Caustic scrubbing is also used to remove CO_2 from small volumes of air where CO_2-free air is required.

A flow sheet of a typical unit for removing CO_2 in trace quantities is shown in **Figure 5-47**. In the process, sodium carbonate is formed in solution by the reaction of dissolved CO_2 with OH^-, and the alkalinity is maintained at a high level by the periodic addition of fresh sodium hydroxide solution. No attempt is made to regenerate the solution, which is discarded or used

402 Gas Purification

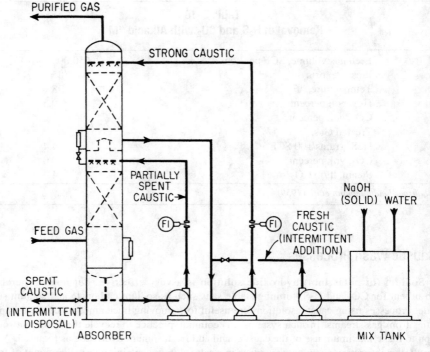

Figure 5-47. Flow diagram of caustic wash process to remove traces of CO_2 from air or other gases.

elsewhere when its alkalinity is reduced to the point where it is no longer effective for CO_2 removal. Some increase in caustic utilization efficiency can be realized by using two or more stages of absorption as shown in the flow diagram. This type of installation has been found to be capable of reducing the CO_2 content of air to as low as 5 ppm from about 300 ppm.

Hydrogen Sulfide Removal

The use of sodium hydroxide for the removal of hydrogen sulfide from gas streams is cost effective only when very small quantities of H_2S are involved and some means is available to dispose of the resulting sulfide solution. The process has been particularly useful for remote oilfield locations where small streams of sour gas must be purified to protect compression and collection equipment. Kent and Abid (1985) point out that a number of these small scrubbing systems employed simple cone-bottom tanks full of caustic solution with a gas sparger at the bottom. These authors also describe a more sophisticated approach developed by the Dow Chemical Company.

The Dow process incorporates a low residence time absorber to minimize co-absorption of CO_2 with the H_2S. Selective absorption avoids problems of equipment plugging due to precipitation of sodium carbonate and reduces the amount of caustic consumed by reaction with CO_2.

Alkaline Salt Solutions for Acid Gas Removal   403

Development of the Dow system has been described by Hohlfield (1979), and is summarized in **Table 5-19.** A flow sheet is shown in **Figure 5-48.** Static mixers are recommended as contactors although other devices, such as venturis, spray towers, and even a section of piping, may be adequate.

The three units listed in **Table 5-19** were operated on a feed gas containing about 1,000 ppm H_2S and 3.5% CO_2. It was desired to reduce the hydrogen sulfide concentration to 100 ppm or less to prevent fouling of compression equipment. The final production unit used a 4-in. in-line static mixer with a gas residence time of 0.02 sec and a pressure drop of 2 psi. It reduced the H_2S concentration to about 90 ppm at an L/G ratio of 4 gal/Mscf.

A relatively high pH (nominally 10) is required to assure effective H_2S absorption. However, an excessively high pH (over 12) indicates that either caustic is being wasted or the contactor is inefficient.

The key reactions that may occur in the contactor are

$$NaOH + H_2S = NaHS + H_2O \tag{5-39}$$

$$2NaOH + H_2S = Na_2S + 2H_2O \tag{5-40}$$

$$2NaOH + CO_2 + Na_2CO_3 + H_2O \tag{5-41}$$

The preferred reaction is 5-39 because it represents the maximum NaOH utilization and produces a very soluble product, NaHS. Reaction 5-40 also removes H_2S, but uses twice as

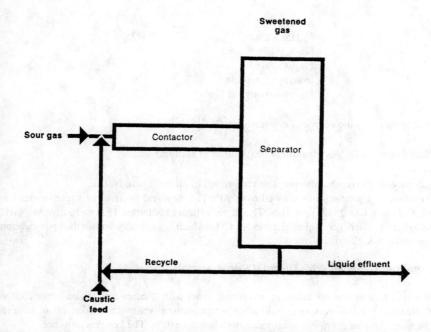

Figure 5-48. Flow sheet of Dow low-residence-time absorber system for selective H_2S removal. (*Hohlfield, 1979*)

Table 5-19
Development of the Dow Low-Residence-Time Scrubber System for Selective H_2S Absorption

	¼-in. Lab	1-in. Pilot	4-in. Production
Gas flow rate, Mcfd	1.0	50	1,000
Contact time, sec	0.02	0.01	0.02
H_2S in gas, ppm	1,000	1,000	925
Maximum absorption, ppm residual	100	80	90
Pressure drop across contactor, psi	2	1	2
Caustic concentration, %	2	1	1

Source: Hohlfield, (1979)

much caustic, and the product salt has limited solubility (about 10% by weight in water). Reaction 5-41 is undesirable. It wastes caustic and produces a salt of limited solubility. The short residence time contactor is intended to minimize its occurrence.

Disposal of the liquid effluent is the main problem with this process. According to Kent and Abid (1985), the effluent can be sold if it meets certain specifications. Typical specifications for marketable sulfide solutions are

Sulfidity	150
NaHS, wt.%	20–35
Na_2S, wt.%	2
Na_2CO_3, wt.%	1.5
Iron, ppm, soluble	300
Hydrocarbons, wt.%	0.2

Sulfidity indicates the degree of conversion to NaHS. It is defined as

Sulfidity = 2 × (moles of sulfur × 100)/(moles of sodium)

A sulfidity of 200 represents complete conversion of all sodium to NaHS.

Tests on an operating unit showed over 90% H_2S removal from a gas stream containing 0.5% CO_2 and 400–2,000 ppm H_2S. The liquid effluent contained 18.6 to 30.0 wt.% NaHS, 0.28 to 0.56 wt.% Na_2S, and negligible Na_2CO_3, which is generally within the specifications (Kent and Abid, 1985).

Mercaptan Removal with Thermal Regeneration

Because mercaptans are acidic in nature they react with alkalies and can be removed from gas streams by a caustic wash. Sodium carbonate, which is a much weaker alkali than the hydroxide, can also react with mercaptans to a limited extent. The key reactions are

$$NaOH + RSH = RSNa + H_2O \tag{5-42}$$

$$Na_2CO_3 + RSH = RSNa + NaHCO_3 \tag{5-43}$$

Both reactions are reversible at elevated temperature (about 250°F).

In the practical application of the process for the purification of large volume gas streams, it is very important that H_2S and CO_2 concentrations be reduced to very low levels before the gas enters the mercaptan absorber. Both of these impurities are stronger acids than mercaptans and consume caustic by forming nonregenerable salts. When the gas stream is processed in a conventional amine plant ahead of the mercaptan removal unit, the H_2S concentration is normally reduced to less than 0.25 grain/100 scf (about 4 ppmv) to meet product specifications; but the carbon dioxide content is often over 100 ppm. As a result modifications to the amine plant may be necessary to provide more complete carbon dioxide removal and make the removal of mercaptans with caustic economical. Alernatively, the mercaptan removal unit should have an upstream caustic wash to remove trace amounts of CO_2 and H_2S.

A plant designed to remove mercaptans from natural gas by absorption in caustic and thermal regeneration has been described by Williams (1964). A flow sheet of the installation is shown in **Figure 5-49.** The plant treats 15 MMscfd of 450–600 psig natural gas containing 30 grains mercaptan/100 scf. The stripping column operates at 28 psia with a feed and bottom temperature of 250°F and a reflux temperature of 140–180°F. Design rates are 7 gpm of circulating absorbent and 10 lb steam per gallon of solution.

Initially, gas entering the caustic scrubber contained at least 10 grain CO_2/100 scf which resulted in excessive caustic consumption. This was reduced to less than 3 grain/100 scf by modifying the amine plant. Because of the presence of CO_2, the circulating solution typically contained 3 to 8 wt.% Na_2CO_3. The NaOH concentration was maintained at 1–3 wt.% by the periodic addition of fresh caustic solution.

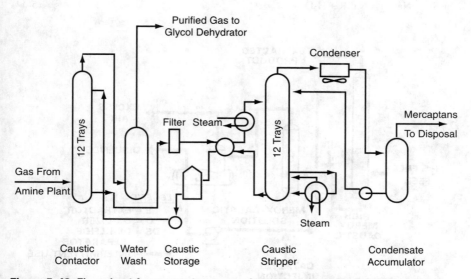

Figure 5-49. Flow sheet for mercaptan removal process with thermal regeneration. (*Williams, 1964*)

406 Gas Purification

Carbon dioxide in the feed gas was not the only cause of excessive caustic consumption. Foaming in the contactor and the reaction of caustic with components in the make-up water also contributed to caustic loss. These problems were reduced by preventing liquid hydrocarbons, which caused the foaming, from entering, or condensing in, the absorber and by using demineralized water for make-up. The combined actions reduced caustic consumption from 24 to less than 10 lb/MMscf of gas treated.

UOP Merox Process

In 1958, UOP introduced a catalyst based process to accelerate the oxidation of mercaptans to disulfides at or near ambient temperature. The process, which is licensed by UOP as the Merox process, is used to remove mercaptans from liquid or gaseous hydrocarbon streams. It operates by either converting the mercaptans to less objectionable disulfides in the flowing stream or removing the mercaptans with a caustic wash and then converting them to disulfides (Staehle et al., 1984). Only the approach involving both caustic absorption and subsequent oxidation to disulfides is applicable to gas purification.

When used for gas treating, the Merox process simply involves the absorption of mercaptans from the gas by countercurrent contact with caustic soda solution, oxidation of the dissolved mercaptans to disulfides by contact with air, and separation of the disulfide liquid from the caustic solution by settling and decanting off the disulfide phase. Note that both CO_2 and H_2S are removed in a caustic prewash prior to contacting the gas with the Merox caustic solution.

A simplified flow diagram of the Merox process as applied to gases or light liquid hydrocarbons is shown in **Figure 5-50**. The key reactions in the process are

Absorption:

$$RSH + NaOH = NaSR + H_2O \tag{5-44}$$

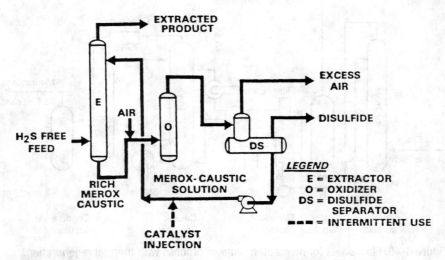

Figure 5-50. Flow sheet of Merox process. (*Staehle et al., 1984*)

Oxidation:

$$4NaSR + O_2 + 2H_2O = 4NaOH + 2RSSR \qquad (5\text{-}45)$$

Overall reaction:

$$\overset{(OH^-)}{4RSH + O_2 = 2RSSR + 2H_2O} \qquad (5\text{-}46)$$

The efficiency of mercaptan removal is highest for light mercaptans (i.e., R is a light hydrocarbon radical) and concentrated caustic solutions. The circulating caustic solution contains a dispersed Merox catalyst which promotes the conversion of mercaptans to disulfides at a relatively low temperature without promoting other reactions and does not affect the formation of mercaptides.

Although the Merox process is applicable to gases, most units have been designed to process liquid hydrocarbons. As of January, 1991, more than 1,450 Merox units of all types had been commissioned (UOP, 1992A).

Merichem Mercaptan Removal Process

The Merichem Company offers a treating system for removing mercaptans from gas (or liquid) streams that employs the same chemical reactions as the Merox process, but uses a unique contactor design. The Merichem contactor, called FIBER-FILM, utilizes a tightly packed bundle of vertical metallic fibers which are preferentially coated by a falling film of the alkaline aqueous phase (Wizig, 1985). A simplified diagram of a typical FIBER-FILM contactor designed for gaseous hydrocarbon application is shown in **Figure 5-51,** and a block diagram of Merichem's treating system for removing mercaptans and other impurities from fuel gas is shown in **Figure 5-52.**

Caustic H_2S and CO_2 Removal System. Gas from the amine plant passes through a strainer, which removes particles greater than 150 microns, and enters the caustic H_2S and CO_2 extraction system where remaining traces of these gases are removed. A FIBER-FILM contactor with recirculated caustic is used for this purpose. Caustic flows downward adhering to the fibers until it enters the pool of aqueous solution where it becomes part of the caustic inventory. Gas flows cocurrent with caustic in the spaces between the fibers and disengages upon exiting the shroud. Carbon dioxide and hydrogen sulfide react with sodium hydroxide according to equations 5-41 and 5-40, respectively.

As caustic in the prewash system reacts with H_2S and CO_2, its alkalinity decreases. Since the alkalinity must be maintained at a high level to prevent CO_2 and H_2S from entering the mercaptan absorber, the caustic solution is removed and replaced when it is only about 40% reacted. To prevent the precipitation of solid salts, it is necessary to control the caustic concentration and temperature. An initial caustic strength no higher than 25° Be and an operating temperature of at least 70°F are recommended.

Mercaptan Absorption System. The mercaptan absorber is also a FIBER-FILM unit, and operates in a manner similar to the CO_2 and H_2S removal unit previously described. Gas enters the top of the unit where it contacts regenerated caustic from the caustic regeneration

408 Gas Purification

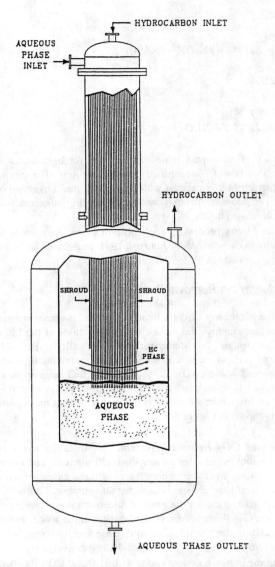

Figure 5-51. Diagram of typical gaseous FIBER-FILM™ contactor installation. (*Wizig, 1985*)

system. Mercaptans react with the caustic in accordance with reaction 5-42. Product gas from the unit goes to compression and drying operations, while the mercaptan-containing caustic is sent to the regeneration system.

Caustic Regeneration System. The caustic regeneration system consists of an oxidizer tower and a settler. Rich caustic solution from the mercaptan absorber containing mercaptide

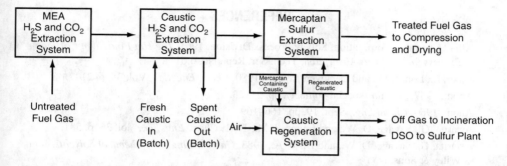

Figure 5-52. Block diagram of Merichem mercaptan removal process. (*Wizig, 1985*)

ions and dissolved catalyst is heated to the optimum temperature for oxidation, about 125°F, and is fed into the bottom of the oxidizer tower. Air is injected into the bottom of the oxidizer tower, causing reaction 5-45 to proceed as the air-solution-catalyst mixture rises through the tower. The liquid mixture is drawn off near the top of the oxidizer tower and flows into a settler, while excess air is removed from the very top of the tower for disposal. Disulfides formed by the oxidation reaction are only sparingly soluble in the caustic and separate in the settler as a light oily phase (disulfide oil or DSO) that is drawn off. The heavier, regenerated caustic phase is pumped back to the mercaptan absorber.

Operating Requirements. Operating requirements are given by Wizig (1985) for a plant designed to treat 12.9 MMscfd fuel gas containing 15 ppmw H_2S, 2,113 ppmw methyl mercaptan (as sulfur), and 100 ppmw CO_2 at 156 psig and 100° F. The treated gas specifications are 4 ppmw (max.) H_2S; 35 ppmw (max.) total sulfur; and 5 ppmw (max.) sodium compounds, as Na. The annual plant operating requirements are estimated to be

NaOH (dry basis), lb	92,757
Spent caustic disposal, gal	87,736
Air, MMscf (at 3.28 scfm)	1.65
Electricity, kWh	55,900
Disulfide oil produced, gal	107,450
Steam (50 psig), lb (at 432 lb/h)	3,628,800
Catalyst, lb	175

About 3 scfm of offgas, containing about 88% nitrogen, 12% oxygen, and 0.01% disulfide, are discharged from the oxidizer. In this plant design it is disposed of by incineration with added fuel gas in a sulfur plant thermal oxidizer.

In addition to this process, which is aimed primarily at the removal of mercaptans from a gas stream, Merichem offers a number of other processes using the FIBER-FILM contactor to remove various impurities from gas and liquid hydrocarbon streams. As of 1992, it was reported that 204 units had been installed treating 2.0 MMbpsd (million barrels per stream day) liquid and 39.5 MMscfd gas (Merichem, 1992).

REFERENCES

Allied Chemical Corporation, Solvay Process Division, 1961, *The Hot Potassium Carbonate Process for Acid Gas Absorption,* Tech./Ser. Rept. No. 6.61.

Alper, E., Lohse, M., and Deckwere, W. D., 1980, *Chem. Eng. Sci.*, Vol. 35, p. 2147.

Anon., 1957, *Sulphur* (special number), pp. 21–23.

Anon., 1960, *Chem. Eng.,* Sept. 19, pp. 166–169.

Astarita, G., Savage, D. W., and Longo, J. M., 1981, *Chem. Eng. Sci.*, Vol. 36, p. 581.

Astarita, G., Savage, D. W., and Bisio, A., 1983, *Gas Treating with Chemical Solvents,* John Wiley & Sons, NY.

Astarita, G., 1967, *Mass Transfer with Chemical Reaction,* Elsevier, Amsterdam.

Bähr, H., and Mengdehl, H., 1935, U.S. Patent 1,990,217.

Bähr, H., 1938, *Proc. Am. Petrol. Inst.* Sect. 3, 19M, May, p. 37.

Baker, R. L., and McCrea, D. H., 1981, "The Benfield LOHEAT Process: An Improved HPC Absorption Process," paper presented at AIChE 1981 Spring National Meeting, Houston, TX, April 5–9.

Bartoo, R. K., Gemborys, T. M., and Wolf, C. W., 1991, "Recent Improvements to the Benfield Process Extend its Use," paper presented at Nitrogen '91 Conference, June, Copenhagen, Denmark.

Bartoo, R. K., and Ruzicka, S. J., 1983, "Benfield Process Provides Energy Reduction in New and Existing Ammonia Plants," paper presented at the 7th International Conference of the British Sulphur Corporation, November, London, England.

Bartoo, R. K., 1984, "Benfield Process for Acid Gas Removal," paper presented at AIChE 1984 National Meeting, March, Atlanta, Georgia.

Behrens, E. A. and Behrens, J., 1904, German Patent 162,655.

The Benfield Corporation., 1971, "The Way to Low Cost Scrubbing of CO_2 and H_2S from Industrial Gases."

Benson, H. E., and Parrish, R. W., 1974, *Hydro. Process.*, Vol. 53, No. 4, pp. 81–82.

Benson, H. E., and Field, J. H., 1960, *Petrol. Refiner,* Vol. 39, April, pp. 127–132.

Benson, H. E., Field, J. H., and Jimeson, R. M., 1954, *Chem. Eng. Progr.*, Vol. 50, p. 356.

Benson, H. E., and Field, J. H., 1955, British Patent 725,000.

Benson, H.E., Field, J. H., and Haynes, W. P., 1956, *Chem. Eng. Progr.*, Vol. 52, p. 433.

Benson, H.E., 1961, *Petrol. Refiner,* Vol. 40, April, pp. 107–108.

Bienstock, D., and Field, J. H., 1961, *Corrosion,* Vol. 17, July, pp. 337t–339t.

Bienstock, D., and Field, J. H., 1965, United States Patent 3,181,929.

Bocard, J. P., and Mayland, B. J., 1962, *Hydro. Process.*, Vol. 41, April, pp. 128-132.

Buck, B. O., and Leitch, A. R. S., 1958, *Oil Gas J.*, Vol. 56, Sept. 22, pp. 99–103.

Clayman, M. A., and Clark, J. R., 1980, "Low Energy Natural Gas Purification Using Benfield Processes," *Gas Conditioning Conference Proceedings,* University of Oklahoma, Norman, OK.

Danckwerts, P. V., 1970, *Gas-Liquid Reactions,* McGraw-Hill Book Company, NY.

Eickmeyer and Associates, Inc., 1992, personal communication.

Eickmeyer, A. G., 1962, *Chem. Eng. Prog.*, Vol. 58, No. 4, April, pp. 89–91.

Eickmeyer, A. G., 1958, *Chem. Eng.*, Vol. 65, August, p. 113.

Eickmeyer, A. G., 1971, *Oil Gas J.*, Vol. 69, August 9, pp. 74–75.

Exxon Research and Engineering Co., 1991, *Flexsorb HP Promoted Hot Potassium Carbonate Solvent for CO_2 and H_2S Removal,* Brochure 08069102 GRC, September.

Exxon Research and Engineering Co., 1992, "Flexsorb Solvents," *Gas Processing Handbook '92, Hydro. Process.*, April, p. 100.

Farbenindustrie, I. G., G.m.b.H., 1935, French Patent 787,782.

Ferguson, K. R., and Stutheit, A. G., 1991, *Oil Gas J.*, Dec. 2, p.54.

Ferrell, J. K., Staton, J. S., and Rousseau, R. W., 1987, "Performance and Modeling of a Hot Potassium Carbonate Acid Gas Removal System in Treating Coal Gas," EPA Report No. EPA/600/7-87/023, November.

Field, J. H., Johnson, E. G., Benson, H. E., and Tosh, J. S., 1960, *U.S. Bur. Mines Rep. Invest. No. 5660.*

Furnas, C. C., and Bellinger, F., 1938, *Trans. Am. Inst. Chem. Engrs.*, Vol. 34, No. 251.

Gangriwala, H. A., and Chao, I-Meen, 1985, "The Catacarb Process for Acid Gas Removal," in *Acid and Sour Gas Treating Processes,* Newman, S.A., Ed., p. 170, Gulf Publishing Co., Houston, TX.

Garner, F. H., Long, R., and Pennell, A., 1958, *J. Appl. Chem.*, May 8, pp. 325–336.

Giammarco-Vetrocoke, 1992A, "Giammarco-Vetrocoke," *Gas Processing Handbook '92, Hydro. Process.*, April, p. 103.

Giammarco-Vetrocoke, 1992B, personal communication, L. Tomasi, July 17.

GKT, 1992, "Operational Results of the VACASULF-PROCESS," Gesellschaft fur Kohle-Technologie mbH (GKT) brochure, Essen, January.

Goldstein, A. M., Edelman, A. M., Beisner, W. D., and Ruziska, P. A., 1984, *Oil and Gas J.*, Vol. 82, July 16, p. 70.

Gollmar, H.A., 1945A, *Chemistry of Coal Utilization,* Vol. 2, edited by H. H. Lowry, p. 984, John Wiley, NY.

Gollmar, H. A., 1945B, *Chemistry of Coal Gasification,* Vol. 2, edited by H. H. Lowry, p. 985, John Wiley & Sons, Inc., NY. p. 985.

Gollmar, H. A., 1949, U.S. Patent 2,464,805.

Gollmar, H. A., and Hodge, W. W., 1956, Proc. Ind. Waste Conf., 10th Conf., Vol. 89, March, pp. 35–48.

Grosick, H. A., and Kovacic, J. E., 1981, "Coke Oven Gas and Effluent Treatment," Chapter 18 in *Chemistry of Coal Utilization, 2nd Supplementary Volume,* Elliott, M.A., ed., John Wiley & Sons, NY.

Grover, B. S., and Holmes, E. S., 1987, "MVR Provides Reduced-Energy Benfield CO_2 Removal Process," paper presented at the AIChE Ammonia Safety Symposium, Minneapolis, MN, August.

Grover, B. S., 1987, U.S. Patent 4,702,898.

Harte, C. R., Baker, E. M., and Purcell, H. H., 1933, *Ind. Eng. Chem.*, Vol. 25, No. 528.

Hill, A. E., and Hill, D. G., 1927, *J. Am. Chem. Soc.*, Vol. 49, April 7, p. 968.

Hohlfield, R.W., 1979, *Oil Gas J.*, Oct. 15, p. 129.

ICF Kaiser Engineers, 1992, Personal Communication, C. Lasky, July 21.

Jeffreys, G. V., and Bull, A. F., 1964, *Trans. Inst. Chem. Engrs.*, Vol. 42, p. 118.

Jenett, E., 1962, *Oil Gas J.*, Vol. 60, April 30, pp. 60–72.

Joshi, S. V., Astarita, G., and Savage, D. W., 1981, "Transport with Chemical Reactions," AIChE Symposium Series No. 202, Vol. 77, p. 63.

Kastens, M. L., and Barraclough, R., 1951, *Ind. Eng. Chem.*, Vol. 43, Sept., pp. 1882–1892.

Katell, S. and Faber, J. H., 1960, *Petrol. Refiner,* Vol. 39, March, pp. 187–190.

Kent, V. A., and Abid, R. A., 1985, "Selective H_2S Caustic Scrubber," *Proc. Gas Conditioning Conference,* University of Oklahoma, Norman, OK.

Killeffer, D. H., 1937, *Ind. Eng. Chem.*, Vol. 29, p. 1293.

Koppers Company, Inc., 1978, *Cleaner Gas from Coke Ovens,* Koppers Engineering and Construction brochure (now part of ICF Kaiser Engineers).

Krupp Wilputte, 1992, personal communication, H. Max Hooper.

Kurtz, J. K., 1955, *Problems and Control of Air Pollution,* p. 215, Reinhold Publishing Corporation, NY.

Leuhddemann, R., Noddes, G., and Schwarz, H.G., 1959, *Oil Gas J.*, Vol. 57, Aug. 3, pp. 100–104.

Liedholm, G. E. (Shell Development Company), 1958, personal communication, Feb. 17.

Litvinenko, M. S., 1952, *J. Appl. Chem.* (U.S.S.R.), Vol. 25, pp. 775–794.

Maddox, R. N. and Burns, M. D., 1967, *Oil Gas J.*, Vol. 65, Nov. 3, pp. 122–131.

Mapstone, G. E., 1966, *Hydro. Process.*, Vol. 45, March, pp. 145–148.

Massey, M. Y., and Dunlap, R. W., 1975, *Journal of the Air Pollution Control Association,* Vol. 25, No. 10, pp. 1019–1027.

Merichem Co., 1992, "Mericat/Thiolex/Regen," *Gas Processing Handbook '92, Hydro. Process.,* April, p.120.

Morse, R. J., 1968, *Oil Gas J.*, Vol. 66, April 22.

Mullowney, J. F., 1957, *Petrol. Refiner,* Vol. 36, December, pp. 149

Palo, R. O., and Armstrong, J. B., 1958, *Petrol. Refiner,* Vol. 37, December, pp. 123–128.

Parrish, R. W., and Neilson, H. B., 1974, "Synthesis Gas Purification Including Removal of Trace Contaminants by the BENFIELD Process," Paper presented at the 167th National Meeting of the American Chemical Society, Division of Ind. & Eng. Chem., March, Los Angeles, CA.

Pasternak, R., 1963, *Brennstoff-Chem.,* Vol. 4, No. 44, pp. 105–110.

Pasternak, R., 1962, *Brennstoff-Chem.,* Vol. 3, No. 43, pp. 65–67.

Peck, E. B., 1938, *Proc. Am. Petrol. Inst.*, Sec. 3, 19M:51.

Perry, J. H., ed., 1950, *Chemical Engineers' Handbook.* 3rd Ed., p. 198, McGraw-Hill Book Company, Inc., NY.

Powell, A. R., 1941, U.S. Patent 2,242,323.

Powell, A. R., 1938, *The Science of Petroleum,* Vol. 3., edited by A .E. Dunstan, p. 1804, Oxford University Press, NY.

Quinn, E. L., and Jones, C. L., 1936, *Carbon Dioxide,* American Chemical Society Monograph Series, Reinhold Publishing Corporation, NY.

Redin, V. N., Eliseev, O. I., and Il'n, A.P., 1986, "Improvement of the Process of Regeneration of Alkaline Wash Oil in Vaccum-Carbonate Sulfur Removal," *Koka i Khimiya,* No. 5, p. 30.

Reed, R. M., and Updegraff, N. C., 1950, *Ind. Eng. Chem.*, Vol. 42, November, p. 2269.

Reich, G. T., 1931, *Chem. & Met. Eng.*, Vol. 38, pp. 38, 136–145.

Rib, D. M., Kimura, S. G., and Smith, D. P., 1982, "Performance of a Coal Gas Cleanup Process Evaluation Facility," presented at AIChE 1982 Spring National Meeting, June 9.

Riesenfeld, F. C., and Mullowney, J. F., 1959, *Oil Gas J.*, Vol. 57, May 11, pp. 86–91.

Rosebaugh, T. W., 1938, *Proc. Am. Petrol. Inst.*, Sect. 3, 19M, May, pp. 47–52.

Rosenstein, L., and Kramer, G. A., 1934, U.S. Patent 1,945,16.3.

Roughton, F. J., and Booth, V. H., 1938, *Biochem. Journal.*, Vol. 32, p. 2049.

Sartori, G., and Savage, D. W., 1983, *Ind. Eng. Chem. Fundam.*, Vol. 22, p. 239.

Sartori, G., Ho, W. S., Savage, D. W., Chludzinski, G.R., and Wiechert, S., 1987, *Separation and Purification Methods*, Vol. 16, No. 2, p.171.

Savage, D. W., Sartori, G., and Astarita, G., 1984, *Faraday Discuss. Chem. Soc.*, Vol. 77, p. 17.

Savage, D. W., Astarita, G., and Joshi, S., 1980, *Chem. Eng. Sci.*, Vol. 35, p. 1513.

Say, G. R., Heinzelmann, F. J., Iyengar, J. N., Savage, D. W., Bisio, A., and Sartori, G., 1984, *Chem. Eng. Prog.*, Oct., p. 72.

Scott, B., Daniels, J. R., and Chao, I-Meen, 1987, "Corrosion History of a Hot Pot CO_2 Absorber," presented at AIChE Symposium "Safety in Ammonia and Related Facilities," paper No. 23.

Seidell, A., 1940, *Solubilities of Inorganic and Metal Organic Compounds*, Vol. 1, Van Norstrand Company, Inc., Princeton, NJ.

Shell Development Co., 1992, W. Saulmon, personal communication.

Sherwood, T. K., and Pigford, R. L., 1952A, *Absorption and Extraction*, 2nd ed., pp. 358–364, McGraw-Hill Book Company, Inc., NY.

Sherwood, T. K., and Pigford, R. L., 1952B, *Absorption and Extraction* 2nd ed., p. 337, McGraw-Hill Book Company, Inc., NY.

Shneerson, A. L., and Leibush, A. G., 1946, *J. Appl. Chem.*, U.S.S.R., Vol. 19, No. 9, pp. 869–880.

Shreve, R. N., 1945, *The Chemical Process Industries*, 1st ed., p. 106, McGraw-Hill Book Company, Inc., New York.

Shrier, A. L.,and Danckwerts, P. V., 1969, *Ind. Eng. Chem. Fundam.*, Vol. 8, No. 3, August, p. 415.

Smith, G. A., 1953, *Gas World*, Coking Section, Oct. 3, pp. 65–69.

Sorell, G., 1990, "Corrosion in Natural Gas Processing Plants," Topical Report May 1989–January 1990, prepared for Gas Research Institute (GRI), PB90-205204.

Spector, N. A., and Dodge, B. F., 1946, *Trans. Am. Inst. Chem. Engrs.*, Vol. 42, p. 827.

Sperr, F. W., Jr., and Hall, R. E., 1925, U.S. Patent 1,533,733.

Sperr, F. W., Jr., 1921, *Am. Gas Assoc. Proc.*, Vol. 3, pp. 282–364.

Staehle, B. E., Verachtert, T. A., and Salazer, J. R., 1984, *Merox 1984, 25 Years of Treating Experience*, Publication of UOP Process Division, UOP Inc., October, Des Plaines, IL.

Tepe, J. B., and Dodge, B. F., 1943, *Trans. Am. Inst. Chem. Engrs.*, Vol. 39, p. 255.

Tosh, J. S., Field, J. H., Benson, H. E., and Anderson, R.B., 1960, *U.S. Bur. Mines Rept. Invest. No. 5622.*

Tosh, J. S., Field, J. H., Benson, H. E., and Haynes, W.P., 1959, *U.S. Bur. Mines Rept. Invest. No. 5484.*

Tseng, P. C., Ho, W. S., and Savage, D. W., 1988, *AIChE Journal,* Vol. 34, No. 6, June, p. 922.

UOP Inc., 1992A, "Merox," *Gas Processing Handbook '92, Hydro. Process.*, April, p. 120.

UOP, 1992B, "Benfield," *Gas Processing Handbook '92, Hydro. Process.*, April, p. 90.

UOP, 1993, personal communication.

Wainwright, H. W., Egleson, G. C., Brock, C. M., Fisher, J., and Sands, A. E., 1953, *Ind. Eng. Chem.*, Vol. 45, June, p. 1378.

Weinberg, H. N., Eisenberg, B., Heinzelmann, F. J., and Savage, D. W., 1983, "New Gas Treating Alternatives for Saving Energy in Refining and Natural Gas Production," Presented at the 11th World Petroleum Congress, Aug. 31, London, England.

Wiberg, E. A., Gangriwala, H. A., and Chao, I-Meen, 1982, "Process Design—Heart of an Estimate," presented at the Kansas City Section of the AIChE, Nov. 19, 1992.

Williams, W. W. 1964, *Hydro. Process.*, Vol. 43, No. 7, July, p. 121.

Williamson, R. V., and Mathews, J. H., 1924, *Ind. Eng. Chem.*, Vol. 16, pp. 1157–1161.

Wizig, H. W., 1985, "An Innovative Approach for Removing CO_2 and Sulfur Compounds from a Gas Stream," *Gas Conditioning Conference Proceedings,* University of Oklahoma, Norman, OK.

Zahnstecher, L. W., 1950, *Heat Eng.*, Vol. 25, April, p. 61.

Chapter 6

Water as an Absorbent for Gas Impurities

INTRODUCTION, 416

WATER-WASH PROCESS EQUIPMENT, 418

    Packed Beds, 419
    Wet Cyclones, 420
    Countercurrent Spray Columns, 420
    Venturi Scrubbers, 422
    Jet Scrubbers, 422

CARBON DIOXIDE ABSORPTION IN WATER, 423

    Basic Data, 427
    Design and Operation, 427

HYDROGEN SULFIDE REMOVAL BY ABSORPTION IN WATER, 436

ABSORPTION OF FLUORIDES, 438

    Basic Data, 439
    Absorber Design, 441
    Operating Data, 448
    Materials of Construction, 450
    Disposal of Absorbed Fluoride, 453

HYDROGEN CHLORIDE ABSORPTION, 453

CHLORINE ABSORPTION IN WATER, 458

    Solubility Data, 460
    Absorption Coefficients, 461
    Materials of Construction, 463

REFERENCES, 463

INTRODUCTION

The principal advantage of water as a absorbent for gas impurities is, of course, its availability at a low cost. This factor alone is sufficient to make the use of water worth considering for the removal of gas impurities which are reasonably soluble in it. Water is particularly applicable to the treatment of large volumes of low-pressure exhaust gas for the prevention of air pollution because solvent losses are difficult to avoid in such installations. Organic solvents, in general, have sufficient vapor pressure to cause the occurrence of appreciable losses by vaporization into the purified gas stream. Practically any chemical absorbent other than water requires a tight system and, unless a salable reaction product is formed, a regenerative cycle. Water, on the other hand, can be used in simple scrubbing units with less concern over leakage and frequently on a once-through basis with the rich solution being discarded.

Water may also be applicable to the washing of high-pressure gases where the solubility of an impurity (such as CO_2), which is only sparingly soluble at low pressure, is brought up to an economically high level by the high CO_2 partial pressure.

Although a detailed discussion of particulate removal from gases is beyond the scope of this text, it should be noted that water scrubbing is frequently used for this purpose. In many cases, a water wash is used to remove both particles and vapor phase impurities. Water contact stages are also frequently used to quench high-temperature gases. In some instances, the quench step serves four distinct purposes, i.e., temperature reduction, condensation of liquids from the vapor phase, absorption of soluble impurities, and removal of particulate material. The quenching of the product from coal gasification units with water is an example of such a multipurpose operation. Soluble components absorbed by the water in such processes include ammonia, hydrogen cyanide, phenol, and hydrochloric acid. Heavy hydrocarbons are condensed as a separate liquid phase, and particles of ash and unreacted coal are captured by the liquids.

Gas phase impurities which have been removed by water scrubbing operations include ammonia, sulfur dioxide, carbon dioxide, hydrogen chloride, hydrogen fluoride, silicon tetrafluoride, silicon tetrachloride, chlorine, aldehydes, organic acids, and alcohols. The need to absorb ammonia occurs primarily in connection with the purification of gas streams from the thermal treatment of coal and other nitrogen containing fuels. Since the processes developed for removing ammonia from such gases with water are closely related to those for removing hydrogen sulfide and carbon dioxide, they are discussed together in Chapter 4. Similarly, the absorption of sulfur dioxide in water, or more accurately, in dilute aqueous solutions, is covered in Chapter 7.

Specific impurities considered in this chapter are carbon dioxide, hydrogen sulfide, hydrogen fluoride, silicon tetrafluoride, hydrogen chloride, silicon tetrachloride, and chlorine. All of these form acids in aqueous solution, and the prevention of corrosion is a common problem. They are by no means equally corrosive, however, and also differ in the nature of the absorption processes. Carbon dioxide, hydrogen sulfide, and chlorine are relatively insoluble in water, and their absorption rates are determined largely by the liquid-film resistance. Hydrogen fluoride, silicon tetrafluoride, hydrogen chloride, and silicon tetrachloride, on the other hand, react rapidly with or are very soluble in water, and their absorption rates are generally high and controlled by gas-film resistance.

When any of the previously mentioned impurities are absorbed from a gas stream, small amounts of the primary gaseous components are also absorbed. This effect is particularly noticeable in high-pressure operations and can materially affect the economics of the water-wash process for such cases. Water-solubility data for components, which typically make up the major portion of gases to be purified by the water-wash process, are presented in **Table 6-1** in the form of Henry's law constants at several temperatures. Similar data, at a single temperature (25°C), are given in **Table 6-2** for thirteen organic chemicals which have sufficiently high solubilities in water to be considered candidates for water-wash gas purification processes. The data are from a compilation of Henry's law constants for 362 organic compounds in water that was prepared by Yaws et al. (1991).

Table 6-1
Solubility of Various Gases in Water*

Gas	H, at Temperature, °C					
	0	10	20	30	40	50
Carbon monoxide, CO	3.52×10^4	4.42×10^4	5.36×10^4	6.20×10^4	6.96×10^4	7.61×10^4
Hydrogen, H_2	5.79×10^4	6.36×10^4	6.83×10^4	7.29×10^4	7.51×10^4	7.65×10^4
Methane, CH_4	2.24×10^4	2.97×10^4	3.78×10^4	4.49×10^4	5.20×10^4	5.77×10^4
Nitrogen, N_2	5.29×10^4	6.68×10^4	8.04×10^4	9.24×10^4	10.4×10^4	11.3×10^4
Oxygen, O_2	2.55×10^4	3.27×10^4	4.01×10^4	4.75×10^4	5.35×10^4	5.88×10^4

*Values given are for H in Equation $P = Hx$, where x = mole fraction of solute in the liquid phase and P = partial pressure of solute in the gas phase, atmospheres.
Source: International Critical Tables (1937)

Table 6-2
Henry's Law Constants for Organic Chemicals in Water at 25°C (77°F)

Formula	Name	Mol. Wt.	Henry's Law Constant, atm/mol. fract.
COS	Carbonyl sulfide	60.070	2812E+01
CS_2	Carbon disulfide	76.131	1067E+00
CH_2O_2	Formic acid	46.026	6147E-05
CH_4O	Methanol	32.042	3858E-04
$C_2H_3Cl_3$	1,1,2-trichloroethane	133.405	5337E-02
C_2H_3N	Acetonitrile	41.052	1115E-03
$C_2H_4Cl_2$	1,2-dichloroethane	98.960	6544E-02
C_2H_4O	Acetaldehyde	44.053	5584E-03
$C_2H_4O_2$	Acetic acid	60.052	6644E-05
C_2H_6O	Ethyl alcohol	46.069	4515E-04
$C_3H_4O_2$	Acrylic acid	72.063	2299E-05
C_3H_6O	Allyl alcohol	56.080	3091E-04
C_3H_6O	Acetone	56.080	2376E-03

Data of Yaws et al. (1991)

WATER-WASH PROCESS EQUIPMENT

In the absorption of sparingly soluble gases such as CO_2, H_2S, and Cl_2, the liquid film resistance is normally controlling, very low overall absorption coefficients are observed, and relatively tall countercurrent packed or tray columns are required. The design of such equipment is discussed in detail in Chapter 1.

Quite different conditions and criteria are encountered in the selection and design of water scrubbers to remove highly soluble gaseous impurities such as HF, SiF_4, HCl, and $SiCl_4$ from low pressure exhaust gas streams. Because of the high solubility (and/or the rapid reaction with water) the liquid film resistance is very small, the gas film resistance becomes controlling, and high overall gas absorption rates are realized. Furthermore, the vapor pressure of the dissolved component over the resulting dilute solution is often negligible. As a result, short packed or tray countercurrent columns; cocurrent or cross flow contactors; spray systems; and special devices such as wet cyclones, venturis, and jet scrubbers can be considered. Typically, low gas-side pressure drop, minimum system size, and operating simplicity are key criteria in the selection of exhaust gas scrubbers.

A generalized selection guide for wet scrubbers covering both absorption and particulate collection functions is given in **Table 6-3,** which is based on a review of the subject by Hanf and MacDonald (1975). Typical gas velocities and gas-side pressure drop values for the most common types of wet scrubbers are given in **Table 6-4.** The values given are considered typical for the specified devices in low pressure exhaust gas service and do not represent operat-

Table 6-3
Selection Guide for Wet Scrubbers

Type of Wet Scrubber	Gas Phase Impurities		Entrained Liquids		Particles			
					Fumes $<1\mu$	Dusts $1-5\mu$	Dusts over 5 μ	
	High Solubility	Low Solubility	Mists $<10\mu$	Drops $>10\mu$			Low Loading	High Loading
Countercurrent Packed Tower	E	E	F	E	NR	NR	E	NR
Cocurrent Packed Tower	G	F	F	E	NR	NR	E	NR
Crossflow Packed Scrubber	E	F	F	E	NR	NR	E	NR
Parallel Flow Washer	NR	NR	P	E	NR	NR	NR	NR
Wet Cyclone	F	NR	P	E	NR	NR	G	G
Venturi	F	NR	E	E	E	E	E	E
Spray Tower	F	NR	NR	G	NR	NR	E	F
Eductor Venturi (Jet)	F	NR	E	E	G	E	E	E

Code for collection efficiencies: E = 95–99%, G = 85–95%, F = 75–85%, P = 50–75%,
NR = Not Recommended.
Based on data of Hanf and MacDonald (1975)

Table 6-4
Typical Gas Velocities and Pressure Drops for Wet Scrubbers

Scrubber Type	Typical Gas Velocities, ft/sec	Pressure Drop, in. of water
Packed Contactors		
Countercurrent	5–10	0.2–0.7 (per ft)
Cocurrent	10–60	1–3 (per ft)
Cross Flow	5–10	4 (total)
Wet Cyclone		
Entrance	40–80	
Superficial	10–15	4 (total)
Spray Tower		
Countercurrent	4–8	1–3 (total)
Venturi Scrubber		
Throat	130–180	8–12 (total)

Velocity and pressure drop values based on air at 70°F.
Data from Parsons (1988), Hanson and Danos (1982), and McCarthy (1980)

ing limits. The table does not include jet scrubbers, which provide energy in the liquid stream to move the gas without an overall pressure drop.

Packed Beds

Packed beds used in exhaust gas scrubbers are usually shallow (e.g., 2–6 ft) and often utilize special low-pressure drop, high-efficiency plastic packing. Examples of random packings of this type include Jaeger Tri-Packs (Jaeger Products, Inc.), Intalox Snowflake packing (Norton Chemical Process Products Corp.), Tellerettes (Ceilcote Air Pollution Control), and Maspac (Clarkson Controls and Equipment Co.). Typical mass transfer data for the absorption of water soluble gas impurities in a column packed with 2-in. Jaeger Tri-Packs are given in **Table 6-5.**

Very compact packed absorber systems have been developed for exhaust gas scrubbing applications. An example of one manufacturer's complete package absorption system is illustrated in **Figure 6-1.** This unit provides the packed section, liquid sump, liquid pump and distribution system, and exhaust fan within a single shell. It is available in 33- to 72-in. diameters. A typical unit (60-in. diameter) uses a 15-hp exhaust fan to move 7,850 cfm of gas through the scrubbing section, and a 3-hp pump to circulate 75 gpm of liquid over the packing (Interel Environmental Technologies, Inc., 1992).

When countercurrent action is not required to attain the desired removal efficiency, cocurrent or crossflow packed absorber operation may be advantageous. Cocurrent operation allows the use of very high gas velocities without concern for flooding; however, the high velocities result in a correspondingly high pressure drop, and a gas/liquid separator must be provided. According to Parsons (1988), pressure drops of 1 to 3 in. of water per foot of packing depth are encountered with vertical flow cocurrent exhaust gas scrubbers. A special type

Table 6-5
Mass Transfer Data for the Absorption of Soluble Gases into Water Using 2-in. Jaeger Tri-Packs

Impurity	G lb/hr. ft^2	L lb/hr. ft^2	Temp °F	HTU inches
HCl	1,792	2,048	77	10.6
NH$_3$	512	1,024	68	8.4
NH$_3$	512	4,096	68	5.4
HF	1,844	3,072	77	6.9
CH$_3$COCH$_3$	1,700	860	68	15.2

Data provided by Jaeger Products, Inc. (1990)

of cocurrent packed absorber utilizes an in-line static mixing unit for contacting the gas and liquid followed by a conventional gas/liquid separator. Such systems are described by Rader et al. (1988, 1989). The in-line mixing elements are claimed to have significantly lower pressure drop than conventional random packings such as raschig or Pall rings.

In crossflow scrubbers, as shown in **Figure 6-2,** the gas flows horizontally while the liquid flows downward through one or more relatively thin beds of packing or mesh. A key characteristic of crossflow scrubbers is that the cross-sectional areas for gas and liquid flow are not the same. Usually the area for liquid flow is less than that available for gas flow, so adequate wetting of the packing is possible with a relatively low liquid flow rate. The beds of packing are normally inclined toward the incoming gas. The slope helps compensate for the tendency of the liquid to migrate toward the back of the bed under the influence of the flowing gas stream. A specialized version of the crossflow scrubber uses thin plastic mesh pads that are wetted by sprays applied to the upstream sides. Countercurrent staging can be obtained by collecting the liquid at the bottom of each set of pads and spraying it onto the faces of the pads that are immediately upstream with regard to gas flow.

Wet Cyclones

Wet cyclones are seldom used as gas absorbers, but are important as gas/liquid separators downstream of venturis or other cocurrent contact devices. Cyclonic action is sometimes used in spray chambers to increase the relative velocity of droplets. The liquid may be added as a spray into the feed gas or sprayed from nozzles within the chamber.

Countercurrent Spray Columns

Countercurrent spray columns typically employ a bank of spray nozzles near the top of the tower and often have additional banks of spray nozzles at lower levels. An efficient mist eliminator is required above the upper spray nozzles to capture any small droplets that are entrained in the upflowing gas. Key requirements for efficient absorption in spray columns are the use of high pressure nozzles (e.g., 60 psig or higher) to assure the production of high velocity, small diameter droplets; the use of many small nozzles carefully located to cover the entire scrubber cross-section without depositing excessive liquid on the walls; and the

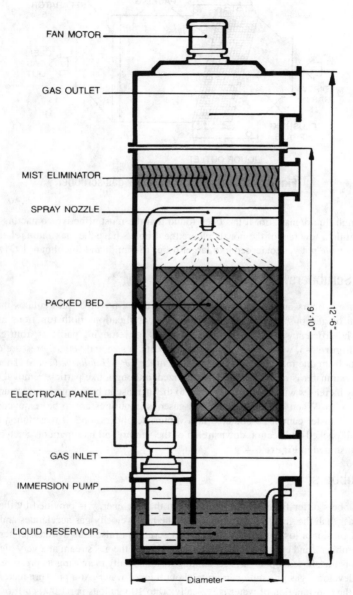

Figure 6-1. Package type packed countercurrent exhaust gas scrubber. *Interel Environmental Technologies, Inc.*

422 Gas Purification

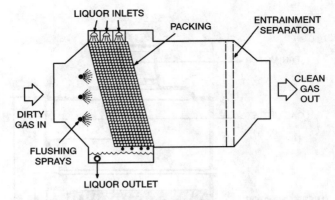

Figure 6-2. Diagram of crossflow gas scrubber.

use of a high liquid/gas ratio. It has been found that the most effective contacting is accomplished within a few feet of the nozzle where the droplets have the maximum relative velocity and they have not yet coalesced into larger drops (Schmidt and Stichlman, 1991).

Venturi Scrubbers

Venturi scrubbers are probably used more for the removal of particulates than for the absorption of gaseous species, although in many applications both functions are accomplished. The efficiency of typical venturi scrubbers for removing particles from gas is illustrated by **Figure 6-3,** which shows the effect of particle size and venturi pressure drop based on the data from one manufacturer as published in the *EPA Handbook* (1986). Their efficiency for mass transfer is not as readily characterized; although, like particle removal efficiency, it normally increases with pressure drop due to the increase in both liquid surface area and relative velocities. When both particulate and gaseous impurities are to be removed and sufficient mass transfer cannot be obtained in a venturi scrubber alone, it is common practice to add a small packed bed section downstream of the venturi and its separator. Such an arrangement is illustrated in **Figure 6-4.**

Jet Scrubbers

Jet scrubbers differ from venturi scrubbers in that the energy is provided by the incoming liquid rather than the gas. In venturi scrubbers the high velocity gas accelerates and breaks up the liquid causing a loss in gas pressure. In jet scrubbers (also called water jet scrubbers, jet venturi scrubbers, and ejectors) the liquid is injected into the gas stream at a very high velocity, its energy is transferred to the gas, accelerating it and actually increasing its pressure. Jet scrubbers can develop a gas pressure gain up to about 8 in. of water. At a pressure increase of 1 in. of water, the consumption of water is typically 50 to 100 gallons per 1,000 cu ft of gas (Perry and Green, 1984). The configuration of a typical jet scrubber unit is shown in **Figure 6-5.** In a complete jet scrubber installation the ejector discharges downward into a separator surge tank.

Jet scrubber units are not very efficient as gas compressors or as gas absorbers, but they are simple to operate and almost maintenance free. They are particularly applicable to relatively small volumes of gas containing corrosive impurities. If necessary, two or more can be

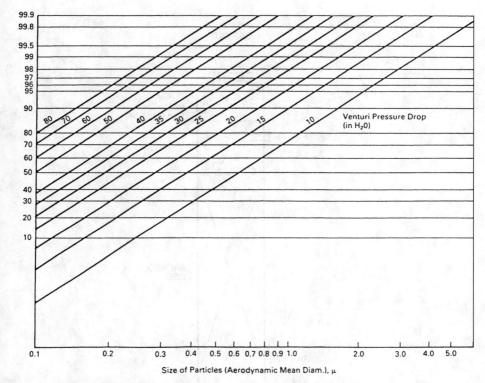

Figure 6-3. Typical venturi scrubber particle collection efficiency curves. *From EPA Handbook, 1986*

placed in series with countercurrent flow of the liquid from stage to stage to provide higher efficiency removal and a more concentrated product liquor. Another common arrangement is to use one or more jet scrubbers to move the gas and provide a preliminary washing stage followed by a short countercurrent packed absorption zone to assure complete removal of soluble gaseous impurities.

CARBON DIOXIDE ABSORPTION IN WATER

The absorption of carbon dioxide in water at elevated pressure was formerly an important industrial process, particularly for the purification of synthesis gas for ammonia production. The process has now generally been replaced by more efficient systems which employ chemical or physical solvents with much higher capacities for carbon dioxide than water. Such systems are described in Chapters 2, 3, 5, and 14. A description of the water wash process for carbon dioxide removal is included in this chapter because of its historical interest, its technical value as a classical liquid film-controlled operation, and the hope that the extensive work done on the process will prove useful in the development of new processes or applications.

(text continued on page 426)

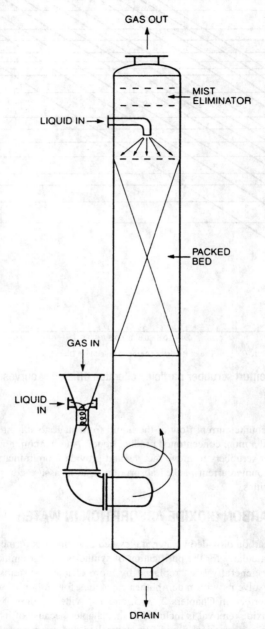

Figure 6-4. Diagram of venturi scrubber-packed absorber combination system. *Croll-Reynolds Co., Inc.*

Water as an Absorbent for Gas Impurities **425**

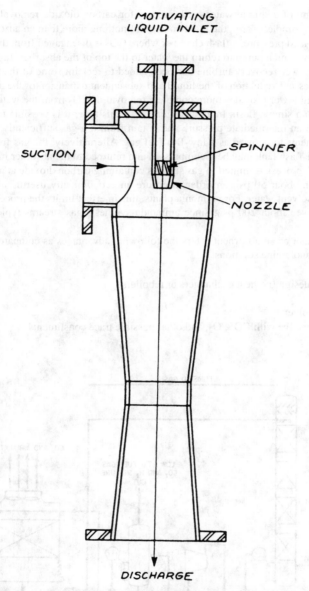

Figure 6-5. Diagram of venturi jet scrubber. This unit is normally installed on top of a separator/liquid storage tank. *Croll-Reynolds Co., Inc.*

(text continued from page 423)

A flow diagram of a simple water-wash process for carbon dioxide removal is shown in **Figure 6-6**. In its simplest form, the plant consists of nothing more than an absorption tower operating at elevated pressure, a flash chamber where CO_2 is disengaged from the water after pressure reduction, and a pump to return the water to the top of the absorber. In the flow diagram shown, a power-recovery turbine has been added to reclaim some of the power available from the pressure-reduction of the liquid and subsequent expansion of the absorbed gas, and a degasifying tower provides more complete removal of CO_2 from the water than could be obtained with a simple flash. In an arrangement of this type, it is possible to operate the flash chamber at an intermediate pressure and obtain from it a gas sufficiently rich in combustible components to be used as a low-Btu fuel gas. Alternatively, the gas from the intermediate pressure flash tank may be recompressed and returned to the absorber inlet.

In general, the process is limited to gas streams containing carbon dioxide at a partial pressure greater than about 50 psia in order to ensure an economically useful carbon dioxide capacity of the solvent water. For ammonia plants, this in effect limits the process to absorption pressures over about 200 psig since ammonia-synthesis gas streams typically contain about 25% CO_2.

The use of water as an absorbent offers the following advantages as compared, for example, to monoethanolamine solutions:

1. Simple plant design (no heat exchangers or reboilers)
2. No heat load
3. Inexpensive solvent
4. Solvent not reactive with COS, O_2, and other possible trace constituents

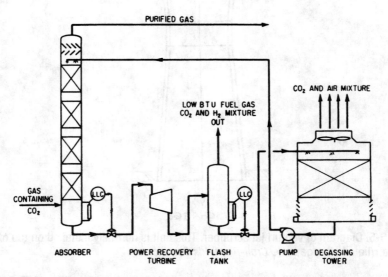

Figure 6-6. Simplified flow diagram of process for absorption of carbon dioxide from gas streams with water.

5. No nitrogenous solvent vapors introduced into the gas stream

The principal disadvantages of the water-wash process are

1. Substantial loss of hydrogen or other valuable constituents of the gas stream
2. Very high pumping load
3. Poor CO_2-removal efficiency
4. Impure by-product CO_2

In view of these disadvantages, the water-wash process is rarely used today for carbon dioxide removal.

Basic Data

The solubility of carbon dioxide in water, considered as a function of pressure and temperature, is presented in **Figures 6-7** and **6-8,** which are based upon the compilation of Dodds et al. (1956). **Figure 6-7,** which covers pressures from 1 to 700 atm, represents smooth curves of the data from numerous investigators. **Figure 6-8,** which covers primarily the region of sub-atmospheric pressures, is based entirely on the data of Morgan and Maass (1931) and the experimental points are therefore shown.

Design and Operation

Packed-Tower Design

The absorption of carbon dioxide in water has been shown to be almost entirely liquid-film controlled—presumably because of the relatively low solubility of carbon dioxide. Considerable research has, therefore, been conducted on the CO_2-H_2O system in connection with both absorption and desorption to determine the liquid-film resistance to mass transfer when various packings are used. Some of the data obtained are directly applicable to the design of commercial installations for carbon dioxide absorption and desorption.

The work of Cooper et al. (1941) is particularly valuable because they employed commercial-size packing (2- by 2- by $\frac{1}{16}$-in. steel raschig rings) and covered the range of very high liquid-flow rates commonly encountered in practice where substantially complete CO_2 removal is required. These investigators found that H_{OL} values for liquid-flow rates greater than about 20,000 lb/(hr)(sq ft) were considerably higher than those indicated by previous data obtained at lower liquid-flow rates. Cooper et al. suggest that this apparent discrepancy may result from the high ratio of liquid- to gas-velocities in a column operating in this manner. Under these conditions, when the ratio of average linear liquid- and gas-velocities is greater than unity, the slower gas stream may be carried downward by the liquid, thus tending to destroy the true countercurrent action and thereby increasing the observed H_{OL}. The results of this investigation are summarized in **Figures 6-9** and **6-10.**

In **Figure 6-9,** lines of equal mG_M/L_M are plotted in which m is the slope of the equilibrium line, y_e/x, G_M is the molar gas velocity, lb moles/(hr)(sq ft), and L_M is the molar liquid velocity, lb moles/(hr)(sq ft). This parameter, which is discussed in detail in Chapter 1, is called the stripping factor. It represents the ratio of the slope of equilibrium line to the slope of the operating line. It is normally less than 1.0 for practical absorption towers, in which essentially

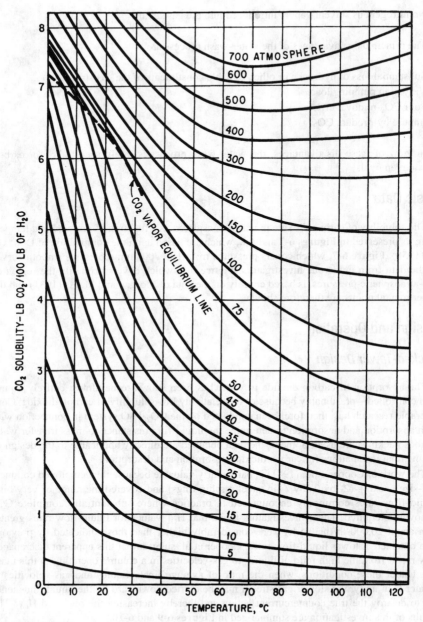

Figure 6-7. Solubility of carbon dioxide in water at pressures of 1 atm and greater. *From Dodds et al. (1956)*

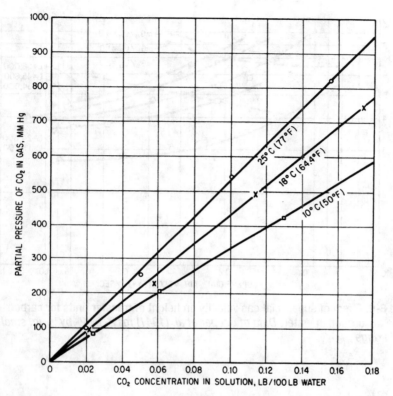

Figure 6-8. Solubility of carbon dioxide in water at low pressure. *Data of Morgan and Maass (1933)*

complete CO_2 removal from the gas is effected with excess water and may be as high as 2.0 for stripping towers, in which almost complete removal of CO_2 from the water is necessary. The data, therefore, indicate that for towers utilizing 2- by 2- by ¹⁄₁₆-in. steel raschig rings, H_{OL} values greater than 7 ft may be expected in the absorber and values from 4 to 5½ ft in the stripper. Data of Sherwood and Holloway (1940) for 2-in. ceramic rings have been included in **Figure 6-10** (which is a cross plot of **Figure 6-9**) for purposes of comparison. The effective V_G (superficial gas velocity) indicated for the ceramic-ring curve represents a value corrected for the difference in free volume between ceramic and steel rings.

The data of Sherwood and Holloway (1940) are of general value in estimating H_L (or $k_L a$) for the lower liquid-flow rate region. Although these data were largely obtained in experiments involving the desorption of oxygen, sufficient data were also obtained with carbon dioxide to indicate that the resulting correlation (equation 6-1), or the corresponding form (equation 6-2), applies equally well to this gas:

$$\frac{k_L a}{D_L} = \alpha \left(\frac{L}{\mu_L} \right)^{1-n} \left(\frac{\mu_L}{\rho_L D_L} \right)^{1-s} \tag{6-1}$$

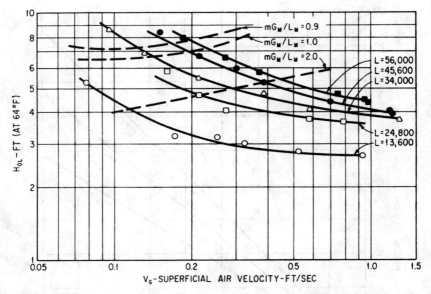

Figure 6-9. Effect of superficial gas velocity on height of transfer units for carbon dioxide absorption in water. *Data of Cooper et al. (1941) for 2- by 2- by ¹⁄₁₆-in. steel raschig rings*

or

$$H_L = \frac{1}{\alpha}\left(\frac{L}{\mu_L}\right)^n \left(\frac{\mu_L}{\rho_L D_L}\right)^s \tag{6-2}$$

Where: $k_L a$ = volume coefficient of mass transfer for liquid phase, lb moles/(hr)(cu ft)(lb mole/cu ft)
H_L = height of an individual liquid-film transfer unit, ft
L = liquid mass velocity, lb/(hr)(sq ft)
μ_L = viscosity of liquid, lb mass/(ft)(hr)
ρ_L = density of liquid, lb/cu ft
D_L = diffusion coefficient for liquid phase, sq ft/hr (approximately 6.6×10^{-5} sq ft/hr for CO_2 at 18°C)
α, n, and s = constants with the values given in **Table 6-6**

The preceding equations (equations 6-1 and 6-2) are intended to provide a means for calculating the liquid-film coefficient, k_L, and height of a transfer unit, H_L, for the general case. If the gas film also presents an appreciable resistance, an overall coefficient (or height) must be calculated by means of the conventional equations for adding the effects of both films. In the case of CO_2 absorption, the gas-film resistance can generally be assumed to be of negligible importance and values calculated by equations 6-1 and 6-2 can be used directly as approximate overall values for $K_L a$ and H_{OL}.

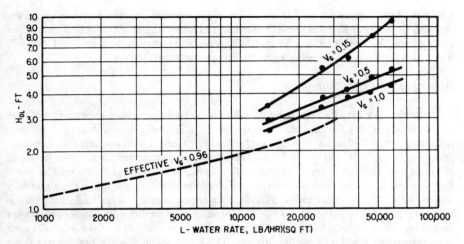

Figure 6-10. Effect of superficial liquid flow rate on height of transfer units for carbon dioxide absorption in water. *Solid line data of Cooper et al. (1941) for 2- by 2- by ¹⁄₁₆-in. steel raschig rings. Dotted line based on data of Sherwood and Holloway (1940) for 2-in. ceramic raschig rings*

	Table 6-6 Values of Constants α, n and s		
Packing	α	n	s
2-in. rings	80	0.22	0.50
1.5-in. rings	90	0.22	0.50
1.0-in. rings	100	0.22	0.50
0.5-in. rings	280	0.35	0.50
⅜-in. rings	550	0.46	0.50
1.5-in. saddles	160	0.28	0.50
1.0-in. saddles	170	0.28	0.50
0.5-in saddles	150	0.28	0.50
3-in. tile	110	0.28	0.50

Rixon (1948) has presented absorption data from a small commercial plant designed to produce relatively pure hydrogen from fermentation gas containing approximately 9.4% CO_2, 50.2% CO, 34.6% H_2, 1.4% CH_4, and 4.4% N_2 (by volume). In addition, he has presented extensive absorption and desorption data obtained during a pilot-plant study which preceded construction of the commercial unit. Results are given for both the absorption and desorption stages of the pilot plant and for the absorption step only of the commercial unit. His results, which are correlated in terms of $K_L a$, are in general agreement with those of Cooper et al. (1941) in showing a decrease in the absorption coefficient with decreased gas rate at the high liquid rates used in the absorber. No effect of gas rate was observed in the stripper. The value of $K_L a$ was found to increase with increased liquid rate in both the stripper and the absorber. Operating data for typical runs on both plants are presented in **Table 6-7**, which also gives physical dimensions for the equipment tested.

Absorption in Plate Columns

Plate-efficiency data for the carbon dioxide-water system have been presented by Walter and Sherwood (1941). For the case of a single plate, 18 in. in diameter with seven 4-in. caps, Murphree plate efficiencies from 1.8 to 2.6% were observed with operating temperatures averaging 10° to 12°C (50° to 54°F) and liquid/gas flow-ratios from 2.2 to 16 moles liquid per mole gas. These data are of interest in illustrating the very low efficiencies obtainable with equipment of this type; however, conventional bubble-cap columns are not normally utilized for the absorption of carbon dioxide in water because they do not lend themselves to the very high ratios of liquid- to gas-rate employed. Several modified tray designs have been developed, however, which appear to offer promise. In one of these (Cooper, 1945) each plate consists of a series of parallel troughs. The liquid which fills the troughs overflows through the spaces between them to form curtains of falling liquid in the space below the plate. The liquid falls into the troughs of the plate below entraining gas bubbles in the liquid. The primary gas flow is horizontal between each plate in a direction parallel to the liquid curtains and upward through spaces at alternate ends of adjacent plates.

In another somewhat similar design, a level of water several inches deep is maintained on each tray, and the bottom of each tray is perforated so that water jets flow downward to the tray below, entraining gas bubbles in the liquid on that tray. As in the previous design, the gas flows horizontally in the space between adjacent trays, then up through an opening at one end of the upper tray and back in the opposite direction in the next tray space. Plates of this type are sometimes referred to as "shed" or "shower" trays.

Operating Data

Typical flow rates and gas analyses have been presented by Yeandle and Klein (1952) who discuss various ammonia-synthesis gas-purification processes. According to these authors, a typical water-absorption unit would handle 5,000 scf/min of synthesis gas at 250 to 275 psig with a flow of 4,000 gpm of water that has been deaerated and cooled. With a conventional packed absorber, the gas leaving the top of the column would contain 0.5 to 0.8% carbon dioxide.

Data are presented in **Table 6-8** for the water-wash section of a large ammonia plant (Walthall, 1958) for two periods representing summer and winter conditions. It will be noted that considerably less water is required in the winter because of the lower ambient temperature, and this results in a reduced loss of hydrogen with the flashed gas. In general, the temperature of the gas entering the absorber is 15° to 30°F higher than that of the feed water. The water at this plant contains 96 ppm dissolved solids and 119 ppm total solids and has a pH of 7.2. The absorber is 10 ft ID by 81 ft high and is packed with raschig rings in three sections as follows:

1. Top 19 ft, 3-in. aluminum rings
2. Middle 17 ft, 3-in. ceramic rings
3. Bottom 17 ft, 2-in. ceramic rings

Table 6-7
Typical Test Data for Absorption and Desorption of CO$_2$ from Water in Packed Towers

Run Variables	Pilot-Plant Absorption, Run No.			Pilot-Plant Desorption, Run No.			Commercial-Plant Absorption, Run No.		
	2	4	6	2	25	10	1	6	11
Physical dimensions:									
Column ID, in.		9¾			15			27	
Packed height, ft		25.5 (25 ft, 5½ in.)			15			18 (two 9-ft sections)	
Packing, stoneware rings		1 × 1-in.			1 × 1-in.			1½ × 1½-in.	
Water rate, lb/(hr)(sq ft)	24,030	14,450	17,400	10,390	7,340	2,445	33,200	33,200	22,400
Water temp, °F	59.0	63.5	54.5	ambient	ambient	ambient	84.2	84.2	78.8
Inlet gas rate, lb/(hr)(sq ft)	134.0	116.0	120.0	136.0	96.0	332.0	76.2	181.0	231.5
Absolute pressure, atm	11.9	11.9	11.9	1.0	1.0	1.0	21.4	21.4	21.4
CO$_2$ in inlet gas, % by vol	61.2	60.6	59.1	neg.	neg.	neg.	9.63	10.12	9.20
CO$_2$ in exit gas, % by vol	0.49	4.97	3.9	3.7*	1.80*	0.17*	0.35	0.63	2.14
CO$_2$ in inlet water, lb/cu ft	3.1 × 10^{-4}	3.1 × 10^{-4}	3.1 × 0^{-4}	4.4 × 10^{-4}	not available	4.4 × 10^{-4}	1.25 × 10^{-4}	1.25 × 10^{-4}	1.37 × 10^{-4}
CO$_2$ in outlet water, lb/cu ft	0.337	0.471	0.404	95.0	3.1 × 10^{-4}	17.3	0.0288	0.0709	0.0993
$K_L a$, lb/(hr)(cu ft)/(lb/cu ft)	Not calculated because of spray-zone effect			95.0	61.0	17.3	20.1	50.0	35.2

*Percent CO$_2$ in air at point 9 ft above bottom of packing.
Source: Data of Rixon (1948)

Table 6-8
Operating Data on Removal of CO_2 from Synthesis Gas by Water Scrubbing

Scrubber Design Variables	July	December
Temperatures, °F:		
Water entering	80	52
Gas leaving	85	55
Pressures, psig:		
Gas entering	254	260
Gas leaving	248	256
Flow rates:		
Feed gas, scfm	10,550	12,150
Product gas (calculated), scfm	8,350	9,800
Water feed, gpm	8,250	5,275
NH_3 production in synthesis section, lb/min	162	189
Gas composition, %:		
Feed gas:		
CO_2	16.7	16.2
CO	2.1	2.8
H_2	59.8	60.1
CH_4	0.3	0.2
N_2 (including argon)	21.1	20.7
Product gas:		
CO_2	0.81	0.72
CO	2.70	2.87
Flashed gas:		
CO_2	75.2	79.5
O_2	0.0	0.0
CO	0.7	0.8
H_2	18.0	15.1
CH_4	0.3	0.1
N_2 (including argon)	5.8	4.5

Source: Data courtesy of Tennessee Valley Authority (1958)

Water scrubbing has been proposed for the removal of CO_2 from methane produced by anaerobic digestion. A pilot unit for this type of process was first put into operation in 1978 at a wastewater treatment plant in Modesto, California (Henrich, 1981). The process, which is called the Binax system, uses an absorption tower operating in the range of 100 to 500 psig and an atmospheric pressure air-blown regeneration tower. About 90% of the methane is recovered from feed gas to the absorber, which contains 30 to 50% carbon dioxide and trace amounts of hydrogen sulfide. The product gas is 98% methane. Plans to construct a Binax plant to produce approximately 650,000 scfd of pipeline quality gas from digester gas containing about 35% CO_2 at Baltimore's Back River Wastewater Treatment Plant have been reported (Henrich and Ross, 1983). Water scrubbing is attractive for this type of application because of its relatively low capital cost in small sizes, simplicity of operation and maintenance, and use of a readily available nonhazardous absorbent.

Corrosion

Unlike the amine- and alkali-salt-process solutions or essentially anhydrous organic solvents which are generally basic or neutral, water becomes quite acid when appreciable quantities of carbon dioxide are absorbed. As would be expected, this results in corrosion problems in water-wash CO_2-removal plants. A compensating factor is that, unlike the other processes, the water system operates at or near ambient temperatures throughout the cycle. The low temperature level is, of course, a favorable factor with regard to corrosion, and the absence of heat exchangers reduces the amount of corrosible metal exposed.

The corrosion of steel by carbon dioxide and other dissolved gases has been studied on a laboratory scale by Watkins and Kincheloe (1958) and Watkins and Wright (1953). These investigators found that the presence of oxygen greatly accelerates the rate of corrosion by carbon dioxide, while hydrogen sulfide in small quantities inhibits carbon dioxide corrosion. Comparative data taken from their corrosion curves are presented in **Table 6-9**.

The tests were conducted in a dynamic test apparatus with water containing dissolved gas flowing at a constant rate past the specimens. Although such tests are of value in illustrating comparative effects, extreme caution should be used in extrapolating to plant-equipment design. The authors found, in fact, that the mere substitution of a different lot of mild steel changed the corrosion rate by a factor of three.

In commercial operations, corrosion is minimized by the addition of inhibitors such as potassium dichromate to the water, the use of stainless steel in areas of high turbulence, and the application of protective coatings to the interior of the absorber and other vessels. Wooden cooling towers are adaptable to the stripping operation where countercurrent contact with air is required. Conventional water treatment to control algae may also be required in installations where the water is exposed to light. Foaming is not usually a problem; however, when it does occur, due to the presence of oil or other impurities in the water, conventional foam-inhibitors of the high-molecular-weight alcohol (e.g., ocenol) or silicone types have been found to be effective.

Table 6-9
Corrosion of Mild Steel by Carbon Dioxide and Other Gases in Water*

Gas Concentration, ppm		Corrosion of Mild Steel, mils/year	
O_2	H_2S	A: CO_2 conc, 200 ppm	B: CO_2 conc, 600 ppm
8.8	0	28	60
4.3	0	18	44
1.6	0	12	34
0.4	0	17	27
<0.5	35	6	6
<0.5	150	15	16
<0.5	400	17	21

*Temperature 80°F, exposure time 72 hr.
Source: Data of Watkins and Kincheloe (1958) and Watkins and Wright (1953)

HYDROGEN SULFIDE REMOVAL BY ABSORPTION IN WATER

Although hydrogen sulfide is appreciably more soluble in water than carbon dioxide, the use of water for removing this impurity from gas streams is not a commercially important process. Several attempts were made to commercialize the process, especially for treating natural gases of very high hydrogen sulfide content. The fact that no heat is required for acid-gas desorption, and, consequently, no heat exchange equipment is necessary, suggested the possibility of substantial savings in capital and operating costs over the conventionally used ethanolamine processes. However, the savings resulting from the reduced energy requirements are not as significant as expected because large amounts of steam are generated in the conversion of hydrogen sulfide to elemental sulfur in Claus type units which are necessary adjuncts of desulfurization plants, especially when large amounts of hydrogen sulfide are involved. In many cases sufficient steam is produced to drive the rotating equipment and satisfy the heat requirements for solution regeneration in an amine plant. Furthermore, the relatively high hydrocarbon content of the acid gas evolved from the water leads not only to appreciable losses of product gas, but also to a considerable increase in the size of the Claus unit, and to operational difficulties due to cracking of hydrocarbons on the Claus catalyst. Lastly, design improvements made in ethanolamine units have resulted in a substantial reduction of energy requirements. These considerations and the fact that water washing alone cannot produce pipeline quality gas have eliminated the process from large-scale industrial use.

In spite of the above considerations, the absorption of H_2S in water may be economical in special cases, and it has been deomonstrated to be a technologically feasible operation. Pilot-plant experiments were conducted in Canada (Burnham, 1959), and a rather large commercial water wash installation, treating about 60 million cu ft per day of natural gas containing 15% hydrogen sulfide and 10% carbon dioxide at a pressure of 1,100 psig was operated in Lacq, France, by Societe Nationale des Petroles d'Aquitaine (S.N.P.A.) (Barbouteau and Galoud, 1954). This unit was intended to remove the bulk of the hydrogen sulfide and carbon dioxide from the sour gas prior to final purification with diethanolamine. However, the plant has been converted to the S.N.P.A.-DEA process (see Chapter 2), and, at present, no commercial units using water washing for acid-gas removal from high-pressure natural gas are known to be in operation.

The solubility of hydrogen sulfide in water at moderate pressures has been determined by Wright and Maass (1932) and the behavior of the hydrogen sulfide-water system evaluated in more detail by Selleck et al. (1952) up to a pressure of 5,000 psig. Fortunately, the data from both of these investigations indicate that Henry's law holds reasonably well for the system at conditions which would normally be encountered in gas-purification operations. Equilibrium gas and liquid compositions can readily be calculated from Henry's law coefficients such as those presented in **Table 6-10.**

Experimentally determined vapor-liquid equilibrium data for the system hydrogen sulfide-carbon dioxide-methane-water at pressures ranging from atmospheric to 1,014 psia and temperatures from 85° to 115°F have been reported by Froning et al. (1964). These authors found that the equilibrium constants K can be represented by the following equations:

$$K_{Methane} = 306,000/P + 2.19t + 3,910 \ t/P - 145.0 \ AG - 121.6R \tag{6-3}$$

$$K_{CO_2} = -3,500/P + 0.12t + 360.0 \ t/P + 8.30 \ AG - 5,825 \ R/P \tag{6-4}$$

$$K_{H_2S} = 4.53 - 1,087/P + 110.0 \ t/P + 4.65 \ AG \tag{6-5}$$

Table 6-10
Solubility of H_2S in Water*

Pressure, atm	H at Various Temperatures, °C						
	5	10	20	30	40	50	60
1	3.12×10^2	3.64×10^2	4.78×10^2	6.04×10^2	7.35×10^2	8.65×10^2	9.81×10^2
2	3.19×10^2	3.69×10^2	4.80×10^2	6.06×10^2	7.39×10^2	8.77×10^2	10.02×10^2
3	3.26×10^2	3.72×10^2	4.83×10^2	6.09×10^2	7.42×10^2	8.83×10^2	10.11×10^2

*Values given are for H in Equation $P = Hx$, where x = mole fraction of solute in the liquid phase and P = partial pressure of solute in the gas phase, atmospheres.
Source: Based on data of Wright and Maass (1932)

Where: K = mole fraction in gas phase/mole fraction in water phase
P = system pressure, psia
t = system temperature, °F
AG = mole fraction $CO_2 + H_2S$ in gas phase
R = mole fraction, H_2S/AG

Comparison of calculated K values using the above equations, with those reported in the literature for the binary systems methane-water, carbon dioxide-water and hydrogen sulfide-water shows quite good agreement.

In its simplest form, the water-wash-process plant consists of nothing more than an absorber, flash vessel, and recycle pump. As justified by the economics, additional equipment may be added to reduce hydrocarbon losses, recover energy from the high-pressure water, or control operating temperature. A process-flow diagram of a pilot plant used to study water absorption for natural-gas purification is shown in **Figure 6-11**. The pilot-plant study was made by the Shell Oil Company to investigate the feasibility of absorbing H_2S and CO_2 from the gas of a Canadian field which analyzed about 45% acid gas. Results of the study indicated that 80 to 90% of the H_2S (initial concentration 35%) and about 60% of the CO_2 (initial concentration about 10%) could be removed by water washing at 500 to 700 psig. The pilot unit purified about 300,000 scf/day of gas with a water circulation of about 50 gpm (Anon., 1955).

An extensive study of corrosion was made in Shell's pilot-plant investigation inasmuch as this was expected to be one of the major operating problems. The results of the corrosion study have been presented in detail by Bradley and Dunne (1957). It was found that carbon steel in the absorber and rich-solution line was corroded rapidly at first (up to 100 mils per year (mpy) during the first 2 days), but corrosion slowed down appreciably (to 10 to 30 mpy after about 10 days) as a result of the formation of a heavy protective iron sulfide scale. The protective film was found to form more slowly on steel samples in the flash vessel. Hydrogen blistering and sulfide-corrosion cracking were found to be potential problems requiring the use of resistant materials and stress relieving. A corrosion inhibitor of the water-dispersible amine type was found to reduce all forms of corrosion markedly when used in a concentration of 0.1% in the absorption water. In general, the data indicate that stress-relieved carbon steel could be used for vessels and major piping. Special alloys, e.g., type

304, type 316, or K-Monel, are preferred for critical components such as pump impellers, Bourdon tubes, thermometer wells, orifice plates, and relief-valve springs.

The commercial unit described by Barbouteau and Galaud used a more complex flow scheme than is shown on **Figure 6-11**. The sour gas was first contacted with water in a column, provided with bubble-cap trays, at about 1,100 psig and from there passed to the diethanolamine unit for final acid-gas removal. The water leaving the high-pressure absorption column passed through a power recovery turbine to a flash drum maintained at 220 psig where most of the hydrocarbon gas dissolved in the water, together with substantial amounts of hydrogen sulfide and carbon dioxide were flashed off. The flash gas was washed with water in a secondary absorption tower and then sent to the fuel gas system.

The effluent water from the first flash drum together with the wash water from the secondary absorber was flashed, through another power recovery turbine, to essentially atmospheric pressure and stripped in a regenerating column. To facilitate expelling of acid gases, a small amount of deoxygenated flue gas was injected at the bottom of the regeneration column. The acid gases were fed to a Claus type sulfur plant.

Corrosion studies conducted in the unit gave similar results as those reported by Bradley and Dunne (1957).

ABSORPTION OF FLUORIDES

The removal of fluorine compounds from industrial exhaust gases is of increasing importance as an air-pollution-control measure. Fluorides may be emitted from many processes and actual vegetation damage or cattle fluorosis have reportedly resulted from the operation of plants producing aluminum, phosphate fertilizer, iron, enamel, and bricks. The two opera-

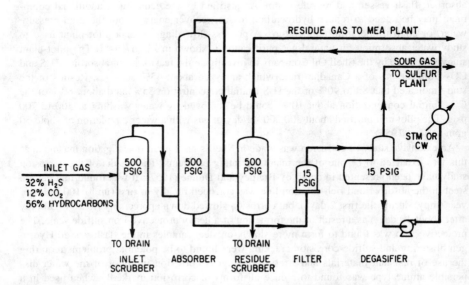

Figure 6-11. Simplified flow diagram of pilot plant used to study H_2S removal from natural gas by water absorption. *Data of Bradley and Dunne (1957)*

tions of major importance from the standpoint of quantity of fluorine compounds liberated are phosphate-rock treatment and aluminum-metal reduction.

The chemical reactions which lead to the evolution of gaseous fluorine compounds in the above operations are not completely understood; however, the resulting compounds are usually silicon tetrafluoride, SiF_4, and hydrogen fluoride, HF. Particulate matter is also usually present in the exhaust-gas streams and this may contain other nonvolatile fluorine compounds. Available data on the mechanism by which fluorides are liberated have been reviewed by Semrau (1956). He concludes that the principal mechanism in high-temperature operation is pyrohydrolysis resulting in the formation of HF, while the formation of SiF_4 is limited to low-temperature operations in which fluosilicates or other fluoride compounds are decomposed with acids. In practice, aluminum-plant exhaust gases are usually found to contain primarily HF with some particulate fluorides such as NaF and AlF_3; gases from the acidulation of phosphate rocks contain primarily SiF_4. Other phosphate-rock processing operations which involve heat, e.g., those using nodulizing kilns and calcium phosphate furnaces, evolve primarily hydrogen fluoride.

Fortunately, both hydrogen fluoride and silicon tetrafluoride are very soluble in water, and most commercial installations for the control of fluoride emission make use of this fact. Since large volumes of low-pressure gas must be handled, the design of exhaust-gas contacting equipment is governed to a large extent by the requirement of a minimum pressure drop. Low investment and operating costs are also important with equipment of this type because the resulting acid solutions do not usually represent an economically recoverable by-product. A further factor influencing the design of fluoride scrubbers is the presence of solid particulate matter in the gas stream and the formation of solids by reactions occurring in the scrubbing liquid. As a result of the considerations discussed above, fluoride-absorption units usually are based on water-spray systems or relatively open grid packing. The effluent solution may be recycled to build up the acid concentration, treated with lime slurry to precipitate the fluoride ion, or discarded without further treatment.

Basic Data

The vapor pressure of hydrogen fluoride over aqueous solutions is shown in **Figure 6-12**, which is based on the data of Brosheer et al. (1947). As can be seen, HF is quite soluble in water at relatively low temperatures. A gas containing 130 ppm (by volume) of HF, for example, representing a vapor pressure of 0.1 mm Hg, could theoretically produce a scrubber-effluent solution containing almost 5 percent HF. In commercial installations, however, a very large excess of water is usually used and the discharged solutions seldom contain over 0.1 percent HF. When the data of **Figure 6-12** are plotted as the logarithm of the partial pressure versus the reciprocal of the absolute temperature, essentially straight lines are obtained, the slope of which indicates that the heat of vaporization is in the range of 10.8 to 11.2 Cal/g mole (19,800 Btu/lb mole) for solution concentrations from 2 to 30% HF.

Boiling point curves and vapor-liquid compositions for the binary system $HF-H_2O$ are reported by Munter et al. (1949), and the data have been extended to the anhydrous HF portion of the system by Ohmi (1990).

Although hydrogen fluoride in solution is toxic and very corrosive to most materials, it is actually a relatively weak acid. It has an ionization constant of 7.4×10^{-4} at 25°C; this makes it only slightly stronger than acetic acid.

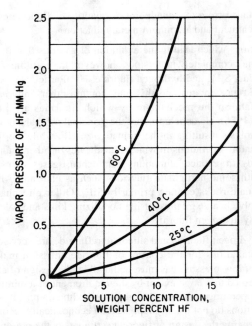

Figure 6-12. Vapor pressure of hydrofluoric acid over dilute aqueous solutions. *Data of Brosheer et al. (1947)*

When silicon tetrafluoride is absorbed by water, it reacts to form fluosilicic acid. The mechanism of the absorption process has been studied by Whynes (1956), who suggests that the reaction probably occurs in steps, as represented by equations 6-6 and 6-7:

$$SiF_4 + 2H_2O = SiO_2 + 4HF \tag{6-6}$$

$$2HF + SiF_4 = H_2SiF_6 \tag{6-7}$$

The simple fluosilicic acid probably reacts with additional SiF_4 or SiO_2 to form a more complex form of this compound. The second reaction (equation 6-7) is reversible to the point that solutions of fluosilicic acid exert a definite vapor pressure of HF and SiF_4.

Whynes has presented data on the concentration of fluosilicic acid in equilibrium with SiF_4 vapors; these are reproduced in **Figure 6-13.** The curves indicate that an exhaust gas containing 0.3% silicon tetrafluoride can produce a solution containing about 32% fluosilicic acid at temperatures below 70°C. This has been confirmed in a plant wash-tower by continuously recirculating the solution for several hours.

Tail gas scrubbers for phosphate operations typically utilize very dilute aqueous solutions. Data on the vapor pressure of HF and SiF_4 over solutions containing less than 2.5 percent H_2SiF_6 are given in **Figures 6-14** and **6-15** which are based on Russian data (Illanionev, 1963) as presented by Hansen and Danos (1980).

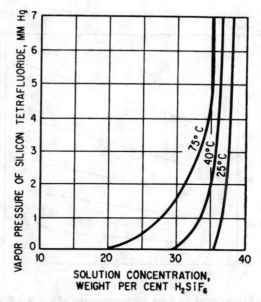

Figure 6-13. Vapor pressure of silicon tetrafluoride over aqueous solutions of fluosilicic acid. *Data of Whynes (1956)*

Fluosilicic acid solutions exhibit very low pH values, as indicated by the data in **Table 6-11** presented by Sherwin (1955) for industrial acid samples.

Absorber Design

As both hydrogen fluoride and silicon tetrafluoride are very soluble in water, the gas-film resistance would be expected to be the controlling factor in their absorption. This has generally been veified by experimental evidence although Whynes (1956) found the SiF_4 absorption to be complicated by a tendency to form mist particles in the gas; there is also a tendency of the silica (produced by reaction with water) to form a solid film on the outside of the water droplets, thus hindering absorption.

Absorbers which have been used or proposed for the removal of fluoride vapors from gas streams with water include the following general types (see also Chapter 1 and the first section of this chapter):

1. Spray chambers (vertical and horizontal)
2. Counterflow packed towers (typically low-pressure drop packing)
3. Venturi scrubbers, in which high gas velocities cause atomization of the water
4. Ejectors (jet scrubbers) in which high-velocity water jets are used to scrub and exhaust the gas
5. Cross-flow packed scrubbers

442 Gas Purification

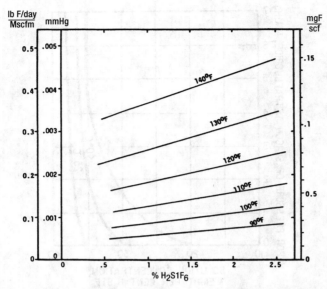

Figure 6-14. Vapor pressure of HF over dilute solutions of H_2SiF_6. *Russian data (Illanionev, 1963) as presented by Hansen and Danos (1980)*

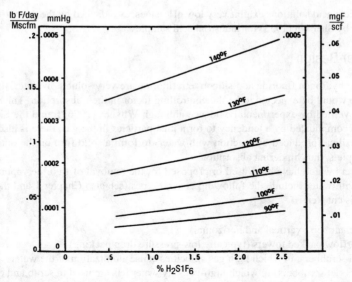

Figure 6-15. Vapor pressure of SiF_4 over dilute solutions of H_2SiF_6. *Russian data (Illanionev, 1963) as presented by Hansen and Danos (1980)*

Table 6-11
pH values of Fluosilicic Acid Solutions

Concentration of H_2SiF_6, % by wt	pH
1.0	1.4
0.1	2.2
0.01	3.0
0.001	3.8

A wide variety of spray-chamber designs has been employed for fluoride removal from tall cylindrical towers to horizontal vessels with cross- or cocurrent flow of water and gas. In typical spray chambers, water is introduced at several points from nozzles operating with water pressures of 15 to 60 psig.

Gas velocities are generally low, in the range of 4 to 5 ft per second, to avoid excessive entrainment of liquid droplets. This results in the requirement for large-diameter vessels; however, it also results in very low gas-side pressure drop, typically less than 3 in. of water.

Packed towers are generally more efficient than spray chambers, but suffer from the disadvantages of a somewhat higher pressure drop and a greater tendency to clog. The most successful packings are grid types, which present less pressure drop to the gas than conventional raschig rings and berl saddles; special low pressure drop plastic shapes which are available from several manufacturers; and plastic mesh pads.

Venturi scrubbers are very efficient in providing atomization of the water and good contact between the gaseous and liquid phases; however, they require a relatively high gas-side pressure drop and, consequently, a significant amount of power. This type of scrubber is probably in more widespread use for removing particulate matter than for absorbing soluble vapors.

Ejectors offer the advantage of simplicity in that they serve as exhausters for the gas as well as contactors for absorption of the soluble vapors. Their principal drawback is a relatively large power requirement because of the comparative inefficiency of ejectors, when used for moving gas, as compared to well-designed blowers. The power is expended in pumping the liquid.

Crossflow scrubbers are normally designed with one or more relatively shallow panels of packing slightly inclined toward the incoming gas. An open packing such as woven plastic mesh has reportedly given good results and can be cleaned readily with a high-pressure water hose (Hansen and Danos, 1980). During operation, the packing must be uniformly wetted, and this is usually accomplished by the use of low-pressure spray nozzles aimed at the face of the bed.

Wet cells composed of beds of wetted saran fibers have been proposed for fluoride absorption and tested on a pilot scale (Berly et al., 1954). However, no commercial installations are known.

Because of the basically different operating principles of the various types of contacting equipment employed, they cannot logically be compared on the basis of conventional mass-transfer coefficients or heights of transfer units. In fact, the volume coefficient of absorption K_Ga, has little value as a means of correlating spray equipment because the area for mass transfer, a, varies with liquid rate, nozzle design, liquid pressure, distance from nozzle, and other factors. Performance of different fluoride-vapor absorbers may, however, be compared on the basis of the number of transfer units in a given piece of equipment, or more simply, on the basis of percentage removal efficiency.

An interesting approach to the problem of correlating data from equipment of this type has been suggested by Lunde (1958), who attempted to relate the number of transfer units to the total power requirements of full-scale equipment.

For the case of a dilute gas stream, the number of transfer units based on an overall gas-phase driving force is defined as follows:

$$N_{OG} = \int_{y_2}^{y_1} \frac{dy}{y - y_e} = \frac{K_G a P h}{G_M} \qquad (6\text{-}8)$$

Where: N_{OG} = number of transfer units (overall gas)
y = concentration of vapor in gas stream, mole fraction
y_e = vapor concentration in equilibrium with liquid, mole fraction
$K_G a$ = overall mass transfer coefficient based on the gas, lb moles/(hr)(cu ft)(atm)
P = total pressure, atm
h = tower height, ft
G_M = gas-flow rate, lb moles/(hr)(sq ft)

In HF and SiF$_4$ absorption units, where a large excess of water is used, the concentration of acid in the solution is very low and y_e can be neglected. The equation is thus simplified to

$$N_{OG} = \int_{y_2}^{y_1} \frac{dy}{y} = \ln \frac{y_1}{y_2} \qquad (6\text{-}9)$$

Since the absorption efficiency (E) is directly related to y_1 and y_2 [$E = (y_1 - y_2)/y_1 \times 100$] the number of transfer units can be calculated directly from the following equation:

$$N_{OG} = \ln \left(\frac{1}{1 - E/100} \right) \qquad (6\text{-}10)$$

In accordance with this equation, an absorption efficiency of 95% requires three transfer units, while 99% efficiency requires about five units.

Lunde suggests that the number of transfer units for a packed tower is primarily dependent upon the height of the tower and is only slightly affected by the power introduced in the gaseous or liquid phase in accordance with the following general relationship:

$$N_{OG} :: \frac{P_L^{0.1-0.5} h^{0.6-1.0}}{P_G^{0.1}} \qquad (6\text{-}11)$$

Where: P_L = power introduced in the liquid
P_G = power introduced in the gas
h = tower height

With spray towers, on the other hand, the proposed equation is

$$N_{OG} :: \frac{P_L^{-1}}{P_G^{0.1}} \qquad (6\text{-}12)$$

That is, the number of transfer units is expected to be approximately proportional to the power introduced in spraying the liquid. This power would, of course, increase with an increase in either the quantity of water or the pressure drop across the nozzles.

No equation is proposed for venturi or jet scrubbers; however, it should be noted that, with the former, most of the power must be supplied with the gas stream, which atomizes the liquid introduced; in the latter type, all of the power is introduced with the water, which provides atomization and moves the gas stream.

Data presented by Lunde for the absorption of HF in several types of equipment are reproduced in **Table 6-12** and plotted in **Figure 6-16**. He concludes that for a given requirement of HF or SiF_4 removal efficiency, grid-packed towers require the least power and venturi scrubbers the most, with spray towers in between. His analysis of the data of Berly et al. (1954) and First and Warren (1956), for the absorption of HF in wetted beds of fiber (wet cells), indicates that the power consumption of this type of equipment is intermediate between those of the spray and venturi scrubbers. Data on the absorption of SiF_4 in spray, packed, and jet scrubbers, presented by Pettit (1951 A, B) and Sherwin (1954), were found to show no correlation with power introduced, although the spray-scrubber points fall reasonably well on the line for spray scrubbers handling HF, as shown in **Figure 6-16**.

Sherwin (1954) also presents data relating to the effect of nozzle design, number of nozzles, and water-flow rate on the performance of spray towers for SiF_4 absorption. The three types of nozzles tried, which included one solid-cone and two hollow-cone spray-pattern designs, did not appear to give appreciably different performances, nor did any improvement result from the substitution of as many as nine nozzles for one centrally located one. As would be expected, however, the water-flow rate was found to have a very strong effect. Using the average performance values for all nozzles, N_{OG} has been calculated for several water-flow rates from Sherwin's study, and the results are shown in **Table 6-13**.

Data on which the table is based were obtained in the first of six towers of a commercial plant. Gas and liquid flowed cocurrently downward through the unit, which was 7 ft square by 28 ft high. Approximately 10,000 cfm of gas containing about 0.6% SiF_4 entered the unit at a temperature of about 60°C (140°F). Sufficient data are not available to permit plotting the points on **Figure 6-16**; however, if power input is assumed to be proportional to water-flow rate, the data would form a line closely paralleling that plotted for spray scrubbers.

Sherwin's data on the relative efficiencies of each of the six spray towers operating in series are also of interest. The six units were arranged for downward gas flow in the first, then alternately upward and downward flow, with the gas leaving the top of the last section. Water was fed to the system through 24 spray nozzles (four per tower), at a pressure of 8 psig and a rate of 3.8 gpm per section. Each unit was 7 ft square by 28 ft high. Pertinent results of this study are tabulated in **Table 6-14**.

The high apparent efficiency of the final stage is believed due to the presence of some packing which was installed in this section for the purpose of removing silica particles from the gas.

446 *Gas Purification*

Table 6-12
Hydrogen Fluoride Absorption

Installation	Type of Absorbing Equipment	Absorbent	Gas Rate, lb/(hr) (sq ft)	Liquid Rate, lb/(hr) (sq ft)	Horsepower Expended per 1,000 cu ft gas/min.			$K_G a$ lb mole/(hr)-(cu ft)(atm)	N_{OG}
					Gas	Liquid	Total		
A	Crossflow spray	Water	2,110	72	~0.0067	0.0089	~0.016	~11	~0.33
A	Crossflow spray	Water	1,880	72	~0.0056	0.0098	~0.015	~12	~0.38
A	Crossflow spray	Water	2,080	103	~0.0067	0.0067	~0.013	~12	~0.25
A	Crossflow spray	Water	1,830	84	~0.0067	0.017	~0.024	~15	~0.62
A	Crossflow spray	Water	1,400	92	~0.0061	0.017	~0.023	~25	~1.09
B	Crossflow spray	Lime water	2,050	105	~0.006	0.013	~0.019	~35	~1.50
C	Counterflow spray	Water	2,000	800	~0.2	0.10	~0.3	9	~5.85
D	Parallel flow spray	Lime water	13,800	3,800	~0.23	0.017	~0.25	51	2.58
E	Counterflow spray	Water	2,000	380	0.24	0.02	0.26	~4.3	2.5
F	Venturi	Water	76,000*	42,000*	4.7	0.071	4.8	......	2.9
G	Venturi	Water	~70,000*	~40,000–65,000*	2.1	0.074	2.2	......	2.0
G	Venturi	Water	~70,000*	~40,000–65,000*	2.4	0.11	2.5	......	2.7
G	Venturi	Water	~70,000*	~40,000–65,000*	2.9	0.071	3.0	......	2.3
G	Venturi	Water	~70,000*	~40,000–65,000*	3.5	0.095	3.6	......	3.0
G	Venturi	Water	~70,000*	~40,000–65,000*	4.0	0.12	4.1	......	3.9
G	Venturi	Water	~70,000*	~40,000–65,000*	4.5	0.13	4.6	......	2.3

*Based on throat cross section. Source: Data of Lunde (1958)

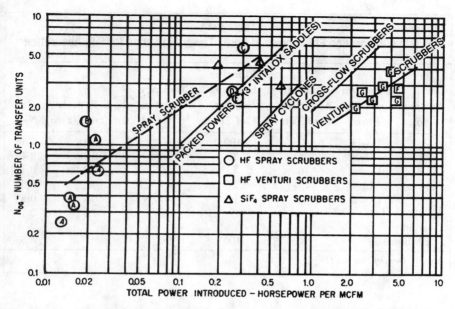

Figure 6-16. Correlation of number of transfer units with total power consumed by HF and SiF$_4$ scrubbers. Letters in HF spray and venturi scrubber points refer to installations listed in Table 6-12. *Spray and venturi data from Lunde (1959), other data from Hansen and Danos (1982)*

Table 6-13
Effect of Water-Flow Rate on SiF$_4$ Absorption in Cocurrent Spray Tower

Water-Flow Rate, gpm	Approx. Absorption Efficiency, %	N_{OG}
3	30	0.36
9	45	0.60
15	60	0.92
21	72	1.27
27	81	1.66

It is interesting to note the small difference between countercurrent and cocurrent operation in the various sections. This is due to the fact that solution concentration has little effect on the absorption rate in the ranges encountered. In a separate experiment to evaluate this factor, solution from the first two towers was recirculated continuously over the first tower to build up the fluosilicic acid concentration. The fluoride-removal efficiency was found to be relatively constant (60 to 80%) for solution concentrations from 0 to 15% fluosilicic acid and then to drop rapidly to about 35% as a concentration of 25% acid was approached.

A detailed compilation of engineering design methods for wet scrubbers is presented by Calvert et al. (1972). The study includes a sample calculation for the design of a water scrub-

**Table 6-14
Relative Absorption Efficiencies of Spray Towers in Series for SiF$_4$ Absorption**

Section No.	Direction of Gas Flow	Fluorine Absorbed, % of Total	Fluorine Entering Section, % of Total	Efficiency of Section, %	K_Ga, lb moles/ (hr)(cu ft)(atm)	N_{OG}
1	Down	30	100	30	0.31	0.36
2	Up	25	70	36	0.38	0.45
3	Down	16	45	36	0.38	0.45
4	Up	13	29	45	0.51	0.60
5	Down	7	16	44	0.49	0.58
6	Up	8	9	89	1.87	2.19
Over-all performance		99	100	99		4.63

Source: Data of Sherwin (1954)

ber to remove SiF$_4$ from gases evolved from a phosphoric acid manufacturing operation. The calculation is based on processing gas from the digestion of 2,000 T/day of phosphate rock in sulfuric acid to produce 600 T/day P$_2$O$_5$ equivalent. Approximately 400 lb/hr of SiF$_4$ are emitted from the reactor and enter the scrubber in approximately 220,000 ft^3/hr of gas saturated with water at 144°F. In order to obtain a removal efficiency of 99.8%, it is concluded that two scrubber stages are required. The first stage is simply a vertical duct approximately 3 ft diam. by 33 ft high, with cocurrent downward flow of gas and water. Approximately 80 gpm of water are sprayed in at the top of the duct. The second stage is a 3 ft diam. by 12 ft high cyclonic spray scrubber employing 22 nozzles, each operating with 6 gpm of water and a water pressure of 60 psig. An efficiency of 91.8 percent fluoride removal is predicted for the first stage and over 99.8 percent for the overall system. Detailed discussions of the design of wet scrubbers for phosphate plants are also presented by Hansen and Danos (1980, 1982).

Operating Data

In the production of superphosphate fertilizer from phosphate rock, the rock is normally pulverized, mixed with sulfuric acid, and discharged into a den where the reaction between rock and acid proceeds. The fresh superphosphate is then removed from the den by an elevator and conveyed to storage for curing. Exhaust gases containing fluorine compounds are drawn from the mixing and den operations and, in some instances, from the elevator and other units.

Typical Florida pebble phosphate rock contains about 3.6% fluoride expressed as elemental fluorine, and approximately 32% of this is released during the acidulation process (Pettit, 1951). Almost all of the fluoride vapors are evolved in the mixer and den although a slight evolution of vapor occurs during subsequent handling and storage operations. The composition of mixer gases and total flue gases from a plant handling Morocco rock is presented in **Table 6-15**. The fluorine evolved from this rock corresponds to approximately 1% of the weight of the rock or 25% of the fluorine originally present.

Table 6-15
Composition of Exhaust Gases from Continuous Superphosphate Acidulation Process

Exhaust Conditions	Mixer Gases	Flue Gas (mixer plus den)
Gas temperature, °C	78 (172°F)	60 (140°F)
Gas quantity, lb moles/hr	146.8	792.4
Analysis, mole percent:		
Air	34.5	78.9
CO_2	16.1	3.0
SiF_4	2.0	0.6
H_2O	47.4	17.5

Source: Data of Sherwin (1954)

Comparative data on three types of scrubbers used in superphosphate plants are presented by Pettit (1951 A, B). Although no conclusions are drawn in the study, the data, which are summarized in **Table 6-16,** indicate that the horizontal scrubber offered the highest efficiency with the lowest water-flow rate. The high efficiency of this unit probably resulted from the use of high-pressure nozzles and the long tortuous path which the gas stream was forced to follow.

Data on 13 scrubbers handling superphosphate-den gas have been presented by Sherwin (1954). Ten of these are more or less conventional spray-tower systems, one is a packed tower, and one is a jet-exhauster system. The spray towers show values for K_Ga ranging from 0.62 to 2.65; the packed tower, a K_Ga of 3.7, and the jet exhauster, a K_Ga of 15.6. The volume for the jet exhauster is based on the volume of the "tower" which would enclose the vertical venturi pipe from the jet level to the level of the liquid in the tank below. The system obviously gives a very high volume-coefficient of performances; however, power consumption was reported high and overall fluorine-removal efficiency for two units in series was not as high as that of the majority of the spray installations. A portion of the data on these units is summarized in **Table 6-17.**

It will be noticed that the two-stage spray tower (installation 4) gives slightly better performance than the six-stage spray tower. The two major reasons for this appear to be (a) the lower gas velocity which allows the mist formed to settle out and (b) the appreciably higher water-circulation rate. Silica-deposition problems generally favor the use of a spray tower for this service over the more compact packed tower.

Hansen and Danos (1982) report on experience with a large (18 ft × 8½ ft × 46 ft) cross-flow scrubber in a phosphoric acid plant. The scrubber consisted of a spray chamber followed by multiple packed beds of plastic woven mesh. With regard to the spray chamber section of the scrubber, they conclude that a spray nozzle pressure over 60 psig is required to attain 80% fluoride removal efficiency (1.5 transfer units); the amount of spray chamber water should be about 20 to 30 gpm/1,000 acfm; and full cone spray nozzles directed countercurrent to the gas flow are preferred. The plastic woven mesh may be irrigated with low-pressure co-current sprays; however, the nozzles should be mounted so that they are equidis-

Table 6-16
Comparative Performance Data on Superphosphate Plant Exhaust-Gas Scrubbers

Design Variables	Vertical-Spray Towers*	Ejector Unit+		Horizontal-Spray Scrubber++
		Sprays	Ejector	
Water-flow rate, gpm	292	60	300	60
Water pressure, psig	32	30	60	62
No. of nozzles (or ejectors)	12	24	1	80
Exhaust-gas rate (max), cfm	13,000	7,500		14,000
Plant size, tons superphosphate/hr	15	15		40
Water-use rate, gpm/1,000 cfm gas	24.3	48		4.3
Apparent average efficiency of fluoride removal, %	95	98		99

*Two cylindrical towers 42 in. in diameter by 40 ft high and one tower 42 in. in diameter by 20 ft high. Gases pass down through first tower, up through second, and down through third. First two towers are fitted with six spray nozzles, each pointed against the gas flow. Last tower is dry.
+Ejector has gas-suction and discharge connections 30 in. in diameter and a 3-in. water connection. Nozzle opening is 30 cm. Unit discharges into covered sump containing 20 spray nozzles mounted on top. Four spray nozzles are also mounted in base of stack. To prevent entrainment, wooden baffles and a thin bed of 1-in. raschig rings are installed in the stack.
++Horizontal scrubber is 33 ft 3¼ in. long, 4 ft 9¼ in. wide, and 11 ft 10 in. high. Unit is divided into eleven compartments so that gas goes up through first, then down through second compartment, and continues alternately up and down throughout the unit. The first 10 compartments contain 8 spray nozzles each; the last is dry.
Source: Data of Pettit (1951 A, B)

tant from the packing face and should be designed so that, when partially plugged, they do not create a single jet of water that can wear holes in the woven mesh.

Data on a commercial water spray installation for removing HF and other fluoride compounds from the exhaust gases of a nodulizing kiln have been reported in considerable detail by Magill et al. (1956) and are reproduced in **Table 6-18**. Limited data are also available on a large jet scrubber operating at a TVA installation manufacturing high analysis superphosphate (Anon., 1962). The unit is used to pull and scrub 12,500 scfm of air containing silicon tetrafluoride vapor and entrained phosphate dust and to develop a suction head of minus 1 in. of water. The ejector has a 36-in. diameter suction chamber and is almost 16 ft high. The spray nozzle has a 5-in. diameter bronze spiral insert covered with 3/16 in. thick neoprene. The scrubber discharges downward into a brick-lined concrete sump. Liquor is recycled to the nozzle by means of a centrifugal pump at a rate of 744 gpm, a pressure of 60 psig, and a maximum temperature of 135°F.

Materials of Construction

As would be expected with chemicals as corrosive as hydrofluoric and fluosilicic acids, the selection of suitable construction materials is an important design consideration. As a general principle, it can be stated that unprotected carbon steel is not suitable for aqueous

Table 6-17
Summary of Data for Typical Commercial Superphosphate Den-Gas Scrubbers

Scrubber Variables	Spray Units A	Spray Units B	Packed Tower	Jet System
Installation number	1	4	12	13
Stages in series	6	2	3	2
Size of dens, tons superphosphate/hr	28	9	6.5	25
SiF_4 in feed gas, % by volume	0.29	0.54	0.17	0.88
SiF_4 in exit gas, ppm by volume	38	50	110	200
Gas flow, cfm/(ton superphosphate) (hr)	250	134	462	120
Liquid flow, gpm/(ton superphosphate) (hr)	0.75	1.42	—	12.0
Gas velocity, ft/sec	24	0.77	6.3	
Fluoride-removal efficiency, %	98.6	99.1	93.1	97.8
Effective height, ft (height × no. of passes)	170.0	35.0	45	49
Tower volume, cu ft/(ton superphosphate) (hr)	293.0	101.0	56	4.8
Liquid circulation rate gal/1,000 cu ft of gas	3.0	10.7	10	100
Gas contact time, sec	70	46	7.1	0.2
K_Ga	0.616	1.04	3.68	15.6
N_{OG}	3.8	4.7	2.7	3.8

Source: Data of Sherwin (1954)

Table 6-18
Operating Data on Water-Spray Tower Treating Nodulizing-Kiln Exhaust Gas

Design Variables	
Tower diameter, in.	18
Tower height, ft	80
No. of spray injection-points	6
Gas inlet temperature, °C	300
Gas exit temperature, °C	72
Gas inlet rate, cfm	52,000
Gas exit rate, cfm	26,500
Rate of inlet HF, lb F_2/day	4,000
Rate of exit HF, lb F_2/day	97
Fluorine removed (as HF), %	97.6
Rate of inlet NaF as dust, lb F_2/day	340
Rate of exit NaF as dust, lb F_2/day	14
Removal of F_2 in dust, %	95.9
Water flow, gpm	700
Hydrated lime consumption, lb/day	24,000

Source: Data from Air Pollution Handbook (1956)

solutions of these acids at any concentration and, therefore, corrosion-resistant alloys, organic materials (wood or polymers), concrete, or brick must be employed.

Exhaust gas containing hydrofluoric acid is normally produced at an elevated temperature, and carbon-steel ducts may be employed to convey the gas to the purification unit provided its temperature is well above the dew point. It is usually necessary to precool the gas before it enters the actual absorption equipment in order that the latter can be constructed from, or lined with organic materials. The gas can be most readily cooked by the use of water sprays within the ducts, and the spray section of the conveyor duct should be constructed of stainless steel or other material which can stand both high temperatures and corrosive liquids. After cooling, the gas and dilute aqueous hydrofluoric acid can be handled in equipment constructed or lined with resistant materials such as polyvinyl chloride, polyethylene, Kynar and neoprene. For large aluminum-plant scrubbers, hydrofluoric acid-absorption towers have been constructed of clear heart-redwood staves with internal piping of polyvinyl chloride and brass, stainless-steel, or Monel nozzles. New scrubber vessels are more commonly constructed of FRP with a Dynel shield to protect glass fibers from the fluorides.

Ejectors for hydrofluoric acid service may be constructed of cast iron or steel and lined with a protective film such as neoprene or Kynar, or they may be fabricated entirely from a hydrofluoric acid-resistant material. Asplit "F," a modified phenolic resin fortified with an inert carbon filler, has proved to be satisfactory for this service (Brown and Tomlinson, 1952).

In silicon tetrafluoride operations, the gas is usually available at moderate temperature (from phosphate-rock acidulation); however, relatively high fluoride concentrations may be developed. Depending on whether recirculation of fluosilicic acid solution is practiced, the fluosilicic acid concentration in the liquid may range from a fraction of 1% to over 10%. Because of the wide range of operating conditions and performance requirements of fertilizer-plant exhaust-gas purifying systems, a considerable variety of construction materials have been utilized.

Experience in Great Britain (Sherwin, 1954) indicates that good-quality engineering brick without frogs, bonded with a latex hydraulic-type cement, is the most satisfactory type of massive construction material for silicon tetrafluoride-absorption towers. In the United States, the towers are more commonly constructed of FRP or wood, with or without a protective organic coating. Tower basins and sumps are usually constructed of conventional portland-cement concrete. This material is apparently protected from severe attack by the precipitation of silica and other compounds in the pores as a result of the initial reaction between fluosilicic acid and constituents of the cement.

Monel metal and high chromium-molybdenum-nickel stainless steel are generally the most resistant materials of construction for equipment in contact with fluosilicic acid. However, they are very expensive, and their use is generally reserved for precision parts in very severe service, such as impellers and nozzles, particularly where recirculation is practiced. Common brass has proved quite satisfactory for exterior piping and, in some cases, for spray nozzles. Lead is not satisfactory for handling fluosilicic particles.

In addition to causing erosion of solution-handling equipment, silica particles, which precipitate from both gaseous and liquid phases, complicate the design of lines and vessels. The silica coats the inside of ducts and columns, plugs packings, and settles out of the solution stream wherever its velocity is reduced. The problem is best handled by designing the equipment so that all surfaces in the absorber are flushed with adequate quantities of water and the resulting solution is kept moving as rapidly as possible in lines and collection basins. In spite of such precautions, it is common practice to flush out silicon tetrafluoride-absorption systems with high-pressure water once every 2 or 3 days and to physically scrape deposited silica from the walls and ducts several times a year.

Disposal of Absorbed Fluoride

In both HF and SiF_4 absorption operations, the product water may be too acid and toxic for direct disposal to sewers. It is, therefore, common practice to neutralize the effluent with limestone (or lime) in a separate tank. The fluoride ion is precipitated as calcium fluoride, and where other components such as silica and iron are present, these may also be precipitated as a result of the pH change. The solids can be readily separated by settling or filtration and disposed of separately, although occasionally the neutralized mixture, which is relatively innocuous, can be disposed of as a dilute slurry.

Efforts have been made in some installations to recover the fluorine evolved in phosphate-rock processing as cryolite or other marketable products. The quantity of fluorine evolved in such operations is very large so that steps to recover it would appear warranted. A process developed by the Tennessee Valley Authority (TVA) is claimed to accomplish this (Anon., 1957). In the TVA process, the water used to scrub nodulizing-kiln exhaust gas is maintained at a pH of 5 to 6 by the continuous addition of ammonia, and the resultant rich liquor is treated to precipitate impurities and yield a valuable NH_4F solution. The absorber solution is recirculated to bring its fluorine content up to about 35 g/liter. The rich solution is then treated with sufficient ammonia to raise the pH to 9, thus precipitating iron, silica, and part of the phosphorus. The precipitate is filtered off, and the solution is then used to make cryolite (by adding sodium sulfate and alum at a pH of 6) or aluminum fluoride.

Additional studies conducted by TVA on the removal of fluorine from aqueous scrubber effluents and the production of useful fluorine compounds have been reported by Tarbutton et al. (1958) and by Barber and Farr (1970). The latter study describes cryolite recovery from effluents from phosphorus furnaces.

HYDROGEN CHLORIDE ABSORPTION

The absorption of hydrogen chloride in water (or in dilute HCl) is a very important operation in the manufacture of commercial hydrochloric acid. It is also important in the purification of exhaust gas from chemical operations which form HCl as a by-product, and from refuse incineration where HCl is formed by the combustion of chlorine-containing plastics and solvents. Where HCl vapor is present as an impurity in gas streams, it can readily be removed by washing with water. The only complicating factors are the extreme corrosiveness of the resultant solution and the problem of its disposal.

Vapor-pressure data for the hydrogen chloride-water systems are presented in **Table 6-19**. As can be seen, the vapor pressure of hydrogen chloride over dilute aqueous solutions is extremely low although it increases appreciably with increased temperature. The heat of solution is considerable: about 240 Btu/lb of 35% hydrochloric acid produced at room temperature. Therefore, heat removal is necessary if it is desired to effect very complete removal of hydrogen chloride from a concentrated gas stream or to produce a solution of maximum concentration. This may be accomplished by using cooled absorbers or by recycling the acid through a cooler and back to the absorption unit.

Because of the very high solubility of HCl in water and the rapidity with which the reaction with water occurs, the absorption is completely gas-film controlled. With concentrated gas streams, in which the low concentration of inert gas permits rapid diffusion through the

Table 6-19
Partial Pressure of Hydrogen Chloride Gas Over Hydrochloric Acid Solutions in Water (Variation with Temperature)

Conc. of HCl, lb per 100 lb H_2O	Partial pressure HCl, mm Hg, at			
	0°C	20°C	50°C	110°C
78.6	510			
66.7	130	399		
56.3	29.0	105.5	535	
47.0	5.7	23.5	141	
38.9	1.0	4.90	35.7	760
31.6	0.175	1.00	8.9	253
25.0	0.0316	0.25	2.12	83
19.05	0.0056	0.0428	0.55	28
13.64	0.00099	0.0088	0.136	9.3
8.70	0.000118	0.00178	0.0344	3.10
4.17	0.000018	0.00024	0.0064	0.93
2.04	—	0.000044	0.00140	0.280

Source: Data from Perry and Green (1984)

gas stream, absorption is extremely rapid and very simple devices such as externally cooled, wetted-wall columns may be used. With less concentrated gases, the absorption rate is reduced somewhat, and multi-stage scrubbers or conventional ceramic-packed towers may be required.

A two-stage jet scrubber system is reported to be capable of recovering 15% HCl solution from the waste gas of a refuse incineration plant containing as little as 0.005 kmol HCl per kmol air (0.5 vol%) with a scrubbing liquid temperature as high as 70°C. (Wiegand Evaporator Division, GEA Food and Process Systems, 1992). The absorption operation is shown diagramatically in **Figure 6-17,** which is based on batch transfer of the liquid from stage to stage.

Referring to the figure, the second stage is started with fresh water (point 4) which is continuously recirculated until its composition reaches that indicated by point 5 on the chart. At this time, the liquid is transferred to the 1st stage tank from which it is recirculated until its composition reaches that indicated by point 6 (15% HCl). At the end of a cycle, gas from the first stage will be approximately in equilibrium with 15% HCl as represented by point 2 on the chart, while cleaned gas from the second stage will approach equilibrium with the liquid in that stage (point 3).

Although the absorption of concentrated HCl vapors in the manufacture of hydrochloric acid is beyond the scope of this book, the removal of traces of HCl from the tail gas of such a process is a gas-purification problem. This type of operation is particularly important in the manufacture of HCl as a by-product of hydrocarbon-chlorination reactions, because the hydrogen chloride is associated with appreciable volumes of inert gas in the effluent from such processes. An excellent review of hydrochloric acid manufacture is presented in *Encyclopedia of Chemical Technology* (1974). A standard design tail-gas absorber, as described by this text, is filled with stoneware packing and takes liquid loadings of 1 gal/(min)(sq ft) or higher and gas velocities from 1 to 3 ft/sec. A weak acid containing 20% or less HCl is produced in the

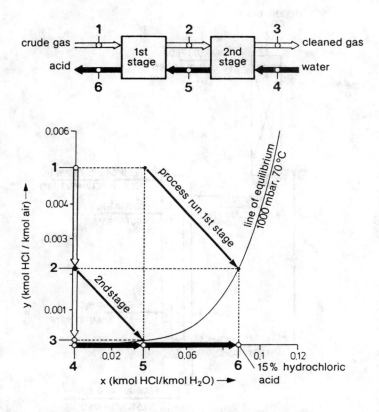

Figure 6-17. Absorption diagram for removal of HCl from a dilute gas stream by two stages of jet scrubbing. *Wiegand Evaporator Division, GEA Food and Process Systems*

tower, and the HCl concentration in the gas vented is reduced to the range of 0.1 to 0.3%. The towers are usually operated under suction with a fan located in the vent-gas stream.

Experimental data on the operation of a complete hydrogen chloride-absorption pilot plant, including both the primary cooler-absorber and the tails tower, have been presented by Coull (1949). The unit was constructed of Karbate (trademark of National Carbon Company, now Union Carbide Corp., for impervious carbon and graphite products) and consisted of a vertical water-jacketed tube through which the HCl gas and liquid flowed cocurrently for production of strong acid, plus a packed column through which unabsorbed tail gas was passed countercurrent to a stream of water. A schematic diagram of the unit together with data from one typical run are presented in **Figure 6-18**.

The cooler-absorber contained a Karbate tube, ⅞-in. ID by 1¼-in. OD and 108 in. long, which served as a cooled, wetted-wall column, and the tails tower consisted of a 4-in. ID Karbate tube packed to a depth of 4 ft 5 in. with ½-in. carbon raschig rings. Losses of hydrogen chloride in the vent gas were found to be negligible, and, in general, the absorption of hydrogen chloride in the tails tower appeared to take place in the lower portion, as indicated by the height of the hot zone. The run used to illustrate the schematic diagram of **Figure 6-18**

456 Gas Purification

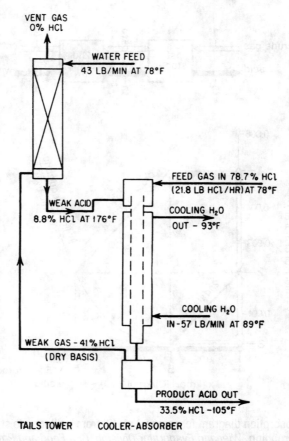

Figure 6-18. Schematic diagram of hydrochloric acid—absorption pilot plant with operating data from a typical run. *From Coull et al. (1949)*

was not typical of this regard, however, as the hot zone extended 40 in. up from the bottom of the 5-ft high tails tower.

A similar installation was employed by Dobratz et al. (1953), who evaluated three different absorption tubes in the cooler-absorber (1½-in. ID Karbate, 1-in. ID tantalum, and 0.88-in. ID stainless steel) and investigated the production of 36 to 40% hydrochloric acid solutions. These authors present complete data, including heat-transfer and absorption-coefficient correlations for the cooler-absorber, but give no data on the performance of their 4-in. by 4-ft packed tail-gas scrubbing tower.

Kantyka and Hincklieff (1954) have shown that a single-tower adiabatic absorber is capable of recovering a 28% hydrochloric acid of commercial quality from by-product gas produced during batch chlorination of organic compounds.

An evaluation of wet-fiber filters for use in the absorption of hydrogen chloride vapors by water in connection with an unusual gas-purification problem was presented by First and Warren (1956). The hydrogen chloride-containing gas in this case was produced as the result of an operation used to make very pure silica by the combustion of silicon tetrachloride. The gas contained a white fume of silicon dioxide and some unburned silicon tetrachloride as well as the hydrogen chloride. The gas absorber consisted of two stages of wetted, glass-fiber mats plus a third-stage dry mat which served as a mist eliminator. The mats were constructed of curly glass fibers of 50-μ diameter. The wet stages were each 4 in. thick, and the dry pad was 2 in. thick. Gas entered at a temperature of 350°F and was precooled by a spray nozzle in the inlet duct. Absorbent was sprayed on the fiber mats at a rate of 3.8 gpm/sq ft for a gas-flow rate of 216 scfm/sq ft. Pressure drop through the unit was found to be 4 in. of water gauge.

With water as the absorbent, the HCl content of the gas was reduced from 3.15 mg/cu m to 0.0025 mg/cu m, a removal efficiency of 99.9%. Other tests made, using a 5% aqueous solution of sodium carbonate as the absorbent, revealed essentially the same removal efficiency. Although the apparatus proved very effective for HCl removal, little or no decrease in the quantity of fine silica fume was effected by the system.

A commercial installation that uses jet scrubbers in combination with packed columns to remove hydrochloric acid vapors from the exhaust gas of a chemical manufacturing plant is described by Papamarcos (1990). A schematic diagram of this installation is shown in **Figure 6-19.** The jet scrubbers are designated "Jet Venturis" by the manufacturer, Croll-Reynolds Co. Inc.; however, they are not venturi scrubbers. The system consists of two jet scrubbers in series which use water to recover 98% of the HCl, followed by two short packed columns in series which use a dilute caustic solution for final purification to about 2

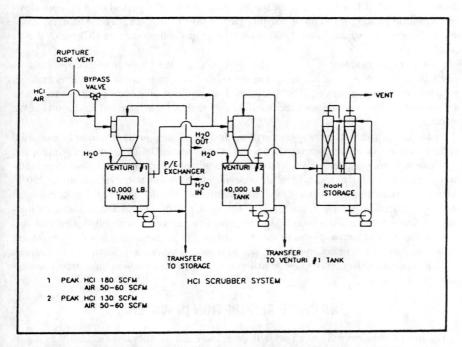

Figure 6-19. Schematic diagram of HCl scrubber system employing two Jet Venturi scrubbers and two packed columns. *From Papamarcos, (1990)*

458 Gas Purification

ppm HCl in the product gas. Two packed columns are used instead of one tall one because of height limitations.

The system illustrated in **Figure 6-19** is designed to handle a variable gas flow rate from 60 to 240 scfm as well as a variable gas composition with regard to HCl content. Each jet scrubber discharges into a 5,000 gallon tank. When the solution in the first tank reaches 32% by weight the solution is transferred to product storage. The contents of the second tank, at a concentration of about 10%, is then transferred to the first, and the second tank is filled with fresh water. The heat of solution of HCl, which is generated primarily in the first jet scrubber stage, is removed by a plate and frame heat exchanger in the solution recirculation line of this stage.

In this installation, the jet scrubber units are constructed of fiber reinforced plastic (FRP) with PVC spinners in the spray nozzles. The tail-gas scrubber towers, tank, and ductwork are also constructed of FRP with polypropylene packing and mist eliminators.

According to Croll-Reynolds (1992), a combination jet scrubber-packed tower system has also proved effective for absorbing silicon tetrachloride produced as a byproduct in the production of pure zirconium metal. The $SiCl_4$ is captured and held in storage tanks; however, a scrubber system is required to absorb small quantities which may escape during plant maintenance operations. Water is a suitable absorbent because $SiCl_4$ reacts rapidly with water to form highly soluble HCl and nonvolatile SiO_2.

The scrubber system consists of two contact devices in series: a jet scrubber for the initial stage and a packed tower for final cleanup. The jet scrubber was selected as the primary contactor because it has no moving parts or complex internals that might be corroded by the acid or plugged by the sticky silica gel formed. The second stage is a packed tower designed to absorb traces of HCl or $SiCl_4$ that escape the jet scrubber. It is packed with an open, low-pressure drop packing made up of polypropylene loops, called Spiral Pac. The overall system is designed to handle 3,000–4,000 acfm of inlet gas and provide an HCl removal efficiency of 99.9%.

A jet scrubber installation used for the absorption of HCl vapor from exhaust gas in a small chemical company is shown in **Figure 6-20**. The system consists of three small jet scrubbers operating in parallel. The scrubbers discharge into individual separator tanks and the product gas flows out through the manifold and vertical pipe in the left foreground of the picture.

Hydrochloric acid is extremely corrosive to most metals, and great care must be exercised in the choice of construction materials. Special alloys such as Durichlor, the Chlorimets, and the Hastelloys are suitable for hydrochloric acid solutions under some conditions. Pure tantalum is inert to hydrochloric acid at all concentrations and at temperatures up to 350°F. Among the suitable nonmetals are fiberglass-reinforced plastic, acidproof brick, chemical stoneware, chemical porcelain, glass, glass-lined steel, rubber (natural and synthetic for low-temperature service), plastics (vinyl chloride, polyethylene, polystyrene, filled phenolics, and fluorocarbons), and various forms of carbon. Carbon and graphite have found extensive use in the handling of wet or dry hydrogen chloride at temperatures up to about 750°F. Karbate, a resin-impregnated carbon or graphite material, is widely used for heat-transfer and absorption equipment.

CHLORINE ABSORPTION IN WATER

The need for the removal of chlorine from gas streams occurs most frequently as a gas-purification problem in connection with the manufacture, liquefaction, transfer, and storage of elemental chlorine. The problem also occurs to some extent as a result of magnesium

Figure 6-20. Jet Venturi scrubber system for removing HCl vapor from exhaust gas. *Croll-Reynolds Co., Inc.*

chloride electrolysis, hydrocarbon chlorination, and other chlorine-producing or -consuming operations.

The principal source of chlorine-containing gas in caustic-chlorine plants is the liquefaction step where noncondensables are vented from chlorine condensers as "sniff" gas containing 30 to 40% chlorine by weight. Dilute gas may be collected at other points in the operation; this gas also requires purification before it can be vented to the atmosphere. A number of processes have been developed to recover the chlorine from the vent-gas streams, including its use for the manufacture of bleach. Where the demand for bleach does not justify this operation, a regenerative recovery system is necessary, and one of the simplest of these involves absorption in water. The absorption of chlorine gas in water is also an important step in the manufacture of certain types of wood pulp. In this application, the process is intended primarily to provide a source of concentrated bleaching solution; however, design data which have been obtained for the absorption step are equally applicable to gas-purification or chlorine-recovery operations.

A schematic diagram of the water-absorption process for recovering chlorine from caustic-chlorine process "sniff" gas developed by the Hooker Electrochemical Company (Anon., 1957) is shown in **Figure 6-21**. In this process, chlorine-containing noncondensable gas from the liquefaction stage of chlorine manufacture is scrubbed countercurrently with water in a packed absorber. The resulting chlorine-free gas can be vented to the atmosphere, and the

460 Gas Purification

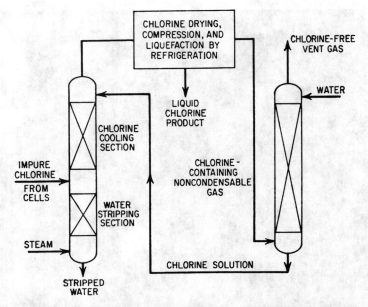

Figure 6-21. Simplified diagram of water-absorption process for removing chlorine from waste gases of electrolytic caustic chlorine plants. *Hooker Electrochemical Company process (Anon., 1957)*

aqueous chlorine solution is used to cool impure chlorine from the electrolytic cell by direct contact in a second packed tower. Chlorine remaining in the water after contact with the hot-cell gas is removed by steam stripping in the lower portion of the same tower, and the stripped water is discarded.

For cases where very small quantities of chlorine are present and essentially complete removal is required, it is common practice to use a dilute solution of sodium hydroxide as the scrubbing agent. The caustic increases the rate of absorption and decreases the vapor pressure of chlorine over the solution; both effects contribute to improved chlorine removal efficiency.

Solubility Data

Solubility data for the chlorine-water system have been published by Whitney and Vivian (1941). These authors point out that the two-phase system may be visualized as consisting of chlorine gas at such a partial pressure that it is in equilibrium with molecular chlorine in the solution—a relationship which can be assumed to follow Henry's law. In addition, the dissolved molecular chlorine will be in equilibrium with hypochlorous acid and hydrogen and chloride ions in the solution in accordance with the following reaction:

$$Cl_{2(aq)} + H_2O = HOCl + H^+ + Cl^- \tag{6-13}$$

Water as an Absorbent for Gas Impurities

For the case of chlorine in pure water, the total dissolved chlorine can be expressed as

$$C = \frac{p}{H'} + \left(\frac{K_e p}{H'}\right)^{1/3} \qquad (6\text{-}14)$$

Where: C = total concentration of chlorine in water, lb moles/cu ft
p = partial pressure of chlorine vapor over the solution, atm
K_e = Equilibrium constant for reaction of dissolved chlorine and water given above
H' = Henry's law coefficient for equilibrium between gaseous chlorine and dissolved but unreacted chlorine in water (atm)(cu ft)/lb mole

Values for the equilibrium and Henry's law constants given by Vivian and Whitney (1947) are presented in **Table 6-20**.

Table 6-20
Henry's Law and Reaction Equilibrium Constants for Chlorine in Water

Temperature, °F	H', (atm)(cu ft)/lb mole	K_e, (lb moles/cu ft)^2
50	141	7.10×10^{-7}
59	171	8.55×10^{-7}
68	213	10.7×10^{-7}
77	256	12.8×10^{-7}

Source: Data of Vivian and Whitney (1947)

Absorption Coefficients

An extensive investigation of the absorption of chlorine by water in packed columns was made by Vivian and Whitney (1947), who confirmed earlier findings (Adams and Edmunds, 1937; Whitney and Vivian, 1940) that the absorption coefficients for this system are much lower than would be predicted from liquid-film correlations. Vivian and Whitney proposed an explanation of the low coefficients based on the relative rates of hydrolysis and of diffusion and suggested the use of a pseudocoefficient using an unhydrolyzed-chlorine driving force for correlation of absorption data for this system.

The normal overall liquid-film absorption coefficient $K_L a$ was found to increase with liquid rate in accordance with the relationship $K_L a \propto L^{0.6}$. This exponent is somewhat lower than the value observed by Sherwood and Holloway (1940) for other sparingly soluble gases. Gas-flow rate was found to have no appreciable effect on the value of the coefficient over a tenfold range of gas rates—further confirming the liquid-film-controlled nature of the absorption.

The normal overall absorption coefficient was found to increase with increased temperature. The data plotted as a straight line on semilog coordinates indicated a relationship of the following form:

$$K_L a = m e^{nt} \qquad (6\text{-}15)$$

462 Gas Purification

With t in °F, n was found to have the value 0.0115. As an alternate correlation, Vivian and Whitney noted that the absorption coefficient is approximately proportional to the sixth power of the absolute temperature over the relatively narrow range studied (35° to 90°F).

Vivian and Whitney obtained data in two towers (4-in. and 14-in. ID). However, for convenience of operation, the major portion of the runs was made in the 4-in. column. The results for both towers are correlated for a single typical temperature in **Figure 6-22.** The 4-in. tower was packed to a height of 2 ft with 1-in. tile raschig rings while the 14-in. tower was packed to a height of 4 ft with similar rings. As can be seen, the towers show different performance characteristics when calculated as normal overall liquid-film coefficients, but this difference disappears when pseudocoefficients are used. The pseudocoefficient values are observed to be closer to predicted liquid-film coefficients (based on the oxygen-water

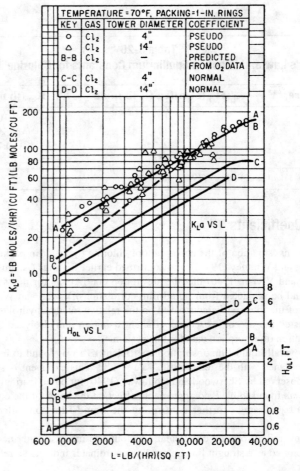

Figure 6-22. Effect of liquid-flow rate on $K_L a$ and H_{OL} for chlorine absorption in water. *Data of Vivian and Whitney (1947)*

system) at high liquid rates, while the normal overall coefficients are closer to predicted values at very low liquid rates. This is explained on the basis of the relatively slow chemical reaction between chlorine and water. At low liquid rates (and, therefore, low absorption rates) the reaction has time to proceed almost to equilibrium in the liquid film and a conventional coefficient applies. At high liquid rates, on the other hand, chlorine molecules are absorbed faster than they can react with water in the film zone so that an appreciable portion of the dissolved chlorine may penetrate into the body of the liquid in the molecular state. If this occurs, the driving force which is available to produce diffusion is the difference in molecular-chlorine concentration between the interface and the body of the liquid rather than the difference in total chlorine concentration.

The use of pseudocoefficients as proposed by Vivian and Whitney represents a procedure by which chlorine absorbers can be designed on the basis of liquid-film coefficients obtained with other systems. However, if actual chlorine-absorption data, such as those presented by Vivian and Whitney, are available for the desired packing, the normal liquid-film coefficients can probably be employed in the design of commercial equipment with reasonable assurance.

Materials of Construction

The chlorine-water system is both an acid and a strong oxidizing agent and, therefore, constitutes an extremely corrosive environment. Materials normally resistant to oxidizing conditions, such as stainless steels, may be attacked by aqueous Cl_2 solutions because the oxidation products formed are soluble in water and cannot protect the underlying metal. Carbon steel is rapidly attacked. However, rubber is resistant to the solutions at moderate temperatures so that rubber-lined steel equipment can be employed. Rubber-lined absorbers are used, for example, in the Hooker "sniff"-gas chlorine-recovery process previously described (Anon, 1957). Ceramics and glass are unaffected by chlorine solutions so that glass pipe and stoneware packing are excellent construction materials for this system. Among the metals, the best resistance is shown by nickel, silver, tantalum, Chlorimet 3, and Durichlor.

REFERENCES

Adams, S. W., and Edmonds, R. G., 1937, *Ind. Eng. Chem.*, Vol. 29, p. 447.

Anon., 1955, *Chem. Eng.*, Vol. 62, February, p. 110.

Anon., 1957, *Chem. Eng.*, Vol. 64, Part 1/June, p. 154.

Anon., 1962, *Southern Power and Industry,* June, p. 18.

Anon., 1957, *Chem. Eng. News,* Vol. 35, Part 3/Sept. 23, p. 81.

Barber, J. C., and Farr, T. D., 1970, *Chem. Eng. Progr.*, Vol. 66, November, pp. 56–62.

Barbouteau, L., and Galaud, R., 1964, *Gas Purification Processors,* Chap. 7, Edited by G. Nonhebel, George Newnes Ltd., London.

Berly, E. M., First, M. W., and Silverman, L., 1954, *Ind. Eng. Chem.*, Vol. 46, pp. 1769–1777.

Bradley, B. W., and Dunne, N. R., 1957, *Corrosion,* Vol. 13, April, p. 238t.

Brosheer, J. C., Lenfesty, F. A., and Elmore, K. L., 1947, *Ind. Eng. Chem.*, Vol. 39, March, p. 423.

Brown, C. R., and Tomlinson, R. W., 1952, *Air Pollution and Smoke Prevention Assoc. Am. Proc.*, Vol. 45, p. 69–74.

Burnham, J. G., 1959, *The Refining Eng.*, February, pp. C-15–C-19.

Calvert, S., Goldshmid, J., Leith, D., and Mehta, D., 1972, *Wet Scrubber System Study, Vol. 1, Scrubber Handbook*, EPA-R2-72-118a, August, Distributed by NTIS as PB-213016.

Cooper, C. M., Christl, R. J., and Peery, L. C., 1941, *Trans. Am. Inst. Chem. Engrs.*, Vol. 37, p. 979.

Cooper, C. M., 1945, U.S. Patent 2,398,345.

Coull, J. C., Bishop, H., and Gaylord, W. M., 1949, *Chem. Engr. Progr.*, Vol. 45, August, pp. 525–531.

Croll-Reynolds Co., Inc., 1992, Bulletin CR-666-E, "Silicon Tetrachloride Fume Problem Solved."

Dobratz, C. J., Moore, R. J., Barnard, R. D., and Meyer, R. H., 1953, *Chem. Engr. Progr.*, Vol. 49, November, p. 611.

Dodds, W. S., Stutzman, L. F., and Sollami, B. J., 1956, *Ind. Eng. Chem.*, Data Series, Vol. 1, No. 1, p. 94.

EPA, 1986, Handbook, *Control Technologies for Hazardous Air Pollutants*, EPA/625/6-86/014, September.

First, M., and Warren, R., 1956, *J. Air Pollution Control Assoc.*, Vol. 6, pp. 32–34.

Froning, H. R., Jacoby, R. H., and Richards, W. L., 1964, *Hydro. Process.*, Vol. 43, April, pp. 125–130.

Hanf, E. W., and MacDonald, J. W., 1975, *Chem. Eng. Progr.*, Vol. 71, p. 3, March, pp. 48–52.

Hansen, A. O., and Danos, R. J., 1980, "The Design and Selection of Scrubbers for Granulation Plants," presented at the Central Florida Section, AIChE Annual Meeting, Clearwater, FL, May 24.

Hansen, A. O., and Danos, R. J., 1982, *Chem. Eng. Prog 78*, No. 3, March, p. 40.

Henrich, R. A., 1981, *Biocycle*, May/June, p. 27.

Henrich, R. A., and Ross, B., 1983, "Landfill and Digester Gas Purification by Water Extraction and Case Study of Commercial System, Baltimore's Back River Wastewater Treatment Plant," presented at the IGT Symposium on Energy from Biomass and Waste, Disney World, FL, Jan. 27.

Illanionev, W. V. et al., 1963, Zhumal Prikladnoi Khimll, Vol. 36, No. 2, February, p. 237.

Interel Environmental Technologies, Inc., 1992, Brochure, "Gas Absorption/Desorption Systems," PBS-908-B.

International Critical Tables, 1937, Vol. 4, McGraw-Hill Book Company, Inc., NY.

Jaeger Products, Inc., 1990, Product Bulletin 600, "Plastic Jaeger Tri-Packs," JPI 3M/11-90.

Kantyka, T. A., and Hincklieff, H. R., 1954, *Trans. Inst. Chem. Engrs.*, Vol. 32, pp. 236–242, London.

Kirk, R. E., and Othmer, D. F., 1974, *Encyclopedia of Chemical Technology,* 3rd edition, John Wiley & Sons, NY.

Lunde, K. E., 1958, *Ind. Eng. Chem.*, Vol. 50, March, p. 293.

Magill, P. L., Holden, F. R., and Ackley, C., 1956, *Air Pollution Handbook,* p. 96, McGraw-Hill Book Company, Inc., NY.

McCarthy, J. E., 1980, *Chem. Eng. Progr.*, May, p. 58.
Morgan, O. M., and Maass, O., 1931, *Can J. Research,* Vol. 5, p. 162.
Munter, P. A., Aepli, O. T., and Kossatz, R. A., 1949, *Ind. Eng. Chem.*, Vol. 41, p. 1504.
Ohmi, T., 1990, *J. Electochem. Soc.*, Vol. 137, No. 3, March, p. 787.
Papamarcos, J., 1990, *Plant Services,* September, p. 45.
Parsons, R. A. ed., 1988, *American Society of Heating, Refrigerating and Air Conditioning Engineers (ASHRAE) Handbook,* Equipment Volume, Chap. 11, "Industrial Gas Cleaning and Air Pollution Control."
Perry, R. H., and Green, D. W., Eds., 1984, *Chemical Engineers' Handbook,* 5th edition, McGraw-Hill Book Company, NY.
Pettit, A. B., 1951A, *Air Pollution and Smoke Prevention Assoc. Am. Proc.*, Vol. 44, p. 98.
Pettit, A. B., 1951B, *Chem. Eng.*, Vol. 58, No. 8, pp. 250–252.
Rader, R. G., Musakis, M., Grosz-Roell, F., and Maugweiler, W., 1989, *Chem. Eng.*, July, p. 137
Rader, R. G., Musakis, M., Grosz-Roell, F., and Maugweiler, W., 1988, *Co-current Absorption/Scrubbing with Static Mixing Units,* Publication CCAS-4 of Koch Engineering Co., Inc.
Rixon, F. F., 1948, *Trans. Inst. Chem. Engrs.,* London, Vol. 26, pp. 119–130.
Schmidt, B., and Stichlman, J., 1991, *Chem. Eng. Technol.*, Vol. 14, p. 162.
Selleck, F. T., Carmichael, L. T., and Sage, B. H., 1952, *Ind. Eng. Chem.*, Vol. 44, September, p. 2219.
Semrau, K. T., 1956, *Emission of Fluorides from Industrial Processes.*, paper presented at American Chemical Society Air Pollution Symposium, Sept. 17–19, Atlantic City, NJ.
Sherwin, K. A., 1954, *Trans. Inst. Chem. Engr.*, London, Vol. 32 (Suppl. 1, pp. S129–140.
Sherwin, K. A., 1955, *Chem & Ind.*, London, Vol. 41, pp. 1274–1281.
Sherwood, T. K., and Holloway, F. A. L., 1940, *Trans. Am. Inst. Chem. Engrs.*, Vol. 36, Feb. 25, p. 21.
Tarbutton, G., Farr, T. D., Jones, T. M., and Lewis, H. T., Jr., 1958, *Ind. Eng. Chem.*, Vol. 50, No. 10, pp. 1525–1528.
Vivian, J. E., and Whitney, R. P., 1947, *Chem. Eng. Progr.*, Vol. 43, December, p. 691.
Walter, J. F., and Sherwood, T. K., 1941, *Ind. Eng. Chem.*, Vol. 33, p. 493.
Walthall, J. H. (Tennessee Valley Authority), 1958, personal communication, Oct. 29.
Watkins, J. W., and Wright, J., 1953, *Petrol. Engr.*, Vol. 25, November, pp. B-50–57.
Watkins, J. W., and Kineheloe, G. W., 1958, *Corrosion,* Vol. 14, No. 7, pp. 55–58.
Whitney, R. P., and Vivian J. E., 1941, *Ind. Eng. Chem.*, Vol. 33, p. 741.
Whitney, R. P., and Vivian, J. E., 1940, *Paper Trade J.*, Vol. 110, No. 20, p. 29.
Whynes, A. L., 1956, *Trans. Inst. Chem. Engr.*, London, Vol. 34, p. 118.
Wiegand Evaporator Division, GEA Food and Process Systems, 1992, GEA Wiegand Brochure P 07 E (562-07-90.09/2000) "Gas Scrubbing Plants."
Wright, R. H., and Maass, O., 1932, *Can. J. Research,* Vol. 6, p. 94.
Yaws, C., Yang, H. C., and Pan, X., 1991, *Chem. Eng.*, November, p. 179.
Yeandle, W. W., and Klein, G. F., 1952, *Chem. Eng. Progr.*, Vol. 48, July, pp. 349–352.

CHAPTER 7
Sulfur Dioxide Removal

REQUIREMENTS FOR SO_2 REMOVAL, 469
    United States Regulations, 470
    Japanese Regulations, 472
    European Regulations, 473
    Canadian Regulations, 474
    Regulatory Benefits, 474

FORMATION OF SULFUR DIOXIDE AND SULFUR TRIOXIDE FROM COMBUSTION OF FUEL, 474
    Sulfur Dioxide, 474
    Sulfur Trioxide, 475

PROCESS CATEGORIES AND ECONOMICS, 476

SELECTION CRITERIA, 491

ALKALINE EARTH PROCESSES, 496
    Limestone/Lime Process, 496
    Magnesium Oxide Process, 539

ALKALI METAL PROCESSES, 544
    Non-Regenerable, Alkali Metal-Based Processes, 544
    Double Alkali Process, 546
    Alkali Metal Sulfite-Bisulfite (Wellman-Lord) Process, 554
    ELSORB Process, 559
    Electrodialysis Regeneration (SOXAL) Process, 560
    Electrolytic Regeneration (Stone & Webster/Ionics) Process, 561
    Zinc Oxide Process, 562
    Citrate Process, 563

AMMONIA PROCESSES, 564

Basic Data, 565
Heat-Regenerative Ammonia Process, 568
Cominco Sulfur Dioxide-Recovery Process, 569
Ammonia-Ammonium Bisulfate (ABS) Process, 573
Catalytic/IFP/CEC Ammonia Process, 573
Walther Process, 575
ATS Technology™ Process, 578
General Electric Ammonium Sulfate Process, 579
Ammonia-Lime Double Alkali Process, 581
Ammonia-Calcium Pyrophosphate, 582

AQUEOUS ALUMINUM SULFATE PROCESS, 582

Dowa Dual Alkali Process, 582

FERROUS SULFIDE WITH THERMAL REGENERATION, 585

Sulf-X Process, 585

SULFURIC ACID PROCESSES, 585

Chiyoda Thoroughbred 101 Process, 585
ISPRA Bromine-Based Process, 588
Noell-KRC Peroxide-Based Process, 589

AMINE PROCESSES WITH THERMAL REGENERATION, 589

Basic Data, 590
Sulphidine Process, 591
ASARCO Process, 593
CANSOLV Process, 595
Dow Process, 596

SEAWATER PROCESSES, 597

Fläkt-Hydro Seawater Process, 599
Bechtel Seawater Process, 600
Bischoff Seawater Process, 601

PHYSICAL SOLVENT PROCESSES, 601

Solinox Process, 602

MOLTEN SALT PROCESS, 603

SPRAY DRYER PROCESSES, 604

    Aqueous Carbonate Process, 606
    Lime Slurry Spray Dryer Processes, 607
    Duct Spray Dryer Process, 614
    Gas Suspension Spray Dryer Process, 614
    Spray Dryer Byproduct Disposal and Use, 615

DRY SORPTION PROCESSES, 616

    Injection of Dry Alkaline Solids, 617
    Dry Lime/Limestone Processes, 618
    Dry Soda Processes, 624
    Dry Sorption Byproduct Disposal and Use, 626
    Dry Metal Oxide Processes (Regenerative), 627
    Alkalized Alumina Processes, 631

ADSORPTION PROCESSES, 634

    Basic Data, 634
    Carbon Adsorption Process with Water Wash Regeneration, 634
    Carbon Adsorption Process with Thermal Regeneration, 637
    Other Activated Coke Processes, 640

CATALYTIC OXIDATION PROCESSES, 641

    SNOX Process, 642
    Cat-Ox Process, 643
    Catalytic Oxidation/Electrochemical Process, 645

GAS PHASE RADIATION-INDUCED CHEMICAL REACTION PROCESSES, 645

    Ebara E-Beam Process, 645

ELECTROCHEMICAL CONVERSION TO SULFUR, 646

GAS PHASE REDUCTION, 646

    Parsons Flue Gas Cleanup (FGC) Process, 646

REFERENCES, 647

Sulfur dioxide removal from flue gases has probably been the subject of more research than any other gas purification operation. This research had very little commercial impact until air pollution control regulations sparked explosive growth in the number of installed Flue Gas Desulfurization (FGD) systems. Since the vast majority of SO_2 emissions are from fossil fuel-fired boilers at power stations, these sources have been widely controlled. Extensive application of wet FGD occurred in the United States and Japan in the 1970s and in Germany and the rest of Europe in the 1980s. Partly due to a lack of understanding of fundamental chemical reactions and their impacts on process and equipment design, early U.S. wet FGD systems earned a reputation for being troublesome and expensive to operate. As a result, the simpler and less expensive spray dryer FGD processes were rapidly adopted for the low sulfur coal applications that dominated the U.S. utility FGD market in the 1980s. Dry injection processes also emerged as a viable option for applications where removal requirements were modest. But, at the same time, the wet limestone/lime FGD technology matured. Many of the problems experienced with the early wet scrubbers were solved, and U.S. and overseas designs incorporated numerous advances. Present designs have close to 100% reliability and high SO_2 removal efficiency (90 to 98%) (Saleem, 1991B). Recent installations incorporate many improvements, such as standardized designs, in situ oxidation, reduced or no outlet flue gas reheat, and single train absorber systems, resulting in lower cost and improved reliability.

FGD process selection is usually based on securing the lowest life-cycle cost consistent with the least risk of failure to meet reliability and performance requirements. Processes that make marketable byproducts have been selected for those locations where byproduct disposal costs are high, where the law offers virtually no alternative (e.g., in Germany where gypsum is produced), or where there is a market for gypsum (e.g., in Japan, where there is no natural gypsum). The requirement for low risk has dictated the selection of simple processes that have undergone exhaustive development and that have progressed to commercial operation through successively larger demonstrations and multiple full-size systems. This extreme conservatism, which is forced upon the power industry by its regulated nature, creates a tremendous obstacle to the introduction of new FGD technology. Nonetheless, the large potential market continues to lure process developers. The most successful of these have been the Japanese and German variations of the conventional wet limestone/lime technology. This chapter provides information to help in the selection and application of suitable FGD processes from the large number of possible alternatives. Consistent with this large number of alternatives is an enormous volume of published information, some of which is reviewed in this chapter.

REQUIREMENTS FOR SO_2 REMOVAL

One of the first precedents for establishing limits on sulfur dioxide discharge in terms of ground level concentration was set in connection with the operation of smelters in the Salt Lake district of Utah in 1920. This resulted in the imposition of a regime which limited the sulfur dioxide concentration to 1 ppm (for an hourly average) at the level of vegetation during the growing season (Katz and Cole, 1950; Swain, 1921).

The Trail, Canada smelter of the Consolidated Smelting and Mining Company of Canada, Ltd., was the subject of a prolonged international investigation which resulted in the establishment of an operating regime setting the maximum permissible discharge in terms of tons per hour under certain weather conditions and restricting sulfur dioxide emissions in terms of ground level concentration and duration (Dean and Swain, 1944). The first known instance where sulfur dioxide removal was a legal requirement for the operation of a large power plant was at the Battersea Station of the London Power Company, which was constructed in 1929 (Hewson et al., 1933). More recently, blanket restrictions in many areas have limited the quantities and concentrations of SO_2 that can be emitted.

Environmental regulations are the driving force behind the need for and selection of FGD systems and dictate many design criteria. For example, they limit the amounts of the pollutants which can be discharged to the atmosphere and to any waterway. They also place limits on the concentration of toxic metals and other chemicals in landfilled byproduct, which can significantly affect the FGD process selection. Landfill material characteristics, such as leachate composition, permeability, and compressive strength and the availability of a suitable landfill site can also be important. Expected future regulations on traces of toxic substances and fine particulate may also affect the selection of a sulfur dioxide removal process.

Regulations are often written in terms of thermal megawatts consumed, MW_t, because of the variations in plant efficiency. In a typical coal-fired power plant, the efficiency of converting heat to electricity is 33 to 35.5%, making the net electric megawatts produced, MW_e, approximately one-third of the thermal megawatts consumed. In co-generation plants, part of the steam produced can be used for purposes other than generating electricity, or electricity is generated by a gas turbine as well as a steam turbine with consequent higher heat utilization, causing the relationship between MW_e and MW_t to vary significantly. For comparative purposes, in a typical coal-fired power plant one MW_e produces about 2,400 scfm (standard cubic feet per minute) of flue gas under normal operating conditions for many coals. However, the MW_e equivalent volume can be 10% higher for some coals, such as those from the Powder River Basin. Actual gas volumes are affected by the boiler exit gas temperature, which typically ranges between 220°F and 350°F for many coal-fired plants, and the gas pressure, which varies due to the altitude of the plant and the fans in the gas system.

IEA Coal Research's *FGD Handbook: Flue Gas Desulphurization Systems* tabulates SO_2 emission standards, guidelines, and proposed regulations in 21 countries (Klingspor and Cope, 1987). Ellison (1993) also gives ranges of national emission standards for many countries and SO_2 emission ceilings/targets in the European Community for large existing combustion plants.

United States Regulations

In the United States, the Air Quality Act of 1967 required the establishment of air quality criteria for pollutants such as SO_2, the monitoring of the pollution levels of all metropolitan areas, and the selection of air quality regions. It also required a report on the technology that could be used to control these pollutants. The Clean Air Act of 1970 dictated a revised strategy with emphasis on attainment of the clean air standards by 1975 (Engdahl, 1973). The failure of the Act to achieve its goals prompted its replacement by the Clean Air Act of 1977. Because of this law, some power plants that were both high SO_2 emitters and located in areas that failed to attain the air quality criteria for SO_2 were retrofitted with FGD systems. All new power plants were required to reduce SO_2 emissions, which in this era meant FGD systems.

The Clean Air Act Amendments of 1990 built upon the SO_2 control requirements of 1977 by expanding the clean air provisions in the revised Title I, adding acid rain provisions, and capping SO_2 emissions at 8.9 million tons of SO_2 in the new Title IV.

Both Titles I and IV affect new SO_2-emitting facilities, but Title I is the primary regulation. It continues the requirement that new facilities meet a criterion called New Source Performance Standards (NSPS). Refer to **Table 7-1**. With today's emission controls, this standard is considered to be relatively straightforward to meet, but even with low sulfur coal, uncontrolled emission of SO_2 is not permitted. In SO_2 attainment areas (areas where the ambient air quality standards are being met), new facilities must continue to go through the process called Prevention of Significant Deterioration (PSD). PSD does not involve a set level of emission or a set degree of reduction, but requires use of the Best Available Control Technology (BACT). A "Top Down" BACT analysis, in which the applicant must use the top (most effective) emission control system that cannot be shown to be technically, economically, or environmentally infeasible, is required. BACT is often more stringent than NSPS. In SO_2 non-attainment areas, new facilities are allowed, but they must satisfy a requirement called Emission Offset Policy. This policy requires reducing emissions from one

Table 7-1
Federal New Source Performance Standards

Source Category	Affected Facilities	Maximum Emissions
Fossil-Fueled Steam Generators	Coal- and Oil-Fired Boilers	Solid Fuel: 1.2 lb $SO_2/10^6$ Btu* Liquid Fuel: 0.8 g $SO_2/10^6$ Btu
Sulfuric Acid Plants	Process Equipment	2 kg SO_2/metric ton (mton) (4 lb SO_2/ton H_2SO_4) and 0.075 kg acid mist/mton H_2SO_4 (0.15 lb acid mist/ton H_2SO_4)
Petroleum Refiners	Refinery Process Equipment including waste-heat boilers and fuel gas combustors	Fuel gas max H_2S: 230 mg/dry std m^3 (0.10 grain/dry std ft^3)
Primary Copper Smelters	Roaster, Smelting Furnace Copper Converter	0.065% SO_2 by vol
Primary Zinc Smelters	Roaster, Sintering Machine	0.065% SO_2 by vol
Primary Lead Smelters	Sintering Machine, Dross Reverberatory Furnace, Electric Smelting Furnace, and Converter	0.065% SO_2 by vol
Petroleum-Refinery Sulfur Recovery Plants	Claus Plant oxidation or reduction with incineration reduction without incineration	0.025% SO_2 by vol dry at 0% excess air 0.030% by vol reduced sulfur compounds dry at 0% excess air and 0.0010% by vol H_2S dry at 0% excess air

*Maximum emissions allowed vary with the sulfur content of the fuel and other factors, but in no event exceed 1.2 lb/10^6 Btu.
Source: CFR (1990)

or more existing facilities to offset all the new SO_2 emissions, and using Lowest Achievable Emission Rate (LAER) technology to minimize the new emissions. Unlike BACT, LAER does not accept high cost as an argument against the use of a technology.

Title IV's SO_2 emissions cap means that all new SO_2-emitting facilities must purchase SO_2 emission allowances freed up by reduction of SO_2 emissions at existing facilities. Thus, construction of a new facility may require SO_2 emission controls at existing facilities that would not require them for any other reason. No matter how tightly emissions are controlled, an emission allowance must be obtained for every ton of SO_2 emitted over the life of the facility. These allowances can be obtained from within the company, or from other sources. However, ultimately they are obtained only through emissions reductions at existing facilities. Market conditions in 1996 indicate that other provisions are forcing such rapid emission reductions that allowances are currently inexpensive. However, when developing a new SO_2-emitting facility, allowances may create a financial incentive to reduce emissions even lower than Title IV would require.

For existing facilities, the Title I PSD provision is not new, so it would be unusual for new measures to be required at a facility simply because it is located in a sulfur dioxide nonattainment area. Acid rain is an issue apart from breathable air quality, so it affects facilities although the air around them attains the clean air standard. Title IV requires Phase I units (a specific group of 261 units) to reduce SO_2 emissions in two steps, with first step compliance by the Phase I date (January 1, 1995), and second step compliance by the Phase II date (January 1, 2000). The structure of the law is such that the utilities can decide which units to control, when, and by how much. The utilities can choose interim solutions of modest control such as coal switching or dry sorbent injection for all their units to meet the Phase I date, with more costly measures reserved to meet the Phase II date, or they can implement highly effective (and costly) measures on a few units in Phase I, and on the rest by the Phase II date. There are incentives for early compliance. A larger group than the Phase I group, all other units 75 MW_e and larger, called Phase II units, must also comply by the Phase II date. However, these units should be able to comply with relatively low-cost methods.

In general, the U.S. federal regulations do not require any particular technology, and it is very difficult to generalize about allowable emission levels or required percent reduction from uncontrolled emission levels. While state and local regulations generally reiterate the federal requirements, these agencies do have the regulatory option to impose more stringent emission limits. Few areas, however, actually implement tighter requirements.

Japanese Regulations

Japan began regulation of SO_2 emissions with the Air Pollution Control Law in 1968. Japanese law began differently than U.S. law, setting emission limits for each plant, rather than air quality objectives for every area. This was amended in 1970, but it failed to deal with heavily industrialized areas. In 1974, a further amendment initiated the concept of regulation to achieve air quality in each area, similar to the U.S. approach in Title I.

Japanese limits vary with the region and the stack height and are in terms of cubic meters of SO_2 per hour. The permissible SO_2 emissions vary between about the equivalent of 8 ppmv and 190 ppmv. Early FGD systems were usually installed on units that burned high-sulfur oil, but in the 1980s many new coal-fired units were equipped with FGD systems.

European Regulations

FGD in Europe began as local authorities in some German states required the retrofit of SO_2 scrubbers to handle a portion of the flue gas from certain boilers that were particularly problematic. By 1983, some 14 units had been installed. The environmental movement built rapidly in Germany, and a law known as GFAVO passed in June 1983. It established the federal limits shown in **Table 7-2,** and required compliance by June 1988. In spite of the feverish activity this schedule required, the industry responded and nearly all of the some 150 affected units complied on schedule. However, the cost and disruption it caused probably influenced the framers of the flexible, two-phase acid rain law enacted in the U.S. in November 1990.

In Germany, state and local authorities continue to enforce stricter SO_2 limits (200 mg/Nm^3 is common) where special problems exist.

The European Community (EC) has enacted legislation that asks its members to adopt limits similar to the German GFAVO. FGD systems were installed rapidly in Austria, Denmark, Holland, and Turkey. Italy and the U.K. followed shortly thereafter. Even before the

Table 7-2
German Federal SO$_2$ Emission Regulations for Coal-Fired Units

		Units Larger Than 300 MW$_t$ (Approx. 105 MW$_e$)	Units 100 to 300 MW$_t$ (Approx. 35 to 105 MW$_e$)	Units 50 to 100 MW$_t$ (Approx. 17 to 35 MW$_e$)
New Units	Fluidized Bed Combustion		400 mg/Nm^3 (between 0.3 and 0.8 lb/10^6 Btu) and 75% removal	
	Pulverized Coal Firing or Stoker Firing	400 mg/Nm^3 (between 0.3 and 0.8 lb/10^6 Btu) and 85% removal*	2,000 mg/Nm^3 (between 1.5 and 3.9 lb/10^6 Btu) and 60% removal	2,000 mg/Nm^3 (between 1.5 and 3.9 lb/10^6 Btu) and use low-sulfur coal
Existing Units	Remaining Service Life over 30,000 hours (approximately 4 years base-loaded)	2,500 mg/Nm^3 (between 1.9 and 4.8 lb/10^6 Btu)		
	Remaining Service Life 10,000 hours to 30,000 hours (1½ to 4 years base-loaded)			
	Remaining Service Life less than 10,000 hours	Existing Limits		

*If very high SO$_2$ content or widely variable SO$_2$ content indicates 400 mg/Nm^3 is unattainable with BACT, 650 mg/Nm^3 (0.5 to 1.3 lb/10^6 Btu) applies.
Source: Siegfriedt and Ludwig (1984)

dissolution of the Eastern Bloc and the Soviet Union, activity had begun there. The EC now sees scrubbing in Eastern Europe as much more cost-effective than further measures within member nations, so FGD systems are being installed in the Czech Republic, Slovenia, Poland, and Russia, generally with funds from Western governments and without any regulatory requirements. A number of the Eastern European countries have enacted regulations with compliance scheduled in the late 1990s.

Canadian Regulations

In Canada, the Federal Ministry of the Environment has no specific SO_2 removal requirements, but encourages the provinces to regulate SO_2 emissions. Each province has applied its own approach, depending on the mix of emissions found there. For example, most of Ontario's emissions are from power plants that belong to provincially-owned Ontario Hydro, so the province prepared a plan specific to that utility. British Columbia has very little pollution from power plants, so it targets industrial sources. The differences in approach even extend to the means of measurement, where some provinces choose the European method (mg/Nm^3) while others choose the U.S. method (lb/10^6 Btu). The Canadian attitude is also unusual, evidenced by a greater spirit of cooperation between the regulatory agencies and industry and by industry's seeming willingness to comply early and/or beyond the letter of the regulations.

Regulatory Benefits

Sulfur dioxide emissions in the United States in 1975 were officially estimated at 33 million tons per year. The goal of the Clean Air Act Amendments of 1977 was to reduce that to 28 million tons per year by 1990 (Bauman and Crenshaw, 1977). However, low growth in electrical demand, coupled with aggressive state programs and significant reductions in industrial emissions, brought the total to 18.9 million tons by 1990. The goal of the Clean Air Act Amendments of 1990 is to achieve 8.9 million tons per year total emissions by 2001 (Public Law 101-549, 1990).

FORMATION OF SULFUR DIOXIDE AND SULFUR TRIOXIDE FROM COMBUSTION OF FUEL

Sulfur Dioxide

Flue gases from combustion processes normally contain less than 0.5 vol % sulfur dioxide. The relationship between the sulfur content of the fuel and the sulfur dioxide content of the resulting flue gas is shown in **Table 7-3**. This table gives the sulfur dioxide content of combustion gases from several typical fuels.

Stack gas from smelters handling sulfur ores, on the other hand, can have very high sulfur dioxide concentrations. Therefore, the economics of recovering sulfur values from such gases can be much more favorable. Of course, the problems of discharging such gases without sulfur dioxide removal are also much more acute.

Table 7-3
Sulfur Dioxide Concentrations in Typical Combustion Flue Gases[1]

Fuel	Sulfur in Fuel, wt %	With Excess Air			With Air Heater Leakage		
		H_2O in Flue Gas, vol %	SO_2 in Flue Gas, ppmw	SO_2 in Flue Gas, ppmv	H_2O in Flue Gas, vol %	SO_2 in Flue Gas, ppmw	SO_2 in Flue Gas, ppmv
Coal: Missouri Beever (Raw)	5.9	9.5	10,972	5,059	8.5	9,541	4,381
Coal: Illinois No. 6	3.5	9.5	6,507	2,997	8.5	5,658	2,596
Coal: Pennsylvania	1.7	8.0	2,671	1,236	7.2	2,323	1,070
Coal: New Mexican	0.5	8.4	830	383	7.5	722	332
Coal: Powder River Basin	0.35	13.3	818	371	11.8	711	322
Oil: California No. 2	0.27	12.2	307	138	10.8	267	120
Oil: Low Sulfur No. 6	0.21	11.9	234	105	10.6	203	91

Notes:
1. These SO_2 concentrations assume no sulfur in the ash and no conversion of SO_2 to SO_3. In the U.S., credit for coal cleaning is allowed, but seldom practiced.
2. Gas concentrations are based on 20% excess air for coal and on 15 and 18% excess air, respectively, for No. 2 and No. 6 oil. Air heater leakage is 15% for all cases. Air heater leakage (%) = (lb wet air leakage X 100%)/(lb wet flue gas) per American Society of Mechanical Engineers (ASME) Power Test Code (PTC) 4.1.
3. SO_2 concentrations in flue gases from the combustion of natural gas are typically extremely low due to the removal of the sulfur bearing compounds from the gas prior to sale and are not included in this table for that reason.
4. Gas concentrations are usually expressed in terms of ppmv by convention.

Many large smelting operations that produce very high concentrations of sulfur dioxide feed the gas stream directly into a sulfuric acid plant. The design and operation of acid plants of this type are not discussed in this text, as they are considered to represent a separation and chemical manufacturing operation, not a gas purification process. On the other hand, the removal of sulfur dioxide from dilute smelter off-gas streams and the recovery of unconverted sulfur dioxide from the acid plant tail gas constitutes gas purification problems and are reviewed in this chapter.

Sulfur Trioxide

Most combustion gases that contain sulfur dioxide also contain a small, but significant amount of sulfur trioxide (or its reaction product with water, sulfuric acid). This component is of considerable importance because of its highly corrosive nature, its effect on the chemistry of many sulfur dioxide recovery processes, and its suspected critical role in air pollution problems. The amount of sulfur trioxide emitted to the atmosphere is a function of combustion air/fuel ratio, fuel composition, combustion temperature, time at temperature, the presence or absence of a catalyst, electrostatic precipitator conditioning with ammonia, and the type of flue gas desulfurization system. The equilibrium concentrations of the principal sulfur species in the combustion gas from a typical fuel oil at several air/fuel ratios have been calculated by Pebler (1974). The results show that in excess air mixtures at equilibrium, SO_2 is the most stable compound above 1,000°K; SO_3 is the predominant sulfur compound between 900 and 600°K; while, on further cooling, H_2SO_4 gains dominance over SO_3.

Sulfuric acid condenses below 400°K. Fortunately, equilibrium conditions are not attained in conventional combustion processes. However, the presence of catalytically active material, such as vanadium in oil and iron pyrites in coal, can increase SO_3 formation. In the absence of actual analytical data for specific cases, a rough estimate of the SO_3 concentration expected in combustion gases from coal and oil may be obtained from **Table 7-4,** which presents data compiled by Pierce (1977).

The sulfuric acid dew point of combustion gases, which is the key parameter with regard to stack corrosion, has been studied by a number of investigators. The available data have been reviewed and correlated by Pierce (1977), who presented the results in graphical form. Selected points from his correlations are given in **Table 7-5. Tables 7-4** and **7-5** can be used to estimate the sulfuric acid dew point of typical combustion gases in the absence of any other data. Flue gases should be kept above the estimated dew point to prevent corrosion of metals (e.g., carbon steel) by sulfuric acid. Mixtures of acids with higher dew point temperatures may condense; however, only limited information is available on the phenomena. Berger et al. (1984) review the corrosivity of various flue gas condensates. For the cases studied, sulfate was the predominant anion contributor to the acidity of the condensate, while chlorides and fluorides contributed to a lesser extent.

PROCESS CATEGORIES AND ECONOMICS

A great many processes have been proposed for removing sulfur dioxide from gas streams. Relatively few processes have attained commercial status; and, of those that have, many have not found a significant place in the U.S. market. In the U.S., only the limestone/lime wet FGD systems, predominantly with spray, tray, or packed tower absorbers; and the lime spray dryer systems are widely accepted today. Users have opted for low cost, proven systems that operate with high availability (Schwieger and Haynes, 1985). Today, the FGD impact on utility system equivalent forced outage rates is usually less than 1% (NERC, 1991). The reliance of utility companies on extremely mature technologies makes it difficult for suppliers of new technologies to bid on scrubber contracts (McGraw-Hill, 1991).

Although the primary emphasis of this text is on commercial processes, other processes can provide valuable background data pertaining to the development of new improved processes. For this reason, some developing processes that appear to have the potential for future commercialization are presented. Also, some processes that are no longer considered viable, but once represented major developmental efforts or commercial operations, are reviewed. With the new legislation mandating control of NO_x from many sources, there is a renewed interest in combined SO_2/NO_x technologies by both developers and users due to the potential cost savings. The NO_x control aspects of a number of these processes are covered briefly in this chapter. More information on some of these processes and NO_x only control processes is provided in Chapter 10.

To organize the many FGD processes that have been developed, the authors have departed from the categorization as either regenerable or non-regenerable commonly used by many,

Table 7-4
Estimate of Sulfur Trioxide in Combustion Gases

Fuel	Excess Air wt %	Oxygen in Gas, vol %	H$_2$O in Gas, vol %	Sulfur Trioxide Expected in Gas, ppmv, with Fuel Sulfur Content of:					
				0.5 wt %	1.0 wt %	2.0 wt %	3.0 wt %	4.0 wt %	5.0 wt %
Coal	25	4	8–13	3–7	7–14	14–28	20–40	27–54	33–66
Oil	25	4	11	12	15	18	22	26	30
Oil	17	3	12	10	13	15	19	22	25
Oil	11	2	13	6	7	8	10	12	14
Oil	5	1	13	2	3	3	4	5	6

Source: Pierce (1977) except for H$_2$O in gas, which are calculated values for typical fuel compositions

Table 7-5
Sulfuric Acid Dew Point of Typical Combustion Gases

Water Vapor in Gas vol %	Acid Dew Point °C (°F) for Sulfur Trioxide Concentrations of:						
	0.5 ppmv	1 ppmv	5 ppmv	10 ppmv	25 ppmv	100 ppmv	500 ppmv
6	104 (219)	110 (230)	125 (258)	132 (270)	142 (288)	157 (315)	177 (350)
10	110 (229)	116 (240)	131 (267)	137 (279)	147 (296)	162 (323)	180 (357)
14	114 (236)	120 (247)	134 (274)	141 (286)	150 (302)	165 (328)	183 (361)

Source: Pierce (1977)

e.g., by IEA Coal Research in *FGD Handbook: Flue Gas Desulphurization Systems* (Klingspor and Cope, 1987). Here, FGD processes are categorized based on the initial SO$_2$ removal step, which reveals the greatest commonality among the processes. A detailed categorization of FGD processes following this procedure is given in **Figures 7-1a** and **7-1b**. Specific sulfur dioxide removal processes are described in the subsequent sections of this chapter, which generally follows the sequence of **Figures 7-1a** and **7-1b**.

Table 7-6 provides a comprehensive list of FGD processes identified by name (or developer) and categorized in accordance with the methodology of **Figures 7-1a** and **7-1b**. The table includes commercially important processes as well as processes that have been abandoned or not yet fully developed. Data are also given on the U.S. supplier and status of maturity of each process. **Table 7-7** provides similar information on FGD byproduct treatment processes.

Not all process types indicated in **Figures 7-1a** and **7-1b** are represented by commercial processes. In fact, a list of U.S. power plant FGD systems operational, under construction, or planned as of December 1987, includes only 11 different processes (See **Table 7-8**). Between 1977 and 1983, the number of operating plants increased from 29 to 114, while the number of processes employed increased by only one (Pedco, 1977; Laseke et al., 1983).

478 Gas Purification

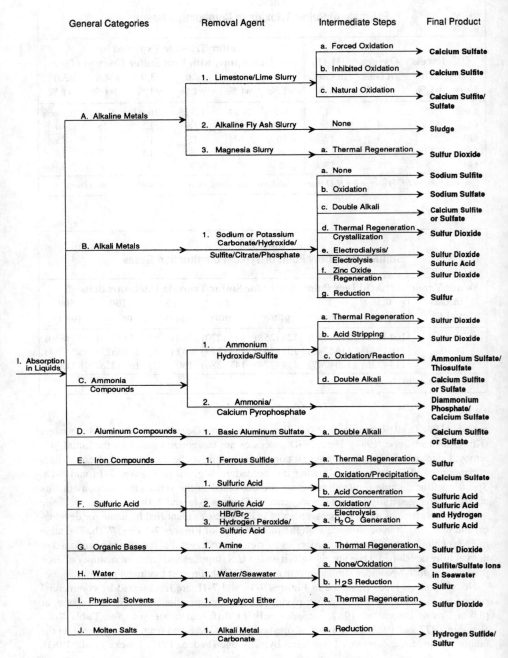

Figure 7-1a. Categorization of wet sulfur dioxide removal processes.

Figure 7-1b. Categorization of dry sulfur dioxide removal processes.

Between 1983 and 1987, the number of operating plants increased from 114 to 149, and the number of processes increased by two (PEI Associates, 1989B). In mid-1977, Japan had almost 1,000 operational FGD plants utilizing about 15 basically different types of processes (Ando, 1977). However, the growth of FGD capacity in Japan was slow after 1977 (37 plants were built in the 1978–83 time period), and virtually no plants involving processes other than limestone/lime were built (Ando, 1983). A comprehensive evaluation and status report covering 189 flue gas desulfurization processes and 24 subsystems with regard to their applicability to power plants has been published by the Electric Power Research Institute (EPRI) (Behrens et al., 1984).

(text continued on page 489)

Table 7-6
FGD Process Suppliers

The process categories and process category numbers in this table correspond to those of **Figures 7-1a** and **7-1b**.

This list has been prepared as a preliminary guide to the reader. The large number of developers and suppliers makes it impossible to be completely accurate, and the constant flux in the marketplace will assure that the list is promptly out of date. The reader is advised to use registers and regularly published lists of pollution control equipment suppliers (such as published by some trade magazines) and to consult suppliers for more accurate information.

The ownership/license status of the processes listed varies widely. Some of the processes are not considered proprietary in which case the listed suppliers have the expertise, experience, and/or willingness to offer the process commercially. Before proceeding with a project, the reader should ascertain the ownership status of the desired process. U.S. suppliers have been identified where known.

Key to Status

D	Developmental	S	Development Apparently Stopped
C	Commercial	U	Unknown
I	Commercialized, but Apparently Inactive		

Process Categories and Processes	**U.S. Supplier**	**Status**
ABSORPTION IN LIQUIDS		
IA1a **Limestone/Lime with Forced Oxidation**		
ABB Environmental Systems (C-E, Peabody)	ABB Envir. Systems	C
AirPol	AirPol	C
Babcock & Wilcox	Babcock & Wilcox	C
Bechtel (Pressure Hydrated Dolomite)	Bechtel	C
Bischoff	Joy	C
Chiyoda CT-121	Chiyoda International	C
Chuba-MKK		S
Clean Gas Systems	Clean Gas Systems	C
Deutsche Babcock	Riley Environmental	C
GEESI (Chemico based)	GEESI	C
Kawasaki (Limestone to Gypsum)	Mayernick & Associates	C
Kawasaki (Mg enhanced)		C
Kobe Steel Kobelco (CaCl) (also part of Cal-NO$_x$)		I
Mitsubishi	Pure Air	C
Mitsui (Chemico based)		C
Nippon Kokan (NKK)		C
Nippon Steel (slag sorbent)		U
Noell/KRC	Research-Cottrell	C
Procedair	Procedair	C
Riley/Saarberg-Hölter (S-H-U) (lime/clear solution)	NaTec	I

Table 7-6 (Continued)
FGD Process Suppliers

Process Categories and Processes	U.S. Supplier	Status
ABSORPTION IN LIQUIDS (Continued)		
Saarberg-Hölter (S-H-U) (limestone/ formic acid)	NaTec	C
Showa-Denko		I
Steinmüller (Chemico based)		C
Sumitomo-Fujikasui (Moretana) (ClO_2 for NO_x)		D
Thioclear [to gypsum & $Mg(OH)_2$]	Dravo	D
Thyssen		C
Ube		C
IA1b Limestone/Lime with Inhibited Oxidation[1]		
ABB Environmental Systems (C-E, Peabody)	ABB Envir. Systems	C
Babcock & Wilcox	Babcock & Wilcox	C
Riley/Deutsche Babcock	Riley Environmental	C
GEESI	GEESI	C
Research-Cottrell	Research-Cottrell	C
Thiosorbic	Dravo	C
UOP	Wheelabrator	C
IA1c Limestone/Lime with Natural Oxidation[1]		
ABB Environmental Systems (C-E, Peabody)	ABB Envir. Systems	C
AirPol	AirPol	C
American Air Filter		I
Anderson 2000	Anderson 2000	C
Bischoff	Joy	C
Babcock & Wilcox	Babcock & Wilcox	C
Riley/Deutsche Babcock	Riley Environmental	I
GEESI (Chemico)	GEESI	C
Lewis Rotary Drum	Dravo	I
Mitsubishi	Pure Air	C
Nippon Kokan (NKK)		U
Procedair	Procedair	C
Research-Cottrell	Research-Cottrell	C
Thiosorbic	Dravo	C
Thioclear [Gypsum & $Mg(OH)_2$]	Dravo	D
Thyssen (CEA)		I
UOP	Wheelabrator	C
Weir (Magnesium-enhanced)	Kellogg	I
IA2 Alkaline Fly Ash Slurry		
ABB Environmental Systems (Peabody)	ABB Environmental	C
Bechtel	Bechtel	C
CEA/ADL		I
Simmering-Graz-Pauker AG (Vienna)		C

[1]Produces sludge that may require treatment for disposal.

Table 7-6 (Continued)
FGD Process Suppliers

Process Categories and Processes	U.S. Supplier	Status
ABSORPTION IN LIQUIDS (Continued)		
IA3a Magnesium Oxide[2]		
Bischoff (Acid Stripping and Thermal Regeneration)		C
Clean Gas Systems	Clean Gas Systems	C
GEESI (Chemico) (Thermal Regeneration)	GEESI	I
Grillo (Thermal Regeneration)		I
Kawasaki Magnesium Hydroxide to $MgSO_4$ Discharge		C
Magnesium Hydroxide Process (MHP) to $MgSO_4$	IHI	C
Mitsui Mining (Thermal Regeneration)		U
Onahama-Tsukishima (Thermal Regeneration)		U
Raytheon (formerly United Engineers and Constructors) (Thermal Regeneration)	Raytheon	C
Ube Magnesium Hydroxide to $MgSO_4$ Discharge		C
IB1a Sodium Carbonate/Sodium Hydroxide		
ABB Environmental	ABB Environmental	C
Arthur D. Little (ex-CEA/ADL)	Arthur D. Little	C
AirPol	Airpol	C
Anderson 2000	Anderson 2000	C
Clean Gas Systems	Clean Gas Systems	C
IHI-TCA		U
Kawasaki		C
Kurabo		I
Kureha		U
Oji		U
Ontario Hydro (FMC)[1]	Ontario Hydro	I
Procedair	Procedair	C
Showa Denko		U
Tsukishima-Bachco		I
UOP	Wheelabrator	I
IB1b Sodium/Potassium Salt with Oxidation		
Passamaquoddy Recovery Scrubber	Passamaquoddy	C
Sumitomo-Fujikasui (Moretana) (ClO_2 for NO_x)		C
IB1c Sodium Salt with Lime or Limestone Double Alkali		
Arthur D. Little (ex-CEA/ADL) (lime)	Arthur D. Little	C
Arthur D. Little (ex-CEA/ADL) (limestone)	Arthur D. Little	S
AirPol	AirPol	C
Anderson 2000 (lime)	Anderson 2000	C
Asahi (limestone/oxidation; chelate for NO_x)		D
Buell (lime)	GEESI	I

[1]Licensed for industrial applications to Advanced Air Technology.
[2]Separate process converts SO_2 to sulfur or acid.

Table 7-6 (Continued)
FGD Process Suppliers

Process Categories and Processes	U.S. Supplier	Status
ABSORPTION IN LIQUIDS (Continued)		
Kawasaki		U
Kureha (sodium acetate/lime)		I
Kureha-Kawasaki (limestone/oxidation)		C
Ontario Hydro (FMC) (lime)[1]	Ontario Hydro	I
Ontario Hydro (FMC) (limestone)[1]	Ontario Hydro	S
Showa Denko-Ebara (limestone/oxidation)		U
Tsukishima (limestone/oxidation)		U
Zurn (lime)	Zurn	
IB1d Sodium Salt with Thermal Regeneration[2]		
Aquaclaus Phosphate (sodium phosphate)	Stauffer	I
ELSORB (sodium phosphate)	Elsorb	D
Tung (organic extraction)	Raycon R & D	D
Wellman-Lord (sodium carbonate)	John Brown (Davy)	C
IB1e Sodium Salt with Electrolytic/Electrodialysis Regeneration[1]		
IONICS	Ionics	S
SOXAL (Allied/Aquatech)	Allied	D
IB1f Sodium Salt with Zinc Oxide Regeneration[1]		
Zinc Oxide (soda/zinc oxide)		D
IB1g Citrate with Reduction		
Fläkt-Boliden (sodium citrate)	ABBFläkt	I
U.S. Bureau of Mines (sodium citrate)	Stauffer	I
U.S. Bureau of Mines (potassium citrate)	Pfizer	S
IC1a Ammonia with Thermal Regeneration[1]		
Cominco Exorption	Cominco	I
IFP/Stackpol 150		S
Simon-Carves		S
IC1b Ammonia with Acid Stripping[1]		
ABS	TVA	S
Cominco	Olin	C
IC1c Ammonia/Ammonium Sulfate, Sulfite & Thiosulfate		
ATS Technology	Coastal Chem, Inc.	C
GEESI	GEESI	D
Nippon Kokan (NKK)		C
NONOX (multiple byproducts)	Intelligent Resources	D
Tampella (sulfite liquor)		C
Ube		I
Walther		C
IC1d Ammonia, Lime Double Alkali		
Kurabo		S
Nippon Kokan (NKK)		D
SCRA		S

[1]Separate process converts SO_2 to sulfur or acid.

Table 7-6 (Continued)
FGD Process Suppliers

Process Categories and Processes	U.S. Supplier	Status
ABSORPTION IN LIQUIDS (Continued)		
IC2 Ammonia, Calcium Pyrophosphate		
Pircon-Peck (IIT)		D
ID1a Aluminum Sulfate/Lime or Limestone Double Alkali		
Dowa		I
ICI Steam Stripping		I
IE1a Ferrous Sulfide with Thermal Regeneration		
Sulf-X	PENSYS	U
IF1a Sulfuric Acid with Gypsum Byproduct		
Chiyoda CT-101 (CT-102 uses same SO_2 section)		I
Showa Denko		U
IF2a Bromine with Sulfuric Acid & Hydrogen Byproducts		
ISPRA	Ferlini/General Atomics	D
IF3a Hydrogen Peroxide to Sulfuric Acid		
Noell KRC	Noell	D
IG1a Amine with Thermal Regeneration[1]		
Asarco	Asarco	C
Cansolv (Union Carbide Canada)	Union Carbide	D
Dow	Dow	I
Lurgi Sulphidine		C
NOSOX (Ethanolamine Glutarate)	Monsanto	S
TVA (melamine)		D
UCAP	UOP	D
IH1a Seawater		
Fläkt-Hydro	ABB	C
Bechtel	Bechtel	D
Bischoff	Joy	D
II1a Physical Solvent		
Linde	Lotepro	C
IJ1a Molten Salts		
Rockwell	ABB Fläkt	S
SPRAY DRYER (Spray Dryer and Duct Spray Dryer)		
IIA1a Soda to Sodium Sulfite/Sulfate		
Aqueous Carbonate Open-Loop (rotary or nozzle)	ABB	I
Babcock & Wilcox (nozzle)	Babcock & Wilcox	D

[1]Separate process converts SO_2 to sulfur or acid.

Table 7-6 (Continued)
FGD Process Suppliers

Process Categories and Processes	U.S. Supplier	Status
SPRAY DRYER (Spray Dryer and Duct Spray Dryer) (Continued)		
Koch		C
Niro Atomizer (rotary)	Joy	I
Teller (nozzle)	Research-Cottrell	C
IIA1b Soda with Molten Carbonate Reduction to H_2S^1		
Aqueous Carbonate Closed-Loop	ABB	I
IIA2 Lime to Calcium Sulfite/Sulfate		
Spray Dryer		
ABB (Fläkt, Rockwell, Carborundum, C-E) (rotary)	ABB	
Babcock & Wilcox (nozzle)	Babcock & Wilcox	C
Buell/Anhydro (rotary)	GEESI	C
Joy/Ecolaire (rotary or nozzle)	Joy	I
Lodge-Cottrell (low-speed rotary)	Lodge-Cottrell	C
Procedair	Procedair	C
Niro Atomizer (rotary)	Joy	C
Research-Cottrell/Komline-Sanderson (rotary)	Research-Cottrell	C
Wheelabrator (rotary or nozzle)	Wheelabrator	C
Duct Spray Dryer		
CZD	Bechtel	D
E-SO$_x$	Babcock & Wilcox/EPA	D
HALT (Hydrate Add'n @ Low Temp)	Dravo Lime, U.S. DOE	D
HYPAS	EPRI	U
In-Duct	GEESI	D
Gas Suspension Absorption		
Air Pol GSA	Airpol	C
DRY SORPTION		
IIIA1 Dry Lime/Limestone		
Furnace Sorbent Injection (FSI), Lime/Limestone Injection with Multiple Burners (LIMB), Lime Injection and Hydrated Lime Injection in Upper Furnace (LI)		
ABB C-E	ABB	C
ARA (with ash reactivation for precipitator)		D
Babcock & Wilcox	Babcock & Wilcox	C
Babcock-Hitachi	Babcock & Wilcox	D
DISCUS (Inland Steel)	Research-Cottrell	C
Fossil Energy Research Corp. (lime and urea)		D
R-SO$_x$ (LI)	Electric Power Services	D

[1] *Separate process converts H_2S to sulfur or acid.*

486 Gas Purification

Table 7-6 (Continued)
FGD Process Suppliers

Process Categories and Processes	U.S. Supplier	Status
DRY SORPTION (Continued)		
LIFAC (Tampella)	Kaiser	C
Ontario Hydro	Ontario Hydro	D
Österreichische Draukraftwerke		C
PromiSox/Intevep	Energy & Envir Rsch	D
Riley/Research-Cottrell (lime and sodium bicarbonate)		D
Sonox (Ontario Hydro R&D)	Research Cottrell	D
Steinmüller (with polisher)		D
TAV Trocken (LIMB & LI)		D
Economizer Injection		
EI (Economizer Injection)	Research-Cottrell	D
Duct Injection		
ADVACATE (silicate enhanced)	ABB Fläkt	D
Babcock & Wilcox (also part of SOx-NOx-Rox-Box)	Babcock & Wilcox	D
Babcock-Hitachi	Babcock & Wilcox	D
Coolside	Consol/Babcock & Wilcox	D
LILAC	MHI	D
Synergistic Reactor	Aerological Resources	D
Fluidized Bed		
Babcock-Hitachi	Babcock & Wilcox	D
Lin (catalytic SO$_3$)		D
Lurgi CFB	Environmental Elements	C
Procedair	Procedair	C
Wulff		U
IIIA2 Dry Soda (Duct Injection)		
Sodium bicarbonate/nahcolite, i.e., natural sodium bicarbonate		
ABB Environmental (Fläkt)	ABB	C
EPRI (Arm & Hammer/Church & Dwight, Kerr-McGee, Multi Mineral, Natrona Resources)		C
NaTec	NaTec	C
Solvay		U
Wheelabrator	Wheelabrator	C
Sodium sesquicarbonate (Na$_2$CO$_3$ • NaHCO$_3$ • H$_2$O)/trona, i.e., natural sodium sesquicarbonate		
EPRI (FMC)	FMC	D
EPRI (Cominco)	Cominco	D
IIIA3a Dry Metal Oxide[1]		
Dept. of Energy Solid Oxide (copper oxide)		D
Exxon (copper oxide)		S
Houdry (Air Products) (copper & silica oxides)		S

[1]Separate process converts SO$_2$ to sulfur or acid.

Table 7-6 (Continued)
FGD Process Suppliers

Process Categories and Processes	U.S. Supplier	Status
DRY SORPTION (Continued)		
Sorbtech (formerly Sanitech) (magnesium oxide)	Sorbtech	D
Shell (copper oxide)	UOP	S
Showa-Denko (activated magnesia)		S
IIIA4a Alkalized Alumina[1]		
NOXSO	NOXSO	D
U.S. Bureau of Mines		S
IIIB1 Carbon Adsorption[1]		
Babcock-Hitachi	Babcock & Wilcox	I
Chemibau-Reinluft	Gilbert	S
EPDC Activated Char		D
Hugo Peterson (Steinmüller)		C
Mitsui/Bergbau-Forschung (Carbo Tech)	GEESI	C
Steag	Conradus Assoc., Ltd	C
Sulfacid	Lurgi	C
Sulfocarbon		D
Sumitomo		C
Takeda		D
Unitika		S
Westvaco	Westvaco	S
IIIB2a Non-Reactive Adsorbent with Thermal Regeneration		U
GAS PHASE CONVERSION		
IVA1a Catalytic Oxidation		
Bayer Double Contact (Farbenfabriken)		C
Cat-Ox	Monsanto Enviro-Chem	C
DESONOX	Degussa	D
Electrochemical Membrane	U of GA	D
SNPA	Topsoe	C
WSA-SNOX (Haldor Topsøe)	ABB Environmental	C
IVA1b Ammonia to Ammonium Sulfate, Nitrate		
E-Beam (radiation-induced chemical reaction)	Ebara	D
ENEL Pulse-Energization		D
Karlsruhe Electron Streaming Treatment		D
IVB1a Electrochemical Conversion to Sulfur		
IGR/Helipump Electrochemical Cell	IGR Enterprises	D
IVB2a Catalytic Reduction		
Parsons FGC	Ralph M. Parsons	D
Chevron	Chevron	S
Ontario Research Foundation		D
Ruthenium/Alumina		D

[1]Separate process converts SO_2 to sulfur or acid.
Source: Siegfriedt et al. (1993)

Table 7-7
FGD Byproduct Treatment Process Suppliers

Process Categories and Processes	U.S. Supplier	Status
SULFUR DIOXIDE TO SULFURIC ACID		
Allied		I
CIL		C
Monsanto	Monsanto	C
Nitrosyl-Sulfuric Acid		C
SULFUR DIOXIDE REDUCTION TO SULFUR		
Allied Chemical (natural gas)	Allied Chemical	C
ASARCO (natural gas)	ASARCO	S
RESOX (coal)	Foster-Wheeler	D
Texas Gulf Sulfur (methane)		S
Topsøe	Haldor Topsøe	C
CALCIUM SULFITE TO GYPSUM		
Sulfuric Acid Digestion (Japan) (Double Alkali)		U
Kawasaki		C
CALCIUM SULFITE TO SULFUR		
Process Calx (converts sludge to CaS or FeS)		D
CALCIUM SULFITE TO HYDROGEN AND GYPSUM		
Hark (electrolysis)	Ecolomics	D
CALCIUM BYPRODUCT TREATMENT		
Amax Resource Recovery Systems, Inc.		D
Calcilox	Dravo Lime	I
Cefill	Cementa (Sweden)	U
Chemfix Controls	Nat'l Environmental	C
Environmental Technology Corp.		D
Marston Associates		D
Ontario Liquid Waste Disposal Limited		D
Poz-O-Tec	Conversion Systems	C
Stabilisat	Steinmüller (Germany)	U
TERRA • CRETE	Sludge Fixation Tech.	D
Werner Pfleiderer Corporation		U
Wirtschaftsgut	Knauf (Germany)	U
SODIUM BYPRODUCT TREATMENT		
Fersona (with ash to form briquettes)		U
Sinterna	Industrial Resources	U
(Na_2SO_4 with acidic ferric solution to form very insoluble double salt)		

Source: Siegfriedt et al. (1993)

Table 7-8
U.S. Power Plant Flue Gas Desulfurization Systems as of December 1987

Type System	Operational	Under Construction	Planned
Limestone, Wet (Non-Regenerative)	67	4	22
Lime, Wet (Non-Regenerative)	30	—	4
Lime/Alkaline Fly Ash, Wet	13	—	—
Lime, Spray Dryer	14	2	6
Lime, Hydrated—Duct Injection	1	—	—
Magnesium Oxide (Regenerable)	3	—	—
Sodium Carbonate, Wet (Non-Regenerative)	7	1	1
Sodium/Lime Double Alkali (Non-Regenerable)	6	—	—
Sodium Sulfite (Wellman-Lord, Regenerative)	7	—	—
Sodium Carbonate Spray Dryer	1	—	—
Trona Duct Injection	—	—	1
Undecided	—	—	19
Total Number of Plants	149	7	53

Source: PEI Associates (1989B)

(text continued from page 479)

Present trends in selecting FGD processes are indicated by the data in **Table 7-9.** This table shows the recent FGD technology selections (as well as some key design data for the systems) resulting from implementation of Phase I of the Clean Air Act Amendments of 1990. None of these systems are included in **Table 7-8,** which predates the 1990 Clean Air Act Amendments.

IEA Coal Research's *FGD Installations on Coal-Fired Plants* (Vernon and Soud, 1990) compiles data gathered from around the world on over 500 FGD installations. The report shows that wet scrubbers using calcium-based sorbents are the most widely used. Increasingly, processes that produce gypsum are favored. Use of spray dryers and sorbent injection is growing in the U.S. and Europe, especially on small units, although their non-usable byproduct may hinder further expansion. Despite their potentially high-value byproducts, regenerable processes have achieved only limited use. New processes, especially those combining SO_2 and NO_x removal, are continually being developed. However, experience indicates that only a small portion of these technologies will achieve widespread commercial use. IEA Coal Research's *FGD Handbook* (Klingspor and Cope, 1987) also provides information on the major types of FGD systems in use and planned.

The economics of FGD systems are site-specific and should be evaluated on a case-by-case basis. Nevertheless, some idea of the overall economics of flue gas desulfurization can be gained from **Table 7-10** which provides data on 34 different FGD processes compiled from two reports issued by the Electric Power Research Institute. This organization has emerged as the major compiler of cost data for FGD systems. Care should be taken in using these data since costs are strongly affected by the assumed bases, including scope of items

Table 7-9
FGD Technology Selections for Phase I of the 1990 Clean Air Act Amendments (As of June 1992)

	Number of Stations/Units	No. of Stations @ Design SO$_2$ Removal Efficiency, % (Refer to note 1)	Total MW$_e$	Number of Units with Absorber Module Size in % of Unit Size and in MW$_e$ per Module	Absorber Material
Wet Limestone with Forced Oxidation	11/20	With O.A. 1 @ 98% 1 @ 97% 1 @ >95% 6 @ 95% 2 @ 93% Without O.A. 7 @ 95% 2 @ 93% 1 @ 92% 1 @ 90%	10,941	8 w/ one 100% module each 2 w/ two 67% modules each 3 w/ two 50% modules each 2 w/ five 50% modules total 1 w/ three 50% modules total 2 w/ three 40% modules each 2 w/ three 33% modules each (100 to 650 MW$_e$ each)	Refer to note 2.
Wet Limestone with Inhibited Oxidation	1/1	91	650	1 w/ two 67% modules total (436 MW$_e$ each)	Ferralium 225 Duplex Stainless Steel
Wet Lime, Magnesium-Enhanced (Thiosorbic Lime)	2/5	95 & 98	4,520	3 w/ one 100% module each (640 MW$_e$ each) 2 w/ six 20% modules each (217 MW$_e$ each)	317L & 317LM Stainless Steels

Notes:
1. O.A. stands for organic acid. Organic acids are specified for four stations and increase the removal efficiency by up to 6%.
2. Various materials are specified for the absorbers of these wet limestone systems: four stations propose to use unclad alloys (three 317L, LM or LMN stainless steels and one Hastelloy C276), two clad steel (one with Hastelloy C276 and one unspecified), two carbon steel wallpapered with Hastelloy C276, one a combination of solid and wallpapered Hastelloy C276, and two rubber lined carbon steel.
3. While the above Title IV/Phase I units account for much of the FGD business for U.S. suppliers during this period, there were a significant number of awards for new and

included in capital costs, unit costs (and credits) assumed for calculating operating costs, assumed on-stream time, and the time frame of the estimate. Many recent U.S. awards for limestone systems have been considerably below the EPRI database values due to the use of single 100% absorber modules, high velocity absorbers, no reheat, and no bypass, and simplified byproduct disposal in addition to very low profit margin, all of which differ from the EPRI database. It is also difficult to draw conclusions about new processes by comparing costs with those of proven systems because of the large uncertainties in cost estimates for processes still in the developmental stage.

A wealth of information is available for the prediction of the costs of large FGD systems. This information has been mostly developed/funded by EPRI and the EPA. The following references provide capital and operating cost information:

- EPRI CS-3342 (Keeth et al., 1983) and EPRI GS-7193 (Keeth et al., 1991B, 1992) provide the cost data that are summarized in **Table 7-10.** Costs from the earlier report are higher than the more recently reported data, and costs from both reports are high based on marketplace activity in the late 1990s.
- EPRI CS-3696 (Shattuck et al., 1984) provides a manual procedure for calculating retrofit FGD system costs.
- EPRI CS-5408-CCM (Stearns, 1987) covers a computerized version of the procedure outlined by Shattuck et al. (1984).
- EPRI GS-7525-CCML (Keeth et al., 1991A) describes the computer model used in preparing the data in EPRI GS-7193 (Keeth et al., 1991B, 1992).
- Sopocy et al. (1991) assembles a number of EPRI computer programs into a single package that can evaluate the applicability of various SO_2 control technology options for a given application, determine costs of the technologies (fifteen in 1991), and evaluate proposals. It can also simulate wet limestone/lime processes.
- EPA/600/S7-90/022 (Maibodi et al., 1991) presents a computer model developed by the U.S. Environmental Protection Agency to estimate costs and performance of coal-fired utility boiler emission control systems. The model, which is based on user supplied data, generates a material balance and an equipment list from which capital investment and revenue requirements are estimated. The model covers a number of conventional and emerging technologies.
- EPA/600/S7-90-91 (Emmel and Maibodi, 1991), PB91-133322 (Emmel and Maibodi, 1990), and EPA/600/7-88/014 (Radian, 1988) provide EPA cost estimates for specific coal-fired electric generating plants.
- Czahar et al. (1991) provide a handbook to aid in least-cost planning of emission control and acid rain compliance measures required by utilities.

SELECTION CRITERIA

Factors considered in the selection of sulfur dioxide removal systems vary with the type of system, but some important parameters are

- gas flow rate (size)
- inlet and outlet sulfur dioxide concentrations
- installed cost
- types, quantities, qualities, and availabilities of sorbents, water, steam, and power

(text continued on page 494)

492 Gas Purification

Table 7-10
Comparative Economics of Processes for Desulfurizing Power Plant Flue Gas

	Capital Costs, $/kW				Levelized Busbar Costs, mills/kWh (Current Dollars)[1]			
	A	B	C	D	A	B	C	D
	Dec 1982	Dec 1982	Jan 1990	Jan 1990	Dec 1982	Dec 1982	Jan 1990	Jan 1990
	2 @ 500 MW$_e$	2 @ 500 MW$_e$	300 MW$_e$	300 MW$_e$	2 @ 500 MW$_e$	2 @ 500 MW$_e$	300 MW$_e$	300 MW$_e$
	4.0% S	0.48% S	2.6% S	2.6% S	4.0% S	0.48% S	2.6% S	2.6% S
	90% Rem	70% Rem	90% Rem	90% Rem UON	90% Rem	70% Rem	90% Rem	90% Rem UON
Ref. Data Base								
Process	New	New	New	Retrofit	New	New	New	Retrofit
Limestone—F.O. (Disposal Grade)	177	—	166.2	216.2	15.8	—	11.5	13.0
Limestone—F.O. (Wallboard Grade)	—	—	183.7	243.4	—	—	11.5	—
Limestone—I.O.	—	—	169.2	234.6	—	—	11.2	—
Limestone—DBA	—	—	163.5	211.8	—	—	11.3	—
Limestone—Formic Acid (S-H-U)	130	—	145.5	189.2	15.7	—	10.3	—
Limestone—N.O.	175	110	—	—	17.9	8	—	—
Limestone—N.O. Dual Alkali	162	—	152.8	198.8	15.6	—	10.5	—
Limestone—Dowa	173	—	—	—	14	—	—	—
Chiyoda Thoroughbred 121	139	—	146.4	190.2	13.5	—	9.6	—
Pure Air Cocurrent	—	—	146.6	195.9	—	—	10.2	—
Limestone—F.O. Bischoff	—	—	176.5	229.4	—	—	11.8	—
Limestone—F.O. Noell-KRC/RC	—	—	182.0	236.6	—	—	11.9	—
Limestone F.O. NSP Bubbler	—	—	176.7	229.7	—	—	10.9	—
Lime—Magnesium-Enhanced	—	—	145.8	189.8	—	—	11.6	—
Lime—Dual Alkali	147	—	149.7	203.0	17.1	—	11.3	—
Lime—Conventional	163	111	—	—	19.5	7.4	—	—
Lime Spray Dryer	—	—	129.8	166.3	—	—	10.5	11.3
Tampella LIFAC	—	—	—	221.2 @80% Rem	—	—	—	13.9 @80% Rem
Lurgi CFB	—	—	—	137.1	—	—	—	9.6

Table 7-10 (Continued)
Comparative Economics of Processes for Desulfurizing Power Plant Flue Gas

	Capital Costs, $/kW				Levelized Busbar Costs, mills/kWh (Current Dollars)[1]			
Ref. Data Base	A Dec 1982 2 @ 500 MW$_e$ 4.0% S 90% Rem	B Dec 1982 2 @ 500 MW$_e$ 0.48% S 70% Rem	C Jan 1990 300 MW$_e$ 2.6% S 90% Rem	D Jan 1990 300 MW$_e$ 2.6% S 90% Rem UON	A Dec 1982 2 @ 500 MW$_e$ 4.0% S 90% Rem	B Dec 1982 2 @ 500 MW$_e$ 0.48% S 70% Rem	C Jan 1990 300 MW$_e$ 2.6% S 90% Rem	D Jan 1990 300 MW$_e$ 2.6% S 90% Rem UON
Process	New	New	New	Retrofit	New	New	New	Retrofit
Advacate Moist Duct Injection	—	—	—	84.4	—	—	—	7.0
Furnace Sorbent Injection	—	—	—	85.9 @ 50% Rem	—	—	—	8.3 @ 50% Rem
Economizer Injection	—	—	—	88.0	—	—	—	8.3
Duct Sorbent Injection	—	—	—	91.0 @ 50% Rem	—	—	—	8.5 @ 50% Rem
HYPAS Dry Injection	—	—	—	128.7 1.5% S @ 60% Rem	—	—	—	8.8 1.5% S @ 60% Rem
Duct Spray Dryer	—	—	—	74.4 @ 50% Rem	—	—	—	6.8 @ 50% Rem
Nahcolite Injection	—	27	—	—	—	8.3	—	—
Trona Injection	—	26	—	—	—	7	—	—
Wellman-Lord	274	—	217.0	278.3	25.5	—	—	13.3
Magnesium Oxide	269	—	217.8	283.1	19.4	—	—	13.7
Sulf-X	295	—	—	—	20	—	—	—
Fläkt-Boliden	391	—	—	—	28.7	—	—	—
Aqueous Carbonate	401	—	—	—	29.5	—	—	—
Conosox	431	—	—	—	44.8	—	—	—
Passamaquoddy Potassium	—	—	370.3	481.4	—	—	19.2	—

Notes:
1. Operating costs include normal operating expenses, capital charges, and credits for byproducts where applicable. Costs are for 65% capacity factor, 30-year (new) and 15-year (retrofit) levelized basis. The cost basis is per the study date. Note that the levelized costs are on a current basis; that is, they include the effect of interest and inflation.
2. Abbreviations: Rem stands for Removal (of sulfur dioxide), UON stands for Unless Otherwise Noted, F.O. stands for Forced Oxidation, I.O. for Inhibited Oxidation, N.O. for Natural Oxidation, and DBA for dibasic acid.
3. Data sources: A & B—EPRI Report CS-3342 (Keeth et al., 1983); C & D—EPRI Report GS-7193 (Keeth et al., 1991B, 1992)

(*text continued from page 491*)

- types, quantities, characteristics, and disposal options for solid and liquid byproducts/wastes
- gas side pressure drop
- operating and maintenance labor and material
- space and sparing requirements
- ease and time of installation
- new vs. retrofit
- materials of construction

For salable byproduct processes, byproduct characteristics and purity are significant considerations. For disposable byproduct processes, the availability of disposal sites, byproduct structural properties, and the landfill leachate properties are important factors. The need, or the potential need, to remove NO_x and other pollutants should also be considered in the selection process. Some processes have the capability to remove NO_x or other pollutants or can be modified to remove them. Failure to define the characteristics and availabilities of potential sorbents and wastes early in the project can lead to higher cost or poorer performance than expected. The use of a single, large module versus multiple modules or the inclusion of a spare module are also important considerations. In view of the steady improvement in scrubber reliability, there has been a growing trend toward the use of larger modules with no spares. The need for a quencher/pre-scrubber for temperature reduction, the requirement to handle failure of such a system, and/or the need to remove chlorides or particulate matter are other early considerations.

For disposable byproduct processes, an early decision must often be made regarding the use of wet limestone/lime vs. spray dryer systems. Inlet and outlet SO_2 concentrations are usually important in making such a decision. Many spray dryer FGD systems now operate or have been proposed to operate in the range of 90–95% SO_2 removal, while wet limestone/lime scrubbers are capable of removing about 98% SO_2. If the regulatory requirements escalate to 97% or 98% SO_2 removal, the FGD selection could change from a spray dryer to a wet scrubber type.

Frank and Hirano (1990) survey the potential for the production and consumption of alternative, usable, commercial byproducts in conjunction with a major reduction in national emissions of SO_2 and NO_x. They conclude that the potential byproduct yields from the U.S. acid rain control program greatly exceed available markets for the chemical products. Byproducts evaluated in the study include gypsum, sulfuric acid, ammonium sulfate, ammonium sulfate/nitrate, and nitrogen/phosphorous fertilizer. Henzel and Ellison (1990) present a review of past, present, and potential future disposal practices and commercial FGD byproduct utilization. They indicate that the only discernable trend is the production of usable gypsum by wet FGD systems. The 1990 Clean Air Act Amendments may create a need for disposal sites, which tend to be expensive and scarce and which could in themselves be environmental problems. Systems that produce usable byproducts are expected to become more important in the future as the disposal option becomes less viable.

The SO_2 gas produced by many regenerable processes can be converted in an auxiliary plant into any of several byproducts, including liquid SO_2, H_2SO_4, and elemental sulfur. The marketability of these products depends on local demand and economic factors. Transportation distances and transportation methods, i.e., pipeline, rail or road, are important factors in an economic analysis (Giovanetti, 1992A). Usually sulfuric acid is much more marketable than elemental sulfur, and elemental sulfur is more marketable than liquid sulfur dioxide. However, elemental sulfur is the least costly to store and transport. *EPRI Report CS-3696* provides a decision logic approach for selecting a process for retrofit situations that takes many of the above factors into account (Shattuck et al., 1984). **Table 7-11** gives typical quantities of sorbents required and byproducts produced on a pound of SO_2 removed basis for several FGD processes.

It should be noted that FGD systems are not the only method of controlling flue gas SO_2 emissions. Other potential methods are fuel cleaning, switching, and blending; unit retirement; purchase of SO_2 emission allowances (in the U.S.); and the use of other technologies such as atmospheric fluidized bed combustion, pressurized fluidized bed combustion, gasification with fuel gas clean-up, etc. Only FGD systems are discussed in this chapter.

Table 7-11
Typical Quantities of Sorbents and Byproducts for Various FGD Processes[1]
(Pounds per pound of SO_2 removed)

Quantities of Sorbent Required		Quantities of Byproduct Produced	
Limestone (Wet Process)[2]	1.83	Gypsum Byproduct[8]	3.15
Lime (Wet Process)[3]	1.02	Inhibited Oxidation Byproduct[9]	2.88
Hydrated Lime (Wet Process)[3]	1.35	Natural Oxidation Byproduct[10]	4–7+
Mg-Enhanced Lime (Wet Process)[4]	1.08	Lime Spray Dryer Byproduct[11]	3.24–3.69
Lime (Spray Dryer Process)[5]	1.20–1.47	Sulfuric Acid, 98.5%	1.55
Soda Ash (Wet Process)[6]	1.66	Elemental Sulfur	0.50
Sodium Hydroxide (Wet Process)[7]	2.50	Sodium Sulfite/Sulfate[12]	2.03
Ammonia	0.53	Ammonium Sulfate	2.03

Notes:
1. *Quantities of sorbent and byproduct are calculated based on stoichiometric equations and assumptions given below. Most sorbents contain some water that has not been included. Quantities will vary with the quality of the sorbent, the presence of other acid species such as HCl and HFl in the flue gas, and other factors.*
2. *Dry limestone with 6% inerts and 1.10 Ca/S ratio.*
3. *Dry lime with 10% inerts and 1.05 Ca/S ratio.*
4. *Dry magnesium-enhanced lime with 10% inerts, 5% magnesium oxide, and 1.05 Ca/S ratio.*
5. *Dry lime with 5% inerts and 1.3–1.6 Ca/S ratio.*
6. *Dry 99.8% pure soda ash with aqueous sodium salts as the byproduct.*
7. *50% concentration sodium hydroxide (water included) with aqueous sodium salts as the byproduct.*
8. *With 6% inerts in limestone and 10% moisture, (no fly ash or lime).*
9. *With 6% inerts in limestone and 20% moisture, without fly ash and lime added.*
10. *Naturally oxidized, wet, from limestone, without fly ash and lime added.*
11. *With 5% inerts and 20% moisture, 1.3–1.6 Ca/S ratio, fly ash omitted.*
12. *Dry salts only. Typical byproduct from the wet soda processes is a dilute solution, about 10–15% salt.*

ALKALINE EARTH PROCESSES

Limestone/Lime Process

The limestone/lime process is currently predominant in power plant flue gas scrubbing and has been the subject of numerous studies and publications. The large number of reports by EPRI covering both limestone/lime systems as well as other FGD systems are particularly valuable.

In the limestone/lime process, flue gas is contacted with an aqueous slurry of limestone or lime. Sulfur dioxide in the gas reacts with the slurry to form calcium sulfate (and sulfite). These compounds are collected as a relatively inert byproduct for disposal, and the purified gas is discharged to the atmosphere (after passing through a mist eliminator). A flow diagram is shown in **Figure 7-2**. For purposes of this discussion, the use of slurries of alkaline fly ash or special reactive lime-based sorbents will be considered as modifications of the basic limestone/lime process.

The process is believed to have originated in the U.S. with Eschellman (1909), who patented a method of purifying burner gases using milk of lime (a slurry of lime in water).

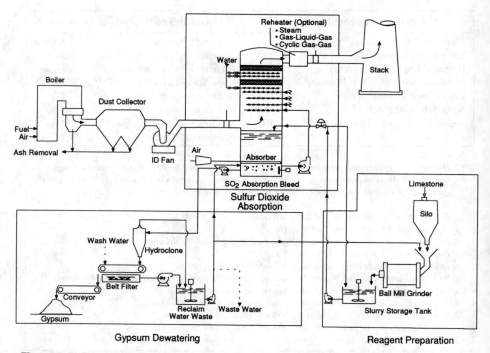

Figure 7-2. Flow diagram, SO_2 removal by limestone process with forced oxidation. (*Kneissel et al., 1989*)

The history of the development of the process through the early British plants in the 1930s to large, modern installations has been traced by Marten (Anon., 1977).

The earliest commercial applications of the limestone/lime process were in London, England. The first unit at the Battersea Power Station was put into operation in 1931. This was followed by improved units at Bankside, Swansen, and Fulham. The initial process was primarily a once-through water wash, using a small amount of chalk slurry added to the natural alkalinity of the Thames River. In 1938, the Fulham Power Plant was the first to use recycle in the process. This process is basically the same as the modern limestone/lime systems.

A modification of the process, which was installed on several United States power plants in the early 1970s, involves the injection of limestone into the furnace followed by the scrubbing of flue gas with a lime slurry (Miller, 1976; Jonakin and McLaughlin, 1969). This process appeared to offer the advantage of in situ calcination of limestone to lime and two-stage contacting (dry plus wet). Early operations in the U.S., however, encountered numerous operational problems. A simple limestone or lime wet scrubber technique **(Figure 7-2)** has been generally preferred for more recent installations.

Another early process "improvement" which has not been favored in recent plants is the concept of combining fly ash particulate removal with the SO_2 removal scrubber. This scheme offers a very large potential for cost savings by eliminating the need for an electrostatic precipitator or fabric filter. However, it also introduces a number of problems. First, the draft fan is located downstream of the wet scrubber where the fan is subject to corrosion and solids deposition (balance) problems. Second, the chemical and physical operating characteristics of the wet scrubber system are affected by the addition of the fly ash particulate. A further advantage of a separate dry ash removal step is that the dry ash can be sold or mixed with the wet byproduct to produce a product more suitable for disposal. It is not likely that there will be a change from the present usage of electrostatic precipitators (ESPs) and fabric filters for particulate removal ahead of the FGD system due to the increasingly stringent particulate regulatory requirements. High particulate removal efficiencies cannot be achieved economically by venturis or other wet scrubbers, but can be met by electrostatic precipitators and fabric filters. In addition, a very high pressure drop is required across a high efficiency venturi, which means a large power requirement for the fan.

Unlike U.S. utilities, many German power suppliers favor the use of wet, vertical, axial fans for induced draft (I.D.) service. There are several reasons for this: duct and fan arrangements are less costly and ductwork is simplified and takes less space, which can be particularly helpful in retrofit situations. To protect the fans from corrosion, the stator is covered with soft rubber, and the wetted surfaces of blades and the impeller are made of corrosion resistant material. Impellers may also be coated with hard rubber (McIlvaine, 1989).

Process Description

Today, there are three main variations of the wet limestone/lime FGD process: limestone, lime, and magnesium-enhanced lime. Wet limestone and lime systems simply use limestone and lime slurries, respectively, as the absorbents. Magnesium-enhanced lime systems use a special lime that can be made by calcining either a limestone naturally containing sufficient magnesium to produce 3 to 8% magnesium oxide in the lime, or by adding magnesium carbonate or dolomite to a limestone that does not contain the necessary amount of magnesium carbonate, and then calcining the mixture. The calcining process that converts calcium carbonate to calcium oxide (lime) also converts the magnesium carbonate to magnesium oxide, which "enhances" the lime.

There are three sub-categories for these main categories, which are identified according to the byproduct formation step. The sub-categories are referred to as natural, forced, and inhibited oxidation. Oxidation in wet limestone or lime FGD systems refers to the conversion of calcium sulfite hemihydrate (called calcium sulfite) to calcium sulfate dihydrate, i.e., gypsum (called calcium sulfate). Natural oxidation occurs if nothing is done to force or inhibit oxidation of the "naturally" formed mixture of calcium sulfite and calcium sulfate. In limestone systems, forced oxidation is accomplished by sparging (bubbling) air into the slurry either in the bottom of the absorber vessel (in situ oxidation) or in an external tank (ex situ oxidation) to oxidize the calcium sulfite to calcium sulfate. The forced oxidation byproduct is mostly calcium sulfate. Inhibited oxidation is accomplished by adding chemical additives to minimize the formation of calcium sulfate, causing most of the byproduct to be calcium sulfite. It has been found that both forced and inhibited oxidation offer advantages over natural oxidation. The primary advantage of both is a decrease in gypsum scaling in the absorber. Another benefit can be better dewatering with consequent improved landfill properties. **Table 7-12** identifies the types of oxidation used with limestone and lime scrubbers.

Natural oxidation levels can range from 10% for high sulfur coals to almost 100% for very low sulfur coals (Saleem, 1991A). High levels of SO_2 in the flue gas tend to decrease percent oxidation; whereas, high levels of oxygen increase oxidation. High fly ash concentrations increase oxidation due to the catalytic effect of some metallic oxides present in the fly ash.

At oxidation levels above about 15–20%, gypsum scaling normally occurs and can cause plugging of vessel internals (Babcock & Wilcox, 1992B). Below this range, co-precipitation of calcium sulfite and calcium sulfate limits the gypsum scaling as described in the Basic Chemistry discussion which follows this section. Above about 65–70% oxidation, the slurry is subsaturated in gypsum and gypsum scaling does not generally occur. Calcium sulfite has a hard crystalline structure and forms what is called "hard" scale that is difficult to remove. Calcium sulfate (gypsum) forms a soft scale that is easier to remove. Gypsum scaling has probably been the largest cause of absorber downtime in U.S. FGD applications. Gypsum precipitation occurs when the scrubbing liquid becomes sufficiently supersaturated with respect to gypsum, and can occur through either crystal growth or nucleation. At the higher relative saturations and in the absence of an adequate supply of seed crystals, nucleation dominates, causing uncontrolled formation of nuclei, resulting in scaling on whatever surfaces are available. As long as a high concentration of gypsum solids is maintained in the slurry, crystal growth dominates, and scaling does not occur (Moser and Owens, 1991).

Forced oxidation (in limestone systems) can increase oxidation to calcium sulfate to well over 95%. In fact, one supplier of forced oxidation systems guarantees 99.5+% conversion to calcium sulfate and considers operation at 95% a result of process chemistry imbalances (Klingspor, 1993). The calcium sulfate so formed is precipitated in the absorber sump/reaction tank as gypsum, provided sufficient time and seed crystals are available. This reduces the amount of dissolved sulfite returned to the absorber, and minimizes the possibility of sulfite oxidation and sulfate scale deposition on equipment surfaces.

Forced oxidation is not required for scale control with lime systems (Gogineni and Maurin, 1975) since sulfate scaling is controlled either by the naturally occurring seed crystals or by co-precipitation of sulfate with sulfite as described in the Basic Chemistry section. Forced oxidation may also be used to oxidize calcium sulfite when alkaline fly ash is the primary source of alkali. The large commercial FGD system at the Colstrip Power Station in Montana, which uses a mixture of fly ash and lime as sorbent, has consistently produced an efflu-

Table 7-12
Status of Oxidation Modes in Limestone/Lime FGD Systems[1]

	Type of FGD System		
Oxidation Mode	**Limestone**	**Lime**	**Magnesium-Enhanced Lime**
Forced	Most widely used combination in the world today.	No large systems of this type were found. Several large systems of this type by S-H-U were converted to limestone based on economics. There are smaller systems, e.g., on municipal solid waste incinerators.	In the U.S., a process called Thioclear, has been pilot-plant tested by Dravo. Gypsum and magnesium hydroxide are produced. In Japan, Kawasaki has a similar process.
Inhibited	Newer and used to a lesser extent than forced oxidation.	Newer and not widely used.	Used on only two plants. Wider usage is expected in the future as inhibited oxidation has been tested on many magnesium-enhanced lime FGD systems.
Natural	Widely used with early FGD systems, but no longer widely used.	Not widely used. At one station, lime/fly ash systems produce a highly oxidized byproduct without the use of forced oxidation.	Use has been primarily regional in the U.S. Ohio Valley. The majority of the magnesium-enhanced lime systems operate in the natural oxidation mode.

Note:
1. Based on mid-1994 market conditions.

ent approaching 100% oxidation (Grimm et al., 1978). This is accomplished by maintaining a low pH (less than 5.6), operating with a high level of suspended solids in solution (12 to 15% by weight), and providing a long residence time for slurry in a stirred tank external to the scrubber (8 to 10 hours, based on bleed rate). The Colstrip byproduct has been observed to increase in pH from about 5, as discharged from the absorber loop, to about 8, after 10 to 20 hours in the settling pond. This is believed to contribute to the self-hardening characteristic that slowly occurs with the Colstrip byproduct.

The limestone forced oxidation system is the most widely used wet scrubbing system in the world today, comprising roughly one-third of the wet scrubbing systems. Limestone is used because it is inexpensive, while forced oxidation is preferred because it reduces the scaling potential within the scrubber, enhances dewatering capability, and produces a byproduct suitable for either landfill or sale as gypsum. Gypsum is usable in wallboard plants, agriculture, and the cement industry. Although it is not typical to sell the gypsum in the U.S., the scale-controlling benefits of forced oxidation permit greater scrubber availability, and the enhanced dewatering capability reduces the waste disposal area requirement, which is attractive in congested areas (Telesz et al., 1990).

Converting from natural to forced oxidation has been reported to increase limestone utilization and/or SO_2 removal efficiency or possibly lower L/G (Burke et al., 1990A). Others, however, have found little effect (Klingspor, 1993).

Inhibited oxidation also has attractions. Its principal benefit is controlling the deposition of gypsum scale by minimizing the formation of calcium sulfate. A side benefit of low oxidation levels is the possible growth of larger calcium sulfite crystals, thus yielding better dewatering. Therefore, the inhibited oxidation system enjoys the benefits of low waste disposal quantity as well as scale control. Although inhibited oxidation systems are limited to the production of a disposable byproduct, they have lower power consumption and only moderately higher chemical consumption than forced combustion systems (Telesz et al., 1990).

Most of the large lime systems in the U.S. are the magnesium-enhanced type due to the particularly beneficial effects of magnesium on lime systems. Magnesium-enhanced lime scrubbers have not enjoyed the worldwide popularity of the limestone forced oxidation system, but they have been quite popular in the U.S. Ohio Valley, where the lime is often delivered by barge. There are over 8,000 MW_e of magnesium-enhanced lime scrubbers in operation or start-up. Most are located in a beltway from Pittsburgh, Pennsylvania, to Evansville, Indiana, but there are also magnesium-enhanced lime scrubbers operating on Units 1–3 at the Four Corners Plant of Arizona Public Service. The Ohio Valley magnesium-enhanced lime scrubbers use a reagent naturally containing approximately 5% MgO. The Four Corners Units 1, 2, and 3 use a locally blended lime product to achieve the same results (Telesz et al., 1990).

Several patents relating to the use of magnesium oxide as an additive to lime scrubbing systems have been obtained by the Dravo Corporation, Pittsburgh, Pennsylvania (Selmeczi, 1975A, C). Their proprietary system is offered as the Thiosorbic Flue Gas Desulfurization process. Pilot plant and commercial experience with the process are described by Selmeczi and Stewart (1978).

In magnesium-enhanced lime FGD systems, magnesium sulfite in solution enhances SO_2 removal and the system can operate at high efficiencies with much lower liquid-to-gas ratios than a limestone system. For such systems, the L/G is typically less than 40 gpm/1,000 acfm vs. 67 to greater than 100 for a typical limestone system. Additional benefits are that absorber scaling is minimal, clarifier overflow can be used for washing mist eliminators, and the system is easy to control. On the negative side, magnesium-enhanced lime systems produce a slurry with very fine calcium sulfite crystals, which, relative to gypsum, are more difficult to dewater. Large thickeners and filter installations are required, and the filtercake tends to liquify. Typical cake solids content is 45% to 50%; in some cases, there is not enough fly ash available to make the mixed fly ash and filtercake dry enough for easy handling (Laslo and Bakke, 1984). It is therefore often necessary to add large amounts of lime to the filter cake to accelerate the pozzolanic reaction and make the material suitable for landfill. However, some improvement in dewatering characteristics can be obtained by the use of inhibited oxidation.

Forced oxidation will not resolve these problems. Also, in-loop forced oxidation will oxidize the magnesium sulfite used to enhance SO_2 removal and thereby destroy the benefits of using magnesium-enhanced lime (Bakke, 1985). To alleviate these problems, Dravo is developing a new magnesium lime process called the Thioclear process. This process uses magnesium hydroxide as the sorbent and produces gypsum and magnesium hydroxide byproducts with an external oxidation process (Benson et al., 1990).

Typically, the magnesium-enhanced lime process has a lower capital cost, but higher operating cost than limestone forced oxidation. This leads to an advantage for the magnesium-enhanced lime process under any conditions that reduce the total sorbent cost. These conditions include low sulfur coal, short remaining plant life, low capacity factor, low lime cost, and small unit size (Ireland and Ogden, 1991).

The selection of a limestone or a lime FGD system depends on many factors in addition to cost. **Table 7-13** gives a comparison of features that are relevant to the selection of limestone or magnesium-enhanced lime as the sorbent. Note that the comparison is based on magnesium enhanced lime rather than straight lime due to the advantages of magnesium enhancement.

The basic components of the limestone/lime system are an absorber vessel for contacting the gas with the absorbent slurry, a mist eliminator to remove entrained moisture from the cleaned gas, a reaction tank (either external or part of the absorber sump) for the slurry where the chemical reactions can proceed, a dewatering system to remove byproduct solids from the liquid absorbent, and a fresh sorbent feed system. Usually, a particulate removal system is required upstream of the absorber. In Japan and Europe, reheaters are frequently used to raise the temperature of the cleaned stack gas exiting the system above the dew point. Early U.S. practice was to use steam or bypass-gas reheat systems. Currently, the usual U.S. approach is to use a "wet stack," rather than a reheater. Pumps, blowers, and controls are, of course, necessary accessories to the process. **Figures 7-3** and **7-4** are simplified drawings of a spray tower absorber and a tray tower absorber, respectively. Many of the major system components are visible in the photograph **(Figure 7-5)** of a scale model of a large limestone process FGD system.

Key items which must be evaluated in the design of limestone/lime systems include limestone vs. lime, byproduct oxidation (forced, natural, or inhibited), sorbent selection, use of additives, water balance including chloride concentration (open- or closed-loop water balance), need for and scope of a waste water treatment system (especially with FGD systems capable of good dewatering), absorbent cycle design (slurry concentration, recycle rate), absorber type, byproduct dewatering (e.g., thickener vs. hydrocyclones, vacuum filters vs. centrifuges), byproduct processing and handling techniques, disposal options, mist eliminator design and operation, wet stack/reheater design, and materials of construction. These items and typical FGD process problems are discussed in subsequent sections of this chapter.

Basic Chemistry

When sulfur dioxide dissolves in water, a portion of it ionizes according to the following equations:

$$SO_{2(g)} = SO_{2(aq)} \tag{7-1}$$

$$SO_{2(aq)} + H_2O = H^+ + HSO_3^- \tag{7-2}$$

$$HSO_3^- = H^+ + SO_3^{2-} \tag{7-3}$$

The solubility of sulfur dioxide in pure water in equilibrium with pure gases is given in **Figure 7-6**, which is based on the data of Parkison (1956). As indicated by the equations, the amount of sulfur dioxide absorbed by an aqueous system can be increased by reducing the

(text continued on page 506)

Table 7-13
Comparison of Wet Limestone/Magnesium-Enhanced Lime System Key Features

Feature	Wet Limestone	Wet Magnesium-Enhanced Lime
Sorbent Cost and Quantity Required	Limestone costs significantly less than lime. Ca/S ratio can be 1.1 or less.	Lime contains more calcium per pound of sorbent than limestone with correspondingly lower transportation and storage costs. Ca/S ratio can be 1.05 or less.
Sorbent Storage Requirements	Limestone can be stored in open piles, enclosed to prevent materials handling problems where freezing can occur.	Lime is shipped and stored as a powder. Lime can absorb water forming calcium hydroxide, giving off heat and caking. Transportation must be in closed containers and storage must be in covered silos.
Sorbent Preparation Requirements	Limestone must be finely pulverized to react at a reasonable rate: from 80% of the particles passing through a 200 mesh to 95% through a 325-mesh screen. A milling system with a high power requirement is needed to grind the limestone.	Lime must be slaked.
Process Control	Process response to changes in SO_2 concentration is slower than with a Mg-enhanced lime system due to the time it takes $CaCO_3$ to dissolve.	Process response to SO_2 concentration changes is more flexible and forgiving than for a limestone system due to the buffering effect of the magnesium sulfite.
Absorption	With natural oxidation, gypsum scaling in the absorber is a concern. Use of forced oxidation or the newer inhibited oxidation reduces scaling. Chlorides slow dissolution of limestone, which tends to reduce dissolved alkalinity, causing SO_2 removal efficiency and pH to drop. Forced oxidation helps to offset both effects. Organic acid additives can increase SO_2 removal and decrease L/G. (Bakke, 1985)	There is little, if any, tendency for scaling in the absorber. Chlorides reduce the dissolved alkalinity, requiring more magnesium to be added to the process. Chlorides can cause corrosion. Lime is more chemically reactive than limestone and, therefore, has a higher dissolution rate, a higher operating pH, and a lower L/G ratio. The absorbers and recycle pumps are therefore smaller. Higher removal efficiencies are possible.
Dewatering	With natural oxidation, dewatering is poor. With forced oxidation, dewatering to 90+% solids is now common. With inhibited oxidation, dewatering is enhanced. Good dewatering in most cases requires a purge stream to prevent chloride build-up that would cause corrosion and, for wallboard gypsum, poor salability. This could necessitate a waste water treatment system. Chloride removal systems are very expensive. Chemical treatment of the waste water to reduce trace metals and suspended solids is also expensive.	With natural oxidation, dewatering is to 45–50% solids range. The dewatering equipment must be larger than for a limestone system as the smaller calcium sulfite crystals settle more slowly in a thickener and do not filter as effectively. Inhibited oxidation improves dewatering. (Magnesium enhancement degrades the dewatering properties of a lime system.)

Table 7-13 (Continued)
Comparison of Wet Limestone/Magnesium-Enhanced Lime System Key Features

Feature	Wet Limestone	Wet Magnesium-Enhanced Lime
Byproduct Disposal	Mixing natural oxidation byproduct with fly ash and lime, forced oxidation, and possibly inhibited oxidation can make the byproduct suitable for landfill. Treatment of byproduct has a further advantage where fly ash cannot be sold in that it renders the ash essentially non-leachable. Gypsum may be salable. If gypsum is produced, fly ash is not needed for byproduct treatment, and may be salable.	Mixing natural oxidation byproduct with fly ash and lime or producing an inhibited oxidation byproduct is necessary to make the product suitable for landfill.

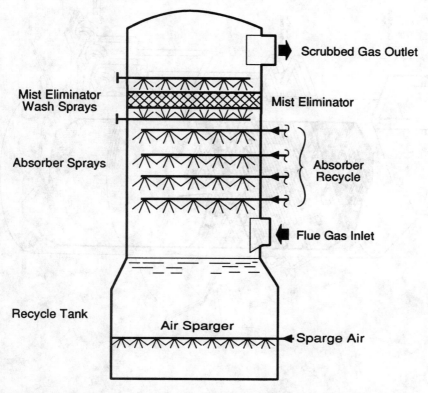

Figure 7-3. Schematic of spray type absorber tower for *in situ* forced oxidation. (*GEESI, 1992*)

504 *Gas Purification*

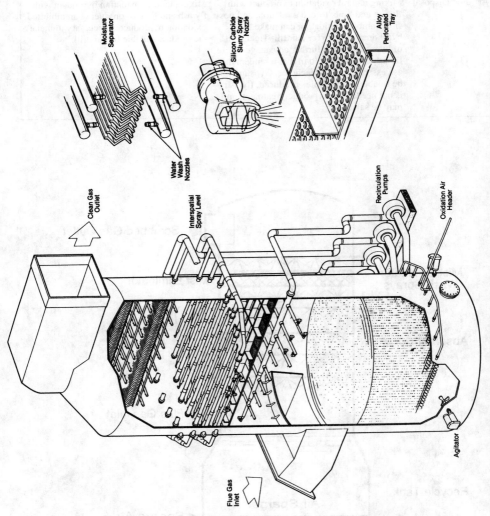

Figure 7-4. Schematic of tray tower absorber. (Babcock and Wilcox, 1992A)

Figure 7-5. Photograph of a scale model of the limestone process FGD plant on Unit 2 of Alabama Electric Cooperative's Tombigbee Station. Components from left to right are inlet ductwork; I.D. fans; common recycle tank, spray tower absorbers, one behind the other; slurry preparation tank; limestone silo; and limestone ball mill (in enclosure). *Courtesy of Peabody Process Systems, Inc.*

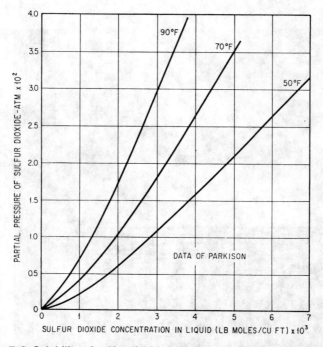

Figure 7-6. Solubility of sulfur dioxide in pure water. *Data of Parkison (1956)*

(text continued from page 501)

hydrogen ion concentration or by removing HSO_3^- or SO_3^{2-}. The addition of calcium oxide or carbonate to the system accomplishes both of these actions. In the presence of lime and limestone, the following reactions occur:

Lime Dissolution:

$$Ca(OH)_{2(s)} = Ca(OH)^+ + OH^- \tag{7-4}$$

$$Ca(OH)^+ = Ca^{2+} + OH^- \tag{7-5}$$

Limestone Dissolution:

$$CaCO_{3(aq)} + H^+ = Ca^{2+} + HCO_3^- \tag{7-6}$$

Reaction with Dissolved SO_2:

$$Ca^{2+} + HSO_3^- = CaSO_{3(aq)} + H^+ \tag{7-7}$$

$$Ca^{2+} + SO_3^{2-} = CaSO_{3(aq)} \tag{7-8}$$

$$CaSO_{3(aq)} + \tfrac{1}{2}H_2O = CaSO_3 \cdot \tfrac{1}{2}H_2O_{(s)} \tag{7-9}$$

Oxidation:

$$HSO_3^- + \tfrac{1}{2}O_2 = SO_4^{2-} + H^+ \tag{7-10}$$

$$Ca^{2+} + SO_4^{2-} = CaSO_{4(aq)} \tag{7-11}$$

$$CaSO_{4(aq)} + 2H_2O = CaSO_4 \cdot 2H_2O_{(s)} \tag{7-12}$$

Coprecipitation:

$$Ca^{2+} + (1-x)SO_3^{2-} + xSO_4^{2-} + \tfrac{1}{2}H_2O = Ca(SO_3)_{1-x}(SO_4)_x \cdot \tfrac{1}{2}H_2O_{(s)} \tag{7-13}$$

Liberation of CO_2 (from limestone):

$$CO_3^{2-} + H^+ = HCO_3^- \tag{7-14}$$

$$HCO_3^- + H^+ = H_2CO_{3(aq)} \tag{7-15}$$

$$H_2CO_{3(aq)} = CO_{2(g)} + H_2O \tag{7-16}$$

Equilibrium conditions for reactions 7-2 and 7-3 (sulfite/bisulfite distribution) and for reactions 7-14 and 7-15 (carbonate/bicarbonate distribution) are defined by the curves of **Figure 7-7** (Head, 1977). Equations for calculating equilibrium constants for reactions 7-1 through 7-5 and 7-8 as a function of temperature, at zero ion strength, have been compiled by Pasiuk-Bronikowska and Rudzinski (1991).

Solubility Product Constants. The dissolution and precipitation of solid species involved in equations 7-9, 7-12, and 7-13 are governed primarily by solubility although supersaturation, which can cause gypsum scaling, occurs under some conditions. For relatively insoluble salts, solubility is best expressed in terms of the solubility product, which, for salts containing one cation and one anion, is simply the product of the activities of the ions in the saturated solution expressed as moles per 1,000 grams of solution. At the low concentrations of interest, the activity is approximately equal to the molar concentration, moles per liter (M). Typical solubility product values are given in the **Table 7-14.**

The solubility product constants are useful in evaluating the effect of changing the concentration of other components in the solution. For example, increasing the concentration of sulfite ion by adding a soluble sulfite salt will cause the concentration of calcium ions in the solution to decrease (at saturation) in order to maintain a constant solubility product. As indicated by **Figure 7-7,** decreasing pH has the effect of decreasing the concentrations of SO_3^{2-} and CO_3^{2-} in solution, thereby increasing the concentration of calcium ions at saturation. As a result, the solubility of calcium sulfite, for example, varies from about 0.001 M at a pH of 6 to over 0.1 at a pH of 4.5 (Hudson, 1980).

SO_2 Absorption Mechanism. The individual steps involved in the removal of SO_2 from gas streams by the limestone/lime process may be summarized as follows:

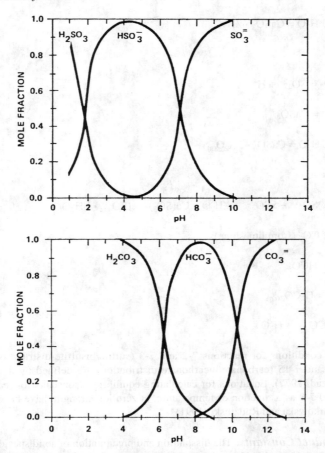

Figure 7-7. Bisulfite-sulfite and bicarbonate-carbonate distributions as a function of pH.

Salt	Solubility Product at Temperature Noted, M^2
$CaSO_3 \cdot \frac{1}{2}H_2O$	2.76×10^{-7} (40°C)
	6.8×10^{-8} (50°C)
$CaSO_4 \cdot 2H_2O$ (gypsum)	1.20×10^{-6} (40°C)
	2.20×10^{-5} (50°C)
$CaCO_3$ (calcite)	0.99×10^{-8} (15°C)
	0.87×10^{-8} (25°C)
$MgSO_3 \cdot 3H_2O$	5.07×10^{-5} (50°C)
$MgSO_3 \cdot 6H_2O$	8.14×10^{-5} (50°C)

Table 7-14
Typical Solubility Product Values

Source: Faist et al. (1981) and Hudson (1980)

1. Transfer of SO_2 in the gas phase to the gas/liquid interface
2. Dissolving SO_2 into water at the interface
3. Ionization of dissolved SO_2 (Note: Hydrolysis of dissolved SO_2 to form sulfurous acid molecules in solution is often included as a step, but there is no strong evidence of its existence (Jolly, 1966; Nannen et al., 1974)
4. Transfer of H^+, HSO_3^-, and SO_3^{2-} ions from the interface into the liquid interior
5. Dissolving and ionization of $Ca(OH)_2$ or $CaCO_3$ to form Ca^{2+}
6. Reaction of Ca^{2+} with SO_3^{2-} and HSO_3^- to form $CaSO_3$ in solution
7. Precipitation of $CaSO_3 \cdot \frac{1}{2}H_2O$
8. Dissolving O_2 in water at the interface
9. Transfer of dissolved O_2 from the interface into the liquid interior
10. Oxidation of sulfite ions to sulfate ions
11. Reaction of Ca^{2+} with SO_4^{2-} to form $CaSO_4$ in solution
12. Precipitation of $CaSO_4 \cdot 2H_2O$
13. Co-precipitation of $Ca(SO_3)_{1-x}(SO_4)_x \cdot \frac{1}{2}H_2O$

Research indicates that the rate controlling mechanisms for SO_2 absorption are usually step 1, gas phase mass transfer of SO_2; step 4, liquid phase mass transfer; and step 5, dissolving $CaCO_3$. In the ideal case, steps 4 and 5 are sufficiently fast that gas phase mass transfer is controlling. There is evidence that this case is approached at low flue gas SO_2 concentrations and relatively high solution pH. High SO_2 removal efficiencies are attained under these conditions.

The efficiency of SO_2 removal is generally lower at high SO_2 concentrations in the gas (greater than about 3,000 ppm) because, at high concentrations, the quantity of SO_2 in solution at the gas-liquid interface exceeds the amount of available alkalinity, causing a decrease in pH. It is apparent from equation 7-2 that a decrease in pH (increase in hydrogen ion concentration) will drive the reaction to the left, increasing the amount of dissolved SO_2 in solution and decreasing the rate of SO_2 absorption.

Absorption Enhancement. The problem of pH reduction near the interface can be mitigated by the use of an additive, which buffers the solution so that its pH is not drastically reduced by the addition of an acid former (in this case SO_2). Typical buffers that have been employed in wet limestone FGD plants are weak organic acids such as DBA (a mixture of dibasic acids containing adipic, glutaric, and succinic acids) and formic acid. DBA is more effective on a molar basis than formic acid because it is dibasic with buffering pHs of 4.3 and 5.5. Formic acid buffers at a single pH of 3.75 (Stevens et al., 1991).

Jankura et al. (1991) give the following equations for the buffering action of a typical organic acid (adipic):

Dissociation:

$$\text{H-Ad-H} + 2OH^- = Ad^{2-} + 2H_2O \tag{7-17}$$

Sulfur Dioxide Absorption:

$$2SO_2 + 2H_2O = 2H^+ + 2HSO_3^- \tag{7-18}$$

Buffering Effect:

$$Ad^{2-} + 2H^+ = H\text{-}Ad\text{-}H \tag{7-19}$$

The symbol Ad represents the adipate radical. Adipate ions in solution combine with hydrogen ions to form adipic acid molecules that remain in solution but are not ionized, thereby preventing the hydrogen ion concentration (acidity) from increasing as SO_2 is absorbed and improving the SO_2 removal efficiency.

Formic acid acts in a similar manner in stabilizing the pH; however, it differs from other carboxylic acids used as buffering agents in several important respects: (1) it is less expensive than other pure acids on a weight or molar equivalent basis, (2) it can be purchased, stored, and added as a neutral salt such as sodium formate, and (3) it has the unique ability to inhibit sulfite oxidation while improving sulfur dioxide absorption efficiency (Moser et al., 1990).

Magnesium has been found to have a significant effect on sulfur dioxide absorption, particularly in lime-based systems. According to Benson (1985), this effect is due to high alkalinity in magnesium-enhanced scrubbing liquors caused by both an increased concentration of SO_3^{2-} and the formation of $MgSO_3°$ ion pairs. The increased concentration of SO_3^{2-} is the result of the much higher solubility product of magnesium sulfite compared to calcium sulfite. See **Table 7-14**. An additional benefit of magnesium in lime-based systems with moderate oxidation is a reduction in gypsum scaling potential. This occurs because the increased concentration of SO_3^{2-} decreases the concentration of Ca^{2+} (to maintain a constant solubility product for calcium sulfite), thereby decreasing the product of the concentrations of Ca^{2+} and SO_4^{2-} to a level below that required to precipitate gypsum scale.

Solids Deposition. The precipitation of insoluble salts is a key chemical reaction in limestone-lime scrubbing. It affects scaling of equipment, dewatering of byproduct, and marketability/disposability of the byproduct. Gypsum scaling has been a major industrial problem in the past; but, according to Moser and Owens (1991), the current understanding of process chemistry should be adequate to essentially eliminate scaling as a problem in future operations. These authors describe the following approaches to minimize scaling in limestone-lime systems:

1. Forced oxidation to convert most of the sulfite to sulfate in conjunction with maintaining an adequate inventory of gypsum seed crystals in the slurry.
2. Inhibition of oxidation to minimize the formation of sulfate to a level that can be removed by coprecipitation of calcium sulfate with calcium sulfite.
3. Modification of crystal formation by the use of an additive that alters the supersaturation at which gypsum nucleation occurs.

The first two approaches are widely used, while only one commercial application of the third approach is reported.

Gypsum is known to form supersaturated solutions in water. At concentrations above 1.3 to 1.4 times saturation, nucleation occurs resulting in scale deposition on any available surfaces. When large quantities of gypsum crystals are suspended in the solution (above about 70% oxidation level), supersaturation does not occur, and gypsum preferentially deposits as crystal growth on the existing crystals. This characteristic can be utilized to minimize scaling by injecting air into the slurry and by recycling of slurry containing gypsum crystals. Injecting air forces oxidation of sulfite to sulfate ensuring an adequate supply of calcium sulfate, and recycling slurry provides an excess of crystal surfaces as sites for deposition.

An alternative approach for the prevention of calcium sulfate scale deposition is to operate with a solution less than saturated with respect to $CaSO_4$. This can be accomplished by continuously removing $CaSO_4$ from solution as a co-precipitate with calcium sulfite at a rate which is equal to that at which it is formed by oxidation. Up to about 15 mol % sulfate can be incorporated into the calcium sulfite crystals as a solid solution (Radian Corp., 1976). Therefore, if oxidation can be maintained at a rate less than 15% of the rate of formation of new sulfite, absorber operation free of gypsum deposition is possible. A convenient technique to assure this mode of operation is oxidation inhibition.

Oxidation Inhibition Chemistry. The most commonly used additive to inhibit oxidation is thiosulfate. This ion was initially added to FGD systems in the form of sodium thiosulfate solution. However, in late 1987, tests at an operating utility plant demonstrated that elemental sulfur could be used at a cost which is only about 20% of the cost of adding sodium thiosulfate (Moser et al., 1990). The sulfur is converted to thiosulfate by the following reaction. Conversion efficiency is on the order of 50%:

$$S + SO_3^{2-} = S_2O_3^{2-} \tag{7-20}$$

Lee et al. (1990) found that alkaline hydrolysis of sulfur under slaker/lime tank conditions in lime-based wet FGD plants resulted in much more effective conversion of sulfur to thiosulfate than is possible in limestone-based systems.

Thiosulfate is believed to inhibit sulfite oxidation by reacting with free radicals generated in the chain reactions involved in sulfite oxidation. The chain reactions are catalyzed by transition metal ions such as Fe_3^+, and the use of a chelating agent such as ethylenediaminetetraacetate (EDTA) to remove the metal ions has been shown to augment the oxidation inhibition properties of thiosulfate (Maller et al., 1990).

Chloride Effects. The effect of a high chloride ion concentration in the scrubbing liquor is of increasing importance because of the trend toward tightly closed water loops in FGD systems. The chloride may originate from the coal (and be introduced as HCl in the flue gas) or enter with the make-up water. The former source ends up as calcium chloride in the scrubber liquor, while the make-up water is more apt to contribute sodium chloride. Technically, both sources provide the same anion (chloride) but different cations. Because of its widespread presence in scrubbing liquors, extensive experimental work has been performed on the effects of chloride on SO_2 absorption (Rader et al., 1982; Downs et al., 1983; Laslo et al., 1983; Laslo and Bakke, 1983; and Chang, 1984).

The 1984 Chang report is quite comprehensive and presents the results of a series of tests conducted in a 0.1 MW_e FGD pilot plant which employs a three-stage turbulent contact absorber (TCA). In addition to experimental data, the report includes a discussion of the chemical reactions and mass transfer phenomena involved. The theory shows, for example, that increasing the Cl^- ion concentration from 200 to 100,000 ppm (in the form of $CaCl_2$) reduces the total alkalinity (HCO_3^-, $CaHCO_3^+$, SO_3^{2-}, and $CaSO_3$) by over 40%. The decreased alkalinity would be expected to decrease the rate of reaction of dissolved SO_2, causing a decrease in the liquid phase mass transfer rate and, therefore, a reduction in SO_2 removal efficiency.

The experimental results are in general agreement with the theoretical model. The observed effects of chloride concentration on SO_2 removal efficiency for a limestone-based system scrubbing a gas containing about 2,500 ppm SO_2 are summarized in **Figure 7-8.** The results show that the cation associated with the chloride has a major influence on system performance. Calcium chloride has the most severe effect on SO_2 removal efficiency in either the natural or forced oxidation modes. Magnesium and sodium chlorides have various effects as shown in the figure. However, both have slightly beneficial or zero effect at moderate chloride ion concentrations.

In addition to decreasing SO_2 removal efficiency, the accumulation of calcium chloride in the limestone slurry causes a decrease in pH, an occasional decrease in slurry settling rate, and an increase in gypsum scaling potential. DBA was found to be very effective in counteracting the effect of calcium chloride on SO_2 removal efficiency. However, as the concentration of calcium chloride increases in the slurry, higher DBA concentrations are needed to maintain a constant SO_2 removal efficiency.

Calcium chloride was also found to reduce the SO_2 removal efficiency and system pH in a lime-based slurry scrubbing high SO_2 concentration gas (about 2,500 ppm), but had no significant effect on low concentration gas (500 ppm SO_2) (Chang, 1984).

Computer Models. Numerous computer programs have been developed to simulate limestone/lime FGD processes and aid in the design of plants. Some of the programs are propri-

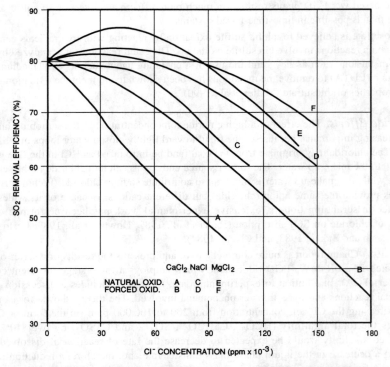

Figure-7-8. Comparison of SO_2 removal efficiency as a function of the slurry chloride ion concentration for natural and forced oxidation conditions and three sources of chloride in a limestone-based FGD system. *Data of Chang (1984)*

etary and are used for the design of licensed versions of the process. An early database model, the Bechtel-Modified Radian Equilibrium Program, (BMREP), provides activity coefficients and equilibrium constants for FGD species (Henzel et al., 1982). This program calculates solution equilibria in an overall Slurry Scrubber Model that predicts scrubber performance and absorber sump/reaction tank compositions based on process chemistry. Gage and Rochelle (1990) describe successful tests of the Slurry Scrubber Model versus a pilot plant study involving various types and grinds of limestone.

Probably the most comprehensive, available program is FGDPRISM (Flue Gas Desulfurization Process Integration and Simulation Model) recently developed by the Electric Power Research Institute (Noblett et al., 1990; Noblett et al., 1991). The model can simulate spray and tray type absorbers as well as complete systems; perform material balances and equilibrium calculations; and evaluate various alternatives such as: natural and forced oxidation, magnesium enhancement of lime systems, and dual-loop limestone systems. The program can be used as an aid to design new systems. It can also be used as a tool to analyze laboratory and pilot scale data; to evaluate proposed changes to an existing full scale system (chemical and/or mechanical modifications); and to assess vendor bids with regard to SO_2 removal capability, scaling potential, and flexibility. To obtain accurate results it is necessary to calibrate the model with existing plant data or a database from similar designs.

Sorbent Selection

The choice between lime or limestone is largely an economical one, as both reagents can normally accomplish the required SO_2 removal efficiency. **Table 7-13** compares the advantages and disadvantages of lime vs. limestone. As indicated in **Table 7-10,** limestone process operating costs are generally lower than lime and because of this and other factors, limestone is usually the preferred sorbent. The use of a fly ash slurry, on the other hand, requires that the coal being burned have an ash with a relatively high concentration of alkaline metal oxides (e.g., CaO or MgO) and a relatively low sulfur content. Generally, the coals best suited to alkaline fly ash scrubbing are the western lignites and subbituminous coals (Kaplan and Maxwell, 1977).

Limestones may be characterized as (1) high-calcium limestones containing at least 95% $CaCO_3$; (2) high-magnesium limestones which are predominately $CaCO_3$ but contain over 5% $MgCO_3$; and (3) dolomitic limestones containing $CaCO_3$ and $MgCO_3$ in approximately equal molar concentrations. The first of these categories is the best for use as uncalcined limestone, the second is less effective, and the third is generally not suitable. $MgCO_3$ is relatively unreactive and can also render some of the calcium unreactive in the form of $MgCO_3 \cdot CaCO_3$ (dolomite) crystals. However, when limestone containing $MgCO_3$ is calcined, the magnesium content is converted to a soluble, reactive form. As previously discussed, the presence of magnesium ions in the scrubbing solution greatly improves process operation.

There are no universally accepted models for choosing a particular limestone. Key factors in the limestone selection decision are delivered cost, limestone hardness, and composition. Because limestone cost is a major operating expense of a wet limestone FGD system, it is an area where substantial cost savings are possible. Recent ongoing work under EPRI sponsorship has shown that selection of a limestone or limestones for a given application can involve the following (Jarvis et al., 1991):

- Preliminary screening of limestone supplies to identify candidate limestones and to determine delivered costs and availabilities.

- Chemical analysis to determine the limestone composition: calcium carbonate, solid solution (soluble) magnesium carbonate, dolomite, and inerts (acid insolubles), etc. This information is used to make a preliminary estimate of limestone and disposal costs, to determine the impact of the inerts and dolomite on the gypsum byproduct composition, and to make a preliminary estimate of magnesium availability. Solid solution magnesium (magnesium not present as dolomite) generally improves FGD system performance. Dolomite is relatively insoluble, decreases carbonate utilization, and can render a gypsum byproduct unsuitable for wallboard manufacture. Species such as iron, silicon, and aluminum can also adversely affect scrubber operation under some circumstances.
- Grindability testing to estimate the Grindability Index and the Bond Work Index allows estimation of the equipment size, grinding energy costs, and the limestone particle distribution. It may not always be possible or desirable to grind all limestones to the same degree. Increasing the fineness of the grind results in a higher limestone utilization at constant pH. Alternatively, increasing the fineness can be used to provide higher SO_2 removal at constant limestone utilization (with increasing absorber feed pH) (Burke et al., 1990B). A grindability test procedure is described in EPRI Report CS-3612 (EPRI, 1988B).
- Reactivity measurement for input into a process performance model. Reactivity is a measure of how quickly the limestone dissolves. Dissolved sulfite, magnesium, and aluminum fluoride complexes can inhibit limestone dissolution. The effect of sulfite can be particularly significant since it is a major constituent in wet limestone slurry. Aluminum fluoride, though not often present, can be significant in quantities as low as a few ppm (Farmer et al., 1987). Its presence in a wet limestone/lime FGD system is usually associated with upsets in the operation of the upstream particulate control system. EPRI has a licensable procedure for measuring limestone reactivity (Jarvis et al., 1991).
- Magnesium availability measurement to determine the available magnesium and the total available carbonate for input to the performance model.
- Process performance modeling using EPRI's FGDPRISM computer model (or another model) and the limestone characteristics to define the SO_2 removal, the limestone utilization, and the byproduct composition.
- Economic analysis to rank the limestones.

Magnesium-enhanced lime is usually acceptable as a sorbent if the magnesium oxide content is between 3% and 8% and inerts are less than 5% with the balance being calcium oxide. Lesser amounts of magnesium oxide may be suitable with inhibited oxidation (Roden, 1992).

Additives

Additives are compounds added to the slurry of a limestone or lime FGD system to improve performance, reliability, or operating flexibility. They can reduce costs for both new and existing FGD systems. Achieving high SO_2 removal efficiencies and reduction of gypsum scaling are the two main benefits obtained with additives. Improvement in solids byproduct characteristics by inhibiting oxidation is being demonstrated at one plant and may be commercially available in the near future. Possible negative aspects of additive use in FGD systems are the effects on air, water, and solid waste quality (Moser and Owens, 1991). **Table 7-15** summarizes the uses of additives in wet limestone and wet lime FGD systems.

Table 7-15
Uses of Additives in Wet Limestone/Lime FGD Systems

Additives	Type of FGD System	
	Limestone	Lime
Organic Acids: Adipic Acid & Dibasic Acid (DBA), Badische EP-306 & EP-501, Formic Acid, Citric Acid, and Acetic Acid; Sodium & Calcium Formate.	Buffer pH	Not Used
Magnesium Oxide, Dolomitic Lime	Increase liquid-phase alkalinity	Increase liquid-phase alkalinity (Magnesium enhanced lime)
Sodium & Ammonium Thiosulfates, Sulfur, Sulfides & Polysulfides	Inhibit oxidation	Inhibit oxidation

Source: Moser and Owens (1991), Blythe et al. (1991), and Babcock & Wilcox (1992B)

Organic acids increase SO_2 removal efficiency and produce other performance enhancements in a limestone FGD system by buffering the pH of the scrubbing slurry. Organic acids are used in relatively low concentrations, e.g., adipic acid is used in concentrations of 200–1,500 ppm, DBA in concentrations of 200–2,000 ppm, and formate in concentrations of 500 to 5,000 ppm (Blythe et al., 1991; Moser and Owens, 1991). The concentrations of organic acids used in actual applications are often in the lower parts of the above ranges. Citric and acetic acids were used in the past; however, their use is no longer common.

Until recently, the major impetus for the use of organic additives has been to boost the performance of FGD systems that fail to achieve SO_2 compliance. Increases in SO_2 removal efficiency of 5 to 20% have been reported. However, in addition to promoting high SO_2 removal efficiency, organic acids may reduce the quantity of limestone required, limestone grinding requirements, the volume of byproduct produced (less solid), and the required L/G ratio. Possible secondary benefits include reduction in mist eliminator plugging, ability to operate the absorber at lower gas velocities resulting in less slurry carryover, i.e., by bypassing part of the untreated gas, and reduction in the amount of water required by lowering limestone stoichiometry. This can also lead to improved thickener performance and reduced maintenance (Babcock & Wilcox, 1992B).

Magnesium oxide and dolomitic lime (which contains magnesium oxide) cause the scrubbing liquor of a lime FGD system to hold a larger inventory of liquid-phase alkalinity in the form of sulfite ions. This allows the design liquid-to-gas ratio to be reduced for the same SO_2 removal (Moser and Owens, 1991). Magnesium sulfite concentrations of 4,000 to 10,000 ppm are used (Roden, 1992). Thiosulfate and similar additives inhibit oxidation of sulfite to sulfate in a limestone or lime system, which decreases gypsum scaling. Thiosulfate is used in relatively low concentrations, about 200 to 2,000 ppm (Moser et al., 1990).

Limestone systems are limited by the rate at which the solid dissolves. They are said to be "liquid-phase" limited. Certain species, such as magnesium (or sodium), which form more soluble sulfite salts than calcium, allow the liquor to hold a higher concentration of liquid-phase alkalinity in the form of sulfite ions. The magnesium suppresses the calcium in solution and the liquid phase alkalinity is from the $MgSO_3$. Magnesium (or sodium) allows higher SO_2 removal efficiencies. Because lime is more soluble than limestone, wet lime FGD systems can operate at a higher pH, where the buffering effects of the small amounts of sulfite present are more effective. Also, these systems operate at lower L/Gs. In the U.S., nearly all of the lime systems on utility boilers use naturally occurring magnesium-enhanced lime as the sorbent, and in a sense contain their own built-in additive. Benson (1985) provides a more detailed explanation of the role of magnesium in increasing SO_2 removal and improving reliability in magnesium-enhanced FGD systems.

At Associated Electric Cooperative's Thomas Hill Unit 3, 12,000 to 14,000 ppm of magnesium in the limestone slurry was used to meet a 1.2 lb SO_2/million Btu emissions limit, but availability was unacceptable with this additive. The additive was switched to DBA and later to DBA and sodium formate (Moser and Owens, 1991; Roden, 1992). As of 1993, Public Service of Indiana's Gibson Unit 5 was the only plant known to be using a magnesium additive (dolomitic lime) to improve the operation of a limestone system. Sulfur is used with the dolomitic lime to inhibit oxidation.

Prior to the mid-1980s, the primary technique utilized in the U.S. for combating gypsum scaling was forced oxidation. In effect, sparge air was the additive (Moser and Owens, 1991). More recently, other additives have been successfully used to combat gypsum scaling. The most common has been thiosulfate, which reduces the formation of gypsum by inhibiting the oxidation of sulfite to sulfate. If forced oxidation is not employed, use of thiosulfate/sulfur is desirable to minimize gypsum scaling (Babcock & Wilcox, 1992B). In 1992, it was reported that three full-scale, magnesium-enhanced lime FGD systems were using sulfur to inhibit oxidation for crystal modification and consequent improved dewatering (Roden, 1992).

Thiosulfate can be created in situ at reduced cost from emulsified sulfur. Emulsified sulfur is being used on about two dozen FGD systems in the U.S. A refinement is the use of small quantities of a chelating agent, ethylenediaminetetraacetate (EDTA), in conjunction with thiosulfate. This combination is being demonstrated on one full-scale FGD system (Moser and Owens, 1991; Blythe et al., 1991). An added benefit of thiosulfate addition is that, at sufficiently low oxidation levels, significant improvement in byproduct dewatering occurs due to the improvement in calcium sulfite particle morphology. At Arizona Public Service's Four Corners Station, Units 4 and 5 (1,600 MW_e total) have wet lime FGD systems. By adding thiosulfate, it was found that oxidation could be reduced from about 10% to 3–5%, allowing bypassing of the vacuum filters at virtually all conditions. Another apparent benefit associated with thiosulfate is a higher pH level for a given sorbent stoichiometry. This can result in either higher SO_2 removal efficiency or reduced sorbent consumption (Babcock & Wilcox, 1992B).

Currently, site-specific tests are often conducted to estimate the consumption and effects of additives (Moser and Owens, 1991). Some significant major process variables to examine when considering use of an additive are additive concentration, pH of the absorber feed slurry, and dissolved calcium concentration (Moser et al., 1990). Other important factors to consider in additive use are pH meter maintenance and operation, thickener operation, absorber and mist eliminator plugging, foaming (caused by soap-like impurities) in the recirculation tanks, system water balance, and corrosion of the additive feed system (Babcock & Wilcox, 1992B).

Capital costs for additive feed systems typically run less than 1% of the total FGD system capital cost, and operating costs for additives amount to about 1 to 2% of total operating costs. Benefits can partially or, in many cases, completely offset these additional costs (Moser and Owens, 1991).

Absorber System Design

The key factors affecting performance of the absorber system are the sorbent used (limestone or lime); liquid/gas ratio; gas velocity; slurry pH; SO_2 concentration in the gas; absorber design; solids concentration in the slurry; concentration of other components such as magnesium ions, chloride ions, and additives; and the degree of oxidation of the slurry. The stoichiometric ratio (moles of limestone or lime added per mole of SO_2 absorbed) is often considered to be an important design parameter, but becomes a dependent variable if the pH of the entering slurry and other design factors are fixed. Comparative values for these and other performance factors are given in **Table 7-16**.

Table 7-16
Comparative Operating Conditions for SO_2 Absorption by Limestone and Lime Slurries in a Spray Tower

Absorbent	Limestone with F.O. (Disposal Grade Gypsum)	Limestone with DBA[2], F.O. (Disposal Grade Gypsum)	Limestone with I.O. (Byproduct Fixated)	Lime with I.O. (Byproduct Fixated)
SO_2 in Feed Gas, ppmw	3,600	3,600	3,600	3,600
SO_2 Removal, %	90	90	90	90
Solids in Slurry, %	15	15	15	6
Gas Velocity, ft/sec	10	10	10	10
L/G, gal/1,000 cf	96	67	100	40
pH of Feed Slurry	5.3	5.3	5.8	6.5
Absorber Sump/Reaction Tank Retention Time, min.	6	6	6	6
Stoichiometric Ratio	1.1	1.1	1.1	1.1
Magnesium Concentration (effective ppm)	0	0	0	14,500
Chloride Concentration (ppm)	0	0	0	0
$CaSO_4 \cdot 2H_2O/(CaSO_3 \cdot 0.5H_2O + CaSO_4 \cdot 2H_2O)$, lb/100 lb	100	100	11.6	1.60
Final Byproduct Solids, wt %	85	85	80	63

Notes:
1. Abbreviations: F.O. stands for Forced Oxidation, I.O. for Inhibited Oxidation, DBA for dibasic acid.
2. DBA concentration is 1,270 ppm in the recycle slurry, or about 20 lb/ton of SO_2 removed.
3. This table should be used for comparative purposes only.
Source: Keeth et al. (1991B)

518 *Gas Purification*

In general terms, individual parameters that increase SO_2 removal are

- Increasing the liquid flow rate
- Increasing the pH of the feed slurry
- Increasing the effective magnesium ion concentration
- Increasing the chloride ion concentration if pH is held constant
- Using an additive

Effective magnesium ion concentration is defined here as the concentration in excess of that required to neutralize chloride ions. Increasing the chloride ion concentration with pH held constant is effective because it results in the addition of more limestone or lime to maintain the pH. The increase in SO_2 absorption attributed to chloride is actually the result of increasing stoichiometry. Increasing the gas flow rate and the inlet SO_2 concentration may either increase or decrease the SO_2 removal, depending on the type of system and the sorbent (Head, 1977).

Numerous scrubber designs have been proposed for the limestone/lime process. The principal requirements are low pressure drop, freedom from plugging problems, adequate contact efficiency to meet the SO_2 removal requirements, and low cost. Currently, the leading scrubber types are the spray and tray tower. Rod trays and static packed beds are used in some applications; however, venturi and mobile bed designs are no longer popular. The *FGD & DeNO$_x$ Newsletter* (1992A) and the *FGD and DeNO$_x$ Manual* (1978) provide some information on the design and use of rod scrubbers. A number of designs in current use are covered in a later section of this chapter titled "Process Variations."

While computer programs such as FGDPRISM are widely used, a simplified approach for sizing FGD absorbers is the use of an overall volumetric gas film mass transfer coefficient and a logarithmic mean driving force for the transfer of SO_2 from the gas phase to the liquid. The model assumes a linear operating line, negligible equilibrium partial pressure of SO_2 over the slurry, and atmospheric pressure operation of the absorber. The use of this approach for a GEESI limestone system spray tower is described by Saleem (1991B). The rate of SO_2 absorption in the tower is defined as follows:

$$R = G(Y_1 - Y_2) = K_g a \cdot V \cdot (Y_1 - Y_2)/\ln(Y_1/Y_2) \tag{7-21}$$

Where: R = rate of SO_2 absorption, moles/hr
G = flue gas flow rate, moles/hr
Y_1, Y_2 = absorber inlet and outlet SO_2 concentrations, respectively, moles SO_2/mole flue gas
V = effective volume of the absorber, ft^3
$K_g a$ = overall volumetric mass transfer coefficient, moles/hr-ft^3

GEESI has refined and calibrated the model on numerous commercial spray towers. Equation 7-21 can be written in terms of the SO_2 removal efficiency (E):

$$K_g a = (G/V) \cdot \ln[1/(1-E/100)] \tag{7-22}$$

Where: $E = (1 - SO_{2\,out}/SO_{2\,in}) \cdot 100$
$SO_{2\,in}$ = SO_2 in inlet gas, lb/10^6 Btu or other consistent unit
$SO_{2\,out}$ = SO_2 in outlet gas, lb/10^6 Btu or other consistent unit

Equation 7-22 can be used to predict the efficiency when the overall mass transfer coefficient for a given absorber is known. The overall mass transfer coefficient is experimentally determined from pilot plant and full-sized units. For the GEESI open spray tower, the overall mass transfer coefficient, $K_g a$, has been correlated to three variables: gas velocity, liquid density, and inlet SO_2 concentration:

$$K_g a = C \cdot U^m \cdot L^n / Y_1^p \tag{7-23}$$

Where: C = proportionality constant
U = gas velocity, fps
L = liquid spray density, gpm/ft³ of tower
m, n, p = correlation coefficients

With this model, pH and droplet size are not variables in the equations. The pH is maintained in a given range to produce a negligible SO_2 back pressure for efficient absorption. Droplet size, which is somewhat indeterminate due to significant collision and agglomeration in the spray zone, is maintained by other means (Saleem, 1991B). The effects of magnesium, chlorides, and other impurities must also be taken into account.

Limestone systems must be appropriately designed to be free of plugging and scaling. This involves a number of considerations. The reaction tank size must be selected large enough and the liquid-to-gas ratio high enough to avoid supersaturation of calcium sulfite and calcium sulfate, which can cause uncontrolled precipitation. Sufficient seed crystals must be present in the slurry to enhance the precipitation rate and to provide host sites for preferential precipitation. Retention time in the absorber sump/reaction tank must be adequate for precipitation to occur. The absorber must be designed to accommodate scaling that can occur under upset conditions. A high sorbent utilization rate is important to prevent scaling of the mist eliminators. Scaling in mist eliminators occurs as a result of excess limestone in the entrained slurry droplets reacting with residual SO_2. The reaction product forms deposits on the mist eliminator surfaces and can obstruct mist eliminator gas passages. This can lead to higher velocity through the mist eliminator, decreased mist eliminator performance, and increased pressure drop (Saleem, 1991B).

Process controls are important to maintain pH, slurry density, and water balance. SO_2 removal efficiency is maintained by pH control. Scaling and crystal growth are regulated by control of the slurry density. The rates of blowdown and maximum mist eliminator washing are functions of water balance control. These controls are normally automated and feed-forward control can be incorporated (Saleem, 1991B).

FGD system reliability involves sparing of components such as pumps, the sorbent preparation system, the dewatering system, pH and density meters, etc. With absorber modules, sparing involves use of complex flue gas path manifolds and troublesome dampers. As a result, there is a growing trend toward single absorber modules, which greatly simplifies the flue gas system. A number of very large single absorber train systems have been built, and the use of single absorber modules is common practice outside the United States (Saleem, 1991B).

Slurry Processing

Slurry from the SO_2 absorption step flows into the sump of the absorber vessel or one or more reaction tanks, where the following occur:

1. Dissolution of the added limestone or lime
2. Reaction to form and precipitate calcium sulfite and co-precipitate calcium sulfite and sulfate
3. Absorption of oxygen (in forced oxidation systems)
4. Precipitation of calcium sulfate

In wet limestone FGD systems, the complete oxidation to calcium sulfate normally requires the introduction of a step that is optimized for oxygen absorption. In this step, air is bubbled into the absorber sump/reaction tank to provide forced oxidation. Due to the low solubility of oxygen in water, forced oxidation is aided by extended liquid residence time. Increased residence time generally also increases limestone utilization (Head, 1977). Many newer limestone/lime FGD systems operate with 6 to 10 minutes residence time in the reaction tank. This is lower than used in some early systems where residence times were 10 to 40 minutes. The main difference is that current designs are based on a better understanding of the FGD chemistry.

During the 1970s, forced oxidation for gypsum production was exclusively carried out in a separate oxidation vessel. It was believed that the low pH required for efficient oxidation could best be generated in an external vessel by addition of sulfuric acid or use of flue gas. Therefore, the process of SO_2 absorption and gypsum production were considered incompatible due to the relatively high pH needed for efficient SO_2 absorption and the low pH needed for efficient oxidation. A large number of FGD plants were, therefore, built using the external forced oxidation process (Saleem, 1991A).

A better understanding of oxidation kinetics has led to the development of *in situ* forced oxidation in which both SO_2 absorption and oxidation steps are carried out in the same vessel without the need for external oxidizers and sulfuric acid addition. The sump of the absorber is the reaction tank. Because of simplicity and lower costs, *in situ* forced oxidation has become the worldwide standard for gypsum production.

Borgwardt (1977) and Hudson (1980) give the results of extensive experimental studies on oxidation in FGD systems. These results and other data from operating systems lead to the following conclusions on the requirements for effective forced oxidation by air sparging:

1. The pH in the reaction tank must be low (usually in the range of 5–5.8) to ensure nearly 100% oxidation to sulfate. The exact pH used depends on the process and site specific factors. High chloride limestone systems can have a pH as low as 4.8, and some fly ash systems a pH as low as 4. At low pH, the sulfite crystals dissolve and the resulting ions are oxidized to form gypsum. Complete sulfite oxidation results in a near zero SO_2 equilibrium vapor pressure. Therefore, SO_2 removal can be high in spite of the low pH. High limestone utilization is required to achieve the low pH. In any event, gypsum specifications usually call for less than 5% impurities in the gypsum byproduct. Since limestone usually contains a significant amount of impurities, the limestone utilization must be high to meet the gypsum purity specification.
2. The air sparging rate is usually about 3 times the stoichiometric amount to ensure high conversion of SO_3^{2-} to SO_4^{2-}.
3. Increased slurry depth increases air bubble residence time in the absorber sump/reaction tank and thereby oxygen utilization. However, the power required to inject air increases with depth. Increased depth with constant tank diameter also increases slurry residence time, which may increase oxygen utilization and gypsum crystal size.

4. As discussed in the Basic Chemistry section, it is necessary to keep the calcium sulfate relative saturation less than about 1.35 to induce precipitation on seed crystals rather than on equipment.
5. The need for agitation of the absorber sump/reaction tank contents varies with individual designs.

Suspended solid particles, which consist primarily of calcium sulfate, calcium sulfite, and unreacted absorbent, are continuously withdrawn from the absorber sump/reaction tank and processed to concentrate the solids to produce the final byproduct. With natural oxidation, the solids are typically removed by gravity settling in a thickener forming a thick product containing 35% to 45% solids. This slurry can be pumped to storage ponds or further processed by secondary dewatering with vacuum filters or centrifuges. The byproduct may be further treated to improve handling and ultimate disposal. With forced oxidation, the gypsum crystals are relatively large and primary dewatering can be accomplished by hydrocyclones followed by secondary dewatering with filters or centrifuges. Ponding or stacking is also possible. Salable gypsum, however, requires washing during the secondary dewatering step to remove soluble salts such as chlorides, thus resulting in the need for a water blowdown to purge these salts from the system (Saleem, 1991B). With inhibited oxidation, the use of thiosulfate as an additive at the wet lime FGD systems at Four Corners Station Units 4 and 5 was found to reduce oxidation of sulfite to sulfate and to also improve dewatering characteristics sufficiently to allow bypassing of the vacuum filters at virtually all conditions (Babcock & Wilcox, 1992B).

Byproduct Disposal

The characteristics of the final byproduct from wet limestone/lime FGD systems can vary widely, depending primarily on the degree of oxidation to gypsum, the method of removal from the absorption circuit, and subsequent handling procedures. **Tables 7-17** and **7-18** give some typical characteristics of the byproduct from forced and natural oxidation limestone/lime FGD plants. Insufficient published data are available to provide comparable information on byproduct from inhibited oxidation FGD systems. **Table 7-17** also includes data on the characteristics of lime spray dryer byproduct (which is generally highly oxidized) for comparison with forced oxidation wet limestone/lime FGD byproduct.

The moisture content of the filtered and stacked byproduct from a forced oxidation system is 35% (dry basis), as shown in **Table 7-17**. This high value is reported to be due to rainfall accumulation, and is not typical, as many forced oxidation systems produce a filtered and stacked byproduct containing 13–17% moisture (dry basis). For wallboard applications, low moisture content is required to minimize the amount of energy needed to evaporate the water. FGD suppliers typically claim the capability of producing 6–12% moisture content material to meet this requirement.

The forced oxidation byproduct, which is primarily gypsum, has better structural and landfill properties than the natural oxidation byproduct, as indicated by the data of **Tables 7-17 and 7-18**, and it probably also has better properties than the inhibited oxidation byproduct. Tests have shown that conversion to gypsum eliminates the liquefiable property of the naturally oxidized byproduct and permits the generation of filtercakes of very low water content. Scanning electron microscope photographs show that the oxidation process converts the crystal structure from flat plate-like $CaSO_3 \cdot \frac{1}{2}H_2O$ crystals to dense monoclinic $CaSO_4 \cdot 2H_2O$ (gypsum) crystals (Goodwin, 1978).

Table 7-17
Typical FGD Forced Oxidation (and Lime Spray Dryer) Byproduct Characteristics

Characteristic/Property	Wet Limestone/Lime FGD Byproduct (Oxidized to Gypsum)				Lime Spray Dryer with Western Coals
	Untreated Settled (Ponded)	Filtered & Stacked	Blended with Fly Ash	Stabilized	
Densities (lb/ft^3 max):					
Bulk Aerated (Poured)	—	—	—	—	36 to 60
Bulk Settled, Dry	60.5	81.5	94.0	99.2	49 to 78
Bulk Settled, Wet	96.6	110.4	108.1	115.4	71 to 104
Particle	—	—	—	—	143 to 175
Moisture (lb moisture/100 lb dry byproduct)	60.2	35.1	15.0	16.3	16 to 38 (Note 2)
Permeability, (cm/sec)	8.0×10^{-5}	1.0×10^{-5}	1.0×10^{-6}	5.0×10^{-9}	3.1×10^{-9} to 1.6×10^{-7}
Unconfined Compressive Strength, Average (psi)	12	31	66	278 @ 1 mo. 1364 @ 4 mo.	140 to 900
Leachate pH	2.8 to 12.8	2.8 to 12.8	2.8 to 12.8	2.8 to 12.8	9.7 to 12.8
Toxicity: RCRA Extraction Procedure (EP):	Note 3	Note 3	Note 3	Note 3	Non-Hazardous

Notes:
1. *Wet limestone/lime byproduct values are from Physical Evaluation of FGD Byproducts by Smith (1992A) of Conversion Systems, Inc. (CSI) except for the leachate pH and toxicity data, which are from EPRI Report FP-977, p. 7-7 & 7-8 (Duval et al, 1979). Spray dryer byproduct values are from the EPRI (1988A) Report CS-5782.*
2. *Range of optimum values for compaction. Actual optimum value depends on byproduct properties.*
3. *The determination of whether or not a particular byproduct is hazardous under the EPA definition should be made on a case-by-case basis (EPRI Report FP-977 p. 7-2 & 7-7). (Duval et al., 1979)*

At Georgia Power's Yates plant, slurry from the Chiyoda CT-121 scrubber (a forced oxidation system) is continuously withdrawn, and the crystals are separated from the water using a "stacking method" first developed in the phosphate fertilizer industry. With this technique, a dragline digs naturally sedimented gypsum from the bottom of the stack. This material is then used to build a dike around the edge of the stack. Gypsum-containing slurry flows to the diked area where gravity causes the gypsum solids to settle while clear liquid flows to a recycle water pond (*FGD & DeNO$_x$ Newsletter,* 1993B).

Byproduct from natural oxidation FGD systems tends to form a dilute sludge which presents a difficult disposal problem if not treated. The material has a high viscosity at low shear rates, which causes plugging, and tends to remain liquefied when the shear force is removed unless it is further dewatered. This makes land disposal somewhat complicated (Rossoff and Rossi, 1974; Jones, 1977). In addition, the high water content of naturally oxidized byproduct increases both the FGD system water consumption and the volume of

Table 7-18
Typical FGD Natural Oxidation Byproduct Characteristics

Characteristic/Property	Wet Limestone/Lime FGD Byproduct (Naturally Oxidized)			
	Untreated Settled (Ponded)	Filtered	Blended with Fly Ash	Stabilized
Reference (note 1)	A	B	C	D
Densities (lb/ft³ max):				
Bulk Aerated (Poured)	—	—	—	—
Bulk Settled, Dry	22 to 33	—	—	52 to 73
Bulk Settled, Wet	71 to 73	—	—	90 to 107
Particle	—	—	—	—
Moisture (lb moisture/100 lb dry byproduct)	122 to 184	100 to 150	25 to 82	47 to 73
Permeability, (cm/sec) Immediate	1×10^{-4} to 1×10^{-5}	Free Draining	7×10^{-5}	4×10^{-6} to 7×10^{-6}
After Curing	—	—	7×10^{-5}	5×10^{-6} to 1×10^{-7}
Unconfined Compressive Strength, Average (psi)	0	0	0 to 10	20 to 1140
Leachate pH	Note 2	Note 2	Note 2	Note 2
Toxicity: RCRA Extraction Procedure (EP):	Note 3	Note 3	Note 3	Note 3

Notes:
1. A and D values are previously unpublished data from Conversions Systems, Inc. (CSI) based on testing for 20 clients (Smith, 1992B); B and C values are from CSI, Sludge Disposal by Stabilization—Why? (Smith, 1977)
2. Leachate pH information is not available.
3. The determination of whether or not a particular byproduct is hazardous under the EPA definition should be made on a case-by-case basis (EPRI Report FP-977 p. 7–2 & 7–7). (Duval et al., 1979)

byproduct requiring disposal. Physical instability of this byproduct is primarily caused by small platelet or needle-like crystals of calcium sulfite hemihydrate, which settle very slowly and trap water even when filtered. Four methods are used for improving dewatering: (1) adding dry solids (i.e., fly ash), (2) using more effective dewatering equipment, (3) controlling crystal size, and (4) oxidizing to obtain gypsum (Jones, 1977). Many operators are opting for forced oxidation due to disposal difficulties with the naturally oxidized byproduct and other benefits from forced oxidation, such as reduced scaling in the absorber and possibly greater tolerance for chlorides.

There are a number of proprietary processes used to treat the naturally oxidized byproduct so that it can be used as landfill or disposed of without containment. Treatment is usually accomplished by the addition of fly ash and/or lime. Usually, both lime and fly ash are employed. Some fly ashes, such as those from lignite, naturally contain sufficient lime to adequately treat the byproduct. These treatment processes generally harden the material to improve stability at disposal sites and bind the soluble constituents to minimize leaching. In addition to improving structural properties, treatment may eliminate the need to line the disposal pit with clay (Hilton, 1991). Encapsulation of the fly ash in a low permeability material by these processes may also be of value.

Many older systems pond the natural byproduct (and coal ash). Usually, the slurried solids are pumped to the pond at a solids concentration less than 50%. The slurry either is the thickener underflow or comes directly from the scrubber, with the pond acting as the thickener/clarifier. Clarified liquor is usually returned to the scrubber (Jones, 1977). The two major environmental considerations with ponding are the potential water pollution problem associated with the soluble material and the land-degradation potential of non-settling or physically unstable solids. If the disposal site is to be reclaimed when the pond is retired, the retention of water in the byproduct is a serious concern.

Most new wet limestone/lime FGD plants in the U.S. dispose of the byproduct as landfill. It might be thought that chloride-laden water could be purged from the FGD system by raising the amount of water in the discarded byproduct. However, leachate from the landfill is usually returned to the scrubber as a means of leachate disposal. If it is not recycled, the high chloride water can become a pollution problem. In either case, there is no net benefit in increasing the amount of water in the byproduct (Smith 1992B).

The water pollution potential of FGD byproduct depends primarily on its chemical properties, but is also affected by the physical properties (e.g., permeability) related to leachate generation. These properties vary widely with the coal, type of scrubber (limestone, lime, dual alkali), amount of unreacted sorbent in the byproduct, amount of ash in the byproduct, degree of dewatering, and degree of oxidation to sulfate. Potential pollutants are soluble metals, chemical oxygen demand (COD) from the sulfites, excessive total dissolved solids (TDS), and excessive levels of other major chemical species (sulfate, chloride, magnesium, sodium). The concentrations tend to diminish with time due to the flushing out of the solubles. The long-range pollution potential, after this initial flushing takes place, is based primarily on the solubility of calcium sulfate. Calcium sulfate is significantly more soluble than calcium sulfite in neutral pH water. Lining the disposal pit can (1) attenuate the pollution migration into the ground via ion exchange or adsorption, and/or (2) greatly reduce the rate of pollution migration due to the low permeability of the liner. Compacted soil, clay or a special liner may be used (Jones, 1977). The common use of impervious membranes (landfill liners) to block and return waste liquid serves to generate a substantial quantity of collected leachate that must be recycled to the FGD system or chemically treated before discharge (*FGD & DeNO$_x$ Newsletter,* 1993B). Another concern is contact of calcium sulfite with acidic water (e.g., acidic mine drainage), which can result in SO$_2$ off-gassing.

Regardless of the method of disposal, a number of important questions must be answered in the design of a byproduct disposal site—Is the byproduct considered to be hazardous by federal, state, or local authorities? Are landfill liners and/or monitoring wells required? Is a leachate collection system required? Will any processing of the byproduct be needed at the landfill or elsewhere? Is it necessary to treat the byproduct? Valuable information on the disposal of wet FGD byproducts is given in several publications. Knight et al. (1980) provide comprehensive information on byproduct disposal. Duval et al. (1978) give information that is still current on the treatment of FGD byproduct. Garlanger and Ingra (1983A, B) evaluate engineering properties and wet stacking for the FGD byproduct from a TVA Widow's Creek unit. Smith (1977) and Samanta (1977) present information on naturally oxidized byproduct comparing it to forced oxidation product, and Yu (1991) presents information on evaluating disposal methods for gypsum.

Uses for FGD byproduct include construction of structural landfills; liners for liquid waste ponds; road bases; parking lots; and the manufacture of structural shapes, synthetic aggregates, and artificial reefs. Minnick (1983), EPRI (1987), Smith and Rau (1981), Henzel and Ellison (1990), and Smith (1992C) provide information on FGD byproduct uses. The usual U.S. practice of landfill disposal of FGD byproduct is in contrast to the practice in foreign countries. In Japan, due to the lack of natural gypsum, the gypsum produced by FGD systems is used in wallboard manufacture. In Europe, waste disposal restrictions also favor the production of usable byproducts.

Mist Elimination

The elimination of fine droplets of entrained slurry from the gas leaving the absorber has proven to be a difficult problem, primarily due to the tendency of the recovered liquid to deposit solid material on the mist eliminator surfaces causing plugging and inefficient operation. A number of points relevant to mist eliminator design and operation follow:

1. The highest gas velocities and the best mist removal efficiencies are obtained with horizontal gas flow (vertically configured mist eliminator profiles). Removal of droplets down to 8 to 15 microns in size is possible, although 20 microns is more typical. The higher pressure drop associated with the high velocities used with this type mist eliminator increases operating cost.
2. Vertical gas flow mist eliminators can provide adequate performance provided the projected net face velocity is kept at least 10% below the droplet break-through velocity. Removal of droplets down to 38 to 40 microns in size is possible with this design. Mist particles generated by the slurry sprays and mist eliminator wash system are generally in the 50–100 micron range, therefore, low carryover rates (0.01–0.05 grains/scf) are achievable. Most mist-eliminator systems are designed for vertical gas flow, including those for most scrubbers designed for compliance with Phase I of the Clean Air Act's Title IV Acid Rain provisions.
3. Actual gas velocity through the mist-eliminator section should be considered. This velocity can be as much as one-third higher than the superficial velocity in the scrubber vessel due to the cross-sectional area blanked off by mist-eliminator supports and other obstructions.
4. The gas velocity distributions entering and exiting the mist eliminator should be as uniform as possible. A good rule-of-thumb is for the maximum flow velocity to not vary from the arithmetic mean by more than ±25%. Model studies to ensure that gas velocities do not exceed the maximum should be conducted for every installation. Absorbers with trays tend to have a slightly more uniform velocity distribution entering the mist eliminator. However, the recycle sprays in both spray and tray type absorbers make the greatest contribution towards a uniform mist-eliminator inlet velocity distribution (Saleem, 1991B). Also, the severe change in gas flow direction from vertical to horizontal ahead of horizontal flow mist eliminators can make obtaining uniform velocity distributions difficult (Van Buskirk, 1992).

5. Two-stage mist eliminators (with the first stage having wider between-vane spacing to minimize scaling) are preferred. The first stage is referred to as the bulk entrainment separator; the second as the fine entrainment separator. The first stage captures most of the mist, and the second captures the entrainment from mist eliminator washing and residual droplets.
6. Chevron vanes (and to a lesser extent slat equivalent arrangements) are commonly used. Currently, various blade profiles have 1 to 4 passes (changes in gas flow direction). The majority of Acid Rain Phase I scrubbers use three- and four-pass designs and most are three-pass designs.
7. Water washing of the mist eliminator reduces scaling. Two scale reduction mechanisms have been proposed: (1) water on the mist eliminator surfaces dilutes collected slurry droplets below the saturation concentration, and (2) water dissolves/washes away previously deposited soft scale (gypsum). Full implementation of the first mechanism would require continuous or frequent washing of mist eliminators. Normally this is not possible because water low in dissolved solids (usually fresh water) is required to prevent plugging from the wash-water. This amount of water can exceed the system make-up water requirement. Regarding the second scale reduction mechanism, orienting the nozzles for impingement on all internal surfaces is difficult, and hard-scale (calcium sulfite) deposits can only be removed by manual, high-pressure water washing. This latter operation has proven to be very detrimental to the mist-eliminator profiles, especially if the vanes are plastic.
8. Wash-water systems are critical to reliable operation. Sequential washing of all mist eliminator sections should be performed on both the inlet and outlet faces (except the outlet face of the last stage) at least once each hour during operation. One manufacturer recommends washing each zone a minimum of 1–2 minutes with water sprayed at 30–35 psig pressure.
9. A wash-water spray on the exit side of the second (last) stage mist eliminator is mandatory to maintain high availability, but should operate only once per day since mist carry-over increases during this washing.
10. Full cone spray nozzles should be used with overlapping spray patterns arranged so that no areas are left unwashed.
11. Wash-water collection devices are required with vertical (horizontal gas flow) or tilted design mist eliminators. They are not typically needed for horizontal (vertical gas flow) mist-eliminators as the wash-water falls naturally into the absorber.
12. Perforated plates or other precollection devices are no longer widely applied as integral parts of mist eliminator systems.
13. Both fiberglass reinforced plastic (FRP) and thermoplastics (polypropylene) are used for mist eliminators. FRP (0.125 to 0.140 in. thick) was widely used for the U.S. Phase I scrubbers due to its ability to withstand temperature excursions in the 300° to 350°F range. In Europe, polypropylene is used with elastomeric-lined absorbers where the flue gas temperature is limited to about 190°F. Glass or talc-filled polypropylene has better properties than the polypropylene without filler. Compared to FRP, polypropylene is homogeneous throughout and not as susceptible to degradation if cracking occurs. Alloy mist-eliminator vanes have been incorporated in many installations. However, halogens in the entrained liquid can cause rapid corrosion on the normally very thin (20-gauge, approximately 1 mm, or thinner) vanes.

Liquid concentrations in the gas exiting mist eliminators are often higher than quoted by manufacturers (Jones et al., 1991). EPRI has studied the operating limits of several mist eliminator designs and has developed a mist eliminator troubleshooting guide (Rhudy, 1990). Some information on the design and selection of mist eliminators has been accumulated in the *FGD and DeNO$_x$ Manual* (1979). Hanf (1992) notes that forced oxidation in conjunction with engineered mist eliminator wash systems and washable designs results in high mist eliminator availability. In Europe, the availability for this combination has been as high as 98% for more than 35,000 hours of operation.

Wet Stacks, Flue Gas Reheat, Cooling Tower Discharge, and Condensing Heat Exchangers

Limestone/lime FGD plants generally have either a "wet stack" or an outlet flue gas reheat system. With a "wet stack," the outlet ductwork from the absorbers to the stack and the stack itself are designed to promote drop-out and collection of the water droplets from the gas stream. In a reheat system, as many droplets as possible are evaporated before they reach the stack (Fink, 1992). Many new wet limestone/lime FGD systems use a wet stack to avoid the considerable expense and problems associated with flue gas reheating.

A well-designed "wet stack" system has minimum droplet carry-over from the mist eliminators, uses sloped duct floors, has properly located and amply sized liquid collectors and drains in the outlet ductwork and stack, limits the gas velocity particularly in the stack, and has smooth outlet ductwork and stack walls to allow entrained liquid collected on the walls to flow to the drains. Most of the entrained liquid is removed in the outlet ductwork. Usually, reduced-scale flow model studies are performed to evaluate the best locations and sizes for liquid collectors and drains. EPRI's *Entrainment in Wet Stacks* (Maroti and Dene, 1982) provide some guidelines which aid in evaluating and solving wet stack entrainment problems.

With wet stacks, reentrainment can cause H_2SO_4 droplet fallout near the chimney and consequent corrosion of the plant facilities. To prevent reentrainment, velocities in the wet stack liner should be in the 30 to 60 fps range. This is substantially lower than the traditional dry stack exit velocities of 60 to 90 fps or higher. With both wet and dry stacks, the plume must be ejected clear of the top of the chimney to preclude downwash of the plume at the chimney outlet and corrosion of the chimney exterior shell. A choke can be used to increase the discharge velocity up to 110 fps to reduce the occurrence of downwash. For concrete chimneys, the liner should extend above the top of the shell to prevent the plume from contacting and damaging the concrete shell in high wind conditions. In the wet stack case, a choke liquid collector should be installed which is specially designed for the selected choke geometry.

A flue gas reheat system represents an alternative to the wet stack approach. Flue gas reheat reduces condensation and consequent corrosion of downstream equipment, improves rise and dispersion of the stack gas, and suppresses the formation of a visible plume. Flue gas reheat of 25° to 50°F is typical. Reheaters may be any one of several types: combustion, steam/water, regenerative, and bypass. Combustion reheaters and steam- or water-to-gas reheaters may be in-line, indirect, or recirculated exhaust flue gas type. Various combinations have been used such as bypass with indirect steam reheat. Regenerative reheaters can be two water-coil type heat exchangers with a pumped heat transfer medium or basket type gas-to-gas heat exchangers. The gas-to-gas reheaters may employ pre-heating of the desulfurized flue gas to prevent corrosion of the heat exchanger.

The two most commonly used reheater designs are (1) in-line heat exchange employing steam and (2) direct combustion of gas or oil with injection of the hot combustion products into the flue gas exiting the mist eliminator. A less common approach uses a steam-to-air heat exchanger to heat air to mix with the outlet flue gas before it enters the stack. This design eliminates the possibility of entrained droplets fouling the heat exchanger surfaces, but is relatively expensive to operate. Perhaps the simplest approach is bypass reheat where part of the flue gas is bypassed around the absorber. This technique, however, can only be employed when the product gas purity requirements are met by scrubbing only part of the flue gas. This usually means low sulfur coal applications. With the cap on SO_2 emissions and the consequent financial incentives to sell SO_2 allowances, bypass reheat is not likely to be widely used in the future.

Most U.S. users of wet FGD systems currently favor wet stack designs over reheat due to the problems associated with reheaters (Smigelski and Maroti, 1986). Corrosion and plugging of reheater tubes exposed to the flue gases are major problems. As dilute sulfuric acid droplets evaporate, the acid is concentrated, and the acid concentration passes through the range where its corrosivity is the greatest. In some cases, even the more corrosion resistant alloys have proven unsatisfactory. Many reheater problems have been attributed to mist eliminator inefficiency and inadequate reheater soot blowers (Bielawski, 1992). Other important reheater design considerations are the prevention of fouling, the provision of an adequate number of spare tubes, the ability to replace individual tubes, the use of adequate fouling factors, and the availability of maintenance access. EPRI Report CS-5980 (Krause et al., 1988) provides some valuable information on reheaters.

In another approach, which has been used in Europe, flue gas from the scrubber is discharged through the plant's large natural draft cooling tower. This avoids many of the problems associated with flue gas reheat. Saarberg Holter-Lurgi (SHL) has patented the process and as of 1989 had 16 FGD systems equipped or retrofitted for cooling tower discharge. The biggest utility in Europe (RWE) has indicated significant savings in operating and maintenance costs on 14 applications. The updraft from the cooling tower air has a volume of up to 25 times greater than that of the flue gas and effectively disperses the treated flue gas into the atmosphere. Also, the concentration of the pollutants is reduced by a factor of ten or more (compared with conventional discharge via a stack) as a result of the stack gas being mixed with the air from the cooling tower. To protect the cooling tower shell from acid attack, the inside shell surface above the fill and the top 20% of the outside shell surface must be epoxy-coated (McIlvaine, 1989; Glamser, et al., 1989).

A new concept still under development is low-temperature, water-saturated stack gas discharge. With this scheme the flue gas is cooled in a condensing heat exchanger below the normal wet scrubber exit temperature and consequently contains much less water vapor. The process uses teflon-covered heat exchanger tubes to minimize corrosion. Cooling and reducing the moisture content of the gas can greatly reduce liquid condensation from the stack discharge and the plume, if visible, would be extremely short. While the low stack temperature would reduce buoyancy and yield essentially no stack plume rise, an elevated stack exit velocity could be achieved with a choke device at the top of the stack. Another benefit that could accrue from the use of a condensing heat exchanger is the collection of sub-micron particles (condensation of moisture on the surface facilitates capture), trace heavy metals (which are concentrated in the fine particulate), SO_3, and volatile metals (condensed due to the lower temperature). Also, the heat removed from the flue gas by the condensing heat exchanger could be used to improve overall thermal efficiency by heating boiler feedwater. A 30 MW_e plant using this concept was scheduled to go into operation in late 1994 (Ellison, 1991; Heaphy et al., 1993; Ellison et al., 1994; Johnson et al., 1994).

Materials of Construction

The selection of appropriate materials of construction is a major factor in designing a maintainable wet limestone/lime FGD system. The materials selections typically vary with the operating environment, e.g., temperature range, pH, chemistry, and abrasiveness. The pH and the chloride and fluoride concentrations are the primary factors in corrosion (Bacha, 1992). Chloride levels in the system are raised significantly by closed-loop operation and by the production of salable gypsum, which requires good dewatering. Operating and maintenance practices are also important, e.g., encrustation if not removed can promote corrosion, and poor maintenance of coatings and linings can aggravate corrosion problems. Field experience and costs are usually the principal bases for materials selection. The type of construction: new, retrofit or existing, can also affect material selections (Rosenberg et al., 1991).

Typical materials used in limestone/lime FGD systems are identified in **Table 7-19.** Material selections vary considerably within a system. Due to the high cost of corrosion resistant materials, the consequences of selecting inadequate materials, and new materials becoming available, a thorough analysis is required for each specific plant design. The components of particular concern are prescrubbers, absorber inlets, outlet ductwork, stack liners, and in-line reheaters (Koch and Beavers, 1982). The wet/dry interfaces, i.e., absorber inlets and bypass-to-outlet duct connections are the most troublesome areas. In some installations chloride concentrations at the absorber inlet have exceeded 100,000 ppm due to water evaporation, providing an extremely corrosive environment (Nischt et al., 1991). Liquid that clings to the walls of the absorber above the mist eliminator can continue to absorb SO_2, lowering the pH, and necessitating the use of a different material at this location than in the lower parts of the absorber.

Carbon steel is generally suitable where the flue gas is substantially above the acid dew point and/or where alkaline conditions exist. Bypass ducts, rarely used on new plants in the U.S. now that regulations create an incentive for maximum SO_2 removal, are usually made of carbon steel, except at the connection to the outlet duct. Carbon steel bypass ducts must be used continuously or isolated when not in use to control corrosion due to flue gas damper leakage, gas cooling, and condensation.

Stainless steels, e.g., types 316L and 317LMN; nickel-based alloys, e.g., 625, C276, and C22; and, in a few cases, titanium are used depending on the aggressiveness of the environment. Extremely high chloride concentrations demand the use of more corrosion resistant materials. Where higher grade alloy materials are required, clad materials or $\frac{1}{16}$-in.-thick wallpaper materials are sometimes cost effective. An alloy system has high initial cost, but, when selected and installed properly, has the longest life and lowest maintenance cost (Bacha, 1992). To reduce costs, alloy thickness may be varied with location. Alloys fail generally by pitting or crevice corrosion, but general corrosion, erosion-corrosion, and stress corrosion cracking can also occur (Rosenberg et al., 1991; Mathay, 1990). **Table 7-20** summarizes data on pitting corrosion for some alloys in typical wet limestone/lime FGD environments. C276 and C22 are currently widely recommended for wet/dry interfaces. Data in **Table 7-20** should be used with caution as they apply only for the test conditions.

Organic linings, e.g., plastics and elastomers (rubbers), are suitable for some locations, can provide low initial cost, and sometimes have surprisingly low annual maintenance costs. However, complete replacement of these materials every 10 to 15 years is to be expected and flue gas temperature excursions are always a concern. In addition, routine inspection and

(text continued on page 532)

Table 7-19
Typical Materials of Construction for Wet Limestone/Lime FGD Systems[1]

Component	Typical Materials Used[6]		
	Metals	**Organics**	**Nonmetallic Inorganics**
Gas Side:			
Inlet Ducts, Inlet Dampers: Frame & Blades	Carbon Steel[2]		
Prescrubbers, Absorber Wet/Dry Interface[3]	316L, 317LMN, 625, C276, C22 Titanium Grades 2 & 7		Hydraulic Cement, Chemically Bonded Concrete or Brick with Backup Membrane
Absorber[4]			
Sump/Recycle Tank	316L, 317LMN, 625, C276, C22		
Vessel, Trays	316L, 317LMN, 625, C276, C22	Rubber Lined CS, Plastic Lined CS	
Spray Nozzles	Stellite		Ceramics such as Silicon Carbide
Mist Eliminators		Thermoplastic, e.g., Glass Coupled or Talc Filled Polypropylene, FRP	
Reheaters, In-Line & Gas Mixing Zones[5]	316L, 317LMN, 625, C276, C22 Titanium Grades 2 & 7		
Outlet Ducts	316L, 317LMN, 625, C276, C22	FRP, Plastic Lined CS	Hydraulic Concrete, Borosilicate Block
Stack Liner	316L, 317LMN, 625, C276, C22	FRP, Plastic Lined CS	Borosilicate Block
Liquid Side:			
Slurry Piping: Internal and External	316L, 317LMN, 625, C276, C22	Rubber Lined CS, FRP	
Pumps	High Chrome	Rubber Lined	
Fresh (Alkaline) Sorbent Storage Tanks	Carbon Steel		
Limestone/Lime System: Storage Silos	Carbon Steel		

Notes:
1. Materials identified are not suitable for all applications.
2. Upstream of the wet/dry interface.
3. Combinations of materials are sometimes used in prescrubbers.
4. The corrosion potential of the absorber environment varies, and sometimes materials are varied to accommodate this. The inlet wet/dry interface has a severe corrosion potential, the sump to the liquid level a mild to moderate corrosion potential, the zone above the sump to the mist eliminator a moderate corrosion potential, and the area above the mist eliminator a severe to moderate corrosion potential.
5. The area downstream of reheaters is similar to wet-dry interfaces and is particularly corrosive (Bacha, 1992).
6. Abbreviations: CS stands for carbon steel, FRP for flakeglass-/fiberglass-reinforced plastic. Alphanumeric designations are ANSI classifications, and 625 is a manufacturer's designation for a nickel alloy material, Inconel 625.

Source: Rosenberg et al. (1991), DoVale et al. (1991), Mathay (1990), and Bacha (1992)

Table 7-20
Corrosion of Alloys in an FGD Environment Data on Pitting Occurrence, Density and Rate[1] at pH of 5.2

Alloy	Position	10,000 ppm Cl			20,000 ppm Cl			30,000 ppm Cl		
		BM	W	HAZ	BM	W	HAZ	BM	W	HAZ
Inconel 625	U									
	M									
	B									
Hastelloy G	U									
	M		• 20			• 28		20 •		
	B					• 48		48 •		
Alloy 904L	U	•	188 •	156 •••	40 •		••		24 •••	•••
	M	••••	••	20 •••	••	12 •		48 •••	••	•••
	B									
SS Type 317 LM	U	•	32 •••		•	48 •••	•••	52 •	•••	•••
	M	••••		•	20 •••	20 ••	••	80 •••	••	•
	B	•	4	•	12 •	12		100 •		
SS Type 316L	U	200 ••••	•	•••	200 •••	•••	•••	116 •••	•••	•••
	M	56 ••••	•	•		•••	64 ••••	40 ••	••	•••
	B	••••		8 •••	24 •			4 •		

Coupon Position
U = Above slurry
M = At slurry/vapor interface
B = Immersion in slurry

Condition
BM = Base Metal
W = Weld
HAZ = Heat Affected Zone

Pit Density
• = <10/in²
•• = 10–1,000/in²
••• = >1,000/in²
•••• = locally > 1,000/in²

Note 1. Numerical values indicate estimated pitting rate in mils/yr based on linear projections from 90 tests.
Source: Saleem (1991B)

(text continued from page 529)

touch-up on a 6-month to 1-year cycle is normal. Various plastic resins are used, including polyesters, vinyl esters, and epoxies; but the most common is polyester. Plastics often contain flakes of glass filler, and are referred to as flakeglass reinforced plastic. Failure of plastics can be caused by high and low temperature excursions, chemical attack (e.g., acid induced hydrolysis of the polyester), abrasion, improper application to the metal surface, or permeation of substances that can attack the substrate (Rosenberg et al, 1991). The success of rubber linings depends to a large extent on good application and the absence of microporosity in the elastomer (Siegfriedt et al., 1990) and chlorbutyl and layered rubber have been successfully used (Nischt et al., 1991).

Nonmetallic inorganic materials include various ceramic blocks, bricks, tiles, and other shapes and hydraulically and chemically bonded concrete and mortars. Borosilicate foam blocks are usually bonded with urethane asphalt. The combination of these two materials results in a lining that is resistant to permeation by aggressive condensate. The thermal insulating properties of the blocks and the flexibility of the urethane material contribute to the good performance of the lining. However, the borosilicate glass is easily damaged by maintenance personnel, and failures can occur due to abrasion, puncturing, and cracking. Also, scale that forms on the lining cannot be easily removed. These problems can be minimized by applying a topcoat of chemically-bonded, abrasion-resistant mortar.

Acid-resistant brick is also commonly used for stack liners. Free-standing brick chimney liners have been found to lean due to differential moisture expansion of the bricks (Rosenberg et al., 1991). Concern has been expressed about brick chimney liners in severe seismic activity areas (Mathay, 1990). Acid-resistant brick linings can also shrink in service creating mechanical stress. Alumina bricks are more abrasion resistant than acid-resistant bricks and are used where abrasion is a consideration. Hydraulically bonded concretes, used in pre-scrubbers, outlet ducts, and stacks, have high permeability and will experience dissolution of the calcium aluminate in sulfuric acid solutions if the pH is less than 4.

EPRI has developed *Guidelines for FGD Material Selection and Corrosion Protection* (TR-100680) (Rosenberg et al., 1993), which includes a materials experience database at operating U.S. FGD plants. They have also developed software for estimating material costs (CS-3628, Cloth, 1984).

FGD Problems and Potential Solutions

Some common wet limestone/lime FGD system problems and potential causes are listed in **Table 7-21**. EPRI has issued a report discussing a variety of FGD problems and their solutions. This report is entitled, *Investigation of Flue Gas Desulfurization Chemical Process Problems,* Report GS-6930 (Radian, 1990).

Process Variations

For large, coal-fired utility boilers worldwide, the most common FGD system utilizes limestone in an open spray tower with countercurrent flow (gas up, spray down). However, significant process variations exist. Some of the more important variations are: limestone vs. lime vs. magnesium-enhanced lime, spray tower vs. tray tower, gas flow cocurrent vs. countercurrent vs. multi-pass, organic acid enhancement, reaction tank separate vs. integral with the absorber, forced vs. natural vs. inhibited oxidation, and *in situ* vs. *ex situ* forced oxida-

Table 7-21
Wet Limestone/Lime FGD Process Problems and Causes

Process Problem	Potential Causes
Insufficient SO_2 Removal	Low liquid phase alkalinity Low mass transfer surface area and/or inadequate contact time between the flue gas and the scrubbing slurry Low L/G Low limestone reactivity
Poor Reagent Utilization	Inadequate pH measurement and/or control system Poor limestone particle size distribution (grind) SO_3 blinding Aluminum/fluoride inhibition of the process
Mist Eliminator Scaling and Plugging	Mist collected or wash water has high gypsum relative saturation Poor limestone utilization Infrequent washing of the mist eliminators Low wash intensity
Scaling of Absorber Packing	Localized areas of high gypsum relative saturation Poor distribution of the slurry over the surface of the packing
Poor Solids Dewatering in Thickener[1] or Vacuum Filter[2]	Inadequate oxidation/small crystal size Inadequate operating procedures Line plugging Filter cloth blinding Filtercake discharge problems
Positive Water Balance	Use of make-up water for limestone grinding High pump and agitator seal water usage Excessive fresh water usage for mist eliminator washing Bypassing flue gas Insufficient surge capacity in the FGD system
H_2S Formation	Anaerobic sulfur-reducing bacteria in the thickener due to FGD system being idle for a long period of time

Notes:
1. Thickener problems include poor solids dewatering, cloudy thickener overflow, low density underflow, and high density underflow (high rate torque or unpumpable underflow).
2. Vacuum filter problems include poor solids dewatering and cloudy filtrate.
Source: (Moser et al., 1988)

tion. Some major features of the systems offered by several suppliers are described in the following section.

General Electric (GEESI) In Situ Forced Oxidation Process. The General Electric Environmental Services Inc. (GEESI) process is a limestone FGD process with *in situ* forced oxidation. A typical configuration is illustrated in **Figure 7-2** and a typical absorber tower in

534 *Gas Purification*

Figure 7-3. An open spray tower with a high turndown ratio is used. Limestone is typically ground to 90% minus 325 mesh in a ball or tower mill. Candidate limestones are evaluated by GEESI during the design phase. Limestone slurry is added to the scrubber sump based on feed-forward control using calculated inlet SO_2 flow with pH trim override. Oxidation air is blown into the sump about 18 ft below the liquid surface. The typical oxygen/sulfur mole ratio is about 3.0. A bleed stream from the single stage hydrocyclone overflow is returned to the sump. The hydrocyclone underflow, which consists primarily of gypsum, goes to dewatering. Drum or belt filters or centrifuges may be used depending on the gypsum moisture specifications.

The spray tower is equipped with three to six spray levels depending upon the sulfur content of the coal and the required SO_2 removal efficiency. The spray levels are about five ft apart. The gas side system pressure drop is about 4 to 6 in. of water of which 1.5 to 2.5 in. is across the spray tower. A prescrubber is used when the incoming flue gas has a high ash content and an ultra pure white gypsum is required. The need for a waste water treatment system is site-specific and depends upon the discharge water quality requirements.

This system has been installed in countries worldwide including the United States, Germany, the UK, Finland, the Netherlands, Japan, Taiwan, Poland, and Austria. Several systems employ a single absorber module with absorber diameters ranging from 40 to 62 ft. The largest single absorber module (approximately 62 ft diameter) designed by GEESI treats flue gas from a 700 MW_e power plant boiler (Kneissel et al., 1989).

Babcock & Wilcox Process. The Babcock & Wilcox FGD process can be designed to utilize either limestone or lime as the reagent and either forced or inhibited oxidation. As of late 1992, Babcock & Wilcox had designed and supplied over 15,000 MW_e of wet FGD systems, of which over 9,000 MW_e were in operation and 6,000 MW_e were scheduled to begin operation by 1997. System sizes range from 60 MW_e to 1,370 MW_e. The 1,370 MW_e installation is at Cincinnati Gas and Electric's Zimmer station and is the largest FGD system in the world.

The absorber, which is a countercurrent spray/tray tower type, is illustrated in **Figure 7-4**. The absorber tower has an integral reaction tank, patented absorber tray, and symmetrical reducing outlet. The integral reaction tank reduces the equipment footprint. The absorber tray causes flue gas to be evenly distributed in the absorption zone, effectively utilizing the slurry distribution from the recycle spray nozzles. In a system where liquid phase diffusion is important, the absorber tray provides liquid hold-up through which the gas passes prior to entering the absorption spray zone. This not only provides a significant amount of SO_2 removal, but increases reagent utilization. The absorber tray is compartmentalized by a grid of baffles that prevents migration of the liquor (and thus the gas) across the absorber tower cross section. The gas velocity through the perforations is typically several times the superficial velocity in the absorber tower. A vigorous frothing action results, which helps maintain absorber tray cleanliness. Another unique feature of the Babcock & Wilcox absorber tower is the patented interspacial header arrangement. This feature allows two spray headers to be located at the same elevation which permits the use of a shorter tower than required for earlier designs.

The absorber tower symmetrical reducing outlet provides a uniform velocity distribution at the front face of the horizontal mist eliminator. Experience has shown that poor flue gas velocity distribution can cause excessive mist carryover. The two-stage, triple-washed mist-eliminator system provides both a "roughing" chevron for capture of large drops of slurry and a "finishing" chevron to capture smaller drops of slurry and wash-water. Wash headers on both the lower and upper surface of the first stage mist eliminator and the lower surface of the second stage mist eliminator minimize pressure drop and overall particle carryover by reducing plugging of mist-eliminator passages (Martinelli, 1992).

Saarberg-Hölter (S-H-U) Process. The S-H-U process is a limestone wet scrubbing process with formic acid enhancement. The S-H-U absorber has both cocurrent and countercurrent sections. The flue gas enters the absorber at the top of the cocurrent flow section and flows downward past several levels of spray nozzles where the pH drops rapidly. The scrubber slurry collects in the sump. The flue gas then turns upward into the second scrubbing stage, the countercurrent flow section, where the final increment of SO_2 removal occurs. The flue gas exits either through a combination of a vertical flow and a horizontal flow mist eliminator or vertical flow mist eliminators.

The formic acid buffer permits optimum SO_2 removal in the pH range of 4.2 to 5.2. The advantages of formic acid buffering are

- Increased SO_2 removal capability
- Improved ability to respond to operating load and SO_2 variations
- High reagent utilization (Ca/S = 1.01 to 1.03)
- Reduced L/G required for a given SO_2 removal resulting in a lower power consumption
- High oxidation efficiency to allow production of wallboard or disposable quality gypsum
- Lower maintenance requirement and greater availability due to reduced potential for scaling and plugging
- High tolerance for chlorides

Forced oxidation ensures an oxidation level of 99.9%, which is achieved by injecting air into the absorber sump. The byproduct gypsum can be disposal grade or commercial grade. A disposal grade gypsum can be produced by dewatering with a rotary drum filter to an 80% solids filtercake. Alternatively, commercial grade gypsum can be produced by washing for chloride control and dewatering with a centrifuge or a hydrocyclone/belt filter system.

Between 1974 and 1992 S-H-U systems were installed on 30 coal-, oil-, and lignite-fired boilers ranging in size from 125 to 900 MW_e (8,000 MW_e total). As of 1992, S-H-U had 1,500 MW_e under contract, including a system for a 316 MW_e unit at Milliken Station in the U.S. (Schütz et al., 1989; Glamser et al., 1989; Glamser, 1992).

Noell KRC Double-Loop Limestone Process. The current Noell KRC FGD process has evolved from the Research Cottrell FGD system of the 1970s. Variations of the design have been used in the U.S., Europe, and Canada. As of 1992, total systems installed, in design, or under construction exceeded 27,000 MW_e in combined capacity. A single 450-MW_e absorber module unit is being erected at New Brunswick Power's Belledune Station in Canada. It is designed for 95% SO_2 reduction with 4.5% sulfur coal (Majdeski, 1992). The system is only suitable for limestone. A forced oxidation byproduct is the usual choice, although an inhibited oxidation byproduct has also been produced. Additives have not been found to be cost-effective.

The Double-Loop FGD process has two separate loops in a single tower: a lower quench loop and an upper absorber loop. Flue gas from the particulate collector is first quenched in the lower quench loop with a recirculating slurry of calcium sulfite, gypsum, and fresh calcium carbonate. The quenched flue gas then flows up through a gas/liquid separator and the upper absorber loop before exiting the absorber through a mist eliminator. The gas/liquid separator prevents mixing of the slurry between the two loops.

In the upper absorber loop, recirculating limestone slurry, at a pH of 5.8 to 6.0, absorbs the bulk of the SO_2. A number of spray levels, and sometimes a wetted film contactor, are used. The system design keeps the dissolved chloride level low in this loop, avoiding the detrimental effect of high chlorides on SO_2 removal efficiency. Also, the corrosion potential is minimized by the low chloride and relatively high pH values, allowing the use of more cost effective materials of construction in this section of the tower.

The quench loop operates at a pH of about 4.0. It receives overflow from the absorber loop feed tank containing some unreacted limestone, which dissolves quickly at the low pH while absorbing 25 to 30% of the SO_2 from the incoming gas. External air is injected through a series of sparger pipes in the quench loop sump for oxidation to gypsum. The quench loop operates in a closed-loop mode using water returned from dewatering. No additives are used and salable gypsum is produced. The formation of gypsum crystals in the sump reduces the scaling potential by providing suspended gypsum particles for crystal growth and by maintaining suitably low calcium sulfite levels. The system provides for independent limestone feed to each loop for pH control. This minimizes the potential for pH variation, thus further reducing the scaling potential.

Typically, a bleed stream from the quench loop, containing 15 to 18% solids, flows to the dewatering system. Most plants require two stages of dewatering—primary and secondary. For gypsum, hydrocyclones are preferable to thickeners as the primary dewatering device because of lower cost, compact size, and adequate particle size separation. Thickeners would produce too dense an underflow when processing gypsum. When the system is operated in the inhibited oxidation mode, the unoxidized calcium sulfite is usually first dewatered in a thickener. Vacuum filters are used for secondary dewatering, producing an end product with typically 60% solids for calcium sulfite and 85–90% solids for gypsum. The sulfite cake is usually blended with fly ash (and possibly lime) to produce a landfillable byproduct of low permeability. Gypsum byproduct can usually be used for commercial wallboard or cement production (Dhargalkar and Tsui, 1990; Majdeski, 1992).

Bischoff Process. The Bischoff limestone FGD process (licensed in the U.S. by Joy) uses large capacity spray nozzles (300 to 600 gpm per nozzle), in situ forced oxidation, large pumps (e.g., 33,000 gpm), a sump with no mechanical agitation, and a unique mist eliminator.

In the absorber, limestone slurry is contacted with the flue gas by four to eight levels of spray nozzles. Each spray level has up to eight spray nozzles. The lower nozzles spray both cocurrent and countercurrent to the flue gas flow. The upper levels are countercurrent only. The flue gas exits the spray tower horizontally through vertical mist eliminators arranged around the perimeter of the top of the spray tower. This design reportedly has the advantage that water drainage does not affect gas flow, and the design results in more even gas flow distribution over the surface of the mist eliminator with reduced carry-over.

The scrubber sump is designed for uniform slurry flow from top to bottom. Oxidation takes place in the upper portion of the sump. Reacted slurry enters at the top at a low pH (4 to 5). The calcium sulfite reacts with oxygen in the air bubbled up through the slurry to form calcium sulfate. An oxygen/sulfur mole ratio of 2 to 3 is maintained in the forced oxidation zone. The slurry flows down to the crystallization zone in the lower portion of the sump. A special grid prevents back-mixing of slurry from the crystallization zone, which is maintained at a pH of 5.5. Make-up limestone slurry is added in the crystallization zone. Five to eight minutes residence time causes precipitation and growth of the large gypsum crystals.

Recycle is from the bottom of the sump. A bleed stream from the lower portion of the crystallization zone goes to byproduct recovery, consisting of hydrocyclones, and a filter or centrifuge. The gypsum byproduct contains 5–10% moisture depending on requirements. Recirculation pumps convey the limestone slurry from the conical bottom sump to the various spray levels.

The Bischoff wet FGD process utilizes either limestone or lime as the sorbent. Bischoff's early scrubber experience was with lime and natural oxidation. With the increased demand for gypsum as the preferred byproduct and the use of sea water for at least a portion of the process water, Bischoff has evolved their wet FGD technology in the direction of limestone with forced oxidation. Limestone is the preferred sorbent when seawater is used as the process water.

Joy Technologies Inc. has sold two 500 MW_e Bischoff FGD systems in North America to Ontario Hydro for their Lambton Station Units 3 and 4. They are also engineering Bischoff scrubbers for four 500 MW_e boilers at Taiwan Power's Taichung Thermal Power Station, Units 1 through 4 (Hegeman and Kutemeyer, 1989; Stearns, 1985; Krippene, 1992).

Pure Air/Mitsubishi (MHI) Process. Pure Air is a general partnership of Air Products and Chemicals and Mitsubishi Heavy Industries America, Inc. The partnership was formed to design and build sulfur dioxide removal systems for the utility marketplace using the MHI FGD process and alternatively to own/operate and maintain these systems under long term contracts. More than 100 systems of this type have been sold worldwide. A system at Northern Indiana Public Service Company's Bailly Generating Station treating flue gas from Units 7 and 8 (615 MW_e total) began commercial operation in June 1992. This system employs a single absorber module with no spare, utilizes direct injection of dry limestone into the absorber modules, a wastewater evaporation system, as well as other innovations.

The MHI FGD system process is a cocurrent flow, *in situ* oxidation system where quenching, particulate removal, and SO_2 removal occur simultaneously. While a pre-scrubber to remove chlorides and some fly ash is included in Japanese and German systems which produce a wallboard grade gypsum byproduct, a pre-scrubber has not been necessary in U.S. applications.

Distinguishing features of the process are the high velocity (15–20 ft/sec) open absorption grid, the cocurrent design, and the air rotary sparger. The absorption grid provides a high gas/liquid contact area and uniform distribution of flue gas flow. The cocurrent flow of the flue gas and the scrubbing liquid reduces mist carryover by as much as 95% while allowing a flue gas velocity increase of almost 100% over current designs. The sump contents are agitated and oxidized simultaneously by Pure Air's rotary air sparger. Traditional fixed air spargers and mixers can also be used with the MHI system (Ashline et al., 1989; Camponeschi, 1992).

Chiyoda Thoroughbred 121 Process. This process differs from conventional limestone/lime systems in that the absorbent is maintained at a pH between 4 and 5 compared with 5.5–6 for most other processes. Also, a single reactor is used to remove SO_2, oxidize the reaction products, and form calcium sulfate. No recycle pumps and no thickener or hydrocyclones are required as with conventional FGD systems. A large gypsum crystal is produced due to the retention time in the reactor and the absence of recycle pumps. SO_2 removal efficiencies of 95% over a wide range of inlet SO_2 loadings are possible (Anders and Torstrick, 1981; Mirabella, 1992B). By 1991, there were 12 operating units worldwide and five others under construction (Mirabella, 1992A).

538 Gas Purification

A key feature of the process is the special reactor design, which is illustrated in **Figure 7-9**. The unit is called a Jet Bubbling Reactor (JBR). This design provides a large inventory of absorbent in the reaction zone, which eliminates the need for liquid recycle. The low pH favors the oxidation of sulfite to sulfate, so complete conversion of the solid product to gypsum is feasible.

In the process, as operated by the Mitsubishi Petrochemical Company at their Yokkaichi, Japan, complex, flue gas from a boiler burning 3 to 4% sulfur fuel oil is treated to remove 97% of the SO_2 (Classen, 1983; Kaneda et al., 1983). Flue gas from the air preheater is first blown into a precooler where a fine spray of recirculated water humidifies and cools the flue gas while removing particulate and other impurities such as chlorides.

The humidified gas flows into the JBR where it bubbles through a shallow zone of absorbent. Sulfur dioxide is absorbed, oxidized, and reacted with calcium ions to precipitate calcium sulfate and form a gypsum slurry. Make-up limestone is added as 20% slurry, air is blown into the bottom reactor zones of the JBR to enhance the oxidation reactions, and the product gypsum is continuously withdrawn as a slurry containing about 15% $CaSO_4 \cdot 2H_2O$.

The gypsum slurry is dewatered in a solid bowl decanter centrifuge. The byproduct gypsum, which is essentially dry, is taken to storage, and the centrate is recycled. A bleed stream of the precooler water is continuously circulated through a thickener to concentrate captured fly ash. The fly ash slurry is mixed with a bleed stream of centrate, neutralized with limestone slurry, and filtered. The resulting solids are discarded. A portion of the filtrate is also discarded to maintain the water balance and dispose of soluble salts while the main stream is recycled to the absorption circuit.

According to Kaneda et al. (1983), operation is smooth and trouble-free. SO_2 removal efficiencies are in the 97–99% range with inlet SO_2 concentrations from 1,000–2,000 ppm. Limestone utilization is greater than 99%, and the gypsum, which is sold to cement and wallboard manufacturers, typically contains about 99.2 wt % $CaSO_4 \cdot 2H_2O$.

In the U.S., a Chiyoda CT-121 system has been installed on Georgia Power's Yates Unit No. 1, a 100 MW_e coal-fired unit, as part of the Department of Energy's Clean Coal Tech-

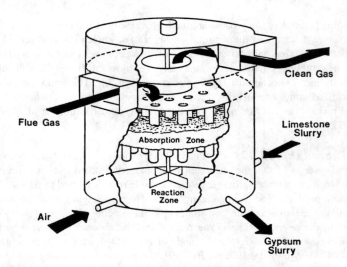

Figure 7-9. Jet bubbling reactor (JBR). *From Kaneda et al. (1983)*

nology Program. Start-up was in October 1992. SO_2 removal is 93–98%, more than 90% of the particulate leaving the electrostatic precipitator is captured in the scrubber, and reliability and availability are 98%. Limestone utilization approaches 100% (PETC, 1993).

Magnesium Oxide Process

Regenerable magnesium oxide processes were developed in the late 1960s by the Grillo Company of Hamborn, Germany; the Chemico Construction Company in the United States (Anon., 1977); and by Showa Denko in Japan. A commercial-size demonstration unit was installed on a 150-MW_e oil-fired steam-generating boiler at the Mystic Power Station of the Boston Edison Company. The unit was started up in 1972 and operated intermittently until June 1974. It was designed by Chemico (now GEESI) and incorporated a single-stage venturi module. Numerous operating problems were encountered. However, the plant successfully demonstrated that the basic concepts were sound. Ninety percent SO_2 removal could be accomplished, magnesia could be regenerated and recycled, and high-quality sulfuric acid could be recovered from the SO_2 removed (Koehler and Dober, 1974).

Philadelphia Electric Company (PECo) and United Engineers and Constructors (now Raytheon) initiated a study in 1971 which led to construction of a demonstration unit at PECo's Eddystone Unit 1 by 1975 and culminated in the installation and startup of three commercial systems in 1982. These were retrofitted to Eddystone Units 1 and 2 (335 and 355 MW_e respectively) and Cromby Unit 1 (160 MW_e). Emission compliance tests conducted shortly after startup showed that all units met or exceeded state requirements for SO_2 and particulate control (MacKenzie et al., 1983).

According to Ando (1977), a magnesium oxide process developed by Chemico-Mitsui is operating in Japan on an oil-fired boiler. The plant provides 90% SO_2 removal efficiency with 100% operability. The process is also used at a number of pulp and paper plant boilers for SO_2 recovery and recycle of the $MgSO_3/Mg(HSO_3)_2$ to the process.

In the U.S., the ability to dispose of limestone/lime scrubber byproducts close to the plant has limited the need for a regenerable system and inhibited the use of the magnesium oxide process on a large scale. At this time, the process is being marketed in Eastern Europe. It is expected that future units will be 200 MW_e or greater in size based on economic considerations. The presence of a nearby acid plant, where the magnesium salts can be regenerated and the SO_2 product utilized, enhances the economics of this process. Also, the very high sulfur dioxide removal efficiency is a feature with increasing attractiveness.

Process Description

Magnesium oxide is not as suitable for a disposable byproduct process as calcium oxide because magnesium sulfate is quite soluble and magnesium sulfite is several times as soluble as the equivalent calcium compound. Furthermore, magnesium compounds are generally more expensive than similar calcium materials. However, the principal product of the absorption of SO_2 by magnesium oxide, magnesium sulfite, decomposes at relatively modest temperatures. This characteristic makes it suitable for a regenerative cycle using a calcination step to release the absorbed SO_2.

Simplified flow diagrams representing the magnesium oxide process are shown in **Figures 7-10** and **7-11**. These flow diagrams are based on the initial Eddystone No. 1 installation. An aqueous slurry containing magnesium oxide is used as the absorbent in a scrubbing step sim-

(text continued on page 542)

540 *Gas Purification*

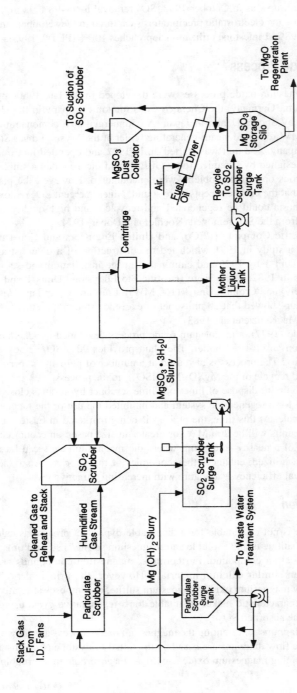

Figure 7-10. Flow diagram of magnesium oxide process, flue gas scrubbing section.

Sulfur Dioxide Removal 541

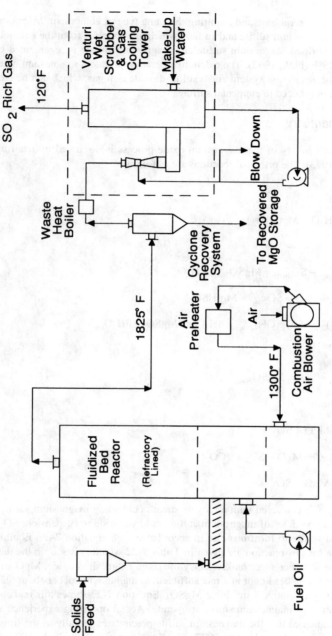

Figure 7-11. Flow diagram of magnesium oxide process, absorbent regeneration section.

(text continued from page 539)

ilar to lime slurry scrubbing and requiring the same type of equipment. Magnesium oxide is converted to magnesium sulfite and sulfate, which are removed from the solution and dried. The mixture of dried magnesium sulfite and magnesium sulfate is regenerated by calcining at about 1,825°F (Fink, 1992). The calcining operation produces magnesium oxide, which is returned to the absorption system and a sulfur-dioxide rich gas, which can be fed to a sulfuric acid plant or reduced to elemental sulfur.

Process Chemistry

The basic chemistry of the magnesium oxide process is discussed in detail by McGlammery et al. (1973). The principal reactions are as follows:

Slaking:

$$MgO_{(s)} + H_2O = Mg(OH)_{2(aq)} \tag{7-24}$$

Absorption:

$$Mg(OH)_{2(aq)} + SO_{2(aq)} = MgSO_{3(aq)} + H_2O \tag{7-25}$$

$$MgSO_{3(aq)} + H_2O + SO_{2(aq)} = Mg(HSO_3)_{2(aq)} \tag{7-26}$$

$$Mg(HSO_3)_{2(aq)} + Mg(OH)_{2(aq)} + 4H_2O = 2MgSO_3 \cdot 3H_2O_{(s)} \tag{7-27}$$

Oxidation:

$$MgSO_{3(aq)} + \tfrac{1}{2}O_2 = MgSO_{4(aq)} \tag{7-28}$$

Regeneration:

$$MgSO_3 = MgO + SO_2 \tag{7-29}$$

$$MgSO_4 + \tfrac{1}{2}C = MgO + SO_2 + \tfrac{1}{2}CO_2 \tag{7-30}$$

$$MgSO_4 = MgO + SO_3 \tag{7-31}$$

The flue gas is contacted with a recycle stream containing magnesium sulfite and sulfate in solution and particles of magnesium sulfite and hydroxide in suspension. SO_2 reacts with the dissolved sulfite to form bisulfite in accordance with equation 7-26. Equilibrium vapor pressure data for this reaction are given in **Table 7-22** which is based on the data of Pinaev (1963). In the scrubber surge tank, a slurry of slaked magnesium oxide, MgO and $Mg(OH)_2$, is added to the scrubber liquor at a rate sufficient to maintain the pH at about 6.3, converting the $Mg(HSO_3)_2$ to relatively insoluble $MgSO_3$. Equation 7-27 shows this reaction leading to the precipitation of magnesium sulfite trihydrate. Actual operating experience at the Eddystone Plant indicated that the magnesium sulfite precipitates initially as the trihydrate in the surge tank (Gille and Mackenzie, 1977; MacKenzie et al., 1983; Fink, 1992). In the pilot plant phase of the work, it was thought that it would be necessary to reduce the $MgSO_4$ with carbon to produce SO_2 as shown by equation 7-30. However, it was found that by heating

Table 7-22
Vapor Pressure of Sulfur Dioxide over Solutions Containing Magnesium Sulfate and Magnesium Sulfite-Bisulfite
(Active Mg refers to magnesium associated with sulfite-bisulfite ions)

Temp. °C	Solution Composition			Mole Ratio SO_2/Mg	P_{SO_2} mmHg
	$MgSO_4$ Concentration g/l	Active Mg g/l	SO_2 g/l		
30	54.75	2.01	7.04	1.31	0.030
30	52.00	2.16	11.0	1.49	0.041
30	51.67	4.22	18.2	1.62	0.052
30	51.30	5.53	25.0	1.69	0.073
60	51.75	3.77	11.9	1.18	0.060
60	53.65	3.48	12.7	1.37	0.069
60	50.20	6.32	23.6	1.40	0.111
60	51.75	8.12	31.6	1.46	0.141
30	106.50	4.15	18.1	1.63	0.073
60	105.50	6.23	23.6	1.38	0.133

Source: Pinaev (1963)

the salt to 1,850°F, SO_3 was driven off as shown by equation 7-31. This being suitable for the production of sulfuric acid, the carbon reduction step was eliminated (Fink, 1992).

Design and Operation

The Eddystone Unit 1 plant employs three venturi rod-type absorbers. The Eddystone Unit 2 plant uses two spray towers (plus one spare). Both plants handle about 1×10^6 acfm of flue gas and are designed to remove 92% of the incoming SO_2. The Cromby Unit 1 plant utilizes two spray towers (plus one spare) for 416,000 acfm of flue gas and is designed to remove 95% of the SO_2. A comparison of actual vs. design performance for one Cromby absorber is given in **Table 7-23**.

All of the magnesium oxide plants, including the one in Japan, encountered numerous operating problems. According to Ando (1977), the problems at the Chemico-Mitsui plant were mainly in the regeneration steps. The problems were solved, and the plant operates quite satisfactorily with almost 100% availability.

Operating problems encountered in the U.S. plants included pipe and pump corrosion; plugging of systems for feeding MgO powder, slaking MgO, and feeding MgO slurry; caking in the $MgSO_3$ dryer; dust production in the dryer; solids handling malfunctions; regenerator product gas filter plugging; and high MgO losses.

It should be noted that many of the operating problems were satisfactorily resolved at the plants where they were encountered. None are considered to represent fundamental barriers to the successful application of the magnesium oxide process for SO_2 removal. Moreover, the PECo units have operated continuously for more than 15 years.

Table 7-23
Comparison of Actual and Design Performance for Magnesium Oxide Scrubbing at PECo Cromby Unit 1 Plant

Parameter	Actual	Design
Gas Flow, m^3/s (acfm)	104 (220,000)	98.2 (208,100)
L/G, l/m^3 (gal/1,000 acf)	5.4–6.7 (40–50)	6.7 (50)
Pressure Drop, Pa (in W.C.)	500–623 (2.0–2.5)	1,245 (5.0)
pH	6.7–6.9	6.8
% Solids	15–20	15
Crystal Form	Trihydrate	Trihydrate
SO$_2$ Removal, %	96–98	95

Source: MacKenzie et al. (1983)

ALKALI METAL PROCESSES

Many processes have been developed based on SO$_2$ removal by absorption in an aqueous solution of a soluble alkali metal compound. Sodium compounds are preferred over potassium or the other alkali metals strictly on the basis of cost. In its simplest form, the process consists of contacting the gas with a solution of sodium carbonate (or sodium hydroxide) to form sodium sulfite, followed by disposal of the spent absorbent solution as waste or as a raw material for some other industrial process. More complicated forms have been proposed to reduce the costs of active absorbent make-up and spent absorbent disposal. These processes incorporate a variety of steps to regenerate the absorbent and produce a byproduct that is readily disposable or salable.

Non-Regenerable, Alkali Metal-Based Processes

The non-regenerable, alkali metal-based processes are particularly applicable to situations involving relatively small quantities of SO$_2$, where the advantage of a simple system with low capital cost outweighs the operating cost penalties associated with use of an expensive chemical and disposal of soluble waste. The processes are also applicable to special situations where the resulting sodium sulfite has a significant market value. This is the case in Japan where a large number of such plants exist, producing sodium sulfite or sodium sulfate (Ando, 1977). EPA/600/7-85/040, *Recent Developments in SO$_2$ and NO$_x$ Abatement Technology for Stationary Sources in Japan,* lists 262 Na$_2$SO$_3$/Na$_2$SO$_4$ FGD plants (Ando, 1985). About 80% of the Japanese plants produce sodium sulfite for paper mills and the rest oxidize the sulfite by air-bubbling to sulfate, which is either used in the glass industry or purged in waste water. Operating data for a typical plant employing NaOH as the absorbent and producing Na$_2$SO$_3$ as a byproduct are given by Ando (1977). The plant treats 190,000 Nm^3/hr (112,000 scfm) of flue gas containing 1,400 ppm SO$_2$ in a packed tower and produces an outlet gas containing only 6 ppm SO$_2$. An L/G of 1.2 l/Nm^3 (~9 gal/1,000 scf) is used with a liquor pH of 6.5.

A large number of non-regenerative sodium-based absorption units are also in use in the U.S. for industrial boilers and other applications. Several large units were placed in service on coal-fired utility boilers between 1974 and 1986. The first three units were designed and installed by Combustion Equipment Associates in association with Arthur D. Little, Inc. (Pedco, 1977) at the Reid Gardner Power Station of the Nevada Power Company. This type of system was economically attractive at the Reid Gardner Station because of a combination of factors, including the availability of sodium carbonate, the use of relatively low-sulfur coal (0.5 to 1.0% sulfur), and the plant location in a warm, arid zone where the effluent can be evaporated in ponds without the use of equipment or energy for byproduct drying.

In each of the first three Reid Gardner units (110 MW_e each) approximately 473,000 acfm of 350°F flue gas is processed. The gas first flows through a venturi scrubber where it is cooled to about 119°F, then through an absorber containing a single sieve tray, and finally through a radial mist eliminator before flowing to the stack. Heated air, added to the gas after it exits the absorber, reheats the gas. About 5,000 gpm of recycled solution is pumped to the venturi, and about 900 gpm to the absorber. Make-up of alkali and water is provided by 53 gpm (maximum) of concentrated sodium carbonate solution and 129 gpm of water from the ash pond. Sulfur dioxide removal is reported to be 85%. The FGD system of Reid Gardner Unit 4 (initially 250 MW_e, upgraded to 302 MW_e) is preceded by a fabric filter, so a venturi scrubber is not required. Otherwise, the unit is similar to the other three units at Reid Gardner. This unit was furnished by Thyssen/CEA (PEI Associates, 1989A; Day, 1991).

Five other large, wet soda once-through systems, all on electric generating units, have been constructed—four 500 MW_e systems at the Jim Bridger Station and one 330 MWe system at the Naughton Station, all in Wyoming. These employ waste soda liquor from soda producing plants in the area and the waste is evaporated in ponds. These units were furnished by Universal Oil Products and Babcock & Wilcox (Echols, 1992; Tolman, 1992).

To avoid the cost and disposal problems of once-through processes employing alkali metal compounds, a considerable amount of research and development effort has been expended on techniques for regenerating this type of absorbent. Processes used employ precipitation of insoluble compounds (double alkali), and thermal decomposition (Wellman-Lord and Elsorb). Processes under development or which have been investigated include precipitation of insoluble compounds (zinc oxide), low-temperature reduction of sulfite (citrate and potassium formate processes), high-temperature reduction (aqueous carbonate process), electrodialysis (SOXAL), and electrolytic (Stone & Webster/Ionics Process). Descriptions of some of these processes are provided in subsequent sections.

Oxidation to Sulfate

Oxidation of alkali metal sulfites to sulfates occurs to some extent in all alkali metal-based SO_2 absorption processes that treat flue gases containing free oxygen. As previously noted, about 20% of the nonregenerable sodium hydroxide/carbonate-based plants in Japan continue the oxidation to essentially 100% sodium sulfate by bubbling air through the solution.

Oxidation of potassium compounds to the sulfate form occurs in the Recovery Scrubber process licensed by Passamaquoddy Technology, which was originally developed for cement plant applications. The process uses waste cement kiln dust, which contains limestone, alkali, and calcium sulfate, in an aqueous slurry as a flue gas scrubbing reagent. The first commercial application of the process began operating at the Dragon Products Company cement plant in Thomaston, Maine in December 1990 (Morrison, 1991).

The Passamaquoddy Recovery Scrubber process offers several advantages for cement plant use: it recovers a major fraction of the waste cement kiln dust in a form suitable for reuse; it removes SO_2 from the flue gas; and it produces a salable byproduct, potassium sulfate. Because of the high temperature attained in cement kilns, the most volatile species, such as potassium compounds, are selectively vaporized. The dust collected from the flue gas is therefore rich in potassium (and also sulfate) and is not suitable for reuse in the kiln without purification.

In the process, the cement kiln dust is collected in a recycling stream of slurry which contacts the flue gas. Carbon dioxide is absorbed in the alkaline solution and reacts with calcium sulfate and calcium hydroxide to precipitate calcium carbonate and release sulfate ions. Sulfur dioxide is absorbed and is then oxidized by oxygen from the flue gas stream to form additional sulfate ions. The sulfate remains in solution as the very soluble potassium sulfate.

As a result of these reactions, the recirculating slurry contains primarily potassium sulfate in solution and particles of calcium carbonate in suspension. A portion of the slurry is continuously treated to remove solids (which are washed and recycled to the kiln) and recover potassium sulfate (by evaporation and crystallization). In the initial operation of the Dragon Products Co. plant the process achieved 92% sulfur dioxide removal efficiency.

Other applications of the Recovery Scrubber process have been proposed by Morrison (1991). In general, these are based on the use of biomass ash from pulp and paper plants and other wood burning facilities. Biomass ash is rich in potassium and very alkaline, so its use presents the possibility of both SO_2 removal and potassium sulfate recovery.

Oxidation of sulfite to a soluble alkali metal sulfate is also a feature of the Moretana process, also known as the Sumitomo-Fujikasui process. The process was developed for simultaneous SO_2 and NO_x removal. It uses an aqueous solution of NaOH to absorb SO_2 and NO_x and ClO_2 to oxidize SO_2 to sulfate and NO_x to nitrate in the solution. Byproducts of the reactions are Na_2SO_4, $NaNO_3$, and $NaCl$.

According to Ando (1985), five small commercial Moretana process plants have been operated in Japan. At most of the plants, 99% of the sulfur dioxide and 90% of the nitrogen oxides were removed. Two disadvantages of the process are the high cost of ClO_2 consumed and the presence of chlorides in the spent solution, which creates a disposal problem.

Double Alkali Process

In the double alkali (or dual alkali) process for flue gas desulfurization, the gas is contacted with a solution of soluble alkali, such as sodium sulfite or sodium hydroxide, which absorbs the SO_2. The resulting solution is then reacted with a second alkaline material (normally lime or limestone) to precipitate the absorbed SO_2 as insoluble calcium sulfite and regenerate the absorbent solution. Several alkali combinations are possible; however, this discussion is limited to the sodium/calcium case.

Process Description

The overall effects of the double alkali process are identical to those of the limestone/lime slurry processes—SO_2 is removed from the gas, lime or limestone is consumed, and a calcium sulfite or sulfate byproduct is produced. The intermediate steps, however, are quite different and result in a complete separation of the SO_2 absorption and byproduct precipitation reactions. This approach permits the gas to be contacted with a clear solution of highly soluble salts, thereby minimizing scaling, plugging, and erosion problems in the absorbent circuit. The use of a clear reactive solution instead of a slurry also offers the potential for a higher SO_2 absorption rate because the SO_2 removal reaction is not limited by the rate of dissolution of solid particles.

The double alkali process was first described in a patent issued in 1918 to Howard and Stantial (1918). In the proposed process, a 2.5% solution of sodium hydroxide or sodium carbonate was used as the first alkali, followed by a lime solution as the second. Interest in the process was revived in the 1960s and early 1970s as a result of the serious operational problems then being encountered with early limestone/lime slurry systems. General Motors initiated pilot plant work on the process in 1969 that culminated in the construction of a commercial plant at GM's Chevrolet Parma Plant near Cleveland, Ohio. This plant, which operates with a dilute absorbent solution, was started in 1974 (Dingo, 1974). Arthur D. Little Inc. performed early work in developing the double alkali process (Lunt and Shah, 1973). Another version of the process employing a more concentrated absorbent solution was pioneered by the FMC Corporation (Legatski et al., 1976). A system of this type was placed into operation at their Modesto, California plant in December 1971. Concentrated mode processes were also developed in Japan by Kureha Chemical Industry Company, Ltd., working with Kawasaki Heavy Industries, Ltd., and by Showa Denko KK jointly with Ebara Manufacturing Company (Kaplan, 1976). Both of the Japanese processes feature the use of limestone for regeneration, sulfuric acid for sulfate removal, and a separate oxidizer to convert precipitated calcium sulfite to gypsum.

The double alkali process has not been as widely accepted as limestone/lime slurry scrubbing. However, a significant number of commercial units have been installed in Japan and the United States. Ando lists 47 indirect limestone/lime flue gas desulfurization plants operational in Japan at the end of 1977 (mostly sodium/calcium double alkali systems). He concludes that such processes are about equal to direct limestone/lime processes with regard to SO_2 removal efficiency, power consumption, and operability (Ando, 1977). In the U.S., six relatively small double alkali plants were operational and two large power plant installations were under construction in mid-1977 (Pedco, 1977; Kaplan, 1976). By mid-1983, six large double alkali systems had been sold, all for Midwest power plants burning high-sulfur coals. Data on the four plants that were in operation during 1983 are given by Glancy et al. (1983). All consistently met their SO_2 removal performance criteria, and availabilities were found to be generally higher than for direct limestone scrubbing.

Two independent cost studies published in 1983 indicate that the cost of owning and operating a limestone double alkali system would be less than that of conventional limestone scrubbing for relatively high-sulfur fuels (Reisdorf et al., 1983; Hollinden et al., 1983B). In spite of this, no new large double alkali systems have been built beyond those previously described. The principal supplier of the technology, FMC, closed its office in December 1988, and sold its lime and limestone dual alkali patents to Ontario Hydro. Ontario Hydro has indicated intent to use the technology on their own units and willingness to license the technology, but for a number of reasons they have not been able to use the technology themselves. In the U.S., Advanced Air has the license for industrial projects. The technology is essentially dormant at this time for large utility power plants (Schneider, 1992; Taylor, 1992); however, a number of industrial systems continue to be built (Lunt, 1993).

Figure 7-12 shows the absorber section of a large double alkali plant during installation. A generalized flow sheet for the double alkali process is shown in **Figure 7-13**. The gas is contacted with a clear solution containing sodium sulfite, sodium sulfate, sodium bisulfite, and, in some cases, sodium hydroxide or carbonate. Sodium sulfite is the principal reactive component and is converted to bisulfite by the absorbed SO_2. A sidestream of the recycling absorbent solution is removed and treated with lime (or limestone), which reacts with the

Figure 7-12. Venturi scrubber system being installed in a double alkali FGD plant. The unit shown handles 80,000 CFM of stack gas from the combustion of coal containing 3.2% sulfur. *FMC Corporation*

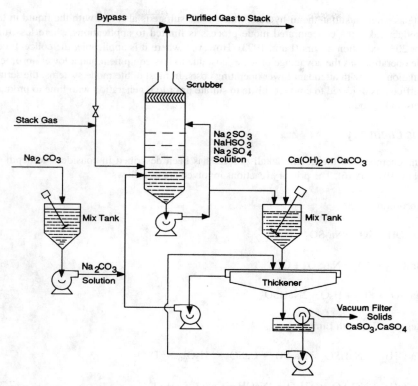

Figure 7-13. Flow diagram for double alkali flue gas desulfurization process.

bisulfite solution to form insoluble calcium sulfite and soluble sodium sulfite (and hydroxide). The insoluble precipitate is removed by settling in a thickener followed by filtration of the byproduct and washing of the filtercake. The clear liquor from the settling and filtration steps is returned to the absorber. A small amount of sodium ion make-up is required. This is typically added as sodium carbonate (or sodium hydroxide) solution to the thickener or absorber circuit.

Two distinct types of sodium/calcium-based double alkali processes have evolved, generally designated as "concentrated" and "dilute" modes. The distinction is based on the concentration of active alkali, NaOH, Na_2CO_3, $NaHCO_3$, Na_2SO_3, and $NaHSO_3$, in the circulating solution. (Salts such as Na_2SO_4 and NaCl, which are not directly involved in the SO_2 absorption reactions, are considered to be inactive.) A system is operating in the concentrated mode when the concentration of dissolved sulfite is so high that calcium is precipitated as calcium sulfite rather than sulfate (gypsum). This occurs at an active Na^+ concentration greater than about 0.15 M and is caused by the very low solubility of calcium sulfite relative to calcium sulfate. High concentrations of sodium sulfite cause calcium to be precipitated so completely that the solubility product for the more soluble calcium sulfate cannot be exceeded. Although $CaSO_4 \cdot 2H_2O$ will not normally precipitate from a concentrated mode operation, up to about 17% of the total moles of calcium removed can co-precipitate with calcium sulfite as calcium sulfate hemi-hydrate. Since some sulfate is also lost with the liquid in the byproduct cake, the concentrated mode process is limited to applications where less than about 20% oxidation occurs (Lunt, 1993). However, where it is applicable, the concentrated mode operation has the advantage of lower cost due to less equipment and a lower absorbent circulation rate with attendant lower operating costs. In most dilute mode systems, the scrubber effluent is oxidized to convert sulfite to sulfate prior to regeneration with lime to produce a gypsum byproduct.

Basic Chemistry

The chemistry of the double alkali process has been described in considerable depth by Kaplan (1974, 1976). The primary reactions involved are as follows:

Absorption:

$$2NaOH + SO_2 = Na_2SO_3 + H_2O \tag{7-32}$$

$$Na_2CO_3 + SO_2 = Na_2SO_3 + CO_2 \tag{7-33}$$

$$Na_2SO_3 + SO_2 + H_2O = 2NaHSO_3 \tag{7-34}$$

Regeneration with Lime:

$$Ca(OH)_2 + 2NaHSO_3 = Na_2SO_3 + CaSO_3 \cdot \tfrac{1}{2}H_2O_{(s)} + \tfrac{3}{2}H_2O \tag{7-35}$$

$$Ca(OH)_2 + NO_2SO_3 + \tfrac{1}{2}H_2O = 2NaOH + CaSO_3 \cdot \tfrac{1}{2}H_2O_{(s)} \tag{7-36}$$

Regeneration with Limestone:

$$CaCO_3 + 2NaHSO_3 = Na_2SO_3 + CaSO_3 \cdot \tfrac{1}{2}H_2O_{(s)} + \tfrac{1}{2}H_2O + CO_2 \tag{7-37}$$

Oxidation:

$$HSO_3^- + \tfrac{1}{2}O_2 = SO_4^{2-} + H^+ \tag{7-38}$$

$$SO_3^{2-} + \tfrac{1}{2}O_2 = SO_4^{2-} \tag{7-39}$$

Sulfate Removal:
Dilute Oxidized Mode:

$$Na_2SO_4 + Ca(OH)_2 + 2H_2O = 2NaOH + CaSO_4 \cdot 2H_2O_{(s)} \tag{7-40}$$

Concentrated Mode ($x \leq 0.17$):

$$xNa_2SO_4 + (1-x)Na_2SO_3 + Ca(OH)_2 = 2NaOH + \tfrac{1}{2}H_2O + Ca(SO_3)_{(1-x)}(SO_4)_x \cdot \tfrac{1}{2}H_2O_{(s)} \tag{7-41}$$

Dilute Oxidized Mode and Concentrated Mode:

$$Na_2SO_4 + 2CaSO_3 \cdot \tfrac{1}{2}H_2O + H_2SO_4 + 3H_2O = 2NaHSO_3 + 2CaSO_4 \cdot 2H_2O_{(s)} \qquad (7\text{-}42)$$

Softening (Dilute Mode Only):

$$Ca_2^+ + Na_2CO_3 = 2Na^+ + CaCO_{3(s)} \qquad (7\text{-}43)$$

$$Ca_2^+ + CO_2 + H_2O = 2H^+ + CaCO_{3(s)} \qquad (7\text{-}44)$$

$$Ca_2^+ + Na_2SO_3 + \tfrac{1}{2}H_2O = 2Na^+ + CaSO_3 \cdot \tfrac{1}{2}H_2O_{(s)} \qquad (7\text{-}45)$$

The absorption reactions are relatively straightforward. The principal reaction, 7-34, is readily reversible. This fact is utilized in the Wellman-Lord Process described in a subsequent section. Equilibrium data relating to this reaction are given in **Figure 7-14**.

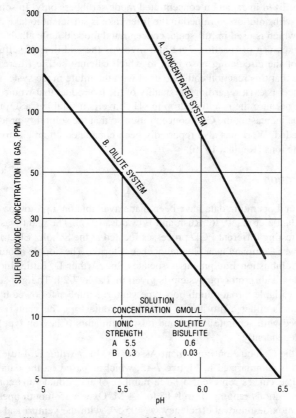

Figure 7-14. Equilibrium concentration of SO_2 in gas over sodium sulfite/bisulfite/sulfate solutions as a function of pH at 1 atm pressure and 130°F. Concentrated system: FMC laboratory data, dilute system: calculated. *Data of Legatski et al. (1976)*

The process configuration is controlled to a considerable extent by the regeneration reactions, which are complicated by the presence of both sulfite and sulfate. In dilute mode, sodium sulfate can be regenerated by reaction 7-40 only when the concentration of SO_3^{2-} is quite low because this ion forms a calcium compound that is much less soluble than gypsum. In addition, the OH^- concentration must be low and the SO_4^{2-} concentration relatively high. In actual practice, General Motor's dilute mode process is maintained at 0.1 M OH^- and 0.5 M SO_4^{2-} (Dingo, 1974), although other dilute mode systems operate at higher concentrations. Many dilute mode industrial plants operate at much higher concentrations: e.g., 0.15 M OH^- and 1.0 M SO_4^{2-} (Lunt, 1993). Because of the equilibrium in causticizing Na_2SO_4 by reaction 7-40, a high residual calcium level results. This causes sufficient calcium ions to be present in the regenerated solution (approximately 800 ppm) to produce scaling in the scrubber. As a result, dilute mode process designs typically employ a softening step in which sodium carbonate and carbon dioxide are added to precipitate calcium by reactions 7-43 and 7-44.

A number of techniques have been developed to remove sulfate from systems employing solutions that are too concentrated to permit the precipitation of gypsum. As previously noted, some calcium sulfate will coprecipitate with calcium sulfite by reaction 7-41. This reaction has been utilized in Japan on concentrated mode scrubbers only. In some cases, this plus normal losses with solution entrained in the filter cake is sufficient to take care of oxidation. A technique, which is used in full-scale, concentrated mode double alkali systems operating in Japan, makes use of reaction 7-42 and involves the addition of sulfuric acid to a small slip stream of the circulating absorbent to which calcium sulfite filtercake is added. While it is possible that this reaction could be used with the dilute mode system, this has not been done, and concerns exist regarding the quality of the byproduct that would be produced. The process works because the resulting drop in pH converts insoluble $CaSO_3$ to more soluble $Ca(HSO_3)_2$, thus increasing the Ca^{2+} concentration so that the solubility product for calcium sulfate is exceeded. Work has also reportedly been conducted on an electrolytic process for removing sulfate ions (Kaplan, 1976).

Design and Operation

Detailed design and operating data have been made available on a prototype double alkali FGD System (LaMantia et al., 1976; Rush and Edwards, 1977). The plant was one of three prototype systems testing different FGD processes located at the Scholtz Electric Generating Station of Gulf Power Company near Chattahoochee, Florida. The double alkali process unit was provided by Combustion Equipment Associates, Inc./Arthur D. Little, Inc. The design basis for the double alkali prototype system is given in **Table 7-24.** The gas-scrubbing system consisted of a variable throat plumb bob–type venturi scrubber followed by an absorber designed to operate as either a tray tower or a spray contactor. The venturi scrubber and tower were equipped with separate sumps and mist eliminators (Chevron type) so that they could be tested independently.

Performance of the Combustion Equipment Associates, Inc./Arthur D. Little, Inc. plant at the Scholtz Station is summarized in **Figure 7-15,** which is based on the data of Rush and Edwards (1977). The curves represent a large number of individual runs conducted with active alkali concentrations ranging from 0.15 to 0.4 M. Over a 15-month operating period, the average sulfur dioxide removal efficiency was 95.5% with the venturi and tray tower in operation, and 90.7% with the venturi alone. Efficiencies over 95% were readily attainable with liquid-to-gas ratios (L/G) on the order of 25 gal/1,000 acf in the venturi (primarily for dust removal) and 5 to 7 gal/1,000 acf in the tray tower. These results are generally consis-

Table 7-24
Design Basis for Combustion Equipment Associates, Inc./Arthur D. Little, Inc. Prototype Double Alkali System

Flue Gas, Inlet:	
Flow rate, acfm	75,000
Temperature, °F	275
O_2 concentration, % dry basis	6.5 (max)
Particulate loading, gr/scf dry	0.02 (from precipitator)
SO_2 concentration, ppm (dry)	1,800–3,800
Design Performance:	
SO_2 removal, %	90 (min)
Maximum SO_2 removal rate, lb/hr	1,530
Particulate, gr/scf (dry)	0.02

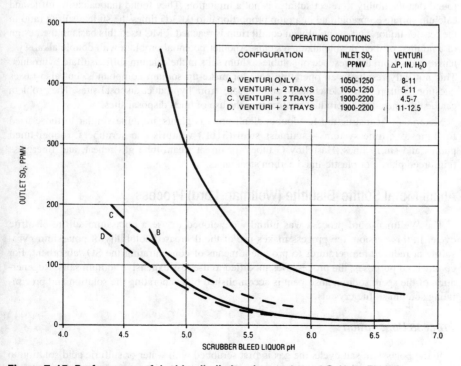

Figure 7-15. Performance of double alkali absorber system at Scholtz Electric Generating Station.

554 *Gas Purification*

tent with reported data from other concentrated mode double alkali plants. FMC, for example, utilizes a scrubbing solution with a pH of approximately 6.5 and typically observes SO_2 collection efficiencies in excess of 90% with a relatively low pressure drop scrubber (Legatski et al., 1976).

On the basis of the Scholtz Station test results, Rush and Edwards (1977) concluded that the overall performance of a properly designed and operated double alkali system should be superior to that of direct limestone and lime systems because

1. The system is highly resistant to upset and the potential for scaling is eliminated except in extreme upset conditions.
2. The handling of slurries in the absorption section is eliminated.
3. The most important control parameter, pH, has a wide acceptable range of operation.

The principal limitation that they observed relates to the inability of concentrated mode double alkali systems to reject large amounts of sulfate. This characteristic tends to limit the application of this process to plants burning fuels containing at least 1% sulfur. For example, for fuels containing less than 1% sulfur, a concentrated mode system cannot be operated at the excess air levels typical of pulverized coal boilers without an intentional purge of sodium sulfate. Above 2% sulfur, the operation is excellent. In the 1% to 2% range, successful operation is highly dependent on combustion air rate and other factors. However, LaMantia et al. (1977) found that the ability to reject sulfates is not a limitation. They found that calcium sulfite and calcium sulfate co-precipitate in a ratio proportional to 0.0365 times the sulfate/sulfite ratio in the reactor liquor, and eventually an equilibrium is reached. FMC used this basis in the design of a number of their double alkali systems. Another potential problem with double alkali systems results from the presence of soluble sodium salts in the calcium sulfite/sulfate byproduct. This not only results in an operating cost for make-up sodium compounds, but also raises questions with regard to leaching of soluble salts from byproduct disposal sites. The problem can be minimized by byproduct treatment or the use of lined disposal sites.

Materials of construction for double alkali process plants are quite similar to those used for limestone/lime systems—stainless steel (316L) venturis and scrubbers, rubber-lined pumps and slurry lines, Hastelloy G tubes for direct steam tube gas reheat, and fiberglass reinforced plastic or plastic-lined carbon steel tanks.

Alkali Metal Sulfite-Bisulfite (Wellman-Lord) Process

The Wellman-Lord process was initially developed to use a potassium sulfite-bisulfite cycle. In this version, the process makes use of the decreased solubility of potassium pyrosulfite at reduced temperatures to provide a means of concentrating the SO_2 absorbent. For commercial purposes, the process was modified to use less expensive sodium salts. Regeneration of the spent sodium absorbent is accomplished by evaporating the solution and precipitating sodium sulfite crystals.

Process Description

In the potassium salt cycle, the gas is first scrubbed with water or sulfuric acid solution to remove particulate and sulfur trioxide. It is then contacted with a potassium sulfite solution which removes SO_2 by the following reaction:

$$K_2SO_3 + SO_2 + H_2O = 2KHSO_3 \tag{7-46}$$

A bleed stream of the absorbent liquid is cooled to about 40°F to convert the bisulfate to pyrosulfite and cause crystallization of this less soluble form:

$$2KHSO_3 = K_2S_2O_5 + H_2O \tag{7-47}$$

The $K_2S_2O_5$ crystals are removed, mixed with water to form a slurry, and fed to a steam stripper. Separation of the $K_2S_2O_5$ as a solid provides maximum concentration of the SO_2-containing compound. In the stripper, the slurry is heated to about 250°F, which causes the crystals to dissolve and form bisulfite. The latter is decomposed to sulfite and SO_2 by the following reaction:

$$2KHSO_3 = K_2SO_3 + H_2O + SO_2 \tag{7-48}$$

Under optimum conditions, each pound of SO_2 produced requires 4 to 4.5 lb of steam. The vapor leaving the top of the stripper is a saturated mixture of steam and SO_2. Most of the steam is condensed on cooling the mixture to about 110°F (Chemical Construction, 1970). The condensate, which is a saturated solution of SO_2 in water, is returned to the stripper. The SO_2 may be fed to an acid plant, marketed as liquid SO_2, or reduced to elemental sulfur.

Initial pilot tests were conducted at Tampa Electric Company's Gannon Station and indicated a 90% plus removal of SO_2. Another unit rated at 56,500 cfm was installed at the Crane Station of the Baltimore Gas & Electric Co. (Besner, 1970). It is reported that this pilot unit was not fully successful (Farthing, 1971). A significant problem in the original process was loss of expensive potassium salts. As a result, the process was modified to use sodium sulfite as the active absorbent. Such a cycle was incorporated into a commercial unit installed on an Olin Corporation sulfuric acid plant in Paulsboro, New Jersey (Martinez et al., 1971).

The chemistry of the sodium cycle is extremely simple and may be represented by the following reaction for both absorption and regeneration:

$$Na_2SO_{3(aq)} + H_2O_{(l)} + SO_{2(g)} = 2NaHSO_{3(aq)} \tag{7-49}$$

Following successful operation of the Paulsboro Plant, the Wellman-Lord process was applied commercially in a considerable number of installations in Japan, Europe, and the U.S. As of 1992, thirty-eight Wellman-Lord plants had been built with gas flows up to 2.4 million scfm. Of these, 26 were still operating; four had been shut down due to plant closures, etc.; two were demonstration units and were shut down after completion of the program; three were shut down due to fuel conversions; two were converted to other processes; and the fate of one was not determined. Applications include coal-, oil- and coke-fired boilers as well as off-gases from chemical processes, Claus plants, and sulfuric acid plants (John Brown E&C, 1992). The smallest plant recovers 600 lb/hr of SO_2 from a sulfuric acid plant tail gas stream, and the recovered SO_2 is returned to the acid plant. An existing Claus plant at the site can make the process suitable for small applications if the Claus plant can accept the additional SO_2. Inlet SO_2 concentrations in commercial applications have varied between 640 ppmv and 17,750 ppmv, and outlet concentrations have been as low as 65 ppmv. An outlet concentration of 65 ppmv is said to be the lower limit of the technology (Giovanetti, 1992C).

The concentrated (typically 97% wt) SO_2 gas produced by the Wellman-Lord process can be converted into any of several byproducts in an auxiliary plant: e.g., liquid SO_2, H_2SO_4, or elemental sulfur. The concentrated sulfuric acid produced is very high quality, approaching electrolytic grade. Make-up water into the process must be condensate quality. Make-up into the prescrubber can be general service water (Giovanetti, 1992C).

A flow diagram of the Wellman-Lord process as applied to utility installations is shown in **Figure 7-16**. This diagram and the process description are based on a paper by Bailey (1974) of Davy Powergas, Inc. (now John Brown E&C). Gas from the power plant at a temperature of 250° to 300°F first enters a gas-saturation prescrubber unit. This unit serves to remove fly ash and halides and reduce the gas temperature to the 120° to 130°F range. An acidic water-fly ash slurry is recirculated through the prescrubber, and a bleed stream is withdrawn continuously to remove fly ash to the disposal pond. A mist eliminator is provided in the lower section of the absorber tower to minimize carry-over of fly ash slurry into the main absorbing circuit. After flowing through the mist eliminator, the flue gas passes into the sulfur dioxide absorption section of the tower where it is contacted with sodium sulfite/bisulfite solution. The active component in the solution is primarily in the form of sulfite when it enters the top of the column and becomes progressively richer in bisulfites as it proceeds downward. Typically the column will contain three to five sieve or valve trays.

In order to provide the necessary amount of liquid to hydraulically load each tray and maintain countercurrent operation in the overall tower, it is necessary to recirculate liquid over each tray by using an external pump. This detail is not shown in the flow diagram. The product gas passes through a conventional mist eliminator and reheater (50°F reheat is typical) and is then vented to the stack. The rich solution from the bottom of the absorption section is pumped to a storage tank, which provides feed to the regeneration portion of the plant.

Regeneration of the sodium bisulfite-rich solution is accomplished in a forced-circulation vacuum evaporator (single or double effect). The increased temperature and steam stripping vapor cause decomposition of the bisulfite to sulfite, which crystallizes from solution to form a slurry. Steam and sulfur dioxide are carried overhead to a series of condensers. In large plants, the first condenser is the heat exchanger of a second-effect evaporator. The SO_2 and steam are ultimately cooled to as low a temperature as possible to reduce the load on the vacuum pump and provide relatively pure SO_2 to the next operation. Normally, the product gas will contain approximately 85% SO_2. The regenerated absorbent (slurry) leaving the evaporators is combined with steam-stripped condensate from the partial condensers. The resulting solution is pumped to the absorbent solution storage tank for use in the absorber.

In the absorbent cycle of the Wellman-Lord Process, as in all alkali-based SO_2 removal processes, some oxidation to sulfate occurs due to oxygen in the gas stream. In addition, at the temperature of regeneration, disproportionation is possible by the following reaction:

$$2NaHSO_3 + 2Na_2SO_3 = 2Na_2SO_4 + Na_2S_2O_3 + H_2O \qquad (7\text{-}50)$$

Both of these mechanisms result in the formation of inactive salts, which must be purged from the recirculating absorbent. In the Wellman-Lord Process, this is accomplished by continuously removing a small slip stream of the absorbent solution for disposal or processing. One processing technique consists of a fractional freeze crystallization operation, which produces solids containing approximately 70% sodium sulfate and 30% sodium sulfite from a typical absorber solution containing 7.1% sulfate, 5.7% sulfite, and 21% bisulfite (Radian, 1977). Although the solid effluent produced by this operation represents a considerable

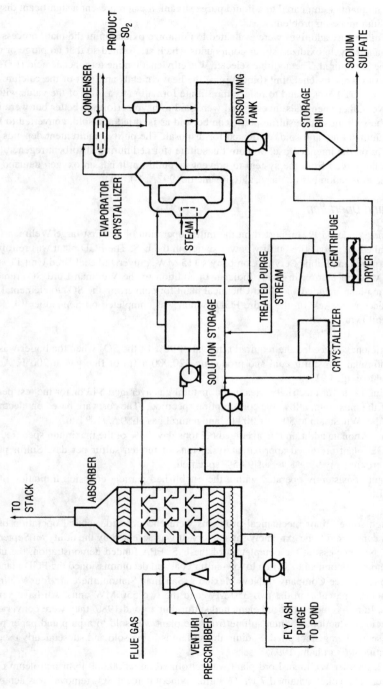

Figure 7-16. Flow diagram of the Wellman-Lord SO_2 removal process.

reduction in quantity compared to a liquid purge stream, it can represent a significant disposal and sodium make-up problem.

As a result, many additives were evaluated to minimize oxidation in the main process and to reduce and recycle oxidized sulfur compounds which are formed so that no purge or only a small one is required. The additive selected was ethylenediamine tetra-acetic acid (EDTA). It was said to work by chelating (binding-up) the heavy metals and part of the calcium and magnesium. EDTA was found to reduce the sulfate formation to a third of the value without the additive. Other chemicals investigated were said to have performed better but were hazardous. The sodium sulfate/sulfite purge can be sold as is, dried and sold, converted to high purity sodium sulfate and sold, or treated for disposal. The purity requirement depends to a large extent on the intended use. Sodium thiosulfate (formed during absorbent regeneration) must also be purged from the system to prevent its build-up. It has an oxygen demand that requires neutralization before disposal (Giovanetti, 1992A, B).

Design and Operation

Link and Ponder (1977) reported on the initial operation of the first large Wellman-Lord plant installed on a coal-fired utility power station in the U.S. The FGD plant was retrofitted to Northern Indiana Public Service Company's 115 MW_e pulverized coal-fired Unit 11 at the Dean H. Mitchell Station in Gary, Indiana. In addition to the Wellman-Lord SO_2 removal system, the plant included an Allied Chemical plant for converting the SO_2 to elemental sulfur. The acceptance test on the entire FGD system was completed on September 14, 1977, with the following results:

1. SO_2 efficiency exceeded the required removal of 90% of the SO_2 when the boiler was firing a nominal 3% sulfur coal and producing 320,000 acfm of flue gas at 300°F. Actual removal averaged 91%.
2. Operating costs for electricity, steam, and natural gas averaged $43/hr for the test period, 77% of the maximum allowable consumption specified. (The costs are based on electricity at 7 mills/kWh, steam at $0.50/1,000 lb, and natural gas at $0.55/$10^6$ Btu.)
3. Sodium carbonate addition rate averaged 6.2 tons/day, 94% of the maximum specified.
4. The FGD plant produced approximately 25 tons of molten sulfur per day. Sulfur purity was approximately 99.9% versus 99.5% specified.
5. The plant consistently operated within the established particle emission limitation of 0.1 lb/10^6 Btu.

Although some minor mechanical problems were encountered, general operation of the plant was considered to be extremely satisfactory. It was accepted by the utility on September 16, 1977. Nonetheless, at the completion of the U.S. EPA-funded demonstration, the utility found it more economical to switch to low-sulfur coal and decommissioned the FGD plant.

At Public Service Company of New Mexico's, San Juan Station, there are four Wellman-Lord systems, one each on the two 350 MW and the two 550 MW units. All have venturi scrubbers. Initially, they produced elemental sulfur, but around 1980, they were converted to sulfuric acid production. Sodium sulfate from the plant was sold to a pulp and paper plant, but is presently being stockpiled. Sodium thiosulfate is being sold and subsequently used for water treatment (Burnham, 1993).

In 1992, another Wellman-Lord plant was retrofitted on a 500,000 lb/hr petroleum coke-fired boiler. The coke contained 7 wt. % sulfur. Ninety percent SO_2 removal was achieved, and both sulfuric acid and sodium sulfate were produced (Rossi, 1993). A number of mechanical and process control problems were experienced on two units installed in Germany (325 MW_e and 350 MW_e in size). The *FGD & $DeNO_x$ Newsletter* (1991A) reviews the problems and their solutions.

Materials of Construction

The materials of construction used in four Wellman-Lord process plants were reviewed by Bailey and Heinz (1975). They report that absorber vessels constructed of either carbon steel lined with flake glass polyester or 316L stainless steel hold up quite well. The preferred materials for absorber internals are 316L stainless steel supports, fiberglass reinforced polyester (FRP) liquid distributors, polypropylene packing, and polypropylene mist eliminators. Piping for the absorbing solutions may be FRP, but an internal resin-rich layer of polyester with Dynel is recommended for improved wear resistance. Stainless steel (316L) is also generally suitable for piping and is the alloy of choice for pumps. Storage tanks for the absorbent solution may be of FRP or flake glass polyester-lined carbon steel. In the evaporation section of the plant, 316L stainless steel is required almost exclusively because of the high temperatures involved and the presence of highly corrosive SO_2 vapors. Due to its susceptibility to accelerated attack by chloride, conventional stainless steel is not recommended for installations where high chloride concentrations are encountered. In one such plant, Incoloy 825 was successfully used for the evaporator and Durimet 20 was used for wetted parts of pumps instead of 316L stainless steel. The *FGD & $DeNO_x$ Newsletter* (1991A) reviews some materials selection problems on the two Wellman-Lord systems in Germany.

ELSORB Process

The ELSORB process, offered by Elkem Technology of Norway, is similar to the Wellman-Lord process in that it utilizes an aqueous solution of sodium salts for SO_2 absorption. However, the salts are sodium salts of phosphoric acid, which are claimed to buffer the solution, thereby reducing the amount of recirculation and the size of the regeneration equipment. Oxidation of sulfite to sulfate is said to be less than 0.5%–1.0% (with 8–12% O_2 in the gas stream). Regeneration by heating and evaporation is similar to the Wellman-Lord process, but there is very little crystal formation, so there are no solids in the circulation loop and no encrustation in the evaporator. The Elsorb regeneration process operates at 120°C and atmospheric pressure. Corrosion problems have not been reported, and 316 stainless steel is used in the absorber and evaporator (Ulset and Erga, 1991; Peterson, 1992).

The capital cost is projected to be $200/kW based on a 300 MW_e power plant burning 2.6% sulfur coal and 95% SO_2 removal. This capital cost is claimed to be 10–15% less than a comparable Wellman-Lord plant because of the smaller evaporator and other equipment. An optimistic 20–30% lower operating cost is projected due mainly to less oxidation loss and correspondingly lower chemical consumption. About 10% less steam consumption is claimed (Ulset and Erga, 1991; Peterson, 1992).

The process has been tested in a mobile pilot plant capable of treating 200 m³/h of gas over a wide range of SO_2 concentrations. This unit has been operated on a coal-fired boiler in Eastern Europe. A 3,400 Nm³/h commercial unit is planned for a Claus tail gas application at an Esso (Exxon) refinery in Norway (Ulset and Erga, 1991; Peterson, 1992).

Electrodialysis Regeneration (SOXAL) Process

Aquatech Systems, a business unit of Allied-Signal, Inc., has patented the SOXAL process, which is a process for regenerating the spent scrubbing solution of an alkaline sodium salt scrubber using electrodialysis cell stacks (electrolytic cells with ion-selective membranes) (Byszewski and Hurwitz, 1991).

Either a single-stage or a two-stage regeneration process is used. The first stage regenerates sodium bisulfite solution to sodium sulfite (which is recycled to the scrubber) and a concentrated stream of SO_2. The optional second stage regenerates the sodium sulfate generated in the scrubber to caustic (which is recycled to the scrubber) and dilute sulfuric acid (which must be disposed of). The first stage consists of an electrodialysis cell stack and a steam stripper to regenerate the sulfurous acid produced by the cell to SO_2 and water. If the second stage is not used, the sodium sulfate from the stripper must be purged and disposed of since it does not react with SO_2. The SO_2 stream from the first stage can be further processed by additional equipment to produce sulfuric acid or elemental sulfur.

The second stage consists of an electrodialysis cell stack, which converts sodium sulfate into a dilute solution of sodium hydroxide that can be recycled to the absorber and sulfuric acid. Use of the second stage minimizes the sodium consumption of the system. Proposed methods for disposal of the dilute acid stream are to form gypsum by reaction with limestone, use the acid to regenerate ion-exchange resin, or use it to clean process units. Chlorides are purged from the single-stage configuration as sodium chloride along with the sodium sulfate and, from the two-stage configuration, as hydrochloric acid along with the dilute sulfuric acid stream (Byszewski and Hurwitz, 1991).

The process can also remove NO_x, and it is presently being proposed in this configuration. With this scheme, the NO_x is first controlled in the boiler by conventional injection of urea to reduce NO_x to N_2 followed by injection of methanol to oxidize residual NO to NO_2, which can be removed in the scrubber. The methanol also reacts with the residual ammonia to reduce ammonia emissions. NO_2 reacts with the scrubbing solution to produce N_2, sodium sulfate, sodium nitrate, and hydrogen. The nitrates are purged from the system either with the sodium sulfate stripper bottoms in the single-stage configuration or with the dilute sulfuric acid in the two-stage configuration. To date the NO_x removal configuration has not been tested (Byszewski and Hurwitz, 1991).

Removal efficiencies of up to 99% for SO_2 and greater than 90% for NO_x are claimed for the combined processes. The electrodialysis cell stack is described as being energy efficient in comparison to electrolytic cells. Condensate quality water is required for make-up (Byszewski, 1992).

A pilot plant test program at the Niagara Mohawk Dunkirk Station on Unit No. 4 boiler was completed in August 1993. The pilot plant treated a flue gas slip stream equivalent to about 3 MW_e. Only the first stage cell was tested. Demonstration of the second stage cell, which splits the sodium sulfate, was not considered necessary, as it has been used commercially. NO_x removal was simulated by injection of NO_x into the flue gas (Byszewski, 1992; Hurwitz, 1993).

Although not yet commercial, this process is considered promising. A study by the Electric Power Research Institute indicated it to be potentially the lowest cost regenerable process and to be competitive with the lowest cost limestone FGD processes (Byszewski and Hurwitz, 1991).

Electrolytic Regeneration (Stone & Webster/Ionics) Process

Stone & Webster Engineering Corporation and Ionics, Inc. joined forces to carry out development of this process for removing sulfur dioxide from power plant flue gases (Humphries and McRae, 1970). The process is based on the absorption of sulfur dioxide in an aqueous solution of sodium hydroxide to form sodium sulfite and bisulfite. These compounds are reacted with sulfuric acid to release sulfur dioxide, which is evolved as a pure gas, and form sodium sulfate in solution. The key to the process is in the reconversion of the inert sodium sulfate to an active absorbent for sulfur dioxide. This is accomplished in an electrolytic cell which generates both sodium hydroxide and sulfuric acid for reuse in the process. The principal chemical reactions involved in the process are

$$2NaOH + CO_2 = Na_2CO_3 + H_2O \tag{7-51}$$

$$Na_2CO_3 + SO_2 = Na_2SO_3 + CO_2 \tag{7-52}$$

$$Na_2SO_3 + \tfrac{1}{2}O_2 = Na_2SO_4 \tag{7-53}$$

$$Na_2SO_3 + SO_2 + H_2O = 2NaHSO_3 \tag{7-54}$$

$$Na_2SO_3 + H_2SO_4 = Na_2SO_4 + SO_2 + H_2O \tag{7-55}$$

$$2NaHSO_3 + H_2SO_4 = Na_2SO_4 + 2SO_2 + 2H_2O \tag{7-56}$$

$$Na_2SO_4 + 3H_2O \xrightarrow{electrolysis} 2NaOH + H_2SO_4 + H_2 + \tfrac{1}{2}O_2 \tag{7-57}$$

Reactions 7-51 through 7-54 occur in the absorber. It is desirable that reaction 7-54 be maximized in order to capture as much SO_2 as possible per unit of regenerated sodium hydroxide. Reaction 7-53 is an undesirable but unavoidable side reaction. However, oxidation of sulfite to sulfate is not as serious in this process as in many other aqueous systems since it does not interfere with the process chemistry or result in a loss of absorbent. Reactions 7-55 and 7-56 represent the SO_2 release step. Both reactions result in the formation of sulfur dioxide gas and sodium sulfate in solution. The final reaction, 7-57, depicts the overall result of electrolysis. Sodium hydroxide and hydrogen are produced at the cathode, while sulfuric acid and oxygen are produced at the anode. The sodium hydroxide is recycled to the absorber, and the sulfuric acid is used to liberate SO_2 from the rich solution. Sulfur dioxide represents the principal product of the process.

The process was tested in a 2,000-acfm pilot plant at Wisconsin Electric Power Company's Valley Station in Milwaukee (Meliere et al., 1974). The test program showed the process to be technically feasible and demonstrated that process reliability can be designed into the system as required by the power generation industry. The average SO_2 removal was 85 to 95% with inlet SO_2 concentrations from about 1,000 to 3,600 ppm. The effluent gas contained 200 to 300 ppm SO_2.

Oxidation of SO_2 in the absorber varied from 7 to 25%. As a result of the oxidation, it is necessary to have two types of electrolytic cells. A three compartment (A type) cell is the basic design that converts sodium sulfate into caustic soda and an impure sulfuric acid solution suitable for reacting with sodium bisulfite. The second type of cell, a four compartment (B type) unit, produces substantially pure 10% sulfuric acid. This acid can be withdrawn from the system for sale or disposal and represents the net production of sulfate ion by oxidation. The amount of sulfate formed in the absorber determines the number of B-type cells required. In the pilot plant, 36% of the electrolyzer cells were of the four-compartment B-type.

The majority of the operating problems experienced at the site were mechanical in nature. After they were resolved, the process had an overall operational availability greater than 90%. One early problem was caused by the presence of precipitate-forming impurities in the feed to the electrolyzer cells. This was resolved by installation of an adequate solution cleanup system. The pilot plant program included the preliminary design of a 75-MW$_e$ prototype. However, no commercial units have been placed in operation to date. According to Elyanow (1992), Ionics is no longer pursuing this process.

Zinc Oxide Process

The zinc oxide process was developed by Johnstone and Singh (1940) at the University of Illinois. Although development was utility sponsored, the process has not been used commercially. However, a considerable amount of pilot-plant work was conducted, and features of the process design were worked out in considerable detail. The process is illustrated in **Figure 7-17**. The flue gas is contacted with a solution of sodium sulfite and bisulfite and sulfur dioxide is absorbed, thus causing an increase in bisulfite content. The solution is next passed into a clarifier, in which particulate matter removed from the gas stream is separated, and finally into a mixer in which it is treated with zinc oxide. At this point, the original ratio of sulfite to bisulfite is restored, and zinc sulfite is precipitated in accordance with the following reactions:

$$ZnO + NaHSO_3 + 2\tfrac{1}{2}H_2O = ZnSO_3 \cdot 2\tfrac{1}{2}H_2O + NaOH \tag{7-58}$$

$$NaOH + NaHSO_3 = Na_2SO_3 + H_2O \tag{7-59}$$

After agitation to promote crystal growth, the precipitate is removed by settling and filtration, and the filtercake is dried and calcined. Calcining of the zinc sulfite results in a gas containing 70% water and 30% sulfur dioxide, which may be cooled, dried, and compressed to produce a nearly pure liquid sulfur dioxide as the final product. Zinc oxide obtained in the calciner is recycled to the process.

As in most processes for recovery of sulfur dioxide from flue gas, oxidation of sulfur dioxide to sulfate introduces a complication. In this case, the sulfate is removed as calcium sulfate, which is formed by treatment with lime.

Lime is added to a clarified side stream of the solution. This results in the precipitation of insoluble calcium sulfite to form a slurry, which is added to the main solution-stream leaving the gas washer. The resulting thin slurry is passed into a clarifier. The calcium sulfite and any fly ash which may have been picked up are then removed as slurry. This slurry is acidified by contacting it with a portion of the product sulfur dioxide. Acidification results in conversion of the calcium sulfite to the more soluble bisulfite form and reaction of the dissolved calcium ions with any sulfate present in the solution to form calcium sulfate, which is relatively insoluble under these conditions. Precipitated calcium sulfate and undissolved ash are removed together in a small filter. The resulting desulfated solution containing dissolved calcium bisulfite is then treated with lime to form the slurry, which is recycled to the process.

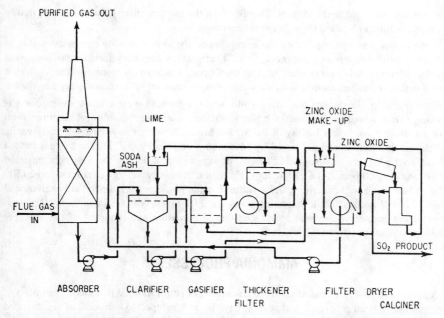

Figure 7-17. Flow diagram of zinc oxide process for sulfur dioxide recovery.

As the zinc oxide process has not been applied commercially since the presentation of complete data in 1940, and recent economic evaluations have not shown it to be economically competitive with other processes, no detailed design data are given. However, it should be noted that this process has been developed very thoroughly with regard to chemical-engineering design data. Much of the work may be useful in connection with other systems, particularly the studies of tower packings (Johnstone and Singh, 1937) and the use of wet cyclone scrubbers (Johnstone and Kleinschmidt, 1938).

Citrate Process

The citrate process was developed specifically for the removal of sulfur dioxide from smelter gases by the Salt Lake City Metallurgy Research Center of the U.S. Bureau of Mines (Rosenbaum et al., 1973). The absorbent is an aqueous solution containing approximately 190 g of citric acid and 80 g of sodium carbonate per liter, and is capable of absorbing 10 to 20 g of sulfur dioxide per liter.

Smelter gases containing 1 to 3% sulfur dioxide are first freed of particulate matter, then cooled to about 120°F, and subsequently contacted countercurrently with the citrate solution in an absorption tower. The loaded solution is reacted with hydrogen sulfide in a stirred, closed vessel, and elemental sulfur is precipitated. The sulfur slurry flows to a thickener, and the thickener underflow is centrifuged to separate the sulfur from regenerated citrate solution which, together with the thickener overflow, is returned to the absorption tower. The sulfur-cake is heated in an autoclave. Liquid sulfur is separated from the residual citrate solution, which is also returned to the system. Two-thirds of the molten sulfur product is converted to hydrogen sulfide to be used for solution regeneration.

The process was operated (with the exception of the conversion of elemental sulfur to hydrogen sulfide) in a pilot plant processing 400 cfm of reverberatory furnace gas, located at the San Manuel, Arizona, smelter of Magma Copper Company. Operating data collected over a period of several months indicate sulfur dioxide removal efficiencies exceeding 90%.

Subsequent to the smelter plant tests, pilot scale operations were conducted on flue gas discharge from a coal-fired industrial boiler simulating a utility application. A demonstration unit was built at the boiler facility of the St. Joe Minerals smelter in Monaca, Pennsylvania. A detailed description was presented by Madenburg and Kurey (1977). The system treated 156,000 scfm of flue gas from a 120 MW_e power station, reducing the SO_2 concentration from 2,000 to 200 ppm (90% efficiency). These authors also give cost data for a large (500 MW_e) plant firing coal containing 2.5% sulfur. Capital costs for such a plant are estimated at $76/kW (1977 dollars) while operating costs are 2.07 mils/kWhr (1978 dollars). Development of the citrate process was abandoned due to corrosion and plugging of the absorber packing with sulfur (Berad, 1992).

AMMONIA PROCESSES

A number of processes based upon the absorption of sulfur dioxide in aqueous solutions of ammonia have been proposed, and several have been developed to commercial or advanced pilot-plant operations. The processes differ primarily in the method of removing the sulfur dioxide from the ammonia-containing solution. Techniques used include steam or inert-gas stripping, oxidation to sulfate, reduction to elemental sulfur, and displacement by a stronger acid. Three processes do not remove the sulfur dioxide from the ammonia-containing solution, but rather produce ammonium-based fertilizer.

The possibility of using an aqueous solution of ammonia to absorb sulfur dioxide was considered as early as 1883 (Ramsey, 1883), and the use of countercurrent washing in stages was disclosed in 1929 (Hansen, 1929). This latter patent also described the use of sulfuric (or other strong acid) to release the absorbed sulfur dioxide. The use of a cyclic ammonia system to concentrate sulfur dioxide, which is later reduced to elemental sulfur with hot carbon, was disclosed in 1934 by Gleason and Loonam (1934) in a patent assigned to Guggenheim Bros. Pilot-scale development work on the Guggenheim process was conducted by the American Smelting and Refining Company at its Garfield, Utah plant (Fleming and Fitt, 1950). However, this work was terminated without construction of a commercial plant.

H.F. Johnstone of the University of Illinois made important contributions to the early development of the ammonia process, particularly with regard to systems employing heat regeneration. Patents covering certain aspects of the operation were obtained as a result of this work (Johnstone, 1937, 1938).

Commercialization of the ammonia process was pioneered by the Consolidated Mining & Smelting Company, Ltd. (Cominco), which operated a 3 ton/day sulfur-producing pilot unit at their Trail plant in 1934 and placed a 40-ton/day commercial plant in operation in 1936 (King, 1950). The sulfur dioxide recovered in these early units was reduced to elemental sulfur. Later changes in the market picture made it more economical to use the concentrated sulfur dioxide streams as feed to sulfuric acid plants. Sulfur dioxide-absorption processes using both heat and acid neutralization were developed at Trail. Present operations use the neutralization process.

Sulfur Dioxide Removal

Processes based on the absorption of SO_2 in ammonia solutions have been commercialized in Japan, Russia, and Germany, while development work on advanced concepts has been conducted in France and the United States (Breed and Hollinden, 1974). As of 1993, there were four commercial ammonia scrubbing processes: the Walther, ATS Technology™, General Electric (GEESI), and the Nippon Kokan (NKK) processes. The ABS and the Cominco processes, described in the next section, have seen no development since 1979 and 1985, respectively (Patterson, 1992; Meyer, 1992).

Basic Data

Vapor Pressure of Sulfur Dioxide and Ammonia

Equilibrium partial vapor pressures over solutions of the ammonia-sulfur dioxide-water system have been reported by Johnstone (1935). His data cover temperatures from 35° to 90°C as well as concentrations in the range likely to be encountered in a cyclic process in which the solution is regenerated by distillation. Johnstone proposed the following equations to predict the partial pressure of sulfur dioxide and ammonia over aqueous solutions:

$$p_{SO_2} = \frac{M(25-C)^2}{(C-S)} \tag{7-60}$$

$$p_{NH_3} = \frac{NC(C-S)}{(2S-C)} \tag{7-61}$$

Where: C = concentration of ammonia, moles/100 moles H_2O
         S = concentration of sulfur dioxide, moles/100 moles H_2O, and
         M, N = empirical constants which vary with temperature according to the equations:

$$\log M = 5.865 - 2,369/T \text{ (T in° K)} \tag{7-62}$$

$$\log N = 13.680 - 4,987/T \tag{7-63}$$

In ordinary operations on waste gases, oxidation occurs to form sulfate ions, which tie up some of the ammonia as ammonium sulfate. If this occurs, equations 7-60 and 7-61 become

$$p_{SO_2} = \frac{M(2S-C+nA)^2}{C-S-nA} \tag{7-64}$$

$$p_{NH_3} = \frac{N(C)(C-S-nA)}{2S-C+nA} \tag{7-65}$$

Where: A = concentration of sulfuric acid (or other strong acid)
         n = valence of acid ion (2 for sulfate)

Comparison of experimental data with values calculated from the equations shows good agreement except near the bisulfite ratio where S approaches C. In this region, small analytical errors are greatly magnified. **Figures 7-18, 7-19,** and **7-20** present vapor-pressure data for typical solutions in graphical form.

The vapor pressure of water over the sulfur dioxide-ammonia-water system was found to follow Raoult's law quite well and can be estimated from the following relationship:

$$p_{H_2O} = p_w \left(\frac{100}{100 + C + S} \right) \qquad (7\text{-}66)$$

Where: p_w = vapor pressure of pure water at the same temperature

Measurements of the solution pH are also presented by Johnstone (1935), and the following empirical equation matches the observed data within 0.1 pH unit over the range of concentrations studied:

$$pH = -4.62(S/C) + 9.2 \qquad (7\text{-}67)$$

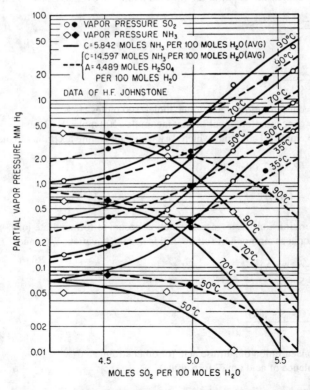

Figure 7-18. Vapor pressures of sulfur dioxide and ammonia over a solution containing 5.842 moles ammonia/100 moles water and over a solution containing an equivalent concentration of free ammonia plus dissolved ammonium sulfate. *Data of Johnstone (1935)*

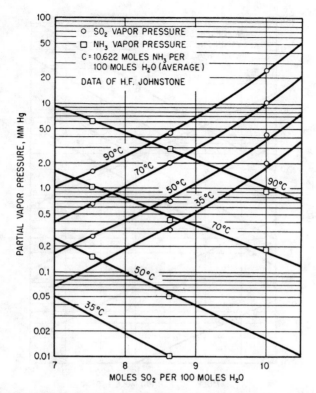

Figure 7-19. Vapor pressures of sulfur dioxide and ammonia over a solution containing 10.622 moles ammonia/100 moles water. *Data of Johnstone (1935)*

The equation cannot be extrapolated entirely to the bisulfite ratio (at which the pH is approximately 4.1).

Heats of absorption for both sulfur dioxide and ammonia in the dilute aqueous solutions considered have been estimated from slopes of vapor pressure curves on the log P versus 1/T scale. The values obtained for sulfur dioxide vary from −9,500 to −11,500 cal/mole, while those of ammonia vary from −19,400 to −22,900 cal/mole.

Oxidation of Absorbed Sulfur Dioxide

In all of the processes utilizing aqueous solutions of ammonia, some of the sulfur dioxide may be oxidized to sulfur trioxide (or in solution to the sulfate ion). The reactions may be represented by the following equations:

$$2SO_2 + O_2 = 2SO_3 \tag{7-68}$$

$$3SO_2 = S + 2SO_3 \tag{7-69}$$

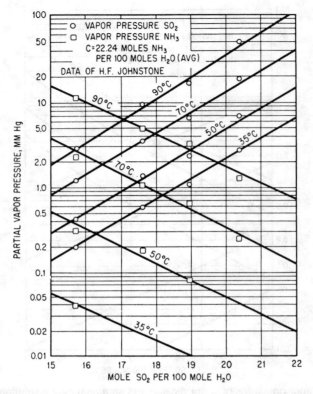

Figure 7-20. Vapor pressures of sulfur dioxide and ammonia over a solution containing 22.24 moles ammonia/100 moles water. *Data of Johnstone (1935)*

Both reactions, of course, result in the formation of ammonium sulfate in the absorbing solution. Reaction 7-68 results from the presence of oxygen in the gas being treated and is normally the most rapid, while reaction 7-69, which can occur without added reactants, is accelerated by the presence of various catalysts (e.g., arsenious oxide and reduced sulfur compounds [Howat, 1940]). The problem of sulfur dioxide oxidation in aqueous ammonium sulfite and bisulfite has been studied by Wartman (1937), who found that the reactions could be inhibited by gallic acid, tannin, pyrogallol, and certain other reducing agents. Processes that have been developed, however, have generally been designed to accept or enhance the oxidation that occurs rather than inhibit it.

Heat-Regenerative Ammonia Process

The heat-regenerative process for recovering sulfur dioxide from gas streams with ammonia solutions was developed on a pilot scale by the American Smelting and Refining Company, but first commercialized by the Consolidated Mining & Smelting Company, Ltd., at their Trail, Canada, plant as their "exorption" process. This operation was subsequently converted to regeneration with sulfuric acid, and no commercial installations of the process are now known. The process is based upon the following reversible reaction:

$$2NH_4HSO_3 = (NH_4)_2SO_3 + SO_2 + H_2O \tag{7-70}$$

The two principal problems of this process are oxidation of sulfur dioxide to form ammonium sulfate and loss of ammonia by vaporization. The oxidation problem can be alleviated by acidifying a portion of the circulating absorbent to release sulfur dioxide and produce ammonium sulfate solution or by carefully controlling the ammonium sulfate concentration in the circulating stream so that the required amount can be removed by a crystallization step. At the Trail plant, the first operation was used because ammonia acidification units were available elsewhere in the plant. Control of the ammonia-vapor-loss problem requires maintenance of minimum temperatures in the absorbers and careful adjustment of solution concentrations.

Design of Distillation Unit

The design of a distillation system for regenerating dilute ammonia solutions containing absorbed sulfur dioxide is complicated by the fact that compound formation takes place in the liquid phase, although all three components are volatile. A method for calculating the number of theoretical plates and quantity of steam required for the cases of distillation by direct steam addition and indirect heating was developed by Johnstone and Keyes (1935). The method is based upon a tray-by-tray calculation of the equilibrium composition at each theoretical contact starting with the desired lean-solution composition. Results of such calculations for regeneration with direct steam addition and indirect heating of the regenerator are shown in **Figure 7-21**. The plot is based upon use of a solution containing 22.27 moles ammonia per 100 moles water, which, according to the authors, is approximately the highest concentration that can be used in this system. It will be noted that the most efficient steam utilization is obtained with direct steam addition over most of the range of solution compositions. The use of 10 rather than 5 theoretical trays produces only a small decrease in the quantity of steam required. More than 10 theoretical trays yields no significant improvement. The feed solution composition is a function of absorber conditions and can be estimated on the basis of the equilibrium vapor-pressure composition curves of **Figures 7-18** to **7-20.** As shown by the equilibrium curves, a reduction in the sulfur dioxide concentration of the solution during stripping will result in a decrease in sulfur dioxide concentration and an increase in ammonia concentration in the vapor. Obviously, a point will be reached where no further enrichment of the vapor with regard to sulfur dioxide occurs. Johnstone and Keyes (1935) note that this limiting sulfur dioxide concentration lies somewhat below the point where the sulfur dioxide and ammonia have the same vapor pressures.

Cominco Sulfur Dioxide-Recovery Process

The process developed at the Trail, Canada smelter of the Consolidated Mining & Smelting Company, Ltd. to absorb sulfur dioxide from exhaust-gas streams produced by their metallurgical operations and sulfuric acid plant is known as the Cominco sulfur dioxide-recovery process. The process is based upon the absorption of the sulfur dioxide in an aqueous solution of ammonium sulfite and the liberation of absorbed sulfur dioxide by the addition of sulfuric acid to the solution, forming ammonium sulfate as a byproduct. The process has also been applied to acid-plant tail gases by the Olin-Mathieson Chemical Corporation at their Pasadena, Texas plant. Olin-Mathieson has acquired the rights to license this process in the United States. A flow sheet of the process as employed by Olin-Mathieson is shown in **Figure 7-22.**

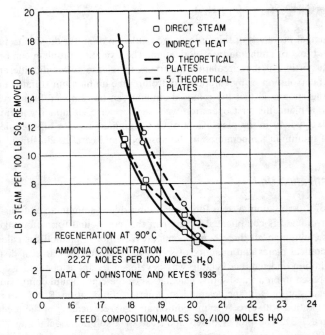

Figure 7-21. Effect of feed composition and number of plates on the quantity of steam required for regenerating a relatively concentrated ammonia solution. *Data of Johnstone and Keyes (1935)*

Absorption Step

Operating data for four absorption systems of plants utilizing the Cominco process are presented in **Table 7-25.** Observed sulfur dioxide removal efficiencies vary from 85 to 97%. The degree of sulfur dioxide removal attainable in a system of this type is obviously dependent upon a large number of variables. Chief among these are

1. Height (and type) of packing in each stage
2. Number of stages
3. Solution-circulation rate in each stage
4. Gas-flow rate
5. Solution composition (with respect to both ammonia and sulfur dioxide) in each stage
6. Temperature

Wood-slat packing is used in all of the absorbers described in **Table 7-25.** Packing of the lead-sinter plant and zinc-roaster plant gas-absorbers is described (Ontario, 1947) as 2- by 6-in. boards on edge, 2 in. apart, with each layer arranged at right angles to the one below it. At intervals, 2- by 8-in. boards are used in place of 2- by 6-in. boards so that the alternate layers are about 2 in. apart, permitting lateral flow of gas.

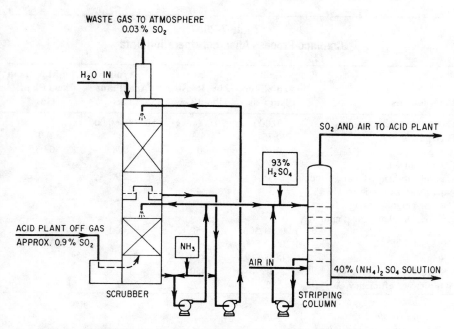

Figure 7-22. Flow diagram of Cominco sulfur dioxide-recovery process.

Aqueous ammonia of approximately 30% concentration is used as make-up in the Trail absorbers. Where several absorption stages are used, the fresh ammonia additive is divided so that a portion goes to the circulating stream of each tower to maintain the proper pH for optimum absorption with minimum ammonia loss. The pH values of the solution in the various absorption units range from about 4.1 to 5.4. The low figure represents the richest solution with regard to sulfur dioxide; this circulates in the first stage of the zinc-roaster system and contacts gas containing 5.5% sulfur dioxide. The pH of solutions in the last stages (with respect to gas) of the Trail absorption systems are approximately 5.1 for the lead-sinter plant unit, 5.2 for the zinc-roaster plant unit, and 5.4 for the acid-plant pretreatment tower (single stage).

Temperatures of absorption must be kept as low as possible to minimize ammonia loss and maintain a favorable equilibrium for absorption of sulfur dioxide. The heat of reaction is removed from the sulfur dioxide-absorption units handling smelter gases by passing the circulating streams of solution through aluminum-tube coolers so that the final gas-contact temperature is no higher than about 35°C (95°F). Temperature control is greatly simplified for the absorbers handling acid-plant tail gas, as this gas stream is so dry that evaporation of water to saturate it provides ample cooling if the gas does not contain more than about 1% sulfur dioxide. A heat balance for a typical case is presented by Burgess (1956) and is based upon the following overall heat of reaction for the absorption of sulfur dioxide in a circulating solution to which 28% ammonia solution is added:

$$SO_{2(g)} + NH_{3(28\% \text{ aq})} + H_2O_{(l)} = NH_4HSO_{3(aq)} \tag{7-71}$$

$\Delta H = -42,750$ Btu/lb mole

Table 7-25
Cominco Process Absorber Operating Data

Plant Factors	Trail Lead Sinter Plant Gas	Trail Zinc Roaster Plant Gas	Trail Acid Plant Gas	Olin-Mathieson Acid Plant Gas
Gas volumetric flow, scfm	150,000	20,000(avg)	50,000–60,000	
Feed gas, % SO_2	0.75	5.5	1.0	0.9
Tail gas, % SO_2	0.10	<0.2	0.08	0.03
SO_2 in rich sol., g/l	500	400–550		
Gas velocity (superficial), ft/sec	4.0	1.7	2.9	
No. of stages in series	3	4	1	2
Packing height per stage, ft.	17	17	25	
Approx. circulation rate, gpm:				
Stage 1**	1,200–1,500	450	1,000	
Stage 2	1,200–1,500	450		900***
Stage 3	600–800	450		
Stage 4		450		
SO_2 removal efficiency, %	85	97	92	97

*Per unit.
**Gas feed to stage 1.
***For stages 1 and 2 combined.

Simple calculations show that for a gas containing 1% SO_2, the heat generated by reaction 7-71 is approximately equal to the heat required for the evaporation of water into vapor at 25°C (77°F), assuming that equilibrium is attained with regard to water, and that the vapor pressure of water over the solution is about 80% that of pure water.

Burgess (1956) points out the importance of feeding a clean gas into the absorber to minimize ammonia losses. The presence of an acid fog in the gas stream from H_2SO_4 plants can cause formation of an ammonium sulfate aerosol, which is not recovered by the scrubber solution, and can result in a tenfold increase in ammonia losses. Too high a pH in the absorbing solution can also cause a fogging condition due to the formation of ammonium sulfite in the gas stream.

As previously noted, the pH of the solution in the Cominco operation ranges from 4.1 to 5.4 with the lowest value for the most concentrated gas stream. Hein et al. (1955) found that, with a gas containing about 0.3% sulfur dioxide, essentially no absorption took place when the pH was 5.6 or less. Their work was conducted using a 2 ft inside diameter pilot-plant scrubber packed with 2-in. ceramic rings. As would be expected, increasing the pH gave increased SO_2 recovery, but also increased ammonia loss. To attain 80% SO_2 recovery with 8 ft of packed height, for example, a pH of about 6.4 was required, and an ammonia loss of about 5% occurred. The ammonia was found to be recoverable, however, by introducing a second-stage absorber.

Stripping Operation

At the Trail plant, a quantity of solution equivalent to both the ammonia added and the sulfur dioxide absorbed is continuously withdrawn from the base of the absorbers and pumped to the acidifiers. These are steel tanks lined with acid-proof brick, 8 ft in diameter and 10 ft high. Only one tank is used at a time. In this vessel, 93% sulfuric acid is added to convert the ammonia to ammonium sulfate and to free the sulfur dioxide. After neutralization, the solution is saturated with sulfur dioxide, and it is necessary to remove the last traces by stripping with steam or air. A tall packed tower is used for this purpose and serves to reduce the sulfur dioxide content of the solution to below 0.5 g/l. The air-sulfur dioxide mixture from the top of the stripper contains about 30% sulfur dioxide and is used along with the pure sulfur dioxide from the neutralizer as feed to the acid plants.

At Olin-Mathieson's plant in Pasadena, Texas, 93% sulfuric acid is added to the solution withdrawn from the scrubber in a Karbate mixing tee. The mixture passes into a lead-lined, steel bubble-plate column through which air is blown. Acid-feed rate to the stripper is regulated by a controller operating on the pH of the neutralized solution. The air-sulfur dioxide mixture is forced into the drying tower of the sulfuric acid plant, and the ammonium sulfate solution is processed in an adjacent fertilizer plant.

Ammonia-Ammonium Bisulfate (ABS) Process

Development work on this process was conducted by TVA at its Colbert Power Plant (Breed and Hollinden, 1974; TVA, 1974). The concept is closely related to the Cominco process in that the SO_2 is absorbed in an ammonium sulfite solution and then liberated by acidulation of the solution. It differs, however, in the technique employed for acidulation.

Spent solution from the SO_2 absorber, containing ammonium bisulfite and ammonium sulfite, is reacted with ammonium bisulfate according to the following equations:

$$NH_4HSO_3 + NH_4HSO_4 = (NH_4)_2SO_4 + H_2O + SO_2 \tag{7-72}$$

$$(NH_4)_2SO_3 + 2NH_4HSO_4 = 2(NH_4)_2SO_4 + H_2O + SO_2 \tag{7-73}$$

The resulting liquor is stripped with air or steam to remove the SO_2 and is then fed to a crystallizer for the production of ammonium sulfate crystals. The crystals are then decomposed by heating to approximately 700°F. The decomposition reaction produces ammonium bisulfate for acidulation and ammonia for recycle to the process in accordance with the following reaction:

$$(NH_4)_2SO_4 \xrightarrow{700°F} NH_3 + NH_4HSO_4 \tag{7-74}$$

In a commercial application of this process, the concentrated SO_2 stream produced by the acidulation and stripping operations would typically be fed to a sulfuric acid plant or sulfur production unit.

Catalytic/IFP/CEC Ammonia Process

A major problem that has been encountered with ammonia scrubbing systems is the appearance of a characteristic "blue plume." The plume is caused by the precipitation of ammonium salts from the vapor phase as extremely small solid particles. A research program was carried out by Catalytic, Inc., and its parent company (at the time), Air Products &

Chemicals, Inc., to develop an understanding of the mechanism of blue plume formation and techniques for avoiding it (Quackenbush et al., 1977; Ennis, 1977). This work led to the establishment of "fumeless" design criteria, which were subsequently patented (Spector and Brian, 1974). The essence of the Air Products/Catalytic fume avoidance technology as presented by Quackenbush et al. (1977), is shown in **Figure 7-23.** The data upon which the figure is based indicate that any combination of vapor pressures and temperatures representable as a point below the lower curve is not conducive to fume formation, while any point above the upper curve represents conditions which are likely to result in the generation of a visible plume. The region between the curves represents a transition zone and provides a margin of safety when the lower curve is used for design purposes.

The proposed fume avoidance correlation is based on the precipitation of ammonium bisulfite by the following reaction:

$$NH_{3(vap)} + SO_{2(vap)} + H_2O_{(vap)} = NH_4HSO_{3(solid)} \tag{7-75}$$

The presence of chlorides in the gas phase can also cause the formation of a plume (NH_4Cl). In order to avoid this occurrence, Quackenbush et al. (1977) recommend that inlet chlorides be kept below 10 ppm by efficient aqueous scrubbing of the gas before it enters the SO_2 absorber. They also note the importance of operating the entire absorption system within the safe zone defined by the curves. For example, the direct contact of ammoniacal solu-

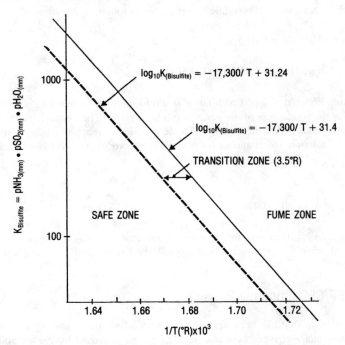

Figure 7-23. Criteria proposed by Air Products and Chemicals, Inc. for fumeless operation of SO_2 absorbers utilizing ammoniacal solutions. *Data of Quackenbush et al.*

tion with hot dry gas or the presence of localized hot or cold spots within the absorber must be avoided. It is also necessary to remove ammonia from the product gas to prevent the formation of fumes outside the scrubber. This can best be accomplished by contacting the gas with slightly acidic water followed by an efficient mist elimination step prior to discharging the product gas into the atmosphere.

Catalytic, Inc. has proposed a complete flue gas purification system which combines their ammonia scrubber technology with the liquid Claus sulfur recovery system developed by Institut Français du Pétrole (IFP) and, if desired, the NO_x removal technology of Chisso Engineering Company (CEC). The IFP process has reportedly operated successfully in a 30 MW_e scale test installation in France, and the CEC process has been piloted in Japan (Quackenbush et al., 1977). The CEC process requires that catalysts (EDTA and ferrous ion) be present in the scrubbing solution. These promote the absorption of NO, forming an adduct. The additives also cause the SO_2 to form a dithionate $((NH_4)_2S_2O_2)$ so a decomposition step for this compound is needed in addition to the standard ammonium sulfite process steps. The NO absorption reaction actually produces additional ammonia by the following overall reaction:

$$2NO + 5SO_2 + 8NH_3 + 8H_2O = 5(NH_4)_2SO_4 \tag{7-76}$$

In the combined process, ammonium sulfate formed by oxidation of sulfite is concentrated to a slurry, then decomposed at 600° to 700°F in a step similar to that used in the ABS process. The liberated SO_2 is reduced in an H_2S generating unit which can utilize reducing gas from a coal gasifier. The generation of H_2S is controlled to produce a two-to-one ratio of H_2S to SO_2 for feed to the IFP liquid Claus reactor. The products of this unit are molten sulfur and ammonia. The ammonia is condensed from the Claus plant tail gas, concentrated, and recycled to the absorber (Radian, 1977).

Walther Process

The Walther process is an aqueous ammonia FGD process that produces ammonium sulfate fertilizer granulate. Ammonia loss due to vaporization into the clean flue gas is controlled by the use of a second absorber vessel with a special mist elimination system specifically designed to remove the ammonia aerosol from the flue gas (Reijnen, 1990).

Chemical reactions governing the process are

Reaction in the Scrubbing Process:

$$(NH_4)_2SO_3 + SO_2 + H_2O = 2NH_4HSO_3 \tag{7-77}$$

Oxidation:

$$NH_4HSO_3 + NH_3 = (NH_4)_2SO_3 \tag{7-78}$$

$$(NH_4)_2SO_3 + \tfrac{1}{2}O_2 = (NH_4)_2SO_4 \tag{7-79}$$

Generation of Aerosols:

$$\{2NH_3 + SO_3 + H_2O\}_{gaseous} = (NH_4)_2SO_4 \,_{solid} \tag{7-80}$$

$$\{NH_3 + HCl\}_{gaseous} = NH_4Cl_{solid} \qquad (7\text{-}81)$$

$$\{NH_3 + HF\}_{gaseous} = NH_4F_{solid} \qquad (7\text{-}82)$$

Three Walther FGD systems had been built as of 1991: the first a $45 million, 750,000 m³/h unit on a power plant in Mannheim, Germany; the second a 191 MW_t, 475,000 m³/h unit on the Karlsruhe municipal heating and power plant Boiler No. 3, also in Germany; and the third, a 40,000 m³/h demonstration unit in Porto Vesme, Sardinia (Hüvel, 1990; *FGD & $DeNO_x$ Newsletter,* 1991B).

The first large commercial Walther plant, at the Mannheim power plant, experienced difficulties that resulted in it being shut down even though it complied with the environmental regulations (Anon., 1988). Aerosols generated in the scrubbing section, i.e., ammonium salt particles with a diameter of up to 1 μm, could not be removed from the clean gas sufficiently to overcome local objections to the white plume. The granulating process consisting of: evaporation of the scrubbing solution in a spray dryer using hot, untreated flue gas; collection of the dust in an electrostatic precipitator; and pelletizing to produce granular ammonium sulfate fertilizer; proved to be troublesome. Also, the fertilizer granulates did not reach the required degree of hardness and bulk density (Hüvel, 1990).

The Walther process was selected for the Karlsruhe system because no landfill was available for disposal, no waste water discharge was permitted, limited space was available for the scrubbing stage, and economics were favorable. The byproduct had to be salable over the long term, i.e., meet ammonia sulfate fertilizer granulate specifications. The flue gas maximum inlet concentrations of SO_2 and NO_x were 2,600 mg/m³ and 1,650 mg/m³, respectively, while the maximum allowable outlet concentrations of SO_2 and NO_x were 200 mg/m³ each (Hüvel, 1990).

The process configuration used at Karlsruhe evolved from the configuration at Mannheim and is depicted in **Figure 7-24.** The initial configuration of the first and second stage scrubbers, as well as the oxidizer and the aerosol removal stage in the second stage scrubber, are similar to those of the Mannheim unit, except for the countercurrent washing in the second stage scrubber.

A number of problems were initially experienced with this system, and the nature of these problems and the corrective actions are described by Hüvel (1990). One problem was insufficient liquid separation from the flue gas passing from the first to the second stage scrubber. This upset the chemistry of the second stage scrubber which in turn caused a high aerosol concentration to exit the scrubber. This problem was corrected by installation of a second droplet separator between the first and second stage scrubbers. Another problem was poor aerosol removal in the second stage scrubber mist eliminator (44% vs. 97% expected). Poor aerosol removal fouled the graphite tubes of the downstream regenerative heat exchanger with ammonium salts and was responsible for a visible plume when the flue gas was vented to the atmosphere. Modifications to the mist eliminator were tried but were not entirely successful in eliminating the plume. The aerosol problem was finally resolved by recirculating oxidized scrubbing solution from the oxidizer into the scrubbing solution of the first stage scrubber, increasing the concentration of the oxidized HSO_4^- and SO_4^{2-} ions, and thereby decreasing the ammonia partial pressure in the first stage scrubber. This decreased the aerosol content downstream of the second stage scrubber to 10 mg/Nm³. In March 1989, tests were conducted on the system, and all guarantees were met with the normal 0.7% sulfur coal.

A subsequent change to a 1.2% sulfur coal resulted in an aerosol content of 40 mg/Nm³ exiting the system and a visible stack plume. This aerosol increase was due to an overload of

Sulfur Dioxide Removal 577

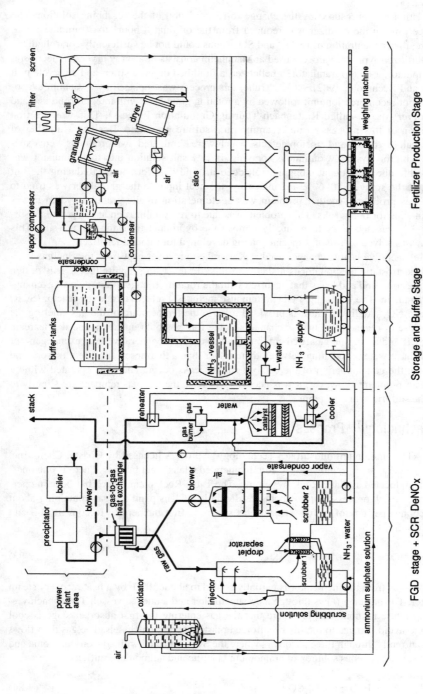

Figure 7-24. The Walther Ammonia FGD System as used at Karlsruhe. (Hüvel, 1990)

the oxidation section caused by the change to recirculation of the scrubbing solution. The residence time in the oxidizer was reduced from the original 8 hours to 20 minutes; and, therefore, the concentration of HSO^{3-} and SO_3^{2-} ions could not be sufficiently controlled.

The fertilizer byproduct process used at Mannheim consists of a spray dryer, an electrostatic precipitator, a mixer/granulator, a pelletizer, a fluid bed dryer, a sifter, and a recycle mill (Martin and Bechthold, 1982). At Karlsruhe, this process was replaced by a vacuum evaporation-vapor recompression unit followed by a rotating granulator with integral fluid bed and rotary dryer (the so-called Kaltenbach-Thuring Granulation Process). This was the first application of this process to make ammonium sulfate granulate. Besides a number of mechanical, corrosion, and erosion defects, which were eliminated, two problems were experienced with this system. Continuous spraying of 75% salt solution into the granulator was not possible due to frequent clogging, salt deposits, and dust formation. Reducing the salt content of the spray to 50–60% and increasing the heat input to the granulator was tried to correct this problem. Another problem was agglomeration of the granulate when it was stored in silos for 6–8 weeks. This problem was due to recrystallization at the granulate surface, and it was necessary to manually remove the byproduct from the storage silo. This problem was solved by use of a special coating developed for mineral fertilizer.

The regenerative heat exchanger used at Karlsruhe to recover heat from the flue gas leaving the electrostatic precipitator is a shell-and-tube design with graphite tubes. This design avoids the corrosion and leaks that occurred with the rotating Ljungstrom type heat exchanger at Mannheim. The Karlsruhe FGD system also includes a selective catalytic reactor (SCR) to remove NO_x. Additional heat is provided to raise the temperature of the gas entering the SCR by use of regenerative heat exchangers in the flue gas both upstream and downstream of the SCR and an in-line gas-fired combustion heater. The regenerative heat exchangers for the SCR have water circulating between them rather than a thermo-oil for safety reasons and to prevent the contamination of the catalyst should leaks occur. The SCR operated without significant problems; however, it was necessary to raise the exit gas temperature slightly to avoid the sulfuric acid dew point.

ATS Technology™ Process

The ATS (ammonium thiosulfate) Technology™ process licensed by Coastal Chem, Inc. (Salt Lake City, Utah) removes SO_2 from incinerated Claus plant tail gas. There is one operating plant, located at Table Rock, Wyoming. The Table Rock plant, which has been in operation since 1979, processes up to about 15 MMscfd of Claus plant tail gas containing about 5,000 pounds per day of sulfur. The tail gas stream is first incinerated to convert H_2S and traces of sulfur to SO_2:

$$2H_2S + 3O_2 = 2H_2O + 2SO_2 \tag{7-83}$$

The hot gas from the incinerator is partially cooled to about 500°F by a heat recovery steam generator. The gas then flows through a de-superheater and a quench vessel. The quench vessel temperature is monitored to ensure that the gas entering the first absorber stage is cool enough to avoid damage to the fiber reinforced plastic absorption vessels. The gas then flows countercurrently through three absorbers where the SO_2 is reacted with aqueous ammonia and converted to an aqueous solution of ammonium bisulfite and ammonium sulfite:

$$2NH_3 + SO_2 + H_2O = (NH_4)_2SO_3 \tag{7-84}$$

$$(NH_4)_2SO_3 + SO_2 + H_2O = 2NH_4HSO_3 \tag{7-85}$$

At the Table Rock plant, SO_2 is reduced to less than 30 ppmv in the flue gas exiting the vent stack.

The ammonium bisulfite/ammonium sulfite solution from the absorbers is reacted with more H_2S and NH_3 to convert the solution to ammonium thiosulfate:

$$NH_4HSO_3 + (NH_4)_2SO_3 + 2H_2S + NH_3 = 2(NH_4)_2S_2O_3 \tag{7-86}$$

The ammonium thiosulfate is then concentrated in an evaporator, and the concentrated solution is pumped to a storage tank for sale. It takes about one-half pound of anhydrous ammonia to react with every pound of H_2S, and this produces approximately 2.18 pounds of ammonium thiosulfate on a dry basis or about 3.6 pounds of ammonium thiosulfate solution.

The ATS process is claimed to be competitive with existing Claus tail gas technology for removing and recovering sulfur compounds. However, the economics must be calculated for each specific location since the cost of ammonia and the selling price for the ammonium thiosulfate solution vary widely.

Fiber reinforced plastic is used for the absorbers, while 316 stainless steel is used for the ATS reactor, the vent stack, and wherever equipment failure could release H_2S. Alloy 20 is used in selected locations (Rooney, 1993; Anon., 1980; Zey et al., 1980).

General Electric Ammonium Sulfate Process

General Electric has developed an ammonia-based process and tested it on a 10,000 acfm pilot plant at the Great Plains Synfuels Plant in Beulah, North Dakota. A 200,000 tpy plant is scheduled to go into operation in March 1997 (GEESI, 1994). The process, which is described by Saleem et al. (1993), uses ammonium sulfate solution for SO_2 absorption and employs *in situ* forced oxidation and low pH. Commercially proven limestone/gypsum process type equipment is used. In fact, the pilot plant ran in both a limestone forced oxidation mode and an ammonium sulfate mode. Sulfur dioxide removal efficiencies of over 99% were achieved with inlet SO_2 levels of up to 6,100 ppm, while ammonium sulfate of more than 99.5% purity was produced. The high-purity, high-market value ammonium sulfate crystals were successfully compacted into a premium granular byproduct. This process has the potential to lower SO_2 scrubbing costs below those of limestone-forced oxidation depending on the market for ammonium sulfate and the cost of ammonia.

Ammonia slip at the pilot plant was less than 3 ppm, and no increase in plume opacity, as a result of the treatment, was observed. The low NH_3 concentration in the outlet gas is attributed to the forced oxidation, which converts ammonium sulfite to essentially non-volatile ammonium sulfate, and the low pH, which suppresses the vapor pressure of ammonia.

Ammonia has a strong affinity for SO_2, thus permitting a compact absorber with a very low liquid-to-gas ratio. However, a low liquid-to-gas ratio requires a relatively high pH and a high ammonium sulfite level in the scrubbing solution to provide the buffering capacity for SO_2 absorption. Both conditions generate high ammonia vapor pressure and, therefore, unavoidable ammonia slip into the gas stream. These conditions, however, can be avoided by designing scrubbers with relatively large liquid-to-gas ratios and *in situ* oxidation of sulfite in the scrubbing solution. The large liquid-to-gas ratio of this process permits low pH operation and, therefore, negligible ammonia vapor pressure over the scrubbing solution. *In situ* forced oxidation further ensures that the scrubbing solution is mostly ammonium sulfate, which is a very stable salt.

The overall reactions of SO_2 absorption and oxidation with aqueous ammonia can be summarized as follows:

$$SO_2 + 2NH_3 + H_2O = (NH_4)_2SO_3 \tag{7-87}$$

$$(NH_4)_2SO_3 + \tfrac{1}{2}O_2 = (NH_4)_2SO_4 \tag{7-88}$$

The actual chemical mechanisms are more complex and include sulfite-bisulfite and sulfate-bisulfate reactions. Oxidation occurs through chain reactions initiated by free radicals, and is significantly influenced by the catalytic activities of transition metal ions as well as the inhibiting effect of oxygen scavenging compounds such as polythionates.

Much fundamental work was performed to develop this process, including work on the oxidation, crystallization, and agglomeration of ammonium sulfate. Most of the published data on the oxidation of ammonium sulfite are not useful because the data are for very dilute solutions and are inconsistent due to the strong catalytic effect of many transition metal ions. Therefore, data on oxidation rates of ammonium sulfite were generated both with and without oxidation catalysts. The oxidation rate was found to be highly dependent on the concentration of ammonium sulfate in the solution, being highest at low concentrations, which indicated that oxidation and crystallization should be separate operations.

After fly ash removal, the hot flue gas flows into a prescrubber vessel where it is co-currently contacted (gas flow down) with saturated ammonium sulfate slurry. The flue gas is cooled close to adiabatic saturation, and ammonium sulfate is crystallized by the evaporation of water from the slurry. Thus, the prescrubber acts as an evaporator/crystallizer in which the waste heat of the flue gas is used for the production of crystals without the use of an expensive external heat source. The prescrubber slurry is recycled from the agitated prescrubber sump. No ammonia is added to the slurry, hence the pH drops to less than 2, effectively preventing any significant amount of SO_2 absorption in the prescrubber.

The cooled, saturated gas leaving the prescrubber is first passed through a vertical mist eliminator and then the countercurrent (gas flow up) SO_2 absorber. In the absorber, SO_2 is removed from the flue gas with sprays of dilute ammonium sulfate solution. This solution is recycled to the absorber sprays from the absorber sump, which also acts as the oxidation reactor. Air is sparged into the base of the absorber sump to oxidize the absorbed SO_2. Anhydrous ammonia is also introduced via the air sparger to maintain the pH of the absorber solution at the desired value and to react with absorbed SO_2 and oxygen to form ammonium sulfate. The clean gas is finally passed through a horizontal mist eliminator to remove any droplets and is then vented to the atmosphere through a stack. If necessary, the clean flue gas can be reheated prior to discharge.

All of the process make-up water is added to the absorber. This ensures that the absorber solution is always dilute and, therefore, readily oxidizable. The exact concentration of the absorber solution is a function of the inlet flue gas temperature and the amount of SO_2 removed. Even at a very high inlet SO_2 concentration of 6,100 ppm, the ammonium sulfate concentration in the absorber solution is less than 30%. Under the more normal high sulfur coal condition of 3,000 ppm SO_2, the ammonium sulfate concentration is about 15%.

The absorber sump liquid level is controlled by the automatic addition of make-up water. The absorber bleed is pumped into the prescrubber as make-up via the vertical mist eliminator. In this way, deposits on the vertical mist eliminator are minimized.

A portion of prescrubber slurry containing ammonium sulfate crystals is automatically withdrawn for dewatering and separation of byproduct based on density control. This slurry bleed is dewatered in a hydrocyclone followed by a centrifuge to produce ammonium sulfate cake containing about 2% moisture. All supernatant liquor recovered from the hydrocyclone and centrifuge is returned to the prescrubber.

The 98% solids centrifuge cake is processed in a drying/compacting system to generate granular ammonium sulfate byproduct containing less than 0.5% moisture. Alternatively, the centrifuge cake can be dried directly to generate a fine ammonium sulfate powder. However, for blending with other fertilizers, a granular byproduct is necessary. In either case, the dried ammonium sulfate byproduct is easily handled, transported, and stored in weather-protected storage facilities.

When dealing with flue gases from oil- or coal-fired boilers, impurities such as fly ash and chlorides are also captured in the prescrubber and thus can accumulate in the byproduct. Should these impurities be undesirable, they can be separated from the ammonium sulfate crystals during dewatering. Since the ammonium sulfate crystals are large and the impurities are either very fine or in solution, the impurities stay with the supernatant liquor. The centrifuge cake has about 2% moisture and thus retains very little of the impurities. If necessary, the cake can be backwashed with clean saturated ammonium sulfate solution for further purification. A small slipstream of the supernatant liquor is filtered to purge out any suspended impurities and returned to the prescrubber. If the removal of any dissolved impurities such as chlorides is desired, the small supernatant slipstream can be further treated in a de-ammoniator to recover ammonia by the liming process. The de-ammoniator is simply a stirred, aerated tank in which the supernatant slipstream is mixed with milk of lime to liberate ammonia, which is then recycled to the absorber. The de-ammoniated slipstream containing calcium chloride, calcium sulfate, etc., can be disposed of after suitable treatment (Saleem et al., 1993).

Ammonia-Lime Double Alkali Process

Several processes have been developed based on the use of ammonium rather than sodium compounds as the active components of the absorbent solution of a double alkali process. As in the sodium-based double alkali process, absorbed sulfur dioxide is precipitated as an insoluble calcium salt to regenerate the absorbent. The Kurabo process represents one form of the technology in which lime is used to remove sulfur from solution and the precipitate is calcium sulfate (gypsum). This process has reportedly been employed in five oil-fired industrial boilers in Japan. Another version of the technology, the SCRA process, uses limestone instead of lime to remove the sulfur compounds from solution. This process has been tested in a small pilot plant, but apparently has not been used commercially (Behrens et al., 1984).

In the Kurabo process, the absorbent solution is maintained at a low pH (3–4) and contains primarily ammonium sulfate. The sulfate is formed from absorbed SO_2 by continuously recycling the solution through a separate oxidation step where it is contacted with air. The low pH greatly limits the solubility of SO_2 in the liquid, so a large L/G ratio is required in the absorber (50–60 gallons per Mscf). However, the low pH suppresses the vapor pressure of ammonia so the formation of an ammonium salt plume at the stack is avoided. A second advantage of this approach is that gypsum rather than calcium sulfite is produced.

The ammonia-based double alkali process has the same advantage as the sodium-based system, compared to wet lime/limestone processes, of using a clear solution in the absorption step. However, both double alkali processes have the disadvantage of greater complexity than lime/limestone scrubbers, which generally causes them to have a higher capital cost. The ammonia-based solution is somewhat easier to regenerate by the precipitation of gypsum than the sodium-based system; however, ammonia is more difficult to handle than sodium compounds. According to Ando (1985), double alkali processes became less attractive as the wet lime/limestone process was improved, and no new ammonia- or sodium-based double alkali plants were built in Japan between 1980 and the time of his report (1985).

Ammonia-Calcium Pyrophosphate

Work on this process has been conducted by the Illinois Institute of Technology (ITT) with the objective of removing SO_2 and fly ash from flue gas while producing a valuable fertilizer byproduct. The process is based on the use of a scrubbing liquid containing calcium and ammonium pyrophosphates in water. Two columns in series are used: the first serving primarily to concentrate the scrubbing solution while absorbing part of the sulfur dioxide, and the second to provide final gas cleanup. The byproduct, consisting mainly of diammonium phosphate, calcium sulfate, and calcium sulfite would, preferably, be sold in slurry form for fertilizer use. The process has been tested in a pilot plant, but no larger installations have been reported (Chi et al., 1982).

AQUEOUS ALUMINUM SULFATE PROCESS

Dowa Dual Alkali Process

The Dowa Dual Alkali Process uses a solution of basic aluminum sulfate to absorb SO_2, air injection to oxidize sulfite to sulfate, and limestone to precipitate the resulting excess sulfate in the form of gypsum.

The process was developed by the Dowa Mining Company of Japan in the early 1980s; and, by 1983, ten commercial systems were operating on a variety of smelters, sulfuric acid plants, and one oil-fired boiler (Nolan and Seaward, 1983). The process was demonstrated in the United States by UOP at TVA's Shawnee Steam Plant (Hollinden et al., 1983A).

A flow diagram of the Dowa process is shown in **Figure 7-25**. A solution of basic aluminum sulfate is used to absorb SO_2 from the gas in a packed contactor. The resulting rich liquor is then pumped through an oxidation tower where air is injected to achieve essentially 100% conversion of sulfite to sulfate. Most of the liquor is recycled to the absorber to provide a sufficiently high L/G ratio for efficient absorption. A slip stream is continuously removed and passed through an external neutralization loop where it is first used to redissolve precipitated aluminum hydroxide then neutralized with limestone to regenerate basic aluminum sulfate solution and precipitate gypsum. The gypsum is removed from the slurry by conventional settling and filtration techniques, and the clear solution is returned to the absorption loop. The basic chemical reactions of the process can be identified as follows:

Absorption:

$$Al_2(SO_4)_3 \cdot Al_2O_3 + 3SO_2 = Al_2(SO_4)_3 \cdot Al_2(SO_3)_3 \tag{7-89}$$

Oxidation:

$$Al_2(SO_3)_3 \cdot Al_2(SO_4)_3 + \tfrac{3}{2}O_2 = Al_2(SO_4)_3 \cdot Al_2(SO_4)_3 \tag{7-90}$$

Sulfur Dioxide Removal 583

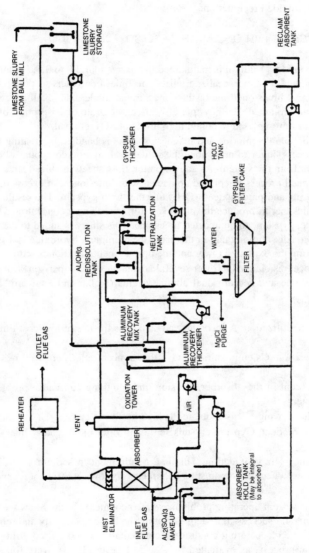

Figure 7-25. Dowa process flow diagram.

Neutralization:

$$Al_2(SO_4)_3 \cdot Al_2(SO_4)_3 + 3CaCO_3 + 2H_2O = Al_2(SO_4)_3 \cdot Al_2O_3 + 2CaSO_4 \cdot 2H_2O + CO_2 \quad (7\text{-}91)$$

Aluminum Hydroxide Precipitation:

$$Al_2(SO_4)_3 + 3CaCO_3 + 2H_2O = 2Al(OH)_3 + 3CaSO_4 \cdot 2H_2O + 3CO_2 \quad (7\text{-}92)$$

The pH of the process solution is maintained in the range of 3.0 to 3.5. Because of the low pH and the complete oxidation realized, all of the limestone reacts to precipitate gypsum. Unlike conventional limestone/lime or the concentrated mode dual alkali processes, there is no calcium sulfite produced. The low pH also affects the rate of absorption of SO_2 and an efficient contactor must be used to obtain high levels of SO_2 removal.

The removal of soluble chloride and magnesium salts is handled by treating a small bleed stream of the process liquor to precipitate the valuable aluminum compounds before purging the clean solution from the system. As shown on the flow sheet, the bleed stream is removed from the regenerated process liquor and contacted with limestone. An excess of limestone is used to raise the pH and precipitate $Al(OH)_3$ together with gypsum. The resulting mixture is separated in a thickener to produce the clean solution purge and a concentrated slurry of aluminum hydroxide, gypsum, and unreacted limestone. This slurry is added to the main process solution loop in the dissolution tank. In this tank, the aluminum hydroxide is dissolved by the acidic solution from the oxidation tower and the primary neutralization reactions are initiated.

The Dowa process is claimed to offer several advantages in comparison with both conventional limestone/lime and sodium dual alkali systems (Nolan and Seaward, 1983). These include

1. 100% limestone utilization. A secondary advantage is the use of low-cost limestone rather than lime.
2. Scale-free operation. Calcium sulfate and sulfite concentrations are well below saturation levels.
3. No slurry problem in the absorber. Erosion and plugging can cause problems in limestone/lime systems.
4. Tolerance to load swings. This is due to the high buffering capacity of the solution.
5. Stable gypsum product. Gypsum crystals settle and filter more readily than sulfite/sulfate mixtures.
6. No requirement to limit oxidation. This is a requirement in sodium dual alkali systems where the sulfate concentration must be kept relatively low for efficient regeneration.

Although some system operating problems were encountered in the Shawnee tests, the test program was generally successful in demonstrating the basic operability and reliability of the process over a range of operating conditions representative of coal-fired utility boilers. The economics of the process, as evaluated by Reisdorf et al. (1983), were found to be very favorable when compared to the more conventional (and more thoroughly developed) processes for desulfurizing the flue gas from a power plant burning high-sulfur coal. However, according to Anazawa (1984), there is no interest on the part of Dowa Mining Corporation in marketing this process in the U.S. or elsewhere.

FERROUS SULFIDE WITH THERMAL REGENERATION

Sulf-X Process

The Sulf-X process, which is the only known system using ferrous sulfide for SO_2 removal, has not been developed to commercial status; however, the chemistry is sufficiently novel to warrant at least a brief mention. A more detailed description and economic evaluation has been prepared by Stearns-Rogers Engineering Corporation for EPRI (Keeth et al., 1983), and is the basis for the discussion that follows.

The Sulf-X process was developed by Pittsburgh Environmental and Energy Systems, Inc. (Pensys). In the process, the flue gas is scrubbed with an aqueous slurry containing FeS and $Fe(OH)_2$ at a pH of about 6.2. The FeS acts as the primary sorbent of sulfur dioxide; while the $Fe(OH)_2$ aids in stabilizing the pH. As a result of the absorption reactions, more complex iron sulfur compounds such as Fe_xS_y and $FeSO_4$ are formed. Prior to regeneration, the absorbent slurry is treated with Na_2S to convert the $FeSO_4$ to Na_2SO_4 and FeS.

Spent slurry from the absorption step is dewatered, and the solids are dried. Regeneration is accomplished by heating the dried solids to about 1,400°F in the presence of coke. The heat decomposes Fe_xS_y (which approximates FeS_2) to FeS and sulfur, and the coke reduces Na_2SO_4 to Na_2S. Elemental sulfur leaves the regenerator as vapor and is condensed as liquid sulfur, the primary byproduct of the process. The FeS and Na_2S are reused in the process.

The Stearns-Rogers study concludes that the Sulf-X process has the potential for lower capital and operating cost than the Wellman-Lord process; however, it has several potential problem areas, including the operation of a high temperature calciner; the handling of three different solid reagents; pyrites, coke, and sodium sulfate; and the need to circulate a complex slurry through much of the process.

SULFURIC ACID PROCESSES

Chiyoda Thoroughbred 101 Process

Information on the Chiyoda Thoroughbred 101 (CT-101) process is included for historical purposes only. The process is no longer commercially available as it has been replaced by the CT-121 process (Mirabella, 1992B).

This process represents another alternative to the double alkali process. Sulfur dioxide is absorbed in dilute sulfuric acid, oxidized to sulfate by air blowing, then precipitated as gypsum by the addition of limestone. The oxidation rate is increased by the use of iron as a catalyst in the circulating acid and is also enhanced by the low pH of the solution. The process, which was developed by Chiyoda Chemical Engineering and Construction Company, Ltd., of Yokohama, has been used quite extensively in Japan. Fourteen plants were reportedly in operation at the end of 1977 (Ando, 1977).

A flowsheet of the Chiyoda Thoroughbred 101 (CT-101) process is shown in **Figure 7-26**. This flowsheet is based on the unit treating one half of the flue gas from a 500-MW_e boiler at the Toyama-Shinko Power Plant of the Hokuriku Electric Power Company, Ltd. Detailed operating data have been made available for this plant by Tamaki (1975). Major process equipment items are listed in **Table 7-26**.

The chemistry of the process is defined by the following equations:

Table 7-26
Key Components in the Chiyoda CT-101 Process Unit at Toyama-Shinko Power Plant

Component and Quantity	Description	Material
Prescrubber (2)	Venturi type	Titanium upper section, 316L lower section
Absorber/Oxidizer (1)	69-ft diameter × 79-ft height	
Absorber section	Packed annulus: diameter 69 and 31 ft × 30-ft height	316L
Oxidizer section	Flooded perforated tray column 30-ft diameter × 30-ft height	316L
Crystallizer (1)	Cylindrical, inner circulation type, 17,000 cu ft, 22 kVA	316L
Clarifier (1)	Cylindrical type, 10,600 cu ft	Rubber lined
Absorbent recycle pump (3, 2 + 1 spare)	Centrifugal, 48,400 gpm, 66-ft head, 800 KVA	Rubber lining and 316L
Air blower (2)	Two stage turbo-fan; 4,400 scfm, 28 psig, 300 kVA	Carbon steel
Centrifuge (3)	Basket type, automatic 1.7 tph cake, 37 kVA each	316L
Mist Eliminator (1)	Impingement type, two stage, 95% mist removal efficiency	Rubber lining Polypropylene plates

Absorption and Oxidation:

$$2SO_2 + O_2 + 2H_2O = 2H_2SO_4 \tag{7-93}$$

$$2FeSO_4 + SO_2 + O_2 = Fe_2(SO_4)_3 \tag{7-94}$$

$$Fe_2(SO_4)_3 + SO_2 + 2H_2O = 2FeSO_4 + 2H_2SO_4 \tag{7-95}$$

Crystallization:

$$H_2SO_4 + CaCO_3 + H_2O = CaSO_4 \cdot 2H_2O + CO_2 \tag{7-96}$$

When the Toyama-Shinko boiler burns 1% sulfur fuel oil, the flue gas contains approximately 450 ppm SO_2 and 2.5 to 4% oxygen. The gas passes through an electrostatic precipitator, which reduces its particulate concentration to about 0.012 gr/scf, and is then fed to the FGD unit at a rate of 467,000 scfm. It flows first through a venturi prescrubber, which cools it from 284° to 140°F, then through a packed absorber where it contacts dilute sulfuric acid, and finally through a mist eliminator before being reheated and released to the stack.

The plant has obtained a 90% desulfurization efficiency, using 2.3% sulfuric acid concentration and a liquid-to-gas ratio of 210 gal/1,000 scf (97,000 gpm).

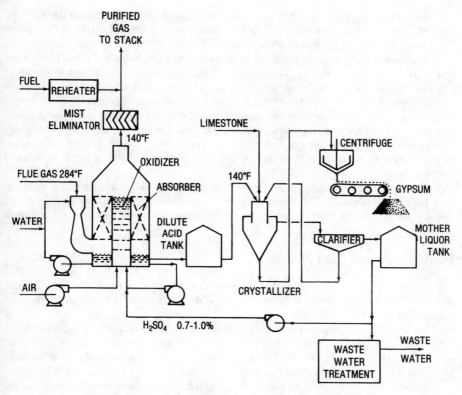

Figure 7-26. Flow diagram of the Chiyoda (CT-101) process.

The solution is oxidized by contact with air in a flooded perforated tray column located in the center of the absorber. Spent dilute sulfuric acid is pumped from the absorber sump to the bottom of the oxidizer and flows upward through this unit cocurrently with 1,900 scfm of air. The oxidized absorbent overflows from the oxidizer to a distributor system, then percolates down through the absorber, which is packed with 3-in. Tellerettes.

A portion of the recirculating H_2SO_4 stream is continuously withdrawn to a crystallizer where the acid is neutralized by the addition of limestone to a concentration of 0.7 to 1.0%. Approximately 30.6 tons/day of limestone are required when processing gas from 1% sulfur fuel oil. The product of the crystallizer is a slurry of gypsum crystals in dilute acid which is sent to a basket-type centrifuge. Gypsum cake from the centrifuge is in the form of a relatively dry powder containing less than 10% free water. This material is widely used in Japan for the manufacture of wallboard and as a retardant for Portland cement.

The liquid from the centrifuge flows into a clarifier from which settled particles are returned to the crystallizer, and clarified liquor (containing 0.7 to 1.0% sulfuric acid) is returned to the absorber/oxidizer circuit. The flow rate of the return stream is approximately 500 gpm. A small amount of the dilute acid is continuously purged from the system to prevent build-up of chlorides that could cause corrosion of stainless steel equipment. A chloride limit of 200 ppm has been specified for the plant. The catalyst, iron sulfate, is not a major cost item. It is estimated that 1,700 lb/day would be required for a CT-101 plant designed for 90% SO2 removal from flue gas of a 250 MW_e boiler burning 3% sulfur fuel oil. The same plant would need about 116 tons/day of limestone (Tamaki, 1975).

A prototype CT-101 process plant was built and tested at the Scholtz Electric Generating Station of Gulf Power Company. The objective of this program was to establish the applicability of the process to a coal-fired steam generator. Detailed operating data for a 25-month test period from 1975 to 1977 have been presented by Rush and Edwards (1977). Although several mechanical problems were encountered, they concluded that the overall performance of a properly designed and operated CT-101 system should be superior to that of direct limestone and lime systems because of its resistance to upsets, freedom from scaling, elimination of slurry handling in the absorption section, and wide latitude with regard to pH or concentration control.

Possible drawbacks of the process are the high corrosivity of the absorbent, the need for very high liquid-to-gas ratios brought about by the low solubility of SO_2 in acid solutions, and the requirement for careful water management to avoid chloride build-up and pollution problems.

ISPRA Bromine-Based Process

This process was developed by the European Community's Institute of Environmental Sciences (ISPRA) located in Italy. Ferlini, an Italian company, is commercializing the process in Europe.

The ISPRA process removes SO_2 with an aqueous solution containing approximately 15 wt% H_2SO_4, 15 wt% HBr, and 0.5 wt% bromine and produces two salable byproducts: commercial quality, 95 wt% H_2SO_4 and H_2 gas. The process has been tested at the 10 MWe level at a pilot plant at an Italian refinery in Sardinia. Greater than 95% desulfurization is said to be possible. The capital cost is said to be comparable to that of a wet limestone/lime system with substantially lower operating costs (Ferlini, 1991). The process is based on the following two chemical reactions:

Oxidation of SO_2 to Sulfuric Acid:

$$SO_2 + Br_2 + 2H_2O = H_2SO_4 + 2HBr \tag{7-97}$$

Regeneration by Electrolysis of the Hydrobromic Acid:

$$2HBr = Br_2 + H_2 \tag{7-98}$$

Before removal of the SO_2, hot flue gas is used to remove essentially all of the HBr from the absorbent and concentrate the sulfuric acid solution. This is accomplished by passing the flue gas through two evaporative concentrators into which mixed, dilute solution is sprayed. The sulfuric acid concentration is first brought up to 70 wt% in the pre-concentrator and then to 95 wt% in the final concentrator. The flue gas is also cooled in these vessels. Further inlet flue gas cooling takes place as the gas passes through a regenerative heat exchanger used to reheat the treated gas before discharge to the atmosphere. After passing through these two concentrators and the heat exchanger, the flue gas passes through the reactor where the bromine solution is sprayed into the gas to remove SO_2. The cleaned flue gas then flows to a spray scrubber where mist is liminated, then through the regenerative heat exchanger where it is heated, and finally to the atmosphere through a stack. A slip stream of scrubbing solution flows from the reactor to the electrolyzer. Hydrogen gas is evolved at the cathode and bromine at the anode. The bromine is added back into the process, and the hydrogen becomes a byproduct. Tests have shown that hydrochloric acid and NO_x in the flue gas do not affect the performance of the process (Caprioglio et al., 1991).

Noell-KRC Peroxide-Based Process

Noell-KRC Umwelttechnik GmbH offers a process for recovering both hydrochloric acid and sulfuric acid from municipal solid waste disposal boiler flue gas (Noell-KRC, 1992). Heavy metals are also removed from the flue gas. The overall process includes the following steps:

1. Gas from the particulate removal equipment is quenched with water. This serves to cool the gas and precipitate heavy metals into the scrubbing liquid.
2. The cooled gas is passed through a two stage HCl scrubber which produces a dilute solution of HCl. This is distilled in a separate unit to make a 31% HCl product. Gas from the HCl scrubber passes through a demister before entering the next step.
3. The HCl-free gas is treated in a sulfur dioxide scrubber where SO_2 is absorbed and continuously oxidized to sulfuric acid. Complete oxidation is assured by recycling a portion of the absorbent through an electrolytic cell that generates hydrogen peroxide in-situ. The hydrogen peroxide quickly oxidizes absorbed sulfur dioxide.

Purified gas from the sulfur dioxide scrubber passes through an efficient demister to remove fine droplets of sulfuric acid. It is then vented through a stack, without reheat, at a temperature of 60–70°C. The process makes commercial grade sulfuric acid (Noell-KRC, 1992).

AMINE PROCESSES WITH THERMAL REGENERATION

Four processes that remove sulfur dioxide from gases using amines are described here. These are the Sulphidine, ASARCO, Dow, and Union Carbide processes. The Sulphidine process was operational prior to World War II while the ASARCO process has been commercial since the late 1940s. Both the Dow and Union Carbide's CANSOLV processes are more recent developments which use proprietary amines.

Processes for the recovery of sulfur dioxide based on chemical absorption in amines, particularly xylidine mixtures (called xylidine here) and dimethylaniline (DMA), have been applied commercially for purifying smelter fumes; however, no applications on gases containing less than about 3.5% sulfur dioxide are known. The use of aromatic amines to absorb sulfur dioxide was disclosed in 1932 in British Patent 371,888, which specifically claimed aniline and its homologues. The first commercially successful process of this type was the Sulphidine process (Weidmann and Roesner, 1935; Roesner, 1937). This process used xylidine. DMA was first used commercially in a sulfur dioxide-absorption plant at the Falconbridge Nickel Company plant in Kristiansand, Norway (Fleming and Fitt, 1950). Later, the American Smelting and Refining Company (now ASARCO) developed a novel flow system based upon DMA (Fleming and Fitt, 1946A; 1946B) and, in 1947, installed a 20 ton/day sulfur dioxide plant at their Selby, California, smelter. In 1974, ASARCO installed a 250 ton/day plant of a modified design at their Tacoma, Washington, smelter. The novel features of the ASARCO process are based upon the flow pattern rather than the solvent. Under some circumstances, it could be advantageous to use xylidine rather than DMA in the proposed flow cycle.

Basic Data

Properties of three aromatic amines that have been proposed for sulfur dioxide absorption are presented in **Table 7-27**. As can be seen, DMA boils at a somewhat lower temperature than either xylidine or toluidine and has a correspondingly higher vapor pressure under the conditions in the absorber. Because of this, losses of DMA by vaporization (or chemical costs to recover it from the gases) may be higher than those of the other amines.

However, xylidine (which is apparently preferable to toluidine) also has disadvantages. Its sulfate is only sparingly soluble in water so that if oxidation of sulfur dioxide occurs in the solution or if sulfur trioxide is present in the gas stream, precautions must be taken to prevent the formation of crystals and subsequent plugging of equipment. The solubility of xylidine sulfite in the solvent is also not as high as would be desirable. At 20°C, for example, crystallization occurs when the concentration of sulfur dioxide reaches 108 g/l (Pastnikov and Astasheva, 1940). Because of this, xylidine is normally used in a mixture with water. Xylidine sulfite is quite soluble in water so that crystallization is avoided, and when sufficient sulfur dioxide has been absorbed, the xylidine-water mixture becomes a single phase.

A comparison of the sulfur dioxide-carrying capacities of DMA and various xylidine-water mixtures is presented in **Figure 7-27**. Data for pure xylidine follow closely the curve for the 2:1 mixture. It will be noted that at low SO_2 concentrations, all of the xylidine-water mixtures have appreciably higher capacities than DMA, while at high sulfur dioxide concentrations, anhydrous DMA appears to have the advantage. Solubility data for xylidine are based upon the work of Pastnikov and Astasheva (1940). These authors found that at 40°C a maximum solubility of sulfur dioxide in xylidine-water mixtures occurred at a ratio of seven parts xylidine to one part water (by volume). The solubility of sulfur dioxide at this ratio, about 455 g/l of mixture (in equilibrium with pure sulfur dioxide), corresponds to the following compound:

$$C_6H_3(NH_2)(CH_3)_2 \cdot H_2SO_3$$

Table 7-27
Properties of Aromatic Amines *

Property	Dimethyl-aniline (DMA)	Xylidine (ortho, para, and meta mixture)	Toluidine (ortho)
Formula	$C_6H_5N(CH_3)_2$	$(CH_3)_2C_6H_3NH_2$	$CH_3C_6H_4NH_2$
Molecular wt	121.18	121.8	107.15
Boiling pt., °C	193	212–223	200
Solubility in water	Very slight	Very slight	Slight
Vapor pressure at 20°C, mm Hg	0.35 mm	0.20 mm	
Temperature at which vapor pressure = 1 mm, °C	29.5	50	44
Specific gravity, 20°/4°	0.956	0.97–0.99	1.0

*Commercial grades

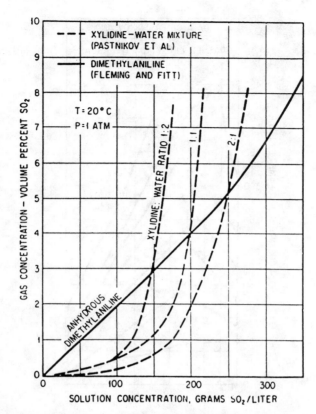

Figure 7-27. Solubility of sulfur dioxide in anhydrous dimethylaniline and various xylidine-water mixtures. *Data of Pastnikov and Astasheva (1940) and Fleming and Fitt (1950)*

The melting point of this compound was found to be 53°C.

The effects of temperature and xylidine/water ratio on the solubility of sulfur dioxide in xylidine-water mixtures are shown in **Figure 7-28.** This figure is based upon pure sulfur dioxide gas. According to Roesner (1937), the heat of reaction for the absorption of sulfur dioxide in xylidine is 4.7 kilocal/g mole (132 Btu/lb) sulfur dioxide absorbed.

Sulphidine Process

The Sulphidine process was developed in Europe by the Gesellschaft für Chemische Industrie in Basel and the Metallgesellschaft, A.G., of Frankfurt (Weidmann and Roesner, 1936A, B). One licensor of this process, Lurgi, has not supplied any of these systems since 1955 because of environmental problems with xylidine and toluidine (Silerberg, 1992). Both xylidine and toluidine, as well as ammonia, are presently classified as air toxics under Title III of the Clean Air Act Amendments of 1990.

592 *Gas Purification*

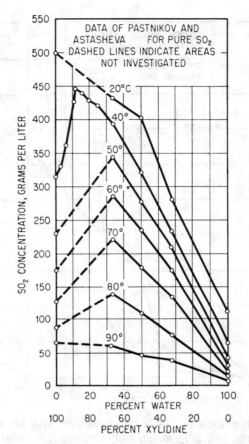

Figure 7-28. Effect of temperature and xylidine/water ratio on solubility of pure SO_2 gas at 1 atm pressure in xylidine/water mixtures. *Data of Pastnikov and Astasheva (1940)*

In the Sulphidine process, the gas feed to the sulfur dioxide-recovery plant is first cleaned in electrostatic precipitators, then passed through two packed absorbers in series where it is contacted with the xylidine-water absorbent. Xylidine vapors are recovered from the gas stream by washing it with dilute sulfuric acid before venting it to the atmosphere. The cleaned gas from the system contains 0.05 to 0.10% sulfur dioxide.

The absorbent used in this process is a mixture of xylidine and water (approximately 1:1). The mixture fed into the top of the absorber consists of two immiscible liquids, but during the absorption of sulfur dioxide, water-soluble xylidine sulfite is formed. The liquid from the bottom of the absorber therefore consists of an aqueous solution of xylidine sulfite.

The SO_2-rich absorbent from the bottom of the absorber, carrying 130 to 180 g/l of SO_2, is pumped to the top of a Raschig ring-packed stripping column in which the SO_2 is removed by heating. The reboiler is heated indirectly by steam, and a temperature of 95° to 100°C is attained. From the top of the stripping column, SO_2-laden vapor is first passed through a cooler where water and xylidine vapor are condensed, then through a water-wash column to further reduce the xylidine content. From this column and from the condenser, water that is saturated with sulfur dioxide and contains some xylidine is returned to the stripping column.

From the bottom of the stripping column, the stripped xylidine-water mixture is passed to a separator in which excess water is removed to purge the system of Na_2SO_4. Xylidine and water in the proper proportions are pumped through a cooler to the top of the second absorber. An aqueous solution of sodium carbonate is added periodically to the circulating liquid stream in the second absorber. The added sodium carbonate is converted by the free sulfur dioxide to sodium sulfite and carbon dioxide. The latter passes out of the column with the waste gas. The sodium sulfite reacts with sulfate ions, which may be formed by oxidation, and the resulting sodium sulfate is removed from the system with the wastewater stream.

ASARCO Process

Process Description

This process developed by the American Smelting and Refining Company (now ASARCO) for the absorption of sulfur dioxide from smelter gases represents an improvement over the Sulphidine process with regard to steam consumption and operating labor requirements. Although the process is reported to be applicable to either dimethylaniline (DMA) or xylidine, DMA has been used in all commercial installations of the process. The principal advantage of DMA is that it does not require water to dissolve the sulfur dioxide compound formed. In addition, as shown by **Figure 7-27,** at high concentrations of sulfur dioxide in the feed gas, DMA can absorb larger quantities of sulfur dioxide than xylidine. Note, however, that with low gas-concentrations, the use of xylidine may have an economic advantage. ASARCO has used the DMA process in their own plants at Selby, California and Tacoma, Washington, and has licensed its use at other locations worldwide. However, no ASARCO amine FGD systems have been built since 1979 (Fay, 1992), and no plants were known to be operating in 1992. The units at the Selby and Tacoma smelters were shut down in 1970 and 1985, respectively, when these facilities closed. About ten of these plants were built worldwide, and the process continues to be available for license. The Tacoma design is recommended by ASARCO, although the Selby design or combinations of the two designs are also available. The processes at Selby and Tacoma produced 500 ppm and 1,000 ppm outlet SO_2 concentrations, respectively.

A flow diagram of the process for the Tacoma plant is shown in **Figure 7-29.** Flue gas containing sulfur dioxide from copper smelting is first thoroughly cleaned and cooled, then contacted with anhydrous DMA in the lower portion of the absorption tower. Gas from this section of the absorber, which contains DMA vapor and a small percentage of sulfur dioxide, next passes through several trays where it is contacted with a 100 g/l ammonium sulfate solution. In the lower trays of this section, sulfur dioxide from the DMA absorber dissolves in water to form sulfurous acid, which captures DMA as DMA sulfite, from which both DMA and SO_2 can be recovered. The top trays of this section are designed to be used alternatively for sulfuric acid absorption of DMA, with some improvement in DMA recovery.

Rich DMA solution from the bottom of the absorber is heated by indirect exchange with hot, lean DMA and is then fed near the top of the stripping column where it is stripped of its sulfur dioxide content by steam. The resulting hot, lean DMA (with condensed steam) is passed through the exchanger, cooled further, and pumped to a separator from which lean DMA is withdrawn as liquid feed to the absorber. The aqueous stream from the sulfurous acid section of the absorption tower is neutralized with ammonia in this separator, releasing

594 *Gas Purification*

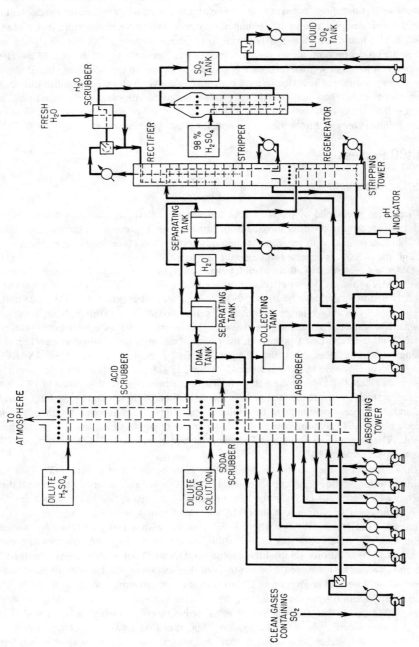

Figure 7-29. Flow diagram of the ASARCO process for sulfur dioxide recovery.

DMA to the lean DMA stream. The aqueous phase from the separator is transferred to the lowest section of the stripping tower where, heated by low-pressure steam injection, it boils to provide the steam for the DMA-stripping operation. Sulfites are decomposed to release SO_2 product; sulfates formed by oxidation (or by sulfuric acid scrubbing in the absorption tower) remain as ammonium sulfate in the stripper aqueous phase. This solution is used as liquid feed to the upper section of the absorber, and a portion is bled from the system to eliminate sulfate.

Sulfur dioxide liberated in the stripping section of the stripping tower is passed through a rectifier where steam is condensed and residual DMA vapors are absorbed. The sulfur dioxide gas is finally dried by countercurrent washing with 98% sulfuric acid, compressed, and condensed as product liquid sulfur dioxide.

Design and Operating Data

Henderson and Yavorsky (1977) describe the Tacoma plant operation. Earlier papers by Fleming and Fitt (1950) and by Henderson and Pfeiffer (1974) describe the Selby plant operation and an initial design of the Tacoma plant based on the Selby plant. The Selby plant used sodium carbonate scrubbing of the gas from the amine absorber to remove additional sulfur dioxide, followed by sulfuric acid scrubbing of the gas to remove vaporized DMA. Removal of the Selby sodium carbonate step and its accompanying neutralization requirement in the Tacoma design reduced sorbent costs and consumption and aqueous bleed requirements. Design and operating data for the Tacoma plant are given in **Table 7-28**.

For the Tacoma plant, the absorption section of the absorption tower provides eight trays, with the bottom five equipped with water coolers. Tail gas from the scrubber contains 500–1,000 ppm SO_2. The rich DMA solution contains 140–160 g/l SO_2. The sulfurous acid scrubbing section contains fifteen trays; the fifth tray from the top is equipped with total draw-off for optional sulfuric acid scrubbing. The stripping tower is divided into three sections: the regenerator where the neutralized acid-wash stream is boiled (containing nine trays), the stripping section where SO_2 is released from the rich DMA stream (containing ten trays), and the rectifying section where steam is condensed (containing five trays).

CANSOLV Process

The Union Carbide CANSOLV process is a relatively new FGD process that produces SO_2. It utilizes a proprietary, thermally regenerable organic amine-based solution, UCAR-SOL Absorbent LH-201, which is non-volatile, stable oxidatively and thermally, and designed to meet applicable health and safety standards. A pilot plant at Suncor Inc.'s oil sands plant at Fort McMurray, Alberta, Canada, treated 7,000 acfm of flue gas from boilers burning 7% sulfur petroleum coke (Hakka and Barnett, 1991).

The process consists of a gas cooling and prescrubbing section, a sulfur dioxide scrubbing section, and a regeneration and solvent purification section. The flue gas cooling and prescrubbing equipment, usually downstream of the particulate removal equipment, reduces the flue gas temperature, removes most of the strong acids and particulate matter, and saturates the gas with water. The SO_2 is then absorbed from the gas in a countercurrent multi-stage scrubber. The scrubber utilizes air atomizing nozzles to take advantage of the absorbent's high reactivity and SO_2 capacity to achieve up to 99% SO_2 removal.

The absorber has interstage solvent collectors and a mist eliminator downstream of the absorption section. The regenerator, which may be trayed or packed, is equipped with a

**Table 7-28
Data from Operations of Tacoma, Washington Sulfur Dioxide Plant, 1978–85**

Plant capacity, design, tons/day	250
Feed gas, avg. SO_2, % by vol	4.5
Recovery of SO_2, %	96.5 to 98.8
Dimethylaniline consumed, lb/ton SO_2 produced	2.0
Sodium carbonate consumed, lb/ton SO_2 produced	10.0
Sulfuric acid, 98%, consumed, lb/ton SO_2 produced	5.0
Steam required, tons/ton SO_2 produced	0.95
Power, kWh/ton SO_2 (including compression of SO_2)	125
Cooling water at 65°F, gpm	6,150
Operating labor:	
Foreman/supervisor, day shift only	1
Shift operators	1/shift
Shipper, day shift only	1

Source: Fay (1992); Henderson and Yavorsky (1977)

steam heated reboiler to regenerate the amine and a vacuum pump to ensure that regeneration occurs at low enough temperatures to suppress the disproportionation of regenerable SO_2 into non-regenerable SO_3. The SO_2 from the regenerator is then dried and may be further processed into sulfuric acid or sulfur.

Almost all of the sulfur trioxide is removed from the process in the prescubber and subsequently neutralized by water treatment. SO_3 absorbed in the scrubber forms stable amine salts, which are removed in a proprietary unit which processes a 1% slipstream of the total solution. This is significant because it is claimed that competitive units have higher slipstreams—up to 30% of the total circulation (Barnett, 1992). Softened water should be suitable for make-up to the absorption circuit, and raw water is used in the prescubber. The bulk of the make-up water goes to the prescubber (Barnett, 1992).

An economic study by an independent engineering firm, commissioned by Union Carbide, compared the CANSOLV FGD process to five other commercial processes. The processes compared were the co-current wet limestone, jet bubbler, countercurrent wet limestone, dry lime, and Wellman-Lord processes. The study showed that the economics for the CANSOLV process can be very favorable when compared with limestone processes in high sulfur applications (Hakka et al., 1991).

Although initial results were promising, Union Carbide has discontinued the development of the CANSOLV process based on their assessment that the current market does not justify further development and because of the significant investment that would be required (Barnett, 1993).

Dow Process

The Dow process is also a relatively new, regenerable, amine-based process with the ability to preferentially recover SO_2 from flue gas. Based on laboratory studies, the process can remove SO_2 to very low levels from gas streams having up to 50,000 ppm of SO_2 (Anon.,

1991). The proprietary absorbent molecule is claimed to have been designed and synthesized to react reversibly with SO_2 and not with other acid gases present. The absorbent is also said to have a very high boiling point, be very stable, and have the EPA designation of "essentially non-toxic." A 1 MW_e-size pilot unit went into operation in June 1991.

In the Dow system, flue gas exiting the particulate removal system is first quenched and scrubbed with water in a prescrubber. The flue gas then passes through a mist eliminator, the SO_2 absorber, another mist eliminator, and then exits to the atmosphere. The SO_2 rich absorbent flows from the absorber to a heat exchanger where it is heated by hot, lean absorbent. The cooled, lean absorbent re-enters the absorber, while the SO_2 is desorbed from the heated, rich absorbent in the SO_2 stripper. The SO_2 proceeds to the byproduct recovery system.

Two waste liquid streams are produced—one from the quench prescrubber and one from the proprietary process that treats the absorber effluent. The stream from the prescrubber contains most of the halides, some of the SO_3, and some of the particulate matter. The balance of these materials is either removed in the scrubber or passes through the system. The prescrubber liquid effluent has a fairly low pH, and it contains the same components as the ash pond feed. It is therefore compatible with this stream.

As the absorbent circulates in the SO_2 absorber, it accumulates impurities that need to be removed. These include fine ash particles, heat-stable salts, and other soluble compounds. Filters are used to remove the fly ash particles. Sulfates in the scrubbing solution, which result from SO_3 and O_2 in the gas stream, as well as other heat stable salts, are removed from a slipstream of lean absorbent using a proprietary process. The waste stream from the slipstream treating process is an absorbent-free, slightly alkaline, aqueous salt solution. For most applications, potable water should be suitable for make-up to the process, and general service water for make-up to the prescrubber (Kirby, 1992).

Three byproduct recovery alternatives have been evaluated—production of sulfuric acid, elemental sulfur, and liquid SO_2 (Kirby et al., 1991). The evaluation shows that sulfuric acid production is highest in capital cost, elemental sulfur production is highest in operating cost, and liquid SO_2 production has the lowest combined capital and operating cost. However, the relatively small market for SO_2 limits the potential application of the latter alternative.

The process is available for license and development. However, development of this process has been discontinued by Dow due to the large investment required to scale up the process, the risks and uncertainties involved in selling to the electric utility market, and the lack of time to adequately prepare for the Acid Rain Phase II market (Whitley, 1993).

SEAWATER PROCESSES

In England, flue gas desulfurization using seawater discharged from the plant cooling system was first implemented in the 1930s (Abrams et al., 1988). This approach was, in part, a replacement for freshwater scrubbing, which offered low buffering capacity and required the addition of chalk to increase SO_2 removal efficiency. Scrubbing flue gas with seawater has been practiced in smelter, refinery, and industrial and utility boiler applications. There are currently over 6.5 million Nm^3/h of flue gases being scrubbed by seawater with guaranteed SO_2 removal efficiencies of up to 99% (Nyman and Tokerud, 1991; Oxley et al., 1991; Ellestad, 1992). Some properties of seawater are given in **Table 7-29** and **Figure 7-30**.

There are two basic seawater FGD process concepts: one uses the natural alkalinity of the seawater to neutralize absorbed SO_2; the other uses added lime. All commercial seawater FGD processes rely on the alkalinity of the bicarbonate in the seawater to neutralize the SO_2

598 Gas Purification

Table 7-29
Typical Seawater Properties

Property		References
Sulfur Concentration, mg/l	500–900	(Oxley et al., 1991; Nyman and Tokerud, 1991)
Magnesium Concentration, mg/l	1,300	(Abrams et al., 1988)
pH	7.5–8.5	(Krippene, 1992)
Alkalinity, milli-equiv/l	2.2–2.4	(Abrams et al., 1988)

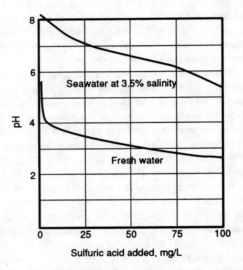

Figure 7-30. The effect of acid addition on the pH of seawater and fresh water. *Reproduced with permission from Oil and Gas Journal, July, 1991, copyright Pennwell Publishing Co. (Nyman and Tokerud, 1991)*

thereby producing sulfite or sulfate, the latter a natural constituent of seawater. A lime-based seawater FGD process has been proposed by Bechtel.

Environmental effects are the major consideration with all seawater FGD processes. Suppliers of these processes claim that the effluent does not endanger the marine environment. This is supported by a number of independent studies. For example, a recent study of the discharge from the Fläkt-Hydro SO_2 scrubber at the Mongstad refinery in western Norway concludes (Botnen et al., 1992): "No harmful impact on the benthos was observed after the outlet was deployed, and the content of organic material and heavy metals, except for lead, remains within the natural range of marine sediment. The environment conditions in the area were good before the outlet was deployed and continue to be so after 18 months of continuous use." Another study was conducted where samples of the bottom fauna and sediment were taken from the discharges of three separate Fläkt-Hydro scrubbers and analyzed. No evidence of harmful impact to the marine bottom fauna was found. Although the sulfate content experienced peaks during start-up of the unit, the sulfur content and all metal concentrations in the sediment were within natural variations (Nyman and Tokerud, 1991). The environmental impact of the Bechtel seawater FGD process effluent, which contains low concentrations of gypsum, fly ash, and non-leachable trace metals, was extensively studied using EPA-800, *Toxicity Test Methods for Aquatic Organisms,* test procedures. Several species of marine organisms were subjected to the effluent, and the effects of seasonal variations were included. It was concluded that no detrimental impact from the effluent discharge is foreseen and that the diluted seawater scrubber system is not detrimental to marine environments (Abrams et al., 1988; Nyman and Tokerud, 1991). With regard to overall ocean contamination issues, Nyman and Tokerud (1991) note that the oceans contain a very large amount of sulfur as sulfate. If this sulfur in the sea were spread out as an even layer, the total ocean area of the world would be covered by a 5-foot thick layer of sulfur. If all the sulfur in all the known oil and coal reserves were added to this layer, the thickness would only increase by the thickness of a sheet of paper.

Environmental requirements often dictate the design of seawater FGD systems. In the United States, EPA coastal water quality standards specify an initial mixing zone (IMZ) where the discharge at the IMZ boundary shall not vary more than ±0.2 units from the natural pH value. Initial mixing is defined to be completed when the momentum-induced velocity of the discharge ceases to produce significant mixing of the effluent (Nyman and Tokerud, 1991). If a seawater FGD process is being considered, an environmental assessment of the local receiving waters should be made. This assessment should include evaluations of depth profiles, currents, and tidal variations, water quality, effluent dilution and dispersion conditions, existing stationary and mobile marine life, and impact of the installation (Ellestad, 1992).

Fläkt-Hydro Seawater Process

Since 1968, the Fläkt-Hydro seawater FGD process has been used in applications ranging from 2 to 375 MW_e, including an initial 125 MW_e, module on the 500 MW_e coal-fired Trombay Unit 5 at Tata Electric Company in Bombay, India (Ellestad, 1992). The Fläkt-Hydro process is a once-through process that absorbs the SO_2 by utilizing the natural alkalinity of seawater. A schematic diagram of the process depicting a typical equipment arrangement is shown in **Figure 7-31**. After particulate removal, the flue gas enters a high turn-down absorber via the inlet quencher duct. The quencher protects the absorber from high temperatures while also removing some SO_2. Alternatively, a gas-to-gas heat exchanger may be used for this purpose. The absorber is the countercurrent type with saddle packing. The total L/G typically varies between 30 and 110 gpm/1,000 cfm (Ellestad, 1992). The flue gas flows up through the absorber and is desulfurized and cooled by the seawater. A mist eliminator at the absorber exit removes entrained water droplets, and the flue gas is reheated (if required) prior to discharge. Sulfite-laden water is discharged to the sea as is or treated prior to discharge. Treatment consists of aeration after mixing with fresh seawater to achieve optimum conditions. Aeration oxidizes the sulfite ions to sulfate ions. Oxidation reduces the COD, raises the O_2 content, and increases the pH back to the initial value. Increasing the velocity of the discharge via effluent pumping may also be used to meet the EPA standard for pH at the IMZ boundary. Dilution with additional seawater may be used to adjust effluent properties (Nyman and Tokerud, 1991).

At Mongstad, Norway's state-owned oil refinery, a Fläkt-Hydro seawater FGD system has been in operation since September 1989, and has achieved SO_2 and SO_3 removal efficiencies

600 Gas Purification

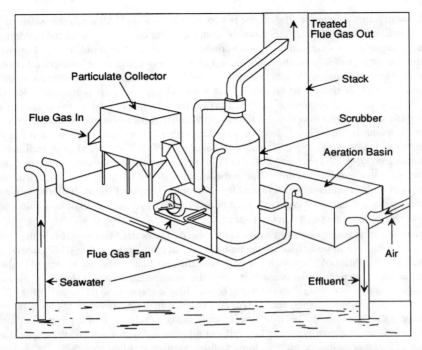

Figure 7-31. Typical Fläkt-Hydro Seawater process schematic. (*FGD and DeNO$_x$ Manual, 1988*)

of 98.8% and 82.8%, respectively. Catalyst fines are removed in an upstream electrostatic precipitator (Nyman and Tokerud, 1991). Although an emergency stack is installed to permit continuous plant operation during an electrostatic precipitator/scrubber shutdown, the unit has been bypassed a maximum of only 98 hours per year, yielding a 98.8% availability. The process uses seawater from the plant's seawater cooling system. Spent seawater flows by gravity, mixes with spent process cooling water, and is returned directly to the sea. Flue gas reheat is accomplished by an auxiliary burner (Nyman and Tokerud, 1991; Glenna and Tokerud, 1991).

Bechtel Seawater Process

Bechtel has developed a conceptual design for a 250 MW$_e$ seawater FGD plant based on their dolomitic lime FGD process as used on Montana Power's Colstrip Units 3 and 4 (Abrams et al., 1988; Shield, 1992). Magnesium hydroxide formed by the reaction of magnesium in the seawater with lime in the regeneration tanks is the primary absorbent.

The system's main component is a top entry absorber incorporating the recycle tank and oxidizer. The flue gas enters through a central downcomer serving as an inlet gas quench chamber where some SO$_2$ is absorbed. The gas turns up and flows through the surrounding outer annular absorption section containing two spray banks and a dual-flow tray. Gas velocity is below 8 fps to maintain uniform gas flow and to provide adequate liquid-to-gas contact. An L/G of 34 gpm/1,000 acfm is proposed. Chevron-type mist eliminators remove entrained liquid, while the need for reheating of cleaned flue gas is case specific (Abrams et al., 1988; Ellestad, 1992).

Slurry from the regeneration tank is pumped to the absorber where magnesium hydroxide absorbs the SO_2, and the resulting magnesium sulfite is oxidized to magnesium sulfate. A portion of the absorber recycle containing the magnesium sulfate returns to the regeneration tank where the magnesium sulfate is converted by reaction with lime to calcium sulfate and magnesium hydroxide. This slipstream is then returned to the absorber recycle loop. A bleed stream from the scrubber sump is diluted with additional seawater (cooling water system discharge in the conceptual design) and discharged to the sea. The calcium sulfate in this stream is in dissolved form when returned to the ocean. The TDS of seawater, typically 33,000–36,000 ppm, is increased by only about one percent, and the natural seawater pH is virtually unchanged (Shield, 1992). Some of the toxics passing through an existing dust removal system are removed in the prescrubber-quencher. The calcium sulfate concentration in the recycle is about 10%.

A number of advantages are claimed for the process, including very high SO_2 removal, the discharge of non-toxic, environmentally acceptable calcium sulfate to the sea in low concentrations, minimal change in TDS, virtually no change in the pH of the seawater effluent, high turn down, capital costs 20–25% less than a typical limestone/lime scrubber system of similar capacity, no scaling due to high solubilities, no aeration requirement, and a slurry recirculation rate about 25% of that required for limestone scrubbing (Shield, 1992).

Bischoff Seawater Process

The Bischoff seawater process is derived from Bischoff's existing limestone/lime FGD process, many applications of which use seawater as the source of slurry water. While several tests have been run using seawater as the absorbent at their existing plants, there are no commercial installations currently in operation. The process basically operates as a once-through, open-water circuit FGD system where the scrubbing water is brought into contact with the flue gas and is then completely discharged. As a result of the contact, bicarbonate in the seawater is replaced by sulfite in solution. The liquid effluent from the scrubber is aerated in the absorber sump to form sulfates before it is diluted with additional seawater to increase the pH and returned to the sea. SO_2 removal rates in excess of 95% have been demonstrated successfully when existing Bischoff limestone/lime wet FGD systems were operated in the seawater process mode (Krippene, 1992).

PHYSICAL SOLVENT PROCESSES

Physical solvent processes are usually most economical when the impurity to be removed is present in a high concentration and/or the gas to be treated is at a high pressure (See Chapter 14). Neither of these conditions is normally present in flue gases requiring desulfurization. However, the solubility of SO_2 is so high in some organic solvents relative to the major components of flue gas (N_2, O_2, and CO_2) that a physical solvent process can be attractive. This is particularly true when the SO_2 concentration in the flue gas is unusually high and/or fluctuates widely. A fluctuating SO_2 concentration can be handled better by a physical solvent than by a chemical solvent because of the different effects of partial pressure on solubility in the two types of solvents. The solubility of SO_2 increases almost linearly with partial pressure in a physical solvent, but it increases only slightly, or not at all, in a chemical solvent when the partial pressure is increased. Therefore a physical solvent will absorb proportionately more SO_2 when the SO_2 concentration in the feed gas increases without requiring a change in the liquid flow rate, while a chemical solvent requires a corresponding increase in liquid rate to handle the added load.

Additional advantages of physical solvents compared to chemical reactive solvents are

- They require less energy to regenerate than thermally regenerative chemical solvents because no heat of reaction is involved.
- They produce a more valuable byproduct (pure sulfur dioxide) than nonregenerative reactive systems such as the limestone/lime slurry processes.
- They operate with a clear, single phase liquid in both absorption and regeneration steps.

According to Becker and Linde (1985), a physical solvent for SO_2 removal must have the following properties:

1. Highly temperature dependent solubility to allow regeneration by thermal means
2. High selectivity for SO_2 relative to N_2, O_2, and CO_2
3. High thermal and chemical stability, and unaffected by impurities in the gas
4. Low vapor pressure
5. Nonpolluting and nonhazardous
6. Readily available at an acceptable price

Few organic solvents meet all of these requirements, and even those that do are not practical for most flue gas applications primarily because of the high liquid flow rates required. However, there appears to be a niche where a physical solvent can be economical; i.e., purifying gases where the SO_2 concentration is relatively high (over about 0.2 vol %), but too low to be an economical feed to a conventional sulfuric acid plant. The economics can, of course, also be affected by other factors such as a fluctuating SO_2 concentration, the presence of other impurities which can be removed simultaneously by the physical solvent, and the need for pure SO_2 locally.

Solinox Process

The Solinox process, developed by Linde A.G. and offered in the U.S. by the Lotepro Corporation, uses the physical solvent tetraethyleneglycol dimethylether to remove sulfur dioxide and other impurities from vent gas (Becker and Linde, 1985). Descriptions of the process and of several commercial applications are given by Sporer (1992) and Hersel and Belloni (1991).

The basic Solinox process employs a typical absorption/desorption cycle with SO_2 removed from the feed gas in a countercurrent absorber, and stripped from the physical solvent in a countercurrent reboiled stripper. In practice, the process is complicated somewhat by the need to water wash the feed gas before it is contacted with solvent to reduce the gas temperature and remove dust and some impurities; and the need to water wash both the purified gas and the stripper off-gas to recover entrained or vaporized solvent. A distinctive feature of the process is its ability to remove hydrocarbons, such as benzene, which are present in some vent gas streams and may require removal to meet air pollution control requirements. Hydrocarbons are generally quite soluble in the solvent. They are absorbed and stripped with the SO_2. The hydrocarbons can be removed from the SO_2 byproduct by a fractionation step or can be destroyed by oxidation during subsequent processing.

The main process steps in a typical Solinox process are

1. The feed gas is cooled and cleaned by contact with water. This can be accomplished in a separate vessel and/or in a short section at the bottom of the absorber.
2. The cooled feed gas is passed upward through the main absorption zone countercurrent to downflowing solvent. Sulfur dioxide and hydrocarbons are absorbed in this step.
3. The purified gas flows upward through a water wash section of the column where traces of solvent are removed from the gas before it is vented to the atmosphere.
4. Rich solution from the absorber, containing the absorbed SO_2 and about 5% water, is heated by indirect heat exchange with hot lean solution and flashed into the stripping column, which operates at a reduced pressure of about 0.5 bar vacuum.
5. The rich solution is stripped of sulfur dioxide and hydrocarbons as it flows downward in the stripping column countercurrent to vapor (primarily water vapor) generated in the reboiler, which is heated by low pressure steam.
6. SO_2-rich vapor from the stripping section of the regenerator flows upward through a reflux section where solvent vapor is removed from the SO_2 fraction.
7. The SO_2 fraction is cooled, compressed, and further processed, as required, for its final disposition.
8. The lean solution from the regenerator is cooled by heat exchange with the rich solution, further cooled by heat exchange with cooling water, then recycled to the absorber.

The basic process may be modified to meet specific requirements. Auxiliary equipment that is sometimes required includes a gas-to-gas heat exchanger to reheat the purified flue gas; an inert gas separator in the rich solution line to reduce the amount of inert gases appearing in the SO_2 byproduct; and a purification unit for the SO_2 byproduct stream.

Commercial plant operating experience described by Sporer (1992) includes data from plants purifying flue gas streams that emanate from lead and zinc smelters, a pulp mill, and a barite ($BaSO_4$) reduction process. The smelter gases fluctuated widely in SO_2 concentration with maximum values of 1.4% for the lead smelter and 2% for the zinc smelter. The pulp mill flue gas averaged about 0.7% SO_2, while the barite reduction process flue gas contained about 0.5% SO_2 and a high loading of dust and other impurities. Although some operating problems were encountered, all of the plants met their design requirements, removing from 95 to 99.3% of the incoming SO_2.

Hersel and Belloni (1991) provide data on the utility consumption of a typical Solinox plant treating 55 MMscfd (55,800 m^3/h) of flue gas. Such a plant would require about 390 kW of electricity (including gas compression); 250 m^3/h of cooling water (10°C temperature rise); 3.7 lt/h of low pressure steam (1.5 bar or 22 psia); and solvent makeup costing $9.30 per hour (U.S. dollars).

MOLTEN SALT PROCESS

The only process in this category that has received significant research and development attention is the Molten Carbonate Process developed by Rockwell International (Oldenkamp and Margolin, 1969; Katz and Oldenkamp, 1969). Although the process has not been commercialized, it is of interest because of the unique technology involved. Its potential advantages are the ability to treat the flue gas at an elevated temperature without adding water vapor and the production of a useful byproduct (sulfur).

The process operates with a closed absorbent cycle in which a molten eutectic mixture of sodium, potassium, and lithium carbonate is circulated to react with the sulfur oxides in the flue gas. The sulfur compounds are absorbed at about 800°F, forming sulfites and sulfates in the melt. The molten salt is next processed in a reducer, operating at about 1,400°F, which uses petroleum coke to convert oxidized sulfur species to the sulfide form. Heat is provided in the reducer by oxidation of a portion of the coke with air.

Molten salt from the reducer is next processed to convert sulfides back to carbonates for recycle to the absorber. This is accomplished in a regeneration column, which operates at about 800°F and uses a mixture of carbon dioxide and water vapor to displace hydrogen sulfide gas from the molten salt. The hydrogen sulfide-rich gas stream from this step is fed directly into a Claus type sulfur plant. Work on the process was terminated after a small demonstration unit developed mechanical problems, including plugging of a mist eliminator at the absorber outlet and corrosion in some lines carrying hot molten salt.

SPRAY DRYER PROCESSES

The use of spray dryers for SO_2 removal has experienced remarkable growth. The first U.S. contract for a spray dryer absorber was awarded in 1977, and by mid-1992 the largest suppliers of these systems had sold about 249 systems. Of these, 69 were for utility applications, 30 for industrial applications, and 150 for waste incinerators (ABB, 1992A, B, C; Joy, 1992).

In spray dryer processes, sulfur dioxide is removed from the flue gas by contact with an atomized spray of reactive absorbent such as lime slurry or sodium carbonate solution. The sulfur dioxide reacts with the absorbent while the thermal energy of the flue gas vaporizes the water in the droplets without saturating the flue gas to produce a fine powder of spent absorbent. The dry product, consisting of sulfite and sulfate salts, unreacted absorbent, and fly ash, is collected in a fabric filter or electrostatic precipitator (ESP). The fabric filter has been shown to be more effective for the collection of the particulate byproduct than a cyclone or ESP in that there is additional reaction of the absorbent with the SO_2. However, some recent work indicates that ESPs can also be effective secondary collectors. Current ESP residence times are much longer (15–25 seconds) than those of the small ESPs used in early investigations. Thus, there is much longer contact time between the absorbent and contaminants than in earlier ESPs. In addition, electric wind blending in the ESP increases contact potential. Proponents claim that problems of dust caking on the discharge electrodes and gas distribution internals have been remedied and that corrosion in high chloride applications can be solved by the use of corrosion resistant materials (*AirTECH News*, 1993).

Spray dryer processes have some significant advantages over the wet scrubber technologies. The gas passes through and exits the spray dryer system well above the adiabatic saturation temperature rather than close to it, so no reheat or wet stack is required. Also, the need for corrosion-resistant materials in the gas path is avoided. The solid byproduct is very dry, facilitating separation from the flue gas and handling and disposal, and there is no waste water discharge stream. Because the gas exiting the system is dry, the draft fans can be located downstream of the system where the gas is cooler and the volumetric flow is less, which reduces the size of the fans. Specific processes, such as the lime spray dryer, have other advantages, including lower capital and operating costs (Niro, 1990). However, other cost studies suggest that levelized busbar costs of the lime spray dryer process may not be significantly lower than those of wet limestone systems. Refer to **Table 7-10**.

Two types of spray dryer processes, both non-regenerative (throw-away), have attained commercial status. The first employs a lime slurry and the second uses sodium carbonate solution. The lime process is by far the most widely used; however, the sodium carbonate system, exemplified by the Aqueous Carbonate Process (ACP), was developed first.

The spray dryer process producing byproducts for disposal was tested in 1977 by Rockwell International and Wheelabrator-Frye, Inc. (supplier of the fabric filter) in a joint program conducted at the Leland Olds Station of the Basin Electric Power Corporation in Stanton, North Dakota. Sodium carbonate, lime, fly ash, and a fly ash/lime mixture were tested as SO_2 reactants. The lime- and fly-ash-containing absorbents were fed to the spray dryer as slurries. Data from the test program have been presented by Estcourt et al. (1978). Typical results for a sodium carbonate solution and a lime slurry are given in **Table 7-30**. For spray dryer and alkaline solids injection processes, it should be noted that the absorbent-to-sulfur ratio (stoichiometric ratio) is usually expressed on the basis of the inlet sulfur dioxide, whereas that of other FGD systems is expressed on the basis of sulfur dioxide removed. For example, at 90% removal, a 1.62 absorbent to sulfur ratio on a sulfur-in basis is equivalent to a 1.80 (1.62/0.90) absorbent to sulfur ratio on a sulfur-removed basis.

As would be expected on the basis of chemical reactivity, lime is appreciably less efficient than sodium carbonate with regard to both SO_2 removal and absorbent utilization when used at similar process conditions. However, by suitable adjustment of spray dryer operating conditions, lime slurry can achieve both high SO_2 removal and high sorbent utilization. Since lime costs less than soda on a weight basis and its equivalent weight is much lower, lime is economically preferred in almost every case (Buschmann, 1993). Sodium carbonate spray dryer units have been used on small systems such as glass furnaces, hazardous waste incinerators, paper mill power boilers, and coke calcining kilns. Sodium carbonate systems generally require larger spray dryers than lime (Saliga, 1990).

Table 7-30
Performance of Spray Dryer/Fabric Filter System for SO_2 Absorption

Stoichiometric Ratio	SO_2 Removal Efficiency, %			Absorbent Utilization, %		
	Spray Dryer	Fabric Filter	Total	Spray Dryer	Fabric Filter	Total
Tests with Sodium Carbonate Solution						
0.5	40	8	48	80	16	96
1.0	82	10	92	82	10	92
1.5	86	12	98	57	8	65
Tests with Lime Slurry						
0.66	35	18	53	53	28	81
0.94	50	19	69	53	21	74
1.21	50	25	75	41	21	62

Notes:
1. SO_2 concentration in feed gas 800–2,800 ppm
2. Fabric filter temperature approximately 200°F
3. Fabric filter performance values are based on feed to the spray dryer, not on feed to the filter

Source: Estcourt et al. (1978)

Aqueous Carbonate Process

The Aqueous Carbonate Process (ACP) was developed by Rockwell International and is now licensed by ABB Fläkt. In this process, the SO_2 is removed by passing the flue gases through a spray dryer where efficient contact with a fine mist of an aqueous sodium carbonate is achieved. The SO_2 reacts with the sodium carbonate (Na_2CO_3) to form sodium sulfite (Na_2SO_3), some of which is further oxidized to sodium sulfate (Na_2SO_4).

Pilot tests of the modified spray dryer absorption unit have been described (Gehri and Gylfe, 1973) which indicate that 90% removal of incoming SO_2 can be realized with liquid to gas ratios of less than 0.4 gal/1,000 scf. In the tests, the $NaCO_3$ absorbent solution was generally maintained as dilute as possible consistent with desired SO_2 removal and the generation of a dry product. Absorbent utilization exceeded 80% in a single pass through the dryer.

Two versions of the ACP have been developed: (1) an open-loop configuration in which the dry spent absorbent is simply removed from the system for disposal and fresh alkali is continuously fed to the spray dryer, and (2) a closed-loop process in which the spent absorbent is regenerated and reused. The open-loop system was installed on the 410 MW_e Coyote Station at Beulah, North Dakota (Botts et al., 1978). The plant used a spray dryer followed by a fabric filter for simultaneous SO_2 and dust removal. As indicated by the flow diagram of **Figure 7-32,** the process is extremely simple. Because the particulate collected on the fabric filter bags remains on the fabric for a period of time, the gas-solids contact time is extended beyond the particulate residence time in the spray dryer so additional absorption of SO_2 by the alkaline material occurs. After this system went commercial in 1981, no other large soda ash spray dryers were built. About 1990, this scrubber was converted from soda ash to lime to eliminate solids build-up in the dryer vessel and to use lower cost lime. The

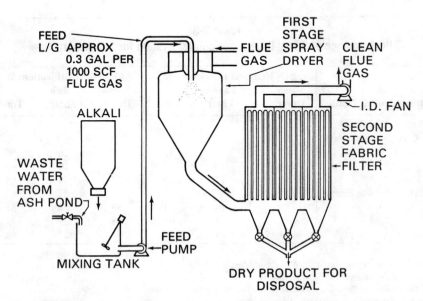

Figure 7-32. Flow diagram of Aqueous Carbonate Process—open loop, two stage system.

problems incurred by the requirement to store soda scrubber waste in a lined disposal pit were also eliminated by conversion to lime (Pozarnsky, 1990).

A 100 MW_e open-loop sodium carbonate spray dryer pilot plant at Jim Bridger Station was tested and evaluated, but a full-scale plant was decided against because the spray dryer outlet gas temperature was too low to adequately dry the product collected in the particulate collector. To raise the outlet temperature, it would have been necessary to raise the inlet temperature, which would have impacted boiler efficiency adversely (Angelovich, 1990).

The closed-loop ACP system using sodium carbonate as the absorbent was selected for a 100 MW_e FGD demonstration plant under a program sponsored by the Empire State Electric Energy Research Corporation (ESEERCO) and the U.S. Environmental Protection Agency (EPA) (Aldrich and Oldenkamp, 1977; Binns and Aldrich, 1977). A flow diagram of the process used in the demonstration plant is shown in **Figure 7-33.** In this application, the spent absorbent particles from the spray dryer are collected in cyclones, with final removal of the remaining particles in an ESP. The design emission rate to the stack is 0.01 gr/scf or less. The gas is at least 50°F above its dew point at the stack inlet so reheat is not required.

The dry spent absorbent is mixed with carbon (petroleum coke or coal) and fed into a refractory-lined reducer vessel that contains a pool of molten sodium carbonate and sodium sulfide at a temperature of about 1,800°F. Air is injected to oxidize part of the carbon to CO and CO_2 in order to provide the heat needed by the endothermic reduction reactions and maintain the overall system at the reaction temperature.

The reduced molten salt mixture containing typically 62% Na_2S, 8% Na_2SO_4, 25% Na_2CO_3, and 5% unreacted carbon and ash is continuously discharged from the reducer vessel and quenched in an aqueous slurry. Soluble constituents of the melt are dissolved in the aqueous medium, which is then filtered to remove unreacted carbon and ash. The clear liquor is reacted with carbon dioxide gas in a series of sieve tray columns to produce, ultimately, a solution of sodium carbonate and a gas stream containing H_2S and CO_2. This gas is fed to a conventional Claus plant where the H_2S is converted to elemental sulfur. The sodium carbonate solution is recycled to the spray dryer as the active absorbent for SO_2.

The demonstration project was terminated for both technical and economic reasons. Among the technical reasons were (1) too great a scale-up, (2) unanticipated technical problems, (3) inability to maintain steady-state conditions long enough to develop a database for a quantitative technical and economic assessment of the process, and (4) lack of adequate sparing of equipment. Among the economic conclusions drawn were (1) such a process is not economically competitive with throw-away processes (and would not be as long as the FGD byproduct continues to be classified as a non-hazardous material), (2) the process is unlikely to be economically competitive with the commercially proven regenerative processes (Wellman-Lord and MgO), which already have limited markets, (3) there is no foreseeable application of the process in New York State (where the interested parties planned to use the process), and (4) the development of new technologies (FBC and IGCC) may be more attractive than a conventional coal combustion plant with regenerative FGD (Stefanski, 1986).

Lime Slurry Spray Dryer Processes

Almost all of the spray dryer FGD systems installed for utility power plant and industrial applications use lime slurry as the absorbent (Palazzolo et al., 1983; Liegois, 1983). Although lime is not as reactive as sodium carbonate, it is usually preferred due to its lower cost and because the spent absorbent can be disposed of more readily than soluble sodium

608   Gas Purification

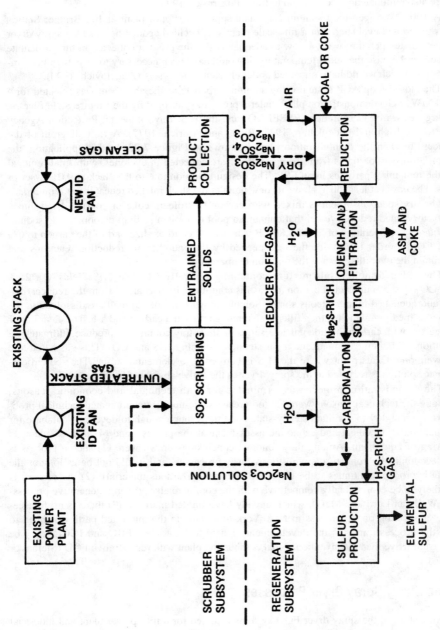

Figure 7-33. Schematic diagram of the Aqueous Carbonate Process.

salts. The chief disadvantages are a relatively low absorbent utilization and the need to use lime instead of the less expensive limestone. Some applications using Western U.S. coals employ the alkalinity of the coal fly ash to reduce lime costs. Limestone has been tested, but with little success to date (Makansi, 1989). While generally thought of as primarily applicable to low sulfur fuel applications, spray dryers are also used in medium and high sulfur coal applications. An early example is Argonne National Laboratories where a spray dryer has been operating since November 1981 with Illinois Basin coal having an average sulfur content of 3.5%. Eighty to eighty-five percent SO_2 removal is achieved. Argonne has also demonstrated that 90% SO_2 removal efficiency can be achieved with 4.2% sulfur Northern West Virginia coal. Another high sulfur coal application is the Salzburg plant in Austria, which has been in operation since 1987 (Farber et al., 1983; Felsvang et al., 1991). Potential factors limiting SO_2 removal efficiency at high inlet SO_2 concentrations appear to be the maximum economic calcium to sulfur ratio, the maximum practical slurry concentration, the maximum available inlet flue gas temperature, the slurry chloride content, the allowable approach to adiabatic saturation, and the ability to recycle sorbent.

Niro has a patent on the recycling of dried byproducts back to the absorbent feed. Recycling reportedly reduces the absorbent consumption by 30 to 50% (Niro, 1990). Enhancers can also reduce spray dryer lime consumption. Brown and Felsvang (1991) have traced the history of enhancer use in lime spray dryer applications. The benefits of chlorides and other deliquescent materials were first recognized by Niro (Hansen et al., 1983). Basic research on the influence of deliquescent materials on lime spray dryer performance was reported by Klingspor (1983). The first full-scale demonstration was by EPA/EPRI at the 100 MW_e Riverside Demonstration Plant (Blythe et al., 1983) where chloride reduced the lime consumption by as much as 30%. In the 1980s, this effect was utilized on full-scale European systems (Felsvang et al., 1988). In the U.S., much research has been performed to determine the influence of chlorides on SO_2 removal with high sulfur coal. Barton et al. (1990) have correlated the effect of SO_2 concentration, byproduct solids chloride content, inlet temperature, lime feed rate, and approach to adiabatic saturation temperature on SO_2 removal.

High chloride levels can necessitate increasing the spray dryer outlet temperature because drying is inhibited by the presence of deliquescent chloride. While raising the outlet temperature decreases the approach to adiabatic saturation, which would normally adversely impact SO_2 removal efficiency, it has been found that there is little or no increase in lime consumption with high chloride/high temperature lime spray dryer operation. Municipal waste incinerators, which are high chloride applications, rely on a high outlet temperature to ensure a free flowing byproduct (Brown and Felsvang, 1991).

Burnett et al. (1991) present important data on the effects of chloride spiking on spray dryer and ESP performance. The testing was performed at a TVA 10-MW_e spray dryer/ESP pilot plant to determine the applicability of spray dryer technology for retrofit with existing ESPs for medium- and high-sulfur coal applications. The results are presented in **Table 7-31,** which gives SO_2 and particulate removal efficiencies for operation with medium and high sulfur coal and with several chloride concentrations in the recycle sorbent. The presence of 0.6% or more chloride in the recycle material (equivalent to about 0.1% or more chloride in the coal) was found to increase the SO_2 removal efficiency from the 75–80% range to the 89–98% range.

With the retrofit of spray dryers to a boiler, the particulate loading to the ESP is much higher, and the ESP must be capable of handling this loading. As shown in **Table 7-31,** the collection efficiency increases greatly due to the addition of chlorides, and this could eliminate the need for upgrading the ESP. The apparent effect of chloride on the ESP performance

Table 7-31
SO$_2$ and Particulate Removal Efficiencies of Spray Dryer and ESP with Chloride Spiking.[1]

Approximate Chloride Levels in Recycle Solids, %	Removal Efficiency, %									
	0.10	0.14	0.24	0.60	0.70	1.10	1.40	1.70	2.50	3.5[2]
SO$_2$ Removal Efficiency with: Medium Sulfur Coal[3]										
(4 lb S/10^6 Btu)[4]	80				89		95			
(5 lb S/10^6 Btu)			89			96		98		
High Sulfur Coal (8 lb S/10^6 Btu)		75–80							95/98[5]	
ESP Particulate Removal Efficiency with: Medium Sulfur Coal[3]										
(4 lb S/10^6 Btu)[4]	99.83		99.80–99.85	99.83				99.95	99.95	99.98
(5 lb S/10^6 Btu)				99.42				99.77		99.96
High Sulfur Coal (8 lb S/10^6 Btu)		99.87[6]								99.96

Notes:
1. Conditions: 320°F inlet gas temperature, 18°F approach to saturation, 1.3 mole Ca(OH)$_2$/mole SO$_2$, except as noted.
2. The 3.5% chloride level in recycle solids corresponds to 0.5% chloride level in the test coal.
3. SO$_2$ removal increased marginally for chloride levels greater than 1.7% for medium sulfur coal cases.
4. 4 lb S/10^6 Btu corresponds to about 2.2% S coal. Differences between the 4 and 5 lb/10^6 Btu values are attributed to ESP upgrades before the lower sulfur coal tests were run.
5. 98% SO$_2$ removal was achieved at 1.6 Ca/S ratio.
6. Chloride level is estimated for this case.

Source: Based on 10 MW$_e$ pilot plant data of Burnett et al. (1991)

is hypothesized to result from the hygroscopic nature of calcium chloride. The moisture makes the particles more cohesive (i.e., "sticky"), so that the particles are better held, both to themselves and to the collection plates, and thus are less likely to become reentrained (Burnett et al., 1991).

Spray dryer/fabric filter systems remove other air pollutants along with the SO$_2$. Heavy metals in particulate form are removed. Mercury, which can be present in several forms, may also be removed depending on the form and other factors. Very high mercury removal can be achieved with the injection of lime slurry containing activated carbon. Hydrochloric and hydrofluoric acids are removed with spray dryers (as they are in wet FGD systems) (Brown and Felsvang, 1991). SO$_3$/H$_2$SO$_4$ are also removed with very high efficiency (but are not removed efficiently in wet FGD systems).

On-line maintenance of the spray dryer atomizers is achieved by locating the draft fans downstream of the atomizer to create a negative pressure in the absorbers. Regular maintenance as frequently as every two weeks may be necessary. In this regard, the emissions averaging period must be long enough to allow the system to catch up for the short, one-half to one hour period when the atomizer is off-line; or, if multiple atomizers are used, there must be enough reserve capacity to compensate for the off-line atomizer. In this latter case, designing for one atomizer out of service could adversely affect the turndown capability or prohibit full load operation during atomizer replacement, both of which are undesirable. In one application, a back-up spray system was used to allow on-line atomizer replacement. In Europe, typical regulations have a short averaging period, but also a yearly exclusion period, such as 72 hours, during which the plant does not have to meet the SO_2 emissions regulations, allowing for atomizer replacement and other maintenance (Zohouralsen, 1992).

A simplified flow diagram for a lime slurry spray dryer system is shown in **Figure 7-34**. This diagram represents the 100 MW_e Demonstration Plant on Units 6 and 7 of the Riverside Station of Northern States Power Company. The plant was used for extensive testing of the effects of various operating variables on a large 46-ft diameter spray dryer absorber (Gutslke et al., 1983).

Spray dryer FGD plants consist of four major subsystems: absorbent preparation, absorption and drying, solids collection, and solids disposal. Because of the quantity of lime required, it is not usually economical to purchase lime in the hydrated form, and pebble lime is normally slaked on site. For installations using less than 2,000–3,000 tons of lime per year, hydrated lime is sometimes preferred because a slaker is not required, though its use can be more troublesome (Potter, 1991; National Lime Assoc., 1982). An evaluation of the available limes is necessary to determine their suitabilities. The National Lime Association (1982) publication, *Lime Handling, Application and Storage,* provides information of value to lime users. Several types of slakers have been tested, including ball mills, paste, and detention types. Ball mill slakers have been used for most large utility applications because they pulverize uncalcined

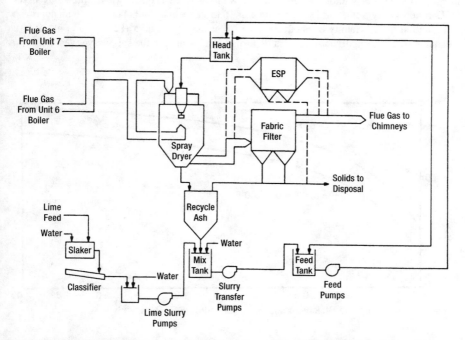

Figure 7-34. Flow diagram of the Lime Slurry Spray Dryer Process as tested at the Riverside Station of Northern States Power Company. (*Gutslke et al., 1983*)

limestone and other solids which the simpler slakers reject as waste material. The water used for slaking is critical. Fresh water is generally required for slaking; however, cooling tower blowdown or another low quality water may be used for dilution (Liegois, 1983). At Laramie River Station, cooling tower blowdown water is used for slaking. The water is softened to remove bicarbonate hardness and prevent calcium carbonate scaling and deposition inside the dual fluid atomizers. This is accomplished by bleeding a small amount of lime slurry into the water softening tank to precipitate calcium carbonate (Larson et al., 1990).

Key factors affecting SO_2 absorption efficiency are the dryer design (e.g., atomizer performance, gas residence time); the absorbent stoichiometry; and the approach to saturation. The general effects of the latter two variables are indicated by **Figure 7-35,** which is based on tests at the Riverside plant. These curves correlate the total SO_2 removed to lime stoichiometry and approach temperature. No other effects such as chloride enhancement are taken into account. The curves were generated by regression analysis of the three parameters using data from 164 tests (Gutslke et al., 1983).

Slurry atomization is generally accomplished by the use of rotary atomizers or two-fluid nozzles. Rotary atomizers appear to be preferred for large installations and nozzles for small units; however, this is not a universal rule. The rotary atomizers have the advantage of high capacity per unit, uniformly fine drop formation, and a lower power requirement. Two-fluid nozzles, using air or steam to provide the atomizing energy, are simpler and easier to maintain. The degree of atomization and chamber design must be coordinated to assure that the droplets are fine enough to provide adequate surface for SO_2 removal, moist enough during most of their flight to be effective absorbents, and sufficiently dry when they strike the walls to flow as a powder without adhering to the walls or to each other. Most lime spray dryer FGD systems have a flue gas residence time of 10 to 12 seconds and operate with an approach to saturation in the vicinity of about 20° to 50°F at the dryer outlet (Palazzolo et al., 1983). Adequate sizing is important to prevent deposits from forming in the spray dryer vessels.

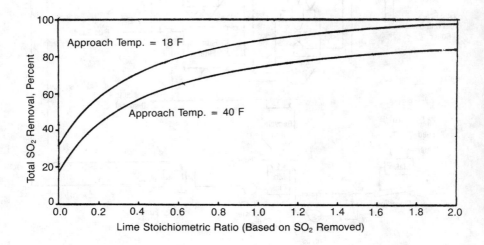

Figure 7-35. SO_2 removal by the Lime Slurry Spray Dryer Process as a function of stoichiometric ratio and approach temperature. (*Gutslke et al., 1983*)

As indicated by **Figure 7-35,** reducing the approach to saturation can result in a significant improvement in SO_2 absorption. Unfortunately, this technique is limited because a margin of safety is required to prevent water condensation in downstream equipment. With given gas inlet conditions (rate, temperature, and humidity), the approach to adiabatic saturation in the spray dryer is established by the amount of water fed in the lime slurry. For typical operating conditions, this generally results in liquid-to-gas ratios ranging from 0.2 to 0.3 gal/1,000 scf.

Absorption efficiency can also be improved by increasing the CaO/SO_2 stoichiometric ratio. This can be accomplished by raising the concentration of lime in the feed slurry. As previously discussed, another approach is to recycle a portion of the spent absorbent using solids that have dropped out in the spray dryer or that have been collected in the fabric filter. Recycle can increase SO_2 removal efficiency and/or absorbent utilization and also increase the utilization of fly ash alkalinity.

The particulate removal equipment represents a key component in spray dryer FGD systems. Not only does it prevent the release of particulate into the atmosphere, it can also provide considerable SO_2 removal capability. This feature is the subject of a U.S. Patent (Gehri et al., 1980). For example, both a fabric filter and an ESP were used in the Riverside Station lime spray dryer tests (Gutslke et al., 1983). The fabric filter typically averaged about 15% SO_2 removal efficiency; whereas, the ESP averaged only 6%. Almost all commercial lime spray dryer FGD installations have employed fabric filters. With systems in which the gas is quenching from very high temperatures, the loss of quench cooling with resulting heat damage to the downstream fabric filter bags is a serious concern.

Combined SO_2/NO_x control is possible in a spray dryer/fabric filter system. However, the temperature of the flue gas at the fabric filter exit must be above about 180°F, which is well above typical adiabatic saturation temperatures. It has been suggested that combined SO_2/NO_x removal might be applicable in retrofit situations where 70% SO_2 removal in conjunction with 40–50% NO_x removal is acceptable. Addition of sodium hydroxide enhances SO_2/NO_x removal. However, the high outlet temperature makes the lime consumption go up appreciably. Significant NO to NO_2 conversion within the fabric filter has been observed under these conditions creating a stack opacity problem (Huang et al., 1988). The high spray dryer exit temperature normally requires a high inlet (boiler outlet) temperature, which adversely affects boiler efficiency.

In the early- to mid-1980s, Joy/Niro tested their lime spray dryer SO_x/NO_x process in Copenhagen, at Argonne National Laboratories, and at Northern States Power's 100 MW_e Riverside Station. In these tests, a small percentage of sodium hydroxide solution was mixed with the milk of lime feed to the atomizer. It was reported that the observed increase in NO_x absorption occurred as a result of a more porous, heavier filter cake deposit on the filter bags at essentially normal fabric filter pressure drop. The solid waste byproduct from the SO_x/NO_x process contains calcium nitrate, which is the end product of NO_x removal. Better NO_x absorption was accomplished at higher O_2 and SO_2 levels in the flue gas and at flue gas temperatures above 190°F at the fabric filter. SO_x/NO_x reduction efficiencies of approximately 85–95%/20–30% were achieved on low sulfur coals, and 85–95%/60–70% on high sulfur coals. The overall stoichiometric ratio for the SO_x/NO_x process was somewhat higher than for a straight dry FGD process. The main disadvantage of the process was the formation of a distinctly yellow stack plume resulting from the relatively high NO_2 vs. NO concentrations contained in the flue gases leaving the system (Krippene, 1992).

Duct Spray Dryer Process

The duct spray dryer process is very similar to the spray dryer process except that the slaked lime slurry is sprayed directly into the ductwork to remove the SO_2. No spray dryer vessel is used. The spent absorbent and the fly ash are captured in a downstream particulate collector. A portion of the collected particulate is recycled and reinjected. Several pilot plant programs are either underway or have been completed (Keeth et al., 1991B).

For example, work on duct spray drying was performed at the converted DOE Duct Injection Test Facility, which operates on a 12 MW_e, 50,000 acfm slipstream. Data, which illustrate the effect of Ca/S ratio on SO_2 removal, are presented in **Table 7-32** (Felix et al., 1991). Testing by Brown et al. (1991) sponsored by the Pittsburgh Energy Technology Center (PETC) on a small 1.7 MW_e pilot plant confirms these results and further indicates that the addition of chlorides to the humidification water and the use of absorbent recycle enhance SO_2 removal, which would be expected due to the similarity of duct absorbent injection to spray drying. Some build-up problems in the duct were experienced, but were in part due to the small size of the duct (Brown et al., 1991).

A duct spray dryer process developed by Bechtel, called the Confined Zone Dispersion (CZD) process, uses a finely atomized slurry of hydrated lime that is injected into the ductwork between the preheater and the electrostatic precipitator. The gas is about 250° to 300°F at the point of injection. A straight run of duct on the order of 50 to 100 ft in length is needed downstream of the injection point to allow slurry droplets to dry before they impact the duct walls or turning vanes. The process has been tested in several pilot scale and proof of concept programs starting in 1986. These tests showed the process to be capable of removing in excess of 50% of the SO_2. More recently, the CZD process has been installed and operated in a demonstration unit at the Seward Station of Pennsylvania Electric Company (Abrams et al., 1987; Abrams and Zaczek, 1991).

Gas Suspension Spray Dryer Process

A process that is closely related to both conventional and duct spray dryer systems is the Gas Suspension Absorption (GSA) process, developed by FLS miljo a/s in Europe and

Table 7-32
Summary of SO_2 Removal with Slurry at the DOE Duct Injection Test Facility

Ca/S Ratio	Approach (°F)	Inlet SO_2 ESP (ppm)	SO_2 Removal, % ESP Inlet	Outlet
1.0			45	50
2.0	20–30	1,200–2,800	60	75
2.5			70	85
1.0			30	40
2.0	50–55	1,200–2,800	50	60
2.5			60	70

Source: Felix et al. (1991)

offered in the U.S. by Airpol, Inc. In the GSA process, the SO_2-containing flue gas from the air preheater flows upward through an empty vertical column (reactor). A freshly slaked lime slurry is sprayed into the bottom of the reactor and flows upward with the gas. SO_2 (and HCl) are absorbed into the droplets of slurry as they dry and react with the lime. As in other spray dryer systems, the drying operation cools the gas stream and produces fine solid particles of spent sorbent.

The mixture of flue gas and dry particles (including fly ash) flows out of the top of the reactor into a cyclone separator. Most (about 90%) of the entrained particles are removed in the cyclone. A major portion of the collected material is recycled to the reactor, while the balance is discharged as byproduct. Gas from the cyclone next passes through a high efficiency particulate collector system then to the stack. Material from this collector is added to the cyclone discharge to form the total byproduct stream. The process has been used in several waste incineration units in Europe, and a 10 MW_e demonstration plant was started up at TVA's National Center for Emissions Research in November 1992 (Airpol, 1993).

Spray Dryer Byproduct Disposal and Use

The byproduct removed from a lime spray dryer/particulate control system is a dry, flowable powder containing calcium sulfite, calcium sulfate, fly ash, and unreacted absorbent. It is usually conveyed pneumatically to a silo for storage prior to disposal and is typically disposed of in a landfill. Water is often added for dust control. This causes pozzolanic reactions to occur resulting in a final byproduct of low permeability and desirable landfill characteristics (Liegois, 1983). **Table 7-17** gives important properties of spray dryer byproduct.

There are a number of EPRI publications that are of value in the design of landfills for lime spray dryer FGD byproduct. EPRI (1988A) Report CS-5782, characterizes eight spray dryer byproducts from western coal applications. It is the source of the spray dryer byproduct data given in **Table 7-17**. EPRI Report CS-5312 (Klimek et al. 1987) gives some guidelines for the design of the byproduct management system for a lime spray dryer. Farber (1988) provides information on the spray dryer byproduct from an eastern coal application, and Cox and Kaplan (1983) describe operating experience for the spray dryer byproduct processing plant at Holcomb Station Unit No. 1.

Lime spray dryer system byproduct has been found to be suitable for some end uses. At many European installations, a large percentage of the fly ash is collected ahead of the spray dryer and is sold. This leads to two categories of applications for spray dryer byproduct: those that can use a low fly ash byproduct and those that can use a high fly ash byproduct. Applications (and potential applications) for the low fly ash byproduct include landfilling, soil sulfation and liming, industrial sludge treatment, cement retardation, cement sulfur content correction, sulfuric acid production (with reductive calcination), and as anhydrite (with thermal treatment) for use in the cement and mining industries. Potential applications that have been studied for the high fly ash byproduct include ocean reef construction, grouts, road construction, liners, membranes, cap materials, autoclaved bricks, sand-lime bricks, plain concrete articles, and ready mixed concrete (Niro, 1988). The additional expense and the complications of such applications have prevented them from becoming popular. In Europe, most older spray dryer applications have permits for disposal of the byproduct in a landfill. The newer plants have fly ash collection ahead of the spray dryer, and the spray dryer byproduct goes to a lined pit. In Europe, the byproduct is now considered a hazardous waste. This has caused many in Europe to prefer wet limestone and lime scrubbers (Buschmann, 1992).

EPRI (1988C) and Cornelissen (1991) provide information on uses for these byproducts. EPRI (1988C) also gives transportation costs, which are important in estimating costs for various disposal options.

For sodium-based spray dryer byproduct disposal and use, refer to the discussion under the heading "Dry Sorption Byproduct Disposal and Use" in the following section. The sodium-based spray dryer byproduct is similar to the sodium-based dry sorption byproduct.

DRY SORPTION PROCESSES

This section describes processes in which the SO_2 is removed from the gas stream by reaction with a dry material. The operation is referred to by the generic name "sorption" to differentiate it from adsorption (which is covered in the next section) and absorption which is usually reserved for liquid-based operations. A considerable amount of development effort has been aimed at dry sorption processes primarily because of their apparent advantage over aqueous absorption systems of permitting stack gas treatment without cooling and saturating the gas with water vapor. Several of the processes involve injection of the dry sorbents into existing boiler plant equipment or ductwork, resulting in a further potential advantage of low capital cost.

As shown in **Table 7-33,** dry sorption processes may be classified as non-regenerable and regenerable, and these two groups may be further categorized on the basis of the type of sorbent used and the contact mechanism employed.

Non-regenerable (also called "throw-away") processes utilize relatively low-cost sorbents such as limestone, lime, and naturally occurring sodium salts, and employ simple flow arrangements. The principal operating costs of non-regenerable dry sorbent systems are the

Table 7-33
Classification of Dry Sorption Processes For Sulfur Dioxide Removal

Contact Mechanism	Type of Process					
	Non-Regenerable[1,2]			Regenerable		
	Lime	Limestone	Soda	Sodium Aluminate	MgO-Based	CuO
Injection						
Lower Furnace	X	X				
Upper Furnace	X	X				
Economizer	X					
Duct	X		X			
Fluidized Bed	X			X		X
Moving Bed					X	X
Fixed Bed						X

Notes:
1. Humidification can be used with many of these processes to enhance collection efficiency.
2. Some processes are hybrids.

purchase of fresh sorbent and the disposal of spent material. Several non-regenerable dry sorption processes have attained commercial status.

Regenerable processes typically utilize more expensive sorbents such as sodium aluminate, magnesium oxide/vermiculite composite, and copper oxide. Also, they require more complex process schemes to move the sorbent between sorption and regeneration operations, and generally a high operating cost for energy and/or reducing agent in the regeneration step. Their main advantage is the elimination of the continuous requirement to procure large quantities of fresh sorbent and dispose of the spent waste material. The production of a salable byproduct (sulfur, sulfur dioxide, or sulfuric acid) seldom represents a significant economic incentive. Although none of the regenerable dry sorbent SO_2 removal processes can be considered fully commercial at this time, development work is still underway, and some of the early results show promise.

In the discussions that follow, dry sorption processes are categorized on the basis of the sorbent utilized: i.e, dry lime/limestone, sodium salt injection, alkalized alumina, dry MgO-based, and copper oxide processes.

Injection of Dry Alkaline Solids

The concept of injecting dry powered alkaline solids into the hot combustion gases in the boiler (or into the exhaust gas downstream of the boiler), then collecting the reacted material together with fly ash, is an extremely simple SO_2 removal technique. As a result, a considerable amount of research and development work has been conducted on this approach. At this time, however, only a few commercial applications exist. The technology is particularly applicable to retrofits because no additional gas train equipment and only a small amount of additional equipment is required. The greatest potential for many of these processes is probably the retrofit of older plants of small to medium size having a moderate SO_2 removal requirement, low plant capacity factor, short remaining life, and limited space available for equipment. All of the technologies rely on the reaction of dry alkali sorbent particles with the SO_2 in the flue gas.

All the dry injection processes share many common features. Smaller sorbent particles achieve higher SO_2 removal because they expose more external surface area to the gas. The initial reaction takes place on these external surfaces very quickly—usually within one second. Particles with many large pores (high internal surface area) remove more SO_2, but the reaction takes longer as the SO_2 must diffuse into the pores, then through the layer of reacted sorbent. The products of reaction can also block the pores as they form. It is believed that the pore structure of a particle is enhanced as a result of the partial decomposition of the particles, i.e., pores are created in the initial phase of the reaction: limestone gives off CO_2 gas when heated, hydrated lime gives off CO_2 in the external calcining process and then H_2O when injected into the furnace, and sodium bicarbonate gives off both CO_2 and H_2O when heated. Humidification ahead of the particulate collector to within about 20°F of adiabatic saturation enhances SO_2 removal. Residence time in the limited reactive temperature range and the SO_2 concentration also affect performance (Bjerle et al., 1990, 1991; Goots et al., 1991). Information on the physical and chemical phenomena associated with duct injection technology encompassing both experimental data and computer modeling is given by Peterson et al. (1989).

Dry Lime/Limestone Processes

Furnace Sorbent Injection Processes

Furnace sorbent injection is a technique in which lime or limestone is injected into wall- and tangentially-fired boilers to react with SO_2. Humidification may or may not be utilized upstream of the ESP for improved sorbent utilization (Keeth et al., 1991B; Princotta, 1990).

The limestone calcination reaction proceeds best at temperatures near 2,300°F. Reaction between sulfur dioxide and calcined limestone particles occurs primarily in the temperature range from about 1,000° to 2,600°F. Temperatures in the vicinity of 3,000°F occur near the bottom of typical boiler furnaces and are high enough to render the limestone inactive if the sorbent is injected at this elevation. As a result, the boiler injection point must be carefully selected. Injection directly with the fuel has resulted in low SO_2 removal efficiencies presumably because of the excessive temperature encountered by the sorbent.

The SO_2 removal efficiency is affected by numerous factors, chief among them being the quantity of sorbent used. In early tests conducted by Combustion Engineering (Plumley et al., 1967), only about 20% of the SO_2 was removed from hot flue gas when a stoichiometric amount of raw dolomite was injected. Similarly, the Wisconsin Electric Co. observed very low SO_2 removal efficiencies with slightly below stoichiometric quantities of injected limestone, but obtained 40 to 50% reduction in SO_2 with 75% excess limestone (Pollock et al., 1967). Other important factors are flue gas temperature distribution in the sorbent injection zone, residence time of the sorbent in the reactive temperature range, sorbent dispersion, and sorbent reactivity (Chughtai et al., 1990).

An extensive furnace sorbent injection demonstration project has been conducted on a wall-fired boiler at Ohio Edison's Edgewater Station. The process is called Lime/Limestone Injection with Multistage Burners (LIMB). Goots et al. (1991) review the process and provide the test results given in **Table 7-34.** This full-scale program characterized the SO_2 removal efficiency for various calcium-based sorbents: calcitic limestone, dolomitic hydrated lime, calcitic hydrated lime, and calcitic hydrated lime with a small amount of added calcium lignosulfonate. Results are presented for the effects of limestone particle size distribution (both with and without humidification to a close approach to saturation upstream of the ESP) and for the effects of injection at different furnace elevations. Humidification to a 20°F approach to saturation ahead of the precipitator was found to enhance removal efficiencies by about 10% over the range of stoichiometries tested. Without humidification, the "ligno" lime gave the highest removal efficiencies. The very fine (100% less than 10 microns) material is not considered a feasible sorbent at this time because the cost is four times as high as that of the other two materials. The higher removal efficiencies of the fine grind materials are attributed in part to the greater particle surface area available for SO_2 sorption. The optimum injection level corresponded to a 2,300°F furnace temperature. None of the sorbents appeared to have any effect on NO_x emissions.

While burning 3.0% sulfur and 10% ash coal during the Edgewater tests, the injection of sorbent at a Ca/S ratio of 2 almost tripled the ash handling system rate. However, humidification of the flue gas was found to increase the ESP's particulate collection capabilities to acceptable levels. The moisture in the gas reduced the electrical resistivity of the ash which improved ESP performance. The humidifier proved to be relatively trouble-free. Due to the cementitious properties of the quicklime and the pozzolanic ash, the byproduct bridged over the ash-handling system's aspirating water jets used to pneumatically convey the ash. The problem was solved by rodding out the jets more frequently. The impact of sorbent injection

Table 7-34
Summary of SO₂ Removal Efficiencies Achieved at the 105 MW$_e$ Edgewater Pulverized Coal Unit 4
(at Ca/S of 2.0 with Injection at the 181-Ft Elevation)

Sorbent and Test Condition	SO$_2$ Removal Efficiency, % Nominal Coal Sulfur, wt%		
	1.6	3.0	3.8
Limestone, Calcitic (80% < 44 μm)			
w/o Close Approach	22	25	NT
with Close Approach	29	NT	NT
Limestone, Calcitic (100% < 44 μm)			
w/o Close Approach	31	NT	NT
Limestone, Calcitic (100% < 10 μm)[4]			
w/o Close Approach	38	NT	NT
Lime, Hydrated Dolomitic			
w/o Close Approach	45	48	52
with Close Approach	62	58	NT
Lime, Commercial Hydrated Calcitic			
w/o Close Approach	51	55	58
with Close Approach	NT	65	NT
Lime, Commercial Hydrated Calcitic, with Calcium Lignosulfonate[5]			
w/o Close Approach	53	63	61
with Close Approach	70	72	71

Notes:
1. The 181-ft injector elevation is approximately opposite the nose of the furnace where combustion temperature is about 2,300°F.
2. w/o stands for without, NT for Not Tested.
3. "Without close approach" is defined as operation at a humidifier outlet temperature sufficient to maintain ESP performance, typically 250° to 275°F. "Close approach" is defined as a 20°F approach to adiabatic saturation (approximately 125°F) measured at the humidifier outlet for the coals used or 145°F. The humidifier is located between the air heater and the ESP.
4. The 10 μm sorbent was considered too expensive.
5. Anomalous results of testing with calcium lignosulfonate appeared to result from unexplained variations in reactivity.

Source: Goots et al. (1991)

on the boiler was dependent on the adequacy of the sootblowing system. Some difficulty was encountered when the byproduct was loaded into trucks for transportation to the landfill because steam generated by the water/quicklime reaction made it impossible for the operator to see how much space remained in a truck bed. However, the steam emanating from the surface of the byproduct presented no sustained problem as it subsided within about 15 minutes. At the ash disposal site, the byproduct posed no significant problems for bulldozers because the pozzolanic reactions did not proceed to any appreciable extent at the relatively low water/byproduct ratios used to ensure that the byproduct could be readily dumped from a truck (Nolan et al., 1990).

Two test landfills containing byproduct from Edgewater have been constructed and instrumented. No problems were encountered with rapid set-up of the wetted byproducts or with dusting of unwetted byproducts. A preliminary chemical characterization of the byproduct indicates that the material is similar to other calcium-injection byproducts with unreacted lime initiating pozzolanic reactions in the wetted byproduct cementing it into a coherent mass. Solidification decreased the permeability of the byproduct markedly (Holcombe et al., 1990).

Leachate from the byproduct is characterized by high pH, low metals content, and high concentrations of common ions, including calcium, potassium, sodium, chloride, and sulfate. These major ions are also present in moderate concentrations in the background groundwater and are unregulated constituents under the primary drinking water standards. Total leachate production is expected to be low, given the low permeability of the weathered solid byproduct. Leachate chemistry shows the effects of equilibrium with cementitious phases in the landfill byproduct. The growth of the cementing phases alters the leaching behavior of the byproduct from that observed in the fresh byproduct as the new mineral phases preferentially incorporate some ions such as calcium (Holcombe et al., 1990).

Furnace sorbent injection using lime/limestone also removes SO_3, HCl, and HF in the boiler. The H_2SO_4 acid dew point is therefore reduced allowing potentially lower boiler outlet temperatures and higher boiler efficiency (Landreth and Smith, 1989; Michele, 1987; Peterson et al., 1991). The removal of chlorides and fluorides may also prove beneficial as HCl and HF are on the EPA's list of hazardous air pollutants.

Tampella Lifac Process. Lifac stands for *L*imestone *I*njection into the *F*urnace and Re*A*ctivation of *C*alcium. With this variation of the furnace sorbent injection process, limestone is injected into the upper part of the furnace, is calcined, and reacts to remove SO_2. The flue gas containing lime, spent lime, and fly ash flows to a vertical activation reactor, located between the air heater and particulate collector. Water, sprayed into the reactor to humidify and cool the gas, and the increased particulate retention time improve SO_2 removal. The dry particulate elutriates from the reactor and flows with the flue gas to a downstream particulate collector. The Lifac process can achieve 80% SO_2 removal efficiency (Keeth et al., 1991B). The largest commercial installation as of 1991 was a 300 MW_e unit. The process is commercial in Finland, the Soviet Union, Canada, and the U.S. (Lifac, 1991; Enwald and Ball, 1991). The Lifac process is being demonstrated on Richmond Power & Light's 60 MW_e Whitewater Valley Unit No. 2 as a part of the Department of Energy's Clean Coal program. Here, the goal is to demonstrate 70 to 80% SO_2 removal at a Ca/S ratio of 2.0. Other furnace injection techniques were unsuccessful here because the boiler has a high heat release rate. The reactor vessel makes the capital cost of the process higher than that of some competing processes, but the use of limestone decreases operating costs (Lifac, 1991).

R-SO_x Process. R-SO_x, a hybrid furnace sorbent injection process developed by Fossil Energy Services International, Inc., is claimed to remove both SO_2 and NO_x. However, only SO_2 removal has been demonstrated. The R-SO_x process uses atmospherically hydrated lime with a proprietary additive to remove SO_2 and would use lime-urea hydrate if NO_x were to be removed. Projected capital and operating costs as of 1990 are $116/kW and 9.8 mills/kWh, respectively, vs. $286/kW and 15.2 mills/kWh for a comparable wet FGD system based on a new 500 MW_e pulverized coal plant burning 0.6% sulfur coal and 95% SO_2 removal efficiency. Limits of the technology are said to be about 95% SO_2 reduction and about 25 ppm outlet SO_2. The process developers claim that the process is particularly well suited for medium and high sulfur applications (Teixeira et al., 1990).

Both the SO_2 removal efficiency and the outlet SO_2 concentration are controlled by varying the recycle and the Ca/S ratios. Most of the process benefit appears to come from recycle, but use of an additive may reduce the sorbent requirement by 30%. Recycle ratios are typically between 1 and 20 with the total particulate flow through the boiler limited to between 2 and 3 times the fly ash flow. Sorbent utilization has been 32% without the additive and 42% with it (Teixeira et al., 1990).

Fresh sorbent mixed with recycled material is injected into the furnace in the region where the flue gas temperature is 1,450°–1,700°F and collected in a downstream particulate collector. Two possible collection equipment configurations have been identified. In the first arrangement, an ESP is operated to collect 90% of the fly ash and 10% of the sorbent with the balance flowing to a downstream fabric filter. This is possible because of the differences in size and resistivities of the ash and sorbent particles. In the second arrangement, all the material is collected together. Collected material is mixed with fresh sorbent and reinjected. In retrofit situations, it is claimed that existing precipitators can be modified to accommodate this process (Teixeira et al., 1990).

Economizer Injection Process

Economizer sorbent injection is identical to furnace sorbent injection except for the injection point location. With economizer sorbent injection, hydrated lime is injected into the boiler economizer inlet. An optimum temperature range (900° to 1,000°F) for SO_2 reaction occurs near this location and 50% SO_2 removal efficiency has been achieved. The process is currently undergoing pilot plant testing (Keeth et al., 1991B).

Duct Injection Process

With duct injection, SO_2 is removed by injecting hydrated lime into the ductwork, typically upstream of the particulate collector. To increase sorbent utilization, water may be injected upstream of the lime injection point to cool and humidify the flue gas, and a portion of the collected particulate may be mixed with fresh sorbent and recycled. SO_2 removal efficiencies of 50% are possible with the basic configuration, but much higher collection efficiencies have been achieved with some variations of the technique. A major concern with this technique is the long term effect of wall wetting and the potential for solids deposition. Programs are underway, including several advanced variations (Keeth et al., 1991B). Some processes are commercial, while others are developmental.

Work on duct injection was performed at the converted DOE Duct Injection Test Facility, which operates on a 12 MW_e equivalent, 50,000 acfm slip stream. Data which illustrate the effect of humidification on duct injection are presented in **Table 7-35** (Felix et al., 1991).

ADVACATE Process. A significant emerging variation of duct sorbent injection is the ADVACATE (Advanced Silicate) process offered by ABB-Fläkt. The first commercial systems are expected to achieve 90% SO_2 removal with a Ca/S ratio of less than 1.2 with costs about 50% of wet FGD. SO_2 removal efficiencies up to 99% have been reported on a pilot plant, including the effects of upstream furnace sorbent injection. Refer to **Table 7-36**. Field evaluation at the 10 MW_e scale by TVA and ABB-Fläkt was completed in October 1992, with results showing up to 90% SO_2 removal at stoichiometric ratios similar to lime spray

622 Gas Purification

Table 7-35
Summary of Most Favorable SO₂ Removal Results Obtained During Dry Hydrated Lime Testing at the DOE Duct Injection Test Facility

Mode of Injection	Ca/S Ratio	Approach (°F)	Inlet SO$_2$ (ppm)	SO$_2$ Removal, % ESP Inlet	ESP Outlet
No Humidification Dry Hydrate Injected Downstream of Water Spray Nozzles	2.32	158	1,350	12	17
Dry Hydrate Injected Upstream of Water Spray Nozzles	3.15	26	1,950	27	37
	2.50	30	1,900	42	53

Source: Felix et al. (1991)

Table 7-36
Edgewater LIMB/ADVACATE Results

ADVACATE Injection Rate (grams/dry standard cubic meter)	Total SO$_2$ Removal (%)	SO$_2$ Removal (%) ADVACATE Only
0	65	0
10	84	54
20	93	80
30	97	91
40	99	97

Note: Testing was on a 0.7 MW$_e$ slip stream, Ca/S = 2.0, and Tout = 63°C (13°C above Tsat).
Source: Princotta (1990)

drying (Buschmann, 1993). Demonstration projects are required and may take 3–7 years from 1992 (Princotta, 1990).

Fläkt Moist Duct Injection (MDI) technology is used with the process. About 25% of the material collected in the dust collector is mixed with lime slurry in a hold tank, pumped through a vertical mill, and reacted for an additional 1.5 to 2 hours at 80° to 90°C. The slurry is then pumped to a mixer mounted on the duct and distributed over the remaining 75% of the dry solids recycle, resulting in a "damp" solid with 30 to 50% moisture. The product solids drop into the duct causing the flue gas to be cooled to within 10°–15°C of saturation within 0.5 seconds. Operation without the formation of duct deposits has been demonstrated in pilot plant tests. The process produces a silicate gel from the mixture of lime and dust collector product. This gel has a high surface area, thin layers of free lime, and substantial moisture allowing for simultaneous in-duct absorption of acid gases and flue gas cooling (Sedman et al., 1991).

Coolside Process. Coolside desulfurization technology, which is offered by Babcock and Wilcox, involves injection of dry hydrated lime into the flue gas and flue gas humidification by water sprays located downstream of the air preheater. SO_2 is captured by reaction with the entrained sorbent particles in the humidifier and with the sorbent collected in the particulate removal system. The humidification water serves two purposes. First, it activates the sorbent to enhance SO_2 removal; and, second, it conditions the flue gas and particulate matter to maintain efficient ESP performance. Spent sorbent is removed from the gas along with the fly ash in the existing particulate collector (ESP or fabric filter). The sorbent activity can be significantly enhanced by dissolving sodium hydroxide or sodium carbonate in the humidification water. Sorbent recycling can be used to improve the sorbent utilization if the particulate collector can handle the resulting increased solids loading.

A demonstration project was conducted on the 104 MW_e Unit No. 4, Boiler 14, at the Ohio Edison Edgewater power plant. Tests using commercial hydrated lime on a once-through basis demonstrated that the Coolside process can routinely achieve 70% SO_2 removal at the design conditions of 2.0 Ca/S and 0.2 Na/Ca molar ratios and 20°F approach to adiabatic saturation temperature. Because of the relatively short test duration, sodium hydroxide was used as the additive in the Edgewater demonstration. Sorbent recycle tests demonstrated the capacity of recycle sorbent to remove additional SO_2 and showed a significant potential for sorbent utilization improvement (Yoon et al., 1991). McCoy et al. (1991) and Nolan et al. (1992) review the economics of the process.

Fluidized Bed Process

Lurgi CFB Process. In this SO_2 removal process, hydrated lime is injected into a circulating fluid bed (CFB) reactor located ahead of the particulate removal system. Water is also injected into the reactor to cool and humidify the gas and increase the SO_2 removal efficiency. A mixture of fly ash and reacted and unreacted sorbent particulate in the gas from the CFB reactor is collected in a special downstream mechanical dust collector "curtain" followed by an ESP. Most of the collected particulate is recycled to the CFB reactor (Keeth et al., 1991B). In 1993, eight systems of this type were operating or were under construction in Europe, and two were under construction in the U.S. (Moore, 1993B).

In this process, both the injected sorbent and waste are dry. Pebble lime is first hydrated (0.5% free moisture) and then conveyed pneumatically to a silo. Hydrated lime from the silo is mixed with recycle material from the ESP and injected into the CFB reactor, forming a fluidized bed. The particulate in the fluidized bed flows with the flue gas upward through the reactor to the downstream mechanical collector and ESP where the particulate is collected. Most of the collected material is recycled via air slides back to the reactor. Recycle rates are very high, e.g., 130 to 1, giving the very high sorbent retention time required for high SO_2 removal efficiency and optimum sorbent utilization. Spray water is added so that the process operates close to the dew point, which further optimizes sorbent utilization. Acid gas component removal efficiency, in decreasing order, is HF, HCl, SO_3, SO_2, and CO_2. Virtually all the strong acids (HF, HCl, and SO_3) are removed and the byproduct typically contains about 8% $CaCO_3$. A fabric filter is not used because of the high inlet grain loading. The gas side pressure drop for the system is typically about 7.5 in. of water with a maximum of 9 in. of water. Velocity through the reactor is an important parameter to assure acceptable operation. Erosion and corrosion at one plant were reported to be unobservable after 3 years operation. The waste, containing about 2% moisture, is landfilled. The process has been used in Germany since 1982 where Lurgi has six installations. The largest German plant treats 585,600 acfm, which correspond to about 160 MW$_e$. This plant has a single CFB reactor (Toher et al., 1991).

Environmental Elements Corporation has the exclusive North American license for the Lurgi CFB process. A 5 MW$_e$ demonstration plant was operated during 1992 for Dakota Gas—formerly the Great Plains Gasification Plant. The pilot plant began operation in the first quarter of 1992. Char from a gasifier plant is burned, so the flue gas is high in CO_2 in comparison to flue gas from coal-fired applications, which could affect sorbent utilization. The total gas flow is 1,200,000 acfm, but only a side stream containing 4,000–6,000 ppm SO_2 was treated (Lanois, 1993).

This process is capable of achieving very high SO_2 removal (>95%) on high sulfur coal (up to 6%) with Ca/S ratios between 1.2 and 1.55 and inlet SO_2 concentrations between 500 and 3,500 ppm. The minimum outlet SO_2 concentration is in the vicinity of 15–20 ppm. Turndown to 30% of design per module is possible based on reactor gas velocities of 6 to 20 feet per second (Hansen and Toher, 1991).

Currently, the combined SO_x/NO_x removal capabilities of the Lurgi CFB process are being investigated, with removal efficiencies of 95%/85%, respectively, being demonstrated. With this variation of the process, a NO_x reduction catalyst is included in the circulating bed material. Fine $FeSO_4 \cdot 7H_2O$ powder (10 µm), without a supporting carrier, is used as the catalyst. Ammonia is the reducing agent. The reactor is located between the economizer and the air preheater and operates at 725°F. No water is used. Reaction products are $CaSO_4$ (anhydrite) and approximately 10% $CaSO_3$. $CaSO_4$ results from the oxidation of SO_2 to SO_3, which is aided by the NO_x catalyst. Usually, this side reaction is minimized in a NO_x control system because it creates corrosion problems. However, in this combined process, it is part of the desulfurization process.

Dry Soda Processes

In this discussion, the term "soda" is used to denote sodium carbonate, sodium bicarbonate, sodium sesquicarbonate, and mixtures of these compounds. Dry soda injection is commercially established in several industries. Four industrial plants installed this technology in the 1987–1992 period. Two are municipal solid waste (MSW) plants in South Carolina and Alaska; another is Ford Motor Glass in Tennessee, where the process is used for control of HCl and HF gas emissions. More recently, Western Slope Refining installed a dry soda system to control SO_2 from their coke calcining operations in Colorado. Public Service Company of Colorado, a pioneer of the technology, has installed a system (25% SO_2 removal) on the 350 MW$_e$ Unit No. 4 of their Cherokee plant and is planning similar systems for other boilers. Wisconsin Electric Power Company plans to install the technology (20–35% SO_2 removal) at their Port Washington Plant on Units 1 and 4 (80 MW$_e$ each) (Bennett, 1992). Pulp and paper mills with kraft recovery boilers equipped with ESPs appear ideal for dry soda injection since they already use sodium sulfate, but this application remains to be demonstrated (Bennett and Nastri, 1990; Hooper, 1990). In this application, control of SO_2 emissions from recovery furnaces would simply increase the sodium sulfate collected in the ESP, and the collected material would be recycled back to the salt-cake mix tank (Hooper, 1990).

Economic evaluations of the dry soda injection technology are available in the EPA IAPCS Economic Model (Version 4) (Maibodi et al., 1991) and have been published in several earlier versions (EPRI, 1986A, B, 1989). These evaluations indicate that the technology has low capital cost compared to most other FGD technologies and is competitive in levelized cost for retrofit applications with small boilers (<150 MW$_e$) firing less than 1% sulfur in their fuel. The technology is also competitive if 70% or less SO_2 removal is required. In a 1991 paper by Bennett and Darmstaedter (1991), a levelized cost of $742/ton of SO_2 removed was calculated for a special case application of 55% SO_2 removal and 15% NO_x removal using the EPA model and taking credit for sodium sulfate ($NaSO_4$) recovery and sale. Process economics are influenced by transportation costs. The major nahcolite (natural sodium bicarbonate) and trona (natural sodium sesquicarbonate) resources are located in Wyoming, Colorado, and California; while sodium bicarbonate is manufactured in Ohio and Georgia (Hooper, 1990).

Salt transport in the process is entirely pneumatic. The dry sorbent is usually delivered by bulk carrier (truck or rail) and off-loaded pneumatically into covered storage silos. From the silos, it is pulverized to 10–20 microns mean mass diameter to optimize both the rate of decomposition and sorbent utilization (Bennett, 1992; Bennett and Nastri, 1990; Hooper, 1990). The pulverized sorbent is conveyed to a day silo, metered into the pneumatic conveying line, and distributed into the flue gas through a system of injection nozzles. The sorbent feed rate can be automatically controlled based on upstream and/or downstream sulfur dioxide gas stream concentrations (Bennett, 1992). The end product (a mixture of sodium sulfate, sodium chloride, sodium nitrate, fly ash, and unreacted soda) is a dry powder easily collected by either a fabric filter or an ESP (Bennett and Nastri, 1990).

The dry soda sorbents are injected as dry powders into the flue gas between the boiler and the particulate collector. Once in the flue gas stream, sodium bicarbonate in the sorbent begins to decompose into sodium carbonate according to the following reaction:

endothermic
$$2NaHCO_3 = Na_2CO_3 + H_2O_{(gas)} + CO_{2\,(gas)} \tag{7-99}$$

In situ decomposition continues to completion, as the SO_2 reacts with the Na_2CO_3. Sorbent injection into the flue gas in the temperature range of 300° to 750°F (375° to 700°F optimum) achieves greater sorbent utilization than injection into 250°F flue gas where decomposition is slow and incomplete (Bennett, 1992).

At the 22 MW_e demonstration of the dry soda injection process at the Public Service Company of Colorado Cameo Unit 1 (Muzio and Sonnichsen, 1982), both nahcolite and trona were used as sorbents and both performed satisfactorily. Pure sodium carbonate was found to provide very limited SO_2 removal (nominally 10%). It was concluded that nahcolite and trona are effective because they thermally decompose to release CO_2 and H_2O, which greatly enhances the porosity and reactive surface area of the residual sodium carbonate. The decomposition of trona occurs at significant rates above 200°F while nahcolite requires temperatures above about 275°F for decomposition. Once thermal decomposition is initiated, SO_2 begins to react with Na_2CO_3 according to the following reaction:

exothermic
$$Na_2CO_{3\,(solid)} + SO_{2\,(gas)} + \tfrac{1}{2}O_{2\,(gas)} = Na_2SO_{4\,(solid)} + CO_{2\,(gas)} \tag{7-100}$$

The solid Na_2SO_4 is collected along with fly ash in an ESP or a fabric filter. Normal particulate emissions and opacity levels have been maintained without flue gas humidification or temperature modification (Bennett, 1992). During the initial testing at the Port Washington Plant, 20 to 35% SO_2 removal was achieved with essentially 100% sorbent utilization at high boiler load. At low boiler loads, utilization decreased to 40–60% presumably because of sorbent fallout in the ductwork. At 74% SO_2 removal, 86% sorbent utilization was achieved (Coughlin et al., 1990).

This process can capture other acid gases, such as HCl from incineration of municipal or medical waste products:

$$NaHCO_{3\,(solid)} + HCl_{(gas)} = NaCl_{(solid)} + CO_{2\,(gas)} + H_2O_{(gas)} \qquad (7\text{-}101)$$

HCl removal efficiencies of greater than 95% have been realized with a system incorporating a fabric filter (Bennett and Nastri, 1990). Removal efficiencies for SO_2/HCl of 70%/90% and for SO_2/NO_x of 90%/70% and 70%/40% have been demonstrated at the 4 MW_e EPRI High-Sulfur Test Center (Bennett and Darmstaedter, 1991). The change in NO removal is directly a function of the SO_2 removal. NO_x removal can only be accomplished with this process as a side reaction to SO_2 collection. But while total NO_x is reduced, NO_2 can increase. Hooper (1988) found that an undisclosed additive improved SO_2 and NO_x removal and reduced NO_2 emissions below the level that caused plume coloration.

Dry Sorption Byproduct Disposal and Use

For the sodium byproduct, potential disposal options include landfilling, disposal in saline environments, dry impoundments, deep-well injection, and ocean dumping. Each of these has drawbacks that must be considered (Eklund and Golden, 1990).

Landfilling has been the standard method to date for byproduct management (Eklund and Golden, 1990). Some states are proposing strict guidelines for the construction of SO_2 control byproduct disposal sites as well as leaching requirements for water which might permeate the site (Bennett and Nastri, 1990). Compacted clay barriers and impervious liners such as high density polyethylene (HDPE) or Claymax (R) can be used. Claymax is a proprietary mat, about ⅛-in. thick, consisting of bentonite clay between layers of geotextile. Geotextiles are woven or non-woven materials that are permeable to water, but not to solids. This technique is approved by some states; however, small cells may be required (Bennett and Nastri, 1990; Hooper, 1990). Undesirable byproduct characteristics are related to the high solubility of sodium salts and include poor leachate quality, high permeability, poor handling properties when mixed with water, and poor landfill strength development. The leachate may have a high sodium concentration, a high total dissolved solids concentration, a high electrical conductivity, a high carbonate concentration, a measurable selenium concentration, and an alkaline pH. High permeability is attributed to the high byproduct sodium content because sodium compounds do not undergo cementitious reactions and can dissolve once the byproduct is compacted and cured. The byproduct becomes fluid at low levels of added moisture, thus creating serious handling problems. Even after curing, the strength may degrade as sodium compounds dissolve (Eklund and Golden, 1990).

Stabilization has been investigated for sodium containing byproducts. In some cases, clay and polymer mixtures have produced permeabilities as low as 10^{-7} cm/sec depending on the ash. The process is sensitive to the ratios of the various components and the requirements for proper mixing and compression, which become difficult when large quantities of byproduct have to be handled (Bennett and Nastri, 1990).

Of the various byproduct utilization options, the most attractive is the separation and recovery of sodium sulfate for sale or reuse (Bennett and Nastri, 1990). Reuse options include recovery and use of sodium compounds in mineral filler, grout, and bricks (Eklund and Golden, 1990). Producing high purity Na_2SO_4 for sale by dissolving the byproduct and recrystallization is in the developmental stage. Although all steps of the process have been used commercially, they have not been used together for this purpose. Development work continues on the removal of nitrates during the dissolution/filtration steps. Another byproduct utilization method is separate collection of fly ash and spent sorbent accomplished by the use of two particulate collectors in series with sorbent injection in between. Air classification and electrostatic powder separation are other techniques under development (Bennett and Nastri, 1990).

For calcium-based dry sorption process byproduct disposal and use, refer to the previous Spray Dryer Byproduct Disposal and Use section. The calcium-based spray dryer byproduct is similar to the calcium-based dry sorption byproduct in many respects. However, byproduct properties can be affected by presence of anhydrite ($CaSO_4$). Anhydrite is present in the byproduct from processes using CaO as the sorbent, but is usually not found in the byproduct from processes using $Ca(OH)_2$. Byproducts with a high anhydrite content may swell significantly when they hydrate, which can cause problems in some situations. This form of calcium sulfate (rather than gypsum, $CaSO_4 \cdot 2H_2O$) is stable in contact with dry (i.e., low relative humidity) gases, which are often present in dry lime/limestone processes.

Dry Metal Oxide Processes (Regenerative)

The oxides of 48 metals were screened by the Tracor Co. in a project conducted for the U.S. National Air Pollution Control Administration to determine which were best suited for the removal of sulfur oxides from flue gases by chemical reaction (Thomas et al., 1969). The screening was accomplished by consideration of the thermodynamic requirements for efficient sulfur oxide removal and product regeneration. Sixteen potential sorbents were selected as a result of this screening process. These were the oxides of titanium, zirconium, hafnium, vanadium, chromium, iron, cobalt, nickel, copper, zinc, aluminum, tin, bismuth, cerium, thorium, and uranium.

The 16 metal oxides were further screened on the basis of their rate of reaction with SO_2 in a flue-gas atmosphere. The oxides were prepared in a kinetically active form by calcining a salt which decomposed to the oxide at a relatively low temperature. The rate data were collected using an isothermal gravimetric technique whereby weight gain of SO_2 was recorded as a function of time. The oxides of copper, chromium, iron, nickel, cobalt, and cerium were found to have economically feasible reaction rates with SO_2. After further evaluation of such factors as sorption reaction stoichiometry, formation of product layers that affect the reaction rate, and SO_3 partial pressure over the sorption product, two materials, copper oxide and iron oxide, were selected as most promising. Finally, preliminary design and economic studies were made for a process employing these oxides with a fluidized bed gas-solid contactor for both sorption and thermal regeneration steps. Copper oxide (CuO) and iron oxide (Fe_2O_3) were found to have promise as potential sorbents for an economically feasible SO_2 removal process.

Copper Oxide Processes

Laboratory scale work on a copper oxide process for SO_2 removal was conducted by the U.S. Bureau of Mines (McCrea et al., 1970). The effort was aimed at developing a dry, regenerable sorbent for SO_2 that would not have the problems of alkalized alumina, i.e., physical degradation, excessive reducing gas consumption, and a high temperature difference between absorption and regeneration steps. A sorbent that appeared to meet these requirements was prepared by the impregnation of copper oxide into porous alumina supports.

The following chemical reactions are important in copper oxide SO_2 removal processes:

$$CuO + SO_2 + \tfrac{1}{2}O_2 = CuSO_4 \tag{7-102}$$

$$CuSO_4 = CuO + SO_3 \tag{7-103}$$

$$CuSO_4 + 2H_2 = Cu + SO_2 + 2H_2O \tag{7-104}$$

$$CuSO_4 + \tfrac{1}{2}CH_4 = Cu + SO_2 + \tfrac{1}{2}CO_2 + H_2O \tag{7-105}$$

Reaction 7-102 goes essentially to completion at temperatures below about 450°C. At higher temperatures, $CuSO_4$ decomposes by either the reverse reaction of 7-102 or reaction 7-103. At about 700°C, regeneration should be complete enough to allow recovery of evolved sulfur oxides and recycle of CuO. Regeneration by the use of reducing gases as in reaction 7-104 and 7-105 can be accomplished at much lower temperatures. The Bureau of Mines work indicated methane to be preferable to hydrogen for regeneration because of the tendency for reduction to proceed all the way to sulfide with hydrogen at low temperature. A conceptual design of a process was proposed as shown in the flow diagram of **Figure 7-36.** Although no pilot or full-scale plant operations of this process have been conducted to date, it appears to offer considerable promise. The major drawbacks are the requirements for a large expensive reactor, which results from a low rate of absorption of SO_2 on the copper oxide-alumina pellets, and the high consumption of reducing gas.

Work on a copper oxide process has also been conducted by Shell in the Netherlands (Doutzenberg et al., 1971). This process, which has been named the Shell Flue Gas Desulfurization (SFGD) process, also uses CuO on an alumina support. It is unique, however, in that it uses fixed beds, and the absorption and regeneration steps are carried out in the same vessel. Two units are used to provide continuous operation. One is used for gas purification, while the other is regenerated. Both steps are accomplished at about the same temperature (400°C). Regeneration is accomplished by the use of a reducing gas such as hydrogen, carbon monoxide, or methane and results in the production of a sulfur dioxide rich gas (reactions 7-104 and 7-105). Hydrogen and CO are preferred as reducing agents for the Shell process because of problems of coke deposition with hydrocarbons. In order to avoid plugging of the fixed bed by soot and ash particles in the flue gas, a novel reactor system was developed in which the gases are made to pass alongside large surfaces of sorbent mass rather than through a particle bed. A pilot plant with a capacity of 21,000 to 35,000 scfh of flue gas was built and operated to demonstrate the parallel passage reactor and other features of the process. The pilot plant operated quite successfully, removing about 90% of the SO_2 from the gas passing through it.

A large commercial unit was started up in 1973 at the Showa Yokkaichi Sekiyo (SYS) refinery in Japan. The system was designed to remove 90% of the SO_2 from the flue gas of a boiler fired with heavy high-sulfur fuel oil. The plant contains two copper-oxide reactors, each sized to handle the entire 125,000 Nm^3/h (77,500 scfm) of flue gas. Off-gas from the unit undergoing regeneration is cooled to condense water and then flows through an absorber-stripper, which damps the cyclic flow and provides a near constant flow of concen-

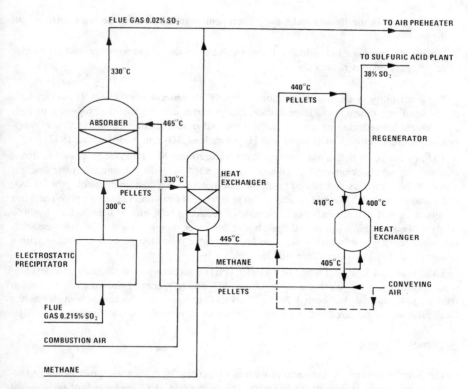

Figure 7-36. Conceptual process flow diagram of SO_2 removal with copper oxide as proposed by Bureau of Mines. (*McCrea et al., 1970*)

trated SO_2 to an existing Claus sulfur plant (Arneson, et al., 1977). The SYS Unit has also been used to demonstrate the capability of the system to remove NO_x, simultaneously with SO_2 (Radian, 1977). A 60 to 70% reduction in NO_x, concentration was obtained by adding NH_3 to the flue gas entering the SFGD System. The ammonia reduces NO to N_2. Carryover of unreacted ammonia with the flue gas was reported to be 2 ppm maximum.

To determine the applicability of the process to flue gas from coal combustion, a small scale (0.6 MW_e) unit was built and placed into operation by UOP, Inc. at Tampa Electric Company's Big Bend Station in North Ruskin, Florida. The pilot plant was started up in 1974 and operated for approximately 2 years. During this time, six runs were made involving over 13,000 acceptance and regeneration cycles with the same acceptor loading. The test program demonstrated

1. The commercial acceptor material has good stability, achieving 90% SO_2 removal across a 4-m bed at a space velocity of 5,000/h after 13,000 cycles.
2. The reactor design is tolerant of high fly ash loadings. Techniques were developed to clean fouled internals *in situ*.

3. Metal oxides in the fly ash and halogen compounds in the flue gas do not interfere with the redox cycle.
4. Mechanical components, including flapper valves, the sequence controller, and the reactor vessel, gave good performance.

The availability of design information on the SFGD process is excellent. However, little commercial experience is available to date. The principal drawback of the process is its high requirement for reducing gas, particularly if the end product is elemental sulfur. Consumption figures reportedly show 6.2 moles of H_2 per mole of SO_2 removed (Radian, 1977).

The process has a potential advantage over conventional SO_2 removal processes in that it can simultaneously remove NO_x. This is accomplished by adding ammonia to the flue gas ahead of the copper oxide bed and using the bed material as a combination acceptor for SO_2 and catalyst for the $NO_x + NH_3$ reaction. An independent evaluation based on pilot-scale test results indicated that the process is capable of reducing NO_x and SO_2 emissions by 90% when applied to a high-sulfur coal-fired boiler, but its economics, for the specific case studied, were less favorable than a separate NO_x catalytic reduction process combined with a conventional FGD system (Burke, 1983).

The Pittsburgh Energy Technology Center (PETC) has conducted research and development on the copper oxide process for combined SO_2 and NO_x control since the late 1960s, but their largest pilot unit has been 1 MW_e. The PETC process uses the same chemistry as the Shell process, but fluidized bed or moving bed reactors are employed (Hoffman et al., 1992).

Sorbtech Process

The Sorbtech (formerly Sanitech) process is a regenerable, dry sorbent process that reportedly can remove over 90% of the SO_2, 10–90% of the NO_x, HCl, and much of the residual fine particulate from flue gas. The process, currently under development, uses new, inexpensive sorbents called Mag*Sorbents made from magnesium oxide (MgO) and an expanded silicate carrier (vermiculite or perlite). A preliminary economic analysis has indicated a cost of $300—$400 per ton (1982 dollars) of SO_2 removed (Nelson and Nelson, 1990).

The best performing Mag*Sorbent consists of 50% wt MgO and 50% wt vermiculite (Sanitech, 1988). The sorbents remove SO_2 by a combination of chemical reaction (MgO and SO_2 react to produce mainly $MgSO_3$ and some $MgSO_4$) and capillary micropore condensation (SO_2 is held by capillary forces within the micropores). The sorbent can be prepared using commercially-available equipment, possibly at the installation site. Long-term storage of the sorbent does not appear to affect sorption performance (Sanitech, 1988).

To increase utilization, flue gas is humidified to within 50°F of the adiabatic saturation temperature by water sprays in the flue gas duct downstream of the ESP. The sorber is a vertical cylindrical vessel with a special radial-panel bed of slowly moving Mag*Sorbent. After humidification, the flue gas enters the sorber through the top, flows downward and then outward horizontally through the sorbent bed to an outer chamber around the sorber perimeter, and exits the unit. Fresh sorbent is fed continuously to the top of the sorber, and the spent sorbent is removed by a rotary feeder at the sorber base (Nelson, 1991).

Regeneration is accomplished by heating in either a reducing or an oxidizing atmosphere. Regeneration in a reducing atmosphere at 750°C for 20 minutes drives off the sorbed SO_2 as elemental sulfur, SO_2, and H_2S. Both the magnesium sulfite and sulfate are decomposed, and NO_x is converted into nitrogen and water. Regeneration in an oxidizing (air) atmosphere at

550°C for 30 minutes releases SO_2, but does not completely decompose the non-reactive sulfates, which remain and reduce sorbent utilization by 5–10% per cycle (Sanitech, 1988).

The regeneration circuit consists of a regenerator, a screening station, and, in the case of regeneration under a reducing atmosphere, a sulfur recovery system. Saturated sorbent is pneumatically conveyed from the sorber to the regenerator, where it is heated to the regeneration temperature. The regenerated sorbent then passes through the screening station, where about 12% of the sorbent is removed and fresh sorbent make-up is added. In the sulfur recovery system, the regenerator off-gas is converted to elemental sulfur by a modified Claus plant or by a new process currently under development at Sorbtech. About 1–4% of the sorbent is degraded to fines during each sorption-regeneration cycle. These fines are removed from the system as part of the 12% of the sorbent removed during screening. There is no other waste stream (Nelson, 1992).

There are a number of potential uses for the spent sorbent, including the manufacture of a slow-release agricultural fertilizer and soil conditioner, pneumatically-injected insulation, wallboard additives, and ingredients for lightweight cement. Premier Services Corporation, the leading U.S. magnesia producer, is assisting Sorbtech in evaluating the spent sorbent's commercial potential (Sanitech, 1988).

Research and development of the process has been ongoing since 1985. As of August 1992, a pilot plant has been in operation on a 2.5 MW_e slipstream of flue gas at Ohio Edison's Edgewater Station in Lorain, Ohio. A prototype system has also been installed on a 100 MW_e equivalent jet engine test facility at the Tyndall Air Force Base in Panama City, Florida. This system, which reduces NO_x by 50–80%, consists of a filter installed atop the test-cell stack (Nelson, 1992).

Alkalized Alumina Processes

Bureau of Mines Process

The Alkalized Alumina process was developed by the U.S. Bureau of Mines and carried through the pilot-scale testing phase (Bienstock et al., 1964; 1967). It has not been applied commercially. The process uses dawsonite [$NaAl(CO_3)(OH)_2$]/sodium aluminate [$NaAlO_2$] as the sorbent. The material is activated at 1,200°F to form a high surface area, high porosity, dry solid, which removes sulfur dioxide from flue gas at temperatures between 300° and 650°F. This process is no longer being pursued due primarily to an excessive sorbent attrition rate.

NOXSO Process

The NOXSO process is a dry regenerable alkalized alumina process that simultaneously removes 90–95% of the SO_2 and 80% of the NO_x from flue gases. Work on the NOXSO process started in 1979. The process is based on the older Bureau of Mines Alkalized Alumina process. A major difference between this process and the former is the use of a sorbent pellet that does not have the uneconomically high attrition rate responsible for the demise of the Bureau of Mines process. (The Bureau of Mines process used solid $NaAlO_2$ pellets that were broken by the mechanical stress of the sorption process.) However, attrition rates are still a significant concern with the NOXSO process. The reported pellet attrition rate is about 0.02% of the fluid bed inventory per hour (Haslbeck 1992). Another difference between the two processes is the sorption temperature: 250°–300°F for the NOXSO process vs. 625°F for the Bureau of Mines process. This lower temperature favors NO_x removal. In order to provide the required low sorption temperature, the process is located downstream of the boiler air heater and particulate collection device rather than upstream of them. In the NOXSO process, a fluidized bed is used as the flue gas sorbent contactor (Haslbeck, 1992; 1984).

Proof-of-concept testing (except for NO_x recycle and the established Claus process) has been conducted on a 5 MW_e slipstream at Ohio Edison's Toronto Station. A demonstration larger than 100 MW_e is planned under a DOE Clean Coal contract (Haslbeck, 1992; Moore, 1993A). A study rated the NOXSO process as the only process with the potential to be equivalent to wet FGD with selective catalytic reduction. The 1991 capital cost is estimated to be $257/kW, and the operating cost to be 11.7 mills/kWh (Cichanowicz et al., 1991).

The process uses pellets of a very porous, high surface area "gamma alumina" (Al_2O_3) substrate with a surface layer of $NaAlO_2$. The layer of $NaAlO_4$ is the active sorbent. The sorbent pellet is prepared by depositing Na_2CO_3 on the alumina pellet. The sodium carbonate and the alumina interact chemically on the first pass through the sorber to form the layer of $NaAlO_2$ (Haslbeck, 1984). Sodium and aluminum are common constituents of fly ash and not expected to cause environmental problems. The pellets presently have an expected life of about 9 months before they must be replaced (Haslbeck, 1992).

The process is depicted in **Figure 7-37.** Flue gas laden with SO_2 and NO_x passes through the fluidized bed sorber where SO_2 and NO_x are simultaneously removed by the sorbent pellets and are converted to sodium sulfate, sodium sulfite, and sodium nitrate. If the flue gas is too hot, it may be necessary to utilize evaporative cooling to achieve the necessary process temperature. Lower temperatures give higher NO_x removal. The upflow of flue gas through the sorber fluidizes the bed of approximately ¹⁄₁₆-in. size sorbent pellets. The pellets leave the sorber primarily by overflowing into a weir pipe and down into a surge bin. Some smaller particles elutriate out of the bed with the flue gas (Haslbeck et al., 1990). Fifty percent of all elutriated particles greater than 50 microns in size are captured in a cyclone and are returned to the process (Haslbeck, 1992).

The sorbent is regenerated in several steps. Material from the sorber surge bin is pneumatically transferred to the sorbent heater. Here, it is heated with air to about 1,150°F to strip the NO_x and a small fraction of the SO_2 (Haslbeck, 1992). The hot air comes from a regenerative sorbent heating-cooling arrangement in which heat from the sorbent leaving the regenerator is transferred to the sorbent entering the regenerator. Additional heat is provided by the burning of natural gas in the air stream ahead of the sorbent heater. The air leaving the sorbent heater is recycled to the boiler as combustion air where about 70–75% of the NO_x is reduced to N_2 by a reburning process (Haslbeck, 1992). As the sorbent reaches the regenerator inlet temperature, it flows via a J-valve to the upper section of the regenerator where the next regeneration step takes place (Haslbeck et al., 1990).

In the upper section of the regenerator, a reducing gas (natural gas) reacts with the sorbent removing about 70% of the sulfur as H_2S and SO_2, which flow to the Claus plant. The inlet J-valve is used to both control the solids feed rate to the regenerator and to isolate the regenerator from the sorbent heater. With the completion of this reaction, the sorbent then drops into the lower section of the regenerator where the final regeneration step takes place (Haslbeck et al., 1990).

In the lower section of the regenerator, the sorbent is contacted with steam to remove the balance of the sulfur as H_2S, which also goes to the Claus plant. From the steam treatment section of the regenerator, the regenerated sorbent then flows to the sorbent cooler via another J-valve arrangement where the sorbent heats the air stream used to heat the sorbent in the sorbent heater. When cooled to about 250°F, the sorbent drops into a surge bin from which it

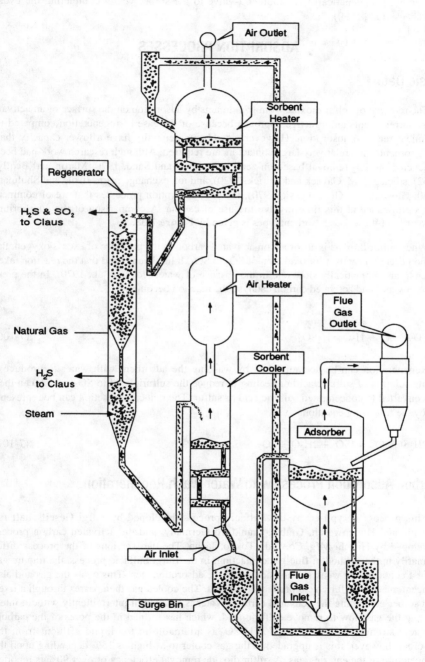

Figure 7-37. Flow diagram of the NOXSO process. (*Haslbeck, 1992*)

is conveyed pneumatically via another J-valve to the sorber vessel completing the cycle (Haslbeck et al., 1990).

ADSORPTION PROCESSES

Basic Data

The removal of sulfur dioxide from gas streams by adsorption on the surface of an activated solid represents an attractive approach because of the ease of regeneration compared to chemical reaction absorption. However, adsorbents generally have a lower capacity than solid sorbents, and relatively large contactors are required. Although research work had been conducted on SO_2 removal based on zeolite (Tamboli and Sand, 1970; Martin and Bently, 1962), silica gel (McGavack and Patrick, 1920), and ion exchange resins (Cole and Shulman, 1960; Glowiak and Gostomczyk, 1970), the only adsorption processes that are of commercial significance at this time involve the use of carbon. A comparison of the equilibrium capacities of these four adsorbent types is given by **Figure 7-38**.

When sulfur dioxide is adsorbed on activated carbon in the presence of excess oxygen, the carbon acts as a catalyst for oxidation of SO_2 to SO_3. It has been found that the reaction takes place at an impractically slow rate in the absence of water (Joyce et al., 1970). In the presence of water and activated carbon catalyst, the reaction becomes

$$SO_2 + \tfrac{1}{2}O_2 + H_2O \xrightarrow{\text{activated carbon}} H_2SO_4 \tag{7-106}$$

Regeneration can be accomplished by washing the adsorbent with water to produce a dilute solution of sulfuric acid or heating to reduce the sulfuric acid to SO_2, which can then be converted to concentrated sulfuric acid or sulfur. The reduction reaction can be represented by the following equation:

$$2H_2SO_4 + C = 2SO_2 + 2H_2O + CO_2 \tag{7-107}$$

Carbon Adsorption Process with Water Wash Regeneration

This process is typified by the Sulfacid process developed by Lurgi Gesellschaft fur Chemie and Huttenwesen, GmbH, Frankfurt, Germany, and the activated carbon process developed by Hitachi Mfg. Co. Ltd., Tokyo, Japan. The two versions of the process differ primarily in the method of flue-gas contacting. In the Lurgi Sulfacid process, the impure gas is first contacted by weak sulfuric acid from the adsorption step. This cools the gas and also concentrates the sulfuric acid product somewhat. The cooled gas then passes through a fixed bed sorber containing the activated carbon. Water is sprayed in intermittently without interrupting the gas flow to remove sulfuric acid, which has formed in the pores of the carbon. An acid strength of only about 7% H_2SO_4 is attainable in the liquid effluent from the absorber; however, this is upgraded in the gas cooler to as high as 15% depending upon the temperature of the entering gas. A sulfur dioxide removal efficiency of over 90% is reported (Dennis and Bernstein, 1968). The process has been used to treat emissions from a sulfuric

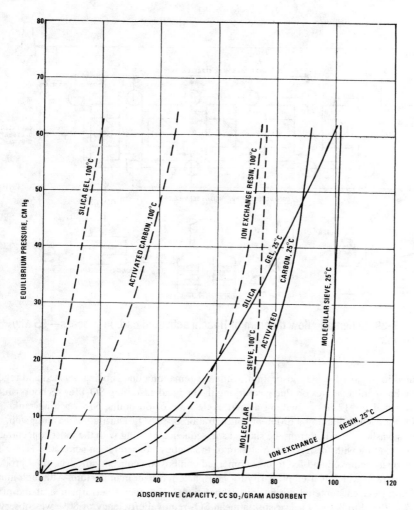

Figure 7-38. Adsorptive capacity for SO_2 of silica gel, activated carbon, ion exchange resin (*Rohm & Haas Co. IRA-400*), and molecular sieve (*Linde Co. Type 5-A*). Data of Cole and Shulman (1960)

acid plant (100,000 cfm) and from a coal-fired unit equivalent to a 2 MW_e power plant (Maurin and Jonakin, 1970).

The Hitachi process, which has been described in considerable detail by Tamura (1970), uses a somewhat more complex gas-contacting arrangement. A schematic flow diagram of the process as employed in a 55 MW_e pilot plant is shown in **Figure 7-39**. The plant processes a portion of the flue gas from a 350 MW_e boiler. A slipstream of gas is removed from the boiler flue duct after the preheater, cleaned of dust, passed through the adsorption beds, and returned to the boiler stack-gas line. At any time in the cycle, four of the carbon beds are

636 *Gas Purification*

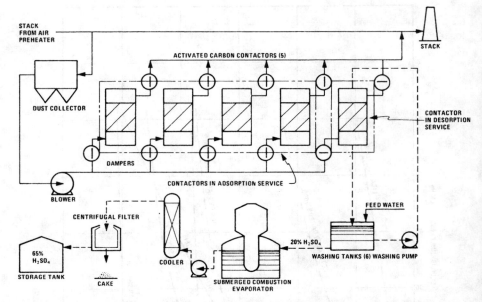

Figure 7-39. Schematic flow diagram of Hitachi activated carbon process—55 Mw_e pilot plant.

operated in parallel for SO_2 adsorption, while the remaining one is being washed and regenerated. Each unit operates on flue gas (drying the bed and adsorbing SO_2) for 48 hours and is then washed for 12 hours making a total cycle time of 60 hours. Six washing tanks are employed, each containing a different concentration of H_2SO_4. During the washing step, the tower is washed with acid from the six tanks in sequence, starting with the most concentrated acid and ending with fresh water. A stream of the highest concentration acid, 20% H_2SO_4, is continuously removed and fed to the submerged combustion concentrator. The final product is 65% H_2SO_4, which is used primarily for phosphate fertilizer manufacture. Although minor problems were encountered during initial operations, the general performance of the plant was considered to be very successful; a mean SO_2 removal efficiency of 80% was observed after 3,000 hours of operation. One plant in Japan operated for over 5 years without appreciable problems, and carbon consumption was very low—about 2% per year. The dilute sulfuric acid product (17%) from this plant was reacted with limestone to produce a salable gypsum (Ando, 1977).

A continuous process has been proposed by Joyce et al. (1970); however, only laboratory-scale tests were conducted to develop process data. In the proposed process, a fixed adsorption bed is used with wash water continuously sprayed on the top surface, and dilute sulfuric acid removed from the bottom of the adsorption vessel. The laboratory data indicated that a maximum H_2SO_4 concentration of 12 to 15% could be obtained from typical flue gas streams. The controlling variables in the adsorption operation were found to be the O_2/SO_3 mole ratio and gas-contact time. The results indicate, for example, that with a 60 second contact time, increasing the O_2/SO_2 ratio from 1.8 to 5.3 increased the conversion from about 70% to almost 100%. With the O_2/SO_2 ratio fixed at 1.8, increasing the contact time from 20 to 80 seconds caused the conversion efficiency to go from less than 40% to over 80%.

The carbon adsorption process with water-wash regeneration appears to be quite simple technically and economically attractive for cases where the dilute acid that is produced is of local value. It is of particular interest for cleaning the tail gas from contact sulfuric acid plants. In such an installation, the weak acid produced could be used instead of make-up water in the acid plant, increasing the total sulfuric acid production while abating air pollution by purifying the tail gas.

Although the water-wash process is simple, it has the drawback of producing a dilute sulfuric acid product, which is difficult to store, ship, or market. Concentration can be troublesome. There appears to be no further development of the Hitachi process (Behrens et al., 1984).

Carbon Adsorption Process with Thermal Regeneration

Carbon adsorption with thermal regeneration offers a number of advantages for flue gas desulfurization. Very high SO_2 removal efficiencies are possible. The process is completely dry; therefore, no saturation of the flue gas occurs, avoiding reheat, visible plumes, etc. The process operates at normal boiler flue gas outlet temperatures so there is no need to heat or cool the flue gas. Carbon adsorption also removes SO_3, and NO_x control can be incorporated by NH_3 addition with the carbon acting as a catalyst at temperatures from 212° to 400°F to convert the NO_x to nitrogen and water. Many other hazardous air pollutants in both gaseous and fine particulate form can also be collected. These include important pollutants such as vapor-phase mercury, dioxins, and furans. Of course, once collected, these pollutants must be disposed of. Some can be returned to the boiler for burning or be decomposed by holding the spent adsorbent at a high temperature for an extended period. This ability to collect hazardous air pollutants has resulted in the technology being applied recently to waste-to-energy plants in Europe either by the use of small amounts of carbon added to spray dryer slurry or by use of a separate carbon adsorption system. The technology has been applied to coal-fired boilers, fluidized catalytic cracking units, waste-to-energy plants, and hospital waste burning plants in Japan and Europe (Makansi, 1992).

The pollution control system configuration varies with the application and the process supplier. If the flue gas contains chlorine, a separate chlorine removal step may be required since chlorine inhibits the adsorption of SO_2. Separate beds may also be provided for different pollutants. For example, when both SO_2 and NO_x are to be controlled, it is necessary to first reduce the SO_2 concentration in the gas in a separate bed. This minimizes ammonium sulfate formation which would increase ammonia consumption and prevent effective denitrification. Also, an initial separate bed is sometimes provided for the removal of heavy metals, organic compounds, and fine particulate; and a final separate guard bed is sometimes provided as a buffer to remove any remaining trace pollutants.

The main drawbacks of the thermal regeneration process are the loss of the sorbent during regeneration and sorbent attrition. Attrition and sorbent loss are initiated during regeneration when some of the carbon is converted to CO_2. This carbon loss leads to loss of mechanical strength and consequent breakage of the sorbent granules. To reduce sorbent cost, the process uses low-cost forms of activated carbon, such as activated coke. Various materials have been used for the activated coke feedstock, including bituminous and anthracite coals. Recently, lignite and petroleum coke have been found by Mitsui to be suitable starting materials (Tsuji and Shiraishi, 1991). Other problems include sorber fires, a narrow temperature range for high NO_x reduction, flow maldistribution leading to deposition in the sorber, high sorber pressure drop due to difficulty in controlling the amount of coke in the beds, handling of fines, abra-

638 Gas Purification

sion of pneumatic piping, corrosion and plugging in the regenerator, and unstable coke combustion (Makansi, 1992). Disposal of waste sorbent fines must also be considered.

Reinluft and Bergbau-Forschung Processes

Reinluft GmbH and Chemiebau Dr. A. Zieren GmbH pioneered the development of the carbon adsorption with thermal regeneration process for SO_2 removal. The process was called the Reinluft (Clean Air) Process (Furkert, 1970; Anon., 1967). Bergbau-Forschung GmbH (Research Center of the German Coal Mining Industry) developed another version of this process known as the Bergbau-Forschung or BF process (Juntzen, et al., 1970). The Bergbau-Forschung process differs from the Reinluft process in the use of low-temperature coke and the use of hot sand instead of hot gas to provide heat for regeneration.

A flow diagram of Chemiebau's design of a Reinluft plant is shown in **Figure 7-40**. Flue gas, at a temperature in the range of about 200° to 320°F, enters the bottom of the adsorption column in which it contacts downflowing streams of granular adsorbent (coke). The size of individual coke particles is between 0.1 and 1.3 in. The purified flue gas leaves the top of the sorber at approximately its original temperature. Ash entrained in the entering flue gas is retained by coke in the lower portion of the sorber and has little effect on its activity. Coke containing adsorbed sulfuric acid is removed from the bottom of the adsorption column, passed over a screen which removes fine particles of coke and ash, and conveyed to the desorber.

The regeneration step consists of reducing adsorbed H_2SO_4 to SO_2 (consuming some of the coke in the process) and stripping the resulting SO_2, H_2O, and CO_2 from the bed. This is

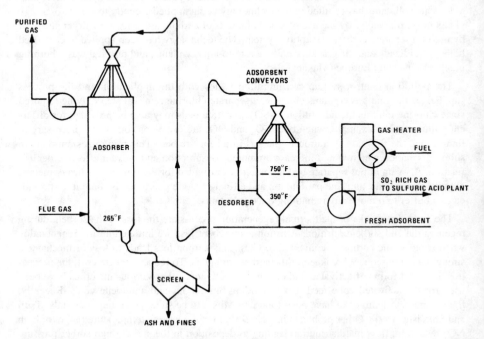

Figure 7-40. Schematic diagram of Reinluft Process—Chemiebau design. (*Anon., 1967*)

accomplished by passing the spent coke downward in the top section of the desorber, countercurrent to a hot gas stream. In this part of the desorber, the high-temperature gas strips off the reaction products and is cooled by the downflowing coke, which is heated to about 720°F in the process. The cooled gas is withdrawn from the top of the desorber and readmitted to the bottom of the lower section. In this section it again undergoes heat exchange with the coke, the upflowing gas being heated and the coke being cooled. The regenerated and cooled coke is then conveyed to the top of the sorber with added fresh material to make up for the amount lost by reaction or attrition. The gas is removed from the top of the lower section of the desorber, a portion of it is withdrawn from the system for conversion of SO_2 to sulfur or sulfuric acid, and the balance is passed through the heater and recycled to the desorber.

The key to the economic potential of the Reinluft process is the low-cost adsorbent used. Instead of employing activated charcoal as makeup adsorbent, the Reinluft process uses low-temperature coke. The relatively inexpensive material is made by carbonizing geologically young fuels such as peat, lignite, and, in some cases, coal at temperatures below 1,300°F. The coke is not activated before use, but becomes activated by the adsorption-regeneration cycle. After 3 to 10 cycles, the coke activity reaches a maximum comparable to gas-adsorption charcoal.

Four Reinluft process plants or pilot plants have been constructed and operated. Although a number of operating problems were encountered, it is reported that the plants adequately demonstrated the basic processes, and a new design has been developed to overcome the observed shortcomings. The principal operating problems were fires in the coke beds and corrosion of equipment by sulfuric acid. The new design (**Figure 7-40**) avoids these problems by completely separating the sorber and desorber portions, maintaining constant (and low) coke temperature at critical locations, and improving the coke-feed and discharge systems to provide uniform distribution and movement of coke particles.

The Bergbau-Forschung (BF) process has been tested in two prototype systems: a 35 MW_e unit at Lunen, West Germany, and a 20 MW_e installation at Gulf Power Company's Scholtz Electric Generating Station (Rush and Edwards, 1977). The latter installation included a RESOX reactor engineered by Foster Wheeler Energy Corporation to reduce the SO_2 to elemental sulfur. Neither the mechanical reliability nor the process operability of the Scholtz Station unit was considered acceptable. The major problems were hot spots in the sorber, poor reliability of the char/sand separator, and plugging of the RESOX system condenser with a mixture of sulfur and carbon particles. Tests at the Lunen facility were reportedly more satisfactory. However, the Lunen plant incorporated a modified Claus process for conversion of sulfur dioxide to sulfur instead of the RESOX process.

Mitsui-BF Process

Bergbau-Forschung licensed the previously described BF process to Mitsui, who further developed and improved the technology. The resulting process is called the Mitsui-BF process. Mitsui in turn has licensed its process to Uhde in Europe and GEESI in the U.S. Mitsui claims four commercial applications of the process worldwide. The largest is the Arzsberg power plant, built in 1987, where two systems treat a total of 1.1 million m^3/h of flue gas. Ninety-eight percent sulfuric acid is produced by an acid plant (Makansi, 1992). According to Oxley et al. (1991), the Mitsui-BF carbon adsorption process is used in a refinery in Chita, Japan, to treat flue gas from a fluid catalytic cracker CO boiler. Design SO_2 and NO_x removals are 90 and 60%, respectively. **Table 7-37** gives capital costs for Mitsui-BF systems (Tsuji and Shiraishi, 1991).

Table 7-37
Capital Costs for Mitsui-BF Systems

Process	Capital Cost, $/kW
SO_x & NO_x	220–240
SO_x only	140–160
NO_x only	70–80

Note: Costs for first two cases apply to two 500 MW_e units. SO_2 at 2,000 ppm, NO_x at 326 ppm. SO_x/NOx removal efficiencies are >90%/>80%. Equipment to produce byproduct elemental sulfur or H_2SO_4 is included. The NO_x only removal case applies to SO_x free gases. For this case, the basis is a 350 MW_e unit with NO_x at 250 ppm and SO_x at 50 ppm and NO_x removal efficiency at 80%. All cases exclude civil work and foundations.
Source: Tsuji and Shiraishi (1991)

In the Mitsui-BF process, activated coke made from bituminous coal is used for dry adsorption of the SO_2. Lignite and petroleum coke are also suitable feedstocks (Tsuji and Shiraishi, 1991). In a typical Mitsui-BF system, flue gas from the boiler passes through a particulate collector and is quenched by water spray before entering the activated coke reactor. The reactor consists of two compartments, one above the other. Both reactor compartments are filled with granular activated coke. Flue gas enters the lower reactor compartment where SO_2 is adsorbed and catalytically oxidized to sulfuric acid on the surface of the activated coke granules. Ammonia is then injected into the gas stream between the two compartments. The SO_2-free gas passes to the upper reactor compartment, where the NO_x is reduced to N_2 and H_2O by reaction with ammonia at 212°–400°F. The coke slowly moves from the upper to the lower reactor compartment and is conveyed by bucket elevator to the regenerator. In the regenerator, the sulfuric acid-loaded coke is first heated by internal coils to 300°–500°C. The adsorbed sulfuric acid is reduced to SO_2, consuming part of the coke in the process. The regenerated coke is then cooled by internal coils before being conveyed by bucket elevator back to the top reactor compartment. Any ammonium sulfate that forms by the reaction of SO_2 and NH_3 is decomposed into N_2, SO_2, and H_2O. Screening removes fines from the process at the bottom of the regenerator column. The SO_2-rich gas from the regenerator is processed into sulfuric acid (Tsuji and Shiraishi, 1991).

Other Activated Coke Processes

Other commercial variations of this process are currently being offered by the Hugo Peterson subsidiary of Steinmüller, Steag, and Sumitomo. Hugo Peterson has installed coke adsorbers on two waste-to-energy (WTE) plants at Garth, eight coal-fired plants at Lausward, and two coal-fired boilers at Flingren, all in Germany. Steag has various applications, including WTE and hospital waste burning plants. Sumitomo installed a 90 MW_e pilot plant at the Matsuhima Power plant in 1987.

RESOX SO₂ Reduction Process

The RESOX process, developed by Foster Wheeler, uses anthracite coal to reduce gaseous SO_2 to elemental sulfur. Development work began on the process in the late 1960s. It was tested at the Scholtz Plant (a 42 MW_e demonstration unit) in the 1970s and at the Lunen plant (a pilot plant) in Germany in 1974 (Rush and Edwards, 1977). The process or variations of it have been used at three plants in Japan (Steiner, 1992).

In the RESOX process, the SO_2 rich gas from the Mitsui-BF regenerator is sent to a reactor where it contacts anthracite coal. The carbon in the coal reacts with the SO_2 to produce elemental sulfur according to the following equation:

$$C_{(solid)} + SO_{2(gas)} = S_{(gas)} + CO_{2(gas)} \tag{7-108}$$

The reaction takes place at 500°–800°F, depending on the coal type and SO_2 inlet concentration. Ash and unreacted coal are removed from the bottom of the reactor, and can be burned in the boiler or disposed of with the boiler ash. The reactor off-gas contains elemental sulfur, H_2S, CO_2, COS, CS_2, and H_2. A downstream condenser removes the elemental sulfur. The uncondensed gas, typically containing 6% H_2S and 4% SO_2, is sent to an incinerator, and the incinerator off-gas is returned to the FGD system. Ninety to ninety-five percent conversion of the SO_2 to elemental sulfur and 99.95% sulfur purity have been achieved (Behrens et al., 1984).

Westvaco Process

The Westvaco process is a dormant, experimental, activated carbon process that has a removal step somewhat similar to that of the Mitsui-BF process, but a regeneration step that does not consume carbon. The sulfuric acid adsorbed on the carbon is reacted with hydrogen sulfide to produce elemental sulfur and water. Part of the sulfur produced is reacted with hydrogen to produce the necessary hydrogen sulfide. Development was discontinued due to the high cost of activated carbon relative to activated coke (Radian, 1977; Behrens et al., 1984; Spears, 1992).

CATALYTIC OXIDATION PROCESSES

The removal of SO_2 from dilute flue gas streams by catalytic oxidation represents an adaptation of the contact catalytic process used in the manufacture of sulfuric acid. Sulfur dioxide is oxidized to sulfur trioxide by reaction with oxygen in the flue gas in the presence of a catalyst. The resulting SO_3 combines with water vapor in the flue gas to form sulfuric acid, which is condensed by cooling and separated from the gas stream. Some of the special problems encountered in adapting the well-known contact sulfuric acid process to flue-gas treating are (1) preventing plugging or poisoning of the catalyst bed by impurities in the gas, (2) maintaining the catalyst at an optimum conversion temperature, and (3) avoiding corrosion by the acid produced. A considerable amount of effort has gone into resolving these problems and developing commercial processes for purifying flue gas by catalytic oxidation.

642 Gas Purification

SNOX Process

The SNOX process is also known as the WSA-SNOX process when both SO_2 and NO_x are removed; or, if only SO_2 is removed, as the WSA process where WSA stands for Wet Sulfuric Acid. This process catalytically reduces both the SO_2 and the NO_x in flue gases by more than 95% and, with integration of the recovered heat from the WSA condenser, is reported to have lower operating costs than conventional technologies. No chemical or additive is required other than ammonia for optional NO_x reduction. Sulfuric acid at 93.2% concentration is produced that is said to meet or exceed U.S. Federal Specifications. The SO_2 conversion catalyst can tolerate up to 50% water vapor and several hundred ppm of chlorides. CO and hydrocarbon emissions are said to be low (Collins et al., 1991).

Two commercial SNOX systems were commissioned in 1991: a 30 MW_e unit in Gela, Italy, and a 305 MW_e plant in Vodskov, Denmark. A 35 MW_e demonstration unit was commissioned in the United States in 1991 (Collins et al., 1991). A number of small WSA plants have been operating since 1980. The denitrification part of the system has been in operation on various small facilities since 1987 (Haldor Topsøe, 1992).

The inclusion or exclusion of denitrification capability does not alter the sulfur dioxide removal process significantly, although the equipment configuration must be different to accommodate the denitrification equipment. The primary difference is the addition of a selective catalytic reactor (SCR) using ammonia to reduce the NO_x ahead of the SO_2 catalyst (Collins et al., 1991).

The SNOX process consists of the following components: particulate collector, gas-to-gas heat exchanger, NO_x SCR, SO_2 converter, sulfuric acid condenser, and acid conditioning unit. **Figure 7-41** shows a typical SNOX process flow diagram for a boiler application. Par-

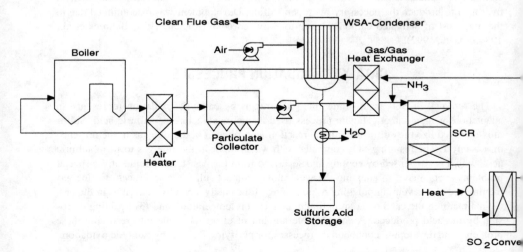

Figure 7-41. Flow diagram of the SNOX integrated process. (*Collins et al., 1991*)

ticulate is removed from flue gas leaving the boiler air heater. The flue gas then passes through the gas-to-gas heat exchanger, which raises the gas temperature above 700°F. A mixture of ammonia and air is added to the gas, and the gas passes through the SCR where the nitrogen oxides are reduced to free nitrogen and water. The flue gas leaves the SCR, is heated to 770°F, and passes through the SO_2-to-SO_3 converter. The SO_3-laden gas passes through the other side of the gas-to-gas heat exchanger where it is cooled to 300°F and then enters a WSA falling film sulfuric acid condenser where it is further cooled with ambient air to below the sulfuric acid dew point before exiting the system at about 210°F. In the condenser, sulfuric acid condenses out of the gas on borosilicate glass tubes and is collected, cooled, conditioned, and stored. The heated cooling air leaves the WSA condenser at over 390°F and is used for furnace combustion air after further heating in the preheater (Collins et al., 1991).

Highly efficient particulate removal is required ahead of the SNOX system since virtually all of the remaining particulate is retained in the SO_2 converter. Build-up of particulate on the sulfur dioxide catalyst causes the pressure drop through the SO_2 converter to rise, necessitating drawdown, screening, and reloading of the catalyst. About 2–3% of the bed material is lost each time the catalyst is processed, and the catalyst is usually adequate for about 10 changes. A high efficiency fabric filter is usually recommended to reduce the particulate load to an acceptable level (Larson, 1990).

Steam production can increase on the order of 1% per each percent of sulfur in the fuel. At 2–3% sulfur, the energy requirements for the WSA process are balanced by the heat generated by the process, making the use of high sulfur fuels attractive (Collins et al., 1991).

Cat-Ox Process

The Cat-Ox process, developed by Monsanto EnviroChem, was originally tested in a small pilot plant built and operated under a joint effort by Pennsylvania Electric, Air Preheater, Research-Cottrell, and Monsanto. This was followed by a larger prototype unit built and operated by Monsanto and Metropolitan Edison. The unit was designed to treat 24,000 scfm or approximately 6% of the total flue gas from a 250-MW_e unit of Metropolitan Edison's Portland, Pennsylvania, generating station (Stites and Miller, 1969; Stites et al., 1969). A flow diagram of the process is shown in **Figure 7-42**.

Flue gas is taken from the boiler at about 950°F and passed through a high-temperature electrostatic precipitator designed to remove over 99.5% of the fly ash from the gas stream. The hot gas then passes through a bed of catalyst in the converter where oxidation of SO_2 to SO_3 occurs. The gas is then cooled by passing first through a tubular economizer and then a regeneration type air preheater. In order to avoid serious corrosion at this point, it is necessary to limit cooling to about 450°F, which is above the dew point of sulfuric acid in the gas stream. The gas then passes through a packed absorber in which it is contacted with a cool stream of sulfuric acid. The gas is cooled to about 225°F, and the resulting hot acid is recycled through a shell and tube heat exchanger in which the heat is transferred to cooling water. Excess acid is drained off as product and further cooled to 110°F for storage.

Gas from the absorber contains some sulfuric acid mist formed during the cooling step plus a small amount of liquid entrained from the absorption tower. Removal of this mist is the key step in the process. In the Monsanto Cat-Ox prototype plant, the mist is removed by a fiber-packed cartridge type eliminator that produces an effluent gas containing less than 1.0 mg 100% H_2SO_4/scf. This amounts to about 10 ppm of acid as mist. The vapor pressure of sulfuric acid at 225°F contributes another 11 ppm so the total quantity of sulfuric acid in the

644 Gas Purification

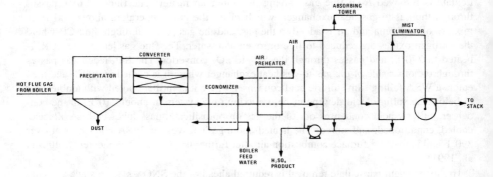

Figure 7-42. Flow diagram of the Cat-Ox Process.

exit stack is about 20 ppm. This is less than the amount of sulfuric acid in the flue gas as it comes from the boiler. The sulfur dioxide content of the product gas is only about 10% of its original value.

Operation of the prototype plant at Metropolitan Edison's Portland Station provided several years of process operating experience. Unfortunately, the high-temperature precipitator and special heat-recovery system made the arrangement tested impractical for all but new stations. As a result, Monsanto developed a "back fit" cycle in which the flue gas is reheated to the 800° to 900°F required by the catalyst (Farthing, 1971). This allows existing low temperature precipitators to be utilized.

A 103-MW$_e$ commercial demonstration plant was installed on a coal-fired boiler at Illinois Power's Wood River Power Station, Unit 4 (Jamgochian and Miller, 1974). Since this was a retrofit application, it was necessary to include equipment for reheating the flue gas to 850°–900°F before feeding it to the sulfuric acid converter. After start-up in September 1972, natural gas was used as fuel for the in-line heaters to reheat the feed gas. The plant operated successfully for 444 hours.

Natural gas became unavailable in October 1972, and it was necessary to substitute oil as fuel for the heaters. Following modification of the burners, an additional period of satisfactory operations was experienced, bringing the total operating time to 602 hours by July 1973. It became apparent, however, that the in-line combustion of oil would cause rapid deterioration of the catalyst. As a result, a major plant modification was undertaken to change to an "external" oil-fired reheat system.

During subsequent attempts to start-up and operate the plant, a number of additional mechanical and structural problems developed. According to a report issued at the time, the unit would require refurbishing before additional operations could be conducted (Radian, 1977). This experience apparently dampened enthusiasm for the process.

The power industry has been reluctant to accept sulfuric acid as a byproduct; however, further progress has been made in commercializing the system. In 1991, a sulfur producing company in Canada started up a Cat-Ox system to reduce the SO_2 emissions from a Claus plant incinerator. While some early material problems were experienced, the system readily met the target of 90% SO_2 reduction and achieved 95% SO_2 reduction (Grendel, 1992).

Catalytic Oxidation/Electrochemical Process

The catalytic oxidation/electrochemical membrane process consists of an upstream commercial sulfuric acid catalyst to convert SO_2 to SO_3 followed by a molten salt electrochemical cell using a sulfur oxide selective membrane. Removal efficiencies of 95% have been simulated. Projected economics for a 500 MW_e power plant burning 3.5% sulfur coal are $96/kW capital cost and 3.24 mills/kWh operating cost. Capital cost includes the catalytic converter and oleum plant and assumes cell replacement twice over a 30-year life (McHenry and Winnick, 1991). The process is in a very early stage of development, and no commercial or demonstration operations have been reported.

The electrochemical cell removes the strongest electron acceptors (Lewis acids) in the gas mixture. Sulfur oxides are the most acidic of all the species present in the flue gas and are therefore theoretically separable by this technique. In this process, the flue gas first passes through a catalytic reactor at 400°C to convert SO_2 to SO_3. The flue gas then passes through channels at the cathodes of the electrochemical cells where SO_3 is selectively removed from the flue gas. The SO_3 is converted to the sulfate ion, which migrates from the cathode to the anode. At the anode, the sulfate ion is converted to SO_3. The SO_3 is removed in a sweep gas stream and used to produce sulfuric acid or oleum (McHenry and Winnick, 1991).

The electrochemical cell consists of an electrolyte-filled membrane sandwiched between two porous, gas-diffusion electrodes, which impose the required voltage gradient across the cell and initiate the required electrochemistry. Near 100% electrochemical efficiency has been achieved (McHenry and Winnick, 1991). Scott (1985) identified the presently used low-melting point electrolyte, potassium pyrosulfate ($K_2S_2O_7$) with small additions of K_2SO_4 and V_2O_5. The electrolyte is retained in the interstices of an inert ceramic matrix. The electrolyte melts near 285°C, depending on the level of V_2O_5 present, and is stable at temperatures in excess of 400°C. The usable temperature range corresponds to temperatures downstream of a boiler economizer.

GAS PHASE RADIATION-INDUCED CHEMICAL REACTION PROCESSES

Ebara E-Beam Process

The Ebara Electron Beam (E-Beam) process utilizes radiation-induced chemical reactions to simultaneously remove both SO_2 and NO_x from a gas stream and produce fertilizer compounds (Frank, 1992). Projected capital costs for power plant applications are $400/kW, and projected operating costs are 13 mills/kWh (Hoffman et al., 1992). With zone irradiation, the capital cost is reported to be halved, and levelized costs are reduced to about 8 to 10 mills/kWh. The process is being tested at five facilities. The largest is a 20,000 m^3/hr flue gas stream from a coal burning plant. It is also being used on a 50,000 m^3/hr automobile tunnel ventilation system for NO_x control. (Frank, 1992). Removal efficiencies of 90% and 80% for SO_2 and NO_x, respectively, have been demonstrated in a 5 MW_e pilot plant (Livengood and Markussen, 1992). Removal efficiencies are said to be a function of the electric power input.

Prior to electron beam irradiation, the gas must be treated to remove particulate, humidified, and cooled to about 158°F. Ammonia is injected into the flue gas ahead of the E-Beam chamber. The flue gas then enters the E-Beam chamber and is irradiated to produce OH radicals and oxygen atoms that react with the SO_2, NO_x, and H_2O in the gas to produce sulfuric and nitric acids. These acids in turn react with the injected NH_3 to form solid particles of ammonium sulfate, $(NH_4)_2SO_4$, and ammonium sulfate-nitrate, $(NH_4)_2SO_4 \cdot 2(NH_4NO_3)$, which are collected in a downstream particulate collector (Frank, 1992). Soluble heavy metals must be reduced to acceptable levels by an upstream particulate removal system if the byproduct is to be used as a fertilizer. The byproduct is a powder that must be converted into a granular form or a liquid solution to be acceptable for agricultural use (McKnight et al., 1985).

ELECTROCHEMICAL CONVERSION TO SULFUR

There are no commercial or near-commercial processes in this category; however, it has been the subject of considerable bench scale research, and is of interest because of the sophisticated technology involved. The process was studied in detail by the Helipump Corp. under a U.S. DOE contract (Cook et al., 1990). It is based on the use of a solid state oxygen-ion-conducting electrochemical reactor that reduces SO_2 to elemental sulfur and NO_x to nitrogen. At the temperature of the reactor, typically 450–500°C, the elemental sulfur is generated as a vapor and can be recovered in a downstream condenser.

The DOE program covered the experimental determination of NO_x and SO_2 reduction rates using yttria, stabilized ceria, and zirconia solid electrolytes with gold electrodes. The use of an electrocatalyst on the surface to enhance NO_x and SO_2 reactivity relative to O_2 was also studied. The presence of fly ash in the flue gas was not found to affect performance.

An economic study of the process was made based on the experimental data. The study indicated that, with small improvements in surface area, electrode cost, and oxygen selectivity, the process could be economically competitive for gases containing no more than 1% oxygen. However, major improvements would be required for the process to be competitive with typical power plant flue gases, which contain about 3% oxygen.

GAS PHASE REDUCTION

Parsons Flue Gas Cleanup (FGC) Process

The Parsons Corporation and a consortium of cosponsors have developed the Parsons FGC process through bench and pilot scale steps. The process, which removes sulfur oxides, nitrogen oxides, and free oxygen from the flue gas, consists of the following steps (Kwong et al., 1991):

1. Simultaneous catalytic reduction of SO_x to H_2S, NO_x to N_2, and O_2 to H_2O in a hydrogenation reactor
2. Selective removal of H_2S from the hydrogenation offgas
3. Conversion of H_2S to elemental sulfur

When the Parsons FGC process was proposed, all three of the key process steps had been practiced commercially in other applications, but they had not yet been adapted and combined for boiler plant stack gas cleanup.

The hydrogenation step is an extension of the technology of the Beavon Sulfur Recovery (BSR) process used for Claus plant tail gas cleanup (see Chapter 8). However, because of the more dilute sulfur dioxide concentration in boiler plant flue gas and the presence of oxygen and fly ash, it was necessary to develop a new catalyst system to achieve virtually complete reduction of SO_x, NO_x, and O_2.

Removal of hydrogen sulfide from the hydrogenation reactor off-gas in the presence of carbon dioxide is accomplished by use of a selective amine such as Exxon's Flexsorb SE Plus or Union Carbide's UCARSOL HS-103 (see Chapters 2 and 3). Regeneration of the selective amine produces a hydrogen sulfide-rich gas which is sent to a sulfur recovery unit. In the Parsons FGC process, the Recycle Selectox process (see Chapter 8) is used for sulfur recovery.

The Parsons FGC process has the advantage of removing both SO_x and NO_x from flue gas and producing a valuable byproduct. Its principal disadvantage is a high fuel requirement since sufficient reducing gas must be provided to react with the residual oxygen in the flue gas as well as the SO_x and NO_x.

REFERENCES

ABB, 1992A, "ABB Dry Flue Gas Desulfurization Experience List for Coal-fired Boilers," ABB Fläkt, Knoxville, TN, July.

ABB, 1992B, "List of Installations Fläkt Dry Absorption System for Aluminum Reduction Furnaces," ABB Fläkt, Knoxville, TN, July.

ABB, 1992C, "ABB Environmental Systems, Gas Absorption Systems for Waste Incinerators," ABB Fläkt, Knoxville, TN, July.

Abrams, J. Z., Blake, J. H., Pennline, R.W., 1987, "The Confined Zone Dispersion Process for Flue Gas Desulfurization—Proof of Concept Tests," presented to the Industrial Gas Cleaning Institute, Arlington, VA, Sept. 22.

Abrams, J. Z., Zaczek, S. J., Benz, A. D., Awerbach, L., and Haidinger, J., 1988, "Use of Seawater in Flue Gas Desulfurization—A New Low-Cost FGD System For Special Applications," *J. of the Air Pollut. Cont. Assoc.*, Vol. 38, No. 7, July, pp. 969–974.

Abrams, J. Z., and Zaczek, S. J., 1991, "The Confined Zone Dispersion (CZD) Demonstration in a Commercial Unit for Flue Gas Desulfurization," *Proceedings: Acid Rain Retrofit Seminar: The Effective Use of Lime,* sponsored by the National Lime Assoc. (Arlington, VA), Philadelphia, PA, Jan. 9–10, pp. 241–260.

AirPol, Inc., 1993, "AirPol Gas Suspension Absorption Demonstration Plant," *AirPol/TVA/DOE Quarterly Project Summary,* January.

AirTECH News, 1993, "ESP and Fabric Filter Conference," Silver Springs, MD, April, p. 46.

Aldrich, R. G., and Oldenkamp, R. D., 1977, "100 MW Second Generation SO_2 Removal Plant for New York State Utilities," paper presented at the 39th Annual Meeting, American Power Conference, Chicago, IL, April 18.

Anazawa, M. (Dowa), 1984, personal communication, Feb. 7.

Anders, W. L., and Torstrick, R. L., 1981, *Computerized Shawnee Lime/Limestone Scrubbing Model Users Manual,* EPA-600/8-81-008 (NTIS PB-178963), U.S. EPA.

Ando, J., 1977, "Status of SO_2 and NO_x Removal Systems in Japan," paper presented at the EPA Symposium on Flue Gas Desulfurization, Hollywood, FL, Nov. 8–11.

Ando, J., 1983, "Status of SO_2 and NO_x Removal in Japan," paper presented at the EPA/EPRI Symposium on Flue Gas Desulfurization, New Orleans, LA, Nov. 1–4.

Ando, J., 1985, *Recent Developments in SO_2 and NO_x Abatement Technology for Stationary Sources in Japan,* EPA/600/7-85/040, U.S. EPA, Research Triangle Park, NC, September.

Angelovich, J. (Utah Power & Light), 1990, personal communication, August.

Anon., 1967, Chem. Eng., October, p. 94.

Anon., 1977, *J. of the Air Pollution Control Assoc.* 27 (October) pp. 948–961, condensed from "The Status of Flue Gas Desulfurization Applications in the United States, A Technological Assessment," Federal Power Commission (Historical Research by J. C. Marten).

Anon., 1980, "Tail Gas Process Makes Fertilizer," *Oil & Gas J.*, Oct. 20, p. 132.

Anon., 1988, "A Once-Faltered Desulfurization Process Will Get Another Tryout," *Chem. Eng.*, June 20, pp. 21–22.

Anon., 1991, "A New Technique to Sock SO_x," *Chem. Eng.*, May p. 17.

Arneson, A. D., Nooy, F. M., and Pohlenz, J. B., 1977, "The Shell FGD Process, Pilot Plant Experience at Tampa Electric," paper presented at the EPA Symposium on Flue Gas Desulfurization, Hollywood, FL, Nov. 8–11.

Ashline, P. M., Camponeschi, B. S., Kitamori, T., and Roth, T. J., 1989, "Advanced On-Site Flue Gas Desulfurization Project," paper presented at 82nd Annual Meeting and Exhibition of the Air and Waste Management Association, Anaheim, CA, June 25–30.

Babcock & Wilcox, 1992A, *Steam, Its Generation and Use,* 40th edition, Barberton, OH, pp. 35–40.

Babcock & Wilcox, 1992B, *Experience with Chemical Additives, Attachment #1,* Barberton, OH, December.

Bacha, F. T. (Fluor Daniel, Chicago, IL), 1992, personal communication, July 20.

Bailey, E. E., 1974, "Continuing Progress for the Wellman-Lord SO_2 Process," *Proceedings: Symposium on Flue Gas Desulfurization,* Atlanta, GA, Nov. 4–7, Vol. II, EPA 650/2-74-126b, December, pp. 745–760.

Bailey, E. E., and Heinz, R. W., 1975, *Chem Eng. Prog.*, Vol. 71, No. 3, pp. 64–68.

Bakke, E., 1985, "Cost Effective Wet FGD Systems on Medium to High Sulfur Coals," paper presented at the 1985 Joint Power Generation Conference, Milwaukee, WI, Oct. 20–24.

Barnett, A. (Union Carbide), 1992, personal communication, IL, June 8.

Barnett, A. (Union Carbide), 1993, personal communication, Oct. 12.

Barton, R. A., Dawson, C. W., Burnett, T. A., Hollinden, G. A., Wertz, K. L., Blythe, G. M., and Rhudy, R. G., 1990, "SO_2 Removal Performance Improvements by Chloride Addition at the TVA 10-MW Spray Dryer/ESP Pilot Plant," paper presented at the EPRI/EPA 1990 SO_2 Control Symposium, New Orleans, LA, May 8–11.

Bauman, R. D., and Crenshaw, J. O., 1977, "Paper 779099—The Air Quality Implications of the Move Toward Coal," *Proceedings of the 12th Intersociety Energy Conversion Engineering Conference (IECEC),* Washington, D.C., Aug. 28–Sept. 2, pp. 629–636. (Presented by American Nuclear Society in cooperation with IECEC Steering Committee, LaGrange Park, IL)

Becker, H., and Linde, G., 1985, "The Linde Solinox Process for Flue-Gas Purification," *Linde Reports on Science and Technology,* Vol. 40, Linde Aktiengesellschaft.

Behrens, G. P., Jones, G. D., Messerole, N. P., Seames, W. S., and Dickerman, J. C., 1984, *The Evaluation and Status of Flue Gas Desulfurization Systems,* EPRI CS-3322, Electric Power Research Institute, Palo Alto, CA, January.

Bennett, R. P. (NaTec Resources, Inc.), 1992, personal correspondence, May 14.

Bennett, R. P., and Darmstaedter, E., 1991, "Sodium Bicarbonate In-duct Injection with Sodium Sulfate Recovery for SO_2/NO_x Control," paper presented at the Fifth Symposium on Integrating Environmental Controls and Energy Production (ASME Conference), New Orleans, LA, March 4–5.

Bennett, R., and Nastri, A., 1990, "The Commercial Status of Sodium Bicarbonate Injection Technology for Dry FGD Applications," paper presented at the 1990 World Environmental Engineering Congress, Atlanta, GA, Oct. 9–12.

Benson, L. B., 1985, "The Role of Magnesium in Increasing SO_2 Removal and Improving Reliability in Magnesium-Enhanced FGD Systems," paper presented at the Second Annual Pittsburgh Coal Conference, Pittsburgh, PA, Sept. 16–20.

Benson, L., Hicks, R., and Johnson, H., 1990, "Advanced Mg-Enhanced Lime SO_2 and NO_x Control Pilot Plant," paper presented at the EPRI/EPA 1990 SO_2 Control Symposium, New Orleans, LA, May 8–11.

Berad, G. (Zinc Corporation of America), 1992, personal communication, Sept. 14.

Berger, D. M., Kieffer, J. K., Trewella, R. J., Wummer, C. J., and Mathay, W. L., 1984, "Determining Corrosivity of Flue Gas Condensate," Power Engineering, October, pp. 56–58.

Besner, D., 1970, *Control of the Emissions of Sulfur Oxide,* Ralph McElroy Co., Austin, TX, p. 102.

Bielawski, G. (Babcock & Wilcox), 1992, personal communication, April 23.

Bienstock, D., Field, J. H., and Myers, J. G., 1964, J. of Eng. for Power Transmission, American Society of Mechanical Engineers, Series A, Vol. 86, pp. 457–464.

Bienstock, D., Field, J. H., and Myers, J. G., 1967, *Process Development in Removing Sulfur Dioxide from Hot Flue Gases,* U.S. Department of Interior, Bureau of Mines, R.I. 7021, July.

Binns, D., and Aldrich, R. G., 1977, "Design of the 100 MW Atomics International Aqueous Carbonate Process Regeneration FGD Demonstration Plant," paper presented at the EPA Symposium on Flue Gas Desulfurization, Hollywood, FL, Nov. 8–11.

Bjerle, I., Xu, F., Karlsson, A., and Gustafsson, L., 1990, "Studies of the SO_2 High Temperature Reactivity of Super Fine Limestone Particles in Laboratory and Pilot Plant Scale," paper presented at the EPRI/EPA 1990 SO_2 Control Symposium, New Orleans, LA, May 8–11.

Bjerle, I., Ye, Z., and Xu, F., 1991, "Studies of the Initial Stage of High Temperature CaO-SO_2 Reaction," paper presented at the EPRI/EPA/DOE 1991 SO_2 Control Symposium, Washington, D.C., Dec. 3–6.

Blythe, G. M., Burke, J. M., Brna, T. G., and Rudy, R. G., 1983, "Field Evaluation of Utility Dry Scrubbing Systems," *Proceedings: Eighth Symposium on Flue Gas Desulfurization,* CS-3706, Vol. 2, Electric Power Research Institute, Palo Alto, CA, November, pp. 10–109 and 10–130.

Blythe, G. M., Slater, T. J., and Moser, R. E., 1991, "Full-Scale Demonstration of EDTA and Sulfur Addition to Control Sulfite Oxidation," paper presented at the EPRI/EPA/DOE 1991 SO2 Control Symposium, Washington, D.C., Dec. 3–6.

Borgwardt, R. H., 1977, *Sludge Oxidation in Limestone FGD Scrubbers,* EPA-600/7-77-061, U.S. Environmental Protection Agency, Research Triangle Park, NC, June.

Botnen, H. B., Johannessen, P. J., and Tvodten, Ø., 1992, *Monitoring the Effect of a Seawater Scrubber Outlet on the Benthic Community of the Marine Recipient,* Dept. of Fisheries and Marine Biology—University of Bergen, Norway, January.

Botts, W. V., Fockler, R. B., and Phelan, J. H., 1978, "Dry Scrubber Systems," paper presented at American Public Power Association Workshop, San Francisco, CA, Feb. 28–March 2.

Breed, C. E., and Hollinden, G. A., 1974, "TVA-EPA Pilot Plant Study of the Ammonia Absorption/Ammonium Bisulfate Regeneration Process," *Proceedings: EPA Symposium on Flue Gas Desulfurization,* Atlanta, GA, Nov. 4–7, Vol. II, EPA-650/2-74-126b, Dec., pp. 1069–1108.

Brown, B., and Felsvang, K., 1991, "High SO_2 Removal Dry FGD Systems," paper presented at the EPRI/EPA/DOE 1991 SO_2 Control Symposium, Washington, D.C., Dec. 3–6.

Brown, C. A., Maibodi, M., and McGuire, L. M., 1991, "1.7 MW Pilot Results for the Duct Injection FGD Process Using Hydrated Lime Upstream of an ESP," paper presented at the EPRI/EPA/DOE 1991 SO_2 Control Symposium, Washington, D.C., Dec. 3–6.

Burgess, W. D., 1956, *Chem. in Can.*, June, pp. 116–120.

Burke, J. M., 1983, *Shell NO_x/SO_2 Flue Gas Treatment Process: Independent Evaluation,* EPA-600/7-82-064 (PB 83-144816), March.

Burke, J. M., Stohs, M., Price, T. J., and Moser, R. E., 1990A, "Results of Sodium Formate Addition Tests at EPRI's High Sulfur Test Center and Associated Electric Cooperative's Thomas Hill Unit 3 FGD System," paper presented at the EPRI/EPA 1990 SO_2 Control Symposium, New Orleans, LA, May 8–11.

Burke, J. M., Owens, D. R., and Moser, R. E., 1990B, "Results from EPRI's High Sulfur Test Center: Factors Affecting Limestone Use in Wet FGD Systems," paper presented at the EPRI/EPA 1990 SO_2 Control Symposium, New Orleans, LA, May 8–11.

Burnett, T. A., Puschaver, E. J., Barton, R. A., Dawson, C. W., Rudy, R. G., Blythe, G., Heineken, K. L., and Durham, M. D., 1991, "Results of Medium- and High-Sulfur Coal Tests on the TVA 10-MW SD/ESP Pilot Plant," EPRI/EPA/DOE 1991 SO_2 Control Symposium, Washington, D.C., Dec. 3–6.

Burnham, R., 1993, (Public Service of New Mexico), personal communication, Oct. 12, 1993.

Buschmann, John (ABB Fläkt), 1992/1993, personal communications.

Byszewski, C. (Aquatech Systems), 1992, personal correspondence, IL, April 30.

Byszewski, C., and Hurwitz, D. (Aquatech Systems), 1991, "Combined SO_x/NO_x Control Via SOXAL, a Regenerative Sodium Based Scrubbing System," paper presented at the EPRI/EPA/DOE 1991 SO_2 Control Symposium, Washington, D.C., Dec. 3–6.

Camponeschi, B. (Pure Air), 1992, personal communication, Sept. 11.

Caprioglio, G., Langenkamp H., and vanVelzen, D., 1991, *The ISPRA Mark 13A Flue Gas Desulfurization Process Design Status and By-Products Utilization,* Ferlini—General Atomics Development Corp., San Diego, CA.

CFR (Code of Federal Regulations), 1990, Title 40, Part 60, *Standards of Performance for New Stationary Sources,* U.S. Government Printing Office, Washington, D.C., July.

Chang, J. C. S., 1984, *Pilot Plant Tests of Chloride Ion Effects in Wet FGD System Performance,* EPA-600/7-84-039, March.

Chemical Construction Corp., 1970, *Engineering Analysis of Emissions Control Technology for Sulfuric Acid Manufacturing Process,* Vol. 1., PB 190 393.

Chi, C. V., Peck, R. E., Tavakili, F., and Wassan, D. T., 1982, *Chem. Eng. Prog.*, June, p. 78.

Chughtai, Y. M., Linneweber, K. W., and Schmid, C., 1990, "Direct Desulfurization in Combination with Polishing Reactor," paper presented at the EPRI/EPA 1990 SO$_2$ Control Symposium, New Orleans, LA, May 8–11.

Cichanowicz, J. E., Dene, C. E., DePriest, W., Gaikwad, R., and Jarvis, J., 1991, "Engineering Evaluation of Combined NO$_x$/SO$_2$ Controls for Utility Application," paper presented at the EPRI/EPA/DOE 1991 SO$_2$ Control Symposium, Washington, D.C., Dec. 3–6.

Classen, D. D., 1983, "Status of the Chiyoda Thoroughbred 121 Flue Gas Desulfurization Process," *Proceedings of Seminar on FGD sponsored by Canadian Electrical Association,* Ottawa, Canada, Sept. 19–20.

Cloth, E. H., 1984, *Economic Analysis of Construction Materials Used in FGD Systems,* CS-3628, Electric Power Research Institute, Palo Alto, CA, August.

Cole, R., and Shulman, H. L., 1960, *Ind. Eng. Chem.*, Vol. 52, October, p. 859.

Collins, D. J., Ricci, R., Speth, C.H., and Bolli, R.E., 1991, "Initial Operating Experience of the SNOX Process," paper presented at the EPRI/EPA/DOE 1991 SO$_2$ Control Symposium, Washington D.C., Dec. 3–6.

Cook, W. J., Cornell, L. P., Keyvani, M., Neyman, M., and Helfritch, D.J., 1990, "Simultaneous Particulates, NO$_x$, SO$_x$ Removal from Flue Gas by All Solid State Electrochemical Technology," Final Report, DOE Contract No. DE-AC22-87PC79855 for period July 18, 1987–Jan. 31, 1990. Prepared for Pittsburgh Energy Technology Center, April 17.

Cornelissen, H. A. W., 1991, "Spray Dry Absorption in Concrete Products," in *Waste Materials in Construction,* Elseiveir Science Publishers B.V., pp. 499-506.

Coughlin, T., Schumacher, P., Andrew, D., and Hooper, R., 1990, "Injection of Dry Sodium Bicarbonate to Trim Sulfur Dioxide Emissions," paper presented at the EPRI/EPA 1990 SO$_2$ Control Symposium, New Orleans, LA, May 8–11.

Cox, G. H., and Kaplan, S. M., 1983, "Start-up & Operation of the Dry FGD System at Holcomb Station Unit No. 1," *Conference Proceedings: Effective Use of Lime,* sponsored by the National Lime Assoc. (Arlington, VA), Denver, CO, Sept. 27–28, p. 229.

Czahar, R., Perrin, L., and McCartney, W., Silver, M.H., 1991, *Least-Cost Utility Planning Handbook for Emissions Control and Acid Rain Compliance,* prepared by Economic and Technical Analysis Group and Independent Power Corporation for National Association of State Utility Consumers' Advocates, Washington, D.C., April.

Day, J. (Nevada Power Company), 1991, personal communication, Sept. 26.

Dean, R. S., and Swain, R. E., 1944, *U.S. Bur. Mines, Bull. No. 453.*

Dennis, R., and Bernstein, R. H., 1968, *Engineering Study of Removal of Sulfur Oxides from Stack Gas,* Report prepared by GCA Corporation for American Petroleum Institute, (GCA-TR-68-15-G).

Dhargalkar, P. H., and Tsui, T. I., 1990, "Recent Operating Experience of Double-Loop Limestone FGD Systems on High Sulfur Coal," paper presented at the EPRI/EPA 1990 SO$_2$ Control Symposium, New Orleans, LA, May 8–11.

Dingo, T. T., 1974, "Initial Operating Experience with a Dual Alkali SO$_2$ Removal System, Part I: Process Performance with a Commercial Dual-Alkali SO$_2$ Removal System" *Proceedings: Symposium on Flue Gas Desulfurization,* Atlanta, GA, November, Vol. I, EPA-650/2-74-126a, December, pp. 519–538.

Doutzenberg, F. M., Naber, J. E., and van Ginneken, A. J. J., 1971, "The Shell Flue Gas Desulfurization Process," paper presented at AIChE 68th National Meeting, Houston, TX, Feb. 28–March 4.

DoVale, A. J., Krause, G. D., and Murphy, G. L., 1991, "Acid Rain FGD System Retrofits," paper presented at the EPRI/EPA 1991 SO_2 Control Symposium, Washington, D.C., Dec. 3–6.

Downs, W., Johnson, D. W., Aldred, R. W., Tonty, L. V., Robards, R. F., and Runyan, R. A., 1983, "Influence of Chlorides on the Performance of Flue Gas Desulfurization," presented at the EPA/EPRI Symposium on Flue Gas Desulfurization, New Orleans, LA, Nov. 1–4.

Duval, Jr., W. A., Gallagher, W. R., Knight, R. G., Kolarz, C. R., McLaren, R. J., and Morasky, T. M., 1978, *State-of-the-Art Sludge Fixation*, EPRI FP-671, Electric Power Research Institute, Palo Alto, CA, January.

Duval, Jr., W. A., Atwood, R. A., Gallagher, W. R., Knight, R. G., McLaren, R. J., and Golden, D. M., 1979, *FGD Sludge Disposal Manual*, EPRI FP-977, Electric Power Research Institute, Palo Alto, CA, January.

Echols, L. (Pacific Power & Light), 1992, personal communication, May 27.

Eklund, A. G., and Golden, D. M., 1990, "Laboratory Characterization of Dry Sodium and Calcium In-Duct Injection By-Products," paper presented at the EPRI/EPA 1990 SO_2 Control Symposium, New Orleans, LA, May 8–11.

Ellestad, A. (Fläkt-Hydro), 1992, personal communications, August & September.

Ellison, W., 1991, *Introductory Technical and Applications Assessment of Condensing Heat Exchanger Corporation (CHX)*, Ellison Consultants, Monrovia, MD, Sept. 27.

Ellison, W., 1993, "Regulations on Air Pollution Control," *International Atomic Energy Agency Research Co-ordination Meeting and Seminar on Radiation Processing of Combustion Flue Gases*, Warsaw-Zakopane, Poland, May 24–28.

Ellison, W., Butcher, T. A., Carbonara, J. C., and Heaphy, J.P., 1994, "Heat Recovery and Pollution Cleanup from Low Grade Fuels," Conference on Ozone Control Strategies for the Next Decade (Century), Air and Waste Management Association, San Francisco, CA, April 11–12.

Elyanow, D. (Ionics), 1992, personal communication, May 26.

Emmel, T., and Maibodi, M., 1991, *Retrofit Costs for SO_2 and NO_x Control Options at 200 Coal-Fired Plants, Project Summary*, EPA/600/S7-90-91, March.

Emmel, T., and Maibodi, M., 1990, *Retrofit Costs for SO_2 and NO_x Control Options at 200 Coal-Fired Plants*, Vol. 1–5, PB91-133322, November.

Engdahl, R. B., 1973, *J. of the Air Pollution Control Assoc.*, Vol. 23, May, pp. 364–375.

Ennis, C. E., 1977, "Sulfur Dioxide Removal with Ammonia: A Fresh Perspective," Second Pacific Chemical Engineering Congress (PACHEC 1977), Denver, CO.

Enwald, T., and Ball, M. E., 1991, "Lifac Demonstration at Poplar River," paper presented at the EPRI/EPA/DOE 1991 SO_2 Control Symposium, Washington, D.C., Dec. 3–6.

EPRI (Electric Power Research Institute), 1986A, *Economic Evaluation of Dry-Injection Flue Gas Desulfurization Technology*, EPRI CS-4373, Electric Power Research Institute, Palo Alto, CA, January.

EPRI (Electric Power Research Institute), 1986B, *Transportation Cost Estimates for Sodium Compounds*, EPRI CS-4764, Electric Power Research Institute, Palo Alto, CA, September.

EPRI (Electric Power Research Institute), 1987, *Utilization Potential of Advanced SO_2 Control By-Products,* EPRI CS-5269, Palo Alto, CA, June.

EPRI (Electric Power Research Institute), 1988A, Laboratory Characterization of Advanced SO_2 Control By-Products: Spray Dryer Wastes, EPRI CS-5782, Palo Alto, CA, May.

EPRI (Electric Power Research Institute), 1988B, *FGD Chemistry and Analytical Methods Handbook,* Vol. 2: Chemical and Physical Test Methods, Rev. 1, EPRI CS-3612, Electric Power Research Institute, Palo Alto, CA, November.

EPRI (Electric Power Research Institute), 1988C, *Advanced SO_2, Control By-Product Utilization (Laboratory Evaluation),* EPRI CS-6044, Electric Power Research Institute, Palo Alto, CA, September.

EPRI (Electric Power Research Institute), 1989, *Sodium Injection Waste Management Guidelines,* EPRI GS-6486, Electric Power Research Institute, Palo Alto, CA, September.

Eschellman, G., 1909, U.S. Patent 900,500.

Estcourt, V. F., Grutle, R. O. M., Gehri, D. C., and Peters, H. J., 1978, "Tests of a Two-Stage Combined Dry Scrubber/SO_2 Absorber Using Sodium or Calcium," paper presented at the 40th Annual Meeting, American Power Conference, Chicago, IL, April 26.

Faist, M. B., Riese, C. E., and Gevirtzman, L., 1981, *Species Distribution Model: A General Computer Program to Calculate the Distribution of Chemical Species Among Several Multicomponent Phases,* submitted to U.S. D.O.E., Morgantown Energy Technology Center, October. (Cited by Benson, 1985.)

Farber, P. S., 1988, "Emissions Control through Dry Scrubbing," *Environ. Prog.,* August, pp. 168–183.

Farber, P. S., Livengood, C. D., Thani, M. I. B., Koch, B. J., McCoy, D. C., and Statnich, M., 1983, "Performance of Spray Dryer Flue Gas Cleaning System on High Sulfur Coal," *Coal Technology '83,* Houston, TX, Nov. 15–17.

Farmer, R. W., Jarvis, J. B., and Moser, R., 1987, "Effects of Aluminum/Fluoride Chemistry on Wet Limestone Flue Gas Desulfurization," 1987 Spring National AIChE Meeting, Houston, TX, March 29–April 2.

Farthing, J. G., 1971, *Electrical World,* May 15, pp. 34–39.

Fay, J. E. (ASARCO), 1992, personal correspondence, May 6.

Felix, L. G., Gooch, J. P., Merritt, R. L., Klett, M. G., Hunt, J. E., and Demain, A. G., 1991, "Scaleup Tests and Supporting Research for the Development of Duct Injection Technology," paper presented at the EPRI/EPA/DOE 1991 SO_2 Control Symposium, Washington, D.C., Dec. 3–6.

Felsvang, K., Hahn, P., Necker, P., Novak, M., Qing, Y., Ruffelsberger, N., Schumacher, E., and Spannbauer, H., 1988, "Update on Spray Dryer FGD Experience in Europe and the People's Republic of China," *Proceedings: First Combined FGD and Dry SO_2 Control Symposium,* EPRI GS-6307, Electric Power Research Institute, Palo Alto, CA, April.

Felsvang, K., Brown, B., and Horn, R., 1991, "Dry Scrubbing Experience with Spray Dryer Absorbers in Medium to High Sulfur Service," *Proceedings: Acid Rain Retrofit Seminar: The Effective Use of Lime,* sponsored by the National Lime Assoc. (Arlington, VA), Philadelphia, PA, Jan. 9–10, pp. 139–186.

Ferlini/General Atomics, 1991, *Brochure: ISPRA: An Advanced Regenerative Flue Gas Desulfurization Process,* San Diego, CA.

FGD & DeNO$_x$ Newsletter, 1991A, "Wellman-Lord Systems Operating in FRG," McIlvaine Company, Northbrook, IL, March, pp. 3–4.

FGD & DeNO$_x$ Newsletter, 1991B, "Walther Plant Operating," McIlvaine Company, Northbrook, IL, August, p. 7.

FGD & DeNO$_x$ Newsletter, 1992A, "Russian FGD Retrofits Will be Different," McIlvaine Company, Northbrook, IL, May.

FGD & DeNO$_x$ Newsletter, 1992B, "Just Focus on the Choices," McIlvaine Company, Northbrook, IL, June.

FGD & DeNO$_x$ Newsletter, 1993A, "20,000-40,000 TPY Reduction in Fine Particulate," McIlvaine Company, Northbrook, IL, September, pp. 4–5.

FGD & DeNO$_x$ Newsletter, 1993B, "Gypsum Has Its Problems" (based on a paper by W. Ellison presented at the Third International Conference on FGD and Chemical Gypsum in Toronto in September), McIlvaine Company, Northbrook, IL, November, pp. 3–4.

FGD and DeNO$_x$ Manual, 1978, "Limestone Scrubbing—Prominent Installation, Central Illinois Light Company," Vol. 1, Chapter 2, McIlvaine Company, Northbrook, IL, April, pp. 139.1–139.2.

FGD and DeNO$_x$ Manual, 1979, "Mist Eliminators," Vol. 1, Chapter 1, Section 2.4, McIlvaine Company, Northbrook, IL, January, pp. 124.481–124.486.

FGD and DeNO$_x$ Manual, 1988, "Fläkt-Hydro Seawater Process," McIlvaine Company, Northbrook, IL, February, pp. 210.1–210.9.

Fink, R. (United Engineers & Constructors, Inc.—now Raytheon), 1992, personal communication, April 24.

Fleming, E. P., and Fitt, T. C., 1946A, U.S. Patent 2,399,013.

Fleming, E. P., and Fitt, T. C., 1946B, U.S. Patent 2,295,587.

Fleming, E. P., and Fitt, T. C., 1950, "Liquid Sulfur Dioxide From Waste Smelter Gases," *Ind. Eng. Chem.*, Vol. 42, No. 11, pp. 2253–2258.

Frank, N. W. (Ebara Environmental Corp.), 1992, personal correspondence, April 27.

Frank, N. W., and Hirano, S., 1990, "Utilization of FGD Solid Waste in the Form of By-Product Agricultural Fertilizer," paper presented at the EPRI/EPA 1990 SO$_2$ Control Symposium, New Orleans, LA, May 8–11.

Furkert, H., 1970, "The Reinluft (Clean Air) Process for the Purification of Flue Gas from Power Stations," *Proceedings of the American Power Conference,* Vol. 32, No. 673.

Gage, C. L., and Rochelle, G. T., 1990, "Modeling of SO$_2$ Removal in Slurry Scrubbing as a Function of Limestone Type and Grind," EPRI/EPA 1990 SO$_2$ Control Symposium, New Orleans, LA, May 8–11.

Garlanger, J. E., and Ingra, T. S., 1983A, *Evaluation of Engineering Properties and Wet Stacking Disposal of Widows Creek FGD Gypsum-Flyash Waste, Addendum 1 EP Toxicity and Column Leaching Tests,* TVA/OP/EDT-84/23, December.

Garlanger, J. E., and Ingra, T. S., 1983B, Evaluation of Engineering Properties and Wet *Stacking Disposal of Widows Creek FGD Gypsum-Flyash Waste,* TVA/OP/EDT-84/20, December.

GEESI (G.E. Environmental Systems, Inc.), 1992, *In Situ Forced Oxidation Spray Tower Absorber,* SP0018 (Illustration), Lebanon, PA, August.

GEESI (G.E. Environmental Systems, Inc.), 1994, "Press Release: GE and MK Awarded $60 Million Contract for an Innovative Flue Gas Desulfurization System," Lebanon, PA, April 12.

Gehri, D. C., and Gylfe, J. D., 1973, "The Atomics International Aqueous Carbonate Process for SO_2 Removal—Process Description and Pilot Test Results," paper No. 73-306, presented at the APCA Meeting in Chicago, IL.

Gehri, D. C., Adams, R. L., and Phelan, H., 1980, U.S. Patent 4,197,278, April 4.

Gille, J. A., and Mackenzie, J. S., 1977, "Philadelphia Electric's Experience with Magnesium Oxide Scrubbing," paper presented at the EPA Symposium on Flue Gas Desulfurization, Hollywood, FL, Nov. 8–11.

Giovanetti, A. 1992A, "Flue Gas Desulfurization with the Wellman-Lord Process," unpresented technical paper, John Brown E&C, Houston, TX, April 21.

Giovanetti, A. (John Brown E&C), 1992B, personal communication, June 8.

Giovanetti, A. (John Brown E&C), 1992C, personal correspondence, April 21.

Glamser, J. (NaTec Resources), 1992, personal communication, Sept. 8.

Glamser, J., Eikmeier, M., and Petzel, H. K., 1989, "Advanced Concepts in FGD Technology: The SHU Process with Cooling Tower Discharge," *Journal of Air and Waste Management Association,* Vol. 39, No. 9, September.

Glancy, D. L., Grant, R. L., Legatski, L. K., Wilhelm, J. H., and Wrobel, B. A., 1983, "Utility Double Alkali Operating Experience," paper presented at EPA/EPRI Symposium on Flue Gas Desulfurization, New Orleans, LA, Nov. 1–4.

Gleason, G. H., and Loonam, A. C., 1934, U.S. Patent 1,972,883 (to Guggenheim Bros.).

Glenna, K. and Tokerud, A. (ABB Fläkt, Oslo, Norway), 1991, "Unique FGD Process Uses Seawater as Absorbent—Desulfurization of Flue Gas Without Chemicals in a Norwegian Oil Refinery," *ABB Review,* April.

Glowiak, B., and Gostomczyk, A., 1970, "Sulfur Dioxide Sorption on Anion Exchangers," paper presented at Second International Clean Air Congress of the International Union of Air Pollution Prevention Association, Washington, D.C., Dec. 6–11.

Gogineni, M. R., and Maurin, P. G., 1975, *Combustion,* Vol. 47, October, pp. 9–15.

Goodwin, R. W., 1978, *J. of the Air Pollution Control Assoc.*, Vol. 28, January, pp. 35–39.

Goots, T. R., DePero, M. J., Purdon, T. J., Nolan, P. S., Hoffman, J. L., and Arrigoni, T. W., 1991, "Results from LIMB Extension Testing at the Ohio Edison Edgewater Station," paper presented at the EPRI/EPA/DOE 1991 SO_2 Control Symposium, Washington, D.C., Dec. 3–6.

Grendel, R. W. (Monsanto Enviro-Chem), 1992, personal communication, May 6.

Grimm, C., Abrams, J. Z., Leffmann, W. W., Raben, I. A., and LaMantia, C., 1978, *Chem. Eng. Prog.*, Vol. 74, February, pp. 51–57.

Gutslke, J. M., Morgan, W. E., and Wolf, S. H., 1983, "Overview and Evaluation of Two Years of Operation and Testing of the Riverside Spray Dryer System," paper presented at the EPA/EPRI Symposium on Flue Gas Desulfurization, New Orleans, LA, Nov. 1–4.

Hakka, L. E., Birnbaum, R. W., and Singleton, M., 1991, "Pilot Testing of the CANSOLV System FGD Process," paper presented at the EPRI/EPA/DOE 1991 SO_2 Control Symposium, Washington, D.C., Dec. 3–6.

Hakka, L. E., and Barnett, A. B., 1991, "The CANSOLV System for Removal of SO_2 from Gas Streams," unpublished Union Carbide paper, Danbury, CT, December.

Haldor Topsøe, 1992, *Reference List for Desulfurization and Denitrification Plants,* Haldor Topsøe, Houston, TX.

Hanf, E. (Munters Corp.), 1992, personal communication, Oct. 15.

Hansen, C., 1929, U.S. Patent 1,740,342.

Hansen, S. K., and Toher, J. G., 1991, "Commercial Use of the EEC/Lurgi Circulating Fluid Bed (CFB) in FGD and Acid Gas Applications," *Proceedings: Acid Rain Retrofit Seminar: The Effective Use of Lime,* sponsored by the National Lime Assoc. (Arlington, VA), Philadelphia, PA, January 9–10, pp. 187–203.

Hansen, S. K., Felsvang, K. S., Morford, R. M., and Spencer, H. W., 1983, "Status of the Joy/Niro Dry FGD System and Its Future Application for the Removal of High Sulfur, High Chloride and NO_x from Flue Gases," (Paper 83-JPGC-APC-8), Joint Power Generation Conference, Indianapolis, IN, Sept. 27–28.

Haslbeck, J. L., Neal, L. G., Wang, C. J., and Tseng, H., 1984, "Evaluation of the NOXSO Combined NO_x/SO_2 Flue Gas Treatment Process," *1st Annual Pittsburgh Coal Conference,* sponsored by the University of Pittsburgh and the U.S. Dept. of Energy's Pittsburgh Energy Technology Center, Pittsburgh, PA, September, pp. 231–240.

Haslbeck, J. L., Woods, M. C., Harkins, S. M., and Ma, W. T., 1990, "The NOXSO Flue Gas Treatment Process: Simultaneous Removal of NO_x and SO_2 from Flue Gas," *Proceedings 25th Intersociety Energy Conversion Engineering Conference,* AIChE (New York), Reno, NV, August 12–17, pp. 161–170.

Haslbeck, J. L. (NOXSO), 1992, personal communication, July 29.

Head, H. N., 1977, *EPA Alkali Scrubbing Test Facility: Advanced Program, Third Progress Report,* EPA-600/7-77-105, September.

Heaphy, J. P., Carbonara, J. C., and Ellison W., 1993, "Integrated Flue Gas Treatment by Wet FGD Operating in a Water-Condensing Mode," paper presented at the EPRI/EPA/DOE 1993 SO_2 Control Symposium, Boston, MA, August 24–26.

Hegeman, R. R., and Kutemeyer, P., 1989, "The Bischoff Flue Gas Desulfurization Process," Proceedings: First Combined Flue Gas Desulfurization and Dry SO_2 Control Symposium, EPRI GS-6307, Electric Power Research Institute, Palo Alto, CA, April.

Hein, L. B., Phillips, A. B., and Young, R. D., 1955, Problems and Control of Air Pollution, edited by F. S. Mallette, Reinhold Publishing Corporation, NY, p. 55.

Henderson, J. M., and Pfeiffer, J. B., 1974, "How ASARCO Liquifies SO_2 Off-Gas at Tacoma Smelter," Mining Engineering, Vol. 26, No. 11, November, pp. 36–38.

Henderson, J. M., and Yavorsky, J., 1977, "Liquid SO_2 from Copper Converter Gases—The ASARCO Dimethylaniline Process," paper presented at the 1977 AIME Annual Meeting, Atlanta, GA, March 8.

Henzel, D. S., and Ellison, W., 1990, "Commercial Utilization of SO_2 Removal Wastes in the Application of New Advanced Control Technology," *EPRI/EPA 1990 SO_2 Control Symposium,* New Orleans, LA, May 8–11.

Henzel, D. S., Laeske, B. A., Smith, E. O., and Swenson, D. O., 1982, *Handbook for Flue Gas Desulfurization Scrubbing with Limestone,* Noyes Data Corp., Park Ridge, NJ.

Hersel, M. P., and Belloni, A. E., 1991, "Options Available in the Solinox Vent Gas Purification Process," *Gas Sep. Purif.,* Vol. 5, June, p. 111.

Hewson, G. W., Pearce, S. L., Pollitt, A., and Rees, R. L., 1933, *Soc. Chem. Ind.*, London, Chemical Engineering Group, Proc., Vol. 15, p. 67.

Hilton, R. G. (Conversion Systems), 1991, personal communication, Nov. 21.

Hoffman, J. S., Smith, D. N., Pennline, H. W., and Yeh, J. T., 1992, "Removal of Pollutants from Flue Gas via Dry, Regenerable Sorbent Processes," unpublished paper provided by H.W. Pennline, Dept. of Energy, Pittsburgh Energy Technology Center, Pittsburgh, PA, March.

Holcombe, L., Weinberg, A., Butler, R., and Harness, J., 1990, "Field Study of Wastes from a Lime Injection Technology," EPRI/EPA 1990 SO_2 Control Symposium, New Orleans, LA, May 8–11.

Hollinden, G., Runyan, R., Newton, S., Garrison, F., Pfeffer, S., and Smith, D., 1983A, "Results of the Dowa Technology Tests at the Shawnee Scrubber Facility," *Proceedings of the Symposium on Flue Gas Desulfurization,* CS-2897, Electric Power Research Institute, Palo Alto, CA.

Hollinden, G. A., Stephenson, C. D., and Stensland, J. G., 1983B, "An Economic Evaluation of Limestone Double Alkali Flue Gas Desulfurization Systems," paper presented at the EPA/EPRI Symposium on Flue Gas Desulfurization, New Orleans, LA, Nov. 1–4.

Hooper, R. G., 1988, "Full Scale Demonstration of Flue Gas Desulfurization by the Injection of Dry Sodium Bicarbonate Upstream of an Electrostatic Precipitator," EPA/EPRI Seventh Symposium: Transfer and Utilization of Particulate Control Technology, Nashville, TN, March 22–25.

Hooper, R. G., 1990, "Abatement of Acidic Emissions by Dry Sodium Bicarbonate Injection," paper presented at 1990 TAPPI Environmental Conference, Seattle, WA, April 8–11.

Howard, H., and Stantial, F. G., 1918, U.S. Patent 1,271,899, July 9.

Howat, D. D., 1940, *Chem. Age* (London), Vol. 43, Nov. 30, p. 249.

Huang, H., Allen, J.W., and Livengood, C. D., 1988, *Combined Nitrogen Oxides/Sulfur Dioxide Control in a Spray-Dryer/Fabric-Filter System,* ANL/ESD TM-8, Argonne National Laboratory, Argonne, IL, November.

Hudson, J. L., 1980, *Sulfur Dioxide Oxidation in Scrubber Systems,* EPA-600/7-80-083, U.S. Environmental Protection Agency, Research Triangle Park, NC, April.

Humphries, J. J., Jr., and McRae, W. A., 1970, *Proceedings of the American Power Conference,* Vol. 32, p. 663.

Hurwitz, D. (Aquatech Systems), 1993, personal communication, Dec. 14.

Hüvel, B., 1990, "Combination of an Ammonia Wet Scrubbing Desulfurization Process (Walther-Process) with a SCR-Denox-System as a Retrofit Measure behind a Coal Wet Bottom Boiler of 191 MW Firing Capacity," *Proceedings: GEN-UPGRADE 90, International Symposium on Performance Improvement, Retrofitting, and Repowering of Fossil Fuel Power Plants,* Volume 3: Policy, Economics, and Environmental Upgrades, EPRI GS-6986, Vol. 3, Washington, D.C., March 6–9, pp. 13-4-1 to 13-4-15.

Ireland, P., and Ogden, G., 1991, "FGD Economics: A Comparison of Magnesium-Enhanced Lime and Limestone Forced Oxidation Systems," *Proceedings: Acid Rain Retrofit Seminar, The Effective Use of Lime,* sponsored by the National Lime Assoc. (Arlington, VA), Pittsburgh, PA, Jan. 9–10, pp. 123–137.

Jamgochian, E. M., and Miller, W. E., 1974, *Proceedings: Symposium on Flue Gas Desulfurization,* Atlanta, GA, November, Vol. II, EPA-650/2-74-126b, December, pp. 762–806.

Jankura, B. J., Milobowski, M. G., Hallstrom, R. U., and Novak, J. P., 1991, "Organic Acid Buffer Testing at Michigan South Central Power Agency's Endicott Station," paper presented at the EPRI/EPA/DOE 1991 SO_2 Control Symposium, Washington D.C., Dec. 3–6.

Jarvis, J. B., Roothaan, E. S., Meserole, F. B., and Owens, D. R., 1991, "Factors Involved in the Selection of Limestone Reagents for Use in Wet FGD Systems," paper presented at the EPRI/EPA/DOE 1991 SO_2 Control Symposium, Washington, D.C., Dec. 3–6.

John Brown E&C, 1992, "Wellman-Lord Installation List," Houston, TX.

Johnson, D. W., Rowley, D. R., Schulze, K. H., and Carrigan, J. F., 1994, "Integrated Flue Gas Treatment and Heat Recovery Using a Condensing Heat Exchanger," paper presented at the American Power Conference, Chicago, IL, April 25–27.

Johnstone, H. F., 1935, *Ind. Eng. Chem.*, Vol. 27, p. 587.

Johnstone, H. F., 1937, U.S. Patent 2,082,006.

Johnstone, H. F., 1938, U.S. Patent 2,134,481.

Johnstone, H. F., and Keyes, D. B., 1935, *Ind. Eng. Chem.*, June, Vol. 27, p. 659.

Johnstone, H. F., and Kleinschmidt, R. V., 1938, *Trans. Am. Inst. Chem. Eng.*, April 25, Vol. 34, p. 181.

Johnstone, H. F., and Singh, A. D., 1937, *Ind. Eng. Chem.*, March, Vol. 29, p. 286.

Johnstone, H. F., and Singh, A. D., 1940, *Univ. Illinois Bulletin Engineering Experimental Station Bulletin No. 324,* Dec. 31.

Jolly, W. J., 1966, *The Chemistry of the Non-Metals,* Prentice Hall, Englewood Cliffs, NJ, p. 67.

Jonakin, J., and McLaughlin, J. F., 1969, *Proc. American Power Conference,* Vol. 31, pp. 543–552.

Jones, J. W., 1977, "Disposal of Flue-Gas-Cleaning Wastes," *Chem. Eng.*, Feb. 14, 1992, pp. 79–85.

Jones, A. F., Rhudy, R. G., and Bowen, C. F. P., 1991, "Results of Mist Eliminator System Testing on an Air-Water Pilot Facility," paper presented at the EPRI/EPA/DOE 1991 SO_2 Control Symposium, Washington, D.C., Dec. 3–6.

Joy Environmental Technologies, Inc., 1992, *Reference List for the Joy/Niro Spray Dryer Absorber Air Quality Systems (SDA AQS),* March.

Joyce, R. J., Lynch, R. T. Sutt, R. F., and Tobias, G. J., 1970, "Effective Recovery of Dilute SO_2," paper presented at Third Joint Meeting of AIChE and Instituto Mexicani de Ingenieros Quimicos, Denver, CO, Aug. 30–Sept. 2.

Juntzen, H., Knoblauch, K., and Peters, W., 1970, "SO_2 Removal from Flue Gases by Special Carbon," paper presented at the Second International Clean Air Congress of the International Union of Air Pollution Prevention Assoc., Washington, D.C., Dec. 6–11.

Kaneda, S., Nichimura, M., Wakui, H., Kuwahara, I., and Classen, D. D., 1983, "Operating Experience with the Chiyoda Thoroughbred 121 FGD Systems," paper presented at the EPA/EPRI Flue Gas Desulfurization Symposium, New Orleans, LA, Nov. 1–4.

Kaplan, N., 1974, *Proceedings: Symposium on Flue Gas Desulfurization,* Atlanta, GA, November, Vol. I, EPA-650/2-74-126a, December, pp. 445–515.

Kaplan, N., 1976, *Proceedings: Symposium on Flue Gas Desulfurization,* New Orleans, March, Vol. I, EPA-600/2-76-136a, May, pp. 387–422.

Kaplan, N., and Maxwell, M. A., 1977, *Chem. Eng.*, Vol. 84, Oct. 17, pp. 127–135.

Katz, M., and Cole, R. J., 1950, *Ind. Eng. Chem.*, Vol. 42, No. 11, pp. 2258–2269.

Katz, B., and Oldenkamp, R.D., 1969, paper presented at the ASME Winter Annual Meeting, Los Angeles, CA, Nov. 11–20, ASME Publ. 69-WA/APC-6.

Keeth, R. J., Miranda, J. E., Reisdorf, J. B., and Scheck, R. W., 1983, *Economic Evaluation of FGD Systems*, Vol. 1, 2 & 3, EPRI CS-3342, Electric Power Research Institute, Palo Alto, CA, Dec.

Keeth, R. J., Baker, D. L., and Radcliffe, P., 1991A, *FGDCOST User's Manual*, GS-7525L, Electric Power Research Institute, Palo Alto, CA.

Keeth, R. J., Baker, D. L., Tracy, P. E., Ogden, G. E., and Ireland, P. A., 1991B, *Economic Evaluation of Flue Gas Desulfurization Systems*, GS-7193, Vol. 1, Electric Power Research Institute, Palo Alto, CA, February.

Keeth, R. J., Baker, D. L., Tracy, P. E., Ogden, G. E., Ireland, P. A., and Von Thun, C. N., 1992, *Economic Evaluation of Flue Gas Desulfurization Systems*, GS-7193, Vol. 2, Electric Power Research Institute, Palo Alto, CA, January.

King, R. A., 1950, *Ind. Eng. Chem.*, Vol. 42, No. 11, pp. 2241–2248.

Kirby, L. (Dow Chemical), 1992, personal communication, June 9.

Kirby, L. H., Kuhr, R. W., Sims, C., and Gullet, D., 1991, "Application of Dow Chemical's Regenerable Flue Gas Desulfurization Technology to Coal-Fired Power Plants," paper presented at the EPRI/EPA/DOE 1991 SO_2 Control Symposium, Washington D.C., Dec. 3–6.

Klimek, A. P., Lees, M. G., McMeekin, E. H., Stewart, M. M., and Golden, D., 1987, *Calcium Spray Dryer Waste Management: Design Guidelines*, EPRI CS-5312, Electric Power Research Institute, Palo Alto, CA, September.

Klingspor, J., 1983, *Kinetics and Engineering Aspects on the Wet-Dry FGD Process*, Department of Chemical Engineering, Lund's Institute of Technology, Sweden, June.

Klingspor, J. S. (ABB Environmental), 1993, personal communication, Dec. 20.

Klingspor, J. S., and Cope, D. R., 1987, *FGD Handbook, Flue Gas Desulfurization Systems* (ICEAS/B5), IEA Coal Research, London, May.

Kneissel, P. J., Staudinger, G., Weitzer, M., and Iyers, R., 1989, "Design, Start-up, and Operating Experience of FGD System with Gypsum By-Product at Riedersbach No. 2 in Austria," *Proceedings: First Combined Flue Gas Desulfurization and Dry SO_2 Control Symposium*, EPRI GS-6307, Electric Power Research Institute, Palo Alto, CA, April.

Knight, R. G., Rothfuss, E. H., Yard, K. D., and Golden, D. M., 1980, *FGD Sludge Disposal Manual*, Second Edition, EPRI CS-1515, Electric Power Research Institute, Palo Alto, CA, September.

Koch, G. H., and Beavers, J. A., 1982, "Laboratory and Field Evaluation of Materials for Flue Gas Desulfurization Systems," paper presented at the EPA/EPRI Symposium on Flue Gas Desulfurization, Hollywood, FL, May 17–20.

Koehler, G. R., and Dober, E. J., 1974, *Proceedings: Symposium on Flue Gas Desulfurization*, Atlanta, GA, November, Vol. II, EPA 650/2-74-126b, December, pp. 673–708.

Krause, H. H., Ireland, P. A., and Charles, G. A., 1988, "Materials and Cleaning Options for Cyclic Reheat Systems: Midterm Report," CS-5980, prepared for the Electric Power Research Institute, Palo Alto, CA, September.

Krippene, B. (Joy Environmental Equipment), 1992, personal communications, August, October and December.

Kwong, K. V., Meissner, R. E., Ahmad, S., and Wendt, C. J., 1991, *Environmental Progress*, Vol. 10, No. 3, August, p. 211.

LaMantia, C. R., Lunt, R. R., Rush, R. E., Frank, T. M., and Kaplan, N., 1976, "Operating Experience—CEA/ADL Dual Alkali Prototype System at Gulf Power/Southern Services, Inc.," *Proceedings: Symposium on Flue Gas Desulfurization*, New Orleans, March, Vol. 1, EPA-600/2-76-136a, May, pp. 423–470.

LaMantia, C. R., Lunt, R. R., Oberholtzer, J. E., Field, E. L., Valentine, J. R. and Kaplan, N., 1977, *Final Report: Dual Alkali Test and Evaluation Program Vol. II, Laboratory and Pilot Plant Programs*, EPA-600/7-77/050b, Arthur D. Little, Inc., Cambridge, MA, May.

Landreth, R. J., and Smith, P. V., 1989, "Retrofit of Sorbent Injection Technology on an Older Coal-Fired Boiler," 82nd Annual Meeting & Exposition, Air and Waste Management Association, Anaheim., CA, June 25–30.

Lanois, G. D. (Environmental Elements), 1993, personal communication, Jan. 4, 1993.

Larson, John (Haldor Topsøe), 1990, personal communication, Nov. 29.

Larson, B. D., Doyle, J. B., and Wolfe, B. A., 1990, "Dry Scrubber 10 Years Later," paper presented to Association of Rural Electric Generating Co-Operatives Annual Meeting, Baton Rouge, LA, June 25–27.

Laseke, B. A. Jr., Melia, M. T., and Kaplan, N., 1983, "Trends in Commercial Applications of FGD," paper presented at the EPA/EPRI Symposium on Flue Gas Desulfurization, New Orleans, LA, Nov. 1–4.

Laslo, D., and Bakke, E., 1983, "The Effect of Dissolved Solids on Limestone FGD Scrubbing Chemistry," paper presented at ASME Joint Power Generation Conference, Indianapolis, IN, Sept. 27.

Laslo, D., and Bakke, E., 1984, "Limestone/Adipic Acid FGD and Stack Opacity Reduction, Pilot Plant Tests at Big Rivers Electric Corp.," paper presented at the 77th Annual Meeting of the APCA, San Francisco, CA, June 25.

Laslo, D., Chang, J. C. S., and Mobley, J. D., 1983, "Pilot Plant Tests on the Effects of Dissolved Salts on Lime/Limestone FGD Chemistry," presented at the EPA/EPRI Symposium on Flue Gas Desulfurization, New Orleans, LA, Nov. 1–4.

Lee, Y. J., Benson, L. B., and O'Hara, R. D., 1990, "Laboratory Study of Thiosulfate Production and Application in Lime-Based Wet FGD," paper presented at the EPRI/EPA 1990 SO_2 Control Symposium, New Orleans, LA, May 8–11.

Legatski, L. K., Johnson, K. E., and Lee, L. Y., 1976, *Proc. Symposium on Flue Gas Desulfurization*, New Orleans, LA, March, Vol. 1, EPA-600/2-76-136a, May, pp. 471–502.

Liegois, W. A., 1983, "Status of Spray Dry Flue Gas Desulfurization," *Conference Proceedings: Effective Use of Lime for Flue Gas Desulfurization*, sponsored by National Lime Association (Arlington, VA), Denver, CO, Sept. 27–28, pp. 61–79.

Lifac North America, 1991, *Lifac Flue Gas Desulfurization: Clean Coal Demonstration Project*, Regina, Saskatchewan, Canada, December.

Link, F. W., and Ponder, W. H., 1977, "Status Report on the Wellman-Lord/ Allied Chemical Flue Gas Desulfurization Plant at Northern Indiana Public Service Company's Dean H. Mitchell Station," paper presented at EPA Flue Gas Desulfurization Symposium, Hollywood, FL, Nov. 8–11.

Livengood, C. D., and Markussen, J. M., 1992, "Emerging NO_x/SO_2 Control Technologies," NO_x Control V Conference, sponsored by the Council of Industrial Boiler Owners, Long Beach, CA, Feb. 9–11.

Lunt, R. R. (Arthur D. Little, Inc.), 1993, personal communication, August.

Lunt, R. R., and Shah, I. S., 1973, "Dual Alkali Process for SO_2 Control," 66th Annual Meeting of American Institute of Chemical Engineers, Philadelphia, PA, November.

MacKenzie, J., Bove, H., and Bitsko, R., 1983, "Operating Experience with the United/PECo Magnesium Oxide Flue Gas Desulfurization Process," Proceedings of Seminar on FGD sponsored by Canadian Electrical Association, Ottawa, Canada, Sept. 19–20.

Madenburg, R. S., and Kurey, R. A., 1977, "Citrate Process Demonstration Plant—A Progress Report," paper presented at the EPA Symposium on Flue Gas Desulfurization, Hollywood, FL, Nov. 8–11.

Maibodi, M., Blackard, A. L., and Page, R. J., 1991, *Integrated Air Pollution Control System*, Version 4.0, EPA/600/S7-90/022, February.

Majdeski, H. (Research-Cottrell), 1992, personal correspondence, Oct. 6, 29.

Makansi, J., 1989, "Dry Scrubbers Aim for High-Sulfur Coal Applications," *Power*, January, pp. 28–29.

Makansi, J., 1992, "Activated Coke Emerges as Cleaning Agent for Stack Gas," *Power*, August, pp. 60–62.

Maller, G., Meserole, F. B., and Moser, R. E., 1990, "Use of EDTA and Thiosulfate to Inhibit Sulfite Oxidation in Wet Limestone Flue Gas Desulfurization Processes: Results of Laboratory-Scale and Bench Scale Testing," EPRI/EPA 1990 SO_2 Control Symposium, New Orleans, LA, May 8–11.

Maroti, L. A., and Dene, C. E., 1982, *Entrainment in Wet Stacks*, EPRI CS-2520, Electric Power Research Institute, Palo Alto, CA, August.

Martin, J. R., and Bechthold, H., 1982, "An Alternative Technology for Flue Gas Desulfurization: Walther Ammonia Process," *Proceedings of the American Power Conference*, April 26–28, Chicago, IL, pp. 413–422.

Martin, D. A., and Bently, F. E., 1962, U.S. Bureau of Mines RI 6321.

Martinelli, R. (Babcock & Wilcox), 1992, personal communication, Sept. 9.

Martinez, J. L., Earl, C. B., and Craig, T. L., 1971, "The Wellman-Lord SO_2 Recovery Process—A Review of Industrial Operation," paper presented at the Environmental Quality Conference for the Extractive Industries of the American Institute of Mining, Metallurgical, and Petroleum Engineers, Inc., Washington, D.C., June 7–9.

Mathay, W. L., 1990, "Current FGD Materials Experience in the United States," *Sixth International Seminar: Solving Corrosion Problems in Air Pollution Control Equipment*, sponsored by the National Association of Corrosion Engineers, Louisville, KY, Oct. 1–19.

Maurin, P. G., and Jonakin, J., 1970, *Chem. Eng.*, Vol. 77 (Deskbook Issue), pp. 173–180.

McCoy, D. C., Statnick, R. M., Stouffer M. R., Yoon H., and Nolan, P. S., 1991, "Economic Comparison of Coolside Sorbent Injection and Wet Limestone FGD Process," paper presented at the EPRI/EPA/DOE 1991 SO_2 Control Symposium, Washington, D.C., Dec. 3–6.

McCrea, D. H., Forney, A. J., and Meyers, J. G., 1970, *J. Air Pollution Control Assoc.*, Vol. 20, December, pp. 819–824.

McGavack, J., and Patrick, N. A., 1920, *J. of the American Chemical Society*, Vol. 42, p. 946.

McGlammery, G. G., Torstrick, R. L., Simpson, J. P., and Phillips, J. F., Jr., 1973, *Sulfur Oxide Removal from Power Plant Stack Gas—Magnesia Scrubbing—Regeneration and Production of Sulfuric Acid,* EPA-R-2-73-244, May.

McGraw-Hill, 1991, "EPRI Official: Few Utilities Willing to Try New SO_2 Scrubber Technologies," *Utility Environment Report,* Dec. 13, p. 7.

McHenry, D. J., and Winnick, J., 1991, "Electrochemical Membrane Process for Flue Gas Desulfurization," unpublished paper provided by J. Winnick, Univ. of Georgia, August.

McIlvaine, R. W., 1989, "New Developments in European Power Plant Air Pollution Control," *Power Engineering,* February, pp. 33–35.

McKnight, R. A., Antonetti, J. O., Mattick, D., Schwieger, R. G., and Elliot, T. C., 1985, "A Utility's View of the Ebara E-Beam NO_x/SO_x Removal Process," paper presented at *Power* Magazine's Second International Conference on Acid Rain, Washington, D.C., March 26, pp. VI-23 to VI-35.

Meliere, K. A., Gartside, R. J., McRae, W. A., and Seamans, T.F., 1974, *Proceedings: Symposium on Flue Gas Desulfurization,* Atlanta, GA, November, EPA-650/2-74-126b, December, pp. 1109–1126.

Meyer, M. (Cominco), 1992, personal communication, April 22.

Michele, H., 1987, "Purification of Flue Gases by Dry Sorbents—Possibilities and Limits," *International Chemical Engineering,* Vol. 27, No. 2, April, pp. 183–196.

Miller, D. M., 1976, *Proceedings: Symposium on Flue Gas Desulfurization,* New Orleans, LA, May, Vol. 1, EPA-600/2-76-136a, pp. 373–385.

Minnick, L. J. 1983, "The Role of Lime in Disposal and Utilization of Utility Wastes," *Conference Proceedings: Effective Use of Lime,* sponsored by the National Lime Assoc. (Arlington, VA), Denver, CO, September 27–28, pp. 99–120.

Mirabella, J. E., 1992A, "List of Chiyoda FGD Plants," Chiyoda International Corp., Seattle, WA, May 1.

Mirabella, J. E. (Chiyoda International Corp.), 1992B, personal correspondence, May.

Moore, M. (NOXSO), 1993A, personal communication, Oct. 11.

Moore, S. (Environmental Elements Corp.), 1993B, personal communication, Oct. 11.

Morrision, G. L., 1991, "Recovery Scrubber—Cement Application Operating Results," paper presented at the EPRI/EPA/DOE 1991 SO_2 Control Symposium, Washington, D.C., Dec. 3–6.

Moser, R. E., Colley, J. D., and Jones, A. F., 1988, "Troubleshooting Utility FGD Chemical Process Problems," *Energy Technology: Proceedings of the Energy Technology Conference,* Vol. 5, EPRI, Palo Alto, CA, Feb. 17–19, pp. 167–178.

Moser, R. E., and Owens, D. R., 1991, "Overview on the Use of Additives in Wet FGD Systems," paper presented at the EPRI/EPA/DOE 1991 SO_2 Control Symposium, Washington D.C., Dec. 3–6.

Moser, R. E., Burke, J. M., Owens, D. R., and Stohs, M. S., 1990, "The Use of Additives to Improve Performance in Wet FGD Systems: HSTC Research Results," *Proceedings: GEN-UPGRADE 90, International Symposium on Performance Improvement, Retrofitting, and Repowering of Fossil Fuel Power Plants,* Volume 3: Policy, Economics, and Environmental Upgrades, EPRI GS-6986, Vol. 3, Washington, D.C., March 6–9, pp. 15-2-1 to 15-2-15.

Muzio, L. J., and Sonnichsen, T. W., 1982, *Demonstration of SO_2 Removal on a 22-MW Coal-Fired Utility—Boiler by Dry Injection of Nahcolite*, Final Report, EPRI RP-1682-2, Electric Power Research Institute, Palo Alto, CA, April.

Nannen, L. W., West, R. E., and Kreith, F., 1974, *J. of the Air Pollution Control Assoc.*, Vol. 24, January, pp. 29–39.

National Lime Association, 1982, *Lime Handling, Application and Storage, Bulletin 213*, 4th ed., Arlington, VA.

Nelson, S. G., 1991, "Sanitech's 2.5 MW_e Magnesia Dry-Scrubbing Demonstration Project," paper presented at the EPRI/EPA/DOE 1991 SO_2 Control Symposium, Washington, D.C., Dec. 3–6.

Nelson, S. G. (Sorbtech), 1992, personal communication, August.

Nelson, S. G., and Nelson, B.W., 1990, "Combined SO_2 and NO_x Control with a New Sorbent," *Proceedings: GEN-UPGRADE 90, International Symposium on Performance Improvement, Retrofitting, and Repowering of Fossil Fuel Power Plants*, Volume 3: Policy, Economics, and Environmental Upgrades, EPRI GS-6986, Vol. 3, Washington, D.C., March 6–9, pp. 13-8-1 to 13-8-14.

NERC (North American Electric Reliability Council), 1991, *Impact of FGD Systems, Availability Losses Experienced by Flue Gas Desulfurization Systems*, North American Electric Reliability Council, Princeton, NJ, July, pp. 1–35.

Niro Atomizer, 1988, *Summary Report on SDA End Product*, APC-EJ-RM/LIB, March 15.

Niro Environmental Protection, 1990, *Sales Brochure: Flue Gas Desulfurization by Spray Dryer Absorption—The Process for the 1990s*, Joy Environmental Equipment, Monrovia, CA.

Nischt, W., Johnson, D. W., and Milobowski, M. G., 1991, "Economic Comparison of Materials of Construction of Wet FGD Absorbers and Internals," paper presented at the EPRI/EPA/DOE 1991 SO_2 Control Symposium, Washington, D.C., Dec. 3–6.

Noblett, J. G. Jr., Hebets, M. J., and Moser, R. E., 1990, "EPRI's FGD Process Model (FGDPRISM)," paper presented at the EPRI/EPA 1990 SO2 Control Symposium, New Orleans, LA, May 6–11.

Noblett, J. G. Jr., DeKraker, D. P., and Moser, R. E., 1991, "FGDPRISM, EPRI's FGD Process Model—Recent Applications," paper presented at the EPRI/EPA/DOE 1991 SO_2 Control Symposium, Washington, D.C. Dec. 3–6.

Noell-KRC, 1992, *Sales Brochure: Hydrochloric Acid/Sulfuric Acid Resources Recovery*, Noell-KRC Umwelttechnik GmbH, Würzturg, Germany.

Nolan, P. S., and Seaward, D. O., 1983, "The Dowa Process Dual-Alkali Flue Gas Desulfurization with a Gypsum Product," *Proceedings of Seminar on Flue Gas Desulfurization sponsored by the Canadian Electrical Association*, Ottowa, Ontario, Canada, September.

Nolan, P. S., Purdon, T. J., Peruski, M. E., Santucci, M. T., DePero, M. J., Hendriks, R. V., and Lachapelle, D. G., 1990, "Results of EPA LIMB Demonstration at Edgewater," paper presented at the EPRI/EPA 1990 SO_2 Control Symposium, New Orleans, LA, May 8–11.

Nolan, P. S., DePero, M. J., and Goots, T. R., 1992, "Technical and Economic Evaluations of the LIMB and Coolside Processes," paper presented at the International Joint Power Generation Conference, Atlanta, GA, Oct. 18–22.

Nyman, G. B., and Tokerud, A., 1991, "Seawater Scrubbing Removes SO_2 From Refinery Flue Gases," *Oil & Gas J.*, July, pp. 52–54.

Oldenkamp, R. D., and Margolin, E. D., 1969, "The Molten Carbonate Process For Sulfur Oxide Emissions," *Chem. Eng. Prog.*, Vol. 65, November, pp. 73–76.

Ontario Research Foundation, 1947, *The Removal of Sulfur Gases from Smelter Fumes*, Province of Ontario: Department of Mines.

Oxley, J. H., Rosenberg, H. S., and Barrett, R. E., 1991, "Sulfur Dioxide and Nitrogen Oxides Control Technologies for the Petroleum Industry," Paper 62A, AIChE 1991 Spring National Meeting, Houston, TX, April 7–11.

Palazzolo, M. A., Kelly, M. E., and Brno, T. G., 1983, "Current Status of Dry SO_2 Control Systems," paper presented at the EPA/EPRI Symposium on Flue Gas Desulfurization, New Orleans, LA, Nov. 1–4.

Parkison, R. V., 1956, TAPPI, Vol. 39, May, pp. 517–519.

Pasiuk-Bronikowska, W., and Rudzinski, K. J., 1991, "Absorption of SO_2 Into Aqueous Systems," *Chemical Engineering Science*, Vol. 46, No. 9, (Printed in Great Britain), pp. 2281–2291.

Pastnikov, V. F., and Astasheva, A. A., 1940, *J. Chem. Ind.*, U.S.S.R., Vol. 17, No. 3, pp. 14–19.

Patterson, J. (Tennessee Valley Authority), 1992, personal communication, May 12.

Pebler, A. R., 1974, *Combustion*, Vol. 46, August, pp. 21–23.

Pedco Environmental Specialists, Inc., 1977, *Summary Report—Flue Gas Desulfurization Systems*—June–July, prepared for U.S. EPA.

PEI Associates, Inc. for U.S. Department of Energy, 1989A, *Utility FGD Survey, January–December 1987, Design and Performance Data for Operating FGD Systems*, DOE/OR/21400-H14, Volume 2, Parts 1, 2 and 3, June.

PEI Associates, Inc., 1989B, *Project Summary, Utility FGD Survey, January–December 1987*, DOE/OR/21400-H13, U.S. Department of Energy, Washington, D.C., June, pp. 1–28.

PETC, 1993, "PETC Project Facts: Demonstration of Innovative Applications of Technology for the CT-121 Flue Gas Desulfurization (FGD) Process," published by Department of Energy, Pittsburgh Energy Technology Center, Pittsburgh, PA, 2 sheets.

Peterson, B. (Elkem), 1992, personal communication, Aug. 19.

Peterson, J. R., Durham, M. D., and Vlachos, N. S., 1989, *Topical Report No. 1, Literature Review, Fundamental Investigation of Duct/ESP Phenomena*, DOE/PC/88850-T1, prepared for DOE Pittsburgh Energy Technology Center, Radian Corporation, Austin, TX, May 9.

Peterson, J. R., Maller, G., Burnette, A., and Rhudy, R. G., 1991, "Pilot-Scale Evaluation of Sorbent Injection to Remove SO_3 and HCl," paper presented at the EPRI/EPA/DOE 1991 SO_2 Control Symposium, Washington, D.C., Dec. 3–6.

Pierce, R. R., 1977, "Estimating Acid Dewpoints in Stack Gases," *Chem. Eng.*, Vol. 84, April 11, pp. 125–128.

Pinaev, V. A., 1963, "SO_2 Pressure over Magnesium Sulfite-Bisulfite-Sulfate Solution," *J. of Applied Chemistry*, USSR, Vol. 36, October, p. 2049.

Plumley, A. L., Whiddon, O. D., Shukto, F.W., and Jonakin, J., 1967, *Proceedings of the American Power Conference*, p. 29.

Pollock, W. A., Tomany, J. P., and Frieling, G., 1967, *Mech. Eng.*, August, p. 21.

Potter, T. L., 1991, "The Lime Industry in the Acid Rain Market," *Proceedings: Acid Rain Retrofit Seminar: The Effective Use of Lime,* sponsored by the National Lime Assoc. (Arlington, VA), Philadelphia, PA, Jan. 9–10, pp. 19–30.

Pozarnsky, D. (Montana-Dakota Utility), 1990, personal communication, July 26.

Princotta, F. T., 1990, "SO_x Technologies for Acid Rain Control," paper presented at the EPRI/EPA 1990 SO_2 Control Symposium, New Orleans, LA, May 8–11.

Public Law 101-549, 1990, "1990 Clean Air Act Amendments," U.S. Government Printing Office, Nov. 13.

Quackenbush, V. C., Polek, J. R., and Agarwal, D., 1977, "Ammonia Scrubbing Pilot Activity at Calvert City," paper presented at EPA Symposium on Flue Gas Desulfurization, Hollywood, FL, Nov. 8–11.

Rader, P. C., Borsare, D. C., and Frabotta, D., 1982, "Process Design of Lime/Limestone FGD Systems for High Chlorides," presented at Coal Technology '82, Houston, TX, Dec. 7–9.

Radian Corp., 1976, *Experimental and Theoretical Studies of Solid Solution Formation in Lime and Limestone SO_2 Scrubbers—Vol. 1,* Final Report, EPA-600/2-76-273a, October.

Radian Corp., 1977, *Evaluation of Regenerable Flue Gas Desulfurization Procedures,* Vol. I, EPRI FP-272, Electric Power Research Institute, Palo Alto, CA, January.

Radian Corp., 1988, *Ohio/Kentucky/TVA Coal-fired Utility SO_2 and NO_x Control Retrofit Study,* EPA/600/7-88/014, August.

Radian Corp., 1990, *Investigation of Flue Gas Desulfurization Chemical Process Problems,* EPRI GS-6930, Electric Power Research Institute, Palo Alto, CA, August.

Ramsay, W., 1883, British Patent 1247.

Reijnen, H. C., 1990, "Removing Aerosols from Flue Gas Desulfurization Systems," *Filtration & Separation,* May/June, pp. 200–202.

Reisdorf, J. B., Keeth, R. J., Miranda, J. E., Scheck, R. W., and Morasky, T. M., 1983, "Economic Evaluation of FGD Systems," paper presented at the EPA/EPRI Symposium on Flue Gas Desulfurization, New Orleans, LA, Nov. 1–4.

Rhudy, R., 1990, *Mist Eliminator System Troubleshooting Guide,* EPRI GS-6984, prepared by Radian, Southern Company, and United Engineers and Constructors, Inc. (now Raytheon), Electric Power Research Institute, Palo Alto, CA, October.

Roden, R. (Dravo Lime), 1992 & 1993, personal communications.

Roesner, G. O., 1937, *Metall u. Erz,* Vol. 34, p. 5.

Rooney, J. (Coastal Chem), 1993, personal communication, July.

Rosenbaum, J. B, George, D. R., Crocker, L., Nissen, W. J., May, S. L., and Beard, H. R., 1973, paper presented at AIME Environmental Quality Conference, Washington, D.C., June 7–9.

Rosenberg, H., Davis, G. O., Hindin, B., Agrawal, A., Sheppard, W., and Koch, G., 1993, *Guidelines for FGD Materials Selection and Corrosion Protection,* TR-100680, Electric Power Research Institute, Palo Alto, CA, April.

Rosenberg, H. S., Davis, G. O., Hindin, B., Radcliffe, P. T., and Syrett, B. C., 1991, "Guidelines for FGD Materials Selection and Corrosion Protection," paper presented at the EPRI/EPA/DOE 1991 SO_2 Control Symposium, Washington, D.C., Dec. 3–6.

Rossi, R. A., 1993, "Refinery Byproduct Emerges as a Viable Powerplant Fuel," *Power,* Aug., pp. 16–22.

Rossoff, J., and Rossi, R. C., 1974, *Disposal of Byproducts from Non-Regenerable Flue Gas Desulfurization Systems: Initial Report,* EPA-650/2-74-037a, May.

Rush, R. E., and Edwards, R. A., 1977, "Operating Experience with Three 20 MW Prototype Flue Gas Desulfurization Processes at Gulf Power Company's Scholtz Electric Generating Station," paper presented at EPA Flue Gas Desulfurization Symposium, Hollywood, FL, Nov. 8–11.

Saleem, A., 1991A, "GE's Worldwide Experience with IFO Based Gypsum Producing Flue Gas Desulfurization Systems," presented at the Second International Conference on FGD and Chemical Gypsum (sponsored by ORTECH), Toronto, Canada, May 12–15.

Saleem, A., 1991B, "Design and Operation of Single Train Spray Tower FGD Systems," paper presented at the EPRI/EPA/DOE 1991 SO_2 Control Symposium, Washington, D.C., Dec. 3–6.

Saleem, A., Janssen, K. E., and Ireland, P. A., 1993, "Ammonia Scrubbing of SO_2 Comes of Age with *In situ* Forced Oxidation," paper presented at the EPRI/EPA/DOE 1993 SO_2 Control Symposium, Boston, MA, Aug. 24–26.

Saliga, J. J. (Fluor Daniel, Chicago, Illinois), 1990, "Telephone Survey of Soda and High Temperature Spray Dryer Applications," Unpublished Report, August.

Samanta, S. C., 1977, "Physical and Chemical Characteristics of Stabilized SO_2 Scrubber Sludges," paper presented at the Sixth Environmental Engineering and Science Conference, University of Louisville, Louisville, KY, Feb. 28.

Sanitech, Inc., 1988, *Phase I Final Report,Study of the Regenerability of a Unique New Sorbent that Removes SO_2-NO_x from Flue Gases,* EPA SBIR Contract 68-02-4484, March 3.

Schneider, F. (Ontario-Hydro), 1992, personal communication on the status of the FMC technology, April 21.

Schütz, M., Eikmeier, M., and Glamser, J., 1989, "Operating Experience with Advanced FGD Technology in West Germany," *Proceedings: First Combined Flue Gas Desulfurization and Dry SO_2 Control Symposium,* EPRI GS-6307, Electric Power Research Institute, Palo Alto, CA, April.

Schwieger, R., and Haynes, A., 1985, "Reliability Concerns, Regulations Lead to Virtual Standardization of Air-Pollution-Control Systems," *Power,* April, pp. 81–93.

Scott, K. D., 1985, *Electrochemical Flue Gas Desulfurization,* Ph.D. Thesis, Georgia Tech.

Sedman, C. B., Maxwell, M. A., and Hall, B., 1991, "Pilot Plant Support for the ADVACATE/MDI Commercialization," paper presented at the EPRI/EPA/DOE 1991 SO_2 Control Symposium, Washington, D.C., Dec. 3–6.

Selmeczi, J. G., 1975A, U.S. Patent 3,914,378.

Selmeczi, J. G., 1975B, U.S. Patent 3,919,393.

Selmeczi, J. G., 1975C, U.S. Patent 3,919,394.

Selmeczi, J. G., and Stewart, D. A., 1978, *Chem. Eng. Prog.,* Vol. 74, February, p. 41–45.

Shattuck, D. M., Ireland, P. A., Keeth, R. J., Mora, R. R., Scheck, R. W., Archambeault, J. A., Rathbon, G. R. and Morasky, T. M., 1984, *Retrofit FGD Cost-Generating Guidelines,* CS-3696, prepared by Stearns Catalytic for Electric Power Research Institute, Palo Alto, CA, October.

Shield, G. (Bechtel), 1992, personal communications, April and August.

Siegfriedt, W. E., and Ludwig, M., 1984, "Desulfurization Processes in West Germany—An Overview," paper presented at American Power Conference, Chicago, IL, April.

Siegfriedt, W. E., Glamser, J. H., and Hannemann, H., 1990, "German FGD Technology: Transfer of Materials Experience to the U.S. Market," *Sixth International Seminar: Solving Corrosion Problems in Air Pollution Control Equipment*, sponsored by National Association of Corrosion Engineers, Louisville, KY, Oct. 17–19.

Siegfriedt, W. E., Wong, A., and Saliga, J. J., 1993, "FGD and FGD Byproduct Treatment Process Suppliers," Unpublished Report, Fluor Daniel, Chicago, IL.

Silerberg, A. N. (Lurgi Corporation), 1992, personal communication on the Sulphidine process, May 19.

Smigelski, J. E., and Maroti, L. A., 1986, "Design and Operation of a Wet Process Based Flue Gas Desulfurization System Without Reheat," Tenth Symposium on Flue Gas Desulfurization, Electric Power Research Institute, Palo Alto, CA, Nov. 18–21.

Smith, C. L., 1977, "Sludge Disposal by Stabilization—Why?," paper presented at The Second Pacific Chemical Engineering Congress, Denver, CO, Aug. 28–31.

Smith, C. L., 1992A, "Physical Aspects of FGD Byproducts," *International Journal of Environmental Issues on Minerals and Energy*, Vol. 1, No. 1, A.A. Balkema, Rotterdam, Netherlands, September, pp. 37–46. (Published by Conversion Systems, Inc. as *Physical Evaluation of FGD Byproducts*.)

Smith, C. L. (Conversion Systems, Inc.), 1992B, personal communication on FGD Byproduct Material Characteristics, June 2.

Smith, C. L., 1992C, *Case Histories in Full Scale Utilization of Fly Ash-Fixated FGD Sludge*, Conversion Systems, Inc., Horsham, PA.

Smith, C. L., and Rau, E., 1981, "Stabilized FGD Sludge Goes to Work," paper submitted for presentation at Coal Technology '81, Houston, TX, Nov. 17–19.

Sopocy, D. M., DePriest, W., Kalanik, J. B., Maurer, A., and Rhudy, R., 1991, "Clean Air Technology (CAT) Workstation," paper presented at the EPRI/EPA/DOE 1991 SO_2 Control Symposium, Washington, D.C., Dec. 3–6.

Spears, P., 1992, personal communication (Westvaco), Aug. 20.

Spector, M. L., and Brian, P. L. T., 1974, U.S. Patent 3,843,789.

Sporer, J., 1992, "The Linde Solinox Process: Gypsum-Free Flue-Gas Desulfurization," *Gas Sep. Purif.*, Vol. 6, No. 3, p. 133.

Stearns Catalytic Corp., 1985, Economic Evaluation of FGD Systems, CS-3342, Vol. 4, Electric Power Research Institute, Palo Alto, CA, July.

Stearns Catalytic Corp., 1987, *Retrofit FGD Cost Estimating Guidelines: Computer Users Manual*, CS-5408-CCM, Electric Power Research Institute, Palo Alto, CA, September.

Stefanski, A., 1986, *Advanced Flue Gas Desulfurization Project, Final Summary Report*, Research Report EP6-9, Empire State Electric Energy Research Corporation, New York, NY, November.

Steiner, P. (Foster Wheeler), 1992, personal communication, Aug. 24

Stevens, G.E., Sitkiewitz, S. D., Phillips, J. L., and Owens, D. R., 1991, "Results of High SO_2 Removal Efficiency Tests at EPRI's High Sulfur Test Center," paper presented at the EPRI/EPA/DOE 1991 SO_2 Control Symposium, Washington, D.C., Dec. 3–6.

Stites, J. G., Jr., and Miller, J. G., 1969, *Proceedings of the American Power Conference*, Vol. 31, p. 553.

Stites, J. G., Jr., Horlacher, W. R., Bachofer, J. L., Jr., and Bartman, J. S., 1969, *Chem. Eng. Prog.*, Vol. 65, October, p. 74.

Swain, R. E., 1921, *Chem. & Met. Eng.*, Vol. 24, pp. 463–465.

Tamaki, A., 1975, *Chem. Eng. Prog.*, Vol. 77, Nov. 5, pp. 55–58.

Tamboli, J. K., and Sand, L. B., 1970, "SO_2 Sorption-Properties of Molecular Sieve Zeolites," paper presented at the Second International Clean Air Congress of the International Union of Air Pollution Prevention Association, Washington, D.C., Dec. 6–11.

Tamura, Z., 1970, "Stack Gas Desulfurization Method by Activated Carbon," paper presented at Second International Clean Air Congress of the International Union of Air Pollution Prevention Association, Washington, D.C., Dec. 6–11.

Taylor, H. (Advanced Air), 1992, personal communication on the status of the former FMC double alkali technology, April 8.

Teixeira, D. P., Muzio, L. J., and Lott, T. A., 1990, "R-SO_x: Recycle Dry Boiler Injection Technology for Flexible SO_2 Control," paper presented at the EPRI/EPA 1990 SO_2 Control Symposium, New Orleans, LA, May 8–11.

Telesz, R. W., Owens, F. C., and Cline, J. R., 1990, "Comparison between Forced Oxidation Limestone and Magnesium Enhanced Lime FGD Systems," paper presented at Power-Gen '90, Orlando, FL, Dec. 4–6.

Thomas, A. D., Davis, D. L., Parsons, T., Schroeder, G. D., and DeBerry, D., 1969, *Applicability of Metal Oxides to the Development of New Processes for Removing SO_2 from Flue Gases,* PB 185 562, July 31.

Toher, J. G., Lanois, G. D., and Sauer, H., 1991, "High Efficiency, Dry Flue Gas SO_x, and Combined SO_2/NO_x Removal Experience with the Lurgi Circulating Fluid Bed Dry Scrubber—A New, Economical Retrofit Option for U.S. Utilities for Acid Rain Remediation," paper presented at the EPRI/EPA/DOE 1991 SO_2 Control Symposium, Washington, D.C., Dec. 3–6.

Tolman, J. (Utah Power & Light), 1992, personal communication on the Naughton Unit 3 Scrubber, May 27.

Tsuji, K., and Shiraishi, I. (Mitsui Mining Co.), 1991, "Mitsui-BF Dry Desulfurization and Denitrification Process Using Activated Coke," paper presented at the EPRI/EPA/DOE 1991 SO_2 Control Symposium, Washington, D.C., Dec. 3–6.

TVA (Tennessee Valley Authority), 1974, *Pilot Plant Study of an Ammonia Absorption-Ammonium Bisulfate Regeneration Process, Topical Report Phases I and II,* EPA-650/2-74-049a, June.

Ulset, T., and Erga, O., 1991, "The ELSORB Process: A New Regenerable FGD Process for Utility Boilers." paper presented at IGCI Forum '91, Washington, D.C.

Van Buskirk, G. (Koch Engineering, Divmet Division), 1992, personal communication, Sept. 11.

Vernon, J. L., and Soud, H. N., 1990, *FGD Installations on Coal-Fired Plants,* IEACR/22, International Energy Association Coal Research, London, UK, April.

Wartman, F. C., 1937, *U.S. Bureau of Mines, Report Invest. No. 3339.*

Weidmann, H., and Roesner, G., 1935, *Metallges.*, Periodic Rev. 11.

Weidmann, H., and Roesner, G., 1936A, *Metallges.*, Periodic Rev. 11, February, pp. 7–13.

Weidmann, H., and Roesner, G., 1936B. *Ind. Eng. Chem.* (News Ed.), Vol. 14, March 20, p. 105.

Whitley, F. (Dow Chemical), 1993, personal communication, Aug. 11.

Yoon, H., Statnick, R. M., Withum, J. A., and McCoy, D. C., 1991, "Coolside Desulfurization Process Demonstration at Ohio Edison Edgewater Power Station," 84th Annual Meeting of the Air & Waste Management Association, Vancouver, British Columbia, Canada, June.

Yu, W. C., 1991, "Evaluation of Disposal Methods for Oxidized FGD Sludge," paper presented at the EPRI/EPA/DOE 1991 SO_2 Control Symposium, Washington, D.C., Dec. 3–6.

Zey, A., White, S., and Johnson, D., 1980, "The ATS Claus Tail Gas Clean-Up Process," *Chem. Eng. Prog.*, Oct., pp. 68–76.

Zohouralsen, O. (Joy Environmental Equipment), 1992, personal communication, June 24.

Chapter 8
Sulfur Recovery Processes

THE BASIC CLAUS PROCESS, 670

    Background, 670
    Basic Data, 671
    Process Description, 674
    Design and Operation, 678

CLAUS PROCESS MODIFICATIONS, 689

    Oxygen-Based Claus Processes, 689
    Isothermal Reactor Concepts, 696
    Developmental Sulfur Recovery Processes, 697
    Miscellaneous Processes, 697

CLAUS PLANT TAIL GAS TREATMENT PROCESSES, 698

    Sub-Dewpoint Claus Processes, 699
    Direct Oxidation of H_2S to Sulfur, 708
    Sulfur Dioxide Reduction and Recovery of H_2S, 717

REFERENCES, 724

THE BASIC CLAUS PROCESS

Background

    The Claus process is not a gas-purification process in the true sense of the word, as its principal objective is recovery of sulfur from gaseous hydrogen sulfide or, more commonly, from acid gas streams containing hydrogen sulfide in high concentrations. Typical streams of this type are the acid gases stripped from regenerable liquids, e.g., the alkanolamine solutions or physical solvents used for the purification of sour gases, such as natural and refinery gases. The effluent gases from the Claus plant are without value and are vented to the atmosphere or are directed to a tail gas treatment system. But whether the Claus plant has a tail gas unit or not, the final effluent gas is usually incinerated to oxidize any residual sulfur to sulfur dioxide. Air pollution control regulations existing in most industrialized countries prohibit the discharge of large amounts of sulfur compounds to the atmosphere; therefore, Claus plants with tail gas treatment units are often mandatory adjuncts to gas-desulfurization installations, and, consequently, the Claus process is of considerable significance within the general scope of gas-purification technology. Furthermore, the Claus process yields sulfur of extremely good quality and thus is a source of a valuable basic chemical.

With growing air pollution concerns, sulfur recovery in Claus type units is increasing to the point where units which normally would not be considered economical are being installed, strictly for the purpose of air pollution control. In addition, the recovery efficiency of Claus type plants is continuously being improved by better plant operation, better design methods, and developments of the process technology. Unfortunately, complete conversion of hydrogen sulfide to elemental sulfur under Claus plant operating conditions is precluded by the equilibrium relationships of the chemical reactions upon which the process is based. As a result of this limitation, the basic Claus process is, in many instances, not adequate to reduce atmospheric emission of sulfur compounds to the level required by air pollution control regulations. In these cases, the basic Claus process has to be supplemented with another process specifically designed to remove residual sulfur compounds from the Claus plant tail gas. Processes of this type, which are usually referred to as "tail gas cleanup" or "tail gas treating" processes, are discussed later in this chapter.

Since the disclosure of the process by Claus in 1883, it has undergone several modifications. The most significant modification was that made by I.G. Farbenindustrie A.G. in 1936 which introduced the process concept currently in use, which consists of a thermal conversion step followed by a catalytic conversion step. As presently used, most process configurations are similar in their basic concept and differ only in the design and arrangement of the equipment.

The literature describing the theoretical as well as design and operational aspects of the Claus process is quite voluminous. In view of this extensive coverage, the following discussion will be directed primarily toward current technology on the design and operation of plants that provide high efficiency sulfur recovery and low emission of sulfurous pollutants to the atmosphere.

Basic Data

The basic chemical reactions occurring in the Claus process are represented by equations 8-1, 8-2, and 8-3, with reactions 8-1 and 8-2 taking place in the thermal stage (reaction furnace) and reaction 8-3 in the catalytic stage (catalytic converters). The thermodynamics and kinetics of the reactions were first rigorously investigated by Gamson and Elkins (1953), who developed a chart of theoretical conversion for pure H_2S as a function of temperature at one atmosphere pressure. **Figure 8-1** was compiled by Paskall (1979) and includes the results of Gamson and Elkins, which were based on a limited number of sulfur species and 1909 thermodynamic data. Paskall (1979) updated the results of Gamson and Elkins using current thermodynamic data. Paskall's results are shown in curve 1 of **Figure 8-1** for all sulfur species (S_1, S_2, S_3, S_4, S_5, S_6, S_7, S_8) and for the same species considered by Gamson and Elkins (S_2, S_6, S_8) in curve 2. Curve 3 depicts the results of Gamson and Elkins. The unusual shape of the conversion curve in **Figure 8-1** is caused by the existence of several sulfur species in the gas phase. The average molecular weight of sulfur vapor increases with decreasing temperature and with increasing sulfur partial pressure. For example, at a sulfur partial pressure of 0.05 atm, and at temperatures below 700°F, sulfur vapor is predominately S_6 and S_8, while at the same partial pressure, but at temperatures above 1,000°F, the sulfur is mostly S_2. Since the equilibrium constant for reaction 8-3 decreases with temperature when S_6 or S_8 is formed and increases with temperature when S_2 is the product, the conversion curve slopes downward with increasing temperature at low temperature, but changes direction at about 1,000°F, and slopes upward at higher temperatures.

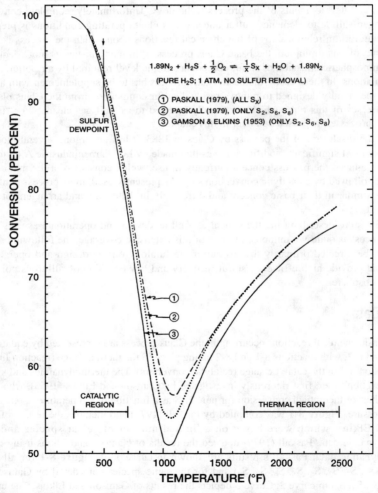

Figure 8-1. Theoretical equilibrium conversion of hydrogen sulfide to sulfur vapor. (*Paskall, 1979*)

$$H_2S + \tfrac{1}{2}O_2 = H_2O + S \tag{8-1}$$

$$H_2S + \tfrac{3}{2}O_2 = H_2O + SO_2 \tag{8-2}$$

$$2H_2S + SO_2 = 2H_2O + 3S \tag{8-3}$$

It can be seen from **Figure 8-1** that the process can be separated into two stages, i.e., a thermal stage, above about 1,700°F, and a catalytic stage, usually between 700°F and a temperature somewhat above the sulfur dew point of the gas mixture. The lower the catalytic stage temperature, the more complete the conversion that can be attained; however, operation at a temperature below the sulfur dew point is not normally feasible because of problems caused by sulfur deposition in the catalyst bed. It is, therefore, advantageous to have several catalytic stages in series with condensation and removal of the sulfur formed after each stage.

The molar heat of reaction for the combination of reactions 8-2 and 8-3, i.e., reaction 8-4, is 262,000 to 311,000 Btu/lb mole with 223,000 to 248,000 Btu/lb mole for reaction 8-2 and 38,000 to 63,000 Btu/lb mole for reaction 8-3 at typical sulfur plant conditions (Gamson and Elkins, 1953).

$$3H_2S + \tfrac{3}{2}O_2 = 3H_2O + 3S \tag{8-4}$$

Since equation 8-3 represents the catalytic stage, it is seen that the temperature rise in the catalyst beds is relatively small, permitting operation at comparatively low temperatures and, consequently, the attainment of a high conversion to sulfur. Since reaction 8-3 is reversible and water is a reaction product, removal of water between catalytic stages would increase conversion. However, attempts to accomplish this have so far failed, primarily because of the corrosiveness of the aqueous condensate and plugging of the equipment with solid sulfur. Thus, the presence of water vapor in the reaction gases throughout the plant imposes a definite limitation on the degree of conversion.

A further limitation on conversion is the occurrence of a number of side reactions, due to the presence of carbon dioxide and light hydrocarbons in the feed gas, resulting in the formation of carbonyl sulfide and carbon disulfide in the thermal stage of the process. These compounds are quite stable and may pass unchanged through the catalytic converters unless provisions are made for their conversion to hydrogen sulfide and carbon dioxide. Formation of carbonyl sulfide and carbon disulfide is a significant consideration in the design and operation of high efficiency Claus plants as the sulfur loss associated with these compounds may amount to an appreciable percentage of the total sulfur loss. For example, Sames and Paskall (1984) show that as much as 17% of tail gas sulfur emissions can be due to COS and CS_2, and Luinstra and d'Haêne (1989) report that COS and CS_2 may be as high as 50% of tail gas sulfur losses.

Various reaction mechanisms have been proposed for the formation of carbonyl sulfide and carbon disulfide and for their subsequent hydrolysis to hydrogen sulfide and carbon dioxide (Paskall and Sames, 1992). The plant data available indicate that carbonyl sulfide is formed primarily from the reaction between elemental sulfur and carbon monoxide, which in turn are derived from hydrogen sulfide and carbon oxides present during combustion of the feed gas in the Claus thermal stage. The production of carbon disulfide in the thermal stage is usually attributed to the presence of hydrocarbons in the feed gas because carbon disulfide is produced commercially by reacting elemental sulfur with saturated hydrocarbons. The resulting concentrations of both carbon disulfide and carbonyl sulfide are found to be well above those predicted by the hydrolysis equilibrium reaction at the furnace temperature. This indicates that, in most circumstances, there is insufficient residence time within the reaction furnace to hydrolyze COS and CS_2 after initial formation.

From a practical standpoint, the exact mechanism of carbonyl sulfide and carbon disulfide formation is less important than the technology to convert these compounds to hydrogen sulfide and sulfur. Equilibrium constants for the hydrolysis of COS and CS_2 are given in Chapter 13 (**Table 13-11**). The reaction equilibria are favored by low temperature, but show almost complete conversion at temperatures up to about 700°F. The reaction rate, of course, increases with increased temperature. It is generally assumed that carbonyl sulfide and carbon disulfide undergo rapid hydrolysis to H_2S at temperatures on the order of 600 to 700°F in the presence of an aluminum oxide type catalyst. Thus, by maintaining sufficiently high temperatures in the first catalytic converter and by use of an active catalyst, near equilibrium conversion of carbonyl sulfide and carbon disulfide to H_2S can be obtained. These conclusions have been substantiated by the work of Grancher (1978), who investigated formation and hydrolysis of carbonyl sulfide and carbon disulfide in large Claus units at Lacq, France.

The catalyst used in the Claus process is normally either granular natural bauxite or alumina shaped into pellets or balls. For high-efficiency plants, an alumina catalyst of high activity is usually preferred. Resistance to attrition and to the relatively high temperatures during activity restoration procedures or rejuvenation are also important catalyst properties. Furthermore, since the Claus process is operated at low pressure (5–12 psig), the catalyst shape must be such that an excessive pressure drop is not incurred at typical design space velocities.

Process Description

There are two basic forms of the Claus process, which may be termed the straight-through and the split-flow processes. The primary difference is that in the straight-through configuration all of the acid gas feed flows through the reaction furnace; whereas, in the split-flow arrangement, a major portion bypasses the furnace and is fed directly to the first catalytic reactor. The selection of the best configuration for a specific case, straight-through or split-flow, is based primarily on the acid gas stream composition. If the acid gas stream consists entirely of H_2S and CO_2, (i.e., no hydrocarbon) and there is no significant preheat of the acid gas or air feed streams, the optimum process is determined by the percent H_2S in the acid gas as indicated by **Table 8-1**. If the acid gas contains hydrocarbons, or if the acid gas or air feed streams are preheated, it is possible to operate straight-through sulfur plants with feed gas streams containing less than the 50% H_2S indicated in the table.

Table 8-1
Guidelines Relating Acid Gas Composition and Claus Plant Configuration

Acid Gas Composition Mol% H_2S	Type of Process Recommended
50–100	Straight-through
20–50	Split-flow

The Straight-Through Process

In the straight-through process, which is shown diagrammatically in **Figure 8-2,** the entire acid-gas stream and the stoichiometric amount of air required to burn one-third of the hydrogen sulfide to sulfur dioxide are fed through a burner into the reaction furnace. At the temperatures prevailing in the furnace, typically 1,800 to 2,500°F, a substantial amount of elemental sulfur is formed, with typically 60 to 70% of the H_2S in the feed gas converted to sulfur (see **Figure 8-1**). This sulfur is condensed by cooling the gases first in a waste heat boiler and subsequently in a sulfur condenser. While high-pressure steam can be generated in the waste heat boiler, the sulfur condenser is normally limited to low pressure steam generation because of the low temperature required to obtain maximum sulfur condensation. The reaction gases leaving the sulfur condenser are reheated to a temperature of 450°–540°F and flow through the first catalytic converter where additional sulfur is produced by the reaction of hydrogen sulfide with sulfur dioxide. Gases leaving the sulfur condenser must be reheated sufficiently to maintain the temperature of the reaction gases above the sulfur dew point as they pass through the first catalytic converter because condensation of sulfur on the catalyst leads to rapid catalyst deactivation. The gases leaving the first catalytic converter are again cooled, and sulfur is condensed. The gases are then reheated before entering the second catalytic converter. The process of reheating, catalytically reacting, and sulfur condensing may be repeated, in one, two, or even three additional catalytic stages. As conversion progresses through the catalytic stages, and more and more sulfur is removed from the gas mixture, the sulfur dew point of the reaction gases is lowered, permitting operation at progressively lower temperatures in each succeeding catalytic converter, thus improving overall conversion (see **Figure 8-1**). Typical inlet temperatures to the second and third catalytic stages are 390°–430°F and 370°–410°F, respectively. After leaving the last sulfur condenser, the exhaust gases, which still contain appreciable amounts of sulfur compounds and a small amount of sulfur as vapor

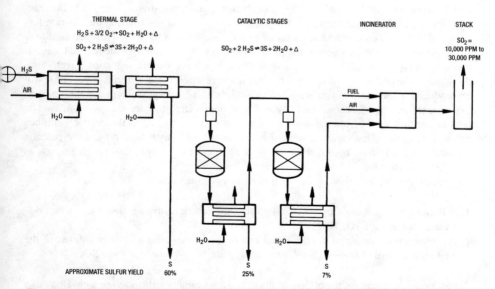

Figure 8-2. Typical flow diagram of two-stage straight-through Claus process plant.

and entrained liquid, are either incinerated, in order to convert all sulfur compounds to sulfur dioxide before venting to the atmosphere, or are further treated in a separate process for removal of residual sulfur compounds. To conserve energy and minimize the loss of sulfur as a vapor in the exhaust gases, the final condenser may function as a boiler feedwater preheater operated so that the effluent gases leave the unit at the lowest practical temperature.

The straight-through process is used for gas streams of high hydrogen sulfide content (typically above about 50 mol% H_2S). Depending on the hydrogen sulfide concentration in the acid gas, 94 to 95% conversion efficiencies can be attained with two catalytic stages and 96 to 97% with three catalytic stages. A fourth catalytic converter is normally not economical as it increases conversion by less than 1%. It should be noted that the quoted conversion efficiencies do not take into account sulfur losses caused by the presence of carbonyl sulfide and carbon disulfide, and that overall conversions must be adjusted downward by the amount of these losses.

As stated earlier, carbonyl sulfide and carbon disulfide hydrolyze fairly readily at temperatures in the range of 600° to 700°F in the presence of water vapor and an active Claus catalyst. It is therefore advantageous to design the first catalytic converter for operation at this temperature level if high conversion efficiency is required and if substantial quantities of carbonyl sulfide and carbon disulfide are present. However, this results in inefficient operation of the first converter with respect to the Claus reaction, and installation of a third converter may be desirable to compensate for this loss in efficiency. If air pollution control regulations require high conversion efficiency, it is usually economical to use only two catalytic converters, and then remove residual sulfur compounds in a tail gas treating unit.

The Split-Flow Process

The split-flow process is used for acid-gas streams containing hydrogen sulfide in such low concentrations that stable combustion, which requires a reaction furnace flame temperature in excess of 1,700°F, could not be sustained if the entire gas stream were fed to the reaction furnace. In this process, one-third or more of the total acid gas is fed to the reaction furnace and sufficient combustion air is added to burn one-third of the total hydrogen sulfide to sulfur dioxide. As a consequence, the production of elemental sulfur in the thermal stage is less than can be accomplished in the straight-through process with the reduction directly proportional to the percentage of acid gas bypassing the reaction furnace. Little or no sulfur will be produced in the reaction furnace when two-thirds of the acid gas (the maximum amount) bypasses the reaction furnace. The hot gases from the reaction furnace are cooled in a waste heat boiler and then combined with the acid gas that has bypassed the reaction furnace before entering the first catalytic stage. Except for the acid gas bypass around the reaction furnace and waste heat boiler, the process is identical to the straight-through process as depicted in **Figure 8-2**.

Operation of the split-flow process is limited by two constraints:

1. Sufficient acid gas must be bypassed so that the reaction furnace flame temperature is greater than about 1,700°F (926°C).
2. The maximum amount of acid gas that can be bypassed is limited to two-thirds of the total, as one-third of the total hydrogen sulfide must be combusted to form SO_2.

Figure 8-3 summarizes the effects of these two constraints on the operating envelope of a split-flow Claus plant (Sames and Paskall, 1985). The maximum bypass (two-thirds of the

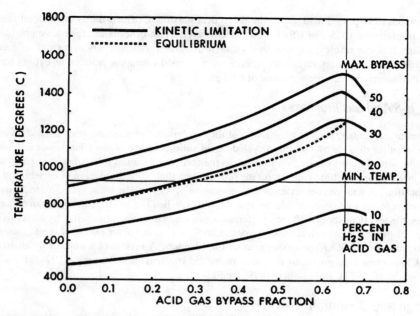

Figure 8-3. Calculated split-flow Claus plant reaction furnace temperatures as a function of feed gas H_2S content and reaction furnace bypass fraction. (*Sames and Paskall, 1985*)

feed gas) is shown as a vertical line and the minimum furnace temperature constraint of 1,700°F (926°C) is shown as a horizontal line. The x-axis is the acid gas bypass fraction; the y-axis is the calculated reaction furnace flame temperature. Furnace temperatures for the solid line curves for 10, 20, 30, 40, and 50 vol% H_2S gas streams are calculated assuming the reactions are kinetically limited. The dotted line for an acid gas stream containing 30 vol% H_2S is based on the assumption that reaction equilibrium is attained within the furnace. Clearly, an acid gas containing 50% H_2S can be easily processed in a split-flow Claus plant; while a 10% H_2S feed would require additional processing steps, such as preheat of the acid gas feed and combustion air or oxygen enrichment. Feed gases containing between 25 and 50% H_2S are suitable feeds for split-flow Claus plants because the furnace temperatures are below the temperature limitations of conventional refractories of 2,900°F (1,600°C) and above the minimum stable furnace temperature of 1,700°F (926°C). For split-flow Claus plants processing feed gases containing 25% or less H_2S, both acid gas and combustion air preheat should be considered to ensure stable operation in case the acid gas flow or composition changes unexpectedly.

The major problem with the split-flow configuration is that feed gas contaminants, which can deactivate the catalyst, have a direct path to the first catalytic converter. COS and CS_2 formation would not be an issue in split-flow plants if equilibrium were attained in the reaction furnace. However, many Claus plant reaction furnaces are kinetically limited and do not achieve equilibrium. Therefore, both COS and CS_2 can be formed in the reaction furnace and it is usually necessary to operate the first converter of a split-flow Claus plant at 620° to 660°F (325° to 350°C) to hydrolyze these compounds (Sames and Paskall, 1985). Increasing the amount of bypassed acid gas to the theoretical maximum of two-thirds of the total feed gas can minimize COS and CS_2 formation. However, these benefits are far outweighed by the harm that can result if oxygen breakthrough from the furnace occurs as a result of feed gas flow or composition variations. Therefore, it is good operating practice to bypass less than the theoretical maximum amount of feed gas.

Low H_2S Acid Gas Processes

Special techniques, such as pre-heating of the combustion air and acid gas streams, generation of sulfur dioxide by burning recycled liquid sulfur, recycle of hot Claus plant tail gas, and addition of light hydrocarbon fuel gas to the feed, are required to process acid gases of very low hydrogen sulfide content, on the order of less than 20%. Replacing or enriching the combustion air with oxygen extends the application of all of these techniques to feed gases with even lower hydrogen sulfide content. However, when the hydrogen sulfide content of the feed is less than about 10%, the preferred approach is to recover the sulfur by direct oxidation of the hydrogen sulfide over a special catalyst (see the descriptions of the Selectox, Superclaus, and MODOP processes later in this chapter). A review of a number of alternative Claus plant configurations has been presented by Beavon and Leeper (1977), Parnell (1985), Chute (1977), and Fischer (1979, 1985).

Design and Operation

Process Design

Process design procedures have been presented by Valdes (1964A, B) and by Opekar and Goar (1966). Since sufficient thermodynamic and kinetic data are available in literature, Claus sulfur plant design methods are well known, and design optimization by the use of computer techniques is commonly practiced. One such method has been proposed by Boas and Andrade (1971). Methods for predicting Claus products by use of a modified Gibbs free energy minimization technique have been presented by several authors. See Maadah and Maddox (1978), Lees (1970), Yen et al. (1985), Pearson and Belding (1985), and Wen et al. (1987). In fact, free energy minimization has become the primary tool for designing Claus sulfur plants. These free energy minimization programs calculate compositions and temperatures using thermodynamic data from the JANAF tables (Stull and Prophet, 1971; Chase et al., 1985). A number of commercial computer programs for simulating Claus plant heat and material balances are available to run on personal computers under the DOS environment, e.g., Sulsim from Western Research (Sames et al., 1985) and TSWEET from Bryan Research and Engineering (BR&E, 1993) and Hysim from Hyprotech Ltd.

The free energy minimization technique represents a significant advance over the reaction equilibrium approach previously used to design Claus plants because it considers many possible chemical species. It is based on the principle that the calculated mixture composition with the minimum Gibbs free energy is at equilibrium. With computer models using the free energy minimization technique, plant configurations can be optimized to a degree that was not possible using hand calculations.

While equilibrium-based calculations provide accurate estimates of catalytic converter performance, these calculation methods do not always match reaction furnace field measurements because many reaction furnaces are kinetically limited and do not reach equilibrium. Some compounds, the most notable ones being carbonyl sulfide and carbon disulfide, do not normally reach equilibrium concentrations in the reaction furnace and waste heat boiler exit gases. There are also strong indications that the temperature of gases exiting the reaction furnace and the concentrations of H_2 and CO in these gases do not equally match equilibrium values as calculated by free energy minimization techniques. For these reasons, some free energy minimization programs (Wen et al., 1987; Sames et al., 1985) supplement the equilibrium analysis with field data or published empirical correlations, such as those of Fischer (1974), Sames and Paskall (1985), and Luinstra and d'Haêne (1989). **Table 8-2** summarizes the correlations of Sames and Paskall (1985) for the concentrations of CO, H_2, COS, CS_2, and S in the outlet gas of a kinetically limited reaction furnace. The correlations are based on plant data taken from more than 300 tests on approximately 100 sulfur trains, including both split-flow and straight-through configurations.

As the previous discussion suggests, the major area of uncertainty in design, and in predicting the performance of existing equipment, is in the reaction furnace-waste heat boiler. Studies comparing kinetically limited and equilibrium-limited reaction furnaces have been published by Sames and Paskall (1985, 1987) and Monnery et al. (1993). Other studies comparing plant data against free energy minimization results have been presented by Sames and Paskall (1987), Sames et al. (1987, 1990), and Monnery et al. (1993). In general, the quantities of sulfur and CO in actual reaction furnace product gas streams are lower than predicted by equilibrium calculations and the quantities of COS and CS_2 are much higher. Field data

Table 8-2
Empirical Correlations for Predicting the CO, H_2, COS, CS_2, and Sulfur Content of a Kinetically Limited Claus Plant Reaction Furnace

CO formation:

R(CO) = fraction of furnace inlet carbon that forms CO
 = $0.002 A^x \exp(4.53A)$, (x = 0.0345).

H_2 formation

R(H_2) = fraction of furnace inlet H_2S that cracks to H_2 and S = 0.056 (standard deviation ±0.024)

COS formation

R(COS) = fraction of furnace inlet carbon that forms COS
 = 0.01 tangent(100A), for $0 \leq A \leq 0.86$
 = 0.143, for A > 0.86

CS_2 formation

R(CS_2) = fraction of hydrocarbon carbon in furnace inlet that forms CS_2
 = $2.6 A^y \exp(-0.965A)$, (y = 0.971)

S formation

R(S) = fraction of furnace inlet H_2S that forms elemental S
 = $1.58^z \exp(-0.73A)$, (z = 1.099)

where, A = mole fraction of H_2S in acid gas feed on a dry basis

Source: Sames and Paskall (1985)

also show that the hydrogen content of the gas exiting actual reaction furnaces is lower than the calculated equilibrium value at feed gas H_2S concentrations greater than 65% and higher than equilibrium values at lower H_2S concentrations (Monnery et al., 1993). There is also very clear evidence that some reactions continue to take place in the waste heat boiler as the reaction gases flow through the tubes (Sames et al., 1990; Dowling et al., 1990; Monnery et al., 1993). Although there is still some uncertainty with regard to the optimum reaction furnace-waste heat boiler modeling approach, the thermodynamic behavior of systems with sufficient residence time can be modeled successfully by assuming equilibrium in the reaction furnace, followed by re-equilibration at waste heat boiler conditions.

Several mechanical designs and arrangements of the major equipment have been reported in literature. The reaction furnace may combine the burners, the combustion chamber, and the waste heat boiler in one integral vessel. This is known as a fire tube design and is the least expensive for unit capacities up to about 30 long tons of sulfur per day. Methods for the design of fire tube reaction furnaces have been reported by Valdes (1965). A different design consisting of a separate combustion furnace and waste heat boiler has been described by Sawyer et al. (1950). This arrangement is usually more economical and practical for capacities in excess of 30 long tons of sulfur per day.

Reaction furnace burner design varies considerably, from the simple coaxial type with the acid gas injected through a central tube and the combustion air through an outer annular space, to the complex, high-intensity type designed for efficient combustion, and used especially when ammonia and hydrocarbons are present in the feed gas (Stevens et al., 1996; Fischer and Kriebel, 1988; Babcock Duiker, 1983; Schalke et al., 1989).

The hot gases leaving the reaction furnace are normally cooled to 500°–600°F by generating 50–600 psig steam in the waste heat boiler. Alternatively, the gases may be cooled in a waste heat exchanger by heat exchange fluids, such as hydrocarbon or synthetic oils and glycol/water solutions.

Reheating of the reaction gases prior to their entry into a catalytic converter may be effected by several methods, i.e., bypassing of hot gases from the waste heat boiler, auxiliary inline acid-gas or fuel gas burners, gas-to-gas heat exchangers, and indirect steam heated or fuel-gas fired heaters. Detailed discussions of reheat systems are given by Grekel et al. (1965), Valdes (1964A, B), Peter and Woy (1969), and Fischer (1979, 1985). The hot gas bypass and inline acid gas burner methods result in slightly lower overall sulfur conversion because some of the acid gas is bypassed around one or more catalytic converters. Both methods have the advantage of low pressure drop and the hot gas bypass has the lowest installation cost of all methods. For a three-bed Claus unit to attain high sulfur recovery, hot gas bypass reheat is not recommended, and in-line acid gas burner reheat is usually restricted to the first catalytic converter. An indirect method of reheat is preferred for the remaining converters. Gas-to-gas heat exchangers and fuel-gas fired indirect heaters are more expensive, and result in higher pressure drops than hot gas bypasses and in-line burners, but their use may be justified if very high conversion is required.

The catalyst beds may be arranged in horizontal or vertical vessels, with more than one bed located in one vessel. Catalyst beds are normally no more than 3–5 feet deep because of pressure drop restrictions. Design space velocities are generally in the range of 1,000 to 2,000 volumes of gas at operating conditions per volume of catalyst per hour. This range of values is intended to address lean to rich variations in the hydrogen sulfide content of acid gas feeds. However, space velocities not exceeding 1,000 volumes of gas at operating conditions per volume of catalyst per hour are preferred for extended run times between catalyst change-out. In very large installations, care has to be taken to assure uniform gas distribution and to avoid channeling.

It is customary to install mist-eliminating devices after the last sulfur condenser to minimize entrainment of sulfur droplets into the incinerator. Installation of mist eliminators after each sulfur condenser is also of value, as catalyst deactivation caused by entrained sulfur may be a problem. Wire mesh pads, located in the outlet channel of the condenser, are usually used for sulfur mist elimination.

When the Claus plant tail gas is discharged to the atmosphere without further purification, it is necessary to assure that all sulfur compounds in the gas are oxidized to sulfur dioxide. Incineration with auxiliary fuel is required as the Claus unit tail gas stream contains insufficient combustibles for self-sustained combustion. Incineration may be accomplished either thermally or catalytically, with thermal incineration being more common. In both cases, the tail gas is heated by fuel gas combustion. Thermal oxidation occurs at temperatures between 1,000° and 1,500°F in the presence of excess oxygen. Catalytic incinerators operate at about 600°–800°F using a controlled amount of air. Catalytic incineration requires significantly less fuel than thermal incineration, but a higher capital expenditure. The resulting gases are then discharged to the atmosphere through a stack. The incinerator may be separate from the stack, or alternatively combined into a single vessel with the stack mounted on the incinerator. Combustion air for a thermal incinerator can be supplied by either natural or forced draft, but a catalytic incinerator requires a forced draft fan due to its positive operating pressure and control requirements.

The mechanical design of small plants is often substantially different from that of large installations, as small units lend themselves to compact packaging. Designs for sulfur plants with capacities less than 50 long tons of sulfur per day have been presented by Grekel et al. (1965). **Figures 8-4** and **8-5** depict large and small Claus plants.

Process Control

The most important control variable in the operation of Claus plants is the ratio of hydrogen sulfide to sulfur dioxide in the reaction gases entering the catalytic converters. Maximum conversion requires that this ratio be maintained constant at the stoichiometric proportion of 2 moles of hydrogen sulfide to 1 mole of sulfur dioxide. Appreciable deviation from the stoichiometric ratio leads to a drastic reduction in conversion efficiency (Valdes, 1964B). The proper ratio is maintained by control of the air flow to the reaction furnace, which can be accomplished simply by automatic air to acid-gas ratio flow control. However, this method is only successful if the hydrogen sulfide content of the acid gas is constant, as it does not compensate for the variations in the actual amount of hydrogen sulfide flowing to the reaction furnace. Nor does this control method account for the presence and flow rate variations of other combustibles in the acid gas, such as hydrocarbons and ammonia. Several methods based on controlling the air flow by continuous analysis of the ratio of hydrogen sulfide to sulfur dioxide in the plant tail gas have been developed, but are not widely used. One such method has been described by Carmassi and Zwilling (1967), Grancher (1978), and Taggart (1980). Most plants use simple flow control for feed forward with reset and/or separate trim air control based on tail gas H_2S/SO_2 analysis. Several analytical instruments based on vapor chromatography and ultraviolet absorption are commercially available. Instruments of the latter type are capable of controlling the air flow rate to within ±0.5% of the optimum value, thereby minimizing the loss of sulfur recovery efficiency.

682 Gas Purification

Figure 8-4. Claus-type sulfur plant producing 875 long tons of sulfur per day. *Courtesy The Parsons Corp.*

Catalyst Deactivation

A serious operating problem in Claus plants is deactivation of the catalyst by deposition of carbonaceous materials and, in some cases, of sulfate. Acid gases usually contain small amounts of hydrocarbons, especially if the sour gas from which the acid gases have been removed is relatively rich in high molecular weight aliphatic and aromatic hydrocarbons, which are somewhat soluble in the absorbent liquids used in gas treating units. When acid gases of high hydrogen sulfide content are processed, the temperature in the reaction furnace is usually high enough to result in complete combustion of all hydrocarbons to carbon dioxide and water, and no carbonaceous material deposition is experienced.

However, at the low combustion temperatures occurring in straight-through plants processing gases containing less than about 40 to 50% hydrogen sulfide, cracking and partial combustion of hydrocarbons produce complex carbonaceous materials that are carried into the catalytic reactors, gradually deteriorating catalyst performance. In addition, hydrocarbons can be fed directly to the first catalytic converter without being burned when a split-flow design is utilized. These hydrocarbons can also cause catalyst deterioration. Catalyst activity can be partially restored by air oxidation of the carbonaceous deposits. However, care must be taken during regeneration not to exceed a temperature of about 1,000°F to avoid thermally induced changes in the catalyst structure. A good catalyst can be regenerated several times, although the activity decreases somewhat with each regeneration.

Figure 8-5. Claus-type sulfur plant producing 7 long tons of sulfur per day. *Courtesy The Parsons Corp.*

Equally as serious as deposition of carbonaceous materials on the catalyst is the gradual accumulation of sulfate, a process known as sulfation, which generally reduces catalyst activity and destroys the capability of the catalyst to hydrolyze carbonyl sulfide and carbon disulfide. This problem has been investigated extensively, and promoted alumina catalyst formulations have been developed that are quite resistant to deactivation by sulfation. For details see Dalla Lana (1973); Pearson (1973, 1978, 1981); Norman (1976); Grancher

(1978); and Dupin and Voirin (1982). Further catalyst research has led to the development of titania (titanium dioxide) based catalysts. These show greater stability to thermal aging and increased resistance to sulfation in comparison to promoted alumina. Titania catalysts also show higher activity with respect to carbonyl sulfide and carbon disulfide hydrolysis. High COS and CS_2 conversions are reported at temperatures on the order of 570°F (Janke, 1990). This permits operating the first catalytic converter at a lower temperature than is practiced with activated alumina, thus gaining an improvement in sulfur recovery.

Another Claus plant operating problem is condensation of sulfur on the catalyst resulting in rapid deactivation. This can be avoided by maintaining the temperature in the catalytic converters above the sulfur dew point of the gas mixture. Should sulfur condense on the catalyst, raising the gas temperature 50°F is usually sufficient to vaporize the condensed sulfur and reestablish catalyst activity (Norman, 1976).

Ammonia Destruction Techniques

Special techniques have to be used for processing gas streams containing appreciable amounts of ammonia such as effluents from refinery sour water strippers. The ammonia must be destroyed in the reaction furnace to avoid deposition of ammonium salts on the catalyst beds. Two methods are available to successfully accomplish this. The first method involves a split-flow reaction furnace design; the second requires a high-intensity reaction furnace burner. It is essential that the ammonia be almost completely destroyed because ammonia concentrations as low as 500 to 1,000 ppmv can cause plugging problems (Anon., 1973).

Split Flow Reaction Furnace. The split-flow reaction furnace design for ammonia destruction is depicted in **Figure 8-6.** In this process, all the combustion air and all the ammonia containing sour water acid gas from the sour water stripper(s) are mixed with a controlled portion of the total amine acid gas stream originating from the amine treating unit(s). As shown in **Figure 8-6,** the combustion of this ammonia rich stream is accomplished in a separate zone (zone 1) of the reaction furnace. Combustion occurs at a temperature of about 2,300° to 2,700°F, which is sufficient to ensure nearly complete combustion of the ammonia with minimum formation of nitrogen oxides and SO_3. Sufficient amine acid gas is diverted to this zone to maintain the required temperature. The remaining amine acid gas is then mixed with the products of combustion from the first zone in zone 2 of the reaction furnace. Typically about 70 to 80% of the amine acid gas bypasses the first combustion zone. For details see Chute (1977), Wiley (1980), Goar (1989), and Beavon (1976, 1977).

Figure 8-7 depicts the temperature in the first zone of a split-flow reaction furnace as a function of the amount of amine acid gas diverted to the second reaction zone. The peak temperature corresponds to stoichiometric combustion of the amine acid gas/sour water acid gas mixture in the first combustion zone. The region of **Figure 8-7** to the left of the peak temperature represents combustion in a reducing atmosphere, while operation to the right corresponds to an oxidizing region. The slope of the reaction furnace temperature versus amine acid gas bypass fraction is less steep than the slope of the curve in the oxidizing region, and temperature control of the split-flow reaction furnace is more stable in the reducing region. Most split-flow reaction furnace designs are designed for reducing atmosphere operation. Problems with the split-flow design include inadequate destruction of hydrocarbons and ammonia that might be contained in the bypassed portion of the amine acid gas and inadequate reaction furnace residence time for thermal Claus sulfur conversion.

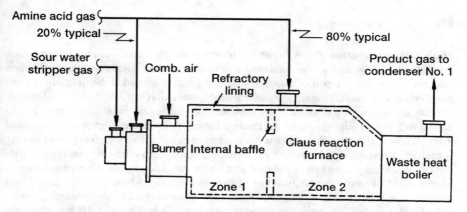

Figure 8-6. Split-flow reaction furnace design for sour water acid gas ammonia destruction. *(Wiley, 1980)*

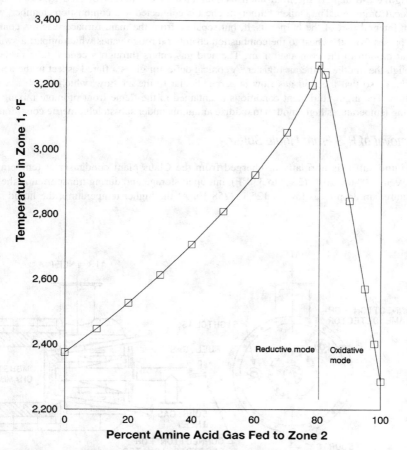

Figure 8-7. Relationship of split-flow reaction furnace temperature to split of amine acid gas to each reaction furnace section. *(Schendel, 1996)*

High-Intensity Burner for Ammonia Destruction. The second ammonia destruction method is based on feeding the combined ammonia-containing stream and the total amine acid gas stream into one port of a specially designed burner located at the front of the reaction furnace, in a similar manner to a conventional Claus unit (Goar, 1989). The total quantity of combustion air is also supplied to the burner. The amine acid gas stream is preheated to 180°–190°F before mixing to raise the reaction furnace temperature to ensure ammonia and hydrocarbon destruction. The burner is designed to ensure that a very high degree of mixing is achieved between the acid gas and air, increasing combustion efficiency and thereby raising the flame temperature. It is claimed that under these conditions the reaction rate of ammonia and oxygen is greater than that of hydrogen sulfide and oxygen. When combined with a reaction furnace of adequate residence time, essentially all of the ammonia is combusted with less than 300 ppmv remaining at the waste heat boiler outlet. Hydrocarbons are also effectively destroyed with this method. Burner and furnace designs utilizing this technology are offered by Comprimo/Goar Allison and Assoc. (Babcock Duiker, 1983; Lagas, 1984; Schalke et al., 1989).

Figure 8-8 depicts the high-intensity Duiker acid gas burner offered by Goar Allison Assoc./Comprimo. The Duiker burner air chest is connected to a combustion chamber, which is an integral part of the burner itself, but separate from the main furnace reaction chamber. Air passes from the chest to the combustion chamber through vanes which impart a swirling, spiral motion to the combustion air. The acid gas enters through a central tube projecting through the middle of the air register. A conical deflector gives a flared aspect to the acid gas discharge so that the acid gas flow is perpendicular to the air flow, which also has a spiral motion. The highly turbulent conditions maintained in the flame front promote mixing, producing temperatures high enough to oxidize ammonia under sub-stoichiometric conditions.

Removal of H_2S From Liquid Sulfur

Liquid sulfur is normally discharged from the Claus plant condenser at temperatures between 140° to 170°C (284° to 338°F), but upon storage and during transportation the temperature can drop to as low as 125°C (258°F). At the higher temperature, the liquid sulfur

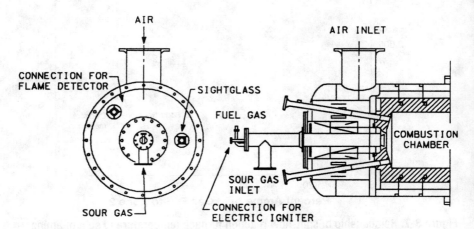

Figure 8-8. Duiker high-intensity burner for ammonia destruction.

product can contain several hundred ppm of H_2S. The H_2S is present as dissolved H_2S and as polysulfide, H_2S_x. When the liquid sulfur is cooled, the polysulfide becomes unstable and slowly dissociates into H_2S and sulfur, causing the emission of H_2S into the sulfur storage vessel vapor space. Accumulation of H_2S, especially in unvented sulfur storage vessels, can lead to a lethal H_2S concentration, and the H_2S concentration in the vapor space may easily exceed the H_2S lower explosive limit.

To prevent hazardous H_2S concentrations, the sulfur must be degassed and the liquid sulfur H_2S content reduced to 10 to 15 ppm (Lagas, 1982). Several commercial methods are described by Watson et al. (1981), Lagas (1982), and Goar (1984). Degassing of the liquid sulfur is typically effected either by vigorous agitation alone, by agitation with air stripping, or by agitation with the addition of a catalyst. Several catalysts, for example, ammonia, have been found to be effective in accelerating the degassing operation (Watson et al., 1981; Lagas, 1982). The Shell process utilizes a stripping column (a box) submerged in the sulfur pit. See **Figure 8-9**. Air is sparged into the box to roll the sulfur, thus accomplishing the nec-

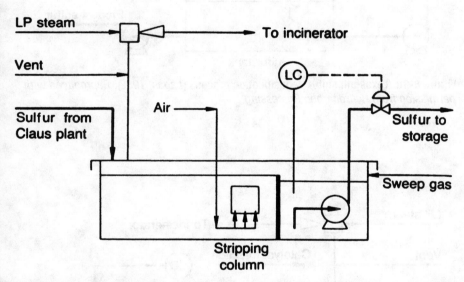

Figure 8-9. Shell sulfur stripper degasification process (*Lagas, 1982*). Reproduced with permission from *Hydrocarbon Processing*

essary agitation, and also acting as a stripping gas. Air is a more effective stripping agent than inert gas, probably because oxygen in the air reacts with H_2S to form elemental sulfur. The Texasgulf process employs a baffle plate column for agitation. The column is installed on the sulfur tank, and sulfur is recycled from the tank to the column. See **Figure 8-10**. The baffle plates agitate the sulfur, releasing the H_2S (Lagas, 1982). The SNEA[P] Aquisulf process, as described by Nougayrede and Voirin (1989), relies on catalyst addition combined with agitation, while the Exxon sulfur degassing process is based on catalyst addition directly to the sulfur pit (Goar, 1984; Watson et al., 1981; Schicho et al., 1985). **Figure 8-11** is a schematic depiction of the SNEA[P] Aquisulf process. Sulfur purity specifications may not be met by degassed sulfur that has been in contact with catalyst and this should be considered when evaluating sulfur degassing options.

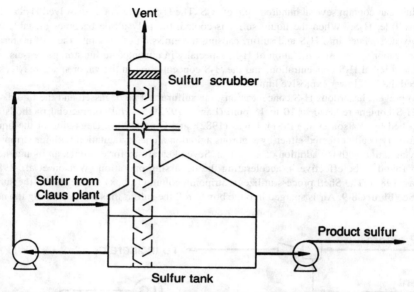

Figure 8-10. Texasgulf sulfur degasification process (*Lagas, 1982*). Reproduced with permission from Hydrocarbon Processing

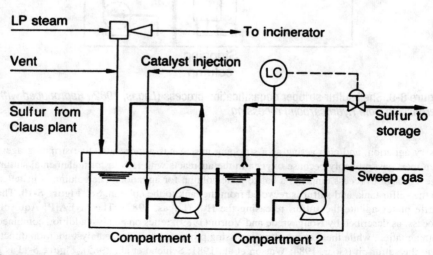

Figure 8-11. SNEA[P] continuous sulfur degasification process (*Lagas, 1982*). Reproduced with permission from Hydrocarbon Processing

CLAUS PROCESS MODIFICATIONS

Oxygen-Based Claus Processes

Attempts to increase acid gas flow through a Claus plant can be limited by the combustion air blower capacity, the practical depth of the condenser sulfur seal legs, or by increased back pressure in the amine stripper. In a normal Claus plant, air is the source of the oxygen required for the reaction. As shown by reaction 8-1, 50 moles of oxygen are required to react with 100 moles of H_2S to produce elemental sulfur. If the oxygen source is air, then 250 moles of air are required to react with 100 moles of H_2S, and the Claus plant tail gas (300 moles total) would contain about 100 moles of water and 200 moles of nitrogen. The tail gas nitrogen content can be reduced or eliminated by replacing some or all of the air with oxygen. For example, if pure oxygen were used instead of air, the tail gas would consist of just 100 moles of water, a volume reduction of 67%. Conversely, 300 moles of H_2S feed could be processed with pure oxygen at the same Claus plant overall pressure drop to give 300 moles of total tail gas, which consists entirely of water vapor.

Oxygen enrichment of Claus plant combustion air has been practiced since the early 1970s (Fischer, 1971). However, the objective of these early applications was to raise the reaction furnace flame temperature when processing acid gases with H_2S contents below 10 vol%. Gray and Svrcek (1981) first demonstrated that oxygen addition had economic advantages. The first commercial installations where oxygen was used to increase plant capacity were by Goar Allison and Assoc./Air Products at the Conoco Lake Charles refinery in 1985 (Goar et al., 1985) and by Lurgi in a European refinery (Fischer, 1985).

These early demonstrations and subsequent plants have proven very successful, and oxygen enrichment has become an established method for providing incremental sulfur plant capacity and for overcoming bottlenecks in existing plants. Oxygen enrichment technologies are now available for license from Goar Allison and Assoc./Air Products, Lurgi, TPA Inc., and BOC Gases. Oxygen enrichment is particularly attractive for debottlenecking existing plants as plant capacities can often be more than doubled for 10 to 15% of the cost of a new air-based Claus plant (Anon., 1995). Oxygen enrichment can also be used to provide standby redundancy. For example, two new parallel Claus trains can be designed so that the oxygen enriched capacity of one train is twice the capacity of an individual train using 100% air (Stevens et al., 1996). This reserve capacity ensures that loss of an individual Claus train will not impact upstream refining operations.

There is some disagreement concerning the intrinsic capability of oxygen enrichment to increase sulfur recovery. Goar et al. (1987) provide examples in which calculated sulfur recoveries for oxygen-based processes exceed those for air-based Claus units by 0.3% for a rich gas feed (3-converter plant), while calculated recoveries for an oxygen-based unit with a lean gas feed are greater by more than 1% (2- or 3-converter plant). Sames and Paskall (1987) have calculated that, on thermodynamic considerations alone, the maximum sulfur recovery increase attributable to oxygen enrichment is 0.1% for 3-converter plants. However, these calculations do not include the effect of oxygen enrichment on the performance of the tail-gas treating unit. In those cases where there is a downstream hydrogenation/selective amine tail gas unit, overall sulfur recovery can be increased because the flow of gas vented to the atmosphere is substantially lower for the oxygen case. Since the concentration of sulfur species in the vented gas will be the same for both the air and oxygen-based designs, the reduction in gas flow means that overall sulfur recovery is higher for the oxygen case.

Oxygen enrichment can be conveniently divided into three categories: low-level, medium-level, and high-level enrichment (Anon., 1995).

1. *Low-level oxygen enrichment* (<28 vol% oxygen)

   The upper safe limit for low-level oxygen enrichment is considered to be 28 vol% oxygen, as special metallurgy and cleaning techniques are required for equipment and piping containing oxygen at higher concentrations. Low-level oxygen enrichment is accomplished by injecting oxygen directly into the Claus furnace combustion air. This technology is not licensed and no Claus plant equipment modifications are required other than providing a tie-in point for oxygen injection. Claus plant capacity increases of about 20 to 25% can be accomplished (Anon., 1995).

2. *Medium-level oxygen enrichment* (28 to 45 vol% oxygen)

   Pure oxygen can be introduced directly into the reaction furnace separately from the air. This can be accomplished with a burner designed specifically for oxygen service (separate oxygen port) or with an independent oxygen lance. Introduction of oxygen in this manner avoids equipment and piping modifications, allowing overall oxygen concentrations greater than 28 vol%.

   The extent of oxygen enrichment is constrained by the maximum temperature limit of the refractory lining in the reaction furnace. Today, refractory is available which is rated at over 3,000°F; however, actual furnace designs rarely go over a calculated temperature of 2,850°F. **Figure 8-12** shows the calculated reaction furnace equilibrium temperature and percent Claus plant capacity increase as a function of the oxygen concentration for a refinery Claus plant feed that contains 92% H_2S in the amine acid gas plus a substantial fraction of sour water acid gas. As indicated in **Figure 8-12,** a typical refinery Claus plant can expect to realize an overall oxygen enrichment of about 45% before running into the reaction furnace refractory temperature limit. According to **Figure 8-12,** an oxygen concentra-

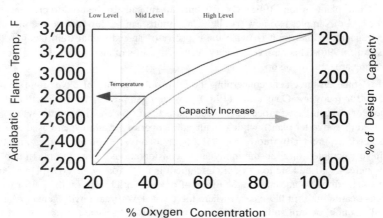

Figure 8-12. Effect of oxygen enrichment on the calculated equilibrium reaction furnace temperature and % Claus plant capacity increase. *Courtesy of BOC Gases*

tion of 45% corresponds to a 75% increase of the original Claus plant design capacity. The maximum allowable oxygen concentration is a function of the acid gas H_2S concentration. Acid gases with low HS concentrations can tolerate higher oxygen enrichment levels without refractory damage. Licensors offering technology for medium-level oxygen enrichment include BOC Gases/Parsons (SURE process), Comprimo/Goar Allison Assoc. (COPE process), Lurgi/Pritchard (OxyClaus process) and TPA Inc. (OxyMax process).

3. *High-level oxygen enrichment* (>than 45 vol% oxygen)

At oxygen concentrations above 45 vol%, it is necessary to provide some means of cooling to protect the reaction furnace refractory from high temperatures. The COPE process (Goar Allison and Assoc./Air Products) provides a recycle blower which quenches the reaction furnace with process gas from the cold side of the first sulfur condenser. See **Figure 8-13.** The BOC Gases SURE Double Combustion process divides the reaction furnace combustion process into two stages with intermediate cooling. See **Figure 8-14.** Both the COPE and the SURE processes can operate at oxygen concentrations up to 100% and, per **Figure 8-12,** can achieve Claus plant capacities over 250% of the original design value. Detailed descriptions of each of these processes are provided in the following discussion.

COPE Process

The COPE process is illustrated in **Figure 8-13.** This Claus process modification includes the addition of a specialized reaction furnace burner and a recycle blower, which are the keys to the effectiveness of the process. Enriching the combustion air with oxygen increases the reaction furnace temperature. For a rich gas feed, the typical maximum temperature of 2,700° to 2,900°F is reached when the oxygen content of the air reaches approximately 45 vol%. To avoid exceeding this temperature limitation, the COPE process recycles cool gas from downstream of the first condenser back to the reaction furnace. As the level of oxygen

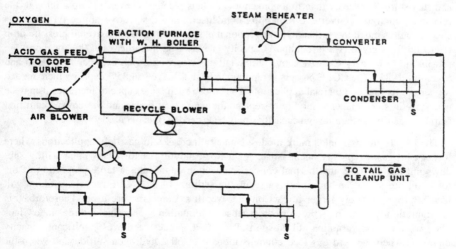

Figure 8-13. COPE sulfur recovery process. (*Goar, 1989*)

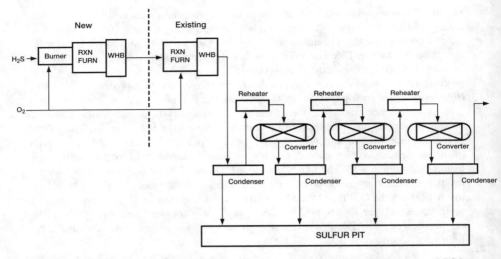

Figure 8-14. BOC Gases/Parsons SURE Double combustion process. *Courtesy of BOC Gases*

enrichment is increased, the amount of recycle is increased accordingly so that the refractory temperature limitation is not exceeded. Enrichment to 100% oxygen is possible with the COPE process. The proprietary Duiker burner, specifically designed for the process, is capable of handling not only the acid gas and air streams, but also the hot recycle gas and oxygen (Hull and Sorensen, 1985; Schalke et al., 1989).

Although the addition of oxygen lowers the overall Claus plant pressure drop, the recycle gas can cause a higher pressure drop through the burner, reaction furnace, waste heat boiler, and first sulfur condenser than in a conventional Claus plant. However, the flow through the reheaters, catalytic converters, remaining condensers, and a selective amine-type tail gas cleanup unit is much lower than in a conventional plant, resulting in a net overall pressure drop reduction for a given H_2S throughput. Capacity increases on the order of 100% (or more) are possible for a refinery Claus/selective amine tail gas unit with a rich acid gas feed (Hull and Sorensen, 1985). The COPE process is reported to have achieved an 85% capacity increase at a refinery in its first commercial application. To achieve such large capacity increases it may be necessary in some plants to modify or replace the waste heat boiler and the downstream first sulfur condenser, as their design capacities may be exceeded (Goar et al., 1987).

The COPE process yields more modest capacity increases in gas field applications where leaner acid gases typically occur. Also, for a two-catalytic reactor Claus plant with a subdewpoint tail gas unit, the thermal conversion section contributes a large proportion of the total unit pressure drop and, for this reason, the addition of the COPE recycle gas loop will have a proportionately larger impact on the overall system pressure drop. Therefore, this configuration will reach its pressure drop capacity limitation sooner than a three-stage Claus plant with a selective amine tail gas unit. The COPE process achieves significant capacity increases when the acid gas feed contains more than 60% hydrogen sulfide and when the thermal conversion section contributes a small fraction of the total pressure drop.

When additional sulfur recovery capacity is required, modifying an existing sulfur recovery unit with a COPE addition has cost advantages compared with the provision of a new sulfur recovery train having the same incremental capacity. Two examples of a Claus plant with a sub-dewpoint tail gas unit are provided by Goar et al. (1987); one having 92.4 mol% and the other 14 mol% hydrogen sulfide in the feed gas. Both examples show substantial savings in favor of the COPE modified units in regards to the incremental cost required to produce additional sulfur. This is principally attributable to the very low capital investment required to convert a conventional Claus plant to the COPE process. The COPE process power and fuel gas costs are also lower than for the combined original and new conventional Claus units. The largest operating cost component for the COPE modified units is the cost of oxygen.

A grassroots facility incorporating the COPE process will have smaller sized equipment downstream of the first condenser due to the reduced process gas flow rates. However, the comparative economics for COPE and conventional air-based Claus plants are much closer for new facilities than for revamps. Goar et al. (1987) compare a grassroots COPE design with a grassroots Claus plant. Both plants have a selective amine tail gas unit. The authors project a cost of sulfur production savings of approximately 9% in favor of the COPE-based unit. A similar study by Hull and Sorensen (1985) indicates identical sulfur production costs. Both examples are based on a rich feed gas. The application of COPE technology to grassroots facilities that process very lean gas streams is generally considered to be uneconomical since the high fraction of inerts, primarily carbon dioxide, in the acid gas does not allow significant downsizing of process equipment.

In 1995, it was reported that 11 COPE units were in operation with an additional six units under license and in various stages of construction (Anon., 1995).

Lurgi OxyClaus Process

The Lurgi OxyClaus process uses a specially designed reaction furnace burner whose operation approaches equilibrium reaction furnace temperatures. The OxyClaus main burner, which is depicted in **Figure 8-15,** consists of a number of individual acid gas burners surrounded by a central start-up burner muffle. Each of the individual acid gas burners consists of three concentric lances: an inner oxygen lance, a middle acid gas lance, and an outer process air lance. This configuration results in a very hot oxygen flame with a core temperature of over 2000°C which is enveloped by a cooler air/acid gas flame. In the very hot oxygen flame, H_2S, CO_2, and H_2O decompose to form H_2, CO, and O_2 according to the following endothermic reactions:

$$H_2S = H_2 + \tfrac{1}{2}S_2 \tag{8-5}$$

$$CO_2 = CO + \tfrac{1}{2}O_2 \tag{8-6}$$

$$H_2O = H_2 + \tfrac{1}{2}O_2 \tag{8-7}$$

These strongly endothermic reactions, which are included in free energy minimization calculations, provide the temperature moderation needed to protect the reaction furnace refractory at oxygen concentrations approaching 45 vol% (Stern et al., 1994; Stevens et al., 1996; Lurgi, 1994). See **Figure 8-12.** The OxyClaus burner can utilize higher oxygen concentrations if the acid gas is leaner.

694 Gas Purification

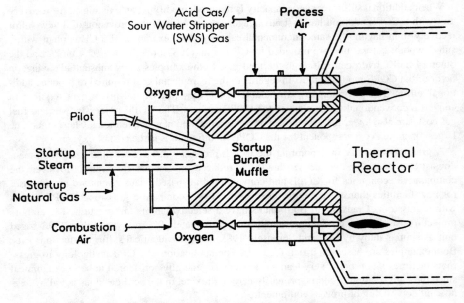

Figure 8-15. Lurgi OxyClaus reaction furnace burner. (*Stevens et al., 1996*)

As of 1995, ten plants using the OxyClaus technology were in operation in either Europe or the United States, and twelve additional units were in design/construction (Anon., 1995).

BOC Gases SURE Process

BOC Gases and the Ralph M. Parsons Co. offer the SURE oxygen-enrichment process. The SURE burner, like the Lurgi OxyClaus burner, is a tip mixed system where the acid gas and oxygen/air are mixed together after leaving the burner. As in the Lurgi OxyClaus burner, the interface between the oxygen stream and the acid gas forms a hot gas envelope within the flame and areas within this region can be 1,800°F (1,000°C) hotter than the average reaction furnace flame temperature. These hot temperatures are very beneficial in destroying contaminants, such as ammonia and heavy hydrocarbons, and lead to endothermic, temperature-moderating reactions (reactions 8-5 through 8-7) which produce hydrogen and carbon monoxide and moderate the reaction furnace temperature, allowing oxygen concentrations up to about 45 vol% without damaging the refractory. BOC has developed a standard range of SURE burners with capacities up to 600 LT/day of sulfur (Anon., 1995).

Figure 8-16 depicts a SURE reaction furnace burner being installed in a North American refinery Claus plant. The SURE burner, like the Lurgi OxyClaus burner, can operate at oxygen concentration levels of 45 vol%, achieving Claus plant capacity increases of about 80%. See **Figure 8-12.** Further details regarding the SURE process are provided by Chen et al. (1995), Hull et al. (1995), and Watson et al. (1995, 1996).

For very high oxygen concentrations, BOC Gases also offers the SURE Double Combustion process. To limit the temperature rise and protect the reaction furnace refractory, the

Figure 8-16. SURE oxygen burner being retrofitted onto a Claus plant reaction furnace in a North American refinery. *Courtesy of BOC Gases*

combustion reactions are carried out in two stages with intermediate cooling as shown in **Figure 8-14.** Acid gases are first subjected to a partial combustion at temperatures well below the safe operating temperature of the refractory, but at temperatures high enough to ensure complete ammonia and hydrocarbon destruction. This first stage of combustion is carried out without attempting to meet the overall stoichiometry requirements or total oxygen demand. The gases are then cooled in a waste heat boiler prior to entering a second reaction furnace where the remainder of the oxygen is introduced.

In the SURE Double Combustion process, there is no sulfur condenser between the first and second waste heat boilers. Also, there is no burner in the second reaction furnace. By design, the gases exiting the first waste heat boiler and entering the second reaction furnace are substantially above the auto ignition temperature. Since the gases entering the second reaction furnace are above the auto ignition temperature, even small quantities of oxygen will react completely, and there is no minimum oxygen flow required to maintain a stable flame. There is also minimal pressure drop through the system because there is just one burner.

The SURE Double Combustion process is particularly attractive in a revamp situation. The existing reaction furnace and waste heat boiler become the No. 2 units; and a new burner, No. 1 reaction furnace, and No. 1 waste heat boiler are added upstream of the existing reaction furnace. It is reported that 100% oxygen enrichment can be achieved with this process. According to **Figure 8-12,** the SURE Double Combustion process can achieve over two and one-half times the original Claus plant design capacity.

As of the end of 1996, there were a total of 12 licensed SURE units with 8 in actual operation. Of the twelve licenses, four were Double Combustion units (Schendel, 1997).

TPA Inc. Oxygen Enrichment Processes

Low-level oxygen enrichment processes (<28 vol% oxygen) offered by TPA Inc. provide Claus plant capacity increases of about 20%. For oxygen enrichment up to about 60 vol%, TPA offers its Oxygen Injection System, which is claimed to achieve capacity increases of 30 to 100% (Anon., 1995). The OxyMax technology for oxygen enrichment levels above 60 vol% is under development and was not commercialized as of 1995 (Anon., 1995). As of 1995, TPA had installed 14 oxygen enrichment systems. The capacity increase for these systems was reported to range from 28 to 100% (Anon., 1995).

Isothermal Reactor Concepts

Linde Clinsulf Process

Linde A.G. of Munich, Germany, has developed an isothermal Claus reactor design in which the heat of reaction is removed directly from the Claus reactor. This concept achieves sulfur recoveries comparable to the conventional adiabatic reactor design, but with fewer reaction stages and less total equipment.

The Linde Clinsulf process, as described by Heisel and Marold (1989), Linde (1988; 1990), and utilizes a vertical reactor with cooling coils installed directly in the reactor bed. Heat is removed by generating steam inside the cooling coils, permitting isothermal operation. Linde has developed a number of different Clinsulf process configurations, each adapted to specific feed compositions and sulfur recovery levels. When one isothermal reactor is included in the process scheme, a sulfur recovery of approximately 94% is claimed, making the process equivalent to a conventional two stage Claus plant. For acid gas feeds containing 20% or more hydrogen sulfide, the Linde Clinsulf process incorporates a conventional reaction furnace, a waste heat boiler, and a first condenser upstream of the isothermal Clinsulf reactor, which contains conventional Claus catalyst. A condenser downstream of the isothermal reactor recovers sulfur from the reactor effluent before the gases are incinerated and discharged to atmosphere. For an acid gas feed that contains 10% or less hydrogen sulfide, sulfur recovery is accomplished by means of a direct oxidation catalyst installed in the isothermal reactor. The feed is preheated in an upstream feed heater, and the sulfur is recovered in a downstream condenser while the tail gases are incinerated.

Increased sulfur recovery can be obtained with a modification of the Clinsulf process that uses two parallel Clinsulf reactors downstream of the second condenser. The first of the two parallel reactors is operated at a temperature below the sulfur solidification point by cooling the reactor internally to less than 100°C at the outlet, while the second reactor is regenerating. The two parallel reactors alternate between reaction/adsorption and regeneration modes of operation. Linde has investigated sub-solid-point operation at its test facility in Höllriegelskreuth and claims that sulfur recoveries up to 99.8% can be achieved (Linde, 1988). In 1994, it was reported that two Clinsulf plants were in operation (Linde A.G., 1994).

BASF Catasulf Process

BASF of Ludwigshafen, Germany, has developed the Catasulf process (Anon., 1992) which utilizes an isothermal tubular reactor. The feed gas containing 5–15 mol% of hydrogen sulfide and a stoichiometric amount of air or oxygen is preheated and then passed through the reactor tubes which contain a direct oxidation catalyst. The heat of reaction evaporates a cooling agent on the shellside of the reactor, which in turn generates medium pressure steam on condensing in a separate heat exchanger. Sulfur is recovered in a downstream condenser. Up to 94% of all sulfur compounds are converted to sulfur. Sulfur conversion is increased to 97.5% when a downstream adiabatic reactor and sulfur condenser are installed. The tail gas is then routed to incineration or a tail gas treatment unit. In 1992, it was reported that one 48 LT/D Catasulf plant was in operation (BASF, 1992).

Developmental Sulfur Recovery Processes

MTE Sulfur Recovery Process

The MTE Sulfur Recovery process, which uses a flowing bed of catalyst particles instead of conventional fixed catalyst beds, has been disclosed by Simek (1991). A test unit has been built and operated. The MTE Sulfur Recovery process feed is the reaction gas from the first sulfur condenser following the reaction furnace and waste heat boiler. The reaction gas is reheated to 350° to 430°F and contacted with sulfur-rich catalyst in the regenerator where additional sulfur is produced while essentially all of the sulfur on the catalyst is vaporized. The sulfur laden gas from the regenerator is partially cooled to condense the bulk of the sulfur, and then further cooled to condense the reaction water and remaining sulfur. A blower circulates the cooled gases into a pipe reactor where the remaining hydrogen sulfide and sulfur dioxide react at a temperature below the sulfur dew point in the presence of regenerated catalyst from the bottom of the regenerator. The now sulfur-rich catalyst is separated from the tail gas in the reactor vessel and flows back to the regenerator. Tail gas is routed out of the unit. Cyclones and electrostatic precipitators are installed in the heads of the regenerator and reactor to prevent catalyst dust from being carried to downstream equipment. Attributes claimed for the process include sulfur recovery in excess of 99%, a small catalyst inventory, and relatively low catalyst consumption.

Richards Sulfur Recovery Process

A modification of the Claus process operating at elevated pressure has been disclosed by Kerr et al. (1982). This process, named the Richards Sulfur Recovery Process (RSRP), was developed jointly by Alberta Energy Company Ltd. and Hudsons Bay Oil and Gas Company Ltd. (now Amoco Canada Petroleum Co. Ltd., of Calgary, Canada). The process was tested in the laboratory and a conceptual design of a commercial plant was proposed. Operation is conducted at pressures ranging from 70 to 300 psig with the catalyst immersed in circulating liquid sulfur which acts as a cooling medium. An interesting feature of the process is that the combustion furnace of a Claus plant is replaced with a catalytic oxidizer, where a portion of the H_2S in the feed gas is partially oxidized with cool liquid sulfur being sprayed over the catalyst bed to maintain the temperature at 700° to 800°F. Conversion of H_2S to sulfur of more than 99% is claimed. There has been no subsequent development of the process, and commercialization is believed to be unlikely.

Miscellaneous Processes

Another modification of the Claus process, operating at much lower temperatures and reportedly usable for H_2S removal from hydrocarbon gases, was developed through the pilot-plant stage in the United States by the Jefferson Lake Sulfur Company (Anon., 1951). In the process, the SO_2 necessary for the reaction is obtained by burning sulfur in an external burner. The gas to be treated is preheated to moderately high temperatures and then contacted with the SO_2 in several chambers containing a special catalyst. The elemental sulfur formed is removed from the gas by condensation. Periodic regeneration is required to maintain high catalyst efficiency. It is claimed that hydrocarbons pass unaffected through the catalyst beds. This process can also be used for acid-gas streams of such low hydrogen sulfide content that combustion could not be sustained even with only one-third of the gas stream.

A process also based on the reaction of H_2S with SO_2 is described by Audas (1951). In this process, which was developed through the pilot-plant stage and is covered by British Patent 653,317, SO_2 is added to the gas in slight excess over the amount theoretically required to react with the H_2S. The gas is then passed through a bed of alumina chips or granules from which it emerges free from sulfur compounds. Depending on the H_2S content of the gas, the operating temperature ranges from 85° to 190°F. Elemental sulfur, water, and excess SO_2 are retained on the alumina. A portion of the alumina is continuously withdrawn from the bottom of the bed, and regenerated alumina is added to the top of the purification vessel. The spent alumina is regenerated by recycling combustion gases at a temperature of 850°F. Sulfur and SO_2 are recovered from these gases. By adjustment of the oxygen content of the circulating gas, the sulfur can be completely converted to SO_2. The advantages claimed for this process are (a) drastic reductions in plant volume and ground area, if compared to an equivalent iron oxide dry-box installation; (b) recovery of relatively pure sulfur and SO_2; and (c) production of extremely dry gas.

Direct catalytic oxidation of hydrogen sulfide to elemental sulfur in the presence of gaseous paraffinic hydrocarbons or other gases has been reported by Grekel (1959). Although pilot-plant work indicated good conversions and no effect on the hydrocarbons, the process has so far not been successfully commercialized.

A process using synthetic zeolites for recovery of sulfur from sour gases under pressure has been described by Haines et al. (1961). In this process, which has been tested in a pilot plant, but which has not been commercialized, hydrogen sulfide is adsorbed on the zeolite which is then regenerated with sulfur dioxide containing gas at high temperatures. The elemental sulfur formed is condensed, and residual hydrogen sulfide and sulfur dioxide are vented to atmosphere.

CLAUS PLANT TAIL GAS TREATMENT PROCESSES

When the Claus sulfur conversion process was first introduced, it was considered to be a means of air pollution control capable of recovering up to about 97% of the sulfur from acid gas streams that would otherwise be burned and vented to the atmosphere. However, as Claus plants became more common, and air pollution control regulations became more stringent, the unrecovered sulfur compounds in Claus plant tail gas streams became the target of further regulation. As a result, many techniques were proposed and/or developed to increase the overall sulfur removal efficiency of sulfur recovery systems and thereby reduce the amount of sulfur escaping into the atmosphere.

Most of the Claus plant tail gas treating processes that have achieved commercial status can be categorized into three basic types: (1) sub-dewpoint Claus processes in which a higher conversion efficiency is obtained for the basic Claus reaction by operating the final catalyst bed(s) of the system at a very low temperature, i.e., below the dew point of sulfur in the gas stream; (2) direct oxidation processes in which the Claus process section of the plant is operated to produce a tail gas containing H_2S, but little or no SO_2, and the H_2S is then oxidized directly to sulfur by use of special highly selective catalyst that promotes only the irreversible oxidation reaction (equation 8-1); and (3) sulfur dioxide reduction processes in which SO_2 and other sulfur compounds in the tail gas are first converted to H_2S, and then removed from the gas stream by any of the well established H_2S removal techniques. Processes representative of these three major categories are described in the following sections.

Sub-Dewpoint Claus Processes

As shown in **Figure 8-1,** the equilibrium conversion of H_2S to sulfur increases with decreasing temperature in the moderate temperature region, and continues to decrease as the temperature is reduced below the sulfur dew point, approaching 100% at a temperature of about 250°F. Conventional Claus process catalyst beds are maintained at a temperature well above the sulfur dew point to avoid the deposition of liquid sulfur on the catalyst, but this precludes the attainment of the high efficiencies possible at lower temperatures.

Sub-dewpoint Claus processes utilize the same chemical reactions as the conventional Claus systems, but operate the catalyst beds at a temperature below the sulfur dew point to take advantage of the high equilibrium conversion possible. The problem of sulfur deposition in the catalyst bed is resolved by using several beds in a cyclic operation, which includes periodic regeneration of each bed by vaporizing deposited sulfur into a hot purge gas. The sub-dewpoint sulfur conversion technique can be used as an add-on to an existing Claus unit or designed into a high efficiency integrated system. Commercial sub-dewpoint processes include the Lurgi/SNEA[P] Sulfreen process, the AMOCO CBA process, and the MCRC sulfur recovery process offered by Delta Projects, Inc.

Sulfreen Process

Introduction. The Sulfreen process was developed jointly by Lurgi Gesellschaft für Wärme und Chemotechnik of Germany and Société Nationale des Petroles d'Aquitaine (now Société National Elf Aquitaine [Production]) of France for the specific purpose of reducing residual sulfur compounds in the tail gases from Claus type sulfur recovery plants. The process is a sub-dewpoint Claus process, as described previously. The reaction between hydrogen sulfide and sulfur dioxide is carried out at lower temperatures than normally used in the Claus process, permitting more complete conversion to elemental sulfur. The sulfur formed by the Claus reaction is adsorbed on the solid catalyst, lowering the sulfur partial pressure in the gas phase and shifting the reaction towards further sulfur formation. The catalyst activity decreases as it becomes progressively laden with sulfur. To maintain catalyst activity at an acceptable level, the sulfur load on the catalyst must be limited. The catalyst is therefore regenerated periodically by thermal desorption of the sulfur.

The catalyst originally used in the Sulfreen process was a specially prepared activated carbon. This catalyst, although highly efficient, requires high temperatures to vaporize the adsorbed sulfur during regeneration. An alumina catalyst similar to that used in Claus units was subsequently developed and all Sulfreen plants since 1972 have been designed to operate with alumina catalyst. Results from an extensive investigation of alumina catalyst performance under conditions representative of the Sulfreen process, i.e., at temperatures below the sulfur dew point, have been reported by Pearson (1976). The study showed that the catalyst retains its activity even when loaded up to 50% by weight with adsorbed sulfur. Use of alumina permits sulfur removal at a relatively low temperature, about 570°F, which not only reduces fuel consumption relative to activated carbon, but also makes it possible to construct the entire plant of carbon steel. In older plants utilizing activated carbon catalyst, stainless steel converters were required to withstand corrosion at the high regeneration temperatures.

Although the Sulfreen process is cyclic, alternating between sulfur adsorption and desorption, the use of several converters in adsorbing, desorbing, and cooling service permits continuous operation.

Process Description. A description of the original form of the Sulfreen process which uses the activated carbon catalyst was presented by Guyot and Martin (1971). The details of the process using alumina catalyst have been described by Cameron (1974), Grancher and Nougayrede (1976), Nougayrede (1976), and Lell and Nougayrede (1991). A typical flow scheme is shown in **Figure 8-17.** A Sulfreen unit includes a group of converters, generally three in larger plants and two in smaller ones, charged with a conventional alumina catalyst. The Claus plant tail gas, containing typically about 1.0 vol% hydrogen sulfide, 0.5 vol% sulfur dioxide, and some sulfur vapor, passes through two converters operating in parallel in adsorption at 260° to 300°F, and about 75 to 85% of the hydrogen sulfide and sulfur dioxide are converted to elemental sulfur. The treated gas from the converters is routed to the incinerator and then discharged to the atmosphere via a stack.

The third converter, which is saturated with sulfur, is offline and undergoing regeneration. This is carried out using a closed loop system and begins by heating the catalyst bed with regeneration gas that has been heated to about 600°F in an indirectly fired heater. The sulfur adsorbed by the catalyst is vaporized into the gas stream and then recovered as a liquid upon cooling in the downstream condenser, where low pressure steam is usually generated. The regeneration gas is then recirculated by a blower through the heater back to the converter. The generation of sulfur vapor peaks at a catalyst bed temperature of about 500°F and then gradually declines to zero at the end of the heating cycle when the bed has reached about 570°F. During the early stages of regeneration, water that has adsorbed onto the catalyst along with the sulfur also desorbs and must be purged from the loop. Constant pressure is maintained in the regeneration loop by means of an equilibration line between the tail gas line and the blower suction.

When regeneration is complete a small quantity of Claus plant acid gas feed is introduced into the loop. The hydrogen sulfide in this gas reduces sulfate that has formed on the alumina catalyst. This sulfate forms on the catalyst during adsorption, when oxygen present in the feed gas reacts with SO_2 to form sulfur trioxide according to reaction 8-8.

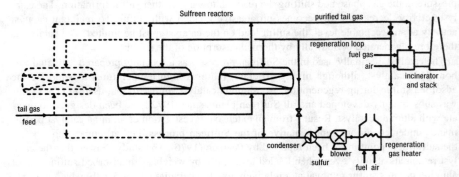

Figure 8-17. Two (or three) reactor Sulfreen unit. (*Lell and Nougayrede, 1991*)

$$SO_2 + \tfrac{1}{2}O_2 = SO_3 \tag{8-8}$$

The sulfur trioxide is adsorbed by the alumina as sulfate, SO_4^{2-}. The sulfate reduction technique, which can be represented by equation 8-9, has proven to be very effective in preserving high catalyst activity over long periods of time.

$$H_2S + SO_3 = S + H_2O + SO_2 \tag{8-9}$$

The regenerated bed is then cooled by a slipstream of treated gas from the adsorbing beds. After leaving the bed, the cooling gas flows through the condenser, blower, and regeneration heater before being routed to the incinerator. An alternative cooling method circulates regeneration gas through the reactor (and condenser) but not through the regeneration heater. When cooling is complete, the regenerated converter is switched to adsorption and the saturated converter in adsorption is switched to regeneration. The various steps of the process cycle are automatic, according to a fixed program sequence.

Performance of alumina catalyst in Sulfreen units is discussed in a study presented by Grancher (1978). The problem of catalyst sulfation and the use of hydrogen sulfide as a sulfate reducing agent are described. **Figure 8-18** shows the effect of H_2S on the efficiency of regenerated catalyst. Another approach reported by Grancher (1978) is the use of a special proprietary catalyst named AM catalyst, which reportedly removes oxygen and is also an effective catalyst for the Claus reaction. The AM catalyst consists of an undisclosed metallic salt supported on activated alumina.

In 1994, it was reported that 40 Sulfreen units were in operation treating tail gas from Claus plants having capacities ranging from 50 to 2,200 long tons of sulfur per day, and five more plants were under construction (Société National Elf Aquitaine (Production) and Lurgi

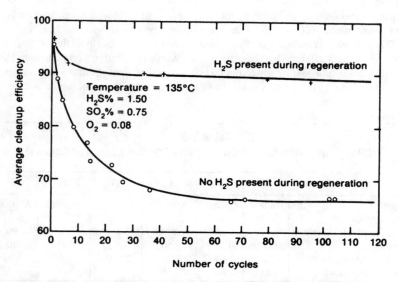

Figure 8-18. Effect of H_2S in regeneration gas on cleanup efficiency of alumina catalyst. (*Grancher, 1978*). *Courtesy of Hydrocarbon Processing*

GmbH, 1994). Operating data for several of these plants have been reported by Guyot and Martin (1971), Cameron (1974), Grancher and Nougayrede (1976), and Nougayrede (1976). It is claimed that with a properly designed and operated plant, better than 99% overall sulfur recovery is attainable with the combination of the Claus-Sulfreen processes. As with a Claus unit, optimum conversion in a Sulfreen unit depends on maintaining the stoichiometric ratio of two moles of hydrogen sulfide to one mole of sulfur dioxide in the Claus tail gas, which requires close control of the Claus plant operation.

Utility consumption data for six Sulfreen units, ranging in capacity from 0.32 to 7.3 MMscf/hour of tail gas from Claus units producing from 90 to 2,100 long tons of sulfur per day, are provided in **Table 8-3** (Nougayrede, 1976). A photograph of a Sulfreen unit at Lacq, France, is shown in **Figure 8-19**.

To meet progressively stricter air pollution regulations, the process licensors have made improvements to the basic Sulfreen process resulting in several new versions. These are described by Lell and Nougayrede (1991). The Two-Stage Sulfreen process uses two adsorption stages in series with intermediate cooling. Sulfur recovery is improved by removing the heat of reaction generated in the first sub-dewpoint converter before the gas enters the second. Overall sulfur recoveries in excess of 99.5% are claimed.

The sulfur recovery efficiency of the basic Sulfreen process is partly limited by the carbonyl sulfide and carbon disulfide content in the Claus plant tail gas. These compounds typically account for 30 to 50% of the overall sulfur losses from a conventional Sulfreen unit, as they do not hydrolyze to any significant degree at the process operating temperature. In the Hydrosulfreen process, the tail gas is pretreated to hydrolyze carbonyl sulfide and carbon disulfide to less than 50 ppm as sulfur, while simultaneously excess hydrogen sulfide is directly oxidized to sulfur. These reactions are performed by contacting the tail gas with an activated titanium dioxide catalyst (CRS 31) at about 570°F. The oxidizing air admitted to the hydrolysis/oxidation reactor is controlled to maintain a ratio of two moles of hydrogen sulfide to one mole of sulfur dioxide at the downstream side, prior to entering the Sulfreen converters. An overall sulfur recovery of 99.4 to 99.7% is claimed for this process. The process licensor reports that the investment for this form of the Sulfreen process is 40 to 55% of the associated, upstream Claus unit cost, as compared to 30 to 45% for the basic Sulfreen process (Anon., 1992).

The Oxysulfreen process involves three pretreatment steps upstream of the Sulfreen reactors. The first step is catalytic hydrogenation and hydrolysis of sulfur compounds to hydrogen sulfide. This is followed by condensation and removal of water from the process gas in a quench tower. Finally, the process gas is heated and passed with air through a reactor con-

Table 8-3
Utility Requirements for Sulfreen Units Treating Claus Tail Gases

Plant	A	B	C	D	E	F
Catalyst	Carbon	Carbon	Alumina	Alumina	Alumina	Alumina
Tail Gas, MMscf/hr	4.6	7.3	7.3	7.0	6.7	0.32
Claus Capacity, LTD Sulfur	1,200	2,000	2,100	1,800	1,600	90
Electricity, h.p.	1,000	1,700	1,500	1,500	1,450	40
Steam Production, lbs/hr	12,000	24,000	20,000	23,000	23,000	850
Fuel Gas, Mscf/hr	25	40	27	33	33	1.25

Source: Nougayrede (1976)

Figure 8-19. Sulfreen plant treating tail gases from a 1,000 long tons per day, Claus-type sulfur plant at Lacq (France). *Lurgi Gesellschaft für Wärme and Chemotechnik and Société Nationale des Petroles d'Aquitaine*

taining CRS 31 catalyst. Approximately 50 to 70% of the hydrogen sulfide in the gas is oxidized directly to sulfur. As in the Hydrosulfreen process, the flow of oxidizing air to the Oxysulfreen reactor is controlled to maintain a ratio of two moles of hydrogen sulfide to one mole of sulfur dioxide in the outlet upstream of the Sulfreen converters. Depending on the initial hydrogen sulfide content of the acid gas and on the quench tower overhead temperature, overall sulfur recoveries of 99.7 to 99.9% are claimed. The process licensor reports that the capital cost for this process is about 90 to 110% of the Claus unit (Anon., 1988).

AMOCO Cold Bed Adsorption (CBA) Process

Introduction. This process, which was developed by AMOCO Canada Petroleum Company, Ltd. in the early 1970s, has been described by Goddin et al. (1974), Nobles et al. (1977), Reed (1983), and Lim et al. (1986). The Cold Bed Adsorption (CBA) process is a "sub-dewpoint" sulfur recovery process and is quite similar in principle to the Sulfreen process, as the Claus reaction is also carried out at sufficiently low temperatures to cause condensation of sulfur on the catalyst. However, unlike the Sulfreen process, where a closed loop of inert gas is used to desorb the adsorbed sulfur, a hot gas slipstream from the Claus unit is used for this purpose.

Process Description. Amoco has numerous configurations for the CBA process (Berman, 1992), involving from two to four total catalytic converters. The total number of converters utilized and their split between conventional Claus operation and sub-dewpoint operation is dependent upon the sulfur recovery desired and the acid gas hydrogen sulfide concentration. Two common configurations are four converters with the third and fourth converters cycling through the adsorption, regeneration, and cooling steps; and three converters with the second and third converters cycling. It is claimed that the first configuration is capable of achieving 99.0 to 99.2% sulfur recovery, and the latter is capable of achieving 98.5 to 99.0%, when recovery is averaged over the entire cycle. Other CBA process configurations may achieve even higher recovery. The process may be combined with the Claus unit, or added downstream as a tail gas treater.

The flow of gases through a four-converter CBA process unit during the regeneration and cooling periods is shown in **Figures 8-20** and **8-21.** The process up to the first Claus converter is identical to a conventional Claus plant. During the regeneration period, shown in **Figure 8-20,** the first converter effluent at 600° to 650°F is fed to CBA converter No. 1. The hot gas vaporizes the sulfur that has condensed on the catalyst, thereby reactivating it. Heating of the catalyst bed is accomplished not only by the sensible heat of the feed gas, but also by the heat of the Claus reaction. The hot, sulfur laden gas is cooled in condenser No. 2 to recover liquid sulfur, and then reheated prior to entering conventional Claus converter No. 2. The sulfur formed in this converter is recovered by cooling the gas to about 260°F in condenser No. 3. The gas is sent at this temperature to CBA converter No. 2, which is in the adsorption mode. The effluent from this converter is sufficiently low in sulfur content so that it can be sent directly to the incinerator.

When regeneration of CBA Converter No. 1 is completed, the valves are switched to start cooling the hot converter as shown in **Figure 8-21.** During the cooling period, tail gas from Condenser No. 3 is diverted to CBA Converter No. 1 and then to CBA Converter No. 2 via condenser No. 4. The two CBA converters are operating in series during this phase, which benefits sulfur recovery. When cooling is completed, the effluent from CBA Converter No. 1 is sent to incineration and regeneration of CBA Converter No. 2 is started.

The sequence of operations for the three-converter configuration, with the second and third converters cycling, is similar to the four-converter sequence. However, there is no second Claus converter, with upstream reheater and downstream condenser, and the final condenser is operated continuously upstream of the adsorbing converter.

The catalyst used in the CBA converters is alumina, the same as that used in the Claus converters. A discussion of catalyst performance in low temperature Claus processes has been presented by Pearson (1976).

The pressure drop through a CBA unit depends to a large extent on the system design. As an add-on to an existing two reactor Claus unit, the addition of two CBA converters, a condenser, and the switching valves can add 1 to 3 psi to the overall plant pressure drop. However, converting an existing three-converter Claus unit to a three-converter CBA unit will have minimal impact on plant pressure drop.

The process licensor claims that the capital cost of a three-converter CBA unit is between 95 and 125% of that of a conventional three-converter Claus unit. This CBA configuration has one less condenser and two less converter reheaters than the Claus unit, but requires the purchase of switching valves. A four-converter CBA unit is claimed to have a capital cost between 120 and 150% of that of a conventional three-converter Claus unit. In 1992, the CBA process was in operation in about 20 facilities worldwide both in gas processing plants and refineries (Berman, 1992).

The sulfur recovery of the CBA process can be increased when used in conjunction with Amoco's ULTRA (Ultra Low Temperature Reaction and Adsorption) or ELSE (Extremely Low Sulfur Emissions) processes. The intent of the ULTRA process (Berman, 1992; Lee et

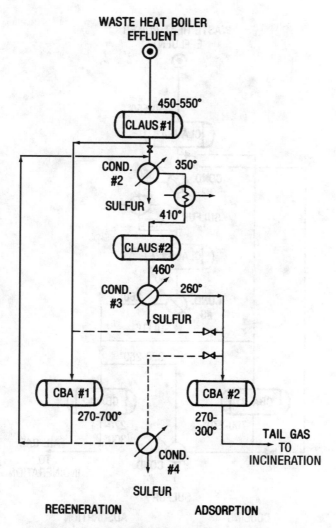

Figure 8-20. AMOCO-CBA process, regeneration.

al., 1984) is to operate the sub-dewpoint converter at a lower temperature than is normally practiced (as low as 180°F) to enhance sulfur recovery by a further improvement to the Claus reaction equilibrium. However, prior to admission to the adsorption bed, the process stream must be hydrogenated to convert sulfur vapor, sulfur dioxide, and organic sulfur compounds to hydrogen sulfide. Also, water vapor must be removed to prevent adsorption on the catalyst and to improve the reaction equilibrium, and one-third of the hydrogen sulfide generated during the hydrogenation step must be oxidized back to sulfur dioxide. Sulfur recoveries on the order of 99.5 to 99.7% were reported during pilot plant testing.

706 *Gas Purification*

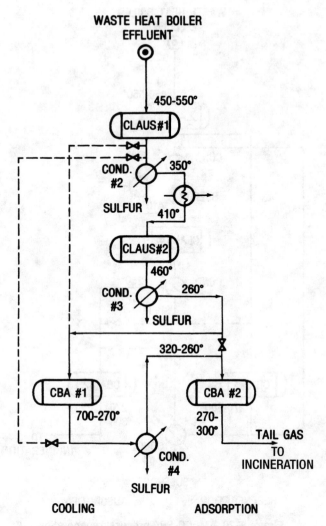

Figure 8-21. AMOCO-CBA process, cooling.

The ELSE process (Lee and Reed, 1986; Berman, 1992) utilizes a solid zinc oxide sorbent to remove residual sulfur compounds from the tail gas of a Claus or Claus/CBA plant. The process is comprised of two ELSE reactor beds, one always in absorption and the other in regeneration. When the absorbing bed has become deactivated through conversion of the zinc oxide to the sulfide, it is regenerated by oxidation at 1,100°F with a dilute air stream. During pilot plant testing the effluent gas from the process was reported to contain less than 50 ppm of sulfur dioxide, corresponding to a sulfur recovery greater than 99.95%. As of mid-1996 there had been no commercial scale ULTRA or ELSE process units built.

MCRC Sulfur Recovery Process

Process Description. The MCRC (Maximum Claus Recovery Concept) process, which was described by Heigold and Berkeley (1983) and Heigold (1991), is licensed by Delta Projects Inc., of Calgary, Canada, and is quite similar to the Amoco CBA process. The process, shown in **Figure 8-22** for the three-bed design, is a sub-dewpoint Claus type, with the equip-

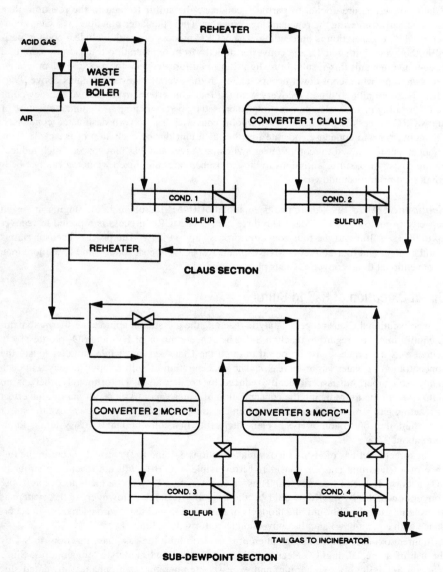

Figure 8-22. Three-converter MCRC unit including Claus section. (*Heigold, 1991*)

ment and piping preceding the second converter the same as for a conventional Claus plant. The second and third converters alternate automatically between sub-dewpoint reaction/adsorption and regeneration modes of operation. Four-converter MCRC units are also available, with three of the converters in cyclic operation. This configuration always has two converters in adsorption with the process gas flowing through them in series.

There are several operational and configurational differences that distinguish the MCRC process from the CBA process. With the MCRC process, the gases leaving the first converter (Claus converter) are cooled to partially condense the sulfur formed in the reaction, then reheated before entering the regenerating converter. Thus, the latter functions as a Claus converter while regeneration is proceeding. The MCRC process does not include a cooling period in the cycle, but switches the converter directly from regeneration to adsorption. With a three-converter unit this results in a slight dip in sulfur recovery for a period of time after bed switching while cool-down occurs, although the overall recovery remains above 99%. There is essentially no dip in recovery with the four-converter unit as the two sub-dewpoint beds operating in series dampen any impact caused by bed switching. A distinctive feature of the MCRC process is that each sub-dewpoint converter has its own designated condenser. The converter and condenser operate together as a unit during each step of the cycle. This arrangement is claimed to require fewer switching valves and less plot space, which reduces piping costs. Also, sulfur emissions to the atmosphere are minimized when the catalyst beds are purged prior to shutdown.

Economics. As of 1993, there had been a total of 16 MCRC plants built, ranging in design capacity from 13 to 525 LT/day. The three-converter MCRC process is reported to recover up to 99.3% sulfur and the four-converter process up to 99.5%. The process licensor claims that the three- and four-converter MCRC plants respectively cost about 7 and 23% more than a conventional three-converter Claus plant.

Direct Oxidation of H_2S to Sulfur

In conventional Claus process catalytic reactors, the gas stream approaches thermodynamic equilibrium with regard to reaction 8-3. The concentrations of H_2S and SO_2 progressively decrease as the process gases proceed through the Claus system; while simultaneously the concentration of water vapor increases. The concentration of sulfur vapor increases in each catalyst chamber, but is periodically reduced by condensation. Unfortunately, there is no simple technique for reducing the concentration of water vapor. As a result of both the effect of reacting gas composition and the requirement to maintain the temperature above the sulfur dew point, the conversion of H_2S to elemental sulfur in a conventional Claus system is limited to about 97%.

The direct reactions of H_2S with oxygen (equations 8-1 and 8-2) normally occur in the furnace of a Claus unit and are considered irreversible, i.e., H_2O will not react with sulfur or SO_2 to form H_2S and free oxygen. Thus, if only reaction 8-1 can be made to occur, 100% conversion of H_2S to sulfur would be theoretically possible. However, if the system is allowed to attain equilibrium, the products of reaction 8-1 will react with each other in accordance with equation 8-3 and the conversion to sulfur will be limited.

In direct oxidation processes, an attempt is made to limit the gas phase reactions to 8-1 by the use of a special, highly selective oxidation catalyst and moderate temperatures. Since reaction 8-1 is highly exothermic, and excessive temperature can enhance unwanted side reactions, the technique is applicable only to dilute H_2S-containing gas streams. However,

Claus plant tail gas is normally quite dilute, and more concentrated H_2S streams can be made dilute by recycling product gas. When direct oxidation is used in conjunction with a conventional Claus plant it is necessary to assure that essentially all sulfur in the tail gas is in the form of H_2S. This can be accomplished by operating the Claus system with slightly less than the stoichiometric amount of air or by providing a separate hydrogenation/hydrolysis step to convert sulfur compounds to H_2S before direct oxidation. Commercial direct oxidation processes include the Superclaus process offered by Comprimo, the Selectox process offered by Unocal/Parsons, Mobil's MODOP process, and the previously discussed Catasulf process.

Superclaus Process

Introduction. The Superclaus process was jointly developed in the Netherlands by Comprimo BV, VEG-Gasinstituut, and the University of Utrecht and has been described by Lagas et al. (1988A, B, 1989, 1994). The process was developed to increase the sulfur recovery capabilities of the Claus process by reducing its inherent thermodynamic limitations. The Superclaus process increases sulfur recovery by replacing the reversible Claus reaction (equation 8-3) with the direct oxidation reaction between hydrogen sulfide and oxygen (equation 8-1), which is considered irreversible. This is accomplished through the use of a new catalyst in the last converter of the Claus unit. The catalyst consists of an alpha-alumina or silica substrate supporting iron and chromium oxides (Goar, et al., 1991). The catalyst is reported to be highly selective to direct hydrogen sulfide oxidation and will convert more than 85% of the H_2S to elemental sulfur. Formation of sulfur dioxide is low, even in the presence of excess air, and there is little reverse Claus reaction reactivity due to the low sensitivity of the catalyst to water. There is also no oxidation of carbon monoxide and hydrogen and no formation of carbonyl sulfide or carbon disulfide. Two versions of the process have been developed: the Superclaus 99 and Superclaus 99.5 processes.

Process Description. The Superclaus 99 process, shown in **Figure 8-23,** consists of a conventional Claus plant thermal stage, followed by two or three converters charged with standard Claus catalyst, and a final converter charged with the new Superclaus catalyst. Each converter is provided with an upstream reheater and an outlet sulfur condenser. The air for the oxidation of the hydrogen sulfide is supplied to two locations. The major portion is introduced into the reaction furnace burner, and the remainder is added to the process gas stream upstream of the Superclaus reactor. In the thermal stage, the acid gas is burned with a substoichiometric amount of air so that an excess of hydrogen sulfide remains in the gas leaving the last Claus converter, thus suppressing the sulfur dioxide concentration. The concentration of hydrogen sulfide is controlled at this point to 0.8–3.0 vol% typically. Maintaining the ratio of two moles of hydrogen sulfide to one mole of sulfur dioxide is no longer necessary, as the hydrogen sulfide remaining in the process gas is directly oxidized to sulfur over the Superclaus catalyst in the presence of excess air. The excess air affords the flexibility in process control required for acid gas feed flow rate and compositional changes. The sulfur produced at each conversion stage is condensed in the immediate downstream condenser. Carbonyl sulfide and carbon disulfide produced in the thermal stage must still be hydrolyzed in the first Claus converter, as they are unaffected by the new catalyst.

In cases where substoichiometric combustion in the reaction furnace is not possible due to minimum flame temperature requirements, the 0.8–3.0% hydrogen sulfide concentration

710 Gas Purification

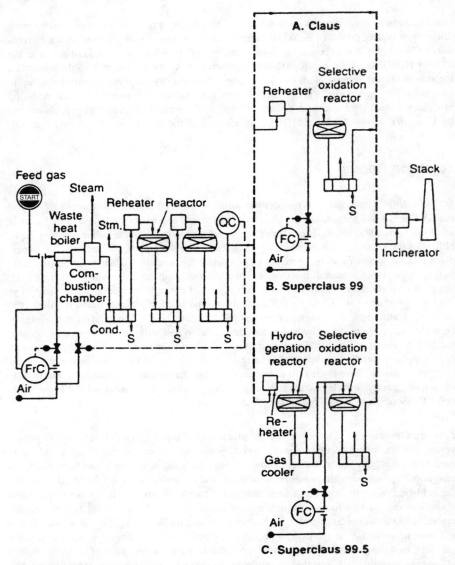

Figure 8-23. Superclaus 99 and 99.5 processes (*Lagas et al., 1989*). Reproduced with permission from Hydrocarbon Processing

is maintained by routing a portion of the reaction furnace acid gas feed downstream of the first condenser.

With one thermal stage and two catalytic Claus stages upstream of the Superclaus selective oxidation stage, an overall sulfur recovery of 99.0% is claimed for a rich feed gas. The process licensor reports that the capital cost of this configuration is approximately 5% greater than that of a typical three-stage Claus plant. With the addition of a further Claus stage, a minimum overall sulfur recovery of 99.3% is claimed, at a capital cost about 15% greater than that of a typical three-stage Claus plant. Overall recoveries for the Superclaus 99 process, however, are subject to the performance of the upstream Claus unit.

The Superclaus 99.5 process introduces a hydrogenation stage between the last Claus reactor and the Superclaus reactor. As all sulfur values are converted to hydrogen sulfide in the hydrogenation reactor over a cobalt/molybdenum catalyst, there is no longer a requirement to operate the Claus unit with excess hydrogen sulfide. The normal ratio of two moles of hydrogen sulfide to one mole of sulfur dioxide is required, but is less critical than for normal Claus plant operation.

In the Superclaus 99.5 process, the gas leaving the hydrogenation reactor is cooled to the optimum inlet temperature for the Superclaus catalyst. As this catalyst is not sensitive to water, there is no need to condense water by quenching, as is practiced in other selective oxidation tail gas processes. In the Superclaus 99.5 process, excess air for selective oxidation of the hydrogen sulfide is added just upstream of the Superclaus reactor. The final condenser recovers the sulfur formed in this stage.

With a rich feed gas and two Claus catalytic stages, the Superclaus 99.5 process is claimed to be capable of a minimum overall sulfur recovery of 99.2%. The process licensor claims that this is obtained with a capital investment that is about 20% greater than that required for a typical three-stage Claus plant. A minimum overall sulfur recovery of 99.4% is claimed when three Claus reactors are included, for an investment reported to be about 30% greater than that for a typical three-stage Claus plant.

The Superclaus process can be used in combination with other Claus process modifications, such as Recycle Selectox and COPE, to achieve higher overall sulfur recovery, and with tail gas units such as BSR/Stretford and SCOT to save utility costs by reducing the sulfur load.

As of 1994, a total of 18 units were under license (Comprimo B.V., 1994). A typical Superclaus retrofit of a Claus unit is described by Nasato et al. (1991).

Selectox Process

Introduction. This process, which was developed by the Union Oil Company of California (now Unocal Corporation) and The Parsons Corp., utilizes a proprietary catalyst (Selectox 33) for the oxidation of relatively low concentrations of hydrogen sulfide to elemental sulfur in a one-step operation. Three applications of the process are shown in **Figure 8-24**.

In the BSR/Selectox version, the process is used for hydrogen sulfide removal from Claus tail gas after hydrogenation in a BSR process hydrogenation section. (The BSR process is described in a subsequent section.) About 99.5% overall sulfur recovery, including the Claus unit, is attainable. Even higher recovery can be achieved if the effluent from the Selectox reactor is treated in a final Claus stage.

The Once Through Selectox process is suitable for dilute acid gas streams containing up to about 5% H_2S, such as geothermal offgas, while the Two Stage Selectox process with recycle can be used for gas streams containing more than 5% hydrogen sulfide. About 80% conversion of hydrogen sulfide to sulfur is reported for the Selectox reactor in these two versions of the process (Beavon et al., 1980).

The principal advantage of the Selectox process, if used in conjunction with the BSR process, is that substantial capital cost savings can be realized by replacing a Stretford system with a Selectox reactor. In addition, problems with liquid effluents are eliminated. The advantage of the other two versions of the process is that they are suitable for the treatment

712 Gas Purification

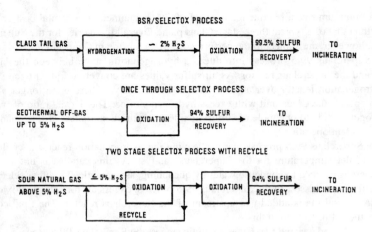

Figure 8-24. Selectox applications. (*Dalla Lana, 1973*)

of very dilute H_2S containing gas streams which cannot be processed in conventional Claus units. The operation is reported to be simple and reliable and the conversion remarkably high (Beavon et al., 1980; Goar, 1982).

Process Description. The process and its first commercial applications have been described by Beavon et al. (1979), Beavon et al. (1980), Hass et al. (1981, 1982), Goar (1982), Warner (1982), and Delaney et al. (1990). In the BSR/Selectox process, which is shown schematically in **Figure 8-25,** the Claus tail gas passes first through the hydrogenation reactor and a two-stage cooling step, where a substantial portion of the water vapor contained in the gas is condensed. After addition of a carefully controlled amount of air to the cooled gas, it enters the Selectox reactor where hydrogen sulfide is catalytically oxidized according to reaction 8-1.

The effluent from the reactor is cooled and sulfur is condensed. When required, the cooled gas may be further processed in a final Claus stage. The gas leaving the final condenser is incinerated either thermally or catalytically before discharge to the atmosphere. Typical gas composition changes in the course of the process are shown in **Table 8-4.**

Inspection of **Table 8-4** shows that the Claus plant tail gas typically contains 30–35% water vapor. Although the selective oxidation reaction (8-1) is not reversible, its products (H_2O and S) can react by the reverse of reaction 8-3 to reduce the net conversion to sulfur. Since the reverse reaction is favored by the presence of water vapor, reducing the water vapor concentration in the gas from 30 to 35% to 1.5 to 4% by cooling aids in obtaining a high conversion. About 80–90% of the H_2S entering the Selectox reactor is converted to sulfur, the conversion being limited primarily by the increase in temperature due to the heat of reaction.

The Selectox 33 catalyst is reported to be highly selective for the oxidation of hydrogen sulfide to sulfur, without formation of SO_3 and without oxidation of either hydrogen or of saturated hydrocarbons. It is claimed to be highly active and stable and to retain its activity over long periods of time without regeneration, when operating at temperatures similar to those encountered in Claus reactors (Beavon et al., 1979; 1980).

The first commercial BSR/Selectox plant started operations in 1978 and reportedly is attaining consistent overall recovery efficiencies of 98.5 to 99.5%, even though the preceding Claus unit recovery efficiency varies between 93 to 96%.

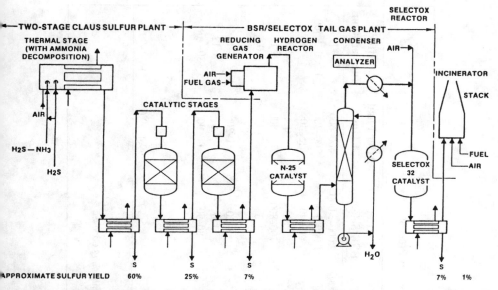

Figure 8-25. BSR/Selectox process for Claus plant emission control. (*Beavon et al., 1980*)

The Once Through Selectox Process consists of the Selectox reactor and sulfur condenser shown in **Figure 8-23**. The allowable concentration of hydrogen sulfide in the gas stream to be treated is limited by the fact that the entire heat of reaction is liberated in the reactor, resulting in excessively high temperature levels if the H_2S concentration exceeds about 5%. At the 5% level, the reactor outlet temperature is about 700°F, which is considered acceptable (Beavon et al., 1980).

The Two Stage Selectox Process with Recycle, which is reportedly suitable for the treatment of gas streams containing up to 40% H_2S (Beavon et al., 1980), is shown schematically in **Figure 8-26**. To overcome the temperature effect due to heat of reaction, a portion of the effluent from the first sulfur condenser is recycled and mixed with the feed gas to the Selectox reactor. In this manner the H_2S concentration in the feed is adjusted to about 5%. The remaining portion of the gas leaving the first sulfur condenser flows to a Claus stage using either alumina or Selectox catalyst. About 82% of the hydrogen sulfide in the feed is recovered as sulfur after the Selectox reactor and an additional 12% is recovered after the Claus stage.

In 1994 it was reported that more than 10 Recycle Selectox Plants were in operation (Unocal Science Div. and the Parsons Corp., 1994). The commercial operation of a Recycle Selectox plant has been described by Goar (1982) and Delaney et al. (1990). This unit is reported to recover about 20 LTPD of sulfur from an acid gas stream containing 13% H_2S (balance mostly CO_2) with a sulfur recovery efficiency of better than 95.0%.

Further variations of the Selectox process, including packaged units and three-stage units with recycle, have been reported by Hass et al. (1981).

MODOP Process

Introduction. A process similar to the Selectox process, the Mobil Oil Direct Oxidation Process (MODOP), has been developed by Mobil Oil AG of Celle, Germany, the German

714 Gas Purification

Table 8-4
Gas Composition Changes in BSR/Selectox Process

Component	Unit	Claus Tail Gas	After Hydrogenation	After Cooling	After BSR/Selectox	After Final Claus	After Incineration
H_2S	ppmv	4,000–10,000	10,000–15,000	12,000–20,000	2,000–3,000	400–600	*1
SO_2	ppmv	3,000–6,000	*0	*0	1,000–1,500	200–300	1,000–1,500
COS	ppmv	300–5,000	10–30	15–40	15–40	15–40	*1
CS_2	ppmv	300–5,000	*0	*0	*0	*0	*0
S†	ppmv	700–1,000	*0	*0	700–800	700–800	*1
H_2	vol %	1–3	2–3	3–4	2–3	2–3	0
CO	vol %	0.5–1	*0	*0	*0	*0	0
CO_2	vol %	1–15	1–15	1–20	1–20	1–20	1–15
H_2O	vol %	30–35	30–35	1.5–4	3–6	3–6	8–12
N_2	vol %	60–70	80–90	80–90	80–90	80–90	80–90
Cumulative percent of Claus feed recovered		93–96	93–96	93–96	98.5–99	99.4–99.6	99.4–99.6

*Approximate.
†As S_1.
Reprinted with permission from Oil and Gas Journal, March 12, 1979, Copyright PennWell Publishing Company
Source: Beavon et al. (1979)

Sulfur Recovery Processes 715

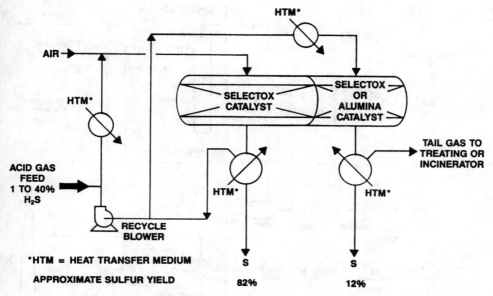

Figure 8-26. Two-stage Selectox process with recycle. (*Beavon et al., 1980*)

affiliate of the Mobil Oil Corporation. The MODOP process recovers hydrogen sulfide from Claus plant tail gas and amine regenerator offgas streams. This process also directly converts hydrogen sulfide to elemental sulfur via the catalytic reaction of equation 8-1. The reaction occurs over a commercially available titanium dioxide (TiO_2) based catalyst. Suitable feed concentrations range from 0.5 to 8% H_2S. One to three reactor stages may be employed depending on the feedstock H_2S concentration and the required overall sulfur recovery.

Process Description. The process configuration for a typical MODOP process for Claus tail gas clean-up is shown in **Figure 8-27.** The tail gas is first heated in a reducing gas generator to about 280°C, then the sulfur components of the tail gas are catalytically converted to hydrogen sulfide in a hydrogenation stage. The gas is then cooled in three stages, initially to recover heat and finally to condense water in the quench column. After the addition of air, the hydrogen sulfide is converted to sulfur by direct oxidation over a TiO_2 catalyst at a temperature above 160°C. About 90% of the H_2S is converted to elemental sulfur, which is recovered as a liquid on cooling the gas in a condenser (Mobil, 1994). If a higher recovery is required, additional oxidation stages may be added as indicated in the figure. The vent gas leaving the final condenser is incinerated before it is released to the stack.

Although the selective oxidation reaction (8-1) is not reversible, the products of reaction 8-1 (S and H_2O) can react by the reverse of reaction 8-3 to reduce overall sulfur recovery. Therefore, in the MODOP process, water is removed from the gas stream in the quench tower to improve the conversion of hydrogen sulfide to sulfur. Conversion can be further increased by removal of additional water in a glycol dehydration tower after the quenching

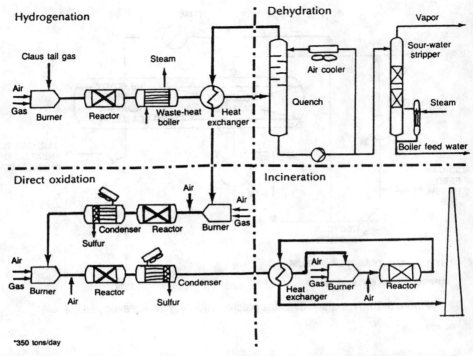

Figure 8-27. Schematic flow diagram of the MODOP process applied to Claus plant tail-gas treatment (*Kettner and Liermann, 1988*). *Reprinted with permission from Oil & Gas Journal, Jan. 11, 1988, copyright PennWell Publishing Co.*

stage. Water recovered from the dehydration of the gas stream may be treated in a sour water stripper and used as boiler feed water make-up.

When treating Claus unit tail gas, the process is capable of overall sulfur recoveries of up to 99.6% The number of reactors included is dependent on the feed gas concentration and the required sulfur recovery. Catalyst selectivity is maximized at a MODOP reactor outlet temperatures of 250°–270°C (482°–518°F). As the oxidation reaction is highly exothermic, additional stages must be employed to limit the reactor outlet temperature to below 320°C (608°F) at high concentrations of H_2S in the feed gas.

The MODOP process may be used to directly treat low hydrogen sulfide content amine regenerator off gas. In this application, the hydrogenation and quench equipment are generally not required due to the absence of non-H_2S sulfur species and the low water content of the feed gas.

The first MODOP unit was commissioned at the Norddeutsche Erdgas Aufbereitungs-Gesellschaft (NEAG) gas processing plant at Voigten, Germany, in 1983. A second unit was installed at the same plant in 1987, and a third MODOP unit was started up in 1991 in Mobile, Alabama (Mobil, 1994). The commercial operation of the MODOP process has been described by Kettner and Liermann (1988) and Kettner et al. (1988).

BASF Catasulf Process

The BASF Catasulf process is a direct oxidation sulfur removal process that uses an isothermal reactor for heat removal. See the prior discussion in this chapter under the Isothermal Reactor Concepts section for more information.

Sulfur Dioxide Reduction and Recovery of H_2S

Although a well designed Claus plant is capable of converting as high as 97% of the hydrogen sulfide in the feed gas to elemental sulfur, the tail gas from such a unit contains sufficiently high concentrations of sulfur compounds to require further treatment to satisfy air pollution control requirements in many jurisdictions. The sulfur compounds that represent the remaining 3% or more of the feed sulfur are in the form of hydrogen sulfide, sulfur dioxide, carbonyl sulfide, carbon disulfide, and elemental sulfur vapor. Since no simple process exists that can remove all of those components simultaneously, techniques have been considered to convert the various forms of sulfur to a single compound that can be removed effectively. The obvious choices are complete oxidation to SO_2 and complete reduction (and/or hydrolysis) to H_2S.

Several processes have been developed based on the oxidation approach, but, with some exceptions, such as the Linde Clintox process (Heisel and Marold, 1992) these have not been commercial successes because of the complexity of SO_2 recovery processes and their comparatively low SO_2 removal efficiencies. Hydrogen sulfide, on the other hand, can be removed at very high efficiency by a number of processes, including the selective amine processes discussed in Chapter 2. Processes that involve the conversion of Claus plant tail gas sulfur compounds to H_2S and the removal of this H_2S are discussed in the following sections. Processes reviewed include Parson's Beavon Sulfur Removal (BSR) process, Shell's SCOT process, FB and D Technologies' Sulften process, the TPA Resulf process, and the Exxon Flexsorb SE Plus process.

Beavon Sulfur Removal (BSR) Process

Introduction. The Beavon Sulfur Removal (BSR) process, as reported by Beavon and King (1970) and Beavon and Vaell (1971), is capable of reducing the total sulfur content of Claus unit tail gases to less than 250 ppm by volume (calculated as sulfur dioxide) and thus of attaining an overall conversion of more than 99.9% of the hydrogen sulfide fed to the Claus unit. The residual sulfur compounds from the Beavon process consist almost entirely of carbonyl sulfide, with only traces of carbon disulfide and hydrogen sulfide. The effluent gas is practically odorless and can often be vented directly to the atmosphere, obviating the need for incineration and the attendant consumption of fuel.

The Beavon Sulfur Removal (BSR) process was developed jointly by the Parsons Corp. of Pasadena, California, and the Union Oil Company of California (now Unocal Corporation of Los Angeles, California). The term Beavon process refers to a group of processes utilized for the removal of residual sulfur compounds from Claus plant tail gases. This family of processes has in common an initial hydrogenation and hydrolysis unit to convert all residual sulfur compounds to H_2S. Individual processes within this family differ from each other in the technology used to remove the H_2S from the Claus tail gas stream. Process improvements and operating experience have been reported by Andrews and Kouzel (1974), Fenton et al. (1975), Beavon and Brocoff (1976), and Kouzel et al. (1977).

In the Beavon Sulfur Removal Process, all sulfur compounds (other than hydrogen sulfide) contained in the tail gas are catalytically converted to hydrogen sulfide, which is subsequently removed by any convenient method. If complete removal of hydrogen sulfide is required, chemical absorption in a methyldiethanolamine solution (BSR/MDEA Process), or oxidation to sulfur in the liquid phase by a redox solution (BSR/Stretford and BSR/Unisulf Processes), may be used. However, if partial removal is adequate, the effluent from the catalytic hydrogenation reactor may be treated, after cooling and condensation of the bulk of the water, either in a final Claus reactor or in a Selectox reactor (BSR/Selectox Process), which may be followed by a Claus reactor (see previous Selectox section). The overall conversions attainable by these two versions of the process range from 98 to more than 99%. In 1994, it was reported that more than 19 Beavon-MDEA plants were operating in the U.S. and Japan, and 2 Beavon-Selectox plants were operating in the U.S. and Germany (Unocal Science and Technology Div. and the Ralph M. Parsons Co., 1994).

Process Description. The Beavon process H_2S conversion step is carried out at elevated temperatures over a cobalt-molybdate catalyst and involves hydrogenation and hydrolysis of sulfur compounds according to equations 8-10 through 8-15.

Hydrogenation reactions (in the presence of hydrogen gas):

$$CS_2 + 2H_2 = C + H_2S \tag{8-10}$$

$$COS + H_2 = CO + H_2S \tag{8-11}$$

$$SO_2 + 3H_2 = H_2S + 2H_2O \tag{8-12}$$

$$S_2 + 2H_2 = 2H_2S \tag{8-13}$$

Hydrolysis reactions (in the presence of water vapor):

$$CS_2 + 2H_2O = CO_2 + 2H_2S \tag{8-14}$$

$$COS + H_2O = CO_2 + H_2S \tag{8-15}$$

Although it is probable that carbonyl sulfide and carbon disulfide are converted to H_2S primarily by hydrolysis, especially since the tail gas contains about 30% water vapor, it is conceivable that hydrogenation also takes place, although to a minor extent.

A schematic flow diagram of the process is presented in **Figure 8-28.** In this version of the process, the hydrogen sulfide formed in the catalytic step is removed by the Stretford or Unisulf process (see Chapter 9), both of which have been demonstrated to be very effective for reducing the hydrogen sulfide content of the hydrogenated gas to less than 10 ppm. A photograph of a BSR process plant incorporating a Stretford unit for H_2S removal is shown in **Figure 8-29.**

By following the flow diagram, it is seen that the Claus plant tail gas is first heated to the temperature required for the catalytic reaction by adding a hot stream of gas resulting from partial combustion of hydrocarbon gas in a line burner. This gas not only supplies the necessary heat, but also enough hydrogen to satisfy the hydrogen demand for the hydrogenation reactions.

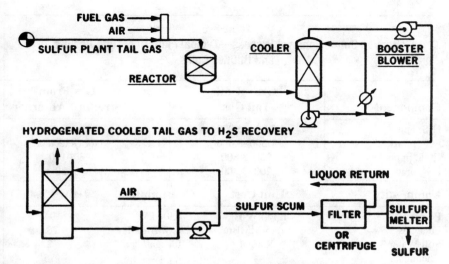

Figure 8-28. Typical process flow diagram—Beavon sulfur removal process using Stretford process for H_2S removal.

After passing through the reactor, the gas is cooled to ambient temperature by direct contact with water. For better heat economy, especially in larger installations, a steam generator may be placed ahead of the direct contact water cooler. The cooled gas, which contains primarily nitrogen, carbon dioxide, hydrogen sulfide, and traces of carbonyl sulfide, is then treated for hydrogen sulfide removal in the Stretford unit. The water condensed from the gas in the direct contact cooler is stripped of hydrogen sulfide in a sour water stripper and then discarded. The stripped hydrogen sulfide is either fed to the hydrogen sulfide removal section or, when permissible, incinerated.

Economics. Typical results from plant operations and utility costs for a BSR/Stretford plant processing tail gas from a 100 long tons per day Claus unit are shown in **Table 8-5**.

Use of an aqueous methyldiethanolamine (MDEA) solution as a selective solvent (see Chapter 2) for hydrogen sulfide removal in tail gas cleanup has been described by Meissner (1983). With this solvent an overall conversion, including the Claus unit, of 99.9% of the H_2S fed to the Claus unit is attainable. However, the treated gas contains some residual H_2S, which may require incineration before discharge to the atmosphere.

Shell Claus Off-Gas Treating (SCOT) Process

Introduction. The SCOT process, which was developed by Shell International Petroleum Maatschappij, The Hague, The Netherlands, is similar in basic principle to the Beavon Sulfur Removal/MDEA Process. It also relies on catalytic conversion of sulfur compounds, other than hydrogen sulfide, contained in the Claus tail gas to hydrogen sulfide, which is then selectively absorbed in an alkanolamine solution with only partial absorption of carbon dioxide. The acid gas is stripped from the amine solution and recycled to the Claus unit. Although it is reported that the process is capable of producing a purified gas stream contain-

Table 8-5
Beavon Sulfur Removal Process—Typical Plant Operating Results and Utilities Costs

Component	Claus Unit Tail Gas	Gas From Stretford Absorber
H_2S, ppmv	3,000–6,000	1
SO_2, ppmv	1,500–3,000	Not Detectable
COS, ppmv	200–3,300	30–100
CS_2, ppmv	200–3,300	9–20

Commodity	Unit Cost	Consumption	Daily Cost ($)
Power	1.3¢/kW-Hr.	300 kW	94
Fuel Gas	$2.25/MMBtu	125 MMBtu/day	281
Soft Water	20¢/gal	10M gal/day	2
Chemicals	—	—	75
Catalysts	—	—	8
Total Costs			460
Less Credit for 50 psig Steam	$2.50/M lb.	2,500 lb/hr.	(150)
Less Credit for Recovered Sulfur	$50/LT	4.9 LT/day	(245)
Net Cost			65
Net Cost/Ton of Sulfur Recovered from Tail Gas			13

Note: Utility costs for unit treating tail gas from 100 LT/day Claus unit operating at 95% conversion.
Source: Kouzel et al. (1977)

ing 10 to 400 ppmv of total sulfur using MDEA or DIPA as the absorbent (measured as sulfur dioxide after incineration), there is usually enough hydrogen sulfide left in the effluent from the amine absorber to require incineration before venting to the atmosphere. The process has been described, and operating experience has been reported by Groenendaal and Van Meurs (1972), Naber et al. (1973), and Harvey and Verloop (1976).

Process Description. A schematic flow diagram of the process is shown in **Figure 8-30**. The flow of gas and liquids in the process is quite similar to that of the Beavon Sulfur Removal/MDEA process. The Claus tail gas is first heated in a line gas heater that may have the additional purpose of supplying the reducing gas required in the subsequent catalytic step. However, reducing gas may also be furnished from an outside source. The hot gas then flows to the reactor where reduction of sulfur compounds occurs over a cobalt-molybdenum or cobalt-nickel catalyst at about 300°C (572°F). The effluent from the reactor is subsequently cooled in two steps, first in a waste heat boiler where low pressure steam is produced, and then by water washing in a cooling tower. In this step, most of the water vapor contained in the gas is condensed. Excess water from the cooling tower, which contains a small amount

Sulfur Recovery Processes **721**

Figure 8-29. Beavon sulfur removal process plant. *Courtesy The Parsons Corp.*

of hydrogen sulfide, is treated in a sour water stripper where hydrogen sulfide is expelled and returned to the Claus unit.

After cooling, the gas enters the amine absorber where essentially all of the hydrogen sulfide is removed, but only a portion of the carbon dioxide is coabsorbed. The rich amine solution is stripped of acid gas in the regenerator by application of indirect heat supplied by a steam heated reboiler, and the acid gas is returned to the Claus unit. This portion of the process is quite similar to the conventional selective ethanolamine processes discussed in Chapter 2.

Selective absorption is based on the fact that the rate of absorption of hydrogen sulfide in alkanolamines is substantially more rapid than that of carbon dioxide. This phenomenon is more pronounced with secondary and tertiary amines than with primary amines (see Chapter 2). Consequently, appreciable selectivity may be attainable by proper selection of the amine and by designing the absorber for short gas and amine solution contact times. Depending on gas composition, choice of amine, and absorber design, co-absorption of carbon dioxide can be limited to about 10 to 40% of the carbon dioxide contained in the absorber feed gas (Naber et al., 1973). In most applications, methyldiethanolamine (MDEA) is the preferred

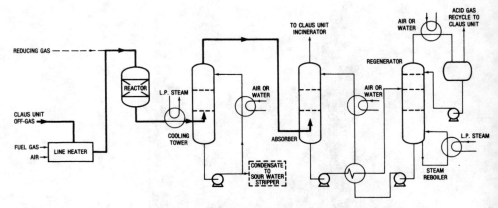

Figure 8-30. Schematic flow diagram—SCOT process.

solvent; however, diisopropanolamine (DIPA), which is less selective, was used in the first SCOT units (Harvey and Verloop, 1976).

The SCOT process is reported to be quite flexible and not very sensitive to upsets in the upstream Claus unit. Relatively wide variations in the H_2S/SO_2 ratio in the Claus plant, as well as fluctuations in the feed gas volume, can be tolerated in a properly designed SCOT unit.

A SCOT unit can be integrated with the desulfurization unit supplying the acid gas to the Claus plant by using the same amine, for example DIPA, in both the desulfurization and the SCOT unit, and stripping the combined rich solutions from the two absorbers in a single regenerator. This scheme is best applied in petroleum refineries where the ADIP process, employing DIPA as the active agent, is used for refinery gas desulfurization.

One rather important advantage of the SCOT process is the absence of liquid effluents that could cause pollution problems. Possible disadvantages are the use of appreciable amounts of energy for amine regeneration and fuel for reducing gas production and purified gas incineration, and the recycle of hydrogen sulfide which results in an increase of about 7–8% in the required capacity of the Claus unit (Naber et al., 1973).

Economics. Capital costs for SCOT units as reported by Harvey and Verloop (1976) are about the same as those for the preceding Claus unit if the SCOT unit is added to an existing Claus unit. The overall costs are lower if a common amine regenerator is used for the desulfurization and SCOT units, or for a grassroots Claus/SCOT installation, where a third Claus catalyst bed may be omitted at the expense of a slightly larger SCOT unit. Further savings are possible by cascading amine solution partially loaded with acid gas from the SCOT unit to the higher pressure absorbers in the desulfurization unit (Wallace and Flynn, 1983). Operating costs reported by Harvey and Verloop (1976) are about $16 per long ton of sulfur produced in the SCOT plant. In 1994, it was reported that 130 SCOT units were in design or operation worldwide on sulfur plants ranging in size from 3 to 2,100 tpd sulfur capacity (Shell International Research Mij B.V. and Shell Oil Co., 1994).

Design and operation of a rather large SCOT unit have been described by Herfkens (1982). This unit treats tail gas from two Claus units with a combined capacity of 1,225 tons of sulfur per day and a required overall recovery of 98.5%. Aqueous MDEA is used for H_2S

removal after hydrogenation, and the acid gas is recycled to the Claus units. The MDEA unit produces a treated tail gas containing less than 500 ppmv of H_2S with only about 10% of the CO_2 being coabsorbed.

Sulften Process

The Sulften process was jointly developed by FB&D Technologies (formerly Ford, Bacon & Davis, Inc.) of Salt Lake City, Utah, and Union Carbide Corporation of Tarrytown, New York. The process is similar to the BSR/MDEA and SCOT processes, relying on catalytic hydrolysis and hydrogenation of sulfur compounds (other than hydrogen sulfide) in the Claus unit tail gas to hydrogen sulfide and subsequent recovery of the H_2S by a selective amine. The Sulften process is unique, however, in that the reaction vessel is divided into two separate sections containing distinct catalysts. The first compartment contains a hydrolysis reactor bed to convert carbon disulfide and carbonyl sulfide to H_2S and carbon dioxide via reactions 8-10 and 8-11. The second compartment contains a conventional hydrogenation catalyst to convert sulfur dioxide and elemental sulfur to H_2S via reactions 8-12 and 8-13. The reducing gas required for the reaction may be supplied either from an external source or by the substoichiometric combustion of fuel gas in the in-line gas preheater.

The Sulften solvent is a specially formulated, proprietary, MDEA solution offered by Union Carbide, known as UCARSOL HS-103, with a high H_2S removal efficiency. The treated gas has a composition of less than 10 ppmv by volume H_2S, which in many jurisdictions eliminates the need for vent gas incineration to comply with environmental regulations, resulting in significant savings of fuel gas. The chemistry of highly selective solvents, such as UCARSOL HS-103, is discussed in Chapter 2. As of 1991, there were eight Sulften units in operation with capacities from 7 to 300 tons per day of sulfur feed to the Claus unit. The commercial operation of a Sulften unit has been described by Tragitt et al. (1986) and Kroop et al. (1985).

Resulf Process

The Resulf process, which is licensed by TPA, Inc., is very similar to the BSR/MDEA, SCOT, and Sulften processes. The Claus plant tail gas is heated; mixed with a reducing gas; passed through a catalyst bed where sulfur vapor, SO_2, COS, and CS_2 are converted to H_2S; cooled; and passed through a selective amine absorber. Acid gas from the amine stripper is recycled to the Claus plant. It is claimed that a high efficiency version of the process, Resulf-10, produces a vent gas from the amine absorber containing a maximum of 10 ppm H_2S (TPA Inc., 1994).

Exxon Flexsorb SE Plus Process

While other amine-based selective tail gas treating processes utilize MDEA, Exxon's Flexsorb SE Plus tail gas treating process is based on sterically hindered amines. Sterically hindered amines are more selective than MDEA and, in tail gas treating applications, recycle less CO_2 back to the Claus plant. Also, sterically hindered amines require less amine circulation for a given level of H_2S removal than comparable MDEA-based units, and therefore have lower utilities consumption. This advantage is somewhat offset by a higher amine solvent cost. The treated gas from a Flexsorb SE Plus unit contains less than 10 ppmv H_2S, which in many jurisdictions eliminates the need for tail gas incineration, resulting in significant fuel gas savings. The Flexsorb processes have been described in some detail by Goldstein (1983), Weinberg et al. (1983), and Chludzinski et al. (1986).

REFERENCES

Andrews, E. J., and Kouzel, B., 1974, "Beavon Sulfur Removal Process Eliminates Sulfur in Claus Plant Tail Gas," paper presented at 53rd GPA Annual Convention, Denver, CO, March 25–27.

Anon., 1951, *Oil and Gas J.*, Vol. 50, No. 4, p. 59.

Anon., 1973, "NPRA's Panel Views Processes," *Hydro. Process.*, April, p. 151.

Anon., 1988, "Oxygen enrichment in Claus process sulfur recovery units," *Sulfur*, No. 195 (March–April), pp. 24–30.

Anon., 1992, *Hydro. Process,* Vol. 71, No. 4, April, p. 126.

Anon., 1995, "Oxygen enrichment enhances plant performance," *Sulphur*, November–December, pp. 45–54.

Audas, F. G., 1951, *Coke and Gas,* Vol. 13, p. 229.

Babcock Duiker, 1983, *Babcock Duiker INFO BULLETIN*, No. 83.013, March.

BASF, 1992, "Catasulf" in *Gas Process Handbook '92, Hydro. Process,* April, p. 91.

Beavon, D. K., 1976, "Process for the Production of Sulfur from Mixtures of Hydrogen Sulfide and Fixed Nitrogen Compounds," U.S. Patent No. 3,970,743.

Beavon, D. K., 1977, U.S. Patent 4,038,036.

Beavon, D. K., and Brocoff, J.C., 1976, "Recent Advances in the Beavon Sulfur Removal Process and the Stretford Process," paper presented at the Second International Conference of the European Federation of Chemical Engineers, University of Salford, England, April 6–8.

Beavon, D. K., and Cameron, D. J., 1970, *Gas Process. (Can.)*, Vol. 62, May/June, pp. 16–21.

Beavon, D. K., and King, F.W., 1970, *Can. Gas J.*, September/October, pp. 22–26.

Beavon, D. K., and Leeper, J. E., 1977, *Sulphur,* No. 133, November/December, pp. 35–39.

Beavon, D. K., and Vaell, R. P., 1971, "Prevention of Air Pollution by Sulfur Plants," paper presented at the AIChE Eighth Annual Technical Meeting, Southern California Section, April 20.

Beavon, D. K., Hass, R. H., and Muke, B., 1979, "High recovery, lower emissions promised for Claus plant tail gas," *Oil and Gas J.*, March 12, p. 76.

Beavon, D. K., Hass, R. H., Kouzel, B., and Ward, J.W., 1980, "Developments in Selectox Technology," paper presented at 7th Canadian Symposium of Catalysis, Edmonton, Alberta, October 19–22.

Berman, M. Y., Amoco Production Company, Tulsa, OK, 1992, personal communication, April 28.

Blohm, C. L., 1952, "Processing Sulfur Bearing Gases," *Petrol. Engr.*, April, pp. C68–C72.

Boas, A. H., and Andrade, R. C., 1971, "Simulate Sulfur Recovery Plants," *Hydro. Process.*, Vol. 50, March, pp. 81–84.

BR&E, 1993, *BR&E Reference Manual 93.0,* Bryan Research and Engineering, Bryan, TX.

Cameron, L. C., 1974, "Aquitaine improves Sulfreen process for more H_2S recovery," *Oil and Gas J.*, Vol. 72, No. 25, June 24, pp. 110–118.

Carmassi, M. J., and Zwilling, J. P., "How SNPA Optimizes Sulfur Plant," 1967, *Hydro. Process.*, Vol. 46, April, pp. 117–121.

Chase, M. W., Davies, C.A., Downey, J. R., Frurip, D. J., McDonald, R. A., and Syverud, A.N., 1985, *JANAF Thermochemical Tables, 3rd Edition,* pub. by the Amer. Chem. Soc. and the Amer. Inst. of Physics for the National Bureau of Standards.

Chen, J. K., Chow, T. K., Cross, F., Hull, R. L., and Watson, D., 1995, "SURE process for Claus sulfur plant oxygen enrichment," paper presented at Sulphur '95, Abu Dhabi, United Arab Emirates, October.

Chludzinski, G. R., Stogryn, E. L., and Weichert, S., 1986, "Commercial Experience with Flexsorb SE Absorbent," paper presented at the 1986 AIChE Spring National Meeting, New Orleans, LA.

Chute, A. E., 1977, "Tailor sulfur plants to unusual conditions," *Hydro. Process.*, Vol. 56, No. 4, pp. 119–124.

Comprimo, B. V., 1994 "Superclaus" in *Gas Processes '94, Hydro. Process.*, April, p. 108.

Dalla Lana, J. G., 1973, Gas Processing/Canada, March–April, pp. 20–25.

Delaney, D. D., Bertram, R. V., and Gowdy, H. W., 1990, "The Selectox Process: Commercial Experience," paper presented at the AIChE 1990 Summer National Meeting, San Diego, CA, August 19–22.

Dowling, N. I., Hyne, J. B., Brown, D. M., 1990, "Kinetics of the Reaction between Hydrogen and Sulfur under High-Temperature Claus Furnace Conditions," *Ind. Eng. Chem. Res.*, Vol. 29, No. 12, pp. 2327–2332.

Dupin, T., and Voirin, R., 1982, "Catalyst enhances Claus operations," *Hydro. Process.*, Vol. 61, No. 11, p. 189.

Fenton, D. M., Woertz, B. B., Brocoff, J. C., and Jirus, E. J., 1975, "Tail Gas Clean-up With the Beavon Sulfur Removal Process," paper presented at NPRA Annual Meeting, San Antonio, TX, March 23–25.

Fischer, H., 1971, "Here's How the Modified Claus Process Treats Low Sulphur Gas," *Oil and Gas International,* Vol. 11, No. 7, July.

Fischer, H., 1974, "Burner/fire box design improves sulfur recovery," *Hydro. Process.*, Vol. 53, No. 10, October, pp. 125–130.

Fischer, H., 1979, "Sulfur costs vary with process selection," *Hydro. Process.*, Vol. 58, No. 3, pp. 125–129.

Fischer, H., 1985, "Claus plants prove flexible," *Hydro. Process,* April, pp. 79–81.

Fischer, H., and Kriebel, M., 1988, "Boosting Throughput in Claus Plants," paper presented at ACHEMA, Frankfurt, Germany, June 6.

Gamson, B. W., and Elkins, R. H., 1953, *Chem. Eng. Prog.*, Vol. 49, No. 4, pp. 203–215.

Goar, B. G., 1982, "World's First Recycle Selectox Sulfur Recovery Unit," *Gas Conditioning Conference Proceedings,* University of Oklahoma, Norman, OK, March 10.

Goar, B. G., 1984, "Sulfur Forming and Degassing Processes," *Gas Conditioning Conference Proceedings,* University of Oklahoma, Norman, OK.

Goar, B. G., Hegarty, W. P., Davis, R., and Kammiller, R., 1985, "Claus plant capacity boosted by oxygen-enrichment process," *Oil and Gas J.*, September 30, pp. 39–41.

Goar, B. G., Hull, R., Bixler, A. D., and Vines, H. L., 1987, "Improving Sulfur Plant Operations with the COPE Process," *Proceedings 66th Annual Gas Processors Association Conference,* Denver, CO, March 16–18, pp. 129–134.

Goar, B. G., 1989, "Emerging New Sulfur Recovery Technologies," Laurance Reid Gas Conditioning Conference Proceedings, University of Oklahoma, Norman, OK, March 6–8.

Goar, B. G., Lagas, J. A., Borsboom, J., and Heijkoop, G., 1991, "Superclaus Updates: How the Process is Performing, Worldwide," paper presented at British Sulphur's 19th International Conference, New Orleans, LA, November 17–20.

Goddin, G. S., Hunt, E. B., and Palm, J. W., 1974, "CBA process ups Claus recovery," *Hydro. Process.,* Vol. 53, No. 10, pp. 122–124.

Goldstein, A. M., 1983, "Commercialization of a New Gas Treating Agent," paper presented at Petroenergy '83 Conference, Houston, TX, September 14.

Grancher, P., 1978, "Advances in Claus technology, Part 1: Studies in reaction mechanics," *Hydro. Process.,* Vol. 57, No. 7, pp. 155–160, "Advances in Claus technology, Part 2: Improvements in industrial units operating methods," *Hydro. Process.,* Vol. 57, No. 9, pp. 257–262.

Grancher, P., and Nougayrede, J., 1976, *Pollution Atmospherique,* No. 70, pp. 129–134.

Gray, M. R., and Svrcek, W. Y., 1981, "Oxygen Use in Claus Sulfur Plants," Gas Conditioning Conference Proceedings, Univ. of Oklahoma, Norman, OK.

Grekel, H., 1959, *Oil and Gas J.,* 57, July 20, pp. 76–79.

Grekel, H., Kunkel, L. V., and McGalliard, R., 1965, *Chem. Eng. Prog.,* Vol. 61, No. 9, pp. 70–73.

Groenendaal, W., and Van Meurs, H. C. A., 1972, *Petroleum and Petrochemical International,* Vol. 12, No. 9, pp. 54–58.

Guyot, G., and Martin, J. E., 1971, "The Sulfreen Process," paper presented at the Canadian Natural Gas Processors Association Meeting, Edmonton, Canada, June.

Haines, H. W., Van Wielingen, G. A., and Palmer, G. H., "Recover Sulfur with Zeolites," 1961, *Petrol. Refiner,* Vol. 40, April, pp. 123–126.

Harvey, C. G., and Verloop, J., 1976, "Experience Confirms Adaptability of the SCOT Process," paper presented at Second International Conference of the European Federation of Chemical Engineers, University of Salford, England, April 6–8.

Hass, R. H., Ingalls, M. N., Trinker, T. A., Goar, B. G., and Purgason, R. S., 1981, "Process meets sulfur recovery needs," *Hydro. Process.,* Vol. 60, No. 5, pp. 104–107.

Hass, R. H., Fenton, D. M., Gowdy, H. W., and Bingham, F. E., 1982, "Selectox and Unisulf: New Techniques for Sulfur Recovery," paper presented at International Sulphur '82 Conference, London, England, November 14–17.

Heigold, R. E., 1991, "MCRC, Its Status and the Status of Sub-Dewpoint Processing," paper presented at British Sulphur's 19th International Conference, New Orleans, LA, November 17–20.

Heigold, R. E., and Berkeley, D. E., 1983, "The MCRC Sub-Dewpoint Sulfur Recovery Process," *Gas Conditioning Conference Proceedings,* University of Oklahoma, Norman, OK.

Heisel, M. P., and Marold, F. J., 1989, "Tests indicate sulfur-recovery process improves on Claus-unit performance," *Oil and Gas J.,* Vol. 87, No. 33, pp. 37–40.

Heisel, M. P., and Marold, F. J., 1992, "How new tail gas treater increases Claus unit thruput," *Hydro. Process.*, March, pp. 83–85.

Herfkens, A. H., 1982, "One company's experience with TGT," *Hydro. Process.*, Vol. 61, No. 11, pp. 199–203.

Hull, R. L., and Sorensen, J. N., 1985, "Oxygen-Based Sulphur Recovery Technology," paper presented at Canadian Gas Processors Association Quarterly Meeting, Calgary, Canada, September 22.

Hull, R., Watson, R., Schendel, R., and Chow, T. K., 1995, "Expand Sulfur Plant Capacity," Fuel Reformulation, September/October, pp. 54–57.

Janke, S. L., 1990, "Catalyst Selection Criteria for Claus Plant Optimization," paper presented at Can Energy 90, Calgary, Canada, May 23.

Kerr, R. K., Jagodzinzki, R. F., and Dillon, J., 1982, "The RSRP: A New Sulfur Recovery Process," Gas Conditioning Conference Proceedings, University of Oklahoma, Norman, OK, March 8–10.

Kettner, R., and Liermann, N., 1988, "New Claus tail-gas process proved in German Operation," *Oil and Gas J.*, January 11, pp. 63–66.

Kettner, R., Luebcke, T., and Sternfels, E. A., 1988, "Experience with the MODOP Tail Gas Treating Process," paper presented at 38th Canadian Chemical Engineering Conference, Edmonton, Canada, October 4.

Kouzel, B., Fuller, R. H., Jirus, E. J., and Woertz, B. B., 1977, "Treat Low Sulfur Gases with Beavon Sulfur Removal Process and the Improved Stretford Process," *Gas Conditioning Conference Proceedings,* University of Oklahoma, Norman, OK, March 7–8.

Kroop, L., Sigmund, P. W., and Taggart, G. W., 1985, "The Sulften™ Process—Advanced Tailgas Treating," *Proceedings Laurance Reid Gas Conditioning Conference,* University of Oklahoma, Norman, OK.

Lagas, J. A., 1982, "Stop emissions from liquid sulfur," *Hydro. Process.*, Vol. 61, No. 10, pp. 85–89.

Lagas, J. A., 1984, "Reduce Operating Problems and Emission From Claus Plants," *Proceedings Sulphur-84,* 3rd International Conference, Calgary, Alberta, pp. 279–289.

Lagas, J. A., Borsboom, J., and Berben, P. H., 1988A, "The Superclaus Process," *Laurance Reid Gas Conditioning Conference Proceedings (Addendum),* University of Oklahoma, Norman, OK, March 7–9.

Lagas, J. A., Borsboom, J., and Berben, P. H., 1988B, "Superclaus-The Answer to Claus Plant Limitations," paper presented at Canadian Chemical Engineering Conference, Edmonton, Canada, October 2–5.

Lagas, J. A., Borsboom, J., and Heijkoop, G., 1989, "Claus process gets extra boost," *Hydro. Process.*, Vol. 68, No. 4, pp. 40–42.

Lagas, J. A., Borsboom, J., and Goar, B.G., 1994, "Superclaus 5 Years Operating Experience," Laurance Reid, Gas Conditioning Conference Proceedings, University of Oklahoma, Norman, OK.

Lee, M. H., Petty, L. E., Wilson, R. H., and Galvin, C., 1984, "ULTRA Tail Gas Cleanup Process," *Chem. Eng. Prog.*, Vol. 80, No. 5, pp. 33–38.

Lee, M. H., and Reed, R. L., 1986, "ELSE Tail Gas Clean-Up Process," Proceedings 65th Annual Gas Processors Association Convention, San Antonio, TX, March 10–12, pp. 109–115.

Lees, R. S., 1970, "Generalized Computer Design and Simulation of Sulphur Plants," M. S. Thesis, University of Alberta, Edmonton, Canada.

Lell, R., and Nougayrede, J. B., 1991, "Reducing Claus plant sulphur emissions with Sulfreen," *Sulphur,* No. 213, March/April, pp. 39–45.

Lim, Y. C., Fukumoto, W., and Pendergraft, P. T., 1986, "Optimized CBA Configuration Improves Gulf Strachan Sulfur Recovery," paper presented at AIChE 1986 Spring National Meeting, New Orleans, LA, April 6–10.

Linde A. G., 1988, "The Clinsulf Sub-Dew-Point Process for Sulphur Recovery," reprint from *Linde Reports on Science and Technology,* No. 44, Linde, A.G.

Linde A. G., 1990, "Linde Processes for Recovery of Sulphur from Gases Containing H_2S," reprint from *Linde Reports on Science and Technology,* No. 47, Linde, A.G.

Linde A. G., 1994 "Clinsulf" in *Gas Processes '94, Hydro. Process.,* April, p. 74.

Luinstra, E. A., and d'Haêne, P. E., 1989, "Catalyst added to Claus furnace reduces sulfur losses," Hydro. Process., Vol. 68, No. 7, July, pp. 53–57.

Lurgi, 1994, "Claus oxygen technology," *Hydro. Process.,* April, p. 74.

Maadah, A. G., and Maddox, R. N., 1978, "Predict Claus products," *Hydro. Process.,* Vol. 57, No. 8, p. 143.

Meissner, R. E., 1983, "Claus Tail Gas Treating with MDEA," paper presented at Union Oil Company of California/The Ralph M. Parsons Company Third BSR and Selectox Users Conference, June 1–2.

Mobil, 1994, "Modop" in *Gas Processes '94, Hydro. Process.,* April, p. 94.

Monnery, W. D., Svrcek, W. Y., and Behie, L. A., 1993, "Modeling the Modified Claus Process Reaction Furnace and the Implications on Plant Design and Recovery," *Can. J. of Chem. Eng.,* Vol. 71, October, pp. 711–724.

Naber, J. E., Wesselingh, J. A., and Groenendaal, W., 1973, "Reduce Sulphur Emission with the Scot Process," *Chem. Eng. Progr.,* Vol. 69, No. 12, pp. 29–34.

Nasato, E., Goar, B. G., and Borsboom, J., 1991, "Superclaus Retrofit, Mobil Oil Canada, Lone Pine Creek Gas Plant," *Laurance Reid Gas Conditioning Conference Proceedings,* University of Oklahoma, Norman, OK, March 4–6.

Nobles, J. E., Palm, J. W., and Knudtson, D.K., 1977, "Design and Operation of First AMOCO CBA Unit," Gas Conditioning Conference Proceedings, University of OK, Norman, OK, March 7–9.

Norman, W. S., 1976, "There are ways to smoother operation of sulfur plants," *Oil and Gas J.,* Vol. 74, No. 46, November 15, pp. 55–60.

Nougayrede, J., 1976, *Sulphur,* No. 127, pp. 37–41.

Nougayrede, J., and Voirin, R., 1989, "Liquid catalyst efficiently removes H_2S from liquid sulfur," *Oil and Gas J.,* Vol. 87, No. 29, July 17, pp. 65–69.

Opekar, P. C., and Goar, B. G., 1966, "This Computer Program Optimizes Sulfur Plant Design and Operation," *Hydro. Process.,* Vol. 45, No. 6, pp. 181–185.

Parnell, D., 1985, "Look at Claus unit design," *Hydro. Process.,* Vol. 64, No. 9, pp. 114–118.

Paskall, H. G., 1979, *Capability of the Modified Claus Process,* Alberta/Canada Energy Resources Research Fund, Alberta, Canada, March.

Paskall, H. G., and Sames, J. A. (editors), 1992, *Sulphur Recovery,* BOVAR/Western Research, Calgary, Canada.

Pearson, M. J., 1973, "Developments in Claus catalysts," *Hydro. Process.*, Vol. 52, No. 2, pp. 81–85.

Pearson, M. J., 1976, *Energy Processing/Canada,* July–August, pp. 38–42.

Pearson, M. J., 1978, "Determine Claus Conversion from catalyst properties," *Hydro. Process.*, Vol. 57, No. 4, pp. 99–103.

Pearson, M. J., 1981, "Special catalyst improves C-S compounds conversion," *Hydro. Process.*, Vol. 60, No. 4, p. 131.

Pearson, M. J., and Belding, W. A., 1985, "Claus Process Modelling to Optimize Sulphur Recovery," Energy Processing/Canada, Vol. 77, No. 5, pp. 31–34.

Peter, S., and Woy, H., 1969, *Chemie Inginieur Technik,* Vol. 41, No. 1 + 2, pp. 1–6.

Reed, R. L., 1983, "Amoco CBA Tail Gas Process Updated," paper presented at Canadian Gas Processors Association Quarterly Meeting, September 14.

Sames, J. A., and Paskall, H. G., 1984, "So you don't have a COS/CS_2 problem, do you?" *Sulphur,* No. 172, May–June.

Sames, J. A., and Paskall, H. G., 1985, "Simulation of reaction furnace kinetics for split-flow sulphur plants," paper presented at Sulphur '85 Conference, London, UK, November.

Sames, J.A., and Paskall, H. G., 1987, "Can Oxygen Enrichment Replace Tail Gas Clean Up in Sulphur Recovery?" paper presented at Sulphur '87 Conference, Houston, TX, April.

Sames, J. A., Ritter, R. A., Paskall, H. G., 1985, "PC Sulphur Plant Simulation," paper presented at the 35th Canadian Chemical Engineering Conference, Calgary, Alberta, Canada, October.

Sames, J. A., Dale, P. R., and Wong, B., 1987, "Evaluation of Reaction Furnace Variables in Modified-Claus Plants," *Laurance Reid Gas Conditioning Conference Proceedings,* University of Oklahoma, Norman, OK.

Sames, J. A., Paskall, H. G., Brown, D. M., Chen, S. K., and Sulkowski, D., 1990, "Field Measurements of Hydrogen Production in an Oxygen-Enriched Claus Furnace," paper presented at Sulphur's 18th International Conference, Cancun, Mexico, April 1–4.

Sawyer F. G., Hader, R. N., Herndon, L. K., and Morningstar, E., 1950, "Sulfur from Sour Gases," *Ind. Eng. Chem.*, Vol. 42, No. 10, October, pp. 1938–1950.

Schalke, P., Godschalk, P., and Goar, B. G., 1989, "Optimum Burner Performance Essential for Optimum Operation of Sulfur Recovery Systems," *Laurance Reid Gas Conditioning Conference Proceedings,* University of Oklahoma, Norman, OK.

Schendel, R., 1997, Consultant, personal communication, March 7.

Schicho, C. M., Watson, E. A., Clem, K. R., and Hartley, D., 1985, "A New, Safer Method of Sulfur Degassing," *Chem. Eng. Prog.*, October, pp. 42–44.

Shell International Research Mij. B. V., and Shell Oil Co., 1994 "Scot" in *Gas Processes '94, Hydro. Process.*, April, p. 98.

Simek, I. O., 1991, "Sulfur unit circulates catalyst," *Hydro. Process.*, Vol. 70, No. 4, p. 45.

Société Nationale Elf Aquitaine (Production), and Lurgi GmbH., 1994, "Sulfreen" in *Gas Processes '94, Hydro. Process.*, April, p. 108.

Stern, L. H., Stevens, D. K., and Nehb, W., 1994, "Sulfur Recovery in the 90's: Oxygen-Use Technology," *Seventy-Third Annual Convention Proceedings Gas Processors Association,* pp. 141–144.

Stevens, D. K., Stern, L. H., and Nehb, W., 1996, "Oxyclaus Technology for Sulfur Recovery," *Laurance Reid Gas Conditioning Conference Proceedings,* University of Oklahoma, Norman, OK.

Stull, D. R., and Prophet, H, 1971, *JANAF Thermochemical Tables, 2nd Edition,* U.S. Dept. of Commerce National Bureau of Standards, U.S. Govt. Printing Office, Report No. 16955.

Taggart, G. W., 1980, "Optimize Claus control," *Hydro. Process.,* Vol. 59, No. 4, pp. 133–137.

TPA Inc., 1994, "Resulf," in *Gas Processes '94, Hydro. Process.,* April, p. 98.

Tragitt, G. N., Armstrong, T. R., Bourdon, J. C., Sigmund, P. W., 1986, "SULFTEN System commercialized," *Hydro. Process.,* February, pp. 27–29.

Unocal Science and Technology Div. and the Ralph M. Parsons Co., 1994, "Beavon-others" in *Gas Processes '94, Hydro. Process.,* April, p. 72.

Valdes, A. R., 1964A, "New Look at Sulfur Plants Part 1: Design," *Hydro. Process.,* Vol. 43, No. 3, pp. 104–108.

Valdes, A. R., 1964B, "New Look at Sulfur Plants Part 2: Operations," *Hydro. Process.,* Vol. 43, No. 4, pp. 122–124.

Valdes, A. R., 1965, "New Way to Design Firetube Reactors," *Hydro. Process.,* Vol. 44, No. 5, May, pp. 223–229.

Warner, R. E., 1982, "Save with Selectox," paper presented at the Canadian Gas Processors Association Quarterly Meeting, Calgary, Alberta, Canada, September 8.

Watson, E. A., Hartley, D., and Ledford, T. H., 1981, "Catalytically degas Claus sulfur," *Hydro. Process.,* Vol. 60, No. 5, pp. 102–103.

Watson, R.W., Hull, R., and Sarssam, A., 1995/96, "The successful use of oxygen in Claus plants," *Hydrocarbon Technology International,* Quarterly:Winter 1995/96, pp. 95–101.

Weinberg, H.N., Eisenberg, B., Heinzelmann, F.J., and Savage, D.W., 1983, "New Gas Treating Alternatives for Saving Energy in Refining and Natural Gas Processing," paper presented at 11th World Petroleum Congress, London, England, August 31.

Wen, T. C., Chen, D. H., Hopper, J.R., and Maddox, R.N., 1987, "Claus Simulation with Kinetics," Energy Processing/Canada, July–August, pp. 27–31.

Wiley, S., 1980, "Off-gas aids Claus operations," *Hydro. Process.,* Vol. 59, No. 4, pp. 127–129.

Yen, C., Chen, D. H., and Maddox, R. N., 1985, "Simulating Various Schemes of the Claus Process," paper presented at the AIChE Spring National Meeting, Houston, TX, March 24–28.

Chapter 9
Liquid Phase Oxidation Processes for Hydrogen Sulfide Removal

INTRODUCTION, 732

PROCESSES OF HISTORICAL INTEREST, 733

    Polythionate Solutions, 734
    Iron Oxide Suspensions, 736
    Iron Cyanide Solutions and Suspensions, 744

THIOARSENATE PROCESSES, 748

    Thylox Process, 748
    Giammarco-Vetrocoke Process, 754

QUINONE AND VANADIUM METAL PROCESSES, 759

    Perox Process, 762
    Takahax Process, 765
    Stretford Process, 769
    Hiperion Process, 794
    Sulfolin Process, 797
    Unisulf Process, 802

CHELATED IRON SOLUTIONS, 803

    Cataban Process, 804
    LO-CAT Process, 805
    Sulfint Process, 823
    SulFerox Process, 825

SULFUR DIOXIDE PROCESSES, 840
   Townsend Process, 841
   IFP Clauspol 1500 Process, 843
   Wiewiorowski Process, 846
   UCBSRP Process, 846

MISCELLANEOUS PROCESSES, 850
   Fumaks Process, 850
   Konox Process, 851
   EIC Copper Sulfate Process, 853
   Permanganate and Dichromate Solutions, 855

REFERENCES, 856

INTRODUCTION

In the mid-nineteenth century, dry iron oxide boxes replaced aqueous calcium hydroxide scrubbers for the removal of sulfur compounds from sour gas streams. Although the new dry method was superior to the prior technology, the process had some inherent disadvantages. The main drawbacks of the dry purification process were (1) large ground-space requirements, (2) high labor costs, and (3) production of low quality sulfur. Through the beginning of this century, the search for more efficient methods for hydrogen sulfide removal from industrial gases continued, and quite naturally turned to purification methods employing liquids in regenerative cycles capable of yielding pure elemental sulfur.

Processes based on the absorption and oxidation of hydrogen sulfide to elemental sulfur in a liquid system have the advantage over absorption-stripping cycles of selectivity for hydrogen sulfide in the presence of carbon dioxide. A high ratio of carbon dioxide to hydrogen sulfide in the feed gas to a typical absorption-stripping type process (e.g., ethanolamine) can result in an acid gas stream that is difficult to process in a Claus plant. Potential drawbacks of liquid phase oxidation processes are the relatively low capacities of the solutions for hydrogen sulfide and oxygen which can result in large liquid flow rates, the difficulty of separating precipitated sulfur from the liquid mixture, and the requirement to dissipate the heat of hydrogen sulfide oxidation at a low temperature level rather than at a level suitable for the generation of steam as in the Claus process.

Since most of the early gas purification work was related to the processing of manufactured and coke-oven gas streams, which contain ammonia as well as hydrogen sulfide, attempts were made to develop combination processes that would remove both impurities while producing marketable products such as ammonium sulfate and elemental sulfur. The first processes of this type were based on a recirculating solution of ammonium polythionate. They proved technically feasible, but quite complex, and were not commercially successful.

Another logical step in the development of continuous hydrogen sulfide removal processes was the utilization of the well-known iron oxide dry box chemistry in a liquid form. This involved the use of a slurry of iron oxide particles in a mildly alkaline aqueous solution. Several processes of this type were developed in Europe and the United States beginning with the pioneering work of Burkheiser shortly before World War I and ending with the once widely used Ferrox process.

The next developmental stage involved the use of more complex inorganic molecules, such as iron-cyanide radicals and thio-arsenates, which exist in more than one form and change from one form to another as they react sequentially with hydrogen sulfide and oxygen. The iron-cyanide systems proved to be somewhat unwieldy; however, the arsenic-based methods were quite effective, and led to the rapid commercial development of the Thylox process just before World War II, and the Giammarco-Vetrocoke process in the 1950s. In recent years, the arsenic based systems have lost market acceptance because of mounting concern about the toxicity of the scrubbing liquor.

Processes using quinones in a redox cycle for H_2S removal were introduced in the 1950s. The Stretford process, which combined a quinone with vanadium salts in the absorbent solution, proved highly successful, and was the dominant process of this type during the 1970s.

In the mid-1970s, another major development took place. Single catalyst systems based on chelated iron were introduced in the marketplace. Although the basic technology behind these processes has been well known for many years, progress in the suppression of catalyst degradation in iron-based systems, coupled with environmental concerns about the presence of vanadium in the blowdown of Stretford plants, have recently made iron chelate based processes like LO-CAT and SulFerox strong contenders in the hydrogen sulfide removal market.

Processes using sulfur dioxide as oxidant instead of molecular oxygen, several of which were developed in the 1960s, have not become a factor in either natural gas or geothermal applications. With the exception of the IFP Clauspol process, which was tailored to Claus tail gas applications, none of these methods has been commercially successful.

In this chapter, most of the liquid redox processes that have attained commercial status for desulfurization of sour gas are reviewed. The processes have been divided into several major groups based on solution chemistry. The major groups and specific processes in each group, which are described in this chapter, are listed in **Table 9-1**. The process descriptions follow the organization of this table.

The polythionate, iron oxide, and iron-cyanide processes are primarily of historical interest and only an abridged discussion of the technology is given. Thioarsenate processes are covered in somewhat greater detail because of the relative importance of these processes in the past. Special attention is given to the quinone and/or vanadium and the iron-chelate processes because they include contemporary processes of major commercial significance. Finally, the sulfur dioxide and miscellaneous processes, which are of lesser importance, are included for their technical interest.

PROCESSES OF HISTORICAL INTEREST

The liquid redox gas desulfurization processes grouped in this section are selected mainly for their historical value. They comprise three separate broad categories and are differentiated by the scrubbing solution chemistry. These process categories are 1) polythionate solutions, 2) iron oxide suspensions, and 3) iron cyanide solutions. Of the three types, only the iron oxide suspension processes were used extensively at one time. Except in very rare cases, most of these plants have since been shut down and replaced with more modern gas desulfurization units.

Table 9-1
Categorization and Commercial Status of Liquid Phase Oxidation Processes for Hydrogen Sulfide Removal

Process Type	Process Name	Developer	Date*	Status
Polythionate	Koppers C.A.S.	Koppers	1945	I
Iron oxide	Ferrox	Sperr/Koppers	1926	C
	Gluud	Gluud	1927	I
	Burkheiser	Burkheiser	1953	I
	Manchester	Manchester Gas	1953	I
Iron-cyanide	Fischer	Fischer/Mueller	1931	I
	Staatsmijnen-Otto		1945	I
	Autopurification		1945	I
Thioarsenate	Thylox	Gollmar/Koppers	1929	I
	Giammarco-Vetrocoke	Giammarco-Vetrocoke	1955	C
Naphtho-quinones and/or vanadium	Perox	Krupp Koppers	1950	C
	Takahax	Tokyo Gas Co., LTD	1964	A
	Stretford	Nicklin/British Gas	1963	A
	Hiperion	Hasebe/Ultrasystems	1986	C
	Sulfolin	Weber/Linde AG	1985	A
	Unisulf	Unocal	1981	C
Iron-chelate	Cataban	Rhodia Inc.	1972	I
	LO-CAT	U.S. Filter Engineered Systems	1978	A
	Sulfint	Integral Eng.	1980	A
	SulFerox	Shell/Dow Chemical	1986	A
Sulfur dioxide	Townsend	Stearns-Roger	1965	I
	IFP Clauspol 1500	IFP	1969	A
	Wiewiorowski	Freeport Sulfur	1969	I
	UCBSRP	UCB	1986	I
Misc. oxidizer	Fumaks	Sumitomo Metals	1970	I
	Konox	Sankyo	1975	I
	Cuprosol	EIC Corp.	1980	I

Status:
I = Inactive or developmental
C = Commercial but not currently being promoted
A = Actively marketed
* = Date denotes the approximate year the process was either commercialized or described in the literature.

Polythionate Solutions

The early development of liquid oxidation processes using polythionate solutions for the removal of hydrogen sulfide from gases derived from coal is to a large extent identified with the work of Feld in Germany. Feld started his studies before the outbreak of World War I, and his principal aim was to devise a process by which hydrogen sulfide and ammonia could be removed simultaneously from coal gas and subsequently converted to ammonium sulfate and elemental sulfur. The chemical basis for this type of process can be illustrated schematically by either one of the following overall equations depending on whether oxygen or sulfur dioxide is used to oxidize ammonium sulfide to ammonium sulfate:

$$(NH_4)_2S + 2O_2 = (NH_4)_2SO_4 \tag{9-1}$$

$$(NH_4)_2S + 2SO_2 = (NH_4)_2SO_4 + 2S \tag{9-2}$$

It should be observed that the simplicity of these two equations is very misleading, as the actual process chemistry is quite complex, with a large number of reaction pathways possible within each system.

The cyclic redox process proposed by Feld involves the absorption of hydrogen sulfide and ammonia in aqueous solutions containing a mixture of ammonium tri- and tetrathionate, together with a small amount of ammonium sulfite. Most of the H_2S and NH_3 are converted to ammonium thiosulfate, ammonium sulfide, and sulfur. The spent solution is regenerated by addition of sulfur dioxide, which reacts with the ammonium thiosulfate forming again tri- and tetrathionate plus ammonium sulfate. The regenerated solution, which gradually accumulates an increasing concentration of ammonium thiosulfate and ammonium sulfate, is recycled for further absorption of hydrogen sulfide and ammonia. When the thiosulfate concentration reaches 30 to 45%, it is oxidized once more with SO_2 to regenerate the polythionates, and then heated. The polythionates are converted to ammonium sulfate, SO_2, and elemental sulfur. Unconverted thiosulfate remaining in the solution reacts with polythionate and is decomposed into ammonium sulfate and elemental sulfur.

A great deal of difficulty was encountered with this process primarily because of the numerous chemical reactions involved. Proper functioning was not only dependent on the concentration ratio between H_2S and NH_3, but also on the maintenance of close temperature control because of the complicated solubility relationships of the various salts present in the system. Feld also experimented with solutions containing zinc and iron polythionates, and later Terres et al. (1954) and Terres (1953) described a process using manganese polythionates and manganese sulfate. The complicated chemistry of the various polythionate processes has been described in some detail by Terres (1953).

Other processes based on the original work of Feld are the Gluud combination process and the Koppers C.A.S. process described by Gollmar (1945). In the Gluud combination process (not to be confused with the Gluud iron oxide process), the hydrogen cyanide-free gas is washed with a solution containing a mixture of thio-compounds and iron sulfate. The spent solution is first saturated with sulfur dioxide and then aerated in a tall tower where the hydrosulfides and the sulfites are converted to thiosulfates, which are then used for further absorption of hydrogen sulfide. During the cyclic operation, thiosulfates accumulate, and a portion of the solution is withdrawn continuously for conversion of the thiosulfates to ammonium sulfate and elemental sulfur.

In the Koppers C.A.S. process—symbolizing cyanogen, ammonia, sulfur—the hydrogen cyanide is first removed with a recirculating solution containing ammonia and elemental sulfur, with the formation of ammonium thiocyanate. Hydrogen sulfide and ammonia are then removed from the gas by contact with a solution containing ammonium polythionate, ammonium sulfite, ammonium thiosulfate, and some iron compounds. This solution is circulated in four towers operated in series, and ammonia is injected into the solution at various points of the system to obtain complete H_2S removal.

Ammonium sulfate and elemental sulfur are produced during solution regeneration. To regenerate the solution, a portion of the solution is withdrawn from all four towers and divided into three parts. One part is regenerated by aeration during which iron sulfide and ammonium hydrosulfide oxidize to form elemental sulfur. The second part is treated with sulfur dioxide, and the sulfur formed is recovered by filtration. The third part is also treated with sulfur dioxide, having first been decanted from the solids. The resulting liquid is then combined with the filtrate from the second part of the solution and heated under pressure at 350°F. In this step the thiocyanate and the other thio-compounds are converted to ammonium sulfate and elemental sulfur.

Over time, many variations of polythionate processes were developed and tried in Europe, but for the most part these processes remained impractical in commercial operation. An excellent review of polythionate processes is presented by Hill (1945).

Iron Oxide Suspensions

A logical step in the development of processes employing liquids in regenerative cycles was the utilization of the reaction between iron oxide and hydrogen sulfide followed by conversion of iron sulfide to iron oxide and elemental sulfur. Several processes using iron oxide suspended in alkaline aqueous solutions were developed in Europe and the United States, beginning with the work of Burkheiser shortly before the first world war. During the 1920s, the Ferrox process was introduced by the Koppers Company of Pittsburgh, Pennsylvania, and an almost identical process was disclosed by Gluud in Germany. Later, a modification of the Ferrox process was developed in England and became known as the Manchester process. The Burkheiser process was not used extensively on a commercial scale because of problems with the ammonia removal cycle. Numerous Ferrox plants were built in the United States, but almost all have been replaced by installations using more efficient processes. The Gluud process was still sparingly used in Europe in the 1980s. The Manchester process, which for a while enjoyed some popularity in Great Britain, has been generally replaced by the more efficient Stretford process, discussed later in this chapter.

In addition to iron oxide, other metals oxides were tried. Nickel, in particular, proved to be an active agent for hydrogen sulfide removal. However, nickel forms soluble salts with hydrogen cyanide from which it cannot be easily regenerated. Because of this problem, and its relatively high price compared to iron, nickel was never used on a large scale.

The iron oxide suspension family of processes is of historical interest because they constitute transitional processes that evolved from the traditional iron-oxide dry-box processes and were forerunners of the chelated-iron based processes that currently share the U.S. market for liquid redox processes together with the vanadium based Stretford process.

Basic Chemistry

The chemistry of all iron oxide suspension processes is based on the reaction of H_2S with an alkaline compound, either sodium carbonate or ammonia, followed by the reaction of the hydrosulfide with iron oxide to form iron sulfide. Regeneration is effected by converting the iron sulfide to elemental sulfur and iron oxide by aeration. This portion of the cycle involves essentially the same reactions as those occurring in dry-box purifiers. The following equations represent the reaction mechanism:

$$H_2S + Na_2CO_3 = NaHS + NaHCO_3 \tag{9-3}$$

$$Fe_2O_3 \cdot 3H_2O + 3NaHS + 3NaHCO_3 = Fe_2S_3 \cdot 3H_2O + 3Na_2CO_3 + 3H_2O \tag{9-4}$$

$$2Fe_2S_3 \cdot 3H_2O + 3O_2 = 2Fe_2O_3 \cdot 3H_2O + 6S \tag{9-5}$$

Besides the main reactions, several side reactions (mostly leading to the formation of undesirable sulfur compounds) occur in the process. These side reactions depend on the operating conditions and the composition of the gas to be treated. Usually a certain amount of thiosulfate formation is inevitable. In some cases it may even be desirable to operate these processes so that hydrogen sulfide is quantitatively converted to thiosulfate according to the following equations:

$$2NaHS + 2O_2 = Na_2S_2O_3 + H_2O \tag{9-6}$$

$$Na_2S + 1\tfrac{1}{2}O_2 + S = Na_2S_2O_3 \tag{9-7}$$

A further side reaction in which sulfur is also converted to an undesirable product is caused by the absorption of hydrogen cyanide in the alkaline material. The hydrogen cyanide is first converted to sodium cyanide and then reacts with elemental sulfur to form thiocyanate:

$$HCN + Na_2CO_3 = NaCN + NaHCO_3 \tag{9-8}$$

$$NaCN + S = NaSCN \tag{9-9}$$

Normally only a small portion of the hydrogen cyanide reacts in this manner as most of the absorbed hydrogen cyanide is stripped from the solution by the regeneration air.

The presence of hydrogen cyanide in the gas to be purified leads to still another side reaction that may have considerable influence on the operation of the process. It was observed (Sperr, 1926) that very noticeable color changes occurred in the solution when gas containing relatively large amounts of hydrogen cyanide (approximately 10% of the hydrogen sulfide) was treated. In this case, the oxidized solution displays a blue coloration indicating the presence of ferric-ferrocyanide complexes, such as Prussian blue, while the fouled solution becomes pale yellow in color. In addition, it was also noticed that, while the reaction between hydrogen sulfide and iron oxide is quite slow, the presence of blue complexes results in rapid conversion of hydrogen sulfide to elemental sulfur. Although the exact chemical nature of the blue and yellow compounds is not known, it is hypothesized that the reactions responsible for the color shift involve oxidation of H_2S to elemental sulfur by conversion of the ferric-ferrocyanide complex to ferrous ferrocyanide. In the regeneration step the ferric-ferrocyanide complex is re-established. The reactions occurring in the cycle can be represented schematically by the following equations:

$$2H_2S + Fe_4[Fe(CN)_6]_3 + 2Na_2CO_3 = 2Fe_2Fe(CN)_6 + Na_4Fe(CN)_6 + 2H_2O + 2CO_2 + 2S \tag{9-10}$$

$$2Fe_2Fe(CN)_6 + Na_4Fe(CN)_6 + O_2 + 2H_2CO_3 = Fe_4[Fe(CN)_6]_3 + 2Na_2CO_3 + 2H_2O \tag{9-11}$$

It is likely that under these conditions hydrogen sulfide does not react at all with iron oxide and that reactions 9-10 and 9-11 are the only ones occurring. The iron, which in most cases is added to the solution as soluble iron sulfate, serves only to replenish the iron-cyanide compounds lost with the sulfur.

Burkheiser Process

This process, which was developed in Germany at approximately the same time as Feld was conducting his polythionate work, is described in some detail by Terres (1953). As in Feld's work, the process objective was the removal of H_2S and NH_3 from coal gas. The principal difference between the two methods is that in the Feld process, H_2S and ammonia are absorbed simultaneously, while in the Burkheiser process they are removed in two consecutive steps. However, only the H_2S-removal portion of the Burkheiser process was successful and applied on a commercial scale.

In the Burkheiser H_2S removal stage, hydrogen sulfide and hydrogen cyanide are absorbed in an aqueous solution containing ammonia, iron oxide, and elemental sulfur. The spent solution leaving the absorber is a slurry containing a mixture of elemental sulfur and iron sulfide solids. To separate the iron sulfide, the slurry is introduced into a "sulfur dissolving" vessel where the suspended free sulfur is converted to soluble ammonium polysulfide by the action of gaseous ammonia and hydrogen sulfide. Subsequently, the iron sulfide solids are removed from the solution by filtration and regenerated by contact with atmospheric oxygen. The iron oxide and elemental sulfur formed in this operation are resuspended in an aqueous solution of ammonia and recycled for further absorption of hydrogen sulfide. The filtrate, which contains a mixture of ammonium polysulfide, ammonium cyanide, ammonium thiocyanate, and ammonia, is heated to approximately 200°F, and the ammonium polysulfide is decomposed into ammonia, hydrogen sulfide, and elemental sulfur. The gaseous ammonia and hydrogen sulfide are absorbed by the spent solution in the "sulfur dissolving" vessel and reused to convert elemental sulfur to ammonium polysulfide as previously discussed.

The reprecipitated free sulfur is separated from the filtrate, and the residual solution, containing cyanides and thiocyanates, is treated with a suspension of calcium hydroxide. The precipitated calcium cyanide and calcium thiocyanate are filtered and added to the coal used in gas manufacturing. During gasification the cyanogen compounds are converted to hydrogen sulfide and ammonia. The advantage of this rather complicated process is that the only end products of gas purification are elemental sulfur and ammonia.

Ferrox Process

The Ferrox process was disclosed by Sperr (1926, 1932) of the Koppers Company, Pittsburgh, PA, and subsequently used on a fairly extensive scale. Ferrox was one of several gas purification processes developed by Koppers in the 1920s. The first was the Seaboard process, which was shortly followed by the Ferrox process in 1926 and the Thylox process in 1929. The Ferrox process was superior to the Seaboard process because more complete hydrogen sulfide removal was obtained, while only small amounts of carbon dioxide were removed at the same time. Over the years, Ferrox plants have been replaced by installations using more efficient processes, but at least one plant was reported to be in operation in 1992.

The Ferrox process represented a marked improvement over dry-box purification because the plants occupied only a fraction of the ground area necessary for dry boxes treating equivalent volumes of gas. In addition, the labor cost was reduced appreciably, and the initial installation cost was somewhat lower than that for dry-box purifiers. A contemporary factor that bolstered commercialization of the Ferrox process was the development of flotation techniques for recovering elemental sulfur after the oxidation step. The principal disadvantage of the process was that complete removal of H_2S could not be obtained as readily and regularly as by the use of dry boxes.

Process Description. A schematic flow diagram of the Ferrox process is shown in **Figure 9-1**. The solution, normally containing 3.0% sodium carbonate and 0.5% ferric hydroxide, is pumped to the top of the absorber where it is countercurrently contacted with the gas fed into the bottom of the vessel. The hydrogen sulfide-containing solution flows from the bottom of the absorber to the thionizer, or regenerator, where elemental sulfur is formed by contact of the solution with air. The sulfur accumulates on the liquid surface as a froth, enters the slurry tank, and is pumped from there to a filter where excess liquid is removed. The regenerated solution is pumped from the thionizer through a heater to the absorber, thus completing the cycle. The liquid obtained in the filter is usually discarded, thus providing a means for continuously purging the system of undesirable salts.

Design and Operation. The absorber used in Ferrox installations has a lower section, or saturator, and an upper section, the absorber proper. The saturator contains a continuous liquid phase, several feet high, through which the raw gas is bubbled before it enters the upper section. The function of the saturator is to provide sufficient contact time to complete the reaction between sodium hydrosulfide and ferric oxide before regeneration of the solution. If essentially complete reaction is achieved, thiosulfate formation in the regenerator is kept at a minimum. The upper part of the absorber contains sprays and wooden hurdles similar to those used in the Seaboard process and usually has a total height of 60 ft (Sperr, 1926).

The thionizer consists of long shallow tanks, each containing several compartments arranged so that the solution can be transferred from one compartment to another. The com-

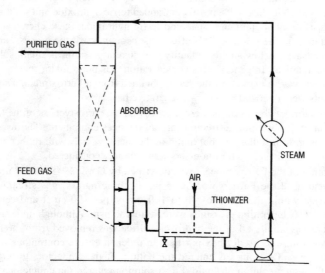

Figure 9-1. Typical flow diagram of Ferrox process.

partments contain directional baffles which ensure proper flow of the solution along the total length of the tanks. The depth of the liquid in the thionizers is approximately 42 in. Air is admitted to the bottom of the thionizer and dispersed into fine bubbles by passage through cloth-covered tubes. Typical cloth tubes are about 5 in. in diameter by 10 ft in length and are mounted on steel pipes which are manifolded above the thionizer tank (Sperr, 1926).

The liquid is circulated at such a rate that a two- to threefold excess of ferric hydroxide over the stoichiometric quantity necessary for the complete reaction with hydrogen sulfide is present. Gollmar (1945) states that the process can be operated with less than the stoichiometric concentration of iron oxide and interprets the function of the iron as a catalytic oxygen carrier. Available historical data from several plants indicate that operation with an excess of iron oxide over the stoichiometric amount was commonly practiced. This excess seems to be required for complete removal of hydrogen sulfide and, also, to minimize thiosulfate formation in the thionizer. For a coal gas plant with a 10 MMscf/day capacity and a hydrogen sulfide removal rate of 400 grains/100 scf, the chemical requirements are approximately 3,500 lb/day of sodium carbonate and 2,800 lb/day of iron.

The air requirements for oxidation of ferric sulfide, assuming complete removal of H_2S, are 300 cu ft/1,000 cu ft of gas. This is equivalent to a ratio of about 10 moles of oxygen to 1 mole of hydrogen sulfide for a gas containing 400 grains hydrogen sulfide per 100 cu ft (Sperr, 1926). Air requirements depend to a large extent on proper oxygen utilization in the thionizer. Since theoretically only ½ mole of oxygen is required per mole of H_2S, it is evident that improved thionizer design should permit the use of appreciably smaller quantities of air.

The efficiency of hydrogen sulfide removal by this process varies from 85 to practically 100%. Sperr (1926) reports that hydrogen sulfide concentrations in the treated gas, sufficiently low to satisfy the U.S. Bureau of Standards lead acetate test, can be achieved in a single absorber. However, when high-purity gas is required, two-stage absorption is recommended. Depending on the hydrogen cyanide content of the gas and the rate of thiosulfate formation, 70 to 80% of the hydrogen sulfide can be recovered as elemental sulfur.

The solids obtained in the filters contain from 30 to 50% elemental sulfur, approximately 50% moisture, and 10 to 20% salts, mostly entrapped ferric hydroxide and sodium carbonate. Because of this loss of sodium carbonate and ferric hydroxide, these chemicals have to be added continuously to the solution. This affects the economics of the process, and for plants where the gas capacity is low and the quantity of recoverable sulfur is small, solution regeneration is uneconomical. In general, this consideration was not a major factor in the United States since there was no market for the sulfur obtained from a Ferrox plant. Therefore, most American plants were operated on a non-regenerative basis.

One of the major drawbacks of the Ferrox process is the corrosiveness of the treating solution which causes fairly rapid destruction of carbon-steel equipment. The use of alloys is uneconomical in most installations, but lining of the major vessels with rubber and the possible use of coated wooden tanks for thionizers was routinely considered.

A modification of the Ferrox process was described by Gard (1948) and Bailey (1966) of Unocal. Unocal used the Ferrox process to purify three natural gas streams containing hydrogen sulfide within the range of 40 to 100 grains per 100 cu ft and carbon dioxide between 4 and 26%. Operating pressures were 80 to 160 psig. Although under normal operating conditions essentially all of the hydrogen sulfide was removed from the feed gas, the Ferrox unit was followed by dry-box purifiers in order to ensure continuous production of pipeline quality gas. The sour gas entered the bottom of the contactors through a sparger and bubbled through a column of liquid. Fresh solution entered the contactor with the feed gas. Spent solution was withdrawn through a draw-off tray and sent to regeneration troughs where it was aerated, and the elemental sulfur formed was skimmed from the liquid. The regenerated liquid was then returned to the contactors. A photograph of this installation is shown in **Figure 9-2.**

Gluud Process

The Gluud process was introduced in Germany in 1927 independently of the American Ferrox process (Gluud and Schoenfelder, 1927). The chemical reactions in this process are the same as those of the Ferrox process with the exception that a dilute solution of ammonium carbonate is used instead of a sodium carbonate solution. The principal difference between the two processes is that the Gluud process employs taller regenerators which permit much more efficient oxygen utilization. This advantage, however, is somewhat offset by the need to compress air against a higher pressure head. For instance, while a typical Ferrox installation requires 300 cu ft air/1,000 cu ft gas, the equivalent plant using the Gluud process requires only 30 cu ft air/1,000 cu ft gas for solution regeneration (Gluud and Schoenfelder, 1927). For a typical coal gas containing 400 grains hydrogen sulfide per 100 cu ft, this air demand is equivalent to 200% of the stoichiometric requirement, or about one

Figure 9-2. Ferrox plant used for removal of hydrogen sulfide from natural gas. *Courtesy of Unocal*

mole of oxygen per mole of hydrogen sulfide. A schematic flow diagram of the Gluud process is shown in **Figure 9-3**.

Manchester Process

A modification of the Ferrox process was developed in England in the 1950s at the Rochdale Works of the Manchester Corporation Gas Department and is known as the Manchester process. This process, which is covered by British Patents 550,272 and 611,917, was subsequently used in several British gas works installations. A large plant, with a capacity to handle 80 million cu ft air/day, was also built to remove hydrogen sulfide from the exhaust air in a viscose cellulose manufacturing plant. In this installation, the hydrogen sulfide content of the air was reduced from 255 to approximately 5 ppm (Roberts and Farrar, 1956).

Process Description. The principal difference between the Manchester and Ferrox processes is the use of multistage treatment in the Manchester process. Fresh solution is fed to each washing stage in the Manchester process, while a single contact is used in the Ferrox process. To ensure completion of the reaction between hydrogen sulfide and iron oxide, the Manchester process provides a separate delay vessel between the absorbers and the regenerators. The regenerators are tall vessels providing relatively long contact times between the solution and the air, which is introduced by means of several rotary or turbo diffusers.

Design and Operation. The design and operation of a typical Manchester plant located in Linacre, Liverpool, is described in detail by Townsend (1953). In this plant, which processes about 3 MMscf gas/day, the gas passes consecutively through six cylindrical absorption towers that are 7 ft 6 in. in diameter by 25 ft in overall height. The towers are packed with wooden boards, and the liquid is distributed in the first two and last two towers by rotating distributors and in the middle two towers by serrated troughs. Fresh solution is pumped into each

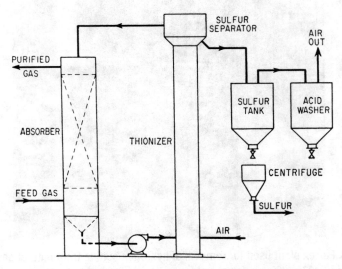

Figure 9-3. Typical flow diagram of Gluud process.

tower and collected in a flooded manifold from which it flows into the delay tanks. The delay tanks are of sufficient size to provide a residence time of 7 to 10 min, which appears to be necessary to drive the reaction between sulfide ion and iron oxide to completion. From the delay tanks, the solution flows to the oxidizers that are 10 ft in diameter by 25 ft in overall height. The effective height of the oxidizers is approximately 16 ft because of the arrangement of the diffusers in the bottom of the vessels. The elemental sulfur liberated in the regenerators is removed as a froth from the top of the vessels and collected in sludge-receiving tanks. From there, the sulfur is processed further and finally recovered in 90 to 95% pure form, the contaminants being primarily iron oxide, sodium carbonate, some thiosulfate, and water. **Figure 9-4** shows a photograph of the absorber section of a Manchester plant treating 12 million scf/day of coal gas. Typical operating data reported for two different plants, each treating about 3 million cu ft gas daily, are shown in **Table 9-2.**

The Manchester process absorber design underwent several modifications, and its configuration varies depending on the operating conditions in each plant. In addition to packed cylindrical vessels, rectangular vessels are also used as coal gas absorbers (R. & J. Dempster, 1957). For rayon plants, where the large volume of exhaust air contains a relatively small amount of H_2S, horizontal spray absorbers with semicircular cross sections are also used (Anon., 1957B).

Figure 9-4. Manchester process plant treating 12 million scf/day of coal gas. Absorbers shown in foreground. *Southwestern Gas Board, England, and W.C. Holmes Co., Ltd.*

Table 9-2
Operating Data for Manchester Process

Design Variables	Plant A*	Plant B†
Gas-flow rate, MSCF/hr	125	140
Liquid-flow rate, gpm	2,400	1,700
Air-flow rate to oxidizers, MSCF/hr	54	21
Temperatures, °F:		
Inlet gas	....	80
Outlet gas	....	86
H_2S, grains/100 scf:		
In	738	705
After 1st stage	328	190
After 2nd stage	102	90
After 3rd stage	17	15
After 4th stage	1	trace
After 5th stage	0.11	nil
After 6th stage	0.04	nil
Suspended iron in soln., % Fe_2O_3	0.046	0.033
Soluble iron in soln., % Fe	0.16	
Alkalinity, N	0.23	0.30
Thiosulfate, N	0.18	
"Blue," %	0.11	
Reagents used, lb/MMscf		
Copperas ($FeSO_4 \cdot 7H_2O$)	344	
Sodium Carbonate	207	

Sources: *Plant A (Townsend, 1953); † Plant B (R. & J. Dempster, 1957)

Iron Cyanide Solutions and Suspensions

Several processes utilizing iron-cyanide complexes as oxidation agents were developed in Europe shortly before or during World War II. All of these processes are identical with respect to the basic chemistry involved in the absorption-regeneration cycle. Trivalent iron is reduced to divalent iron in the absorption step, and the divalent iron is reoxidized during regeneration. Hydrogen sulfide is oxidized to elemental sulfur, which is recovered by flotation and slurry filtration. The primary reaction chemistry of the iron cyanide solutions is similar to that of modern processes based on chelated iron solutions. The reactions are symbolized in the following equations:

Absorption:

$$2Fe^{3+} + H_2S = 2Fe^{2+} + S + 2H^+ \quad (9\text{-}12)$$

Regeneration:

$$2Fe^{2+} + \tfrac{1}{2}O_2 + H_2O = 2Fe^{3+} + 2OH^- \quad (9\text{-}13)$$

The differences between the individual iron-cyanide complex processes stem from the type of complex selected and the method of regeneration. In two processes, that of the Gesellschaft fur Kohlentechnik and the Fischer process, alkaline aqueous solutions of potassium ferricyanide and ferrocyanide are used, and regeneration is carried out by contact with air and electrolysis, respectively. The other two processes of this category, the Staatsmijnen-Otto and the Autopurification processes, employ suspensions of complexed ferric-ferrocyanide compounds in alkaline solutions that are regenerated by air contact. The latter two processes are essentially identical, although they were developed independently in 1945 in the Netherlands and in England, respectively.

Like the more modern iron-complex based processes, these processes claimed two main advantages over prior technology: (1) the capability to completely remove hydrogen sulfide and (2) the production of high purity sulfur. In addition, in many coke-oven gas applications where hydrogen cyanide was already present in the feed gas, sufficient hydrogen cyanide could be obtained from the gas, and chemical make-up requirements consisted solely of the occasional addition of iron sulfate to replace iron losses.

The iron-cyanide processes have not been employed extensively on an industrial scale, and only a few publications describing their operation have appeared. The Fischer (1932) process has been described by Mueller (1931) and Thau (1932). The Staatsmijnen-Otto process has been discussed in detail, especially with respect to its complex chemistry, by Pieters and van Krevelen (1946). The operation of an Autopurification process plant has been reviewed by Craggs and Arnold (1947).

Fischer Process

This process, which was developed and commercialized at the gas works at Hamburg, Germany, and patented by Fischer (1932), utilizes an aqueous solution containing about 20% potassium ferrocyanide and 6% potassium bicarbonate. The solution is subjected to electrolysis, which converts a portion of the ferrocyanide to ferricyanide and an equivalent amount of the bicarbonate to carbonate, releasing at the same time a proportionate volume of hydrogen. The presence of both ferricyanide and carbonate enables the solution to absorb H_2S rapidly and to convert it immediately to elemental sulfur. The reactions involved can be expressed by the following equations:

Electrolysis:

$$2K_4Fe(CN)_6 + 2KHCO_3 = 2K_3Fe(CN)_6 + 2K_2CO_3 + H_2 \tag{9-14}$$

Absorption:

$$2K_3Fe(CN)_6 + 2K_2CO_3 + H_2S = 2K_4Fe(CN)_6 + 2KHCO_3 + S \tag{9-15}$$

The overall reaction, indicating the decomposition of hydrogen sulfide to hydrogen and sulfur, can be written in a simplified form:

$$H_2S = H_2 + S \tag{9-16}$$

In practice, the process as described by Mueller (1931) and Thau (1932) operates as follows. The sour gas is contacted with the regenerated solution in a centrifugal contactor, and the H_2S is converted to sulfur of very small particle size. A centrifugal contactor was chosen because of plugging difficulties experienced with equipment of conventional design. The spent solution is pumped to a settling tank and from there flows to a filter press where the sulfur is removed. If a large settling tank is used, only a portion of the solution has to be filtered with the rest of the liquid being decanted. The combined sulfur-free liquids are pumped to a specially designed electrochemical cell where regeneration is achieved by electrolysis. The regenerated solution is then recycled to the contactor.

A review of the operating data for the Hamburg gasworks plant shows that the regenerated solution contains about 2 to 5% potassium ferricyanide. Approximately 2,000 gal/hr are circulated to treat an hourly volume of about 100,000 cu ft of water-gas containing 175 grains H_2S/100 cu ft. The reported electric power consumption of about 1.8 kwhr/lb of sulfur is rather high (twice the theoretical requirement) and may be one of the reasons for the failure of the process to gain acceptance despite the quality of the sulfur produced.

Staatsmijnen-Otto and Autopurification Processes

Process Description. In these processes, which are practically identical, hydrogen sulfide is removed—primarily from coal gases which also contain ammonia and hydrogen cyanide—by contact with an ammoniacal solution containing suspended ferric-ferrocyanide complexes, usually referred to as "iron blue." In addition, the solution contains ammonium salts which are necessary to stabilize the cyanide complexes. After long periods of use, thiocyanate and thiosulfate also accumulate in the solution. Regeneration of the spent solution is carried out by contact with compressed air in a tall aerating tower. Because of the exothermic nature of the oxidation, reaction heat is liberated in the regenerator and causes an increase in the solution temperature. After leaving the regenerator, the oxidized solution flows through a cooler and from there is returned to the absorber.

Elemental sulfur separates from the solution during regeneration and is collected as a froth at the top of the regenerator. The sulfur froth, which contains 70% water, some ammonia, and a small amount of "blue," is filtered and washed for removal of the bulk of the impurities. The sulfur is then heated with water in an autoclave at a temperature above its melting point and recovered as fairly pure molten sulfur. An extremely pure product may be obtained by further heating of the sulfur at about 600°F—at which temperature organic impurities are decomposed—followed by distillation.

The chemistry of the process is rather complicated, primarily because of the complex behavior of the "blue" and the many side reactions which may occur, depending on the gas composition and operating conditions. A very extensive study of process variables was made by Pieters and van Krevelen (1946). Basically, during absorption, hydrogen sulfide reacts with ammonia to form ammonium hydrosulfide. In the regeneration portion of the process, the hydrosulfide is oxidized to elemental sulfur by reduction of the "blue," which acts as an oxygen carrier. The reduced "blue" is then reoxidized. It is obvious that this interpretation is an oversimplification of the mechanism actually occurring, especially if it is kept in mind that the solution contains dissolved iron-cyanide salts that can react directly with the hydrogen sulfide.

Process Operation. It is claimed that H_2S is removed quantitatively and for the greater part converted to elemental sulfur. However, a certain amount of thiosulfate formation does take place, especially at high pH and low concentrations of "blue." In some instances, it may even be desirable to operate the process so that the H_2S is completely converted to thiosulfate.

Any hydrogen cyanide contained in the gas is absorbed quantitatively and converted to ammonium thiocyanate. In order to avoid losses of sulfur caused by this reaction, the hydrogen cyanide may be removed from the gas before it enters the desulfurization plant, and converted to alkali ferrocyanide. This procedure is economically advantageous, since the ferrocyanide can be used as makeup to replace the "iron-blue" lost with the wet sulfur.

The hydrogen cyanide is commonly removed from the feed gas by contact with a solution of alkali carbonate in a cast-iron vessel filled with iron filings. If the absorption temperature is about 200°F, HCN reacts rapidly with the iron and H_2S and CO_2 are not absorbed. A portion of the solution is withdrawn from the vessel at regular intervals, and the ferrocyanide is salted out by adding alkali carbonate. The crystals are removed and the remaining solution is returned to the absorber.

Typical operating data for the Staatsmijnen-Otto process are presented in **Table 9-3**. These data were obtained at a plant operated by the Societe Carbochimique at Tertre, Belgium. The plant consists of two parallel installations, each composed of a contactor and regenerator, and one final-purification installation, also containing a contactor and regenerator.

Operating experience with a small demonstration plant utilizing the Autopurification process was reported by Craggs and Arnold (1947). The plant, located at Billingham, England, processed 6,000 cu ft/h of coke-oven gas. Operational tests conducted over a period of 4 years resulted in the conclusion that although the plant was capable of producing gas containing less than 1 ppm of H_2S, the operation was erratic, with the treated gas containing as much as 200 ppm H_2S. Laboratory experiments conducted in parallel with the plant tests revealed a fairly good correlation between the ammonia content of the solution and the H_2S

Table 9-3
Typical Operating Data of Staatsmijnen-Otto Process

Gas flow rate, MSCF/hr	1,400
Solution flow rate, gpm	3,500–5,300
Air flow rate, MSCF/hr	56
Cooling-water flow rate, gpm	1,750
Gas contents, grains/100 scf:	
H_2S in inlet gas	170–210
H_2S after first contactor	0.5
H_2S after second contactor	0
HCN in inlet gas	15–26
HCN after first contactor	0
NH_3 in inlet gas	200–250
Solution composition, g/liter:	
Total solids	300–400
"Blue"	1.0 (min.)
Chemical consumption:	
Ferrous sulfate, lb/day	330
Sulfur recovery, %	30–60

Source: Pieters and van Krevelen (1946)

content of the outlet gas. It was found that the treating efficiency of the process is much more sensitive to the ammonia content than to the total iron and cyanide contents of the solution. As a result of these studies, a solution containing 2.0 g/liter of total iron, 2.8 g/liter of cyanide, and 4.0 g/liter of ammonia was specified for optimum performance.

Another shortcoming of the process was the large amount of air required for solution regeneration. The plant is equipped with a tall regenerating tower, and an air volume equal to about 15% of the gas volume treated is required for adequate regeneration. However, the most serious finding of this study came from a comparative cost analysis which indicated that, at the time of the study, the Autopurification process was not competitive with dry-box purification.

THIOARSENATE PROCESSES

There are two main processes in the thioarsenate category. The first, the Thylox process, was developed in the late twenties and used in the U.S. for many years, especially for the purification of coke-oven and other manufactured gases. The second, the Giammarco-Vetrocoke (G-V) process, is a dual H_2S/CO_2 gas sweetening process introduced in Italy in 1955, which was extensively used in Europe and Asia, especially in applications where the H_2S concentration in the feed gas was relatively low.

The Thylox process is no longer used commercially, while the arsenic-based version of the Giammarco-Vetrocoke process is still supported by the licensor in countries where the use of arsenical solutions is permitted. A non-arsenic, evolutionary modification of the G-V process is currently being offered for CO_2 removal, and is claimed to be very competitive. This form of the process can not be classified as a liquid redox process, and is discussed in Chapter 5.

Although the presence of arsenic salts in the treating solution suggests a strong similarity between the two thioarsenate processes, their chemistry and operating characteristics are fundamentally different. The Giammarco-Vetrocoke process, unlike the Thylox process, relies on the use of relatively inert monothioarsenate salts. This practice limits the extent of undesirable side reactions, and makes arsenic sulfide precipitation problems less likely. In spite of the toxicity of the solution, the combination of these two factors has made the arsenic-based Giammarco-Vetrocoke solution somewhat more environmentally acceptable.

Thylox Process

The Thylox process was disclosed by Gollmar (1929A, B) and Jacobsen (1929) and commercialized by the Koppers Company, Inc. A partial list of Thylox plants operating in 1945 showed a daily gas volume of 266 million cu ft, resulting in the production of about 60 long tons of sulfur. After 1950, the Thylox process gradually lost its prominence because of the advent of large-scale use of natural gas and the development of more efficient processes for the desulfurization of natural and manufactured gases. In recent years, the use of the Thylox process for gas desulfurization has been completely abandoned, primarily because of environmental concern due to the toxicity of the scrubbing solution.

The early applications of the Thylox process were aimed at about 80 to 90% hydrogen sulfide removal, but essentially near total hydrogen sulfide removal was eventually achieved through subsequent process improvements. One variation of the process, the so-called "modified Thylox process," was applicable in cases where complete purification of gases containing small amounts of hydrogen sulfide was required.

A neutral or slightly alkaline solution, containing sodium or ammonium thioarsenate as the active ingredient, is used in the Thylox process. Hydrogen sulfide is converted to elemental sulfur of sufficiently high purity to be usable in agriculture, mainly as a fungicide (Sauchelli, 1933). Since the sulfur contains less than 0.5% arsenic, it could also be used as a raw material for the manufacture of various chemicals.

The Thylox process offered some economic advantages over processes previously discussed in this chapter. The consumption of alkali due to thiosulfate formation was reduced markedly, and the sulfur was produced in a much more valuable form. Estimated operating requirements for a plant treating 5 million cu ft/day of refinery gas containing 1,000 grains hydrogen sulfide/100 cu ft were reported by Dunstan (1938), and are shown in **Table 9-4**.

Basic Chemistry

The chemistry of the Thylox process is described in detail by Gollmar (1934). The actual reaction mechanisms taking place during the various phases of the process are quite complicated because of the possible existence of a large variety of ionic species. The principal reactions involve replacement of one atom of oxygen by one atom of sulfur in the thioarsenate molecule during absorption, and the reverse during regeneration. The reactions can be symbolized by the following equations:

Absorption:

$$Na_4As_2S_5O_2 + H_2S = Na_4As_2S_6O + H_2O \qquad (9\text{-}17)$$

Regeneration:

$$Na_4As_2S_6O + \tfrac{1}{2}O_2 = Na_4As_2S_5O_2 + S \qquad (9\text{-}18)$$

These two reactions are quite rapid and are undoubtedly the main reactions occurring under most operating conditions. In cases where gas containing very high concentrations of hydrogen sulfide is treated, or when long contact-times are provided, other, much slower reactions, equations 9-19 and 9-20, may also take place to some extent.

Table 9-4
Operating Requirements of Thylox Process

Type of gas:	Refinery or natural
Plant capacity, MMscf/day	5.0
H_2S content, grains/100 scf	1,000
Sulfur removal, %	98
Gas pressure, psig	60
Labor (operating and maintenance), hr/day	14
Power, kw-hr/day	1,200
Steam, lb/day	15,000
Sodium carbonate, lb/day	600
Arsenic trioxide, lb/day	150
Credit: sulfur recovered, tons/day	3

Source: Dunstan (1938)

Absorption:

$$Na_4As_2S_6O + H_2S = Na_4As_2S_7 + H_2O \tag{9-19}$$

Regeneration:

$$Na_4As_2S_7 + \tfrac{1}{2}O_2 = Na_4As_2S_6O + S \tag{9-20}$$

Fresh Thylox solution is prepared by dissolving arsenic trioxide and sodium carbonate in water, in the proportion of 2 moles of sodium carbonate to 1 mole of arsenic trioxide. The resulting solution contains sodium carbonate and bicarbonate, sodium arsenite, and arsenious acid.

The sodium arsenite reacts readily with hydrogen sulfide to yield a $Na_4As_2S_5$ solution having a straw-yellow color. The formation of sodium thioarsenite can be represented as follows:

$$2Na_2HAsO_3 + 5H_2S = Na_4As_2S_5 + 6H_2O \tag{9-21}$$

If the pH of the thionizer solution is kept at a value of 7.5 or higher, the thioarsenite formed in equation 9-21 will absorb oxygen from air and be converted to thioarsenate. This oxidation reaction occurs rapidly, and the change in arsenic valence from +3 to +5 is accompanied by a change in solution appearance from yellow to colorless. This reaction is shown in equation 9-22:

$$Na_4As_2S_5 + O_2 = Na_4As_2S_5O_2 \tag{9-22}$$

During arsenite oxidation the pH decreases and carbon dioxide is gradually expelled. When the solution is completely oxidized, it contains practically no carbon dioxide. In preparing the solution, it is important to maintain a ratio of a least 2 atoms of sodium to 1 atom of arsenic. If insufficient sodium is present, the pH can drop below 6.7, and part of the arsenic in the thio-arsenate can revert to its lower valence. Under these conditions, a mixture of arsenic trisulfide and sulfur precipitates. For these reasons, the process operates best within a pH range of 7.5 to 8.0. Ammonia can be substituted for the sodium carbonate without changing the characteristics of the process.

As in all liquid-redox processes, part of the sulfur is converted to thiosulfate, although the rate of formation is appreciably lower in the essentially neutral Thylox solution than in more alkaline solutions used in other processes. Hydrogen cyanide, which is absorbed in the absorber, reacts readily with the sulfur formed in the thionizer to yield sodium thiocyanate. Because of these side reactions, the active thioarsenate has to be replenished continuously by addition of arsenic oxide and sodium carbonate.

The amount of carbon dioxide present in typical feed gas, ranging between 0.5 and 8.0%, does not interfere with the removal of hydrogen sulfide because the Thylox solution is not sufficiently alkaline to absorb carbon dioxide to an appreciable extent.

Process Description

A basic flow diagram of the Thylox process is shown in **Figure 9-5.** The gas enters at the bottom of the absorber and is washed countercurrently with solution entering at the top of the vessel. Essentially all of the hydrogen sulfide and hydrogen cyanide are removed in this

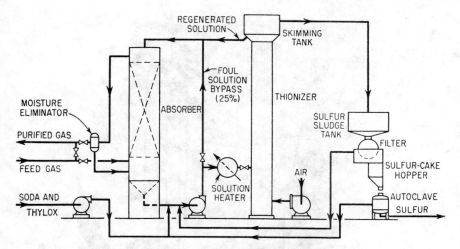

Figure 9-5. Basic flow diagram of the Thylox process.

operation. The foul solution is pumped from the bottom of the absorber through a heat exchanger where it is heated to approximately 110°F. The warm solution enters the bottom of the thionizer and is regenerated by a cocurrent stream of air as it moves upwards through the column. The air not only releases the sulfur, but also acts as a flotation agent for the sulfur that collects at the surface of the solution as a froth. The liquid level in the thionizer is maintained below a weir over which the sulfur froth flows to the sulfur-sludge tank. The regenerated liquid returns by gravity flow to the top of the absorber. The sulfur sludge leaves the sludge tank and is passed through the filters, whence the sulfur-cake is further processed for final product preparation. The sulfur can be recovered as a wet paste, a dry powder, or as cast crude sulfur. Pure sulfur can be obtained by distillation of the crude product. Various means for sulfur recovery are discussed extensively by Gollmar (1945). The filtrate is returned to the foul-solution stream, or it can be partially or totally discarded, thus serving as a system purge. Normally a small portion of the solution is continuously withdrawn from the system to prevent the accumulation of thiosulfates and thiocyanates. The arsenic may be reclaimed from this liquid by adding acid and filtering the arsenic sulfide that is formed. The arsenic sulfide is dissolved in sodium carbonate solution and then returned to the system. A photograph of a Thylox plant is shown in **Figure 9-6.**

Various modifications of the basic process were developed, including 1) two-stage absorption for more complete hydrogen sulfide removal and 2) bypassing a portion of the foul solution (approximately 25%) around the thionizer. In the two-stage absorption process, freshly prepared Thylox solution is used in the second absorber. A plant using two-stage absorption is described by Powell (1936).

Design and Operation

The typical absorber is a steel tank packed with wooden hurdles. Since most Thylox plants operate at essentially atmospheric pressure, and the mass transfer efficiency of this design is relatively poor, a rather large vessel is required. The standard dimensions are commonly 90 ft high and up to 20 ft in diameter, depending on the gas throughput. The scrubbing solution

752 Gas Purification

Figure 9-6. Thylox gas purification plant for removing hydrogen sulfide and producing sulfur. *Koppers Company (Koppers is now part of ICF Kaiser Engineers)*

is usually fed to a distributor ring located at the top of the tank, which directs and partitions the flow through spray nozzles to facilitate adequate distribution of the solution in the absorber.

The thionizer, which operates under slight pressure, is usually a tall empty vessel of much smaller diameter and a height of 120 ft. Air is supplied by compressors, and blown upward through the liquid filled vessel, lifting the sulfur particles to the top surface, where they form a froth. A somewhat different and shorter thionizer design featuring two concentric shells is reported by Powell (1936).

Operating results from three Thylox plants are reported by Powell (1936), Denig (1933), Farquhar (1944), and McBride (1933). The plant described by McBride employs a solution containing ammonium thioarsenate instead of sodium thioarsenate. Operating results from three of these installations are presented in **Table 9-5.** Other operating data for two Thylox plants treating coal gas and blue-water gas are reported by Foxwell and Grounds (1939).

The circulation rate of the treating solution is set so that a considerable excess of arsenic oxide is maintained, over the stoichiometric quantity required to react with the hydrogen sulfide. This is necessary because of the incomplete regeneration obtained in the thionizers. In general, 4 to 5 moles of thioarsenate (measured as As_2O_3) per mole of hydrogen sulfide are circulated through the absorber.

Table 9-5
Operating Data of Thylox Plants

Design Variables	Plant A	Plant B*	Plant C
Gas-flow rate, MSCF/hr	210	900	2,200
Solution-flow rate, gpm:			
Primary solution	280	1,350†	5,000‡
Secondary solution		850	
Air-flow rate, MSCF/hr	16.4	34.0	12.0
Steam rate, lb/hr	1,100		
Temperature, °F:			
Inlet gas	73		75
Outlet gas	83		100
Solution to absorber	97	95	
Composition:			
H_2S content, grains/100 scf:			
Inlet gas	316	61	160
Outlet gas:			
Primary absorber	9.0	3.5	3.0
Secondary absorber		0.25	
HCN content, inlet gas, grains/100 scf	12.0		
As_2O_3 content, %:			
Primary solution	0.7	0.3	
Secondary solution		0.8	
Sulfur recovered, %	64.0		

*Plant B was operating at less than 50% of rated capacity when the operating data were obtained.
†315 gpm bypassed around thionizer.
‡25% bypassed around thionizer.
Sources:
A) (Farquhar, 1944); B) (Powell, 1936); C) (McBride, 1933)

From the data in **Table 9-5** it can be seen that approximately 95% hydrogen sulfide removal can be easily obtained with a single absorber. When a two stage absorber configuration is employed, and the system is properly operated, it is possible to produce a purified gas containing 0.2 to 0.3 grain hydrogen sulfide per 100 cu ft gas.

Reactivation of the solution is carried out using an excess of oxygen over the stoichiometric requirement. The rate of air flow reported in the literature varies considerably from plant to plant, but it appears that about 5 moles of oxygen per mole of hydrogen sulfide are required for proper solution reactivation.

The considerable heat of reaction generated in the oxidation of hydrogen sulfide to elemental sulfur is dissipated by evaporation of water in the thionizer. As long as solutions of low arsenic concentration are used and, consequently, large volumes of liquid are circulated, there is no appreciable solution temperature increase.

754 *Gas Purification*

Thylox solution is somewhat corrosive; therefore, stainless-steel internals are specified for pumps and valves, and stainless-steel clad tube sheets and stainless-steel tubes are normally recommended in solution heaters. Other equipment is made of carbon steel with sufficient corrosion allowance to compensate for the severity of the operating environment.

A process used in Russia, which is practically identical with the Thylox process, is reported by Jegorov et al. (1954). This paper contains a rather detailed description of process equipment, design criteria, and a sample calculation for a typical plant for the treatment of coke oven gas.

The absorbers used in Russian installations are cylindrical vessels packed with wood slats. Linear gas velocities of 2 to 4 ft/sec and pressure drops of 2 to 3 in. of water are used for the design of typical absorbers operating at essentially atmospheric pressure. The treating solution, which contains 8 g/liter of As_2O_3 and 10 g/liter of Na_2CO_3, is circulated at a rate equivalent to a 50% excess over the stoichiometric requirement.

The regenerator described by Jegorov et al. (1954) consists of a liquid filled tower, 100 to 120 ft high, containing several metal screens for redistribution of the air. The volume of the vessel is based on a solution residence time of 40 to 50 min. and the diameter on an air velocity of 500 to 800 cu ft/(hr)(sq ft) of tower cross section. The air-flow rate is equivalent to a ratio of 2.5 moles of oxygen per mole of hydrogen sulfide.

An interesting feature of the Russian process is the two-step method employed for the complete recovery of arsenic from solution waste-streams. In the first step, which is similar to the recovery method used in the Thylox process, the solution is heated to 70°C (158°F), and arsenic sulfide is precipitated by the addition of 75% sulfuric acid. The precipitate is separated from the liquid by filtration, dissolved in aqueous sodium carbonate, and returned to the circulating solution-stream. The clear liquid is then passed to the second step where it is made alkaline with sodium carbonate solution and treated with a solution of ferric sulfate. In this operation the small amount of arsenic remaining in the solution after the first step is fixed and precipitated as ferric arsenite and arsenate. The precipitate is finally removed by filtration, and the filtrate, which contains about 10 to 20 ppm of arsenic, is either discarded or processed for recovery of thiosulfate. Wooden tanks lined with acid-resistant materials are used in both steps of the arsenic-recovery operation. Each tank is sized for a solution residence time of 4 hr and provided with a mechanical agitator.

Giammarco-Vetrocoke Process

The Giammarco-Vetrocoke (G-V) process originally consisted of two different processing cycles: one for H_2S, and one for CO_2. A combination of two cycles in sequence; first for H_2S, and then second for CO_2; or else in single integrated cycle—can be used for acid gas sweetening (Jenett, 1962). The combined process is very versatile, and can be easily adapted to the particular needs of any individual plant depending on the feed gas composition and operating conditions.

The chemistry of the general process is based on alkali carbonate solutions, activated by arsenic salts, or certain organic compounds (Anon., 1960B; Maddox and Burns, 1968). Initially, the alkali carbonate solution was activated by inorganic arsenic in both cycles. Later, the carbon dioxide cycle was made more attractive environmentally by introducing the use of non-toxic organic activators, most commonly glycine. The arsenic based activators enter into the removal reactions, but when organic activators are used they act essentially as catalysts for CO_2 absorption, and the basic gas removal mechanism becomes equivalent to that of alkali carbonate solutions alone (see Chapter 5).

While the use of organic activators instead of arsenic compounds in the CO_2 cycle does not significantly affect the capacity or performance of the plant, the arsenate is essential to the H_2S cycle because it stabilizes the sulfur in a form that is unaffected by the presence of CO_2 in the feed gas. According to Jenett (1962), the feasibility of using the G-V process primarily for hydrogen sulfide removal is limited to applications where the inlet gas hydrogen sulfide concentration does not exceed 1.5%, and where the total sulfur removal capacity requirements are less than 15 tons per day. The process is reported to be capable of producing purified gas containing less than 1 ppm of hydrogen sulfide even when operated at absorption temperatures up to 300°F and in the presence of substantial concentrations of carbon dioxide in the gas to be treated.

Rigorous limitations on the use of the G-V H_2S cycle became increasingly widespread due to the toxic nature of the arsenical solution, and related problems of plant effluent non-compliance with local environmental regulations. For these reasons, as of 1992, it was reported that even though the H_2S removal process was still being offered for licensing, no new plants had been built anywhere in the world for several years (Tomasi, 1992).

By contrast, the application of the G-V process for carbon dioxide removal has no limitations as to inlet gas composition, quantity of carbon dioxide to be removed, or inlet gas temperature. While many older operating plants that employ G-V technology are still arsenic based single CO_2 cycles, most modern plants are dual activated (glycine plus amine), with two-pressure level regeneration. Many of the early licensees are revamping their units, and converting to the low energy operation mode and the use of nontoxic organic activators (Tomasi, 1992). The non-arsenic based G-V process is discussed in Chapter 5.

The general arsenic based process was first disclosed by Giammarco (1955, 1956, 1957) and later described in some detail in the technical literature (Jenett, 1962, 1964; Anon., 1960A, B; Riesenfeld and Mullowney, 1959). It was originally used in Europe, chiefly for coke-oven gas and synthetic gas treating, but the first commercial operation in the U.S., in 1960 in Fort Stockton, Texas, was a high pressure natural gas sweetening application. The plant was built for Transwestern Pipeline Co., and was designed to treat 180 MMscfd of natural gas containing 28% carbon dioxide and 2 grains per 100 cu ft of hydrogen sulfide, at a design pressure of 1,050 psig (Jenett, 1962).

Basic Chemistry

The chemistry of the H_2S removal cycle is uniquely based on an arsenic-activated potassium carbonate solution, and is quite complex. The overall reaction mechanism of the absorption-regeneration cycle can be represented in a simplified form by the following equations:

Absorption:

$$KH_2AsO_3 + 3H_2S = KH_2AsS_3 + 3H_2O \tag{9-23}$$

Digestion:

$$KH_2AsS_3 + 3KH_2AsO_4 = 3KH_2AsO_3S + KH_2AsO_3 \tag{9-24}$$

Acidification:

$$KH_2AsO_3S = KH_2AsO_3 + S \tag{9-25}$$

Oxidation:

$$2KH_2AsO_3 + O_2 = 2KH_2AsO_4 \tag{9-26}$$

The overall reaction is essentially the Claus reaction: the oxidation of hydrogen sulfide to elemental sulfur as shown in equation 9-27:

$$H_2S + \tfrac{1}{2}O_2 = S + H_2O \tag{9-27}$$

The H_2S absorption step as represented by equation 9-23 is rapid, and the rate of absorption is favored by an excess of arsenite.

In the second step, thioarsenite reacts with arsenate forming monothioarsenate (equation 9-24). The stoichiometry indicates that the presence of 1 mole of pentavalent arsenic is required for each mole of hydrogen sulfide absorbed. This reaction is relatively slow, requiring ample residence time, but the reaction rate can be made faster by increasing the concentration of arsenate and by raising the reaction temperature. In general, in order to drive the reaction to completion, an excess of arsenate is recommended. The digestion reaction is a critical step in the process because it stabilizes the sulfur as monothioarsenate which is very stable towards oxygen and carbon dioxide. The stability of the sulfur towards oxygen suppresses the formation of thiosulfate.

The extremely low equilibrium vapor pressure of H_2S over the monothioarsenate (KH_2AsO_3S) ensures a negligible concentration of H_2S in the treated gas, so that it is possible to produce treated gas of very high purity even at elevated absorption temperatures.

In the third step, represented by equation 9-25, monothioarsenate is decomposed into arsenite and elemental sulfur. This is achieved by lowering the pH of the solution. There are two different methods that can be used to precipitate sulfur depending on the pH of the starting solution. If the process operates at elevated temperatures, and the pH of the circulating medium is relatively high, the solution is commonly treated with high pressure carbon dioxide. The treatment converts essentially all of the carbonate to bicarbonate, resulting in sufficient lowering of the pH to precipitate elemental sulfur. On the other hand, with solutions in which the starting pH is already low, this procedure is unnecessary, and the slight increase in acidity resulting from the formation of arsenate during the oxidation step is usually sufficient to achieve the precipitation of elemental sulfur.

The final step of the H_2S removal cycle shown in equation 9-26 is the reoxidation of trivalent to pentavalent arsenic, usually by contact with air. The rate of the oxidation reaction is quite slow, but may be markedly increased by addition of certain catalysts.

Process Description

A typical flow diagram of the Giammarco-Vetrocoke process is shown in **Figure 9-7.** The diagram depicts the basic form of the process where acidification of the solution and oxidation of trivalent to pentavalent arsenic take place simultaneously in the regenerator or oxidation vessel. In the other version where carbon dioxide is used for solution acidification, a separate vessel is located between the "digester" and the regenerator. Typically CO_2 recovered from the G-V plant CO_2 removal flash drum is fed to the bottom of the acidification drum and countercurrently contacted with the cooled digester effluent.

In the basic diagram shown in **Figure 9-7,** the gas enters the bottom of the absorber where it is contacted countercurrently with lean solution entering at the top of the vessel. Essential-

Liquid Phase Oxidation Processes for Hydrogen Sulfide Removal 757

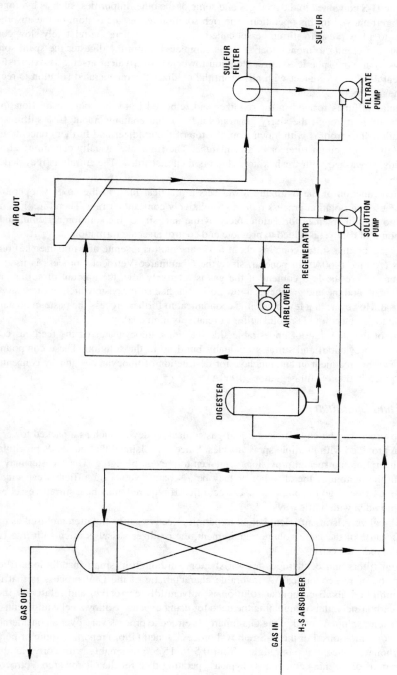

Figure 9-7. Typical flow diagram of Giammarco-Vetrocoke process for H_2S removal.

ly all of the H_2S contained in the feed gas and some of the other impurities, such as hydrogen cyanide, are removed in this operation. The rich solution leaves the bottom of the absorber and flows to a low pressure stirred vessel called a "digester," where the relatively slow conversion of thioarsenite to monothioarsenate is completed. From the digester the spent solution flows to an atmospheric oxidizer flotation tower where trivalent arsenic is oxidized to pentavalent arsenic by contact with air. From the oxidizer the regenerated solution is recycled to the absorber.

Elemental sulfur is formed in the oxidizer and removed from the solution by flotation. The air bubbling through the slurry generates a froth that contains about 10% sulfur and 90% solution. The froth is withdrawn from the top of the oxidizer and further concentrated either in a rotary vacuum filter or in a centrifuge. The filtercake, usually containing about 50% solids, is washed with fresh water which is then discarded. The filtrate is recycled to the process.

The dual function of the oxidizing tower sets a practical limit to the air flow operating range because the flotation process requires a relatively constant air rate. Therefore, it is not practical to control arsenate formation solely by the intensity of the solution aeration, and a small amount of catalyst is added to promote and control arsenate formation.

After washing, the sulfur contains about 0.3% arsenic (as arsenite and thioarsenate) on a dry basis (Anon., 1960A). In general, since the Giammarco-Vetrocoke process is used in applications where the H_2S content of the gas is relatively low, the amount of sulfur produced is small, and the end product is a low purity washed sulfur paste which must be stored for disposal. However, in a few cases it is economical to further process the paste in an autoclave, and to produce liquid or solid sulfur of relatively high quality.

In spite of the slow rate of undesirable side reactions, thiocyanates (if the feed gas contains hydrogen cyanide) and sulfates gradually build up in the solution. These compounds are removed by treatment in an autoclave for destruction of thiocyanates and by concentration and precipitation of sulfate (Anon., 1960A).

Design and Operation

The absorber may be any efficient liquid-gas contacting device such as a packed tower or a column provided with multiple spray nozzles. Since it is claimed that no solids precipitate during absorption, no special provisions are taken to prevent plugging. The digester may be located in the bottom of the absorber or may be a separate vessel of sufficient capacity to allow the reaction to go to completion. The acidifier, if required, may be a stirred vessel or a column provided with baffle trays.

Cylindrical vessels of sufficient height to permit effective sulfur flotation are used as oxidizers. Because of the large volume requirement, the oxidizer vessel is often split into two columns to reduce its size.

Vacuum filters and centrifuges are satisfactory, although in small installations filters appear to be more economical. A favorable characteristic of the G-V process is that the arsenic inhibited alkali carbonate solution is only mildly corrosive, and relatively inert towards carbon steel, thus eliminating the need for using expensive alloy steel equipment.

The treating solutions used in the Giammarco-Vetrocoke process vary over a considerable range of concentrations. For the H_2S removal process, Jenett (1962) reports sodium or potassium carbonate concentrations ranging from 0.5 to 15%, presumably with corresponding concentrations of arsenic compounds. Typical operating data for the Giammarco-Vetrocoke process reported by Jenett (1962) are shown in **Table 9-6**.

Table 9-6
Typical Operating Data for Giammarco-Vetrocoke Process

Alkali	Na or K
Alkali concentration, wt%	0.5–15
Solution capacity, cu.ft. H_2S/gal.	0.15–2.8
Air requirements, cu.ft./1,000 grains H_2S	15–20
Maximum absorption temperature, °F	300
H_2S in treated gas, ppm	0.01–1.0

Source: Jenett (1962)

A comparison of operating characteristics of the Giammarco-Vetrocoke process with those of other early liquid oxidation processes is given in **Table 9-7** (Riesenfeld and Mullowney, 1959).

Table 9-7
Comparison of Typical Operating Data for H_2S Removal Processes

Process	Giammarco-Vetrocoke	Thylox	Ferrox	Manchester
Pressure, psig.	atm	atm	atm	atm
Inlet gas temp.,°F	100–300	100	100	100
H_2S in, grains/100 scf	300–500	300–500	300–500	500–1,000
H_2S out, grains/100 scf	<0.1*	10†	5	0.25‡
Solution capacity, grains/gal.	600	40	70	10

*One-stage absorption.
†One-stage absorption; 0.25 grain/100 scf for two-stage absorption.
‡Six-stage absorption with fresh solution to each stage.
Source: Riesenfeld and Mullowney (1959)

When first introduced, the H_2S absorption capacity of the Giammarco-Vetrocoke solution was much higher than that of other liquid-redox processes available during the same period. In addition, even though amine solutions had higher capacity ratings, the amine solution capacity for H_2S absorption decreased with increased CO_2 concentration in the inlet gas. The net result was that for many gases whose composition included a high content of CO_2, the hydrogen sulfide absorption capacity of the G-V process compared favorably with amine processes.

QUINONE AND VANADIUM METAL PROCESSES

This section reviews three groups of processes each with two representatives. The first group contains the Perox and Takahax processes, which utilize the liquid redox potential of

760 Gas Purification

organic quinone solutions. The two processes in the second group were derived from the first, and employ a combination of quinone compounds and metal salts, usually vanadium. This group is highlighted by the Stretford process, and contains, in addition, the Hiperion process, a process of less commercial significance. The third group, consisting of the Unisulf and Sulfolin processes, evolved directly from Stretford technology, and is characterized by the use of vanadium salts in conjunction with organic additives other than quinones. The Sulfolin process has achieved moderate commercial success.

The quinone-based processes utilize the redox cycle illustrated in **Figure 9-8** to convert hydrogen sulfide to elemental sulfur. In these processes hydrogen sulfide is absorbed into an aqueous solution containing a quinone in the oxidized state. The absorbed hydrogen sulfide is then oxidized to elemental sulfur by the quinone, which is reduced to hydroquinone in the reaction. The hydroquinone is reoxidized to quinone by contact with air in a separate step to complete the cycle.

The chemical formulas for four of the quinones most commonly used historically in hydrogen sulfide removal liquid redox processes are shown in **Figure 9-9** (Douglas, 1990B). The redox potentials of some typical quinone compounds are given in **Table 9-8**.

The first liquid redox quinone process was the Perox process, developed in Germany in 1950. It employs p-benzoquinone, and was a forerunner to other quinone based processes. The second process discussed in this section is the naphthoquinone-based Takahax process, developed in Japan by the Tokyo Gas Company. This process, which achieved significant market penetration in Japan, was introduced in the same time frame as the Stretford process, and constituted an attempt to eliminate the use of heavy metals that is characteristic of Stretford type processes.

The most extensive discussion pertains to the Stretford process, the ADA-vanadium process developed by Nicklin in 1963 for British Gas, which became the most dominant liquid redox H_2S removal technology in the seventies and early eighties.

The section closes with a discussion of three new processes introduced in the 1980s; they include the Unisulf, Sulfolin, and Hiperion processes. Unisulf and Sulfolin are vanadium-based processes that were designed to minimize or eliminate the need for spent solution

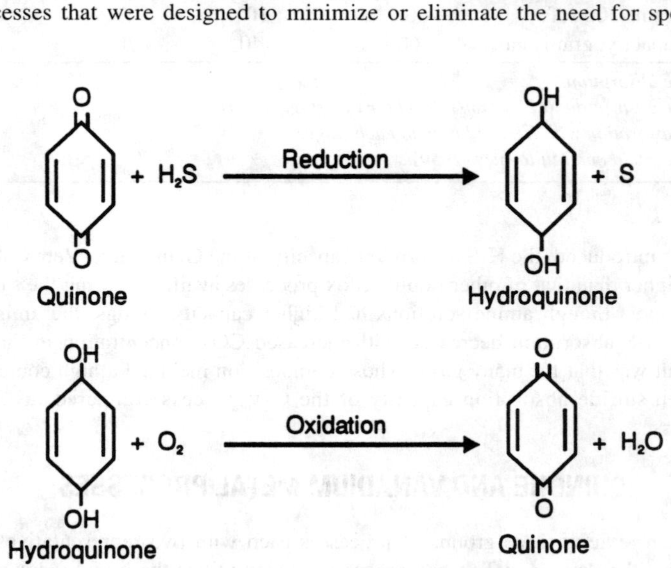

Figure 9-8. Redox cycle of quinone.

p-Benzoquinone

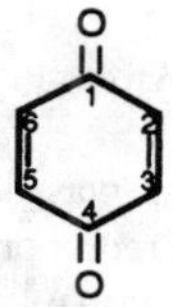

Duroquinone (2, 3, 5, 6 Tetramethylquinone)

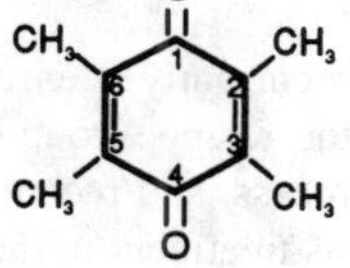

1,4 Naphthoquinone - 2 - sulfonic acid

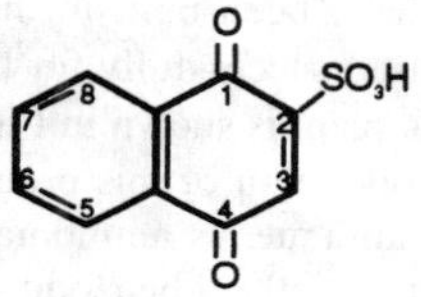

9, 10 Anthraquinone - 2, 7 di-sulfonic acid (ADA)

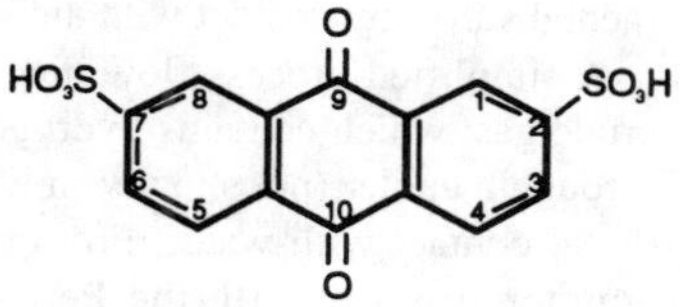

Figure 9-9. Structural formulas of quinones used in hydrogen sulfide removal processes. (*Trofe et al., 1987*)

Table 9-8
Typical Quinone Redox Potentials ($E°$ at 25°C)

	Volts
o - Benzoquinone	0.787
p - Benzoquinone	0.699
3,4 Phenanthrequinone-1-sulfonic acid	0.677
1,2 Naphthoquinone-4-sulfonic acid	0.625
1,4 Naphthoquinone-2-sulfonic acid	0.535
9,10 Anthraquinone-2,7 di-sulfonic acid (ADA)	0.187

Source: Douglas (1990A, B)

purging. They are essentially modifications of the Stretford process in which the formation of thiosulfate salts is abated through the elimination of side-reactions caused by hydrogen peroxide generation. These side reactions are eliminated by abandoning the use of ADA as an oxygen carrier. Hiperion, on the other hand, is a modification of the Takahax process. It reportedly incorporates chelated iron which accelerates the reoxidation rate of the naphthoquinone salt used in the Takahax process. This makes the operation of the oxidizer more efficient and reduces capital expenditure.

Perox Process

The Perox process was developed in Germany after World War II, and was the first process to utilize the redox properties of quinone compounds in the removal of hydrogen sulfide from coke-oven gas. As early as 1956, it was reported that three commercial units with a combined capacity of approximately 30 million cu ft per day were operating in Germany (Reinhardt, 1956).

The process is currently licensed in the United States by Krupp Wilputte, a North American affiliate of the Krupp Group of Essen, Germany. However, Krupp is no longer promoting the Perox process, and recommends instead their vacuum potash process, VACASULF, for coke-oven gas treatment in the steel industry (see Chapter 5). As of 1993, no new Perox plants have been built in the U.S. for the last five years. Nonetheless, the process is still being used successfully for the purification of coal gas in Germany. A picture of an existing German plant is shown in **Figure 9-10.**

The operation of this process is quite simple, and consists of absorption of hydrogen sulfide in an aqueous ammonia solution containing 0.3 gram per liter of an organic oxidation catalyst, usually p-benzoquinone, followed by oxidation of ammonium hydrosulfide to elemental sulfur by contact with air.

A simplified process flow diagram is given in **Figure 9-11** (Krupp Wilputte, 1988). The crude gas, which contains hydrogen sulfide, hydrogen cyanide, and ammonia, is first passed through a cooler (not shown) in which the temperature and ammonia content are adjusted by direct contact with water. From there the gas flows to the contactor in which it is washed countercurrently with the Perox solution and practically all of the H_2S and HCN are absorbed and converted to $(NH_4)_2S$ and NH_4CN.

The spent solution is reactivated in the oxidizer by contact with compressed air and returned to the contactor. The three principal reactions taking place in the regenerator are as follows:

$$\text{hydroquinone} + \tfrac{1}{2}O_2 = \text{p-benzoquinone} + H_2O \tag{9-28}$$

$$\text{p-benzoquinone} + (NH_4)_2S + 2H_2O = \text{hydroquinone} + \tfrac{1}{2}S_2 + 2NH_4OH \tag{9-29}$$

Liquid Phase Oxidation Processes for Hydrogen Sulfide Removal **763**

Figure 9-10. Perox plant in Germany. *Courtesy of Krupp-Wilputte*

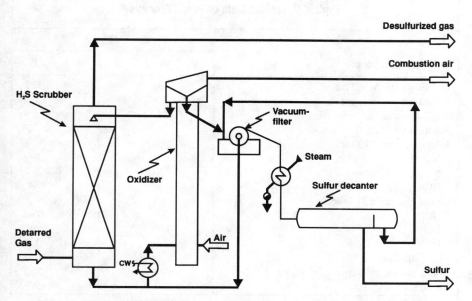

Figure 9-11. Simplified flow diagram of Perox process. (*Krupp-Wilputte, 1988*)

$$NH_4CN + \tfrac{1}{2}S_2 = NH_4SCN \tag{9-30}$$

During regeneration, the hydroquinone derivative is first oxidized to the p-benzoquinone form according to equation 9-28. The organic quinone acts as an oxygen carrier and reacts with the ammonium sulfide formed in the contactor. Equation 9-29 shows the oxidation of ammonium sulfide to elemental sulfur. This reaction also releases the ammonia from the ammonium sulfide and restores most of the ammonia content of the solution. A portion of the elemental sulfur generated reacts with the ammonium cyanide to yield ammonium thiocyanate by equation 9-30. Therefore, of the total sulfur present in the feed as H_2S, the amount converted to elemental sulfur will depend on the HCN concentration in the feed gas.

The sulfur froth and the regenerated scrubbing solution are separated at the top of the oxidizer. The sulfur froth is filtered off and the resulting filter cake is melted to produce sulfur. The filtrate is returned to the solution circuit. Surplus air can be reused as combustion air.

The oxidation of hydrogen sulfide to sulfur results in a net production of water. The overall water balance of the system is maintained by cooling the inlet gas to a temperature somewhat below that of the circulating solution so that the surplus water is continuously carried away by the purified gas.

Excessive accumulation of side-reaction products in the solution, such as thiosulfates and thiocyanates, is prevented by naturally occurring losses. Since these solution losses also entail continuous loss of catalyst, periodic additions of the catalytic compound are required. Operating results from two Perox plants are shown in **Table 9-9** (Pippig, 1953; Anon., 1957A; Brommer and Luhr, 1956).

Table 9-9
Typical Operating Data of Perox Process

	Plant	
Design variables	A	B
Gas flow rate, MSCF/hr	300	519
Solution-flow rate, gpm	1,300	2,650
Air-flow rate, MSCF/hr	12	31.8
Solution temp., °F	72	71
Inlet gas, grains/100 scf:		
H_2S	240	399
NH_3	230	372
HCN	45	54
Outlet gas, grains/100 scf:		
H_2S	0.5	0.06
NH_3	150	295
HCN	2.5	1.8
pH of solution	8.78	8.83
Solution loss, gal/day	390	550
Catalyst loss, lb/day	2.0	9.9

Sources: A) (Pippig, 1953); B) (Anon., 1957); (Brommer and Luhr, 1956)

Takahax Process

The Takahax process was developed by Hasebe (1970) of the Tokyo Gas Company Ltd., to replace the Thylox and Giammarco-Vetrocoke processes (Swaim, 1972). The first commercial application of the Takahax process was to desulfurize coal gas in a revamped 1.4 MMscfd Thylox plant operated by Kamaishi Gas. About a hundred Takahax units were in operation at one time in Japan, primarily in gas works, steel plants, and chemical plants. The process is still extensively used in Japan for the desulfurization of coke oven gas.

The main objectives of the Takahax development program were to devise a liquid redox hydrogen sulfide removal process comparable to the Stretford process and to eliminate the use of heavy metals to oxidize the hydrogen sulfide. The new process also sought to minimize the periodic liquid purging that is commonly used in liquid redox systems to eliminate the build-up of undesirable contaminants in the scrubbing solution.

The Takahax process utilizes naphthoquinone (NQ) compounds as the oxygen carrier. The preferred solutions contain salts of 1, 4-naphthoquinone-2-sulfonic acid dissolved in an alkaline aqueous media within a pH range of 8 to 9 (Hasebe, 1970). The original Takahax process was based on sodium carbonate; however, either sodium carbonate or ammonia are now used as the solution alkaline component. The quinone compound used in the Takahax process has a redox potential which is more than double that of anthraquinone di-sulfonic acid, also known as ADA, the quinone compound used in the Stretford process. The greater redox potential of this naphthoquinone compound promotes a rapid conversion of H_2S to sulfur without the addition of vanadium.

A major drawback of the Takahax process is the slow rate of reoxidation of the reduced hydro-naphthoquinone sodium salt. This increases the regenerator residence time requirements and the capital cost. It is reported that the process is capable of producing treated gas containing less than 10 ppm of hydrogen sulfide even when the raw gas contains substantial quantities of carbon dioxide. In addition, 85 to 95% of the HCN in the feed is removed (GEESI, 1981).

Typical operating conditions for coke-oven gas desulfurization, reported by Hasebe in 1970, are shown in **Table 9-10**.

Table 9-10
Typical Operating Conditions of Takahax Process

Gas composition:	
H_2S, volume %	0.4
CO_2, volume %	5–10
Solution composition:	
Na_2CO_3, grams/liter	40
Catalyst, gram mol/liter	0.0015–0.0020
Volume ratio, gas-liquid:	22
Volume ratio, air-liquid:	1.9
Chemical consumption:	
Na_2CO_3, lb/lb of sulfur recovered	0.4
Quinone, gram mol/lb of sulfur recovered	0.225

Source: Hasebe (1970)

The flow scheme of the Takahax process is quite similar to that of the Perox process, and, as in the Perox process, the oxidation of hydrosulfide to elemental sulfur in the absorber occurs almost instantaneously. Therefore, there is no need for a delay tank downstream of the absorber to complete the reaction. The process requires no steam and operates at ambient pressure.

In the Takahax process the precipitated sulfur is very fine and not amenable to flotation. Therefore, when elemental sulfur recovery is desired, the sulfur recovery technique is based on continuous recirculation of a sulfur slurry of relatively high solids content and removal of sulfur from a slip stream in a filter press.

Improvements made to the process by Nippon Steel during the early 1970s were described by Kozumi et al. (1977). One important new feature was the possible use of ammonia as the alkaline component of the solution in addition to sodium carbonate. This led to the development of two parallel technologies: the Ammonia-Takahax and the Sodium-Takahax processes. The application of the Ammonia-Takahax process is preferred when the feed gas contains sufficient ammonia, and ammonium sulfate is the desired byproduct. The Sodium-Takahax process is preferred when either elemental sulfur or sulfuric acid is the desired byproduct (GEESI, 1981).

Other modifications to the original process were aimed at developing new waste liquor treatment methods to make the Takahax process a closed-loop system.

The Nippon Steel coke-oven gas desulfurization technology provides four different options for handling Takahax waste solution (GEESI, 1981). These are

Type A: Wet oxidation to Ammonium Sulfate
Type B: Incineration to Sulfur Dioxide
Type C: Sulfuric Acid Production
Type D: Sulfur Production

The first two options are available with the Ammonia-Takahax process, and the last two with the Sodium-Takahax process.

In the U.S., the Takahax process has been used in combination with a wet oxidation Hirohax unit (Type A), where the thiosulfate and thiocyanate ammonium salts dissolved in the bleed stream are oxidized to fertilizer grade ammonium sulfate at high temperature (480°–535°F), and pressure (1,000–1,280 psig). In Japan, both wet oxidation units (Type A), and sulfur production units (Type D), are frequently selected, the choice depending primarily on the feed gas composition.

Process Chemistry

The principal reactions in the Takahax process (Barry and Hernandez, 1990) are

Absorber reactions:

$$NH_4OH + H_2S = NH_4HS + H_2O \tag{9-31}$$

$$NH_4OH + HCN = NH_4CN + H_2O \tag{9-32}$$

$$NQ + NH_4HS + H_2O = NH_4OH + S + H_2NQ \tag{9-33}$$

Regeneration reactions:

$$H_2NQ + \tfrac{1}{2}O_2 = H_2O + NQ \tag{9-34}$$

$$2NH_4HS + 2O_2 = (NH_4)_2S_2O_3 + H_2O \tag{9-35}$$

$$S + NH_4CN = NH_4SCN \tag{9-36}$$

$$NH_4OH + NH_4HS + 2O_2 = (NH_4)_2SO_4 + H_2O \tag{9-37}$$

The absorption reactions are those of H_2S and HCN combining with ammonia to form ammonium bisulfide and ammonium cyanide, followed by the oxidation of the ammonium bisulfide by the naphthoquinone (NQ) to form elemental sulfur and the naphthohydroquinone (H_2NQ). The reactions with oxygen take place in the regenerator and lead to the formation of NQ, thiocyanate, thiosulfate, and sulfate. Part of the elemental sulfur reacts with the available NH_4CN in the oxidizer to produce NH_4SCN.

Commercial Applications in the United States

The Takahax process is licensed in the United States by General Electric Environmental Services (GEESI), of Lebanon, PA. The first major installation of the Takahax process in the U.S. was in 1979 at Kaiser Steel's Fontana, CA, plant (GEESI, 1981) followed in 1981 by a second Takahax unit at the by-products plant of Republic Steel's coke plant, located in Chicago, IL (Williams et al., 1983). Both of these installations are Type A Takahax units.

Figure 9-12 is a simplified process flow diagram for a Type A plant (Takahax-Hirohax configuration) (GEESI, 1981). Per descriptions of this process provided by Williams et al. (1983), GEESI (1981), Araki et al. (1990), and Barry and Hernandez (1990), coke oven gas (COG) is cooled to 75°–95°F for tar and naphthalene removal. The cooled COG is contacted countercurrently—in three to four packed stages—with regenerated Takahax solution to simultaneously remove NH_3, H_2S, and HCN. The H_2S and HCN react with ammonia to form ammonium bisulfide and ammonium cyanide. In the absorber, the redox catalyst—1,4-naphthoquinone, 2-sulfonic acid—oxidizes the absorbed H_2S to elemental sulfur and ammonium cyanide reacts with the sulfur to form thiocyanate. The spent solution is pumped from the absorber bottom. To prevent the buildup of thiocyanate, elemental sulfur and sulfur salts, and permit closed loop operation, a slipstream of the spent solution (line 1, **Figure 9-12**) is sent to the Hirohax waste treatment unit. The rest of the solution is cooled to remove the heat of reaction and control the absorbent solution temperature so that sulfur and cyanide absorption are not adversely affected. The cooled solution is then directed to the regenerator. In the regenerator the solution is cocurrently contacted with compressed air. A novel air injection nozzle is reported to increase oxygen utilization from 30% achieved with the conventional air diffusion pipe to 50% (Araki et al., 1990). The redox catalyst is oxidized back to the quinone form, and thiosulfate, sulfate, and thiocyanate are formed from the elemental sulfur. The amount of thiosulfate, elemental sulfur, and sulfate in the oxidized solution leaving the regenerator is controlled by monitoring the solution redox potential (Araki et al., 1990). The regenerated solution flows by gravity back to the absorption column. In most cases, carbon steel can be used for piping and equipment in the Takahax section of the plant (Swaim, 1972). Usually, the absorber and regenerator are epoxy lined.

768 *Gas Purification*

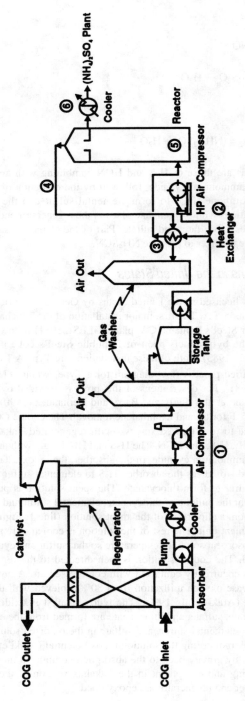

Figure 9-12. Simplified flow diagram of Type A Takahax plant. *(GEESI, 1981)*

The Takahax purge slipstream (line 1, **Figure 9-12**) is directed to a storage tank for conditioning before treatment in the Hirohax unit. Passivators are added to protect the titanium metallurgy used in the hot, high pressure sections of the Hirohax unit (Williams et al., 1983). Ammonia is also added if the slipstream does not contain sufficient ammonia to produce ammonium sulfate from all sulfur compounds. The liquid from the tank is pumped to about 1,000 to 1,280 psig, mixed with high pressure air (line 2, **Figure 9-12**), and heated by heat exchange (line 3, **Figure 9-12**) with reactor offgas (line 4, **Figure 9-12**) to about 390° to 480°F. The heated solution is sent to the oxidation reactor (line 5, **Figure 9-12**) where all salts are oxidized to ammonium sulfate. Reactor temperatures range from 480 to 535°F. The oxidized solution, containing 33 to 37 wt% ammonium sulfate with a free acid content of 2 to 4 wt%, is cooled, ammonia is added to reduce the free acid content, and the solution is then directed to a crystallizer. Offgas from the reactor heats the reactor feed and is washed before being vented to atmosphere.

Another application of the Takahax-Hirohax combination process for the desulfurization of coke-oven gas, at the LTV Steel Co. in Chicago, IL, was reported by Barry and Hernandez (1990). This paper describes in detail the relationship between the redox potential of the regenerated Takahax solution and sulfur deposition problems experienced at this plant in the Hirohax unit heat exchangers. The article also provides a complete outline of the methodology used in calculating the sulfur, ammonia, and water material balance of the Takahax unit. Typical operating conditions at the LTV Takahax plant are summarized in **Table 9-11**. The absorber liquor concentrations are taken at the top of the absorber. The bottom concentrations are essentially the same except for the concentration of HS^-, which decreases towards the bottom. **Figure 9-13** depicts the Takahax plant at Nippon Steel's Nagoya Works.

Stretford Process

The Stretford process was jointly developed in the 1950s in the UK by the North Western Gas Board (now the British Gas Corporation) and the Clayton Aniline Company, Ltd. It was originally conceived to replace the iron oxide boxes used for the removal of hydrogen sulfide from coke-oven gas. However, the process proved to be equally suitable for desulfurization of a variety of other gas streams, such as refinery gas, geothermal vent gas, natural gas, and

Table 9-11
Typical Operating Conditions at LTV Takahax Plant

Component concentration	Inlet gas	Outlet gas	Absorber liquor
H_2S, grains/scf	3.50	0.15	—
HCN, grains/scf	0.80	0.10	—
NH_3, grains/scf	5.50	3.85	>7.0 g/L
HS^-, ppmw			1,000
SCN^-, g/l			27–30
$S_2O_3^{2-}$, g/l			50–70
SO_4^{2-}, g/l			10–40

Source: Barry and Hernandez (1990)

Figure 9-13. Takahax plant—two absorbers in foreground; regenerator behind them on left side. *Courtesy of GE Environmental Services, Inc. (GEESI)*

synthetic gas. Another important application of the Stretford process was its use as the hydrogen sulfide removal step of the Beavon Sulfur Removal Process (see Chapter 8).

The process was initially licensed in Europe by W. C. Holmes Ltd., Huddersfield, UK. Holmes developed and introduced several proprietary improvements in the areas of gas pretreatment, sulfur recovery, and waste-liquor treatment. This branch of the process, used primarily in coke-oven and town gas applications, became known as the Holmes-Stretford process.

In the U.S., the Parsons Corporation of Pasadena, CA, was the original American licensor for British Gas. Another licensor in the U.S. domestic market was the Pritchard Corp., which built Stretford plants for natural gas and refinery fuel gas treating applications.

The Peabody Co. of Stamford, CT, acquired Holmes Ltd. in 1973, and gained access to the Holmes-Stretford technology. After the acquisition, the Holmes-Stretford process was licensed in Europe by Peabody-Holmes and was applied mainly to coke-oven gas plants. In the U.S., the Peabody Co. adapted the Holmes-Stretford technology to the American market, placing greater emphasis on coal gasification and geothermal gas clean-up. In 1986, the Peabody Co. was purchased by Fläkt Ltd. of Sweden, which later became part of ASEA Brown Boveri (ABB) Ltd. of Switzerland. Since the latest acquisition, the Stretford-Holmes technology has been licensed exclusively in the U.S. by ABB Environmental Systems, located in Norwalk, CT.

At its peak in the 1970s, Stretford technology was offered by more than a dozen companies worldwide, including Kobe Steel Ltd., British Gas Corporation, Sim-Chem Ltd., and the previously mentioned companies.

The literature on the Stretford process is extensive. The general outline of the process has been described by Nicklin and Holland (1963A, B); Thompson and Nicklin (1964); Nicklin et al. (1973); Ellwood (1964); Ludberg (1980); Miller and Robuck (1972); Moyes and Wilkinson (1973A, B, 1974); Moyes et al. (1974); Carter et al. (1977); Penderleith (1977); Smith and Mills (1979); Wilson and Newell (1984); Vasan and Willett (1976); and Beavon (1973).

The application of the Stretford process to the purification of natural gas has been discussed by Nicklin et al. (1973) and by Moyes and Wilkinson (1974). The Holmes-Stretford process has been described by Moyes and Wilkinson (1973B) and by Vasan (1979). Other applications of the Holmes-Stretford technology to a geothermal unit and to direct treatment of gasification gas have been reported by Vasan (1978) and Mills and Mosher (1987), respectively. Vancini has described two applications: a unit treating a high-CO_2 geothermal noncondensable gas (Vancini and Lari, 1985), and a unit purifying high-pressure gasification gas (Vancini, 1985). The Beavon-Stretford combination process for Claus plant tail gas treating has been described by Beavon and Brocoff (1976), and coke-oven gas applications have been described by England (1975) and Carter et al. (1977). A unit operating upstream of a Sulfinol plant has been described by Lindsey and Wadleigh (1979) and Kresse et al. (1981).

A total of about 170 Stretford units had been built worldwide by 1987 with an average design capacity of roughly 8.1 long tons of sulfur per day (LTPD) (Trofe et al., 1987). Of the total, half have a capacity of less than 3.5 LTPD. The largest plant built to date is at the SASOL II plant in South Africa with a capacity of 310 long tons sulfur per day. In the U.S., the largest plant was the 105 LTPD Great Plains plant in Beulah, ND. Both of these plants, which treat Rectisol offgas, were later converted to the Sulfolin process to minimize corrosion and chemical costs. As of 1992, more than 100 Stretford plants were still in operation worldwide, with an estimated total design sulfur removal capacity of 400,000 long tons per year. Many of the Stretford plants that have been shut-down are coke-oven gas treating plants (Vancini, 1992).

Process Chemistry

Initial Process Chemistry. As initially conceived and described by Nicklin and Brunner (1961A), the process utilized an aqueous solution containing sodium carbonate and bicarbonate in the proportion of about 1:3, and sodium salts of the 2,6 and 2,7 isomers of anthraquinone disulfonic acid (ADA). The postulated reaction mechanism involved four steps:

1. Absorption of hydrogen sulfide in alkali
2. Reduction of ADA by addition of hydrosulfide ion to a carbonyl group
3. Liberation of elemental sulfur from reduced ADA by interaction with oxygen dissolved in the solution
4. Reoxidation of the reduced ADA (by air)

Although this form of the process was tested successfully in commercial installations, it was soon found that certain inherent features imposed serious limitations on its economic development. The process was based on the oxidation of the hydrogen sulfide to sulfur by the ADA, and the oxidation reaction relied on the dissolved oxygen present in the solution. At the prevailing pH of 8.5 to 9.5, and at ambient temperature, the saturation concentration of dissolved oxygen in the ADA solution is approximately 10 milligrams per liter of solution. The small amount of soluble oxygen limited the maximum HS^- loading to 40 milligrams per liter of solution.

Furthermore, hydrogen peroxide, H_2O_2, was formed in the reoxidation step as a side-product of the reaction between ADA and oxygen. The hydrogen peroxide could oxidize hydrogen sulfide to thiosulfate or colloidal sulfur:

$$2H_2O_2 + 2HS^- = 2H_2O + (\tfrac{1}{4}) S_8 + 2OH^- \tag{9-38}$$

$$4H_2O_2 + 2HS^- = 5H_2O + S_2O_3^{2-} \tag{9-39}$$

As the rates of these reactions are comparable, free sulfur and sodium thiosulfate could be produced in roughly similar proportions. Hence, depending on the availability of hydrogen peroxide, any HS^- loading in the ADA solution in excess of the maximum amount of 40 milligrams per liter could be converted to either free sulfur or thiosulfate.

The need to employ dilute solutions to prevent the formation of thiosulfate resulted in very large circulation rates and considerable power consumption. Furthermore, the formation of elemental sulfur was slow from an industrial standpoint, requiring large reaction tanks and large liquid inventories. In addition, to obtain satisfactory rates of hydrogen sulfide absorption when treating gas streams containing appreciable amounts of carbon dioxide, partial decarbonation of the solution was required before recycle to the absorber.

Final Process Chemistry. To improve the economics of the process, a number of compounds were screened as possible additives to increase both the solution capacity for hydrogen sulfide and the rate of conversion of hydrosulfide to elemental sulfur. The compounds selected were vanadium salts (Nicklin and Holland, 1963A, B) since the combination of ADA plus vanadium proved to be remarkably effective. Hydrosulfide was reduced quite rapidly by vanadate to elemental sulfur, thus reducing the size of the reaction tanks. Also, by introducing vanadate as the oxidant, it was no longer necessary to rely on the oxygen dissolved in the solution, thus permitting substantially higher solution loadings, while less than 2% of the sulfur was converted to thiosulfate. Moreover, the vanadium-ADA system could work without loss of washing efficiency at a lower pH than the process with ADA alone.

This was a distinct advantage when treating gas containing high concentrations of carbon dioxide, because it eliminated the need for decarbonation of the working solution.

Although vanadate reacts readily with hydrosulfide to produce sulfur, a solution containing vanadate alone cannot be regenerated by blowing with air. However, in the presence of ADA, complete oxidation of reduced vanadate can be achieved, and the reduced ADA is readily reoxidized by contact with air.

The overall oxidation reaction to convert H_2S to sulfur can be represented as follows:

$$H_2S + \tfrac{1}{2}O_2 = S + H_2O \tag{9-40}$$

The reaction is exothermic, with the heat of reaction amounting to 3,580 Btu per pound of sulfur. The overall reaction can be subdivided into three separate consecutive steps. These steps are: a) hydrogen sulfide absorption, b) conversion of hydrogen sulfide to elemental sulfur, and c) vanadium reoxidation.

The chemistry of the sequential process can be represented by the following simplified molecular reactions:

$$H_2S + Na_2CO_3 = NaHS + NaHCO_3 \tag{9-41}$$

$$4NaVO_3 + 2NaHS + H_2O = Na_2V_4O_9 + 2S + 4NaOH \tag{9-42}$$

$$Na_2V_4O_9 + 2NaOH + H_2O + 2ADA = 4NaVO_3 + 2ADA\ (reduced) \tag{9-43}$$

$$2ADA\ (reduced) + O_2 = 2ADA + 2H_2O \tag{9-44}$$

Hydrogen Sulfide Absorption. In the Stretford process, the wash liquor in circulation typically contains a total of 30 grams per liter of sodium carbonate and sodium bicarbonate, whose proportions are established by the partial pressure of carbon dioxide above the solution.

The first step in the process is the absorption of the hydrogen sulfide in the aqueous solution, and is represented by equation 9-41. The hydrogen sulfide dissolves in water as hydrogen sulfide, hydrosulfide ion, and/or sulfide ion. The concentration of each species depends on the pH of the solution **(Figure 9-14)**.

When the pH value reaches nine, practically all of the molecular hydrogen sulfide becomes dissociated in the solution. Since the hydrogen sulfide in the gas phase is in equilibrium only with the undissociated H_2S in solution, the capacity of the solution for dissolving the H_2S becomes limited only by the diffusion of the gas into the bulk of the solution. To maintain this high pH, an alkali, usually sodium carbonate, is added to the water.

Although the rate of absorption is favored by high pH, the rate of conversion of the absorbed hydrogen sulfide to elemental sulfur is adversely affected by pH values above 9.5. The process is therefore best operated within a pH range of 8.5 to 9.5.

Conversion of Hydrogen Sulfide to Elemental Sulfur. The form of sulfur that is thermodynamically stable at ambient conditions is the rhombic form, S_8. One of the most plausible sulfur precipitation mechanisms involves the formation of polysulfides (S_x^{2-}, where x = 2–5) along with elemental sulfur. The S_4^{2-} and S_5^{2-} polysulfide species are likely to predominate at the pH and HS^- concentration typical of the Stretford liquor (Giggenbach, 1972).

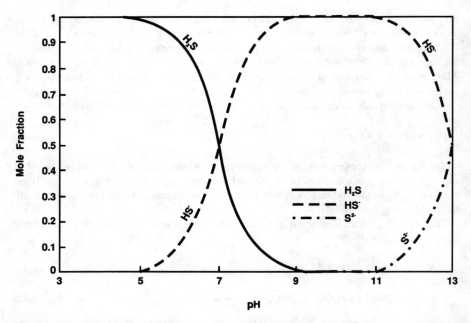

Figure 9-14. Hydrogen sulfide species distribution diagram at 25°C. (*Trofe et al., 1987*)

Kinetic studies conducted by DeBerry (1989) suggest that the polysulfide formation reaction mechanism can be represented by the following equation:

$$4HS^- + 6V^{5+} = S_4^{2-} + 4H^+ + 6V^{4+} \tag{9-45}$$

The polysulfides can then react further with the pentavalent vanadium to form elemental sulfur:

$$S_4^{2-} + 2V^{5+} = \tfrac{1}{2}S_8 + 2V^{4+} \tag{9-46}$$

Because the oxidation of bisulfide ion (equation 9-45) is faster than the oxidation of polysulfides (equation 9-46), the rate at which polysulfides are formed is faster than their disappearance rate, and the Stretford solution should always contain unreacted polysulfides at equilibrium. This trend is also favored by the fact that polysulfide species can be formed by the direct reaction of elemental sulfur immersed in solutions containing bisulfide ions (Schwarzenbach and Fisher, 1960).

Regardless of the actual mechanism, polysulfides play a major role in the sulfur formation in the Stretford process using pentavalent vanadium as the primary oxidant. A more detailed study of the reaction mechanism and kinetics of Stretford reactions was reported by Ryder and Smith (1962) and Phillips and Fleck (1979).

The conversion of hydrosulfide ion to elemental sulfur can be represented in molecular form as described in equation 9-42. The overall reaction is quite rapid and is essentially a function of the vanadate concentration in the solution as shown in **Figure 9-15** (Nicklin and

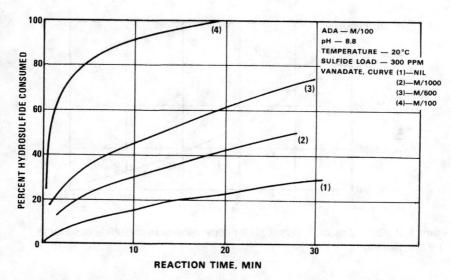

Figure 9-15. Effect of vanadate concentration on rate of reaction in Stretford solution.

Holland, 1963A). According to Thompson and Nicklin (1964), the reaction rate can be approximated by a second order expression as follows:

$$t = [1/k(a - b)] \times \ln[b(a - x)/a(b - x)] \tag{9-47}$$

Where: t = time in minutes

and k = specific reaction rate constant, l/[moles/l)(hr)]
a = initial concentration of vanadium in moles/l
b = initial hydrosulfide concentration in moles/l
x = moles/l of either hydrosulfide or vanadium reacted

Numerical values for the rate constant as a function of pH are shown in **Figure 9-16**. The reaction rate decreases exponentially with increasing pH. The rate of reaction increases with the solution temperature, but high temperatures cause reduced absorption and reduced dissociation of dissolved H_2S to HS^-. In addition, increased temperature and pH accelerate the production of thiosulfate from side reactions. For these reasons, the best operating temperature for the Stretford process is around 95°F.

A general correlation for estimating the time required for conversion is given in **Figure 9-17**. The values shown are based on a reaction-rate constant of 6,000 (pH about 8.5) and a vanadate concentration of 0.01 mole per liter. Such values are commonly found in industrial practice. For other values of the rate constant and molar concentrations, the reaction time is obtained by dividing the reaction time given on the graph by the value of the rate constant multiplied by the molar concentration (Thompson and Nicklin, 1964).

Equation 9-42 shows that 2 moles of vanadate are required for each mole of hydrogen sulfide. In industrial practice, an excess of vanadate is used to avoid overloading the solution

776 Gas Purification

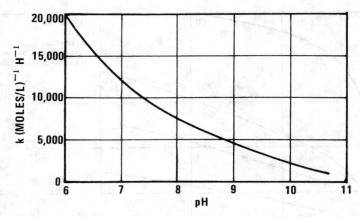

Figure 9-16. Rate of reaction versus pH for conversion of hydrosulfide to sulfur in Stretford solution.

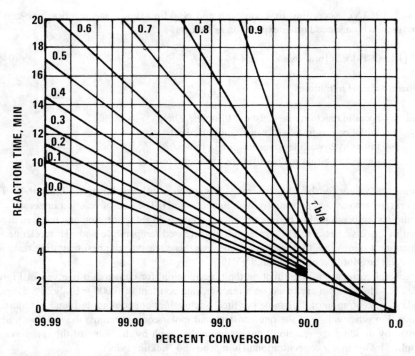

Figure 9-17. Time of reaction versus percent conversion of hydrosulfide to sulfur in Stretford solution.

with sulfide and subsequent formation of thiosulfate during solution regeneration. A concentration of V^{5+} of 0.3–0.5 grams per liter over that stoichiometrically required for the HS^- absorption is commonly used. If the amount of vanadium in solution is insufficient to oxidize all the absorbed hydrogen sulfide, some vanadium may precipitate as a black vanadium-oxygen-sulfur complex. In a properly formulated solution, this problem is completely prevented by adding a small amount of chelating agents to the solution (Nicklin and Brunner, 1961B).

When the oxidizing stage of the Stretford process is functioning correctly, the vanadium in solution is solely in its pentavalent form. The main advantage of using pentavalent vanadium as the oxidant is the increase in HS^- loadings that can be achieved, since the HS^- loading is no longer dependent upon the dissolved oxygen concentration. Theoretically, HS^- liquor loadings in the range of 100 to 500 milligrams per liter can be reached. From a practical standpoint, the operational HS^- loading limit is set by the tendency towards sulfur plugging. It is known that the probability of sulfur plugging problems increases almost exponentially when the HS^- loading exceeds 300 milligrams per liter of solution.

Vanadium Reoxidation. In the oxidizer, reduced vanadate is oxidized by ADA according to equation 9-43. Unlike pentavalent vanadium, tetravalent vanadium is largely insoluble in Stretford liquor. This insolubility is partially overcome through the use of chelating agents containing carboxylic groups, such as sodium citrate. Boron has also been reported to be effective in controlling the formation of vanadium sulfide by forming a boron-vanadium complex (Trofe et al., 1987).

Reoxidation of quadrivalent vanadium in the oxidizer follows a complex mechanism. The reduced V^{4+} formed in the absorber can not be regenerated to the V^{5+} oxidation state by direct contact with air. It is the ADA redox cycle that produces the hydrogen peroxide capable of oxidizing the vanadium, whereas the reduced ADA is easily reoxidized with air (Moyes and Wilkinson, 1973A). **Figure 9-18** is a graphic interpretation of how the redox circuits of vanadium, ADA, and oxygen are linked together in the Stretford process (Zwicky and Mills, 1980).

Reoxidation of ADA by contact with air is fairly rapid. The oxidation rate is controlled by the diffusion of oxygen in the liquid. The oxygen diffusion increases with temperature, solution pH, and contact time. The rate of oxidation can also be appreciably accelerated by the presence of small amounts of iron salts kept in solution by a chelating agent (Thompson and Nicklin, 1964).

Because the ADA acts as an oxygen carrier, its concentration is not directly related to the vanadium molarity, but rather to the hydrosulfide ion loading and the maintenance of sufficient pentavalent vanadium in solution to satisfy the required oxidation rate in the absorber.

Some of the ADA present in the Stretford solution is lost by an oxidative degradation process involving hydroxyl free radicals. ADA degradation losses of 0.18 wt% per day are typical. A method for controlling ADA consumption is to allow the concentration of thiosulfate produced in the absorber to reach a minimum value of 3 wt%. Thiosulfate is an effective reducing agent, and acts as an oxidation inhibitor by reacting with the hydroxyl radicals.

Secondary Reactions. In addition to the primary Stretford reactions, many secondary reactions take place under varying operating conditions. The most serious side reactions are the conversion of hydrosulfide to thiosulfate, and those involving hydrogen cyanide (when present in the feed gas).

The Stretford liquor, when entering the absorber, contains dissolved oxygen taken from the air used in the oxidizer to regenerate the solution. Additional oxygen may be present in

778 Gas Purification

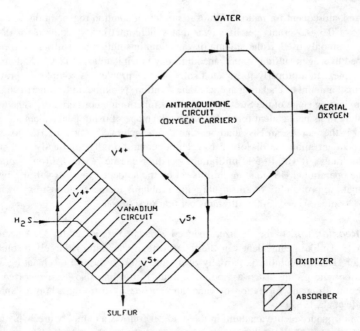

Figure 9-18. Stretford process oxidation circuits (*Zwicky and Mills, 1980*).

the sour gas. This oxygen reacts with unreacted HS⁻ and unconverted polysulfides to produce thiosulfate and sulfate anions.

The principal oxygen-based reactions leading to the formation of thiosulfate in the absorber are (Trofe and DeBerry, 1991)

$$2HS^- + 2O_2 = S_2O_3^{2-} + H_2O \qquad (9\text{-}48)$$

$$S_4^{2-} + O_2 + OH^- = S_2O_3^{2-} + HS^- \qquad (9\text{-}49)$$

Also, any HS⁻ slip to the oxidizer can be converted to thiosulfate by the combined action of oxygen and hydrogen peroxide. Therefore, the system must have sufficient hold-up for the hydrosulfide to be oxidized to elemental sulfur before the solution is brought into contact with air in the oxidizer:

Yet another source of thiosulfate ions is disproportionation of the precipitated sulfur in the alkaline solution:

$$S_8 + 8OH^- = 2S_2O_3^{2-} + 4HS^- + 2H_2O \qquad (9\text{-}50)$$

Liquid Phase Oxidation Processes for Hydrogen Sulfide Removal   779

The rate of formation of thiosulfate is not only dependent upon the degree of conversion of the hydrosulfide to sulfur prior to contacting with oxygen, but also on the pH of the solution, the operating temperature, and the concentration of dissolved solids in the circulating Stretford liquor. The effects of temperature and pH are shown in **Figures 9-19** and **9-20**

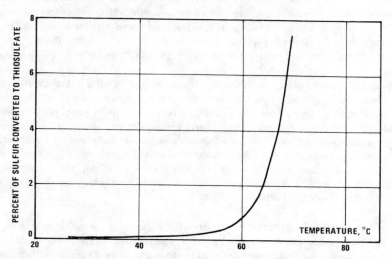

Figure 9-19. Effect of temperature on thiosulfate formation in Stretford solution.

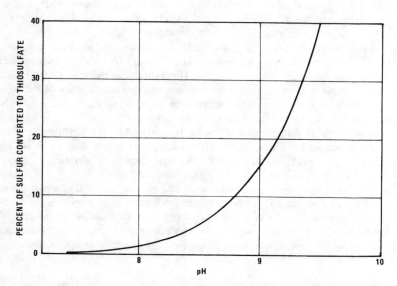

Figure 9-20. Effect of pH on thiosulfate formation in Stretford solution.

(Nicklin and Holland, 1963A). The effect of increased alkalinity on thiosulfate formation is rather significant above a pH of 8.8. This effect is more pronounced in the reactions of hydrosulfide with dissolved oxygen. For instance, at an oxygen partial pressure of 7 psia in the sour gas, a rise in pH from 8.3 to 8.8 triples the amount of sulfur converted to thiosulfate (from 3 to 9%) (Nicklin and Holland, 1963B). Likewise, increasing the concentration of dissolved solids in the liquor from 7 to 20% (by weight), doubles the sulfate formation rate.

The conversion of elemental sulfur to thiosulfate (equation 9-50) increases as the temperature of the solution goes up, becoming quite significant above approximately 120°F. This route to thiosulfate formation is particularly burdensome in Stretford units that melt the sulfur directly without first separating the froth liquor (the froth contains only 5–8% sulfur) by filtration.

In a properly operated plant, thiosulfate formation can be controlled at less than 1% of the sulfur in feed gas (Ludberg, 1980). The maximum concentration of sodium sulfate and thiosulfate in the Stretford liquor is usually maintained between 25 to 30 grams per liter. Associated with each of these sulfur anions is an equivalent amount of sodium cation that must be added to maintain ionic balance. Thus, the formation of soluble sulfur species increases alkali makeup (usually soda ash). Also, the accumulation of sulfate, thiosulfate, and sodium ions in the solution makes it necessary to discard part of the solution periodically to maintain control of the concentration of dissolved salts. This blowdown contributes to the loss of vanadium, ADA, and sodium carbonate.

Effect of Carbon Dioxide Partial Pressure. The presence of carbon dioxide in the sour gas has two major effects in the Stretford process. The first effect is to lower the pH of the solution. The second and more important effect is to decrease the hydrogen sulfide absorption mass transfer rate.

When carbon dioxide is present in the sour gas at a partial pressure greater than about 0.2 psia, it is partially absorbed by the Stretford liquor with the formation of bicarbonate and a consequent lowering of the pH. The lowering of the pH, unless buffered, would result in a lower reaction rate in the oxidation reaction of hydrosulfide to sulfur.

The most significant effect of carbon dioxide is a decrease in the absorber mass-transfer rate for H_2S. In the absence of CO_2, the H_2S mass transfer rate depends primarily on the partial pressure of hydrogen sulfide in the sour gas. However, when the sour gas being treated contains a high concentration of carbon dioxide, the absorption efficiency of the solution may be sufficiently lowered to require an appreciable increase in the absorber height (Nicklin and Holland, 1963A). These effects are described in **Figure 9-21.**

The decrease in the overall absorber efficiency is mitigated by the fact that the Stretford liquor has a high absorption selectivity for H_2S as compared to CO_2 (Garner et al., 1958). **Figure 9-22** (Vancini, 1985) illustrates how selectivity varies with operating pressure in a cross-flow absorber as a function of the HS^- loading in the Stretford liquor. The graph shows that if the absorber is operated at a typical load of 200 mg HS^-/liter and at a pressure of 50 psig, the quantity of absorbed CO_2 amounts to one-tenth of the total present in the feed gas. The low CO_2 absorption capacity of Stretford liquors is one of the distinct advantages of the Stretford process. By utilizing a cross-flow contactor configuration, the absorption of carbon dioxide can usually be kept at acceptable levels unless the process is operated at elevated pressure.

Effect of Inlet Gas Contaminants. Some components and trace contaminants present in industrial gases may cause serious operating problems in Stretford plants. Some contami-

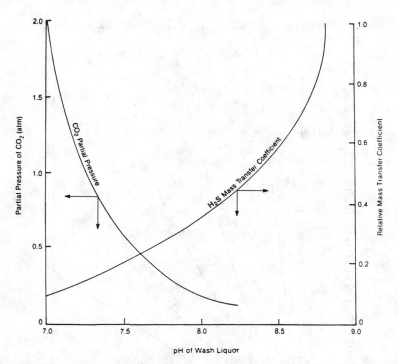

Figure 9-21. Effect of CO_2 partial pressure on Stretford solution pH and H_2S mass transfer coefficient (*Moyes and Wilkinson, 1974*). *Reprinted with permission from Chemical Engineer, copyright 1974, Institution of Chemical Engineers*

nants may produce undesirable secondary reactions, which require increasing chemical makeup. Others may collect on the sulfur particles, causing discoloration and filtration rate reduction. Finally, a few components may physically hinder the regular operation of the Stretford plant, or may cause the formation of undesirable froth at liquid-gas interfaces. Recognized trace contaminants that may cause one or more of the foregoing difficulties are

1. Organic sulfur compounds, which include carbonyl sulfide, carbon disulfide, mercaptans, and thiophene
2. Condensable complex heavy hydrocarbons, all toxic to humans in various degrees (tar and oil). These may include substituted benzenes, naphthalenes, and furanes
3. Solid particulates (e.g., fly ash, unburned coal, or rock dust)
4. Metals (alkali and heavy)
5. Acid gases, including HCN, HCl, and SO_x
6. Reactive gases, including oxygen and ammonia
7. Foam producing compounds

Organic sulfur compounds that may be present in industrial gas include carbonyl sulfide, carbon disulfide, mercaptans (mostly methylmercaptan), and thiophene. These contaminants

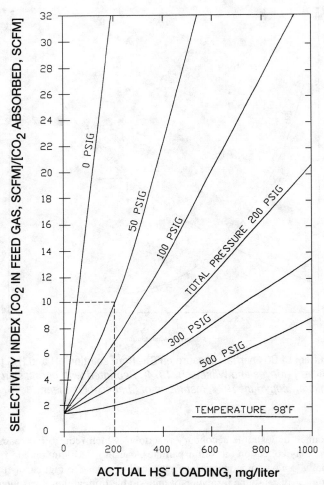

Figure 9-22. Selectivity of a crossflow Stretford absorber for CO_2 as a function of HS^- concentration and total pressure. (*Vancini, 1985*)

are to be found particularly in gas from petroleum refinery and petrochemical plants, gas from coal processing such as carbonization and gasification, and tail-gas from the Claus sulfur recovery process. Coke-oven gas, for example, typically contains 300–400 ppmv of organic sulfur, of which about 80% is CS_2, 15% COS, and 5% thiophene (Fenton and Gowdy, 1979).

Carbonyl sulfide and carbon disulfide pass more or less intact through the Stretford absorber, while about half of the thiophene is typically removed. Methylmercaptan is largely oxidized to dimethyl disulfide, DMDS, which is largely incorporated into the sulfur cake. Higher mercaptans are partially oxidized to disulfide, but most are released to the atmosphere in the oxidizer, in some cases causing severe odor problems, which may require incineration of the oxidizer vent stream.

For any gas containing a high quantity of organic sulfur, a preliminary hydrolysis/hydrogenation step may be necessary before introducing the gas into the Stretford unit. With this pretreatment, all organic sulfur species are converted almost quantitatively to H_2S, which is then removed by the Stretford unit (Kouzel et al., 1977; Moyes et al., 1975).

If the prevailing temperature of the working liquor is lower than the gas temperature, tar and oils in the sour gas to the Stretford absorber may condense in significant amounts. This situation can be prevented by fractional condensation, i.e., cooling the gas in several stages arranged in series, with condensate product removal after each cooler. Often, a recirculated oil medium, like kerosene, is injected into the raw gas ahead of the condensers to make tars more fluid. This type of quenching also collects a major part of the fly ash, soot, and heavy metals, many of which can form hard deposits during the initial gas cooling (Vancini, 1985). In coke-oven gas treating systems, a final electrostatic detarrer is usually installed at the end of the pretreatment step before introduction of the cold sour gas into the Stretford unit.

When present, alkyl benzenes tend to form droplets in suspension in the wash liquor. In the absorber, these droplets associate with precipitated sulfur particles and make it difficult to float-off the sulfur in the oxidizer. Whenever the alkylbenzene concentration of the sour gas exceeds 2 grams per actual cubic meter, the gas must be fractionally condensed and passed through an electrostatic detarrer (Moyes and Wilkinson, 1973A; Smith et al., 1976). The benzene and heavier aromatics (BTX) content of the sour gas should be reduced in the same way to ensure good quality froth in the oxidizer and limit foam formation inside the absorber. Holding the concentration of BTX below 5 grams per actual cubic meter of gas will usually be sufficient to prevent foaming problems.

Phenolic compounds, polycyclic aromatic hydrocarbons, and tars, even in minute amounts, can produce discoloration of the sulfur, or reduce its dewatering ability. A maximum content of 10 milligrams of tar per actual cubic meter of gas is usually acceptable. Naphthalene levels lower than 0.1 gram per actual cubic meter can be safely tolerated before absorber plugging becomes a problem (Moyes and Wilkinson, 1973A).

Solid particulate materials and alkali metals in the treated gas can cause severe corrosion/erosion problems in gas turbine applications. Solid particulates in the sour gas having a mean diameter greater than 10 microns are effectively removed by the absorber and end-up in the sulfur product. This may cause discoloration of the sulfur, and an increase in its ash content. The recommended maximum allowable particulate content is 40 milligrams of dust per actual cubic meter of sour gas feed. When an electrostatic detarrer is employed, the concentration of solid particulate is usually well below this value.

Of the heavy metals that may be present as vapors in the inlet sour gas, mercury is the most significant. Mercury is sometimes present in gasification gas, but is more prevalent in geothermal noncondensable gas. The mercury content of geothermal gas can reach concentrations of several milligrams per standard cubic meter (dry basis). Any mercury introduced to a Stretford plant ends up in the recovered sulfur, which may become a hazardous nonmarketable waste product (sulfuric acid must contain less than 0.1 ppm of mercury to be marketable). Removal of mercury before the sour gas enters the Stretford unit can be accomplished by adsorption in a bed of activated carbon (Coolidge, 1927; Otani et al., 1988; Habashi, 1978). Activated carbon catalyzes the oxidation of some H_2S to sulfur, which remains adsorbed in the carbon bed. This sulfur reacts quantitatively with the mercury vapor content in the inlet gas forming HgS (Lovett and Cunniff, 1974). A residual concentration of less than 0.01 ppm mercury in the sour gas can be obtained by this method.

Hydrogen cyanide, hydrogen chloride, and sulfur oxides are quantitatively absorbed and decomposed by the Stretford liquor, forming sodium thiocyanate, sodium chloride, and sodium sulfate. These contaminants accumulate in the circulating liquor in ionic form and increase the blowdown frequency. The need for frequent system purging increases the volume of hazardous waste by-product generated. In addition, excessive blowdown increases chemical make-up costs because more fresh chemicals are required to maintain solution strength. A more detailed review of various methods available to cope with HCN is provided later in the Stretford liquid waste treatment discussion.

The solubility of inlet gas oxygen in the Stretford liquor is a function of the oxygen partial pressure and the process temperature. Absorbed oxygen is undesirable since it initiates the oxidation of hydrosulfide ions to sulfate and thiosulfate. Some geothermal noncondensable gas may contain over 10 vol% of oxygen as a result of excessive air leakage into the turbine surface condenser, which is operated under vacuum. In this extreme situation, it has been reported that soda ash consumption is two to three times greater than normal.

Ammonia in the inlet gas can react with solid sulfur to produce polysulfides (Penderleith, 1977). The practice of introducing ammonia to remove sulfur deposits in the absorber (Steppe, 1986) can result in the release of ammonia in the oxidizer causing odor problems. The maximum allowable concentration of ammonia is about 0.1 grams of ammonia per standard cubic meter of feed gas (Moyes and Wilkinson, 1973B).

When a fresh charge of chemical makeup reagents (vanadium, ADA, and possibly chelating agent) is introduced into a Stretford unit, some foaming may take place. After a period of a few hours, however, the foam usually subsides. Foam produced by the presence of saponifiable organics in the inlet gas (Smith et al., 1976) is more persistent. The primary corrective measure to combat this situation is the addition of nonionic surfactants having separate functional groups, part soluble in water and part soluble in the organics (for instance, Percol, Primafloc, Pluronics). Care must be exercised not to use excessive amounts of antifoaming agents because they may hinder sulfur flotation in the oxidizer. For this reason their concentration is seldom allowed to exceed a few milligrams per liter of wash liquor. Borax on occasion can also reduce foaming (Vancini, 1986).

Process Description

A schematic flow diagram of a typical low-pressure Stretford unit is shown in **Figure 9-23**. The sour gas is contacted countercurrently with the Stretford solution in the absorber, where practically all the hydrogen sulfide is removed. The treated gas may contain less than 1 ppm of hydrogen sulfide. The rich solution flows from the absorber to a delayed reaction tank, where the conversion of hydrosulfide to elemental sulfur is completed, and the sulfur precipitates in the form of fine particles. The delay tank may be the bottom of the absorber or a separate vessel. From the delay tank, the solution flows to the oxidizer, where it is regenerated by intimate contact with air. In the oxidizer, the sulfur is separated from the solution by flotation, and is removed at the top as a froth containing about 5–8% solids. The regenerated solution, containing typically less than 0.5 wt% suspended sulfur, is recycled to the absorber.

The sulfur froth is collected in a slurry tank and subsequently processed in a filter or centrifuge to remove the solution remaining in the froth. It is usually necessary to wash the sulfur-cake with at least 2 pounds of water per pound of sulfur to recover most of the chemicals contained in the solution and to produce relatively pure sulfur. To maintain the system water balance, wash water and water produced by the reaction have to be evaporated, either with the vent gas in the oxidizer or in a separate evaporator, depending on the quantity of water involved. The sulfur-cake, which contains about 50 to 60% solids, may be further processed by melting in an autoclave. In this manner high grade liquid or solid sulfur is produced. If

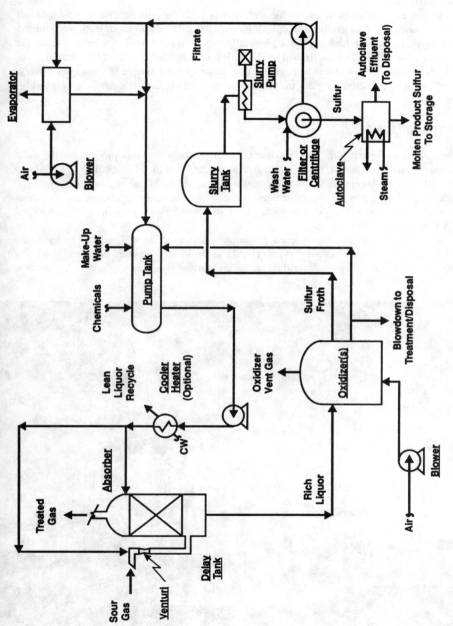

Figure 9-23. Schematic flow diagram of a typical Stretford process.

marketable sulfur is not produced, the sulfur-cake from the filter must be disposed of as a toxic waste.

The process is also applicable to the treatment of high pressure gas, such as the product from a pressurized coal gasification unit (Vancini, 1985). The major modification required is the inclusion of a flash tank to release physically dissolved gas from the solution leaving the absorber. The rich solution is flashed to about 10 psig. The flash gas does not contain a significant amount of H_2S, and in large plants may be sent to an expander for energy recovery. A photograph of a Stretford plant treating geothermal noncondensable gas is shown in **Figure 9-24.**

Design and Operation

Absorber Configuration. The fundamental steps in the absorption process are the mass transfer of H_2S from the gas to the liquid and the oxidation of the HS^- to sulfur (Vancini and Lari, 1985). The absorption of hydrogen sulfide can take place in one or more contact stages, with the gas washed by the Stretford liquor either countercurrently or cocurrently. A subsequent delay tank allows enough residence time to convert major portions of the absorbed H_2S to elemental sulfur.

Under steady-state conditions, each gas-liquid contact device will reduce the hydrogen sulfide in the gas from a partial pressure $[H_2S]_{in}$ at the inlet to a partial pressure $[H_2S]_{out}$ at the outlet. Given the very small residual partial pressure of H_2S over the Stretford liquor, the

Figure 9-24. Stretford plant on geothermal, noncondensable gas. *Courtesy of ABB Environmental Systems*

expression for the Number of Transfer Units, $[NTU]_{req}$, required to reduce an inlet concentration $[H_2S]_{in}$ to an outlet concentration $[H_2S]_{out}$ in the absorber becomes

$$[NTU]_{req} = \ln\{[H_2S]_{in}/[H_2S]_{out}\} \qquad (9\text{-}51)$$

In industrial practice, the absorber may be any efficient gas-liquid contacting device, and its configuration depends on the sour gas H_2S concentration and the desired removal efficiency. Several designs can be effectively used:

1. Packed tower
2. Spray tower
3. Venturi
4. Plate tower

The design choice is normally the approach that provides the most convenient and economical method of obtaining the required number of transfer units $[NTU]_{req}$. Generalized expressions for the number of transfer units attainable with various types of contactors used in the Stretford process have been reviewed by Vancini and Lari (1985).

Packed towers are used primarily in applications that require high removal efficiency, $[NTU]_{req} > 3$, but where the hydrogen sulfide inlet gas concentration is relatively low. When the hydrogen sulfide concentration of the sour gas is higher, and plugging of the packed tower may be a problem, a spray tower can be installed ahead of the packed column to lower the concentration of H_2S reaching the inlet of the packed section of the absorber. In this type of application, the HS^- loading of the wash liquor is kept low by increasing the liquid/gas ratio in the absorber.

Venturi absorbers are also used when there is a high concentration of H_2S in the feed gas. Venturi absorbers (Rothery, 1986) are typically of the jet-type. That is, the liquor is introduced through a single nozzle at a pressure sufficiently high to drive the gas through the throat section of the venturi without significant pressure drop. In venturi-jets, the mass transfer between gas and liquid occurs across the boundary layer on the surface of the droplets produced by the spray nozzle. Venturi absorbers are less efficient than packed columns, and are used in applications where the required number of transfer units $[NTU]_{req}$ ranges from 1.5 to 2.5.

Plate towers can be utilized in high removal efficiency applications, even at medium H_2S concentrations. The preferred design for Stretford applications uses weeping trays, thus avoiding the chance of downcomer plugging. For low-CO_2 gases, each tray is reported to be equivalent to 0.6 transfer units.

Oxidizer Configuration. Several types of oxidizers have been employed (Thompson and Nicklin, 1964). The simplest and most commonly used is a vertical tank containing an air distribution device at the bottom. The liquid and air flow concurrently upward, and the sulfur froth is removed by overflowing a weir at the top of the vessel. The liquor, which is essentially sulfur free, is withdrawn at a point located below the top of the oxidizer.

There are two basic types of air distribution devices: (1) circular perforated plates, and (2) eductor mixers.

The air needed to oxidize the vanadium and froth the sulfur can be introduced through circular perforated plates located at the base of the oxidizer. Two-millimeter holes are typical, with a pressure drop across the holes of 4–6 inches of water (Moyes and Wilkinson, 1973A). The total air pressure must, of course, be sufficient to overcome the head of liquid in the oxidizer as well as the pressure drop across the plate. Each plate is approximately 16–32 inches in diameter. The number and arrangement of the plates, as well as the hole pattern within each plate, are designed to obtain good bubble distribution in the oxidizer cross section.

Eductor mixers introduce the air through tiny holes in the throat of small venturis installed on separate pipes transferring the liquor from the delay tank to the oxidizer. The large shear force at the air-liquor interface provides a high mass-transfer coefficient between the oxygen and the liquor. The air then bubbles upward through the oxidizer to accomplish the sulfur flotation (Vancini and Lari, 1985).

The design of an oxidizer is fully defined by three basic parameters: (1) the required air flow, (2) the oxidizer diameter, and (3) the useful liquid volume. The required air flow is directly proportional to the amount of sulfur being produced and inversely proportional to the useful height of liquid in the oxidizer. The value of the latter proportionality constant corresponds to an oxygen utilization of 0.6–0.7% per foot of useful liquid level in the oxidizer.

The oxidizer diameter is directly proportional to the required air flow and inversely proportional to the square root of the air specific gravity (taken at the bottom of the oxidizer at the liquor temperature). To obtain good sulfur flotation, the air flow rate is 2.5–3.0 actual cubic feet per minute per square foot of cross-sectional oxidizer area (Nicklin and Holland, 1963B).

The volume of liquid retained in the oxidizer is directly proportional to the liquid flow rate (and therefore the amount of sulfur being produced) and the required residence time. The required residence time depends on the number of oxidizers. When one oxidizer is used, the residence time should be around 45 minutes. With two oxidizers in series, the total residence time in both oxidizers should not exceed 30 minutes.

The multi-oxidizer configuration provides better efficiency for oxygen mass transfer. The clear liquor from the first oxidizer is fed to the second oxidizer and sulfur froth is removed from the top of both oxidizers. Often the air volume is apportioned so that a higher air flow rate is introduced into the first oxidizer (to increase oxygen mass transfer through greater turbulence) and a lower air flow rate to the second oxidizer (to facilitate sulfur flotation).

Reagent Consumption. The main chemical reactants present in the Stretford solution—vanadium, ADA, and chelating agents—are continuously lost with the produced sulfur. This chemical loss is a function of the quantity of sulfur recovered, the amount of liquor retained by the filtercake, the concentration of each chemical in the Stretford solution, and the washing efficiency. With regard to carbonate alkalinity, the amount removed from the system depends not only on the factors already enumerated, but also on the extent of secondary oxidation reactions.

In summary, the required chemicals make-up is a function of

- Chemicals physically removed by the wet sulfur-cake
- Chemicals lost by decomposition in the redox environment (especially ADA)
- Chemical inventory change in the system needed to achieve the optimum liquor composition

It is common practice to adjust the Stretford solution composition when the gas flow rate or composition differs from the design values. **Table 9-12** gives an example of composition adjustment following the reduction of hydrogen sulfide loading in a typical Stretford plant.

High reagent consumption in a Stretford system can be experienced when some of the following conditions exist:

Table 9-12
Comparison Between Design and Optimum Operating Conditions for a Typical Stretford Plant

Liquor Composition	Design Conditions at 100% Load	Optimum Conditions at 40% Load
Vanadium, g/L	1.34	0.86
ADA, g/L	2.14	1.93
Na Alkalinity, g/L	5.50	4.50

Source: Vancini and Althem (1985)

- Solution temperature greater than 120°F
- pH greater than 8.8
- Insufficient ADA or vanadium levels
- Unbalanced Stretford liquor composition
- Excess oxidizer residence time
- Absorber overloading with HS^-
- Insufficient residence time in the delay tank
- Excess levels of suspended sulfur and suspended solids in the liquor

Materials of Construction. Stretford plant vessels are constructed primarily of carbon steel, with inert linings (for example, cold-cured epoxy resins) for the oxidizer and the sulfur slurry tanks. Stainless steel impellers are recommended for circulating pumps wetted by Stretford liquor and sulfur slurry pumps. Care should be taken to avoid sulfur deposits on unprotected metal faces (Sundström, 1979; Edwards, 1979). Piping is usually carbon steel, with the exception of the absorber inlet and outlet gas lines, which are fabricated of 316 stainless steel or fiberglass. Corrosion studies of Stretford plants were performed by Wessels (1980) and Hanck et al. (1981).

Sulfur Recovery. The sulfur froth that is collected in the slurry tank consists of Stretford liquor containing 5–8 wt% sulfur. Stretford process sulfur recovery begins with an initial dewatering stage to separate the liquor remaining in the sulfur froth from the sulfur particles. The equipment most commonly used for this purpose is a filter, but in large plants centrifuges are sometimes used instead of filters (Moyes et al., 1975).

Most plants utilize a vacuum rotary-drum filter. The vacuum is applied internally and provides the driving force to draw the solution through the filtercake. Removal of the cake from the drum is critical (Herpers and Korsmeier, 1979; Grande, 1987). The use of cutting knives has been found to be impractical. Instead, cake removal is generally accomplished by reversing the pressure profile across part of the filter and blowing air intermittently through the filtration surface in an outward direction. This procedure dislodges the filtercake from the filter cloth and helps to prevent filter cloth blinding by small particles in the Stretford sulfur (typical mean particle size equals 30 microns) (Foxall, 1986).

Washing of the cake is performed with demineralized water, but typically only two or three liquid displacements can be obtained due to the inherently small wash area available on a rotary-drum filter. The washed filtercake (50 wt% sulfur) can be melted batchwise in a pressurized (50–70 psig) stainless-steel autoclave. Internal steam coils heat the sulfur plus the interstitial liquid to 280°F to reduce its viscosity without reaching the boiling point of the mixture. The total melting operation can take up to 20 hours, with half of the time needed for sulfur melting, and the remaining time required to separate the molten sulfur from the residual liquor. The supernatant liquor liquid phase, which contains almost all the Stretford solution chemicals, is flashed into a knock-out drum, and then flows by gravity to the oxidizer vessel.

The molten sulfur phase is discharged from the bottom of the autoclave into a heated sulfur pit. The purity of the recovered molten sulfur obtained with this method depends primarily on the quality of the wet sulfur-cake introduced in the autoclave. If the cake is produced with one filtration, the sulfur purity may still exceed 99.8%. Usually the ash content will range from 0.02 to 0.04 wt%, and the vanadium content will be less than 7 ppmw (Trofe et al., 1987).

The purity of the sulfur produced can be further upgraded (Vancini, 1988) by employing two consecutive filtrations and using an intermediate reslurry step with demineralized water. This method reduces the ash and heavy metals content of the sulfur to less than 0.005% ash and less than 1 ppmw vanadium, which is usually well below mandated environmental regulation limits. This may enable the Stretford plant operator to produce marketable sulfur if so desired.

Herpers and Korsmeier (1979) have reviewed the methods used to purify Stretford sulfur. Typical specifications of sulfur for sulfuric acid manufacture are reported in **Table 9-13** (Sundström, 1979).

Sulfur can be shipped in the molten state in tank cars or trucks, but the safest and most convenient means of transportation is in solid form. This is most commonly accomplished by pelletizing the sulfur in the shape of pastilles (Kessler and Kwong, 1987). The typical bulk specific gravity of sulfur in pastille form is close to 1.40.

When it is not economically attractive to produce marketable sulfur, the filtercake must be considered a hazardous waste and disposed of accordingly (Vancini, 1988). In general, this is the case for plants whose production capacity is below half a ton of sulfur per day (dry basis).

Bacterial Growth. One problem encountered in many Stretford plants treating gas streams other than coke-oven gas is contamination of the system by bacterial growth (Sundström, 1979; Bromel, 1986). The problem can also be present but is quite rare in coke-oven gas or geothermal applications. The absence of bacterial contamination in these two types of Stretford plants is due to the presence of thiocyanate in the former, and boron compounds in the latter (boron in geothermal steam originates from radioactive decay within the earth's mantle).

Bacteria in the Stretford solution were first discovered in 1978 in a solution sample from a petroleum refinery in Illinois (Wilson and Newell, 1984). The British Gas Corporation insti-

Table 9-13
Typical Sulfur Specifications For Sulfuric Acid Manufacture, Wt% Impurity

Sulfur, min.	99.5–99.7
Ash, max.	0.005–0.02
Carbon, max.	0.05–0.10
Chloride, max.	0.005–0.006
Water, max.	0.05

Source: Sundström (1979)

tuted an in-depth investigation to elucidate this phenomenon. Several strains of bacteria were identified, belonging to the *Pseudomonas* and *Thiobacillus* groups.

The problems created by bacteria build-up in Stretford plants vary in their degree of severity, but usually fall into several general categories:

- Increased consumption of alkali due to lowering of pH
- Uncontrollable foaming rather than normal sulfur froth
- Sulfur that is sticky and tends to adhere to all surfaces
- Absorber plugging
- Corrosion of carbon steel equipment
- Sulfur contamination as determined by carbon analysis
- Increasing occurrence of secondary reactions, with oxidation of thiosulfate to sulfate and sulfuric acid

To combat these problems a number of biocides have been developed whose efficacy varies from plant to plant depending on the type of bacterial strain present and plant operating conditions. Of the many biocide compounds tested, those releasing bromine or formaldehyde have been found to be the most effective. The required dosage is about 1,000 ppm or less. Thiocyanate addition can also be used to hinder the growth of bacteria, although like all other biocides, it undergoes gradual decomposition and thus requires constant make-up.

Characteristics and Treatment of Waste Streams

Several Stretford process waste streams require careful consideration. These waste streams include

- Gas streams: treated gas when vented, oxidizer vent gas, evaporator off-gas
- Liquid streams: Stretford liquor bleed, HCN unit bleed, sulfur melting water-phase bleed
- Solid waste: sulfur-cake, when not marketed or when hazardous

Gas Waste Streams. Treated gas can be a product, like a combustible gas, or an emission to be disposed of. An example of the latter is the noncondensable gas emitted from geothermal power plant gas purification units. This stream is a treated waste gas and is vented to the atmosphere after being diluted with cooling-tower ventilation air.

Oxidizer vent gas is usually vented directly to the atmosphere. It consists of air (depleted of oxygen), water vapor, and traces of gas stripped from the Stretford liquor, typically CO, CO_2, ammonia, hydrocarbons, and organic sulfur vapors. Of these gas components, mercaptans are the most objectionable because of their strong odor. Incineration, although expensive, is the most reliable way to cope with odor problems.

The hydrosulfide oxidation reaction produces water, which accumulates in the Stretford liquor. Other major sources of water are the sulfur wash water and the dilution water associated with the fresh chemicals makeup solution. Water vapor in the oxidizer vent gas and other evaporation losses are normally insufficient to water balance the process, and a small evaporator is commonly used to remove additional water from the system. This evaporator takes a side stream from the circulating liquor loop and evaporates any excess water with air. The evaporator is rather small and evaporator off-gas emission contributions are usually negligible.

Liquid Waste Streams. Liquid waste streams may create disposal problems in Stretford systems (Tallon et al., 1984). Secondary reactions produce a limited quantity of thiosulfate and sulfate soluble salts. The concentration of dissolved salts in the circulating liquor increases until an equilibrium is reached between the amount of salts being formed and the amount of salts being removed in the sulfur-cake. If conditions are such that the dissolved salt concentration in the circulating solution reaches 25 wt%, some liquor has to be bled from the system.

The typical composition of a Stretford bleed stream is as follows:

Alkali (as Na_2CO_3)	12 grams per liter
Vanadium (as $NaVO_3$)	5 grams per liter
ADA	2 grams per liter
$Na_2S_2O_3$ (anhydrous)	240 grams per liter
Na_2SO_4	60 grams per liter

This bleed stream is a toxic substance and must be treated before disposal. Several treatment processes have been used with various degrees of success (Yan and Espenscheid, 1980). These treatment methods can be broadly grouped into two generic categories: biological and chemical treatment.

Biological treatment uses several species of bacteria to convert thiosulfate to a non-toxic form. The bleed stream is first diluted so that the maximum $S_2O_3^{2-}$ concentration does not exceed 0.5 grams/liter (Moyes and Wilkinson, 1973A). This treatment process operates at a pH below 7.5 and a maximum temperature limit of 75°F. A minimum residence time of 12 hours in the bacterial tank is also required. Biological treatment can also be used for the liquid bleed from HCN removal units. In this application, a concentration of 1 gram of SCN^- per liter of waste solution is the maximum thiocyanate level tolerated by the bacteria.

Biological treatment, like other proposed treatments (multiple-effect evaporation, high-temperature incineration) does not recover either the salts or the Stretford reagents in the liquid waste stream. A chemical method that was proposed to recover all inorganic compounds as reusable reagents was the fixed salts reductive incineration method developed by Peabody-Holmes and described by Smith and Mills (1979) and Carter et al. (1977). Unfortunately, the first commercial plant could not overcome recurrent operating problems and was eventually shut down. A similar regeneration process, based on a modification of the same basic concept, has since been proposed by Nittetu Chemical Engineering Ltd. (Mitachi et al., 1981).

Another approach to regenerating Stretford solutions has been proposed by the British Gas Corporation (Wilson and Newell, 1984). However, this process is only applicable to plants treating gas containing HCN in no more than trace amounts. The process involves the acidic decomposition of thiosulfate to produce elemental sulfur and sulfate. During the process, the sodium carbonate in the solution is neutralized by sulfuric acid. The ADA is not affected by the acidification process. The sulfur produced by the British Gas process is very coarse and granular and can be separated by known techniques. After separation of the sulfur, sodium sulfate decahydrate is crystallized at low temperature to prevent destruction of ADA. If desired, the crystals can be further processed to anhydrous sodium sulfate. The first commercial application of this process was in 1990 by Global Sulfur Systems Inc., at a Northern California Power Agency geothermal power plant near Healdsburg, CA. In the test, the thiosulfate content of the spent Stretford solution was reduced from 370 to 95 g/liter, and the treated solution was reused in the Stretford process.

An integrated system to remove ADA and vanadium has been proposed and tested by Dow (Pack, 1986). The Dow Stretford Chemical Recovery Process is a patented process for purging thiosulfate while recovering ADA and vanadium for recycle back to the Stretford unit. The process consists of three unit operations: filtration, activated carbon adsorption, and ion exchange, run as a semi-batch cycle (Hammond, 1986).

A schematic diagram of this process is shown in **Figure 9-25**. A slipstream of the main Stretford solution is pumped through porous tubular filter units. The filtration section removes sulfur particles that could interfere with the performance of the activated carbon or the ion exchange resin. A small fraction of the filtered slipstream is then passed through activated carbon and ion exchange beds in series. The activated carbon recovers the ADA from the purge stream, while the ion-exchange resin removes the vanadium. After the activated carbon and ion exchange beds are fully loaded, they are simultaneously regenerated, thus recovering the ADA and vanadium. Four percent caustic soda is used to remove the vanadium from the ion exchange resin. The effluent from the resin bed is then heated and used to strip the ADA from the activated carbon. A system flush follows to lower the pH in the resin bed and to cool the activated carbon before the bed can be reloaded.

One major difficulty often encountered in the commercial application of the Stretford process to coke-oven gas plants is caused by the presence of hydrogen cyanide in the feed gas. In the Stretford process, hydrogen cyanide is co-absorbed with other acid gases and converted to thiocyanate during the regeneration of the spent solution. High concentrations of thiocyanate reduce the effectiveness of the treating solution, and continuous discharge of a side stream and addition of fresh chemicals is required to maintain solution strength. Because of the toxicity and high chemical oxygen demand of the waste stream, its disposal can create a serious pollution problem.

A review of methods used to cope with the cyanide problem in coke-oven gas desulfurization was presented by Massey and Dunlap (1975). One approach is pretreatment of the

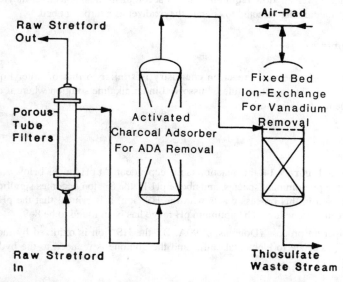

Figure 9-25. Loading step in the Dow Stretford chemical recovery process. (*Pack, 1986*)

gas prior to its entry into the desulfurization unit, either by washing with water or with a polysulfide solution containing suspended sulfur. Detailed operating experience with polysulfide scrubbers for hydrogen cyanide removal has been reported by Penderleith (1977), Carter et al. (1977), and Donovan and Laroche (1981). Two hydrogen cyanide removal units used by Dofasco, Inc. at Hamilton Ontario, Canada, have been described by Donovan and Laroche (1981). One unit used a polysulfide scrubbing solution and the other unit utilized an aqueous ammonia scrubbing liquor. In the first unit, the purge stream was processed in a Fixed Salts Recovery Plant, and in the second unit, the effluent spent solution was processed for chemicals recovery in a Zimpro wet air oxidation process. See the "Effect of Feed Contaminants" section in the SulFerox process discussion for additional information on cyanide removal options.

Hiperion Process

The Hiperion process is a modified version of the Takahax process. Prior to 1994, the process was licensed in the U.S. by Ultrasystems Engineers & Contractors of Irvine, CA. In the Hiperion process, the active catalyst is a combination of naphthoquinone with chelated iron, which is claimed to considerably reduce the reoxidation residence time requirement when compared to the Takahax process (Dalrymple and Trofe, 1989). Since the volume of liquid needed in the oxidizer, and consequently the vessel size, is proportional to the residence time requirement, the change in solution chemistry is intended to reduce capital costs and make the process more attractive economically.

Besides altering the chemical properties of the scrubbing solution, the creators of the Hiperion process also sought to develop an oxidizer configuration that, while inhibiting sulfur plugging, would provide a better gas-liquid mass transfer coefficient than a conventional packed bed. The improved mass transfer coefficient should result in more effective oxygen utilization and lower air flow requirements. To accomplish their objective, they designed a new packing medium and elected to operate the oxidizer as a turbulent bed.

Basic Chemistry

The first step in the basic reaction chemistry is similar to that of other liquid redox processes. That is, hydrogen sulfide is dissolved in an alkaline solution where it dissociates to a bisulfide ion and a proton:

$$H_2S = HS^- + H^+ \tag{9-52}$$

As shown in **Figure 9-14,** this reaction is pH dependent. At pH values below 7, undissociated H_2S is the predominant species, and above pH 12 the S^{2-} ion becomes significant. Since the oxidant used in this process reacts with the HS^- ion, it is critical that the pH be maintained between these values. The optimum pH range has been found to be 8–9.

In the Hiperion process (Douglas, 1990A, B), the HS^- ion is oxidized by the naphthoquinone (NQ) chelate to elemental sulfur and the quinone is reduced to the hydroquinone form (HNQ):

$$NQ:Chelate + 2HS^- = HNQ:Chelate + 2S \tag{9-53}$$

The hydroquinone chelate is subsequently reacted with oxygen in atmospheric air to form the quinone chelate and hydrogen peroxide:

$$HNQ:Chelate + O_2 = NQ:Chelate + H_2O_2 \qquad (9\text{-}54)$$

The reoxidized quinone chelate is returned to the absorber column thus completing the cycle.

In addition, since hydrogen peroxide is an extremely active oxidation agent, it reacts readily with residual unreacted bisulfide ion to form sulfur and water:

$$H_2O_2 + HS^- = H_2O + OH^- + S \qquad (9\text{-}55)$$

An additional benefit of the hydrogen peroxide produced by reaction 9-54 is its effectiveness as a biocidal agent, thus suppressing the growth of biological organisms without the use of other bacteria-prevention chemicals.

Turbulent Bed Mass Transfer Packing

The reactions taking place in the absorber and the reoxidizer are claimed to be almost instantaneous. In the absorber, the driving force for the absorption of H_2S into the liquid is the concentration of H_2S in the feed gas. Theoretically, the concentration of H_2S in the treated gas can be effectively reduced to near zero if sufficient mass transfer area is provided. Thus, effective mass transfer equipment is essential to the operation of the process.

In addition to providing effective contact between gas and liquid, the absorber must also be resistant to sulfur plugging. Solid elemental sulfur has a tendency to adhere to and coat commercially available random packing. Because of this sulfur property, the performance of a conventional packed bed in the oxidizer vessel, where the sulfur slurry is fed to the top of the column, would be questionable.

The creators of the Hiperion process sought to develop a packing medium which would provide effective mass transfer and inhibit sulfur plugging. They chose to design both the absorber and oxidizer as turbulent bed contactors, operating in a high gas velocity regime, under incipient fluidization conditions. To meet these conflicting specifications, a packing was developed which consisted of open, internally finned balls. This characteristic shape increased the available contact surface by making the internal surface of the spheres available for mass transfer.

The packing configuration also enhances fluidization efficiency by breaking up large gas bubbles into smaller ones. One additional feature of the packing elements is the ability to promote turbulent motion in the bed. This effect was achieved by supporting the balls' internal fins on vanes that impart a rotational motion by gas and liquid impingement. The turbulent movement of the packing elements is claimed to inhibit sulfur deposition.

Process Description

The process flow diagram for a typical Hiperion H_2S removal unit is shown in **Figure 9-26** (Douglas, 1990A). The main equipment consists of two reaction vessels—an absorber and an oxidizer. The sour gas enters the bottom of the absorber column and flows upward through a stacked series of beds where it is contacted by downflowing catalyst solution. The H_2S is oxidized to elemental sulfur which forms a slurry in the catalyst solution. The overhead

796 Gas Purification

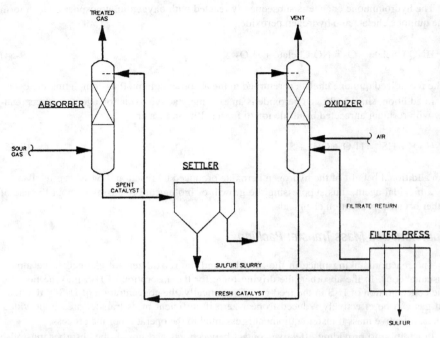

Figure 9-26. Process flow diagram of the Hiperion process H$_2$S removal system. (*Douglas, 1990A*)

sweet gas flows through a knock-out drum equipped with a mist eliminator (not shown) and is recovered as clean gas.

The spent solution from the bottom of the absorber flows to a settler where elemental sulfur is separated from the solution by gravity settling. The spent solution, free of elemental sulfur, overflows a weir and is collected in a separate compartment of the settler before being pumped to the top of the oxidizer column. In the oxidizer, the spent catalyst solution is contacted with air over beds of the specially designed packing. The spent solution is thereby oxidized back to the naphthoquinone chelate form and is then pumped back to the absorber. Vent gas from the oxidizer can be discharged directly to atmosphere or used as combustion air in a furnace.

The sulfur slurry that accumulates in the bottom of the settler vessel is periodically withdrawn and pumped through a plate and frame filter press where the sulfur is removed and the filtrate is returned to the process. The filtercake is then water washed and air blown to minimize catalyst losses (not shown). The wash water is returned to the system where it serves as a primary source of make-up water to replace that lost by evaporation. The water wash and air blow cycle are the determining factors in the chemical makeup requirements of the process. A small amount of catalyst concentrate is added after each filtration cycle to replace filtercake losses. It is claimed that no catalyst degradation occurs under normal operating conditions. The only other chemical make-up requirement consists of a small amount of alkaline solution for pH control.

The dried filtercake consists of 95–98 wt% sulfur. The sulfur product contains no heavy metals or hazardous constituents and is suitable for disposal in a Class II landfill or for further purification and sale.

Commercial Applications

The first commercial application of the Hiperion process in the U.S. was at the Huntway Oil Company Refinery in Benicia, CA (Douglas, 1987). The process was installed in an asphalt unit where it is part of an indirect method for removing hydrogen sulfide from heavy gas oil. In this application, heavy gas oil is stripped of hydrogen sulfide in a trayed column by a recirculating nitrogen stream. The H_2S laden nitrogen is then fed to the Hiperion absorber column where the H_2S is converted to sulfur. The treated nitrogen stream is then recirculated back to the vacuum gas oil stripper to be reused as stripping gas.

The unit was designed for a nitrogen flow rate of 200 scfm at 20 psig with an H_2S concentration in the sour gas of 4,000 ppmv. However, after being put into service, the H_2S concentration of the nitrogen stream increased to between 10,000 and 20,000 ppmv. After some design modifications, the unit has operated satisfactorily at the new sulfur loading and removes over 40 tons of sulfur annually.

A second Hiperion unit was installed in 1989 in a California refinery (Douglas, 1989). The treated stream is a vacuum distillation column offgas, produced at low volume, but with a relatively high H_2S concentration. Because of the low flow rate of the sour gas (less than 50 scfm), and the high design basis H_2S concentration (greater than 10 wt%), this unit was designed with a recycle stream of treated gas which mixes with the incoming sour gas. This approach limits the concentration of H_2S in the feed gas entering the absorber column and reduces the tendency of sulfur to deposit in the absorber. The treated gas from this unit, which is required to contain less than 100 ppmv of H_2S, is used as fuel gas in a refinery furnace.

Sulfolin Process

The Sulfolin process is an aqueous phase oxidative H_2S removal process closely related to the Stretford technology. It was developed by Linde AG for gas streams having a relatively low hydrogen sulfide concentration. The first commercial application of the Sulfolin process was in 1985 at the Sasol II sulfur recovery plant in Secunda, South Africa. The unit consisted of one train designed for 171,000 scfm (275,000 Nm^3/hr) of feed gas and 110 metric TPD of sulfur production (Heisel and Marold, 1987).

The process is characterized by the use of a scrubbing liquor that is reported to be highly stable, easily regenerable, and only mildly corrosive when compared to Stretford liquor. In addition, the process provides a chemical environment that is reportedly less favorable to secondary reactions thus resulting in low by-product formation.

As of 1994, Sulfolin has been employed in six plants. Four of these are installed in South Africa for desulfurizing Rectisol off-gas generated by coal gasification. In addition, there is a Sulfolin unit operating in the U.S. and a smaller plant in West Germany. These last two plants also recover sulfur from CO_2 rich Rectisol off-gas (Heisel et al., 1990). The four South African Sulfolin plants and the U.S. Sulfolin plant were originally Stretford units, but were converted to Sulfolin operation to reduce chemical costs, corrosion, and sulfur plugging problems.

Currently, Sulfolin is offered commercially in the U.S. by the Lotepro Corp., which has offices in Houston, TX, and Valhalla, NY. It is also licensed directly by Linde AG in Hoellriegelskreuth, Germany.

Basic Chemistry

Primary Reactions. The basic chemistry used in the Sulfolin process parallels that employed in Stretford solutions, except for the use of an organic nitrogen vanadium promoter instead of the ADA oxygen carrier (Heisel and Marold, 1987).

First, the H_2S in the feed gas is absorbed by and reacts with the alkaline washing liquor in the absorber by the following reaction:

$$H_2S + Na_2CO_3 = NaHS + NaHCO_3 \tag{9-56}$$

Next, the NaHS formed in the absorber is oxidized by sodium vanadate to elemental sulfur:

$$2NaHS + 4NaVO_3 + H_2O = Na_2V_4O_9 + 4NaOH + 2S \tag{9-57}$$

After oxidation of H_2S to elemental sulfur, the reduced vanadium (V^{4+}) must be reoxidized. As in the Stretford process, direct oxidation with air would be too slow for a viable commercial process. In the Stretford process the ADA redox cycle produces hydrogen peroxide, which rapidly oxidizes vanadium to V^{5+}. In the Sulfolin process the oxidation rate is accelerated by the addition of a promoter, which increases the reactivity of the reduced vanadium. The promoter is described as an organic nitrogen compound, but the exact composition is proprietary.

The regeneration reaction is represented by the expression:

$$Na_2V_4O_9 + 2NaOH + O_2 = 4NaVO_3 + H_2O \tag{9-58}$$

The pH of the Sulfolin solution is determined by the balance between $NaHCO_3$ and Na_2CO_3 in the wash liquor. When CO_2 is present in the feed gas, it reacts in the absorber according to the reaction:

$$CO_2 + Na_2CO_3 + H_2O = 2NaHCO_3 \tag{9-59}$$

The bicarbonate reacts with the caustic produced in equation 9-57 to restore the soda ash content of the solution:

$$NaHCO_3 + NaOH = Na_2CO_3 + H_2O \tag{9-60}$$

Because the rate of the hydrosulfide oxidation reaction increases as the pH of the wash liquor drops, a high CO_2 feed gas concentration is favorable to the kinetics of the process; however, a low pH adversely affects H_2S absorption.

During aeration, some of the $NaHCO_3$ is converted back to soda ash by the stripping of CO_2:

$$2NaHCO_3 = Na_2CO_3 + H_2O + CO_2 \tag{9-61}$$

Because of the higher Na_2CO_3 content, the pH of the oxidized wash liquor is normally higher than that of the reduced solution leaving the absorber.

Secondary Reactions. Besides elemental sulfur, soluble sulfate and thiosulfate salts will accumulate in the liquor due to the following four reactions:

$$2NaHS + 2O_2 = Na_2S_2O_3 + H_2O \qquad (9\text{-}62)$$

$$Na_2S_2O_3 + Na_2CO_3 + 2O_2 = 2Na_2SO_4 + CO_2 \qquad (9\text{-}63)$$

$$2S + Na_2CO_3 + O_2 = Na_2S_2O_3 + CO_2 \qquad (9\text{-}64)$$

$$S + Na_2CO_3 + 3/2O_2 = Na_2SO_4 + CO_2 \qquad (9\text{-}65)$$

The rate of byproduct formation in the Sulfolin process is lower than in the Stretford process because of the absence of hydrogen peroxide. Hydrogen peroxide is formed during the oxidation of quinone containing solutions and can participate in side reactions. As in the Stretford process, mercaptans, carbonyl sulfide, and carbon disulfide will pass through the process almost unchanged.

Process Description

The process flow diagram for a typical Sulfolin plant is shown in **Figure 9-27.** The process consists of a scrubbing cycle and a sulfur/liquor separation stage. The H_2S-containing feed gas enters the scrubber and is contacted by downflowing scrubbing liquor. The rich solution then passes into a reaction tank in which the oxidation of hydrosulfide to elemental sulfur, which started in the absorber, is completed.

The largest vessel in the process is the oxidizer where the scrubbing solution is regenerated with air. The air also floats the elemental sulfur particles suspended in the solution to the

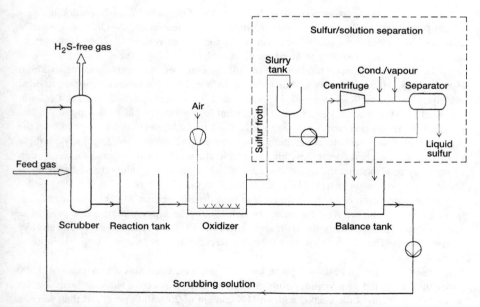

Figure 9-27. Typical Sulfolin plant process flow diagram. (*Weber et al., 1988*). Reprinted with permission from Linde Reports on Science and Technology

top surface. In this manner the sulfur is concentrated from 0.1 wt% at the outlet of the reaction tank to 4–6 wt% in the sulfur froth leaving the top of the oxidizer vessel. The froth flows over a weir and is collected in a slurry storage tank. The regenerated solution flows to an intermediate balance tank from which it is pumped back to the scrubber unit.

The sulfur froth in the slurry tank is pumped to a decanter centrifuge which concentrates the froth to a sulfur cake of about 30–40 wt% sulfur. The recovered scrubbing liquid is sent to the balance tank with the main stream of regenerated solution.

Condensate is used to wash the sulfur cake and dilute the residual scrubbing liquor. The washed cake is fed into a separator/autoclave and steam is injected to melt the sulfur. The liquefied sulfur is held at a temperature of 135°C and continuously separated from the residual scrubbing liquor in the separator/autoclave. The recovered aqueous phase is recycled back to the balance tank, and the reclaimed sulfur (purity > 99.7%) is the final product.

Commercial Applications

The SASOL I sulfur recovery plant initially used a conventional Stretford unit to desulfurize Rectisol regeneration off-gas. Operational problems in the SASOL I Stretford plant led to a decision to switch to the Sulfolin process in the SASOL I, SASOL II, and SASOL III plants.

The following problems were encountered in SASOL I with the Stretford process:

1. Excessive accumulation of sulfate and thiosulfate salts in the Stretford liquor
2. Excessive vanadium and ADA make-up requirements
3. Extreme corrosivity of the solution toward carbon steel
4. Plugging problems due to sulfur deposition on the absorber towers and in the piping system

The basic modifications introduced by the chemistry of the Sulfolin wash solution mitigated the first three problems. The absorber plugging experienced at SASOL I was initially countered by replacing the packed absorber with a three-stage venturi scrubber. However, as had been previously observed in Stretford plants, when the CO_2 content of the feed gas is high, the absorber H_2S mass transfer rate in Sulfolin units becomes too low to achieve high H_2S scrubbing efficiencies. It was thus found at SASOL I that the venturi scrubber provided inadequate contact efficiency for H_2S removal, and it was replaced with a packed column equipped with the necessary mechanical equipment for periodic in-place cleaning of the sulfur deposits. In this proprietary cleaning method, the packed-bed column is flooded with the Sulfolin liquor, and mechanical agitation then removes the deposited solid sulfur. With these modifications the plant operated within acceptable chemical loss limits, and the generation of waste by-products met the required specifications. Operating results relative to chemical losses and corrosion are provided in **Table 9-14** (Weber et al., 1988).

Another application of the Sulfolin process is at the Rheinbraun AG HTW coal gasification plant near Cologne, West Germany (Heisel, 1989). The raw synthesis gas is purified in various steps, and eventually the H_2S, and part of the CO_2, are removed in a non-selective Rectisol wash. The H_2S/CO_2 stream from the Rectisol unit is treated using the Sulfolin process.

In the Rheinbraun AG Sulfolin plant, the feed gas, which contains a maximum of 3,500 ppmv of H_2S, is fed to a venturi scrubber device where it is contacted with lean Sulfolin liquor. The gas leaves the venturi with an H_2S content of 200–500 ppmv, and then enters a packed tower containing three different packings where it is further contacted with Sulfolin solution. The outlet gas contains less than 2 ppmv of hydrogen sulfide. The pressure drop

Table 9-14
Sulfolin Chemical Losses and Corrosion Data

Chemical losses:	
Vanadium, g/liter per day	<0.01
Promoter, % per day	<0.10
Side reactions: $Na_2S_2O_3 + Na_2SO_4$, g/liter per day	<0.10
Corrosion: rate, mm/year	<0.02

Source: Weber et al. (1988)

through the packed column, at full load (17,000 Nm^3/h), increases from 40 mbar at start of run to a maximum of 100 mbar for a sulfur charged column ready for cleaning.

During absorber cleaning, additional sulfur froth is formed in the packed column and is routed to the oxidizer. Sufficient buffer capacity in the sulfur make-up section, especially in the slurry tank, is required to handle the peak sulfur production experienced during absorber cleaning.

Process Economics

Gas streams with large flow rates and high H_2S concentrations are usually treated with a conventional absorption process followed by the Claus process. Typically, the Sulfolin process is best suited for a gas feed containing less than 5 tpd sulfur as H_2S. The ranges of economical choices between the Sulfolin process and the amine absorption + Claus system, based on operating and investment costs, are shown in **Figure 9-28** (Heisel and Marold, 1987).

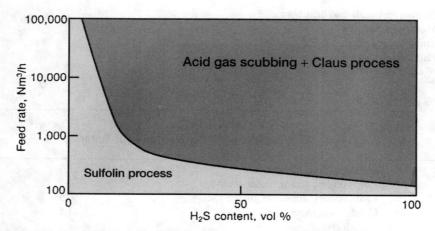

Figure 9-28. Approximate economic ranges of Sulfolin versus acid gas scrubbing + Claus as a function of feed gas rate and H_2S content. (*Heisel and Marold, 1987*). *Reprinted with permission from Linde Reports on Science and Technology*

Design parameters that tend to make the Sulfolin process less attractive economically include (Heisel and Marold, 1987):

1. High feed gas pressure, which requires a high scrubber-tower operating pressure
2. High acid content or high concentration of dissolved oxygen in the feed gas, which causes by-products formation in the scrubbing liquor

Typical utility requirements per metric ton of sulfur for the Sulfolin process are given in **Table 9-15** (Heisel et al., 1990).

Unisulf Process

The Unisulf process, a homogeneous liquid redox catalytic process for oxidizing H_2S to sulfur, was developed by Unocal in 1981 and was licensed jointly by Unocal and The Parsons Corporation. It was intended for clean-up treatment of Claus tail gas, Rectisol or Selexol off-gas, and oil shale retort off gas. Because of problems encountered in commercial operations, the process is no longer offered by Unocal (Bingham, 1992).

As described by Fenton and Gowdy (1981) and by Hass et al. (1982), the process is capable of recovering more than 99.9% of the H_2S from gas streams containing less than 10 mole % H_2S. The treated off-gas typically contains less than 10 ppmv H_2S. It is claimed that the process chemistry minimizes solution degradation and eliminates the need for a purge stream.

The Unisulf process is quite similar to the Stretford process, as is apparent from a cursory inspection of the flow diagram shown in **Figure 9-29**. The principal difference is the absence of quinone compounds in the Unisulf solution. Instead, carboxylated complexing agents (for example, sodium 1-hydroxybenzene-4-sulfonate and sodium 8-hydroxyquinoline-5-sulfonate) are used (Fenton and Gowdy, 1981).

The overall composition of the aqueous solution employed in the Unisulf process includes the following generic components:

- Sodium carbonate and bicarbonate
- Vanadium complex
- Thiocyanate ions
- Carboxylated complexing agent
- A water soluble aromatic compound

Table 9-15
Sulfolin Utility Requirements per Metric Ton of Sulfur

Electric power, MWh	1.0
Low-pressure steam, metric tons	0.3
Medium-pressure steam, metric tons	0.5
Condensate, metric tons	3.5
Chemicals:	
Sodium hydroxide, kg	15.0
Diesel oil, kg	1.0

Source: Heisel et al. (1990)

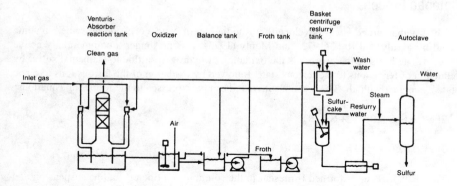

Figure 9-29. Schematic flow diagram of Unisulf process (*Fenton and Gowdy, 1981*). *Hydrocarbon Processing, April, 1982*

Because of this unique composition, the chemical solution was claimed to have longer life than Stretford liquor, lower chemical consumption, and less corrosivity. In effect, since the process made no provision for a purge stream, the only chemical consumption, except during startup, was associated with entrainment losses in the sulfur filtercake (Hass et al., 1982).

Unocal built a total of four Unisulf plants. The first unit treated Rectisol off-gas and had a sulfur recovery capacity of 26 TPD. The second Unisulf unit was commissioned as part of a now shutdown oil-shale synthetic fuels plant located at Parachute Creek, CO. The first application in a petroleum refinery was a 6 TPD unit, constructed at Unocal's Santa Maria, CA, refinery in 1986. This unit processed the tail gas from two 60-ton/d Claus plants. The fourth and last plant was a Beavon-Stretford unit conversion.

CHELATED IRON SOLUTIONS

The iron-chelate systems have been gaining popularity in the last decade relative to other liquid redox technologies primarily because of the non-hazardous nature of the scrubbing solution. Two processes are currently competing for control of the market in the U.S. These are the Wheelabrator Clean Air Systems (now U.S. Filter Engineered Systems) LO-CAT and the Dow Chemical/Shell SulFerox processes. Even though the basic chemistry of the processes is comparable, enough technological differences exist between the two systems to provide a basis for making a process selection. Other iron chelate processes are the Cataban and Sulfint processes.

The economics of these processes depend primarily on the following factors:

1. The effect of solution circulation rate on equipment capital costs
2. The power requirements of the circulation pump and the regeneration air blower
3. The chemical cost due to chelant and iron losses

These factors are influenced by the iron concentration in the solution, the design of the oxidizer vessel, the chemical composition of the solution, and the plant operating conditions. The exact chemical composition of the solution is proprietary for all processes. Chelant degradation rates are quoted by the licensors in most cases, but they are seldom expressed on a common basis, i.e., lb of EDTA/ton of recovered sulfur.

Cataban Process

The Cataban process was developed by Rhodia Inc. of New York, and described in some detail by Meuly and Ruff (1972) and Meuly (1973). It is no longer a commercially active process. The basis of the process is the oxidation of hydrogen sulfide to elemental sulfur by reduction of ferric ions to the ferrous state, followed by oxidation of the ferrous ions to ferric state by contact with air. The reactions involved can be expressed in the following equations:

$$2Fe^{3+} + H_2S = 2Fe^{2+} + S + 2H^+ \quad (9\text{-}66)$$

$$2Fe^{2+} + \tfrac{1}{2}O_2 + H_2O = 2Fe^{3+} + 2OH^- \quad (9\text{-}67)$$

The elemental sulfur formed is present in the solution as a fine crystalline suspension and can be removed by mechanical means such as decantation, filtration, or separation as molten sulfur. Mercaptans contained in the gas stream are oxidized to disulfides, which are insoluble in the solution and can be removed by skimming.

Cataban, the catalytic agent, is a clear orange-brown aqueous solution containing 2 to 4 wt% ferric ion in chelated form. It is reportedly stable over a pH range from 1.0 to 11.0, and over a temperature range from below room temperature to at least 260°F (Meuly, 1973). The reaction between the hydrogen sulfide and ferric ion is extremely fast. Reoxidation of the ferrous ions is also quite fast, but limited by the amount of oxygen dissolved in the solution. Since oxygen is only sparingly soluble in aqueous solutions, rapid oxidation requires highly efficient air-liquid contact.

Complete removal of hydrogen sulfide requires at least the stoichiometric equivalent of the Cataban reagent, or 14 pounds of Cataban reagent per scf of hydrogen sulfide. For a 5 wt% Cataban solution this is equivalent to a circulation rate of 40 gallons per minute to remove 1,000 ppm of hydrogen sulfide from 1,000 cfm of gas. However, in practice about twice this amount is used. (Note that as the Cataban agent contains 2 to 4 wt% ferric ion, a 5 wt% Cataban solution would contain 1,000 to 2,000 ppmw iron. Also, the stoichiometry of equation 9-66 requires about 900 ppmw ferric ion in a 40 gpm solution treating 1,000 cfm of gas containing 1,000 ppm hydrogen sulfide.)

The effectiveness of the process is considerably influenced by the pH of the solution. Hydrogen sulfide removal is incomplete when the pH is below 7.0, but no significant adverse effect of pH is noted within the pH range of 7.0 to 10.0. Above pH 11.5, iron hydroxide is precipitated and the solution becomes ineffective.

Air requirements for solution regeneration depend on the effectiveness of air-liquid contact. However, a four-fold excess over the theoretical quantity of oxygen is recommended (Meuly, 1973). The process is not particularly temperature sensitive. Data reported by Meuly (1973) indicate only minor effects over a temperature range of 80° to 250°C.

Gas impurities such as CO_2 and CO are not removed by the Cataban solution, and their presence is not injurious to the Cataban reagent. Hydrogen cyanide reacts with the Cataban reagent forming ferrocyanide complexes. However, since such complexes are in themselves capable of removing hydrogen sulfide, it is claimed that their presence does not deactivate the Cataban solution. It is reported that hydrogen cyanide has little effect on the Cataban system at concentrations up to 100 to 200 ppm in the gas to be treated (Meuly, 1973).

The Cataban process is primarily useful for the removal of small amounts of hydrogen sulfide from exhaust gas streams where recovery of sulfur is not the prime objective. The process is quite simple, requiring essentially two contact vessels for absorption and regeneration, and auxiliary equipment such as pumps, an air blower, and sulfur filter. If the gas to be purified contains oxygen, the reduction-oxidation reactions take place simultaneously in one vessel.

LO-CAT Process

Background

The LO-CAT process is a liquid redox H_2S removal process that uses an aqueous solution of ferric iron. The iron is held in solution by organic chelating agents. The ferric iron oxidizes the hydrogen sulfide ions absorbed in the solution, converting them to elemental sulfur while the ferric iron is reduced to the ferrous state. The spent ferrous solution is then circulated to an oxidizer where it is regenerated with air. The LO-CAT process is offered by Wheelabrator Clean Air Systems, Inc. (now U.S. Filter Engineered Systems) (Wheelabrator, 1994).

The genesis of the LO-CAT process is found in the early work by Humphreys and Glasgow in London. The initial British process, named the C.I.P. (Chelated Iron Process), used Ethylene Diamine Tetra Acetic Acid (EDTA) to keep the iron in solution. The first application of the process in an oil refinery in Landarcy, Wales, in 1964, was unsuccessful because of the instability of the chelated iron solution. The instability problem was due to a combination of two factors: The precipitation of hydrated iron oxides at elevated pH, and the degradation of the chelating agent over time (Meuly and Ruff, 1972; Meuly, 1973).

The second generation LO-CAT process, based on a dual chelate system, was developed by Thompson in the late 1970s for Air Resources Inc. (ARI), Palatine, IL. ARI was later acquired by Wheelabrator, Inc. (Wheelabrator, 1994). Thompson's invention was described in two patents (Thompson, 1980A, B). Thompson claimed to have overcome the iron precipitation instability problem by the addition of a polyhydroxylated sugar to the EDTA complexing agent. The degradation problem was partially solved in 1982 by ARI with the introduction of a new proprietary ARI-310 catalytic reagent. The ARI-310 solution, when first introduced, was furnished as a liquid concentrate with 18,000 ppm of chelated iron content. It was then diluted with water in the field by about a 40 to 1 ratio, to yield a final working solution containing approximately 500 ppm of total iron.

A significant advantage of the LO-CAT process is that the catalyst solution is non-toxic. The process operates at ambient temperature and requires no heating or cooling of the solution. It is also very efficient in H_2S removal (up to 99.99%), thus virtually eliminating any air pollution due to the presence of hydrogen sulfide in sour gases.

The LO-CAT process has enjoyed considerable commercial success. As of 1994, a total of 105 LO-CAT units had been licensed worldwide and 15 were in the design and construction stage. Of the plants built, 63 were in operation (Wheelabrator, 1994).

The LO-CAT process is offered in several configurations including, in chronological order: The LO-CAT, the Autocirculation LO-CAT, and the LO-CAT II processes. The LO-CAT II process is offered in both "conventional" and Autocirculation versions. The LO-CAT and Autocirculation LO-CAT processes use dilute iron chelate solutions (typically 500 to 2,000 ppmw and 250 to 500 ppmw iron, respectively) and circulate twice the stoichiometric iron requirement required to convert the H_2S to elemental sulfur. The LO-CAT II process, on the other hand, typically circulates less than the stoichiometric iron requirement and uses a special oxidizer design with staged controlled circulation. The special oxidizer design incorporates features that minimize thiosulfate and sulfate formation and regeneration air power requirements. LO-CAT II scrubbing solution iron concentrations are reportedly similar to those used in the LO-CAT process (Reicher, 1995). Wheelabrator/ARI now regard the

LO-CAT II process as their standard offering and would only provide a LO-CAT or Autocirculation LO-CAT design under very special circumstances (Eaton, 1992).

Basic Chemistry

The LO-CAT process overall reaction consists of two primary reactions. The first reaction, which takes place in the absorber, is the oxidation of bisulfide ions to sulfur:

$$2Fe^{3+} + HS^- = 2Fe^{2+} + S + H^+ \tag{9-68}$$

Bisulfide ions are formed when H_2S is absorbed into the aqueous LO-CAT solution. As the LO-CAT solution typically has a pH of 8.0 to 8.5, bisulfide ion is the predominant species. See **Figure 9-14**.

Per equation 9-68, two moles of Fe^{3+} are required per mole of HS^-, or, equivalently, per mole of absorbed H_2S. In the LO-CAT and Autocirculation LO-CAT processes, the molar ratio of iron in the scrubbing solution to H_2S in the feed gas is typically 4:1, twice the stoichiometric requirement (Hardison, 1992). In the LO-CAT II process (both the "conventional" and Autocirculation versions), this ratio is typically less than 2:1.

In the second reaction, the reduced iron is oxidized by dissolved oxygen in the oxidizer:

$$2Fe^{2+} + \tfrac{1}{2}O_2 + H_2O = 2Fe^{3+} + 2OH^- \tag{9-69}$$

In liquid redox systems, thiosulfate or sulfate formation can also take place due to the oxidation of bisulfide ions:

$$2HS^- + 2O_2 = H_2O + S_2O_3^{2-} \tag{9-70}$$

$$S_2O_3^{2-} + \tfrac{1}{2}O_2 = 2SO_4^{2-} \tag{9-71}$$

These side-reactions can occur either when unreacted bisulfide ion is carried over into the regenerator or when the lean solution fed to the absorber contains a large amount of dissolved oxygen. In the LO-CAT process, where the iron concentration is greater than the stoichiometric requirement, it is claimed that the high activity of the catalyst makes the oxidation of bisulfide ions to sulfur almost instantaneous and prevents bisulfide ions from reaching the regenerator. In the LO-CAT II process, which is reported to operate with substoichiometric iron concentrations, bisulfide ions reach the oxidizer. However, thiosulfate and sulfate formation are minimized by preferential baffling in the oxidizer, which causes oxidized solution to mix with spent solution thereby oxidizing bisulfide ions to elemental sulfur in the absence of dissolved oxygen.

The oxygen content of the regenerated solution depends on its pH and the oxidation state of the iron as measured by the solution electropotential. The relationship between the solution oxygen content and its electropotential is shown in **Figure 9-30** (Hardison, 1984). To minimize the formation of thiosulfate by-product in the absorber, the LO-CAT process is designed to operate with less than 1 ppm of dissolved oxygen in the re-oxidized solution. To maintain this low dissolved oxygen level when operating at a typical LO-CAT pH range of 8.0–8.5, the solution leaving the oxidizer must have an electropotential value of approximately −150 millivolts or less. As indicated in **Figure 9-30,** if the electropotential of the

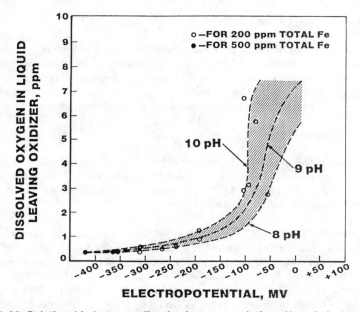

Figure 9-30. Relationship between dissolved oxygen, solution pH, and electropotential in a LO-CAT process solution. (*Hardison, 1984*)

regenerated solution is allowed to become much higher than −150 millivolts (less negative electropotential), the concentration of dissolved oxygen in the system increases rapidly.

As **Figure 9-30** indicates, at a given electropotential the dissolved oxygen concentration is pH dependent. If the pH of the solution is allowed to drift above its normal working range of 8.0–8.5, the oxygen content of the solution will rise, and thiosulfate formation will increase. If the pH is too low, H_2S absorption and reaction are adversely affected. The formation of thiosulfate and sulfate ions according to equations 9-70 and 9-71 tends to reduce the pH of the scrubbing solution, and, consequently, the H_2S removal efficiency. Maintaining good pH control is an essential feature of the LO-CAT process.

The LO-CAT solution electropotential is also related to the solution Fe^{3+}/Fe^{2+} concentration ratio. The higher the Fe^{3+}/Fe^{2+} ratio at constant pH, the higher (less negative) the redox potential. At a constant Fe^{3+}/Fe^{2+} ratio, the lower the pH, the higher the redox potential. To minimize thiosulfate formation and chelant loss, the LO-CAT process is monitored daily to ensure that operation stays within solution electropotential and pH limits (Eaton, 1992).

To ensure that the process operates within the desired pH range and that neither the lower nor the upper pH limit specifications are exceeded, the scrubbing medium must be neutralized or buffered. Potassium or sodium hydroxide is commonly used. In most applications KOH is the preferred neutralizing agent because potassium salts purged from the system with the sulfur product are a more tolerable contaminant when the ultimate use of the sulfur product is in fertilizers.

When KOH is the selected neutralizing agent, it reacts with H_2S in the presence of oxygen to form the corresponding thiosulfate salt according to the following reaction:

$$2KOH + 2H_2S + 2O_2 = K_2S_2O_3 + 3H_2O \tag{9-72}$$

As already indicated, the resulting potassium thiosulfate is purged from the system by the continuous removal of solution contained in the sulfur filtercake. In addition to thiosulfate, two other by-products, $KHCO_3$ and K_2CO_3, are formed. Their relative concentration depends on the CO_2 concentration of the sour gas and the operating pH. Both of these compounds are highly soluble in water and do not normally cause salt precipitation problems.

Process Configurations

As previously noted, the LO-CAT process has been offered in four basic configurations: the LO-CAT, the Autocirculation LO-CAT, and the "conventional" and Autocirculation LO-CAT II processes. At this time, the LO-CAT II process is Wheelabrator's standard offering. Descriptions of each configuration follow.

LO-CAT Process. The LO-CAT configuration was the first commercial offering of Wheelabrator/ARI. It is distinguished by the use of dilute iron chelate solutions (about 500 to 2,000 ppmw) where the Fe^{3+} concentration is typically twice the stoichiometric amount required to oxidize the absorbed H_2S to elemental sulfur. A flow diagram for low pressure gas treating applications of the LO-CAT process is provided in **Figure 9-31** (Hardison, 1984).

The design of the contactor in the LO-CAT (and the LO-CAT II) process depends on the volumetric gas flow rate and the required H_2S removal efficiency. The following contactor configurations, used singly or in combination, have been used commercially and are generally considered to be acceptable (Hardison and McManus, 1987; Eaton, 1992):

- liquid-filled sparged vessels
- packed countercurrent absorbers
- venturi scrubbers
- static mixers
- spray chambers
- mobile bed absorbers
- combinations of the above

In small gas flow applications, where sufficient gas pressure is available to overcome the liquid head, and, when the concentration of H_2S in the feed gas is above 1.5%, liquid-filled sparged vessels are usually the absorber configuration of choice as they are very resistant to sulfur plugging and have high H_2S removal efficiencies. At higher gas volumes, the selected absorber configuration depends to a large extent on the hydrogen sulfide concentration of the feed gas. If the H_2S concentration is low, mobile beds are preferred. If the hydrogen sulfide concentration is high, then venturis alone or followed by a mobile bed absorber are generally preferred (Reicher, 1995).

In general, the absorption of oxygen into the aqueous scrubbing solutions is more difficult than the absorption of H_2S and requires longer residence times. Packed towers have better mass transfer coefficients than liquid-filled vessels, but have a built-in potential for plugging if the circulating solution has high sulfur loadings. Even though the LO-CAT process operates at low sulfur loadings (typically between 0.2 to 0.6 wt% sulfur), the commercial oxidizer design is based on liquid-filled vessels to minimize plugging risks (Niemiec, 1995). Because the LO-CAT process employs dilute iron solutions, oxidizer vessels become rather large at

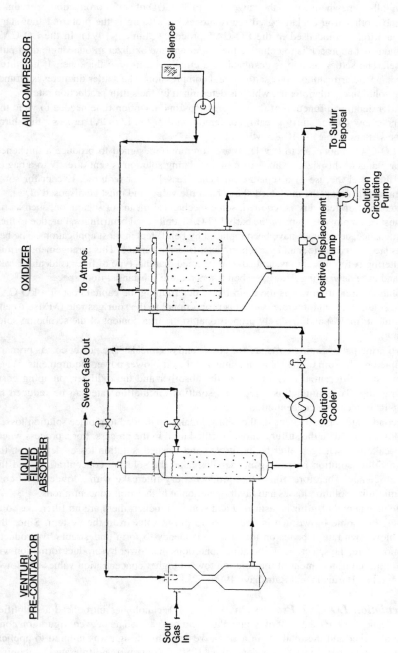

Figure 9-31. Conventional process flow diagram for LO-CAT process treating low-pressure gas. *(Hardison, 1984)*

high sulfur production rates, which tends to place an economic upper limit on train size. This limit often falls somewhere within the range of 15 to 30 LT/D of sulfur production per train.

As in most other iron chelate-based redox processes, settling is the preferred first-stage sulfur separation method used in the LO-CAT process (Quinlan, 1991). In the LO-CAT process, the usual approach is to combine the settler and the oxidizer in one single dual purpose vessel. The settler section is designed as a circular-basin type thickener. It has three operating zones: clarification, zone settling, and compression. The settler diameter is a function of the solution circulation rate, which is determined by the sulfur production rate.

The settler depth is a function of the compression-zone retention time needed to yield the desired underflow (sulfur slurry stream) concentration. For the LO-CAT process, the slurry leaving the settler is typically 10–15 wt% sulfur.

Small LO-CAT plants (up to 0.5 TPD) are commonly designed to produce a thickened slurry. For plants of this size, a slip stream of circulating solution is sent directly to a bag or pressure filter, and the use of a separate thickener vessel is not required. Except for some agricultural applications, the thickened slurry has little value, and must be disposed of.

If greater solution recovery is required, some mechanical means of separation, such as filtration, must be employed. Some of the early LO-CAT units used centrifuges, but, due to their high maintenance, centrifuges have been replaced by belt filters in most applications. The belt filter units are top fed devices and use a horizontal filtering surface. This arrangement permits gravity filtering before vacuum is applied. It is claimed that this type of filter is ideal for cake washing and cake dewatering. However, belt filters require substantial floor space.

The solution circulation rate is the main factor determining the capital cost of a LO-CAT plant. The solution circulation rate (GPM) is proportional to the sour gas rate (MMscfd) and its H_2S content (mol%) and inversely proportional to the iron content of the scrubbing solution (ppmw).

The two principal operating costs are for power and chemicals. The power cost is primarily determined by the sum of the solution pump and the air blower power requirements. When a large pressure differential exists between the absorber and the oxidizer, pumping costs increase rapidly. To counter this effect, the solution circulation rate can be reduced by increasing its iron chelate concentration.

The second major operating cost is the catalyst make-up needed to replace solution losses. Catalyst losses vary with the sulfur removal method used in the process. For a process where the sulfur-cake is discharged after being filtered and washed, solution losses depend on the amount of dilute solution remaining in the filtercake, the solution concentration, and the washing efficiency. Therefore, filters that produce a dry filtercake with a low catalyst content will minimize solution losses and favor operation at higher iron concentrations.

In systems where the sulfur is reslurried and sent to a melter, there are no filtercake solution losses, but some blowdown is necessary to purge salts from the system. Since the required blowdown rate depends on the process tendency to form undesirable by-products, mainly thiosulfate, the selection of operating conditions that lower by-product formation will push the optimum iron content of the solution toward higher concentration values and lower circulation rates (Hardison and Ramshaw, 1990).

Autocirculation LO-CAT Process. In 1983, ARI Technologies introduced a simplified LO-CAT design for the treatment of amine plant acid off-gases. The new configuration combined the absorber and the oxidizer in a single vessel. This design was tailored to applications that required very low iron concentrations (250–500 ppmw) and tolerated contamination of the treated gas with air.

The compact design of the Autocirculation process permits the circulation of large volumes of dilute catalyst solution without the use of pumps, thus offering major power cost savings. However, the absorber in these units operates in an oxygen rich environment and, as a result, conditions are more favorable for thiosulfate production. The increased formation rate of salt by-products requires higher blow-down rates and, in spite of the low solution concentration, tends to drive up chemical costs. This offsets, to some extent, the operating cost benefits achieved through the reduction in power costs.

In spite of this, it is claimed that the combination of a selective amine unit operated at high pressure and an Autocirculation LO-CAT unit operated at atmospheric pressure has been found to be an economic alternative to the amine-Claus process in natural gas treating plants handling relatively low quantities of sulfur. An application of this type was reported by Owens (1991) at the Texas Indian Rock Plant. This plant began operation in 1984 and was the first Autocirculation unit built by ARI. It employs a conventional amine unit and a LO-CAT Autocirculation unit to treat the off-gas from the amine plant regeneration unit.

A schematic of the Autocirculation LO-CAT arrangement is shown in **Figure 9-32**. The sour gas is bubbled through the chelated iron catalyst solution from a distributor located in the centerwell of the liquid-filled absorber/oxidizer vessel. The H_2S is converted to elemental sulfur in the centerwell, and the sulfur particles combine and fall to the conical bottom of the vessel. Regeneration air is injected into an annular space between the outside shell of the vessel and the centerwell. The driving force to maintain catalyst circulation is the density differential between the catalyst solution in the centerwell and the lighter, highly aerated catalyst solution outside the centerwell in the annular space. This natural convection circulation

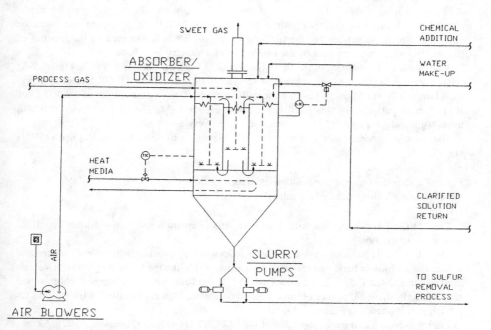

Figure 9-32. Schematic diagram of LO-CAT Autocirculation process. *Courtesy of Wheelabrator Clean Air Systems*

812 Gas Purification

is assisted by the momentum imparted by the air flow, which forces the catalyst solution to move upward through the annular space before it spills over the open-ended centerwell.

As originally designed, the prototype plant experienced severe foaming problems caused by frothing of the sulfur particles at the top of the vessel. The froth would build on top of the catalyst solution until it reached the vent gas stack and developed enough head to block the flow of gas. This problem was ultimately solved by the continuous addition of a foam inhibitor.

Another problem was the plugging of the feed gas header with sulfur particles. This problem was caused by swings in gas flow rate, which resulted in backflow of the catalyst solution into the header during low flow operation periods. This allowed conversion of H_2S to solid sulfur to take place inside the header. This problem was solved by installing an air sparge line at the inlet of the feed gas header and periodically blowing the header.

LO-CAT II Process. In 1990, ARI introduced the AQUA-CAT process to remove H_2S from waste water. In this process the chelated iron chelate solution was continuously added to the waste water feed and discarded with the treated water. As the iron chelate solution was lost, the economics of the process hinged on whether or not the concentration of chelated iron could be held at substoichiometric levels. It was found that the process was still effective when the concentration of chelated iron was well below a 2:1 ferric ion to H_2S stoichiometric ratio. This ratio was the minimum level that the LO-CAT process historically had tried to maintain.

The LO-CAT II concept was born from this observation (Hardison and Ramshaw, 1991), and prompted ARI to claim a significant breakthrough in liquid redox chemistry. Although ARI has not made public all the details of their new insight into the chemistry of the modified LO-CAT process, they maintain that in the LO-CAT II process configuration, the amount of solution circulated to the high-pressure absorber is determined by the kinetics of the absorption reaction, and not by the iron/H_2S stoichiometry (Hardison, 1990).

A simplified flow diagram of the first commercial LO-CAT II plant is shown in **Figure 9-33** (Hardison, 1992). The unit was placed in service in 1991 at the Mobil Oil Exploration and Development North Midway gas field near Bakersfield, CA. A photograph of the LO-CAT II equipment package for this plant is shown in **Figure 9-34.** The Mobil LO-CAT II unit treats wellhead gas produced in a steamflood oil recovery operation.

The principal characteristics of the LO-CAT II process as described by Hardison (1991) are

- Use of a staged, controlled circulation oxidizer
- Use of substoichiometric Fe/H_2S ratios

The LO-CAT II process is available in both Autocirculation and "conventional" configurations (Reicher, 1995).

The LO-CAT II oxidizer vessel has been redesigned to provide better mass transfer and reduce the air blower head and capacity requirements. Unlike the vertical cylindrical vessels used in the LO-CAT design, the new LO-CAT II oxidizers are flat-bottomed, shorter, and rectangular in shape. A picture of the oxidizer box and air blowers installed in the Mobil LO-CAT II unit is shown in **Figure 9-35.**

Figure 9-36 is a schematic depiction of the LO-CAT II oxidizer (Hardison, 1992). As previously noted, the LO-CAT II process operates with less than the stoichiometric amount of iron. Consequently, bisulfide ions are present in the spent solution from the absorber. To pre-

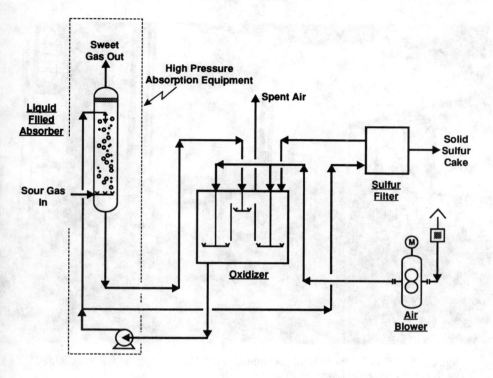

Figure 9-33. Simplified flow diagram of first commercial LO-CAT II plant (Hardison, 1992)

vent the bisulfide from being oxidized to thiosulfate in the oxidizer, the spent solution is mixed with completely oxidized solution in a special reaction chamber. Chelated Fe^{3+} in the oxidized solution converts the bisulfide ions in the spent solution to elemental sulfur in this reaction chamber. As dissolved oxygen is absent, thiosulfate formation is minimized. See **Figure 9-36.** The LO-CAT II oxidizer vessels are also equipped with staged air injection, where air is injected into a series of baffled compartments through which the spent solution flows sequentially. This design permits shorter oxidizer vessels and reduces air compressor head and horsepower requirements. It is reported that LO-CAT II oxidizer vessels can be shop fabricated in sizes up to 10 LTPD of sulfur (Reicher, 1995).

Absorbers for LO-CAT II units can be of several alternative designs. As in the LO-CAT process, liquid-filled absorbers are generally preferred for small gas flows, as they are very resistant to plugging and have high H_2S removal efficiencies. A typical example is the Mobil LO-CAT II unit, where the combination of low gas flow, relatively high efficiency requirements, and high CO_2 partial pressure, led to the selection of a liquid-filled absorber.

In the LO-CAT II process, external scrubbing solution circulation rates are set to obtain good absorber performance and minimize pump horsepower. LO-CAT II circulation rates have been greatly reduced by using less than the stoichiometric iron requirement. It is reported that ratios of chelated iron to H_2S as low as 20% of the stoichiometric requirement have been used. However, because there are unreacted bisulfide ions in the spent solution leaving

Figure 9-34. LO-CAT II equipment package at North Midway gas field, Bakersfield, CA. *Courtesy Wheelabrator Clean Air Systems*

the absorber, it is necessary to ensure that substoichiometric operation does not impose a vapor-liquid-equilibrium (VLE) limit on the H_2S content of the treated gas. It is claimed that VLE limits are not a problem, particularly at high absorber operating pressures, where cost savings due to reduced scrubbing solution circulation are most significant (Hardison, 1992).

Sulfur handling has also been upgraded. In the Mobil unit, which has a sulfur capacity of 1.5 LT/D, a slipstream of the external circulation stream is taken to a dual bag filter system where the sulfur is recovered and stored in 1,000 kg capacity filter bags (Hardison, 1992; Niemiec, 1995). LO-CAT II units equipped with melters have intermediate filtering devices to improve the quality of the sulfur product.

A comparison of design parameters for a conventional amine/LO-CAT combination unit and a direct LO-CAT II plant, is presented in columns #1 and #2 of **Table 9-16** (Hardison, 1991). Column #3 of this table provides equivalent information for the SulFerox process (Kwan and Childs, 1991).

The large gas flow difference between the two LO-CAT options can be explained by the fact that a LO-CAT II plant is designed to handle the full gas flow directly, while in the Amine/LO-CAT combination plant, the conventional LO-CAT unit is sized to treat a much smaller volume of amine plant acid gas.

The table is of limited value since the Fe concentration recommended for the LO-CAT II unit is not included in the data. However, it is interesting that the solution circulation rate for

Liquid Phase Oxidation Processes for Hydrogen Sulfide Removal **815**

Figure 9-35. Oxidizer box and blowers on LO-CAT II plant at North Midway gas field, Bakersfield, CA. *Courtesy of Wheelabrator Clean Air Systems*

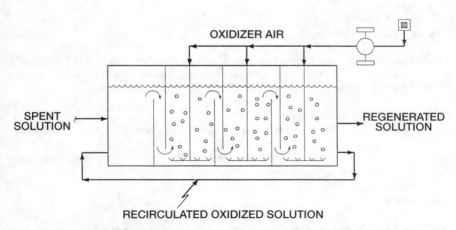

Figure 9-36. Process schematic of LO-CAT II oxidizer. (*Hardison, 1992*)

Table 9-16
Comparison of LO-CAT and SulFerox Design Parameters

Design Basis	Amine/LO-CAT*	LO-CAT II	SulFerox
Gas flow, scfm	176	8,350	8,333
Gas pressure, psia	24	264	264
Temperature, °F	110	97	97
H_2S content, vol%	46	0.98	0.98
Solution circulation rate, gpm	N/A	240	220
Fe concentration, ppmw	500	N/A	40,000
Regenerator air, scfm	1,975	1,580	600

Note:
*Data are for LO-CAT unit operating on acid gas from amine plant.
Source: Kwan and Childs (1991) and Hardison (1991)

SulFerox, which is based on the use of a concentrated iron chelate solution, and that of LO-CAT II, are approximately the same.

The regenerator air requirements for LO-CAT II process are significantly higher than those of SulFerox. This is not surprising since, in addition to reoxidizing the iron, LO-CAT II uses air to maintain autocirculation in the oxidizer, and to evaporate water for heat and water balance. SulFerox, on the other hand, requires an additional reverse osmosis system to maintain the water balance in the system, thus trading off savings in air blower operating costs versus the cost of the power required to run the RO unit.

As of 1995, twenty-four LO-CAT II units were either in operation or under license (Niemiec, 1995). Applications include

- Natural gas direct treatment
- Natural gas amine acid gas
- Coke-oven gas
- Refinery fuel gas
- Chemical plant vent gas

Even though the process licensor intends to replace the conventional LO-CAT system with the new LO-CAT II technology whenever possible, there are some applications where the LO-CAT II design is not suitable. These applications fall into two categories:

1. High pressure CO_2 feed gas applications
2. Units with large gas volumes and very low H_2S gas loadings, typically aerobic odor control

Some typical applications that fall into these categories are wastewater treating, odor control plants, and food grade CO_2 purification.

Applications

CO_2 Rich Oilfield Production Gas. The first application of the LO-CAT process to treat oilfield gas was at the Sable CO_2 Recovery Plant near Plains, TX (Price et al., 1986; Price,

1987). This plant was a LO-CAT configuration with a dilute iron chelate solution and a higher than stoichiometric Fe^{3+} concentration.

Three process schemes for H_2S removal were evaluated to determine the lowest cost approach: (1) an amine unit equipped with an incinerator, (2) a metal oxide slurry process, and (3) the LO-CAT process. A comparison of the projected capital and operating costs is summarized in **Table 9-17** (Price et al., 1986). These cost data are in 1986 dollars and are based on 5 MMscfd of gas with an H_2S concentration of 770 ppmv.

A simplified schematic of the Sable Plant LO-CAT unit is shown in **Figure 9-37**. Gas flowing at a nominal rate of 5 MMscfd is fed to a two-stage absorber operating at a pressure of 20–25 psig. In the first stage, the gas enters a stainless steel venturi-mixer where it is mixed with 150 gpm of regenerated catalyst solution under highly turbulent conditions. About 70% of the H_2S is converted to elemental sulfur in this first stage contactor.

Table 9-17
Cost Analysis for the Sable CO_2 Recovery Plant

Process	Capital Cost, M$	Operating Cost, M$/yr
Amine Treating*	1,080	224
Metal Oxide Slurry	300	269
LO-CAT	700	65

*Processing plant at $980,000 plus an incinerator stack at $100,000.
Source: Price et al. (1986)

The semi-sweet gas and the catalyst solution containing suspended sulfur particles are discharged from the venturi into the second stage packed-bed absorber. The gas rises through the column which is packed with 10 feet of polypropylene packing. Another 150 gpm of regenerated catalyst solution is sprayed on top of the packed bed. The conversion in the packed section is about 99%, resulting in an overall H_2S conversion of about 99.7%. The sweet gas leaving the absorber overhead contains no more than 10–20 ppmv of H_2S.

The dilute sulfur slurry leaving the absorber is piped to the top of the oxidizer-settler. This vessel is operated at atmospheric pressure and is almost full of liquid. Details of this vessel are shown in **Figure 9-38**. The downward flow of slurry is regenerated by exposure to air bubbles rising from an air sparger. The moisture laden, partly spent air vents directly to the atmosphere. This effluent air is virtually free of any H_2S contamination.

Regenerated catalyst solution is extracted via a nozzle located about midway on the oxidizer-settler vessel. This solution is recirculated back to the absorber column. The portion fed to the venturi-mixer is first heated to 120°F to ensure that the absorber is operated above the hydrocarbon dew point of the inlet gas.

The lower half of the settler is a low turbulence zone that allows the sulfur to settle on the bottom cone and gradually drain to the clarifier/filter system through a diaphragm control valve. The clarifier/filter system separates the sulfur product from the residual catalyst solution in the slurry. Most of the catalyst solution is decanted via the clarification mode. The primary use of the filters is to attain final dewatering after the clarifier tank is full of sulfur. The filtered solution is pumped back to the oxidizer through a surge tank, while the damp sulfur-cake is mechanically removed from the filter and stored.

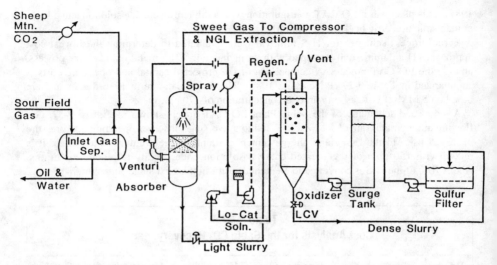

Figure 9-37. Simplified flow diagram of LO-CAT process system at the Sable CO_2 recovery plant. (*Price et al., 1986*)

Table 9-18 provides design and operating data for the Sable Plant LO-CAT unit (Price et al., 1986). Sulfur plugging problems at the Sable LO-CAT plant were reviewed by Price et al. (1986) and Price (1987). Hydrocarbon mist in the feed gas reportedly coated newly formed sulfur particles, lowering the particle density to the point where they floated rather than settled in the oxidizer-settler. This floating tendency disrupted the sulfur settling process, overloaded the settler, and caused plugging problems throughout the LO-CAT unit. Use of a special surfactant to wash the oil coating from the sulfur particles was reported to solve the problem. However, Price et al. (1986) recommend that a coalescing aerosol filter be installed upstream of any LO-CAT unit to remove hydrocarbon mist and minimize sulfur plugging problems.

Geothermal Gas Cleanup. In January 1988, Pacific Gas and Electric converted one of its Stretford H_2S abatement units at The Geysers geothermal power plant (in Northern California) to the LO-CAT process (Henderson and Dorighi, 1989). This application is a LO-CAT process configuration with the use of a dilute iron chelate solution and a greater than stoichiometric Fe^{3+} concentration. The overlying reason for the conversion to the LO-CAT technology was the increasing cost of treating Stretford process wastes, which contain vanadium, and in California, must be disposed of as a hazardous waste.

The electric generators at The Geysers power plant are powered by turbines driven by geothermal steam. After passing through the turbines, the steam is condensed in surface condensers. Noncondensable gases are removed from the surface condensers by a two-stage gas ejector system. Condenser vent gas contains hydrogen sulfide and must be treated before it is released to atmosphere.

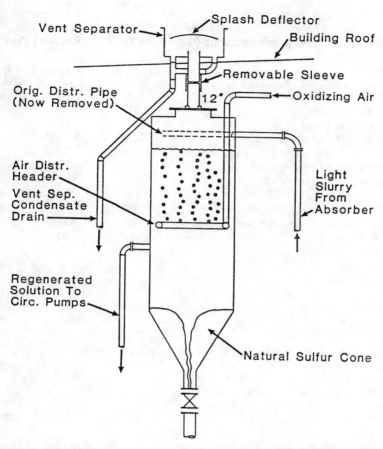

Figure 9-38. Sable CO_2 recovery plant LO-CAT oxidizer-settler (*Price et al., 1986*)

The process design basis is given in **Table 9-19** (Henderson and Dorighi, 1989). The conversion to the LO-CAT process required changing the inventory of treating solution, but no functional changes were initially made to the process equipment. The thiosulfate concentrations given in the table are average reported values without purge.

The new process scheme is shown in **Figure 9-39.** The diagram depicts much of the existing plant equipment used for the operation of the Stretford process prior to the LO-CAT conversion. The revamped process was based on the conventional LO-CAT two-step absorption cycle with a venturi scrubber followed by an absorber column. The sulfur separation method took advantage of the available equipment and followed the standard Stretford technique of skimming off the sulfur froth in the second oxidizer tank. The sulfur slurry from the froth tank, containing about 20% solids, was pumped to a vacuum belt filter where most of the solution was reclaimed through a wash process. The solid sulfur (80% solids content) was then sent to waste disposal.

The conversion project was evaluated in four separate categories:

Table 9-18
LO-CAT Unit Design and Operating Data for the Sable CO_2 Recovery Plant

	Design	2-11-86	4-2-86
Inlet Gas Volumes, Mscfd	410	210	395
CO_2 In Inlet Gas, %	51	7	35
CO_2 Make-Up, Mscfd	—	117	173
Inlet Gas H_2S Content, ppmv	6,200	17,000	10,750
Residue Gas H_2S Content, w/Packing, ppmv	50 max.	15	10
Residue Gas H_2S Content, w/o Packing, ppmv	50 max.	150	—
Sulfur Production, Tons/Mo.*	3.1	4.5	5.5
LO-CAT Catalyst, pH	8.5	8.4	8.9
LO-CAT Catalyst, Redox Potential, mv	−150	−195	−262
LO-CAT Catalyst, Total Iron Content, ppmw	500	1,800	2,467**
Plant Run Time, %	—	95	97
Chemical Usage, 310C, gal/day		5	6.5
310M, gal/day		6	4.0
Surfactant, gal/day		2	3.0
Biostat, gal/day		1	0.3
Anti-foamant, gal/day		rare	rare
KOH, gal/day		10–30	0***

Notes:
*Sulfur production is calculated based upon the H_2S available in the inlet gas and an estimated LO-CAT unit efficiency of 99.9%.
** Abnormally high after test run at 3,000 ppmw.
*** KOH temporarily discontinued while trying to lower pH.
Source: Price et al. (1986)

1. Hydrogen sulfide abatement efficiency
2. Cost of make-up chemicals
3. Need for operating procedure changes
4. Analysis of the effect of the LO-CAT solution on equipment metallurgy

A summary review of the performance of the LO-CAT process yielded the following results (Henderson and Dorighi, 1989):

1. The H_2S abatement efficiency was excellent. The concentration of H_2S in the treated gas averaged about 2 parts per million. This removal was better than Stretford and allowed the unit to operate at full capacity. Previously, when the unit operated in the Stretford mode, absorption efficiency fell off whenever the liquid circulation rate exceeded 75% of pump capacity. This resulted in occasional non-compliance with local emission standards and set a practical operating limit for the unit that was well below the design basis capacity.
2. The main chemical requirements of the LO-CAT process are met by the addition of ARI-310C iron concentrate, ARI-310M chelate, and caustic potash (KOH). To meet the requirements of this application, the iron concentrate was added at a rate of 3.6 lb concen-

Table 9-19
Typical LO-CAT Operating Conditions at The Geysers Unit 15 Geothermal Power Plant

Typical Gas Composition, Mole%	
H_2S	4.0
CO_2	56.0
H_2	16.0
CH_4	14.0
N_2	8.0
O_2	2.0
Process Variables	
Condenser gas flow, scfm	1,500
Sulfur loading, LTPD	3.2
Solution inventory, gal.	130,000
Venturi scrubber rate, gpm	2,250
Packed column rate, gpm	750
Thiosulfate conc., g/L	50–70

Source: Henderson and Dorighi (1989)

trate/100 lb sulfur, and the make-up chelate was added at a rate of 18 lb ARI-310M chelate/100 lb sulfur. These rates were twice the estimated amount. KOH was used as the neutralizing agent and was added at the rate of 0.7 lb KOH/100 lb of sulfur. This rate is equivalent to the amount previously used in the Stretford process.

3. No major changes were required in the operating procedures for the LO-CAT process. However, several adjustments were needed. First, the sulfur removal by flotation required better control of the solution level and vigorous mixing in the balance tank to prevent sulfur settling and unwanted accumulation in that vessel. Second, the reaction tank was not needed for the LO-CAT process since the residence time requirements were much lower than for Stretford operation. Third, the solids content of the LO-CAT sulfur-cake after vacuum dewatering (80–90%) was much higher than that of the Stretford sulfur (50–60%). This facilitated waste removal and reduced operating costs.

4. LO-CAT solution is highly corrosive to carbon steel. All carbon vessel tanks were lined with Ceilcote Flakeline 180 polyester coating, and all piping and process valves were replaced with stainless steel components. The tank liner performed satisfactorily except for some local metal attack where the coating was damaged due to equipment malfunction.

Sulfur Quality

With any LO-CAT unit having a sulfur production capacity greater than about 1.5 LTPD, it may be economical to sell molten sulfur (Quinlan, 1991). The exact breakpoint must be determined on a case by case basis. Also, in recent years, the economic justification for melters has become more difficult as the sulfur product usually demands a low price due to

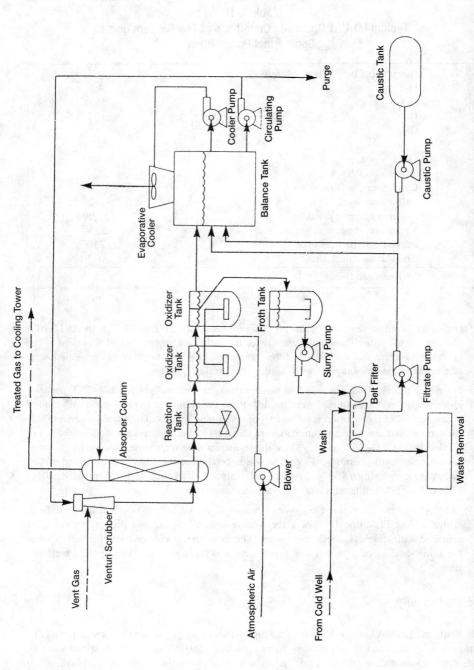

Figure 9-39. Flow diagram of a LO-CAT plant treating vent gas from a geothermal power plant condenser. *(Henderson and Dorighi, 1989)*

worldwide excess capacity. However, melters can sometimes be justified simply to reduce the volume of landfilled material (Niemiec, 1995).

In the older LO-CAT and Autocirculation LO-CAT process configurations, sulfur is separated from the bulk solution by primary settling, which produces a slurry of 10–15 wt% sulfur. See **Figures-9-31** and **9-32**. When a melter is used, this slurry is fed directly to the melter where the temperature is raised above the melting point of sulfur and a liquid/liquid separation is conducted. In both of these older processes, there is usually no intermediate filtering and prewashing step prior to melting. **Table 9-20** summarizes the properties of molten sulfur produced by this technique (Kwan and Childs, 1991).

In the LO-CAT II process, sulfur is recovered by filtration and then washed to remove iron and salts. The washed filtercake is then reslurried with fresh water before melting (Reicher, 1995). As the LO-CAT II and LO-CAT processes utilize similar iron concentrations, the LO-CAT II molten sulfur product should be at least as pure as indicated in **Table 9-20**.

If mercaptans are present in the sour gas, some are absorbed by the LO-CAT solution. Some of the absorbed mercaptans are converted to disulfides and can produce a mild discoloration of the sulfur product. Unconverted mercaptans are desorbed in the oxidizer and vented to the atmosphere, which can cause odor problems if the vent gas is not properly incinerated.

Hydrocarbon solubility in the circulating LO-CAT solution is about the same as in water. Some hydrocarbons end up in the sulfur product and cause the sulfur product to darken. This problem can often be corrected by use of a coalescing filter on the feed gas upstream of the LO-CAT absorber and by keeping the LO-CAT solution temperature above the feed gas temperature. A sulfur product that is dark in appearance can also be an indication of excessive residence time and/or high operating temperature in the sulfur melter (Eaton, 1992).

Sulfint Process

The Sulfint process was developed by Integral Engineering of Vienna, Austria, and is currently licensed by Le Gaz Integral Enterprise of Nanterre, France. The process has been used primarily for sewer gas treatment, and ten plants had been built by 1993—all in Europe (Rossati, 1993). The process is described in considerable detail by Mackinger et al. (1982), and a flow diagram is shown in **Figure 9-40**. The process is based on the use of EDTA as the iron chelating agent. The pH of the wash liquor is slightly alkaline, normally between 7 and 9. The H_2S loading in the absorber and the oxidation rate in the oxidizer are controlled by monitoring the redox potential of the solution.

Table 9-20
Properties of LO-CAT Sulfur

Form	liquid
Color (in solid form)	yellow
Sulfur content, wt%	99.9+
Impurities, ppmw	
Organics	< 400
Ash	< 200
Iron	< 50

Source: Kwan and Childs (1991)

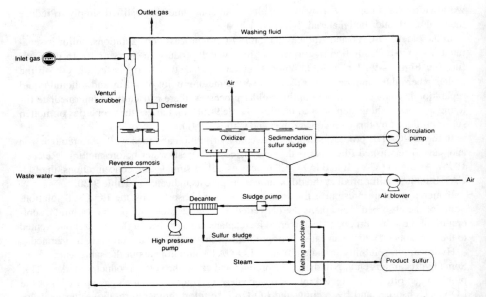

Figure 9-40. Sulfint process flow diagram (*Mackinger et al., 1982*). *Courtesy of Hydrocarbon Processing, March, 1982*

Per **Figure 9-40,** sour gas is brought into intensive contact with the washing fluid in one or more venturi scrubbers where H_2S is absorbed and converted to sulfur. The use of ejector scrubbers reportedly minimizes the formation of sulfur deposits in the absorber. Depending on the required final gas purity, several scrubbers in series may be needed to provide sufficient gas/liquid mass transfer. The treated gas flows out through a demister. The effluent liquid from the scrubber flows to the oxidizer and a sedimentation chamber. The sulfur sludge from the sedimentation chamber is pumped out and dewatered in a centrifugal decanter. The dewatered sulfur is mixed with water and melted in an autoclave to achieve final purification. It is claimed that the addition of a second washing/centrifugal decanting step prior to sulfur melting and subsequent molten sulfur filtration can produce "Claus" purity sulfur.

The washing fluid from the centrifuge may be treated in a reverse osmosis unit to separate the iron chelate from wash water and salts formed by secondary reactions. The chelate complex cannot pass through the membrane and is reclaimed and returned to the main solution circuit as a concentrated solution, thus avoiding catalyst losses. Sulfates, carbonates, and alkalis pass through the membrane and are removed from the treating solution thereby preventing the buildup of contaminant salts in the scrubbing solution (Mackinger et al., 1982). The use of reverse osmosis may be economically attractive if power costs are low, and if the amount of iron chelate in the water lost during sulfur purification is high.

In the oxidizer basin the iron chelate complex is oxidized by finely dispersed aeration. The sulfur entering the oxidizer has a mean diameter size of 10 microns, but it agglomerates and quickly settles to the bottom as a sludge containing about 5 wt% solid matter. The sulfur particle size of the agglomerated sulfur ranges from 0.1 to 0.2 mm (Mackinger et al., 1982).

The design of the Sulfint oxidation and settling tanks reflects their technological origin in biological sewage treatment plants, and the equipment selection is consistent with standard practice in that industry. Most equipment and piping are constructed of stainless steel or glass reinforced plastic (Rossati, 1993).

As in other iron chelate-based processes, the washing solution is non-toxic, and capital costs are comparatively low for small plants. The main drawback of this process is the slow oxidation rate in the oxidizer, which requires the use of large oxidation basins. The original oxidizer designs were not particularly effective and had characteristically low oxygen utilization coefficients. For stoichiometric conversion, this led to high air consumption figures on the order of 15 cubic meters of air per kilogram of H_2S. New improved proprietary designs are reported to have cut the amount of air required from 15 to less than 10 m3/kg sulfur and to have reduced the size of the oxidation basins (Rossati, 1993).

Most environmental regulations in Europe are such that the produced sulfur cannot be landfilled or disposed of easily, and purity requirements limit the marketability of the Sulfint sulfur product. For these reasons, most larger European applications utilize the Claus process (Rossati, 1993). Therefore, the only large scale Sulfint gas purification plant (18 LTPD sulfur) is at a coal gasification plant in Usti, Czech Republic. This plant treats Rectisol off gas. Design data for this plant are given in **Table 9-21** (Mackinger et al., 1982). The other nine Sulfint plants, ranging in capacity from 0.02 to 0.75 LTPD, treat sewer or incinerator off gas (Rossati, 1993).

SulFerox Process

SulFerox is a high iron concentration, chelate-based, liquid redox desulfurization process developed by the Shell Development Co. It is presently licensed under a joint agreement by the Shell Oil Company and the Dow Chemical Co. The first SulFerox unit was installed in 1987 at White Castle, LA, to treat low-pressure natural gas. By late 1995, a total of 37 SulFerox units had been licensed, with over 20 units in operation (Kenny, 1995).

As reported by Fong et al. (1987), the process was designed to achieve three main objectives:

1. High H_2S removal capacity per unit circulation
2. Low chelate degradation rate
3. Simple and compact process design

Table 9-21
Sulfint Plant Data

Feed gas quantity, Nm³/hr	40,000
H_2S in feed gas, g/Nm³	20
H_2S removal efficiency, %	99.9
Sulfur production, metric tons/day	18
Sulfur recovery efficiency, %	94.0
Feed gas temperature, °C	40
Process temperature, °C	20–40
Utilities	
Electricity, kW approx.	1,500
Steam, tons/h	1.2
Waste water, m³/h	10

Source: Mackinger et al. (1982)

SulFerox was the first commercial iron chelate-type liquid redox process to utilize a high iron content treating solution, typically ranging between 1 and 3 wt% iron (Pirtle et al., 1994). Buenger et al. (1991B) report that the reaction with H_2S is, in effect, first order for iron. Therefore, the use of a concentrated iron solution accelerates the kinetics and uncouples the process from the need to have a moderately high pH for efficient H_2S removal. The accelerated kinetics resulting from the higher iron concentration also reduces the lifetime of H_2S, HS^-, and S^{2-} in the process solution. Since thiosulfate formation generally occurs in iron redox systems by regeneration air oxidation of these species, byproduct thiosulfate formation is claimed to be negligible (Allen, 1995). See equation 9-70. Relatively low pH operation also reportedly takes the process out of the region where byproduct thiosulfate is formed. The high iron concentration makes it possible to operate at low circulation rates and high H_2S absorption rates. This reduces equipment size and lowers pumping costs.

The high iron concentration is attained by using a proprietary chelating agent that keeps the iron in solution at a neutral to slightly basic pH (typically 7.0 to 8.0) (Allen, 1995). The key to making SulFerox an economical process was to control the chelant degradation rate to keep chemical costs within acceptable limits.

The nominal chemical costs (in 1994 dollars) for SulFerox process units, based on typical iron and chelant losses, are currently projected at about $235 to $300 per long ton of sulfur (Hammond and Pirtle, 1994). Actual chemical costs experienced in some plants may be considerably higher than this because of unanticipated mechanical losses such as a high frequency of spills, above average leakage due to poor equipment maintenance, or losses due to improper sampling procedures. Additional data on chemical losses for refinery gas treating applications are provided by Day and Allen (1992).

Basic Process Chemistry

The overall SulFerox chemistry, as described by Fong et al. (1987), is rather simple and consists of two sets of chemical reactions, where L denotes the chelant:

Absorption:

$$2Fe^{3+} \bullet L + H_2S = 2Fe^{2+} \bullet L + S + 2H^+ \tag{9-73}$$

Regeneration:

$$2Fe^{2+} \bullet L + \tfrac{1}{2}O_2 + 2H^+ = 2Fe^{3+} \bullet L + H_2O \tag{9-74}$$

The function of the chelant is to prevent the formation of insoluble iron compounds without interfering with the ability of the iron to undergo reduction and oxidation. In the absorption step, the sour gas stream containing hydrogen sulfide comes in contact with a liquid containing a soluble ferric iron chelate, $Fe^{3+} \bullet L$. The hydrogen sulfide forms sulfide and hydrosulfide ions, which are selectively oxidized to form elemental sulfur, and the ferric chelate, $Fe^{3+} \bullet L$, is reduced to the corresponding ferrous chelate, $Fe^{2+} \bullet L$.

In the regeneration step, the ferrous chelate, $Fe^{2+} \bullet L$, is reoxidized with air back to ferric chelate, $Fe^{3+} \bullet L$. The regenerated solution is then recycled to the absorption section.

The second set of chemical reactions summarizes the chelant, free-radical-induced oxidative degradation chemistry and is considerably more complex:

$$2Fe^{2+} \bullet L + O_2 + 2H^+ = 2Fe^{3+} \bullet L + H_2O_2 \tag{9-75}$$

$$Fe^{2+} \cdot L + H_2O_2 = Fe^{3+} \cdot L + \cdot OH + OH^- \tag{9-76}$$

$$Fe^{2+} \cdot L + \cdot OH = Fe^{3+} \cdot L + OH^- \tag{9-77}$$

$$L + \cdot OH = \text{Degradation products} \tag{9-78}$$

In addition, excessive chelant decarboxylation may take place through direct Fe^{3+} oxidation of chelant at high operating temperatures, necessitating a high chemical make-up rate. Consequently, except in special cases, the SulFerox process solution temperature should not exceed about 140°F (Allen, 1995). To prevent hydrocarbon condensation inside the contactor, the SulFerox solution temperature is usually at least 10°F warmer than the feed gas. This limits the feed gas temperature to a maximum of 130°F.

The overall reaction for the SulFerox process can be represented as follows:

$$H_2S + \tfrac{1}{2}O_2 = S + H_2O \tag{9-79}$$

Even though the basic chemistry of the process appears to be deceptively simple, the selection of the right chelant is a critical choice and depends on five major factors:

1. The iron chelant solubility properties
2. The reaction rate of the soluble oxidizing complex, $Fe^{3+} \cdot L$, with H_2S in the absorber
3. The reaction rate of the reduced complex, $Fe^{2+} \cdot L$, with O_2 in the regenerator
4. The degradation rate of the chelant
5. The relative cost of the chelating agent

In general, the reaction rate in the H_2S scrubbing step can be considered to be nearly instantaneous for most iron chelates, but the rate of regeneration of the reduced complex is slower and usually becomes a factor in the design of the process. The regeneration reaction rate for any given chelant increases with the solution Fe^{2+} concentration and with the oxygen partial pressure.

As shown in reaction 9-79, the stoichiometry of the overall reaction calls for one-half mole of oxygen or about 2.4 moles of air per mole of sulfur processed. However, as noted by Buenger et al. (1988), a partially regenerated solution with a high iron concentration may require 2 to 5 times the stoichiometric amount of air. SulFerox regenerators are typically designed for 200% excess air (at 10 psig air pressure) or 3 times the stoichiometric amount (Buenger et al., 1988; Allen, 1995). Actual operating experience indicates that this is a conservative design limit and that for most applications, optimum air regeneration requirements will fall somewhere between 100 and 200% excess air.

One other important factor in the control of the process is that for any given chelant, the iron chelate can exist in many different forms, depending on the chelant and iron concentrations, the solution pH, the temperature, and the overall ionic strength of the solution. Consequently, the process conditions in the regeneration step must be such that the formation of the most stable form of ferric chelate is favored.

Process Description

A schematic flow diagram of a typical SulFerox plant treating sour gas is shown in **Figure 9-41**. The sour gas is contacted cocurrently with the regenerated solution in a sparged tower

828 *Gas Purification*

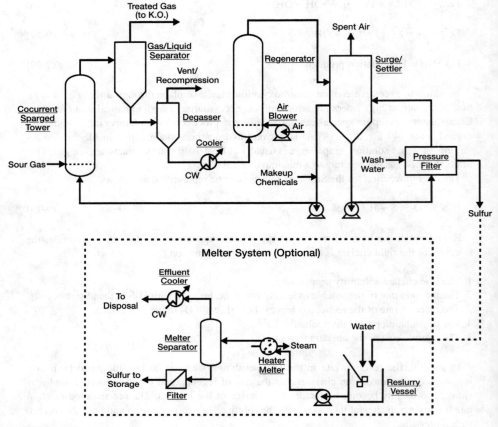

Figure 9-41. SulFerox configuration treating sour gas and producing salable sulfur. *Courtesy of Dow Chemical*

contactor or bubble column where the hydrogen sulfide is immediately converted to elemental sulfur in a very fast, mass transfer limited reaction. Since the reaction is irreversible and not limited by equilibrium, the H_2S can be scrubbed to meet any required specification if enough mass transfer surface area is provided.

The sweetened gas flows from the contactor to a gas/liquid separator, where it is separated from the sulfur-containing spent solution slurry. If the gas is at high pressure, the slurry (typically less than 0.1 wt% solids) is degassed in a flash drum to remove absorbed gases (Allen, 1995). The degassed solution is then cooled to remove the heat of reaction and fed to the regenerator. In the regenerator, the spent solution moves cocurrently with the bottom-sparged air flow, and the Fe^{2+} is oxidized back to Fe^{3+}. The regenerated liquor exits the top of the regenerator and flows into the surge/settler vessel where the spent air is vented. Make-up chemicals are typically added to the suction of the main circulation pump to ensure adequate mixing with the process solution (Allen, 1995).

The elemental sulfur solids precipitated in the contactor are chemically modified to cause particle agglomeration and sinking. The surge/settler vessel acts as a solids thickener and a solution clarifier—the concentrated bottoms are typically on the order of 2.0 wt% solids, while the solids concentration in the circulating solutions is typically on the order of 0.1 wt%. The thickened slurry at the bottom of the surge/settler is fed to a filter press equipped with a cake wash and pressurized diaphragm cake-squeeze system for maximum solution recovery. The recovered liquid is returned to the process. The sulfur-cake produced by this method of operation contains 10 to 25 wt% moisture (Allen, 1995).

If the desired product is molten sulfur instead of wet filtercake, a sulfur melter can be added to the process. The sulfur-cake is reslurried with water and pumped through an exchanger where it is melted using steam. The molten sulfur/water mixture is separated with the aqueous phase going to wastewater treatment. The volume of wastewater is usually less than 2 gallons per minute per long ton per day (LTPD) of sulfur processed (Allen, 1995). This stream is sufficient to purge the small amount of degradation products produced and normally contains less than ½ wt% organics. The molten sulfur is filtered to remove any traces of entrained solids, and then sent to storage.

Figure 9-42 depicts the 0.6 LTPD SulFerox unit at Mobil's Moco Midway-Sunset field near Taft, CA. This plant has a downflowing pipeline contactor to the far left. A gas-liquid separator is to the right of the pipeline contactor, followed by the regenerator and surge drum.

Operating Conditions

1. *Absorber.* Because of the presence of solids in the circulating solution, the contactor is particularly prone to sulfur fouling. Hence, the use of packed towers, with either the gas or the liquid as the continuous phase, is not recommended. Several different fouling-resistant contactor types are used in the SulFerox process: cocurrent upflow sparged towers, a proprietary cocurrent pipeline contactor in either an up or a downflow configuration, and countercurrent spray towers (Anon., 1994). While these contactor types are resistant to sulfur fouling, it is advisable to pay particular attention to the internal finish of vessels and piping during construction to minimize the potential for sulfur accumulation (Al-Mughiery et al., 1992).

   In most cases, the SulFerox process utilizes either a sparged tower or a proprietary pipeline contactor design that is claimed to give high H_2S removal with very short residence times and is reported to be very resistant to sulfur plugging. The contactor type selection for a particular application is based on gas volume (acfm), available pressure drop, percent H_2S removal, and required capital expenditure.

   In general, the diameter of a given contactor is set by the actual volumetric flow rate of the sour gas. Typical superficial design velocities for sparged towers are 0.1 to 1.0 ft/sec, while pipeline contactors range from 5 to 20 ft/sec. Typical pressure drops for sparged towers range from 3 to 10 psi. Downflow pipeline contactors have a pressure drop of about 1.0 psi, and upflow pipeline contactors have pressure drops ranging from 5 to 15 psi, depending on the level of H_2S removal. Spray towers are designed in the 3 to 8 ft/sec superficial vapor velocity range with a pressure drop typically less than 1 psi (Allen, 1995; Pirtle et al., 1994).

   For sparged tower and pipeline type contactors, the contactor length is set by the H_2S removal requirements and is directly related to the number of transfer units (NTU) required for H_2S removal. For a very fast, irreversible reaction, the number of transfer

830 Gas Purification

Figure 9-42. Photograph of 0.6 LTPD SulFerox plant with cocurrent downflow pipeline contactor. *Courtesy of Dow Chemical*

units (NTU) can be expressed as the natural logarithm of the inlet gas H_2S concentration divided by the outlet gas H_2S concentration. See equation 9-51. Commercial SulFerox pipeline contactors have been supplied with NTUs ranging from 2 to 8, while SulFerox sparged tower contactors have NTUs ranging from 1 to 10 (Allen, 1995; Pirtle et al., 1994).

2. *Regenerator.* The size (and capital cost) of the regenerator section, as well as the liquid circulation rate, are primarily dependent on the plant sulfur production rate. According to Buenger and Kushner (1988), for any iron chelate-based process, the circulation rate can be estimated from the following equation:

$$\text{Circulation rate (GPM)} = \frac{1,340,000 \times \text{Sulfur Capacity (LTPD)}}{\text{Iron Concentration (ppmw)}} \quad (9\text{-}80)$$

A comparison of equation 9-80 with equation 9-73 suggests that the recommended SulFerox iron concentration is approximately 200% of the stoichiometric requirement.

The rate of solvent regeneration is affected by the oxygen partial pressure, but the volume of air required must be above the minimum stoichiometric value. At atmospheric pressure, the air volume required to regenerate SulFerox solution is typically in the range of 2 to 3 times stoichiometric (Allen, 1995). Regeneration is usually carried out at low pressure (typically 10 psig) to minimize regeneration air compression costs.

3. *Sulfur Recovery Section.* For plants processing 1 LTPD or less of sulfur, there is no advantage to concentrating the sulfur solids prior to filtration since the required filter surface area is small. In these cases, full stream filtration of the contactor scrubbing solution is possible using an inexpensive plate-and-frame type pressure filter. For plants processing more than 1 LTPD of sulfur, it is generally cost effective to concentrate the sulfur solids in the surge/settler vessel for feed to a filter press (Allen, 1995).

The SulFerox sulfur-cake matrix tends to be quite compressible and reportedly lends itself well to filtration via plate-and-frame filters or automated batch filter presses with recessed membranes for post-filtration sulfur-cake squeezing. It is claimed that only small amounts of wash water are needed to displace the residual process solution from the filtercake. Rotary drum vacuum filters were used in all early SulFerox applications, but variations in feed slurry characteristics due to differing inlet gas contaminants and required additive levels made their performance inconsistent. It is reported that filtration by pressurized-feed filter presses has eliminated the problem of cake quality variation and has substantially reduced iron chelate losses (Anon., 1994). The sulfur filtercake from pressurized-feed filter presses is reported to contain 10 to 25 wt% moisture (Allen, 1995).

The sulfur slurry from the surge/settler vessel should not be melted without an intermediate filtration step because the Fe^{3+} in solution tends to oxidize the chelant at temperatures above 170°F, thus increasing chelant degradation. Water insoluble solid contaminants entrained in the gas stream tend to concentrate in the product sulfur. Simple filtration of the molten sulfur will remove these materials and increase sulfur product quality.

Proper operation of the SulFerox process requires good solution maintenance procedures. It is quite important to maintain the proper levels of iron, pH, additive concentration, and solids in the circulating solution. All these solution parameters must be regularly monitored by utilizing appropriate spectro-photometric, gravimetric, and titration test methods.

Effect of Feed Contaminants

The SulFerox process is highly selective for hydrogen sulfide removal because carbon dioxide does not react chemically with the solution. Carbon dioxide removal is limited by the CO_2 solubility in the circulating solution. When the system is operated at a pH of 8, the solubility of CO_2 in the solution is approximately the same as in pure water. Nevertheless, the effect of CO_2 on the pH of the solution cannot be totally ignored, since the higher the pH of the solution the more CO_2 it will absorb and the more difficult it will be to maintain the desired pH. A small amount of make-up buffering agent (typically NaOH that is converted to Na_2CO_3 and $NaHCO_3$ in the solution) is normally sufficient to prevent significant pH swings in the contactor resulting from CO_2 in the gas stream.

Usually, the higher the CO_2 to H_2S ratio in the sour gas, the more attractive SulFerox will be compared to conventional processes. Good applications include desulfurization of recycled CO_2 streams in enhanced oil recovery projects and desulfurization of sour noncondensable gas from steam in geothermal power production.

Even though it is highly selective for hydrogen sulfide removal, SulFerox can also remove between 50 and 90% of the mercaptans present in the sour gas. The initial reaction step is thought to proceed through the formation of an iron-thiol intermediate complex (Bedell et al., 1988), which leads to the generation of a thiyl radical:

$$Fe^{3+} + RSH = (Fe-SR)^{2+} + H^+ = Fe^{2+} + RS\bullet + H^+ \tag{9-81}$$

Because of the carbon-sulfur bond, oligomerization to polysulfides, which is common with hydrosulfides, is not favored, and the reaction will usually terminate with the formation of the disulfide, RSSR. Some of the disulfides are stripped out in the spent regenerator air. The bulk of the remaining disulfides can normally be decanted from the solution surface, since they are very sparingly soluble in aqueous solutions. If they are not removed, a large proportion of the disulfide impurities formed in the process are adsorbed on the sulfur surface and leave the system in the filtercake. Because this disulfide component also contains unconverted mercaptans, the sulfur produced may have a distinct unpleasant mercaptan odor.

In the SulFerox process, carbonyl sulfide can be hydrolyzed to CO_2 and H_2S that is then reacted to elemental sulfur. However, the hydrolysis reaction rate is quite slow, and consequently the removal rate of COS is limited by the residence time in the contactor. For this reason, it would be expected that a sparged tower absorber would be much more effective at COS removal than the other contactor types. Carbon disulfide in the feed gas is first hydrolyzed to COS and then to CO_2 and H_2S. The net CS_2 removal efficiency is therefore less than that for COS. If the contactor has ample residence time, the SulFerox process may remove as much as 30 to 60% of the carbonyl sulfide present in the sour gas. If the unit has a short residence time contactor, the removal rate for CS_2 and COS may be less than 10%.

At one time it was thought that the SulFerox solution could be "poisoned" by certain components found in refinery gas streams. For example, HCN, NH_3, and SO_2 were considered to have a deleterious effect on the process. In the case of HCN, it was postulated that it would react stoichiometrically with iron, rendering the reacted iron nonregenerable due to the formation of a stable thiocyanate complex. Recent data support the formation of such a thiocyanate complex and high levels of HCN can lead to very high chemical consumption (Allen, 1995). Klanecky et al. (1995) suggest several HCN removal options, including catalytic hydrolysis and scrubbing with polysulfide solutions. Lynch (1982) reviews polysulfide scrubbing for HCN removal, and Huisman (1994) provides data on catalytic hydrolysis of HCN. See the cyanide removal discussion in the Stretford process "Liquid Waste Streams" section for additional information.

Theoretically, if the concentration of ammonia in the feed is sufficiently high, the increase in pH of the SulFerox solution may result in iron precipitation. In practice, this is seldom a problem, and the presence of ammonia simply reduces the amount of additional base required for pH adjustment. The SulFerox process has reportedly treated sour-water stripper gas streams containing up to 30 to 40% ammonia without experiencing pH-induced precipitation problems, although one facility has had to add acid for pH-control instead of the usual caustic (Allen, 1995).

Finally, too high a concentration of SO_2 has an effect somewhat similar to that of HCN, and leads to excessive base consumption and the formation of sulfite, sulfate, or thiosulfate salts. In most cases, a normal solution purge, coupled with good pH control, is sufficient to prevent excessive sulfur salt buildup.

Applications of the SulFerox Process

The SulFerox process can be used for treating either low or high pressure gas. This characteristic, plus its high H_2S selectivity, makes it potentially useful for a large variety of gas treating requirements. Some of the possible applications include the treatment of

1. Sour Natural Gas
2. Amine Plant Acid Gas
3. Low Pressure Casinghead Gas
4. Reduced Claus Tail Gas
5. Geothermal Noncondensable Gas

High Pressure Sour Natural Gas Treatment. An application of the SulFerox process for removing H_2S from high pressure, high CO_2 content natural gas at the No. 1 Mivida plant, located near Pecos, TX, was reported by Iversen et al. (1990).

The original configuration of the No. 1 plant is depicted in **Figure 9-43a.** Water and H_2S are removed from the feed gas by molecular sieve treaters (2×4 bed units). Part of the feed gas and the sour regeneration gas from the molecular sieve treaters are treated in a Sulfinol unit, dried with triethylene glycol (TEG), and combined with the H_2S free stream from the molecular sieve treaters. The combined dry gas stream is sent to a Fluor Solvent unit for CO_2 removal, and the acid gas from the Sulfinol plant is sent to a direct oxidation sulfur recovery unit (SRU). Plant No. 1, as originally configured, was both difficult and expensive to operate. Swings in the molecular sieve treater regeneration gas flow rate led to instabilities in the SRU, and the Sulfinol reboilers ($2 \times 100,000$ lb/hr steam) and the molecular sieve regeneration gas heaters (4×33 MMBtu/hr) were costly to operate.

The addition of a SulFerox unit, as depicted in **Figure 9-43b,** considerably simplified the design. Feed gas containing 1,800 ppmv H_2S is fed directly to the SulFerox unit for H_2S removal. The H_2S free gas is dried and CO_2 is then removed in the Fluor Solvent plant. This simplified configuration eliminates the molecular sieve treaters and their four fired heaters, the Sulfinol unit and two LP steam generators, and the SRU, substantially reducing operating costs.

The Mivida SulFerox unit is designed for about 2 LT/day sulfur production and includes a sulfur melter to process the SulFerox sulfur filtercake for sale as a high quality molten product. A flow diagram of the sulfur melter is shown in **Figure 9-44.** A summary of the projected operating conditions at the Mivida SulFerox plant, which includes a mix of SulFerox pilot plant and design data, is provided in **Table 9-22** (Iversen et al., 1990; Allen, 1995).

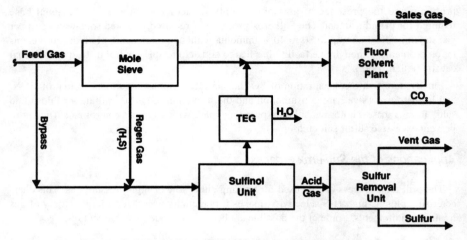

Figure 9-43a. Original configuration of No. 1 Mivida plant.

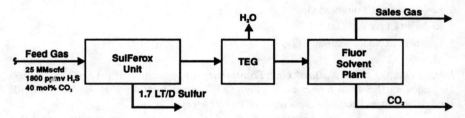

Figure 9-43b. Configuration of No. 1 Mivida plant after addition of a SulFerox unit.

Amine Tail Gas Treating. The use of the SulFerox process for amine tail gas treating is described by Pirtle (1991). The plant is a grassroots installation processing up to 6 MMscfd of wellhead gas at 750–850 psig. The sour gas to the amine treater contains about 4,000 ppmv H_2S, 730 ppmv mercaptans, and 200 ppmv organic sulfur compounds (RSSR, RSR, etc.). The amine treater uses an MDEA (methyldiethanolamine) based solvent. The acid gas from the amine treater is the feed to the SulFerox unit. It contains 46 mol% H_2S, 3.5 mol% mercaptans, and 630 ppmv of other sulfur compounds.

A generic process flow diagram for this type of unit is shown in **Figure 9-45** (Buenger et al., 1991A). In this application, both the treated acid gas and the spent regenerator air are wastestreams. Because of odor problems associated with the presence of mercaptans in the feed, the combined wastestream is incinerated. The sulfur filtercake is collected for disposal.

A high liquid hold-up, sparged absorber is used to improve mercaptan removal. To minimize the presence of "free" mercaptans in the filtercake, buffering agent makeup is used to maintain the lean solution pH at about 7. Mercaptan removal under these conditions is greater than 90%. However, if the sulfur were melted, the release of mercaptan-based compounds could create operating problems.

The complete stream molar flow balance of the plant is shown in **Table 9-23.** From these data it appears that SulFerox partitions most of the disulfides produced by oxidation of mercaptans to the sulfur product, with little going to the treated gas or spent regenerator air streams.

Table 9-22
Projected Operating Conditions for the SulFerox Unit at the No. 1 Mivida Plant

Feed Gas Analysis, Vol%	
Methane	58.5
Ethane	0.3
Propane plus	0.4
CO_2	40.0
H_2S (ppmv)	1800
Pressure, psig	960
Temperature, °F	80
Gas Volume, MMscfd[1]	~25
H_2S out, ppmv	<4.0
Sulfur Recovery, LTPD	~1.7
Solution pH	6.5–7.3
Utility Requirements	
Process Water, gpm	4.0
Electric Power, kVa	300

Notes:
1. Data from Allen (1995).
Source: Iversen et al. (1990) and Allen (1995)

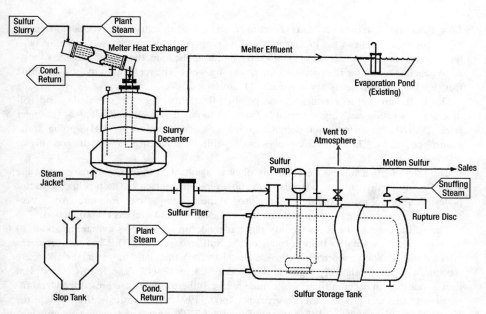

Figure 9-44. Process flow diagram of sulfur melter processing reslurried SulFerox sulfur filtercake. (*Iverson et al., 1990*)

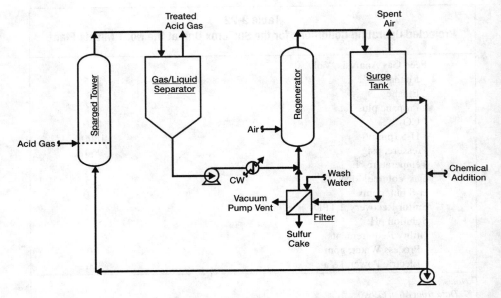

Figure 9-45. Flow diagram of SulFerox plant used for treating amine plant tail gas. (*Buenger et al., 1991A*)

Low Pressure Casinghead Gas Treating. The use of the SulFerox process to remove H_2S from the casinghead gas in a steam-flood oil recovery project expansion in Kern County, CA, is reviewed by Buenger et al. (1991B). In this application, the treated gas is combusted to generate steam for injection into the producing wells. The removal of hydrogen sulfide from the fuel gas was necessary to reduce SO_2 emissions.

One of the process requirements was operating flexibility, since the gas quality and volume at this location can vary appreciably from day to day as shown in **Table 9-24** (Buenger et al., 1991B). The air pollution permit requires 90% overall reduction of H_2S going to the steam generators. The SulFerox process is capable of achieving 97% hydrogen sulfide removal.

The flow diagram for the Kern County plant is identical to **Figure 9-45** except that the Kern County facility uses a cocurrent, downflow pipeline contactor. The pipeline contactor is designed for a 2 psi maximum pressure drop at the design gas rate. Pressure drop is very important in this system because backpressure in the casinghead restricts oil production.

The sulfur-cake at the Kern County plant is made up of particles averaging about 20 microns, which is smaller than normal for the SulFerox process. This has not seriously affected the filtration properties of the elemental sulfur particulate; however, the filtercake moisture content is relatively high, on the order of 25 to 35 wt%. The Kern County plant uses a rotary drum vacuum filter, while more recent SulFerox plants use pressurized feed filter presses that can provide drier filtercakes (Anon., 1994). The washed sulfur-cake at the Kern County plant contains 1,000 ppmw of iron and 7,500 ppmw of organic contaminants; however, the produced sulfur qualifies for agricultural use.

Table 9-23
Material Balance for Amine Plant and SulFerox Unit Treating the Amine Plant Tail Gas

Stream	1 Sour Gas to Amine Treater	2 Sweet Gas from Amine Treater	3 Flash Gas from Amine Treater	4 Acid Gas to SulFerox Unit	5 Treated Acid Gas	6 Spent Regenerator Air
Component, lb mole/hr						
H_2S	1.316	0.000	0.003	1.317	0.022	0.000
N_2/Air	14.410	14.291	0.000	0.008	0.008	9.932
CO_2	2.862	1.591	0.010	1.286	1.158	0.025
Methane	270.723	263.706	6.779	0.081	0.083	0.000
Ethane	22.799	22.121	0.645	0.011	0.018	0.000
Propane	9.804	9.517	0.273	0.003	0.013	0.000
Butane+	7.073	6.952	0.089	0.045	0.025	0.000
CH_3SH	5.27E-02	2.61E-02	5.20E-04	2.66E-02	6.95E-05	9.99E-05
C_2H_5SH	9.22E-02	5.41E-02	9.10E-04	3.80E-02	1.31E-03	4.29E-03
C_3H_7SH	7.90E-02	5.41E-02	7.80E-04	2.49E-02	1.60E-03	4.49E-03
C_4H_9SH	1.65E-02	3.82E-03	1.62E-04	1.26E-02	4.55E-04	9.99E-05
Misc. Sulfur	6.68E-02	6.50E-02	6.59E-04	1.79E-03	4.81E-03	5.29E-03
Press., psig	780	89	50	5	2	2
Temp., °F	41			58	108	108
Flow, MMscfd	3					
Flow, lb mole/hr	329.29	318.51	7.80	2.86	1.34	9.99

Sulfur Removal

Component	% Removal Across Amine Treater	% Removal Across SulFerox Contactor	% Recovered Across SulFerox Unit Including Spent Air as "Unrecovered"
CH_3SH	50.4	99.7	99.4
C_2H_5SH	41.3	96.6	85.3
C_3H_7SH	31.5	93.5	75.5
C_4H_9SH	76.8	96.4	95.6
Misc. Sulfur	2.7	−168.7 (net gain)	−464.2 (net gain)
H_2S		98.3	98.3
Total	100.0		96.8

Source: Buenger et al. (1991A)

Table 9-24
Comparison Between Design and Actual Conditions at the Kern Co. SulFerox Unit

Process Conditions	Design	Actual
Volume, MMscfd	3	1.0–1.9
H_2S inlet, ppmv	2,000	6,000–19,000
CO_2, vol%	80	85
Temperature,°F	80	45
Pressure, psig	5	5
Sulfur, Avg. LTPD	0.23	0.40–0.80

Source: Buenger et al. (1991B)

Geothermal Gas Cleanup. The selection of a primary abatement system for a geothermal power plant is influenced by several factors, including (1) the level of H_2S in the steam and the level of abatement needed, (2) the partition of H_2S between the noncondensable gases and the condensate, and (3) the volume and quality of sulfur that can be produced (Kenny et al., 1988).

The applicability of the SulFerox process to geothermal gas clean up depends on the partition of H_2S between the condensate and noncondensable gases, which depends on the level of ammonia in the steam and the condenser design. As the ammonia level rises, the amount of H_2S dissolved in the steam condensate increases. This makes the SulFerox process less attractive because separate condensate treatment can be required. Since the noncondensable gases are contacted with much less water with indirect cooling than with direct contact condensation, less H_2S is dissolved and a major fraction remains in the gas phase. Therefore, operating plants that have been retrofitted with surface condensers are more likely to yield improved economics when the SulFerox process is used to treat the noncondensable vent gas.

In a Geysers application described by Kenny et al. (1988), the noncondensable gas flow is 4.2 MMscfd with an H_2S content of 1.2 mole%. The SulFerox unit was designed to meet a 20 ppmv H_2S emission standard on the purified noncondensable gas. The amount of sulfur-cake produced is 2 LTPD on a dry basis. The economics for this application, based on 304/316 stainless steel construction, are summarized in **Table 9-25**.

Table 9-25
Process Economics for Geothermal Application of SulFerox Technology

Capital Cost	$883,000
Annual Operating Costs	
Chemicals	$105,000
Power	$30,600

Notes:
1. Unit treats 4.2 MMscfd of gas containing 1.2 mol% H_2S.
2. Costs are given in 1988 constant dollars and include the license fee for the technology.

Source: Kenny et al. (1988)

Sulfur Quality

The SulFerox process converts hydrogen sulfide into elemental sulfur, which is separated from the slurry by filtration. Sulfur quality can be influenced by contaminants in the gas stream. However, filtration, followed by melting, can usually provide a yellow sulfur product with less than 100 ppmw of iron, 500 ppmw of ash, and 1,000 ppmw of carbon. A comparison between Claus-produced sulfur and the filtercake and molten sulfur produced in the SulFerox process is shown in **Table 9-26** (Van Kleeck and Morisse-Arnold, 1990). As indicated earlier, direct melting without concentration of the sulfur slurry is usually not a viable process option due to contamination of the molten sulfur and high chemical losses caused by thermal degradation.

Table 9-26
Comparison Between SulFerox and Claus Sulfur Quality

	Claus	SulFerox	
Type (typical)	Molten	Filter Cake	Molten
Sulfur, wt%	99.5	80	>99.5
Moisture, wt%	none	<18	none
Carbon, ppmw	500	10,000	1,000
Ash, ppmw	30	3,400	500
Iron, ppmw		4,000	100
Heavy metals, ppmw	20		none
Chlorides, ppmw	10		100
Color (yellow)	Bright	Pale	Dark

Source: Van Kleeck and Morisse-Arnold (1990)

Filtration of the aqueous slurry permits good solution recovery and yields a high-quality filtercake. Automated batch filter-press systems provide continuous sulfur removal with good turndown capability (Anon., 1994). As **Figure 9-46** indicates, the SulFerox filtercake residual iron content is a strong function of the filtercake moisture content and relatively unaffected by the wash-water ratio for ratios above 1 lb water to 1 lb sulfur (Van Kleeck and Morisse-Arnold, 1990). For this reason, there is an economic incentive to select filters that minimize the filtercake moisture content and related iron-chelate losses. Experience has shown that mechanical compression filters provide drier, cleaner filtercakes than other filter choices (Anon., 1994).

Carbon contamination of the produced sulfur can result from liquid hydrocarbon droplets in the feed gas or by condensation of feed gas hydrocarbons. This problem is best avoided by operating the process about 10°F above the feed gas hydrocarbon dew point and by removing any hydrocarbon aerosol in the feed gas (compressor lube oil, etc.) with a coalescing filter (Allen, 1995). Yet another potential form of sulfur contamination can occur when the feed gas contains a significant amount of mercaptans. Any disulfides formed will tend to coat the surface of the sulfur particles. If the sulfur is to be recovered as a salable product, special attention needs to be paid to the design of the overall system to avoid production of low-quality, disulfide-contaminated sulfur. Elimination of mercaptans and heavy hydrocar-

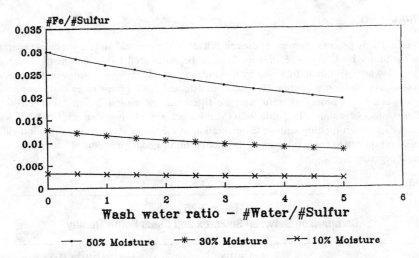

Figure 9-46. Effect of wash-water/sulfur ratio on iron content of SulFerox sulfur filtercake. (*VanKleeck and Morisse-Arnold, 1990*)

bons from the sour gas before it enters the process is an effective technique for reducing disulfide and carbon contamination of the product sulfur.

The ash content of the sulfur normally depends on the amount of inorganic salts present in the filtercake. The best way to reduce the ash content of the sulfur, other than process water pretreatment and careful control of chemical additive make-up rates, is to melt the filtercake. Molten sulfur produced by the SulFerox process is much closer to "Claus" grade sulfur, but still has ash levels several times higher than Claus sulfur specifications. See **Table 9-26**.

SULFUR DIOXIDE PROCESSES

The processes falling within this category are, in essence, extensions of the Claus process discussed in Chapter 8. The reaction involved is the oxidation of hydrogen sulfide to elemental sulfur and water using sulfur dioxide as the oxidant, according to equation 9-82.

$$2H_2S + SO_2 = 3S + 2H_2O \tag{9-82}$$

The first sulfur dioxide-based process was the Townsend process, developed in 1958. This process never advanced beyond the pilot-plant stage due to mechanical and corrosion problems. It was followed by the IFP Clauspol 1500 and the Wiewiorowski processes in 1969. The IFP process is closely related to the Townsend process, but is restricted in application to the treatment of Claus tail gas. The low cost and simplicity of the IFP process has attracted some commercial interest; however, the Wiewiorowski process was never commercialized.

In 1986, the University of California at Berkeley conducted extensive bench scale research on a process similar to the original Townsend process, but using a different physical solvent and a new catalyst formulation. This process has been named the UCBSRP process, but in spite of its scientific interest, it has not attracted the funding needed to be scaled up to the pilot-plant stage.

Townsend Process

The Townsend process, which was disclosed by Reid and Townsend (1958) and Townsend and Reid (1958), and described in a patent granted to Townsend (1965), was proposed as a method for high-pressure natural gas desulfurization and elemental sulfur production in one operation, thus combining the conventional process of absorbing hydrogen sulfide in an aqueous alkaline solution (e.g., ethanolamine), followed by processing the stripped H_2S in a Claus-type sulfur plant. However, the process is claimed to be equally applicable to the treatment of acid-gas streams, such as the effluents from ethanolamine plant regenerators. In this application, the Townsend process would be a substitute for the Claus plant.

A schematic flow diagram of the Townsend process, as applied to high-pressure natural gas treating, is shown in **Figure 9-47**. The sour natural gas enters the base of the reactor column at atmospheric temperature and contacts countercurrently a concentrated stream of di- or triethylene glycol (typically 98% glycol, 2% water) containing dissolved sulfur dioxide. The reaction of hydrogen sulfide with sulfur dioxide is rapid, with the water present acting as a catalyst. The glycol circulation is large enough to maintain a water concentration not exceeding 5% by weight after absorption of both the water contained in the feed gas and the water formed in the reaction.

The treated gas leaving the reactor column is washed in a high-pressure column with concentrated glycol to remove sulfur dioxide carried by the gas out of the reactor. The effluent gas from this column is claimed to be within pipeline specifications with respect to hydrogen sulfide and to be free of sulfur dioxide.

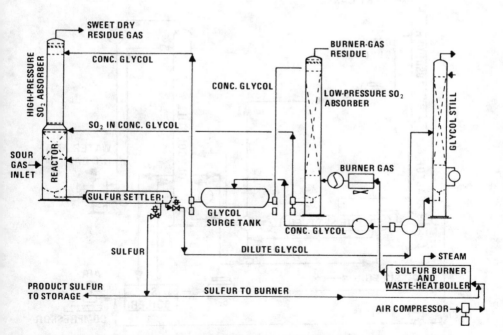

Figure 9-47. Typical flow diagram of Townsend process for high-pressure natural gas treating.

A slurry of elemental sulfur in glycol flows from the bottom of the reactor to a settling tank where the mixture is heated to 250°–275°F. Excess sulfur dioxide is stripped from the solution and returned to the reactor. Liquid sulfur is withdrawn from the bottom of the settling tank. A portion of the liquid sulfur flows to a sulfur burner where it is burned to supply the sulfur dioxide required in the process.

The dilute glycol flows from the settling tank to the glycol still where water is removed and the glycol is reconcentrated. The concentrated glycol then flows to a surge tank where it is split into two streams. One stream flows to the top of the high-pressure sulfur dioxide absorber to remove sulfur dioxide from the purified gas. The other stream flows to a low-pressure column where it absorbs sulfur dioxide from the sulfur burner flue gases and is then recycled to the top of the reactor column.

Application of the process to the treatment of acid-gas streams is shown schematically in **Figure 9-48.** In this version, the sulfur dioxide is supplied by burning one-third of the acid gas in the same manner as in the split-flow Claus process (see Chapter 8). The glycol reactor is the equivalent of the catalytic converters of the Claus plant.

The process was tested on a pilot plant scale in Canada. Severe mechanical and corrosion problems were encountered in the operation of the pilot plant and hindered the continuing development of the process to such extent that it was never commercialized.

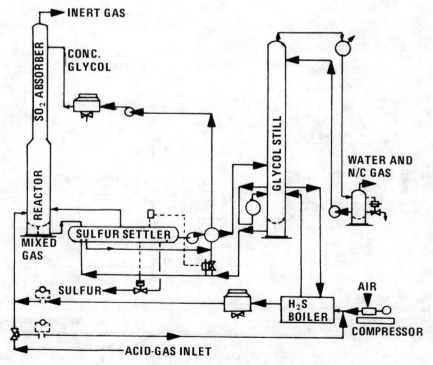

Figure 9-48. Flow diagram of Townsend process for acid gas treating.

IFP Clauspol 1500 Process

A process, very similar to the Townsend process, was disclosed by Renault (1969) and Barthel et al. (1971) of Institut Francais du Petrole (IFP). This process was named Clauspol 1500 and developed specifically for the removal of hydrogen sulfide and sulfur dioxide from Claus unit tail gases.

The catalytic reaction takes place in an essentially anhydrous liquid medium that acts as a common solvent for the H_2S, the SO_2, and the catalyst. The process is operated at temperatures above the melting point of sulfur, but below the sulfur dew point of the gas mixture to be treated. The solvent first proposed (Renault, 1969) was tributyl orthophosphate containing an alkaline substance as the catalyst. However, polyethylene glycol soon became the solvent of choice because of its good thermal and chemical stability, low vapor pressure, low cost, and availability. Additional advantages are the low solubility of sulfur in the solvent and of the solvent in sulfur.

The process gained wide acceptance soon after being proposed, and in 1976 twenty-seven plants were either in operation or in various stages of design and construction (Andrews et al., 1976). It is claimed that among all the processes capable of reducing sulfur compounds in Claus plant tail gases to 1,000 ppm of sulfur dioxide after incineration, the IFP process requires the lowest capital investment and the least expenditure of energy and labor.

The ready acceptance of the Clauspol process in Europe was not duplicated in the U.S. Only four plants were built in the U.S. and only one was still in operation in 1992 (at the Phillips refinery, in Borger, TX). The IFP Clauspol 1500 process is relatively inexpensive and easy to operate; however, the SO_2 emissions (1,000 ppm SO_2) are high in comparison to other competing Claus plant tail gas processes.

Process Description and Operation

The flow scheme of the IFP process, shown in **Figure 9-49**, is extremely simple. The tail gas from the Claus plant, containing hydrogen sulfide and sulfur dioxide, is contacted countercurrently with the solvent in a packed tower. The liquid sulfur formed separates readily by gravity and is withdrawn from a sump located at the bottom of the tower. Since the solvent is only slightly soluble in liquid sulfur, no further purification of the sulfur product is required. The sulfur-free solvent is drawn from the side of the column, and recycled to the top of the packed tower in a pumparound loop. Steam condensate is injected into the recycle stream and vaporized to remove the heat of reaction. The temperature of the solvent entering the tower is maintained between 260° and 280°F to keep the sulfur in a molten state without experiencing excessive glycol losses in the overhead gas. The condensate and the water formed in the reaction, are evaporated and carried out of the tower by the purified gas.

An auxiliary heat exchanger is provided in the circulation loop for startup to raise the solvent/catalyst medium to reaction temperature before Claus gas injection. There is no water build-up in the system, and, because the process is carried out at a temperature above the boiling point of water, the solvent is non-corrosive and the equipment is fabricated entirely of carbon steel.

Proper operation depends primarily on the ratio of hydrogen sulfide to sulfur dioxide and on the total content of hydrogen sulfide and sulfur dioxide in the feed gas. For maximum conversion, it is necessary to maintain the ratio of hydrogen sulfide to sulfur dioxide in the feed gas within 5% of the stoichiometric proportion of two to one. This requires reliable control instruments, such as in-line gas chromatographs or ultraviolet spectrophotometers, to ensure that the composition of the Claus unit tail gas is relatively constant.

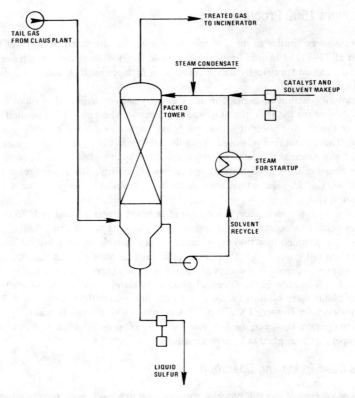

Figure 9-49. Flow diagram of IFP process for Claus-plant tail gas treating. (*Barthel et al., 1971*)

The effect of total sulfur content (in the form of hydrogen sulfide and sulfur dioxide) in the feed gas on conversion is shown in **Table 9-27** (Barthel et al., 1971). It should be noted that sulfur compounds such as carbonyl sulfide and carbon disulfide, which may be present in Claus unit tail gases in appreciable concentrations, are unaffected by the solvent. The term "conversion" therefore applies only to hydrogen sulfide and sulfur dioxide.

Table 9-27
Effect of Total Sulfur on Conversion for the IFP Clauspol 1500 Process

Volume % ($H_2S + SO_2$)	% Conversion ($H_2S + SO_2$)
0.4–0.8	80
0.8–1.5	90
1.5	95

Source: Barthel et al. (1971)

The presence of COS and CS_2 in the Claus tail gas has an adverse effect on the IFP unit sulfur recovery. To minimize the concentration of COS and CS_2 in the Claus burner effluent, it is recommended to run the first Claus converter hotter than normal, and to replace the bauxite catalyst with a more active catalyst. As a general rule, only one tower is required in the IFP unit for a three-stage Claus plant with a sulfur recovery of less than 200 LTPD.

The IFP process may be combined with a Claus unit in a variety of ways as shown in **Tables 9-28** and **9-29**. Since operation of the process is based on physical solubility (i.e., partial pressure) of hydrogen sulfide and sulfur dioxide in the solvent, the solvent flow rates and tower dimensions are essentially the same for a relatively wide range of hydrogen sulfide and sulfur dioxide concentrations in the feed gas. This enables the process to compensate for Claus plant upsets.

**Table 9-28
Sulfur Recoveries for Various Claus/IFP Clauspol 1500 Process Configurations**

Scheme	Sulfur Recovery (% H_2S + SO_2 Conversion)		
	Claus unit	I.F.P. unit	Overall
Furnace + 3 converters + IFP unit	96.8	80.0	99.36
Furnace + 2 converters + IFP unit	94.4	90.0	99.44
Furnace + 1 converter + IFP unit	86.1	95.0	99.30

Source: Barthel et al. (1971)

**Table 9-29
Application of IFP Clauspol 1500 Process for Treating Claus Unit Tail Gas**

Process Conditions	Effluent From		
	1st Converter	2nd Converter	3rd Converter
Tail gas composition, mole %			
H_2S	1.84	0.59	0.34
SO_2	0.74	0.29	0.17
S	1.26	0.14	0.13
H_2O	28.58	29.96	30.25
N_2, CO_2, misc.	67.94	69.02	69.11
Conditions:			
Temperature, °F	260	260	260
Pressure, psig	0.5	0.5	0.5
Sulfur recovery:			
H_2S + SO_2 conversion, %	95	90	80
Production, lb/hr	247	81	43
Treated gas (H_2S + SO_2), ppm	1,100	900	1,000

Source: Barthel et al. (1971)

Wiewiorowski Process

In this process, which was disclosed by Wiewiorowski (1969) of the Freeport Sulfur Company (Anon., 1970), the reaction between hydrogen sulfide and sulfur dioxide takes place in molten sulfur within the temperature range of 240°–320°F (low-viscosity range of liquid sulfur). A basic nitrogen compound, such as ammonia or an amine, in concentrations of 1 to 5,000 ppm is added as the catalyst.

The operation of the process is analogous to that of the IFP process, except that only one liquid phase, i.e., molten sulfur, is present. Any efficient gas-liquid contacting device, such as a packed tower or a turboreactor, is claimed to be suitable for carrying out the reaction (Anon., 1970). As in the IFP process, it is important to maintain the ratio of hydrogen sulfide to sulfur dioxide close to the stoichiometric proportion of two to one in order to achieve maximum conversion to elemental sulfur. It is stated that the process will operate with feed gases containing from 1 to 67% hydrogen sulfide and from 0.5 to 34% sulfur dioxide, and that essentially complete conversion to elemental sulfur can be obtained (Wiewiorowski, 1969). This process has never achieved commercial operation.

UCBSRP Process

In the University of California Berkeley Sulfur Recovery Process (UCBSRP), hydrogen sulfide and sulfur dioxide are dissolved in an organic liquid phase and reacted in the presence of a catalyst at temperatures below the melting point of sulfur. When the sulfur formed in the reaction exceeds its solubility in the solvent medium, it crystallizes from solution and is recovered at high purity values.

The solvent is a polyglycol ether whose exact chemical composition and properties vary with the type of application. The preferred catalyst is 3-pyridyl-carbinol (3-PC) (Lynn et al., 1991). Tertiary aromatic amines such as N,N-dimethyl aniline (DMA) can also be utilized as catalysts.

The choice of solvent medium is very important because the specificity of the gas absorption operation depends on the composition of the sour gas and the selectivity of the solvent for the specific gas component to be removed. In general, most polyglycol ethers are extremely effective in the coabsorption of hydrocarbon gases.

The solubility of four gases in five different polyglycol ethers at infinite dilution is shown in **Table 9-30** (Lynn et al., 1991). The gases are H_2S, CO_2, propane, and N-butane. The solubility is given in moles of solute gas per kg solvent per MPa partial pressure at 25°C. The table also gives the selectivity of the solvents for H_2S relative to the same gas components, expressed as the ratio of the solubility of H_2S to the solubilities of the respective gases in the various solvents.

The solubility of SO_2 is not included in the table, but it should be noted that sulfur dioxide is over ten times more soluble in DGM than hydrogen sulfide (Lynn et al., 1991).

Sciamanna and Lynn (1988) proposed the application of the UCBSRP process for the treatment of sour natural gas to pipeline sales specifications. For this application, tetraglyme (tetraethylene glycol dimethyl ether) was chosen as the solvent. The selection of tetraglyme as the solvent of choice was made because of its low volatility and low vapor pressure (B.P. = 275°C). The solvent vapor pressure was reduced even more by cooling the solvent to 10°C. The catalyst chosen was 3-pyridyl-carbinol (3-(hydroxy-methyl) pyridine). This is a basic, homogeneous, liquid-phase catalyst which has a boiling point of 266°C, is miscible with water at ambient temperature, and has a negligible volatility.

Table 9-30
Gas Solubilities and Solvent Selectivity for Various UCBSRP Solvents

Solvent*	Solubility [mol/kg MPa]				Selectivity for H_2S relative to:		
	H_2S	CO_2	C_3	$n\text{-}C_4$	CO_2	C_3	$n\text{-}C_4$
Diglyme	11.74	2.03	3.46	10.36	5.80	3.39	1.13
Triglyme	10.93	1.65	2.58	7.72	6.62	4.24	1.42
Tetraglyme	10.65	1.50	2.10	6.22	7.10	5.07	1.71
Dowanol DM (DGM)	9.87	1.29	1.45	3.93	7.65	6.81	2.51
Dowanol TBH	7.15	1.17	2.17	6.37	6.11	3.29	1.12

*The chemical names of the five glycol ethers listed in the table are as follows:
Diglyme   - Diethylene glycol dimethyl ether
Triglyme   - Triethylene glycol dimethyl ether
Tetraglyme   - Tetraethylene glycol dimethyl ether
Dowanol DM (DGM) - Diethylene glycol methyl ether
Dowanol TBH   - Diethylene glycol t-butyl ether
Source: Lynn et al. (1991)

Bench scale tests of the application of the UCBSRP process to recycle gas from a crude oil residuum hydrotreater and to a coal gasification stream were reported, respectively, by Lynn et al. (1987) and by Neumann and Lynn (1986).

The natural gas processing scheme proposed by Sciamanna and Lynn (1988) was developed with the aid of the UCBSRP flow-sheet simulator (Neumann, 1986). The process flow diagram presented in this paper is rather complex for a process of this type, reflecting greater emphasis on technical optimization than on commercial feasibility. An illustration of a simplified, updated generic flow diagram for the UCBSRP selective H_2S process is shown in **Figure 9-50** (Lynn et al. 1991).

Per **Figure 9-50**, natural gas entering the plant is first desulfurized in a primary H_2S absorber where the H_2S, water, and the residual heavy hydrocarbon components are removed. The primary absorber is divided into two sections. Chilled sour gas enters the bottom of the first section and is contacted countercurrently at 50 bar with cold recycle solvent containing a low concentration of sulfur dioxide. The driving force for H_2S removal is provided by the simultaneous absorption and reaction of the H_2S with the SO_2 in the recycle solvent. Although the dominant mechanism is physical absorption, the reaction enhances the H_2S absorption rate by lowering its bulk liquid phase concentration. The H_2S content in the vapor exiting the first section of the absorber is less than 4 ppm. The top section of the column is fed with a small amount of cold, lean solvent that serves to reabsorb any SO_2 that has been stripped from solvent in the lower section. The top of the absorber is maintained at a sufficiently low temperature to keep the solvent content of the exiting gas below 10 ppm.

The H_2S laden solvent stream exiting the bottom of the primary absorber is combined with SO_2 rich solution originating in the SO_2 absorber and is fed to the primary reactor/crystallizer. In this vessel, the warm, saturated solution is cooled, and most of the H_2S reacts to form elemental sulfur which precipitates as predominantly bright yellow rhombic crystals. Sufficient SO_2 is added to react with all but 5% of the total amount of H_2S entering the process.

Clarified overflow from the reactor/crystallizer is the main absorbent medium and is pumped back to the primary absorber. This solvent stream is H_2S free, but it is still saturated

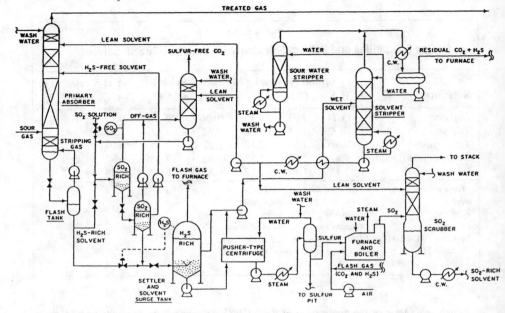

Figure 9-50. Flow diagram of UCBSRP H_2S selective gas purification process (*Lynn et al., 1991*). *Reprinted with permission from the Proceedings of the 1991 GRI Liquid Redox Sulfur Recovery Conference, copyright 1991, Gas Research Institute*

with respect to the other sour gas components. This limits the net co-absorption in the absorber of these other gas components and enhances the solvent H_2S selectivity.

The underflow solvent stream from the reactor/crystallizer contains both dissolved and precipitated sulfur plus the water formed in the reaction between H_2S and SO_2. After flashing to remove part of the residual dissolved H_2S and some of the hydrocarbons, the liquid is cooled to 10°C, and sent to a sulfur settler. The flow to the settler is adjusted to keep the water content of the solvent from exceeding 5% with the flow rate typically at about 10% of the total flow of solvent through the primary absorber. The sulfur settler is a dual-purpose vessel. Its primary function is to separate and recover the precipitated sulfur particles. Its secondary role is to serve as a solvent surge tank.

The 50 wt% slurry of sulfur in solvent from the bottom of the sulfur settler is fed to a pusher-type centrifuge. The sulfur crystals are freed of solvent and washed with water in the centrifuge. The sulfur crystals (about 250 microns in size) are then reslurried in water (33 wt% solids), preheated in a heat exchanger, and pumped to a melter/decanter. The pressure in the decanter is high enough to prevent the vaporization of water when the slurry is heated above the melting point of sulfur.

The wet solvent from the centrifuge is combined with the overflow stream from the surge tank and is pumped to a solvent stripper, where the water (3.5 wt%) is stripped from the solvent to maintain the water balance in the system. The water-vapor flow in the stripper removes unreacted H_2S and most of the other gases, such as CO_2, co-absorbed in the solvent. Most of the lean solvent exiting the bottom of the solvent stripper is used in the SO_2 scrubber to absorb the SO_2 from the combustion gas leaving the furnace. The overhead vapor is condensed in a partial condenser. Approximately 65% of the condensate is fed to a sour-water stripper for reclamation of the water. A portion of the recovered water is reused as wash water in the centrifuge, another portion is supplied to the SO_2 absorber, and the remaining excess water is sent to disposal.

Roughly two-thirds of the molten sulfur produced is sent to sulfur storage in the sulfur-pit, and one-third is burned in a furnace to generate the SO_2 required for the process. Sulfur is fed to the furnace in excess of the stoichiometric amount for oxidation to SO_2 to assure that there is no formation of SO_3. The excess sulfur is condensed in an economizer and recycled to the furnace. The heat of combustion is recovered in a waste-heat boiler. Combustion gas from the furnace is fed to the bottom of the SO_2 absorber. The acid gas is scrubbed with lean solvent, as indicated earlier. The SO_2 content of the furnace stack gas is reduced to less than 1 ppm while generating an 8.2 wt% solution of SO_2.

A reference plant design for the process has been prepared on the basis of a natural gas stream (containing sour condensate) flowing at the rate of 100 MMscfd. The design basis for the plant is presented in **Table 9-31** (Sciamanna et al., 1988).

As the process utilizes a physical solvent, it is relatively stable to variations in composition and gas flow. In addition to H_2S, other gases may be coabsorbed, including propane or higher

Table 9-31
Design Basis for a Plant Utilizing the UCBSRP Process to Treat 100 MMscfd of Natural Gas

DESIGN FEED GAS	
Pressure, bar	50
Temperature, °C	35
Flow rate, kmol/hr	4,986
Composition, mol%	
Methane	70.00
Ethane	9.64
Propane	8.40
Butane	4.50
Pentane and heavier	3.30
Carbon dioxide	3.50
Hydrogen sulfide	0.60
Water	0.06

PRODUCT SPECIFICATIONS	
Treated sales gas:	
Hydrogen sulfide, ppm (max)	4
Water, ppm (max)	147
Carbon dioxide, % (max)	5
Hydrocarbons, dew point at 50 bars (max) °C	4
S purity, wt% (min)	99.98
H_2S or SO_2 content of wastewater, ppm	<1
Stack gas, SO_2 content, ppm (max)	100

Source: Sciamanna et al. (1988)

hydrocarbons. The absorption selectivity can be varied by changing the solvent flow rate. The optimum ratio between the solvent flow rate and the gas flow rate in the absorber is determined by the solvent solubility of the least soluble component that is to be recovered from the sour gas stream, i.e., propane, and not by the mole fraction of that component in the feed.

The reaction between H_2S and SO_2 dissolved in glycol ether at temperatures below 100°C is irreversible and proceeds rapidly to completion when suitably catalyzed. The reaction products appear to be limited to elemental sulfur and water with no significant secondary reactions leading to the formation of undesirable byproducts.

The UCBSRP process is claimed to have significantly lower capital and operating costs than the combination of an ethanolamine absorber/stripper unit plus a Claus plant plus a SCOT tail gas unit. Most of the energy consumed by the process is connected with the recovery and fractionation of propane and heavier hydrocarbons. Approximately 92% of the electrical power usage and 89% of the cooling requirements are associated with hydrocarbon recovery and separation (Sciamanna et al., 1988). The heat generated by the sulfur furnace more than offsets the heat demand required by the desulfurization of the natural gas stream.

Corrosion studies by Crean (1987) indicate that carbon steel and 300 series stainless steels are satisfactory materials of construction for the UCBSRP process.

MISCELLANEOUS PROCESSES

This section describes four very different processes. The Fumaks and Konox processes have been widely used in Japan for the desulfurization of coke-oven gas. The Fumaks process was first developed at Osaka Gas and was later commercialized by Sumitomo Metal Industries Ltd. It utilizes an alkaline solution containing a small amount of picric acid. The Konox process was developed by Sankyo Gas as a replacement for the Ferrox process and employs a ferrate salt as the active oxidizing agent. The Cuprosol process was developed by the EIC Corporation of Newton, MA, for the removal of hydrogen sulfide from geothermal gas, but despite favorable results obtained in a demonstration unit built at the Geysers field in Northern California, it has not matured into a commercial process. Permanganate and dichromate solutions have been used extensively in the past to remove traces of H_2S from gas streams. Both processes are non-regenerative, and their use is declining due to difficulties in disposing of the spent solution.

Fumaks Process

This process, originally developed by Osaka Gas to remove H_2S and HCN from coke-oven gas, was later commercialized by Sumitomo Metals Industries Ltd. of Japan (Hamamura, 1970).

In the Fumaks process, H_2S is removed by scrubbing the gas with an alkaline aqueous solution (NaOH, Na_2CO_3, NH_4OH) containing 0.1% picric acid, which acts as a redox agent (Brodovich et al., 1976). If the raw gas contains ammonia, the ammoniacal solution is preferred. The absorbed H_2S is oxidized to sulfur by the picric acid, and the resulting reduced form of picric acid is then regenerated by atmospheric oxygen.

The Fumaks process removes approximately 95% of the H_2S content of coke-oven gas in a single scrubbing stage. Ammonia is a better reagent than soda ash for the process since it produces less byproduct thiosulfate. The sulfur recovery yields are 85% with ammonia and only 70% with soda ash. The sulfur is precipitated as tiny (1–2 μ in diam.) rhombic sulfur crystals, which favors the formation of ammonium polysulfide. The polysulfide then reacts with the HCN present in the sour gas to form thiocyanate that remains in solution.

Prior to 1976, as many as 28 units were operating in Japan on coke-oven gas and other gas streams. Chemical and utility consumption factors are given in **Table 9-32** (Brodovich et al., 1976):

Konox Process

The Konox process is a liquid-phase oxidation process based on the use of an iron complex. The process was developed in Japan in 1975 by Sankyo to replace the Ferrox process in the treatment of coke oven gas.

The primary process reactions are

H_2S Absorption:

$$4Na_2FeO_4 + 6H_2S = 4NaFeO_2 + 4NaOH + 4H_2O + 6S \qquad (9\text{-}83)$$

Regeneration:

$$4NaFeO_2 + 4NaOH + 3O_2 = 4Na_2FeO_4 + 2H_2O \qquad (9\text{-}84)$$

The active iron is present as a ferrate, Na_2FeO_4, a strong oxidizing agent which reacts almost instantaneously with hydrogen sulfide. The relative ability of Konox solution to absorb hydrogen sulfide compared to that of water and quinone-based H_2S scrubbing liquors is shown in **Figure 9-51** (Kasai, 1975). The high capacity of the Konox solution for hydrogen sulfide permits the use of relatively small circulation rates when compared to other processes. Regeneration is accomplished by blowing the spent Konox solution with air to oxidize the iron from the trivalent (ferrite) state back to the hexavalent (ferrate) state.

It is claimed that side reactions, such as the formation of thiosulfate and sulfate ions from hydrosulfide ions, are minimized because the rapid reaction between H_2S and iron ferrate prevents the buildup of hydrosulfide ions in the solution. This is a distinct advantage over processes where the purging and disposal of thiosulfate salts and the requirements for chemical makeup can represent significant cost items.

Table 9-32
Chemical Consumption/Utility Requirements of the Fumaks Process

Catalyst (picric acid), gm/kg H_2S	15
Alkali (Na_2CO_3), kg/kg H_2S	0.47
Alkali (Na_2CO_3), kg/kg HCN	1.96
NH_3 from gas, kg/kg H_2S	0.15
NH_3 from gas, kg/kg HCN	0.63
Electricity, kWh/1,000 m^3 of gas	7
Process water, tons/1,000 m^3 of gas	0.02

Source: Brodovich et al. (1976)

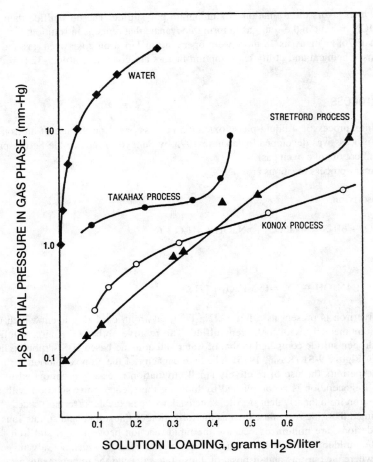

Figure 9-51. Hydrogen sulfide solution loading in various wet oxidation processes (*Kasai, 1975*). Reproduced with permission from Hydrocarbon Processing, February, 1975

The sodium ferrite/ferrate solution is very alkaline and tends to absorb other acid gases such as HCN, CO_2, and SO_2. HCN is a weak acid that reacts with the alkaline solution to form NaCN by a reversible acid/base reaction. Since it is not destroyed (as is H_2S) the NaCN builds up in the solution until the vapor pressure of HCN over the solution is high enough to impede absorption. At this point most of the HCN in the feed gas leaves the absorber with the product gas. A small fraction of the HCN will react with solution components to form NaSCN and ferric ferrocyanide (Prussian blue). This ferrocyanide complex is identical to the oxygen carrier employed in the Staatsmijnen-Otto process, and contributes to the oxidation of hydrogen sulfide to elemental sulfur.

The presence of carbon dioxide in the sour gas increases the concentration of sodium carbonate and bicarbonate in solution, according to equations 9-85 and 9-86 (Kasai, 1975; Brodovich et al., 1976):

$$2NaOH + CO_2 = Na_2CO_3 + H_2O \qquad (9\text{-}85)$$

$$Na_2CO_3 + CO_2 + H_2O = 2NaHCO_3 \qquad (9\text{-}86)$$

This has a deleterious effect on the process because the contact time required for regeneration is a function of the concentration of iron and sodium carbonate in the spent solution. The higher the concentration of sodium carbonate, the longer the time needed to regenerate the Konox solution.

Sulfur dioxide is a stronger acid than either CO_2 or HCN and is absorbed as sodium sulfite. This salt is oxidized in the regenerator to form sodium sulfate, which is extremely stable and nonvolatile. The presence of a significant amount of SO_2 in the feed gas can therefore result in high costs for chemical makeup and solution purge disposal.

Elemental sulfur formed in the Konox process is suspended in the solution as fine particles varying from 1 to 4 microns in size. The sulfur is normally separated from the spent Konox solution by continuous filtration. When the solution contains insufficient active iron, or when the alkalinity becomes too low, the sulfur particles grow in size and float to form a froth at the top of the absorber vessel.

When the process was first introduced, it was claimed that utility and chemical requirements for the Konox process compared favorably with those of contemporary competing processes. A comparison of operating data for several well established H_2S removal processes of that period is given in **Table 9-33** (Kasai, 1975).

EIC Copper Sulfate Process

This process, which was designated Cuprosol by its developer, the EIC Corporation of Newton, MA, specifically targeted the removal of hydrogen sulfide from geothermal steam at relatively high temperatures and pressures.

The main features of the Cuprosol process are described in a patent granted to Harvey and Makrides (1980). **Figure 9-52** is a simplified block diagram of the process as given by

Table 9-33
Comparison of Konox Operating Data With Other Commercial Liquid Redox Processes

Process	Ferrox	Perox	Giammarco-Vetrocoke	Stretford	Takahax	Konox
Gas flow rate (Nm³/hr)	11,700	8,300	11,800	7,080	9,500	9,000
H_2S in (gram/Nm³)	9.16	5.53	5.5	5.73	2–3	4.54
H_2S out (gram/Nm³)	1.38	0.01	0.003	0.01	0.03	0.003
Air rate (Nm³/hr)	3,500	318	450	—	350	320
Soln. flow rate (m³/hr)	—	300	—	455	300	63
Makeup alkali (kg/kg-H_2S)	0.61	0.34	0.10	0.15	0.30	0.10
Power (kwh/kg-H_2S)	0.74	0.83	0.60	2.01	—	0.63

Source: Kasai (1975)

854 Gas Purification

Dagani (1979). A more detailed process flow diagram is included in a DOE report prepared by Brown et al. (1980).

A brief description of the process given by Brown and Dyer (1980), together with **Figure 9-52,** provide a basic understanding of the EIC Cuprosol process, which can be visualized as consisting of four key steps:

1. *Scrubbing.* In this step the dry geothermal steam is contacted with a sulfuric acid solution of copper sulfate. Hydrogen sulfide and ammonia are absorbed to form a precipitate of copper sulfide and a solution of ammonium sulfate and sulfuric acid. Other impurities in the raw steam are also removed by the scrubbing operation.
2. *Liquid-Solids Separation.* A slip stream of the circulating scrubbing liquid is sent to a decanter where it is separated into a concentrated slurry of copper-sulfide solids and a diluted purge stream of clear, acidic copper and ammonium sulfate solution.
3. *Regeneration.* The slurry of copper sulfide solids is pressure leached with air to oxidize the precipitate to soluble copper sulfate. The resulting solution is then recycled back to the scrubber.
4. *Copper Recovery.* The purge stream of clear liquid from Step 2 is neutralized with ammonia, and then treated with iron and hydrogen sulfide to recover dissolved copper for recycle to the process.

Roasting can be used instead of pressure leaching to regenerate the copper sulfate solution in Step 3. If roasting is selected, batch centrifugation instead of decantation is required to concentrate the copper sulfide slurry to effect a cleaner liquid-solids separation. Step 4, copper recovery, can also be accomplished by ion exchange.

The results of demonstration tests that ended in early 1980 were presented in a U.S. DOE report (Pacific Gas and Electric Co., 1980). In the demonstration testing, a stream of approximately 100,000 lb/hr of steam, enough to supply a 5 MW power generating plant, was treated in a crossflow sieve-tray scrubber. The column was 6 ft in diameter by 30 ft in height,

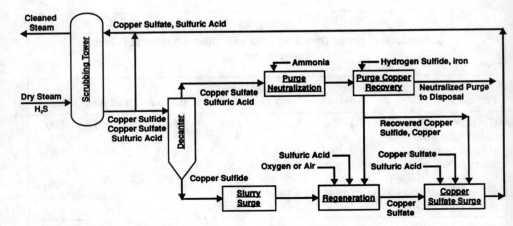

Figure 9-52. Block flow diagram of EIC Cuprosol process. (*Dagani, 1979*)

contained a single sieve tray and downcomer in the contacting section, and was equipped with mesh-type demister pads in the disengagement section.

The dry steam had a nominal inlet H_2S concentration of 220 ppm. Hydrogen sulfide removal efficiency averaged 97% during the 6-month test campaign and occasionally exceeded 99%. Approximately 80% of the ammonia and boron in the feed stream was removed. The other process steps, including regeneration of copper sulfide by pressure leaching, were also demonstrated. One result that was not anticipated was that about 10% of the hydrogen sulfide in the feedstream was oxidized to elemental sulfur, instead of forming copper sulfide.

Prior to the time period in which the EIC method was developed, hydrogen sulfide releases to the atmosphere at geothermal power plants were abated by adding iron salts to the circulating condensate water and using the Stretford process on vent gases from the condenser. The combination of these two methods although effective, was cumbersome, and difficult to operate. This situation sparked the effort to develop a reliable dry steam scrubbing technology. Better condensate treatment methods developed in the 1980s, based on oxidation of H_2S with hydrogen peroxide, and improvements in the application of the Stretford technology to condenser vent gas scrubbing, later reduced the perceived need for the development of dry steam scrubbing methods. In recent years, the improvement of condensate/vent gas purification technology has continued with some of the Stretford units being replaced by the simpler iron-chelate based liquid redox processes.

At its inception, the EIC process appeared to offer the advantages of treating a single low-volume gas stream (high-pressure steam); purifying steam before it entered the turbine (and thus assuring that steam that might be vented from the turbine would be clean); removing other harmful impurities before they entered the turbine; and eliminating the need for operating the condensate system as a slurry of suspended iron compound particles in water (Dagani, 1979). It now appears that these advantages did not give the EIC process a competitive edge, and as a result, the process, despite its technical interest, has failed to achieve market acceptance.

Permanganate and Dichromate Solutions

Buffered aqueous solutions of potassium permanganate and sodium or potassium dichromate are sometimes used to completely remove traces of hydrogen sulfide from industrial gases. Processes employing such solutions are nonregenerative and, because of the high cost of the chemicals used, are only economical when very small amounts of hydrogen sulfide are present in the gas. Permanganate solutions were once used quite extensively for the final purification of carbon dioxide in the manufacture of dry ice.

A schematic flow diagram of the process is shown in **Figure 9-53**. The gas is contacted with the solution in two packed towers operating in series. The solution, which in the case of the permanganate process contains about 4.0% potassium permanganate and 1.0% sodium carbonate, is circulated until approximately 75% of the permanganate in either tower is converted to manganese dioxide. At that point the spent solution is discarded and fresh solution is pumped into the tower. This can be accomplished without interruption of the gas flow if sufficient active permanganate is available in each contact stage to ensure complete H_2S removal. Dichromate solutions usually contain 5 to 10% potassium dichromate, zinc sulfate, and borax. In addition to hydrogen sulfide, traces of organic compounds such as amines are also removed quantitatively in these processes. Use of these processes is declining due to difficulties in disposing of the spent solution.

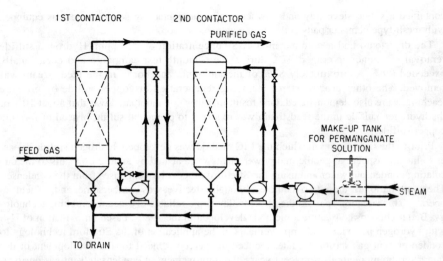

Figure 9-53. Flow diagram of permanganate and dichromate processes.

REFERENCES

Allen, M. A., 1995, personal communication (Dow Chemical), Sept. 14.

Al-Mughiery, S. S., Mills, C., and Bowman, D., 1992, "The SulFerox Process for High Pressure Natural Gas Treating," paper presented at the 1992 Liquid Redox Sulfur Recovery Conference, Austin, TX, October 4–6.

Andrews, J. W., Bonnifay, P., Cha, B. Y., Barthel, Y., Deschamps, A., Frankowiak, S., and Renault, P., 1976, *J. of Air Pollution Control Assoc.*, Vol. 26, July, pp. 664–667.

Anon., 1957A, *Coke and Gas,* Vol. 19, p. 412.

Anon., 1957B, *Chem. Eng.*, Vol. 64, No. 141.

Anon., 1960A, *Sulphur,* Special Issue, pp. 35–38.

Anon., 1960B, "Sweet-Gas Process Makes U.S. Debut," *Chem. Eng.*, Vol. 67, pp. 166–169.

Anon., 1970, *Chem. Eng. News,* Vol. 48, April 27, pp. 68–69.

Anon., 1994, "New equipment for the SulFerox process," *Sulfur,* May–June, pp. 42–45.

Araki, T., Kobayashi, J., Horibe, K., and Miyasaka, N., 1990, "Nippon Steel's Coke Oven Gas Desulfurization Technology," *1990 Ironmaking Conference Proceedings,* pp. 173–182.

Bailey, E. J., 1966, Proc. Western Gas Processors and Oil Refiners Assoc., October, pp. 6–7.

Barry, J., and Hernandez, H., 1990, "Controlling Sulfur Deposition in Wet Oxidation Processes," *1990 Iron Making Conference Proceedings,* pp. 535–546.

Barthel, Y., Bistri, Y., Deschamps, A., Renault, P., Simandoux, J. C., and Dutriau, R., 1971, "Treat Claus tail gas," *Hydro. Process.*, Vol. 50, May, pp. 89–91.

Beavon, D. K., 1973, "The Beavon Sulfur Removal Process," *Proceedings International Conference on Control of Gaseous Sulphur Compound Emissions,* University of Salford (Brit.), Vol. 1, April 10–12.

Beavon, D. K., and Brocoff, J. C., 1976, "Recent Advances in the Beavon Sulfur Removal Process and the Stretford Process," Proceedings Second International Conference on Control of Gaseous Sulphur Nitrogen Compound Emissions, University of Salford, Salford (Brit.), Vol. 2, April 10.

Bedell, S. A., Kirby, L. H., Buenger, C. W., and McGaugh, M. C., 1988, "Chelates' role in gas treating," *Hydro. Process.*, Vol. 2, pp. 63–66.

Bingham, F. E., 1992, (Unocal) personal communication, April 27.

Brodovich, A. I. Shteinberg, E. A., Dolzhanskaya, Y. B., and Sekirna, V. T., 1976, "H_2S Removal from Coke-Oven Gas in Other Countries," *Koksikyhimiya,* No. 9, pp. 39–42.

Bromel, M. C., 1986, "Microbiology of Stretford Solutions," *Proceedings of the 1986 Stretford Users' Conference,* Austin, TX, p. 176.

Brommer, H., and Luhr, W., 1956, *Stahl u. Eisen,* Vol. 76, No. 7, p. 402.

Brown, F. C., and Dyer, W. H., 1980, *Geothermal Research Council Transactions,* Vol. 4, September, p. 667.

Brown, F. C., Harvey, W. W., and Warren, R. B., 1980, *Control of Hydrogen Sulfide Emission from Geothermal Power Plants, U.S. Department of Energy Report, Geothermal Energy,* Contract #EY-76-C-02-2730, Final Report Vol. 2.

Buenger, C. W., Bedell, S. A., and Kirby, L. H., 1988, "Chelates in H_2S Removal," *Proceedings Laurence Reid Gas Conditioning Conference,* Univ. of Oklahoma, Norman, OK.

Buenger, C. W. and Kushner, D. S., 1988, "The Sulferox Process—Plant Design Considerations," *Proceedings of the Gas Processors Association 67th Annual Convention,* pp. 56–62.

Buenger, C. W., Morrisse-Arnold, D., and Hammond, C. A., 1991A, "Performance of the SulFerox Process with Mercaptans," paper presented at the AIChE Spring National Meeting, Houston, TX, April.

Buenger, C. W., Lanning, R. A., and Day, K. K., 1991B, "SulFerox Update 1991—An Operating History from the Kern County Unit," *Proceedings 1991 GRI Liquid Redox Sulfur Recovery Conference,* Austin, TX, May 8–9, pp. 201–209.

Carter, W. A. P., Rodgers, P., and Morris, L., 1977, "Gas Desulphurization by the Stretford Process and the Development of a Process to Treat Stretford Effluent," *Proc. of the Symposium on Treatment of Coke-Oven Gas,* McMaster University, Hamilton, Ontario, Canada, May 26.

Coolidge, A. S., 1927, "The Absorption of Mercury Vapor by Charcoal," *J. Am. Chem. Soc.*, Vol. 49, Part 2, pp. 1949–1952.

Craggs, H. C., and Arnold, M. H. M., 1947, *Chem. & Ind.* (London), Vol. 58, p. 571.

Crean, D. J., 1987, Thesis, Department of Chemical Engineering, University of California, Berkeley.

Dagani, R., 1979, "Cleaning of geothermal steam simplified," *Chem. and Eng. News,* Vol. 75, Dec. 3, pp. 29–30.

Dalrymple, D. A., and Trofe, T. W., 1989, "An Overview of Liquid Redox Sulfur Recovery," *Chem. Eng. Prog.*, March, pp. 43–49.

Day, K. A., and Allen, M. C., 1992, "Experiences Using the Sulferox Process in Refinery Gas Treating," paper presented at the NPRA Annual Meeting, New Orleans, LA, March.

DeBerry, D. W., 1989, "Electrochemical Investigation of the Reactions of Sulfur Species with Vanadium (V)," *Proceedings of the 1989 GRI Liquid Redox Sulfur Recovery Conference,* Austin, TX, p. 10.

Denig, F., 1933, *Gas Age-Record,* Vol. 71, p. 593.

Donovan, J. J., and Laroche, K. J., 1981, "Two Approaches to Effluent Treatment of a Coke-Oven Gas Desulfurization Process," paper presented at 64th Chemical Conference and Exhibition, Chemical Institute of Canada, May 31–June 3.

Douglas, R. A., 1987, "Applications of the Hiperion Sulfur Removal Process," *Proceedings of the 1987 Stretford Conference,* Austin, TX, p. 272.

Douglas, R. A., 1989, "Hiperion Process Update—Commercial and Pilot Plant Experience," *Proceedings of the 1989 GRI Liquid Redox Sulfur Recovery Conference,* Austin, TX, p. 181.

Douglas, R. A., 1990A, "Hydrogen Sulfide Oxidation by Naphthoquinone Complexes—The Hiperion Process," paper presented at the 199th American Chemical Society Meeting, Boston, MA, April.

Douglas, R. A., 1990B, "Direct Oxidation of Hydrogen Sulfide in a Regenerable, Naphthoquinone-Based Process," paper presented at the August AIChE Summer National Meeting, San Diego, CA.

Dunstan, A. E., ed., 1938, *The Science of Petroleum,* New York: Oxford University Press, p. 1807.

Eaton, R. F., 1992, (ARI Technologies) personal communication, Dec. 22.

Edwards M. S., 1979, "H_2S-Removal Processes for Low-Btu Coal Gas," ORNL/TM-6077, Oak Ridge National Lab., Oak Ridge, TN.

Ellwood, P., 1964, "Meta-Vanadates Scrub Manufactured Gas," *Chem. Eng.*, July 20, pp. 128–130.

England, T., 1975, "Commissioning and Operation of the Stretford Process," in *The Coke-Oven Managers' Yearbook,* pp. 255–267.

Farquhar, N. G., 1944, "Sulfur Removal and Recovery from Coke Oven Gas," *Chem. & Met. Eng.,* Vol. 51, July, pp. 94–96.

Fenton, D. M., and Gowdy, H. W., 1979, "The Chemistry of the Beavon Sulfur Removal Process," *Environment International,* Vol. 2, No. 3, p. 183.

Fenton, D. M., and Gowdy, H. W., 1981, U.S. Patent 4,283,379.

Fischer, F., 1932, U.S. Patent 1,891,974.

Fong, H. L., 1987, "SulFerox, A Unique, High Capacity, Iron Chelate Redox Process for Natural Gas Desulfurization," *Proceedings of the 1987 Stretford Conference,* Austin, TX, p. 253.

Fong, H. L., Kushner, D. S., and Scott, R. T., 1987, "Gas Desulfurization Using Sulferox™," Proceedings Laurence Reid Gas Conditioning Conference, University of Oklahoma, Norman, OK, March 2–4.

Foxall, B., 1986, "Sulfur Froth Filtration and Washing Prior to Melting," *Proceedings of the 1986 Stretford User's Conference,* Austin, TX, May 5, p. 134.

Foxwell, G. E., and Grounds, A., 1939, *Chem. & Ind.* (London), August, Vol. 58, p. 163.

Gard, C. D., 1948, *Calif. Oil World,* Vol. 41, December, p. 3.

Garner, F. H., Long, R., and Penell, A., 1958, "The Selective Absorption of Hydrogen Sulphide in Carbonate Solutions," *J. Appl. Chem.* (Brit.), Vol. 8, pp. 325–336.

GEESI (General Electric Environmental Systems, Inc.), 1981, *Coke Oven Gas Desulfurization Systems,* Technical Bulletin COGD 0.7M 9/81R.

Giammarco, G., 1955, Italian Patent 537,564.

Giammarco, G., 1956, Italian Patent 560,161.

Giammarco, G., 1957, Italian Patent 565,320.

Giggenbach, W., 1972, "Optical Spectra and Equilibrium Distribution of Polysulfide Ions in Aqueous Solution at 20°," *Inorganic Chemistry,* Vol. 11, No. 6, pp. 1201–1207.

Gluud, W. and Schoenfelder, R., 1927, *Chem. & Met. Eng.,* Vol. 34, No. 12, p. 742.

Gollmar, H. A., 1929A, U.S. Patent 1,719,762.

Gollmar, H. A., 1929B, U.S. Patent 1,719,177.

Gollmar, H. A., 1945, *Chemistry of Coal Utilization,* Edited by H.H. Lowry, New York: John Wiley & Sons, Inc., Chap. 26.

Gollmar, H. A., 1934, "Chemistry of the Thylox Gas-Purification Process," *Ind. Eng. Chem.,* Vol. 26, February, pp. 130–132.

Grande, M., 1987, "Experiences with Rotary-Drum Vacuum Filters," *Proceedings of the 1987 Stretford Conference,* Austin, TX, p. 185.

Habashi, F., 1978, "Metallurgical Plants: How Mercury Pollution is Abated," *Environ. Sci. & Technol.,* Vol. 12, No. 13, pp. 1372–1376.

Hamamura, K., 1970, *The Operating Progress of Fumaks Desulfurization Facilities,* Sumitomo Metal Industries, Ltd., Japan.

Hammond, C. A., 1986, "The Dow Stretford Chemical Recovery Process," *Environ. Prog.,* Vol. 5, No. 1, pp. 1–4.

Hammond, A., and Pirtle, L., 1994 "Process and Product Comparison: SulFerox and Claus," paper presented at Sulfur '94, Tampa, FL, Nov. 6–9.

Hanck, J. H., Nekoksa, G., and Chhatre, R. M., 1981, "Corrosion Studies on H_2S Abatement at the Geysers Power Plant," *Mater. Perform.,* May, pp. 9–12.

Hardison, L. C., 1984, "Applications of the LO-CAT Process to Sweetening Natural Gas," *Proceedings Gas Conditioning Conference,* University of Oklahoma, Norman, OK, March 5–7.

Hardison, L. C., and McManus, D., 1987, "Chelated Iron Catalyzed Conversion of Gaseous Sulfur Compounds—What to Expect in the 1990s," paper presented at the AIChE Summer National Meeting, Minneapolis, MN, Aug. 16–19.

Hardison, L. C., 1990, "Recent Developments in Acid Gas Treatment Using the Autocirculation LO-CAT Approach," paper presented at the AIChE Spring National Meeting, Houston, TX, April 7–11.

Hardison, L. C., 1991, "LO-CAT II A Big Step Forward in Iron Redox Chemistry," Paper presented at the GRI Liquid Redox Sulfur Recovery Conference, Austin, TX, May 5–7.

Hardison, L. C., and Ramshaw, D. E., 1990, "Economics of Redox Process Selection for Natural Gas H_2S Removal," paper presented at the AIChE Summer National Meeting, San Diego, CA, Aug. 19–22.

Hardison, L. C. and Ramshaw, D. E., 1991, "New Developments in the LO-CAT Hydrogen Sulfide Oxidation Process," *Proceedings Laurence Reid Gas Conditioning Conference,* University of Oklahoma, Norman, OK, March 4–6.

Hardison, L. C., 1992, "Update on LO-CAT Process Developments and Commercial Experience," paper presented at the 1992 GRI Liquid Redox Sulphur Recovery Conference, Austin, TX, Oct. 4–6.

Harvey, W. W., and Makrides, A. C., 1980, U.S. Patent 4,192,854, March 11.

Hasebe, N., 1970, "The Takahax Wet Desulfurization Process," *Chemical Economy Engineering Review*, March, p. 27.

Hass, R. H., Fenton, D. M., Gowdy, H. W., and Bingham, F. E., 1982, "Selectox and Unisulf: New Technologies for Sulfur Recovery," paper presented at the International Sulfur '82 Conference, London, England, Nov. 14–17.

Heisel, M. P., and Marold, F. J., 1987, "New gas scrubber removes H_2S," *Hydro. Process.*, April, pp. 35–37.

Heisel, M. P., Marold, F., and Schrader, U., 1990, "Linde Processes for Recovery of Sulphur from Gases Containing H_2S," *Linde Reports on Science and Technology*, Vol. 47.

Heisel, M. P., 1989, "Operating Experience with the Direct Scrubber Using Sulfolin Liquor at Rheinbraun AG/Berrenrath," *Proceedings of the 1989 GRI Liquid Redox Sulfur Recovery Conference*, Austin, TX, p. 146.

Henderson, J. M., and Dorighi, G. P., 1989, "Operating Experiences of Converting a Stretford to a LO-CAT H_2S Abatement System at Pacific Gas and Electric Company's Geysers Unit 15," Geothermal Resources Council, Transactions, Vol. 13, Oct., pp. 593–595.

Herpers, E. T., and Korsmeier, H., 1979, "The Purification of Stretford Sulfur(Improvements in Filtration of Sulfur Derived from the Stretford Process)," *Inst. Chem. Eng. Symp. Ser. Vol. 57, Control of Sulfur and Other Gaseous Emissions*, Salford (Brit.), April, Inst. of Chem. Engr., pp. Z1–Z37.

Hill, W. H., 1945, *Chemistry of Coal Utilization*, Edited by H.H. Lowry, New York: John Wiley & Sons, Inc., Chap. 27.

Huisman, H. M., 1994, "The Hydrolysis of Carbonyl Sulfide, Carbon Disulfide and Hydrogen Cyanide on Titania Catalysts," Ph.D. Dissertation, Univ. of Utrecht, The Netherlands, October.

Iversen, B., Allen, M., and Bacon, T., 1990, "New High Pressure Process Reduces Sulfur Emissions at Oasis Pipeline's Mivida Plant," *Proceedings Laurance Reid Gas Conditioning Conference*, Univ. of Oklahoma, Norman, OK.

Jacobson, D. L., 1929, U.S. Patent 1,719,180.

Jenett, E., 1962, "Six cases throw light on the Giammarco-Vetrocoke Process," *Oil & Gas J.*, Vol. 60, April 30, pp. 72–77.

Jenett, E., 1964, "An Assessment of the Giammarco-Vetrocoke Process in U.S. Gas Treating Operations—Theory and Practice," *Proceedings of the Gas Conditioning Conference*, University of Oklahoma, Norman, OK.

Jegorov, N. N., Dimitriev, M. M., and Sikov, D. D., 1954, *Desolforazione dei gas*, Milan, Ed. Ulrico Hoepli.

Kasai, T., 1975, "Konox process removes H_2S," *Hydro. Process.*, February, pp. 93–95.

Kenny, K. A., 1995, personal communication (Dow Chemical), Sept. 14.

Kenny, K. A., Bacon T. R., and Kirby, L. H., 1988, "Site Specific Needs Affect Process Choices for H_2S Abatement in Geothermal Power Plants," Paper presented at the Annual Geothermal Resource Council (GRC) Meeting, San Diego, CA, Oct. 10.

Kessler, R., and Kwong, V., 1987, "Sulfur Handling," *Proceedings of the 1987 Stretford Conference,* Austin, TX, p. 167.

Klanecky, D. A., Mamrosh, D. L., Worley, C. M., Arnold, D. K., and Bedell, S. A., 1995, "Effects of Hydrogen Cyanide on Iron Chelate Based H_2S Removal Processes," paper presented at the AIChE Spring National Meeting, Houston, TX, March 23, paper 54d.

Kouzel, B., Woertz, B. B., Fuller, R. H., and Jirus, E. J., 1977, "Treat Low Sulfur Gases with Beavon Sulfur Removal Process and the Improved Stretford Process," *Proc. of the Gas Cond. Conf.,* University of Oklahoma, Norman, OK, March 7–9, p. H-1.

Kozumi, T., Idzutsu, W., Swaim, C. D., Tsurok, H., and Tsuchiya, T., 1977, "Coke Oven Gas Desulfurization by the Takahax Process," paper presented at McMaster Symposium on Treatment of Coke-Oven Gas, McMaster Univ., Hamilton, Ontario, Canada, May 26.

Kresse, T. J., Lindsey, E. E., and Wadleigh, T., 1981, "Stretford plants proving reliable," *Oil & Gas J.,* Jan. 12, pp. 82–87.

Krupp Wilputte, 1988, "H_2S Perox Scrubbing Process," Technical Bulletin, KWC 6.641e.

Kwan, R. K., and Childs, A. M., 1991, "Technical and Economic Comparison of Two Liquid Redox Processes: LO-CAT and Sulferox," *Proceedings 1991 Liquid Redox Sulfur Recovery Conference,* Austin, TX, p. 149.

Lindsey, E. E., and Wadleigh, T., 1979, "The Operational Experiences of Natural Gas Pipeline's Two Stretford Plants," *Proceedings of the Gas Conditioning Conference,* University of Oklahoma, Norman, OK.

Lovett, W. D., and Cunniff, F. T., 1974, "Air Pollution Control By Activated Carbon," *Chem. Eng. Prog.,* Vol. 70, No. 5, pp. 43–47.

Ludberg, J. E., 1980, "Removal of Hydrogen Sulfide from Coke-Oven Gas By the Stretford Process," paper presented at the 73rd Annual Meeting of Air Pollution Control Association, Montreal, Quebec, June 22–27.

Lynch, P. A., 1982, *Iron and Steel Engineer,* December, p. 29.

Lynn, S., Neumann, D. W., Sciamanna, S. F., and Vorhis, F. H., 1987, *Environ. Prog.,* Vol. 6, pp. 257–266.

Lynn, S., Hix, M., Neumann, D. W., Sciamanna, S. F., and Stevens, C. A., 1991, "The UCB Sulfur-Recovery Process," *Proceedings 1991 GRI Liquid Redox Sulfur Recovery Conference,* Austin, TX, pp. 169–180.

Mackinger, H., Rossati, F., and Schmidt, G., 1982, "Sulfint process," *Hydro. Process.,* March., pp. 169–172.

Maddox, R. N. and Burns, M. D., 1968, "Liquid absorption-oxidation processes," *Oil & Gas J.,* Vol. 66, June 3, pp. 90–95.

Massey, M. J., and Dunlap, R. W., 1975, "Economics and Alternatives for Sulfur Removal from Coke Oven Gas," *J. of Air Pollution Assoc.,* Vol. 25, No. 10, pp. 1019–1027.

McBride, R. S., 1933, *Chem. & Met. Eng.,* Vol. 40, August, p. 399.

Meuly, W. C., 1973, "Cataban Process for the Removal of Hydrogen Sulfide from Gaseous and Liquid Streams," Paper presented at the Twelfth Annual Purdue Air Quality Conference, Purdue Univ., Nov. 7–8.

Meuly, W. C., and Ruff, C. D., 1972, *Paper Trade Journal,* May 22.

Miller, S. G., and Robuck, R. D., 1972, "The Stretford Process at East Wilmington Field," *J. Petrol. Technol.,* Vol. 24, May, pp. 545–548.

Mills, B., and Mosher, D. R., 1987, "Direct Application of the Holmes-Stretford Process to the Desulfurization of Low-Btu Gas Derived from the KILnGAS Gasification Process," paper presented at the 1987 Stretford Conference, Austin, TX, Oct. 4–6.

Mitachi, K., Murakami, K., and Sato, J., 1981, "Thermal Decomposition/Regeneration of Desulfurization Liquor," *Chem. Eng. Prog.*, April, pp. 56–61.

Moyes, A. J., Wilkinson, J. S., and Mills, B., 1974, "The Desulphurization of Coke-Oven Gas by the Holmes-Stretford Process," Paper presented at the Midland Sect. of the Coke-Oven Managers' Assoc., Sheffield, Brit., Feb. 14.

Moyes, A. J., and Wilkinson J. S., 1973A, "Development of the Stretford Process," Paper presented at the North-Western Branch of the Institution of Chemical Engineers (Brit.) and at the International Conference for the Control of Gaseous Sulphur Compound Emissions, Salford University (Brit.), April 10–12, (see also *The Chem. Eng.* (Brit.), February 1974, pp. 84–90).

Moyes, A. J., and Wilkinson J. S., 1973B, "High Efficiency Removal of H_2S from Fuel Gases and Process Gas Streams," *Process Eng.*, September, pp. 101, 103, 105.

Moyes, A. J., and Wilkinson J. S., 1974, "Development of the Holmes-Stretford Process," *Chem. Eng.* (Brit.), February, pp. 84–90.

Moyes, A. J., Mills, B., Rothery, E., and Brown, A., 1975, "Practical Engineering Aspects of the Holmes-Stretford Process for the Desulphurization of Coke-Oven Gas," in *Proc. Eng. Aspects Pollut. Control Met. Ind.*, Metals Society, London, November, pp. 101–106.

Mueller, H., 1931, *Gas-u Wasserfach,* Vol. 74, No. 28, p. 653.

Neumann, D. W., 1986, Ph.D. Dissertation, Department of Chemical Engineering, Univ. of California, Berkeley, CA.

Neumann, D. W., and Lynn, S., 1986, "Kinetics of the Reaction of Hydrogen Sulfide and Sulfur Dioxide in Organic Solvents," *Ind. Eng. Chem. Process Des. Dev.*, Vol. 25, pp. 248–251.

Nicklin, T., and Brunner, E., 1961A, *Inst. Gas Engrs.*, British, Pub. 593.

Nicklin, T., and Brunner, E., 1961B, "How Stretford Process is Working," *Hydro. Proc. and Pet. Ref.*, December, pp. 141–146.

Nicklin, T., and Holland, B. H., 1963A, "Removal of Hydrogen Sulfide from Coke-Oven Gas by the Stretford Process," European Symposium on Coke-Oven Gas Cleaning, Saarbrucken, Germany, March (see also *Dechema Monogr.*, 1963, No. 48, pp. 243–271).

Nicklin, T. and Holland B.H., 1963B, "Further Development in the Stretford Process," *Gas World,* Sept. 7, pp. 273–278.

Nicklin, T., Riesenfeld, F. C., and Vaell, R. P., 1973, "The Application of the Stretford Process to the Purification of Natural Gas," Paper presented at the 12th World Gas Conference, Nice, France, June.

Niemiec, W., 1995, personal communication (Wheelabrator Clean Air Systems), Sept. 25.

Otani, Y., Kanaoka, C., Emi, H., Uchijima, I., and Nishino, H., 1988, "Removal of Mercury Vapor from Air with Sulfur Impregnated Adsorbents," *Environ. Sci. & Technol.*, Vol. 22, No. 6, pp. 708-711.

Owens, M., 1991, "LO-CAT Operating Experiences at the Tejas Indian Rock Plant," *Proceedings 1991 Liquid Redox Sulfur Recovery Conference,* Austin, TX, p. 225.

Pacific Gas and Electric Co., 1980, "Demonstration of EIC's Copper Sulfate Process for Removing Hydrogen Sulfide and Other Trace Contaminants from Geothermal Steam at Turbine Inlet Temperature and Pressures," *DOE/RA/27181-01 Final Report,* May.

Pack, G. E., 1986, "Dow Stretford Chemical Recovery Process," *Proceedings of the 1986 Stretford Users' Conference,* Austin, TX, pp. 149–158.

Penderleith, Y., 1977, "Stretford Plant Operating Experience at DOFASCO," *Proc. of the Symposium on Treatment of Coke-Oven Gas,* McMaster University, Hamilton, Ontario, Canada, May 26.

Phillips, J. R., and Fleck, R. N., 1979, "Reaction Kinetics in a Stretford-Process Absorber," paper presented at the American Institute of Chemical Engineers, 72nd Annual Meeting, San Francisco, CA, Nov. 25–29.

Pieters, A. J., and van Krevelen, D. W., 1946, *The Wet Purification of Coal Gas and Similar Gases by the Staatsmijnen-Otto Process,* Amsterdam: Elsevier Publishing Co.

Pippig, H., 1953, *Gas-u. Wasserfach,* Vol. 94, p. 62.

Pirtle, L., 1991, "Startup and Operational Experiences at the Scandia SulFerox Plant," *Proceedings 1991 Liquid Redox Sulfur Recovery Conference,* Austin, TX, May 5–7, p. 231.

Pirtle, L. L., Allen, M. C., Hammond, C. A., and Anderson, K. D., 1994, "SulFerox Design Criteria with Special Considerations for Offshore Applications," paper presented at the 6th Gas Research Institute Sulfur Recovery Conference, Austin, TX, May 15–17.

Powell, A. R., 1936, *Chem. & Met. Eng.,* Vol. 43, July, p. 307.

Price, G. S., Price, B. C., and Hardison, L. C., 1986, "The Use of LO-CAT in the Sable San Andres CO_2 Miscible Flood Project," paper presented at the AIChE Spring Meeting, New Orleans, LA, April 9.

Price, G. S., 1987, "Designing and Operating a LO-CAT® Plant in CO_2 Rich Oil Field Gas Service," *Proceedings of the Sixty-Sixth GPA Annual Convention,* pp. 122–128.

Quinlan, M. P., 1991, "Characteristics & Handling Options for Sulfur from Liquid Redox Processes," *Proceedings 1991 Liquid Redox Sulfur Recovery Conference,* Austin, TX, p. 317.

R. & J. Dempster, Ltd., 1957, personal communication, July 12, Manchester, England.

Reicher, M., 1995, personal communication (Wheelabrator Clean Air Systems), Sept. 22.

Reid, L. S., and Townsend, F. M., 1958, *Oil and Gas J.,* Vol. 56, Oct. 13, p. 120.

Reinhardt, K., 1956, Energietechnik, Vol. 6, No. 10, p. 454.

Renault, P., 1969, U.S. Patent No. 3,441,379.

Riesenfeld, F. C., and Mullowney, J. F., 1959, *Petrol. Refiner,* Vol. 38, May, p. 161.

Roberts, C. B., and Farrar, H. T., 1956, "The treatment of gaseous and liquid effluents attendant in producing viscose cellulose film," *Roy. Soc. Promotion Health J.,* Vol. 76, pp. 36–44.

Rossati, F., 1993, (Le Gaz Integral) personal communication, Jan. 7.

Rothery, E., 1986, "Venturi Absorbers for the Stretford Process," *Proceedings of the 1986 Stretford Users' Conference,* Austin, TX, p. 40.

Ryder, C., and Smith, A. V., 1962, "Application of the Stretford Process to the Removal of Hydrogen Sulphide at High Pressure," *Gas J.,* Dec., p. 348; (also see *Inst. Gas Eng. J.,* 1963, Brit., Vol. 3, No. 6, pp. 283–295).

Sauchelli, V., 1933, "Flotation Sulfur in Agriculture," *Ind. Eng. Chem.*, Vol. 25, April, pp. 363–366.

Schwarzenbach, G., and Fischer, A., 1960, *Helv. Chim. Acta,* Vol. 43, pp. 1365–1390.

Sciamanna, S. F., and Lynn, S., 1988, "An Integrated Process for Simultaneous Desulfurization, Dehydration, and Recovery of Hydrocarbon Liquids from Natural Gas Streams," *Ind. Eng. Chem. Res.*, Vol. 27, No. 3, pp. 500–505.

Smith, C. R., and Mills, B., 1979, "Cost-Effective Improvements to the Holmes-Stretford Process," *Inst. Chem. Eng. Symp. Ser., No. 57,* April, Salford, Brit., published by the Inst. Chem. Eng., Warwickshire, England, pp. U1–U15.

Smith, F., Smith, C., and Daniels, J. D., 1976, "Technical Advances in the Purification and Utilization of Coke Oven Gas," Paper presented at the Southern Section Coke Oven Managers' Assoc. (Brit.), March.

Sperr, F. W., 1926, Gas Age-Record, Vol. 58, p. 73.

Sperr, F. W., 1932, U.S. Patent 1,841,419.

Steppe, R., 1986, "Causes and Prevention of Sulfur Deposition in the Stretford System," *Proceedings of the 1986 Stretford Users' Conference,* Austin, TX, p. 62.

Sundström, O., 1979, "Mercury in Sulphuric Acid," *Sulphur (Brit.),* Jan./Feb., p. 37.

Swaim, C. D., 1972, "Ford, Bacon & Davis Texas—Takahax Desulfurization Process," *Proceedings of the Gas Conditioning Conference,* University of Oklahoma, Norman, OK.

Tallon, J. T., Rittmeyer, R. W., and Maruhnich, E. D., 1984, Analysis of Options for Management of Spent Stretford Solution, Report No. DOE/MC/19392-1721, Dept. of Energy, Washington, D.C., June.

Terres, E., 1953, *Gas-u. Wasserfach,* Vol. 94, No. 9, p. 260.

Terres, E., Buscher, H., and Matroff, G., 1954, *Brennstoff-Chem.,* Vol. 35, No. 9/10, p. 144.

Thau, A., 1932, *Gas World,* Vol. 97, p. 144.

Thompson, R. B., 1980A, U.S. Patent No. 4,189,462, "Catalytic Removal of Hydrogen Sulfide from Gases."

Thompson, R. B., 1980B, U.S. Patent No. 4,218,342, "Composition for Catalytic Removal of Hydrogen Sulfide from Gases."

Thompson, R. J. S, and Nicklin, T., 1964, "Le Procede Stretford," Paper presented at Congress of Association Technique de l'Industrie du Gaz en France.

Tomasi, L., 1992, Giammarco-Vetrocoke, personal communication, July 30.

Townsend, F. M., and Reid L. S., 1958, "Newest Sulfur Recovery Process," *Petrol. Ref.,* November, pp. 263–266.

Townsend, F. M., 1965, U.S. Patent No. 3,170,766.

Townsend, L. G., 1953, "Operation of the Manchester liquid purification plant at Linacre, Liverpool," *Inst. Gas Engrs., Commun.*, p. 429.

Trofe, T. W., Dalyrymple, D. A., and Scheffel, F. A., 1987, *Stretford Process Status and R&D Needs,* Gas Research Institute, Chicago, IL, Contract No. 5083-253-0936.

Trofe, T. W., and DeBerry, D. W., 1991, "Results of Bench Scale Studies on the Formation of Sulfur By-Products in the Stretford Process," *Proceedings 1991 Liquid Redox Sulfur Recovery Conference,* Austin, TX, p. 63.

Ultrasystems Engineers & Contractors, 1989, "The Hiperion Process," Bulletin 5189 P&A, Irvine, CA.

VanKleeck, D., and Morisse-Arnold, D., 1990, "Quality of SulFerox Produced Sulfur," paper presented at the AIChE Summer National Meeting, San Diego, CA, August.

Vancini, C. A., and Althem, H. R., 1985, "Optimization of Peabody-Stretford Process and Operation," paper presented at the 1985 International Symposium on Geothermal Energy, Kailua Kona, HI, Aug. 26–30. (See also *Geothermal Resources Council*, Vol. 9, Pt. 2, Davis, CA, pp. 273–278.)

Vancini, C. A., 1985, "The Peabody-Stretford Process on High-Pressure Gasification Gas," paper presented at the 8th International Coal and Solid Fuel Utilization Conference, Pittsburgh, PA, Nov. 4. (See also *Coal Technol.* (Houston), Vol. 8, No. 5–6, pp. 259–281.)

Vancini, C. A., 1986, "Gas Stream Clean-up for the Stretford Process," *Proceedings of the 1986 Stretford Users' Conference*, Austin, TX, p. 2.

Vancini, C. A., 1988, "Quality and Usability of By-Product Stretford Sulfur from Geothermal Steam Incondensables," paper presented at the Eleventh Annual Energy-Source Technology Conference and Exhibit, New Orleans, LA, Jan. 10–13. (See also *Geothermal Energy Symposium Proceedings*, 1988, publ. by ASME, NY, NY, pp. 391–393.)

Vancini, C. A., and Lari R., 1985, "The Peabody-Stretford Process on Geothermal Incondensable Gas," paper presented at the AIChE Spring National Meeting, Houston, TX, March 27.

Vancini, C. A., 1992, (ABB Environmental) personal communication, July 7.

Vasan, S., 1978, "Holmes-Stretford Process offers economic H_2S removal," *Oil & Gas. J.*, Jan. 2, pp. 78–80.

Vasan, S., 1979, "The Holmes-Stretford Process for Desulfurization of Tail-Gases from Acid-Gas Systems," Paper presented at the Ammonia-from-Coal-Symposium, Muscle Shoals, AL, May 8.

Vasan, S., and Willett, P., 1976, "Economics of Desulfurization of Coal-Gas vs. Flue-Gas Desulfurization," 3rd Symposium on Coal Utilization, Nat. Coal Assoc., Louisville, KY, Oct. 12.

Weber, Bucki, and Hofer, 1988, "Sulfolin-Development of an Oxidative Hydrogen Sulphide Scrubbing Process," *Linde Reports on Science and Technology*, Vol. 44.

Wessels, G. F. S., 1980, "Korrosie in die swawelwaterstof-absorbeerders van die Stretfordproses," *Corrosion and Coatings*, South Africa, October, p. 15.

Wheelabrator, 1994, "ARI LO-CAT II," Gas Processes '94, *Hydro. Process.*, April, p. 71.

Wiewiorowski, T. K., 1969, U.S. Patent No. 3,447,903.

Williams, A. E., Smith, L. O., Koenig, K. A., and Basciani, K. A., 1983, "Design, construction and start-up of a modern coke plant," *Iron and Steel Engineer*, May, pp. 45–53.

Wilson, B. M., and Newell, R.D., 1984, "H_2S Removal by the Stretford Process—Further Development by the British Gas Corporation," paper presented at the National AIChE Meeting, Atlanta, GA, March 13 (see also *Chem. Eng. Prog.*, No. 10, 1984, pp. 40–47).

Yan, T. Y., and Espenscheid, W. F., 1980, "Removal of Thiosulfate/Sulfate from Spent Stretford Solution," Environ. Science and Technol., Vol. 14, No. 6, June, pp. 732–735.

Zwicky, J. F., and Mills B., 1980, "Desulphurization of coke-oven gas by the Holmes-Stretford process," *Iron Steel Eng.*, December, pp. 37–42.

Chapter 10
Control of Nitrogen Oxides

NO_x FORMATION MECHANISMS, 868

REQUIREMENTS FOR NO_x CONTROL, 868
    U.S. Regulations, 868
    Japanese Regulations, 876
    European Regulations, 876

PROCESS CATEGORIES AND COSTS, 878

PRE-COMBUSTION NO_x CONTROL PROCESSES, 879

COMBUSTION NO_x CONTROL PROCESSES, 880
    Low-Excess Air Firing, 882
    Burners Out of Service and Fuel Biasing, 882
    Overfire Air, 883
    Flue Gas Recirculation, 884
    Low-NO_x Burners, 884
    Fuel Reburning, 885
    Water/Steam Injection, 885
    Dry Low-NO_x Combustors, 886

POST-COMBUSTION NO_x CONTROL PROCESSES, 887
    Selective Non-Catalytic Reduction (SNCR) Processes, 888
    Selective Catalytic Reduction (SCR) Processes, 904
    Combined NO_x/SO_2 Post-Combustion Processes, 928

REFERENCES, 936

Efforts to reduce nitrogen oxides (NO_x) air emissions have been underway for many years because of the detrimental effects of NO_x on health and the environment. NO_x denotes the two nitrogen oxide compounds usually associated with air pollution: nitric oxide (NO) and nitrogen dioxide (NO_2). NO converts to NO_2, a yellowish-brown gas, in the atmosphere and in the presence of light. The direct health effects of excessive NO_2 concentrations in the air include bronchitis and pneumonia. In addition, NO_2 and volatile organic compounds (VOCs) in the lower atmosphere interact to form ozone, which causes eye irritation, crop damage, and the breakdown of rubber and plastics (Mark et al., 1978). NO_2 is also an acid rain precursor, leading to damage to trees, lakes, and buildings. VOC is the more direct ozone precursor and, initially, VOC controls were emphasized with the establishment of requirements for large- and mid-sized sources. The remaining VOC sources are small and numerous, so it is very difficult to achieve large reductions in their emissions. Therefore, NO_x emissions have been targeted for control where ozone concentrations are a concern.

However, recent studies have shown that a moderate reduction in NO_x concentration may actually increase the ozone concentration, and a large NO_x reduction may be necessary to reduce ozone significantly. Maximum ozone concentrations do not normally occur near the NO_x emission sources due to the time it takes for the reactions to proceed; in fact, these concentrations may occur in adjacent states (Wolff, 1994).

Fossil fuel combustion also produces small amounts of nitrous oxide (N_2O) in addition to NO_x, and this source accounts for about 1 to 3% of the global emissions of N_2O. The percentage is uncertain due to lack of balance between estimated total global N_2O emissions and N_2O buildup and destruction rates in the atmosphere (Hofmann et al., 1993; Levine, 1991). Other large sources of nitrous oxide emissions to the atmosphere are the soils of some forests, oceans, biomass burning, and fertilization. In the atmosphere, N_2O acts as a greenhouse gas as do carbon dioxide, methane, and water vapor. N_2O is also believed to contribute to the chemical destruction of stratospheric ozone (Turco, 1985). N_2O is not a contributor to tropospheric ozone or acid rain and is not currently a regulated species in the U.S. (Elkins, 1990). However, it is regulated in other countries, and U.S. regulators may consider it as NO_x in future control requirements.

Man-made NO_x emissions are produced from both mobile and stationary sources. Stationary sources, which account for approximately 53% of total NO_x emissions, can be broken down as follows:

Power Stations	53%
Internal Combustion Engines	20
Industrial Boilers	14
Process Heaters	5
Combustion Turbines	2
Other	6
Total	100%

NO_x emissions from power stations vary with the type of boiler as well as with other factors such as fuel type, boiler heat release rate, excess air, and mode of operation (e.g., cyclic vs. base load). Coal-fired boilers typically produce more NO_x than oil- or gas-fired units. Wall-fired wet bottom and cyclone-fired boilers usually have the highest NO_x emissions, while tangentially-fired and arch-fired boilers have the lowest.

In internal combustion engines, the lowering of excess air to reduce NO_x emissions increases CO and hydrocarbon emissions. In combustion turbine installations, reduction of NO_x by injection of steam or water into the combustor increases CO emissions.

NO_x FORMATION MECHANISMS

There are two main types of NO_x formed during combustion: fuel NO_x and thermal NO_x. NO_x formed by the oxidation of nitrogen bound in the fuel is called "fuel NO_x." NO_x formed by the fixation of nitrogen in the combustion air at the high temperatures associated with combustion is called "thermal NO_x." Most of the NO_x formed in combustion processes is NO.

The type of NO_x formed varies with the fuel. Natural gas produces mostly thermal NO_x since this fuel is very low in nitrogen. Coal, on the other hand, produces mostly fuel NO_x, resulting from nitrogen in the coal.

When coal is combusted there are two sources of fuel NO_x: volatile nitrogen (nitrogen that evolves during fuel devolatilization) and char nitrogen (nitrogen that remains with the residual solid matter following devolatilization). During the initial heating of the coal, devolatilization occurs quickly (within a few microseconds for pulverized coal). Oxygen concentration is the primary factor that controls NO_x during the devolatilization process, and reducing conditions during devolatilization favor the formation of N_2 instead of NO_x. The residual char left after the initial devolatilization steps burns more slowly than the volatiles, and the corresponding slow rate of char nitrogen conversion to NO is difficult to control; however, less char nitrogen is converted than volatile nitrogen (Booher, 1993).

Fuel NO_x is most effectively controlled by reducing the oxygen in the flame zone early in the combustion process. The formation of fuel NO_x depends on the following factors:

- Nitrogen content of the fuel
- Excess air in the flame zone
- Primary/secondary air ratio
- Ratio of fixed carbon to volatile matter in fuel

Thermal NO_x formation becomes significant at around 2,700°F and increases exponentially with temperature from that point. Thermal NO_x is most effectively controlled by reducing temperatures in the combustion zone below 2,700°F as quickly as possible (Booher, 1993). The formation of thermal NO_x depends on the following factors:

- Amount of oxygen available
- Temperature profile in the combustion zone
- Pressure (with some types of gas turbine combustors and some other pressurized systems)
- Residence time in the combustion zone

Fundamental research into NO_x formation was performed by Zel'dovich et al. (1947), who developed equations governing the formation of NO_x. Since then many have studied the subject, including Boardman and Smoot (1989), Booher (1993), Bozzuto (1991), Burch et al. (1991), Haussmann and Kruger (1989), Knill and Morgan (1989), Lightly et al. (1989), Shaw (1973), Toqan et al. (1989), and Wendt et al. (1989).

REQUIREMENTS FOR NO_x CONTROL

U.S. Regulations

The effort to reduce emissions of pollutants from mobile and stationary sources has been a subject of expanded research and investigation since the passage of the first Clean Air Act in

1963 and the formation of the U.S. Environmental Protection Agency (EPA). Since then, electric utilities and industrial sources have been faced with increasingly more stringent air quality emission control requirements for both new and existing fossil fuel power plants. The existing legislation allows the EPA to continue to escalate requirements for some time to come. Where NO_x or ozone exceeds standards (non-attainment areas), the law requires the EPA to escalate NO_x control until attainment is achieved. Thus, NO_x control will require utility and industrial attention for at least the next 20 years.

The U.S. initiated the development of NO_x control technology, but followed the lead of Japan and Germany in adopting strict NO_x control measures. The U.S. program provides for a much more gradual implementation than did the programs in these countries. Title I of the Clean Air Act Amendments of 1990 requires state agencies to implement programs to improve the air quality in those regions where the air quality fails to meet the stated standards (Mansour et al., 1991). Excessive NO_x concentrations and ground level ozone (which results in part from NO_x emissions) are covered in Title I. Title IV places restrictions on units with high NO_x emissions. **Table 10-1** gives an overview of the regulations and their applicability for both existing and new boilers prior to November 23, 1994.

The 1995 compliance deadline for the Phase I boilers (called Group I boilers—second column in **Table 10-1**) has been delayed by court action. On November 23, 1994, in *Alabama Power Company et al. v. U.S. EPA,* a District of Columbia Circuit Court of Appeals held that the EPA had improperly interpreted the term "low NO_x burner technology" when promulgating the Title IV Phase I Group I (1995 compliance date) utility NO_x emission limits. As a result, the EPA must repropose the regulations. The EPA had defined "low NO_x burner technology" to include overfire air. The court ruled that the term encompassed only low NO_x burners, not overfire air.

The ruling could also affect the Phase I and Phase II boilers that must comply by the year 2000 (called Group II boilers—last two columns in **Table 10-1**) since, according to the Clean Air Act Amendments of 1990, NO_x emission limits for these boilers should result in costs comparable to the Phase I Group I NO_x control costs.

Regulations are often written in terms of thermal megawatts consumed, MW_t, because of the variations in plant efficiency. In a typical coal-fired power plant, the efficiency of converting heat to electricity is 33 to 35.5% making the net electric megawatts produced, MW_e, approximately one-third of the thermal megawatts consumed. In co-generation plants, part of the steam produced is used for purposes other than generating electricity, or electricity is generated by a gas turbine as well as a steam turbine with consequent higher heat utilization, causing the relationship between MW_e and MW_t to vary significantly. In a pulverized coal-fired plant, for comparative purposes, one MW_e corresponds to about 2,400 scfm of flue gas under normal operating conditions for many coals, but the flue gas volumetric flow, on a MW_e basis, can be 10% higher for coals such as those from the Powder River Basin. Actual gas volumes are affected by the boiler exit gas temperature, which typically ranges between 220° and 350°F for many coal-fired plants, and the pressure, which varies due to the altitude of the plant and other factors.

Existing Plants in NO_x or Ozone Non-Attainment Areas

Many of the combustion power plants in the U.S. will be affected by NO_x and ozone air quality standards. The new regulations for ozone non-attainment areas will affect virtually all existing power plants located in urban areas, particularly in Southern California's South Coast Air Quality Management District (SCAQMD) and throughout the Northeast. The

(text continued on page 872)

Table 10-1
Overview of NO$_x$ Rules, Affected Plants, and Controls

	1992 Rulemaking 1995 Compliance (Delayed) Acid Rain Phase I Dry-Bottom Boilers	1993 Rulemaking 1993–2010 Compliance Existing Boilers in Ozone Non-Attainment Area	1993 Rulemaking 1993 & Beyond Compliance All New Units and Existing Units Classified as New Units	1997 Rulemaking 2000 Compliance Acid Rain Phase I Wet-Bottom Boilers	1997 Rulemaking 2000 Compliance Acid Rain Phase II Dry- & Wet-Bottom Boilers
Approximate numbers and types of affected boilers	179 Coal-Fired Boilers: 94 Tangentially-Fired, 85 Wall-Fired	About 1,000 Coal-, Oil-, & Gas-Fired Boilers of All Types	All new units and existing units classified as new units due to major life extension modifications (i.e., "WEPCO" court order) and repowering.	77 Coal-Fired Boilers: 43 Cyclones, 18 Cell-Burners, 16 Wall-Fired	801 Coal-Fired Boilers: All Types-698 Dry-Bottom 103 Wet-Bottom
Size of affected boilers	60–950 MW$_e$	All sizes	All sizes	115–1,300 MW$_e$	Over 25 MW$_e$
Vintage of affected boilers	1950–1979	All ages	New	1953–1977	All ages
Number of states affected	19	37	All	15	43
NO$_x$ control regulatory requirements	Tangentially-Fired: 0.45 lb NO$_x$/10^6 Btu Wall-Fired: 0.50 lb NO$_x$/10^6 Btu If these limits cannot be met with low-NO$_x$ burners (LNB) and overfire air (OFA), alternative emissions limits (AEL) can be obtained provided several requirements are satisfied.	To be determined by SIP's; Refer to **Tables 10-2** and **10-3** for regulatory requirements in ozone non-attainment areas.	All new units: New Source Review—New NSPS & SIP. Refer to **Tables 10-4a, b,** and **c** for NSPS requirements. Non-Attainment Areas: LAER & Emission Offsets	Limits are to be defined and result in costs comparable to those for Phase I dry-bottom boilers. This could mean a different limit for each type of boiler. Tangentially-Fired: 0.38 lb NO$_x$/10^6 Btu (EPA estimate) Wall-Fired: 0.43 lb NO$_x$/10^6 Btu (EPA estimate) Early compliance by Phase II boilers is encouraged through a voluntary grandfathering procedure. Phase II boilers can meet Phase I limits up until eight years after the 2000 compliance deadline. This allows utilities the flexibility of using planned outages for hardware

Table 10-1 (Continued)
Overview of NO$_x$ Rules, Affected Plants, and Controls

NO$_x$ control regulatory requirements (continued)	For AELs, a 15-month extension is possible if the applicable control technology is not in adequate supply. Bubbling (or averaging of emissions from multiple units) is allowed. Phase I units that comply with their SO$_2$ limits by using scrubbers do not have to comply with their NO$_x$ limits until 1997.	However, regional studies indicate that moderate reductions in NO$_x$ concentrations may increase ozone concentrations and that large NO$_x$ reductions may be necessary to reduce ozone significantly. This could affect both the probable control limits and the acceptable control technologies given in **Tables 10-2** and **10-3**. The direction of the change is indeterminate.	Attainment Areas: Existing PSD & BACT. Requirements are becoming more strict. changes. Boilers can be opted out of the early election program or can be terminated from it after a year of noncompliance, but can't be reinstated. Bubbling (or averaging of emissions from multiple units) is allowed.
NO$_x$ control technology	LNB on all boilers; OFA on all tangentially-fired boilers; OFA on wall-fired boilers only if needed to meet the limits; SNCR, etc. is permissible if reduction from baseline emissions is 65% or greater, but no control technology is necessary if these limits can be met without them.	Refer to **Table 10-2** for probable measures in ozone non-attainment areas.	Non-Attainment Areas: LAER (i.e., SCR & SNCR) & Emission Offsets Attainment Areas: BACT Probably combustion controls, possibly SNCR for stoker-fired boilers, but probably not SCR for any boilers unless the bubble concept is employed. Using the bubble concept, very stringent controls could be employed on some boilers to avoid any controls on others.

Note: BACT stands for Best Available Control Technology, LAER for Lowest Achievable Emission Rate, NSPS for New Source Performance Standard, PSD for Prevention of Significant Deterioration, SCR for Selective Catalytic Reduction, SIP for State Implementation Plan, SNCR for Selective Non-Catalytic Reduction, and WEPCO for Wisconsin Electric Power Company.
Source: Kantor and Siegfriedt (1992) and Blaszak (1994)

(text continued from page 869)

SCAQMD is the only NO_x non-attainment area in the U.S. Virtually all boilers in the SCAQMD are affected and must reduce NO_x by an average of 90%. Selective Catalytic Reduction (SCR) has already been retrofitted on over 4,000 MW_e of boiler capacity in the SCAQMD. In the Northeast, NESCAUM (Northeast States for Coordinated Air Use Management), which is made up of representatives from eight northeastern states: Connecticut, Maine, Massachusetts, New Hampshire, New Jersey, New York, Rhode Island, and Vermont, was formed to solve the non-attainment problems of this ozone transport area and has proposed Reasonably Available Control Technology (RACT) values and other rules stricter than those of the EPA, including SCR for NO_x control.

Title I has changed the basis for setting NO_x emission limits to RACT. RACT requirements are defined by each state with consideration of the EPA Guidance Documents, but a reasonable estimate of RACT requirements is given in **Table 10-3**. The U.S. EPA requires each state to develop a SIP (State Implementation Plan) to implement the requirements of the law. The new operating permits required by Title V of the amendments for existing plants, issued under the SIP, will define the RACT NO_x limit for each source. While this affects plants in all areas, it has a special impact on those in ozone non-attainment areas. For those

Table 10-2
Probable Measures in Ozone Non-Attainment Areas

Ozone Area Class	Attainment Date	Major Source Size, tpy	Probable Measures for Stationary Sources
Marginal	November, 1993	≥100	RACT[1] for existing control technology guidelines (CTG) categories through November 15, 1990.
Moderate	November, 1996	≥100	RACT[1] for existing and new CTG categories through 1996. Case-specific RACT for major sources.
Serious	November, 1999	≥50	RACT[1] for existing and new CTG categories through 1999. Case-specific RACT for major sources.
Severe	November, 2005 or 2007	≥25	RACT[1] for existing and new CTG categories through 2005. Case-specific RACT for major sources. BACT for modifications of ≥100 tpy sources. (SNCR and/or SCR will likely be required for retrofit for some sources.)
Extreme	November, 2010	≥10	RACT[1] for existing and new CTG categories through 2010. Case-specific RACT for major sources. LAER for modifications of ≥100 tpy sources. (SNCR and SCR will be required for retrofit on utility units and some other sources.)

Note 1: NO_x control techniques to meet federally proposed RACT limits for boilers are given in Table 10-3.
Source: Blaszak (1992)

Table 10-3
Federally Proposed NO$_x$ RACT for Boilers[1]

Firing Mode	Proposed Limit[2] lb/10^6 Btu	Assumed NO$_x$ Control Technologies
Pulverized Coal, Dry Bottom, Tangentially-Fired	0.45	Burners, overfire air
Pulverized Coal, Dry Bottom, Wall-Fired	0.50	Burners
All Other Coal-Fired	0.70	Technology driven. Refer to Note 3.
Oil/Gas, Tangentially-Fired	0.20	Burners, overfire air
Oil/Gas, Wall-Fired	0.30	Burners
All Other Oil/Gas-Fired	0.55	Technology driven. Refer to Note 3.

Notes:
1. *This table represents only one proposal. It is not clear that this is likely to be the rule. In fact, the rule may defer to regional or state agencies to set limits.*
2. *For comparison, 1 lb NO$_x$/10^6 Btu is approximately equivalent to 1,230 mg/Nm$_3$ or 598 ppm NO$_x$ for coal at 6% O$_2$, 1,540 mg/Nm$_3$ or 748 ppm NO$_x$ for oil at 3% O$_2$, and 1,590 mg/Nm$_3$ or 775 ppm NO$_x$ for gas at 3% O$_2$. Conversions are approximate since they depend on the chemical analysis of the fuel and operating variables. For all fuels, 1 lb NO$_x$/10^6 Btu is exactly equal to 430 nanograms/J.*
3. *Technology for this group does not yet exist or is developing. The framers of the regulations are "forcing," i.e., causing technology to be developed to meet these regulations.*

Source: Fotis (1992)

units for which RACT technology is emerging, the permit may stipulate installation of the NO$_x$ control equipment at a later date (Public Law 95–95, 1977). Further, RACT compliance today is no assurance of RACT compliance in the future.

New Plants

The Clean Air Act requires that new facilities meet a criterion called New Source Performance Standards (NSPS) for combustion sources, regardless of the plant's location (Public Law 101-549, 1990). NSPS set minimum national requirements for new NO$_x$ emitting combustion sources. **Tables 10-4a, b,** and **c** present NO$_x$ emission NSPS (current as of August 1995) for various combustion source types and sizes. Modern combustion sources usually meet these standards easily. Title IV of the Clean Air Act Amendments of 1990 requires the EPA to promulgate revision of the NSPS for NO$_x$. The EPA is likely to issue new, more stringent NSPS based on the improved control technologies. Title I of the 1990 amendments sets a maximum NO$_x$ concentration for breathable air quality. In areas that attain the NO$_x$ standard, an analysis must be performed to satisfy Prevention of Significant Deterioration (PSD) of the air quality rules. PSD does not involve a set emission concentration or limit or a set reduction, but requires use of the Best Available Control Technology (BACT). The required process for determining the technology is a "Top Down" BACT analysis, in which the applicant must agree to use the top (most effective) emission control technology that cannot be shown to be technically, economically, or environmentally infeasible. Recent permits

Table 10-4a
NSPS for Electric Utility Steam Generating Units Greater than 250×10^6 Btu/h
(Current as of August 1995)[5]

Fuel Type	NO_x Emission Limit in Terms of Heat Input[1] lb/10^6 Btu (ng/J)[4]
Gaseous fuels:	
Coal-derived fuels	0.50 (215)
All other fuels	0.20 (86)
Liquid fuels:	
Coal-derived fuels	0.50 (215)
Shale Oil	0.50 (215)
All other fuels	0.30 (129)
Solid fuels:	
Coal-derived fuels	0.50 (215)
Any fuel containing more than 25% by weight coal refuse	Refer to Note 2.
Any fuel containing more than 25% by weight lignite if the lignite is mined in North Dakota, South Dakota, or Montana, and is combusted in a slag tap furnace. (Refer to note 3)	0.80 (344)
Any fuel containing more than 25% by weight lignite not subject to the 340 ng/J heat input emission limit. (Refer to note 3)	—
Subbituminous coal	0.50 (215)
Bituminous coal	0.60 (258)
Anthracite coal	0.60 (258)
All other fuels	0.60 (258)

Notes:
1. *NO_x is expressed as NO_2.*
2. *Exempt from NO_x standards and NO_x monitoring requirements.*
3. *Any fuel containing less than 25% by weight lignite is not prorated but its percentage is added to the percentage of the predominant fuel.*
4. *Conversion is 1 lb/10^6 Btu = 430 ng/J (nanograms per joule).*
5. *CFR, Title 40, Part 60, Subparts D & Da, 1992 (CFR, 1992).*
Source: CFR (1992)

resulting from PSD analyses have required low-NO_x burners to produce less than 0.50 lb NO_x per 10^6 Btu fired (approximately 615 mg/Nm^3, burning coal at 6% O_2). It should be noted that in most areas where new power plants are located, more stringent local and state rules will also apply because most urban areas fail to attain the ozone standard.

In areas not meeting NO_x or ozone air quality standards, new plants are subject to a new source review and must use Lowest Achievable Emission Rate (LAER) emissions controls. Cost of the controls is not a consideration with LAER. This means post-combustion controls,

Table 10-4b
NSPS for Industrial-Commercial-Institutional Steam Generating Units between 100 and 250 × 10^6 Btu/h
(Current as of August 1995)[2]

Fuel/Steam Generating Unit Type	NO_x Emission Limit in Terms of Heat Input[1] lb/10^6 Btu/(ng/J)
(1) Natural gas and distillate oil, except (4):	
(i) Low heat release rate	0.10 (43)
(ii) High heat release rate	0.20 (86)
(2) Residual oil:	
(i) Low heat release rate	0.30 (129)
(ii) High heat release rate	0.40 (172)
(3) Coal:	
(i) Mass-feed stoker	0.50 (215)
(ii) Spreader stoker and fluidized bed combustion	0.60 (258)
(iii) Pulverized coal	0.70 (301)
(iv) Lignite, except (v)	0.60 (258)
(v) Lignite mined in North Dakota, South Dakota, or Montana and combusted in a slag tap furnace	0.80 (344)
(vi) Coal-derived synthetic fuels	0.50 (215)
(4) Duct burner used in a combined cycle system:	
(i) Natural gas and distillate oil	0.20 (86)
(ii) Residual oil	0.40 (172)

Notes:
1. NO_x is expressed as NO_2.
2. CFR, Title 40, Part 60, Subpart Db, 1992 (CFR, 1992).

such as SCR or possibly Selective Non-Catalytic Reduction (SNCR), will likely be required in a growing number of cases. New plants in these areas will also be required to obtain offsets of their NO_x emissions from existing sources (Public Law 101-549, 1990). Offsets are reductions in NO_x from other sources in the area that are at least equal to the total potential NO_x emissions from the proposed facility.

In states such as California and New Jersey, which have severe pollution problems, BACT (applicable in attainment areas) has been interpreted as either SCR or SNCR for certain applications. In other states that do not have a severe pollution problem, combustion control technologies have been acceptable. New sources must undergo a case-by-case review, and this has resulted in a trend toward lower NO_x emission limits. Technology limits are not yet fully defined.

Other Pending Provisions

Inter-pollutant trading would be very complex and is not yet attractive to owners or to the EPA. However, this concept is being studied by the EPA. National market-based NO_x emission

Table 10-4c
NSPS for Combustion Turbines (Current as of August 1995)[5]

Combustion Turbine Size and Use	NO_x Emission Limit[1] (ppm)
Electric utility stationary gas turbines with heat input greater than 100×10^6 Btu/h[2, 3]	75[4]
Stationary gas turbines with heat input between 10×10^8 and 100×10^6 Btu/h (about 1,000 hp to 10,000 hp)[2, 3]	150[4]
Stationary gas turbines with a manufacturer's rated base load at ISO conditions of 30 megawatts or less except electric utility stationary gas turbines[2, 3]	150[4]
Stationary gas turbines with heat input less than 10×10^6 Btu/h; those using water or steam for NO_x control when ice fog is a traffic hazard; those used for emergency, military facilities (except garrison facilities), military training facilities, fire fighting, research and development (on a case-by-case basis); those located where (and when) mandatory governmental water restrictions apply, certain ones covered by other regulations, and those with heat input greater than 10×10^6 Btu/h and fired by natural gas when firing emergency fuel, and regenerative cycle units with heat input less than or equal to 100×10^6 Btu/h	None

Notes:
1. NO_x is expressed as NO_2.
2. Based on lower heating value of the fuel fired.
3. At 15% oxygen on a dry basis.
4. Adjustments to the NO_x emission limits are allowed for unit efficiency and fuel-bound nitrogen.
5. CFR, Title 40, Part 60, Subpart GG, 1992 (CFR, 1992).

allowances are still only a concept, but could be adopted if SO_2 allowances are traded successfully. This is the only trading concept that the EPA is considering seriously at this time.

Japanese Regulations

Government regulations for NO_x emissions were first issued in Japan in 1973 (Yamamura and Suyama, 1988). For new large plants, regulations have become increasingly stringent and appear to have stabilized at the current limits as summarized in **Table 10-5**.

European Regulations

In the 1980s, there were significant developments in European NO_x control legislation and policies. The impetus was concern about the effects of NO_x emissions on forests. Although vehicles produce the majority of NO_x emissions, stationary sources were the main targets of legislation.

Table 10-5
NO$_x$ Regulations in Japan

Plant Type	Limit (ppm)
Coal	200 @ 6% O$_2$
Oil	130 @ 4% O$_2$
Gas	60 @ 5% O$_2$
Diesel Engine Cogeneration	950 (Refer to Note 1)
Gas Turbine Cogeneration	70 (Refer to Note 2)

Notes:
1. Typical limit; actual value can be as low as 50 ppm in some areas.
2. Typical limit; limits set by local regulations could be lower.
Source: Yamamura and Suyama (1988)

Emission standards for NO$_x$ in Europe vary a great deal from country to country. In addition to national limits, local regulations may be applied in many countries. Standards have been set for various fuels such as coal and other solid fuels, oil, natural gas, and municipal waste.

Tables 10-6 and **10-7**, respectively, list the plant capacities covered by NO$_x$ regulations and NO$_x$ emission standards for coal-fired plants in Europe. Most countries with coal stan-

Table 10-6
Plant Capacities Covered by NO$_x$ Control Regulations in Europe

European Country	New Plants Current Regulations	Existing Plants Current Regulations	Planned Regulations
Austria	>50 MW$_t$		>50 MW$_t$
Belgium	>50 MW$_t$		
Denmark	>50 MW$_t$		
Federal Republic of Germany	>1 MW$_t$	>50 MW$_t$ > MW$_t$ (Refer to Note 1)	
Italy	>100 MW$_t$	>400 MW$_t$	
Netherlands	(Refer to Note 2)	(Refer to Note 2)	
Sweden	(Refer to Note 2)		(Refer to Note 2)
Switzerland	>1 MW$_t$	>1 MW$_t$	
United Kingdom	>700 MW$_t$		
European Community	>50 MW$_t$		

Notes:
1. The "greater than 1 MW$_t$" requirement applies only to circulating fluidized bed type boilers.
2. No minimum plant size is stated in the regulation.
3. MW$_t$ stands for thermal megawatts. Electrical megawatts (MW$_e$) are about ⅓ of MW$_t$ values.
Source: Hjalmarsson and Vernon (1989)

Table 10-7
Range of NO$_x$ Emission Standards for Coal Combustion Plants in Europe

European Country	New Plants		Existing Plants	
	Current Regulations mg/m^3 (mg/MJ)	Planned Regulations mg/m^3	Current Regulations mg/m^3	Planned Regulations mg/m^3 (mg/MJ)
Austria	800			200–600 from 1994
Belgium	650–800	200–400 from 1996		
Denmark[2]	1,150 (400			
Federal Republic of Germany	200–500		200–1,300	
Italy	650		1,200	
Netherlands	400–800		1,100	
Sweden[2]	140 (50)			140–280 (50–200) from 1995
Switzerland	200–300		200–500	
United Kingdom	670 (330)			
European Community	650–1,300			

Notes:
1. The standards in mg/m^3 are based on a temperature of 0°C, a pressure of 1 atmosphere, and 6% O$_2$ in dry flue gas, except for some types of combustion technology in the Federal Republic of Germany where O$_2$ is between 5 and 7%.
2. Danish and Swedish regulations are in terms of mg/MJ. Conversions are approximate.
Source: Hjalmarsson and Vernon (1989)

dards also have standards for gas- and oil-fired plants. These standards are summarized in **Table 10-8** (Hjalmarsson and Vernon, 1989).

Germany and Austria are the only countries with national NO$_x$ emission limits for waste incinerators. The limit in Germany is 500 mg/m^3 for all plants (at 11% O$_2$ for feed rates >750 kg waste/h, 17% O$_2$ for feed rates <750 kg waste/h), and in Austria the limit is 100 mg/m^3 for plants larger than 750 kg waste/h (Hjalmarsson and Vernon, 1989).

PROCESS CATEGORIES AND COSTS

NO$_x$ control technologies for combustion sources can be conveniently divided into three main categories:

- Pre-combustion
- Combustion
- Post-combustion

Costs for the various NO$_x$ reduction techniques are given in several tables in this chapter. Two computer programs are available for estimating the costs of NO$_x$ control techniques. These are EPA's Integrated Air Pollution Control System (IAPCS) model for coal-fired

Table 10-8
Range of NO$_x$ Emission Standards for Oil and Natural Gas Combustion Plants in Europe

European Country	Oil-Fired Plants		Natural Gas-Fired Plants	
	New Plants mg/m³ (mg/MJ)	Existing Plants mg/m³ (mg/MJ)	New Plants mg/m³ (mg/MJ)	Existing Plants mg/m³ (mg/MJ)
Austria	450	150–450	350	150–300
Belgium	450, but 150 after 1995		350, but 100 after 1995	
Denmark	—	—	—	—
Federal Republic of Germany	250–450	250–700	200–350	200–500
Italy	650	1,200		
Netherlands	300	700	200	500
Sweden[2]	140–560 (50–200)	140–560 (50–200) from 1995	140–560 (50–200)	140–560 (50–200)
Switzerland	150–450		100–200	
United Kingdom	564			
European Community	450		350	

Notes:
1. The standards in mg/m³ are based on standard temperature and pressure at 3% O$_2$.
2. Swedish regulations are based on mg/MJ. Conversions are approximate.
Source: Hjalmarsson and Vernon (1989)

power plants and the Electric Power Research Institute's CAT Compliance Planning Workstation (Kaplan, 1993; Sopocy et al., 1991). The Integrated Air Pollution Control System model is useful in examining the cost sensitivity of a technology to a number of variables. The CAT workstation can be used to assess a multitude of compliance options and scenarios, especially for multi-unit systems. It uses best available correlations of NO$_x$ production with fuel, boiler/burner type, and other combustion parameters. It can estimate the emissions from individual boilers, plants, and utility systems, identify NO$_x$ controls to meet emission reduction targets, and estimate the cost of NO$_x$ reduction retrofits.

PRE-COMBUSTION NO$_x$ CONTROL PROCESSES

Pre-combustion NO$_x$ control is aimed at reducing the formation of fuel NO$_x$ and consists primarily of switching to a fuel with lower nitrogen content. Gas combustion generates less NO$_x$ than oil, and oil less than coal.

Virtually all experts concede that, unlike sulfur, it is impractical to remove nitrogen from fuels to reduce NO$_x$ emissions and no process for removing fuel bound nitrogen is on the commercial horizon. With coal, factors such as coal chemistry, including volatile species, oxygen, and moisture content, appear to control the formation of NO$_x$ more than the nitrogen content. For all fuels, NO$_x$ control focuses on minimizing NO$_x$ formation during the combustion process and/or reducing its concentration in the flue gas in a downstream treatment process (Makansi, 1991).

COMBUSTION NO$_x$ CONTROL PROCESSES

The formation of *thermal* NO$_x$ in a combustion zone can be reduced by decreasing (1) available oxygen, (2) peak temperature, and/or (3) time at high-temperature, high-oxygen conditions. The formation of *fuel* NO$_x$ is relatively independent of temperature and occurs only when the fuel contains nitrogen compounds. Fuel NO$_x$ formation can be decreased by providing reducing conditions and sufficient residence time in the initial combustion zone to assure that N$_2$ rather than NO$_x$ is the final product of nitrogen compound oxidation. Unfortunately, the conditions that favor low thermal and fuel NO$_x$ formation are not optimal for high fuel combustion efficiency, so compromises are necessary in the design of combustion NO$_x$ control processes.

Commercial combustion NO$_x$ control technologies utilize one or more of the above basic principles in the following concepts:

1. Fuel staging
2. Combustion air staging
3. Total combustion air reduction
4. Flame quenching

Fuel staging consists of establishing two distinct combustion zones; part of the fuel is mixed with all of the air in the first stage, which operates at a reduced temperature because of the large amount of excess air, and the balance of the fuel is added in the second stage. The second stage also operates at a reduced temperature (compared to single stage combustion) because of heat loss from the first stage and the dilution effects of inerts and combustion products in the gas from the first stage. Fuel reburning represents a special case of fuel staging in which auxiliary fuel is added downstream of the main combustion zone in the furnace.

Combustion air staging also involves the use of two combustion zones; however, it differs from fuel staging in that the first stage operates under substoichiometric rather than excess air conditions. In combustion air staging, only a portion of the required air (and all of the fuel) is added to the first stage to maintain a relatively low temperature and reducing conditions. The balance of the air is introduced further in the furnace to complete the combustion (Kuhr et al., 1988; Casiello, 1991). Overfire Air (OFA), Burners Out of Service (BOOS), and fuel biasing utilize the basic concept of combustion air staging.

Both fuel staging and combustion air staging may be accomplished by special burner designs or by modifying the operation of the boiler furnace. Total combustion air reduction is an operating technique in which combustion is accomplished with very little excess air. This reduces NO$_x$ formation by reducing the amount of oxygen available to react with nitrogen. Flame quenching consists of injecting either steam or water (for combustion turbine applications) into the combustion zone. These thermal diluents reduce the flame temperature and thus the formation of thermal NO$_x$.

NO$_x$ combustion control technologies using these concepts are identified in **Table 10-9**. The development status of these technologies as well as general performance and cost information are also summarized. The NO$_x$ reductions indicated in **Table 10-9** are representative only. The reported, potential NO$_x$ reductions with each technique vary widely depending on the source of the data. Using these technologies either alone or in combination, the average

Table 10-9
Features of Combustion NO$_x$ Control Technologies

Technology	Developmental Status	Approximate NO$_x$ Reduction, %	Retrofit Cost, $/kW
Boilers:			
Low-Excess Air Firing (LEA)	This marginal technique has been widely used for more than a decade.	10–30	0–10
Burners Out of Service (BOOS)	This technique has been successful on gas-fired units, but has had poor results on coal.	10–50	0–10
Fuel Biasing	Results have been similar to BOOS.	10–30	0–10
Overfire Air (OFA)	In use for more than 30 years. Very effective in combination with sub-stoichiometric combustion techniques.	10–30	5–15
Flue Gas Recirculation (FGR)	This technique is a viable technique for natural gas. It was ineffective with coal and was withdrawn from the market.	20–60	4–12
Low-NO$_x$ Burners (LNB)	Technology is in commercial operation.	20–65	6–25+
Fuel Reburning	Reburning with gas (in gas- or coal-fired units) has shown some promise, but is not yet widely commercial. Reburning with coal requires demonstration of coal micronizing technology except on cyclone boilers where 40–60% reductions have been demonstrated (McKinney, 1993).	40–60	30–50
Combustion Turbines:			
Water/Steam Injection[1]	Commercially available.	60–80	Note 2
Dry Low-NO$_x$ Combustors[1]	Commercially available from some manufacturers.	>90	Note 2

Notes:
1. Applicable to combustion turbines only.
2. Users have not been required to retrofit combustion turbines with these technologies.

Source: Pacer (1992), Brusger et al. (1990), Katzberger and Sloat (1990), Pachello (1993), Makansi (1988), and Campobenedetto (1994)

boiler can achieve 10 to 65% NO_x reduction, while combustion turbines can achieve greater than 60%. Depending on the fuel fired and the burner design, possible impacts include increased CO, particulate, and unburned hydrocarbons emissions; loss of combustion efficiency (i.e., increased carbon in the flyash); increased slagging, fouling, and corrosion; changes in furnace heat-release rates; unstable or even unsafe furnace operation; reduced turndown capability; added complexity of controls; and increased maintenance. **Table 10-10** indicates the technologies that are applicable to various types of coal-fired boilers.

Table 10-10
Combustion NO_x Control Technologies for Coal-Fired Boilers

Firing Mode	NO_x Technology
Pulverized Coal: Wall-Fired Dry Bottom	LNB, OFA
Pulverized Coal: Tangentially-Fired Dry Bottom	LNB, OFA
Pulverized Coal: Wet Bottom	Fuel reburning
Pulverized Coal: Cell Burners	Low-NO_x cell burners or conventional LNB, with or without OFA. Use of inverted OFA/LNB cells.
Pulverized Coal: Burners in Arch Arrangement	New burner design, OFA, fuel reburning
Cyclone	Fuel reburning
Stoker	Fuel reburning, additional OFA

Source: Pacer (1992)

Table 10-11 indicates projects intended to demonstrate some of these combustion modification technologies for commercial feasibility or demonstrate applicability where they have not been used previously. The degree of NO_x reduction achievable while maintaining acceptable boiler performance is an important element of these test programs. As of mid-1994, all of the programs had been completed except the TVA micronized coal reburning program.

Low-Excess Air Firing

Low-Excess Air (LEA) firing is an operational control strategy. Since high excess air causes high NO_x emissions, the idea is to operate the boiler with the lowest excess-air that is safe, efficient, and practical. The technology should be incorporated into any efficient low-NO_x combustion system. It has been reported that 10–20% NO_x reduction can be achieved per percentage point reduction in boiler O_2. This strategy requires process control modifications rather than new or modified hardware. It is the least expensive to install and can increase boiler efficiency. LEA is also applicable to modern waste-to-energy facilities (Makansi, 1988).

Burners Out of Service and Fuel Biasing

Off-stoichiometric conditions are achieved by modifying the primary combustion zone stoichiometry, i.e., the air/fuel ratio. Burners Out of Service (BOOS), which is an operational

Table 10-11
NO$_x$ Emission Control Technology Demonstrations

Technology	Sponsors	Project Site
Overfire Air	Southern Co. (CCT)	Hammond
	Southern Co. (CCT)	Smith
	EERC (CCT)	Cherokee
Low-NO$_x$ Burners	Southern Co. (CCT)	Hammond
	Southern Co. (CCT)	Smith
	EERC (CCT)	Cherokee
Low-NO$_x$ Cell Burners	B&W (CCT)	JM Stuart
Coal Reburning	B&W (CCT)	Nelson Dewey
Gas Reburning	EERC (CCT)	Hennepin, Lakeside
	EERC (CCT)	Cherokee
	EPA, DOE, GRI	Niles
Micronized Coal Reburning	TVA (CCT)	Shawnee

Note: These projects generally have several participants in addition to the indicated sponsors. All Clean Coal Technology (CCT) projects are co-funded by Department of Energy.
Source: U.S. DOE (1992) and Flora et al. (1991)

technique, achieves reductions in NO$_x$ emissions from multiple-burner level oil- and gas-fired boilers by blocking the fuel flow to an upper level of burners allowing only air to pass through them. The fuel that would have gone to them is redirected to the burners in service, providing a crude form of air staging. Significantly higher O$_2$ imbalances and CO emissions have resulted using this technique, but both can be controlled by implementation of the proper BOOS pattern. BOOS is not regarded as a suitable technique for coal-fired boilers (Makansi, 1988; Kuhr et al., 1988; Wood, 1994). With Fuel-Biasing, the furnace is divided into upper and lower combustion zones. The lower zone is operated fuel-rich to control NO$_x$ formation, and the upper zone is operated fuel-lean to complete the combustion. The technique is proven only on oil- and gas-fired utility boilers (Makansi, 1988). These techniques are generally applicable only to large, multiple-burner combustion devices (Wood, 1994).

Overfire Air

With Overfire Air (OFA), fuel burns initially with minimal air and sometimes at a deficiency of air (sub-stoichiometrically) with additional air introduced as overfire air. OFA ports are located above the highest elevation of burners or above the grate of a stoker-fired boiler. Ten to twenty percent of the total combustion air flow is introduced above the burner zone. It is important that adequate mixing of the OFA with the primary combustion products occurs. When combined with low-NO$_x$ burners, this technology is very effective. The points to consider in OFA design are port location, number, and geometry; boiler dimensions; fuel type; boiler operation; fan capacity; port spacing; boiler pressure drop; and the structural integrity of the boiler walls (Makansi, 1988). A variation of this technique, lance air, consists of installing air tubes around the periphery of each burner to supply staged air. The technique is generally applicable only to large, multiple-burner combustion devices (Wood, 1994).

Flue Gas Recirculation

With Flue Gas Recirculation (FGR), flue gas is introduced with the combustion air and acts as a thermal diluent to reduce the combustion temperature. Usually, the amount of flue gas recirculated corresponds to 10–20% of the combustion air. FGR reduces only thermal NO_x. It is suitable only for oil- and gas-fired boilers. Results with coal have been generally disappointing. In coal-fired stoker units, FGR provides better grate cooling. FGR has been successfully applied on industrial solid fuel-fired units and is considered appropriate for waste-to-energy plants. Retrofit modifications include new ductwork, gas recirculation fan(s), flue gas/air mixing devices and controls (Makansi, 1988; Wood, 1994). Gas recirculation fans can be troublesome.

A closely related technique, reduction of air preheat temperature, reduces NO_x formation because it also reduces peak flame temperature. But a substantial energy penalty results, about 1% efficiency loss for each 40°F reduction in preheat. In some cases, this may be offset by adding or enlarging the existing economizer (Wood, 1994).

Low-NO_x Burners

Low-NO_x burners (LNB) are expected to be widely used to meet the NO_x control requirements of Title IV. The major boiler, combustion turbine, and burner companies have developed low-NO_x burners for both retrofit and new boilers, turbines, and heat recovery steam generators (HRSGs). Each manufacturer has its own design, but all generally incorporate features to slow the rate of air-fuel mixing, reduce oxygen availability in the critical NO_x formation zones, and reduce the peak flame temperature. Many boiler vendors have incorporated OFA and/or staged combustion into their burner designs. Most of the burner manufacturers have proven low-NO_x burners for gas and oil fuel, but only a few have low-NO_x burners for pulverized coal applications.

All low-NO_x burners reduce NO_x emissions by reducing the formation of thermal and fuel NO_x in the combustion area. This is accomplished by reducing flame temperatures by staging and controlling secondary air. Air staging involves removing some of the combustion air from the burner and introducing it evenly into the flame later to complete char burnout. The amount of oxygen in the fuel-rich primary or devolatization zone is reduced, inhibiting NO_x formation. Secondary air is then injected evenly into the flame to complete the combustion at lower temperatures.

Older pulverized coal burners typically have no moving parts and receive secondary air from a common windbox. No effort is made to distribute the air evenly to the burners except by manual adjustment of registers during commissioning. Early research on low-NO_x, pulverized coal burners led to compartmentalizing the windboxes, since the open windbox resulted in significant imbalances in secondary air flow to the burners. Compartmentalization proved to be extremely expensive and involved significant physical windbox and air duct changes.

The major U.S. manufacturers of coal-fired burners have put significant effort into design improvements (particularly in materials). Results with low-NO_x, coal-fired burners have been mixed. NO_x reduction is a function of boiler geometry, fuel properties, and boiler operation. Not all existing boilers can be economically retrofitted with low-NO_x, coal-fired burners. The retrofit can be complicated and may require windbox redesign. Operation and maintenance of low-NO_x, coal-fired burners requires more operator attention and upgrading to microprocessor-based burner controls is advisable. Also, low-NO_x, coal-fired burners may increase CO and unburned carbon emissions.

Garg (1994) reviews the specification of low-NO_x burners for furnaces. He identifies five major requirements for these burners that apply to all types of fuels:

1. Significant reduction in NO_x formation
2. A flame pattern compatible with the furnace geometry
3. Easy maintenance and accessibility
4. A stable flame at turndown conditions
5. The ability to handle a wide range of fuels

Low-NO_x, coal-fired burners potentially affect electrostatic precipitators if unburned carbon emissions increase. Increases of up to 2 times baseline values have been reported. Carbon re-entrains easily in an electrostatic precipitator, potentially increasing emissions because it has very low resistivity (Piepho et al., 1992; Baublis and Miller, 1992), particularly if it is greater than 10% of the fly ash. Carbon can also adsorb SO_3. The adsorption of SO_3 can cause the electrostatic precipitator to lose its conditioning from either the natural SO_3 in the flue gas or from injected SO_3. Potential long-term effects are still not known. Other effects resulting from a low-NO_x, coal-fired burner conversion, such as more fly ash due to less slag formation in the boiler and more flue gas volume, appear to be case specific. When excessive particulate emissions result from a combined low-NO_x, burner conversion and a coal switch, flue gas conditioning can be used to assure compliance (Kumar, 1993). In one case when low-NO_x, coal-fired burner conversion caused high carbon in the ash, dual flue gas conditioning was used successfully (Krigmont and Coe, 1990). A retrofit combining fuel reburning with low-NO_x, coal-fired burners may also be used to reduce unburned carbon.

Fuel Reburning

Fuel reburning involves diverting 10–20% of a boiler's fuel input to create a secondary combustion zone downstream of the primary zone. In the secondary or reburn zone, NO_x from the primary zone is reduced to elemental N_2. Additional air is added higher up in the furnace to complete the combustion. Mixing of the reburned fuel, additional overfire air, and combustion gases is critical. The important characteristics of the reburn fuel are low nitrogen and high volatility. Natural gas and low-nitrogen fuel oils meet these criteria. Coal may be applicable if it is low in nitrogen, can be dispersed evenly in the furnace, and can be burned out within the available residence time. Demonstrations of gas and coal reburning are part of U.S. Department of Energy Clean Coal Technology Projects and are being tried elsewhere as well. Reburn configurations are customized specifically for existing boilers, and in some instances may be the only applicable in-furnace NO_x control solution (Makansi, 1988; LaFlesh and Borio, 1993; Yagiela, 1993).

Water/Steam Injection

The injection of water or steam into the combustion zone is usually applied to combustion turbines only. While application to oil- and gas-fired boilers is feasible, it is rarely practiced (Makansi, 1988). When water is injected (usually as a fine spray) into a turbine combustor, heat from the burning fuel vaporizes the water and brings the resulting mixture of fuel, air, water vapor, and combustion products to a lower temperature than would occur otherwise. Since the residence time of air in the combustion zone is unchanged by water injection, the lower rate of thermal NO_x formation resulting from the lower flame temperature causes a decrease in NO_x emissions. (Water injection rates of 0.75 to 1.2 lb water per lb fuel have typically been used.) Water injection limitations are set by the onset of flame instability, high CO emissions, increased unburned hydrocarbon emissions, severely increased wear rates of combustion hardware, and possibly by too close of an approach to the compressor flow surging region (Schreiber, 1991).

Steam injection can also be used to cool and dilute the flame. Since the heat of vaporization is provided by a heat source external to the turbine combustor, there is less flame cooling per unit of steam injected. Steam injection is also less prone to cause the flame instability that leads to dynamic pressure pulsation and to higher combustor component wear (Schreiber, 1991).

In a combined cycle application, steam rather than water may be the desirable choice as the diluent to maximize the thermodynamic efficiency of the plant. However, the decision to use steam or water injection is usually an economic choice dependent on whether the steam is more valuable to generate electricity or to control NO_x. For a natural gas-fired machine, approximately 60% more steam than water is needed to control NO_x to a given concentration. However, the energy to generate the steam comes from gas turbine exhaust heat rather than turbine fuel. The heat rate penalty is generally less for steam than for water (since additional energy is needed to vaporize the water) and the impact of steam on pressure oscillations and CO generation is less than that of water, especially at higher flow rates. Still there is a heat rate penalty of more than 2% when steam is used to control NO_x emissions from a natural gas-fired combined cycle plant to achieve less than 42 ppmvd (ppm volumetric and dry) NO_x in the exhaust (Schorr, 1992). Where regulatory requirements dictate lower emission levels, operating hour limits or post combustion treatment may be needed for NO_x compliance.

Dry Low-NO_x Combustors

All of the dry low-NO_x combustors currently available for combustion turbines utilize the lean pre-mix principle of operation, which creates a homogeneous, fuel-lean mixture of fuel and air prior to combustion. This mixture is then introduced to the combustion zone of the combustion chamber at a controlled velocity sufficiently higher than the local speed of flame propagation to prevent flashback into the pre-mix zone. However, the pre-mixture velocity must be low enough to avoid blowing the whole flame downstream (Schreiber, 1991).

When burning in this mode, all parts of the fuel-air mixture are at or above the stoichiometric air/fuel ratio. Therefore, there is no diffusion flame front (plane along which combustion starts, or base of the flame) where high temperature exists. The resulting flame is lower in temperature, since the energy released by combustion must heat a greater mass of air that is in contact with it at the moment of combustion. Since the burning rate is a function of the temperature, the cooler flame associated with the lean pre-mix mode of operation requires more time for full burnout of fuel than the hotter diffusion flame. Therefore, dry low-NO_x combustors are more complex than diffusion type combustors in that they require precise control of local velocities and sequencing of fuel/air ratios during transients, starts, and stops. To prevent lean blowout of the main flame, pilot diffusion flames generating NO_x at a high rate (but at a low total mass flow) may be employed.

This technology is relatively new and is still being actively developed and refined. Siemens and ABB both offer dry low-NO_x silo type combustors. The General Electric Company offers a can-annular combustor and Westinghouse is developing a can-annular system. The combustors are generally capable of operation in two modes: dry low-NO_x operation firing gaseous fuel only and operation in the diffusion flame mode firing oil with steam- or water-injection for NO_x reduction.

POST-COMBUSTION NO$_x$ CONTROL PROCESSES

Post-combustion processes, by definition, are located downstream of the combustion zone. Post-combustion NO$_x$ control processes can be classified as

- Selective non-catalytic reduction (SNCR)
- Selective catalytic reduction (SCR)
- Combined NO$_x$/SO$_2$ processes
- Other NO$_x$-only processes

SNCR and SCR use ammonia or urea to reduce NO$_x$ in the combustion gases to elemental nitrogen. A significant common consideration with both of these processes is ammonia slip, i.e., unused ammonia in the downstream flue gas. However, ammonia slip, which exists whether ammonia or urea is the reagent, can be effectively controlled. The acceptable amount of ammonia slip is usually dictated by fouling limitations and by permit requirements. Regulators often set 10 to 15 ppm at the stack as the maximum allowable ammonia concentration.

Several detrimental side reactions occur when ammonia slip is high. Of prime concern are the reactions between excess ammonia, SO$_3$, and water in the flue gas to form ammonium sulfate and ammonium bisulfate. These reaction products, if present in sufficiently large quantities, contribute to fouling and corrosion. Depending on the concentration of SO$_3$ in the gas, ammonium sulfate can form as a powdery substance at temperatures between 500° and about 580°F. At lower temperatures, typically between 200°–350°F, ammonium sulfate converts to ammonium bisulfate if sufficient SO$_3$ is present. The deposition temperature of ammonium bisulfate as a function of ammonia and sulfur trioxide concentrations in the flue gas is given by Pease (1984). Temperatures between 500° and 580°F occur upstream of the air preheater, and drop into the lower range within the air preheater. At air preheater intermediate zone temperatures, in particular, ammonium bisulfate is a corrosive, sticky liquid, which has an affinity for dust and soot and promotes the buildup of deposits. Limiting air preheater fouling and corrosion to low levels requires that the ammonia slip be maintained below acceptable limits at all times. To mitigate the effects of ammonia salts on downstream equipment, typically allowable values of ammonia slip are 5 ppm for hot-side/high-dust SCR applications, less than 2 ppm for hot-side/low-dust SCR applications, 2–5 ppm for cold-side SCR applications, and 5–10 ppm for oil-fired SCR applications (Robie et al., 1991A). For natural gas-fired applications, the ammonia slip value is usually dictated by environmental constraints because there is little or no sulfur in the gas (Cho and Dubow, 1992). Higher slip values are possible for high-dust applications since the abrasive action of the ash will remove ammonia salts. For SNCR applications, ammonia slip values as high as 20 ppm have been used. These values are largely based on experience and each case requires individual consideration. Tests with oil-fired boilers (Hurst, 1981) have shown that the ammonium salts that deposit on ducts and equipment can be removed by periodic water washing.

The ammonia salts can also contaminate the fly ash and flue gas desulfurization (FGD) wastewater, create a visible stack plume (at about 20 ppm), and add another pollutant to the flue gas. Where ash has been sold in Germany, the ammonia slip has been limited to 2 ppm. This value was determined based on experience, and depends in part on the quantity of ash. As the flue gas cools further, any HCl present in the flue gas and the excess ammonia can combine to form ammonium chloride which, when present in sufficient quantities, causes a visible stack plume. Under some conditions, ammonium nitrate can form. Ammonia slip, by combining with the sulfur trioxide in the flue gas, can reduce the resistivity of the fly ash, which could necessitate additional sulfur trioxide injection into the flue gas ahead of an electrostatic precipitator to maintain its performance.

Ammonia-related plume formation and resulting opacity problems are often overlooked during design with potential costly consequences. As an example, at Biogen Power 1, an 18 MW_e fluidized bed boiler firing waste coal fines, it was found that chlorine concentrations as low as 0.01% in the fuel caused visible plume/opacity problems. At 300°F, NH_3 and HCl are gases; at 200°F they combine to form solid ammonium chloride which can form a detached plume. To mitigate the problem, it was proposed that the NO_x limit be raised in the winter when ozone concentrations are the lowest and plume opacity problems are the greatest. This recommendation, based on modeling studies, would reduce the ammonia slip and virtually eliminate the plume formation problem (Lange et al., 1994).

With SNCR systems, high NO_x reduction usually means high ammonia slip. Transient thermal conditions under cycling conditions can also cause high ammonia slip and prevent slip from being reliably guaranteed in SNCR applications. Pilot plant tests and kinetic modeling by Muzio et. al. (1991) also established that SNCR processes based on the use of ammonia, urea, and cyanuric acid all produce some N_2O. Ammonia injection produced the lowest levels, while cyanuric acid produced the highest levels.

Table 10-12 compares SNCR and SCR technologies. Effective temperature ranges, typical NO_x reductions, typical ammonia slips, reagents, reagent utilizations, enhancers, reagent dilution/carrier gas requirements, injection method, and energy losses are summarized. Although not considered here, some companies offer combined SNCR/SCR processes.

Selective Non-Catalytic Reduction (SNCR) Processes

Selective Non-Catalytic Reduction is a post-combustion NO_x control method that reduces NO_x via injection of ammonia or a urea-based reagent into the upper furnace and/or convection section of the boiler. As SNCR requires an elevated temperature, it is sometimes referred to as thermal reduction. The process is particularly applicable to older furnaces with longer residence times in the SNCR reaction temperature range, where combustion controls are very expensive or not possible, where the boiler has a low capacity factor or short remaining life, or where system-wide averaging can be used.

NO_xOUT Process

General. The NO_xOUT process is a commercial SNCR process for the reduction of NO_x using NO_xOUT A (a proprietary mixture of urea and small amounts of chemicals that inhibit urea precipitation and scaling) (Fuel Tech, 1989A). In some applications, additional chemicals, called enhancers, are injected into the boiler to improve process performance. The process emerged from research on the use of urea to reduce nitrogen oxides initially conducted in 1976 by the Electric Power Research Institute (EPRI). EPRI obtained the first patent on the fundamental urea process in 1980 (Comparato et al., 1991). Since 1990, Nalco Fuel Tech, EPRI's exclusive commercial licensing agent, has commercialized the technology as the NO_xOUT process (Fuel Tech, 1989B). Other U.S. companies have licensed the technology from Nalco Fuel Tech including Research-Cottrell, Foster Wheeler, Wheelabrator, RJM, and Todd Combustion (Hofmann, 1994). The process has been demonstrated in over 60 tests

(text continued on page 892)

Table 10-12
SNCR vs. SCR for Boilers

	SNCR Using Urea	SNCR Using Ammonia	SCR Using Ammonia
Effective Temperature Range	1,650–2,200°F without enhancers, down to 1,400°F with enhancers. As low as 1,000°F based on flame tube data, but reaction time may be a concern. Upper temperature limit is determined by oxidation to NO_x and depends on gas composition and other factors.	1,600–2,000°F without enhancers, down to 1,300°F with enhancers. The optimum temperature is about 1,750°F depending on the application. At about 1,830°F, NH_3 oxidization to NO_x in the gas stream becomes significant, and this reaction becomes predominant above 2,200°F.	Temperature range depends on the catalyst formulation. For base metal catalysts, 450–1,050°F possible, 600–700°F typical. Upper limit is typically established by catalyst sintering. The lower limit is typically established to prevent ammonium sulfate deposition. Ammonium sulfate formation depends on the fuel sulfur content: typical temperatures are 570–580°F with coal and 500°F with natural gas (no sulfur) at full load. Temperatures less than 400°F are possible for low load operation under some conditions. Usually, the ammonia is shut off and the ammonia is allowed to desorb from the catalyst before temperature is ramped down into this range. Catalyst activity rates and poisoning are important considerations particularly at low load. Precious metal catalysts (with a lower temperature range) and zeolites (suitable to about 960°F) are typically not used for boilers.
Typical NO_x Reductions, %,[1,2,3]	40–70 (80% with circulating fluidized bed boilers)		Coal: 50-85 Oil: 50–90 Gas: 50–95
Typical Ammonia Slip, ppmv	<5 to 20. Low slip is not likely at high NO_x reductions. Slip depends on the boiler and the residence time.		Coal: 2–5 Oil: 3–10 Gas: 10
Reagent	50% aqueous urea with inhibitors to reduce scaling caused by the dilution water, etc.	Aqueous ammonia (≈28% concentration) or pressurized anhydrous ammonia. Aqueous ammonia is increasingly preferred due to safety considerations.	
Typical Reagent Utilization Neglecting Ammonia Slip, %	25–60. Reagent utilization tends to be low for high NO_x reduction with low ammonia slip. Reagent utilization tends to be high for high inlet NO_x values and low for low inlet NO_x values.		Usually ≈ 100%. Above 850–900°F, NH_3 starts to catalytically oxidize to NO_x with some catalysts and on metal (stainless steel or carbon steel) surfaces. Some catalysts sinter at these temperatures.

(table continued on page 890)

Table 10-12 (Continued)
SNCR vs. SCR for Boilers

	SNCR Using Urea	SNCR Using Ammonia	SCR Using Ammonia
Enhancers Used	Typically oxygenated hydrocarbons to increase reagent utilization, increase NO_x reduction efficiency, widen temperature range, vary optimum reaction temperature, and reduce ammonia slip. Not often used.	Hydrogen to extend effective temperature range. Not often used.	No enhancers are injected.
Enhancer Quantity, mole/mole reagent	When used, 0.05–0.10 moles/mole of urea at lower temperatures.	When used, 0.2–0.5 moles/mole of ammonia.	Not applicable.
Reagent Dilution/Carrier Gas Requirements	Diluted with water to 5–25% concentration. Air is the carrier gas. (Steam is no longer used.) Typically, 4 scfm of plant air per injector is currently used for atomization and for injector cooling.	Not diluted with water. Air or steam at 1–2% of the flue gas flow rate is the carrier gas. This gas requirement is reduced by the amount of water vaporized if aqueous ammonia is vaporized.	Not diluted with water. Carrier gas is air, flue gas, or possibly steam at about 2 psig. Ammonia/carrier gas volumetric ratio is a minimum of about 1 to 20 for anhydrous ammonia to keep the ammonia concentration in the air below the lower explosive limit. Lower concentrations may be used based on injection diffusion patterns, the ammonia vaporization heat requirement, minimum flow control ranges, and other factors.
Injection Method	Wall injectors and, if required for distribution, lances. Typically few injection points.	Wall injectors and, if required for distribution, lances. Typically many injection points.	Injection lances/grids.

Table 10-12 (Continued)
SNCR vs. SCR for Boilers

	SNCR Using Urea	SNCR Using Ammonia	SCR Using Ammonia
Energy Losses	Energy requirements of the storage, handling, and delivery systems for the injectors/lances. Energy to compress the plant air that atomizes the diluted urea and cools injectors. Flue gas heat to evaporate the liquid injected. At $\geq 22.6\%$ concentrations, the heat released by the urea disassociating is $\geq$ the heat required to evaporate the dilution water, and no flue gas heat is needed. Gas side pressure drop does not increase unless ammonium salts cause pluggage of downstream equipment.	With aqueous ammonia, the energy losses include the heat added to vaporize and raise the temperature of the ammonia. With anhydrous ammonia, the heat added to vaporize and raise the temperature of the ammonia. Energy to deliver the carrier gas. Flue gas heat to raise the temperature of the injected ammonia and air to the boiler exit temperature. Gas side pressure drop does not increase unless ammonium salts cause pluggage of downstream equipment.	With aqueous ammonia, the energy losses include the heat added to vaporize and raise the temperature of the ammonia. With anhydrous ammonia, the heat added to vaporize and raise the temperature of the ammonia. Energy to deliver the carrier gas. Flue gas heat to raise the temperature of the injected ammonia and air to the boiler exit temperature. Energy to overcome the gas side pressure drop through the SCR—typically 2–3 in. wg (1–6 range). Pressure drop does not increase above these values unless ammonium salts or fly ash cause pluggage of downstream equipment. For cold side/low dust SCR, the additional heat added by the fired heater to raise the flue gas to the operating temperature of the catalyst and the additional pressure drop through the regenerative air heater and the fired heater.

Notes:
1. NO_x reductions are based on typical inlet NO_x concentrations of 200–2,500 ppm for coal; 100–400 ppm for oil; and 50–150 ppm for natural gas. While the range of inlet NO_x concentrations is the same for both SNCR and SCR, possible NO_x reduction with SNCR varies with the inlet NO_x concentration, being higher for higher inlet NO_x concentrations. For SCR, this correlation does not exist, as the amount of SCR catalyst can be varied to suit a particular application. Reliability and ammonia slip are greater concerns with NO_x control efficiencies above 80–90%. The SNCR temperature and residence time cannot be varied beyond what optimization of injection locations can accomplish. (The inlet NO_x concentrations given here represent those to which the technologies have been applied. Some boilers can produce higher NO_x concentrations than indicated. For example, some oil-fired boilers can produce 1,000 ppm of NO_x, and some gas-fired boilers 300–400 ppm.) Stringent ammonia slip control may limit the NO_x reduction achievable in some cases.
2. Air pollution regulatory requirements heavily influence the NO_x removal requirements.
3. NO_x reduction efficiencies above and below the specified ranges are possible depending on residence time and temperatures in the boiler/SCR, ammonia slip, etc.

Source: Clark and Comparato (1994), McIntyre, (1994), and Pritchard (1994)

892 Gas Purification

(text continued from page 888)

and commercial installations. Applications include a wide range of combustors and fuels as shown in **Table 10-13**. The largest boiler application of the process was on an 850 MW$_e$ oil-fired unit, where 50% NO$_x$ reduction was achieved. This was at Niagara Mohawk's Oswego Station where the process was used while flue gas recirculation fans were out of service.

It is reported that the process can achieve NO$_x$ reductions of 30 to 85%. Inlet NO$_x$ concentrations as high as 2,000 ppm have yielded NO$_x$ reduction efficiencies of 50–80%. If required, ammonia slip under 2 ppm has been demonstrated, but not necessarily at higher NO$_x$ reductions.

Capital costs for some recent U.S. urea-based SNCR systems are reported in **Table 10-14**. Chemical costs account for a large portion of the total operating costs. Compared to ammonia-based SCR, the higher price of urea is claimed to be offset, in some cases, by eliminating the costs for SCR catalyst replacement and the energy needed to overcome the pressure drop through the catalytic reactor (Fuel Tech, 1988).

Basic Chemistry. In the NO$_x$OUT process, an aqueous solution containing about 50% of urea-based reducing agent, called NO$_x$OUT A, is diluted to 5–25% concentration and injected into the flue gas to reduce the NO$_x$ (Fuel Tech, 1989A). The process parameters that are most important in determining performance are the flue gas temperature distribution, residence time in the reaction temperature range, distribution/mixing of the chemicals with the flue gas, inlet NO$_x$ concentration, and amount of NO$_x$ reduction required for a specific application. There is a benefit in operating on the high side of the reaction temperature range in that NH$_3$ formation is suppressed while NO$_x$ reduction is only slightly diminished (Hofmann et al., 1989). The urea reacts with the NO$_x$ to form molecular nitrogen, carbon dioxide, and water. The exact reaction mechanism is uncertain, owing partially to the complexities of urea pyrolysis and also to the wide reaction temperature range (Epperly et al., 1988A). Based on

	Table 10-13 Applicability of the NO$_x$OUT Process
Combustor Types	Boilers: Pulverized coal wall-, corner-, tangentially-, cyclone-, and cell-fired; package; Volund grate-fired; grate stoker; CO; waste heat; circulating fluidized bed
	Combustion turbines configured with downstream duct burners
	Petroleum and Petrochemical: Ethylene crackers, catalytic crackers, bottom-fired process heaters
	Miscellaneous: waste incinerators, glass furnaces, calciners, sludge combustors
Fuel Types	Coal: bituminous, brown
	Oil: No. 6
	Gas: natural, methane, refinery, organic
	Pulp and paper: wood, wood waste, bark, paper sludge, black liquor, hog fuel
	Miscellaneous: municipal solid waste, tires, coke, contaminated solids
Source: Mincy (1992)	

Table 10-14
Capital Costs of Some Recent U.S. Urea-Based SNCR Retrofits

Facility	Fuel	NO_x Reduction Efficiency	Cost	Notes
Long Island Lighting Company's Port Jefferson Station	No. 6 Oil	40%	$10–15/kW $800/ ton NO_x	Demonstration only. Tangentially-fired unit. N_2O increased 10–15 ppm over baseline conditions.
Wisconsin Electric Power Company's Valley Station	Coal	Up to 50% (Refer to note 1)	$8/kW $400–600/ton NO_x	Demonstration only Front wall-fired dry bottom boiler.
New England Power's Salem Harbor Station	Coal	At high load: 50–60% At low loads: 60–75% (Refer to note 2)	$15/kW $900/ton NO_x	Being converted for long-term testing. Front wall-fired dry bottom boiler. N_2O emissions were consistent with previously reported results.

Notes:
1. *While gas phase analysis for ammonia slip generally showed very low or non-detectible concentrations at Valley Station, samples of the fly ash showed concentrations of ammonia significantly above the no-effect industry standard for reuse of 80 mg/kg.*
2. *The ammonia slip was between about 10 to 20 ppm at 60% NO_x reduction depending on the coal.*

Source: Wax (1993B), Hofmann et al. (1993), and Shore et al. (1993)

Nalco Fuel Tech field measurements for boiler applications, over 95% of the NO_x is in the form of NO, so that the predominant overall reaction is between urea and NO. The reaction of NO with urea is temperature dependent, typically takes place within 30 milliseconds to one or more seconds, and can be expressed as

$$CO(NH_2)_2 + 2NO + \tfrac{1}{2}O_2 = 2N_2 + CO_2 + 2H_2O \tag{10-1}$$

This chemical reaction indicates that one mole of urea is required to react with two moles of NO. Test results indicate that greater than stoichiometric quantities must be injected to achieve a desired NO_x reduction. In actual use, urea utilization is typically 20 to 60%. Excess urea degrades to nitrogen, carbon dioxide, and small amounts of ammonia. Possible byproducts due to poor process conditions (i.e., improper mixing, low residence time, lack of proper temperature, etc.) are CO, high NH_3, and N_2O. There is no urea emission. The reaction of NO_x with urea is effective over a temperature range of 1,650° to 2,200°F, below which ammonia is formed and above which NO_x emissions actually increase (Mincy, 1992). **Figure 10-1** shows this temperature relationship. Within this range, urea decomposes and reacts with NO_x. If the temperature falls below this window, then the urea/NO_x reaction rate decreases, causing increased ammonia slip and reduced NO_x control.

In some installations, small amounts of proprietary chemicals called enhancers (typically oxygenated hydrocarbons) may be injected to increase urea utilization, increase NO_x reduction efficiency, widen the reaction temperature range, vary the optimum reaction tempera-

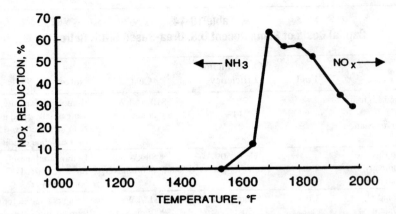

Figure 10-1. Reduction of NO_x with urea as a function of temperature. (*Fuel Tech, 1989B*)

ture, reduce the ammonia slip, and/or reduce scaling by the dilution water (Hofmann et al., 1989; Epperly et al., 1988B). Typical situations where enhancers may be of value are where a boiler experiences frequent load swings (shifting reaction temperature zone), where chemical residence time is short, or where the boiler burns multiple fuels (Lin et al., 1991). Although the chemical composition of the enhancers is considered proprietary, it is claimed that they are readily available, require no special safety precautions, and produce no secondary byproducts for disposal. However, enhancers add to the operating cost of the system. Enhancers are believed to moderate the free radical pool available to sustain NO-reducing reactions and can increase the effective temperature range of the process to 1,400°–2,200°F (Epperly et al., 1988B; Lin et al., 1991). Refer to **Figure 10-2.** Note that the data in **Figures 10-1 and 10-2** are based on flame tube testing. The symbols marking points on the curves

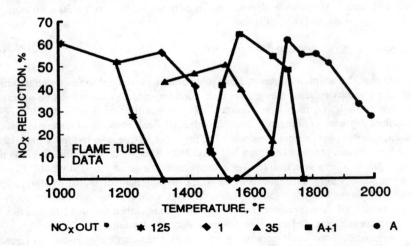

Figure 10-2. Reduction of NO_x with urea and with urea plus $NO_x OUT^{tm}$ enhancers as a function of temperature. (*Fuel Tech, 1989B; Mincy, 1992*)

and listed under **Figure 10-2** identify the trade names of the various NO_xOUT enhancers tested. The enhancer/urea (E/U) ratio is a significant process variable because it affects NO_x reduction and is particularly important in the control of ammonia slip. NO_xOUT technology is claimed to control NH_3 slip to low concentrations by carefully selecting the quantity of enhancer. It is reported that enhancers are used on only a small percentage of the installations, and improved injection techniques have minimized their use (Hofmann et al., 1993).

Ammonia slip generally increases with increased NO_x reduction and, at some NO_x reduction level, unacceptably high ammonia slip results. **Figure 10-3** illustrates this phenomenon. Ammonia slip tends to increase with (Hofmann et al., 1989)

- decreasing temperature
- increasing reagent-to-NO_x mole ratio (increasing NO_x reduction)
- decreasing enhancer/urea (E/U) ratio

Figure 10-3 is based on a series of tests on a brown coal fired boiler. The lower curve is based on injection at one level. The upper curve is based on injection at two levels and optimization of the amounts of urea and enhancer at each of the two levels. Following this optimization, it was found possible to further increase NO_x reduction by an additional 6% with no increase in ammonia slip by injecting enhancer at a third level (not shown in the figure) (Epperly et al., 1988B).

N_2O generation is a potential concern with urea injection since it can increase N_2O emissions 10 to 15 ppm over baseline conditions. These values can correspond to about 5 to 15%

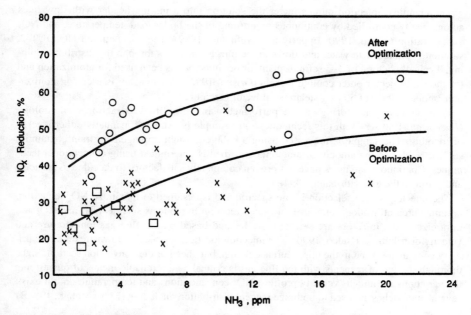

Figure 10-3. Control of ammonia slip for the NO_xOUT^{tm} process (□ and x = two sets of test data obtained before optimization, o = data obtained after optimization). (*Epperly et al., 1988B*)

of the NO_x reduced. The amount of N_2O formed is related to the amount of NO_x reduced, temperature, and the use of enhancers. CO emissions can also increase, but can be maintained at less than a 10 ppm increment (Shore et al., 1993; Hofmann et al., 1993; Berg et al., 1993; Rini and Cohen, 1993; Hunt et al., 1993; Kwan et al., 1993; Teixeira et al., 1993; Mincy, 1992). Nalco Fuel Tech has installed systems which meet N_2O regulations in Sweden, Germany, and Italy. N_2O formation is controlled primarily by the location and profile of the injection (Mincy, 1992).

Process Description. The process is shown in **Figure 10-4.** The system includes (Mincy, 1992; Grisko, 1992)

- The NO_xOUT A (urea solution) storage tank and skid-mounted module with solution recirculation pumps and heat exchanger for maintaining the solution at 80°F. (Since urea is non-toxic, urea storage requirements are simple.) Separate enhancer storage tanks (unheated) are required if enhancers are used.
- One skid-mounted module for metering and mixing, consisting of the concentrated NO_xOUT A solution metering pumps, dilution water booster pumps, and static mixer. If enhancers are used, concentrated enhancer pumps are required.
- One or more levels of distribution "modules," with each module consisting of NO_xOUT A wall-injectors and all associated piping, including the piping to the injectors for the diluted NO_xOUT A solution, the enhancer (if required), atomizing air or steam, and cooling air. Air is typically used for atomizing because it is easier to handle.

The chemical injection pumps meter the solution into a mixing header with flow rates automatically controlled by continuous monitoring of the boiler load, temperature, and NO_x concentration (Nalco, 1990; Epperly and Hofmann, 1989). The concentrated NO_xOUT A solution is diluted with water and flows to the injector modules through the distribution piping. Typically, a total of 5–7 scfm of plant air per injector has been used for atomization and for the wall-injector outer cooling jackets (Grisko, 1992). When steam is used for atomization, about 50 lb/hr of 50 psig steam per injector is required.

Ammonia slip monitoring can be performed using simplified extraction methodology (*Fuel Tech,* 1989A). With this technique, a gas sample is extracted manually either weekly or monthly. The sample is processed using a bubble-through impingement train containing sulfuric acid, and the amount of ammonia absorbed is determined using conventional techniques. Continuous monitors have not proven completely satisfactory to date due to particles and sulfur in the gas (Hofmann, 1994).

The injector system selection is site specific and is designed to maximize NO_x reduction, optimize urea utilization, minimize unwanted reaction byproducts such as NH_3 and CO, and minimize costs. Injectors are selected and located based on the flue gas temperature and velocity distributions (Nalco, 1990). The injection location is typically downstream of the primary combustion zone in the upper furnace region, but ahead of the convection section. Since thorough mixing of reagents with the flue gas is critical, and a reliable means of quantifying the variations in flue gas velocity profiles, NO_x concentrations and temperature do not exist; several approaches are used to optimize reagent distribution, including (Epperly et al., 1988B)

- Designing and locating injectors for proper reagent mixing with the flue gas based on boiler geometry
- Spacing reagent injectors consistent with spray patterns

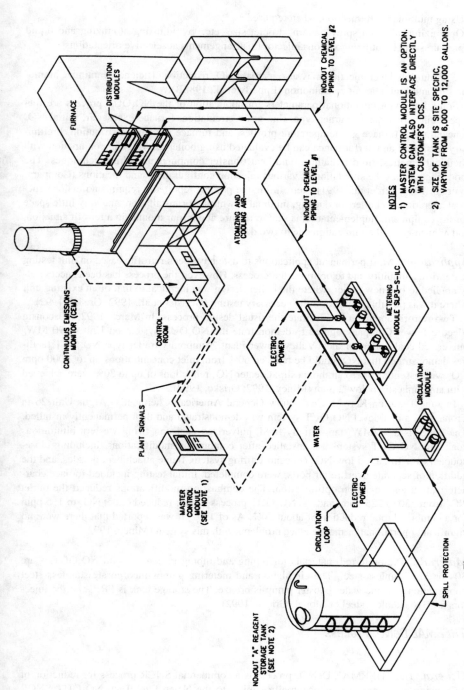

Figure 10-4. Schematic diagram of the NO_xOUT^{tm} process. *Courtesy of Nalco Fuel Tech*

- Using multiple level injection, when required
- Optimizing injection spray pattern, droplet size, etc., by adjusting atomizing and liquid pressures and by utilizing various injector tip arrangements at selective orientations

Use of multiple level injection may also increase NO_x reduction efficiency, minimize ammonia slip, and limit reagent consumption (Epperly et al., 1988B).

Some important operational parameters for application of the NO_xOUT process are fuel type, boiler load range, burner type, flue gas recirculation, furnace excess air, initial NO_x concentration, furnace gas temperature profiles, and furnace gas flow distribution. Preliminary feasibility and performance can be evaluated using boiler drawings in conjunction with chemical kinetics, fluid dynamics, and heat transfer computer modeling techniques. The modeling enables the simulation of various injection configurations and locations (Comparato et al., 1991; Lin et al., 1991). Retrofitting the process reportedly requires no major structural alterations to boilers or ducting, minimal downtime for installation, and very little space for the equipment. Implementation is said to require about four months in a retrofit situation, and adjustment and optimization one to two days.

Applications. Most permanent applications to date have been retrofits based on unit testing to confirm suitability and to optimize the process. However, the process has been successfully applied to one new boiler, where the boiler design was identical to that of an existing unit where it was possible to perform the necessary testing (Geissler et al., 1992; Grisko, 1992).

Two retrofit applications illustrate the usual design process. In March 1992, Wisconsin Electric Power Company agreed to demonstrate the NO_xOUT system on Unit 4 (90 MW_e pulverized coal boiler) of their Valley Power Plant. Temporary trailer-type NO_xOUT facilities demonstrated that NO_x could be reduced 60% from inlet concentrations up to 2,000 ppm NO_x with essentially zero ammonia slip. Greater NO_x reductions of up to 80% were achieved with ammonia slip below 2 ppm (Mincy, 1992; Grisko, 1992).

In June 1989, on Kerr-McGee's (now General American Chemical's) Argus Unit 26 in Trona, CA, a dual level NO_xOUT system was demonstrated and then permanently installed. The boiler is a 75 MW_e tangentially-fired, pulverized coal unit firing western bituminous coal. The NO_xOUT system was installed after combustion modifications, including close-coupled over-fire air, a low-NO_x concentric firing system, flame attachment nozzles, and the addition of separate overfire air ports, were completed. Initial testing included furnace characterization and injection optimization. The combustion modifications reduced the outlet NO_x from 330 to 225 ppm, and the NO_xOUT process further reduced outlet NO_x to 165 ppm for a combined total reduction of about 50%. As of 1992, it was reported that there were no significant operational or maintenance problems with this system (Mincy, 1992).

Materials. The preferred materials for piping and tubing in contact with NO_xOUT A are 304 or 316 stainless steel. The circulation and metering pumps incorporate stainless steel construction, and the water booster pump is bronze. The storage tank is FRP, and the injectors are 316 stainless steel or Monel (Grisko, 1992).

THERMAL DeNO$_x$ Process

General. The THERMAL DeNO$_x$ process is a commercial SNCR process for reduction of NO_x using ammonia. It is conceptually similar to the Nalco Fuel Tech NO_xOUT SNCR process and operates within approximately the same temperature range as the NO_xOUT process. It differs from the NO_xOUT process in the following ways:

- Ammonia gas is injected rather than urea solution. Either an aqueous or liquid anhydrous ammonia storage system is acceptable.
- Enhancers are not usually used; although hydrogen was used in some early applications.
- Either steam or air can be used as the carrier gas.
- The process can be designed for both new and retrofit applications without testing.

The process was developed by Exxon Research and Engineering Company (ER&E), and is available under license from ER&E (Exxon, 1989). The first commercial application was in 1974. The process has been demonstrated in over 130 applications. As shown in **Table 10-15,** these include a wide range of combustors and fuels. The largest boiler application of the process to date has been on Steag AG's Herne Unit IV, a 383 MW_e (plus district heating steam) coal-fired (1.5% sulfur) unit, where 55% NO_x control was achieved. SNCR proved effective only during steady-state operation. During load changes, the furnace's cross-sectional temperature profile became too unpredictable to maintain adequate NO_x reductions (Makansi, 1992B). To reduce ammonia slip from 10 ppm to 5 ppm at full load, this SNCR system was replaced by an SCR system (McIntyre, 1994). The SNCR process has not been applied to combustion turbines since combustion turbine exhaust temperatures are below the effective temperature window of the process.

Basic Chemistry. The THERMAL $DeNO_x$ process achieves gas phase NO_x reduction by injection of a small quantity of ammonia (by means of either air or steam carrier gas) into the boiler. The injected ammonia, in the presence of excess oxygen, selectively reacts with the NO_x and reduces it to nitrogen. The overall chemical reaction in the THERMAL $DeNO_x$ process is best described by the following reaction:

$$6NO + 4NH_3 = 5N_2 + 6H_2O \qquad (10\text{-}2)$$

Table 10-15
Applicability of the THERMAL $DeNO_x$ Process

Combustor Types	Boilers: Pulverized coal, cyclone, stoker, CO, waste heat, circulating fluidized bed
	Petrochemical applications
	Miscellaneous: waste incinerators, glass furnaces, industrial furnaces, industrial heaters
Fuel Types	Coal, up to 1.5% sulfur
	Oil: refined, crude
	Gas: natural, CO, waste vapors
	Pulp and paper: wood, wood refuse
	Miscellaneous: municipal solid waste, coke (up to 3% sulfur), sludge, hazardous waste, tires, nut shells

Source: Exxon (1981–1992)

In order for reaction 10-2 to occur in the required temperature range, oxygen must be present in the flue gas.

The reaction between NO_2 and NH_3 is not known for certain. However, field observations and calculations indicate that flue gas NO_2 concentrations are typically less than 5% of the total NO_x. Therefore, NO_2 is not a major concern based on present NO_x reduction levels (Medock, 1990; Muzio and Arand, 1976).

The selective NO_x reduction in the THERMAL DeNO$_x$ process strongly depends upon the temperature of the combustion products, as **Figure 10-5** shows. NO_x reduction occurs in the temperature range of approximately 1,600° to 2,000°F, with the peak reductions occurring at about 1,750°F. The effectiveness of the reaction can be extended from 1,600°F down to about 1,300°F (as would occur in a load-following unit) by injecting hydrogen along with the ammonia (Exxon, 1982).

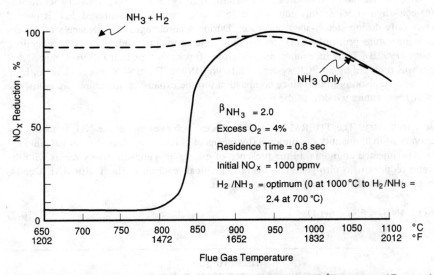

Figure 10-5. Temperature dependency of the THERMAL DeNO$_x$^{tm} process. (*Exxon, 1982*)

NO_x reduction becomes less and less effective as the temperature increases beyond the optimum temperature of about 1,750°F. In fact, at about 1,830°F, ammonia starts oxidizing to NO according to the following reaction:

$$4NH_3 + 5O_2 = 4NO + 6H_2O \tag{10-3}$$

At temperatures above 2,200°F, this reaction becomes predominant (Hurst and White, 1986).

For the THERMAL DeNO$_x$ process, **Figure 10-6** shows the relationship between NO_x reduction and the amount of ammonia injected in terms of the molar ratio of injected ammonia to initial NO, β_{NH3}. NO_x reduction efficiency increases rapidly with increasing β_{NH3} up to a β_{NH3} of about 2, but increases marginally with larger values. This figure also shows that the NO_x reduction at any given β_{NH3} is practically independent of the initial NO_x concentra-

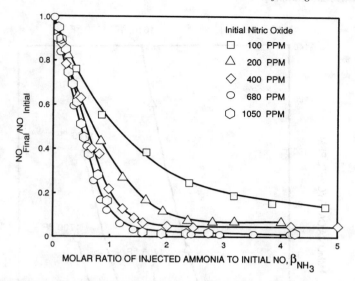

Figure 10-6. Nitric oxide reduction as a function of ammonia injection rate and initial NO_x concentrations (1,760°F, 2% excess oxygen). (*Muzio and Arand, 1976*)

tion when it is greater than 400 ppm. Below 400 ppm, however, lower NO_x reduction is achieved at lower initial NO_x concentrations (Muzio and Arand, 1976). Conversely, the data show that a higher value of β_{NH_3} is required to attain the same fractional reduction of NO when the initial NO concentration decreases. In practice, the NH_3/NO_x ratio is varied depending on the initial NO_x concentration. An NH_3/NO_x molar ratio on the order of 1.5 is common for initial NO_x concentrations of 200 ppm or less. As the initial NO_x concentration increases, the ratio is reduced toward 1.0 (Hurst, 1981).

Figure 10-7 (Muzio and Arand, 1976) gives the ammonia slip as a function of temperature and ammonia injection rate as measured in laboratory tests. Ammonia slip depends strongly upon temperature. For all practically useful values of β_{NH_3}, the ammonia slip increases with lower temperatures and higher β_{NH_3}.

The THERMAL $DeNO_x$ process is claimed to have no measurable effect on CO, CO_2, or SO_x emissions (Haas, 1992B).

Process Description. The process is shown in **Figure 10-8.** The system includes (Haas, 1992B; Medock, 1990; Hurst, 1983)

- Either an anhydrous or an aqueous ammonia storage and supply system. Both ammonia systems consist of an ammonia storage tank, vaporizer, control valves, mixing module, and associated piping. The anhydrous tank is pressurized, whereas the aqueous tank is atmospheric. The anhydrous system vaporizer is the once-through type. In the aqueous ammonia system, an ammonia stripping system or a hot air vaporization system can be used instead of a vaporizer.
- Carrier gas supply system consisting of either low pressure steam supply lines (if steam is the carrier gas) or an air compressor (if air is the carrier gas), control valves, and associated piping.

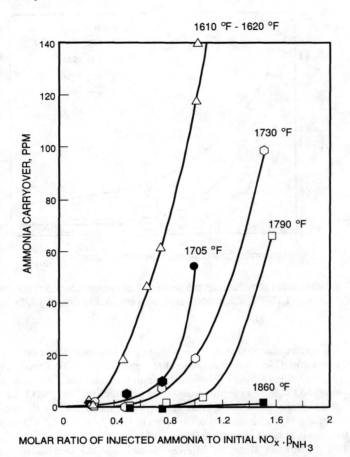

Figure 10-7. Ammonia carryover as a function of ammonia injection rate and temperature (4% excess oxygen, initial NO_x 300 ppm). (*Muzio and Arand, 1976*)

- One or more sets of wall-mounted injectors and associated distribution piping.
- Control system consisting of programmable process controllers to inject controlled quantities of ammonia into the flue gas stream.

The process is equally amenable to either anhydrous or aqueous ammonia feed. The decision to use either anhydrous or aqueous ammonia is usually based on local regulations. The decision to vaporize or strip the aqueous ammonia is based primarily on economics, which favor stripping with larger systems. Vaporization reduces carrier gas requirements (Haas, 1992A). Ammonia storage and handling systems are covered in greater detail in the SCR section of this chapter.

The mixed vaporized ammonia and carrier gas (air or steam) is injected into the boiler by wall-nozzles (injectors). The carrier gas pressure is maintained constant, and the flow rate depends on the number of injector zones in service.

Control of Nitrogen Oxides   903

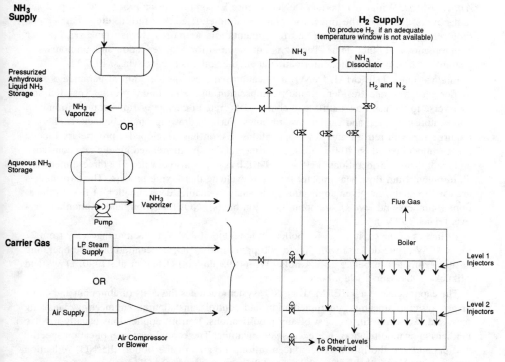

Figure 10-8. THERMAL DeNO$_x$^{tm} process flow diagram. (*Haas, 1992A; Hurst, 1983*)

Flow control valves meter the mixture into the boiler based on feed-forward measurement of the fuel rate and/or feedback of the measured NO$_x$ emission. Suitable logic is built into the control system to optimize injection rates and to require operator intervention in the event that the THERMAL DeNO$_x$ system experiences large changes without corresponding changes in boiler operating parameters.

The injector system configuration is based on site-specific parameters and is designed to optimize the process as is the case with the NO$_x$OUT process. Sets of wall injectors are used, consisting of large jets located at or near boundary walls of the injection zone(s). The wall injectors have several advantages over injection grids: lower cost, improved load-following capability, simplicity of installation, lower fouling tendency, and no cooling requirements. Their obvious disadvantage is the difficulty of achieving the required mixing, especially across large injector planes. Virtually all applications today are based on using wall injectors. Better load-following capabilities with the use of injectors, multiple levels of injection, and other refinements have also eliminated the use of hydrogen. The carrier gas is used to achieve adequate mixing with the flue gas, since the amount of ammonia required for the process is quite small. The carrier gas requirements are typically 1 to 2% of the flue gas flow rate (Hurst, 1981). For a municipal solid waste facility, the amount of carrier gas can be 50–100 pounds of air per pound of ammonia injected.

Applications. With each thermal DeNO$_x$ license Exxon provides a process design package that covers design of the injectors (number, diameter, configuration, etc.), carrier gas requirements, suggested piping and instrumentation, ammonia storage tank size, power requirements, and other items. Testing is not required for NO$_x$ reduction and ammonia slip guarantees; however, detailed information about the boiler is needed (Haas, 1992A).

Engineering a THERMAL DeNO$_x$ application requires consideration of parameters such as flue gas flow rate (residence time); composition and temperature profiles (parallel and transverse to the gas flow path); load change data (rate of change of flow, composition, and temperature); mixing; and carrier gas attributes. Furnace drawings and tube bank details are required for both retrofit and new applications. Exxon has developed a proprietary kinetic process model based on 40 reactions, experimental data on the related kinetic rate constants, and theoretical considerations (Fellows, 1990). Exxon has also developed a three-dimensional, turbulent fluid flow/heat transfer model to evaluate the mixing process. The two models are used to analyze a given application. The analysis enables determination of injector locations (numbers and levels), ammonia injection rate(s), NO$_x$ reduction, and ammonia carryover (Haas, 1992A).

Estimates of the capital cost for boilers in the 30 to 500 MW$_t$ capacity range vary from 22 to 38 $/kW (1992 dollars) with 300 to 600 ppmv initial NO$_x$ concentrations. The operating cost is estimated to be 1.5 to 2.4 mils/kWh (1992 dollars) (Khan et al., 1990; Davis and Mikucki, 1989; Hurst, 1983; Fellows, 1989).

The capital cost of a THERMAL DeNO$_x$ system includes the costs of ammonia and carrier gas systems, injectors, instrumentation and controls, installation, engineering, licensing fees, and, for a retrofit, the cost of any modifications. Retrofit applications usually involve neither major modifications nor excessive downtime. The capital cost is application specific and depends on the initial NO$_x$ concentration, the size of the boiler, the NO$_x$ reduction required, the load following capability, and the ammonia carryover limitations.

The operating cost consists of the costs of the ammonia and carrier gas, electrical power consumption, and normal maintenance. If the boiler has to operate at higher-than-otherwise flue gas exit temperatures (e.g., to curtail ammonium salts emissions and/or low temperature air heater corrosion), then the cost of the lower boiler efficiency due to the higher boiler exit temperature should also be considered in calculating the operating costs.

The NO$_x$ reduction achieved in the THERMAL DeNO$_x$ process in a given application depends on the acceptable ammonia slip. In general, 40 to 70% NO$_x$ reduction with <5 ppm to 20 ppm ammonia slip can be obtained.

The process is suited to retrofit applications because of its low capital cost, relatively short installation time, and capability to meet newer (and tougher) regulations at low costs when used in combination with combustion modifications.

Materials. The injector is made of stainless steel. Additional information on ammonia storage and handling systems is covered in the SCR section of this chapter.

Selective Catalytic Reduction (SCR) Processes

General

Background. Selective Catalytic Reduction (SCR) uses a catalyst to increase the rate of selective chemical reactions between NO$_x$ and ammonia to produce nitrogen and water. This process has the highest NO_x reduction capability (greater than 90%) of any available NO_x control technology and is the most widely commercialized post-combustion control technology today.

The SCR process was originally invented and patented in the U.S. in 1959; however, its first applications were primarily limited to nitric acid plant pollution control (Anderson et al., 1961). Initial development of SCR technology included laboratory studies, which led to the evaluation of many catalyst materials and formulations, life test evaluations, and optimum process design (Campobenedetto and Murataka, 1990). The first commercial utility application of the process was in Japan in 1978 (Campobenedetto and Murataka, 1990), and commercialization in Germany began in 1985 (Gouker and Brundrett, 1991).

Due to the safety concerns associated with ammonia transportation and storage of ammonia near populated areas, the use of urea instead of ammonia has been investigated. Noroozi (1993) concludes that based on economics, anhydrous ammonia is the SCR reactant of choice. Should safety hazards or regulations dictate a ban on anhydrous ammonia, aqueous ammonia can be used. If even more restrictive safety considerations are present, the use of urea is theoretically possible, although it has not yet been used at an operating plant.

Status. The Japanese began the application of the SCR technology to boilers in the mid- to late-'70s. By 1985, there were more than 200 Japanese commercial applications with NO_x reductions ranging from 40 to 60% (Ando and Sedman, 1987). Among the many important contributions the Japanese have made to SCR development is the switch from noble metal to base metal oxide catalysts. In addition, the Japanese developed honeycomb and plate type metal catalysts to optimize geometry and to avoid plugging and erosion due to fly ash in coal-fired services. These improvements reduced the size and cost of the SCR reactors by a factor of two. About 80% of the boilers that went into operation in Japan between 1979 and 1990 were equipped with SCR systems (Kunimoto et al., 1990). Current development in Japan is focused on understanding the effects of fly-ash compounds on catalyst life (Gouker and Brundrett, 1991) and on achieving NO_x reductions greater than 95% (Pritchard, 1994).

SCR technology development shifted from Japan to Germany in the mid-1980s. German utilities and industries had more stringent regulatory standards to contend with, and the German SCR systems had to be designed for much higher NO_x reduction efficiencies (up to 90%). In addition to dry bottom units, as in Japan, SCR was retrofitted to slag tap boilers. Inlet NO_x concentrations, and coal sulfur and ash contents were also much higher in Germany than in Japan. Significant technical advances mitigated the loss of catalyst activity due to excessive fly ash, including catalyst poisoning, physical fouling, bulk plugging, and erosion. Other advances included the use of soot blowers, flow straightening vanes, and dummy catalyst layers. Even with high sulfur/high ash coal, catalyst life in German systems surpassed predictions based on Japanese experience (Gouker and Brundrett, 1991). By 1991, SCR had been installed on more than 32 gigawatts of power plant capacity in Germany, mostly in coal-fired plants (Iskandar, 1990). SCR has also been applied to many other power plants in Europe.

In the U.S., commercial SCR usage has been largely limited to gas- and oil-fired boilers, turbines, and industrial applications such as process heaters (Gouker and Brundrett, 1991). Pilot plant testing on coal-fired boilers, led by EPRI and the DOE, began in 1990 at six utility sites (Gulian et al., 1991). Three full-scale, coal-fired plants with SCR were in operation in 1994. These are at the Chambers, Keystone, and Indiantown plants (Franklin, 1994). Information on the first three SCRs to be used on coal-fired plants in the U.S. is provided in **Table 10-16.**

906 *Gas Purification*

The German and Japanese coal-fired power plant SCR experience, though significant, can not be directly applied in the U.S. due to differences in coal characteristics and the variable load operation of U.S. plants. According to Gouker and Brundrett (1991), the five key issues related to high ash/high sulfur U.S. coals must be addressed before SCR gains widespread acceptance in the U.S. (see page 907).

Table 10-16
Comparison of the First Three SCRs on Coal-Fired Boilers in the U.S.

	Chamber's Works Cogeneration Plant (Carney's Point, NJ)	Keystone Cogeneration Plant (Logan Township, NJ)	Indiantown Cogeneration Plant (Indiantown, FL)
Type of Boiler	Pulverized Coal	Pulverized Coal	Pulverized Coal
Plant Capacity	2 at 140 MW_e each	230 MW_e	330 MW_e
Start-Up Date	October 1993	July 1994	January 1995
Flue Gas Flow Rate, lb/hr	1,346,500	2,010,000	3,191,000
Flue Gas Temperature, Operating, °F	748	721	690
Inlet NO_x Concentration, ppmvd[1]	196	197	214
NO_x Removal Efficiency, Nominal, %	63[2]	63	50
Outlet NO_x Concentration, ppmvd[1]	73[2]	73	107
Outlet NO_x Concentration, lb $NO_x/10^6$ Btu	0.10[2]	0.10	0.15
Catalyst Type and Pitch	Ceramic honeycomb, 7.5 mm pitch	Plate, 6 mm pitch	Plate, 6 mm pitch
Flow Direction	Vertical down	Vertical down	Vertical down
Pressure Drop, Initial Operating, in. wc[3]	2.5	2	1.5
High-Dust or Low-Dust	High-dust	High-dust	High-dust
Ammonia Slip at SCR Exit, ppmvd[1]	5[4]	5	5
Coal Type	Eastern Bituminous	Eastern Bituminous	Eastern Bituminous
Coal Sulfur Content, %	2	1.09	1.09
Catalyst Life Guarantee, years[5]	10	10	3

Notes:
1. *At 7% O_2, ppmv (dry).*
2. *Permit allows higher emissions under some conditions.*
3. *With initial catalyst charge.*
4. *Based on a thirty-day rolling average.*
5. *Catalyst supplier furnishes at no additional cost to owner all catalyst required for the specified number of years of operation.*

Source: Franklin (1994)

1. Catalyst space velocity, superficial velocity, reaction temperature, and ammonia slip as a function of the NO_x reduction level;
2. Acceptable ammonia slip under high SO_2 and SO_3 conditions;
3. Catalyst performance and deactivation rate in flue gas and fly ash from U.S. coals;
4. Performance of the air preheater when exposed to high SO_3 concentrations and subsequently high concentrations of ammonium sulfate and ammonium bisulfate, and
5. Adhesion characteristics of fly ash from U.S. coals to SCR catalyst, the effectiveness of soot blowing, and the effects of residual fly ash on catalyst activity.

Despite these outstanding issues, SCR is emerging as the post-combustion technology of choice in many applications due to a favorable combination of reasonable cost, available guarantees, high NO_x reduction capability, low ammonia slip, and predictability of performance.

Process Description

Boiler Applications. For boiler applications, the SCR system consists of an ammonia storage, handling, and injection system; a catalytic reactor; controls; and instrumentation. Depending upon the location of the catalyst reactor in the flue gas path, the SCR system can be classified into one of the following basic categories (refer to **Figure 10-9**):

1. *Hot-side, high-dust system.* The SCR catalyst is located downstream of the economizer and upstream of the air preheater and any emission control equipment. This is the most commonly used configuration, as the required reaction temperature is usually available and no auxiliary heat input (and associated capital and energy expenditure) is required. Disadvantages of this configuration are that a large-pitch catalyst must be used to accommodate the heavy dust loading, and the reactivity of the catalyst must be limited to minimize the oxidization of SO_2 to SO_3. Although the presence of ash ultimately increases the required catalyst volume, it has been observed in the field that fly-ash particles actually assist in removing ammonium bisulfate and sulfate deposits from catalyst and air preheater elements. The space required for the SCR equipment is also a consideration with this configuration.
2. *Hot-side, low-dust system.* The SCR catalyst is located downstream of a hot-side electrostatic precipitator (ESP), and upstream of the air preheater and the FGD system. The major advantage of this configuration is the absence of significant fly ash in the flue gas, which minimizes catalyst erosion, required volume, and deactivation. Also, fly ash removal upstream of the ammonia injection point avoids NH_3 contamination of the fly ash and reduces the potential for catalyst poisoning. Disadvantages include the increased costs and problems associated with hot-side precipitators and the greater susceptibility of the catalyst to plug with fine dust particles and ammonium salts.
3. *Cold-side, low-dust system.* The SCR catalyst is located downstream of the air preheater, ESP, and FGD system. This configuration is also known as a tail-end system. Since the fly ash and SO_x are removed upstream, this SCR system requires the least catalyst volume and can utilize a smaller pitch and a more active catalyst. However, in order to achieve NO_x reduction performance in this location, it is necessary to include an auxiliary heat source to maintain the appropriate reaction temperature. The costs associated with the auxiliary heat source are usually greater than the cost savings resulting from minimized catalyst volume. This system is usually applied only to boilers where hot-side configurations can not be applied due to space limitations or to waste-to-energy plants where it is prudent to remove the acidic gases prior to entering the SCR.

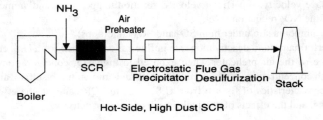

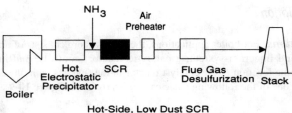

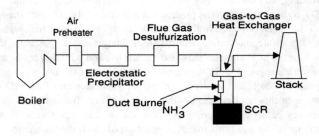

Figure 10-9. Basic SCR configurations. (*Robie et al., 1991A*)

Only the three basic SCR configurations: hot-side, high dust; hot-side, low dust; and cold-side, low dust are discussed in detail in this chapter, although other process designs are available for boilers. For example, a selective catalytic reduction air heater (CAT-AH) system based on a Ljungstrom type rotary air heater has been developed by Kraftanlagen AG Heidelberg (KAH). Since this type of air heater is designed with a large surface area in a small volume and the heat transfer surface provides intimate contact with the flue gas, moderate NO_x reduction can be achieved by applying catalyst material to the hot end surface of the air heater. This technology has been applied commercially in coal- and oil-fired plants in Germany and is currently in use at a 215 MW gas- and oil-fired plant in California (Reese et al., 1991).

The use of SCR in boiler applications with Western U.S. coal may prove difficult because of the cementitious properties of the coal ash. While SCRs normally operate above the dew point of the flue gas, they may go through the dew point, during startups and shutdowns, which could cause blinding of the catalyst by the ash.

Combustion Turbine Applications. The first application of SCR to combustion turbines for NO_x control was in 1980 at the Kawasaki plant of the Japanese National Railway. The first application of SCR to a utility combined-cycle powerplant for NO_x control was at Tokyo Electric's 2,000 MW_e Futtsu Power Station. This plant has 14 GE Frame 9 combustion turbines in a combined-cycle configuration (Schorr, 1992). As of 1990, the total installed U.S. SCR combustion turbine capacity included over 3,600 MW_e in 110 units (May et al., 1991).

Unique technical issues must be addressed with SCR turbine applications. Based on experience to date, SCR works best in base-loaded, combined-cycle combustion turbine applications where the fuel is natural gas. The reasons for this are the extreme dependency of the catalytic NO_x-ammonia reaction and the catalyst life on temperature, and the major problems associated with the use of sulfur bearing liquid fuels. (However, newer catalyst formulations that limit SO_2 to SO_3 conversion can reduce this concern.) In simple-cycle configurations, full scale, commercially proven, base metal oxide catalyst generally cannot be used because the exhaust gas temperature is above the operating range for SCR catalyst. In combined-cycle applications, the catalyst is located in the Heat Recovery Steam Generator (HRSG), **Figure 10-10,** where the temperature can be controlled in the optimum range of 600°–750°F, rather than in the combustion turbine exhaust, where the temperature typically exceeds 1,000°F (Schorr, 1992). In natural gas applications, operating temperatures of 450°–500°F or

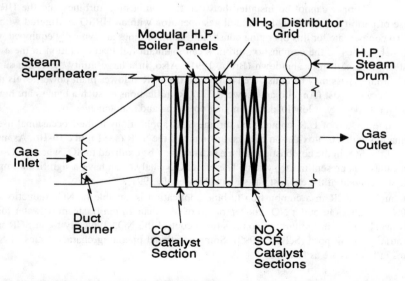

Figure 10-10. Schematic diagram of an NO_x SCR and a CO oxidation reactor in a heat-recovery steam generator (HRSG). *(Schorr, 1992)*

910 Gas Purification

lower are acceptable with some catalysts (Cohen, 1993, Campobenedetto, 1994). At lower temperatures, catalyst efficiency is poor; however, when the lower temperature is caused by a reduced load, the required NO_x removal efficiency may still be met because of the increased gas residence time in the catalyst bed.

In HRSG applications, the base metal oxide catalyst is normally positioned after several rows of tubes in the superheater and/or evaporator where the temperature of the exhaust gases at the catalyst is within the optimum range of operation. Maintaining the catalyst in the narrow operating temperature window over the entire operating load range is necessary to maintain a high NO_x reduction (e.g., 90%). Substantial temperature variations can occur at any given HRSG location if the combined-cycle plant is required to follow load, or if the HRSG is supplementary-fired under some, but not all load conditions. The temperature variation in such cases may make it impractical to find a single location for the catalyst that is always in the acceptable range (Schorr, 1992).

Steam production may be more difficult to control if the SCR is installed in the middle of the HRSG, because a portion of the turbine exhaust can no longer be diverted around the HRSG to the stack (Bretz, 1991). Incorrect placement or variations in the HRSG temperature outside the design range can result in deviations in the operating temperature for the catalyst, which in turn can lead to unnecessary increases in ammonia usage, reduced catalyst performance, and premature catalyst replacement. For proper placement, the following require evaluation (May et al., 1991):

- anticipated combustion turbine load swings
- shifts in HRSG steam demand
- impact of duct-firing on HRSG temperature profile
- changes in HRSG performance with time

A bypass damper cannot be installed between the combustion turbines and the HRSG because of permit requirements for natural gas operation with an HRSG configured with an SCR. To compensate for this operating constraint, steam turbines are typically equipped with full-flow bypasses so the combustion turbines can maintain full load operation in the event of a steam turbine trip or shutdown (Bretz, 1991). Also, in a large utility system, a superheater partial bypass may be considered to operate the SCR at lower plant power levels than would otherwise exist (Angello and Lowe, 1989). Ramifications of such a bypass are higher cost, heat rate penalty at low loads, and higher operating and maintenance costs.

A unique feature of U.S. combustion turbine SCR applications is their occasional use in combination with a CO-oxidation catalyst upstream of the SCR (see **Figure 10-10**). As much as half of the SO_2 in the combustion turbine exhaust may be oxidized to SO_3 when a CO-oxidation catalyst is present in the system. Therefore, CO catalyst can have a significant impact on the SO_3 content of the exhaust gas stream (May et al., 1991).

Although an SCR in a combustion turbine application is capable of NO_x reductions of 80–90% or higher (an outlet NO_x concentration of less than 10 ppm), steam or water injection is usually used in the turbine combustor to control the NO_x, typically to an SCR inlet concentration of 42 ppm (Schorr, 1992). Some additional operating characteristics of SCR systems in HRSGs are as follows:

- For new combustion turbine applications, the ammonia slip is typically guaranteed to not exceed 10 ppm over the life of the catalyst (Cohen, 1993).

- Typical space velocities for combustion turbine (clean fuel) applications are between 12,000 and 22,000 hr^{-1} for ceramic honeycomb catalysts (Pritchard, 1994), but lower for the plate-type catalysts (Campobenedetto, 1994). (Note: In coal-fired applications for both types of catalysts, the space velocities are lower and their ranges closer because wider openings are required in the ceramic honeycomb type catalyst.) For platinum catalysts, which are less widely used in these applications, higher space velocities are possible.
- Pressure drops across the SCR system range from 1 to 6 in. wg, which reduces the power generating capacity and increases the heat rate of the turbine (May et al., 1991; Pritchard, 1994). Typical pressure drop values are 2–3 in. wg (Cohen, 1993).

There is very limited experience in the U.S. firing distillate oil in combustion turbines equipped with SCRs (May et al., 1991) primarily because sulfur in the oil can cause operating problems. The fuel sulfur problem is generally related to ammonium salts formation. Ammonium bisulfate forms in the lower temperature section of the HRSG where it deposits on the walls and heat transfer surfaces. The surface deposits result in increased pressure drop, reduced heat transfer and power output, and lower cycle efficiency (Schorr, 1992). If significant oil-firing is expected, designing the system for 5 ppm or less ammonia slip can minimize these effects (Cohen, 1993). Similar problems can occur in SCR systems when sulfur-containing refinery gas is used as the primary fuel.

Basic Chemistry

In the selective catalytic reduction process, NO_x in flue gas is selectively reduced to nitrogen and water vapor by reaction with HH_3 in the presence of a catalyst. The SCR process is described by the following overall chemical reactions (Blair et al., 1981):

$$4NO + 4NH_3 + O_2 = 4N_2 + 6H_2O \tag{10-4}$$

$$2NO_2 + 4NH_3 + O_2 = 3N_2 + 6H_2O \tag{10-5}$$

$$6NO + 4NH_3 = 5N_2 + 6H_2O \tag{10-6}$$

$$6NO_2 + 8NH_3 = 7N_2 + 12H_2O \tag{10-7}$$

The catalyst lowers the activation energy of the NO_x reduction reactions, thereby allowing the reactions to proceed at flue gas temperatures lower than those required for non-catalytic NO_x reduction processes. With most catalysts, the upper operating temperature for the SCR process is limited to about 750°F to prevent catalyst sintering and consequent deactivation, while the lower operating temperature is between about 500° and 580°F. The lower operating temperature limitation depends on the SO_3 content of the flue gas. If the operating temperature is too low, SO_3 in the flue gas can cause deposition of ammonium sulfate on the catalyst surface and resulting catalyst deactivation until removed. Some permanent plugging of pores may occur with long and repeated cycling to excessively low temperatures. Temperature cycling and lower temperatures can occur at part load or during startup.

The reactions involving oxygen comprise the dominant path for NO_x reduction; as a practical minimum, 1% O_2 by volume is required. NO_x in flue gas is predominantly NO, and the SCR process has a molar ratio of NO-reduced to NH_3-injected close to the theoretical value of 1.0 for reaction 10-4. **Figure 10-11** illustrates a typical relationship between the NO_x reduction efficiency, NH_3 slip, and the NH_3/NO_x molar ratio. As the figure depicts, the relationship between the NO_x reduction efficiency and NH_3/NO_x molar ratio is nearly linear for efficiencies up to 85 to 90%. A low space velocity is required to achieve high NO_x reduction efficiency with low ammonia slip.

In general, the molar ratio is always higher than unity due to ammonia slip and the decomposition of small quantities of ammonia according to the following reaction:

$$4NH_3 + 3O_2 = 2N_2 + 6H_2O \tag{10-8}$$

In addition to the NO_x reactions, the catalyst can also oxidize SO_2 to SO_3:

$$2SO_2 + O_2 = 2SO_3 \tag{10-9}$$

SO_3 can react subsequently with H_2O in the flue gas to form sulfuric acid, which can then combine with the unreacted ammonia to form finely divided ammonium sulfate and ammonium bisulfate particles. The SO_2 to SO_3 conversion percentage across the catalyst, K_{SO_2}, is defined in equation 10-10.

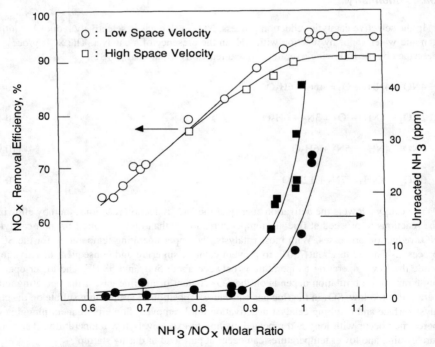

Figure 10-11. Effect of the NH_3(feed)/NO_x(inlet) molar ratio on NO_x removal efficiency and unreacted ammonia in an SCR (data based on an NO_x concentration of 350 ppm and a reactor temperature of 660°F). (*Itoh and Kajibata, 1981*)

$$K_{SO_2} = (C_{SO3D} - C_{SO3U}) \cdot (100)/C_{SO_2U} \tag{10-10}$$

Where: C_{SO_3U} = the SO_3 concentration upstream of the catalyst,
        C_{SO_3D} = the SO_3 concentration downstream of the catalyst, and
        C_{SO_2U} = the SO_2 concentration upstream of the catalyst.

The SO_2 to SO_3 conversion is significantly affected by (Jaerschky et al., 1991; Cohen, 1993):

- The chemical composition of the catalyst, particularly the vanadium pentoxide content.
- The area velocity, which is the ratio of the flue gas volumetric flow rate to the surface area of the catalyst.
- The flue gas conditions, particularly the flue gas temperature. The conversion to SO_3 increases exponentially with the flue gas temperature.

Figure 10-12 shows the relationship between NO_x reduction, flue gas temperature, and SO_2 to SO_3 conversion for a given catalyst composition and volume. The SO_2 to SO_3 conversion is typically maintained below 3% for many applications and below 1% for coal-fired applications (Cohen, 1993).

System Design

Site-Specific Considerations. The proper specification of an SCR system requires consideration of site specific information. This information includes type of fuel, flue gas flow rate and temperature, initial NO_x concentrations, ash quantity and characteristics, SO_2 and SO_3 concentrations in the flue gas, water vapor concentration, excess oxygen concentration, plant mode of operation, emission compliance requirements, catalyst life requirements, ammonia slip requirements, and data on the existing equipment in the flue gas train.

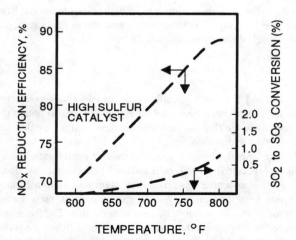

Figure 10-12. NO_x reduction efficiency and SO_2 conversion vs. SCR temperature. *(Behrens et al., 1991A)*

Catalyst Location and Arrangement. Practical and economic considerations determine the optimum catalyst location. Additional concerns affecting the location and arrangement of the SCR catalyst involve proper mixing of ammonia and the flue gas, the amount of ash in the flue gas, and the ash characteristics. The catalyst arrangement should be flow model-tested for uniform gas flow distribution through the catalyst and mixing of the ammonia with the flue gas upstream of the catalyst. Effects of the SCR system arrangement on the downstream equipment should also be considered.

Catalyst Volume. The performance of an SCR reactor depends significantly upon the diffusion surface area (or specific surface area), which includes the surface area of both macro-pores and micro-pores in the catalyst. SCR performance increases with the specific surface area. Since this area is catalyst-specific, it is not a convenient design parameter (Cho and Dubow, 1992). A widely used parameter is space velocity, which is usually defined as the volumetric flue gas flow rate in normal ft^3/hr (at 32°F, 1 atm) divided by the bulk volume of the catalyst.

NO_x reduction efficiency, N, is defined as the quantity of NO_x reduced divided by the quantity of NO_x in the inlet stream. NO_x reduction efficiency may be correlated (for a 1.0 molar ratio) in terms of the space velocity, SV, and the NH_3/NO_x molar ratio, m, as follows:

$$N = m(1 - \exp(-K/SV)) \tag{10-11}$$

where K is a constant, which depends on many factors, including the catalyst composition, diffusion characteristics of ammonia and NO_x in the gas stream and catalyst, O_2 and H_2O concentrations, gas temperature and velocity, and catalyst age (Cho and Dubow, 1992). In general, for a given configuration, higher values of the molar ratio and lower values of space velocity, i.e., more ammonia and longer residence time, cause the NO_x reduction efficiency to be higher. Lower space velocity values greatly reduce ammonia slip at NO_x/NH_3 molar ratios over about 0.85 (see Figure 10-11).

In a specific application, the inlet NO_x and the desired NO_x emission concentrations define N. Given the value of N, m is calculated from the simple material balance equation by fixing the value of the NH_3 slip, S and assuming a reaction stoichiometry of 1 mole NH_3 per mole of NO_x and no NH_3 destruction (Cho and Dubow, 1992):

$$m = N + S/\text{inlet } NO_x \tag{10-12}$$

From the values of m and N, catalyst-specific curves or tables (dependent upon reaction rates and deterioration expectancy) can be used with equation 10-11 to obtain the required value of SV. Space velocity and the flue gas flow rate are then used to calculate the volume of catalyst required for a given application.

The required space velocity depends upon the catalyst configuration and properties (such as activity, stability, expected life), NO_x inlet and outlet concentrations, flue gas temperature, flue gas SO_3 and SO_2 concentrations, fly ash content and composition, and the desired ammonia slip. Space velocities can vary widely depending on these factors. Comparisons based on space velocity should therefore be used with caution. The fly ash concentration and composition dictate the allowable pitch, which in turn establishes the surface area per unit catalyst volume. **Figure 10-13** (Robie et al., 1991A) shows the typical relationship between the NO_x reduction efficiency and the flue gas temperature at various space velocities. As the figure shows, higher space velocities (less catalyst volume) at higher temperatures can

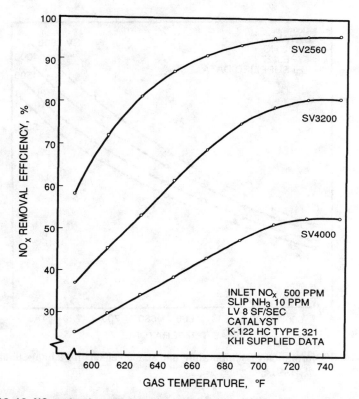

Figure 10-13. NO_x reduction efficiency vs. temperature and space velocity for a low-sulfur catalyst. *(Robie et al., 1991A)*

achieve the same NO_x reduction as low space velocities at low temperatures. Higher flue gas temperatures and lower space velocities increase SO_3 formation, which can have a detrimental effect on the downstream equipment. This relationship for a high-sulfur catalyst is shown in **Figure 10-14** (Robie et al., 1991A). In both **Figures 10-13** and **10-14,** the space velocity (SV) is expressed as normal cubic feet (32°F, 1 atm.) of flue gas per hour per cubic foot of catalyst. The linear velocity (LV), as used in **Figure 10-13,** represents the value obtained when the volumetric flow rate of the gas in normal cubic feet per second is divided by the cross-sectional area of the catalyst in the catalyst bed. Its units are normal feet per second (SF/SEC). Data in the figures are based on Kawasaki Heavy Industries' (KHI) catalyst.

Catalyst Selection. Because catalyst volume is determined on the basis of catalyst-specific space velocity curves, the choice of a catalyst type, composition, and configuration establishes the overall SCR reactor design. In general, the SCR systems today are based on modular-type catalyst configurations using either the parallel plate or honeycomb matrix design. In the parallel plate design, the catalyst material is generally attached to a stainless steel or ceramic substrate. Refer to **Figure 10-15.** Honeycomb catalysts can be homogeneous or composite. The honeycomb matrix is an extrusion of either homogeneous catalyst or a sub-

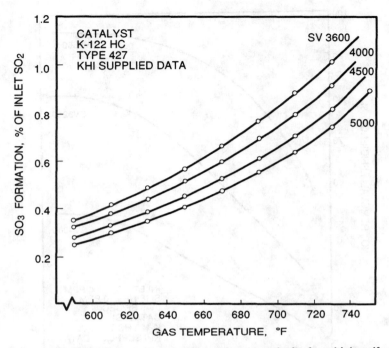

Figure 10-14. SO_3 formation vs. temperature and space velocity for a high-sulfur catalyst. (*Robie et al., 1991A*)

strate, a strong, thin-walled ceramic support to which a thin layer of catalytic material is bonded to form a composite structure. Spalling and erosion of the catalyst layer are concerns with composite catalysts. The parallel plate type is illustrated conceptually in **Figure 10-15** and the honeycomb types in **Figures 10-16** and **10-17**. Commercial honeycomb catalysts have many more gas passages than the few shown in **Figure 10-16.**

The most commonly used SCR catalysts are composed of metal oxides such as titania and vanadia. These are called base metal oxide catalysts to distinguish them from catalysts containing precious metals such as platinum. In base metal oxide catalysts, the vanadium controls the reactivity, but it also catalyzes SO_2 oxidation. Therefore, for high-sulfur applications, the vanadium content of the catalyst elements should be minimized (Behrens et al., 1991A). By reducing the residence time in the catalyst, i.e., increasing the area velocity, SO_2 oxidation to SO_3 can be reduced. At the lower gas flows associated with lower loads, less SO_3 is often generated because the lower gas temperatures decrease SO_3 production more than the lower area velocities increase it (Cohen, 1993).

Recently, high-temperature zeolite catalysts have been introduced, which are reported to be effective and stable over a wider temperature range than base metal oxide or precious metal catalysts. However, the temperature range is zeolite type specific. For example, a naturally occurring mordenite zeolite can operate between 430°–970°F depending upon the formulation. A synthetic zeolite, ZSM-5, has a narrower operating temperature range of 570°–900°F. Another synthetic zeolite, which is coated on a ceramic honeycomb structure, is claimed to be operational at temperatures between 675°–1,075°F (Schreiber, 1991). By com-

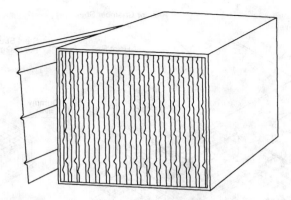

Figure 10-15. Plate-type catalyst unit. (*Campobenedetto and Murataka, 1990*)

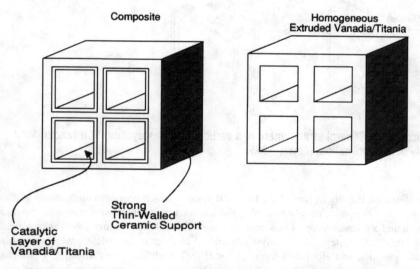

Figure 10-16. Portions of honeycomb-type catalyst blocks (composite and extruded types). (*Speronello et al., 1991*)

parison, with vanadia-titania catalyst, ammonia begins to thermally decompose above 800°–850°F and, at higher temperatures, to oxidize to NO_x, which is counterproductive. This is not usually the case with zeolite catalysts, and this fact must be considered when evaluating the applicability of a zeolite catalyst or whether SCR itself is even applicable. By far the greatest advantage of the high-temperature zeolite catalyst is its potential for large scale commercial use in turbine exhaust applications, especially in simple-cycle plants. Also, in combined-cycle plants the extended temperature range of zeolite catalysts will ease the problem of providing suitable SCR catalyst locations within the HRSG.

A Norton Corporation zeolite catalyst has been operating since December 1990, on the exhaust duct of a 4 MW_e Solar Centaur Gas Turbine Cogeneration Plant installed at Uno-

918 *Gas Purification*

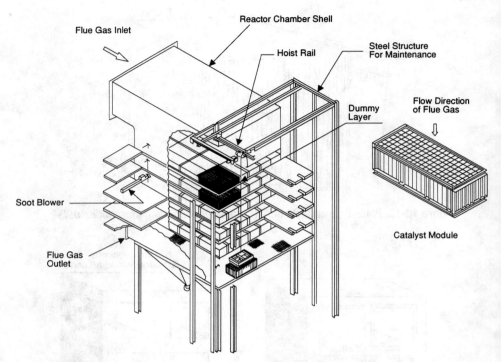

Figure 10-17. General arrangement of a vertical flow, honeycomb SCR reactor for a utility boiler application. (*Lowe, 1985*)

cal's Research Facility in Brea, CA. The SCR system, which operates continuously at about 960°F, achieves approximately 80% NO_x reduction on the average with ammonia slip under 20 ppm and a pressure drop of less than 3 in. wg (Anon., 1992). **Figure 10-18** schematically shows this SCR system. The catalyst is installed in front of the HRSG, which simplifies HRSG operation and eliminates the need for HRSG modification by allowing a one-piece construction versus two separate sections. Zeolite catalysts are expected to be less of a disposal problem than base metal oxide catalysts because they contain no heavy metals and reportedly qualify as a non-hazardous waste (Robie et al., 1991A; Schreiber, 1991).

Cormetech has developed a titania-tungsten catalyst that should operate up to approximately 1,050°F and that is claimed to be competitive with zeolite catalyst. The first application of this catalyst was scheduled to go into operation in February 1995, on a simple-cycle gas turbine unit (Pritchard, 1994).

Recently, a precious metal-based (platinum) catalyst was applied commercially in the U.S. This catalyst reportedly can operate within a temperature window of 425°–525°F, with an optimum operating temperature of about 475°F. The low-temperature range allows placement of the catalyst outside of the high-pressure section of the HRSG, upstream of the economizer and stack. The limitations of this type of catalyst are (1) above 525°F, the catalyst oxidizes NH_3 producing additional NO_x; and (2) the catalyst is limited to low sulfur fuel applications because it oxidizes SO_2 to SO_3 (May et al., 1991). Some catalyst manufacturers

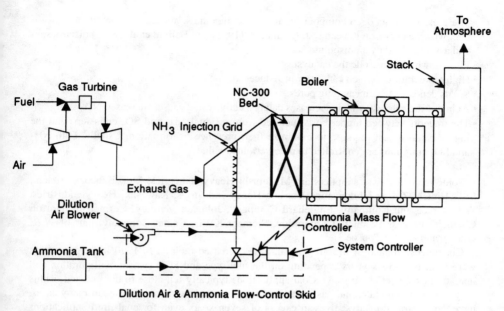

Figure 10-18. SCR process using high-temperature zeolite catalyst (NC-300) in a turbine exhaust cogeneration application. (*Anon., 1992*)

offer a catalyst that they claim destroys excess ammonia, and at least one offers a precious metal catalyst that not only reduces NO_x but also controls both NH_3 and CO emissions (Schorr, 1992). There are also catalysts that are reported to destroy N_2O.

A goal in catalyst design is large numbers of both macro-pores (that gas can diffuse into) and micro-pores (to support the reduction reactions) (Behrens et al., 1991A). These pores are much smaller than the gas passages through the catalyst. For extruded catalysts, the gas passages are the openings made by the extrusion dies; for plate-type catalysts, they are the openings created by the bends in the plates. Smaller passages through the catalyst result in more surface area (more macro-pores and micro-pores) for the same catalyst volume, which increases NO_x reduction efficiency. However, smaller passages also increase the flue gas pressure drop and the sensitivity of the catalyst to plugging by ash particles. The size of the flue gas passages can be selected to accommodate a range of flue gas ash concentrations. Large passages are used in high ash concentration applications. Typical pitches for coal applications range from 6.5 to 7.5 mm for honeycomb catalysts. Plate-type catalysts have lower area/volume ratios than honeycomb-extrudate types due to plate spacing limitations that increase their required volume (Franklin, 1993). The parallel plate design typically offers low flue gas pressure drop, and minimizes catalyst erosion and masking problems attributable to high ash concentrations (Robie et al., 1991A). With all catalysts, thermal stress must be considered. Temperature ramp rates when temperatures are below the dew point must be low enough to allow any liquid in the catalyst to evaporate (Pritchard, 1994).

Catalyst selection must also consider the following operating conditions:

- The presence of halogen compounds, alkaline compounds, arsenic, and ammonium bisulfate, which can poison the catalyst. Chen et al. (1991) and Balling et al. (1991) discuss several aspects of catalyst poisoning.
- Fly ash, which can erode the catalyst.
- High temperatures, which can sinter or reduce catalyst pore volume.
- Solids deposits, which can plug pores.
- Aging of the catalyst, which decreases catalyst activity and causes increased NH_3 slip.
- The tradeoff between higher costs incurred from residual NH_3 and SO_3 emissions and the higher catalyst cost required to minimize their presence or formation. Usually, both NH_3 and SO_3 emissions are controlled to predetermined values.

Some major catalyst suppliers are Mitsubishi Heavy Industries, Ltd., Babcock Hitachi, Ltd., Hitachi-Zosen, Ishikawajima-Harima Heavy Industries, Kawasaki Heavy Industries, W.R. Grace and Company, Englehard Company, Johnson Matthey, Cormetech, Norton Chemical Process Products Corporation, Siemens/KWU, Camet Company (part of W.R. Grace), and Haldor Topsøe.

Guarantees on catalyst life, in general, have increased rapidly. In 1990/1991, guarantees were from two to six years depending on the fuel type and SCR system configuration and operation (Robie et al., 1991A). Coal applications typically fell close to the low end of this range, oil in the middle, and gas at the upper end. Actual field applications, in many cases, have demonstrated catalyst lives in excess of seven years even for coal-fired applications (Campobenedetto and Murataka, 1990). Based on the lives of operating catalysts and the very few cases of catalyst poisoning being reported, much longer lives are currently being guaranteed.

SCR system life can be extended by using a catalyst management/replacement program. These programs are typically based on monitoring the NH_3 slip, and include rotation, replacement, and/or addition of catalyst layers at scheduled intervals. At the Chambers Works Cogeneration Plant (refer to **Table 10-16**), compliance was guaranteed for ten years based on the use of a catalyst addition/replacement program (Franklin, 1993).

Reactor Design. Individual catalyst elements are assembled into blocks or modules that are then arranged in horizontal or vertical layers within the reactor on engineered support structures. In large systems, there are several layers with space between each layer for access. **Figure 10-17** depicts a general arrangement of a vertical flow, honeycomb SCR reactor. The reactor housing can be oversized to provide for additional catalyst layers if this is required by the catalyst management program. The SCR reactor houses and supports the individual catalyst units and includes a sealing system to prevent flue gas from bypassing the catalyst modules (Campobenedetto and Murataka, 1990).

The cross-sectional area and the number of catalyst layers are selected to optimize (within allowable space limitations) gas velocity, pressure drop, and gas flow distribution for the required NO_x reduction (Behrens et al., 1991B). In the presence of fly ash, it is preferred to direct the flue gas vertically downward to facilitate the passage of fly ash through the catalyst. This is especially pertinent with load-following units, where the ash carrying velocity may not be maintained otherwise (Robie et al., 1991A). The maximum allowable linear velocity (flue gas velocity entering the catalyst based on the overall cross-sectional area of the catalyst bed) depends on the catalyst type and the application. For hot-side/high-dust applications, the linear velocity should be kept below 20 ft/sec to mitigate catalyst erosion. In low-dust applications, the erosion potential is minimal, and the linear velocities are based on maintaining laminar flow through the catalyst (Robie et al., 1991A; Cho and Dubow, 1992). In either case, significant penalties in pressure drop and catalyst life can occur if linear velocity is too high. Typical pressure drop values (attributable to the entire SCR system) are from less than 4- up to 6-in. wg for hot-side SCR systems and much higher for cold-side systems due to the additional equipment in the gas path (Franklin, 1993; Cohen, 1993).

There are several peripheral considerations required in SCR reactor design. Structural steel support is designed to include platforms for operation and maintenance of the catalyst, and expansion joints are provided to compensate for three dimensional thermal expansion with temperature. Hoist or monorail systems are often provided to facilitate catalyst installation, removal, and replacement. In high-dust applications, soot blowers are often provided to alleviate catalyst masking and plugging. Also, for high-dust applications, a top layer of "dummy" catalyst is sometimes installed to act as a sacrificial leading edge, thus protecting the active catalyst from erosion (Lowe, 1985). The layer is often harder than the catalyst to withstand erosion, and the additional pressure drop that it creates tends to improve gas distribution. Straightening vanes, turning vanes, and baffles should also be considered for promoting proper flue gas distribution and minimizing both turbulence and potential catalyst masking (Cho and Dubow, 1992).

When necessary, an SCR gas bypass is provided for startup and shutdown. Use of the bypass to permit operation (full or partial) during catalyst removal/replacement is not recommended. SCR bypasses offer protection when the flue gas temperature is below the dew point, such as during startup and shutdown. However, most applications do not have SCR bypasses, since routines are used during startup and shutdown which preclude their need (Cho and Dubow, 1992), and regulations sometimes prohibit their use. Also, experience in Japan and Germany has shown them to be costly and not required to prevent damage due to low-load oil firing, thermal gradients, and other conditions. However, an economizer bypass is required on most boiler applications to avoid ammonium salt deposition at the low operating temperatures associated with low loads (Campobenedetto, 1994).

Process Control

A simplified SCR process control logic diagram is shown in **Figure 10-19.** The NH_3 injection requirements of an SCR process are controlled based on the NO_x concentration at the inlet of the SCR. Measurement of the NO_x concentration at the SCR inlet combined with the flue gas flow yields the NO_x flow signal. The NH_3 control module multiplies this signal by the NH_3/NO_x molar ratio setpoint to obtain the NH_3 flow demand signal. The NH_3 demand signal is compared with the actual NH_3 flow, and the NH_3 flow is adjusted to meet the demand. The ammonia slip is measured downstream of the SCR reactor and recorded as is the NO_x concentration downstream of the SCR reactor bypass. Although not shown on this diagram, the NO_x analyzer downstream of the SCR reactor can be used to reset the NH_3/NO_x molar ratio setpoint in the NH_3 control module.

Ammonia feed is shut off by a low temperature switch when the operating flue gas temperature drops below the minimum recommended value. This prevents deactivation of the catalyst from ammonium bisulfate deposition. This control feature is also applied during system startup and shutdown. An economizer bypass is used to maintain the flue gas temperature above the minimum recommended SCR operating temperature and an SCR bypass is shown (though not often provided). The SCR bypass is used to protect the SCR catalyst during startup and shutdown when the flue gas temperature can be below its dew point. Economizer bypasses are used at the Chambers, Indiantown, and Keystone power plants (Franklin, 1993).

922 Gas Purification

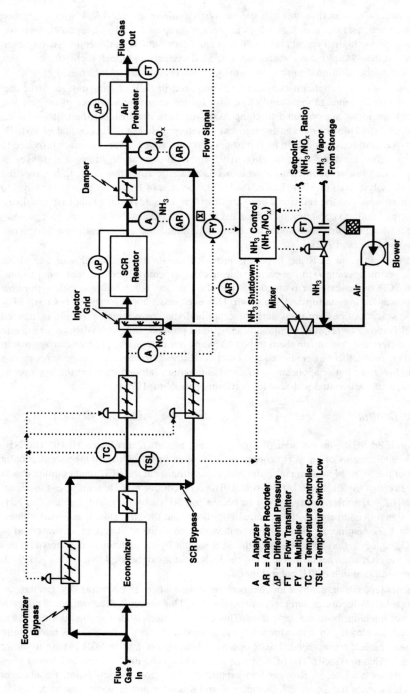

Figure 10-19. Simplified SCR process control logic diagram.

Emissions Control Systems

Continuous NO_x monitoring is required for SCR process control and may be required by environmental regulations. Ammonia slip monitoring is not required as part of the SCR process control scheme; however, it may be required by permit. An ammonia concentration monitor can be used to avoid or predict downstream process equipment operating problems. It is also utilized, in conjunction with NO_x monitoring, for determining the effectiveness or activity of the SCR catalyst. Currently, reliable continuous NH_3 monitors are not available for accurate, low concentration ammonia determination, and manual sampling techniques, based on wet chemical analysis of flue gas samples, are usually used for such measurements. However, at the Chambers Works Cogeneration Plant, an ammonia monitor is being used to report ammonia emissions on a continuous basis (Franklin, 1993).

Ammonia Storage and Handling System

Ammonia is a common industrial chemical, which is hazardous to health and is highly explosive under some conditions. It is an irritant to skin, eyes, and respiratory and mucous membranes, and it can cause death at concentrations greater than 5,000 ppm. The Occupational Safety and Health Administration (OSHA) in 29 CFR 1910.1000 has set a short term (15 minute) exposure limit of 35 ppm for ammonia. The American Conference of Governmental Industrial Hygienists' required limit for an eight-hour time-weighted average exposure is 25 ppm, and for a 15-minute short-term exposure the limit is 35 ppm (ACGIH, 1991). The National Institute for Occupational Safety and Health recommends breathing apparatus at 500 ppm, because this concentration represents Immediate Danger to Life and Health (IDLH) (NIOSH, 1990). Ammonia is on the EPA list of extremely hazardous substances under the TITLE III, section 302 of the Superfund Amendments and Reauthorization Act of 1986 (SARA) (Schorr, 1992). It is potentially explosive in air at atmospheric pressure in concentrations of 16–25% and at temperatures as low as 1,200°F (Robie et al., 1991A).

Ammonia is shipped by rail or truck, rail car capacity being 160,000 pounds and truck capacity 35,000 pounds. Anhydrous ammonia is generally stored in pressure vessels rated at 250 psig, while aqueous ammonia can be stored at atmospheric pressure (Robie et al., 1991A; Franklin, 1992). Because of its availability, only one to two weeks of ammonia storage capacity is usually provided. Ammonia safety systems are provided at the storage area to detect and protect against leakage.

In the U.S., the current trend is to use aqueous ammonia. This is based primarily on safety concerns, although aqueous ammonia also has safety concerns associated with its use. Fire marshalls and local governing or regulatory agencies may also limit the use of anhydrous ammonia. There have been a number of conversions from anhydrous to aqueous systems in the U.S. In Europe, ammonia storage requirements vary widely. The strictest European requirements have mandated, for anhydrous systems, double-walled tanks, underground piping, and vaults below grade to contain spills (Lewis, 1993; Siegfriedt and Pacer, 1991). Other design features for ammonia systems may include (Franklin, 1992; Siegfriedt and Pacer, 1991)

- Ammonia tanks and piping protected from local traffic, potential coal mill explosion, and potential steam turbine failure by remote location, shielding, or the use of underground vaults
- Flame backflash valves with bypasses for cleaning

- A deluge system that is activated above maximum allowable ammonia concentrations
- Protective safety suits and oxygen tanks for personal use
- Dedicated ammonia piping and venting from each storage vessel
- Basin curbs at grade for containment of spills

In addition, all applicable codes and standards for ammonia safety and storage requirements must be followed. ANSI (American National Standards Institute) K61.1, *Storage and Handling of Anhydrous Ammonia,* is widely used for the design of anhydrous systems. There is no comparable code for aqueous systems.

Choosing between aqueous and anhydrous ammonia systems involves considerations such as safety, required storage capacity, availability, and regulations. Aqueous ammonia is generally considered safer so that diking, a deluge system, and ammonia leakage monitoring are usually not required. Also, cleanup is relatively easy in case of a major leak. Anhydrous ammonia, by comparison, is considered a hazardous material, and is more difficult to permit. On the positive side, anhydrous ammonia requires much less physical space for storage. Also, it may not require a separate heat source for vaporization because if the required ammonia flow is small, enough heat may be absorbed through the tank surface for adequate vaporization. Most cases, however, require ammonia flow beyond the capacity of the heat absorbed through the tank surface, and an electric vaporizer is usually provided. Without a vaporizer, the ammonia gas is nearly saturated and condensation can occur (Franklin, 1992, 1993; Cohen, 1993).

Ammonia injection systems may use either anhydrous ammonia, aqueous ammonia that is completely vaporized, or ammonia that is stripped from the aqueous solution. Usually, the most economical system is the anhydrous system, and the least economical is the aqueous ammonia system with complete vaporization. If the ammonia is stripped from the aqueous solution, the ammonia-laced stripped water must be disposed of and the complexities of stripping the ammonia from the water must be dealt with. Concerns about ammonia-containing wastewater disposal and safety have prevented widespread use of ammonia stripping (Lewis, 1993).

Figure 10-20 depicts an anhydrous ammonia system and an aqueous ammonia system with vaporization. In the anhydrous ammonia system, an evaporator is used to maintain the vapor pressure in the ammonia storage tank. The gaseous NH_3 flows to an NH_3/carrier gas mixing chamber where the proper quantity of gaseous ammonia is mixed with the carrier gas. The carrier gas, either compressed air or low pressure steam, is mixed at a minimum NH_3/carrier gas volumetric ratio of approximately 1 to 20 (Campobenedetto and Murataka, 1990).

In an SCR system using aqueous ammonia, the ammonia solution, about 28 wt% concentration, is stored at atmospheric pressure. It is pumped from the storage tank, metered, and sprayed into a vaporizer vessel. Ambient air provided by the blower is typically heated to 600°–850°F and introduced to the vaporizer. In the vaporizer, the cold aqueous ammonia and the hot air mix together, vaporizing the liquid ammonia and water (Cho and Dubow, 1992).

The injection system design is common to both types of ammonia systems. Injection grids are designed to provide an even distribution of ammonia from the evaporator or vaporizer into the flue gas. Proper mixing is ensured by using external, manually-set control valves to balance the NH_3/air mixture to match the flue gas flow distribution pattern and by using the appropriate number and arrangement of injection grid nozzles/orifices. The minimum distance required for NH_3 to mix with the flue gas is about 30 feet from the point of injection. If space restraints limit the available mixing length, the distance can be shortened to 10 to 12

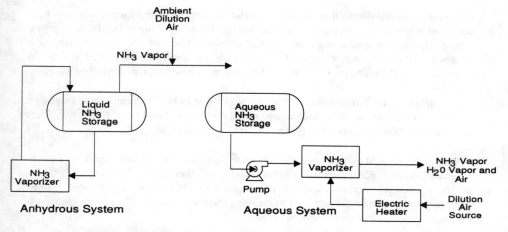

Figure 10-20. Schematic diagrams of ammonia storage and handling systems.

feet by installing a mixing device (Campobenedetto and Murataka, 1990). Makansi (1992A) also discusses ammonia storage, handling, and injection.

Materials. Material requirements vary with the type of system. Both anhydrous and aqueous storage tanks can be carbon steel, although aluminum and stainless steel have been used for aqueous systems to minimize the corrosion that occurs when the aqueous solution is exposed to air. Some aqueous ammonia storage tanks operate at 15–25 psig to minimize the amount of air entering the tank, and a few tanks use nitrogen blanketing. Stress-relief of carbon steel tanks is not necessary for ammonia storage tanks when the ammonia contains more than 0.2% water, which is usually the case for the anhydrous ammonia used in NO_x control applications. However, stress relief of both anhydrous and aqueous ammonia tanks is often performed as a safety precaution (Wehner, 1994).

In anhydrous systems, carbon steel is acceptable for all other components due to the non-corrosive nature of anhydrous ammonia. In aqueous systems, stress-relieved carbon steel or 304/304L stainless steel is required for piping, the vaporizer, and any equipment exposed to aqueous ammonia. Small bore piping can be 316L stainless steel; however, carbon steel (not stress-relieved) is sometimes the preferred material due to cost considerations. Piping 2 inches and larger can be schedule 80 carbon steel. Carbon steel is also acceptable downstream of the vaporizer. Valves can be carbon steel with 316 stainless steel trim and Buna-N or Teflon for other parts.

The injection grid can be carbon steel, but near the injectors, stainless steel may be used due to the corrosive effects of the combination of ammonia and fly-ash buildup (Cohen, 1993). The entire grid is sometimes made of stainless steel (Pritchard, 1994). Copper, galvanized steel, and viton should not be used in ammonia systems.

Catalyst Management Program

Eventually, a catalyst reaches the end of its useful life and strategies must be developed for recycling or for economical, environmentally safe disposal. The base metal oxide catalysts are of primary concern due to their hazardous metal content, including metals that may have deposited in the catalyst, such as lead and arsenic (zeolite catalysts are claimed to be non-hazardous requiring no special disposal). Typically, responsibility for recycling or disposal is assumed by the SCR catalyst vendor. Options for catalyst disposal are reviewed in detail by Robie et al. (1991A). The main options available for spent base metal oxide catalyst follow.

Disposal as Waste/Hazardous Waste. As of October 1994, this material was listed as a hazardous waste (due to the vanadium and nickel content) in the state of California. There were no restrictions on disposal in the rest of the U.S. at that time.

Reuse. Spent catalyst may be used as a substitute for a commercial product or as an ingredient in an industrial process to make another product, and, thereby may be removed from regulation as a federal hazardous waste. For example, the spent catalyst may be used as a raw material for cement manufacture or as a concrete aggregate. The possible uses for spent catalyst must be determined on a case-by-case basis.

Cascading. Catalyst with moderate remaining activity may be used in another SCR application where the lower activity is suitable, creating a cascade effect. The catalyst remains exempt from hazardous waste classification until it can no longer be effectively utilized.

Regeneration. In the U.S., spent catalyst has been partially regenerated by sandblasting it to clear the catalyst passages. (Effective restoration cannot be obtained when heat sintering occurs.)

Reclamation. Metals in the spent catalyst, such as vanadium, tungsten, and titanium, may be reclaimed for use in other products.

SCR Economics

SCR capital cost estimates and levelized cost estimates are summarized in **Table 10-17**. It is important to note that there are large differences in the estimates. Early estimates are much higher than recent awards and quotes. In particular, note the significantly lower SCR capital costs provided by Wax of the Institute of Clean Air Companies (ICAC) based on 1993 market activity. More recent costs are lower due to the effects of competition, technical advances, operating experience, and fewer changes to the balance of plant than in the early estimates. Also, early SCR units were oversized as suppliers underestimated catalyst activity. Moreover, in the U.S., the price of catalyst has dropped dramatically since the process was first introduced (Wax, 1993A). Cochran et al. (1993) also discuss some of the factors behind the reduction in SCR costs experienced in the U.S. market.

Levelized costs have also dropped because guaranteed catalyst life is now typically seven-years vs. the two years used several years ago for coal-fired applications. The seven-year catalyst life is based on about 10 SCRs in Japan with Hitachi catalyst that operated over five years with no catalyst replacement. Half of these SCRs operated over seven years with no catalyst replacement.

The cost variability in **Table 10-17** is due to many factors, including date of the estimate, SCR size and type, initial and controlled NO_x concentrations, fuel characteristics, catalyst life/replacement considerations, and modifications to downstream equipment. Operating cost variability depends on the following factors (Robie et al., 1991B):

Table 10-17
SCR Costs

SCR Application	Capital Cost, $/kW	Levelized Cost, $/ton of NO_x reduced	Levelized Busbar Cost, mills/kWh	References
New coal-fired unit	78–93 50–65	3,300–3,800 Not given	5.3–6.1 Not given	Robie et al., 1991A[1] Cochran et al., 1993
Retrofit to conventional furnaces (wall- or tangential-fired)	96–105 75–100[2] 30–70[3]	2,750–4,250 Not given Not given	5.9–6.5 Not given Not given	Robie et al., 1991A Cochran et al., 1993 Wax, 1993A
Retrofit to cyclone-fired boilers	125–140	1,100–1,250	8.2–9.1	Robie et al., 1991A
Retrofit to oil-fired boilers	90–98	3,600–7,500	4.2–4.6	Robie et al., 1991A
Retrofit to oil- & gas-fired boilers	15–30	Not given	Not given	Wax, 1993A
Retrofit as post-FGD SCR	140	2,850	6.8[4]	Robie et al., 1991B[1]
Combustion turbine installations (3.5 to 80 MW)	30–100[5]	—	—	May et al., 1991

Notes:
1. Capital and levelized cost estimates by Robie et al. (1991A) are based on constant 1990 dollars and 400–550 MW_e gross unit capacity. Robie et al. (1991B) cost estimates are based on December 1989 dollars and a 536 MW_e capacity unit. EPRI's Technical Assessment Guide (TAG) was used as a basis for estimating these costs.
2. For the most difficult retrofit high-dust SCR installations.
3. Refer to **Table 10-18** for a cost breakdown.
4. A reheat credit of 0.93 mills/kWh is assumed for the post-FGD SCR case. It would not be applicable to sites where a wet stack is used.
5. The costs by May et al. (1991) are based on data collected on an SCR-site field program. Combustion turbine sizes ranged from 3.5 to 80 MW_e. The capital costs given represented from 5 to 25% of the total installed cost of a combined cycle combustion turbine system. Levelized cost data were not given.

- SCR catalyst replacement frequency
- Ammonia consumption
- Ammonia vaporization steam
- Incremental sootblowing steam/air for the catalyst
- Dilution air blower fan horsepower consumption
- Incremental ID fan/booster fan horsepower consumption
- Incremental FD fan horsepower consumption
- Incremental ESP power consumption
- Water treatment chemicals
- Air heater efficiency loss
- Increased air heater leakage
- Incremental FGD reheat steam consumption
- SCR catalyst disposal
- Incremental fly-ash disposal
- Natural gas consumption (cold-side SCR reheat fuel)

928 Gas Purification

The following observations can be made regarding SCR costs:

- The levelized cost per ton of NO_x removed decreases with increasing inlet NO_x concentration.
- Capital costs for SCR retrofits can be substantially higher than for new applications because SCR retrofits often require costly modifications to existing equipment.
- SCR capital costs are higher for cyclone-fired boilers because their higher NO_x emissions require more catalyst volume (Robie et al., 1991B).
- The levelized costs of hot-side SCR systems are usually lower than the costs of cold-side (post FGD) systems due primarily to the additional expense for the external heat source and supplemental natural gas firing needed by cold-side units.

Wax (1993A) gives a cost breakdown for hot-side, high-dust SCR retrofit applications, which is presented in **Table 10-18**. Wax (1993C) has also compiled capital cost data for a number of SCR installations, which are presented in **Table 10-19**.

Combined NO_x/SO_2 Post-Combustion Processes

The control of both SO_2 and NO_x concentrations in flue gas by a single process has obvious advantages over the use of two separate processes, and considerable effort has gone into attempts to develop an economical combined NO_x/SO_x control process. Unfortunately, no combined process has yet achieved widespread acceptance, although numerous approaches have been proposed and a few have reached commercial status. Combined NO_x/SO_x control processes that have been used commercially, or are believed to be in an advanced stage of development, are described in Chapter 7, "Sulfur Dioxide Removal." To avoid repetition, this section is limited to comparative evaluations of available combined processes and very brief descriptions of selected technologies.

Several organizations have conducted reviews and evaluations of combined NO_x/SO_x control processes, including EPRI (DePriest et al. 1989; DePriest et al., 1990; Cichanowicz et al., 1991); Pittsburgh Energy Technology Center (PETC) (Pan, 1991); Argonne National Laboratory (ANL) (Livengood, 1989); PETC and ANL (Livengood and Markussen, 1993 and 1994), and Tennessee Valley Authority (TVA) (Faucett et al., 1978). General survey papers on the subject have also been released by Davis and Mikucki (1989), Livengood and Markussen (1991), Makansi (1990), and Kuhr et al. (1988).

The purpose of the EPRI study was to identify any processes (expected to be commercially available around 1997) that might offer such significant benefits over advanced FGD/SCR that they would merit R&D support. The project initially developed a list of 70 processes. This list was reduced to about 25 processes in the first screening, and 19 of these were finally selected for more detailed technical and economic evaluation. These 19 have been classified into 6 groups as indicated in **Table 10-20** (Cichanowicz et al., 1991).

The Tennessee Valley Authority study (Faucett et al., 1978), although very thorough, is outdated. It provides an excellent review of the state-of-art for NO_x control technology as of 1977. Davis and Mikucki (1989) used a modified form of the TVA study approach for classifying processes as "bench scale," "pilot scale," and "demonstration/commercial scale." They were concerned primarily with pollution control from industrial boilers and concluded

Table 10-18
Cost Breakdown for a Retrofit Hot-Side, High-Dust SCR[1]

Item	Typical Value $/kW[2]	Notes
Ducting	3.00	Ducting cost can be highly variable. This cost assumes that the SCR reactor is located over the air heater.
Fan Upgrade/Replace	0.00	Usually, the existing boiler draft fans will have enough capacity to handle the additional load of an SCR. On a pressurized boiler, the combined air side pressure drop increase due to retrofit of LNB and gas side pressure drop to retrofit of SCR might be large enough to necessitate fan modifications. Increased air heater leakage due to increased differential pressure across it also affects fan capacity requirements.
Structural	4.00	Site-work and foundations costs are excluded. This item can be significant in a difficult retrofit.
Storage & Distribution	2.00	Storage and distribution cost is based on use of anhydrous ammonia.
Reactor/Catalyst	15.00	Reactor/catalyst cost is based on 6,000 ft^3 of catalyst and $400/ft^3 cost with added allowance for the cost of the reactor. Current market conditions are assumed.
Controls	1.00	Controls cost includes two NO$_x$ analyzers, one ammonia analyzer, and control hardware.
Air Heater	1.50	Air heater cost is for replacement of the air heater baskets. Given the low ammonia slip values, this may not be necessary.
Installation	10.60	40% of the purchased equipment cost.
Total Process Capital	37.10	
Fees & Contingency (45%)	16.70	These costs are for general facilities, fees, and contingencies. The indicated value may err on the high side as no general facilities may be required, and engineering fees may only be 10%.
Total Plant Cost	53.80	

Notes:
1. Based on a 200 MW$_e$ power plant
2. 1991 current dollars
Source: Wax (1993A)

Table 10-19
Capital Costs of Some Recent U.S. SCR Retrofits

Facility	Fuel	Design NO_x Reduction Efficiency	Cost	Notes
Los Angeles Department of Water and Power Haynes Station	Natural gas with low-sulfur No. 2 oil for backup	92% with natural gas 71% with fuel oil	Two 230 MW_e units: $25/kW Two 330 MW_e units: $24/kW	Costs exclude ammonia storage and supply systems and inlet and outlet NO_x continuous monitoring systems.
Southern California Edison Los Alamitos Station	Natural gas	87%	Two 480 MW_e units: $30/kW	
Southern California Edison Ormond Beach Station	Natural gas	87% & 93%	Two 750 MW_e units: $25/kW	

Source: Wax (1993C)

Table 10-20
Combined NO_x/SO_2 Processes Identified by EPRI for Technical Evaluation

Process Category	Process Name
Solid Sorption/Regeneration	UOP/PETC Fluidized bed CuO Rockwell Moving bed CuO NOXSO Mitsui/BF Activated Coke Sumitomo/EPDC Activated Char Sanitech Nelsorbent Lehigh Univ. Low Temp. Battelle ZnO Spray Dryer
Irradiation of the Flue Gas	Ebara E-Beam ENEL Pulse-Energization
Wet Scrubbing	Argonne/Dravo ARGONNOX Dow Electrochemical Regen.
Gas/Solid Catalytic	Haldor Topsøe WSA-SNOX Degussa Catalytic B&W SNRB Parsons Flue Gas Cleanup
Dry Injection Additives	Argonne High-Temp. Spray Dry PETC Mixed Alkali Spray Dry
Electrochemical	IGR/Helipump

Source: Cichanowicz et al. (1991)

- Processes that produce salable byproducts should be considered for application at sites where chemicals such as sulfuric acid are already being produced. Modified forms of the CuO and NOXSO processes are specifically recommended.
- Throwaway processes are preferred for nonindustrial sites. Some form of fluidized bed combustion appears to be the leading candidate.

The combined NO_x/SO_2 control process categories and specific processes listed in **Table 10-20** are described briefly in the balance of this section in the order given in the table.

Solid Sorption/Regeneration Processes

The processes categorized as Solid Sorption/Regeneration in **Table 10-20** all operate by contacting the gas with particles; however, the mechanisms involved in the various processes differ markedly.

Copper Oxide. Copper oxide based processes remove NO_x from the gas by selective catalytic reduction with ammonia using the copper oxide as a catalyst. They remove SO_2 from the gas stream by sorption and reaction with the copper oxide to form copper sulfate. The sorbent is regenerated by contact with a reducing gas, producing a concentrated sulfur dioxide stream that can be used as feed to a sulfuric acid plant. The process, as developed by UOP and PETC, utilizes a fluidized bed of sorbent in the main contactor. A variation proposed and tested by Rockwell International uses a moving bed of sorbent in a cross-flow panel arrangement. Additional details on the history and basic technology of the process are given in Chapter 7 under the heading "Copper Oxide Process."

NOXSO Process. Unlike the copper oxide process, the NOXSO process uses a solid sorbent that actually captures and retains NO_x as well as SO_2. The sorbent contains about 3.8% sodium on the surface of a gamma-alumina substrate (Livengood, 1989). The reaction mechanisms are not completely understood, but the spent sorbent contains a wide variety of compounds, including sodium sulfate and sodium nitrate. Regeneration of the sorbent requires two stages: a thermal stage (about 600°C) where the nitrogen compounds are decomposed to produce an NO_x-rich stream that can be recycled back to the boiler in the combustion air, and a reducing stage where a gas, such as methane, is used to produce a concentrated stream of SO_2 and H_2S, which is used as feed to a Claus sulfur plant. Additional details on the process are given in Chapter 7 in the section titled "NOXSO Process."

Activated Coke or Char. Processes based on active forms of carbon resemble the copper oxide processes in that they remove NO_x by selective catalytic reduction with ammonia using the sorbent as a catalyst. However, the mechanism for SO_2 removal is entirely different. Sulfur dioxide is adsorbed on the active carbon, which also acts as a catalyst for the oxidation of adsorbed SO_2 to SO_3. In the presence of moisture, the sulfur trioxide forms sulfuric acid on the char. Regeneration of the sulfuric acid-laden char is accomplished in a separate vessel where the sorbent is heated to about 400°C. At this temperature, the sulfuric acid reacts with a portion of the carbon, forming a gas phase containing sulfur dioxide, carbon dioxide, moisture, and various impurities. This gas stream is further processed to produce sulfuric acid or elemental sulfur, and the remaining char is recycled, with makeup, to the contactor. Several variations of the basic process are discussed in Chapter 7 under the broad heading "Adsorption Processes."

Sorbtech Process. This process resembles the NOXSO process in the use of a sorbent that removes and retains both SO_2 and NO_x from the flue gas and is regenerated for reuse. The sorbent is composed of magnesia and expanded vermiculite. It removes SO_2 and NO_x from the gas by a combination of reaction with the magnesia and capillary pore condensation. The sorbent is regenerated by heating to 700°C in a slightly reducing environment. Although field tests have been conducted on portions of the process (Nelson, 1990), it is still in the very early stages of development. Additional data on the process are given in Chapter 7 under the heading "Sorbtech Process."

Lehigh University Low Temperature Process. Bench scale tests on a variety of catalyst/sorbents have been conducted at Lehigh University. The object of the program is to evaluate the ability of various catalyst compositions and configurations to remove SO_2 and NO_x. The work is in a very early stage of development; however, the construction and field testing of a small rotating reactor is planned (Makansi, 1990).

Battelle ZnO Spray Dryer. The Battelle process utilizes a slurry of zinc oxide injected into a spray dryer for the initial flue gas contact. The gas and dry particles then flow to a cloth filter, which removes the partially spent sorbent and also provides additional gas-sorbent contact time. The spent sorbent discharged from the filter contains zinc sulfite, zinc sulfate, and nitrogen compounds. It is thermally decomposed to drive off SO_x and NO_x, which are sent to a processing system for the recovery of sulfuric and nitric acids. The process is in a very early stage of development; however, a patent has been granted to Battelle covering certain aspects of the technology (Rosenberg, 1987).

Flue Gas Irradiation Processes

The original work on the control of SO_2 and NO_x in flue gas using electron beams was conducted in Japan in 1970 by Ebara International Corporation (Ebara). More recently, pilot scale tests of the Ebara process have been conducted in the U.S. (Livengood, 1989). The ionizing radiation produces numerous ions and active species in the gas resulting in the conversion of SO_2 and NO_x to strong acid forms. Two versions of the process have been tested. In one version, ammonia is injected into the gas, and an ammonium sulfate-nitrate byproduct is produced. In the other, lime is used instead of ammonia to neutralize the acids. The lime is fed as a slurry into a spray dryer, and the spent lime, containing calcium sulfite, calcium sulfate, and calcium nitrate, is considered to be a throw-away waste as in conventional spray drying FGD processes. Work on a similar process has been reported by ENEL of Italy (Kuhr et al., 1988). Additional details of the E-Beam process are given in Chapter 7 under the heading "Ebara E-Beam Process."

Wet Scrubbing Processes

Metal Chelate Addition. Metal chelates, such as ferrous ethylene-diamine-tetra-acetate (Fe(II) • EDTA), enhance the absorption of NO_x into aqueous solutions by reacting quickly with dissolved NO to form the complex Fe(II) • EDTA • NO. The coordinated NO can react with sulfite and bisulfite ions, forming hydroxylamine-N-disulfonates (HADS) and releasing the ferrous chelate to react with additional NO. When the aqueous solution is a sodium or calcium-based SO_2 absorbent, the addition of ferrous chelate results in a combined NO_x/SO_x control process.

An early problem with this chelate system was the oxidation of ferrous ions to the inactive ferric form. Argonne National Laboratory has been conducting an extensive study to develop additives that inhibit this oxidation (Mendelsohn et al., 1991). Further development of the additive approach is being carried out under an agreement between Argonne and Dravo Lime Co. Another approach is under development by The Dow Chemical Co. A key feature of the Dow process is the use of an electrochemical cell to reduce ferric ions to the ferrous state (Livengood, 1989).

Sodium Chlorite Addition. Another process, offered by Belco Technologies for commercial application, uses sodium chlorite as an additive to the SO_2 scrubbing solution. The sodium chlorite converts NO to NO_2, and dissolved sulfur dioxide reduces the NO_2 to nitrogen (AirTECH News, 1993).

Gas/Solid Catalytic Processes

Haldor Topsøe WSA-SNOX. The WSA-SNOX process was developed by Haldor Topsøe A/S, Denmark, and is offered in the U.S. by ABB Combustion Engineering, Inc. The process is one of the most highly developed of the combined NO_x/SO_2 FGD systems. Several industrial systems are in operation in Europe, and a demonstration unit has been operated in the U.S. (Kingston et al., 1990).

In the WSA-SNOX process, the flue gas passes first through a conventional SCR unit where NO_x is reduced to N_2 by ammonia. The flue gas is then heated slightly and passed through a second catalyst where SO_2 is oxidized to SO_3. The SO_3 is hydrated to form sulfuric acid and concentrated to 95 wt% acid in an air-cooled falling-film condenser constructed of glass. Ammonia slip from the SCR reactor is oxidized in the SO_2 converter, eliminating a problem encountered in conventional SCR processes. A more detailed description of the process is given in Chapter 7 in the section titled "SNOX Process."

Degussa DESONOX Process. The DESONOX process, conceived by the German firm Degussa and being developed jointly by Stadtwerke Munster, Lentjes, and Lurgi, is closely related, but not identical, to the WSA-SNOX process. It also reduces the NO_x to nitrogen in an SCR step using ammonia, and oxidizes the SO_2 to SO_3 over a second catalyst. However, in the DESONOX process both catalysts are in the same chamber and operate at about the same temperature. The SCR catalyst is zeolite based, and the oxidizing catalyst contains precious metals. Both are ceramic honeycomb structures. A large demonstration plant was constructed in Germany. Its initial operation is described by Brand et al. (1989).

SO_x-NO_x-Rox-Box (SNRB). This process is based on the well known technologies of in-duct injection of a calcium- or sodium-based sorbent for SO_2 removal and SCR with ammonia for NO_x control. Its novelty is in conducting the two operations, plus fly ash collection, in the same system. The sorbent is injected into the flue gas either before or after the economizer and reacts with SO_2 in both the ducts and the filtercake on the cloth filter bags. Ammonia is injected into the flue gas and reacts with NO_x on a catalyst suspended within the filter bags. The filter is operated hot (425°–470°C) to accelerate the reactions; and as a result, a special high-temperature, woven-ceramic filter bag material is required. Low exit SO_2 and SO_3 concentrations may permit lower air preheater exit temperatures and greater system thermal efficiency. A 5 MW process demonstration was completed in May 1993, at Ohio Edison's R.E. Berger plant under the second round of the DOE Clean Coal Technology program. SO_2 removal efficiencies greater than 80% were achieved using lime at Ca/S ratios of 1.8/2.0. Testing with sodium bicarbonate showed that SO_2 removals above 90% were attainable at a normalized stoichiometry (Na_2/S) of 2.0 over a fabric filter temperature range of 220°/470°C. NO_x reductions greater than 90% were achieved at NH_3/NO_x molar ratios of 0.85/0.90. A minimum life of three years is currently being projected for the NO_x catalyst. An economic evaluation using EPRI guidelines estimates process capital costs of $240/kW and an annual levelized cost of $509/ton SO_2 + NO_x removed for a 500 MW plant burning 2.5 wt% S coal (1993 dollars) assuming a 15-year plant life (Livengood and Markussen, 1991; 1994).

Parsons Flue Gas Cleanup Process. The Parsons Flue Gas Cleanup (FGC) process consists of three steps:

1. Simultaneous catalytic reduction of SO_2 to H_2S and NO_x to N_2 in a hydrogenation reactor.
2. Recovery of H_2S by selective absorption from the gas stream.
3. Conversion of H_2S to elemental sulfur.

These three technologies have been applied in commercial applications, but their use in a single process for flue gas treatment is a relatively new development. Potential development issues with this application are excessive reducing of gas consumption due to oxygen in the flue gas and catalyst plugging by fly ash (Livengood and Markussen, 1994). During 1990, pilot scale tests of the hydrogenation step with a boiler flue gas produced consistent 99%+ SO_2 conversion and 92% NO_x conversion with the two best catalyst systems (Kwong et al., 1991).

Dry Additive Injection

Sulfur dioxide removal by the injection of alkaline solids (or slurries) into the gas stream and separation of spent particles by filtration or electrostatic precipitation is a well established technology. As would be expected, attempts have been made to modify this technology to accomplish combined NO_x/SO_2 control. Although the approach is still developmental, it offers the promise of process simplicity. Two basic mechanisms have been investigated:

1. Injection of a mixture of an SO_2 sorbent, such as lime, and a selective reducing agent, such as urea, into the furnace at a temperature region where both thermal reduction of NO_x and sorption of SO_2 can occur.
2. Injection of a reactive sodium compound either with or without a calcium sorbent. It is theorized that SO_2 is first absorbed and the resulting sodium sulfite acts as a catalyst for the oxidation of NO to NO_2. The NO_2 then reacts with the solids to form nitrogen, nitrates, nitrites, and sulfur-nitrogen compounds.

A furnace injection process was studied by Himes et al. (1990). They conducted tests using a dry lime-urea hydrate injected into a high-temperature gas stream simulating the upper furnace region of a boiler. The tests demonstrated simultaneous SO_2 and NO_x reductions of 65% at mole ratios of 3 and 1.5, respectively, for Ca/S and N/NO_x. Gullett et al. (1991) report on tests with a slurry of lime in urea-based solution. These tests, which were conducted in a natural gas reactor, achieved 80% reduction of both impurities at reactant/pollutant stoichiometric ratios of 2/1 and 1/1 for SO_2 and NO_x, respectively.

In conventional spray-drying FGD systems using lime slurry for SO_2 removal, NO_x in the flue gas is relatively unaffected. Small scale research at the Pittsburgh Energy Technology Center (PETC) indicated that operation at higher than normal temperature with NaOH as an additive to the lime resulted in significant NO_x reduction. The process was studied in full scale tests at Argonne National Laboratory (Brown et al., 1988). The ANL tests demonstrated that

- A temperature of at least 180°F was required for NO_x reduction.
- Most NO_x reduction occurred in the filtercake and reduction was strongly dependent on filtercake thickness.
- NO_x reduction increased with SO_2/NO_x ratio.
- Caustic soda increased NO_x reduction.
- Reduction efficiencies of 70% for SO_2 and 35% for NO_x were easily attained.

Duct injection represents a third feed alternative. In one concept, sodium bicarbonate is injected into the flue gas duct at about 150°C in addition to hydrated lime injected into the convective section of the boiler (Helfritch et al., 1991). In another concept, recommended for small plants, sodium bicarbonate, without lime, is injected into the duct upstream of a cloth filter (Darmstaedter, 1990). A general problem with the use of sodium compounds is the release of NO_2, which can cause a visible yellow plume at the stack. The production of NO_2 tends to increase with temperature, but can be minimized by the injection of urea or ammonia. Disposal of the solid product, which contains water-soluble sodium salts, can also be a problem.

Electrochemical Processes

Several organizations have conducted research on the use of electrochemical processes for flue gas purification, but the technology has not yet progressed beyond the bench scale stage. A process under development by Helipump, Inc. has shown some promise for removing both SO_2 and NO_x. The technology is derived from work on high-temperature solid-oxide fuel cells using ceramic electrolytes coated with proprietary electrocatalysts (Kuhr et al., 1988). The cell operates in the temperature range of 850°–950°F and reduces SO_2 and NO_x to elemental sulfur and nitrogen, which remain in the flue gas. In an operating plant the sulfur would be condensed and recovered downstream of the electrochemical cell.

Miscellaneous NO_x Removal Processes

Tri-Mer TRI-NO_x Process. This is a commercial, proprietary NO_x control process based on multistage scrubbing of the gas stream. It uses an undisclosed solvent that is claimed to also remove HCl, SO_2, Cl_2, HNO_3, HF, and other residual inorganic gases simultaneously with the NO_x. Over 50 installations were reported to be in operation in 1992, the largest treating 54,000 cfm of gas (Tri-Mer Corporation, 1992). The liquid effluent from the scrubber contains soluble salts, which in many cases can be disposed of in a municipal sewer system. If this is not acceptable, a water treatment package must be added. The process is said to be applicable to all ratios of NO to NO_2 and is claimed to be capable of reducing NO_x concentrations to below 10 ppm on a continuous basis.

Nissan Permanganate Process. The Nissan Permanganate process was developed by Nissan Engineering. This process was originally designed to treat HNO_3 tail gas; but, with the installation of an FGD system prior to the denitrification system, the process could be adapted to fully treat power-plant stack gas. A chloride prescrubber may be required for a coal-fired power plant. The NO_x is absorbed in a scrubber using a solution of KOH and $KMnO_4$. The liquid effluent from the scrubber is regenerated, and HNO_3 is produced using an electrolysis unit, a "manganate reactor," and an electrolytic oxidizer. Little information and no economic data have been published for this process. In 1978, it was reported that since 1972 four small plants (100–2,000 Nm^3/hr) had been built (Faucett et al., 1978).

Tokyo Electric-Mitsubishi Heavy Industries Process. The Tokyo Electric-Mitsubishi Heavy Industries (MHI) process is a wet oxidation-absorption system developed jointly by Tokyo Electric Power and MHI. This process was developed for treating clean flue gas (no SO_x). The NO_x is oxidized by O_3 to N_2O_5, absorbed in H_2O to form 8–10% HNO_3, and then concentrated to a 60% acid solution in the byproduct recovery section. A calcium sulfite scrubber removes any excess ozone from the flue gas. In December 1974, a 33.3 MW_{equiv} plant began operating. No operating results or costs have been published (Faucett et al., 1978).

REFERENCES

ACGIH, 1991, *1991–1992 Threshold Limit Values for Chemical Substances and Physical Agents and Biological Exposure Indices,* American Conference of Governmental Industrial Hygienists for Chemical Substances and Physical Agents and Biological Exposure Indices, Cincinnati, OH.

AirTECH News, 1993, "Wet Scrubber/Filter Used to Remove NO_x from Flue Gas," Silver Springs, MD, May, pp. 59–60.

Anderson, H. C., Green, W. J., and Steele, D. R., 1961, "Catalytic Treatment of Nitric Acid Plant Tail Gas," *Industrial Engineering Chemistry,* Vol. 53, No. 3, pp. 199–204.

Ando, J., and Sedman, C. B., 1987, "Status of Acid Rain and SO_2 and NO_x Abatement Technology in Japan," *Proceedings: Tenth Symposium on Flue Gas Desulfurization—EPRI/EPA,* EPRI CS-5167, Electric Power Research Institute, Palo Alto, CA, May, pp. 2-45 to 2-66.

Angello, L., and Lowe, P., 1989, "Gas Turbine Nitrogen Oxide (NO_x) Control: Current Technologies and Operating Experiences," paper presented at the 1989 Joint Symposium on Stationary Combustion NO_x Control—EPA/EPRI, published as EPRI GS-6423, San Francisco, CA, March 6–9.

Anon., 1992, "Cogen Plant First to Use New High-Temperature De-NO_x Process," *Oil & Gas J.,* April 13, p. 53.

Balling, L., Sigling, R., Schmelz, H., Hums, E., and Spitznagel, G., 1991, "Poisoning Mechanism in Existing SCR Catalytic Converters and Development of a New Generation for Improvement of the Catalytic Properties," paper presented at the 1991 Joint Symposium on Stationary Combustion NO_x Control—EPA/EPRI, Washington, D.C., March 25–28.

Baublis, D. C., and Miller, J. A., 1992, "A Utility Perspective: NO_x Control for Six Tangentially-Fired Boilers," *Power-Gen '92,* Vol. XI, Orlando, FL, Nov. 17–19, pp. 297–308.

Behrens, E. S., Ikeda, S., Yamashita, T., Mittelbach, G., Yanai, M., 1991A, "SCR Operating Experience on Coal-Fired Boilers and Recent Progress," paper presented at the 1991 Joint

Symposium on Stationary Combustion NO_x Control—EPA/EPRI, Washington, D.C., March 25–28.

Behrens, E. S., Ikeda, S., Yamashita, T., Mittelbach, G., and Yanai, M, 1991B, "SCR Cuts NO_x Emissions Successfully at Coal-fired Plants," *Power Engineering,* September, pp. 49–52.

Berg, M., Bering, H., and Payne, R, 1993, "NO_x Reduction by Urea Injection in a Coal Fired Utility Boiler," paper presented at the 1993 Joint Symposium on Stationary Combustion NO_x Control—EPRI/EPA, Bal Harbour, FL, May 24–27.

Blair, J. B., Massey, G., and Hill, H., 1981, "Refinery Catalytically Cuts NO_x Emissions," *Oil & Gas J.,* Jan. 12, pp. 99–104.

Blaszak, T. P. (Fluor Daniel), 1992 and 1994, personal communication.

Boardman, R. D., and Smoot, L. D., 1989, "Prediction of Fuel and Thermal NO in Advanced Combustion Systems," *Proceedings of the 1989 Symposium on Stationary Combustion Nitrogen Oxide Control,* GS-6423, Electric Power Research Institute, Palo Alto, CA, July, pp. 6B-20—6B-21.

Booher, J. H., 1993, "NO_x Formation and Control via Furnace Cleanliness Management," paper presented at the Western Fuels Conference (sponsored by Diamond Power Specialty Group), Kansas City, MO, Aug. 23–24.

Bozzuto, C., 1991, "Review of NO_x Formation Basics," paper presented at the Council of Industrial Boiler Owners (CIBO) Fourth Annual NO_x Control Conference, Concord, CA, Feb. 11–12.

Brand, R., Engler, E., Honnen, W., Kleine-Mollhoff, P., and Koberstein, E., 1989, "First Operating Experience with the DESONOX Process for Simultaneous Removal of NO_x and SO_2 from Flue Gas," paper presented at the 82nd Annual Meeting & Exhibition of the Air and Waste Management Association, Anaheim, CA, June 25–30.

Bretz, E. A., 1991, "Gas-Turbine-Based Combined Cycle Plants," *Electrical World,* August, pp. 41–47.

Brown, K. A., Huang, H., Allen, J. W., and Livengood, C. D., 1988, "Combined Nitrogen Oxides/Sulfur Dioxide Control in a Spray-Dryer/Fabric-Filter System," ANL/ESD/TM-8, Argonne National Laboratory, Argonne, IL, November.

Brusger, E., Wayland, R., and Hewson, T., 1990, *Clean Air Response: A Guidebook to Strategies,* GS-7105, Part II, Chapter III, Electric Power Research Institute, Palo Alto, CA, December.

Burch, T. E., Tillman, F. R., Chen, W. Y., Lester, T. W., Conway, R. B., and Sterling, A. M., 1991, "Partitioning of Nitrogenous Species in the Fuel-Rich Stage of Reburning," *Energy and Fuels,* American Chemical Society, March/April, pp. 231–237.

Campobenedetto, E. J. (Babcock & Wilcox), 1994, personal communication.

Campobenedetto, E. J., and Murataka, T., 1990, "Application of Selective Catalytic Reduction Systems for NO_x Control," *Proceedings of the American Power Conference,* Chicago, IL, April 23–25, pp. 589–594.

Casiello, F., 1991, "NO_x Control Techniques, Limitations & Costs: Gaseous and Liquid Fuels," paper presented at the Council of Industrial Boiler Owners (CIBO) Fourth Annual NO_x Control Conference, Concord, CA, Feb. 11–12.

CFR (Code of Federal Regulations), 1992, Title 40, Part 60, U.S. Government Printing Office, Washington, D.C., July.

Chen, J., Yang, R. T., and Cichanowicz, J. E., 1991, "Poisoning of SCR Catalysts," *1991 Joint Symposium on Stationary Combustion NO_x Control—EPA/EPRI,* Washington, D.C., March 25–28.

Cho, S. M. and Dubow, S. Z., 1992, "Design of a Selective Catalytic Reduction System for NO_x Abatement in a Coal-Fired Cogeneration Plant," *Proceedings of the American Power Conference,* Chicago, IL, April 13–15, pp. 717–722.

Cichanowicz, J. E., Dene, C. E., DePriest, W., Gaikwad, R., and Jarvis, J., 1991, "Engineering Evaluation of Combined NO_x/SO_2 Controls for Utility Application," paper presented at the EPRI/EPA/DOE 1991 SO_2 Control Symposium, Washington, D.C., Dec. 3–6.

Clark, G., and Comparato, J. (Nalco Fuel Tech), 1994, personal communications, May.

Cochran, J. R., Gregory, M. G., and Rummenhohl, V., 1993, "The Effect of Various Parameters on SCR System Cost," paper presented at Power-Gen '93, Dallas, TX, Nov. 17–19.

Cohen, M. (ABB Combustion), 1993, personal communication, Nov. 3.

Comparato, J. R., Buchs, R. A., Arnold, D. S., and Bailey, L. K., 1991, "NO_x Reduction at the Argus Plant Using the NO_xOUT® Process," paper presented at the 1991 Joint Symposium on Stationary Combustion NO_x Control—EPA/EPRI, Washington, D.C., March 25–28.

Darmstaedter, E., 1990, "Sodium Bicarbonate Injection: A Small Plant SO_2/NO_x Option," *Power Engineering,* December, pp. 25–27.

Davis, M. L., and Mikucki, W. J., 1989, "Choosing a Technology for Simultaneous Control of NO_x/SO_x from Industrial Boilers," paper presented at the 82nd Annual Meeting & Exhibition of the Air & Waste Management Association, Anaheim, CA, June 25–30.

DePriest, W., Jarvis, J. B., Cichanowicz, J. E., and Dene, C. E., 1990, "Engineering Evaluation of Combined NO_x/SO_x Removal Processes: Second Interim Report," paper presented at the EPRI/EPA 1990 SO_2 Control Symposium, New Orleans, LA, May 8–11.

DePriest, W., Jarvis, J. B., and Cichanowicz, J. E., 1989, "Engineering Evaluation of Combined NO_x/SO_2 Removal Processes: Interim Report," *Proceedings of the 1989 Symposium on Stationary Combustion Nitrogen Oxide Control,* GS-6423, Vol. 2, Electric Power Research Institute, Palo Alto, CA, July, pp. 7A-67–7A-85.

Elkins, J., 1990, "Current Uncertainties in the Global Atmospheric N_2O Budget," European Workshop on the Emission of Nitrous Oxide, Lisbon, Portugal, June 6–8, 1990. (From Hofmann, J. E., Johnson, R. A., Schumacher, P. D., Sload, A., Afonso, R., 1993, "Post Combustion NO_x Control for Coal-Fired Utility Boilers," paper presented at the 1993 Joint Symposium on Stationary Combustion NO_x Control—EPRI/EPA, Bal Harbour, FL, May 24–27.)

Epperly, W. R., and Hofmann, J. E., 1989, "Control of Ammonia and Carbon Monoxide Emissions in SNCR Technologies," paper prepared for presentation at the AIChE 1989 Summer National Meeting, Aug. 20–23.

Epperly, W. R., Broderick, R. G., and Peter-Hoblyn, J. D., 1988A, "Control of Nitrogen Oxides Emissions From Stationary Sources," *Proceedings of the American Power Conference,* Chicago, IL, April 20, pp. 911–915.

Epperly, W. R., Hofmann, J. E., O'Leary, J. H., and Sullivan, J. C., 1988B, "The NO_xOUT® Process for Reduction of Nitrogen Oxides," paper presented at the 81st A.P.C.A. Annual Meeting and Exhibition, Dallas, TX, June 23.

Exxon, 1981–1992, Applications Experience Reports, Exxon Research and Engineering Company, Florham Park, NJ.

Exxon, 1982, "The Non-Catalytic Denitrification Process (Thermal DeNO$_x$) for Glass Melting Furnaces," Exxon Research and Engineering Company, Technology Licensing Division, Florham Park, NJ, paper presented at the Society of Glass Technology Symposium, England, May.

Exxon Research and Engineering, 1989, *THERMAL DeNO$_x$ Process,* Technology Licensing Division, Exxon Research and Engineering Company, Florham Park, NJ.

Faucett, H. L. Maxwell, J. D., and Burnett, T. A., 1978, "Technical Assessment of NO$_x$ Removal Processes for Utility Applications," AF-568, Electric Power Research Institute, Palo Alto, CA, March.

Fellows, W. D., 1989, "Application of the THERMAL DeNO$_x$ Process to Glass Melting Furnaces," GPI/CMP Workshop, Pittsburgh, PA, Sept. 25.

Fellows, W. D., 1990, "Experience with the Exxon Thermal DeNO$_x$ Process in Utility and Independent Power Production," paper presented at the 40th Canadian Chemical Engineering Conference, Halifax, N.S., Canada, August.

Flora, H., Barkley, J., Janik, G., Marker, B., and Cichanowicz, J. E., 1991, "Status of 1 MW SCR Pilot Plant Tests at Tennessee Valley Authority and New York State Electric & Gas," paper presented at the 1991 Joint Symposium on Stationary Combustion NO$_x$ Control—EPRI/EPA, Washington, D.C., March 25–28.

Fotis, S. C. (Van Ness, Feldman & Curtis), 1992, memorandum to Ben Yamagata (Van Ness, Feldman & Curtis) on Report on NO$_x$ Electric Utility Meeting at EPA, Washington, D.C., May 29. (Provided to members of the Clean Coal Technology Coalition, Washington, D.C.)

Franklin, H. (Foster Wheeler Energy Corp.), 1992/1993/1994, personal communication.

Fuel Tech, 1988, "First Order for NO$_x$OUT Process Received," *News Bulletin 001/5-88,* Fuel Tech, Inc., Stamford, CT.

Fuel Tech, 1989A, Brochure: "Fuel Tech is a Dynamic, High Technology Company," Fuel Tech, Inc., Stamford, CT.

Fuel Tech, 1989B, "The NO$_x$OUT Process is a New Chemical and Mechanical System For Cost Effective NO$_x$ Reduction From Fossil Fueled Combustion Sources," Fuel Tech, Inc., Stamford, CT.

Garg, A., 1994, "Specifying Better Low-NO$_x$ Burners for Furnaces," *Chemical Engineering Progress,* January, pp. 46–49.

Geissler, P., Grisko, S. E., and O'Leary, J. H., 1992, "NO$_x$OUT® System at the Exeter Energy Tire Burning Facility," paper 92-43.05 presented at the AWMA Annual Meeting and Exhibition, Kansas City, MO, June 21–26.

Gouker, T. R., and Brundrett, C. R., 1991, "SCR Catalyst Developments for the U.S. Market," paper presented at 1991 Joint Symposium on Stationary Combustion NO$_x$ Control—EPA/EPRI, Washington, D.C., March 25–28.

Grisko, S. E. (Nalco Fuel Tech), 1992, personal communication, September.

Gulian, F. J., Gouker, T. R., and Burndrett, C. P., 1991, "Slipstream Testing of SCR Catalysts in U.S.," paper presented before the Division of Petroleum Chemistry, Inc., American Chemical Society, Atlanta, GA, April 14–19.

Gullett, B. K., Bruce, K. R., Hansen, W. F., Hofmann, J. E., 1991, "Furnace Slurry Injection for Simultaneous SO$_2$/NO$_x$ Removal," paper presented at the EPRI/EPA/DOE 1991 SO$_2$ Control Symposium, Washington, D.C., Dec. 3–6.

Haas, G. A. (Exxon Research and Development), 1992A, personal communication, Nov. 20.

Haas, G. A., 1992B, "Selective Non-Catalytic Reduction (SNCR): Experience with the Exxon THERMAL DeNO$_x$ Process," paper presented at the NO$_x$ Control V Conference, Council of Industrial Boiler Owners, Long Beach, CA, Feb. 10–11.

Haussmann, G. J., and Kruger, C. H., 1989, "Evolution and Reaction of Fuel Nitrogen During Rapid Coal Pyrolysis and Combustion," *Proceedings of the 1989 Symposium on Stationary Combustion Nitrogen Oxide Control,* GS-6423, Electric Power Research Institute, Palo Alto, CA, July, pp. 6B-61–6B-74.

Helfritch, D. J., Bortz, S. J., and Beittel, R., 1991, "Dry Sorbent Injection for Combined SO$_2$ and NO$_x$ Control," *Seventh Annual Coal Preparation, Utilization, and Environmental Control Contractors Conference—Proceedings,* Pittsburgh, PA, July 15–18, pp. 223–230.

Himes, R. M., Muzio, L. J., and Thompson, R. E. 1990, "Combined SO$_2$/NO$_x$ Control with Lime-Urea Hydrate," paper presented at the EPRI/EPA 1990 SO$_2$ Control Symposium, New Orleans, LA, May 8–11.

Hjalmarsson, A. K., and Vernon, J., 1989, "Policies for NO$_x$ Control in Europe," paper presented at the 1989 Joint Symposium on Stationary Combustion NO$_x$ Control—EPA/EPRI, EPRI GS-6423, San Francisco, CA, March 6–9.

Hofmann, J. E. (Nalco Fuel Tech), 1994, personal communication, February.

Hofmann, J. E., von Bergmann, J., Böekenbrink, D., and Hein, K., 1989, "NO$_x$ Control in a Brown Coal-Fired Utility Boiler," paper presented at the EPRI/EPA 1989 Joint Symposium on Stationary Combustion Nitrogen Oxide Control, EPRI GS-6423, Vol. 2, Electric Power Research Institute, Palo Alto, CA, July, pp. 7A.53–7A.66.

Hofmann, J. E., Johnson, R. A., Schumacher, P. D., Sload, A., and Afonso, R., 1993, "Post Combustion NO$_x$ Control for Coal-Fired Utility Boilers," paper presented at the 1993 EPRI/EPA Joint Symposium on Stationary Combustion NO$_x$ Controls, Miami Beach, FL, May 23–27.

Hunt, T., Schott, G., Smith, R., Muzio, L., Jones, D., and Steinberger, J., 1993, "Selective Non-Catalytic Operating Experience Using Both Urea and Ammonia," paper presented at the 1993 Joint Symposium on Stationary Combustion NO$_x$ Control—EPRI/EPA, Bal Harbour, FL, May 24–27.

Hurst, B. E., 1981, "Applicability of THERMAL DeNO$_x$ to Oil Field Steam Generators," Exxon Research & Engineering Company, Florham Park, NJ, June.

Hurst, B. E., and White C. M., 1986, "THERMAL DeNO$_x$: A Commercial Selective Non-Catalytic NO$_x$ Reduction Process for Waste-to-Energy Applications," paper presented at the ASME 12th Biennial National Waste Processing Conference, Denver, CO, June 2.

Hurst, B. E., 1983, "Improved THERMAL DeNO$_x$ Process for Coal-Fired Utility Boilers," paper presented at the 11th Annual Stack Gas/Coal Utilization Meeting, Paducah, KY, Oct. 6. (Sponsored by Battelle Memorial Institute, Columbus, OH.)

Iskandar, R. S., 1990, "NO$_x$ Removal by Selective Catalytic Reduction," Cormetech, Inc., Corning, NY, April.

Itoh, H. and Kajibata, U., 1981, "Countermeasures for Problems in NO$_x$ Removal Process for Coal-Fired Boilers," Joint EPA/EPRI Symposium on Stationary Combustion Control, May. (Taken from *Technical Feasibility and Cost of Selective Catalytic Reduction (SCR) NO$_x$ Control,* EPRI GS-7266, prepared by United Engineers and Constructors, Inc. (C. P. Robie and P. A. Ireland) for the Electric Power Research Institute, Palo Alto, CA, May, 1991.)

Jaerschky, R., Merz, A., and Mylonas, J., 1991, "SO_3 Generation—Jeopardizing Catalyst Operation?," paper presented at the 1991 Joint Symposium on Stationary Combustion NO_x Control—EPA/EPRI, Washington, D.C., March 25–28.

Kantor, B. L., and Siegfriedt, W. E., 1992, "The NO_x Control Mandate: Demystifying the Requirements of the Clean Air Act," 92-JPGC-Pwr-22, paper presented at the ASME/IEEE International Joint Power Generation Conference, Atlanta, GA, Oct. 18–22.

Kaplan, N., 1993, "NO_x Control Costs in the IAPCS Model," paper presented at the 1993 Joint Symposium on Stationary Combustion NO_x Control—EPRI/EPA, Bal Harbour, FL, May 24–27.

Katzberger, S. M. and Sloat, D. G., 1990, "Options for Compliance with Acid Rain Legislation," ASME Paper No. 90-JPGC/Pwr-22, paper presented at ASME/IEEE International Joint Power Generation Conference, Boston, MA, Oct. 21–25.

Khan, S. R., Desai, M. S., and Gawin, A. F., 1990, "NO_x Control Technologies: An Overview of Effectiveness for Stationary Combustion Sources," *Proceedings of the 52nd American Power Conference,* Chicago, IL, April, pp. 595–604.

Kingston, W. H., Cunninghis, S., Evans, R. J., and Speth, C. H., 1990, "Demonstration of the WSA-SNOX Process Through the CCT Program," paper 90-JPGC/FACT-17 presented at the Joint ASME/IEEE Power Generation Conference, Boston, MA, October 21–25.

Knill, K. J., and Morgan, M. E., 1989, "The Effect of Process Variables on NO_x and Nitrogen Species Reduction in Coal Fuel Staging," *Proceedings of the 1989 Symposium on Stationary Combustion Nitrogen Oxide Control,* GS-6423, Electric Power Research Institute, Palo Alto, CA, July, pp. 6B-75–6B-92.

Krigmont, H. D., and Coe, E. L., 1990, "Experience with Dual Flue Gas Conditioning of Electrostatic Precipitators," *Proceedings: Eighth Particulate Control Symposium,* EPRI GS-7059, Vol. 1, Electric Power Research Institute, Palo Alto, CA, November, pp. 30.1–30-15.

Kuhr, R. W., Wedig, C. P., and Davidson, L. N., 1988, "The Status of New Developments in Flue Gas NO_x and Simultaneous NO_x/SO_x Cleanup," paper presented at the 1988 Joint Power Generation Conference, Philadelphia, PA, Sept. 26–28.

Kumar, K. S. (Research Cottrell), 1993, personal communication, Feb. 9.

Kunimoto, T., Toyoda, T., Kaneko, S., Imamoto, Y., and Iida, K., 1990, "The State-of-the Art SCR Technologies and R&D Trend Update," presented at *Gen Upgrade '90 International Symposium on Performance Improvement, Retrofitting, and Repowering of Fossil Fuel Power Plants, Vol. 3, Policy, Economics, and Environmental Upgrades,* EPRI GS-6986, (Sponsored by International Energy Agency, U.S. Department of Energy and Electric Power Research Institute, March 6–9, pp. 14-6-1 to 14-6-13.

Kwan, Y., Mansour, M. N., Carnevale, J. J., Iseri, K. A., and Garcia, A. R., 1993, "Experience on Automated Urea Injection for NO_x Reduction at Scattergood Unit 1," paper presented at the 1993 Joint Symposium on Stationary Combustion NO_x Control—EPRI/EPA, Bal Harbour, FL, May 24–27.

Kwong, K. V., Meissner III, R. E., Ahmad, S., and Wendt, C. J., 1991, "Application of Amines for Treating Flue Gas from Coal-Fired Power Plants," Environmental Progress, Vol. 10, No. 3, August, pp. 211–215.

LaFlesh, R., and Borio, R., 1993, "ABB C-E Services' Experience with Reburn Technology—Utility Demonstrations and Future Applications," TIS 8608, ABB C-E Services, Windsor, CT.

Lange, H. B., Reddy, V., and DeWitt, S. L., 1994, "Plume Visibility Related to Ammonia Injection for NO_x Control: A Case History," *Seventh Annual NO_x Control Conference—Proceedings* (sponsored by the Council of Industrial Boiler Owners), Oak Brook, IL, May 2–4, pp. 273–282.

Levine, J. S., 1991, "The Global Atmospheric Budget of Nitrous Oxide," paper presented at the EPA/EPRI 1991 Joint Symposium on Stationary Combustion NO_x Control—EPA/EPRI, Washington, D.C., March 25–28.

Lewis, E. (Babcock & Wilcox), 1993, personal communication, January.

Lightly, J. S., Gordon, D. L., Pershing, D. W., Owens, W. D., Cundy, V. A., Leger, C. N., 1989, "The Effect of Fuel Nitrogen on NO_x Emissions from a Rotary-Kiln Incinerator," *Proceedings of the 1989 Symposium on Stationary Combustion Nitrogen Oxide Control,* GS-6423, Electric Power Research Institute, Palo Alto, CA, July, pp. 5B-45–5B-64.

Lin, M. L., Diep, D. V., and Dubin, L., 1991, "Unique Features of Urea-Based NO_xOUT Process For Reducing NO_x Emissions," paper presented at the 8th Pittsburgh Coal Conference, Pittsburgh, PA, Oct. 14–18.

Livengood, C. D., 1989, "Combined SO_2/NO_x Control Technologies" paper presented at the Energy/Chemistry Conference, sponsored by the Organization for Energy Planning, Cairo, Egypt, April 1–2. Published as *Department of Energy Report DE 8904197* (Conf-8904197-2).

Livengood, C. D., and Markussen, J. M., 1991, "Recent Developments in Combined Control of SO_2 and NO_x," paper presented at NO_x Control Technologies and Methods, sponsored by Council of Industrial Boiler Owners, Concord, CA, Feb. 10–12.

Livengood, C. D., and Markussen, J. M., 1994, "FG Technologies for Combined Control of SO_2 and NO_x," *Power Engineering,* January, pp. 38–42.

Livengood, C. D., and Markussen, J. M., 1993, "Status of Flue-Gas Treatment Technologies for Combined SO_2/NO_x Reduction," (work supported by the U.S. Department of Energy under contract W-31-109-ENG-38), paper presented at the Sixth International Symposium on Integrated Energy and Environmental Management, sponsored by AWMA, EPRI, and ASME, New Orleans, LA, March 10–12.

Lowe, P. A., 1985, "Utility Operating Experience with Selective Catalytic Reduction of Flue Gas NO_x," *Proceedings Power Magazine's 2nd International Conference on Acid Rain,* Washington, D.C., March, McGraw Hill Publ. Co., NY, NY, pp. VI.1–VI.22.

Makansi, J. K., 1988, "Reducing NO_x Emissions: Boilers, Gas Turbines, and Engines," *Power,* September, pp. S1-S12 & S26.

Makansi, J. K., 1990, "Will Combined SO_2/NO_x Processes Find a Niche in the Market?," *Power,* September, p. 26.

Makansi, J. K., 1991, "How Fuel Quality Affects NO_x Formation—and CAA Compliance," Power, July, pp. 54–55 & 81.

Makansi, J. K., 1992A, "Ammonia: It's Coming to a Plant Near You," *Power,* May, pp. 16–22.

Makansi, J. K., 1992B, "Herne Sets Example for German District Heating," *Electric Power International,* McGraw-Hill, Inc., NY, NY, March, pp. 49–51.

Mansour, M. N., Nass, D. W., Brown, J., and Jantzen, T. M., 1991, "Integrated NO_x Reduction Plan to Meet SCAQMD Requirements for Steam Electric Power Plants," *Proceedings of the American Power Conference,* Chicago, IL, April 29–May 1, pp. 964–970.

Mark, H. F., Othmer, D. F., Overburger, C. G., Seaborg, G. T., Grayson, M., and Eckroth, D., 1978, "Air Pollution," *Kirk-Othmer Encyclopedia of Chemical Technology,* Vol. 1, 3rd Ed., John Wiley & Sons, NY, pp. 639–641.

May, P. A., Campbell, L. M., and Johnson, K. L., 1991, "Environmental and Economic Evaluation of Gas Turbine SCR NO_x Control," paper presented at the 1991 Joint Symposium on Stationary Combustion NO_x Control—EPA/EPRI, Washington, D.C., March 25–28.

McIntyre, A. (Exxon) 1994, personal communication, May.

McKinney, K. (Metzler & Associates), 1993, personal communication, Feb. 19.

Medock, M., 1990, "An Overview of Non-Catalytic NO_x Control," *Solid Waste & Power,* February, pp. 46–51.

Mendelsohn, M. H., Livengood, C. D., and Harkness, J. B. L., 1991, "Combining SO_2/NO_x Control Using Ferrous•EDTA and a Secondary Additive in a Lime-Based Aqueous Scrubber System," paper presented at the EPRI/EPA/DOE 1991 SO_2 Control Symposium, Washington, D.C., Dec. 3–6, pp. 133–145.

Mincy, J. E. (Nalco Fuel Tech), 1992, personal communication, September.

Muzio, L. J., and Arand, J. K., 1976, "Homogeneous Gas Phase Decomposition of Oxides of Nitrogen," EPRI FP-253, Project 461-1, Final Report, KVB, Inc., August.

Muzio, L. J., Montgomery, T. A., Quartucy, G. C., Cole, J. A., and Kramlich, J. C., 1991, "N_2O Formation in Selective Non-Catalytic NO_x Reduction Processes," paper presented at the 1991 Joint Symposium on Stationary Combustion NO_x Control—EPA/EPRI, Washington, D.C., March 25–28.

Nalco Fuel Tech, 1990, Bulletin NFT-10, *NO_xOUT—The NO_xOUT Process is a New Chemical and Mechanical System for Cost-Effective NO_x Reduction from Fossil Fueled Combustion Sources,* Nalco Fuel Tech, June.

Nelson, S. G., 1990, "Field Testing of a New SO_2 and NO_x Sorbent" presented at the EPRI/EPA 1990 SO_2 Control Symposium, New Orleans, LA, May 8–11.

NIOSH, 1990, *Pocket Guide to Chemical Hazards,* National Institute for Occupational Safety and Health, Public Health Service, Center for Disease Control, U.S. Department of Health and Human Services, Washington, D.C., June. (Available from NIOSH, Cincinnati, OH.)

Noroozi, S., 1993, "Urea Enhances Safety in SCR Applications," *Power Engineering,* December, pp. 28–29.

Pacello, V. (ABB Combustion), 1993, personal communication on gas turbines, January.

Pacer, D. W. (Fluor Daniel), 1992, personal communication, June.

Pan, Y.S., 1991, *Recent Advances in Flue Gas Desulfurization Technologies,* report DOE/PETC/TR-91/4, Pittsburgh Energy Technology Center, Pittsburgh, PA.

Pease, R. R., 1984, *Status Report on Selective Catalytic Reduction for Gas Turbines,* Southern California Air Quality Management District Engineering Division Report, July.

Piepho, J., Cioffi, P., LaRue, A., and Waanders, P., 1992, "Seven Different Ways Low-NO_x Strategies Move from Demonstration to Commercial Status," paper presented at Power-Gen '92, Orlando, FL, Nov. 19.

Pritchard, S. (Cormetech, Inc.), 1994, personal communication, June.

Public Law 95-95, 1977, *Clean Air Act Amendments of 1977,* U.S. Government Printing Office, Washington, D.C., Aug. 7.

Public Law 101-549, 1990, *Clean Air Act Amendments of 1990,* U.S. Government Printing Office, Washington, D.C., Nov. 15.

Reese, J. L., Mansour, M. N., Mueller-Odenwald, H., Johnson, L. W., Radak, L. J., Rundstrom, D. A., 1991, "Evaluation of SCR Air Heater for NO_x Control on a Full-Scale Gas- and Oil-Fired Boiler," paper presented at the 1991 Joint Symposium on Stationary Combustion NO_x Control—EPA/EPRI, Washington, D.C., March 25–28.

Rini, M. J., and Cohen, M. B., 1993, "Evaluating the SNCR Process for Tangentially-Fired Boilers," paper presented at the 1993 Joint Symposium on Stationary Combustion NO_x Control—EPRI/EPA, Bal Harbour, FL, May 24–27.

Robie, C. P., Ireland, P. A., and Cichanowicz, J. E., 1989, "Technical Feasibility and Economics of SCR NO_x Control in Utility Applications," *Proceedings of the 1989 Symposium on Stationary Combustion Nitrogen Oxide Control,* GS-6423, Electric Power Research Institute, Palo Alto, CA, July, pp. 6A-05–6A-124.

Robie, C. P., Ireland, P. A., and Cichanowicz, J. E., 1991A, "Technical Feasibility and Cost of Selective Catalytic Reduction (SCR) NO_x Control," EPRI GS-7266, Electric Power Research Institute, Palo Alto, CA, May.

Robie, C. P., Ireland, P. A., and Cichanowicz, J. E., 1991B, "Technical Feasibility and Cost of SCR for U.S. Utility Application," 1991 Joint Symposium on Stationary Combustion NO_x Control—EPA/EPRI, Washington, D.C., March 25–28.

Rosenberg, H. S., 1987, U.S. Patent 4,640,825, Feb. 3.

Schorr, M. M., 1992, "NO_x Emission Control for Gas Turbines: A 1992 Update," paper presented at the Council of Industrial Boiler Owners NO_x Control V Conference, Long Beach, CA, Feb. 10–11.

Schreiber, H., 1991, "Combustion NO_x Controls for Combustion Turbines," paper presented at the 1991 Joint Symposium on Stationary Combustion NO_x Control—EPA/EPRI, Washington, D.C., March 25–28.

Shaw, H., 1973, "The Effects of Water, Pressure, and Equivalence Ratio on Nitric Oxide Production in Gas Turbines," American Society of Mechanical Engineers, Paper 73-WA/GT-1.

Shore, D. E., Buening, H. J., Prodan, D. A., Teetz, R. D., Muzio, L. J., Quartucy, G. C., Sun, W. H., Carmignani, P. G., Stallings, J. W., and O'Sullivan, R. C., 1993, "Urea SNCR Demonstration at Long Island Lighting Company's Port Jefferson Station, Unit 3," paper presented at the 1993 Joint Symposium on Stationary Combustion NO_x Control—EPRI/EPA, Bal Harbour, FL, May 24–27.

Siegfriedt, W. E., and Pacer, D. W., 1991, German SCR/SNCR Trip Report (unpublished report), Fluor Daniel, Chicago, IL, September.

Sopocy, D. M., DePriest, W., Kalanik, J. B., Maurer, A., Rhudy, R., 1991, "Clean Air Technology (CAT) Workstation," paper presented at the EPRI/EPA/DOE 1991 SO_2 Control Symposium, Washington, D.C., Dec. 3–6.

Speronello, B. K., Chen, J. M., Durilla, M., and Heck, R. M., 1991, "Application of Composite NO_x SCR Catalysts in Commercial Systems," paper presented at the 1991 Joint Symposium on Stationary Combustion NO_x Control—EPRI/EPA, Washington, D.C., March 25–28.

Teixeira, D. P., Lin, C. I., Muzio, L. J., Jones, D. G., and Okazaki, S., 1993, "Selective Noncatalytic Reduction (SNCR) Demonstration in a Natural Gas-Fired Boiler," paper presented at the 1993 Joint Symposium on Stationary Combustion NO_x Control—EPRI/EPA, Bal Harbour, FL, May 24–27.

Toqan, M. A., Teare, J. D., Beer, J. M., Weir, Jr., A. J., and Radak, L. J., 1989, "NO_x Reduction in Fuel-Rich Natural Gas and Methanol Turbulent Diffusion Flames," *Proceedings of the 1989 Symposium on Stationary Combustion Nitrogen Oxide Control,* GS-6423, Electric Power Research Institute, Palo Alto, CA, July, pp. 6B-21–6B-38.

Tri-Mer Corporation, 1992, sales literature, *Tri-Mer Corporation Air Pollution Control Systems,* Owosso, MI.

Turco, R. P., 1985, "The Photochemistry of the Stratosphere," *The Photochemistry of Atmospheres,* J. S. Levine, editor, Orlando: Academic Press, Inc., pp. 77–128. Referenced by Joel S. Levine, in "The Global Atmospheric Budget of Nitrous Oxide," paper presented at the 1991 Joint Symposium on Stationary Combustion NO_x Control—EPA/EPRI, Washington, D.C., March 25–28.

U.S. Department of Energy, 1992, *Clean Coal Technology Demonstration Program—Program Update 1991,* Publication No. DOE/FE-0247P, February.

Wax, M. J. (Institute of Clean Air Companies, Washington, DC), 1993A, letter to Dr. Praveen K. Amar, Northeast States for Coordinated Air Use Management (NESCAUM), Boston, MA, April 14. (Used with permission.)

Wax, M. J. (Institute of Clean Air Companies, Washington, DC), 1993B, letter to Dr. Praveen K. Amar, Northeast States for Coordinated Air Use Management (NESCAUM), Boston, MA, May 26. (Used with permission.)

Wax, M. J. (Institute of Clean Air Companies, Washington, DC), 1993C, letter to Dr. Praveen K. Amar, NESCAUM (Data compiled by Wax based on telephone communications with personnel at LADWP and Southern California Edison and from Loen, H., Freitas, T., and Wilkinson, J., "Full-Scale SCR Retrofit the First at a U.S. Steam Plant," *Power,* April 1993, pp. 77–86.), Boston, MA, June 29. (Letter used with permission.)

Wehner, M. (LaRouche Industries), 1994, personal communication, March 16.

Wendt, J. O. L., Bose, A. C., and Hein, K. R. G., 1989, "Fuel Nitrogen Mechanism Governing NO_x Abatement for Low and High Rank Coals," work supported by the U.S. Department of Energy, Pittsburgh Energy Technology Center, Contract DE-AC22-84PC70771, Pittsburgh, PA.

Wolff, G. T., 1994, "NO_x-VOC-O_3 Relationship," *Seventh Annual NO_x Control Conference—Proceedings,* (sponsored by the Council of Industrial Boiler Owners), Oak Brook, IL, May, pp. 113–136.

Wood, S. C., 1994, "Select the Right NO_x Control Technology," *Chemical Engineering Progress,* January, pp. 32–38.

Yagiela, A., 1993, "Cyclone Reburn & Low-NO_x Cell™ Burners," Technical Information Exchange—NO_x Reduction Solutions, (sponsored by Babcock & Wilcox), March, Chicago, IL.

Yamamura, M., and Suyama, K., 1988, "Operation Experience of Selective Catalytic NO_x Removal Systems for Miscellaneous Flue Gas," paper presented at the 81st Annual Meeting of APCA, Dallas, TX, June 19–24.

Zel'dovich, Y. B., Sadovnikov, P. Y., and Frank-Kamenetskii, D. A., 1947, *The Oxidation of Nitrogen with Combustion,* Izd. Akad. Nauk SSSR (Academy of Sciences, U.S.S.R., Institute of Chemical Physics), Moscow-Leningrad, trans. by M. Shelf.

Chapter 11

Absorption of Water Vapor by Dehydrating Solutions

INTRODUCTION, 946

    Water Content of Saturated Gas, 947
    Available Dehydration Processes, 952

GLYCOL DEHYDRATION PROCESSES, 953

    Glycol Selection, 953
    Process Description, 955
    Basic Design Data, 964
    Dehydration Plant Design, 972
    Dehydration Plant Operation, 988

INJECTION SYSTEMS, 997

    Process Description, 997
    Injection Process Design Data, 1002
    Injection Process Design, 1002

DEHYDRATION WITH SALINE BRINES, 1007

    Calcium Chloride for Gas Dehydration, 1008
    Lithium Halides for Air Dehydration, 1010

REFERENCES, 1017

INTRODUCTION

Water vapor is probably the most common undesirable impurity in gas streams. Usually, it is not the water vapor itself that is objectionable, but rather the liquid or solid phase that may precipitate from the gas when it is compressed or cooled. Liquid water almost always accelerates corrosion, and ice (or solid hydrates) can plug valves, fittings, and even gas lines. To prevent such difficulties, essentially all fuel gas which is transported in transmission lines must be at least partially dehydrated. Compressed air used to operate automatic valves and instruments in refineries and chemical plants must also be thoroughly dry. There are occasionally other reasons why gas streams must be dehydrated as, for example, in catalytic processes where water constitutes a catalyst poison or source of undesirable side reaction and in the air conditioning field, where dehumidification is frequently a requirement.

Water Content of Saturated Gas

The quantity of water in saturated natural gas at various pressures can be estimated from **Figure 11-1,** which is based on the correlation of McKetta and Wehe (1958). This chart provides essentially the same data as the frequently used correlation of McCarthy et al. (1950), but has the advantage of including corrections for gas specific gravity and water salinity. The corrections are used as simple multipliers for water content values shown on the main chart. For example, if the gas has a molecular weight of 26 and is in equilibrium with an aqueous phase containing 3% salt, the correction factors would be $C_G = 0.98$ and $C_S = 0.93$. For this case, and conditions of 150°F and 3,000 psia, the gas would have a water content of (0.93)(0.98)(104) = 95 lb/MMscf.

Figure 11-1 also shows a hydrate formation line for a 0.6 specific gravity gas. To the left of this line solid hydrates will form when saturated gas is cooled. For example, if 2,000 psia, 0.6 specific gravity, saturated gas is cooled below about 69°F, hydrates will form. At pressures below about 150 psia, on the other hand, cooling to 32°F is necessary to precipitate a solid phase, and in this case, ordinary ice will form. The hydrates form more readily (i.e., at a higher temperature or lower pressure) with gases of greater density and less readily with very light gases. For example, at a pressure of 1,000 psia, hydrates form at about 60°F in natural gas of 0.60 specific gravity, while they form at 67° and 71°F, respectively, in gases of 0.75 and 1.00 specific gravity (Arnold and Pearce, 1961). **Figure 11-2** is a simple chart showing the hydrate point vs. gas pressure for three values of specific gravity (Pearce and Sivalls, 1993). A more detailed discussion of the conditions that cause the formation of hydrates is given in the *GPSA Engineering Data Book* (1987).

In **Figure 11-1** the lines below and to the left of the hydrate/ice formation line represent a meta-stable equilibrium between water vapor in the gas phase and supercooled liquid water. The actual equilibrium with solid ice or hydrate is at a lower water content. The effect is depicted in **Figure 11-3,** which also extends the water content scale of **Figure 11-1** down to 0.1 lb water/MMscf. The data on equilibrium water contents in the 0.1 to 1.0 lb water/MMscf range are necessary for the design of the recently developed "superdehydration" processes. Water content data down to as low as 0.001 lb/MMscf are plotted by Bucklin et al. (1985). Such extremely low values are of interest in the design of natural gas turboexpander plants.

The water vapor content of saturated natural gas is also affected by gas composition. The presence of substantial concentrations of CO_2 or H_2S, for example, increases the equilibrium water concentration, particularly at pressures above 1,000 psia. This effect has been correlated by Robinson et al. (1978) for gas pressures from 1,000 to 10,000 psia and combined acid

(text continued on page 950)

948 Gas Purification

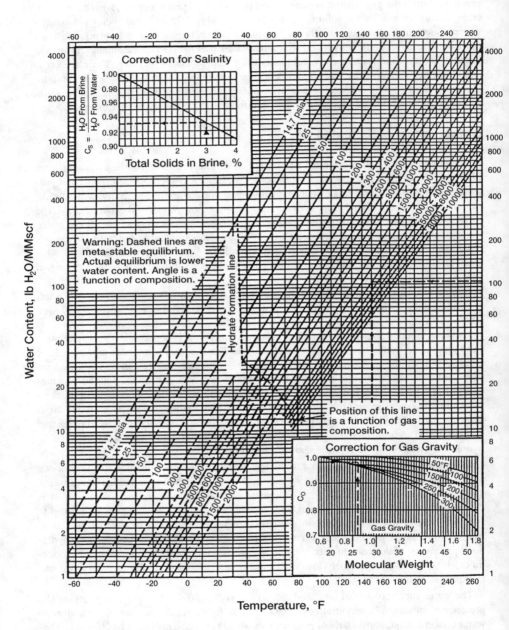

Figure 11-1. Water content of saturated natural gas. *Data of McKetta and Wehe (1958)*

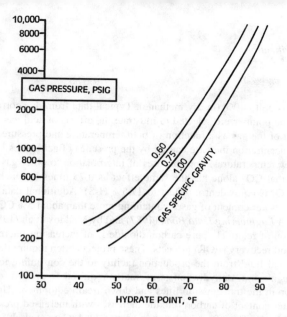

Figure 11-2. Conditions for hydrate formation in natural gas. *From Pearce and Sivalls (1993)*

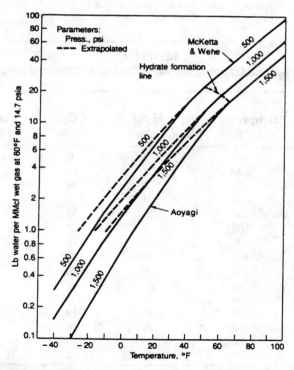

Figure 11-3. Equilibrium water content of natural gas below the hydrate point. *From Hicks and Senules (1991), based on Aoyagi et al. (1979). Reproduced with permission from Hydrocarbon Processing, April 1991*

(*text continued from page 947*)

gas concentrations up to 40% (in dry methane). Typical data from this correlation are given in **Table 11-1**. The points were selected to illustrate the effects of acid gas concentration on the water content of the gas as a function of both temperature and pressure. It will be noted that the water concentration increase caused by the presence of acid gases is greatest at high pressures and low temperatures. For purposes of interpolation to other gas compositions, it can be assumed that CO_2 alone has the same effect as 0.75 times as much H_2S (e.g., 10% H_2S plus 10% CO_2 is equivalent in effect to 17.5% H_2S). Additional data and correlations for estimating the water content of gases containing more than about 5% CO_2 and/or H_2S are given in the *GPSA Engineering Data Book* (1987) and by Maddox et al. (1988).

The dehydration of relatively pure carbon dioxide is of increasing interest because of its use in enhanced oil recovery (EOR) projects. These projects often require the transmission of CO_2 as supercritical fluid from the production facility to the consuming locations. As with natural gas transmission, dehydration is normally required to prevent corrosion and/or hydrate formation in the transmission lines and downstream equipment. Unlike natural gas, the saturated water content of carbon dioxide increases with increased pressure at pressures above about 1,000 psia. This effect is shown in **Figure 11-4** (Case et al., 1985). A much more detailed discussion of the variation of water content of CO_2-rich gas streams with temperature, pressure, and composition is given by Diaz et al. (1991).

The water content of saturated air at pressures from 1 to 1,000 atm is given in **Figure 11-5**, which is based on the data of Landsbaum et al. (1955). The water content of air at atmos-

Table 11-1
Effect of H_2S and CO_2 on Water Vapor Content of Saturated Natural Gas

Pressure, psia	Temperature, °F	H_2S, Vol%	CO_2, Vol%	Water Concentration, lb/MMscf
1,000	100	0	0	58.9
1,000	100	10	10	63.9
1,000	100	20	20	71.9
1,000	200	0	0	630
1,000	200	20	20	733
6,000	100	0	0	23.1
6,000	100	10	10	38.5
6,000	100	20	20	73.6
6,000	200	0	0	197
6,000	200	20	20	397
10,000	100	0	0	19.9
10,000	100	10	10	36.1
10,000	100	20	20	71.8
10,000	200	0	0	159
10,000	200	20	20	378

Data of Robinson et al. (1978)

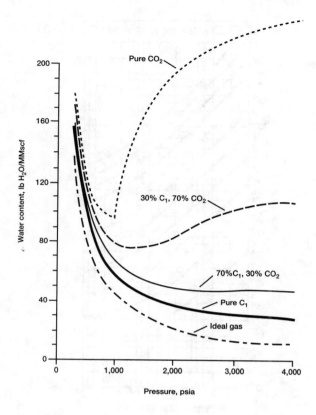

Figure 11-4. Water vapor content of saturated carbon dioxide-rich gas at 100°F. *From Case et al. (1985). Reproduced with permission from Oil & Gas Journal, May 13, 1985, copyright PennWell Publishing Co.*

pheric pressure and various degrees of saturation is most conveniently estimated from psychrometric charts, which are reproduced in most standard air conditioning and chemical engineering texts.

Figures 11-1, 11-3, and **11-4** follow the common practice of stating the water content of natural gas in terms of lb/MMscf and that of air in terms of lb water/lb dry air. Another useful method of indicating the water content of any gas is in terms of the water dew point. This is defined as the temperature to which a gas must be cooled (at constant water content) in order for it to become saturated with respect to water vapor (i.e., attain equilibrium with liquid water). Since dehydration is frequently practiced to prevent the precipitation of liquid water from gases when they are cooled, the dew point is a more direct indication of the dehydration effectiveness than the absolute water content. If a dew point of 40°F is desired, for example, a natural gas stream would require dehydration to 62 lb/MMscf at 100 psia or 9 lb/MMscf at 1,000 psia. Since the water vapor pressure over dehydrating solutions nor-

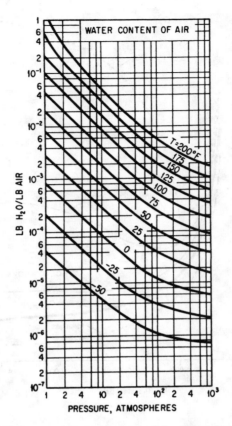

Figure 11-5. Water vapor content of saturated air. *After Landsbaum et al. (1955)*

mally varies with temperature in approximately the same manner as the vapor pressure of pure water, the effectiveness of a given solution can be evaluated in terms of the difference between the dew point of the dehydrated gas and the contact temperature. This difference is known as the "dew-point depression" and is roughly constant for a given dehydration system (i.e., solution strength and contact efficiency) over a fairly wide range of temperatures and pressures.

Available Dehydration Processes

Commercial processes for removing water vapor from gas streams can be classified as follows:

1. Absorption by hygroscopic liquids (or reactive solids)
2. Adsorption by activated solid desiccants
3. Condensation by cooling and/or compression

Only the first two methods are covered in detail, in this and the following chapter, as the third is primarily mechanical and is normally economical only under special circumstances.

Glycerol was one of the first liquids used for drying fuel gas, and the design of a plant utilizing it for city gas was described by Tupholme (1929). Calcium chloride solution was reportedly the first liquid used for dehydrating natural gas. This was employed during the early 1930s (Swerdloff, 1957). Diethylene glycol (DEG) was first used to dehydrate natural gas in fall 1936. This material and its close relative, triethylene glycol (TEG), proved to be very effective. In 1957, it was estimated that there were at least 5,000 glycol-type natural gas dehydration plants in the U.S. and Canada (Polderman, 1957); it is now estimated that this number has grown to over 20,000 plants.

Triethylene glycol has become the industry standard for natural gas; however, other glycols are employed where they are able to meet the requirements at lower cost. In addition to DEG and TEG, tetraethylene glycol (T_4EG) and glycol blends have found application for the dehydration of natural gas. The blends represent impure products of the manufacturing process and are, therefore, available at a lower cost than highly purified DEG, TEG, or T_4EG. However, the blends are significantly less effective than the pure compounds (Grosso, 1978). For air dehumidification, triethylene glycol and lithium chloride solutions are the only two liquid systems in common use.

For processes in which the liquid agent is injected directly into the gas stream to inhibit the formation of hydrates downstream of the injection point, low viscosity at the operating temperature is a more important property than dehydration ability. Ethylene glycol or methanol are most commonly used for such applications with methanol preferred for cryogenic systems operating below about −40°F and ethylene glycol for moderate temperature applications. The higher glycols are also occasionally used in injection processes, particularly where high temperature operation can lead to excessive vaporization of ethylene glycol.

For the special case of dehydrating supercritical carbon dioxide, Shell Oil Company has developed the technology of using glycerol (Diaz and Miller, 1984). Detailed discussions of the process are given by Diaz et al. (1991) and Wallace (1985). At subcritical conditions the conventional glycols, DEG and TEG, are effective for dehydrating CO_2-rich gas streams; however, at supercritical conditions the CO_2-rich fluids can dissolve substantial amounts of the glycols. Under these conditions, glycerol is an attractive desiccant. At 95°F and 1,200 psia, for example, the solubility of glycerol in carbon dioxide is only 2 lb/MMscf compared to 76, 150, and 150 lb/MMscf for EG, DEG, and TEG, respectively. The drying capability of glycerol is roughly similar to TEG. Operation of the Shell Oil Company Glycerol process in a Hungarian enhanced oil recovery (EOR) process is described by Udvardi et al. (1990).

Sulfuric acid is an excellent dehydrating agent, but because of its extreme corrosiveness, it is now used only for special applications, such as the drying of gas streams in sulfuric acid plants. Many other liquids possess dehydrating properties, including solutions of sodium or potassium hydroxide and the halides of several metals; however, none of these materials is in widespread use for gas dehydration, primarily because they are difficult to handle.

GLYCOL DEHYDRATION PROCESSES

Glycol Selection

Data on the physical properties of four glycols are given in **Table 11-2**. Diethylene glycol and triethylene glycol are the principal glycols used for gas dehydration, with triethylene glycol applications predominating. Diethylene glycol is preferred for applications below

Table 11-2
Properties of Glycols

Property	Ethylene Glycol (EG)	Diethylene Glycol (DEG)	Triethylene Glycol (TEG)	Tetraethylene Glycol (T_4EG)
Formula	CH_2-OH | CH_2-OH	CH_2CH_2-OH / O \ CH_2CH_2-OH	CH_2-O-CH_2CH_2-OH | CH_2-O-CH_2CH_2-OH	CH_2CH_2-O-CH_2CH_2-OH / O \ CH_2CH_2-O-CH_2CH_2-OH
Molecular Weight	62.1	106.1	150.2	194.2
Boiling Point @ 760 mm Hg	197.6°C (387.7°F)	245.8°C (474.4°F)	288.0°C (550.4°F)	314.0°C (597.2°F)
Initial Decomposition Temp, °F	329	328	404	—
Density @ 77°F (25°C), g/ml	1.110	1.113	1.119	1.120
Freezing Point	−12.7°C (9.1°F)	−7.8°C (17.6°F)	−7.2°C (19.04°F)	−5.6°C (22°F)
Viscosity, abs, cp @77°F (25°C) @140°F (60°C)	16.5 5.08	28.2 7.6	37.3 9.6	39.9 10.2
Surface Tension @ 25°C, dyne/cm	47	44	45	45
Specific Heat @ 77°F (25°C), Btu/lb°F	0.58	0.55	0.53	0.52
Heat of Vaporization (760 mmHg), Btu/lb	364	232	174	—
Heat of Solution of Water in Infinite Amount of Glycol (approx. 80°F) Btu/lb	—	58	86	—
Flash Point, °F (C.O.C.)	240	280	320	365

Note 1: C.O.C. = Cleveland Open Cup method
Source: Union Carbide, 1971; Worley, 1966; Gallaugher and Hibbert, 1937

about 50°F because of the high viscosity of TEG in this temperature range. Tetraethylene glycol is recommended for contact temperatures above about 120°F to minimize vapor losses. Additional physical property data on the two principal glycols are given in **Table 11-3**. The factors that have led to the widespread use of glycols for gas dehydration are their unusual hygroscopicity, their excellent stability with regard to thermal and chemical decomposition, their low vapor pressures, and their ready availability at moderate cost. Photographs of typical glycol dehydration plants are presented in **Figures 11-6** and **11-7**.

Process Description

Basic Process

A simplified flow diagram of a typical triethylene glycol dehydration plant for natural gas service is shown in **Figure 11-8** (Bentley, 1991). The process is quite simple. After flowing through a separator (not shown in the diagram) and through a knockout section to remove entrained liquid, the feed gas then flows up through a chimney tray into the absorption section of the contactor, which typically contains between four and ten bubble-cap trays. The concentrated glycol, normally containing 0.5 to 2% water, is fed to the top of the contactor and absorbs water from the gas while flowing downward through the column. The dried gas leaves the top of the contactor and is used to cool the glycol feed. As indicated on the flow diagram, it may then pass through a scrubber which removes any entrained glycol droplets before the product gas enters the pipeline.

Rich glycol flowing out of the bottom of the contactor typically contains 3 to 7% water and must be reconcentrated before it can be reused for water absorption. The rich glycol is often used to provide cooling and condense water vapor at the top of the reconcentrator. This raises the temperature of the rich glycol, which then may be further heated by heat exchange with hot lean glycol. It then enters a reduced pressure flash tank where dissolved hydrocar-

Table 11-3
Effect of Temperature on Glycol Properties

Glycol	Temp., °F	Sp. Gr.	Viscosity, cp.	Sp. Ht., Btu/(lb) (°F)	Thermal Conduct., Btu/(h) (ft) (°F)
DEG	50	1.127	72	0.53	0.146
	100	1.107	18	0.56	0.135
	150	1.089	7.3	0.58	0.125
	200	1.064	3.6	0.60	0.115
	300	1.021	1.3	0.66	—
TEG	50	1.134	88	0.485	0.140
	100	1.111	23	0.52	0.132
	150	1.091	8.1	0.55	0.125
	200	1.068	4.0	0.585	0.118
	300	1.022	1.5	0.65	—

Source: Union Carbide, 1971

Figure 11-6. Large glycol natural gas dehydration plant. *Courtesy of Southern Counties Gas Co. of California*

bon gases are released. The released gases are recovered and used for fuel or other purposes. After flashing, the rich glycol passes through a filtration system and a second glycol/glycol heat exchanger where it is further heated, and finally enters the reconcentrator column above a short packed or tray section. Efficient recovery of heat from the reconcentrator products is necessary to minimize fuel consumption in the reboiler.

Because of the extreme difference in the boiling points of glycol and water, a very sharp separation can be achieved with a relatively short column. Water reflux must be provided at the top of the column to effect rectification of the vapors and minimize glycol losses in the overhead vapor stream. This is normally provided by condensing a portion of the overhead vapor. The amount of reflux is held at the minimum consistent with good plant operation because it directly affects the quantity of heat required in the reboiler. Typically, a condenser heat duty of about 25% of the reboiler duty will provide sufficient reflux to limit glycol losses to less than 2 lb glycol per MMScf of feed gas (GPSA, 1987). In the flow diagram of **Figure 11-8,** the extent of vapor condensation is controlled by a simple bypass valve in the rich solution line. More effective control can be attained by using a three-way valve in the solution line to apportion solution between the reflux coil and the bypass line in response to a temperature sensor located at the top of the still column (Bucklin, 1993).

Heat for the distillation is provided by a direct fired reboiler. Hot lean glycol leaves the reboiler and flows to a surge tank (which often contains cooling coils) and is pumped through the glycol/glycol heat exchangers and back to the contactor.

Figure 11-7. Small package-type glycol natural gas dehydration plant. *Courtesy of BS&B Engineering Co., Inc.*

Enhanced Stripping Processes

The degree of dehydration that can be attained with a glycol solution is primarily dependent on the extent to which water is removed from the solution in the reconcentrator. The operation of atmospheric pressure distillation units for water removal is limited by the maximum temperature that can be tolerated without excessive decomposition of the glycol (about 400°F for TEG). Concentration of TEG to 98.5 to 99.0% is attainable in a simple atmospheric pressure still. When significantly higher concentrations are needed to meet stringent gas dehydration requirements, the use of an enhanced stripping technique is necessary.

The flow diagram of **Figure 11-8** shows an optional stripping gas column that operates on the hot lean glycol flowing from the reboiler to the surge tank. When this type of stripping is used, a small stream of dry natural gas is fed into the bottom of the stripping gas column to reduce the partial pressure of water vapor in the gas phase. The gas aids in removing water from the glycol and finally leaves the primary stripping column with the vented water vapor. A simpler but less effective technique is to inject the inert gas directly into the glycol in the reboiler. According to Wieninger (1991), a concentration of 99.5% can be obtained by injecting stripping gas into the reboiler, and a concentration as high as 99.9% can be obtained with a separate stripping gas column between the reboiler and the surge tank (see **Figure 11-8**).

958 *Gas Purification*

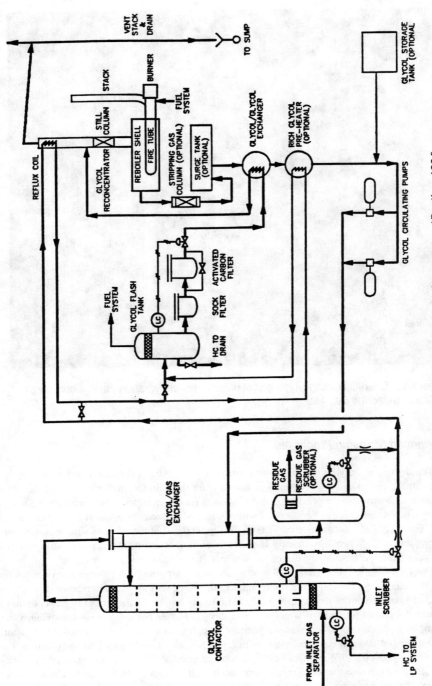

Figure 11-8. Flow diagram of typical glycol dehydration process. *(Bentley, 1991)*

The reboiler gas injection method has the additional advantage of stirring the glycol which helps avoid overheating of glycol in contact with the hot firetube surface.

An alternate approach for the enhancement of reconcentrator performance is the use of an azeotrope former as shown in **Figure 11-9** (Smith, 1990). In this approach a volatile hydrocarbon liquid is fed into the glycol regeneration system. The hydrocarbon increases the volatility of water in the solution and, after vaporization, acts as stripping gas in the lean glycol stripper. Mixed vapors from this stripper flow through the glycol reboiler and the rich glycol stripper, which is refluxed with aqueous condensate to minimize glycol losses. Vapors from the rich glycol stripper are totally condensed and collected in a separator. Condensed hydrocarbon liquids form an immiscible phase which is recycled to the regeneration system. Liquid water is discarded.

The process was originally developed by Dow Chemical Company in the mid 1970s and called the Drizo or Super Drizo process. OPC Engineering acquired an exclusive license in 1986 and has made several improvements to the process. The process has the advantages over the use of noncondensible stripping gas of permitting the recovery of BTX components (as excess liquid hydrocarbon) and avoiding the consumption of valuable sales gas; however, it does require payment of a license fee (OPC Engineering, Inc., 1992; Fontenot et al., 1986; Frazier and Force, 1982; Fowler, 1975; Pearce et al., 1972).

Smith and Skiff (1990) reported that this type of process can achieve concentrations of over 99.99% with triethylene glycol, resulting in potential product gas water dew points in the −100° to −140°F range. In more recent papers, Smith (1993) and Smith and Humphrey (1995) cite upper limit glycol concentrations of 99.997 to 99.999% based on use of the Drizo technology. They report experience with one plant producing treated gas with an indicated

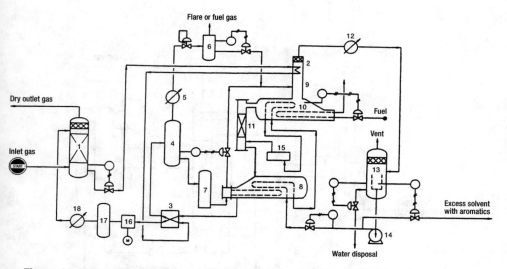

Figure 11-9. Flow diagram of Drizo process showing (1) glycol contactor, (2) reflux condenser, (3) glycol-glycol plate exchanger, (4) flash tank, (5) solvent recovery condenser, (6) recovered solvent drum, (7) glycol filter, (8) surge tank/exchanger, (9) rich stripper, (10) glycol reboiler, (11) lean stripper, (12) solvent-water condenser, (13) solvent-water separator, (14) solvent pump, (15) solvent superheater, (16) glycol pump, (17) acoustical filter, and (18) glycol cooler. (*Smith, 1990*)

water dew point of −139°F. The ability to produce dew points near −150°F can make glycol dehydration units applicable to turboexpander plants, which now commonly use the more expensive molecular sieve adsorbent dehydration systems.

Enhanced stripping can also be attained by operating the reconcentrator under vacuum. BS&B Engineering, Inc., offers the process shown in **Figure 11-10** and claims that hundreds of plants have been installed since it was first introduced in 1957 (BS&B Engineering, Inc., 1992). In the process illustrated, the rich glycol is concentrated to approximately 99% by atmospheric pressure stripping in a conventional reconcentrator. The partially regenerated glycol, at a temperature of about 400°F, is flashed across a throttling valve to subatmospheric pressure, reheated to 400°F, and fed into a vacuum drum. Vapors from the vacuum drum are partially condensed and pumped into the atmospheric pressure regenerator. Glycol from the vacuum drum, at a concentration as high as 99.9%, is cooled and recycled to the absorber. In an alternate vacuum regeneration system described by Polderman (1957), a single-stage reconcentrator is operated under vacuum with the entire vapor output condensed. Data on ten vacuum regeneration glycol dehydration plants are given by Polderman.

Another interesting technique for enhancing glycol concentrator performance has been disclosed by Reid (1975). The method, called "COLDFINGER" is illustrated in **Figure 11-11**. A cooling coil (the cold finger) is placed in the vapor space above hot glycol taken from the reboiler of a conventional regenerator, and a trough is placed below the coil to catch condensate. Since the vapor above concentrated glycol is much richer in water vapor than in glycol vapor, its condensation removes water from the system and additional water is vaporized from the hot glycol. The liquid captured in the trough contains some glycol which is recovered in the plant regenerator. More than 25 "COLDFINGER" units were in operation in 1975.

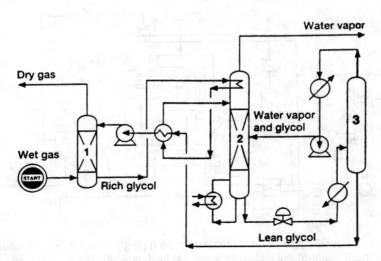

Figure 11-10. Flow diagram of glycol-dehydration plant with vacuum dehydration showing (1) glycol contactor, (2) atmospheric pressure still and reboiler, and (3) vacuum drum. (*BS&B Engineering Company, Inc., 1992*). *Reproduced with permission from Hydrocarbon Processing, April 1992*

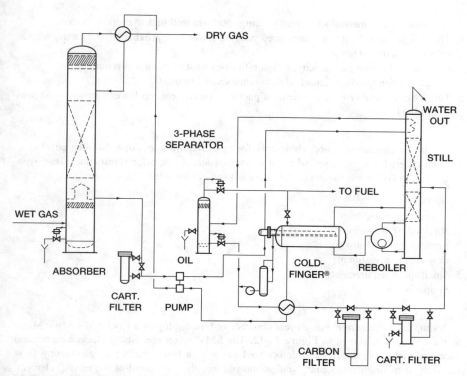

Figure 11-11. Flow diagram of "COLDFINGER" glycol dehydration process. (*Reid, 1975*)

Alternate Absorber Designs

In the basic glycol dehydration plant design, the typical contactor is described as a countercurrent column containing four to ten bubble-cap trays. Although this description is probably appropriate for the majority of operating glycol units, there is a growing trend to consider other types of contacting devices. Chief among these are high efficiency structured packings and co-current contactors.

Random-packed absorption columns have seldom been used for large commercial-size dehydrators because of the very low liquid flow rates usually employed. As pointed out in Chapter 1, a low liquid flow rate normally leads to the selection of a bubble-cap column. The recent trend toward extremely low dew-point requirements (−40°F and below) has resulted in a reexamination of contactor designs to attain more theoretical stages at a reasonable cost. Structured packing, which is offered by several vendors, is claimed to offer greatly increased throughput, lower pressure drop, and lower column height than bubble-cap trays for the same service.

In order to investigate the applicability of structured packing in TEG dehydrator service, ARCO Oil and Gas Co. conducted tests on 11 different commercially available structured packings in a field-operated pilot plant (Kean et al., 1991). They concluded

962 *Gas Purification*

1. A number of commercial structured packings perform well in dehydration service.
2. The design gas velocity in a structured packing tower is approximately twice the normal velocity in a trayed tower.
3. Because of the high gas velocity, a high efficiency mist eliminator is required.
4. A high efficiency drip-point liquid distributor should be used.
5. Turndown capability of the structured packing is excellent, no lower capacity limit was determined.

Co-current contacting is applicable only for cases requiring less than 1.0 theoretical contact; however, for those cases where it is appropriate, it can offer significant advantages. According to Baker and Rogers (1989) these advantages include

1. Minimal pressure drop
2. No flooding problem
3. Broad turndown ratio
4. Low capital investment
5. Small space requirements
6. Simple operation

A simplified diagram of a co-current absorber system employing a Koch SMV Static Mixer as the contactor is shown in **Figure 11-12.** The SMV mixer consists of stacked corrugated sheets of metal, plastic, or ceramic oriented to create a large number of intersecting flow channels or mixing cells. Lean glycol is sprayed into the line ahead of the mixing element, which divides and distributes the liquid while providing an extended liquid surface for contact. The results of tests on a static mixer co-current contactor of this type have been reported by Pyles and Rader (1989). They conclude that the system provides highly efficient water removal and close approach to equilibrium over a wide gas flow range in a short pipe length.

Air Dehydration

In low-pressure dehydration plants such as those used for air conditioning, the pressure drop through the absorber becomes a major design factor, and it is common practice to use spray nozzles in conjunction with a minimum of low-pressure-drop packing in the absorption zone. A schematic diagram of a unit of this type is shown in **Figure 11-13.** In this design, cooling coils are installed in the absorber to remove the latent heat of condensation of water. These also serve as packing. Cooling is required when low-pressure gas or air is dehydrated, because the relatively large amount of water in such gas streams otherwise causes an appreciable temperature increase. The increased temperature can reduce the dehydration efficiency and increase the loss of glycol by vaporization. In the design of **Figure 11-13,** the regenerator is also a spray column, and air is used as a stripping vapor in conjunction with the heating coils. Reflux is provided by condensing a portion of the water from the regenerator air stream on cooling coils located above the glycol-feed point. Glycol from the absorber is used as coolant in the coils before it enters the regenerator. The solution-flow arrangement of this unit is such that a portion of the liquid pumped from the basin is passed through the regenerator as a slip stream operating in parallel to the absorber instead of in series, as is the practice in high-pressure gas-dehydration plants.

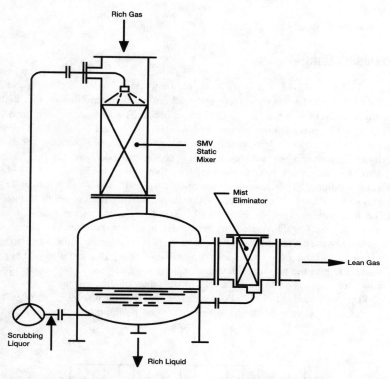

Figure 11-12. Diagram of vertical SMV co-current absorber with liquid recirculation followed by a gas-liquid separator. (*Pyles and Rader, 1989*)

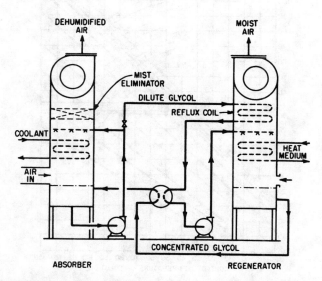

Figure 11-13. Diagram of glycol dehydration unit used in air conditioning service. (*Gifford, 1957*)

Basic Design Data

Figures 11-14 and **11-15,** which are based on the work of Polderman (1957) and Parrish et al. (1986), present data on the water dew points of gases in equilibrium with diethylene and triethylene glycol solutions at various temperatures. The TEG curves are based on the Parrish et al. data rather than the widely used Worley (1967) data because the Parrish et al. data cover a wider range of dew-point depressions and TEG concentrations; are thermodynamically consistent; and generally result in more conservative designs.

The Parrish et al. results agree fairly well with those given by Rosman (1973) for most dehydrating conditions. They also agree reasonably well with the Worley data at TEG concentrations of 90 and 99.97%, but indicate dew points ranging from about 6° to 14°F higher for TEG concentrations of 98, 99, 99.5 and 99.9% (at 100°F contact temperature).

Hicks and Senules (1991) suggest that the Parrish et al. dew-point curves were generated for low-pressure systems and may not be applicable at high pressure. However, Parrish et al. point out that unless the gas contains a large fraction of acid gases and is at a high pressure above about 500 psig the pressure effects are usually insignificant. Earlier studies had indicated dew-point depression to be relatively independent of pressure up to pressures of at least 2,000 psia (Swerdloff, 1957; Polderman, 1957; Townsend, 1953); however, more recent studies have shown that the pressure effect can be important. Manning and Wood

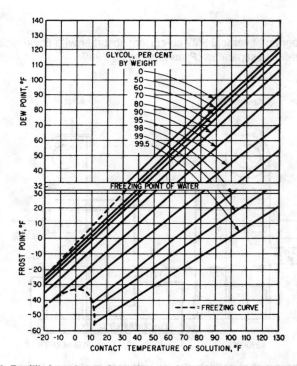

Figure 11-14. Equilibrium dew points of gases in contact with diethylene glycol solution. *Data of Polderman (1957)*

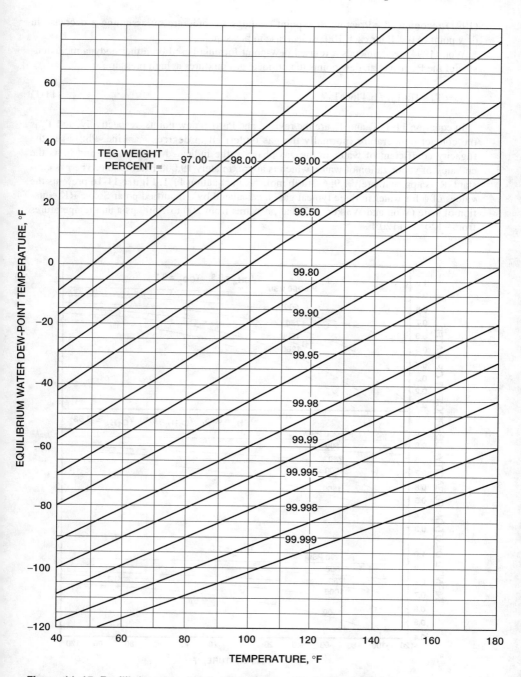

Figure 11-15. Equilibrium dew points of gases in contact with triethylene glycol solution. *From Bucklin and Won (1987), based on Parrish et al. (1986)*

(1991) recommend correcting dew-point readings for pressure by assuming a decrease in dew point of 0.9°F for each 100 psi increase in pressure.

Won (1994) recommends a revised dew-point formula based on a thermodynamic extension of the Parrish et al. dew points to non-ideal gas mixtures at high pressures:

$$T_r = T_d + [R(T_d)^2/ h_v] \ln(\phi T_d/\phi_{Tg}) \tag{11-1}$$

In equation 11-1, T_r and T_d are the revised and Parrish dew points, respectively, and T_g is the contact temperature (normally the gas outlet) all in degrees Rankine. Phi (ϕ) is the fugacity coefficient of water in the gas phase at the indicated temperature, and h_v is the enthalpy of vaporization of water, which is approximately 19,000 Btu/lb mole. The gas constant, R, is approximately 1.986 Btu/lb mole °R in equation 11-1. **Figure 11-16** presents the values of ϕ for water in three typical gases, calculated by a modified polar Soave RK equation of state (Won and Walker, 1979) at pressures from 200 to 1,500 psia and temperatures from −120° to +120°F.

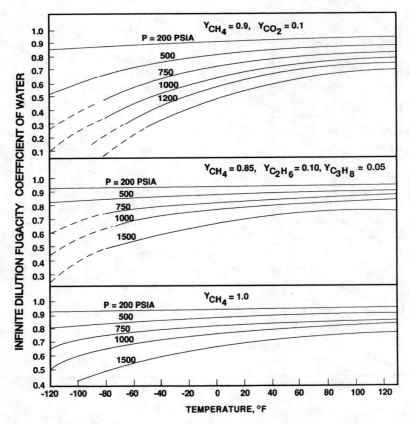

Figure 11-16. Effect of temperature on the infinite dilution fugacity coefficient of water in typical natural gases. (*Won, 1994*)

The revision reduces the dew-point temperature predicted by the Parrish et al. correlation. As an example, the equilibrium water dew point of a methane gas containing 10 mole percent carbon dioxide contacted with 99.9 wt% TEG at 100°F and 1,500 psia is estimated to be −33°F based on the Parrish correlation, **Figure 11-15.** The revised dew point, based on equation 11-1 and **Figure 11-16** is about −48°F, which is 15 degrees lower than the original prediction. If the gas consists primarily of methane, a smaller correction is indicated (about six degrees for these conditions).

At water dew points below the hydrate formation temperature, the vapor phase is actually in equilibrium with solid hydrate rather than liquid water. Correlations such as **Figure 11-1** for the water content of saturated natural gas assume that the gas is in equilibrium with subcooled liquid water at temperatures below the normal ice or hydrate formation point. The dew-point charts, **Figures 11-14** and **11-15,** are based on this same assumption, so the two charts are consistent. However, as indicated by **Figure 11-3,** the actual water content of saturated gas in equilibrium with hydrate is slightly less than shown in **Figure 11-1,** and conversely, for a given water content, the actual temperature at which water deposits from gas (as hydrate in the hydrate formation region) may be higher than that indicated on the dew-point chart. This difference can be significant in the design of cryogenic plants since the hydrate point can be as much as 20°F higher than the metastable water dew point at temperatures below about −100°F (Bucklin et al., 1985).

The activity coefficient for water in concentrated glycol solutions is useful in the calculation of the equilibrium constant for water in the water-glycol-natural gas system. The equilibrium constant is necessary for calculating the required number of theoretical stages by commonly used design correlations. Activity coefficients suggested by Hubbard (1989) based on the Parrish et al. (1986) data are given in **Figure 11-17.**

Figures 11-18 and **11-19** are useful for the design of regenerators. The vertical lines drawn at 340°F for diethylene glycol and 375°F for triethylene glycol represent approximate

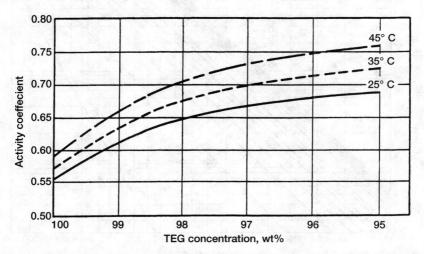

Figure 11-17. Activity coefficients for water in triethylene glycol. *From Hubbard (1989). Reproduced with permission from Oil & Gas Journal, Sept. 11, 1989, copyright PennWell Publishing Co.*

968 Gas Purification

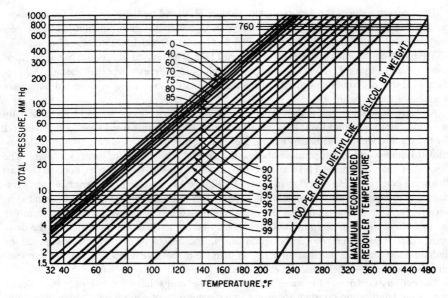

Figure 11-18. Total vapor pressure of various diethylene glycol solutions vs. temperature. *Dow Chemical Company Data (1956)*

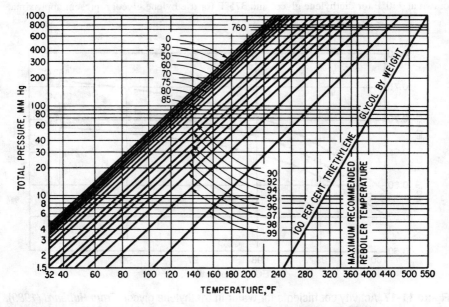

Figure 11-19. Total vapor pressure of various triethylene glycol solutions vs. temperature. *Dow Chemical Company Data (1956)*

maximum design temperatures for these materials. The intersection of these lines with the various boiling point curves marks the recommended regeneration pressure for each solution concentration (in the absence of stripping gas). Where maximum dehydration is required, triethylene glycol reboilers may be operated at temperatures as high as 400°F. With atmospheric pressure operation and no stripping vapor in the regenerator, 400°F operation results in a lean TEG concentration in the range of 98.5 to 99.15%. Recent reports indicate that 340°F may be somewhat high for DEG because, according to Smith (1993), DEG will start to degrade rapidly at a reboiler temperature as low as 325°F.

Important physical property data for diethylene glycol and triethylene glycol are presented in **Figures 11-20** through **11-28. Figure 11-20** shows the effect of temperature and glycol concentration on the specific gravity of triethylene and diethylene glycol solutions. To estimate the specific gravity of glycol solutions at a temperature other than 60°F, it is safe to assume that the specific gravity-temperature curves will be approximately parallel to those for 100% glycols as long as the solutions are relatively concentrated. The vapor pressure chart, **Figure 11-23,** may be used as a basis for vapor loss estimates by assuming Raoult's law holds for glycol in the concentrated solution employed. The freezing point diagrams of glycol-water systems shown in **Figure 11-14** are useful in the design of glycol injection systems to avoid the formation of solid phases.

(*text continued on page 972*)

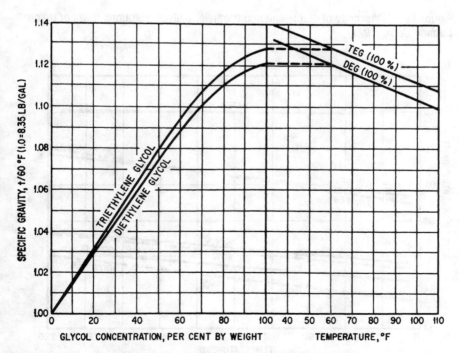

Figure 11-20. Specific gravity of diethylene and triethylene glycol solutions at 60°F, and effect of temperature on specific gravity of the pure glycols. *Data of Union Carbide Corp. (1971)*

970 Gas Purification

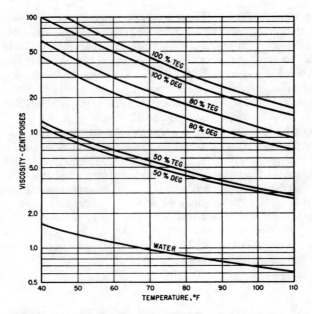

Figure 11-21. Viscosity of diethylene and triethylene glycol solutions. *Data of Union Carbide Corp. (1971)*

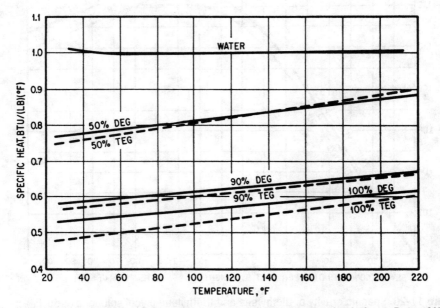

Figure 11-22. Specific heat of diethylene and triethylene glycol solutions. *Data of Union Carbide Corp. (1971)*

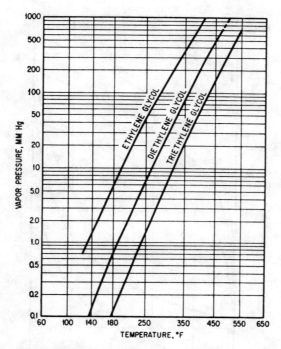

Figure 11-23. Vapor pressure of pure glycols. *Data of Union Carbide Corp. (1971)*

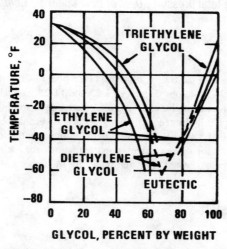

Figure 11-24. Freezing points of ethylene, diethylene, and triethylene glycol solutions. *Data of Union Carbide Corp. (1971)*

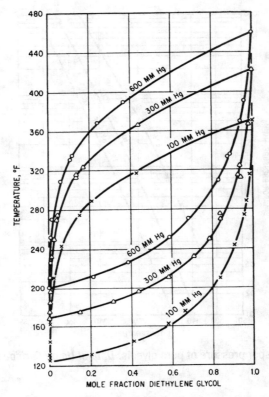

Figure 11-25. Vapor/liquid composition diagrams for diethylene glycol and water at 100, 300, and 600 mm Hg total pressure. *Data of Dow Chemical Company (1956)*

(*text continued from page 917*)

The vapor/liquid composition diagrams, **Figures 11-25** and **11-26,** are applicable to the design of vacuum regeneration columns. **Figures 11-27** and **11-28** give data on the solubility of natural gas and carbon dioxide in glycols. Additional data on the solubility of carbon dioxide and data on the solubility of hydrogen sulfide in glycols are presented by Jou et al. (1988).

Since triethylene glycol is the most widely used dehydrating agent, and equations are frequently preferable to graphs for design calculations, equations for estimating several key properties of TEG solutions, excerpted from the paper by Manning and Wood (1991), are given in **Table 11-4.**

Dehydration Plant Design

Inlet Separator

An efficient inlet separator is an essential component in a glycol dehydration system. The primary function of the separator is to remove liquid water from the feed gas stream; howev-

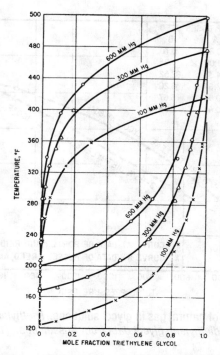

Figure 11-26. Vapor/liquid composition diagrams for triethylene glycol and water at 100, 300, and 600 mm Hg total pressure. *Data of Dow Chemical Company (1956)*

er, it also removes other impurities that may contaminate the glycol or cause operating problems. These include salt (in the liquid water), liquid hydrocarbons, compressor lubricants (if the gas has been compressed), well treating chemicals, and solid particles such as sand, rust, and iron sulfide. The inlet separator may be a free-standing vertical or horizontal vessel or integral with the contactor. If separate, it should be close to the contactor so that further condensation of liquid water or hydrocarbons does not occur in the connecting line.

Typically, the separator is a vertical cylindrical vessel equipped with vanes and/or mesh to remove fine droplets of liquid from the gas. Tables for estimating the required diameter of separators for various gas flow rates and conditions are given in several publications, including *The API Specification for Glycol Type Gas Dehydration Units* (1990) and a paper by Pearce and Sivalls (1993). The tables are based on the Souders-Brown correlation, which is quite simple to use directly, i.e.,

$$V = K((d_L - d_G)/d_G)^{0.5} \qquad (11\text{-}2)$$

Where: V = Allowable superficial gas velocity, ft/s
d_L = Density of the liquid, lb/ft^3

(*text continued on page 976*)

974 Gas Purification

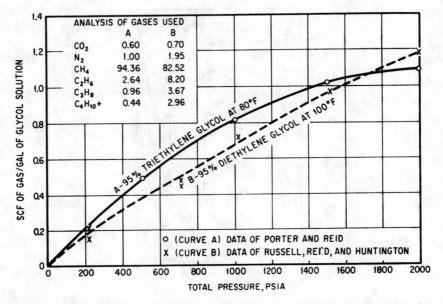

Figure 11-27. Solubility of natural gas in glycol solutions. *Triethylene glycol data from Porter and Reid (1950), diethylene glycol data from Russell et al. (1945)*

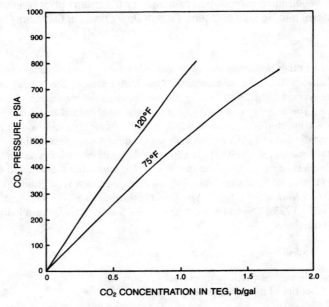

Figure 11-28. Solubility of carbon dioxide in pure triethylene glycol. *Data of Takahashi and Kobayashi (1982)*

Table 11-4
Equations for Estimating Properties of Triethylene Glycol Solutions

Specific Gravity (at t/60°F) = A + Bx + Cx^2
Where x = wt % TEG

t, °F	A	B	C
−50	1.0502	1.8268 E-3	−5.2009 E-6
0	1.0319	1.7455 E-3	−4.8304 E-6
50	1.0121	1.5247 E-3	−2.8794 E-6
100	0.9920	1.7518 E-3	−5.4955 E-6
200	0.9627	1.4068 E-3	−3.5089 E-6
300	0.9177	1.2511 E-3	−2.0848 E-6

Specific Heat Btu/(lb)(°F) = A + Bt + Ct^2
Where t = temperature, °C

wt % TEG	A	B	C
0	1.00540	−2.7286 E-4	2.9143 E-6
20	0.92490	2.0429 E-4	2.4524 E-6
40	0.83229	6.2286 E-4	1.3714 E-6
60	0.72200	9.4000 E-4	8.0000 E-7
80	0.60393	1.2043 E-3	2.8571 E-7
100	0.48614	1.3929 E-3	−5.7140 E-8

Thermal Conductivity Btu/(hr)(ft)(°F) = A + Bt
Where t = temperature, °C

wt % TEG	A	B
0	0.33667	7.1667 E-4
20	0.29000	4.0000 E-4
40	0.25133	3.3333 E-4
60	0.20933	−1.6667 E-4
80	0.17267	−2.8333 E-4
100	0.14133	−3.1667 E-4

Viscosity (Cp) = (A)(10)B^x
Where x = wt % TEG

t, °F	A	B
0	1.53010	2.9967 E-2
50	1.09200	1.9348 E-2
100	0.58916	1.5763 E-2
150	0.37045	1.3481 E-2
200	0.27371	1.1731 E-2
300	0.14026	8.1319 E-3

Source: Manning and Wood, 1991

(*text continued from page 973*)

    d_G = Density of the gas, lb/ft^3
    K = An empirical constant varying from about 0.12 for highly loaded gas to 0.35 for gas streams containing small amounts of liquid, according to the API Specification (1990)

The separator height must be sufficient to house the separator elements and provide surge capacity for collected liquids. If both liquid water and liquid hydrocarbons are present in the feed gas, a three-phase separator must be provided with adequate residence time to allow the two liquid phases to separate from each other. Pearce and Sivalls (1993) suggest vertical separator shell heights of 5 to 7.5 ft for 16 to 60-in.-diameter two-phase units and 7.5 to 10 ft for three-phase units of the same diameter range. The heights are based on providing one-minute residence time for liquid in the two-phase separator and 5 minutes for each of the two liquids in the three-phase unit.

Contactor

Tray Columns. The contactor, or absorber, is typically a countercurrent column containing an integral scrubber at the bottom, a central trayed or packed absorption section, and a demister at the top. The basic design correlations for absorption columns are described in Chapter 1 and only matters specific to dehydration contactors are covered in the following paragraphs.

Most glycol dehydration contactors contain 4 to 10 bubble-cap trays with 24-in. tray spacing. Sieve or valve trays are occasionally used where relatively constant gas flow conditions are anticipated. Large towers (i.e., over 4 ft in diameter) often utilize a 30-in. tray spacing to provide accessibility for maintenance, while very small towers (i.e., 12-in. or less diameter) typically use random packing such as ceramic saddles or stainless steel Pall rings (Manning and Wood, 1991). There is a growing trend to consider structured packing for use where bubble-cap trays would normally be specified because of its higher allowable gas velocity and lower height requirement.

After establishing design requirements and conditions, a typical procedure for designing a bubble-cap column for a TEG dehydration plant includes the following steps:

1. Convert the required product gas water content to a dew-point temperature by use of either **Figure 11-1** or **11-3**.
2. Use **Figure 11-15** to select a TEG concentration that shows an equilibrium dew point about 20°F below the required product gas dew point at the expected contact temperature.
3. Select a TEG flow rate in the range of 2 to 6 gallons of lean glycol per pound of water absorbed. Three gallons per pound of water is a typical rate for small dehydrators, while lower rates (e.g., 2–2.5 gallons per pound) are more economical for large plants.
4. Make a material balance around the column using **Figures 11-1** and **11-3** to determine the water content of the feed and product gas streams and the selected TEG flow rate to calculate the water content of the rich TEG.
5. Make a heat balance around the absorber and calculate the temperature of the exit streams.
6. Estimate the number of trays required using precalculated charts, a McCabe-Thiele diagram, or a design correlation such as the Kremser equation, and, where necessary, a tray efficiency based on experience.

7. Calculate the required column diameter using the Souders-Brown or other appropriate correlation.
8. Repeat steps 2 through 7 as required to optimize performance and economics taking into account regeneration requirements and the overall plant design.

Charts showing the effects of TEG concentration and flow rate on the dew-point depression attainable with 4, 6, and 8 actual trays at 100°F contact temperature are given in the *GPSA Engineering Data Book* (1987), based on the data of Worley (1967). Similar charts for 1.0, 1.5, 2.0, and 2.5 equilibrium stages at 80°F and 100°F contact temperature and 600 psia pressure are given by Manning and Wood (1991), based on the Parrish et al. (1988) equilibrium data. Three of the latter charts are reproduced as **Figures 11-29, 11-30,** and **11-31.**

A clear picture of gas and liquid composition changes on the trays of an absorber can be gained from **Figure 11-32,** which represents a McCabe-Thiele type tray diagram for a typical natural gas dehydration plant. This type of diagram is particularly useful for unusual or difficult dehydration cases where a large number of trays or a close approach to equilibrium may be required. For a given dehydration problem, the diagram can be used to estimate the required number of trays, solution rate, or solution concentration. Several combinations, which will give the desired dehydration, can be worked out and the most economical selected.

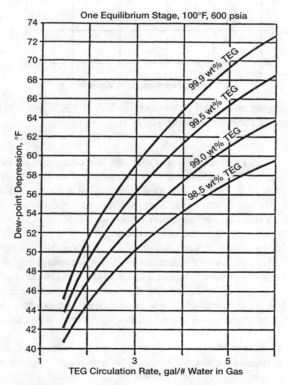

Figure 11-29. Predicted dew-point depression for TEG at 100°F and 600 psia, one equilibrium stage. (*Manning and Wood, 1991*)

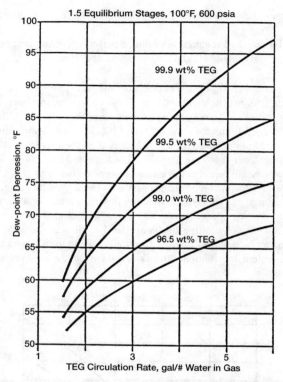

Figure 11-30. Predicted dew-point depression for TEG at 100°F and 600 psia, 1.5 equilibrium stages. (*Manning and Wood, 1991*)

For the case shown in **Figure 11-32,** it is assumed that a natural gas stream is saturated with water at 500 psia and 90°F and that it is desired to dehydrate this gas to a water content of 10 lb/MMscf (dew point 28°F). With triethylene glycol a concentration of 98.5% can readily be attained with simple atmospheric pressure regeneration. The dew-point chart, **Figure 11-15,** shows an equilibrium dew point of about 15°F for this glycol concentration, equivalent to a 13°F approach at the top of the column.

Assuming that 4 gal glycol/lb of water absorbed are circulated, the solution will be diluted from 98.5 to about 95.9%. These two liquid compositions in combination with the inlet- and outlet-gas compositions (as estimated from **Figure 11-1** or **11-3**) are used to establish an operating line on the diagram. The equilibrium line is obtained by converting dew-point data from **Figure 11-15** to water content of the gas for the specific temperature and pressure considered. To simplify the analysis, it is assumed that temperature is constant over the length of the column.

With the operating and equilibrium lines drawn on the chart as indicated, it can be seen that about 1.5 theoretical stages are required. Assuming an individual Murphree tray efficiency of 40%, the number of actual trays can be estimated by using vertical steps on the tray diagram that extend 40% of the distance from the operating line to the equilibrium line at each tray point. The use of this procedure indicates that at least six actual trays should be used. The

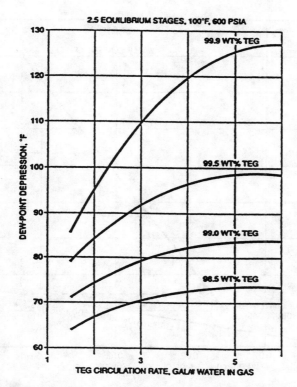

Figure 11-31. Predicted dew-point depression for TEG at 100°F and 600 psia, 2.5 equilibrium stages. (*Manning and Wood, 1991*)

same result is obtained by assuming an overall column efficiency of 25%, a value that is commonly recommended for bubble-cap columns in glycol dehydrator service. A slightly higher overall efficiency (i.e., 33%) is usually used in the design of valve tray columns.

Examination of **Figure 11-32** will reveal that the glycol solution can be permitted to become much more dilute in passing through the column without approaching equilibrium with the gas at any point (i.e., a liquid rate less than 4 gal/lb water can be used). However, the use of a lower liquid rate will require more trays. The optimum design must, therefore, take into account the cost of additional column height vs. costs associated with a higher liquid flow rate.

In the foregoing analysis, it is assumed that the temperature is constant through the absorption column. For a more exact evaluation, heat effects must be considered. It is good practice to design the system with the glycol feed at a slightly higher temperature than the inlet gas (typically about 10° to 20°F hotter). Also, an amount of heat is liberated in the absorber equal to the heat of condensation of the water absorbed plus the heat of solution of this water in the glycol. In general, the gas stream has a considerably higher total heat capacity than the liquid stream, so that the liquid leaving the bottom of the absorber will be at approximately the temperature of the entering gas. The exit gas temperature can therefore be estimated by a heat balance around the column. In the case of high-pressure gas absorption,

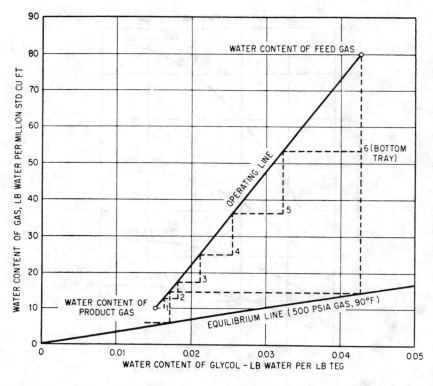

Figure 11-32. Tray diagram for absorber design; high-pressure natural gas dehydration with triethylene glycol solution.

the net effect is usually a small increase in gas temperature on passing through the column (1° or 2°F), and this is normally of little significance.

The heat of water absorption is of importance in low-pressure gas dehydration and air dehumidification. In such applications, it is common practice to use cooling coils in the absorption zone to remove heat which is liberated. In units used for air conditioning dehumidification, the cooling coils may also serve as packing in the air-glycol contact zone as illustrated in **Figure 11-13.**

The diameter of tray columns can be calculated by conventional column sizing techniques (see Chapter 1); however, a conservative gas velocity should be used because of the tendency of glycol solutions to foam under some conditions. Swerdloff (1957) presented a nomograph for calculating the constant in the Souders-Brown correlation (equation 11-2) which takes into account the tray spacing and the acceptable glycol loss. For the typical case of a 24-in. tray spacing and an acceptable glycol loss of 1 lb/MMscf, the correlation gives a value for the factor K of 0.11. More recent practice is to use an efficient mist eliminator above the tray section and much higher K values. Manning and Wood (1991) suggest a K value of 0.167 for equation 11-1 when applied to bubble-cap columns. *The API Specification for Glycol-Type Gas Dehydration Units* (1990) specifies a K factor of 0.16 for 24-in. tray spacing and 0.12 for 18-in. spacing. Kean et al. (1991) claim that a K factor of 0.18 is typically used for bubble-cap contactors, and in many applications the columns have been pushed to a K factor as high as 0.22.

Charts for estimating the required internal diameter of natural gas dehydration columns using triethylene glycol and equipped with valve-type trays have been presented by Camerinelli (1970). One of these charts, for the case of 24-in. tray spacing, is reproduced in **Figure 11-33**. The chart is based on a constant gas inlet temperature of 120°F; however, for pressures up to the maximum shown, 1,200 psia, gas temperature was found to have very little effect. The column size obtained from the chart is based on a design gas capacity of about 70% of flooding.

Structured Packing. The design of dehydration columns packed with high-efficiency structured packing is discussed by Kean et al. (1991). They tested 11 commercial structured packings and found their HETP values to range from 3.7 to 5.5 ft per theoretical stage at a gas flow factor, F_s, of 3 and a liquid flow rate of 0.3 gpm/ft². These results can be compared with bubble-cap trays which, at 2 ft spacing and 25% overall tray efficiency, require about 8 ft of column height per theoretical stage.

The Kean et al. study showed that a design point of $F_s = 3$ is reasonable for the more efficient structured packings. For applications not requiring very high efficiency (e.g., product gas water content of 7 lb/MMscf), higher capacity packings can be used with design rates up to $F_s = 3.5$. The F_s factor, which is commonly used in the design of structural packing installations, is defined as

$$F_s = V_s \, d_G^{0.5} \tag{11-3}$$

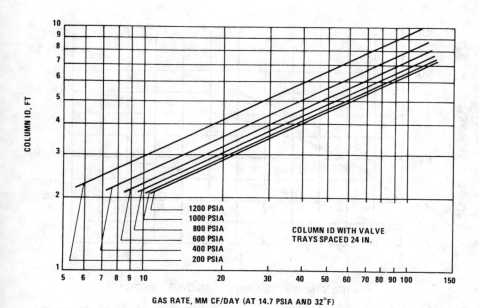

Figure 11-33. Chart for estimating the internal diameter of triethylene glycol absorption columns equipped with valve-type trays. *Data of Camerinelli (1970)*

Where: V_s = Superficial gas velocity, ft/s
d_G = Gas density, lb/ft³

An F_s factor of 3 equates to a K factor of about 0.36 in the Souders-Brown correlation (equation 11-2), which is about twice that normally recommended for tray columns.

Two special requirements for the effective use of structured packing are efficient mist eliminators and uniform liquid distribution. Kean et al. (1991) recommend a vane type eliminator with a 3- to 4-in. mesh pad face to minimize carry-over. A simple wire-mesh mist eliminator is not considered adequate at the high gas velocities employed with structured packing. They also recommend that a high-efficiency drip-type distributor be used with typical matrix dimensions of 4 in. by 4 in. and the peripheral drip points located no more than 2 in. from the wall. The distributor must be installed as close as possible to the packing to ensure good distribution and minimize carry-over.

Cocurrent Contactors. With cocurrent absorption, the product gas approaches equilibrium with the rich solution rather than with the lean solution as in countercurrent operations. As a result, the process is more sensitive to liquid flow rate (which determines the rich solution composition) and is not generally applicable when extremely low dew points are required. However, this type of contactor is capable of a closer approach to equilibrium than is possible on the top tray of a countercurrent unit, and it is much smaller than a conventional tray or packed column.

The design of cocurrent contactors employing in-line mixing elements is normally handled by the equipment supplier. **Figure 11-34,** from Baker and Rogers (1989), depicts dew-

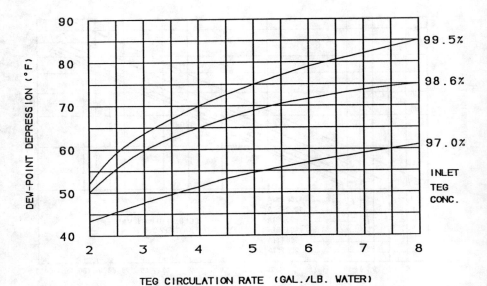

Figure 11-34. Calculated dew-point depression vs. circulation rate for triethylene glycol solutions in a cocurrent contactor; basis 98% approach to equilibrium. *From Baker and Rogers (1989)*

point depressions attainable with a Koch Engineering Co. cocurrent SMV contactor designed to achieve a 98% approach to equilibrium. This chart can be used for any contact temperature and pressure because, as previously discussed, the dew-point depression for a given system is relatively independent of these factors. Pressure drops are said to range from 3 to 5 psi when the unit is used in a vertical downflow configuration.

Pyles and Rader (1989) describe a program conducted by Sun Exploration and Production Co. to evaluate the Koch cocurrent contactor for possible application to glycol dehydration units on offshore platforms or at other sites where its small size and other characteristics make it an attractive alternative to a conventional column. For the test application, a 6-in.-diameter contactor less than 3 ft long was designed to handle 10 to 50 MMscfd of 1,000 psig natural gas with an inlet moisture content of over 20 lb water/MMscf and an outlet moisture content below the pipeline specification of 7 lb water/MMscf. The results confirmed the design; the outlet gas and rich glycol were very close to equilibrium over a wide range of gas flow rates. With a lean TEG concentration of 98.6% and a gas outlet temperature of about 100°F, the product gas moisture content was consistently below 7 lb water/MMscf.

Flash Tank

Most glycol dehydration plants employ a flash tank to remove hydrocarbons from the rich glycol. The amount of hydrocarbons present depends primarily upon absorber conditions (pressure and temperature), feed gas composition, and whether or not a glycol-powered pump is used. The solubility of a typical natural gas in TEG and DEG is shown in **Figure 11-27**. A commonly used value for natural gas solubility in TEG is 1 scf/gal at 1,000 psig and 100°F (Manning and Wood, 1991). A higher quantity is absorbed if the gas contains appreciable amounts of heavy hydrocarbons, particularly aromatics.

When a glycol-powered pump is used, the separator serves to remove the off-gas from the pump, which typically amounts to 3 scf/gal with 500 psig absorption and 6 scf/gal with 1,000 psig absorption (Sivalls, 1976). According to Ballard (1986), the separator usually works best in the temperature range of 140°F to 160°F. He recommends at least an 8 minute retention time for a two-phase separator and 10 to 45 minutes for a three-phase separator. The three-phase type is required if liquid hydrocarbons are present in the rich solution. Manning and Wood suggest a 10 minute retention time for a two-phase separator and 20 minutes (at 150°F) for a three-phase unit to provide ample time for breaking any oil-glycol emulsion. *The API Specification for Glycol-Type Gas Dehydration Units* (1990) recommends 5 minutes retention time for two-phase separators and 10 to 30 minutes for three-phase units. The *GPSA Engineering Databook* (1987) states that only a 3 to 5 minute retention time in the flash drum is required for degassing.

When high-pressure gas streams containing very high concentrations of CO_2 are dehydrated, a simple flash tank may not be adequate for removing dissolved gas due to the high solubility of CO_2 in glycols at high CO_2 partial pressure. It is desirable to remove dissolved CO_2 from the solution entering the regeneration system to minimize corrosion in the still and reboiler, reduce the heat load, and limit vapor traffic in the still. This problem has been studied by Glaves et al. (1983) who developed the design for a plant to dehydrate 550 MMscfd of 1,080-psia gas containing 71.7% CO_2. The original design was reevaluated on the basis of more recently published test data (Takahashi and Kobayashi, 1982) (**Figure 11-28**). It was concluded that an intermediate pressure stripper was needed to remove CO_2 from the rich glycol before it enters the regeneration column. Calculations based on the new data showed that operation of the intermediate pressure stripping column at 450 psia with 1.5 MMscfd of stripping gas (containing 19% CO_2) would result in a reduction of CO_2 from 1.60 lb CO_2 per lb of glycol in the liquid leaving the absorber to 0.29 lb/lb in the liquid leaving the intermediate stripper. Although greatly reduced by this stripper, the amount of CO_2 in the glycol feed to the regenerator in this case would still be greater than the amount of water, significantly increasing the size of the regeneration column required. One positive aspect of the dissolved CO_2 is its action as a stripping vapor to assist in removing water from the glycol.

Regeneration System Design

The regeneration of diethylene or triethylene glycol generally requires only the simple distillation of a binary mixture, the two components of which have widely differing boiling points and do not form azeotropes. About the only difficulty in this otherwise straightforward engineering-design problem is that excessive decomposition may occur if the temperature reaches too high a level. Conservative operating temperatures are about 325°F for DEG and 375°F for TEG; however, satisfactory operation has been observed with appreciably higher reboiler temperatures. TEG reboilers are reported to operate successfully with a glycol outlet temperature of 400°F. To overcome the temperature limitation when very concentrated solutions are required, the distillation process may be modified by the use of vacuum, and inert stripping gas, or a liquid hydrocarbon azeotrope former.

Stripping Column. Because of the wide difference in boiling points, the separation of water from diethylene or triethylene glycol requires very few equilibrium stages. The number of stages for a specific case can readily be estimated by means of a McCabe-Thiele diagram as described by Townsend (1985). Such an analysis usually shows that the separation can be accomplished with two or three theoretical trays, one of which is the reboiler. The stripping column height is more commonly established on the basis of practical considerations, and since the vapor and liquid quantities are normally small, sizing can be quite generous.

For relatively small plants, the regenerator column is frequently installed directly on top of the reboiler and packed with 1 or 1½-in. ceramic saddles or stainless steel Pall rings. A minimum packing height of 4 ft is used for small columns and up to 15 ft for larger units. For very large plants where a regenerator column diameter of 24 in. or more is required, bubble-cap columns are commonly used. The number of actual trays used in commercial columns ranges from 10 to 20 with the solution inlet located somewhat below the midpoint. The apparent large excess of trays is used to minimize the loss of glycol with the overhead vapor. Because of the very low liquid loading on trays above the feed point, care should be taken that the trays are well sealed and weep holes are sufficiently small to prevent draining during operation.

Reflux to the top of the regenerator column may be supplied by a number of alternate systems. The simplest is to install an uninsulated or finned section at the top of the column to provide cooling and condense a portion of the water vapor, which then flows back down as reflux. This system is used on a large number of small dehydrators, but is difficult to control under adverse weather conditions. A tubular water- or glycol-cooled condenser may be used either on top of the tower for gravity return of reflux or in a separate vessel. In the latter case a reflux pump must be provided. The use of rich glycol as coolant, as illustrated in **Figure 11-8,** has the advantage of recovering heat which would otherwise be lost.

The use of a water- or glycol-cooled condenser provides good control, but for many small plants some first-cost savings can be realized by utilizing steam condensate or fresh water as reflux. The water is introduced directly to the top plate of the column. The principal problems with this arrangement are the possible introduction of salts if process water is used, and the need for accurate flow control because of the small quantities of reflux required. Typically, the amount of reflux is one-fourth to one-third the amount of water evaporated from the glycol.

The diameter of the stripping column is determined on the basis of the largest cross-sectional area required to handle gas and liquid flows at any point in the column. This is normally at the bottom of the column immediately above the reboiler. The vapor load consists of the water and glycol vapors from the reboiler plus any hydrocarbons which may be desorbed from the glycol or added as stripping vapor. The liquid load consists of the glycol stream plus reflux water.

Reboiler. With regenerators employing kettle-type reboilers equilibrium can be assumed to be attained in the reboiler. The temperature-pressure-composition relationships can be obtained from **Figures 11-18** and **11-19**. The vapor-pressure values given on these charts represent the total pressure of water and glycol, which is equal to the total reboiler pressure if no inert stripping gas is added. In the latter case, the partial pressure of the added inert gas must be subtracted from the total reboiler pressure to give the solution vapor-pressure to be used in reading the charts.

The maximum recommended temperature shown on **Figure 11-19** is considered to be quite conservative and can be increased to 400°F by very careful design of the reboiler system to limit retention time and metal wall temperatures. According to **Figure 11-21**, a 400°F reboiler temperature will result in a glycol concentration of about 98.5%. In practice, concentrations as high a 99.1% are frequently obtained due to the stripping effect of dissolved hydrocarbons and, in some cases, operation of the system at elevations above sea level.

The reboiler heat load for conventional systems can be estimated roughly by use of the following equation proposed by Sivalls (1976):

$$Q = 2,000L \tag{11-4}$$

Where: Q = Total heat duty of the reboiler in Btu/hr
L = Glycol flow rate in gal/hr

The above equation assumes typical values for the following items:

- Sensible heat required to raise the glycol temperature
- Heat of vaporization of water from glycol solution
- Heat required to revaporize the reflux
- Heat losses from reboiler and stripping column

A more accurate determination of reboiler heat load requires that each of these items be calculated for the specific plant conditions. Pearce and Sivalls (1993) suggest that Item 4 can be estimated by the following equation:

$$Q_4 = 0.24\,(A)(T_F - T_A) \tag{11-5}$$

Where: Q_4 = Heat loss from reboiler and column, Btu/hr
A = Total exposed surface area of reboiler and column, ft^2
T_F = Temperature of fluid in vessel, °F
T_A = Minimum ambient air temperature, °F
0.24 = Approximate heat loss from large insulated surfaces, Btu/hr ft^2 °F

Heat for the regeneration step is normally provided by direct combustion of natural gas in tubes in the reboiler or by steam. Hot oil and waste heat sources have also been proposed. The use of exhaust gases from engines driving gas compressors has been described by Carmichael (1964). In his example, the exhaust gases leave the engine at an average temperature of 1,260°F. They lose about 60°F enroute to the glycol regenerator and leave that unit at a temperature between 400° and 450°F. The advantages compared to a direct-fired reboiler burning part of the natural gas are claimed to be lower cost, reduced maintenance because of lower temperatures in the heater tubes, and improved safety as a result of eliminating the open-flame system.

When steam is used as the source of reboiler heat, it is often necessary to reduce the steam pressure upstream of the reboiler steam coil to limit the metal wall temperature and thereby minimize thermal degradation of the glycol. If the tube wall temperature is limited to 430°F, as suggested by the API Specification (1990) for fire-tube designs, a maximum steam-side temperature on the order of 460°F (450 psig steam) is indicated. For fire-tube reboilers the API Specification gives a normal range for heat flux of 6,000 to 10,000 Btu/hr ft^2. Manning and Wood (1991) recommend that the heat flux be limited to 6,000 Btu/hr ft^2 when the reboiler temperature is 400°F and up to 8,000 Btu/hr ft^2 when the reboiler temperature is set at 360°F.

Enhanced Stripping Systems. Stripping can be enhanced without exceeding the decomposition temperature of the glycol by any mechanism that lowers the partial pressure (or fugacity) of water vapor in the gas phase over the glycol solution. Three approaches have been developed: gas injection, azeotrope formation, and vacuum stripping. The "COLDFINGER" system apparently reduces the partial pressure of water vapor in the gas phase of a separate stripping chamber, thereby increasing the concentration of TEG in the solution that is in the chamber.

The use of stripping gas is a simple and effective technique for improving performance of glycol regeneration systems and is the most frequently used stripping enhancement method. Originally the gas was injected directly into the reboiler, which increased the maximum TEG concentration attainable with 400°F operation from 99.1% to about 99.6%. The process was improved by Stahl (1963), who proposed using a short packed column in the downcomer between the reboiler and the surge tank for countercurrent contact of the injected gas with hot semi-lean glycol from the reboiler. Since very little additional water is vaporized from the glycol in the stripping gas column, a small amount of inert gas is able to cause a significant reduction in the gas phase water vapor partial pressure.

The effect of stripping gas quantity on the predicted regenerated glycol concentration is shown in **Figure 11-35** (Parrish et al., 1986). The chart is based on the assumption that the reconcentrator located above the reboiler is equivalent to two theoretical stages. The chart also shows the relative efficiency of injecting the gas directly into the reboiler vs. injecting it at the bottom of a countercurrent stripping gas column located below the reboiler (see **Figure 11-8**). **Figure 11-35** also shows the effect of varying the number of equilibrium stages in the stripping gas column. Typically, the stripping gas column is a 2- to 4-ft-long packed section in the downcomer between the reboiler and surge tank, containing ceramic saddles or Pall rings. Structured packing has been used in some installations to provide a maximum number of theoretical trays in the limited vertical space available. If the packed section is equivalent to two theoretical plates, and a typical stripping gas rate of 2 scf/gal TEG is used, a glycol concentration of about 99.85% is predicted.

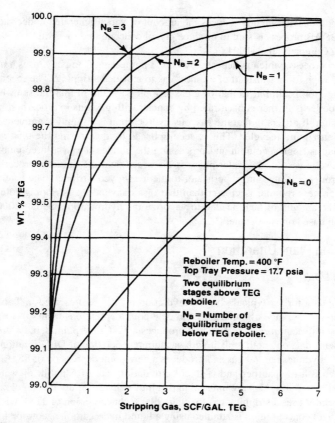

Figure 11-35. Effect of stripping gas on triethylene glycol concentration attainable, as a function of the number of equilibrium stages in the column below the reboiler. (*Parrish et al., 1986*)

Vacuum stripping can be accomplished with a stripping column-reboiler combination that is essentially identical to one used for atmospheric pressure operation, but operates at a reduced pressure. The only significant difference is the requirement to totally condense the overhead vapor stream and to provide a vacuum pump to remove noncondensible gases. **Figures 11-26** and **11-27** provide vapor-liquid composition diagrams for DEG and TEG with water at various subatmospheric pressures. These can be used to construct McCabe-Thiele type diagrams for determining the number of plates required to perform the required separation at selected subatmospheric pressures.

The vacuum regeneration system offered by BS&B Engineering Company (see **Figure 11-10**) limits the vacuum system to a second stage of regeneration. Most of the water contained in the rich glycol is removed in a conventional atmospheric pressure still-reboiler combination. The semi-lean glycol from the reboiler, at a concentration of about 99%, is flashed into a vacuum drum, with the addition of heat to compensate for the latent heat of vaporization. This arrangement avoids the requirement to design the still and reboiler for vacuum conditions and greatly reduces the amount of vapor generated under vacuum conditions. The BS&B process is said to produce glycol concentrated to as high as 99.9% (BS&B Engineering Company, Inc., 1994).

The Drizo process, which is licensed by OPC Engineering Company, uses a liquid hydrocarbon azeotrope former instead of an inert gas to enhance stripping. The regeneration system configuration is similar to that of a gas stripping system in that the semi-lean solution from the reboiler is further concentrated by contact with gaseous hydrocarbon in a countercurrent column. It differs in that the gaseous hydrocarbon is a liquid at ambient temperature and is condensed and recycled. This is accomplished by condensing the total vapor stream from the main stripping column and processing the condensate in a three-phase separator. The gas phase is vented, the water phase is discarded, and the liquid hydrocarbon phase is recycled through a heater and superheater to the lean-glycol stripper. The heat requirement of a Drizo process reboiler is greater than that of a conventional stripping column-reboiler system by the amount of heat needed to vaporize the hydrocarbon: about 1,000 Btu/gal of hydrocarbon based on iso-octane.

Dehydration Plant Operation

Operating Data

Operating data for six typical glycol dehydration plants are presented in **Table 11-5**. The plants cover a wide range of conditions and dew-point depressions from 40° to 132°F. Plant D operates with vacuum regeneration and produces a 77°F dew-point depression. Polderman (1957) presents data on another plant with vacuum regeneration of DEG solution (at 175 mm Hg) which dehydrates natural gas and produces a dew-point depression of 85° to 100°F. This plant uses an 8-tray absorber and a glycol circulation rate of 5 gal/lb of water removed. Plants E and F of **Table 11-4** are unusual in that they dehydrate pure carbon dioxide. In Plant E the regenerator uses a stripping gas to attain 99.0% concentration of TEG; while in Plant F the same glycol concentration is attained using a liquid hydrocarbon azeotrope former (Drizo process). The absorber of Plant F operates very close to the critical point for carbon dioxide, which is the reason for the indicated high loss of glycol. Both of the CO_2 dehydration plants in the table produce a gas containing about 10 lb water/MMscf. However, since this water concentration represents a lower dew point for carbon dioxide than it would for natural gas at the same pressure, the plants show an unusually high dew-point depression.

The reboiler temperatures of plants listed in **Table 11-5** are well below the recommended maximums; however, higher temperatures are often used. Worley (1967) reported that up to 10 years experience with several hundred TEG units utilizing a reboiler temperature of 400°F has failed to indicate any evidence of measurable losses by degradation. Dew-point depressions in excess of 100°F are reported for these units. A dew point of –95°F is reported by Smith and Skiff (1990) for a plant employing only 1.5 gal TEG/lb water removed. This represents a dew-point depression of at least 150°F (assuming a minimum contact temperature of 55°F), and was attained by using the Drizo process to regenerate the TEG to 99.999% concentration (10 ppm H_2O).

Glycol Loss

Glycol loss constitutes one of the most important operating problems of dehydration units. Most of this loss occurs as carry-over of solution with the product gas, although a small amount of glycol is lost by vaporization into the gas stream. An additional small amount is

Table 11-5
Glycol Dehydration Plant Operating Data

Plant Variables	A	B	C	D	E	F
Gas rate, MMscfd/day	50*	60	10.8	73.5	14.7	16.5
Absorber pressure, psig	1,000	750	385	720	590	1,080
Type of gas	Nat.	Nat.	Nat.	Nat.	CO_2	CO_2
Solution rate, gpm	5–10	6	2.27	15.6	9.7	6.0
Glycol used	DEG	DEG	TEG	DEG	TEG	TEG
Lean Sol. Conc., % Glycol	95	95	98.25	97.8	99.0	99.0
Rich Sol. Conc., % Glycol	90	90	96.95	96.5	97.9	95.5
Absorber:						
Diameter, in	50	36	36			
Height	33 ft 7 in.	28 ft	12 ft 6 in.			
No. of trays	4	4	4	7	10	8
Regenerator:						
Diameter, in	26	18	12¾			
Height	29 ft 10 in.	35 ft	6 ft 6 in.			
No. of trays	20	15	Rings			
Feed tray	15 (from top)	—	Rings			
Reboiler:						
Temp., °F	—	310	352	290	388	375
Pressure, psia	Atm.	20	Atm.	7.4	Atm.	Atm.
Temp. of feed gas, °F	78	60–68	55	84	100	102
Temp of product gas, °F	—	—	—	86		
Dew point of product gas, °F	+38	+10, +15	−4	+9	−10	−30
Dew-point depression†, °F	40	50, 53	59	77	110	132
Glycol loss, lb/MMscf	—	—	—	0.28		23

*Rate to each of three absorbers.
†Depression based on feed-gas temperature if product-gas temperature not given.
Data sources: Plant A, Hull (1945) and Senatoroff (1945); Plant B, Love (1942); Plant C, Peahl (1950); Plant D, Polderman (1957); Plants E and F, Zabrik and Frazier (1984)

always lost through mechanical leakage, and some may be lost with the vapors leaving the regenerator. By careful plant operation, total glycol losses can be maintained below 0.5 lb/MMscf of gas treated; however, a loss of 1 lb/MMscf is sometimes considered acceptable.

Since the major glycol loss is by entrainment, any design or operating action which reduces this item can result in a considerable improvement in plant economics. Neal et al. (1989) describe a case in which loss from a TEG contactor was reduced from about 9 lb/MMscf to about 0.009 lb/MMscf by replacing a standard 4-in.-thick mesh pad made of 0.011-in.-diameter 304 stainless steel wire with a density of 9 lb/ft³ by a 6-in.-thick pad containing fine multifilament dacron thread knitted in with the wire and incorporating a draining feature.

The major causes of glycol losses experienced with 1,200 package-type skid-mounted glycol units operated in the San Juan Basin were reported to be (1) overloading of the glycol absorber when shut-in wells were placed on line and (2) operational failure of inlet separator dump valves. An improvement developed for these units to minimize inlet separator malfunctioning was the use of pipe coils inside the separator for the circulation of hot glycol. This prevented freezing of water collected in the separator during cold weather and helped to break the heavy viscous oil emulsions produced in some wells (Fowler, 1957).

Foaming

Excessive entrainment can usually be traced to foaming in the contactor. It has been found that foaming can result from contamination of the glycol with hydrocarbons, finely divided solids, or salt water brought in with the feed gas. It is important, therefore, that the incoming gas be passed through an efficient separator before it contacts the glycol and that the circulating stream of glycol be maintained in a clean condition. When foaming does occur, it can usually be brought under control by the addition of a foam inhibitor. Several proprietary brands are on the market. The use of trioctyl phosphate has been described (Swerdloff, 1957). In this case the inhibitor, used in a concentration of 500 ppm, reduced the loss from as high as 15 lb glycol/MMscf to less than 0.5 lb/MMscf. Pearce and Sivalls (1993) report that a silicone-type antifoam agent is usually successful in the 25–150 ppm range.

Corrosion

Corrosion can be a serious problem in the operation of glycol dehydration plants. Since the pure glycol solutions are themselves essentially non-corrosive to carbon steel, it is generally believed that the corrosion is accelerated by the presence of other compounds that may come from the oxidation or thermal decomposition of the glycol, or enter the system with the gas stream. The rate of corrosion will, of course, be influenced by the temperature of the solution, velocity of the fluid, and other factors. In general, the principles that have been employed in combating corrosion are

- The use of corrosion-resistant alloys
- The use of corrosion inhibitors
- The prevention of solution contamination
- The use of process-design modifications to minimize temperatures and velocity

The principal chemical factors involved in glycol-plant corrosion are believed to be the oxidation of glycols to form organic acids and the absorption of acidic compounds, principally H_2S and CO_2, from the gas stream. The oxidation of diethylene glycol has been studied by Lloyd (1956) with regard to oxidation product and rate-governing factors. He found that oxidation of diethylene glycol resulted in the formation of an organic peroxide as an intermediate product and formic acid and formaldehyde in copious quantities. The oxidation rate was found to increase with increased oxygen partial-pressure and increased temperature and to be accelerated by the presence of acid.

Lloyd and Taylor (1954) investigated the effect of glycol deterioration products and of various added chemicals on the rate of corrosion by diethylene glycol solutions. The corrosion tests were conducted by heating the glycol solution in a flask, with samples immersed in the solution and suspended in the vapor space. The vapor-phase samples, which were wetted with condensate, showed by far the most serious corrosion, and conclusions with regard to the corrosiveness of the various solutions were based on these samples. These conclusions were

- Glycol solutions which had been made acid, either by auto-oxidation or by the addition of acetic acid, were consistently more corrosive than the neutral samples.
- Low concentrations of neutral salts did not affect corrosion rates.
- The glycol solutions that were made alkaline showed fairly low corrosion rates.

It is postulated that corrosion in this system is caused by the presence of a volatile acid that vaporizes and condenses with water on the mild-steel corrosion coupons. Alkaline buffers, such as potassium phosphate, combine with the free organic acids and reduce their vapor pressure to a negligible value. Organic alkaline materials, such as monoethanolamine, may also act by vaporizing with the organic acids and neutralizing them when condensation occurs.

A number of corrosion inhibitors have been successfully applied in commercial plants; these include monoethanolamine (Kruger and Mozelli, 1952) and sodium mercaptobenzothiazole (Pearce and Sivalls, 1993). The use of the latter material in a plant that had previously encountered very severe corrosion has been described by Swerdloff and Duggan (1955). The plant operated on a gas containing 58 grains H_2S/100 scf, 54 grains mercaptans/100 scf, 1.36% CO_2, and, at times, traces of oxygen. After 2 years of operation, the contactor trays were found to be severely corroded, and, after 3 years, the dried-gas line blew out, about 80 ft downstream from the contactor. When the corrosion was first noticed, the pH of the solution was found to range from 4.1 to 5.0. Steps taken to remedy the situation included the installation of stainless-steel lining and trays in the contactor, the addition of sodium mercaptobenzothiazole to the glycol, cooling of the gas stream from 100° to 80°F prior to its contact with the solution, and installation of a system to decrease entry of oxygen into the gas stream. The net result of all of these changes was a very great reduction in the corrosion rate of any iron in contact with the glycol, as measured by coupons. The inhibitor was used as a 45% solution of sodium mercaptobenzothiazole in water and added directly to drums of glycol used as makeup. Concentration in the dehydration solution amounted to about 1%.

Two additional steps were taken to minimize corrosion downstream of the glycol dehydrator unit. One was the use of a product-gas scrubber to minimize glycol entrainment in the gas stream, and the other was the use of a second inhibitor that was added to the dried-gas stream. This material was of the polyethanolrosinamine-type consisting of 70% solution in isopropyl alcohol of a mixture of 90% ethoxylated rosinamine (11 moles ethylene oxide with each mole of rosinamine) and 10% free rosinamine. This inhibitor was injected at the rate of 0.1 gal/day for a gas volume of 60 MMscf/day. As a result of the above measures, the corrosion rate of coupons suspended in the dried-gas line decreased from almost 30 mils/year to as low as 0.2 mil/year during about 4 years of testing.

Solution Maintenance

Proper maintenance of the glycol solution is critical for trouble-free operation of the dehydration process. Chemical analysis of the solution can often prevent or identify the cause of operating problems. Analyses are typically made for water, pH, lower glycols, hydrocarbons, foaming tendency, and inorganic salts (particularly chlorides). Methods of conducting analysis of glycol are described in detail by Grosso et al. (1979). **Table 11-6,** based on the data of Bentley (1991), provides a guide to acceptable ranges for several analysis items.

Ballard (1986) suggests checking the glycol pH periodically and keeping it in the range of 7.0 to 7.5 by the addition of borax, ethanolamine (usually triethanolamine), or other alkaline chemicals. Too high a pH (e.g., over 8.0–8.5) is undesirable because it can increase the tendency of the solution to foam and form emulsions with hydrocarbons. **Figure 11-36,** from

Table 11-6
Acceptable Ranges for Lean Glycol Analyses

Hydrocarbon content, %	1.0
Salt content, ppm	200–300
Solids content, ppm	100
pH	7.0–7.5
Iron content, ppm	10–15 (100 ppm max)
Foam, time to break, seconds	3.5

Data of Bentley, 1991

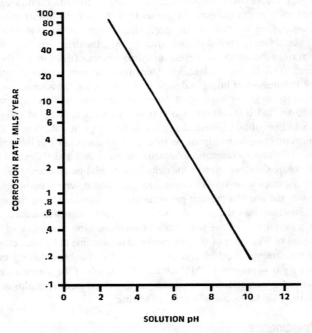

Figure 11-36. Relationship between diethylene glycol pH and rate of corrosion. *From Pearce and Sivalls (1993)*

Pearce and Sivalls (1993), graphically illustrates the effect of pH on corrosivity of DEG solution on carbon steel.

A technique that is used at many plants to minimize the entry of oxygen into the solution is to gas-blanket the glycol storage tank. Since some gas streams contain free oxygen, this precaution will not always prevent oxidation of the glycol from occurring. In air dehydration, of course, there is no hope of excluding oxygen from the system, and the use of inhibitors is desirable.

Filtration

In addition to causing corrosion, contamination of the glycol solution can result in fouling of heat-exchanger surfaces and loss in operating efficiency. The solution may become contaminated with oxidation products as previously described, by corrosion products (usually iron oxide or iron sulfide), and by solid or liquid particles brought in with the gas stream. Solid contaminants are objectionable in that they settle out in tanks, contactor and still trays, heat exchangers, and other vessels. They may also be a factor in accelerating corrosion (or erosion). The use of some means for removing suspended particles is therefore usually justified. Filters of the common waste-pack or cartridge-type have proved quite successful and are usually located in the line carrying the rich glycol solution from the contactor. Ballard (1977, 1986) recommends the use of filter elements designed to remove particles over 5 microns in size. He also recommends cloth fabric filter elements in favor of paper or fiberglass and suggests that the solids content in the glycol be held below 0.01 wt% (100 ppm).

Activated carbon is also employed to remove impurities from glycol solutions. It is particularly effective for removing nonfilterable heavy hydrocarbons. The activated carbon bed should be located downstream of the rich glycol filter to avoid plugging with solids. Coal-based carbons are reportedly more widely used than wood-based. Carbon particle sizes typically range from 8 by 30 to 5 by 7 mesh (Ballard, 1988).

The carryover of carbon particles can also be a problem. Simmons (1981) cites one case in which the top two trays of a glycol contactor were found to be plugged with a mixture of heavy oils and carbon fines. He recommends the use of a solids filter downstream of the carbon unit to prevent carbon fines from entering the system. According to Pearce and Sivalls (1993), activated carbon filters usually operate on a 10% slipstreams and are sized for a glycol flow of 1–5 gpm/ft^2 of cross-sectional area.

The testing and operation of a crossflow-membrane microfiltration system is described by Meadows (1989). The system uses a filter module consisting of polypropylene tubes with 0.2 micron pores. The "dirty" fluid is pumped into the tube side at one end of the module. The pressure is greater inside the tubes than in the shell chamber, which causes clean fluid to permeate through the tube walls leaving particles inside the tubes to flow out the opposite end as concentrate. The system reduced the solids content of glycol, containing on the order of 30,000 ppm of solids less than 0.5 microns in size, to an average suspended solids content of less than 50 ppm.

Salt Removal

Contamination of the glycol with sodium chloride and/or calcium chloride is a common problem. The best solution is an efficient separator in the gas feed line; however, no separator is 100% efficient and some brine contamination is inevitable when brine is present in the feed gas stream. Sodium chloride decreases in solubility with increased temperature and therefore precipitates out on hot surfaces (e.g., in the fire-tube reboiler). Calcium chloride has the more common characteristic of being less soluble at low temperatures than at high temperatures. It, therefore, tends to come out of solution in low temperature portions of the system. For example, 95% TEG solution will hold about 4% NaCl at 100°F and less than 2% at 300°F; while the same glycol solution will hold 1% CaCl$_2$ at 100°F and about 35% of this salt at 300°F.

Salt that deposits on heat exchange surfaces can be removed continuously by the use of scraped surface heat exchangers in conjunction with centrifuges to remove scrapings from the product liquid. Such a system has been used on a solution that was supersaturated with respect to calcium chloride and also contained some sodium chloride (Pearce and Sivalls, 1993). The removal of salt (and coke) from fire-tube walls is normally accomplished by mechanical means during shutdown. Salt can be removed directly from the glycol solution by vacuum distillation in a reclaimer or by ion exchange using a strong acid/strong base ion exchange resin (Grosso et al., 1979).

Glycol Reclaiming

According to Simmons (1981) there are only four reasons for replacing the glycol in a dehydration unit. These are

1. Inability to maintain proper dew-point depression with the existing glycol charge
2. Fire-tube failure due to fouling
3. Foaming that cannot be controlled by an antifoam agent
4. Excessive corrosion that cannot be controlled by corrosion inhibitors

The choice between purchasing new glycol or reclaiming the old is purely a matter of economics. Frequently the cost of disposal of waste glycol is an important consideration. If it is determined that glycol reclamation is cost effective, a second decision must be made with regard to using a commercial reclamation service or procuring equipment for on-site reclamation. Only a very large dehydration plant can justify a permanently installed reclaimer; however, a single reclaimer serving several plants or a portable unit may be an attractive approach.

The design and operation of a trailer-mounted vacuum reclaimer which was used to process the TEG of a number of dehydration units by the La Vaca Gathering Company, is described by Armstrong (1979). The unit was designed to process 0.5 gpm of glycol, but actually handled up to 2.0 gpm. The reclaimer operates in a semi-batch mode, i.e., feed is continuous at 0.5–2.0 gpm until sufficient non-volatile material accumulates in the heater vessel to raise the boiling point to 400°F at 28 in. of mercury vacuum. The reclaimer is then shut down and cleaned out.

The reclaimer operates on a slipstream of the glycol dehydration plant, so shutdown of the dehydrator is not required while the glycol is being reclaimed. The time required for processing varies from a few days to three or four weeks, depending on the size of the dehydration plant. According to Armstrong, the system has proved outstandingly effective in eliminating salt deposition and fire-tube failures. The unit was constructed primarily from surplus equipment; however, he estimated that a new unit would cost on the order of $50,000 in 1979.

Aromatics Absorption

The glycols used to remove water from high-pressure natural gas streams also absorb a small amount of hydrocarbons and are particularly effective in absorbing aromatic hydrocarbons such as benzene, toluene, ethylbenzene, and the xylene isomers (BTEX). Some of the absorbed hydrocarbons are flashed off in the flash tank and recycled to the gas stream or used as fuel; however, a significant fraction reaches the regeneration system and appears as vapor in the stripping column offgas.

The Clean Air Act Amendments of 1990 regulate the emissions of 189 hazardous (or toxic) air pollutants, including BTEX compounds, from major and area sources. A major source is defined as a stationary source (or group of sources) that emits more than 10 tons per year of any one pollutant or 25 tons per year total. Many glycol dehydration plants fall into this category. All other stationary sources are considered to be area sources. Several states already have regulations restricting the emissions of hydrocarbons from dehydration units, and widespread imposition of restrictions is expected.

Although BTEX is of major concern, other potential emissions must also be considered in the design and operation of glycol dehydration plants. These include volatile organic compounds (VOCs), ethylene glycol (one of the 189 listed pollutants), and glycol breakdown products. Combustion products from the reboiler fire-tube must also be considered, but these can usually be controlled by good burner design and burning low-sulfur fuel. A general discussion of emission controls for glycol dehydration equipment is given by Sams (1990).

The amount of BTEX absorbed in the contactor is a function of its solubility in the glycol used, concentration in the feed gas, absorption pressure and temperature, number of theoretical trays, and glycol circulation rate. The Henry's law constant for benzene in TEG at 1,000 psia is plotted in **Figure 11-37,** which presents values calculated by Fitz and Hubbard (1987)

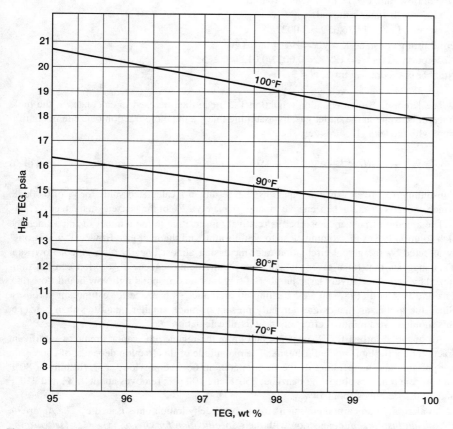

Figure 11-37. Henry's law constant for benzene in triethylene glycol solutions at 1,000 psia. *From Fitz and Hubbard (1987). Reproduced with permission from Oil & Gas Journal, Nov. 23, 1987, copyright PennWell Publishing Co.*

using previously published data on the solubility of benzene in DEG as a basis. The solubility of benzene in TEG is estimated to be 69% greater than in DEG at the same conditions. The TEG-benzene curves can be adjusted to 500 psia by multiplying by 0.83, and to 1,500 psia by multiplying by 1.20.

Fitz and Hubbard (1987) suggest that the amount of benzene absorbed can be estimated by the use of a McCabe-Thiele type diagram or the Kremser-Brown equation. They illustrate the use of the Kremser-Brown approach for a typical case as follows:

Pressure, P = 1,000 psia
Temperature, T = 80°F
Feed gas benzene content, y_{n+1} = 0.1 mole %
Gas flow rate, V = 10 MMscfd = 26,350 lb mole/day
Glycol concentration (average) = 98 wt % TEG
Glycol molecular weight (average) = 130.9
Glycol flow rate, L = 71 lb mole/day (based on 3 gal
 TEG per lb water removed)
Flow ratio, L/V = 71/26,350 = 0.0027
Equilibrium constant, K = y/x = H/P = 0.012
Absorption factor, A = L/VK = 0.0027/0.012 = 0.225
Number of theoretical trays, N = 3

The Kremser-Brown equation (equation 1-20) can then be used to calculate y_1, the mole fraction benzene in the outlet gas, by assuming that y_o, the mole fraction benzene in equilibrium with the lean glycol, is zero, i.e.,:

$$y_1 = y_{n+1} - y_{n+1}((A^{N+1} - A)/(A^{N+1} - 1)) \tag{11-6}$$

Substituting the given values in equation 11-6 results in a calculated value for y_1 of 0.00078, which converts to 452 lb benzene/day or approximately 22% of the benzene in the feed.

The amount of benzene absorbed is relatively insensitive to the number of trays since the rich solution leaves the column close to equilibrium with the inlet gas. However, it is strongly affected by the glycol circulation rate, increasing almost linearly with glycol flow rate. This suggests that for a given dehydration requirement, benzene absorption can be minimized by using a relatively large number of trays and a correspondingly low liquid flow rate. Fitz and Hubbard (1987) suggest treating all aromatics as benzene for estimation purposes. Since the higher aromatics are normally present in much smaller quantities than benzene, this should have a minimal effect on the overall calculation.

All of the absorbed BTEX does not end up in the regenerator vapor stream; a significant fraction may flash off in the flash tank. The magnitude of this fraction depends, of course, on the pressure and temperature of the flash operation. For the previous case, flashing at 80°F and 30 psia removes almost no benzene; 350°F and 100 psia removes about 23%; and 350°F and 30 psia removes about 74%.

An alert on hydrocarbon emissions from glycol dehydration units issued by *The American Petroleum Institute,* in connection with their *Specification for Glycol-Type Gas Dehydration Units* (1990), contains a mass balance for a 10 MMscfd TEG unit operating at 800 psia and 130°F with a feed gas containing 100 ppm benzene. The study indicates that 10% (3 tons per year) of the benzene is absorbed and discharged in the regenerator vapor stream. In this example, with flashing at 100 psia and 185°F, only a trace of the benzene is removed in the flash tank.

The emission of BTEX from the regeneration system of glycol dehydration plants can be reduced by several means, including

1. Minimizing the amount absorbed (e.g., low liquid flow rate)
2. Maximizing the amount flashed (e.g., low-pressure, high-temperature flash conditions)
3. Providing a separator or skimmer to remove liquid hydrocarbons from the rich glycol
4. Completely burning regenerator offgases (e.g., flare or incinerator)
5. Efficient cooling and condensation of regenerator vapor stream

Condensation of water and condensible hydrocarbons in the regeneration system offgas is a very effective means of preventing the emission of BTEX and VOCs. If no stripping gas is used, the offgas can be almost totally condensed and separated into aqueous and hydrocarbon liquid phases. Part of the liquid hydrocarbon can be recycled to the regenerator as in the Drizo process, or all of it can be recovered as a marketable product.

Unfortunately, the aqueous condensate may contain sufficient dissolved BTEX and other organics to constitute a disposal problem. To avoid this problem and provide complete recovery of BTEX compounds, the Radian Corporation, under contract to GRI, is developing a control technology, The R-BTEX process (Rueter et al., 1992). The process, which is shown in **Figure 11-38**, includes the following steps:

1. Using the hot regenerator offgas in a BTEX stripper to remove organic compounds from aqueous condensate, which is produced in the next step
2. Cooling and condensing the offgas from step 1 by indirect heat exchange with glycol and cold water to produce an aqueous phase that is fed to step 1, a noncondensible phase that is vented or burned, and a liquid hydrocarbon phase that represents a marketable product
3. Cooling the stripped water from the BTEX stripper in a small cooling tower where it is further stripped and cooled by contact with air
4. Using the cold water from the cooling tower as final coolant in step 2

The developers expect the R-BTEX process to be applicable to either new or retrofit applications and to achieve greater than 95% recovery of BTEX as a salable liquid.

INJECTION SYSTEMS

Process Description

An operation that is closely related to glycol dehydration, but is aimed more at preventing the formation of solid hydrates than at removing water from the gas is the injection of a hydrophilic liquid into the gas stream. Several liquids have been used or proposed for injection. The principal ones are listed in **Table 11-7**. The injection process is particularly applicable in conjunction with liquid hydrocarbon recovery by refrigeration and, because of the low freezing temperatures of their aqueous solutions, methanol and ethylene glycol are the most commonly used hydrate inhibitors. Diethylene and triethylene glycol are used primarily to prevent hydrate formation in lines leading to conventional dehydration units using the same glycol.

998 Gas Purification

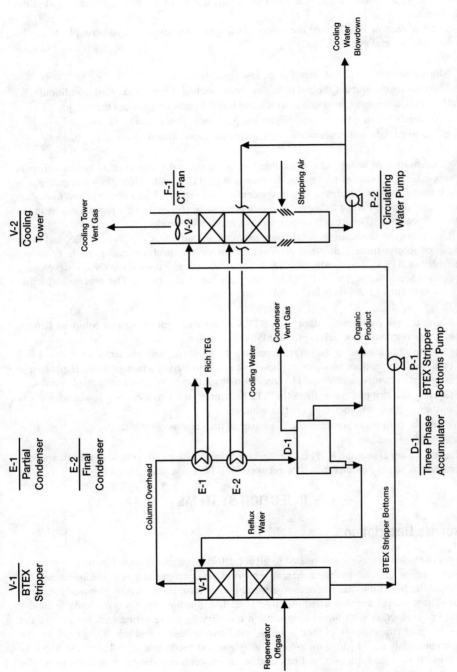

Figure 11-38. Flow diagram of R-BTEX process. *From Rueter et al. (1992)*

Table 11-7
Hydrate Suppression by Inhibitor Injection

Inhibitor	Minimum Temperature	Remarks
Methanol	−140°F, possibly to −160°F	Significant vapor loss above −20°F.
Ethylene Glycol	−45°F	Lower vapor losses and less solubility in liquid hydrocarbons than methanol. Significant vapor losses above 30°F.
Diethylene and Triethylene glycol	−10°F	Useful for warm low-pressure systems when EG losses are significant or where system also includes a DEG or TEG dehydrator.

Glycol Injection

In the glycol injection process, as illustrated in **Figure 11-39,** the gas is first passed through an inlet separator where liquid water and/or hydrocarbons are removed. Glycol is sprayed into the gas after it leaves the inlet separator and before it undergoes any cooling that can reduce the temperature below the hydrate point. It is common practice to spray the glycol on the tube sheet of the gas-to-gas or gas-to-refrigerant heat exchanger to assure good distribution of the glycol into all of the tubes. Tubes that are not wetted with glycol may plug with hydrates as the wet gas is cooled. After passing through the heat exchanger, the gas-glycol mixture is further cooled by expansion through a valve or orifice or by mechanical refrigeration. The cooling causes both water and liquid hydrocarbons to condense. The water is absorbed into the glycol, while the liquid hydrocarbons form a separate phase. The chilled mixture of gas, hydrocarbon liquids, and glycol-water solution flows into the primary separator where the liquid and gas phases are separated. The liquid mixture from the primary separator is generally warmed in a fired heater or by heat exchange with the sales gas to reduce liquid viscosities. The mixture tends to form emulsions, which are quite stable at low temperature, but separate more rapidly as the temperature increases. The mixture is then fed to the oil-glycol separator, the oil is removed as product, and the glycol is sent to the regenerator where it is reconcentrated for reuse.

In cold oil plants operating in the −40°F range, the separation of the glycol-water mixture from the hydrocarbon condensate may be accomplished at the low contact temperature. The low level refrigeration available from the cold condensate can then be used in the high-pressure demethanizer column of the hydrocarbon fractionation train.

Ethylene glycol is generally preferable to diethylene or triethylene glycol for this type of operation because it is less soluble in liquid hydrocarbons, and liquid hydrocarbons are less soluble in ethylene glycol than in the higher glycols. Ethylene glycol is also more effective on a weight basis for hydrate inhibition, and separates more readily from liquid hydrocarbons because of its lower viscosity. On the other hand, the two higher glycols have lower vapor pressure, which results in comparatively lower vaporization losses. To minimize the possibility of solid-phase formation in the glycol solution, compositions near the eutectic (i.e., 60 to 80 wt% glycol in water) are commonly employed, compared to 95% or higher concentrations used in more conventional dehydrator designs. The dilute aqueous solutions have the further advantage of low solubility in liquid hydrocarbons.

In glycol injection systems, the glycol provides some dehydration, but its primary function is to act as an "antifreeze" agent in suppressing the formation of solid hydrates. The high degree of dehydration attained in the case illustrated in **Figure 11-39** may be attributed more to the cooling that occurs than to the effect of the glycol solution. Many different flow schemes are used for glycol injection systems. In its simplest form, the process is very similar to conventional glycol dehydration with a section of the gas line serving as the gas-liquid contactor. Details of the design and operation of glycol injection systems are given by Arnold and Pearce (1961, 1966); Robirds and Martin (1962); and Sheilan (1991).

Methanol Injection

For temperatures lower than about −40°F, glycol injection is impractical because of the high viscosity of glycol solutions at such low temperatures. Methanol's low viscosity and other favorable characteristics make it the fluid of choice for hydrate inhibition in very low

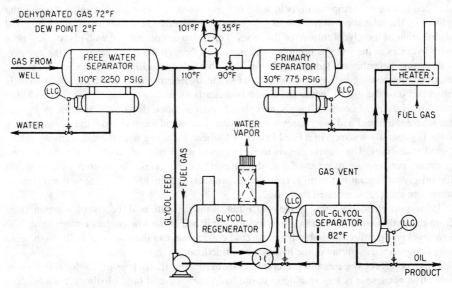

Figure 11-39. Flow diagram of process employing glycol injection and cooling to dehydrate natural gas. *Based on data of Sullivan (1952)*

Absorption of Water Vapor by Dehydrating Solutions 1001

temperature applications such as turboexpander gas refrigeration plants for LPG recovery. This application of the methanol injection process is described by Herrin and Armstrong (1972); Nelson and Wolfe (1981); and Nielsen and Bucklin (1983). A typical flow diagram as given by Nielsen and Bucklin is reproduced in **Figure 11-40.**

After passing through a free-water knockout separator, the gas is cooled in a gas-to-gas heat exchanger in which methanol is sprayed on the tubesheets to inhibit the formation of solid hydrates. A methanol-water solution condenses in the heat exchanger and subsequent chiller and is removed from the gas stream in the separator. The aqueous solution is flashed to remove dissolved gas, filtered, and distilled to recover methanol. A significant amount of methanol dissolves in the liquid hydrocarbon product and is recovered by washing either the total hydrocarbon stream or the propane fraction with water from the methanol still.

Nielsen and Bucklin compared the cost of a methanol injection plant for 600 psig natural gas to costs of solid-bed dehydration. They concluded that the methanol injection plant would have lower capital and operating costs than an activated alumina system for low water content gas (4 lb/MMscf) and a lower capital cost, but about the same operating cost as a molecular sieve system for high water content feed gas (23 lb/MMscf).

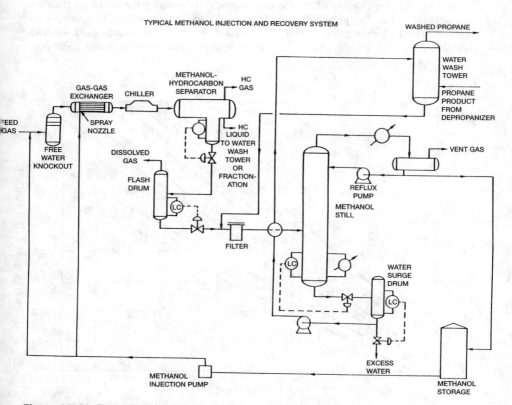

Figure 11-40. Flow diagram of typical methanol injection and recovery system. *From Nielsen and Bucklin (1983)*

Injection Process Design Data

Figures 11-41 through **11-45** provide data applicable to the design of ethylene glycol and methanol injection systems. **Figure 11-41** shows the water vapor dew point for gas in equilibrium with ethylene glycol solutions. A phase diagram for the water-methanol system is given in **Figure 11-42,** and the density of water-methanol solutions is given in **Figure 11-43.** A method for predicting hydrate formation temperatures in the presence of ethylene glycol or methanol solutions has been developed by Maddox et al. (1991) and is described in the Injection Process Design section that follows. This method requires activity coefficients for water in ethylene glycol and methanol solutions. These are given in **Figures 11-44** and **11-45.**

Injection Process Design

The design of glycol and methanol injection systems is determined to a large extent by the overall gas handling system which establishes the gas quantities, pressures, temperatures, and compositions at the injection and separation points. The minimum quantity of inhibitor required can be determined by a material balance between the water removed from the gas and the amount that can be absorbed by the inhibitor within its hydrate prevention range.

The first step in the design of an injection system is determining the minimum inhibitor concentration required to prevent hydrate formation. Hammerschmidt (1934) developed an equation based on experimental data for methanol that has been widely used for both methanol and glycol, but is applicable only at low inhibitor concentrations (below about 20%

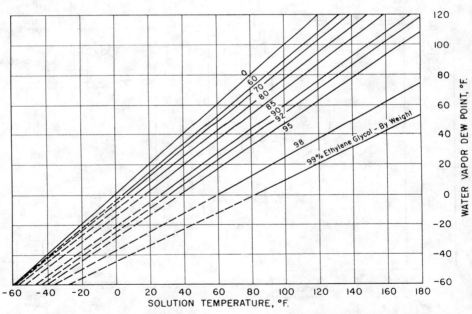

Figure 11-41. Water vapor dew point over aqueous ethylene glycol solutions. *Data of Arnold and Pearce (1966)*

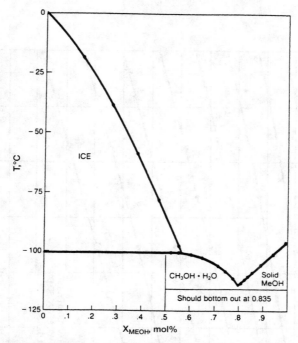

Figure 11-42. Water-methanol system phase diagram. *From Nielsen and Bucklin (1983)*

by weight). Nielsen and Bucklin (1983) describe a modified Hammerschmidt approach applicable to methanol injection systems using more concentrated solutions and operating at lower temperatures. This technique indicates, for example, that 90% methanol provides a hydrate depression of 233.5°F (the maximum possible with methanol). However, the technique is not recommended for temperatures below −160°F.

Maddox et al. (1991) propose a technique for calculating the hydrate temperature of natural gas streams in the presence of hydrate inhibitors that gives results in excellent agreement with experimental data over a wide range of temperatures and concentrations. The experimental data available for comparison include temperatures from ambient to −100°F, methanol concentrations from 20 to 85%, and ethylene glycol concentrations from 20 to 50%.

The Maddox et al. procedure is based on equations describing the thermodynamics of hydrate formation and applies to either methanol or ethylene glycol. A hand calculation version of the technique includes several simplifying assumptions, but appears to give results that are comparable to the computer version and much closer to experimental data than the previously proposed calculation methods. The hand calculation model for hydrate formation is

$$\ln(x_w a_w) = (-2,063)(1/T - 1/T_0) \tag{11-7}$$

Where: x_w = mole fraction water in solution
a_w = activity coefficient of water in solution
T = hydrate temperature in presence of inhibitor, °R
T_0 = hydrate temperature in absence of inhibitor, °R

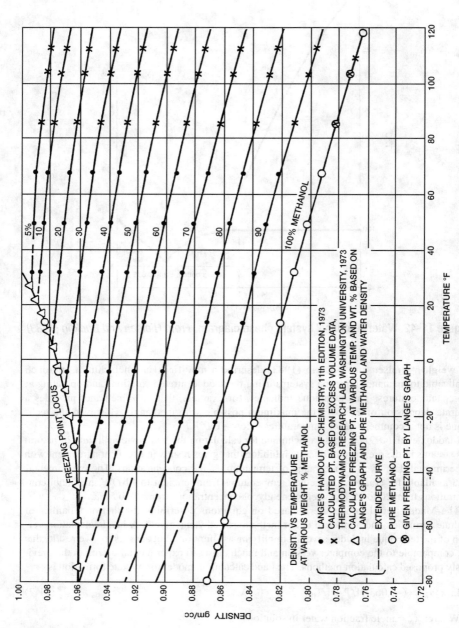

Figure 11-43. Density of methanol-water solutions. *From Nielsen and Bucklin (1983)*

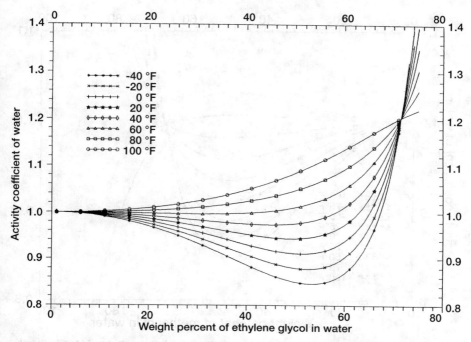

Figure 11-44. Activity coefficient of water in ethylene glycol solutions. *Data of Maddox et al. (1991)*

The activity coefficient for water as determined by experimental inhibitor hydrate temperatures is plotted in **Figures 11-44** and **11-45** for glycol and methanol respectively. The equation and the curves are used in a trial and error calculation to estimate the hydrate temperature.

As an example, consider the case of 20% by weight methanol in 0.6 specific gravity gas at a pressure of 2,000 psia. From **Figure 11-1**, T_0 is 69°F (529°R). From **Figure 11-45**, the activity coefficient, a_w, is 0.963 for 20% methanol (0.1232 mole fraction) at 69°F. Substituting in the equation 11-7:

$$\ln(0.8768 \times 0.963) = -2063(1/T - 1/529)$$

$$T = 507°R = 47°F$$

Repeating the process using 47°F as the assumed value of T_0 does not significantly change the result. Therefore, the hydrate temperature of natural gas in contact with 20% methanol is 47°F. The procedure can be used to estimate the required final inhibitor concentrations. This, coupled with a mass balance, gives the amount and concentration of inhibitor to be injected.

In the design of ethylene glycol injection systems, it is good practice to use about twice the theoretical minimum glycol circulation rate to allow for imperfect distribution of glycol in the heat exchanger tubes. Bucklin (1993) recommends the use of 80% ethylene glycol as feed with dilution to about 75% on an overall basis for systems operating in the –40°F range. The refrigerant-side tube wall temperature is normally limited to about –50°F due to materials limitations. The glycol solution is quite viscous at the cold end (in the 500 cp range); however,

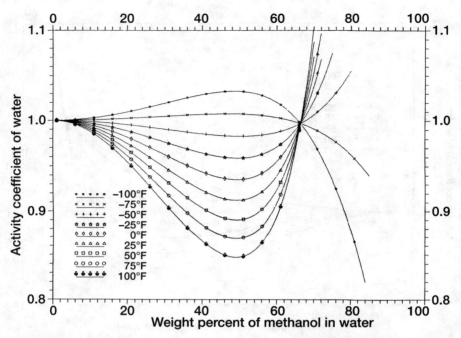

Figure 11-45. Activity coefficient of water in methanol solutions. *Data of Maddox et al. (1991)*

heat transfer is not significantly affected by the presence of the viscous glycol. Robirds and Martin (1962) recommend somewhat more dilute solutions of ethylene glycol, i.e., a practical design target of 60% glycol returning to the regenerator, and regeneration back to about 70%. Herrin (1990) points out that the most important consideration is the concentration of glycol in the rich solution not in the injected feed solution. He cites a case in which excessive amounts of hydrocarbon were lost from the glycol flash tank (operating at −20°F) due to poor hydrocarbon-glycol separation. The problem was solved by reducing the glycol concentration in the feed sufficiently to cause the rich glycol concentration to drop from about 85% to 69% with a corresponding drop in viscosity from 220 cp to 90 cp. With methanol it is common practice to set the concentration of methanol leaving the primary separator at 90%, as this gives the maximum hydrate-point depression (Nielsen and Bucklin, 1983).

The required rate of inhibitor injection is equal to the sum of (1) inhibitor in the water condensate, (2) inhibitor in the liquid hydrocarbon condensate, and (3) inhibitor vapor in the gas phase. In the methanol injection plant described by Nelson and Wolfe (1981), more than half the injected methanol and essentially all of the water contained in the inlet gas are recovered in the aqueous phase of the high-pressure separator. This liquid is fed to a 30-tray distillation column, where 99% pure methanol is recovered overhead.

Precise data on the equilibrium distribution of ethylene glycol and methanol between aqueous solutions and liquid hydrocarbons are not available. Nielsen and Bucklin (1983) present a compilation of several sources of data for methanol and suggest using the most conservative data set, which can be represented by a line with the approximate equation:

$$\log S = -4{,}140/T + 10.81 \tag{11-8}$$

Where: S = methanol solubility in liquid hydrocarbon, mole/100 moles
T = absolute temperature, °R

They further suggest adding a 20% safety margin to the calculated solubility. Thus for a temperature of −100°F, equation 11-8 indicates a solubility of 0.2 moles methanol/100 moles hydrocarbon, and a value of 0.24 would be used to calculate the amount of methanol in the hydrocarbon liquids. This methanol is reclaimed by water washing the appropriate cut during distillation of the hydrocarbons.

The amount of methanol in the natural gas product stream is normally quite small relative to the other two phases, particularly at very low contact temperatures (lower than −100°F). Nielsen and Bucklin (1983) present a series of charts for estimating methanol loss in the vapor phase as a function of solution concentration, pressure, and temperature. Typical points from these charts are given in **Table 11-8**.

Inhibitor losses due to solubility in the liquid hydrocarbon phase and vaporization into the gas stream are generally less significant with ethylene glycol than with methanol. Arnold and Pearce (1966) suggest using a value of 0.01% by weight for the solubility of glycol from a typical injection solution (70–85% concentration) into a typical liquid hydrocarbon (specific gravity 4.5 lb/gal). Vaporization losses for ethylene glycol may be estimated by use of **Figure 11-23** with the assumption that Raoult's law holds for the concentrations employed.

DEHYDRATION WITH SALINE BRINES

The use of calcium chloride brine for water absorption is quite old. However, for most applications, it is gradually being replaced by the glycols and by more efficient salts, such as lithium bromide and lithium chloride. The lithium salts are used to a growing extent in air conditioning installations for dehumidification.

Table 11-8
Loss of Methanol in Natural Gas

Temperature, °F	Methanol Loss in Gas Phase, lb methanol loss/(MMscf) (mole fraction methanol in the aqueous phase)		
	200 psia	400 psia	600 psia
−40	16.0	11.4	11.3
−70	2.9	2.7	3.1
−100	0.57	0.53	0.83
−140	0.031	0.045	>0.05
gas analysis:	93–95% CH_4 5% C_2H_6 0–2% CO_2		

Data of Nielsen and Bucklin, 1983

Calcium Chloride for Gas Dehydration

The equilibrium dew point of gases in contact with aqueous solutions of calcium chloride is shown in **Figure 11-46,** which is based on the data of Brockschmidt (1942). This author presents a comparison of operating data for a plant using a 35% calcium chloride solution and essentially the same unit employing a 95% solution of diethylene glycol. In order to permit glycol to be used in the plant, it was necessary to replace the calcium chloride-solution reboiler with a 13-plate regeneration column and to add heat exchangers and solution preheater. The comparison shows that the glycol gave a dew-point depression averaging about 45°F as compared with a dew-point depression averaging only 19°F for the calcium chloride solution. During comparison periods of about 7 months' operation with each liquid, the glycol was found to have removed about twice the quantity of water as the calcium chloride solution. In view of such poor performance, coupled with operating problems and corrosion, it is no wonder that conventional dehydration units utilizing calcium chloride solutions have been almost entirely replaced by glycol systems for natural-gas dehydration.

More recently, however, a novel application of calcium chloride to natural-gas dehydration has been introduced and a number of small units installed. A schematic diagram of this

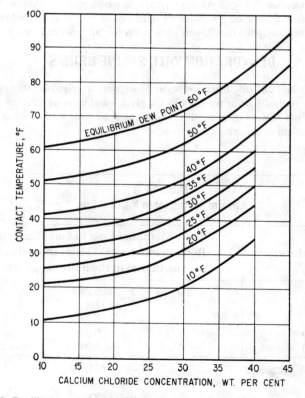

Figure 11-46. Equilibrium dew point of gases in contact with calcium chloride solutions. *Data of Brockschmidt (1942)*

type of dehydration unit is shown in **Figure 11-47,** which is based on a description by Fowler (1957). The unit contains a bed of ⅜- to ¾-in. calcium chloride pellets and five specially designed brine-circulating trays. Gas enters the bottom of the column, passes up through the brine trays where it contacts progressively more concentrated calcium chloride solutions, then continues upward through the bed itself, where additional water is absorbed on the surface of the pellets, forming concentrated brine which drips down onto the trays continuously. The design of the trays is such that the gas aspirates liquid upward to provide circulation and thereby maintains sufficient liquid on each tray without the need for a pump. The concentrated calcium chloride brine dripping from the bed of pellets has a specific gravity of 1.40, and this is reduced to approximately 1.15 to 1.20 by the time it reaches the bottom of the column. This brine is considered expendable and is dumped into a pit along with any free water produced by the well. The units are recharged with calcium chloride pellets periodically, bringing the calcium chloride bed to a depth of 8 ft. As the chemical is used up, the bed settles; however, the efficiency of the unit is not appreciably reduced as long as the level is above 24 in. It is reported that units of this type can give a dew point of as low as 7°F with a bed-depth as low as 2 ft and a gas temperature of 127°F. Up to 3.5 lb H_2O/lb $CaCl_2$ can be absorbed in units equipped with trays vs. 1.1 lb H_2O/lb $CaCl_2$ for units which use only dry $CaCl_2$ for dehydration (*GPSA Engineering Data Book,* 1987).

Since the frequency of recharging and the cost of chemicals is proportional to the water content of the gas, it is desirable that the water content of feed gas be maintained as low as possible by operating at a temperature close to that of the hydrate freezing point. A plot of

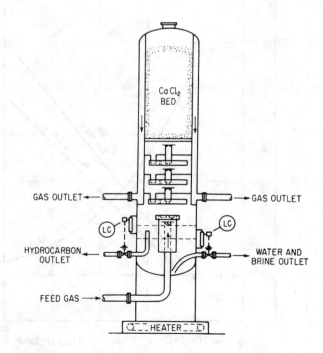

Figure 11-47. Schematic diagram of natural gas dehydration unit employing calcium chloride pellets. *Data of Fowler (1957)*

chemical costs, recharging days, and chemical consumption vs. the flowing temperature of the gas, assuming 500 lb pressure, is presented in **Figure 11-48**. The principal operating difficulties of this unit have been caused by the freezing of brine on the trays.

The results of field tests on 250 units over an 18-month period have been summarized by Fowler (1964). Calcium chloride has also been proposed for hydrate prevention in natural-gas gathering lines. In this application an aqueous solution is injected near the well head and collected at the downstream point after gas cooling has occurred. In tests reported in Russian literature (Andryushchenko and Vasilchenko, 1963), the process was found to be very effective. However, purging of the solutions with natural gas prior to injection was found necessary to reduce their corrosive action.

Lithium Halides for Air Dehydration

Data on the two lithium salts which are useful for air dehydration are presented in **Table 11-9** and **Figures 11-49** and **11-50**. As shown in the figures, lithium bromide is considerably more soluble in water, and the saturated solutions of this salt have a lower vapor pressure than lithium chloride solutions at the same temperature and thus can provide a greater degree of dehydration. For most operations, however, the degree of dehydration provided by lithium

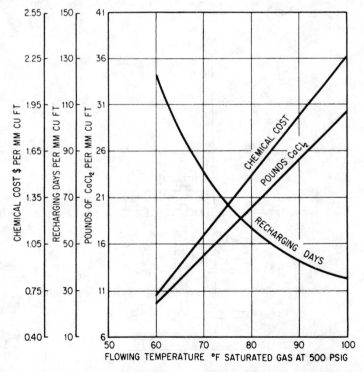

Figure 11-48. Effect of gas temperature on operation of calcium chloride pellet dehydration unit. *Data of Fowler (1957)*

Table 11-9
Properties of Lithium Salts Used for Dehydration

Property	Lithium Chloride	Lithium Bromide
Formula	LiCl	LiBr
Molecular weight	42.40	86.86
Melting point, °C	614	547
Solubility in water, g/100g	63.7 at 0°C 130.0 at 95°C	145 at 4°C 254 at 90°C
pH of 1% Sol	6.4	6.8
Heat of fusion, Cal/mole	4	5

Physical-property data from *Chemical and Physical Properties of Lithium Compounds*, publication of the Foote Minerals Company, Philadelphia, PA (1956)

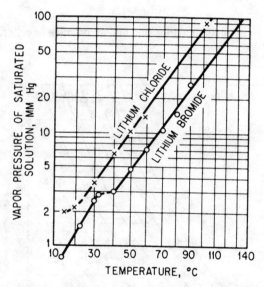

Figure 11-49. Vapor pressure of saturated solutions of lithium chloride and bromide. Data of Foote Minerals Company (1956)

chloride is adequate and, because of its somewhat lower cost, it is, therefore, the preferred compound. Dehydration units designed to employ lithium-halide solutions are essentially the same as those which use triethylene glycol. The principal difference between the halide solutions used for dehydration and the glycols is that the active component in the halide solution has essentially zero vapor pressure, and, therefore, no rectification section is necessary in the regenerator.

The dew-point depression theoretically attainable with three lithium chloride solutions is presented in **Figure 11-51**, which is plotted on a psychrometric chart to illustrate the applica-

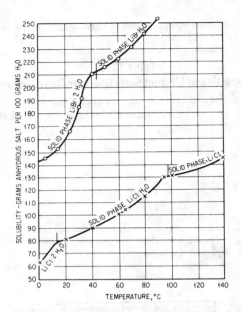

Figure 11-50. Solubility of lithium chloride and lithium bromide in water. *Data of Foote Minerals Company (1956)*

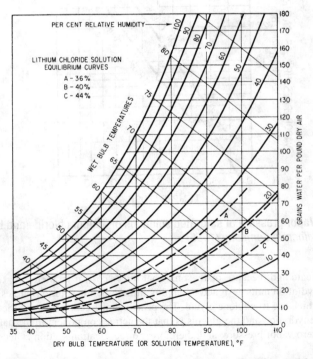

Figure 11-51. Equilibrium water content of air in contact with lithium chloride solutions. *Data of Gifford (1957)*

tion of the data to air conditioning problems. The solution of a typical air conditioning dehumidification problem is shown in **Figure 11-52** as presented by Gifford (1957). In this problem, it is assumed that air is available at 95°F, 75°F wet bulb (99 grains water per pound dry air), and it is desired to determine the degree of dehydration attainable with 44% lithium chloride solution. Two cases are considered. In Case A, it is assumed that the absorbent solution can be cooled to 80°F by the available cooling medium. In Case B, it is assumed that a solution temperature of 60°F can be maintained. Further assumptions are that dilution effects are negligible and that, in both cases, the equipment design is such that a 90% approach to equilibrium can be attained. By moving 90% of the distance along the line from the point representing the inlet-air condition to the point on the solution equilibrium curve corresponding to the solution temperature, it is seen that, for Case A, the air temperature can be reduced to 81°F (dry bulb) and 56½°F wet bulb. This corresponds to a 35°F dew point or a water con-

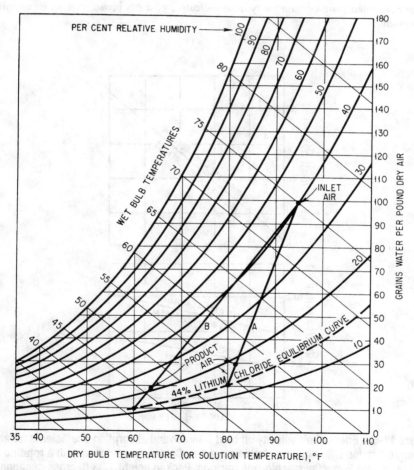

Figure 11-52. Solution of typical air conditioning dehydration problem employing lithium chloride solution. *Data of Gifford (1957)*

1014   *Gas Purification*

tent of 30 grains/lb dry air. In Case B, with a 60°F solution, the air can be dried and cooled to 64°F dry bulb and 46¼°F wet bulb, corresponding to a 24°F dew point or a water content of 19 grains/lb dry air.

Absorption and heat-transfer data for the dehumidification of air with lithium chloride solutions in a short column packed with 2-in. clay raschig rings (stacked) are given in **Figure 11-53** as presented by Bichowsky and Kelley (1935). The data are based on commercial-type work and are believed to be dependable within about 5%. Solution concentrations are not given; however, the authors report that, in the range of concentrations generally used, the coefficients are not found to be sharp functions of the concentration. Data for one of the experiments on which the curves of **Figure 11-53** are based are given in **Table 11-10**.

The absence of a concentration effect was also observed by Tohata et al. (1964A, B) for wetted wall and perforated plate columns. In the wetted wall column study, lithium chloride solutions in the range of 18.7 to 28.4% were employed. The results showed the gas-phase resistance to be controlling. Very high-stage efficiencies (up to 90%) were observed in the perforated plate column study, also indicative of a gas-phase resistance controlled absorption.

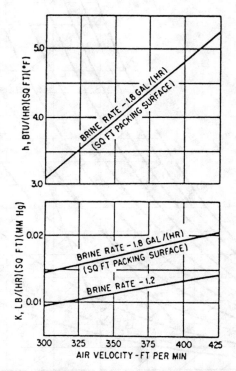

Figure 11-53. Effect of air velocity on heat transfer and absorption coefficients for air dehumidification with lithium chloride solutions. Column is equipped with a rotating distributor and 2-in. clay raschig-ring packing. Packed height 1.33 ft, cross-sectional area 2.4 sq ft, packing surface area 27 sq ft/cu ft. *Data of Bichowsky and Kelley (1935)*

Table 11-10
Sample Data for Experiment Using LiCl Brine and Absorber Packed with Raschig Rings

Variable	Numerical Value
Air rate, cu ft/min.	655
Tower cross section, sq ft	2.4
Packing height, ft	1.33
Packing surface, sq ft/cu ft	27
LiCl solution rate, gpm	2.75
LiCl solution, specific gravity	1.20
Temperatures, °F:	
Solution in	70
Solution out	78.2
Air in, dry bulb	82.8
Air in, wet bulb	66.0
Air out, dry bulb	79.0
Air out, wet bulb	61.2
Pressures, mm Hg:	
Partial pressure of water in inlet air	11.4
Partial pressure of water in outlet air	8.6
Vapor pressure of inlet brine	5.4
Vapor pressure of outlet brine	7.4
Absorption coefficient K, lb/(hr) (sq ft) (mm Hg)	0.0224
Heat transfer coefficient h, Btu/(hr) (sq ft) (°F)	4.6

Source: Bichowsky and Kelley, 1935

Figure 11-54 is a photograph of a commercial Kathabar air-dehydration unit manufactured by the Kathabar Systems Division of Somerset Technologies, Inc. The Kathene solution used in this unit is a solution of lithium chloride with appropriate additives. Kathabar humidity control systems are offered by the Kathabar Systems Division of Somerset Technologies, Inc.

A flow diagram of a duplex air-dehydration installation utilizing a 44 to 45% aqueous solution of lithium chloride is shown in **Figure 11-55**. This particular unit is installed in a penicillin-processing plant. Its job is to remove 30 gal/hr of water from 3,500 cfm of air, reducing the water content to 9 grains/lb so the humidity level in the plant can be maintained at 16 grains/lb to prevent moisture from damaging the hygroscopic penicillin (Anon., 1951). As shown in the diagram, outside air is drawn through unit A; this is cooled by circulating water at 85°F, and its moisture content is reduced from 122 grains to about 36 grains/lb. This partially dehydrated fresh air is combined with 2,850 cfm of recirculated air, and the mixture is passed through the second dehydration unit, which utilizes Freon at 38°F as coolant. In this unit the moisture content is reduced to 9 grains/lb. In both absorbers, the principal contact surface is the outside of fin-type heat exchangers through which the coolant is circulated. About 90% of the lithium chloride solution from the basin is recycled in the absorbers,

1016   *Gas Purification*

Figure 11-54. Commercial lithium chloride air-dehydration unit (Kathabar system). *Courtesy of Kathabar Systems Division, Somerset Technologies, Inc.*

and the remainder is bypassed to a regenerator. The regenerator is heated by low-pressure steam to about 230°F, which is well below the boiling point of the solution. Regeneration is accomplished at this temperature by the use of air as stripping vapor to sweep evaporated water out of the regenerator. The regenerated solution then flows to the sump of the first absorber; here it is cooled by dilution and the sensible heat is ultimately removed by the circulating water in the cooling coil.

The use of lithium chloride as a liquid desiccant in a heating, ventilating, and air conditioning system for a large office building is described by Meckler (1979). The system is unique in its integration of energy producing and using systems to minimize utility energy costs. The lithium chloride (Kathabar) system is required only during the summer for air dehumidification. When it is in use, the latent heat of absorption is rejected to a cooling tower via water coils in the contactor. A regenerator concentrates the lithium chloride solution by contacting it with exhaust air from the building. Heat for regeneration is supplied by a hot water circuit at 150°–200°F, which extracts waste heat from the jacket and exhaust of a diesel engine and also obtains heat, when available, from a flat-plate solar collector.

Although they are appreciably less corrosive than calcium chloride solution, lithium chloride and bromide brines are somewhat corrosive, particularly in the presence of impurities (especially copper) and it is desirable to utilize inhibitors. One such inhibitor is lithium chromate, which is particularly useful in these solutions as it does not introduce a foreign cation.

An incidental benefit obtained with lithium chloride units is the appreciable degree of sterilization of the air treated. Research at the University of Toledo on lithium chloride solutions

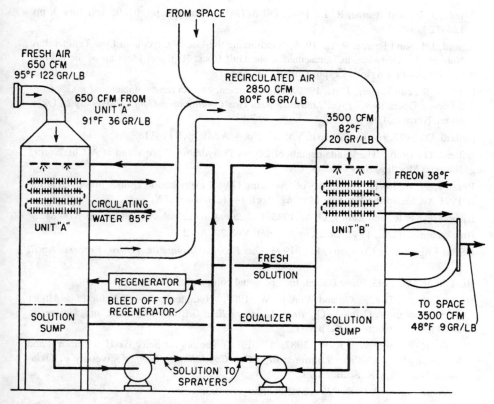

Figure 11-55. Flow arrangement and design specifications for air-dehumidification plant employing lithium chloride solution. (*Anon., 1951*)

reportedly indicated 97% removal of air-borne microorganisms from the air processed (Gifford, 1957). This feature is of particular importance in the evaluation of air conditioning systems for hospitals and food-processing plants.

REFERENCES

American Petroleum Institute (API), 1990, "Specifications for Glycol-Type Gas Dehydration Units," *API Spec. 12 GDU,* First Edition, Dec. 15, 1990.

Andryushchenko, F. K., and Vasilchenko, U. P., 1963, *Neft i Gas Prom. Nauchn.-Tekhn. Sb.*, (4).

Anon., 1951, *Chem. Eng. News,* Feb. 26, p. 819

Aoyagi, K., Song, K. Y., Sloan, E. D., Dharmawardhona, P. B., and Kobayashi, R., 1979, "Improved Measurement and Correlation of Methane Gas in Equilibrium with Hydrate," *Proc. 58th Annual Convention, Gas Processors Association,* San Antonio, TX.

Armstrong, R. A., 1979, *Oil & Gas J.*, Oct. 8, p. 83.

Arnold, J. L., and Pearce, R. L., 1961, *Oil & Gas J.*, June 19, pp. 92–95 and July 3, pp. 125–129.

Arnold, J. L., and Pearce, R. L., 1966, "Optimizing the Use of Glycols in Low Temperature Natural Gas Dehydration," presented at the Gulf Coast Regional Meeting of the Natural Gas Processor's Assoc., Corpus Christi, TX, Feb. 18.

Baker, J. R., and Rogers, J. A., 1989, "High Efficiency Co-current Contactors for Gas Conditioning Operations," *Proc. Laurance Reid Gas Conditioning Conf.*, University of Oklahoma, Norman, OK, p. 163.

Ballard, D., 1977, *Hydro. Process.*, Vol. 56, No. 4, April, pp. 111–118.

Ballard, D., 1986, "The Fundamentals of Glycol Dehydration," presented at AIChE Spring National Meeting, New Orleans, LA.

Bentley, M. T., 1991, "The Basics of Operating Glycol Dehydration Units," presented at the 1991 AIChE Spring National Meeting, April 1–11, Houston, TX.

Bichowsky, F. R., and Kelley, G. A., 1935, *Ind. Eng. Chem.*, Vol. 27, August, pp. 878–882.

Brockschmidt, C. L., 1942, *Gas*, Los Angeles, Vol. 28, April, p. 28.

BS&B Engineering Company, Inc., 1994, *Gas Process Handbook, Hydro. Process*, April, p. 798.

Bucklin, R. W., 1993, Fluor Daniel, Inc., personal communication.

Bucklin, R. W., Toy, K. G., and Won, K. W., 1985, "Hydrate Control of Natural Gas Under Arctic Conditions Using TEG," *Proc. Laurance Reid Gas Conditioning Conf.*, University of Oklahoma, Norman, OK, p. B-1.

Bucklin, R. W., and Won, K. W., 1987, "HIGEE Contactors for Selective H_2S Removal and Superdehydration," *Proc. Laurance Reid Gas Conditioning Conf.*, University of Oklahoma, Norman, OK, Section D.

Camerinelli, I., 1970, *Hydro. Process.*, February, p. 103.

Carmichael, C. J., 1964, *Oil & Gas J.*, Nov. 2, pp. 72–74.

Case, J. L., Ryan, B. F., and Johnson, J. E., 1985, *Oil & Gas J.*, May 13, p. 103.

Diaz, Z., and Miller, J. H., 1984, "Drying Substantially Supercritical CO_2 with Glycerol," United States Patent No. 4,478,612, Oct. 23.

Diaz, Z., Nasir, P., and Wallace, C. R., 1991, "Fundamentals of CO_2 Dehydration," presented at the 1991 AIChE Spring National Meeting, April 1–11, Houston, TX.

Dow Chemical Company, 1956, *Properties and Uses of Glycols*.

Fitz, C. W., and Hubbard, R. A., 1987, *Oil & Gas J.*, Nov. 23, p. 72.

Fontenot, C. E., Perry, L. A., Smith, R. S., and Zabicik, D. J., 1986, "Economic Comparison of Two Enhanced Glycol Dehydration Processes," *Proc. Laurance Reid Gas Conditioning Conf.*, University of Oklahoma, Norman, OK, p. F-1.

Foote Minerals Company, 1956, *Chemical and Physical Properties of Lithium Compounds*.

Fowler, A., 1975, "Super-Drizo, The Dow Dehydration Process," *Proc. Gas Conditioning Conf.*, University of Oklahoma, Norman, OK.

Fowler, O. W., 1957, *Oil & Gas J.*, Vol. 55, April 29, p. 188.

Fowler, O. W., 1964, *Oil & Gas J.*, Vol. 62, No. 31, p. 123.

Frazier, C. W., and Force, J. E., 1982, "Drizo Protects Turbo Expander Plant," *Proc. Gas Conditioning Conf.*, University of Oklahoma, Norman, OK.

Gallaugher, A. F., and Hibbert, H., 1937, *J. Am. Chem. Soc.*, Vol. 59, p. 2524.

Gas Process Supplier's Assoc. (GPSA) 1987, *Engineering Data Book* Vol. II, Sec. 20, Tenth edition, Tulsa, OK.

Gifford, E. W., 1957, *Heating, Piping, Air Conditioning J.*, Sect. 29, April, pp. 156–159.

Glaves, P. S., McKee, R. L., Kensell, W. W., and Kobayashi, R., 1983, *Hydro. Process.*, Vol. 62, No. 11, November, p. 213.

Grosso, S., 1978, *Oil & Gas J.*, Feb. 13, pp. 106–110.

Grosso, S., Pearce, R. L., and Hall, P. D., 1979, *Oil & Gas J.*, Part 1, Sept. 24, p. 176, Part 2, Oct. 1, p. 56.

Hammerschmidt, E. G., 1934, *Ind. Eng. Chem.*, Vol. 26, pp. 851–855.

Herrin, J. P., 1990, "Solving Glycol Dehydration Unit Operating Problems by Use of Process Fundamentals," *Proc. Laurance Reid Gas Conditioning Conf.*, University of Oklahoma, Norman, OK, p. 1.

Herrin, J. P., and Armstrong, R. A., 1972, "Methanol Injection and Recovery in a Turbo Expander Plant," *Proc. Gas Conditioning Conference,* University of Oklahoma, Norman, OK.

Hicks, R. L., and Senules, E. A., 1991, *Hydro. Process.*, Vol. 70, No. 4, April, p. 55.

Hubbard, R. A., 1989, *Oil & Gas J.*, Sept. 11, p. 47.

Hull, R. H., 1945, *Calif. Oil World,* Vol. 38, August, pp. 4–9.

Jou, F. Y., Mather, A. E., and Otto, F. D., 1988, "Acid Gas Solubilities in Glycols," *Proc. Laurance Reid Gas Conditioning Conf.*, University of Oklahoma, Norman, OK, p 163.

Kean, J. A., Turner, H. M., and Price, B. C., 1991, *Hydro. Process.*, Vol. 4, April, p. 47.

Kruger, H. O., and Mazelli, J. R., 1952, *Proc. Pacific Coast Gas Assoc.*, Vol. 43, p. 179.

Landsbaum, E. M., Dodds, W. S., and Stutzman, L. F., 1955, *Ind. Eng. Chem.*, Vol. 47, January, p. 101.

Lloyd, W. G., 1956, *J. Am. Chem. Soc.*, Vol. 78, p. 72.

Lloyd, W. G., and Taylor, F. C. Jr., 1954, *Ind. Eng. Chem.*, Vol. 46, November, pp. 2407–2416.

Love, F. H., 1942, *Petrol. Engr.*, Vol. 13, No. 13, p. 46.

Maddox, R. N., Lilly, L. L., Moshfeghian, M., and Elizondo, E., 1988 "Estimating Water Content of Sour Natural Gas Mixtures," *Proc. Laurance Reid Gas Conditioning Conf.*, University of Oklahoma, Norman, OK, p. 75.

Maddox, R. N., Moshfeghian, M., Lopez, E., Tu, C. H., Shariat, A., and Flynn, A. J., 1991, "Predicting Hydrate Temperature at High Inhibitor Concentration," *Proc. Laurance Reid Gas Conditioning Conf.*, University of Oklahoma, Norman, OK, p. 272.

Manning, W. P., and Wood, H. S., 1991, "Design Guidelines for Glycol Dehydrators," presented at the 1991 AIChE Spring National Meeting, April 1–11, Houston, TX.

McKetta, J. J., and Wehe, A. H., 1958, *Petroleum Refiner, Hydro. Process.*, Vol. 37, No. 8, August, p. 153.

McCarthy, E. L., Boyd, W. L., and Reid, L. S., 1950, *J. Petrol. Tech.*, Vol. 189, p. 241.

Meadows, R. E., 1989, *Oil & Gas J.*, May 15, p. 47.

Meckler, G., 1979, *Specifying Engineer,* April.

Neal, R., Franke, S., and Patel, K., 1989, *Pipeline & Gas Journal,* July, p. 35.

Nelson, K. D., and Wolfe, L., 1981, "Methanol Injection and Recovery in a Turbo Expander Plant," *Proc. Gas Conditioning Conference,* University of Oklahoma, Norman, OK.

Nielsen, R. B., and Bucklin, R. W., 1983, *Hydro. Process.*, Vol. 62, No. 4, April, p. 71.

OPC Engineering, Inc., 1992, *Gas Process Handbook '92, Hydro. Process.*, April, p. 98.

Parrish, W. P., Won, K. W., and Baltatu, M. E., 1986, "Phase Behavior of the Triethylene Glycol-Water System and Dehydration/Regeneration Design for Extremely Low Dew Point Requirements," presented at 65th Annual GPA Convention, March 10–12, San Antonio, TX.

Peahl, L. H., 1950, *Oil & Gas J.*, Vol. 49, July 13, p. 92.

Pearce, R. L., and Sivalls, C. R., 1993, "Fundamentals of Gas Dehydration Design and Operation with Glycol Solutions," *Proc. Laurence Reid Gas Conditioning Conference,* University of Oklahoma, Norman, OK.

Pearce, R. L., Protz, J. E., and Lyon, G. W., 1972, "Drizo—Improved Regeneration of Glycol Solutions," *Proc. Gas Conditioning Conference,* University of Oklahoma, Norman, OK.

Polderman, L. D., 1957, *Oil Gas J.*, Vol. 55, Sept. 23, pp. 107–112.

Porter, J. A., and Reid, L. S., 1950, *J. Petrol. Tech.*, July, p. 189.

Pyles, S., and Rader, R. G., 1989, "Single Stage Co-Current Contactor Replaces Trayed Column on Offshore Platform for Dehydration," presented at January 1989 Production Technology Symposium, 12th Annual Energy-Sources Technology Conference, Houston, TX.

Reid, L. S., 1975, "Coldfinger, and Exhauster for Removing Trace Quantities of Water from Glycol Solutions used for Gas Dehydration," *Proc. Gas Conditioning Conference,* University of Oklahoma, Norman, OK.

Robinson, J. N., Wichert, E., and Moore, R. G., 1978, *Oil & Gas J.*, Feb. 6, pp. 76–78.

Robirds, K. D., and Martin, J. C. III, 1962, *Oil & Gas J.*, Vol. 60, April 30, p. 85.

Rosman, A., 1973, *Soc. Petrol. Eng. J.*, October, p. 297.

Rueter, C. O., Thompson, P. A., Lowell, P. S., Nelson, T. P., Evans, J. M., and Gamez, J. P., 1992, "Research on Emissions of BTEX and VOC from Glycol Dehydrators," *Proc. Laurance Reid Gas Conditioning Conference,* University of Oklahoma, Norman, OK, p. 187.

Russell, G. F., Reid, L. S., and Huntington, R. I., 1945, *Petrol. Refiner,* Vol. 24, December, p. 137.

Sams, G. W., 1990, "Emission Control for Glycol Dehydration Equipment," *Proc. Laurance Reid Gas Conditioning Conference,* University of Oklahoma, Norman, OK, p. 23.

Senatoroff, H. K., 1945, *Oil & Gas J.*, Vol. 44, December, pp. 98–108.

Sheilan, M. "Optimizing Glycol Injection Refrigeration Plants," 1991, *Proc. Laurance Reid Gas Conditioning Conference,* University of Oklahoma, Norman, OK, p. 96.

Simmons, C. V. Jr., 1981, *Oil & Gas J.*, Sept. 28, p. 313.

Sivalls, C. R., 1976, "Glycol Dehydration Design Manual," *Proc. Gas Conditioning Conf.*, University of Oklahoma, Norman, OK.

Smith, R. S., 1990, *Hydro. Process.*, February, p. 75.

Smith, R. S., 1993, "Custom Glycol Units Extend Operating Limits," *Proc. Laurance Reid Gas Conditioning Conf.*, University of Oklahoma, Norman, OK, p. 101.

Smith, R. S., and Humphrey, S. E., 1995, "High Purity Glycol Design Parameters and Operating Experience," *Proc. Laurance Reid Gas Conditioning Conference,* University of Oklahoma, Norman, OK, p. 142.

Smith, R. S., and Skiff, T. B., 1990, "Drizo Gas Dehydration, Solution for Low Dew Points/Aromatics Emissions," *Proc. Laurance Reid Gas Conditioning Conf.*, University of Oklahoma, Norman, OK, p. 61.

Stahl, W., 1963, "Method and System for Reconcentrating Ethylene Glycols," U.S. Patent 3,105,748, October.

Sullivan, J. H., 1952, *Oil & Gas J.*, Vol. 50, March 3, p. 70.

Swerdloff, W., 1957, *Oil & Gas J.*, Vol. 55, April 29, pp. 122–129.

Swerdloff, W., and Duggan, M., 1955, *Petrol. Refiner*, Vol. 34, March, p. 208.

Takahashi, S., and Kobayashi, R., 1982, GPA TP-9, Gas Processors Association, December, Tulsa, OK.

Tohata, H., Yamada, T., Nakada, T., and Itorgaki, H., 1964A, *Kagaku Kogaku*, Vol. 28, No. 10, pp. 832–836.

Tohata, H., Yamada, T., Nakada, T., and Sasu, A., 1964B, *Kagaku Kogaku*, Vol. 28, No. 2, pp. 155–158.

Townsend, F. M., 1953, "Vapor Liquid Equilibrium Data for Diethylene Glycol Water and Triethylene Glycol Water in Natural Gas Systems," *Proc. Gas Hydrate Control Conf.*, University of Oklahoma, Norman, OK, May 5–6.

Townsend, F. M., 1985, "Glycol-Water Distillation," *Proc. Laurance Reid Gas Conditioning Conf.*, University of Oklahoma, Norman, OK, p. L-1.

Tupholme, 1929, *Gas Age-Record*, Vol. 63, pp. 311–313.

Udvardi, G., Gerecs, L., Ouchi, Y., Nagakura, F., Thoes, E. A., and Wallace, C. B., 1990, *Oil & Gas J.*, Oct. 22, p. 74.

Union Carbide Corp., 1971, Glycols.

Wallace, C. B., 1985, *Oil & Gas J.*, June 24, p. 98.

Wieninger, P., 1991, "Operating Glycol Dehydration Systems," *Proc. Laurance Reid Gas Conditioning Conference,* University of Oklahoma, Norman, OK, p. 23.

Won, K. W., 1994, "Thermodynamic Basis of the Glycol Dew-Point Chart and its Application to Dehydration," presented at the 73rd Gas Processors Association Annual Convention, March 7–9, New Orleans, LA.

Won, K. W., and Walker, C. K., 1979, *Advances in Chemistry Series,* Am. Chem. Soc., No. 182, p. 35.

Worley, M. S., 1966, "Twenty Years of Progress with TEG Dehydration," presented at CNGPA meeting, Calgary, Alberta, Canada, Dec. 2.

Worley, M. S., 1967, "Super-dehydration with Glycols," *Proc. Gas Conditioning Conf.*, University of Oklahoma, Norman, OK.

Zabrik, D. J., and Frazier, C. W., 1984, "Dehydration of CO_2 with TEG, Plant Operating Data," *Proc. Gas Conditioning Conference,* University of Oklahoma, Norman, OK.

Chapter 12
Gas Dehydration and Purification by Adsorption

INTRODUCTION, 1023

    Adsorption Cycles, 1024
    General Design Concepts, 1026

WATER VAPOR ADSORPTION, 1030

    Desiccant Materials, 1034
    Mechanism of Water Adsorption, 1044
    Dehydration System Design Approach, 1048
    Operating Practices, 1069

USE OF MOLECULAR SIEVES FOR GAS PURIFICATION, 1070

    Basic Data, 1072
    Inert Gas Purification, 1074
    Carbon Dioxide and Water Removal from Ethylene, 1076
    Carbon Dioxide Removal from Cryogenic Plant Feed Gas, 1076
    Removal of Sulfur Compounds, 1078
    Hydrogen Purification by Pressure Swing Adsorption (PSA), 1081

HYDROCARBON RECOVERY WITH SILICA GEL, 1086

ORGANIC VAPOR ADSORPTION ON ACTIVE CARBON, 1087

    Properties of Gas Adsorption Carbons, 1088
    VOC Removal and Solvent Recovery with Active Carbon, 1093
    Hydrocarbon Recovery with Active Carbon, 1109
    Fluidized, Moving, and Rotating Bed Processes, 1109
    Canister and Panel Systems for Air Purification, 1117

BIOFILTERS FOR ODOR AND VOC CONTROL, 1124

    Soil Beds, 1124
    Peat Beds, 1125

IMPREGNATED ADSORBENT APPLICATIONS, 1126
Sulfur Compound Removal with Impregnated Carbon, 1126
Mercury Removal with Impregnated Carbon, 1127

MISCELLANEOUS APPLICATIONS OF ADSORPTION, 1128
Hydrochloric Acid on Alumina, 1128
Radioactive Isotope Adsorption, 1128
Adsorption of Iron and Nickel Carbonyls, 1128

REFERENCES, 1129

INTRODUCTION

The unit operation of adsorption is of increasing importance in gas purification and forms the basis for commercial processes that remove water vapor, organic solvents, odors, and other vapor-phase impurities from gas streams. In adsorption, materials are concentrated on the surface of a solid as a result of forces existing at this surface. Since the quantity of material adsorbed is directly related to the area of surface available for adsorption, commercial adsorbents are generally materials that have been prepared to have a very large surface area per unit weight. For gas purification, the adsorbent particles may be irregular granules or preformed shapes, such as tablets or spheres, and the gas to be purified is passed through a bed of the material. The gas-phase impurity is selectively concentrated on the internal surfaces of the adsorbent while the purified gas passes through the bed.

The nature of the forces that hold certain molecules at the solid surface is not thoroughly understood and numerous theories have been proposed to explain the phenomenon. The most familiar theory is that of Langmuir (1916, 1918) who proposed that the forces acting in adsorption are similar in nature to those involved in chemical combination. Sites of residual valency are assumed to exist on the surface of solid crystals. When an adsorbable molecule from the gaseous phase strikes a suitable unoccupied site, the molecule will remain instead of rebounding into the gas. As in the evaporation of liquids, the adsorbed molecule may leave the surface when suitably activated; however, other molecules will continually adhere. When adsorption is first started, a large number of active sites exist, and the number of molecules adhering exceeds the number of those leaving the surface. As the surface becomes covered, the probability of a molecule in the gas finding an unoccupied space is decreased, until finally the rate of condensation equals the rate of evaporation, which represents the condition of equilibrium. In accordance with the Langmuir theory, the adsorbed material is held onto the surface in a layer only one molecule deep, although it is recognized that these adsorbed molecules may have their force fields shifted in such a manner that they can attract a second layer of molecules, which could have some attraction for a third layer, and so on. The adsorption mechanism postulated by Langmuir would require that the equilibrium quantity of a compound adsorbed from a gas increase with increased gas pressure, but at a constantly decreasing rate. Equilibrium isotherms of this shape are referred to as Langmuir isotherms.

The forces holding adsorbed molecules to the surface may be quite weak, resembling those that cause molecules to coalesce and form the liquid phase, or so strong that the adsorbed material cannot be removed without a chemical change taking place. The weaker—physical or van der Waals—forces are apparently responsible for most adsorption phenomena; this explains why, in general, compounds that have low vapor pressures are adsorbed in greater quantity than relatively noncondensable gases. The chemical type of adsorption, which has been given the name "chemisorption," is of less importance in industrial adsorption processes. An example of chemisorption is the adsorption of oxygen on charcoal at temperatures above 0°C. When an attempt is made to desorb the oxygen by elevating the temperature, it is released as an oxide of carbon.

When a vapor- (or liquid-) phase component concentrates on a solid by adhering to the solid surfaces, even though the surfaces may consist of the interior of submicroscopic pores, the phenomenon is known as "adsorption." If, on the other hand, penetration of the solid or semisolid structure occurs and produces a solid solution or a chemical compound, the phenomenon is termed "absorption." The general term "sorption" has been proposed to cover both cases.

Although adsorption can be practiced with many solid compositions, the great majority of gas-purification and dehydration adsorbents are based on some form of silica, alumina (including bauxite), carbon, or certain silicates, the so-called molecular sieves. The silica and alumina-base adsorbents are primarily used for dehydration, while activated carbon has the specific ability of adsorbing organic vapors and is very important for this purpose. The molecular sieves have very unusual properties with regard to both dehydration and the selective adsorption of other compounds.

Whether the process involves the removal of water vapor or of some other gas-phase impurity, the basic concepts involved in the design of the installation are similar. The gas must be passed through a bed of the adsorbent material at a velocity consistent with pressure drop and other requirements and under conditions that will allow the required material transfer to occur. The bed will eventually become loaded with the impurity and must then either be discarded, removed for reclaiming, or regenerated in place.

When regeneration is practiced, it is normally accomplished by increasing the temperature, reducing the total pressure, reducing the partial pressure with a stripping gas, displacing the adsorbate with a more strongly adsorbed species, or by using a combination of these operations. Most adsorption systems use fixed beds; however, processes have also been developed which use moving or fluidized bed concepts.

This chapter is intended as a practical guide to selecting, designing, and operating adsorption systems for the removal of vapor-phase impurities from gas streams. It does not cover the theoretical aspects of adsorption processes that are aimed more at separation than purification, or liquid phase processes. Comprehensive coverage of the general field of adsorption is available in books by Yang (1987), Wankat (1986), Ruthven (1984), and Liapis (1987). Yang's book is of particular interest because it is limited to the processing of gases and covers both the theory of adsorption and the modeling of adsorption cycles. The Liapis book is a collection of papers presented at a conference on adsorption fundamentals. It provides a good review of ongoing research, but little in the way of practical design information. Useful summaries of adsorption principles and general design techniques are presented in chapters on the subject in two handbooks on separation processes (Keller et al., 1987; Kovach, 1988).

Adsorption Cycles

Regenerative adsorption processes are normally based on the use of one or more of the following basic cycles:

Temperature Swing Adsorption (TSA) Cycle

This is the cycle commonly used for gas dehydration and the removal of organic impurities with activated carbon. The gas to be purified is passed through a bed of the adsorbent at a relatively low temperature until the bed is essentially saturated with the impurity at this temperature. The bed temperature is then raised, and more gas is passed through the bed until equilibrium is attained at the higher temperature. The difference between the loading at the low temperature and that at the high temperature represents the net removal capacity, sometimes called the "delta loading."

TSA cycles require a considerable amount of heat to raise the temperature of the adsorbent and vessel as well as to supply the heat of adsorption. The heating and cooling cycle is time consuming, typically over an hour. As a result, the TSA processes are used primarily for the removal of small concentrations of impurities from gases. Frequently, an inert gas purge is used in combination with increased temperature to assure complete regeneration.

Inert Purge Adsorption Cycle

In this cycle, the gas to be purified is passed through the bed until the adsorbent is essentially saturated with the adsorbate at its partial pressure in the feed. A nonadsorbing gas containing very little or none of the impurity is then passed through the bed, reducing the partial pressure of adsorbate in the gas phase so that desorption occurs.

The heat of adsorption causes the temperature of the gas and adsorbent to rise during the adsorption phase and decrease during desorption. Since the increase in temperature limits the amount of material that can be captured in the bed during the adsorption step, the process is, in general, limited to low concentration changes in the adsorbent bed. However, since bed heat-up and bed cool-down are not required, very short cycle times can be employed, typically a few minutes. The simple inert gas purge cycle is used primarily for separating hydrocarbons, not for gas purification.

Displacement Purge Cycle

In this process, regeneration is effected by the use of a purge gas (desorbent) that is more strongly adsorbed than the component removed during the adsorption step. Typically the heat of adsorption of the desorbent is approximately the same as that of the original adsorbate. As a result, the process operates in an essentially isothermal manner and relatively large net loadings can be realized. Like the inert purge adsorption cycle, the process can be operated with short cycle times, on the order of a few minutes, and is more commonly used for separating hydrocarbons than purifying gases. A problem with displacement surge systems is the need to separate the desorbent from both the product and purge streams.

The use of active carbon beds to adsorb organic vapors with steam regeneration is primarily a TSA cycle; however, desorption is aided by the displacement of organic compounds by water. The adsorbed water then serves to limit temperature rise in the bed during the next organic vapor adsorption step.

Pressure Swing Adsorption (PSA) Cycle

The PSA concept is the most rapidly developing adsorption cycle. According to Yang (1987), the original idea for a PSA cycle was disclosed by Guerin de Montgareuil and Domine in France in 1957, and in a somewhat different form by Skarstrom in the U.S. in 1958.

1026 Gas Purification

The PSA cycle makes use of the simple fact that the partial pressure of adsorbate in the gas phase can be reduced by lowering the total pressure. Pressure reduction can thus be used to regenerate adsorbent that has been loaded with adsorbate at an elevated pressure. Since it is not necessary to heat or cool the bed between or during the adsorption and desorption steps, very rapid cycling is possible. The process is now widely used for hydrogen purification, air separation, hydrocarbon separation, and air drying, and new applications are under development.

General Design Concepts

Design data and procedures that have been developed for gas dehydration with solid desiccants, organic vapor removal with activated carbon, and other important adsorption applications are presented in later sections of this chapter under the specific application. Design concepts that are applicable to solid adsorption systems in general follow.

Mass Transfer Zone (MTZ)

Adsorption in a typical fixed-bed system can be visualized by reference to **Figure 12-1**, which is a plot of X, the concentration of adsorbate on the adsorbent, versus L, the length of

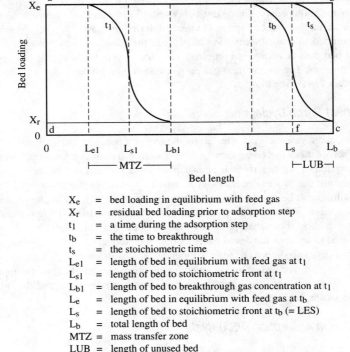

X_e = bed loading in equilibrium with feed gas
X_r = residual bed loading prior to adsorption step
t_1 = a time during the adsorption step
t_b = the time to breakthrough
t_s = the stoichiometric time
L_{e1} = length of bed in equilibrium with feed gas at t_1
L_{s1} = length of bed to stoichiometric front at t_1
L_{b1} = length of bed to breakthrough gas concentration at t_1
L_e = length of bed in equilibrium with feed gas at t_b
L_s = length of bed to stoichiometric front at t_b (= LES)
L_b = total length of bed
MTZ = mass transfer zone
LUB = length of unused bed

Figure 12-1. Schematic diagram showing adsorbate loading of bed as a function of bed length at various times in the adsorption cycle.

bed from the gas inlet. At the start of the cycle, the adsorbate concentration in the bed is X_r, representing the residual concentration after regeneration. At an intermediate time into the cycle, t_1, the bed nearest to the gas inlet is in equilibrium with the feed gas and has an adsorbate concentration, X_e. In front of this equilibrium section of the bed, from L_{e1} to L_{b1}, a transition zone exists where the bed loading varies in an S-shaped curve from the equilibrium concentration to the residual concentration. Since mass transfer of adsorbate from the gas phase to the adsorbent is actively occurring in this zone, it is commonly called the mass transfer zone or MTZ.

Length of MTZ

The length of the MTZ is determined by the rate of mass transfer of adsorbate from the gas phase to its ultimate location on solid surfaces within the pores of the adsorbent. When the mass transfer rate is extremely high, the MTZ reduces to a plane, commonly called the "stoichiometric front," which is located on the chart at L_{s1} at time, t_1. The location of the stoichiometric front at any time can be readily calculated by a simple material balance equating the amount of adsorbate removed from the gas in the given time with the amount picked up by the adsorbent in a bed of length, L, when its loading goes from X_r to X_e. Although the mass transfer rate is seldom high enough in real systems to make the MTZ coincide with the stoichiometric front, the location of the hypothetical plane representing the front is a very useful concept in some shortcut design procedures.

The overall rate of mass transfer of adsorbate in the bed is affected by external transport from the bulk of the gas to the external surfaces of the adsorbent particles, axial dispersion and backmixing in the gas phase, and internal transport within the pores. External transport can be correlated by equations similar to those used for mass transfer in packed absorption columns, such as the Ranz-Marshall (1952) equation:

$$Sh = 2.0 + 0.6 \, Sc^{1/3} \, Re^{1/2} \qquad (12\text{-}1)$$

where Sh, Sc, and Re refer to the Sherwood, Schmidt and Reynolds numbers, respectively.

Wakao and Funazkri (1978) reanalyzed available test data and proposed the following revised version of the equation:

$$Sh = 2.0 + 1.1 \, Sc^{1/3} \, Re^{0.6} \qquad (12\text{-}2)$$

The new equation includes the effects of axial dispersion. It is applicable at low Reynolds numbers (down to about 3) where equation 12-1 cannot be used. At high Reynolds numbers (over about 100), the two equations give similar results.

The estimation of internal mass transfer is very complex. The subject is covered in depth by Yang (1987), who points out that it involves four mechanisms: gaseous diffusion, convective flow due to diffusion, surface flow on the pore wall, and viscous flow.

In comparing external versus internal mass and heat transfer, Yang (1987) concludes that the major resistance is within the pores for mass transfer and in the gas film for heat transfer. However, the heat transfer rate is not normally used in absorber designs; the gas is simply assumed to be at the same temperature as the particles with which it is in contact.

Because of the complexity of fundamental approaches for predicting the length of the MTZ, it is common practice to estimate it on the basis of operating or test data with similar systems, use simple empirical correlations developed for specific applications, or neglect it entirely in determining the required bed length.

The following factors are listed by Kovach (1988) as affecting the length and rate of movement of the MTZ:

1. Type of adsorbent
2. Particle size
3. Bed depth (or length)
4. Gas velocity
5. Temperature
6. Concentration of adsorbate in the gas
7. Concentration of other gas components
8. Pressure
9. Required removal efficiency
10. Possible decomposition or polymerization of contaminants on the adsorbent

The adsorbent type, particle size, and gas velocity are the principal factors. Gas velocity has a strong effect on both the length and rate of movement of the MTZ. Its rate of movement is directly proportional to the carrier gas velocity. The MTZ decreases with particle size; therefore, it is advantageous to use particles that are as small as possible, consistent with pressure drop limitations. The pressure drop limitations also affect bed depth, which should generally be as large as possible since the fraction of the bed loaded to the equilibrium capacity increases with increased bed depth.

Length of Unused Bed (LUB)

As the MTZ moves through the bed, the concentration of adsorbate in the outlet gas eventually rises to an arbitrarily set "breakthrough" concentration, typically the minimum detectable or the maximum allowable adsorbate concentration. The time required for this to occur is called the "breakthrough time," which is indicated by curve "t_b" on **Figure 12-1**. The overall bed loading at this time is designated by the area "abcd" under the loading curve and represents the maximum useful bed capacity. Since the curve "bc" is symmetrical, the maximum useful bed capacity can also be represented by the rectangle "aefd" where the line "ef" lies on the stoichiometric front. The bed length "L_s" is called the length of equivalent equilibrium section (LES). The remaining bed area "egcf" represents the unused bed capacity, and the length of bed from the stoichiometric front "L_s" to the end of the bed "L_b" at breakthrough is called the length of unused bed (LUB). As shown on the chart, LUB is equal to one-half the length of the MTZ.

The total length of bed required to adsorb a given amount of material can be estimated by simple addition of LES, calculated from equilibrium loading data, and LUB, estimated on the basis of available correlations or test data. Weight units can, of course, be used instead of length units in a calculation of this type; however, length is convenient in that it leads directly to the vessel size required.

The key test data required for the estimation of LUB are the time required for breakthrough, t_b, and the time required for the stoichiometric front to reach the end of the bed, t_s, in a system in which a stable mass transfer zone has developed. The equation for calculating LUB as given by Keller et al. (1987) is

$$LUB = (1 - t_b/t_s)L_b \qquad (12\text{-}3)$$

Where: LUB = length of unused bed
t_b = time to breakthrough
t_s = stoichiometric time
L_b = total length of bed

Thermal Effects

The adsorption process is always exothermic, and, as would be expected, adsorption capacity decreases with increased temperature. These effects cause more of a problem in separations where a large fraction of the feed gas may be adsorbed than in gas purification where, generally, only a small amount of the feed gas is adsorbed.

In gas dehydration the effect of heat of adsorption is primarily a function of operating pressure (assuming that the feed gas is saturated with water). With high-pressure gas (over about 500 psig), the temperature increase during the adsorption step is only about 2° to 4°F and can usually be neglected. With low-pressure gas (or air), the large amount of water associated with each pound of gas causes a significant temperature rise that must be considered in the design. The details of dehydrator design are discussed in a subsequent section of this chapter.

During adsorption most of the heat is generated in the MTZ. When the heat capacity of the gas stream is high relative to that of the bed, the gas will carry the heat forward ahead of the MTZ. This effect can be quantified by use of a "crossover ratio," R, defined by the following equation (Keller et al. 1987):

$$R = C_{pg}(X_i - X_r)/C_{ps}(Y_i - Y_o) \tag{12-4}$$

Where: C_{pg} = gas heat capacity, Btu/mole °F
C_{ps} = solid heat capacity, Btu/lb °F
X_i = adsorbent loading at inlet, mole/lb
X_r = residual adsorbent loading, mole/lb
Y_i = molar ratio of adsorbate to carrier gas at gas inlet, mole/mole
Y_o = molar ratio of adsorbate to carrier gas at gas outlet, mole/mole

When R is much greater than 1, the heat evolved is carried away in the gas, and the system operates in an essentially isothermal manner. When R is less than 1, the heat front lags behind the MTZ, and the temperature of the equilibrium section is increased causing a decrease in equilibrium loading.

For cases in which the heat is carried out of the MTZ by the product gas, a heat balance can be used to estimate the temperature of the bed downstream of the MTZ, and, therefore, of the product gas. The following heat balance equation is given by Kovach (1988):

$$QG = (GC_{pa} + mC_{ps} + VC_v)(\Delta t) \tag{12-5}$$

Where: Q = heat of adsorption, Btu/lb
G = weight of adsorbed adsorbate, lb
C_{pa} = adsorbate heat capacity, Btu/lb °F
C_{ps} = adsorbent heat capacity, Btu/lb °F
C_v = carrier gas heat capacity, Btu/scf °F
m = weight of adsorbent, lb
V = volume of carrier gas processed, scf
Δt = temperature rise of the bed, °F

The equation is based on the assumption that all of the gas processed, the entire bed of adsorbent, and the adsorbate end up at the same temperature. Actually, the cool incoming gas reduces the temperature of the inlet portion of the bed and is warmed in the process. This causes the temperature of the outlet gas to be slightly higher than that calculated by equation 12-5.

Pressure Drop

Pressure drop is a critical factor in the design of adsorption systems, as it normally determines the allowable gas velocity and, therefore, the bed cross-sectional area. Although much work has been done on the subject, no completely satisfactory general correlation has been developed that takes into account the shapes of individual particles, size distribution, void fraction, and aging effects, as well as the more readily characterized gas properties and conditions. It is, therefore, common practice to use experimental and operating data and semi-empirical correlations aimed at specific adsorbent types and applications. Typical data and correlations are presented in subsequent sections covering dehydration with solid desiccants and organic vapor adsorption on activated carbon.

WATER VAPOR ADSORPTION

A large number of solid materials will take up water vapor from gases, some by actual chemical reaction, others through formation of loosely hydrated compounds, and a third group by adsorption as described above. Desiccants that operate by adsorption are of primary importance for commercial gas dehydration. The types most commonly used for this purpose are

1. *Silica-based adsorbents.* This group includes pure, activated silica gel and special formulations containing a small percentage of other components such as alumina.
2. *Alumina-based adsorbents.* These include impure, naturally occurring materials such as bauxite and high purity activated aluminas derived from gels or crystalline minerals.
3. *Molecular sieves.* This category covers a large family of synthetic zeolites characterized by extremely uniform pore dimensions.

With the possible exception of the molecular sieves, which normally require somewhat higher regeneration temperatures, the equipment and process-flow arrangements for all of the adsorbents are essentially identical and need be described only once. In many cases, the adsorbents themselves are interchangeable and equipment designed for one can be operated quite effectively with another.

In its simplest form, a plant for removing water vapor from gases by adsorption will consist of two vessels filled with granular desiccant together with sufficient auxiliary equipment so that one bed of desiccant can be regenerated while the other is being used for dehydration. Regeneration is accomplished by passing hot gas through the bed. When the first bed is spent and the second completely regenerated, their effective positions in the flow pattern are reversed by suitable valving. The complete cycle is repeated periodically so that the process is in effect continuous with regard to gas dehydration. The principal difference between various adsorption-dehydration processes is the means of providing heat for regeneration. To make the process truly continuous, some work has been done on units in which the beds are made to move from the regeneration zone to the adsorption zone rather than switching the gas-feed point by valving. However, the moving-bed units have been industrially important only for low pressure applications such as air-conditioning dehydration.

Photographs of large field installations for solid-desiccant dehydration of high-pressure natural gas are shown in **Figures 12-2** and **12-3,** and a small package-type unit for drying instrument air is shown in **Figure 12-4.**

Figure 12-2. A typical natural gas dehydration plant employing dry desiccant in vertical vessels. *Courtesy of United Gas Pipe Line Company*

Figure 12-3. A large field installation for drying high-pressure natural gas with a solid desiccant. These adsorbers are of the horizontal compartmental type with four compartments per vessel. The gas is introduced through manifolds inside the units to provide even distribution. Regeneration gas is heated in a salt-bath indirect-fired heater. *Courtesy of Black, Sivalls & Bryson, Inc.*

Figure 12-4. Package-type solid-desiccant dryer. Unit shown has a capacity of 1,000 scf/min. of air at 100-psig operating pressure. It is designed for an 8-hour tower reversal cycle with 250-psig steam reactivation and product air with a −40°F dew point. *Courtesy of C.M. Kemp Manufacturing Company*

Many gas-dehydration problems can be solved by using either a solid desiccant or a liquid system. However, the principal areas of application of the dry-desiccant processes are

1. Cases where essentially complete water removal is desired
2. Installations (usually small) in which the operating simplicity of the granular-desiccant system makes it attractive

In the dehydration of relatively large volumes of high-pressure natural gas, liquid dehydrating systems (diethylene glycol or triethylene glycol) are usually more economical if dew-point depressions of 40° to 140°F are required. If higher dew-point depressions, up to about 180°F, are necessary, either type may be selected on the basis of intangible factors. If dew-point depressions consistently higher than about 180°F are required, solid-desiccant dehydration is generally specified.

In general, where a simple triethylene glycol unit (atmospheric-pressure regeneration) is applicable, it is more economical from both an initial and operating-cost standpoint than a typical dry-desiccant system. A comparison of approximate equipment costs for natural-gas dehydration by dry-desiccant and glycol processes is presented in **Figure 12-5.** Operating costs for dry-desiccant systems are typically 20 to 30% higher than simple glycol dehydration units.

In comparison to liquid systems, solid-desiccant dehydration plants offer the following advantages:

1. Ability to provide extremely low dew points
2. Insensitivity to moderate changes in gas temperature, flow rate, pressure, etc.
3. Simplicity of operation and design of units
4. Relative freedom from problems of corrosion, foaming, etc.
5. Adaptability to dehydration of very small quantities of gas at low cost

The process has the following disadvantages:

1. High initial cost
2. Generally higher pressure drop
3. Susceptibility to poisoning or breakup
4. Relatively high heat requirement

In some cases dry bed dehydration is actually less costly than glycol absorption. Such a case, as described by Alexander (1988), is the dehydration of natural gas at an offshore platform. A glycol plant for the proposed service would be uneconomical because of the requirement for a complex system to recover vapors from the glycol still to avoid air pollution problems. Silica gel was selected for this plant in favor of alumina to avoid the catalytic effect of

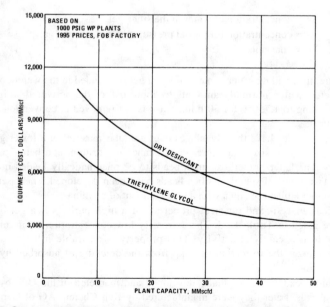

Figure 12-5. Comparative initial equipment costs for typical dry-desiccant and glycol dehydration processes in high-pressure natural gas service.

alumina on the COS formation reaction. Molecular sieve was not used because of its higher cost and its requirement for high temperature regeneration.

Desiccant Materials

Important physical properties of typical desiccant materials are listed in **Table 12-1**. Equilibrium water-capacity data are presented in **Figure 12-6**.

Silica Gel

Silica gel is commercially available as a powdered, granular, and spherical bead material of various size-ranges. The individual particles have a hard, glassy appearance resembling quartz. The material may be represented by the formula $SiO_2 \cdot nH_2O$. It is produced by reacting sodium silicate with sulfuric acid, coagulating the mixture into a hydrogel, washing to remove sodium sulfate, and drying the hydrogel to produce the commercial adsorbent. The end product is highly porous, with pores estimated to average 4×10^{-7} cm in diameter.

A typical chemical analysis of commercial silica gel is given in **Table 12-2** (W.R. Grace & Co., 1992). The analysis is on a dry basis. The loss on ignition at 1,750°F is reportedly 6.0 percent maximum.

The equilibrium partial pressure of water vapor over silica gel containing various concentrations of adsorbed water is shown in **Figure 12-7**. This figure is based upon the data of Taylor (1945) and Hubard (1954), extrapolated to cover the very low water range by application of the Freundlich relationship:

$$W = KC^n \tag{12-6}$$

Where: W = concentration of water in the silica gel
C = concentration (or partial pressure) of water in the gas phase
K and n = constants

In this figure, the residual water content of silica gel is included in the weight of the desiccant. This water, which normally amounts to about 6% of the activated weight, can be removed by heating to 1,750°F for 30 min., but it is not removed at conventional regeneration temperatures.

Bead type adsorbents have the advantages over granular material of a lower gas pressure drop for the same mesh size range, greater resistance to attrition, and better particle flow characteristics. The latter property permits beads to be pneumatically loaded into dehydration vessels. Two types of spherical silica beads have been developed; one essentially pure silica and the other a blend of silica with a small amount of alumina.

The chemical analysis and physical properties of a high purity silica gel bead (Grace Bead Gel) are given in **Table 12-3**. The extremely low concentration of alumina in these beads results in low catalytic activity. This property is desirable in some applications because it minimizes the formation of coke from the cracking of adsorbed hydrocarbons during regeneration.

Silica/alumina beads were originally developed and marketed in the U.S. by Mobil Oil Corp. as Sovabeads. Later they were manufactured by Kali-Chemie AG of Germany under

(text continued on page 1038)

Table 12-1
Important Physical Properties of Typical Desiccant Materials

Physical properties	Type of desiccant (typical commercial products)					
	1. Silica gel (Davison 03)	2. Silica-base beads (Sorbead R)	3. Activated alumina (Alcoa grade F-1)	4. Alumina-base beads (LaRoche A-201)	5. Activated bauxite	6. Molecular sieve 4A & 5A
True specific gravity	2.1–2.2	—	3.3	3.1–3.3	3.40	—
Bulk density, lb/cu ft (4–8 mesh)	45	49	52–55	48	50–52	40–45
Apparent specific gravity	1.2	—	1.6	—	1.6–2.0	1.1
Average porosity, %	50–65	—	51	65	35	—
Specific heat, Btu/(lb)(°F)	0.22	0.25	0.24	—	0.24	0.2
Thermal conductivity, Btu(in.)/(ft²)(hr)(°F)	1.0	1.37	1.0(100°F); 1.45(200°F)	—	1.09(360°F; 4–8 mesh)	—
Water content (regenerated), %	4.5–7	4–6	6.5	6.0	4–6	varies
Reactivation temperature, °F	250–450	300–450	350–600	350–850+	350+	300–600
Particle shape	granular	spheroidal	granular	spheroidal	granular	cylindrical pellets
Surface area, sq meter/g	720–760	650	210	325	—	—
Static sorption at 60% RH, %	29	33.3	14–16	20	10	22

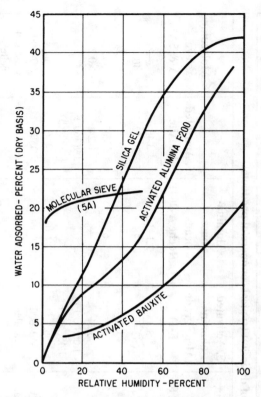

Figure 12-6. Equilibrium water capacity of solid adsorbents versus relative humidity based on air at atmospheric pressure and 77°F. *Silica gel data from W.R. Grace & Co. (1992); activated alumina data from Alcoa (1991); activated bauxite data from Amero et al. (1949); molecular sieve data from UOP (1990)*

Table 12-2 Typical Chemical Analysis of Commercial Silica Gel, Percent	
Silica (SiO$_2$)	99.71
Iron as Fe$_2$O$_3$	0.03
Aluminum as Al$_2$O$_3$	0.10
Titanium as TiO$_2$	0.09
Sodium as Na$_2$O	0.02
Calcium as CaO	0.01
Zirconium as ZrO$_2$	0.01
Trace elements	0.03

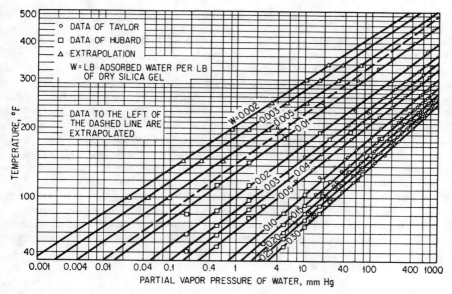

Figure 12-7. Equilibrium partial pressure of water vapor over silica gel containing various amounts of adsorbed water. Residual water, which cannot be removed by conventional regeneration, is included in the weight of the desiccant. *Based on data of Taylor (1945) and Hubard (1954)*

Table 12-3 Properties of Grace Bead Gel	
Chemical Analysis	
SiO_2, %	99.80
Al_2O_3, %	0.029
Fe_2O_3, ppm	<100
pH	4.5
Physical Properties	
Pore Volume, cm^3/g	0.45
Average Pore Diameter, $A°$	22
Surface Area, m^2/g	800
Total Volatile at 1,750°F, %	5.5
Bulk Density, lb/ft^3	48
Specific Heat, Btu/(lb)(°F)	0.22
Thermal Conductivity, Btu(in.)/(ft^2)(h)(°F)	1

Data from W.R. Grace & Co., Davison Chemicals Division (1993)

1038 Gas Purification

(*text continued from page 1034*)

license from Mobil and distributed in the U.S. by Weskem, Inc., under the name of Sorbead (Kali-Chemie, 1979; Weskem, Inc., 1982). Although Weskem is no longer promoting Sorbead, a very similar product, Natrasorb, is offered by Multiform Desiccants. Both Sorbead and Natrasorb are in the form of uniform beads, about 0.14 in. (3.5 mm) in diameter. Two basic types are used in gas dehydration; Sorbead R (Natrasorb T), a very active form used in the bulk of the bed, and Sorbead W (Natrasorb TW), a less active form, used where contact with liquid water may occur. The Natrasorb T and TW desiccants are reported to contain 3% and 10% alumina respectively, the balance being silica (Multiform Desiccants, 1984).

The very active form is susceptible to damage if contacted with liquid water; in fact, regenerated beads shatter if dropped into water. To avoid deterioration of beads by the action of water droplets that may be entrained in the feed gas, a buffer desiccant is recommended for the portion of the bed nearest the feed-gas entry. This may be any solid desiccant that is not susceptible to damage by liquid water, including the less active silica/alumina beads.

The water capacity of Sorbead R is indicated by **Figure 12-8,** which is based on atmospheric pressure, air, and isothermal adsorption. The equilibrium capacity shown on the chart

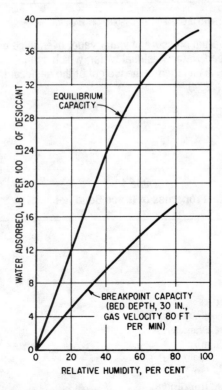

Figure 12-8. Effect of relative humidity on capacity of Sorbead R desiccant. Data obtained with air at atmospheric pressure and under essentially isothermal conditions (77°F). *Data of Mobil Oil Corp. (1971)*

was obtained by passing air at 77°F and various known humidities through the desiccant until no more moisture could be absorbed and the air emerged at the same humidity at which it entered. The break-point capacity was obtained by passing the air through the bed until the humidity of the air leaving the bed started to rise sharply. In either case, the capacity value was then determined by noting the increase in the weight of the bed due to water adsorption and relating this to the original weight of the dried material. The gas velocity and bed depth must be specified when reporting break-point capacity values because these are a function of operating conditions. The effects of adiabatic versus isothermal operation and other variables on break-point capacity are discussed in a later section.

Activated Alumina

The first commercially available activated alumina was a granular material designated F-1 by the manufacturer, the Aluminum Company of America (Alcoa). This material was introduced in the 1930s and was an important solid desiccant for several decades. A spherical form of activated alumina, H-151 Alumina Gel, was later introduced by Alcoa and proved quite successful. Production of H-151 by Alcoa ceased in 1985, and it was replaced by newer spherical products such as H-152 and F-200. The granular F-1 is still available, but is not considered to be an important solid desiccant for gas dehydration (Woosley, 1990). Both granular and spherical activated alumina desiccants have been introduced by other manufacturers and several of these are listed in **Table 12-4.**

Goodboy and Fleming (1984) point out that the term "alumina" is a misused expression for aluminum oxides and hydroxides that exist in at least five thermodynamically stable phases and many more metastable transition forms. The commercial activated alumina desiccants are generally manufactured by a process involving the following steps (Woosley, 1990):

1. Aluminum trihydroxide, $Al(OH)_3$, (gibbsite) is thermally decomposed at a temperature in the range of 375° to 450°C to release water vapor and form a partially dehydrated Al_2O_3. Complete dehydroxylation is not allowed to occur. The product is a powder still capable of losing 4 to 7% on ignition at high temperature.
2. The activated powder is bonded using water as the sole or primary binding medium to form spheres or granules of the desired size and shape.
3. The formed particles are given a second thermal treatment to solidify the bond and reactivate the adsorbent, producing a final adsorbent with a loss on ignition of 4 to 8%.

The chemical analysis and physical properties of a typical activated alumina desiccant are given in **Table 12-5.** The data are for a spherical product, A-201, produced by LaRoche Chemicals (LaRoche is the successor company to Kaiser Chemicals, which produced the activated alumina products prior to mid-1988).

The water adsorptive properties of several commercially available alumina desiccants are given in **Table 12-6.** As indicated by the data in this table, and also by **Figure 12-6,** high quality activated aluminas typically have the capacity to adsorb 35–40 pounds of water per 100 pounds of regenerated alumina when equilibrated with 90% relative humidity (RH) air.

Activated Bauxite

Activated bauxite used for gas dehydration usually appears as hard, brownish-red granules. The material is made from naturally occurring bauxite by heating it under controlled

Table 12-4
Properties of Commercial Activated Alumina Desiccants

Type	Supplier	Form	Size, mesh, in., or mm	Bulk Density, lb/cu ft	Crush Strength lbf	Equilib. Capacity @ 60% RH, wt %
A-2	LaRoche	gran.	8 × 14	42	—	18
			12 × 32			
A-201	LaRoche	spher.	½" × ¼"	48	—	20
			3 × 6	—	—	—
			5 × 8	—	35	—
			7 × 12	—	—	—
A-204	LaRoche	spher./	½" × ¼"	50	—	18
		gran.	3 × 6	—	—	—
			5 × 8	—	20	—
			7 × 12	—	—	—
			7 × 14	—	—	—
			14 × 28	—	—	—
A	Rhone-Poulenc	spher.	2 × 5 mm	48	29	22
			5 × 10 mm	—	55	20
F-200	Alcoa	spher.	⅛"	48	30	19
			3/16"	—	55	—
			¼"	—	70	—
H-152	Alcoa	spher.	—same as F-200—			22
DD420	Discovery Chemical	spher.	¼" × 5 mesh	52	40	20
			5 × 8	—	—	—
			8 × 14	—	—	—
DD430	Discovery Chemical	spher.	¼" × 5 mesh	56	45	22
			5 × 8	—	—	—
			8 × 14	—	—	—

LaRoche product data from LaRoche Chemicals (1988), other data from Woosley (1990)

conditions to vaporize water from the hydrated alumina. The major component of the bauxite ore is usually alumina trihydrate. After activation, a typical activated bauxite (Florite) has the composition given in **Table 12-7**.

The principal advantage of activated bauxite is its low cost as compared to that of synthetic desiccants. It has the additional advantage of resisting breakup in the presence of liquid water and, in spite of its low cost, it is capable of providing extremely low dew points. Its principal disadvantage is a somewhat smaller capacity for water than that of the other desiccants. The total capacity of Florite activated bauxite as a function of relative humidity is shown in **Figure 12-6**, which is based on air at atmospheric pressure. The break-point capacity for air, at room temperature and a 55°F dew point, flowing at a rate of 0.141 cu ft/min through 4- to 8-mesh Florite in a column 1.3 in. in diameter and 36 in. deep, is indicated to be about 6½% by weight according to the data of Amero et al. (1949). These authors recom-

Table 12-5
Chemical Analysis and Properties of LaRoche A-201 Activated Alumina

Chemical Analysis	wt %
SiO_2	0.02
Fe_2O_3	0.02
TiO_2	0.002
Na_2O	0.35
Loss on Ignition	6.0
Al_2O_3	93.6
Physical Properties	
Surface Area, m^2/g	325
Pore Volume, cm^3/g	0.50
Static Sorption (at 60% RH)	20
Bulk Density, packed, lb/ft^3	48
Crushing Strength, lb force	35

Notes:
1. Static sorption determined on 5×8 mesh spheres.
2. Crushing strength is for 5 mesh (4 mm) spheres.
Data from LaRoche Chemicals (1990)

Table 12-6
Water Adsorption Properties of Activated Alumina Desiccants

Type	Supplier	Loss on Ignition wt %	Equilibrium Capacity at		
			10% RH, wt %	60% RH, wt %	90% RH, wt %
A (2×5 mm)	Rhone-Poulenc	3–5	8	22	35
F-200	Alcoa	6.5	7	19	40
H-200	Alcoa	4.0	12	22	40

Notes: 1. All percentage values are based on the weight of the activated material.
2. Loss on ignition values are based on loss in weight between 250°C and 1,100°C.
Data from Woosley (1990)

Table 12-7
Composition of Activated Bauxite

Component	Wt %
Al_2O_3	70–75
Fe_2O_3	3–4
SiO_2	11–12
TiO_2	3–4
Volatile (water)	4–6

mend the use of 5% capacity for gases at temperatures less than 100°F, having a relatively high humidity, and containing no constituents that will cause undue fouling.

Molecular Sieves

Although naturally occurring molecular-sieve adsorbents have been known for many years, this type of material did not become of commercial importance until the introduction of synthetic molecular sieves by Union Carbide Corporation's Linde Division in 1954. The molecular sieves differ from conventional adsorbents primarily in their ability to adsorb small molecules while excluding large ones, so that separations can be made based on molecular size differences. They have the additional property of relatively high adsorption capacity at low concentrations of the material being adsorbed and have an unusually high affinity for unsaturated and polar-type compounds.

In August 1988, the Catalysts, Adsorbents, and Process Systems (CAPS) business of Union Carbide was merged into a new venture operating under the name UOP. Molecular sieves offered by UOP are basically the same materials as those formerly provided by Union Carbide. Several other companies now offer molecular sieves, including Zeochem and the Davison Chemicals Division of W.R. Grace & Company.

The commercial molecular sieves generally belong to the zeolite class of minerals, i.e., hydrated alkali metal or alkaline earth aluminosilicates, which are activated by heat to drive off the water of crystallization. The crystals have a robust cubic structure, which does not collapse on heating, so that activation results in a geometric network of cavities connected by pores. The pores are of molecular dimensions and cause the sieving action of these materials.

About 40 different zeolite structures have been discovered; however, only three synthetic types are widely used in commercial applications: Zeolite A, Faujasite, and Pentasil. The pores of Zeolite A are restricted by 8-membered oxygen rings. The free aperture for this structure is about 3.0 A° for the K^+ form (3A), 3.8 A° for the Na^+ form (4A), and 4.3 A° for the Ca^{++} form (5A). Faujasite is represented by two forms, X and Y, with pores restricted by 12-membered oxygen rings. The pores of these materials are relatively large with a free aperture of about 8.1 A°. The X and Y zeolites differ from each other only with regard to the Si/Al ratio which controls cation density, and therefore, affects adsorptive properties. The Pentasil-type zeolite has pores restricted by a 10-membered oxygen ring, which gives a free aperture of about 6.0 A°. This zeolite was first proposed by Mobil in the early 1970s and named ZSM-5. At about the same time, a pure silica analog with the same framework structure was prepared by Union Carbide and named Silicalite (Ruthven, 1988).

Typical properties of the four most widely used molecular sieves are given in **Table 12-8.** All of the sieves are excellent desiccants for dehydrating gas; however, as indicated in the table, their special properties make certain types preferable for specific applications. In addition to the primary types listed, special sieves are available for dehydration applications requiring resistance to acids in the gas being treated and for applications where COS formation may be catalyzed by sieve components.

Molecular sieves are generally available as cylindrical pellets, spherical beads, and powder. Pellets and beads are most commonly used for dehydration and gas purification applications. The pellets are formed by extrusion and usually have a fixed diameter of $\frac{1}{16}$ or $\frac{1}{8}$ in., and a variable length equal to 1 to 4 times the diameter. The bead size is characterized by screen cut, which identifies the Tyler screen size through which all of the beads pass and the size that retains all of the beads, in that order, with the two sizes separated by an "x." The two most commonly used screen cuts are 4×8 and 8×12.

Table 12-8
Basic Types of Commercial Molecular Sieves

Basic type	Nominal pore diameter, Angstroms	Bulk density of pellets, lb/cu ft (1)	H_2O capacity, (%/wt) (2)	Molecules adsorbed (typical) (3)	Molecules excluded	Typical applications
3A	3	47	20	H_2O, NH_3	Ethane and larger	Dehydration of unsaturated hydrocarbons
4A	4	45	22	H_2S, CO_2, SO_2, C_2H_4, C_2H_6, C_3H_6	Propane and larger	Static desiccant in refrigeration systems, etc. Drying saturated hydrocarbons
5A	5	43	21.5	$n\text{-}C_4H_9OH$,	Iso compounds, 4 carbons rings and larger	Separates n-paraffins from branched and cyclic hydrocarbons
13X	10	38	28.5	Di-n-propyl-amine	$(C_4F_9)_3N$ and larger	Coadsorption of H_2O, H_2S, and CO_2

Notes:
1) Bulk density for 1/8" pellets.
2) Pounds $H_2O/100$ lb activated adsorbent at 17.5 mm Hg partial pressure and 25°C, adsorbent in pellet form.
3) Each type adsorbs listed compounds plus those of all preceding types.
Data from UOP (1990)

A relatively new shape for molecular sieve pellets is the TRISIV adsorbent offered by UOP. This is an extrusion with a trilobal (cloverleaf) cross-section. The diameter of each lobe is typically 1/16 in.; however, since the lobes are joined to give a much larger effective diameter for the extrusion, the pellet length is about 3 to 4 times the lobe diameter. The impact of particle shape on performance was analyzed by Ausikaitis (1983). He compared TRISIV with beads and pellets for the same dehydration service and concluded that, for the specific conditions and assumptions of the study, the TRISIV design offered a small but significant economic advantage.

A major application of molecular sieves is for gas drying ahead of cryogenic processing, where extremely low dew points are required. Molecular sieve units are widely used prior to the cryogenic extraction of helium from natural gas, cryogenic air separation, liquefied natural-gas production, and deep ethane recovery from natural gas using the cryogenic turboexpander process. Although the molecular sieves are somewhat more expensive than other adsorbers, they offer the following advantages:

1. They provide good capacity with gases of low relative humidity.
2. They are applicable to gases at elevated temperatures.
3. They can be used to adsorb water selectively.
4. They can be used to remove other selected impurities together with water.
5. They can be used for adiabatic drying.
6. They provide extremely low dew points.
7. They are not damaged by liquid water.

The equilibrium capacity of molecular sieve, type 5A, for water vapor at 25°C is presented in **Figure 12-6** with curves for other adsorbents, which are included for purposes of comparison. The effect of temperature on the equilibrium adsorption-capacity of molecular sieve, type 5A, activated alumina, and silica gel is shown in **Figure 12-9** for a water-vapor partial pressure of 10 mm. These data show that at 200°F, for example, the molecular-sieve capacity is 15 percent by weight, while the other adsorbents have an almost negligible capacity at this temperature. The extremely high drying efficiency of molecular sieves makes them useful for "trimmer" beds in conjunction with silica-gel or alumina desiccants. In this type of installation, a small bed of molecular sieves following the conventional adsorbent removes the last traces of moisture from the gas stream, permitting the primary bed to be loaded to a higher capacity and producing a lower product-gas dew point.

Mechanism of Water Adsorption

When a gas containing water is passed through a bed of freshly regenerated adsorbent, the water is adsorbed first near the inlet portion of the bed, and the dehydrated gas passes through the rest of the bed with only a small amount of additional drying taking place. As the adsorbent nearest the inlet becomes saturated with water at the condition of the feed gas, the zone of rapid water adsorption moves inward and ultimately progresses through the entire bed. When this "adsorptive wave" reaches the outlet end, the water content of the product gas is observed to rise rapidly, signifying the "break point" for the particular operating conditions.

The adsorption of water results in the evolution of heat in the active adsorption zone. With high-pressure gas (above about 500 psig), the large weight of gas associated with each pound of water picks up the heat released with only a small temperature rise, on the order of 2° to 4°F. With low-pressure gas or air, on the other hand, there are fewer pounds of gas per pound of water vapor so that a much greater temperature rise is possible. The heat is actually

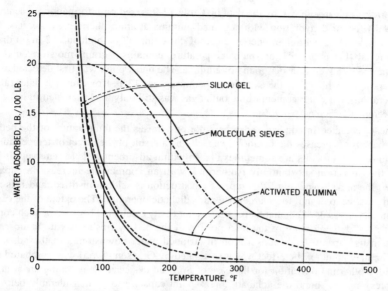

Figure 12-9. Comparison of the effects of temperature on the capacity of molecular sieve, type 5-A, silica gel, and activated alumina in equilibrium with water vapor at 10-mmHg partial pressure. The dotted lines show the effect of 2% residual water at the start of adsorption. *Data of UOP (1990)*

Table 12-9
Temperature Changes Within an Adiabatic Dehydration Bed

Distance from air inlet, in.	Temperature, °F, for operation times given					
	0 hr	0.5 hr	2.5 hr	4.5 hr	7.0 hr	11 hr
0 (inlet)	75	75	75	75	75	75
3	75	165	105	95	90	85
12	75	95	245	140	115	100
21	75	92	145	235	160	120
30 (outlet)	75	90	93	185	213	170

Source: Data of Derr for air dehydration with activated alumina (1938)

liberated inside the adsorbent particles, as a result of water condensation and adsorbent wetting. If no cooling coils are provided in the bed, the heat is transferred to the gas stream in the active adsorption zone. On leaving this zone, however, the hot gas encounters cool (and dry) adsorbent, and heat transfer occurs in the reverse direction, warming the bed downstream to the adsorption zone and recooling the gas stream, which emerges from the bed only slightly warmer than it was upon entering. The high-temperature wave thus progresses through the bed somewhat in advance of the adsorptive wave, so that the exit-gas temperature starts to rise well before the dehydration break point occurs.

This effect is clearly shown by the data in **Table 12-9,** based on information presented by Derr (1938) for air dehydration with activated alumina at atmospheric pressure. The table presents temperatures within an uncooled bed of desiccant, 32 in. high and 12 in. in diameter, drying air that is at 75°F dry-bulb temperature, contains 9 grains moisture/cu ft, and flows at a rate of 5.2 cu ft/hr per pound of alumina. The break point was first detected after 7 hr of operation at which time the exit-air temperature was 213°F. However, 1 hr later (after 8 hr of operation), the water content of the outlet gas was still only 0.34 grains/cu ft, indicating a water-removal efficiency of 96%.

The heat generated in adiabatic adsorbers not only raises the temperature of the bed and the gas, but also decreases the operating capacity as a result of the effect of temperature on equilibrium. Cooling coils are sometimes placed within adsorbent beds to remove this heat, making the operation substantially isothermal, with an appreciable increase in capacity. However, the added expense of this type of construction is seldom justified, and it is more common practice to design for a larger bed and adiabatic operation. The extent of the capacity reduction as a result of adiabatic operation is not easily calculated because of such complicating factors as the cooling of the inlet portion of the bed by fresh gas with a subsequent increase in its capacity and readsorption of stripped water downstream of the adsorptive wave. The magnitude of the effect was investigated by Grayson (1955) who compared adiabatic with isothermal operation for the case of Sorbead desiccant dehydrating air at atmospheric pressure. He found the adiabatic break-point capacity to be considerably below the isothermal break-point capacity and, under some conditions, to decrease with increased water content of the feed gas. This effect is illustrated in **Table 12-10** by examples of Grayson's data, which were obtained with air, at atmospheric pressure and 80°F dry-bulb temperature, passed through a bed of Sorbead 36 in. deep.

The effect of gas velocity on break-point capacity is also indicated by the data of **Table 12-10.** As can be seen, increasing the gas velocity from 30 to 60 ft/min. decreases both the adiabatic and isothermal break-point capacities with a given quantity of adsorbent. Grayson also investigated the effects of humidity and bed depth. As indicated in the table, increased humidity generally (although not always) decreased the capacity at break point. Increasing the bed depth was found to increase the break-point capacity (on a unit weight basis) appreciably, as a result of the cooling effect of the inlet gas on the first portion of the bed contacted.

Equations describing nonisothermal adsorption in large fixed beds have been developed by Leavitt (1962). His analysis indicates that for the case of a feed gas containing one adsorbable and one nonadsorbable component in a large diameter bed, two separate transfer zones tend to form. The equations permit calculation of concentrations, loadings, and temperatures in both zones and in the interzone region as well as the speeds and lengths of the zones as they move through the adsorber.

As pointed out above, an increase in gas velocity decreases the break-point capacity for a given bed of desiccant. However, if the bed is made deeper to compensate for the increased gas velocity and provide the same cycle time as with lower gas rate, the average capacity of the bed at the break point is increased. This occurs because the mass transfer zone (MTZ) is a smaller fraction of the deeper bed.

As in other mass-transfer operations, the rate of transfer of water vapor from the gaseous phase is a function of the gas flow rate, the size and shape of the desiccant particles, and the properties of the gaseous and adsorbate phases. If the mass-transfer coefficient is very high, the mass transfer zone will be quite short, i.e., complete dehydration will be obtained until the break point is reached; at this time the water content of the product gas will rise very rapidly. If the mass-transfer coefficient is quite low, on the other hand (or if the bed is very shallow),

Table 12-10
Adiabatic and Isothermal Break-point Capacities for Sorbead Desiccant Dehydrating Air at Atmospheric Pressure

Air Velocity, ft/min.	30		60	
Air Wet-bulb temp.,* °F	35	60	35	60
Break-point Capacity Adiabatic, wt %	7.4	5.6	5.8	4.2
Isothermal, wt %	11.0	20.0	8.0	16.8

*Inlet air.
Source: Grayson, 1955

some water vapor may pass through with the gas from the start, and the water content of the exit gas will increase slowly as the entire bed becomes saturated. Most commercial installations fall between these two extremes in that there is first a period of maximum dehydration, then after a definite break point, the water content of the product gas is observed to rise at a moderate rate. Break-point capacity curves of this type for Activated Alumina H-151 in high-pressure natural-gas service are shown in **Figure 12-10**. The data illustrated were obtained in a twin-tower dehydration unit, each tower 16 ft high and 3 ft in diameter. Natural gas, at approximately 850 psig, was dehydrated using downflow during dehydration and upflow (with 360°F gas) during regeneration. One tower required 3,900 lb H-151 Activated Alumina Gel Balls, ¼ in. to 6 mesh, and handled 16 to 19 MMscf/day of gas.

In the dehydration of air at atmospheric pressure for air conditioning, the pressure drop through the desiccant must be kept to an absolute minimum, and very shallow beds are

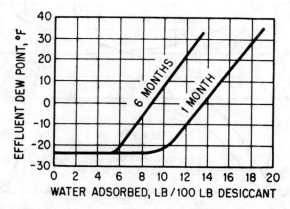

Figure 12-10. Capacity test curves for Activated Alumina H-151 after 1 month and 6 months of service, dehydrating natural gas at approximately 850 psig and 80°F; gas flow rate, 16 to 18 million scf/day; tower size, 16 ft high, 3 ft diameter. *Data of Getty et al. (1953)*

1048 Gas Purification

employed. Because of this, the initial break point occurs very quickly, and the dew point of the product gas climbs steadily during most of the adsorption cycle. This type of operation has been analyzed quite thoroughly by Ross and McLaughlin (1951) for silica-gel adsorption. The results of typical adsorption and desorption tests by these authors are shown in **Figure 12-11**.

Dehydration System Design Approach

The normal approach for the design of fixed bed dehydration systems involves the following steps:

1. Select the adsorbent material based on the required product gas water concentration, the composition of the feed gas, projected capital and operating costs, and other factors.

	NOMINAL	ADSORPTION	DESORPTION
BED THICKNESS (IN.)	4		
ROOM DB (°F)	70	70.0	70.1
ROOM RH (%)	35	34.8	34.6
AIR FLOW (CFM)	110-112.5	110.6	113.2
FACE VEL (FPM)	55-56.25	55.3	56.6
WATTS		4500	4531
CFM/WATT		0.025	0.025

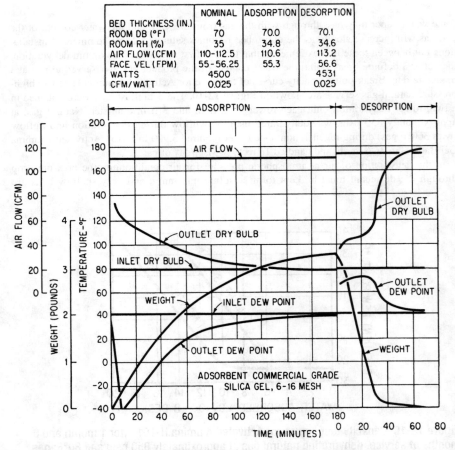

Figure 12-11. Characteristic operation curves for a solid adsorbent dehumidifier. *Data from Ross and McLaughlin (1951)*

2. Determine the allowable gas superficial velocity based on a specified pressure drop limitation, e.g., 7 psi for high pressure operation.
3. Calculate the required bed cross-sectional area and/or the number of beds of a given diameter based on the superficial velocity selected in step 2.
4. Select a cycle time (typically 8 to 12 hours) and calculate the quantity of adsorbent required based on an estimated end-of-life equilibrium capacity.
5. Determine the total bed height based on the volume of adsorbent calculated in step 4 and an estimated length of the mass transfer zone (MTZ).
6. Calculate the total heat required during the regeneration step to heat and desorb the adsorbate, heat the adsorbent, heat the vessel and internals, heat the exit gas, and provide for heat losses through the vessel walls.
7. Establish a heating/cooling time schedule based on the cycle time selected in step 4, and calculate the regeneration gas flow rate required to provide the heat needed in the time available.

Detailed design procedures, which generally follow these steps are described by Polley (1981), Campbell (1974) and the *GPSA Engineering Data Book* (1987). The GPSA approach is highly simplified, but widely used for preliminary design studies of natural gas dehydration plants. Data and correlations used in the GPSA procedure are included in the following discussions where appropriate.

Desiccant Selection

A chart developed by Coastal Chemical Company, Inc., to aid in the selection of solid desiccants is reproduced in **Figure 12-12** (Veldman, 1991). The chart considers the four widely used desiccants: activated alumina (F-200), silica gel beads (Sorbead R and Sorbead W), alumina gel beads (H-152), and molecular sieves (3A or 4A).

Molecular sieves are the most expensive, but provide the lowest dew point. They are normally used to dehydrate natural gas feed streams for cryogenic hydrocarbon recovery units. Type 4A is the type most commonly employed, but 3A is gaining favor because it is less active catalytically and can be regenerated at a lower temperature than 4A. A disadvantage of molecular sieves for gas dehydration is that they can be fouled by impurities in the gas, such as amine, caustic, chlorides, glycol, and liquid hydrocarbons. However, fouling problems can be minimized by the installation of a buffer layer of one foot or more of activated alumina (for amine, caustic, and chlorides) or activated carbon (for glycol or hydrocarbon liquids) on top of the molecular sieve bed (*Veldman, 1991*).

Activated alumina is the least expensive of the manufactured desiccants and has a high water capacity at high feed-gas saturation levels. It has the further advantage of stability in the presence of alkaline impurities. Its two main disadvantages are the co-adsorption of hydrocarbons, which reduces its capacity for water and can lead to the loss of valuable hydrocarbon components to the fuel gas, and rehydration, which destroys its activity. Rehydration is caused by contact with high levels of water vapor at elevated temperatures. Rehydration can be alleviated by using dry gas for regeneration; however, this results in the requirement for larger vessels because the unit being used for dehydration must process gas for both product and regeneration streams.

The main advantages of silica gel are that it has a high capacity for water, can be regenerated at a low temperature, and is not catalytic for sulfur conversion reactions. Silica gel (including silica-base beads) also has a high capacity for pentane and higher hydrocarbons

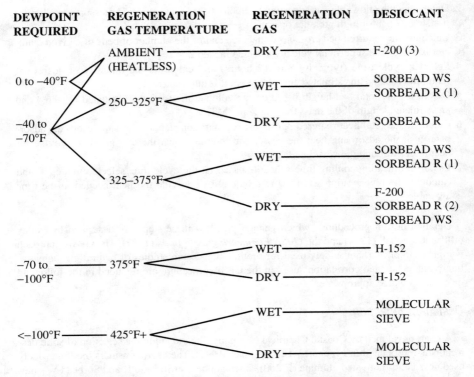

Figure 12-12. Coastal Chemical Company, Inc. dehydration desiccant selection chart. *Data of Veldman (1991)*

and can be used for combined dehydration/hydrocarbon recovery processes. A problem with silica gel is its tendency to shatter when contacted with liquid water. This problem can be prevented by providing a layer of water-resistant desiccant, such as Sorbead W, on top of the bed. Silica gel deteriorates in the presence of alkaline materials such as amines, but is quite resistant to acidic compounds such as H_2S and SO_2.

As indicated by the data in **Figure 12-12,** desiccants that provide lower product gas dew points tend to require higher regeneration temperatures. They also generally have higher heats of adsorption. The average heats of water adsorption for the three basic types of solid desiccants are silica gel—1,400, activated alumina—1,600, and molecular sieves—1,800 Btu/lb H_2O (Veldman, 1991). As a result of both the temperature of regeneration and the heat of adsorption, the energy consumption of solid-desiccant dehydration systems generally increases with increased dew-point depression requirements.

When both H_2S and CO_2 are present in the gas, the adsorbent may catalyze the following reaction to form COS:

$$H_2S + CO_2 = COS + H_2O \tag{12-7}$$

This phenomenon has been identified as the simultaneous adsorption of H_2S and rate-limited catalytic reaction with CO_2 (Turnock and Gustafson, 1972). An investigation of the problem as it relates to the removal of low concentrations of H_2S (less than 160 ppmv) from natural gas has been reported by Cines et al. (1976). They found sieves of the 5A type to be least catalytic, the type 4A to be of intermediate activity, and the 13A sieve to be the most catalytic. The COS formed is not strongly adsorbed and therefore appears in the treated gas long before H_2S appears.

The problem of COS formation can also be encountered in molecular sieve beds performing primarily as dehydrators. McAllister and Westerveld (1978) report that a large cryogenic plant in the Middle East produced a raw gas liquid containing 800 ppm COS as a result of the reaction occurring in the molecular sieve dehydrator. They recommend removal of acid gas prior to dehydration to avoid the problem.

The COS formation reaction can cause two problems in natural gas dehydration plants: (1) the formation of H_2O by the reaction affects the degree of dehydration attainable, and (2) the release of COS affects the product gas purity. When liquefied hydrocarbons are recovered from the dehydrated natural gas, the COS concentrates in the propane fraction requiring the installation of a propane treater.

The problem of COS formation has been greatly reduced by the development of molecular sieves that are poor catalysts for reaction 12-7. Davison Chemical's SZ-5 sieve, for example, has been shown to produce about one-fourth as much COS as the standard 5A sieve under the same operating conditions. Zeochem's Z3-04 is a special grade of 3A sieve recommended for cases where COS formation may be a problem; but a 3A sieve is otherwise applicable. The practical application of Z3-04 sieve is described by Trent et al. (1993). In one plant the dehydrator feed gas contained 0.70 ppm H_2S and 0.0 ppm COS and entered the system at 80°F and 750 psig. With the original charge of 4A adsorbent, the product gas contained 0.40 ppm H_2S and 0.25 ppm COS, indicating an appreciable conversion of H_2S to COS. After changing to type Z3-04 sieve, the exit gas was found to contain 0.38 ppm H_2S and 0.0 ppm COS, indicating that the conversion was very low. Several applications of Union Carbide's COS-minimizing molecular sieve are described by Markovs (1990).

The formation of COS is increased by high levels of H_2S and CO_2 in the gas, low concentration of water vapor, high temperature, extended time at reaction conditions, and alkaline conditions. The Davison SZ-5 material is specially formulated to substitute calcium for sodium to reduce sodium alkalinity compared to the standard 5A sieve. According to Alexander (1988), the COS formation reaction is catalyzed by both sodium and aluminum compounds. As a result, he concludes that silica-based desiccants are superior in this regard to either alumina or molecular sieves.

Pressure Drop

In the design of solid-desiccant dehydration plants, it is important that the pressure drop through the bed be estimated as accurately as possible, because the work required to overcome this pressure drop can represent a major operating cost.

Generalized correlations for estimating pressure drop of gas flowing through a bed of granular particles have been proposed by Rose (1945), Brownell and Katz (1947), Leva (1947), and Ergun (1952). A modified form of the Ergun equation is suggested by UOP for use with molecular sieve beds (1991B). This equation, which appears to be quite suitable for general use with all fixed-bed adsorbents, is

$$\Delta P/L = (f_t C_t G^2 / \rho D_p) 10^{-10} \tag{12-8}$$

Where: C_t = pressure-drop coefficient $(ft)(hr^2)(in.^2)$
D_P = effective particle diameter, ft
f_t = friction factor
G = superficial mass velocity, $lb/(hr)(ft^2)$
L = bed depth, ft
ΔP = pressure drop, psi
ρ = fluid density, lb/ft^3
$\Delta P/L$ = the pressure drop per unit length of bed, psi/ft

The value of D_P for cylindrical pellets is given by equation 12-9.

$$D_P = D_c/[\tfrac{2}{3} + \tfrac{1}{3}(D_c/L_c)] \tag{12-9}$$

Where: D_c = particle diameter, ft
L_c = particle length, ft

The friction factor, f_t, and the pressure-drop coefficient, C_t, are determined from **Figure 12-13.** The friction factor is plotted against a modified Reynolds number in which

μ = fluid viscosity, $lb/(hr)(ft)$

The pressure-drop coefficient, which takes into account the packing density and also includes required conversion factors and constants, is plotted against the external void fraction of the bed. Suggested values for D_P, the effective particle diameter, and ε, the external void fraction, are given in **Table 12-11** for a number of desiccants based on the data of Fair (1969) and manufacturers' bulletins.

Figures 12-14 and **12-15** present quick pressure-drop estimation techniques for the two conditions most frequently encountered, i.e., high-pressure natural gas and air at atmospheric pressure. **Figure 12-15,** which is based on the data of Allen (1944), has been extended to include two grades of silica gel. Available data for pressure drop through beds of Sorbead R and LaRoche A-201 Active Alumina (¼ in. × 8 mesh) indicate that both of these ball-shaped desiccants would be represented quite closely by the 4–8 mesh curve of **Figure 12-15.**

A simpler modification of the Ergun equation is suggested by the *GPSA Engineering Data Book* (1987) (and also by W.R. Grace & Co., 1992) for molecular sieve adsorbents:

$$\Delta P/L = B\mu V + CdV^2 \tag{12-10}$$

Where: ΔP = pressure drop across the bed, psi
L = bed depth, ft

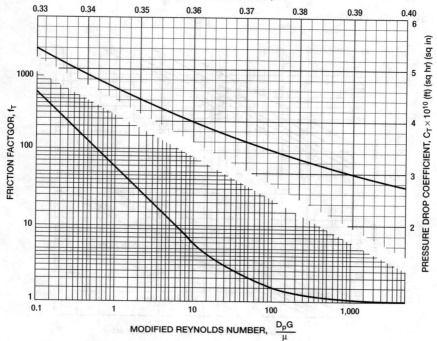

Figure 12-13. Correlation for estimating friction factor and pressure drop coefficient for use in calculation of pressure drop of gas flowing through beds of granular adsorbents. *Data of UOP (1991B)*

Table 12-11
Desiccant Properties for Pressure Drop Calculation

Desiccant	Particle form	Mesh size	Bulk density, lb/ft³	Effective diameter D_P, ft	External void fraction, ε
Silica Gel	Granules	3 × 8	45	0.0127	0.35
"	"	6 × 16	45	0.0062	0.35
"	Spheres	4 × 8	50	0.0130	0.36
Alumina	Granules	4 × 8	52	0.0130	0.25
"	"	8 × 14	52	0.0058	0.25
"	"	14 × 28	54	0.0027	0.25
"	Spheres	¼ in.	52	0.0208	0.30
"	"	⅛ in.	54	0.0104	0.30
Molecular Sieves	Granules	14 × 28	30	0.0027	0.25
"	Pellets	⅛ in.	45	0.0122	0.37
"	"	1/16 in.	45	0.0061	0.37
"	Spheres	4 × 8	45	0.0109	0.37
"	"	8 × 12	45	0.0067	0.37

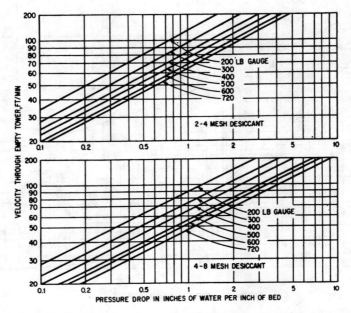

Figure 12-14. Pressure drop of gas flowing through adsorbent beds at elevated pressures. Specific gravity of gas, 0.677. *Data of Cappell (1944)*

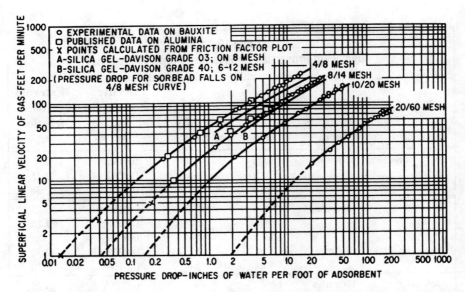

Figure 12-15. Pressure drop for air at atmospheric pressure and ambient temperature flowing through beds of granular adsorbents. *Silica gel data from W.R. Grace & Co. (1992); other data from Allen (1944)*

μ = gas viscosity, centipoise
V = superficial velocity, ft/min.
d = gas density, lb/cu ft

B and C are constants that take into account the intrinsic properties of the bed and have the following values for the particles listed:

Particle type	B	C
⅛ in. bead	0.0560	0.0000869
⅛ in. extrudate	0.0722	0.000124
1/16 in. bead	0.152	0.000136
1/16 in. extrudate	0.238	0.000210

A similar equation is suggested by LaRoche Chemicals (1990) for use with their spherical A-201 alumina. The following values of the constants are recommended:

A-201 bead size	B	C
½" × ¼"	0.00275	0.0000375
3–6 mesh	0.00933	0.0000689
5–8 mesh	0.0227	0.000108
7–12 mesh	0.0632	0.000180

Pressure drops calculated by these equations and charts are for the desiccant beds only and do not include vessel entrance and exit losses.

For molecular sieve dehydrators, the *GPSA Engineering Data Book* (1987) suggests setting P/L equal to 0.333 psi/ft and using a design pressure drop through the bed of 5 to 8 psi. Damron (1977) suggests a maximum pressure drop of 9 to 10 psi to prevent bed crushing and sieve dusting in dehydration plants that are designed to dry high-pressure gas prior to cryogenic processing.

Since some bed-settling and particle-breakage may occur, it is not unusual for the pressure drop through desiccant beds to increase appreciably with time. The pressure-drop record of a silica-gel dehydration unit is shown in **Figure 12-16** (Herrmann, 1955). The data for this figure were obtained in a twin-tower unit, each bed 15 ft high and 38 in. in diameter. The unit handled 15 MMscfd of natural gas at about 1,000 psig and 95°F inlet temperature. Data presented by Harrell (1957) also show increased pressure drop versus time for both silica-gel beads and alumina pellets. His data are summarized in **Table 12-12** for the case of 3.0 MMscfh of natural gas at 570 psig and 72°F. The dehydrator beds are 6 ft 6 in. in diameter and 20 ft deep.

Bed Capacity

Although the equilibrium capacity is of interest in comparing desiccants, the design capacity is always lower for the following reasons:

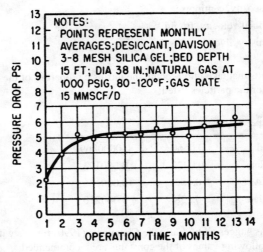

Figure 12-16. Effect of time in service on pressure drop through silica gel desiccant handling high-pressure natural gas. *Data of Herrmann (1955)*

Table 12-12
Pressure-drop Changes in Desiccant Beds with Length of Service

Length of service, months	Pressure drop, psi	
	Silica-gel beads	Alumina pellets
8	8.5	1.8
18	9.0	2.6
35	10.5	2.9

1. As shown in **Figure 12-10,** which illustrates a typical drying period, water vapor appears in the product gas well before the average moisture content of the bed reaches the value which it would have if equilibrium were attained with the inlet gas.
2. Since the break point represents the limit of capacity with maximum dehydration, it is common design practice to provide a factor of safety and end the dehydration period before the break occurs. (Where a higher dew point is allowable, units are sometimes designed to exceed the break-point capacity.)
3. An appreciable decrease in capacity normally occurs with extended operation. It is, therefore, necessary to design for the minimum capacity expected before desiccant replacement is required.

For molecular sieve dehydrators, the *GPSA Engineering Data Book* (1987) suggests a base design capacity of 13 pounds of water per 100 pounds of sieve based on equilibrium with saturated gas at 75°F. This value can be corrected for gas that is not completely saturated with water and for different temperatures by multiplying 13 by correction factors obtained from **Figures 12-17** and **12-18**. The corrected design capacity is then used to calculate the pounds of molecular sieve required in the equilibrium zone, and this is converted to volume on the basis of a bulk density of 42 to 45 lb/cu ft for spherical particles and 40 to 44 lb/cu ft for extruded cylinders.

Figure 12-19 illustrates the decrease in adsorbent capacity with time in service for several adsorbents. All of the data in this figure are for units dehydrating natural gas; however, the operating conditions and gas analyses vary considerably so that curves for different adsorbents are not strictly comparable. The curves are typical, however, and indicate the basis for establishing design capacities. Recommended design capacities for adsorbents in high-pressure natural-gas service, assuming the gas to be clean and essentially saturated with water,

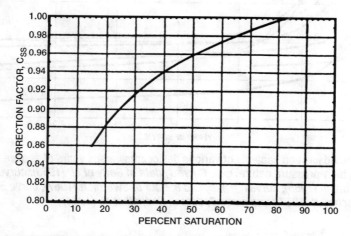

Figure 12-17. Molecular sieve capacity correction factor for unsaturated inlet gas. *From GPSA Engineering Data Book (1987). Reproduced with permission, copyright 1987, Gas Processors Supplier's Association*

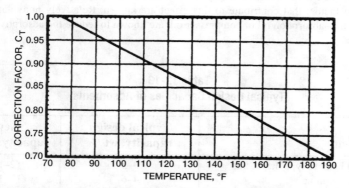

Figure 12-18. Molecular sieve capacity correction factor for adsorption at temperatures above 75°F. *From GPSA Engineering Data Book (1987). Reproduced with permission, copyright 1987, Gas Processors Supplier's Association*

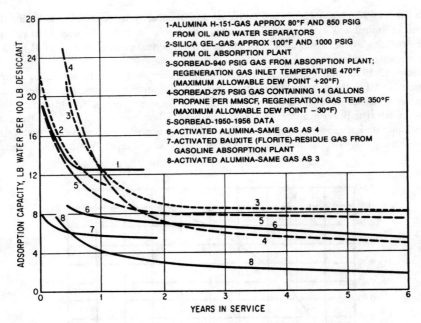

Figure 12-19. Adsorption capacity of various desiccants versus years of service dehydrating high-pressure natural gas. *Curve 1, data of Getty et al. (1953); curve 2, data of Herrmann (1955); curves 3, 4, 6, and 8, data of Swerdloff (1967); curve 5, data of Hammerschmidt (1957); and curve 7, data of Cappell et al. (1944)*

are presented in **Table 12-13**. For comparison, the drying capacities of the adsorbents when new are included. The latter data are based on air of 70% humidity (at 80°F), dehydrated at a rate of 15 cu ft/hr per pound of adsorbent, to a dew point of −50°F.

The phenomenon of decreased capacity with service is not fully understood; however, some of the factors that contribute to this effect are known. Relatively nonvolatile compounds, such as absorption oil or compressor oil, are known to reduce the adsorptive capaci-

Table 12-13
Typical Design Capacities of Adsorbents

Adsorbent	Typical design capacity, wt %	New capacity, wt %
Activated bauxite-type desiccants	3–5	12–13
Activated Alumina	4–6	13–15
Silica-gel and silica-base beads	5–8	18–21
Alumina-gel balls	5–12	20–24
Molecular Sieve Type A	7–14	14–18

ty. It has also been shown that, even without contamination, the adsorption capacity is gradually reduced by the effect of water and heat during regeneration (Hammerschmidt, 1945).

It has been observed that the desiccant at the top of the bed (gas inlet) decreases in capacity most rapidly. This is not unexpected inasmuch as the top portion of the bed contacts the wet inlet gas during the entire adsorption period and, in addition, is more apt to adsorb small quantities of heavy hydrocarbons or other desiccant poisons that may be present. During regeneration, which is normally conducted with upflow of heated gas, all of the steam generated in the lower portion of the bed must pass upward through the top section. Hydrocarbons adsorbed near the top are thus carried out without contacting the lower portion of the bed. Data relative to the decline in capacity of bauxite desiccant as a function of position in the adsorber have been presented by Hammerschmidt (1957). These data show bauxite, with an original capacity of 8% water, to have the capacity shown in **Table 12-14** after 2½ years of service.

Table 12-14
Capacity of Bauxite after 2½ Years Service

Position in Adsorber	Capacity, wt %
Tray 1 (gas inlet)	3.2
Tray 2	3.9
Tray 3	4.6
Tray 4	5.7
Tray 5 (gas outlet)	5.7

As indicated by the equilibrium-capacity data for the various desiccants, the quantity adsorbed is a function of the vapor pressure of water in the gas stream (and thus of the relative humidity of the feed-gas stream). Capacity values used for design must also be adjusted downward if the gas being dehydrated is less than 100% saturated. Data on two desiccants used to dehydrate natural gas containing a maximum of 7 lb water/million cu ft at 72°F and 570 psig are presented by Harrell (1957). Very complete gas dehydration is required at this plant, since the gas stream is chilled to −105°F, and it is necessary to prevent ice formation in the chillers. Several desiccants were tested, and silica-gel beads and alumina-base pellets were selected as most favorable. Both of these showed capacities of 7% immediately after installation and decreased in capacity as shown in **Table 12-15**.

The drying of cracked-gas streams poses a particular problem because of polymerization of the olefins on the desiccant. Molecular sieves are affected less than silica or alumina in this type of service because of their capability to exclude large molecules. A comparison of four desiccants for drying a cracked gas stream was conducted at Union Carbide Corporation's Texas City olefins plant (Pierce and Stieghan, 1966). For both high- and low-density alumina, the capacity was found to decrease from about 5 lb H_2O/100 lb desiccant after 25 cycles to about 1 lb/100 lb after 200 cycles. Silica gel's capacity dropped from about 10 to 2 lb/100 lb in the same period, while the capacity of molecular sieve 3A decreased by only about 25% from 13.7 to 10.2 lb H_2O/100 lb desiccant, in the period between 25 and 200 drying cycles. As a result, all of the adsorption units were recharged with Type 3A molecular sieve (replacing alumina). Because of its greater capacity for water, a depth of about 45% of the original bed depth was used.

Table 12-15
Capacity of Desiccants Used in Dehydrating Gas Containing a Maximum of 7 lb H_2O/MMscf at 72°F and 570 psig

Age of desiccant, months	Capacity, wt%		No. of reactivations	
	Silica-gel beads	Alumina pellets	Silica-gel beads	Alumina pellets
0	7.0	7.0	1	1
8	5.0	3.9	145	145
18	2.7	2.5	272	283
27	2.0	2.2	425	436
35	2.2	2.1	587	598

Source: Harrell, 1957

At Phillips Petroleum Company's plant in Sweeny, Texas, considerable cost savings were effected by converting two dehydration units treating cracked gas streams from activated alumina to Type 3A molecular sieves. In addition to replacing the bed material, the system was modified from a lead-trim mode to a parallel mode, and a compound bed consisting of both ⅛-in. and ¹⁄₁₆-in. pellets was used. In the lead-trim mode of operation, two beds are used for dehydration in series while the third is being regenerated. This is advisable with activated alumina to prevent leakage of moisture as the material loses capacity due to accelerated aging. In the parallel mode, two beds are used for dehydration in parallel while the third is regenerated. This reduces pressure drop and therefore compression energy in the main gas stream and permits the use of smaller pellets. The overall effect of the changes was an estimated savings of $300,000 per year (Ezell and Gelo, 1982).

According to Silbernagel (1967), butadiene and acetylene (including methyl acetylene and propadiene) cannot usually be dried at all with silica or alumina desiccants because of rapid fouling of the bed by adsorbed hydrocarbons. With Type 3A molecular sieve, on the other hand, design water loadings of 7 to 14 lb H_2O/100 lb desiccant can be used with a 2- to 4-year molecular sieve life. A 380° to 425°F purge-gas temperature is normally used for regeneration of the adsorbent in this service. Although the Linde 3A molecular sieve is able to exclude all molecules larger than methane and thereby resists fouling due to polymerization of adsorbed hydrocarbons, it is apparently quite sensitive to attack by acids, including H_2S and CO_2, when present in high concentrations. This problem has been resolved by the development of acid-resistant molecular sieves. In one plant employing an acid-resistant Type A sieve to dehydrate natural gas containing 26% H_2S and 5 percent CO_2 at 2,200 psig, a normal service life of 2 years per desiccant charge has been achieved (Kraychy and Masuda, 1966).

Drying Tower Dimensions

For the typical case of a vertical, cylindrical dehydrator, the diameter is selected to give a superficial gas velocity that results in an acceptable pressure drop as described in the previous section entitled "Pressure Drop." The length is then established by the quantity of adsorbent required to remove water from the gas before breakthrough occurs. Estimation of this

Gas Dehydration and Purification by Adsorption

required quantity is complicated because only part of the bed attains equilibrium with the gas; the remainder near the exit end of the adsorber varies from the equilibrium concentration to essentially zero concentration at the very end. Active adsorption is still proceeding in this section when the adsorption cycle is stopped.

As discussed in the previous section of this chapter entitled "General Design Concepts," the zone of active adsorption is called the mass transfer zone (MTZ). The length of this zone is a measure of how rapidly adsorption is occurring. A short MTZ is desirable in that it enables a large fraction of the total bed to approach equilibrium with the feed gas before breakthrough occurs. Although equations have been developed to define the shape of the concentration-vs.-bed length curve, and thus the length of the MTZ in terms of mass transfer units and system parameters (Tan, 1980 and 1984, Vermeulen et al., 1973), simpler approaches are generally used for solid bed dehydrator design.

In one simplified approach, the total bed is viewed as the sum of two sections: the length of equivalent equilibrium section (LES) and the length of unused bed (LUB). LES is equivalent to the position of the stoichiometric front at breakthrough, while LUB is one-half the length of the MTZ. When breakthrough occurs, LUB is defined by equation 12-11:

$$LUB = L_o - L_s \qquad (12\text{-}11)$$

Where: LUB = length equivalent to unused bed, ft
L_o = total bed length, ft
L_s = position of stoichiometric front in bed, ft

The position of the hypothetical stoichiometric front at any time is given by the following equation:

$$L_s = 100G/\rho_b \times (\Delta Y/\Delta X)t \qquad (12\text{-}12)$$

Where: G = the gas-feed rate, lb mole/hr ft^2
ρ_b = adsorbent bulk density, lb/ft^3
$\Delta X = Xe - Xo$ where Xe and Xo are the adsorbate loadings in equilibrium with the feed gas and on the regenerated adsorbent, respectively, lb mole/100 lb activated adsorbent
$\Delta Y = Ye - Yo$ where Ye and Yo are the concentrations of adsorbable component in the gas feed and in the gas in equilibrium with regenerated adsorbent, respectively, lb mole/lb mole
t = time from the start of adsorption, hrs

The equations are best used in conjunction with actual test data. L_s is calculated from equation 12-12 by substituting t_b, the time at which breakthrough occurs, for t and using adsorption isotherms to establish ΔX and ΔY. LUB is then determined by equation 12-11. This LUB value can then be applied to the design of a plant with an entirely different length of equilibrium section, provided factors affecting the mass transfer zone are the same. A detailed review of adsorption column design using the mass-transfer-zone concept has been presented by Lukchis (1973).

An alternative correlation for estimating LUB is equation 12-3 (page 1028), which requires data on t_b, the time to breakthrough, and t_s, the time required for the stoichiometric front to reach the end of the bed. These time values can be obtained on a test unit operating at similar conditions.

1062 Gas Purification

In the GPSA (1987) procedure, the water adsorbed in the MTZ is neglected for the purpose of estimating the amount of sieve required for adsorption, but the length of the MTZ is included in calculating the total bed depth. The length of the MTZ is estimated by the following equation:

$$L_{MTZ} = (V/35)^{0.3}(Z) \qquad (12\text{-}13)$$

Where: L_{MTZ} = length of mass transfer zone, ft
V = superficial gas velocity, ft/min.
Z = 1.70 for ⅛-in. sieve
0.85 for ¹⁄₁₆-in. sieve

The total bed depth required can thus be calculated as either the sum of LUB and the stoichiometric length at breakthrough or the sum of the MTZ length and the equilibrium length at breakthrough. In high-pressure natural gas service, about five to six feet are added to the total calculated bed depth to get the required tangent-to-tangent dehydration vessel length. The additional space is required for support balls (or structures), gas distribution space, nozzle connections, and the possible future addition of more desiccant.

The final selection of vessel dimensions requires consideration of gas pressure drop, the cost of vessel construction, and the need for good gas distribution. According to Amero et al. (1949), such considerations usually results in a height:diameter ratio between 2:1 and 5:1 and a gas velocity between 20 and 60 ft/min, based on the empty vessel.

In a sample design calculation for a unit to dehydrate 190 MMscfd of natural gas at 700 psig and 80°F, Polley (1991) suggests design criteria of 6 ft for vessel diameter, 15 ft minimum for bed depth, and 7 psi maximum for pressure drop. These values are typical, but not limiting; in many cases it is economical to use larger diameter vessels, up to the maximum that can be transported to the site. Height:diameter ratios as low as 1:1 have been used successfully.

Where deep beds are indicated, it is common practice to install intermediate support trays at intervals of 4 or 5 ft to minimize the load on particles at the bottom of the bed and aid in gas distribution. Even in high pressure gas streams, pressure drop is important and several design modifications have been utilized to minimize it, including the use of horizontal rather than vertical vessels (see **Figure 12-3**) and the use of radial flow from a central core to an outer annulus in vertical vessels. In low pressure gas and atmospheric pressure air service, pressure drop is of extreme importance, and large diameter shallow beds with a height:diameter ratio of 1:1 or less are often employed.

For a given total gas flow and desiccant-bed volume, a deep bed is more effective than a shallow bed because a deep bed permits the desiccant to attain a higher average loading. This advantage is gained, however, at the expense of pressure drop because the deep bed must operate at a higher gas velocity.

Allowable gas velocities during both adsorption and regeneration are limited by considerations of particle entrainment and bed agitation as well as pressure drop. This problem has been analyzed by Ledoux for the case of up-flow of gas, and the following semiempirical equation has been proposed (1948):

$$G^2/d_g d_a Dg = 0.0167 \qquad (12\text{-}14)$$

Where: G = mass velocity of gas, lb/(sec)(sq ft)
d_g = gas density (flow conditions), lb/cu ft

d_a = adsorbent bed density, lb/cu ft
D = average particle diameter, ft
g = acceleration due to gravity, ft/(sec)^2

As the equation is dimensionless, any other consistent set of units could be used. Actual design velocities should be somewhat lower than might be indicated by the equation because of uncertainties inherent in the use of an average particle size and bed density for commercial adsorbents.

An alternative approach mentioned by Wunder (1962) for establishing the maximum permissible downflow gas velocity in beds of granular alumina makes use of a momentum concept:

$$N_M = V \times M \times P \leq 30{,}000 \qquad (12\text{-}15)$$

Where: N_M = a design parameter proportional to the gas momentum
V = superficial gas velocity, ft/min.
M = molecular weight of gas
P = system pressure, atm

Wunder claims that desiccant attrition should not be a problem if N_M is equal to or less than 30,000. This value is based on granular alumina, but is also recommended for silica gel.

The following gas velocities are given by Barrow (1983) as typical design values that will give an acceptable pressure drop for ⅛-in. molecular sieve in gas dehydration service.

Pressure (psia)	Velocity (fpm)
200	55–71
600	32–48
1,000	25–40

It is recommended that the velocity of the gas being dehydrated be limited to a value that will result in a pressure drop no higher than 7 psi. During regeneration, the gas velocity must be high enough to create a pressure drop of about 0.1 psi/ft. This leads to velocities in the range of 5–8 fpm for ⅛-in. molecular sieve and 2–3 fpm for ¹⁄₁₆-in. material.

Desiccant Regeneration

Regeneration (or reactivation) of the desiccants is accomplished by taking advantage of the fact that the capacities of all of the desiccants decrease with increased temperature. Usually a stripping gas is also employed to flush the released water vapor from the bed and reduce the partial pressure of water vapor in the gas to the lowest point possible during regeneration. Sufficient heat is required to provide the latent heat of vaporization of the adsorbed water, and to raise the temperature of the bed and associated equipment to the final regeneration temperature. Theoretically, the heat of wetting must also be provided; however, this is quite small relative to the heat of vaporization and is usually neglected. The required heat is generally supplied by passing a stream of preheated gas through the bed as pointed out above, the preferred flow direction for the regeneration gas being the reverse of that taken by the gas during dehydration. A regeneration-gas temperature of about 350° to 400°F is typical for all of the adsorbents, except molecular sieves. These materials require regener-

1064   *Gas Purification*

ation at a temperature in the range of 500° to 700°F if they are to provide maximum capacity and minimum dew point (−120° to −150°F). However, they reportedly will still provide one and one-half to two times the capacity, at dew points 15° to 40°F lower than conventional desiccants, when regenerated in the same manner as the conventional desiccants (Clark, 1958). The regeneration pressure for all adsorbents is usually the same as that used for dehydration; however, lower pressures are sometimes used to improve the stripping of adsorbed hydrocarbons. Low pressure regeneration is seldom used in natural gas dehydration systems because it requires the addition of carefully controlled depressurization and repressurization steps in the cycle, and it usually produces more low pressure effluent gas than can be used locally for fuel.

A regeneration cycle for a natural gas dehydration unit employing a molecular sieve desiccant is shown in **Figure 12-20**. In this unit, gas at 600°F is used for regeneration. It will be noted that the temperature of the gas leaving the unit rises rapidly at first, then more slowly as the exit gas temperature approaches that of the inlet. This effect is even more pronounced when a low regeneration gas flow, and therefore a longer regeneration period, is used; in fact, in some cases a temperature plateau is observed, marking a period when the bulk of the water is evaporated.

The regeneration period of about five hours shown in **Figure 12-20** is fairly typical. When the dehydrator vessels are cycled every 8 hr, for example, the regeneration gas quantity and temperature are usually set to provide complete regeneration in 4½ hr, with cooling in about 3 hr, leaving about ½ hr for switching and standby before the vessel is brought on stream as a dehydrator.

Bed cooling is most commonly accomplished at the same gas flow rate as regeneration with the regeneration heater bypassed. The cooling gas is usually passed through the bed in the same direction as the heating gas, i.e., upflow. This is particularly advantageous when dry gas is used for cooling (the most common case), because upflow cooling insures that the bottom of the bed is always dry. However, if wet feed gas is used for cooling, the gas flow direction should be reversed from upflow during heating to downflow during cooling, so that any water deposited by the cooling gas will be at the top of the bed and, therefore, will not be picked up by the product gas during the adsorption cycle.

The *GPSA Engineering Data Book* (1987) design procedure offers a very simplified, but useful set of equations for regeneration calculations. Two assumptions in the procedure are that 10% of the heat required to evaporate the adsorbed water and raise the temperature of the vessel and contents to the hot gas temperature will be lost, and that one-half of the heat put into the regeneration gas will be utilized. The latter assumption implies that the area under the exit gas curve of **Figure 12-20** is equal to one-half the area under the inlet gas curve during the heating period, which appears to be approximately the case for the example illustrated.

The quantities of heat energy required for a molecular sieve system are

$$Q_w = (1{,}800 \text{ Btu/lb})(\text{lb of water in the bed}) \tag{12-16}$$

$$Q_{si} = (\text{lb of sieve})(0.22 \text{ Btu/lb°F})(T_{rg} - T_i) \tag{12-17}$$

$$Q_{st} = (\text{lb of steel})(0.12 \text{ Btu/lb°F})(T_{rg} - T_i) \tag{12-18}$$

$$Q_{hl} = (Q_w + Q_{si} + Q_{st})(0.10) \tag{12-19}$$

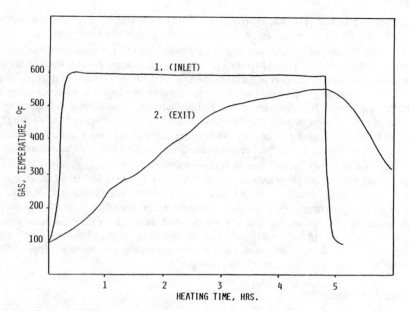

Figure 12-20. Regeneration cycle for molecular sieve, natural gas dehydration unit. *Data of Cummings (1983)*

$$Q_{tr} = (2)(Q_w + Q_{si} + Q_{st} + Q_{hl}) \qquad (12\text{-}20)$$

Where: Q_w = heat needed to desorb water, Btu
Q_{si} = heat needed to raise sieve temperature, Btu
Q_{st} = heat needed to raise steel temperature, Btu
Q_{hl} = heat loss, Btu
Q_{tr} = total heat needed for regeneration, Btu
T_{rg} = temperature of regeneration gas, °F
T_i = temperature of inlet gas, °F

The calculated value of the total heat required for regeneration is used to determine the amount of regeneration gas that must be passed through the bed by the following equation:

$$R_g = Q_{tr}/(C_p)(T_{rg} - T_i)(\text{heating time hours}) \qquad (12\text{-}21)$$

Where R_g is the required amount of regeneration gas, lb/h, C_p is the heat capacity of the gas, Btu/lb °F, and the other symbols are as given above. After calculating the required amount of regeneration gas, its superficial velocity through the bed and pressure drop should be estimated. Damron (1977) suggests that the superficial velocity should be in the range of 5 to 10 ft/min. (for high pressure natural gas dehydrators). If it is below about 5 ft/min. hot gas channeling may occur and the gas rate should be increased, which normally results in a shorter heating cycle. The *GPSA Data Book* recommends that the gas velocity be maintained high enough to produce a pressure drop of 0.10 psi/ft. This results in a minimum gas velocity of about 5 ft/min. for ⅛-in. beads, but lower acceptable velocities for finer particles.

In a process developed by Maloney Steel of Calgary, Canada, the customary cooling step is eliminated (Palmer, 1977). The Maloney process typically employs three beds of molecular sieve adsorbent. Regeneration is accomplished by a continuous stream of inlet gas that is heated first in a gas-to-gas heat exchanger by regeneration gas from the adsorber tower and finally in a gas-fired salt bath heater to 500°F. Each adsorber is in service two-thirds of the time and in regeneration one-third of the time. Since no time is required for cooling, the adsorbent is actively drying gas for a greater fraction of the overall cycle period than in conventional designs and, as a result, the process requires a smaller volume of adsorbent.

The principal drawback of the process is the possibility of producing poorly dried gas for a brief period when a hot, regenerated bed is first put on stream for dehydration service. However, operating experience of more than a year with a commercial unit did not show evidence of such occurrences (possibly due to instrumentation limitations). In theory, the heat transfer zone moves through the bed several hundred times faster than the mass transfer zone. As a result, except for a brief period at the beginning of the cycle, the heat transfer zone moves rapidly ahead of the mass transfer zone, providing an adequate supply of cool adsorbent to pick up water contained in the feed gas.

Process Flow Systems

A typical flow diagram for a high-pressure natural-gas dehydration plant is shown in **Figure 12-21.** In this arrangement, the regeneration gas is taken from the main wet-gas stream ahead of a pressure-reducing valve, which maintains sufficient pressure drop to enable the regeneration gas to flow through a heater, dehydrator, cooler, and separator and back into the wet-gas feed stream. In an alternate arrangement that has been proposed, the cooled regeneration gas reenters the system at a mid-point of the tower in dehydrating service. This uses the pressure drop across the first half of the bed instead of the pressure drop across a valve to provide the driving pressure for the regeneration gas and has the effect of reducing the required pressure drop across the entire dehydration plant. In the unit shown, steam at 386°F is used to heat the regeneration gas to 360°F. When regeneration of the bed is completed, as evidenced by a rise in the exit-gas temperature to approximately that of the inlet gas, the steam to the heater is shut off (or the heater may be bypassed), and the slip stream of wet feed gas is used to cool the regenerated bed. When the bed temperature approximates that of the dehydration-plant feed, it is ready for dehydration service. Switching the two dehydrators from dehydration service to regeneration heating and then to regeneration cooling is accomplished by the use of 12 valves, which are numbered on the drawing. The valve positions for the various operations are given in **Table 12-16.**

In the arrangement shown, the gas flow is downward during dehydration, when the gas velocity is highest, in order to avoid disturbing the bed. It is upward during regeneration in order that the bottom of the bed will be thoroughly regenerated, as it is the last point of contact with the gas being dehydrated; and it is downward during cooling in order that any water deposited from the cooling gas will be at the top of the bed where it cannot be revaporized into the gas stream during dehydration.

A more common arrangement for large natural gas dehydration plants is depicted in **Figure 12-22.** In this system, the gas used for regeneration is recompressed and recycled to the wet feed gas stream. Less energy is required to recompress the small stream of regeneration gas than is lost by reducing the pressure of the main gas stream through a valve. The **Figure**

Gas Dehydration and Purification by Adsorption **1067**

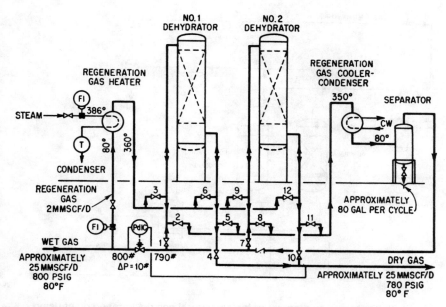

Figure 12-21. Process flow diagram of typical natural gas dehydration plant using an alumina or silica desiccant and wet gas regeneration.

Table 12-16
Positions of Valves in Dehydration Plant of Figure 12-21 During Operating Cycles*

Valve No.	Bed No. 1 on dehydrating service		Bed No. 2 on dehydrating service	
	Bed No. 2, regenerating	Bed No. 2, cooling	Bed No. 1, regenerating	Bed No. 1, cooling
1	O	O	C	C
2	C	C	O	C
3	C	C	C	O
4	O	O	C	C
5	C	C	C	O
6	C	C	O	C
7	C	C	O	O
8	O	C	C	C
9	C	O	C	C
10	C	C	O	O
11	C	O	C	C
12	O	C	C	C

*C = closed; O = open.

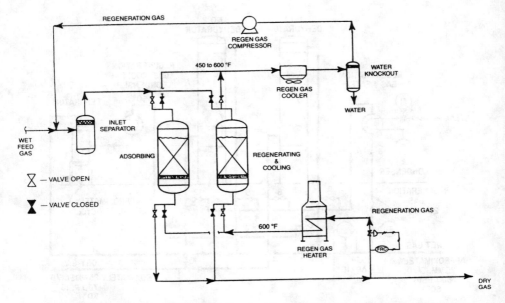

Figure 12-22. Process flow diagram of typical natural gas dehydration plant using molecular sieve desiccant and dry gas regeneration. *From GPSA Engineering Data Book (1987). Reproduced with permission, copyright 1987, Gas Processors Supplier's Association*

12-22 system has the further advantage of permitting dry gas to be used for regeneration and cooling. The dry gas is more effective in removing water during regeneration and does not add water to the bed during the cooling cycle. It can be fed to the bottom of the bed during both regeneration and cooling, reducing the number of valves required compared to the system of **Figure 12-21.** The two flow diagrams also differ with regard to regeneration gas heating. In **Figure 12-21,** this gas is heated by steam to a temperature of 360°F, which is typical of silica and alumina desiccant systems; while in **Figure 12-22,** the regeneration gas flows through a direct-fired heater, which raises its temperature to 600°F—typical of a molecular sieve system.

The length of the dehydration period may be varied within wide limits. Shorter periods are more economical with regard to equipment size and desiccant charge; however, operating costs may be higher because of shorter desiccant life and more frequent valve operation. Obviously, sufficient time must be allowed for the complete regeneration and cooling of one tower while the other is in service; however, the regeneration time requirement may also be varied by adjusting the quantity and temperature of the regeneration gas stream. In general, the length of the drying cycle is established on the basis of operating labor schedules, resulting in the usual specification of 8-, 12-, 16-, and 24- hour cycles. Eight to twelve hour cycles are most typical. It is common practice to switch beds on the basis of the established time schedule rather than to take advantage of the full capacity of the desiccant. The desiccant bed capacity decreases with age, however, and ultimately a time is reached when the bed just barely accomplishes the desired dehydration at the end of the drying cycle and must be replaced.

Cummings (1983) points out the economic advantage of adjusting the dehydration cycle to take advantage of the higher capacity of the adsorbent at the beginning of its life cycle. He provides the following breakdown of heat requirements for regeneration of molecular sieve systems:

	Percent of Heat Input
Sensible heat to desiccant	9
Sensible heat to vessels and internals	15
Sensible heat to water plus the heat of desorption	16
Sensible heat carried out in exit regeneration gas	60

Since only about 16% of the heat input is actually used to remove water, a doubling of the water load in the bed increases the heat consumption by only about 16%. This results in a significant fuel savings during the early life of the desiccant charge.

To take advantage of the fuel savings possible with a maximum length dehydration cycle, it is necessary to know when breakthrough is imminent. This can be accomplished by installing a gas sample tap 12–18 in. from the bottom of the bed. The presence of moisture in gas from the sample tap indicates that the breakpoint is approaching the end of the bed and marks the appropriate cycle time for the specific operating conditions.

A process flow system using three towers packed with silica gel at the top and molecular sieve at the bottom is often selected to provide a combination of natural gas dehydration, control of hydrocarbon dew-point temperature, and recovery of hydrocarbon liquids (Badger Engineers, 1984). At any one time, two towers are in adsorption service while the third is being regenerated or cooled. The feed gas is first cooled and passed through a liquid separator before it flows to the top of the adsorber towers. The silica gel in the top portion of the bed adsorbs both water and heavy hydrocarbons and also serves as a guard for the molecular sieve, which does the final dehydration. Natural gas heated to 500–600°F is used for regeneration. Adsorbed water and heavy hydrocarbons are recovered from the regeneration gas by passing it through a cooler/condenser and separating the products. A portion of the dried product gas is used for cooling before the adsorber is placed back in adsorption service. Gas used for regeneration and cooling is recycled to the active adsorber.

Operating Practices

As previously noted, the capacity of solid desiccants decreases with time, and ultimately the beds must be replaced. In order to predict the replacement time well in advance, it is good practice to determine periodically the maximum capacity of the beds. This is accomplished by allowing the beds to remain on stream past the normal cycle time and then monitoring the dew point of the product gas to detect the break point and the point at which it reaches the maximum allowable value.

Since the decrease in capacity is believed to be due to a considerable extent to heavy hydrocarbons that are deposited on the bed, all possible precautions should be followed to prevent such materials from entering the unit. An efficient scrubber, ahead of the dehydration plant, is recommended for removing liquid hydrocarbons, water, and other particulate impurities that may be present in the feed-gas stream.

1070 Gas Purification

Liquid water is particularly objectionable with high-capacity gel-type adsorbents because it can cause particle breakage. To minimize this effect, it is common practice to guard susceptible desiccants, such as Sorbead R, with a layer of liquid-water-resistant desiccant, such as Sorbead W or Activated Alumina. Another precaution is to insulate the tower surfaces and connecting pipelines so as to prevent the condensation of water on the cooled metal surfaces and avoid the possibility of such water reaching the desiccant.

Although regeneration can be accomplished at 300°F for all desiccants except molecular sieves, higher temperatures (up to 400°F) provide lower dew points and may actually increase the desiccant's useful life by providing more complete removal of adsorbed hydrocarbons. Periodic regeneration at a higher than normal temperature is also occasionally of value in reducing the rate of capacity decline.

Hydrogen sulfide in the feed-gas to solid-bed dehydrators can cause difficulties, particularly if the gas also contains a trace of oxygen. In the latter case, the desiccant acts as a catalyst for the oxidation of hydrogen sulfide to elemental sulfur, which deposits on the particles. Some of this is vaporized during regeneration and can cause plugging of cooler condensers. Wilkinson and Sterk (1950) present data on the dehydration of a very sour gas, containing 1,800 to 2,000 grains. H_2S/100 scf (and presumably no oxygen), with Sorbead and Activated Alumina. They found the Activated Alumina to be inactivated rapidly while the Sorbead was relatively unaffected by the H_2S.

The presence of H_2S in the gas to be dehydrated is particularly serious when bauxite that contains iron oxide is used as the desiccant. The iron oxide reacts with the hydrogen sulfide to form iron sulfide, which changes the characteristics of the bauxite; this results in deactivation and disintegration of the particles.

An operating precaution that may seem rather obvious, but which nevertheless is frequently overlooked, is the prevention of sudden pressure surges. If the dehydration vessel is vented rapidly, for example, very high, localized gas velocity may occur in the bed, causing bed motion, attrition, and even entrainment of the desiccant particles in the gas stream. Ballard (1978) recommends that the pressurization rate be limited to less than 50 psi per minute and that the depressurization rate be held below 30 psi per minute to avoid desiccant breakage and damage to the vessel internals.

Uniform distribution of gas entering the bed is very important to prevent gas channeling and desiccant damage. Neither the wet feed gas nor the regeneration gas should impinge directly on the bed. A perforated basket or screen-wrapped slotted pipe, with the feed gas exiting radially at a low velocity, provides better feed gas distribution than a perforated plate above the bed (Ballard, 1978).

Two types of bed supports are commonly used: (1) several layers of screen on a horizontal grid supported by beams, and (2) inert alumina balls that completely fill the bottom head of the vessel. The upper screen must be smaller in mesh size than the desiccant particles and is often covered with a few inches of small inert alumina balls to minimize contact with the desiccant. When the bed is supported entirely on inert balls, they are graduated in size, with large (typically ¾ in.) balls filling the bottom head, then a layer of ½-in. balls, and finally a layer of small (e.g., ¼ in.) balls directly under the desiccant. The upper layers of balls are typically 6 to 12 in. in depth. A layer of large (e.g., 2 in.) balls is often placed on the top of the desiccant bed to improve gas distribution and minimize bed particle movement (Ballard, 1978).

USE OF MOLECULAR SIEVES FOR GAS PURIFICATION

In addition to their use for dehydration as described in a previous section of this chapter, the molecular-sieve adsorbents are finding application for many other gas-purification problems. Most of these take advantage of the fact that the molecular sieves show a high adsorptive selectivity for polar and unsaturated compounds. Polar compounds, such as water, carbon dioxide, hydrogen sulfide, sulfur dioxide, ammonia, carbonyl sulfide, and mercaptans, are very strongly adsorbed and can readily be removed from such nonpolar systems as natural gas or hydrogen.

In most gas-purification cases, water vapor is also present in the impure gas and is removed by the molecular-sieve adsorbent along with the other impurities. Since water is adsorbed more strongly than any of the other components, it concentrates initially at the inlet portion of the bed where it displaces the other impurities that had previously been adsorbed. These desorbed impurities are then readsorbed farther down the column, and this impurity adsorption zone moves through the bed in advance of the water adsorption zone. This phenomenon is illustrated graphically in **Figure 12-23** for the case of mercaptan adsorption from natural gas (Conviser, 1965). As indicated by the illustration, a much smaller bed would be required, for the same adsorption period, if dehydration alone were desired. In that case a small amount of mercaptan sulfur would be removed with the water, but, at the water breakthrough point, the sulfur content of the product gas would be the same as that of the feed. In a similar manner, the presence of any more strongly adsorbed component will affect capacity and breakthrough point for a given impurity.

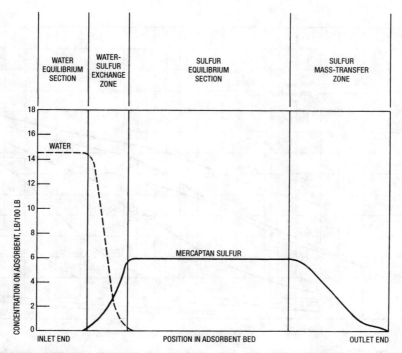

Figure 12-23. Adsorption zones in a molecular sieve bed, adsorbing both water vapor and mercaptans from natural gas. *Data of Conviser (1965)*

Basic Data

Equilibrium capacity data for molecular sieves and the gases CO_2, H_2S, SO_2, and NH_3 are presented in **Figures 12-24, 12-25, 12-26,** and **12-27,** respectively. Data for silica gel are included on the NH_3 chart for comparison purposes. As pointed out above, the capacity attainable in plant operations involving multicomponent gas streams cannot be estimated directly from equilibrium data obtained with a single component. However, such data are useful as an indication of maximum possible capacities and as a base for more complex design analyses.

One approach to the correlation and prediction of multicomponent adsorption equilibria for molecular sieves has been proposed by Yon and Turnock (1971). This approach uses a Loading Ratio Correlation (LRC), which makes use of constants that are derived from pure-component behavior. According to the authors, a single set of constants can characterize adsorption equilibria over broad ranges of temperature, pressure, and concentration. Although the technique is not excessively cumbersome to use, a detailed description is beyond the scope of this text, and the reader is referred to the original paper for additional information (Yon and Turnock, 1971). For the specific case of natural gas purification (H_2S and water removal) using a 5A molecular sieve, a detailed design calculation procedure has been presented by Chi and Lee (1973). The proposed design method has been compared with experimental data over a wide range of conditions and found to adequately predict hydrogen

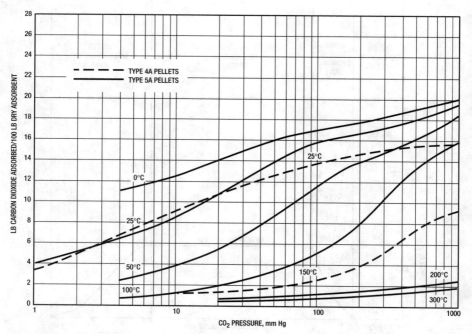

Figure 12-24. Equilibrium isotherms for carbon dioxide adsorption on molecular sieves, types 4A and 5A. *Data of UOP (1993A)*

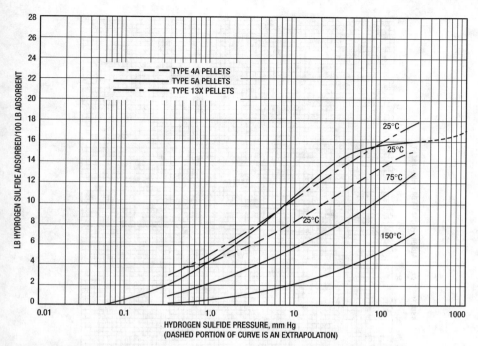

Figure 12-25. Equilibrium isotherms for hydrogen sulfide adsorption on molecular sieves, types 4A, 5A, and 13X. *Data of UOP (1993A)*

sulfide break time. It is based on the use of a dynamic H_2S saturated capacity, which is correlated with H_2S and CO_2 partial pressures, and a "Lost Bed Height" (similar to LUB), which correlates with H_2S flux (lb/ft^2 hr) and the ratio of CO_2 to H_2S partial pressures.

Process cycles used for gas purification are generally similar to those used for dehydration. Fixed beds of adsorbent are employed, with gas typically flowing downward during the adsorption cycle, upward during regeneration and downward during cooling. Once the bed capacity has been established from pilot-plant tests or fundamental data, the general process design techniques described in previous sections of this chapter can be applied. An exception is pressure swing adsorption which employs pressure reduction instead of a hot purge gas for bed regeneration. Another nontypical molecular sieve process, which makes use of both the catalytic and adsorptive properties of molecular sieves, is the Pura Siv-N process developed by Union Carbide Corporation for the removal of NO_x from nitric acid plant tail gas (Farnoff, 1971). The process utilizes two beds of a special acid-resistant molecular sieve. When tail gas from the acid plant is passed through one of the beds, the NO is catalytically converted to NO_2, which is immediately adsorbed as an equilibrium mixture of NO_2 and N_2O_4. The operation is cycled to the other bed prior to NO_2 breakthrough, and the NO_2 in the spent bed is desorbed by heating to about 600°F. Additional specific applications of molecular sieves for gas purification are described in the following sections.

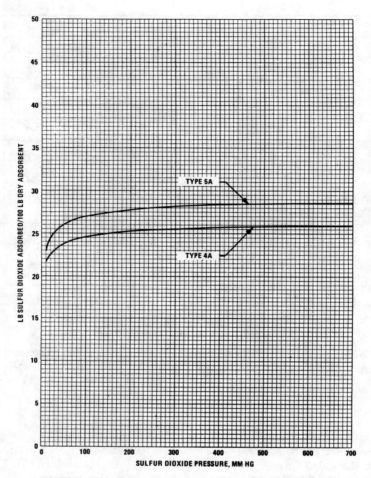

Figure 12-26. Equilibrium isotherms for sulfur dioxide adsorption at 25°C (77°F) on molecular sieves, types 4A and 5A. *Data of UOP (1993A)*

Inert Gas Purification

Inert gas, for use in annealing and other operations requiring a nonreactive atmosphere, is customarily made by removing carbon dioxide and water vapor from a combustion gas made under carefully controlled conditions. A flow diagram of the purification process employing molecular sieves is shown in **Figure 12-28.** Natural gas is burned with air in near stoichiometric proportions to give a product gas containing approximately 89% nitrogen and 11% carbon dioxide plus the water originally present in the air and that formed during the combustion. The combustion gas is cooled by heat exchange with regeneration air and then by passage through a water-cooled exchanger. The cooled gas is then passed through one of the three molecular-sieve beds for water and CO_2 removal. During this time, the second molecu-

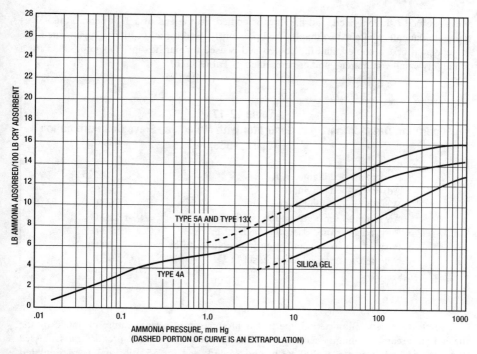

Figure 12-27. Equilibrium isotherms for ammonia adsorption at 25°C (77°F) on molecular sieves, types 4A, 5A, and 13X, and on silica gel. *Data of UOP (1993A)*

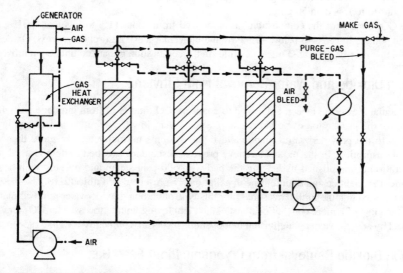

Figure 12-28. Molecular sieve process for purification of inert gas used in annealing operations. *From Clark (1958)*

lar-sieve bed is regenerated, and the third is cooled. At the end of a 1-hr period, the valves are automatically switched so that the bed previously used for adsorption is regenerated, the regenerated bed is cooled, and the cooled bed is used for purification. Operating data for a relatively small commercial unit are presented in **Table 12-17**.

Table 12-17
Operating Data for Inert-gas Generator with Molecular-Sieve Gas Purification (Clark, 1958)

Adsorbent-bed diameter, ft	2.25
Adsorbent-bed height, ft	3.5
Weight of adsorbent per bed, lb	600
Air required for regeneration, scf/hr	12,500
Purified gas from adsorption bed, scf/hr	2,200
Purified gas used as purge during cooling, scf/hr	200
Net inert-gas production rate, scf/hr	2,000

Source: Clark, 1958

During the regeneration period, the bed is brought up to a temperature of 450°F by the heated air stream. Cooling is accomplished in two parts—for 5 min., product gas is passed through the bed at a rate of 200 scf/hr and vented; and for 55 min., gas is recirculated through the bed and a cooler at the rate of 12,500 scf/hr. However, even during the 55-min. period, fresh product gas is added and recirculated gas vented at the rate of 200 scf/hr. At the end of the 1-hr cooling and purging period, the oxygen content in the recirculation gas has been reduced to less than 50 ppm.

The CO_2 content of the combustion gas is reduced from about 11% to less than 0.01%, and the dew point is lowered to below $-75°F$ by the molecular-sieve adsorption process. Oxides of nitrogen that may be present in trace quantities in the combustion gas are completely removed.

Carbon Dioxide and Water Removal from Ethylene

Molecular sieves have been employed to coadsorb CO_2 and H_2O from ethylene gas which is used for the production of polyethylene plastic. A very low CO_2 content is required in the feed gas to the polymerization plants, and liquid systems normally require more than one stage of absorption. In the molecular-sieve process, a standard two-bed system is used with regeneration accomplished by heating the bed with 600-psig steam and purging with heated methane. Pertinent design data for one installation are presented in **Table 12-18**.

It is reported that the initial investment for this type of carbon dioxide-water removal plant is lower than that of a more conventional installation employing liquid processes for CO_2 removal followed by adsorptive dehydration and that operating costs are competitive (Clark, 1958).

Carbon Dioxide Removal from Cryogenic Plant Feed Gas

Molecular sieves are used in many installations for the simultaneous removal of carbon dioxide and water from gas streams which undergo low-temperature liquefaction in a subsequent operation. Even relatively low concentrations of carbon dioxide and water can cause

Table 12-18
Design Data for Molecular-Sieve Ethylene Gas Purification Plant

Ethylene-gas feed rate, MMscf/day	1.75
Feed-gas pressure, psig	430
Adsorber diameter (ID), in.	60
Adsorber height, ft	32
Adsorbent charge (per vessel), lb	15,000
	(Linde type 5-A molecular sieve)
Steam pressure, psig	600
Steam rate, lb/day	11,400
Cooling-water rate, gal./day	57,200
Ethylene in purged gas (recycled to furnace) MMscf/day	0.05
Methane in purged gas, MMscf/day	0.06
Inlet-gas CO_2 content, ppm	3,000
Product-gas CO_2 content, ppm	<1
Product-gas dew point, °F	<−100°F

Source: Clark, 1958

the formation of frost, or "rime," on the surfaces of cryogenic heat exchanger fins. One of the earliest CO_2 removal applications of molecular sieves was for purifying the air feed to cryogenic air separation plants. In this service, continuous product gas purities of less than 1 ppm CO_2 and 1 ppm H_2O are readily attained.

The design of molecular sieve front end purification (FEP) systems for cryogenic air separation plants is described by Kerry (1991). In a typical design, the feed air is compressed to about 6–8 atm and cooled to about 8°C (46°F) before it enters the bed of molecular sieve. The adsorption step is stopped before breakthrough of CO_2 occurs. Since propane, ethylene, acetylene, and higher hydrocarbons as well as H_2O are more strongly adsorbed than CO_2, only the inlet portion of the bed will be near saturation with regard to these impurities at the end of the adsorption step so they will be removed from the gas with very high efficiency.

The regeneration of molecular sieve beds in FEP systems is accomplished by flowing heated nitrogen through the bed, in the reverse direction to adsorption, at approximately atmospheric pressure. Regeneration temperatures are modest, 90° to 100°C (194° to 212°F), but an amount of nitrogen equal to about 20 to 30% of the inlet air is required. The bed is cooled with low temperature nitrogen before it is ready for the adsorption step.

A growing field of applications is front-end feed purification for natural-gas liquefaction plants. **Figure 12-29** is a simplified flow diagram of a three-tower molecular-sieve adsorption plant, and **Figure 12-30** is a photograph of a plant of this type. This arrangement has one tower drying and purifying gas, another tower being cooled, and a third tower being regenerated by a hot purge at the same time. The towers are manifolded to common inlet and outlet headers. and are automatically switched from adsorption to heating to cooling in sequence. The same gas is used for cooling one bed and heating another so purge gas requirements are about one-half that of a dual bed system. In the flow diagram shown, the purge gas is withdrawn from the purified gas stream, and after use it is cooled to knock out most of the water and is then returned to the pipeline. Boil-off gas from the liquefaction plant is also frequently used for purging.

1078 Gas Purification

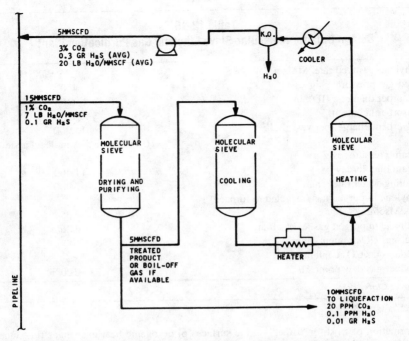

Figure 12-29. Simplified flow diagram of typical three-tower molecular sieve adsorption system for removing carbon dioxide and water from natural gas prior to liquefaction. *From Thomas and Clark (1967)*

Removal of Sulfur Compounds

In natural-gas processing, the mercaptans are generally the most strongly adsorbed impurities (other than water), followed by H_2S with CO_2 being the most weakly adsorbed compound in this series. However, all mercaptans are not adsorbable by type 4A or 5A sieves because pore size limitations prevent adsorption of any but the lightest members of the family. As a consequence, 13X is the preferred adsorbent for complete sulfur removal from natural-gas streams. Existing commercial sulfur removal units treat natural-gas streams at flow rates ranging from 2 to over 200 MMscfd. Data on a large plant are given in **Table 12-19.** The presence of a trace of glycol, glycol degradation products, and absorber oil in the feed gas is noted as these heavy molecules affect both capacity (by coadsorption) and bed life (due to coke formation). It is claimed that a properly designed molecular sieve, sulfur removal system will perform for 3 to 5 years before new adsorbent is required (Conviser, 1965).

Molecular sieves are very effective for H_2S removal and can produce gas containing extremely low levels of this impurity. Since H_2S is adsorbed more strongly than CO_2, the molecular sieves also offer the capability for selectively removing H_2S from gas streams that contain both impurities. However, the process has not gained general acceptance because of the problem of disposing of the sulfur-rich regeneration gas. The problem has been bypassed in one process scheme that is applicable to cases where it is desired to leave part of the CO_2 in the gas stream while removing essentially all of the H_2S and water. This flow arrangement

Figure 12-30. Molecular-sieve gas purification plant. Automated three-tower system removes carbon dioxide and water from natural gas prior to liquefaction. *Courtesy of Memphis Gas, Light and Water Division*

is shown diagrammatically in **Figure 12-31** for a typical case. The gas feed to the dry bed and liquid absorption units is ratioed in such a manner that the final mixed product meets pipeline CO_2 specifications. The increased gas volume that results from leaving 3% CO_2 in the product gas provides a credit for gas sold on a volume basis. Use of the dry-bed adsorption process on a portion of the gas has the further advantage of permitting the liquid absorber system to operate closer to maximum capacity, as a product gas which is slightly off specification with regard to H_2S, and water can be tolerated for dilution with the very pure adsorber product.

Molecular sieves have also found application for desulfurization of natural-gas feed to ammonia plants. Removal of all types of sulfur compounds ahead of these plants is desirable because sulfur acts as a temporary poison to steam-hydrocarbon reforming catalysts and a permanent poison to expensive low-temperature shift conversion catalysts. An installation employing a standard dual bed adsorption system has been described by Lee and Collins (1968). The authors also describe comparative tests of a molecular sieve and a commercial grade of impregnated activated carbon in a dual-bed mobile pilot unit. The test results indicated that the molecular sieve could treat 2 to 4 times as much gas per unit volume of adsorbent as the carbon. The commercial plant consistently provided gas to the primary reformer containing less than 0.3 ppm (vol) peak total sulfur from a feed gas averaging about 0.6 ppm

Table 12-19
Design and Operating Data for Large Mercaptan Removal Plant

Natural gas feed rate, MMscf/day	200
Feed gas pressure, psig	750
Adsorber description	(two horizontal vessels)
Diameter, ft	6
Length, ft	36
Bed depth, ft	3
Adsorbent charge, (per vessel) lb	25,000 (Type 13X)
Inlet gas composition	
Mercaptan and heavy sulfur, grains/100 scf as H_2S	2
Water, lb/Mscf	<7
Glycols, glycol degradation products, and absorber oil	trace
H_2S and CO_2	nil
C_4 hydrocarbons, % vol.	0.03
C_1, C_2, C_3, and N_2	balance
Product gas, total sulfur, grains/100 scf as H_2S	<0.06
Operating Cycle:	

Step	Flow Direction	Time
1. Purify	Down	12 hours
2. Hot purge (600°F)	Up	8 hours
3. Cool purge	Down	4 hours

Source: Conviser, 1965

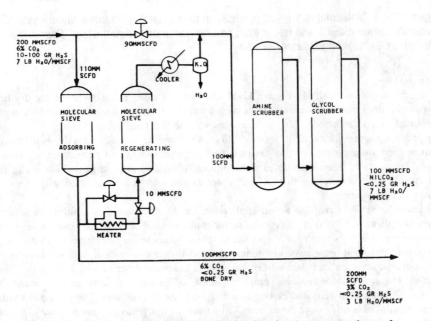

Figure 12-31. Combination molecular sieve-liquid absorbent process scheme for purifying carbon dioxide-rich natural gas. *Data of Thomas and Clark (1967)*

H$_2$S, 0.7 ppm mercaptans, 1.6 ppm sulfides, and 1.2 ppm residual sulfides. In this application, the high-sulfur purge gas can be fed into the plant fuel system so that the sulfur is finally vented to the atmosphere as dilute sulfur dioxide. Although this disposal method is preferable to venting the original sulfur compounds, it is not broadly acceptable and constitutes a significant drawback to the process.

Hydrogen Purification by Pressure Swing Adsorption (PSA)

Although the pressure swing cycle is not limited to hydrogen purification or to molecular sieve adsorbents, this combination represents the principal current gas purification use of the process. Air separation, another important application, is not considered a purification process and is, therefore, not treated here; however, it is adequately described in other texts, e.g., Yang (1987).

The PSA cycle operates between two pressures, adsorbing impurities at the higher pressure and desorbing them at the lower pressure with no temperature change except that due to the heat of adsorption and desorption. The absence of a heat requirement leads to a simple installation compared to a thermal cycle. This advantage is counterbalanced, however, by a greater gas loss due to the venting and low-pressure purge operations.

PSA gas purification and separation operations can be categorized as equilibrium-based or diffusion-rate-based. Equilibrium-based separations, which are the most common, depend on the ability of the adsorbent to adsorb a greater quantity of "heavy" species than of "light" species at equilibrium. Diffusion-rate-based operations separate components because one species diffuses more rapidly than the other components. The performance of equilibrium-based systems can be predicted by relatively simple models, while the diffusion-rate-based systems require more complicated models to account for the effects of mass transfer within the individual adsorbent particles.

Equilibrium-based models are described by Knaebel (1991), and Knaebel and Hill (1985). Diffusion-rate-based models are described by Farooq and Ruthven (1991) and Shin and Knaebel (1987). Equilibrium-based separations are generally capable of producing a higher purity product and are, therefore, used for gas purification applications; while PSA operations based on diffusion rates are more often applicable to bulk separations such as the production of oxygen and nitrogen from air.

The principle of the PSA process is illustrated by **Figure 12-32** and **Table 12-20**. The partial pressures of adsorbed impurities are lowered by reducing the adsorber pressure from the high pressure of the feed gas to the low pressure of the tail gas, then further lowered by the use of a high-purity hydrogen purge. The relative strength of adsorption of typical impurities is given in the table. If adsorption is continued to breakthrough, the lighter impurities appear first, followed by the intermediate, and finally the heavier impurities.

A minimum pressure ratio of approximately 4:1 between adsorption and desorption pressures is normally required for hydrogen purification. According to Miller and Stoecker (1989), the optimum feed pressure for hydrogen purification units in refinery applications is 200–400 psig. The product hydrogen from the PSA unit is available at pressures about 10 psi lower than the feed. The hydrogen recovery in PSA units is typically 80 to 92% at optimum conditions and as low as 60% when the tail gas is delivered at a relatively high pressure (e.g., 80 psig). Typically, the tail gas is used locally as low-pressure fuel.

The performance of a simple PSA plant designed to produce high purity hydrogen from demethanizer off-gas is given in **Table 12-21**. The data indicate that a very high purity

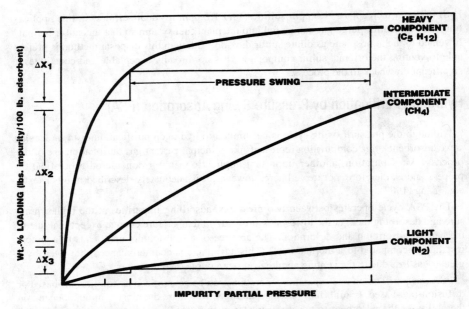

Figure 12-32. Generalized adsorption isotherms for heavy, intermediate, and light components showing effect of pressure swing on net adsorbate loading. *From Haun et al. (1990)*

Table 12-20
Relative Strength of Adsorption of Typical Impurities

Non-adsorbed	Light	Intermediate	Heavy
H_2	O_2	CO	C_3H_6
He	N_2	CH_4	C_4H_{10}
	Ar	C_2H_6	C_5^+
		CO_2	H_2S
		C_3H_8	NH_3
		C_2H_4	BTX
			H_2O

hydrogen is obtained, but less than 70% of the hydrogen in the feed is recovered as product. The waste gas can be recompressed and used as fuel, however.

Hydrogen is commonly produced by steam reforming natural gas or naphtha in combination with processes for purifying the reformer off-gas. Typically, the reformed gas is subjected to a water-gas shift reaction to produce crude hydrogen containing 75–80% H_2, 0.1–1% CO, 15–25% CO_2, 1–5% CH_4, and a trace of N_2 (dry basis) at a pressure of 150–350 psig. The use of PSA technology to remove the listed impurities plus water vapor has become the

Table 12-21
Performance of Pressure Swing Adsorption Process

	Feed	Product	Waste
Flow, MMscfd	1.2	0.5	0.7
Pressure, psig	200	198	2.0
Composition, mol %			
H_2	61.1	99.999+	33.3
CH_4	36.7	1 ppm	63.0
C_2H_4	0.6	1 ppm	1.0
C_2H_6	0.1	1 ppm	0.1
CO	1.1	1 ppm	1.9
N_2	0.4	1 ppm	0.7

Source: Stewart and Heck, 1967

state-of-the-art gas purification method. With a layer of activated carbon at the feed end of the adsorber to remove H_2O and CO_2 and a layer of 5A zeolite to remove CO and CH_4 at the product end, a PSA system can produce 99.999% pure H_2 with a recovery of 76.2% from a typical reformer process gas (Sircar and Kratz, 1988).

The high purity of the hydrogen produced in a PSA system can result in side benefits in an overall hydrogen plant by, for example, reducing the need for gas recycle and purging. Hiller et al. (1987) performed an economic analysis of various approaches for producing hydrogen for refinery hydroprocessing applications. They conclude that substantial investment and operating cost savings are possible in a hydrocracker plant using 99.9% hydrogen from a PSA system compared to one using 97% hydrogen purified in an amine unit/methanation system.

Early work on the use of PSA processes for the purification of hydrogen was conducted by the Union Carbide Corporation, and the advanced technology that they developed is now licensed by UOP as the Polybed PSA Process. The process utilizes a relatively large number of adsorption beds (3 to 12) in a specific sequence of steps to provide improved separation efficiency (Fuderer and Rudelstorfer, 1976). The basic steps for hydrogen purification are described as follows (UOP, 1991A):

1. *Adsorption:* High pressure feed gas is passed through the adsorber, impurities are retained, and pure hydrogen is withdrawn from the other end.
2. *Cocurrent depressurization:* Feed is stopped and the unit is partially depressurized by removing gas through the product port. Hydrogen removed during this step is used to purge and partially pressurize another unit.
3. *Counter-current depressurization:* The unit is further depressurized to its lowest pressure by removing gas through the inlet port. Part of the adsorbed impurities are removed during this step, and the gas removed represents a portion of the tail gas.
4. *Purge:* Hydrogen in counter-current flow is used to purge the remaining impurities from the bed at low pressure. The effluent gas becomes part of the tail gas stream.
5. *Repressurization:* Purified hydrogen, flowing in the counter-current direction, is used to repressurize the bed to the feed gas pressure level, completing the cycle and preparing the bed for step 1.

The first U.S. Polybed PSA unit was started up in 1978 and gave an on-stream factor of 0.999 during its first 14 months of operation (Heck, 1980). A comparison of the Polybed PSA process with a "conventional" process for producing hydrogen by naphtha reforming, reported by Heck and Johanson (1978), is summarized in **Table 12-22.** The indicated production costs are a strong function of the assumptions used in the evaluation, and would be less favorable for PSA if, for example, the cost of fuel is assumed to be significantly less than the cost of naphtha feedstock and if less credit is given for export steam. Of probably greater significance is the extremely high purity of the hydrogen produced by the adsorption process.

The Polybed PSA process has been an outstanding success. The wide variety of applications and the number of units operating on each feed source as of September 1989 are listed in **Table 12-23.** According to a report in the *Gas Process Handbook* (UOP, 1996), over 500 Polybed PSA units were in operation or under construction at that time. A photograph of a large UOP Polybed PSA unit is shown in **Figure 12-33.**

Numerous modifications to the basic PSA process have been proposed. In most cases, the changes consist of adding additional depressurization and pressurization steps to increase the recovery of hydrogen. A modification aimed at the simultaneous production of high purity hydrogen and carbon dioxide has been developed by Air Products and Chemicals (Sircar, 1979). The process uses six parallel adsorbers (A beds) designed to selectively remove CO_2

Table 12-22
Comparison of Polybed PSA (Pressure Swing Adsorption) Hydrogen Plant with "Conventional" Design

	Polybed PSA	Conventional*
Costs:		
Total Installed Cost (1977 basis), $ Million	16.5	16.0
Production Cost**, $/1,000 scf H_2	1.30	1.40
Design Basis:		
Capacity, MMscfd contained H_2	50	50
Product Pressure, psig	350	350
Hydrogen Purity, %	99.999	97
Feedstock	Naphtha	Naphtha
Feed and Utilities:		
Feedstock, MM Btu/h (low heat value)	796	533
Fuel, MM Btu/h (low heat value)	50	306
Boiler Feedwater, lb/h	157,500	63,800
Cooling Water, gpm (25°F rise)	4,120	7,480
Electric Power, (kW)	730	920
Export Steam, lb/h	70,400	11,700

*"Conventional" plant uses a liquid process for CO_2 removal followed by methanation.
**Based on feed and fuel, $2.50/MM Btu (low heat value); boiler feedwater, $0.20/1,000 lb; cooling water, $0.05/1,000 gal; electric power, $0.03/kWh; superheated HP steam, $3.50/1,000 lb; capital charges, 32%/year of investment; and maintenance, 2%/year of investment for PSA plant and 3% for conventional plant.
Source: Heck and Johanson, 1978

Table 12-23
Hydrogen Purification by Polybed PSA, Applications, September 1989

Feed Source	H_2 in Feed, %	No. Units
Steam reformer	64–96	123
Refinery streams	45–92	64
Ethylene offgas	35–95	50
H_2, CO cold box	72–99	16
Partial oxidation	54–94	11
MeOH offgas	58–84	7
Chlorine offgas	98–99	9
NH_3 plant uses	51–88	10
Styrene offgas	79–88	9
Coke oven gas	50–59	10
CO-SORB offgas	43–93	6
Misc. offgas	60–98	30
Total no. of units		345

From UOP (1991A)

Figure 12-33. Large UOP Polybed PSA Unit. *Courtesy of UOP*

and H_2O from the feed gas and three parallel adsorbers (B beds) packed with an adsorbent that selectively removes CO_2, CO, CH_4, and N_2. One A bed and one B bed are operated in series during the adsorption step, but they undergo separate depressurization, rinse, purge, and repressurization cycles during subsequent operations. The exact cycle of six A bed steps and seven B bed steps is described by Sircar and Kratz (1988). Pilot plant tests indicated that the process is capable of producing ultra pure hydrogen (99.999%) at a recovery of 86–88% simultaneously with pure carbon dioxide (99.4%) at a recovery of 90+%. Since most of the CO_2 is recovered as a pure gas, the fuel gas byproduct is rich in CO, CH_4, and H_2 and has a higher heating value than the fuel gas produced by "conventional" PSA plants.

A non-hydrogen application of PSA technology is UOP's NITREX process for removing nitrogen from natural gas. The process is suggested for small to moderate gas volumes containing 5 to 50% nitrogen. The product gas contains less than the 3% nitrogen required by some pipelines. Carbon dioxide and water vapor pass through the process and leave with the sales gas (UOP, 1993B).

HYDROCARBON RECOVERY WITH SILICA GEL

Silica gel can be used to recover hydrocarbons from natural gas; however, this use is declining in favor of continuous processes such as cryogenic condensation and absorption. The adsorption process is particularly applicable to lean gas streams containing relatively low concentrations of hydrocarbons heavier than propane. Dehydration is also accomplished during the process because of the greater affinity of silica gel for water compared to hydrocarbons. However, the hydrocarbons are normally present in higher concentrations than water; therefore, the silica gel bed adsorbs considerably more hydrocarbon liquids than water during each cycle. In view of its declining importance and the fact that hydrocarbon recovery is not considered to be primarily a gas purification operation, only a brief description of the silica gel adsorption process is presented.

Since the hydrocarbons in natural gas are present in a continuous series with only slightly different volatilities between adjacent members of the series, no clear-cut breakthrough point is observed. As a result, somewhat more complex design procedures are required to optimize the bed size and cycle time than with single component adsorption units. Empirical design methods for both the adsorption and condensation steps have been developed by Humphries (1966). His correlations are based on a considerable amount of data obtained with a small three-tower test unit processing gas from the Lolita Field in Jackson County, Texas. Most of the test data were obtained with Sorbead H adsorbent. This adsorbent is a spherical silica gel base material very similar to Sorbead R (see dehydration), but with considerably larger pores to provide a greater capacity for liquid hydrocarbons.

Silica gel adsorption plants are frequently operated with extremely short cycle times (as short as 20 minutes) in order to provide maximum hydrocarbon recovery. The effect of cycle time on recovery is illustrated by the data in **Table 12-24** for a plant processing a gas containing 0.192 gal./Mscf of pentanes and higher hydrocarbons (Mobil Oil Corp., 1971).

The operation of short-cycle units has been described in detail by Ballard in several papers (1963, 1965A, B). Many of the operating precautions that he mentions are applicable to solid-bed adsorption plants generally.

1. An adequate inlet scrubber should be provided to remove all solids and free liquids from the incoming gas stream.
2. Inlet gas nozzles and/or baffle plates should be designed to prevent bed agitation at the top of the tower.

**Table 12-24
Effect of Cycle Time on Recovery**

Cycle time, hr	Hydrocarbon Recovery, gal./MMscf
32	10
16	20
8	30
4	40

3. A buffer layer equal to about 5% of the total bed volume should be used on top of adsorbents which fracture in the presence of free liquids. The buffer particle density should be similar to that of the main bed.
4. Gas flow through the bed should be held to about 50% of the design capacity for the first 10 or 12 cycles to let the adsorbent cure properly.
5. Rapid pressure changes should be avoided. Pressure reduction in particular can cause adsorbent breakage.

ORGANIC VAPOR ADSORPTION ON ACTIVE CARBON

The removal of organic vapors from air and other gases by active carbon is probably second in importance to dehydration as an industrial application of adsorption. Active carbon is the preferred adsorbent in this application because of its selectivity for organic compounds.

The process is of particular importance in the removal and recovery of volatile organic compounds (VOCs) and the removal of obnoxious odors or other trace impurities from air streams in connection with air pollution control. It is used to a lesser extent to recover hydrocarbons from natural, manufactured, and coke oven gases, although this was formerly a major application.

The use of active carbon to control VOC emissions is of increasing importance. Over 700 plant locations employing adsorption systems are listed in a 1983 EPA report (Troxler et al., 1983). This report covers full-scale activated-carbon vapor-phase adsorption applications and gives specific flow rates, chemicals adsorbed, and sources of emissions. A more recent report (EPA, 1988) presents detailed test data from 12 different sites that remove a wide variety of organic compounds. The report concludes that continuous removal efficiencies over 95% are achievable with the process.

A comprehensive study of the costs of controlling emissions from 102 plants by carbon adsorption or catalytic incineration was conducted by Du Pont (Kittleman and Akell, 1978). Only sources emitting more than 3 lb/hr were reported, and carbon adsorption was the method selected for 60% of the sources in the small (3 to 8 lb/hr) and large (over 15 lb/hr) categories and for 90% of the intermediate size sources. The study showed that the cost of abatement increased very rapidly (per lb of organic material recovered) as the size of the source decreased below about 100 lb/hr. It was concluded that about 75% of the inventoried emissions came from sources that emit more than 500 lb/hr, and these could be controlled (80% removal) for only about 10% of the cost required to control all of the inventoried sources.

The term "active carbon" covers a multitude of carbon-based materials that possess adsorptive power. Many manufacturers of these materials refer to their products as "activated carbon," and the two terms are considered synonymous and used interchangeably. The term "activated charcoal" is generally reserved for those carbons that are derived from woody materials.

The gas-adsorptive properties of wood charcoal were recognized as early as 1773 by Scheele (Deitz, 1944); however, it was not until the middle of the nineteenth century that the property was utilized commercially. This application consisted of the use of charcoal air-filters in the ventilation and disinfection of sewers (Stenhouse, 1961). The next major step in the development of active carbon for gas adsorption occurred during World War I, when the use of poison gas by the Germans made it necessary to develop a hard, granular carbon for gas-mask use.

The principal operation in the manufacture of active carbon is the heating of carbon-containing material so that volatile components, either originally present or formed during the heating operation, are distilled off, leaving a highly porous structure. Source materials that have been used for the production of active carbon include bones, coal, coconut shells, coffee grounds, fish, fruit pits, kelp, molasses, nutshells, peat, petroleum coke, rice hulls, sawdust, and wood of all types. The amount of activity developed by simple heating is dependent upon the composition and properties of the raw materials. With many substances, additional steps are required to obtain a highly active carbon. Two such methods, which have been used extensively, are the incorporation of chemical additives (particularly metallic chlorides) into the pulverized carbonaceous material before heating and the controlled oxidation of the char by using suitable oxidizing gases at elevated temperatures.

Properties of Gas Adsorption Carbons

In addition to a high degree of activity, special requirements must be met by gas-adsorption carbons. They must possess sufficient strength to withstand abrasion, they must be available in granular form to provide low-pressure-drop beds, and they must be as dense as possible to minimize the adsorber-space requirement.

Of the many carbonaceous materials that form active charcoal, only relatively few—coconut shells, fruit pits, and cohune and babassu nutshells—readily yield chars with all of the properties desired for gas-adsorbent use. Because of the limited supply of these materials, special preparatory treatments have been developed to enable other base materials to be used for gas-adsorbent carbon. In its most common form, the pretreatment consists of pulverizing carbonaceous material, incorporating a suitable binder, and pelleting or extruding to form a dense, compressed material. The pellets or spaghetti-like extrusions are then carbonized at temperatures from 700° to 900°C. Various types of wood and coal have been found to be suitable base materials, and materials such as sugar, tar, and lignin can be used as binders.

As can be realized from the above discussion, active carbon is not a single, standardized adsorbent. Not only do carbons from different sources differ in appearance and adsorptive capacity, but also in selectivity for various gases. In general, however, for compounds of similar molecular structure, the quantity retained on any active carbon will increase with increased molecular weight or critical temperature. These effects are illustrated in **Figures 12-34** and **12-35,** which present adsorption isotherms for several hydrocarbons on a coconut-shell carbon (Columbia G grade activated carbon) and on a coal-base material (Lewis et al. 1950). The figures show the coconut-shell carbon to have a higher capacity, although somewhat less selectivity, for propane over ethane. Several exceptions to the molecular-weight, critical-temperature rule are noted. Propylene is adsorbed more strongly than propane and 1-butene more

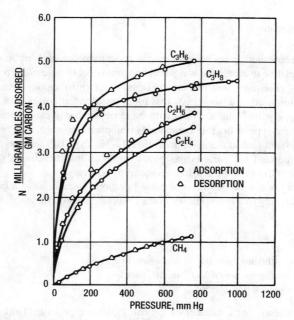

Figure 12-34. Adsorption isotherms for hydrocarbons on activated coconut-shell carbon at 25°C (77°F). *Data of Lewis et al. (1950)*

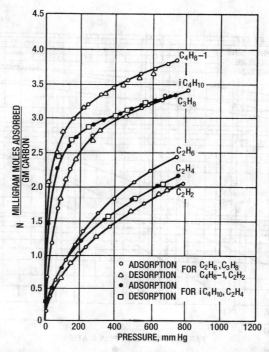

Figure 12-35. Adsorption isotherms for hydrocarbons on coal-based activated carbon at 25°C (77°F). *Data of Lewis et al. (1950)*

1090 Gas Purification

strongly than isobutane, both cases being in reverse order to molecular weight. Acetylene is in the correct order relative to molecular weight, although it is out of place with regard to critical temperature. These exceptions illustrate an additional rule (which applies even more strongly to silica gel and some other adsorbents), that more unsaturated compounds are generally adsorbed more strongly than saturated ones with the same number of carbon atoms.

Isotherms for three typical organic compounds (thiophene, toluene, and hexane) on Calgon type BPL activated carbon are given in **Figure 12-36**. Type BPL carbon is a coal-based adsorbent recommended for various gas phase applications such as solvent recovery, removal of organic sulfur compounds, and hydrocarbon recovery.

Isotherms for organic compounds on activated carbon typically have a convex upward shape and approximately fit a curve known as the Freundlich isotherm, which has the following equation:

$$w_e = k\, P^m \tag{12-22}$$

Where: w_e = equilibrium capacity, lb adsorbate/lb carbon
P = partial pressure of VOC in gas, psia
k, m = empirical parameters

Freundlich parameters for a number of typical VOCs are given in **Table 12-25**. These parameters are reported by Vatavuk et al. (1990) and are based on Calgon data (1980). They apply only to the indicated partial pressure ranges. Equilibrium capacity data outside these ranges and for compounds not included in the table can usually best be obtained from activated carbon suppliers.

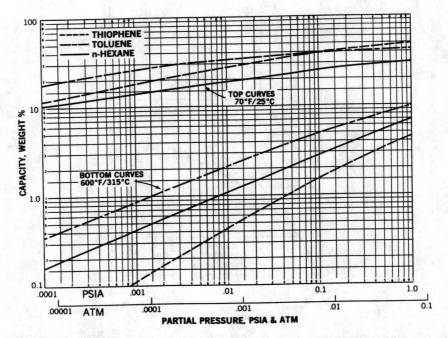

Figure 12-36. Adsorption isotherms for three organic compounds in air at 70°F and 600°F on Calgon BPL carbon. *From Calgon Carbon Corp. (1987)*

Table 12-25
Freundlich Equation Parameters for Selected Adsorption Isotherms

Adsorbate	Adsorption Temp. (°F)	Isotherm Parameters k	m	Range of Isotherm (psia)
Benzene	77	0.597	0.176	0.0001–0.05
Chlorobenzene	77	1.05	0.188	0.0001–0.01
Cyclohexane	100	0.508	0.210	0.0001–0.05
Dichloroethane	77	0.976	0.281	0.0001–0.04
Phenol	104	0.855	0.153	0.0001–0.03
Trichloroethane	77	1.06	0.161	0.0001–0.04
Vinyl Chloride	100	0.200	0.477	0.0001–0.05
m-Xylene	77	0.708	0.113	0.0001–0.001
	77	0.527	0.0703	0.001–0.05
Acrylonitrile	100	0.935	0.424	0.0001–0.015
Acetone	100	0.412	0.389	0.0001–0.05
Toluene	77	0.551	0.110	0.0001–0.05

Notes:
1. Data from Vatavuk et al. (1990).
2. Based on Calgon type BPL carbon.
3. Equations should not be extrapolated beyond the indicated ranges.

A new form of carbon, activated carbon fiber (ACF), has recently been commercialized and shows great promise for VOC removal (Kenson, 1990). Beds of the material consist of mats of extremely fine, microporous, carbon filaments. The individual micropores extend inward from the fiber surface; whereas, in granular carbon the micropores are extensions of much larger macropores, which are in contact with the surface. Since solvent molecules do not have to diffuse through macropores en route to the final site of adsorption in the micropores in ACF, adsorption and desorption are much faster in ACF than in granular carbon. This allows the use of very short regeneration cycles for ACF (about 10 minutes) versus regeneration time of 30 to 60 minutes for granular carbon beds.

ACF is claimed to be superior to granular activated carbon for adsorbing reactive compounds such as cyclohexanone and styrene for the following reasons:

1. The low metals content of ACF results in low catalytic activity reducing the rate of reaction of the solvent heel with air or steam.
2. The thorough regeneration possible with ACF leaves only a small heel of solvent in the bed to react and build up heat.
3. The small fiber diameter and low fiber density allow efficient removal of exothermic reaction heat from the bed.

The ability to operate with very short adsorption and desorption times allows ACF systems to use small beds and short cycles and also makes it suitable for use in rotary devices where the bed moves from adsorption to desorption zones. A rotating adsorber using ACF, called the KPR system, is described in a later section of this chapter.

Standardized tests have been developed to aid in the evaluation of active carbons with regard to capacity for organic vapors, retentivity of the adsorbed vapors, service life, hardness, and other factors. The following tests are of particular value.

Carbon Tetrachloride Activity. The numerical value obtained from this test indicates the adsorptive capacity of the carbon for concentrated organic vapors. It is obtained by measuring the quantity of carbon tetrachloride vapor adsorbed, at 25°C and 760 mm Hg, from air that has been saturated with carbon tetrachloride vapors at 0°C, and is expressed as a percentage of the original charcoal weight.

Carbon Tetrachloride Retentivity. This factor represents the ability of a carbon to retain a previously adsorbed vapor. Its numerical value indicates the percent by weight of carbon tetrachloride remaining in the carbon (based on the weight of activated carbon) after blowing dry air at 25°C for 6 hr through the bed saturated from the activity test.

Minute Service. This is a measure of the length of time during which a specified thin bed of active carbon will completely adsorb an organic vapor, preventing any breakthrough of the vapor through the carbon. The service life is measured with chloropicrin vapors under standardized test conditions.

Hardness (C.W.S. ball-abrasion method). This is a measure of the resistance of activated carbon to mechanical breakage during handling and use. The value represents the percentage of 6- to 8-mesh carbon remaining on a 14-mesh screen after shaking with steel balls under specified conditions for 30 min.

Typical test values and other properties of several grades of commercial activated carbon are presented in **Table 12-26**.

Table 12-26
Typical Properties of Commercial Activated Carbons

Manufacturer	Calgon	Barnebey & Sutcliffe		Westates	
Carbon Designation	BPL	1261	AC	KP 601	VOCarb
Base	coal	coconut shell	coconut shell	coal	coconut shell
Total Surface Area, BET, m^2/g, (typ.)	1,050–1,150	—	—	1,100	1,250
Packed Apparent Density, g/cc (typ.)	0.48	0.48–0.50	0.50–0.54	0.40–0.41	0.47
Carbon Tetrachloride Activity, %	62 (min.)	60 (min.)	55–65	65 (typ.)	65 (typ.)
Hardness, Ball Abrasion, %	—	97 (min.)	95 (typ.)	95 (min.)	97 (min.)
Moisture, % (max.)	—	5	5	2	2 (max.)
Total Ash, % (max.)	—	5	5	—	2 (max.)
Specific Heat, at 100°C	0.25	—	—	—	—

Data provided by Calgon Carbon Corporation; Barnebey & Sutcliffe Corporation; and Westates Carbon, Inc.

VOC Removal and Solvent Recovery with Active Carbon

In many industrial processes, relatively volatile organic solvents are used as carrier liquids and dissolving agents. During certain processing steps, these solvents are vaporized into the air. In many cases, the removal of vaporized solvent from the air and its recovery for reuse is an economic necessity. In other cases, its removal is desirable to prevent air pollution. Typical commerical units are illustrated in **Figures 12-37** and **12-38**. Solvent-recovery processes are important in plants manufacturing cellulose acetate rayon, plastic-coated paper or cloth, plastic films, rubber products, and smokeless powder. They are also used in connection with such operations as solvent extraction, high-speed printing, painting and varnishing, and degreasing of metal parts.

Some form of activated carbon is used in these processes rather than silica- and alumina-base adsorbents, because of carbon's selectivity for organic vapors in the presence of water. Typical solvents, which can be recovered from air streams by activated carbon, include hydrocarbons such as naphtha or petroleum ether; methyl, ethyl, isopropyl, butyl, and other alcohols; chlorinated hydrocarbons such as carbon tetrachloride, ethylene dichloride, and propylene dichloride; esters such as methyl, ethyl, isopropyl, butyl, and amyl acetate; acetone and other ketones; ethers; aromatic hydrocarbons such as benzene, toluene, and xylene; carbon disulfide, and many other compounds.

Figure 12-37. Automatically operated package-type activated-carbon unit for gas purification or solvent recovery. *Courtesy of Barnebey & Sutcliffe Corp.*

Figure 12-38. Large adsorption unit installed to recover acetone from acetate-yarn manufacturing operations. Outdoor installation was possible in this case because of the mild climate. This unit was preassembled in New Jersey, then dismantled and shipped in sections for reassembly at the site (Ocotian, Mexico). *Courtesy of Union Carbide Corp.*

Spivey (1988) points out that activated carbons are generally suitable for the following ranges of VOCs:

1. VOCs with molecular weights between roughly 50 and 200 corresponding to boiling points between about 67° and 350°F.
2. All aliphatic and aromatic hydrocarbons that fulfill the requirements of item 1, i.e., carbon numbers between about C_4 and C_{14}.
3. Most common halogenated solvents (subject to item 1), including carbon tetrachloride, ethylene dichloride, methylene chloride, perchlorethylene, and trichloroethylene.
4. Most common ketones (e.g., acetone, methylethylketone) and some esters (e.g., butyl and ethyl acetate).
5. Common alcohols (ethanol, propanol, butanol).

Several types of organic compounds are not suitable for adsorption in conventional activated carbon systems. These can be categorized as follows (Spivey, 1988):

1. Compounds that react with carbon or with the steam used for regeneration, i.e., organic acids such as acetic acid, aldehydes such as formaldehyde, some ketones such as cyclohexanone, some easily hydrolyzed esters such as methyl acetate, and some easily hydrolyzed halogenated hydrocarbons such as ethyl chloride.

2. Compounds that polymerize on the carbon.
3. High molecular weight compounds that are difficult to remove such as plasticizers, resins, heavy hydrocarbons, phenols, glycols, and amines.

In some cases, process modifications can be made to accommodate the compounds listed previously as unsuitable. These include the use of vacuum regeneration to remove high molecular weight compounds from the carbon and the use of a hot inert gas instead of steam to permit the adsorption of readily hydrolyzable compounds. Another approach is to use special adsorbents, such as activated carbon fibers, which are less subject to the listed problems.

Many carbon adsorption systems for VOC removal are installed for the specific purpose of air pollution abatement rather than solvent recovery. Examples of such systems are adsorbers that purify the exhaust air from sewage treatment units and those that purify the exhaust from stripping columns that use air to remove VOCs from water. In these systems the spent carbon may be discarded, returned to the manufacturer for high temperature regeneration, or regenerated without solvent recovery. It is not uncommon for the activated carbon to be regenerated by purging with hot air to produce a VOC-containing gas, which is much more concentrated than the original exhaust stream. The concentrated exhaust can then be incinerated at lower cost than incineration of the entire original exhaust stream.

Process Description

A flow diagram of a typical solvent recovery plant is shown in **Figure 12-39**. As in dehydration, two adsorbers are normally used, with one adsorbing while the other is being regenerated. Regeneration is accomplished by passing low-pressure steam upward through the bed. The steam raises the temperature of the bed, reducing its equilibrium capacity for the adsorbed vapors, provides latent heat of vaporization for the solvent, and acts as a stripping vapor to reduce the partial pressure of solvent in the vapor phase. The steam passes out of the vessel with the desorbed solvent, and both are then condensed to permit solvent recovery.

In the flow diagram shown, steam and solvent vapors discharged from the adsorber being regenerated are condensed in a water-cooled heat exchanger, the condensate is collected in a separate vessel, and the solvent, which is insoluble in water, is decanted and returned to the process. When the solvent is partially or completely soluble in water, a more elaborate separation step is required. This step, which usually includes distillation, must, of course, be designed for the specific separation involved.

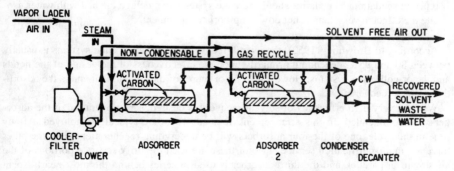

Figure 12-39. Flow diagram of typical solvent recovery plant.

Some steam is condensed and adsorbed in the bed during the stripping operation, and this quantity of water must be removed from the bed before the next regeneration cycle in order to prevent a buildup of water in the adsorption bed. With carbons that have a high, selective adsorption capacity for the solvent in the presence of moisture, this excess water can be removed during the adsorbing period by vaporization into the air that passes through the bed. If the solvent is not adsorbed strongly in the presence of water, it is necessary to dry the bed between regeneration and adsorption cycles. This may be accomplished by passing dry air through the bed for a short period of time.

The vapor-laden air feed to the system is usually filtered at the inlet of the recovery plant to prevent dust or other particulate matter from entering the activated carbon bed. The concentration of solvent vapors in the air entering most solvent recovery units ranges from about ¼ to 2 lb/cu ft. With flammable solvents, fire-hazard considerations limit the maximum allowable solvent concentration. This should be less than 50% of the lower explosive limit (LEL) when continuous monitoring is used, and less than 25% of the LEL when intermittent monitoring is employed (Parsons, 1988). Data on LEL for organic compounds are available from the Underwriters' Laboratories, Inc. Values for common solvents in air are given in Chapter 13 (**Table 13-4**).

Special care must be taken in the design and operation of adsorption units handling ketone solvents, such as methyl ethyl ketone (MEK), methyl isobutyl ketone (MIBK), and cyclohexanone, because of their reactivity in the presence of activated carbon. Traces of metal compounds in the carbon act as catalysts for decomposition reactions, which are further accelerated by the presence of heat and moisture. The decomposition reactions are exothermic and if allowed to continue can cause "hot spots" in the bed and ultimately fires. According to Collins (1988), ketone adsorption and recovery can be accomplished safely if these guidelines are followed:

1. Humidity levels should be maintained in the range of 40–60% to provide sufficient moisture in the carbon to dissipate ketone reaction heat. Extended adsorption cycles, which might dry the bed excessively, should be avoided.
2. Air and steam should be uniformly distributed to avoid the accumulation and retention of solvent in one portion of the bed.
3. A separate cooling cycle should be used to cool and condition the carbon bed after regeneration with steam.
4. Instrumentation should be installed to provide an alarm signal in case of high bed temperature. Carbon monoxide monitoring of the effluent gas is also useful to detect the occurrence of oxidation. The alarms should activate systems that deluge the bed with water, isolate the affected vessel, and shut down appropriate equipment.

According to Browning (1952), the overall recovery efficiency of such systems is usually between 80 and 95%, depending primarily upon the effectiveness of the design of the hoods and vapor-collection systems. However, 99% (or more) of the solvent that enters the adsorption plant is recovered.

Where a very high efficiency of solvent recovery is desired, series operation of the adsorbers is recommended. In this arrangement, four adsorbers are typically employed. At any time in the cycle, one of the four adsorbers will be undergoing regeneration while the other three are in air-purification service. Two of these are used to purify separate portions of the air stream in parallel, and the third adsorber is used in series behind the most nearly spent adsorber. This permits each adsorber to be operated until it actually passes some solvent and thus acquires a higher solvent charge than would be possible if it were taken off stream before any solvent appeared in the product air. This type of series operation reportedly results in a solvent-recovery efficiency of 99.7 to 99.8% of the solvent in the entering air (Browning, 1952).

Two adsorption process modifications have been proposed by Mattia (1970A) for application to contaminated air streams in which the organic vapor concentrations are too low to be recovered economically by conventional means. Both techniques involve passing the contaminated air through a bed of active carbon as the primary gas-purification step. They differ from conventional systems, however, in the carbon regeneration step. In one version, called the Zorbcin Process (Mattia, 1969), regeneration is accomplished by blowing hot air through the bed. A blowdown stream of the regeneration air is passed to an incinerator (either thermal or catalytic) where the stripped organic vapor is burned and a hot flue gas is produced. A regulated amount of this hot flue gas is returned to the regeneration cycle to provide heat for regeneration. A small amount of natural gas is also burned in the incinerator to provide the initial heat requirements.

In the second process modification (Mattia, 1970B), solvents contained in the stripping gas from the primary adsorbers are recovered by passing the gas through a secondary adsorber. In this case, heat is provided for regeneration of the primary adsorbers by passing a portion of the recirculating gas stream through a fired heater. The secondary adsorber is regenerated by passing steam through the bed, and the solvents are recovered by condensing the steam-solvent mixture and separating the solvent by decantation or distillation.

A third process modification, based on the use of vacuum to aid regeneration, has also been described (Parmele et al., 1979); (Kenson, 1979). Vacuum regeneration can be cost effective when conventional steam injection causes a corrosion problem or when the absorbate may undergo polymerization or reaction due to the high temperature or presence of water during steam regeneration. Vacuum-stripped carbon adsorption systems have been used for a wide variety of industrial emission sources, including plants manufacturing household products, magnetic tape, pharmaceuticals, and polymers. In one system, controlling emissions from a pill-coating operation, the bed is heated to ~65°C by a combination of convective heating using a preheated gas stream and conductive heating from elements embedded in the carbon, and regeneration is accomplished by subjecting the heated bed to a 1 mm Hg vacuum. The desorbed organics are removed from the adsorber by a closed-loop purge gas stream that is passed through a refrigerated condenser and returned to the adsorber. The condensed organic liquid is collected in a condensate recovery tank.

In many solvent recovery systems, adsorption represents only one step in a complex series of chemical engineering operations. The design of a complete system for recovering methylene chloride and methanol from air emitted from a dryer in a resin processing plant has been described by Drew (1975). The overall solvent recovery system includes a water scrubber to remove resins and cool the air to 100°–110°F; a standard 2-bed carbon adsorber unit designed for 95% solvent removal efficiency; a condenser and decanter to handle the vapors that are stripped from the carbon by steam; an extraction column in which water is used to remove the water soluble methanol from the methylene chloride phase; a stripping column to remove dissolved methanol and methylene chloride from the waste water; and a drying column to remove water from the recovered methylene chloride. These items of equipment and operations are representative of those required for complete solvent recovery systems; however, each system must, of course, be tailored to the properties of the specific solvent involved.

Design Approach. Design of a fixed-bed adsorption system for removing VOCs from an exhaust gas normally includes the following steps, although not necessarily in the order given:

1. Select an adsorbent. This can best be accomplished on the basis of operating experience with similar systems and vendor recommendations for the specific VOCs and conditions involved.
2. Calculate the required bed area. This is based on the volume of gas to be treated and a selected superficial gas velocity. A lower limit of 20 ft/min. is suggested to assure proper air distribution (EPA, 1988). Typical velocities are between 50 and 100 ft/min. Several beds operating in parallel may be required to provide the calculated area.
3. Select an appropriate bed depth. Typical depths are in the range of 1.5 to 3.0 ft. An adequate depth is needed to assure that the MTZ, which is usually on the order of 3 in. deep (EPA, 1988), occupies only a small fraction of the total bed. The maximum bed depth of 3.0 ft is based on pressure drop considerations and much higher values can be used if required.
4. Establish an overall adsorption-desorption cycle. Steps 2 and 3 give a preliminary value for the volume of adsorbent required in the adsorption step. The time available for adsorption is equal to the net working capacity of this quantity of adsorbent (lb adsorbate) divided by the amount to be removed from the gas (lb adsorbate per hour). Typical adsorption times range from 1 to 12 hours. In a two-bed system, the minimum adsorption time is that required for regenerating the other bed. The regeneration period must normally include time for heatup, steam purging, cooldown and drying, and standby to allow for valve switching and a margin of safety. The total time required is typically 1 to 2 hours.

   In a multi-bed system, the relationship between the number of beds adsorbing (N_a); the number desorbing (N_b); and the times required for each operation (t_a and t_b, respectively) is as follows (Vatavuk et al., 1990):

$$t_d \leq t_a (N_d/N_a) \tag{12-23}$$

   For example, in a system of seven beds (six adsorbing and one being regenerated), if an eight-hour adsorption time is desired, it is necessary that the bed being regenerated complete the entire procedure (desorption, cooling, and standby) in less than 1⅓ hour (8 × ⅙). Since the frequency of regeneration affects the life of the adsorbent, it has been suggested that the optimum regeneration frequency for systems treating gas streams with moderate to high VOC loadings is once every 8 to 12 hours (Vatavuk et al. 1990).
5. Adjust the preliminary bed area and depth estimates to provide acceptable vessel sizes and cycle times.
6. Estimate the steam requirement for regeneration. Typical steam requirements are in the range of about 1.5 to 6 pounds of steam per pound of adsorbate removed, although values below 1.0 and above 20 have been noted (Nelson et al., 1984). The amount of steam required depends primarily on the size and loading of the carbon bed. Large beds with high loading require the least steam per pound of adsorbate. However, it also depends upon the type of VOC adsorbed, the degree of regeneration desired, and other factors. Vatavuk et al. (1990) recommend that a value of 3.5 lb steam/lb of adsorbed VOC be used for preliminary estimates.
7. Design the overall system, including vessels, internals, fans, pumps, condenser, decanter, instrumentation, ductwork, and piping.

Adsorption Vessels. Activated carbon beds are normally contained in horizontal, cylindrical vessels for large plants and in vertical vessels for smaller installations. Downflow of the gas being purified is employed to minimize movement of the adsorbent (which is relatively light). The granular carbon is normally supported by structural members and a screen. The screen size recommended by one manufacturer for various mesh sizes of carbon is presented in **Table 12-27**.

The quantity of carbon required depends on the cycle time and the working capacity of the adsorbent. The working capacity is the net amount of VOC adsorbed per pound of carbon per cycle. It is essentially the equilibrium capacity at adsorber inlet conditions corrected for: (1) the "heel" of adsorbate remaining on the carbon after regeneration, (2) the loss of capacity with time until the carbon is replaced, and (3) the portion of the bed at the exit end, which cannot be allowed to attain equilibrium loading due to the presence of the MTZ and any margin of safety. Values for the working capacity are best obtained from operating experience with similar systems; however, if no data are available, Vatavuk et al. (1990) suggest using a value of 50% of equilibrium capacity at the adsorber inlet.

The capacity of activated coconut-shell charcoal for several common VOCs is indicated by the data of **Table 12-28**. The values given, which are based on the data of Lamb and Coolidge (1920), do not represent complete equilibrium, but are very close to it for the conditions 0°C (32°F) and 10 mm Hg pressure of solvent vapor over the carbon. More detailed data for one of the solvents (benzene) are presented in **Figure 12-40** to illustrate the typical effects of the partial pressure of the solvent and the temperature on the quantity adsorbed. The two upper curves represent adsorber operating conditions and the two lower curves represent conditions during regeneration. Adsorption isotherms for other VOCs have generally similar shapes as discussed in the previous section entitled "Properties of Gas-Adsorption Carbons."

When the gas contains several species of VOC, the more strongly adsorbed will displace less strongly adsorbed compounds during the adsorption cycle, and, as a result, several mass transfer zones will move through the bed. It is necessary to size the bed on the basis of its capacity when the leading MTZ reaches the end of the bed and breakthrough of this component would occur.

In most installations, the beds are switched at the end of each cycle by automatically operated valves. The cycle time may be based on a fixed period, which is known to be adequate

**Table 12-27
Screens Recommended for Support of Granular Activated Carbon**

Carbon mesh size	Screen specifications		
	Mesh	Wire diameter, in.	Opening, in.
4 × 8	10	0.035	0.065
6 × 10	12	0.028	0.055
8 × 12	14	0.025	0.046
10 × 20	24	0.014	0.0277
12 × 30	35	0.012	0.0166

Courtesy of Barnebey & Sutcliffe Corp.

Table 12-28
Carbon Capacity and Integral Heats of Adsorption for Organic Compounds on Activated Coconut-shell Charcoal

Compound	Formula	Solvent adsorbed (0°C and 10 mm Hg vapor pressure)		Integral heat of adsorption (1 g-mole of solvent on 500 g of carbon at 0°C†), cal
		ml gas/g*	lb/100 lb carbon	
Ethyl chloride	C_2H_5Cl	109	31	12,000
Carbon disulfide	CS_2	125	43	12,500
Methyl alcohol	CH_3OH	110	16	13,100
Ethyl bromide	C_2H_5Br	92	45	13,900
Ethyl iodide	C_2H_5I	106	74	14,000
Chloroform	$CHCl_3$	92	49	14,500
Ethyl formate	$HCOOC_2H_5$	97	32	14,500
Benzene	C_6H_6	97	34	14,700
Ethyl alcohol	C_2H_5OH	—	—	15,000
Carbon tetrachloride	CCl_4	78	54	15,300
Diethyl ether	$(C_2H_5)_2O$	84	28	15,500

*Gas volume measured at standard condition (0°C, 760 mm Hg).
†Equivalent to a solvent concentration of 44.6 ml gas/g carbon.
Source: Based on data of Lamb and Coolidge (1920)

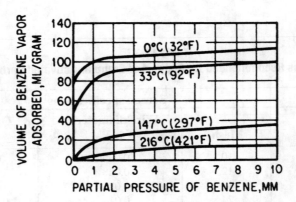

Figure 12-40. Equilibrium capacity of activated coconut-shell charcoal for benzene. Gas volume measured in ml at standard conditions (0°C, 760 mm Hg); charcoal outgassed at 550°C before tests. *Data of Coolidge (1924)*

to prevent VOC carry-through with the air stream, or may be varied to accommodate differing conditions. In one arrangement, a vapor detector is installed to monitor air discharged from one portion of the carbon bed so that a cycle change can be made when the bed is fully charged with adsorbate (Ray and Logan, U.S. Patent 2,211,162). Where an occasional trace of VOC can be tolerated in the outlet gas stream, a continuous monitor of the product gas can be used to take the adsorber off line when a specified concentration is reached. The advantage of using continuous impurity monitors to determine cycle times is that they permit full utilization of bed capacity and result in less frequent regeneration with a corresponding increase in bed life.

Air Velocity and Pressure Drop. The cost of power to move air through the system constitutes one of the major operating expenses of active-carbon, solvent-recovery installations. In view of this, it is imperative that pressure drop through the beds be kept to a minimum, and adsorbent depths greater than about 3 ft are seldom used. The pressure drop can be estimated by the use of generalized correlations (see **Figure 12-13,** page 1001); however, since these units are normally operated with the same gas (air), at approximately the same temperature and pressure (atmospheric), specific pressure-drop correlations can be conveniently used. **Figure 12-41** presents pressure-drop data for activated carbon of various particle-size grades (based on Tyler-screen cuts) with atmospheric air at 70°F, flowing downward. Pressure drop data for natural gas flowing through Calgon BPL, 4 × 10 mesh carbon at 70°F, and various pressures are given in **Figure 12-42.** Recommended maximum downflow and upflow velocities for natural gas flowing through Calgon BPL, 4 × 10 mesh carbon, as a function of pressure, are given in **Figures 12-43** and **12-44.** Curves are presented for operation at 70°F, representing adsorption conditions, and 600°F, representing regeneration with a hot inert gas (Calgon Carbon Corporation, 1987). The maximum allowable velocity to prevent attrition is sometimes estimated from the correlation of Ledoux (1948) (equation 12-14,

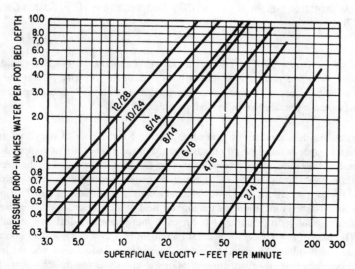

Figure 12-41. Pressure drop through dry-packed activated-carbon beds for air flowing downward at 1 atm pressure and 70°F. *Union Carbide Corp.*

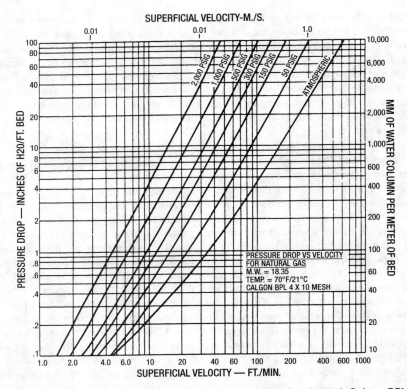

Figure 12-42. Pressure drop vs. velocity for natural gas flowing through Calgon BPL 4 × 10 carbon. Molecular weight of gas = 18.35; temperature = 70°F. *Calgon Carbon Corp. (1987)*

page 1010) for the case of upflow gases. This equation can also be used as a guide for the downflow case. A more practical limitation is to control the velocity so that the pressure drop through the bed does not create a pressure at the bottom layer of adsorbent greater than that required to cause breakage of individual particles.

Heat Effects. As described in the preceding section for the case of dehydration, heat is evolved when a vapor is taken up by an adsorbent. For the case of organic-vapor adsorption on active carbon, this heat of adsorption cannot be assumed to be equal to the heat of condensation from vapor to liquid phase, although in general, compounds with high heats of condensation (or vaporization) also have relatively high heats of adsorption. The initial portion of a vapor adsorbed on carbon has the highest heat of adsorption, and thereafter the heat of adsorption normally drops somewhat for each increment as additional quantities are adsorbed. The heat of adsorption for a small increment of vapor at a constant concentration in the carbon is called the differential or instantaneous heat of adsorption, and the heat quantity evolved between two adsorbate concentrations is known as the integral heat of adsorption. This latter value is of more interest for engineering calculations and is most commonly measured with an initial adsorbate concentration of zero. Typical integral heat-of-adsorption

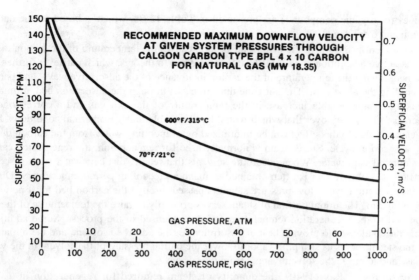

Figure 12-43. Recommended maximum downflow velocity as a function of pressure for natural gas flowing through Calgon BPL 4 × 10 carbon at 70°F and 600°F. *Calgon Carbon Corp. (1987)*

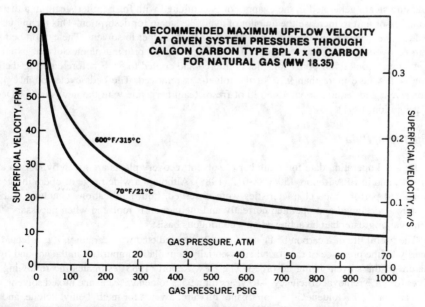

Figure 12-44. Recommended maximum upflow velocity as a function of pressure for natural gas flowing through Calgon BPL 4 × 10 carbon at 70°F and 600°F. *Calgon Carbon Corp. (1987)*

values for organic compounds are presented in **Table 12-28,** which is based on the data of Lamb and Coolidge (1920).

As solvent is adsorbed, the heat of adsorption causes the temperature of the carbon to rise in the active adsorption zone. This heat is transferred to the gas, which carries it further into the bed, raising the temperature of the carbon in advance of the adsorptive wave, and partially cooling the air stream. The hot zone thus proceeds through the adsorption bed, ultimately causing an appreciable increase in the temperature of the exit gas and a corresponding decrease in capacity over that which would be obtainable under isothermal conditions. Loss in capacity due to this effect can be minimized by evaporating water from the carbon during the adsorption cycle. Since steam is normally used for regeneration, the regenerated carbon is saturated with water. When adsorbing solvents that readily displace water from carbon, a sufficient cooling effect is often obtained to make a precooling period unnecessary. During the regeneration cycle, low-pressure steam is passed through the carbon bed for a period of 15 to 60 min. The initial portion of steam serves primarily to raise the temperature of the carbon bed to the regeneration temperature and is condensed in the process. Some additional steam is condensed to provide heat of desorption for the adsorbed solvent, and the remainder of the steam serves as stripping vapor to lower the partial pressure of the solvent in the vapor phase and to purge the solvent vapors from the system.

According to Ray (1940), the quantity of steam required for regeneration in a well-designed system ranges from 3 to 5 lb steam/lb solvent recovered. Steam consumption represents a major cost in the operation of activated carbon adsorption systems for solvent recovery, and techniques have been developed to reduce steam usage. In one large plant, for example, the hot mixture of steam and solvent vapor, which comes off the bed during the desorption step, is used as the heat source in a low-pressure falling film evaporator. Low-pressure steam generated in the evaporator is combined with fresh boiler steam in a thermocompressor unit to increase the amount of available steam for desorption. This technique has reduced net steam consumption from 4.5 to 2.7 lb steam per lb solvent. The carbon bed system recovers about 6,000 tons/year of toluene at a large printing plant. Solvent-laden air, averaging about 1,500 ppmv toluene, is filtered and cooled to 85°F before entering the carbon beds where more than 95% of the toluene is removed. The beds cycle roughly 4,400 times/year and require about 7,466 lb of fresh steam per cycle with the new system (Dedert Corporation, 1991).

Operating Data

Detailed operating data for a number of solvent recovery plants are given by Nelson et al. (1984), and in EPA Report 450/3-88-012 (1988). Although a wide range of operating conditions and performance characteristics were observed, both publications conclude that if a system is designed and operated correctly and the carbon is replaced when necessary, 95% removal or greater can be achieved on a continuous basis.

The useful life of a carbon bed cannot usually be predicted with certainty. It is affected primarily by the presence of dust or other essentially nonvolatile impurities in the gas and by the nature of the solvent being adsorbed. The EPA Report (1988) estimates the following bed lives for 98% removal efficiency—10 years for a 95% toluene/5% hexane mixed solvent, over 6 years for a 50% toluene/50% isopropanol mixture, 5 years for methyl ethyl ketone, and less than 5 years for a mixture containing cyclohexanone and methyl ethyl ketone.

Nelson et al. (1984) present operating data for six adsorption systems (selected from more than fifty) which represent a variety of industries and solvents. Most of the sites were visited (and tested) twice to observe the effect of age on performance. Ten sets of data are presented. Selected data from four of the data sets (two sites) are summarized in **Table 12-29**.

In most cases, the changes in performance observed between the first and second tests were attributed to factors other than carbon degradation. However, the observed change for plant 2 is believed to be due to degradation because the efficiency increased to over 98% when the carbon was replaced immediately after the second test. The results, therefore, indicate a useful life of about 4 years for carbon operating under the conditions of plant 2.

The observed steam consumption in the tests of **Table 12-29** range from 5.2 to 15 lb steam per lb solvent recovered, which is considered somewhat high for well-designed and operated plants. In the full series of tests reported by Nelson et al., steam/solvent ratios ranged from slightly less than 1.0 to about 20 lb/lb. Steam temperatures ranged from 230° to 300°F, and steaming periods from 25 to 55 minutes.

The steam consumption and other utility requirements of a commercial acetone-recovery installation are given by Ray (1940). In this plant, 4 lb of steam are used per pound of acetone recovered, and an acetone-water condensate is obtained which averages 20 to 33% acetone. A portion of the steam is required to distill this mixture. This plant also uses 8.75 gal. of cooling water (at 55°F) and 0.082 kwhr of electricity per pound of solvent. Moffett (1943) reports that 3.5 to 4.3 lb of 5-psig steam are required, per pound of recovered solvents, in a plant recovering solvents from printing operations at the *New York Daily News* Rotogravure printing plant. This steam consumption apparently also includes the amount necessary to recover the solvent dissolved in the water layer discharged from the decanter. The plant recovers over 300,000 gal./year of ink solvent. In addition to the steam requirement, the plant uses approximately 10 gal. of cooling water (at 70°F) and 0.13 to 0.17 kwhr of electricity per pound of recovered solvent.

**Table 12-29
Performance of Typical Carbon Adsorption Systems for Solvent Recovery**

Plant number	2		5	
Test	Orig.	Repeat	Orig.	Repeat
Solvent	50THF/ 50Tol.	75THF/ 25Tol.	Tol.	Tol.
Inlet air flow, scfm	9,800	9,500	9,100	7,800
Inlet solvent conc., ppmv	1,470	1,140	1,940	896
Adsorption time, min.	64	65	150	200
Desorption time, min.	32	32	50	40
Steam/solvent, lb/lb	5.2	6.5	8.2	15
Carbon age, years	2.0	3.8	5.0	6.5
VOC control efficiency, %	99.7	95.2	98.6	80.5
Capital cost, $/scfm	51		24	
Total cost, $/lb solvent	0.28		0.09	

Notes: THF = Tetrahydrofuran, Tol. = Toluene, lb solvent = pounds of solvent recovered.
 Data from Nelson et al. (1984)

1106 Gas Purification

At the Ford Motor Company's spark plug plant in Fostoria, Ohio, trichloroethylene vapor in the air from degreaser units is recovered by adsorption on activated carbon. Two 72-in.-diameter adsorbers are used, each containing 1,500 lb of carbon pellets. The system is reported to be capable of recovering 400 to 450 gal. of liquid trichloroethylene per day with a collection efficiency of over 90%. Operating costs were only about 3% of the value of the recovered solvent (Anon., 1969).

Data on European practice for solvent recovery in the printing industry are presented by Benson and Courouleau (1948), who discuss the Acticarbone process of the Parisian concern Carbonisation et Charbons Actifs. A flow diagram for a typical application of this process, employed to recover a mixed solvent containing 40% toluene, 14% butyl acetate, and 46% ethanol from printing-plant exhaust air is shown in **Figure 12-45**. Operating data for the installation and typical design ranges are given in **Table 12-30**.

Waste treatment systems can emit significant quantities of VOCs to the atmosphere and activated carbon is often used to control such emissions. The results of a pilot plant program conducted on air emissions from a covered activated-sludge unit are reported by Keener et al. (1988). The tests showed that carbon loading decreases with increasing relative humidity, increasing gas velocity, and decreasing bed height. Operating ranges included bed heights

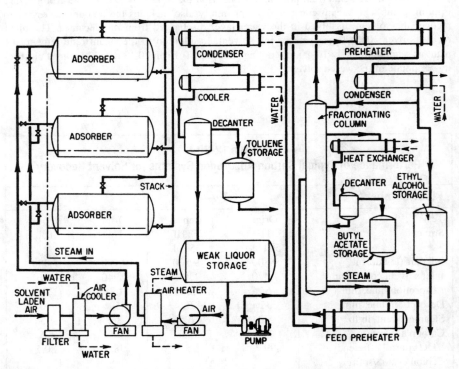

Figure 12-45. Flow diagram for Acticarbone process of Carbonisation et Charbons Actifs (a Parisian concern). This installation recovers a mixed solvent containing toluene, butyl acetate, and ethanol from a printing-plant air stream. *From Benson and Courouleau (1948)*

Table 12-30
Typical Operating Data and Ranges of Process-design Factors for Acticarbone Process Solvent-adsorption Plants

Operating data:	
Air rate to adsorbers (2 in parallel), scf/min.	11,000
Solvent content of air feed, lb/Mscf	0.43
Air temperature at adsorber inlet, °F	90
Steam rate to adsorber being regenerated, lb/hr	1,800
Steam/solvent ratio	3.5:1
Time cycle for each adsorber:	
Adsorption, hr	2
Steaming, min	30
Drying [with hot (220°F] air), min.	15
Cooling (with cold air), min.	15
Typical range of design factors:	
Solvent adsorbed by activated carbon per cycle, % by wt	10 to 14
Solvent concentration in air stream, lb/Mscf	0.3 to 0.5
Power required, kwhr/lb of solvent recovered	0.10 to 0.15
Cooling water, gal./lb of solvent recovered	7 to 10
Steam, lb/lb of solvent recovered	3.5 to 5
Activated carbon, lb/ton of solvent recovered	0.5 to 1

Source: Benson and Courouleau, 1948

from 8 to 24 inches, superficial gas velocities of 50 to 100 fpm, relative humidities of 40 to 80%, temperatures of 65° to 80°F, and inlet impurity concentrations of 21 to 83 ppm (total VOCs). Removal efficiencies were generally over 90% and breakthrough times were as long as 140 hours; however, VOC loading on the carbon was quite low, normally less than 10 lb/100 lb.

Activated carbon is also used to control emissions from stripping columns in which air is used to remove hydrocarbons from ground water. Typically, such exhaust air streams contain less than 100 ppmv VOCs and are saturated with water vapor. It is necessary to reduce the relative humidity before effective adsorption is possible. According to Valentin (1990), the air or gas should usually have a relative humidity not significantly above 80% for the adsorption of odor-causing organics. If it is higher, slight heating will bring down the relative humidity to an acceptable level without significantly reducing bed capacity. The operating temperature should not be above about 100°F and preferably should be below 80°F.

The major operating problems encountered in activated-carbon solvent-recovery processes are contamination of the activated carbon and corrosion of equipment. In improperly designed systems, attrition and plugging of the bed may also constitute problems. Contamination can occur as a result of resinous or polymerizable substances in the air stream that remain on the carbon during the regeneration cycle and reduce its activity. Certain types of contaminants that cannot normally be removed and recovered satisfactorily with active carbon can be tolerated in very small amounts, as they accumulate in the top portion of the bed, which is first contacted with the air and partially removed during the regenerative cycle. Since the major portion of the bed remains in good condition, a reasonable service life can be realized, and the carbon may occasionally be returned to the manufacturer for reactivation. Some contaminants must be removed from the vapor-laden air stream prior to its entry into the adsorbent bed. Phenolic materials, for example, which are encountered in operations designed to recover alcohol from certain resin-impregnating operations, can be removed by scrubbing with caustic solution in a packed or spray-type scrubber. Traces of polymerizable or very heavy compounds can also be removed in a guard chamber, which is placed in the gas stream ahead of the main adsorbers. The carbon or other adsorbent in the guard chamber is permitted to become contaminated in order to protect the main carbon-charge.

Although most dry solvents are not particularly corrosive, corrosive conditions may occur in the adsorption equipment as a result of the steaming operation. Many solvents hydrolyze in the presence of liquid water or steam at elevated temperatures, and activated carbon may act as a catalyst for this and other decomposition reactions. Esters such as ethyl acetate are particularly corrosive because hydrolysis results in the formation of organic acids. When electrolytes are present or are formed during the process, corrosion of iron and steel can be further accelerated by galvanic action between the carbon and the metal screen or vessel. Solvents such as hydrocarbons that do not decompose can usually be handled in plain steel equipment. Special construction materials, e.g., fiber reinforced plastic (FRP), copper, Everdur, Monel, and stainless steel, may be required for other solvents. Special adsorption-vessel designs may also be employed to minimize corrosion by preventing the wet steam-solvent mixture from contacting the vessel shell.

Oxidizing gases such as oxygen and chlorine may react with carbon under some conditions, and contact with these gases at high temperatures should be avoided. Activated carbon can be used with high-velocity air at temperatures up to about 150°C; with low air velocity, somewhat higher temperatures are permissible.

Attrition of activated carbon can be minimized by proper design of the adsorption vessels. The air-flow rate should be below 100 ft/min., preferably below 60 ft/min., and good air distribution to the top of the bed should be provided. Plugging of activated carbon beds can be prevented by eliminating any sources of carbon attrition and providing an adequate screen or filter ahead of the adsorption vessel.

Costs of VOC Recovery Systems

Correlations are given by Vatavuk et al. (1990) that are useful in developing a preliminary cost estimate for a proposed system. The total adsorption system equipment cost is

$$C_A = R_C [C_C + (N_A + N_D)C_V] \tag{12-24}$$

Where: C_A = Total adsorber equipment cost including carbon, vessels, fans, pumps, decanters, instrumentation, and piping, fall, 1989 dollars
R_C = Ratio of C_A to the cost of carbon and vessels
  = $5.82\ Q^{-0.133}$
Q = Feed gas flow rate, acfm (in the range of 4,000 to 500,000)
C_C = Cost of carbon, $, (a typical cost of $2.00/lb in fall, 1989 dollars is suggested)
N_A = Number of beds adsorbing at a time
N_D = Number of beds desorbing at a time
C_V = Cost of each vessel, $
  = $271\ S^{0.778}$ (FOB vendor in fall, 1989 dollars, for 304 stainless steel, cylindrical, low-pressure vessels)
S = External surface area of each vessel, ft^2 (in the range of 97 to 2,110)

The cost of any required external ductwork and other nonstandard auxiliary equipment must be added to C_A, and the total multiplied by 1.08 to cover sales taxes and freight. The product is the total purchased equipment cost, PEC. The PEC must be multiplied by 1.61 to cover direct and indirect installation costs. The resulting installed equipment cost plus the cost of site preparation and buildings, as required, represents the total capital investment for the fixed-bed adsorption system.

Hydrocarbon Recovery with Active Carbon

The removal of aromatics and relatively heavy hydrocarbons from gas streams with fixed beds of activated carbon is essentially the same process as solvent recovery, and similar adsorbents and equipment are used. The principal differences are that in hydrocarbon recovery the feed is typically a natural gas or other combustible gas stream rather than air, and adsorption is usually (but not always) conducted at elevated pressure. The basic design approach for hydrocarbon recovery systems follows the same general logic as that described for solvent recovery systems. A brief outline of the key design steps is given in the Calgon Carbon Corporation bulletin, *Heavy Hydrocarbon Removal or Recovery from Gas Streams* (1987).

Adsorption isotherms for several of the hydrocarbons typically removed from gas streams are given in the previous section of this chapter entitled "Properties of Gas Adsorption Carbons." Additional data can be obtained from the carbon manufacturers. A set of equilibrium data for the hydrocarbons and natural gas sulfur compounds on activated carbon (Pittsburgh BPL) has been presented by Grant et al. (1962, 1964). These data and some additional experimental data for isobutane and carbon dioxide adsorption on Columbia NXC 8 × 10 carbon have been used by Hasz and Barrere (1964) as a basis for predicting isotherms for all of the common constituents of natural gas using the Polyani potential theory of adsorption (1914).

Field test data showing the effects of natural gas composition and flow rate are given by Enneking (1966). The use of adsorption to recover hydrocarbons from gas streams at refineries and petrochemical plants is described by Cantrell (1982). The recovery of benzol (a benzene-rich light oil) from manufactured and coke-oven gas streams was formerly an important application of activated carbon adsorption, but is no longer considered significant. A typical benzol removal installation is described by Howell (1943) and Walker et al. (1944). Hydrocarbon recovery processes are not described in detail herein because of their similarity to solvent recovery processes and the fact that they are intended primarily for recovery, not gas purification.

Fluidized, Moving, and Rotating Bed Processes

Hypersorption

Many attempts have been made to develop continuous adsorption processes, but until recently they have met with only limited success. One of the earliest versions, developed by the Union Oil Co. (now Unocal), used a moving bed of activated carbon particles to recover ethylene and other hydrocarbons from gas streams. Six large plants were built between 1947 and 1949; however, commercial utilization of the process was discontinued shortly thereafter, apparently due to excessive attrition and other solids handling problems (Berg, 1951).

1110 Gas Purification

Courtaulds' Fluidized Bed Process

A later development was a fluidized bed process for removing carbon disulfide vapors from air leaving a viscose (rayon) manufacturing plant. The process was developed by Courtaulds, Ltd., and installed at its viscose plant in Holzwell, Wales.

A detailed description of the process and its development is provided by Rowson (1963). A 250,000-cfm air stream containing 1,000 ppm (vol) CS_2 and 20 to 30 ppm H_2S required treatment. Preliminary process studies indicated that a fixed-bed plant would be uneconomical. A large cross-sectional area of bed would be required (about 6,000 ft^2) to provide a reasonable air velocity. Because of the low concentration of CS_2, fixed-bed adsorption would be inefficient and the break point would occur early. The decision was therefore made to develop a fluidized bed process. Such a process offers the potential advantages of fully continuous operation, countercurrent contacting with attendant high and uniform solvent loadings on the carbon, and high gas velocities leading to small vessel diameters.

Commercial operation of the Courtaulds, Ltd., fluidized bed plant commenced in 1960, about 8 years after initiation of development work on the project. It is reported that the plant—code named Landmark—cost $1.7 million (Anon., 1963). A flow diagram of the process is shown in **Figure 12-46**. Design data and typical operating conditions are listed in **Table 12-31**.

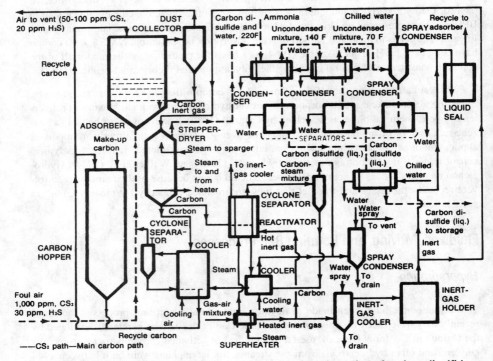

Figure 12-46. Flow diagram of fluidized bed process for adsorption of carbon disulfide on active carbon.

Table 12-31
Design and Operating Data for Fluidized Bed Carbon Disulfide Adsorption Plant
(Rowson, 1963; Anon., 1963)

Air flow rate	250,000 cfm
Inlet concentration	1,000 ppm CS_2, 30 ppm H_2S
Outlet concentration	50–100 ppm CS_2, 20 ppm H_2S
CS_2 recovery rate	1.2 tons/hr
CS_2 adsorption efficiency	85–95%
Adsorbent	6–14 mesh coconut charcoal
Adsorbent circulation rate	50,000 lb/hr
CS_2 loading on adsorbent	
lean	0.5%
rich	7.0%
Stripping steam required	12,000 lb/hr
Adsorber design	
size	38 ft dia. × 45 ft high
number of beds	5
bed depth	2–3 inches (each)
pressure drop	3 psi

Source: Rowson, 1963; Anon., 1963

Air from the viscose plant is first contacted by an alkaline ferric oxide suspension in a spray scrubber (not shown in the flow diagram) to remove the bulk of the H_2S. Removal is necessary because hydrogen sulfide is catalytically oxidized by air, in the presence of active carbon, to elemental sulfur, which is extremely difficult to strip from the carbon.

The air is forced through the fluidized bed adsorber and a dust collection system by means of a 1,000-hp fan. Carbon is conveyed to the top of the adsorber column and on to the top tray. Carbon level on the tray is maintained by a weir which it flows over into a downcomer leading to the tray beneath. After passing over a total of five trays, the carbon flows downward through seal legs into the stripper vessel. The seal legs are purged with inert gas to prevent steam and CS_2 from entering the adsorber.

The adsorbent is stripped of CS_2 by countercurrent contact with 300°F steam while passing downward in plug flow through an 18-ft-diameter vessel. The bottom half of the vessel contains a number of vertical, externally finned tubes. High-pressure steam inside the tubes heats and dries the carbon which flows downward on the outside. Steam liberated from the carbon flows upward and augments the stripping steam. The hot dry carbon is split into two streams; 10% is reactivated and the remainder is cooled to 160°F by contact with fresh air. This is accomplished in a single 7-ft-diameter fluidized bed.

Reactivation of a continuous bypass stream of carbon serves to remove elemental sulfur, sulfuric acid, and other low volatility compounds that cannot be removed by normal stripping. The operation is conducted by passing low-pressure, highly superheated steam through the carbon in a single fluidized bed.

The carbon disulfide is recovered from the exit gases leaving the stripper by a four-stage condensation train. About 2 lb/hr of ammonia is added to the exit gas before it enters the condensers to protect the aluminum tubes from corrosion. The first two stages serve to condense the bulk of the water (shown as a single unit in the flow diagram). The third stage condenses most of the CS_2, and final cleanup is obtained in the fourth stage—a spray tower using 35°F water. CS_2 and water are virtually immiscible so the heavier CS_2 can be separated from the water readily by decantation.

Operating experience with the Courtaulds plant has shown the recovery efficiency to be at least as high as for a fixed-bed plant and steam consumption a little lower. Carbon attrition has proven to be somewhat of a problem as the accumulation of fines in the bed increases the residence time, which in turn leads to a greater adsorption of water at the expense of CS_2. This problem can be partially resolved by drawing off fines collected from the exit gas instead of returning them to the bottom adsorber tray as shown in the flow diagram. However, this solution leads to an increased requirement for makeup carbon.

Purasiv HR Process

The Purasiv HR process is an improved fluidized bed, activated carbon process based on the use of a spherical beaded adsorbent developed by Kureha Chemical of Japan. The process was introduced in the United States and Canada by Union Carbide Corporation in 1978 (Union Carbide Corp., 1983). The unique form of the carbon adsorbent represents the key feature of the process. The beads, which are about 0.7 mm in diameter, create a homogeneous fluidized bed in the adsorption section and a free-flowing dense bed in the desorption section while providing a much higher resistance to attrition than conventional granular or pelletized material. The beads are produced by a proprietary process that involves shaping molten petroleum pitch into spherical particles which are subsequently carbonized and activated under controlled conditions.

A simplified flow diagram of the Purasiv HR system is given in **Figure 12-47.** The solvent-laden air is fed into the bottom of the adsorption section and passes upward countercurrent to the carbon beads, which are fluidized on a series of perforated trays. The beads flow from each tray to the one beneath it by an overflow weir arrangement and finally flow from the bottom tray into a seal pot where they form a dense-packed bed. The carbon then flows downward as a moving dense bed through a secondary adsorber section where it removes hydrocarbons from the nitrogen used for regeneration; a desorption section where it is heated while being stripped of hydrocarbons by a rising stream of nitrogen; and finally a cooler where the carbon is cooled to ambient temperature. The regenerated and cooled carbon is then conveyed to the top tray of the adsorber column. High temperature solvent-laden nitrogen from the desorption section passes through a water-cooled condenser and a separator which removes liquid solvent. Nitrogen from the separator, which is cool but still saturated with solvent, is passed through the secondary adsorber section and is then recycled to the bottom of the desorption section.

This description and the flow diagram refer to the nitrogen-stripping technique, designated Type N, which is particularly suitable for the recovery of water-soluble solvent. The process can also be used with a more conventional steam-stripping technique, designated Type S, which is useful for the recovery of chlorinated hydrocarbons and other water-insoluble solvents. The Type S system is similar to the Type N except that the secondary adsorber and provisions to recycle the stripping vapor are not required. Commercial Purasiv HR systems recovering ketones, tetrahydrofuran, toluene, isopropyl alcohol, acetone, kerosene, methyl-

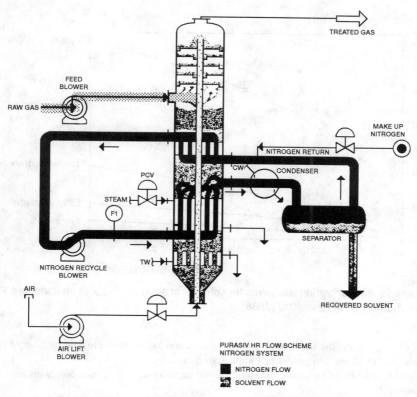

Figure 12-47. Flow diagram of Purasiv HR process. *Union Carbide Corp.*

ene chloride, and various solvent mixtures were reported to be in operation in the United States in 1983 (Union Carbide Corp. 1983).

Polyad FB Process

This process was developed by a division of Nobel Industries, a major chemical corporation in Sweden. The heart of the process is a unique adsorbent called Bonopore 110, which is described as a microporous, crosslinked polymer in the form of spherical particles with a narrow size distribution (Heinegard, 1988). The particles have a pore diameter of approximately 80 Å, a specific surface area of about 800 sq m per gram, and a diameter of about 0.5 mm. Adsorption isotherms for four typical organic compounds and Bonopore 110 are given in **Figure 12-48.** The polymer adsorbent is claimed to have the following advantages over activated carbon:

1. Easier regeneration; activated carbon has a higher percentage of very fine pores that hold adsorbate molecules more tightly.

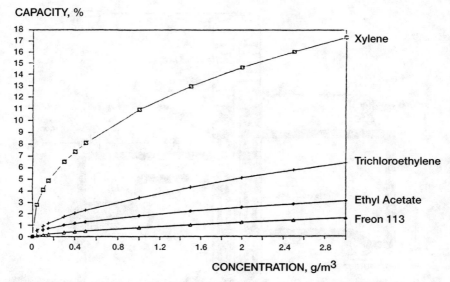

Figure 12-48. Adsorption isotherms for solvents in air at 77°F (25°C) on Bonopore 110 adsorbent. *Data of Heinegard (1988)*

2. Longer lifetime; the small pores of activated carbon can become clogged by heavy hydrocarbons resulting in degradation of adsorption capacity.
3. Less sensitivity to humidity; activated carbon capacity decreases markedly with increased humidity.
4. No catalytic effect; can be used to adsorb solvents that have a tendency to oxidize, polymerize, or hydrolyze.

Fixed bed and rotor configurations were evaluated for use with the new polymer adsorbent, but a fluidized bed process was selected as optimum to utilize its unique properties of fast adsorption kinetics, excellent abrasion resistance, and good fluidization characteristics. In the fluidized bed process, the adsorbent is transported between an adsorption section, where the incoming air is purified, and a regeneration section, where desorption occurs.

The adsorption section contains one or more beds of polymer particles maintained in a fluidized state by the flow of gas being purified. Regenerated polymer particles are continuously added to the top bed, and saturated particles are continuously removed from the bottom. The saturated adsorbent is pneumatically transported to the top of the desorber. The particles flow down through the desorber where they are heated, stripped by contact with a small stream of air or inert gas, and cooled. The cooled and regenerated adsorbent is transported back to the adsorber.

Solvent vapors are condensed out of the purge gas stream, and the condensate flows into a collection tank. The uncondensed gas is returned to the desorber at a point above the heater. The cool adsorbent in this zone removes most of the remaining solvent vapors from the purge gas, which exits the top of the desorber and is recycled to the feed gas entering the adsorber.

An economic comparison was made of the Polyad FB process, an activated-carbon fixed-bed plant, and a thermal incinerator system for purifying 6,300 scfm of air from a gravure printing plant containing 510 ppmv toluene. The study indicated the Polyad FB plant to have a capital cost about 17% less, and an operating cost about 50% less than the activated carbon adsorption plant. Its economic advantage over thermal incineration was even greater. Several full-scale Polyad FB plants have been installed in Sweden and the process is being marketed in the U.S. by Weatherly, Inc. (Heinegard, 1988).

KPR Rotating Adsorber System

The KPR system incorporates two novel concepts: the use of a rotor to move the adsorbent between adsorption and regeneration zones, and the use of activated carbon fiber (ACF) in the form of corrugated paper as the adsorbent. The process was developed in Japan and is offered in the U.S. by Metro-Pro Corporation (Kenson and Jackson, 1989; Benamy, 1987).

The rotor may be in the form of a disk or cylinder. Diagrams of the cylindrical type showing adsorbing and desorbing operations are given in **Figures 12-49** and **12-50**. The KPR system is designed to concentrate dilute VOC gas streams before final treatment by a conventional carbon adsorption system or incineration. The rotor, consisting of a honeycomb structure of the ACF corrugated paper, is slowly rotated through adsorption and desorption sectors of the apparatus. The VOC-containing feed gas and the hot air used for regeneration flow in opposite directions through the tubular paths of the honeycomb structure in their respective sectors. VOCs are removed from the gas in the adsorption sector and desorbed into the air in the regeneration sector. The regeneration off-gas contains the desorbed VOCs

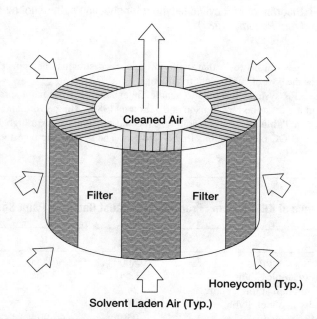

Figure 12-49. Diagram of KPR cylinder-type rotor showing path of air being purified. *From Benamy (1987)*

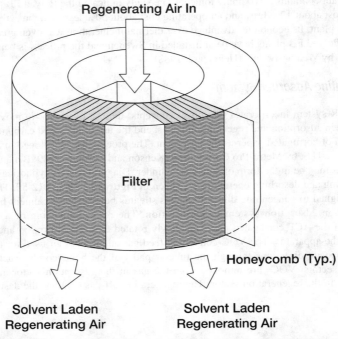

Figure 12-50. Diagram of KPR cylinder-type rotor showing path of hot air used for regeneration. *From Benamy (1987)*

in a volume of gas only ⅕ to ⅟₁₅ that of the original feed streams. The concentration effect greatly reduces the cost of subsequent treatment.

Three large KPR systems installed to handle the exhaust air from paint spray booths at an aerospace firm are described by Kenson and Jackson (1989). The performance of these units is summarized in **Table 12-32**. As would be expected, the units show a higher removal efficiency at higher VOC concentrations in the feed gas. The system can be effectively coupled

Table 12-32
Performance of KPR Systems Processing Exhaust Gas from Paint Spray Booths

System	A	B	C
System capacity, scfm	105,000	70,000	35,000
VOC quantity at inlet, lb/hr	23.04	3.668	221
VOC concentration at inlet, ppmv	24	5.8	204
VOC quantity at outlet, lb/hr	1.197	0.3623	11.1
Removal efficiency, %	95	90	95

Data of Kenson and Jackson (1989)

to a catalytic incinerator, with heat from the incinerator providing the heat required to raise the temperature of the stripping air in a gas-to-gas heat exchanger. Kenson (1991) suggests that the KPR system would also be economically attractive for control of BTX emissions from petrochemical plants.

Munters Zeol Rotating Zeolite Adsorber

A process offered by Sweden's Munters Zeol employs a large, disk-shaped ceramic rotor that is honeycombed with channels impregnated with a hydrophobic, all-silicon zeolite. Although it does not use carbon, the process is mentioned here because it is aimed at removing low concentrations (below 1,000 ppm) of VOCs from contaminated air streams. When used for this purpose, the impure air flows through one section of the rotor's channels, where the zeolite adsorbs the VOCs, while heated air or inert gas flows through another section to desorb material previously adsorbed. The assembly rotates continuously so that each channel passes through adsorption and desorption zones alternately. Adsorption typically occurs at 80°–100°F, desorption at 360°F. Reportedly, the process was first offered in the U.S. in 1991 after more than a dozen units had been sold in Europe (Anon., 1991).

Moving Bed Process

A moving bed process has been tested for use in removing VOCs from the exhaust air of an aircraft painting facility (Larsen and Pilat, 1991); however, no commercial applications are known. The test system employed a 4-in.-thick, down-flow moving bed of Westvaco BX-7540 activated carbon pellets through which contaminated air flowed horizontally, cross-flow to the carbon. The cross-section of moving bed through which the air passed was roughly the area of a 6-in.-diameter duct. The moving bed was found to have a VOC collection efficiency in the range of 77.1 to 99.6% with most of the tests showing efficiencies over 90%. The reason for the fluctuations in removal efficiency was not determined; however, the efficiency appeared to be unaffected by carbon flow rate in the range of 5 to 8 lb/hr; air flow rate over the range of 27 to 185 ft/min.; or inlet VOC concentration over the range of 0.2 to 16 ppm (as MEK). The investigators conclude that a moving bed system of this type may have application for the control of VOCs, especially at low concentrations; however, additional research and development is needed for the design of full-scale systems.

Canister and Panel Systems for Air Purification

The problem of odor removal differs from that of solvent recovery primarily in that the impurities are present in much lower quantities and that usually no attempt is made to recover the adsorbed compounds. Large volumes of atmospheric-pressure air must be handled in air-conditioning types of equipment, and a very low pressure drop is mandatory. The pressure-drop requirement makes the use of very thin beds of activated carbon desirable. Fortunately, most odorous vapors are compounds of relatively high molecular weight, which are adsorbed readily, and can therefore be removed in shallow beds. Commercial equipment for the removal of odors from air streams is designed to give maximum face-area for the passage of air in a minimum space. The equipment is unitized for ease of replacement of spent carbon, and the units are commonly in the form of cylindrical canisters or larger unit cells containing flat or corrugated beds.

1118 Gas Purification

Principal requirements of carbon for use in odor-removal equipment are a high capacity and low pressure drop. The material must also be dustless. Specifications for activated carbon used in air purification are presented in **Table 12-33**. Typical commercial equipment is illustrated in **Figure 12-51, 12-52,** and **12-53**.

Process Design

Canister and panel systems are typically integrated with a building's heating, ventilating, and air conditioning (HVAC system), and may treat 100% "fresh" air, 100% recirculated air, or some combination of the two. In cold or hot climates, it is common practice to recirculate indoor air through the carbon panels to remove cigarette smoke and other internally generated pollutants while conserving the energy required to heat or cool outside air. In mild but smoggy climates, the air treated in a carbon adsorber may consist entirely of outside air.

Standardized canister and panel adsorption system designs are offered by the manufacturers. Typically, thin beds ½ to 2 in. thick are used to minimize pressure drop. Face velocities are in the range of about 30 to 60 ft/min. A common design for large systems uses many rectangular panels 1 to 2 ft on each side, arranged in a housing to create an accordion-folded, extended sur-

Table 12-33
Specifications for Activated Carbon Used in Air Purification

Property	Specification, Minimum
Activity for carbon tetrachloride	50%
Retentivity for carbon tetrachloride	30%
Apparent density	0.40 g/ml
Hardness (ball abrasion)	80%
Size distribution	6 to 14 mesh (Tyler-sieve series)

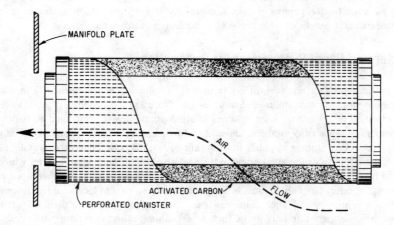

Figure 12-51. Diagram of canister-type activated-carbon air-purification unit.

Figure 12-52. Air-purification unit employing cylindrical canisters, showing typical method of installation.

face of carbon bed. An example of an activated carbon panel is a unit 18″ × 18″ × 1½″ thick, which is offered as model PAA by Barnebey & Sutcliffe Corporation (1990). When operated at its nominal flow rate of 83 cfm air, the panel has a pressure drop of approximately 0.35 in. of water. A single housing may contain more than 200 of these individual panels.

The activated carbon in HVAC systems can usually be returned to the manufacturer for regeneration after it becomes saturated with pollutants. This can be done by removing and returning either the individual units or the bulk carbon. The carbon may last as long as 4 or 5 years before replacement or regeneration is required, depending, of course, on the concentration of adsorbable impurities in the air. Extremely long cycles are possible because even relatively severe odor problems are usually caused by only very small concentrations of odorous compounds, and a good grade of carbon can pick up 0.1 to 0.2 lb of odorant per pound of carbon before regeneration is required. The upper limit of odorant concentration for the practical use of replacement cartridge-type activated carbon adsorption is reported to be about 2 to 5 ppm (Turk, 1958). Above this level the service life of the cartridges is greatly diminished, necessitating frequent reactivation.

According to Graham and Bayati (1990), regeneration of spent carbon requires a temperature of about 1,200°F and proceeds through the following steps:

1120 Gas Purification

Figure 12-53. Corrugated bed-type activated charcoal air-purification cell. This unit contains 35 lb of 50-min. activated coconut-shell charcoal and handles 1,000 cfm of air with a pressure drop of 0.25 in. of water. *Courtesy of Barnebey & Sutcliffe Corp.*

1. Flash desorption of VOCs
2. Pyrolysis of residual hydrocarbons remaining on the carbon surfaces
3. Re-establishment of the carbon pore structure via the steam-carbon reaction
4. Cool down of regenerated carbon

VOCs released in the process are partially destroyed in the regeneration furnace and finally oxidized at high efficiency by an afterburner.

Turk (1955) proposed the following formula to predict the service life of an adsorption system between regenerations:

$$t = \frac{6.43\,(10^6)\,SW}{eQ_r MC_v} \tag{12-25}$$

Where: t = service life, hr
S = ultimate proportionate saturation of carbon (retentivity, fractional)
W = weight of carbon, lb
e = adsorption efficiency (fractional)
Q_r = quantity of air processed by adsorption equipment, cfm
M = average molecular weight of contaminants
C_v = concentration of contaminants, ppm

For a commercial cell, containing 45 lb carbon, and rated at 700 cfm air at an operating efficiency of about 95%, the service life in hours (t) is calculated to be 4.35×10^5 S/MC$_v$. Constants of some important atmospheric odorants for use in this equation are given in **Table 12-34**. The table also gives calculated values for service life based on a concentration of impurity equal to the odor-threshold value. Additional odor-threshold concentration values are given by Wark and Warner (1976). Data obtained on the adsorption of contaminants from air streams in which they are present in low ppm concentrations indicate that generally reported retentivities represent maximum values (Stankovich, 1969). Appreciably lower capacities can be expected when contaminants are present in concentrations near the odor threshold.

The relative capacity (or retentivity, S, of **Table 12-34**) of a good grade of activated coconut-shell charcoal (50 min. by chloropicrin test), for a number of vapors (including some which may be classed as odors) has been presented by Barnebey (1958), and a portion of his data is reproduced in **Table 12-35**. Most materials encountered in odor-removal problems fall in classes 3 or 4, which means that the weight of carbon required will be three to six times the weight of odorant present in the air purified during the period of operation. If specific chemical odorants are not identified in the air to be purified, Barnebey suggests that the required quantity of carbon be estimated on the basis of

1. Possible odor sources (e.g., number of occupants of a room)
2. Type of occupancy (e.g., laboratories, toilets, offices, etc.)
3. Consideration of the amount of outside air that normal air-conditioning-design practice would call for to eliminate odors and stuffiness
4. Use of a preliminary test with a small adsorber
5. Assumption of an average odor composition that, for example, will be just perceptible at a concentration of 0.01 lb of odorant per million cubic feet of air
6. Rules of thumb and experience

Typical yearly requirements in pounds of 50-min. activated coconut-shell charcoal for various occupancies have been given as follows (ASHRAE, 1970):

Residences or offices—total per person
Smokers 2 lbs
NonSmokers 1
Hotels, per person average occupancy 3
Bar or tavern, per occupant 5

Table 12-34
Characteristics of Some Atmospheric Odorants

Substance	Molecular weight, M	Retentivity, S	Odor-threshold concentration, ppm	Service life for odor-threshold concentrations, hr*
Acrolein (heated fats)	56	0.15	1.8	650
Hexane (hydrocarbon)	86	0.16	almost odorless	
Phenol (carbolic acid)	94	0.30	0.28	5,000
Valeric acid (body odor)	102	0.35	0.00062	2.4×10^6

*Calculated value.
Source: Turk, 1958

Table 12-35
Capacity of 50-min. Activated Coconut-shell Charcoal for Vapors

Vapor	Capacity Index No.	Vapor	Capacity Index No.	Vapor	Capacity Index No.
Acetaldehyde	2	Carbon dioxide	1	Hydrogen sulfide	3
Acetic acid	4	Carbon monoxide	1	Isopropyl alcohol	4
Acetone	3	Carbon tetrachloride	4	Masking agents	4
Acrolein	3	Chlorine	3	Mercaptans	4
Alcohol	4	Chloropicrin	4	Ozone	4
Amines	2	Cigarette smoke	4	Perfumes, cosmetics	4
Ammonia	2	Cresol	4	Perspiration	4
Animal odors	3	Diesel fumes	3	Phenol	4
Anesthetics	3	Disinfectants	4	Propane	2
Benzene	4	Ethyl acetate	4	Pyridine	4
Body odors	4	Ethylene	1	Ripening fruits	4
Butane	2	Essential oils	4	Smog	4
Butyl alcohol	4	Formaldehyde	2	Solvents	3
Butyric acid	4	Gasoline	4	Stuffiness	4
Cancer odor	4	Hospital odors	4	Toluene	4
Caprylic acid	4	Household smells	4	Turpentine	4

The capacity index numbers have the following meanings:
4: High capacity for all materials in this category. One pound takes up about 20 to 50% of its own weight—averaging about ⅓ (33⅓%). This category includes most of the odor-causing substances.
3: Satisfactory capacity for all items in this category. These constitute good applications, but the capacity is not as high as for category 4. Adsorbs about 10 to 25% of its weight—averaging about ⅙ (16.7%).
2: Includes substances that are not highly adsorbed, but that might be taken up sufficiently to give good service under the particular conditions of operation. These require individual checking.
1: Adsorption capacity is low for these materials. Activated charcoal cannot be satisfactorily used to remove them under ordinary circumstances.
Source: Barnebey, 1958

ASHRAE Standard 62 (1989) requires 20 cfm/person of outdoor air for office buildings to maintain a high-quality internal atmosphere. Internally generated pollutants can also be controlled by recirculating the air through an activated carbon adsorber. The rate of removal of pollutants by the adsorber is the product of the air flow rate, the pollutant concentration in the air, and the removal efficiency of the adsorption unit. Because of this relationship, low efficiency adsorbers can achieve adequate pollution control if a sufficiently high recirculation rate is used. For example, in a building operating with the addition of 5 cfm/person of outdoor air, a 15% efficiency adsorber with 100 cfm/person recirculation rate, can reduce the indoor concentration of a targeted pollutant by a factor of four compared to the level attained by dilution alone with the 5 cfm/person of outdoor air (Liu, 1991).

The use of adsorptive air purification in air conditioning applications can frequently show an economic advantage over the use of a large volume of fresh outside air to reduce odor and stuffiness because of the high cost of heating and cooling the outside air to the desired interior temperature.

Operating Data

Data on the performance of panels of granular coconut-shell activated carbon treating air in six buildings in the Los Angeles, California, area are given by Bayati and Graham (1991).

The results of their study are summarized in **Table 12-36**. The adsorption systems generally consisted of banks of 24″ × 24″ × 1″ panels treating from less than 25% to as high as 100% outside air. The results show that activated carbon can adsorb and retain significant amounts of volatile matter from ambient air. Material balances based on the amount of volatile matter adsorbed, the amount of air processed, and the assumption of 100% efficiency, indicate that the inlet air streams contained 10 to 150 ppb VOCs. The highest levels were observed in geographical areas known to have more polluted air (i.e., Pasadena, San Marino, and downtown Los Angeles).

Activated carbon filters have been used successfully for many years at the District of Columbia's refuse transfer station to remove garbage odors from air and prevent contamination of the surrounding atmosphere (Enneking and Todd, 1969). The filters operate only about 2 hours per day during periods when the collection trucks return to the station to transfer their loads. A fan capacity of 85,300 cfm is provided on the basis of seven air changes per hour. The air is first passed through cyclone type collectors to remove large dust particles, then through glass fiber filters to remove the remaining dust, and finally through 3420 canister type carbon filters for odor removal. The canisters are each 10½ in. long and 4⅜ in. outside diameter with a ¾-in.-thick annular bed of carbon. Pressure drop through the canisters is 0.15 in. of water at the rated throughput of 25 cfm each. The canisters have an initial odor removal efficiency of 95% and require replacement every 4 years.

An example of the commercial application of active-carbon adsorption in the control of air pollution by removing odorous materials from the exhaust air of a chemical plant has been described by Lorentz (1950). The odors in this case arise during steps in the synthesis of Vitamin B-1. These steps involve the production and handling of sodium dithioformate,

Table 12-36
Performance of Activated Carbon Panels in HVAC Service

Facility/ Location	Air Flow, cfm	Carbon, lb (1)	Time, days (2)	V.M., wt % (3)
Hospital/ Newport Beach	30,000	1,872	943	6.27
Hospital/ Pasadena	6,000	270	930	10.04
Law Library/ Downtown L.A.	56,000	3,495	234	5.92
Hospital/ Upland	48,000	2,995	739	12.44
Library/ San Marino				
Admin. Area	5,900	343	430	14.08
Gallery	12,200	790	430	4.40

Notes: (1) Total amount of activated carbon.
(2) Number of days carbon was in service, assuming 24 hours per day operation.
(3) Volatile matter in carbon after indicated days of operation. Carbon was not necessarily saturated.
Data of Bayati and Graham (1991)

and the vapors have a tenacious garlic-sulfide stench. The odor-containing air is drawn from exhaust vents of the chemical-process equipment and adjacent work area at a rate of 12,000 cfm at a suction pressure of ⅜ in. of water. Air from the fan is first passed through air-conditioning-type dust filters with a total cross-sectional area of about 33 sq ft and then through a bank of 503 Dorex Type 42 carbon canisters. Pressure drop through the canisters is 0.14 in. of water at an air-flow rate of 24 cfm per unit. Approximately 800 lb of carbon are contained in the bank of canisters and between 300 and 400 lb of air-borne material is collected in the carbon by the end of the 6-month cycle. In addition to the odorous compounds, the carbon picks up some water and traces of organic solvents from the air stream, which leaves the unit entirely free from objectionable odor. The spent carbon is returned to the suppliers for reactivation, and about ¾ of the original carbon is reclaimed and returned. The carbon bed in the canister is about 1-in. thick, and a total cross-sectional area of about 1,000 sq ft of carbon surface is exposed to the flowing air stream. Total pressure drop through the system is about ¾ in. of water.

An interesting application that combines the features of odor-removal and solvent-recovery plants is described by Larkin and Davis (1957). This installation removes solvent vapors from air from a silk-screening operation. The principal objective is air purification rather than solvent recovery as the vapors were found to be extremely irritating to workers. The unit utilizes a single large canister, 7½-ft long by 5-ft OD, containing a 6-in.-thick cylindrical bed of carbon. The charge consists of 1,100 lb of 50-min. activated coconut-shell carbon. Regeneration is accomplished *in situ* by passing low-pressure steam through the unit, for a short period of time, once a month. Steam and solvent vapors are condensed; however, the removed mixed solvents have little value. The carbon has been found to last for approximately 5 years in this installation.

BIOFILTERS FOR ODOR AND VOC CONTROL

Where very large volumes of air must be treated for the removal of traces of organics, as in odor control, an effort has been made to find a less costly alternative to activated carbon adsorption. One such alternative is the biofilter, in which a bed of porous soil or a bed of peat, tree bark, or similar biomass is used to contact the polluted air. In these processes, microorganisms in the soil or peat assist adsorption by causing bio-oxidation of adsorbed organic compounds. Surface catalysis by minerals in the bed may also be a factor in enhancing oxidation of relatively nonbiodegradable molecules. Typical biofilters are about 3 ft deep and last between one and seven years. Early designs were open to the atmosphere at the top, but more recent concepts utilize sealed containers, which provide better control of the process and permit the exit gas to be monitored more accurately.

A total of 14 companies (all European) is listed in a December 1992 report as offering biofilter and bioscrubber technology (Fouhy, 1992). One of these, Clair Tech B.V., is responsible for more than 53 biofilter installations, including a unit that treats 140,000 m^3/hr of air, reducing the concentration of hydrocarbon solvents from 500 mg/m^3 to less than 150 mg/m^3.

Soil Beds

The design and performance of soil beds are described by Bohn and Bohn (1988). Basically, the system consists of a horizontal network of perforated pipe placed about 2 ft below ground level. The impure air is forced through the pipe network and upward through the soil before it enters the atmosphere. Contaminants in the feed air are adsorbed on particles of soil, then rapidly oxidized by microbiological processes or surface catalysis, while nitrogen, oxygen, carbon dioxide, and water vapor are largely unaffected and are discharged to the atmosphere. The oxidation continuously renews the bed's capacity for adsorption.

The removal efficiency of soil beds for organic compounds varies from 80–90% for highly stable light gases, such as methane and propane, to about 99% for more complex compounds such as aldehydes and organic acids. Inorganic impurities such as SO_2, NO_x, and H_2S are also removed with high efficiency by most soils; however, the effectiveness will decrease if sufficient acid gas is present to neutralize the soil's natural alkalinity. When soil pH drops to about 3, SO_2 is no longer removed, and NO_x removal stops at a pH of about 2. The life of the soil bed for removing acidic inorganic gases may be extended by the addition of lime. The life for removing organic compounds is virtually unlimited.

Soils vary widely in composition and permeability, but the gas permeability of soils is always much lower than that of activated carbon beds, and as a result, much lower gas velocities must be employed. Typical air flow rates are in the range of 4–10 m^3/hr m^2 (13–33 ft^3/hr ft^2) with pressure drops of about 2–3 in. of water.

The capacity of a soil bed for gas purification is generally limited by gas permeability rather than the rate of oxidation of organic compounds. For example, with an input gas containing 1,000 ppm of organic pollutants flowing 40 hours per week, the amount of organic matter added to the soil per year will be only about 2 Kg/m^2 (9 tons per acre). This annual rate is about the same as the natural rate of addition of organic matter to soils in temperate regions. Much higher organic loadings can be destroyed by soil beds when necessary (Bohn and Bohn, 1988).

Peat Beds

The use of peat beds is a relatively recent extension of soil bed technology. According to Valentin (1990), the process was first used in 1975 for the control of livestock odors and is now in widespread use, particularly in Germany, Scandinavia, the Netherlands, the UK, and Ireland.

A peat bed facility, as described by Valentin, consists of a concrete-lined rectangular pit, a raised perforated floor to support the peat bed, and an underfloor distribution area for the polluted air feed. The peat bed is typically a mixture containing about one-half sphagnum or trichophorum peat and about one-half woody heather or brush wood, which is added to minimize consolidation of the peat. The bottom layer is entirely heather to aid in air distribution, and the entire bed is about 1 m (39 in.) deep. A plastic membrane is used at the walls to prevent air bypassing.

It is important that the peat bed be thoroughly wetted when installed and kept moist during operation. This can be accomplished by the use of a spray system over the bed, to be used as required, plus a water spray scrubber for the feed air upstream of the peat bed.

Bed operation is quite simple; in addition to keeping the bed moist, it is necessary to control the temperature in the range of 10°–45°C (about 50°–110°F). Bed pH will normally be between 4 and 7.5. After about one year of operation the bed must be turned to compensate for consolidation, and new material added as needed. Complete replacement of the peat and heather is required after five years or more. The superficial velocity of air through the bed can be as high as 200 m/hr (11 ft/min.) for low odor concentrations, decreasing to about 110–130 m/hr (6–7 ft/min.) for strong odors.

The typical capital cost for a peat bed system in the UK is given as $550 per square meter of cross-section area, including associated fans, scrubber, and local ducting. At a flow rate of 125 m^3/hr m^2, this is about $4.50 per m^3/hr ($7.65 per scfm) of capacity (Valentin, 1990).

This estimate can be compared with a capital cost range for a proprietary biofilter system of $5.50 to $30 per m^3/hr ($9.40 to $50 per scfm) reported in 1992 (Fouhy, 1992) and an estimate for soil beds of about $5 to $6 per m^3/hr ($8.50–$10 per scfm), which was published four years earlier (Bohn and Bohn, 1988).

IMPREGNATED ADSORBENT APPLICATIONS

Impregnated adsorbents provide a combination of adsorption and chemical reaction to remove traces of specific impurities from gas streams. The process is used to remove impurities that are adsorbable; the presence of a reactive chemical in the pores of the adsorbent serves to lock adsorbed material in place, increasing the capacity of the bed and decreasing the vapor pressure of the adsorbate to essentially zero.

Sulfur Compound Removal with Impregnated Carbon

Two types of impregnating agents are used to increase the effectiveness of carbon for the removal of traces of H_2S and low molecular-weight sulfur compounds. The types are alkaline materials such as sodium hydroxide or sodium carbonate, which hold the sulfur compounds by acid-base reaction, and metal oxides, which hold the sulfur as metal sulfides or sulfates. Examples of the two types are Calgon Carbon Corporation's Type IVP (Improved Vapor Phase), which is reported to be impregnated with sodium hydroxide (Bourke and Mazzoni, 1989), and Type FCA, which is stated to be a metal-oxide impregnated carbon (Calgon, 1978).

Laboratory studies reported by the Calgon Carbon Corporation (1977) show Type IVP carbon to attain a 20–25% loading by weight of H_2S from an air stream containing 1% by volume of this compound, compared to 10% loading for conventional carbon under the same conditions. The impregnated carbon adsorbed 9.5% methyl mercaptan by weight from a gas stream containing 1.3% of the impurity, compared to a loading of 4.1% for conventional carbon.

A laboratory study conducted in Japan showed an even greater effect (Ikeda et al., 1988). The work covered the adsorption of H_2S and methyl mercaptan (CH_3SH) on activated carbon impregnated with NaOH or Na_2CO_3 and the adsorption of trimethylamine [$(CH_3)_3N$] on activated carbon that had been sulfonated by contact with sulfuric acid. The results for H_2S and CH_3SH showed that impregnation with NaOH or Na_2CO_3 increased the capacity as much as 40 to 60 times that of the original carbon. The adsorption of trimethylamine was helped by sulfonation of the carbon, particularly at low concentrations in the gas (less than 10 ppmv). At higher concentrations most of the amine is adsorbed by physical adsorption.

Impregnated carbon beds are used at many petrochemical plants to recover traces of sulfur compounds, which can poison catalysts used in subsequent operations (e.g., ammonia and methanol synthesis). A typical application is described by Bourke and Mazzoni (1989). The system treats 6,900 scfm of natural gas at 190 psig and 49°F containing up to 10 ppmv H_2S. The adsorption system uses two vessels in series, each of which contains 8,450 pounds of Calgon FCA carbon. When the H_2S concentration in the outlet gas from an adsorber reaches 0.22 ppm, the vessel is regenerated with steam containing a small percentage of oxygen. The oxygen serves to convert H_2S to elemental sulfur in place, releasing the alkali to react with additional hydrogen sulfide during the next adsorption period. The regenerated bed is then placed in second position to "polish" the purified gas leaving the bed in first position. Because of the low level of H_2S in the feed, the time to regeneration is on the order of six months. H_2S is not detectable in the final product gas from the system.

Caustic impregnated carbon is widely used to remove H_2S and other odorous compounds from exhaust air streams at municipal sewage disposal facilities. Hydrogen sulfide is normally the predominant odor contributor, but it occurs with a wide variety of volatile organic compounds. The impregnated carbon is effective because it can physically adsorb organic species while chemically reacting the H_2S. Rabosky and Matta (1985) suggest the following design parameters for sewage plant applications of Calgon Type IVP, 4×6 mesh, impregnated carbon:

1. 6 to 30 air changes per hour
2. 50- to 100-ft per minute face velocity
3. 18- to 36-in. bed depth

They report that adherence to these values will result in H_2S capacities in excess of 0.12 grams H_2S/cc of carbon.

In another commercial application, which is described by Rowley and Lawlor (1984), Calgon Type IVP carbon is used to remove traces of hydrogen sulfide and other sulfur compounds from gases vented from tanks and storage pits of a Stretford natural gas treating plant. The impregnated adsorbent bed is 3 ft deep and is operated in the upflow mode.

Mercury Removal with Impregnated Carbon

Mercury occurs in some natural gas streams and can result in severe corrosion of aluminum (e.g., brazed aluminum heat exchangers in LNG and turboexpander plants). The use of sulfur-impregnated activated carbon is a proven commercial process for removing the mercury. Sulfur is the active ingredient, securely fixing mercury, as sulfide, in the microporous structure of the carbon (Leeper, 1980).

An example of a commercial application of impregnated carbon for mercury removal is described by Bourke and Mazzoni (1989). The adsorber treats 30 MMscfd of natural gas at 1,100 psig and 110°F. The inlet gas contains over 50 micrograms of mercury/Nm^3, and no mercury is detected in the product gas (i. e., less than 0.001 micrograms/Nm^3). The bed contains 1,500 pounds of Calgon Type HGR impregnated carbon and is expected to remain effective for many years. Thirty Calgon impregnated granular activated carbon units were reported to be in operation in 1994, removing mercury from natural gas, hydrogen, cracked gas, and air (Calgon Carbon Corp., 1994).

Non-impregnated carbon can also be used for adsorbing mercury vapor; however, the capacity is quite low. Mahwar (1986) examined the proposed use of carbon as a regenerable adsorbent to capture mercury in hydrogen gas produced in a chlor-alkali plant. The carbon was found to reduce the mercury concentration of dry hydrogen to non-detectable levels; however, the capacity was only 0.374 grams Hg/kg carbon. Complete regeneration was accomplished by purging the carbon bed with 100°C air for 18 hours. The mercury is recoverable from the air by cooling and condensation. A related process, offered by UOP, is based on the chemisorption of mercury on a special molecular sieve-based adsorbent, HgSIV. This adsorbent may be used in conjunction with a conventional desiccant in a unit that simultaneously dries the gas and removes mercury. The HgSIV adsorbent is regenerated during the drier regeneration cycle. In 1994, seven HgSIV units were reported to be in operation treating LPG and natural gas (UOP, 1994).

MISCELLANEOUS APPLICATIONS OF ADSORPTION

Hydrochloric Acid on Alumina

Traces of hydrochloric acid can be removed effectively from gas streams by the use of activated alumina adsorbent. Although the acid is very strongly adsorbed, the rate of adsorption is relatively low, apparently due to slow diffusion within the pores. The low rate produces an extended MTZ which results in the requirement of a deep bed to attain an acceptable capacity prior to breakthrough. The use of the smallest practical particle size aids in reducing the length of the MTZ, and, therefore, increases the working capacity of alumina beds. Typical design capacities for a properly designed system are 6–8 lb chloride/100 lb alumina for a standard alumina (Kaiser Aluminum & Chemical Corp., Type A-201) and 12–14 lb/100 lb for a promoted alumina (Kaiser Aluminum & Chemical Corp., Type A-203 Cl) (Pearson and Janke, 1986).

Radioactive Isotope Adsorption

A highly specialized application of adsorbents is the removal of radioactive isotopes from exhaust air or other gas streams. Ruthven (1988) reports that Kr^{85} can be effectively adsorbed by silicalite or de-aluminized hydrogen mordenite (molecular sieves) in a chromatographic-type column. Other unusual molecular sieve zeolites (silver cation type X or silver cation morganite) are effective adsorbents for I^{129} and can be regenerated thermally.

Radioactive iodine resulting from nuclear reactor operations is commonly adsorbed on activated carbon. Problems encountered in this application include weathering by humid air, the presence of a portion of the iodine in compounds such as methyl iodide which are not as readily adsorbed, and the effects of radiation. At the Savannah River Plant (SRP) activated carbon beds must be replaced after 3 to 5 years to ensure that iodine adsorption efficiency meets the required design level (99.85% removal (Milham, 1968). In this plant, Barnebey-Cheney Type 416 carbon is used to purify a flow of 100,000 cfm of air. The effect of radiation on the adsorption of iodine and methyl iodide is described by Jones (1968).

Adsorption of Iron and Nickel Carbonyls

Iron and nickel carbonyls are volatile compounds that are present in many industrial gas streams, particularly those with high concentrations of carbon monoxide which have been in contact with ferrous alloys at elevated temperature. A typical gas that may contain the metal carbonyls is the product gas from an oxygen blown gasifier. Since the carbonyls can poison catalysts used in subsequent operations, their removal is usually required. Golden et al. (1991) screened five commercial adsorbents for effectiveness in removing iron carbonyl. Two of these, Calgon BPL activated carbon and Linde H-Y zeolite showed the highest capacity for the iron compound and were further tested for nickel carbonyl adsorption. The carbon showed higher capacity than the zeolite for both carbonyls, but poorer regeneration characteristics with the iron compound. It is concluded that an effective adsorption bed could consist of a zeolite layer on top to remove iron carbonyl followed by a layer of carbon to remove the nickel compound. The combination bed would be regenerated by purging with nitrogen at about 120°C (248°F).

REFERENCES

Alcoa, 1991, "F-200 Activated Alumina for Adsorption Applications," Alcoa Industrial Chemicals Division, Chemicals Product Data Sheet, SEP 941, November.

Alexander, R. A., 1988, "An Offshore Dehydration System for the Production of the Norphlet Sour Gas in Mobil Bay," presented at the 20th Annual Offshore Technology Conference, Houston, TX, May 2–5.

Allen, H. V., Jr., 1944, *Petrol. Refiner.,* Vol. 23, July, pp. 93–98.

Aluminum Company of America, 1969, *Activated and Catalytic Aluminas,* July 14.

Amero, R. C., Moore, J. W., and Cappell, R. G., 1949, *Chem. Eng. Progr.,* Vol. 43, July, p. 349.

Anon., 1963, *Chem. Eng.,* April 15, pp. 92–93.

Anon., 1969, *Filtration Engineering,* December.

Anon., 1991, *Chem. Eng.,* Vol. 98, No. 5, May, p. 17.

ASHRAE Guide and Data Book, 1970, p. 463.

Ausikaitis, J. P., 1983, "TRISIV Adsorbent—The Optimization of Momentum and Mass Transport via Adsorbent Particle Shape Modification," presented at the International Conference on the Fundamentals of Adsorption, Klais, Upper Bavaria, West Germany, May 6–11.

Badger Engineers, Inc., 1984, *Hydro. Process.,* April, p. 55.

Ballard, D., 1963, "Discussion of Short-Cycles, Solid Desiccant Hydrocarbon Recovery Units," paper presented at the Natural Gas Processor's Association Meeting, Lafayette, LA, Jan. 8.

Ballard, D., 1965A, *Oil Gas J.,* July 12, pp. 101–110.

Ballard, D., 1965B, *Hydro. Process.,* Vol. 44, April, pp. 131–136.

Ballard, D., 1978, "How to Improve Cryogenic Dehydration," prs. at the 1978 Gas Conditioning Conference, University of Oklahoma, Norman, OK.

Barnebey, H. L., 1958, *Heating, Piping Air Conditioning, J. Sec.,* Vol. 153, March.

Barnebey & Sutcliffe Corporation, 1990, "Series FJS and FRS Side-Loading Activated Carbon Adsorbers," Brochure T-827, October.

Barrow, J. A., 1983, *Hydro. Process.,* January, p. 117.

Bayati, M. A., and Graham, J. R., 1991, "The Use of Activated Carbon for the Removal of Trace Organics in the Control of Indoor Air Quality and Climate," Toronto, Canada, February.

Benamy, R. L., 1987, "New Technology for VOC Control," presented at the 1987 Winter Meeting of the Am. Soc. of Heating, Refrigerating, and Air-Conditioning Engineers, Inc., New York, published in *ASHRAE Transactions,* Vol. 93, Part 1.

Benson, R. E., and Courouleau, P. H., 1948, *Chem. Eng. Progr.,* Vol. 44, June, p. 459.

Berg, C., 1951, *Chem. Eng. Progr.,* Vol. 47, No. 11, November, p. 585.

Bohn, H., and Bohn, R., 1988, *Chem. Eng.,* Vol. 95, No. 6, April 25, p. 23.

Bourke, and Mazzoni, 1989, "The Roles of Activated Carbon in Gas Conditioning," presented at the 1989 Laurence Reid Gas Conditioning Conference, University of Oklahoma, Norman, OK, March 6–9.

Brownell, L. E., and Katz, D. E., 1947, *Chem. Eng. Progr.,* Vol. 43, pp. 537–549.

Browning, F. M., 1952, *Chem. Eng.,* Vol. 59, October, p. 158.

Calgon Carbon Corporation, 1977, *Chem. Eng.,* Feb. 14, p. 47.

Calgon Carbon Corporation, 1978, "Sulfur Control with Metal Oxide-Impregnated Carbon," Bulletin 23-67a.

Calgon Carbon Corporation, 1980, *Adsorption Handbook.*

Calgon Carbon Corporation, 1987, "Heavy Hydrocarbon Removal or Recovery from Gas Streams," Bulletin 23-66b.

Calgon Carbon Corporation, 1994, "Mercury Removal," *Gas Processes '94, Hydro. Process,* April, p. 92.

Campbell, J. M., 1974, *Gas Conditioning and Processing,* Vol. II, 3rd edition, Norman, OK.

Cantrell, C. J., 1982, *Chem. Eng., Progr.,* October, p. 56.

Cappell, R. G., Hammerschmidt, E. G., and Derschner, W.W., 1944, *Ind. Eng. Chem.,* Vol. 36, September, p. 779.

Chi, C. W., and Lee, H., 1973, *Gas Purification by Adsorption, AIChE Symposium Series No. 134,* Vol. 69, pp. 96–101.

Cines, M. R., Haskell, D. M., and Houser, C. G., 1976, *Chem. Eng. Progr.,* Vol. 72, No. 8, August, pp. 89–93.

Clark, E. L., 1958, *Proc. Gas Conditioning Conf.,* University of Oklahoma, Norman, OK.

Collins, P. J., 1988, "Theory of Carbon Adsorption and Distillation Technology as Applied to VOC Control in the Packaging Rotogravure Industry," TAPPI, *Proceedings of the 1988 Environmental Conference,* Charleston, SC, p. 5.

Conviser, S. A., 1965, *Oil & Gas J.,* Vol. 63, Dec. 6, pp. 130–135.

Coolidge, A. S., 1924, *J. Am. Chem. Soc.,* Vol. 46, March, p. 596.

Cummings, W. P., 1983, "Dry Bed Dehydration," *Proc. Laurence Reid Gas Conditioning Conference,* University of Oklahoma, Norman, OK.

Damron, E., 1977, "Cryogenic Plant Design, An Operator's Viewpoint," presented at the Gas Processors Association Meeting, March 23.

Dedert Corporation, 1991, *Chem. Eng.,* Vol. 98, No. 12, December, p. 137.

Deitz, V. R., 1944, *Bibliography of Solid Adsorbents,* Washington, D.C.: U.S. Cane Sugar Refiners and Bone Char Manufacturers and National Bureau of Standards, p. ix.

Derr, R. B., 1938, *Ind. Eng. Chem.,* Vol. 30, April, p. 384.

Drew, J. W., 1975, *Chem. Eng. Progr.,* Vol. 71, No. 2, February, pp. 92–99.

Enneking, J. C., 1966, *Hydro. Process.,* Vol. 45, October, pp. 189–192.

Enneking, J. C., and Todd, H. H., Jr., 1969, *ASHRAE J.,* July.

EPA, 1988, "Carbon Adsorption for Control of VOC Emissions Theory and Full Scale System Performance," Report EPA-450/3-88-012, PB91-182006, June.

Ergun, S., 1952, *Chem. Eng. Progr.,* Vol. 48, pp. 89–94.

Ezell, E. L., and Gelo, G. F., 1982, *Hydro. Process.,* May, p. 191.

Fair, J. R., 1969, *Chem. Engr.,* July 14, pp. 90–102.

Farooq, S., and Ruthven, D. M., 1991, *Chem. Eng. Sci.,* Vol. 46, p. 2213.

Fornoff, L. L., 1971, "The Pura Siv-N Process," presented at the AIChE 64th Annual Meeting, San Francisco, CA, Nov. 28.

Fouhy, K., 1992, *Chem. Eng.,* Vol. 99, No. 12, December, p. 41.

Fuderer, A., and Rudelstorfer, E., 1976, U.S. Patent No. 3,986,849, Oct. 19.

Gas Processors Supplier's Assoc. (GPSA), 1987, *Engineering Data Book,* Vol. II, Sec. 20, Tenth edition, Tulsa, OK.

Getty, R. J., Lamb, C. E., and Montgomery, W. C., 1953, *Petrol. Eng.,* Vol. 25, August, p. B-7.

Golden, T. C., Hsuing, T. H., and Snyder, K. E., 1991, *Ind. Eng. Chem. Res.,* Vol. 30, No. 3, p. 502.

Goodboy, K. P., and Fleming, H. L., 1984, *Chem. Eng. Progr.,* November, p. 63.

Graham, J. R., and Bayati, M. A., 1990, "Removal of VOC Contaminants using GAC Technology," *Proc. Third Annual Hazardous Materials Management Conference and Exhibition,"* Chicago, IL, March.

Grant, R. J., and Manes, M., 1964, *Ind. Eng. Chem. Fundamentals,* Vol. 3, No. 3, pp. 221–224.

Grant, R. J., Manes, M., and Smith, S. B., 1962, *AIChE J.,* Vol. 8, No. 3, pp. 403–406.

Grayson, H. G., 1955, *Ind. Eng. Chem.,* Vol. 47, January, p. 41.

Hammerschmidt, E. G., 1945, *Gas (Los Angeles),* Vol. 21, January, pp. 32–33.

Hammerschmidt, E. G., 1957, CEP-57-7, presented at Production Conference, Operating Section, American Gas Association, Bal Harbor, FL, May 20–22.

Harrell, A. G., 1957, *Oil & Gas J.,* Vol. 55, Oct. 28, p. 121.

Hasz, J. W., and Barrere, C. A., Jr., 1964, *Chem. Eng. Progr. Symposium Series,* Vol. 96, No. 65, pp. 48–56.

Haun, E. E., Anderson, R. F., Kauff, D. A., Millder G. Q., and Stoeker, J., 1990, "The Efficient Refinery, Hydrogen Management in the 1990s," presented at the UOP Spring 1990 Technology Conference.

Heck, J. L., and Johanson, T., 1978, Abstract of Paper Presented at Western Gas Processors and Oil Refiners Association Meeting, Long Beach, CA, published in *Oil & Gas J.,* Jan. 9, pp. 91–92.

Heck, J. L., 1980, *Oil & Gas J.,* Feb. 11, p. 122.

Heinegard, C., 1988, "The Polyad FB Process for VOC Control and Solvent Recovery," Presented at the 81st Annual Meeting of the APCA, Dallas, TX, June 19–24.

Herrmann R. H., 1955, *Oil & Gas J.,* Vol. 53, April 18, pp. 144–147.

Hiller, M. H., Lacatena, J. J., and Miller, G. Q., 1987, "Hydrogen for Hydroprocessing Operations," presented at the 1987 NPRA Annual Meeting, March 29–31, San Antonio, TX.

Howell, A. K., 1943, *Gas J.,* Vol. 242, Sept. 15, p. 337.

Hubard, S. S., 1954, *Ind. Eng. Chem.,* Vol. 46, February, p. 356.

Humphries, C. L., 1966, *Hydro. Process.,* Vol. 45, December, pp. 88–96.

Ikeda, H., Asaba, H., and Takeuchi, Y., 1988, *Journal of Chemical Engineering of Japan,* Vol. 21, No. 1, p. 91.

Jones, L. R., 1968, *Proc. 10th AEC Air Cleaning Conf.* (CONF-680821), pp. 204–215.

Kali-Chemie, 1979, KC-Trockenperlen, Brochure P2000 12.79.

Keener, T. C., Henderson, G., and Rickabaugh, J., 1988, "The Use of Activated Carbon for the Control of Volatilized Organic Emissions from a Covered Activated Sludge Basin," presented at the 81st Annual Meeting of the APCA, Dallas, TX, June 19–24.

Keller, G. E., Anderson, R. A., and Yon, C. M., 1987, "Adsorption," Chapter 12 in *Handbook of Separation Technology,* R. W. Rousseau editor, John Wiley & Sons, New York.

Kenson, R. D., 1979, *Pollution Engineering,* July, p. 38.

Kenson, R. E., and Jackson, J. F., 1989, "Case History of an Aerospace Plant Paint Spray Booth VOC Emission Control Installation," presented at the 82nd Annual Meeting and Exhibition, Air and Waste Management Association, Anaheim, CA, June 25–30.

Kenson, R. E., 1990, "Recovery of Reactive Monomers from Polymer Reactor Exhaust Using Activated Carbon Adsorbents," presented at the AIChE Summer Meeting, San Diego, CA, Aug. 19–22.

Kenson, R. E., 1991, "Comparison of Carbon Adsorption, Adsorption, Condensation, and Incineration for Control of Dilute Air Toxics Emissions," presented at the AIChE Summer National Meeting, Pittsburgh, PA, Aug. 20.

Kerry, F. G., 1991, *Chem. Eng. Progr.,* Vol. 87, No. 8, August, p. 48.

Kittleman, T. A., and Akell, R. B., 1978, *Chem. Eng. Progr.,* Vol. 74, No. 4, April, pp. 87–91.

Knaebel, K. S., and Hill, G. B., 1985, *Chem. Eng. Sci.,* Vol. 40, p. 2351.

Knaebel, K. S., 1991, "Pressure Swing Adsorption Processes," presented at the AIChE National Meeting, Pittsburgh, PA, August.

Kovach, J. L., 1988, "Gas Phase Adsorption", Section 3.1 in *Handbook of Separation Techniques for Chemical Engineers,* second edition, P. A. Schweitzer editor, McGraw-Hill Book Company, New York.

Kraychy, P. N., and Masuda, A., 1966, *Oil & Gas J.,* Vol. 64, Aug. 8, p. 66.

Lamb, A. B., and Coolidge, A. S., 1920, *J. Am. Chem. Soc.,* Vol. 42, p. 1146.

Langmuir, I., 1916, *J. Am. Chem. Soc.,* Vol. 38, p. 2207.

Langmuir I., 1917, *J. Am. Chem. Soc.,* Vol. 34, p. 1883.

Langmuir, I., 1918, *J. Am. Chem. Soc.,* Vol. 40, p. 1361.

Larkin, S. C., and Davis, W., 1957, *Industry Power,* May.

LaRoche Chemicals, 1988, "Alumina Products and Technology," Brochure 1C 107, R9/88.

LaRoche Chemicals, 1990, "Fixed-Bed Dehydration with Activated Aluminas," Brochure 1C-235.

Larsen, E. S., and Pilat, M. J., 1991, *J. Air & Waste Management Assoc.,* Vol. 41, No. 9, September, p. 1199.

Leavitt, F. W., 1962, *Chem. Eng. Progr.,* Vol. 58, August, pp. 54–59.

Ledoux, E., 1948, *Chem. Eng.,* Vol. 55, March, p. 118.

Lee, M. N. Y., and Collins, J. J., 1968, "Ammonia Plant Feed Desulfurization with Molecular Sieves," paper presented at the Tripartite AIChE Meeting, Montreal, Canada, Sept. 25.

Leeper, J. E., 1980, *Hydro. Process.,* November, p. 239.

Leva, M., 1947, *Chem. Eng. Progr.,* Vol. 43, pp. 549–554.

Lewis, W. K., Gilliland, E. R., Chertow, B., and Cadogan, W. P., 1950, *Ind. Eng. Chem.,* Vol. 42, pp. 1326–1332.

Liapis, P. C., 1987, *Fundamentals of Adsorption,* Engineering Foundation (distributed by AIChE), New York.

Liu, R. -T., 1991, "Control of VOC Using Activated Carbon Adsorbers," presented at the AIChE National Meeting, Pittsburgh, PA, Aug. 18–21.

Lorentz, F., 1950, *Chem. Eng. Progr.,* Vol. 46, Aug., p. 377.

Lukchis, G. M., 1973, *Chem. Eng.,* Vol. 80, No. 13, June 11, p. 111–116.

Mahwar, R. S., 1986, *Chemical Age of India,* Vol. 37, No. 12, p. 853.

Markovs, J., 1990, "Sour Gas Drying and Purification with Molecular Sieves," Proc. of the Laurence Reid Gas Conditioning Conference, University of Oklahoma, Norman, OK, March 15.

Mattia, M. M., 1969, U.S. Patent 3,455,089.

Mattia, M. M., 1970A, *Chem. Eng. Progr.,* Vol. 66, December, p. 74–79.

Mattia, M. M., 1970B, U.S. Patent 3,534,529.

McAllister, W. S., and Westerveld, W. W., 1978, *Oil & Gas J.,* Jan. 16, p. 66–77.

Milham, R. C., 1968, *Proc. 10th AEC Air Cleaning Conf.* (CONF-680821), p. 167–169.

Miller, G. Q., and Stoecker, J., 1989, "Selection of a Hydrogen Separation Process," presented at the 1989 NPRA Annual Meeting, March 19–21, San Francisco, CA.

Mobil Oil Corporation, 1971, *Mobil Sorbead Desiccants.*

Moffett, T. F., 1943, *Heating and Ventilating,* Vol. 40, April, p. 33–36.

Multiform Desiccants, 1984, "Natrasorb Silica Gel Beads, T, TW, TH, TI," Multiform Desiccants, Inc., Brochure.

Nelson, T. F., Blacksmith, J. R., and Randall, J. L., 1984, "Field Evaluation of Volatile Organic Compound Removal Efficiency for Full-Scale Carbon Adsorption Systems," EPA-600/D-84-211, PB 84-238690, August.

Palmer, G. H., 1977, *Hydro. Process.,* April, pp. 103–106.

Parmele, C. S., O'Connell, W. L., and Basdekis, H. S., 1979, *Chem. Eng.,* Dec. 31, p. 59.

Parsons, R. A., editor, 1988, *ASHRAE Handbook,* Chapter 11, Industrial Gas Cleaning and Air Pollution Control."

Pearson, M. J., and Janke, S. L., 1986, *Oil & Gas J.,* May 12, p. 64.

Pierce, J. E., and Stieghan, D. L., 1966, (*Hydro. Process.*) *Petrol. Refiner.,* Vol. 45, March, pp. 170–172.

Polley, F. F., 1981, "Natural Gas Dehydration—A Solid Desiccant Engineering Clinic," Proc. of the 1981 Gas Conditioning Conference, University of Oklahoma, Norman, OK.

Polyani, M., 1914, *Verh. Dtsch. Phys. Ges.,* Vol. 16, p. 1012.

Rabosky, J. G., and Matta, S. R., 1985, "Effective Odor Control with Granular Activated Carbon Systems," presented at the 57th Annual Conference of the Water Pollution Control Association of Pennsylvania, Aug. 7–9.

Ranz, W. E., and Marshall, W. R., 1952, *Chem. Eng. Progr.,* Vol. 48, No. 173.

Ray, A. B., 1940, *Chem. & Met. Eng.,* Vol. 47, May, pp. 329–332.

Rose, H. E., 1945, *Inst. Mech. Engs.,* (London), Vol. 153, pp. 141–168.

Ross, W. L., and McLaughlin, E. R., 1951, *Refrig. Eng.,* Vol. 59, February, p. 167.

Rowley, P., and Lawlor, L., 1984, *Chemical Processing,* June, reprint provided by Calgon Carbon Corporation.

Rowson, H. M., 1963, *Brit. Chem. Eng.,* March.

Ruthven, D. M., 1988, *Chem. Eng. Progr.,* Vol. 84, No. 2, February, p. 42.

Ruthven, D. M., 1984, *Principles of Adsorption and Adsorption Processes,* John Wiley & Sons, New York.

Shin, H. S., and Knaebel, K. S., 1987, *AIChE Journal,* Vol. 33, p. 634.

Silbernagel, D. R., 1967, *Chem. Eng. Progr.,* Vol. 63, No. 4, pp. 99–102.

Sircar, S., 1979, U.S. Patent 4,171,206.

Sircar, S., and Kratz, W. C., 1988, *Separation Science and Technology,* Vol. 23, Nos. 14 & 15, p. 2397.

Spivey, J. J., 1988, *Environmental Progress,* Vol. 7, No. 1, February, p. 31.

Stankovich, A. J., 1969, "The Capacity of Activated Charcoals Under Dynamic Conditions for Selected Atmospheric Contaminants in the Low Parts per Million Range," *ASHRAE Symposium Bulletin, Odors and Odorants: An Engineering View.*

Stenhouse, J., 1961, *Chem. News,* Vol. 3, p. 78.

Stewart, H. A., and Heck, J. L., 1967, "Hydrogen Purification by Pressure Swing Adsorption," paper presented at the AIChE 64th National Meeting, New Orleans, LA, March 16–20.

Swerdloff, W., 1967, "Dehydration of Natural Gas," *Proc. Gas Conditioning Conf.,* University of Oklahoma, Norman, OK.

Tan, K. S., 1980, *Chem. Eng.,* March 24, p. 117.

Tan, K. S., 1984, *Chem. Eng.,* Dec. 24, p. 57.

Taylor, R. K., 1945, *Ind. Eng. Chem.,* Vol. 37, July, p. 649.

Thomas, T. L., and Clark, E. L., 1967, *Oil & Gas J.,* Vol. 65, No. 12, pp. 112–115.

Trent, R. E., Craig, D. F., and Coleman, R.L., 1993, "The Practical Application of Special Molecular Sieves to Minimize the Formation of Carbonyl Sulfide During Natural Gas Dehydration," proc. of the 1993 Laurance Reid Gas Conditioning Conference, University of Oklahoma, Norman, OK, March 3.

Troxler, W. L., Parmele, C. S., Barton, C. A., and Hobbs, F. D., 1983, "Survey of Industrial Applications of Vapor-Phase Activated Carbon Adsorption for Control of Pollutant Compounds from Manufacture of Organic Compounds," EPA-600/2-83-035, August, (PB83-200618).

Turk, A., 1955, *Ind. Eng. Chem.,* Vol. 47, p. 966.

Turk, A., 1958, *Ind. Wastes,* Vol. 3, Jan./Feb., pp. 9–13.

Turnock, P. H, and Gustafson, K. J., 1972, "Advances in Molecular Sieve Technology for Natural Gas Sweetening," *Proc. of the Gas Conditioning Conference,* University of Oklahoma, Norman, OK.

Union Carbide Corporation, 1983, "Purasiv HR," Brochures F-48668A5M and F-48668A15M.

UOP, 1990, "UOP Molecular Sieves," UOP Brochure F-1979J, 1/90.

UOP, 1991A, "Polybed PSA Systems," Publication of UOP Process Plants and Systems, EP3015B, March.

UOP, 1991B, "Fixed-Bed Pressure Drop Calculations," Brochure F-4183b, reprinted 3/91.

UOP, 1996, "Hydrogen (Polybed PSA)," *Gas Processes, '96, Hydro. Process.,* April, p. 108.

UOP, 1993A, "Molecular Sieves Non-Hydrocarbon Materials Data Sheets," Brochure XF-19.

UOP, 1993B, "NITREX," Brochure GP 5121.

UOP, 1994, "Mercury Removal," *Gas Processes '94, Hydro. Process.,* April, p. 92.

Valentin, F. H. H., 1990, *Chem. Eng.,* January, p. 112.

Vatavuk, W. M., Klotz, W. L., and Stallings, R. L., 1990, "Carbon Adsorbers," Chapter 4 in *OAQPS Control Cost Manual,* 4th ed., EPA 450/3-90-006, January.

Veldman, R., 1991, "Improving Solid Desiccant Performance in Natural Gas Dehydration," presented at the AIChE National Meeting, Houston, TX, April 11.

Vermeulen, T., Klein, G., and Hiester, N. K., 1973, Section 16 in *Chemical Engineers' Handbook,* 5th edition, editors, R. H. Perry and C. H. Chilton, McGraw Hill, New York.

W.R. Grace & Co., 1992, "Davison Silica Gels," Davison Chemical Division, Technical Data.

W.R. Grace & Co., 1993, "Bead Gel Spherical Silica Gel Adsorbents," Davison Chemical Division, Brochure 1C-5-1082.

Wakao, N., and Funazkri, T., 1978, *Chem. Eng. Sci.,* Vol. 33, p. 1375.

Walker, C. R., Applebee, H. C., and Howell, A. K., 1944, *Gas J.,* Vol. 243, pp. 310–346, 379.

Wankat, P. C., 1986, *Large Scale Adsorption and Chromatography,* CRC Press, Boca Raton, FL.

Wark, K., and Warner, K., 1976, *Air Pollution, Its Origin and Control,* Harper & Row, New York, NY.

Weskem, Inc., 1982, "Sorbead R/H/W," Product Data Sheet.

Wilkinson, E. P., and Sterk, B. J., 1950, *Petrol. Engr., Reference Annual,* p. D-30.

Woosley, R. D., 1990, "Activated Alumina Desiccants," in *Alumina Chemicals, Science and Technology Handbook,* L. D. Harte, editor, The American Ceramic Soc., Inc., Westeville, OH.

Wunder, J. W. J., 1962, *Oil & Gas J.,* Aug. 6, pp. 137–148.

Yang, R. T., 1987, *Gas Separation by Adsorption Processes,* Butterworth Publishers, Stoneham, MA.

Yon, C. M., and Turnock, P. H., 1971, "Multicomponent Adsorption Equilibrium on Molecular Sieves," paper presented at the AIChE 68th National Meeting, March 1.

Chapter 13
Thermal and Catalytic Conversion of Gas Impurities

INTRODUCTION, 1136

THERMAL CONVERSION OF GAS IMPURITIES, 1137
    Thermal Oxidation of VOCs and Odors, 1137

CATALYTIC CONVERSION OF GAS IMPURITIES, 1145
    Catalytic Oxidation of VOCs and Odors, 1148
    Catalytic Oxidation of Sulfur Compounds to Sulfur Oxides, 1162
    Conversion of Organic Sulfur Compounds to H_2S, 1165
    Conversion of Carbon Monoxide to Carbon Dioxide (Shift Conversion), 1172
    Conversion of Carbon Oxides to Methane (Methanation), 1177
    Selective Hydrogenation of Acetylenic Compounds, 1180

REFERENCES, 1183

INTRODUCTION

The thermal and catalytic conversion processes described in this chapter involve the chemical reaction of one or more gas phase species, including the impurities to be removed, to form new species, which remain in the gas. In thermal processes the desired reaction occurs at an acceptably high rate as a result of operating the reactor at an elevated temperature. In catalytic processes the reaction rate is accelerated at a low or moderate temperature by the presence of an active catalyst.

The new compounds formed by the conversion reactions are either unobjectionable and therefore permitted to remain in the gas, or more readily removed than the initial impurities. In the first case, the thermal or catalytic conversion process constitutes the entire purification operation while, in the second case, additional steps, such as absorption or adsorption, are required. Examples of the two cases are (1) oxidation of volatile organic compounds (VOCs), where the resulting carbon dioxide and water vapor remain in the gas and (2) catalytic conversion of organic sulfur compounds to H_2S, which is then removed in a separate step.

Processes described in this chapter include (1) thermal oxidation of VOCs and odors, (2) catalytic oxidation of VOCs and odors, (3) catalytic oxidation of sulfur compounds to sulfur oxides, (4) catalytic conversion of organic sulfur compounds to hydrogen sulfide, (5) conversion of carbon monoxide to carbon dioxide (shift conversion), (6) conversion of oxides of carbon to methane (methanation), and (7) conversion of acetylene to ethylene (selective hydrogenation).

Thermal and catalytic conversion processes that are not covered in this chapter include (1) processes associated with the removal of SO_2, which are covered in Chapter 7; (2) processes used for converting H_2S to elemental sulfur, including treatment of sulfur plant tail gas, which are covered in Chapter 8; and (3) processes for reducing NO_x, which are covered in Chapter 10.

THERMAL CONVERSION OF GAS IMPURITIES

Only two noncatalytic thermal conversion processes are important for gas purification. These are (1) the reaction of urea and/or ammonia with NO_x to remove the NO_x from hot combustion exhaust gas and (2) the reaction of oxygen with combustible impurities to oxidize them to less noxious species. NO_x removal is discussed in Chapter 10, so thermal oxidation of VOCs and other impurities is the only process of this type described in this chapter. The discussion of reactor design for thermal conversion processes is, therefore, directed entirely toward this application, and is included in the section that follows.

Thermal Oxidation of VOCs and Odors

Oxidation is a very effective technique for removing small concentrations of VOCs and odor-causing compounds from exhaust gas streams before they are released to the atmosphere. The process has the advantages over adsorption of being applicable to essentially all organic compounds, providing high efficiency at very low impurity concentration, and disposing of, rather than concentrating, the impurities. Carbon adsorption has an advantage over oxidation when the recovered organic compounds have significant value.

Oxidation of VOCs in a gas stream with an excess of air (often called incineration) produces primarily carbon dioxide and water vapor, which are relatively innocuous. When the gas impurities are chlorine- or sulfur-containing compounds, the products of combustion will normally include HCl and SO_2. A significant concentration of these compounds in the incinerator off-gas can result in the requirement for an auxiliary scrubber or other removal system. When nitrogen compounds are present in the feed, they typically form elemental nitrogen; however, under some conditions the presence of nitrogen compounds in the feed can result in the emission of nitrogen oxides in the exhaust.

Oxidation of the gaseous impurities may be accomplished in a thermal or catalytic incinerator. In the thermal incinerator a burner raises the temperature of the feed gas to a level high enough to cause a rapid reaction between combustible impurities and oxygen in the gas. The temperature required for thermal oxidation is typically over 1,200°F. In the catalytic incinerator a catalyst is used to increase the rate of the oxidation reactions at a lower temperature, typically about 600°F. Because of its higher operating temperature, a thermal incinerator usually requires more fuel with a resulting higher operating cost than a catalytic unit; however, it is simpler in design and typically has a lower capital cost.

Process Description

Thermal oxidation is probably the simplest and most efficient method of removing VOCs and odors from exhaust gas streams before they are released to the atmosphere. The process is typically applied to exhaust air streams where the quantity or type of organic material present does not justify recovery.

In its simplest form, a thermal oxidation system consists of a burner and a refractory-lined chamber or afterburner. The burner provides a stable flame that raises the temperature of the impure feed gas to the required level, and the chamber provides the necessary residence time for the reactions to proceed. The burner/afterburner system must also be designed to provide rapid and complete mixing so that all portions of the feed gas are subjected to the required time/temperature conditions.

In most cases the gas to be processed is an exhaust air stream and contains an adequate concentration of oxygen, but insufficient fuel, to support combustion. It is therefore necessary to use an auxiliary fuel, such as natural gas or oil, in the burner. In the few cases where the gas to be processed is low in oxygen (i.e., less than about 16%), additional air must be added.

Since the gas leaves the afterburner chamber at an elevated temperature, it is usually economically attractive to recover some of the heat energy. This can be accomplished by using the hot gas to preheat the gas to be treated (primary heat recovery) or using it in an external system such as a waste heat boiler (secondary heat recovery). Feed-gas preheating is the most common heat recovery technique, and two types of systems are employed: recuperative and regenerative. Simplified flow diagrams of thermal incineration systems using recuperative and regenerative heat exchangers, respectively, are shown in **Figures 13-1** and **13-2**.

Recuperative systems use conventional shell-and-tube or plate-type heat exchangers. Plate-type units are reported to be more economical for low to moderate temperature service (to about 1,000°F), while shell and tube are preferable for higher temperatures (van der Vaart et al., 1990). Because of the temperature limitations of conventional heat exchangers, recuperative systems are normally designed to heat the feed gas to no more than 1,200°F. A conventional shell-and-tube recuperative system can typically recover 60 to 80% of the available energy.

When the gas contains a significant concentration of organic compounds, their oxidation can contribute to the heat requirement of the system and, with sufficient heat recovery, it is sometimes possible to operate without the use of any auxiliary fuel. The relationship between the amount of heat recovery required, and the concentration of VOCs in the gas (expressed as percent of Lower Explosive Limit, LEL) for a typical case is shown in **Figure 13-3**. As indicated in the figure, slightly less heat recovery is required for 1,400°F than for 1,500°F operation. The dotted line represents a thermal oxidizer treating 200°F exhaust air containing solvents equivalent to 20% of the LEL at an oxidation temperature of 1,400°F, and shows that about 58% heat recovery would be required for self-sustaining operation.

In regenerative systems, the hot gas from the combustion chamber is passed through a packed bed of temperature- and chemical-resistant material such as chemical porcelain, which is heated while the gas is cooled. The gas leaves the bed at a low temperature, having given up most of its sensible heat to the ceramic. At the same time the feed gas stream, with auxiliary air, if required, is passed through a second, previously heated, packed bed which heats it to the preheat temperature and is cooled in the process. When the first bed is fully

Thermal and Catalytic Conversion of Gas Impurities **1139**

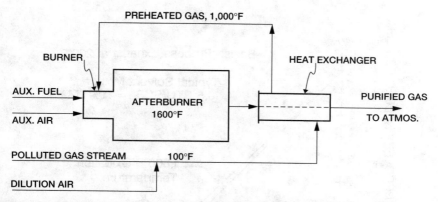

Figure 13-1. Simplified flow diagram of typical thermal oxidation system with recuperative heat recovery.

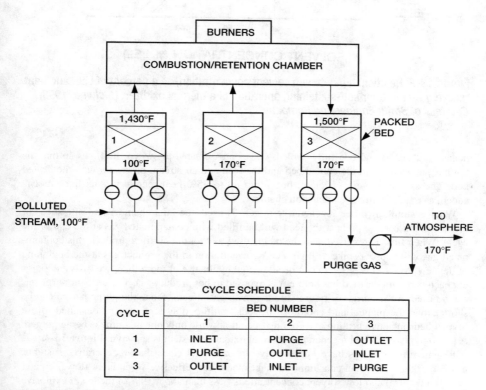

Figure 13-2. Simplified flow diagram of three-bed thermal oxidation system with regenerative heat recovery and positive purge.

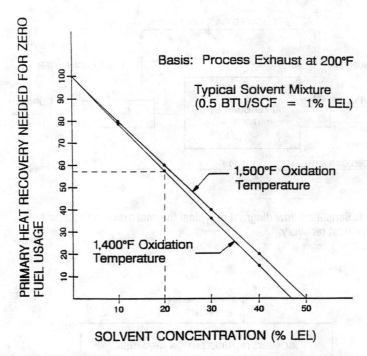

Figure 13-3. Relationship between solvent concentration, as a percent of LEL, and heat recovery required for self-sustaining operation of a thermal oxidizer. (*McIwee, 1993*). *Courtesy of Smith Engineering Company*

heated and the second bed fully cooled, valves are actuated to switch the flows so that the hot gas passes through the cooled bed and the feed and air stream passes through the heated bed. The system is very efficient, providing up to 95% heat recovery. Auxiliary fuel is added, as required, in the combustion chamber.

When a simple two-bed regenerative system is used, the inlet chamber and interstices of the bed being used to preheat the feed will be filled with contaminated gas at the end of the cycle. When flow is reversed, and the bed is used to cool the purified product, this contaminated gas will cause a spike of high VOC concentration in the product gas at the beginning of the new cycle. This problem can be alleviated by the use of more than two beds and purging each bed with purified gas before it is exposed to hot product gas. The effluent purge gas can be used in the burner or returned to the main gas stream. Since the purge period can be shorter than the preheat and cooldown periods, multi-bed systems can be designed to use several beds in active heating and cooling operation while only one at a time is being purged. A photograph of a 5-bed regenerative thermal oxidation system is shown in **Figure 13-4.**

Regenerative systems have appreciably higher capital costs than recuperative systems; however, their lower fuel consumption makes them cost effective when large flows (greater than 10,000 cfm) and low solvent concentrations (less than about 10% of the lower explosive limit) are present (Renko, 1990). The regenerative systems have other potential advantages. They can operate at temperatures up to about 2,000°F, while recuperative systems are limit-

Figure 13-4. A 145,000-scfm five-bed regenerative oxidizer with 86% heat recovery at a paper and film coating facility. *Courtesy of Smith Environmental Corporation*

ed to hot gas temperatures of about 1,500°F. Temperatures above 1,500°F are sometimes necessary to attain the required destruction efficiency. Also, the regenerative systems are free of leakage and corrosion problems associated with the heat exchanger design; however, they do require large, gas-tight cycling valves, which are unnecessary in recuperative designs.

Design Approach

The basic elements in the design of thermal incinerators for gaseous impurities are the combustion chamber temperature and the time the reacting gas is held at the temperature. The two are interrelated, i.e., if the residence time is reduced, the same destruction efficiency can often be realized by increasing the temperature. The destruction efficiency (or the required time/temperature combination) is also influenced by the type of compounds being destroyed and the degree of mixing in the combustion chamber.

Table 13-1 gives the theoretical combustion chamber temperatures required to destroy 99.99% of the listed compounds with a one-second residence time under excess air conditions (van der Vaart et al., 1990). Actual units are typically designed to attain significantly lower destruction efficiencies by providing somewhat less than 1 second reaction time at a temperature in the range of 1,200°–2,000°F.

On the basis of several studies, based on actual field test data, van der Vaart et al. (1990) report that commercial incinerators should generally be run at 1,600°F with a nominal residence time of 0.75 seconds to provide 98% destruction of non-halogenated organics. Valentin (1990) states that highly effective destruction of odors is obtained with a residence time of 0.3 to 0.5 seconds in the mixing chamber plus 0.1 to 0.3 seconds in the afterburner and an after-

Table 13-1
Theoretical Temperatures Required for 99.99% Destruction by Thermal Incineration in One-Second Residence Time

Compound	Temperature, °F
Acrylonitrile	1,344
Allyl Chloride	1,276
Benzene	1,350
Chlorobenzene	1,407
1, 2-dichloroethane	1,368
Methyl chloride	1,596
Toluene	1,341
Vinyl chloride	1,369

From van der Vaart et al. (1990)

burner temperature in the range of about 1,200° to 1,500°F. These values average 0.6 seconds total residence time at 1,350°F. For VOCs, McInnes et al. (1990) suggest total residence times of 0.5 to 1.0 seconds at 1,200° to 1,600°F. Renko (1990) gives typical residence times of 0.33 to 0.5 seconds for such service. Residence times and temperatures suggested in the *EPA Handbook* (1991) for 98 and 99% destruction efficiency are given in **Table 13-2**. In some states, residence time and temperature requirements are specified for thermal incinerators as these parameters can be monitored more readily than the outlet gas composition.

When a regenerative heat exchanger with a thermal energy recovery in the 80 to 95% range is used, the temperature of the preheated VOC-containing gas will generally be higher than the ignition temperature of typical VOCs. For example, with 95% thermal energy recovery and a 1,500°F combustion chamber temperature, the preheated gas will be at approximately 1,430°F. As indicated by the data in **Table 13-3**, this is well above the ignition temperature for common VOCs, and some oxidation of VOCs will begin to occur (by design) in the packed bed before the gas enters the combustion chamber (Seiwert, 1993).

Once a residence time and a combustion chamber temperature are selected, the design is relatively straightforward and includes the following steps:

Table 13-2
Design Parameters for Thermal Incinerators

Destruction Efficiency, %	Nonhalogenated Temperature, °F	Time, sec.	Halogenated Temperature, °F	Time, sec.
98	1,600	0.75	2,000	1.0
99	1,800	0.75	2,200	1.0

Note: Values are conservative and assume adequate mixing of gases in the incinerator.
From EPA Handbook (1986)

Table 13-3
Ignition Temperatures for Typical Volatile Organic Compounds (VOCs)

VOC	Ignition Temperature, °F
Acetone	870
Acetonitrile	970
Isobutyl Alcohol	780
Isopropyl Alcohol	750
Methanol	878
Methyl Ethyl Ketone	759
Methylene Chloride	1,030
Toluene	896
Xylene	867

Data of Seiwert (1993). Courtesy of Smith Engineering Company

1. Determine if dilution air is required to assure that the concentration of combustibles in the feed gas to the incinerator system is less than 25% of the Lower Explosive Limit (LEL), or less than 50% if continuous monitoring is employed. LEL and heat of combustion data for a number of compounds are given in **Table 13-4**. If dilution air is required, calculate the amount and the total quantity of feed.

2. Assume a heat exchanger efficiency and calculate the feed stream temperature after heat exchange. The heat exchanger efficiency, E, is defined as the percent of the sensible heat in the flue gas between the combustion chamber outlet temperature and the unheated feed gas temperature that is transferred to the feed gas. E is approximately equal to the ratio (as a percent) of the temperature rise of the feed gas to the potential temperature rise if it were heated to the combustion chamber outlet temperature, i.e., $E = 100 (T_{F2} - T_{F1})/(T_{E1} - T_{F1})$, where T_{F1} is the temperature of the unheated feed gas; T_{F2} is the temperature of the heated feed gas; and T_{E1} is the temperature of the combustion chamber exit gas. This relationship can be used to estimate the temperature of the heated feed gas because the temperatures of the unheated feed and the combustion chamber outlet gas are known and the heat exchanger efficiency can be estimated with reasonable accuracy. The *EPA Handbook* (1986) suggests assuming a value of 50% for E if no other information is available; however, efficiencies as high as 80% for recuperative and 95% for regenerative heat exchangers are feasible.

3. Calculate the amount of supplementary fuel that must be added to attain the desired combustion chamber temperature. In most cases the oxygen content of the feed gas will be high enough (over about 16%) to provide an adequate supply of oxygen for combustion of both fuel and impurities, and the concentration of combustible impurities will be so low that their heat of combustion can be neglected. When precise calculations are required, or when the feed gas contains a solvent concentration over 2–3% of the LEL, it is necessary to take the heat of combustion of the gas impurities into account. The amount of fuel required can be estimated using a simple heat balance around the combustion chamber. The calculated amount of fuel should be increased as required to account for 5–10% heat losses from the incinerator. If the heat balance indicates that excess heat is available from combustion of the VOCs, a lower efficiency heat recovery unit can be used or, with a

**Table 13-4
Lower Explosive Limit (LEL) and Heat of Combustion of Selected Compounds**

Compound	LEL (ppmv)	Lower Heat of Combustion (Btu/scf)
Methane	50,000	892
Ethane	30,000	1,588
Propane	21,000	2,274
n-Butane	16,000	2,956
Isobutane	18,000	2,947
n-Pentane	15,000	3,640
Isopentane	14,000	3,631
Neopentane	14,000	3,616
n-Hexane	11,000	4,324
Ethylene	27,000	1,472
Propylene	20,000	2,114
n-Butene	16,000	2,825
1-Pentene	15,000	3,511
Benzene	13,000	3,527
Toluene	12,000	4,196
Xylene	11,000	1,877
Acetylene	25,000	1,397
Naphthalene	9,000	5,537
Methyl alcohol	60,000	751
Ethyl alcohol	33,000	1,419
Ammonia	160,000	356
Hydrogen sulfide	40,000	583

Note: Heats of combustion based on 70°F and 1 atm.
Data from EPA Handbook (1991).

regenerative system, part of the gas can be permitted to bypass the packed beds. If no auxiliary fuel is required, it is still customary to maintain a pilot burner in operation to assure stable operation of the combustion chamber.

4. Calculate the combustion chamber volume. This requires that the actual volumetric flow rate of hot gas passing through the combustion chamber be known. If the auxiliary fuel is natural gas, the actual flow rate of hot gas will be equal to the flow rate of feed gas plus the flow rate of natural gas corrected to the combustion chamber temperature. The hot gas flow rate in the combustor must also include any air added to dilute the VOC concentration to a safe value, raise the oxygen content of the gas stream, or support combustion of fuel in the burner. The required combustion chamber volume is calculated from the total gas volumetric flow rate, at the actual operating temperature and pressure, and the required residence time. The calculated volume is customarily increased by 5% to allow for fluctuations in operating conditions (*EPA Handbook,* 1991).

5. Determine the size of the heat exchanger required to heat the feed gas and cool the exit gas from the combustion chamber to the temperatures defined in step 2. For shell-and-tube recuperative heat exchangers this is a standard heat exchanger calculation. An overall heat transfer coefficient in the range of 2 to 8 Btu/h-ft °F is typical for this type of heat exchanger (*EPA Handbook,* 1986). The design of regenerative heat exchanger systems is somewhat more complex, but is described in standard engineering texts (e.g., Perry and Green, 1984).

Economics

Capital cost curves for thermal incinerators (based on combustion chamber volume) and recuperative heat exchangers (based on surface area) are given in the *EPA Handbook* (1986). Equations and curves for estimating capital costs are also given by van der Vaart et al. (1990). Typical equipment costs read from the curves are given in **Table 13-5**. The tabulated costs represent equipment only. The total capital cost, including all direct and indirect costs of installation, can be approximated by multiplying the equipment cost by a factor of 1.9. The result is in 1988 dollars and does not include any costs required for site preparation or buildings. According to McInnes et al. (1990), the total capital cost for a typical system capable of handling 20,000 scfm of polluted exhaust gas ranges from $380,000 to $480,000. Operating costs for a system of this size can range from $25,000 to $225,000 per year, depending primarily on the degree of heat recovery, the VOC concentration in the gas, and the operating schedule.

CATALYTIC CONVERSION OF GAS IMPURITIES

The field of chemical catalysis is extremely complex, and any attempt to discuss its theoretical aspects, even in the most elementary form, would be beyond the scope of this book. Detailed discussions of the subject are available in a number of texts, (e.g., Perry and Green, 1984; Lee, 1985; Carberry, 1976; Fogler, 1986; Smith, 1981; and Campbell, 1988).

Table 13-5
Equipment Costs for Thermal Incinerator Systems

Heat Recovery, % Heat Exchanger	0 none	50 recup.	70 recup.	95 regen.
Plant size, Mscfm		Equipment cost, M$		
3	68	130	150	—
10	90	170	200	300
50	120	260	310	800
100	—	—	—	1,360

Notes:
1. Costs in April 1988 dollars.
2. Costs are for equipment only, F.O.B, not installed.

Data from van der Vaart et al. (1990)

Catalysis is based on the observation that the rate of certain reactions is influenced by the presence of substances—called catalysts—which are not changed by the reactions. Catalysts may increase (positive catalysis) or decrease the rate of reaction (negative catalysis) or direct the reaction along a specific path. Depending on the phase relationship between the catalyst and the initial reactants, catalytic processes may involve homogeneous or heterogeneous catalysis. Homogeneous catalysis is characterized by the fact that the catalyst and the reactants are in the same phase, while in heterogeneous catalysis the catalyst and the reactants are in different phases. Catalytic gas purification processes utilize positive catalysis and, since the catalysts are generally solids, involve heterogeneous catalysis.

In heterogeneous catalysis the contact between the catalyst and the reactants at the phase boundary is of utmost importance. This contact is established by adsorption of the reactants on the catalyst surface, resulting in increased concentration of the reactants. Adsorption takes place in localized areas called "active centers." The overall mechanism of heterogeneous catalysis involves (a) mass transfer of the reactants from the fluid to the catalytic surface and of the products from the catalytic surface to the fluid; (b) diffusion of reactants and reaction products into and out of the pores of the catalyst; (c) activated adsorption of reactants and desorption of products at the interface; and (d) surface reactions of adsorbed reactants to form chemically adsorbed products. Considering that each of these steps is quite complex in itself, the highly complicated nature of the overall mechanism can be appreciated.

Catalysts used in most gas-purification processes are metals or metal compounds, usually supported on an inert carrier of large suface area. However, unsupported catalysts are also used. Typical carriers are alumina, ceramic (e.g., cordierite), bauxite, activated carbon, and metal wires. In cases where very highly active catalysts are required, the catalytic surface is "activated" by special procedures. It is not uncommon to use two or more catalytic materials in one preparation because the activity of one catalytic material may frequently be increased beyond that expected by the simple additive effect of the additional components. Additives of this type are known as promoters.

Since catalysts are not changed or consumed during the conversion process, they are theoretically usable for an indefinite period of time. In practice, however, most catalysts deteriorate or are gradually deactivated during operation and have to be replaced or regenerated periodically. Loss in catalyst performance may be due to any of the following mechanisms:

1. *Poisoning* is a loss in activity due to chemical reactions between the catalyst materials and substances present in the gas stream with the formation of stable reaction products.
2. *Masking* is the gradual accumulation of heavy organic molecules, carbon, corrosion products, metal oxides, or other impurities on the catalyst surface restricting access to active sites by gaseous reactants.
3. *Sintering* is the agglomeration or densification of the catalyst or support material, which reduces the effective area of catalyst exposed to the process stream.
4. *Attrition* is the breakup of catalyst shapes to form fine particles, which can result in increased pressure drop through the catalyst bed or excessive loss of catalyst by entrainment of small particles in the gas stream.

Much of the progress in catalytic conversion processes has been in the development of catalysts, systems, and operating procedures that minimize the effects of these phenomena. Commercial catalysts must not only have the required activity and resistance to deactivation, but must also be available in shapes that provide effective contact with the gas at low pressure drop and meet installation and operating requirements such as recharging and regeneration.

The heart of a typical catalytic gas purification process is a contacting or reaction vessel, often called a reactor or converter, which contains the catalyst. The catalyst may be in the form of pellets, spheres, or granules arranged in single or multiple fixed beds or in tubes, or it may be in the form of a monolithic honeycomb panel, supported in a frame. Typical catalyst shapes used in fixed beds are shown in **Figure 13-5**. A commercial honeycomb configuration is illustrated in **Figure 13-6**.

The size of the converter required for a given duty is determined primarily by the design space velocity for the specific reaction conditions. Space velocity is an indirect measure of the contact time between the gas and the catalyst. It is usually defined as the volume of gas at standard conditions passing through a unit volume of catalyst per unit of time. However, space velocity is occasionally expressed in terms of actual flowing gas volume per volume of catalyst per unit of time, and it is extremely important that the applicable definition be known before a space velocity value is used for design.

Figure 13-5. Typical catalyst shapes used in fixed bed converters. *Courtesy of Haldor Topsøe, Inc.*

Figure 13-6. Honeycomb type, platinum metal group catalyst. *Courtesy of Engelhard Corp.*

The design space velocity establishes the volume of catalyst required for a given conversion; however, other factors must be considered in establishing the bed depth and cross-sectional area. The chief factors are pressure drop, attrition (or particle entrainment in the case of upflow reactors), and channeling. Low depth-to-diameter ratios are indicated when pressure drop is an important consideration (as in low-pressure, high-volume, exhaust air processing). High ratios are used with high-pressure systems where the actual volume of gas is relatively small, and the cost of pressure containment is more important than the cost of moving the gas through the bed. Correlations for the calculation of pressure drop through beds of pelleted or granular materials are given in Chapter 12. Pressure drop data for special catalyst shapes and honeycomb monoliths are normally available from the suppliers.

Catalytic Oxidation of VOCs and Odors

Catalytic oxidation is similar to thermal oxidation in that the gaseous impurities to be destroyed are reacted with oxygen at elevated temperature to form primarily carbon dioxide and water vapor. It differs in that the reactions are made to occur at a relatively low temperature by the use of a solid catalyst. The process is widely used for destroying VOCs and odorous compounds in exhaust air streams and for eliminating carbon monoxide from flue gas streams produced by combustion processes. It is also applicable for reacting hydrogen and oxygen in gas streams where either of these constitutes an impurity and the other is present in excess.

Process Description

Although the required operating temperature is much lower for catalytic than for thermal oxidation processes, the polluted gas stream must still be heated to a temperature high enough to initiate the reaction. Typical minimum temperatures for the initiation of catalytic oxidation of various organic compounds are given in **Table 13-6.** Both sets of data are based on the use of platinum metal group (PMG) catalysts. The data of Suter (1955) were obtained under typical commercial operating conditions at hydrocarbon concentrations corresponding to approximately 10% of the lower limit of flammability.

Typically, the reaction temperature is in the range of 300° to 900°F, and is determined on the basis of the material being oxidized, the catalyst used, and the conversion desired. The temperatures required for 99% conversion of a large number of gaseous pollutants over Haldor Topsøe metal oxide catalysts are given in **Table 13-7.**

Recovery of heat from the gas leaving the reaction chamber is usually economical and, because of the relatively low temperatures involved, conventional metal gas-to-gas recuperative heat exchangers are used.

A generalized flow diagram of a catalytic oxidation system is shown in **Figure 13-7.** The gas to be purified is preheated to the desired temperature by mixing with the combustion products of a burner and/or by indirect heat exchange with the hot product gas or other heat source. The preheated gas is then passed through a bed of catalyst which accelerates the rate of reaction between oxygen and combustible impurities in the gas. Typically about 80 to 99% of the impurities are destroyed. The hot product gas is usually cooled by heat exchange with the impure feed gas then vented to the atmosphere.

Regenerative heat exchange can be used with catalytic as well as thermal oxidation. Such a system is offered by Haldor Topsøe A/S as the REGENOX process (Haldor Topsøe, 1990). In this process the catalyst also acts as the heat transfer medium for regenerative heat

Table 13-6
Catalytic Ignition Temperatures of Typical VOC's and Hydrocarbons

Component	Catalytic Ignition Temp., °F	°C
Benzene	400	204
Toluene	400	204
Formaldehyde	300	149
Methylethylketone	475	246
Carbon monoxide	320	160
Methane	760	404
Ethane	680	360
Propane	650	343
n-Butane	570	299
n-Pentane	590	310
n-Hexane	630	332
n-Heptane	580	304
n-Octane	490	254
n-Decane	500	260
n-Dodecane	540	282
n-Tetradecane	550	288
o-Xylene	470	243

Note:
Values given are ignition temperatures; operating temperatures are always higher.
Data on benzene, toluene, formaldehyde, methylethylketone, and carbon monoxide from Engelhard Corp. (1991)
Balance of data from Suter (1955)

exchange. Two beds are used with a burner between them. The feed gas is preheated by passage through the first bed, further heated, if necessary, by the burner, then cooled while it passes through the second bed. After the second bed is thoroughly heated, the gas flow is reversed.

Catalysts

Many catalyst chemical formulations and geometric shapes are used to promote oxidation-reduction reactions. Chemical types used for VOC oxidation include platinum, platinum alloys, copper chromate, copper oxide, cobalt oxide, chromium oxide, manganese oxide, and nickel. The catalysts are often categorized as platinum metal group (PMG) and base metal (or metal oxide) types. The active catalyst is often supported on an inert carrier such as gamma alumina. Catalyst forms include metal ribbons, mesh and gauze; honeycomb monoliths; and small beads or particles that can be used in a fixed, fluidized, or moving bed.

Catalytic oxidation systems are subject to poisoning or masking by halogens, phosphorus, lead, bismuth, arsenic, antimony, mercury, iron, tin, zinc, sulfur, alkali metals, and silica.

Table 13-7
Lowest Catalytic Combustion Temperature at Which Over 99% Conversion is Obtained

Compound	°C	°F
Ethylene	300	572
Butane	290	554
Butene	230	446
Heptane	275	527
Benzene	300	572
Toluene	270	518
Xylene	280	536
Naphthalene	270	518
Methanol	190	374
Formaldehyde	190	374
Ethanol	210	410
1-Propanol	210	410
1-Pentanol	200	392
Cresol	240	464
Diisobutylketone	210	410
Methylethylketone	240	464
Butyric acid	200	392
Ethylene oxide	190	374
Phthalic acid anhydride	270	518
Maleic acid	200	392
Dibutylphthalate	275	527
Pyridine	250	482
Tributylamine	200	392
Dimethylformamide	230	446
Toluene diisocyanate	285	545
Chlorobenzene	380	716
Chloroform	350	662
Acetonitrile	220	428
Acrylonitrile	220	428
Offset solvents (boiling point 230–270°C)	260	500
Carbon monoxide	180	356
Ammonia	250	482
Hydrogen cyanide	250	482
Ill-smelling exhaust air from:		
Onion/garlic processing	200	392
Coffee roasting	180	356
Meat destruction	200	392

Notes: 1. Temperatures are bed inlet temperatures.
    2. Catalyst is Topsøe CK-302 metal oxide.
Source: Haldor Topsøe A/S (1991A, 1993)

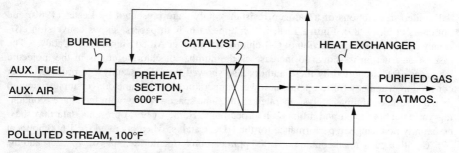

Figure 13-7. Simplified flow diagram of fixed bed catalytic oxidation system with recuperative heat recovery.

Based on the poison susceptibility of a Johnson Matthey platinum-rhodium catalyst oxidizing ethane at 480°C, the severity of poisons tested, on a molar basis, is in the following order (Jung and Becker, 1987):

As > Fe > Na > P > SiO_2

In these tests the silica was believed to act as a masking agent rather than a catalyst poison.

In addition, solid particles in the feed stream can cause plugging of fixed beds of catalyst. In processing exhaust air streams containing VOCs, the presence of halogens is normally the most significant problem. Particulates, if present, can be filtered from the gas stream before they reach the catalyst, and the other poisons listed are seldom present in significant quantities in such streams.

Because of the widespread use of chlorinated hydrocarbons in industry, the presence of halogens in exhaust gas impurities is very common. This has caused problems, particularly with platinum group metal (PGM) catalysts, which make up about 75% of the catalysts used for VOC destruction. Reportedly, the problem lies with the catalyst carrier or wash coat rather than with the catalytic metals. Alumina, which is usually used as the carrier, reacts with chlorine compounds to form aluminum chloride, which can cover or otherwise degrade the metal catalysts. According to Parkinson (1991), alternate carriers are being developed to avoid this problem.

Several catalyst systems have recently been developed that resist poisoning by chlorine. An example is the ARI International Econ-Abator system, which uses chromia-alumina catalyst granules in a shallow fluidized or moving bed (ARI International, 1992). The catalyst is formulated by impregnating chromia throughout a porous gamma alumina substrate. The catalyst particles are kept clean by the scouring action of the fluidized bed, which continuously exposes fresh active catalyst to the gas stream (Parkinson, 1991).

According to van der Vaart et al. (1990), gases containing chlorinated compounds have been successfully oxidized over metal oxide catalysts such as chromia/alumina, cobalt oxide, and copper oxide/manganese oxide; while sulfur-containing VOCs can be oxidized by the use of platinum-based catalysts, which are rapidly deactivated by chlorides.

Allied Signal offers two poison-resistant oxidation catalysts: Halocarbon Destruction Catalyst (HDC), which is claimed to be capable of reducing emissions of chlorinated hydrocarbons by 95%, and Poison Resistant Oxidation (PRO) catalyst, which is designed to resist poisoning by phosphorus, silicon, and heavy metals (Shearman, 1992).

Detailed descriptions of a halogen resistant catalyst are presented by Lester (1989) and Summers et al. (1989). **Figure 13-8,** which appears in both papers, shows conversion efficiency curves for the destruction of C-1 chlorocarbons over Allied Signal HDC Catalyst. The ease of destruction is shown to increase as the number of chlorine atoms in the molecule increases. Tests reported by these authors also showed the HDC Catalyst to provide higher conversion efficiency (or to operate at lower temperature for the same conversion efficiency) than a chromia-alumina catalyst operated at the same space velocity with feed gases containing various chlorinated and aromatic hydrocarbons. Herbert (1991) presents data that show essentially no change in performance for the HDC Catalyst when operated for 1,600 hours at 375°C with a gas stream containing 1,000 ppm carbon tetrachloride, 9.8 vol.% water, and the balance air.

The development of Allied Signal's poison-resistant catalyst is described by Summers et al. (1989) and Lester and Summers (1988). The work was conducted because of the rapid deactivation observed for a conventional PMG type catalyst used to oxidize VOCs in the exhaust air from dryers of lithographic web offset presses. The problem was found to be due to phosphorous introduced in additives to a solution used in the printing process. A new catalyst, designated PZ-1157, was developed specifically for such applications and tested in parallel with the conventional catalyst. During the test period, samples of the catalyst were periodically removed from the inlet portion of the beds (where the highest concentration of poison is encountered) and tested for activity using n-heptane as the model hydrocarbon. Although both catalysts lost activity, the slopes of the conversion-vs.-time curves indicated that the new catalyst would have a useful life at least double that of the old material. The performance of the PZ-1157 catalyst in commercial graphic arts applications has shown

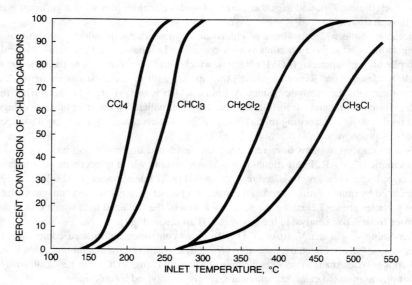

Figure 13-8. Destruction of C-1 chlorocarbons over Allied Signal HDC Catalyst. Conditions: H_2O concentration 1.5%; space velocity 15,000 hr^{-1}; chlorocarbon concentrations, CCl_4 900 ppmv, $CHCl_3$ 500 ppmv, CH_2Cl_2 800 ppmv, CH_3Cl 600 ppmv. (*Summers et al., 1989*)

hydrocarbon conversion efficiencies over 95% for material in service as long as 40 months (Summers et al., 1989).

A comprehensive description of PMG honeycomb monolithic catalysts of the type shown in **Figure 13-6** is given by Heck et al. (1988). They point out that the honeycomb design provides the following advantages compared to packed beds of granules:

1. Low pressure drop
2. Fast warm up
3. Low volume and weight
4. No attrition, dimensionally stable
5. No limitation on reactor orientation
6. Good shock and vibration resistance
7. High geometric surface area
8. Resistant to plugging
9. No channeling
10. No dusting
11. Efficient utilization of precious metal
12. Cleanable

Commercial honeycombs are available containing 10 to 600 cells per square inch (CPSI). Comparison of a 200 CPSI honeycomb with a particle bed of equal geometric surface area at equal space velocity indicates that the particle bed has 50 to 100 times higher pressure drop per foot of depth (Heck et al., 1988).

The microstructure of a ceramic honeycomb structure wall is shown schematically in **Figure 13-9**. The ceramic wall provides physical support, gas flow paths, and a high geometric

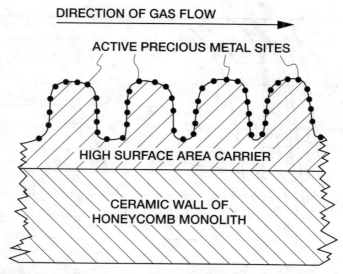

Figure 13-9. Schematic diagram of wall structure of honeycomb type, platinum metal group catalyst. (*Based on Heck et al., 1988*)

1154 *Gas Purification*

surface area; the carrier provides a stable, high surface area for catalyst dispersion; and the catalyst provides the active sites required to promote the oxidation reactions.

Honeycomb structures are also available that use a metal rather than a ceramic substrate (Johnson Matthey, 1992). According to Jung and Becker (1987), the most common metal substrate design is a stack of alternate layers of flat and corrugated foil strips. The resulting metal monolith, with a cell density of 100, 200, or 400 CPSI, is covered with a washcoat of high surface area alumina that serves as a support for the active PGM catalyst. It is claimed that the metal honeycomb design gives a lower pressure drop for a given amount of catalyst coating and provides a further advantage over ceramic-based designs in situations where thermal shock is experienced (Jung and Becker, 1987).

A catalyst that has been used effectively for oxidizing traces of acetylene in the air feed to low temperature separation plants and for providing hydrocarbon-free air for instruments and other special applications consists of a mixture of manganese dioxide and copper oxide, called Hopcalite. This catalyst also converts traces of carbon monoxide to carbon dioxide and traces of ozone to oxygen. The basic composition is about 60% manganese dioxide and 40% copper oxide; however, relatively small additions of silver as a promoter have been found to make the catalyst particularly effective for most air streams (Rushton, 1954). The degree of conversion of several hydrocarbons over a commercial Hopcalite catalyst as a function of temperature is shown in **Figure 13-10**.

The effectiveness of a catalyst system can be decreased by several mechanisms, including mechanical vibration, particulate abrasion, poisoning, masking, and thermal aging. Poisoning occurs when certain materials such as phosphorus, halogens, and heavy metals combine chemically with the catalyst forming inactive compounds. Masking is the result of the accumulation of solid material on the surface of the catalyst. It may be caused from entrained dust, metal oxides, or organic char formed by operating the reactor at too low a temperature. Thermal aging of precious metal catalysts is caused by the agglomeration of extremely small

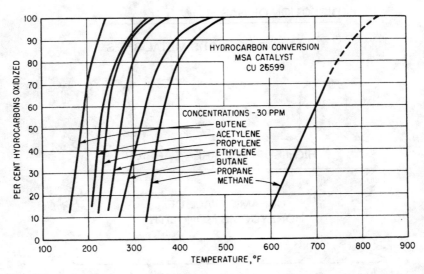

Figure 13-10. Oxidation of hydrocarbons over commercial Hopcalite as a function of temperature. (*Mine Safety Appliance Company*)

catalyst particles to form larger particles reducing the available surface area. With metal oxide catalysts, thermal aging can be caused by sintering of the carrier or of the metal oxide catalyst particles.

Careful design of the system can minimize the rate of catalyst deactivation by these phenomena. PMG catalyst that is deactivated by masking can often be regenerated by techniques that destroy or remove the foreign materials. Processes employed to regenerate such catalysts and typical results obtained in practice are described by Heck et al. (1988). Thermal aging can be minimized by operating at the lowest possible temperature and can be accounted for in the design by the use of a suitable thermal aging model. The development of such a model for PMG catalysts is discussed by Cordonna et al. (1989).

System Design

The key elements in the design of catalytic oxidation systems are space velocity and temperature. Space velocity, as used in the design of VOC oxidation catalyst beds, is usually defined as standard cubic feet of gas per cubic foot of catalyst per hour. It has the units of reciprocal hours. Space velocity is not simply the reciprocal of residence time because (1) the gas volume is measured at standard not actual conditions, and (2) the catalyst volume is the total volume, including both solid catalyst and spaces for gas flow.

Design space velocities for VOC oxidation are generally in the range of 10,000 to 60,000 hr^{-1} for precious metal catalysts and 5,000 to 15,000 hr^{-1} for base metal oxide catalysts (van der Vaart et al., 1990). Much higher space velocities can be used for carbon monoxide oxidation over platinum-based honeycomb catalyst systems. Design values in the range of 100,000 to 150,000 hr^{-1}, depending upon temperature and conversion requirements, are suggested by Engelhard (1987), and successful test operation at even higher space velocities has been reported (Chen et al., 1989).

Typically, the relationship between space velocity, temperature, and conversion efficiency is determined experimentally for the specific catalyst and gas composition involved, or is based on operating data from similar systems. Examples of space velocity/temperature/conversion test data are given in the following section. In selecting the space velocity and temperature values to be used for design, it is important to consider the possible effects of aging, poisoning, feed gas composition changes, and other factors on conversion efficiency during the life of the catalyst.

Once design values for space velocity and operating temperature have been selected, the design procedure resembles that for thermal incineration. If heat recovery is to be used, the heat exchanger efficiency is assumed, and the temperature of the feed gas to the preheat chamber is estimated. A heat balance is made around the burner, preheat chamber, and catalyst bed to determine the amount of auxiliary fuel that must be burned to produce the design temperature in the catalyst bed. Heat produced by oxidation of the VOCs should be included in the heat balance. The heat of combustion of typical VOCs is approximately 50 Btu/scf of exhaust gas when the VOC concentration is 100% of the LEL. This results in a heat of combustion of about 13 Btu/scf when the VOC content is 25% of the LEL (*MECA Guidebook,* 1992).

The following items should be considered in the thermal design of a catalytic oxidation unit:

1. The temperature of the gas entering the catalyst bed must be above the catalytic ignition temperature. This temperature varies with catalyst and VOC, and is typically in the range of 300° to 900°F. Catalytic ignition temperatures for several compounds on precious metal catalysts are given in **Table 13-6**.

2. The catalyst bed temperature must be high enough to give the desired conversion based on the design space velocity for the specific impurities and type of catalyst. Typical minimum temperature requirement data are given in **Table 13-7.**
3. The catalyst bed temperature must not exceed the level that can cause deterioration of the catalyst or support. The maximum allowable continuous operating temperature ranges from 1,200° to 1,400°F depending on the type of catalyst.
4. The auxiliary fuel burner should be operated to provide a minimum of about 5% of the total energy input regardless of heat exchanger capability to stabilize the flame in the preheat combustion chamber.
5. The auxiliary fuel should not contain impurities (such as vanadium or sulfur) which can cause poisoning or masking of the catalyst.

These limitations may result in the requirement to add dilution air to prevent excessive temperatures in the catalyst. In general, the heat of combustion of the gas entering the preheat chamber should be less than that of a gas with a VOC content about 20% of the LEL. These limitations may also affect the amount of heat recovery that can be utilized.

Typical catalytic combustion unit design parameters are listed in **Table 13-8** (*EPA Handbook,* 1991). The values given are considered quite conservative. The minimum inlet temperature of 600°F assures ignition of any likely organic compound in the gas, and the minimum outlet temperature of 1,000°F assures an adequate overall reaction rate, while the indicated maximum outlet temperature of 1,200°F is intended to prevent thermal degradation of the catalyst. The space velocities given in the table represent typical ranges for monolithic catalysts. Beds of pelletized or granular catalysts normally operate at lower space velocities.

After establishing a heat and material balance around the burner/preheat chamber/catalyst system, the total volumetric flow rate (scfh) of gases passing through the catalyst bed is calculated, including the flow rates of feed gas, diluent air, and combustion products. This total flow rate is used with the design space velocity to estimate the volume of catalyst required (volume of catalyst = total gas flow rate at standard conditions/space velocity).

The optimum configuration for the required volume of catalyst (i.e., bed thickness and face area) is determined on the basis of pressure drop, gas distribution, and vessel size considerations. Pressure drop is normally the most important factor, and is limited to a few inches of water, due to the high cost of compressing large volumes of atmospheric pressure exhaust gas; however, the pressure drop must be high enough to provide good distribution of gas passing

Table 13-8
Typical Catalytic Oxidation Unit System Design Parameters

Parameter	Typical Value
Destruction efficiency, %	95
Catalyst inlet temperature, °F	600
Catalyst outlet temperature, °F	1,000–1,200
Space velocity, l/hour:	
Base metal	10,000–15,000
Precious metal	30,000–40,000

Data from EPA Handbook (1991)

through the catalyst bed. The pressure drop for hot gas flowing through platinum based honeycomb catalyst at 20 ft/sec is shown as a function of space velocity in **Figure 13-11.** The pressure drop through beds of granular catalyst particles can be calculated by the use of standard packed bed correlations (see Chapter 12).

Operating Data

PGM catalysts are used extensively for controlling carbon monoxide emissions from gas turbine cogeneration systems and data on the development of this application are given by Chen et al. (1989). A schematic diagram of a carbon monoxide catalyst installation in a gas turbine system is given in **Figure 13-12.** A plot of carbon monoxide conversion vs. temperature for three space velocities with an Engelhard PMG honeycomb catalyst is reproduced in **Figure 13-13.** The data are for a typical honeycomb structure containing about 200 cells per square inch (CPSI). Reducing this value to 64 CPSI reduces conversion from over 80% to about 55% at the same temperature (350°C) and space velocity (300,000 hr^{-1}). Of course, reducing the number of cells per square inch also reduces the pressure drop (from about 0.6 to 0.2 in. of water under the above conditions (Chen et al., 1989).

The flow arrangement and operating conditions of a large catalytic oxidation system for the packing industry are shown in **Figure 13-14.** The installation, a Haldor Topsøe CATOX design, employs both a recuperative heat exchanger and a heat recovery system. The concentration of VOC (acetone) in the exhaust gas is as high as 10 g/Nm^3 (3,865 ppmv), and its

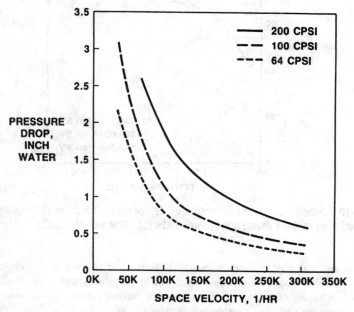

Figure 13-11. Pressure drop vs. space velocity for honeycomb type catalysts with 64, 100, and 200 cells per square inch in carbon monoxide oxidation service at 350°C (662°F) and 25 ft/sec face velocity. (*Chen et al., 1989*)

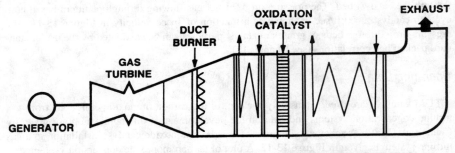

Figure 13-12. Schematic diagram of catalytic carbon monoxide oxidation system installed on a gas turbine. (*Chen et al., 1989*)

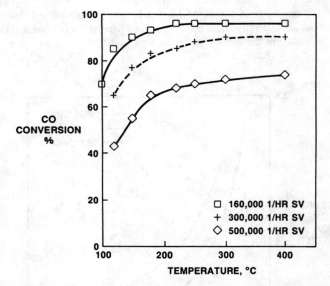

Figure 13-13. Carbon monoxide conversion vs. temperature for three different space velocities with a precious metal honeycomb catalyst. (*Chen et al., 1989*)

heat of combustion is used to produce steam and heat water. A total of 1.8 million kcal/h (7.1 million Btu/h) is recovered with more than 75% in the form of high pressure steam. It is reported that the value of heat recovered can pay back the installation costs in about one year under West European economic conditions (Andreasen, 1988).

Figure 13-15 is a photograph of a large fixed bed catalytic oxidation unit that cleans 55,000 Nm^3/h (approximately 34,000 scfm) of exhaust air from a phthalic anhydride plant. The largest piece of equipment is the vertical gas-to-gas heat exchanger located in the left-

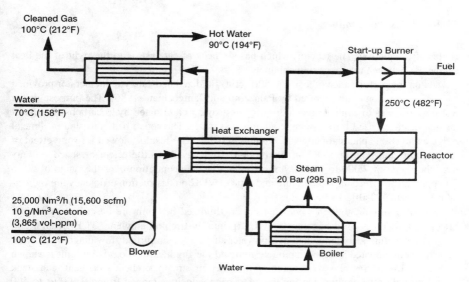

Figure 13-14. Flow diagram of CATOX catalytic oxidation system with recuperative heat exchange and two heat recovery units. (*Andreasen, 1988*)

Figure 13-15. Large CATOX catalytic oxidation unit cleaning 55,000 Nm^3/hr (34,200 scfm) of exhaust air from a phthalic anhydride plant. The large vessel to the left of center is a vertical heat exchanger. The converter, a six-bed unit, is to right of center, and the fuel tank is at the far right. The horizontal start-up burner is in the foreground. *Courtesy of Haldor Topsøe, Inc.*

center of the picture. The reactor, which has six beds of catalyst, is to the right of the heat exchanger between two large vertical ducts.

Data on the performance of a small (500 scfm) fluidized bed catalytic incinerator provided by ARI International are reported by Palazzolo and Jamgochian (1986). The purpose of the test program was to evaluate techniques for destroying chlorinated hydrocarbons in exhaust air that has been used for stripping contaminated water. The active bed consisted of a metal oxide catalyst on spherical pellets of aluminum oxide. The results showed a strong effect of temperature and type of VOC on destruction efficiency, but little effect of space velocity over the range of 7,000 to 10,500 hr^{-1} or inlet VOC concentration over the range of 50 to 200 ppmv. The effects of temperature and type of VOC on destruction efficiency are indicated by the data in **Table 13-9.**

Tests of a similar, though somewhat larger, fluidized bed unit are described by Hylton (1991). This unit is used to remove VOCs (primarily trichloroethylene, TCE) from 1,200 scfm of air exiting a contaminated groundwater stripper. A simplified flow diagram showing both the contaminated water stripping system and the fluidized bed oxidation unit is shown in **Figure 13-16.** Samples were taken of the inlet air stream, preheater effluent, and stack over a period of 19 months. During the test, space velocities ranged from 10,600 to 15,500 h^{-1} due primarily to the effects of catalyst loss and replenishment on the bed height. In addition to increasing the space velocity, catalyst loss had the effect of decreasing the pressure drop through the bed, which varied from 1,490 to 2,100 Pa (6.0 to 8.5 in. of water). It was concluded that the unit was capable of achieving 99% destruction of TCE when operated at 370°C and a space velocity of about 11,700 hr^{-1}.

Table 13-9
Destruction Efficiency of a Fluidized Catalyst Bed for Test Compounds

Component	Destruction Efficiency, %	
	650°F	950°F
Cyclohexane	99	99+
Ethylbenzene	98	99+
Pentane	96	99+
Vinyl chloride	93	99
Dichloroethylene	85	98
Trichloroethylene	83	98
Dichloroethane	81	99
Trichloroethane	79	99
Benzene	55	95
Tetrachloroethylene	52	92

Notes:
1. *Values given are mean values based on all destruction efficiencies measured for the compound and are not adjusted for differences in space velocity or other factors.*
2. Ranges of conditions:
   Space velocity   7,000–10,500 hr^{-1}
   Throughput   330–500 scfm
   Inlet concentration   1.0–14.1 ppmv
Data from Palazzolo and Jamgochian (1986)

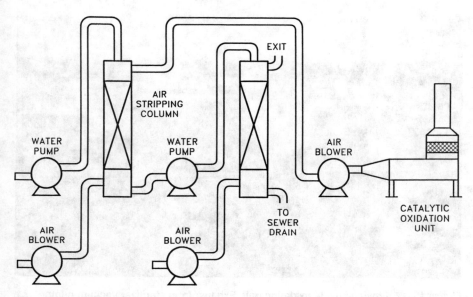

Figure 13-16. Simplified flow diagram of contaminated water stripping system with fluidized bed catalytic oxidation unit. (*Hylton, 1991*)

Typical operating conditions and performance for the REGENOX regenerative heat exchange catalytic oxidation process as provided by Haldor Topsøe (1990) are conversion efficiency: >99%, heat exchange efficiency: 95%, operating temperature in hottest zone: 300°C (572°F). The system uses a metal oxide-based catalyst that is capable of operating at temperatures up to 600°C (1,112°F).

Package type catalytic oxidation units that contain manganese dioxide-copper oxide catalyst are used commercially to destroy ethylene oxide emitted by sterilization systems. **Figure 13-17** is a photograph of such a unit installed on the roof of a building. This unit is unique in that it processes both a concentrated stream of ethylene oxide from the sterilizer and an exhaust air stream containing only traces of ethylene oxide. Inlet air passes through a prefilter, heat exchanger, and heater in series, bringing its temperature up to at least 280°F. The hot air then enters a mixing chamber where ethylene oxide from the sterilizer vacuum pump is introduced through a flame arrestor and distribution manifold. The air-ethylene oxide mixture next passes through two series mounted catalyst cells where ethylene oxide is oxidized to carbon dioxide and water. The guaranteed efficiency is 99.9% destruction of ethylene oxide in the sterilizer exhaust and 99.0% destruction in the low concentration air stream (Donaldson Company, Inc., 1992).

Economics

According to McInnes et al. (1990), catalytic oxidation systems have a capital cost about 40% higher than thermal units, but lower maintenance costs. An operating cost of about $15,000 per year is said to be typical for a 20,000 cfm system, excluding catalyst replacement.

A curve for estimating the cost of a catalytic incinerator as a function of the gas flow rate is given in the *EPA Handbook* (1986). The cost of the heat exchanger, from another curve,

Figure 13-17. Ethylene oxide oxidation unit. Exhaust from sterilizer vacuum pump enters via black line on right. *Courtesy of Donaldson Company, Inc.*

and of the catalyst, from the manufacturer, must be added to obtain the major equipment purchased cost. Equations and curves for estimating capital costs are also given by van der Vaart et al. (1990). Typical equipment costs read from the curves for fixed and fluidized catalytic incineration systems are given in **Table 13-10**. These numbers include the catalyst and all auxiliary equipment F.O.B., but not installation. The complete installed cost can be approximated by multiplying the equipment cost by a factor of 1.9. The result is total cost of the incineration system in April 1988 dollars, but does not include any costs for site preparation or buildings.

A simplified procedure for developing the conceptual design and costs of a catalytic oxidation unit is given in the *MECA Guidebook* (1992). The procedure includes equations for performing the steps outlined in the previous "System Design" section and a simple equation for estimating the equipment cost for processing an exhaust stream containing a low VOC concentration (<25% LEL) at atmospheric pressure and a catalyst bed temperature below 1,200°F. The equipment cost is converted to a total system installed cost by multiplying by an appropriate factor.

Catalytic Oxidation of Sulfur Compounds to Sulfur Oxides

The catalytic oxidation of sulfur compounds in gas streams is the basis for numerous gas purification processes, most of which are covered in other sections of this book. The oxidation of SO_2 to SO_3 as part of a sulfur dioxide removal process is discussed in Chapter 7. The oxidation of H_2S to elemental sulfur by the Claus process and related technologies is covered in Chapter 8. The liquid-phase oxidation of H_2S to elemental sulfur is described in Chapter 9, and the oxidation of H_2S and other sulfur compounds in dry boxes and in other sulfur scavenging processes is included in Chapter 16. An area of technology that is not covered

Table 13-10
Equipment Costs for Catalytic Incinerator Systems

Heat Recovery, %	0		50		70	
Catalyst Bed	Fixed	Fluid	Fixed	Fluid	Fixed	Fluid
Plant size, Mscfm						
3	88	130	110	140	120	150
10	180	220	200	240	240	275
25	280	410	350	480	390	550
50	330	—	500	—	580	—

Notes:
1. Costs in thousands of dollars (April 1988).
2. Fixed bed includes packed and monolith designs.
3. Equipment and catalyst only, F.O.B. not installed.
Data from van der Vaart et al. (1990)

elsewhere is the selective catalytic oxidation of sulfur compounds in combustible gas streams to form sulfur dioxide and trioxide, which are then neutralized. The technology is no longer commercially significant, but it is historically interesting and may find future applicability. Three processes in this category were widely used for the purification of manufactured, coke oven, and synthesis gas during the 1930s and 1940s. They are described in detail in previous editions of this book, but only very briefly in the next section.

Katasulf Process

The Katasulf process was developed in Germany in the early 1930s by Hans Bähr (1938). It is based on the somewhat surprising discovery that oxygen reacts selectively with hydrogen sulfide in the presence of hydrogen at 750°F, with a suitable catalyst, to form sulfur dioxide and water. The primary purpose of the process was to remove hydrogen sulfide and ammonia, present in coke-oven gas, in a single system while producing salable ammonium sulfate and elemental sulfur.

In the basic form of the Katasulf process, the gas was first passed through an electrostatic filter for tar removal and then preheated to 750°F by heat exchange with the gas leaving the catalytic converter. After addition of air, the gas was admitted to the catalyst chamber where the H_2S was converted to SO_2. Because of the exothermic nature of the reaction, the gas temperature increased considerably in the converter. The exit gas was partially cooled by heat exchange with the incoming gas. It was then passed through an absorption tower where it was washed countercurrently with a concentrated aqueous ammonium sulfite-bisulfite solution, which absorbed both the SO_2 and NH_3.

The circulating ammonium sulfite-bisulfite solution was maintained at a salt concentration of 50 to 60% by continuous withdrawal of a side stream. This stream was acidified with sulfuric acid, and the resulting salts converted to ammonium sulfate in solution and liquid elemental sulfur by heating under pressure at 290°F.

The preferred catalyst for the Katasulf process was composed of two metals—one (e.g., iron, nickel or copper) combined with the hydrogen sulfide to form metal sulfide, and the other (e.g., tungsten, vanadium, or chromium) served as an oxygen carrier.

The process was not highly efficient by today's standards. In one commercial plant, the H_2S was reduced from 450 grains/100 scf to about 0.6; the NH_3 was reduced from 250 grains/100 scf to about 0.6; and the organic sulfur was reduced from 23 grains/100 scf to about 12 (Bähr, 1938).

North Thames Gas Board Process for Organic Sulfur Removal

This process was probably an outgrowth of the Carpenter-Evans process described in a later section, as the same catalyst material, nickel subsulfide was used; however, the technology is quite different, involving oxidation rather than reduction. The process was developed as a means of removing organic sulfur compounds from coal gas after hydrogen sulfide removal by conventional iron oxide dry boxes. It has been described by Griffith (1952) and Plant and Newling (1948).

After purification in dry boxes, the gas was pressurized slightly (to 20–30 in. water); preheated in a gas-to-gas heat exchanger to about 430°–570°F; and passed through the bed of catalyst. Oxidation of the sulfur compounds, and some of the hydrogen, caused an increase in gas temperature to about 700°F at the catalyst bed exit. The hot gas was partially cooled in the heat exchanger then further cooled by washing with a dilute solution of sodium carbonate, which also served to remove the SO_2.

The catalyst was made by boiling ¼-in. china pellets in a solution of nickel sulfate followed by drying at 750°F and sulfiding with hydrogen sulfide contained in coal gas. The product contained approximately 8% nickel subsulfide. The kinetics of the oxidation reactions were investigated by Crawley and Griffith (1938) who concluded that carbon disulfide is chemisorbed and oxidized by impact with oxygen, while carbonyl sulfide reacts by collision with chemisorbed oxygen. A space velocity of about 1,000 scf/(h)(cu ft of catalyst) was used.

A major problem of the process was rapid deactivation of the catalyst by the deposition of carbonized and tarry materials. As a result of the deposition, it was necessary to remove the catalyst every 3–4 months and regenerate it by controlled oxidation at 800° to 1,050°F in a separate kiln.

The process was only moderately effective in removing sulfur compounds. It reduced the total organic sulfur content of the gas by about 70% (from 29 to 9 grains/100 scf); however, most of the sulfur in the residual gas was thiophene which is known to be an extremely stable compound.

Soda Iron Process

This process was disclosed in Germany in 1934 and widely used for the removal of organic sulfur compounds from synthesis gas. It represents an extension of the classical iron oxide dry box process. Its basis is the oxidation of organic sulfur compounds to oxides of sulfur at elevated temperatures over a catalyst consisting of hydrated iron oxide and sodium carbonate. The oxides of sulfur react with the sodium carbonate and are retained in the bed as sodium sulfate. The required oxygen is provided by adding a small amount of air to the gas ahead of the catalytic converters. The process, as it was practiced at a German installation, has been described in detail by Sands et al. (1948).

The catalyst consisted of a mixture of natural iron oxide ore with 30 to 70% sodium carbonate. The catalyst was used until about 90% of the sodium carbonate had been converted to sodium sulfate. The process was very effective in removing organic sulfur, but required the use of several converters in series and low space velocities. In one German plant operating at essentially atmospheric pressure with a space velocity of about 90 vol/(vol)(hr) and at temperatures in the converters from 300° to 500°F, the organic sulfur concentration was reduced from about 19 grains/100 scf to 0.05 grain/100 scf. Experiments reported by Sands et al. (1950) showed that by operating at a pressure of 300 psig, a temperature of 480° F, and a space velocity of 450 vol/(vol)(hr), 25 grains of organic sulfur/100 scf could be reduced to less than 0.1 in a simulated synthesis gas.

Although the catalyst was very effective for the conversion of carbonyl sulfide, carbon disulfide, and mercaptans, it was ineffective for thiophene. Other problems with the process were the low space velocities required and the need to remove and dispose of spent catalyst containing all of the collected sulfur as sodium sulfate.

Conversion of Organic Sulfur Compounds to H_2S

The principal organic sulfur compounds present in industrial gas streams are carbonyl sulfide, carbon disulfide, mercaptans of low molecular weight, sulfides such as dimethylsulfide, disulfides such as dimethyldisulfide, and thiophene. The concentrations of these compounds vary over a considerable range and often depend on the prior processing history of the gas stream. For example, the mercaptan content of natural gas may vary from less than 1 grain/100 scf to 5–10 grains/100 scf depending upon the extent of ethane, propane, and butane removal in a preceding natural gas liquids recovery step. Total organic sulfur concentrations in coal and synthesis gases made from high-sulfur fuel range typically from 20 to 50 grains/100 scf with carbonyl sulfide and carbon disulfide as the major ingredients. In refineries, organic sulfur compounds are found primarily in gas streams from delayed cokers, visbreakers, and fluid catalytic cracking units. Concentrations can range up to several hundred grains/100 scf (as sulfur) depending on the process, the sulfur content of the feed, and the presence of contaminants such as flue gas or air in the gas stream.

The organic sulfur compounds are much less acidic than hydrogen sulfide and are therefore not effectively removed by conventional alkaline solution-based hydrogen sulfide removal processes. Physical solvents, however, generally show a very high solubility for organic sulfur compounds (see Chapter 14). The absorbed organic sulfur compounds are not destroyed in the process, but appear in the acid gas stream, which is usually fed to a Claus sulfur plant where they are converted to elemental sulfur along with the hydrogen sulfide.

The catalytic conversion of organic sulfur compounds in coal gas to be used as fuel is no longer a commercially significant operation. However, the technology is important in the purification of gas streams used in the synthesis of ammonia, synthetic natural gas, methanol, and other chemicals. It is also a necessary part of Claus type sulfur plant technology; however, this application is covered in Chapter 8.

Basic Chemistry

The principal chemical reactions taking place in the catalytic conversion of organic sulfur compounds to hydrogen sulfide can be expressed by the following equations:

Hydrogenation reactions:

$$CS_2 + 2H_2 \rightleftharpoons C + 2H_2S \qquad (13\text{-}1)$$

$$COS + H_2 = CO + H_2S \tag{13-2}$$

$$RCH_2SH + H_2 = RCH_3 + H_2S \tag{13-3}$$

$$C_4H_4S + 4H_2 = C_4H_{10} + H_2S \tag{13-4}$$

Hydrolysis (in the presence of water vapor):

$$CS_2 + 2H_2O = CO_2 + 2H_2S \tag{13-5}$$

$$COS + H_2O = CO_2 + H_2S \tag{13-6}$$

Besides the reactions shown in equations 13-1 to 13-6, a number of other reactions, involving oxygen, hydrogen, hydrogen sulfide, hydrocarbons, nitrogen compounds, and carbon monoxide may also take place, depending on the feed-gas composition and the operating conditions. For example, hydrogenation of carbon disulfide may be represented by the following mechanism:

$$CS_2 + 4H_2 = 2H_2S + CH_4 \tag{13-7}$$

$$CS_2 + H_2O + H_2 = CO + 2H_2S \tag{13-8}$$

Side reactions occurring in the treatment of coal gas with different catalysts have been investigated by Key and Eastwood (1946). These authors found that significant amounts of mercaptans may be formed from hydrogen sulfide and unsaturated hydrocarbons, depending on the temperature and catalysts used. Wedgewood (1958) gave a detailed study of side reactions occurring in a plant using the Holmes-Maxted process, which is described later in this chapter. Side reactions resulting in the formation of polymeric materials, which deactivate the catalyst, are especially undesirable.

Equilibrium constants for the hydrolysis of carbon disulfide according to equation 13-5 have been determined experimentally by Terres and Wesemann (1932). Since the reaction takes place in two steps, the equilibrium constants for the two intermediate reactions—shown in equations 13-9 and 13-10—were determined, and the constant for the overall reaction was obtained by simple multiplication:

$$H_2S + CO_2 = H_2O + COS \tag{13-9}$$

$$COS + H_2S = CS_2 + H_2O \tag{13-10}$$

As can be seen, the reaction in equation 13-9 is the same as that in equation 13-6, which represents the hydrolysis of carbonyl sulfide. The first reaction, equation 13-9, was studied within the temperature range of 350° to 600°C (662° to 1,112°F) and the second reaction, equation 13-10, at 700° to 900°C (1,292° to 1,652°F). The experimental data were extrapolated over the range of 20° to 1,000°C (68° to 1,832°F) by conventional thermodynamic calculations. The equilibrium constants thus obtained for the reactions in equations 13-5 and 13-6 and the opposite of the reaction in equation 13-10 are shown in **Table 13-11.**

Key and Eastwood (1946) determined equilibrium constants for the hydrogenation of carbonyl sulfide (equation 13-2) at 300°C (572°F) and 600°C (1,112°F). A value of 15.0 was

Table 13-11
Equilibrium Constants for the Hydrolysis of Carbonyl Sulfide and Carbon Disulfide

Temperature, °C	$K_1 = \dfrac{(H_2S)(CO_2)}{(COS)(H_2O)}$	$K_2 = \dfrac{(COS)(H_2S)}{(CS_2)(H_2O)}$	$K_3 = \dfrac{(H_2S)^2(CO_2)}{(CS_2)(H_2O)^2}$
20	7.25×10^5	7.15×10^9	5.21×10^{15}
100	3.16×10^4	3.84×10^7	1.21×10^{12}
200	2.75×10^3	7.15×10^5	1.91×10^9
300	5.55×10^2	5.28×10^4	2.9×10^7
400	1.85×10^2	8.33×10^3	1.5×10^6
500	81.0	2.22×10^3	1.8×10^5
600	43.1	7.94×10^2	3.4×10^4
700	26.1	3.45×10^2	9.2×10^3
800	17.3	1.74×10^2	3.0×10^3
900	12.3	100.0	1.2×10^3
1,000	4.5	63.0	2.8×10^2

Source: Data of Terres and Wesemann (1932)

found for the equilibrium constant at both temperatures. Although these authors do not claim high accuracy for their experiments, calculation of the constants for the hydrolysis reaction (equation 13-6), from the experimental values and accepted values for the equilibrium constants of the water-gas reaction, gives good agreement with the experimental data of Terres and Wesemann.

The equilibrium constant for other reactions can be calculated with fairly good accuracy from thermodynamic data such as those presented by Kelley (1937). Calculated equilibrium constants for the reactions in equations 13-1 and 13-2 are shown in **Table 13-12**.

Table 13-12
Equilibrium Constants for the Hydrogenation of Carbonyl Sulfide and Carbon Disulfide

Temperature, °K	°C	$K = \dfrac{(H_2S)(CO)}{(COS)(H_2)}$	$K = \dfrac{(H_2S)^2(C)}{(CS_2)(H_2)^2}$
500	227	7.55	6.99×10^{11}
600	327	10.9	1.03×10^9
700	427	13.8	1.27×10^7
800	527	16.0	2.72×10^5
900	627	17.8	1.68×10^4
1,000	727	19.0	1.81×10^3

Source: Calculated from data of Kelley (1937)

Historical Background

Carpenter-Evans Process. The first catalytic process used commercially for the hydrogenation of organic sulfur compounds was developed in England by Carpenter (1913) and Evans (1915). The catalyst used in this process was a sulfide of nickel, shown by Evans and Stanier (1924) to be nickel subsulfide with the formula Ni_3S_2. The catalyst was prepared by soaking firebrick in a solution of nickel chloride, heating to 930°F to obtain nickel oxide, and reacting the product with hydrogen sulfide in crude coal gas to form the active nickel subsulfide.

The catalyst was found to be most active at 800° to 850°F converting about 80% of the organic sulfur in the gas to hydrogen sulfide. During operation, carbon deposited on the catalyst as a result of the carbon disulfide hydrogenation reaction (reaction 13-1), and periodic regeneration was required. In one plant the catalyst was regenerated every 30–35 days. This was accomplished by blowing air through the bed at a temperature somewhat lower than the normal operating temperature. In commercial applications on coal gas, the hydrogen sulfide formed was normally removed by the use of iron oxide dry boxes.

Holmes-Maxted Process. The most extensively used commercial process for removing organic sulfur compounds from coal-derived fuel gases was invented by E. B. Maxted (1937) and developed by W.C. Holmes & Company, Ltd. of England. The process has been described in considerable detail by Wedgewood (1958), Maxted and Priestly (1946), Maxted and Marsen (1946), Priestly and Marris (1947), Priestly (1957), and Priestly (1958). In 1958 it was reported that the process had been used in more than 50 installations ranging in capacity from 5,000 to 4,000,000 scf/d of gas (W.C. Holmes & Co., 1958).

The Holmes-Maxted process catalyst was a normal thiomolybdate with the formula, for a bivalent metal, M, of $MMoS_4$. M was typically copper, iron, zinc, cobalt, or nickel. A commercial catalyst, e.g., copper thiomolybdate, was made by impregnating granular bauxite with a solution containing copper molybdate, ammonium sulfate, and excess ammonia; heating to about 750°F to drive off ammonia and decompose ammonium sulfate; and finally forming the thiomolybdate by contacting the catalyst with gas containing hydrogen sulfide at a temperature between 570° and 750°F (Maxted and Priestly, 1946).

The process used a special converter design which allowed catalyst to be added and removed semicontinuously. This was necessary because of the continuous formation of carbon on the catalyst. Fresh catalyst was added at a rate of about 1 cu ft/MMscf of gas, and an equivalent amount, containing about 10 wt% carbon, was removed. The spent material was regenerated in a separate vessel by contact with a mixture of air and flue gas at a controlled temperature in the range of 480° to 660°F.

Results from one year of continuous operation of a plant treating about 500,000 scf/d of coal gas are given in **Table 13-13.** Removal efficiencies for individual sulfur compounds in the same installation are given in **Table 13-14.** The data show that the total organic sulfur removal efficiency of the process depended strongly on the type of sulfur compounds present. The process was quite efficient for carbon disulfide removal; much less so for carbonyl sulfide; and very poor for thiophene.

Current Practice

Most ammonia plants utilize natural gas or light hydrocarbons as feedstock and include three major catalytic operations: steam reforming, shift conversion, and ammonia synthesis.

Table 13-13
Typical Operating Results of Holmes-Maxted Process (one year's operation)

Operating variables	
Total gas volume treated, cu ft	198,105,000
Average daily gas volume treated, cu ft	542,000
Fuel gas used for heating converter, cu ft	755,000
Fuel gas used for catalyst regeneration, cu ft	126,000
Catalyst discharged from converter, lb	12,310
Catalyst regenerated, lb	10,800
Catalyst temperature, °F	680–725
Space velocity, vol/(vol)(hr)	1,300–2,000
Pressure drop through catalyst, in. H_2O	2.8
Organic sulfur, grains/100 scf:	
Inlet	23.1 (average)
Outlet	6.7 (average)
Oxygen, %:	
Inlet	1.1 (average)
Outlet	0
Carbon on spent catalyst, %	7.5

Source: Priestly, 1957

Table 13-14
Typical Removal of Individual Organic Sulfur Compounds (one year's operation)

Compound	Conc., grains/100 cu ft		Removal, %
	Inlet	Outlet	
Thiophene	4.5	3.6	20
Carbon disulfide	14.5	1.3	91
Carbonyl sulfide	4.1	1.8	56
Total organic sulfur	23.1	6.7	71

Source: Priestly, 1957

The synthesis catalysts are extremely sensitive to sulfur, and it is necessary to reduce the concentration of all sulfur compounds to very low levels prior to the synthesis step. The preferred purification practice with natural gas or light hydrocarbon feed is to remove the bulk of the sulfur from the hydrocarbon feedstock in an absorption or adsorption unit and the final traces of sulfur in a nonregenerative zinc oxide bed (see Chapter 16).

The zinc oxide bed is not very effective in capturing organic sulfur compounds at its normal operating temperatures. When such compounds are present, it is therefore necessary to convert them to H_2S before the gas enters the zinc oxide bed. This is usually accomplished by hydrogenation in a bed of cobalt-molybdenum catalyst, which may be located in a separate vessel or installed as a layer on top of the zinc oxide.

The catalysts are composed of cobalt and molybdenum oxides on an active alumina base. Typical chemical and physical properties of a commercial hydrotreating catalyst recommended for converting organic sulfur compounds to hydrogen sulfide are given in **Table 13-15**. This catalyst must be sulfided prior to use. Its general range of operating conditions are temperature 500°–750°F, pressure 100–500 psig, and space velocity 500 to 1,500 cu ft/(cu ft)(hr) (United Catalysts, Inc., 1985).

Catalysts used for the hydrogenation of organic sulfur compounds in hydrocarbon gas streams are typically operated in the temperature range of about 650°–700°F. Temperatures above about 750°F cannot be tolerated because of potential damage to the gamma alumina substrate. At typical operating pressures, a hydrogen concentration of at least 2% (or a hydrogen partial pressure of at least 10 psia) is required in the gas stream to drive the hydrogenation reactions at a satisfactory rate (Huber, 1993).

Since sulfur compounds are normally present at very low concentrations (1–10 ppm), their hydrogenation does not cause a very significant temperature change within the bed. When olefins are present in the feed stream, however, they will also be hydrogenated in the presence of an active cobalt-molybdenum catalyst. Since the hydrogenation of olefins is highly exothermic, it may be necessary to use a feed temperature considerably below the 650°–700°F operating range to avoid exceeding the upper temperature limit. The feed temperature must, of course, be high enough to initiate the olefin hydrogenation reaction—typically about 450°F.

When the feed to a synthesis operation is derived from the gasification of coal or heavy oil rather than from the reforming of natural gas, it will normally contain COS as well as H_2S. If hydrogen sulfide is removed from the gas and the gas is then passed through a shift conversion catalyst, the COS will be converted to H_2S (at approximately its equilibrium concentration), and the H_2S produced can then be removed in a later step. On the other hand, if the gas

Table 13-15
Properties of Hydrodesulfurization Catalyst

Catalyst designation:	United Catalysts, Inc. C49-1-4.0
Catalyst form:	⅛ in. extrusions
Chemical composition, wt%	
CoO	3.0–4.0
MoO_3	9.0–11.0
Al_2O_3	83–88
Other heavy metals	<0.10
Bulk density, lb/cu ft	35 ± 3
Surface area, m^2/g	200–300
Pore volume, cc/g >29.2 A°	0.55–0.65
Crush strength, lb (min. avg.)	15
Attrition, wt% loss	<5

Data provided by United Catalysts, Inc. (1993)

is not first treated for hydrogen sulfide removal, or if some of it does not contact shift conversion catalyst, it is necessary to include a separate COS conversion step before hydrogen sulfide removal. A flow sheet of such a system as applied to methanol synthesis is shown in **Figure 13-18** (Udengaard and Berzins, 1985).

COS can be converted to H_2S by hydrogenation (reaction 13-2) or hydrolysis (reaction 13-6). Hydrogenation is normally accomplished with a cobalt-molybdenum catalyst, which is also active for shift conversion. If shift conversion is not desired during the COS removal step, a hydrolysis catalyst, which is not active for shift conversion, can be utilized. One such catalyst is Topsøe CKA, which consists mainly of activated alumina and is available as 3×10^{-3} or 6×10^{-3} m (⅛ or ¼ in.) extrudates. **Table 13-16** shows operating conditions and COS converter performance for five cases. The first four are projected values for plants designed to produce synthetic natural gas (SNG), methyl alcohol (MeOH), fuel for a power plant, and ammonia. The fifth case presents actual operating data from a coal-to-ammonia plant. The COS concentrations in the product gas streams represent values close to equilibrium for the composition and temperature conditions in the converters.

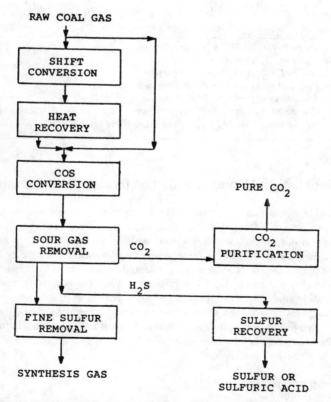

Figure 13-18. Flow diagram for the production of synthesis gas for methanol or synthetic natural gas (SNG) from coal gasification product gas. (*Udengaard and Berzins, 1985*). *Reproduced with permission from Acid and Sour Gas Treating Processes, edited by S. A. Newman, copyright 1985, Gulf Publishing Company*

Table 13-16
Typical COS Conversion Unit Performance On Coal-Based Plants

Case	1	2	3	4	5
Pressure, atm	31.2	24.2	24.2	31.6	31.2
Temperature, °C	224	191	235	204	214
Gas comp., wt.% dry					
H_2	47.9	44.6	35.8	53.2	52.4
CO	16.0	22.3	44.6	2.9	2.8
CO_2	32.8	29.8	18.0	42.2	42.6
H_2S	0.1	0.7	1.0	1.0	0.8
Inerts	3.3	2.6	0.6	0.7	1.4
H_2O/dry gas, mol ratio	0.27	0.26	0.68	0.41	0.42
COS conversion data					
Inlet, ppmv	70	400	500	51	22
Outlet, ppmv	3	8	6	10	7
Conversion, %	96	98	99	80	68

Notes: Case 1. Synthesis Gas for SNG, Projected.
Case 2. Synthesis Gas for Methanol, Projected.
Case 3. Gas for Combined Cycle Power, Projected.
Case 4. TVA Ammonia from Coal Plant, Projected.
Case 5. TVA Ammonia from Coal Plant, Op. data.
Data of Udengaard and Berzins (1985). Reproduced with permission from Acid and Sour Gas Treating Processes, edited by S. A. Newman, Copyright 1985, Gulf Publishing Co.

Conversion of Carbon Monoxide to Carbon Dioxide (Shift Conversion)

Formation of hydrogen and carbon dioxide by the reaction of carbon monoxide with water vapor in the presence of a catalyst is one of the earliest industrial applications of catalysis. The technology of this process, which is known as shift conversion, has been highly developed since its early inception, and, at present, it is used extensively for the production and purification of hydrogen. The process is applicable to the treatment of hydrogen produced in partial oxidation units, steam-hydrocarbon reformers, water-gas generators, and steam-iron reactors. It is also suitable for the adjustment of the H_2/CO ratio in synthesis gases and the purification of gases produced in controlled-atmosphere generators. Since the process is aimed more at the production of hydrogen than at its purification and has been adequately covered in the literature, the following discussion is limited to a general description of the process, its characteristics, and applications.

Process Description

A typical process system is illustrated by the schematic flow diagram of **Figure 13-19,** which is based on the treatment of gas produced by the reforming of natural gas hydrocarbons with steam. The 1,500° to 1,600°F gas mixture emerging from the reformer, which contains mainly hydrogen, carbon monoxide, and carbon dioxide, is used to generate high pressure steam and is cooled to about 700°F in the process. For hydrogen plants using a reactive or physical solvent for carbon dioxide removal, steam may be added to the cooled reformer effluent as indicated in **Figure 13-19**. The added steam forces the shift reaction (equation 13-11) to the right to produce higher purity hydrogen. Plants using Pressure Swing Adsorption (PSA) for hydrogen purification do not normally use steam addition prior to the high temperature shift.

The gas is then passed through the bed of catalyst in the first (high temperature) conversion stage. About 80 to 95% of the carbon monoxide is converted to carbon dioxide, and a quantity of hydrogen is produced, which is equivalent to that of the carbon monoxide reacted. A typical temperature rise across this reactor is 100°F. The first conversion stage is primarily intended for the production of hydrogen and is, therefore, not considered to be a gas purification step.

The hot gas leaving the first converter is cooled to about 400°F by generating steam and/or preheating boiler feed water. The gas is then passed through the second (low temperature) shift conversion stage with a temperature rise of about 40°F. While most plants utilizing a reactive or physical solvent for carbon dioxide removal have a low temperature shift converter, most PSA-based plants do not. In plants that do have a low temperature converter, the effluent from this unit is often cooled by heating boiler feedwater and, finally, by heat exchange with air or cooling water before carbon dioxide removal by a conventional technique, such as adsorption in a PSA unit or absorption in a reactive or physical solvent. Carbon dioxide removal processes are described in detail in Chapters 2, 3, 5, 12, and 14.

The flow scheme depicted in **Figure 13-19** may be modified in many ways. Two of the most common modifications are (a) the use of the outlet gas from the low temperature converter to provide heat to the amine plant reboiler, and (b) the addition of a methanator for conversion of the last remaining traces of carbon monoxide and carbon dioxide. The shift

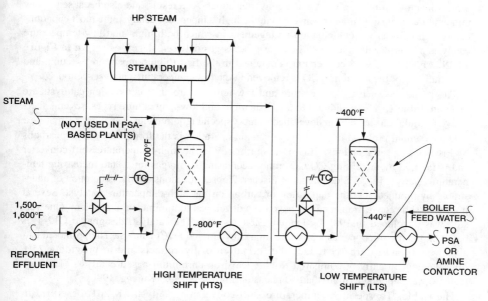

Figure 13-19. Schematic flow diagram of a shift conversion plant.

conversion process may be operated over a considerable range of pressures and temperatures, depending primarily upon the conditions of the gas source and requirements for the product. Pressures from 30 to over 600 psig are not uncommon. Typical operating temperatures range from about 400° to 500°F for low temperature catalysts, and from about 600° to 950°F for high temperature catalysts. Hydrogen plants utilizing PSA produce hydrogen that is up to 99.99% pure; while plants using amine, hot potassium carbonate, or physical solvent systems produce 96–97% pure hydrogen.

Catalysts

Two basic types of catalyst are normally used for shift conversion, depending on the temperature at which the reaction is carried out. The high temperature catalyst is typically chromium-promoted iron oxide, which is active within the temperature range of about 650° to 950°F. Typical preparations contain 70 to 80% ferric oxide and 5 to 15% chromic oxide and are available as tablets of various sizes (e.g., ¼ × ⅛ in. and ⅜ × 3⁄16 in.). This catalyst is relatively insensitive to sulfur compounds.

Low temperature catalysts typically contain copper and zinc and are effective over a temperature range of about 350° to 650°F. They are used where very low carbon monoxide concentrations are required in the product gas. These catalysts are extremely sensitive to poisoning by sulfur compounds, and the feed gas has to be thoroughly desulfurized before contacting the catalyst. Chlorides and silica from entrained boiler feedwater in the steam can also poison the low temperature catalyst. Since the equilibrium for the shift reaction is more favorable at lower temperatures, almost complete conversion of carbon monoxide—on the order of 99%—is possible with low temperature catalysts.

Since the shift conversion reaction is exothermic, the gas increases in temperature as it passes through the bed when a significant amount of CO is reacted. This has an adverse effect on equilibrium conditions at the converter outlet. As a result, and also because the low temperature catalyst is more expensive than the high temperature catalyst, the most economical system is usually a combination arrangement in which both high- and low-temperature catalysts are used with intercooling of the process gas between stages, as shown in **Figure 13-19.** About 80 to 95% conversion is obtained in the first (high temperature) section, and essentially all of the remaining CO is removed in the low-temperature catalyst section.

The properties of high-temperature and low-temperature shift conversion catalysts are given in **Table 13-17.** In addition to the basic high- and low-temperature types, several alternative formulations have been developed to meet special requirements.

One potential problem in the shift conversion operation is the formation of hydrocarbons by Fischer Tropsch reactions. This occurs primarily in the high-temperature shift converter and is favored by increasing CO/CO_2 ratio, decreasing steam/gas ratio, and increasing temperature (Hartye, 1985). Side effects of Fischer Tropsch reactions are the generation of undesirable by-products and the deposition of carbon on the high-temperature catalyst. Several improved shift conversion catalysts have been developed to minimize the problem. One version was developed in 1985 by United Catalysts, Inc. The catalyst, designated C12-4, is based on the use of copper as a promoter in a conventional Fe/Cr catalyst. The promoted catalyst was found to greatly reduce the formation of byproducts and to provide a 25% improvement in activity compared to classical iron-chromium based catalysts. The new material was supplied to more than 50 plants during 1987–1992 (Kujang, 1992).

The Haldor Topsøe high-temperature shift-conversion catalyst, SK-201, described in **Table 13-17,** also contains a small amount of copper as a promoter. This formulation is

Table 13-17
Properties of Shift Conversion Catalysts

High temperature:	
Catalyst designation	SK-201
Catalyst composition	
Iron, wt%	59
Chromium, wt%	6
Graphite, wt%	4
Sulfur, ppmw	<150
Oxygen, as metal oxide	balance
Normal operating temperature, °F	610–930
Optimal inlet temperature, °F	610–660
Normal operating pressure, psig	0–700
Low temperature:	
Catalyst designation	LK-801
Catalyst composition	
Base	copper, zinc, aluminum
Copper content, lb/cu ft (approx.)	19
Activation	Reduction by H_2 in a carrier gas at 300°–445°F for 20 hr
Normal operating temperature, °F	375–525
Normal inlet temperature, °F	>375; 25–35 above dew point
Poisons	sulfur, chlorine, silica

Haldor Topsøe, Inc. (1989, 1992)

reported to practically eliminate the formation of hydrocarbons at typical steam/carbon ratios and to provide an activity appreciably higher than conventional chromia-promoted iron oxide catalysts (Haldor Topsøe, Inc., 1992). A copper-promoted high-temperature shift converter catalyst is also offered by ICI.

A highly active sulfur-tolerant shift-conversion catalyst has been developed for use with high sulfur concentration synthesis gas. This catalyst utilizes mixed metal sulfides and requires the presence of sulfur compounds in the gas to remain active (Nielsen and Hansen, 1981). The sulfur tolerant catalyst can operate over a wide range of temperatures (390°–890°F), and offers the advantage of allowing sulfur and carbon dioxide removal to be accomplished in one step (Haldor Topsøe, Inc., 1991B).

More complete information on shift conversion catalyst availability and properties, as well as design methods for shift converters employing both high- and low-temperature catalysts, are available from catalyst manufacturers.

Basic Chemistry

Carbon monoxide reacts exothermally with steam at elevated temperatures according to equation 13-11.

$$CO + H_2O = CO_2 + H_2 \qquad (13\text{-}11)$$

The heats of reaction and equilibrium constants, within the range of 500° to 1,000°K (439° to 1,341°F), are given in **Table 13-18**. The reaction results in no change in gas volume. The equilibrium is, therefore, not significantly affected by pressure. Since the reaction is exothermic, the gas temperature increases as it passes through the catalyst bed—about 13°F per dry mole percent CO converted.

Design and Operation

The important design variables for a shift converter are temperature, pressure, space velocity, steam/gas ratio, and the carbon monoxide content of the inlet gas. Since these variables interact, some working in opposite directions, it is not possible to define generally valid optimum operating conditions, and, therefore, each case has to be analyzed individually. However, certain generalizations with respect to individual variables can be stated.

The activity of conventional high-temperature shift conversion catalysts increases markedly with temperature and is sufficiently high at 650°F for commercial operation at pressures above 100 psig. Somewhat higher operating temperatures are normally required at near-atmospheric pressures. Higher temperatures may also be required to account for the reduction of catalyst activity as it ages. Kujang (1992) gives a plot of high temperature catalyst activity vs. time onstream that shows a drop to about 55% of its original value after five years. He suggests gradually raising the catalyst inlet temperature from its initial value of 650°–680°F to a final value of 730°–750°F to compensate for loss of activity due to aging.

The more active copper-promoted, high-temperature shift catalysts can be operated at somewhat lower temperatures. Haldor Topsøe (1992) recommends 610°–660°F as the optimum inlet temperature for their SK-201 copper-promoted iron/chromia-based catalyst and report that it can be operated at temperatures down to 570°F.

The space velocity used in commercial installations depends to a considerable extent on other operating conditions. Typical values for the high-temperature converter range from about 1,000 to 3,000 vol/(vol)(hr). Somewhat higher space velocities may be used in the sec-

Table 13-18
Heats of Reaction and Equilibrium Constants for Shift-Conversion Reaction

Temperature, °K	°C	Heat of Reaction ΔH, cal/g mole	$K = \dfrac{(CO_2)(H_2)}{(CO)(H_2O)}$
500	227	−9.520	134
600	327	−9.294	27.6
700	427	−9.051	8.95
800	527	−8.802	4.11
900	627	−8.553	2.20
1,000	727	−8.311	1.37

Source: Data of Wagman et al. (1945); Nielsen and Hansen (1981)

ond- and third-stage converters where the carbon monoxide concentration in the inlet gas is quite low.

The steam/gas ratio required for optimum conversion varies with both temperature and pressure. For a given set of operating conditions, with respect to pressure and temperature, conversion first increases with an increasing steam/gas ratio and after reaching an optimum, decreases upon further addition of steam. This behavior is due to the favorable effect on the equilibrium of high concentrations of water vapor and the unfavorable effect of decreased contact time.

Steam/dry gas ratios ranging from 1:1 to 5:1 vol/vol are typical for first stage conversion, and 0.5:1 to 1:1 vol/vol are normally used in the second and third stages. There can be an economic advantage to operating the ammonia plant reformer section at a lower steam/gas ratio, which results in a low steam/gas ratio in the feed to the first shift conversion stage. According to Hartye (1985), several hundred ppm of hydrocarbons may be produced if the high-temperature converter is operated with a steam/gas ratio below 0.6 vol/vol with inlet CO/CO_2 levels above 1.6. The production of hydrocarbons can be minimized by (1) adding more steam at the first stage converter inlet, (2) operating the converter at a lower temperature, or (3) using a special high-temperature catalyst formulation that inhibits by-product formation.

The carbon monoxide content of the inlet gas is not an important design variable with regard to the attainable approach to equilibrium. Concentrations ranging from a few percent to more than 50% have very little effect on the percentage of conversion obtained in a properly designed unit.

Conversion of Carbon Oxides to Methane (Methanation)

Catalytic hydrogenation of carbon monoxide and carbon dioxide to methane is normally used to eliminate small quantities of these compounds remaining in gas streams after bulk removal by other techniques. A typical application of the process is in the removal of small amounts of residual CO and CO_2 from hydrogen, following shift conversion and bulk removal of CO_2 by absorption. The methanation process is suitable for the purification of gas streams containing the oxides of carbon at a maximum concentration of about 2.5 mole %, and is particularly advantageous if the presence of methane in the treated gas is not objectionable in processes following the methanation step. Under proper operating conditions, the reaction goes almost to completion, and exit gases containing only a few ppm of the oxides of carbon are obtained.

Removal of carbon monoxide and carbon dioxide by methanation is required for the protection of certain hydrogenation and ammonia synthesis catalysts against rapid deactivation. It is also necessary when the hydrogen is used in hydroprocessing operations because CO_2 and CO can lead to temperature excursions and catalyst damage in the reactors. Furthermore, methanation is an essential step in the reaction systems associated with the Fischer-Tropsch synthesis and with the production of synthetic natural gas from liquid hydrocarbons and coal. Grayson (1956) presented a detailed discussion of methanation in connection with the Fischer-Tropsch process.

The methanation process used in the production of synthetic natural gas is substantially different from the purification application described in this section. It involves the methanation of high concentrations of carbon oxides and is conducted at high temperatures. The heat evolved during the reaction is usually recovered by the generation of high-pressure steam.

1178 Gas Purification

One version, the RM process, in which methanation is carried out in several stages, is described by White et al. (1974).

A description of methanator design and operation is given by Allen and Yen (1973). These authors point out the danger of forming nickel carbonyl (a highly toxic gas) by the reaction of elemental nickel in the catalyst with CO in the gas. The reaction is favored by high pressure and low temperature. To avoid the formation of nickel carbonyl, it is recommended that the operating temperature be maintained above 400°F whenever the pressure is above 300 psig.

An alternative technique for removing CO and CO_2 from synthesis gas is a two-step process involving methanolation (the conversion of the carbon oxides to methanol) to remove most of the impurities, followed by methanation to remove the last traces. The production of methanol has the potential advantage over methane of using less hydrogen and producing a potentially useful byproduct. The methanol is removed from the gas stream by water scrubbing. The resulting solution may then be distilled for recovery of a commercially pure methanol byproduct or recycled to the reformer to recover the hydrogen content (Sogaard-Andersen and Hansen, 1991).

Process Description

Although many types of catalysts have been investigated for the hydrogenation of carbon monoxide and carbon dioxide to methane, catalysts of high nickel content are most commonly used for gas purification. Typical commercial preparations contain 20 to 35% nickel and are available in the form of tablets, extruded pellets, rings, and small spheres of various sizes. The catalyst does not require regeneration, and has a typical service life on the order of ten years when operated with a poison-free feed gas. The properties of a typical methanation catalyst are given in **Table 13-19**.

Methanation catalysts are easily poisoned by sulfur, and, therefore, sulfur compounds must be removed from the gas stream before it enters the methanator. The flow scheme of

Table 13-19
Properties of Methanation Catalyst

Catalyst designation	PK-5
Composition	
Carrier	Alumina
Catalyst, wt % Ni in reduced catalyst	27
Surface area, m^2/g	250
Configuration	¼-in. OD extruded rings
Normal inlet temperature, °F	about 570
Maximum contin. operating temperature, °F	840
Temperature increase across methanator,	
°F per 1% CO	135
°F per 1% CO_2	108
°F per 1% O_2	315

Data of Haldor Topsøe, Inc. (1988)

the process is quite simple. A flow diagram typical of the final purification step in a hydrogen plant using either a physical or reactive solvent is depicted in **Figure 13-20.** A heat exchanger using high-temperature shift-converter effluent is used to start the methanator by heating the feed gas to about 600°F. When the methanator is operating, the temperature rise across the reactor is approximately 50°F, resulting in an outlet temperature of about 650°F. With this temperature rise, the methanator feed/effluent heat exchanger can be designed to supply all the heat required for feed preheat during normal operation, and the heat exchanger using the high-temperature shift effluent for heating the methanator feed is often not required after steady-state operation has commenced. Commercial methanation reactors normally contain a single bed of catalyst with a minimum depth to diameter ratio of 1:1.

Basic Chemistry

The reactions involved in methanation of carbon monoxide and carbon dioxide can be represented by equations 13-12 and 13-13:

$$CO + 3H_2 = CH_4 + H_2O \tag{13-12}$$

$$CO_2 + 4H_2 = CH_4 + 2H_2O \tag{13-13}$$

The heats of reaction and equilibrium constants within the range of 400° to 1,000°K (261° to 1,341°F) are shown in **Table 13-20.**

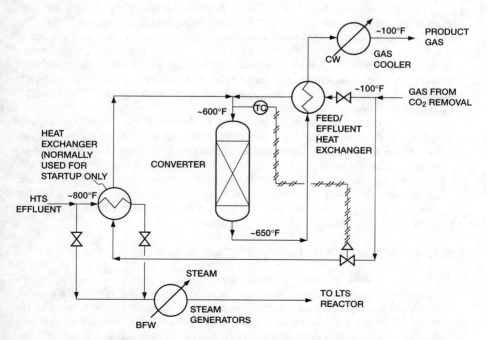

Figure 13-20. Schematic flow diagram of a methanation unit.

Table 13-20
Heats of Reaction and Equilibrium Constants for Methanation of Carbon Monoxide and Carbon Dioxide

Temperature, °K	°C	Heat of Reaction ΔH cal/g mole Eq. (13-12)	Eq. (13-13)	$K = \dfrac{(CH_4)(H_2O)}{(CO)(H_2)^3}$	$K = \dfrac{(CH_4)(H_2O)^2}{(CO_2)(H_2)^4}$
400	127	−50.353	−40.643	4.009×10^{15}	2.709×10^{12}
500	227	−51.283	−41.763	1.148×10^{10}	8.712×10^{7}
600	327	−52.061	−42.768	1.980×10^{6}	7.310×10^{4}
700	427	−52.703	−43.652	3.726×10^{3}	4.130×10^{2}
800	527	−53.214	−44.412	27.97	6.032
900	627	−53.610	−45.058	0.131	0.2888
1,000	727	−53.903	−45.592	0.0265	0.0365

Source: Data of Wagman et al. 1945

Design and Operation

The process can be operated over a considerable range of temperatures and pressures. Inlet gas temperatures may vary from 350° to 750°F; however, because of the unfavorable equilibrium at elevated temperature, the exit gas temperature should not exceed 825°F. The strongly exothermic nature of the reaction results in a temperature rise during conversion of 135°F for each 1% CO; 108°F for each 1% CO_2; and 315°F for each 1% O_2. These values can be used to estimate the required inlet temperature which will result in the desired final temperature. The large amount of heat generated by the reaction is a primary reason that the process is considered unsuitable for the treatment of gases containing more than about 2.5 mole % carbon oxides.

The pressure range over which the process can be operated extends from essentially atmospheric pressure to 12,000 psig. Space velocities of 1,000 to 2,000 volumes per volume per hour are typical for operation at atmospheric pressure. Much higher space velocities—up to 20,000 volumes per volume per hour—can be used at elevated pressures. Operation at very high pressure usually results in carbon-oxide concentrations below detectable limits. Typical operating conditions for a commercial installation are given in **Table 13-21**. The data indicate that no significant change in catalyst activity or pressure drop occurred over a period of more than four years.

Selective Hydrogenation of Acetylenic Compounds

Most olefinic gas streams obtained from refinery off-gases or from the cracking of saturated hydrocarbons contain acetylenic compounds in concentrations ranging from a few tenths of 1% to about 2%. These compounds must be removed if the olefins are to be used for the production of certain petrochemicals. Although liquid purification processes employing selective solvents have been developed, selective catalytic hydrogenation is usually the pre-

Table 13-21
Methanator Operating Data for 1,150 t/d NH$_3$ Plant

Catalyst age, months	33	86
Temperature, °F		
Inlet	601	597
Midpoint	621	642
Bottom of bed	621	643
Pressure, psig		
Inlet	402	401
Outlet	397	396
CO Concentration		
Inlet, %	0.22	0.38
Outlet, ppmv	<1	—

Data of Haldor Topsøe, Inc. (1988)

ferred technique, especially if the acetylenic compounds are present in relatively low concentrations and gas of very high purity is required.

Process Desicription

Removal of acetylene may be effected at several points during the olefin purification process. The selection of the most economical arrangement depends on a great many factors and has to be evaluated for each specific case. Basically, the choice is between treating the raw, cracked gas or the purified olefin stream. Acetylene concentrations of less than 10 ppmv in the final product gas can be obtained by treatment at either point in the process. In most commercial processes, acetylene is removed from the cracked gas, usually after removal of aromatics and acid gases. In some installations selective catalytic hydrogenation is used to eliminate small concentrations of acetylene and its homologues from purified olefin streams. A detailed discussion of the advantages and disadvantages of acetylene removal at various points of the olefin recovery and purification system was presented by Reitmeier and Fleming (1958).

Catalysts suitable for selective hydrogenation of acetylenic compounds in cracked gas streams contain elements of group VI and VIII of the periodic table. An early catalyst was molybdenum sulfide supported on activated alumina (Key and Eastwood, 1946). This was followed by the development of cobalt molybdate and nickel based catalysts (Giaro, 1956; Barry, 1950). Modern catalysts for impure (sulfur-bearing) cracked gas streams typically contain nickel, cobalt, and chromium on a silica-alumina base (United Catalysts, 1993).

Palladium and promoted palladium catalysts are used for the hydrogenation of small amounts of acetylenic compounds from purified olefin streams. These catalysts are sensitive to sulfur poisoning and, therefore, the feed gas must be free of sulfur prior to hydrogenation. The noble metal catalysts are more active than the base metal types and can therefore operate at lower temperatures—as low as 80°F—and are capable of producing effluent purities below 1 ppmv acetylene under optimum conditions. An improved catalyst, consisting of promoted palladium on alumina, was introduced by United Catalysts, Inc., in 1988. The catalyst is reported to remove acetylene selectively while increasing ethylene yields, and to provide a longer than normal life cycle compared to previously available catalysts (Parkinson and Johnson, 1989).

Commercial plants for selective acetylene hydrogenation consist usually of two or three reactors containing the catalyst in fixed beds. One or two beds (operating in parallel) are in operation, while the third is being regenerated. The gas usually flows downward through the catalyst bed and is not cooled during the reaction. Regeneration, which is necessary to remove polymeric materials from the catalyst, is accomplished by the use of superheated steam, or a mixture of steam and air, which flows upward through the catalyst. Steam may also be added to the inlet gas to control selectivity and to remove particulate matter.

Basic Chemistry

Hydrogenation of acetylene and methylacetylene can be represented by the following equations:

$$C_2H_2 + H_2 = C_2H_4 \tag{13-14}$$

$$C_2H_2 + 2H_2 = C_2H_6 \tag{13-15}$$

$$C_3H_4 + H_2 = C_3H_6 \tag{13-16}$$

$$C_3H_4 + 2H_2 = C_3H_8 \tag{13-17}$$

The heats of reaction and equilibrium constants for equations 13-14 and 13-15 within the range of 300° to 1,000°K (80° to 1,341°F) are given in **Table 13-22**.

Table 13-22
Heats of Reaction and Equilibrium Constants for Hydrogenation of Acetylene to Ethylene and Ethane

Temperature, °K	°C	Heat of Reaction ΔH cal/g mole Eq. (13-14)	Eq. (13-15)	$K = \dfrac{(C_2H_4)}{(C_2H_2)(H_2)}$	$K = \dfrac{(C_2H_6)}{(C_2H_2)(H_2)^2}$
300	27	−41.711	−74.451	3.37×10^{24}	1.19×10^{42}
400	127	−42.368	−75.553	7.63×10^{16}	2.65×10^{28}
500	227	−42.911	−76.485	1.65×10^{12}	1.31×10^{20}
600	327	−43.311	−77.211	1.19×10^{9}	3.31×10^{14}
700	427	−43.645	−77.767	6.50×10^{6}	3.10×10^{10}
800	527	−43.676	−78.157	1.28×10^{5}	2.82×10^{7}
900	627	−44.014	−78.432	5.88×10^{3}	1.17×10^{5}
1,000	727	−44.099	−78.584	2.23×10^{2}	1.46×10^{3}

Source: Data of Wagman et al. (1945), Kilpatrick et al. (1946), and Prosen et al. (1945)

Design and Operation

Reactors most commonly used in the process consist of cylindrical vessels containing the catalyst in an adiabatic fixed bed with a maximum depth of about 10 ft. The bed depth-to-diameter ratio is normally less than 1:1. In cases where the concentration of acetylene in the feed gas is sufficiently high to cause an excessive temperature increase during the conversion operation, isothermal tubular reactors are used, with the catalyst inside the tube and a coolant on the outside.

Process conditions depend on the type of gas being treated and the catalyst employed. Acetylene removal from cracked gas streams containing sulfur, carbon monoxide, and a large excess of hydrogen can be carried out with nickel/cobalt/chromium/sulfur-based catalysts at 300° to 500°F and 50 to 500 psig. Such an operation will provide a product stream containing less than 10 ppmv acetylene.

Sulfur-free cracked gas streams can be purified with a palladium base catalyst at a lower temperature (100° to 350°F), which is capable of reducing the acetylene concentration to less than 3 ppmv (and under some conditions less than 1 ppmv). Palladium and promoted palladium based catalysts are also used for the removal of traces of acetylenic compounds from concentrated olefin streams, and are the preferred catalysts in such applications.

A mathematical analysis of optimum operating conditions for the catalytic removal of acetylene from ethylene with a palladium catalyst has been presented by Huang (1979). Parameters investigated include temperature, space velocity, and feed gas composition. The results of the analysis show that, for the case studied, temperature is the most sensitive parameter. Optimum operating conditions, yielding the lowest acetylene concentration (1 ppmv) and the lowest ethylene loss, include a reactor temperature of 240°F and a space velocity of 7,000 (vol)/(vol)(hr).

REFERENCES

Allen, D. W., and Yen, W. H., 1973, *Chem. Eng. Progr.,* Vol. 69, No. 1, January, p. 75.

Andreasen, J., 1988, "Air Pollution Control by Catalytic Processes," presented at the World Bank Seminar, Washington, D. C., March, reprint provided by Haldor Topsøe, Inc.

ARI International, 1992, *Chem. Eng.,* Vol. 99, No. 8, August, p. 92.

Bähr, H., 1938, *Chem. Fabrik,* Vol. 11, No. 1/2, p. 10

Barry, A. W., 1950, U.S. Patent 2,511,453.

Campbell, I., 1988, *Catalysis at Surfaces,* Chapman and Hall, London, England.

Carberry, J. J., 1976, *Chemical and Catalytic Reaction Engineering,* McGraw-Hill, New York, NY.

Carpenter, C., 1913, *J. Gas Lighting,* Vol. 122, p. 1010.

Chen, J. M., Heck, R. M., Burns, K. R., and Collins, M. F., 1989, "Commercial Development of Oxidation Catalyst for Gas Turbine Cogeneration Applications," presented at the 82nd Annual Meeting of the Air & Waste Management Assoc., Anaheim, CA, June 25–30.

Cordonna, G. W., Kosanovich, M., and Becker, E. R., 1989, "Gas Turbine Emission Control," *Platinum Metals Review,* Vol. 33, No. 2, April.

Crawley, B., and Griffith, R. H., 1938, *J. Chem. Soc.,* p. 720.

Donaldson Company, Inc., 1992, "Eto-Abator Ethylene Oxide Emission Control System," Brochure PB 500 SCFM ETO.

Engelhard Corporation, 1987, Data sheet entitled "Oxidation Catalyst System-CO Abatement," EC-3628 Rev. 11/87.

Engelhard Corporation, 1991, Brochure entitled "Engelhard's Catalyst Technology—for VOC Air Pollution Abatement that Meets Everyone's Approval," EC-5249 Rev. 4/91.

Environmental Protection Agency (EPA), 1991, *Handbook, Control Technologies for Hazardous Air Pollutants,* Report No. EPA/625/6-91/014, June.

Environmental Protection Agency (EPA), 1986, *Handbook, Control Technologies for Hazardous Air Pollutants,* Report No. EPA/625/6-86/014, September.

Evans, E. V., 1915, *J. Soc. Chem. Ind.,* (London), Vol. 34, pp. 9–14.

Evans, E. V., and Stanier, H., 1924, *Proc. Roy. Soc.* (London), Vol. 105A, p. 626.

Fogler, H. S., 1986, *Elements of Chemical Reaction Engineering,* Prentice-Hall, Englewood Cliffs, NJ

Giaro, J. A., 1956, U. S. Patent 2,735,897.

Grayson, M., 1956, *Catalysis,* Vol. 4, Edited by P. H. Emmett, Reinhold Publishing Co., NY.

Griffith, R. H., 1952, *Ind. Eng. Chem.,* Vol. 44, No. 5, p. 1011.

Haldor Topsøe, Inc., 1988, "Topsøe Methanation Catalyst PK-5," Brochure PK-5/5/88.

Haldor Topsøe, Inc., 1989, "Topsøe Low Temperature Shift Catalysts," Brochure LK-801/LK-821/9/89.

Haldor Topsøe A/S, 1990, "Topsøe REGENOX Regenerative Catalytic Combustion," Brochure Eqp. 9/90.

Haldor Topsøe A/S, 1991A, "Topsøe CATOX Catalytic Combustion Technology for Air Purification," Brochure E. Comb. 01.02/91.

Haldor Topsøe, Inc., 1991B, "Topsøe Catalytic Program," Brochure Cat. Progr. 5/91.

Haldor Topsøe, Inc., 1992, "Topsøe High Temperature Shift Catalyst SK-201," Brochure SK-201 11/92.

Haldor Topsøe A/S, 1993, Personal Communication, March 31.

Hartye, R. W., 1985, "Effect of Low Steam on Water Gas Shift Operation," presented at the AIChE 1985 Ammonia Symposium, Safety in Ammonia Plants and Related Facilities, Seattle, WA, August.

Heck, R. M., Durilla, M., Bouney, G., and Chen, J. M., 1988, "Air Pollution Control—Ten Years Operating Experience with Commercial Catalyst Regeneration," presented at the 81st APCA Annual Meeting & Exhibition, Dallas, TX, June 19.

Herbert, K. J., 1991, "Catalysts for Volatile Organic Compound Control in the 1990s," presented at Georgia Power Company, VOC Control Workshop, Atlanta, GA, June 25, Reprint provided by Allied-Signal, Inc.

Huang, W., 1979, *Hydro. Process.,* Vol. 58, No. 10, October, p. 131.

Huber, R. J., 1993, United Catalysts, Inc., Personal Communication.

Hylton, T. D., 1991, "Evaluation of the TCE Catalytic Oxidation Unit at Wurtsmith Air Force Base," presented at the AIChE Summer National Meeting, Pittsburgh, PA, Aug 20.

Johnson Matthey, 1992, "DeNOx and CO Replacement Catalysts," Johnson Matthey Environmental Products, Catalytic Systems Division, Bulletin 92–95.

Jung, H. J., and Becker, E. R., 1987, "Emission Control for Gas Turbines," *Platinum Metals Review,* Vol. 31, No. 4, October.

Kelley, K. K., 1937, U. S. Bur. Mines, Bull., No. 406.

Key, A., and Eastwood, A. H., 1946, Forty-fifth Report of the Joint Research Committee of the Gas Research Board and the University of Leeds, London, The Gas Research Board (presently the Gas Council), Publ. GRB 14/4.

Kilpatrick, J. E., Prosen, E. J., Pitzer, K. S., and Rossini, F. D., 1946, *J. Research Nat. Bur. Standards,* Vol. 36, pp. 599–612.

Kujang, P. T., 1992, "New Developments in High Temperature Shift Catalyst," presented at the United Catalysts, Catalysts Seminar, Bali, Indonesia, May 26–27.

Lee, H. H., 1985, *Heterogeneous Reactor Design,* Butterworth Publishers, Stoneham, MA.

Lester, G. R., 1989, "Catalytic Destruction of Hazardous Halogenated Organic Chemicals," presented at the Air and Waste Management Association, 82nd Annual Meeting and Exposition, Anaheim, CA, June 25–30.

Lester, G. B., and Summers, J. C., 1988, "Poison-Resistant Catalyst for Purification of Web Offset Press Exhaust," presented at the Air Pollution Control Association, 81st Annual Meeting, Dallas, TX, June 19–24.

Maxted, E. B., 1937, British Patent 490,775.

Maxted, E. B., and Marsen, A., 1946, *J. Soc. Chem. Ind.* (London), Vol. 65, p. 51.

Maxted, E. B., and Priestly, J. J., 1946, *Gas J.,* Vol. 247, pp. 471, 515, 556, 593.

McIlwee, R. J., 1993, "The Basics of VOC & Odor Emissions Control," unpublished report, Smith Engineering Company, Ontario, CA.

McInnes, R., Jelinek, S., and Putsche, V., 1990, *Chem. Eng.,* Vol. 97, No. 9, September, p. 108.

MECA, 1992, *Catalytic Control of VOC Emissions,* a guidebook provided by MECA, Manufacturers of Emission Controls Association, Washington, DC.

Nielsen, H., and Hansen, J. B., 1981, "Catalysts and Processes for the Water Gas Shift Reaction," paper provided by Haldor Topsøe A/S, Copenhagen, Denmark.

Palazzolo, M. A., and Jamgochian, C. L., 1986, "Destruction of Chlorinated Hydrocarbons by Catalytic Oxidation", U. S. EPA-600/2-86-079, September.

Parkinson, G., 1991, *Chem. Eng.,* Vol. 98, No. 7, July, p. 37.

Parkinson, G., and Johnson, E., 1989, *Chem. Eng.,* Vol. 96, No. 9, September, p. 30.

Perry, R. H., and Green, D. W. (editors), 1984, *Perry's Chemical Engineer's Handbook,* 6th edition, McGraw-Hill, NY.

Plant, J. H. G., and Newling, W. B. S., 1948, "The Catalytic Removal of Organic Sulfur Compounds from Coal Gas," Inst. Gas Engrs, Commun. 344.

Priestly, J. J., 1957, *Gas Times,* Vol. 91, October/November, p. 640.

Priestly, J. J., 1958, *Gas in Ind.,* Vol. 1, pp. 21–28.

Priestly, J. J., and Marris, H. Q., 1947, *Gas Times,* Vol. 50, pp. 396–399.

Prosen, E. J., Pitzer, K. S., and Rossini, F. D., 1945, *J. Research Nat. Bur. Standards,* Vol. 34, pp. 403–411.

Reitmeier, R. E., and Fleming, H. W., 1958, *Chem. Eng. Progr.,* Vol. 54, No. 12, December, p. 48.

Renko, R. J., 1990, *Chem. Eng. Progr.,* Vol. 86, No. 10, October, p. 47.

Rushton, J. H., 1954, *Advances in Catalysis,* Vol. 2, p. 107, edited by W. G. Frankenburg et al., Academic Press, Inc., NY.

Sands, A. E., Wainwright, H. W., and Schmidt, L. D., 1948, *Ind. Eng. Chem.,* Vol. 40, No. 4, p. 607.

Sands, A. E., Wainwright, H. W., and Egleson, G. C., 1950, U. S. Bur. Mines Rept. Invest. 4699.

Seiwert, J. J., 1993, "Advanced Regenerative Thermal Oxidation Technology for Air Pollution Control", unpublished report, Smith Engineering Company, Ontario, CA.

Shearman, J., 1992, *Chem. Eng.,* Vol. 99, No. 3, March, p. 59.

Smith, J., 1981, *Chemical Engineering Kinetics,* 3rd edition, McGraw-Hill, NY.

Sogaard-Andersen, P., and Hansen, O., 1991, "Methanolation in Ammonia Plants" presented at the 1991 AIChE Symposium, Safety in Ammonia Plants and Related Facilities, Los Angeles, CA, Nov. 17–20.

Summers, J. C., Frost, A. C., and Sawyer, J. E., 1989, "Volatile Organic Compounds Emission Control by Catalysts," presented at the STAPPA/ALAPCO Technical Briefing, New Orleans, LA, October, reprint provided by Allied Signal, Inc.

Suter, H. R., 1955, *J. Air Pollution Control Assoc.,* Vol. 5, No. 3, p. 173.

Terres, E., and Wesemann, H., 1932, *Angew. Chem.,* Vol. 45, pp. 795–801.

Udengaard, N. R., and Berzins, V., 1985, "Catalytic Conversion of COS for Gas Cleanup," Chapter 22 in *Acid and Sour Gas Treating Processes,* edited by S. A. Newman, Gulf Publishing Co., Houston, TX.

United Catalysts, Inc., 1985, "C49 Product Bulletin," 06-85.

United Catalysts, Inc., 1993, Personal Communication.

van der Vaart, D. R., Spivey, J. J., Vatavuk, W. M., and Wehe, A., 1990, "Thermal and Catalytic Incinerators," Chapter 3 in *OAQPS Control Cost Manual,* Report No. EPA 450/3-90-006, January.

Valentin, F. H. H., 1990, *Chem. Eng.,* Vol. 97, No. 1, January p. 112.

W.C. Holmes & Co., 1958, Personal Communication.

Wagman, D. D., Kilpatrick, J. E., Pitzer, K. S., and Rossini, F. D., 1945, *J. Research Nat. Bur. Standards,* Vol. 35, pp. 467–496.

Wedgewood, W., 1958, *Inst. Gas Engrs.,* Publ. No. 525, London.

White, G. A, Roszkowski, T. R., and Stanbridge, D. W., 1974, "The RM Process," *Proceedings 168th National Meeting,* American Chemical Society, Division of Fuel Chemistry, Atlantic City, NJ, Sept. 8–13.

Chapter 14
Physical Solvents for Acid Gas Removal

BACKGROUND, 1188

PROCESS DESCRIPTION, 1191

PROCESS DESIGN, 1192
- Solvent Circulation Rate, 1193
- Recycle Gas Rate, 1194
- Water Disposition, 1195

PROCESS SELECTION, 1196

SIMPLE PHYSICAL SOLVENT PROCESSES, 1198
- Fluor Solvent Process, 1198
- Selexol Process, 1202
- Sepasolv MPE Process, 1210
- Purisol Process, 1210
- Rectisol Process, 1215
- Ifpexol Process, 1223
- Estasolvan Process, 1224
- Methylcyanoacetate Process, 1225

MIXED PHYSICAL/CHEMICAL SOLVENT PROCESSES, 1225
- Sulfinol Process, 1225
- Amisol Process, 1231
- Selefining Process, 1232

REFERENCES, 1234

BACKGROUND

When the acid gas impurities make up an appreciable fraction of the total gas stream, the cost of removing them by heat regenerable solvents may be out of proportion to the value of the treated gas. This has provided the major impetus for the development of processes that employ nonreactive organic solvents as the treating agents. These materials physically dissolve the acid gases, which are then stripped without the application of heat by merely reducing the pressure.

Early efforts to employ water as a physical solvent met with limited success (see Chapter 6), but the solubilities of CO_2 and H_2S in water are too low for water wash to be a practical commercial process. The earliest commercial process based on an organic physical solvent, methanol, was the Rectisol Process, which has been used for synthesis gas applications where the removal of other impurities in addition to CO_2 and H_2S and the production of treated gas containing only ppm levels of CO_2 and H_2S is required. This process operates at very low temperatures (to minus 100°F) and is quite complex compared to other physical solvent processes. As a result, the Rectisol process is not considered applicable to most gas treating services, although it continues to find application in purifying synthesis gases derived from the gasification of heavy oil and coal.

This trend to physical solvents accelerated in 1960 with the introduction of the Fluor Solvent Process, which was followed by several other physical solvent processes. More recently, a new class of process based on the use of a mixed absorbent, containing both a physical and a chemical solvent, has been commercialized. Both simple physical solvent and mixed solvent processes are described in this chapter.

A listing of the major physical solvent gas purification processes that have been or are currently offered for commercial use and the solvents used by each is provided in **Table 14-1**. Many more solvents have been proposed and evaluated in the past, and the search for superior solvents is continuing. A screening study to optimize physical solvent processes for the purification of gases at high pressure has been presented by Zawacki et al. (1981). In this work a large number of physical solvents were screened and, after selection of two solvents (the dimethyl ether of tetraethylene glycol and N-formyl morpholine), process schemes were proposed for a variety of applications.

A key parameter in the screening of potential solvents is the solubility of the gaseous impurities to be absorbed. Techniques for evaluating the solubilities of gases in polar solvents of the type used as physical solvents are described by England (1986) and Sweeney et al. (1988). The latter authors propose a technique which is particularly effective for evaluating a series of functionally related solvents. The need for experimentation is not eliminated, but is greatly reduced. A comparison of calculated versus experimental data for six solvents and three gases presented by Sweeney et al. is summarized in **Table 14-2**.

The calculated values for the first five solvents in the table are based on experimental data for n-Methyl-2-pyrrolidone (NMP), a solvent used in the Purisol process. Calculated values for the Selexol solvent are based on experimental data for the compound tetraethylene glycol dimethyl ether ($CH_3O(CH_2CH_2O)_4CH_3$). The Selexol solvent is reported to be a mixture of polyethylene glycol dimethyl ethers ($CH_3O(CH_2CH_2O)_xCH_3$), where x ranges from 3 to 9, with an average molecular weight of about 272 (Sweeney et al., 1988).

Although many organic solvents appear to be suitable for use as physical solvents, their actual number is limited by certain criteria that must be fulfilled to make them acceptable for economic operation. In order to be practical, the solvents must have an equilibrium capacity for acid gases several times that of water, coupled with a low capacity for the primary con-

Table 14-1
Physical Solvent Processes

Process Name	Solvent	Process Licensor
Simple Physical Solvents		
Fluor Solvent	Propylene carbonate (PC)	Fluor Daniel
SELEXOL	Dimethyl ether of polyethylene glycol (DMPEG)	Union Carbide
Sepasolv MPE	Methyl isopropyl ether of polyethylene glycol (MPE)	Badische (BASF)
Purisol	N-Methyl-2-pyrrolidone (NMP)	Lurgi
Rectisol	Methanol	Lurgi and Linde AG
Ifpexol	Methanol	Institut Français du Pétrole (IFP)
Estasolvan	Tributyl phosphate	IFP/Uhde
Methylcyanoacetate	Methylcyanoacetate	Unocal
Mixed Physical/Chemical Solvents		
Sulfinol	Sulfolane and DIPA or MDEA	Shell Oil/SIPM
Amisol	Methanol and secondary alkylamine	Lurgi
Selefining	Undisclosed physical solvent and tertiary amine	Snamprogetti

Table 14-2
Comparison of Calculated with Experimental Solubility Data

	Solubility, Mol% at 1 atm and 298°K					
	CO_2		H_2S		COS	
Solvent	Calc.	Exp.	Calc.	Exp.	Calc.	Exp.
Dimethyl-2-pyrrolidone	2.31	1.99	22.7	21.1	7.30	5.85
n-Methyl-2-piperidone	2.39	1.68	24.0	22.5	7.52	5.95
n-Methyl-caprolactam	1.88	1.62	21.3	21.7	6.13	5.26
n-Methyl-4-piperidone	2.83	2.25	24.6	24.4	8.62	7.11
n-Ethyl-pyrrolidone	1.91	1.30	21.2	15.6	2.02	3.98
Selexol solvent	3.39	3.56	29.6	22.4	10.3	9.65

Data of Sweeney et al. (1988)

stituents of the gas stream, e.g., hydrocarbons and hydrogen. In addition, they must have low viscosity and low or moderate hygroscopicity. They must be noncorrosive to common metals as well as nonreactive with all components in the gas and, preferably, have a very low vapor pressure at ambient temperature. Finally, they must be available commercially at a reasonable cost.

The simplest version of a physical solvent process involves the regeneration of the solvent by flashing to atmospheric pressure or vacuum, or by inert gas stripping. This approach produces a treated gas that still contains small amounts of acid gas. If H_2S is present at only very low concentrations or is entirely absent, this flow scheme is usually applicable since CO_2 concentrations as high as 2 or 3% can often be tolerated in the product gas. Where H_2S is present in significant amounts, however, thermal regeneration has generally been necessary to accomplish the thorough stripping of the solvent needed to reach stringent H_2S purity requirements. Heat requirements are usually far less for physical solvents than for reactive solvents, such as amines, since the heat of desorption of the acid gas for the physical solvent is only a fraction of that for reactive solvents. The circulation rate of the physical solvent may also be less, particularly when the acid gas partial pressure is high.

Physical solvent processes are used primarily for acid-gas removal from high-pressure natural-gas streams and for carbon dioxide removal from crude hydrogen and ammonia synthesis gases produced both by partial oxidation and steam-hydrocarbon reforming. Since solvent processes are most efficient when operated at the highest possible pressure, carbon dioxide removal from reformer effluents is best carried out after compression of the process gas to the ultimate pressure required for such processes as ammonia synthesis or hydrocracking. Under these circumstances, the molecular weight of the CO_2-rich gas is sufficiently high to permit use of relatively inexpensive centrifugal compressors to reach the required discharge pressure.

As shown in Table 14-2, most organic solvents have an appreciably higher solubility for hydrogen sulfide than for carbon dioxide, and a certain degree of selective hydrogen sulfide removal can be attained. This feature is of special significance when the ratio of carbon dioxide to hydrogen sulfide in the crude gas is so high that the acid gas stream resulting from complete removal cannot be processed in a Claus sulfur recovery unit. By removing essentially all of the hydrogen sulfide and only a portion of the carbon dioxide, this ratio can often be lowered sufficiently to permit normal processing in a Claus plant. The selectivity displayed by physical solvents has been applied to excellent advantage in many instances and is described in more detail later in the chapter.

Minor gas impurities such as carbonyl sulfide, carbon disulfide, and mercaptans are quite soluble in most organic solvents, and these compounds are removed to a large extent together with the acid gases. The solubility of hydrocarbons in organic solvents increases with the molecular weight of the hydrocarbon. Consequently, hydrocarbons above ethane are also removed to a large extent and flashed from the solvent together with the acid gas. Although special designs for the recovery of these compounds have been proposed, physical solvent processes are generally not economical for the treatment of hydrocarbon streams that contain a substantial amount of pentane-plus hydrocarbons. Aromatic hydrocarbons are especially difficult to deal with. Even in trace amounts they require a special step to separate them from the solvent because they are very strongly absorbed by most of the solvents used in these processes and tend to accumulate in the solvent.

In another class of process usually referred to as mixed solvent processes, an amine is blended with a physical solvent so that the bulk removal capabilities of the physical solvent are combined with the amine's ability to achieve very low residual acid-gas specifications in a single treating step. These processes are typified by Shell's Sulfinol Process.

PROCESS DESCRIPTION

In their simplest form, physical solvent processes require little more than an absorber, an atmospheric flash vessel, and a recycle pump. No steam or other heat source is required. After the absorbed gases are desorbed from the solution by flashing at atmospheric pressure, the lean solution contains acid gas in an amount corresponding to equilibrium at 1 atm acid-gas partial pressure; and this, therefore, represents the theoretical minimum partial pressure of acid gas in the purified-gas stream. To obtain a higher degree of purification, vacuum or inert gas stripping or heating of the solvent must be employed. Other process modifications are used to minimize loss of valuable gas components, provide a relatively low temperature of operation, and otherwise improve process economics.

Figure 14-1 presents three different configurations of physical solvent processes differing in the method by which the solvent is stripped. The first illustrates regeneration by simple flashing. One or more flash steps may be employed. **Figure 1(A)** shows a two stage flash system with the final flash to atmospheric pressure. An alternate to this case that is not shown is to conduct the final flash at vacuum conditions. **Figure 1(B)** shows the use of inert gas stripping to lower the acid gas content of the lean solvent. Nitrogen and fuel gas have been used as stripping agents. Air may also be used, but is not feasible in systems where sulfur species occur due to the formation and accumulation of elemental sulfur and to other side reactions. **Figure 1(C)** shows thermal regeneration of the solvent, a method that is widely practiced with physical solvents to achieve pipeline quality gas.

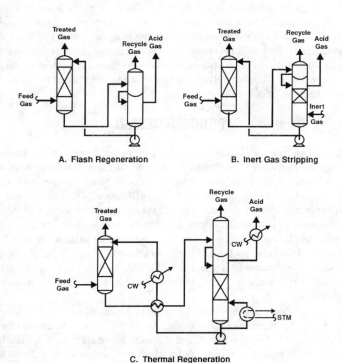

Figure 14-1. Simplified flow diagrams of physical solvent processes showing three basic methods of solvent regeneration.

The absorption step is the same with all three types of regeneration. The gas enters the bottom of the absorber, which contains either packing or trays, and is washed by a descending stream of regenerated solvent. The rich solvent leaves the bottom of the absorber and flows to the regeneration system. When multiple flashes are used, the gases flashed at the highest pressure level will contain most of the dissolved non-acidic gases, and are usually recompressed and returned to the absorber inlet to minimize losses of product gas. Quite frequently, solvent regeneration by pressure reduction is adequate for attaining the required gas purity, and the solvent leaving the lowest pressure flash is directly recycled to the top of the absorber. If product gas of higher purity is required, the residual acid-gas content of the solvent can be further reduced by stripping with an inert gas. Complete removal of the last remaining acid gas can be effected by heat regeneration and reboiling of the solvent as shown in **Figure 14-1(C).**

Although not shown in **Figure 14-1,** the rich solvent is frequently expanded through hydraulic turbines that are used to supply about half of the lean solvent pump horsepower. This has the dual purpose of reducing energy requirements and cooling the regenerated solvent before reuse, since less net shaft work is introduced into the system through the lean solvent pump. Since the absorption capacity of the solvents for acid gases increases as the temperature is lowered, it is advantageous to operate at the lowest possible temperature. Often, sufficient autorefrigeration is available in the system to make outside refrigeration unnecessary. In other cases the inclusion of auxiliary mechanical (or absorption) refrigeration results in a more efficient and less costly plant.

With most processes, solvent recovery from the effluent streams is not required since the vapor pressures of the solvents are sufficiently low at operating conditions. With some processes, however, such as the Purisol process, effluent streams are water washed.

Most commercially used solvents are very stable and reclaiming is not required. However, water removal by distillation is sometimes necessary to maintain the water concentration of the solvent at low levels. Alternatively, the feed gas can be dehydrated by glycol injection or other means to keep water out of the system.

PROCESS DESIGN

The design of absorption and stripping columns for physical solvents is relatively straightforward (compared to reactive absorbents) because no chemical reactions occur in the liquid phase. Design equations and simulation models commonly used for hydrocarbon separations are generally applicable to physical solvent gas purification. The key requirement is adequate liquid/vapor equilibrium data covering all components and conditions encountered in the process. Thermal and physical property data are also necessary for complete designs.

Computer models based on a tray-by-tray heat and material balance are the best approach for final designs; however, preliminary studies can make use of the Kremser equation which is described in Chapter 1 (equation 1-20). An application of this equation to the design of a Selexol system for selectively absorbing H_2S in the presence of CO_2 is described by Sweny (1985).

The Kremser equation correlates three factors: (1) the fraction of a given gas component that is absorbed, (2) the number of theoretical plates in the column, N, and (3) the absorption factor, A. A is defined as L/KV, where L and V are the liquid and gas flow rates in moles per unit time, and K is the equilibrium constant for the given component, y/x. The symbols y and x have their usual meaning of mole fraction of the given component in the gas and in the liquid, respectively, at equilibrium.

For any compound, the fraction absorbed increases as A is increased or as the number of trays is increased. However, complete removal of a component from the gas cannot be attained unless A is greater than 1, no matter how many trays are used. Any desired removal efficiency can be obtained over a wide range of values for A and N. For example, the Kremser equation shows that 98% removal of a given component is possible with either 25 trays at A = 1.25 or 7 trays at A = 1.5.

These considerations are very important in the design of physical solvent absorbers because the conditions selected to meet the requirements for the key component (e.g., H_2S) may have a major effect on column performance with regard to other, less soluble components (e.g., CO_2, COS, and CH_4). If maximum selectivity is desired, a large number of trays should be used, with the lowest possible value of A that will permit attainment of the desired removal efficiency for the key component. "A" will then be slightly over 1.0 for H_2S and much lower than 1 for the less soluble gases, which will, therefore, not be efficiently removed. On the other hand, if it is desired to remove a selected less soluble component (e.g., COS) more efficiently, it is necessary to increase A for that component (normally by increasing the liquid rate). At values of A significantly below 1.0, increasing A has a much greater effect on the fraction removed than increasing the number of trays. As a result, it is often possible to meet the requirements for both H_2S and COS removal with a high liquid rate and a relatively short column.

The selectivity of a physical absorption process can be enhanced by the use of more than one stripping and absorption stage. For example, if H_2S is absorbed selectively relative to CO_2, the rich solution from the primary absorber will contain a higher H_2S/CO_2 ratio than the feed gas. If the rich solution is stripped (or partially stripped), and the released gas is subjected to selective reabsorption, the rich solution from the reabsorber will contain an even higher H_2S/CO_2 ratio. This type of flow scheme can be useful in providing an acid gas stream with a high enough H_2S content for use in a standard Claus plant. Many other flow schemes have been developed to meet specific requirements and take advantage of the properties of specific solvents. Several of these alternative flow schemes are described in subsequent sections of this chapter in connection with commercial physical solvent processes.

Since the equilibrium constant is a function of temperature, absorber and stripper designs must take into account temperature changes that occur as a result of absorption (or desorption) of components. Zawacki et al. (1981) report using heats of absorption values of 150, 180, and 1,000 Btu/lb for CO_2, H_2S, and H_2O, respectively, for calculating the temperature rise due to absorption in two different solvents (N-formyl morpholine and the dimethyl ether of tetraethylene glycol). They also assumed tray efficiencies of 15 to 20% for absorption in these solvents, and estimated that 50 to 60 actual trays would represent practical column heights.

Solvent Circulation Rate

As in other types of gas treating processes, probably the most important process factor that dictates capital cost of the plant is the solvent circulation rate. This is particularly true for bulk removal plants, but also applies to those plants that use thermal regeneration of the solvent. The reason for this is obvious, since circulation rate affects the size and the cost of virtually every piece of equipment, including the absorber, piping, circulation pumps, and flash drums.

Every effort should be made to minimize solvent circulation rate. For a given solvent this can be accomplished most readily by reducing the contact temperature since the solvent's capacity for absorbing acid gases increases as the temperature is decreased. Factors which reduce solvent temperature include the following:

1194 Gas Purification

- The Joule-Thomson cooling effect of expanding CO_2 through the plant—from feed gas pressure to atmospheric pressure—cools the solvent. No additional equipment is required to realize this benefit.
- The use of heat exchangers to recover refrigeration from process streams such as the absorber overhead and flash gas streams is a common practice.
- A cooling effect is achieved if the rich solvent is directed through hydraulic turbines to provide part (up to almost 50%) of the solvent pumping energy. Without these turbines, motor or steam driven turbines would be needed and additional shaft work would be added to the system.
- In plants where the feed gas is very high in CO_2 content, i.e., 30% or higher, a portion of the absorbed CO_2 can be flashed from the solvent at an intermediate pressure, then directed through expansion turbines which are used to drive circulation pumps. The flashed gas is cooled during expansion through the turbine and can then be used in a heat exchanger to remove heat from the circulating solvent.
- The use of external mechanical or absorption refrigeration to further reduce solvent temperature has often been used. For bulk removal plants where the lean solvent is not heated, applying the refrigeration at the warmest point in the cycle (the rich solvent leaving the absorber) is preferred since the capital and energy requirements for the refrigeration system are the lowest. In plants where thermal regeneration is used, the refrigeration is applied to the lean solvent leaving the stripper.

An additional benefit derived from chilling the solvent and reducing its flow rate is that the amount of light hydrocarbon (or H_2 + CO in a syngas application) absorbed in the solvent may be significantly reduced. This benefit is possible because the solubilities of CO_2 and H_2S generally increase significantly when the operating temperature is reduced, while the solubilities of CH_4, H_2, and CO show little change with temperature. **Figure 14-2** illustrates this phenomenon for methanol and the Sepasolv MPE solvent. The solubility of CH_4 in these solvents is seen to change much less with temperature than the solubilities of H_2S, COS, and CO_2. In some solvents, CH_4, H_2, and CO may actually become less soluble as the temperature is reduced. The dimethyl ether of tetraethylene glycol is an example of such a solvent as indicated by the data in **Table 14-3** from Zawacki et al. (1981).

The temperature to which a solvent may be cooled is limited primarily by its increased viscosity and the resulting decrease in solvent heat and mass transport capabilities. The cost of providing the required cooling is, of course, also a factor. The minimum acceptable temperature depends on the solvent used. Methanol, for example, can be used at temperatures as low as $-95°F$ (Knapp, 1968), while propylene carbonate is limited to temperatures above $0°F$ (Freireich and Tennyson, 1977).

Recycle Gas Rate

In every physical solvent process, a portion of the most valuable constituent of the process gas—ordinarily either CH_4 or H_2—is unavoidably dissolved in the rich solvent. With CH_4 this potential loss can be as high as 10% of the total CH_4. Standard practice is to flash the rich solvent to an intermediate pressure, ½ to ⅓ of the absorber pressure, and to recompress the flash gas and recycle it to the feed gas. This additional step can reduce CH_4 losses to typically 2 or 3% of the CH_4 present in the feed gas. The CH_4 (or H_2) loss is clearly a function of the pressure at which the flash is conducted and standard optimization procedures can be followed to determine this pressure. In some cases, several intermediate flashes, rather than a single flash, are used to reduce compression horsepower. Another method of reducing compression horsepower that has been proposed is to scrub CO_2 from the flash gas in a "reab-

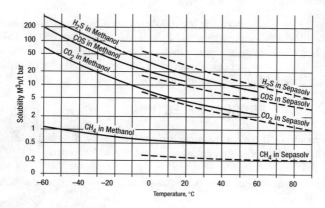

Figure 14-2. Effect of temperature on the solubility of gases in methanol and Sepasolv MPE solvent (*Ranke and Mohr, 1985; Wölfer, 1982*). *Reproduced with permission from "Acid and Sour Gas Treating Processes," S. A. Newman, Ed., copyright 1985, Gulf Publishing Company*

Table 14-3
Effect of Temperature on the Solubility of Various Gases in the Dimethyl Ether of Tetraethylene Glycol

Gas	Solubility at:	
	80°F	40°F
Methane	0.43	0.37
Carbon monoxide	0.12	0.05
Hydrogen	0.08	0.05
Carbon dioxide	3.61	4.15
Hydrogen sulfide	20.67	25.85

Solubility values are in the units of volumes of gas (60°F, 1 atm) per volume of solvent (60°F, 1 atm).

Data source: Zawacki et al. (1981)

sorber." This is especially applicable when the feed gas contains a high concentration of CO_2, since the flash gas will be even more concentrated with CO_2.

Water Disposition

The feed gas, whether it is natural or synthesis gas, will almost always contain water. Typically, the gas will be water-saturated at feed gas conditions. Another source of water entering the system may be stripping gas used to regenerate the solvent. Some water will be discharged from the plant in the gas streams leaving the system; however, water will accumulate in the solvent until the amounts entering and leaving the system are equal. Although most processes can tolerate several percent water in the solvent, it is sometimes necessary to control water buildup. This can be accomplished by distilling water from a solvent slipstream or dehydrating the feed gas and/or stripping gas.

PROCESS SELECTION

A number of proven physical solvent processes are available for most applications. In addition to capital cost, the following factors must be included in any comparison:

- Process performance in terms of treated gas purity, and acid gas composition (e.g., suitability as Claus plant feed)
- Loss of light and heavy hydrocarbon (or other valuable constituents)
- Experience and ingenuity of the designer in adapting the process to the case at hand
- Experience and method of dealing with impurities that may be present, such as COS, NH_3, aromatic hydrocarbons, etc.
- Experience with regard to corrosion, foaming, or other operating problems
- Cost of initial solvent charge
- Cost of replacement solvent—as influenced by process temperatures, vapor pressure of solvent, solvent stability, and fugitive losses
- Energy and/or stripping gas requirements
- Process royalty cost

Depending upon the extent of process data available, a comparison may be made of the carrying capacities of the various solvents. **Table 14-4,** which is based on the data of Bucklin and Schendel (1985), presents gas solubility data for three different processes: Fluor Solvent (propylene carbonate), Selexol (polyethylene glycol dimethyl ether), and Purisol (n-methyl-2-pyrrolidone). Solubilities of CO_2 and H_2S as well as hydrocarbons and other gases are shown, with all data collected at 25°C. The solubilities shown are single component data. In real systems there can be substantial interactions between the solutes, the net effect of which is usually to decrease the CO_2 and H_2S solubilities and to increase the hydrocarbon solubilities. Also, since all processes do not operate at the same temperature, the relationship between solubility and temperature must also be included in the evaluation.

The comparative data in **Table 14-4** show that, at equal temperatures, all three solvents have nearly the same CO_2 capacity. For H_2S, however, Selexol and Purisol have capacities about three times that of Fluor Solvent. For this reason, it is clear that Selexol or Purisol would be preferred to Fluor Solvent for cases where the feed gas contains a substantial concentration of H_2S or where selective H_2S removal is required. However, when CO_2 dictates the design, i.e., when H_2S is absent or present only in trace amounts, Fluor Solvent has an inherent advantage since hydrocarbon solubilities are substantially less in it than in the other two solvents.

Table 14-5 provides some key physical property data for DMPEG, NMP, PC, and methanol. Vapor pressure versus temperature data for DMPEG, NMP, methanol, and MPE are plotted in **Figure 14-3.** Methanol has by far the highest vapor pressure and requires very low temperature operation. NMP has a much higher vapor pressure than DMPEG, PC, or MPE and normally requires water washing of the process offgases to limit solvent losses. The Selexol, Fluor Solvent, and Sepasolv MPE processes require no water wash steps. The

Table 14-4
Gas Solubility Data for Selexol, Purisol, and Fluor Solvent Process Absorbents.
Volume Gas/Volume Liquid @ 25°C and 1 Atm.

Gas	Selexol (DMPEG)	Purisol (NMP)	Fluor Solvent (Propylene Carbonate)
H_2	0.047	0.020	0.027
N_2	—	—	0.029
CO	0.10	0.075	0.072
C_1	0.24	0.26	0.13
C_2	1.52	1.36	0.58
CO_2	3.63	3.57	3.41
C_3	3.70	3.82	1.74
iC_4	6.79	7.89	3.85
nC_4	8.46	12.4	5.97
COS	8.46	9.73	6.41
iC_5	16.2	—	11.9
NH_3	17.7	—	—
nC_5	20.1	—	17.0
H_2S	32.4	36.4	11.2
nC_6	39.9	—	46.0
CH_3SH	82.4	121	92.7

Data from Bucklin and Schendel (1985)

Table 14-5
Solvent Comparative Data

Process Solvent	Selexol (DMPEG)	Purisol (NMP)	Fluor Solvent (PC)	Rectisol (Methanol)
Vapor Pressure, mm Hg @ 25°C	$.073 \times 10^{-2}$	40×10^{-2}	8.5×10^{-2}	—
Viscosity, cp @ 25°C	5.8	1.65	3.0	0.6
Maximum feasible operating temperature, °C	175	—	65	—
Density, kg/m^3 @ 25°C	1,030	1,027	1,195	785
Boiling point, °C	240	202	240	65
Freezing point, °C	−28	−24	−48	−92
Molecular weight	280	99	102	32
Specific heat @ 25°C, Btu/(lb)(°F)	0.49	0.40	0.339	0.556
Thermal conductivity, Btu/(hr)(ft^2)(°F/ft)	0.11	0.095	0.12	0.122

Data from Bucklin and Schendel (1985) and Ranke and Mohr (1985)

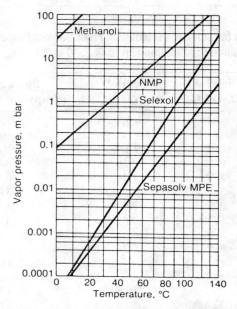

Figure 14-3. Vapor pressure of various physical solvents (*Wölfer, 1982*). *Courtesy of Hydrocarbon Processing*

Selexol solvent has the highest viscosity, which can significantly affect mass and heat transfer as the solvent temperature is reduced. Selexol, however, is suitable for operation at temperatures up to 175°C, while, for stability reasons, Fluor Solvent (propylene carbonate) is limited to a maximum operating temperature of about 65°C.

SIMPLE PHYSICAL SOLVENT PROCESSES

Fluor Solvent Process

The Fluor Solvent process, which is licensed by Fluor Daniel, Inc., was introduced in 1960 (Kohl and Buckingham, 1960). Although several solvents were covered by U.S. patents, only propylene carbonate has been used commercially. The process has been applied in 13 commercial installations—nine processing natural gas, two ammonia synthesis gas, and two hydrogen.

Basic Data

Selected physical properties of propylene carbonate are tabulated in **Tables 14-5** and **14-6.** Data on the solubility of various gases in propylene carbonate have been reported by several investigators (Dow Chemical Company, 1962; Schmack and Bittrich, 1966; Makranczy et al., 1965; Bucklin and Schendel, 1985). The equilibrium solubilities of hydrogen sulfide and carbon dioxide as a function of pressure are shown in **Figure 14-4.** Although there is some scattering of points, it is evident that the solubilities of both acid gases follow Henry's law

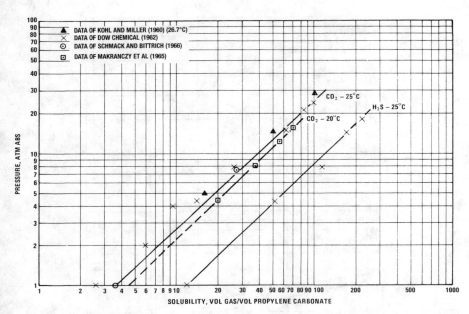

Figure 14-4. Effect of pressure on the solubility of carbon dioxide and hydrogen sulfide in propylene carbonate (gas volumes at 0°C and 760 mm Hg).

Table 14-6
Physical Properties of Propylene Carbonate

Empirical formula	$C_4H_6O_3$
Molecular weight	102.09
Vapor Pressure at 60°F	0.046 mm Hg
Viscosity at 60°F	2.6 centistokes
at 0°F	6.4 centistokes
Water Solubility in Solvent @ 25°C	94 g/l
Solvent solubility in water @ 25°C	236 g/l
CO_2 solubility in solvent @ 25°C, 1 atm	0.455 ft³/US gal

Sources: Dow Chemical Canada, Ltd. (1962); Fluor Daniel (1993); Bucklin and Schendel (1985)

up to a pressure of about 20 atm. The effect of temperature is shown for carbon dioxide and hydrogen in **Figure 14-5**. It is interesting to note that the solubility of hydrogen increases with increasing temperature.

No data appear to be available in the open literature on the specific effects of dissolved carbon dioxide, hydrogen sulfide, and other gases on the solubility of the individual compo-

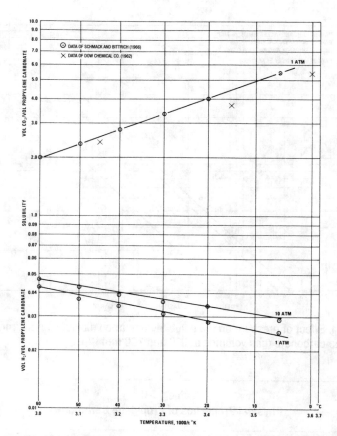

Figure 14-5. Effect of temperature on the solubility of hydrogen and carbon dioxide in propylene carbonate (gas volumes at 0°C and 760 mm Hg).

nents. However, the sketchy information available on gas mixtures containing carbon dioxide and methane indicates that at high partial pressures of methane the solubility of carbon dioxide is somewhat reduced, while the solubility of methane is appreciably increased by the presence of dissolved carbon dioxide (Dow Chemical, 1962; Makranczy et al., 1965). In view of the low solubility of hydrogen in propylene carbonate, it is reasonable to assume that its presence has no significant effect on the solubility of carbon dioxide or hydrogen sulfide. Solubilities of several gases, in terms of Bunsen coefficients (volume of gas at 0°C and 760 mm Hg per volume of liquid), are shown in **Table 14-7**. Additional solubility data are included in **Table 14-4**.

Process Design and Operation

The Fluor Solvent process uses flash regeneration as depicted in **Figure 14-1(A)**. Commercial plants utilize several flashes of the rich solvent at decreasing pressure levels with recycling of the gas evolved in the high pressure flash. Lean solvent temperatures as low as

Table 14-7
Solubility of Gases in Propylene Carbonate

Solute	Bunsen Coefficient	Reference
H_2	0.03	A
O_2	0.09	B
N_2	0.09	B
CO	0.06	B
COS	6.25	B
C_2H_2	8.6	C
CH_4	0.13	B
C_2H_6	0.58	B
CO_2	3.22	D
H_2S	10.6	D

Data sources: A) Schmack and Bittrich (1966); B) Fluor Daniel (1993); C) Dow Chemical Co. (1962); D) Bucklin and Schendel (1985)

0°F have been used in this process. A modification of the process resulting in an appreciable reduction of hydrocarbon gas loss has been described by Freireich and Tennyson (1977). In this scheme, the gas from the second intermediate pressure flash is washed with solvent in a small absorber. The overhead from this absorber is recompressed and recycled to the plant inlet. It is claimed that the value of the recovered gas amply compensates for the cost of the additional equipment, and that a payout as low as three months is realized with a fuel cost of $2 per MMbtu.

The operation of the Fluor Solvent process in the initial commercial plant has been described by Buckingham (1961). This author has also reported performance data for four plants treating natural gas (Buckingham, 1964). Data from these plants are given in **Table 14-8**.

An economic study comparing the Fluor Solvent process with the activated hot potassium carbonate process for the removal of carbon dioxide from synthesis gas for the production of ammonia and urea has been reported by Cook and Tennyson (1969). The authors concluded that the process is more economical than activated hot potassium carbonate in all cases stud-

Table 14-8
Operating Data of Fluor Solvent Process Plants

Plant	Design feed gas rate MMscfd	Feed gas composition		Absorption pressure, psig	Acid gas partial pressure in feed, psia	Sales gas specifications	
		CO_2	H_2S			CO_2	H_2S
A	220	53%	3 gr/100 scf	850	CO_2: 458	2%	0.25 gr/100 scf
B	10	17%	—	450	CO_2: 79	5%	—
C	20	22.8%	—	800	CO_2: 186	1%	—
D	28	10–30%	5–15%	1,000	$CO_2 + H_2S$: 250–300	0.1%	0.8 gr/100 scf

Source: Buckingham (1964)

1202 Gas Purification

ied, which included production of ammonia and urea by steam reforming of natural gas and naphtha and by partial oxidation.

A comparison of Fluor Solvent to other physical solvent processes (Bucklin and Schendel, 1985) illustrates that for cases where CO_2 removal requirements dictate the plant design, i.e., where little or no H_2S is present, the Fluor Solvent process enjoys a significant advantage over the other processes due to lower solubilities of the gas being purified, light hydrocarbons in the case of natural gas and H_2 in the case of synthesis gas. This advantage is evidenced by either a higher percent hydrocarbon or hydrogen recovery or by lower compression requirements for gas flashed from the rich solvent at intermediate pressures.

A photograph of a Fluor Solvent plant processing natural gas is shown in **Figure 14-6.**

Selexol Process

The Selexol process uses a physical solvent and, as a result, is generally similar to other processes discussed in this chapter. The process was originally developed by Allied Chemical Corporation. In 1982, the Norton Company purchased the rights to the process; in 1990,

Figure 14-6. Fluor Solvent process plant treating high pressure natural gas. *Fluor Daniel and El Paso Natural Gas Company*

Union Carbide acquired the process from Norton; and, in 1993, UOP acquired the rights to the Selexol process and is now responsible for process licensing.

The Selexol process has found a very wide range of applications. It was originally used to remove CO_2 from an ammonia plant in Nebraska, followed soon after by H_2S and CO_2 removal from natural gas in the U.S. and in Europe. Other applications include desulfurization and CO_2 removal from synthesis gas derived from the partial oxidation of heavy petroleum stocks and from coal gasification. Natural gas treating applications include several, where in addition to production of pipeline specification gas, a relatively pure stream of carbon dioxide is produced for reinjection into oil formations, so-called enhanced oil recovery or EOR. A relatively new use for the process dating back to 1979 is the purification of landfill gas drawn from the biological degradation of municipal waste in sanitary landfills. This application is characterized by the occurrence of chlorinated and aromatic hydrocarbons as impurities in the landfill gas.

Union Carbide reported that as of 1992, a total of 53 Selexol plants had been installed. These comprise 10 for CO_2 removal from various synthesis gases, 12 for CO_2 removal from natural gas, 15 for selective H_2S removal (with or without CO_2 removal), 8 for desulfurization of synthesis gas, and 8 for landfill gas purification (Epps, 1992A).

This process has been described quite extensively in the literature (Hegwer and Harris, 1970; Sweny and Valentine, 1970; Sweny, 1973; Valentine, 1974; Clare and Valentine, 1975; Valentine, 1975; Sweny, 1976; Raney, 1976; Van Deraerschot and Valentine, 1976; Judd, 1978; Swanson, 1978; Sweny, 1980; Hernandez and Huurdeman, 1989; Epps, 1992B).

Basic Data

The treating solution used in the Selexol process is a mixture of homologues of the dimethylether of polyethylene glycol. This material has been shown to be chemically stable as well as non-toxic and biodegradable. It has a very low vapor pressure as well as a high capacity for various impurities, including H_2S, CO_2, COS, mercaptans, and others.

The relative solubilities of various gases in Selexol solvent as compared to methane are shown in **Table 14-9** (Shah and McFarland, 1988; Epps, 1992B). The data in this table indicate that H_2S is almost 9 times as soluble as CO_2. This high ratio facilitates the use of Selexol for the selective removal of H_2S from gas streams also containing CO_2. The data also show that hydrocarbons are quite soluble and their solubility increases with increasing molecular weight. Propane, for example, has a relative solubility of about 15.4, which is similar to that of CO_2; while hexane, with a relative solubility of 167, is somewhat more soluble than H_2S. Water is more soluble than any of the listed compounds except HCN, and liquid water is miscible with the Selexol solvent. These solubility characteristics have led to the development of Selexol solvent applications for removing hydrocarbons and/or water from natural gas, e.g., simultaneous hydrocarbon and water dew point control (Epps, 1994). Actual solubility values for a number of components in the Selexol solvent (and in two other solvents) are given in **Table 14-4** in terms of volume of gas per volume of liquid at 25°C and 1 atm partial pressure. The effect of partial pressure on solubility is shown in **Figure 14-7** for several gases. Physical properties of the Selexol solvent are given in **Table 14-5**. The solvent's flash point is reported to be about 304°F (Sweny and Valentine, 1970).

Process Design and Operation

The basic flow scheme of the process is very simple, requiring only an absorption stage and regeneration by flashing at successively decreasing pressure levels as depicted in **Figure**

Table 14-9
Relative Solubilities of Various Gases in SELEXOL Solvent

Component	$R = \dfrac{K' \text{ Methane}}{K' \text{ Component}}$	Component	$R = \dfrac{K' \text{ Methane}}{K' \text{ Component}}$
H_2	0.20	NH_3	73
N_2	0.30	nC_5	83
CO	0.43	H_2S	134
C_1	1.0	C_6	167
C_2	6.5	CH_3SH	340
C_2H_4	7.2	C_7	360
CO_2	15.2	CS_2	360
C_3	15.4	SO_2	1,400
iC_4	28	C_6H_6	3,800
COS	35	CH_2Cl_3	5,000
nC_4	36	C_4H_4S	8,200
iC_5	68	H_2O	11,000
C_2H_2	68	HCN	19,000

$K' = y/x'$ where y is the mole fraction of the component in the vapor phase and x' is the mole fraction of the component in the liquid phase considering only the solvent and the component.
Sources: Shah and McFarland (1988); Epps (1992B)

14-1(A). If the objective is bulk removal of carbon dioxide from a gas stream, successive flashes are all that is required for regenerating the solution. More extensive regeneration can be achieved by vacuum flashing, by stripping with air or an inert gas, or by application of heat. In the case of hydrogen sulfide removal, stripping with inert gas or heat is the usual procedure.

If selective removal of sulfur compounds is required, the Selexol plant consists of an absorber, a flash, and a steam-heated stripping column [see **Figure 14-1(C)**]. The flashed gas is compressed and recycled to the absorber inlet, and the regenerator effluent is processed for the production of elemental sulfur.

In cases where both selective sulfur removal and complete carbon dioxide removal are required, such as with coal-derived substitute natural gases, two successive independent absorption-regeneration cycles are used as shown in **Figure 14-8.** Various flow schemes are discussed in some detail by Van Deraerschot and Valentine (1976) and Sweny (1980). These authors claim that, with proper plant design, removal of sulfur compounds to concentrations as low as a few parts per million can be achieved with relatively low co-absorption of carbon dioxide.

Operating results from three plants treating natural gas containing various amounts of hydrogen sulfide and carbon dioxide (Hegwer and Harris, 1973) are summarized in **Table 14-10.**

In Plant A, the bulk of the solvent is regenerated by flashing at three pressure levels (400 psig, 200 psig, and atmospheric), and a side stream is further regenerated by an additional atmospheric flash at elevated temperature. A split-flow circuit is used, with the semistripped solvent and the completely stripped solvent being fed to the absorber at different points.

Physical Solvents for Acid Gas Removal **1205**

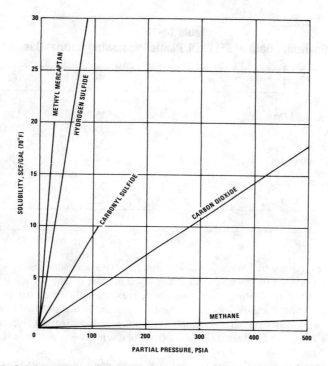

Figure 14-7. Solubility of gases in SELEXOL solvent. *Data of Sweny and Valentine (1970)*

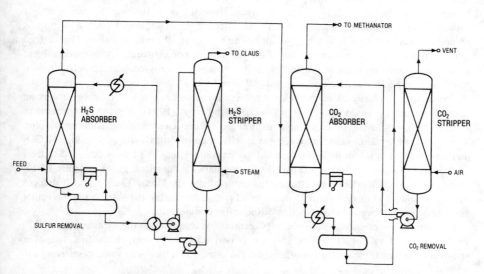

Figure 14-8. Flow diagram of SELEXOL process for selective hydrogen sulfide removal and complete carbon dioxide removal.

Table 14-10
Operating Data of SELEXOL Plants Processing Natural Gas

Plant	A	B	C
Feed gas:			
Volume, MMscfd	275	250	130
Pressure, psig	1,000	1,000	1,000
CO_2,%	43	3.5	18
H_2S, gr/100 scf	1	8	8
Treated gas:			
CO_2,%	3.5	3.0	2.5
H_2S, gr/100 scf	0.25	0.25	0.25

Data of Hegwer and Harris (1970)

Extensive power recovery by hydraulic turbines and gas expanders supplies the total pumping energy required. A single stream circuit is employed in Plant B. The solvent is regenerated by flashing at 300 psig, 175 psig, and atmospheric pressure and then by stripping with air at elevated temperature. The flow scheme of Plant C is similar to that of Plant B, except that a split-flow cycle for semistripped and completely stripped solvent is used. Solvent flashing is carried out at 250 psig, 190 psig, and atmospheric pressure. Power recovery is also practiced in Plants B and C.

A Selexol unit that is integrated into the design of an ammonia plant located in the Netherlands is described by Hernandez and Huurdeman (1989). Since minimum steam consumption is a key requirement, air stripping of the Selexol solvent is employed. The flow diagram is depicted in **Figure 14-9**. In this arrangement, three stages of flashing, including one at subatmospheric pressure, are used prior to air stripping. Lean solution from the stripper is cooled by heat exchange with a refrigerant before it is pumped to the absorber. The air feed to the stripper is cooled using a portion of the partially stripped solvent. The warmed solvent heats the stripper vent gas which increases its enthalpy, thereby conserving refrigeration. Some data on plant operation are provided in **Table 14-11**.

Commercial applications of the Selexol solvent for simultaneous hydrocarbon dew-point control and natural gas dehydration are described by Epps (1994). A plant design used in several European installations pretreats natural gas before it enters a molecular sieve unit. The design is intended to meet a treated gas specification of a maximum of 0.50 mole% CO_2 and a maximum of 6.5 mole% ethane and heavier components. A plant is designed to treat 26 MMscfd of gas at 32°F and 603 psia. Operating data for this plant, given in **Table 14-12**, show that it meets the CO_2 and ethane-plus removal specifications. The plant also reduces the water content of the gas from 75 ppmv to 12 ppmv, decreasing the load on the molecular sieve unit, and removes a major fraction of the sulfur components.

Commercial experience with the Selexol technology as applied to the treatment of landfill gas is described by Epps (1992B). This gas is derived from municipal landfills and, after removal of potentially harmful impurities, the gas is used as a fuel or in some cases is sold to natural gas distributors.

Typical landfill gas compositions cited by Epps (1992B) are shown in **Table 14-13**. Oxygen is present because air is drawn into the landfill due to the slight vacuum that occurs

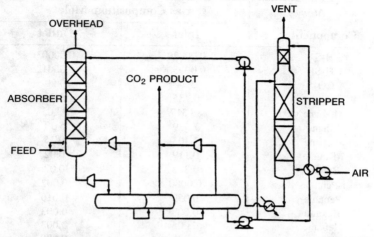

Figure 14-9. Flow diagram of SELEXOL process for ammonia synthesis gas treating (*Hernandez and Huurdeman, 1991*)

Table 14-11
Ammonia Plant SELEXOL Unit Operating Data

	Design	Operating
Plant capacity, NH_3 t/d	1,360	1,440
Feed gas CO_2 content, mol%	18.2	18.1
Treated gas CO_2 content, ppmv	1,000	440
CO_2 product recovery, % of feed	81	82
CO_2 product purity, mol%	99.0	99.4
CO_2 product temperature, °F	63	43
CO_2 product pressure, psia	21.6	23.3
Solvent losses, lbs/year		11,000

Data of Hernandez and Huurdeman (1989)

within the gas gathering system. The primary breakdown products of the municipal waste, largely paper, cardboard, and other organic materials, are methane and carbon dioxide. Small but significant amounts of H_2S are also formed.

Heavy and aromatic hydrocarbons as well as chlorinated hydrocarbons are generated by the municipal waste due to biodegradation and vaporization. If not removed from the landfill gas, these impurities can result in corrosion, scaling, and other problems and would also present environmental problems if discharged to the atmosphere.

As depicted in **Figures 14-10** and **14-11,** the Selexol process used for landfill gas purification consists of two stages. In Stage 1 all impurities except CO_2 are removed completely and medium BTU gas is produced. This gas is suitable for local use as fuel for boilers or gas tur-

Table 14-12
Inlet and Outlet Gas Compositions for SELEXOL Unit Pretreating Natural Gas

Component	Gas Composition, Mole%	
	Inlet	Outlet
H_2S	0.00020	0.00003
CH_3SH	0.00050	0.00012
CO_2	2.44	0.29
N_2	0.785	0.88
Methane	88.317	93.02
Ethane	7.539	5.33
Propane	2.403	0.35
i-Butane	0.119	0.02
Butane	0.303	0.07
i-Pentane	0.0080	0.002
Pentane	0.059	0.016
Hexane	0.008	0.003
Heptane	0.005	0.003
n-Decane	0.002	0.000
Water	0.008	0.001

Data of Epps (1994)

Table 14-13
Typical Landfill Gas Composition

Component	Range	Median
Methane, Mole%	41.2–60	53.4
Carbon Dioxide, Mole%	30.9–46.1	40.7
Nitrogen, Mole%	0.2–22.6	4.5
Oxygen, Mole%	0.01–5.2	0.35
Hydrogen Sulfide, ppmv	4.0–99	30.0
Impurities*, ppmv	72.0–2,000	181.0

*Impurities

Methylmercaptan	Decane	1,1 - Dichloroethane
Ethylmercaptan	Undecane	1,2 - Dichloroethane
Carbonyl Sulfide	Dodecane	Benzene
Carbon Disulfide	1, 1 - Dichloroethylene	Toluene
Methylene Chloride	Trichloroethylene	o,m,p - Xylene
Ethane	Perchloroethylene	Acetonitrile
Propane	Tetrachloroethylene	1,2 - Dibromomethane
Butene	Chlorobenzene	Benzyl Chloride
Pentanes	Ethylbenzene	Chloromethane
Hexane	Vinyl Chloride	1,2 - Dichloroethane
Heptane	Ethylene Dibromide	1,1,1 - Trichloroethane
Octane	o,m,p - Dichlorobenzene	
Nonane		

Data of Epps (1992B)

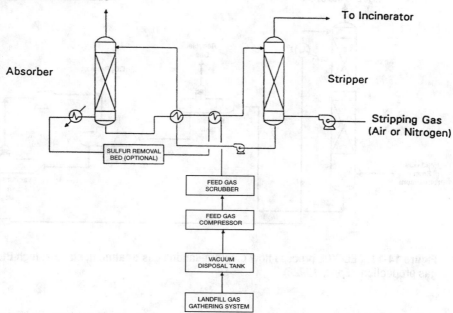

Figure 14-10. SELEXOL process flow diagram, landfill gas treatment; stage 1, medium Btu gas production. (*Epps, 1992B*)

bines. If high-BTU gas is required to meet pipeline specifications, it is necessary to remove the CO_2 as well as the other impurities. This is accomplished in Stage 2.

Figure 14-10, which represents Stage 1 purification, shows an optional fixed bed sulfur removal unit, such as iron sponge, to remove H_2S from the gas before it contacts the Selexol solvent. The solvent removes the heavy and chlorinated hydrocarbons since their solubilities in the solvent are exceedingly high, in some cases over an order of magnitude higher than the H_2S solubility. Air is used to strip absorbed hydrocarbons from the solvent and the resulting impure air is incinerated. If the H_2S is not removed ahead of the absorber, it will dissolve in the solvent and be oxidized in the air stripper, producing elemental sulfur and causing operating problems within the system. If steam stripping is used instead of air, it is possible to eliminate the fixed bed H_2S removal step, in which case most of the H_2S is stripped from the solvent by the steam and must be disposed of by incineration or other means. If the feed gas contains oxygen, some of the H_2S may be oxidized to elemental sulfur, even when steam is used for stripping. The first stage serves to remove H_2S, heavy hydrocarbons, and chlorinated compounds from the feed gas. When medium BTU gas is desired, no further treatment is required.

When high BTU gas is required, the gas is usually compressed to a pressure in the range of 150 to 350 psig and subjected to a second stage of absorption for CO_2 removal. This is a conventional Selexol unit in which solvent regeneration is accomplished by flashing, using a final vacuum flash if required to achieve higher CO_2 removal. Direct discharge to the atmosphere of the CO_2 from the second stage is ordinarily practiced. Flash gas from the highest

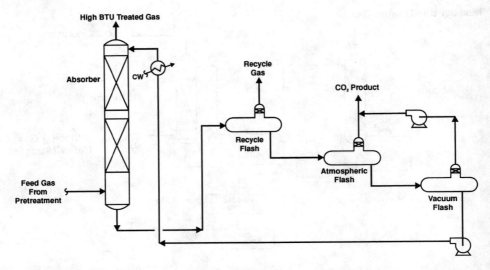

Figure 14-11. SELEXOL process flow diagram, landfill gas treatment; stage 2, high Btu gas production. (*Epps, 1992B*)

pressure flash step is recycled to improve overall methane recovery. This scheme is shown in **Figure 14-11**.

Sepasolv MPE Process

This process, which was developed by BASF of West Germany, is quite similar to the Selexol process, both with respect to the solvent used and the mode of operation. It was initially developed primarily for the selective removal of H_2S from natural gas, but reportedly is also suitable for CO_2 removal from synthesis gases. Two commercial Sepasolv MPE operating plants are described in the literature (Wölfer et al., 1980), but BASF (1992) reports that this process is no longer licensed.

The Sepasolv MPE solvent is described as a mixture of polyethylene glycol methyl isopropyl ethers with a mean molecular weight of about 316. Physical properties of the solvent are given in **Table 14-14**. Bucklin and Schendel (1985) and Wölfer et al. (1980) present gas solubility data that show MPE gas solubilities are similar to those of the Selexol solution.

Purisol Process

This process has been developed and commercialized by Lurgi GmbH of Frankfurt, Germany, and as of 1996 seven units were in operation or under construction (Lurgi Öl-Gas-Chemie GmbH, 1996). Discussions of the basic features of the process and of its application to the purification of natural gas, hydrogen, and synthesis gases are provided in several references: (Hochgesand, 1968; Hochgesand, 1970; Stein, 1969; Kapp, 1970; Beavon and Roszkowski, 1969; Lurgi, 1978; Lurgi, 1988A; Kriebel, 1989).

The solvent used in the Purisol process is N-methyl-2-pyrrolidone (NMP), a high boiling liquid, which has an exceptionally high solubility for hydrogen sulfide. NMP is particularly

Table 14-14
Physical Properties of Sepasolv MPE

Molecular Weight	316
Density, 20°C, g/cc	1.002
Specific Heat, 0°C, KJ/kgK	1.94
Specific Heat, 100°C, KJ/kgK	2.18
Viscosity, 20°C, in Pa. S	7.2
Viscosity, 0°C, in Pa. S	15.0
Freezing Point, °C	−25

Data of Wölfer (1982)

suitable for selective hydrogen sulfide absorption in the presence of carbon dioxide, since, as shown in **Table 14-4,** H_2S solubility is almost 9 times that of CO_2. NMP is also used in a number of other chemical processing applications, including acetylene recovery from pyrolysis gases, butadiene recovery from C_4 fractions, and as an extractive agent for aromatics recovery from oil refinery stocks. The Purisol process is reportedly capable of yielding gas streams containing less than 0.1 percent carbon dioxide and a few parts per million of hydrogen sulfide (Lurgi, 1978).

Basic Data

Physical properties of NMP are given in **Table 14-5** and in **Figure 14-12**. Equilibrium solubility data for carbon dioxide, hydrogen sulfide, methane, and propane are presented in **Table 14-15**. The data of Boston and Schneider (1971) were obtained with a gas mixture containing 97.10% methane, 1.02% propane, 1.17% carbon dioxide, and 0.71% hydrogen sulfide at a total pressure of 850 psia and a temperature of 74°F. Additional gas solubility data are given in **Table 14-4**.

Process Operation

As with other physical solvents, the optimum method for regenerating the solvent depends upon the purity required for the treated gas. In order of increasing severity, the applicable regeneration methods are flashing to atmospheric pressure, vacuum flashing, inert gas stripping, and regeneration by heating.

Grünewald (1989) reports that Purisol is effective in removing COS and that the removal can be enhanced by raising the absorption temperature and increasing the water concentration in the solvent. Both of these measures serve to promote the hydrolysis of COS to CO_2 and H_2S. The highest COS removal is obtained by the use of an undisclosed catalyst dispersed in the solvent and by increasing the residence time of the solvent in the absorber to further promote COS hydrolysis. NMP is somewhat more volatile than other physical solvents so water washing of gaseous effluents is required to minimize solvent losses.

The Purisol process is particularly well suited to the purification of high-pressure, high CO_2 synthesis gas for gas turbine integrated gasification combined cycle (IGCC) systems because of Purisol's high selectivity for H_2S. Extreme purity with regard to sulfur com-

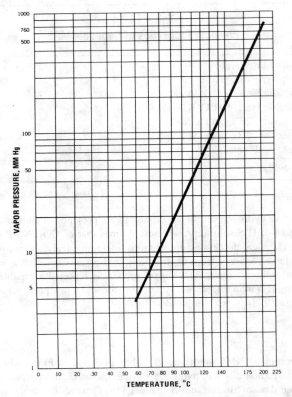

Figure 14-12. Vapor pressure of N-methyl-2-pyrrolidone. *Data of General Aniline and Film Corp.*

pounds is not normally required for such fuel gas, and carbon dioxide in the purified gas expands through the gas turbine to provide additional power.

Grünewald (1989) suggests the flow arrangement shown in **Figure 14-13** as optimum for the processing of gas turbine fuel gas for IGCC applications. Hydrogen sulfide and some carbon dioxide are removed from the raw gas in the H_2S absorber. The rich solution is flashed to a lower pressure and the flashed gas is passed through a reabsorber (for selective removal of H_2S) before it is recompressed and recycled to the main gas stream. The solution is then completely stripped of acid gas by a second (hot) flash step followed by a reboiled stripping operation. Gas from the hot flash step is purified and recycled, while the reboiled stripper offgas, which has a high H_2S concentration, is sent to an oxygen-blown Claus unit. The Claus unit is operated so that its tail gas contains H_2 and CO, as well as H_2O and CO_2 and any unconverted H_2S and SO_2. The tail gas is passed through a hydrogenation reactor where all sulfur compounds are converted to H_2S. It is then cooled to remove water and the remaining gas, consisting primarily of CO_2 with small amounts of H_2 and H_2S, is recycled through the reabsorber to the fuel gas stream. The net result of the process is selective hydrogen sulfide removal from the fuel gas and conversion of the H_2S to elemental sulfur without

Table 14-15
Equilibrium Solubility of Gases in N-Methyl-2-Pyrrolidone

Gas	Temperature, °C	Partial pressure, atm. abs.	Solubility, vol/vol*	Reference
CO_2	20	1.0	3.95	A
CO_2	23.5	0.67	2.0	C
CO_2	35	1.0	3.0	B
CO_2	35	10.0	32.0	B
H_2S	20	1.0	48.8	A
H_2S	23.5	0.41	14.3	C
H_2S	35	1.0	25.0	B
CH_4	20	1.0	0.28	A
CH_4	23.5	56	12.2	C
C_3H_8	23.5	0.59	1.9	C

*Gas volume at 0°C and 760 mm Hg.
Data sources: A) Stein (1969); B) Hochgesand (1970); C) Boston and Schneider (1971)

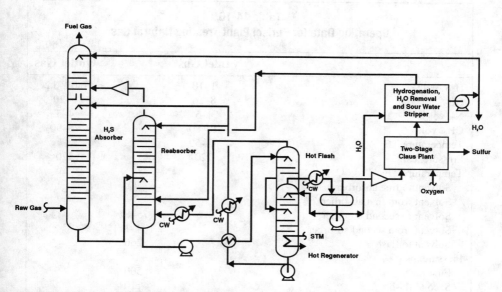

Figure 14-13. Proposed optimum configuration for combined cycle fuel gas desulfurization, incorporating a Purisol unit, a two-stage oxygen-blown Claus unit, and a Claus tail gas hydrogenation unit. (*Grünewald, 1989*)

the production of a Claus plant offgas stream or the use of a Claus tail gas unit with its associated tail gas incinerator.

Operation of a Purisol plant processing about 40 MMscfd of natural gas containing hydrogen sulfide and carbon dioxide has been described by Stein (1969). The principal objective in this installation is to remove a maximum of hydrogen sulfide with only partial removal of carbon dioxide. Absorption of the acid gases is carried out in two columns, operating in series, the first provided with trays and the second with packing. The solvent passes from the second to the first absorber and is then regenerated by flashing at three different pressure levels, augmented by stripping with inert gas at elevated temperature in the last flash. Cooling of the solvent is required between the two absorption columns and, of course, after regeneration. The gas released in the first flash is recompressed and returned to the inlet of the first absorber. The acid gases are processed in a Claus type sulfur recovery unit. Typical operating data from this installation are given in **Table 14-16**.

Examples of process conditions proposed by Lurgi (1978) for three cases involving (a) essentially complete carbon dioxide removal from high-pressure gases with high carbon dioxide content, (b) bulk removal of hydrogen sulfide from natural gas, and (c) selective removal of hydrogen sulfide from natural gas are presented in **Table 14-17**. The first case involves solvent regeneration by flashing and inert-gas stripping, see **Figure 14-1(B)**. In the second case, the solvent is simply flashed at three pressure levels. In the third case, requiring complete solvent regeneration, flashing and high temperature regeneration with reboiling are employed, see **Figure 14-1(C)**.

Table 14-16
Operating Data for Purisol Plant Treating Natural Gas

	Inlet Gas	Outlet Gas
H_2S, vol %	1–10	0.02–0.2
CO_2, vol %	8–26	6–20
N_2, vol %	4–5	4–5
CH_4, vol %	70–80	75–90
Temperature, °F	32–59	77
Pressure, psig	720	570
Temperature, °F:		
Solvent to first absorber	70	
Solvent from first absorber	86	
Solvent to second absorber	75–82	
Solvent from second absorber	104–111	
Solvent to flash	266	
Pressure, psig:		
First flash	500	
Second flash	210	
Third flash	7.5	

Data from Stein (1969)

Table 14-17
Typical Process Conditions for Acid-Gas Removal with Purisol Process

Case:		1	2	3
Feed gas:				
Volume, MMscfd		100	100	100
Pressure, psig		1,070	510	1,070
Temperature, °F		110	80	80
H_2	vol.%	64.53	—	—
CO_2	vol.%	33.15	1.0	15.0
H_2S	vol.%	—	34.0	6.0
CO	vol.%	1.50	—	—
CH_4	vol.%	0.44	63.7	75.0
C_2H_6-C_4H_{10}	vol.%	—	1.1	—
$C_6H_{12}^+$	vol.%	—	0.2	—
N_2	vol.%	0.38	—	4.0
Treated gas:				
H_2	vol.%	96.44	—	—
CO_2	vol.%	0.10	1.2	13.6
H_2S	vol.%	—	2.0	2 ppm
CO	vol.%	2.24	—	—
CH_4	vol.%	0.59	95.4	82.0
C_2H_8-C_4H_{10}	vol.%	—	1.4	—
$C_6H_{12}^+$	vol.%	—	—	—
N_2	vol.%	—	0.63	4.4
Utilities:				
Power, kW*		2,100	1,600	1,100
Steam, (45 psig), lb/hr		3,750	1,500	13,000 (60 psig)
Cooling water (75°F), gpm		1,300	750	820
Condensate, lb/hr		2,850	2,000	2,200
NMP loss, lb/hr		6.5	11	9

*Without power recovery.
Data source: Lurgi GmbH (1978)

Rectisol Process

The Rectisol process, which uses methanol as its solvent, was the first physical organic solvent process. Although it is a true physical solvent process, it has a number of features that set it apart from the other physical solvent processes. First, Rectisol has the demonstrated ability to separate troublesome impurities that are produced in the gasification of coal or heavy oil, including hydrogen cyanide, aromatics, organic sulfur compounds, and gum-forming hydrocarbons. The use of methanol also facilitates dehydration and the prevention of ice and hydrate formation at the low temperatures used in the process. The second distinguishing

1216 Gas Purification

feature is that the Rectisol process operates at much lower temperatures than other physical solvent processes—with operating temperatures as low as −75° to −100°F. At these temperatures, methanol still has a low viscosity so that mass and heat transfer are not significantly impaired, and the solvent's carrying capacity for both CO_2 and H_2S becomes very high, considerably higher than that of other physical solvents at their typical operating temperatures. These features lead to the ability to achieve very sharp separations, with H_2S concentrations of typically 0.1 ppm and CO_2 concentrations of just a few ppm in the treated gas. Likewise, Rectisol can achieve concentrated H_2S streams suitable for Claus plant feeds and CO_2 off-gases essentially free of H_2S.

Operation at very low temperatures with very sharp separations results in relatively complex flow schemes. This, combined with the need for low level refrigeration, leads to high plant costs. As a result, most applications of the Rectisol process represent relatively difficult gas treating conditions where other gas treating processes are not suitable for one reason or another. Typical applications are the purification of gas streams in the heavy oil partial oxidation processes of Shell and Texaco and the Lurgi coal gasification process, as used at the Sasol plants in South Africa.

The Rectisol process was initially developed in Germany by Lurgi GmbH. It was developed further jointly with Linde AG (Kriebel, 1989) and is now offered by both firms. The major use of the process is in coal- and heavy oil-based facilities to produce ammonia, methanol, hydrogen, SNG, Fischer-Tropsch liquids, and oxo alcohols. In 1996 it was reported that more than 100 units were in operation or under construction (Lurgi Öl-Gas-Chemie GmbH and Linde AG, 1996).

A large industrial Rectisol plant used for the purification of gas obtained by coal gasification in Lurgi gasifiers at the first Fischer-Tropsch plant of South African Oil, Coal, and Gas Corporation (SASOL) in South Africa has been described (Hoogendoorn and Solomon, 1957; Ranke, 1973). This plant, which was installed in the 1950s, consists of three identical purification units with a total capacity of 164 MMscfd of gas and a common regeneration section. In the 1970s, two much larger coal gasification facilities were installed by Sasol, with each Rectisol purification system treating over 1 billion standard cubic feet per day of gas. Each of these Rectisol facilities consists of four independent trains. Another large Rectisol facility was installed at the Great Plains gasification plant in North Dakota and was designed to produce 137.5 MMscfd of substitute natural gas (SNG) from lignite.

Basic Data

Physical property data for methanol are given in **Table 14-5.** The solubility of carbon dioxide in methanol as a function of temperature at a partial pressure of one atmosphere is shown in **Figure 14-14** (Herbert, 1956). The effect of partial pressure on the equilibrium solubility at two temperatures is shown in **Figure 14-15** (Hochgesand, 1968). Equilibrium solubilities of H_2S and CO_2 at two temperatures are reported by Hochgesand (1970) and are presented in **Table 14-18.** Additional data on the effects of temperature on the solubilities of gases in methanol are provided by **Figure 14-2.** The vapor pressure of methanol (Dreisbach, 1952) is given in **Figure 14-16.** The high vapor pressure of methanol results in relatively high vapor pressure losses and correspondingly high methanol makeup requirements.

Process Description and Operation

The Rectisol process is available in a wide variety of configurations to meet specific requirements and feed conditions. Systems can be designed to (1) remove all impurities, pro-

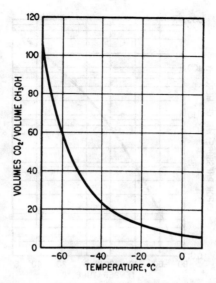

Figure 14-14. Solubility of carbon dioxide in methanol, partial pressure of carbon dioxide = 1 atm. *Data of Herbert (1956)*

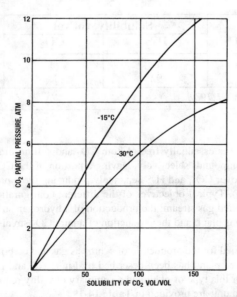

Figure 14-15. Effect of partial pressure on solubility of carbon dioxide in methanol. *Data of Hochgesand (1968)*

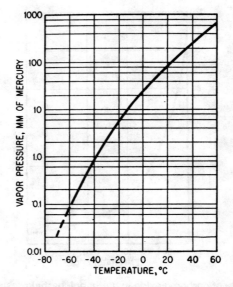

Figure 14-16. Vapor pressure of methanol. *Data of Dreisbach (1952)*

**Table 14-18
Equilibrium Solubilities of H_2S and CO_2 in Methanol**

Temperature, °C	Solubility, vol/vol		Selectivity H_2S/CO_2
	H_2S	CO_2	
−10	41	8	5.1
−30	92	15	6.1
Source: Data of Hochgesand (1970)			

ducing a treated gas that is essentially free of both CO_2 and sulfur compounds; (2) remove H_2S and other sulfur compounds selectively, leaving a portion of the CO_2 in the treated gas; and (3) remove and recover CO_2, and H_2S separately producing three product streams (purified gas, CO_2, and H_2S). Typical objectives of the different configurations include the production of a highly purified gas stream; the production of a hydrogen sulfide-rich gas stream suitable as feed to a Claus plant; and the production of a pure CO_2 stream that can be used in the synthesis of urea.

A Rectisol unit designed for the production of synthesis gas and carbon dioxide, which are suitable as feeds to a urea plant, has been described by Linde AG and Lurgi GmbH (1992). The material balance for this type of plant and a summary of utility requirements based on a 1,000 tons/day ammonia unit are provided in **Table 14-19.**

Two basic configurations of the Rectisol process are the nonselective standard process and the selective version. In the nonselective or single-step process the CO_2 and H_2S are

Table 14-19
Rectisol Process Material Balance and Utility Requirements

Material Balance, Mol%

	Feed	Purif. Gas	CO_2	Tail Gas	H_2S Stream
H_2	62.47	95.27	0.84	0.07	0.31
N_2 + Ar	0.51	0.76	0.08	25.32	4.36
CO + CH_4	2.67	3.97	0.33	0.04	0.08
CO_2	34.10	20 ppm	98.75	74.57	68.01
H_2S + COS	0.25	0.1 ppm	2 ppm	5 ppm	27.24

Conditions

Flow, kmol/h	6,021.1	3,936.1	1,283.9	1,003.6	54.2
Pressure, bar	78	76	1.8	1.1	2.5

Utility Requirements

Power (shaft, without power recovery), kW	1,100
Steam (6 bar), lb/h	14,300
Refrigerant (at 235°K), Btu/h	6.75×10^6
Cooling water (Δt = 10°C), gpm	1,320
Stripping gas (4 bar), kmol/h	256.7
Methanol loss, lb/h	66

Basis: 1,000 ton/d ammonia plant (Linde AG and Lurgi GmbH, 1992)

absorbed simultaneously; whereas, in the selective or two-step design H_2S is removed in the first step followed by CO_2 removal in the second step. The majority of commercial applications employ the standard configuration.

Many of the operating details of the commercial facilities are held confidential by the licensors. However, good general descriptions are available in the previously cited articles, and in a non-confidential DOE document (Miller and Lang, 1985) describing the Great Plains SNG-from-coal plant in North Dakota. The description which follows is taken from this latter reference.

Standard Rectisol Plant

In the North Dakota plant, lignite is gasified using Lurgi technology. After removal of heavy impurities such as coal tar and particulate solids from the gasifier effluent and a water gas shift conversion step, the Rectisol process is used to remove naphtha, H_2S, CO_2, and other contaminants such as HCN, mercaptans, COS, and organic sulfur impurities. The process gas then undergoes methanation, which produces methane from H_2 and carbon oxides to yield pipeline specification gas referred to as SNG (synthetic or, alternatively, substitute natural gas).

The Great Plains Rectisol unit consists of two identical absorption and regeneration trains with a common naphtha extraction and methanol recovery train. The basic flow scheme is shown in **Figure 14-17**. The composition of the feed gas to the Rectisol plant at the Great Plains plant is 39% CO_2 and 0.35% H_2S. Feed gas pressure to the Rectisol unit is close to 400 psig, while the combined feed gas flow for both Rectisol trains is 556 MMscfd.

The feed gas is first chilled by heat exchange with process off-gas streams and then by ammonia refrigeration to a temperature between −30° and −55°F. The feed gas then passes to the prewash and absorber column. The prewash and absorber column is divided into three major sections: the bottom prewash, the middle main absorption, and the top final absorption sections. The middle main absorption section is divided into upper and lower segments by a chimney tray. The bottom or prewash section removes naphtha and other relatively heavy impurities, while H_2S, CO_2, and most of the lighter impurities are absorbed in the middle main absorption sections. The top final absorption section removes the residual traces of CO_2 and H_2S from the product gas. Methanol is used in all three sections to accomplish these separations.

The leanest methanol, from the hot regenerator column, is fed to the top final absorption section of the prewash and absorber column. Most of the methanol from the flash regenerator is returned to the upper main absorption section. Methanol from this upper main absorption section is withdrawn from the prewash and absorber column and chilled by ammonia refrigeration. Most of this methanol is recycled back to the top tray of the lower main absorption section. This removes the heat of absorption and permits higher solution loadings. A small slipstream of the rich, chilled methanol is used for heavy impurity removal in the bottom prewash section.

The treated gas with H_2S and CO_2 concentrations reduced to 0.2 ppm and 1.5 volume percent, respectively, is used to precool the feed gas, and then passes to the battery limits enroute to methanation. With the design feed gas, both Rectisol trains produce 355 MMscfd of product gas which, after methanation, yields 137 MMscfd of SNG. The rich methanol from the main absorption section of the prewash and absorber column is chilled by ammonia refrigeration and flows to a series of six flash regeneration steps, A through F. In these six flash regeneration stages, the rich solvent is flashed at successively lower pressures with the final flash at vacuum conditions. Offgas from the first flash, containing significant amounts of H_2, CO, and CH_4, is recycled to recompression facilities for eventual recovery of these valuable components. Flash gas from the other flashes is sent to a Stretford unit for removal of H_2S. Gas from the three lowest pressure flashes requires compression to 8 psig before Stretford treating.

Some of the methanol from the flash regenerator is regenerated in the hot regenerator column where H_2S is stripped to very low residual levels. This tower is reboiled with low pressure steam; while the overhead gas stream containing acid gases is washed with steam condensate for methanol removal, then combined with the flash regenerator offgas, and directed to the Stretford plant for H_2S removal.

Methanol from the prewash section of the prewash and absorber column, which contains naphtha and other impurities, is flashed in the prewash flash vessel for removal of dissolved gases, which also pass to the Stretford unit. The flashed methanol then flows to the naphtha extractor along with several aqueous streams. When the aqueous streams and the prewash methanol are mixed, the methanol preferentially dissolves in the aqueous phase and the naphtha is recovered as an immiscible layer in the extractor. The naphtha is then distilled in the naphtha stripper for removal of dissolved gases and small amounts of water.

The methanol-water cut from the naphtha extractor is distilled in the azeotrope column where a methanol-water-naphtha azeotrope is taken overhead and then recycled to the naphtha extractor. The bottom product from the azeotrope column contains methanol and water

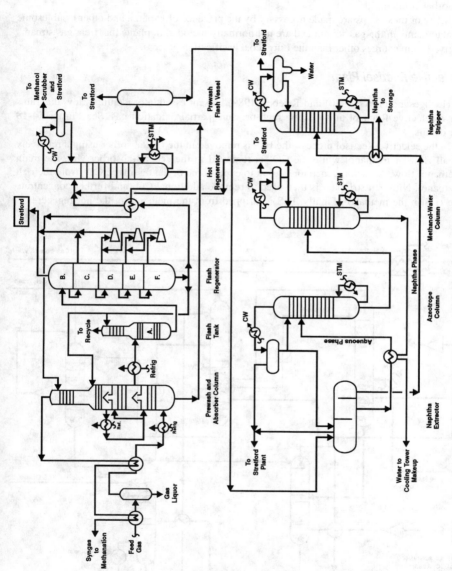

Figure 14-17. Simplified process flow diagram of Rectisol process system at the SNG-from-coal plant in North Dakota. *(Miller and Lang, 1985)*

1222   *Gas Purification*

and is sent to the methanol-water column. In this tower, the water is rejected from the bottom and used off-site for cooling tower makeup. The methanol overhead stream is recycled to the hot regenerator column for eventual reuse in the final absorption section of the prewash and absorber column.

Many of these steps are made necessary by the presence of naphtha and other troublesome contaminants in the gas stream and are not normally included in plants that treat gas streams derived from sources other than the Lurgi coal gasification process.

Selective Rectisol Plant

Hochgesand (1968, 1970) and Kriebel (1989) present essentially identical flow diagrams for a selective Rectisol unit treating high-pressure partial oxidation gas. See **Figure 14-18** for details.

In the selective Rectisol process, the H_2S is removed in the first absorber using a relatively small flow of methanol while the CO_2 is removed in the second absorber with the main methanol flow. A flash regenerator is used to expel some of the dissolved CO_2 from the rich methanol. Nitrogen stripping is then used to remove additional CO_2 and further concentrate the H_2S in the methanol. Finally, H_2S is stripped from the methanol in the hot regenerator,

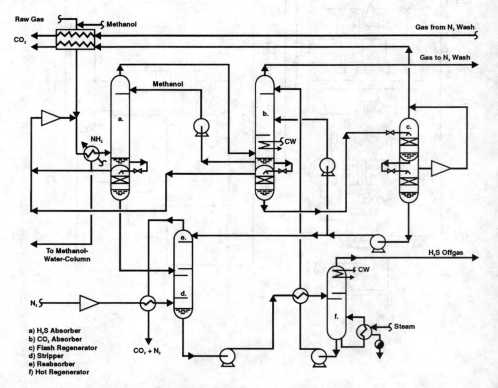

Figure 14-18. Simplified process flow diagram of selective Rectisol process. (*Kriebel, 1989*)

which is a conventional trayed distillation tower with reboiler and condenser. An additional tower, which is not shown, is also used to remove water accumulating in the circulating methanol stream.

Ifpexol Process

This process has recently been introduced by Institut Français du Pétrole (IFP). The process employs methanol as the sole agent for treating natural gas to achieve (1) dehydration, (2) natural gas liquids recovery, and (3) acid gas removal. These steps are integrated into an overall processing system. The first Ifpexol unit was constructed in 1992. In 1996, it was reported that six Ifpexol process plants were in operation and five more were in various stages of design. Capacities range from 10 to 350 MMscfd (Institut Français du Pétrole, 1996). The technology is described by Minkkinen and Levier (1992). The overall system involves two separate processes that may be used independently or in combination. IFPEX-1 removes condensable hydrocarbons from the feed gas, while IFPEX-2 accomplishes acid gas removal and recovery.

Figure 14-19 is a flow diagram of the combined Ifpexol concept. In the first step water is separated from the natural gas to allow low-temperature processing in the subsequent steps. This is accomplished by contacting the natural gas feed with a methanol-rich stream in a conventional countercurrent contactor. Methanol, since it is more volatile than water, is vaporized and taken overhead, while the water stream containing as little as 50 ppm of methanol is withdrawn as bottoms. Control of this column to achieve the desired splits of water and methanol while minimizing methanol losses in the water stream is critical to proper operation.

Liquid hydrocarbons and a water-methanol liquid stream are condensed from the gas as it is chilled for low temperature acid gas removal by the IFPEX-2 process with the water-methanol cut being recycled to the dehydration step. The chilled mixture is passed through a separator from which the liquid water-methanol cut is recycled to the dehydration column, the liquid hydrocarbon cut is removed for recovery, and the gas phase is sent to the IFPEX-2

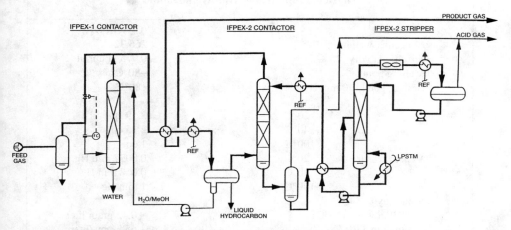

Figure 14-19. Flow diagram of Ifpexol process, consisting of IFPEX-1 and IFPEX-2 units in series. (*Menkkinen and Levier, 1992*)

process absorber. In the IFPEX-2 section, acid gases are removed from the natural gas stream in a conventional absorber-stripper system, much as in the Rectisol process. IFP states that hydrocarbon removal in this acid gas removal step can be controlled within limits by adjusting absorption parameters, primarily the water content of the methanol.

Selective removal of H_2S or removal of essentially all the acid gas can be achieved. Pipeline H_2S specifications of 0.25 grains per 100 SCF and CO_2 levels of 1% are claimed to be feasible (Institut Français du Pétrole, 1992).

Estasolvan Process

The Estasolvan process was announced jointly by Institut Français du Pétrole of France and Friedrich Uhde GmbH of West Germany (Franckowiak and Nitschke, 1970). The process was demonstrated in two pilot plants, but it has recently been reported that further promotion of the process has been halted, and the process is now commercially unavailable (Uhde, 1992).

The Estasolvan process was based on the use of tri-n-butyl phosphate as a solvent. Selected physical properties of this material are listed in **Table 14-20**. The process development work focused primarily upon the selective removal of H_2S from natural gas that also contained CO_2. The solubility of CO_2 in the solvent is significantly lower than in most other physical solvents used for gas treating, but this, of course, is beneficial in accomplishing selective removal of H_2S.

One proposed variation of the process incorporated simultaneous absorption of acid gases and liquefied petroleum gases (LPG) in the solvent, followed by separation of the absorbed components by fractional distillation. High percentage removal of both COS and mercaptans was also claimed.

Table 14-20
Physical Properties of Tributylphosphate

Formula		$(C_4H_8)_3PO_4$
Molecular weight		266.32
Specific gravity (25°C)		0.973 gram/ml
Melting point		−80°C
Boiling point (30 mm Hg)		180°C
Viscosity		
	20°C	3.19 cp
	40°C	2.15 cp
	100°C	1 cp
Vapor pressure	20°C	0.0037 mm Hg
	40°C	0.018 mm Hg
	100°C	1 mm Hg
Solubility		
TBP in water (25°C)		0.42 gram/liter
Water in TBP (25°C)		65 grams/liter

Data source: Franckowiak and Nitschke (1970)

Methylcyanoacetate Process

The Methylcyanoacetate process, disclosed by the Union Oil Company of California (Woertz, 1971), was tested in a pilot plant, but has not been used commercially. The solvent, methylcyanoacetate (MCA), is reported to be stable and to have high capacity for acid gases. An interesting feature of the process is that the solvent is appreciably more selective for acid gases contained in hydrocarbon gas streams than the solvents used in commercial processes of this type. For example, it is reported by Woertz (1975) that under comparable operating conditions, propylene carbonate removes about 50% of the propane and 100% of the butanes present in a hydrocarbon feed gas, while MCA removes only about 30% of the propane and 75% of the butanes. Vapor-liquid equilibrium data for a number of gases in methylcyanoacetate have been reported (Woertz, 1975).

MIXED PHYSICAL/CHEMICAL SOLVENT PROCESSES

Sulfinol Process

In contrast to the processes described previously in this chapter, the Sulfinol process employs a mixture of a chemical and a physical solvent. This is often referred to as a "mixed solvent process." In a mixed solvent process, the physical solvent removes the bulk of the acid gas while the chemical solvent (an alkanolamine in the Sulfinol process) purifies the process gas to stringent levels, all in a single step. Although the process flow sheet resembles that of a conventional amine treating unit, the presence of the physical solvent enhances the solution capacity, especially when the gas stream to be treated is at high pressure and the acidic components are present in high concentrations. The Sulfinol process also has demonstrated its ability to achieve high efficiency removal of other impurities, namely COS, mercaptans, and other organic sulfur compounds.

The Sulfinol process is licensed by the Shell Oil Company in the U.S. and by Shell International Petroleum Maatschappij (SIPM) in the Netherlands. The process has found wide application in the treatment of natural, refinery, and synthesis gases. The Sulfinol process can meet the requirement for deep CO_2 removal to 50 ppm for LNG plants, as well as the opposite extreme of bulk CO_2 removal using flash regeneration. **Table 14-21** shows the ranges of feed gas composition and conditions and of treated gas purity specifications of licensed Sulfinol plants (Shell, 1992). In 1996, more than 180 commercial units were reported to be in operation or under construction (Shell, 1996).

Basic Data

The Sulfinol solvent consists of sulfolane (tetrahydrothiophene dioxide) and an alkanolamine, usually diisopropanolamine (DIPA) or methyldiethanolamine (MDEA), and water. The solvent with DIPA is referred to as Sulfinol-D or simply as Sulfinol, and the solvent with MDEA is referred to as Sulfinol-M. Typically, Sulfinol-D is used when essentially complete removal of both hydrogen sulfide and carbon dioxide and deep removal of carbonyl sulfide is desired. Sulfinol-M is used for the selective removal of hydrogen sulfide over carbon dioxide and the partial removal of carbonyl sulfide (Nasir, 1990). Both Sulfinol solvents are reported to be capable of removing mercaptans and alkyl sulfides to very low levels.

The equilibrium solubility of hydrogen sulfide in sulfolane and in the Sulfinol-D solvent, as a function of partial pressure, is shown in **Figure 14-20** (Dunn et al., 1964). For compari-

Table 14-21
Ranges of Feed Gas Compositions, Conditions, and Treated Gas Purity For Licensed Sulfinol Plants

	Feed Gas	Treated Gas Purity
Pressure, psia	22 to 1,330	
Acid Gas Content, %vol		
H_2S	0 to 53.6	1 to 8 ppmv
CO_2	2.6 to 43.5	50 ppmv to 3.7%m+
COS	0 to 1,000 ppmv	3 to 160 ppmv
RSH	0 to 3,000 ppmv	4 to 160 ppmv
Acid Gas Partial Press, psia		
H_2S	0 to 713	
CO_2	2.7 to 396	
$H_2S + CO_2$	12 to 748	
H_2S/CO_2 Ratio, Volume	0 to 20.4	

Source: (Shell, 1992)

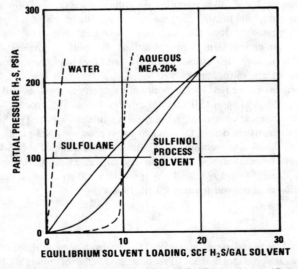

Figure 14-20. Solubility of hydrogen sulfide in Sulfinol solvent. *(Dunn et al., 1964)*

son, the solubility of hydrogen sulfide in water and in 20% aqueous monoethanolamine solution is also shown. It should be noted that, from the standpoint of solvent capacity, the Sulfinol solvent is inferior to aqueous monoethanolamine at low hydrogen sulfide partial pressures. However, as the partial pressure of hydrogen sulfide becomes higher, the capacity of the Sulfinol solvent continues to increase, while that of the monoethanolamine solution remains almost constant because the stoichiometry of the chemical reaction between hydrogen sulfide and the amine limits the capacity of the monoethanolamine solution.

Process Description

A typical flow diagram of a Sulfinol unit is shown in **Figure 14-21.** The scheme is identical to that of a typical alkanolamine system as discussed in Chapter 2, with the exception of the flash tank, which is optional in aqueous ethanolamine systems treating low pressure gases, but almost a necessity in a Sulfinol unit. Because of the relatively high solubility of hydrocarbons in the solvents, omission of the flash tank would lead to high concentrations of hydrocarbons in the acid gas and possibly to operational difficulties in Claus units where the hydrogen sulfide is converted to elemental sulfur. The flash gas from the flash tank can either be recompressed and recycled to the absorber inlet, or, as usually practiced in natural gas-treating plants, treated and used as plant fuel.

Process Operation

Operating data from commercial units using Sulfinol-D as reported by Dunn et al. (1965) and Frazier (1970) are presented in **Table 14-22.** Pilot-plant data obtained with an East Texas natural gas are shown in **Table 14-23** (Dunn et al., 1964). For comparison, performance data of the same pilot plant with an aqueous 20% monoethanolamine solution are also given. Additional pilot-plant data obtained in processing four different Canadian natural-gas streams are given in **Table 14-24** (Dunn et al., 1964).

A Sulfinol-M plant in Rankin County, Mississippi, which treats 100 MMscfd of natural gas containing 34% H_2S and produces pipeline quality gas and 1,275 LT/d of elemental sulfur, is described by Christensen (1979). The plant also removes 97% of the COS, which is present in the feed gas at a concentration of 705 ppm.

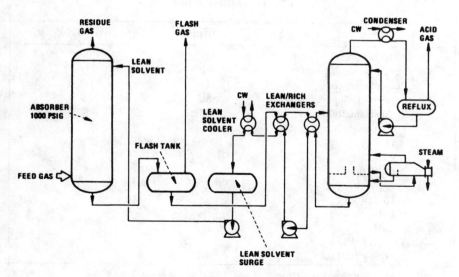

Figure 14-21. Typical flow diagram of Sulfinol process.

1228 *Gas Purification*

Table 14-22
Commercial Operating Data for the Sulfinol-D Process

Plant	A	B	C	D
Feed gas:				
Volume, MMscf/d	32	150	50	150
Pressure, psig	1,000	1,000	1,000	1,000
H_2S, gr/100 scf	1.60(%)	6.35	27.00	19.00
CO_2, %	6.90	3.30	9.16	6.81
COS, ppm	7	—	—	—
RSH, gr/100 scf	19(ppm)	0.75	0.20	—
Total hydrocarbon, %	91.00	96.33	89.88	91.40
Treated gas:				
CO_2, %	—	0.30	0.30	0.30
H_2S, gr/100 scf	<0.1–0.6	0.25	0.25	0.25
RSH, gr/100 scf	—	0.25	—	—
Solvent rate, gpm	315–335	—	—	—
Solvent loading, scf/gal	6.0	—	—	—
Reboiler duty, MMBtu/MMscf gas	10.1	—	—	—

Data sources: Plant A–Frazier (1970); Plants B, C, and D–Dunn et al. (1965)

Table 14-23
Pilot-Plant Data for Sulfinol-D Process and Aqueous MEA
(East Texas Natural Gas)

	Sulfinol	20% MEA
Absorber pressure, psig	1,000	1,000
Feed gas rate, scf/min	40.7	23.6
Solvent rate, gpm	1.0	1.0
Solvent loading, scf/gal	8.5	4.9
Total steam, lb/lb acid gas*	1.42	3.80
Feed gas:		
H_2S, %		15.00
CO_2, %		6.00
COS, ppm		60
N_2, %		7.50
CH_4, %		57.69
C_2H_6, %		6.24
C_3H_8 +, %		7.57
Treated gas:		
H_2S, gr/100 scf	<1	<1

*Combined duty of preheater and reboiler
Data of Dunn et al. (1964)

Table 14-24
Pilot-Plant Data for Sulfinol-D Process
(Canadian Natural Gas)

Gas Stream	A	B	C	D
Pressure, psia	995	935	935	715
Feed gas:				
H_2S, %	26.40	16.20	15.60	3.40
CO_2, %	5.20	5.40	5.50	1.30
COS, %	0.03	0.01	0.01	—
RSH, gr/100 scf	0.75	0.14	0.14	—
Treated gas:				
H_2S, gr/100 scf	0.4	0.13	0.1	0.15
CO_2, gr/100 scf	<1	<1	<1	<1
RSH, gr/100 scf	<0.01	—	—	—
COS, ppm	2.0	1.0	1.0	—
Acid gas:				
Hydrocarbon, %	1.9	0.8	0.7	3.6
Steam, lb/lb acid gas	0.9	1.0	1.0	1.3

Data of Dunn et al. (1964)

A summary of plant operation is given in **Table 14-25**. The operation of a Sulfinol-M plant in The Netherlands that treats 300 MMscfd of natural gas containing 0.15 to 0.44% H_2S and 2.87 to 4.25% CO_2 is described by Taylor and Hugill (1991). This plant is designed to reject CO_2 during the absorption step and also to increase the H_2S content of the feed to the Claus unit by a "flash enrichment" step. **Figure 14-22** is a simplified flow diagram showing the integration of the Claus plant and its tail gas treater and a molecular sieve dryer with the Sulfinol unit. Also shown is the flash enrichment step. **Table 14-26** summarizes the overall performance of the plant.

Studies of hydrocarbon solubility in the Sulfinol solvent conducted during the early pilot plant tests indicate that aliphatic hydrocarbons up to pentane are largely rejected by the solvent. However, aromatics are absorbed quite efficiently (Dunn et al., 1964). Sulfinol solvent has proved to be very stable. For Sulfinol-D, losses due to degradation are generally less than 5 lb per MMscf of raw gas. No significant losses or degradation of the solvent have been experienced with Sulfinol-M. Corrosion is, in general, no problem in Sulfinol units, and carbon steel is a satisfactory material of construction (Dunn et al., 1964). However, isolated cases of absorber corrosion, particularly in the lower section of the vessel, have been reported (Schmeal et al., 1978). Blisters and "ring pits" were observed on trays, downcomers, and vessel walls. The corrosion was ascribed to boiling and flashing of acid gas. It was especially pronounced in units processing gases where the acid gases are present in low ratios of carbon dioxide to hydrogen sulfide. Lowering of temperatures in the absorber bottom and protection of that area with stainless steel liners were found to be effective remedies.

In addition to its effectiveness for acid-gas removal, the Sulfinol solvent shows excellent capability for the removal of carbonyl sulfide and mercaptans. In one test, about 96% removal

Table 14-25
Operating Data from Rankin County, Mississippi Sulfinol-M Plant

	Design	Actual
Feed Gas Flow Rate, MMscfd	100	5 to 106
Temperature, °F	100	135–150
H_2S, mol %	34	34.4
CO_2, mol %	9	7.7
COS, ppm	500	705
Residue Gas		
H_2S, ppm	8	1
CO_2, mol %	Nil	0.02
COS, ppm	70	15
Sulfur		
Production, LT/d	1,275	120–1,434
Recovery efficiency, %	97.7	98.6–96.5

Data source: Christensen (1979)

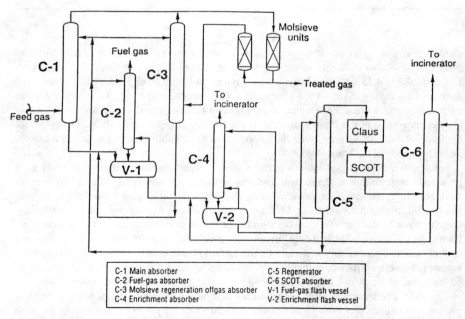

C-1 Main absorber
C-2 Fuel-gas absorber
C-3 Molsieve regeneration offgas absorber
C-4 Enrichment absorber
C-5 Regenerator
C-6 SCOT absorber
V-1 Fuel-gas flash vessel
V-2 Enrichment flash vessel

Figure 14-22. Simplified flow diagram of Sulfinol unit integrated into an overall process including a Claus sulfur recovery unit and a tail gas treatment system. (*Taylor and Hugill, 1991*)

Table 14-26
Performance of Sulfinol-M Desulfurization Unit at Emmen, The Netherlands

	Design	Actual Performance	
		High H_2S	Low H_2S
Feed Gas, MMscfd	150*	150	150
H_2S, mol %		0.44	0.15
CO_2, mol %		4.25	2.87
Pipeline Gas			
H_2S, ppm	<3.5	2.6	2.2
CO_2 slip, %	>60	62.4	60.8
Reboiler Duty, MMBtu/h	<51	49	46

*All data are for one of two trains.
Data source: Taylor and Hugill (1991)

of methyl mercaptan was reported (Dunn et al., 1964). When a Sulfinol absorber is designed to reduce the H_2S concentration in the gas to 4 ppm, it will typically remove about 70% of the other sulfur compounds (Shell Oil Co., 1992). The removal efficiency for other sulfur compounds in a given column can, of course, be increased by increasing the solvent flow rate.

Amisol Process

The Amisol process is similar to the Sulfinol process in that it uses a combination of a physical and chemical solvent for acid gas removal. It was developed by Lurgi GmbH and employs methanol as the physical solvent, as does the Rectisol process, which was co-developed by Lurgi (Bratzler and Doerges, 1974). The first plants practicing the Amisol process used alkanolamines (MEA and DEA) as the chemical solvents; however, the more recent plants, where selective H_2S removal was an objective, have used alphatic alkylamines (Kriebel, 1985, 1989).

The Amisol process can be used for either selective desulfurization or complete removal of CO_2, H_2S, COS, and other organic sulfur compounds. Reportedly sulfur can be removed to less than 0.1 ppm and CO_2 to less than 5 ppm (Lurgi GmbH, 1988B). Through 1993, there were six plants employing the Amisol process, five of which removed H_2S and other impurities from gases derived from the gasification of coal, peat, or heavy oil, while the sixth removed CO_2 from reformer effluent and recycle gases (Lurgi GmbH, 1993).

Basic Data

The specific alkylamines—diisopropylamine (DIPAM) and diethylamine (DETA)—differ from MEA and DEA in that they have (a) greater chemical stability, (b) higher acid gas loading, (c) high H_2S selectivity, (d) easier regeneration (including a lower reboiler temperature), and (e) higher volatility. Like the alkanolamines, they are soluble in water. Properties of the alkylamines are shown in **Table 14-27**. Properties of DEA are included for comparison.

**Table 14-27
Properties of Amines Used in the Amisol Process**

	DIPAM	DETA	DEA
Chemical Formula	$[(CH_3)_2CH]_2 NH$	$(C_2H_5)_2 NH$	$(HOC_2H_4)_2 NH$
Molecular Weight	101.2	73.14	105.14
Boiling Point, °C	84	56	269
Vapor Pressure, mmbar			
at 20°C	100	253	0.1
Specific gravity, at 20°C	0.716	0.704	1.092

Source: Kriebel (1985)

Kriebel (1985, 1989) presents solubility data for Amisol systems using DIPAM and DETA showing that the carrying capacities of the Amisol mixed solvents for H_2S are significantly greater than both DEA and MDEA at H_2S partial pressures above about one bar.

Process Operation

Amisol plants are in use in Europe and China to purify various synthesis gases derived from coal, peat, and heavy oil and fuel gas derived from coal. Feed gas compositions and conditions for three different coal gasification cases described by Kriebel (1985) are presented in **Table 14-28**.

The flow diagram for one of the cases (selective desulfurization of gas from an air-blown VEW entrained bed gasifier) is shown in **Figure 14-23**. This plant uses the newer version of the Amisol process that employs DETA as the chemical solvent.

A prewash to remove impurities such as HCN and NH_3 is contained in the bottom of the absorber. For other gas streams, such as that obtained from the British Gas/Lurgi slagging gasifier, a more elaborate prewash is required. Because of the high volatility of the solvent components, both the absorber, which operates at ambient temperature, and the regenerator, which operates at about 80°C, require water wash stages at the top of the columns. The water absorbs vaporized solvent and the resulting aqueous solution is fed to a distillation column. Methanol and amine vapors from the top of the still are condensed to provide heat for the regenerator and the condensate is added to the main circulating solvent stream. Solvent-free water from the bottom of the still is reused as wash water.

The absorber/regenerator system shown on the left of the diagram is conventional, but the additional reabsorber, shown on the right, is not commonly used in amine plants. It serves to selectively remove H_2S from the CO_2-rich vapor stream extracted from the regenerator, concentrating the H_2S offgas to a level that can be sent to a Claus plant. Enrichment schemes such as this have usually been limited to pure physical solvent systems.

Selefining Process

This process was developed by Snamprogetti SpA of Milan, Italy, for selectively removing H_2S from natural and synthesis gases also containing CO_2. The Selefining solvent consists of a tertiary amine mixed with an undisclosed physical solvent. The physical solvent

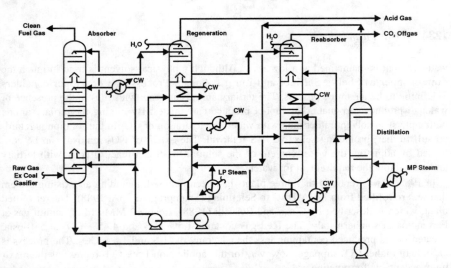

Figure 14-23. Simplified flow diagram of Amisol process for selective desulfurization of gas from an air-blown entrained-bed coal gasifier. (*Kriebel, 1985*)

Table 14-28
Composition of Amisol Plant Feed Gas Streams from Coal Gasifiers

	RBW Hydro Gasifier	VEW Entrained Bed Gasifier	BGL Slagging Gasifier
Major Components, Vol %			
CO_2	1.8	4.17	3.96
CO	7.2	22.36	60.65
H_2	53.1	11.46	25.15
N_2	0.8	61.48	0.87
CH_4	36.3	0.2	8.11
C_2H_6	0.6		0.33
C_2H_4	—	—	0.14
C_3 hydrocarbons	—	—	0.16
C_4 hydrocarbons	—	—	0.04
C_5 hydrocarbons	—	—	0.01
C_6H_6	0.05	—	—
H_2S	0.1	0.21	0.49
Trace Components, ppmv			
COS	5	244	200
RSH	134	—	—
CS_2	—	14	200
HCN	9	66	600
NH_3	9	34	100
Thiophene	—	—	10
Naphthalene	—	—	7
Pressure, Bar	51	19	55
Temperature, °C	40	60	40

RBW = Rheinische Braunkohlenwerke, A. G. Kohn
VEW = Vereinigte Electrizitatswerke Westfalen, Dortmun
BGL = British Gas Corporation, London
Source: Kriebel (1985)

water content is maintained at a low level. Although a physical solvent is used in much the same manner as Shell's Sulfinol and Lurgi's Amisol Process, the developer considers Selefining to be an amine process. The tertiary amine can react with CO_2 in the presence of water to produce carbonate, but cannot produce carbamates. By mixing the tertiary amine with a physical solvent rather than with water, the hydration of CO_2 in the solution is retarded still further, promoting selective H_2S removal. The extent of CO_2 removal can be controlled by adjusting the water content of the solution. H_2S can react directly with tertiary amines without the requirement for water.

In 1988, it was reported that three plants had been licensed, including an existing system that was retrofitted from MEA-DEG to Selefining (Snamprogetti SpA, 1988). The retrofitted plant, which is described by Gazzi and Rescalli (1988), treats 12 MMscfd of natural gas at Ferrandina in southern Italy. The feed gas contains 1.5% H_2S and 4.45% CO_2 at 580 psia. Treated gas is produced containing less than 1 ppmv of H_2S and 1.5% CO_2. The process is capable of higher CO_2 slippage; however, for this application 1.5% CO_2 represents the maximum acceptable for pipeline use.

REFERENCES

BASF, 1992, private communication, Dr. Hefner, May 14.

Beavon, D. K., and Roszkowski, T. R., 1969, "Purisol removes carbon dioxide from hydrogen, ammonia syngas," *Oil & Gas J.,* Vol. 67, April 14, pp. 138–142.

Boston, F. C., and Schneider, M. L., 1971, *Proceedings Gas Conditioning Conference,* University of Oklahoma, Norman, OK.

Bratzler, K., and Doerges, A., 1974, "Amisol Process Purifies Gas," *Hydro. Process.,* April, pp. 78–80.

Buckingham, P. A., 1961, "Fluor Solvent Process Demonstrated in EPNG Plant," paper presented at NGAA Regional Meeting, Odessa, TX, May.

Buckingham, P. A., 1964, "Fluor Solvent Process Plants: How They Are Working," *Hydro. Process.,* Vol. 43, April, pp. 113–116.

Bucklin, R. W., and Schendel, R. L., 1985, "Comparison of Physical Solvents Used for Gas Processing," in *Acid and Sour Gas Treating Processes,* S. A. Newman, editor, Gulf Publishing Co., Houston, TX, pp. 42–79.

Christensen, R. E., 1979, "Shell solves sour-gas plant problems," *Oil & Gas J.,* Jan. 15, pp. 126–129.

Clare, R. T., and Valentine, J. P., 1975, "Acid Gas Removal Using the Selexol Process," paper presented at Second Quarterly Meeting of the Canadian Gas Processors Association, Edmonton, Alberta, Canada, June 5.

Cook, T. P., and Tennyson, R. N., 1969, "Improved Economics in Synthesis Gas Plants," *Chem. Eng. Progr.,* Vol. 65, No. 11, pp. 61–64.

Dow Chemical of Canada, Ltd., 1962, *Gas Conditioning Fact Book.*

Dreisbach, R. R., 1952, *Pressure-Volume-Temperature Relationships of Organic Compounds, 3rd ed.,* Chart 20, New York: McGraw-Hill Book Company, Inc.

Dunn, C. L., Freitas, E. R., Goodenbour, J. W., Henderson, H. T., and Papadopoulos, M. N., 1964, "New Pilot Plant Data on Sulfinol Process," *Hydro. Process.,* Vol. 43, March, pp. 150–154.

Dunn, C. L., Freitas, E. R., Hill, E. S., and Sheeler, J. E. R., 1965, "First Plant Data From Sulfinol Process," *Hydro. Process.,* Vol. 44, April, pp. 137–140.

England, C., 1986, *Chem. Eng.,* No. 8, April 28, p. 63.

Epps, R., 1994, "Use of SELEXOL Solvent for Hydrocarbon Dewpoint Control and Dehydration of Natural Gas," *Laurence Reid Gas Conditioning Conference Proceedings,* University of Oklahoma, Norman, OK, Feb. 27–March 2.

Epps, R., 1992A, Union Carbide Chemicals and Plastics Company, Inc., personal communication, May 6.

Epps, R., 1992B, "Processing of Landfill Gas for Commercial Applications," paper presented at ECO WORLD '92, Washington, D.C., June 15.

Fluor Daniel unpublished data, 1993, personal communication, P. Buckingham, May 20.

Franckowiak, S., and Nitschke, E., 1970, "Estasolvan: New Gas Treating Process," *Hydro. Process.,* Vol. 49, May, pp. 145–148.

Frazier, J., 1970, "How's the Sulfinol Process Working?" *Hydro. Process., Vol.* 49, April, pp. 101–102.

Freireich, E., and Tennyson, R. N., 1977, "Increased Natural Gas Recovery from Physical Solvent Gas Treating Systems," *Proceedings Gas Conditioning Conference,* University of Oklahoma, Norman, OK, March 7–9.

Gazzi, L., and Rescalli, C., 1988, "An MEA-DEG Plant Successfully Retrofitted to Selenfining," *Energy Prog.,* Vol. 8, No. 2, June, pp. 113–117.

Grünewald, G., 1989, "Selective Physical Absorption Using the Purisol Process," paper presented at 6th Continental Meeting of the European Chapter of the GPA, Bremen, Germany, May 19.

Hegwer, A. M., and Harris, R. A., 1970, "Selexol Solves High H_2S/CO_2 Problem," *Hydro. Process.,* Vol. 49, April, pp. 103–104.

Herbert, W., 1956, *Erdol u. Kohle,* Vol. 9, No. 2, pp. 77–81.

Hernandez, R. J., and Huurdeman, T. L., 1989, "Solvent Unit Cleans Synthesis Gas," *Chem. Eng.,* February, pp. 154, 156.

Hernandez, R. J. and Huurdeman, T. L., 1991, "SELEXOL Solvent Process in Ammonia Synthesis Gas Treating, Use and Experience," paper SC-1683A, Union Carbide Chemicals and Plastics, Inc. May 3.

Hochgesand, G., 1968, *Chemie-Ing.-Techn.,* Vol. 40, No. 9/10, pp. 432–440.

Hochgesand, G., 1970, "Rectisol and Purisol," *Ind. Eng. Chem.,* Vol. 62, No. 7, pp. 37–43.

Hoogendoorn, J. C., and Solomon, J. M., 1957, *Brit. Chem. Eng.,* Vol. 2, May, pp. 238–244.

Institut Français du Pétrole, 1992, "Ifpexol," *Hydro. Process., Gas Process Handbook '92,* April, p. 112.

Institut Français du Pétrole, 1996, "Ifpexol," *Hydro. Process., Gas Process '96,* April, p. 124.

Judd, D. K., 1978, "Selexol unit saves energy," *Hydro. Process.,* Vol. 57, No. 4, April, pp. 122–124.

Kapp, E., 1970, *Erdöl and Kohle-Erdgas-Petrochemie Vereinigt mit Brennstoff Chemie,* Vol. 23, No. 9, pp. 566–571.

Knapp, H., 1968, "Low Temperature Absorption—the Rectisol Process," *Proceedings of the 1968 Gas Conditioning Conference,* University of Oklahoma, Norman, OK.

Kohl, A. L., and Miller, F. E., 1960, U.S. Patent 2,926,751, March 1.

Kohl, A. L., and Buckingham, P. A., 1960, "Fluor Solvent CO_2 Removal Process," *Petr. Refiner,* Vol. 39, May, pp. 193–196.

Kriebel, M., 1985, "Improved Amisol Process for Gas Purification," in *Acid and Sour Gas Treating Processes,* S. A. Newman, editor, Gulf Publishing Co., Houston, TX, pp. 112–130. (See also *Energy Prog.,* Vol. 4, No. 3, September, pp. 143–146).

Kriebel, M., 1989, *Ullmann's Encyclopedia of Industrial Chemistry, Gas Production,* VCH Verlagsgesellschaft mbH. Weinheim, pp. 253–258.

Linde AG, and Lurgi GmbH, 1992, *Hydro. Process., Gas Process Handbook '92,* April, p. 125.

Lurgi GmbH, 1978, "Purisol for Gas Treating," Brochure 1163/6.78.

Lurgi GmbH, 1988A, "Purisol," *Hydro. Process., 1988 Gas Process Handbook,* April, p. 69.

Lurgi GmbH, 1988B, "Amisol," *Hydro. Process., 1988 Gas Process Handbook,* April, p. 53.

Lurgi GmbH, 1993, "References: Gas and Synthesis Technology," Brochure 1569e/9.93/4.10.

Lurgi Öl-Gas-Chemie GmbH, 1996, "Purisol," *Hydro. Process, Gas Processes '96,* April, p. 133.

Lurgi Öl-Gas-Chemie GmbH, and Linde AG, 1996, "Rectisol," *Hydro. Process, Gas Processes '96,* p. 134.

Makranczy, J., Szeness, M. M., and Rusz, L., 1965, *Veszpremi Vegyipari Egyetem Kozlemenyei,* Vol. 9, pp. 95–105 (University of Chemical Industries, Institute of General and Inorganic Chemistry, Veszprem, Hungary).

Miller, W. R., and Lang, R. A., 1985, *Great Plains Coal Gasification Plant Public Design, Volume I,* DOE/CH/10088-1874, Technical Information Center, U.S. DOE.

Minkkinen, A., and Levier, J. F., 1992, "Ifpexol: Complete Gas Treatment with a Basic Single Solvent," *Laurance Reid Gas Conditioning Conference Proceedings,* University of Oklahoma, Norman, OK.

Nasir, P., 1990, "A Mixed Solvent for a Low Total Sulfur Specification," paper presented at the AIChE National Meeting, San Diego, CA, Aug. 19.

Raney, D. R., 1976, "Remove carbon dioxide with Selexol," *Hydro. Process.,* Vol. 55, No. 4, pp. 73–75.

Ranke, G., 1973, *Linde Reports on Science and Technology,* Vol. 18.

Ranke, G., and Mohr, V. H., 1985, "The Rectisol Wash—New Developments in Acid Gas Removal from Synthesis Gas," in *Acid and Sour Gas Treating Processes,* S. A. Newman, editor, Gulf Publishing Company, Houston, TX, pp. 80–111.

Schmack, P., and Bittrich, H. J., 1966, *Wissenschaftl. Zeitschrift,* Vol. 8, No. 2/3, pp. 182–186.

Schmeal, W. R., MacNab, A. J., and Rhodes, P. R., 1978, "Corrosion in Amine/Sour Gas Treating Contactors," *Chem. Eng. Progr.,* Vol. 74, No. 3, pp. 37–42.

Shah, V. A., and McFarland, J., 1988, "Low cost ammonia and CO_2 recovery," *Hydro. Process.,* March, pp. 43–46.

Shell Oil Company, 1992, private communication, D. Lancaster, April 21.

Shell Oil Company, and Shell International Research Mij. B.V., 1996, "Sulfinol," *Hydro. Process, Gas Processes '96,* April, p. 142.

Snamprogetti SpA, 1988, *Hydro. Process., 1988 Gas Process Handbook,* April, p. 71.

Stein, W. H., 1969, *Erdöl-Erdgas-Zeitschrift,* Vol. 85, pp. 467–470.

Swanson, C. G., 1978, "Carbon Dioxide Removal in Ammonia Synthesis Gas by Selexol," paper presented at 71st AIChE Meeting, Miami Beach, FL, Nov. 12–16.

Sweeney, C. W., Ritter, T. J., and McGinley, E. B., 1988, "A Strategy for Screening Physical Solvents," *Chem. Eng.*, Vol 95, No. 9, June 20, pp. 119–125.

Sweny, J. W., 1973, "Synthetic Fuel Gas Purification by the SELEXOL Process," paper presented at 165th National Meeting of the American Chemical Society, Division of Fuel Chemistry, Dallas, TX, April 8–12.

Sweny, J. W., 1976, "The SELEXOL Process in Fuel Gas Treating," paper presented at 81st National Meeting of the American Institute of Chemical Engineers, Kansas City, MO April 11–14.

Sweny, J. W., 1980, "High CO_2-High H_2S Removal with Selexol Solvent," paper presented at 59th Annual GPA Convention, Houston, TX, March 17–19.

Sweny, J. W., and Valentine, J. P., 1970, "Physical Solvent Stars in Gas Treatment/Purification," *Chem. Eng.*, Vol. 77, No. 19, Sept. 7, pp. 54–56.

Sweny, J. W., 1985, "Gas Treating with a Physical Solvent" in *Acid and Sour Gas Treating Processes*, S. A. Newman, editor, Gulf Publishing Company, Houston, TX, pp. 1–41.

Taylor, N. A., and Hugill, J. A., 1991, "Sulfinol-M Provides the Solution to a Tough Treating Challenge," *Laurance Reid Gas Conditioning Conference Proceedings*, Univ. of Oklahoma, Norman, OK, March.

Uhde GmbH, 1992, private communication, Dr. Mundo, April 23.

Valentine, J. P., 1974, "New solvent process purifies crude, coal acid gases," *Oil & Gas J.*, Vol. 72, No. 46, pp. 60–62.

Valentine, J. P., 1975, "Economics of the SELEXOL Solvent Gas Purification Process," paper presented at the 79th National Meeting of the American Institute of Chemical Engineers, Houston, TX, March 19.

Van Deraerschot, R., and Valentine, J. P., 1976, "The SELEXOL Solvent Process for Selective Removal of Sulfur Compounds," paper presented at 2nd International Conference on the Control of Gaseous Sulphur and Nitrogen Emission, Salford University, England, April 6–8.

Woertz, B. B., 1971, *J. of Pet. Tech.*, April, pp. 483–490.

Woertz, B. B., 1975, *Soc. Pet. Eng. J.*, Feb. 7–12.

Wölfer, W., Schwartz, E., Vodrazka, K., and Volkamer, K., 1980, "Solvent shows greater efficiency in sweetening of gas," *Oil & Gas J.*, Jan. 21, pp. 66–70.

Wölfer, W., 1982, "Helpful hints for physical solvent absorption," *Hydro. Process.*, Vol. 61, No. 11, November, pp. 193–197.

Zawacki, T. S., Duncan, D. A., and Macriss, R. A., 1981, "Process optimized for high pressure gas cleanup," *Hydro. Process.*, Vol. 59, No. 4, April, pp. 143–149.

Chapter 15
Membrane Permeation Processes

INTRODUCTION, 1238

History and Status, 1239

PROCESS TECHNOLOGY, 1242

Transport Mechanisms, 1242
Design and Operating Considerations, 1245
Membrane and Module Configurations, 1246
Flow Arrangements, 1250
Simulation and Design Calculations, 1252
Fields of Application, 1258

APPLICATION CASE STUDIES, 1259

Hydrogen, 1259
Carbon Dioxide, Hydrogen Sulfide, and Water Removal, 1270
Helium Removal from Natural Gas, 1281
Air Separation, 1282
Solvent Vapors, 1288

REFERENCES, 1291

INTRODUCTION

Membrane technology, as applied to gases, involves the separation of individual components on the basis of the difference in their rates of permeation through a thin membrane barrier. The rate of permeation for each component is determined by the characteristics of the component, the characteristics of the membrane, and the partial pressure differential of the gaseous component across the membrane. Since separation is based on a difference in the rates of permeation rather than on an absolute barrier to one component, the recovered component that flows through the membrane (the permeate) is never 100% pure. Also, since a finite partial pressure differential is required as the driving force, some portion of the permeating component remains in the residue gas, and 100% recovery is not possible. As these generalizations would suggest, the process is particularly suitable for bulk removal operations rather than for the removal of trace impurities from gas streams. It should be noted, however, that relatively high product purities and high recoveries are possible with membrane systems (at increased cost) by the use of multiple stages and recycle systems or when used in combination with other technologies.

The residue gas product normally leaves the unit at a pressure close to that of the feed, while the permeate product, which must pass through the membrane, leaves at a much reduced pressure. The principal (and/or highest purity) product may be either the permeate (e.g., the production of hydrogen from dilute gas streams) or the residue gas (e.g., the purification of natural gas by the removal of excess carbon dioxide from high pressure feed), and the process may be considered either separation or purification.

Gas purification and separation by membrane permeation has many advantages, including

- Low capital investment
- Ease of operation. Process can be operated unattended
- Good weight and space efficiency
- Ease of scale up. However, there is little economy of scale (see disadvantages below)
- Minimal associated hardware
- No moving parts
- Ease of installation
- Flexibility
- Minimal utility requirements
- Low environmental impact
- Reliability
- Ease of incorporation of new membrane developments. Users can install the next generation of membranes into existing equipment at the scheduled membrane replacement time (Schell, 1983)

The principal disadvantages are

- A clean feed is required. Particulates, and in most cases entrained liquids, must be removed. Filtration to remove particles down to one micron in size is preferred.
- Because of their modular nature, there is little economy of scale associated with larger membrane installations.
- Because membranes use pressure as the driving force of the process, there may be a considerable energy requirement for gas compression.

History and Status

Since the 19th century it has been known that certain polymer membranes can separate gases by permeation. As early as 1831, Mitchell reported that different gases permeate membranes at different rates. Graham, in 1866, discussed the mechanism of permeation and demonstrated experimentally that mixtures of gases can be separated using rubber membranes. In 1950, Weller and Steiner reported on permeation processes of industrial importance, the separation of oxygen from air and the recovery of helium from natural gas. However, the selectivity and production rates of the membranes available at the time were poor and the large membrane areas required made membrane permeation economically unattractive. In 1960, Loeb and Sourirajan developed a technique to cast cellulose acetate into a film that had an active thin surface layer and a highly porous supporting layer. These asymmetric membranes provided increased permeation rates while retaining their selectivity for specific gases. The development of these membranes improved the economics of membrane applications and led to increased interest in membrane technology.

During the 1970s, considerable research and developmental work was devoted to membranes. Many potential applications were identified, but commercialization was slow. In 1977, Monsanto demonstrated its first full scale membrane separator at Texas City, Texas, in a hydrogen/carbon monoxide ratio adjustment application (Burmaster and Carter, 1983). In 1979, Monsanto commercialized its hollow fiber membrane module as the Prism separator. From 1979 to 1982 Prism separators were evaluated in several refinery hydrogen purification applications (Bollinger et al., 1982). The success of these pilot tests established the commercial viability of gas separation with membranes. The first large scale commercial CO_2 membrane separation project was the installation of two membrane separation facilities at the Sacroc tertiary oil recovery project in West Texas in 1983. Up to 80 MMscfd of gas has been processed in these facilities (Parro, 1984).

Since the early 1980s, membrane technology has advanced rapidly and continues to advance. In addition to cellulose acetate and polysulfone, the polymers used in making gas separation membranes include polyimides, polyamides, polyaramid, polydimethylsiloxane, silicon polycarbonate, neoprene, silicone rubber, and others. Today membranes can be designed to withstand a 2,000 psi pressure differential. Membranes used in hydrogen or carbon dioxide applications operate at temperatures up to 200°F, while those used in solvent applications can operate at temperatures up to about 400°F (Baker, 1985).

Improvements in manufacturing methods have resulted in improved membrane performance and economics. A flux increase of 5% and a separation factor increase of 20% have resulted from improved manufacturing methods (Hamaker, 1991), while during the mid-1980s, air separation membranes became from two to four times more efficient. Not only are the polymers rapidly changing, but capital costs are also coming down. Advances in membranes reduced the installed cost of a membrane plant by about 40% during the 1980s (Spillman, 1989). Because of the rapid changes in technology and costs, economic studies and cost data become outdated soon after publication, and the reader should use caution in using cost data presented later in this chapter as technical advances are continuing to improve performance and reduce costs.

The developmental and commercial successes of the early 1980s, and the perceived large market, attracted many companies into the field. As the technology matured and the market became extremely competitive, some companies dropped out and others changed ownership. Companies offering commercial scale membrane systems are listed in **Table 15-1**. The table also identifies the principal areas of application for each company's products.

In the field of air separation, the improving economics of membrane-based processes have encouraged large industrial gas suppliers to join forces with membrane suppliers. This trend is pointed out by Prasad et al. (1994) who provide the following chronology:

1985: Union Carbide Industrial Gases, Inc. and Albany International Membrane Venture form a joint venture.
1986: Innovative Membrane Systems (formerly Albany International Membrane Venture) becomes a wholly owned subsidiary of Union Carbide Industrial Gases, Inc. (now Praxair, Inc.).

Table 15-1
Commercial-Scale Membrane Suppliers

Company	CO$_2$	H$_2$	Air O$_2$	Air N$_2$	Other*
A/G Technology (AVIR)	X		X	X	
Air Products (Permea)	X	X	X	X	X
Asahi Glass (HISEP)			X	X	
Cynara (Dow)	X				
Dow (Generon)			X	X	
DuPont		X			
Grace Membrane Systems	X	X			X
Hoescht Celanese (Separex)	X	X			X
International Permeation	X				X
Membrane Technology and Research					X
Nippon Kokan K.K.					X
Osaka Gas			X		
Oxygen Enrichment Co.			X		
Perma Pure					X
Techmashexport (USSR)			X		
Teijin Ltd.			X		
Toyobo			X		
Ube Industries		X			X
Union Carbide (Linde)		X	X	X	
UOP/Union Carbide			X		

Note:
*Includes solvent vapor recovery, dehumidification, and/or helium recovery membranes.
Source: Spillman (1989) and Prasad (1994)

1988: British Oxygen Co. (BOC) and Dow/Generon Membrane Systems form a joint venture.
1989: L'Air Liquide and DuPont form a joint venture.
1990: Medal AL becomes a wholly owned subsidiary of L'Air Liquide, Inc.
1991: Permea, Inc. becomes a wholly owned subsidiary of Air Products and Chemicals, Inc.

The commercial application of membrane-based hydrogen processes also expanded rapidly because of the high permeability of selected membranes for hydrogen, the high value of pure hydrogen, and the wide range of hydrogen containing gas streams in refinery, petrochemical, and industrial chemical operations. According to Koros and Fleming (1993), the major suppliers of membrane systems for hydrogen applications are Permea, Medal, UOP, Ube, and Separex. In 1996, Medal reported that more than 60 of their hydrogen recovery systems were in operation or under construction (Medal, 1994).

PROCESS TECHNOLOGY

Transport Mechanisms

It is generally agreed that a solution-diffusion mechanism governs the transport of gases through all commercially important nonporous membranes. The mechanism involves the following: (a) adsorption of the gas at one surface of the membrane, (b) solution of the gas into the membrane, (c) diffusion of the gas through the membrane, (d) release of the gas from solution at the opposite surface, and (e) desorption of the gas from the surface. Since these steps are not necessarily independent, the term permeation is used to describe the overall transport of gases through a membrane.

In its simplest form, the solution-diffusion model considers only steps b, c, and d. This model is based on two assumptions: (1) the concentration of a component in a membrane at its surface is directly proportional to the partial pressure of the component in the gas phase adjacent to the surface, and (2) the rate at which a component passes through a membrane is proportional to the concentration gradient (concentration/distance) in the membrane. These two assumptions represent Henry's law and Fick's first law of diffusion, respectively, and can be stated as follows:

$$c_i = k_i p_i \quad \text{(Henry's law)} \tag{15-1}$$

$$J_i = -D_i(dc_i/dx) \quad \text{(Fick's law)} \tag{15-2}$$

Where: c_i = local concentration of i in the membrane
J_i = steady state flux of i
k_i = solubility coefficient
p_i = partial pressure of i in the gas
D_i = local diffusivity
x = distance through active membrane

Combining and integrating equations 15-1 and 15-2 over the full membrane thickness across the membrane, yields

$$J_i = P_i \Delta p_i / l \tag{15-3}$$

Similarly, for a hollow tubular membrane, such as a hollow fiber, the steady state rate of gas permeation is (Stern, 1986)

$$J_i = P_i(2\pi L \Delta p_i / \ln(R_o/R_i)) \tag{15-4}$$

Where: $P_i = k_i D_i$, the permeability coefficient
$\Delta p_i = p_{i(feed)} - p_{i(permeate)}$ (the partial pressure difference across the membrane)
l = membrane thickness ($x = l$)
R_o = effective outer radius of the tube
R_i = effective inner radius of the tube
L = length of tube

Any consistent set of units may be used in equations 15-1–15-4. Permeability coefficient data are often given in Barrers. One Barrer = $10^{-10}(cm^3\{STP\})(cm)/(cm^2)(sec)(mmHg)$. Since commercial membranes normally consist of a very thin active layer on a thicker porous substrate, the effective thickness may not be accurately known, and it is more convenient to use the permeation rate, P_i/l, for the overall membrane, as a correlating factor. Typical engineering units for the permeation rate are $(scf)/(ft^2)(hr)(100\ psi)$. One $(scf)/(ft^2)(hr)(100\ psi) = 1.55 \times 10^{-5}\ (cm^3)\{STP\}/(cm^2)(sec)(cmHg)$. The units of flux, J, are $(scf)/(ft^2)(hr)$.

The Henry's law/Fick's law model previously described is a simplification of the actual permeation mechanism, and more complex models have been proposed. The dual system model, for example, is a more precise representation for many cases. It assumes that gas molecules which dissolve in the dense regions of the membrane surface follow Henry's law; while molecules that adsorb on the walls of microscopic cavities in the membrane surface follow Langmuir's adsorption isotherms. Equations based on the dual system model have been developed and presented by Lee et al. (1988). Additional discussions of transport mechanisms are provided by Lacey and Loeb (1972) and Stern (1986).

Permeation rates for specific components and membranes are not constants. They vary with temperature, pressure, and the presence of other components in the gas. Detailed permeation rate data for commercial membranes are normally considered proprietary; however, some comparative data have been published. **Table 15-2** lists typical permeation rate data for a number of membranes and gases. The cellulose acetate data are from Mazur and Chan (1982), and the other data are based on a paper by Tomlinson and Finn (1990).

The permeation data given in **Table 15-2** are relative values based on a permeation rate of 1.0 for oxygen in cellulose acetate and polysulfone. However, data presented by Schell and Hoernschemeyer (1982) for the permeation rates of various gases in cellulose acetate indicate that actual values, expressed as $(scf)/(ft^2)(hr)(100\ psi)$, are close to the relative values listed in the table. For example, they show the permeation rate for oxygen, measured at room temperature and 100 psig, to actually be about 1.0 $(scf)/(ft^2)(hr)(100\ psi)$, and the actual permeation rates for other gases in cellulose acetate to be similar to those listed.

Table 15-2
Permeation Rates of Gases Through Membranes

Membrane	Relative Permeation Rate[1]									
	H_2	N_2	O_2	CH_4	CO_2	H_2O	He	H_2S	CO	C_2H_6
Polysulfone	13	0.2	1	0.22	6					
Cellulose Acetate	12	0.18	1	0.2	6	100	15	10	0.3	0.1
Polyamide	9	0.05	0.5	0.05						
Dow Product	136	8	32		93					
Permea Product	22	0.4	2.3	0.4	9					
PDMS[2]	649	281	604							

Notes:
1. Permeation rates are relative, based on 1 for oxygen in polysulfone and cellulose acetate.
2. PDMS = poly(dimethyl siloxane); silicone rubber.
Sources: Tomlinson and Finn (1990) and Mazur and Chan (1982).

Lee et al. (1995) evaluated available data on the permeability of gases in cellulose acetate to obtain values for use in designing systems for removing CO_2 from natural gas. They concluded that the following would be reasonable values for computer simulation of the process:

Permeability rate for $CO_2 = 9 \times 10^{-5}$ $(cm^3)\{STP\}/(cm^2)(sec)(cmHg)$
  (or approximately 5.8 $(scf)/(ft^2)(hr)(100\ psi)$
Selectivity for $CO_2/CH_4 = 20$
Selectivity for $N_2/CH_4 = 1$
Selectivity for $(C_2+)/CH_4 = 0.4$

Additional permeability data for cellulose acetate membranes are provided by Li et al. (1990), Donohue et al. (1989), and Ettouney et al. (1995).

Gases with high permeability rates are often called "fast" gases. The data in **Table 15-2** indicate that low molecular weight and highly polar gases tend to be fast gases; while the slower gases are nonpolar and/or higher molecular weight. Detailed data on organic solvent permeation rates are given by Baker et al. (1986).

The separation factor of a membrane, α_{ij}, is defined as:

$$\alpha_{ij} = P_i/P_j \tag{15-5}$$

and is an indication of a membrane's ability to separate species i and j. A separation factor greater than 1 indicates that, at equal partial pressures, component i permeates through the membrane faster than j. Very high (or very low) separation factors result in easy separations. No separation is possible if $\alpha_{ij} = 1$.

Separation factors are influenced by membrane materials, feed composition, temperature, and pressure. Separation factors for various systems are presented in **Table 15-3.** Note that the separation factor for oxygen and nitrogen is considerably lower than those for other component pairs generally considered viable for commercial membrane separation. Oxygen and nitrogen are very similar in molecule size and solubility, which makes their separation via

Table 15-3
Separation Factors for Various Component Systems in Commercial Membranes

Component System	Ranges of Separation Factors	Typical Separation Factors in Cellulose Acetate
CO_2/CH_4	10–50	25
O_2/N_2	3–12	
H_2/CH_4	45–200	45
H_2/CO	35–80	
H_2/N_2	45–200	45
H_2S/CH_4	40–60	50
He/CH_4	60–100	60
CO_2/C_2H_6	44–52	50

Sources: Stookey et al. (1986) and Spillman et al. (1988).

membranes one of the more difficult processes. However, due to the abundance of "free" feedstock a high recovery is not required.

The separation factors for solvent/air processes have a very wide range depending on the membrane material. For example, the separation factor for toluene/nitrogen can be as low as 40 for nitrile rubber or as high as 10,000 for neoprene (Baker et al., 1986).

Design and Operating Considerations

The separation efficiency of a membrane for a given gas mixture will depend on the gas composition, the pressure difference between the feed and the permeate, and the separation factor for the two components at the specific operating conditions. The higher the separation factor, the greater the selectivity of the membrane and the higher the product purity.

The gas composition and pressure differential become very important when the more permeable gas in the feed has a low concentration. Since the partial pressure of the fast component on the permeate side cannot exceed its partial pressure on the feed side, high feed gas pressures and low permeate pressures are required to obtain efficient separations even with high separation factors. The differential pressure across the membrane relates directly to the membrane area required. Compression costs on the other hand are a function of pressure ratio. Therefore, operation at high pressure with a substantial pressure differential across the membrane but with a reasonably low pressure ratio, is economically advantageous where recompression of the permeate is required.

The general effects of varying key operating factors, while holding other conditions constant, for a typical single-stage membrane permeation system, can be summarized as follows:

1) Increasing the overall differential pressure across the membrane leads to an increase in permeate flow rate and a decrease in the concentration of the fast gas in the permeate stream. The differential pressure at any point in a module can be affected by pressure drop in either the feed or permeate flow channels. If the permeate flow rate is sufficiently high it can generate a back pressure that reduces the differential pressure and rate of permeation.
2) Increasing the feed gas flow rate decreases the percent recovery of the fast gas as permeate and decreases the purity of the residue. However, increasing the feed rate increases the purity of the permeate and increases the percent recovery of the slow gas in the residue.
3) Decreasing the feed gas flow rate below a critical value decreases the separation efficiency due to a boundary layer effect. The concentration of the fast gas is depleted in the feed gas adjacent to the membrane surface, reducing its partial pressure and therefore its rate of permeation. Since the rate of permeation of the slow gas is not affected (or is increased) this reduces permeate purity. The critical flow rate is determined by the degree of mixing at the membrane surface, and this is a function of gas velocity; gas properties such as viscosity, density, and diffusivity; and module design.
4) Increasing the temperature raises most permeabilities by about 10 to 15% per 10°C and has little effect on separation factors (Schell and Hoernschemeyer, 1982).
5) Increasing the membrane area increases the purity of the residue; while decreasing the membrane area increases the purity of the permeate.

In hydrocarbon membrane systems, permeation of fast gases (e.g., H_2 or CO_2) increases the concentration of heavy components (e.g., C_3+ hydrocarbons) in the remaining gas. The increased concentration of readily condensible components, as well as the decrease in temperature that frequently accompanies permeation, may cause the condensation of liquid hydrocarbons within the membrane unit, interfering with its operation. Typical remedies for this problem are preheating the gas and removing heavy hydrocarbons from the gas ahead of the membrane unit. Even where membranes can physically tolerate the condensate, performance will normally suffer. The condensed liquid (usually made up primarily of slow permeating components) can cover the membrane surface, forming an additional barrier to permeation. Alternatively, the liquid may wet the membrane creating a gas leakage path, which can result in permeate contamination.

Membrane life is an important factor affecting process economics. Membranes typically require replacement every three to seven years. It is therefore necessary to consider factors that may shorten or possibly extend their life. Funk et al. (1986) investigated the effect of impurities in the gas on a cellulose acetate membrane in acid gas service. They found that various components in the gas affect the permeability, the tensile strength, and the elastic modulus of the membrane. Some of the adverse effects of individual components can be mitigated by proper design (e.g., temperature control), but other impurities may require removal prior to membrane processing. Therefore, the feed gas should be evaluated for the presence of particulates, entrained liquids, oil mist (compressors), condensible components, all trace compounds, and water. Pretreatment systems that may be needed include:

1) A high efficiency separator to remove particles and oil mist
2) A liquid knock-out drum to remove liquid hydrocarbons and water
3) Preheat or reheat steps to raise the gas temperature sufficiently above its water and hydrocarbon dew point to prevent condensation in the module
4) Component removal steps to eliminate compounds, such as solvents, BTX, or ammonia, that may cause membrane deterioration

The design of pretreatment equipment must also account for both normal operating and upset conditions.

Since many of the process variables are highly interdependent, the overall design of membrane systems requires consideration of all requirements and conditions. This is illustrated in **Figure 15-1,** which shows the effects of percent hydrogen in the permeate, percent hydrogen recovery, and permeate pressure on the relative costs of recovering hydrogen in a typical application. These three factors, as well as others, must be evaluated to optimize overall economics for each specific case. Most membrane system suppliers have computer programs to optimize the system design (Poffenbarger and Gastinne, 1989; Spillman, 1989; MacLean et al., 1983).

Membrane and Module Configurations

Membranes

The key requirements for a membrane to be used in an economical gas purification or separation process are:

1) High permeability for the component to be removed
2) High selectivity for the component to be removed in relation to other components
3) High membrane stability in the presence of all gas components which will come into contact with the membrane
4) Uniformity—freedom from pinholes or other defects

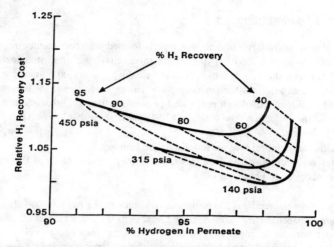

Figure 15-1. Effects of hydrogen permeate purity, permeate pressure, and percent hydrogen recovery on relative cost of hydrogen recovery from a 75% H_2, 815 psia refinery gas stream. (*Heyd, 1986*)

5) Low effective thickness of the active portion of the membrane to ensure a high permeation rate
6) Physical strength to withstand the required operating conditions

Items 1 through 3 are related primarily to the polymers used in fabricating the membrane; while items 4 through 6 relate to the fabrication method. A key breakthrough in the development of fabrication methods was provided by Loeb and Sourirajan (1960). They developed a technique for casting asymmetric cellulose acetate membranes with a uniform, very thin (0.1–1.0 micron) skin on a strong porous substrate (100–200 microns thick). Since this original work, which was actually aimed at the development of reverse osmosis membranes, the basic approach has been applied to a variety of polymeric materials and to sheet and hollow fiber configurations for use in both gas and liquid phase operations.

Development work has also investigated alternative asymmetric membrane systems, including (a) an ultra thin nonporous film laminated to a much thicker microporous backing (which may be a different material) and (b) a very thin nonporous film applied as a coating to a thicker microporous substrate (Stern, 1986). A complex membrane structure reportedly used in the Monsanto Prism separator is a "skinned" asymmetric hollow fiber of polysulfone coated with a thin film of silicone rubber (about 1 micron thick). The polysulfone skin (about 0.1 micron thick) is the active separator, while the silicone rubber serves to seal any defects in the base membrane without affecting the intrinsic permeability of the membrane (Koros and Chern, 1987).

Koros and Fleming (1993) present a comprehensive review of membrane-based gas separation with emphasis on membrane materials, formation techniques, and module designs. The most popular module configurations are the hollow fiber and spiral-wound designs due to their high packing density.

Spiral-Wound Configuration

In the spiral-wound configuration an envelope is formed with two membrane sheets separated by a porous support material. Typically the module consists of several such envelopes. The material between the membranes (permeate channel spacer) supports them against the operating pressure and defines the permeate flow channel. The envelope is sealed on three sides. The fourth side is sealed to a perforated permeate collection tube, and the envelope is wrapped around the collection tube with a net-like spacer sheet that has two functions:

1) It keeps adjacent membranes apart to form a feed channel.
2) It promotes turbulence of the feed gas mixture as it passes through the module, thus reducing concentration polarization.

During operation, the feed gas mixture enters one face of the module, travels axially along the feed channel spacer and membrane surface, and exits the other face as a residue or retentate. The more permeable gases pass through the membranes and travel in a spiral path inward within the envelope through the permeate channel spacer until they reach the perforated collection tube and finally exit as permeate (**Figure 15-2**). The feed channel spacer is a key feature of the spiral-wound module and, as shown by Da Costa et al. (1991, 1994), its design significantly affects module performance. Typically the modules have about 1,000 square feet of surface per cubic foot of volume and are 4–12 inches in diameter by 36–42 inches long. Up to six modules may be housed in a single pressure vessel shell.

Hollow Fiber Configuration

The hollow fiber configuration consists of thousands of hollow fibers packaged in bundles mounted in a pressure vessel resembling a shell and tube heat exchanger. For high pressure applications the fiber diameter is usually on the order of 100 µm ID and 150–200 µm OD. The bundles are capped on one end and have an open tube sheet on the other end

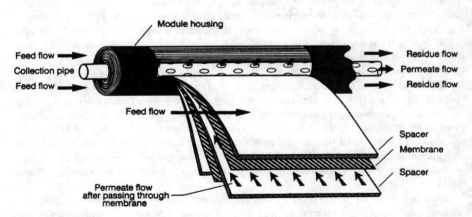

Figure 15-2. Diagram of a spiral-wound membrane permeation element. (*Courtesy of Membrane Technology and Research, Inc.*)

(**Figure 15-3**). The bundles typically have 3,000 square feet of membrane surface per cubic foot of module volume and the module dimensions range from 4–12 in. in diameter by 4–20 ft long. The feed gas is introduced on the shell side because hollow fibers are much stronger under compression than expansion. The faster permeating gases migrate into the fiber bore and exit via the open end of the bundle. For low pressure applications the fibers have a diameter greater than 400 μm and the feed gas enters the bore side while the permeate exits via the shell side. This configuration reduces pressure drop on the feed side.

Not all membrane materials can be made into a thin selective layer on a porous substrate in a hollow fiber form. Consequently, spiral-wound membranes, which can be made from a wider range of materials, usually have higher permeation rates. However, this is offset by the much higher packing density of hollow fiber modules, resulting in similar overall productivity per unit module volume for the two configurations. This situation could change if developments in polymer science lead to more effective thin films in a hollow fiber form.

Numerous mechanical designs of modules have been developed for both the spiral-wound and hollow-fiber concepts. The various designs are aimed at optimizing such features as: membrane area per unit volume, gas flow distribution and pressure drop, seals and fastenings, and assembly technology. With either spiral-wound or hollow-fiber systems, large

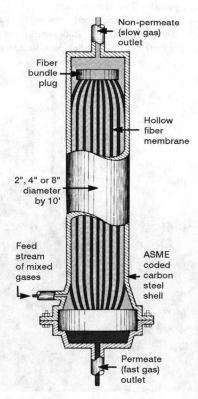

Figure 15-3. Diagram of a hollow-fiber membrane permeation element. (*Courtesy of Permea*)

1250 Gas Purification

commercial installations normally require a large number of individual modules. This is evident in **Figure 15-4,** which shows a UOP Advanced Membrane System for removing carbon dioxide from natural gas.

A third module design—plate and frame—utilizes a stack of disk shaped membranes separated by sheets of porous filter paper and membrane support plates. The flow pattern is very much like that of a plate and frame filter. This design is not competitive for large commercial applications because of the relatively small membrane area per unit volume attainable. However, plate and frame designs are used for producing oxygen enriched air in small medical applications.

Flow Arrangements

To obtain the specified flowrates and purities, the optimum arrangement of the modules needs to be determined. Series flow, **Figure 15-5,** provides for high recoveries for a given feed rate. The permeate purity varies from module to module, which makes it possible to produce multiple permeate products. The substantial velocity changes that occur when large portions of the feed are recovered as permeate can be accommodated by using successively smaller elements. The series flow arrangement has a slightly higher pressure drop between the feed and the residue than a parallel flow arrangement.

Parallel flow, **Figure 15-6,** allows for higher feed rates for the same recovery. Turn down is accomplished by taking elements out of service. In designing parallel flow systems, particular care must be paid to the gas distribution systems. Non-uniform gas flow to the elements

Figure 15-4. Large commercial installation of UOP Advanced Membrane System for removing carbon dioxide from natural gas. (*Courtesy of UOP*)

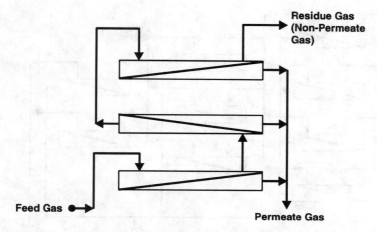

Figure 15-5. Series flow configuration.

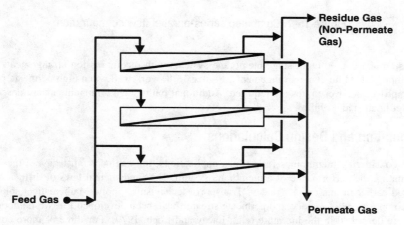

Figure 15-6. Parallel flow configuration.

can result in declines in product purity and recovery. High recovery in one element can cause liquid formation.

The combined series-parallel flow sequence, **Figure 15-7,** provides for good turndown ratios while accommodating high feed rates with low residue flow rates. A reduction in the number of modules in parallel accommodates the reduction in flow as permeate is removed.

As stated earlier, separation by membrane permeation is not an absolute separation. Each species has a finite permeability through the membrane and the enrichment is achieved due to relative permeabilities, not zero permeability for one of the species. As higher purities are approached, recovery of product declines rapidly and single-stage systems become increasingly inefficient. Therefore, single-stage processes frequently are applicable to bulk removal of a species or the concentration of feeds to other purification processes.

1252 Gas Purification

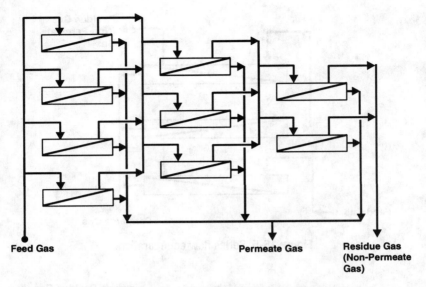

Figure 15-7. Combined series-parallel flow configuration.

As an alternative, multi-stage membrane systems with recompression arrangements may be employed. Multi-stage arrangements, **Figure 15-8,** work well when high permeate purity or improved recovery rates are required. Additional multi-stage arrangements are described by Spillman et al. (1988).

Simulation and Design Calculations

In considering membranes for various applications, the engineer must have a means of estimating the performance of a membrane system. The fundamental laws of diffusion discussed earlier in this chapter under "Transport Mechanisms" apply. Mathematical solutions for compositions of residue and permeate streams have been developed for two components and are described in the literature (e.g., Lacey and Loeb, 1979). For three or more compo-

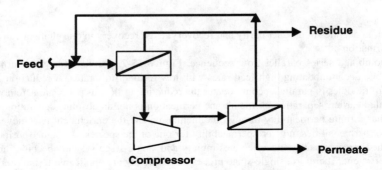

Figure 15-8. Multi-stage flow configuration.

nents a strict mathematical solution to the differential equations has not been found. However, a numerical solution is possible, and with the prevalence of computers, this approach is not only practical but widely used.

The basic equation defining the rate of flow of a component through a membrane is equation 15-3 which can be written in the form:

$$J_i = R_i (p_{i,feed} - p_{i,perm}) \tag{15-6}$$

Where: J_i = steady state flux of i, moles/(time)(area) = M_i/A
M_i = flow rate of i through a given area, moles/time
$R_i = \dfrac{k_i D_i}{l} = \dfrac{P_i}{l}$
p_i = partial pressure of i in the gas
A = area
k_i = solubility coefficient for i
D_i = local diffusivity for i
l = membrane thickness
P_i = permeability coefficient for i

Any consistent set of units may be used. Gas volume (e.g., scf) is often used instead of moles.

The partial pressure of each individual component changes as the separation is carried out because different components are being removed from the feed (high pressure) side at different rates. Because the partial pressure of individual components is a function of position along the membrane surface, mathematical integration of equation 15-6 over the entire length of a membrane surface is an interesting exercise for two components, but not practical for three or more components.

Commercial membranes today are asymmetric membranes. This means that the active membrane surface is only a very thin layer on top of a porous substrate. Components which permeate the membrane must travel out of the porous membrane substrate before entering the bulk permeate stream. Therefore, the effective partial pressure and mole fraction on the permeate side of the active membrane surface are only a function of the material passing through the membrane, not a function of the bulk permeate stream.

A computer simulation of the process can be made by considering small incremental areas of the membrane individually. A permeation analysis and a material balance are performed on the first incremental area. The residue gas from this area is treated as the feed to the next area and the operation is repeated. The analysis continues, adding areas until the residue gas meets the product purity requirement or other process requirements are attained.

If the assumed areas are small enough, the feed and residue gas compositions for each ΔA are similar enough that the driving force for permeation can be based on the composition of the feed rather than on an average of the feed and residue compositions. The permeate composition for each incremental area must take into account the permeation rates of all components, but is not affected by permeate from the other areas.

Figure 15-9 is a simple diagram showing gas flows adjacent to and through an increment of area, ΔA. The computer simulation is based on this diagram and the following equations:

For a small increment of membrane area, equation 15-6 becomes

$$M_i = R_i (p_{i,feed} - p_{i,perm}) \Delta A \tag{15-7}$$

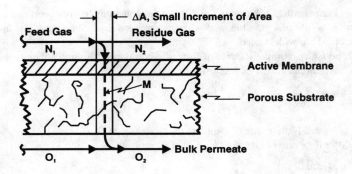

N₁ = Flow of Feed to ΔA
N₂ = Flow of Residue from ΔA
   = Flow of Feed to Next ΔA
O₁ = Bulk Flow of Permeate Prior to ΔA
O₂ = Bulk Flow of Permeate After ΔA
M = Flow of Permeate Through ΔA

Figure 15-9. Diagram of gas flow pattern at small increment of membrane area for computer simulation model.

Substituting for partial pressures:

$$M_i = R_i (y_{i,feed} \pi_{feed} - y_{i,perm} \pi_{perm}) \Delta A \tag{15-8}$$

The mole fraction of any component, i, in the permeate is

$$y_{i,perm} = M_i / \Sigma M_i \tag{15-9}$$

Similarly, the mole fraction of component i in the residue from increment area ΔA is

$$y_{i,residue} = N_{2i} / \Sigma N_{2i} \tag{15-10}$$

Combining equations 15-8 and 15-9 yields

$$M_i = \frac{R_i \Delta A \, y_{i,feed} \, \pi_{feed}}{\left[1 + \dfrac{R_i \Delta A \, \pi_{perm}}{\Sigma M_i}\right]} \tag{15-11}$$

This is the key equation in the computer simulation program. Other important equations define the increase in bulk permeate and decrease in residue gas flow at each ΔA:

$$O_{2,i} = O_{1,i} + M_i \tag{15-12}$$

$$N_{2,i} = N_{1,i} - M_i \tag{15-13}$$

In equations 15-7 through 15-13,

ΔA = incremental area
M_i = flow of component i through ΔA, moles/time
ΣM_i = flow of all components through ΔA
N_1 = flow of feed gas to ΔA
N_2 = flow of residue gas from ΔA
O_1 = bulk flow of permeate from prior ΔA's
O_2 = bulk flow of permeate after ΔA
R_i = permeability rate for i, moles/(time)(area)(partial pressure differential)
y_i = mole fraction i in gas
π = total pressure

The permeation rate for each component i must be known, and the total feed pressure and permeate pressure must also be known.

A reasonable ΣM_i is assumed for the first increment of area, ΔA. It is then corrected by subsequent iterations, until a solution is found where M_i calculated for each component equals (within the specified tolerance) the M_i used in the previous iteration. The sum of all M_i from each iteration is also used for the calculation of the new ΣM_i. At the first increment of membrane area, $y_{i,feed}$ is simply the mole fraction of component i in the feed.

After the solution for M_i for each component has converged, the residue gas molar flow, $N_{2,i}$, for each component is calculated by subtracting M_i, the permeate flow, from the feed flow $N_{1,i}$. This residue flow then becomes the feed flow to the next small increment of membrane area.

The procedure is repeated until the desired product specifications are met. After each increment of area is calculated, the permeate flow for each component, $O_{1,i}$, is increased by the permeate flow for that incremental area, M_i. Similarly, at each increment of area, the residue gas for each component, N_1, is reduced by the permeate flow passing through that increment of area, M_i. The sum of all $N_{2,i}$'s from the final increment of membrane area is the residue gas for the system. Each increment of area is also accumulated so that the total area required is known as well as the gas compositions and quantities of the residue and permeate streams.

A simplified flow sheet for the required calculations is shown in **Figure 15-10.**

Table 15-4 summarizes computer output from a program using the calculation logic outlined in **Figure 15-10.** The data are from a computer simulation of a spiral-wound cellulose acetate membrane unit recovering CO_2 from 7.4 MMscfd (812 mol/hr) of a high CO_2 content gas, rich in hydrocarbons, as might be found in a CO_2 flood enhanced oil recovery (EOR) project. Input to the program consisted of the feed composition, flow rate, pressure, and temperature; the permeate pressure; and the specification that the residue gas contain 40 vol% CO_2.

In addition to the flow rates and compositions of the residue gas and permeate, the output indicates the approximate membrane surface area required, and the number of increments of membrane area used for the calculation. At 1,000 sq ft of membrane surface area/cubic foot of membrane module volume, approximately 13.1 cu ft of membrane module volume is required. A membrane module 8 in. in diameter and 36 in. long occupies approximately 1 cu ft and contains approximately 1,000 sq ft of membrane surface area. Therefore, the simulation indicates that 13 modules (8 in. diam × 36 in.) are required for the separation.

1256 *Gas Purification*

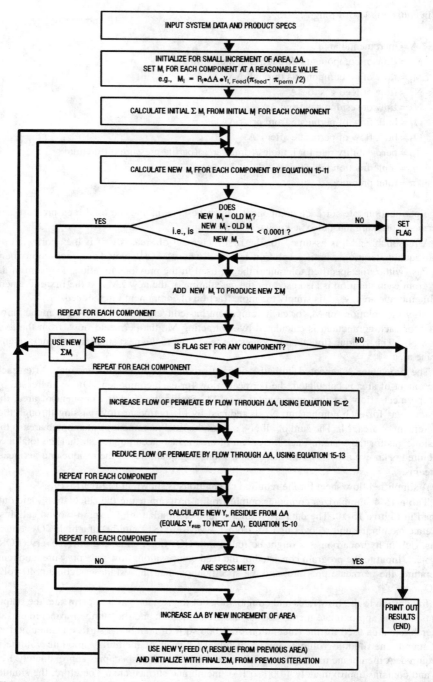

Figure 15-10. Calculation flowsheet for computer simulation of membrane permeation.

Table 15-4
Computer Simulation of CO_2 Recovery Using a Single-Stage, Spiral-Wound Cellulose Acetate Membrane Unit

Component	Feed MOL/HR	Feed MOL FRAC.	Residue MOL/HR	Residue MOL FRAC.	Permeate MOL/HR	Permeate MOL FRAC.
Carbon dioxide	735.58	.905887	44.29	.399763	691.29	.985849
Nitrogen	1.32	.001626	1.12	.010074	.20	.000291
Methane	40.24	.049557	33.17	.299400	7.07	.010084
Ethane	11.74	.014458	10.65	.096103	1.09	.001559
Hydrogen sulfide	.10	.000123	.00	.000006	.10	.000142
Propane	12.41	.015283	11.63	.104938	.78	.001119
Isobutane	2.28	.002808	2.14	.019279	.14	.000206
Butane	4.71	.005800	4.41	.039827	.30	.000425
Isopentane	1.41	.001736	1.32	.011923	.09	.000127
Pentane	1.34	.001650	1.26	.011331	.08	.000121
Hexane	.46	.000567	.43	.003890	.03	.000041
Heptane	.41	.000505	.38	.003467	.03	.000037
TOTAL	812.00		110.78		701.22	

Area = 13,110.9 Square feet, 125 Increments used

Note:
Operating Conditions: Feed at 500.0 psia, 140.0 degrees F, permeate 100.0 psia.
Source: Schendel (1995)

The permeation rate, R_i, is not necessarily a constant. For example, it is known to vary with temperature and pressure, and corrections for these factors are readily included in the calculations. Because changes in composition from increment to increment are normally small, even component interference effects can be included in the calculations if required.

Gas composition effects can be significant. The concentration of carbon dioxide in natural gas, for example, can affect its rate of permeation (per unit of partial pressure differential). Data provided by Hogsett and Mazur (1983) indicate that increasing the CO_2 concentration from 15 to 50% in 200 psia natural gas increases the CO_2 permeation rate from about 4.0 to 5.0 (scf)/(ft^2)(hr)(100 psi) with GASEP membranes.

Hogsett and Mazur (1983) also suggest a simplified approach for estimating the approximate membrane area required for a multicomponent system. The approach avoids the need to use a computer simulation model, but is reportedly accurate to only about ± 20% of the actual required area. The method is based on the following steps:

1) Sort all components into two groups (fast and slow permeators), with the fast permeators typically having permeation rates 15 times those of the slow permeators.
2) Calculate weighted average permeation rates for the two groups.
3) Use equations developed for two-component systems to calculate the membrane area required to meet fast permeator removal requirements.

The required two-component equations and details of the calculation procedure are given by Hogsett and Mazur (1983).

Ettouney et al. (1995) compared two-component and four-component models for the analysis of natural gas well enrichment. In the flow system evaluated, gas from the well is recycled through a permeation unit and back to the well. Removal of a CO_2 and H_2S-rich permeate results in an increase in purity of the well contents with time. Results of the study showed that proper design of this system requires that the models take into consideration flow patterns, the presence of more than one species, and permeability rate functions that include the effects of both composition and pressure. The use of a simplified two-component model with constant permeabilities gave large deviations from the more detailed models.

Fields of Application

Membrane systems are now available that are economically attractive for many applications, and the fields of application are growing steadily. The main current commercial applications of membrane-based gas purification and separation are:

1) Hydrogen recovery from nitrogen-bearing gases, e.g., ammonia synthesis purge gas
2) Hydrogen removal and recovery from hydrocarbons (e.g., methane) and other slower permeating gases (e.g., carbon monoxide)
3) Removal of carbon dioxide, hydrogen sulfide, and water vapor from methane and other hydrocarbon gases (e.g., upgrading natural gas to meet pipeline specifications)
4) Oxygen or nitrogen separation from air (e.g., producing nitrogen for inert blanketing)
5) Helium removal and recovery from natural gas
6) Solvent vapor removal from exhaust gases

The small molecular size of hydrogen and helium allow them to diffuse rapidly through membranes. Therefore, these elements are readily separable from larger molecule gases such as methane and heavier hydrocarbons. The acid gases, carbon dioxide and hydrogen sulfide, and water have larger molecules and diffuse more slowly than hydrogen or helium. However, they are much more soluble in polymers used for the manufacture of membranes. Since permeability is the product of solubility and diffusivity, highly soluble gases can have permeation rates comparable to those of lighter gases; while gases such as methane, which has both a relatively large molecule and low solubility, have low permeation rates.

The separation of oxygen and nitrogen is difficult because the size and shape (and hence the diffusivity) of the molecules are quite similar. In addition, the solubility and diffusivity of the faster gas (oxygen) are generally quite low, resulting in a low rate of permeation and the requirement for a large membrane area. However, because of the industrial importance of this separation and the scarcity of simple alternatives, it has been the subject of extensive research and development work. In addition, the feedstock is "free." Therefore, high recovery is not a requirement for this separation. Low recovery operation can be used to improve the separation efficiency (i.e., the permeate purity).

Organic solvents typically exhibit low diffusivity rates but very high solubilities in appropriate membranes. As a result, satisfactory permeation rates can be obtained relative to the other components in exhaust gases. The partial pressure differential can be improved by operating the permeate side under vacuum. This is normally more economical than compressing large volumes of atmospheric pressure exhaust air. However, the process is not effective in the extremely low partial pressure range required for removing traces of organic solvents from air, and is used primarily for the bulk removal of solvents from relatively concentrated exhaust air streams.

APPLICATION CASE STUDIES

Hydrogen

Ammonia Synthesis Purge

Ammonia is produced by reacting hydrogen with nitrogen over a catalyst. The hydrogen is usually produced in a steam-methane reformer (SMR) and the nitrogen comes from the air supplied to a secondary reformer. Not all the methane (natural gas) is converted to hydrogen in the SMR nor is the conversion of nitrogen and hydrogen to ammonia complete in the synthesis reactor. The reactor product gases, after removal of the ammonia, are recycled back to the reactor feed to improve yields. Inert gases, such as argon from the air and methane, build up in this closed loop and reduce the nitrogen-hydrogen reaction rate. Therefore, a continuous gas purge to the fuel gas system is maintained to keep these inerts at a manageable level. However, this purge gas contains valuable hydrogen. Currently many ammonia plants recover hydrogen from the purge stream. This is a popular application for membranes because of the high permeation rate of hydrogen and the high pressure of the purge gas.

Shirley and Borzik (1982) reported on the installation of a hydrogen recovery membrane system in a 1,000 ton per day ammonia plant that resulted in a 5% overall capacity increase. MacLean et al. (1988) reported the installation of a membrane unit in a 600 ton per day ammonia plant. The unit provided an 89% hydrogen stream with an 86% hydrogen recovery, representing a 4% ammonia plant capacity increase.

A comparison of membrane separation versus cryogenic separation in a typical large ammonia plant was made by Schendel et al. (1983). For the case study described in this paper, the operating conditions were modified to increase the methane content at the entrance to the synthesis loop to about double that allowed without hydrogen recovery. The pertinent process variables for the 15 MMscfd feed stream are summarized in **Table 15-5**. To prevent densification of the membrane or formation of an insoluble phase in the cryogenic system, the ammonia in the feed to both systems is reduced to very low levels in a water scrubber. To prevent formation of a solid phase in the cryogenic unit, molecular sieves are used to remove the water picked up in the scrubber.

The inerts concentration in the synthesis loop is held constant. Since the cryogenic system produces a slightly purer recycle hydrogen stream, the purge rate required to hold the inerts at a fixed level is less in this system. Hydrogen permeates through the membrane and is recovered as a low pressure product. The recompression requirements of the membrane system are, therefore, considerably greater than those of the cryogenic unit. To reduce the recompression costs, the membrane unit is operated in two stages.

The utility requirements are also presented in **Table 15-5**. The external power requirement for the compressor horsepower is that required to integrate the recovery unit into the total plant. The primary advantage of the cryogenic unit is the lower recycle recompression requirements. The higher on-skid electrical costs for the cryogenic unit reflects the use of the molecular sieve to dry the feed stream to the recovery unit. The cryogenic unit also requires a small purge stream of nitrogen.

The estimated capital equipment costs are virtually identical, approximately $1.35 million (1983 dollars). The membrane unit consists of two skids, while the cryogenic unit requires

Table 15-5
Comparison of Membrane and Cryogenic Separation Units for Hydrogen Recovery in an Ammonia Plant

	Membrane Type, Monsanto PRISM	Cryogenic Type, Petrocarbon S-2000
Feed Composition, mole%		
H_2	60.8	60.8
N_2	20.0	20.0
CH_4	12.1	12.2
Ar	3.2	3.1
NH_3	3.9	3.9
Feed Gas Quantity, lb moles/hr	1,767	1,503
Pressure at Separator Inlet, psig	1,973	1,000
Hydrogen Recovery, %	95.7	94.6
Hydrogen Purity, mole%	87.8	92.5
Ammonia Recovery, %	99.8	98.7
Recycle Product to High Stage Compressor, %	49.8	100
Recycle Product to Low Stage Compressor, %	50.2	—
Electricity, kWh/h	30	80
Cooling water, gpm	200	225
Steam (600 psig), lbs/hr	1,910	1,760
Nitrogen, scfh	Startup only	180
Instrument air, scfh	2,100	1,800
Turbine condensate, gpm	12	Minor make-up
External power differential, kWh/h	470	—

Source: Schendel et al. (1983)

four. When installation and maintenance costs are factored in, the two processes are considered to be competitive.

Oxo-alcohol Synthesis Gas

In the production of oxo-alcohols, carbon monoxide is reacted with hydrogen at a one-to-one ratio to form an aldehyde. The aldehyde is then reacted with pure hydrogen to form the desired oxo-alcohol product. The hydrogen and carbon monoxide synthesis gas is made by steam reforming of natural gas or by partial oxidation of hydrocarbons. The raw synthesis gas has a hydrogen to carbon monoxide ratio range of 3:1 to 2:1. Before the synthesis gas can be used, the ratio of hydrogen to carbon monoxide must be adjusted. Cryogenics, molecular-sieves, pressure swing adsorption (PSA), and membranes are used to make the ratio adjustment. The following case histories indicate how membranes have been used to increase capacity and/or efficiency in achieving the correct ratio of hydrogen/carbon monoxide for synthesis.

MacLean and Graham (1980) reported on a membrane system used to debottleneck Monsanto's Texas City plant where the cold box did not have enough capacity to supply carbon monoxide feedstock for acetic acid production. The cold box had three product streams, a pure carbon monoxide stream for acetic acid, a 1.3:1 hydrogen/carbon monoxide stream for oxo-alcohols, and a hydrogen stream for methanol. The solution was to install a membrane separation system in parallel with the cold box. The 3.1:1 H_2/CO feed was split between the membrane unit and the cold box. The cold box was then operated to produce hydrogen and the pure carbon monoxide required for acetic acid production. The membrane unit produced the 1.3:1 H_2/CO ratio stream required for oxo-alcohol production while recovering 93% of the carbon monoxide. The permeate stream, consisting of 96% pure hydrogen, was fed to the methanol plant. As an indication of membrane aging, the separation efficiency declined less than 10% during the first three years of operation.

A membrane unit was also installed on the Texas City methanol plant purge gas stream (Burmaster and Carter, 1983). The membrane unit recovered about one half of the hydrogen and carbon dioxide that was previously sent to the fuel gas system. The methanol plant capacity was increased by 2.6%, and except for compressor limitations, could have been increased by 3.9%.

The integration of a membrane system with pressure swing adsorption (PSA) at an aldehyde synthesis plant was studied by Doshi et al. (1989). Synthesis gas ratio adjustment using PSA alone produces high purity hydrogen as well as the proper syngas ratio. However, the drawback of the system is the relatively expensive tail gas compression that is required. PSA tail gas is produced at low pressure and requires a large compressor to boost the pressure to the aldehyde synthesis pressure. The use of a membrane system alone will produce syngas at the proper hydrogen/carbon monoxide ratio, but the purity of the hydrogen permeate stream is low. The syngas produced is at approximately the same pressure as the feed gas and can be used for aldehyde synthesis without additional compression. The hydrogen permeate stream generally has to be upgraded, which usually results in the loss of carbon monoxide from the system.

An integrated system that uses both membranes and PSA on the product gas from a partial oxidation (POX) unit, minimizes the drawbacks of either system when it is used alone. In the integrated system, **Figure 15-11,** the feed gas enters the membrane unit where the residue stream, enriched in carbon monoxide, is produced at approximately feed pressure. The lower pressure hydrogen permeate stream is fed to a PSA unit where the carbon monoxide is adsorbed and a high purity hydrogen stream is produced. The low-volume, CO-rich PSA tailgas stream is compressed and combined with the residue stream from the membrane unit. The study shows the integrated system to have lower capital and operating costs than either of the individual systems and to provide 100% recovery of hydrogen and carbon monoxide. An economic comparison of a stand-alone PSA and an integrated membrane–PSA system is presented in **Table 15-6.**

Catalytic Reforming

Catalytic reforming is a process in which hydrocarbon molecules are structurally rearranged to higher octane forms. The reforming process is a net producer of hydrogen that, if recovered, can be used in hydroprocessing. The following case history of a demonstration membrane system for the recovery of hydrogen from a catalytic reformer unit (CRU) off-gas was described by Yamashiro et al. (1985).

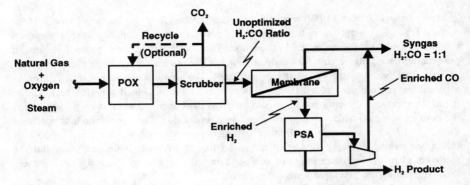

Figure 15-11. Integrated membrane plus PSA system in a natural gas partial oxidation (POX) plant. (*Doshi et al., 1989*)

Table 15-6
Comparative Economics of PSA Alone vs. a Membrane/PSA Integrated System for Producing Hydrogen and Aldehyde Syngas

Basis:	
Raw Feed Rate, MMscfd	20.0
Synthesis Gas Rate, MMscfd	14.4
Hydrogen Product Rate, MMscfd	5.6
Raw H_2/CO Gas Pressure, psig	420
Aldehyde Reaction Pressure, psig	400
Hydrogen Use Pressure, psig	150

COMPOSITIONS, MOL-%	RAW GAS H_2/CO	H_2 PRODUCT	SYN. GAS
H_2	63.4	99.999	49.1
CO	35.4	10 ppm	49.1
Ar/N_2	0.4	trace	0.6
CH_4	0.8	trace	1.2
CO_2	trace	trace	trace
H_2O	Sat'd	Dry	Sat'd

	PSA ONLY	MEMBRANE + PSA
Compression Required, BHP	1080	415
Separation Equipment Cost, $MM US	1.425	1.375
Installation Cost, $MM US	0.175	0.225
Installed Compressor Cost, $MM US	0.864	0.332
Compressor Operating Cost, $MM US (3 yr–8000 hr/yr–5¢/kWh)	0.966	0.371
Total Capital Cost + 3 yrs Operation, ($MM US)	3.430	2.303

Source: Doshi et al. (1989)

The 850 Mscfd unit was installed in the No. 2 Reformer at the Cosmo Oil Refinery in Chiba, Japan. The process flow for the unit is shown in **Figure 15-12.** The reformer effluent is cooled, the liquid and gas separated, and the gas fed to an absorber to remove the heavy components. Gas from the absorber, containing approximately 80% hydrogen and 20% methane and saturated with absorption oil components at a dew point of 95°F and a pressure of 398 psia, is fed to a filter separator to remove any residual liquids. It is then preheated to a temperature above the hydrocarbon dew point of the residual gas leaving the membrane unit and fed to the membrane separator. Heating the gas prevents condensation of the heavy hydrocarbons as the gas dew point increases with hydrogen removal.

The 97% hydrogen permeate gas is of sufficient purity and pressure that it can be fed directly into the 256 psia hydrogen supply system. Typical operating conditions are provided in **Table 15-7.** The unit was operated at a low pressure ratio to deliver hydrogen to the refinery hydrogen supply system without recompression. This low pressure ratio resulted in a low hydrogen recovery of about 30%.

During startup, the temperature of the feed was varied from 104° to 180°F. At higher temperatures the gas permeation rate increased by more than a factor of two. A corresponding decrease of membrane selectivity was also noted, but was not great enough to alter the system performance significantly. The permeate gas pressure was also varied while maintaining constant feed pressure. It was shown that membrane selectivity and hydrogen permeation rate are independent of differential pressure and pressure ratio for the range of conditions studied (Schell and Houston, 1985).

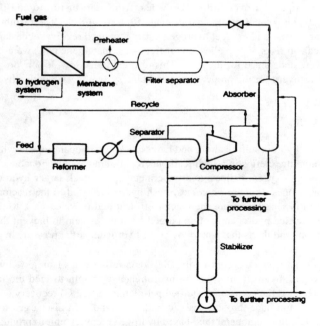

Figure 15-12. Flow diagram of reformer system with membrane hydrogen recovery unit on offgas (*Yamashiro et al., 1985*). Reproduced with permission from Hydrocarbon Processing, February 1985

Table 15-7
Operating Conditions for Catalytic Reformer Offgas Hydrogen Recovery Membrane System

	Feed	Permeate	Residue
Flow, Mscfd	850		
Composition, vol %:			
H_2	80	97	74
CH_4	20	3	26
Dew Point, °F	95		
Inlet Pressure, psig	380	260	
Inlet Temperature, °F	140		
Hydrogen Recovery, %	30		

Source: Yamashiro et al., 1985

Another example of membranes being used for the recovery of hydrogen from a CRU was reported by Lane (1983). The unit feed was the offgas from a CRU that formerly went to a hydrogen plant. The feed gas flow rate was nominally 4 MMscfd, but varied from 2 to 10 MMscfd. The membrane unit typically produced 98% hydrogen at 250 psig with 36% recovery. The system operated over full CRU cycles, from start-of-run to end-of-run, which caused the hydrogen content of the feed to range from 62 to 87%. The membrane unit percent hydrogen recovery and product hydrogen purity remained relatively constant.

To lessen scaleup risks, the smallest commercial-size membrane units were installed even though they were undersized for the flow. This resulted in the low 36% hydrogen recovery. The predicted recovery for the proper size membrane system is 93% with a hydrogen purity of 94%.

Hydroprocessing

Hydrotreating, hydrodesulfurization, and hydrocracking are operations in which hydrogen is used to saturate the olefins in a hydrocarbon stream; remove objectionable elements such as sulfur, nitrogen, oxygen, halides, and trace metals; and crack larger hydrocarbon molecules into smaller ones. In these processes, fresh hydrogen is fed to the reactor, and unconsumed hydrogen is separated from the reactor effluent and recycled back to the reactor. A portion of the recycled hydrogen is often purged from the system to prevent the build up of light hydrocarbons and inerts that would lower the hydrogen partial pressure in the reactor.

Hydrotreating. At Conoco's Ponca City Oklahoma refinery, a 71 mole % hydrogen high-pressure purge gas stream from a gas-oil hydrotreater was split to feed the light-cycle oil hydrodesulfurizer and the cryogenic liquified petroleum gas (LPG) recovery unit. The purge stream was used on a once-through basis in both units and then discharged to the fuel gas system. The installation of a membrane-based hydrogen recovery unit to produce high purity hydrogen from this purge stream was described by Shaver et al. (1991). A schematic of the system is presented in **Figure 15-13.** The membrane unit is designed to produce 7 MMscfd of 95 mole % hydrogen from the 12 MMscfd hydrotreater high pressure purge with a 75%

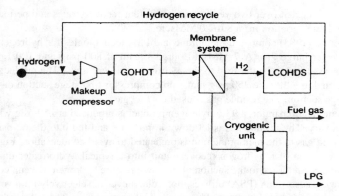

Figure 15-13. Flow diagram of membrane system installed to purify gas oil hydrotreater (GOHDT) offgas to provide hydrogen for light-cycle oil hydrodesulfurizer (LCOHDS) (*Shaver et al., 1991*). *Reproduced with permission from Hydrocarbon Processing, June 1991*

hydrogen recovery. The high-purity permeate hydrogen is sent to the hydrodesulfurizer. The available pressure drop from 1,050 psig at the hydrotreater to 430 psig at the hydrodesulfurizer provides the driving force for the membrane separation. Because of the high purity of this permeate hydrogen, the hydrodesulfurizer offgas stream is still 90 plus percent hydrogen, and can be recycled back to the gas oil hydrotreater. The residue stream from the membrane unit is fed to the cryogenic unit for recovery of LPG. The pretreatment of the feed to the membrane unit consists of a knock out drum, a feed preheater, and a dry gas filter. The economics of this system is presented in **Table 15-8** and indicates a 1.7-year payback period.

Table 15-8
Economics of Membrane Hydrogen Recovery System on Hydrotreater Offgas

Investment, $	662,000
Debits, $/yr	
Steam consumption	22,000
Lost hydrogen fuel value	75,000
Maintenance and overhead	29,000
Total	126,000
Credits, $/yr	
Gas oil HDT product upgrade	396,000
LCO HDS product upgrade	74,000
Reduced power consumption	42,000
Total	512,000
Earnings, $/yr	386,000
Simple payback period, yr	1.7

Source: Shaver et al. (1991)

1266 Gas Purification

Operating data collected over two years show consistent recovery rates and permeate purity even with feed rate variation from 8 to 20 MMscfd.

Schendel et al. (1983) compared membrane and PSA technologies for hydrogen recovery from hydrotreater purge gas. A 7 MMscfd purge stream at 800 psig and 100°F with 72% hydrogen was assumed to be the feed to the hydrogen recovery units. The product hydrogen could be returned to either the 250 psig make-up compressor first stage suction or to the 450 psig inter-stage suction. The residue gas is used for fuel gas.

For the membrane separator, a 93% hydrogen product is produced at 250 psig with an 81% hydrogen recovery. The residue gas is let down in pressure and fed into the 60 psig fuel gas system. The feed gas to the membrane unit is preheated to avoid condensation. **Figure 15-14** shows the integration of a hydrogen recovery unit into a typical hydrotreater unit with the recovered hydrogen returned to the suction of the low-pressure hydrogen makeup compressor.

Pressure swing adsorption (PSA) utilizes molecular sieves to selectively remove hydrocarbons and other impurities to produce a high purity hydrogen stream. The greater the pressure swing (between the high pressure of adsorption and the low pressure of desorption), the greater the unit capacity and product recovery.

The PSA system evaluation was based on two different operating scenarios: product at 450 psig and residue gas at 60 psig, and product at 250 psig and residue gas at 5 psig. In the 450 psig product case, the hydrogen is returned to the compressor interstage suction and the residue gas is fed to the 60 psig fuel gas system. In the 250 psig product case, the hydrogen is returned to the 250 psig compressor suction and the residue gas is sent to a low pressure burner. The hydrogen recovery was much greater for the low pressure case.

The operating conditions and the operating and capital costs for the three cases are presented in **Table 15-9.** As can be seen from the cost data, the membrane system shows a

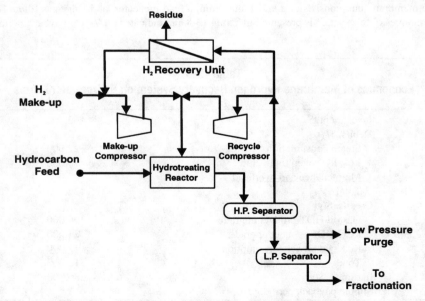

Figure 15-14. Integration of hydrogen recovery unit into typical hydrotreater system. (*Schendel et al., 1983*)

Table 15-9
Operating Conditions and Economics for Hydrogen Recovery from Hydrotreater Purge Gas by Membrane and PSA Systems

	Membrane	PSA	PSA
Waste Gas Pressure, psig	(1)	60	5
Feed H_2, %	72	72	72
Feed Pressure, psig	800	800	800
Feed Temperature, °F	100	100	100
Feed Flow Rate, MMscfd	7	7	7
High Purity H_2, %	93	99.5	99.9
High Purity H_2, psig	250	450	250
High Purity H_2, MMscfd	4.42	3.04	4.05
Waste Gas H_2, %	36	51	34
Waste Gas Flow Rate, MMscfd	2.58	3.96	2.95
H_2 Recovery, %	81	60	80
Capital Costs, M$			
Equipment	530	1,050	875
Installation	100	175	150
Total Cost	630	1,225	1,025
Operating Costs, M$/yr[2]			
Steam for Preheat	10	—	—
Compression to Reactor Press.	140	46	130
Contrib. to H_2 Cost, $/Mscf	0.11	0.05	0.09
Total H_2 Cost, $/Mscf[3]	0.20	0.29	0.24

Notes:
1. Membrane system residue gas produced at operating pressure and, after pressure reduction, fed into the fuel gas system.
2. Utility costs based on 5¢/kwh and $5/MM Btu.
3. H_2 cost based on 8,000 hr/yr operation for a five-year span.

Source: Schendel et al. (1983)

lower capital cost and lower total hydrogen cost than either of the PSA cases. The membrane and the low-pressure PSA case incur a significant cost for recompressing the hydrogen to reactor pressure. The low-pressure PSA case has the better economics of the two PSA cases. However, it may be difficult to find a use for the 5 psig waste gas. Also, depending on the sensitivity of the hydrotreater to hydrogen partial pressure, the higher purity of the PSA gas may have definite advantages.

Tonen Technology K.K. and UBE Industries (1990) reported on the installation of a hydrogen recovery facility at the Wakayama refinery of Tonen Company, Ltd. Polyimide membrane modules, arranged in eight rows of two trains each, were designed to produce 5,153 scfm of product hydrogen with a minimum purity of 95%. The feed is filtered to remove hydrocarbon mist and then preheated to 160°–170°F prior to entering the modules. The hydrogen-rich permeate is collected and fed to the hydrogen system. The residue gas is collected, cooled to 150°F, and discharged to the fuel gas system. Plant operating data are presented in **Table 15-10**.

Table 15-10
Operating Data for Hydrogen Recovery Membrane System at Wakayama Refinery

	Feed	Product	Offgas
Flow Rate, scfm	15,490	10,095	5,395
Pressure, psig	343	124	102
Temperature, °F	90	155	99
Composition, mol %			
H_2	77.6	98.2	39.1
C_1	17.0	1.6	45.8
C_2	4.4	0.2	12.2
C_3	0.3		.09
C_4+	0.7		20
H_2S, ppm	400	300	590
BTX, ppm	300		
Mw	6.30	2.3	13.81
H_2 Recovery, %		82.5	

Source: Tonen Technology and UBE Industries (1990)

Hydrocracking. Hydrocrackers typically operate at higher pressures than hydrotreaters or hydrodesulfurization (HDS) units. Bollinger et al. (1984) performed a study to optimize hydrogen recovery from hydrocracker purge gas streams. Various membrane separation operating options were studied. Operating options included constant recycle purity, constant purge rate, constant make-up compressor horsepower, and constant hydrogen make-up rate.

The optimized system, which includes the recovery of hydrogen from both the high- and low-pressure purge streams, is shown in **Figure 15-15,** which includes material balance data for the system. As indicated in the material balance, hydrogen is consumed during the hydrocracking reaction (chemical hydrogen) and some of the unreacted hydrogen is discharged in purge streams from the high- and low-pressure separators. The optimized design depicted in **Figure 15-15** recovers hydrogen from the two purge streams thereby minimizing hydrogen losses. When compared to the hydrocracker without membrane units, the optimized system results in a 10% increase in the hydrogen partial pressure of the recycle gas leaving the high-pressure separator, a slight reduction in make-up hydrogen, and an increased chemical hydrogen consumption (26.7 to 31.8 MMscfd). Assuming that chemical hydrogen consumption per barrel of feed remains constant, the increase in chemical hydrogen consumption corresponds to an increase in hydrocracker throughput of 19%. While the make-up hydrogen flow for the optimized system decreases from 40 to 38.7 MMscfd, the total flow through the make-up hydrogen compressor increases from 40 to 59.0 MMscfd as both permeate streams must be compressed in addition to the make-up hydrogen.

"Butamer" Offgas. In the UOP licensed "Butamer" process, normal butane is catalytically isomerized to isobutane. The process produces isobutane and hydrogen streams. Even with high hydrogen recycle rates, some feedstock is cracked into methane, ethane, and propane. To maintain high reactor efficiency, some of the recycle gas is purged, usually to the fuel gas

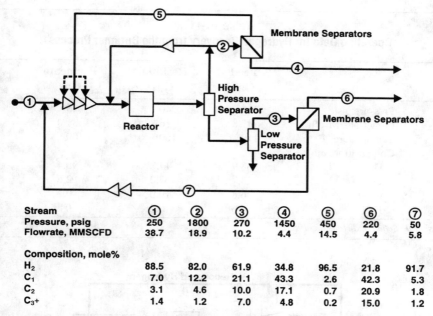

Figure 15-15. Optimized flow scheme for recovery of hydrogen from hydrocracker purge streams (*Bollinger et al., 1984*). *Reproduced with permission from Chemical Engineering Progress, copyright 1984, American Institute of Chemical Engineers*

system. This stream contains not only hydrogen and the cracking products, but also some butanes.

Membranes have been considered for the recovery of hydrogen from Butamer units. However, to maintain catalyst activity, small amounts of an organic chloride are introduced into the feed stream. The chlorides are converted to HCl by the catalyst, and the purge gas from the Butamer unit contains traces of HCl. The effect of HCl on membranes is a concern. A simplified process schematic for a "Butamer" unit with hydrogen recovery is shown in **Figure 15-16**. The membrane unit is located downstream of a caustic wash used to remove HCl from the purge gas. Schell and Houston (1982) described the integration of a membrane unit with a "Butamer" unit, designed to process 47,800 scfh of feed gas. The flow rates and purities are shown in **Table 15-11.** They report that under bone-dry feed conditions, the cellulose acetate membrane, which was located upstream of the caustic wash in this plant, was not affected by HCl; however, special materials and adhesives were used to ensure resistance to the HCl. Cooley and Dethloff (1985) reported on a demonstration unit installed initially upstream of the caustic wash unit. They found that HCl concentrations of 2,000 and 4,000 ppm in the purge gas impaired the membrane performance. The unit was relocated downstream of the caustic wash as shown in **Figure 15-16** and the performance improved. The unit was designed for a feed rate of 971 scfh at 295 psig with 61% hydrogen in the feed gas.

1270 Gas Purification

**Table 15-11
Operating Data for Hydrogen Recovery from the Butamer Process**

	Feed Gas	Residual Gas	Permeate Gas
Pressure, psia	265	240	15
Flow Rate, Mscfh	47.8	16.8	31.0
Temperature, °F	110	100	100
Composition, mol %			
H_2	68.9	17.8	96.4
C_1	23.7	63.0	2.6
C_2+	6.8	19.0	0.2
HCl	0.6	0.2	0.8

Source: Schell and Houston (1982)

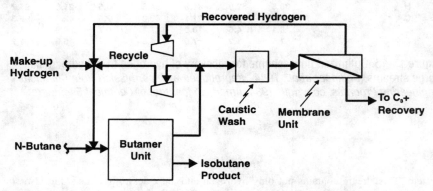

Figure 15-16. Membrane system for recovering hydrogen from Butamer system purge gas. (*Cooley and Dethloff, 1985*)

Other Hydrogen Applications

MacLean and Narayan (1982) have described other applications of membrane systems for hydrogen separation, including toluene hydrodealkylation and coal liquefaction processes.

Carbon Dioxide, Hydrogen Sulfide, and Water Removal

Enhanced Oil Recovery (EOR)

When CO_2 is injected into an oil reservoir at sufficient pressure, it dissolves in the oil present in the substrata reducing its viscosity, allowing it to flow more freely, and thereby increasing oil production. When the oil is brought to the surface and its pressure reduced, the injected CO_2 is released from the oil and discharged with the associated gas.

In CO_2 flood EOR projects, the CO_2 is typically recovered from the associated gas and recycled back into the oil producing formation. With this continual recycle of CO_2, both the volume and CO_2 content of the associated gas progressively increase. In the design of CO_2 flood EOR systems, the objective is to minimize capital expenditure when the associated gas volume and CO_2 content are low, but have enough design flexibility so that the system will be operable in the future when the associated gas volume and CO_2 content are high. In many ways, the modular nature of membranes makes them ideally suited for this application. In the initial phases of the EOR project, capital costs can be minimized by adding the minimum number of membrane modules and CO_2 re-injection compressors. Additional membrane modules and CO_2 compression can be added later when the associated gas volume and CO_2 content are higher. Other CO_2 recovery processes do not have this flexibility.

The first commercial scale membrane installation used with CO_2 flood EOR was the Sacroc project in West Texas. Injection of CO_2 into the field at volumes up to 200 MMscfd began in 1972. To handle the anticipated increase in the associated gas CO_2 concentration, Sacroc installed three CO_2-removal facilities. The plants were installed in conjunction with three existing processing facilities operated by Sun Exploration & Production Co., Chevron U.S.A., and Monsanto. The Sun and Chevron facilities use the Benfield hot potassium carbonate process, and the Monsanto facility uses the monoethanolamine (MEA) process. The Sun hot potassium carbonate plant was designed to reduce the CO_2 content of 160 MMscfd of associated gas from 24 to 0.5% CO_2, while the Chevron plant was designed to reduce the CO_2 concentration of 46 MMscfd of associated gas from 24 to 1.0% CO_2. The Monsanto MEA plant was designed to treat 16.5 MMscfd of the 24% CO_2 feed gas, reducing its CO_2 concentration to 0.01% (Parro, 1984).

In the late 1970s, Sacroc realized that the CO_2 content of the gas produced from the field would peak for a few years at levels greater than the CO_2-removal plants' capacity. The CO_2 content of the field gas had gradually increased from 0.5% prior to injection to 40% CO_2. Sacroc contracted with The Cynara Company to build and operate two membrane units. The new membrane units were installed and operated integrally with the Sun and Chevron hot potassium carbonate plants.

The Sun membrane unit was designed to recover 50 MMscfd of CO_2 at 520 psig with an allowable pressure drop of 40 psig. The Chevron membrane unit was designed to recover 20 MMscfd of CO_2 at 480 psig with an allowable pressure drop of 40 psig. The membrane separation process flow is given in **Figure 15-17.** The two important features of this design are the dehydration of the inlet gas and the operation of the membranes at reduced temperatures. The inlet gas is first cooled by cross exchange with the CO_2 and hydrocarbon product gas streams. This reduces both the moisture and hydrocarbon content of the feed gas and the size of the dehydration equipment. After dehydration, the feed gas is cooled by cross exchange with the residue gas and finally refrigerated before being directed to the membrane modules. The low-temperature feed results in a higher separation factor and a reduced volume of gas. The membranes are single stage and are arranged in parallel.

The amount of feed gas processed depends on the field production rate, and changes in flow were handled by increasing or decreasing the number of membrane modules on line. When the plant was shut down, depressurized, and restarted; the permeate gas flow was found to be 5 to 15% lower than previous levels. The permeate flow gradually improved over a few days of operation, but did not recover entirely. Some loss of flux remained (with a slightly better separation factor), and additional modules were required to maintain the same permeate flow rate (Marquez and Hamaker 1986).

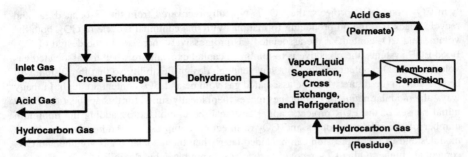

Figure 15-17. Membrane system for treating gas at Sacroc enhanced oil recovery (EOR) project in West Texas. (*Cutler and Johnson, 1985*)

Additional field test data on the use of membrane permeation to provide a concentrated CO_2 stream for EOR were reported by Russell (1984) and Mazur and Chan (1982). Estimated capital costs for processing a 800 psig, 20 MMscfd feed gas stream containing varying amounts of CO_2 were developed by Coady and Davis (1982) and are presented in **Figure 15-18**. In all cases, the permeate stream contains 5% hydrocarbons while the residue gas contains 2% CO_2. To obtain a 95% CO_2 permeate stream, a two-stage membrane system with interstage recompression is required for feed gas CO_2 concentrations less than 75 vol%. At feed gas CO_2 concentrations greater than 75%, the second-stage membrane and interstage recompression can be eliminated.

Low permeate pressure results in the least amount of membrane area required to achieve the desired separation. However, this is at the expense of recompression horsepower required for reinjection. Marquez (1991) reported on a series of tests performed to determine membrane performance with high-permeate back pressures. The tests covered feed gas pressures from 313 to 363 psig (333 psig average) and permeate back pressures from 189 to 250 psig (221 psig average). The permeate composition averaged about 97% CO_2 and 0.13% H_2S, representing removal of about 37% of the CO_2 and 40% of the H_2S from the feed gas. It was concluded that permeation into a high back pressure system is feasible and can result is considerable recompression cost savings. New membrane modules were used for the tests. Permeation rate measurements showed that the flux stabilized after a month of operation at 79% of the initial rate.

In work sponsored by the U.S. Department of Energy (1989), the cost was developed for a membrane unit installed upstream of an amine unit. The membrane unit was designed to process 170 MMscfd of feed gas containing 17% H_2S and 45% CO_2. It was estimated that 280,000 ft² of membrane would be required to remove 70% of the acid gas. At an installed first cost of $20 per ft² of membrane, the cost of the unit would be $5.6 million. The estimated annual steam savings in the amine plant were $5 million to $10 million based on 0.8 to 1.6 pounds of steam per pound of acid gas removed and a cost of $2.28 per 1,000 pounds of steam. The net annual savings, including membrane replacement, were $4.4 million to $9.4 million.

Goddin (1982) compared several methods for recovering CO_2 from a CO_2-flood project associated gas stream. In this study, the associated gas hydrocarbon and nitrogen flow rates were held constant while the CO_2 content increased up to 90 vol%. This simulates the change in associated gas composition over the EOR project life. The following CO_2 recovery cases were evaluated:

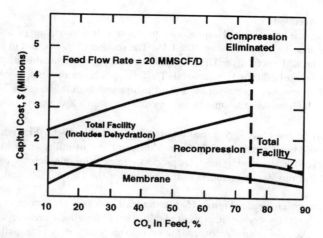

Figure 15-18. Estimated capital costs for carbon dioxide recovery from a 20 MMscf/d gas stream (*Coady and Davis, 1982*). Reproduced with permission from Chemical Engineering Progress, copyright 1982, American Institute of Chemical Engineers

1) *Conventional Amine*

The amine unit is based on a 30% DEA solution. The feed gas is compressed to 285 psia and pretreated to remove heavy hydrocarbons. After passing through the DEA absorber, the sweet offgas is compressed to 650 psia and sent to the existing gas plant. The sour CO_2 stream from the DEA stripper, at 20 psig, is compressed to 450 psia and sent to a Selexol sweetening process designed to reduce the H_2S content of the CO_2 product gas to less than 100 ppm.

2) *Cryogenic Fractionation (Ryan-Holmes Process)*

The feed to the cryogenic unit is compressed to 625 psia, dehydrated, and chilled. The overhead gas from the first cryogenic column, which contains the methane and lighter components, is sweet pipeline gas. The bottom product from this column flows to the CO_2 column, where the sweet CO_2 leaves in the overhead. Lean oil is added to the top of the first column to prevent the CO_2 from freezing, and also is added to the top of the second column to break up a CO_2/ethane azeotrope. The bottoms from the CO_2 column are sent to a lean oil recovery unit. The propane and lighter components are processed in a DEA unit to remove the CO_2 and H_2S. The acid gases removed here are processed in a Claus unit for sulfur recovery. Two cases were considered. Case A assumes that all of the ethane recovered has the value of liquid hydrocarbon. Case B assumes that only a portion of the ethane has the high liquid hydrocarbon value since 80% of the hydrocarbons would be recovered in the existing gasoline plant. Credit was given for the differential value of ethane in natural gas liquids (NGL) versus fuel gas.

3) *TEA Bulk CO_2 Removal*

In the TEA bulk removal process, a TEA solution is used to remove the H_2S and CO_2 from the feed. The CO_2 and H_2S are then removed from the TEA solution by flashing it to about 20 psia. The TEA absorber overhead stream, containing about 20% CO_2 and some H_2S, is sent to a DEA unit for final clean up. The acid gas streams from the DEA unit and the TEA flash tower are compressed to 450 psia and sent to a Selexol unit for H_2S removal.

4) *Membrane Permeation*

The membrane unit was designed to produce a permeate with a maximum of 5% hydrocarbons and a residue stream containing 20% CO_2. The residue stream is sent to a DEA unit for cleanup. The acid gas from the DEA unit and the sour permeate streams are compressed and sent to a Selexol unit for H_2S removal. To develop a range of costs, two cases were considered. The low cost case assumed a short membrane module life and a high permeation rate. The high cost case assumed a long module life with a lower permeation rate.

A summary of capital costs versus feed rate is given in **Table 15-12. Figures 15-19** and **15-20** present the cost of removing CO_2 for two different sets of utility costs. Also plotted are the cost curves for sweetening the CO_2 product stream using the Selexol process. The cost of the Selexol process is included in the curves for the DEA, TEA, and membrane processes.

For the low energy cost case (**Figure 15-19**), the least cost system, over most of the CO_2 concentration range, is the cryogenic process with full credit for ethane recovery. When only partial credit is taken for ethane recovery, the cryogenic system is not competitive. Membrane permeation is more economical than DEA and TEA over the entire range, and has an increasing advantage over DEA at CO_2 concentrations above 20%. The effect of higher energy cost (**Figure 15-20**) is to make TEA and the membrane process more economical than any of the others over the entire CO_2 concentration range.

In the mid 1980s, several large CO_2-flood projects were initiated based on the availability of naturally occurring CO_2 brought in by pipeline to West Texas. At that time crude prices were high and, significantly, NGL prices were also high. Overall economics were quite similar to the scenario presented by Goddin in **Figure 15-19**. Three of the large projects: Amerada Hess's Seminole Plant (2 × 85 MMscfd, 77% CO_2) (Schaffert et al., 1986; Wood et al., 1986), Shell's Wasson Denver unit (275 MMscfd, 93% CO_2) (Flynn, 1983; Youn et al., 1987), and Arco's Willard unit (72.9 MMscfd, 86% CO_2) (Price and Gregg, 1983), and a smaller plant, the Mitchell Alvord South CO_2 plant (7.5 MMscfd, 85% CO_2) (McCann et al., 1987) used cryogenic distillation (the Ryan Holmes process). All of these projects recovered NGL and cryogenic distillation was chosen because economics favored NGL production.

Table 15-12
Summary of CO_2 Removal Costs for DEA, Cryogenic, TEA, and Membrane Facilities

Percent CO_2	20	40	60	80	90
Flow, MMscfd	18.2	24.3	36.5	73.3	148.0
Capital, $MM					
DEA	9.7	16.4	25.5	54.5	103.6
Cryo A	16.2	20.8	28.3	42.6	73.5
Cryo B	16.2	20.8	28.3	42.6	73.5
TEA-DEA	—	15.0	21.6	36.9	65.0
Perm-Low	—	13.8	18.8	29.8	47.0
Perm-High	—	16.6	23.4	37.9	60.5

Source: Goddin (1982)

Membrane Permeation Processes **1275**

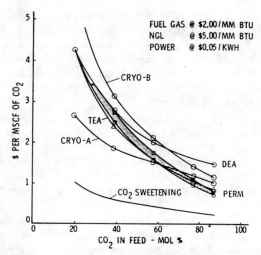

Figure 15-19. Estimated costs of carbon dioxide removal for a CO_2-flood EOR project—low utility cost case (*Goddin, 1982*). *Reproduced with permission from Proceedings of the 61st Annual Convention of the GPA, copyright 1982, Gas Processors Association*

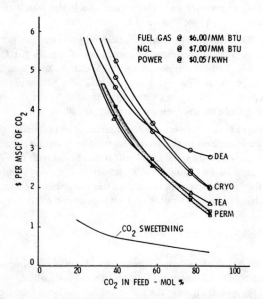

Figure 15-20. Estimated costs of carbon dioxide removal for a CO_2-flood EOR project—high utility cost case (*Goddin, 1982*). *Reproduced with permission from the 61st Annual Convention of the GPA, copyright 1982, Gas Processors Association*

Chevron's Sacroc project (Parro, 1984; Schendel and Nolley, 1984) and Amoco's Central Mallet unit (100 MMscfd, 85% CO_2) (Anon., 1985) used membranes. However, the Sacroc project predates the Ryan Holmes process, and Amoco was under contractual obligation to supply gas with NGL's still present to a downstream NGL extraction plant. Therefore in the mid-1980s, the market generally sustained the conclusion of Goddin (1982) that, when NGL was recovered, economics favored cryogenic distillation over the use of either membranes or amine treating in CO_2-flood EOR projects.

Interest in CO_2-flood EOR fell with the drop in oil prices that occurred in the early 1980s, but seems to be reviving. Recent CO_2-flood EOR projects using membranes include Mobil's Salt Creek project (64-110 MMscfd, 70% CO_2), which started up in 1992; Shell's McCarney, Texas plant, which began operation in 1993 (11 MMscfd, 70% CO_2); and Amoco's Mallet plant in West Texas (30–100 MMscfd, 80% CO_2), which started up in 1994 (Cynara, 1995). One of the key factors favoring membranes in these more recent projects is the ability to delay capital expenditures.

Schendel (1984) proposed the integration of a membrane process and cryogenic distillation. Feed gas from the field passes through a hydrocarbon dew point control unit and then through a membrane unit. This first membrane unit removes the bulk of the CO_2. The residue is fed to the cryogenic unit. Rather than suppress the CO_2/ethane azeotrope with a hydrocarbon recycle stream, the CO_2-rich azeotrope is allowed to go overhead to a second membrane unit where CO_2 is removed as the permeate. The primary advantage of this scheme is that the cryogenic unit can be sized for the relatively constant hydrocarbon rate, not the increasing CO_2 rate over time. As the CO_2 rate increases, additional front-end membrane modules can be added as required. The disadvantage of this concept is that the CO_2 product is produced at low pressure, requiring recompression. Some cryogenic distillation process configurations can produce the CO_2 product as a liquid which can be easily pumped to high pressure (Wood et al., 1986; Ryan and Schaffert, 1984).

Boustany et al. (1983) compared the performance and economics of hot potassium carbonate, cryogenic separation, and membrane permeation processes for CO_2 removal. They assumed a 100 MMscfd feed stream containing 80% CO_2 available at 25 psig and 100°F. The gas is assumed to be sweet. The cryogenic unit is similar to that previously described by Goddin (1982) except that the CO_2 recovered from the DEA unit is compressed and combined with that coming from the cryogenic column overhead.

The membrane unit consists of two sections in series. The permeate from the second section is compressed and recycled to the feed. The residue from the second stage is fed to a DEA unit. The CO_2 recovered from the DEA unit is combined with that from the first membrane section and compressed to 500 psig. A summary of operating conditions and cost data is presented in **Table 15-13**. It is concluded that the membrane permeation units (PRISM separation) offer both capital and operating cost savings over the other systems.

Well Fracturing

In this process, which is applicable to certain types of wells, high pressure CO_2 is injected into a reservoir to fracture the formation. A slurry of sand is then fed into the well to fill the fractures and provide a porous channel for gas or oil flow. The associated gas flow after the fracture contains a high amount of CO_2, which decreases to pipeline transmission levels over a relatively short period of time. Membranes work well for treating the gas resulting from the fracturing process. Because of their modular nature and portability, they can be used to

Table 15-13
Summary of Data for Comparison of Cryogenic, Hot Potassium Carbonate, and Membrane Processes

	Hot Potassium Carbonate	Cryogenic	Membrane
CO_2 Product Stream			
CO_2 Recovery, %	96.9	93.0	96.9
CO_2 Purity, %	99.8	95.1	95.5
Hydrogen Recovery, %	99.5	81.0[1]	83.5
Capital Costs, $MM			
CO_2 Removal	21.1	24.2	16.1
DEA Treating	—	4.9	4.9
Feed Compression			
25 to 250 psig	11.1	11.1	12.1
250 to 500 psig	0.9	6.2	0.9
Hydrocarbon Liquid Recovery	4.0	—	—
CO_2 Compression 10 to 500 psig	7.9	—	5.1
Total Capital	45.0	46.4	43.1
Operating Costs, $MM	11.78	8.46	6.46

Notes:
1. *Assumes CO_2 column operation based on ethane rejection/propane recovery.*
Source: Boustany et al. (1983)

maintain pipeline quality gas until the CO_2 levels return to normal and can then be removed and used elsewhere (Schendel and Seymour, 1985).

Pipeline Natural Gas

Typical specifications for pipeline gas are CO_2 less than 2%, H_2S less than 4 ppmv, and water less than 7 lb/MMscf. Water vapor is a very fast gas and, therefore, dehydration to pipeline specifications can be achieved by membrane systems while also removing the slower CO_2 or H_2S acid gases. Membrane permeation systems can remove CO_2 to the required 2% level; however, at low CO_2 concentrations, the CO_2 partial pressure driving force is reduced and a significant amount of hydrocarbon (primarily methane) is lost with the CO_2-rich permeate. The problem becomes even more severe when trying to remove substantial quantities of H_2S to meet a 4 ppmv specification. Normally, with membranes only modest adjustments to H_2S concentration in the ppm range are economically feasible. However, subsequent treatment with other processes to meet the H_2S specification is possible.

The Gas Research Institute (GRI) performed an extended field evaluation of a membrane unit operating on low quality natural gas (Meyer and Gamez, 1995; Lee et al., 1995). The unit was a standard two-tube Grace Membrane Systems (GMS) test system designed to treat 0.5 MMscf/d of 750 psig natural gas containing 6.0% carbon dioxide. Two types of asymmetric cellulose acetate membrane (standard and higher density) were tested. During the

573-day test period the unit operated smoothly, reducing the CO_2 content of the gas to meet the pipeline specification of less than 2.0% carbon dioxide. The system also provided gas dehydration to 7 lb H_2O/MMscf or less.

GRI evaluated process economics based on the test results. For the case of a plant treating 37 MMscf/d at 725 psig, the study indicated that the membrane process would be competitive with DEA or MDEA systems if the gas contained less than 20% CO_2 and appreciably less expensive than the amine processes if the gas contained over 20% CO_2 (Meyer and Gamez, 1995).

Bhide and Stern (1993) conducted a study to optimize the process configuration and operating conditions and to assess the economics of membrane processes for removing CO_2 from natural gas. The base case was a 35 MMscf/d natural gas feed stream at 800 psia with CO_2 concentrations in the range of 5 to 40 mole%. The optimum configuration was found to be a three-stage system consisting of a single permeation stage in series with a two-stage permeation cascade with recycle.

For the base case operating conditions, membrane properties, and economic parameters assumed in the study, membrane processes were found to be more economical than DEA over the entire range of CO_2 concentrations considered (with no H_2S in the feed). When H_2S is also present, the results showed that the cost of meeting product gas specifications (<2 mole% CO_2 and <4 ppmv H_2S) increases with increased H_2S concentration. For example, with feed containing 1,000 ppm H_2S, the membrane process was found to be more economical than DEA only when the total acid gas in the feed exceeded about 16 mole%.

Fournie and Agostini (1987) studied the removal of CO_2 from gas that was to be sold as liquified natural gas (LNG) and from gas that was to be sold in the gaseous phase. For gas to be sold as LNG, with a 100 ppm CO_2 specification, they concluded that membranes were 1.1 to 2.5 times more capital intensive than amine units. Three feed CO_2 concentration ranges were investigated for gas to be sold as a gaseous product containing less than 2% CO_2. For a feed concentration range of 5 to 20% CO_2, membrane systems were estimated to be one-half the investment cost of diethanolamine processes, but were not recommended because of the low hydrocarbon recovery and the problem of finding a use for the low pressure permeate. The predicted investment cost for CO_2 concentrations ranging from 20 to 40% was about the same as that for conventional processes, but the hydrocarbon losses were significant. For CO_2 concentrations greater than 40%, the estimated investment was about one-half that of conventional processes. Their overall conclusion is that membranes are most effective for high CO_2 concentrations, especially when the CO_2-rich permeate can be used for EOR applications.

The capital costs of membrane systems to remove H_2S to meet a sales specification of 4 ppm were found to be about two times the cost of conventional systems. The use of membranes for bulk removal of H_2S down to about 2% combined with conventional processing to bring the concentration down to 4 ppm resulted in capital costs about the same as conventional processes, but with a much lower overall energy consumption (Fournie and Agostini, 1987).

The use of natural gas as the circulating fluid drilling gas in a deep gas formation (rather than air or conventional muds) was reported by Cooley and Dethloff (1985). The gas available for this use was saturated with water and contained 6% CO_2 and 3,000 ppm H_2S. The CO_2 and H_2S levels were considered too high so a gas purification system based on membrane permeation was installed. The membrane unit reduced the CO_2 content to 2%. The H_2S content was reduced to 600 ppm, which then could be economically treated with an iron sponge unit. Operating data are presented in **Table 15-14**.

Some of the early tests demonstrating the feasibility of using membranes to remove CO_2 from natural gas to produce a sales grade gas have been reported by Mazur and Chan (1982)

Table 15-14
Operating Data for Drilling Gas Application of Membrane Process

	Inlet	Product Gas
Flow, MMscfd	1.22	1.07
Pressure, psia	900	900
Temperature, °F	100	90
CO_2 Content, mol %	6.05	2.0
H_2S Content, ppm	3000	600

Source: Cooley and Dethloff (1985)

and Schell and Houston (1982). Pipeline gas with less than 5% CO_2 was produced from wellhead gas containing up to 30% CO_2. Inlet pressures ranged from 250 psig to 800 psig.

Another economic comparison of membranes and amine treatment was presented by Babcock et al. (1988). They considered CO_2 feed gas levels ranging from 10 to 60 vol% as well as various operating pressures and flow rates. The base case feed gas contained less than 0.5% H_2S and had a flow rate of either 37.2 or 60 MMscfd at 725 psig and 100°F. The product gas specification called for CO_2 and H_2S concentrations to be less than 2% and 4 ppm, respectively.

The conventional DEA and MDEA split-flow/flash amine systems operated at 900 psia. The circulating solutions were 30% DEA and 50% MDEA. Acid gas in both systems was released at 25 psia. The multistage membrane system consisted of an initial membrane bulk removal unit followed by a two-stage permeation system with recycle.

Some of the estimated capital and operating costs are summarized in **Table 15-15**. Processing costs include operating expenses, lost product gas, and capital charges. The lost product gas includes gas lost in the permeate and gas used for fuel for recycle recompression. The processing cost comparison is presented graphically in **Figure 15-21**. The plotted results indicate that membrane processes are more economical than conventional DEA over the entire feed CO_2 concentration range. They are more economical than two-stage MDEA at CO_2 concentrations below about 15% and above about 42%; while the MDEA system is slightly more economical in the range between these two concentrations.

McKee et al. (1991) prepared an economic comparison of membrane, amine, and membrane/amine hybrid processing systems to aid in determining when the hybrid system would have an economic advantage. In the hybrid system, the bulk of the CO_2 is removed with a small area membrane system and the amine system is used to remove the residual CO_2 to meet specification. Feed gas CO_2 concentrations from 4 to 27% and flow rates between 5 MMscfd and 75 MMscfd were used to generate cost data. The amine system design parameters were 35% DEA, 70% approach to equilibrium, and a reboiler steam rate of 0.7 pounds per gallon of DEA circulated. The membrane was designed as a single stage with a 20 psig permeate pressure. Feed gas values of $0.50/MMBtu to $2.00/MMBtu were used.

The cost comparisons at $1.50/MMBtu feed gas value show that

1) Amine systems are preferred when flow rates are high (over about 20–30 MMscfd) and CO_2 concentrations are low (below about 16 mole%).

Table 15-15
Capital and Operating Costs for Amine and Membrane Systems

CO_2 in Feed, %	10	20	25	30	50	80
Flow Rate, MMscfd	37.2	37.2		37.2	37.2	37.2
DEA						
Capital, $MM	4.54	6.21		7.50	9.50	11.80
Processing, $/Mscf	0.24	0.36		0.46	0.64	0.87
Multistage Membrane						
Capital, $MM	3.33	3.69		3.37	2.45	1.44
Processing, $/Mscf	0.19	0.25		0.27	0.24	0.14
Single-Stage Membrane						
Processing, $/Mscf	0.26	0.32		0.32	0.28	0.15
Flow Rate, MMscfd	60.0		60.0			60.0
MDEA						
Capital, $MM	7.32		7.60			8.76
Processing, $/Mscf	0.21		0.23			0.30

Notes:
1. Feed gas conditions are 725 psig and 100°F for all cases. All feed gases have less than 0.5 vol% H_2S.
2. In all instances product specifications are $CO_2 < 2$ vol%, $H_2S < 4$ ppm.

Source: Babcock et al. (1988)

2) Membrane systems have an advantage at low feed rates (below about 20–30 MMscfd) over the entire range of feed gas compositions.
3) The hybrid system is favorable at high flow rates (above about 30 MMscfd) and high CO_2 concentrations (above about 16 mole %).

Membrane units have a smaller plot space requirement and weigh less than most other process units. These factors, as applied to an offshore installation, were evaluated by Cooley (1990). A comparison was made between an amine/glycol unit and a membrane unit designed to treat 96 MMscfd of natural gas at 1,000 psig. The systems were designed to reduce the CO_2 content of the gas from 15.8 to <2% and dehydrate the gas to North Sea specifications. The results of the comparison are presented in **Table 15-16**.

Landfill Gas Purification

Landfill gas is produced at low pressure (near atmospheric), has a high CO_2 content, and contains numerous trace contaminants. The operation of a membrane system producing approximately 800 Mscfd of high Btu gas at 500 psig, containing less than 3.5% CO_2 was described by Houston and Schell (1986). The feed gas from the collection system is first compressed to 525 psig. The gas is then treated in a carbon bed adsorber to remove organics, fed to a filter/separator to remove condensed liquids, and preheated to 130°F to prevent water condensation inside the membrane. The gas is then fed to a two stage membrane unit

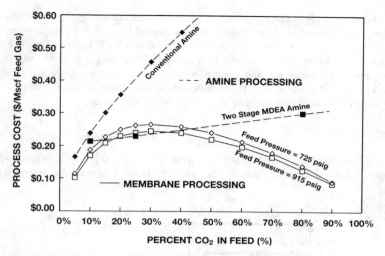

Figure 15-21. Cost comparison between amine and membrane processes for purification of 60 MMscf/d of natural gas (*Babcock et al., 1988*). *Reproduced with permission from Energy Progress*

Table 15-16
Plot, Cost, and Weight Requirements for Membrane and Amine/Glycol Dehydration Acid Gas Removal Units

	Amine Plant	Membrane Plant
Weight, metric tons	1,605	503
Installed Cost, $MM	30.5	18.7
Deck Area, sq. meters	600	214

Source: Cooley (1990)

with recycle. The CO_2-rich permeate from the first stage is vented and the second-stage permeate is compressed and recycled to the first stage to improve methane recovery. Advantages claimed for membrane systems in this application are low capital and operating costs, simplicity of operation and maintenance, compact size, and modular construction. The unit also dehydrates the gas.

Helium Removal from Natural Gas

Helium occurs naturally as a minor constituent (<5%) of some natural gas streams. It is usually recovered by a two-step process: the first recovers a crude helium concentrate (about 50%) from the natural gas, and the second purifies the helium to a high purity product (>99.99% for Grade A Helium). Several hybrid processes for accomplishing both steps of the helium recovery and purification operation have been evaluated by Choe et al. (1988).

1282 Gas Purification

The most economically attractive process is a hybrid system that combines cryogenic, PSA, and membrane units to produce 99.99% helium from dilute natural gas. The process is shown in **Figure 15-22.** The feed is natural gas containing 2.1% helium. The feed is first cooled to −60°F to condense the heavier hydrocarbons. Then the gas is cooled to −240°F to condense most of the methane and some nitrogen. At this point the gas contains 30–35% helium. The crude helium is then fed to a two-stage membrane unit that produces a 95% helium stream. PSA is used to upgrade this stream to Grade A purity.

Air Separation

The initial attempts to separate air with membranes were aimed primarily at the production of oxygen, and date back over a century (Prasad et al., 1994). However, the oxygen-selective

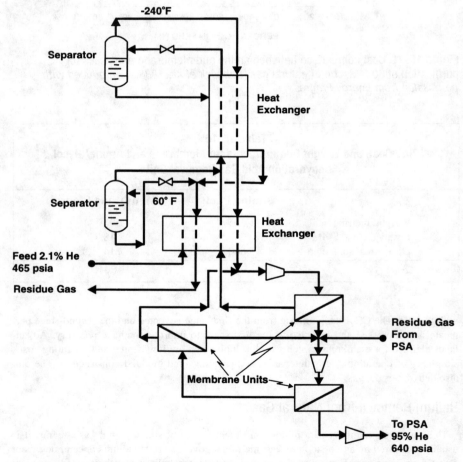

Figure 15-22. Cryogenic/Membrane hybrid process for removing helium from natural gas. (*Choe et al., 1988*)

nature of currently available polymeric membranes makes them more suitable for the production of nitrogen. In a typical operation, a single permeation stage can produce relatively pure nitrogen as residue gas at elevated pressure; while the permeate is oxygen-enriched air at atmospheric pressure. As a result of the process advantages of nitrogen production, and the rapidly developing commercial market for nitrogen, the development and use of membrane systems for nitrogen production have expanded rapidly.

Nitrogen

In a paper prepared in 1993, it was estimated that the production of nitrogen from air had reached well over 2,000 tons per day, and more than 1,000 commercial units had been installed worldwide (Prasad et al., 1994). The practical economic limit for nitrogen purity has risen from 95–97 to 99.9%, with the ability to produce 99.9995% nitrogen from compressed air (Rice, 1990). However, extremely high purities require the use of a hybrid system employing a chemical deoxygenation step following the membrane unit. Nitrogen recoveries have gone from 22 to 30% (Anon, 1990). The major constituents in air, other than nitrogen and oxygen, are water vapor and carbon dioxide. Both gases are faster gases than oxygen and are concentrated in the oxygen permeate stream. Therefore, it is possible to obtain nitrogen with a dew point less than 40°F and a carbon dioxide concentration less than 20 ppm using membrane separators.

Nitrogen is presently produced via cryogenic, PSA, and membrane systems. It is produced on site or delivered as liquid or as gas in cylinders. **Figure 15-23** shows the typical economic areas of application of various nitrogen supply systems based on the quantity and purity of nitrogen required.

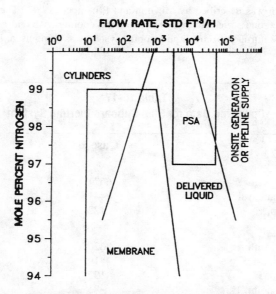

Figure 15-23. Economic areas of applicability of nitrogen supply systems (*Beaver et al., 1988*). *Reproduced with permission from AICHE Symposium Series, copyright 1988, American Institute of Chemical Engineers*

1284 *Gas Purification*

The primary use for nitrogen produced by membrane separation is inert gas blanketing. Most flammable gases require a minimum of 10–12 vol% oxygen to sustain combustion. Since high purity nitrogen is not required for this application, 95–97% nitrogen from a membrane unit is satisfactory for inerting storage tanks.

A membrane system to produce nitrogen for inerting the fuel tanks of military aircraft was described by Bhat and Beaver (1988) and Gollan and Kleper (1987). Depending on the type of aircraft and the flight conditions, bleed air from the jet engine compressors is available at 25 to 150 psig and temperatures up to 450°F. The air is cooled to 70°–160°F depending on the membrane limitations, filtered to remove particulates, and then fed to the membrane unit. The inert gas produced has a nitrogen concentration of 88 to 97%.

Crude and LNG tankers began inert blanketing of cargo tanks for safety in the early 1970s. Chemical tankers require a dry, non-reactive gas for purging lines and for preventing condensation inside cargo tanks. Metzger et al. (1984) reviewed the design of a system installed onboard two ships. The system was designed to produce a 95 to 99% nitrogen stream. Operating data for the system are presented in **Table 15-17**. Dry instrument gas can also be provided by attaching a connection to one or more of the membrane modules.

Bhat and Beaver (1988) also reported on the use of membranes to control the composition of the atmosphere in the storage and transportation of fruit and vegetables. To slow respiration, the oxygen content of the atmosphere needs to be in the 1 to 5% range. However, two metabolic products, carbon dioxide and ethylene, also must be controlled to prevent accelerated ripening and decay. Membranes can provide primary control of the oxygen and secondary control of the carbon dioxide and ethylene. Although the resultant gas mixture may not be the optimum, the cost is less than that required to produce an exact optimum mixture of the gases.

The use of a membrane system to replace delivered liquid nitrogen in a powdered metal sintering operation was described by McGinn and Pfitzinger (1988). The sintering furnace consisted of three zones: presintering, high heat, and cooling. A blend of 20% dissociated ammonia and 80% nitrogen from a liquid storage system was being used. In the new system,

Table 15-17
Operating Data for Ship Onboard Inerting System

	Case 1	Case 2
Input		
Flow, scfh	6674	4026
Pressure, psig	441	441
Temp, °F	122	122
Inert Gas		
Flow, scfh	3320	989
Press, psig	419	419
Temp, °F	122	122
Dew point, °F	−85	−85
% Nitrogen	95	99

Source: Metzger et al. (1984)

500 scfh of membrane-separated nitrogen was used in the presintering and cooling zones and dissociated ammonia was used in the high heat zone. The membrane-separated nitrogen contained a small amount of oxygen, which proved to be an advantage in the presintering zone by aiding in the removal of lubricants. The removal of the lubricants with the membrane-separated gas was far more effective than with the previously used nitrogen atmosphere. In addition to improved quality, a reduction in operating costs was observed. Delivered liquid nitrogen cost, including storage was $0.53 per hundred scf, while membrane gas production cost was only $0.28 per hundred scf.

Beaver et al. (1986) reported the results of a feasibility study for generating 600 MMscfd of nitrogen on a ship moored to an offshore platform. The nitrogen generated was to be injected into an offshore oil reservoir to arrest the sinking of the seabed and the production platform fixed to the seabed. In the proposed process, turbine exhaust gases are cooled in a heat recovery boiler and an exhaust gas cooler. The gases are then preheated to prevent condensation and fed to a membrane unit. The nitrogen-rich residue stream from the membrane unit contains 2% oxygen and about 1% CO_2. The oxygen is reduced to less than 5 ppm by catalytic reduction. A glycol dehydration system is used to remove water down to a $-4°F$ dew point. Product nitrogen is produced at 68°F and 900 psia. The system has three identical and independent process trains, which include nine gas turbines, nine heat-recovery units, six compressor systems, and six membrane-separation units. The heat recovery units produce about 600,000 pounds per hour of steam. It was concluded that the membrane process would be an energy efficient, low cost method of generating nitrogen.

A comparison of the capital and operating costs of a membrane system versus PSA for producing nitrogen was developed by Gollan and Kleper (1986). The results are presented in **Table 15-18.** The analysis shows that the cost per standard cubic foot of nitrogen produced is the same for both systems, but the capital cost of the PSA system is one-third higher.

The integration of membranes and catalytic oxygen removal to produce 99.9995% nitrogen was described by Beaver et al. (1988). Compressed air at 125 psig and 100°F is fed to a membrane separator that produces a 99.5% nitrogen stream. The nitrogen stream, at 110 psig, is then fed to a catalyst module where the nitrogen is mixed with a small amount of hydrogen and passed over a palladium catalyst. The catalytic reaction of oxygen and hydrogen to form water reduces the nitrogen stream oxygen content to less than 5 ppm.

Beaver et al. (1988) also described a small scale application that integrates membranes and cryogenics to produce 5 liters per day of liquid nitrogen. A 100 psig compressed and filtered air stream is fed to a membrane unit that produces a 99% nitrogen stream. This stream is then sent to a cold head that is located inside a liquid nitrogen Dewar flask. The membrane module, made of hollow fibers, is two inches in diameter by three feet long. The unit can be reconfigured to produce liquid oxygen by changing the internal piping. High purity oxygen can be produced by adding a compressor and a second-stage membrane module.

McReynolds (1985) described a commercially available package unit producing up to 99% nitrogen and up to 40% oxygen. The unit is sized to deliver up to 15,900 cfh of 95% nitrogen at 105 psi and 77°F. Depending on the operational mode and capital depreciation option selected, the total operating cost ranges between $0.071 to $0.15 per hundred cubic feet, assuming $0.05/kWh electrical cost.

Oxygen

Oxygen is presently produced via cryogenic distillation, PSA, vacuum swing adsorption (VSA), and membrane systems. It is produced on site or delivered as liquid or as gas in

Table 15-18
Operating Cost Comparison for 95% Nitrogen Generation

	Membranes	PSA
Capacity, ton/day	3	3
Capital Cost, FOB	$75,000	$100,000
Installation, %	20	20
Installed Cost, $	$90,000	$120,000
Operating Costs, $/day		
Membrane Replacement	16.01	0.00
Power Cost	35.10	40.50
Cooling Water Cost	0.43	0.43
Operator Labor Cost	4.30	4.30
Maintenance & Taxes	3.89	5.18
Capital Charges	32.93	43.90
Depreciation	12.96	17.28
Total Operating Costs		
$/day	105.61	111.59
$/ton	35.20	37.20
$/100 SCF	0.13	0.13

Note: Membrane performance and costs based on A/C Technology Units.
Source: Gollan and Kleper (1986)

cylinders. **Figure 15-24** shows the market economics relative to size and purity. Note that the use of oxygen begins at 21%, its atmospheric concentration, while applications using nitrogen require concentrations beginning at 90%.

A single-stage membrane yields oxygen concentrations up to 40%. Multi-stage systems can produce oxygen concentrations up to 60%; however, the economics are much less attractive. Oxygen passes through the membrane in the N_2/O_2 separation. Therefore, the feed gas (air) must be compressed, and the oxygen enriched air is produced at low pressure.

Oxygen-enriched air produced by membranes has applications that include industrial combustion efficiency improvement, fermentation efficiency improvement, respiratory care, and pulp bleaching. Although these applications hold promise, commercial implementation has been slow. The major potential market is combustion enhancement, but progress has been limited by combustion system development. There is also limited use in respiratory care, particularly in Japan (Spillman, 1989). One advantage membranes have in medical applications is elimination of the humidification step. Oxygen produced via adsorption (PSA or VSA) or cryogenics is dry and requires humidification before patient use. Since water is a fast gas, it permeates with the oxygen in a membrane system.

An application that integrates membranes with PSA is an extension of a previous nitrogen application—fuel tank blanketing on board military aircraft. The oxygen-rich permeate is fed to a PSA unit to produce 90% oxygen for crew breathing (Beaver et al., 1988).

A comparison of oxygen-enriched air (OEA) produced by membranes and PSA was performed by Gollan and Kleper (1986). Their data are presented in **Table 15-19**. Membranes show a significant advantage in capital cost and power consumption.

Table 15-19
Operating Cost Comparison for 35% Oxygen Enriched Air Generation

	Membranes	PSA
Capital Cost, FOB	$250,000	$480,000
Installation, %	15	15
Installed Cost	$287,500	$552,000
Operating Costs, $/day		
Membrane Replacement	38.11	0.00
Power Cost	85.50	130.50
Operator Labor Cost	8.60	8.60
Maintenance & Taxes	9.91	19.02
Capital Charges	105.18	201.95
Depreciation	33.03	63.41
Total Operating Costs		
$/day	280.33	423.49
$/ton	28.03	42.35

Source: Gollan and Kleper (1986)

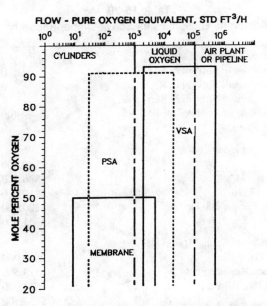

Figure 15-24. Economic areas of applicability of oxygen supply systems (*Beaver et al., 1988*). *Reproduced with permission from AICHE Symposium Series, copyright 1988, American Institute of Chemical Engineers*

Solvent Vapors

Although permeability data on organic vapors and air were available in the early 1970s (Spangler, 1975; Rogers et al., 1972) process development did not begin until the mid-1980s. Operating conditions for membranes used in vapor recovery applications are quite different from the conditions seen in hydrogen or carbon dioxide applications. Inlet pressures are typically less than 125 psig, and a vacuum pump is frequently used on the permeate side to increase the pressure ratio. The permeate is often compressed for recovery of the solvents by condensation.

Peinemann et al. (1986) described the use of membranes to recover solvent from an industrial oven. Hot solvent-laden air from the oven is fed to a blower to boost its pressure to 17.6 psia and then to a membrane system. A vacuum pump is used to maintain a low pressure on the permeate side of the membrane. The solvent-rich permeate is compressed and the solvent is condensed and recovered. The hot-solvent depleted residue gas is recycled back to the oven to reduce energy consumption. The operating and cost data for this system are presented in **Table 15-20**.

In the manufacture of polyvinyl chloride (PVC), an offgas stream is produced that contains unreacted vinyl chloride monomer. This stream is usually compressed and condensed to recover as much monomer as possible. However, the vent stream from the condenser still contains a significant amount of monomer. Baker et al. (1991) described an installation that recovers 100 to 200 lb/hr of monomer from a condenser vent stream. The process vent stream is compressed to 65 psig and sent to a condenser operating at 14°F. The condenser vent stream, containing 50% vinyl chloride monomer, is sent to the membrane unit. The

Table 15-20
Membrane Economics for Solvent Vapor Recovery

Plant Operating Characteristics	
Concentration of Solvent in Feed Air	0.5 vol%
Concentration of Solvent in Residue	0.1 vol%
Concentration of Solvent in Permeate	4.4 vol%
Volume Flow of Feed Air @ 17.6 psi	1400 scfm
Pressure Ratio	0.05
Capital Cost of Plant	
Membranes	$40,500
Other	$40,500
Blower, 17 Hp	$30,000
Vacuum Pumps, 200 Hp	$220,000
Total	$331,000
Operating Costs (Annual Basis)	
Annual Fixed Costs	$48,000
Membrane Replacement	$13,500
Utilities ($0.05/kwh)	$71,000
Total	$132,500

Note: 1,000 L/day solvent recovery, 0.5 vol% solvent in feed.
Source: Peinemann et al. (1986)

monomer in the vent gas is reduced to less than 1%. The discharge of the vacuum pump on the permeate stream is recycled back to the inlet compressor suction.

Baker et al. (1991) also described the installation of a membrane system on a CFC-11 and CFC-113 drum filling line. The unit is designed to process 10 scfm of saturated vent gas from the drum filling operation. The gas passes through a drier and then to a condenser operating at 5°F. The condenser reduces the CFC concentration in the vent gas from 65 to 21%. A single-stage membrane is then used to reduce the CFC-11 concentration to 1.2%. Operating data and system costs are presented in **Table 15-21,** and the unit is pictured in **Figure 15-25.** The unit recovers approximately two pounds of CFC per 55 gallon drum filled.

Table 15-21
Vapor Recovery Membrane System Performance and Costs

System Performance	CFC-11	CFC-113
Condenser inlet solvent concentration	65%	25%
Condenser temperature	–15°C	–15°C
CFC removal:		
Condenser only	82%	68%
Membrane separator + condenser	98%	96%
System Costs		
Capital cost	$ 45,000	
Operating costs	$ 8,500/yr	
	$0.01/lb of CFC recovered	

Source: Baker et al. (1991)

Figure 15-25. CFC vapor recovery system. *Courtesy of Membrane Technology and Research, Inc.*

1290 Gas Purification

The use of membranes to recover HCFC-123 from the drying chamber of a film coating operation was described by Baker et al. (1992). A flow diagram of the system is shown in **Figure 15-26**. The vapor stream exiting the drying chamber is compressed to 125 psig, dried to prevent icing in the condenser, and cooled to 5°F where most of the HCFC-123 is condensed out. The condenser vent is routed to a membrane unit where the HCFC-123 concentration is reduced to less than 100 ppm. The unit recovers 15 pounds per hour and will pay for itself in 3,000 hours of operation based on $5/lb for HCFC-123. The system is pictured in **Figure 15-27**.

Other solvent vapor applications for membranes include recovery of refrigerants from the purge stream of low temperature chiller refrigeration systems (Wijmans et al., 1991) and reducing emissions from industrial sterilizers (Baker and Wijmans, 1991), hydrocarbon storage tanks, dry cleaning operations, and printing and coating processes (Baker, 1985).

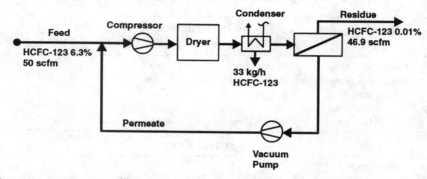

Figure 15-26. Flow diagram of membrane system for the recovery of HCFC-123 vaporized in a film drying operation. (*Baker et al., 1992*)

Figure 15-27. HCFC vapor recovery system. *Courtesy of Membrane Technology and Research, Inc.*

REFERENCES

Anon., 1990, "Membranes Shoot for the Big Time," *Chem. Eng.*, April, pp. 37–43.

Anon., 1985, "Amoco Starts Up CO_2 Recovery Plant in Big West Texas Field," Sept. 9, p. 80.

Babcock, R. E., Spillman, R. W., Godden, C. S., and Cooley, T. E., 1988, "Natural Gas Cleanup: A Comparison of Membrane and Amine Treatment Processes," *Energy Prog.*, September, pp. 135–142.

Baker, R. W., and Wijmans, J. G., 1991, "Process for Reducing Emissions from Industrial Sterilizers," U.S. Patent No. 5,069,686, December.

Baker, R. W., Noriaki, Y., Mohr, J. M., and Kahn, A. J., 1986, "Separation of Organic Vapors from Air," Presented at the American Chemical Society Regional Meeting, Denver, CO, June 8–12.

Baker, R. W., Kaschemekat, J., Wijmans, J. G., and Simmons, V. L., 1992, "Membrane Vapor Separation Systems for the Recovery of VOCs," Presented at the Air & Waste Management Association Annual Meeting, Kansas City, MO, June 21–26.

Baker, R. W., 1985, "Process for Recovering Organic Vapors from Air," U.S. Patent No. 4,553,983, Nov 6.

Baker, R. W., Kaschemekat, J., Simmons, V. L., and Wijams, J. G., 1991, "Membrane Pervaporation and Vapor Separation Systems for the Control of VOCs," Presented at the Ninth Annual Membrane Technology/Planning Conference, Nov. 6.

Beaver, E. R., Graham, T. E., Johannessen, T., and Kvivik, H., 1986, "Inert Gas Generation Systems for Offshore Platforms," *Energy Prog.*, September, pp. 149–154.

Beaver, E. R., Bhat, P. V., and Sarcia, D. S., 1988, "Integration of Membranes with Other Air Separation Technologies," *AIChE Symposium Series,* No. 261, Vol. 84, pp. 113–123.

Bhat, P. V., and Beaver, E. R., 1988, "Innovations in Nitrogen Inerting Using Membrane Systems," *AIChE Symposium Series,* No. 261, Vol. 84, pp. 124–129.

Bhide, B. D., and Stern, S. A., 1993, "Membrane Processes for the Removal of Acid Gases From Natural Gas; I. Process Configurations and Optimization of Operating Conditions; II. Effects of Operating Conditions, Economic Parameters, and Membrane Properties," *J. Memb. Sci.,* Vol. 81, Part I, pp. 209–237; Part II, pp. 239–252.

Bollinger, W. A., Long, S. P., and Metzger, T. R., 1984, "Optimizing Hydrocracker Hydrogen," *Chem. Eng. Prog.,* May, pp. 51–57.

Bollinger, W. A., MacLean, D. L., and Narayan, R. S., 1982, "Separation Systems for Oil Refining and Production," *Chem. Eng. Prog.,* October, pp. 27–32.

Boustany, K., Narayan, R. S., Patton, C. J., and Stookey, D. J., 1983, "Economics of Removal of Carbon Dioxide from Hydrocarbon Gas Mixtures," *Proceedings of the Sixty-Second Gas Processors Association Annual Convention,* pp. 146–149.

Burmaster, B. M., and Carter, D. C., 1983, "Increased Methanol Production Using Membrane Separators," paper presented at the AIChE National Meeting, Houston, TX, March 27–31.

Choe, J. S., Agrawal, R., Auvil, S. R., and White, T. R., 1988, "Membrane/Cryogenic Hybrid Systems for Helium Purification," *Proceedings of the Sixty-Seventh Gas Processors Association Annual Convention,* pp. 251–255.

Coady, A.B., and Davis, J. A., 1982, "CO_2 Recovery by Gas Permeation," *Chem. Eng. Prog.,* October, pp. 44–49.

Cooley, T. E., and Dethloff, W. L., 1985, "Field Tests Show Membrane Processing Attractive," *Chem. Eng. Prog.,* October, pp. 45–50.

Cooley, T.E., 1990, "The Use of Membranes for Natural Gas Purification," Presented at the Gas Processors Association Meeting, European Chapter, Biarritz, France, May 17 and 18.

Cutler, G., and Johnson, J., 1985, "Large-Scale CO_2 Recovery with Membranes," *Proceedings of the 1985 Laurence Reid Gas Conditioning Conference Proceedings,* Univ. of Oklahoma, Norman, OK, pp. F1–F11.

Cynara, 1995, "Gas Membrane CO_2 Separations," Brochure: Experience List for CO_2 Membrane Applications, Cynara Co., Houston, TX, Rev. 8/95 jah, Sept. 6.

DaCosta, A. R., Fane, A. G., Fell, C. J. D., and Franken, A. C. M., 1991, "Optimal Channel Spacer Design for Ultrafiltration," *J. Memb. Sci.,* Vol. 62, pp. 275–291.

DaCosta, A. R., Fane, A. G., and Wiley, D. E., 1994, "Spacer Characterization and Pressure Drop Modelling in Spacer-Filled Channels for Ultrafiltration," *J. Memb. Sci.,* Vol. 87, pp. 79–98.

Donohue, M. D., Minhas, B. S., and Lee, S. Y., 1989, "Permeation Behavior of Carbon Dioxide—Methane Mixture in Cellulose Acetate Membranes," *J. Memb. Sci.,* Vol. 42, p. 197.

Doshi, K. J., Werner, R. G., and Mitariten, M. J., 1989, "Integration of Membrane and PSA Systems for the Purification of Hydrogen and Production of Oxo-Alcohol Syngas," *AIChE Symposium Series No. 272,* Vol. 85, pp. 62–67.

Ettouney, H. M., Al-Enezi, G., and Hughes, R., 1995, Modeling of Enrichment of Natural Gas Wells by Membranes," *Gas Sep. Purif.,* Vol. 9, No. 1, pp. 3–11.

Flynn, A. J., 1983, "Wasson Denver Unit—CO_2 Treatment," *Proceedings of the Sixty Second Annual Convention of the Gas Processors Assoc.,* pp. 142–145.

Fournie, F. J. C., and Agostini, J. P., 1987, "Permeation Membranes Can Efficiently Replace Conventional Gas Treatment Processes," *J. of Pet. Tech.,* June, pp. 707–712.

Funk, E. W., Kulkarni, S. S., and Swamikannu, A. X., 1986, "Effect of Impurities on Cellulose Acetate Membrane Performance," *AIChE Symposium Series,* No. 250, Vol. 82, pp. 27–34.

Goddin, C. S., 1982, "Comparison of Processes for Treating Gases with High CO_2 Content," *Proceedings of the Sixty-First Annual Gas Processors Association Annual Convention,* pp. 60–68.

Gollan, A., and Kleper, M. H., 1986, "Membrane-Based Air Separation," *AIChE Symposium Series,* No. 250, Vol. 82, pp. 35–47.

Gollan, A., and Kleper, M. H., 1987, "State-of-the-Art: Gas Separation," *Proceedings of the Fifth Annual Membrane Technology/Planning Conference,* Boston, MA, October, pp. 145–160.

Graham, T., 1866, *Philos. Mag.,* Vol. 32, p. 401.

Hamaker, R. J., 1991, "Evolution of a Gas Separation Membrane 1983–1990," Presented at the International Conference on Effective Industrial Membrane Processes—Benefits and Opportunities, Edinburgh, Scotland, March 19–21.

Heyd, J., 1986, "Hydrogen Recovery Using Membranes in Refining Applications," Presented at the National Petroleum Refiners Association Annual Meeting, Los Angeles, CA, March 23–25.

Hogsett, J. E., and Mazur, W. H., 1983, *Hydro. Process.,* August, p. 52.

Houston, C. D., and Schell, W. J., 1986, "Recovery of Methane from Landfill Gas by the SEPAREX Membrane Process," *Proceedings of Energy from Biomass and Wastes X, Institute of Gas Technology,* Chicago, IL, April 7–10.

Koros, W.J., and Fleming, G. K., 1993, "Review, Membrane-Based Gas Separation," *J. Memb. Sci.,* Vol. 83, pp. 1–80.

Koros, W. J., and Chern, R. T., 1987, "Separation of Gaseous Mixtures Using Polymer Membranes," Chapter 20 In *Handbook of Separation Process Technology,* R.W. Rouseau, ed., New York: John Wiley and Sons, pp. 862–953.

Lacey, R. E., and Loeb, S., 1979, *Industrial Processing with Membranes,* New York: Robert E. Krieger Publishing Company, pp. 279–339.

Lane, V. O., 1993, "Plant Permeation Experience," *Hydro. Process.,* August, p. 56.

Lee, A. L., Feldkirchner, H. L., Stern, S. A., Houde, A. Y., Gamez, J. P., and Meyer, H. S., 1995, "Field Tests of Membrane Modules for the Separation of Carbon Dioxide from Low-Quality Natural Gas," *Gas Sep. Purif.,* Vol. 9, No. 1, pp. 35–45.

Lee, S. Y., Minhas, B. S., and Donohue, M. D., 1988, "Effect of Gas Composition and Pressure on Permeation Through Cellulose Acetate Membranes," *AIChE Symposium Series,* No. 261, Vol. 84, pp. 93–101.

Li, K., Acharya, D. R., and Hughes, R., 1990, "Membrane Gas Separation with Permeate Purging," *Gas. Sep. Purif.,* No. 4, p. 81.

Loeb, S., and Sourirajan, S., 1960, University of California, Los Angeles, Report No. 60-60.

MacLean, D. L., and Narayan, R. S., 1982, "PRISM Separator Applications—Present and Future," In *Hydrogen Energy Progress IV,* Vol. 2. T. N. Verizoglu, W. D. Van Vorst, and J. H. Kelly, eds., New York: Pergamon Press, pp. 837–847.

MacLean, D. L., Prince, C. E., and Chae, Y.C., 1988, "Energy—Saving Modifications in Ammonia Plants," *Chem. Eng. Prog.,* March, pp. 98–104.

MacLean, D. L., Stookey, D. J., and Metzger, T. R., 1983, "Fundamentals of Gas Permeation," Hydro. Process., August, pp. 47–51.

MacLean, D. L., and Graham, T.E., 1980, "Hollow Fibers Recover Hydrogen," *Chem. Eng.,* Feb. 25, pp. 54–55.

Marquez, J. J., 1991, "Application of Hollow Fiber Membranes with High Permeate Pressures," *Proceedings of the Seventieth Gas Processors Association Annual Convention,* San Antonio, TX, March 11–12, pp. 212–221.

Marquez, J. J., and Hamaker, R. J., 1986, "Development of Membrane Performance During SACROC Operations," Presented at the AIChE Spring National Meeting and Petro. Expo '86, New Orleans, LA, April 6–10.

Mazur, W. H., and Chan, M. C., 1982, "Membranes for Natural Gas Sweetening and CO_2 Enrichment," *Chem. Eng. Prog.,* October, pp. 38–43.

McCann, P., Ryan, J. M., and O'Brien, J. V., 1987, "The Mitchell Alvord South CO_2 Plant Propane Recovery—Ryan Holmes," paper 74A presented at the AIChE 1987 Spring National Meeting, Houston, TX, March 29–April 2.

McGinn, K. S., and Pfitzinger, M. S., 1988, "Producing a Sintering Atmosphere Through a Hollow Fiber Membrane," *Proceedings of the International Powder Metallurgy Conference,* Orlando, FL, pp. 229–241.

McKee, R. L., Changela, M. K., and Reading, G. L., 1991, "CO_2 Removal: Membrane Plus Amine," *Hydro. Process.,* April, pp. 63–67.

McReynolds, K. B., 1985, "A New Air Separation System," *Chem. Eng. Prog.*, June, pp. 27–29.

MEDAL, (An Air Liquide group company), 1996, "Hydrogen (Medal)," *Gas Processes '96, Hydro. Process.*, April, p. 119.

Metzger, T. R., Handermann, A. C., and Stookey, D. J., 1984, "Shipboard Inert Gas Generation," Presented at the AIChE 1984 Summer National Meeting, Philadelphia, PA, Aug. 22.

Meyer, H. S., and Gamez, J. P., 1995, "Gas Separation Membranes: Coming of Age for Carbon Dioxide Removal from Natural Gas," *Proceedings of the 45th Annual Laurance Reid Gas Conditioning Conference*, Norman, OK, Feb. 26–March 1, pp. 284–306.

Mitchell, J. K., 1831, *J. Roy. Inst.*, Vol. 2, No. 101, p. 307.

Parro, D., 1984, "Membrane CO_2 separation proves out at Sacroc tertiary recovery project," *Oil & Gas J.*, Sept. 24, pp. 85–88.

Peinemann, K. V., Mohr, J. M., and Baker, R. W., 1986, "The Separation of Organic Vapors from Air," *AIChE Symposium Series*, No. 250, Vol. 82, pp. 19–26.

Poffenbarger, G. L., and Gastinne, P., 1989, "Hydrogen Membrane Applications and Design Considerations," Presented at the AIChE National Meeting, Houston, TX, April 2–6.

Prasad, R., Notaro, F., and Thompson, D. R., 1994, "Evolution of Membranes in Commercial Air Separation," *J. Memb. Sci.*, Vol. 94, pp. 225–248.

Price, B. C., and Gregg, F. L., 1983, "CO_2/EOR: from source to resource," *Oil and Gas J.*, August 22, pp. 116–122.

Rice, A. W., 1990, "Process for Capturing Nitrogen from Air Using Gas Separation Membranes: Deoxygenation," U.S. Patent No. 4,894,068.

Rogers, C. E., Fels, M., and Li, N. N., 1972, "Separation by Permeation Through Polymeric Membranes," In *Recent Developments in Separation Science*, Vol. II. Cleveland: Chemical Rubber Company, pp. 107–155.

Russell, F. G., 1984, "Applications of the DELSEP Membrane System," *Chem. Eng. Prog.*, October, pp. 48–52.

Ryan, J. M., and Schaffert, F. W., 1984, "CO_2 Recovery by the Ryan/Holmes Process," *Chem. Eng. Prog.*, October, pp. 53–56.

Schaffert, F. W., Wood, N. V., and O'Brien, 1986, "The Seminole San Andres Unit CO_2 Recovery Plant—Meeting the Challenges of EOR Gas Processing," *Proceedings of the Sixty-Fifth Annual Convention of the Gas Processors Assoc.*, pp. 86–89.

Schell, W. J., 1983, "Membrane Use/Technology Growing," *Hydro. Process.*, August, pp. 43–46.

Schell, W. J., and Houston, C. D., 1985, "Refinery Hydrogen Recovery with SEPAREX Membrane Systems," Presented at the National Petroleum Refiners Association Annual Meeting, San Antonio, TX, March 24–26.

Schell, W. J., and Hoernschemeyer, D. L., 1982, "Principles of Gas Separation," Presented at the *AIChE Symposium*, Anaheim, CA, June 7–10.

Schell, W. J., and Houston, C. D., 1982, "Spiral-Wound Permeators For Purification and Recovery," *Chem. Eng. Prog.*, Oct., pp. 33–37.

Schendel, R. L., 1995, personal communication (consultant), Sept. 6.

Schendel, R. L., and Seymour, J., 1985, "Take Care In Picking Membranes to Combine with Other Processes for CO_2 Removal," *Oil & Gas J.*, Feb. 18, pp. 84–86.

Schendel, R. L., 1984, "Using Membranes for the Separation of Acid Gases and Hydrocarbons," *Chem. Eng. Prog.,* May, pp. 39–43.

Schendel, R. L., Mariz, C. L., and Mak, J. Y., 1983, "Is Permeation Competitive?," *Hydro. Process.,* August, pp. 58–62.

Schendel, R. L., and Nolley, E., 1984, "Commercial Practice in Processing Gases Associated with CO_2 Floods," paper presented at the World Oil and Gas Show and Conference, Dallas, TX, June 4–7.

Shaver, K. G., Poffenbarger, G. L., and Grotewold, D. R., 1991, "Membranes Recover Hydrogen," *Hydro. Process.,* June, pp. 77–80.

Shirley, J., and Borzik, D., 1982, "Hollow Fiber Gas Separator Boosts NH_3 Output by 50 tpd," *Chem. Process.,* January, pp. 30–31.

Spangler, G. E., 1975, "Analysis of Two Membrane Inlet Systems on Two Potential Trace Vapor Detectors," *Amer. Lab.,* Vol. 7, No. 7, pp. 36–8, 40–5.

Spillman, R. W., 1989, "Economics of Gas Separation Membranes," *Chem. Eng. Prog.,* January, pp. 41–62.

Spillman, R. W., Barrett, M. G., and Cooley, T. E., 1988, "Gas Membrane Process Optimization," Presented at the AIChE National Meeting, New Orleans, LA, March 9.

Spillman, R. W., and Cooley, T. E., 1989, "Membrane Gas Treating," *Proceedings of the Sixty-Eighth Gas Processors Association Annual Convention,* San Antonio, TX, pp. 186–194.

Stern, S. A., 1986, "New Developments in Membrane Processes for Gas Separations," In *Synthetic Membranes,* M.B. Chenoweth, ed., New York: MMI Press.

Stookey, D. J., Patton, C. J., and Malcom, G. L., 1986, "Membranes Separate Gases Selectively," *Chem. Eng. Prog.,* November, pp. 36–40.

Tomlinson, T. R., and Finn, A. J., 1990, "H_2 Recovery Processes Compared," *Oil & Gas J.,* Jan. 15, pp. 35–39.

Tonen Technology, and UBE Industries, 1990, "UBE Polyimide Membrane System for Hydrogen Recovery and Purification of Hydrofiner Offgas," Unpublished report provided by UBE Industries, March 16.

U.S. Dept. of Energy, 1989, "Membrane Separation Processes for Liquid Hydrocarbons and Gases in the Petrochemical Industry," DOE Contract No. DE-AC01-87CE40762, Report DOE/CH10093-57; DE89009464, September.

Weller, S., and Steiner, W. A., 1950, "Engineering Aspects of Separation of Gases—Fractional Permeation Through Membranes," *Chem. Eng. Prog.,* November, pp. 585–590.

Wijmans, J. G., Kaschemekat, J., and Baker, R.W., 1991, "A Membrane Process for the Recovery of Volatile Organic Compounds from Process and Vent Streams," Presented at the Air & Waste Management Association 84th Annual Meeting and Exhibition, Vancouver, British Columbia, June 16–21.

Wood, N. V., O'Brien, J. V., and Schaffert, F.W., 1986, "Seminole CO_2 flood gas plant has successful early operations," *Oil and Gas J.,* Nov. 17, pp. 50–54.

Yamashiro, H., Hirajo, M., Schell, W. J., and Maitland, C.F., 1985, "Plant Uses Membrane Separation," *Hydro. Process.,* February, pp. 87–89.

Youn, K. C., Poe, W. A., Sattler, J. P., and Inlow, H. L., 1987, "CO_2-recovery plant key to West Texas EOR operations," *Oil and Gas J.,* Nov. 23, pp. 76–78.

Chapter 16
Miscellaneous Gas Purification Techniques

SULFUR SCAVENGING PROCESSES, 1297

Available Technologies, 1297
Dry Sorption Processes, 1298
Liquid Absorption Processes, 1309
Comparison of H_2S Scavenging Processes, 1318

HIGH-TEMPERATURE, REGENERATIVE SULFUR REMOVAL SYSTEMS, 1320

Background, 1322
Zinc Ferrite and Titanate Sorbents, 1324
Haldor Topsøe Tin Oxide-Based Process, 1328
Z-Sorb Promoted Zinc Oxide-Based Process, 1329

ABSORPTION AND CONDENSATION PROCESSES FOR EXHAUST GAS PURIFICATION, 1330

Absorption of Impurities by Reactive Solvents, 1330
Physical Absorption Processes, 1332
Condensation, 1332

LOW-TEMPERATURE GAS PURIFICATION PROCESSES, 1337

Hydrogen Purification, 1338
The CNG Process, 1342
The Ryan/Holmes Processes, 1342
Nitrogen Removal from Natural Gas, 1344

CARBON MONOXIDE REMOVAL WITH COPPER COMPOUNDS, 1346

Aqueous Copper Ammonium Salt Process, 1347
COSORB Process, 1357

HYDROCARBON REMOVAL BY OIL ABSORPTION, 1359

Light Oil Removal from Coal Gas, 1359
Naphthalene Removal from Coal Gas, 1366

REFERENCES, 1369

SULFUR SCAVENGING PROCESSES

The expression "Sulfur Scavenging" is used to describe a group of processes which operate in a nonregenerative manner to remove small quantities of sulfur compounds, usually H_2S, from gas streams. The sulfur quantity may be small, and a scavenging process applicable, because of either a low volume of gas or a very low concentration of sulfur in a larger volume of gas. Typical applications are the treatment of natural gas at wells that are remote from gas processing facilities (usually less than 10 MMscfd containing 10 to 1,000 ppm H_2S) and the final purification of feed gas to ammonia or other petrochemical processes following bulk sulfur removal in a regenerative system.

Sulfur scavenging processes are typically batch operations. They make use of materials that capture and retain sulfur compounds, but have a finite effective capacity. When this capacity is reached, the spent sorbent (liquid or solid) must be removed and replaced with fresh material. The spent sorbent is normally disposed of as waste, and the production of an environmentally acceptable waste has become a key factor in the selection of scavenging processes for specific applications. Other important considerations are operating safety and reliability (unattended operation is often required) and, of course, capital and operating costs.

Available Technologies

Sulfur scavenging processes can generally be categorized as follows:

A. Dry Sorption Processes

1. *Iron Oxide.* This is probably the most widely used process. Iron oxide captures hydrogen sulfide by forming iron sulfide. Bed life can be extended by admitting air to oxidize some of the sulfide to elemental sulfur and regenerate the iron oxide; however, this is not common practice as it interferes with bed porosity and removability.
2. *Zinc Oxide.* Zinc oxide operates by reacting with hydrogen sulfide to form zinc sulfide, which is quite stable. It is widely used to remove traces of hydrogen sulfide from hot synthesis gas streams.
3. *Alkaline Solids.* Strongly alkaline solids such as sodium hydroxide/lime granules are effective reagents for removing hydrogen sulfide (and other acid gases) by a simple acid-base reaction.
4. *Adsorbents.* Molecular sieves, activated carbon, and impregnated activated carbon are effective adsorbents for hydrogen sulfide and higher molecular weight sulfur compounds and are often used in scavenger-type operations. Adsorption, including adsorption with impregnated adsorbents, is discussed in Chapter 12.

B. Liquid Phase Absorption Processes

1. *Iron Oxide Slurries.* The chemistry of this type of process is essentially identical to that of the dry iron oxide process. The principal advantage is the ease of removing and replacing the spent reactant.
2. *Zinc Oxide Slurries.* These are more reactive (and more expensive) than iron oxide slurries. Typically a weak acid anion, which forms a soluble zinc salt, is added to increase the activity of zinc in solution.
3. *Oxidizing Solutions.* Many oxidizing agents, including dichromate, permanganate, peroxide, nitrite, chlorite, and oxygen have been proposed. In most cases, dissolved hydrogen sulfide is oxidized to elemental sulfur.
4. *Aldehydes.* Aldehydes, such as formaldehyde, react rapidly with hydrogen sulfide to form relatively nonvolatile organic sulfur compounds. A major problem with this type of process is the objectionable odor of the final product.
5. *Alkylamine/Aldehyde Condensation Products.* Compounds of this type are used in liquid form and react with hydrogen sulfide to form products that remain in the liquid state. The products have a very objectionable odor, but are reported to have corrosion inhibition properties.
6. *Triazines.* Certain triazines, such as 1,3,5 tri-(2-hydroxyethyl)-hexahydro-S-triazine, react rapidly with hydrogen sulfide to form water soluble liquid byproducts. A claimed advantage for this type of process is the corrosion inhibition properties of the byproducts.
7. *Alkaline Solutions.* Processes of this type usually employ a water solution of sodium hydroxide. The operation normally absorbs all acid gases, although some selectivity for hydrogen sulfide relative to carbon dioxide is possible. The general use of sodium hydroxide solutions for purifying gases is described in detail in Chapter 5 under the heading "Caustic Scrubbing."

Specific processes, representing examples of several of the categories previously listed, are described in the following sections.

Dry Sorption Processes

This section covers processes which scavenge H_2S and related sulfur compounds from gas streams by reaction with a solid. It does not cover adsorption or combined adsorption-reaction using impregnated adsorbents, both of which are discussed in Chapter 12. With the exception of the iron oxide dry box process, which can be considered to be partially regenerative, the processes are not regenerative. Closely related processes that use dry solids to remove relatively large quantities of sulfur at elevated temperature in a regenerative scheme are not included in this section, but are discussed in the section that follows under the heading "High-Temperature Sulfur, Regenerative Removal Systems."

Iron Oxide Process

The iron oxide gas purification process is one of the oldest methods used for the removal of objectionable sulfur compounds from industrial gases. Around the middle of the nineteenth century, the iron oxide process was introduced in England to replace a wet purification process utilizing calcium hydroxide as the active agent. The iron oxide process gained widespread use in Europe and the U.S. for the treatment of coal gases, but began to lose favor near the middle of the twentieth century as natural gas replaced coal gas, and efficient liquid processes, such as those based on the alkanolamines, were introduced (see Chapter 2).

The process is currently used almost exclusively for the treatment of relatively small volumes of gas, where the amount of sulfur removed does not justify the expense and complexity of a regenerative process with sulfur recovery. However, because of the widespread use of the process in the past, and the extensive scientific and engineering base that was developed to support it, a brief description of the background and technology of the process follows. More detailed descriptions of several iron oxide-based processes, design approaches, and operating experiences are given in previous editions of this book.

Historical Background. The first installations utilized a simple form of the iron oxide (or dry-box) process. In this form of the process, hydrogen sulfide was removed completely by reaction with hydrated ferric oxide, resulting in the formation of ferric sulfide. After removal from the box and exposure to atmospheric oxygen, the ferric sulfide oxidized to elemental sulfur and ferric oxide. The oxidized mixture was reloaded into the box and used to react with additional hydrogen sulfide. The cycle could be repeated several times before the material lost activity due to the presence of excessive amounts of elemental sulfur.

Later, it was determined that a much simpler and more economical process resulted when the iron oxide was revivified *in situ*, by addition of small amounts of air to the gas at the inlet of the purification system. Another method of *in situ* revivification consisted of the circulation of oxygen-containing gas through the spent bed after it had been used to purify oxygen-free gas. *In situ* oxidation of the bed was found to result in large savings in labor costs for loading and unloading the dry boxes.

Two forms of iron oxide were used in the early dry box processes: unmixed oxides and mixed oxides. Unmixed oxides included modified iron ores containing up to 75% iron oxide and residue from the purification of bauxite, called "Lux" in Europe, which contained 25 to 50% iron oxide. Natural bog ore found in Denmark and Holland was also widely used.

The mixed oxides, or sponges, were prepared by supporting the finely divided oxide on media such as wood shavings. The advantage of mixed oxides is that the bulk density, the iron oxide content, the moisture, and the pH can be controlled better than in unmixed oxides. In addition, mixed oxides have less tendency to cake, and free passage of the gas and higher final sulfur loading can be achieved. Iron sponge (iron oxide supported on wood chips) is the most common form used today. Typical composition ranges for two types of iron oxide materials used in the 1950s are given in **Table 16-1** (Moignard, 1952).

During its period of widespread use for the purification of coal gas, the iron oxide process was subject to numerous modifications and improvements. The major forms of the process can be categorized as follows:

1. *Conventional dry box purifiers.* A simplified flow diagram of the basic iron-oxide purification process as used for low-pressure coal gas treatment is shown in **Figure 16-1**. A purifying plant consisted of one or more series of four to six boxes with the piping arranged in such a manner that the inlet gas could enter each box first, thus allowing changes in the order of the boxes. Each box contained one or several layers of oxide, each about 2 ft deep, supported on wooden or steel grids (Ward, 1964; Taylor, 1950; Dent and Moignard, 1950)
2. *Tower Purifiers.* Tower purifiers were developed to reduce the ground area required for iron oxide boxes and to simplify the loading and unloading operations. Several tower designs were employed. In general, the tower consisted of a carbon steel shell enclosing

Table 16-1
Composition of Purifying Materials

Characteristic	Bog Ore	Oxide Supported on Wood Shavings
Moisture, %	45–55	38–42
Loss on ignition (dry basis), %	25–35	36–41
Fe_2O_3 (dry basis), %	45–55	46–48
Bulk density, lb/cu ft	—	19–21

Source: Moignard, 1952

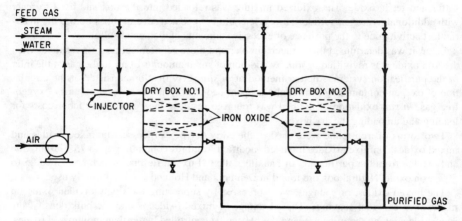

Figure 16-1. Schematic flow diagram of basic iron-oxide purification process as employed for low-pressure manufactured gas.

up to 14 superimposed removable baskets, each basket containing one or two shallow layers of iron oxide. The baskets were loaded and unloaded by use of traveling cranes. (Guntermann and Schnurer, 1957; Bairstow, 1957).

3. *Continuous Processes.* In an attempt to reduce the size and operating cost of dry purification plants and to obtain more efficient utilization of the oxide, work on continuous iron-oxide processes was undertaken in Europe after World War II. Probably the most successful was the "Gastechnik" process developed in Germany, which has been described by Moore (1956) and Sexauer (1959). More than 60 Gastechnik plants were in operation in Germany in 1956 and several were installed in England. The process used cylindrical towers filled with a deep bed of roughly spherical oxide pellets. Pellets were periodically added to and removed from the bed through air locks without interruption of the gas flow. After removal, the pellets were contacted with perchlorethylene for the extraction of sulfur, then returned to one of the towers.

4. *High-Pressure Purifiers.* Although the widespread use of iron-oxide purifiers for high-pressure gas is a relatively recent development, a plant treating 15 MMscfd of natural gas at

325 psig was described as early as 1944 (Turner). Typical high-pressure installations have used iron sponge in deep beds (e.g., 10 ft deep). Additional details of high-pressure iron-oxide purifiers are given in a later section entitled "Current Applications of Iron Sponge."

Basic Chemistry. The basic chemistry of the process can be represented by the following equations:

$$2Fe_2O_3 + 6H_2S = 2Fe_2S_3 + 6H_2O \tag{16-1}$$

$$2Fe_2S_3 + 3O_2 = 2Fe_2O_3 + 6S \tag{16-2}$$

Combining equations 16-1 and 16-2,

$$6H_2S + 3O_2 = 6H_2O + 6S \tag{16-3}$$

Reactions 16-1 through 16-3 represent a simplified mechanism for the process, and, depending on operating conditions, a number of other reactions may occur. Principal variables affecting the reaction mechanisms are temperature, moisture content, and pH of the purifying material.

Iron oxide is also capable of removing mercaptans by the reaction (Zapffe, 1963):

$$Fe_2O_3 + 6RSH = 2Fe(RS)_3 + 3H_2O \tag{16-4}$$

When used for scavenging traces of sulfur from natural gas in the absence of air, the principal reactions are 16-1 and 16-4.

There are several forms of iron oxide known, but two, namely α $Fe_2O_3 \cdot H_2O$ and τ $Fe_2O_3 \cdot H_2O$, are considered to be particularly useful as purifying materials (Griffith, 1954; Ward, 1964). Some Fe_3O_4 (equivalent to $Fe_2O_3 \cdot FeO$) may also be present. Extensive tests have shown that the iron oxide must be present as a true hydrate, not just wetted iron oxide, to be effective in removing sulfur compounds from the gas (Anerousis and Whitman, 1984). The hydrated iron oxides react readily with hydrogen sulfide, and the resulting ferric sulfide is easily reoxidized to an active form of ferric oxide. The cycle proceeds most satisfactorily at moderate temperatures, approximately 100°F, and in an alkaline environment. At temperatures above 120°F and in neutral or acid environments, ferric sulfide loses its water of crystallization and changes into a mixture of FeS_2 and Fe_8S_9. These sulfides are not readily reconverted to hydrated ferric oxide, but oxidize slowly to ferrous sulfate and polysulfides, neither of which are useful for hydrogen sulfide removal.

Excellent discussions on the properties of purification materials have been presented by Smith (1957) and Ward (1964). The physical mechanism of hydrogen sulfide absorption by iron oxide was investigated by Avery (1939) and Dent and Moignard (1950). The studies showed that under proper conditions of temperature, moisture content, and pH, the sulfur formed on the oxide particle is continuously displaced by iron oxide, which migrates from the particle center to the surface, thus exposing fresh oxide for further reaction with hydrogen sulfide and oxygen.

Moignard (1952, 1955) recommended the use of supported iron oxide of a coarse and open texture, low bulk density (on the order of 20 to 25 lb/cu ft, dry), and the highest possible moisture content (30 to 50%) consistent with free gas flow. An optimum pH range of 8 to 8.5 is also recommended. Grades of iron sponge currently available contain nominally 6.5, 9,

Table 16-2
Typical Specifications for 15-lb Iron Sponge

Water Content (Loss on Drying, wt%)

Iron Sponge Product	30.60
Iron Oxide Particulates	17.70

Size Distribution of Iron Oxide Particulates, wt%

Retained on mesh	
16	4.22
30	54.62
60	32.72
100	4.49
140	1.58
200	0.79
325	1.06
400	0.26
Smaller than 400 mesh	0.26

Chemical Analysis of Dried Iron Oxide Particulates, wt%

Iron as Fe_2O_3	58.67
Iron as Fe_3O_4	20.40
Sulfur as S	0.49
Copper as Cu	0.11
Zinc as Zn	0.01
Lead as Pb	0.01
Silicon as Si	1.02
Aluminum as Al	0.02
Phosphorus as P	0.02
Balance primarily wood substrate material	
Flooded pH (1)	10.2
Leachable pH (2)	7.88
Weight of Iron Oxide, lb/bushel	17.61

Notes: 1. Flooded pH is determined by soaking a bushel of iron sponge in an excess of distilled water for 24 hours, then measuring the pH of a representative sample of the water.
2. Leachable pH is determined by recycling 100 ml of distilled water over 1 g of sponge, then measuring the pH of the water.
Source: Anerousis and Whitman, 1984.

and 15 lb of active iron oxide per bushel (one bushel is 1.25 cu ft) (Anerousis and Whitman, 1984). Typical specifications for 15-lb sponge are given in **Table 16-2**.

The two most important criteria for the selection of purifying materials are capacity and activity. Theoretically 1 lb of ferric oxide absorbs 0.64 lb of hydrogen sulfide when completely transformed to iron sulfide. In operations without oxygen addition this value is not attainable, although capacities as high as 0.56 lb sulfur/lb ferric oxide have been reported (Taylor, 1956; Perry, 1970). In general, it is assumed that about 50% of the theoretical sulfiding capacity is attainable in the first cycle of operation and the capacity decreases progressively with additional cycles. When oxygen is present in the feed gas, much higher capacities can be obtained. Final sulfur contents as high as 2.5 lb sulfur/lb iron oxide have been reported (Taylor, 1956).

Current Applications of Iron Sponge. While iron sponge is no longer used for treating large volumes of natural gas, it is still widely used for removing hydrogen sulfide from small volumes of moderately sour natural gas at sites remote from gas processing plants. The sorbent material consists of wood chips or shavings impregnated with hydrated ferric oxide, Fe_2O_3, and sodium carbonate to control pH. It is used as a fixed bed in a vertical, cylindrical vessel with downflowing gas. Spent material is periodically removed from the vessel and

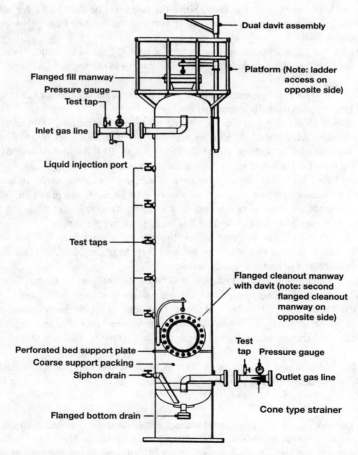

Figure 16-2. Typical iron sponge contactor for elevated pressure natural gas service. (*Anerousis and Whitman, 1984*).

1304 *Gas Purification*

replaced with fresh sponge. A diagram of a typical iron sponge contactor for natural gas purification is shown in **Figure 16-2.**

Key mechanical features of iron sponge (and other solid bed) contactors are

1. Top and bottom manways for loading and unloading sorbent material.
2. A porous bed support strong enough to carry the full weight of the bed plus the calculated gas pressure drop, and designed to retain fine particles of sorbent.
3. Pressure sensors on the feed and outlet gas lines to indicate possible bed plugging or collapse.
4. Gas sample taps at the bed inlet and outlet and at several intermediate points in the bed to permit monitoring of the sour gas front position and avoid breakthrough.

Two iron-sponge beds are often used in series, with a valve and manifold system to permit the lead unit to be removed from service, emptied, recharged, and placed back in service as the final unit without interrupting gas flow. Since the bed must be kept moist for effective operation, many iron sponge vessels have a water spray above the bed or a liquid injection line in the inlet gas line.

The following design and operating guidelines for iron sponge scavenging systems are based on the publications of Schaack and Chan (1989) except as otherwise indicated.

1. Operation at 65°–115°F is recommended to minimize wax deposition on the chips and achieve effective hydrogen sulfide removal.
2. If regeneration with air is not practiced, the use of sponge containing 15-lb active iron oxide per bushel is recommended. With air regeneration, 9 lb/bushel sponge is preferred. Two additional grades of sponge are available: 6.5 and 20 lb/bushel. When properly packed in the vessel, a bushel of sponge occupies about one cu ft.
3. A bed height of 10 ft is recommended for general use; greater depth results in bed compaction and increased pressure drop unless tray or grid supports are provided. Maddox (1974) recommends a minimum bed height of 4 to 10 ft. Anerousis and Whitman (1984) recommend 10 ft minimum for H_2S and 20 ft if mercaptans are present.
4. A maximum superficial gas velocity of 10 ft/min is recommended. The gas velocity should be limited to a value that results in no more then 15 grains of sulfur deposition per minute per square foot of bed cross-sectional area.
5. The moisture content should be maintained at 40% by weight, plus or minus 15%. Manning and Thompson (1991) recommend a moisture content of at least 20%.
6. Manning and Thompson (1991) state that about one pound of sodium carbonate should be included per bushel of sponge to assure an alkaline environment. During operation, the pH of water in the bed should be maintained between 8 and 10 by the addition of sodium carbonate.
7. A design loading of about 0.42 lb sulfur/lb iron oxide is recommended if regeneration with air is not practiced. For the case of 15 lb/bushel sponge this is equivalent to a loading of about 6.2 lb sulfur/cu ft of sponge.

Although iron sponge is not considered hazardous, it must be handled with care, and unnecessary contact should be avoided. Precautions must be taken during the removal of spent material to prevent fires. If the material is allowed to dry out in the presence of air, oxidation of the highly reactive iron sulfide can generate enough heat to ignite the wood chips. To prevent this, it is recommended that sufficient water be sprayed on spent material exposed to air to maintain its temperature below 120°F until controlled oxidation has occurred (Schaack and Chan, 1989).

SulfaTreat Process. The SulfaTreat process offered by The SulfaTreat Co. is similar in operation to the iron sponge technique for removing hydrogen sulfide and mercaptans from gas streams. It also uses one or more fixed beds of granular, iron oxide-containing material in simple pressure vessels with downflow of gas. The principal difference is that, instead of iron sponge, the process uses a proprietary granular material called SulfaTreat. The process and bed material are described by Samuels (1990) and Wendt (1991).

According to the manufacturer, SulfaTreat is a substantially dry, free-flowing material, composed of a unique, proprietary iron compound mixed with supplemental chemicals, which has been granulated to facilitate ease of handling. The insoluble particles range in size between 4 and 30 mesh, and have a bulk density of 70 lb/cu ft (Gas Sweetener Associates, Inc., 1991).

The iron oxide in SulfaTreat is reportedly present in two forms: Fe_2O_3 and Fe_3O_4. Hydrogen sulfide reacts with both forms to produce a mixture of iron sulfides. The conversion efficiency in commercial operations has been found to range between 0.55 and 0.716 lb H_2S reacted/lb of iron oxide (Samuels, 1990). This is somewhat higher than the value of 0.42 lb sulfur (or 0.45 lb H_2S) /lb iron oxide recommended for iron sponge bed design.

Some of the advantages claimed for SulfaTreat over iron sponge are

1. No risk of fire during change-out (non-pyrophoric).
2. Easier change-outs because the material does not become cemented.
3. Low pressure drop without gas channeling because of uniform porosity. Pressure drop can be estimated by use of the Ergun equation (see Chapter 12).
4. Control of pH and the presence of liquid water are not required; however, the gas must be saturated or nearly saturated with water for good efficiency.

According to Samuels (1990), replacing iron sponge with SulfaTreat in a Canadian installation resulted in significant improvements in operation and economics. Although material costs were higher for SulfaTreat, the overall costs were lower due primarily to fewer required change-outs. For a six-month period in 1988, treating costs, including labor, material, and disposal, averaged \$6.16/lb of H_2S removed with iron sponge in the units. During the same months in 1989, with SulfaTreat, costs averaged \$5.37/lb of H_2S removed. In 1994, it was reported that over 400 SulfaTreat plants were in operation or planned (The SulfaTreat Co., 1994).

Zinc Oxide-Based Processes

High-Temperature Nonregenerative Processes. For many years zinc oxide has been the preferred sorbent for removing traces of hydrogen sulfide from natural gas feed to steam-hydrocarbon reformers producing synthesis gas for the production of ammonia and other petrochemicals. The zinc oxide is typically in the form of cylindrical extrudates 3–4 mm in diameter and 4–8 mm in length. Several forms are available for operation at temperatures from about 400° to 750°F.

In cases where the natural gas contains stable organic sulfur compounds, it is necessary to use a hydrogenation catalyst, such as cobalt-molybdenum, ahead of the zinc oxide bed to

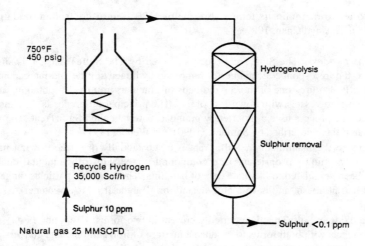

Figure 16-3. Typical feedstock desulfurization system for a 1,000 ton/d ammonia plant. (*Carnell, 1986*)

convert the organic compounds into the more reactive hydrogen sulfide. A typical flow arrangement for an ammonia plant is shown in **Figure 16-3.** The hydrogenation catalyst operates at essentially the same temperature as the zinc oxide and serves to convert organic sulfur compounds into the more reactive hydrogen sulfide. (The catalytic conversion of organic sulfur compounds to hydrogen sulfide is discussed in Chapter 13.)

Hydrogen sulfide reacts with zinc oxide by the following reaction:

$$ZnO + H_2S = ZnS + H_2O \qquad (16\text{-}5)$$

The equilibrium constant for this reaction is

$$K_p = P_{H_2O}/P_{H_2S} \qquad (16\text{-}6)$$

where P_{H_2O} and P_{H_2S} are the partial pressures of water vapor and hydrogen sulfide, respectively, in the gas phase.

As shown in **Figure 16-4,** the equilibrium constant decreases rapidly with temperature. At the high operating temperature used in these reactors, equilibrium is closely approached, and, therefore, the product gas hydrogen sulfide concentration can be estimated accurately with equation 16-6, if the temperature and water vapor concentration are known.

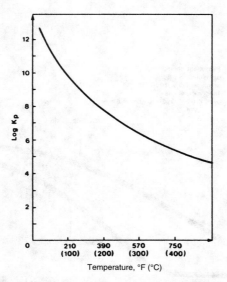

Figure 16-4. Equilibrium constant, K_p, for the reaction: $ZnO + H_2S = ZnS + H_2O$, where $K_p = P_{H_2O}/P_{H_2S}$. *(Haldor Topsøe, 1986)*

The design of zinc oxide units for high-temperature hydrogen sulfide removal is relatively straightforward and based on extensive operating experience. Because of the uniformity of commercial shapes, pressure drop can be calculated on the basis of conventional correlations such as the Ergun equation (see Chapter 12) or manufacturers' data. The maximum sulfur loading is in the range of 30–40 lb sulfur/100 lb sorbent, dependent on the specific material used. Because of the extremely low levels of hydrogen sulfide involved, beds are usually designed to last over a year.

In operation, a reaction zone gradually moves through the bed with completely reacted zinc sulfide behind this zone and essentially unreacted zinc oxide ahead of it. The length of the reaction zone normally represents a portion of the bed that cannot be fully utilized, and it is therefore desirable to make it as short as possible. Increasing the temperature helps by increasing the rate of diffusion within the solid particles. Almost complete conversion of zinc oxide can be obtained by using two beds in series, installed so that each bed can be used as the first in the series before it is changed out.

Low-Temperature Zinc Oxide (ICI) Process. Imperial Chemical Industries (ICI) of Britain reported the development of a form of zinc oxide that performs more efficiently than conventional forms at low temperatures (Carnell, 1986). The ICI material has a slightly reduced density and greatly increased porosity and surface area. The performance of this product is compared with a conventional high-temperature zinc oxide in **Figure 16-5.** The figure indicates that the new material is significantly more effective than standard zinc oxide in the temperature range of about 100° to 400°F.

Carnell (1986) provides information on the first commercial application of the ICI zinc oxide sorbent on a North Sea oil platform. Although the material is capable of removing hydrogen sulfide from gases at low temperature, its performance does improve with tempera-

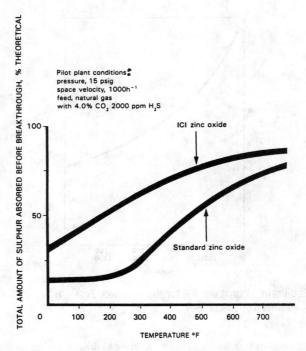

Figure 16-5. Accelerated breakthrough tests for new ICI and standard zinc oxide using natural gas. (*Carnell, 1986*)

ture, so it was installed downstream of a compressor to take advantage of the heat of compression. This provided an inlet gas at about 300°F. A simplified flow diagram of the installation is shown in **Figure 16-6**. Operating experience with the system over an 18-month period showed it to be completely reliable and predictable. The spent bed, containing in excess of 20 wt% sulfur, is discharged by gravity flow and may be reprocessed for metal recovery.

ICI Katalco offers the low-temperature zinc oxide process as part of a family of fixed-bed processes under the trademark, Puraspec. The Puraspec processes employ combinations of catalysts, catalytic absorbents, and regenerable adsorbents to remove a wide variety of impurities from gases and hydrocarbon liquids (ICI Katalco, 1991). Each process is custom designed to meet the specific requirements of the purification problem. The composition of the materials used is considered proprietary.

The operation of a Puraspec plant that removes hydrogen sulfide from natural and associated gas is described by Siemek (1993). The plant was commissioned in April 1990. It consists of two fixed-bed reactors in series, with interconnecting piping to allow either reactor to be operated in the lead position. In a test operation shortly after startup, the unit reduced the hydrogen sulfide concentration of 51,000 cu m/h (46 MMscfd) of gas from 53 mg/cu m to 0.6 mg/cu m. The absorbent in the lead bed was replaced in October 1991. The spent material was found to contain more than 200 kg sulfur/cu m. The reactor was recharged with new sorbent and placed in the lag position. The sorbent in the second bed was similarly discharged and replaced in August 1992. Arrangements were made for reprocessing the spent bed material for metal recovery by ICI Katalco through its Catalyst Care Program (Siemek, 1993).

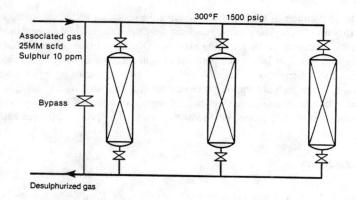

Figure 16-6. Three-bed zinc oxide system for desulfurizing associated gas on an offshore platform. (*Carnell, 1986*)

Alkaline Solids

Sofnolime RG. This process, which is licensed by Molecular Products Ltd., UK, uses fixed beds of an alkaline granular solid called Sofnolime RG. The material is claimed to be a synergistic mixture of hydroxides that reacts with acid gases such as H_2S, COS, CO_2, SO_2, and RSH. Typical reactions are

$$2NaOH + H_2S = Na_2S + 2H_2O \qquad (16\text{-}7)$$

$$Ca(OH)_2 + CO_2 = CaCO_3 + H_2O \qquad (16\text{-}8)$$

The sofnolime RG granules are sandwiched between a ceramic ball support layer beneath the bed and a top layer of ceramic balls to retain the bed. Gas (or liquid) flow is upward. As of early 1994, there were 15 installations worldwide (Molecular Products Ltd., 1994).

Liquid Absorption Processes

Iron Oxide Slurry

Slurrisweet Process. The Slurrisweet process was developed by Gas Sweetener Associates as a process that would utilize the same chemistry as the iron sponge process, but avoid some of the difficulties of loading and unloading the contactor and of disposing of the spent material (Sivalls, 1982). The process was later developed further and offered commercially; however, the process is no longer being marketed (Bucklin, 1994). The technology is based on the use of a slurry of iron oxide particles in water to contact the gas and react with the hydrogen sulfide. According to Schaack and Chan (1989) air must be injected into the sour gas stream to attain high absorption efficiency for H_2S and stabilize the solution. They also report that problems of foaming, short batch life, rapid settling of the slurry, and corrosion were encountered in early plants, but were later resolved by process changes and the use of additives.

Zinc Oxide Slurry

Chemsweet Process. Chemsweet is Natco's trademark for a white powder formulation containing inorganic zinc compounds that is mixed with water to form the absorbent for this process. According to Manning et al. (1981), the powder contains zinc oxide, zinc acetate, and a dispersant to keep the zinc oxide particles in suspension. One part of powder is mixed with five parts of water to give the desired slurry density. The zinc acetate dissolves to give a controlled source of zinc ions. These react instantly with dissolved hydrogen sulfide to form insoluble zinc sulfide and acetic acid. The acid causes additional zinc to go into solution. The chemical reactions can be represented by

$$ZnAc_2 + H_2S = ZnS + 2HAc \qquad (16\text{-}9)$$

$$ZnO + 2HAc = ZnAc_2 + H_2O \qquad (16\text{-}10)$$

Equation 16-11 represents the overall reaction:

$$ZnO + H_2S = ZnS + H_2O \qquad (16\text{-}11)$$

The process is carried out in a simple, vertical bubble contactor as shown in **Figure 16-7**. The contactor is equipped with inlet and outlet gas nozzles, slurry feed and overflow nozzles, a drain, a manway, a mist eliminator, and a gas distributor. Gas bubbles rise from the distributor and provide the agitation to keep the zinc oxide particles in suspension.

The pH of the solution is maintained low enough to avoid carbon dioxide absorption, but high enough to minimize vessel corrosion. Additives buffer the slurry pH above 6.0 (Schaack and Chan, 1989). The inlet gas must be saturated with water and free of liquid contaminants. An efficient inlet separator is recommended (Manning, 1979).

In the design of a zinc oxide slurry contactor, an expansion factor must be applied to the liquid volume at rest to account for the presence of gas bubbles during operation. Schaack and Chan (1989) suggest an expansion factor of 1.25; while Manning and Thompson (1991) recommend using 1.10 to 1.20. The latter authors point out that the bubbles rising through the slurry quickly attain a terminal velocity of about 48 ft/min. The superficial velocity, V_s, is therefore related to the expansion factor, f, by the following relationship:

$$V_s = 48(f-1)/f \qquad (16\text{-}12)$$

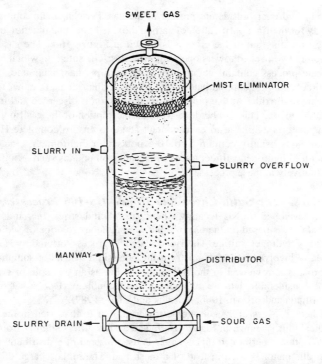

Figure 16-7. Bubble tower proposed for Chemsweet process. (*Manning, 1979*)

If, for example, an expansion factor of 1.15 is selected, a superficial velocity of 6.23 ft/min can be used to calculate the required contactor diameter.

In calculating the required contactor height, the liquid depth is determined on the basis of the desired time between charges, typically 15 to 90 days, and increased by the anticipated expansion. An adequate disengagement space must be allowed between the top of the expanded liquid and the mist eliminator.

The Chemsweet process has been used successfully in many commercial installations ranging in capacity from 5 Mscfd to 12 MM scfd. No difficulty has been experienced meeting the standard gas purity requirement of ¼ grain H_2S/100 scf, and the process is also reported to remove RSH and COS quantitatively (Houghton and Bucklin, 1994). The principal disadvantage of the process is the high cost of the reactant, which has generally limited the applicability of the process to cases where low quantities of sulfur must be removed, and has caused many of the original installations to be replaced by other processes (Bucklin, 1994).

Oxidizing Solutions

Permanganate and Dichromate Solutions. Buffered aqueous solutions of potassium permanganate and sodium or potassium dichromate are used to remove completely traces of hydrogen sulfide from industrial gases. Processes employing such solutions are non-regenerative and, because of the high cost of the chemicals used, are only economical when very

1312 Gas Purification

small amounts of hydrogen sulfide are present in the gas. Permanganate solutions are used quite extensively for the final purification of carbon dioxide in the manufacture of dry ice.

A schematic flow diagram of the process is shown in **Figure 16-8.** The gas is contacted with the solution in two packed towers operating in series. The solution, which in the case of the permanganate process contains about 4.0% potassium permanganate and 1.0% sodium carbonate, is circulated until approximately 75% of the permanganate in either tower is converted to manganese dioxide. At that point, the spent solution is discarded and fresh solution is pumped into the tower. This can be done without interruption of the gas flow if sufficient active permanganate is available in each contact stage to ensure complete H_2S removal. Dichromate solutions usually contain 5 to 10% potassium dichromate, zinc sulfate, and borax. In addition to hydrogen sulfide, traces of organic compounds such as amines are also removed quantitatively in these processes.

Nitrite Solutions—The Sulfa-Check and Hondo HS-100 Processes. The Sulfa-Check process, developed by NL Treating Chemicals/NL Industries, Inc. and marketed by Exxon Chemical Co., is based on the use of a buffered aqueous solution of sodium nitrite to absorb and destroy hydrogen sulfide. The Hondo HS-100 process, offered by Hondo Chemicals, is believed to involve similar chemistry using potassium instead of sodium nitrite. The Sulfa-Check process is reviewed in the following section as an example of a nitrite-based system because considerable data are available on the technology (Dobbs, 1986; Bhatia and Allford, 1986; Bhatia and Brown, 1986; Schaack and Chan, 1989).

The Sulfa-Check process utilizes a bubble tower similar to that used in the Chemsweet process. A typical unit is shown in **Figure 16-9.** The design differs slightly from that shown in **Figure 16-7** in that a perforated pipe sparger is used instead of a distribution plate and a vane-type mist eliminator is used instead of a mesh pad. The initial charge of Sulfa-Check solution is gradually converted into a slurry containing particles of elemental sulfur and other precipitated solids as the solution absorbs hydrogen sulfide. When essentially all of the active nitrite has been consumed, the spent slurry is discharged and replaced with fresh solution.

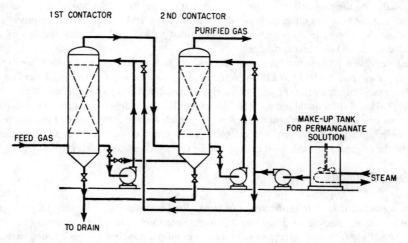

Figure 16-8. Flow diagram of permanganate and dichromate processes.

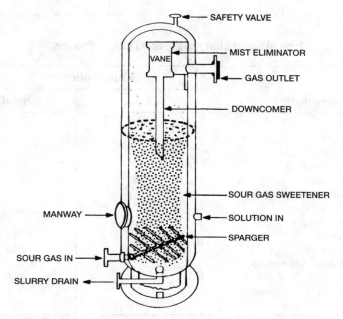

Figure 16-9. Bubble tower proposed for Sulfa-Check process. (*Dobbs, 1986*)

The chemistry of the Sulfa-Check process is quite complex, but the overall reaction can be represented by the following simple equation:

$$3H_2S + NaNO_2 = NH_3 + 3S + NaOH + H_2O + \text{some } NO_x \tag{16-13}$$

In the presence of CO_2, the NaOH is neutralized to produce sodium carbonate and bicarbonate in solution. Sodium bicarbonate is sparingly soluble and may appear with the solid phase in the slurry. The composition of a typical spent slurry is given in **Table 16-3**.

The process has been widely used. In 1986, it was reported that more than 100 units were in operation in the U.S. and Canada (Bhatia and Brown, 1986). Advantages claimed for the process include

1. The spent slurry is classified as a nonhazardous waste.
2. The slurry is noncorrosive to mild steel.
3. Change-out of spent absorbent is simple.
4. Equipment is simple and low in cost.

The scrubbing vessel contains only two internals: a gas sparger at the bottom and a mist eliminator at the top. The vessel size is determined by the feed gas flow rate and pressure, the hydrogen sulfide content of the gas, and the desired turnaround time. Design guidelines include

1. Optimum efficiency is in the temperature range of about 75° to 110°F.
2. Maximum superficial velocity of gas in the vessel should be below 0.16 ft/sec.
3. The quantity of solution required is 0.0473 gal/MMscf per ppm of H_2S in the gas.

Table 16-3
Typical Spent Slurry of the Sulfa-Check Process

Component	Composition, wt %	
	Liquid	Solid
Water	36.7	
Nitrite	1.0	
Nitrate	0.9	
Bicarbonate	8.2	
Carbonate	1.2	
Sulfate	1.7	
Tetrathionate	2.5	
Ammonium	0.3	
Sodium	7.3	
Sulfur		20.2
Sodium Bicarbonate		17.7
Sodium Sulfate		0.7
Sodium Tetrathionate		1.6
Total	59.8	40.2

Note: The spent slurry had a pH of 8.0 and contained no heavy metals, cyanides, sulfides, or chlorinated organic compounds.
Data of Bhatia and Allford (1986)

4. The liquid height, in feet, required to effectively remove hydrogen sulfide from the feed gas is equal to 2.5 times the logarithm of the H$_2$S concentration, in ppm. For example, the removal of 10 ppm requires a liquid height of at least 2.5 ft while 100 ppm requires at least 5.0 ft.

After selecting a desired turn-around time, these criteria can be used to estimate the required contactor diameter and height. The overall height must include a margin of safety in the liquid depth, room for liquid expansion during operation, droplet disengaging space, and a mist-eliminator zone.

As indicated by equation 16-13, the Sulfa-Check process produces a small amount of NO$_x$. Normally only NO is produced; however, the presence of air and other impurities appears to accelerate the formation of NO and cause a portion to be oxidized to NO$_2$. When the carbon dioxide concentration in the feed is low, the release of ammonia into the product gas is also enhanced (Schaack and Chan, 1989). As a result of these considerations, the Sulfa-Check process may not be applicable when

1. Air is present or injected into the gas stream.
2. The feed gas contains less than about 0.1% CO$_2$.
3. Sweetened, undiluted gas is sold directly to a utility.

Aldehyde Processes

A number of hydrogen sulfide scavenging processes have been developed that make use of the rapid reaction between an aldehyde, such as formaldehyde, and H$_2$S. Typically, the absorbent is a mixture of formaldehyde and methanol. Schaack and Chan (1989) list the following formaldehyde-methanol-based processes: Scavinox; Prohib-196; Champion Exp Cortron JS-49-4; Baker Magnatreat M118W (formerly Dichem E118W); Techniscav; and Couger H-925. Several of these processes were experimental or used for a limited number of applications. Others have been used quite extensively; however, it is believed that use of this technology is decreasing.

Considerable design and operating data for the Scavinox process are given by Schaack and Chan, and these data form the principal basis for the following discussion. The chemistry of the hydrogen sulfide-formaldehyde reaction is quite complex, producing cyclic carbon-sulfur compounds and, to a lesser extent, mercaptans. The main reaction path can be represented by

$$3HCOH + 3H_2S = \begin{array}{c} S \\ CH_2 \diagup \diagdown CH_2 \\ | \quad \quad | \\ S \diagdown \diagup S \\ CH_2 \end{array} + 3H_2O \quad (16\text{-}14)$$

(formaldehyde)   (trithiane)

The formaldehyde and reaction products, particularly the reaction products with mercaptans, give the spent chemicals a strong, objectionable odor. This problem and the difficulty in disposing of waste material have probably been the principal reasons aldehyde-based processes have not found wider application.

The liquid reactant can be used either by injection into producing wells or flowlines or, preferably, as the absorbent in a bubble tower gas-liquid contactor similar to those previously described for the Chemsweet and Sulfa-Check processes. An advantage of the formaldehyde-methanol mixture over water-based solutions or slurries is its low freezing point. The typical composition is about 60% formaldehyde and 40% methanol which has a freezing point in the −58°F range. Reducing the contact temperature reduces the reaction rate; however, in one Canadian installation, Dome Petroleum Ltd. was able to attain 100% hydrogen sulfide removal with the gas temperature as low as 32°F (Schaack and Chan, 1989).

Potential problems with formaldehyde-based processes are the objectionable odor of both the fresh and spent solutions and the tendency of the solutions to polymerize and flocculate. The latter phenomena can cause plugging of the contact equipment, make clean-out difficult, and, if deep well injection is used for waste disposal, cause plugging of the formation. Flocculation can be avoided by diluting the mixture with two or three parts water, but this increases the volume of each charge required for a given change-out schedule. Deposits in the equipment can be cleaned out by use of dilute caustic; however, polymer deposits in the formation cannot be removed readily and may rule out this type of disposal.

Alkylamine/Aldehyde Condensation Products

Solutions of alkylamine/aldehyde condensation products have been used in several sulfur-scavenging processes. The spent solution is free of sulfur particles, and no nitrogen oxide gases are formed; however, it has a very obnoxious odor. It is reported to have corrosion-inhibition properties and to be approved for Class 1, non-hazardous, deep-well disposal (Houghton and Bucklin, 1994). Examples of processes in this category are Gas Treat 114, Gas Treat 115, and Gas Treat

1316 Gas Purification

117, which are developments of Champion Technologies, Inc., and Sulfa-Check 6138 and Sulfa-Check 6139, which were developed by Exxon Chemical Co. (Houghton and Bucklin, 1994).

Polyamine Solutions

The Sulfa-Scrub process, developed by Quaker Chemical Corp. and marketed by Petrolite, and the Sulfa-Guard process, offered by Coastal Chemicals, are examples of polyamine-based processes. A description of the Sulfa-Scrub process, based on papers by Dillon (1991 A, B), follows.

The Sulfa-Scrub process utilizes a cyclic polyamine of the triazine class to absorb and react with hydrogen sulfide. The chemical formulas for a specific triazine, 1,3,5 tri-(2 hydroxyethyl)-hexahydro-5-triazine, and products of its reaction with H_2S are shown in **Figure 16-10**. It is reported that the process will also remove mercaptans, but not COS.

Advantages claimed for the Sulfa-Scrub process include

1. The absorbent is a liquid.
2. It is selective for H_2S in the presence of CO_2.
3. The reaction products are water-soluble liquids.
4. The reaction products are excellent corrosion inhibitors.
5. The spent material is nonhazardous and has a low toxicity.

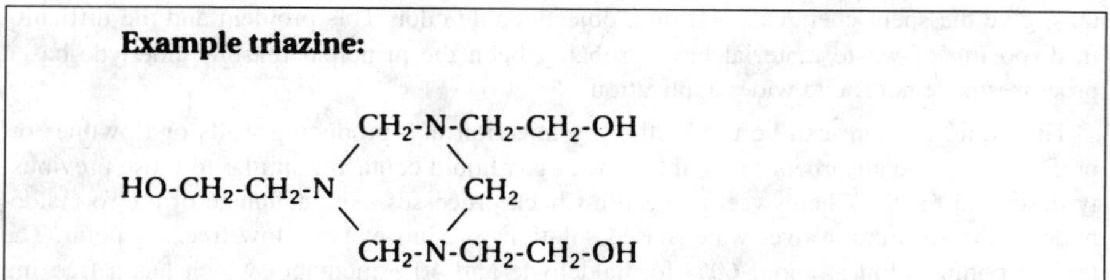

Example triazine:

```
                    CH₂-N-CH₂-CH₂-OH
                   /    \
    HO-CH₂-CH₂-N         CH₂
                   \    /
                    CH₂-N-CH₂-CH₂-OH
```

1,3,5 tri – (2hydroxyethyl) – hexahydro – 5 – triazine

Example products:

```
                    CH₂-S              S-CH₂
                   /     \            /     \
    HO-CH₂-CH₂-N           CH₂-S-S-CH₂        N-CH₂-CH₂-OH
                   \     /            \     /
                    CH₂-S              S-CH₂
```

Bis-dithiazine

```
            S-CH₂
           /     \
    CH₂           N-CH₂-CH₂-OH
           \     /
            S-CH₂
```

Dithiazine

Figure 16-10. Chemical formulas of triazine and typical reaction products. (*Dillon, 1991B*)

Since the absorbent and waste products are liquids, the process can be operated in either a batch or continuous mode. Data from treating trials at two plants are given in **Table 16-4**. Case A was operated as a batch process using a conventional bubble tower. Case B was a continuous operation with the triazine injected into the gas ahead of a converted iron sponge tower. In both cases, complete hydrogen sulfide removal was observed. Triazine product consumption ranged from 0.04 to 0.06 gal/MMscf per ppm of H_2S in the feed gas.

Caustic Scrubbing

The general subject of caustic scrubbing to remove acidic impurities from gas is discussed in Chapter 5. As pointed out in that discussion, hydrogen sulfide can be absorbed selectively in the presence of carbon dioxide if a short-residence-time contactor is employed. The Koch Engineering Company offers such a system based on the use of a Koch static mixer as the short-residence-time contactor (Koch, 1986). The mixer is installed in the line leading to a high-efficiency separator that removes entrained droplets of spent caustic from the gas. When used for H_2S scavenging, the reaction product is sodium bisulfide, NaHS (or NaSH), and the process is sometimes called the NaSH process.

According to Manning and Thompson (1991), two or three mixer settler contact stages are needed to make pipeline-quality gas. However, a single stage may be adequate when high purity gas is not required, as in the production of fuel gas for local use in field heater sys-

**Table 16-4
Sulfa-Scrub Plant Operating Tests**

Type of Contactor	Bubble Tower	Line Injection Followed by Converted Iron Sponge Tower
Tower Design		
Diameter, ft	3.0	3.5
Height, ft	30	20
Capacity, cu ft	184	192
Mist eliminator	pad	pad
Feed Gas		
Rate, MMscfd	4.0	14.5
Temperature, °F	85	95
Pressure, psig	700	780
Superficial vel., ft/s	0.15	0.45
H_2S, ppm	32	35
Results,		
Treated gas H_2S, ppm	0	0
Run time, days	45	—
Triazine used, gal/d	8.0	20.3
gal/ppm H_2S/MMscf	0.06	0.04

Data of Dillon (1991A)

tems. Typically, a single stage will reduce the hydrogen sulfide content of the gas from 7,500 ppm to about 600 ppm.

The caustic is usually received as a 50% solution and diluted to 20% for use in the process. Softened water is desirable as the diluent to avoid scaling problems. In most oil field installations, the spent solution is diluted with production water and disposed of by injection into the formation. The overall process is more complex than most batch systems. In addition to the static mixers and separators, the process needs storage tanks for concentrated and dilute caustic and for spent solution; recycle and makeup pumps; and an instrumentation/control system. The process has several advantages over other scavenging systems, including the use of a readily available, low-cost chemical; compact size; and the production of a waste product, NaHS, which has the potential for reuse in other industrial processes such as paper manufacture.

Comparison of H_2S Scavenging Processes

Several attempts have been made to evaluate and compare hydrogen sulfide scavenging processes based on economic analyses. Unfortunately, such studies are of limited value because they are highly dependent on the specific conditions and assumptions used in the study. Also, other factors, such as the environmental acceptability of the reactant and its waste products, operator acceptance of the process, and winterizing requirements, may be more important than cost in selecting a process. The very high rate of process modification, new process development, and removal of existing processes from the market also makes comparative evaluation difficult in this field.

Schaack and Chan (1989) made both economic and qualitative comparisons of six different scavenging processes. **Table 16-5** gives the results of their qualitative evaluation, which is intended to include both economics and operational characteristics. The Scavinox process scored highest in this analysis, but is no longer marketed under this name. A related process, Di-Chem, which also appears to be commercially unavailable, scored third in the analysis, but appreciably below the second process, Sulfa-Check. The commonly used iron sponge process produced the lowest (worst) score.

An economic evaluation of four scavenging processes was presented by Trahan and Manning (1992), and in 1994 Houghton and Bucklin reported the results of an economic study covering nine different scavenging processes and four sets of conditions. **Table 16-6** presents a comparison of operating costs for all three studies. The costs are given as dollars per ton of sulfur removed in order to minimize the effects of differences in plant size and sulfur concentration on the results. The Schaack and Chan results are in 1987 Canadian dollars. The others are in U.S. dollars as of the approximate date of the references. The Schaack and Chan iron sponge-process data reported in the table are for a two-bed system. They also give data on a single-bed iron sponge unit, but the results show an unreasonably high operating cost due in part to the inclusion of a cost of lost gas production for a 48-hour period while the single bed is being recharged.

The results compiled in **Table 16-6** appear to indicate

1. Operating costs per ton of sulfur are not strongly affected by plant size or H_2S concentration.
2. Caustic scrubbing has the lowest operating cost.
3. The Hondo HS-100, iron sponge, and Sulfatreat processes generally show the second-lowest operating costs (if the Schaack and Chan result for iron sponge is neglected).

Table 16-5
Scavenging Process Qualitative Comparison

Process	Selection Index
Scavinox	22.9
Sulfa-Check	22.8
Di-Chem	16.9
Slurrisweet	16.5
Zinc oxide slurry	15.8
Iron sponge (one tower)	15.1
Iron sponge (two towers)	15.1

Notes:
Selection index = $C + O + P + 0.2W + 0.5E + 0.4A$
Where C = Capital cost; O = Operating cost; P = Process reliability; W = Winterization constraints; E = Ease of Operation; and A = Operator acceptance

Data from Schaack and Chan (1989)

A comparison of capital cost estimates from the two economic studies that provide such information is given in **Table 16-7**. The data are given as dollars per million standard cubic feet per day (MMscfd) plant capacity. Results of the two studies are not directly comparable because the Schaack and Chan estimates are based on a grass-roots plant and include such incidental items as heated process buildings for water-based processes, truck loading dock and shelter where needed, and a water well with pump. The Houghton and Bucklin estimates (U.S. dollars) are based on the principal items of equipment only and a more recent time period.

Results of the capital cost comparisons generally show the caustic scrubbing and amine-based processes to be the least expensive followed by the other liquid processes. The solid bed processes have the highest capital costs.

Houghton and Bucklin (1994) included an evaluation of the ICI Low Temperature Zinc Oxide Process; however, the operating costs proved excessive, so only a single, low sulfur concentration case was studied. The chemical cost alone was about $40,000 per ton of sulfur removed. They concluded that practical applications of the zinc oxide process are generally limited to gases with very low sulfur content. The process may be particularly useful where space requirement is an important consideration, as on an offshore oil and gas production platform.

Both operating and capital costs are highly dependent on the design bases and assumptions. The data given in **Tables 16-6** and **16-7** from reference A (Schaack and Chan, 1989) are based on a 30-day change-out period. They also considered a 60-day change-out period. This greatly reduced the operating costs, but increased the capital costs proportionately. Reference B (Trahan and Manning, 1992) based their study on a variable change-out period, but a fixed-vessel diameter (30 inches). The data of reference C (Houghton and Bucklin, 1994) are based on both a variable change-out period and a variable-vessel diameter to provide a near optimum design for each assumed case.

Houghton and Bucklin (1994) provide a summary of the organic sulfur removal capabilities of several commercial nonregenerative H_2S removal processes. This information is

1320 Gas Purification

Table 16-6
Operating Costs for Sulfur Scavenging Processes

Reference	A	B	C			
Case	—	—	1	2	3	4
Feed Gas Flow, MMscfd	5	0.5	6	6	6	0.5
Feed Gas H_2S content, ppmv	100	400	10	400	800	400
Feed Gas Pressure, psig	1,000	200	510	510	510	200
	Operating Costs, Dollars per ton of Sulfur Removed					
Iron Oxide, Dry Bed						
Iron Sponge	24,270	6,235	3,771	3,771	3,771	4,845
Sulfatreat	—	6,588	3,714	3,630	3,625	3,998
Iron Oxide, Slurry						
Slurrisweet	14,860	—	—	—	—	—
Zinc Oxide, Dry Bed						
ICI, ZnO	—	—	41,724	—	—	—
Zinc Oxide, Slurry						
Chemsweet	15,290	—	17,982	15,376	15,368	16,065
Nitrite Solution						
Hondo HS-100	—	—	4,422	3,841	3,820	4,036
Sulfa-Check	16,550	9,467	8,792	8,077	8,029	8,315
Aldehyde Solution						
Di-Chem	23,060	—	—	—	—	—
Scavinox	13,080	—	—	—	—	—
Amine Solution						
Sulfa-Guard	—	8,823	8,177	7,214	7,158	7,990
Sulfa-Scrub	—	—	13,859	12,895	12,840	13,672
Caustic Scrubbing						
NaSH	—	—	3,242	1,401	1,346	2,179

References: A) Schaak and Chan, 1989; B) Trahan and Manning, 1992; C) Houghton and Bucklin, 1994

given in **Table 16-8,** which also identifies the basic process type and supplier. Data on the Sofnolime RG process provided by Molecular Products, Ltd. (1992) are also included.

HIGH-TEMPERATURE, REGENERATIVE SULFUR REMOVAL SYSTEMS

The primary incentive for developing a process that can remove hydrogen sulfide from gas streams at high temperatures is its potential for improving the thermal efficiency of integrated-gasification-combined-cycle (IGCC) power plants. Conventional pulverized coal boiler systems have efficiencies in the range of 30–35%. The IGCC system with hot gas cleanup has the potential to provide efficiencies in the range of 43–46% (Gupta and Gasper-Galvin,

Table 16-7
Capital Costs of Sulfur Scavenging Processes

Reference	A	C			
Case	—	1	2	3	4
Feed Gas Flow, MMscfd	5	6	6	6	0.5
Feed Gas H_2S Content, ppmv	100	10	400	800	400
Feed Gas Pressure, psig	1,000	510	510	510	200
	Capital Cost, Dollars per MMscfd Feed Gas Capacity				
Iron Oxide, Dry Bed					
Iron Sponge	144,000	50,300	109,000	151,000	312,000
Sulfatreat	—	50,300	95,400	115,000	347,000
Iron Oxide, Slurry					
Slurrisweet	132,000	—	—	—	—
Zinc Oxide, Dry Bed					
ICI, ZnO	—	27,000	—	—	—
Zinc Oxide, Slurry					
Chemsweet	161,000	38,800	74,500	105,000	141,000
Nitrite Solution					
Hondo HS-100	—	38,800	63,800	74,500	249,000
Sulfa-Check	68,000	38,800	63,000	92,400	249,000
Aldehyde Solution					
Di-Chem	84,000	—	—	—	—
Scavinox	67,000	—	—	—	—
Amine Solution					
Sulfa-Guard	—	16,600	28,200	31,500	125,000
Sulfa-Scrub	—	15,500	28,200	31,500	125,000
Caustic Scrubbing					
NaSH	—	15,500	30,600	45,300	136,000

References: A) Schaak and Chan, 1989; C) Houghton and Bucklin, 1994

1991). Purification of the fuel gas at a temperature near that at which it is generated avoids the energy losses associated with cooling it to a conventional gas purification process temperature (typically about 100°F) and reheating it to the required gas turbine unit feed temperature. It is estimated that operating the gas purification system at an elevated temperature can increase the efficiency of an IGCC system by about 6% (Furimsky and Yumura, 1986).

Although the development of a high-temperature desulfurization process can be justified on the basis of its effect on the IGCC system efficiency, it also offers potential benefits for other applications such as the purification of feed gases for high-temperature fuel cells and catalytic synthesis operations. Because of the potential importance of the process, the status of development of high-temperature desulfurization processes is reviewed, although there are no significant commercial applications at this time.

Table 16-8
Impurity Removal Capabilities of Non-Regenerable Sulfur-Scavenging Processes

Process Type and Trade Name	Licensor	H_2S	RSH	COS	CO_2
Dry Iron Oxide					
Iron Sponge	Various	Yes	Yes	No	No
Sulfatreat	Gas Sweetener Assoc.	Yes	Yes	No	No
Dry Zinc Oxide					
Puraspec	ICI Katalco	Yes	Yes	Yes	No
Alkaline Solids					
Sofnolime RG	Molecular Products	Yes	Yes	Yes	Yes
Zinc Oxide Slurry					
Chemsweet	Natco	Yes	Yes	Yes	No
Alkali Nitrite					
Hondo HS-100	Hondo Chemical	Yes	No	No	Some
Sulfa-Check	Exxon Chemical	Yes	No	No	No
Alkylamine/Aldehyde					
Gas Treat-114	Champion Technol.	Yes	Yes	No	No
Sulfa-Check 6138 & 6139	Exxon	Yes	Yes	No	No
Polyamine					
SulfaGuard	Coastal Chemical	Yes	Yes	No	No
Enviro-Scrub	Quaker Pet. Chem Co.	Yes	Yes	No	No
Sulfa-Scrub	Petrolite	Yes	Yes	No	No
Caustic Solution					
NaSH	Koch Engineering	Yes	Yes	Yes	Some

Sofnolime RG data from Molecular Products, Ltd (1992), all other data from Houghton and Bucklin (1994)

Background

Two basic approaches have been considered for high-temperature sulfur removal from gases: contact with reactive molten salts and contact with reactive solids. The molten salt approach has encountered several problems, including containment materials corrosion, salt vaporization, and salt solidification on cooling. As a result, recent efforts have concentrated on the use of reactive solids, and only this approach is discussed.

The reactive solids concept is, in principle, very simple. The hydrogen sulfide is removed from the gas by reacting with a metal oxide to form a stable, non-volatile metal sulfide. The metal sulfide is then subjected to a regeneration step in which it reacts with oxygen and/or water vapor to remove the sulfur (usually as sulfur dioxide) and produce reusable metal oxide. Both the sorption and regeneration steps are carried out at elevated temperatures.

The development of commercially viable processes is not nearly as simple. Some of the problems encountered in developing a suitable sorbent are particle breakup due to chemical changes and mechanical attrition; vaporization of key components such as zinc; and side reactions such as the formation of metal sulfates. Even with an ideal sorbent there are major problems in developing the overall system and hardware. Fixed-bed systems require high temperature, high-pressure valves and produce a regeneration gas stream that varies in composition and quantity during the cycle. Fluidized and moving bed systems are very susceptible to particle attrition problems and require complex solids transport systems. Finally, all of the commercial and near-commercial, high-temperature, regenerative hydrogen sulfide removal processes produce sulfur dioxide, which presents a much more difficult disposal problem than the concentrated hydrogen sulfide stream produced by most low-temperature regenerative processes. Sulfur dioxide can be reduced to elemental sulfur, but the process requires a reducing gas, and only one commercial process is currently available (the Allied Process). Sulfur dioxide can also be sold as the pure compound or oxidized to sulfuric acid; however, the markets for these two chemicals are limited and/or site dependent.

The state-of-the-art, as of about 1984, of hot desulfurization of coal gas streams is summarized by Jalan (1985). A comprehensive review of potential solid sorbent materials for hot gas desulfurization is presented by Furimsky and Yumura (1986). A number of metals known to have a strong affinity for sulfur are evaluated. The materials are classified into two groups: alkaline earth metal compounds such as calcium oxide, limestone, dolomite, and calcium silicate; and transition metal compounds such as iron oxide, zinc oxide, zinc ferrite, and manganese oxide. The key reactions of several metal oxides with hydrogen sulfide and the logarithms of the equilibrium constants for the reactions are given in **Table 16-9**. The positive values for log K generally indicate that all of the metal oxides except magnesium have a strong affinity for hydrogen sulfide.

Table 16-9
Log K Values for the Reaction of Metal Oxides with H_2S at 1,000°K

Reaction		Log K
$CaO + H_2S$	$\rightarrow CaS + H_2O$	3.36
$MgO + H_2S$	$\rightarrow MgS + H_2O$	−1.79
$FeO + H_2S$	$\rightarrow FeS + H_2O$	2.29
$FeO + 2H_2S$	$\rightarrow FeS_2 + H_2O + H_2$	0.58
$\frac{1}{2}Fe_2O_3 + \frac{1}{2}H_2S$	$\rightarrow FeS + \frac{3}{2}H_2O + \frac{1}{4}S_2$	2.40
$\frac{1}{2}Fe_2O_3 + 2H_2S$	$\rightarrow FeS_2 + \frac{3}{2}H_2O + \frac{1}{2}H_2$	1.77
$\frac{1}{3}Fe_3O_4 + \frac{2}{3}H_2S$	$\rightarrow FeS + \frac{4}{3}H_2O + \frac{1}{6}S_2$	1.99
$\frac{1}{3}Fe_3O_4 + 2H_2S$	$\rightarrow FeS_2 + \frac{4}{3}H_2O + \frac{2}{3}H_2$	0.99
$MnO + H_2S$	$\rightarrow MnS + H_2O$	2.80
$ZnO + H_2S$	$\rightarrow ZnS + H_2O$	3.88
$MoO_2 + 2H_2S$	$\rightarrow MoS_2 + 2H_2O$	5.2*
$MoO_3 + 3H_2S$	$\rightarrow MoS_2 + 3H_2O + \frac{1}{2}S_2$	8.7*
$CoO + H_2S$	$\rightarrow \frac{1}{9}Co_9S_8 + H_2O + \frac{1}{18}S_2$	3.10
$NiO + H_2S$	$\rightarrow NiS + H_2O$	4.1*
$NiO + H_2S$	$\rightarrow \frac{1}{3}Ni_3S_2 + H_2O + \frac{1}{6}S_2$	2.76

*extrapolated values
Data of Furimsky and Yumura (1986)

Iron oxide- and zinc oxide-based sorbents are used commercially for hydrogen sulfide removal from gases at low and moderate temperatures, and have therefore been considered for high-temperature applications. Iron oxide has the disadvantage of forming several stable oxides (e.g., FeO, Fe_2O_3, Fe_3O_4) and transforms from one to another (and even to elemental iron) depending on the gas composition and other factors. Such transformations can accelerate sorbent attrition and the production of fines. All of the oxides can react with hydrogen sulfide; however, the reaction rates vary. Regeneration of iron sulfide with air and/or steam can be difficult because some part of the FeS appears to be inaccessible to the oxidant, and the regeneration offgas may contain elemental sulfur as well as sulfur dioxide, particularly when steam is present.

Zinc oxide is very effective as a nonregenerable scavenger for hydrogen sulfide at low and moderate temperatures. Unfortunately, the pure oxide is difficult to regenerate completely because it undergoes sintering at a temperature of about 700°C (1,292°F); and, under some conditions, elemental zinc may form and be vaporized (Westmoreland et al., 1977).

Zinc Ferrite and Titanate Sorbents

Sorbent Development

In order to overcome the problems of iron and zinc oxides when used alone, consideration was given to combination sorbents. The sorbent, zinc ferrite ($ZnFe_2O_4$) was originally developed by the U.S. Department of Energy/Morgantown Energy Technology Center (DOE/METC) (Grindley and Steinfeld, 1981). The sorbent contains an equimolar mixture of zinc oxide and iron oxide calcined at 800° to 850°C (1,372° to 1,562°F) with a suitable binder (Gupta et al., 1992).

Although zinc ferrite proved to be superior to either iron oxide or zinc oxide alone, it was found to have some limitations. According to Gupta and Gasper-Galvin (1991), zinc ferrite sorbents have a temperature limitation of about 649°C (1,200°F) due to iron oxide reduction, zinc vaporization, and particle sintering. They are also limited to the treatment of gases containing sufficient steam to inhibit reduction. As a result of these limitations, the development emphasis has shifted to a more stable zinc compound, zinc titanate.

A key difference in the various zinc-based formulations is the rate of zinc loss due to vaporization. This is illustrated by **Figure 16-11** from a paper on a DOE/METC program aimed at developing suitable sorbents for use in a high-temperature, high-pressure fluidized-bed reactor (Gangwal, et al., 1991A). The data were obtained in a thermogravimetric reactor (TGR) at atmospheric pressure and 650°C (1,202°F) with gas simulating the product of a KRW coal gasifier, but without H_2S. Sorbents tested included Fe_2O_3, ZnO, $ZnFe_2O_4$, TiO_2, and three zinc titanate formulations described as follows:

Designation	Molar Ratio, ZnO/TiO_2	Bentonite %	Fabrication Technique
ZT-2	0.8	2	granulation
ZT-5	1.5	2	granulation
L-3758	1.5	2	undisclosed

The results showed that the pure oxides—Fe_2O_3 and TiO_2—exhibited no weight loss with time (after an initial stabilization period) while ZnO lost weight rapidly. Zinc ferrite also

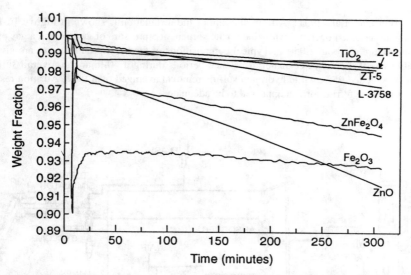

Figure 16-11. Zinc loss rates for various sorbents in a thermogravimetric reactor. (*Gangwal, et al., 1991A*)

showed a significant rate of weight loss, while the three zinc titanate formulations showed very low losses. The best sorbent, designated ZT-2, lost about 1% of its weight after five hours.

The DOE/METC report concludes that both zinc ferrite and zinc titanate sorbents have the capability to reduce the hydrogen sulfide concentration in coal gas to less than 20 ppm in a fluidized bed reactor. However, the zinc ferrite sorbents were found to suffer structural weakening and losses due to attrition and vaporization at temperatures above about 550°C (1,022°F); whereas, the zinc titanate sorbents, prepared by a proprietary granulation technique, showed excellent sulfur capacity, regenerability, attrition resistance, and zinc vaporization characteristics at temperatures up to about 650°C (1,202°F).

The development of sorbents for a moving bed system is described by Ayala et al. (1991). It was found that ellipsoidal pellets made by a rounding and densification process were more attrition resistant, but somewhat less reactive than extruded cylindrical pellets. Zinc titanate formulations were generally more attrition resistant than the corresponding zinc ferrite cases.

Reactor Designs

Zinc ferrite and zinc titanate sorbents have been extensively evaluated for fixed-bed reactors (Gangwal et al., 1989; Grindley and Goldsmith, 1987; Grindley and Steinfeld, 1985). However, according to Gupta et al. (1992), the fixed-bed reactor system has limited practical potential because of (1) the need for high-temperature, high-pressure valves; (2) the difficulty of handling the heat released during regeneration; and (3) the nonuniform composition of the regeneration offgas during the cycle period. The DOE/METC program was therefore directed toward moving and fluidized bed systems.

In order to demonstrate a moving-bed concept, the General Electric Company (GE) constructed a 24-ton per day pilot plant incorporating an air-blown fixed-bed coal gasifier; high-temperature cyclones for particulate removal; and a moving-bed metal oxide desulfurization system (Furman et al., 1991). A flow diagram of the installation is shown in **Figure 16-12.** The system has been operated with zinc ferrite sorbent at a pressure of 20 atm and a gas outlet temperature of about 1,050°F. Typical compositions of feed and product gases are given in **Table 16-10.** For selected material balance periods during a 60-hour run, overall sulfur removal exceeded 95%, while hydrogen sulfide removal averaged over 98%. Attrition resistance of the zinc ferrite sorbent appeared to be adequate.

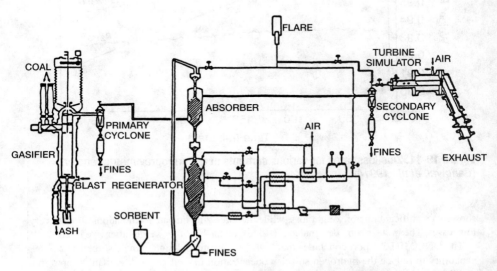

Figure 16-12. GE moving-bed hot gas cleanup test system. (*Furman et al., 1991*)

Table 16-10
Typical Gas Compositions for Tests of Zinc Ferrite Sorbent with Illinois No. 6 Coal Feed to Gasifier

Component	Reactor Inlet, Vol. %		Reactor Outlet, Vol. %	
	Wet	Dry	Wet	Dry
Nitrogen	33.9	45.2	35.5	44.4
Carbon Dioxide	11.6	15.5	14.1	17.6
Carbon Monoxide	10.4	13.9	8.8	11.0
Hydrogen	15.8	21.1	18.5	23.1
Methane	3.1	4.1	3.1	3.9
Hydrogen Sulfide	0.23	0.31	0.003	0.004
Carbonyl Sulfide	0.03	0.04	0.000	0.000
Water	25.0		20.0	

Source: Furman et al., 1991

Sulfur Recovery from Regenerator Off-Gas

When zinc ferrite, zinc titanate, or a similar sorbent is used in a hot gas desulfurization process, regeneration is normally accomplished by contacting the spent material with a dilute air stream. In fixed-bed systems the regenerator offgas typically contains 1 to 3% sulfur dioxide, while the moving bed system being developed by GE produces a gas containing up to 13% sulfur dioxide (Dorchak et al., 1991).

A comprehensive evaluation of processes capable of recovering sulfur from hot regeneration offgas was conducted by The Parsons Corporation, under contract to DOE (O'Hara et al., 1987). The study considered numerous processes and process combinations for the treatment of a gas containing 1.5 to 5.0 volume % SO_2 and 95% nitrogen. Economic data were developed for the four most promising combinations based on the regeneration offgas from a unit treating fuel gas derived from high-sulfur midwestern coal for use in a 100 MW_e molten carbonate fuel cell power plant. The two cases showing the lowest 10-year levelized busbar costs (L_{10}) were

1. Ten percent of the regenerator offgas is purified in a Wellman-Lord unit, and the concentrated sulfur dioxide from this unit is added to the remaining 90% of the gas which is then processed in a BSRP system. The Wellman-Lord process is described in Chapter 7. The BSRP system consists of a Beavon Sulfur Removal (BSR) unit to convert all sulfur compounds to hydrogen sulfide (see Chapter 8) followed by a Stretford unit to convert the hydrogen sulfide to elemental sulfur (see Chapter 9).
2. The regeneration offgas is fed directly to an augmented Claus unit and the Claus plant tail gas is treated in a BSRP unit before venting to the atmosphere. In the augmented Claus unit the sulfur dioxide-containing stream is partially reduced by contact with a gas-containing hydrogen and carbon monoxide at about 2,000°F, and the reduced stream, containing hydrogen sulfide amd sulfur dioxide in a mole ratio of 2:1, is then fed to a conventional Claus unit.

An alternative to the augmented Claus process, called the Direct Sulfur Recovery Process (DSRP) has been developed by the Research Triangle Institute (RTI) (Gangwal et al., 1990, 1991B; Dorchak et al., 1991;). The process is based on the catalytic conversion of sulfur dioxide directly to elemental sulfur without first reducing a portion of it to hydrogen sulfide. A small slip stream of the coal gas can be used as the reducing agent. Typical reactions occurring are

$$2H_2 + SO_2 = 2H_2O + 1/n\ S_n \tag{16-15}$$

$$2CO + SO_2 = 2CO_2 + 1/n\ S_n \tag{16-16}$$

$$SO_2 + 3H_2 = H_2S + 2H_2O \tag{16-17}$$

$$SO_2 + 2H_2S = 2H_2O + 3/n\ S_n \tag{16-18}$$

The first two reactions occur rapidly in the presence of the catalyst and are not thermodynamically limited. If no other reactions occur, these two reactions are capable of producing a very high yield of sulfur. As indicated by the stoichiometry, they require two moles of reducing gas per mole of SO_2. The third reaction is not thermodynamically limited, but its rate may be controllable by proper selection of catalyst and conditions. It does not represent a desirable reaction path. The fourth reaction is the Claus reaction, which is limited by equilibrium and proceeds at a rate determined by the catalyst and the operating conditions. In small scale tests of the process, conversions were at times greater than 96%; although attainment of equilibrium would result in a conversion of only 50 to 60%. Increased pressure was found to improve the sulfur yields. The process appears promising, but has not yet been tested on a commercial scale.

Haldor Topsøe Tin Oxide-Based Process

A process has been developed by Haldor Topsøe that treats coal gas in the temperature range from 350° to 500°C (662°F to 932°F) and delivers a product gas that is environmentally acceptable as gas turbine fuel with regard to the common impurities: sulfur, nitrogen compounds, and halogens (Nielsen and Rudbeck, 1993). The overall gas purification scheme includes halogen removal, water-gas shift conversion, ammonia adsorption, and sulfur removal.

The halogen removal unit consists of a fixed bed of solid sodium carbonate. This reacts quantitatively with halogens present in gas from the coal gasification unit. The most common halogen impurity, HCl, is neutralized by the sodium carbonate to form NaCl which remains in the bed. The fixed bed of sodium carbonate also acts as a filter to remove traces of dust that may have passed through the gasifier offgas dust removal unit.

Water-gas shift conversion is included primarily as a means of reducing the concentration of water vapor in the gas. A low water concentration improves both ammonia adsorption efficiency and sulfur conversion in subsequent steps. Ammonia is removed by adsorption on an appropriate solid such as an acid zeolite and may be recovered by desorption at a higher temperature.

The key process step is the removal of hydrogen sulfide by reaction with a tin oxide-based sorbent. The process is based on the reversible reaction:

$$H_2 + H_2S + SnO_2 = 2H_2O + SnS \tag{16-19}$$

The equilibrium constant for this reaction is

$$K = (y_1^2)/(y_2)(y_3) \tag{16-20}$$

where y_1, y_2, and y_3 are the mole fractions of steam, hydrogen, and hydrogen sulfide, respectively, in the gas phase. At 400°C (752°F), K is approximately 400. Reaction 16-20 is favored by high hydrogen and low water vapor concentrations; however, the ratio of water to hydrogen must be above about 1.5 (at 400°C) to avoid the reduction of tin oxide to elemental tin.

A flow sheet of the Haldor Topsøe process using steam regeneration is shown in **Figure 16-13**. Regeneration involves the reverse of the absorption reaction 16-19. Superheated, medium-pressure steam is used in an amount equal to about 30 times the amount of sulfur to be removed.

With gas from high-sulfur coals, the use of steam would be uneconomical and an alternative regeneration mode is suggested (Nielsen and Rudbeck, 1993). In this scheme, the sulfur removal system is integrated with a Claus plant and hot tail gas from the Claus plant is used for regeneration of the SnS. The resulting dilute H_2S gas stream is fed directly to the Claus plant.

The Haldor Topsoe process, with steam regeneration, offers the advantage over the zinc ferrite or zinc titanate processes of recovering the sulfur as an almost 100% hydrogen sulfide stream, which may be economically converted to sulfur or sulfuric acid by conventional

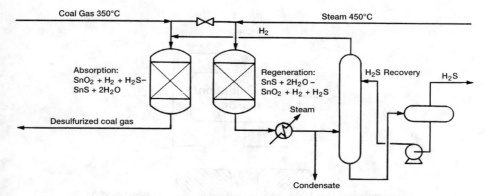

Figure 16-13. Haldor Topsøe tin oxide-based process with steam regeneration. *(Nielsen and Rudbeck, 1993)*

means. It has the disadvantage of operating at temperatures in the range of 350° to 500°C (662° to 932°F), which is considerably below the outlet temperature of most gasifiers. No commercial installations have been reported.

Z-Sorb Promoted Zinc Oxide-Based Process

The Z-Sorb process, developed by Phillips Petroleum Company appears to be closely related to the zinc ferrite and titanate processes developed under DOE sponsorship. It uses an extruded sorbent containing about 50% zinc oxide, a promoted metal oxide, which also reacts with hydrogen sulfide to form a metal sulfide; and a matrix material to provide pore volume and aid in prolonging sorbent life. The sorbent is used in fixed beds that operate at about 700K (800°F) during absorption and 810–920K (1,000°–1,200°F) during regeneration.

A schematic diagram of the Z-Sorb process is given in **Figure 16-14.** The figure represents an actual commercial operation of the process at a gasoline plant where the Z-Sorb beds are used to remove H_2S from the tail gas of a sulfur plant; however, other applications are possible.

In the application illustrated, acid gas containing about 17% hydrogen sulfide is fed to a combination direct oxidation/Claus system which produces elemental sulfur and an exhaust stream containing 1.7% hydrogen sulfide. In the sulfur plant, a portion of the hydrogen sulfide is oxidized directly to sulfur, while another portion reacts with recycled sulfur dioxide by the conventional Claus reaction. The amount of oxidizing gas fed to the sulfur plant is controlled to assure that sulfur in the tail gas is in the form of hydrogen sulfide, not sulfur dioxide.

The H_2S-containing gas from the sulfur plant is passed through one of the Z-Sorb beds where hydrogen sulfide reacts with zinc oxide to form zinc sulfide at about 800°F. At the same time, diluted air is passed through the second bed which attains a temperature in the 1,000°–1,200°F range due to the oxidation reactions. The SO_2-containing regeneration gas is fed to the sulfur plant with the inlet feed. In the 5 ton/day commercial installation, total sulfur removal exceeded the required 96%. It is estimated that efficiencies over 99.9% could be obtained with relatively minor modifications to the equipment and flow arrangement (Brinkmeyer and Delzer, 1990).

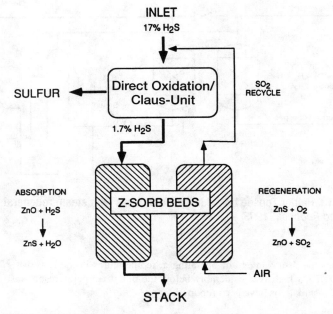

Figure 16-14. Schematic diagram of system for incorporating Z-Sorb beds into a sulfur plant. (*Brinkmeyer and Delzer, 1990*)

ABSORPTION AND CONDENSATION PROCESSES FOR EXHAUST GAS PURIFICATION

This section does not cover the purification of combustion process exhaust gas, which is the subject of Chapters 7 and 10, but rather the exhaust, usually polluted air, which is discharged from numerous chemical and industrial operations. The impurities to be removed are typically volatile organic compounds (VOCs) and noxious inorganic compounds such as hydrochloric acid and ammonia. Several techniques are available for removing the impurities, including adsorption, catalytic and thermal oxidation, absorption, and condensation. Some generalized guidelines for selecting an appropriate technique are given in Chapter 1. Adsorption is discussed in Chapter 12. Catalytic and thermal oxidation are covered in Chapter 13. This section covers the remaining approaches—absorption and condensation.

Absorption of Impurities by Reactive Solvents

The use of reactive solvents has a significant advantage over physical solvents, where applicable, because normally the impurity is destroyed or altered by the reaction and does not exhibit a significant vapor pressure over the solution once absorbed. When the liquid phase reaction is irreversible and rapid, which is often the case, a countercurrent contactor is not necessary, and simple venturi or spray scrubbers can provide high efficiency. The design of such equipment is discussed in Chapter 1.

Neutralization Reactions

In the handling of liquefied or compressed toxic gases, important safety features are the storage of the gas in tanks within a tight enclosure and the capability to remove the gas from enclosure air in the event of leakage. Somerville (1990) recommends chemical scrubbing for the removal of toxic gases from air and provides guidelines for the design of toxic gas scrubbing systems. Important criteria for the hardware are the ability to respond instantly after a long idle period and the absence of valves or low points in the liquid absorbent lines. The use of eductor-venturi scrubbers is recommended because they provide excellent gas-liquid contact while inducing air flow without the need for mechanical blowers. Scrubber design factors for many of the commonly encountered acidic and basic toxic gases are given in **Table 16-11**.

The table is based on the release and vaporization of 2,000 lb/h of toxic gas into the air of a containment building and evacuation of the air through a two-stage eductor-venturi scrubber system. Each stage of the system contains a stoichiometric quantity of the reagent. Sodium hydroxide solution is used as the reagent for all acid gases except hydrofluoric acid, which requires potassium hydroxide because of the low solubility of sodium fluoride. Sulfuric acid is recommended for absorbing ammonia because of its low cost and ready availability and the high solubility of ammonium sulfate. The indicated minimum airflows will keep the first stage scrubber temperature below 160°F when the ambient temperature is 90°F.

Absorption with Oxidation

Absorption of impurities into an aqueous solution in which the impurities are oxidized is a technique that is frequently used for odor control. Sodium hypochlorite is probably the most widely used absorbent because of its low cost and effectiveness in oxidizing most organic compounds. Since odorous air streams typically contain a wide variety of compounds, some

Table 16-11
Venturi Scrubber Design Factors for Toxic Gases

Toxic Gas	Scrubber Liquid	Reagent Req'd., lb	Heat Release, Btu lb gas	Min. Air Flow, acfm	Min. Vent. Size, in.	Liquid Flow, gpm	Noz. Pres., psig
HCl	20% NaOH	2,200	1,580	2,750	16	125	75
H_2S	20% NaOH	2,350	720	820	12	75	75
Cl_2	20% NaOH	2,360	630	2,400	16	125	75
NH_3	35% H_2SO_4	5,760	2,550	4,600	20	250	80
SO_2	20% NaOH	2,500	1,100	1,600	12	75	75
$COCl_2$	20% NaOH	3,250	2,000	3,300	18	175	75
HF	20% KOH	5,600	5,600	3,800	18	175	80
HCN	20% NaOH	3,000	420	420	10	50	80

Design Basis: 2,000 lb/hr toxic gas to the scrubber.
Data from Somerville (1990). Reproduced with permission from Chemical Engineering Progress, Copyright 1990, American Institute of Chemical Engineers

of which are not readily identifiable, it is good practice to conduct small scale tests of the proposed absorption/oxidation system.

According to Dunson (1991), the effective removal of VOCs with sodium hypochlorite solution requires operation under conditions that produce a residual chlorine concentration of several parts per million (volume) in the product gas. Removal of the chlorine and further removal of VOCs can be accomplished by use of a second scrubbing step with hydrogen peroxide solution. As an alternative to a two-stage system (sodium hypochlorite plus hydrogen peroxide) a single stage of sodium sulfite scrubbing may be used. Deodorization of exhaust air by scrubbing with sodium sulfite solution in a venturi scrubber is reported to be comparable in cost and more effective than the two-stage system.

A problem with sodium hypochlorite is its ability to chlorinate some compounds to form volatile reaction products. Ammonia and alkylamines, for example, can be chlorinated to form chloramines. The problem can be avoided by removing alkaline nitrogen compounds from the gas stream with an acid wash prior to the sodium hypochlorite scrubber. An alternative approach described by Valentin (1990) is the use of a catalyst in the scrubber liquid to promote oxidation. Hydrated nickelic oxide, which is formed in solution from nickel sulfate and sodium hypochlorite at a pH of 9 to 10.5, is recommended. A nickel concentration of 50–75 ppm is said to be effective, making removal of up to 95% of the odorous compounds possible in a single stage.

Recommended design parameters for sodium hypochlorite scrubbers are packing height, 2.5–4 m (or more) of high-efficiency packing; superficial gas velocity, 1.6–2.5 m/s; and solution flow rate, 15–22 m^3/h m^2 (Valentin, 1990).

Physical Absorption Processes

Physical absorption using water as the solvent is most commonly used for removing highly soluble inorganic impurities such as ammonia and hydrochloric acid from exhaust streams being vented to the atmosphere. Ammonia absorption (and the use of aqueous ammonia for absorption) are covered in Chapter 10. Most other water wash applications are covered in Chapter 6. The design of absorbers for physical absorption processes is discussed in some detail in Chapter 1. A highly simplified design procedure for packed absorbers for use in air pollution control systems is provided in the *U.S. EPA Handbook* (1991).

The use of physical absorption to remove VOCs from exhaust air streams has found limited application. In Germany, where the limits for solvents in air were reduced to 20 mg/m^3 as of Jan. 1, 1989, one plant has been built to absorb 10–200 g/m^3 dichloromethane from a 30 m^3/h air stream at ambient temperature using tetraethylene glycol dimethyl ether as the absorbent. Absorption is accomplished in a countercurrent spray column. Desorption occurs in an electrically heated column at 100°–130°C and 80–100 millibar pressure. The operation requires 11 kW power and 5 m^3/h cooling water (Anon., 1989).

Condensation

Condensation is frequently used in the chemical industry to recover valuable products from gas streams. In many cases, it serves as the initial gas purification step by removing the bulk of the organic compounds from a gas stream before it undergoes final cleanup. As indicated in **Figure 1-1,** a high percentage removal of impurities by condensation can only be achieved when the inlet concentration is high (over about 5,000 ppmv). Removal efficiencies of more than about 95% are seldom attainable with condensation because the VOC content of the product gas cannot be reduced below the equilibrium vapor pressure of the condensate at the operating temperature.

Two basic types of condensers are used: surface and direct contact. Surface condensers are generally shell-and-tube heat exchangers with coolant flowing inside the tubes and vapor condensing from the gas stream on the shellside. The condensing vapors form a liquid film on the outside of the tubes, which flows to the bottom of the condenser and is drained to a collection tank. Direct contact condensers use a spray of cooled liquid (typically water) to contact and cool the gas. The surface of the spray droplets provides the extended surface area required for heat transfer, and the condensing vapor joins the spray of liquid. Although the direct contact condenser avoids the cost of metal heat transfer surface, it has the disadvantage of requiring a very high liquid flow rate. Another disadvantage is that in correcting an air pollution problem, the direct contact condenser may create a water pollution problem. Treatment of the organic condensate-water coolant mixture can be expensive (Corbitt, 1990).

For small volume gas streams containing high concentrations of VOCs, it is often economical to compress the gas before it enters the condenser. This causes the vapor pressure of the condensed organics to represent a smaller fraction of the total pressure, and thus the concentration of VOCs in the product gas to be lower than with atmospheric pressure condensation. According to Hill (1990), for the case of vapors released when gasoline is loaded into ships, compression followed by cooling and condensation is competitive with compression followed by absorption in hydrocarbon liquid and less expensive in both capital and operating costs than condensation by refrigeration. Because of the high cost of compression, systems that require the gas to be compressed are uneconomical for large, dilute exhaust gas streams.

Condensation System Design

For purposes of this discussion, the most commonly encountered conditions are considered, i.e., a large dilute exhaust gas stream at essentially atmospheric pressure cooled in the shell side of a shell-and-tube exchanger. A simplified schematic diagram of such a system is given in **Figure 16-15**.

The gas leaving the condenser can be assumed to be in equilibrium with condensate at the gas outlet temperature, although the liquid may be further cooled before it leaves the vessel. For a single organic compound, the required partial pressure in the product gas can be calculated readily from the inlet concentration and the desired percent removal:

$$P_p = y_o P_t \tag{16-21}$$

$$y_o = y_i(1 - E)/(1 - Ey_i) \tag{16-22}$$

Where: P_p = partial pressure of VOC at outlet, mm Hg
P_t = total system pressure, mm Hg
y_o = mole fraction at outlet (ppmv $\times 10^{-6}$)
y_i = mole fraction at inlet (ppmv $\times 10^{-6}$)
E = removal efficiency, expressed as a fraction

The required condensation temperature can then be obtained from a vapor pressure-temperature chart for the specific organic compound. Vapor pressure data for four typical VOCs are given in **Figure 16-16**. For example, if the gas contains 13,000 ppmv styrene (y_i = 0.013) and 90% removal is desired (E = 0.9), the above equations yield a value for y_o of

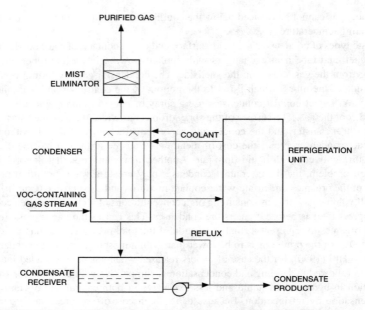

Figure 16-15. Schematic diagram of refrigerated condenser system with condensate reflux to prevent ice formation on heat transfer surfaces.

0.001315, which corresponds to a partial pressure of 1.0 mm Hg. The required condensation temperature indicated by **Figure 16-16** is 20°F, and based on **Table 16-12,** the appropriate coolant is a brine solution. Additional data for this example, which is described in more detail in the *U.S. EPA Handbook* (1991), are given in **Table 16-13.**

The design becomes somewhat more complex when a mixture of compounds is involved. However, the components of the mixture are usually mutually soluble in the liquid phase, and, as a first approximation for related solvents, it can be assumed that the mixture follows Raoult's law, i.e., the partial pressure of each component in the product gas will be equal to the vapor pressure of the pure component at the gas outlet temperature times its mole fraction in the liquid phase. For more precise calculations and more complex liquid mixtures, it is necessary to use vapor-liquid equilibrium (VLE) data for the specific system. The estimation and correlation of VLE data are discussed in various chemical engineering texts, such as *Perry's Handbook* (Perry et al., 1984), Reid and Sherwood (1966), and Prausnitz (1969).

When the design condensation temperature is near or below 32°F, water vapor in the feed gas can condense as ice on the heat-transfer surfaces, interfering with system operation. Forrester and Le Blanc (1988) point out three possible approaches for avoiding ice problems: (1) using two separate condenser coils and passing hot gas from the compressor through each one alternately to melt ice deposits; (2) washing the coils with a reflux stream of the condensed phase (this system is shown in **Figure 16-15,** and is applicable if water is soluble in the condensed liquid and the resulting solution has a sufficiently low freezing point); and (3) cooling the gas feed in two steps: first to about 35°F to condense the bulk of the water as liquid, then to the desired condensation temperature to remove the organic vapors.

After establishing the required exit gas temperature and the amount of organic compounds to be condensed, it is necessary to determine the quantity of heat that must be removed (sen-

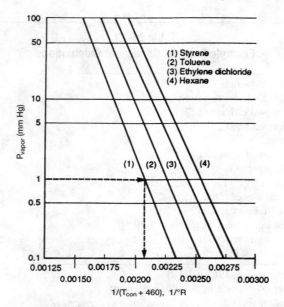

Figure 16-16. Vapor pressures of four typical volatile organic compounds; P_{vapor} = vapor pressure of compound (mm Hg), T_{con} = condensation temperature (°F). *(U.S. EPA Handbook, 1991)*

**Table 16-12
Selection of Coolant for Surface Condenser Systems**

Required Condensation Temperature, °F	Typical Coolant
60 to 80	Water
45 to 60	Chilled Water
−30 to 45	Brine (e.g., calcium chloride or ethylene glycol solution)
−90 to −30	Refrigerant (e.g., Freon 12)

Note: Coolant temperature will typically be about 15°F lower than the required condensation temperature.
Based on data from U.S. EPA Handbook (1991)

sible plus latent) in order to size the heat exchanger and refrigeration system. The detailed design of heat exchange equipment and refrigeration units is beyond the scope of this book; however, it should be noted that heat transfer coefficients for condensing vapors from an inert gas are generally much lower than for condensing pure vapors, but somewhat higher than cooling (or heating) a noncondensible gas. The *U.S. EPA Handbook* (1991) suggests an overall coefficient of 20 Btu/(hr)(sq ft)(°F) for a typical condenser with a liquid coolant and a temperature approach of 15°F at the gas outlet end of the heat exchanger.

Table 16-13
Example Case for Condenser Calculations

Max. inlet gas flow rate, scfm	2,000
Impurity (Styrene) concentration, ppmv	13,000
Required removal efficiency, %	90
Condensation temp., °F	20
Inlet gas temp., °F	90
Total heat removal load, Btu/hr	239,000
Coolant inlet temp., °F	5
Coolant outlet temp., °F	30
Overall ht. trans. coef., Btu/(hr)(ft^2)(°F)	20
Condenser surface area, sq ft	370
Coolant flow rate (29% CaCl$_2$ soln.), lb/hr	14,700
Refrigeration unit capacity, tons	20
Recovered product, lb/hr	373
Costs for above case (1988 dollars)	
Condenser (8-ft tube length, fixed tubesheet)	34,800
Refrigeration unit (20 tons, 20°F condensation temp.)	107,000
Total	141,800

Notes: Total cost includes auxiliary equipment, instrumentation, sales tax, freight, and installation costs, but not site preparation or buildings.
Data from U.S. EPA Handbook (1991)

Applications

The application of refrigerated condensation to remove hydrocarbons from the gases vented during the transfer of liquids to and from marine vessels is discussed with other techniques by Hill (1990). The study concludes that refrigerated condensation is considerably more expensive than the other techniques considered for this application, including compression-cooling-condensation; compression-absorption; adsorption; absorption-adsorption-absorption; and combustion, but has the following advantages:

1. It is a stand-alone unit with no circulating absorption fluid.
2. The recovered products can be pumped directly to storage.
3. It can handle compounds that may not readily desorb from carbon or that may cause temperature excursions in activated carbon systems.

According to Forrester and Le Blanc (1988), refrigerated condensation has been used most often for handling the vent gas from storage tanks, but has also been used successfully at barge and tank-car loading and transfer stations. It has been used with such volatile organic liquids as gasoline, alcohols, ketones, and nitriles, and at condensation temperatures down to about −100°F. The condensers are often used in combination with absorbers, adsorbers, and incinerators to reduce the load on the auxiliary systems.

In the design of condensation systems for vent gas streams, it is necessary to take into account the large differences in operating conditions between summer and winter. These differences are evident in the data of **Table 16-14,** which compares January and July operating conditions for a system designed to remove 95% of the hydrocarbons from a vent gas stream at all times of the year. Since the gas entering the unit in the winter is colder and contains less hydrocarbon vapor, it is necessary that the condenser operate at a much lower temperature to provide the same percent removal (95%) as in the summer (Forrester and Le Blanc, 1988).

LOW-TEMPERATURE GAS PURIFICATION PROCESSES

Separation of gaseous mixtures by condensation and distillation at low temperatures has been practiced extensively since Carl von Linde disclosed his air-liquefaction process at the beginning of the twentieth century. In most applications, these processes are not aimed at the removal of low concentrations of impurities from gases, but rather at the separation and recovery of pure components from mixtures. However, the technology is quite versatile and may also be used for separations that can be classified as gas purification such as: (1) the purification of hydrogen for ammonia synthesis and other uses, (2) the removal of nitrogen from natural gas, and (3) the removal of high concentrations of carbon dioxide and hydrogen sulfide from natural gas.

Although improvements continue to be developed in the field of low-temperature separation, it can be considered to be a mature technology. For some applications it is being supplanted by newer approaches such as pressure swing adsorption (PSA) and membrane permeation (see Chapters 12 and 15). However, low-temperature processes remain important for many cases. They are generally favored over PSA for removing components that are weakly adsorbed such as carbon monoxide, methane, and nitrogen, or that are needed at high pressure such as carbon dioxide for enhanced oil recovery (EOR) projects. They are usually more economical than membrane processes for large capacity operations where they benefit from the economies of scale. When the operating temperature is below about $-100°F$, the technology is called cryogenics. A general review of the principles and applications of cryogenics is given by Springmann (1985).

Since the compounds being separated are not normally reactive with each other and also because chemical reactions are extremely slow at cryogenic temperatures, cryogenic separation operations are purely physical. They follow the same vapor-liquid equilibria principles and system design technologies as commonly used for separating hydrocarbons that are liq-

Table 16-14
Seasonal Variations in Operating Conditions of Condensation System

	January	July
Inlet temp. of vent gas, °F	50	90
Hydrocarbon equilibrium temp., °F	32	83
Outlet gas temp., °F	−50	−15
Condensation rate, lb/hr	41	160
Energy load, MBtu/hr	31	79

Data of Forrester and Le Blanc (1988)

uid at ambient temperatures. The key special requirements of cryogenic systems that differentiate them from conventional condensation/distillation operations are (1) a refrigeration system capable of reducing temperatures to the required cryogenic level, and (2) highly efficient insulation to minimize heat entry into the cold system.

Cryogenic refrigeration is normally provided by one or more of three basic technologies. They are (1) expansion of compressed gas through a throttling valve (the Joule-Thompson effect), (2) expansion of compressed gas through an engine which does external work, and (3) vaporization of a liquefied gas. The Joule-Thompson effect is not applicable to hydrogen, which has a negative Joule-Thompson coefficient. The most commonly used approach is the expansion of compressed gas through a turboexpander.

Efficient insulation is mandatory for an economical cryogenic system. This requirement normally results in the installation of all (or groups of) cryogenic equipment items in a "cold box," which is basically a modular frame packed with insulation material.

The discussions that follow include some processes that are clearly cryogenic and some that are merely "low temperature"; however, all involve liquefaction of normally gaseous components, require refrigeration systems, and operate below 0°F.

Hydrogen Purification

Separation of hydrogen from coke-oven gas by liquefaction of the higher-boiling components was first practiced on an industrial scale in France and Germany shortly after the end of World War I. Since then, this technique has been applied successfully to the separation of pure hydrogen from a variety of industrial gas streams, such as petroleum refinery offgases and the product streams from steam hydrocarbon reforming and natural gas partial oxidation. The basic process consists of cooling the gas stream to condense out all components except hydrogen. The removal of carbon monoxide can be enhanced by washing the gas with liquid methane or liquid nitrogen. The use of nitrogen is particularly effective when the product gas is to be used for ammonia synthesis since a 75% hydrogen, 25% nitrogen mixture is desired.

Process Arrangements

Basic types of cryogenic processes used for producing purified hydrogen from impure gas streams are listed in **Table 16-15.** Simple condensation processes operate by liquefying CO (and higher boiling impurities) in several cooling steps to produce a gaseous product containing approximately 98% hydrogen and 2% carbon monoxide. The liquid CO contains all of the methane present in the original gas stream and can be purified by distillation to produce a 99.8% pure CO gaseous product and a CO/CH_4 bottoms product suitable for recycle to the syngas generation step. If higher purity hydrogen is desired, it can be further purified by pressure swing adsorption (PSA) or by shifting the CO to CO_2, removing the CO_2, and methanating the remaining carbon oxides to CH_4. The PSA process will produce 99.9% pure hydrogen; while the water gas shift system is capable of producing about 99.7% purity (Goff and Wang, 1987).

The use of a methane wash system results in the production of hydrogen containing only parts per million carbon monoxide, but 1 to 2% methane. The wash column bottoms is typically flashed to remove dissolved hydrogen, then distilled in a CO/CH_4 splitter to separate the carbon monoxide and methane. The carbon monoxide is recovered as 98% pure byproduct in the splitter overhead.

Table 16-15
Cryogenic Processes for Purifying Hydrogen-Containing Gas Streams

Type of Process	Products	Purity %	Recovery %
Condensation	Hydrogen	98	90–95
Condensation and CH_4/CO separation	Hydrogen	98	90–95
	Carbon Monoxide	99.8	95
Methane Wash	Hydrogen	99	99
	Carbon Monoxide	98	
Nitrogen Wash	3:1 $H_2:N_2$ Mix	99	99

Data of Springmann (1985)

A nitrogen wash system is normally used only when the hydrogen is to be used in ammonia synthesis. Some nitrogen is allowed to pass overhead with the purified hydrogen. Additional gaseous nitrogen is then added as required to give a final product containing the required 3:1 hydrogen to nitrogen ratio. The product gas contains only traces of carbon monoxide and less than 1% total impurities.

Cryofining

The Cryofining process for recovering high purity hydrogen and liquid byproducts from refinery fuel gas is licensed by Air Products and Chemicals, Inc. The process is applicable to refinery offgases and purge streams containing 30 to 70% hydrogen and up to 30% liquefiable hydrocarbons (e.g., ethane, propane, and olefins).

A simplified flow diagram of the process is shown in **Figure 16-17.** The gas at a pressure in excess of about 400 psia is first cooled to a temperature slightly above the hydrate point to

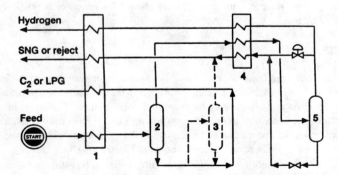

Figure 16-17. Simplified flow diagram of Cryofining process for recovering high purity hydrogen and liquid products from refinery fuel gas (*Air Products and Chemicals, Inc. 1996*). *Reproduced with permission from Hydrocarbon Processing, April 1996*

condense liquid water, then dried to less than 1 ppm water by a conventional dehydration unit (these steps are not shown on the diagram). The dry gas is cooled by heat exchange with the cold product streams in heat exchanger (1) to condense liquids, which are removed from the gas in separators (2, 3). The liquid from the separators represents an ethane or liquefied petroleum gas (LPG) byproduct. The gas is further cooled in heat exchanger (4) to condense additional liquids which are removed in separator (5). Liquid product from this separator is passed through a throttle valve to a lower pressure at which vaporization provides refrigeration for the system. Gas from the separator, after use in the heat exchangers, represents the final hydrogen product.

The performance of this system for a feed gas at 600 psia containing 33.7% hydrogen is reported to be: hydrogen product purity, 95%; hydrogen recovery, 85%; and hydrocarbon liquids recovery, 76%. The balance of the feed is available as fuel. The economics of cryogenic hydrogen recovery are favored by moderately high impurity levels (30 to 70%), hydrogen purity requirements in the 90 to 98% range, and the concurrent need for hydrocarbon liquids recovery (Air Products and Chemicals, Inc., 1996).

Ammonia Purge Purification

A somewhat similar process scheme has been proposed for recovering pure hydrogen and ammonia from the purge gas of an ammonia synthesis loop (Shaner, 1978). The purge is drawn off to keep inerts—i.e., methane and argon—from building up in the synthesis loop. The gas is first prepurified to remove most of the ammonia by water washing, dehydrated by molecular sieve adsorption, then passed into the cryogenic unit.

In the cryogenic unit, the dry gas at a pressure of about 1,000 psig is cooled by heat exchange with product streams to condense methane and argon. The condensate is separated from the gas, throttled to a lower pressure to provide refrigeration, then mixed with a portion of the hydrogen product to give an adequate temperature difference in the heat exchanger. After use as a coolant, the mixture is used as fuel gas. A portion of the hydrogen gas from the separator is used to regenerate the molecular sieve so that adsorbed ammonia is returned to the synthesis loop with the hydrogen.

In this system, the fuel gas pressure has a strong influence on efficiency. For example, if 92% product hydrogen purity is desired, a recovery of about 92% of the hydrogen is possible when the fuel gas pressure is 8 psig; while only about 82% recovery is possible with a fuel gas pressure of 60 psig.

Design Data

The fact that hydrogen boils at a considerably lower temperature than all the other components present in the gas streams treated in low-temperature processes permits essentially complete removal of most impurities by fractional condensation. However, to obtain complete separation it is necessary to cool the gas mixture to a temperature considerably below the boiling point of the constituent to be removed. This temperature can be estimated from the vapor pressures of the pure constituents shown in **Figure 16-18**.

The refrigeration required in low-temperature processes can be estimated from the specific heats and the latent heats of vaporization (or condensation) of the constituents of the gas mixture. Values for the most frequently encountered gases are given in **Tables 16-16** and **16-17**.

While the sensible heat of the incoming gas can be removed by heat exchange with cold product streams, the latent heat required for condensing impurities is much greater and can

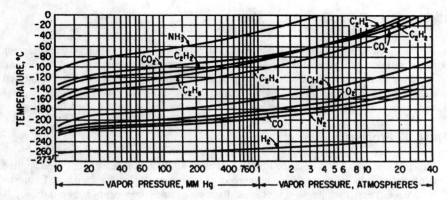

Figure 16-18. Vapor pressure of various gases at low temperatures; left half in mm Hg, right half in atm. *Data of Stull (1947)*

Table 16-16
Boiling Points and Latent Heats of Vaporization of Gases

Gas	Boiling Point, °C	Latent heat of vaporization at normal boiling point and 1 atm, calories/g mole	Reference
Hydrogen	−249.4	206	U.S. NBS (1955)
Oxygen	−183.0	1,630	U.S. NBS (1952)
Nitrogen	−195.8	1,333	Kirk-Othmer (1952)
Carbon monoxide	−191.5	1,443	Din (1956)
Methane	−161.5	1,955	U.S. NBS (1952)
Ethane	−88.6	3,517	U.S. NBS (1952)
Ethylene	−103.8	3,237	Egan and Kempt (1937)

Table 16-17
Specific Heats of Gases at Constant Pressure* (cal/g mole)

	Temperature, °C					
Gas	−50	−80	−100	−140	−200	Reference
Hydrogen	6.654	6.472	6.309	5.864	5.120	U.S. NBS (1955)
Oxygen	6.990	6.994	7.004	7.051		U.S. NBS (1955)
Nitrogen	6.978	6.974	6.994	7.034		U.S. NBS (1955)
Carbon monoxide	6.99	7.01	7.05	7.19		Din (1956)
Methane	—	7.97 (−74°C)	7.21 (−115°C)	7.24 (−131°C)		Millar (1923)

One atmosphere.

most conveniently be supplied by a boiling liquid. Since the entropy of the system increases with the difference between the condensation temperature of the impurity and the boiling temperature of the refrigerant, it is advantageous to use liquids boiling close to the desired condensation temperature. Typical liquid refrigerants used in the process are methane, carbon monoxide-nitrogen mixtures, and nitrogen.

The CNG Process

This process, which was described by Hise et al. in 1982, has apparently not been applied commercially. It is mentioned here because of the unusual concepts incorporated into the flow scheme. These include the use of crystallization of one component (carbon dioxide) as a means of stripping another component (hydrogen sulfide) from solution, and the use of a slurry of carbon dioxide crystals in an organic solvent to condense and absorb additional carbon dioxide. The process was a joint development of CNG Research Corporation, the U.S. Department of Energy, and Helipump Corporation. It was proposed for use in the removal of acid gas from coal-derived gas streams at pressures above 300 psig.

The feed gas, containing over 25% carbon dioxide, is cooled to a temperature near the carbon dioxide triple point, and liquid carbon dioxide is condensed. The remaining gas is contacted with pure liquid carbon dioxide in one absorber to remove hydrogen sulfide and other impurities and then with a slurry of carbon dioxide crystals in an organic liquid in a second absorber to remove additional carbon dioxide. Hydrogen sulfide is stripped from the liquid carbon dioxide stream by adiabatic flashing, which cools the liquid and causes the formation of pure carbon dioxide crystals. The crystals are vaporized to form pure carbon dioxide gas. Organic liquid from the second absorber is regenerating by flashing off carbon dioxide gas, causing the formation of solid crystals of carbon dioxide and re-forming the slurry for recycle.

The Ryan/Holmes Processes

The Ryan/Holmes processes represent a family of related techniques for recovering carbon dioxide and hydrocarbon liquids from carbon dioxide-rich gas streams. They are licensed by Process Systems International, Inc. The processes incorporate low-temperature distillation with a circulating natural gas liquid (NGL) additive. Their primary field of application has been in the processing of gas from enhanced oil recovery (EOR) projects. The technology has been described by Process Systems International, Inc. (1992); Wood et al. (1986); Ryan and O'Brien (1986); Schaffert and Ryan (1985); and Ryan and Schaffert (1984).

Process Description

A simplified flow diagram of one version of the Ryan/Holmes technology is shown in **Figure 16-19.** In a typical application of the process for carbon dioxide-rich EOR gases, the feed gas stream is first passed through a separator (1) to remove any liquid water or hydrocarbons, then compressed to 360 psig (2), and dehydrated in a conventional glycol or molecular sieve unit (3).

In the four-column fractionation section, the dehydrated feed is fed first to the ethane recovery column (4) where ethane and heavier components (and H_2S, if present) are recovered in the bottom product. The column is operated as an extractive distillation unit with a recycled stream of C_4+ hydrocarbons as the additive. The additive acts as an absorption

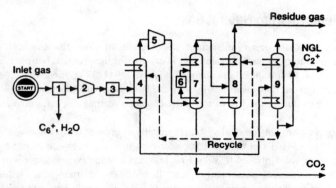

Figure 16-19. Simplified flow diagram of Ryan/Holmes process for removing and recovering hydrocarbons, H_2S, and CO_2 from natural gas or associated gas from enhanced oil recovery projects (*Process Systems, International, 1992*). Reproduced with permission from Hydrocarbon Processing, April 1992

medium to recover C_2+ hydrocarbons and also breaks the ethane-carbon dioxide azeotrope. The overhead stream contains primarily carbon dioxide, methane, nitrogen, and any unrecovered higher hydrocarbons.

Gas from the ethane recovery column is compressed to about 650 psig (5), cooled, and partially condensed by providing reboil heat and by indirect contact with a refrigerant in heat exchanger (6), and then fed into the carbon dioxide recovery column (7).

The carbon dioxide recovery column operates as a simple distillation unit without the use of an additive. The overhead gas is primarily methane containing 15 to 30% carbon dioxide. The bottoms product is liquid carbon dioxide containing a small amount of ethane. This liquid stream is typically pumped to a pressure of about 2,000 psig for use in EOR injection.

Overhead gas from the carbon dioxide recovery column flows directly into the demethanizer column (8), which operates with additive. This column serves to reduce the carbon dioxide content of the gas stream to the specified level for use as sales or fuel gas (typically less than 2%). Bottoms from the demethanizer join the stream of additive being fed to the ethane recovery column, and the total liquid stream from that column is fed to the additive recovery column (9).

The additive recovery column is basically a stripper to remove hydrogen sulfide and light hydrocarbons, primarily ethane and propane, from the circulating additive stream. Excess C_4+ hydrocarbons that accumulate in the stripped additive are removed as net product, which can be combined with the overhead product as indicated in the flow diagram or handled separately. When the plant feed contains hydrogen sulfide, this will appear in the overhead of the additive recovery column and may be removed from this stream by amine treating (not shown on the diagram).

Applications

In 1992, it was reported that there were nine licensees for the Ryan/Holmes process for plants ranging in size from slightly over 1 to 290 MMscfd (Process Systems International, Inc., 1992).

Nitrogen Removal from Natural Gas

Nitrogen occasionally occurs in natural gas in concentrations that are high enough to require its removal to meet heating value specifications for the product. Nitrogen is also normally present in the gas produced in enhanced oil recovery (EOR) projects which employ the injection of compressed flue gas or nitrogen. In the latter case, the nitrogen content of the gas gradually increases during the life of the EOR project, creating a special challenge for the nitrogen-removal process.

Since cryogenic nitrogen-removal systems typically operate at temperatures as low as 115 to 80K (−250° to −310°F) in their coldest sections, it is necessary to remove impurities which might form solid phases to very low levels. Gaseous components which require removal, with their freezing points and typical allowable concentrations, are listed in **Table 16-18**. Particulate impurities, such as dust from a molecular sieve unit and entrained compressor lubrication oil, must also be removed to a very low level to minimize fouling of extended surface heat exchangers.

Several cryogenic process flow schemes for purifying gas streams of different nitrogen concentrations are described by Harris (1980) and Vines (1986). Two process cycles suitable for removing nitrogen from natural gas are shown in **Figures 16-20** and **16-21**. According to Vines (1986), both of these cycles are capable of adapting to widely varying inlet gas conditions.

The single-column heat pumped cycle shown in **Figure 16-20** uses a fractionation tower, T-1, in which both the condensing and reboiling duties are provided by a closed-loop methane heat pump system. Natural gas feed to the system is precooled in heat exchanger E-1 using both the overhead nitrogen and bottoms methane products as coolants. This cycle has the advantage of producing high-pressure nitrogen, but at the expense of heat-pump energy consumption.

The double column cycle shown in **Figure 16-21** is particularly efficient for handling high-pressure gas containing a relatively high concentration of nitrogen, where the rejected nitrogen is not needed at elevated pressure. In this case, a significant portion of the refrigeration is provided by the expansion of the nitrogen.

Table 16-18
Components Requiring Removal Prior to Cryogenic Nitrogen Rejection Processes

Component	Freezing Point, K	°F	Allowable Conc., ppm
Water	273	32	0.1
Carbon Dioxide	217	−70	10 to 1,000
Methanol	176	−144	1.0
Benzene	279	42	0.1
Hydrogen Sulfide	188	−122	50 to 500
Ethylene Glycol	257	3	1.0

Data of Vines (1986). Reproduced with permission from Chemical Engineering Progress, Copyright 1986, American Institute of Chemical Engineers

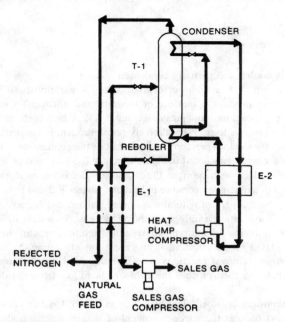

Figure 16-20. Diagram of single-column heat pumped cycle for removing nitrogen from natural gas (*Vines, 1986*). *Reproduced with permission from Chemical Engineering Progress, Copyright 1986, American Institute of Chemical Engineers*

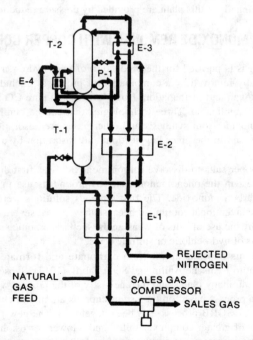

Figure 16-21. Diagram of double-column heat-pumped cycle for removing nitrogen from natural gas (*Vines, 1986*). *Reproduced with permission from Chemical Engineering Progress, Copyright 1986, American Institute of Chemical Engineers*

The feed gas is cooled and partially condensed in heat exchanger E-1 and flashed into high pressure column, T-1, which performs a partial separation between nitrogen and methane. The bottom product, a mixture of methane and nitrogen, is subcooled in heat exchanger E-2 then flashed into low pressure column, T-2, which performs the final separation. The overhead product from T-1, relatively pure nitrogen, is condensed in E-4, which serves as a reboiler for the low-pressure column. Part of the condensate is used as reflux for T-1 while the remainder is subcooled in heat exchanger E-3 and used as reflux for the low-pressure column. Product methane from the bottom of T-2 is raised in pressure by pump, P-1, and rewarmed to ambient temperature in heat exchangers E-2 and E-1.

In many cases, the recovery of natural gas liquids (ethane and higher hydrocarbons) can be profitably integrated with the nitrogen-removal system. According to Vines (1986), the recovery of 70–85% of the ethane typically requires temperatures in the range of 190 to 155K (−120° to −180°F). One effective arrangement for integrating the natural gas liquids recovery and nitrogen removal functions is to partially condense the feed gas and pass it through a demethanizer, then process the cold overhead gas from the demethanizer in a nitrogen-removal system.

A commercial nitrogen-removal plant operating in France has been described by Streich (1970). The process reduces the nitrogen content of natural gas from about 14% to about 2.5% using a two-column system. The product gas leaves the plant in two streams, one containing about 70% of the methane feed at 355 psig and the other, containing the balance of the methane, at 16 psig. Both streams contain approximately 2% nitrogen. The waste nitrogen contains less than 0.5% methane. Essentially all of the separation work, energy losses, and refrigeration demands of this plant are provided by pressure reduction of the gas.

CARBON MONOXIDE REMOVAL WITH COPPER COMPOUNDS

When synthesis gas is purified for the production of pure hydrogen or ammonia, carbon monoxide is generally removed by a combination of processes, including shift conversion, carbon dioxide removal, and methanation. However, when a pure CO byproduct is desired, this approach is not applicable. Three technologies that can be employed to remove and recover carbon monoxide from synthesis and other gases are adsorption (see Chapter 12); cryogenics, as discussed in the preceding section; and absorption by a liquid, which is discussed next.

The ability of copper salts to dissolve carbon monoxide was first discovered by Leblanc (1850). Early interest in the phenomenon centered around its use in gas analysis for the determination of carbon monoxide. The first aqueous solutions were generally acid and unsuited for use in conventional equipment because they caused severe corrosion. This problem was mitigated by the use of salts of weak acids, such as carbonic and formic, rather than previously used salts of hydrochloric or sulfuric acids.

The use of aqueous cuprous-ammonium carbonate and formate solutions was first described in a patent to Badische Anilin and Soda Fabrik (1914). Subsequently, the aqueous process was utilized in many plants and the chemistry of the process was studied extensively. More recently, a process based on the use of cuprous aluminum chloride in an aromatic solvent carrier (the COSORB process) has been developed. The new process is reported to have the advantages of greater complex stability and a lower corrosion rate. Both the aqueous and non-aqueous processes are discussed in the following sections. Although the aqueous process is now seldom used, it is included in the discussion because of its historical and technical interest. A more detailed discussion of the aqueous process is given in previous editions of this book. The non-aqueous approach is currently being used in several plants; however, it too appears to be losing ground relative to pressure swing adsorption and cryogenic processes.

Aqueous Copper Ammonium Salt Process

In this process, which is illustrated diagramatically in **Figure 16-22,** carbon monoxide at high pressure is absorbed by an aqueous solution of a copper-ammonium salt in a countercurrent contactor with the formation of a cuprous-ammonium-carbon monoxide complex. The preferred compounds for this service are copper-ammonium formate, carbonate, and acetate, although other salts of weak acids have been proposed. The solution is regenerated by the application of heat, which destroys the complex to liberate almost all of the absorbed carbon monoxide. Because of its mildly alkaline nature, the solution also absorbs carbon dioxide, which is also liberated during regeneration. The process is complicated somewhat by side reactions such as the reduction of cupric to cuprous ions by carbon monoxide, and the auto-oxidation of cuprous to cupric ions with the precipitation of elemental copper. To prevent the latter reaction, it is necessary to provide oxygen to the system to maintain a sufficiently high concentration of cupric ions.

Chemical Reactions

Reaction 16-23 has been proposed by van Krevelen and Baans (1950) as the principal carbon monoxide absorption-desorption mechanism.

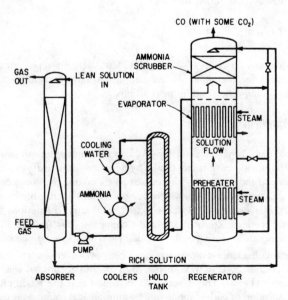

Figure 16-22. Simplified flow diagram of copper-ammonium-salt process for carbon monoxide removal from gases.

$$Cu(NH_3)_2^+ + CO + NH_3 = Cu(NH_3)_3(CO)^+ \tag{16-23}$$

Secondary reactions—occurring in the absorber and reversed to some extent in the regenerator—which result in the absorption of carbon dioxide are

$$2NH_4OH + CO_2 = (NH_4)_2CO_3 \tag{16-24}$$

$$(NH_4)_2CO_3 + CO_2 + H_2O = 2NH_4HCO_3 \tag{16-25}$$

In addition to the above absorption-desorption reactions, certain oxidation-reduction reactions can occur in this system. One of the most important of these is the reduction of cupric to cuprous ion by dissolved carbon monoxide as follows:

$$2Cu^{++} + CO + 4OH^- = 2Cu^+ + CO_3^- + 2H_2O \tag{16-26}$$

The reaction in equation 16-26 depletes the supply of cupric ions in the solution, shifting the equilibrium of the following reaction (equation 16-27) to the right:

$$2Cu^+ = Cu + Cu^{++} \tag{16-27}$$

To avoid the precipitation of elemental copper by the foregoing reaction, it is necessary to maintain about ⅕ of the copper in the cupric state. This is accomplished by injecting air into the system to oxidize cuprous ion as follows:

$$4Cu^+ + O_2 + 2H_2O = 4Cu^{++} + 4OH^- \tag{16-28}$$

Note that in equations 16-26 to 16-28, Cu^+ and Cu^{++} actually represent complex copper-ammonium ions; however, the ammonia and acid groups have been left out of the equations as they do not play an important part in the reactions.

Vapor-Liquid Equilbria

The vapor-liquid equilibrium relationships for copper-ammonium salt solutions containing dissolved carbon monoxide have been studied by a number of investigators. Hainsworth and Titus (1921) measured the vapor pressure of carbon monoxide over copper-ammonium carbonate solutions. Experimental data on formate solutions were obtained by Larson and Teitsworth (1922). Zhavoronkov and Reshchikov (1933) studied solutions of chlorides, formates, lactates, and acetates. Zhavoronkov and Chagunava (1940) made a detailed study of formate-carbonate mixtures, including the solution from an operating plant.

A method of estimating carbon monoxide vapor pressures over aqueous copper-ammonium salt solutions, under conditions outside the range of experimental data, was proposed by van Krevelen and Baans (1950). These authors concluded that the basic chemical reaction involved in the absorption of carbon monoxide by ammoniacal solutions is expressed in equation 16-26. By considering equilibrium relationships in this reaction, they derived an equation relating the carbon monoxide partial pressure to the solution composition as follows:

$$\frac{KH}{\gamma_t} = \frac{m}{(m-1)(mB-A)p_{co}} = C_{eq} \qquad (16\text{-}29)$$

Where: K = equilibrium constant for the reaction in equation 16-23
H = Henry's law constant in the equation $(CO) = Hp_{co}$
γ_t = the overall activity coefficient
m = moles of monovalent copper ion per mole of carbon monoxide absorbed
A = total cuprous salt concentration, g moles/liter
B = total free ammonia concentration before carbon monoxide absorption, g mole/liter
p_{co} = partial pressure of carbon monoxide, atm
C_{eq} = apparent equilibrium constant

Equation 16-29 can be used for calculating CO vapor pressures providing that C_{eq} is known. Van Krevelen and Baans have developed satisfactory equations relating C_{eq} to temperature and ionic strength of the solution for copper-ammonium chloride solutions (based on their own experiments) and copper-ammonium formate solutions (based on the data of Larson and Teitsworth). These equations follow.

For copper-ammonium chloride:

$$\log_{10} C_{eq} = \frac{11,900}{2.3RT} - 0.040I - 8.790 \qquad (16\text{-}30)$$

and for copper-ammonium formate,

$$\log_{10} C_{eq} = \frac{13,500}{2.3RT} - 0.040I - 9.830 \qquad (16\text{-}31)$$

Where: T = the absolute temperature, K
I = ionic strength of the solution
$= \frac{1}{2} \Sigma c_i z_i^2$
c_i = concentration of the ion, i, g moles/liter
z_i = valence of the ion i

As an example of the application of equations 16-30 and 16-31, consider the case of a solution with the following analysis:

Cu^+ .... 0.94 g atoms/liter
Cu^{++} .... 0.02 g atoms/liter
$HCOO^-$ .... 1.72 g equiv/liter
NH_3 (total)............. 7.01 g moles/liter

If it is assumed that each cuprous ion is associated with two molecules of ammonia and one formate ion, while each cupric ion is associated with four molecules of ammonia and two formate ions, the following concentrations can be deduced:

NH$_3$ (as uncomplexed ammonium formate) = 1.72 − 0.94 − 2(0.2)
= 0.74

NH$_3$ (free + complex) = 7.01 − 0.74 = 6.27

NH$_3$ (free) = 6.27 − 2(0.94) − 4(0.02) = 4.31

For this system the ionic strength, I, is equal to

½[0.94 + (0.02)(4) + 1.72 + 0.74] = 1.74

For the case of T = 295K (22°C, 71.6°F) and a carbon monoxide concentration in the solution equivalent to 0.777 moles CO/mole Cu$^+$ (m = 1.29), C_{eq} is calculated to be 1.25 by equation 16-31 and p_{co} is calculated to be 0.78 atm by equation 16-29. This may be compared with an experimental value of about 0.84 atm obtained by van Krevelen and Baans. They show that the equations also fit the data of Larson and Teitsworth remarkably well.

A much simpler, though not as rigorous, relation between carbon monoxide pressure and solution composition is presented by Zhavoronkov and Chagunava (1940), and a nomograph for quickly determining the approximate quantity of carbon monoxide absorbed in various copper-ammonium carbonate solutions at 25°C and 1 atm carbon monoxide pressure has been presented by Egalon et al. (1955).

In addition to the carbon monoxide vapor pressure, the vapor pressures of other solution components are of interest in assisting the calculation of possible losses and heat requirements. Copper-ammonium carbonate solutions have an appreciably higher vapor pressure than solutions of the formate or salts of stronger acids because of the relatively high decomposition pressure of ammonium carbonate. The total vapor pressures of a number of solutions have been measured by Zhavoronkov and Reshchikov (1933) and Zhavoronkov and Chagunava (1940).

The vapor pressures of the individual components of typical plant solutions have been determined by Zhavoronkov (1939). One set of his data is presented in **Figure 16-23** as a plot of pressure versus temperature on a log p versus 1/T scale.

Heat Effects

The molar heats of vaporization of ammonia, carbon dioxide, and water from a typical copper-ammonium-salt solution (mixed formate and carbonate) have been calculated by Zhavoronkov (1939) using the Clausius-Clapeyron equation and the slopes of the log p versus 1/T lines as plotted in **Figure 16-23**. For the solution illustrated in this figure, he obtained the results in **Table 16-19**.

As indicated in **Table 16-19,** heat is required to release any of the volatile components from the solution, and conversely heat is liberated if they are absorbed. The absorption of carbon dioxide can, in fact, account for a considerable portion of the heat generated in the absorber.

The absorption of carbon monoxide is also exothermic, and its heat of absorption can be estimated by a similar analysis of available vapor-pressure data. Zhavoronkov and Reshchikov (1933) and Zhavoronkov and Chagunava (1940) have done this for the solutions that they studied and some of their results are presented in **Table 16-20**.

The calculated values for the differential heat of solution do not appear to differ significantly for formate, carbonate, or lactate solutions. A value of about 11,000 cal/g mole of CO

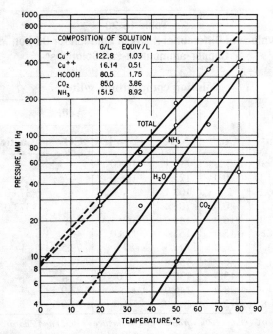

Figure 16-23. Vapor pressures of the individual components of a typical copper-ammonium formate solution. *Data of Zhavoronkov (1939)*

	Table 16-19 Approximate Heats of Vaporization of Copper-solution Components	
	Calculated heat of vaporization	
Component	**cal/g mole**	**Btu/lb mole**
Ammonia	9,280	16,700
Carbon dioxide	15,100	27,200
Water	12,610	22,700
Mixed vapors	10,500	18,900
Source: Mixed formate and carbonate solution, data of Zhavoronkov (1939)		

absorbed (19,800 Btu/lb mole) is believed to be a reasonable engineering approximation. Similar calculations based on the data of Hainsworth and Titus (1921) for a copper-ammonium carbonate solution indicate a heat absorption of the same magnitude.

Calculation of heat effects also requires a knowledge of the heat capacity of the solution. For engineering estimates, this can be taken to be 0.8 Btu/(lb)(°F).

Table 16-20
Average Values of Differential Heats of Solution for Carbon Monoxide in Copper-ammonium-salt Solutions*

Solution	Solution composition, g/liter						Diff. heat of sol. cal/g mole
				Acid			
	Cu^+	Cu^{++}	NH_3	Formic	CO_2	Lactic	
1	119.0	2.5	117.1	166.0	—	—	12,000
2	104.1	20.0	123.9	166.0	—	—	9,600
3	106.2	2.5	103.1	—	—	373.0	11,600
4	88.7	20.0	106.4	—	—	373.0	10,600
5	123.1	16.0	145.2	162.5	—	—	12,200
6	88.2	15.0	163.4	128.0	40.7	—	9,600
7	103.1	26.0	141.3	87.5	79.5	—	10,500
8	101.3	40.0	156.8	37.5	47.0	—	9,200
9	94.3	13.0	136.8	0.0	85.4	—	13,100
10	131.4	28.0	145.9	158.7	83.0	—	9,400
11	91.8	12.0	148.9	71.0	52.0	—	13,200

*Temperature, 0–80°C. Data of Zhavoronkov and Reshchikov (1933) and Zharoronkov and Chagunava (1940)

Solution Composition

The optimum solution composition is determined by a delicate balance of many factors including capacity for CO, stability, and cost. Since the primary active ingredient is the cuprous ion, it is desirable to have the solution as concentrated in respect to this component as possible. Ammonia increases copper solubility and also increases the effectiveness of the cuprous ion for absorbing carbon monoxide; however, the allowable ammonia concentration is limited by ammonia vapor-pressure considerations. The acid ion is necessary to maintain the copper in solution. The most inexpensive acid is carbonic acid, but carbonate solutions have higher ammonia and CO_2 vapor pressures and cannot hold as much copper as do solutions of organic acids such as formic, acetic, or lactic. The organic acids are subject to decomposition and loss; however, their advantages make their use justifiable in many cases. The cupric ion is also a necessary component of the solution. However, too high a concentration is uneconomical as this ion is inactive as a carbon monoxide absorbent.

A study of copper solubility and the stability of carbonate and formate solutions was made by Pavlov and Lopatin (1947). These authors found the maximum possible concentration of copper to be 160 g/liter in carbonate solutions and about 210 g/liter in formate solutions. However, these values require uneconomically high ammonia concentrations. They recommend concentrations within the ranges given in **Table 16-21**. The lower portions of the ranges are preferred to minimize ammonia loss.

Chemical analysis of solution samples from several operating formate plants showed total Cu concentrations from 103 to 150 g/liter; HCOOH concentrations from 55 to 88 g/liter; and

Table 16-21
Recommended Composition Range for Copper-Ammonium Formate Solutions

Component	Concentration Range, g/l
Total Cu	135–175
Cu^{++}	15–40
HCOOH	90–120
CO_2	60–70
NH_3	110–150

Data of Pavlov and Lopatin (1947)

NH_3 concentrations from 125 to 151 g/liter. Data on acetate solution composition are given with other operating data in **Table 16-22,** which is based on a plant formerly operated by the Tennessee Valley Authority.

Plant solutions are normally prepared by dissolving metallic copper in a mixture of ammonia, acid, and water. The use of distilled water is desirable since chlorides or sulfates, which may be introduced with less-pure water sources, can result in corrosion. It is necessary to blow air into the dissolver to oxidize the copper to the cupric state, and the cupric copper that is formed is capable of dissolving additional elemental copper by oxidizing it to the soluble cuprous form. The reaction is also of value as a means of controlling the cupric/cuprous ion ratio in the solution as it is made. Careful control of ammonia and acid concentrations is also required. An excess of acid relative to ammonia can cause the solution to become corrosive, while insufficient ammonia or acid can result in the precipitation of copper compounds.

Absorption

According to van Krevelen and Baans (1950), the reaction of CO with solution components is very fast. Since the physical solubility of CO in water is very low, the reaction zone in the liquid coincides with the liquid-gas interface. The rate of absorption, which is quite rapid, is determined entirely by the mass transfer of reactants and reaction products to and from the interface.

Yeandle and Klein (1952) reported that a typical CO-absorption unit operating at 1,800 psig will reduce the CO content of the gas stream from about 3.5% to less than 25 ppm, using about 75 gpm of copper-ammonium formate solution for 4,000 scf/min of feed gas. They claim that in practical plant experience with a solution inlet temperature of 0° to 5°C (32° to 41°F), about 70% of the theoretical CO/Cu^+ ratio of 1:1 can be attained.

Detailed operating data on the absorption stages of a plant utilizing a solution containing acetic acid (as well as CO_2 and some formic acid) are presented in **Table 16-22.** The absorber in this plant is 3-ft in diameter by 61-ft high and is packed with two 21-ft sections of 2-in. steel raschig rings. It is followed by a secondary scrubber, 2-ft in diamter by 47-ft high, which is packed with two 17-ft sections of 1-in. steel raschig rings. The secondary scrubber employs a solution fortified with excess ammonia to remove the last traces of CO_2. The solution from this column is recirculated until it begins to lose CO_2-absorption capacity;

Table 16-22
Typical Plant-Operating Data for Absorption of CO and CO_2 in Copper-ammonium-salt Solution

Operating Variables	July	December
Temperatures, °F:		
Gas entering primary scrubber	106	87
Gas leaving secondary scrubber	46	36
Solution entering primary scrubber	32	30
Solution leaving primary scrubber	69	65
Pressures:		
Gas entering primary scrubber, psig	1,635	1,750
Gas leaving secondary scrubber, psig	1,535	1,710
Flows:		
Gas entering primary scrubber, cfm*	8,350	9,800
Gas leaving secondary scrubber, cfm*	8,050	9,425
Gas dissolved in solution (calculated), cfm†	300	375
Solution entering primary scrubber, gpm	117	119
Solution entering secondary scrubber, gpm	7	7
Ammonia addn. to copper solution, lb/min	0.8	0.7
Ammonia production in synthesis section, lb/min	163	189
Compositions:		
Gas†† entering primary scrubber:		
CO_2, % §	0.81	0.72
CO, % §	2.70	2.87
Gas entering secondary scrubber, CO, ppm	<5	<5
Gas leaving secondary scrubber:		
CO_2, ppm §	<5	<5
CO_2 + CO, ppm §	<5	<5
H_2, % ¶	74.6	74.7
CH_4, % ¶	0.2	0.2
N_2, % ¶	25.2	25.1
Solution entering primary scrubbers:		
Temp. of sample when sp. gr. was measured, °F	88	82
Specific gravity	1.153	1.157
Cu^{++}, g/liter	22.2	21.2
Cu^{+}, g/liter	84.4	90.3
Total Cu, g/liter	106.6	111.5
Cu^{+}/Cu^{++}, g/liter	3.80	4.26
NH_3, g/liter	150.1	147.8
HCOOH, g/liter	8.5	10.9
CO_2, g/liter	100.2	92.7
CH_3COOH, g/liter	51.1	55.0
H_2O, g/liter	736.2	739.1
Free NH_3, eq/liter	3.24	3.33

Table 16-22 (Continued)
Typical Plant-Operating Data for Absorption of CO and CO₂ in Copper-ammonium-salt Solution

Operating Variables	July	December
Free NH_3/total Cu, eq/liter	1.60	1.59
Ammonia activity §	3.0	3.0
Solution entering secondary scrubber:		
Temp. of sample when sp. gr. was measured, °F	77	72
Specific gravity	1.091	1.092
Cu^{++}, g/liter	45.7	28.0
Cu^{+}, g/liter	22.9	63.0
Total Cu, g/liter	68.6	91.0
Total NH_3, g/liter	212.3	208.1
CO_2, g/liter	96.3	76.9
Free NH_3, mole %	17.5	18.0

*Gas volumes are reduced to 1 atmosphere and 60°F (calculated flows).
†This gas contains some H_2, CH_4, and N_2.
††Gas entering the primary scrubber contains approximately 0.1 percent oxygen. Dissolved oxygen in water used in CO_2 scrubber is the source of this oxygen.
§These analyses were made according to the procedures described by Brown and coworkers (1945)
¶This analysis was made with an Orsat gas analyzer.
Source: Walthall (1958)

at that time it is drained to the regeneration system and replaced with regenerated solution (Walthall, 1958).

Solution Regeneration

Regeneration of the copper-ammonium-salt solution is accomplished primarily by pressure reduction and the application of heat. Unfortunately, these operations have other effects, e.g., the vaporization of ammonia and the production of side reactions, which result in the requirement for a somewhat complex regenerator design. The regeneration temperature should be below 180°F in order to minimize the vaporization of ammonia and occurrence of side reactions. The pressure of regeneration should be as low as economically feasible. Operation at 1 atm is probably most common.

The temperature limitation of regeneration (below 180°F) satisfactorily prevents undue losses of ammonia and solution decomposition but, unfortunately, also limits the degree of carbon monoxide removal that is attainable by simple evaporation. The last traces of carbon monoxide can be removed from the solution, however, by permitting the reaction in equation 16-26 to proceed. By this reaction, carbon monoxide is oxidized to carbonate by the cupric ion, which is itself reduced to the cuprous state. This reaction is very slow at room temperature but quite rapid at 170° to 180°F, so that it can be made to proceed most satisfactorily while the solution is still at regeneration temperature.

1356 Gas Purification

According to Egalon et al. (1955), a solution hold-time of 15 to 20 minutes is sufficient to ensure that the last traces of carbon monoxide will be oxidized. A French regeneration-system design, described by these authors, is shown in **Figure 16-22**. In this design, the temperature at the top of the scrubbing section is 40°C (104°F). The solution flows out of the bottom of this section at 65°C (149°F) and into the bottom of the lower section of the column. The solution flows upward through the tubes of the first heat-exchanger section, which serves as a preheater, then through the tubes of the second steam-heated exchanger, which serves as an evaporator. Vapors from the evaporator pass through a chimney tray and directly up through the scrubber zone while the solution overflows at a temperature of 72°C (161.6°F) into an insulated holding tank, where removal of the last traces of carbon monoxide is achieved by oxidation with cupric ions. This reaction is slightly exothermic, and the solution leaves the holding tank at approximately 77°C (170.6°F). No provision is made in this design for the addition of air to adjust the cupric/cuprous ion-ratio because, in the plant for which the system was designed, sufficient air was present in the gas feed to the absorber to reoxideize the necessary amount of cuprous ion. In other designs, air is added at the base of the regenerator column or in a separate vessel if dilution of the carbon-monoxide stream is undesirable.

Operating Problems

Principal operating problems of the copper-ammonium-salt CO-removal process may be listed as follows:

1. Loss of active solution components
2. Control of cuprous/cupric ion-ratio
3. Recovery of pure carbon monoxide or a carbon monoxide-hydrogen mixture
4. Formation of precipitates
5. Corrosion

Loss of ammonia with the purified gas from the absorber can be minimized by operating at a sufficiently high pressure and with a low-temperature solution. The loss that is incurred in this operation can be estimated from partial-pressure data such as those presented in **Figure 16-23**. Ammonia losses can be more serious in the regeneration step because of the higher temperatures and lower pressures involved, and, as mentioned in the preceding section, several process schemes have been developed to minimize losses from this operation. When ammonia is recovered by washing the exhaust gases with the rich copper solution, it is important that this solution be as cool as possible. Since a temperature rise occurs in the absorber, this in turn means that the lean solution fed to the absorber should be at the lowest possible temperature. Plants generally operate with the lean-solution temperature in the range of 32° to 68°F with perhaps 40°F a preferred value.

When copper-ammonium carbonate is employed in the solution, the carbonate content equilibrates at a satisfactory operating level as a result of the absorption of CO_2 from the gas being purified and the stripping of an equal quantity in the regeneration system. With copper-ammonium formate, on the other hand, a gradual loss of formic acid occurs due to oxidation to carbon dioxide. In the presence of cupric ion, it would be expected that the reaction would proceed as follows:

$$HCOOH + 2Cu^{++} + 2(OH)^- = CO_2 + 2H_2O + 2Cu^+ \qquad (16\text{-}32)$$

If formic acid is lost more rapidly than ammonia or if it is not replaced at the rate it is lost, the resulting excess of ammonia reacts with carbon dioxide so that the solution gradually approaches the copper-ammonium carbonate composition. This problem can be minimized by the use of more stable organic acids such as acetic or lactic. Cuprous lactate has also been reported to have the advantage of a high stability to reduction by CO (Gump and Ernst, 1940).

The cuprous/cupric ion-ratio can generally be controlled at the desired value (about 5:1 on a weight basis) by adding the proper amount of oxygen with the feed gas or as air to the regenerator. In some cases where excess oxygen is present, control of the ratio by reduction of Cu^{++} with CO is required.

In a conventional regeneration system, the CO produced is contaminated with CO_2 and also with nitrogen (if air is added to the regenerator). If pure CO or a $CO-H_2$ mixture is desired for recycle to the shift converter, special steps are required. The extent to which these are warranted depends on the relative value of the recovered CO. Some increase in purity can be realized by adding air in a separate vessel (similar to the hold tank in **Figure 16-22**), so that the gas from the primary regenerator is at least nitrogen-free. A relatively CO_2-free stream of $CO-H_2$ can be obtained by stripping in stages.

The copper-ammonium-salt solutions are generally not corrosive to mild steel; however, the gases evolved during regeneration can be quite corrosive due to the presence of carbon dioxide. To prevent corrosion in the vapor zones, it is good practice to use stainless steel (vessels or liners) in the exhaust-gas scrubber section and above the liquid in the evaporator section.

COSORB Process

The COSORB process, developed by Tenneco Chemicals, Inc. in the early 1970s, was initially aimed at the recovery of high-purity carbon monoxide from nonconventional gas stream sources (Haase et al., 1982). However, since its introduction, the process has been licensed for carbon monoxide removal and recovery from a wide variety of gas streams, including those from hydrocarbon-steam reforming, natural gas partial oxidation, coke ovens, and coal gasification. KTI became the owner and exclusive licensor of the process in 1983 and it is now offered by KTI COSORB Ltd., Inc. of Houston, Texas. The process is reported to be capable of removing more than 99.5% of the carbon monoxide from gas streams while producing a carbon monoxide product that is over 98% pure (KTI, 1994).

The COSORB process resembles the copper-ammonium salt process described in the preceding section in that it also uses a copper compound that forms a complex with absorbed CO. It is significantly different, however, in that the absorbent is nonaqueous. The active component is cuprous aluminum chloride ($CuAlCl_4$) dissolved in an aromatic base, toluene (Haase, 1975). The absorption-desorption reaction can be represented by the following equation:

$$CuAlCl_4C_7H_8 + CO = CuAlCl_4CO + C_7H_8 \tag{16-33}$$

The reaction proceeds to the right during absorption and to the left when the complex is heated during regeneration.

The absorbent is inert to gases such as hydrogen, carbon dioxide, methane, and nitrogen normally found in synthesis gas, although these compounds exhibit some degree of physical solubility in the toluene base. Water in the feed gas reacts quantitatively with the active ingredient in the absorbent to form HCl gas and a waste product that is soluble in the solvent. Because of this reaction, it is necessary to predry the feed gas to a COSORB unit, preferably to less than 1 ppm H_2O. Other gaseous impurities, which may be present in the feed gas such as hydrogen sulfide, sulfur dioxide, ammonia, methanol, and olefins can also react irreversibly with the solvent and should be reduced to low levels before the gas enters the COSORB process absorber.

The COSORB process has the advantage over cryogenic and adsorption processes for CO removal and recovery of being based on chemical selectivity rather than physical property differences. It is capable of producing a purer CO product than cryogenic processes when the feed gas contains nitrogen because nitrogen and carbon monoxide are quite difficult to separate by distillation. Similarly, it produces a purer CO product than adsoprtion (PSA) when the feed gas contains methane because methane is adsorbed even more strongly than carbon monoxide. However, when the principal objective is hydrogen purification, cryogenic and adsorption processes are usually more economical.

A simplified flow diagram of a COSORB unit is shown in **Figure 16-24.** The overall process generally includes: (1) feed gas preparation, (2) carbon monoxide absorption (complexing), (3) carbon monoxide desorption (decomplexing), (4) aromatic solvent recovery from effluent gas streams, and (5) compression of CO product stream. Steps (1) and (5) are not shown on the flow sheet.

The feed gas preparation section typically includes refrigeration and molecular sieve adsorption for water removal and activated carbon adsorption for removal of sulfur compounds to acceptable levels. The pretreated feed gas enters the bottom of the absorber where it is contacted countercurrently with the active solvent. Gas leaving the top of the absorber, with a residual carbon monoxide concentration in the ppm range, is treated for solvent recovery.

The flow diagram shows a preflash system, which is considered optional in the process. Rich solution from the absorber is flashed to a lower pressure in this unit to remove physically dissolved gases. The released gases are washed with lean solvent to reabsorb any carbon monoxide and may be further treated for aromatic solvent recovery before exiting the process.

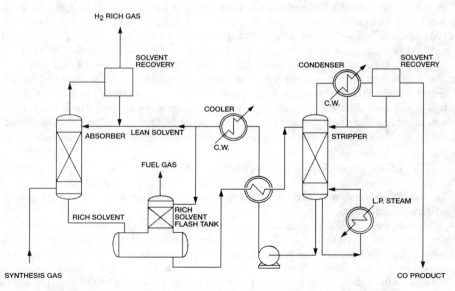

Figure 16-24. Simplified diagram of the COSORB process. (*KTI, 1994*)

The rich solvent is next heated by heat exchange with regenerated solvent and fed to the top of the stripper. This column operates with a steam heated reboiler and a water cooled condenser. The solvent is heated to about 135°C (275°F) in the stripper. Carbon monoxide offgas passes through the condenser then through an aromatic solvent recovery unit before release as product. Hot lean solvent from the stripper is pumped through the heat exchanger and a cooler before returning to the top of the absorber.

Because of the high volatility of the aromatic solvent, it is necessary to include aromatics recovery operations on all product gas streams. This is accomplished for the absorber and stripper off gas streams by a combination of compression, refrigeration, and treatment with activated carbon. Chilling is generally adequate to recover aromatics from the small amount of gas produced by the preflash system.

It is reported that mild steel is generally applicable as the material of construction for COSORB units; however, some specifically selected stainless steels, copper nickel alloys, and brass are also acceptable (Haase et al., 1982).

HYDROCARBON REMOVAL BY OIL ABSORPTION

The removal of hydrocarbon vapors from gas streams by absorption in liquid oils is an important part of many industrial operations. In some cases, such as natural gasoline recovery, the absorption step is but a portion of a refining process that produces several commercial products, and the gas purification aspects of the operation are of little importance. In other cases, such as the removal of aromatics from coke-oven gas, the absorption process serves to improve the value of the product gas. Processes for the removal of light oil (primarily benzene) and naphthalene from coke-oven gas are described in this section.

In general, the design of hydrocarbon absorption systems is straightforward. Since mass transfer is not complicated by the occurrence of chemical reactions, conventional absorption coefficient, theoretical plate, and absorption factor concepts can be used for design calculations (see Chapter 1). Basic data for such calculations, including the thermodynamic properties of compounds found in coke-oven gas and equilibrium data for several gas-coal liquid systems, are given in the U.S. DOE *Coal Conversion Systems Data Book* (1982) and other hydrocarbon data compendia.

Light Oil Removal from Coal Gas

The principal sources of coal gas are coke ovens operated to produce coke for iron and steel-producing operations. The light oil content of the gas depends upon both the manner of carbonization and the nature of the coal. Typically, it accounts for about 1% (by volume) of the raw gas stream, and the recovered liquid has a composition as indicated by **Table 16-23**.

As indicated by the analysis, benzene is by far the most abundant compound in the light oil. For this reason the total liquid recovered is sometimes referred to as benzol. The marketable benzene products of specified purity are also referred to as benzol (e.g., 1° benzol, industrial benzol, etc.), and the word "toluol" has the same significance with regard to toluene. Carbon disulfide can also be removed from the gas stream during the light oil recovery operation, and, in some installations, which are specifically designed for this service, carbon disulfide removal is of equal importance to benzol recovery.

A simplified flow diagram of a coke by-product recovery plant is shown in **Figure 16-25**, which is based on the paper of Kroll and Barry (1991). A flow diagram of a typical light oil recovery section is given in **Figure 16-26**. Light oil removal is accomplished by washing the

Table 16-23
Typical Composition and Yield of Crude Light Oil from Coke-oven Gas Operation

Constituent	Original Conc., gal/ton coal	Conc. in Crude light oil, %
Benzene	1.85	57.8
Toluene	0.45	14.1
Xylenes and light solvent naphtha	0.30	9.4
Unsaturated hydrocarbons, etc.	0.16	5.0
Naphthalene and other heavy hydrocarbons	0.24	7.5
Wash oil	0.20	6.2
Total crude light oil	3.20	100.0

Source: Porter, 1924

gas with lean oil in a countercurrent absorption tower. Rich oil from the bottom of the absorber is pumped through heat exchangers in which it is heated first by the vapors leaving the top of the still, then by the lean solution leaving the bottom of the still, and finally by steam, before it is fed into the still (stripping column) near the top. In the process illustrated, the still is operated under vacuum, although atmospheric pressure operation is more conventional. Steam is admitted into the bottom of the still to provide stripping vapor for the light oils and thereby to minimize the temperature required for adequate stripping.

Lean oil from the still is pumped through a heat exchanger and cooler and back to the absorber. Vapors from the still are cooled with partial condensation by indirect heat exchange with the rich oil. The condensate is passed to a separator, which removes water, and the liquid hydrocarbons are returned to the still as reflux. The partially cooled vapors are then passed to the benzol condenser where relatively complete condensation is obtained by the use of cooling water. In this particular plant, the condensate and noncondensed vapors are then passed through a vacuum pump to a separator. From this, noncondensed gases are returned to the fuel gas main, water condensate is removed for disposal, and the hydrocarbon layer is transferred to the crude-benzol storage tanks.

Recent national standards for hazardous air pollutants have placed strict controls on benzene emissions from coke-oven facilities. These regulations result in the requirement to enclose and seal all process vessels and tanks and to direct all vent streams to a collection system where benzene can be recovered and destroyed. The design of vapor control systems for coke-oven byproduct recovery plants to minimize the emission of benzene to the atmosphere is discussed by Kroll and Barry (1991).

Wash Oils

Although a large number of hydrocarbon liquids have been proposed for absorbing the light oil fraction of coal gas, the most commonly used materials are petroleum liquids (gas, oil, and straw oil) and coal-tar fractions (creosote oil). The coal-tar oils have the obvious advantage of being available at any coke plant. However, they have several disadvantages, chief among these being their tendency to become more viscous with use. Because of this

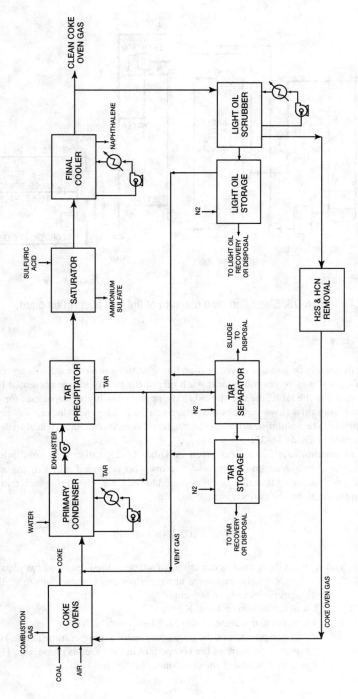

Figure 16-25. Simplified flow diagram of a coke-byproduct recovery plant. *(Kroll and Barry, 1991)*

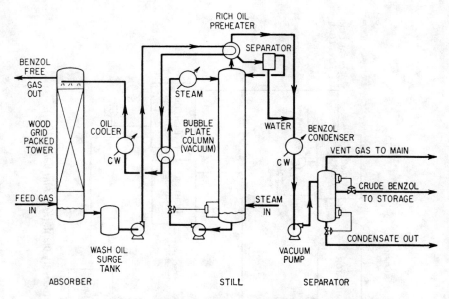

Figure 16-26. Simplified flow diagram of light oil absorption plant.

factor, and frequently because of cost considerations, petroleum-base wash oils are generally favored. The properties of several typical wash oils of the two types are presented in **Table 16-24**. As shown in the table, the coal-tar oils have appreciably higher capacities for benzene than the petroleum oils. However, this advantage can be lost by the "thickening" phenomenon noted above. The solubilities of important light oil constituents in a typical hydrocarbon oil are presented in **Table 16-25**.

The use of partition coefficients (as given in **Table 16-25**) rather than equilibrium constants was quite common among early investigators who produced much of the available data on coal gas systems. If Raoult's law is assumed to hold, i.e., p = xP for each component, the partition factor, k, can be defined as follows:

$$k = c_L/c_G = (62,300 \, s \, T)/(MP)$$

Where: c_L and c_G = equilibrium concentrations of solute in the liquid and gas phases, respectively, expressed in weight per unit volume units (e.g., lb/cu ft)
s = specific gravity of solvent
T = absolute temperature, K
M = molecular weight of solvent
P = vapor pressure of the pure component at temperature T, mm Hg
p = partial pressure of the component in the gaseous phase, mm Hg
x = mole fraction of the component in the liquid

Table 16-24
Properties of Typical Petroleum and Coal-tar Oils Used for Benzol Recovery

Wash Oil (fresh)	Distillation Range, °C	Specific Gravity (20/20°C)	Viscosity, 20°C, cp	Volatility (loss to gas) at 20°C		Absorptive capacity, concentration of benzene in liquid in equilibrium with benzene vapor at 6.5 mm Hg pressure	
				g/cu m	gal/million cu ft	g/100 g solution	vol./100 vol. absorbent
Creosote oils:							
Light fraction (free from tar acids)	205–265	0.9535	25	2.8	18.3	4.55	5.15
Light (tar acids 17.5% by volume)	200–300 (95%)	1.0135	35	1.9	11.7	3.95	4.75
Medium (tar acids 20% by volume)	200–350 (90%)	1.031	51	2.0	12.1	3.6	4.35
Petroleum-gas oils:							
Light fraction, (a)	214–285 (95%)	0.8295	36	1.7	13.0	3.2	3.1
Fraction (b)	210–400 (95%)	0.8635	70	1.75	12.5	2.9	2.95
Fraction (c)	260–365 (95%)	0.849	92	0.9	6.7	2.8	2.8

Source: Hoffert and Claxton, 1938

Measured partition coefficients can be used to calculate apparent molecular-weight values for the solvent by application of equation 16-34; this has been done for the data of **Table 16-25.** Since the apparent molecular-weight values do not vary over a very wide range, they can be interpolated readily to provide a reasonably accurate basis for estimating partition coefficients at intermediate temperatures or for alternate components (of a similar nature).

In addition to capacity, the following factors are of importance in the selection of a suitable wash oil:

1. The specific gravity should be far enough removed from that of water to permit satisfactory settling without the formation of an emulsion.
2. The viscosity should be as low as possible.
3. The initial boiling point of the oil should be as high as possible to minimize vaporization into the purified gas and to permit separation of the light oil and wash oil to be accomplished readily.

Table 16-25
Solubilities of Light-oil Constitutents in Typical Petroleum Oil*

Solute	Solubility (Partition coefficient, k) lb per cu ft in liquid/ lb per cu ft in gas			Apparent molecular wt. of oil		
	25°C (77°F)	80°C (176°F)	130°C (266°F)	25°C (77°F)	80°C (176°F)	130°C (266°F)
Carbon disulfide	235	56	22	187	161	143
Benzene	650	114	36	260	215	199
Toluene	2,030	280	79	272	228	193
Xylene	7,570	716	170	248	222	200
Naphthalene	171,000	8,620	1,320	321	276	251

*Designated "Benzol Absorption Oil," specific gravity at 20°C (68°F), 0.871; viscosity at 20°C (68°F), 12.4 cp; boiling point: first drop 270°C (518°F), 50 percent 331°C (627.8°F), 75 percent 380°C (716°F).
Source: Data of Silver and Hopton, 1942

Absorber Design

In view of the long history of light oil recovery (dating back to before 1889), it is not surprising that a great number of absorber designs have been developed. These include horizontal multi-chamber scrubbers and mechanical washers as well as conventional countercurrent columns. Currently used absorbers are generally packed or spray columns as described in Chapter 1. Wood grid packing was formerly used in packed towers, but it has been largely displaced by newer low-pressure drop, metal-packing designs.

Since the equilibrium curves for light-oil components are reasonably close to linear and the operating line is relatively straight in typical light-oil absorption operations, simplified design equations such as equation 1-11, based on K_Ga; equation 1-17, based on transfer units; and equation 1-18, based on theoretical plates, can be employed. Modified forms of these equations based upon weight/volume partition factors (see **Table 16-25**) rather than mole fractions have been used by a number of investigators. In these units, the parameter L_M/mG_M of equations 1-17 and 1-18 becomes Lk/G, where the gas and liquid rates are based on volume and k is the partition coefficient c_L/c_G. As with the parameter L_M/mG_m, an Lk/G value of 1.0 represents the minimum liquid/gas ratio which can result in complete removal of the component in a tower of infinite height. In actual practice, Lk/G values on the order of 1.3 are typical for light oil absorption. The required flow ratio of oil to gas for obtaining essentially complete removal of carbon disulfide, benzene, toluene, xylene, or napththalene with the typical petroleum fraction used to obtain the data in **Table 16-25** can be calculated directly by dividing the assumed Lk/G value (e.g., 1.3) by the partition coefficients given in the table. Such calculations show that an oil rate of approximately 35 gal/1,000 cu ft of gas is required to provide substantially complete removal of carbon disulfide (at 77°F) as compared to 13 gal/1,000 cu ft for benzene removal and only 4 gal/1,000 cu ft for toluene removal.

Stripper Design

The rich "benzolized" wash oil is stripped of its light oil content by fractional distillation—most commonly in a tray column operated at substantially atmospheric pressure with direct-steam addition to the bottom. The number of theoretical trays required for the separation can be calculated by conventional techniques and the actual requirement estimated on the basis of an assumed tray-efficiency. Overall efficiencies of 45 to 70% have been reported as typical for light oil stripper stills (Silver and Hopton, 1942; Glawacki, 1945), and 10 to 15 actual trays are commonly employed below the feed.

Raschig-ring packed stripping stills are also used to some extent and data on several such columns have been presented by Silver and Hopton (1942). For purposes of design, these authors recommend a K_La for benzene of between 5 and 8 lb/(hr)(cu ft)(lb/cu ft) [or lb moles/(hr)(cu ft)(lb mole/cu ft)] for 1½-in. rings at oil rates from 100 to 150 gal/(hr)(sq ft) and superficial steam velocities of from 0.2 to 0.3 fps. Both liquid- and steam-flow rates are based on the empty column, and the steam-velocity values assume that no other vapors are present.

Operating Data

Operating data from four plants were presented by Silver and Hopton (1942). A summary of data on two of these plants, which illustrates the performance of commercial installations designed primarily for benzol removal and sulfur removal, respectively, is presented in **Table 16-26**.

Table 16-26
Light-Oil Recovery Plant Operating Data

Plant Variables	Plant A*	Plant B†
Gas-flow rate, scf/hr	45,700	264,500
Oil circulation rate:		
gal/hr	510	9,000
gal/1,000 cu ft	11.2	34.1
Mean wash temperature, °F	68.9	63.5
Steam-flow rate:		
lb/hr	201	1,080
lb/gal of oil	0.39	0.12
Mean still temperature, °F	239.9	176
Still pressure, psia	15.4	4.5
Light-oil recovery:		
gal/10,000 cu ft	2.23	2.60
percent	88	87
Carbon disulfide in feed, grains/100 cu ft	18.7	22.8
Carbon disulfide recovery, %	52	91

*Designed primarily for benzol recovery.
†Designed primarily for carbon disulfide recovery.
Source: Silver and Hopton, 1942

The principal operating difficulties in light oil recovery plants are wash-oil deterioration, as evidenced by an increase in viscosity or the formation of sludge, and corrosion—which has been shown to be due primarily to the presence of ammonium thiocyanate in the oil (Cawley and Newall, 1945). The formation of sludge is believed to be due to oxidation and/or polymerization of certain coal-gas constituents such as indene or styrene. The sludge deposits interfere with plant operation by coating heat-transfer surfaces and tower packings. The sludge problem can be minimized by prior treatment of the gas with an electrostatic precipitator to remove sludge-forming components that are present in particulate form; however, a more practical approach is to regenerate the wash oil. One successful method of accomplishing this involves the use of steam distillation of a small slip stream of the wash oil which is bypassed into a separate vessel. A relatively large amount of steam must be passed through the oil to vaporize the wash-oil components. However, the steam and wash-oil vapors are passed directly into the main stripper so that their heat content is not lost. Nonvolatile sludge residue is periodically discarded from the regenerator vessel.

Naphthalene Removal from Coal Gas

The removal of naphthalene from coal gases is of considerable importance because, unlike other hydrocarbons present in the coal gases, naphthalene condenses as a solid when the temperature of the gas is lowered, resulting in the plugging of processing equipment and distribution lines. Naphthalene is formed in varying quantities during coal carbonization, depending on the carbonization temperature. In typical high-temperature carbonization operations, the gas may contain as much as 250 grains of naphthalene per 100 cu ft, while less than 100 grains/100 cu ft may be formed under low-temperature carbonization conditions. Because of its high boiling point (218°C), most of the naphthalene condenses with the tar in the primary gas coolers. However, appreciable amounts, typically 15 to 50 grains/100 cu ft, remain in the gas after the primary cooling step. In order to protect distribution lines against deposition of solid naphthalene, it is customary to remove the naphthalene so that the final concentration is about 2 grains/100 cu ft (Hopton, 1953).

Process Description

A closed system for removing naphthalene from coke-oven gas in the final cooler is described by Kroll and Barry (1991). In this system, recycled water and fresh water are sprayed into the process gas at high flow rates to cool and saturate the gas with water. Naphthalene is condensed and collected with the water. The naphthalene is removed from the water by contact with liquid tar in a liquid-liquid extraction step at the bottom of the spray contactor. After naphthalene removal the water is pumped through an indirect heat exchanger (cooler) and recycled with fresh water makeup to the spray nozzles.

Naphthalene is adequately removed in the oil-washing operation for benzol recovery, and a separate naphthalene removal step is not required if benzol is recovered. This type of operation is shown in the overall coke byproduct recovery flow diagram (**Figure 16-25**) where residual naphthalene in the gas from the final cooler is removed by the light oil scrubber. However, in some cases it is advantageous to remove naphthalene in a separate step before the gas is processed for the removal of benzol and other impurities. This is usually accomplished by oil washing.

The types of oils used for light-oil absorption are also suitable for naphthalene removal. Petroleum oils which are free of naphthalene are preferable from the standpoint of maximum naphthalene removal. However, coal-tar oils have a somewhat higher absorptive capacity.

A schematic flow diagram of a naphthalene-removal installation is shown in **Figure 16-27**. The gas is washed with the oil in an absorber of special design, permitting efficient contact between a large volume of gas and a small quantity of liquid. Because of the similarity of the operation, the same absorber designs are used for naphthalene removal as for benzol absorption. Countercurrent towers containing two stages of packing are normally used. In the lower section of the tower, where the bulk of the naphthalene is removed, the gas is washed with partially rich oil which is recirculated at a high rate. The partially purified gas enters the upper section of the tower where the naphthalene content is reduced to the desired level by contact with a small stream of fresh oil. The fresh oil is required at such a low rate that continuous countercurrent addition would result in poor contact between the gas and the liquid, and the fresh oil is, therefore, often injected intermittently at a sufficiently high rate to wet the packing of the upper tower section. The oil leaving the upper section of the absorber joins the recirculating oil stream in the lower section, thereby providing fresh oil for bulk naphthalene absorption. Venturi scrubbers are sometimes used as naphthalene absorbers. Rich oil is continuously withdrawn from the system at the same rate at which fresh oil is added; it is then stripped of naphthalene by steam distillation and reused, as shown in **Figure 16-27**. A convenient way of disposing of the stripped naphthalene consists of returning the stripper overhead vapors to the gas-collecting main ahead of the primary cooler in order to condense the naphthalene with the tar in the primary cooler. However, the rich oil is not always regenerated and may be processed in a variety of ways, depending on the overall economics of the installation.

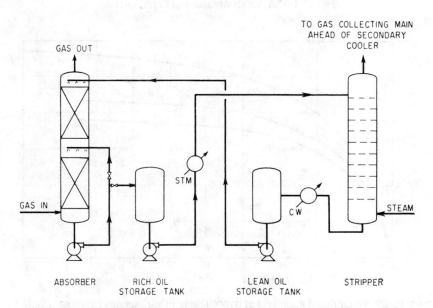

Figure 16-27. Simplified flow diagram of naphthalene-removal installations.

Process Design

Absorption columns can be designed by the use of simplified equations as described in the preceding section for light oil absorption. Because of the relatively low vapor pressure of naphthalene and its high solubility in various oils (see **Table 16-25**), very low oil rates are required to obtain a high removal efficiency. The vapor pressure of naphthalene as a function of temperature is shown in **Figure 16-28** and vapor/liquid equilibria for naphthalene and three different petroleum oils are presented in **Figure 16-29** (Speece, 1925). In general, oil-

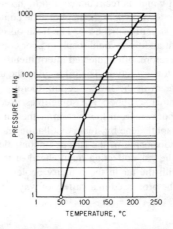

Figure 16-28. Vapor pressure of naphthalene.

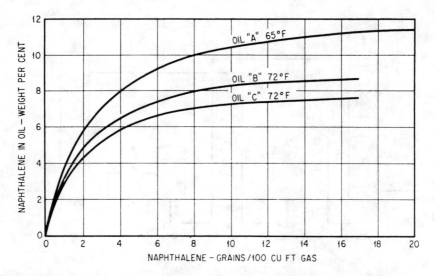

Figure 16-29. Vapor liquid equilibria of naphthalene in various oils (*Speece, 1925*). **Oil A:** Crude oil saturated with benzene. **Oil B:** residuum gas oil. **Oil C:** crude oil.

circulation rates of about 0.01 to 0.1 gal/1,000 cu ft of gas are required to reduce the naphthalene content of typical coal gases to less than 2 grains/100 cu ft. An oil-circulation rate of 0.035 gal/1,000 cu ft of gas has been calculated by Hopton (1953) for an Lk/G value of 1.3 by using partition factors for petroleum oil given in **Table 16-25**.

Stripping columns are also designed in the same manner as those used in light-oil recovery plants. Hopton (1953) recommends a steam rate of 16.5 lb/gal of oil for operation at 248°F and 1 psig pressure. This value is based on G/Lk value of 1.3 and the use of petroleum oil.

REFERENCES

Air Products and Chemicals, Inc., 1996, "Hydrogen (Cryofining)" in *Gas Processes '96, Hydro Process.,* April, p. 104.

Anerousis, J. P., and Whitman, S. K., 1984, "An Updated Examination of Gas Sweetening by the Iron Sponge Process," SPE 13280, September, Houston, TX.

Anon., 1989, *Chem. Eng.,* Vol. 96, No. 3, March, p. 17.

Avery, H. B., 1939, *Chem. & Ind.* (London), Vol. 58, p. 171.

Ayala, R. E., Gal, E., and Jain, S. C., 1991, "Development of Durable Sorbents for the GE Moving-Bed Hot-Gas Desulfurization Process," presented at the 1991 Summer National Meeting of the AIChE, Pittsburgh, PA, Aug. 18–21.

Badische, Anilin, and Soda, Fabrik, 1914, German Patent 289,694.

Bairstow, R., 1957, *Gas J.,* Vol. 289, p. 243.

Bhatia, K., and Allford, K. T., 1986, *Oil & Gas J.,* Oct. 20, p. 44.

Bhatia, K., and Brown, T. M., 1986, "Examination of Field Data From a New Gas Sweetening Process," presented at the 61st Annual Technical Conference and Exhibition of the Society of Petroleum Engineers, New Orleans, LA, Oct. 5–8, SPE 15456.

Brinkmeyer, F. M., and Delzer, G. A., 1990, "Z-Sorb—A New Sulfur Removal Process," presented at the 1990 AIChE Summer National Meeting, San Diego, CA, Aug. 19–22.

Brown, E. H., Cline, J. E., Felger, M. M., and Howard, R. B., Jr., 1945, *Ind. Eng. Chem., Anal. Ed.,* Vol. 17, pp. 280–282.

Bucklin, R. W., 1994, Bob Bucklin and Associates, personal communication.

Carnell, P. J. H., 1986, "Gas Sweetening with a New Fixed Bed Absorbent," *Proc. Laurance Reid Gas Conditioning Conf.,* March 3–5, University of Oklahoma, Norman, OK.

Cawley, C. M., and Newall, H. E., 1945, *J. Soc. Chem. Ind.,* (London), Vol. 64, p. 285.

Corbitt, R. A., Ed., 1990, *Handbook of Environmental Engineering,* McGraw-Hill, NY.

Dent, F. J., and Moignard, L. A., 1950, *The Purification of Town Gas by Means of Iron Oxide,* Gas Research Board, Commun. 52 (Gr. Brit.).

Dillon, E. T., 1991A, *Hydro. Process.,* December, p. 65.

Dillon, E. T., 1991B, "Gas Sweetening With a Novel and Selective Alkanolamine," *Proceedings of the Seventieth GPA Annual Convention.*

Din, F. (Ed.), 1956, *Thermodynamic Functions of Gases,* Vol. 1, Butterworth Scientific Publications, London, p. 164.

Dobbs, J. B., 1986, "Gas Sweetening with an Effective One-Step Process," *Proc. 1986 Gas Conditioning Conference,* Univesity of Oklahoma, Norman, OK.

Dorchak, T. P., Gangwal, S. K., and McMichael, W. J., 1991, "Development of the Direct Sulfur Recovery Process," presented at the 1991 Summer National Meeting of the AIChE, Pittsburgh, PA, Aug. 18–21.

Dunson, J. B., Jr., 1991, "Chemical Scrubbing for Control of Air Emissions," presented at AIChE Summer Meeting, Pittsburgh, PA, Aug. 18–21.

Egalon, R., Vanhille, R., and Willemyns, M., 1955, *Ind. Eng. Chem.,* Vol. 47, May, pp. 887–899.

Egan, C. J., and Kempt, J. D., 1937, *J. Am. Chem. Soc.,* Vol. 59, p. 1264.

Forrester, J. S. and Le Blanc, J. G., 1988, *Chem. Eng.,* Vol. 95, No. 8, May 23, p. 145.

Furimsky, E., and Yumura, M., 1986, *Erdol and Kohle—Erdgas—Petrochemie vereinigt mit Brennstoff-Chemie,* Vol. 39, No. 4, April, p. 163.

Furman, A. H., Kimura, S. G., Ayala, R. E., Cook, C. S., and Gal, E., 1991, "High Temperature Removal of Coal Gas Contaminants," presented at the 1991 Summer National Meeting of AIChE, Aug. 18–21.

Gangwal, S. K., Harkins, S. M., Woods, M. C., Jain, S. C., and Bossart, S. J., 1989, *Environ. Progr.,* Vol. 8, No. 4, November, p. 265

Gangwal, S. K., Gupta, R. P., and Jain, S. C., 1991A, "Enhanced Durability of Desulfurization Sorbents for Fluidized-bed Applications," Report on contract DE-AC21-88MC25006 at Eleventh Contractor's Review Meeting on Gasification and Gas Stream Cleanup, DOE/METC, August.

Gangwal, S. K., McMichael, W. J., and Dorchak, T. P., 1991B, *Environ. Progr.* Vol. 10, No. 3, August, p. 186.

Gangwal, S. K., McMichael, W. J., and Dorchak, T. P., 1990, "The Direct Sulfur Recovery Process," Presented at AIChE Summer National Meeting, San Diego, CA, Aug. 19–20.

Gas Sweetener Associates, Inc. 1991, *Operations and Technical Manual,* June.

Glowacki, W. M., 1945, *Chemistry of Coal Utilization* Vol. 2, Edited by H. H. Lowry, New York: John Wiley & Sons, Inc., p. 1195.

Goff, S. P., and Wang, S. I., 1987, *Chem. Eng. Progr.,* Vol. 83, No. 8, August, p. 46.

Griffith, R. H., 1954, *Gas J.,* Vol. 280, pp. 254–256.

Grindley, T., and Goldsmith, H., 1987, "Development of Zinc Ferrite Desulfurization Sorbents for Large-Scale Testing," Presented at AIChE Annual Meeting, New York, NY, Nov. 15–20.

Grindley, T., and Steinfeld, G., 1981, *Development and Testing of Regenerable Hot Coal-Gas Desulfurization Sorbents,* pp. 45–1125.

Grindley, T., and Steinfeld, G., 1985, "Desulfurization of Hot Coal Gas by Zinc Ferrite," in *Acid and Sour Gas Treating Processes,* Gulf Publishing Co., Houston, TX, p. 419.

Gump, W., and Ernst, I., 1940, *Ind. Eng. Chem.,* Vol. 22, April, p. 382.

Guntermann, W., and Schnurer, F., 1957, *Gas-u. Wasserfach,* Vol. 100, No. 25, pp. 643–649.

Gupta, R., Gangwal, S. K., and Jain, S. C., 1992, Preprint of paper for publication in *Energy and Fuels,* January.

Gupta, R. P., and Gasper-Galvin, L. D., 1991, "Screening of Supported Metal Oxides for Hot-Gas Desulfurization," presented at the 1991 AIChE Annual Meeting, Los Angeles, CA, Nov. 17–21.

Haase, D. J., 1975, *Chem. Eng., Vol. 82,* No. 16, Aug. 4, p. 52.

Haase, D. J., Duke, P. M., and Cates, J. W., 1982, *Hydro. Process.,* March, p. 103.
Hainsworth, W. R., and Titus, T. Y., 1921, *J. Am. Chem. Soc.,* 43-1-11.
Haldor Topsøe, Inc., 1986, *Topsøe Sulfur Absorption Catalysts, HYZ,* Brochure HTZ-10/86.
Harris, R. A., 1980, *Hydro. Process.,* Vol. 59, No. 8, August, p. 111.
Hill, J., 1990, *Chem. Eng.,* Vol. 97, No. 5, May, p. 133.
Hise, R. E., Massey, L. G., Adler, R. J., Brosilow, C. B., Garner, N. C., Brown, W. R., Cook, W. J., and Petrik, M., 1982, "The CNG Process, A New Approach to Physical Absorption Acid Gas Removal," paper presented at AIChE Annual Meeting, Nov. 14–18, 1992.
Hoffert, W. H., and Claxton, G., 1938, *Motor Benzole: Its Production and Use,* 2nd ed., London, The National Benzole Association.
Hopton, G. U., 1953, *The Cooling, Washing and Purification of Coal Gas,* London: North Thames Gas Board.
Houghton, J. E., and Bucklin, R. W., 1994, "Nonregenerable H_2S Scavenger Update," presented at 44th Annual Laurance Reid Gas Conditioning Conference, University of Oklahoma, Norman, OK, Feb. 27–March 2.
ICI Katalco, 1991, *Puraspec Purification Processes,* Brochure DH/991/1/Pur 9015.
Jalan, V., 1985, "High Temperature Desulfurization of Coal Gases," in *Acid and Sour Gas Treating Processes,* S. A. Newman, Editor, Gulf Publishing Co., Houston, TX.
Kirk, R. E., and Othmer, D. F. (Eds.), 1952, *Encyclopedia of Chemical Technology,* Vol. 9, Interscience Publishers, Inc., NY, p. 404.
Koch Engineering Company, Inc., 1986, *Static Mixing Application Report, Oil and Gas Production,* Brochure A-F1, Ps-248 (10/86).
Kroll, P. J., and Barry, J. M., 1991, "Control of Benzene Emissions from Coke By-product Plants," presented at the AIChE Summer National Meeting, Pittsburgh, PA, Aug. 18–21.
KTI (Kinetics Technology International Corporation), 1994, personal communication and brochure, *The KTI-COSORB Process.*
Larson, A. T., and Teitsworth, C. S., 1922, *J. Am. Chem. Soc.,* Vol. 78.
Leblanc, F., 1850, *Compt. Rend.,* Vol. 30, p. 483.
Maddox, R. N., 1974, *Gas and Liquid Sweetening,* Campbell Petroleum Series, Norman, OK.
Manning, F. S., and Thompson, R. E., 1991, *Oilfield Processing of Petroleum, Volume One: Natural Gas,* Pennwell Publishing Company, Tulsa, OK.
Manning, W. P., 1979, "Chemsweet—A New Gas Sweetener," *Proc. 1979 Gas Conditioning Conference,* University of Oklahoma, Norman, OK (also published in *Oil & Gas J.,* Oct. 15, 1979, p. 122).
Manning, W. P., Rehm, S. J., and Schmuhl, J. L., 1981, "Method of Removing Hydrogen Sulfide from Gas Mixtures," U.S. Patent 4,276,271, June 30.
Millar, R. W., 1923, *J. Am. Chem. Soc.,* Vol. 45, p. 874.
Moignard, L. A., 1952, *The Performance of Oxide Purifiers,* Gas Council (Gt. Brit.), Research Commun. GC7.
Moignard, L. A., 1955, *Gas World,* Vol. 142, pp. 816–818.
Molecular Products, Ltd, 1994, "Sofnolime RG," *Gas Processes '94, Hydro. Process.,* April, p. 102.

Moore, D. B., 1956, *Gas World,* Vol. 143, pp. 153–155.

Nielsen, P. E. H., and Rudbeck, P., 1993, *Hot Gas Cleaning in IGCC Power Plants,* Power Generation Technology, Published by Sterling Publications, International Limited, London, England.

O'Hara, J. B., Chow, T. K., and Findley, J. E., 1987, *Sulfur Recovery from Hot Coal Gas Desulfurization Processes,* DOE/MC/21097—2338, DE87 006477, NTIS.

Pavlov, K. F., and Lopatin, K. I., 1947, *J. Appl. Chem.* (U.S.S.R.), Vol. 20, No. 12, pp. 1223–1234.

Perry, C. R., 1970, "A New Look at Iron Sponge Treatment of Sour Gas," Paper presented at Gas Conditioning Conference, University of Oklahoma, Norman, OK, March 31–April 1.

Perry, R. H., Green, D. W., and Mahoney, J. O., Editors, 1984, *Perry's Chemical Engineers' Handbook,* 6th edition, McGraw Hill, NY.

Porter, H. C., 1924, *Coal Carbonization,* Reinhold Publishing Corporation, NY.

Process Systems International, Inc., 1992, "Ryan/Holmes" in *Gas Process Handbook, 1992, Hydro. Process.,* April, p. 126.

Prausnitz, J. M., 1969, *Molecular Thermodynamics of Fluid-Phase Equilibria,* Prentice-Hall, Englewood Cliffs, NJ.

Reid, W. T., and Sherwood, T. K. 1966, *The Properties of Gases and Liquids,* 2nd ed., McGraw-Hill, NY.

Ryan, J. M., and O'Brien, J. V., 1986, *Oil & Gas J.,* Oct. 6, p. 62.

Ryan, J. M., and Schaffert, F. W., 1984, *Chem. Eng. Progr.,* Vol. 84, No. 10, October, p. 53.

Samuels, A., 1990, *Oil & Gas J.,* Feb. 5, p. 44.

Schaack, J. P., and Chan, F., 1989, *Oil & Gas J.,* Part 1—Jan. 23, p. 51; Part 2—Jan. 39, p. 81; Part 3—Feb. 20, p. 45.

Schaffert, F. W., and Ryan, J. M., 1985, *Oil and Gas J.,* Jan. 28, p. 133.

Sexauer, W., 1959, *Gas-u. Wasserfach,* Vol. 100, No. 27, pp. 687–690.

Shaner, R. L., 1978, *Chem. Eng. Progr.,* Vol. 74, No. 5, May, p. 47.

Siemek, P., 1993, "Gas Treatment Enhances Austria's Energy Supply," *Pipeline and Gas Journal,* March.

Silver, L., and Hopton, G. U., 1942, *J. Soc. Chem. Ind.,* London, Vol. 61, March, p. 37.

Sivalls, C. R., 1982, "Slurrisweet Acid Gas Treating Process" *Proc. Gas Conditioning Conference,* University of Oklahoma, Norman, OK.

Smith, B., 1957, "Dry Methods for Removing Hydrogen Sulfide from Gases," *Trans. Chalmers Univ. Technol.,* Gothenburg, No. 184.

Somerville, R. L., 1990, *Chem. Eng. Progr.,* Vol. 84, No. 12, December, p. 64.

Speece, J. E., 1925, *Proc. Pacific Coast Gas Assoc.,* Vol. 16, p. 345.

Springmann, H., 1985, *Chem. Eng.,* Vol. 92, No. 10, May 13, p. 58.

Streich, M., 1970, *Hydro. Process.,* Vol. 49, No. 5, April, p. 86.

Stull, D. R., 1947, *Ind. Eng. Chem.,* Vol. 39, p. 517.

SulfaTreat Co., 1994, "SulfaTreat," Gas Processes '94, *Hydro. Process.,* April, p. 104.

Taylor, D. K., 1956, *Oil & Gas J.,* Vol. 54 (79):125; (81):260; (83):133; (84):147.

Taylor, H. F., 1950, *Gas J.,* Vol. 264, p. 421.

Trahan, D. O., and Manning, W. P., 1992, "Batch Gas Sweetening Processes" presented at the GRI Liquid Sulfur Recovery Conference, Oct. 4-6.

Turner, C. F., 1944, *Oil & Gas J.,* (Annual Pipeline Number), Vol. 42, Sept. 23, pp. 192, 195, 199.

U.S. Department of Energy, 1982, *Coal Conversion Systems Technical Data Book,* HCP/T2286-01, Section I, "Properties of Process Materials."

U.S. Environmental Protection Agency, 1991, *EPA Handbook, Control Technologies for Hazardous Air Pollutants,* EPA/625/6-91/014, June.

U.S. Nat. Bur. Standards, 1952, Circ. No. 500.

U.S. Nat. Bur. Standards, 1955, Circ. No. 564.

Valentin, F. H. H., 1990, *Chem. Eng.,* Vol. 97, No. 1, January, p. 112.

Van Krevelen, D. W., and Baans, C. M. E., 1950, *J. Phys. & Colloid Chem.,* Vol. 54, pp. 370–390.

Vines, H. L., 1986, *Chem. Eng. Progr.,* Vol. 82, No. 11, November, p. 40.

Walthall, J. H. (Tennessee Valley Authority), 1958, personal communication, Oct. 29.

Ward, E. R., 1964, *Gas Purification Processes,* edited by G. Nonhebel, London, Georges Newnes Limited, Chap. 8.

Wendt, R. P., 1991, "Reaction of Hydrogen Sulfide with SulfaTreat," presented at the AIChE Spring National Meeting, Houston, TX, April 7–11.

Westmoreland, P. R., Gibson, J. B., and Harrison, D. P., 1977, *Environ. Sci. Techn.,* Vol. 11, No. 5, p. 488.

Wood, N. V., O'Brien, J. V., and Schaffert, F. W., 1986, *Oil & Gas J.,* Nov. 17, p. 50.

Yeandle, W. W., and Klein, G. F., 1952, *Chem. Eng. Progr.,* Vol. 48, July, pp. 349–352.

Zapffe, F., 1963, *Oil & Gas J.,* Aug. 19, p. 103.

Zhavoronkov, N. M., 1939, *J. Chem. Ind.,* (U.S.S.R.), Vol. 16, No. 10, pp. 36–37.

Zhavoronkov, N. M., and Chagunava, V. T., 1940, *J. Chem. Ind.* (U.S.S.R.), Vol. 17, No. 2, pp. 25–29.

Zhavoronkov, N. M., and Reshchikov, P. M., 1933, *J. Chem. Ind.* (U.S.S.R), Vol. 10, No. 8, pp. 41–49.

Appendix
Units and Conversion Factors

SI UNITS

Quantity	SI Unit	Symbol
Length	meter	m
Mass	kilogram	kg
Time	second	s
Temperature	kelvin	K
Amount of substance	mole	mol
Force	newton	$N = kg\ m/s^2$
Work or energy	joule	$J = N\ m$
Power	watt	$W = J/s$
Pressure	pascal	$Pa = N/m^2$
Dynamic viscosity		$Pa\ s$ or $N\ s/m^2$
Kinematic viscosity		m^2/s
Diffusivity		m^2/s
Surface tension		$J/m\ s$ or N/m
Enthalpy		J/kg
Entropy		$J/kg\ K$
Heat capacity		$J/kg\ K$
Thermal conductivity		$W/m\ K$
Heat transfer coefficient		$W/m^2\ K$
Mass transfer coefficient		m/s

FREQUENTLY USED CONVERSION FACTORS

Length	1 in.		$= 2.54$ cm
	1 ft		$= 0.3048$ m
Area	1 in.^{2}		$= 6.4516$ cm^2
	1 ft^2		$= 0.092903$ m^2
Volume	1 in.^{3}		$= 16.387$ cm^3
	1 ft^3	$= 7.4805$ gallons (U.S.)	$= 0.02832$ m^3
	1 gallon (U.S.)		$= 3785.4$ cm^3
	1 barrel (oil)	$= 42$ gallons (U.S.)	$= 0.15987$ m^3
Mass	1 lb	$= 7,000$ grains	$= 0.45359$ kg
	1 ton (long)	$= 2,240$ lb	$= 1016.06$ kg
Force	1 lbf		$= 4.4482$ N

	1 dyn		$= 10^{-5}$ N
Temperature difference	1°F (°R)		= 5/9 °C (°K)
Energy (work, heat)	1 cal (I.T.)		= 4.1868 J
	1 erg		$= 10^{-7}$ J
	1 Btu (I.T.)	= 252 cal (I.T.)	= 1,055.056 J
Heating value (volumetric)	1 Btu/ft^3		= 37.259 kJ/m^3
Velocity	1 f/s		= 0.3048 m/s
Volumetric flow	1 ft^3/s		= 0.028316 m^3/s
	1 U.S. gal/min (gpm)		= 63.09 cm^3/s
Mass flow	1 lb/h		= 0.1260 g/s
Density	1 lb/in.^{3}		= 27.680 g/cm^3
	1 lb/ft^3		= 16.019 kg/m^3
Pressure	1 atm (std)	= 14.696 lbf/in.^{2}	= 101.3250 kPa
	1 lbf/in.^{2} (psi)		= 6.8948 kPa
	1 mm Hg (0°C)	= 1 torr	= 133.32 Pa
Power	1 hp (British)		= 745.70 W
	1 Btu/h		= 0.29307 W
	1 ton of refrigeration		= 3516.9 W
Viscosity, dynamic	1 P (poise)		= 0.1 N s/m^2
	1 lb/ft s		= 1.4882 N s/m^2
Viscosity, kinematic	1 St (stoke)		$= 10^{-4}$ m^2/s
Surface tension	1 erg/cm^2	= 1 dyn/cm	$= 10^{-3}$ N/m
Mass transfer coefficient	1 ft^{3}h/ft^2	= 1 ft/h	= 0.084667 m/s
Heat transfer coefficient	1 Btu/h ft^2°F		= 5.6783 W/m^2 K
Enthalpy	1 Btu/lb		= 2.326 kJ/kg
Heat capacity	1 Btu/lb°F		= 4.1868 kJ/kg K

ABBREVIATIONS, CONSTANTS, AND USEFUL NUMBERS

For SI units m = milli, 10^{-3}; C = centi, 10^{-2}; k = kilo, 10^3; and M = mega, 10^6
scf = standard cubic feet of gas (60°F, 1 atm)
Mscf = thousand standard cubic feet of gas
MMscf = million standard cubic feet of gas
s.t.p. = standard temperature and pressure (273.15 K and 1.013×10^5 Pa)
1 grain per 100 scf = 24.19 mg/m^3 (s.t.p.)
$1/4$ grain H$_2$S per 100 scf $\approx$ 4 ppmv H$_2$S
Volume of 1 lb mol of ideal gas at 60°F and 1 atm = 379.5 scf
Volume of 1 kmol of ideal gas at s.t.p. = 22.41 m^3
Gas constant, R = 1.986 Btu/lb mol°R = 8.314 J/mol K

Index

Absorbers,
  design of, 12–35
  for ethanolamine processes, 111
  for exhaust gas purification, 1329–1332
  for fluoride removal, 441–448
  for glycol dehydration process, 976–983
  for limestone/lime sulfur dioxide removal process, 501–504
  for liquid hydrocarbon treating, 166–170
  selection of, 6–8
  for water wash process, 418–448
  (see also Columns; Contactors; Towers)
Absorption, definition of, 2
Absorption coefficients,
  for ammonia in water, 299–302
  for carbon dioxide, in ethanolamine solutions, 115
    in hot potassium carbonate solutions, 349, 351–353
    in sodium carbonate-bicarbonate solutions, 379, 380
    in sodium hydroxide-carbonate solutions, 380
    in water, 427–432
  for carbon monoxide in copper ammonium salt solutions, 1253
  for chlorine in water, 461–463
  for fluorides in water, 444–448
  for hydrogen sulfide in ethanolamine solutions, 115
Acetylene,
  hydrogenation of, 1180–1183
    catalysts for, 1181
    equilibrium constants for, 1182
    heats of reaction for, 1182
  vapor pressure of, 1341
Acticarbone process, 1106
Activated alumina,
  capacity for water of, 1036
  as catalyst for sulfur conversion, 674
  gas dehydration with, 1039
  physical properties of, 1035, 1040, 1041
  removal of hydrochloric acid with, 1128
Activated bauxite, 1039, 1040
  capacity for water, 1036

  dehydration with, 1039
  physical properties of, 1035, 1041
Activated Benfield process, 362
Activated carbon,
  adsorption of organic vapors on, 1087–1124
  adsorption isotherms for hydrocarbons on, 1089, 1090
  air purification with, 1117–1124
  capacity of, for organic compounds, 1089, 1090, 1100, 1122
  gas velocity through beds of, 1103
  heat of adsorption of organic compounds on, 1100
  pressure drop through beds of, 1101, 1102
  properties of, 1088–1093
  solvent recovery with, 1093–1117
  sulfur dioxide removal with, 634–641
  support screens for beds of, 1099
  tests for evaluation of, 1092
  VOC removal with, 1093–1109
Activated carbon filters for amine solutions, 250–255
  sizing criteria for, 252
  pressure drop through, 253, 254
Activated charcoal, 1088 (see also Activated carbon)
Active alumina (see Activated alumina)
Active carbon (see Activated carbon)
Activity coefficient,
  of water in ethylene glycol, 1005
  of water in methanol, 1006
Adip process, 53
Adip solutions,
  specific gravity of, 102
  specific heat of, 107
  viscosity of, 105
  (see also Diisopropanolamine solutions)
Adipic acid, 514–517
Adsorbents, dehydration, 1034–1044
  capacity for water of, 1036–1038
  design capacity of, 1055–1060
  pressure drop through beds of, 1051–1055
  properties of, 1035, 1037, 1040, 1043
  regeneration of, 1063–1066

selection of, 1049–1051 (see also specific adsorbents)
Adsorption, 1023–1129
for air purification, 1117–1124
cycles, 1024–1026
definition, 2
for dehydration (see Dehydration)
design methods, 1026–1030, 1048–1063
fluidized bed, 1109–1115
of gas impurities on molecular sieves, 1018–1086
heat effects during, 1029, 1102
of hydrochloric acid, 1128
of iron and nickel carbonyls, 1128
mechanism of, 1023, 1024, 1044–1048
of mercury, 1127
for odor removal, 1117–1127
of organic vapors on active carbon, 1087–1117
pressure swing cycle, 1081–1086
of radioactive isotopes, 1128
of traces of sulfur, 1126
of sulfur dioxide, 634–641
of water vapor, 1030–1070 (see also Dehydration)
Adsorption vessels, design of, 1060–1063
Air,
water content of, in contact with lithium chloride solutions, 1012
water content of saturated, 952
Air dehydration,
by Kathabar system 1015–1017
with lithium chloride solutions, 1010–1017
with glycol solutions, 962, 963
Air pollution by sulfur dioxide, 469–474
Air purification,
by adsorption, 1077, 1117–1124
by catalytic oxidation, 767–783
for low temperature separation, 1077
Air separation by membrane permeation, 1282–1287
Aldehyde process for sulfur scavenging, 1315
Alkacid process, 397–400
solution "dik," 399
solution "M," 400, 401
vapor pressure of hydrogen sulfide over solutions, 400
Alkali carbonate solutions for acid gas removal, 334–393
Alkali metal compounds for sulfur dioxide removal, 544–564

Alkali metal sulfite-bisulfite process, 554–559
Alkaline earth compounds for sulfur dioxide removal, 496–544
Alkalized Alumina process, 631
Alkanolamines, 40–277 (see also Ethanolamines)
Alumina (see Activated alumina)
Aluminum sulfate solution for sulfur dioxide removal, 582–584
Amine process for hydrogen sulfide and carbon dioxide removal (see Ethanolamine process, and specific amines)
Amine Guard process, 42
Amine recovery, water wash for, 58–59
Amines, aromatic, sulfur dioxide recovery with, 589–597 (see also Dimethylaniline; Toluidine; Xylidine)
Amisol process, 1231–1233
Ammonia,
in coal gas, 280, 292
distillation of, 298
heat of vaporization of, from copper-ammonium-salt solutions, 1350, 1351
heats of reaction of, with hydrogen sulfide and carbon dioxide, 320
heat of solution of, 291–294
Henry coefficient for, in pure water, 285
losses of from copper-ammonium-salt solutions, 1356
reactions with hydrogen sulfide and carbon dioxide, 319–321
recovery of, as ammonium salts, 308–318
use in Katasulf process, 1163
vapor pressure of, 283–292, 1341
over ammonia-sulfur dioxide-water solution, 565–567
over aqueous solutions, 283
Ammonia absorption by water,
in packed towers, 299–302
in spray towers, 301–302
Ammonia-Ammonium Bisulfate (ABS) process, 573
Ammonia-calcium pyrophosphate process, 582
Ammonia-lime double alkali process, 581, 582
Ammonia removal, 292–302
by Chevron WWT process, 314–318
by direct process, 308
by indirect process, 308–310
by Phosam process, 311
by semidirect process, 308–310

Ammonia solutions,
  boiling point diagram of, 284
  carbon dioxide removal with, 319
  equilibrium with coke oven gas, 292
  heat of absorption of sulfur dioxide in, 567
  hydrogen sulfide removal with, 318–326
  specific gravity of, 282
  sulfur dioxide removal with, 564–582
  vapor pressure of ammonia, hydrogen sulfide, carbon dioxide and water over, 283–292
Ammonia-sulfur dioxide-water solutions,
  pH of, 566
  vapor pressure of ammonia and sulfur dioxide over, 565–567
Ammonium sulfate,
  production by Katasulf process, 1163
  saturator for production of, 310
  solubility of, in water, 295
  physical properties of solutions of, 295
Ammonium sulfite solutions, removal of sulfur dioxide with, 564–582
AMOCO cold bed absorption process, 703–706
Aqueous Carbonate process, 606, 607
Aqueous Aluminum Sulfate process, 582–584
Aromatics, absorption of in glycol solutions, 994–997
Arsenic trioxide, use of in Giammarco-Vetrocoke process, 371–378
ASARCO process, 593–595
ATS Technology process, 578, 579
Autopurification process, 746–748

Babcock and Wilcox FGD process, 534–535
BASF Catasulf process, 696, 697
BASF (MDEA) process for carbon dioxide removal, 60
Battelle Zinc Oxide spray dryer process, 932
Bauxite (see Activated bauxite)
Beavon Sulfur Removal process, 717–719
Bechtel seawater process, 600–601
Benfield process (see Hot potassium carbonate process)
Benzene,
  capacity of activated carbon for, 1100
  heat of adsorption on activated carbon, 1100
  solubility of,
    in petroleum oil, 1363
    in triethylene glycol, 995
Benzol removal from coal gas, 1359–1366

Bicarbonate-carbonate distribution vs. pH, 508
Binax process, 434
Biofilters, 1124–1127
Bischoff FGD process, 536, 537
Bischoff seawater process, 601
Bisulfite-sulfite distribution vs. pH, 508
Bog ore, use of in iron oxide process, 1299
Boiling points of gases, 1341
British Oxygen Corporation SURE process, 694–696
BTEX absorption in glycols, 994–997
Burkheiser process, 738

Calcium chloride,
  dehydration with, 1008–1010
  dew point of gases in contact with solutions of, 1008
  pellets, dehydration with, 1009
CANSOLV process, 595, 596
Carbon, activated (see Activated carbon)
Carbon dioxide,
  adsorption isotherms on molecular sieve, 1072
  catalytic conversion of, 1177–1180
  corrosion of steel by, in ethanolamine process, 199–210
  heat of reaction of,
    with ammonia, 320
    with ethanolamines, 91–99
    with potassium carbonate, 358
  heat of vaporization of, from copper-ammonium-salt solutions, 1350–1351
  ionization constant of, 331
  rate of absorption of,
    in sodium carbonate solutions, 1350–1351
    in potassium carbonate solutions, 349, 351–353
    in sodium carbonate, bicarbonate, and hydroxide, 379–381
  reactions with ammonia, 319, 320
  solubility of,
    in methanol, 1217, 1218
    in triethylene glycol, 974
    in water, 379, 423, 427–429
  vapor pressure of, 1341
    over ammonia solutions, 283–292
    over ethanolamine solutions, 62–91
    over potassium carbonate solutions, 341–343, 370
    over sodium carbonate solutions, 379
  water vapor content of, 951

Carbon dioxide absorption,
    by ethanolamine solutions, 113–120
    by monoethanolamine solutions, in plate columns, 116–118
    by water in packed columns, 427–432
        in plate columns, 432
Carbon dioxide removal,
    with alkali carbonate solutions, 334–393
    with ammonia solutions, 319
    by BASF (MDEA) process, 60
    by Binax process, 434
    by Catacarb process, 363–368
    by CNG process, 1342
    with copper-ammonium-salt solutions, 1353–1355
    by Estasolvan process, 1224
    by ethanolamine process, 40–277
    by Flexsorb HP process, 369–371
    by Fluor Solvent process, 1198–1202
    by Giammarco-Vetrocoke process, 371–378
    by Ifpexol process, 1223–1224
    by membrane permeation, 1270–1281
    by Methylcyanoacetate process, 1225
    by molecular sieves, 1076–1078
    by organic solvents, 1188–1234
    by potassium carbonate solutions, 334–362
    by Purisol process, 1210–1214
    by Rectisol process, 1215–1223
    by Selexol process, 1202–1210
    by Sepasolv MPE process, 1210
    with sodium hydroxide solutions, 401–402
    by Sulfinol process, 1225–1231
    with water, 423–435
Carbon disulfide,
    capacity of activated carbon for, 1200
    effect on ethanolamines, 49, 50, 239–241
    heat of adsorption on activated carbon, 1200
    hydrogenation of, 1165–1171
    solubility of, in petroleum oil, 1364 (see also Organic sulfur)
Carbon disulfide removal,
    from air, 1110–1112
    from coal gas, 1364–1365
    by catalytic conversion, 1165–1172
    by hot potassium carbonate process, 351
Carbon monoxide,
    catalytic conversion of, 1172–1180
    heat of solution of, in copper-ammonium-salt solution, 1348–1351
    latent heat of, 1341
    solubility of,
        in copper-ammonium-salt solution, 1350–1351
        in water, 417
    specific heat of, 1350
    vapor pressure of, 1350
Carbon monoxide removal,
    by condensation, 1338
    by Cosorb process, 1357–1359
    with copper-ammonium-salt solutions, 1347–1359
    by methanation, 1177–1181
    by nitrogen wash, 1338
    by shift conversion, 1172–1177
Carbon tetrachloride activity of active carbons, 1092
Carbon tetrachloride retentivity of active carbons, 1092
Carbonyl sulfide,
    effect of on ethanolamines, 49–53, 239–241
    formation of, by molecular sieve adsorbents, 1051
    hydrogenation of, 1165–1171
    in Claus process, 673, 674 (see also Organic sulfur)
Carbonyl sulfide removal,
    with hot carbonate solutions, 357
    from hydrocarbon gases, 1170
    from liquid hydrocarbons, 173, 174
    from synthesis gases, 1168–1172 (see also Organic sulfur)
Carl Still process, 322
Carpenter-Evans process, 1168
Cataban process, 804, 805
Catacarb process, 363–368
Catalysis in gas purification, 1145–1183
Catalysts,
    for acetylene hydrogenation, 1181
    for ethylene oxide destruction, 1161
    for carbon monoxide oxidation, 1157, 1160
    for Claus process, 674
    for hydrogen sulfide oxidation, 1164
    for methanation, 1178, 1179
    for organic sulfur hydrogenation, 1170
    for organic sulfur oxidation, 1164
    for shift conversion, 1174, 1175
    for VOC and odor oxidation, 1149–1155
Catalytic conversion,
    of carbon oxides to methane, 1177–1180
    of carbon monoxide to carbon dioxide, 1172–1177
    of organic sulfur to hydrogen sulfide, 1165–1171

Catalytic hydrogenation of acetylenic compounds, 1180–1183
Catalytic/IFP/CEC Ammonia scrubbing process, 573–575
Catalytic oxidation for sulfur dioxide removal, 641–645
Catalytic oxidation of VOCs and odors, 1148–1163
  combustion temperatures for, 1150
  economics of, 1161–1163
  ignition temperatures for, 1149
Catalytic reactors, design of, 1146–1148
CATOX catalytic oxidation system, 1157–1159
Cat-Ox process for sulfur dioxide removal, 643–644
Caustic wash, for carbon dioxide removal, 401, 402
  for hydrogen sulfide removal, 402–404
  for mercaptan removal, 404–406
Chelated iron solutions for hydrogen sulfide removal, 803–840
Chelates for nitrogen oxide control, 932
Chemical conversion, definition, 2 (see also Catalytic conversion)
Chemsweet process, 1310, 1311
Chevron WWT process, 314–318
Chiyoda Thoroughbred 101 process, 585–588
Chiyoda Thoroughbred 121 process, 537–539
Chlorine,
  absorption of, in water, 458–463
  reaction of, with water, 460
  solubility of, in water, 461
Citrate process, 563, 564
Claus process, 670–724
  ammonia destruction in, 684–686
  catalyst deactivation, 682–684
  isothermal reactor concepts for, 696–697
  oxygen-based systems, 689–696
  sub-dewpoint systems, 699–708
  tail gas purification for, 698–724
  theoretical conversion for, 676
CNG process, 1342
Coal gas,
  impurities in, 281
  light oil removal from, 1359–1366
  naphthalene removal from, 1366–1369
  organic sulfur removal from, 1164–1172

Cobalt-molybdate catalysts, 1170
Cocurrent absorption, for amine process, 61
  for glycol dehydration, 962, 963
Coke oven gas, purification of, 297–299, 318–326, 1359–1367
Coldfinger dehydration process, 960, 961
Columns,
  absorption, design of, 12–35
  packed,
    absorption of fluorides in, 443
    absorption of acid gases by ethanolamines in, 113–120
    absorption of carbon dioxide by water in, 427–432
    flooding correlations for, 27–29
  plate,
    absorption of carbon dioxide by ethanolamine solutions in, 116–118
    absorption of carbon dioxide by water in, 432
  relative costs of, 8
  selection of, 6–8
  spray,
    for selective hydrogen sulfide removal, 320
    for fluoride absorption, 273, 281
  types of, 6–11, 111 (see also Absorbers; Contactors; Towers)
Combustion control for NO_x reduction, 880–886
Cominco process, 569–573
Condensation, definition, 2
  for exhaust gas purification, 1332–1337
Contactors, selection of, 6–8 (see also Absorbers; Columns; Towers)
Coolside process, 623
COPE process, 691–693
Copper, solubility of in copper-ammonium-salt solutions, 1352
Copper-ammonium-salt solutions,
  carbon monoxide removal with, 1346–1359
  corrosion in, 1357
  heat capacity of, 1351
  heat of solution of carbon monoxide in, 1352
  regeneration of, 1355–1356
  solubility of carbon monoxide in, 1351, 1352
Copper oxide process,
  for sulfur dioxide removal, 627–630

Index **1381**

for combined sulfur dioxide/nitrogen
  oxide removal, 931
Corrosion,
  in ammonia solutions for acid gas
    removal, 326
  in chlorine absorption by water, 463
  in copper-ammonium-salt process, 1357
  in ethanolamine process, 188–224
  in fluoride removal, 450–453
  in glycol dehydration process, 990, 991
  in hydrogen chloride absorption, 458
  in lime/limestone FGD systems, 529–532
  in lithium halide solutions, 1016
  in Thylox process, 754
  in water wash process,
    for carbon dioxide removal, 435
    for hydrogen sulfide removal, 437
  inhibitors in ethanolamine systems, 223
  inhibitors in Catacarb process, 366
Cosorb process, 1357–1359
Cost,
  of carbon dioxide removal, 1274, 1275,
    1277, 1280, 1281
  of Chevron WWT process, 317
  of enriched air generation, 1287
  of membrane carbon dioxide removal
    systems, 1273
  of membrane hydrogen recovery systems,
    1265–1267
  of membrane solvent recovery systems,
    1288, 1289
  of natural gas dehydration equipment, 1033
  of nitrogen generation, 1286
  of sulfur dioxide removal systems,
    492–495
  of sulfur scavenging processes, 1320, 1321
  of thermal incineration, 1145
  of VOC removal by adsorption, 1108, 1109
Cryofining process, 1339, 1340
Cuprous-ammonium-salt solutions (see
  Copper-ammonium-salt solutions)

Dehumidification (see Dehydration)
Dehydration,
  by adsorption, 1030–1069
    adsorbent regeneration, 1063–1066
    adsorbents for, 1034–1044
    dehydrator design for, 1060–1063
    flow systems for, 1066–1069
    operating practices in, 1069–1070
    pressure drop in, 1051–1055
  of air, by Kathabar process, 1015–1017

  with calcium chloride, 1008–1017
  with glycols, 953–997 (see also Glycol
    dehydration)
  by Ifpexol process, 1223
  with lithium halide solutions, 1010–1017
  by Rectisol process, 1215
  with saline brines, 1007–1017
Desiccants, solid (see Adsorbents;
  Dehydration)
Dew point of gases in contact with,
  calcium chloride solutions, 1008
  diethylene glycol solutions, 964
  triethylene glycol solutions, 965
Dew-point depression, 952
  in commercial glycol dehydrators, 989
Diammonium phosphate, 292, 310
  heat of formation of, 292
DIAMOX process, 324, 325
Dichromate solutions, hydrogen sulfide
  removal with, 855
Diethanolamine, 50, 51
  degradation of, 232–242
  irreversible reaction with carbon dioxide
    of, 235–238
  reaction with carbonyl sulfide, 239
  thermal reclaiming of, 261–263
  vapor pressure of, over aqueous
    solutions, 231
Diethanolamine solutions,
  corrosion in, 200, 212, 213
  heat capacity of, 105
  heat of reaction with acid gases, 99
  specific gravity of, 101
  vapor pressure of, 93
  vapor pressure of acid gases over, 70–76
  viscosity of, 103
Diethylene glycol,
  properties of, 954
  vapor pressure of, 971
Diethylene glycol solutions,
  dehydration with, 953–997
  dew point of gases in contact with, 964
  freezing points of, 971
  solubility of natural gas in, 615
  specific gravity of, 969
  specific heat of, 970
  vapor/liquid equilibrium diagrams for, 972
  vapor pressure of, 968
  viscosity of, 970
Diglycolamine, 51
  irreversible reaction with carbon dioxide,
    238, 239

reaction with COS, 240–241
thermal reclaiming of, 258, 259
Diglycolamine solutions,
  density of, 101
  heat capacity of, 106
  heat of reaction with acid gases, 94, 99
  vapor pressure of, 94
  vapor pressure of acid gases over, 72–78
  vapor/liquid equilibrium diagrams for, 260
  viscosity of, 104
Diisopropanolamine, 53
  irreversible reaction with carbon dioxide, 238
  use of in Sulfinol process, 1225 (see also Adip)
Diisopropanolamine solutions,
  heat of reaction with acid gases, 99
  thermal reclaiming of, 261, 262
  vapor pressure of, 96
  vapor pressure of acid gases over, 79–81
Dimethylaniline,
  properties of, 590
  solubility of sulfur dioxide in, 591
  sulfur dioxide recovery with, 593–595
Direct process for ammonia recovery, 296, 297
Double alkali process, 546–554
Dow process for sulfur dioxide removal, 596, 597
Dowa Dual Alkali process, 582–584
Drizo process, 959
Dry-box process (see Iron oxide process)
Dry soda process for sulfur dioxide removal, 624–626
  byproduct disposal for, 626
Duct spray dryer process for sulfur dioxide removal, 614

Ebara-E-Beam process, 645, 646, 982
EIC Copper Sulfate process, 853, 855
Econamine process, 52
Economics (see Costs)
Ejectors, absorption of fluorides in, 441–448
Electrochemical process for nitrogen oxide removal, 935
Electrodialysis, 264–266, 560
Electrolytic Regeneration process, 561, 562
Elsorb process, 559
Equilibrium constants,
  for hydrogenation,
    of acetylene, 1182
    of carbon disulfide, 1167
    of carbonyl sulfide, 1167
  for hydrolysis,
    of carbon disulfide, 1167
    of carbonyl sulfide, 1167
  for methanation, 1180
  for shift conversion, 1176
Equilibrium vapor pressure (see Vapor pressure)
Estasolvan process, 1224, 1225
Ethane,
  boiling point of, 1341
  latent heat of, 1341
  vapor pressure of, 1341
Ethanolamine processes, 40–277
  Absorber,
    diameter of 112–114
    height of, 113–120
    selection of, 111
    temperature profile of, 120–123, 139
  chemical losses in, 231–242
  cocurrent absorption in, 61
  corrosion in, 188–244
  design of, 103–144
  flow schemes for, 57–60
  operating data for, 144–151
  stripping system design for, 123, 141–144
  water wash for amine recovery in, 58, 59
Ethanolamines,
  cost of, 49
  effect of gas impurities on, 239–242
  heats of reaction of hydrogen sulfide and carbon dioxide with, 91–99
  pH of during neutralization, 47
  physical properties of, 49
  (see also Diethanolamine; Monoethanolamine; Methyldiethanolamine; Triethanolamine; Diglycolamine; Diisopropanolamine; Adip)
Ethanolamine solutions,
  concentration of, 56
  degradation of, 232–242
  foaming of, 224–231
  organic sulfur removal with, 151–155
  purification of, 48–56
  reactions of hydrogen sulfide and carbon dioxide with, 44–48
  solubility of methane in, 267, 268
  vapor pressure of, 91–97
  vapor pressure,

of carbon dioxide over, 62–91
of hydrogen sulfide over, 62–91
Ethylene,
 boiling point of, 1341
 injection, for hydrate control, 999, 1000
 latent heat of, 289
 vapor pressure of, 289
Ethylene glycol,
 dehydration with, 999, 1005
 physical properties of, 954
 vapor pressure of, 971
Ethylene glycol solutions,
 freezing points of, 971
 water vapor dew point over, 1002
Ethylene purification with molecular sieve, 1076

Ferrous sulfide for sulfur dioxide removal, 585
Ferrox process, 738–741
Fiber-Film contactor, 407–409
Film coefficients, 15–21
Filtration,
 of ethanolamine solutions, 252–255
 of glycol solutions, 993
Fire hazard properties of organic solvents, 1096
Fischer process, 745, 746
Flakt-Hydro seawater process, 599, 600
Flexitray plates, 8
Flexsorb SE Plus process, 723–724
Flexsorb HP process, 369–371
Flooding of packed towers, 27–29
Flue gas,
 desulfurization (FGD) processes, 466–647
 sulfur dioxide concentration in, 475
Fluidized-bed active carbon adsorption, 1109–1115
Fluor Econamine process, 52
Fluorides,
 removal with water, 438–453
  absorbers for, 441–448
  disposal of absorbed, 453
  rate of absorption of, 444–448
  materials of construction for, 450–453
Fluosilicic acid solutions,
 pH of, 443
 vapor pressure of silicon tetrafluoride over, 271
Fluor Solvent process, 1198–1202
Foaming,
 of ethanolamine solutions, 224–231
 of glycol solutions, 990
Freezing points,
 of glycols, 971
 of ethanolamine solutions, 108
Fumaks process, 850, 851

Gas suspension spray dryer process, 614, 615
GAS/SPEC amine process, 42
Gastechnik process for hydrogen sulfide removal, 1300
General Electric Ammonium Sulfate process, 579
General Electric (GEESI) process, 533, 534
Geothermal gas cleanup,
 by LO-CAT process, 818–821
 by SulFerox process, 838
Giammarco-Vetrocoke process,
 for carbon dioxide removal, 371–378
 for hydrogen sulfide removal, 754–759
Girbotol process (see Ethanolamine processes)
Gluud Combination process, 735
Gluud iron oxide process, 736
Glycol-Amine process, 50
Glycol dehydration process, 953–997
 absorber design for, 961, 962, 976–983
 aromatics absorption in, 994–997
 in air conditioning, 962, 963
 corrosion in, 990, 992
 dew-point depression in, 964–967
 filters for, 993
 flash tank design for, 983, 984
 flow schemes for, 955–963
 foaming in, 990
 glycol losses in, 988, 989
 inlet separators for, 972–976
 plant operation in, 988–997
 regenerator design for, 984–988
Glycol injection process, 997–1007
Glycols,
 properties of, 954
 reclaiming of, 994

Heat of adsorption,
 of benzene on activated carbon, 1100
 of carbon disulfide on activated carbon, 1100
 of organic compounds on activated carbon, 1100
Heat capacity (see also Specific Heat)
 of copper-ammonium-salt solutions, 1351
 of diethanolamine solutions, 105

1384 *Gas Purification*

of monoethanolamine solutions, 105
of potassium carbonate solutions, 346
Heat of formation of diammonium
  phosphate, 292
Heat of fusion of lithium halides, 104
Heat of reaction,
  of acetylene hydrogenation, 1182
  of carbon dioxide with potassium
    carbonate, 358
  of hydrogen sulfide and carbon dioxide
    with ammonia, 320
  of hydrogen sulfide and carbon dioxide
    with ethanolamines, 91–99
  of hydrogen sulfide with oxygen, 673
  of methanation, 1180
  of sulfur dioxide with xylidine, 591
  of shift conversion, 1176
Heat of solution,
  of ammonia in water, 291–294
  of carbon monoxide in copper-
    ammonium-salt solutions, 1350
  of water in glycols, 954
Heat of vaporization,
  of components from copper-ammonium-
    salt solutions, 1351
  of glycols, 954
  of hydrofluoric acid, 439
Heat transfer coefficients for air dehydration
  with lithium chloride solutions, 1014
Height of transfer units (see Transfer units)
Helium removal from natural gas, 1281, 1282
Henry's law constant,
  for benzene in triethylene glycol, 995
  for chlorine in water, 461
  for inert gases in water, 417
  for organic chemicals in water, 417
Hiperion process, 794–797
Hi-Pure process, 339
Hitachi FGD process, 634
Holmes–Maxted process, 1168
Hondo HS-100 process, 1312–1314
Hopcalite catalyst, 1154
Hot activation process (see Vacuum
  carbonate process)
Hot potassium carbonate process, 334–362
  corrosion in, 359–362
  economics of, 362
  materials of construction for, 361–362
  solution regeneration, 357–359
  (see also Potassium carbonate)
Hydrate suppression by inhibitor injection,
  997–1007

Hydrates, 947–949
Hydrocarbon recovery,
  with silica gel, 1086–1087
  with active carbon, 1109
  with oil absorption, 1359–1369
Hydrocarbons,
  adsorption isotherms for on activated
    carbon, 1089–1090
  catalytic oxidation of, 1148–1162
Hydrochloric acid, removal of,
  with water, 453–458
  by adsorption on alumina, 1128
Hydrochloric acid solutions,
  heat of solution of hydrochoric acid in, 453
  materials of construction for, 458
  vapor pressure of hydrogen chloride
    over, 454
Hydrofluoric acid,
  heat of vaporization of, 439
  ionization constant of, 439
  removal of, with water, 438–453
  vapor pressure of, over solutions, 440
  (see also Fluorides)
Hydrogen,
  boiling point of, 1341
  latent heat of, 1341
  purification of, by cryogenic processes,
    1338–1342
  recovery of, by membrane process,
    1259–1270
  solubility of, in water, 417
  specific heat of, 1341
  vapor pressure of, 1341
Hydrogen chloride (see Hydrochloric acid)
Hydrogen cyanide, in coal gases, 280, 281
Hydrogen cyanide removal,
  by Perox process, 762–764
  by Seaboard process, 389
  by Staatsmijnen-Otto process, 746–748
  by Stretford process, 783, 784
  by Thylox process, 750
  by vacuum carbonate process, 387, 388
Hydrogen fluoride (see Hydrofluoric acid;
  Fluorides)
Hydrogen sulfide,
  absorption of,
    in ethanolamine solutions, 113–120
    in potassium carbonate solutions,
      353–357
  adsorption isotherms on molecular
    sieves, 1021
  conversion to sulfur, 670–724

corrosion of steel by, in ethanolamine
   process, 199–210
heat of reaction,
   with ammonia, 320
   with ethanolamines, 91–99
   with oxygen, 673
ionization constant of, 331
rate of absorption of,
   in diethanolamine solutions, 115
   in potassium carbonate solutions, 356
   in sodium carbonate solutions,
      387, 388
reactions of, with ammonia, 319, 320
solubility of, in water, 436, 437
species in water, distribution of, 774
vapor pressure,
   over Alkacid solutions, 400
   over ammonia solutions, 283–292
   over ethanolamine solutions, 62–91
   over potassium carbonate solutions,
      347–349
   over tripotassium phosphate
      solutions, 395
Hydrogen sulfide removal,
   by aldehyde processes, 1315
   by Alkacid process, 397–400
   by alkanolamines, 40–277
   by alkylamine/aldehyde condensation
      products, 1315, 1316
   by Amisol process, 1231–1232
   with ammonia solutions, 318–326
   by Autopurification process, 746–748
   by Burkheiser process, 738
   by Cataban process, 804, 805
   by Catacarb process, 363–368
   by caustic scrubbing, 402–404, 1317, 1318
   by CBA process, 703–706
   by chelated iron solutions, 803–840
   by Chemsweet process, 1310–1311
   by DIAMOX process, 324–325
   with dichromate solutions, 855, 1311, 1312
   by EIC copper sulfate process, 853–855
   by Estasolvan process, 1224
   by ethanolamine processes, 40–277
   by Ferrox process, 738–741
   by Fischer process, 745, 746
   by Fumaks process, 850, 851
   by Gastechnik process, 1300
   by Gas/Spec solvents, 41
   by Giammarco-Vetrocoke process,
      754–759
   by Gluud combination process, 735

by Gluud process, 736
by Hiperion process, 794–797
by Hi-Pure process, 339
by Hondo HS-100 process, 1312–1314
by IFP Clauspol 1500 process, 843–845
by Ifpexol process, 1223, 1224
by iron chelate solutions, 803–840
by iron cyanide complexes, 744–748
by iron oxide process, 1298–1305
by iron oxide suspensions, 734–736
by Katasulf process, 1163, 1164
by Konox process, 851–853
by Koppers C.A.S. process, 735, 736
by Krupp Wilputte process, 324
by liquid oxidation processes, 731–856
by Lo-Cat process, 804–823
by Manchester process, 742–744
by membrane permeation, 1270–1281
by Methylcyanoacetate process, 1225
with organic solvents, 1188–1233
by oxidation to oxides of sulfur,
   1162–1165
with permanganate solutions, 855,
   1311, 1312
by Perox process, 762–764
with polyamine solutions, 1316
with polythionate solutions, 734–736
by Purisol process, 1210–1215
by Rectisol process, 1215–1223
by Seaboard process, 381–383
selective,
   with ammonia solutions, 321
   by low-residence-time caustic
      scrubbing, 402, 403
   with methyldiethanolamine, 46,
      149, 150
   by Selexol process, 1202–1210
   by Sepasolv MPE process, 1210
   by tripotassium phosphate process,
      393–396
by Selectox process, 711–713
by Selexol process, 1202–1210
by Sepasolv MPE process, 1210
by Slurrisweet process, 1309, 1310
with sodium hydroxide solutions,
   402–404, 1317, 1318
by Sofnolime RG process, 1309
by Sodium phenolate process, 396, 397
by Staatsmijnen-Otto process, 746–748
with sterically hindered amines, 48, 56
by Still Otto process, 322–324
by Stretford process, 769–794

1386 *Gas Purification*

by Sulfa-Check process, 1312–1314
by SulfaTreat process, 1305
by Sulferox process, 825–840
by Sulfinol process, 1225–1231
by Sulfint process, 823–825
by Sulfolin process, 797–802
by Sulfreen process, 699–703
by Takahax process, 765–769
with thioarsenate solutions, 748–759
by Thylox process, 748–754
by tin oxide, 1328, 1329
by Townsend process, 841–842
by tripotassium phosphate process, 393–396
by UCARSOL process, 42
by UCBSRP process, 846–850
by Unisulf process, 802, 803
by Vacasulf process, 392, 393
by vacuum carbonate process, 383–392
with water, 436–438
by Wierwiorowsky process, 846
by zinc ferrite and titanate, 1324–1328
by zinc oxide, 1305–1309, 1329, 1330
by Z-Sorb process, 1329
Hydrogenation,
of acetylene, 1180–1183
catalysts for, 1181
heats of reaction in, 1182
of carbon disulfide, 1165–1171
of carbonyl sulfide, 1165–1171
Hydrolysis,
of carbon disulfide, 1165–1171
of carbonyl sulfide, 1165–1171

IFP Clauspol 1500 process, 843–845
Ifpexol process, 1223, 1224
Ignition temperatures of VOCs, 1143
Impregnated adsorbents, 1126, 1127
Indirect process for ammonia recovery, 296, 297
Inert gas purification with molecular sieves, 1074–1076
Intalox saddles, 22, 29
Injection,
of liquid inhibitors for hydrate control, 997–1007
of dry sorbent for sulfur dioxide removal, 617–623
Ion exchange for amine solution purification, 264
Ionization constant,
of carbon dioxide, 331

of hydrofluoric acid, 439
of hydrogen sulfide, 331
Iron carbonyl removal by adsorption, 1128
Iron chelate solutions for hydrogen sulfide removal, 803–840
Iron-cyanide complexes for hydrogen sulfide removal, 744–748
Iron oxide process, 1298–1305
basic chemistry of, 1301, 1303
continuous, 1300
design guidelines for, 1304
high pressure operation of, 1301, 1304–1305
operation of, 1304
purifying materials for, 1300
tower purifiers for, 1299
Iron oxide suspensions for hydrogen sulfide removal, 736–744
Iron sponge specifications, 1302
ISPRA bromine-based FGD process, 588, 589

Katasulf process, 1163, 1164
Kathabar air dehydration systems, 1015–1017
Konox process, 851–853
Koppers C.A.S. process, 735, 736
KPR rotating adsorber, 1115–1117
Krupp Wilputte process, 324

Latent heat of gases, 1341
Lehigh University process, 932
Light oil,
composition of, 1360
solubility of constituents in wash oils, 1364
Light oil removal from coal gas, 1359–1366
partition coefficients for, 1364
properties of wash oils for, 1363
Lime slurry spray dryer process, 607–614
Limestone/Lime injection process, 617–621
Limestone/Lime FGD process, 496–539
absorber design for, 517–519
additives for, 514–517
basic chemistry of, 501, 506–513
byproduct disposal in, 521–525
materials of construction for, 529–532
mist elimination in, 525–527
slurry processing in, 519–521
sorbent selection for, 513, 514
Linde Clinsulf process, 696
Liquid hydrocarbon treating, 156–174
absorber design for, 166–171

removal of COS by, 173, 174
settlers and coalescers for, 172, 173
water wash for, 171, 172
Liquid/liquid equilibrium data for acid gases in liquid hydrocarbons and amine solutions, 157–163
Lithium bromide, solubility of, in water, 1011, 1012
Lithium bromide solutions, vapor pressure of, 1011
Lithium chloride, solubility of, in water, 1011, 1012
Lithium chloride solutions,
air dehydration with, 1010–1017
rate of absorption of water vapor in, 1014
vapor pressure of, 1011
water content of air in contact with, 1012
Lo-Cat process, 805–823
Low temperature processes for gas purification, 1337–1346
Lower explosive limit (LEL), 1144
LPG treating, 156–174 (see also Liquid hydrocarbon treating)
Lurgi OxyClaus process, 693, 694

Magnesium oxide FGD process, 539–544
Manchester process, 742–744
Mass transfer, in gas absorbers, 15–27
in solid bed dehydration, 1046
Mass Transfer Zone (MTZ), 1026–1028
Materials of construction,
for absorption of chlorine in water, 463
for ethanolamine process, 189–190, 216, 225
for fluorides removal, 450–453
for hot potassium carbonate process, 361, 362
for hydrogen chloride absorption, 455–458
for hydrogen sulfide removal with water, 437, 438
for hydrogen sulfide removal with ammonia solutions, 326
for limestone/lime process, 529–532
for sour water strippers, 308
for Thylox process, 754
for vacuum carbonate process, 390
for carbon dioxide removal with water, 435
for Wellman-Lord process, 559
MCRC sulfur recovery process, 707, 708
Membrane permeation,
definition of, 2

design calculations for, 1252–1258
flow arrangements for, 1250–1252
processes for, 1238–1290
equipment suppliers, 1241
transport mechanisms for, 1242–1245
Mercaptans, removal of,
by adsorption, 1078–1081, 1126
by catalytic conversion, 1166
by caustic wash process, 404–410
by ethanolamine solutions, 155, 156
by hot potassium carbonate process, 357
by Merichem process, 407–410
by UOP Merox process, 406, 407
Mercaptans (see also Organic Sulfur)
Merichem process, 407–410
Mercury removal with impregnated carbon, 1127
Methanation for removal of carbon oxides, 1177–1179
catalysts for, 1178
equilibrium constants for, 1180
heats of reaction in, 1180
Methane,
boiling point of, 1341
latent heat of, 1341
removal of, by condensation, 1340
solubility of,
in ethanolamine solutions, 267, 268
in water, 417
specific heat of, 1341
vapor pressure of, 1341
Methanol,
acid gas removal with, 1215–1223 (see also Rectisol and Ifpexol processes)
injection, for hydrate control, 997–1007
solubility of carbon dioxide in, 1217
vapor pressure of, 1218
Methanol-water,
phase diagram, 1003
solution density, 1004
Methylacetylene, removal of, 1182
Methylcyanoacetate process, 1225
Methyldiethanolamine, 53
Methyldiethanolamine solutions,
heat capacity of, 106
selective removal of hydrogen sulfide with, 46, 149, 150
vapor pressure of, 97
vapor pressure of acid gases over, 79, 82–85
viscosity of, 104

Methylene chloride, removal from air, 1097
Minute service of active carbons, 1092
Mitsui-BF process, 639, 640
Mixed amines, 54, 55
MODOP process, 713–716
Molecular sieves,
    capacity for water of, 1036, 1043, 1056–1058
    carbon dioxide and water removal with, 1076–1078
    dehydration with, 1042–1044
    ethylene purification with, 1076
    equilibrium isotherms,
        for carbon dioxide on, 1072
        for hydrogen sulfide on, 1073
        for sulfur dioxide on, 1074
        for ammonia on, 1075
    gas purification with, 1018–1076
    hydrogen purification with, 1076–1081
    inert gas purification with, 1074–1076
    natural gas purification with, 1077–1081
    physical properties of, 1035, 1043
    sulfur removal with, 1078–1081
Molten carbonate process, 603, 604
Monoethanolamine, 49, 50
    degradation of, 232–242
    irreversible reaction of, with carbon dioxide, 233–235
    reaction of, with carbonyl sulfide, 239–241
    reaction of, with oxygen, 232, 233
    reactions of, with hydrogen sulfide and carbon dioxide, 44–48
    vapor pressure of, over aqueous solutions, 231
Monoethanolamine solutions,
    heat capacity of, 105
    heats of reaction with acid gases, 99
    purification of, 242–266
    solubility of, methane in, 267, 268
    specific gravity of, 49
    stripping of, 123–133
    viscosity of, 49, 102
    vapor/liquid equilibrium composition diagram, 257
    vapor pressure of acid gases over, 62–70
Monoethanolamine-glycol mixtures, 50
MTE sulfur recovery process, 697
Munters Zeol adsorber, 1117
Murphree plate efficiency, 24

Naphthalene,
    effect of on vacuum carbonate process, 390
    removal of, from coal gas, 1366–1369
    solubility in petroleum oil, 1364
    vapor pressure of, 1368
Natural gas,
    solubility of,
        in amine solutions, 267
        in glycols, 974
    water vapor content of, 947–951
Nickel carbonyl removal by adsorption, 1128
Nitric oxide in coal gas, 280
Nitrogen,
    boiling point of, 1341
    in carbonization products, 280
    latent heat of, 1341
    solubility of, in water, 417
    specific heat of, 1341
    vapor pressure of, 1341
Nitrogen compounds in coal gas, 280
Nitrogen oxide emission control, 866–936
    by activated carbon, 931
    by Battelle ZnO spray dryer process, 932
    by burners out of service (BOOS), 882
    by combined NO_x/SO_x postcombustion processes, 928–936
    by combustion control processes, 880–886
    by copper oxide processes, 931
    by DESONOX process, 933
    by dry low-NO_x combustors, 886
    by electochemical process, 935
    by flue gas irradiation, 932
    by Lehigh University low temperature process, 932
    by low excess air firing (LEA), 882
    by overfire air, 883
    by flue gas recirculation (FGR), 884
    by fuel reburning, 885
    by low-NO_x burners (LNB), 884, 885
    by metal chelates, 932
    by Nissan permanganate process, 936
    by NO_xOUT process, 888–898
    by NOXSO process, 931
    by Parsons Flue Gas Cleaning (FGC) process, 934
    by postcombustion processes, 887–936
    by precombustion processes, 879–880
    requirements for, 868–879
    by selective catalytic reduction (SCR), 904–930
    by selective non-catalytic reduction (SNCR), 888–904
    by sodium chlorite, 932
    by Sorbtech process, 932

by SO_x-NO_x-ROX-BOX (SNRB) process, 933
by Thermal $DeNO_x$ process, 898–904
by Tokyo Electric-MHI process, 936
by TRI-NO_x process, 935
by water/steam injection, 885
by WSA-SNO_x process, 933
Nitrogen oxide formation mechanisms, 868
Nitrogen removal from natural gas, 1344–1346
N-methyl-2 pyrrolidone,
  physical properties of, 1211
  use of, in Purisol process, 1210–1215
Noell KRC double loop limestone process, 535, 536
Noell KRC peroxide-based process, 589
North Thames Gas Board process, 1164
NOxOUT process, 888–898
NOXSO process, 631–634, 931

Odorants, characteristics of, 1121
Odor removal,
  by adsorption, 1117–1127
  by biofiltration, 1124–1127
  by thermal oxidation, 1137–1145
Organic compounds,
  capacity of activated carbon for, 1089, 1090
  heat of adsorption of, on activated carbon, 1100
  (see also Volatile organic compounds, VOCs)
Organic solvents, acid gas removal with, 1198–1234
Organic sulfur,
  catalytic conversion of, 1165–1172
  content of gases, 1165
  content of liquid hydrocarbons, 163–164
Organic sulfur removal, by amine solutions, 151–156
  by Carpenter Evans process, 1168
  by catalytic conversion, 1165–1171
  by Holmes-Maxted process, 1168
  by North Thames Gas Board process, 1164
  by Rectisol process, 1215–1219
  by soda-iron process, 1164
  from synthesis gases, 1165–1171
Organic vapor adsorption on active carbon, 1087–1117
Oxygen,
  boiling point of, 1341
  latent heat of, 1341
  solubility of, in water, 417
  specific heat of, 1341
  vapor pressure of, 1341

Packed towers (see Absorbers; Columns; Towers)
Pall rings, 9, 22, 29
Parsons Flue Gas Cleaning (FGC) process, 646, 934
Particulate removal by wet scrubbers, 418
Partition coefficients for light oil absorption, 1364
Permanganate solutions,
  for hydrogen sulfide removal, 855, 856, 1311, 1312
  for nitrogen oxide removal, 936
Perox process, 762–764
pH,
  of ammonia-sulfur dioxide-water solutions, 566
  of bisulfate-sulfite solutions, 508
  of bicarbonate-carbonate solutions, 508
  of bisulfide-sulfide solutions, 774
  control in iron oxide purifiers, 1301
  of ethanolamine solutions, 47
  of fluosilicic acid solutions, 443
Phenol removal with ammonia solutions, 308
Phosam and Phosam W processes, 311–314
Physical solvents for acid gas removal, 1188–1234
Plate efficiency, 24–26
  for carbon dioxide absorption in MEA solutions, 117, 118
  for water vapor absorption in glycol solutions, 978
Plates, design of, 21
Polyad FB process, 1113–1115
Polyethylene glycol dimethylether,
  properties of, 1197
  solubility of gases in, 1197
  use of, in Selexol process, 1202–1210
Polythionate solutions for hydrogen sulfide removal, 734–736
Potassium bicarbonate, solubility of, 341
Potassium carbonate,
  heat of reaction of with carbon dioxide, 358
  solubility of, 341
Potassium carbonate solutions,
  carbon dioxide removal with, 334–362
  heat capacity of, 346

hydrogen sulfide absorption in, 353–357
regeneration of, 357–359
specific gravity of, 345
vapor pressure,
   of carbon dioxide over, 341–343
   of hydrogen sulfide over, 347
   of water over, 344
(see also Hot Potassium Carbonate process)
Potassium N-dimethylglycine (see Alkacid process)
Potassium permanganate, 855, 856, 936, 1311, 1312
Potassium sulfite-bisulfite cycle, 554, 555
Precombustion processes for nitrogen oxide control, 879, 880
Pressure drop of gas,
   through activated carbon, 1101, 1102
   through solid desiccant beds, 1051–1055
Pressure drop of ethanolamine solutions through activated carbon filters, 253, 254
Pressure swing adsorption, 1081–1086
Propylene carbonate, physical properties of, 1197, 1199
Purasiv HR process, 1112, 1113
Pura Siv-N process, 1073
Pure Air/MHI FGD process, 537
Purisol process, 1210–1215
Pyridine bases, in coal gases, 280, 281

Quinones, 759–797

Radioactive isotope adsorption, 1128
R-BTEX process, 997, 998
Reclaimers for amine solutions, 255–264
Rectisol process, 1215–1223
Reinluft process, 638, 639
Resulf process, 723
RESOX process, 641
R-SO_x process, 620, 621
Richards sulfur recovery process, 697
Ryan/Holmes process, 1342, 1343

Saarberg-Hölter (SHU) FGD process, 535
Saline brines, dehydration with, 1007–1017
Saturator for ammonium sulfate production, 309, 310
SCOT process, 719–723
Seaboard process, 381–383
Seawater FGD process, 597–601

Selective catalytic reduction (SCR) of NO_x, 904–930
Selective noncatalytic reduction (SNCR) of NO_x, 888–904
Selection of processes for gas impurity removal, 2–5
Selectox process, 711–713
Selexol process, 1202–1210
Semidirect process for ammonia recovery, 296, 297
Sepasolv MPE process, 1210
Shed trays, 9
Shift conversion, 1172–1177
   catalysts for, 1174, 1175
   equilibrium constants for, 1176
   heats of reaction for, 1176
Silica-base beads, capacity for water of, 1036
   gas dehydration with, 1034–1039
   physical properties of, 1035, 1036
Silica gel,
   capacity for water of, 1036, 1037
   chemical analysis of, 1036
   equilibrium partial pressure of water over, 1037
   gas dehydration with, 1034–1039
   hydrocarbon recovery with, 1086, 1087
   physical properties of, 1035
Silicon tetrafluoride, vapor pressure of, over fluosilicic acid solutions, 440, 441 (see also Fluorides)
Slurrisweet process, 1309, 1310
SNPA-DEA process, 51
SNOX process for FGD, 642, 643
Soda injection process for FGD, 624–626
Soda-iron process, 1164
Sodium alanine, 397–400
   (see also Alkacid process)
Sodium carbonate, use of, in soda–iron process, 1169
Sodium carbonate solutions,
   carbon dioxide removal with, 378–381
   hydrogen sulfide removal with, 387–393
   rate of absorption of carbon dioxide in, 379–380
   vapor pressure of carbon dioxide over, 379
Sodium chlorite, for nitrogen oxide removal, 933
Sodium dichromate, hydrogen sulfide removal with, 1311
Sodium hydroxide solutions,
   carbon dioxide removal with, 401, 402

hydrogen sulfide removal with, 402–404, 1317, 1318
    rate of absorption of carbon dioxide in, 380, 381
Sodium phenolate process, 396, 397
Sodium sulfite-bisulfite cycle for FGD, 555–559
Sofnolime RG process, 1309
Solid-desiccant dehydration process, 1030–1069 (see also Adsorption, Dehydration)
Solinox process for FGD, 602, 603
Solubility,
    of ammonium sulfate in water, 295
    of Benzene,
        in petroleum oil, 1364
        in triethylene glycol, 995
    of carbon dioxide,
        in methanol, 1217, 1218
        in water, 423, 427–429
        in triethylene glycol, 974
    of carbon disulfide in petroleum oil, 1364
    of carbon monoxide in copper-ammonium-salt solution, 1348
    of gases in N-methyl-2 pyrrolidone, 1213
    of gases in water, 417, 428, 429, 436, 437, 440, 441
    of gases in potassium carbonate solution, 346
    of light oil components in wash oils, 1364
    of lithium halides in water, 1012
    of naphthalene in petroleum oil, 1364
    of natural gas in glycols, 974
    of potassium carbonate/bicarbonate mixtures in water, 341
    of sulfur dioxide,
        in dimethylaniline, 591
        in water, 341
        in xylidine/water mixtures, 592
    of toluene and xylene in petroleum oil, 1364
Solubility product for calcium and magnesium salts, 508
Solvent recovery,
    by Acticarbone process, 1106
    with activated carbon, 1093–1124
    by KPR rotating adsorber, 1115–1117
    by membrane permeation, 1288–1290
    by Polyad FB process, 1113–1115
    by Purasiv HR process, 1112, 1113
Solvents, organic, fire-hazard properties of, 1096

Sorbead, 1038, 1039
Sorbtech process, 630, 631, 932
Sour water stripping, 302–308
SO_x-NO_x-ROX-BOX (SNRB) process, 933
Space velocity, 1147
    for methanation, 1180
    for shift conversion, 1176, 1177
    for soda-iron process, 1165
    for VOC oxidation, 1155
Specific gravity,
    of Adip solutions, 102
    of adsorbents, 1035
    of ammonia solutions, 282
    of alkanolamine solutions, 98, 100, 101
    of glycol solutions, 969
    of glycols, 954, 955
    of potassium carbonate solutions, 345
Specific heat,
    of adsorbents, 1035
    of alkanolamine solutions, 105–107
    of gases, 1341
    of glycols, 954
    of glycol solutions, 970
Sponge iron oxide, 1302
Spray contactors, design of, 34–35
Spray dryer FGD process, 604–616
    disposal of byproduct from, 615, 616
Staatsmijnen-Otto process, 746–748
Sterically hindered amines, 48, 56
Still-Otto process, 322
Stone & Webster Ionics process, 561
Stretford process, 769–794
    materials of construction for, 789
    treatment of waste streams from, 791–794
Stripping systems,
    for ammonia solutions, 322
    for ethanolamine plants, 123, 141–144
    for glycol dehydration plants, 984–988
    for sour water, 302–308
Structural packing for glycol dehydrators, 981, 982
SulfaTreat process, 1305
SulFerox process, 825–840
Sulfidine process, 591–593
Sulfinol process, 1225–1231
Sulfint process, 823–825
Sulfolane, 1225–1231
Sulfolin process, 797–902
    economics of, 801
Sulfreen process, 699–703
Sulften process, 723

Sulfur compound removal with impregnated carbon, 1126–1127
Sulfur recovery,
  by AMOCO cold bed adsorption (CBA) process, 703–706
  by British Oxygen Corporation SURE process, 694–696
  by BASF Catasulf process, 696, 697
  by Claus process, 670–724
  by COPE process, 691–693
  by direct oxidation of hydrogen sulfide, 708–717
  by Linde Clinsulf process, 696
  by Lurgi OxyClaus process, 693–694
  by MCRC sulfur recovery process, 707–708
  by MODOP process, 713–716
  by MTE sulfur recovery process, 697
  by Selectox process, 711–713
  by Richards process, 697
  by Sulfreen process, 699–703
  by Superclaus process, 709–711
  by TPA, Inc., oxygen enrichment process, 696
Sulfur, removal of hydrogen sulfide from, 686–688
Sulfur dioxide,
  adsorption isotherms for, 635, 1074
  concentration in flue gas, 475
  heat of absorption of, in ammonia solutions, 567
  heat of reaction of, with xylidine, 591
  oxidation of, in ammonia solutions, 567, 568
  reactions of,
    in limestone/lime process, 501–513
    in magnesium oxide process, 542
  reduction to hydrogen sulfide, 717–723
  regulations governing emissions of,
    in Canada, 474
    in Europe, 473
    in Japan, 472
    in the United States, 470–472
  requirements for removal, 469–474
  solubility of,
    in dimethylaniline, 591
    in pure water, 506
    in xylidine-water mixtures, 591
  vapor pressure of, over ammonia-sulfur dioxide-water solutions, 565–567
Sulfur Dioxide removal, 466–647
  by adsorption, 634–641
  by ADVACATE process, 621, 622
  with alkali metal compounds, 544–564
  by alkali metal sulfite-bisulfite process, 554–559
  with alkaline earth compounds, 496–544
  by Alkalized Alumina process, 631
  by Ammonia-Ammonium Bisulfate (ABS) process, 573
  by Ammonia-Calcium Pyrophospate process, 582
  by Ammonia-Lime double alkali process, 581, 582
  with ammonia solutions, 564–582
  by Aqueous Aluminum Sulfate process, 582–584
  by Aqueous Carbonate process 606, 607
  with aromatic amines, 589–597
  by ASARCO process, 593–595
  by ATS Technology process, 578, 579
  by Babcock and Wilcox process, 534, 535
  by Bechtel seawater process, 600, 601
  by Bischoff seawater process, 601
  by Bischoff process, 536, 537
  by CANSOLV process, 595, 596
  by carbon adsorption, 634–640
  by catalytic oxidation, 641–645
  by catalytic oxidation/electrochemical process, 645
  by Cat-Ox process, 643–644
  by Chiyoda Thoroughbred 101 process, 585–588
  by Chiyoda Thoroughbred 121 process, 537
  by Citrate process, 563, 564
  by Cominco process, 569–573
  by Coolside process, 623
  by copper oxide process, 627–630
  by Dow process, 596, 597
  by Dowa process, 582–584
  by double alkali process, 546–554
  by dry soda injection process, 624–626
  by duct spray dryer process, 614
  by Ebara E-Beam process, 645, 646
  by electrochemical reduction tosulfur, 646
  by electrolytic regeneration process, 561, 562
  by ferrous sulfide, 585
  by General Electric ammonium sulfate process, 579–581
  by General Electric (GEESI) limestone process, 533, 534
  by Hitachi activated carbon process, 634

by ISPRA bromine based process, 588, 589
by LIMB process, 618–620
by lime slurry spray dryer process, 607–614
by limestone/lime injection, 618–623
by limestone/lime process, 496–539
by magnesium oxide process, 539–544
by Mitsui BF process, 639, 640
by Noell KRC double loop limestone process, 535, 536
by Noell KRC peroxide based process, 589
by NOXO process, 631–634
by Parsons Flue Gas Cleanup (FGC) process, 646, 647
by Pure Air/Mitsubishi (MHI) process, 537
by Reinluft process, 638, 639
by RESOX process, 641
by R-SOX process, 620, 621
by Saarberg-Hölter (S-H-U) process, 535
by seawater processes, 597–601
by SNOX process, 642, 643
by Solinox process, 602, 603
by Sorbtech process, 630, 631
by spray dryer process, 604–616
by Sulfacid process, 634
by Sulf-X process, 585
by Tampella Lifac process, 620
by Walther process, 575–578
by Wellman Lord process, 554–559
by zinc oxide process, 562–563
Sulfur dioxide removal processes,
categorization of, 476–479
cost comparison of, 492–495
selection of, 491–495
status of, 480–488
suppliers of, 480–488
Sulfur scavenging processes, 1297–1320
Sulfuric acid, 569, 573
Sulfur trioxide formation, 475, 476
Sulf-X process, 585
Superclaus process, 709–711
Super Drizo process, 959
Superphosphate plant exhaust gases (see Fluorides)
Synthesis gas,
removal of organic sulfur from, 1165–1172
removal of oxides of carbon from, 1177–1180

Takahax process, 765–769
Tampella Lifac process, 620

Tetraethylene glycol, properties of, 954
TEXTREAT amine process, 42
Thermal conductivity of adsorbents for gas dehydration, 1035
Thermal $DeNO_x$ process, 898–904
Thermal oxidation of VOCs and odors, 1137–1145
economics of, 1145
Thermal reclaiming of amine solutions, 255–264
Thioarsenate solutions, hydogen sulfide removal with, 748–759
Thiophene, removal of by Holmes-Maxted process, 1169 (see also Organic sulfur)
Thiosulfate formation,
in Stretford process, 772, 777–780
in Thylox process, 750
Thylox process, 748–754
materials of construction for, 754
Tin oxide process, 1328, 1329
Tokyo Electric-MHI process, 936
Toluene, solubility of, in petroleum oil, 1364
Toluidine, properties of, 590
sulfur dioxide recovery with, 591–593
Towers, absorption, design of, 12–35 (see also Absorbers; Columns)
adsorption, design of, 1060–1063 (see also Adsorption; Adsorption vessels)
Townsend process, 841, 842
TPA, Inc., oxygen enrichment process, 696
Trail, Canada, sulfur dioxide recovery plants at, 569–573
Transfer units,
height of, 18
for ammonia absorption in packed towers, 299–302
for carbon dioxide absorption in water, 427–432
number of, 18, 19
for ammonia absorption in spray towers, 301–302
for fluoride absorption, 444–448
Tray diagram,
for carbon dioxide absorption in monoethanolamine solution, 118
for carbon dioxide stripping, 119
for water absorption in glycol, 980
Trays (see Plates)
Tributyl phosphate,
physical properties of, 1224
use of in Estosolvan process, 1224

Triethanolamine, 41, 45, 49
Triethanolamine solutions,
  vapor pressure of 95
  vapor pressure of acid gases over, 79, 85, 86
Triethylene glycol,
  dehydration with, 953–997
  physical properties of, 954, 975
  solubility of carbon dioxide in, 974
  solubility of natural gas in, 974
  vapor pressure of, 971
Triethylene glycol solutions,
  dew point of gases in contact with, 965, 977–979
  freezing points of, 971
  specific gravity of, 969
  specific heat of, 970
  vapor/liquid equilibrium diagrams for, 975
  viscosity of, 970
TRI-NO$_x$ process, 935
Tripotassium phosphate process, 393–396
  steam consumption in, 395, 396
Tripotassium phosphate solution, vapor pressure of hydrogen sulfide over, 395
Turbogrid plates, 7–9

UCARSOL process, 42
UCBSRP process, 846–850
Uniflux plates, 7
Unisulf process, 802, 803
UOP Merox process, 406, 407
USS Phosam process, 311–314

Vacasulf process, 392, 393
Vacuum carbonate process, 383–392
  operating cost, 390–392
  operating problems, 389–390
  solution regeneration, 389
Valve plates, 7–9
Vapor/liquid composition diagrams,
  for Diglycolamine solutions, 260
  for monoethanolamine solutions, 257
Vapor pressure,
  of ammonia,
    over ammonia–sulfur dioxide-water solutions, 565–567
    over aqueous solutions, 283–292
  of ammonium sulfate solution, 295
  of carbon dioxide,
    over potassium carbonate, 341–343
    over sodium carbonate-bicarbonate solutions, 380, 381
  of carbon dioxide and hydrogen sulfide
    over ammonia solutions, 283–292
    over ethanolamine solutions, 62–91
  of carbon monoxide over copper-ammonium-salt solutions, 1348–1350
  of copper-ammonium-salt solutions
  of diethanolamine solutions, 93
  of ethanolamine solutions, 91–97
  of ethanolamines, 49
  of ethylene dichloride, 1335
  of gases at low temperature, 1341
  of glycol solutions, 968
  of hexane, 1335
  of hydrofluoric acid solutions, 440
  of hydrogen chloride over hydrochloric acid solutions, 454
  of hydrogen sulfide,
    over Alkacid solutions, 400
    over potassium carbonate solutions, 347, 349
    over tripotassium phosphate solutions, 395
  of lithium halide solutions, 1011
  of methyldiethanolamine solutions, 97
  of monoethanolamine solutions, 92
  of naphthalene, 1368
  of organic compounds, 1335
  of silicon tetrafluoride over fluosilicic acid solutions, 440, 441
  of sulfur dioxide,
    over ammonia-sulfur dioxide-water solutions, 565–567
    over magnesium salt solutions, 543
  of toluene, 1335
  of triethanolamine solutions, 95
  of water,
    over ammonia solutions, 283–292
    over ammonia-sulfur dioxide-water solutions, 565–567
    over potassium carbonate solutions, 344, 351, 352
Venturi scrubbers, 422
  absorption of fluorides in, 441, 443, 447
Viscosity,
  of Adip solutions, 105
  of ammonium sulfate solutions, 295
  of DGA solutions, 104
  of diethanolamine solutions, 103
  of glycol solutions, 970
  of glycols, 955
  of monoethanolamine solutions, 102

of potassium carbonate solutions, 345
VOC removal,
   by active carbon, 1087–1124
      cost of, 1108–1109
   by biofilters, 1124–1127
   by catalytic oxidation, 1148–1163
   by KPR rotating adsorber, 1115–1117
   by Munters Zeol adsorber, 1117
   by moving bed process, 1117
   by thermal oxidation, 1137–1145
   by various processes, efficiency of, 5
VOCs, ignition temperatures of, 1143
Volatile organic compound (see VOC)

Water,
   absorption,
      of ammonia by, 299–302
      of carbon dioxide by, 427–432
      of chlorine by, 458–463
      of fluorides by, 438–453
      of hydrogen chloride by, 454–458
      of hydrogen sulfide by, 436–438
   capacity of adsorbents for, 1036
   solubility of,
      carbon dioxide in, 428, 429
      of gases in, 417 (see also individual gases)
      of hydrogen sulfide in, 437
      of sulfur dioxide in, 506

Water/methanol phase diagram, 1003
Water vapor,
   adsorption of, 1030–1069
   in saturated air, 952
   in saturated natural gas, 947–951
   removal of (see Dehydration)
Water wash for amine recovery, 58, 59
Water wash processes, 415–462
   equipment for, 418–422
Wellman Lord process, 554–559
Wet scrubbers, 418
Wiewiorowski process, 846
WSA-SNOX process, 933
   (see also SNOX process)

Xylene, solubility of, in petroleum oil, 1364
Xylidine,
   heat of reaction with sulfur dioxide, 591
   properties of, 592
   sulfur dioxide recovery with, 589–593
Xylidine-water mixtures, solubility of sulfur dioxide in, 591

Zinc ferrite and titanate sorbents, 1324–1328
Zinc oxide, for hydrogen sulfide removal, 1305–1309
   for sulfur dioxide removal, 562, 563
Zinc oxide slurry process, 1310, 1311
Z-Sorb process, 1329

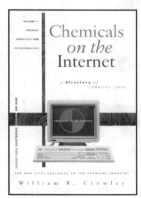

RULES OF THUMB FOR CHEMICAL ENGINEERS
Carl Branan, Editor

Packed full of useful information, this volume helps solve field engineering problems with its hundreds of common sense techniques, shortcuts, and calculations. The safety chapter covers lower explosive limit and flash, flammability, and static charge, while the chapter on energy conservation addresses process efficiency, heat recovery, and equivalent fuel values.
1994. 368 pages, figures, tables, charts, appendixes, index, large-format flexible cover.
ISBN 0-88415-162-X
#5162 $79   £63

FLUID CATALYTIC CRACKING HANDBOOK
Design, Operation, and Troubleshooting of FCC Facilities
Reza Sadeghbeigi

In clear language, this manual provides practical information covering all key areas of fluid catalytic cracking (FCC).
The book focuses on FCC process technology, FCC feedstock properties, and the advantages and disadvantages of different FCC catalysts.
Enhance unit reliability with step-by-step techniques for effective troubleshooting.
1995. 322 pages, figures, tables, appendixes, glossary, index, 6" x 9" hardcover.
ISBN 0-88415-290-1
#5290 $89   £72

Visit Your Favorite Bookstore

Or order directly from:
Gulf Publishing Company
P.O. Box 2608 • Dept. KU
Houston, Texas 77252-2608
713-520-4444 • FAX: 713-525-4647
Send payment plus $9.95 ($11.55 if the order is $75 or more, $15.95 for orders of $100 or more) shipping and handling or credit card information. CA, IL, PA, TX, and WA residents must add sales tax on books and shipping total.

Price and availability subject to change without notice.

Thank you for your order!

CHEMICALS ON THE INTERNET
A Directory of Industry Sites
Volumes 1 and 2
William R. Crowley, Editor

This time-saving, two-volume directory gives you each chemical or mineral industry site's URL address, describes what is to be found, and reveals additional hypertext links. Each volume comes with a free HTML browser disk.

Volume 1
Organic Chemicals and Petrochemicals
1997. 244 pages, 770 Web sites, index, free computer disk, 7" x 10" paperback.
ISBN 0-88415-139-5
#5139 $69   £55

Volume 2
Inorganic Chemicals and Minerals
1997. 244 pages, 650+ Web sites, index, free computer disk, 7" x 10" paperback.
ISBN 0-88415-140-9
#5140 $69   £55

THE NEW WEIBULL HANDBOOK
Second Edition
Robert B. Abernethy

This second edition includes the author's latest research findings, most notably, which numerical method is best and when. Special methods, such as Weibayes, are presented with actual case studies.
1996. 252 pages, figures, tables, 8½" x 11" lay-flat paperback.
ISBN 0-88415-507-2
#5507 $89   £73